T0391698

THE OXFORD HANDBOOK OF

DEVELOPMENTAL
COGNITIVE
NEUROSCIENCE

OXFORD LIBRARY OF PSYCHOLOGY

THE OXFORD HANDBOOK OF AUTISM AND CO-OCCURRING
PSYCHIATRIC CONDITIONS
Edited by Susan W. White, Brenna B. Maddox, and Carla A. Mazefsky

THE OXFORD HANDBOOK OF ADOLESCENT SUBSTANCE ABUSE
Edited by Robert A. Zucker and Sandra A. Brown

THE OXFORD HANDBOOK OF ACCURATE
PERSONALITY JUDGMENT
Edited by Tera D. Letzring and Jana S. Spain

THE OXFORD HANDBOOK OF EVOLUTIONARY PSYCHOLOGY
AND RELIGION
Edited by James R. Liddle and Todd K. Shackelford

THE OXFORD HANDBOOK OF LIFELONG LEARNING, 2E
Edited by Manuel London

THE OXFORD HANDBOOK OF EVOLUTIONARY PSYCHOLOGY
AND PARENTING
Edited by Viviana A. Weekes-Shackelford and Todd K. Shackelford

THE OXFORD HANDBOOK OF CULTURAL NEUROSCIENCE AND
GLOBAL MENTAL HEALTH
*Edited by Joan Y. Chiao, Shu-Chen Li, Robert Turner, Su Yeon Lee-Tauler,
and Beverly Pringle*

THE OXFORD HANDBOOK OF THE POSITIVE HUMANITIES
Edited by Louis Tay and James O. Pawelski

THE OXFORD HANDBOOK OF INFIDELITY
Edited by Tara DeLecce and Todd K. Shackelford

THE OXFORD HANDBOOK OF EEG FREQUENCY
Edited by Philip Gable, Matthew Miller, and Edward Bernat

THE OXFORD HANDBOOK OF EVOLUTIONARY PSYCHOLOGY
AND ROMANTIC RELATIONSHIPS
Edited by Justin K. Mogilski and Todd K. Shackelford

THE OXFORD HANDBOOK OF ACCEPTANCE AND
COMMITMENT THERAPY
Edited by Michael P. Twohig, Michael E. Levin, and Julie M. Petersen

THE OXFORD HANDBOOK OF SELF-DETERMINATION THEORY
Edited by Richard M. Ryan

THE OXFORD HANDBOOK OF PSYCHOLOGY AND LAW
Edited by David DeMatteo and Kyle C. Scherr

THE OXFORD HANDBOOK OF PSYCHOLOGY AND
SPIRITUALITY, 2E
Edited by Lisa J. Miller

THE OXFORD HANDBOOK OF HUMAN SYMBOLIC EVOLUTION
Edited by Chris Sinha, Andy Lock, and Nathalie Gontier

THE OXFORD HANDBOOK OF COGNITIVE ARCHAEOLOGY
Edited by Thomas Wynn, Karenleigh A. Overmann, and Frederick L. Coolidge

THE OXFORD HANDBOOK OF

DEVELOPMENTAL COGNITIVE NEUROSCIENCE

Edited by

KATHRIN COHEN KADOSH

OXFORD
UNIVERSITY PRESS

Great Clarendon Street, Oxford, OX2 6DP,
United Kingdom

Oxford University Press is a department of the University of Oxford.
It furthers the University's objective of excellence in research, scholarship,
and education by publishing worldwide. Oxford is a registered trade mark of
Oxford University Press in the UK and in certain other countries

Published in the United States of America by Oxford University Press
198 Madison Avenue, New York, NY 10016, United States of America

British Library Cataloguing in Publication Data
Data available

Library of Congress Control Number: 2023947801

ISBN 978–0–19–882747–4

DOI: 10.1093/oxfordhb/9780198827474.001.0001

Printed and bound by
CPI Group (UK) Ltd, Croydon, CR0 4YY

For Jonathan and Itamar, who have never lost that early magic

Contents

Contributors

Laura Albantakis, Physician, Psychiatrist in training for adult psychiatry and psychotherapy and for child and adolescent psychiatry and psychotherapy, Department of Psychiatry and Psychotherapy, LMU University Hospital, Munich

Nicholas B. Allen, Ann Swindells Professor of Clinical Psychology, Department of Psychology, University of Oregon

Nadja Althaus, Lecturer, School of Psychology, University of East Anglia

Michael L. Anderson, Rotman Canada Research Chair in Philosophy of Science; Professor, Department of Philosophy, Western University

Morteza Ansarinia, Postdoctoral Fellow, Research Unit Education, Culture, Cognition & Society (ECCS), University of Luxembourg and Institute of Cognitive Science and Assessment

Alexander L. Anwyl-Irvine, Data Science Manager, Behavioural Science, Lloyd's Banking Group, UK

Salomi S. Asaridou, Departmental Lecturer, Department of Experimental Psychology, University of Oxford

Sikoya M. Ashburn, Postdoctoral Fellow in Psychology and Neuroscience, University of North Carolina at Chapel Hill

Duncan E. Astle, Professor, MRC Cognition and Brain Sciences Unit & University of Cambridge

Orianna Bairaktari, PhD Student, Institute of Education, Multilanguage and Cognition Lab, University College London

Marjolein E. A. Barendse, Postdoctoral Student, Department of Psychology, University of Oregon

Ryan Barry-Anwar, Postdoctoral Fellow, Brain Cognition and Development Lab, University of Florida

Patricia J. Bauer, Asa Griggs Candler Professor of Psychology, Emory University

Daphne Bavelier, Professor of Cognitive Neuroscience, Psychology and Education Science (FPSE), Geneva University

Giacomo Bignardi, Postdoctoral Fellow, Department of Psychology, University of Cambridge

Daniel Brady, PhD Student, School of Psychology, University of Surrey

Philip Brandner, Postdoctoral Fellow, Erasmus University Rotterdam, the Netherlands

Peter Bright, Professor of Psychology, Anglia Ruskin University

Sarah M. Burke, Research Scientist, Faculty of Medical Sciences, University of Groningen, The Netherlands

Pedro Cardoso-Leite, Associate Professor, Research Unit Education, Culture, Cognition and Society (ECCS), University of Luxembourg and Institute of Cognitive Science and Assessment

Kevin Champagne-Jorgensen, Postdoctoral Fellow, Neuroscience Graduate Program, McMaster University, Canada; Department of Immunology, University of Toronto, Canada

Kathrin Cohen Kadosh, Associate Professor, Developmental Cognitive Neuroscience, University of Surrey

Roi Cohen Kadosh, Professor of Cognitive Neuroscience, School of Psychology, University of Surrey

Fallon Cooke, Postdoctoral Fellow, Infant Sleep Australia; Department of Paediatrics, University of Melbourne; Intergenerational Health, Murdoch Children's Research Institute

Hannah Corcoran, Paediatric Fellow, The Royal Children's Hospital

Eveline A. Crone, Professor of Developmental Neuroscience in Society, Leiden University

Edwin S. Dalmaijer, Assistant Professor, University of Bristol

Kathy T. Do, PhD Student, Department of Psychology and Neuroscience, University of North Carolina

Jessica A. Dugan, Postdoctoral Fellow in Psychology, Emory University

Guinevere F. Eden, Professor of Pediatrics, Georgetown University

Olivia M. Elvin, Clinical Psychologist, School of Applied Psychology, Griffith University

Roberto Filippi, Professor, Institute of Education, Multilanguage and Cognition Lab, University College London

Leon Fonville, Senior MRI Scientist, Invicro, A Konica Minolta Company

Alexander Fraser, Postdoctoral Research Associate, Department of Experimental Psychology, University of Oxford

Bryce L. Geeraert, Postdoctoral Fellow, Brain Create Centre, University of Calgary

Judit Gervain, Full Professor of Developmental Psychology, Department of Developmental and Social Psychology, University of Padua

Amanda L. Griffiths, Head of Sleep Medicine, Respiratory and Sleep Paediatrician, Department of Respiratory and Sleep Medicine, Royal Children's Hospital

Suzanne van de Groep, Postdoctoral Researcher, Erasmus University

Olaf Hauk, Senior Investigator Scientist, MRC Cognition and Brain Sciences Unit, University of Cambridge

Teresa Iuculano, Associate Research Professor, Department of Psychology, Centre National de la Recherche Scientifique & Université Paris Cité, La Sorbonne

E. J. H. Jones, Professor, Centre for Brain and Cognitive Development and School of Psychological Sciences, Birkbeck, University of London

Anna R. Kauer, PhD Student, School of Psychology, University of Surrey

Rebecca C. Knickmeyer, Associate Professor in the Department of Pediatrics and Human Development, Michigan State University

Anna Kolesnik-Taylor, Associate Professor, School of Psychological Sciences, Birkbeck College, University of London

Enikő Ladányi, Postdoctoral Fellow, Department of Linguistics, University of Potsdam

Jennifer Y. F. Lau, Professor of Youth Resilience and Co-Director of the Youth Resilience Unit

Catherine Lebel, Professor of Radiology, University of Calgary

Hayley Leonard, Research Director, National Centre for Social Research

Klaus Libertus, Assistant Professor, University of Pittsburgh

Sarah Lloyd-Fox, Principle Research Associate, Department of Psychology, University of Cambridge

Anna A. Matejko, Assistant Professor, University of Durham

Claire M. Matthews, Postdoctoral Fellow, Toronto Metropolitan University

Whitney I. Mattson, Research Scientist, Senior, The Research Institute of Nationwide Children's Hospital

Daphne Maurer, Distinguished University Professor of Psychology, Neuroscience & Behaviour, McMaster University

Cameron C. McKay, Data Science Manager, Center for the Study of Learning, Department of Pediatrics, Georgetown University

Vinod Menon, Professor of Psychiatry and Behavioral Sciences and of Neuroscience, and Director of the Stanford Cognitive and Systems Neuroscience Laboratory in the Stanford University School of Medicine

Evelyne Mercure, Assistant Professor, Department of Psychology, Goldsmiths University

Kathryn L. Mills, Assistant Professor, Department of Psychology, University of Oregon

Kathryn Modecki, Associate Professor Centre for Mental Health and School of Applied Psychology and Menzies Health Institute Queensland, Griffith University

Josephine Mollon, Postdoctoral Fellow, Boston Children's Hospital, Harvard Medical School

Catherine J. Mondloch, Professor, Psychology Department, Brock University

Michele A. Morningstar, Assistant Professor, Department of Psychology, Queen's University

Victoria Mousley, Postdoctoral Researcher, Centre for Brain and Cognitive Development, Birkbeck, University of London

Eric E. Nelson, Principal Investigator and Director of Neuroimaging Research, Abigail Wexner Research Institute, Nationwide Children's Hospital

Karen-Anne McVey Neufeld, Research Associate, Brain-Body Institute, McMaster University

Aarthi Padmanabhan, Postdoctoral Fellow, Cognitive and Systems Neuroscience Laboratory, Stanford University

Eric Pakulak, Associate Professor, Department of Psychology, Stockholm University

Andriani Papageorgiou, Independent Researcher, Institute of Education, Multilanguage and Cognition Lab, University College London

Sabina Pauen, Professor, Developmental Psychology, University of Heidelberg

Marcie Penner, Associate Professor of Psychology, King's University College at Western University

Eva Periche-Tomas, Postdoctoral Fellow, Brain Research Imaging Centre, Cardiff University

Stefanie Peykarjou, Junior Research Group Leader, Developmental and Biological Psychology, University of Heidelberg

Jennifer H. Pfeifer, Professor, Department of Psychology, University of Oregon

Victoria Pile, Lecturer in Clinical Psychology, Psychology Department, King's College London

Mitchell J. Prinstein, John Van Seters Distinguished Professor of Psychology and Neurosciences, University of North Carolina at Chapel Hill

Vicente Raja, Research Fellow, Department of Philosophy, Universidad de Murcia

Monika B. Raniti, Postdoctoral Research Fellow, Medicine, Dentistry and Health Sciences University of Melbourne

Jess E. Reynolds, Canadian Institutes of Health Research (CIHR) Postdoctoral Fellow, Department of Radiology, Cumming School of Medicine, University of Calgary

Tracy Riggins, Professor of Psychology, University of Maryland

Katherine M. Ryan, Research Fellow, School of Applied Psychology, Griffith University

Gaia Scerif, Professor of Developmental Cognitive Neuroscience, Department of Experimental Psychology, University of Oxford

Leonhard Schilbach, Head, Department of General Psychiatry 2, LVR-Klinikum Düsseldorf

Christian Schmahl, Medical Director, Department of Psychosomatic Medicine and Psychotherapy, Central Institute of Mental Health

Emmanuel Schmück, Research Unit Education, Culture, Cognition & Society (ECCS) University of Luxembourg and Institute of Cognitive Science and Assessment

Flora Schwartz, Postdoctoral Fellow, Université Toulouse Jean Jaurès

Lisa S. Scott, Professor, Department of Psychology, University of Florida

Maurizio Sicorello, Postdoctoral Fellow, Central Institute of Mental Health Clinic for Psychosomatic Medicine and Psychotherapy, Heidelberg University

Paul T. Sowden, Professor of Psychology, Cognition and Creativity, University of Winchester

Nikolaus Steinbeis, Professor of Developmental Neuroscience, Division of Psychology and Language, University College London

Courtney Stevens, Professor of Psychology, Department of Psychology, Willamette University

Christian K. Tamnes, Professor, Department of Psychology, University of Oslo

Eva H. Telzer, Professor, Department of Psychology and Neuroscience, University of North Carolina

Hanna Thaler, Postdoctoral Fellow, Independent Max Planck Research Group for Social Neuroscience, Max Planck Institute of Psychiatry

Gabrielle-Ann Torre, Postdoctoral Fellow, Georgetown University

John Trinder, Professor Emeritus, University of Melbourne

Lucina Q. Uddin, Professor, Department of Psychiatry and Biobehavioral Sciences, University of California Los Angeles

Allison M. Waters, Deputy Head of School (Research), School of Applied Psychology, Griffith University

Kate E. Watkins, Professor of Cognitive Neuroscience, Department of Experimental Psychology, University of Oxford

CHAPTER 1

INTRODUCTION TO DEVELOPMENTAL COGNITIVE NEUROSCIENCE

Current state-of-the-art and new frontiers

KATHRIN COHEN KADOSH

Und jedem Anfang wohnt ein Zauber inne

[There is magic in every beginning]

Hermann Hesse

WHEN I was a student of Psychology at the Johann Wolfgang Goethe-University in Frankfurt, Germany in the early 2000s, I was taught that the science was working towards a better understanding of how brain structure and function can explain human behavior. Different behaviors were associated with specific brain areas and all attempts at improving or enhancing behavior focused on changing responsiveness in these brain modules (Cohen Kadosh and Johnson 2007). This scientific approach soon seemed too limited; as certain conditions such as neurodevelopmental disorder eluded localization to a particular brain region. Only when I came across the work on cortical specialization in development by Mark H. Johnson (Johnson 2011; Johnson et al. 2002; Johnson et al. 2015) and Annette Karmiloff-Smith (Johnson et al. 2002; Karmiloff-Smith 1998) did I grasp that working to understand brain development from matured adult brains was more of a hindrance than a help. To understand the human brain structure and function and its influence on behavior, it is important to start at the beginning.

The aim of this handbook is to share my passion for the "magical beginning" of human development. It brings together the voices of the leading developmental cognitive neuroscientists in the field that work on understanding human development and the complex interplay of genetic, environmental, and brain maturational factors that shape social and cognitive functioning in development (Cohen Kadosh 2011; Johnson et al. 2009; Johnson et al. 2002). It also includes a number of chapters on new, emerging research areas that show promise for understanding both brain and behavior in development, such as nutrition and the microbiome gut-brain axis (Cohen Kadosh et al. 2021b) and sleep (Anderson et al. 2017; Orchard et al. 2020; Smith et al. 2019). Looking beyond early developmental changes, this handbook places importance on the period of adolescence. Adolescence is an important developmental juncture, where increasing peer orientation and changes in social behavior, alongside ongoing cognitive improvements and brain maturation, all work together to allow for considerable opportunity of growth (Blakemore 2008; Burnett et al. 2011; Tamnes et al. 2017), but also increased risk of mental health problems (Keshavan et al. 2014).

The developmental cognitive neuroscience (DCN) research approach intersects nature and nurture and considers both health and disease models. DCN focuses on understanding the complexity of human development, necessitating a multi-level and multi-factor research approach to grasp change and plasticity which, by definition, is multidisciplinary. By assuming complexity from the outset, the DCN research approach can provide much needed insights into both the initial set-up of brain networks and cognitive mechanisms, but also into adaptability across the developmental trajectory. This is relevant not only for scientists studying typical and atypical development, but also for interventional work looking for critical or sensitive periods where interventions would be most effective. This new handbook, which is suitable for both lay readers and experts, aims to provide the reader with a comprehensive overview of the current research in DCN and it highlights current gaps and directions for future research.

As Karmilloff-Smith, a giant in the field of DCN once said: "if you want to get ahead, have a theory" (Karmiloff-Smith and Inhelder 1975), so in the first chapter of this handbook Raja et al. critically evaluate current theoretical frameworks in DCN and explain how the neural re-use hypothesis could provide a unifying framework for a number of theories. The chapter also offers guidance for the interpretation of the studies presented in later chapters. The handbook also provides a comprehensive description and update on key DCN methodologies, such as the chapter by Mills and Tamnes on functional magnetic resonance imaging, which focuses on design and analysis approaches for longitudinal structural and functional brain development in childhood and adolescence, and which still represent the gold standard for mapping (functional) brain development. The chapter by Geeraert, Reynolds, and Lebel on diffusion tensor imaging studies of white matter tract development provides a comprehensive overview of current progress and also links the developmental research closely to reading acquisition. The chapter by Stevens and Pakulak takes a different approach by looking at the effects of early risk and adversity on the developing brain. The chapters on encephalography, event-related potentials, and steady-state visual-evoked potentials by Barry-Anwar, Riggins, and Scott, magnetoencephalography (Dalmaijer et al.), near-infrared spectroscopy (Lloyd-Fox), and eye-tracking (Althaus) offer a comprehensive guide to these popular methodologies (including a useful checklist in the Barry-Anwar et al. chapter for designing and analyzing DCN experiments), and discuss the unique challenges posed by testing developing populations. These insights should be of interest to all researchers working with special populations, but specifically those working with patients with atypical brain function and damage, as well as participants with neurodevelopmental disorders.

The handbook also introduces new approaches to understanding cognitive and brain function in development, including the microbiome gut-brain axis. Over two chapters (Knickmeyer; Champagne-Jorgensen and McVey Neufeld), the reader is introduced to the emerging research field of the microbiome gut-brain axis in development, and how gut-based bacteria shape and influence brain function throughout development. The first chapter by Knickmeyer focuses on the critical period of infancy and the early years during which the gut microbiome is first established, whereas the second chapter by Champagne-Jorgensen and McVey Neufeld focuses on the period of adolescence, which has been described as a second sensitive period for the fine-tuning of the microbiome gut-brain axis (Ezra-Nevo et al. 2020). This new-found knowledge is prime for DCN research as gut microbiome changes coincide with critical brain maturation, as well as increased peer-orientation and a changing social environment (Cohen Kadosh et al. 2021a; McVey Neufeld et al. 2016).

The handbook also offers dedicated chapters on developmental cognitive neuroscience research in a number of cognitive domains. For language development, the chapter by Ladányi and Gervain provides a comprehensive overview of neurocognitive bases of language acquisition in infancy, whereas the chapter by Filippi et al. looks at the special case of multi-language acquisition and cognitive development, such as attention abilities, as well as neurodevelopmental disorders like autism. The chapter by Mercure and Mousley summarizes the evidence of experience-dependent plasticity in brain activation for language acquisition, such as in hearing children of deaf parents or in infants from varying socio-economic backgrounds. Lastly, the Asaridou and Watkins chapter comprehensively reviews the neural basis of speech and language impairments and particularly developmental language disorder, whereas the chapter by Matejko et al. discusses the neurocognitive evidence of reading acquisition and developmental dyslexia and how dyslexia relates to other disorders, for instance dyscalculia or neurodevelopmental disorders such as attention deficit hyperactivity disorder.[*]

In the numerical domain, Cohen Kadosh and Iuculano wrote a comprehensive chapter on numerical and mathematical abilities in children, whereas the chapter by Menon, Padmanabhan, and Schwartz focuses on dyscalculia and mathematical learning disabilities from a systems neuroscience perspective. For developmental changes in the domain of attention, Fraser and Scerif provide an overview of the current research into changes in infancy and childhood and discuss some implications for research into neurodevelopmental disorders, such as autism and Fragile X, while Elvin et al. describe attentional biases in children and adolescents within the context of psychopathology and bias-modification-based interventions. The chapter by Kauer and Sowden looks at the role of attention in the development of creativity, setting out evidence highlighting that the development of attention is as integral to the development of creativity as it is to other aspects of cognition. Lastly, the chapter by Cardoso-Leite et al. provides a comprehensive overview of the literature on the behavioral and neurocognitive bases of video games and how they can be used to shape and enhance developmental outcomes in attention.

For perceptual development, the chapter by Matthews, Maurer, and Mondloch summarizes neurocognitive development of face perception and particularly facial identity processing in infancy and childhood and the chapter by Pauen and Peykarjou describes the development of categorical thinking for both the visual and auditory domain, with a strong focus on optimal paradigms and methodologies.

For the domain of memory, Bauer and Dugan's chapter describes episodic and autobiographical memory and the neurodevelopmental changes in the supporting brain networks.

Exploring social cognitive abilities, four chapters review developmental changes in childhood and adolescence. The Steinbeis chapter focuses on the neurocognitive mechanisms supporting prosocial behavior, whereas the chapter by Burke et al. provides an overview of the ongoing developmental changes in trust and reciprocity in adolescence. The chapter by Do, Prinstein, and Telzer looks at peer pressure, particularly within the context of risk and reward processing in adolescence and from a neurodevelopmental disorder perspective. Finally, the adolescence chapter by Barendse and Pfeifer concentrates on the key role of puberty hormones on brain maturation and functioning.

From a neurodevelopmental perspective, the chapter by Kolesnik-Taylor and Jones provides a comprehensive overview of the DCN evidence on autism in infancy and childhood, while the chapter by Thaler, Albantakis and Schilbach describes social cognitive abilities in autism at later developmental stages and into early adulthood. Moving on from cognitive domains, two chapters focus on motor development, the chapter by Libertus summarizes current progress and makes a strong case for the important contribution that motor development has on perceptual and cognitive development. The chapter by Brady and Leonard looks at developmental coordination disorder and discusses current theories and neurocognitive models for an under-researched neurodevelopmental disorder, with excellent pointers for future research.

A number of chapters focus on mental health in development, which presently is prominent in public discourse. As a result, there is demand for scientific guidance not only for understanding the root causes for mental health but also for new intervention approaches that take into account the considerable developmental changes that children and young people experience (Burnett et al. 2011; Cohen Kadosh et al. 2013). The chapter by Morningstar, Mattson and Nelson on affective disorders in children and adolescents describes deficits in social information processing and the underlying neural correlates. Sicorello and Schmahl synthesize evidence on the neurobiology and developmental psychopathology of borderline personality disorder. Fonville and Mollon provide a comprehensive overview of the literature on early psychotic experiences in development and the evidence of changes in the underlying neurocognitive bases. The chapter by Corcoran, Cooke, and Griffiths approaches life-long (mental) health from a different angle by discussing the importance of sleep in development, which also holds important implications for understanding cognitive functioning and brain maturation. This is extended by the chapter by Raniti, Trinder, and Allen which reviews current evidence on anxiety, depression, and sleep in adolescence. And finally, Pile and Lau review the efficacy of cognitive, and particularly attention bias-modification-based, intervention in children and young people with depression.

The story of human development and of the different factors that shape and influence the individual across the life-span is an age-old one, as it not only allows us to understand ourselves and how we have come to live together in the way that we do, but also comprehend the brilliant diversity of human interactions and life. A better understanding of our past and present will also help us with making decisions for the future and for the future of our many wonderful research participants.

REFERENCES

Anderson, J. R., Carroll, I., Azcarate-Peril, M. A., Rochette, A. D., Heinberg, L. J., Peat, C., et al. (2017). A preliminary examination of gut microbiota, sleep, and cognitive flexibility in healthy older adults. *Sleep Medicine*, *38*, 104–107. doi:10.1016/j.sleep.2017.07.018

Blakemore, S. J. (2008). The social brain in adolescence. *Nature Reviews Neuroscience*, *9*(4), 267–277. doi:10.1038/nrn2353

Burnett, S., Sebastian, C., Cohen Kadosh, K., and Blakemore, S. J. (2011). The social brain in adolescence: Evidence from functional magnetic resonance imaging and behavioural studies. *Neuroscience & Biobehavioral Review*, *35*(8), 1654–1664. doi:10.1016/j.neubiorev.2010.10.011

Cohen Kadosh, K. (2011). What can emerging cortical face networks tell us about mature brain organisation? *Developmental Cognitive Neuroscience*, *1*(3), 246–255.

Cohen Kadosh, K., Basso, M., Knytl, P., Johnstone, N., Lau, J. Y. F., and Gibson, G. R. (2021a). Psychobiotic interventions for anxiety in young people: A systematic review and meta-analysis, with youth consultation. *Translational Psychiatry*, *11*(1), 352. doi:10.1038/s41398-021-01422-7

Cohen Kadosh, K. and Johnson, M. H. (2007). Developing a cortex specialized for face perception. *Trends in Cognitive Science*, *11*(9), 367–369. doi:10.1016/j.tics.2007.06.007

Cohen Kadosh, K., Linden, D. E., and Lau, J. Y. (2013). Plasticity during childhood and adolescence: innovative approaches to investigating neurocognitive development. *Development Science*, *16*(4), 574–583. doi:10.1111/desc.12054

Cohen Kadosh, K., Muhardi, L., Parikh, P., Basso, M., Jan Mohamed, H. J., Prawitasari, et al. (2021b). Nutritional support of neurodevelopment and cognitive function in infants and young children—an update and novel insights. *Nutrients*, *13*(1), 199. doi:10.3390/nu13010199

Ezra-Nevo, G., Henriques, S. F., and Ribeiro, C. (2020). The diet-microbiome tango: How nutrients lead the gut brain axis. *Current Opinion in Neurobiology*, *62*, 122–132. doi:10.1016/j.conb.2020.02.005

Johnson, M. H. (2011). Interactive specialization: A domain-general framework for human functional brain development? *Developmental Cognitive Neuroscience*, *1*(1), 7–21.

Johnson, M. H., Grossmann, T., and Cohen Kadosh, K. (2009). Mapping functional brain development: Building a social brain through interactive specialization. *Developmental Psychology*, *45*(1), 151–159. doi:10.1037/a0014548

Johnson, M. H., Halit, H., Grice, S. J., and Karmiloff-Smith, A. (2002). Neuroimaging of typical and atypical development: A perspective from multiple levels of analysis. *Development and Psychopathology*, *14*(3), 521–536. Retrieved from https://www.ncbi.nlm.nih.gov/pubmed/12349872

Johnson, M. H., Jones, E. J., and Gliga, T. (2015). Brain adaptation and alternative developmental trajectories. *Development and Psychopathology*, *27*(2), 425–442. doi:10.1017/S0954579415000073

Karmiloff-Smith, A. (1998). Development itself is the key to understanding developmental disorders. *Trends in Cognitive Sciences*, *2*, 389–398.

Karmiloff-Smith, A. and Inhelder, B. (1975). If you want to get ahead, get a theory. *Cognition*, *3*(3), 195–212.

Keshavan, M. S., Giedd, J., Lau, J. Y., Lewis, D. A., and Paus, T. (2014). Changes in the adolescent brain and the pathophysiology of psychotic disorders. *Lancet Psychiatry*, *1*(7), 549–558. doi:10.1016/S2215-0366(14)00081-9

McVey Neufeld, K. A., Luczynski, P., Seira Oriach, C., Dinan, T. G., and Cryan, J. F. (2016). What's bugging your teen?—The microbiota and adolescent mental health. *Neuroscience Biobehavioral Review*, *70*, 300–312. doi:10.1016/j.neubiorev.2016.06.005

Orchard, F., Gregory, A. M., Gradisar, M., and Reynolds, S. (2020). Self-reported sleep patterns and quality amongst adolescents: Cross-sectional and prospective associations with anxiety and depression. *Journal of Child Psychology and Psychiatry*, *61*(10), 1126–1137. doi:10.1111/jcpp.13288

Smith, R. P., Easson, C., Lyle, S. M., Kapoor, R., Donnelly, C. P., Davidson, E. J., et al. (2019). Gut microbiome diversity is associated with sleep physiology in humans. *PLoS One*, *14*(10), e0222394. doi:10.1371/journal.pone.0222394

Tamnes, C. K., Herting, M. M., Goddings, A. L., Meuwese, R., Blakemore, S. J., Dahl, R. E., et al. (2017). Development of the cerebral cortex across adolescence: A multisample study of inter-related longitudinal changes in cortical volume, surface area, and thickness. *Journal of Neuroscience*, *37*(12), 3402–3412. doi:10.1523/JNEUROSCI.3302-16.2017

CHAPTER 2

THE NEURAL REUSE HYPOTHESIS

VICENTE RAJA, MARCIE PENNER, LUCINA Q. UDDIN, AND MICHAEL L. ANDERSON

INTRODUCTION

DEVELOPMENTAL Cognitive Neuroscience (DCN) is the interdisciplinary study of the biological underpinnings of cognition and behavior as these change over the course of development from infancy through adulthood. It is concerned with understanding both typical and abnormal development. Neural reuse is an evolutionary and developmental process whereby parts of the brain—at various spatial scales—are placed into new functional partnerships, being by such a process reused for different purposes. The neural reuse hypothesis, then, is that neural reuse is a crucial complement to Hebbian neuroplasticity, and a necessary and important part of the overall biological mechanisms for learning and cognitive development. The study of neural reuse is still in its early stages—its own underlying mechanisms are, for instance, very poorly understood—but the evidence for the importance of this process is mounting, and its implications have begun to change both the methodological and inferential practices in the cognitive neurosciences.

In this chapter, we will describe neural reuse in more detail, and outline some of the evidence for its importance to and impact on brain development. We will then compare the neural reuse hypothesis with some other prominent theories in DCN and suggest that reuse might offer a unifying framework for the field. Finally, we will explore some of the implications of neural reuse for the study of math cognition, brain dynamics, and neurodevelopmental disorders.

The Neural Reuse Hypothesis

Two central questions regarding the activity of the brain are how neural resources are deployed to accomplish cognitive tasks and how that shapes the structural and functional organization of the system. The neural reuse hypothesis aims to answer these questions by considering the nature of brain dynamics while attending to evolutionary and developmental considerations. In doing so, the neural reuse hypothesis engages with the common issues of the field of DCN. In this section we will introduce the neural reuse hypothesis, outline its fundamental tenets, discuss some predictions that follow from it, and consider its position among the different frameworks of contemporary DCN.

What Is Neural Reuse?

Neural reuse is an organizational principle for the functions of the central nervous system (CNS hereafter) that holds that, wherever possible, new functions are accomplished by reusing existing circuits, often in novel combinations. As a principle, neural reuse has been presented in the form of different hypotheses—e.g., the shared circuits model (Hurley, 2005, 2008), the neural exploitation hypothesis (Gallese and Lakoff, 2005), the massive redeployment hypothesis (Anderson, 2007), and the neuronal recycling hypothesis (Dehaene, 2005, 2009); see Anderson (2010) for discussion.[1] More generally, neural reuse is the basis of a framework for the psychological sciences that brings together aspects from biology, embodied cognition, and evolution (Anderson, 2014). From a broad point of view, the framework based on neural reuse is built on two basic assumptions. First, the CNS is fundamentally a control system and other functions must find their space within the boundaries and constraints imposed by such a system. The primary and original task of the brain is to control the action of the organism when it faces a challenging environment and any other function (e.g., "higher" order cognitive capacities such as language or imagination) is usually developed by using the existing neural resources.[2] Second, some form of neural plasticity is the key to understanding how the existing resources of a control system may be used for processes and behaviors not directly related to control itself.

The core tenet of neural reuse is that different parts of the CNS are *used* and *reused* to accomplish different functions. Reuse occurs at multiple spatial scales, but, crucially, at the scale of neural networks or neural regions. Reuse can be considered an evolved principle of neural resource allocation that unfolds over developmental time (Anderson and Finlay 2014).[3] Due to the principle of reuse, neural networks that develop early (e.g., networks developed for basic perceptual tasks or for control of action) are reused in functions that develop later (e.g., language or decision making). As is detailed in Anderson (2014), the phenomenon of neural reuse is characterized by the soft-assembling of different neural networks in different functional systems as required by the cognitive task such that "rather than developing new structures de novo, resource constraints and efficiency considerations dictate that whenever possible neural, behavioral, and environmental resources should have been reused and redeployed in support of any newly emerging cognitive capacities" (p. 7).

Taking neural reuse as an organizational principle, Anderson (2014, 2016) suggests a cognitive architecture based on plastic networks. Such an architecture consists of a set of interactive neural networks (or regions), with no clear boundaries between them (see Pessoa, 2012), that actively soft-assemble in new patterns of connectivity, or actively soft-assemble in the form of already established connectivity patterns, to accomplish various tasks, via a process that has been dubbed *interactive differentiation and search*. In this sense, new cognitive functions are not generally achieved by the generation of new brain regions or by the functional development of existing ones in isolation. The flourishing of new cognitive functions and the participation of different brain regions in supporting them is not a matter of the isolated activity of a new or existing region, but comes from the way different non-task-specific regions are soft-assembled to enable the given task. Such large-scale networks are known as TALoNS (Transiently Assembled Local Neural Subsystems).

As an organizational principle, neural reuse stands against both componential (also known as modular) and holist accounts of the structural and functional organization of the brain and their basic commitments. Contemporary cognitive neuroscience is mostly based on a strongly componential version of the computational approach to cognitive science (see Edelman, 2008; Gallistel and King, 2009) and so it is entangled with the idea of the brain as a collection of different components or modules, each of them devoted to performing a specific function (see Anderson, 2014: xix–xx). The most extended alternative to such a componential account of the structural and functional organization of the CNS is holism. In a holist architecture there are no brain areas that behave as function-specific modules. On the contrary, all brain areas participate in all brain functions as observed, for example, in classic connectionist networks (McClelland, and the PDP Research Group, 1986). Unlike both modular and holist accounts, neural reuse allows for some degree of specialization of brain regions—they all have their own functional biases or *neural personalities*—but they need not have fixed, modular functions; instead they participate in supporting a range of different behaviors and cognitive functions depending on their soft-assembled interconnections with other regions.

Predictions

Neural reuse promotes a departure from the classic understanding of the CNS both as componential structure of function-specific modules and as a general network of functionally equivalent components. Such a departure and, in particular, the plausibility of accepting a neural reuse-based architecture as a paradigm for the functional organization of the brain is based in part on the empirical support of three predictions that follow from the hypothesis.

The first—and most obvious—prediction is that the same local neural networks participate in different cognitive tasks. These networks play different functional roles in the larger soft-assembled groups of networks (TALoNS) in which they are used and reused. This prediction is supported by several studies. Some of these studies involve the meta-analysis of thousands of functional neuroimaging results to explore the functional diversity of different regions of the brain along with their connectivity patterns (Anderson, Kinnison, and Pessoa, 2013; see also Anderson and Penner-Wilger, 2013; Bolt et al., 2017). The underlying idea of these studies is that, if neural reuse is right, we should find activation of the same brain regions, classically understood as function-specific local neural networks or modules, across different task domains (e.g., action execution, attention, vision, emotion, reasoning, or memory, among others). A measurement of diversity borrowed from ecology was used to *quantify* the functional diversity of each region of the brain, and the results showed that the vast majority of brain regions are extremely diverse, averaging over 0.75 using a scale running from 0 (dedicated to a single task) to 1 (active across all measured tasks) (Anderson, Kinnison, and Pessoa, 2013; Anderson, 2016). Interestingly, the fulfillment of this prediction also constitutes empirical evidence against the componential framework of cognitive neuroscience: if brain regions participate in different task domains, the idea of brain regions as modules (i.e., function-specific local neural networks) is not justified.

The second prediction is that the functional differences between cognitive tasks are reflected in the diverse patterns of interaction between many of the same local neural networks. In other words, if neural reuse is right, the CNS enables different cognitive tasks in virtue of the way different local neural networks interact with each other, and not in virtue of their intrinsic activity alone. The same local neural network may be playing two different roles in two different TALoNS, so the aim of neuroscience should be to understand how a given local network is integrated in larger structures and the ways it interacts with the other components of those structures. A way to examine the differential interaction between local neural networks across task domains is to investigate in which tasks different local neural networks are more likely to work together (see Ciric et al., 2017 for the related idea of "contextual connectivity").

Anderson et al. (2010; Anderson and Penner-Wilger, 2013) conducted a coactivation analysis on a large set of neuroimaging data investigating the brain-bases of several different cognitive functions. They looked for brain regions that were more likely than chance (as predicted by their general activation likelihood) to be active at the same time given different tasks. When the observed coactivation likelihood was significantly greater than chance, this is taken to reflect a functional interaction or connection between the brain regions. As in the case of the first prediction, this second prediction was corroborated (Anderson et al., 2010; Anderson and Penner-Wilger, 2013). Individual regions of the brain were active across multiple tasks but cooperated with different partners in each condition. These results appear to offer empirical evidence against a holistic framework of brain activity. For holism, brain regions are expected to activate to some degree no matter the cognitive task and, thus, the likelihood of finding differential coactivation is low.

The third prediction is that later developing cognitive functions (or behaviors, or abilities) should be enabled by larger groups of local neural networks more broadly distributed along the brain. The rationale behind this prediction is that the more recent functions find a larger set of neural networks to be potentially used and reused and, given that, there is no reason to assume that the functions will be highly localized. Anderson (2010, 2014, 2016) offers evidence to support this prediction. More recent cognitive functions, like language, are enabled by larger groups of local neural networks than older cognitive functions, like visual perception. Crucially, these groups of neural networks (TALoNS) are not only larger in the case of recent cognitive functions, but also more broadly scattered over the brain.

The empirical support for these three predictions along with other studies at different neural scales—e.g., studies on the ubiquity of neuromodulation at different scales (Bargmann, 2012; Hermans et al., 2011), or studies on mixed selectivity in activation of single neurons (Cisek, 2007; Rigotti et al., 2013)—point to the plausibility of neural reuse as the organizational principle for the functions of the central nervous system. It also depicts the emergence of new cognitive functions through evolution and development in terms of the soft-assembling of existing local neural networks such that the different interactions between the networks underwrite the functional diversity and flexibility of the CNS. This is the sense in which the neural reuse hypothesis can provide new insights for DCN and even can become a general framework for the field.

Neural Reuse and Other Frameworks

As a field located in the intersection between developmental psychology and developmental neuroscience, DCN aims to account for the emergence of the functions and the structure of the adult human brain in terms of ontogenesis. On the one hand, DCN aims to understand how different psychological functions (e.g., visual categorization, language, reasoning) emerge in the process of development. On the other hand, DCN addresses the way these psychological functions are implemented in the CNS and its activity. To achieve such a twofold aim, DCN researchers have normally worked under three different frameworks, each of them based on a different understanding of cognitive development and leading to different ways of analyzing the empirical data in the field. These three frameworks are the *maturational viewpoint, interactive specialization*, and the *skill learning* framework (Johnson, 2013). In this section, we succinctly review these frameworks and analyze the position of the neural reuse hypothesis among them.

According to the maturational viewpoint, the emergence of new cognitive functions through growth is related to the maturation of particular brain areas. The process of development of the CNS and the different psychological functions it enables is based on the neuroanatomical maturation of specific brain regions. In other words, new functions such as language or reasoning are based on local neural networks that get more complex and more mature through development. In this sense, the maturational view is characterized by three interrelated theses.

The first thesis is that the CNS is a modular/componential system. Cortical areas are independent function-specific regions that, moreover, have their own developmental schedule and timing. This assumption is needed for two reasons. On the one hand, for development to be a product of maturation of cortical areas, those areas must be in charge of different psychological functions. On the other hand, if these cortical areas have their own developmental schedule and timing, the fact that different cognitive functions emerge at different moments of the developmental process is easy to accommodate.

The second thesis is that there is a fixed mapping between brain regions and their functions during development. This assumption is related to the previous one: if cortical areas are function-specific, there is no way to observe plasticity beyond local neural networks. Also, the idea of a fixed mapping between structure and function in the CNS is expected from a maturational viewpoint insofar as development is due to the maturation of local neural networks and not to the diversification of their functionality.

The last thesis has to do with the restricted nature of neural plasticity and with some biological determination of the development of cortical areas. According to the proponents of the maturational viewpoint, neural plasticity is usually confined to local neural networks. Cases in which neural plasticity is observed across or between broader regions are generally due to extreme conditions caused by injuries or other neural problems. That plasticity would mainly be a local phenomenon fits well with both the idea of genetically determined maturation of these local networks, and with the idea that psychological functions map onto specific local neural networks (modules). Modules are easy to understand as genetically determined and, thus, on this view the structure and function of the CNS do not depend on epigenetic factors or on interactions at the neural and the agential levels.

Another theoretical framework of DCN is known as interactive specialization (Johnson, 2000, 2001, 2010). According to the proponents of interactive specialization, the functional development of the CNS consists of the organization and stabilization of patterns of interactions between neural regions. Unlike the maturational viewpoint, interactive specialization understands development in terms of the constitution of stable functional states composed by different local networks. In this sense, interactive specialization rests on three different theses.

The first thesis of interactive specialization is that development is coordinated by the interaction of densely interconnected local neural networks. In this sense, interactive specialization stands against the idea of the modularity or the componential character of the CNS. Cognitive functions develop through the establishment and refinement of interactive networks of different brain regions and not by the genetically determined maturation of function-specific local neural networks. Indeed, the properties of the activity of local brain regions are constituted by the patterns of connectivity they hold with different regions.

The second thesis of interactive specialization is that the CNS is a developmentally plastic system. Unlike the maturational viewpoint, which holds that local function is genetically determined, interactive specialization holds that the final functional specialization depends crucially on the experiences of the organism, and the way experience drives specific neural interactions. Such plasticity is crucial since interactive specialization depicts the brain of a newborn as an initially non-specialized, general purpose system with some anatomic, but few functional constraints on how patterns of interaction will shape local networks. Such a depiction, combined with the idea of specialization based on the refinement of local cortical networks via large-scale interaction, requires plasticity at different levels of the CNS to explain developmental changes. Without plasticity, there would be no mechanism for shaping the activation profile of local neural networks via their interactions with other networks, nor for the emergence of new combinations of cortical networks that support new cognitive functions.

As a consequence of the two previous theses, the third thesis of interactive specialization is that different behaviors and cognitive functions are supported by combinations of cortical networks that are constituted by local neural networks that have refined their activation patterns due to their interactions with each other and their context. This thesis suggests that when new cognitive functions or behaviors are acquired, there is a constitution of a new, or at least newly refined, combination of local neural networks. Moreover, any such new combination also affects the proper activity of its constituents. The emergence of new combinations of cortical networks refines the activation profile of local neural networks, such that they transition from non-specific activation—i.e., activation across a broad domain of contexts—to a state of specific activation—i.e., activation in concrete domains and contexts.

Finally, some researchers in the field of DCN defend skill learning as a theoretical framework. According to these researchers, the same brain regions that participate in the development of early cognitive functions (e.g., attention, motor control) also participate in the development of later, more complex functions (e.g., mathematical knowledge, expertise in playing football). Thus, the acquisition of new skills is always presented in terms of the same patterns through the history of the development of the CNS (Gauthier and Nelson, 2001; Poldrack, 2002).

As in the case of the maturational viewpoint and interactive specialization, the starting point of skill learning as a framework for DCN is a set of theses. The first thesis is that there are specific cortical areas that are specialized in the development of basic cognitive functions (mostly perceptual and motor) and have been since birth. These are the brain regions in charge of the early development of visual attention or posture control, for example, and become the seed for the activity of the CNS in further developmental processes.

The second thesis, based on the previous one, is that later development and learning of complex skills, both in terms of new functions and expertise in existent ones, is based on the skills acquisition that occurred in the early stages of development. In other words, the cortical activity that enables the acquisition of basic cognitive functions shapes the way more complex skills are acquired. Following this thesis, the third is that plasticity plays a restricted role in development. Plasticity is needed as far as the activity brain regions that enable early development shape the activity of their own activity and the activity of other regions in later stages of development. However, the CNS does not undergo dramatic changes in its structure to function mapping.

The three theoretical frameworks for DCN just reviewed offer a very broad and useful understanding of the neurodevelopmental principles that could allow for the acquisition of adult-level skills and neural architecture. However, each suffers from some shortcomings. For example, the maturational viewpoint is unable to accommodate empirical research that suggests the participation and change of more than one local neural network in the development of individual cognitive functions (Luna et al., 2001; Supekar et al., 2009), nor research supporting the ways in which specific experience can dramatically change local network architecture and function (Von Melcher, Pallas, and Sur, 2000). In the case of interactive specialization, the problem is the opposite: the fact that some brain regions (e.g., auditory cortex) seem to be highly specialized from birth goes against the assumption that large-scale neural interactions are always necessary for local specialization. Finally, skill learning is based in a parallel activity of brain regions in the early and the later stages of development. However, such a parallelism does not seem to be enough to account for the stronger claim of a common mechanism (or a common shaping) for the acquisition of skills at different stages of the developmental process.

The neural reuse hypothesis can provide a way to maintain the most important insights of the previous theoretical frameworks for DCN while bringing them together under a common developmental umbrella. First, the neural reuse hypothesis can accommodate the thesis that local neural networks are, to some extent, function-specific without fully embracing a modular/componential understanding of the CNS. Neural reuse allows for functional biases in local neural networks:[4] local neural networks are more likely to activate in some cognitive tasks than in other ones. For example, the visual cortex tends to activate during visual events. However, this does not mean that the visual cortex *only* activates during visual events or that it is the only cortical region that activates during these events. Neural reuse also allows for a role to be played by genetic factors (albeit not deterministic ones) in specifying functional biases (Anderson and Finlay, 2014). In this sense, neural reuse keeps the best insight of the maturational viewpoint and is also able to accommodate the empirical data regarding the diversity of functions supported by most brain regions.

Second, the neural reuse hypothesis can accommodate the thesis that cognitive functions are supported by combinations of cortical areas that may play different roles within each combination, depending on their interactions with each other. In this sense, neural reuse is compatible with the best insight of interactive specialization while maintaining the idea that there can be strong functional biases, accounting for the fact that some areas, like the auditory cortex, seem to have a clear function-specific role throughout development. On the other hand, neural reuse adds a thesis that is incompatible with one of the theses of interactive specialization. According to the latter, brain regions transit from non-specific activation in many contexts to concrete activation profiles in specific contexts. Neural reuse, however, suggests that brain regions, by virtue of being used and reused, remain active in a wider range of contexts. Indeed, because local function can change depending on the network of which it is transiently a part, many regions never acquire a single specialization at all.

Third, the neural reuse hypothesis can accommodate the thesis that brain regions at early stages of development shape the activity of the brain regions implicated in later stages of development, incorporating the central insight of the skill learning framework. Moreover, neural reuse offers the mechanism by which such shaping occurs. Since brain regions that are active during the acquisition of basic cognitive functions during early stages of development are reused for the acquisition of more complex functions in later stages of development, it is not surprising that the activity of the regions at a previous stage shapes, to some extent, their activity at later stages.

These are just three examples of the benefits of taking the neural reuse hypothesis as a possible candidate to be the theoretical framework for DCN. Neural reuse seems to be able to accommodate most of the central theses of other frameworks without suffering from their shortcomings. In the following, we analyze concrete instances of the neural reuse hypothesis applied to developmental events and issues.

NEURAL REUSE AND MATHEMATICS

We provide two instances of how the application of the neural reuse hypothesis has provided insights into the nature of, and processes involved in, the development of mathematics: examining the existence of a number modules, and the relation between finger representations and number representations.

Number Module

It has been proposed that the horizontal segment of the intraparietal sulcus (hIPS) houses a domain-specific number module (Dehaene et al., 2003; Hubbard et al., 2005). Penner-Wilger and Anderson (2014; in preparation) examined this proposal by performing a meta-analysis of neuroimaging data (N = 2071 experiments) using techniques driven by the framework of neural reuse, including cross-domain modeling (Penner-Wilger and Anderson, 2011), and functional diversity analysis (Anderson, Kinnison, and Pessoa, 2013). To thoroughly evaluate this claim of domain specificity, it is important to determine both the *activation consistency* of the hIPS—the probability of hIPS activation given that a numerical task is being performed—and the *functional specificity* of the hIPS—the probability that a numerical task is being performed given hIPS activation. Their findings showed that, although the hIPS clearly participate in numerical processing, hIPS activation is not ubiquitous for numerical tasks; only 17.6 percent of numerical tasks showed activation in the left hIPS and 20.5 percent in the right hIPS. Moreover, only 15 percent of left hIPS and 18 percent of right hIPS activations were for mathematics tasks (16 percent and 20 percent when numerical tasks were more broadly construed). As shown in Figure 2.1, hIPS activation is observed across a wide range of non-numerical tasks in diverse domains, including perception, action, cognition, and emotion. When comparing the functional diversity of the hIPS to domain-general regions that contribute to number processing—the left angular gyrus (AG) and the bilateral posterior superior parietal lobule (PSPL)—it was shown that the right and left hIPS are significantly more diverse than the whole-brain average, and significantly more diverse than the domain general left AG and bilateral PSPL. Thus, employing the framework and associated techniques of neural reuse provided results that address a debate in numerical cognition (Ansari, 2008) and show that the hIPS does not house a domain-specific number module. This knowledge, and the application of a neural reuse framework, can guide the understanding of typical mathematical development as well as developmental dyscalculia.

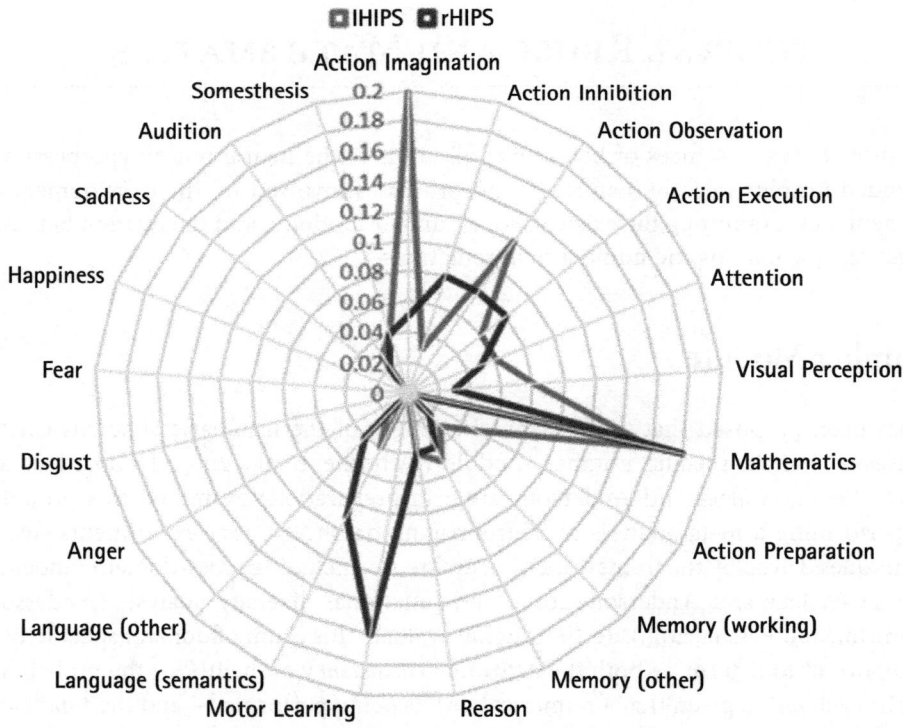

FIGURE 2.1 Radial axis represents the proportion of observed activations within the lhIPS (light gray) and rhIPS (dark gray) that fall into each domain and each vertex corresponds to one of the domains investigated.

Fingers and Number

Finger gnosis, the ability to mentally represent one's fingers, without visual feedback, is related to math performance (Fayol, Barrouillet, and Marinthe, 1998; Noël, 2005; Penner-Wilger et al., 2007, 2009, 2014, 2015). Finger gnosis is commonly assessed using a finger localization task (Baron, 2004; Noël, 2005), wherein the participant's hand is occluded from their view while a finger, or two fingers, are touched. The participant is then asked to indicate the touched finger(s). Performance is measured in terms of number of fingers correctly identified. Finger gnosis ability predicts performance on a variety of math measures in children, both concurrently and longitudinally (β's range from 0.22 to 0.36; Fayol et al., 1998; Noël, 2005; Penner-Wilger et al., 2007, 2009). Finger gnosis ability also predicts performance on a variety of math measures in adults (β's range from 0.21 to 0.30; Penner-Wilger et al., 2014, 2015).

Neural reuse provides a productive framework for understanding the relation between finger gnosis and number representations. On the *redeployment view* (Penner-Wilger and Anderson, 2008, 2013), the behavioral link between finger and number representations is at least partially explained by neural reuse (specifically redeployment, that is the reuse of the same local region performing the same underlying operation or "working"). Thus, one (or more) local brain regions, over evolutionary and/or developmental time, have come to perform the same operation in support of both uses. Zago et al. (2001) pinpointed a region of overlap between finger and number representation in the left precentral gyrus. In support of the redeployment view, regions associated with finger gnosis are activated during tasks requiring the representation of number (Andres, Michaux, and Pesenti, 2012; Dehaene et al., 1996; Zago et al., 2001), rTMS and direct cortical stimulation disrupt both finger gnosis and tasks requiring the representation of number (Rusconi, Walsh, and Butterworth, 2005; Roux et al., 2003), and there is interference between tasks involving finger gnosis and tasks requiring the representation of number (Brozzoli et al., 2008).

Penner-Wilger and Anderson (2011) conducted a meta-analysis of imaging data (sixty-five studies showing post-subtraction activation within the region of interest in the left precentral gyrus) to determine the full complement of tasks, across domains, that this region of interest (ROI) was implicated in (i.e., cross-domain modeling). The goal of this work was to identify common requirements across tasks/uses to guide structure-function mapping. In addition to number and finger representation tasks, the ROI was implicated in generation, inhibition, and order tasks. Common requirements across these uses were identified, including ordered storage and mapping, and a candidate working that could implement both these requirements was proposed—*an array of pointers*. Next, a computational model was built and tested that performs both finger gnosis and number comparison tasks (i.e., determining which of two presented numbers is larger) using the proposed shared working implemented in artificial spiking neurons (Stewart, Penner-Wilger, Waring, and Anderson, 2016). The performance of the model was compared to human performance data (accuracy and response times) and showed a good fit on both tasks. Given that the model could successfully perform both tasks using the same operation, and that the model performance mirrored that of human participants, it lends support for the view that the observed neural overlap between finger and number representation reflects neural reuse, specifically redeployment (same ROI, same working). This work offers a concrete instance of the reuse of a basic operation, specifically one grounded in sensorimotor systems, in a high-level, abstract cognitive task. Further behavioral evidence that mathematics is grounded in sensorimotor systems is found in the research of Landy and colleagues, which shows that the perception and solution of arithmetical and algebraic equations demonstrate perception-action phenomena, and suggests people interact with equations using strategies similar to those employed in the manipulation of physical objects (Landy and Goldstone, 2007; Landy and Goldstone, 2009; Landy and Goldstone, 2010; Landy and Linkenauger, 2010; for further discussion see Anderson, 2014).

Neural Reuse and Brain Dynamics

Recent developments in the analysis of intrinsic functional brain networks suggest that an additional layer of complexity exists when the time domain is considered in functional neuroimaging investigations. While earlier studies examined averaged correlation values across the duration of fMRI data collection to create an overall index of functional connectivity between brain regions, it is now widely acknowledged that functional connectivity metrics can vary as a function of time; a phenomenon first described by Chang and colleagues and now referred to as "dynamic functional connectivity" (Chang and Glover, 2010). One approach to quantifying brain dynamics is the "sliding-window" approach. Here, functional connectivity is computed on the order of seconds rather than the more conventional approach averaging across minutes. This approach allows for the quantification of "dwell time" (the amount of time the brain spends in a given functional network connectivity state) and "frequency of occurrence" (the number of times a given functional network connectivity state occurs) (Allen et al., 2014). These metrics can be used, in principle, to track developmental differences in brain dynamics and the emergence of adult-like patterns of functional connectivity. Another approach for quantifying dynamic functional connectivity involves the identification of critical time points when the blood oxygenation level-dependent (BOLD) signal intensity surpasses a certain threshold, giving rise to multiple stable spatial patterns (co-activation patterns) that can be obtained by clustering of the critical time frames (Chen, Chang, Greicius, and Glover, 2015). Other approaches for assessing dynamics in fMRI time series are described in Preti, Bolton, and Van De Ville (2017).

As it is increasingly recognized that *dynamic* interactions between brain regions constitute the basis of cognition, analysis of brain dynamics is taking center stage (Uddin, 2014).[5] Initial analyses of brain dynamics across development revealed that increasing age is associated with greater variability of functional connection strengths across time

(Hutchison and Morton, 2015). Analysis of time-varying changes in functional coupling between brain regions promises to be a fruitful avenue for future work in DCN (Uddin and Karlsgodt, 2018).

The rapidly growing and increasingly fruitful field of fMRI dynamic connectivity represents a concrete empirical validation of a central thesis of the neural reuse framework: a local neural network is not to be understood in terms of its role in a 'static' network, but rather in terms of its time-varying participation in soft-assembled groups of networks (TALoNS; Anderson, 2014).

Neural Reuse and Neurodevelopmental Disorders

That the focus in understanding the neural bases of cognitive development should be on dynamic connectivity is increasingly borne out by the empirical literature on neurotypical individuals; but it is equally becoming clear that this focus is crucial to our understanding of deviations from normal development (Hadders-Algra, 2018; Uddin and Karlsgodt, 2018). Neural development is marked not just by the sculpting of local circuit architectures (Johnson, 2001, 2011), especially important in the first 3 months post-term (Hadders-Algra, 2018), but also by significant changes in overall patterns of both physical and functional connectivity, including the creation of new functional partnerships as new skills and cognitive capacities emerge in development (Anderson and Finlay, 2014). Especially important is the trend, starting around 3 months post-term and accelerating after around 1 year, of the increasing dominance of long-range and global connectivity over local connectivity. These changes are crucial to normal development, and their disruption or alteration has been consistently associated with neurodevelopmental disorders including cerebral palsy, autism spectrum disorder (ASD), attention deficit hyperactivity disorder (ADHD), and schizophrenia (Hadders-Algra 2018).

In addition to the—perhaps at this juncture in the history of the field, obvious—point that dynamic connectivity should be a crucial focus of future research, neural reuse has other implications for the study of neurodevelopmental disorders. One important such implication is the need to represent the behavioral profiles of brain regions and networks in a nuanced, multi-dimensional way that better captures the complexity of the underlying causal dispositions of these structures. Task-bound functional attributions such as "face processing," "visual word form recognition," and the like, reflect at best one small facet of the functional contribution of a typical brain region; functional fingerprints, such as those offered in the third section, above, represent a better option.

This recommendation holds not just for capturing the behavioral repertoire of structures in the neurotypical adult brain, but also for the brain's functional development. Capturing neurodevelopment as a trajectory through a high-dimensional behavioral space can lead to more sensitive diagnostic practices, if it becomes possible to quantify with precision deviations from the expected developmental trajectory.

The realization that individual regions of the brain in fact participate in multiple functional coalitions may also lead us to explore clinical interventions that move beyond directly targeting lost skills, and toward a mixture of direct and indirect exercise of damaged neural tissues. Knowing the healthy functional fingerprint of a given neural region may suggest multiple possible paths to rehabilitation following a brain injury.

The importance of neural reuse in development also suggests some hypotheses regarding the differential effects of different sorts of deficits or injury. For instance, a neonatal stroke that primarily affects local grey matter will make that region (and its attendant functional bias) unavailable for inclusion in emerging neural partnerships (leading to what may initially present as a focal deficit). In contrast, a connectivity disorder like periventricular leukomalacia may instead slow the process of building the neural coalitions themselves. In the first case, it may be possible for neighboring regions to acquire a similar functional bias—as when the somatosensory cortex reorganizes after an amputation—and serve instead of the damaged region, leading to near or full recovery. In contrast, connectivity disorders would appear to be relatively more likely to cause more widespread behavioral deficits and developmental delays.

Age is also an important factor. As brain regions become incorporated into different neural coalitions they become "functionally burdened" in the sense that any changes to local structure will impact multiple functional partnerships and therefore be difficult to implement (especially insofar as existing partnerships reinforce existing connectivity). The existence of critical periods, along with the slowing of learning rates and increased recovery times observed across the lifespan, may have as much to do with the difficulties of overcoming such functional burden as with any changes in the underlying biological mechanisms of neuroplasticity. More generally, neural reuse invites increased attention to order effects in both normal and abnormal development.

Conclusion

In this chapter, we have defended neural reuse as a unifying framework for DCN. It has been our contention that neural reuse provides all the benefits of the most prominent theories of DNC (maturational viewpoint, interactive specialization, and skill learning) while avoiding their limitations. To defend our proposal, first, we have described the neural reuse hypothesis and reviewed three main predictions regarding the functional organization of the CNS that follow from it and several studies that fulfill them.

Second, we have reviewed the most prominent theories of DCN, and explained why neural reuse may serve as a unifying framework for the field, able to accommodate the advantages of the maturational viewpoint, interactive specialization, and skill learning while avoiding some of their limitations. In brief, neural reuse allows for functional biases in local neural networks and also embraces the idea that cognitive functions are supported by combinations of cortical areas playing different functional roles. In this way, neural reuse accommodates the maturational viewpoint and interactive specialization. Moreover, neural reuse is a way to understand how the same neural regions can participate in the acquisition of different skills at different developmental stages, as suggested by the skill learning theory.

Finally, we have succinctly examined three instances of the application of the neural reuse hypothesis to the study of three different developmental issues: numerical and mathematical learning, the role of dynamics in development, and neurodevelopmental disorders. We take these three examples to show that neural reuse is not just a high-level theoretical proposal regarding DCN, but that it also provides concrete ways to approach development issues and to generate and test hypotheses in the field.

In fact, research into neural reuse as a developmental process remains in its infancy, and there is still very much to be learned. An obvious implication of neural reuse for which there is suggestive but not conclusive evidence is that learning should generally involve not just the shaping of a local network (as Dehane's neural recycling hypothesis might be taken to imply; Dehaene, 2009; Dehaene and Cohen, 2007) but also changes in functional connectivity patterns, i.e., the patterns of cooperation between networks. For instance, it has been observed that learning to control a prosthetic limb via a brain-machine interface induces not just changes in local neural tuning, but also large-scale changes in the patterns of functional/activation correlation *between* cells (Lebedev and Nicolelis, 2006). Similarly, Bassett et al. (2011) found that "network flexibility," operationalized as the number of times a region changed its functional partnerships during learning, predicted the degree of learning gain. Much more work investigating such phenomena is needed.

Moreover, we currently know very little about the range of typical and atypical developmental pathways leading to adult functional configurations; as measured both by the changing functional fingerprints of individual regions and the changing patterns of functional cooperation between regions. Movements like "population neuroscience" (Paus, 2010) are poised to shed light on these questions, but *only* if the complex functional profiles of individual parts of the brain and their changes over time are actually captured; but this is rarely done. Things are more hopeful in the area of network neuroscience, as the field has now moved decisively to a thoroughly network-oriented view of brain function (e.g., van Essen et al., 2013).

Finally, just to offer one more area of future research (but, see Anderson, 2014 for a discussion of dozens of open questions), although we have uncovered some of the functional and architectural traces left by the operation of neural reuse in development, we know next to nothing about the underlying biological mechanisms. An important area of investigation will be how new neural partnerships are formed and stabilized, and how the several uses of individual brain regions are coordinated. Surely neuromodulation has an important role to play (e.g., Bargmann, 2012), but its role in large-scale network changes is underexplored. Not incidentally, this example highlights the ways in which the neural reuse framework can bring focus to issues across spatial scales, thus requiring the concerted efforts of scientists from specialties—from neurogenetics to large-scale connectomics—that still too rarely interact. Understanding human development in general, and neurodevelopment in particular, will of course require the efforts of many, and a good deal of time.

Notes

1. Concisely, the neural exploitation and shared circuits model builds on the basic notion that conceptual content is grounded in sensorimotor experience, and language understanding and other higher cognitive functions therefore utilize some of the neural circuits underpinning basic sensorimotor functions. In contrast, the neuronal recycling and massive redeployment hypotheses suggest that reuse of existing brain circuitry is a universal principle of the brain's functional architecture, with neuronal recycling emphasizing the importance of local plasticity in adapting evolved regional function to novel cultural achievements like reading and mathematics, and massive redeployment based on the notion that local function can often be retained; novel capacities come largely via putting the same building blocks together in new combinations.
2. Note that as stated, this view ignores the *action shaping* or *action creation* role of the CNS— the very coordination of bodily resources that makes coherent action possible—in favor of a sensorimotor control view (see Keijzer et al., 2013; Keijzer, 2015). A more complete framework would include this crucial role.
3. In this sense, the theory of reuse should be thought of as thoroughly evo-devo in spirit.
4. As well as accepting that there is a *range* of functional diversity in the brain; some regions are highly specialized, others barely at all (Anderson, Kinnison, and Pessoa, 2013).
5. Note that there remain outstanding questions with regard to the origin of these dynamics (e.g., Laumann et al., 2017).

References

Allen, E. A., Damaraju, E., Plis, S. M., Erhardt, E. B., Eichele, T., and Calhoun, V. D. (2014). Tracking whole-brain connectivity dynamics in the resting state. *Cerebral Cortex*, 24(3), 663–676.

Anderson, M. L. (2007). Massive redeployment, exaptation, and the functional integration of cognitive operations. *Synthese*, 159(3), 329–345.

Anderson, M. L. (2010). Neural reuse: A fundamental organizational principle of the brain. *Behavioral and Brain Sciences*, 33(4), 245–266.

Anderson, M. L. (2014). *After Phrenology: Neural Reuse and the Interactive Brain*. Cambridge, MA: MIT Press.

Anderson, M. L. (2016). Précis of after phrenology: Neural reuse and the interactive brain. *Behavioral and Brain Sciences* 39 , e120.

Anderson, M. L., and Finlay, B. (2014). Allocating structure to function: the strong links between neuroplasticity and natural selection. *Frontiers in Human Neuroscience*, 7: 918.

Anderson, M. L., and Penner-Wilger, M. (2013). Neural reuse in the evolution and development of the brain: Evidence for developmental homology? *Developmental Psychobiology*, 55(1), 42–51.

Anderson, M. L., Brumbaugh, J., and Suben, A. (2010). Investigating functional cooperation in the human brain using simple graph-theoretic methods. In W. Chaovalitwongse, P. Pardalos, and P. Xanthopoulos (eds.), *Computational Neuroscience* (pp. 31–45). New York: Springer.

Anderson, M. L., Kinnison, J. and Pessoa, L. (2013). Describing functional diversity of brain regions and brain networks. *NeuroImage 7,* 50–58.

Andres, M., Michaux, N., and Pesenti, M. (2012). Common substrate for mental arithmetic and finger representation in the parietal cortex. *NeuroImage, 62*(3), 1520–1528.

Ansari, D. (2008). Effects of development and enculturation on number representation in the brain. *Nature Reviews Neuroscience, 9,* 278–291.

Bargmann, C. I. (2012). Beyond the Connectome: How neuromodulators shape neural circuits. *Bioessays 34*(6), 458–465.

Baron, I. S. (2004). *Neuropsychological Evaluation of the Child.* New York, NY: Oxford University Press.

Bassett, D. S., Wymbs, N. F., Porter, M. A., Mucha, P. J., Carlson, J. M., and Grafton, S. T. (2011). Dynamic reconfiguration of human brain networks during learning. *Proceedings of the National Academy of Sciences, 108*(18), 7641–7646.

Bolt, T., Nomi, J. S., Yeo, B. T., and Uddin, L. Q. (2017). Data-Driven Extraction of a Nested Model of Human Brain Function. *Journal of Neuroscience, 37*(30), 7263–7277.

Brozzoli, C., Ishihara, M., Göbel, S. M., Salemme, R., Rossetti, Y., and Farnè, A. (2008). Touch perception reveals the dominance of spatial over digital representation of numbers. *Proc. Natl. Acad. Sci. U.S.A., 105,* 5644–5648.

Chang, C., and Glover, G. H. (2010). Time–frequency dynamics of resting-state brain connectivity measured with fMRI. *Neuroimage, 50*(1), 81–98.

Chen, J. E., Chang, C., Greicius, M. D., and Glover, G. H. (2015). Introducing co-activation pattern metrics to quantify spontaneous brain network dynamics. *Neuroimage, 111,* 476–88.

Ciric, R., Nomi, J. S., Uddin, L. Q., and Satpute, A. B. (2017). Contextual connectivity: A framework for understanding the intrinsic dynamic architecture of large-scale functional brain networks. *Scientific Reports, 7*(1), 6537.

Cisek, P. (2007). Cortical mechanisms of action selection: The affordance competition hypothesis. *Philosophical Transactions of the Royal Society B: Biological Sciences, 362*(1485), 1585–1599.

Dehaene, S. (2005). Evolution of human cortical circuits for reading and arithmetic: The "neuronal recycling" hypothesis. In S. Dehaene, J. R. Duhamel, M. D. Hauser, and G. Rizzolatti (eds.), *From Monkey Brain to Human Brain: A Fyssen Foundation Symposium* (pp. 133–158). Cambridge, MA: MIT Press.

Dehaene, S. (2009). *Reading in the Brain.* New York: Penguin Viking.

Dehaene, S., and Cohen, L. (2007). Cultural recycling of cortical maps. *Neuron, 56*(2), 384–398.

Dehaene, S., Piazza, M., Pinel, P., and Cohen, L. (2003). Three parietal circuits for number processing. *Cognitive Neuropsychology, 20,* 487–506.

Dehaene, S., Tzourio, N., Frak, V., Raynaud, L., Cohen, L., Mehler, J., and Mazoyer, B. (1996). Cerebral activations during number multiplication and comparison: A PET study. *Neuropsychologia, 34,* 1097–1106.

Edelman, S. (2008). *Computing the Mind: How the Mind Really Works.* Oxford, UK: Oxford University Press.

Fayol, M., Barrouillet, P., and Marinthe, C. (1998). Predicting arithmetical achievement from neuro-psychological performance. *Cognition, 68,* B63–B70.

Gallese, V., and Lakoff, G. (2005). The brain's concepts: The role of the sensory-motor system in conceptual knowledge. *Cognitive Neuropsychology*, 22(3–4), 455–479.

Gallistel, C. R., and King, A. P. (2009). *Memory and the Computational Brain: Why Cognitive Science will Transform Neuroscience*. Oxford, UK: Wiley-Blackwell.

Gauthier, I., and Nelson, C. (2001). The development of face expertise. *Current Opinion in Neurobiology*, 11, 219–224.

Hadders-Algra, M. (2018). Early human motor development: From variation to the ability to vary and adapt. *Neuroscience & Biobehavioral Reviews*, 90, 411–427.

Hermans, E. J., van Marle, H. J., Ossewaarde, L., Henckens, M. J., Qin, S., van Kesteren, M. T., Schoots, V. C., Cousijn, H., Rijpkema, M., Oostenveld, R., and Fernández, G. (2011). Stress-related noradrenergic activity prompts large-scale neural network reconfiguration. *Science*, 334(6059), 1151–1153.

Hubbard, E. M., Piazza, M., Pinel, P., and Dehaene, S. (2005). Interactions between number and space in parietal cortex. *Nature Reviews Neuroscience*, 6, 435–448.

Hurley, S. L. (2005). The shared circuits hypothesis: A unified functional architecture for control, imitation and simulation. In S. Hurley and N. Chater (eds.), *Perspectives on Imitation: From Neuroscience to Social Science* (pp. 76–95). Cambridge, MA: MIT Press.

Hurley, S. L. (2008). The shared circuits model (SCM): How control, mirroring, and simulation can enable imitation, deliberation, and mindreading. *Behavioral and Brain Sciences*, 31(1), 1–58.

Hutchison, R. M., and Morton, J. B. (2015). Tracking the brain's functional coupling dynamics over development. *Journal of Neuroscience*, 35(17), 6849–6859.

Johnson, M. H. (2000). Functional brain development in infants: Elements of an interactive specialization framework. *Child Development*, 71, 75–81.

Johnson, M. H. (2001). Functional brain development in humans. *Nature Reviews Neuroscience*, 2, 475–483.

Johnson, M. H. (2010). *Developmental Cognitive Neuroscience*, 3rd ed. Chichester, UK: Wiley-Blackwell.

Johnson, M. H. (2011). Interactive Specialization: A domain-general framework for human functional brain development? *Developmental Cognitive Neuroscience*, 1(1), 7–21.

Johnson, M. H. (2013). Theories in developmental cognitive neuroscience. In J. Rubenstein and P. Rakic (eds.), *Neural Circuit Development and Function in the Brain* (pp. 191–206). London, UK: Elsevier Academic Press.

Keijzer, F. (2015). Moving and sensing without input and output: Early nervous systems and the origins of the animal sensorimotor organization. *Biology and Philosophy*, 30(3), 311–331.

Keijzer, F., van Duijn, M., and Lyon, P. (2013). What nervous systems do: Early evolution, input–output, and the skin brain thesis. *Adaptive Behavior*, 21(2), 67–85.

Landy, D., and Goldstone, R. L. (2007). How abstract is symbolic thought? *Journal of Experimental Psychology. Learning, Memory, and Cognition*, 33(4), 720–733.

Landy, D., and Goldstone, R. L. (2009). How much of symbolic manipulation is just symbol pushing? In N. A. Taatgen and H. van Rijn (eds.), *Proceedings of the 31st Annual Conference of the Cognitive Science Society* (pp. 1072–7). Austin, TX: Cognitive Science Society.

Landy, D., and Goldstone, R. L. (2010). Proximity and proceedings in arithmetic. *Quarterly Journal of Experimental Psychology*, 63(10), 1953–1968.

Landy, D., and Linkenauger, A. (2010). Arithmetic notation … now in 3D! In A. Ohlsson and R. Catrambone (eds.), *Proceedings of the 32nd Annual Conference of the Cognitive Science Society* (pp. 2164–9). Austin, TX: Cognitive Science Society.

Laumann, T. O., Snyder, A. Z., Mitra, A., Gordon, E. M., Gratton, C., Adeyemo, B., ... and McCarthy, J. E. (2017). On the stability of BOLD fMRI correlations. *Cerebral Cortex*, *27*(10), 4719–4732.

Lebedev, M. A., and Nicolelis, M. A. (2006). Brain–machine interfaces: past, present and future. *TRENDS in Neurosciences*, *29*(9), 536–546.

Luna, B., Thulborn, K. R., Munoz, P. D., et al. (2001). Maturation of widely distributed brain function subserves cognitive development. *NeuroImage*, *13*(5), 786–793.

Noël, M. -P. (2005). Finger gnosia: A predictor of numerical abilities in children? *Child Neuropsychology*, *11*, 413–430.

Paus, T. (2010). Population neuroscience: Why and how. *Human Brain Mapping*, *31*(6), 891–903.

Penner-Wilger, M., and Anderson, M. L. (2008). An alternative view of the relation between finger gnosis and math ability. In B. C. Love, K. McRae, and V. M. Sloutsky (eds.), *Proceedings of the 30th Annual Conference of the Cognitive Science Society* (pp. 1647–1652). Austin, TX: Cognitive Science Society.

Penner-Wilger, M., and Anderson, M. L. (2011). The relation between finger gnosis and mathematical ability: Can we attribute function to cortical structure with cross-domain modeling? In L. Carlson, C. Hoelscher, and T.F. Shipley (eds.), *Proceedings of the 33rd Annual Conference of the Cognitive Science Society* (pp. 2445–2450). Austin, TX: Cognitive Science Society.

Penner-Wilger, M., and Anderson, M. L. (2013). The relation between finger gnosis and mathematical ability: Why redeployment of neural circuits best explains the finding. *Frontiers in Psychology*, *4*, 877.

Penner-Wilger, M., and Anderson, M. L. (2014). Functional diversity of the intraparietal sulcus: Evidence against a domain-specific number module. *Canadian Journal of Experimental Psychology*, *68*, 259.

Penner-Wilger, M., Fast, L., LeFevre, J., Smith-Chant, B. L., Skwarchuk, S., Kamawar, D., and Bisanz, J. (2007). The foundations of numeracy: Subitizing, finger gnosia, and fine-motor ability. In D. S. McNamara and J. G. Trafton (eds.), *Proceedings of the 29th Annual Cognitive Science Society* (pp. 1385–1390). Austin, TX: Cognitive Science Society.

Penner-Wilger, M., Fast, L., LeFevre, J., Smith-Chant, B. L., Skwarchuk, S., Kamawar, D., and Bisanz, J. (2009). Subitizing, finger gnosis, and the representation of number. In N. A. Taatgen and H. van Rijn (eds.), *Proceedings of the 31st Annual Conference of the Cognitive Science Society* (pp. 520–25). Austin, TX: Cognitive Science Society.

Penner-Wilger, M., Waring, R. J., and Newton, A. T. (2014). Subitizing and finger gnosis predict calculation fluency in adults. In M. Bello P., Guarini M., McShane M. & Scassellati B. (Eds.) *Proceedings of the 36th Annual Conference of the Cognitive Science Society* (pp. 1150–55). Austin TX: Cognitive Science Society.

Penner-Wilger, M., Waring, R. J., Newton, A. T., and White, C. (2015). Finger gnosis and symbolic number comparison as robust predictors of adult numeracy. [Abstract]. In D. C. Noelle, R. Dale, A. S. Warlaumont, J. Yoshimi, T. Matlock, C. D. Jennings, and P. P. Maglio (eds.), *Proceedings of the 37th Annual Conference of the Cognitive Science Society* (p. 2963). Austin TX: Cognitive Science Society.

Pessoa, L. (2012). Beyond brain regions: Network perspective of cognition-emotion interactions. *Behavioral and Brain Sciences*, *35*, 158–159.

Poldrack, R. A. (2002). Neural systems for perceptual skill learning. *Behavioral and Cognitive Neuroscience Reviews*, *1*(1), 76–83.

Preti, M. G., Bolton, T. A., and Van De Ville, D. (2017). The dynamic functional connectome: State-of-the-art and perspectives. *Neuroimage*, *160*, 41–54.

Rigotti, M., Barak, O., Warden, M. R., Wang, X. J., Daw, N. D., Miller, E. K., and Fusi, S. (2013). The importance of mixed selectivity in complex cognitive tasks. *Nature*, *497*(7451), 585–590.

Roux, F.-E., Boetto, S., Sacko, O., Chollet, F., and Tremoulet, M. (2003). Writing, calculating, and finger recognition in the region of the angular gyrus: A cortical study of Gerstmann syndrome. *Journal of Neurosurgery*, *99*, 716–27.

Rumelhart, D. E., McClelland, J. L., and the PDP Research Group (1986). *Parallel Distributed Processing* (Vol. 1 & 2). Cambridge, MA: MIT Press.

Rusconi, E., Walsh, V., and Butterworth, B. (2005). Dexterity with numbers: rTMS over left angular gyrus disrupts finger gnosis and number processing. *Neuropsychologia*, *43*(11), 1609–24.

Stewart, T. C., Penner-Wilger, M., Waring, R. J., and Anderson, M. L. (2017). A common neural component for finger gnosis and magnitude comparison. In G. Gunzelmann, A. Howes, T. Tenbrink, and E. J. Davelaar (eds.), *Proceedings of the 39th Annual Conference of the Cognitive Science Society* (pp. 1150–55). Austin, TX: Cognitive Science Society.

Supekar, K. S., Musen, M. A., and Menon, V. (2009). Development of large-scale functional brain networks in children. *PLoS Biology*, *7*(7), e1000157.

Uddin, L. Q. (2014). Dynamic connectivity and dynamic affiliation: Comment on "Understanding brain networks and brain organization" by Pessoa. *Physics of Life Reviews*, *11*(3), 460.

Uddin, L. Q., and Karlsgodt, K. H. (2018). Future directions for examination of brain networks in neurodevelopmental disorders. *Journal of Clinical Child & Adolescent Psychology*, *47*(3), 483–497.

Van Essen, D. C., Smith, S. M., Barch, D. M., Behrens, T. E., Yacoub, E., Ugurbil, K., and Wu-Minn HCP Consortium. (2013). The WU-Minn human connectome project: An overview. *Neuroimage*, *80*, 62–79.

Von Melchner, L., Pallas, S. L., and Sur, M. (2000). Visual behaviour mediated by retinal projections directed to the auditory pathway. *Nature*, *404*(6780), 871.

Zago, L., Pesenti, M., Mellet, E., Crivello, F., Mazoyer, B., and Tzourio-Mazoyer, N. (2001). Neural correlates of simple and complex mental calculation. *NeuroImage*, *13*, 314–327.

CHAPTER 3

ELECTROPHYSIOLOGY IN DEVELOPMENTAL POPULATIONS

Key methods and findings

RYAN BARRY-ANWAR, TRACY RIGGINS,
AND LISA S. SCOTT

INTRODUCTION

ELECTROPHYSIOLOGICAL recordings of human brain activity were first recorded by Hans Berger in 1929 (Berger, 1929). This technique was rapidly adapted for use in developmental populations, with the earliest studies examining infant brain activity during sleep (Smith, 1938) and during seizures (Harris and Tizard, 1960). Over the next few decades, electroencephalogram (EEG) methods were adapted to examine the development of visual and auditory sensory processes (e.g., Kurtzberg and Vaughan, 1979; Shelburne, 1973; Shulman-Galambos and Galambos, 1978). Since this time, developmental EEG studies have provided a wealth of information not only about sensory and perceptual systems but also attention, learning, cognitive, and social abilities.

The EEG method is ideally suited for examining development. First, EEG is relatively inexpensive compared to other neuroimaging techniques. Second, EEG does not require motor or verbal responses from participants. Third, EEG data can be compared across a variety of ages. Fourth, given the appropriate set-up, a large amount of data (100–200+ trials) can be acquired quickly (e.g., 10–15 minutes). Finally, EEG recordings are less sensitive to movement artifacts relative to other neuroimaging methods.

Although EEG methods continues to advance, today EEG is primarily used to study development using three different techniques: EEG, Event-Related Potentials (ERP), and steady-state Visual Evoked Potential (ssVEP). All of these techniques record electrophysiological activity from electrodes placed on the scalp while participants are in a quiet attentive state or in response to stimuli. The signal recorded is a measure of the electrical potential between two electrodes on the scalp. Specifically, responses from each scalp electrode are compared, during recording, to a reference electrode (often Cz, located at the top center of the head). These measurements are later re-referenced off-line in order to maximize their potential (see Data Processing, Analysis, and Publication Criteria for details).

The EEG signal is composed of multiple sine waves of different frequencies and can be examined in both the frequency and time domains (see Figure 3.1). EEG and ssVEP typically examine frequencies of interest (e.g., alpha, mu, theta, gamma, or a specific stimulated frequency; see EEG section below). ERPs, however, focus on the time domain (on the order of milliseconds). ERPs and ssVEPs are similar in that they are evoked and time-locked to presented stimuli. Whereas ssVEPs are recorded in response to visual stimuli, ERPs are recorded to visual, auditory, or tactile stimuli as well as in response to behavior (e.g., button press). ssVEPs are evoked when a visual stimulus is presented periodically, at a specific frequency (e.g., 6 Hz or one stimulus every 167 ms). The resulting ssVEP can be seen in the frequency domain as a peak in the spectra at the stimulating frequency (e.g., 6 Hz).

The EEG reflects postsynaptic activity of populations of synchronously active cortical pyramidal cells oriented in the same direction relative to the scalp. Figure 3.2 shows an excitatory neuronal response leading positive ions to flow into the postsynaptic cell, creating a negative extracellular charge near the dendrites of the postsynaptic cell and a positive charge near the cell body, creating a dipole. If a large population of pyramidal cells are aligned within the cortex and simultaneously active, this dipole sums and quickly propagates to the scalp where it can be recorded by electrodes. Measuring the EEG can provide millisecond temporal resolution and is especially useful for understanding the timing of neural processing. Measuring where the scalp signals are coming from is a more complicated problem because the number and orientation of neural sources is difficult to determine. However, advances in source localization techniques have led to a number of investigations using a combination of EEG measures, MRI images, and mathematical modeling to estimate the sources of EEG responses

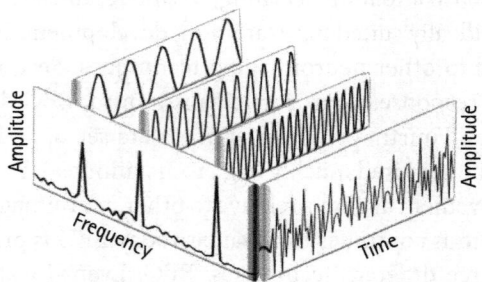

FIGURE 3.1 Depiction of the relation between the EEG time and frequency domains.

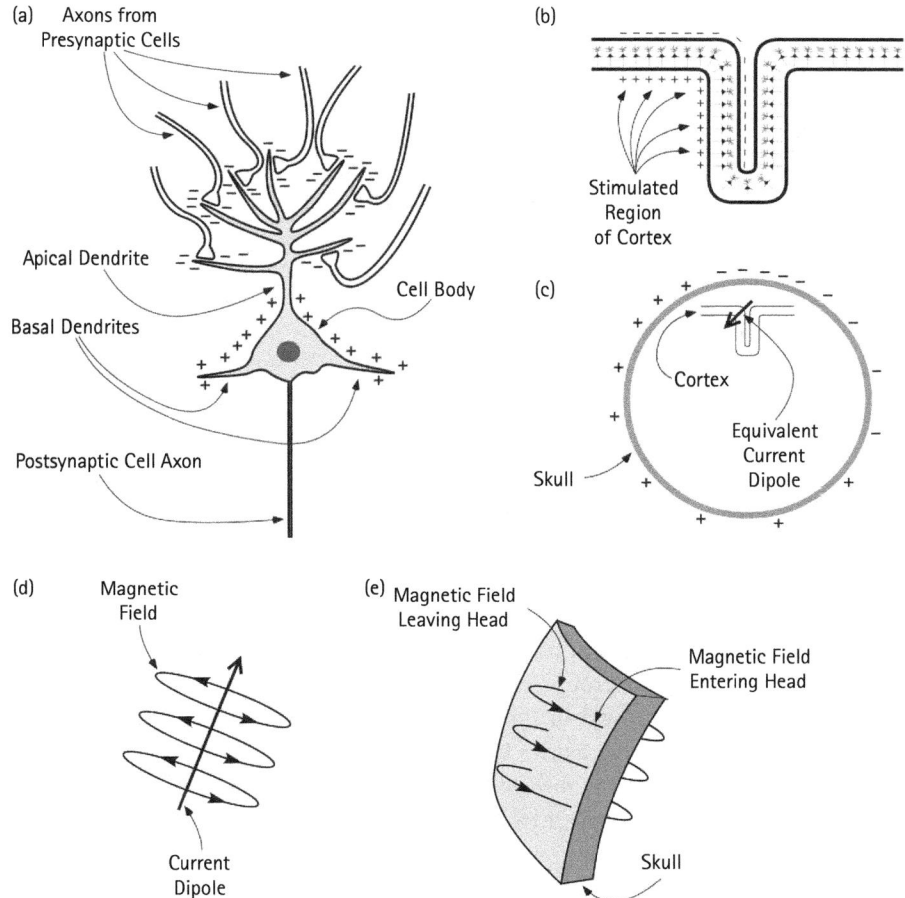

FIGURE 3.2 Principles of ERP generation. (a) Schematic pyramidal cell during neurotransmission. An excitatory neurotransmitter is released from the presynaptic terminals, causing positive ions to flow into the postsynaptic neuron. This creates a net negative extracellular voltage (represented by the - symbols) in the area of other parts of the neuron, yielding a small dipole. (b) Folded sheet of cortex containing many pyramidal cells. When a region of this sheet is stimulated, the dipoles from the individual neurons summate. (c) The summated dipoles from the individual neurons can be approximated by a single equivalent current dipole shown here as an arrow. The position and orientation of this dipole determine the distribution of positive and negative voltages recorded at the surface of the head. (d) An example of a current dipole with a magnetic field traveling around it. (e) An example of the magnetic field generated by a dipole that lies just inside the surface of the skull. If the dipole is roughly parallel to the surface, the magnetic field can be recorded as it leaves and enters the head; no field can be recorded if the dipole is oriented radially.

(e.g., Hämäläinen et al., 2011; Lenartowicz et al., 2016; Ortiz-Mantilla et al., 2012; Reynolds and Richards, 2005; Reynolds et al., 2010; Xie et al., 2018).

METHODOLOGICAL OVERVIEW AND KEY FINDINGS

EEG

To quantify the ongoing EEG signal, segments of interest are often decomposed into frequency bands using spectral analyses, such as Fourier transformation. Spectral resolution depends on the length of the segments analyzed, with longer segments resulting in higher resolution. For example, a 40 second trial provides a resolution of 0.025 Hz (1/40s), whereas a 2 second trial provides a resolution of 0.5 Hz (1/2s). Fourier analysis estimates spectral power at each frequency but is often averaged across specific frequency bands (e.g., 6–9 Hz) for analysis. Power and coherence are two measures derived from an EEG that are often used to index brain development and functional connectivity. Power is the level of mean squared microvolts; an increase in power is interpreted as an increase in activity of groups of synchronously active neurons. Coherence is a measure of phase synchrony within a frequency band between two electrodes or groups of electrodes and is calculated as the frequency-dependent squared cross correlation between these two locations (Nunez, 1981). Coherence is often interpreted as an index of functional connectivity between two neuronal populations (for a review: Thatcher, 2012). In addition to using previous research to guide analysis of frequency bands, there are two common methods for determining appropriate frequency bands in early development (Pivik et al., 1993). In the wide-band method, all frequencies with evidence of power are analyzed. In the small-band method frequency bands are defined by peaks seen in individual spectra (for review: Bell and Cuevas, 2012).

When recording EEG, neural responses during an active state (i.e., during stimulus presentation) are compared to a quiet attentive or baseline state. In developmental research, baselines often range between 1–2 seconds and 1–2 minutes. To record a quiet attentive or baseline period, researchers often show events such as an experimenter blowing bubbles, colorful balls moving inside a bingo wheel, or an animated fixation cross/object. These stimuli are thought to minimally engage attention but, depending on the question of interest and task design, researchers may consider using less engaging stimuli. The activity recorded during baseline is then either analyzed on its own to examine resting state differences in development or is compared to activity recorded during a task. This comparison to a task allows researchers to quantify event-related synchronization or desynchronization. Synchronization occurs when the power is greater during the task than at baseline, whereas desynchronization occurs when the power is greater at baseline than during the task. Regional patterns of synchronization and desynchronization may arise from activation in one area of the cortex that is paired with inhibition of the surrounding areas (Neuper and Pfurtscheller, 2001; Orekhova et al., 2001). In addition to power synchronization, oscillations can also be synchronized in phase (see Yordanova and Kolev, 2009).

EEG has been used to examine the development of attention (Xie et al., 2018), working memory (Bell, 2012; Wolfe and Bell, 2004), memory(Cuevas et al., 2012), attention to social cues and stimuli (Hoehl et al., 2014; Michel et al., 2015), object processing (Gliga et al., 2010; Kaufman et al., 2003; 2005), motor development (Marshall et al., 2011; Southgate et al., 2009), and brain development generally (Bell and Fox, 1992; Cuevas and Bell, 2011). The most studied EEG oscillations in developmental populations include alpha (Marshall et al., 2002; Orekhova et al., 1999; Stroganova et al., 1998), mu (Marshall et al., 2011; Southgate et al., 2009), gamma (Gliga et al., 2010; Grossmann et al., 2007), and theta (Bosseler et al., 2013; Michel et al., 2015; Orekhova et al., 1999). For a detailed review of EEG bands see Saby and Marshall (2012). Oscillations in high frequency bands are thought to reflect the activity of small neuronal networks (or short distance connections), whereas slower oscillations are thought to reflect the activity of much larger networks of neurons (or long-distance connections; Buzsáki and Draguhn, 2004).

Posterior Alpha

In adults, alpha includes frequencies between 8–12 Hz and is thought to reflect general alerting and ongoing visual information processing (Adrian and Matthews, 1934; Klimesch et al., 1998) that may play a role in selective attention (Foxe et al., 1998; Foxe and Snyder, 2011). Posterior alpha is typically recorded over occipital regions. Traditionally, suppression was thought to be the typical alpha response because opening the eyes suppressed alpha-band activity (Pfurtscheller et al., 1996).

Infant alpha includes frequencies between 6–9 Hz, recorded over posterior brain regions. Like the adult alpha, the infant alpha is reduced in amplitude during visual attention and is maximal at occipital sites during darkness (Stroganova et al., 1999). Multiple domains (e.g., working memory, inhibitory control, visual attention, joint attention, and emotion processing/regulation) have been examined using infant alpha (Cuevas and Bell, 2011; Hoehl et al., 2014; Michel et al., 2015; Orekhova et al., 1999, 2001; Stroganova et al., 1998; Wolfe and Bell, 2007).

In both adults (Ward, 2003) and infants (Michel et al., 2015; Hoehl et al., 2014; Xie et al., 2018), alpha desynchronization has been associated with increased demands of attention and task load. For example, posterior occipital alpha desynchronization occurs when 4- and 9-month-old infants viewed novel objects during joint attention tasks (Hoehl et al., 2014; Michel et al., 2015) and during sustained attention in 10- and 12-month-old infants (Xie et al., 2018). These findings suggest that alpha desynchronization over occipital regions is a good index of task-related attention during infancy.

Central Alpha/mu Rhythm

Infant mu rhythm/central alpha oscillations includes frequencies between 6–9 Hz recorded over sensorimotor cortex (central regions). *Central alpha/mu* has been related

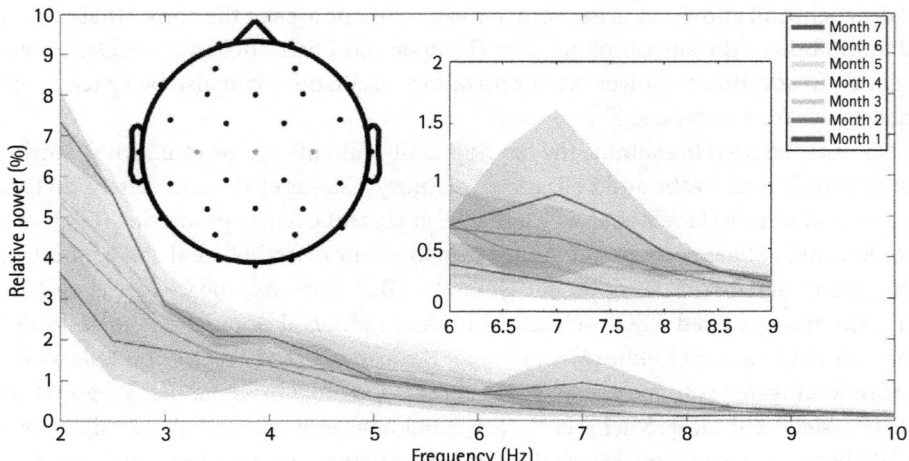

FIGURE 3.3 Monthly spectral profiles at motor cortices. Each color-coded curve presents averaged relative power density from sessions of the same monthly age. Shaded areas show standard deviations for each monthly age with corresponding colors. Left inset shows geodesic locations of each electrode on the scalp, and green ones are selected to evaluate spectral profiles in motor cortices. Right inset provides a zoom-in view of spectral profiles in 6–9 Hz.

Source: Reprinted from Xiao R., Shida-Tokeshi J., Vanderbilt D. L., and Smith B. A. (2018). Electroencephalography power and coherence changes with age and motor skill development across the first half year of life, *PLoS ONE* 13(1): e0190276. https://doi.org/10.1371/journal.pone.0190276

to motor development (Cuevas et al., 2014; Gonzalez et al., 2016; Xiao et al., 2018) and exhibits mirroring properties (Marshall et al., 2011; Southgate et al., 2009). The mu rhythm is desynchronized when an individual performs an action or watches someone perform the same action. For example, mu/central alpha power recorded while an experimenter manipulated an object was not present at 1 month of age, when infants were themselves not capable of manipulating objects, but emerged by 7 months when they were able to do so (Xiao et al., 2018; see Figure 3.3). Guidelines and best practices for recording and reporting the infant mu rhythm have been outlined in Cuevas, Cannon, Yoo, and Fox (2014).

Resting Alpha and Frontal Alpha Asymmetry

In infancy, resting alpha has been defined as 6–9 Hz oscillations recorded over frontal electrodes during a quiet attentive state (Cuevas and Bell, 2011). Resting alpha power increases nearly linearly with development during infancy (Bell and Fox, 1992; Cuevas and Bell, 2011; Marshall et al., 2002), and thus, greater alpha power has been interpreted as a general measure of brain development. Interestingly, higher quality maternal behavior at 5 months has been associated with higher resting alpha power at 10 and 24 months, suggesting that experience and parenting quality impact brain development during infancy (Bernier et al., 2016).

Relative power recorded over right and left frontal regions (e.g., right EEG power—left EEG power) can be used to examine individual differences in affective style and emotion regulation. Greater left frontal activity has been related to withdrawal and avoidance behaviors in both infants and adults (Davidson, 2000; Fox, 1994). Recently, Poole et al. (2018) measured resting frontal alpha longitudinally from 6 to 8 years while children viewed affective videos. At 8 years, the children who showed a stable right frontal asymmetry exhibited more avoidant social behaviors, whereas children who showed a stable left frontal asymmetry displayed more approach-related behaviors.

Gamma

Gamma rhythms, or high-frequency oscillations (30–100 Hz in adults), are involved in object perception (Gruber et al., 2002; Gruber and Müller, 2005), attention (Müller et al., 2000; Pantev et al., 1991; Tiitinen et al., 1993), language (Pulvermüller et al., 1996), and memory retrieval (Engel et al., 2001; C. S. Herrmann et al., 2004, 2010) in adults. In infants, the gamma band includes oscillations in the range of 20–60 Hz. Changes in infant gamma have been linked to changes in perceptual and cognitive abilities including active maintenance of objects in memory in 6-month-olds (Kaufman et al., 2003, 2005) and stronger responses to familiar stimuli than unfamiliar stimuli (e.g., direct vs. averted gaze: Grossmann et al., 2007; objects with known or unknown labels: Gliga et al., 2010). Increased power in gamma frequency bands over frontal and parietal regions was found to be associated with increased executive function and increased verbal IQ in highly disadvantaged 4-year-old girls, suggesting that gamma power may be an important neural marker for cognitive function and an index of resilience in girls (Tarullo et al., 2017).

Theta

Theta oscillations are recorded over frontal regions and include frequencies 4–8 Hz in adults and 3–6 Hz in infants. Theta has been related to processing emotional information (Aftanas et al., 2001; Sammler et al., 2007) and memory encoding in adults (Basar et al., 2000; Klimesch 1999; Pare et al., 2002) and infants (Begus et al., 2015). In infants, theta has also been found to be positively related to attention modulation/executive control of attention (Orekhova et al., 1999; Wolfe and Bell, 2007; Xie et al., 2018). In addition, an increase in theta may also reflect expectations and/or a preparedness to learn in infants (see Begus et al., 2016). In the domain of language, increased theta synchronization in 6- to 12-month-old infants was found in response to infant-directed speech compared to adult-directed speech (Zhang et al., 2011) and an increase in theta power was reported in 6-month-olds for infrequently presented relative to frequently presented native and non-native syllables (Bosseler et al., 2013). However, by 12 months, infants displayed an increase in theta for syllables from the native language, regardless of frequency. These language-related results highlight attention-based changes in response to language during the second half of the first year of life and support previous perceptual narrowing findings (for recent review see Hadley et al., 2014).

ERPs

The following section provides an introduction to the ERP technique, components, and important empirical findings. Those interested in conducting ERP studies with developmental populations are referred to the following materials for specific methodological suggestions: DeBoer et al., 2007; Hoehl and Wahl, 2012; Thierry, 2005; Johnson et al., 2001; Keil et al., 2014; Luck, 2014; Picton and Taylor, 2007.

ERPs are derived from the continuous EEG by averaging time-locked segments of the neural response to an event of interest, such as a specific visual, auditory, or tactile stimulus or motor response, such as a button press. Averaging the EEG, in response to multiple instances of these events results in a reduction of the noise and unrelated neural activity and yields time-locked positive and negative deflections that reflect neural responses to the experimental manipulation or condition. These positive and negative deflections are called components and can be quantified and analyzed. The number of trials included in the average is directly related to the quality of the data, such that signal quality increases as the number of averaged trials increases (Luck, 2014; see Boudewyn et al., 2018 for discussion of trial number; see Snyder et al., 2002 for infant example). In developmental research, investigators often examine peak or mean amplitude, latency, or area under the curve as dependent measures. The time-locked nature of ERPs makes them particularly useful for research questions aimed at understanding sequence or timing of perceptual, cognitive, or social processing.

In general, developmental ERP components vary both qualitatively and quantitatively from adult components, as a result of differences in age, task, and data quality. The latencies of components tend to be delayed and amplitudes blunted (or not as peaked) compared to adults. However, voltage responses themselves are larger in amplitude for younger children and infants relative to adults likely due to skull conductivity differences (e.g., Taylor et al., 2004; Kuefner, 2010). Larger voltage responses may result in increased sensitivity of EEG measurements in developmental populations (Wendel et al., 2010). Delayed latency and slow-wave ERP responses are typically interpreted as resulting from a reduction in myelination of axons and synaptic efficiency in developmental populations relative to adults (Dubois et al., 2008; see Picton and Taylor, 2007 for review). However, it is also plausible that delayed peaks and slow wave responses result from increased variability in neural responses across trials and/or across participants. More specifically, increased variability may be due to an increase in intransient brain responses, more variability in phase, or to lower quality data. Figure 3.4 shows examples of how variability in the latency or phase of the response can result in different averaged responses and delayed peaks.

Although an exhaustive summary of the developmental ERP literature is beyond the scope of this review, this section describes several ERP components commonly observed in developmental populations. ERPs have been used to study many aspects of development including: auditory processing and language acquisition (Friedrich and Friederici, 2006; Junge et al., 2012; Mills et al., 2004; Stevens et al., 2006), visual

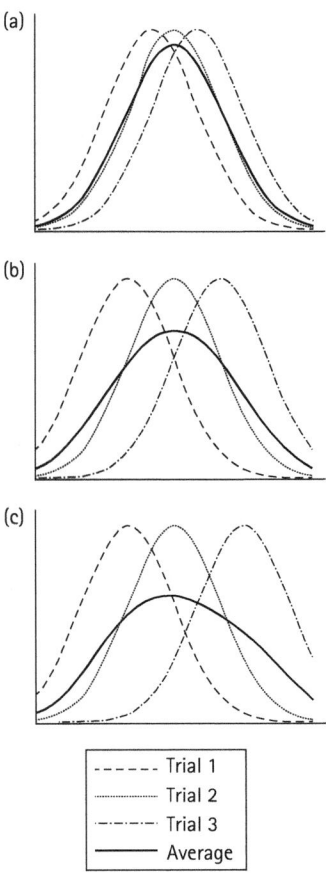

----- Trial 1
.... Trial 2
--·--·-- Trial 3
——— Average

FIGURE 3.4 Example of how the latency or phase of a response impacts the averaged response. Each graph shows three waveforms and their average. All graphs show the same three single-trial waveforms. The latencies vary by a common factor in (b) causing the averaged waveform to be reduced in amplitude (b) compared to (a). The latencies vary randomly in (c) causing the averaged waveform to be both reduced in amplitude and peaks earlier than (a).

recognition memory (Carver et al., 2000; Nelson, 1994; Riggins et al., 2009, 2013; Riggins and Rollins, 2015; Webb et al., 2005), attention (Pickron et al., 2018; Reynolds et al., 2010; Richards, 2003), categorization (Grossmann et al., 2009; Marinović et al., 2014), face processing (de Haan et al., 2003; Scott and Monesson, 2010; Scott and Nelson, 2006; Scott, Shannon et al., 2006), object processing (Pickron et al., 2018; Scott, 2011), emotion processing (Hoehl and Striano, 2008; Leppänen et al., 2007; Nelson and de Haan, 1996; Peltola et al., 2009; Vogel et al., 2012), executive functioning (e.g., cognitive control, performance monitoring, inhibition; Donkers and Van Boxtel, 2004; Jodo and Kayama, 1992), and reading (Coch 2014; Coch and Meade, 2016; Hasko et al., 2013; Heldmann et al., 2017; Stites and Laszlo, 2017).

P1/P100 Component

The P1/P100 is recorded over occipital scalp locations. In adults the P1 peaks around 100 ms and is sensitive to low-level stimulus properties including color, shape, contrast, and luminance (Rossion and Caharel, 2011; Taylor, 2002)and is thought to originate from striate and extrastriate visual areas (Clark and Hillyard, 1996; Di Russo et al., 2002). A meta-analysis (Batty and Taylor, 2006; Itier and Taylor, 2004a, 2004b; Taylor et al., 2001) of four studies that recorded ERPs in response to faces in children suggest large decreases in P1 amplitude and smaller decreases in P1 peak latency across childhood (Taylor et al., 2004).

Recently, reports have also suggested the P1 may be involved in face and object perception. For example, P1 was reported to be slightly larger in response to scrambled faces than intact faces but equal for faces and cars relative to scrambled faces in children ages 4 years to adults (Kuefner, 2010). In addition, in infants, de Haan and Nelson (1999) found a larger P1 in response to faces compared to objects, and Scott (2011) reported a P1 inversion effect after infants learned to name pictures of strollers with nonsense words. Face category priming was also shown to modulate both P1 amplitude and latency in 9-month-old infants, suggested sensitivity to primed categories (Peykarjou et al., 2014, 2016). However, low-level stimulus properties may be driving these differences in P1 response. In line with this, one study found scrambled images with similar mean luminance and hues to intact images but containing increased high contrast borders and greater image area evoked larger P1 responses than intact faces and bodies in 3-month-old infants (Gliga and Dehaene-Lambertz, 2005).

N170 (Children) and N290/P400 (Infants) Components

In adults, the N170 is a negative peaked component that is most prominent over right posterior temporal brain regions (Bentin et al., 1996; for a review: Rossion and Jacques, 2008) and is robustly generated in response to faces and sensitive to some aspects of face processing. Although previously thought to be specific to faces, the N170 has also been implicated in non-face perceptual expertise (Jones et al., 2018; Scott et al., 2006, 2008) and categorization processes (Carmel and Bentin, 2002; Rossion et al., 2000).The topography of the N170 in children is similar to that in adults, but latencies have been reported to decrease as a function of age, with adult levels reached in middle childhood or adolescence (see Figure 3.6a and b). In a recent investigation, 4- to 6-year-olds exhibited adult-like N170 components (peaking at around 220 ms) that distinguished upright and inverted faces (Hadley, Pickron, et al., 2015). However, the adult-like effect (inverted greater than upright) was only found in children who participated in an early infant learning study in which faces or objects were labeled with individual-level names for 3 months. These results suggest that the development of the N170 component is in part impacted by experience individuating faces or objects. Taylor and colleagues (2001) were the first to report a bifurcated N170 response in children that suggested a qualitatively different N170 response in children relative to adults. However, recently it was proposed that the bifurcated appearance of the N170 in children (Kuefner, 2010; see

Figure 3.6c) is due to the merging of two negative peaks, the N170 and the N250. The N250 is a component that is implicated in face and object expertise, repetition, and individuation in adults (Jones et al., 2018; Pierce et al., 2011; Schweinberger et al., 2002; Schweinberger et al., 2004; Scott et al., 2008, 2006). Importantly, although the morphology, magnitude, and timing of the N170 changes across development, face-related condition differences did not differ for all nine ages tested (Kuefner et al., 2010).

In infants there is no empirical evidence of an adult-like N170, however there are two reported developmental analogs. Both the occipital temporal N290 and the P400 components have been found to respond to and distinguish between faces and objects, inverted and upright faces/objects, and familiar and unfamiliar faces/objects (e.g., de Haan et al., 2002; Halit et al., 2003; Scott et al., 2006; for review see de Haan et al., 2003). However, the majority of evidence suggests that the N290 is the primary developmental equivalent of the N170 component. The N290 has been reported in infants as young as 3 months of age (Halit et al., 2003) and source localization suggests it is primarily related to activity in occipital-temporal brain regions (Guy et al., 2016; Johnson et al., 2005).

de Haan and Nelson (1999) first reported the P400 and found it to be faster in response to faces compared to objects (but see Guy et al., 2016 who report no P400 latency difference between faces and objects). The P400 is prominent over lateral occipital-temporal brain regions and peaks between 350–450 ms after stimulus onset. The P400 distinguishes between upright and inverted faces (de Haan et al., 2002; McCleery et al., 2009; Scott and Monesson, 2010), human and monkey faces (Halit et al., 2003), familiar and unfamiliar faces and face groups (Scott, Shannon et al., 2006; Vogel et al., 2012), and featural and configural face changes by hemisphere (Scott and Nelson, 2006). In a source analysis of the P400 and Nc components, Guy and colleagues (2016) suggested that these two components overlap and the P400 may be the positive expression of the source(s) generating the Nc component. If supported with future work, these findings suggest that the P400 may be more impacted by attention-related processes than face or object perception.

In one recent investigation, 9-month-olds exhibited marginally greater N290 amplitude and reduced P400 amplitude to faces paired with labels relative to faces paired with a non-speech noise (Barry-Anwar, Hadley, and Scott, 2019; see Figure 3.5). However, 6-month-olds only differentiated these two conditions in the second half of trials for N290 amplitude, indicating that infant ERP responses are impacted by age and experience and that processing may shift from one component to another during learning and across development.

Mismatch Negativity (MMN)/Mismatch Response

The mismatch negativity response is evoked when an occasional novel or mismatching stimulus is embedded within a stream of matching stimuli. For example, when infants hear the sound da imbedded within a stream of seventy-five presentations of the sound ba, a mismatch response should be present if infants can differentiate between the two sounds. This response is thought to reflect automatic change detection and is often used in studies of auditory or language development (Brannon et al., 2008; Cheour et

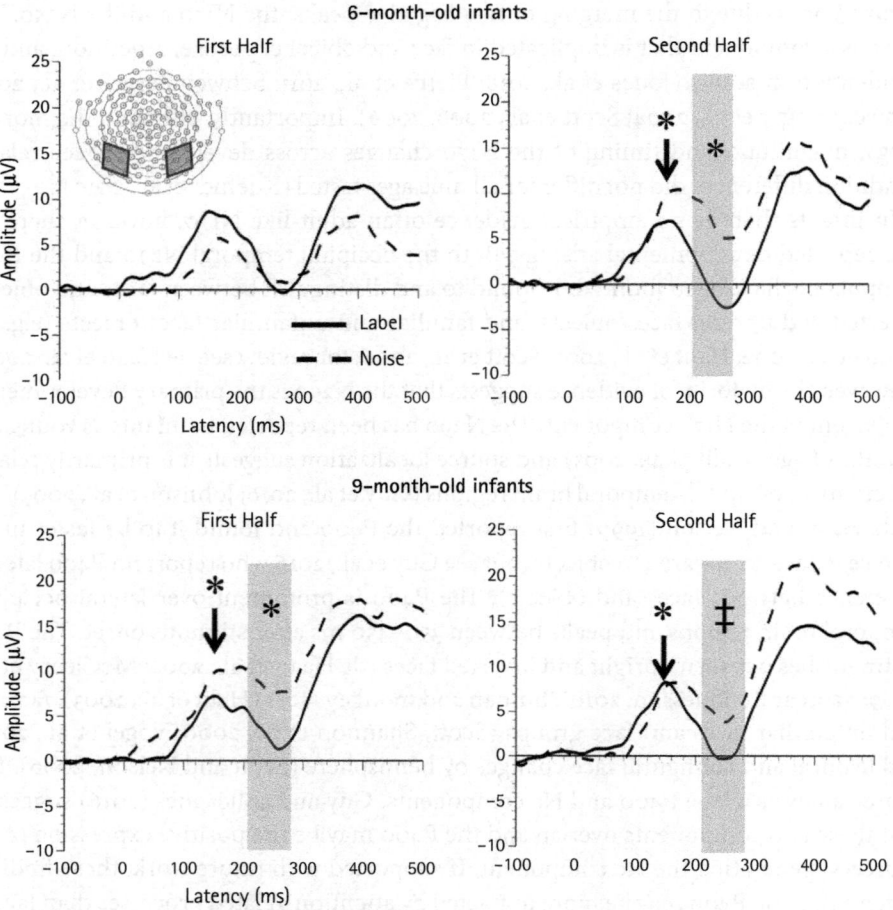

FIGURE 3.5 Waveforms of ERP responses for 6- and 9-month-olds for the first and second halves of the training period, collapsed across hemispheres. The responses in both the label and noise conditions are shown. Significant P1 latency differences are marked with an arrow and an asterisk and significant N290 amplitude differences are marked with a box and asterisk. Indicates p <0.05, ‡ indicates p = 0.05.

Source: Reprinted from *Neuropsychologia*, *108*(8), Ryan Barry-Anwar, Hillary Hadley, Stefania Conte, Andreas Keil, and Lisa S. Scott, The developmental time course and topographic distribution of individual-level monkey face discrimination in the infant brain, 25–31, https://doi.org/10.1016/j.neuropsychologia.2017.11.019, © 2017 Elsevier Ltd. All rights reserved.

al., 2000; Dehaene-Lambertz and Gliga, 2004; Kushnerenko et al., 2002; Marshall et al., 2009; Novitski et al., 2007; for review: Garrido et al., 2009).

Negative Central (Nc) Component

In infants and children, the Nc (negative central) component is a prominent negative component recorded from frontal and central scalp locations and occurs between approximately 400–700 ms after stimulus onset. In infants, the Nc is sensitive to stimulus

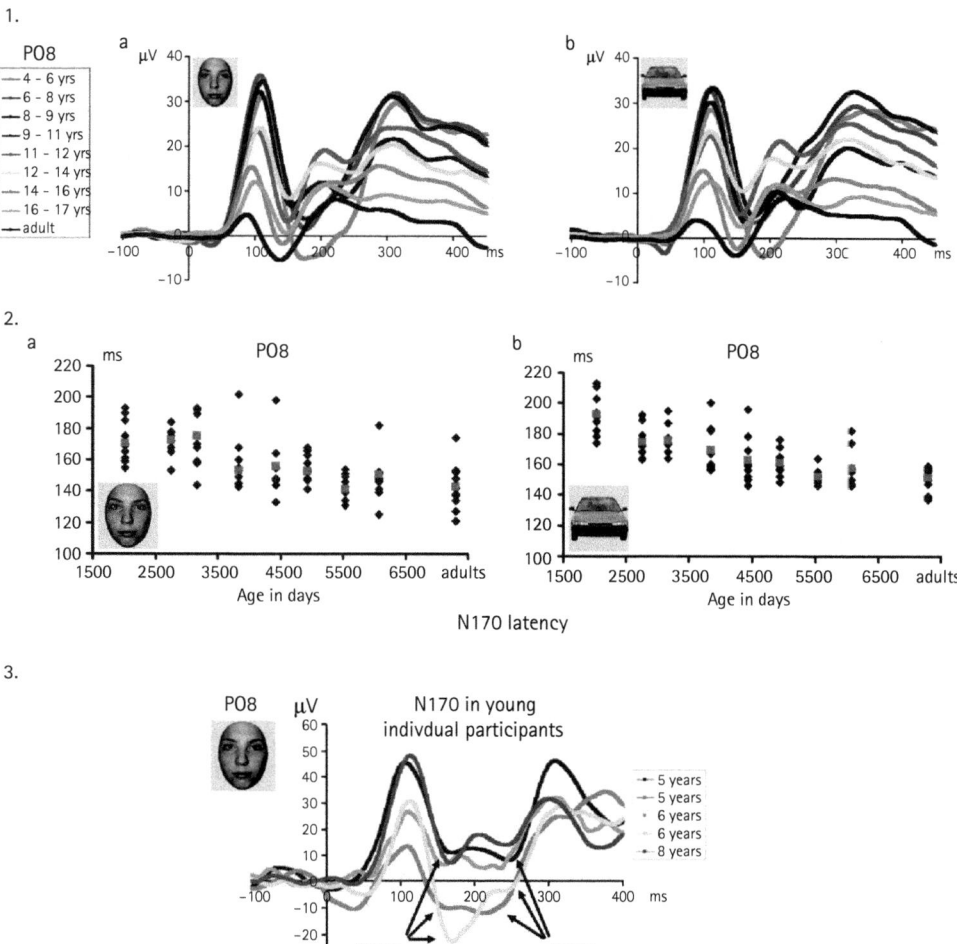

FIGURE 3.6 (1) Grand-average waveform in response to (a) faces and (b) cars in all age groups and adults (in black) illustrating the dramatic differences of amplitude, latency and width of the N170 component. Note that, especially evident for the face stimuli, in the three youngest age groups the component is particularly wide due to the merging in grand-averaged data of the N170 with a late negative deflection. This phenomenon can be clearly observed on individual data (Figure 3.6). A 30-Hz low-pass filter has been applied to each waveform in this figure. (2) Peak latency of the N170 in response to pictures of faces (a) and cars (b). Individual data points are plotted according to the mean age (in days) of each age group. Mean latency for each age group is shown in red. The overall latency delay for cars was found across all age groups, and the developmental decrease of latency appears to follow a similar pattern for faces and cars. Note the longer latency of the N170 in the three youngest age groups compared to the others, but also the generally smaller variability (25 ms at most). The N170 latency appears to reach an adult-like level at 14–16 years old, both for faces and cars. (3) Illustration of young individual participant waveforms showing the two negative deflections which are merged on grand averaged data. The second peak may correspond to the N250 in older populations. A 30-Hz low-pass filter has been applied to each waveform in this figure.

Source: Reprinted from Dana Kuefner, Adélaïde de Heering, Corentin Jacques, Ernesto Palmero-Soler, and Bruno Rossion (2010) Early visually evoked electrophysiological responses over the human brain (P1, N170) show stable patterns of face sensitivity from 4 years to adulthood.

Kuefner, de Heering, Jacques, Palmero-Soler, and Rossion. *Frontiers in Human Neuroscience* 3:67. https://doi.org/10.3389/neuro.09.067.2009 © 2010. Licensed under CC BY-ND 4.0.

probability (Ackles and Cook, 1998, 2009; Courchesne et al., 1981; Karrer and Ackles, 1987; Karrer and Monti, 1995), is greater during periods of increased attention (Nelson and Collins, 1991, Pickron et al., 2018, Reynolds and Richards, 2005, 2009; Webb et al., 2005), and is larger in response to preferred (e.g., mother's face or own toy) or salient (e.g., novel) stimuli (Bauer et al., 2003; Carver et al., 2000; de Haan and Nelson, 1997; de Haan and Nelson, 1999; Snyder et al., 2002).

Nc amplitude is larger during periods of attention compared to inattention, has been previously linked to attention orienting and sustained attention (Reynolds, Courage, et al., 2013; Reynolds, Zhang, et al., 2013; Reynolds and Richards, 2005, 2019; Richards, 2003; for review see Reynolds et al., 2010), and has been localized to the anterior cingulate and regions within the prefrontal cortex (Guy et al., 2016; Reynolds and Richards, 2005; Richards et al., 2010). Moreover, variations in attention indexed by the Nc have also been shown to be modulated by memory in infants (Bauer et al., 2003; Carver et al., 2000) and children (Riggins and Rollins, 2015; Robey and Riggins, 2016).

Late Slow Wave Components

Slow wave activity often follows the Nc (600–900 ms after stimulus onset), does not have a distinct peak, and is more widely distributed across the scalp. In infants, slow wave activity can be either positive or negative in amplitude. Negative slow wave (NSW) has been related to novelty detection, whereas positive slow wave (PSW) has been interpreted as indexing the updating of a partially encoded stimulus or context in working memory (Nelson, 1994; Nelson et al., 1998; see DeBoer et al., 2007 for a review). The dependent measure is often the area under the curve or mean amplitude. The cortical source of PSW in infants is thought to be in temporal regions and a separate (but possibly overlapping) anterior source is thought to generate the NSW (Reynolds and Richards, 2005).

A late slow wave (LSW) is often elicited in memory paradigms that present old/familiar items and new/novel items with equal probabilities to children. LSW responses tend to be larger for old compared to new items (e.g., Marshall et al., 2009), but this pattern depends on participant age and the task (Riggins and Rollins, 2015). In studies of 4- to 5-year-olds requiring memory for items and details of these items (i.e., a recollection effect), LSW has been reported to differentiate items with correct source judgments from old items remembered without details and new items (Riggins et al., 2013; see also Robey and Riggins, 2016). Additionally, the magnitude of the LSW has been shown to be positively related to performance on memory tasks (Geng et al., 2018; Riggins et al., 2013).

N200 (N2a, N2b, N2c, N2pc)

The N200 is a negative-going peaked component reaching its maximum voltage approximately 200 ms after successful inhibition of a stimulus (e.g., Go/No-go, Stop-signal, Stroop or Flanker tasks). The N200 is thought to reflect cognitive control processes that are essential for inhibitory control and interference suppression (Jodo and Kayama, 1992) and/or the detection and monitoring of conflict (Donkers and Van

Boxtel, 2004). Larger N200 amplitudes result for less frequent responses, inhibiting prepotent responses, and reconciling response conflict (Azizian et al., 2006; Bruin and Wijers, 2002). Neural generators of the N200 are thought to include frontal and superior temporal cortex as well as anterior cingulate cortex (Huster et al., 2010; Lamm et al., 2006).

The N200 has been referred to as the N2a, N2b, N2c, or N2pc depending on the particular paradigm used. For example, the N2pc has been observed in tasks with distractors that need to be suppressed for successful task completion, and is proposed to reflect a combination of attention selection and distractor suppression (Mazza et al., 2009; Shimi et al., 2015). Some developmental studies of N200 report decreases in amplitude and latency with age (e.g., Davis et al., 2003; Johnstone et al., 2005; Lewis et al., 2006), however other studies observe no effects of age but rather effects of task performance (e.g., Lamm et al., 2006). These differences may be due to variations in tasks, participant ages and/or aptitude, which can alter neural generators (Lamm et al., 2006; see Downes et al., 2017 for discussion). Fewer studies have examined N200 subcomponents, although Couperus and Quirk (2015) report that N2pc can be detected in children as young as 9 years of age.

P300 (P3, P3a, P3b)

The P300 is a positive-going peaked component that reaches it maximum voltage approximately 300 ms after stimulus onset in adults. A P300 response is commonly elicited during tasks engaging attention, memory, and/or problem solving (Polich, 2012). Although the exact cognitive function(s) reflected by the P300 remain debated, there seems to be general agreement that the P300 reflects information processing and updating of working memory (Polich, 2012). In contemporary adult research, the P300 is often further subdivided into P3a and P3b subcomponents. The P3a (also referred to as the novelty P300), has a fronto-central topography that is elicited in response to novel stimuli, and is not thought to require an active response from the participant. In contrast, the P3b has a parietal topography and is typically observed in tasks that involve intentional or conscious discrimination of novel stimuli. Common paradigms used to elicit a P300 response include the manipulation of auditory or visual stimulus presentation probabilities, with one stimulus presented frequently (e.g., 80 percent of the time) and another presented infrequently (e.g., 20 percent).

Developmental studies suggest that the P300/P3b can be detected in children (albeit at longer latencies) and appears to reflect similar cognitive processes to those in adults (see Downes et al., 2017, Riggins and Scott, 2020). With age, latency and amplitude have been shown to decrease (van Dinteren et al., 2014). Many continuities appear to exist in the P300 component across development, yet two striking developmental differences have been documented (Riggins and Scott, 2020). The first is the lack of a canonical P300 in infancy, especially in the visual domain (e.g., Courchesne et al., 1981), and the second is that ERP responses to novel, unexpected stimuli differ dramatically across development and modality (auditory versus visual; e.g., Courchesne, 1977).Finally, the P300 appears reduced in amplitude in atypically developing groups (e.g., children with autism, Courchesne et al., 1984; attention deficit hyperactivity disorder (ADHD), Doehnert et al., 2010; and adolescents at risk for alcoholism, Hill and Shen, 2002).

N400

The N400 is a negative-going waveform that peaks around 400 ms after stimulus onset and is typically maximal over centro-parietal electrode sites. The N400 is most commonly associated with lexical-semantic processes. For example, N400 amplitude is larger for pseudowords than for words and larger for words that are semantically incongruous to a given context than words that are congruous (e.g., He spread the warm bread with butter versus He spread the warm bread with socks; Kutas and Hillyard, 1980; for reviews see Friederici, 2006; Kutas and Van Petten, 1994; Kutas and Federmeier, 2000). However, the N400 has also been used to examine semantic memory and recognition memory and is elicited by a variety of stimuli, including pictures, music, and videos (see Kutas and Federmeier, 2011) and emotional/social processing (Rueschemeyer et al., 2014). Studies have identified a distributed set of brain regions that contribute to the N400, including: the anterior medial temporal lobe, middle and superior temporal areas, inferior temporal areas, and prefrontal areas (Kutas and Federmeier, 2011; Lau et al., 2008; Van Petten and Luka, 2006).

The neural architecture supporting N400 processing appears to be available around the end of the first year, as N400 effects have been reported in 9-month-olds while viewing unexpected events in action sequences (Reid et al., 2009) and when the mother presents incongruent objects and words pairs (Parise and Csibra, 2012). In the domain of language, picture/word congruity N400 effects are often reported in children over a year of age (Friedrich and Friederici, 2010) and semantic congruity effects in sentences have been reported in infants (Asano et al., 2015). Mills, Conboyand, and Paton (2005) report similar latency N400 components from 13 months to 3 years. In an investigation with 9-month-olds, infants showed evidence of integrating word-picture pairings across the infant N200 and N400 components (Junge et al., 2012). N400 amplitudes have also been shown to vary with real-world language skills, such as productive vocabulary (Friedrich and Friederici, 2006, 2010)and are predictive of language skill at 30 months (Friedrich and Friederici, 2006).

In older children (6–11 years), the N400 is modulated by emotional content of a vocal stimulus such that the N400 was attenuated to angry relative to happy and neutral voices (Chronaki et al., 2012). The N400 continues to be examined throughout childhood and adolescence to investigate multiple aspects of language, memory and social processing (e.g., reading development, see Coch, 2014; rhyme decisions, see Welcome and Joanisse, 2018; language within a social context, Westley et al., 2017).

Error-Related Negativity (ERN, Ne) and Error-Related Positivity (Pe)

The ERN is a negative-going deflection after an erroneous response during various speeded-response tasks, such as Flanker or Go/No-go, even when the participant is not explicitly aware of making the error. The ERN is typically observed ~100 ms after an erroneous response and is maximal over frontal and central electrode sites and is thought to be generated by the anterior cingulate cortex (ACC; Ito et al., 2003). The ERN is generally thought to index detection of errors and conflicts associated with selecting an incorrect response (Hajcak, 2012). The ERN can be recorded in children as early as 3 years of age (Grammer et al., 2014), however, amplitude increases with age, with the greatest changes found in adolescence (Davies et al., 2004; A. Meyer et al., 2014; Taylor et al., 2018). Similar to findings in adults (Olvet and Hajcak, 2008), the ERN has been shown to increase in clinically anxious children (Meyer et al., 2012), but this relation may vary with age (Torpey et al., 2013). During adolescence, enhanced ERN has been associated with externalizing symptoms, such as impulsiveness (Taylor et al., 2018).

The ERN is often followed by a positivity, known as the Pe. The Pe is a positive deflection with a centro-parietal distribution that occurs 200–500 ms after an erroneous response if the participant is aware of their error (Falkenstein et al., 1991). Pe amplitude is thought to reflect the conscious awareness of the error, with greater awareness related to greater amplitude Pe's (Nieuwenhuis et al., 2001; Overbeek et al., 2005). Pe has been shown to be generated by posterior cingulate cortex (M. J. Herrmann et al., 2004; Overbeek et al., 2005; Vocat et al., 2008). In contrast to the ERN, Pe is thought to show an earlier maturational profile (Downes et al., 2017). Specifically, age-related differences have been observed in young children (e.g., 3–7 years) but not in mid-childhood to adolescence (Davies et al., 2004). This lack of developmental change in Pe in childhood may result from the superposition of different components during the Pe time window (Arbel and Donchin, 2011) or low signal, as the Pe is not observed in all trials (Downes et al., 2017).

Feedback-Related Negativity (FRN)

The FRN is a negative deflection that occurs approximately 250–350 ms after presentation of feedback to a response that is maximal over fronto-central electrodes and larger to negative compared to positive feedback (Holroyd and Coles, 2002; Miltner et al., 1997). The FRN has been localized to the ACC (Luu et al., 2003). Some researchers who are interested in separating feedback regarding gains vs. losses may refer to this component as the Reward Positivity (RewP) to highlight that the gains, and not the losses, are the construct of interest (e.g., Proudfit, 2015).

Developmental studies report that the FRN response can be detected in children as young as 2.5 years (M. Meyer et al., 2014). Generally, larger FRN amplitudes and greater latencies are observed in children and more adult-like FRN responses emerge across adolescence (Crowley et al., 2013). In addition, some studies suggest FRN may be less differentiated between reward compared to loss conditions in younger individuals (e.g., Hammerer et al., 2011; Zottoli and Grose-Fifer, 2012).The FRN has been widely used as a marker of risk-taking and impulsiveness in adolescence (Downes et al., 2017), and differences in FRN have been associated with increased likelihood of conduct problems (Gao et al., 2016) and antisocial behavior (Sheffield et al., 2015).

ssVEPs

The ssVEP is an exogenous ERP generated in response to a periodic stimulus. The response in the brain is at the exact frequency of presentation. Although the ssVEP technique has been used to study low-level processing in infants for nearly three decades (e.g., Hamer and Norcia, 1994; Norcia et al., 1990), the technique has become more popular recently for the study of covert attention (Christodoulou et al., 2017; Robertson et al., 2012) as well as higher-level visual processing in infants (e.g., face processing, Barry-Anwar et al., 2018; Peykarjou et al., 2017; and object categorization, Farzin et al., 2012; de Heering and Rossion, 2015).

When recording ssVEPs, it is necessary to reduce low-level stimulus differences. This can be accomplished by: presenting images at a random luminance at each presentation, presenting images at a random size at each presentation, averaging dark and light pixel contrasts across all images used (for example using the SHINE Matlab toolbox; Willenbockel et al., 2010), or presenting stimuli on a noise background that has been standardized to the mean of the stimuli. A stimulation frequency anywhere from 1 Hz–12 Hz is common in infancy. In addition to responses occurring at the presentation frequency, harmonic responses occur because the brain is a nonlinear system. Because of this, responses are also often seen at exact multiples of the frequency. For example, when stimuli are presented at 6 Hz, brain responses may be observed at 6 Hz, 12 Hz, 18 Hz, and so on. Commonly a sinusoidal contrast modulation is used (see Figure 3.7) for presenting stimuli because it generates fewer harmonic responses than a square wave presentation (Norcia et al., 2015; Regan, 1989).

When analyzing ssVEPs, it is necessary to include an integer number of presentations. At least one full period of the response of interest is necessary. For example, if you are interested in a 6 Hz response, the segments transformed to the frequency domain should be a multiple of 167 ms (1/6 Hz). This can be accomplished by removing the first and/or last portion of a trial if necessary. Researchers analyze the amplitude of the response, the signal to noise ratio (SNR), or both. The noise used for the SNR has ranged anywhere from ±0.5 Hz to ±1 Hz. The SNR is calculated by dividing the amplitude at the frequency of interest by the average of the surrounding noise bins. A similar metric, the baseline corrected amplitude (BCA) is sometimes used in place of the SNR. When calculating BCA, the average noise amplitude is simply subtracted from the amplitude at the frequency of interest. Thus, the signal equal to noise level is 1 when using SNR but 0 when using BCA.

Important Paradigms and Findings

Early work measuring ssVEPs in infants used the sweep VEP paradigm. In this paradigm, a stimulus is parametrically varied over a range of values rather than being presented at a constant value. The sweep VEP is commonly used to measure spatial acuity, contrast sensitivity (e.g., Norcia et al., 1990), and motion sensitivity (Hamer and Norcia, 1994). The stimuli are presented at a constant rate and the amplitude and phase of the ssVEP response at the presentation frequency are analyzed.

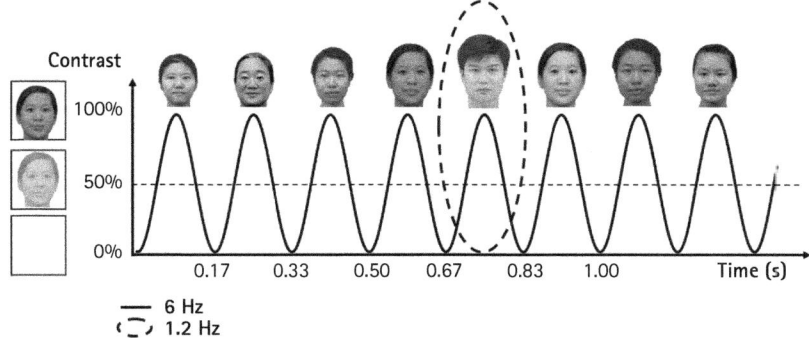

FIGURE 3.7 Fast periodic visual stimulation (FPVS) oddball task. Example shows female and male faces presented at 6 Hz. Every fifth face (circled) is male. Thus, male faces were presented at a rate of 1.2 Hz (6 Hz/5). Faces were presented at a random luminance and size at each presentation in order to reduce neural responses to low-level stimulus differences.

Recent work has validated the use of ssVEP measures for the study of overt and covert attention in infants using frequency tagging (Christodoulou et al., 2017; Robertson et al., 2012). In this paradigm, multiple stimulation frequencies are used simultaneously to measure evoked responses to different stimuli/conditions. This method measures relative attention to simultaneously presented stimuli, conditions, or parts of stimuli. In one study using this technique, 12-week-old infants were found to covertly attend to objects prior to an overt shift in looking (Robertson et al., 2012). In another investigation, frequency tagging allowed researchers to quantify attention to complex versus sparse images in 4-month-olds (Christodoulou et al., 2017). Although little work has been done using this technique with developmental populations, frequency tagging simultaneously presented stimuli is a useful tool for measuring the development of attention.

Recent research examining face processing in infants and young children has used fast periodic visual stimulation (FPVS; (Liu-Shuang et al., 2014; Rossion, 2014; see Figure 3.7) in the form of an oddball paradigm (Barry-Anwar et al., 2018; de Heering and Rossion, 2015; Lochy et al., 2019; Peykarjou et al., 2017) or a pattern onset/offset design (Farzin et al., 2012). In the oddball paradigm, an infrequently presented oddball stimulus is presented at a fixed interval in a periodic train of standard stimuli. It is common to use a standard presentation rate of 6Hz and present the oddball stimuli as every fifth stimuli. A response at 1.2 Hz (6 Hz/5) indicates that the stimuli have been discriminated, whereas a response at 6 Hz indicates that the stimuli have simply been processed by the visual system (see Figure 3.8). An oddball task can be used to study categorization of stimuli or discrimination between stimuli. De Heering and Rossion (2015) examined categorization of faces and objects in 4- to 6-month-old infants and found a right-lateralized categorization response, similar to findings in adults. Lochy and colleagues (2019) extended this study and added a sample of 5-year-old children.

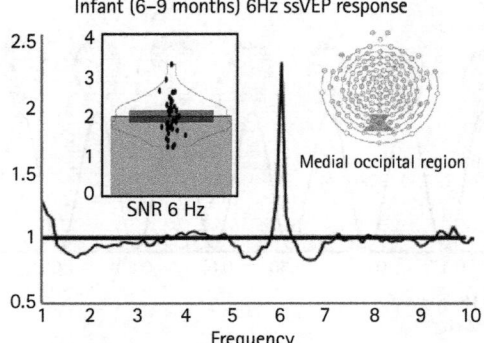

FIGURE 3.8 The SNR to the 6 Hz base stimulation frequency for 6–9-month-old infants (n = 30). The bar graph is a plot of the mean (dark line), 95 percent CI (shaded box), and individual data points averaged across the five channels (marked on the channel map) used to make up the medial occipital region. The frequency graph is the SNR at the medial occipital region and shows a peak at the 6 Hz stimulation frequency.

Source: Data from *Neuropsychologia*, *108*(8), Ryan Barry-Anwar, Hillary Hadley, Stefania Conte, Andreas Keil, and Lisa S. Scott, The developmental time course and topographic distribution of individual-level monkey face discrimination in the infant brain, 25–31, https://doi.org/10.1016/j.neuropsychologia.2017.11.019, 2017.

Although a categorization response was present, the response was bilateral over occipitotemporal regions. Categorization of different types of faces has also been examined (Peykarjou et al., 2017). Nine-month-old infants were presented with streams of both ape faces that contained human oddball faces and streams of human faces that contained ape oddball faces. A 1.2 Hz categorization response was found over occipital regions in both conditions. Barry-Anwar et al. (2018) used an oddball paradigm to examine individuation of monkey faces in 6- and 9-month-old infants. A 1.2 Hz individuation response was found in both ages, but the topography differed such that the 9-month-old showed a more right-lateralized occipitotemporal response while 6-month-olds showed a more medial occipital response.

In pattern onset/offset paradigms, stimuli that are equated on one low-level attribute (the one the experimenter hopes to control for) but vary on a high-level attribute that does not need to be controlled for (the one the experimenter is interested in examining) are alternated. Farzin et al. (2012) used a pattern onset/offset design to present 4- to 6-month-old infants with alternating images (6 Hz) of intact (3 Hz) and scrambled faces and objects. The response to the low-level changes are symmetric for the on and off phases of the pattern, leading the response to be evident in the even harmonics (e.g., 6 Hz, 12 Hz, 18 Hz, etc.). The response to the high-level global structure changes, however, differ for the on and off phases of the pattern, leading the higher-level response to be evident in the odd harmonics (e.g., 3 Hz, 9 Hz, 15 Hz, etc.). Responses to the global structure of objects was evident over the medial occipital region, however, responses to faces was found over left, right, and medial occipital regions.

DATA COLLECTION AND PROCESSING

The following section details considerations and recommendations for the collection, processing, analysis and publication of developmental electrophysiological data. This section ends with a publication checklist adapted for developmental research (Keil et al., 2014).

Data Collection Systems

Several commercial high- and low-density montage EEG data collection systems are available for use with developmental populations. These systems differ based on whether they have active versus passive electrodes, the type and size of recording net/cap used, the software/hardware components, and the preparation time required. All of these factors are important to consider when purchasing an EEG system for developmental research. Active electrodes include built-in readout circuitry that amplifies the neural signal and buffers the EEG signals at the electrode, prior to cabling, and then also send the signals to an amplifier, making the signal more robust to environmental interference. Passive electrodes, on the other hand, send the scalp recorded signal to an amplifier without pre-amplification. Although some argue that active electrodes may provide a better signal and require fewer trials, the density of the recording montage may be limited due to the increased weight of active electrodes on the scalp and the preparation time can be increased relative to some commercially available high-density passive-electrode systems. Net electrode systems allow for high-density recording (between 64–256 electrodes) and are typically easier and faster to apply than cap systems that require gels and scalp abrasion at each electrode site (about 30–45 minutes for 64 electrodes). However, net systems may have increased low-frequency noise from skin potentials (due to high-electrode impedance), may increase electrode bridging, and may be more susceptible to movement artifacts (Johnson et al., 2001; Luck, 2014). High-impedance recordings also may require more trials to get the same level of statistical significance because of their greater susceptibility to skin potentials (Kappenman and Luck, 2010).

Experimental Design Considerations

As with most methods, experimental design should be driven by theory and a priori predictions that are grounded in previous research. Several EEG/ERP/ssVEP tasks have been used successfully with developmental populations and a useful place to start is to identify tasks that have previously resulted in successful data collection and good quality data in the age group/population of interest. A second strategy is to adapt tasks used with adult populations for use with infant/child populations. Task demands are important to consider for each age group of interest and pilot testing is typically necessary in order to modify tasks so that they are not too taxing for developmental participants resulting in increased attrition and low-quality data.

Artifacts and Artifact Reduction/Correction

Artifacts are present in EEG data in adults and developmental populations and trials, time segments, or the artifacts themselves need to be removed or corrected. Typical artifacts include ocular artifacts (blinks, saccades), movement or muscle artifacts, skin potentials, electrocardiography, and electrical noise (i.e., 60 Hz line noise). Artifact detection and correction parameters used for adults typically identify an upper voltage threshold and mark data that goes above that limit within a certain time window as an artifact. Trials with artifacts are then either removed or corrected using various techniques. Artifact detection and correction techniques used with adults are often ineffective in developmental populations, and the parameters used in children at one age may not be appropriate at another age and so many researchers adjust the parameters used for different ages and/or manually edit each trial or data segment for artifacts. In adults, eye artifacts are often the largest artifact source present in the EEG data. However, in infants, ocular artifacts have less of an impact on the EEG response. Although not fully understood, the smaller ocular eye artifacts may be because infants blink less often and/or their blinks do not impact their neural activity (see DeBoer et al., 2007; Nelson, 1994). One useful, but underutilized, method for identifying artifacts is to use Independent Component Analysis (ICA) (Gao et al., 2010; Vigario et al., 2000). ICA decomposes the EEG into separable and independent components and allows for eye-movement and eye-blink artifacts to be identified and either removed manually or through an automated process (for methods, see Chaumon et al., 2015; Mognon et al., 2011; Nolan et al., 2010). However, the applicability and reliability of ICA for artifact identification in infant populations is not well documented.

Attention Getters, Animation, Breaks, and Background Images

Developmental research typically requires that researchers incorporate games, attention getters, animation of stimuli, breaks, and background patterns in order to retain attention during tasks. For infants, animated cartoons or images are typically presented in between trials/blocks to redirect attention to the screen. Some researchers use short video clips or novel toys to entertain infants during net/cap/electrode placement. It is also important to try to work with parents and caregivers to schedule participants for testing during times when they are typically awake and happy. Breaks should be incorporated into paradigms so that parents can feed or change their infants if needed without removing the net/cap. Stimuli can also be presented on a black and white or colorful patterned background to help maintain attention, although this background should not vary by condition or group. Having a researcher monitor attention to the screen through a live video feed and only presenting trials during periods of attention is an effective way of increasing data quality. Methodology that integrates data acquisition and eye-tracking and uses gaze-contingent attention-getters and stimulus presentation increases both the quality and quantity of infant/child EEG data and also allows for grouping of trials based on fixation data (Maguire et al., 2014). In some infant visual studies, a researcher is often seated next to the infant and lightly taps on the screen or snaps her fingers in between trials in order to redirect infant attention. Although this method should be reduced as much as possible and the location of the researcher should be counterbalanced across participants, person redirection is sometimes necessary for maintaining infant attention to the screen and increasing retention. In toddlers and children, tasks can be presented within a game or fun context (spaceship, underwater diving adventure, etc.). For example, Hadley et al. (2014) had 4-to 5-year-olds complete a match-to-sample task with faces and objects within the context of a Finding Nemo game.

Attrition

An ideal goal should be to create paradigms that allow for retaining 75 percent or more of tested participants. Piloting is useful for determining what tasks can be used. For example, with ERP tasks, high-quality data can typically be obtained in 75 percent or more of infants (e.g., ages 6 –12 months) when the task includes two conditions of 50–75 1000–1200 ms trials. When the number of conditions is increased to four, attrition increases depending on age and the nature of the stimuli. Children, on the other hand, can complete a greater number of conditions and trials, especially if the task is framed within a fun context. Longitudinal studies that follow infants and children over time and clinical or special population investigations likely have an increased attrition rate depending on age and task used. Even without recording EEG, retention rates in these investigations is decreased. In these cases, it is necessary to weigh the importance of the question being asked against the risk of high attrition. Researchers unexpectedly finding themselves with high attrition rates should systematically examine and explicitly state reasons for attrition and discuss how reduced data retention may impact the results and interpretation of findings. Recording EEG and ssVEPs in developmental populations can be done more quickly than ERP and so attrition rates tend to be lower. In practice, it appears that this is true, as ssVEP retention rates range from 60–95 percent (Barry-Anwar et al., 2018; Christodoulou et al., 2017; Farzin et al., 2012; de Heering and Rossion, 2015; Peykarjou et al., 2017; Robertson et al., 2012) and EEG retention rates range from 88–100 percent (Bell and Cuevas, 2012)and ERP retention rates are approximately 50 percent (Stets et al., 2012).

DATA PROCESSING, ANALYSIS, AND PUBLICATION CRITERIA

After collection, EEG data is run through several processing steps in order to analyze it. Data processing steps and parameters vary across laboratories, ages tested, and task/study designs. Typical data processing steps for EEG, ERP, and ssVEP are shown in Table 3.1. EEG measures the difference between the electric potentials at one location and in the location of the reference electrode (Nunez, 1981). There is some debate about best practices for referencing and re-referencing but the reference choice impacts the data substantially and should be considered carefully (see Lei and Liao, 2017; Luck, 2014). Commonly, online recording references are the vertex/Cz, the nose, or one or both of the mastoids. Once the data is recorded online it is typically filtered and then can be re-referenced to any other electrode or the mean of any group of electrodes (e.g., the average of all electrodes for the average reference, which is often used in developmental research). The spatial resolution of the electrode montage is important to consider when choosing an offline reference (Dien, 1998; Ryynänen et al., 2006). In adults 64 electrodes or more are needed in order to use the average reference without distorting the data. Moreover, source localization techniques require both a high-density montage (64+), low measurement noise, and may be improved with lower skull resistivity (as is found in infants and young children; Ryynänen et al., 2006). Before using a filter, it is important to understand how these different types of filters work, what filter parameters do and how filtering may distort or bias EEG data (see Luck, 2014; Widmann et al., 2015). Online filtering is necessary during data acquisition and typically includes an 80–100 Hz low-pass filter. The cut-off value for the online filter should be one third the sampling rate or less (Luck, 2014). Offline filtering includes: high-pass/low-pass/band-pass/band-stop; finite/infinite impulse response, FIR/IIR filters, and filter parameters include: cut-off frequencies, filter order and roll-off, ripple, delay, and causality.

As described in Artifacts and artifact reduction/correction above, artifacts in EEG data need to be accurately detected and either removed or corrected. Several algorithms exist for detecting and removing artifacts, but the parameters used in adults may not be appropriate for developmental populations. When looking at condition differences across ages, blocks, or trials it is necessary to segment and average the data, although single trial analyses are becoming prevalent in studies with adults (Delorme et al., 2015; Ouyang et al., 2015; Rousselet et al., 2011; Williams et al., 2015) and may be useful for better characterizing the time course of electrophysiological developmental processes. For EEG and ssVEP, the data is transformed from the time domain to the frequency domain, and for ERP, the data is averaged across conditions and the baseline response is subtracted from the data. Dependent measures of interest, including power, coherence (EEG), mean or peak amplitude, area under the curve, latency (ERP), SNR, baseline corrected amplitude, or z-scores (ssVEP) are then computed. Measurements of peak

Table 3.1 Typical EEG data–processing steps for EEG, ERP, and ssVEP data.

EEG	Filtering (low-pass, high-pass, bandpass, or notch filters, see Luck, 2014 detailed recommendations)	Re-referencing (Typically from the online reference Cz or one of the mastoids to the average reference or mastoid reference)	Artifact Detection and Removal/ Correction/ Interpolation	Averaging trials within conditions/ participants	Fourier Transform from time domain to frequency domain	Computing power and/or coherence and compare to conditions of interest and/or baseline recordings
ERP		Segmenting or epoching data into trials	Artifact Detection and Removal/ Correction/ Interpolation	Averaging trials within conditions/ participants	Baseline Correction (typically 100–200 ms)	Compute mean or peak amplitude, latency, or other DV
SsVEP	Segmenting or epoching data into trials	Re-referencing (Typically from the online reference Cz or one of the mastoids to the average reference or mastoid reference)	Interpolation	(1) Averaging trials within conditions/ participants (2) Fourier Transform from time domain to frequency domain	(1) Fourier Transform from time domain to frequency domain (2) Computing SNR individually for each trial	(1) Computing SNR/BCA/ Z scores (2) Averaging the SNR over trials within conditions/ participants

amplitude in ERP datasets have been criticized and means (or trimmed means, Wilcox, 2005) are preferred as functional differences often occur between component peaks. These dependent measures can be analyzed using a variety of parametric statistical tools (for review, see DeBoer et al., 2007), non-parametric analysis (Maris and Oostenveld, 2007; in adults, see Blair and Karniski, 1993; Guan et al., 2018; in infants, see Kouider et al., 2015), Bayesian analyses (in infants, see Peykarjou et al., 2017), current source density and source localization analyses (in infants, see Guy et al., 2016; Hämäläinen et al., 2011; Reynolds et al., 2010; Reynolds and Richards, 2005), and multivariate pattern analysis methods (MVPA; in adults, see Cauchoix et al., 2014).

Although not specific to EEG research, null hypothesis significance testing has limitations (Wagenmakers, 2007), and if it is important to demonstrate the absence of an effect, Bayesian statistics may be more useful. It is also useful for individual differences to be considered. Significant effects of group means do not show how many infants/children exhibit that effect and so including figures with individual data points is recommended. In addition, the nature of EEG recording often leads to the use of multiple comparisons in analyses and so it is important to apply appropriate corrections (e.g., Bonferroni correction or see Pernet et al., 2015 for other methods).

The field of human developmental electrophysiology has previously been remiss in not including all the methodological details necessary for direct replication and meta-analyses (see Stets et al., 2012, who reported that 47 percent of published studies could not be included in a published meta-analysis because of missing methodological information). In 2014, *Psychophysiology* published a committee report that outlines publications guidelines and recommendations for studies using EEG (Keil et al., 2014). Rather than reiterating the suggestions put forward by this committee, we have republished (with permission) an adapted version of this table for developmental research and refer the reader to Keil and colleagues' (2014) paper for a more detailed description of these suggestions (Table 3.2). All studies using EEG methodology should incorporate this checklist into their laboratory publication procedures.

FUTURE OF EEG/ERP/ssVEP DEVELOPMENTAL RESEARCH

The future of electrophysiology for the study of human development is promising. Electrophysiological techniques are becoming increasingly accessible for researchers and methodological advances are allowing investigators to apply electrophysiological methods to a variety of perceptual, cognitive, social, and clinical domains from early in infancy through adolescence and into adulthood. There are several exciting areas of research currently being conducted with developmental populations that will likely advance our understanding of development. However, there are also several shortcomings that the field, as a whole, needs to address.

Table 3.2 Publication checklist for developmental EEG/ERP/ssVEP research.

Hypotheses

☐ Specific hypotheses and predictions for the measures are described in the Introduction

Participants

☐ Characteristics of the participants are described, including age, sex, race/ethnicity, parental education and SES levels, and other relevant characteristics

☐ Explanation of how sample size was determined (power analysis, previous research, etc.)

Recording characteristics and instruments

☐ The type of EEG sensor is described, including make and model

☐ All sensor locations are specified, including the online reference electrode(s)

☐ Sampling rate is indicated

☐ Online filters are described, specifying the type of filter and including roll-off and cut-off parameters (in dB, or by indicating whether the cut-off represents half power/half amplitude)

☐ Amplifier characteristics are described

☐ Electrode impedance or similar information is provided

☐ Redirecting techniques and/or attention monitoring procedures are described

Stimulus and timing parameters

☐ Timing of all stimuli, responses, and inter-trial intervals are fully specified including whether intervals are from onset or offset

☐ Characteristics of the stimuli are described such that replication is possible

Description of data preprocessing steps

☐ The order of all data preprocessing steps is included

☐ Re-referencing procedures (if any) are specified, including the location of all sensors contributing to the new reference

☐ Segmentation procedures are described, including epoch length and baseline time period

☐ Artifact rejection procedures and parameters are described, including the type and proportion of artifacts rejected, whether the data was manually edited and whether entire trials or partial trials (e.g., zero padding) were removed or interpolated

☐ Artifact correction procedures are described, including the number of components removed, the interpolation or removal method used, and whether the same correction procedures were used with all participants and age groups

☐ Offline filters are described, specifying the type of filter and including roll-off and cut-off parameters (in dB, or by indicating whether the cut-off represents half-power/half amplitude)

☐ The number of trials used for averaging (if any) is described, reporting the number of trials in each condition and each group of subjects. This should include both the mean number of trials in each cell and the range of trials included

Measurement procedures

☐ Measurement procedures are described, including the measurement technique (e.g., mean amplitude), the time window and baseline period, sensor sites, etc.

(continued)

Table 3.2 Continued

☐ For peak amplitude measures, the following is included: whether the peak was an absolute or local peak, whether visual inspection or automatic detection was used, and the number of trials contributing to the averages used for measurement

☐ An a-priori rationale is given for the selection of time windows, electrode sites, etc.

☐ Both descriptive and inferential statistics are included

Statistical analyses

☐ Appropriate correction for any violation of model assumptions is implemented and described (e.g., Greenhouse–Geisser, Huynh–Feldt, or similar adjustment)

☐ The statistical model and procedures are described and results are reported with test statistics, in addition to p-values

☐ An appropriate adjustment is performed for multiple comparisons

☐ If permutation or similar techniques are applied, the number of permutations is indicated together with the method used to identify a threshold for significance

Figures

☐ Line plots (e.g., ERP waveforms) include the following: sensor location, baseline period, axes at zero points (zero physical units and zero ms), appropriate x- and y-axis tick marks at sufficiently dense intervals, polarity, and reference information, where appropriate

☐ Variability information is included (waveform plots with variation plotted, bar or scatterplot plots include individual subject data or other estimate of variance)

☐ Scalp topographies and source plots include the following: captions and labels including a key showing physical units, perspective of the figure (e.g., front view; left/right orientation), type of interpolation used, location of electrodes/sensors, and reference

☐ Coherence/connectivity and time-frequency plots include the following: a key showing physical units, clearly labeled axes, a baseline period, the locations from which the data were derived, a frequency range of sufficient breadth to demonstrate frequency-specificity of effects shown

Spectral analyses

☐ The temporal length and type of data segments (single trials, averages) are defined

☐ The decomposition method is described, and an algorithm or reference given

☐ The frequency resolution (and the time resolution in time-frequency analyses) is specified

☐ The use of windowing function is described, and its parameters (e.g., window length, type) are given

☐ The method for baseline adjustment or normalization is specified, including the temporal segments used for baseline estimation, and the resulting unit

ssVEP analyses

☐ The noise bins used for signal to noise ratio, baseline corrected amplitude, or z score calculations are stated

☐ Whether or not harmonics were included and procedures used to include or exclude harmonics are clearly explained

☐ An a priori rationale for the selection of electrodes is provided

Table 3.2 Continued

Source–estimation procedures

☐ The volume conductor model and the source model are fully described, including the number of tissues, the conductivity values of each tissue, the (starting) locations of sources, and how the sensor positions are registered to the head geometry

☐ The source estimation algorithm is described, including all user-defined parameters (e.g., starting conditions, regularization)

Principal component analysis (PCA)

☐ The structure of the EEG data submitted to PCA is fully described

☐ The type of association matrix is specified

☐ The PCA algorithm is described

☐ Any rotation applied to the data is described

☐ The decision rule for retaining/discarding PCA components is described

Independent component analysis (ICA)

☐ The structure of the EEG data submitted to ICA is described

☐ The ICA algorithm is described

☐ Preprocessing procedures, including filtering, detrending, artifact rejection, etc., are described

☐ The information used for component interpretation and clustering is described

☐ The number of components removed (or retained) per subject is described

Multimodal imaging

☐ Single-modality results are reported

Current source density and Laplacian transformations

☐ The algorithm used and the interpolation functions are described

Single–trial analyses

☐ All preprocessing steps are described

☐ A mathematical description of the algorithm or a reference to a complete description is provided

Open access and data availability

☐ Links to data, processing scripts, and stimuli are provided if open access

Source: Reprinted from *Psychophysiology 51*(1), Andreas Keil, Stefan Debener, Gabriele Gratton, Markus Junghöfer, Emily S. Kappenman, Steven J. Luck, Phan Luu, Gregory A. Miller, and Cindy M. Yee. Committee report: Publication guidelines and recommendations for studies using electroencephalography and magnetoencephalography, 1–21, https://doi.org/10.1111/psyp.12147, Copyright © 2013 Society for Psychophysiological Research.

Areas in Need of Improvement

First, additional work needs to be done to link adult and developmental EEG techniques as our understanding of the developmental trajectory of EEG is limited. Future work should continue to seek to bridge the apparent gap between developmental and adult ERP components. Second, although the field of psychology is in the midst of an open and replicable science revolution (Chambers, 2017), electrophysiological studies in developmental populations have been slow to join this revolution. The nature of conducting electrophysiological research combined with the difficulties in recruitment, retention, and testing of developmental populations has led to lower sample sizes, decreased power, and in some cases low-quality data. However, these problems are not insurmountable if studies are designed appropriately, power analyses are conducted a priori to determine appropriate sample sizes, tasks, and recording parameters are modified to increase sensitivity, effect sizes are clearly reported, and the extent to which electrophysiological responses are reliable within and across participants and ages are quantified (e.g., see Munsters et al., 2019). Future work should examine the impact of error variance on neural responses across ages, as measured by the SNR, determine the time course and topographical distribution of effect size measures that index perceptual, cognitive, and social processes, and quantify internal consistency (e.g., Cronbach's alpha, Cronbach, 1951; Thigpen et al., 2017). Including this information in publications will allow us to identify robust markers for characterizing both inter- and intra-individual developmental differences. Finally, the field of developmental electrophysiology needs to open its doors to researchers who do not collect EEG data from developmental populations but may be able to advance ways in which it is processed and analyzed. To do this, researchers should deposit raw EEG data sets on open access repositories along with the stimuli, processing scripts/codes, and detailed methods information.

Exciting Current and Future Areas of Developmental Electrophysiological Research

First, studies linking behavioral responses with electrophysiological methods continue to be important for elucidating both behavioral and neural development. Research investigations that examine both diverging and converging behavioral and neural findings can lead to significant advances in our understanding of the timing and mechanisms underlying developmental processes (for review, see Reynolds and Guy, 2012; for example, see Vogel et al., 2012). In addition to linking brain and behavior, concurrent recording of multiple physiological methods including heart rate measures (e.g., Cuevas et al., 2012; Xie and Richards, 2016), galvanic skin response (GSR, Zahn-Waxler et al., 1995) and eye-tracking (e.g., Monroy et al., 2019) allows for the determination of changes in attention and arousal. In addition, investigating the relation between the development of EEG, ERP, and ssVEP or between components and frequencies may yield additional insight into developmental processes. Finally, combining EEG measures with MRI/fMRI or fNIRS may provide additional spatial information and allow for increased source estimation accuracy (Akalin Acar et al., 2016; Richards et al., 2016).

Other exciting areas of emerging research include: 1) examining social learning from live interactions before or during EEG recording between infants and caregivers/experimenters; 2) dual recordings between parents and children or between peers; and 3) the use of brain computer interfaces (BCIs) to allow for neural responses to be recorded, analyzed, and translated into commands that can be relayed to a variety of output devices in real time (Carver and Vaccaro, 2007; Hirotani et al., 2009; Kopp and Lindenberger, 2011; Leong et al., 2017; Pineda et al., 2014, Striano et al., 2006).

ACKNOWLEDGMENTS

Funding for review work was provided to L. Scott from the National Science Foundation (NSF) (BCS-1560810;1728133) and from the College of Liberal Arts at the University of Florida. The authors thank Andreas Keil and Emily Kappenman for feedback on a previous draft of this chapter, Dara Oyewole, Amanda Wasoff, and Gabriella Silva for editorial assistance, and members of the UF Brain, Cognition and Development Laboratory, including Allison Carr, Ethan Kutlu, Gabriella Silva, and Amanda Wasoff for relevant discussion and feedback.

REFERENCES

Ackles, P. K. and Cook, K. G. (1998). Stimulus probability and event-related potentials of the brain in 6-month-old human infants: A parametric study, *International Journal of Psychophysiology*, 29(2), 115–143. doi:10.1016/S0167-8760(98)00013-0

Ackles, P. K. and Cook, K. G. (2009). Event-related brain potentials and visual attention in six-month-old infants, *International Journal of Neuroscience*, 119(9), 1446–1468. doi:10.1080/00207450802330579

Adrian, E. D. and Matthews, B. H. C. (1934). The interpretation of potential waves in the cortex, *The Journal of Physiology*, 81(4): 440–471. doi:10.1113/jphysiol.1934.sp003147

Aftanas, L., et al., (2001). Event-related synchronization and desynchronization during affective processing: Emergence of valence-related time-dependent hemispheric asymmetries in theta and upper alpha band, *International Journal of Neuroscience*, 110, 197–219.

Akalin Acar, Z., et al., (2016). High-resolution EEG source imaging of one-year-old children, *Conference Proceedings IEEE Eng Med Biol Soc.*, 2016, 117–120. doi:10.1109/EMBC.2016.7590654

Arbel, Y. and Donchin, E. (2011). When a child errs: The ERN and the Pe complex in children, *Psychophysiology*, 48(1), 55–63. doi:10.1111/j.1469–8986.2010.01042.x

Asano, M., et al., (2015). Sound symbolism scaffolds language development in preverbal infants., *Cortex; A Journal Devoted to the Study of the Nervous System and Behavior*, 63, 196–205. doi:10.1016/j.cortex.2014.08.025

Azizian, A., et al., (2006). Beware misleading cues: Perceptual similarity modulates the N2/P3 complex, *Psychophysiology*, 43(3), 253–260. doi:10.1111/j.1469–8986.2006.00409.x

Barry-Anwar, R., et al., (2018). The developmental time course and topographic distribution of individual-level monkey face discrimination in the infant brain, *Neuropsychologia*, *108*(8), 25–31. doi:10.1016/j.neuropsychologia.2017.11.019

Barry-Anwar, R., Hadley, H., and Scott, L. S. (2019). Differential neural responses to faces paired with labels versus faces paired with noise at 6- and at 9-months, *Vision Research*, *157*, 264–273. doi:10.1016/j.visres.2018.03.002

Basar, E., et al., (2000). Brain oscillations in perception and memory., *International journal of psychophysiology*, *35*(2–3), 95–124. doi:10.1016/S0167-8760(99)00047-1

Batty, M. and Taylor, M. J, (2006). The development of emotional face processing during childhood., *Developmental Science*, *9*(2), 207–20. doi:10.1111/j.1467-7687.2006.00480.x

Bauer, P. J., et al., (2003). Developments in long-term explicit memory late in the first year of life: behavioral and electrophysiological indices, *Psychological Science*, *14*(6), 629–635. doi:10.1046/j.0956-7976.2003.psci_1476.x

Begus, K., Southgate, V., and Gliga, T. (2015). Neural mechanisms of infant learning: differences in frontal theta activity during object exploration modulate subsequent object recognition., *Biology Letters*, *11*(5), 20150041. doi:10.1098/rsbl.2015.0041

Begus, K., Gliga, T., and Southgate, V. (2016). Infants' preferences for native speakers are associated with an expectation of information, *Proceedings of the National Academy of Sciences*, *113*(44), 12397–12402. doi:10.1073/pnas.1603261113

Bell, M. A. (2012). A psychobiological perspective on working memory performance at 8 months of age, *Child Development*, *83*(1), 251–265. doi:10.1111/j.1467-8624.2011.01684.x

Bell, M. A. and Cuevas, K. (2012). Using EEG to study cognitive development: Issues and practices, *Journal of Cognition and Development*, *13*(3), 281–294. doi:10.1080/15248372.2012.691143

Bell, M. A. and Fox, N. A. (1992). The relations between frontal brain electrical activity and cognitive development during infancy, *Child Development*, *63*(5), 1142–1163

Bentin, S., et al. (1996). Electrophysiological studies of face perception in humans, *Journal of Cognitive Neuroscience*, *8*(6), 551–65. doi:10.1162/jocn.1996.8.6.551

Berger H. (1929). Uber das elektrenkephalogramm des menschen. *Archiv Psychiat Nervenkr*, *87*, 527–570.

Bernier, A., Calkins, S. D., and Bell, M. A. (2016). Longitudinal associations between the quality of mother-infant interactions and brain development across infancy, *Child Development*, *87*(4), 1159–1174. doi:10.1111/cdev.12518

Blair, R. C. and Karniski, W. (1993). An alternative methods for significance testing of wave-form differences potentials, *Psychophysiology*, *30*, 518–524.

Bosseler, A. N., et al. (2013). Theta brain rhythms index perceptual narrowing in infant speech perception, *Frontiers in Psychology*, *4*(October), 1–12. doi:10.3389/fpsyg.2013.00690

Boudewyn, M. A., et al. (2018). How many trials does it take to get a significant ERP effect? It depends, *Psychophysiology*, *55*(6), 1–16. doi:10.1111/psyp.13049

Brannon, E. M., et al. (2008). Electrophysiological measures of time processing in infant and adult brains: Weber's law holds, *Journal of Cognitive Neuroscience*, *20*(2), 193–203. doi:10.1162/jocn.2008.20016

Bruin, K. J. and Wijers, A. A. (2002). Inhibition, response mode, and stimulus probability: A comparative event-related potential study, *Clinical Neurophysiology*, *113*(7), 1172–1182. doi:10.1016/S1388-2457(02)00141-4

Buzsáki, G. and Draguhn, A. (2004). Neuronal oscillations in cortical networks, *Science*, *304*(5679), 1926–1929. doi:10.1126/science.1099745

Carmel, D. and Bentin, S. (2002). Domain specificity versus expertise: Factors influencing distinct processing of faces, *Cognition*, 83(1), 1–29. doi:10.1016/S0010-0277(01)00162-7.

Carver, L. J., Bauer, P. J., and Nelson, C. A. (2000). Associations between infant brain activity and recall memory, *Developmental Science.* 3(2), 234–246. doi:10.1111/1467-7687.00116.

Carver, L. J. and Vaccaro, B. G. (2007). 12-Month-old infants allocate increased neural resources to stimuli associated with negative adult emotion, *Developmental Psychology*, 43(1), 54–69. doi:10.1037/0012-1649.43.1.54.

Cauchoix, M., et al. (2014). The neural dynamics of face detection in the wild revealed by MVPA, *The Journal of Neuroscience.* 34(3), 846–854. doi:10.1523/JNEUROSCI.3030–13.2014.

Chambers, C. (2017). *The Seven Deadly Sins of Psychology: A Manifesto for Reforming the Culture of Scientific Practice.* Princeton, NJ, US: Princeton University Press.

Chaumon, M., Bishop, D. V. M., and Busch, N. A. (2015). A practical guide to the selection of independent components of the electroencephalogram for artifact correction. *Journal of Neuroscience Methods*, 250, 47–63. doi:10.1016/j.jneumeth.2015.02.025.

Cheour, M., Leppa Ènen, P. H. T., and Kraus, N. (2000). Mismatch negativity (MMN) as a tool for investigating auditory discrimination and sensory memory in infants and children, *Clinical neurophysiology*, 111, 4–16.

Christodoulou, J., Leland, D. S., and Moore, D. S. (2017). Overt and covert attention in infants revealed using steady-state visually evoked potentials. *Developmental Psychology*, 54(5), 803–815. doi:10.1037/dev0000486.

Chronaki, G., et al. (2012). Isolating N400 as neural marker of vocal anger processing in 6-11-year old children, *Developmental Cognitive Neuroscience*, 2(2), 268–276. doi:10.1016/j.dcn.2011.11.007.

Clark, V. P. and Hillyard, S. A. (1996). Spatial selective attention affects early extrastriate but not striate components of the visual evoked potential, *Journal of Cognitive Neuroscience*, 8(5), 387–402. doi:10.1162/jocn.1996.8.5.387.

Coch, D. (2014). The N400 and the fourth grade shift. *Developmental Science*, 18(2), 254–69. doi:10.1111/desc.12212.

Coch, D. and Meade, G. (2016). N1 and P2 to words and wordlike stimuli in late elementary school children and adults. *Psychophysiology*, 53(2), 115–128. doi:10.1111/psyp.12567.

Couperus, J. W. and Quirk, C. (2015). Visual search and the N2pc in children. *Attention, Perception, and Psychophysics*, 77(3), 768–776. doi:10.3758/s13414-015-0833-5.

Courchesne, E, (1977). Event-related brain potentials: Comparison between children and adults., *Science*, 197(4303), 589–592.

Courchesne, E., et al. (1984). Autism: Processing of novel auditory information assessed by event-related brain potentials. *Electroencephalography and Clinical Neurophysiology/Evoked Potentials*, 59(3), 238–248. doi:10.1016/0168-5597(84)90063-7.

Courchesne, E., Ganz, L., and Norcia, A. M. (1981). Event-related brain potentials to human faces in infants. *Child Development*, 52(3), 804–811.

Cronbach, L. J. (1951). Coefficient alpha and the internal structure of tests. *Psychometrika*, 16(3), 297–334. doi:10.1007/BF02310555.

Crowley, Michael J., et al. (2013). A developmental study of the feedback-related negativity from 10-17 years: Age and sex effects for reward versus non-reward. *Developmental Neuropsychology*, 38(8), 595–512. doi:10.1080/87565641.2012.694512.

Cuevas, K., et al. (2012). Electroencephalogram and heart rate measures of working memory at 5 and 10 months of age. *Developmental Psychology*, 48(4), 907–917. doi:10.1037/a0026448.

Cuevas, K., et al. (2014). The infant EEG mu rhythm: Methodological considerations and best practices. *Developmental Review, 34*(1), 26–43. doi:10.1016/j.dr.2013.12.001.

Cuevas, K. and Bell, M. A. (2011). EEG and ECG from 5 to 10 months of age: Developmental changes in baseline activation and cognitive processing during a working memory task. *International Journal of Psychophysiology, 80*(2), 119–128. doi:10.1016/j.ijpsycho.2011.02.009.

Davidson, R. J. (2000). Affective style, psychopathology, and resilience: Brain mechanisms and plasticity. *American Psychologist, 55*(11), 1196–1214. doi:10.1037/0003-066X.55.11.1196.

Davies, P. L., Segalowitz, S. J., and Gavin, W. J. (2004). Development of response-monitoring ERPs in 7- to 25-year-olds. *Developmental Neuropsychology, 25*(3), 355–376. doi:10.1207/s15326942dn2503_6.

Davis, E. P., et al., (2003). The X-trials: Neural correlates of an inhibitory control task in children and adults. *Journal of Cognitive Neuroscience, 15*(3), 432–443. doi:10.1162/089892903321593144.

DeBoer, T., Scott, L. S., and Nelson, C. A. (2007). Methods for acquiring and analyzing infant event-related potentials. In M. de Haan (ed.), *Infant EEG and Event-Related Potentials, Studies in Developmental Psychology* (pp. 5–37). New York, NY: Psychology Press.

de Haan, M., Johnson, M. H., and Halit, H. (2003). Development of face-sensitive event-related potentials during infancy: A review. *International Journal of Psychophysiology, 51*(1), 45–58. doi:10.1016/S0167-8760(03)00152-1.

de Haan, M. and Nelson, C. A. (1997). Recognition of the mother's face by six-month-old infants: a neurobehavioral study. *Child development, 68*(2), 187–210.

de Haan, M. and Nelson, C. A. (1999). Brain activity differentiates face and object processing in 6-month-old infants. *Developmental Psychology, 35*(4), 1113–1121. doi:10.1037/0012-1649.35.4.1113.

de Haan, M., Pascalis, O., and Johnson, M. H. (2002). Specialization of neural mechanisms underlying face recognition in human infants. *Journal of Cognitive Neuroscience, 14*(2), 199–209. doi:10.1162/089892902317236849.

Dehaene-Lambertz, G. and Gliga, T. (2004). Common neural basis for phoneme processing in infants and adults. *Journal of Cognitive Neuroscience, 16*(8), 1375–1387. doi:10.1162/0898929042304714.

de Heering, A. and Rossion, B. (2015). Rapid categorization of natural face images in the infant right hemisphere. *eLife, 4*(JUNE), 1–14. doi:10.7554/eLife.06564.

Delorme, A., et al. (2015). Grand average ERP-image plotting and statistics: A method for comparing variability in event-related single-trial EEG activities across subjects and conditions. *Journal of Neuroscience Methods, 250*, 3–6. doi:10.1016/j.jneumeth.2014.10.003.

Dien, J. (1998). Issues in the application of the average reference: Review, critiques, and recommendations, *Behavior Research Methods, Instruments, & Computers, 30*(1), 34–43. doi:10.3758/BF03209414.

Di Russo, F., et al. (2002). Cortical sources of the early components of the visual evoked potential. *Human Brain Mapping, 15*(2), 95–111. doi:10.1002/hbm.10010.

Doehnert, M., et al. (2010). Mapping attention-deficit/hyperactivity disorder from childhood to adolescence:no neurophysiologic evidence for a developmental lag of attention but some for inhibition, *Biological Psychiatry, 67*(7), 608–616. doi:10.1016/j.biopsych.2009.07.038.

Donkers, F. C. and Van Boxtel, G. J. (2004). The N2 in go/no-go tasks reflects conflict monitoring not response inhibition. *Brain and Cognition, 56*(2), 165–176. doi:10.1016/j.bandc.2004.04.005.

Downes, M., Bathelt, J., and De Haan, M. (2017). Event-related potential measures of executive functioning from preschool to adolescence. *Developmental Medicine and Child Neurology, 59*(6), 581–590. doi:10.1111/dmcn.13395.

Dubois, J., et al. (2008). Microstructural correlates of infant functional development: example of the visual pathways. *Journal of Neuroscience*, 28(8), 1943–1948. doi:10.1523/jneurosci.5145-07.2008.

Engel, A. K., Fries, P., and Singer, W. (2001). Dynamic predictions: Oscillations and synchrony in top–down processing. *Nature Reviews Neuroscience*, 2(10), 704–716. doi:10.1038/35094565.

Falkenstein, M., et al. (1991). Effects of crossmodal divided attention on late ERP components. II. Error processing in choice reaction tasks. *Electroencephalography and Clinical Neurophysiology*, 78(6), 447–455. doi:10.1016/0013-4694(91)90062-9.

Farzin, F., Hou, C., and Norcia, A. M. (2012). Piecing it together: infants' neural responses to face and object structure. *Journal of vision*, 12(13), 1–14. doi:10.1167/12.13.6.

Fox, N. A. (1994). Dynamic cerebral processes underlying emotion regulation. *Monographs of the Society for Research in Child Development*, 59(2–3), 152–166. doi:10.1111/j.1540-5834.1994.tb01282.x.

Foxe, J. J., Simpson, G. V., and Ahlfors, S. P. (1998). Parieto-occipital ~10 Hz activity reflects anticipatory state of visual attention mechanisms. *NeuroReport*, 9(17), 3929–3933. doi:10.1097/00001756-199812010-00030.

Foxe, J. J. and Snyder, A. C. (2011). The role of alpha-band brain oscillations as a sensory suppression mechanism during selective attention. *Frontiers in Psychology*, 2, 1–13. doi:10.3389/fpsyg.2011.00154.

Friederici, A. D. (2006). The neural basis of language development and its impairment. *Neuron*, 52(6), 941–952. doi:10.1016/j.neuron.2006.12.002.

Friedrich, M. and Friederici, A. D. (2006). Early N400 development and later language acquisition. *Psychophysiology*, 43(1), 1–12. doi:10.1111/j.1469-8986.2006.00381.x.

Friedrich, M. and Friederici, A. D. (2010). Maturing brain mechanisms and developing behavioral language skills. *Brain and Language*, 114(2), 66–71. doi:10.1016/j.bandl.2009.07.004.

Gao, J. F., et al. (2010). Automatic removal of eye-movement and blink artifacts from EEG signals. *Brain Topography*, 23(1), 105–114. doi:10.1007/s10548-009-0131-4.

Gao, Y., et al. (2016). Dysfunctional feedback processing in adolescent males with conduct disorder. *International Journal of Psychophysiology*, 99, 1–9.

Garrido, M. I., et al. (2009). The mismatch negativity: A review of underlying mechanisms, *Clinical Neurophysiology*, 120(3), 453–463. doi:10.1016/j.clinph.2008.11.029.

Geng, F., Canada, K., and Riggins, T. (2018). Age- and performance-related differences in encoding during early childhood: Insights from event-related potentials. *Memory*, 26(4), 451–461. doi:10.1080/09658211.2017.1366526.

Gliga, T. and Dehaene-Lambertz, G. (2005). Structural encoding of body and face in human infants and adults. *Journal of Cognitive Neuroscience*, 17(8), 1328–1340. doi:10.1162/0898929055002481.

Gliga, T., Volein, A., and Csibra, G. (2010). Verbal labels modulate perceptual object processing in 1-year-old children. *Journal of Cognitive Neuroscience*, 22(12), 2781–2789. doi:10.1162/jocn.2010.21427.

Gonzalez, S. L., Reeb-Sutherland, B. C., and Nelson, E. L. (2016). Quantifying motor experience in the infant brain: EEG power, coherence, and mu desynchronization. *Frontiers in Psychology*, 7(February), 1–6. doi:10.3389/fpsyg.2016.00216.

Grammer, J. K., et al. (2014). Age-related changes in error processing in young children: A school-based investigation, *Developmental Cognitive Neuroscience*, 9, 93–105. doi:10.1016/j.dcn.2014.02.001.

Grossmann, T., et al. (2007). Social perception in the infant brain: Gamma oscillatory activity in response to eye gaze. *Social Cognitive and Affective Neuroscience*, 2(4), 284–291. doi:10.1093/scan/nsm025.

Grossmann, T., et al. (2009). The neural basis of perceptual category learning in human infants. *Journal of Cognitive Neuroscience*, 21(12), 2276–2286. doi:10.1162/jocn.2009.21188.

Gruber, T. and Müller, M. M. (2005). Oscillatory brain activity dissociates between associative stimulus content in a repetition priming task in the human EEG. *Cerebral Cortex*, 15(1), 109–116. doi:10.1093/cercor/bhh113.

Gruber, T., Müller, M. M., and Keil, A. (2002). Modulation of induced gamma band responses in a perceptual learning task in the human EEG. *Journal of Cognitive Neuroscience*, 14(5), 732–744. doi:10.1162/08989290260138636.

Guan, Y., Farrar, M. J., and Keil, A. (2018). Oscillatory brain activity differentially reflects false belief understanding and complementation syntax processing. *Cognitive, Affective and Behavioral Neuroscience*, 18(1), 189–201. doi:10.3758/s13415-018-0565-9.

Guy, M. W., Zieber, N., and Richards, J. E. (2016). The cortical development of specialized face processing in infancy. *Child Development*, 87(5), 1581–600. doi:10.1111/cdev.12543.

Hadley, H., et al. (2014). A mechanistic approach to cross-domain perceptual narrowing in the first year of life, *Brain Sciences*, 4, 613–634. doi:10.3390/brainsci4040613.

Hadley, H., Pickron, C. B., and Scott, L. S. (2015). The lasting effects of process-specific versus stimulus-specific learning during infancy. *Developmental Science*, 18(5), 842–852. doi:10.1111/desc.12259.

Hajcak, G. (2012). What we've learned from mistakes: Insights from error-related brain activity. *Current Directions in Psychological Science*, 21(2), 101–106. doi:10.1177/0963721412436809.

Halit, H., De Haan, M., and Johnson, M. H. (2003). Cortical specialisation for face processing: Face-sensitive event-related potential components in 3- and 12-month-old infants, *NeuroImage*, 19(3), 1180–1193. doi:10.1016/S1053-8119(03)00076-4.

Hämäläinen, J. A., Ortiz-Mantilla, S., and Benasich, A. A. (2011). Source localization of event-related potentials to pitch change mapped onto age-appropriate MRIs at 6 months of age. *NeuroImage*, 54, 1910–1918. doi:10.1016/j.neuroimage.2010.10.016.

Hamer, R. D. and Norcia, A. M. (1994). The development of motion sensitivity during the first year of life. *Vision Research*, 34(18), 2387–2402. doi:10.1016/0042-6989(94)90283-6.

Hammerer, D., et al. (2011). Life span differences in electrophysiological correlates of monitoring gains and losses during probabilistic reinforcement learning. *Journal of Cognitive Neuroscience*, 23(3), 579–592. doi:10.1162/jocn.2010.21475.

Harris, R. and Tizard, J. P. M. (1960). The electroencephalogram in neonatal convulsions. *The Journal of Pediatrics*, 57(4), 501–520. doi:10.1016/S0022-3476(60)80078-9.

Hasko, S., et al. (2013). The time course of reading processes in children with and without dyslexia: An ERP study. *Frontiers in Human Neuroscience*, 7, 570. doi:10.3389/fnhum.2013.00570.

Heldmann, M., et al. (2017). Development of sensitivity to orthographic errors in children: An event-related potential study. *Neuroscience*, 358, 349–360. doi:10.1016/j.neuroscience.2017.07.002.

Herrmann, C. S., Fründ, I., and Lenz, D. (2010). Human gamma-band activity: A review on cognitive and behavioral correlates and network models. *Neuroscience and Biobehavioral Reviews*, 34(7), 981–992. doi:10.1016/j.neubiorev.2009.09.001.

Herrmann, C. S., Munk, M. H. J., and Engel, A. K. (2004). Cognitive functions of gamma-band activity: Memory match and utilization. *Trends in Cognitive Sciences*, 8(8), 347–355. doi:10.1016/j.tics.2004.06.006.

Herrmann, M. J., et al. (2004). Source localization (LORETA) of the error-related-negativity (ERN/Ne) and positivity (Pe). *Cognitive Brain Research*, 20(2), 294–299. doi:10.1016/j.cogbrainres.2004.02.013.

Hill, S. Y. and Shen, S. (2002). Neurodevelopmental patterns of visual P3b in association with familial risk for alcohol dependence and childhood diagnosis. *Biological Psychiatry*, 51(8), 621–631. doi:10.1016/S0006-3223(01)01301-4.

Hirotani, M., et al. (2009). Joint attention helps infants learn new words: Event-related potential evidence. *NeuroReport*, 20(6), 600–605. doi:10.1097/WNR.0b013e32832a0a7c.

Hoehl, S., et al. (2014). Eye contact during live social interaction modulates infants' oscillatory brain activity, *Social Neuroscience*, 9(3), 300–308. doi:10.1080/17470919.2014.884982.

Hoehl, S. and Striano, T. (2008). Neural processing of eye gaze and treat-related emotional facial expressions in infancy. *Child development*, 79(6), 1752–1760. doi:10.2307/27563590.

Hoehl, S. and Wahl, S. (2012). Recording infant ERP data for cognitive research, *Developmental Neuropsychology*, 37(3), 187–209. doi:10.1080/87565641.2011.627958.

Holroyd, C. B. and Coles, M. G. H. (2002). The neural basis of human error processing: Reinforcement learning, dopamine, and the error-related negativity. *Psychological Review*, 109(4), 679–709. doi:10.1037/0033-295X.109.4.679.

Huster, R. J., et al. (2010). The role of the cingulate cortex as neural generator of the N200 and P300 in a tactile response inhibition task. *Human Brain Mapping*, 31(8), 1260–1271. doi:10.1002/hbm.20933.

Itier, R. and Taylor, M. (2004a). Face inversion and contrast-reversal effects across development: In contrast to the expertise theory. *Developmental Science*, 7(2), 246–260. doi:10.1111/j.1467-7687.2004.00342.x.

Itier, R. and Taylor, M. (2004b). N170 or N1? Spatiotemporal differences between object and face processing using ERPs. *Cerebral Cortex*, 14(2), 132–142.

Ito, S., et al. (2003). Performance monitoring by anterior cingulate cortex during saccade countermanding. *Science*, 302(October), 120–122. doi:10.1126/science.1087847.

Jodo, E. and Kayama, Y. (1992). Relation of a negative ERP component to response inhibition in a Go/No-go task, *Electroencephalography and Clinical Neurophysiology*, 82(6), 477–482. doi:10.1016/0013-4694(92)90054-L.

Johnson, M. H., et al. (2001). Recording and analyzing high-density event-related potentials with infants. Using the Geodesic sensor net. *Developmental neuropsychology*, 19(3), 295–323. doi:10.1207/S15326942DN1903_4.

Johnson, M. H., et al. (2005). The emergence of the social brain network: Evidence from typical and atypical development. *Development and psychopathology*, 17(3), 599–619. doi:10.1017/S0954579405050297.

Johnstone, S. J., et al. (2005). Development of inhibitory processing during the Go/NoGo task. *Journal of Psychophysiology*, 19(1), 11–23. doi:10.1027/0269-8803.19.1.11.

Jones, T., et al. (2018). Neural and behavioral effects of subordinate-level training of novel objects across manipulations of color and spatial frequency. *European Journal of Neuroscience*. doi:10.1111/ejn.13889.

Junge, C., Cutler, A., and Hagoort, P. (2012). Electrophysiological evidence of early word learning. *Neuropsychologia*, 50(14), 3702–3712. doi:10.1016/j.neuropsychologia.2012.10.012.

Kappenman, E.S. and Luck, S. J. (2010). The effects of electrode impedance on data quality and statistical significance in ERP recordings. *Psychophysiology*, 47(5), 888–904. doi:10.1111/j.1469-8986.2010.01009.x.

Karrer, R. and Ackles, P. K. (1987). Visual event-related potentials of infants during a modified oddball procedure. In Johnson, R., Parasuraman, R., and Rohrbaugh, J. W. (eds) *Current Trends in Event-Related Potential Research* (pp. 603–608). Amsterdam: Elsevier.

Karrer, R. and Monti, L. A. (1995). Event-related potentials of 4-7-week-old infants in a visual recognition memory task. *Electroencephalography and Clinical Neurophysiology*, *94*(6), 414–424.

Kaufman, J., Csibra, G., and Johnson, M. H. (2003). Representing occluded objects in the human infant brain. *Proceedings of the Royal Society B: Biological Sciences*, *270*(Suppl 2), S140–S143. doi:10.1098/rsbl.2003.0067.

Kaufman, J., Csibra, G,. and Johnson, M. H. (2005). Oscillatory activity in the infant brain reflects object maintenance. *Proceedings of the National Academy of Sciences of the United States of America*, *102*(42), 15271–15274. doi:10.1073/pnas.0507626102.

Keil, A., et al. (2014). Committee report: Publication guidelines and recommendations for studies using electroencephalography and magnetoencephalography. *Psychophysiology*, *51*(1), 1–21. doi:10.1111/psyp.12147.

Klimesch, W., et al. (1998). Induced alpha band power changes in the human EEG and attention. *Neuroscience Letters*, *244*(2), 73–76. doi:10.1016/S0304-3940(98)00122-0.

Klimesch, W. (1999). EEG alpha and theta oscillations reflect cognitive and memory performance: A review and analysis. *Brain Research Reviews*, *29*(2–3), 169–195. doi:10.1016/S0165-0173(98)00056-3.

Kopp, F. and Lindenberger, U. (2011). Effects of joint attention on long-term memory in 9-month-old infants: An event-related potentials study. *Developmental Science*, *14*(4), 660–672. doi:10.1111/j.1467-7687.2010.01010.x.

Kouider, S., et al. (2015). Neural dynamics of prediction and surprise in infants. *Nature Communications*, *6*, 1–8. doi:10.1038/ncomms9537.

Kuefner, D. et al. (2010). Early visually evoked electrophysiological responses over the human brain (P1, N170) show stable patterns of face-sensitivity from 4 years to adulthood. *Frontiers in Human Neuroscience*, *3*(January), 1–22. doi:10.3389/neuro.09.067.2009.

Kurtzberg, D. and Vaughan, H. G. J. (1979). Maturation and task specificity of cortical potentials associated with visual scanning. In Lehmann, D. and Callaway, E. (eds) *Human Evoked Potentials: Applications and Problems; Proceedings. NATO Special Program/Panel on Human Factors* (pp. 185–199). New York, NY: Plenum Press.

Kushnerenko, E., et al. (2002). Maturation of the auditory change detection response in infants: A longitudinal ERP study. *NeuroReport*, *13*(15), 1843–1848.

Kutas, M. and Federmeier, K. D. (2000). Electrophysiology reveals semantic memory use in language comprehension. *Trends in Cognitive Sciences*, *4*(12), 463–4670. doi:10.1016/S1364-6613(00)01560-6.

Kutas, M. and Federmeier, K. D. (2011). Thirty years and counting: Finding meaning in the N400 component of the event-related brain potential (ERP). *Annual Review of Psychology*, *62*(1), 621–647. doi:10.1146/annurev.psych.093008.131123.

Kutas, M. and Hillyard, S. A. (1980). Reading senseless sentences: brain potentials reflect semantic incongruity. *Science*, *207*(4427), 203–205.

Kutas, M. and Van Petten, C. (1994). ERP psycholinguistics electrified: Event-related brain potential investigations. In Gernsbacher, M. A. (ed.) *Handbook of Psycholinguistics* (pp. 83–143), Cambridge, MA: Academic Press. doi:10.1016/B978-012369374-7/50018-3.

Lamm, C., Zelazo, P. D., and Lewis, M. D. (2006). Neural correlates of cognitive control in childhood and adolescence: Disentangling the contributions of age and executive function. *Neuropsychologia*, *44*(11), 2139–2148. doi:10.1016/j.neuropsychologia.2005.10.013.

Lau, E. F., Phillips, C., and Poeppel, D. (2008). A cortical network for semantics: (De) constructing the N400. *Nature Reviews Neuroscience*, *9*(12), 920–933. doi:10.1038/nrn2532.

Lei, X. and Liao, K. (2017). Understanding the influences of EEG reference: A large-scale brain network perspective. *Frontiers in Neuroscience*, *11*(APR), 1–11. doi:10.3389/fnins.2017.00205.

Lenartowicz, A., et al. (2016). Alpha desynchronization and fronto-parietal connectivity during spatial working memory encoding deficits in ADHD: A simultaneous EEG-fMRI study. *NeuroImage: Clinical*, *11*, 210–223. doi:10.1016/j.nicl.2016.01.023.

Leong, V., et al. (2017). Speaker gaze increases information coupling between infant and adult brains. *Proceedings of the National Academy of Sciences*, *114*(50), p. 201702493. doi:10.1073/pnas.1702493114.

Leppänen, J. M., et al. (2007). An ERP study of emotional face processing in the adult and infant brain. *Child Development*, *78*(1), 232–245. doi:10.1111/j.1467–8624.2007.00994.x.

Lewis, M., et al. (2006). Neurophysiological correlates of emotion regulation in children and adolescents. *Journal of Cognitive Neuroscience*, *18*(3), 430–443.

Liu-Shuang, J., Norcia, A. M., and Rossion, B. (2014). An objective index of individual face discrimination in the right occipito-temporal cortex by means of fast periodic oddball stimulation. *Neuropsychologia*, *52*(1), 57–72. doi:10.1016/j.neuropsychologia.2013.10.022.

Lochy, A., de Heering, A., and Rossion, B. (2019). The non-linear development of the right hemispheric specialization for human face perception. *Neuropsychologia*, 126, 10–19. doi:10.1016/j.neuropsychologia.2017.06.029.

Luck, S. J. (2014). *An Introduction to the Event-Related Potential Technique*. 2nd edn. Cambridge, MA: MIT Press.

Luu, P., et al. (2003). Electrophysiological responses to errors and feedback in the process of action regulation. *Psychological Science*, *14*(1), 47–53. doi:10.1111/1467-9280.01417.

Maguire, M. J., Magnon, G., and Fitzhugh, A. E. (2014). Improving data retention in EEG research with children using child-centered eye tracking. *Journal of Neuroscience Methods*, *238*, 78–81. doi:10.1016/j.jneumeth.2014.09.014.

Marinović, V., Hoehl, S., and Pauen, S. (2014). Neural correlates of human-animal distinction: An ERP-study on early categorical differentiation with 4- and 7-month-old infants and adults. *Neuropsychologia*, *60*(1), 60–76. doi:10.1016/j.neuropsychologia.2014.05.013.

Maris, E. and Oostenveld, R. (2007). Nonparametric statistical testing of EEG- and MEG-data. *Journal of neuroscience methods*, *164*(1), 177–90. doi:10.1016/j.jneumeth.2007.03.024.

Marshall, P. J., Bar-Haim, Y., and Fox, N. A. (2002). Development of the EEG from 5 months to 4 years of age. *Clinical Neurophysiology*, *113*(8), 1199–208. doi:10.1016/S1388-2457(02)00163-3.

Marshall, P. J., Reeb, B. C., and Fox, N. A. (2009). Electrophysiological responses to auditory novelty in temperamentally different 9-month-old infants. *Developmental Science*, *12*(4), 568–582. doi:10.1111/j.1467-7687.2008.00808.x.

Marshall, P. J., Young, T., and Meltzoff, A. N. (2011). Neural correlates of action observation and execution in 14-month-old infants: An event-related EEG desynchronization study. *Developmental Science*, *14*(3), 474–480. doi:10.1111/j.1467-7687.2010.00991.x.

Mazza, V., Turatto, M., and Caramazza, A. (2009). Attention selection, distractor suppression and N2pc. *Cortex*, *45*(7), 879–890. doi:10.1016/j.cortex.2008.10.009.

McCleery, J. P., et al. (2009). Atypical face versus object processing and hemispheric asymmetries in 10-month-old infants at risk for autism. *Biological Psychiatry, 66*(10), 950–957. doi:10.1016/j.biopsych.2009.07.031.

Meyer, A., et al. (2012). Additive effects of the dopamine D2 receptor (DRD2) and dopamine transporter (DAT1) genes on the error-related negativity (ERN) in young children. *Genes, Brain, and Behavior, 11*(6), 695–703. doi:10.1111/j.1601-183X.2012.00812.x.

Meyer, A., Bress, J. N., and Proudfit, G. H. (2014). Psychometric properties of the error-related negativity in children and adolescents. *Psychophysiology, 51*(7), 602–610. doi:10.1111/psyp.12208.

Meyer, M., et al. (2014). Neural correlates of feedback processing in toddlers. *Journal of Cognitive Neuroscience, 26*, 1519–1527.

Michel, C., et al. (2015). Theta- and alpha-band EEG activity in response to eye gaze cues in early infancy, *NeuroImage, 118*, 576–583. doi:10.1016/j.neuroimage.2015.06.042.

Mills, D. L., et al. (2004). Language experience and the organization of brain activity to phonetically similar words: ERP evidence from 14- and 20-month-olds. *Journal of Cognitive Neuroscience, 16*(8), 1452–1464. doi:10.1162/0898929042304697.

Mills, D. L., Conboy, B. T., and Paton, C. (2005). Do changes in brain organization reflect shifts in symbolic functioning? In Namy, L. L. (ed.) *Emory symposia in Cognition. Symbol Use and Symbolic Representation: Developmental and Comparative Perspectives* (pp. 123–153). Mahwah, NJ: Lawrence Erlbaum Associates Publishers.

Miltner, W. H. R., Braun, C. H., and Coles, M. G. H. (1997). Event-related brain potentials following incorrect feedback in a time-estimation task: Evidence for a generic neural system for error detection. *Journal of Cognitive Neuroscience, 9*(6), 788–798. doi:10.1162/jocn.1997.9.6.788.

Mognon, A., et al. (2011). ADJUST: An automatic EEG artifact detector based on the joint use of spatial and temporal features *Psychophysiology, 48*(2), 229–240. doi:10.1111/j.1469-8986.2010.01061.x.

Monroy, C. D., et al. (2019). Sensitivity to structure in action sequences: An infant event-related potential study. *Neuropsychologia, 126*, 92–102. doi:10.1016/j.neuropsychologia.2017.05.007.

Müller, M. M., Gruber, T., and Keil, A. (2000). Modulation of induced gamma band activity in the human EEG by attention and visual information processing. *International journal of psychophysiology, 38*(3), 283–299. doi:10.1016/S0167-8760(00)00171-9.

Munsters, N. M., et al. (2019). Test-retest reliability of infant event related potentials evoked by faces. *Neuropsychologia, 126*, 20–26. doi:10.1016/j.neuropsychologia.2017.03.030.

Nelson, C. A (1994). Neural correlates of recognition memory in the first postnatal year of life. In Dawson, G. and Fischer, K. (eds.) *Human Behavior and the Developing Brain* (pp. 269–313). New York, NY: Guilford Press.

Nelson, C. A., et al. (1998). Delayed recognition memory in infants and adults as revealed by event-related potentials. *Int J Psychophysiol, 29*(2), 145–165. doi:10.1016/s0167-8760(98)00014-2.

Nelson, C. A. and Collins, P. F. (1991). Event-related potential and looking-time analysis of infants responses to familiar and novel events: Implications for visual recognition memory. *Developmental Psychology, 27*(1), 50–58. doi:10.1037/0012-1649.27.1.50.

Nelson, C. A. and de Haan, M. (1996). Neural correlates of visual responsiveness to facial expressions of emotion. *Developmental Psychobiology, 29*(September 1995), 1–18. doi:10.1002/(SICI)1098-2302(199611)29.

Neuper, C. and Pfurtscheller, G. (2001). Event-related dynamics of cortical rhythms: Frequency-specific features and functional correlates. *International Journal of Psychophysiology*, 43(1), 41–58. doi:10.1016/S0167-8760(01)00178-7.

Nieuwenhuis, S., et al. (2001). Error-related brain potentials are differentially related to awareness of response errors: Evidence from an antisaccade task. *Psychophysiology*, 38(5), 752–760. doi:10.1017/S0048577201001111.

Nolan, H., Whelan, R., and Reilly, R. B. (2010). FASTER: Fully Automated Statistical Thresholding for EEG artifact Rejection. *Journal of Neuroscience Methods*, 192(1), 152–162. doi:10.1016/j.jneumeth.2010.07.015.

Norcia, A., Tyler, C., and Hamer, R. (1990). Development of Contrast Sensitivity in the Human infant. *Vision Research*, 30(10), 1475–1486. doi:10.1016/0042-6989(90)90028-J.

Norcia, A. M., et al. (2015). The steady-state visual evoked potential in vision research: A review. *Journal of Vision*, 15(6), 4. doi:10.1167/15.6.4.doi.

Novitski, N., et al. (2007). Neonatal frequency discrimination in 250-4000-Hz range: Electrophysiological evidence. *Clinical Neurophysiology*, 118(2), 412–419. doi:10.1016/j.clinph.2006.10.008.

Nunez, P. (1981). *Electrical Fields of the Brain*. New York, NY: Oxford.

Olvet, D. M. and Hajcak, G. (2008). The error-related negativity (ERN) and psychopathology: Toward an endophenotype. *Clinical Psychology Review*, 28(8), 1343–1354. doi:10.1016/j.cpr.2008.07.003.

Orekhova, E. V., Stroganova, T. A., and Posikera, I. N. (2001). Theta synchronization during sustained anticipatory attention in infants over the second half of the first year of life. *International Journal of Psychophysiology*, 32(2), 151–172. doi:10.1016/S0167-8760(99)00011-2.

Orekhova, E. V., Stroganova, T. A. and Posikera, I. N, 2001, Alpha activity as an index of cortical inhibition during sustained internally controlled attention in infants. *Clinical Neurophysiology* 112(5), 740–749. doi:10.1016/S1388-2457(01)00502-8.

Ortiz-Mantilla, S., Hämäläinen, J. A., and Benasich, A. A. (2012). Time course of ERP generators to syllables in infants: A source localization study using age-appropriate brain templates. *NeuroImage*, 59(4), 3275–3287. doi:10.1016/j.neuroimage.2011.11.048.

Ouyang, G., Sommer, W., and Zhou, C. (2015). A toolbox for residue iteration decomposition (RIDE):A method for the decomposition, reconstruction, and single trial analysis of event related potentials. *Journal of Neuroscience Methods*, 250, 7–21. doi:10.1016/j.jneumeth.2014.10.009.

Overbeek, T. J. M., Nieuwenhuis, S., and Ridderinkhof, K. R. (2005). Dissociable components of error processing. *Journal of Psychophysiology*, 19(4), 319–329. doi:10.1027/0269-8803.19.4.319.

Pantev, C., et al. (1991). Human auditory evoked gamma-band magnetic fields. *Proceedings of the National Academy of Sciences of the United States of America*, 88(20), 8996–9000. doi:10.1073/pnas.88.20.8996.

Pare, D., Collins, D., and Pelletier, J. G. (2002). Amygdala oscillations and the consolidation of emotional memories. *Trends in Cognitive Sciences*, 6, 306–314.

Parise, E. and Csibra, G. (2012) Electrophysiological evidence for the understanding of maternal speech by 9-month-old infants. *Psychological Science*, 23(7), 728–733. doi:10.1177/0956797612438734.

Peltola, M. J., et al. (2009). Emergence of enhanced attention to fearful faces between 5 and 7 months of age. *Social Cognitive and Affective Neuroscience*, 4(2), 134–142. doi:10.1093/scan/nsn046.

Pernet, C. R., et al. (2015). Cluster-based computational methods for mass univariate analyses of event-related brain potentials/fields: A simulation study. *Journal of Neuroscience Methods*, *250*, 85–93. doi:10.1016/j.jneumeth.2014.08.003.

Peykarjou, S., et al. (2017). Rapid categorization of human and ape faces in 9-month-old infants revealed by fast periodic visual stimulation. *Scientific Reports*, *7*(1), 12526. doi:10.1038/s41598-017-12760-2.

Peykarjou, S., Pauen, S., and Hoehl, S. (2014). How do 9-month-old infants categorize human and ape faces? A rapid repetition ERP study. *Psychophysiology*, *51*(9), 866–878. doi:10.1111/psyp.12238.

Peykarjou, S., Pauen, S., and Hoehl, S. (2016). 9-Month-old infants recognize individual unfamiliar faces in a rapid repetition ERP paradigm. *Infancy*, *21*(3), 288–311. doi:10.1111/infa.12118.

Pfurtscheller, G., Stancák, A., and Neuper, C. (1996). Event-related synchronization (ERS) in the alpha band - An electrophysiological correlate of cortical idling: A review. *International Journal of Psychophysiology*, 24(1–2), 39–46. doi:10.1016/S0167-8760(96)00066-9.

Pickron, C. B., et al. (2018) Learning to individuate: The specificity of labels differentially impacts infant visual attention. *Child Development*, *89*(3), 698–710. doi:10.1111/cdev.13004.

Picton, T. W. and Taylor, M. J. (2007). Electrophysiological evaluation of human brain development. *Developmental Neuropsychology*, *31*(3), 249–278. doi:10.1080/87565640701228732.

Pierce, L. J., et al. (2011). The N250 brain potential to personally familiar and newly learned faces and objects. *Frontiers in Human Neuroscience*, *5*(October), 1–13. doi:10.3389/fnhum.2011.00111.

Pineda, J. A., et al. (2014). Neurofeedback training produces normalization in behavioural and electrophysiological measures of high-functioning autism. *Philosophical Transactions of the Royal Society B: Biological Sciences*, *369*(1644). Available at: http://rstb.royalsocietypublishing.org/content/369/1644/20130183.abstract.

Pivik, R. T., et al. (1993). Guidelines for the recording and quantitative analysis of electroencephalographic activity in research contexts. *Psychophysiology*, *30*, 547–558.

Polich, J. (2012). Neuropsychology of P300. In Kappenman, E. S. and Luck, S. J. (eds) *The Oxford Handbook of Event-Related Potential Components* (pp. 159–188). New York, NY: Oxford University Press. doi:10.1093/oxfordhb/9780195374148.013.0089.

Poole, K. L., et al. (2018). Trajectories of frontal brain activity and socio-emotional development in children. *Developmental Psychobiology*, *60*(4), 353–363. doi:10.1002/dev.21620.

Proudfit, G. H. (2015). The reward positivity: From basic research on reward to a biomarker for depression. *Psychophysiology*, *52*(4), 449–59. doi:10.1111/psyp.12370

Pulvermüller, F., et al. (1996). High-frequency cortical responses reflect lexical processing: An MEG study. *Electroencephalography and Clinical Neurophysiology*, *98*(1), 76–85. doi:10.1016/0013-4694(95)00191-3.

Regan, D. (1989). *Human Brain Electrophysiology: Evoked Potentials and Evoked Magnetic Fields in Science and Medicine*. Amsterdam: Elsevier.

Reid, V. M., et al. (2009). The neural correlates of infant and adult goal prediction: Evidence for semantic processing systems. *Developmental psychology*, *45*(3), 620–629. doi:10.1037/a0015209.

Reynolds, G. D., Courage, M. L., and Richards, J. E. (2010). Infant attention and visual preferences: Converging evidence from behavior, event-related potentials, and cortical source localization. *Developmental Psychology*, *46*(4), 886–904. doi:10.1037/a0019670.

Reynolds, G. D., Courage, M. L., and Richards, J. E. (2013). The development of attention. In Reisberg, D. (ed.) *The Oxford Handbook of Cognitive Psychology* (pp. 1000–13). New York, NY: Oxford University Press. doi:10.1093/oxfordhb/9780195376746.013.0063.

Reynolds, G. D. and Richards, J. E. (2005). Familiarization, attention, and recognition memory in infancy: An event-related potential and cortical source localization study. *Developmental Psychology*, 41(4), 598–615. doi:10.1037/0012-1649.41.4.598.

Reynolds, G. D. and Richards, J. E. (2009). Cortical source localization of infant cognition, *Developmental Neuropsychology*, 34(3), 312–329. doi:10.1080/87565640902801890.

Reynolds, G. D. and Richards, J. E. (2019). Infant visual attention and stimulus repetition effects on object recognition. *Child Development*, 90(4), 1027–1042. doi:10.1111/cdev.12982.

Reynolds, G. and Guy, M. (2012). Brain–behavior relations in infancy: Integrative approaches to examining infant looking behavior and event-related potentials. *Developmental Neuropsychology*, 37(3), 210–225.

Reynolds, G. D., Zhang, D., and Guy, M. W. (2013). infant attention to dynamic audiovisual stimuli: Look duration from 3 to 9 months of age. *Infancy*, 18(4), 554–577. doi:10.1111/j.1532-7078.2012.00134.x.

Richards, J. E. (2003). Attention affects the recognition of briefly presented visual stimuli in infants: An ERP study. *Developmental Science*, 6(3), 312–328. doi:10.1111/1467-7687.00287.

Richards, J. E., et al. (2016). A database of age-appropriate average MRI templates. *NeuroImage*, 124, 1254–1259. doi:10.1016/j.neuroimage.2015.04.055.

Richards, J. E., Reynolds, G. D., and Courage, M. L. (2010). The neural bases of infant attention. *Current Directions in Psychological Science*, 19(1), 41–46. doi:10.1177/0963721409360003.

Riggins, T., et al. (2009). Consequences of low neonatal iron status due to maternal diabetes mellitus on explicit memory performance in childhood. *Developmental Neuropsychology*, 34(6), 762–779. doi:10.1080/87565640903265145.

Riggins, T. and Rollins, L. (2015). Developmental differences in memory during early childhood: insights from event-related potentials. *Child Development*, 86(3), 889–902. doi:10.1111/cdev.12351.

Riggins, T., Rollins, L., and Graham, M. (2013). Electrophysiological investigation of source memory in early childhood. *Developmental Neuropsychology*, 38(3), 180–196. doi:10.1080/87565641.2012.762001.

Riggins, T. and Scott, L. S. (2020), P300 Development from infancy to adulthood, *Psychophysiology*, 57, e13346. doi:10.1111/psyp.13346.

Robertson, S. S., Watamura, S. E., and Wilbourn, M. P. (2012). Attentional dynamics of infant visual foraging. *Proceedings of the National Academy of Sciences*, 109(28), 11460–11464. doi:10.1073/pnas.1203482109.

Robey, A. and Riggins, T. (2016). Event-related potential study of intentional and incidental retrieval of item and source memory during early childhood. *Developmental Psychobiology*, 58(5), 556–567. doi:10.1002/dev.21401.

Rossion, B., et al. (2000). The N170 occipito-temporal component is delayed and enhanced to inverted faces but not to inverted objects: An electrophysiological account of face-specific processes in the human brain. *Neuroreport*, 11(1), 69–74.

Rossion, B. (2014). Understanding individual face discrimination by means of fast periodic visual stimulation, *Experimental Brain Research*, 232(6), 1599–1621. doi:10.1007/s00221-014-3934-9.

Rossion, B. and Caharel, S. (2011). ERP evidence for the speed of face categorization in the human brain: Disentangling the contribution of low-level visual cues from face perception. *Vision research*, 51(12), 1297–1311. doi:10.1016/j.visres.2011.04.003.

Rossion, B. and Jacques, C. (2008). Does physical interstimulus variance account for early electrophysiological face sensitive responses in the human brain? Ten lessons on the N170., *NeuroImage*, 39(4), 1959–1979. doi:10.1016/j.neuroimage.2007.10.011.

Rousselet, G. A., et al. (2011). Modeling single-trial ERP reveals modulation of bottom-up face visual processing by top-down task constraints (in some subjects). *Frontiers in Psychology*, 2(JUN), 1–19. doi:10.3389/fpsyg.2011.00137.

Rueschemeyer, S. A., Gardner, T., and Stoner, C. (2014). The Social N400 effect: How the presence of other listeners affects language comprehension. *Psychonomic Bulletin and Review*, 22(1), 128–134. doi:10.3758/s13423-014-0654-x.

Ryynänen, O. R. M., Hyttinen, J. A. K., and Malmivuo, J. A. (2006). Effect of measurement noise and electrode density on the spatial resolution of cortical potential distribution with different resistivity values for the skull. *IEEE Transactions on Biomedical Engineering*, 53(9), 1851–1858. doi:10.1109/TBME.2006.873744.

Saby, J. N. and Marshall, P. J. (2012). The utility of EEG band power analysis in the study of infancy and early childhood. *Developmental Neuropsychology*, 37(3), 253–273. doi:10.1080/87565641.2011.614663.

Sammler, D., et al. (2007). Music and emotion: Electrophysiological correlates of the processing of pleasant and unpleasant music. *Psychophysiology*, 44(2), 293–304. doi:10.1111/j.1469-8986.2007.00497.x.

Schweinberger, S. R., et al. (2002). Human brain potential correlates of repetition priming in face and name recognition. *Neuropsychologia*, 40(12), 2057–2073. doi:10.1016/s0028-3932(02)00050-7.

Schweinberger, S. R., Huddy, V., and Burton, A. M. (2004). N250r: A face-selective brain response to stimulus repetitions. *NeuroReport*, 15(9), 1501–1505. doi:10.1097/01.wnr.0000131675.00319.42.

Scott, L. S., et al. (2006). A reevaluation of the electrophysiological correlates of expert object processing. *Journal of Cognitive Neuroscience*, 18(9), 1453–1465. doi:10.1162/jocn.2006.18.9.1453.

Scott, L. S., et al. (2008). The role of category learning in the acquisition and retention of perceptual expertise: A behavioral and neurophysiological study. *Brain Research*, 1210, 204–215. doi:10.1016/j.brainres.2008.02.054.

Scott, L. S. (2011). Mechanisms underlying the emergence of object representations during infancy. *Journal of Cognitive Neuroscience*, 23, 2935–2944. doi:10.1162/jocn_a_00019.

Scott, L. S. and Monesson, A. (2010). Experience-dependent neural specialization during infancy. *Neuropsychologia*, 48(6), 1857–1861. doi:10.1016/j.neuropsychologia.2010.02.008.

Scott, L. S. and Nelson, C. A. (2006). Featural and configural face processing in adults and infants: A behavioral and electrophysiological investigation. *Perception*, 35(8), 1107–1128.

Scott, L. S., Shannon, R. W., and Nelson, C. A. (2006). Neural correlates of human and monkey face processing in 9-month-old infants. *Infancy*, 10(February), 171–86. doi:10.1207/s15327078in1002_4.

Sheffield, J. G., et al. (2015). Reward related neural activity and adolescent antisocial behavior in a community sample. *Developmental Neuropsychology*, 40, 363–378.

Shelburne, S. A. (1973). Visual evoked responses to language stimuli in normal children. *Electroencephalography and Clinical Neurophysiology*, 34(2), 135–143. doi:10.1016/0013-4694(73)90040-0.

Shimi, A., Nobre, A. C., and Scerif, G. (2015). ERP markers of target selection discriminate children with high vs. low working memory capacity. *Frontiers in Systems Neuroscience*, 9, 153. doi:10.3389/fnsys.2015.00153.

Shulman-Galambos, C. and Galambos, R. (1978). Cortical responses from adults and infants to complex visual stimuli. *Electroencephalography and Clinical Neurophysiology*, 45(4), 423–435.

Smith, J. R. (1938). The electroencephalogram during normal infancy and childhood: II. The nature of the growth of the alpha waves. *The Pedagogical Seminary and Journal of Genetic Psychology*, 53(2), 455–469. doi:10.1080/08856559.1938.10533821.

Snyder, K., Webb, S. J., and Nelson, C. A. (2002). Theoretical and methodological implications of variability in infant brain response during a recognition memory paradigm. *Infant Behavior and Development*, 25(4), 466–494. doi:10.1016/S0163-6383(02)00146-7.

Southgate, V., et al. (2009). Predictive motor activation during action observation in human infants. *Biology Letters*, 5(6), 769–772. doi:10.1098/rsbl.2009.0474.

Stets, M., Stahl, D., and Reid, V. M (2012). A meta-analysis investigating factors underlying attrition rates in infant ERP studies. *Developmental Neuropsychology*, 37(3), 226–252. doi:10.1080/87565641.2012.654867.

Stevens, C., Sanders, L., and Neville, H. (2006). Neurophysiological evidence for selective auditory attention deficits in children with specific language impairment. *Brain Research*, 1111(1), 143–152. doi:10.1016/j.brainres.2006.06.114.

Stites, M. C. and Laszlo, S. (2017). Time will tell: A longitudinal investigation of brain-behavior relationships during reading development. *Psychophysiology*, 54(6), 798–808. doi:10.1111/psyp.12844.

Striano, T., Reid, V. M., and Hoehl, S. (2006). Neural mechanisms of joint attention in infancy. *The European Journal of Neuroscience*, 23(10), 2819–2823. doi:10.1111/j.1460-9568.2006.04822.x.

Stroganova, T., Orekhova, E., and Posikera, I. N. (1999). EEG alpha rhythm in infants. *Clinical Neurophysiolog: Official Journal of the International Federation of Clinical Neurophysiology*, 110(6), 997–1012. doi:10.1016/S1388-2457(98)00009-1.

Stroganova, T. A., V. Orekhova, E., and Posikera, I. N. (1998). Externally and internally controlled attention in infants: An EEG study. *International Journal of Psychophysiology*, 30(3), 339–351. doi:10.1016/S0167-8760(98)00026-9.

Tarullo, A. R., et al. (2017). Gamma power in rural Pakistani children: Links to executive function and verbal ability. *Developmental Cognitive Neuroscience*, 26, 1–8. doi:10.1016/j.dcn.2017.03.007.

Taylor, J. B., et al. (2018). The error-related negativity (ERN) is an electrophysiological marker of motor impulsiveness on the Barratt Impulsiveness Scale (BIS-11) during adolescence. *Developmental Cognitive Neuroscience*, 30(January), 77–86. doi:10.1016/j.dcn.2018.01.003.

Taylor, M. J., et al. (2001). Eyes first! Eye processing develops before face processing in children. *Cognitive Neuroscience and Neuropsychology*, 12(8), 1671–1676.

Taylor, M. J. (2002). Non-spatial attentional effects on P1. *Clinical Neurophysiology*, 113(12), 1903–1908. doi:10.1016/S1388-2457(02)00309-7.

Taylor, M. J., Batty, M., and Itier, R. J. (2004). The faces of development: A review of early face processing over childhood. *Journal of Cognitive Neuroscience*, 16(8), 1426–1442. doi:10.1162/0898929042304732.

Thatcher, R. W. (2012). Coherence, phase differences, phase shift, and phase lock in EEG/ERP analyses. *Developmental Neuropsychology*, 37(6), 476–96. doi:10.1080/87565641.2011.619241.

Thierry, G. (2005). The use of event-related potentials in the study of early cognitive development. *Infant and Child Development*, 14(1), 85–94. doi:10.1002/icd.353.

Thigpen, N. N., Kappenman, E. S., and Keil, A. (2017). Assessing the internal consistency of the event-related potential: An example analysis. *Psychophysiology, 54*(1), 123–138. doi:10.1111/psyp.12629.

Tiitinen, H. T., et al. (1993). Selective attention enhances the auditory 40-Hz transient response in humans. *Nature, 364*(6432), 59–60. doi:10.1038/364059a0.

Torpey, D. C., et al. (2013). Error-related brain activity in young children: Associations with parental anxiety and child temperamental negative emotionality. *Journal of Child Psychology and Psychiatry, and Allied Disciplines, 54*(8), 854–862. doi:10.1111/jcpp.12041.

van Dinteren, R., et al. (2014). P300 development across the Lifespan: A systematic review and meta-analysis. *PLoS One, 9*(2), e87347. doi:10.1371/journal.pone.0087347

Van Petten, C. and Luka, B. J. (2006). Neural localization of semantic context effects in electromagnetic and hemodynamic studies. *Brain and Language, 97*(3), 279–293. doi:10.1016/j.bandl.2005.11.003.

Vigario, R., et al. (2000). Independent component approach to the analysis of EEG and MEG recordings. *IEEE Transactions on Bio-Medical Engineering, 47*(5), 589–593. doi:10.1109/10.841330.

Vocat, R., Pourtois, G., and Vuilleumier, P. (2008). Unavoidable errors: a spatio-temporal analysis of time-course and neural sources of evoked potentials associated with error processing in a speeded task. *Neuropsychologia, 46*(10), 2545–2555. doi:10.1016/j.neuropsychologia.2008.04.006.

Vogel, M., Monesson, A., and Scott, L. S. (2012). Building biases in infancy: The influence of race on face and voice emotion matching. *Developmental Science, 15*(3), 359–372. doi:10.1111/j.1467-7687.2012.01138.x.

Wagenmakers, E. (2007). A practical solution to the pervasive problems of p values. *Psychonomic Bulletin & Review, 14*(5), 779–804.

Ward, L. M. (2003). Synchronous neural oscillations and cognitive processes, *Trends in Cognitive Sciences, 7*(12), 553–559. doi:10.1016/j.tics.2003.10.012.

Webb, S. J., Long, J. D., and Nelson, C. A. (2005). A longitudinal investigation of visual event-related potentials in the first year of life. *Developmental Science, 8*(6), 605–616. doi:10.1111/j.1467-7687.2005.00452.x.

Welcome, S. E. and Joanisse, M. F. (2018). ERPs reveal weaker effects of spelling on auditory rhyme decisions in children than in adults. *Developmental Psychobiology, 60*(1), 57–66. doi:10.1002/dev.21583.

Wendel, K., et al. (2010). The influence of age and skull conductivity on surface and subdermal bipolar EEG leads. *Computational Intelligence and Neuroscience, 2010,* 397272. doi:10.1155/2010/397272.

Westley, A., Kohút, Z., and Rueschemeyer, S. A. (2017). 'I know something you dont know': Discourse and social context effects on the N400 in adolescents, *Journal of Experimental Child Psychology, 164,* 45–54. doi:10.1016/j.jecp.2017.06.016.

Widmann, A., Schroger, E., and Maess, B. (2015). Digital filter design for electrophysiological data – a practical approach. *Journal of Neuroscience Methods, 250,* 34–46. doi:10.1016/j.jneumeth.2014.08.002.

Wilcox, R. R. (2005). *Introduction to Robust Estimation and Hypothesis Testing.* New York, NY: Elsevier Academic Press.

Willenbockel, V., et al. (2010). Controlling low-level image properties: The SHINE toolbox. *Behavior Research Methods, 42*(3), 671–684. doi:10.3758/BRM.42.3.671.

Williams, N. J., Nasuto, S. J., and Saddy, J. D. (2015). Method for exploratory cluster analysis and visualisation of single-trial ERP ensembles. *Journal of Neuroscience Methods*, *250*, 22–33. doi:10.1016/j.jneumeth.2015.02.007.

Wolfe, C. D. and Bell, M. A. (2004). Working memory and inhibitory control in early childhood: Contributions from physiology, temperament, and language. *Developmental Psychobiology*, *44*(1), 68–83. doi:10.1002/dev.10152.

Wolfe, C. D. and Bell, M. A. (2007). The integration of cognition and emotion during infancy and early childhood: Regulatory processes associated with the development of working memory. *Brain and Cognition*, *65*(1), 3–13. doi:10.1016/j.bandc.2006.01.009.

Xiao, R., et al. (2018). Electroencephalography power and coherence changes with age and motor skill development across the first half year of life. *PLoS ONE*, *13*(1), 1–17. doi:10.1371/journal.pone.0190276.

Xie, W., Mallin, B. M., and Richards, J. E. (2018). Development of infant sustained attention and its relation to EEG oscillations: An EEG and cortical source analysis study. *Developmental Science*, *21*(3), 1–16. doi:10.1111/desc.12562.

Xie, W. and Richards, J. E. (2016). Effects of interstimulus intervals on behavioral, heart rate, and event-related potential indices of infant engagement and sustained attention. *Psychophysiology*, *53*(8), 1128–1142. doi:10.1111/psyp.12670.

Yordanova, J. and Kolev, V. (2009). Event-related brain oscillations: Developmental effects on power and synchronization. *Journal of Psychophysiology*, 23(4), 174–182. doi:10.1027/0269-8803.23.4.174.

Zahn-Waxler, C., et al. (1995). Psychophysiological correlates of empathy and prosocial behaviors in preschool children with behavior problems. *Development and Psychopathology*, *7*(1), 27–48. doi:DOI:10.1017/S0954579400006325.

Zhang, Y., et al. (2011). Neural coding of formant-exaggerated speech in the infant brain. *Developmental Science*, *14*(3), 566–581. doi:10.1111/j.1467-7687.2010.01004.x.

Zottoli, T. M. and Grose-Fifer, J. (2012). The feedback-related negativity (FRN) in adolescents. *Psychophysiology*, *49*(3), 413–420. doi:10.1111/j.1469-8986.2011.01312.x.

CHAPTER 4

··

LONGITUDINAL STRUCTURAL AND FUNCTIONAL BRAIN DEVELOPMENT IN CHILDHOOD AND ADOLESCENCE

··

KATHRYN L. MILLS AND CHRISTIAN K. TAMNES

INTRODUCTION

··

THE brain undergoes profound development from birth to adulthood.

Developmental changes range from growth in overall size to subtle functional reorganization and can be observed on a microcellular scale as well as by analyzing brain images acquired by magnetic resonance imaging (MRI). The changing nature of the brain during the first two decades of life suggests that environmental inputs could impact developmental trajectories. The concept of brain plasticity refers to the brain's ability and potential to change. This ability is inherent in the brain—that is, the brain is constantly changing in response to the environment. However, the brain is more capable of change during specific periods of life than at others, and it is well known that the developing brain, compared to the adult brain, is most plastic. In this chapter, we first introduce key findings from post-mortem and animal studies on postnatal brain development and discuss selected methodological considerations for MRI studies on human brain development. We then focus on and discuss typical development of brain structure and function from childhood to early adulthood, as well as the cognitive relevance of these changes.

Post-Mortem and Animal Studies

Starting in the womb, and continuing in the first years of life, the brain overproduces synaptic connections between neurons (Huttenlocher, 1984). Neurons are located in several parts of the brain, including the cerebral cortex, which forms the thin outer layer of the brain. The process of overproducing synaptic connections between neurons is intrinsically regulated, and is not thought to be influenced by environmental experiences (Huttenlocher and Dabholkar, 1997). Different regions of the cerebral cortex demonstrate different trajectories of synaptic overproduction within the first few years, correlating to the overall functionality of that brain region. For example, as visual competency increases between the ages of 2–8 months, so does the proliferation of connections between brain cells in the region of the cortex that processes vision (Huttenlocher, 1984).

As the human brain continues to mature, its overabundance of synaptic connections diminishes, leaving the adult human brain with roughly 60 percent of the synaptic connections that were present during infancy (Huttenlocher and Dabholkar, 1997). The elimination of many of these connections can be influenced by environmental factors. For example, a child who develops amblyopia through congenital strabismus can still attain proper vision through intervention, but only if this intervention occurs before the age of 5 years. The synaptic density in the visual cortex diminishes to adult levels around the age of 5 years, which marks the end of a period of heightened plasticity in the visual system (Huttenlocher, 1984). The studies that have identified these sensitive periods where the brain adapts to certain environmental inputs (or the lack of environmental inputs) are largely laboratory studies conducted on animals. While it is possible to measure periods of sensitivity to environmental influences in humans (for example clinical phenomena like interventions for strabismus), the only studies in humans capable of examining the underpinning cellular processes involve collecting samples of post-mortem human brain tissue.

Post-mortem work has also provided evidence that some brain regions continue to gain myelin, a fatty electrically insulating layer around the axons of neurons, into the second and even third decade of life (Benes, 1989; Yakovlev and Lecours, 1967). Yakovlev and Lecours (1967) theorized that protracted changes in white matter development parallel the changes in behaviour that occur in later stages of development. Although constrained by small sample sizes, these foundational studies challenged prevailing ideas that brain development is complete by early childhood, and spurred subsequent work investigating brain changes beyond the first decade of life. Comparative studies of primates add to this picture; the human brain undergoes a more prolonged developmental course compared to other species, including primates (Leigh, 2004). For example, myelination of the human brain continues beyond adolescence, whereas degree of myelination in the chimpanzee brain reaches adult levels at roughly the same time as the animal becomes sexually mature (Miller et al., 2012). Much of the potential—and many of the vulnerabilities—of our brains will depend on transactional processes between this prolonged maturation and our experiences.

In Vivo Neuroimaging

Although post-mortem and animal studies took some of the first steps in our understanding of both the microscopic and macroscopic changes occurring in the brain across development, MRI has, in many contexts, become the instrument of choice to measure changes in both brain structure and brain function. MRI technologies allow us the ability to examine how the brain of living, developing, individuals change across age. Furthermore, compared to other common *in vivo* imaging methods, MRI offers high spatial resolution (typically 1 mm^3 for structural and 2 mm^3 for functional) and decent temporal resolution (down to a measurement every second). While MRI technology cannot measure the number of synaptic connections between brain cells or amount of myelin, we can quantify larger anatomical changes, as well as a proximal measure of brain activity. Thus, with MRI, scientists can link developmental changes in cognition, emotions, and behavior to changes occurring in the brain. For example, the powerful magnetic fields within the MRI machine are able to detect the level of blood flow in the different regions of the brain. Because of this ability to detect blood flow, the parts of the brain that require more energy during a certain cognitive task or behavior can be identified using MRI. However, most MRI studies are unable to specify what brain chemicals (neurotransmitters) are being released.

Early important structural and functional MRI studies quantitatively comparing children and adults, i.e., cross-sectional studies, were published in the 1990s (e.g., Giedd et al. 1996; Jernigan and Tallal, 1990; Reiss et al., 1996; Sowell et al., 1999; Thomas et al.,1999). Since then, a large number of cross-sectional and an increasing number of longitudinal investigations have examined brain developmental trajectories across the first decades of life using different MRI methods. Through these studies we have learned that the human brain undergoes not only a protracted development, with aspects of our brain maturing into the third decade of life, but that this development is tissue-specific, multifaceted, regional, and coordinated.

MAGNETIC RESONANCE IMAGING (MRI) OF LONGITUDINAL BRAIN DEVELOPMENT

MRI Methods

MRI is a collection of imaging techniques that uses strong magnetic fields, electric field gradients, and radio waves to produce high-quality images of living organs. Because tissues differ in their magnetic properties, protocols designed to create anatomical images of the brain are based on signal intensities and contrasts that distinguish between grey matter, white matter, and cerebrospinal fluid, as well as different brain structures. Other protocols, like diffusion MRI (dMRI) which, when modelled e.g. like

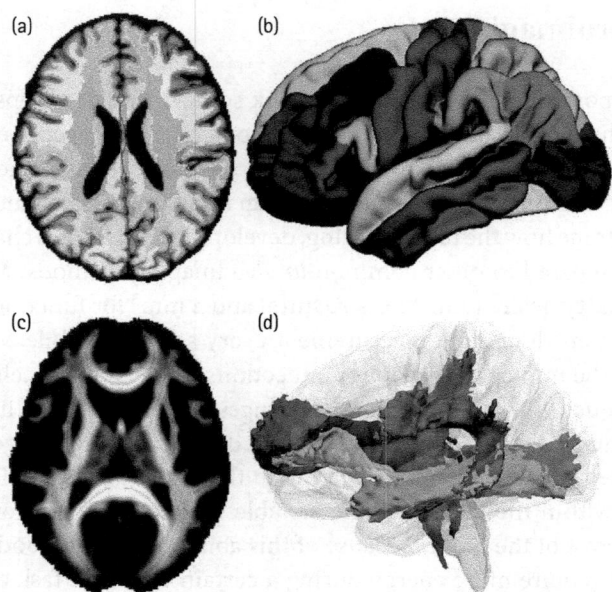

FIGURE 4.1 Methods to examine structural brain changes with MRI.

(a) Horizontal slice of T1 image showing a whole brain segmentation used for volumetric analyses, and (b) a left lateral view of an averaged parcellated cerebral cortex used for surface-based analyses, both from FreeSurfer. (c) Horizontal slice of TBSS mean FA white matter skeleton overlaid on a mean FA map, and (d) a left lateral view of a 3D rendering of probabilistic fiber tracts from the Mori Atlas.

in diffusion tensor imaging (DTI), quantifies preferences of randomly diffusing water molecules to move in different directions, can create images to probe tissue micro-structural properties (see Figure 4.1). Functional MRI (fMRI) techniques on the other hand, can be used to produce images that reflect physiological processes that accompany neural activity. fMRI measures changes in blood flow and oxygenation, which vary based on the processing demands placed on the brain region. fMRI is typically used to investigate blood-oxygenation-level dependent (BOLD) signals that are temporally correlated with experimental conditions or stimuli or specific responses (task-fMRI), or to investigate spontaneous fluctuations in BOLD signals during task-free states, so-called resting-state fMRI.

Challenges and Limitations

A core challenge with neuroimaging studies in general is that it is difficult to ascribe individual differences or longitudinal changes in imaging phenotypes from e.g., ana-tomical MRI, DTI, or fMRI scans to specific cellular and molecular properties or events (Lerch et al., 2017; Zatorre, Fields, and Johansen-Berg, 2012). Concerning brain development in particular, there are several hypotheses about the central involved mechanisms (Crone and Ridderinkhof, 2011; Paus, 2013). One hypothesis is that the early increases and later reductions in grey matter volume may partly relate to initial overproduction and subsequent pruning of synaptic connections, respectively. However, given that synaptic boutons comprise only a fraction of grey matter volume, and even when synapses are particularly dense, they are estimated to represent less than 1.5 percent of cortical volume (Bourgeois and Rakic, 1993), it is unlikely that the marked decreases in cortical volume observed across adolescence are mainly reflective of synaptic pruning. A speculative alternative hypothesis is that synaptic pruning is also accompanied by a reduction in the number of cortical glial cells, and that these events together account for some of the reductions in cortical grey matter volume observed during development (K. L. Mills and Tamnes, 2014). Growth of subcortical white matter and reduction in grey matter likely also relate to continued myelination, both in white matter and intra-cortically, as well as increases in axon diameter and changes in fiber packing density (Paus, 2010). Undoubtedly, there is a myriad of both parallel and interacting cellular processes underlying the observed changes in imaging measured across childhood and adolescence. The relative roles of specific cellular processes for developmental changes in brain structure, microstructure, and activation patterns are likely also age-dependent, with different contributions for instance in infancy and during adolescence.

The current developmental neuroimaging literature largely consists of studies either on infancy or late childhood and adolescence, while few studies have investigated the period between infancy and school age (Brown and Jernigan, 2012). The main reason for this is that young children are likely to move while inside the MRI scanner, which results in poor image quality. In contrast, older children and adolescents can be instructed to stay still, and MRI examinations of infants can, with good timing and some luck, be performed for non-clinical purposes during natural sleep after feeding. Studies of infants and young children, however, have specific major challenges, including image registration, choice of atlases, the large scale of anatomical changes, and the change of image intensity contrasts (Cusack, McCuaig, and Linke, 2018; Phan et al., 2018).

In all developmental neuroimaging studies, preparing the participant and behavioral interventions such as movie watching and friendly feedback during acquisition (Greene et al., 2018) are important in order to reduce motion and optimize data quality. Additionally, quality control procedures, both pre- and post-processing, are of great importance for the resulting data quality (Vijayakumar et al. 2018). It has for instance been shown that post-processing quality control, in the form of exclusion of scans defined as quality control failures on the basis of visual inspection and review of extreme values, has a large impact on identified developmental trajectories for cortical thickness (Ducharme et al., 2016). While the biases introduced by motion-related artefacts have received substantial attention in the developmental fMRI literature (Fair et al., 2012; Power, Schlaggar, and Petersen, 2015), effects of subtle motion on structural measures have also increasingly been recognized as potentially impacting models of structural brain development (Alexander-Bloch et al., 2016; Yendiki et al., 2014). In order to improve reproducibility and effectiveness, well-documented and standardized (Backhausen et al., 2016) or automated (e.g., Esteban et al., 2017; Klapwijk et al., 2019) quality control procedures have recently been developed and are increasingly used.

Finally, there are also a number of considerations and challenges that are specific for longitudinal neuroimaging studies. These issues are associated with different stages of the research process, from study design and participant attrition, to data processing and statistical analysis (King et al., 2018; Tamnes, Roalf et al., 2018; Telzer et al,. 2018; Vijayakumar et al., 2018). While the most commonly used software for structural MRI have developed features specifically for longitudinal data (e.g., FreeSurfer, Reuter et al., 2012; ANTS, Tustison et al., 2017), most available software packages for fMRI analysis do not include longitudinal processing or longitudinal statistical analyses. Hopefully, future programs will close the gap between the modeling capabilities of current software and the models that are necessary to answer longitudinal developmental questions (Madhyastha et al., 2018).

STRUCTURAL BRAIN DEVELOPMENT

Brain and Tissue Volumes

The brain changes throughout life, but it is clear that, besides during the prenatal period, the changes observed during the first few postnatal years qualitatively and quantitatively far exceed the changes seen across the rest of the lifespan (Gilmore, Knickmeyer, and Gao, 2018). A cross-sectional study found that whole brain volume increases by 101 percent in the first year and then an additional 15 percent in the second year (Knickmeyer et al., 2008). A later longitudinal study further mapped these very rapid changes in brain morphology in early infancy by examining a large group of newborns aged 2 to 90 days old (Holland et al., 2014). The results showed that the brain's total volume increases approximately by a staggering 1 *percent per day* in the period immediately after birth (Holland et al., 2014). In contrast to the increase in volume observed in early childhood, a multi-sample longitudinal study found that whole brain volume reduces in size during adolescence (K. L. Mills et al., 2016). When these findings are considered alongside those from a meta-analysis of longitudinal studies, it appears that whole brain volume increases until around age 13 years, then decreases until some point in the early twenties, after which it remains relatively stable until it later begins to decrease again (Hedman et al., 2012).

Importantly, volumetric development of the two main tissue types of the brain, grey matter and white matter, follow distinct developmental trajectories (see Figure 4.2). Grey matter, i.e., the cerebral, cerebellar cortex, and distinct subcortical structures, is composed of neuronal bodies, glial cells, dendrites, blood vessels, extracellular space, and both unmyelinated and myelinated axons. Similar to whole brain volume, cortical

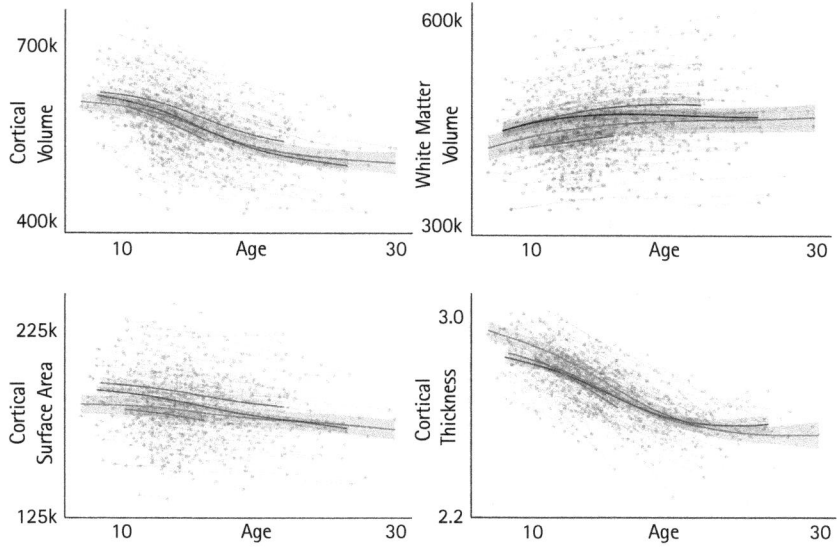

FIGURE 4.2 Structural brain developmental trajectories.

Developmental trajectories for global measures for cerebral cortical volume, cerebral white matter volume, cortical surface area, and mean cortical thickness. The colored lines represent the GAMM fitting while the lighter colored areas correspond to the 95 percent confidence intervals. Pink, NIMH Child Psychiatry Branch, Bethesda, Maryland, USA; purple, Pittsburgh Cohort, Pittsburgh, Pennsylvania, USA; blue, NeuroCognitive Development dataset, Oslo, Norway; green, Braintime dataset, Leiden, The Netherlands.

Adapted from Tamnes et al. (2017) Development of the cerebral cortex across adolescence, a multisample study of inter-related longitudinal changes in cortical volume, surface area, and thickness. *J Neurosci. 37*(12), 3402–12; and Mills et al., Structural brain development between childhood and adulthood, Convergence across four longitudinal samples. *NeuroImage, 141,* 273–81, under creative commons licence (CC BY 4.0).

grey matter volume increases rapidly after birth, approximately doubling in size in the first year of life (Gilmore et al., 2012). It then reaches its greatest volume in childhood, and decreases in late childhood and throughout adolescence, before stabilizing in young adulthood (Lebel and Beaulieu, 2011; K. L. Mills et al., 2016). Unlike earlier characterizations of cortical grey matter volume reaching a "peak" at a specific age in late childhood or early adolescence (Giedd et al., 1999; Lenroot et al., 2007), it is likely that cortical grey matter reaches its highest volume in early childhood and plateaus for a number of years before it begins to decrease in late childhood (Gilmore, Knickmeyer, and Gao, 2018).

In contrast to cortical grey matter volume, cerebral white matter, which occupies almost half of the human brain and consists largely of organized myelinated axons, continues to increase in volume from birth and across the second decade of life before stabilizing at some point in the late teens or early twenties (Knickmeyer et al., 2008; Lebel and Beaulieu, 2011; K. L. Mills et al., 2016). In addition to the tissue-specific patterns described above, there are also regional and component specific differences in brain developmental timing and pace.

Regional Patterns of Development in the Cerebral Cortex

An important addition to the general patterns described above is that the cerebral cortex does not develop uniformly, but with regional patterns. For example, the parietal lobes and lateral occipital cortices show larger volumetric reductions in late childhood and early adolescence, whereas the medial frontal cortex and anterior temporal cortex pick up the pace in the teen years (Tamnes et al., 2013). Development of the cerebral cortex thus appears to generally follow a posterior–anterior pattern, with relatively late development of anterior brain regions in the frontal and temporal cortices.

Because the cerebral cortex morphometrically is a layer of tissue enveloping the cerebrum, it can be measured in terms of its distinct components: thickness and surface area. Importantly, cortical thickness and surface area are phylogenetically (Geschwind and Rakic, 2013) and genetically (Kremen et al., 2013) distinct, and they are increasingly used metrics instead of cortical volume (Winkler et al., 2010). Cortical thickness and surface area are also ontogenetically distinct. In the first two years of life, both cortical thickness and surface area increase over time, although with a much larger increase in the latter (Lyall et al., 2015). From mid-childhood to adulthood, they show different developmental trajectories. Results from recent longitudinal studies suggest a monotonic decline for cortical thickness, while surface area increases in childhood and then slightly decreases during adolescence (K. L. Mills, Lalonde, et al., 2014; Tamnes et al., 2017; Vijayakumar et al., 2016; Wierenga, Langen, Oranje, et al., 2014). Greater cortical volumes and surface areas are, on average, observed in boys compared to girls, but relatively smaller and few sex differences are observed for cortical thickness, or for developmental change in any of these cortical measures (K. L. Mills, Lalonde, et al., 2014; Vijayakumar et al., 2016). While surface area has been found to be the primary determinant of individual differences in cortical volume in adults, and also the main driving factor for cortical volume increases in the early years of life (Lyall et al., 2015), the dominant contributor to cortical volume reductions between late childhood and early adulthood is cortical thinning (Tamnes et al., 2017). The developmental patterns of cortical thinning from childhood to early adulthood are similar between homologous and contiguous regions, as well as between the core nodes of the default mode network: the medial prefrontal cortex and medial posterior cortex (Khundrakpam et al., 2019).

The human cortex is highly convoluted, with approximately one third of the surface exposed on gyri and two-thirds buried within sulci, and developmental changes related to this feature can be measured using a global (whole brain) or local gyrification index. During the first 2 years of life, the global gyrification index increases around 25 percent, with substantial regional heterogeneity (Li et al., 2014). Local changes in gyrification are stronger for association cortices in infancy, and regions located in sensorimotor, auditory, and visual cortices show relatively less change in gyrification (Li et al., 2014). Global gyrification decreases between childhood and young adulthood (Alemán-Gómez et al., 2013; Raznahan et al., 2011). One longitudinal study demonstrated that the cortex "flattens" during adolescence, mostly due to decreases in sulcal depth and increases in sulcal width (Alemán-Gómez et al., 2013). As seen in infancy, changes in gyrification vary across the cortex during childhood and adolescence, with regions in medial prefrontal cortex, occipital cortex, and temporal cortex undergoing little to no change between ages 6–30 years (Mutlu et al., 2013). However, similar to what has been found in whole brain (Raznahan et al., 2011) and lobar-level (Alemán-Gómez et al., 2013) analyses, linear decreases in local gyrification are observed across the majority of the cortex (Mutlu et al., 2013).

Subcortical Structures

The growth rates of subcortical grey matter structures in infancy are similar to that of cortical volume, except that the hippocampus appears to show a somewhat slower growth than other subcortical structures (Gilmore et al., 2012; Gilmore, Knickmeyer, and Gao, 2018). Several subcortical structures and cortical in-folds also show substantial structural change between childhood and young adulthood, although generally at a lower rate than is observed in the cortex (Tamnes et al., 2013). Longitudinal studies of children and adolescents have found that the thalamus, pallidum, amygdala, caudate, putamen, and nucleus accumbens all show significant changes in volume across the second decade of life (Goddings et al., 2014; Herting et al., 2018; Wierenga, Langen, Ambrosino, et al., 2014). The caudate, putamen, and nucleus accumbens decrease at a near linear rate during this time, whereas the amygdala, thalamus, and pallidum demonstrate non-linear increases in volume. Findings for the hippocampus have been inconsistent, but recent studies, including a multisite longitudinal developmental study (Herting et al., 2018) and a large longitudinal study (Tamnes, Bos et al., 2018), demonstrate a non-linear developmental trajectory with subtle increases into mid-adolescence and subsequent decreases. Together, these developmental findings for subcortical structures are in contradiction with hypotheses and developmental models that assume that subcortical structures are mature by adolescence, as it is now clear that these regions are still undergoing structural development throughout the second decade of life. Males on average consistently show larger subcortical volumes than females (Wierenga, Bos, et al., 2018), but findings are less clear for sex differences in their developmental trajectories (Herting et al., 2018).

White Matter Microstructure

In addition to volumetric studies, white matter development is often investigated with dMRI. The most typical quantification of dMRI is achieved in a tensor model and this is referred to as DTI. Several indices can be derived; fractional anisotropy (FA) is used as a measure of the directionality of diffusion, mean diffusivity (MD) reflects the overall magnitude of diffusion, whilst axial (AD) and radial diffusivity (RD) are diffusivity along and across the longest axis of the diffusion tensor, respectively. These indices can be analyzed on a voxel-wise basis or in regions of interest, or with tractography techniques to reconstruct long-range connections, yielding possibilities for inferring patterns of structural connectivity. The major white matter fiber pathways in the brain are present and identifiable at birth, but rapid changes in DTI indices are seen across infancy (Qiu, Mori, and Miller, 2015). A longitudinal study of infants ranging from 2 weeks to 2 years (scanned annually) found that over the first 2 years of life, FA in ten major tracts increased by 16–55 percent, RD decreased by 24–46 percent, and AD decreased by 13–28 percent, with faster changes in the first year than in the second (Geng et al., 2012).

In addition to the early changes in fiber integrity that have been observed, longitudinal DTI studies have documented consistent patterns of continued development of white matter microstructure in later childhood and adolescence. These studies have demonstrated that, with increasing age, FA increases, while MD and RD decrease in most white matter regions, but the results for AD are less consistent (Lebel and Deoni, 2018). In preschool and the early school years, FA showed linear increase over time for most white matter regions, while MD and RD showed linear decreases (Krogsrud et al., 2016). Lebel and Beaulieu (2011) studied a much broader age range, 5–32 years, and found that almost all tracts investigated showed nonlinear developmental trajectories, with decelerating increases for FA and decelerating decreases for MD, primarily due to decreasing RD.

There are also regional differences in white matter microstructure development. Consistent with the posterior–anterior cortical grey matter developmental pattern, a pattern of maturation in which major tracts with frontotemporal connections develop more slowly than other tracts has emerged (Lebel et al., 2008). Of the major fiber bundles, the cingulum, implicated in cognitive control, and the uncinate fasciculus, implicated in emotion and episodic memory, appear to be among those with the most prolonged development (Lebel et al., 2012; Olson et al., 2015).

DTI has also been used to measure changes in brain connectivity across the brain. One longitudinal study found that only a small percentage of connections changed structurally between the ages of 15–19 years, and that these connections were largely between "hub" regions of the brain (Baker et al., 2015). Specifically, connections between subcortical hubs decreased across late adolescence and connections between frontal–subcortical and frontal–parietal hubs increased across late adolescence (Baker et al., 2015). A recent study replicated and extended these findings in two independent longitudinal datasets with participants ranging from ages 4–13 years (Wierenga, van den Heuvel, et al., 2018). As observed in Baker et al., 2015, the hub regions of the brain showed the most substantial changes in connectivity between childhood and early adolescence (Wierenga, van den Heuvel, et al., 2018). Further, this study found that white matter tracts connecting distal regions of the brain demonstrated larger developmental changes in FA than white matter tracts connecting proximal brain regions (Wierenga, van den Heuvel, et al., 2018).

The brain is highly organized, and the layout is often described as highly efficient in regards to signaling (Achard and Bullmore, 2007). A longitudinal study of twins aged 9–15 years demonstrated that the efficiency of the brain's structural connections is highly heritable, and that this efficiency increases across early adolescence (Koenis et al., 2015). A follow-up study of the same cohort demonstrated that global efficiency of the brain's structural connections follows a non-linear trajectory, increasing between 10–13 years before stabilizing between ages 12–18 years (Koenis et al., 2018). This study also found that adolescents with more efficient structural brain networks (both local and global) had higher IQs, and that the coupling between structural network efficiency and IQ increased with age (Koenis et al., 2018). Further, the increased correlation between structural network efficiency and IQ observed with age appeared to be under strong genetic control (Koenis et al., 2018).

Structural Brain Development and Cognition

What are the functional consequences of the prolonged structural development of the human brain? One of the main characteristics of the brain during development is its plasticity, or capacity for change in response to external influences, which can range from concerted interventions to everyday behaviors. There have only been a few studies that have measured the impact of interventions on brain structure during development. One recent study investigated how a reading intervention impacted the integrity of white matter tracts in children aged 7–12 years (Huber et al., 2018). This novel study identified struggling readers and assigned them to either an 8-week reading intervention or school as usual, and the students were scanned four times: once at baseline, after 2.5 weeks of intervention, after 5 weeks of intervention, and after the intervention had been completed (Huber et al., 2018). This intensive longitudinal study allowed the researchers to observe the rapid changes that occurred in white matter tracts, including the left arcuate fasciculus and inferior longitudinal fasciculus—tracts which connect a core network of brain regions involved in reading and word comprehension (Huber et al., 2018).

While intervention studies are crucial to fully understand how environmental factors can impact the developing brain, some characteristics of the environment (e.g., socioeconomic status) cannot be manipulated in experimental studies. There have been a few observational studies investigating the impact of the environment on brain development. One longitudinal study of children aged 11–20 years examined the interacting effects of parenting style and socioeconomic disadvantage on structural brain development (Whittle et al., 2017). In this study, socioeconomic disadvantage measured at the neighborhood level was related to differences in volumetric brain development in the amygdala and temporal lobes, more specifically with greater increases in amygdala volumes and relatively increased cortical thickening longitudinally. Furthermore, positive parenting strategies attenuated the effects of socioeconomic disadvantage on brain development in some regions (Whittle et al., 2017).

There have been several studies investigating how changes in brain structure correlate with changes in cognitive abilities or behaviors. For example, changes in cortical thickness in prefrontal cortical regions during adolescence have been associated with developmental improvements in aspects of cognitive control, including proactive control (Vijayakumar et al., 2014a), and cognitive reappraisal strategies (Vijayakumar et al., 2014b). Further, individual differences in developmental changes in DTI indices relate to cognitive development across ages 8–28 years (Simmonds et al., 2014). In this study, earlier white matter development in adolescence was associated with faster and more efficient behavioral responding and better inhibitory control, whereas later (delayed) white matter development was associated with poorer performance, suggesting that the timing of developmental changes in white matter microstructure is important for cognitive development (Simmonds et al., 2014). A separate longitudinal DTI study examined the relationship between temporal discounting and frontostriatal circuitry in individuals between the ages of 8–26, and found that greater frontostriatal white matter integrity was related to the preference for waiting for larger reward during this age period (Achterberg et al., 2016).

FUNCTIONAL BRAIN DEVELOPMENT

Compared to structural brain development, there have been fewer longitudinal investigations of functional brain development utilizing either task-based or resting-state fMRI techniques, and the majority include only two time points per participant. The current section focuses exclusively on longitudinal fMRI studies and does not cover the large number of cross-sectional fMRI studies that have correlated brain responses, or functional connectivity, with age, cognition, or behavior during developmental periods.

Developmental Changes in Functional Connectivity

The first longitudinal study to investigate developmental changes in the brain's functional organization did not use data acquired during the canonical "resting-state" when participants usually rest quietly with eyes closed or fixated on a cross-hair, but rather examined functional connectivity while participants listened passively to meaningless speech (Sherman et al., 2014). As with previous cross-sectional studies investigating age-related differences in functional brain networks (Fair et al., 2007, 2009, 2008), this study demonstrated that the integration (within-network connectivity) and segregation (between-network connectivity) of the default mode and central executive networks changes between childhood and early adolescence. Specifically, core nodes within each of these networks became more strongly connected to their respective networks, and less connected with the other network, between the ages of 10–13 years (Sherman et al., 2014).

A recent study that made use of publicly available datasets found that the brain's functional organization remained stable from adolescence to early adulthood (Horien et al., 2019). Despite the availability of longitudinal resting state datasets for developing populations, in addition to the datasets that have been acquired but not publicly shared, not much is published on this topic. There are a few studies comparing the development of resting state functional brain connectivity between clinical groups (e.g., Jalbrzikowski et al., 2017; B. D. Mills et al., 2017), but fewer studies investigating typically developmental trajectories of functional brain organization or connectivity between predefined regions.

Developmental changes in functional brain organization can also be related to cognitive processes and behaviors assessed outside of the scanner. A recent longitudinal study demonstrated that individual differences in functional brain organization can account for behavioral changes in temporal discounting with age (Anandakumar et al., 2018). In this study, functional connections that could explain temporal discounting changes above age were first identified in a longitudinal sample of 7–15-year-olds, and then confirmed in an independent cross-sectional sample. For both samples, connectivity strength within and between valuation and cognitive control systems accounted for variance in temporal discounting behavior that was not explained by age (Anandakumar et al., 2018). Specifically, greater connectivity strength between cognitive control regions, as well as between cognitive control and valuation regions, was related to a preference for waiting for a larger reward. In contrast, greater connectivity strength between valuation network regions was related to a preference for taking an immediate, smaller, reward.

In a study that combined three longitudinal datasets spanning ages 6–22 years, functional connectivity strength between regions of the lateral frontoparietal network showed a nonlinear pattern, with minimal change in late childhood but a pronounced change in the transition into adolescence (Wendelken et al., 2017). Further, the functional connectivity strength within this network was associated with reasoning ability in adolescents and adults (Wendelken et al., 2017).

Developmental Changes in Neural Responses During Task fMRI

One of the first longitudinal fMRI investigations of brain development demonstrated that neural responses to social stimuli continue to change between childhood and adolescence (Pfeifer et al., 2011). Between the ages of 10–13 years, children begin to show greater neural activity in the ventral striatum and ventromedial prefrontal cortex while looking at faces in general, whereas the anterior temporal cortex showed greater neural activity in particular for emotional faces (Pfeifer et al., 2011). The longitudinal design of this study also provided the opportunity to understand how changes in neural responsivity in socioemotional situations relate to emotion-regulation capacities. The authors demonstrated that heightened ventral striatal activity across this age period was correlated with decreasing susceptibility (i.e., increasing resistance) to peer influence (Pfeifer et al., 2011).

Another longitudinal study examined non-linear changes in neural recruitment during an inhibitory control task across ages 9–26 years (Ordaz et al., 2013). Inhibitory control continued to improve throughout adolescence, but recruitment of brain regions involved in motor response control did not show developmental changes, suggesting that brain functioning regarding this aspect of inhibitory control matures earlier—in late childhood (Ordaz et al., 2013). However, activation in certain executive control regions decreased with age until adolescence, and error processing activation in the dorsal anterior cingulate cortex showed continued increases all the way into young adulthood and was associated with task performance (Ordaz et al., 2013). By taking a longitudinal approach, the study provides strong evidence that the continued maturation of error-processing abilities underlies the protracted development of inhibitory control over adolescence.

There has been one longitudinal study that has examined the impact of puberty on the capacity to experience (some) fear-evoking experiences as an exciting thrill (Spielberg et al., 2014). This study demonstrated that increases in testosterone over a 2-year period of pubertal maturation was positively related to recruitment of both the amygdala and nucleus accumbens when perceiving stimuli typically associated with threat (Spielberg et al., 2014). The level of recruitment of these regions was further related to greater approach behavior—in this case, responding faster to threatening faces (Spielberg et al., 2014). Overall, the authors suggest that the experience of threat shifts to a more complex process during pubertal maturation, which is specifically associated with changing testosterone levels.

There are developmental increases in the dorsal striatum, but not ventral striatum, during the anticipation of rewards or losses during adolescence, which, interestingly, do not differ between reward and loss conditions (Lamm et al., 2014). A separate study found that a network of brain regions involved in processing reward (including the ventral striatum and medial prefrontal cortex) are consistently recruited in response to rewards over the course of adolescence, and the propensity to engage in risky behavior is positively correlated with reward-related activation within this reward-related network (van Duijvenvoorde et al., 2014). Further, this study did not find evidence for a "peak" in the brain's response to rewards in mid-adolescence (van Duijvenvoorde et al., 2014). However, a subsequent study with a larger sample, which examined longitudinal changes in neural responses to rewards during a laboratory risk-taking task across ages 8–27 years, did find evidence for heightened activity in the nucleus accumbens in response to rewards during adolescence, as compared to in both childhood and adulthood (Braams et al., 2015). All in all, these discrepant findings could be resolved through replication efforts as more datasets examining longitudinal studies of functional brain development are acquired.

CONCLUSIONS AND FUTURE DIRECTIONS

Thanks to two decades of advances in neuroimaging acquisition and analysis and an increasing number of longitudinal studies, we have gained new insight into how the human brain, both in terms of structural architecture and functional activation patterns, develops throughout childhood and adolescence. This fundamental knowledge opens up many novel research questions that are currently being pursued. How are the brain structural and functional developmental changes related, and how do they jointly support development of cognition and emotion (Fengler, Meyer, and Friederici, 2016; Ullman, Almeida, and Klingberg, 2014)? How do early influences and novel social experiences characteristic of the adolescent period affect neurocognitive development (Blakemore and Mills, 2014; McLaughlin, Sheridan, and Nelson, 2017)? What brain measures demonstrate a differential developmental pattern when comparing females and males, and what are the roles of puberty and specific hormonal changes in these differential patterns (Herting and Sowell, 2017; Wierenga, Bos, et al., 2018)? How do the genetic, social and neural dimensions together augment risk or promote resilience for emerging mental health problems observed from childhood to young adulthood (Alnæs et al., 2018; Bos et al., 2018; Kaufmann et al., 2017)?

A critical premise for answering many of these questions is that we need to move beyond a focus on averages and towards studying variability and individual differences (Fisher, Medaglia, and Jeronimus, 2018; Foulkes and Blakemore, 2018). Patterns of structural brain development have been linked to individual differences in psychological traits (Ferschmann et al., 2018), as well as intelligence (Schnack et al., 2015). Several studies have demonstrated that patterns of structural brain development are not the same for everyone within a group, e.g., boys (Wierenga et al., 2018), or children with a specific clinical diagnosis (Alexander-Bloch et al., 2014; Bethlehem et al., 2018), or at an individual level (K. L. Mills, Goddings, et al., 2014). Both structural and functional longitudinal neuroimaging studies indicate substantial individual differences in brain developmental trajectories, but few studies have formally evaluated individual differences in slopes and tried to relate these to other variables of interest (but some are starting to, see Becht et al., 2018). Large longitudinal datasets or datasets with many timepoints, such as the Adolescent Brain Cognitive Development (ABCD) study, and both exploratory and hypothesis-driven studies are needed to achieve this.

ACKNOWLEDGMENTS

We thank all of the families and participants who have been involved in the longitudinal projects described in this chapter, as well as the research staff who kept these projects going throughout the years. We also thank the developers of open source, freely available software that have allowed researchers to characterize brain developmental patterns. Finally, we thank the researchers and institutions who have made their datasets and analytic tools open and available for other researchers to use. This collaborative spirit has benefitted developmental cognitive neuroscience and continues to quicken the pace of discovery. Christian K. Tamnes is funded by the Research Council of Norway.

REFERENCES

Achard, Sophie and Ed Bullmore. (2007). Efficiency and cost of economical brain functional networks. *PLOS Computational Biology*, 3(2), e17. doi:10.1371/journal.pcbi.0030017.

Achterberg, Michelle, Jiska S. Peper, Anna C. K. van Duijvenvoorde, René C. W. Mandl, and Eveline A. Crone. (2016). Frontostriatal white matter integrity predicts development of delay of gratification, A longitudinal study. *Journal of Neuroscience*, 36(6), 1954–1961. doi:10.1523/JNEUROSCI.3459-15.2016.

Alemán-Gómez, Yasser, Joost Janssen, Hugo Schnack, Evan Balaban, Laura Pina-Camacho, Fidel Alfaro-Almagro, Josefina Castro-Fornieles, et al. (2013). The human cerebral cortex flattens during adolescence. *The Journal of Neuroscience, The Official Journal of the Society for Neuroscience*, 33(38), 15004–15010. doi:10.1523/JNEUROSCI.1459-13.2013.

Alexander-Bloch, Aaron F., Liv Clasen, Michael Stockman, Lisa Ronan, Francois Lalonde, Jay Giedd, and Armin Raznahan. (2016). Subtle in-scanner motion biases automated measurement of brain anatomy from in vivo MRI. *Human Brain Mapping*, 37(7), 2385–2397. doi:10.1002/hbm.23180.

Alexander-Bloch, Aaron F., Philip T. Reiss, Judith Rapoport, Harry McAdams, Jay N. Giedd, Ed T. Bullmore, and Nitin Gogtay. (2014). Abnormal cortical growth in schizophrenia targets normative modules of synchronized development. *Biological Psychiatry, Brain Development and Connectivity in Schizophrenia*, 76(6), 438–446. doi:10.1016/j.biopsych.2014.02.010.

Alnæs, Dag, Tobias Kaufmann, Nhat Trung Doan, Aldo Córdova-Palomera, Yunpeng Wang, Francesco Bettella, Torgeir Moberget, Ole A. Andreassen, and Lars T. Westlye. (2018). Association of heritable cognitive ability and psychopathology with white matter properties in children and adolescents. *JAMA Psychiatry*, 75(3), 287–295. doi:10.1001/jamapsychiatry.2017.4277.

Anandakumar, Jeya, Kathryn Mills, Eric Earl, Lourdes Irwin, Oscar Miranda-Dominguez, Damion V. Demeter, Alexandra Walton-Weston, Sarah Karalunas, Joel Nigg, and Damien A. Fair. (2018). Individual differences in functional brain connectivity predict temporal discounting preference in the transition to adolescence. *Developmental Cognitive Neuroscience*, 34, 101–113. doi: 10.1016/j.dcn.2018.07.003.

Backhausen, Lea L., Megan M. Herting, Judith Buse, Veit Roessner, Michael N. Smolka, and Nora C. Vetter. (2016). Quality control of structural MRI images applied using FreeSurfer: A hands-on workflow to rate motion artifacts. *Frontiers in Neuroscience*, 10, 558. doi:10.3389/fnins.2016.00558.

Baker, Simon T. E., Dan I. Lubman, Murat Yücel, Nicholas B. Allen, Sarah Whittle, Ben D. Fulcher, Andrew Zalesky, and Alex Fornito. (2015). Developmental changes in brain network hub connectivity in late adolescence. *Journal of Neuroscience*, 35(24), 9078–9087. doi:10.1523/JNEUROSCI.5043-14.2015.

Becht, Andrik I., Marieke G. N. Bos, Stefanie A. Nelemans, Sabine Peters, Wilma A. M. Vollebergh, Susan J. T. Branje, Wim H. J. Meeus, and Eveline A. Crone. (2018). Goal-directed correlates and neurobiological underpinnings of adolescent identity: A multimethod multisample longitudinal approach. *Child Development*, 89(3), 823–836. doi:10.1111/cdev.13048.

Benes, F. (1989). Myelination of cortical-hippocampal relays during late adolescence. *Schizophrenia Bulletin*, 15(4), 585–593.

Bethlehem, Richard A. I., Jakob Seidlitz, Rafael Romero-Garcia, and Michael V. Lombardo. (2018). Using normative age modelling to isolate subsets of individuals with autism expressing highly age-atypical cortical thickness features. *BioRxiv*, January, 252593. doi:10.1101/252593.

Blakemore, Sarah-Jayne, and Kathryn L. Mills. (2014). Is adolescence a sensitive period for sociocultural processing? *Annual Review of Psychology*, 65, 187–207. doi:10.1146/annurev-psych-010213-115202.

Bos, Marieke G. N., Sabine Peters, Ferdi C. van de Kamp, Eveline A. Crone, and Christian K. Tamnes. (2018). Emerging depression in adolescence coincides with accelerated frontal cortical thinning. *Journal of Child Psychology and Psychiatry, and Allied Disciplines*, 59(9), 994–1002. doi:10.1111/jcpp.12895.

Bourgeois, Jean-Pierre and Pasko Rakic. (1993). Changes of synaptic density in the primary visual cortex of the macaque monkey from fetal to adult stage. *The Journal of Neuroscience, The Official Journal of the Society for Neuroscience*, 13(7), 2801–2820.

Braams, Barbara R., Anna C. K. van Duijvenvoorde, Jiska S. Peper, and Eveline A. Crone. (2015). Longitudinal changes in adolescent risk-taking: A comprehensive study of neural responses to rewards, pubertal development, and risk-taking behavior. *The Journal of Neuroscience, The Official Journal of the Society for Neuroscience*, 35(18), 7226–7238. doi:10.1523/JNEUROSCI.4764-14.2015.

Brown, Timothy T. and Terry L. Jernigan. (2012). Brain development during the preschool years. *Neuropsychology Review*, 22(4), 313–333. doi:10.1007/s11065-012-9214-1.

Crone, Eveline A. and K. Richard Ridderinkhof. (2011). The developing brain: From theory to neuroimaging and back. *Developmental Cognitive Neuroscience*, 1(2), 101–9. doi:10.1016/j.dcn.2010.12.001.

Cusack, Rhodri, Olivia McCuaig, and Annika C. Linke. (2018). Methodological challenges in the comparison of infant fMRI across age groups. *Developmental Cognitive Neuroscience*, 33, 194–205. doi:10.1016/j.dcn.2017.11.003.

Ducharme, Simon, Matthew D. Albaugh, Tuong-Vi Nguyen, James J. Hudziak, J. M. Mateos-Pérez, Aurelie Labbe, Alan C. Evans, Sherif Karama, and Brain Development Cooperative Group. (2016). Trajectories of cortical thickness maturation in normal brain development—the importance of quality control procedures. *NeuroImage*, 125(January), 267–279. doi:10.1016/j.neuroimage.2015.10.010.

Duijvenvoorde, Anna C. K. van, Zdeňa A. Op de Macks, Sandy Overgaauw, Bregje Gunther Moor, Ronald E. Dahl, and Eveline A. Crone. (2014). A cross-sectional and longitudinal analysis of reward-related brain activation: Effects of age, pubertal stage, and reward sensitivity. *Brain and Cognition*, 89(August), 3–14. doi:10.1016/j.bandc.2013.10.005.

Esteban, Oscar, Daniel Birman, Marie Schaer, Oluwasanmi O. Koyejo, Russell A. Poldrack, and Krzysztof J. Gorgolewski. (2017). MRIQC, Advancing the automatic prediction of image quality in MRI from unseen sites. *PloS One*, 12(9), e0184661. doi:10.1371/journal.pone.0184661.

Fair, Damien A., Alexander L. Cohen, Nico U. F. Dosenbach, Jessica A. Church, Francis M. Miezin, Deanna M. Barch, Marcus E. Raichle, Steven E. Petersen, and Bradley L. Schlaggar. (2008). The maturing architecture of the brain's default network. *Proceedings of the National Academy of Sciences*, 105(10), 4028–4032. doi:10.1073/pnas.0800376105.

Fair, Damien A., Alexander L. Cohen, Jonathan D. Power, Nico U. F. Dosenbach, Jessica A. Church, Francis M. Miezin, Bradley L. Schlaggar, and Steven E. Petersen. (2009). Functional brain networks develop from a "local to distributed" organization. *PLOS Computational Biology*, 5(5), e1000381. doi:10.1371/journal.pcbi.1000381.

Fair, Damien A., Nico U. F. Dosenbach, Jessica A. Church, Alexander L. Cohen, Shefali Brahmbhatt, Francis M. Miezin, Deanna M. Barch, Marcus E. Raichle, Steven E. Petersen, and Bradley L. Schlaggar. (2007). Development of distinct control networks through segregation and integration. *Proceedings of the National Academy of Sciences*, *104*(33), 13507–13512. doi:10.1073/pnas.0705843104.

Fair, Damien A., Joel T. Nigg, Swathi Iyer, Deepti Bathula, Kathryn L. Mills, Nico U. F. Dosenbach, Bradley L. Schlaggar, et al. (2012). Distinct neural signatures detected for ADHD subtypes after controlling for micro-movements in resting state functional connectivity MRI data. *Frontiers in Systems Neuroscience*, *6*, 80. doi:10.3389/fnsys.2012.00080.

Fengler, Anja, Lars Meyer, and Angela D. Friederici. (2016). How the brain attunes to sentence processing: Relating behavior, structure, and function. *NeuroImage*, *129*(April), 268–278. doi:10.1016/j.neuroimage.2016.01.012.

Ferschmann, Lia, Anders M. Fjell, Margarete E. Vollrath, Håkon Grydeland, Kristine B. Walhovd, and Christian K. Tamnes. (2018). Personality traits are associated with cortical development across adolescence: A longitudinal structural MRI study. *Child Development*, *89*(3), 811–822. doi:10.1111/cdev.13016.

Fisher, Aaron J., John D. Medaglia, and Bertus F. Jeronimus. (2018). Lack of group-to-individual generalizability is a threat to human subjects research. *Proceedings of the National Academy of Sciences*, *115*(27), E6106–E6115. doi:10.1073/pnas.1711978115.

Foulkes, Lucy and Sarah-Jayne Blakemore. (2018). Studying individual differences in human adolescent brain development. *Nature Neuroscience*, *21*(3), 315–323. doi:10.1038/s41593-018-0078-4.

Geng, Xiujuan, Sylvain Gouttard, Anuja Sharma, Hongbin Gu, Martin Styner, Weili Lin, Guido Gerig, and John H. Gilmore. (2012). Quantitative tract-based white matter development from birth to age 2 years. *NeuroImage*, *61*(3), 542–557. doi:10.1016/j.neuroimage.2012.03.057.

Geschwind, Daniel H. and Pasko Rakic. (2013). Cortical evolution: Judge the brain by its cover. *Neuron*, *80*(3), 633–647. doi:10.1016/j.neuron.2013.10.045.

Giedd, Jay N., J. Blumenthal, N. O. Jeffries, F. X. Castellanos, H. Liu, A. Zijdenbos, T. Paus, A. C. Evans, and J. L. Rapoport. (1999). Brain development during childhood and adolescence: A longitudinal MRI study. *Nature Neuroscience*, *2*(10), 861–863. doi:10.1038/13158.

Giedd, Jay N., J. W. Snell, N. Lange, J. C. Rajapakse, B. J. Casey, P. L. Kozuch, A. C. Vaituzis, et al. (1996). Quantitative magnetic resonance imaging of human brain development: Ages 4–18. *Cerebral Cortex*, *6*(4), 551–560.

Gilmore, John H., Rebecca C. Knickmeyer, and Wei Gao. (2018). Imaging structural and functional brain development in early childhood. *Nature Reviews Neuroscience*, *19*(3), 123–137. doi:10.1038/nrn.2018.1.

Gilmore, John H., Feng Shi, Sandra L. Woolson, Rebecca C. Knickmeyer, Sarah J. Short, Weili Lin, Hongtu Zhu, Robert M. Hamer, Martin Styner, and Dinggang Shen. (2012). Longitudinal development of cortical and subcortical gray matter from birth to 2 years. *Cerebral Cortex*, *22*(11), 2478–2485. doi:10.1093/cercor/bhr327.

Goddings, Anne-Lise, Kathryn L. Mills, Liv S. Clasen, Jay N. Giedd, Russell M. Viner, and Sarah-Jayne Blakemore. (2014). The influence of puberty on subcortical brain development. *NeuroImage*, *88*(March), 242–251. doi:10.1016/j.neuroimage.2013.09.073.

Greene, Deanna J., Jonathan M. Koller, Jacqueline M. Hampton, Victoria Wesevich, Andrew N. Van, Annie L. Nguyen, Catherine R. Hoyt, et al. (2018). Behavioral interventions for reducing head motion during MRI scans in children. *NeuroImage*, *171*(May), 234–245. doi:10.1016/j.neuroimage.2018.01.023.

Hedman, Anna M., Neeltje E. M. van Haren, Hugo G. Schnack, René S. Kahn, and Hilleke E. Hulshoff Pol. (2012). Human brain changes across the life span: A review of 56 longitudinal magnetic resonance imaging studies. *Human Brain Mapping*, *33*(8), 1987–2002. doi:10.1002/hbm.21334.

Herting, Megan M., Cory Johnson, Kathryn L. Mills, Nandita Vijayakumar, Meg Dennison, Chang Liu, Anne-Lise Goddings, et al. (2018). Development of subcortical volumes across adolescence in males and females: A multisample study of longitudinal changes. *NeuroImage*, *172*(May), 194–205. doi:10.1016/j.neuroimage.2018.01.020.

Herting, Megan M. and Elizabeth R. Sowell. (2017). Puberty and structural brain development in humans. *Frontiers in Neuroendocrinology*, *44*, 122–137. doi:10.1016/j.yfrne.2016.12.003.

Holland, Dominic, Linda Chang, Thomas M. Ernst, Megan Curran, Steven D. Buchthal, Daniel Alicata, Jon Skranes, et al. (2014). Structural growth trajectories and rates of change in the first 3 months of infant brain development. *JAMA Neurology*, *71*(10), 1266–1274. doi:10.1001/jamaneurol.2014.1638.

Horien, Corey, Xilin Shen, Dustin Scheinost, and R. Todd Constable. (2019). The individual functional connectome is unique and stable over months to years. *NeuroImage*, *189*, 676–687. doi:10.1016/j.neuroimage.2019.02.002.

Huber, Elizabeth, Patrick M. Donnelly, Ariel Rokem, and Jason D. Yeatman. (2018). Rapid and widespread white matter plasticity during an intensive reading intervention. *Nature Communications*, *9*(1), 2260. doi:10.1038/s41467-018-04627-5.

Huttenlocher, Peter R. (1984). Synapse elimination and plasticity in developing human cerebral cortex. *American Journal of Mental Deficiency*, *88*(5), 488–496.

Huttenlocher, Peter R. and A. S. Dabholkar. (1997). Regional differences in synaptogenesis in human cerebral cortex. *The Journal of Comparative Neurology*, *387*(2), 167–178.

Jalbrzikowski, Maria, Bart Larsen, Michael N. Hallquist, William Foran, Finnegan Calabro, and Beatriz Luna. (2017). Development of white matter microstructure and intrinsic functional connectivity between the amygdala and ventromedial prefrontal cortex: Associations with anxiety and depression. *Biological Psychiatry*, *82*(7), 511–521. doi:10.1016/j.biopsych.2017.01.008.

Jernigan, T. L. and P. Tallal. (1990). Late childhood changes in brain morphology observable with MRI. *Developmental Medicine and Child Neurology*, *32*(5), 379–385.

Kaufmann, Tobias, Dag Alnæs, Nhat Trung Doan, Christine Lycke Brandt, Ole A. Andreassen, and Lars T. Westlye. (2017). Delayed stabilization and individualization in connectome development are related to psychiatric disorders. *Nature Neuroscience*, *20*(4), 513–515. doi:10.1038/nn.4511.

Khundrakpam, Budhachandra S., John D. Lewis, Seun Jeon, Penelope Kostopoulos, Yasser Itturia Medina, François Chouinard-Decorte, and Alan C. Evans. (2019). Exploring individual brain variability during development based on patterns of maturational coupling of cortical thickness: A longitudinal MRI study. *Cerebral Cortex*, *29*(1), 178–188. doi:10.1093/cercor/bhx317.

King, Kevin M., Andrew K. Littlefield, Connor J. McCabe, Kathryn L. Mills, John Flournoy, and Laurie Chassin. (2018). Longitudinal modeling in developmental neuroimaging research: Common challenges, and solutions from developmental psychology. *Developmental Cognitive Neuroscience*, *33*, 54–71. doi:10.1016/j.dcn.2017.11.009.

Klapwijk, Eduard T., Ferdi van de Kamp, Mara van der Meulen, Sabine Peters, and Lara M. Wierenga. (2019). Qoala-T, A Supervised-learning tool for quality control of FreeSurfer segmented MRI data. *NeuroImage*, *189*, 116–129. doi:10.1016/j.neuroimage.2019.01.014.

Knickmeyer, Rebecca C., Sylvain Gouttard, Chaeryon Kang, Dianne Evans, Kathy Wilber, J. Keith Smith, Robert M. Hamer, Weili Lin, Guido Gerig, and John H. Gilmore. (2008).

A structural MRI study of human brain development from birth to 2 years. *The Journal of Neuroscience: The Official Journal of the Society for Neuroscience, 28*(47), 12176–12182. doi:10.1523/JNEUROSCI.3479-08.2008.

Koenis, Marinka M. G., Rachel M. Brouwer, Martijn P. van den Heuvel, René C. W. Mandl, Inge L. C. van Soelen, René S. Kahn, Dorret I. Boomsma, and Hilleke E. Hulshoff Pol. (2015). Development of the brain's structural network efficiency in early adolescence: A longitudinal DTI twin study. *Human Brain Mapping, 36*(12), 4938–4953. doi:10.1002/hbm.22988.

Koenis, Marinka M. G., Rachel M. Brouwer, Suzanne C. Swagerman, Inge L. C. van Soelen, Dorret I. Boomsma, and Hilleke E. Hulshoff Pol. (2018). Association between Structural Brain Network Efficiency and Intelligence Increases during Adolescence. *Human Brain Mapping, 39*(2), 811–836. doi:10.1002/hbm.23885.

Kremen, William S., Christine Fennema-Notestine, Lisa T. Eyler, Matthew S. Panizzon, Chi-Hua Chen, Carol E. Franz, Michael J. Lyons, Wesley K. Thompson, and Anders M. Dale. (2013). Genetics of brain structure: Contributions from the Vietnam era twin study of aging. *American Journal of Medical Genetics. Part B, Neuropsychiatric Genetics, The Official Publication of the International Society of Psychiatric Genetics, 162B*(7), 751–761. doi:10.1002/ajmg.b.32162.

Krogsrud, Stine K., Anders M. Fjell, Christian K. Tamnes, Håkon Grydeland, Lia Mork, Paulina Due-Tønnessen, Atle Bjørnerud, et al. (2016). Changes in white matter microstructure in the developing brain—a longitudinal diffusion tensor imaging study of children from 4 to 11 years of age. *NeuroImage, 124*(Pt A), 473–486. doi:10.1016/j.neuroimage.2015.09.017.

Lamm, C., B. E. Benson, A. E. Guyer, K. Perez-Edgar, N. A. Fox, D. S. Pine, and M. Ernst. (2014). Longitudinal study of striatal activation to reward and loss anticipation from mid-adolescence into late adolescence/early adulthood. *Brain and Cognition, 89*(August), 51–60. doi:10.1016/j.bandc.2013.12.003.

Lebel, Catherine and Christian Beaulieu. (2011). Longitudinal development of human brain wiring continues from childhood into adulthood. *The Journal of Neuroscience: The Official Journal of the Society for Neuroscience, 31*(30), 10937–10947. doi:10.1523/JNEUROSCI.5302-10.2011.

Lebel, Catherine and Sean Deoni. (2018). The development of brain white matter microstructure. *NeuroImage, 182,* 207–218. doi:10.1016/j.neuroimage.2017.12.097.

Lebel, Catherine, M. Gee, R. Camicioli, M. Wieler, W. Martin, and Christian Beaulieu. (2012). Diffusion tensor imaging of white matter tract evolution over the lifespan. *NeuroImage, 60*(1), 340–352. doi:10.1016/j.neuroimage.2011.11.094.

Lebel, Catherine, L. Walker, A. Leemans, L. Phillips, and Christian Beaulieu. (2008). Microstructural maturation of the human brain from childhood to adulthood. *NeuroImage, 40*(3), 1044–1055. doi:10.1016/j.neuroimage.2007.12.053.

Leigh, S. R. (2004). Brain growth, life history, and cognition in primate and human evolution. *American Journal of Primatology, 62*(3), 139–164. doi:10.1002/ajp.20012.

Lenroot, Rhoshel K, Nitin Gogtay, Deanna K. Greenstein, Elizabeth Molloy Wells, Gregory L. Wallace, Liv S. Clasen, Jonathan D. Blumenthal, et al. (2007). Sexual dimorphism of brain developmental trajectories during childhood and adolescence. *NeuroImage, 36*(4), 1065–1073. doi:10.1016/j.neuroimage.2007.03.053.

Lerch, Jason P., André J. W. van der Kouwe, Armin Raznahan, Tomáš Paus, Heidi Johansen-Berg, Karla L. Miller, Stephen M. Smith, Bruce Fischl, and Stamatios N. Sotiropoulos. (2017). Studying neuroanatomy using MRI. *Nature Neuroscience, 20*(3), 314–326. doi:10.1038/nn.4501.

Li, Gang, Li Wang, Feng Shi, Amanda E. Lyall, Weili Lin, John H. Gilmore, and Dinggang Shen. (2014). Mapping longitudinal development of local cortical gyrification in infants from birth to 2 years of age. *Journal of Neuroscience, 34*(12), 4228–4238. doi:10.1523/JNEUROSCI.3976-13.2014.

Lyall, Amanda E., Feng Shi, Xiujuan Geng, Sandra Woolson, Gang Li, Li Wang, Robert M. Hamer, Dinggang Shen, and John H. Gilmore. (2015). Dynamic development of regional cortical thickness and surface area in early childhood. *Cerebral Cortex, 25*(8), 2204–2212. doi:10.1093/cercor/bhu027.

Madhyastha, Tara, Matthew Peverill, Natalie Koh, Connor McCabe, John Flournoy, Kate Mills, Kevin King, Jennifer Pfeifer, and Katie A. McLaughlin. (2018). Current methods and limitations for longitudinal FMRI analysis across development. *Developmental Cognitive Neuroscience, 33*, 118–128. https,//doi.org/10.1016/j.dcn.2017.11.006.

McLaughlin, Katie A., Margaret A. Sheridan, and Charles A. Nelson. (2017). Neglect as a violation of species-expectant experience: Neurodevelopmental consequences. *Biological Psychiatry, 82*(7), 462–471. doi:10.1016/j.biopsych.2017.02.1096.

Miller, Daniel J., Tetyana Duka, Cheryl D. Stimpson, Steven J. Schapiro, Wallace B. Baze, Mark J. McArthur, Archibald J. Fobbs, et al. (2012). Prolonged myelination in human neocortical evolution. *Proceedings of the National Academy of Sciences, 109*(41), 16480–16485. doi:10.1073/pnas.1117943109.

Mills, Brian D., Oscar Miranda-Dominguez, Kathryn L. Mills, Eric Earl, Michaela Cordova, Julia Painter, Sarah L. Karalunas, Joel T. Nigg, and Damien A. Fair. (2017). ADHD and attentional control, impaired segregation of task positive and task negative brain networks. *Network Neuroscience, 02*(02), 200–217. doi:10.1162/netn_a_00034.

Mills, Kathryn L., Anne-Lise Goddings, Liv S. Clasen, Jay N. Giedd, and Sarah-Jayne Blakemore. (2014). The developmental mismatch in structural brain maturation during adolescence. *Developmental Neuroscience, 36*(3–4), 147–160. doi:10.1159/000362328.

Mills, Kathryn L., Anne-Lise Goddings, Megan M. Herting, Rosa Meuwese, Sarah-Jayne Blakemore, Eveline A. Crone, Ronald E. Dahl, et al. (2016). Structural brain development between childhood and adulthood, convergence across four longitudinal samples. *NeuroImage, 141*(November), 273–281. doi:10.1016/j.neuroimage.2016.07.044.

Mills, Kathryn L., François Lalonde, Liv S. Clasen, Jay N. Giedd, and S. J. Blakemore. (2014). Developmental changes in the structure of the social brain in late childhood and adolescence. *Social Cognitive and Affective Neuroscience, 9*(1), 123–131. doi:10.1093/scan/nss113.

Mills, Kathryn L. and Christian K. Tamnes. (2014). Methods and considerations for longitudinal structural brain imaging analysis across development. *Developmental Cognitive Neuroscience, 9*(July), 172–190. doi:10.1016/j.dcn.2014.04.004.

Mutlu, A. Kadir, Maude Schneider, Martin Debbané, Deborah Badoud, Stephan Eliez, and Marie Schaer. (2013). Sex differences in thickness, and folding developments throughout the cortex. *NeuroImage, 82*(November), 200–207. doi:10.1016/j.neuroimage.2013.05.076.

Olson, Ingrid R., Rebecca J. Von Der Heide, Kylie H. Alm, and Govinda Vyas. (2015). Development of the uncinate fasciculus, implications for theory and developmental disorders. *Developmental Cognitive Neuroscience, 14*(Supplement C), 50–61. doi:10.1016/j.dcn.2015.06.003.

Ordaz, Sarah J., William Foran, Katerina Velanova, and Beatriz Luna. (2013). Longitudinal growth curves of brain function underlying inhibitory control through adolescence. *The Journal of Neuroscience, The Official Journal of the Society for Neuroscience, 33*(46), 18109–18124. doi:10.1523/JNEUROSCI.1741-13.2013.

Paus, T. (2010). Growth of white matter in the adolescent brain, myelin or axon? *Brain and Cognition, 72*(1), 26–35. doi:10.1016/j.bandc.2009.06.002.

Paus, T. (2013). How environment and genes shape the adolescent brain. *Hormones and Behavior, 64*(2), 195–202. doi:10.1016/j.yhbeh.2013.04.004.

Pfeifer, Jennifer H., Carrie L. Masten, William E. Moore 3rd, Tasha M. Oswald, John C. Mazziotta, Marco Iacoboni, and Mirella Dapretto. (2011). Entering adolescence, resistance to peer influence, risky behavior, and neural changes in emotion reactivity. *Neuron, 69*(5), 1029–1036. doi:10.1016/j.neuron.2011.02.019.

Phan, Thanh Vân, Dirk Smeets, Joel B. Talcott, and Maaike Vandermosten. (2018). Processing of structural neuroimaging data in young children, bridging the gap between current practice and state-of-the-art methods. *Developmental Cognitive Neuroscience, 33*, 206–223. doi:10.1016/j.dcn.2017.08.009.

Power, Jonathan D., Bradley L. Schlaggar, and Steven E. Petersen. (2015). Recent progress and outstanding issues in motion correction in resting state FMRI. *NeuroImage, 105*(January), 536–551. doi:10.1016/j.neuroimage.2014.10.044.

Qiu, Anqi, Susumu Mori, and Michael I. Miller. (2015). Diffusion tensor imaging for understanding brain development in early life. *Annual Review of Psychology, 66*(January), 853–876. doi:10.1146/annurev-psych-010814-015340.

Raznahan, Armin, Phillip Shaw, Francois Lalonde, Mike Stockman, Gregory L. Wallace, Dede Greenstein, Liv Clasen, Nitin Gogtay, and Jay N. Giedd. (2011). How does your cortex grow? *The Journal of Neuroscience, 31*(19), 7174–7177. doi:10.1523/JNEUROSCI.0054-11.2011.

Reiss, A. L., M. T. Abrams, H. S. Singer, J. L. Ross, and M. B. Denckla. (1996). Brain development, gender and IQ in children. A volumetric imaging study. *Brain, A Journal of Neurology, 119*(Pt 5) (October), 1763–1774.

Reuter, Martin, Nicholas J. Schmansky, H. Diana Rosas, and Bruce Fischl. (2012). Within-subject template estimation for unbiased longitudinal image analysis. *NeuroImage, 61*(4), 1402–1418. doi:10.1016/j.neuroimage.2012.02.084.

Schnack, Hugo G., Neeltje E. M. van Haren, Rachel M. Brouwer, Alan Evans, Sarah Durston, Dorret I. Boomsma, René S Kahn, and Hilleke E. Hulshoff Pol. (2015). Changes in thickness and surface area of the human cortex and their relationship with intelligence. *Cerebral Cortex, 25,* 6, 1608–1617. doi:10.1093/cercor/bht357.

Sherman, Lauren E., Jeffrey D. Rudie, Jennifer H. Pfeifer, Carrie L. Masten, Kristin McNealy, and Mirella Dapretto. (2014). Development of the default mode and central executive networks across early adolescence: A longitudinal study. *Developmental Cognitive Neuroscience 10* (October), 148–159. doi:10.1016/j.dcn.2014.08.002.

Simmonds, Daniel J., Michael N. Hallquist, Miya Asato, and Beatriz Luna. (2014). Developmental stages and sex differences of white matter and behavioral development through adolescence: A longitudinal diffusion tensor imaging (DTI) study. *NeuroImage, 15*(92), 356–368 doi:10.1016/j.neuroimage.2013.12.044.

Sowell, E. R., P. M. Thompson, C. J. Holmes, T. L. Jernigan, and A. W. Toga. (1999). In vivo evidence for post-adolescent brain maturation in frontal and striatal regions. *Nature Neuroscience, 2*(10), 859–861. doi:10.1038/13154.

Spielberg, Jeffrey M., Thomas M. Olino, Erika E. Forbes, and Ronald E. Dahl. (2014). Exciting fear in adolescence: Does pubertal development alter threat processing? *Developmental Cognitive Neuroscience, 8*(April), 86–95. doi:10.1016/j.dcn.2014.01.004.

Tamnes, Christian K., David R. Roalf, Anne-Lise Goddings, and Catherine Lebel. (2018). Diffusion MRI of white matter microstructure development in childhood and adolescence:

Methods, challenges and progress. *Developmental Cognitive Neuroscience, 33,* 161–175. doi:10.1016/j.dcn.2017.12.002.

Tamnes, Christian K., Marieke G. N. Bos, Ferdi C. van de Kamp, Sabine Peters, and Eveline A. Crone. (2018). Longitudinal development of hippocampal subregions from childhood to adulthood. *Developmental Cognitive Neuroscience, 30*(April), 212–222. doi:10.1016/j.dcn.2018.03.009.

Tamnes, Christian K., Megan M. Herting, Anne-Lise Goddings, Rosa Meuwese, Sarah-Jayne Blakemore, Ronald E. Dahl, Berna Güroğlu, et al. (2017). Development of the cerebral cortex across adolescence: A multisample study of inter-related longitudinal changes in cortical volume, surface area, and thickness. *The Journal of Neuroscience, The Official Journal of the Society for Neuroscience, 37*(12), 3402–3412. doi:10.1523/JNEUROSCI.3302–16.2017.

Tamnes, Christian K., Kristine B. Walhovd, Anders M. Dale, Ylva Østby, Håkon Grydeland, George Richardson, Lars T. Westlye, et al. (2013). Brain development and aging: Overlapping and unique patterns of change. *NeuroImage, 68C*(March), 63–74. doi:10.1016/j.neuroimage.2012.11.039.

Telzer, Eva H., Ethan M. McCormick, Sabine Peters, Danielle Cosme, Jennifer H. Pfeifer, and Anna C. K. van Duijvenvoorde. (2018). Methodological considerations for developmental longitudinal FMRI research. *Developmental Cognitive Neuroscience, 33,* 149–160. doi:10.1016/j.dcn.2018.02.004.

Thomas, K. M., S. W. King, P. L. Franzen, T. F. Welsh, A. L. Berkowitz, D. C. Noll, V. Birmaher, and B. J. Casey. (1999). A developmental functional MRI study of spatial working memory. *NeuroImage, 10*(3 Pt 1), 327–338. doi:10.1006/nimg.1999.0466.

Tustison, Nicholas J., Andrew J. Holbrook, Brian B. Avants, Jared M. Roberts, Philip A. Cook, Zachariah M. Reagh, James R. Stone, Daniel L. Gillen, and Michael A. Yassa. (2017). The ANTs longitudinal cortical thickness pipeline. *BioRxiv,* July, 170209. doi:10.1101/170209.

Ullman, Henrik, Rita Almeida, and Torkel Klingberg. (2014). Structural maturation and brain activity predict future working memory capacity during childhood development. *Journal of Neuroscience, 34*(5), 1592–1598. doi:10.1523/JNEUROSCI.0842–13.2014.

Vijayakumar, Nandita, Nicholas B. Allen, George Youssef, Meg Dennison, Murat Yücel, Julian G. Simmons, and Sarah Whittle. (2016). Brain development during adolescence: A mixed-longitudinal investigation of cortical thickness, surface area, and volume. *Human Brain Mapping, 37*(6), 2027–2038. doi:10.1002/hbm.23154.

Vijayakumar, Nandita, Kathryn L. Mills, Aaron F. Alexander-Bloch, Christian K. Tamnes, and Sarah Whittle. (2018). Structural brain development: A review of methodological approaches and best practices. *Developmental Cognitive Neuroscience, 33,* 129–148. doi:10.1016/j.dcn.2017.11.008.

Vijayakumar, Nandita, Sarah Whittle, Murat Yücel, Meg Dennison, Julian Simmons, and Nicholas B. Allen. (2014a). Prefrontal structural correlates of cognitive control during adolescent development, a 4-year longitudinal study. *Journal of Cognitive Neuroscience, 26*(5), 1118–1130. doi:10.1162/jocn_a_00549.

Vijayakumar, Nandita, Sarah Whittle, Murat Yücel, Meg Dennison, Julian Simmons, and Nicholas B. Allen. (2014b.) Thinning of the lateral prefrontal cortex during adolescence predicts emotion regulation in females. *Social Cognitive and Affective Neuroscience, 9*(11), 1845–1854. doi:10.1093/scan/nst183.

Wendelken, Carter, Emilio Ferrer, Simona Ghetti, Stephen K. Bailey, Laurie Cutting, and Silvia A. Bunge. (2017). Frontoparietal structural connectivity in childhood predicts development of functional connectivity and reasoning ability: A large-scale longitudinal investigation. *Journal of Neuroscience, 37*(35), 8549–8558. doi:10.1523/JNEUROSCI.3726–16.2017.

Whittle, Sarah, Nandita Vijayakumar, Julian G. Simmons, Meg Dennison, Orli Schwartz, Christos Pantelis, Lisa Sheeber, Michelle L. Byrne, and Nicholas B. Allen. (2017). Role of positive parenting in the association between neighborhood social disadvantage and brain development across adolescence. *JAMA Psychiatry, 74*(8), 824–832. doi:10.1001/jamapsychiatry.2017.1558.

Wierenga, Lara M., Marieke G. N. Bos, Elisabeth Schreuders, Ferdi vd Kamp, Jiska S. Peper, Christian K. Tamnes, and Eveline A. Crone. (2018). Unraveling age, puberty and testosterone effects on subcortical brain development across adolescence. *Psychoneuroendocrinology, 91*(May), 105–114. doi:10.1016/j.psyneuen.2018.02.034.

Wierenga, Lara M., Martijn P. van den Heuvel, Bob Oranje, Jay N. Giedd, Sarah Durston, Jiska S. Peper, Timothy T. Brown, Eveline A. Crone, and The Pediatric Longitudinal Imaging, Neurocognition, and Genetics Study. (2018). A multisample study of longitudinal changes in brain network architecture in 4-13-year-old children. *Human Brain Mapping, 39*(1), 157–170. doi:10.1002/hbm.23833.

Wierenga, Lara M., Marieke Langen, Sara Ambrosino, Sarai van Dijk, Bob Oranje, and Sarah Durston. (2014). Typical development of basal ganglia, hippocampus, amygdala and cerebellum from age 7 to 24. *NeuroImage, 96*(August), 67–72. doi:10.1016/j.neuroimage.2014.03.072.

Wierenga, Lara M, Marieke Langen, Bob Oranje, and Sarah Durston. (2014). Unique developmental trajectories of cortical thickness and surface area. *NeuroImage, 87*(February), 120–126. https,//doi.org/10.1016/j.neuroimage.2013.11.010.

Wierenga, Lara M., Joseph A. Sexton, Petter Laake, Jay N. Giedd, and Christian K. Tamnes. (2018). A key characteristic of sex differences in the developing brain, greater variability in brain structure of boys than girls. *Cerebral Cortex, 28*(8), 2741–2751. doi:10.1093/cercor/bhx154.

Winkler, Anderson M, Peter Kochunov, John Blangero, Laura Almasy, Karl Zilles, Peter T Fox, Ravindranath Duggirala, and David C Glahn. (2010). Cortical thickness or grey matter volume? The importance of selecting the phenotype for imaging genetics studies. *NeuroImage, 53*(3), 1135–1146. doi:10.1016/j.neuroimage.2009.12.028.

Yakovlev, P.A. and I. R. Lecours. (1967). The myelogenetic cycles of regional maturation of the brain. In *Regional Development of the Brain in Early Life*. Alexandre Minkowski (ed.). Oxford: Blackwell.

Yendiki, Anastasia, Kami Koldewyn, Sita Kakunoori, Nancy Kanwisher, and Bruce Fischl. (2014). Spurious group differences due to head motion in a diffusion MRI study. *NeuroImage, 88*, 79–90. doi:10.1016/j.neuroimage.2013.11.027.

Zatorre, Robert J., R. Douglas Fields, and Heidi Johansen-Berg. (2012). Plasticity in gray and white: Neuroimaging changes in brain structure during learning. *Nature Neuroscience, 15*(4), 528–536. doi:10.1038/nn.3045.

CHAPTER 5

DIFFUSION IMAGING PERSPECTIVES ON BRAIN DEVELOPMENT IN CHILDHOOD AND ADOLESCENCE

BRYCE L. GEERAERT, JESS E. REYNOLDS, AND CATHERINE LEBEL

INTRODUCTION

WHITE matter supports efficient and synchronous communication between brain regions and enables coordinated information processing. Communication between brain regions is essential for the sophisticated cognitive functions and behaviors necessary throughout human life, and thus understanding how white matter develops and how it supports cognitive maturation is of great interest. A comprehensive understanding of healthy white matter development is also essential for identifying deviations from normal maturation that may occur in developmental disorders, diseases, and brain injuries. Identification of early biomarkers of developmental disorders or indicators of prognosis in disease and injury can help inform intervention and treatment approaches. Biological outcomes of these treatments may also be assessed via measurement of structural changes within white matter networks.

White matter in the human brain consists primarily of axons along with oligodendrocytes that form the surrounding myelin sheath, as depicted in Figure 5.1. Bundles of axons connecting cortical regions are referred to as white matter tracts. Important microstructural factors like the density of axons within the tract, the diameter of these axons, degree of myelination, and coherence of axon orientations determine

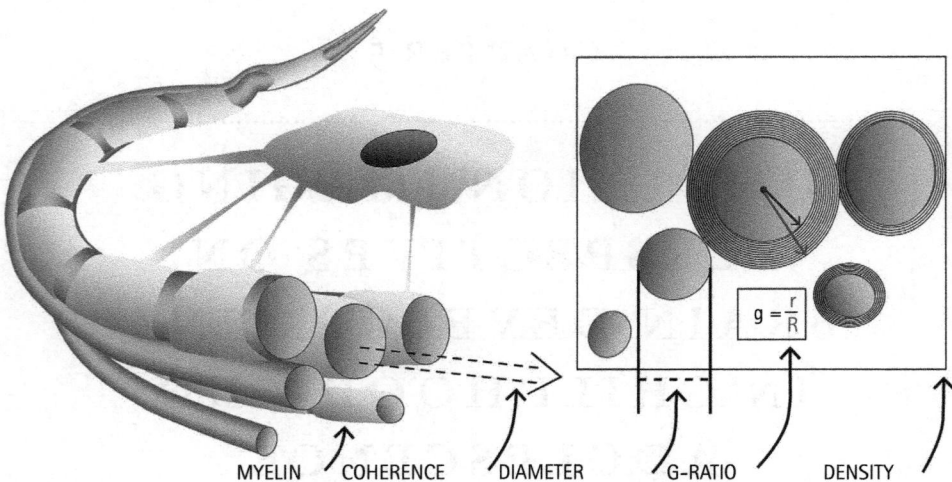

$$g = \frac{r}{R}$$

MYELIN   COHERENCE   DIAMETER   G-RATIO   DENSITY

FIGURE 5.1 Brain white matter tissue consists primarily of bundles of axons (blue) propagating between cortical regions. Key structural features of these bundles include myelin provided by oligodendrocytes, fiber coherence, axon diameter and density, the g-ratio (g), the ratio of axon diameter to total fiber diameter. Each of these features influences the speed and efficiency of communication along white matter fibers.

the communication efficiency between cortical regions. White matter connections are largely in place at birth, but their maturation (including myelination, axon growth, etc.) continues postnatally. Throughout life, white matter undergoes substantial development and is capable of plastic change in order to adapt to environmental influences and refinements in cognitive function.

In this chapter, we present an overview of the current literature on white matter maturation, with a focus on diffusion magnetic resonance imaging (dMRI). Diffusion tensor imaging (DTI) has dominated the literature on white matter development to date, due to its clinical feasibility and sensitivity to white matter microstructure, though other techniques are beginning to see more widespread application. After introducing DTI, we discuss global white matter development trends as well as variance in developmental timing seen between networks and regions. Next, we discuss major genetic and environmental influences on white matter development which may underlie individual variability in trajectories. We then comment on links between white matter development and sophisticated cognitive and behavioral skills, and how atypical white matter development is linked to deficits in these functions. Our discussion of links between white matter structure, behavior, and cognition finishes with commentary on the changes observed in white matter following intervention treatments.

The chapter concludes with commentary on advanced imaging techniques that are becoming more clinically feasible and widely applied to study white matter maturation. These techniques generally support the trends observed in DTI studies, and help to clarify the factors driving white matter maturation. By better understanding the influences and outcomes of behavioral and cognitive conditions, genetics, experience, and environmental factors, better identification, intervention, and preventative methods may be developed to ensure healthy outcomes for children and adolescents.

Early Studies of White Matter Development

Early, foundational studies of brain anatomy used histological methods to investigate tissue structure and physiology. These studies, involving chemical staining of post-mortem tissue, represent some of the most direct information available regarding white matter microstructure. From childhood to early adulthood, increases in total brain volume, cortical growth and pruning, increases in white matter volume and myelination, and reductions in axonal tortuosity have been associated with age in post-mortem samples (Dekaban, 1967; Yakovlev and Lecours, 1967; Huttenlocher, 1979; Benes, 1989; Joosten and Bar, 1999; Landing, Shankle et al., 2002). While fiber pathways are established by birth, increases in axon diameter are rapid and widespread post-natally. Similarly, myelination occurs rapidly throughout the brain during infancy, and histology has shown this process continues in some regions, such as the hippocampus, into adolescence and adulthood (Kinney, Brody et al., 1988; Benes, 1989; Benes, Turtle et al., 1994). In general, these white matter developmental processes follow posterior-to-anterior, central-to-peripheral, and inferior-to-superior patterns of development, where brain regions involved in sensory and motor function (e.g., visual cortex, corticospinal tracts) mature first and fastest, while those involved in higher-order cognitive processing (e.g., prefrontal areas) mature last (Yakovlev and Lecours, 1967; Benes, 1989). Histological studies have shown age-related differences in structural composition in primary motor circuits (pyramidal tracts) into adulthood (Lassek, 1942). This implies that white matter development accompanies both the acquisition of new skills and their continued refinement.

Despite their impressive specificity to white matter structure, histological studies are unable to investigate longitudinal development and are limited by the availability of previously healthy tissue for research, particularly from young populations. Medical imaging methods have been developed to enable the study of in vivo brains and extend histological findings. Magnetic resonance imaging (MRI) has been one of the most useful methods for detailing brain development as it does not involve ionizing radiation, enabling safe longitudinal studies of healthy brain development even in young children. During the past 15–20 years, our understanding of white matter development has greatly improved through MRI research.

Mirroring histological findings, early imaging studies identified increases of white matter volume across childhood and adolescence, while gray matter volume follows a u-shaped trajectory (increases followed by decreases) corresponding to periods of synaptic arborization and pruning (Pfefferbaum, Mathalon et al., 1994; Giedd, Jeffries et al., 1999; Good, Johnsrude et al., 2001). Anatomical imaging studies identified increases of white matter signal intensity, interpreted as white matter "density," which showed continued maturation of the internal capsule and left arcuate fasciculus into adolescence (Paus, Zijdenbos et al., 1999). Studies also showed a rostral-to-caudal wave of growth in the corpus callosum in late childhood and early adolescence (Thompson, Giedd et al., 2000). More recent imaging studies consistently verify these increases of white matter volume across childhood and adolescence (Lenroot, Gogtay et al., 2007; Lebel and Beaulieu, 2011; Walhovd, Westlye et al., 2011), but also use techniques like diffusion MRI to probe white matter microstructure more specifically.

Diffusion Magnetic Resonance Imaging (dMRI)

Diffusion MRI (dMRI) provides more specific and sensitive measures of white matter maturation than the anatomical imaging methods initially used to assess white matter volume changes. Throughout the brain, water molecules are constantly undergoing Brownian motion and, in the absence of barriers, have an equal probability to move in any direction. However, when barriers are present, water diffusion will be restricted. When these barriers are ordered, such as in the case of bundles of axons, the probability of water diffusion along the axon fiber will be greater than the probability of water diffusion across the axonal membrane. Diffusion MRI methods are designed to probe white matter structure by assessing the motion of hydrogen atoms in water in a chosen direction (Le Bihan and Breton, 1985). Diffusion in the chosen direction is detected as a loss of signal proportional to displacement in that direction. Diffusion measured along an axon would produce a large signal loss in the resultant image, while diffusion measured across the axonal membrane would produce a small loss of signal (and a brighter image). By comparing a diffusion-weighted image to a reference image with no diffusion weighting, diffusivity along the chosen direction can be estimated. DTI, one of the most widely used dMRI methods, measures diffusivity in many directions to model diffusion probability as a tensor, a three-dimensional construct often visualized as an ellipsoid (see Figure 5.2). Measurements in at least six non-collinear directions are required to estimate the tensor, but for increased accuracy modern DTI sequences typically measure thirty directions or more (Jones, 2004).

Measures which describe the amount and directionality of diffusion can be calculated from DTI. Fractional anisotropy (FA), the most commonly used DTI measure, describes how directionally restricted water diffusion is (or how elongated the ellipsoid is) in an area. FA ranges from 0–1, with higher FA representing increased anisotropy, or directionality of diffusion. Mean diffusivity (MD) describes the amount of water movement within a region averaged over all directions (or how large the tensor is). In white matter, diffusion is hindered across cellular membranes and myelin, and relatively unrestricted along the axon (Beaulieu, 2002). This results in higher FA in white matter compared to grey matter or cerebrospinal fluid (CSF), where oriented membranes are less prevalent. MD is high in CSF but similar in white matter and gray matter, as both white and gray matter present barriers (e.g., cell membranes) which reduce diffusion. DTI measures are sensitive to important microstructural factors such as the presence of axons, axonal density and diameter, fiber bundle coherence, and myelin (Beaulieu, 2002) (see Figure 5.1). While FA and MD are very sensitive to different microstructural factors, they cannot distinguish among these factors. Axial diffusivity (AD) and radial diffusivity (RD) describe diffusion parallel and perpendicular to the primary axis of the tensor, respectively, and have increased sensitivity to specific aspects of white matter microstructure. In animal models, AD has been linked to axonal degradation, while RD has been linked to demyelination and remyelination (Song, Sun, et al., 2002; Song, Sun et al., 2003). Thus, in human development AD is primarily sensitive to changes in axonal integrity and coherence, while RD has increased sensitivity to myelin and axonal membranes. AD and RD are useful tools to speculate on the microstructural processes driving trends in MD and FA, though they remain sensitive to multiple processes.

Figure 5.2 depicts the typical range of values of the various diffusion parameters in a healthy human brain, and how the tensor manifests in different diffusion conditions.

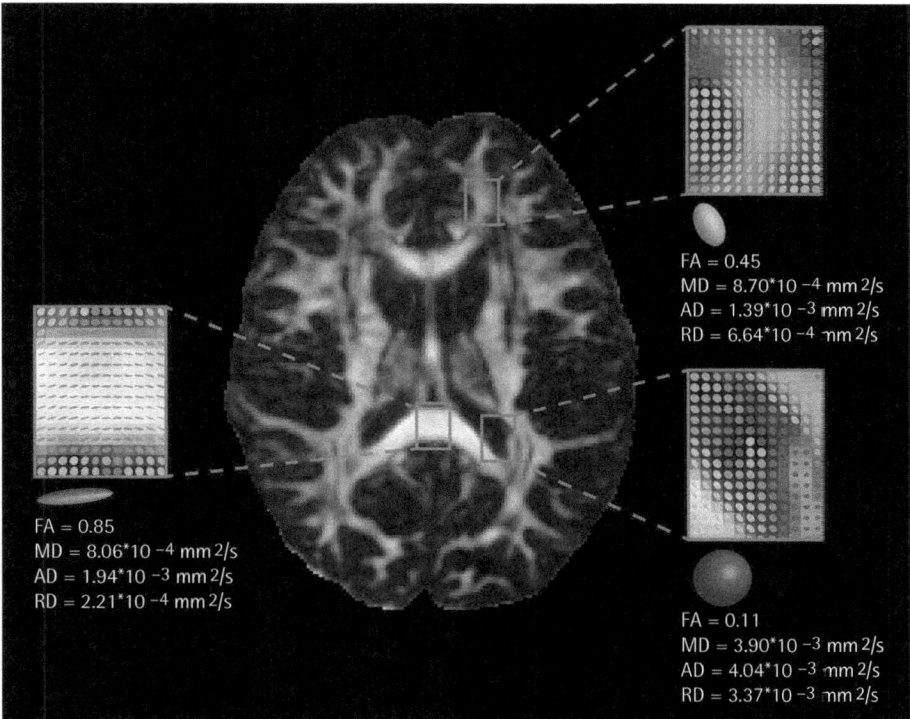

FA = 0.45
MD = 8.70*10 −4 mm2/s
AD = 1.39*10 −3 mm2/s
RD = 6.64*10 −4 mm2/s

FA = 0.85
MD = 8.06*10 −4 mm2/s
AD = 1.94*10 −3 mm2/s
RD = 2.21*10 −4 mm2/s

FA = 0.11
MD = 3.90*10 −3 mm2/s
AD = 4.04*10 −3 mm2/s
RD = 3.37*10 −3 mm2/s

FIGURE 5.2 Tensor descriptions of diffusion are shown in a healthy human brain (11-year-old girl) for three brain regions: the genu (top right), splenium (left), and lateral ventricle (bottom right). The cutouts display tensors for each voxel within the regions shown in red boxes. An enlarged tensor is shown below each box to represent one of the voxels in the region and provide the diffusion parameter values. By convention, tensors are colored by their primary orientation (Green: posterior-anterior orientation, red: medial-lateral, and blue: inferior-superior). FA = fractional anisotropy, MD = mean diffusivity, AD = axial diffusivity, RD = radial diffusivity.

For example, in the central splenium of the corpus callosum (bottom left), axons are highly coherent, densely packed, and well myelinated, resulting in high FA and AD, and low MD and RD. Compared with the splenium, a frontal projection of the forceps minor (top right) has lower FA and AD, and higher RD. This reflects less restricted diffusion, likely caused by less coherently oriented white matter. Finally, a tensor in the CSF of the ventricles (bottom right) shows high MD, RD, and AD, and very low FA, reflecting nearly unrestricted diffusion.

DTI parameters are sensitive to changes in white matter and have provided a wealth of information regarding typical changes with age, and white matter abnormalities in diseases, disorders, and brain injury across the lifespan. However, interpretation of DTI trends or group differences should be done within context of the limitations of the technique. Because diffusion is detected as a loss of signal, increased sensitivity to diffusion also carries greater vulnerability to stochastic machine error (noise) and subject motion. In order to outweigh the influence of noise, researchers can choose imaging parameters such as diffusion sensitivity (also described as the b-value), sequence timing, and image resolution to maximize signal. Furthermore, subject motion is an important consideration, particularly in pediatric studies. Motion can be minimized by using behavioral training or distractions, short protocols, or constraints (e.g., foam padding around the head) (Thieba, Frayne et al., 2018). After acquisition, motion-corrupted images should be removed (Tamnes, Roalf et al., 2017), but even smaller amounts of motion may be a critical covariate to consider during statistical analysis (Roalf, Quarmley et al., 2016). Furthermore, the tensor models one primary fiber direction per voxel, whereas much of the brain contains multiple fiber populations (Jeurissen, Leemans et al., 2013). In cases with overlapping fiber bundles, FA measures may be artificially low, and not accurately reflect the underlying microstructure.

After acquisition, DTI data is typically preprocessed with steps including correction for eddy current distortions, subject motion, signal drift, and Gibb's ringing. Once the tensor and diffusion parameters (FA, MD, etc.) have been calculated for each voxel within the brain, DTI data may be analysed in a number of different ways. A variety of user-friendly software packages are freely available for DTI analysis, each with their own advantages and disadvantages. For example, tract-based spatial statistics (TBSS) (Smith, Jenkinson et al., 2006) provides high local sensitivity to changes by analysing data voxel-by-voxel, and is automated, making it a widely used tool for DTI analysis. However, TBSS is sensitive to registration errors and requires strict multiple comparison correction, which may mask subtle changes. Alternative programs such as ExploreDTI (Leemans, Jeurissen et al., 2009) or TrackVis (Wang, Benner et al., 2007) allow the user to virtually reconstruct whole white matter fibers via tractography, increasing sensitivity to global changes across tracts and avoiding registration errors, but these programs require time and expertise of the user to manually delineate each tract in each participant.

DTI measures are influenced by choice of sequence parameters, analysis methods, and inherent differences between scanners. As multi-site studies become more popular to facilitate larger sample sizes, they often try match protocols as closely as possible and compensate for scanner differences by statistically controlling for site of data acquisition or inclusion of traveling phantoms participants scanned at all sites (Gouttard, Styner et al., 2008; Casey, Cannonier et al., 2018). Still, direct comparisons of diffusion parameters across scanners or sites should be avoided. Even comparison between regions within the same subject must be interpreted carefully, as diffusion measures can vary due to tract architecture rather than differences in "integrity" or "maturity" as is commonly reported. For example, the corticospinal tract shows high FA in inferior portions where fibers are large and myelinated, but decreased FA in superior portions due to intersection with the corpus callosum (Groeschel, Hagberg et al., 2016). The tensor model does not appropriately model regions including multiple fiber populations with unique orientations, which can result in artificially low FA. Finally, one must keep in mind that DTI is sensitive to multiple aspects of white matter microstructure, and it is not possible to identify the relative contributions of axonal packing, myelin, coherence, membrane permeability, and other factors to determine the driving factors behind measured trends.

More advanced dMRI methods, for example neurite orientation dispersion and density imaging (NODDI) (Zhang, Schneider et al., 2012) and diffusion kurtosis imaging (DKI) (Jensen, Helpern et al., 2005), as well as non-diffusion imaging techniques such as myelin water imaging (MWI) (MacKay, Whittall et al., 1994; Deoni, Rutt et al., 2008) and magnetization transfer (MT) (Wolff and Balaban, 1994; Varma, Duhamel et al., 2015) are discussed later in the chapter. These methods are able to provide more specific measures of white matter microstructure than DTI, but have not yet been as widely applied.

Study Design

A mix of research designs have been used to explore white matter microstructural development. Choice of study design influences the sensitivity to different types of relationships and developmental growth trajectories. Early studies of development were primarily cross-sectional, and have provided valuable information on developmental trends across participants. Cross-sectional studies are able to include wide age ranges and large numbers of subjects, but they have limited ability/capacity to explore individual variation or specific growth trajectories. More recently, longitudinal studies have expanded the literature and can assess change within individuals, providing a more sensitive measure of maturation. Longitudinal studies with more time-points enable a more accurate exploration of growth trajectories, which may better capture developmental trends. While longitudinal studies can account for inter-subject variability, it takes years to follow subjects (some of whom will drop out), and longitudinal studies are also sensitive to scanner upgrades. Thus, many studies use accelerated longitudinal designs, where age at intake varies across a wide range and subjects are only followed for a few years, in order to capture and model a more extensive period of development across a relatively short span of data collection. A number of large multi-site longitudinal databases are currently being collected (e.g., the Developing Human Connectome Project (dHCP), HCP Lifespan Development Study (HCP-D), Pediatric Longitudinal

Imaging, Neurocognition, and Genetics (PLING), C-MIND, Adolescent Brain Cognitive Development Study (ABCD®)), and will be invaluable resources in future exploration of white matter maturation.

TRENDS IN WHITE MATTER DEVELOPMENT

Diffusion tensor imaging is a popular and sensitive measure of white matter integrity and has been extensively applied to investigate white matter tissue in healthy and atypical populations. In this section, we outline trends in DTI measures at various scales, beginning with a whole brain perspective, then increasing specificity to discuss white matter networks and regional variation.

Developmental Trends of DTI Measures

Using a range of acquisition and analysis procedures, DTI studies of typical neurodevelopment in childhood, adolescence, and young adulthood consistently demonstrate robust increases in FA and decreases in MD across the brain. The increases of FA and decreases of MD occur nonlinearly, with the steepest slopes in infancy and early childhood (Reynolds, Grohs et al., 2019), and the changes gradually slowing until they reach a plateau in adolescence or early adulthood (Figure 5.3). MD decreases seem to occur just slightly later than increases of FA, with plateaus/minima occurring a few years later than FA plateaus/peaks (Lebel, Gee et al., 2012). Developmental changes in FA and MD have been shown to be driven largely by decreases in RD (Wang, Adamson et al. 2012), suggesting increasing myelination and axonal density, as well as decreasing membrane permeability (Beaulieu, 2002; Mukherjee, Miller et al., 2002; Song, Yoshino et al., 2005; Qiu, Tan et al., 2008). Trends in AD, typically speculated to capture internal axon structure, fiber coherence, and axonal integrity, are less clear. Slight decreases (Lebel, Walker et al., 2008; Kumar, Nguyen et al., 2012; Moura, Kempton et al., 2016; Pohl, Sullivan et al., 2016), no change (Rollins, Vacha et al., 2009), and slight increases of AD (Giorgio, Watkins et al., 2010; Brouwer, Mandl et al., 2012) with age have been reported across adolescence. The lack of consistent findings for AD suggests that axonal integrity plays a relatively minor role in developmental processes, though detection of changes in AD may be confounded by interactions with sex, genetics, or other factors, and may become clearer in larger, longitudinal studies.

Figure 5.3 demonstrates different scales of development from infancy to adolescence by showing trends from three different studies. The nonlinear nature of maturational changes is clear. During infancy, changes are rapid and apparent even over several weeks. During early childhood, developmental changes continue, though they may be obscured by the rapid changes in infancy. Studies of older children and adolescents show continued maturation, but on a much more gradual scale, with changes only apparent

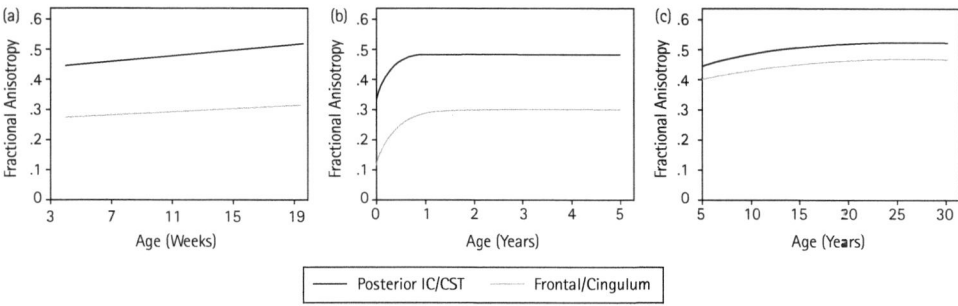

FIGURE 5.3 Developmental trends and regional variation in FA from three different studies. Rapid age-related changes occur in the first weeks of life (a; trend lines from Dubois et al., 2014), with changes apparent even over a few weeks. These rapid changes in infancy appear to level off during the preschool years (b; data from Paydar et al., 2014), though changes in childhood are obscured by the large changes in infancy. The continued development during childhood and adolescence is observed when infants are excluded from the trendlines (c; data from Lebel et al., 2008). Approximately the same amount of change in FA can be seen to occur during the 4-month period from 3–19 weeks (a), as during the 25 years from 5–30 years (c). Projection fibers (posterior internal capsule [IC] or corticospinal tract [CST]) follow similar trajectories to frontal-temporal association fibers, although regional variation can be observed with higher FA in the CST.

over several years. Regional variation (discussed further in Section 3.3) is also apparent, with consistently greater FA in the CST compared to frontal and temporal association fibers. Finally, differences in data acquisition and analysis methods will confound the diffusion parameter values, leading to the discrepancies observed in different datasets shown in Figure 5.3.

DTI has also been used to explore developmental trends in subcortical grey matter structures and white matter tract "regional termination zones" (e.g., basal ganglia, thalamus). Studies demonstrate increases of FA and reductions of MD in subcortical grey matter into adulthood, which may reflect ongoing development (e.g., myelination and/ or increasing density) of tracts emanating and terminating in these grey matter regions (Lebel, Walker et al., 2008; Simmonds, Hallquist et al., 2014; Mah, Geeraert et al., 2017). These changes likely reflect maturation of these grey matter areas, but are difficult to interpret, as they may also be influenced by additional factors such as macromolecular content, neural and glial membranes, and iron deposition (Mukherjee, Miller et al., 2002; Pfefferbaum, Adalsteinsson et al., 2010).

While brain development across childhood and early adulthood has been consistently demonstrated to be complex and non-linear, the specific developmental trajectories are not clear. Different studies posit various models (e.g., quadratic, cubic, exponential), and all are strongly influenced by the dataset characteristics. Factors such as the age range, age distribution across the range, study size, and number of data points per individual can have substantial impacts on the model fits and how appropriately they describe the data (Fjell, Walhovd et al., 2010). It is important to remember that a model used in one dataset may not be the best model for another. However, across models, data from numerous developmental studies converge to show that rapid development of diffusion metrics occurs in early infancy, and changes begin to slow from 1 to 2 years of age (Dubois, Dehaene-Lambertz et al., 2014). More gradual increases in FA and decreases in MD continue to occur until adolescence or early adulthood (~15 to 40 years), depending on the tract (Lebel, Gee et al., 2012). Diffusion measures then remain relatively stable until changes reverse during the aging process, with decreases in FA and increases in MD observed in later adulthood (Moseley, 2002; Lebel, Gee et al., 2012).

Development of White Matter Networks

Brain regions—and brain connections—do not function in isolation, and a network approach to understanding brain development can be informative over and above investigations of individual connections. Typically, graph theory is used to model the whole-brain network using a standard atlas of choice—such as the Automated Anatomical Labeling template (Tzourio-Mazoyer, Landeau et al., 2002). The atlas is used to define regions (nodes), and connections between each pair of nodes can be calculated using dMRI data. These connections (edges) can then be weighted (e.g., by the mean FA of each connection), or unweighted. The connectome (see Figure 5.4 for an example) can then be used to identify hubs and calculate metrics including efficiency, path length, and clustering coefficients (Watts and Strogatz, 1998; Sporns, Tononi et al., 2005). Efficiency and path length describe the ability for information to quickly pass from one location to another, while the clustering coefficient quantifies the amount of connections entering and exiting a node. If desired, functional MRI data can be combined with dMRI to provide insight about functional networks and their connectivity within the graph. Graph theory analysis has provided valuable insights into the architecture of a healthy brain network and changes in the whole brain network throughout development.

Healthy brain networks are typically organized into a highly modular, small world architecture. The small world architecture is characterized by distinct functional networks with many connections between nodes within the same network, and a small "rich club" of highly interconnected hub nodes which connect these distinct functional networks (Fair, Cohen et al., 2009; Cao, Huang et al., 2017). The small world architecture enables coordinated processing within functionally specific networks, facilitates transfer of relevant information between networks along reinforced pathways, and reduces vulnerability to disruptions caused by damage to one part of the network (Watts and Strogatz, 1998; Bullmore and Sporns, 2009; Bullmore and Sporns, 2012). Across typical development, individual networks tend to become more integrated and connections between the rich club of hubs are emphasized. These processes result in increasing modularity, which tends to peak in young adulthood (Vogel, Power et al., 2010) alongside increases in global efficiency (Bathelt, Barnes et al., 2017). Increased modularity has been linked to increased reading performance (Bailey, Aboud et al., 2018), and lower modularity may be predictive of disorders including attention deficit

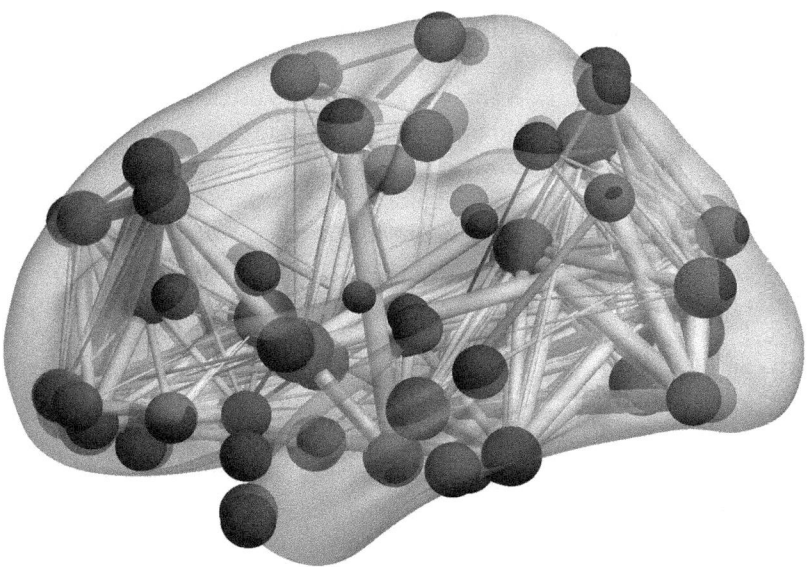

FIGURE 5.4 An example whole brain graph from a healthy 13.9-year-old male. Network nodes are determined by brain regions from a standard atlas (e.g., Automated anatomical labeling (AAL) (Tzourio-Mazoyer, Landeau et al. 2002)), and their connections edges are calculated from diffusion MRI data. In this figure, node size represents degree centrality, with larger nodes having more connections to other brain regions. Edge diameter represents the FA of each connection, approximating tract efficiency between regions. Other useful measures can be computed from the graph to assess ease of information transfer (or path length), clustering of connections, presence of network hubs, and efficiency of the network.

Source: Image created by Xiangyu Long, used with permission.

hyperactivity disorder (Uddin, Kelly et al., 2008). Thus, typical development of white matter impacts development of networks responsible for important cognitive and behavioral functions.

Network-based analyses of white matter structure and development offer a valuable addition to region-based analyses, and may detect otherwise unnoticed effects. For example, children's academic attainment has been linked to global efficiency and clustering coefficient while no links between white matter structure and academic attainment were observed in region-of-interest-based analyses (Bathelt, Gathercole et al., 2018; Bathelt, Scerif et al., 2019). Network-based approaches are also ideal to evaluate links between brain and genetics. For example, mutations within a single gene have been connected to cognitive impairments and decreases in global clustering coefficient and efficiency, with regionally specific effects depending upon gene expression variations (Bathelt, Barnes et al., 2017). Despite regional variation, subjects with the mutation shared a profile of cognitive deficits. This finding highlights that alterations in different tracts within the same network may result in a similar profile of deficits, and is a strong argument for assessment of typical and atypical development at the network level.

Regional Variability in White Matter Development

Brain white matter shows considerable variation in microstructure, as reflected by DTI parameters. In general, central and posterior white matter (e.g., the corpus callosum, especially the splenium; CST) has higher FA values than more peripheral and frontal regions (e.g., frontal white matter); see Figure 5.3. This pattern holds true across infancy, childhood, and adolescence, reflecting differences in microstructural organization (e.g., myelination and axonal packing), rather than maturation differences. Additionally, white matter microstructural developmental trajectories, including the timing of peaks and plateaus, vary across the brain (Lebel and Beaulieu, 2011; Reynolds, Grohs et al., 2019). In line with post-mortem histological studies (Yakovlev and Lecours, 1967; Benes, 1989), DTI studies show that white matter tract maturation follows posterior-to-anterior and central-to-peripheral patterns of development within the brain (Dubois, Hertz-Pannier et al., 2006; Dubois, Dehaene-Lambertz et al., 2008; Löbel, Sedlacik et al., 2009; Qiu, Mori et al., 2015; Krogsrud, Fjell et al., 2016). This is apparent in studies of infants, where all regions experience rapid changes, but FA increases and MD decreases in commissural and projection fibers appear to slow earlier (~1 year) than those in association tracts (~2 years) (Hermoye, Saint-Martin et al., 2006; Geng, Gouttard et al., 2012; Dubois, Dehaene-Lambertz et al., 2014; Paydar, Fieremans et al., 2014). In studies of older children and adolescents, the fornix appears to be one of the earliest developing fibres, with little-to-no changes detected during late childhood or adolescence (Lebel, Gee et al., 2012; Simmonds, Hallquist et al., 2014). Following this, commissural (e.g., genu, body, splenium of the corpus callosum) and projection (e.g., CST, posterior thalamic radiation) fibres reach maturity during late childhood or early adolescence (~10–15 years), and frontal and temporal connections (e.g., cingulum, uncinate, superior longitudinal fasciculus) continue maturing into young adulthood (~20–30+ years) (Tamnes, Ostby et al., 2010; Lebel, Gee et al., 2012; Simmonds, Hallquist et al., 2014). Within the corpus callosum, a posterior to anterior wave of development has been demonstrated in infants, children, and adolescents, with the splenium reaching plateau earlier than the body or genu fibres (Hasan, Kamali et al., 2009; Lebel, Caverhill-Godkewitsch et al., 2010; Dubois, Dehaene-Lambertz et al., 2014; Paydar, Fieremans et al., 2014).

This hierarchical development of white matter roughly follows the timeline of behavioral development. Most tracts that demonstrate earlier maturation are involved in basic processes and lower-level sensorimotor functions. For example, the fornix, consistently demonstrated to show early maturation, has been linked to memory and spatial working memory (Aggleton and Brown, 1999; Ly, Adluru et al., 2016); the posterior thalamic radiation and inferior longitudinal fasciculus (ILF) are among the early maturing projection and association fibers, and both are involved in visual processing (Catani, Jones et al., 2003). The corpus callosum, where development is largely complete by late childhood, supports early motor development and interhemispheric communication (e.g., Aboitiz and Montiel, 2003; Chang, Hung et al., 2015). More prolonged development of fronto-temporal tracts likely reflects later behavioral maturation of higher emotional, cognitive, and executive functioning processes. For example, the superior longitudinal fasciculus, uncinate fasciculus, and cingulum are involved in language, intelligence, and emotion processing (Catani, Jones et al., 2003; Parker, Luzzi et al., 2005; Jung and Haier, 2007).

INFLUENCES ON WHITE MATTER DEVELOPMENT

A look at any scatterplot of individual DTI data points shows substantial variability, and the spread of points often exceeds the developmental changes that occur across the age range. Variation is also present in individual changes over time, as observed by differing slopes. While some scatter in the data is likely due to measurement error, individual variability in white matter microstructure and maturation patterns can also be influenced by factors of interest, including sex, genetics, and the environment. Understanding this variation and its causes can be very informative and is a topic of great interest in the current literature. In this section, we outline current understandings of the interplay between these influences and white matter development.

Sex Differences in Development

Differences between males and females behaviorally and cognitively, as well as in the timing of onset, expression, and prevalence of neurodevelopmental and neuropsychiatric conditions, suggest that there are sex differences in brain development (Giedd, Raznahan et al., 2012). Neurodevelopmental sex differences may provide some insight into the mechanisms of these disorders, and why their timing of onset and/or manifestations vary by sex. Research has consistently demonstrated that males have larger absolute white matter and total brain volume than females (Lenroot, Gogtay et al., 2007; Perrin, Herve et al., 2008). Sex differences in developmental trajectories of brain volumes have also been demonstrated, with females reaching their peak brain volumes approximately 1.5 years before males (Lenroot, Gogtay et al., 2007), coinciding with earlier puberty and growth spurts in females.

However, findings for sex differences in white matter microstructural development are much more mixed. A number of cross-sectional studies did not identify significant sex differences in FA or MD of children, adolescents, or young adults (e.g., Bonekamp, Nagae et al., 2007; Eluvathingal, Hasan et al., 2007; Giorgio, Watkins et al., 2008; Lebel, Walker et al., 2008; Bava, Thayer et al., 2010; Giorgio, Watkins et al., 2010; Uda, Matsui et al., 2015). Other studies report significant sex differences only in certain brain areas, though the location and direction of differences vary and do not produce any consistent findings (Schneiderman, Buchsbaum et al., 2007; Schmithorst, Holland et al., 2008;

Chiang, McMahon et al., 2011). Interestingly, the few studies exploring differences in developmental trajectories between males and females generally show more pronounced age-related changes in diffusion measurements in males (Asato, Terwilliger et al., 2010; Bava, Boucquey et al., 2011; Clayden, Jentschke et al., 2012; Wang, Adamson et al., 2012; Simmonds, Hallquist et al., 2014; Seunarine, Clayden et al., 2016). Because of this apparent age-sex interaction, it is possible that study design and large age ranges contribute to the previous inconsistent findings of sex differences. It is worth noting that these studies have not linked these sex differences in developmental trajectories to behavior. Thus, it remains unclear whether these differences do, in fact, underlie sex differences in neurodevelopmental or neuropsychiatric conditions. Further, studies to date have often conflated the biological effects of sex with the cultural and societal influences of gender and gender norms. To better understand both sex and gender differences during white matter development, and their interaction with age, large longitudinal studies, and studies assessing sex and gender separately, are necessary.

Genetics

Most aspects of brain structure and function show some degree of heritability, including volume and cortical thickness (Posthuma, De Geus et al., 2002; Winkler, Kochunov et al., 2010) and functional brain activation patterns during tasks (Blokland, McMahon et al., 2008; Karlsgodt, Kochunov et al., 2010) and at rest (Glahn, Winkler et al., 2010). A range of genetic single-nucleotide polymorphisms have been demonstrated to influence white matter microstructure in healthy individuals (Kohannim, Jahanshad et al., 2012); genes and alleles that lead to a heightened risk of schizophrenia, depression, and Alzheimer's disease amongst others, have also been associated with specific alterations in white matter (Shaw, Lerch et al., 2007; Tham, Woon et al., 2011; Zhou, Liu et al., 2017).

Genes also interact with age and environment to affect white matter differently across development. In a large study of 705 twins and their siblings, Chiang and colleagues (Chiang, McMahon et al., 2011) showed a reduced influence of genetics with age. During adolescence, 70–80 percent of variation in FA could be contributed to genetic factors, compared to only 30–40 percent in adults (Chiang, McMahon et al., 2011). This decreased heritability with age suggests an increasing influence of environmental factors, such as diet, education, and home environment, with age. However, these genetics-brain structure relationships are complex, and confounded by environmental conditions (Turkheimer, Haley et al., 2003). Chiang and colleagues (Chiang, McMahon et al., 2011) also identified gene-environment interactions on white matter structure, with higher socioeconomic status and IQ related to higher heritability of white matter structure. Thus, genes and the environment interact in a complex non-linear manner that changes throughout development, and the relationships with white matter structure remain unclear. Large studies incorporating both neuroimaging and genetics, such as the Pediatric Longitudinal Imaging, Neurocognition, and Genetics (PLING) study (Jernigan, Brown et al., 2016) will help to disentangle these relationships.

Environment

An increasing influence of environmental factors on brain white matter microstructure across the lifespan is suggested by the decreasing heritability of white matter metrics with age. Prenatal exposure to substances, including alcohol (Treit, Lebel et al., 2013) and illegal drugs (e.g., cocaine (Lebel, Warner et al., 2013)) appear to impact white matter development. Other prenatal factors, such as maternal anxiety (Rifkin-Graboi, Meaney et al., 2015), depression (Lebel, Walton et al., 2016), and obesity (Ou, Thakali et al., 2015) are related to diffusion metrics in infants and/or children. Postnatally, socioeconomic status (Chiang, McMahon et al., 2011), neglect (Bick, Zhu et al., 2015), nutrition (Ou, Andres et al., 2015), parental education (Noble, Korgaonkar et al., 2013; Foulkes and Blakemore, 2018), life experience, culture, and exercise (Chaddock-Heyman, Erickson et al., 2014), have all been identified as related to white matter structure. In general, adverse environments (e.g., alcohol/drug exposure, low socioeconomic status) are associated with lower FA and/or higher MD in children, suggesting delayed development and/or less development overall (Foulkes and Blakemore, 2018). However, exceptions have been noted; for example, prenatal depressive symptoms were associated with lower MD in right frontal regions in young children, suggesting premature brain development (Lebel, Walton et al., 2016). One longitudinal study found steeper decreases of MD in children with prenatal alcohol exposure compared to unexposed controls, perhaps suggesting compensatory plasticity (Treit, Lebel et al., 2013). The mechanisms underlying these associations remain largely unknown, but may include epigenetic effects or altered stress responses. A fruitful avenue for future research may be to try to understand the separate and cumulative effects of different environmental influences, as they rarely act in isolation.

LINKS BETWEEN WHITE MATTER DEVELOPMENT AND COGNITION

DTI enables an exploration of relationships between white matter microstructure and different aspects of cognition. Numerous links between white matter and cognitive abilities have been identified, however there are fewer studies examining how relationships between cognition and white matter change over time. Here, we touch briefly on the broader literature, and use reading as an example to describe links between white matter structure/development and cognitive functions.

Typical and Atypical Cognition

Broadly, better cognitive performance is linked with increased FA and/or decreased MD in children (Nagy, Westerberg et al., 2004; Matejko and Ansari, 2015; Deoni,

O'Muircheartaigh et al., 2016), suggesting more mature patterns of white matter structure in these children. For example, higher FA in the left superior longitudinal, inferior longitudinal, and inferior fronto-occipital fasciculi, superior corona radiata,
and corticospinal tract, white matter connections connecting brain regions involved
in math (e.g., posterior parietal cortex and intraparietal sulcus), has consistently been
linked to better math performance (Li, Hu et al., 2013). Increased plasticity (greater
FA increases) of the left fronto-temporal section of the superior longitudinal fasciculus is associated with greater math improvements during a 2-month tutoring program
(Jolles, Wassermann et al., 2016). Links between white matter and memory have also
been identified in childhood. Increased FA and/or decreased diffusivity, particularly
in fronto-parietal connections, has been linked to better declarative auditory-verbal
memory (Mabbott, Rovet et al., 2009) and to working memory performance (Nagy,
Westerberg et al., 2004, Vestergaard, Madsen et al., 2010; Krogsrud, Fjell et al., 2018).
A longitudinal study of 148 4–11-year-old children demonstrated greater decreases in
MD, AD, and RD in the right inferior longitudinal, inferior fronto-occipital, and uncinate fasciculi, and the forceps major were associated with greater visuospatial working
memory improvements (Krogsrud, Fjell et al. 2018). In another longitudinal study of
6- to 25-year olds, working memory performance was predicted by FA in fronto-parietal
and fronto-striatal white matter that was measured 2 years prior (Darki and Klingberg,
2015). Links between white matter and cognition are dynamic and nuanced, and vary
depending upon how a cognitive function is assessed. To illustrate the dynamic links between white matter and cognition, we turn to the example of reading.

Reading is an essential skill typically learned during childhood that involves attention,
sensory integration, phonological processing, memory, and other components. Brain
regions involved in reading, most notably the left inferior frontal, temporal-parietal,
and temporal-occipital areas (Price, 2012; Martin, Schurz et al., 2015), are connected via
dorsal and ventral white matter pathways including the arcuate, superior, and inferior
longitudinal, inferior fronto-occipital, and uncinate fasciculi (Catani, Jones et al., 2005;
Ben-Shachar, Dougherty et al., 2007; Welcome and Joanisse, 2014). Figure 5.5 provides a
schematic overview of the major white matter connections involved in reading.

DTI studies have also identified specific brain regions related to separate reading
processes. For example, improved phonological processing performance has been
linked with higher FA in the left inferior longitudinal, and fronto-occipital, and uncinate fasciculi, and lower MD in the left arcuate fasciculus (Lebel, Shaywitz et al.,
2013; Welcome and Joanisse 2014; Reynolds, Long et al., 2019), along with higher
FA and lower MD in the arcuate fasciculus (Reynolds, Long et al., 2019), and greater
corpus callosum volume (Welcome and Joanisse, 2014). Improved word identification,
involving both sight reading proficiency and phonological processing, has been linked
to higher FA in more widespread brain regions, including the arcuate fasciculus and ILF
(Yeatman, Dougherty et al., 2012), internal capsule, thalamus, right corona radiata, and
portions of the corpus callosum (Lebel, Shaywitz et al., 2013). Reading fluency involves
comprehension of connected text and narrative memory, and has been linked to higher
FA in the inferior internal capsule (Sullivan, Zahr et al., 2010), right uncinate fasciculus

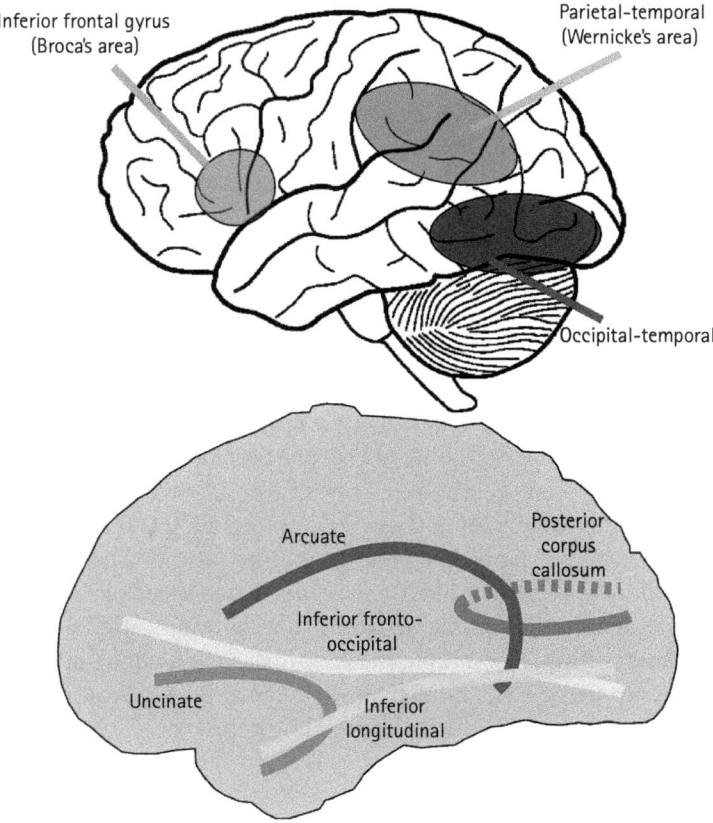

FIGURE 5.5 Schematic of the reading network. Connections between occipital, temporal, and inferior frontal cortices are emphasized through tracts commonly related to reading. The reading network is typically left-lateralized, with a heavy emphasis on leftward cortical regions and tracts.

(Arrington, Kulesz et al., 2017), and corpus callosum (Gordon, Lee et al., 2011). Finally, domain general skills such as attention control and working memory are involved in coordination of reading fluency and comprehension (Arrington, Kulesz et al., 2014) and involve coordination between multiple distributed cognitive networks throughout the brain among many white matter pathways.

While most studies find higher FA and lower MD to relate to better reading performance, notably, several DTI studies have observed opposite findings. A small number of studies have found reading ability to be negatively correlated with FA and positively correlated with MD in the posterior corpus callosum (Dougherty, Ben-Shachar et al., 2007; Odegard, Farris et al., 2009; Vandermosten, Boets et al., 2012). This reduced connectivity between left and right posterior reading areas in good readers may reflect better segregation of left hemisphere reading regions, and thus better specialization of this area; alternatively, increased connectivity in poor readers may reflect a compensatory mechanism.

A small number of longitudinal studies have related reading skills to white matter development over time. In general, children with better reading skills show faster increases of FA than children with poor reading skills (Yeatman, Dougherty et al., 2012; Wang, Mauer et al., 2017; Reynolds, Long et al., 2019). Faster changes in FA or MD are associated with larger gains in reading skills in children with and without developmental disorders (Treit, Lebel et al., 2013; Wang, Mauer et al., 2017). Differences in developmental trajectories in children with and without reading difficulties have also been observed, with dysfluent readers not displaying age-related FA increases in temporal-parietal areas as are seen in fluent readers, suggesting they may have delayed white matter maturation in reading-related tracts (Reynolds, Long et al., 2019). Differences in the relationship between white matter and reading extend into populations with reading disorders, such as dyslexia.

The most consistent brain region affected in dyslexia is a left temporal-parietal region that includes the arcuate fasciculus and corona radiata (Vandermosten, Boets et al., 2012, Vanderauwera, Wouters et al., 2017). Lower FA and/or increased MD here suggest delayed white matter development in these regions, though it is unclear whether this is due to less myelin, reduced axonal packing, or another process. Decreased maturity in reading-related white matter tracts likely reflects decreased speed or synchronicity of information transfer within the reading network. This communication inefficiency could drive reading deficits through multiple mechanisms, such as difficulty distinguishing phonemes when acquiring spoken language in infancy or a delayed transition to the rapid visual processing typical of the mature reading network.

Outcomes of Intervention

Across a range of neurodevelopmental disorders and brain injuries, intervention approaches have exploited the fact that the brain demonstrates plasticity at both neuronal and network levels (e.g., (Nudo, 2006). The use of DTI has enabled researchers to characterize intervention-related changes in brain microstructure. Considering the example of reading, poor readers aged 8–10 years old showed increases of FA and reductions of RD following an intensive reading intervention in the same region (left frontal area) where poor readers differed from controls at baseline, suggesting that the intervention facilitated an increase in myelination in the brain (Keller and Just, 2009). Furthermore, the FA increases in this left frontal area were correlated with the improvements in reading (Keller and Just, 2009). Another study showed increases of FA and reductions of MD in the arcuate and inferior longitudinal fasciculi after an intensive reading intervention, while the structure in the corpus callosum remained stable (Huber, Donnelly et al., 2018). MD of all three of these tracts correlated with reading abilities before the intervention, however only the corpus callosum remained significantly correlated with reading ability following the intervention (Huber, Donnelly et al. 2018). Interestingly, the data suggested that intervention-induced changes in diffusion metrics were not a "normalization" of white matter, but rather increased the differences between the two groups. These intervention studies suggest not only that white matter microstructure changes during intensive instruction, but also that its relationship to reading abilities changes.

White matter plasticity, the extent and capacity for FA and MD changes, has also been observed in other training and intervention programs (for example, increases in FA in visuo-motor related connections when teaching adults to juggle (Scholz, Klein et al., 2009)), and plasticity likely underlies, at least in part, other effective interventions to reduce or prevent behavioral and cognitive deficits in a variety of developmental disorders. Interventions are likely to be particularly effective if sensitive windows can be identified for these treatments. A clearer understanding of typical white matter development has the potential to lead to earlier diagnosis of many childhood difficulties and disorders by identifying deviations from normal; this would allow early intervention at a time when the brain demonstrates maximal capacity for neuroplasticity.

IMAGING WHITE MATTER MICROSTRUCTURE IN NEW DETAIL

DTI measures are sensitive to multiple aspects of white matter micro- and macrostructure, including myelin, axon packing, membrane permeability, axon diameter, white matter volume, and water content (Beaulieu, 2002). However, as discussed through this chapter, their specificity is limited. Additionally, the diffusion tensor is not able to resolve more than one fiber population per voxel, as is the case in much of the brain (Jeurissen, Leemans et al., 2013). This can lead to artificially low FA values in areas of crossing fibers, which may bias results in certain areas.

More advanced imaging techniques can augment DTI findings and inform discussion of the processes driving white matter development. Measuring upwards of sixty diffusion directions (compared to the typical measurement of approximately thirty) and/or incorporating multiple diffusion weightings (b-values) allows use of advanced diffusion models to better represent in-vivo diffusion and provide more specific measures. DKI (Jensen, Helpern et al., 2005) evaluates deviations from Gaussian diffusion observed at high b values. Though the application of DKI to neurodevelopment has been relatively scarce to date, DKI measures appear to be more sensitive to age-related changes than FA and MD (Paydar, Fieremans et al., 2014; Das, Wang et al., 2017; Grinberg, Maximov et al., 2017). Another advanced diffusion model, neurite orientation dispersion and density imaging (NODDI) (Zhang, Schneider et al., 2012), incorporates multiple components of diffusion, to more accurately represent environmental conditions experienced by water within and around white matter than the tensor model. Like DKI, the NODDI-derived neurite density index (NDI) appears to be more sensitive to age-related changes in white matter than FA or MD (Chang, Owen et al., 2015; Genc, Malpas et al. 2017; Mah, Geeraert et al., 2017). The orientation dispersion index (ODI), also derived from

NODDI, shows no changes with age in children and adolescents, suggesting that axon coherence remains stable during development (Chang, Owen et al., 2015; Genc, Malpas et al., 2017; Mah, Geeraert et al., 2017). These methods generally require more imaging time than DTI, which has limited their application in pediatric research. However, improved hardware and software (e.g., multiband imaging) is facilitating more clinically feasible scan times, enabling contemporary studies to apply these advanced methods to measure brain development. As these advanced diffusion models, and others, become more widely used, they will further advance our understanding of the processes underlying normal and atypical white matter development.

Non-diffusion MRI methods such as MT imaging and myelin water imaging also offer increased specificity to myelin compared to diffusion methods. MT imaging is sensitive to macromolecules including myelin, and typically reports the magnetization transfer ratio (MTR) (Wolff and Balaban, 1994; Henkelman, Stanisz et al., 2001). One combined DTI-MT study showed robust increases of FA with no changes of MTR during late childhood (Moura, Kempton et al., 2016); other studies have shown slight decreases of MTR in males but not females during adolescence (Perrin, Leonard et al., 2009; Pangelinan, Leonard et al., 2016). This suggests that myelination is not driving development during this time period, and that there may be sex-specific mechanisms at work, particularly in adolescence. An extension of MT is inhomogeneous MT (ihMT), which appears to be even more specific to myelin within the human brain (Girard, Prevost et al., 2015; Varma, Duhamel et al., 2015; Geeraert, Lebel et al., 2018). IhMT has been applied in a semi-longitudinal cohort of 6–15-year-olds, where no relationships were observed between ihMT and age, similar to previous developmental studies using MT (Geeraert, Lebel et al., 2019).

Myelin water imaging models water exchange between compartments and reports the myelin water fraction (MWF). MWF has been used to demonstrate substantial myelination during infancy and early childhood, following similar trends of posterior-to-anterior, central-to-peripheral patterns of change observed in DTI and post-mortem developmental studies (Deoni, Dean et al., 2012; Dean, O'Muircheartaigh et al., 2016). Age-related changes in MWF appear to happen earlier in females (Dean, O'Muircheartaigh et al., 2015), consistent with DTI findings. Finally, g-ratio is the ratio of axon diameter to outer fiber diameter, and is related to the efficiency of signal conduction along axons. One study has used this to assess brain development in a small sample (n=18), reporting decreases across the brain from birth to age ~7.5 years (Dean, O'Muircheartaigh et al., 2016). Another study combined NODDI, ihMT, mcDESPOT, and g-ratio imaging to describe widespread changes in NDI and MWF during late childhood and adolescence, but no changes in ihMT and only one region with decreases in g-ratio (Geeraert, Lebel et al., 2019). Combining multiple microstructurally sensitive imaging techniques is a promising direction for future studies which aim to clarify white matter development or links to cognition which have been initially described by DTI.

Many of the measures discussed previously have considerable overlap in their contrast. For example, ihMT and myelin water imaging share ~95 percent of their variance in the brains of healthy children, suggesting they are both quite sensitive, and

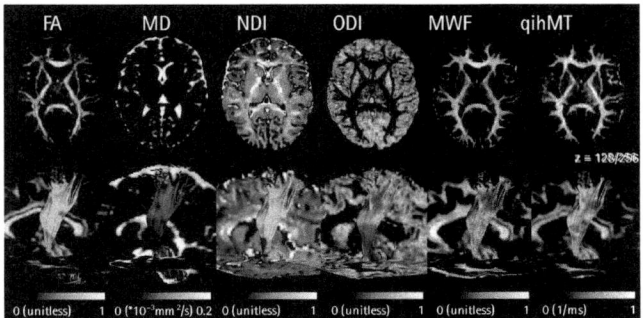

FIGURE 5.6 Imaging measures of white matter structure provide complementary information about the characteristics of white matter tissue and fiber tracts in the brain of an 11-year-old male subject. FA = fractional anisotropy, MD = mean diffusivity, NDI = neurite density index, ODI = orientation dispersion index, MWF = myelin water fraction, qihMT = quantitative ihMT.

relatively specific, to myelin. RD from DTI also shares 70–80 percent variance with these measures, demonstrating reasonable sensitivity, but less specificity, as expected (Geeraert, Lebel et al., 2017). Figure 5.6 shows parameter maps and a reconstruction of the corticospinal tract coloured by parameter value for different white matter metrics in the same healthy brain. FA, NDI, MWF, and qihMT are all sensitive to myelin, as is evident in their shared contrast. These metrics provide variation along the length of the tract, with the highest FA/NDI/MWF/qihMT values apparent in the central portion of it, reflecting higher myelination in these regions. MD values are more uniform across the brain as they are not specific to myelin and instead reflect the presence of any type of diffusion barriers. ODI, a measure of tract coherence, also shows less variation along the length of the tract, with higher dispersion only apparent in the superior portion where the tract begins to fan. In areas of crossing fibers (for example, to each side of the genu of the corpus callosum), artificially low FA values are apparent; MWF and qihMT, neither of which is affected by crossing fibers, remain high in these areas.

Microstructurally sensitive measures have been applied to begin to describe links between white matter microstructure and cognition in new detail. NODDI has been applied to identify greater axonal density in the posterior corpus callosum of poor readers (Huber, Henriques et al., 2019), and decreased NDI and ODI have been linked to intelligence in a sample of 18–40-year-olds (Genc, Fraenz et al., 2018). These findings support the hypothesis that cognitive networks undergo pruning during maturation, resulting in a refined and efficient system. McDESPOT has also been related to cognitive development within young children, demonstrating clear links between myelination and cognitive maturation (Deoni, O'Muircheartaigh et al., 2016). A recent longitudinal MWF study also reports positive associations between white matter structure and cognitive ability during infancy and early childhood (Dai, Hadjipantelis et al., 2019). DKI measures have not been linked to cognitive measures in typical development, but have been linked to processing speed and executive function in studies of older adults with schizophrenia and multiple sclerosis, such that indications of greater integrity were linked to improvements in cognitive performance (Bester, Jensen et al., 2015; Kochunov, Rowland et al., 2016). Beyond mcDESPOT, other myelin-sensitive measures such as MTR and ihMT techniques have not yet been applied to study cognition in healthy development. Some studies have employed MT imaging in studies of older adults, but have found either weak links or no connection between MTR and cognition (Deary, Bastin et al., 2006; Penke, Muñoz Maniega et al., 2012). Overall, many new techniques may be applied in future studies to describe how white matter microstructure supports cognition.

DTI remains a useful technique given its ease of acquisition and the readily available analysis tools, and its sensitivity to changes in white matter throughout childhood and adolescence. Its shared variance with more specific techniques makes it apparent that it is measuring, at least in part, the desired features of white matter microstructure. However, its limited specificity makes detailed interpretations difficult. Advanced white matter imaging techniques are gaining wider application as new technologies make scan times more feasible in children. Together, DTI and advanced techniques will continue to enhance our understanding of white matter development, sex differences, and links with cognitive abilities in the years to come.

Conclusions

White matter development is a complex and nuanced process. Rapid changes in white matter structure and organization occur in the early years of life, dominated by myelination. More gradual remodelling continues throughout adolescence and into adulthood, likely driven by axonal packing, although this remains unclear. White matter develops relatively uniformly within cognitive and behavioral networks, but significant variation is seen between developmental timing of brain regions. For example, central tracts (primary sensorimotor networks) mature more rapidly than association fibers of the brain related to sophisticated cognitive functions (those connecting frontal and temporal regions). The general small-world architecture of adult white matter networks is present at birth, but as white matter becomes mature and myelinated, recruitment of spatially neighboring regions decreases in favor of reinforced long-distance connections within cognitive and behavioral networks.

White matter structure and its development plays a critical role in the healthy function of cognitive networks. As an example of links between white matter development and cognitive abilities, children with better reading skills often exhibit higher FA and lower diffusion in frontal, parietal, and temporal white matter areas associated with reading, generally suggesting a more mature brain in children with better reading performance. Furthermore, the growth and strengthening of white matter connections is linked with development of reading skills, and reading interventions produce changes in white matter.

The majority of literature on white matter development to date has used DTI. DTI is sensitive technique and has greatly enhanced our understanding of white matter maturation from birth to adulthood. Newer, more advanced models and techniques overcome some of the limitations of DTI and are providing new clarity on the developing brain. The application of these newer techniques, particularly in combination with each other and/or DTI, will be advantageous to holistically describe white matter microstructural development and its nuanced variation throughout the brain and in relation to cognitive abilities, genetics, and environmental influences. It is through intelligent application of imaging methods that new light can be shed upon white matter development and its internal and external influences in both typical and atypical conditions.

REFERENCES

Aboitiz, F. and J. Montiel (2003). One hundred million years of interhemispheric communication: The history of the corpus callosum. *Brazilian Journal of Medicine and Biology Research*, 36(4): 409–420.

Aggleton, J. P. and M. W. Brown (1999). Episodic memory, amnesia, and the hippocampal-anterior thalamic axis. *Behavioral Brain Science*, 22(3): 425–444; discussion 444–489.

Arrington, C. N., P. A. Kulesz, D. J. Francis, J. M. Fletcher, and M. A. Barnes (2014). The contribution of attentional control and working memory to reading comprehension and decoding. *Scientific Studies of Reading*, 18(5): 325–346.

Arrington, N. C., P. A. Kulesz, J. Juranek, P. T. Cirino, and J. M. Fletcher (2017). White matter microstructure integrity in relation to reading proficiency. *Brain and Language*, 174: 103–111.

Asato, M. R., R. Terwilliger, J. Woo, and B. Luna (2010). White matter development in adolescence: A DTI Study. *Cerebral Cortex*, 20(9): 2122–2131.

Bailey, S. K., K. S. Aboud, T. Q. Nguyen, and L. E. Cutting (2018). Applying a network framework to the neurobiology of reading and dyslexia. *Journal of Neurodevelopmental Disorders*, 10(1): 37.

Bathelt, J., J. Barnes, F. L. Raymond, K. Baker, and D. Astle (2017). Global and local connectivity differences converge with gene expression in a neurodevelopmental disorder of known genetic origin. *Cerebral Cortex*, 27(7): 3806–3817.

Bathelt, J., S. E. Gathercole, S. Butterfield, C. Team, and D. E. Astle (2018). Children's academic attainment is linked to the global organization of the white matter connectome. *Dev Sci*, 21(5): e12662.

Bathelt, J., G. Scerif, A. C. Nobre, and D. E. Astle (2019). Whole-brain white matter organization, intelligence, and educational attainment. *Trends Neuroscience Education*, 15: 38–47.

Bava, S., V. Boucquey, D. Goldenberg, R. E. Thayer, M. Ward, J. Jacobus, and S. F. Tapert (2011). Sex differences in adolescent white matter architecture. *Brain Research*, 1375: 41–48.

Bava, S., R. Thayer, J. Jacobus, M. Ward, T. L. Jernigan, and S. F. Tapert (2010). Longitudinal characterization of white matter maturation during adolescence. *Brain Research*, 1327: 38–46.

Beaulieu, C. (2002). The basis of anisotropic water diffusion in the nervous system—a technical review. *NMR Biomed*, 15: 435–455.

Ben-Shachar, M., R. F. Dougherty, and B. A. Wandell (2007). White matter pathways in reading. *Current Opinion in Neurobiology*, 17(2): 258–270.

Benes, F. M. (1989). Myelination of cortical-hippocampal relays during late adolescence. *Schizophrenia Bulletin*, 15(4): 585–593.

Benes, F. M., M. Turtle, Y. Khan, and P. Farol (1994). Myelination of a key relay zone in the hippocampal formation occurs in the human brain during childhood, adolescence, and adulthood. *Archives General Psychiatry*, 51(6): 477–484.

Bester, M., J. H. Jensen, J. S. Babb, A. Tabesh, L. Miles, J. Herbert, R. I. Grossman, and M. Inglese (2015). Non-Gaussian diffusion MRI of gray matter is associated with cognitive impairment in multiple sclerosis. *Multiple Sclerosis*, 21(7): 935–944.

Bick, J., T. Zhu, C. Stamoulis, N. A. Fox, C. Zeanah, and C. A. Nelson (2015). Effect of early institutionalization and foster care on long-term white matter development: A randomized clinical trial. *Jama Pediatrics*, 169(3): 211–219.

Blokland, G. A., K. L. McMahon, J. Hoffman, G. Zhu, M. Meredith, N. G. Martin, P. M. Thompson, G. I. De Zubicaray, and M. J. Wright (2008). Quantifying the heritability of task-related brain activation and performance during the N-back working memory task: A twin fMRI study. *Biological Psychology*, 79(1): 70–79.

Bonekamp, D., L. M. Nagae, M. Degaonkar, M. Matson, W. M. Abdalla, P. B. Barker, S. Mori, and A. Horská (2007). Diffusion tensor imaging in children and adolescents: Reproducibility, hemispheric, and age-related differences. *Neuroimage*, 34(2): 733–742.

Brouwer, R. M., R. C. Mandl, H. G. Schnack, I. L. van Soelen, G. C. van Baal, J. S. Peper, R. S. Kahn, D. I. Boomsma, and H. E. Hulshoff Pol (2012). White matter development in early puberty: A longitudinal volumetric and diffusion tensor imaging twin study. *PLoS One*, 7(4): e32316.

Bullmore, E. and O. Sporns (2009). Complex brain networks: Graph theoretical analysis of structural and functional systems. *Nature Reviews Neuroscience*, 10(3): 186–198.

Bullmore, E. and O. Sporns (2012). The economy of brain network organization. *Nature Reviews Neuroscience*, 13(5): 336–349.

Cao, M., H. Huang and Y. He (2017). Developmental connectomics from infancy through early childhood. *Trends in Neuroscience*, 40(8): 494–506.

Casey, B. J., T. Cannonier, M. I. Conley, A. O. Cohen, D. M. Barch, M. M. Heitzeg, M. E. Soules, T. Teslovich, D. V. Dellarco, H. Garavan, C. A. Orr, T. D. Wager, M. T. Banich, N. K. Speer, M. T. Sutherland, M. C. Riedel, A. S. Dick, J. M. Bjork, K. M. Thomas, B. Chaarani, M. H. Mejia, D. J. Hagler, Jr., M. Daniela Cornejo, C. S. Sicat, M. P. Harms, N. U. F. Dosenbach, M. Rosenberg, E. Earl, H. Bartsch, R. Watts, J. R. Polimeni, J. M. Kuperman, D. A. Fair, A. M. Dale, and A. I. A. Workgroup (2018). The Adolescent Brain Cognitive Development (ABCD) study: Imaging acquisition across 21 sites. *Developmental Cognitive Neuroscience*, 32: 43–54.

Catani, M., D. K. Jones, R. Donato, and D. H. Ffytche (2003). Occipito-temporal connections in the human brain. *Brain*, 126(Pt 9): 2093–2107.

Catani, M., D. K. Jones, and D. H. Ffytche (2005). Perisylvian language networks of the human brain. *Annuals of Neurology*, 57(1): 8–16.

Chaddock-Heyman, L., K. I. Erickson, J. L. Holtrop, M. W. Voss, M. B. Pontifex, L. B. Raine, C. H. Hillman, and A. F. Kramer (2014). Aerobic fitness is associated with greater white matter integrity in children. *Frontiers in Human Neuroscience*, 8: 584.

Chang, C.-L., K.-L. Hung, Y.-C. Yang, C.-S. Ho, and N.-C. Chiu (2015). Corpus Callosum and Motor Development in Healthy Term Infants. *Pediatric Neurology*, 52(2): 192–197.

Chang, Y. S., J. P. Owen, N. J. Pojman, T. Thieu, P. Bukshpun, M. L. Wakahiro, J. I. Berman, T. P. Roberts, S. S. Nagarajan, E. H. Sherr, and P. Mukherjee (2015). White matter changes of neurite density and fiber orientation dispersion during human brain maturation. *PLoS One*, *10*(6): e0123656.

Chiang, M.-C., K. L. McMahon, G. I. de Zubicaray, N. G. Martin, I. Hickie, A. W. Toga, M. J. Wright, and P. M. Thompson (2011). Genetics of white matter development: A DTI study of 705 twins and their siblings aged 12 to 29. *Neuroimage*, *54*(3): 2308–2317.

Clayden, J. D., S. Jentschke, M. Muñoz, J. M. Cooper, M. J. Chadwick, T. Banks, C. A. Clark, and F. Vargha-Khadem (2012). Normative development of white matter tracts: Similarities and differences in relation to age, gender, and intelligence. *Cerebral Cortex*, *22*(8): 1738–1747.

Dai, X., A.-O. Hoo, P. Hadjipantelis, J. L. Wang, S. C. L. Deoni, and H. G. Muller (2019). Longitudinal associations between white matter maturation and cognitive development across early childhood. *Human Brain Mapping*, *40*(14).

Darki, F. and T. Klingberg (2015). The role of fronto-parietal and fronto-striatal networks in the development of working memory: A longitudinal study. *Cerebral Cortex*, *25*(6).

Das, S. K., J. L. Wang, L. Bing, A. Bhetuwal, and H. F. Yang (2017). Regional values of diffusional kurtosis estimates in the healthy brain during normal aging. *Clinical Neuroradiology*, *27*(3): 283–298.

Dean, D. C. 3rd, J. O'Muircheartaigh, H. Dirks, B. G. Travers, N. Adluru, A. L. Alexander, and S. C. Deoni (2016). Mapping an index of the myelin g-ratio in infants using magnetic resonance imaging. *NeuroImage*, *132*: 225–237.

Dean, D. C. 3rd, J. O'Muircheartaigh, H. Dirks, N. Waskiewicz, K. Lehman, L. Walker, I. Piryatinsky, and S. C. Deoni (2015). Estimating the age of healthy infants from quantitative myelin water fraction maps. *Human Brain Mapping*, *36*(4): 1233–1244.

Deary, I. J., M. E. Bastin, A. Pattie, J. D. Clayden, L. J. Whalley, J. M. Starr, and J. M. Wardlaw (2006). White matter integrity and cognition in childhood and old age. *Neurology*, *66*: 505–512.

Dekaban, A. S. (1967). Changes in brain weights during the span of human life: Relation of brain weights to body heights and body weights. *Annals of Neurology*, *4*(4): 345–356.

Deoni, S. C., D. C. Dean, 3rd, J. O'Muircheartaigh, H. Dirks, and B. A. Jerskey (2012). Investigating white matter development in infancy and early childhood using myelin water faction and relaxation time mapping. *Neuroimage*, *63*(3): 1038–1053.

Deoni, S. C., J. O'Muircheartaigh, J. T. Elison, L. Walker, E. Doernberg, N. Waskiewicz, H. Dirks, I. Piryatinsky, D. C. Dean 3rd, and N. L. Jumbe (2016). White matter maturation profiles through early childhood predict general cognitive ability. *Brain Structure and Function*, *221*(2): 1189–1203.

Deoni, S. C. L., B. K. Rutt, T. Arun, C. Pierpaoli, and D. K. Jones (2008). Gleaning multicomponent T1 and T2 information from steady-state imaging data. *Magnetic Resonance in Medicine*, *60*(6): 1372–1387.

Dougherty, R. F., M. Ben-Shachar, G. K. Deutsch, A. Hernandez, G. R. Fox, and B. A. Wandell (2007). Temporal-callosal pathway diffusivity predicts phonological skills in children. *PNAS*, *104*(20): 8556–8561.

Dubois, J., G. Dehaene-Lambertz, S. Kulikova, C. Poupon, P. S. Hüppi, and L. Hertz-Pannier (2014). The early development of brain white matter: A review of imaging studies in fetuses, newborns and infants. *Neuroscience*, *276*: 48–71.

Dubois, J., G. Dehaene-Lambertz, M. Perrin, J. F. Mangin, Y. Cointepas, E. Duchesnay, D. L. Bihan, and L. Hertz-Pannier (2008). Asynchrony of the early maturation of white matter bundles in healthy infants: Quantitative landmarks revealed noninvasively by diffusion tensor imaging. *Human Brain Mapping*, *29*(1): 14–27.

Dubois, J., L. Hertz-Pannier, G. Dehaene-Lambertz, Y. Cointepas and D. Le Bihan (2006). Assessment of the early organization and maturation of infants' cerebral white matter fiber bundles: A feasibility study using quantitative diffusion tensor imaging and tractography. *NeuroImage*, *30*(4): 1121–1132.

Eluvathingal, T. J., K. M. Hasan, L. Kramer, J. M. Fletcher, and L. Ewing-Cobbs (2007). Quantitative diffusion tensor tractography of association and projection fibers in normally developing children and adolescents. *Cerebral Cortex*, *17*(12): 2760–2768.

Fair, D. A., A. L. Cohen, J. D. Power, N. U. Dosenbach, J. A. Church, F. M. Miezin, B. L. Schlaggar, and S. E. Petersen (2009). Functional brain networks develop from a "local to distributed" organization. *PLoS Computational Biology*, *5*(5): e1000381.

Fjell, A. M., K. B. Walhovd, L. T. Westlye, Y. Ostby, C. K. Tamnes, T. L. Jernigan, A. Gamst, and A. M. Dale (2010). When does brain aging accelerate? Dangers of quadratic fits in cross-sectional studies. *NeuroImage*, *50*(4): 1376–1383.

Foulkes, L. and S. J. Blakemore (2018). Studying individual differences in human adolescent brain development. *Natural Neuroscience*, *21*(3): 315–323.

Geeraert, B., R. M. Lebel, A. C. Mah, S. C. Deoni, D. C. Alsop, G. Varma and C. Lebel (2018). "A comparison of inhomogeneous magnetization transfer, myelin volume fraction, and diffusion tensor imaging measures in healthy children. *NeuroImage*, *182*: 343–350.

Geeraert, B. G., R. M. Lebel, A. C. Mah, S. C. Deoni, D. C. Alsop, G. Varma, and C. Lebel (2017). A comparison of inhomogeneous magnetization transfer, myelin volume fraction, and diffusion tensor imaging measures in healthy children. *NeuroImage*, *182*: 343–350.

Geeraert, B. L., R. M. Lebel, and C. Lebel (2019). A multiparametric analysis of white matter maturation during late childhood and adolescence. *Human Brain Mapping*, *40*(15): 4345–4356.

Genc, E., C. Fraenz, C. Schluter, P. Friedrich, R. Hossiep, M. C. Voelkle, J. M. Ling, O. Gunturkun, and R. E. Jung (2018). Diffusion markers of dendritic density and arborization in gray matter predict differences in intelligence. *Nature Communications*, *9*(1): 1905.

Genc, S., C. B. Malpas, S. K. Holland, R. Beare, and T. J. Silk (2017). Neurite density index is sensitive to age related differences in the developing brain. *Neuroimage*, *148*: 373–380,

Geng, X., S. Gouttard, A. Sharma, H. Gu, M. Styner, W. Lin, G. Gerig, and J. H. Gilmore (2012). Quantitative tract-based white matter development from birth to age 2 years. *Neuroimage*, *61*(3): 542–557.

Giedd, J. B., J., Jeffries, N. O., Castellanos, F. X., Liu, H., Zijdenbos, A., Paus, T., Evans, A. C., and Rapoport, J. L. (1999). Brain development during childhood and adolescence: A longitudinal MRI study. *National Neuroscience*, *2*(10): 861–863.

Giedd, J. N., A. Raznahan, K. L. Mills, and R. K. Lenroot (2012). Review: Magnetic resonance imaging of male/female differences in human adolescent brain anatomy. *Biology of Sex Differences*, *3*(1): 19.

Giorgio, A., K. E. Watkins, M. Chadwick, S. James, L. Winmill, G. Douaud, N. De Stefano, P. M. Matthews, S. M. Smith, H., Johansen-Berg, and A. C. James (2010). Longitudinal changes in grey and white matter during adolescence. *NeuroImage*, *49*(1): 94–103.

Giorgio, A., K. E. Watkins, G. Douaud, A. C. James, S. James, N. De Stefano, P. M. Matthews, S. M. Smith, and H. Johansen-Berg (2008). Changes in white matter microstructure during adolescence. *NeuroImage*, *39*(1): 52–61.

Girard, O. M., V. H. Prevost, G. Varma, P. J. Cozzone, D. C. Alsop, and G. Duhamel (2015). Magnetization transfer from inhomogeneously broadened lines (iHMT): Experimental optimization of saturation parameters for human brain imaging at 1.5 tesla. *Magnetic Resonance Imaging*, *73*: 2111–2121.

Glahn, D. C., A. Winkler, P. Kochunov, L. Almasy, R. Duggirala, M. Carless, J. Curran, R. Olvera, A. Laird, and S. Smith (2010). Genetic control over the resting brain. *Proceedings of the National Academy of Sciences, 107*(3): 1223–1228.

Good, C. D., I. S. Johnsrude, J. Ashburner, R. N. Henson, K. J. Friston, and R. S. Frackowiak (2001). A voxel-based morphometry study of ageing in 465 normal adult human brains. *NeuroImage, 14*(1 Pt 1): 21–36.

Gordon, E. M., P. S. Lee, J. M. Maisog, J. Foss-Feig, M. E. Billington, J. Vanmeter, and C. J. Vaidya (2011). Strength of default mode resting-state connectivity relates to white matter integrity in children. *Developmental Science, 14*(4): 738–751.

Gouttard, S., M. Styner, M. Prastawa, J. Piven, and G. Gerig (2008). *Assessment of Reliability of Multi-site Neuroimaging Via Traveling Phantom Study.* Berlin, Heidelberg: Springer.

Grinberg, F., Maximov, II, E. Farrher, I. Neuner, L. Amort, H. Thonnessen, E. Oberwelland, K. Konrad, and N. J. Shah (2017). Diffusion kurtosis metrics as biomarkers of microstructural development: A comparative study of a group of children and a group of adults. *NeuroImage, 144*(Pt A): 12–22.

Groeschel, S., G. E. Hagberg, T. Schultz, D. Z. Balla, U. Klose, T. K. Hauser, T. Nagele, O. Bieri, T. Prasloski, A. L. MacKay, I. Krageloh-Mann, and K. Scheffler (2016). Assessing white matter microstructure in brain regions with different myelin architecture using MRI. *PLoS One,* 11(11): e0167274.

Hasan, K. M., A. Kamali, A. Iftikhar, L. A. Kramer, A. C. Papanicolaou, J. M. Fletcher, and L. Ewing-Cobbs (2009). Diffusion tensor tractography quantification of the human corpus callosum fiber pathways across the lifespan. *Brain Research, 1249*: 91–100.

Henkelman, R. M., G. J. Stanisz, and S. J. Graham (2001). Magnetization transfer in MRI: A review. *NMR in Biomedicine,* 14(2): 57–64.

Hermoye, L., C. Saint-Martin, G. Cosnard, S. K. Lee, J. Kim, M. C. Nassogne, R. Menten, P. Clapuyt, P. K. Donohue, K. Hua, S. Wakana, H. Jiang, P. C. van Zijl, and S. Mori (2006). Pediatric diffusion tensor imaging: Normal database and observation of the white matter maturation in early childhood. *NeuroImage, 29*(2): 493–504.

Huber, E., P. M. Donnelly, A. Rokem, and J. D. Yeatman (2018). Rapid and widespread white matter plasticity during an intensive reading intervention. *Nature Communications,* 9(1): 2260.

Huber, E., R. N. Henriques, J. P. Owen, A. Rokem, and J. D. Yeatman (2019). Applying microstructural models to understand the role of white matter in cognitive development. *Developmental Cognitive Neuroscience 36*: 100624.

Huttenlocher, P. R. (1979). Synaptic density in human frontal cortex—developmental changes and effects of aging. *Brain Research, 163*: 195–205.

Jensen, J. H., J. A. Helpern, A. Ramani, H. Lu, and K. Kaczynski (2005). Diffusional kurtosis imaging: The quantification of non-gaussian water diffusion by means of magnetic resonance imaging. *Magnetic Resonance in Medicine 53*(6): 1432–1440.

Jernigan, T. L., T. T. Brown, D. J. Hagler, Jr., N. Akshoomoff, H. Bartsch, E. Newman, W. K. Thompson, C. S. Bloss, S. S. Murray, N. Schork, D. N. Kennedy, J. M. Kuperman, C. McCabe, Y. Chung, O. Libiger, M. Maddox, B. J. Casey, L. Chang, T. M. Ernst, J. A. Frazier, J. R. Gruen, E. R. Sowell, T. Kenet, W. E. Kaufmann, S. Mostofsky, D. G. Amaral, and A. M. Dale (2016). The pediatric imaging, neurocognition, and genetics (PING) data repository. *NeuroImage,* 124: 1149–1154.

Jeurissen, B., A. Leemans, J. D. Tournier, D. K. Jones, and J. Sijbers (2013). Investigating the prevalence of complex fiber configurations in white matter tissue with diffusion magnetic resonance imaging. *Human Brain Mapping, 34*(11): 2747–2766.

Jolles, D., D. Wassermann, R. Chokhani, J. Richardson, C. Tenison, R. Bammer, L. Fuchs, K. Supekar, and V. Menon (2016). Plasticity of left perisylvian white-matter tracts is associated with individual differences in math learning. *Brain Structure and Function, 221*(3): 1337–1351.

Jones, D. K. (2004). The effect of gradient sampling schemes on measures derived from diffusion tensor MRI: A Monte Carlo study. *Magnetic Resonance in Medicine, 51*(4): 807–815.

Joosten, E. A. J. and D. P. R. Bar (1999). Axon guidance of outgrowing corticospinal fibres in rats. *Journal of Anatomy, 194*: 15–32.

Jung, R. E. and R. J. Haier (2007). The parieto-frontal integration theory (P-FIT) of intelligence: Converging neuroimaging evidence. *Behavioral and Brain Sciences, 30*(02): 135.

Karlsgodt, K. H., P. Kochunov, A. M. Winkler, A. R. Laird, L. Almasy, R. Duggirala, R. L. Olvera, P. T. Fox, J. Blangero, and D. C. Glahn (2010). A multimodal assessment of the genetic control over working memory. *The Journal of Neuroscience: The Official Journal of the Society for Neuroscience, 30*(24): 8197–8202.

Keller, T. A. and M. A. Just (2009). Altering cortical connectivity: Remediation-induced changes in the white matter of poor readers. *Neuron, 64*(5): 624–631.

Kinney, H. C., B. A. Brody, A. S. Kloman, and F. H. Gilles (1988). Sequence of central nervous system myelination in human infancy. II. Patterns of myelination in autopsied infants. *Journal of Neuropathology & Experimental Neurology, 47*(3): 217–234.

Kochunov, P., L. M. Rowland, E. Fieremans, J. Veraart, N. Jahanshad, G. Eskandar, X. Du, F. Muellerklein, A. Savransky, D. Shukla, H. Sampath, P. M. Thompson, and L. E. Hong (2016). Diffusion-weighted imaging uncovers likely sources of processing-speed deficits in schizophrenia. *Proceedings of the National Academy of Sciences of the United States of America, 113*(47): 13504–13509.

Kohannim, O., N. Jahanshad, M. N. Braskie, J. L. Stein, M.-C. Chiang, A. H. Reese, D. P. Hibar, A. W. Toga, K. L. McMahon, and G. I. De Zubicaray (2012). Predicting white matter integrity from multiple common genetic variants. *Neuropsychopharmacology, 37*(9).

Krogsrud, S. A.-O., A. M. Fjell, C. K. Tamnes, H. Grydeland, P. Due-Tonnessen, A. Bjornerud, C. Sampaio-Baptista, J. Andersson, H. Johansen-Berg, and K. B. Walhovd (2018). Development of white matter microstructure in relation to verbal and visuospatial working memory—A longitudinal study. *PLoS One, 13*(4).

Krogsrud, S. K., A. M. Fjell, C. K. Tamnes, H. Grydeland, L. Mork, P. Due-Tønnessen, A. Bjørnerud, C. Sampaio-Baptista, J. Andersson, and H. Johansen-Berg (2016). Changes in white matter microstructure in the developing brain—A longitudinal diffusion tensor imaging study of children from 4 to 11 years of age. *NeuroImage, 124*: 473–486.

Kumar, R., H. D. Nguyen, P. M. Macey, M. A. Woo, and R. M. Harper (2012). Regional brain axial and radial diffusivity changes during development. *Journal of Neuroscience Research, 90*(2): 346–355.

Landing, B. H, Shankle, W. R., Hara, J., Brannock, J., and Fallon, J. H. (2002). The development of structure and function in the postnatal human cerebral cortex from birth to 72 months: Changes in thickness of layers II and III co-relate to the onset of new age-specific behaviors. *Pediatric Pathology & Molecular Medicine, 21*: 321–342.

Lassek, A. M. (1942). The human pyramidal tract: V. postnatal changes in the axons of the pyramids. *Archives of Neurology and Psychiatry, 47*(3): 422–427.

Le Bihan, D. and E. Breton (1985). In vivo magnetic resonance imaging of diffusion. *Interventional Cardiology Review, 301*(15): 1109–1112.

Lebel, C. and C. Beaulieu (2011). Longitudinal development of human brain wiring continues from childhood into adulthood. *Journal of Neuroscience, 31*(30): 10937–10947.

Lebel, C., S. Caverhill-Godkewitsch, and C. Beaulieu (2010). Age-related regional variations of the corpus callosum identified by diffusion tensor tractography. *Neuroimage*, 52(1): 20–31.

Lebel, C., M. Gee, R. Camicioli, M. Wieler, W. Martin, and C. Beaulieu (2012). Diffusion tensor imaging of white matter tract evolution over the lifespan. *Neuroimage*, 60(1): 340–352.

Lebel, C., B. Shaywitz, J. Holahan, S. Shaywitz, K. Marchione, and C. Beaulieu (2013). Diffusion tensor imaging correlates of reading ability in dysfluent and non-impaired readers. *Brain Language*, 125(2): 215–222.

Lebel, C., L. Walker, A. Leemans, L. Phillips, and C. Beaulieu (2008). Microstructural maturation of the human brain from childhood to adulthood. *NeuroImage*, 40(3): 1044–1055.

Lebel, C., M. Walton, N. Letourneau, G. F. Giesbrecht, B. J. Kaplan, and D. Dewey (2016). Prepartum and postpartum maternal depressive symptoms are related to children's brain structure in preschool. *Biological Psychiatry*, 80(11): 859–868.

Lebel, C., T. Warner, J. Colby, L. Soderberg, F. Roussotte, M. Behnke, F. D. Eyler, and E. R. Sowell (2013). White matter microstructure abnormalities and executive function in adolescents with prenatal cocaine exposure. *Psychiatry Research: Neuroimaging*, 213(2): 161–168.

Leemans, A., B. Jeurissen, J. Sijbers, and D. Jones (2009). ExploreDTI: A graphical toolbox for processing, analyzing, and visualizing diffusion MR data. In *Proceedings of the International Society for Magnetic Resonance in Medicine*.

Lenroot, R. K., N. Gogtay, D. K. Greenstein, E. M. Wells, G. L. Wallace, L. S. Clasen, J. D. Blumenthal, J. Lerch, A. P. Zijdenbos, A. C. Evans, P. M. Thompson, and J. N. Giedd (2007). Sexual dimorphism of brain developmental trajectories during childhood and adolescence. *Neuroimage*, 36(4): 1065–1073.

Li, Y., Y. Hu, Y. Wang, J. Weng and F. Chen (2013). Individual structural differences in left inferior parietal area are associated with schoolchildrens' arithmetic scores. *Frontiers in Human Neuroscience*, 7(844).

Löbel, U., J. Sedlacik, D. Güllmar, W. A. Kaiser, J. R. Reichenbach, and H.-J. Mentzel (2009). Diffusion tensor imaging: the normal evolution of ADC, RA, FA, and eigenvalues studied in multiple anatomical regions of the brain. *Neuroradiology*, 51(4): 253–263.

Ly, M., N. Adluru, D. J. Destiche, S. Y. Lu, J. M. Oh, S. M. Hoscheidt, A. L. Alexander, O. C. Okonkwo, H. A. Rowley, M. A. Sager, S. C. Johnson, and B. B. Bendlin (2016). Fornix microstructure and memory performance is associated with altered neural connectivity during episodic recognition. *Journal of the International Neuropsychological Society*, 22(2): 191–204.

Mabbott, D. J., J. Rovet, M. D. Noseworthy, M. L. Smith, and C. Rockel (2009). The relations between white matter and declarative memory in older children and adolescents. *Brain Research*, 1294: 80–90.

MacKay, A. L., K. P. Whittall, J. Adler, D. Li, D. Paty, and D. Graeb (1994). In vivo visualization of myelin water in brain by magnetic resonance. *Magnetic Resonance Imaging*, 31: 673–677.

Mah, A., B. Geeraert, and C. Lebel (2017). Detailing neuroanatomical development in late childhood and early adolescence using NODDI. *PLoS One*, 12(8): e0182340.

Martin, A., M. Schurz, M. Kronbichler, and F. Richlan (2015). Reading in the brain of children and adults: A meta-analysis of 40 functional magnetic resonance imaging studies. *Human Brain Mapping*, 36(5): 1963–1981.

Matejko, A. A. and D. Ansari (2015). Drawing connections between white matter and numerical and mathematical cognition: a literature review. *Neuroscience Biobehavioral Review*, 48: 35–52.

Moseley, M. (2002). Diffusion tensor imaging and aging—a review. *NMR in Biomedicine*, 15(7–8): 553–560.

Moura, L. M., M. Kempton, G. Barker, G. Salum, A. Gadelha, P. M. Pan, M. Hoexter, M. A. Del Aquilla, F. A. Picon, M. Anes, M. C. Otaduy, E. Amaro, Jr., L. A. Rohde, P. McGuire, R. A. Bressan, J. R. Sato, and A. P. Jackowski (2016). Age-effects in white matter using associated diffusion tensor imaging and magnetization transfer ratio during late childhood and early adolescence. *Magnetic Resonance Imaging, 34*(4): 529–534.

Mukherjee, P., Miller, J. H., Shimony, J. S., Philip, J. V., Nehra, D., Snyder, A. Z., Conturo, T. E., Neil, J. J., McKinstry, R. C. (2002). Diffusion-tensor MR imaging of gray and white matter development during normal human brain maturation. *American Journal of Neuroradiology, 23*: 1445–1456.

Nagy, Z., H. Westerberg, and T. Klingberg (2004). Maturation of white matter is associated with the development of cognitive functions during childhood. *Journal of Cognitive Neuroscience, 16*(7): 1227–1233.

Noble, K. G., M. S. Korgaonkar, S. M. Grieve, and A. M. Brickman (2013). Higher education is an age-independent predictor of white matter integrity and cognitive control in late adolescence. *Developmental Science, 16*(5): 653–664.

Nudo, R. J. (2006). Plasticity. *NeuroRx, 3*(4): 420–427.

Odegard, T. N., E. A. Farris, J. Ring, R. McColl, and J. Black (2009). Brain connectivity in non-reading impaired children and children diagnosed with developmental dyslexia. *Neuropsychologia, 47*(8–9): 1972–1977.

Ou, X., A. Andres, R. T. Pivik, M. A. Cleves, and T. M. Badger (2015). Brain gray and white matter differences in healthy normal weight and obese children. *Journal of Magnetic Resonance, 42*(5): 1205–1213.

Ou, X., K. M. Thakali, K. Shankar, A. Andres, and T. M. Badger (2015). Maternal adiposity negatively influences infant brain white matter development. *Obesity (Silver Spring, Md.) 23*(5): 1047–1054.

Pangelinan, M. M., G. Leonard, M. Perron, G. B. Pike, L. Richer, S. Veillette, Z. Pausova, and T. Paus (2016). Puberty and testosterone shape the corticospinal tract during male adolescence. *Brain Structure & Function, 221*(2): 1083–1094.

Parker, G. J., S. Luzzi, D. C. Alexander, C. A. Wheeler-Kingshott, O. Ciccarelli, and M. A. Lambon Ralph (2005). Lateralization of ventral and dorsal auditory-language pathways in the human brain. *NeuroImage, 24*(3): 656–666.

Paus, T, Zijdenbos, A., Worsley, K. Collins, D. L., Blumenthal, J., Giedd, J. N., Rapoport, J. L., and Evans, A. C. (1999). Structural maturation of neural pathways in children and adolescents: In vivo study. *Science, 283*(5409): 1908–1911.

Paydar, A., E. Fieremans, J. I. Nwankwo, M. Lazar, H. D. Sheth, V. Adisetiyo, J. A. Helpern, J. H. Jensen, and S. S. Milla (2014). Diffusional kurtosis imaging of the developing brain. *American Journal of Neuroradiology, 35*(4): 808–814.

Penke, L., Muñoz Maniega, S., Bastin, M. E., Valdes Hernandez, M. C. Murray, C., Royle, N. A., Starr, J. M., Wardlaw, J. M., and Deary, I. J. (2012). Brain white matter tract integrity as a neural foundation for general intelligence. *Molecular Psychiatry, 17*: 1026–1030.

Perrin, J. S., P. Y. Herve, G. Leonard, M. Perron, G. B. Pike, A. Pitiot, L. Richer, S. Veillette, Z. Pausova, and T. Paus (2008). Growth of white matter in the adolescent brain: role of testosterone and androgen receptor. *Journal of Neuroscience, 28*(38): 9519–9524.

Perrin, J. S., G. Leonard, M. Perron, G. B. Pike, A. Pitiot, L. Richer, S. Veillette, Z. Pausova, and T. Paus (2009). Sex differences in the growth of white matter during adolescence. *NeuroImage, 45*(4): 1055–1066.

Pfefferbaum, A., E. Adalsteinsson, T. Rohlfing, and E. V. Sullivan (2010). Diffusion tensor imaging of deep gray matter brain structures: Effects of age and iron concentration. *Neurobiological Aging*, 31(3): 482–493.

Pfefferbaum, A., D. H. Mathalon, E. V. Sullivan, J. M. Rawles, R. B. Zipursky, and K. O. Lim (1994). A quantitative magnetic resonance imaging study of changes in brain morphology from infancy to late adulthood. *Archives of Neurology*, 51: 874–887.

Pohl, K. M., E. V. Sullivan, T. Rohlfing, W. Chu, D. Kwon, B. N. Nichols, Y. Zhang, S. A. Brown, S. F. Tapert, K. Cummins, W. K. Thompson, T. Brumback, I. M. Colrain, F. C. Baker, D. Prouty, M. D. De Bellis, J. T. Voyvodic, D. B. Clark, C. Schirda, B. J. Nagel, and A. Pfefferbaum (2016). Harmonizing DTI measurements across scanners to examine the development of white matter microstructure in 803 adolescents of the NCANDA study. *NeuroImage*, 130: 194–213.

Posthuma, D., E. J. De Geus, W. F. Baaré, H. E. H. Pol, R. S. Kahn, and D. I. Boomsma (2002). The association between brain volume and intelligence is of genetic origin. *Nature Neuroscience*, 5(2): 83.

Price, C. J. (2012). A review and synthesis of the first 20 years of PET and fMRI studies of heard speech, spoken language and reading. *NeuroImage*, 62(2): 816–847.

Qiu, A., S. Mori, and M. I. Miller (2015). Diffusion tensor imaging for understanding brain development in early life. *Annual Review of Psychology*, 66: 853–876.

Qiu, D., L. H. Tan, K. Zhou, and P. L. Khong (2008). Diffusion tensor imaging of normal white matter maturation from late childhood to young adulthood: Voxel-wise evaluation of mean diffusivity, fractional anisotropy, radial and axial diffusivities, and correlation with reading development. *NeuroImage*, 41(2): 223–232.

Reynolds, J. E., M. N. Grohs, D. Dewey, and C. Lebel (2019). Global and regional white matter development in early childhood. *NeuroImage*, 196: 49–58.

Reynolds, J. E., X. Long, M. N. Grohs, D. Dewey, and C. Lebel (2019). Structural and functional asymmetry of the language network emerge in early childhood. *Developmental Cognitive Neuroscience*, 39: 100682.

Rifkin-Graboi, A., M. J. Meaney, H. Chen, J. Bai, W. B. R. Hameed, M. T. Tint, B. F. Broekman, Y.-S. Chong, P. D. Gluckman, and M. V. Fortier (2015). Antenatal maternal anxiety predicts variations in neural structures implicated in anxiety disorders in newborns. *Journal of the American Academy of Child & Adolescent Psychiatry*, 54(4): 313–321, e312.

Roalf, D. R., M. Quarmley, M. A. Elliott, T. D. Satterthwaite, S. N. Vandekar, K. Ruparel, E. D. Gennatas, M. E. Calkins, T. M. Moore, R. Hopson, K. Prabhakaran, C. T. Jackson, R. Verma, H. Hakonarson, R. C. Gur, and R. E. Gur (2016). The impact of quality assurance assessment on diffusion tensor imaging outcomes in a large-scale population-based cohort. *NeuroImage*, 125: 903–919.

Rollins, N., B. Vacha, P. Srinivasan, J. Chia, J. Pickering, C. W. Hughes, and B. Gimi (2009). Simple developmental dyslexia in children: Alterations in diffusion-tensor metrics of white matter tracts at 3T. *Radiology*, 251(3): 882–891.

Schmithorst, V. J., S. K. Holland, and B. J. Dardzinski (2008). Developmental differences in white matter architecture between boys and girls. *Human Brain Mapping*, 29(6): 596–710.

Schneiderman, J. S., M. S. Buchsbaum, M. M. Haznedar, E. A. Hazlett, A. M. Brickman, L. Shihabuddin, J. G. Brand, Y. Torosjan, R. E. Newmark, C. Tang, J. Aronowitz, R. Paul-Odouard, W. Byne, and P. R. Hof (2007). Diffusion tensor anisotropy in adolescents and adults. *Neuropsychobiology*, 55(2): 96–111.

Scholz, J., M. C. Klein, T. E. Behrens, and H. Johansen-Berg (2009). Training induces changes in white-matter architecture. *Nature Neuroscience*, 12(11).

Seunarine, K. K., J. D. Clayden, S. Jentschke, M. Munoz, J. M. Cooper, M. J. Chadwick, T. Banks, F. Vargha-Khadem, and C. A. Clark (2016). Sexual dimorphism in white matter developmental trajectories using tract-based spatial statistics. *Brain Connectivity*, 6(1): 37–47.

Shaw, P., J. P. Lerch, J. C. Pruessner, K. N. Taylor, A. B. Rose, D. Greenstein, L. Clasen, A. Evans, J. L. Rapoport, and J. N. Giedd (2007). Cortical morphology in children and adolescents with different apolipoprotein E gene polymorphisms: An observational study. *Lancet Neurology*, 6(6): 494–500.

Simmonds, D. J., M. N. Hallquist, M. Asato, and B. Luna (2014). Developmental stages and sex differences of white matter and behavioral development through adolescence: A longitudinal diffusion tensor imaging (DTI) study. *Neuroimage*, 92: 356–368.

Smith, S. M., M. Jenkinson, H. Johansen-Berg, D. Rueckert, T. E. Nichols, C. E. Mackay, K. E. Watkins, O. Ciccarelli, M. Z. Cader, P. M. Matthews, and T. E. Behrens (2006). Tract-based spatial statistics: Voxelwise analysis of multi-subject diffusion data. *NeuroImage*, 31(4): 1487–1505.

Song, S.-K., S.-W. Sun, M. J. Ramsbottom, C. Chang, J. Russell, and A. H. Cross (2002). Dysmyelination revealed through MRI as increased radial (but unchanged axial) diffusion of water. *NeuroImage*, 17(3): 1429–1436.

Song, S.-K., J. Yoshino, T. Q. Le, S.-J. Lin, S.-W. Sun, A. H. Cross, and R. C. Armstrong (2005). Demyelination increases radial diffusivity in corpus callosum of mouse brain. *NeuroImage*, 26(1): 132–140.

Song, S. K., Sun, S. W., Ju, W. K., Lin, S. J., Cross, A. H., Neufeld, A. H. (2003). Diffusion tensor imaging detects and differentiates axon and myelin degeneration in mouse optic nerve after retinal ischemia. *NeuroImage*, 20: 1714–1722.

Sporns, O., G. Tononi, and R. Kotter (2005). The human connectome: A structural description of the human brain. *PLOS Computational Biology*, 1(4): e42.

Sullivan, E. V., N. M. Zahr, T. Rohlfing, and A. Pfefferbaum (2010). Fiber tracking functionally distinct components of the internal capsule. *Neuropsychologia*, 48(14): 4155–4163.

Tamnes, C. K., Y. Ostby, A. M. Fjell, L. T. Westlye, P. Due-Tonnessen, and K. B. Walhovd (2010). Brain maturation in adolescence and young adulthood: Regional age-related changes in cortical thickness and white matter volume and microstructure. *Cerebral Cortex*, 20(3): 534–548.

Tamnes, C. K., D. R. Roalf, A. L. Goddings, and C. Lebel (2017). Diffusion MRI of white matter microstructure development in childhood and adolescence: Methods, challenges and progress. *Developmental Cognitive Neuroscience*, 33: 161–175.

Tham, M. W., P. S. Woon, M. Y. Sum, T.-S. Lee, and K. Sim (2011). White matter abnormalities in major depression: Evidence from post-mortem, neuroimaging and genetic studies. *Journal of Affective Disorders*, 132(1): 26–36.

Thieba, C., A. Frayne, M. Walton, A. Mah, A. Benischek, D. Dewey, and C. Lebel (2018). Factors associated with successful MRI scanning in unsedated young children. *Frontiers in Pediatrics*, 6: 146.

Thompson, P. M., Giedd, J. N., Woods, R. P., MacDonald, D., Evans, A. C., and Toga, A. W. (2000). Growth patterns in the developing brain detected by using continuum mechanical tensor maps. *Nature* 404: 190–193.

Treit, S., C. Lebel, L. Baugh, C. Rasmussen, G. Andrew and C. Beaulieu (2013). Longitudinal MRI reveals altered trajectory of brain development during childhood and adolescence in fetal alcohol spectrum disorders. *Journal of Neuroscience*, 33(24): 10098–10109.

Turkheimer, E., A. Haley, M. Waldron, B. d'Onofrio, and I. I. Gottesman (2003). Socioeconomic status modifies heritability of IQ in young children. *Psychological Science*, 14(6): 623–628.

Tzourio-Mazoyer, N., B. Landeau, D. Papathanassiou, F. Crivello, O. Etard, N. Delcroix, B. Mazoyer, and M. Joliot (2002). Automated anatomical labeling of activations in SPM using a macroscopic anatomical parcellation of the MNI MRI single-subject brain. *NeuroImage*, 15(1): 273–289.

Uda, S., M. Matsui, C. Tanaka, A. Uematsu, K. Miura, I. Kawana, and K. Noguchi (2015). Normal development of human brain white matter from infancy to early adulthood: A diffusion tensor imaging study. *Developmental Neuroscience*, 37(2): 182–194.

Uddin, L. Q., A. M. Kelly, B. B. Biswal, D. S. Margulies, Z. Shehzad, D. Shaw, M. Ghaffari, J. Rotrosen, L. A. Adler, F. X. Castellanos, and M. P. Milham (2008). Network homogeneity reveals decreased integrity of default-mode network in ADHD. *Journal of Neuroscience Methods*, 169(1): 249–254.

Vanderauwera, J., J. Wouters, M. Vandermosten, and P. Ghesquiere (2017). Early dynamics of white matter deficits in children developing dyslexia. *Developmental Cognitive Neuroscience*, 27: 69–77.

Vandermosten, M., B. Boets, J. Wouters, and P. Ghesquiere (2012). A qualitative and quantitative review of diffusion tensor imaging studies in reading and dyslexia. *Neuroscience & Biobehavioral Reviews*, 36(6): 1532–1552.

Varma, G., G. Duhamel, C. de Bazelaire, and D. C. Alsop (2015). Magnetization transfer from inhomogeneously broadened lines: A potential marker for myelin. *Magnetic Resonance in Medicine*, 73(2): 614–622.

Vestergaard, M., K. S. Madsen, W. F. C. Baaré, A. Skimminge, L. R. Ejersbo, T. Z. Ramsøy, C. Gerlach, P. Åkeson, O. B. Paulson, and T. L. Jernigan (2010). White matter microstructure in superior longitudinal fasciculus associated with spatial working memory performance in children. *Journal of Cognitive Neuroscience*, 23(9): 2135–2146.

Vogel, A. C., J. D. Power, S. E. Petersen, and B. L. Schlaggar (2010). Development of the brain's functional network architecture. *Neuropsychology Review*, 20(4): 362–375.

Walhovd, K. B., L. T. Westlye, I. Amlien, T. Espeseth, I. Reinvang, N. Raz, I. Agartz, D. H. Salat, D. N. Greve, B. Fischl, A. M. Dale, and A. M. Fjell (2011). Consistent neuroanatomical age-related volume differences across multiple samples. *Neurobiology of Aging*, 32(5): 916–932.

Wang, R., T. Benner, A. G. Sorensen, and V. J. Wedeen (2007). Diffusion toolkit: A software package for diffusion imaging data processing and tractography. *Proceedings of the International Society for Magnetic Resonance in Medicine*, 15: 3720.

Wang, Y., C. Adamson, W. Yuan, M. Altaye, A. Rajagopal, A. W. Byars, and S. K. Holland (2012). Sex differences in white matter development during adolescence: A DTI study. *Brain Research*, 1478: 1–15.

Wang, Y., M. V. Mauer, T. Raney, B. Peysakhovich, B. L. C. Becker, D. D. Sliva, and N. Gaab (2017). Development of tract-specific white matter pathways during early reading development in at-risk children and typical controls. *Cerebral Cortex*, 27(4): 2469–2485.

Watts, D. J. and S. H. Strogatz (1998). Collective dynamics of "small-world" networks. *Nature*, 393: 440–442.

Welcome, S. E. and M. F. Joanisse (2014). Individual differences in white matter anatomy predict dissociable components of reading skill in adults. *NeuroImage*, 96: 261–275.

Winkler, A. M., P. Kochunov, J. Blangero, L. Almasy, K. Zilles, P. T. Fox, R. Duggirala, and D. C. Glahn (2010). Cortical thickness or grey matter volume? The importance of selecting the phenotype for imaging genetics studies. *NeuroImage*, 53(3): 1135–1146.

Wolff, S. D. and R. S. Balaban (1994). Magnetization transfer imaging: Practical aspects and clinical applications. *Radiology, 193*(3): 593–599.

Yakovlev, P. I. and A.-R. Lecours (1967). The myelogenetic cycles of regional maturation of the brain. In *Regional Development of the Brain in Early Life,* 3–70. Oxford: Blackwell.

Yeatman, J. D., R. F. Dougherty, M. Ben-Shachar, and B. A. Wandell (2012). Development of white matter and reading skills. *Proceedings of the National Academy of Sciences, 109*(44): E3045.

Zhang, H., T. Schneider, C. A. Wheeler-Kingshott, and D. C. Alexander (2012). NODDI: practical in vivo neurite orientation dispersion and density imaging of the human brain. *NeuroImage, 61*(4): 1000–1016.

Zhou, Y., J. Liu, N. Driesen, F. Womer, K. Chen, Y. Wang, X. Jiang, Q. Zhou, C. Bai, D. Wang, Y. Tang, and F. Wang (2017). White matter integrity in genetic high-risk individuals and first-episode schizophrenia patients: Similarities and disassociations. *BioMed Research International, 2017*: 3107845.

CHAPTER 6

MAGNETOENCEPHALOGRAPHY AND DEVELOPMENTAL COGNITIVE NEUROSCIENCE

EDWIN S. DALMAIJER,
ALEXANDER L. ANWYL-IRVINE,
GIACOMO BIGNARDI, OLAF HAUK,
AND DUNCAN E. ASTLE

INTRODUCTION

WHEN neurons in the brain "fire," their membranes are ever so slightly depolarized. When a large number of neurons in the same area with roughly the same orientation fire in concert, their combined depolarization changes the local electric and magnetic fields. Magnetoencephalography (MEG) is a technique that can measure these perturbations of the magnetic field with sensors around the head, and thus it directly reflects brain activity. Any undergraduate textbook covering neuroimaging methods will include a phrase about the temporal resolution benefits of MEG. This is true enough, MEG and electroencephalography (EEG) measure neural activity on a sub-second scale, and this is a clear benefit relative to more sluggish methods like functional resonance imaging (fMRI). By capturing neuronal processes, the MEG signal can be used to better understand how neurophysiology is changing with time. For example, key developmental changes in the mechanism may reflect changes in the *timing*, rather than extent of activity. Because MEG activity reflects neuronal activity—rather than some down-stream metabolic process like the blood oxygenation level dependent (BOLD) response—it allows researchers to pin their findings to changes and developments in the activity of neuronal processes themselves, rather than vascular changes.

Using MEG to study developmental cognitive neuroscience is not without its complications, but there is an increasing array of good options for analysis. Many analyses can be conducted with reliable "off the shelf" packages like FieldTrip (Oostenveld, Fries, Maris, and Schoffelen, 2011; Popov, Oostenveld, and Schoffelen, 2018), the Amsterdam Decoding and Modelling Toolbox ADAM (Fahrenfort, van Driel, van Gaal, and Olivers, 2018), the Statistic Parametric Mapping Toolbox SPM (Penny, Friston, Ashburner, Kiebel, and Nichols, 2011), and the OHBA Software Library OSL for Matlab; or MNE-Python (Gramfort et al., 2013; Jas et al., 2018) for Python. In our experience, some more complex analyses are likely to require an additional bit of bespoke coding. Nonetheless, MEG provides a window on a set of electrophysiological processes and can provide a child-friendly way of understanding how different brain regions are coordinated. The sensitivity of MEG to neural activity itself allows developmental researchers to link their findings to mechanistic accounts of brain organization and function from the animal literature.

In this chapter we cover the basics of recording MEG from children and then provide a brief overview of the current state of the field in terms of analysis options. Our aim is to provide a guide to developmental cognitive neuroscientists wanting to start using MEG and highlight the unique place of this technique in the methodological repertoire.

MEG: The Basics

Capitalising fully on the potential of MEG requires a basic understanding of how the signal is generated and recorded. When an electric current moves through a wire, it generates a magnetic field around it. The orientation of the field is dependent on the orientation of the wire, and the direction of the current. (Some readers might recall the "right-hand rule" from high school physics classes.) In the brain, electric currents run through neuronal axons by means of ion transport during action potentials. As a consequence, the electric potential and magnetic field around a neuron is subtly altered. If enough neurons with a similar location and orientation fire in concert, their combined electrical activity changes their local field (Figure 6.1). Whilst EEG measures electric potentials that reach the scalp, MEG measures magnetic fields produced by all currents inside the head (Hämäläinen, Hari, Ilmoniemi, Knuutila, and Lounasmaa, 1993). Some properties of neuronal firing that affect the measurable MEG signal are neurons' orientations, specifically because poorly aligned neurons are less likely to produce a unified signal, and sometimes can even cancel each other out. In addition, the frequency at which neurons fire will impact the frequency of the measurable MEG signal. We will cover these concepts in the next paragraphs.

While EEG and MEG share many analysis features and can even be combined, they also have some practical distinctions. Because the magnetic field detectable at the scalp is less than one-billionth of the strength of the Earth's magnetic field, MEG measurement necessitates a magnetically shielded room and highly sensitive sensors. Currently,

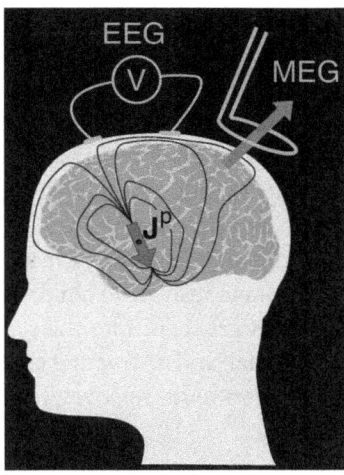

FIGURE 6.1 Patches of electrical cortical activity are modelled as "current dipoles" (arrow). These dipoles represent the so-called primary currents (Jp) that are caused by post-synaptic excitatory potentials. The black lines indicate the so-called volume currents which are a result of passive current flow through the electrically conducting tissues brain, CSF, skull, and scalp. The EEG measures electric voltage differences at the scalp surface produced by volume currents, and the MEG measures the summed magnetic field of all (primary and volume) currents inside the head. Note that the volume currents are affected by tissue conductivity as well as distance to the primary currents.

Source: image from Matti Hämäläinen's talk "MEG/EEG Source Estimation Approaches: A Spectrum of Purpose-Built Optimal Tools," part of the MIT Center for Brains, Minds, and Machines' MEG workshop, © Professor Matti Hämäläinen.

superconducting quantum interference devices (SQUIDs) are the most popular technology in MEG systems, which utilize coils that measure small voltage changes produced by a change in the surrounding magnetic field. To achieve superconductivity, the sensors are housed in a liquid helium helmet dewar. This is currently how most MEG systems work, but the next generation of machines will likely do away with this cumbersome super-cooling technology, and utilize smaller stand-alone optically pumped magnetometers (OPMs) that will allow lightweight and mobile setups more akin to a bulky EEG cap than a static scanner (Boto et al., 2018).

A number of factors influence our ability to detect neural activity with MEG and EEG. It is thought that the main source of MEG and EEG signals are currents that originate from cortical pyramidal neurons because their long and uniformly oriented dendrites are more likely to produce a coherent electromagnetic signal than variably oriented neurons. Magnetic sources within a conducting sphere (like approximately the skull and surrounding cerebrospinal fluid) emit detectable MEG signals when directed tangential to the sphere (Ahlfors, Han, Belliveau, and Hämäläinen, 2010; Melcher and Cohen, 1988). Source depth also weakens sensitivity and accuracy in source localization (Hillebrand and Barnes, 2002; Leahy, Mosher, Spencer, Huang, and Lewine, 1998). MEG is often credited with having better spatial resolution than EEG, not only because whole-head

MEG systems usually contain more sensors than typical EEG systems, but also because the MEG signal is less affected by the geometry and make-up of the head. For example, fatty layers around the skull cause smearing of extra-cranial EEG signal. However, MEG is less sensitive to radial, deep, or spatially extended sources (Ahlfors et al., 2010, 2009; Goldenholz et al., 2009), and therefore a combination of EEG and MEG is thought to provide the most complete information about neural activity (Henson, Mouchlianitis, and Friston, 2009; Molins, Stufflebeam, Brown, and Hämäläinen, 2008; Sharon, Hamalainen, Tootell, Halgren, and Belliveau, 2007). In our lab, we have found it practically challenging to record both MEG and EEG simultaneously from children. The setup time means that children get restless before you even get them into the scanner. That said, in an ideal world collecting both would be optimal, and distracting children with a film or cartoons can drastically improve their willingness to sit (relatively) still for longer periods of time.

MEG Safety

Sensors in MEG scanners are passive: they are sensitive to fluctuations in the magnetic field as they occur naturally. This is in stark contrast with MRI scanners, which generate their own magnetic field, and as such present risks for participants who carry metal in or on their body. MEG scans do not present such risks.

The sensors in a SQUIDs-based MEG scanner require cooling through liquid helium. In older devices, this required regular refilling, while modern devices recycle evaporated helium. If helium leaks as a consequence of improper refilling or poor maintenance, this would replace the oxygen in the scanner environment, with obvious negative health consequences. MEG scanners should thus be operated and maintained by qualified staff and fitted with a measurement device that sound an alarm when oxygen levels dip.

In sum, in properly maintained and operated MEG scanners, the largest risk participants face is that they might trip over a cable or the chair.

Signal Quality

The raw data from the scanner cannot be analysed as is because it typically contains multiple artifacts. The first step of any MEG analysis is to look for system-related artifacts which can contaminate downstream analyses, such as "bad" sensors with high amounts of noise or no signal (Gross et al., 2013). Next, physiological artifacts from the heart, ocular, and face muscles, can also influence MEG signals. Muscle activity is usually easy to spot, as its alteration of the magnetic field generates high-amplitude MEG signal. Movements of the eye also alter the magnetic field, because each eye is effectively a dipole (the retina is more negatively charged than the lens). Crucially, eye movements and blinks can be caused by experimental stimuli and confound MEG comparisons between conditions. Many MEG systems have input channels for recording heart rate (electrocardiography), eye movement (electrooculography), or reference magnetic sensors to cancel environmental magnetic noise. Finally, head movements can reduce the statistical sensitivity of MEG analyses, and can be prevented during data collection and corrected after (Stolk, Todorovic, Schoffelen, and Oostenveld, 2013).

When children are your participants the data may contain a lot of these artifacts. A good way to remove them without losing all of your data is to use a multivariate component analysis (for example an independent component analysis) to identify variance components associated with external noise (Gonzalez-Moreno et al., 2014), or physiological artifacts (Plöchl, Ossandón, and König, 2012) and then regress these from the data. When conducting research with children, not taking this approach to removing contaminated data may leave you with little artifact-free data and can thus force you to exclude a lot of participants. It should be noted that despite the availability of these methods, the best way to avoid artifacts is to prevent movement in the first place. In our experience, captivated children are less likely to fidget. Exciting experiments or playing an engaging video when possible is thus a worthwhile consideration. In addition, soft foam blocks and pads are ideal for wedging a child's head snugly within the MEG helmet. You could even opt for using MEG head-casts (Meyer et al., 2017; Troebinger et al., 2014).

MEG data collection has several practical features that make it appealing to a developmental cognitive neuroscientist. First, MEG machines run quietly, and children can be seated upright in some models, allowing for a much more comfortable experience for young participants. It is even possible for a parent or guardian to join the child in the scanner room. However, children may be more likely to produce movement artifacts. So, if you are used to collecting MEG data from adults, you will need to consider shorter testing bouts and more frequent breaks. For very young children (under the age of 5–6), using smaller MEG helmets allows the sensors to be placed closer to the child's scalp, and provides better control of head movement, permitting more accurate measurement (Adachi and Haruta, 2014). In our experience, a standard setup should work well for children from around age 8, when their heads approach adult size.

It is important to note that head position within the MEG scanner affects its sensitivity, in particular to frontal and anterio-temporal sources (Marinkovic, Cox, Reid, and Halgren, 2004). While most participants will rest their heads on the back of the MEG scanner, researchers with interest in anterior areas (e.g., decision making or language) could consider positioning their participants towards the front, for example with foam cushioning behind the head.

Once you have a clean signal there are many different ways to analyse the data, depending upon the question you are asking. In the subsequent sections we will work through some of these options, gradually increasing in complexity.

FROM SENSORS TO SOURCES

One of the main benefits of using MEG over EEG is the better prospect of moving from sensor space recordings to exploring activity at the level of cortex. Whilst complex, source reconstruction allows the researcher to ask an increasingly sophisticated set of questions about the developing brain—e.g., how does activity in area x influence activity in area y across developmental time? In our experience source reconstruction can also have a practical advantage to studying activity purely at sensor space. In reality each child will be sitting in a slightly different position within the scanner, their head will be in a different location, and there could be differences in the underlying neural architecture producing the signals. A good MEG source reconstruction technique will account for these differences and place all of the data into a common space by using a child's individual MRI scan to create a source model taking into account head position during the scan. This can be vital for identifying clear effects at a group level.

In some experiments, e.g., when we are only looking at early sensory-evoked brain responses, we expect very focal activation with a spatial extent of several square millimeters or a few centimeters (e.g., in primary auditory cortex, or visual area V1). Such a small patch of electrically active cortex can be modelled as a current "dipole," i.e., a point source of electric current at a certain location and with a certain orientation that represents a small patch of active cortex (Scherg and Berg, 1991). When cortical activity must be assumed to be more spatially distributed, e.g., in experiments of higher cognitive function or for very noisy data, brain activity can be approximated by a large number of dipoles (100s or 1000s) distributed across the cortical surface or brain volume (Dale and Sereno, 1993; Michel et al., 2004).

Ideally, we would like to reconstruct the exact distribution of brain activity in the brain that has given rise to our measured signals. Unfortunately, with many more possible dipole sources (infinitely many in the continuous case) but only several dozen or hundred sensors, the best we can hope for is an approximation of the real source distribution. This "inverse problem" is so called because every distribution of measured EEG and MEG signals, no matter how accurate, can in principle be generated by infinitely many source distributions inside the brain. In order to obtain an interpretable solution, we either have to introduce meaningful a priori constraints (e.g., about where we expect likely sources to be, and how many dipoles there are), or use minimal a priori assumptions and accept the limited spatial resolution of our result, i.e. interpret it as a "blurred version" of the true brain activity. Methods of the former type are often called "dipole fits" (Scherg and Berg, 1991), and of the latter type "distributed source solutions" (Dale and Sereno, 1993).

Brain activity in experiments with developing children can be expected to be complex and noisy, relative to their adult counterparts, and therefore distributed source solutions are most appropriate. Examples of these are "minimum-norm"-type methods (Dale and Sereno, 1993; Hauk, Wakeman, and Henson, 2011), "beamformers" (Hillebrand, Singh, Holliday, Furlong, and Barnes, 2005; Sekihara and Nagarajan, 2008), and "multiple sparse priors" (Friston et al., 2008; Henson, Wakeman, Litvak, and Friston, 2011). The combination of EEG and MEG generally improves the spatial resolution of these methods (Henson et al., 2009; Molins et al., 2008; Sharon et al., 2007).

A detailed overview of the large number of source estimation methods is not within the scope of this chapter, but several overviews are available (e.g., Baillet, Mosher, and

Leahy, 2001; Hansen, Kringelbach, and Salmelin, 2010; Michel et al., 2004; Sekihara and Nagarajan, 2015). As a general rule, the choice of source estimation method should be independent of the results obtained in a particular experiment—the methods should follow your assumptions, and not vice versa. When we move on to the more complex aspects of neuronal coupling in source space, there may be advantages and disadvantages to some of these different source reconstruction approaches.

EVENT-RELATED FIELDS

An event-related field (ERF) refers to a change in the average magnetic field in relation to point of interest in an experiment—an event. Typically, MEG scanners can have different types of sensors: magnetometers produce a single measurement of the magnetic field, and gradiometers produce a reading of a gradient in the magnetic field between two points. These sensors can also be aligned differently in relation to the head—axial aligned sensors pick up sources that are radial and planar aligned sensors are more sensitive to gradients across the brain. Readings from multiple magnetometers and/or gradiometers are summed together and baselined at a specific event (time 0) so values represent a change from this point. Over many trials, these time-locked signals are averaged: trial-to-trial random "noise" (activity that is not related to the event) averages out, while the event-related signal remains. Thus, the averaged signal characterizes the activity of populations of neurons across many trials in response to a given event. The ERF is a waveform that contains several peaks and troughs that are typically referred to as "components," usually named after the time or sequence location they occur in (e.g., M100 or M1).

An important question is how many trials one should run to collect enough data to compute an ERF. While the general answer "as many as possible" is perhaps too simplistic, the precise answer depends on many experimental details, including the magnitude of the expected effect. As with any experiment, running a power analysis to compute the number of required trials is a good start. In addition, researchers should account for signal loss due to eye, head, and body movement. In practice, many experiments will require at least 100 trials per condition, and often more. Anyone who has ever been around a 5-year-old will appreciate that it is unlikely they will sustain attention and avoid fidgeting for hundreds of trials. This is why we recommend regular breaks, and to make tasks engaging wherever possible.

Where EEG event-related components are prefixed with "N" or "P" to denote negative or positive peaks due to EEG's ability to record polarity, changes in magnetic field lack polarity, and are therefore suffixed or prefixed with an "M." Figure 6.2 is a cartoon example of two evoked fields: one for trials with auditory stimuli producing clear M100 and M150 components, and a control trial without audio.

Experienced EEG researchers will be familiar with these plots. ERFs are similar to event-related potentials (ERPs) in EEG, where the signal measured is the electric

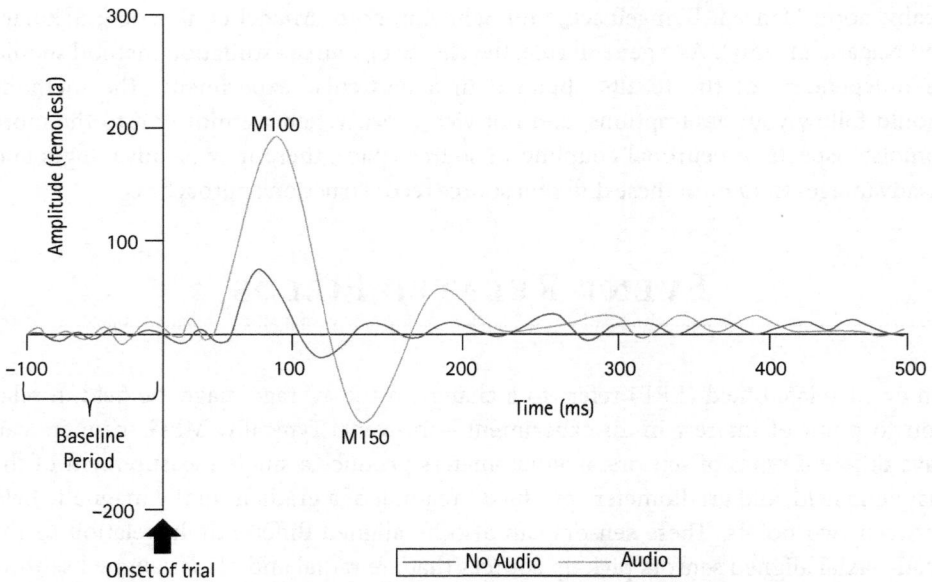

FIGURE 6.2 Cartoon example of two evoked fields: one for trials with auditory stimuli that produce clear M100 and M150 components (light gray line), and a control trial without audio (dark gray line).

potential across the scalp rather than the magnetic field. The key difference between the two methods (beyond measuring electric potential versus magnetic field) is that ERFs represent dipole sources tangentially oriented to the sensors, but less so radial sources—like the top of a gyrus (Kikuchi, Yoshimura, Mutou, and Minabe, 2016). This should be considered when running a study looking for well-known ERP components, as the effect may look somewhat different in MEG, or could even be undetectable.

Averaging across many trials is a good way of getting a rough approximation of population-level neural activity, revealing relatively large effects that are driven by amplitude. However, this will lose any spectral information, or event-related changes in phase (Pfurtscheller and Lopes da Silva, 1999). In our lab we often do a traditional ERF analysis, but this is rarely where our analysis ends. There are so many more possibilities afforded by MEG than simply to calculate the magnetic equivalent of ERPs, but this is often a good starting point. We will cover some of these more complex options in later sections.

Auditory Evoked Fields

ERFs can be characterized using manipulations of a number of different sensory inputs. Auditory Evoked Fields, for example, are generated by an auditory stimulus—usually a tone (Orekhova et al., 2013), but also more complex stimuli such as speech sounds (Yoshimura et al., 2014). Often studies use an "oddball" design, where a standard stimulus is repeated over and over again so that the brain's response to it is attenuated, and occasionally a single deviant sound is played in order to elicit certain components (Donkers et al., 2015). Spatially, MEG is able to distinguish between left- and right-hemisphere auditory evoked responses to tones whereas EEG struggles (Kikuchi et al., 2016). Given that children tend to have smaller brains and therefore a higher source density, this makes MEG a good tool for studying any developmental spatial differences in auditory evoked response.

A field evoked from an unexpected change in the stimulus (e.g., the standard followed by a deviant sound in the oddball task) is termed "mismatch field." Often, these changes lead to a decrease in magnetic field strength in a given population of neurons—termed magnetic mismatch negativity (MMNm). Generally, this is best recorded from frontal sensors (Marco et al., 2012). Auditory mismatch negativity has been reported from a very early age, for instance in neonates who displayed a consistent MNNm response to vowel sounds with an infrequent rising pitch (Kujala et al., 2004). This indicated a potential ability to discriminate between simple vowel sounds from an early age. MMNm has been used to investigate processes in a number of developmental clinical groups. For example, an MMNm component in children with autism has been shown to be delayed relative to controls (Marco et al., 2012; Oram Cardy, Flagg, Roberts, and Roberts, 2005).

Yoshimura et al., 2014 investigated the auditory evoked field over 1–2 years in 2–6 year olds, and found that the P1m component was a marker for language development. Greater P1m amplitude in response to a spoken syllable in an oddball task (where pitch changed in some trials) predicted a higher score on a language task. Across development, auditory evoked responses change. For instance, adults show a P50m/N100 m response in response to a click, whereas children show two distinct components before this—P50m and P100m, which overlapped with the N100m (Orekhova et al., 2013). One possibility is that alterations in auditory evoked field morphology over development arise due to layer-by-layer changes in the auditory cortex (Moore and Linthicum, 2007).

Visual Evoked Fields

ERFs can also be characterized in relation to visual stimuli, these are commonly referred to as "visual evoked fields." Stimuli utilized in these studies can range from visual gratings (Orekhova et al., 2015) to complex stimuli like faces (Deffke et al., 2007). The "primary visual evoked response" is an increase in activity 100 ms after stimulus presentation (Martin et al., 2007). The M100 component's latency is proposed as representing the time-course of neural response, and the magnitude represented by the amplitude (Rojas, 2013). In clinical developmental research, this component has been observed to be delayed in children and adolescents with foetal alcohol spectrum disorder (FASD), which implies that sensorimotor integration problems in FASD may be related to lower-level visual processing deficits (Coffman et al., 2013).

Research in developmental psychology has utilized faces to elicit an evoked response. The M170 component has been shown to be sensitive to faces, but not to other visual stimuli (Liu, Harris, and Kanwisher, 2002). This component is also present in children (He, Brock, and Johnson, 2014; Kylliäinen, Braeutigam, Hietanen, Swithenby, and Bailey, 2006). Unlike adults, children showed a preceding M100 component that was larger and overlapped with the M170, and that could be source-localized to the fusiform gyrus, but importantly requires a custom MEG scanner or helmet to detect (He, Brock, and Johnson, 2015).

Somatosensory Evoked Fields

ERFs are generated by somatosensory stimuli, for example a mechanically induced airpuff (Götz et al., 2011) or a tap on the fingertips (Bast et al., 2007; Demopoulos et al., 2017). These taps can follow different patterns, for instance an oddball structure much like auditory oddball experiments, where uncommon deviant taps are stronger than repeated standard taps. It is worth noting that task requirements change elicited ERFs, for example between oddball and one-back tasks (Götz et al., 2011).

Research into these types of evoked responses in MEG with children are less common. One notable example reported a reduced peak amplitude at 40 ms in the somatosensory evoked field over primary-sensory cortex in a small sample of children with autism relative to healthy controls (Marco et al., 2012). This amplitude was then related to the degree of abnormal tactile behaviour displayed.

TIME-FREQUENCY ANALYSIS

Whilst ERFs are a good starting point, they cannot characterize the oscillatory frequency of activity across a given population of neurons. The data recorded by the MEG sensors often fluctuates in a semi-regular manner, for simplicity this is best thought of having the characteristics of a wave that can vary in frequency. Different populations of neurons are capable of producing different frequencies of magnetic field modulation at different times, and time-frequency analysis aims to characterize these different frequencies in relation to an event or over a fixed time period.

Techniques like the Fourier Transform are used to decompose the relatively complex data into simpler constituent frequencies, and a moving window is then applied to the data which allows us to look at how these constituents vary in power over time. The end result is a spectrum of different frequencies that varies as time progresses. Inferential statistics can then be used to see if this information varies between given conditions, for example if a particular frequency is associated with a certain stimulus relative to another.

Frequencies commonly observed in the brain are often referred using the Greek alphabet: delta (~ 0.2 Hz–3 Hz), theta (~3 Hz–8 Hz), alpha (~ 8 Hz–12 Hz), beta (~ 12 Hz–27 Hz), low gamma (~ 27 Hz–40 Hz), and high gamma (~ 40 Hz).

Resting-State MEG in Developmental Cognitive Neuroscience

MEG has been used to deliver the first system-wide measure of resting-state functional connectivity that reflects neural activity rather than a down-stream metabolic process. If you are going to collect MEG data from children then we would encourage you to also collect a resting-state scan. We usually start our sessions with a 9-minute eyes-closed scan, asking children to think about nothing in particular but to avoid a falling asleep. These data can provide important insight into individual differences in neurophysiology that can be tied to individual differences in cognition across your participants. Unlike task-based measures, resting-state data can provide insight into how a system functions that is unbiased by task choice or any strategy that children might apply. As such it is worth including in your protocol.

Resting-state connectivity analyses in functional magnetic resonance imaging (fMRI) were first published on in 1995 (Biswal, Zerrin Yetkin, Haughton, and Hyde, 1995). It entails the recording of blood-oxygenation level dependent (BOLD) signal during periods of rest, while the participant is not engaged in any task, either with open or closed eyes. Connectivity is assessed by correlating the activity in one voxel with every other voxel. The resulting groups of voxels with correlating activity are considered to be networks, some of which overlap with task-dependent networks. For an overview of identified resting-state networks, see e.g., (Rosazza and Minati, 2011; Van den Heuvel and Hulshoff Pol, 2010).

When using MEG to explore resting-state connectivity it is technically possible to find correlating groups of sensors, but researchers generally prefer to transform the signal from sensor space into source space. The Achilles heel for an analysis like this is signal leakage (Palva et al., 2018). This is the phenomena of erroneously attributing activity from brain area A when reconstructing the activity of brain area B. This will inflate the apparent correlation in activity between these areas or result in spurious functional connectivity. As a result, a beamformer is a popular method for source reconstruction if the end goal is to explore functional connectivity (Astle, Barnes, Baker, Colclough, and Woolrich, 2015; Astle, Luckhoo, et al., 2015; Brookes et al., 2011). This method can actively suppress the activity coming from other sources whilst reconstructing the activity for a particular area (Schoffelen and Gross, 2009), although it should be noted that this feature is not exclusive to beamforming. In addition to software approaches, head casts can improve the reliability of beamformer reconstruction and the following connectivity analysis (Liuzzi et al., 2017).

Once the MEG signal is in source space, the resting-state connectivity analysis has similarities to resting-state fMRI (Figure 6.3). There are many approaches that can be taken. In our lab we typically calculate the signal's frequency-specific amplitude envelope using a Hilbert transform (Luckhoo, Brookes, and Woolrich, 2014), after which the result is down-sampled (Brookes et al., 2011), concatenated across individuals, and

Pipeline for Source–Space Electrophysiological Connectivity Analysis

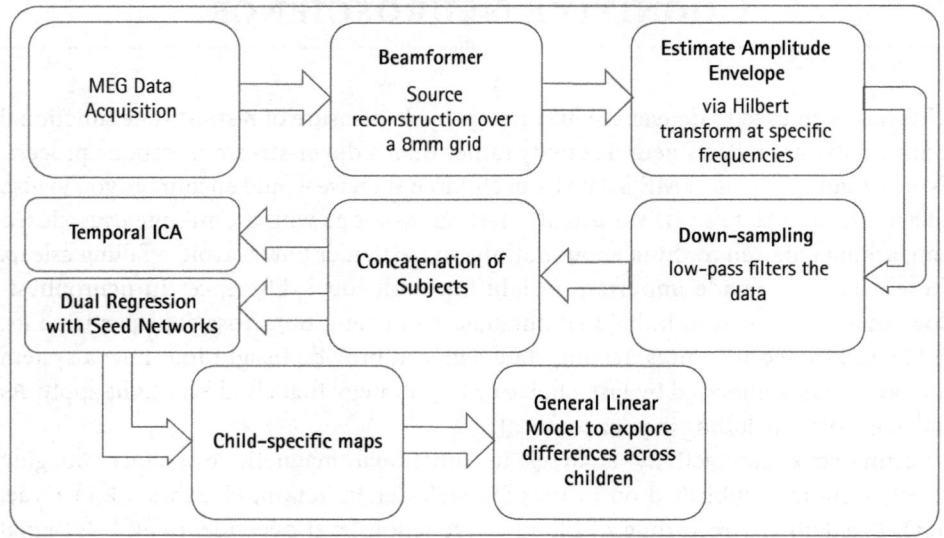

FIGURE 6.3 The MEG acquisition pipeline.

Source: reproduced from Jessica J. Barnes, Mark W. Woolrich, Kate Baker, Giles L. Colclough, and Duncan E. Astle, Electrophysiological measures of resting state functional connectivity and their relationship with working memory capacity in childhood, *Developmental* Science, 19 (1), pp. 19-31, doi:10.1111/desc.12297, Figure 6.1 © 2015 The Authors. Developmental Science Published by John Wiley and Sons Ltd. Licensed under Creative Commons Attribution License (CC BY).

then subjected to a MEG-adapted analysis for resting-state connectivity (Luckhoo et al., 2014). Because this analysis uses the envelope of ongoing spontaneous oscillations, rather than the raw signal itself, it is possible to remove zero-lag correlations in the raw data (which could reflect signal leakage) and still find significant correlations in the envelope. This is one of the analysis steps that is needed in order to have confidence that the networks you identify reflect genuine co-occurring activity across different spatially distributed populations of neurons. (For more methodological details, see Astle, Barnes, et al., 2015; Astle, Luckhoo, et al., 2015; Barnes, Woolrich, Baker, Colclough, and Astle, 2016.)

Barnes and colleagues (Barnes, Woolrich, et al., 2016) collected resting-state MEG data in thirty-one children, aged 8–11 years. Seven networks were identified: cerebellum (evident in the beta band), sensori-motor (evident in the beta band), bi-lateral fronto-parietal (with nodes in theta band in superior-parietal and frontal lobes), right-hemisphere fronto-parietal (evident in the beta band), anterior cingulate (evident in the alpha band), visual (with nodes in the theta band in early areas of visual cortex), and left-hemisphere fronto-parietal (with nodes in beta band in left frontal cortex); see Figure 6.4. Interestingly, these networks partly map on to networks found in adults (Brookes et al., 2011), but with clear omissions such as the default mode network and a posterior-parietal component. This could be evidence for a developmental difference in

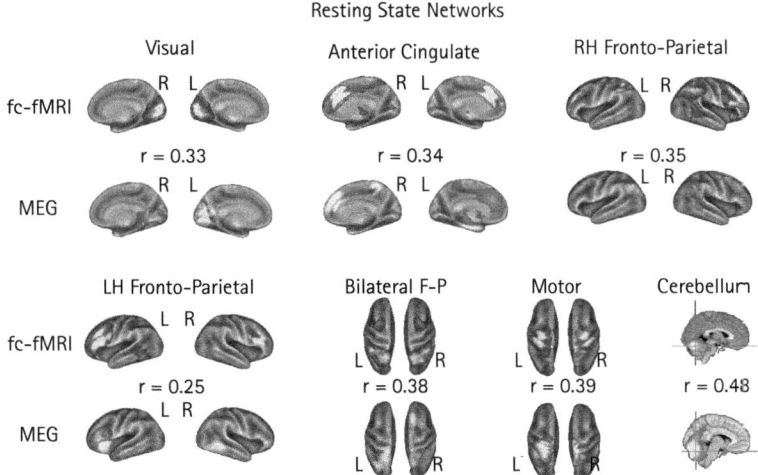

FIGURE 6.4 Resting state networks in MEG during development compared to their counterparts from fMRI in adults.

Source: reproduced from Jessica J. Barnes, Mark W. Woolrich, Kate Baker, Giles L. Colclough, and Duncan E. Astle, Electrophysiological measures of resting state functional connectivity and their relationship with working memory capacity in childhood, *Developmental Science*, 19 (1), pp. 19–31, doi:10.1111/desc.12297, Figure 6.3 © 2015 The Authors. Developmental Science Published by John Wiley and Sons Ltd. Licensed under Creative Commons Attribution License (CC BY).

cortical organization, but it might also reflect a poorer signal in the MEG dataset. Much more work is needed to unpack whether the differences you see in the MEG relative to fMRI counterparts. Spatial working memory correlated positively with the connection strength between the bilateral fronto-parietal network and an area of the brain that spanned from left inferior temporal cortex to posterior cingulate cortex. However, no such result could be found for verbal working memory (Barnes, Woolrich, et al., 2016).

The relationship between fronto-parietal networks identified using resting-state MEG in children and their working memory is further cemented by Astle and colleagues (Astle, Barnes, et al., 2015), who subsequently trained the working memory of these children. Over 4–6 weeks children took part in 20–25 training sessions at home. Each lasted 30–45 minutes and comprised 120 trials from a selection of working memory tasks. In half of the sample, the training was adaptive to participant's performance (adaptive condition), and in the other half it was not (placebo condition). Both groups improved in their composite working memory scores, but the adaptive training group did so much more than the placebo group. Individual pre- and post-training differences in working memory scores correlated with individual differences in functional connectivity strength. Specifically, a bilateral fronto-parietal network showed greater connectivity increases with the left superior parietal lobule, and with left inferior temporal cortex (both appeared in the beta band). The authors concluded that working memory training puts high demands on fronto-parietal networks, and thereby improves their internal and external connectivity.

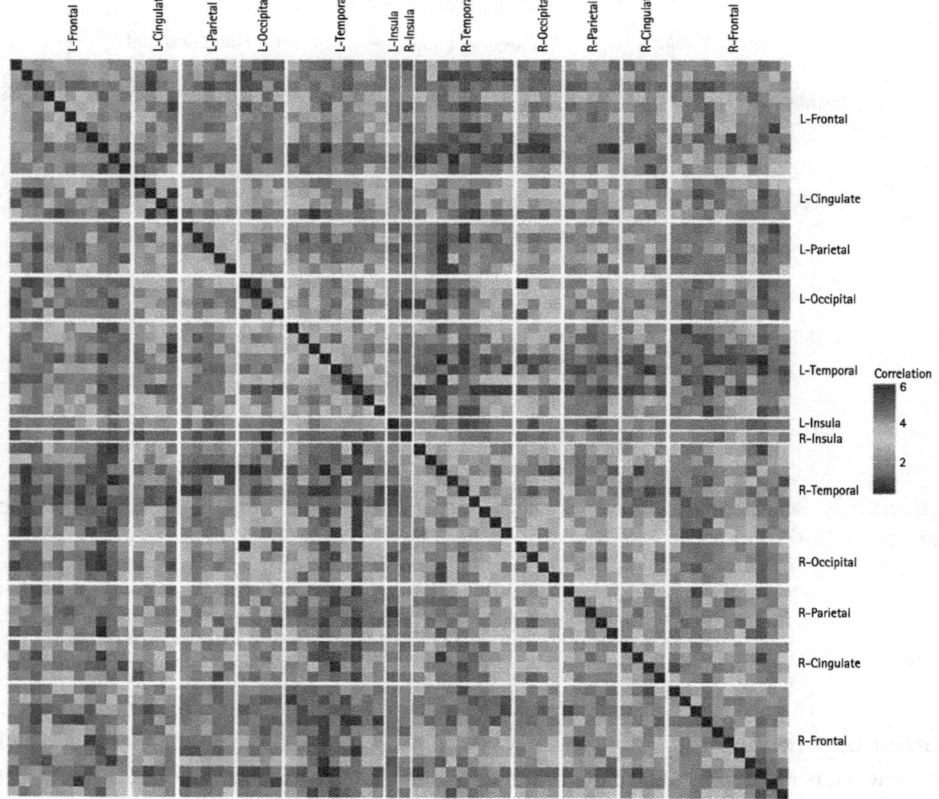

FIGURE 6.5 Functional connectivity in MEG data from a cross-sectional study of eighty children aged 7–10. Each data point reflects the strength of correlation in the time-course of the MEG signal, using resting-state data, whilst controlling for signal spread.

Source: reprinted from Johnson, Amy, "The environment and child development: a multivariate approach," PhD thesis, University of Cambridge, Cambridge, United Kingdom, 2018.

These results may be interesting in their own right, but they also highlight the potential value of including a resting-state scan in your protocol. If you are interested in how the brain develops then knowing what changes occur independent of a task or strategy can be a powerful way of demonstrating and interpreting differences or changes over time. Since the first observations of resting-state functional connectivity using MEG in children outlined above there have been a number of technical advancements. The most prominent of which is the incorporation of algorithms capable of identifying sub-second fluctuations in networks, rather than simply averaging across the whole 9 minute period (Baker et al., 2014; Hawkins and Akarca et al., 2018).

The organizational principles of system-wide neural networks can be captured using graph theory. This mathematical framework provides a formal way of summarizing and comparing networks across individuals. The pipeline of a typical graph theory analysis can take into account functional connectivity data, maps this onto a parcellation of brain areas, and then uses the connectivity matrix to compute whether (and how well) different brain areas are connected. Multiple metrics can be used to provide a comparison across individuals, including how well connected different particular nodes are (number of paths per area), how well particular nodes cluster together (number of within-cluster versus number of between-cluster paths), how efficient a network is (average path length between areas), and how modular a network is. One classic finding is the "small world" brain, which entails that functionally related areas have a high number of connections between them, but also have a few long-range connections to functionally different areas (Bullmore and Sporns, 2009; Rubinov and Sporns, 2010).

It is possible to use EEG and MEG data to derive whole-brain functional connectomes. With EEG data, using electrodes as nodes (with the caveat that this sensor-space analysis is particularly susceptible to aforementioned signal leakage), connectome-derived measures change between the ages of 5 to 7 (N=227, longitudinal study). Specifically, the clustering index (reflecting how well neighbouring nodes connect to each other rather compared to non-neighbouring nodes) increases from 5 to 7 years of age, as does the average path length (reflecting the average shortest path between one node and another), whereas weight dispersion (reflecting the difference between highest and lowest weight in a network, an index of network development) decreases. The authors interpret these findings and the general decrease in connectivity as a reflection of neural pruning: the reduction of connections to make a network less random and chaotic, and more like an efficient small-world network (Boersma et al., 2011).

We have recently used whole-brain source projected MEG data to construct functional connectomes in children aged 7–10 (N = 80; Johnson, 2018; Figure 6.5). This works in principle, but in our experience these MEG-derived functional connectomes can be noisy and to date the methods need to be better developed to yield robust results. Specifically, lower signal-to-noise ratios typically result in lower connectivity estimates. Special consideration is important when designing the statistical approach to compare connectomes across individuals. There are so many connections that even if those connections only reflected noise, they could be combined to significantly predict any outcome variable. Permutation testing or data simulation approaches are the best way to test whether any relationships are genuine. Generating reliable and robust whole-brain functional connectomes that reflect neurophysiological relationships within source space is one of the current challenges facing developmental cognitive neuroscience. In our view, this will require some more serious methodological work, but it is achievable.

FUNCTIONAL NEURAL COUPLING DURING TASK PERFORMANCE

So far, we have covered data collection, basic pre-processing, simple event-related analyses e.g., ERFs), and the current suite of resting-state analyses available to developmental cognitive neuroscientists. In this final section we focus on more complex analyses to MEG data acquired during task performance.

To date, the vast majority of studies exploring neural-cognitive relationships during task performance have used fMRI. In some cases, researchers have used this method to explore how cognitive processes depend upon the coordination of different neural populations, or how the development of a cognitive process is linked to the changing functional relationship between different areas (Casey, Galvan, and Hare, 2005). The same can be done with MEG, but the additional temporal information can be pivotal in understanding the potential function of a particular connection. For example, coordinated activity across fronto-parietal networks has been linked to a wide variety of higher-order cognitive processes like attentional control and working memory (Duncan, 2010). Developmental changes in these networks are thought to be linked to age-related improvements in these cognitive functions (Klingberg, Forssberg, and Westerberg, 2002). Using MEG, researchers were able to show key developmental differences in the dynamics of these networks. In childhood, the coordination of oscillations in these networks in anticipation of memory items was strongly linked to their subsequent retention, but this relationship was not present in adults (Astle, Luckhoo, et al., 2015). In this example, the temporal information about the role of the network is crucial. Were this an fMRI study, which is typically too slow to disentangle anticipatory and stimulus-locked activity, we would likely conclude that we have identified some developmental change in a memory-related process. The temporal resolution shows that this is actually likely to be an attentional process reflecting preparation for the memoranda. Indeed, in this example, the authors then extracted pre-stimulus activity from this network and demonstrated that trial-to-trial fluctuations in its pre-stimulus activity were significantly predictive of trial-to-trial changes in early visual activity when the stimulus was subsequently presented, consistent with the view that this mechanism reflects an anticipatory attentional process. In short, being able to track the connectivity relationships over time provides crucial information about the likely role of any neural process, see Figure 6.6.

As noted at the outset of this chapter, the benefits of using MEG extend beyond just the added temporal information. Different elements of the signal can be used to explore how brain regions interact at a neurophysiological level. For example, the phase of the signal from fronto-parietal areas selectively couples with visual cortex processing a target item, relative to a distracter (Kuo, Nobre, Scerif, and Astle, 2016). This closely mirrors the "communication through coherence" account of large-scale cortical organization proposed on the basis of intracranial recordings with both humans and adults (Bastos, Vezoli, and Fries, 2015; Fries, 2005). This is an important feature of MEG-derived functional coupling metrics: their electrophysiological nature provides a direct link to neuro-psychologically mechanistic accounts of brain function which are typically heavily based on intracranial recordings. This provides the developmental cognitive neuroscientist with a unique opportunity to relate developmental changes in mechanism to basic neuroscience foundations from direct cell recording methods.

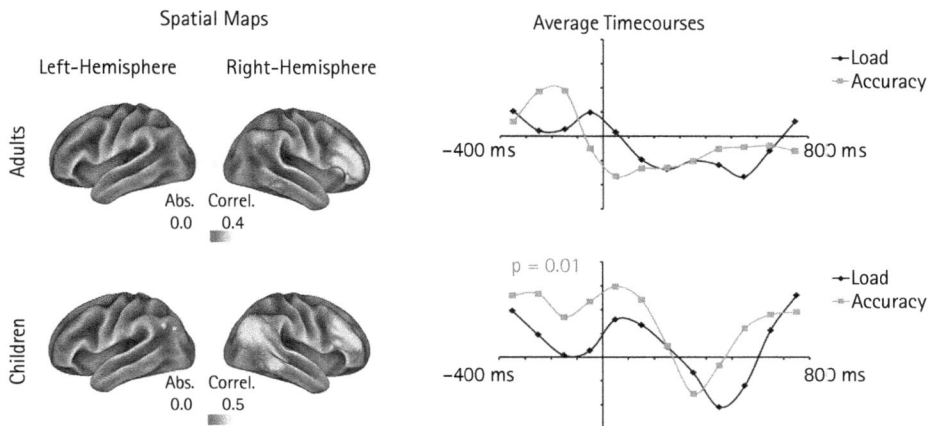

FIGURE 6.6 The right lateralized fronto-parietal identified using an ICA of MEG activity in both adults (upper panel) and children (lower panel). The left-hand images show the spatial extent of the component networks (in terms of the absolute Pearson Correlation values between each brain location and this component); the right-hand images show the time course of a GLM contrast for these networks. The black line reflects the effect of VSTM load, and the light gray line reflects the effect of subsequent accuracy.

Source: reprinted from Duncan E. Astle, Henry Luckhoo, Mark Woolrich, Bo-Cheng Kuo, Anna C. Nobre, and Gaia Scerif. The Neural Dynamics of Fronto-Parietal Networks in Childhood Revealed using Magnetoencephalography, *Cerebral Cortex*, 25(10), pp. 3868–76 doi:10.1093/cercor/bhu271 Copyright © 2014, Oxford University Press. Licensed under Creative Commons CC BY.

Phase amplitude coupling provides a final example of this. Dynamic neural activity in areas of the frontal and parietal lobe is thought to be involved in coding task-critical information (Desimone and Duncan, 1995; Duncan and Owen, 2000). Long-range connections enable these brain areas to organize ongoing sensory processing in a goal-directed fashion. Previous research in human adults (e.g., Voytek et al., 2010), non-human primates (e.g., Haegens, Nacher, Luna, Romo, and Jensen, 2011), and rodents, (e.g., Hyman, Zilli, Paley, and Hasselmo, 2010) provides a neurophysiological framework by which this functional integration may occur. Slow oscillations, putatively expressed in long-range functional networks, modulate the excitability of local neural assemblies (Jensen, Bonnefond, and VanRullen, 2012; Jensen and Colgin, 2007), and the high-frequency activity reflected by these local processes can subsequently become coupled to the slower rhythm (Canolty et al., 2006). This mechanism of cross-frequency phase-amplitude coupling provides a basis by which attentional demands, including spatial location (Sauseng, Klimesch, Gruber, and Birbaumer, 2008; Siegel, Donner, Oostenveld, Fries, and Engel, 2008), timing (Cravo, Rohenkohl, Wyart, and Nobre, 2013; Lakatos et al., 2005), and relevance (Cohen et al., 2009a, 2009b; Siegel, Warden, and Miller, 2009), can influence ongoing sensory, cognitive, or motor processes.

It is possible to measure phase amplitude coupling in children using MEG. We recently demonstrated that the phase of the alpha signal from the fronto-parietal cortical areas became significantly coupled to the amplitude of high-frequency gamma

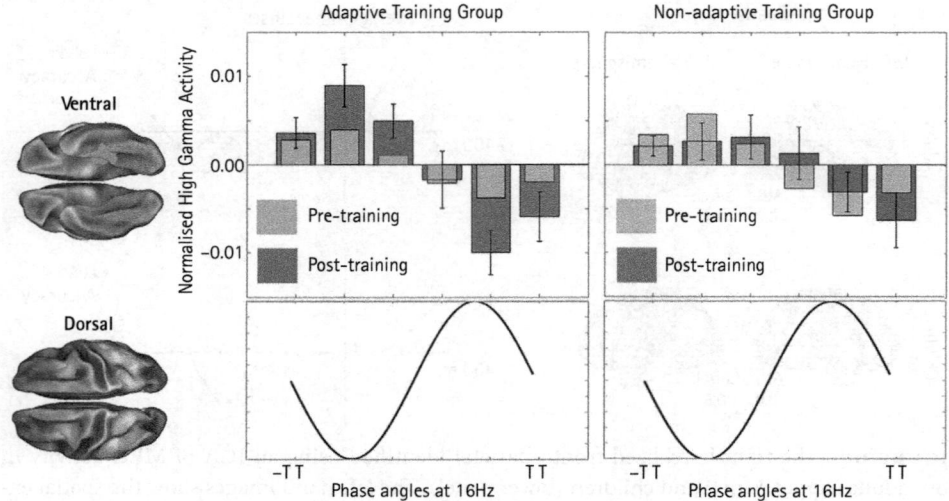

FIGURE 6.7 Phase-ordered amplitude plots. High gamma (83–93 Hz) inferior-temporal cortex replotted according to their position in the cycle of the 16 Hz rhythm from the superior fronto-parietal network. Error bars indicate SEM.

Source: adapted from Duncan E. Astle, Jessica J. Barnes, Kate Baker, Giles L. Colclough, and Mark W. Woolrich, "Cognitive Training Enhances Intrinsic Brain Connectivity in Childhood," *The Journal of Neuroscience*, 35(16), pp. 6277–6283; Figure 6.4, DOI: doi:10.1523/JNEUROSCI.4517-14.2015 (c) 015, The Authors. Licensed under Creative Commons CC BY.

activity within inferior temporal cortex. The strength of this coupling was significantly enhanced by adaptive working memory training (Barnes, Nobre, Woolrich, Baker, and Astle, 2016). At a cellular level, the alpha rhythm provides this temporal structure by modulating the excitability of local neuronal assemblies (Canolty et al., 2006; Canolty, Ganguly, and Carmena, 2012; Canolty et al., 2010; Canolty and Knight, 2010; Fries, 2005). This is possible because these rhythmic changes alter a neuron's output spike timing and its sensitivity to synaptic input (Canolty et al., 2010; Canolty and Knight, 2010), and these local neuronal computations are apparent as ongoing gamma activity (Fries, 2009). MEG provides the developmental cognitive neuroscientist with the rare opportunity to place their findings within what is known about neurophysiological mechanisms, see Figure 6.7.

FINAL SUMMARY

In this chapter, we have covered the basics of how to collect and pre-process MEG in children. We have also worked through various analysis options, from ERFs and time-frequency decomposition, to functional connectivity at rest and during task perform-ance. In our view, MEG provides the developmental cognitive neuroscientist with a way to ask neurophysiological questions about brain development, and how these changing mechanisms link to disorder or cognitive change over time.

References

Adachi, Y. and Haruta, Y. (2014). Whole-head child MEG system and its applications. In S. Supek and C. J. Aine (eds.), *Magnetoencephalography*, (pp. 599–607). New York: Springer. doi:10.1007/978-3-642-33045-2_27.

Ahlfors, S. P., Han, J., Belliveau, J. W., and Hämäläinen, M. S. (2010). Sensitivity of MEG and EEG to source orientation. *Brain Topography*, 23(3), 227–32. doi:10.1007/s10548-010-0154-x.

Ahlfors, S. P., Han, J., Lin, F.-H., Witzel, T., Belliveau, J. W., Hämäläinen, M. S., and Halgren, E. (2009). Cancellation of EEG and MEG signals generated by extended and distributed sources. *Human Brain Mapping*, NA-NA. doi:10.1002/hbm.20851.

Astle, D. E., Barnes, J. J., Baker, K., Colclough, G. L., and Woolrich, M. W. (2015). Cognitive training enhances intrinsic brain connectivity in childhood, *Journal of Neuroscience*, 35(16), 6277–6283. doi:10.1523/JNEUROSCI.4517-14.2015.

Astle, D. E., Luckhoo, H., Woolrich, M., Kuo, B.-C., Nobre, A. C., and Scerif, G. (2015). The neural dynamics of fronto-parietal networks in childhood revealed using magneto-encephalography. *Cerebral Cortex*, 25(10), 3868–3876. doi:10.1093/cercor/bhu271.

Baillet, S., Mosher, J. C., and Leahy, R. M. (2001). Electromagnetic brain mapping. *IEEE Signal Processing Magazine*, 18(6), 14–30. doi:10.1109/79.962275.

Baker, A. P., Brookes, M. J., Rezek, I. A., Smith, S. M., Behrens, T., Probert Smith, P. J., and Woolrich, M. (2014). Fast transient networks in spontaneous human brain activity. *ELife*, 3, e01867. doi:10.7554/eLife.01867.

Barnes, J. J., Nobre, C., Woolrich, M. W., Baker, K., and Astle, D. E. (2016). Training working memory in childhood enhances coupling between frontoparietal control network and task-related regions. *The Journal of Neuroscience*, 36(34), 9001–9011. doi:10.1523/JNEUROSCI.0101-16.2016

Barnes, J. J., Woolrich, M. W., Baker, K., Colclough, G. L., and Astle, D. E. (2016). Electrophysiological measures of resting state functional connectivity and their relationship with working memory capacity in childhood. *Developmental Science*, 19(1), 19–31. doi:10.1111/desc.12297.

Bast, T., Wright, T., Boor, R., Harting, I., Feneberg, R., Rupp, A., ... Baumgärtner, U. (2007). Combined EEG and MEG analysis of early somatosensory evoked activity in children and adolescents with focal epilepsies. *Clinical Neurophysiology*, 118(8), 1721–1735. doi:10.1016/j.clinph.2007.03.037

Bastos, A. M., Vezoli, J., and Fries, P. (2015). Communication through coherence with inter-areal delays. *Current Opinion in Neurobiology*, 31, 173–180. doi:10.1016/j.conb.2014.11.001.

Biswal, B., Zerrin Yetkin, F., Haughton, V. M., and Hyde, J. S. (1995). Functional connectivity in the motor cortex of resting human brain using echo-planar MRI. *Magnetic Resonance in Medicine*, 34(4), 537–541. doi:10.1002/mrm.1910340409.

Boersma, M., Smit, D. J. A., De Bie, H. M. A., Van Baal, G. C. M., Boomsma, D. I., De Geus, E. J. C., ... Stam, C. J. (2011). Network analysis of resting state EEG in the developing young brain: Structure comes with maturation. *Human Brain Mapping*, 32(3), 413–425. doi:10.1002/hbm.21030.

Boto, E., Holmes, N., Leggett, J., Roberts, G., Shah, V., Meyer, S. S., ... Brookes, M. J. (2018). Moving magnetoencephalography towards real-world applications with a wearable system. *Nature*, 555(7698), 657–661. doi:10.1038/nature26147

Brookes, M. J., Woolrich, M., Luckhoo, H., Price, D., Hale, J. R., Stephenson, M. C., ... Morris, P. G. (2011). Investigating the electrophysiological basis of resting state networks using magnetoencephalography. *Proceedings of the National Academy of Sciences*, 108(40), 16783–16788. doi:10.1073/pnas.1112685108.

Bullmore, E. and Sporns, O. (2009). Complex brain networks: Graph theoretical analysis of structural and functional systems. *Nature Reviews Neuroscience*, 10(3), 186–198. doi:10.1038/nrn2575.

Canolty, R. T., Edwards, E., Dalal, S. S., Soltani, M., Nagarajan, S. S., Kirsch, H. E., ... Knight, R. T. (2006). High gamma power is phase-locked to theta oscillations in human neocortex, *Science*, 313(5793), 1626–1628. doi:10.1126/science.1128115.

Canolty, R. T., Ganguly, K., and Carmena, J. M. (2012). Task-dependent changes in cross-level coupling between single neurons and oscillatory activity in multiscale networks. *PLoS Computational Biology*, 8(12), e1002809. doi:10.1371/journal.pcbi.1002809.

Canolty, R. T., Ganguly, K., Kennerley, S. W., Cadieu, C. F., Koepsell, K., Wallis, J. D., and Carmena, J. M. (2010). Oscillatory phase coupling coordinates anatomically dispersed functional cell assemblies, *Proceedings of the National Academy of Sciences*, 107(40), 17356–17361. doi:10.1073/pnas.1008306107.

Canolty, R. T. and Knight, R. T. (2010). The functional role of cross-frequency coupling. *Trends in Cognitive Sciences*, 14(11), 506–515. doi:10.1016/j.tics.2010.09.001.

Casey, B. J., Galvan, A., and Hare, T. A. (2005). Changes in cerebral functional organization during cognitive development. *Current Opinion in Neurobiology*, 15(2), 239–244. doi:10.1016/j.conb.2005.03.012.

Coffman, B. A., Kodituwakku, P., Kodituwakku, E. L., Romero, L., Sharadamma, N. M., Stone, D., and Stephen, J. M. (2013). Primary visual response (M100) delays in adolescents with FASD as measured with MEG: Visual M100 delays in adolescents with FASD. *Human Brain Mapping*, 34(11), 2852–2862. doi:10.1002/hbm.22110.

Cohen, M. X., Axmacher, N., Lenartz, D., Elger, C. E., Sturm, V., and Schlaepfer, T. E. (2009a). Good vibrations: Cross-frequency coupling in the human nucleus accumbens during reward processing. *Journal of Cognitive Neuroscience*, 21(5), 875–889. doi:10.1162/jocn.2009.21062.

Cohen, M. X., Axmacher, N., Lenartz, D., Elger, C. E., Sturm, V., and Schlaepfer, T. E. (2009b). Nuclei accumbens phase synchrony predicts decision-making reversals following negative feedback. *Journal of Neuroscience*, 29(23), 7591–7598. doi:10.1523/JNEUROSCI.5335-08.2009.

Cravo, A. M., Rohenkohl, G., Wyart, V., and Nobre, A. C. (2013). Temporal expectation enhances contrast sensitivity by phase entrainment of low-frequency oscillations in visual cortex. *Journal of Neuroscience*, 33(9), 4002–4010. doi:10.1523/JNEUROSCI.4675-12.2013.

Dale, A. M. and Sereno, M. I. (1993). Improved localization of cortical activity by combining EEG and MEG with MRI cortical surface reconstruction: A linear approach. *Journal of Cognitive Neuroscience*, 5(2), 162–176. doi:10.1162/jocn.1993.5.2.162.

Deffke, I., Sander, T., Heidenreich, J., Sommer, W., Curio, G., Trahms, L., and Lueschow, A. (2007). MEG/EEG sources of the 170-ms response to faces are co-localized in the fusiform gyrus. *NeuroImage*, 35(4), 1495–1501. doi:10.1016/j.neuroimage.2007.01.034.

Demopoulos, C., Yu, N., Tripp, J., Mota, N., Brandes-Aitken, A. N., Desai, S. S., ... Marco, E. J. (2017). Magnetoencephalographic imaging of auditory and somatosensory cortical responses in children with autism and sensory processing dysfunction. *Frontiers in Human Neuroscience*, *11*, 259. doi:10.3389/fnhum.2017.00259.

Desimone, R. and Duncan, J. (1995). Neural mechanisms of selective visual attention. *Annual Review of Neuroscience*, *18*(1), 193–222.

Donkers, F. C. L., Schipul, S. E., Baranek, G. T., Cleary, K. M., Willoughby, M. T., Evans, A. M., ... Belger, A. (2015). Attenuated auditory event-related potentials and associations with atypical sensory response patterns in children with autism. *Journal of Autism and Developmental Disorders*, *45*(2), 506–523. doi:10.1007/s10803-013-1948-y.

Duncan, J. (2010). The multiple-demand (MD) system of the primate brain: Mental programs for intelligent behaviour. *Trends in Cognitive Sciences*, *14*(4), 172–179. doi:10.1016/j.tics.2010.01.004.

Duncan, J. and Owen, A. M. (2000). Common regions of the human frontal lobe recruited by diverse cognitive demands. *Trends in Neurosciences*, *23*(10), 475–83. doi:10.1016/S0166-2236(00)01633-7.

Fahrenfort, J. J., van Driel, J., van Gaal, S., and Olivers, C. N. L. (2018). From ERPs to MVPA using the Amsterdam Decoding and Modeling Toolbox (ADAM). *Frontiers in Neuroscience*, *12*, 368. doi:10.3389/fnins.2018.00368.

Fries, P. (2005). A mechanism for cognitive dynamics: Neuronal communication through neuronal coherence. *Trends in Cognitive Sciences*, *9*(10), 474–80. doi:10.1016/j.tics.2005.08.011.

Fries, P. (2009). Neuronal gamma-band synchronization as a fundamental process in cortical computation. *Annual Review of Neuroscience*, *32*(1), 209–224. doi:10.1146/annurev.neuro.051508.135603.

Friston, K., Harrison, L., Daunizeau, J., Kiebel, S., Phillips, C., Trujillo-Barreto, N., ... Mattout, J. (2008). Multiple sparse priors for the M/EEG inverse problem. *NeuroImage*, *39*(3), 1104–1120. doi:10.1016/j.neuroimage.2007.09.048.

Goldenholz, D. M., Ahlfors, S. P., Hämäläinen, M. S., Sharon, D., Ishitobi, M., Vaina, L. M., and Stufflebeam, S. M. (2009). Mapping the signal-to-noise-ratios of cortical sources in magnetoencephalography and electroencephalography. *Human Brain Mapping*, *30*(4), 1077–1086. doi:10.1002/hbm.20571.

Gonzalez-Moreno, A., Aurtenetxe, S., Lopez-Garcia, M.-E., del Pozo, F., Maestu, F., and Nevado, A. (2014). Signal-to-noise ratio of the MEG signal after preprocessing. *Journal of Neuroscience Methods*, *222*, 56–61. doi:10.1016/j.jneumeth.2013.10.019.

Götz, T., Huonker, R., Miltner, W. H. R., Witte, O. W., Dettner, K., and Weiss, T. (2011). Task requirements change signal strength of the primary somatosensory M50: Oddball vs. one-back tasks: Task requirements affect the SEF M50. *Psychophysiology*, *48*(4), 569–577. doi:10.1111/j.1469-8986.2010.01116.x.

Gramfort, A., Luessi, M., Larson, E., Engemann, D. A., Strohmeier, D., Brodbeck, C., ... Hamalainen, M. (2013). MEG and EEG data analysis with MNE-Python. *Frontiers in Neuroscience*, *7*, 267. doi:10.3389/fnins.2013.00267.

Gross, J., Baillet, S., Barnes, G. R., Henson, R. N., Hillebrand, A., Jensen, O., ... Schoffelen, J.-M. (2013). Good practice for conducting and reporting MEG research. *NeuroImage*, *65*, 349–363. doi:10.1016/j.neuroimage.2012.10.001

Haegens, S., Nacher, V., Luna, R., Romo, R., and Jensen, O. (2011). Oscillations in the monkey sensorimotor network influence discrimination performance by rhythmical inhibition of neuronal spiking. *Proceedings of the National Academy of Sciences*, *108*(48), 19377–19382. doi:10.1073/pnas.1117190108.

Hämäläinen, M., Hari, R., Ilmoniemi, R. J., Knuutila, J., and Lounasmaa, O. V. (1993). Magnetoencephalography—theory, instrumentation, and applications to noninvasive studies of the working human brain. *Reviews of Modern Physics*, 65(2), 413–497. doi:10.1103/RevModPhys.65.413.

Hansen, P. C., Kringelbach, M. L., and Salmelin, R. (2010). *MEG: An Introduction to Methods*. New York: Oxford University Press.

Hauk, O., Wakeman, D. G., and Henson, R. (2011). Comparison of noise-normalized minimum norm estimates for MEG analysis using multiple resolution metrics. *NeuroImage*, 54(3), 1966–1974. doi:10.1016/j.neuroimage.2010.09.053.

Hawkins, E., Akarca, D., Zhang, M., Brkić, D., Woolrich, M., Baker, K., and Astle, D. E. (2018). *Functional Network Dynamics in a Neurodevelopmental Disorder of Known Genetic Origin* [Preprint]. doi:10.1101/463323.

He, W., Brock, J., and Johnson, B. W. (2014). Face-sensitive brain responses measured from a four-year-old child with a custom-sized child MEG system. *Journal of Neuroscience Methods*, 222, 213–217. doi:10.1016/j.jneumeth.2013.11.020.

He, W., Brock, J., and Johnson, B. W. (2015). Face processing in the brains of pre-school aged children measured with MEG. *NeuroImage*, 106, 317–327. doi:10.1016/j.neuroimage.2014.11.029.

Henson, R. N., Mouchlianitis, E., and Friston, K. J. (2009). MEG and EEG data fusion: Simultaneous localisation of face-evoked responses. *NeuroImage*, 47(2), 581–589. doi:10.1016/j.neuroimage.2009.04.063.

Henson, R. N., Wakeman, D. G., Litvak, V., and Friston, K. J. (2011). A parametric empirical Bayesian framework for the EEG/MEG inverse problem: Generative models for multisubject and multi-modal integration. *Frontiers in Human Neuroscience*, 5. doi:10.3389/fnhum.2011.00076.

Hillebrand, A. and Barnes, G. R. (2002). A quantitative assessment of the sensitivity of wholehead meg to activity in the adult human cortex. *NeuroImage*, 16(3), 638–650. doi:10.1006/nimg.2002.1102.

Hillebrand, A., Singh, K. D., Holliday, I. E., Furlong, P. L., and Barnes, G. R. (2005). A new approach to neuroimaging with magnetoencephalography. *Human Brain Mapping*, 25(2), 199–211. doi:10.1002/hbm.20102.

Hyman, J. M., Zilli, E. A., Paley, A. M., and Hasselmo, M. E. (2010). Working memory performance correlates with prefrontal-hippocampal theta interactions but not with prefrontal neuron firing rates. *Frontiers in Integrative Neuroscience*. doi:10.3389/neuro.07.002.2010.

Jas, M., Larson, E., Engemann, D. A., Leppäkangas, J., Taulu, S., Hämäläinen, M., and Gramfort, A. (2018). A reproducible MEG/EEG group study with the MNE software: Recommendations, quality assessments, and good practices. *Frontiers in Neuroscience*, 12, 530. doi:10.3389/fnins.2018.00530.

Jensen, O., Bonnefond, M., and VanRullen, R. (2012). An oscillatory mechanism for prioritizing salient unattended stimuli. *Trends in Cognitive Sciences*, 16(4), 200–206. doi:10.1016/j.tics.2012.03.002.

Jensen, O. and Colgin, L. L. (2007). Cross-frequency coupling between neuronal oscillations. *Trends in Cognitive Sciences*, 11(7), 267–269. doi:10.1016/j.tics.2007.05.003.

Johnson, A. (2018). *The Environment and Child Development: A Multivariate Approach* (Doctoral Dissertation). University of Cambridge, Cambridge, United Kingdom.

Kikuchi, M., Yoshimura, Y., Mutou, K., and Minabe, Y. (2016). Magnetoencephalography in the study of children with autism spectrum disorder: Magnetoencephalography for autism. *Psychiatry and Clinical Neurosciences*, 70(2), 74–88. doi:10.1111/pcn.12338.

Klingberg, T., Forssberg, H., and Westerberg, H. (2002). Increased brain activity in frontal and parietal cortex underlies the development of visuospatial working memory capacity during childhood. *Journal of Cognitive Neuroscience*, *14*(1), 1–10. doi:10.1162/089892902317205276.

Kujala, A., Huotilainen, M., Hotakainen, M., Lennes, M., Parkkonen, L., Fellman, V., and Näätänen, R. (2004). Speech-sound discrimination in neonates as measured with MEG. *NeuroReport*, *15*(13), 2089.

Kuo, B.-C., Nobre, A. C., Scerif, G., and Astle, D. E. (2016). Top–down activation of spatiotopic sensory codes in perceptual and working memory search. *Journal of Cognitive Neuroscience*, *28*(7), 996–1009. doi:10.1162/jocn_a_00952.

Kylliäinen, A., Braeutigam, S., Hietanen, J. K., Swithenby, S. J., and Bailey, A. J. (2006). Face and gaze processing in normally developing children: A magnetoencephalographic study. *European Journal of Neuroscience*, *23*(3), 801–810. doi:10.1111/j.1460-9568.2005.04554.x.

Lakatos, P., Shah, A. S., Knuth, K. H., Ulbert, I., Karmos, G., and Schroeder, C. E. (2005). An oscillatory hierarchy controlling neuronal excitability and stimulus processing in the auditory cortex. *Journal of Neurophysiology*, *94*(3), 1904–1911. doi:10.1152/jn.00263.2005.

Leahy, R. M., Mosher, J. C., Spencer, M. E., Huang, M. X., and Lewine, J. D. (1998). A study of dipole localization accuracy for MEG and EEG using a human skull phantom. *Electroencephalography and Clinical Neurophysiology*, *107*(2), 159–173. doi:10.1016/S0013-4694(98)00057-1.

Liu, J., Harris, A., and Kanwisher, N. (2002). Stages of processing in face perception: An MEG study. *Nature Neuroscience*, *5*(9), 910–916. doi:10.1038/nn909

Liuzzi, L., Gascoyne, L. E., Tewarie, P. K., Barratt, E. L., Boto, E., and Brookes, M. J. (2017). Optimising experimental design for MEG resting state functional connectivity measurement. *NeuroImage*, *155*, 565–576. doi:10.1016/j.neuroimage.2016.11.064

Luckhoo, H. T., Brookes, M. J., and Woolrich, M. W. (2014). Multi-session statistics on beamformed MEG data. *NeuroImage*, *95*, 330–335. doi:10.1016/j.neuroimage.2013.12.026.

Marco, E. J., Khatibi, K., Hill, S. S., Siegel, B., Arroyo, M. S., Dowling, A. F., ... Nagarajan, S. S. (2012). Children with autism show reduced somatosensory response: An MEG study: Reduced somatosensory response in autism. *Autism Research*, *5*(5), 340–351. doi:10.1002/aur.1247.

Marinkovic, K., Cox, B., Reid, K., and Halgren, E. (2004). Head position in the MEG helmet affects the sensitivity to anterior sources. *Neurology and Clinical Neurophysiology: NCN*, *30*.

Martin, T., McDaniel, M. A., Guynn, M. J., Houck, J. M., Woodruff, C. C., Bish, J. P., ... Tesche, C. D. (2007). Brain regions and their dynamics in prospective memory retrieval: A MEG study. *International Journal of Psychophysiology*, *64*(3), 247–258. doi:10.1016/j.ijpsycho.2006.09.010.

Melcher, J. R., and Cohen, D. (1988). Dependence of the MEG on dipole orientation in the rabbit head. *Electroencephalography and Clinical Neurophysiology*, *70*(5), 460–472. doi:10.1016/0013-4694(88)90024-7.

Meyer, S. S., Bonaiuto, J., Lim, M., Rossiter, H., Waters, S., Bradbury, D., ... Barnes, G. R. (2017). Flexible head-casts for high spatial precision MEG. *Journal of Neuroscience Methods*, *276*, 38–45. doi:10.1016/j.jneumeth.2016.11.009.

Michel, C. M., Murray, M. M., Lantz, G., Gonzalez, S., Spinelli, L., and Grave de Peralta, R. (2004). EEG source imaging. *Clinical Neurophysiology*, *115*(10), 2195–2222. doi:10.1016/j.clinph.2004.06.001.

Molins, A., Stufflebeam, S. M., Brown, E. N., and Hämäläinen, M. S. (2008). Quantification of the benefit from integrating MEG and EEG data in minimum ℓ2-norm estimation. *NeuroImage*, 42(3), 1069–1077. doi:10.1016/j.neuroimage.2008.05.064.

Moore, J. K. and Linthicum, F. H. (2007). The human auditory system: A timeline of development. *International Journal of Audiology*, 46(9), 460–78. doi:10.1080/14992020701383019.

Oostenveld, R., Fries, P., Maris, E., and Schoffelen, J.-M. (2011). FieldTrip: Open source software for advanced analysis of MEG, EEG, and Invasive Electrophysiological Data. *Computational Intelligence and Neuroscience*, 2011, 1–9. doi:10.1155/2011/156869.

Oram Cardy, J. E., Flagg, E. J., Roberts, W., and Roberts, T. P. L. (2005). Delayed mismatch field for speech and non-speech sounds in children with autism. *NeuroReport*, 16(5), 521.

Orekhova, E. V., Butorina, A. V., Sysoeva, O. V., Prokofyev, A. O., Nikolaeva, A. Y., and Stroganova, T. A. (2015). Frequency of gamma oscillations in humans is modulated by velocity of visual motion. *Journal of Neurophysiology*, 114(1), 244–255. doi:10.1152/jn.00232.2015.

Orekhova, E. V., Butorina, A. V., Tsetlin, M. M., Novikova, S. I., Sokolov, P. A., Elam, M., and Stroganova, T. A. (2013). Auditory magnetic response to clicks in children and adults: Its components, hemispheric lateralization and repetition suppression effect. *Brain Topography*, 26(3), 410–427. doi:10.1007/s10548-012-0262-x.

Palva, J. M., Wang, S. H., Palva, S., Zhigalov, A., Monto, S., Brookes, M. J., ... Jerbi, K. (2018). Ghost interactions in MEG/EEG source space: A note of caution on inter-areal coupling measures. *NeuroImage*, 173, 632–643. doi:10.1016/j.neuroimage.2018.02.032

Penny, W. D., Friston, K. J., Ashburner, J. T., Kiebel, S. J., and Nichols, T. E. (2011). *Statistical Parametric Mapping: The Analysis of Functional Brain Images*. London: Elsevier.

Pfurtscheller, G. and Lopes da Silva, F. H. (1999). Event-related EEG/MEG synchronization and desynchronization: Basic principles. *Clinical Neurophysiology*, 110(11), 1842–1857. doi:10.1016/S1388-2457(99)00141-8.

Plöchl, M., Ossandón, J. P., and König, P. (2012). Combining EEG and eye tracking: Identification, characterization, and correction of eye movement artifacts in electroencephalographic data. *Frontiers in Human Neuroscience*, 6. doi:10.3389/fnhum.2012.00278.

Popov, T., Oostenveld, R., and Schoffelen, J. M. (2018). FieldTrip made easy: An analysis protocol for group analysis of the auditory steady state brain response in time, frequency, and space. *Frontiers in Neuroscience*, 12, 711. doi:10.3389/fnins.2018.00711.

Rojas, D. (2013). Electrophysiology of Autism. In K. McFadden and M. Fitzgerald (eds.), *Recent Advances in Autism Spectrum Disorders—Volume II*. London: IntechOpen. doi:10.5772/54770.

Rosazza, C., and Minati, L. (2011). Resting-state brain networks: Literature review and clinical applications. *Neurological Sciences*, 32(5), 773–785. doi:10.1007/s10072-011-0636-y.

Rubinov, M., and Sporns, O. (2010). Complex network measures of brain connectivity: Uses and interpretations. *NeuroImage*, 52(3), 1059–1069. doi:10.1016/j.neuroimage.2009.10.003

Sauseng, P., Klimesch, W., Gruber, W. R., and Birbaumer, N. (2008). Cross-frequency phase synchronization: A brain mechanism of memory matching and attention. *NeuroImage*, 40(1), 308–317. doi:10.1016/j.neuroimage.2007.11.032.

Scherg, M. and Berg, P. (1991). Use of prior knowledge in brain electromagnetic source analysis. *Brain Topography*, 4(2), 143–150. doi:10.1007/BF01132771

Schoffelen, J.-M. and Gross, J. (2009). Source connectivity analysis with MEG and EEG. *Human Brain Mapping*, 30(6), 1857–1865. doi:10.1002/hbm.20745

Sekihara, K. and Nagarajan, S. S. (2008). *Adaptive Spatial Filters for Electromagnetic Brain Imaging*. New York: Springer. doi:10.1007/978-3-540-79370-0

Sekihara, K. and Nagarajan, S. S. (2015). *Electromagnetic Brain Imaging: A Bayesian Perspective*. New York: Springer.

Sharon, D., Hamalainen, M. S., Tootell, R. B., Halgren, E., and Belliveau, J. W. (2007). The advantage of combining MEG and EEG: Comparison to fMRI in focally stimulated visual cortex. *NeuroImage*, *36*(4), 1225–1235.

Siegel, M., Donner, T. H., Oostenveld, R., Fries, P., and Engel, A. K. (2008). Neuronal synchronization along the dorsal visual pathway reflects the focus of spatial attention. *Neuron*, *60*(4), 709–719. doi:10.1016/j.neuron.2008.09.010.

Siegel, M., Warden, M. R., and Miller, E. K. (2009). Phase-dependent neuronal coding of objects in short-term memory. *Proceedings of the National Academy of Sciences*, *106*(50), 21341–21346. doi:10.1073/pnas.0908193106.

Stolk, A., Todorovic, A., Schoffelen, J.-M., and Oostenveld, R. (2013). Online and offline tools for head movement compensation in MEG. *NeuroImage*, *68*, 39–48. doi:10.1016/j.neuroimage.2012.11.047

Troebinger, L., López, J. D., Lutti, A., Bradbury, D., Bestmann, S., and Barnes, G. (2014). High precision anatomy for MEG. *NeuroImage*, *86*, 583–91. doi:10.1016/j.neuroimage.2013.07.065.

Van den Heuvel, M. P. and Hulshoff Pol, H. E. (2010). Exploring the brain network: A review on resting-state fMRI functional connectivity. *European Neuropsychopharmacology*, *20*(8), 519–534. doi:10.1016/j.euroneuro.2010.03.008.

Voytek, B., Canolty, R. T., Shestyuk, A., Crone, N. E., Parvizi, J., and Knight, R. T. (2010). Shifts in gamma phase–amplitude coupling frequency from theta to alpha over posterior cortex during visual tasks. *Frontiers in Human Neuroscience*, *4*, 191. doi:10.3389/fnhum.2010.00191.

Yoshimura, Y., Kikuchi, M., Ueno, S., Shitamichi, K., Remijn, G. B., Hiraishi, H., … Minabe, Y. (2014). A longitudinal study of auditory evoked field and language development in young children. *NeuroImage*, *101*, 440–447. doi:10.1016/j.neuroimage.2014.07.034.

CHAPTER 7

···

FUNCTIONAL NEAR INFRARED SPECTROSCOPY (FNIRS)

···

SARAH LLOYD-FOX

INTRODUCTION

···

THE development of non-invasive brain imaging techniques over the last 50 years has led to an exponential growth in our understanding of brain function and structure. Over the last two decades, advances in neuroimaging technology and software have opened a new avenue for research on the developing human brain, allowing us to investigate questions that until recently would have seemed impossible with existing behavioral methods. Importantly, neuroimaging methods do not rely on an overt signal from the participant (such as a point or verbal information), which may be difficult to elicit in certain populations and ages. Furthermore, while young infants may have a restricted behavioral repertoire (for example due to an immature visual system or limited hand-eye coordination) or demonstrate similar patterns of visual attention to differing stimuli, the measurement of underlying cortical processes may reveal the recruitment of substantially different functional networks. Or, importantly, neuroimaging may be able to demonstrate that different behavioral responses utilize common or shared cognitive processes. A further strength of neuroimaging is its adaptability for use across a range of ages enabling us to study developmental change and maturation across both cross-sectional and longitudinal studies. Moreover, brain imaging measurements can be used objectively across a wide range of populations, regardless of the culture or setting.

Neuroimaging methods either detect the direct activation related to electrical activity of the brain (e.g., electroencephalography (EEG), magnetoelectroencephalography (MEG)) or the consequent hemodynamic response (e.g., positron emission tomography (PET), functional magnetic resonance imaging (fMRI), functional near infrared spectroscopy (fNIRS)). Understanding the conceptual, methodological, and statistical challenges that these technologies pose for the study of the developing brain will allow researchers to direct the appropriate method to the question under consideration (Peterson, 2003). A major limitation for developmental brain imaging research is methodological. Many of these techniques, which are well established in adults, have limiting factors restricting or preventing their use in infants and young children (see Figure 7.1). PET requires the use of radioisotopes, whilst fMRI and MEG require the participant to remain very still, usually swaddled or restrained. There has been some infant research published using these techniques (Blasi et al., 2011; Dehaene-Lambertz et al., 2002; Huotilainen et al., 2003; Imada et al., 2006; Shultz et al., 2014; Tzourio-Mazoyer et al., 2002). However, this work has generally been restricted to the study of sleeping or sedated infants or children over the age of 5 years. Studies are rarely undertaken with PET on infants and children unless there is a clinical need (Chugani et al., 1987; Tzourio-Mazoyer et al., 2002) as the use of radioisotopes in healthy developing populations is not encouraged or generally approved by ethical committees. One other technique increasingly used with developmental populations is MEG (for a review see Hari and Salmelin, 2012). MEG has very fine temporal resolution and potentially finer spatial resolution (Cheour et al., 2004). Typically, participants sit in a seat with a MEG-scanner placed around their head and must remain relatively motionless. For these reasons MEG has

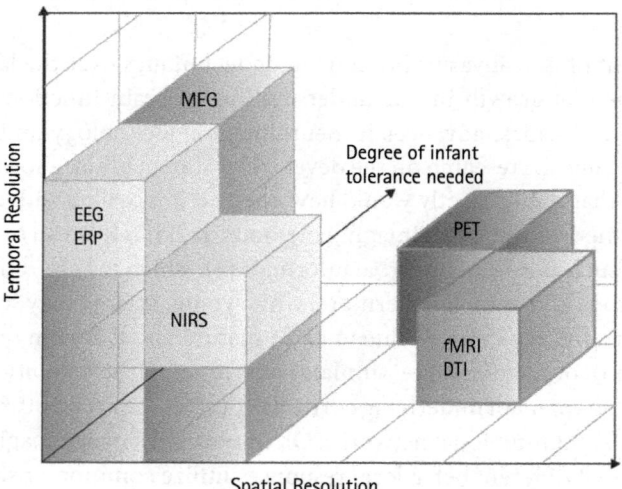

FIGURE 7.1 This figure illustrates the relative degree of tolerance needed from the infant for each method, ranging from light grey (low) to dark grey (high), and the spatial and temporal resolution of NIRS compared with other infant functional neuroimaging methods. EEG, electroencephalography; ERP, event-related potential; MEG, magnetoencephalography; NIRS, near infrared spectroscopy; fMRI, functional magnetic resonance imaging; DTI, diffusion tensor imaging; PET, positron emission tomography.

Source: adapted from S. Lloyd-Fox, A. Blasi, and C.E. Elwell, Illuminating the developing brain: The past, present and future of functional near infrared spectroscopy. *Neuroscience & Biobehavioral Reviews*, 34(3), 269–84, doi:10.1016/j.neubiorev.2009.07.008, Copyright © 2009 Elsevier Ltd. All rights reserved.

largely been restricted to the study of older children and adults rather than with early developmental populations, though recent advances and pioneering work suggests this technique may become more and more suitable (Draganova et al., 2005; Imada et al., 2006; Travis et al., 2011).

To study brain function in infants and young children, the most frequently used methods are EEG and fNIRS. They both use a similar experimental setting, are non-invasive, and easy to administer with this age range. MRI has been used extensively with older children for studies of brain function and structure, and more frequently for the study of brain anatomy in younger children and infants. Given that all these techniques are non-invasive, the ability to take repeated measurements within the same individuals is facilitative, particularly for longitudinal, developmental, and intervention studies.

The advantages and disadvantages of these three most popular neuroimaging methods for use with infants are outlined in Figure 7.1 and Table 7.1. The major advantages of functional near infrared spectroscopy (fNIRS) compared with EEG is that it is less susceptible to data corruption by movement artifacts and offers a more highly spatially resolved image of activation allowing the localization of brain responses to specific cortical regions. In infants and young children the spatial resolution of neighboring fNIRS

Table 7.1 Relative attributes of the three most widely used neuroimaging techniques for the study of infant brain development.

Technique	fNIRS	EEG	fMRI
Type of response measured	Changes in HbO$_2$ and HHb concentration	Neuronal excitation	Changes in BOLD (mainly HHb concentration)
Spatial localization of response	Good	Relatively poor	Very good
Time locking of response	Good	Very good	Relatively poor
Acquisition of signal	Milliseconds	Milliseconds	Seconds
Timing of signal	Seconds	Milliseconds	Seconds
Participant State	Awake/asleep	Awake/asleep	Asleep/immobile
Experimental Setting	Infant on parent's lap/seated or mobile	Infant on parent's lap/seated or mobile	Infant wrapped up on bed in MRI scanner
Freedom of movement of participants	Relatively high	Relatively high	None
Freedom of movement of equipment	Yes	Yes	No
Length of preparation of participants	Short	Short/Medium	Long
Length of experiment	Short	Short	Long
Instrumentation noise	None	None	High—ear protection needed
Cost of study	Fairly low	Fairly low	Relatively high

measurement channels is usually approximately 10–30 mm (with new image reconstruction methods it can be even more specific), while responses in EEG are generally averaged across channels covering several centimeters of tissue—(in contrast to fMRI which can measure responses within voxels of 1–3 mm). For both fNIRS and EEG there is no capacity for measuring brain structure for anatomical reference, therefore MRI will always be the primary choice for obtaining structural information. The temporal resolution of EEG is highest: the precision can reach up to a thousand hertz. Though fMRI and fNIRS measure the same hemodynamic response, generally fMRI techniques have an intrinsically limited acquisition rate usually at a minimum of 1 Hz, whereas fNIRS can acquire data rapidly, generally at 10 Hz (though it can be acquired up to a rate of several hundred Hz), thus providing a more complete temporal picture. Furthermore, in comparison with fMRI, fNIRS can measure both oxy- (HbO_2) and deoxy-hemoglobin (HHb) providing a more complete measure of the hemodynamic response.

When undertaking research with infants, fNIRS headgear can be placed on the head and a study ready to begin within 30 seconds. For EEG, when using saline electrodes administration time is relatively short, and the study can begin within approximately 1 minute. However, the gel electrodes (which provide the advantage of a better signal to noise ratio) can take considerably longer (10–15 minutes) to prepare and so in developmental populations can have a significant impact on compliance. It is recommended with both methods that two experimenters run the study so that one can entertain the participant while the headgear is prepared. For MRI, depending on the age of the population under study, the administration and preparation time can be considerable (i.e., several hours), and participant compliance to allow for artifact-free signals sometimes challenging (Leroux et al., 2013; Kotsoni et al., 2006). A current disadvantage of fNIRS in comparison to fMRI, and to a large extent EEG, is that the lack of standardized (i) headgear placement across commercially available systems, and (ii) approaches to data processing and analyses can make comparisons across research labs more challenging. These issues should be resolved as the fNIRS community continues to grow (Boas et al., 2014).

Despite some of the concerns outlined previously, fNIRS research has made important contributions over the last two decades. The chapter is organized to first provide a basic overview of fNIRS and how it is implemented with infants. It then reviews some of the areas in which fNIRS has made particular contributions to our understanding of early development. Finally, future directions of fNIRS in developmental cognitive neuroscience are discussed.

FUNCTIONAL NEAR INFRARED SPECTROSCOPY (fNIRS)

The use of near infrared light to monitor intact organs began as a discovery at a dinner table with the passage of light being observed through a steak bone at a family supper in the 1970s by Frans Jöbsis-vanderVliet. As he held the cleaned bone up to a light he could clearly discern a shadow of a finger visible through the object and inferred the possible existence of an optical window into the body (Jöbsis-vanderVliet, 1999). This window, from 650–950 nm, takes advantage of the relative transparency of bone to near-infrared wavelengths of light (as well as the advantageous optical properties within the near infrared range of other substances such as water and melanin). Initially, infant work with near infrared spectroscopy (NIRS) took place within a clinical setting investigating cerebral oxygenation in preterm and term neonates (i.e., Brazy et al., 1985). Then, in 1993, four research labs published the first reports on the use of NIRS to detect the hemodynamic response to functional cortical activation (Chance et al., 1993; Hoshi and Tamura, 1993; Kato et al., 1993; Villringer et al., 1993). Since then, the technology has been used to investigate cortical function in a range of age groups including adults (Ferrari and Quaresima, 2012) and children (Nagamitsu et al., 2012). It is relatively recently that researchers have realized the potential of NIRS as an assay of infants' neuronal activity and brain organization (Lloyd-Fox et al., 2010). The term functional NIRS (fNIRS) is used for the application of NIRS for the purpose of functional neuroimaging, which is a rapidly growing field (Boas et al., 2014).

Several attributes make fNIRS ideal for developmental research with infants and young children. fNIRS (i) can be undertaken while the participants are asleep or awake (lying down, sitting on a chair or their parent's lap), (ii) can accommodate some degree of movement, is (iii) non-invasive, (iv) portable and low-cost relative to other neuroimaging methods, (v) involves rapid administration time as the headgear can often be placed on the head and a study ready to begin within one minute, (vi) usually involves short lengths of assessment (5–10 minutes), and (vii) in infants, takes advantage of the light transport properties of tissue which allows light to travel further and interrogate a larger area of the brain at this age.

Basic Principles of fNIRS

Neuronal activation originates in the neurons in the brain as electrical signals are passed between cells. During functional brain activation, the metabolic demands of neurons change, provoking a complex set of changes in oxygen and glucose consumption, local cerebral blood flow, and blood volume. This activation-induced vascular response is known as the hemodynamic response function (see Figure 7.2), so called because of the protein—hemoglobin—in red blood cells that transports oxygen and contains iron. Hemoglobin chromophores come in two forms, oxyhemoglobin (HbO_2) (oxygen carrying) and deoxyhemoglobin (HHb) (without oxygen). The changes in concentration of these chromophores can be used as surrogate markers of brain blood flow and hence provide a means of investigating brain function. To a first approximation, a typical hemodynamic response to cortical neuronal activation in adults drives an increase in local blood flow that is disproportionate to the local oxygen demand, thus leading to an initial dip in HbO_2 and decrease in HHb, followed by a larger increase in HbO_2

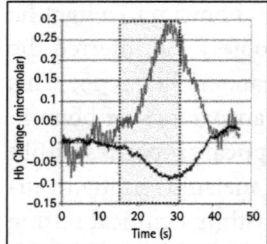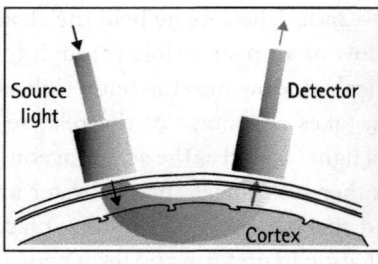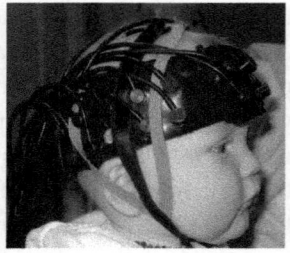

FIGURE 7.2 This figure illustrates a typical hemodynamic response to functional activation showing an increase in HbO$_2$ (light gray) and a decrease in HHb (dark gray) (left panel); the passage of the majority of light that the detector is most likely to measure from an adjacent source in a NIRS array positioned on the head (middle panel); and an infant wearing the Birkbeck-NIRS infant headgear (right panel).

Sources: parts A and B: reproduced from Lloyd-Fox, Sarah, Functional near infrared spectroscopy (fNIRS). In Hopkins, B., Geangu, E., and Linkenauger, S., eds, *The Cambridge Encyclopedia of Child Development*. 2nd Edition, p. 131, Figure 7.2 © Cambridge University Press, 2017.

Part C: © S. Lloyd-Fox.

and decrease as it is displaced from the veins. The shape (latency and magnitude) of the hemodynamic response function (HRF) may vary according to the age of the participant, evoking stimuli (i.e., differences in amplitude are observed between brief and prolonged stimulus presentation) as well as the underlying neural activity (Issard and Gervain, 2018). In general, the initial dip in HbO$_2$ (and increase in HHb) occurs within the first few seconds of stimulation followed by a peak in activation 10–30 seconds later. Overall, these changes result in an increase in total hemoglobin (HbT). This change in local hemoglobin concentration is the basis of fNIRS.

With this optical technique, fNIRS, infants wear lightweight headgear which facilitates the delivery to, and detection of near infrared light from the head. The light migrates from sources located within the headgear and travels through the skin, skull, and underlying brain tissue. Detectors within the same headgear measure light reflecting back to the surface of the head (see Figure 7.2). In comparison to visible light, biological tissue is relatively transparent to light in the near infrared part of the spectrum. Therefore, several centimeters of tissue can be illuminated through the scalp and skull. Chromophores are compounds whose absorption of near infrared light (in the wavelength range 650–1000 nm) is oxygen status dependent in tissue. As the oxygenated and deoxygenated form of the chromophore hemoglobin have different absorption spectra (meaning that the amount of light absorbed by the two chromophores differs), near-infrared spectroscopy methods can differentially measure the changes in concentration of HbO$_2$ and HHb within the blood and tissue (hemodynamics). This is because changes in the amount of the diffuse light reaching the detectors corresponds to changes in the optical properties of the tissue (in particular, absorption properties due to increases/decreases in HbO$_2$/HHb concentration) around the sources and detectors. By arranging the sources and detectors within the headgear into a series of paired channels, spatially localized measurements can be obtained. These channels can form single pairs

(Meek et al., 1998), sequential grid-like patterns (see Figure 7.2), or overlapping patterns allowing for interrogation of signals at overlapping positions (Eggebrecht et al., 2014) and at multiple depths.

In infants, the majority of the light measured by the detector in each channel has interrogated the tissue approximately midway between the source and detector, and half this distance in depth from the scalp surface (Fukui et al., 2003). For example, for an array with source-detector separations of 20 mm, the majority of the light will reach a depth of approximately 10 mm from the skin surface. In infants, the surface of the cortex is reached at a depth of between 4–10 mm (in 5–24-month-olds; (Salamon et al., 1990). Therefore, in infants, a portion of the detected light will have sampled the brain and thus provide a measure of changes in cerebral hemoglobin concentration (HbO$_2$ and HHb) and thus brain function. The light transport properties of tissue, and the thickness of the tissue and skull differ over development (Duncan et al., 1995; Fukui et al., 2003). In preterm infants and neonates—if measurement channels are placed over the whole head—in addition to optical topography (two-dimensional (2D)) representations of the data), it is possible to interrogate both cortical and subcortical brain regions and generate a three-dimensional (3D) image of the head (optical tomography). However, beyond this age, when the head size exceeds approximately 11 cm diameter and light transport properties of tissue alter, this is no longer possible. In contrast to fMRI which can measure the whole brain, the majority of fNIRS studies measure responses within cortical regions near to the surface of the brain. Consequently, arrays of channels are usually placed over targeted brain regions, and measure cortical brain regions only, with regions such as the fusiform gyrus or insula being out of reach beyond around 2 months of age (Jönsson et al., 2018).

Study Design

The design of the study must take into consideration the temporal characteristics of the hemodynamic response (i.e., the vascular response occurs over a scale of seconds rather than milliseconds). As a consequence, while some developmental studies use an event related design (i.e., Emberson et al., 2015, Taga and Asakawa, 2007), the majority of infant fNIRS work has been conducted using a block design for stimulus presentation (Issard and Gervain, 2018). The common method is to present the experimental condition followed by a control or baseline condition typically of longer duration to allow the hemodynamic response initiated during the experimental condition to return to a baseline level. This control condition is usually either of minimal stimulation (e.g., silence during sleeping studies) or designed to cause stimulation to a lesser extent than the experimental condition. Furthermore, to reduce physiological (i.e., oscillations from vasomotion) and anticipatory effects it is common practice to jitter experimental stimulus onset by varying the duration of the control trials so that they do not follow a predictable pattern. Finally, one must establish the number of trial repetitions required for a robust response. Under ideal experimental conditions a single trial would be sufficient to yield a significant response (i.e., Carlsson et al., 2008), however inadequate signal to noise and the presence of motion artifacts typically require the repetition of several trials. Yet, one must also keep in mind the phenomenon of repetition suppression or enhancement whereby the repetition of several trials can result in adaptation effects where neural responses to repeated stimuli can diminish over time (Krekelberg et al., 2006) or enhance for example due to learning (for further discussion see Turk-Browne, Scholl, and Chun, 2008). If the signal to noise ratio is adequate fewer trials could yield more reliable data. It may therefore be possible to design paradigms where relatively few trials are used to test multiple experimental conditions within one study.

Instrumentation

There is a range of commercially produced as well as "in-house" manufactured fNIRS systems available, which use different techniques: continuous wave (CW), time resolved, spatially resolved, and frequency resolved spectroscopy. The choice of which system to use is often driven by the cost and availability of infant- or child-appropriate arrays and headgear. There have been a number of well-written reviews on instrumentation (Minagawa-Kawai et al., 2008; Wolf et al., 2007). The majority of fNIRS infant research has been undertaken using CW systems to measure attenuation changes at two wavelengths, as they provide the simplest approach (for a detailed report see Hebden et al., 2004). CW optical topography systems use arrays of multiple sources and detectors to provide two-dimensional maps of the cortical hemodynamic response. Each detector records the amount of light coming from neighboring sources. To identify the source associated with a given detected signal either the sources are illuminated sequentially or are intensity modulated at unique frequencies. If using the latter method, the signals are then decoded in hardware using lock-in amplifiers or in software using a Fourier transform. In this way, measurements can be taken at a rate of several Hz, enabling the time course of the hemodynamic response to be accurately charted simultaneously across multiple channels.

In addition to CW-NIRS systems, time-resolved (TR-NIRS) and frequency-resolved (FR-NIRS) spectrometers can be used to derive an absolute concentration of HbO_2- and HHb. These techniques are generally impractical for functional infant NIRS studies and have yet to be used routinely in clinical neonatal settings (i.e., Cooper et al., 2014). However, these systems are important for use in determining the spatial and temporal variations in optical pathlength which could influence the measured hemodynamic response (Ijichi et al., 2005; Sakatani et al., 2006). Beyond standard commercialized systems there is an opportunity to measure with multiple wavelengths to enhance spectral resolution not just of the hemodynamic chromophores but also of cytochrome oxidase (oxCCO), a chromophore which can provide a marker of cellular oxygen metabolism (Heekeren et al., 1999). These have been used in studies with both adults and infants (Brigadoi et al., 2017; Kolyva et al., 2012; Siddiqui et al., 2017).

Data Processing and Analysis

When analysing NIRS data it is necessary to have an understanding of the practical issues that may impact upon successful data collection including: 1) the development of the arrays and headgear to reduce the effects of movement of the infant, particularly important when the infant is awake; 2) the design of the study, considering the effects of boredom, anticipation, and the synchronization of systemic/biorhythmic responses; 3) an understanding of the hemodynamic response in infants and how to interpret a significant result; 4) contamination of the cortical signal by vascular changes in other tissue layers such as the skin, and 5) co-registration between the hemodynamic response measured at the surface of the head and the underlying cortical anatomy (for a review of these issues see Aslin, 2012; Lloyd-Fox et al., 2010).

fNIRS measures the quantitative regional hemodynamic change that results from a localized change in brain *activation* state relative to a *control* state. The signal processing steps required to extract useable hemodynamic changes from the optical data typically involve: (i) conditioning or smoothing the data by means of correction of linear trends (very slow fluctuations that could be of physiological origin) and removal of high frequency noise, through detrending methods, low pass and high pass filtering; (ii) conversion of attenuation into changes in hemoglobin oxygenation (Beer-Lambert law; Kocsis et al., 2006); (iii) detection, removal, and sometimes correction of movement artifacts in the signal; (iv) incorporation of an assessment of infant compliance/attention to the stimulus (typically extracted from the video recording of the session); and (v) selection of valid trials for block averaging and statistical analysis of the data with predefined objective criteria and analysis plans. See previous reviews (Lloyd-Fox et al., 2010; Tak and Ye, 2014) and recent research papers (i.e., Di Lorenzo, Pirazzoli et al, 2019) for an overview of recommended data processing and analysis approaches. Furthermore, recent developmental work has begun to address the impact of extra-cerebral factors on the HRF signal. These include investigating the use of short channels which interrogate shallow tissue layers external to the cortex, and used to extract signal changes unrelated to brain function (Gagnon et al., 2014; Yucel et al., 2015).

As described earlier, an increase in HbO_2 and a decrease in HHb concentration represent the standard adult hemodynamic change, however from the very first fNIRS infant study in 1998, it was proposed that there could be a different infant-specific pattern of hemodynamic response. Despite numerous publications since then, controversy still remains regarding the exact nature of this infant response. Further understanding of the infant hemodynamic response could come from infant physiological models of brain circulation and metabolism in accord with those proposed for adults (Banaji et al., 2008). In addition, simultaneous data acquisition with NIRS and fMRI could provide structural localization of the origin of the signal changes observed in infants. For the present, it is paramount that researchers report both HbO_2 and HHb changes in infant studies until the relative functional importance of the two chromophores and the inconsistent HHb changes observed in infants can be explored further.

Recently, researchers have begun to optimize the precision of localizing fNIRS data by using information from structural age-appropriate MRI templates. Alternatively, structural infant MRIs measured during one session have been used in combination with NIRS functional data measured during a second session, to localize responses. This technique is often used in transcranial magnetic stimulation (TMS) studies with adults with the use of MRI templates and provides a more accurate alternative to scalp anatomical landmarks (such as aligning data based on the position of the nasion, inion, and preauricular points), which can only provide a general understanding of the underlying brain regions. For developmental work, it is largely impractical to obtain structural MRI scans for each participant, which is why age-appropriate templates are an essential tool. In recent work the reliability of co-registration of individual MRI-NIRS vs MRI templates has been investigated (Lloyd-Fox et al., 2014b). NIRS channels were co-registered to MRIs from the same infants aged 4–7 months, showing reliable estimates of frontal and temporal cortical regions across this age range with the individual MRI data. Furthermore, similar results were evidenced with the use of age-appropriate MRI templates, demonstrating the utility of this approach when individual MRIs cannot be undertaken. A standardized scalp surface map of fNIRS channel locators was generated for researchers to reliably locate cortical regions for this age range (see Figure 7.3). Consideration of scalp-brain distance and its variability and relation to scalp external landmarks (i.e., the ears, nasion) for accurate localization is an important consideration (i.e., Beauchamp et al, 2011).

Several groups have also used localization techniques such as 3D virtual registration or digitizers to record the location of the NIRS channels in other age groups (Emberson et al., 2017; Tsuzuki and Dan, 2014) in combination with age-appropriate templates or age and head size matched MRI scans to more accurately localize their functional data. These approaches are being incorporated into open source software packages thereby driving forward more standardized methods for fNIRS analyses. These tools will be of great benefit to developmental researchers studying infancy and are highly recommended.

Given the potential to obtain valid data with relatively few trial repetitions, fNIRS holds great potential for the study of individual differences. It is essential, if researchers are to move forward with the use of fNIRS as a measure of individual differences, that the main factors influencing reliability in the measures are identified. In recent work, the replicability of obtaining consistent hemodynamic responses at the group and individual level across a longitudinal study (Blasi et al., 2014) was investigated. This was assessed by measuring (i) how many fNIRS channels showed significant hemodynamic responses during a functional paradigm; (ii) how similar the spatial maps of activation were; and (iii) how replicable the hemodynamic time course was across two time points. Group analyses showed a high degree of correlation in the magnitude and spatial distribution of the response and the shape of the hemodynamic response (see Figure 7.4) across the two measurements (8.5 months apart: at 4–8 and 12–16 months). Similar to fMRI research in adults (Bennett and Miller, 2010), reliability showed greater variability across individuals, but was acceptable when a specific region of interest was targeted.

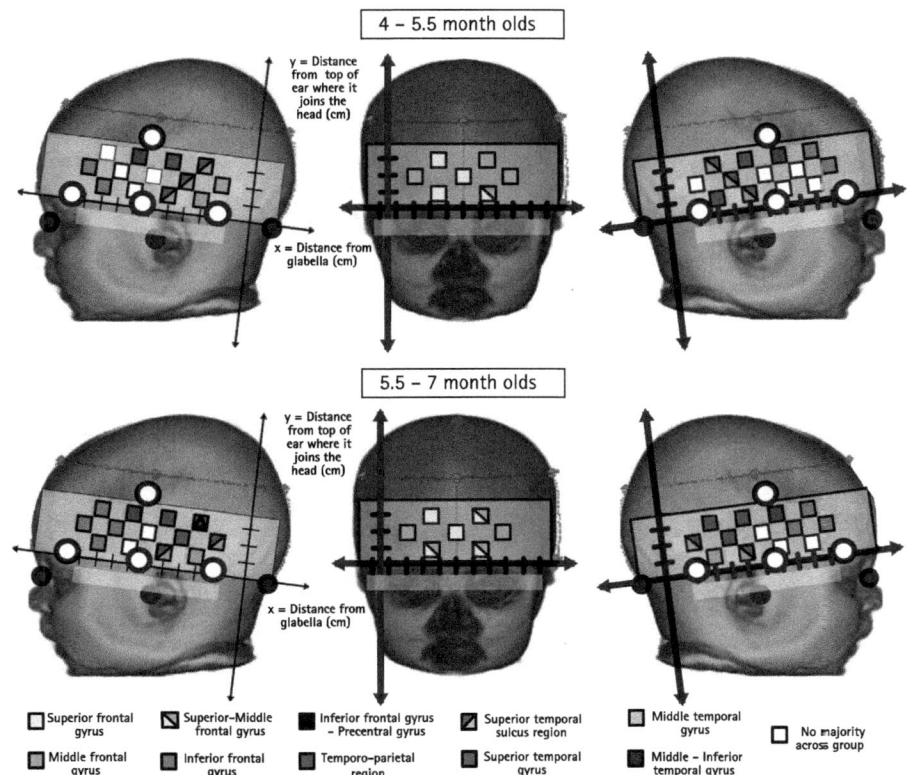

FIGURE 7.3 Reference maps for regions of interest in the frontal, temporal and parietal cortex for placement of fNIRS channels in infants of 4–5.5 months (top row) and 5.5–7 months (bottom row). The regions highlighted were identified during the atlas projections in 75–100% of the 55 infants tested. The distances given are relative to a referential axial curve between the glabella and the point at which the top of the ear joins the head. Note that for the superior temporal sulcus and temporo-parietal locators these are defined as regions as the identity across the group was split between the superior temporal-middle temporal gyri and superior temporal-postcentral gyri respectively (the atlases do not define sulci). The positions with a purple marker are closest/overlapping with the temporo-parietal junction (median distance of channel from TPJ is <2mm). The white markers indicate the position of the nasion, inion and pre-auricular points and the red dashed lines and markers identify the position of the fNIRS headband on the infant head.

Source: reproduced from Sarah Lloyd-Fox, John E. Richards, Anna Blasi, Declan G. M. Murphy, Clare E. Elwell, and Mark H. Johnson, Coregistering functional near-infrared spectroscopy with underlying cortical areas in infants. *Neurophotonics*, 1(2), 025006, Figure 8, doi:10.1117/1.NPh.1.2.025006 © The Authors, 2014.

fNIRS is, therefore, a highly suitable technique for infant studies, and its reliability at the single participant level can be improved further by adopting strategies that reduce signal variability such as accurate positioning of sensor arrays over regions of interest, regression techniques to examine residual signals at the surface of the head, improving resilience of the sensor arrays to signal artifacts, and accounting further for the changes

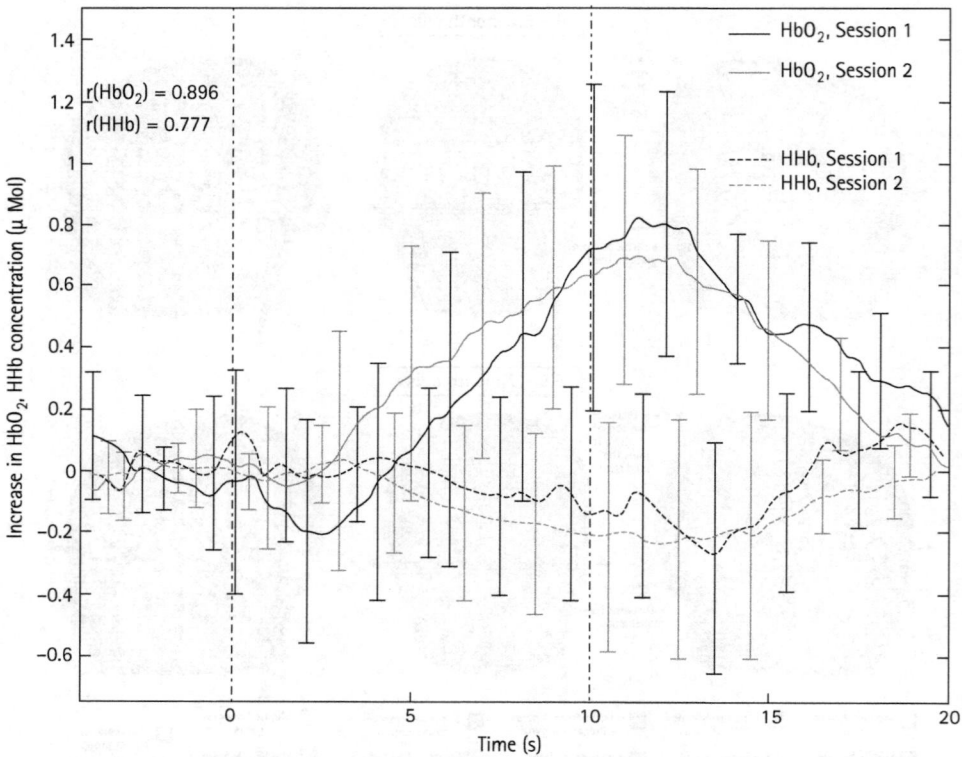

FIGURE 7.4 Test re-test study: Mean time course changes in HbO_2 (solid lines) and HHb (dotted lines) across all channels in the cortical region of interest and across all infants per session. Dash-point vertical lines indicate the start and end of the task presentation; r is the correlation coefficient between the time courses for each chromophore.

Source: adapted from Anna Blasi, Sarah Lloyd-Fox, Mark H. Johnson, and Clare Elwell. Test–retest reliability of functional near infrared spectroscopy in infants. *Neurophotonics*, 1(2) 025005, Figure 7.5, doi:10.1117/1.NPh.1.2.025005 © The Authors, 2014.

in brain morphology in the developing brain. These findings highlight the importance of using standardized and reliable data acquisition parameters.

fNIRS Research with Infants

Due to the relatively transparent properties of the neonate skin, skull, and underlying tissue, initial work with infants took place within a clinical setting investigating cerebral oxygenation in neonates. Work particularly focused on sick and/or preterm babies to study infants at risk for subsequent neurodevelopmental abnormalities (i.e., Brazy et al., 1985; Nicklin et al., 2003; Wyatt et al., 1986). Later, developmental researchers began to apply NIRS to the investigation of functional cortical activation (fNIRS): in 1998 the first study to be published on infants was undertaken by Meek and colleagues (Meek et al., 1998) in London, UK, to assess occipital responses to visual chequerboards.

Since this study, the number of published fNIRS infant studies continues to rise dramatically. Furthermore, whilst once limited to isolated sites across Japan, Europe, and the USA, the number of research laboratories that have now acquired, or are in the process of acquiring, a system for fNIRS continues to expand across the globe at a remarkable rate. The use of NIRS to study functional brain activation in infants is a rapidly increasing research area (Cristia et al., 2013; Gervain et al., 2011; Lloyd-Fox et al., 2010; Minagawa-Kawai et al., 2008). Whereas in early fNIRS studies the main aim was typically to detect the neural response to basic stimuli in primary cortical areas, such as response to acoustic tones in the auditory cortex (Sakatani et al., 1999) or stroboscopic flashing light in the visual cortex (Hoshi et al., 2000; Zaramella et al., 2001), as with EEG, researchers continue to address ever more complex questions about brain and cognition such as the processing of object permanence and identity (i.e., Baird et al., 2002; Wilcox et al., 2010), social communication (i.e., Grossmann, 2013; Minagawa-Kawai et al., 2009), language and voice perception (i.e., Gervain et al., 2008; Lloyd-Fox et al., 2012), human actions (i.e., Farroni et al., 2013; Grossmann et al., 2013; Ichikawa et al., 2010), attention, learning and memory (i.e., Benavides-Varela et al., 2011; Emberson et al., 2015), and social and affective touch (Jönsson et al., 2018; Kida and Shinohara, 2013; Miguel et al., 2017; Pirazzoli et al., 2018). Furthermore, the technique has been successfully applied to the study of functional connectivity in infancy (i.e., Homae et al., 2010; Keehn et al., 2013; Watanabe et al., 2010).

Several research labs have begun to use fNIRS in more ecologically valid scenarios (Hakuno et al., 2018; Lloyd-Fox, Széplaki-Köllőd et al., 2015; Shimada and Hiraki, 2006; Urakawa et al., 2015), some of which highlighted the need for carefully constructed study design and interpretation. For example, researchers have shown live interactive stimuli (such as an actor demonstrating goal directed actions to infants) produced, altered, or enhanced responses in comparison with televized versions of the same stimuli (Shimada and Hiraki, 2006). Others have highlighted the potential for multi-participant data collection. A recent study acquiring simultaneous fNIRS signals from two infants, both wearing fNIRS headgear within the same session, used this setup to investigate responses within a communicative context while both infants interacted with an adult (Lloyd-Fox, Széplaki-Köllőd et al., 2015). In this way they could investigate functional activation to social cues directed both to the target infant, and activation while the target infant observed interaction with another infant. This approach has also been adopted for two-person "dual-brain hyper-scanning" to study social interactions and interactive specialization in adults (Hirsch et al., 2017; Liu et al., 2016). With advances in wireless technology now on the horizon there is great potential for wider application across home and community settings, and therefore the prospect of reaching less socially mobile participants.

A recent shift in the use of fNIRS has been towards the study of the infant brain at the individual level (Lloyd-Fox, Wu et al., 2015; Ravicz et al., 2015; Wilcox et al., 2014). This advance in the application of fNIRS is important as the identification of significant

hemodynamic responses within individuals allows us to look at the relationship between brain function and other variables such as age, demographics, and behavioral
data. For example, in a recent study a relationship between individual infant's patterns of
brain activation to the perception of others' actions and the development of the infant's
own fine motor skills was found (Lloyd-Fox, Wu et al., 2015) as measured by the *Mullen
Scales of Early Learning* (Mullen, 1995, a behavioral assessment of cognitive development). Further, combining neural markers with behavioral assessments allowed the
relative contribution of other factors such as age, gender, and general cognitive development to be assessed.

One way to provide further information about the mechanistic processes behind
the recordings is to provide concurrent information by combining the advantages
of different neuroimaging methods together. This practice has been successfully
implemented in research with adult participants providing simultaneous measurements
from two techniques, e.g., combined fNIRS and EEG (Moosmann et al., 2003; Wallois
et al., 2012), fNIRS and MRI (Steinbrink et al., 2006), and fMRI and EEG (Dale and
Halgren, 2001; Eichele et al., 2005). Recently, researchers have also implemented multimodal measures for the study of infants (Telkemeyer et al., 2009; Verriotis et al., 2016).
In the first of these (Telkemeyer et al., 2009), newborn infants (3.32 ± 1.27 days) were
presented with tonal auditory stimuli that varied in temporal structure and length
(12, 25, 160, and 300 ms). Analysis of the auditory evoked potentials (AEPs) revealed
a focused response in the frontal and central electrodes, however, the EEG measures
did not reveal any significant differences between the four conditions (12, 25, 160, and
300 ms segments). In contrast, the hemodynamic measures (fNIRS) revealed a significant effect for the 25 ms condition demonstrating that newborns are sensitive to
speech-like structures from the first days of life. Within a hypothesis driven context,
this study highlights the importance of multimodal investigation, as the concurrent
measures allowed the significant EEG effect to be investigated in further detail with
the fNIRS. Furthermore, this multimodality approach has the potential to improve
neurodevelopmental research through the study and/or clinical neuro-monitoring of
infants at risk of compromised development due to, for example, a developmental disorder or in preterm and term infants with acute brain injury (Toet and Lemmers, 2009).

Studying Developmental Disorders with fNIRS

Developmental functional and structural imaging studies increasingly help to define the typically developing brain and assist in generating hypotheses about
atypical function and target brain structures for research in developmental psychopathology (Bush et al., 2005). Critically, this knowledge has allowed a shift in the use
of neuroimaging towards the study of the developing brain in situations where this development may be compromised in some way. These may include the impact of acute
brain injury in early infancy, chronic or long-term conditions (such as cerebral palsy
or epilepsy), the impact of environmental factors (such as poverty and nutrition), and

genetically related conditions or developmental disorders (such as Williams syndrome, Down's syndrome, autism spectrum disorders, or attention deficit hyperactivity disorder (ADHD)). In particular, objective measures of brain function within the first few months and years of life could be used to assess how the timing and nature of developmental disorders impact on cognitive development, and to inform and evaluate intervention strategies. As outlined previously, several advantages of fNIRS (i.e., tolerance of movement, use in ecologically valid settings, easy setup, and administration) make it a highly suitable method for the study of individuals with psychiatric or developmental disorders. Indeed, the number of fNIRS studies that have been directed towards psychiatric research questions has increased significantly in recent years (Ehlis et al., 2014). In adults and children these include studies on schizophrenia, eating disorders, affective syndromes, personality disorders, anxiety, and developmental disorders such as ADHD (Ehlis et al., 2014). For example, studies in these areas have found altered hemodynamic responses (during tasks such as the Stroop test, Go/No-Go or working memory paradigms) in the frontal cortex (Jourdan Moser et al., 2009; Negoro et al., 2010; Schecklmann et al., 2008; Xiao et al., 2012). Clinical research into conditions such as epilepsy in infants and young children has also made substantial progress recently (i.e., Roche-Labarbe et al., 2008).

fNIRS has also been increasingly used in the study of infants and children who may be at risk of compromised development due to increased familial likelihood of developing disorders, prematurity, or who may have suffered brain injury at birth. This form of research is particularly important in prospective longitudinal studies of infants at risk as it enables the identification of biomarkers of compromised development prior to the typical onset of observable behavioral markers, which usually become apparent in the second year of life or later. Whilst behavioral measures have been unable to distinguish between infants with low- and high-familial likelihood of developing autism spectrum disorders (ASD), several recent EEG and fNIRS studies have identified differences in brain function in younger infants (Braukmann et al., 2018; Elsabbagh et al., 2012, 2009; Fox et al., 2013; Guiraud et al., 2011; Lloyd-Fox et al., 2018, 2013; McCleery et al., 2009). For example, Lloyd-Fox and colleagues (Lloyd-Fox et al., 2018) used a social visual and auditory paradigm to investigate functional brain responses in infants with increased familial likelihood of developing ASD. Infants were presented with social human movements (i.e., "Peek-a-boo") and non-human static images (i.e., cars, helicopters, trains) while listening to vocal (i.e., laughter, crying, yawning) and non-vocal sounds (i.e., water running, rattles, and bells). This study found that infants who went on to develop ASD at 3 years of age showed evidence of diminished activation in regions of the inferior frontal and temporal lobes (including regions of the superior temporal—temporoparietal cortex), in response to both the visual and auditory social cues when they were just 4–6 months of age (see Figure 7.5).

These findings are in line with EEG and fMRI research in older children finding atypical patterns of brain activation to the perception of social human actions (Pelphrey and Carter, 2008) and auditory stimuli (Čeponiené et al., 2003; Eyler et al.,

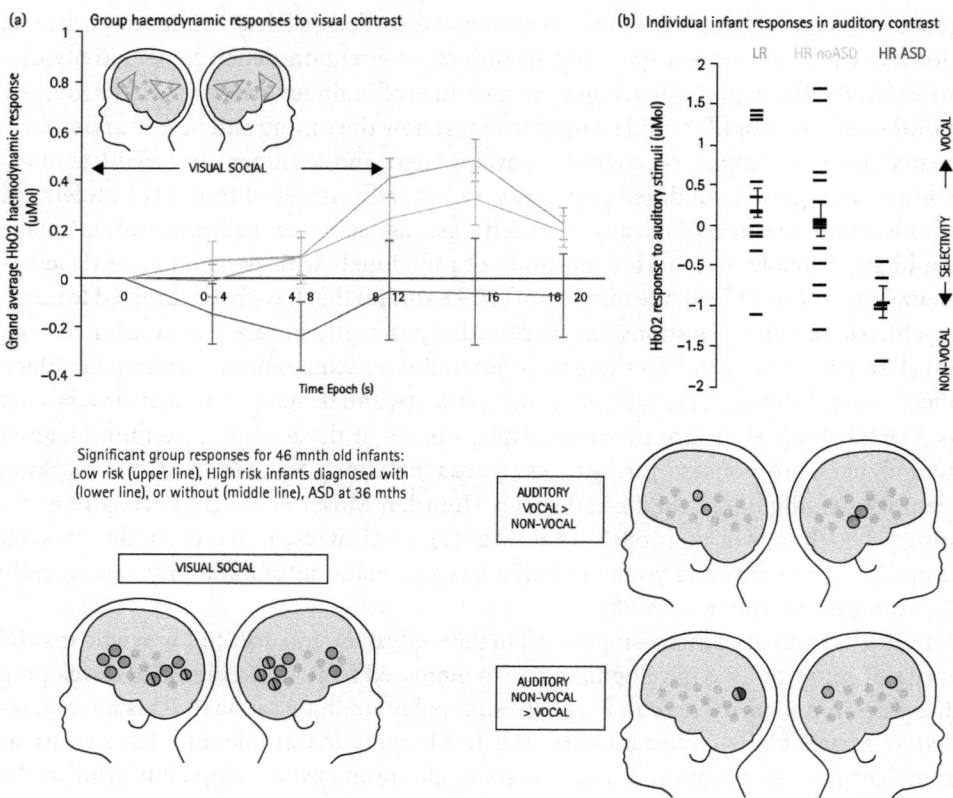

FIGURE 7.5 Significant fNIRS responses to social and non-social stimuli in 4–6 month old infants. Infants with a low familial likelihood of developing ASD are shown as the upper line, those with increased familial likelihood but who did not go on to develop ASD are shown as the middle line, and those infants with increased familial likelihood and who went on to develop ASD at three years of age are shown as the lower line.

Sources: Part A: reproduced from S. Lloyd-Fox, A. Blasi, G. Pasco, T. Gliga, E. J. H. Jones, D. G. M. Murphy, C. E. Elwell, T. Charman, M. H. Johnson, and the BASIS Team. Cortical responses before 6 months of life associate with later autism. *European Journal of Neuroscience*, 47(6), 736–49, Figure 7.2, doi:10.1111/ejn.13757, © 2017 The Authors. European Journal of Neuroscience published by Federation of European Neuroscience Societies and John Wiley & Sons Ltd. Licensed under CC BY.

Part B: reproduced from S. Lloyd-Fox, A. Blasi, G. Pasco, T. Gliga, E. J. H. Jones, D. G. M. Murphy, C. E. Elwell, T. Charman, M. H. Johnson, and the BASIS Team, Cortical responses before 6 months of life associate with later autism. *European Journal of Neuroscience*, 47(6), 736–49, Figure 7.4, doi:10.1111/ejn.13757, © 2017 The Authors. European Journal of Neuroscience published by Federation of European Neuroscience Societies and John Wiley & Sons Ltd. Licensed under CC BY.

2012). Furthermore, fNIRS has been used to evidence altered functional connectivity during the first year of life in infants at risk for ASD (Keehn et al., 2013), in infants with Down's Syndrome (Imai et al., 2014), and in early childhood with chidren with ASD (Zhu et al., 2014). These results are consistent with the view that atypical functioning of the brain may be manifest from the first few months of life. Yet, it is important to be aware of a number of conceptual issues that come with the study of atypical development (Vanderwert and Nelson, 2014). Firstly, there is a substantial need for careful phenotyping of the population in question and without this it may prove very difficult to identify why one group differs from another. Secondly, many of these studies are conducted with small sample sizes (Imai et al., 2014; Lloyd-Fox et al., 2018). Therefore, there is a great need for larger-scale longitudinal research to be undertaken, particularly as there can be such a high degree of individual variability in some developmental disorders. Thirdly, most studies that have investigated developmental disorders with fNIRS thus far have simply reported group differences in the hemodynamic response. Although this is important, such findings do not address the deeper issue of why such differences have arisen, whether there is a difference in neural circuitry, whether particular effects differ across disorders, or whether they are present at birth or arise through infancy and childhood (Vanderwert and Nelson, 2014).

fNIRS Outside of the Research Lab

Evidence of the flexibility of this technology is provided by the recent rapid expansion of fNIRS to a global health platform, with fNIRS infant and early childhood studies now under way in several low- and middle-income countries. The application of fNIRS in such contexts has helped expand the investigation of the effect of complex and interacting socio-economic, psychosocial, and environmental adversities on brain development and children's developmental outcomes (Blasi et al., 2019). To undertake this research it is essential that fNIRS can operate successfully in the field (i.e., rural/urban community/clinical settings) and not just in well-resourced research labs (Katus et al., 2019). In 2013, we were the first research team to do this (Lloyd-Fox et al., 2013) using fNIRS to study infant brain function in a field station in a rural village, Keneba, part of the Medical Research Council Unit The Gambia, at the London School of Hygiene and Tropical Medicine (MRCG at LSHTM). This initiative demonstrated the potential for fNIRS research in resource-poor communities as the technology can be packed and transported, it is easy and quick to set up; with basic training of research assistants, data can be readily collected and quality checks can be performed almost immediately (Blasi et al., 2019). In addition to ongoing work in The Gambia (Lloyd-Fox et al., 2014a; 2017; 2019), fNIRS global health research has expanded to other sites across Africa (Jasińska and Guei, 2018) and Asia (Perdue et al., 2019; Wijeakumar et al., 2019). Thus, this recent research has demonstrated that fNIRS can be used to study development in a range of settings.

CONCLUSIONS

I anticipate that further refinement and application of fNIRS over the next 10 years will contribute significantly to the advancement of our understanding of the developing brain.

Over the past decade major progress has been made in fNIRS: the development of multiple source-detector distance arrays to investigate depth discrimination of the hemodynamic response; an ever-increasing number of channels allowing for a wider coverage of the head; availability of sophisticated open-source software for data analysis; and advances in the design of the headgear providing improved quality of the optical signals (for review see Lloyd-Fox et al., 2010). A further recent advance in data acquisition has been a return of interest in measuring cytochrome c oxidase (oxCCO) with multi-wavelength fNIRS (NIRS was first developed for measuring oxCCO but was largely superseded by functional studies of HbO_2 and HHb). The advantage of measuring oxCCO is that it allows us to understand cellular oxygen metabolism (rather than hemoglobin which is only going to tell us about the carriage of oxygen in the blood), and therefore recent pioneering studies with infants (Siddiqui et al., 2017) hold great promise for future work.

Future innovations on the horizon include (i) the development of wearable, mobile, fNIRS systems that will enable research to be far more ecologically valid with testing done at clinics or even in the home, and the capacity for recordings while the infant or child is fully mobile (i.e., crawling, walking); (ii) combined EEG-fNIRS systems for enhanced data collection of both temporal and spatial information; and (iii) combined commercial multi-wavelength systems for enhanced data collection of both hemodynamic and metabolic markers of activation (HbO_2, HHb, and oxCCO). As for a blue skies concept? I would hope that technological innovation will allow for the development of a truly low-cost, wearable, and mobile fNIRS system, with solar or battery powered capabilities, to enable research to reach a wider spectrum of participants across the globe. As this field of research continues to rapidly expand, it is paramount that the interpretation of, and subsequent claims arising from, fNIRS data are informed by a clear understanding of physiology of the neural and vascular response as well as the complexities of fNIRS measurements. There have been remarkable accomplishments in infant fNIRS research over the last decade encompassing instrument design, probe development, protocol optimization, and data analysis and interpretation. These have set a firm foundation for the continued role of fNIRS as an essential tool for advancing developmental neuroscience research in the future.

References

Aslin, R.N. (2012). Questioning the questions that have been asked about the infant brain using near-infrared spectroscopy. *Cognitive Neuropsychology*, 29, 7–33. doi:10.1080/02643294.2012.654773.

Baird, A.A., Kagan, J., Gaudette, T., Walz, K.A., Hershlag, N., and Boas, D.A. (2002). Frontal lobe activation during object permanence: Data from near-infrared spectroscopy. *NeuroImage*, 16, 1120–1126. doi:10.1006/nimg.2002.1170.

Banaji, M., Mallet, A., Elwell, C.E., Nicholls, P., and Cooper, C.E. (2008). A model of brain circulation and metabolism: NIRS signal changes during physiological challenges. *PLOS Computer Biology*, 4, e1000212. doi:10.1371/journal.pcbi.1000212.

Beauchamp, M.S., Beurlot, M.R., Fava, E., Nath, A.R., Parikh, N.A., Saad, Z.S., Bortfeld, H., and Oghalai, J.S. (2011). The developmental trajectory of brain-scalp distance from birth through childhood: Implications for functional neuroimaging. *PLoS ONE*, *6*, e24981. doi:10.1371/journal.pone.0024981.

Benavides-Varela, S., Gómez, D.M., Macagno, F., Bion, R.A.H., Peretz, I., and Mehler, J. (2011). Memory in the neonate brain. *PLoS ONE, 6*, e27497. doi:10.1371/journal.pone.0027497.

Bennett, C.M. and Miller, M.B. (2010). How reliable are the results from functional magnetic resonance imaging? *Annals of the New York Academy of Sciences*, *1191*, 133–155. doi:10.1111/j.1749-6632.2010.05446.x.

Blasi, A., Lloyd-Fox, S., Johnson, M.H., and Elwell, C. (2014). Test-retest reliability of functional near infrared spectroscopy in infants. *Neurophotonics*, *1*, 025005. doi:10.1117/1.NPh.1.2.025005.

Blasi, A., Lloyd-Fox, S., Katus, L., and Elwell, C.E. (2019). fNIRS for tracking brain development in the context of global health projects. *Photonics*, *6*(3), 89. doi:10.3390/photonics6030089.

Blasi, A., Mercure, E., Lloyd-Fox, S., Thomson, A., Brammer, M., Sauter, D., Deeley, Q., Barker, G.J., Renvall, V., Deoni, S., Gasston, D., Williams, S.C.R., Johnson, M.H., Simmons, A., and Murphy, D.G.M. (2011). Early specialization for voice and emotion processing in the infant brain. *Current Biology*, *21*, 1220–1224. doi:10.1016/j.cub.2011.06.009.

Boas, D.A., Elwell, C.E., Ferrari, M., and Taga, G. (2014). Twenty years of functional near-infrared spectroscopy: Introduction for the special issue. *NeuroImage*, *85*, 1–5. doi:10.1016/j.neuroimage.2013.11.033.

Braukmann, R., Lloyd-Fox, S., Blasi, A., Johnson, M.H., Bekkering, H., Buitelaar, J.K., and Hunnius, S. (2018). Diminished socially selective neural processing in 5-month-old infants at high familial risk of autism. *European Journal of Neuroscience*, *47*, 720–728. doi:10.1111/ejn.13751.

Brazy, J.E., Lewis, D.V., Mitnick, M.H., and Jöbsis-Vander Vliet, F.F. (1985). Monitoring of cerebral oxygenation in the intensive care nursery. *Advanced Experimental Medical Biology*, *131*, 843–848.

Brigadoi, S., Phan, P., Highton, D., Powell, S., Cooper, R.J., Hebden, J., Smith, M., Tachtsidis, I., Elwell, C.E., and Gibson, A.P. (2017). Image reconstruction of oxidized cerebral cytochrome C oxidase changes from broadband near-infrared spectroscopy data. *Neurophotonics 4*(2), 021105. doi:10.1117/1.NPh.4.2.021105.

Bush, G., Valera, E.M., and Seidman, L.J. (2005). Functional neuroimaging of attention-deficit/hyperactivity disorder: A review and suggested future directions. *Biological Psychiatry*, *57*, 1273–1284. doi:10.1016/j.biopsych.2005.01.034.

Carlsson, J., Lagercrantz, H., Olson, L., Printz, G., and Bartocci, M. (2008). Activation of the right fronto-temporal cortex during maternal facial recognition in young infants, *Acta Paediatra*, *97*, 1221–1225. doi:10.1111/j.1651-2227.2008.00886.x.

Čeponienė, R., Lepistö, T., Shestakova, A., Vanhala, R., Alku, P., Näätänen, R., and Yaguchi, K. (2003). Speech–sound-selective auditory impairment in children with autism: They can perceive but do not attend. *Proceedings Of the National Academy of Sciences U. S. A.*, *100*, 5567.

Chance, B., Zhuang, Z., UnAh, C., Alter, C., and Lipton, L, (1993). Cognition-activated low-frequency modulation of light absorption in human brain. *Proceedings of the National Academy of Sciences U. S. A.*, *90*, 3770–3774.

Cheour, M., Imada, T., Taulu, S., Ahonen, A., Salonen, J., and Kuhl, P. (2004). Magnetoencephalography is feasible for infant assessment of auditory discrimination. *Experimental Neurology*, *190*, 44–51. doi:10.1016/j.expneurol.2004.06.030.

Chugani, H.T., Phelps, M.E., and Mazziotta, J.C. (1987). Positron emission tomography study of human brain functional development. *Annals Neurology*, *22*, 487–497. doi:10.1002/ana.410220408.

Cooper, R.J., Magee, E., Everdell, N., Magazov, S., Varela, M., Airantzis, D., Gibson, A.P., and Hebden, J.C. (2014). MONSTIR II: A 32-channel, multispectral, time-resolved optical tomography system for neonatal brain imaging. *Review of Scientific Instruments*, 85, 053105. doi:10.1063/1.4875593.

Cristia, A., Dupoux, E., Hakuno, Y., Lloyd-Fox, S., Schuetze, M., Kivits, J., Bergvelt, T., van Gelder, M., Filippin, L., Charron, S., and Minagawa-Kawai, Y. (2013). An online database of infant functional near infrared spectroscopy studies: A community-augmented systematic review. *PLoS ONE*, 8(3), e58906. doi:10.1371/journal.pone.0058906.

Dale, A.M. and Halgren, E. (2001). Spatiotemporal mapping of brain activity by integration of multiple imaging modalities. *Current Opinion in Neurobiology*, 11, 202–208.

Dehaene-Lambertz, G., Dehaene, S., and Hertz-Pannier, L. (2002). Functional neuroimaging of speech perception in infants. *Science*, 298, 2013–2015. doi:10.1126/science.1077066.

Di Lorenzo, R., Pirazzoli, L., Blasi, A., Bulgarelli, C., Hakuno, Y., Minagawa, Y., and Brigadoi, S. (2019). Recommendations for motion correction of infant fNIRS data applicable to multiple data sets and acquisition systems. *Neuroimage*, 200, 511–527. doi:10.1016/j.neuroimage.2019.06.056.

Draganova, R., Eswaran, H., Murphy, P., Huotilainen, M., Lowery, C., and Preissl, H. (2005). Sound frequency change detection in fetuses and newborns, a magnetoencephalographic study. *NeuroImage*, 28, 354–361. doi:10.1016/j.neuroimage.2005.06.011.

Duncan, A., Meek, J.H., Clemence, M., Elwell, C.E., Tyszczuk, L., Cope, M., and Delpy, D. (1995). Optical pathlength measurements on adult head, calf and forearm and the head of the newborn infant using phase resolved optical spectroscopy. *Physics in Medicine & Biology*, 40, 295.

Eggebrecht, A.T., Ferradal, S.L., Robichaux-Viehoever, A., Hassanpour, M.S., Dehghani, H., Snyder, A.Z., Hershey, T., and Culver, J.P. (2014). Mapping distributed brain function and networks with diffuse optical tomography. *Nature Photonics*, 8, 448–454. doi:10.1038/nphoton.2014.107.

Ehlis, A.-C., Schneider, S., Dresler, T., and Fallgatter, A.J. (2014). Application of functional near-infrared spectroscopy in psychiatry. *NeuroImage*, 85, 478–488. doi:10.1016/j.neuroimage.2013.03.067.

Eichele, T., Specht, K., Moosmann, M., Jongsma, M.L.A., Quiroga, R.Q., Nordby, H., and Hugdahl, K. (2005). Assessing the spatiotemporal evolution of neuronal activation with single-trial event-related potentials and functional MRI. *Proceedings of the National Academy of Sciences U. S. A.*, 102, 17798–17803. doi:10.1073/pnas.0505508102.

Elsabbagh, M., Mercure, E., Hudry, K., Chandler, S., Pasco, G., Charman, T., Pickles, A., Baron-Cohen, S., Bolton, P., and Johnson, M.H. (2012). Infant neural sensitivity to dynamic eye gaze is associated with later emerging autism. *Current Biology*, 22, 338–342. doi:10.1016/j.cub.2011.12.056.

Elsabbagh, M., Volein, A., Csibra, G., Holmboe, K., Garwood, H., Tucker, L., Krljes, S., Baroncohen, S., Bolton, P., and Charman, T. (2009). Neural correlates of eye gaze processing in the infant broader autism phenotype. *Biological Psychiatry*, 65, 31–8. doi:10.1016/j.biopsych.2008.09.034.

Emberson, L.L., Cannon, G., Palmeri, H., Richards, J.E., and Aslin, R.N. (2017). Using fNIRS to examine occipital and temporal responses to stimulus repetition in young infants: Evidence of selective frontal cortex involvement. *Developmental Cognitive Neuroscience*, 23, 26–38. doi:10.1016/j.dcn.2016.11.002.

Emberson, L.L., Richards, J.E., and Aslin, R.N. (2015). Top-down modulation in the infant brain: Learning-induced expectations rapidly affect the sensory cortex at 6 months. *Proceedings of the National Academy Sciences U. S. A.*, 112, 9585–9590. doi:10.1073/pnas.1510343112.

Eyler, L.T., Pierce, K., and Courchesne, E. (2012). A failure of left temporal cortex to specialize for language is an early emerging and fundamental property of autism. *Brain*, *135*, 949–960. doi:10.1093/brain/awr364.

Farroni, T., Chiarelli, A.M., Lloyd-Fox, S., Massaccesi, S., Merla, A., Di Gangi, V., Mattarello, T., Faraguna, D., and Johnson, M.H. (2013). Infant cortex responds to other humans from shortly after birth. *Science Reports*, *3*, 2851. doi:10.1038/srep02851.

Ferrari, M. and Quaresima, V. (2012). A brief review on the history of human functional near-infrared spectroscopy (fNIRS) development and fields of application. *NeuroImage*, *63*, 921–935. doi:10.1016/j.neuroimage.2012.03.049.

Fox, S.E., Wagner, J.B., Shrock, C.L., Tager-Flusberg, H., and Nelson, C.A. (2013). Neural processing of facial identity and emotion in infants at high-risk for autism spectrum disorders. *Frontiers in Human Neuroscience*, *7*(89), 1–18. doi:10.3389/fnhum.2013.00089.

Fukui, Y., Ajichi, Y., and Okada, E. (2003). Monte Carlo prediction of near-infrared light propagation in realistic adult and neonatal head models. *Applied Optics*, *42*, 2881–2887.

Gagnon, L., Yücel, M.A., Boas, D.A., and Cooper, R.J. (2014). Further improvement in reducing superficial contamination in NIRS using double short separation measurements. *Neuroimage*, *85*, 127–135. doi:10.1016/j.neuroimage.2013.01.073.

Gervain, J., Macagno, F., Cogoi, S., Peña, M., and Mehler, J. (2008). The neonate brain detects speech structure. *Proceedings of the National Academy of Sciences*, *105*, 14222–14227. doi:10.1073/pnas.0806530105.

Gervain, J., Mehler, J., Werker, J.F., Nelson, C.A., Csibra, G., Lloyd-Fox, S., Shukla, M., and Aslin, R.A. (2011). Near-infrared spectroscopy: A report from the McDonnell infant methodology consortium. *Developmental Cognitive Neuroscience*, *1*, 22–46.

Grossmann, T. (2013). The role of medial prefrontal cortex in early social cognition. *Frontiers in Human Neuroscience*, *7*(340), 1–6. doi:10.3389/fnhum.2013.00340.

Grossmann, T., Cross, E.S., Ticini, L.F., and Daum, M.M. (2013). Action observation in the infant brain: The role of body form and motion. *Social Neuroscience*, *8*, 22–30. doi:10.1080/17470919.2012.696077.

Guiraud, J.A., Kushnerenko, E., Tomalski, P., Davies, K., Ribeiro, H., and Johnson, M.H. (2011). Differential habituation to repeated sounds in infants at high risk for autism. *NeuroReport*, *22*(16), 845–849. doi:10.1097/WNR.0b013e32834cobec.

Hakuno, Y., Pirazzoli, L., Blasi, A., Johnson, M.H., and Lloyd-Fox, S. (2018). Optical imaging during toddlerhood: Brain responses during naturalistic social interactions. *Neurophotonics*, *5*, 011020. doi:10.1117/1.NPh.5.1.011020.

Hari, R. and Salmelin, R. (2012). Magnetoencephalography: From SQUIDs to neuroscience. *NeuroImage*, *61*, 386–396. doi:10.1016/j.neuroimage.2011.11.074.

Hebden, J.C., Gibson, A., Austin, T., Yusof, R.M., Everdell, N., Delpy, D.T., Arridge, S.R., Meek, J.H., and Wyatt, J.S. (2004). Imaging changes in blood volume and oxygenation in the newborn infant brain using three-dimensional optical tomography. *Physics in Medicine & Biology*, *49*, 1117–1130.

Heekeren, H.R., Kohl, M., Obrig, H., Wenzel, R., von Pannwitz, W., Matcher, S.J., Dirnagl, U., Cooper, C.E., and Villringer, A. (1999). Noninvasive assessment of changes in cytochrome-c oxidase oxidation in human subjects during visual stimulation. *Journal of Cerebral Blood Flow Metabolism*, *19*, 592–603. doi:10.1097/00004647-199906000-00002.

Hirsch, J., Zhang, X., Noah, J.A., and Ono, Y. (2017). Frontal temporal and parietal systems synchronize within and across brains during live eye-to-eye contact. *NeuroImage*, *157*, 314–330. doi:10.1016/j.neuroimage.2017.06.018.

Homae, F., Watanabe, H., Otobe, T., Nakano, T., Go, T., Konishi, Y., and Taga, G. (2010). Development of global cortical networks in early infancy. *Journal of Neuroscience*, *30*, 4877–4882. doi:10.1523/JNEUROSCI.5618-09.2010.

Hoshi, Y., Kohri, S., Matsumoto, Y., Cho, K., Matsuda, T., Okajima, S., and Fujimoto, S. (2000). Hemodynamic responses to photic stimulation in neonates. *Pediatric Neurology*, *23*, 323–327.

Hoshi, Y. and Tamura, M. (1993). Detection of dynamic changes in cerebral oxygenation coupled to neuronal function during mental work in man. *Neuroscience Letters*, *150*, 5–8. doi:10.1016/0304-3940(93)90094-2.

Huotilainen, M., Kujala, A., Hotakainen, M., Shestakova, A., Kushnerenko, E., Parkkonen, L., Fellman, V., and Näätänen, R. (2003). Auditory magnetic responses of healthy newborns. *Neuroreport*, *14*, 1871–1875. doi:10.1097/01.wnr.0000090589.35425.10.

Ichikawa, H., Kanazawa, S., Yamaguchi, M.K., and Kakigi, R. (2010). Infant brain activity while viewing facial movement of point-light displays as measured by near-infrared spectroscopy (NIRS). *Neuroscience Letters*, *482*, 90–94. doi:10.1016/j.neulet.2010.06.086.

Ijichi, S., Kusaka, T., Isobe, K., Okubo, K., Kawada, K., Namba, M., Okada, H., Nishida, T., Imai, T., and Itoh, S. (2005). Developmental changes of optical properties in neonates determined by near-infrared time-resolved spectroscopy. *Pediatric Research*, *58*, 568–573. doi:10.1203/01.PDR.0000175638.98041.0E.

Imada, T., Zhang, Y., Cheour, M., Taulu, S., Ahonen, A., and Kuhl, P.K. (2006). Infant speech perception activates Broca's area: A developmental magnetoencephalography study. *Neuroreport*, *17*, 957–962. doi:10.1097/01.wnr.0000223387.51704.89.

Imai, M., Watanabe, H., Yasui, K., Kimura, Y., Shitara, Y., Tsuchida, S., Takahashi, N., and Taga, G. (2014). Functional connectivity of the cortex of term and preterm infants and infants with Down's syndrome. *NeuroImage*, *85*, 272–278. doi:10.1016/j.neuroimage.2013.04.080.

Issard, C. and Gervain, J. (2018). Variability of the haemodynamic response in infants: Influence of experimental design and stimulus complexity. *Developmental Cognitive Neuroscience*, *33*, 182–193. doi:10.1016/j.dcn.2018.01.009.

Jasińska, K.K. and Guei, S. (2018). Neuroimaging field methods using functional near infrared spectroscopy (NIRS) neuroimaging to study global child development: Rural Sub-Saharan Africa. *Journal of Visualized Experiments*, *132*, 57165. doi:10.3791/57165.

Jöbsis-vanderVliet, F.F. (1999). Discovery of the near-infrared window into the body and the early development of near-infrared spectroscopy. *Journal of Biomedical Optics*, *4*, 392.

Jönsson, E.H., Kotilahti, K., Heiskala, J., Wasling, H.B., Olausson, H., Croy, I., Mustaniemi, H., Hiltunen, P., Tuulari, J.J., Scheinin, N.M., Karlsson, L., Karlsson, H., and Nissilä, I. (2018). Affective and non-affective touch evoke differential brain responses in 2-month-old infants. *NeuroImage*, *169*, 162–171. doi:10.1016/j.neuroimage.2017.12.024.

Jourdan Moser, S., Cutini, S., Weber, P., and Schroeter, M.L. (2009). Right prefrontal brain activation due to Stroop interference is altered in attention-deficit hyperactivity disorder—A functional near-infrared spectroscopy study. *Psychiatry Research*, *173*, 190–195. doi:10.1016/j.pscychresns.2008.10.003.

Kato, T., Kamei, A., Takashima, S., and Ozaki, T. (1993). Human visual cortical function during photic stimulation monitoring by means of near-infrared spectroscopy. *Journal of Cerebral Blood Flow Metabolism*, *13*, 516–520. doi:10.1038/jcbfm.1993.66.

Katus, L., Hayes, N. K., Mason, L., Blasi, A., McCann, S., Darboe, M. K., de Haan, M., Moore, S. E., Lloyd-Fox, S., and Elwell, C. E. (2019). Implementing neuroimaging and eye tracking methods to assess neurocognitive development of young infants in low- and middle-income countries. *Gates Open Research*, *3*, 1113. doi:10.12688/gatesopenres.12951.2.

Keehn, B., Wagner, J.B., Tager-Flusberg, H., and Nelson, C.A. (2013). Functional connectivity in the first year of life in infants at-risk for autism: A preliminary near-infrared spectroscopy study. *Frontiers in Human Neuroscience, 7*, 444. doi:10.3389/fnhum.2013.00444.

Kida, T. and Shinohara, K. (2013). Gentle touch activates the prefrontal cortex in infancy: An NIRS study. *Neuroscience Letters, 541*, 63–66. doi:10.1016/j.neulet.2013.01.048

Kocsis, L., Herman, P., and Eke, A. (2006). The modified Beer–Lambert law revisited. *Physics in Medicine & Biology, 51*, N91.

Kolyva, C., Tachtsidis, I., Ghosh, A., Moroz, T., Cooper, C.E., Smith, M., and Elwell, C.E. (2012). Systematic investigation of changes in oxidized cerebral cytochrome c oxidase concentration during frontal lobe activation in healthy adults. *Biomedical Optics Express, 3*, 2550–2566. doi:10.1364/BOE.3.002550.

Kotsoni, E., Byrd, D., and Casey, B.J. (2006). Special considerations for functional magnetic resonance imaging of pediatric populations. *Journal of Magnetic Resonance Imaging, 23*, 877–886. doi:10.1002/jmri.20578.

Krekelberg, B., Boynton, G.M., and van Wezel, R.J. (2006). Adaptation: From single cells to BOLD signals. *Trends in Neuroscience, 29*, 250–256.

Leroux, G.I., Lubin, A. l, Houdé, O., and Lanoë, C.I., (2013). How to best train children and adolescents for fMRI? Meta-analysis of the training methods in developmental neuroimaging. *Neuroeducation, 2*, 44–70.

Liu, N., Mok, C., Witt, E.E., Pradhan, A.H., Chen, J.E., and Reiss, A.L. (2016). NIRS-based hyperscanning reveals inter-brain neural synchronization during cooperative jenga game with face-to-face communication. *Frontiers in Human Neuroscience, 10*, 82. doi:10.3389/fnhum.2016.00082.

Lloyd-Fox, S., Begus, K., Halliday, D., Pirazzoli, L., Blasi, A., Papademetriou, M., Darboe, M.K., Prentice, A.M., Johnson, M.H., Moore, S.E., and Elwell, C.E. (2017). Cortical specialisation to social stimuli from the first days to the second year of life: A rural Gambian cohort. *Developmental Cognitive Neuroscience, 25*, 92–104. doi:10.1016/j.dcn.2016.11.005.

Lloyd-Fox, S., Blasi, A., and Elwell, C.E. (2010). Illuminating the developing brain: The past, present and future of functional near infrared spectroscopy. *Neuroscience and Biobehavioral Reviews, 34*, 269–284. doi:10.1016/j.neubiorev.2009.07.008.

Lloyd-Fox, S., Blasi, A., Elwell, C.E., Charman, T., Murphy, D., and Johnson, M.H. (2013). Reduced neural sensitivity to social stimuli in infants at risk for autism. *Proceedings of the Royal Society B. Biological Sciences, 280*(1758), 20123026. doi:10.1098/rspb.2012.3026.

Lloyd-Fox, S., Blasi, A., McCann, S., Rozhko, M., Katus, L., Mason, L., Austin, T., Moore, S. E., Elwell, C. E., and the BRIGHT Project Team. (2019). Habituation and novelty detection fNIRS brain responses in 5- and 8-month-old infants: The Gambia and UK. *Developmental Science, 22*, e12817. doi:10.1111/desc.12817.

Lloyd-Fox, S., Blasi, A., Mercure, E., Elwell, C.E., and Johnson, M.H. (2012). The emergence of cerebral specialization for the human voice over the first months of life. *Social Neuroscience, 7*, 317–330. doi:10.1080/17470919.2011.614696.

Lloyd-Fox, S., Blasi, A., Pasco, G., Gliga, T., Jones, E.J.H., Murphy, D.G.M., Elwell, C.E., Charman, T., Johnson, M.H., and the BASIS Team. (2018). Cortical responses before 6 months of life associate with later autism. *European Journal of Neuroscience, 47*, 736–749. doi:10.1111/ejn.13757.

Lloyd-Fox, S., Papademetriou, M., Darboe, M.K., Everdell, N.L., Wegmuller, R., Prentice, A.M., Moore, S.E., and Elwell, C.E. (2014a). Functional near infrared spectroscopy (fNIRS) to assess cognitive function in infants in rural Africa. *Nature Scientific Reports, 4*, 4740. doi:10.1038/srep04740.

Lloyd-Fox, S., Richards, J.E., Blasi, A., Murphy, D.G.M., Elwell, C.E., and Johnson, M.H. (2014b). Coregistering functional near-infrared spectroscopy with underlying cortical areas in infants. *Neurophotonics* 1(2), 025006. doi:10.1117/1.NPh.1.2.025006.

Lloyd-Fox, S., Széplaki-Köllőd, B., Yin, J., and Csibra, G. (2015). Are you talking to me? Neural activations in 6-month-old infants in response to being addressed during natural interactions. *Cortex*, 70, 35–48. doi:10.1016/j.cortex.2015.02.005.

Lloyd-Fox, S, Wu, R., Richards, J.E., Elwell, C.E., and Johnson, M.H. (2015). Cortical activation to action perception is associated with action production abilities in young infants. *Cerebral Cortex*, 25, 289–297. doi:10.1093/cercor/bht207.

McCleery, J.P., Akshoomoff, N., Dobkins, K.R., and Carver, L.J. (2009). Atypical face versus object processing and hemispheric asymmetries in 10-month-old infants at risk for autism. *Biology and Psychiatry*, 66, 950–957. doi:10.1016/j.biopsych.2009.07.031.

Meek, J.H., Firbank, M., Elwell, C.E., Atkinson, J., Braddick, O., and Wyatt, J.S. (1998). Regional hemodynamic responses to visual stimulation in awake infants. *Pediatric Research*, 43, 840–843.

Miguel, H.O., Lisboa, I.C., Gonçalves, Ó.F., and Sampaio, A. (2017). Brain mechanisms for processing discriminative and affective touch in 7-month-old infants. *Developmental Cognitive Neuroscience*, 35, 20–27. doi:10.1016/j.dcn.2017.10.008.

Minagawa-Kawai, Y., Matsuoka, S., Dan, I., Naoi, N., Nakamura, K., and Kojima, S. (2009). Prefrontal activation associated with social attachment: Facial-emotion recognition in mothers and infants. *Cerebral Cortex*, 19, 284–292. doi:10.1093/cercor/bhn081.

Minagawa-Kawai, Y., Mori, K., Hebden, J.C., and Dupoux, E. (2008). Optical imaging of infants' neurocognitive development: Recent advances and perspectives. *Developmental Neurobiology*, 68, 712–728. doi:10.1002/dneu.20618.

Moosmann, M., Ritter, P., Krastel, I., Brink, A., Thees, S., Blankenburg, F., Taskin, B., Obrig, H., and Villringer, A. (2003). Correlates of alpha rhythm in functional magnetic resonance imaging and near infrared spectroscopy. *NeuroImage*, 20, 145–158.

Mullen, E.M. (1995). *Mullen Scales of Early Learning*. Circle Pines, MN: American Guidance Service Inc.

Nagamitsu, S., Yamashita, Y., Tanaka, H., and Matsuishi, T. (2012). Functional near-infrared spectroscopy studies in children. *Biopsychosociology Medicine*, 6, 7. doi:10.1186/1751-0759-6-7

Negoro, H., Sawada, M., Iida, J., Ota, T., Tanaka, S., and Kishimoto, T. (2010). Prefrontal dysfunction in attention-deficit/hyperactivity disorder as measured by near-infrared spectroscopy. *Child Psychiatry and Human Development*, 41, 193–203. doi:10.1007/s10578-009-0160-y.

Nicklin, S., Hassan, I., Wickramasinghe, Y., and Spencer, S. (2003). The light still shines, but not that brightly? The current status of perinatal near infrared spectroscopy. *Archives of Disease in Childhood: Fetal & Neonatal Edition*, 88, F263–F268. doi:10.1136/fn.88.4.F263.

Pelphrey, K.A. and Carter, E.J. (2008). Charting the typical and atypical development of the social brain. *Developmental Psychopathology*, 20, 1081–1102.

Perdue, K., Jensen, S.K.G., and Kumar, S. et al. (2019). Using functional near-infrared spectroscopy to assess social information processing in poor urban Bangladeshi infants and toddlers, *Developmental Science*, 22, e12839. doi:10.1111/desc.12839.

Peterson, B.S. (2003). Conceptual, methodological, and statistical challenges in brain imaging studies of developmentally based psychopathologies. *Developmental Psychopathology*, 15, 811–832.

Pirazzoli, L., Lloyd-Fox, S., Braukmann, R., Johnson, M.H., and Gliga T. (2018). Hand or spoon? Exploring the neural basis of affective touch in 5-month-old infants. *Developmental Cognitive Neuroscience*, 35, 28–35. doi:10.1016/j.dcn.2018.06.002

Ravicz, M.M., Perdue, K.L., Westerlund, A., Vanderwert, R.E., and Nelson, C.A. (2015). Infants' neural responses to facial emotion in the prefrontal cortex are correlated with temperament:

A functional near-infrared spectroscopy study. *Frontal Psychology*, 6. doi:10.3389/fpsyg.2015.00922.

Roche-Labarbe, N., Zaaimi B., Berquin P., Nehlig A. Grebe R., and Wallois, F. (2008). NIRS-measured oxy- and deoxyhemoglobin changes associated with EEG spike-and-wave discharges in children. *Epilepsia*, *49*, 1871–1880. doi:10.1111/j.1528-1167.2008.01711.x.

Sakatani, K., Chen, S., Lichty, W., Zuo, H., and Wang, Y.P. (1999). Cerebral blood oxygenation changes induced by auditory stimulation in newborn infants measured by near infrared spectroscopy. *Early Human Development*, *55*, 229–236.

Sakatani, K., Yamashita, D., Yamanaka, T., Oda, M., Yamashita, Y., Hoshino, T., Fujiwara, N., Murata, Y., and Katayama, Y. (2006). Changes of cerebral blood oxygenation and optical pathlength during activation and deactivation in the prefrontal cortex measured by time-resolved near infrared spectroscopy. *Life Science*, *78*, 2734–2741. doi:10.1016/j.lfs.2005.10.045.

Salamon, G., Raynaud, C., Regis, J., and Rumeau, C. (1990). *Magnetic Resonance Imaging of the Pediatric Brain*. New York, NY: Raven Press.

Schecklmann, M., Ehlis, A.-C., Plichta, M.M., and Fallgatter, A.J. (2008). Functional near-infrared spectroscopy: A long-term reliable tool for measuring brain activity during verbal fluency. *NeuroImage*, *43*, 147–155. doi:10.1016/j.neuroimage.2008.06.032.

Shimada, S. and Hiraki, K. (2006). Infant's brain responses to live and televised action. *NeuroImage*, *32*, 930–939. doi:10.1016/j.neuroimage.2006.03.044.

Shultz, S., Vouloumanos, A., Bennett, R.H., and Pelphrey, K. (2014). Neural specialization for speech in the first months of life. *Developmental Science*, *17*, 766–764. doi:10.1111/desc.12151.

Siddiqui, M.F., Lloyd-Fox, S., Kaynezhad, P., Tachtsidis, I., Johnson, M.H., and Elwell, C.E., (2017). Non-invasive measurement of a metabolic marker of infant brain function. *Nature Scientific Reports*, *7*, 1330. doi:10.1038/s41598-017-01394-z.

Steinbrink, J., Villringer, A., Kempf, F., Haux, D., Boden, S., and Obrig, H. (2006). Illuminating the BOLD signal: Combined fMRI-fNIRS studies. *Magnetic Resonance Imaging*, *24*, 495–505. doi:10.1016/j.mri.2005.12.034.

Taga, G. and Asakawa, K. (2007). Selectivity and localization of cortical response to auditory and visual stimulation in awake infants aged 2 to 4 months. *NeuroImage*, *36*, 1246–1252.

Tak, S. and Ye, J.C. (2014). Statistical analysis of fNIRS data: A comprehensive review. *NeuroImage*, *85*, 1, 72–91. doi:10.1016/j.neuroimage.2013.06.016.

Telkemeyer, S., Rossi, S., Koch, S.P., Nierhaus, T., Steinbrink, J., Poeppel, D., Obrig, H., and Wartenburger, I. (2009). Sensitivity of newborn auditory cortex to the temporal structure of sounds. *Journal of Neuroscience*, *29*, 14726.

Toet, M.C. and Lemmers, P.M.A. (2009). Brain monitoring in neonates. *Early Human Development*, *85*, 77–84. doi:10.1016/j.earlhumdev.2008.11.007.

Travis, K.E., Leonard, M.K., Brown, T.T., Hagler, D.J., Curran, M., Dale, A.M., Elman, J.L., and Halgren, E. (2011). Spatiotemporal neural dynamics of word understanding in 12- to 18-month-old-infants. *Cerebral Cortex*, *21*(8), 1832–1839. doi:10.1093/cercor/bhq259.

Tsuzuki, D. and Dan, I. (2014). Spatial registration for functional near-infrared spectroscopy: From channel position on the scalp to cortical location in individual and group analyses. *NeuroImage*, *85*, 92–103. doi:10.1016/j.neuroimage.2013.07.025.

Turk-Browne, N.B., Scholl, B.J., and Chun, M.M. (2008). Babies and brains: Habituation in infant cognition and functional neuroimaging. *Frontiers of Human Neuroscience*, *2*, 16. doi:10.3389/neuro.09.016.2008.

Tzourio-Mazoyer, N., De Schonen, S., Crivello, F., Reutter, B., Aujard, Y., and Mazoyer, B. (2002). Neural correlates of woman face processing by 2-month-old infants. *NeuroImage*, *15*, 454–461. doi:10.1006/nimg.2001.0979.

Urakawa, S., Takamoto, K., Ishikawa, A., Ono, T., and Nishijo, H. (2015). Selective medial prefrontal cortex responses during live mutual gaze interactions in human infants: An fNIRS study. *Brain Topography, 28,* 691–701. doi:10.1007/s10548-014-0414-2.

Vanderwert, R.E. and Nelson, C.A. (2014). The use of near-infrared spectroscopy in the study of typical and atypical development. *NeuroImage, 85,* 264–271. doi:10.1016/j.neuroimage.2013.10.009.

Verriotis, M., Fabrizi, L., Lee, A., Cooper, R.J., Fitzgerald, M., and Meek, J. (2016). Mapping cortical responses to somatosensory stimuli in human infants with simultaneous near-infrared spectroscopy and event-related potential recording. *eNeuro, 3*(2). doi:10.1523/ENEURO.0026-16.2016.

Villringer, A., Planck, J., Hock, C., Schleinkofer, L., and Dirnagl, U. (1993). Near infrared spectroscopy (NIRS): A new tool to study hemodynamic changes during activation of brain function in human adults. *Neuroscience Letters, 154,* 101–104.

Wallois F, Mahmoudzadeh M, Patil A, and Grebe R. (2012). Usefulness of simultaneous EEG-NIRS recording in language studies. *Brain and Language, 121,* 110–123. doi:10.1016/j.bandl.2011.03.010.

Watanabe, H., Homae, F., and Taga, G. (2010). General to specific development of functional activation in the cerebral cortexes of 2- to 3-month-old infants. *NeuroImage, 50,* 1536–1544. doi:10.1016/j.neuroimage.2010.01.068.

Wijeakumar, S., Kumar, A., Delgado Reyes, L.M., Tiwari, M., and Spencer, J.P. (2019). Early adversity in rural India impacts the brain networks underlying visual working memory, *Developmental Science, 22,* e12822. doi:10.1111/desc.12817.

Wilcox, T., Haslup, J.A., and Boas, D.A. (2010). Dissociation of processing of featural and spatiotemporal information in the infant cortex. *NeuroImage, 53,* 1256–1263. doi:10.1016/j.neuroimage.2010.06.064.

Wilcox, T., Hirshkowitz, A., Hawkins, L., and Boas, D.A. (2014). The effect of color priming on infant brain and behavior. *NeuroImage, 85*(1), 302–313. doi:10.1016/j.neuroimage.2013.08.045.

Wolf, M., Ferrari, M., and Quaresima, V. (2007). Progress of near-infrared spectroscopy and topography for brain and muscle clinical applications. *Journal of Biomedical Optics, 12,* 062104. doi:10.1117/1.2804899.

Wyatt, J.S., Cope, M., Delpy, D.T., Wray, S., and Reynolds, E.O. (1986). Quantification of cerebral oxygenation and haemodynamics in sick newborn infants by near infrared spectrophotometry. *The Lancet, 2,* 1063–6.

Xiao, T., Xiao, Z., Ke, X., Hong, S., Yang, H., Su, Y., Chu, K., Xiao, X., Shen, J., and Liu, Y. (2012). Response inhibition impairment in high functioning autism and attention deficit hyperactivity disorder: Evidence from near-infrared spectroscopy data. *PloS One, 7,* e46569. doi:10.1371/journal.pone.0046569.

Yücel, M.A., Selb, J., Aasted, C.M., Petkov, M.P., Becerra, L., Borsook, D., and Boas, D.A. (2015). Short separation regression improves statistical significance and better localizes the hemodynamic response obtained by near-infrared spectroscopy for tasks with differing autonomic responses. *Neurophotonics, 2,* 035005. doi:10.1117/1.NPh.2.3.035005.

Zaramella, P., Freato, F., Amigoni, A., Salvadori, S., Marangoni, P., Suppjei, A., Schiavo, B., and Chiandetti, L. (2001). Brain auditory activation measured by near-infrared spectroscopy (NIRS). *Journal of Neonates and Pediatric Research, 49,* 213–219.

Zhu, H., Fan, Y., Guo, H., Huang, D., and He, S. (2014). Reduced interhemispheric functional connectivity of children with autism spectrum disorder: Evidence from functional near-infrared spectroscopy studies. *Biomedical Optical Express, 5,* 1262. doi:10.1364/BOE.5.001262.

....

CHAPTER 8

BEHAVIORAL TESTING AND EYE-TRACKING TECHNOLOGY

....

NADJA ALTHAUS

Using infants' looking behavior to gain insight into cognitive processes is one of the most frequently used and easily accessible methods in developmental cognitive science. The infant's visual system develops rapidly, and visual preferences can be found within hours of birth (Braddick and Atkinson 2011). Systems using video cameras to film infants' eyes were in use long before automatic eye-tracking systems, while remote eye trackers that did not require head-mounted gear became widely available for infancy research in the early 2000s.

Infants selectively attend to stimuli and show clear preferences for "interesting" over "uninteresting" material, and demonstrate clearly when a stimulus is surprising rather than expected. This is exploited to answer a plethora of different questions. For some of these, preferential looking provides a direct answer (e.g., is a red, shiny ball more visually salient than a brown, dull piece of cloth?) but for others, preferential looking provides *indirect* insight into cognitive processes (e.g., whether infants form categories of animals or whether infants recognize a word even after hearing just a part of it). While the concept of tracking infants' eye movements to determine what they are interested in, or what they find surprising, is simple to start with, interpreting the data is often not. In this chapter, we will firstly review the relationship between eye movements and underlying cognitive processes (first section). In the second section we will provide a brief characterization of different types of eye movements, fixations, saccades and smooth pursuit. The remainder of the chapter is divided into two parts: in the third section we cover fundamental paradigms of looking time methods, specifically novelty preference procedures and intermodal preferential looking (IPL). Most aspects covered in this section apply to both traditional, manually scored gaze tracking methods as used in many classic infant looking time studies, and to more recently available automatic eye-tracking systems which provide a higher spatial and temporal resolution. The fourth section is specific to automatic eye tracking, addressing more specifically the additional metrics that are available to study cognition using this technology. Lastly, in the final section we will cover how technological developments have advanced "looking time studies" beyond the possibilities of manually scoring gaze from a camera recording—in particular with a view towards techniques such as head-mounted eye tracking, gaze-contingent procedures, and pupillometry.

Eye Movements and Cognitive Processes

The relationship between eye movements and cognitive processes has been established at least since Yarbus' (1967) investigation into an observer's scan patterns during picture viewing (for a historical overview of looking time measures, see Aslin and McMurray 2004). This study involved the participant exploring a picture after having been given different instructions, ranging from "free examination" to "remember positions of people and objects in the room." Scan patterns differed widely depending on those instructions, suggesting that higher-level cognitive functions had influenced eye movements.

Fixations are resting points in the gaze patterns, thought to reflect the focus of overt attention (Kowler et al. 1995). Measuring fixations reveals cognitive processes at a much finer temporal scale than behavioral responses (e.g., button presses) could, assessing even subconscious shifts of attention. Where research with young children or infants is concerned, eye movements are especially valuable. The time spent looking at a stimulus is one of the few measures available to researchers, since preverbal infants cannot be verbally instructed to perform a specific task or be asked to make manual responses.

But what processes are involved in directing gaze at a specific target? Eye movements are triggered towards a region in the visual field in order to actively sample the most "interesting" parts of the visual input, to maximize information gain (e.g., Najemnik and Geisler 2005; Gegenfurtner 2016). At the behavioral level, eye movements may be triggered by bottom-up processes such as the visual saliency of a stimulus (Itti et al. 1998; Kienzle et al. 2009) as well as higher level processes involving object recognition (Einhäuser et al. 2008). Longer-term processes such as task planning can also affect where an observer is likely to look (e.g., Hayhoe and Ballard 2005). Schütz et al. (2011) provide a detailed review of the behavioral evidence concerning the question of what factors predict eye movements. Research on the primate and human visual system has led to a detailed picture of the neurophysiology underlying eye movements, including a complex network of structures such as cerebellum, the superior colliculi and the reticular formation (for a review see Krauzlis et al. 2017).

Before we move on to the question of how eye movements can provide insight into cognitive development, we must first consider specific characteristics of eye movements and gaze.

CHARACTERIZATION OF EYE MOVEMENTS

Gaze patterns are characterized by fixations (during which the eye comes to a rest for a period of time) and saccades (during which the eyes move from one fixation location to the other). Information about the visual environment is acquired during fixations, whereas visual input is suppressed during saccades (Thiele et al. 2002) and changes in a visual scene may go unnoticed if they occur during a saccade (e.g., Grimes 1996; Henderson and Hollingworth 2012). Smooth pursuit, by contrast, allows to track a moving target closely as it travels across the visual scene (Aslin 1981; Westheimer 1954). Vergence movements (to align the foveas of both eyes in order to allow focusing on objects at different distances from the viewer) and vestibulo-ocular movements (to provide a stable view of the world despite head movements) are not covered here in detail, as they are less relevant to measuring developmental cognitive processes.

Fixations

During fixations, a specific location in space (i.e., the center of the visual field) is inspected for a certain period of time. This is when information uptake takes place. The duration of fixations varies considerably, e.g., the Henderson and Luke (2014) report mean fixation durations of approximately 210 ms for adults' reading and approximately 275 ms for scene memoration, whereas Wass et al. (2013) report a mean fixation duration of 521 ms for adults and 604 ms for 12-month-olds in a video viewing task. Contrary to what the name indicates, the eye is in fact not completely still during a fixation. Micro-saccades allow the visual system to remain active, while at the same time attending to a specific location. Without micro-saccades, the input to the visual system would remain identical, and cells in visual cortex would stop responding due to fatigue. Small movements, of which the observer is not aware, cause the physical input to change on a millisecond scale, preventing adaptation from causing the visual image to fade (e.g., Martinez-Conde et al. 2006). During fixations active viewing takes place, such as scene exploration, and this is followed by the encoding of visual information in working memory. Fixation duration is therefore one of the primary variables of interest in infant looking time (see "Eye Tracking in Developmental Science fourth section"), and this can be influenced by both bottom-up information (Itti and Koch 2001) such as visual saliency and top-down factors, for instance task demands (Yarbus 1967; Henderson et al. 1999; Althaus and Mareschal 2012). However, individual differences in infants' fixation duration have also been linked to temperament and attentional control (Papageorgiou et al. 2014).

Saccades

During saccades, the eye moves from one fixation location to the next. The visual system is effectively blind while this movement is carried out (known as saccadic suppression, Matin 1974; Thiele et al. 2002). The purpose of moving the eyes is to change the input to the fovea, i.e., the part of the retina with high spatial resolution. The planning of a saccade takes time, so we know that eye movements occurring after an event (say, the onset of a stimulus) are only "responsive" to this stimulus if the fixation is landed a certain amount of time after the event. The fastest responses, for humans, have been demonstrated in tasks where observers are presented with a display containing an image of an animal and a distracter, and asked to respond as quickly as possible with regard to the location of the animal. Here, saccades are launched as quickly as 120 ms after stimulus onset in adults (Kirchner and Thorpe 2006).

Neonates display saccadic eye movements towards a salient target, especially in the absence of competition. Aslin and Salapatek (1975) investigated saccadic eye movements in one- and two-month-olds (N = 48) by presenting a central fixation target followed by ring-shaped targets at different locations in the visual periphery. Human coders tracked eye movements manually. Saccades were particularly likely if the initially presented central fixation target disappeared, but less so if it remained, i.e., if there was competition between targets. This competition effect was examined by Hood and Atkinson (1993), who systematically varied the temporal gap between the disappearance of the central fixation stimulus and appearance of the peripheral target in a study with infants at one-and-a-half months (N = 19), three months (N = 20) and six months (N = 20). While one-and-a-half -month-olds had generally longer latencies to move their eyes compared to the older age groups, they were particularly slow to disengage when there was a temporal overlap between central and peripheral stimuli. By contrast three- and six-month-olds' performance was similar for the different temporal conditions. This marked improvement in disengaging abilities has been attributed to a shift from subcortical to cortical control mechanisms (Atkinson 1984; Atkinson 2000; Johnson 1990). Gredebäck et al. (2006) measured the development of saccadic latencies using a paradigm that involved targets moving on a straight trajectory but suddenly changing direction. This required infants to produce a saccade in order to foveate on the target. Researchers found considerable improvements in saccadic reaction times between four (mean RT 595 ms) and eight months (mean 442 ms) of age (N = 16, longitudinal study). Yang et al. (2002), however, reported longer RTs compared to adults even at 12 years of age, indicating that there is continuous development of saccadic control throughout infancy and childhood (see Csibra et al. 1998 for neural correlates of saccade planning).

Smooth Pursuit

Smooth pursuit eye movements involve following a small moving target in order to keep this target on the fovea for high-acuity vision. The ability to move the head and eyes in this way develops over the first few months of life (Aslin 1981). The complexity derives in particular from the requirement to predict target motion as processing is otherwise too slow to allow stabilizing on the target (Barnes 2008). At birth, infants only use smooth pursuit eye movements to follow a moving target for a small portion of time (around 15% of the time; Kremenitzer et al. 1979), and even at one month von Hofsten and Rosander (1996) found infants lagging behind the target by about 180 ms. Large improvements in smooth pursuit are found in particular between two and three months, with some infants displaying up to 90% smooth pursuit for slow, sinusoidal motion (calculated as the ratio of eye gain with and without saccades) but performance remains highly variable across individuals (von Hofsten and Rosander 1997). Smooth pursuit gains still develop throughout childhood and are not adult-like until well into adolescence (Salman et al. 2006).

Having reviewed the main types of eye movements, we now turn to an overview of how looking time is used in studies on cognitive development. The next section aims to provide insight into the general techniques in this field, and particularly into two of the most frequently used approaches that have been employed in many variations—novelty preference procedures and IPL.

LOOKING TIME IN DEVELOPMENTAL SCIENCE

Confronted with a population that cannot easily be instructed to respond to a given task or question, researchers in cognitive development have always had to design tasks in which the variable of interest is "naturally occurring" behavior. Gaze is one of the most important behaviors that allow studying young children's mental representations, being both "cognitively inexpensive" in the sense that eye movements require little metabolic energy (e.g., Schütz et al. 2011), therefore offering a highly sensitive method to study cognition, and are "available" for measurement from the earliest stages of life. Looking time studies have been conducted with infants on the day of birth, demonstrating, for instance, a preference for faces over non-face-like stimuli (Johnson et al. 1991).

Eye movement data can be obtained in various ways. Cameras are often used to record the participant's face, sometimes from multiple angles (e.g., Schafer and Plunkett 1998; Fernald et al. 1998; Stager and Werker 1997). The video tape can then be scored offline by a manual coder. High intercoder reliability is typically achieved for such setups, provided the different target displays (usually two) are placed far enough apart to allow an observer to discriminate between looking at the left item or the right item, and looking away. Often, this involves placing one camera right above each display location, so that when the child lands a fixation on one of the pictures they are looking almost directly at the corresponding camera, whereas the camera located above the remaining picture shows them clearly not looking at the camera. Similar set-ups, often used for habituation (see the following section) where infant looking time has to be tracked on-line in order to determine the course of the experiment, involve a human on-line coder, who might observe the child through a small opening near the stimulus display or via camera (Stager and Werker 1997). This procedure involves pressing a button when gaze at a stimulus begins and again when the child looks away. Such set-ups are neither technically complicated nor expensive, and therefore available to a large range of developmental laboratories.

Since the early 2000s, eye trackers have become available for developmental research. The critical innovation was the introduction of so-called remote eye trackers, which no longer require viewers to use a chin-rest or wear a heavy head-mounted device (Aslin and McMurray 2004). Instead, information from the camera mounted above or below the stimulus display uses optical sensors to locate the subject's eye from a distance. The camera then records the pupil center and cornea reflection, taking images every few milliseconds. An algorithm analyzes these in order to determine the gaze location. In contrast to scoring looking time from a video tape, eye tracking offers a much finer spatial (and nowadays temporal) resolution. In many labs, eye trackers are gradually replacing "traditional" video-based systems, both for the fine-grained spatial and temporal detail they provide, but also for ease of data evaluation, considering that a great deal can be achieved completely automatically and no longer requires time-consuming manual scoring.

In the following sections we will look at the most common techniques used in looking time research, highlighting the foundational principles, such as novelty preference and infants' preferential looking at a named target, that drive progress in the area. In this section we will focus on looking time studies in general, i.e., all aspects covered here apply to both manually scored, camera-based systems, and automatic eye tracking. Aspects specific to automatic eye tracking will be covered in the section "Eye Tracking in Developmental Science".

Novelty Preference Procedures

One of the earliest systematic studies using infants' looking time was Fantz' (1964) continuous familiarization procedure. Here, infants (aged one to six months) were repeatedly presented with pairs of complex images, so that the image on one side was the same on every trial, while the image on the other side changed on every trial. Fantz reported that two- to six-month-olds' looking time to the stable image decreased over time and gradually showed a preference for the novel image over the familiar one. This provided evidence that infants could *discriminate* between the familiar image and the novel ones, implying that infants must have encoded the familiar image, and it also demonstrated that *novelty* engages infants' attention. This observation formed the basis of most looking time studies today: preferential looking tasks tap into mental representations by asking which of two images appears more novel to infants, thereby providing a metric of perceived distance to the familiarized items. Novelty preference is used to probe infant working memory in the change detection task (e.g., Ross-Sheehy et al. 2003; Simmering 2012), to assess number processing (e.g., Xu and Spelke 2000; Xu and Arriaga 2007), infants' understanding of the physical world (Spelke et al. 1992), as well as perception and categorization (e.g., Althaus and Mareschal 2012; Eimas and Quinn 1994; Quinn et al. 1993; Oakes and Ribar 2005; Oakes et al. 2009; Sučević et al. 2021; see Rakison and Yermolayeva 2010 for a review).

The central idea underlying most looking time studies is to present a set of target stimuli that are to be viewed in sequence. This may be the same item repeatedly (e.g., to assess discrimination abilities), or different items, for example different members of a category (e.g., Eimas and Quinn 1994), static or dynamic stimuli (e.g., Sučević et al. 2021; Pulverman et al. 2008; Gliga et al. 2008). Presentation may happen in pairs (e.g., Bomba and Siqueland 1983; Eimas and Quinn 1994; Behl-Chadha 1996), which has been shown to facilitate learning (Oakes and Ribar 2005; Kovack-Lesh and Oakes 2007; Oakes, Kovack-Lesh, and Horst 2009). In such a procedure, infants typically display a decrease in looking time with increased number of trials. This has been interpreted as familiarization (Fantz 1964). After this, they are presented with one or more test trials, which involve the presentation of both a familiar stimulus and a novel, contrasting stimulus (alternatively these items are presented in sequence, e.g., Younger and Cohen 1983; Stager and Werker 1997). A preference, i.e., longer looking time, to the contrasting stimulus can be interpreted as successful encoding of the familiarized item(s): infants indicate with their longer looking (reflecting higher interest) that they perceive this novel item as different. If they do not notice the difference, by contrast, they spend equal amounts of time gazing at each test stimulus (because these are perceived as equally interesting). For example, Eimas and Quinn (1994) familiarized three- to four-month-olds ($N = 48$) with cat images, and used several contrasting categories as test items, such as horses, tigers, and female lions. While these young infants showed a clear preference at test for horses and tigers, they did not show a preference for female lions, indicating that they perceived them as the same kind of thing as the cats they saw during familiarization. Infants just a few months older (aged six to seven months, $N = 16$), by contrast, showed a preference even for the female lion, indicating that categorization skills become more refined during the first months of life.

While the idea of infants becoming familiar with a stimulus, or a type of stimulus, during repeated presentations, and exhibiting a preference for a contrasting item afterwards is at the heart of most looking time studies, we can further discriminate between several categories of looking time studies which tap into different cognitive processes.

Habituation Studies

One of the indicators that learning is taking place is a decrease in infants' looking time with increased stimulus exposure, i.e., they *habituate*. Since there may be individual variation in the amount of time or exposure infants require until their looking time drops, studies emphasizing this aspect of learning expose infants to a series of stimuli whose length or duration is not fixed a priori (e.g., Bertenthal et al. 1984; Bushnell and Roder 1985; Cohen and Strauss 1979; Johnson et al. 2004). Instead, the procedure is "infant-controlled" (Horowitz et al. 1972) in the sense that the experimenter merely defines a criterion to end the habituation process, but the actual duration depends on the individual infant's behavior. Typically this criterion involves a decrease in looking time by a specific amount or to a specific fraction of the looking time exhibited on the first trial (e.g., looking time on three subsequent trials is 65% of looking at the first three trials, Pulverman et al. 2008). Once this criterion is reached, the infant is said to have "habituated" to the stimulus. This method may result in infants being exposed to very different numbers of trials across the sample—some infants will maintain looking for longer and take more trials to reach the habituation criterion, whereas others will habituate rapidly. One benefit of the method is that at the offset of the habituation period, infants are at a comparable state in terms of their current level of interest in the habituation items. Following habituation, the test trials are initiated. Testing can then either involve a sequence of individual stimuli so that a preference is established across multiple trials, or involve paired presentations so that a preference can be established as the proportion of looking at one of the stimuli. One theory of habituation is the comparator theory (Sokolov 1963), according to which looking time relates directly to the time it takes to update a mental representation to incorporate the current stimulus. At the beginning of habituation, e.g., to a category of (similar but not identical) objects, establishing this mental representation is more effortful and requires encoding more visual features than towards the end of habituation, when stimuli are perceived as more or less matching the previously seen items and only few visual details need to be encoded during updating. Looking time declines because the mental representation approaches the visual variability encountered in the stimuli (see Colombo and Mitchell 2009, for a discussion of the theoretical account of habituation). Oakes (2010) gives an extensive review of habituation and its use to assess infant cognition.

Familiarization Studies

In *familiarization* studies, infants are instead presented with a stimulus sequence of fixed length (e.g., Behl-Chadha 1996; Bomba and Siqueland 1983; Eimas and Quinn 1994 Gliga et al. 2008; Quinn et al. 1993, 2001), which means infants may not be fully habituated. The advantage is that all infants have had the same amount of exposure by the time they reach the end of familiarization, or at least they had the same opportunity to explore the stimuli, since their actual looking times may still differ. Like in habituation studies, looking time is expected to decrease over the course of the familiarization phase, but in practice due to the fixed-length familiarization period some infants maintain looking (i.e., they are not sufficiently habituated for their looking times to drop by the time the familiarization phase ends). Even if a decrease in looking is not systematically observed across a sample of infants, this does not necessarily indicate a lack of learning. There are many examples where infants do not show a systematic decrease in looking time during familiarization, but demonstrate successful learning through novelty preference at test.

Quinn et al. (1993), Experiment 1, familiarized three- to four-month-old infants ($N =$ 34) with either eight photographs of cats or eight photographs of dogs. Infants' looking time during the familiarization phase showed no significant decrease, but infants clearly preferred an out-of-category item at test (a bird photograph). Crucially, a control experiment established that there is no a priori preference for birds over cats or dogs. In other words, the clear preference for the bird photograph after familiarization must be due to learning from the familiarization exemplars, even though looking time did not decrease throughout that phase. The authors in this case attributed the lack of habituation to category complexity and interestingness, which may have maintained looking time even though learning took place (see Younger 1985 for similar results).

Violation of Expectation

A variant of the previous familiarization/habituation paradigm is the "violation of expectation" paradigm that is typically used to study infants' understanding of physical object properties (e.g., Baillargeon 1986, 2004; Spelke et al. 1992; Jackson and Sirois 2009 for a variant using pupil dilation). The familiarization phase involves the repetition of a specific event, such as the disappearance and reappearance of an object from behind an occluder. At test, an "unexpected" (often physically impossible) event is shown, e.g., an object that has physically changed during occlusion, or reappears later than expected. An increase in looking time is interpreted as "violation of expectation" or surprise that the expected event has not occurred—therefore the child must represent mentally the property that is being tested.

Interpretation of Novelty Preference Paradigms

While there is an expectation that infants who successfully learn during the familiarization phase should exhibit longer looking times for novel items at test, several research groups found contradicting results. Sometimes infants exhibit a preference for the familiar item instead. If this is systematically found across infants, this still reflects successful learning (Houston-Price and Nakai 2004; Cohen 2004). Hunter and Ames (1988) established that there is a systematic progression from initial familiarity preference to later novelty preference. If infants are familiarized with stimuli for a short amount of time they are likely to prefer the familiar item at test. By contrast, if they are familiarized with a longer sequence, they become gradually more likely to exhibit novelty preference. As Hunter and Ames (1988) demonstrated, this shift from familiarity to novelty preference is further affected by aspects such as stimulus complexity, with more complex stimuli requiring more familiarization time before novelty preference emerges.

A particular difficulty in novelty preference procedures is the occurrence of a null-preference, where the averaged novelty preference scores do not differ significantly from 0.5 (i.e., chance level responding). This is extremely difficult to interpret, as such a result could either arise because infants are truly not discriminating between familiarization and test stimuli, i.e., no learning has taken place, but it could equally be due to other reasons. Equal looking times to familiar and novel test items merely indicate that infants at this point in time find both images equally interesting. One reason for a lack of preference could be a priori differences in the interestingness of the two types of stimuli, i.e., the by now familiar category item may be so visually salient that it continues to attract a lot of looking, despite its familiarity. Control conditions can be used to assess prior preferences. Eimas and Quinn (1994), as mentioned previously, investigated infants' learning of animal categories and in their main study (Experiment 1) three- to four-month-olds who were familiarized with cats ($N = 24$) showed a preference for tigers, but not for female lions (preference score for female lions $M = .52$). However, the authors also included a control experiment (Experiment 3) in which they tested whether three- to four-month-olds ($N = 40$) exhibited a preference given any of the investigated category contrasts, without a prior familiarization phase. To examine this, they presented infants with eight pairs of items from the two contrasting categories (e.g., cats vs female lions), and determined the overall mean preference score. Since these infants did not show a preference for cats over female lions, it can be concluded that the null result in Experiment 1 did not occur because cats are inherently more attractive to young infants, i.e. it is not the result of novelty (female lions) competing with higher attractiveness (cats). In particular since the same infants showed a novelty preference when presented with a novel cat vs a tiger, it therefore does seem safe to conclude that the infants in Experiment 1 indeed failed to form a category of cats that was narrow enough to exclude female lions, which are after all more similar looking to cats than tigers.

Another potential reason for a null preference may be fatigue, for instance if the familiarization phase is too long and infants are no longer fully attentive during the test trials. Studies using subtle contrasts at test therefore often include an extra test trial after the critical one that includes a more extreme novel or out-of-category item. If infants are fatigued, they should not show a preference for either the subtle or the extreme test stimuli. Conversely, if they do show a preference for only the extreme test, then it is likely that the null preference on the critical trial indeed represents a failure to discriminate the items. For example, in Eimas and Quinn's (1994) study, the infants who showed no preference for the female lion over a novel cat after familiarization with cat exemplars were also presented with a test trial pairing a cat with a tiger and the clear preference on that trial indicates that infants were not simply too fatigued to exhibit novelty preference for the female lion.

What do infants learn during familiarization? As described in the previous section, a theory of familiarization/habituation is that the infant forms a mental representation over time that corresponds to the familiarization stimuli, and the more accurate this representation becomes, the less surprising the encountered items seem, which leads to less processing time and therefore lower looking time with increasing levels of familiarization. In the context of categorization, the assumption is therefore often that the outcome of familiarization is a representation approximating the "average" of the familiarization items. However, recent results have provided more nuanced insights into this, showing that there are sequence effects (Mather and Plunkett 2011), meaning that the similarity of successive items can have an impact on learning. Mather and Plunkett's work with ten-month-old infants indicated that the same eight stimulus items could lead to significant novelty preference if presented a sequence that maximized variability across successive items, but to no significant novelty preference if presented in a sequence in which successive items were more likely to be similar. Althaus et al. (2020) further demonstrated that in fact memory plays a large role and in fact recency effects can influence novelty preference. In a study with ten-month-olds ($N = 96$) they showed that if the item shown last during familiarization (i.e., immediately prior to the test trial) had high similarity to the familiar category test item, then infants were highly likely to show novelty preference. If, by contrast, the final item shown during familiarization was a category exemplar relatively far from the familiar category test stimulus, then infants showed no significant novelty preference.

Intermodal Preferential Looking

Not all looking time paradigms involve a familiarization period. Preferences that manifest without a learning phase in the laboratory can provide just as much insight into mental representations. One prominent example is the intermodal preferential looking (IPL) paradigm, coined by Golinkoff et al. (1987), which is used to test language comprehension, and in particular word recognition, and has been used in many variations to assess infants' early language skills (Golinkoff et al. 2013). Here, infants are presented with a pair of visual stimuli (e.g., a duck and an elephant) as well as a spoken word (e.g., "duck"). Typically, the pictures are presented on the screen in silence for a certain amount of time, the "pre-naming phase," before the auditory label occurs, so that a change in preference for the named target can be obtained by comparing the pre-naming and the post-naming phase, e.g., Styles and Plunkett (2009a) used a pre- and post-naming phase of 2500 ms each each, leading to a 5000 ms visual stimulus presentation with the target word beginning half-way through the picture presentation. An overall preference for the named target in the post-naming phase is interpreted as successful word recognition (Golinkoff et al. 1987; Houston-Price et al. 2007; Delle Luche et al. 2015). The method is also known as "looking while listening" (Fernald et al. 2008), which was originally introduced as a term to emphasize the analysis of the time course of looking on a frame-by-frame basis, allowing a careful observation of shifts in looking at target vs distracter depending on the occurrence of the target word. This approach has been used to look at processing speed in word recognition (Fernald et al. 1998), the processing of partial words (Fernald et al. 2001), as well as establishing the youngest age at which infants recognize words. In their seminal study on word recognition in six- to nine-month-olds Bergelson and Swingley (2012) used an IPL task in which prerecorded sentences containing the target words were replaced by live utterances from the participant's caregiver. This was achieved by asking the caregiver to repeat a phrase presented to them via headphones while their infant was looking at the visually presented stimuli, either pairs of pictures or scene stimuli containing the target items as well as distracters. Caregivers were wearing a visor so they could not see the target display. Hearing the words spoken by the familiar voice, infants ($N = 33$) demonstrated successful recognition of referents of highly frequent nouns (body parts such as hand, leg, eyes, and food-related items such as banana, milk, spoon). While the preference for the named target was not as strong as in older children, even the younger (six- to seven-month-olds, $N = 20$) infants in Bergelson and Swingley's sample showed an increase in looking at a target if it was named compared to the same item on trials where this item was not named—clearly a lot earlier than what was previously thought.

Another application of the IPL paradigm is in mispronunciation studies, which aim to assess the encoding specificity of early words, i.e., the question of whether early lexical entries represent phonological detail. Thus, Swingley and Aslin (2000, 2002) and Bailey and Plunkett (2002) found that 15- to 24-month-olds were sensitive to mispronunciations of familiar words, indicating phonological detail is encoded. White and Morgan (2008) testing consonant mispronunciations, and Mani and Plunkett (2011) addressing vowels, demonstrated that mispronunciation effects (i.e., the decrease in preferential looking compared to trials with an accurately pronounced target) also scale with the degree of mispronunciation, i.e., mispronunciation effects are larger if the mispronounced item differs from the accurate word form by more phonological features (e.g., "sook," a three-feature mispronunciation of "book," produces a larger mispronunciation effect than "took," a two-feature mispronunciation, or "dook," a one-feature mispronunciation).

IPL trials are also used for testing novel word learning in combination with a familiarization phase (e.g., Schafer and Plunkett 1998; Ballem and Plunkett 2005; Yoshida et al. 2009). Here, the familiarization phase exposes the child to images of novel objects together with novel labels. Schafer and Plunkett (1998), for example, presented 15-month-olds ($N = 29$) with two novel objects and two corresponding novel words ("bard" and "sarl") eight times each, and this familiarization phase was followed by IPL trials that tested whether infants had formed a correct mapping between word and object. The 15-month-olds in their study showed a preference for the accurate referent during the post-naming phase. IPL has shown to provide a highly sensitive metric of infants' mental representation, compared to other ways to assess newly learned word-object mappings. Stager and Werker (1997) used a "switch task" (Werker et al. 1998) to assess 14-month-olds' learning of minimal pairs "bih" vs "dih." In this paradigm, infants are also presented with a familiarization phase pairing two novel objects with corresponding novel labels ("bih"/"dih"). Testing involves a "same" trial that preserves the familiarized pairing of word and object, and a "switch" trial in which an object is presented together with the label that was previously associated with the other item. A difference in looking time between same and switch trials constitutes evidence that infants have formed the correct word-object mapping (Werker et al. 1998). Surprisingly, 14-month-olds showed no evidence for word-object mapping in this set-up with minimal pairs "bih" and "dih," whereas that age group was successful with more distinct words "lif" and "neem." While this provides insight into the relative difficulty that word-object mapping poses for 14-month-olds, Yoshida et al. (2009) were able to demonstrate that this age group is in fact able to learn minimal pairs by using an IPL test trial instead. Presumably the cognitive demand is lower for IPL trials compared to the switch task as target and distracter can be compared directly. More recently Lany (2018) used a similar set-up to investigate the relationship between infants' speed of processing in an IPL task (the time taken to switch to a target on trials where the distracter is fixated at the time the auditory stimulus is heard) with familiar words and their ability to learn novel words. At 17 months ($N = 35$), faster processing speed in IPL with familiar words predicted higher ability to learn novel words, but for 30-month-olds ($N = 34$) such a relationship was only found with a more complex novel-word learning task in which familiarization involved dynamic scenes and variable labeling phrases, and testing also involved videos instead of static images.

Finally, a variant of IPL has been used to study priming effects in the developing lexicon (e.g., Arias-Trejo and Plunkett 2009; Styles and Plunkett 2009b; Mani and Plunkett 2010). To investigate semantic priming, Styles and Plunkett (2009b) preceded a standard IPL trial by a phrase ending in a noun that was semantically related (or unrelated) to the target item (e.g., "Yesterday I saw a cat!" followed by a trial in which the named target is *dog*, i.e., "dog" is presented as an auditory label and the visual display shows an image of a dog and an unrelated distracter). While 18-month-olds' target looking was similar across trials with a related and unrelated prime, 24-month-olds looked longer at the target if it was preceded by a related word. Arias-Trejo and Plunkett (2009) further found that infants at 21 months exhibited inhibition effects for unrelated targets. Together, these studies provide evidence that by 21 months, infants already begin to establish a lexicon structure that incorporates a system of semantic links between different lexical entries. By preceding an IPL trial not with a spoken, related word, but with a picture of a single related (or unrelated) item, Mani and Plunkett (2010) tested phonological priming in 18-month-olds. Their finding of preferential target looking only for trials with a related prime picture demonstrates not only that phonologically unrelated items inhibit target looking, but also that 18-month-olds implicitly generate labels for pictures in the first place.

As this brief overview shows, in-depth analysis of eye movement behavior offers powerful insights into cognitive development over the first months and years of life. Although some of the work cited previously was conducted with automatic eye trackers, the methodological approaches described so far can all, in principle, be carried out with simple camera-based systems and manual scoring of video data. However, automatic eye tracking has not just made looking time studies much easier to carry out in the sense that the data analysis is much faster, it has also led to a number of methodological advances that provide deeper insight into cognitive processes by providing more fine-grained information about infants' gaze. We therefore turn our attention to those aspects of looking time studies that are specific to automatic eye tracking.

Eye Tracking in Developmental Science

Eye-tracking methodology relies on measuring the displacement of the first so-called Purkinje reflection with respect to the pupil. The four Purkinje reflections are created when light enters the eye and is refracted on the different surfaces: inner/outer cornea surface and inner/outer lens surface (Cornsweet and Crane 1973; Duchowski 2007). Only the first reflection is bright, and therefore is the one monitored by eye trackers. The reflection is usually created by an infrared diode in the vicinity of the eye tracker's camera. Image processing algorithms are used to locate both the center of the pupil and the Purkinje reflection. The relative position of both then indicates the subject's direction of gaze: as the eyeball rotates horizontally or vertically, the reflection is displaced with respect to the center of the pupil. Since the relationship between the two features (pupil and reflection) is dependent on the characteristics of the subject's eyes and the relative position of the eye to the camera, a calibration procedure has to be performed before recording eye movements (Aslin and McMurray 2004; Gredebäck et al. 2010).

For infant eye tracking, i.e., "remote" eye tracking, the camera filming the eyes is not placed on the subject's head, as in older systems, but under or in front of the screen on which stimuli are displayed. Not attaching the cameras to the head, however, requires head tracking technology that keeps track of where the subject's head is located in space, and compensates for movement. Eye-tracking systems provide different solutions to head tracking. Some systems require a circular sticker to be placed on the subject's forehead (e.g., SR Research Ltd. 2008). This way, image recognition software can determine distance and angle of the head. Others utilize a dual camera system to determine where the subject's head is located in 3D space (e.g., Tobii Technology AB 2016). Since two cameras deliver subtly different images of the visual scene, distance in space can be estimated on the basis of these differences. Both approaches require the subject to move only within a specific area ("box") in front of the eye tracker, with recovery time needed to relocate the head and eyes after larger movements (for a recent evaluation of remote eye-tracking systems see Niehorster et al. 2018).

The quality of the data recorded in the session is dependent on the quality of the calibration (for a comprehensive review see Holmqvist et al. 2011). Good calibration quality is achieved with a stable head position and the subject reliably shifting their gaze towards a specific screen location. Both of these factors present challenges with infants, as they tend to move a lot and can hardly be instructed to remain still or look at a specific location.

In calibration with adults, small calibration targets are used to indicate the intended gaze location on the screen (e.g., a dot). Once the subject has moved the eyes and landed in a stable position, the eye image, or rather, the pupil and reflection position, are "logged" and future gaze locations are compared by an algorithm to these known locations.

The calibration procedure needs to be adapted to use with children. Researchers typically use colorful, animated calibration targets that loom and are accompanied by attention-grabbing sounds. The number of calibration points is typically reduced, e.g., to five points, i.e., four corners and center of the screen (compared to nine arranged in a three-by-three grid in adults). Schlegelmilch and Wertz (2019) reported that calibration targets with high contrast centers were beneficial for data quality, as well as those with high visual complexity.

As mentioned previously, one of the main aims of using eye tracking is the higher spatial and temporal resolution of the data. The sampling frequency is restricted by the speed of the camera equipment. While scoring video tapes from standard video equipment typically resulted in a 25 Hz resolution, current eye-tracking systems handle 300, 600, or even 1000 Hz, thus allowing gaze tracking at high temporal accuracy. In the following, we will first briefly address the most common metrics eye-tracking studies use that go beyond the looking time measures discussed previously. In the later section we then turn to an overview of the literature, focussing particularly on novel approaches specific to automatic eye tracking.

Metrics of Interest in Developmental Automatic Eye-Tracking Studies

Eye-tracking data is extremely rich. The following section provides a short overview of the different metrics that can be extracted from remote eye tracker systems.

Fixation Location

The tracker takes a picture of the eye every few milliseconds, depending on the tracking frequency. Every single image corresponds to a sample, for which a spatial location is determined by the eye-tracking software. A fixation filter is then used to identify fixations and saccades from the sequence of sample locations and their timestamps (Holmqvist et al. 2011). As data from infants is often so noisy due to movement and/ or calibration quality (Hessels et al. 2015; Wass et al. 2013) that researchers have even suggested correcting the output of an automatic fixation filter manually (Saez de Urubain et al. 2015), robust algorithms are necessary to identify fixations and saccades (Hessels et al. 2017). Fixation targets, often particularly the location of the first fixation (e.g., Gliga et al. 2009), are one of the most important metrics in assessing infants' mental representations, perception of novelty, or preference.

Fixation Duration

The fixation duration is the amount of time the eye remains stable in one fixation before saccading away. The duration of the first fixation on a target is often the longest and par-ticularly informative (e.g., Frankenhuis et al. 2013; Hessels et al. 2016; van Renswoude et al. 2019). Individual differences in fixation duration have been associated with infant temperament (Papageorgiou et al. 2014), and shorter fixation durations have been found in infants at risk for autism and those who would later be diagnosed with autism (Wass et al. 2015).

Latency to Fixation

Just like reaction time, latency to fixation (i.e., the length of time, in milliseconds, that it takes from start of trial to move the eye to the desired fixation point) has been used as an indicator of processing speed. Due to eye movements being metabolically inexpensive (Richardson et al. 2007; Schütz et al. 2011), compared to e.g., reaching behavior or even head turns, the measure is highly sensitive. There is a subtle distinction between "saccadic reaction time" ("saccadic latency"), which measures the time between the onset of a stimulus and the time a saccade is *launched*, and "latency to fixation," which measures the time between stimulus onset and the onset of a fixation, i.e., the *landing* of a saccade. Both are used, depending on which metric is of more interest to the researchers. Latencies are not just useful as reaction times *after* a stimulus presentation (e.g., Franklin et al. 2008; see Gredebäck et al. 2006; and Kenward et al. 2017; for reviews of saccadic reaction time as a measure in infancy). Anticipatory eye movements, i.e., saccades to a target location that are launched before the arrival or appearance of a target at that location, have been used as an indicator of expectation and prediction (Falck-Ytter et al. 2006; Gredebäck et al. 2009; Melzer et al. 2012; Monroy et al. 2017). Other interpretations of anticipatory eye movements have included learning, i.e., after familiarization with a specific event that involves the appearance of an object at a specific location (e.g., Addyman and Mareschal 2010; Albareda-Castellot et al. 2011; Kaldy et al. 2016; McMurray and Aslin 2004).

Looking Time/Dwell Time

A frequently used aggregate measure is total looking time or dwell time, indicating the overall duration of looking spent on a given stimulus. This does not take into consideration the number or duration of individual fixations, but instead reflects the sum of these. While not as detailed as some other metrics, this is useful as an indicator of how interesting a stimulus is, in particular compared to others and often reported as the "proportion of target looking" (PTL). When reported with regard to an entire stimulus, e.g., one picture out of two presented on a screen, this is identical to looking time as reported in studies that involve manually scoring looking directed at a specific stimulus. Dwell time, however, can also be calculated for a spatially confined "area-of-interest" (AOI), which is defined a priori. This way, the high spatial resolution of eye tracking (compared to, e.g., manually scored camera recordings) can be utilized. Many recent studies rely on accurate measurements of e.g., infants' gaze towards specific facial features, object parts, or items in an array or cluttered display (e.g., Althaus and Mareschal 2012; Althaus and Plunkett 2016; Amso et al. 2014; Gliga et al. 2009; Lewkowicz et al. 2012; for detailed examples see "High Spatial Resolution: Making Use of Detailed Areas-of-Interest").

Time Course of Looking

Rather than analyzing individual fixations or dwell time aggregated over the whole trial or part thereof, the time course of looking across the trial can provide more fine-grained information about the temporal characteristics of infant looking behavior. For every time slot (either corresponding to individual eye-tracking samples or longer intervals comprising several samples), the proportion of trials during which gaze is directed at a target is determined. This way, it is possible to inspect how looking at the target varies as a function of time. This type of analysis has been particularly useful in the context of studying word recognition, where the time course of looking at a target vs distracter item can reveal the speed of processing (Bergelson and Swingley 2012; Ackermann et al. 2020). While looking patterns over time have been analyzed even with video-based (manually scored) systems (e.g., Swingley et al. 1999; Swingley and Aslin 2000), this has become much easier to do with eye-tracking systems (Ackermann et al. 2020; Bergelson and Swingley 2012; Chow et al. 2017; Ren et al. 2019; Shukla et al. 2021). Linear mixed effects models or growth curve analyses (Mirman 2014) can reveal differences at a very fine time scale, but do require many trials. They are therefore more frequently encountered in the context of studies where no learning is necessary (e.g., recognition of familiar words or mispronunciations, rather than word learning studies or familiarization-based paradigms where there is essentially a single test trial per participant). For instance, Chow et al. (2017) used a growth curve analysis with a visual world paradigm, an adaptation of the IPL paradigm with four visual items (phonological/semantic competitors of the auditory target and two unrelated distracters), to explore the time course of word recognition in toddlers. This task demonstrated that 24-month-olds ($N = 24$) and 30-month-olds ($N = 24$) move their eyes faster to phonological matches than semantic ones.

Pupil Size

Modern eye trackers also allow measuring the size of the pupil. One of the main causes for a change in pupil size is the pupillary light reflex (PLR), i.e., the pupil constricts in adaptation to an increase in ambient light (or dilates to adapt to a darker environment). While this has been studied in infants at risk for autism, who were found to exhibit a hypersensitive PLR (Nyström et al. 2015), most infant studies using pupillometry are concerned with phasic changes in pupil size. These reflect for example arousal, cognitive load, and memory (Hepach and Westermann 2016; Sirois and Brisson 2014). Pupil dilation caused by these factors is tied to subcortical activation, primarily the locus coeruleus, and therefore allows tapping into the noradrenergic system (Laeng et al. 2012). The pupil response is slow, potentially taking several seconds to full dilation. Some methodological considerations are necessary in order to use pupil size as an index of cognitive processes, mainly to be able to separate the response to changes in luminance from cognitively induced changes. This includes the use of isoluminant stimuli, as well the inclusion of a baseline period just before presentation of the relevant stimulus to determine dilation on a trial-by-trial and subject-by-subject basis. Since measurement is usually not error-free, interpolation and smoothing procedures are needed in order to preprocess the signal (see Sirois and Brisson 2014; Mathôt et al. 2018 for methodological considerations). Pupillometry offers some advantages over looking time, for example the use of a single test stimulus as opposed to paired presentations as in preferential looking, where distracter choice can be an important factor in experimental design (see the following section).

Beyond Preferential Looking: Where Eye Tracking Offers New Opportunities

Eye tracking offers benefits going beyond merely saving the time it takes to manually score video data. High spatial and temporal resolution, as well as the availability of pupil size information have already been mentioned. However, the rapid digital availability of gaze data adds another advantage—the possibility to use gaze patterns as a factor in an experimental manipulation, i.e., gaze-contingent tracking. Another approach that has only become recently possible with eye-tracking equipment becoming smaller is a combination of head-mounted cameras and eye tracking. With head-worn equipment, the child's view of the world can be recorded and simultaneous eye tracking provides insight into which particular objects their overt attention is directed to. In the following sections we review some of the more recent approaches in infant looking time studies, highlighting four areas where automatic eye tracking has provided advances beyond the possibilities of manually scored, video-based systems.

High Spatial Resolution: Making Use of Detailed Areas-of-Interest

Eye tracking allows assessing overt shifts of attention at a very fine spatial resolution. While looking time in more traditional, camera-based set-ups is mostly restricted to two stimulus items (left image, right image), eye-tracking studies can reliably investigate looking patterns in displays with many items. Gliga et al. (2009) showed that six-month-olds' attention was attracted more by faces than by other object categories in a circular display of six items. Other studies analyze children's gaze directed at different targets within a complex natural scene (e.g., Amso et al. 2014; Helo et al. 2017; Kelly et al. 2019; Öhlschläger et al. 2020), and at object parts rather than whole objects (e.g., Althaus and Mareschal 2012, 2014; Althaus and Plunkett 2015, 2016; Hurley and Oakes 2015; Kovack-Lesh et al. 2014). For instance, Helo et al. (2017) investigated to what extent 24-month-olds' attention is attracted by semantic inconsistencies in scene perception, such as a sock hanging on a kitchen rack vs a ladle hanging in the same place. Using manually defined AOIs as well as several visual predictors such as a saliency map analysis, the authors found that 24-month-olds' gaze was more dependent on low-level visual features compared to adults, and only showed an effect of semantic consistency for visually salient items. Highlighting a similar dependence on visual saliency, albeit in a younger age group, Althaus and Mareschal (2012) showed that 12-month-olds spend more time on salient image regions at the beginning of a familiarization phase, but this preference shifted towards category-relevant features (evidenced by an increase in looking at antlers during familiarization with deer) as familiarization progressed. Here, the authors defined AOIs for five different object parts. This was done in a data-driven procedure that centered AOIs on a manually defined spot (e.g., the center of the body, head, and so on) and then gradually expanded a rectangle as long as more gaze points were encountered. This meant that a maximum of recorded gaze points were included in areas of interest and therefore provided a buffer against small measurement errors. The definition of AOIs is a difficult aspect in eye-tracking studies involving complex scenes, and the analysis of sub-object level eye gaze is not unproblematic as the spatial resolution together with uncertainty in calibration may make it difficult to unambiguously attribute a fixation to a specific object part. Other studies investigating attention to object parts (e.g., Althaus and Plunkett 2015, 2016) have avoided the problem of AOIs being too close in space by designing stimuli with individual parts that are clearly separated.

Gaze-Contingent Eye Tracking

The fully automatic processing of eye movement data provides an additional tool for researchers in experimental design, rather than data accuracy. It is much easier to evaluate gaze on the spot and allow for gaze variables to determine the course of the experiment. Traditionally, looking time has been evaluated on-line in order to determine, for instance, the end of a habituation phase (e.g., Cohen and Strauss 1979; Johnson et al. 2004). Experimenters also sometimes monitor the child's attention in order to initiate a trial, which may involve ad hoc redirection of the child's attention to the screen via a microphone (e.g., with phrases like "Look! What's happening?"— e.g., Althaus et al. 2020). These aspects can occur in a fully automatic way using eye tracking, e.g., by including a gaze-contingent attention getter at the beginning of a trial, which consists of a small, animated looking target at the center of the screen, potentially with accompanying sound. Once the eye tracker registers gaze on the target (or rather, within an AOI defined around it), the trial proper is initiated. However, gaze-contingent procedures also allow more specific experimental control. For instance, Sučević et al. (2021) used gaze-contingent AOIs in a categorization study to either trigger a sound file in which the object currently being inspected by the infant is labeled, or to trigger object motion, i.e., the presentation of a video sequence. Here, gaze-contingent elements were used in order to establish a mapping between object and label/motion. The study demonstrated not only that both labels and motion patterns facilitated infants' categorization, but looking patterns also revealed that labels and motion led to similarity-focused exploration (as evidenced by high rates in shifting between items from the same category). However, one open question is to what extent the introduction of gaze as a control mechanism affects the child's performance, i.e., whether infants learn that they are in control of these additional stimuli. Wang et al. (2012) demonstrated that in a simple visual setting even six-month-olds rapidly learn to move their eyes in a way that triggers rewarding stimuli (here, gaze directed at a red dot triggered the appearance of a photograph in a different location on the screen, whereas a different red dot that was equally visually attractive did not have this gaze-contingent function). Anticipatory gaze shifts toward the photograph's location indicated additionally that infants learned to expect the resulting visual change.

As a tool, gaze-contingent tracking offers opportunities for studying social interactions and what social behavior is grounded in. So, for instance, Deligianni et al. (2011) demonstrated that it is possible to elicit orienting behavior from eight-month-old infants by animating objects in a gaze-contingent way. Here, objects on the screen "responded" to infants' gaze by tilting in a specific direction. Similar to gaze following, this caused infants to shift their gaze in the same direction, clearly demonstrating that this be-havior does not depend on the presence of a face or eyes. Investigating social develop-ment in two-year-olds at risk for autism spectrum disorder (ASD), Vernetti et al. (2018) used three gaze-contingent conditions that allowed the toddlers to choose (a) between a social (face) and a nonsocial (toy) stimulus, which each were animated in response to looking behavior, (b) between a more (face turned towards toddler) or less (face turned away) engaging face stimulus, and (c) between an invariant (always "hello") or a variable (different phrases) video being played in response to gaze. Using these three conditions, the researchers were able to distinguish between a preference for social interaction, which was displayed by all children in the study, and a preference for variability of the social interaction, which was exhibited by toddlers not diagnosed with ASD by age three, but not by those who received a diagnosis. Clearly, the use of gaze-contingent tracking as a way for the participant to seek out specific stimuli, thereby choosing what they will be exposed to, is a powerful tool to disentangle complex cognitive processes.

A very different use of gaze-contingent tracking is as a mechanism in attentional con-trol training. A seminal study was conducted by Wass et al. (2011), who used several gaze-contingent tasks for attentional training with 11-month-old infants. Tasks involved for instance maintaining attention on a target object (a butterfly) while ignoring distracters, or re-orienting after distraction to one of several windows that had previ-ously contained a target. On average, participants received 77 minutes of training. The authors found improvements in cognitive control and sustained attention and reduced saccadic reaction times. Importantly, a trend was also found for looking behavior during free play after training in the sense that children shifted looking faster and more often between an object and the caregiver.

While gaze-contingent eye tracking is evidently a powerful tool for "interactive" studies, it is also highly dependent on calibration quality. Depending on how small the gaze-contingent AOI is, a calibration that is even slightly noisy can lead to looks at target not triggering the gaze-contingent event, or conversely, the gaze-contingent event may be triggered "accidentally," without a subject actively gazing at the target. For the analysis of looking time data, gaze-contingent paradigms also bring challenges, given the fact that each participant's experience may be different, depending on where they looked, and when. However, it is clear that gaze-contingent tracking holds a lot of potential to investigate cognitive processes in an eye-movement based scenario where interaction with the participant is required.

Pupillometry

Pupil size, as mentioned previously, provides an additional metric that indexes cogni-tive processes. This allows tapping into a different physiological system to assess infants' attentive state and arousal. Jackson and Sirois' (2009) seminal study using a violation-of-expectation paradigm with eight-month-olds showed how pupil dilation can dis-entangle looking time results, which in the absence of pupil data would be difficult to interpret. After familiarizing infants with video sequences of toy trains of different colors entering a tunnel and re-emerging from it, they were presented with test trials that were either possible but used a novel train color, impossible with a color switch as the train emerged from the tunnel, using familiar colors, or impossible but with a switch involving novel colors. For looking time there was an interaction of plausibility and familiarity with impossible events leading to lower looking times for familiar colors, but longer looking times for novel colors. However, pupil data clearly showed that the time-locked pupil size only increased on presentation of an impossible event when it was combined with a novel color, making it clear that infants only reliably detected the violation in the context of a salient perceptual feature change.

Over the past decade, pupillometry has been used more widely in infancy research. For example, Hellmer et al. (2018) demonstrated that by seven months, the pupil response reflects whether or not a particular image has been seen or not, similar to adults (where four-month-olds do not yet exhibit this effect). Similarly, pupil size indexes performance on the change detection task (Ross-Sheehy and Eschmann 2019).

But using pupillometry in a developing population has also provided novel insights in language development. Byers-Heinlein et al. (2017) showed that bilingual children show decreased preferential target looking for sentences involving code-switching to the nondominant language (e.g., "Find the *chien*!" compared to "Find the dog!"), and used pupil size measurements to demonstrate that cognitive load drove this effect.

One advantage compared to looking time studies is that the methodology does not rely on a preference measure, or paired presentations. This means that in the context of IPL, where normally the use of distracter items is necessary, this can now be avoided. By definition, distracters are often not a variable of interest in experimental design, and yet need to be carefully controlled in order to avoid spurious biases. Using pupillometry can provide a significant advantage here, specifically where subtle effects are to be detected. Thus Tamási et al. (2017) used pupillometry to measure toddlers' graded mispronunciation sensitivity, systematically varying the number of phonological features differing from the accurate pronunciation of the onset sound of a target word (e.g., *pony*, correct—*tony*, one-feature difference—*vony*, two-feature difference—*zony*, three-feature difference). Toddlers' pupil size change after word onset differed systematically with the degree of mispronunciation. Investigating instead the learning of novel words, Nencheva et al. (2021) used pupillometry to examine toddlers' attentional patterns to child-directed speech. They not only demonstrated that child-directed speech leads to more synchronous patterns in two-year-olds' pupil size, but also that the use of specific, attention-attracting contours (so-called hill-shapes, with a rising-then-falling pitch) is directly related to word learning success.

Head-Mounted Eye Tracking

In particular, studies of naturalistic dyadic interactions have recently benefited from the addition of head-mounted eye trackers (see Smith, Yu, Yoshida et al. 2015 for a review). Here, infants are outfitted with a head-mounted camera and eye-tracking device, offering insight into their first-person view and allowing researchers to determine which object in their visual field attracts infants' attention throughout an interaction. Early approaches uncovered that toddlers' visual experience involves different dynamics compared to that of adults, often switching between single objects that are dominant in the first-person view, compared to an adult's view that is more stable, encompassing several objects at a time (Smith et al. 2011). Yu and Smith (2012) established, using this technique, that successful naming scenarios, which result in successful word recognition performance later, are typically scenes in which the caregiver labels an object that is dominant in the child's view at the time. Unsuccessful naming events (which did not result in successful word recognition) by contrast usually did not show the target object in a close-up view. Looking at these dynamics from a more long-term perspective, Slone et al. (2019) found that toddlers who produced more variable views of the same object (by manually rotating and moving the item) at 15 months showed better vocabulary learning over the six following months. This demonstrated not just that variability in visual exposure is an important factor in learning about objects and eventually words, but also that the learning child themselves takes an active role in achieving this.

Researchers have also used head-mounted eye-tracking set-ups to illuminate our understanding of how face perception skills relate to face exposure early in life. Thus, Sugden and Moulson (2017) reported that almost 90% of one- and three-month-olds' exposure to faces is in an upright position during normal everyday life. Jayaraman et al. (2015) reported a marked decline in the amount of exposure to faces over the first year of life, with very young infants receiving a lot of exposure not just in terms of absolute time (about one quarter of recorded video frames), but also in terms of persistent exposure to the same faces (Jayaraman and Smith 2019). After this, the proportion of video frames with faces decreases to less than 10% by 11 months (Jayaraman et al. 2015), with *hands* becoming more prominent instead (Fausey et al. 2016). Examining hand-eye coordination using a similar head-mounted set-up with 11- to 24-month-olds Yu and Smith (2017) demonstrated the emergence of joint attentional events, in which caregiver and child attend to the same object, a crucial basis for word learning.

Head-mounted eye tracking brings its own set of challenges, such as for example the need for scoring looking towards entirely unpredictable, dynamic stimuli where AOIs cannot be identified a priori. However, the advances that have already been made in this area make it clear that head-mounted eye trackers have a large role to play in future research into children's social interactions and curiosity driven exploration of the world.

CONCLUSION

There are more looking time and eye-tracking paradigms for infancy research than this review could cover. The aim of this chapter was to outline the versatility and scope of looking time and eye tracking research. We began the overview with a short introduction on the characteristics of eye movements as such, and moved on to a review of looking time methodologies in developmental science. The chapter first focused on classic looking time studies and the kinds of approaches that were developed with manually scored camera-based systems that offer limited spatial and temporal resolution by comparison to modern eye-tracking systems. These types of studies established the methodological foundations every infant study concerned with gaze behavior builds upon, such as familiarization/habituation and novelty preference procedures. In the final part of the chapter we moved on to those approaches in studying infant gaze that were only made possible by the advent of automatic eye-tracking systems, with higher tracking frequencies, precise spatial localization of the gaze target, and immediately accessible digital data that can be used in gaze-contingent paradigms to change the experimental flow. The very final section covered the first studies using head-mounted eye tracking, an expanding field that without doubt will provide a wealth of insight into young children's experience in the world, as the field of Developmental Science gradually ventures beyond laboratory-based studies and out into the real world. As the sheer variety of methods and areas of application demonstrate, looking time studies hold a firm place as one of the main branches of research in cognitive development.

References

Ackermann, L., Hepach, R., and Mani, N. (2020). Children learn words easier when they are interested in the category to which the word belongs. *Developmental Science*, 23(3), 1–12.

Addyman, C. and Mareschal, D. (2010). The perceptual origins of the abstract same/different concept in human infants. *Animal Cognition*, 13(6), 817–833.

Albareda-Castellot, B., Pons, F., and Sebastián-Gallés, N. (2011). The acquisition of phonetic categories in bilingual infants: New data from an anticipatory eye movement paradigm. *Developmental Science*, 14(2), 395–401.

Althaus, N., Gliozzi, V., Mayor, J., et al. (2020). Infant categorization as a dynamic process linked to memory: Infant categorisation linked to memory. *Royal Society Open Science*, 7(10). https://royalsocietypublishing.org/doi/full/10.1098/rsos.200328

Althaus, N. and Mareschal, D. (2012). Using saliency maps to separate competing processes in infant visual cognition. *Child Development*, 83(4), 1122–1128.

Althaus, N. and Mareschal, D. (2014). Labels direct infants' attention to commonalities during novel category learning. *PLoS ONE*, 9(7), e99670. https://journals.plos.org/plosone/article/citation?id=10.1371/journal.pone.0099670

Althaus, N. and Plunkett, K. (2015). Timing matters: The impact of label synchrony on infant categorisation. *Cognition*, 139, 1–9.

Althaus, N. and Plunkett, K. (2016). Categorization in infancy: Labeling induces a persisting focus on commonalities. *Developmental Science*, 19(5), 770–780.

Amso, D., Haas, S., and Markant, J. (2014). An eye tracking investigation of developmental change in bottom-up attention orienting to faces in cluttered natural scenes. *PLoS ONE*, 9(1), 1–7.

Arias-Trejo, N. and Plunkett, K. (2009). Lexical-semantic priming effects during infancy. *Philosophical Transactions of the Royal Society B: Biological Sciences*, 364(1536), 3633–3647.

Aslin, R. N. (1981). Development of smooth pursuit in human infants. In D. F. Fisher, R. A. Monty, and J. W. Senders (eds.), *Eye Movements: Cognition and Visual Perception* (pp. 31–51). Hillsdale, NJ: Lawrence Erlbaum Associates, Inc.

Aslin, R. N. and McMurray, B. (2004). Automated corneal-reflection eye tracking in infancy: Methodological developments and applications to cognition. *Infancy*, 6(2), 155–163.

Aslin, R. N. and Salapatek, P. (1975). Saccadic localization of visual targets by the very young human infant. *Perception & Psychophysics*, 17(3), 293–302.

Atkinson, J. (1984). Human visual development over the first six months of life. A review and a hypothesis. *Human Neurobiology*, 3, 61–74.

Atkinson, J. (2000). *The Developing Visual Brain*. Oxford: Oxford University Press.

Baillargeon, R. (1986). Representing the existence and the location of hidden objects: Object permanence in 6- and 8-month-old infants. *Cognition*, 23(1), 21–41.

Baillargeon, R. (2004). Infants' physical world. *Current Directions in Psychological Science*, 13(3), 89–94.

Bailey, T. M. and Plunkett, K. (2002). Phonological specificity in early words. *Cognitive Development*, 17(2), 1265–1282.

Ballem, K. D. and Plunkett, K. (2005). Phonological specificity in children at 1;2. *Journal of Child Language*, 32(1), 159–173.

Barnes, G. R. (2008). Cognitive processes involved in smooth pursuit eye movements. *Brain and Cognition*, 68(3), 309–326.

Behl-Chadha, G. (1996). Basic-level and superordinate-like categorical representations in early infancy. *Cognition*, 60(2), 105–141.

Bergelson, E. and Swingley, D. (2012). At 6-9 months, human infants know the meanings of many common nouns. *Proceedings of the National Academy of Sciences*, 109(9), 3253–8.

Bertenthal, B. I., Proffitt, D. R., and Cutting, J. E. (1984). Infant sensitivity to figural coherence in biomechanical motions. *Journal of Experimental Child Psychology*, 37(2), 213–230.

Bomba, P. C. and Siqueland, E. R. (1983). The nature and structure of infant form categories. *Journal of Experimental Child Psychology*, 35(2), 294–328.

Braddick, O. and Atkinson, J. (2011). Development of human visual function. *Vision Research*, 51(13), 1588–1609.

Bushnell, E. W. and Roder, B. J. (1985). Recognition of color-form compounds by 4-month-old infants. *Infant Behavior and Development*, 8(3), 255–268.

Byers-Heinlein, K., Morin-Lessard, E., and Lew-Williams, C. (2017). Bilingual infants control their languages as they listen. *Proceedings of the National Academy of Sciences*, 114(34), 9032–9037.

Chow, J., Aimola Davies, A., and Plunkett, K. (2017). Spoken-word recognition in 2-year-olds: The tug of war between phonological and semantic activation. *Journal of Memory and Language*, 93, 104–134.

Cohen, L. B. and Strauss, M. S. (1979). Concept acquisition in the human infant. *Child Development*, 50(2), 419–424.

Cohen, L. B. (2004). Uses and misuses of habituation and related preference paradigms. *Infant and Child Development: An International Journal of Research and Practice*, 13(4), 349–352.

Colombo, J. and Mitchell, D. W. (2009). Infant visual habituation. *Neurobiology of Learning and Memory*, 92(2), 225–234.

Cornsweet, T. N. and Crane, H. D. (1973). Accurate two-dimensional eye tracker using first and fourth Purkinje images. *Journal of the Optical Society of America*, 63(8), 921–928.

Csibra, G., Tucker, L. A., and Johnson, M. H. (1998). Neural correlates of saccade planning in infants: A high-density ERP study. *International Journal of Psychophysiology*, 29(2), 201–215.

Deligianni, F., Senju, A., Gergely, G., et al. (2011). Automated gaze-contingent objects elicit orientation following in 8-month-old infants. *Developmental Psychology*, 47(6), 1499–1503.

Delle Luche, C., Durrant, S., Poltrock, S., et al. (2015). A methodological investigation of the intermodal preferential looking paradigm: Methods of analyses, picture selection and data rejection criteria. *Infant Behavior and Development, 40*, 151–172.

Duchowski, A. T. (2007). *Eye Tracking Methodology: Theory and Practice*. Springer.

Eimas, P. D. and Quinn, P. C. (1994). Studies on the formation of perceptually based basic-level categories in young infants. *Child Development, 65*(3), 903–17.

Einhäuser, W., Spain, M., and Perona, P. (2008). Objects predict fixations better than early saliency. *Journal of Vision, 8*(14), 18.1–26.

Falck-Ytter, T., Gredebäck, G., and von Hofsten, C. (2006). Infants predict other people's action goals. *Nature Neuroscience, 9*(7), 878–879.

Fantz, R. L. (1964). Visual experience in infants: Decreased attention to familiar patterns relative to novel ones. *Science, 146*(3644), 668–670.

Fausey, C. M., Jayaraman, S., and Smith, L. B. (2016). From faces to hands: Changing visual input in the first two years. *Cognition, 152*, 101–107.

Fernald, A., Pinto, J. P., and Swingley, D. (1998). Rapid gains in the speed and efficiency of word recognition by infants in the second year. *Psychological Science, 9*(3), 228–231.

Fernald, A., Swingley, D., and Pinto, J. P. (2001). When half a word is enough: Infants can recognize spoken words using partial phonetic information. *Child Development, 72*(4), 1003–1015.

Fernald, A., Zangl, R., Portillo, A. L., et al. (2008). Looking while listening: Using eye movements to monitor spoken language. *Developmental Psycholinguistics: On-line Methods in Children's Language Processing, 44*, 97.

Frankenhuis, W. E., House, B., Barrett, H. C., et al. (2013). Infants' perception of chasing. *Cognition, 126*(2), 224–233.

Franklin, A., Drivonikou, G. V., Bevis, L., et al. (2008). Categorical perception of color is lateralized to the right hemisphere in infants, but to the left hemisphere in adults. *Proceedings of the National Academy of Sciences of the United States of America, 105*(9), 3221–3225.

Gegenfurtner, K. R. (2016). The interaction between vision and eye movements. *Perception, 45*(12), 1333–1357.

Gliga, T., Elsabbagh, M., Andravizou, A., et al. (2009). Faces attract infants' attention in complex displays. *Infancy, 14*(5), 550–562.

Gliga, T., Mareschal, D., and Johnson, M. H. (2008). Ten-month-olds' selective use of visual dimensions in category learning. *Infant Behavior and Development, 31*(2), 287–293.

Golinkoff, R. M., Hirsh-Pasek, K., Cauley, K. M., et al. (1987). The eyes have it: Lexical and syntactic comprehension in a new paradigm. *Journal of Child Language, 14*(01), 23–45.

Golinkoff, R. M., Ma, W., Song, L., et al. (2013). Twenty-five years using the intermodal preferential looking paradigm to study language acquisition: What have we learned? *Perspectives on Psychological Science, 8*(3), 316–339.

Gredebäck, G., Johnson, S., and Von Hofsten, C. (2010). Eye tracking in infancy research. *Developmental Neuropsychology, 35*(1), 1–19.

Gredebäck, G., Örnkloo, H., and Von Hofsten, C. (2006). The development of reactive saccade latencies. *Experimental Brain Research, 173*(1), 159–164.

Gredebäck, G., Stasiewicz, D., Falck-Ytter, T., et al. (2009). Action type and goal type modulate goal-directed gaze shifts in 14-month-old infants. *Developmental Psychology, 45*(4), 1190.

Grimes, J. (1996). On the failure to detect changes in scenes across saccades. In K. A. Akins (ed.), *Vancouver Studies in Cognitive Science*, Vol. 5. *Perception*, 89–110. Oxford: Oxford University Press.

Hayhoe, M. and Ballard, D. (2005). Eye movements in natural behavior. *Trends in Cognitive Sciences, 9*(4), 188–194.

Hellmer, K., Söderlund, H., and Gredebäck, G. (2018). The eye of the retriever: Developing episodic memory mechanisms in preverbal infants assessed through pupil dilation. *Developmental Science, 21*(2), 1–10.

Helo, A., Ommen, S. Van, Pannasch, S., Danteny-dordoigne, L., et al. (2017). Infant behavior and development in fluence of semantic consistency and perceptual features on visual attention during scene viewing in toddlers. *Infant Behavior and Development, 49*, 248–266.

Henderson, J. M. and Hollingworth, A. (2012). Global change blindness during scene perception. *Psychological Science, 14*(5), 493–497.

Henderson, J. M. and Luke, S. G. (2014). Stable individual differences in saccadic eye movements during reading, pseudoreading, scene viewing, and scene search. *Journal of Experimental Psychology: Human Perception and Performance, 40*(4), 1390–1400.

Henderson, J. M., Weeks Jr, P. A., and Hollingworth, A. (1999). The effects of semantic consistency on eye movements during complex scene viewing. *Journal of Experimental Psychology: Human Perception and Performance, 25*(1), 210.

Hepach, R. and Westermann, G. (2016). Pupillometry in infancy research. *Journal of Cognition and Development, 17*(3), 359–377.

Hessels, R. S., Andersson, R., Hooge, I. T. C., et al. (2015). Consequences of eye color, positioning, and head movement for eye-tracking data quality in infant research. *Infancy, 20*(6), 601–633.

Hessels, R. S., Hooge, I. T. C., and Kemner, C. (2016). An in-depth look at saccadic search in infancy. *Journal of Vision, 16*(8):10, 1–14.

Hessels, R. S., Niehorster, D. C., Kemner, C., et al. (2017). Noise-robust fixation detection in eye movement data: Identification by two-means clustering (I2MC). *Behavior Research Methods, 49*(5), 1802–1823.

Holmqvist, K., Nyström, M., Andersson, R., et al. (2011). *Eye Tracking: A Comprehensive Guide to Methods and Measures*. Oxford: Oxford University Press.

Hood, B. M. and Atkinson, J. (1993). Disengaging visual attention in the infant and adult. *Infant Behavior and Development, 16*(4), 405–422.

Horowitz, F. D., Paden, L., Bhana, K., et al. (1972). An infant-controlled procedure for studying infant visual fixations. *Developmental Psychology, 7*, 90.

Houston-Price, C., Mather, E., and Sakkalou, E. (2007). Discrepancy between parental reports of infants' receptive vocabulary and infants' behaviour in a preferential looking task. *Journal of Child Language, 34*(4), 701–724.

Houston-Price, C. and Nakai, S. (2004). Distinguishing novelty and familiarity effects in infant preference procedures. *Infant and Child Development: An International Journal of Research and Practice, 13*(4), 341–348.

Hunter, M. A. and Ames, E. W. (1988). *A multifactor model of infant preferences for novel and familiar stimuli*. In C. Rovee-Collier and L. P. Lipsitt (eds.), *Advances in Infancy Research*, Vol. 5 (pp. 69–95). New York: Ablex Publishing.

Hurley, K. B. and Oakes, L. M. (2015). Experience and distribution of attention: Pet exposure and infants' scanning of animal images. *Journal of Cognition and Development, 16*(1), 11–30.

Itti, L. and Koch, C. (2001). Computational modelling of visual attention. *Nature Reviews Neuroscience, 2*(3), 194–203.

Itti, L., Koch, C., and Niebur, E. (1998). A model of saliency-based visual attention for rapid scene analysis. *IEEE Transactions on Pattern Analysis and Machine Intelligence, 20*(11), 1254–1259.

Jackson, I. and Sirois, S. (2009). Infant cognition: Going full factorial with pupil dilation. *Developmental Science*, 12(4), 670–679.

Jayaraman, S., Fausey, C. M., and Smith, L. B. (2015). The faces in infant-perspective scenes change over the first year of life. *PloS One*, 10(5), e0123780.

Jayaraman, S. and Smith, L. B. (2019). Faces in early visual environments are persistent not just frequent. *Vision Research*, 157, 213–221.

Johnson, M. H. (1990). Cortical maturation and the development of visual attention in early infancy. *Journal of Cognitive Neuroscience*, 2, 81–95.

Johnson, M. H., Dziurawiec, S., Ellis, H., et al. (1991). Newborns' preferential tracking of face-like stimuli and its subsequent decline. *Cognition*, 40(1–2), 1–19.

Johnson, S. P., Slemmer, J. A., Amso, D., et al. (2004). Where infants look determines how they see: Eye movements and development of object perception. *Infancy*, 6(8), 185–201.

Kaldy, Z., Guillory, S. B., and Blaser, E. (2016). Delayed match retrieval: A novel anticipation-based visual working memory paradigm. *Developmental Science*, 19(6), 892–900.

Kelly, D. J., Duarte, S., Meary, D., et al. (2019). Infants rapidly detect human faces in complex naturalistic visual scenes. *Developmental Science*, 22(6), e12829.

Kenward, B., Koch, F.-S., Forssman, et al. (2017). Saccadic reaction times in infants and adults: Spatiotemporal factors, gender, and interlaboratory variation. *Developmental Psychology*, 53(9), 1750–1764.

Kienzle, W., Franz, M. O., Schölkopf, B., et al. (2009). Center-surround patterns emerge as optimal predictors for human saccade targets. *Journal of Vision*, 9(5):7, 1–15. https://jov.arvoj ournals.org/article.aspx?articleid=2122672

Kirchner, H. and Thorpe, S. J. (2006). Ultra-rapid object detection with saccadic eye movements: Visual processing speed revisited. *Vision Research*, 46(11), 1762–1776.

Kowler, E., Anderson, E., Dosher, B., et al. (1995). The role of attention in the programming of saccades. *Vision Research*, 35(13), 1897–1916.

Kovack-Lesh, K. A., McMurray, B., and Oakes, L. M. (2014). Four-month-old infants' visual investigation of cats and dogs: Relations with pet experience and attentional strategy. *Developmental Psychology*, 50(2), 402–413.

Kovack-Lesh, K. A. and Oakes, L. M. (2007). Hold your horses: How exposure to different items influences infant categorization. *Journal of Experimental Child Psychology*, 98(2), 69–93.

Krauzlis, R. J., Goffart, L., and Hafed, Z. M. (2017). Neuronal control of fixation and fixational eye movements. *Philosophical Transactions of the Royal Society B: Biological Sciences*, 372(1718), 20160205.

Kremenitzer, J. P., Vaughan, H. G., Kurtzberg, D., et al. (1979). Smooth-pursuit eye movements in the newborn infant. *Child Development*, 50(2), 442–448.

Laeng, B., Sirois, S., and Gredebäck, G. (2012). Pupillometry: A window to the preconscious? *Perspectives on Psychological Science*, 7(1), 18–27.

Lany, J. (2018). Lexical-processing efficiency leverages novel word learning in infants and toddlers. *Developmental Science*, 21(3), 1–12.

Lewkowicz, D. J. and Hansen-Tift, A. M. (2012). Infants deploy selective attention to the mouth of a talking face when learning speech. *Proceedings of the National Academy of Sciences of the United States of America*, 109(5), 1431–1436.

Mani, N. and Plunkett, K. (2010). In the infant's mind's ear: Evidence for implicit naming in 18-month-olds. *Psychological Science*, 21(7), 908–913.

Mani, N. and Plunkett, K. (2011). Does size matter? Subsegmental cues to vowel mispronunciation detection. *Journal of Child Language*, 38(03), 606–627.

Martinez-Conde, S., Macknik, S. L., Troncoso, X. G., et al. (2006). Microsaccades counteract visual fading during fixation. *Neuron*, 49(2), 297–305.

Mather, E. and Plunkett, K. (2011). Same items, different order: Effects of temporal variability on infant categorization. *Cognition*, 119(3), 438–447.

Mathôt, S. (2018). Pupillometry: Psychology, physiology, and function. *Journal of Cognition*, 1(1), 16.

Matin, E. (1974). Saccadic suppression: A review and an analysis. *Psychological Bulletin*, 81(12), 899.

McMurray, B. and Aslin, R. N. (2004). Anticipatory eye movements reveal infants' auditory and visual categories. *Infancy*, 6(2), 203–229.

Melzer, A., Prinz, W., and Daum, M. M. (2012). Production and perception of contralateral reaching: A close link by 12 months of age. *Infant Behavior and Development*, 35(3), 570–579.

Mirman, D. (2014). *Growth Curve Analysis and Visualization using R*. Boca Raton, FL: CRC Press/Taylor & Francis Group.

Monroy, C. D., Gerson, S. A., and Hunnius, S. (2017). Toddlers' action prediction: Statistical learning of continuous action sequences. *Journal of experimental child psychology*, 157, 14–28.

Najemnik, J. and Geisler, W. S. (2005). Optimal eye movement strategies in visual search. *Nature*, 434(7031), 387–391.

Nencheva, M. L., Piazza, E. A., and Lew-Williams, C. (2021). The moment-to-moment pitch dynamics of child-directed speech shape toddlers' attention and learning. *Developmental Science*, 24(1), 1–15.

Niehorster, D. C., Cornelissen, T. H. W., Holmqvist, K., et al. (2018). What to expect from your remote eye-tracker when participants are unrestrained. *Behavior Research Methods*, 50(1), 213–227.

Nyström, P., Gredebäck, G., Bölte, S., et al. (2015). Hypersensitive pupillary light reflex in infants at risk for autism. *Molecular Autism*, 6(1), 1–6.

Oakes, L. M. (2010). Using habituation of looking time to assess mental processes in infancy. *Journal of Cognitive Development*, 11(3), 255–268.

Oakes, L. M., Kovack-Lesh, K. A., and Horst, J. S. (2009). Two are better than one: Comparison influences infants' visual recognition memory. *Journal of Experimental Child Psychology*, 104(1), 124–131.

Oakes, L. M. and Ribar, R. J. (2005). A comparison of infants' categorization in paired and successive presentation familiarization tasks. *Infancy*, 7(1), 85–98.

Öhlschläger, S. and Võ, M. L. (2020). Development of scene knowledge: Evidence from explicit and implicit scene knowledge measures, *Journal of Experimental Child Psychology*, 194, 104782.

Papageorgiou, K. A., Smith, T. J., Wu, R., et al. (2014). Individual differences in infant fixation duration relate to attention and behavioral control in childhood. *Psychological Science*, 25(7), 1371–1379.

Pulverman, R., Golinkoff, R. M., Hirsh-Pasek, K., et al. (2008). Infants discriminate manners and paths in non-linguistic dynamic events. *Cognition*, 108(3), 825–830.

Quinn, P. C., Eimas, P. D., and Rosenkrantz, S. L. (1993). Evidence for representations of perceptually similar natural categories by 3-month-old and 4-month-old infants. *Perception*, 22(4), 463–475.

Quinn, P. C., Eimas, P. D., and Tarr, M. J. (2001). perceptual categorization of cat and dog silhouettes by 3- to 4-month-old infants. *Journal of Experimental Child Psychology*, 79(1), 78–94.

Rakison, D. H. and Yermolayeva, Y. (2010). Infant categorization. *Wiley Interdisciplinary Reviews: Cognitive Science*, 1(6), 894–905.

Ren, J., Cohen Priva, U., and Morgan, J. L. (2019). Underspecification in toddlers' and adults' lexical representations. *Cognition*, 19, 103991.

Richardson, D. C., Dale, R., and Spivey, M. J. (2007). Eye movements in language and cognition. *Methods in Cognitive Linguistics*, 18, 323–344.

Ross-Sheehy, S. and Eschman, B. (2019). Assessing visual STM in infants and adults: Eye movements and pupil dynamics reflect memory maintenance. *Visual Cognition*, 27(1), 78–92.

Ross-Sheehy, S., Oakes, L. M., and Luck, S. J. (2003). The development of visual short-term memory capacity in infants. *Child Development*, 74(6), 1807–1822.

Saez de Urabain, I. R., Johnson, M. H., and Smith, T. J. (2015). GraFIX: A semiautomatic approach for parsing low- and high-quality eye-tracking data. *Behavior Research Methods*, 47(1), 53–72.

Salman, M. S., Sharpe, J. A., Lillakas, L., et al. (2006). Smooth pursuit eye movements in children. *Experimental Brain Research*, 169(1), 139–143.

Schafer, G. and Plunkett, K. (1998). Rapid word learning by fifteen-month-olds under tightly controlled conditions. *Child Development*, 69(2), 309–320.

Schlegelmilch, K. and Wertz, A. E. (2019). The effects of calibration target, screen location, and movement type on infant eye-tracking data quality. *Infancy*, 24(4), 636–662.

Schütz, A. C., Braun, D. I., and Gegenfurtner, K. R. (2011). Eye movements and perception: A selective review. *Journal of Vision*, 11(5), 1–30.

Shukla, M. and de Villiers, J. (2021). The role of language in building abstract, generalized conceptual representations of one-and two-place predicates: A comparison between adults and infants. *Cognition*, 213, 104705.

Simmering, V. R. (2012). The development of visual working memory capacity during early childhood. *Journal of Experimental Child Psychology*, 111(4), 695–707.

Sirois, S. and Brisson, J. (2014). Pupillometry. *Wiley Interdisciplinary Reviews: Cognitive Science*, 5(6), 679–692.

Slone, L. K., Smith, L. B., and Yu, C. (2019). Self-generated variability in object images predicts vocabulary growth. *Developmental Science*, 22(6), e12816.

Smith, L. B., Yu, C., and Pereira, A. F. (2011). Not your mother's view: The dynamics of toddler visual experience. *Developmental Science*, 14(1), 9–17.

Smith, L. B., Yu, C., Yoshida, H., et al. (2015). Contributions of head-mounted cameras to studying the visual environments of infants and young children. *Journal of Cognition and Development*, 16(3), 407–419.

Sokolov, E. N. (1963). *Perception and the Conditioned Reflex*. New York: Macmillan.

Spelke, E. S., Breinlinger, K., Macomber, J., et al. (1992). Origins of knowledge. *Psychological Review*, 99(4), 605.

SR Research Ltd. (2008). *Eye Link 1000 User Manual*. Version 1.4.0. Ontario, Canada: Copyright SR Research Ltd. Mississauga.

Stager, C. L. and Werker, J. F. (1997). Infants listen for more phonetic detail in speech perception than in word-learning tasks. *Nature*, 388(6640), 381–382.

Styles, S. J. and Plunkett, K. (2009a). What is "word understanding" for the parent of a one-year-old? Matching the difficulty of a lexical comprehension task to parental CDI report. *Journal of Child Language*, 36(4), 895–908.

Styles, S. J. and Plunkett, K. (2009b). How do infants build a semantic system? *Language and Cognition*, 1(01), 1–24.

Sučević, J., Althaus, N., and Plunkett, K. (2021). The role of labels and motions in infant category learning. *Journal of Experimental Child Psychology*, 205, 105062.

Sugden, N. A. and Moulson, M. C. (2017). Hey baby, what's "up"? One- and 3-month-olds experience faces primarily upright but non-upright faces offer the best views. *Quarterly Journal of Experimental Psychology*, 70(5), 959–969.

Swingley, D. and Aslin, R. N. (2000). Spoken word recognition and lexical representation in very young children. *Cognition*, 76, 147–166.

Swingley, D. and Aslin, R. N. (2002). Lexical neighborhoods and the word form representations of 14-month-olds. *Psychological Science*, 13(5), 480–484.

Swingley, D., Pinto, J. P., and Fernald, A. (1999). Continuous processing in word recognition at 24 months. *Cognition*, 71, 73–108.

Tamási, K., McKean, C., Gafos, A., et al. (2017). Pupillometry registers toddlers' sensitivity to degrees of mispronunciation. *Journal of Experimental Child Psychology*, 153, 140–148.

Thiele, A., Henning, P., Kubischik, M., et al. (2002). Neural mechanisms of saccadic suppression. *Science*, 295(5564), 2460–2462.

Tobii Technology AB (2016). *Tobii Studio User's Manual*. Version 3.4.5. Copyright Tobii Technology AB, Stockholm, Sweden.

Wang, Q., Bolhuis, J., Rothkopf, C. A., et al. (2012). Infants in control: Rapid anticipation of action outcomes in a gaze-contingent paradigm. *PloS One*, 7(2), e30884.

Wass, S. V., Jones, E. J. H., Gliga, T., et al. (2015). Shorter spontaneous fixation durations in infants with later emerging autism. *Scientific Reports*, 5(Lc), 1–8.

Wass, S., Porayska-Pomsta, K., and Johnson, M. H. (2011). Training attentional control in infancy. *Current Biology*, 21(18), 1543–1547.

Wass, S. V., Smith, T. J., and Johnson, M. H. (2013). Parsing eye-tracking data of variable quality to provide accurate fixation duration estimates in infants and adults. *Behavior Research Methods*, 45(1), 229–250.

Westheimer, G. (1954). Eye movement responses to a horizontally moving visual stimulus. *A.M.A. Archives of Ophthalmology*, 52, 932–943.

White, K. S. and Morgan, J. L. (2008). Sub-segmental detail in early lexical representations. *Memory and Language*, 59, 114–132.

van Renswoude, D. R., van den Berg, L., Raijmakers, et al. (2019). Infants' center bias in free viewing of real-world scenes. *Vision Research*, 154(January 2018), 44–53.

Vernetti, A., Senju, A., the BASIS Team, et al. (2018). Simulating interaction: Using gaze-contingent eye-tracking to measure the reward value of social signals in toddlers with and without autism. *Developmental Cognitive Neuroscience*, 29, 21–29.

von Hofsten, C. and Rosander, K. (1996). The development of gaze control and predictive tracking in young infants. *Vision Research*, 36(1), 81–96.

von Hofsten, C. and Rosander, K. (1997). Development of smooth pursuit tracking in young infants. *Vision Research*, 37(13), 1799–1810.

Werker, J. F., Cohen, L. B., Lloyd, V. L., et al. (1998). Acquisition of word-object associations by 14-month-old infants. *Developmental Psychology*, 34(6), 1289–1309.

Xu, F. and Spelke, E. S. (2000). Large number discrimination in 6-month-old infants. *Cognition*, 74(1), B1–B11.

Xu, F. and Arriaga, R. I. (2007). Number discrimination in 10-month-old infants. *British Journal of Developmental Psychology*, 25(1), 103–108.

Yang, Q., Bucci, M. P., and Kapoula, Z. (2002). The latency of saccades, vergence, and combined eye movements in children and in adults. *Investigative Ophthalmology & Visual Science*, 43(9), 2939–2949.

Yarbus, A. (1967). *Eye Movements and Vision*. New York: Plenum Press.

Yoshida, K. A., Fennell, C. T., Swingley, D., et al. (2009). Fourteen-month-old infants learn similar-sounding words. *Developmental Science, 3*, 412–418.

Younger, B. A. (1985). The segregation of items into categories by ten-month-old infants. *Child Development, 56*(6), 1574–1583.

Younger, B. A. and Cohen, L. B. (1983). Infant perception of correlations among attributes. *Child Development, 54*(4), 858–867.

Yu, C., & Smith, L. B. (2012). Embodied attention and word learning by toddlers. *Cognition, 125*(2), 244–262.

Yu, C. and Smith, L. B. (2017). Hand–eye coordination predicts joint attention. *Child Development, 88*(6), 2060–2078.

CHAPTER 9

RECOGNIZING FACIAL IDENTITY

Prolonged development during infancy and childhood

CLAIRE M. MATTHEWS, DAPHNE MAURER,
AND CATHERINE J. MONDLOCH

FACE DETECTION

ADULTS readily detect faces; faces capture attention when embedded in complex arrays (Bindemann et al., 2005; Langton et al., 2008) and are even detected in the absence of normal facial features, such as stimuli comprised only of black-and-white contours (Mooney faces: Mondloch et al., 2013b; Schwiedrzik et al., 2018), paintings with objects replacing facial features (e.g., Archimbaldo paintings), and pure-noise images (Liu et al., 2014). Such illusory perception, *pareidolia*, shows how exquisitely the adult perceptual system is tuned to face-like stimuli.

Face detection is associated with two neural markers. The N170 (an event-related potential peaking 170 mos after stimulus onset) is larger for upright faces than for most other stimuli (Bentin et al., 1996; see Eimer, 2011; Rossion and Jacques, 2011, for reviews) and fMRI activation in several regions, most notably the fusiform face area (FFA), is larger for faces than for a variety of nonface objects (Kanwisher et al., 1997; McCarthy et al., 1997), even when the stimuli are ambiguous (Liu et al., 2014; Tong et al., 1998; Hasson et al., 2001).

RECOGNITION OF FACIAL IDENTITY

Two distinct approaches provide abundant evidence that adults are experts at recognizing facial identity. In one approach, researchers present *tightly controlled* images in which all photos are taken with the same camera, from the same distance and under identical lighting conditions; hair is typically masked and blemishes removed. This approach is ideal for measuring the ability to discriminate unfamiliar faces or recognize a face despite systematic, controlled changes (e.g., in lighting or viewpoint). In the second approach, researchers examine the ability to recognize facial identity in *ambient* images—images that capture natural within-person variability in appearance (across variable lighting, makeup, viewpoint, expression, hairstyle, and even age). This approach precludes image recognition, thereby better capturing the type of face processing necessary in everyday life.

Adults perform very well when asked to recognize faces in an old/new recognition task or in a delayed match-to-sample task, at least when identical images are presented during the study and test phases (Duchaine and Nakayama, 2006; Nordt and Weigelt, 2017; Proietti et al., 2019). When faces are wholly unfamiliar, performance drops when nonidentical images of a face are presented—even when the study and test images are taken only moments apart (Megreya et al., 2013) or when a simple change in lighting, viewpoint, or facial expression is made to tightly controlled images (Duchaine and Nakayama, 2006; Nordt and Weigelt, 2017). When faces are familiar, adults recognize them in ambient images (Jenkins et al., 2011; reviewed in Burton, 2013) and even when they are distorted (e.g., stretched; Bindemann et al., 2008; Hole et al., 2002)—highlighting a fundamental difference between the recognition of familiar and unfamiliar faces.

Underlying Processes

Adults' ability to recognize and discriminate faces has been attributed to three underlying processes. 1) Adults process faces as a gestalt; perception of each feature (e.g., the eyes) is influenced by the rest of the face—a pattern dubbed holistic processing (Hole, 1994; Tanaka and Farah, 1993; Tanaka et al., 1998; Young et al., 1987). 2) Adults are exquisitely sensitive to differences in feature shape and spacing, allowing them to detect very small differences among both unfamiliar and familiar faces (Freire et al., 2000; Mondloch et al., 2002; Ge et al., 2003). 3) Adults engage in norm-based coding; each face is represented based on how (e.g., crooked nose), and how much (e.g., large eyes), it differs from a prototype (Rhodes et al., 2014; Valentine, 1991; Valentine et al., 2016).

Adults' ability to recognize familiar faces despite tremendous variability in appearance has been attributed to their building a robust and abstract representation of facial identity—analogous to Face Recognition Units in Bruce and Young's (1986) influential model. This mental representation likely includes an average representation of each person's face that excludes nondiagnostic cues unique to individual instances (Burton et al., 2005) but includes idiosyncratic variability characteristic of each person (Burton et al., 2016). Exposure to within-person variability in appearance facilitates learning such that new instances are recognized (e.g., Andrews et al., 2015; Baker et al., 2017; Dowsett et al., 2016; Murphy et al., 2015; Ritchie and Burton, 2017).

In the remainder of this chapter we provide evidence that, despite rudimentary abilities being present during infancy, the ability to recognize facial identity develops slowly during childhood—a pattern that is attributable to the slow development of some, though not all, underlying processes. Finally, we provide evidence that face-specific experience plays an important role in the development of these skills.

INFANCY

All of these skills have been tested during infancy.

Face Detection

Poor acuity and contrast sensitivity limit the information available to newborns from faces. From closeup, newborns can see only the outline and the most contrasty internal features; farther away, only the outline remains visible (Maurer, 2016; von Hofsten et al., 2014) (see Figure 9.1). Perhaps as a result, newborns fixate mainly on the external contour (Haith et al., 1977; Maurer and Salapatek, 1976). Even when viewing simplified drawings with no bold external contour, newborns look at the internal features only about half the time (Maurer, 1983).

Within minutes of birth, newborns orient preferentially toward an oval with three squares located in the positions of eyes and mouth, a stimulus dubbed *config* because of its face-like configuration, in preference to an oval with those squares inverted (Mondloch et al., 1999; Valenza et al., 1996). At 8 months' gestation, the fetus shows the same orienting preference when bright red blobs are projected onto the mother's abdomen (Reid et al., 2017). In a seminal paper, Morton and Johnson (1991) postulated that the early preference for *config* is mediated by a subcortical mechanism, likely involving the superior colliculus. This mechanism called CONSPEC, they argued, guarantees that

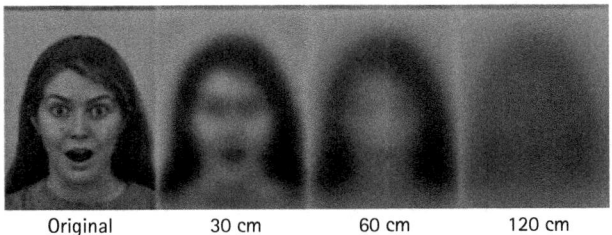

Original   30 cm   60 cm   120 cm

FIGURE 9.1 The information visible to a newborn when viewing a face from various distances. Beyond 30 cm, only the outline remains visible.

Source: used with permission from von Hofsten, O., von Hofsten, C., Sulutvedt, U., Laeng, B., Brennen, T., and Magnussen, S. (2014). Simulating newborn face perception. *Journal of Vision*, 14(13):16, 1–9. Doi: 10.1167/14.13.16

young babies attend to faces, feeding information about facial structure into developing cortical networks that "take over" control around 1–2 months with a mechanism they called CONLERN. Consistent with this theory, when newborns wear a patch over one eye, the preference for *config* is easier to demonstrate in the temporal visual field (e.g., the right visual field for the right eye, which can be mediated without cortical input) than in the nasal visual field (across the nose, which requires input routed through the visual cortex) (Lewis and Maurer, 1992; Maurer et al., 1991; Simion et al., 1998). Subsequent work with adults and animal models suggests that the subcortical mechanism may remain active in adults, supporting fast orienting toward faces in the periphery, especially if threatening, and that the network likely involves the pulvinar and amygdala in addition to the superior colliculus (Johnson et al., 2015).

The neonatal looking preference likely is not the manifestation of an innate face module per se, but rather of more basic preferences that often match a human face. The newborn looks longer at stimuli with an optimal amount of visible energy (Kleiner, 1987; Mondloch et al., 1999), more features in the top than bottom ("top-heaviness") (Macchi Cassia et al., 2004; Simion et al., 2001; Turati et al., 2002; Turati, 2004), and congruency between the widest part of the external contour and of the internal features (Macchi Cassia et al., 2008) (see Figure 9.2). These features match the shape of the newborn's limited visual field—skinny half-pear shapes in front of each eye, cut off by the newborn's nose (Maurer and Maurer, 2019). The looking preference for *config* disappears if the contrast is reversed (Farroni et al., 2005), reflecting a more general bias to attend to dark details against a lighter background that is evident in 3-month-olds (Dannemiller and Stephens, 2001) and adults (Komban et al., 2011) and that matches the modal characteristics of the environment (Maurer and Maurer, 2019).

Improved acuity, contrast sensitivity, and peripheral vision allow babies older than ~ 2 months to receive more information from internal facial features (Lewis and Maurer, 1992; Maurer, 2016). From that age, babies' fixations fall predominantly on the internal features of faces (Hunnius and Geuze, 2004), especially the eyes (Di Giorgio et al., 2013; Haith et al., 1977; Maurer and Salapatek, 1976). At the same time, more complex face

 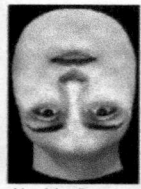 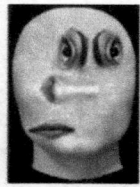   

Upright Face   Upside-Down   Top-Heavy   Bottom-Heavy   Uprigh Face   Top-Heavy
                    Face   Configuration   Configuration   Configuration

FIGURE 9.2 Newborns' "face" preference is mediated, in part, by a bias to look at top-heavy figures. They prefer the stimulus on the left in the left and middle panels, but have no preference between the two stimuli in the right-most panel.

Source: used with permission from: Cassia, V. M., Turati, C., and Simion, F. (2004). Can a nonspecific bias toward top-heavy patterns explain newborns' face preference? *Psychological Science*, 15(6): 379–383. Doi:10.1111/j.0956-7976.2004.00688.x.

preferences develop and override initial structural biases such as top-heaviness: by 3–5 months, babies look at a photograph of a real face in preference to a scrambled image matched in low-level features (Macchi Cassia, Kuefner et al., 2006a; Chien, 2011), a pattern extending to schematic faces (Kleiner, 1987; Mondloch et al., 1999). As in adults, the critical information appears to be in horizontally oriented spatial frequencies (de Heering et al., 2016).

In more complex scenes, older infants orient first and longest toward a face compared to other objects. Though difficult to demonstrate at 3 months, the preference increases over the first year (Kelly et al., 2019). By 6 months, the saliency of faces is so well-established that even in arrays with six stimuli, infants look first and longer toward a face than a body part or an animal (Gluckman and Johnson, 2013; see also Di Giorgio et al., 2012). Inverted faces are as effective as upright faces in capturing first fixations, but not in attracting the longest looking time. Faces with the internal features phase scrambled are ineffective at either (Gliga et al., 2009), suggesting a role for facial structure.

Neural Underpinnings

A variety of techniques have established some overlap between the adult face network and areas active during face detection in older infants. A pioneering study using PET with 2-month-olds found that faces, in comparison to flickering Christmas tree lights, activated a large, distributed network predominantly in the right hemisphere, including the fusiform face area. The network overlapped with the adult face-processing network but extended beyond to include the left superior temporal and inferior frontal gyri, areas that will become specialized for language (Tzourio-Mazoyer et al., 2002).

Only one study has examined the neural basis of newborns' preference for *config*. It did so by slowly turning the stimulus on and off and then examining whether electro-encephalogram (EEG) recordings oscillated at the same frequency. EEG recordings did oscillate at the same frequency when newborns viewed *config*, the standard inverted control, and a top-heavy alternative, but the variation in the amplitude of the EEG was largest for *config*, intermediate for the top-heavy alternative, and smallest for the control stimulus (Buiatti et al., 2019). Source modelling indicated that the *config* response emerged in the ventral stream but extended to other regions. Although much of the activity overlapped with the face network in adults, the authors note that they could not measure subcortical processing and hence could not evaluate whether the response was routed via the superior colliculus, as predicted by Morton and Johnson (1991).

Event-related potentials (ERPs) have revealed changes time-locked to faces versus control images, including a negative wave at ∼ 290 msec and a positive wave shortly after 400 msecs. By 3 months (the youngest age tested), the N290 is larger for human faces than scrambled images, and, from some electrodes, larger than for inverted human faces or monkey faces (Halit et al., 2003; Halit et al., 2004; but see Peykarjou and Hoehl, 2013). Inversion increases the latency of the N290 for human faces but has no effect for monkey faces. By 3 months of age, a later wave, the P400, is slower for monkey than for human faces and for upright than for inverted faces, as well as larger in amplitude for human faces over the right hemisphere (Halit et al., 2003). Because the effect of species and inversion on the amplitude and latency of the N290 and P400 resemble the effects of those variables on the adult N170, they are thought to be its infant precursors (Halit et al., 2003; de Haan et al., 2003). However, some of the effects of inversion and species on the N290 and the P400 also differ from the adult's N170 later during the first year (de Haan et al., 2002; Halit et al., 2003). In addition, at 3 months of age, the N290 and P400 may be mediated by an early primitive mechanism, because 3-month-olds look longer at human faces than at top-heavy scrambled images, but these stimuli elicit comparable N290 responses (Macchi Cassia et al., 2006a, 2006b), Source localization for the N290 suggests a network overlapping, but not identical to, the adult face network, including the (right) fusiform gyrus and surrounding areas from the youngest age tested (3 months), but also the right superior temporal sulcus and surrounding temporal lobe (Conte et al., 2020; Guy et al., 2016; Johnson et al., 2005), as well as prefrontal and parahippocampal clusters (Guy et al., 2016; Johnson et al., 2005). Along with evidence from the P1 and P400 (Conte et al., 2020) these findings suggest that the entire social brain is at least partially active from early infancy (Johnson et al., 2005).

Specialization of the Right vs Left Hemisphere

The role of each hemisphere in infants' face detection has been studied by presenting faces monocularly to only one hemifield (and hence to one hemisphere) in infants 4–10 months old (deSchonen et al., 1993). Either hemisphere can learn to discriminate a schematic face from one with scrambled features. The right hemisphere is better at learning configural changes and the left hemisphere, local changes. Infants demonstrate discrimination of their mother's face from that of a stranger when tested with the right hemisphere but not the left, suggesting that configural processing is being used (deSchonen et al., 1993). Transfer of learning between hemispheres does not seem to occur until the second year of life (Liégeois et al., 2000). (The lack of integration across hemispheres is inconsistent with an explanation of the preference for *config* based on binocular correlation in stimulation of the two eyes, as is the top-heavy bias (Wilkinson et al., 2014).)

Superior learning by the right hemisphere complements findings using near-infrared spectroscopy (NIRS) that show an increase in total hemoglobin, especially in the right hemisphere, when infants 5–8 months old view intact images of internal facial features, compared to a baseline condition with vegetables, scrambled faces, or inverted faces (Honda et al., 2010; Otsuka et al., 2007). Likewise, when 4-month-olds (the only age tested) watched a rapidly changing stream of natural images (6/second) with faces embedded at a specific temporal frequency, the EEG signal over the right hemisphere contained changes locked to that temporal frequency (de Heering and Rossion, 2015). No such signal emerged for scrambled images or in the left hemisphere. In a similar paradigm, 4- to 6-month-olds' EEG contained distinctive responding to faces versus objects, especially in the right hemisphere, with more overlap with findings from adults in the location of the regions responding to faces than those responding to objects (Farzin et al., 2012). The response to faces in such paradigms is enhanced when the baby is exposed to the mother's body odor (Leleu et al., 2019), a modulation indicating that infants' learning about faces is multimodal.

A Special Role for Eyes?

Newborns look longer at faces with eyes open than closed (Batki et al., 2000) and at upright than at inverted faces, but only if the eyes are not occluded (Gava et al., 2008) or the polarity of the contrast is not reversed (Farroni et al., 2005). In both cases, the preferred stimulus is a better match to *config*. They also look longer at a face with direct rather than markedly deviated gaze, but only if the face is upright and oriented straight ahead (Farroni et al., 2002; Farroni et al., 2004; Farroni et al., 2006). After viewing a video of talking strangers, newborns subsequently look longer at the static image of a novel stranger—but only if the eyes are visible with direct gaze (Gava et al., 2008; Guellai and Streri, 2011). By 3 months (but not at birth), infants prefer to look at the face of a chimp or Barbary monkey when the eyes are replaced by human eyes (Dupierrix et al., 2014). Like their preference for *config*, newborns' attention to eyes and preference for direct gaze might be mediated by low-level biases: a clearly visible black-and-white stripe in the top half of an object with dark features on a light background (Maurer and Maurer, 2019).

By 4 months (the youngest age tested), the N290 is larger when babies view upright (but not inverted) faces with direct rather than averted gaze (Farroni et al., 2002), even when the eyes are within an averted head (Farroni et al., 2004). At 6–10 months, the latency of the N290 is shorter for photos with direct than averted gaze and the P1, N290, and P400 all have a larger amplitude when a live observer begins by looking directly at the baby and then shifts gaze to the side than the reverse (Vernetti et al., 2018). Source localization indicates that the N290 modulated by eye gaze originates, among other regions, from right temporal regions and bilateral fusiform gyrus, as it does in adults (Johnson et al., 2005).

FACIAL IDENTITY

After only a few hours, newborns prefer to look at their mother over the mother of another baby, even if they have not seen the mother in the preceding 15 minutes (Pascalis et al., 1995; Bushnell, 2001). The preference is evident within 3 hours of birth if the baby has been able to both see *and* hear the mother (Sai, 2005) and increases with exposure up to 8 hours (the longest tested) (Bushnell, 2001). If both the mother and stranger wear identical scarves, all evidence of recognition disappears (Pascalis et al., 1995)—as would be expected from the newborn's limited scanning of internal facial features.

When tested with tightly controlled images, newborns rapidly learn to recognize a new face and discriminate it from another: after familiarization with one face, they look longer at a new one (Pascalis and de Schonen, 1994; Turati et al., 2006), even after a 2-minute delay (Pascalis and de Schonen, 1994) and even if the faces are filtered to contain only low spatial frequencies (0–0.5 cycle/degree) (de Heering et al., 2008). External features dominate newborns' learning. Despite learning from the internal or external features presented alone, newborns demonstrate recognition of the whole face after seeing only the external features, but not only the internal features (Turati et al., 2006). At 5 weeks, babies demonstrate learning even if the mouth and eyes of the familiarized and novel faces are occluded, but covering the hair eliminates it (Bushnell, 1982). Nevertheless, even newborns are able to generalize habituation between a face directly facing them (*en face*) and the same face turned to a 3/4 view. However, habituation does not generalize between a profile and other viewpoints (Turati et al., 2008), a pattern suggesting that viewing both eyes (as a trigger to *config*) supports learning. By 19 weeks, infants can recognize a learned face even if the hair is covered (Bushnell, 1982). Around this time, they also can learn to recognize an upright face shown during habituation even when it changes in expression (smiling or neutral) or point of view (*en face*, 3/4, or profile) (Pascalis et al., 1998; Turati et al., 2004).

Evidence from ERPs parallels behavior. By 3 months (the youngest age tested), the ERP to a novel face is distinctive between 750 and 1700 msec, even when the familiar face has been presented in various viewpoints (frontal, 3/4, profile) and after a 2-minute delay (Pascalis et al., 1998; see Ichikawa et al., 2019 for similar evidence from functional near-infrared spectroscopy (fNIRs). If additional familiarization is provided by sending home a three-dimensional (3D) model of the face, the P400 is larger for the familiar face than a novel face, especially on frontal central electrodes over the right hemisphere (Moulson et al., 2011). Additional evidence from fNIRS indicates that between 5–8 months, the temporal cortex becomes sensitive to facial identity despite changes in viewpoint or facial expressions and to facial profiles (Ichikawa et al., 2019; Kobayashi et al., 2011; 2014).

Underlying Mechanisms

From birth, infants can discriminate differences in shape (Slater et al., 1983). In contrast, it is not until 5 months that infants demonstrate discrimination between upright faces differing only in the spacing of internal features; 3-months-olds show no such evidence even after habituation to exaggerated differences (Bhatt et al., 2005; Hayden et al., 2007b). The discrimination extends to monkey faces but not houses or inverted faces (Hayden et al., 2007b; Zieber et al., 2013), a pattern suggesting a possible role for experience.

From 3 months, infants' recognition memory for faces includes an average of the faces they have seen recently: after familiarization with four faces, they treat their average as familiar, looking less long at it than a novel face or even one of the four original faces (de

Haan et al., 2001). This is an important step in the development of norm-based coding. By 12 months (youngest age tested), infants look longer at the member of a pair of faces that is closer to average, matching adults' judgments of attractiveness. Infants' preference for the more-average face is limited to human faces, suggesting a role for experience (Damon et al., 2017).

Nevertheless, it is nearly 1 year before infants integrate facial parts into a gestalt. Cashon and Cohen (2004) used a switch paradigm in which babies are habituated to two faces, then tested with a novel face and a composite comprising the external features of one of the habituated faces and the internal features of the other (see Figure 9.3). To adults, the upright composite face looks completely novel because the external and internal features are integrated into a novel gestalt; inverted, it is treated as a collection of familiar pieces. Three-month-olds treat even upright composites as a collection of familiar pieces, showing no recovery of looking. At 4 months, the composite face is treated as novel whether upright or inverted (and whether own- or other-race) (Ferguson et al., 2009). At 5.5 months, the adultlike (upright-only) pattern is shown, but as the baby's perspective changes with sitting up (around 6 months of age), the composite face is again treated as a familiar collection of pieces whether upright or inverted (Cashon and Holt, 2015). At 7 months, the adult-like pattern reemerges and babies also demonstrate

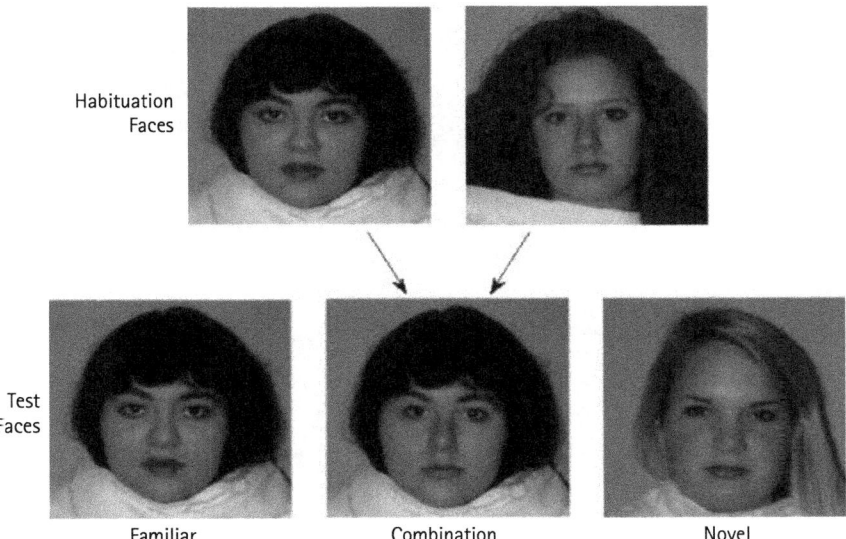

FIGURE 9.3 The switch paradigm for measuring holistic processing. After habituation to two faces, the baby is tested with a combination of the external contour of one and the internal features of the other. If processing piecemeal, the baby should find it as familiar as the face on the left that was shown during habituation. But if processing holistically, the baby should find it novel, like the face on the right.

Source: used with permission from: Cashon, C. H., and Cohen, L. B. (2004). Beyond U-Shaped Development in Infants' Processing of Faces: An Information-Processing Account. *Journal of Cognition and Development*, 5(1): 59–80.

holistic processing of their mother's face with the Composite Face Effect (see the following section) (Nakato et al., 2018).

In sum, by their first birthday, infants show at least rudimentary forms of all the specialized skills adults use to detect faces and process their identity, using neural networks that overlap with those of adults, but which remain to be refined.

RECOGNITION OF FACIAL IDENTITY DURING CHILDHOOD

Research with children after infancy has focused on their ability to recognize newly learned faces and match unfamiliar faces when images incorporate limited, systematic variability in appearance (see Figure 9.4). Children's recognition of previously learned faces shown from a new viewpoint improves between the ages of 5–12 (Croydon et al., 2014), with evidence of additional improvement until age 30 (Germine et al., 2011). The ability to match images of a target across minor changes in lighting, viewpoint, and

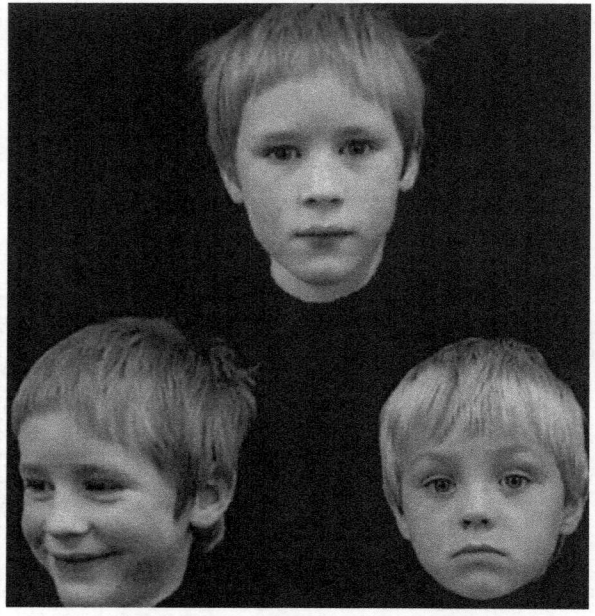

FIGURE 9.4 An example of stimuli used in a match-to-sample task in which children were asked to decide which of the two bottom images depicts the same person as the top image, despite systemic changes in viewpoint and expression.

Source: used with permission from Bruce, V., Campbell, R.N., Doherty-Sneddon, G., Import, A., Langton, S., McAuley, S., and Wright, R. (2000). Testing face processing skills in children. *British Journal of Developmental Psychology*, 18(3): 319–333. Doi: 10.1348/026151000165715

expression also improves across childhood (Bruce et al., 2000; Carey et al , 1980; de Heering et al., 2012; Megreya and Bindemann, 2015; Mondloch et al., 2003a), as does the ability to recognize an unfamiliar face when paraphernalia (e.g., hats, glasses) is added or removed (Diamond and Carey, 1977; Freire and Lee, 2001).

Performance is better with familiar faces. By age 5, there is an advantage for familiar over unfamiliar faces when matching identity in tightly controlled images and, like adults, children rely on internal features to recognize familiar faces, but external features for unfamiliar faces (Bonner and Burton, 2004; Ge et al., 2008; Wilson et al., 2009; but see Newcombe and Lie, 1995). Familiarity also protects 6-year-olds from making errors based on paraphernalia (Diamond and Carey, 1977).

Despite these findings, some researchers have argued that recognition of identity even of unfamiliar faces is *fully quantitatively mature* by the age of 5, with further improvements arising only from general cognitive development (Crookes and Robbins, 2014; McKone et al., 2012; Weigelt et al., 2014). Weigelt and colleagues drew this conclusion based on findings of face-specific improvement in recognition memory throughout childhood (i.e., memory for faces improved more than memory for other categories, such as cars) but no evidence of face-specific perceptual development after 5 years of age (i.e., improvements on perceptual tasks were similar for faces and other categories). In a memory task, Crookes and Robbins (2014) confirmed that error rates increase when newly learned faces are shown from a different versus the same viewpoint. They discounted developmental change because the increase was comparable in 8-year-olds and adults after equating performance in the same-view condition by reducing set size for children. All of these studies used tightly controlled images of unfamiliar (or newly learned) faces with minimal variability, stimuli that are ideal for measuring children's sensitivity to different types of information, but that do not measure the type of expertise adults use to recognize familiar faces across natural variability in appearance (e.g., changes in lighting, hairstyle, expression, age, weight; see Burton, 2013 for discussion). Recognizing identity in ambient images (i.e., those that capture natural variability in appearance, see Figure 9.5) is likely a face-specific skill because no other visual category contains so much variability and, for most observers, only faces are recognized at the individual level (see Baker et al., 2017 for discussion). Moreover, adults' ability to recognize identity in ambient images is limited to familiar faces, but most developmental research has relied on unfamiliar faces. As we will show below, the developmental trajectory for recognizing identity in ambient images is prolonged, especially when faces are unfamiliar.

Adults make errors when matching identity in ambient images of unfamiliar faces (e.g., Bruce et al., 1999; Bruce et al., 2001; Burton et al., 2010; Jenkins et al., 2011; Kemp et al., 1997; Megreya and Burton, 2006, 2008; Megreya et al., 2013) but recognize ambient images of familiar faces with ease (e.g., Bruce, 1982; Bruce et al., 2001; Jenkins et al., 2011), even when image quality is poor (e.g., in closed-circuit television (CCTV) footage, Burton et al., 1999) and after large distortions (Hole et al., 2002). This stark contrast between familiar and unfamiliar face recognition was highlighted in a sorting task by Jenkins and colleagues (2011). Adults sorted 40 photographs (20 images of two

FIGURE 9.5 An example of stimuli that capture natural within-person variability in appearance (i.e., ambient images). All six images are of the same person.

Source: used with permission from Matthews, C. M., and Mondloch, C. J. (2018). Finding an unfamiliar face in a line-up: Viewing multiple images of the target is beneficial on target-present trials but costly on target-absent trials. *British Journal of Psychology*, 109: 758–776. Doi:10.1111/bjop.12301.

people's faces) into piles based on identity. Whereas participants who were familiar with the people correctly sorted the images into two piles, those who were unfamiliar made about seven piles (i.e., perceived seven people to be present). Representations of unfamiliar faces are limited by the instance in which the image was captured; a change in appearance is often perceived as a change in identity. In contrast, representations of familiar faces are robust to variability in appearance, allowing for recognition in novel instances (see, Burton, 2013; Burton et al., 2005; Burton et al., 2011; Hancock et al., 2000; Johnston and Edmonds, 2009). A strong test of fully quantitatively mature face recognition requires comparisons of familiar and unfamiliar faces and the use of ambient images to examine children's ability to form such robust representations—akin to Face Recognition Units (FRUs) in Bruce and Young's (1986) model.

The few studies using ambient images of unfamiliar faces with children have reported age-related improvement and, when adults were tested, significant differences between children and adults (Baker et al., 2017; Laurence and Mondloch, 2016; Matthews et al.,

2018; Neil et al., 2016). Accuracy on Jenkins et al.'s (2011) sorting task with unfamiliar faces increases between the ages of 6–14 years (Neil et al., 2016). This prolonged development is evident even when the task is made child-friendly by reducing cognitive demands (Laurence and Mondloch, 2016; Matthews et al., 2018). The inclusion of control trials showed that children's poor performance was not attributable to generally immature cognitive skills. The developmental pattern for familiar faces is quite different. The only study using ambient images suggests that the ability to recognize a familiar face in ambient images is adultlike by 6 years of age (Laurence and Mondloch, 2016): children aged 6 years and older sorted ambient images of their own teacher (nearly) without error. However, many 4- and 5-year-olds recognized only a subset of their teacher's images, despite knowing her for 3–9 months. Young children's errorless performance on control trials indicates their failure to recognize their teacher cannot be attributed to general cognitive skills. Thus, there appears to be improvement in the ability to recognize familiar faces between 5 and 6 years of age, at which point it is (nearly) adultlike.

Behavioral evidence of prolonged development of identity recognition, at least for unfamiliar faces, is consistent with evidence that face-selective regions in the brain undergo substantial changes during childhood (e.g., Cohen Kadosh et al., 2013a, 2013b; Golarai et al., 2007, 2010, 2017; Scherf et al., 2007). During childhood and adolescence, the fusiform face area (FFA) which, in adults, responds more to faces than nonface objects (Kanwisher et al., 1997) and which is involved in representing invariant aspects of facial identity (i.e., information that does not change across instances; see Haxby et al., 2000), increases in size (Golarai et al., 2007; Peelen et al., 2009; Scherf et al., 2007) and becomes more selective to faces versus nonface objects (Aylward et al., 2005; Golarai et al., 2010; Peelen et al., 2009). This region also becomes increasingly sensitive to facial identity during childhood (Natu et al., 2016; Nordt et al., 2018). For example, when different images of the same person were presented, adults' responsiveness in the FFA decreased across trials, but there was much less change for 7–10-year-olds (Nordt et al., 2018), consistent with children's difficulty recognizing identity across even slight changes in appearance. Collectively, behavioral and neural evidence suggest prolonged development of identity recognition for unfamiliar faces and, in the very few studies to date, faster development for familiar faces.

Underlying Mechanisms

Insights into the source of limitations on children's face processing have been gained from studies of the specific mechanisms used by adults.

Holistic Processing

Holistic processing, a process by which facial features are integrated into a gestalt, is a hallmark of adults' expertise and evident in two experimental tasks. The *composite face*

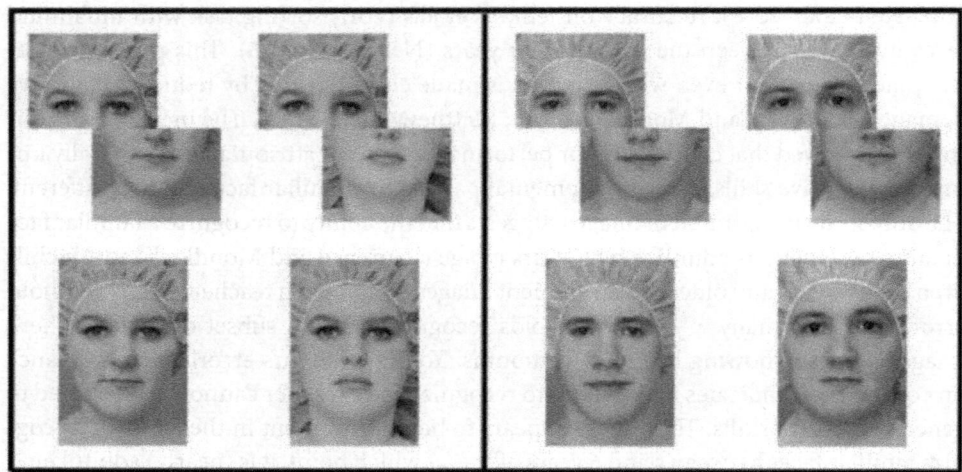

FIGURE 9.6 An example of stimuli from the Composite-face task. The top row depicts faces pairs in the misaligned condition and the bottom row depicts faces pairs in the aligned condition. In the left panel, the top halves are identical in each pair, and in the right panel they are different. The bottom halves are different in all face pairs. Aligning the two halves evokes holistic processing; the halves become fused, and adults perceive a novel identity, making it difficult to decide whether the top halves are the same or different.

Source: used with permission from Le Grand, R., Mondloch, C. J., Maurer, D., Brent, H. P. (2004). Impairment in holistic face processing following early visual deprivation. *Psychological Science*, 15(11): 762–768. Doi: 10.1111/j.0956-7976.2004.00753.x

effect occurs when aligning the top and bottom halves of two different faces impairs recognition of individual parts (see Figure 9.6). When the composite is comprised of familiar faces, adults are slow to recognize the top half, despite being told to ignore the bottom half (Young et al., 1987). When the composite is comprised of unfamiliar faces and the task is to make same/different judgments about the top halves, adults have a high error rate on *same* trials (i.e., when the same top half is aligned with different bottoms, see Hole (1994)). Adults perform better when the two halves are misaligned or the face is inverted, disrupting holistic processing (Hole, 1994; Hole et al., 1999; Le Grand et al., 2004; Young et al., 1987). In the *part/whole task*, participants are asked to recognize facial features when presented in isolation (e.g., Larry's nose, *part* condition) or in the context of the whole face (*whole* condition). Adults more accurately identify the features in the whole than in the part condition (Farah et al., 1998; Goffaux and Rossion, 2006; Pellicano and Rhodes, 2003; Tanaka and Farah, 1993; Tanaka and Sengco, 1997). Both effects are upright face-specific (i.e., not observed for objects such as houses and cars or for inverted faces; Tanaka and Farah, 1993; Macchi Cassia et al., 2009b).

Holistic processing emerges during infancy and is mature by 4 years of age. When making same/different judgments about the top halves of unfamiliar faces, children as young as 3.5 years (Macchi Cassia et al., 2009b) demonstrate a composite face effect and 4- and 6-year-olds show an effect of comparable magnitude to that of adults (de

Heering et al., 2007; Mondloch et al., 2007). At age 6 children also show a composite face effect for familiar faces (Carey and Diamond, 1994). By age 4, children are more accurate in the whole than in the part condition of the part/whole task (Pellicano and Rhodes, 2003; Pellicano et al., 2006; Tanaka et al., 1998). Holistic processing is face-specific during childhood—at least to some degree: children as young as 3.5 years do not show a composite effect for cars (Macchi Cassia et al., 2009b), but 8–10-year-olds do show one for watches, albeit a smaller effect than that observed for faces (Meinhardt-Injac et al., 2017). Similarly, children as young as 5 do not show the part/whole effect with inverted faces (Pellicano and Rhodes, 2003; Tanaka et al., 1998). Collectively these studies suggest that immature holistic processing does not contribute to the slow development of face recognition during childhood (see Crookes and McKone, 2009 for a similar conclusion).

Sensitivity to External Contour, Shape of Features, and their Spacing

To discriminate between identities adults can rely on subtle differences in the shape of facial features, the spacing among them and the shape of the external contour. Systematically isolating each of these cues by replacing the original features with those of another face, by moving the eyes and mouth, or by replacing the original exterior contour with that of another face has repeatedly demonstrated adults' sensitivity to each of these cues (see Figure 9.7). Whether asked to make same/different judgments of unfamiliar faces or detect subtle changes to familiar faces, adults are highly accurate, with > 80% accuracy even when spacing changes are kept within normal limits (Ge et al., 2003; Freire et al. 2000; Mondloch et al., 2002). Sensitivity to feature spacing is tuned by experience: it is reduced for monkey and inverted faces, and for houses (Mondloch et al., 2006b; Robbins et al., 2011).

Sensitivity to each of these cues emerges during infancy but the age at which adult-like performance is achieved varies. Six-year-olds are adultlike when making same/different judgments about unfamiliar faces that differ in external contour, and nearly so when such faces differ in feature shape, with little to no statistical difference at age 10 (Mondloch et al., 2002; Mondloch et al., 2010b). When unfamiliar faces differ in feature spacing, accuracy increases between 6–14 years of age and even 14-year-olds make more mistakes than adults (Mondloch et al., 2002; Mondloch et al., 2003a, see Freire and Lee, 2001 for evidence of age-related improvement in a delayed match-to-sample task). Even when the task is adapted to increase the viewing time and exaggerate feature spacing beyond natural limits, 8-year-olds are still not as accurate as adults (Mondloch et al., 2004), although when featural distortions are extreme, children aged 7–11 perform comparably to adults in a memory task (Gilchrist and McKone, 2003). Threshold sensitivity also decreases: the smallest difference in spacing that children are able to detect decreases between 7 and 11 years of age (Baudoin et al., 2010). Although less sensitive to feature spacing than adults, even 6- to 8-year-olds show face-specificity: an adult-like

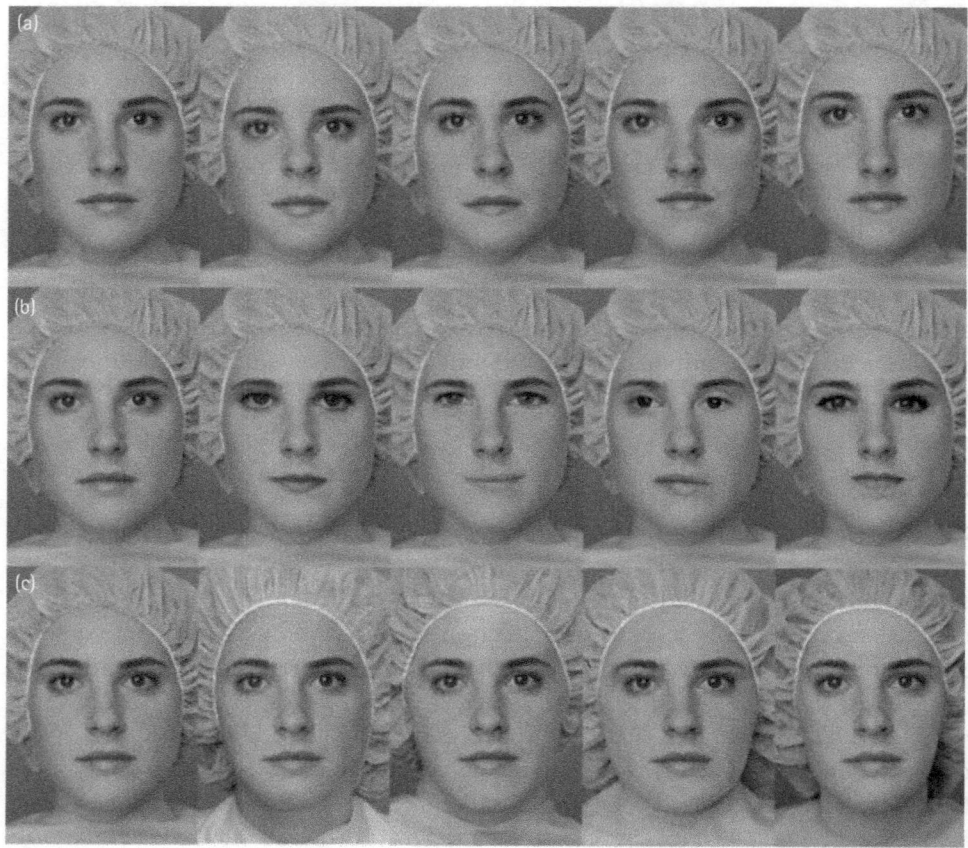

FIGURE 9.7 An example of stimuli created to examine sensitivity to feature shape and spacing, and external contour. The original face is shown as the left-most face in each panel, along with the altered faces in each set. Panel A depicts the spacing set, created by moving the eyes in/out/up/down and moving the mouth up or down. Panel B depicts the feature set, created by replacing the original features with those of another identity. Panel C depicts the external contour set, created by replacing the original exterior contour with that of another face.

Source: used with permission from Mondloch, C. J., Le Grand, R., and Maurer, D. (2002).
Configural face processing develops more slowly than featural face processing.
Perception, 31: 553–566. Doi: 10.1068/p3339

advantage for recognizing feature spacing in upright human over inverted human and monkey faces (Mondloch et al., 2002; Mondloch et al., 2006b).

Although 4-year-olds detect differences in feature spacing under certain circumstances (e.g., when judging distinctiveness, McKone and Boyer, 2006; when asked to recognize facial features in the part/whole task, Pellicano et al., 2006), it is unlikely that they use feature spacing to recognize facial identity. Four-year-olds are at chance at distinguishing familiar(ized) faces from distractors that differ only in feature spacing (Mondloch et al., 2006a), regardless of whether the faces had been learned from a storybook or were personally familiar—best friends from the child's daycare (Mondloch and Thompson, 2008) or even their own face[1] (Mondloch et al., 2006a). Chance performance cannot be attributed to general immature cognition because these same children performed (nearly) without error when the distractor differed in feature shape or external contour. Similarly, 4-year-olds are at chance in a simultaneous match-to-sample task in which they are instructed to find the twins when one member of each triad differed in feature spacing (Mondloch and Thompson, 2008). This is the same age at which children have difficulty recognizing their own teacher in ambient images (Laurence and Mondloch, 2016). Given that sensitivity to feature spacing matures quite late in childhood, while other mechanisms (e.g., holistic processing, sensitivity to feature shape and external contour) are fully developed by age 4–6, immature sensitivity to feature spacing may underlie children's difficulties in recognizing facial identity.

Prototype Formation

One conceptualization of how adults encode individual faces is as unique points in a multidimensional face space. According to Valentine's (1991) influential model, vectors in multidimensional face space represent dimensions along which individual faces vary. The nature of these dimensions remains unspecified, but they might represent features such as the size of the nose or the distance between the eyes (see Turk and Pentland, 1991 for description of eigenfaces). Individual faces are encoded in relation to a prototype—a process known as norm-based coding. This prototype represents the average of all the faces a person has encountered and is constantly updated based on experience. Faces that are located close to the prototype are rated as more typical and attractive than faces located farther away (Potter and Corneille, 2008; Rhodes and Tremewan, 1996; Valentine et al., 2004).

To provide evidence of norm-based coding, researchers use an adaptation paradigm. Repeated exposure (i.e., adaptation) to faces distorted in a similar direction leads to an aftereffect—a temporary shift in the face prototype, influencing the perception of subsequent faces. Researchers have examined figural (i.e., attractiveness) aftereffects by adapting participants to faces with compressed or expanded features. Such distortions lead to the perception of unaltered faces as distorted in the opposite direction and similarly distorted faces as more attractive (Rhodes et al., 2003; Webster and MacLin, 1999). Researchers have examined identity aftereffects by adapting participants to the opposite of an original identity (e.g., "anti-Dan"), derived from a computational face-space; for example, if Dan had widely spaced eyes, anti-Dan's eyes are close together. After repeated exposure to "anti-Dan," participants perceive previously ambiguous faces along the *Dan/anti-Dan* trajectory as more similar to the original identity (i.e., Dan; Anderson and Wilson, 2005; Leopold et al., 2001). These effects are selective such that larger aftereffects are found for opposite faces (e.g., Dan and anti-Dan) than for non-opposite faces (e.g., Dan and anti-Jim) equal in dissimilarity (Rhodes and Jeffery, 2006). The magnitude of identity aftereffects is positively correlated with the ability to recognize newly learned faces (as measured by the Cambridge Face Memory Task), providing evidence of the role of norm-based coding in identity perception (Dennett et al., 2012; Rhodes et al., 2014).

Adults' face space shows evidence of separate prototypes for different face categories (e.g., race, sex, age, upright vs inverted) as shown by opposing (category-contingent) aftereffects. Adapting adults to face categories distorted in opposite directions (e.g., compressed Caucasian and expanded East Asian faces) results in perceptions of normality being shifted in opposite directions for Caucasian versus Asian faces (Jaquet et al., 2008; Little et al., 2008). This suggests that the mature face processing system codes faces relative to category-specific norms; had there been a single prototype, the aftereffects would have cancelled each other out, leading to no shift in the norm.

Children as young as 4 appear to represent faces in a multidimensional face space that shares at least some properties with that of adults (reviewed in Jeffery and Rhodes, 2011). Four-year-olds perceive differences in facial distinctiveness, a judgment that requires reference to a face prototype (McKone and Boyer, 2006). Similarly, 6–10-year-olds rate caricatures (images manipulated to exaggerate differences between a face and the average) as more distinctive than anti-caricatures (images manipulated to enhance similarities between a face and the average; Chang et al., 2002).

By early childhood, children also show characteristics of norm-based coding. Adaptation produces figural aftereffects in children 4–12 years old (Anzures et al., 2009; Hills et al., 2010; Jeffery et al., 2010; Short et al., 2011) and identity aftereffects in children 5–9 years old (Jeffery et al., 2011; Nishimura et al., 2008; Pimperton et al., 2009). As in adults, these identity aftereffects are selective for pairs of opposite faces (e.g., Dan and anti-Dan: Nishimura et al., 2008). The size of children's aftereffects is comparable to that of adults (Jeffery et al., 2010; Nishimura et al., 2008; Pimperton et al., 2009) and, like adults, children experience stronger aftereffects after adaptation to more extreme distortions (Jeffery et al., 2010, 2011, 2013).

Nonetheless, children are less sensitive to deviations from the norm than are adults. When rating unaltered and distorted (e.g., compressed or expanded) faces based on attractiveness, 8-year-olds require more extreme distortions to reliably rate the unaltered faces as more attractive (Anzures et al., 2009; Crookes and McKone, 2009; Jeffery et al., 2010). Children appear to use the same dimensions as adults when coding facial identity but rely on only one of these dimensions at a time (predominately eye color), whereas adults use multiple dimensions simultaneously (Nishimura et al., 2009). Further, 5-year-olds show no evidence of category-specific norms for race, sex, or age (Short et al., 2014). Category-specific prototypes emerge during middle childhood but are not observed for all categories; 8-year-olds show race-specific aftereffects (Short et al., 2011) but even 10-year-olds do not show orientation-specific aftereffects (Robbins et al., 2012), suggesting that separable norms for faces of different categories continue to develop during childhood. Collectively, this evidence suggests that although children engage in norm-based coding, their face space continues to be refined throughout childhood.

Learning to Recognize Ambient Images of Familiar Faces

The ability to recognize faces in ambient images is the true marker of expert recognition—something that is limited to familiar faces in adults. Within-person variability in appearance makes the recognition of a newly encountered face difficult, but exposure to such variability facilitates face learning in adults (e.g., Andrews et al., 2015; Baker et al., 2017; Dowsett et al., 2016; Ritchie and Burton, 2017). Adults' ability to find an identity in a lineup is improved after viewing multiple images of that person (Dowsett et al., 2016; Matthews and Mondloch, 2018); viewing images captured on different days (i.e., high variability) is more beneficial than viewing images captured on the same day (i.e., low variability; Ritchie and Burton, 2017). Variability may be beneficial because adults can represent multiple versions of a person's face (Burton et al., 2016; Young and Burton, 2017) and/or because they form an average representation that contains reliable diagnostic cues but excludes cues specific to a particular instance (Burton et al., 2005; Kramer et al., 2015). Adults extract average representations rapidly and automatically— a process known as ensemble coding (Davis et al., 2020; Kramer et al., 2015; Matthews et al., 2018); after briefly viewing four images of a person's face they report having seen the average of those four images and the images themselves with comparable frequency (see Figure 9.8).

Despite being less accurate than adults at recognizing unfamiliar faces, children aged 6–11, show comparable benefit from viewing multiple images of a newly learned identity in a perceptual task (Matthews et al., 2018). Like adults, they recognize new images of a learned identity more accurately when they had been shown six images of the target identity rather than a single image. Younger children show a slightly different pattern. Although viewing six images increases 4- and 5-year-old's sensitivity to identity, it also leads them to adopt a less conservative response bias (Matthews and Mondloch, 2022). This shift in criterion after viewing six images vs. one image results in children identifying more images of the character but also incorrectly identifying more images of the distractor. This effect of false alarms is not seen in older children or adults, suggesting that the ability to benefit from exposure to multiple images is not fully refined before the age of 6.

Children as young as 6, the youngest age tested, also show comparable ensemble coding to adults (Matthews et al., 2018). After viewing four images of a target identity children recognize the average of those four images (i.e., report the average as having been in the original array) at the same rate at which they recognize the images themselves, regardless of whether the individual instances are viewed simultaneously or sequentially. Formation of such an average was already evident in infancy, from 3 months of age (de Haan et al., 2001).

(a) Trial Presentation   (b) Types of Test Images

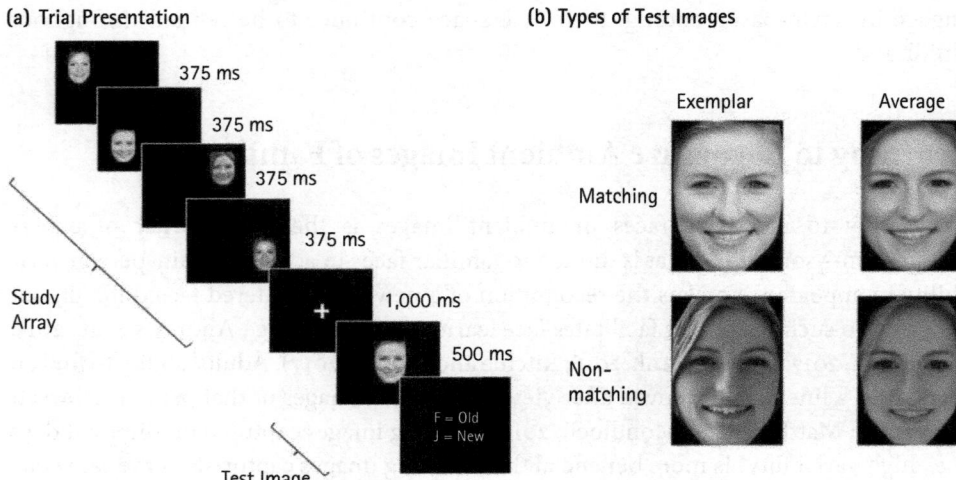

FIGURE 9.8 (a) Trial presentation in the ensemble coding task. On each trial, a study array comprising four images of a single identity is presented (either sequentially or simultaneously; shown here sequentially). Next, a fixation cross is displayed, followed by one of the four types of test images (see panel b). Once the test image disappears, participants are asked to decide whether the test image was present in the study array. (b) Types of test images. Matching Exemplar: a display image from the array; Matching Average: an average of the display images; Non-matching Exemplar: a new image of the identity; Non- Matching Average: an average of four new images of the identity. Ensemble coding is demonstrated when the matching average (panel b) is perceived as having been seen.

Source: used with permission from Davis, E. E., Matthews, C. M., and Mondloch, C. J. (2020). Ensemble coding of facial identity is not refined by experience: Evidence from other-race and inverted faces. *British Journal of Psychology*. Advance online publication. Doi: 10.1111/bjop.12457

In contrast, children are less efficient than adults at learning a newly encountered face when the task involves holding a representation in memory—the challenge faced when meeting new people in daily life. Children aged 6–12 show evidence of learning after watching a video that incorporates high variability in appearance (filmed across 3 days), but unlike adults, show no evidence of learning when the variability is low (filmed on a single day; Baker et al., 2017). Evidence of children's prolonged development in face learning in a memory but not a perceptual task is consistent with Weigelt and colleagues' (2014) conclusion that face memory, but not perception, continues to improve after 5 years of age.

TUNING BY EXPERIENCE

We have reviewed a wealth of data documenting prolonged development of face detection, face recognition, and underlying mechanisms during infancy and childhood.

Although improvements in general cognitive skills (e.g., memory, attention, executive function) certainly contribute to the developmental changes (see Matthews and Mondloch, 2021; McKone et al., 2012), face-specific experience plays a crucial role. In the language of Gottlieb (1976), experience both induces postnatal change and facilitates the refinement of face processing.

What Infants Experience

Infants are exposed to faces more than any other visual stimulus, ~ 25% of the time they are awake over the first 3 months (Jayaraman and Smith, 2018; Sugden et al., 2014), usually within 2 feet, upright, and *en face* (Jayaraman et al., 2015; Jayaraman et al., 2017; Jayaraman and Smith, 2018; Sugden and Moulson, 2017). Even in multiethnic Toronto, the faces predominantly match the baby's (and mother's) ethnicity (96% in one study) and are mostly adults (81%) and female (70%) (Sugden et al., 2014). The primary caretaker (usually the mother) is seen most frequently and longest (Jayaraman and Smith, 2018; Sugden and Moulson, 2018).

By 3 months, the infant has accumulated about 210 (of 800) waking hours with faces directly in front of the face, a number that increases to 620 (of 3100) by 11 months (Jayaraman et al., 2015). Over the next 1–2 years, faces remain prevalent but their prevalence declines to ~ 8% after 18 months (Fausey et al., 2016; Jayaraman et al., 2017).

Tuning by Biased Experience During Infancy

The clearest evidence of experiential tuning comes from comparisons of human faces with a category rarely seen—nonhuman animals. Newborns can discriminate between two monkey faces and they look longer at an upright than an inverted monkey face, perhaps based on the same cues that attract their attention to *config* (Di Giorgio et al., 2012). When structural cues are matched, newborns look equally long at monkey and human faces. A looking preference for the experienced (i.e., human) category develops over the first 3 months (Di Giorgio et al., 2012; Di Giorgio et al., 2013), a preference revealed even with complex arrays at older ages (Jakobsen et al., 2016; Simpson et al., 2020).

Nevertheless, infants continue to be able to discriminate between monkey faces after 3 months of age. For example, at 6 months, 20 seconds of exposure to a human or monkey face is sufficient to elicit a subsequent novelty preference. At 9 months it is sufficient only for the human faces (Pascalis et al., 2002), an effect replicated with different animal species (Pascalis et al., 2005; Scott and Monesson, 2009; Simpson et al., 2011). Thus, monkey faces become harder for infants to discriminate between 6 and 9 months of age, such that they fail the novelty test they "passed" at 6 months. This pattern, a hallmark of biased experience during infancy, has been called perceptual narrowing (Maurer and Werker, 2014).

Longer exposure time to the familiarized monkey face or the insertion of human eyes can prevent the loss of discrimination at 9 and sometimes 12 months (Damon et al., 2015; Fair et al., 2012; Zieber et al., 2013; but see Scott and Monesson, 2009); so can calling attention to the individuality of six monkey faces by giving them different names in a storybook read ~ 30 times between 6 and 9 months (Pascalis et al., 2005; Scott and Monesson, 2009). Mere exposure or a story without individual names is insufficient (Scott and Monesson, 2009). After hearing individual names, the N290 and P400 waves of the ERP also differentiate upright from inverted monkey faces (Scott and Monesson, 2010). Without such training, ERPs at 8–9 months suggest that infants do not process monkey faces at the individual level, are insensitive to whether the monkey's face is upright or inverted, and to whether the monkey's vocalization is congruent with the observed facial movements (Dixon et al., 2019; Grossmann et al., 2012; Scott and Monesson, 2010). Nevertheless, at 9 months, infants' ERPs differentiate repeated from novel monkey faces, with a response even greater than at 6 months, especially in right occipital, right occipitocentral, right central, and left central regions (Barry-Anwar et al., 2018). Those results imply that at the transitional age of 9 months, there is still neural differentiation of individual monkey faces that is not strong enough to be manifest in typical behavioral paradigms.

The results for ethnicity/race[2] are similar and directly correlated with the infant's biased diet of faces. At 3 months, a looking preference for own-race faces emerges despite a continuing ability to discriminate among other-race faces (Bar-Haim et al., 2006; Kelly et al., 2007a; but see Hayden et al., 2007a; Sangrigoli and De Schonen, 2004; and Tham et al., 2015 for evidence of an emerging own-race advantage). By 6–9 months, babies fail to show discrimination with paradigms that they "pass" at younger ages and continue to pass with own-race faces (Kelly et al., 2007b; Kelly et al., 2009; Vogel et al., 2012). This pattern emerges first with female faces (Sugden and Marquis, 2017) and extends to infant faces (Macchi Cassia et al., 2014; Kobayashi et al., 2018). The falloff in discrimination can be prevented at least temporarily by using photographs of expressive faces (with happy or angry expressions; Quinn et al., 2020), encouraging individuation of other-race faces (Anzures et al., 2012; Heron-Delaney et al., 2011; Markant et al., 2016), or living in a multiethnic environment (Tham et al., 2019).

By 8–9 months, babies seem to process other-race faces at the categorical (what is it?) rather than the individual level (whose face is it?) (Anzures et al., 2010; Hayden et al., 2009) and in a piecemeal rather than holistic fashion (Ferguson et al., 2009). They also spend more time fixating the features that differentiate among faces of their own race (for Caucasian faces, the eyes; for Asian faces, the nose) (Liu et al., 2011; Pickron et al., 2017; Wheeler et al., 2011) and shifting fixation between two individual faces of their own race (Fassbender and Lohaus, 2019). ERPs and fNIRs reveal a change in processing coincident with perceptual narrowing, but the use of different paradigms, seemingly inconsistent results (greater for own-race or for other-race) and lack of crossover studies make the findings difficult to interpret (Balas et al., 2011; Timeo et al., 2019: Vogel et al., 2012).

Similar perceptual tuning is seen for age and gender: by 3 months, a looking preference emerges for own-race faces that are adult rather than infant and for own-race faces that are female rather than male (Heron-Delaney et al., 2017; Tham et al., 2015), unless the baby is being raised by a male (Quinn et al., 2002; Quinn et al., 2008). At 9 months, adult faces evoke an increase in oxyhemoglobin in the right temporal area and a decrease in deoxyhemoglobin in the left temporal area, but infant faces do not (Kobayashi et al., 2018). At 3 months, the infant also fails to show discrimination between (own-race) male faces (Quinn et al., 2002), despite continuing to discriminate between faces from unfamiliar species, races, and ages. By 7–9 months, the effect of sex of face has vanished, likely because of increasing exposure to a mixture of male and female faces (Tham et al., 2015). Nevertheless, even at this point, female faces elicit a larger N290, which is also uniquely sensitive to the novelty of the individual female face (Righi et al., 2014).

Persistent Effects of Early Experience

The own-race advantage persists into adulthood on the very tasks that show prolonged development during childhood: recognizing learned unfamiliar faces in memory tasks (e.g., Golby et al., 2001; MacLin and Malpass, 2001; Meissner and Brigham, 2001, Wright et al., 2003, Zhou et al., 2018), matching identity across images of unfamiliar faces (Laurence et al., 2016; Megreya et al., 2011; Meissner et al., 2013; Proietti et al., 2019), and detecting changes in feature shape and spacing (Hayward et al., 2008; Mondloch et al., 2010a; Rhodes et al., 2009; Zhao et al., 2014).[3] Although social cognitive factors might contribute to these effects (Hugenburg et al., 2010; Young et al., 2012) and the other race effect is reduced when perceived threat increases (e.g., with angry expressions; Ackerman et al., 2006), the fact that these effects emerge during infancy (i.e., before developing the classic outgroup homogeneity bias) suggests experience is critical (Anzures et al., 2013; Kelly et al., 2005, 2007b; Tanaka et al., 2013). In contrast, adults do not show an own-race advantage on the tasks that mature early: when recognizing highly familiar faces (Zhou and Mondloch, 2016), ensemble coding (Davis et al., 2020) or learning faces from ambient images in a perceptual matching task (Matthews and Mondloch, 2018).

Exposure to siblings' faces also alters development. By 3 years of age, children without a sibling show superior discrimination and a selective inversion effect for adult relative to child or infant faces. Three-year-olds with a sibling (younger or older) and 6-year-olds with a younger sibling show comparable discrimination and inversion effects for both infant/child and adult faces (Macchi Cassia et al., 2009a, 2012, 2013, 2014). This sibling effect persists into adulthood: only mothers who had a younger sibling show an inversion effect for infant faces (Macchi Cassia et al., 2009a).

The number of faces encountered during development also influences face processing. Adults raised in small towns are less accurate at recognizing faces learned in a memory task and at matching the identity of unfamiliar faces than adults from large towns (Balas and Saville, 2015, 2017). Small-town adults also demonstrate an N170 that is less specific to faces (Balas and Saville, 2015). Similarly, adults who were home-schooled are poorer at sorting ambient images of unfamiliar faces than adults who encountered many more faces by attending school (Short et al., 2017).

When early visual experience is completely lacking —because an infant was born with dense central cataracts in both eyes, the developmental trajectory is altered more profoundly. When first treated during infancy, their face preferences resemble those of newborns and not age mates: they look preferentially toward *config*, even at ages when that preference has disappeared in babies with normal eyes (Mondloch et al., 2013a) and their face preferences change with accumulated experience, rather than their chrono-logical age. Despite the early visual deprivation, as adults (i.e., after years of [nearly] normal visual input following treatment) they perform normally in detecting faces even when local featural cues are eliminated (Mondloch et al., 2013b) and in recognizing identity based on differences in the shape of the external contour or of the internal features (Le Grand et al., 2001a; Le Grand et al., 2001b).

Nevertheless, the early visual deprivation has prolonged effects. As adults, these patients are worse than controls in recognizing upright human faces differing only in the spacing of internal features—a deficit not evident for inverted faces, monkey faces, or the windows and doors of houses (Le Grand et al., 2001a; Le Grand et al., 2001b; Mondloch et al., 2003b; Mondloch et al., 2010b; Robbins et al., 2010). In other words, the deficit is restricted to the category on which adults excel and for which development is protracted. Follow-up experiments indicated that early input to the right hemisphere is critical (Le Grand et al., 2003; see also Dalrymple et al., 2020). Cataract-reversal patients also fail to show evidence of holistic processing of upright human faces: in the composite face task, they are *better* than controls in recognizing that the top halves of two faces are identical when aligned with different bottom halves (Le Grand et al., 2004). Given the lack of these two markers of expert processing, it is not surprising that they also have difficulty recognizing famous faces (de Heering and Maurer, 2014) and learning or discriminating new faces when picture encoding is prevented by adding variable noise or introducing changes in point of view or lighting (Geldart et al., 2002; de Heering and Maurer, 2014; Putzar et al., 2010).

fMRI analyses indicate that cataract-reversal patients use the normal face net-work when processing faces versus objects or houses, but with reduced activity and specialization in both the core and extended face networks and a different pattern of connectivity, particularly with the left fusiform gyrus. The extent of these changes correlates with performance in recognizing facial identity based on the spacing of in-ternal features (Grady et al., 2014). There are also changes in the neural underpinnings of face detection, despite normal behavioral performance. As in controls, the N170 is larger for faces than other categories (scrambled faces, objects, houses), but it has a much greater amplitude, as does the preceding P100, with the size of the amplification correlated with the duration of the visual deprivation during infancy (Mondloch et al., 2013b). That pattern suggests that patients achieve normal accuracy by recruiting additional neural processing. (When the deprivation lasts past infancy, until an average age of 4 years, the N170 no longer differentiates faces from other categories [Röder et al., 2013]).

Of course, these studies do not distinguish between the necessity for *face* experience versus patterned visual input more generally, but a study of monkeys raised in a rich visual environment but without exposure to faces suggests that it is face experience that is critical (Sugita, 2008).

Summary

The ability to detect faces and the mechanisms underlying the recognition of facial identity emerge during infancy. Whereas some face recognition mechanisms mature relatively early in childhood (i.e., holistic processing, sensitivity to feature shape and external contour, norm-based coding, ensemble coding, perceptual learning from ambient images), others show prolonged development (i.e., sensitivity to feature spacing, refinement of multidimensional face space, forming identity representations in memory). The accumulation of experience during infancy and childhood is critical for these developments.

Notes

1. Although Freire and Lee (2001) reported that 4-year-olds were capable of using differences in feature spacing to identify faces in a delayed match-to-sample task, subsequent analyses (McKone and Boyer, 2006) revealed that performance in this age group was not above chance.
2. We use the term "other-race" because it is the traditional label even though it actually refers to an unfamiliar ethnicity since there are no biologically distinct races.
3. The results for holistic processing are mixed (Crookes et al., 2013; Hayward et al., 2013; Michel et al., 2006a, b; Mondloch et al., 2010a; Tanaka et al., 2004).

References

Ackerman, J. M., Shapiro, J. R., Neuberg, S. L., Kenrick, D. T., Becker, D. V., Griskevicius, V., Maner, J. K., and Schaller, M. (2006). They all look the same to me (unless they're angry): From out-group homogeneity to out-group heterogeneity. *Psychological Science*, 17: 836–840. Doi:10.1111/j.1467-9280.2006.01790.x.

Aylward, E. H., Park, J. E., Field, K. M., Parsons, A. C., Richards, T. L., Cramer, S. C., and Meltzoff, A. N. (2005). Brain activation during face perception: Evidence of a developmental change. *Journal of Cognitive Neuroscience*, 17: 308–319. Doi:10.1162/0898929053124884.

Anderson, N. D. and Wilson, H. R. (2005). The nature of synthetic face adaptation. *Vision Research*, 45: 1815–1828. Doi:10.1016/j.visres.2005.01.012.

Andrews, S., Jenkins, R., Cursiter, H., and Burton, A. (2015). Telling faces together: Learning new faces through exposure to multiple instances. *Quarterly Journal of Experimental Psychology*, 68: 2041–2050. Doi:10.1080/17470218.2014.1003949.

Anzures, G., Mondloch, C. J., and Lackner, C. (2009). Face adaptation and attractiveness aftereffects in 8-year-olds and adults. *Child Development*, 80: 178–191. Doi:10.1111/j.1467-8624.2008.01253.x.

Anzures, G., Quinn, P. C., Pascalis, O., Slater, A. M., and Lee, K. (2010). Categorization, categorical perception, and asymmetry in infants' representation of face race. *Developmental Science*, 13(4): 553–564. Doi:10.1111/j.1467-7687.2009.00900.x.

Anzures, G., Quinn, P. C., Pascalis, O., Slater, A. M., Tanaka, J. W., and Lee, K. (2013). Developmental origins of the other-race effect. *Current Directions in Psychological Science*, 22: 173–178. Doi:10.1177/0963721412474459.

Anzures, G., Wheeler, A., Quinn, P. C., Pascalis, O., Slater, A. M., Heron-Delaney, M., Tanaka, J. W., and Lee, K. (2012). Brief daily exposures to Asian females reverses perceptual narrowing for Asian faces in Caucasian infants. *Journal of Experimental Child Psychology*, 112: 484–495. Doi:10.1016/j.jecp.2012.04.005.

Baker, K. A., Laurence, S., and Mondloch, C. J. (2017). How does a newly encountered face become familiar? The effect of within-person variability on adults' and children's perception of identity. *Cognition*, 161: 19–30. Doi:10.1016/j.cognition.2016.12.012.

Balas, B., Westerlund, A., Hung, K., and Nelson, C. A. (2011). Shape, color and the other-race effect in the infant brain. *Developmental Science*, 14: 892–900. Doi:10.1111/j.1467-7687.2011.01039.x.

Balas, B. and Saville, A. (2015). N170 face specificity and face memory depend on hometown size. *Neuropsychologia*, 69: 211–217. Doi:10.1016/j.neuropsychologia.2015.02.005.

Balas, B. and Saville, A. (2017). Hometown size affects the processing of naturalistic face variability. *Vision Research*, 141: 228–236. Doi:10.1016/j.visres.2016.12.005.

Bar-Him, Y., Ziv, T., Lamy, D., and Hodes, R. M. (2006). Nature and nurture in own-race face processing. *Psychological Science*, 17: 159–163. Doi:10.1111/j.1467-9280.2006.01679.x.

Barry-Anwar, R., Hadley, H., Conte, S., Keil, A., and Scott, L. S. (2018). The developmental time course and topographic distribution of individual-level monkey face discrimination in the infant brain. *Neuropsychologia*, 108: 25–31. Doi:10.1016/j.neuropsychologia.2017.11.019.

Batki, A., Baron-Cohen, S., Wheelwright, S., Connellan, J., and Ahluwalia, J. (2000). Is there an innate gaze module? Evidence from human neonates. *Infant Behavior and Development*, 23: 223–229. Doi:10.1016/S0163-6383(01)00037-6.

Baudouin, J. Y., Gallay, M., Durand, K., and Robichon, F. (2010). The development of perceptual sensitivity to second-order facial relations in children. *Journal of Experimental Child Psychology*, 107: 195–206. Doi:10.1016/j.jecp.2010.05.008.

Bentin, S., Allison, T., Puce, A., Perez, E., and McCarthy, G. (1996). Electrophysiological studies of face perception in humans. *Journal Cognitive Neuroscience*, 8: 551–565. Doi:10.1162/jocn.1996.8.6.551.

Bhatt, R. S., Bertin, E., Hayden, A., and Reed, A. (2005). Face processing in infancy: Developmental changes in the use of different kinds of relational information. *Child Development*, 76: 169–181. Doi:10.1111/j.1467-8624.2005.00837.x.

Bindemann, M., Burton, A. M., Hooge, I. T. C., Jenkins, R., and de Haan, E. H. F. (2005). Faces retain attention. *Psychonomic Bulletin and Review*, 12: 1048–1053. Doi:10.3758/BF03206442.

Bindemann, M., Burton, A. M., Leuthold, H., and Schweinberger, S. R. (2008). Brain potential correlates of face recognition: Geometric distortions and the N250r brain response to stimulus repetitions. *Psychophysiology*, 45: 535–544. Doi:10.1111/j.1469-8986.2008.00663.x.

Bonner, L. and Burton, A. M. (2004). 7-11-Year-old children show an advantage for matching and recognizing the internal features of familiar faces: Evidence against a developmental shift. *Quarterly Journal of Experimental Psychology Section A: Human Experimental Psychology*, 57: 1019–1029. Doi:10.1080/02724980343000657.

Bruce, V. (1982). Changing faces: Visual and non-visual coding processes in face recognition. *British Journal of Psychology*, 73: 105–116. Doi:10.1111/j.2044-8295.1982.tb01795.x.

Bruce, V., Campbell, R. N., Doherty-Sneddon, G., Import, A., Langton, S., McAuley, S., and Wright, R. (2000). Testing face processing skills in children. *British Journal of Developmental Psychology*, 18: 319–333. Doi:10.1348/026151000165715.

Bruce, V., Henderson, Z., Greenwood, K., Hancock, P. J. B., Burton, A. M., and Miller, P. (1999). Verification of face identities from images captured on video. *Journal of Experimental Psychology: Applied*, 5: 339–360. Doi:10.1037/1076-898X.5.4.339.

Bruce, V., Henderson, Z., Newman, C., and Burton, A.M. (2001). Matching identities of familiar and unfamiliar faces caught on CCTV images. *Journal of Experimental Psychology: Applied*, 7: 207–218. Doi:10.1037/1076-898X.7.3.207.

Bruce, V. and Young, A. (1986). Understanding face recognition. *British Journal of Psychology*, 77: 305–327. Doi:10.1111/j.2044-8295.1986.tb02199.x.

Buiatti, M., Di Giorgio, E., Piazza, M., Polloni, C., Menna, G., Taddei, F., Baldo, E., and Vallortigara, G. (2019). Cortical route for facelike pattern processing in human newborns. *Proceedings of the National Academy of Sciences USA*, 116: 4625–4630. Doi:10.1073/pnas.1812419116.

Burton, A. M. (2013). Why has research in face recognition progressed so slowly? The importance of variability. *Quarterly Journal of Experimental Psychology*, 66: 1467–1485. Doi:10.1080/17470218.2013.800125.

Burton, A. M., Jenkins, R., Hancock, P. J. B., and White, D. (2005). Robust representations for face recognition: The power of averages. *Cognitive Psychology*, 51: 256–284. Doi:10.1016/j.cogpsych.2005.06.003.

Burton, A. M., Jenkins, R., and Schweinberger, S. (2011). Mental representations of familiar faces. *British Journal of Psychology*, 102: 943–958. Doi:10.1111/j.2044-8295.2011.02039.x.

Burton, A. M., Kramer, R. S. S., Ritchie, K. L, and Jenkins, R. (2016). Identity from variation: Representations of faces derived from multiple instances. *Cognitive Science*, 40: 202–223. Doi:10.1111/cogs.12231.

Burton, A. M., White, D., and McNeill, A. (2010). The Glasgow face matching test. *Behavior Research Methods*, 42: 286–291. Doi:10.3758/BRM.42.1.286.

Burton, A. M., Wilson, S., Cowan, M., and Bruce, V. (1999). Face recognition in poor-quality video: Evidence from security surveillance. *Psychological Science*, 10: 343–248. Doi:10.1111/1467-9280.00144.

Bushnell, I. W. (1982). Discrimination of faces by young infants. *Journal of Experimental Child Psychology*, 33: 298–308. Doi:10.1016/0022-0965(82)90022-4.

Bushnell, I. W. R. (2001). Mother's face recognition in newborn infants: Learning and memory. *Infant and Child Development*, 10: 67–74. Doi:10.1002/icd.248.

Carey, S. and Diamond, R. (1994). Are faces perceived as configurations more by adults than by children? *Visual Cognition*, 1: 253–274. Doi:10.1080/13506289408402302.

Carey, S., Diamond, R., and Woods, B. (1980). Development of face recognition: A maturational component? *Developmental Psychology*, 16: 257–269. Doi:10.1037/0012-1649.16.4.257.

Cashon, C. H. and Cohen, L. B. (2004). Beyond U-shaped development in infants' processing of faces: An information-processing account. *Journal of Cognition and Development*, 5: 59–80. Doi:10.1207/s15327647jcd0501_4.

Cashon, C. H. and Holt, N. A. (2015). Developmental origins of the face inversion effect. *Advances in Child Development and Behavior*, 48: 117–150. Doi:10.1016/bs.acdb.2014.11.008.

Chang, P. P. W., Levine, S. C., and Benson, P. J. (2002). Children's recognition of caricatures. *Developmental Psychology*, 38: 1038–1051. Doi:10.1037/0012-1649.38.6.1038.

Chien, S. H. (2011). No more top-heavy bias: Infants and adults prefer upright faces but not top-heavy geometric or face-like patterns. *Journal of Vision*, 11(6). Doi:10.1167/11.6.13.

Cohen Kadosh, K., Johnson, M. H., Dick, F., Cohen Kadosh, R., and Blakemore, S. J. (2013a). Effects of age, task performance, and structural brain development on face processing. *Cerebral Cortex*, 23: 1630–1642. Doi:10.1093/cercor/bhs150.

Cohen Kadosh, K., Johnson, M. H., Henson, R. N., Dick, F., and Blakemore, S. J. (2013b). Differential face-network adaptation in children, adolescents and adults. *Neuroimage*, 69: 11–20. Doi:10.1016/j.neuroimage.2012.11.060.

Conte, S., Richards, J. E., Guy, M. W., Xie, W., and Roberts, J. E. (2020). Face-sensitive brain responses in the first year of life. *Neuroimage*, 211: 116602. Doi:10.1016/j.neuroimage.2020.116602.

Crookes, K., Favelle, S., and Hayward, W. (2013). Holistic processing for other-race faces in Chinese participants occurs for upright but not inverted faces. *Frontiers in Psychology*, 4: 1–9. Doi:10.3389/fpsyg.2013.00029.

Crookes, K. and McKone, E. (2009). Early maturity of face recognition: No childhood development of holistic processing, novel face encoding, or face-space. *Cognition*, 111: 219–247. Doi:10.1016/j.cognition.2009.02.004.

Crookes, K. and Robbins, R. A. (2014). No childhood development of viewpoint-invariant face recognition: Evidence from 8-year-olds and adults. *Journal of Experimental Child Psychology*, 126: 103–111. Doi:10.1016/j.jecp.2014.03.010.

Croydon, A., Pimperton, H., Ewing, L., Duchaine, B. C., and Pellicano, E. (2014). The Cambridge Face Memory Test for Children (CFMT-C): A new tool for measuring face recognition skills in childhood. *Neuropsychologia*, 62: 60–67. Doi: 10.1016/j.neuropsychologia.2014.07.008.

Dalrymple, K. A., Khan, A. F., Duchaine, B., and Elison, J. T. (2020). Visual input to the left versus right eye yields differences in face preferences in 3-month-old infants. *Developmental Science*, e13029. Doi:10.1111/desc.13029.

Damon, F., Bayet, L., Quinn, P. C., Hillairet de Boisferon, A., Méary, D., Dupierrix, E., Lee, K., and Pascalis, O. (2015). Can human eyes prevent perceptual narrowing for monkey faces in human infants. *Developmental Psychobiology*, 57(5): 637–642. Doi:10.1002/dev.21319.

Damon, F., Méary, D., Quinn, P. C., Lee, K., Simpson, E. A., Paukner, A., Suomi, S. J., and Pascalis, O. (2017). Preference for facial averageness: Evidence for a common mechanism in human and macaque infants. *Scientific Reports*, 7: 46303. Doi:10.1038/srep46303.

Dannemiller, J. L. and Stephens, B. R. (2001). Asymmetries in contrast polarity processing in young human infants. *Journal of Vision*, 1: 112–125. Doi:10.1167/1.2.5.

Davis, E. E., Matthews, C. M., and Mondloch, C. J. (2020). Ensemble coding of facial identity is not refined by experience: Evidence from other-race and inverted faces. *British Journal of Psychology*, 112: 265–281. Doi:10.1111/bjop.12457.

de Haan, M., Johnson, M. H., and Halit, H. (2003). Development of face-sensitive event-related potentials during infancy: A review. *International Journal of Psychophysiology*, 51(1): 45–58. Doi:10.1016/S0167-8760(03)00152-1.

de Haan, M., Johnson, M. H., Maurer, D., and Perrett, D. I. (2001). Recognition of individual faces and average face prototypes by 1- and 3-month-old infants. *Cognitive Development*, 16: 659–678. Doi:10.1016/S0885-2014(01)00051-X.

de Haan, M., Pascalis, O., and Johnson, M. H. (2002). Specialization of neural mechanisms underlying face recognition in human infants. *Journal of Cognitive Neuroscience*, 14(2): 199–209. Doi:10.1162/089892902317236849.

de Heering, A., Goffaux, V., Dollion, N., Godard, O., Durand, K., and Baudouin, J. Y. (2016). Three-month-old infants' sensitivity to horizontal information within faces. *Developmental Psychobiology*, 58(4): 536–542. Doi:10.1002/dev.21396.

de Heering, A., Houthuys, S., and Rossion, B. (2007). Holistic face processing is mature at 4 years of age: Evidence from the composite face effect. *Journal of Experimental Child Psychology*, 96: 57–70. Doi:10.1016/j.jecp.2006.07.001.

de Heering, A. and Maurer, D. (2014). Face memory deficits in patients deprived of early visual input by bilateral congenital cataracts. *Developmental Psychobiology*, 56(1): 96–108. Doi:10.1002/dev.21094.

de Heering, A. and Rossion, B. (2015). Rapid categorization of natural face images in the infant right hemisphere. *Elife*, 4: e06564. Doi:10.7554/eLife.06564.

de Heering, A., Rossion, B., and Maurer, D. (2012). Developmental changes in face recognition during childhood: Evidence from upright and inverted faces. *Cognitive Development*, 27(1): 17–27. Doi:10.1016/j.cogdev.2011.07.001.

de Heering, A., Turati, C., Rossion, B., Bulf, H., Goffaux, V., and Simion, F. (2008). Newborns' face recognition is based on spatial frequencies below 0.5 cycles per degree. *Cognition*, 106: 444–454. Doi:10.1016/j.cognition.2006.12.012.

Dennett, H. W., McKone, E., Edwards, M., and Susilo, T. (2012). Face aftereffects predict individual differences in face recognition ability. *Psychological Science*, 23: 1279–1287. Doi:10.1177/0956797612446350.

de Schonen, S., Deruelle, C., Mancini, J., and Pascalis, O. (1993). Hemispheric differences in face processing and brain maturation. In B. de Boysson-Bardies, S. de Schonen, P. Jusczyk, P. MacNeilage, and J. Morton (eds.), *Developmental Neurocognition: Speech and Face Processing in the First Year of Life*. Dordrecht: Kluwer Academic/Plenum Publishers, pp. 149–163. Doi:10.1007/978-94-015-8234-6.

Diamond, R. and Carey, S. (1977). Developmental changes in the representation of faces. *Journal of Experimental Child Psychology*, 23: 1–22. Doi:10.1016/0022-0965(77)90069-8.

Di Giorgio, E., Leo, I., Pascalis, O., and Simion, F. (2012). Is the face-perception system human-specific at birth. *Developmental Psychology*, 48(4): 1083–1090. Doi:10.1037/a0026521.

Di Giorgio, E., Méary, D., Pascalis, O., and Simion, F. (2013). The face perception system becomes species-specific at 3 months: An eye-tracking study. *International Journal of Behavioral Development*, 37(2): 95–99. Doi:10.1177/0165025412465362.

Di Giorgio, E., Turati, C., Altoè, G., and Simion, F. (2012). Face detection in complex visual displays: An eye-tracking study with 3- and 6-month-old infants and adults. *Journal of Experimental Child Psychology*, 113(1): 66–77. Doi:10.1016/j.jecp.2012.04.012.

Dixon, K. C., Reynolds, G. D., Romano, A. C., Roth, K. C., Stumpe, A. L., Guy, M. W., and Mosteller, S. M. (2019). Neural correlates of individuation and categorization of other-species faces in infancy. *Neuropsychologia*, 126: 27–35. Doi:10.1016/j.neuropsychologia.2017.09.037.

Dowsett, A. J., Sandford, A., and Burton, A. M. (2016). Face learning with multiple images leads to fast acquisition of familiarity for specific individuals. *The Quarterly Journal of Experimental Psychology*, 69: 1–10. Doi:10.1080/17470218.2015.1017513.

Duchaine, B. and Nakayama, K. (2006). The Cambridge Face Memory Test: Results for neuro-logically intact individuals and an investigation of its validity using inverted face stimuli and prosopagnosic participants. *Neuropsychologia*, 44: 576–585. Doi:10.1016/j.neuropsycholo gia.2005.07.001.

Dupierrix, E., de Boisferon, A. H., Méary, D., Lee, K., Quinn, P. C., Di Giorgio, E., Simion, F., Tomonaga, M., and Pascalis, O. (2014). Preference for human eyes in human infants. *Journal of Experimental Child Psychology*, 123: 138–146. Doi:10.1016/j.jecp.2013.12.010.

Eimer, M. (2011). The face-sensitivity of the N170 component. *Frontiers in Human Neuroscience*, 5(119): 1–2. Doi:10.3389/fnhum.2011.00119.

Fair, J., Flom, R., Jones, J., and Martin, J. (2012). Perceptual learning: 12-month-olds' dis-crimination of monkey faces. *Child Development*, 83(6): 1996–2006. Doi:10.1111/j.1467-8624.2012.01814.x.

Farah, M. J., Wilson, K. D., Drain, M., and Tanaka, J. N. (1998). What is "special" about face per-ception? *Psychological Science*, 105: 482–498. Doi:10.1037/0033-295X.105.3.482.

Farroni, T., Massaccesi, S., Pividori, D., and Johnson, M. H. (2004). Gaze following in newborns. *Infancy*, 5(1): 39–60. Doi:10.1207/s15327078in0501_2.

Farroni, T., Csibra, G., Simion, F., and Johnson, M. H. (2002). Eye contact detection in humans from birth. *Proceedings of the National Academy of Sciences USA*, 99(14): 9602–9605. Doi:10.1073/pnas.152159999.

Farroni, T., Johnson, M. H., and Csibra, G. (2004). Mechanisms of eye gaze percep-tion during infancy. *Journal of Cognitive Neuroscience*, 16(8): 1320–1326. Doi:10.1162/0898929042304787.

Farroni, T., Johnson, M. H., Menon, E., Zulian, L., Faraguna, D., and Csibra, G. (2005). Newborns' preference for face-relevant stimuli: effects of contrast polarity. *Proceedings of the National Academy of Sciences USA*, 102(47): 17245–17250. Doi:10.1073/pnas.0502205102.

Farroni, T., Menon, E., and Johnson, M. H. (2006). Factors influencing newborns' preference for faces with eye contact. *Journal of Experimental Child Psychology*, 95(4):Doi:10.1167/12.13.6. 298–308. Doi:10.1016/j.jecp.2006.08.001.

Farzin, F., Hou, C., and Norcia, A. M. (2012). Piecing it together: Infants' neural responses to face and object structure. *Journal of Vision*, 12(13): 6. Doi:10.1167/12.13.6.

Fassbender, I. and Lohaus, A. (2019). Fixations and fixation shifts in own-race and other-race face pairs at three, six and nine months. *Infant Behavior and Development*, 57: 101328. Doi:10.1016/j.infbeh.2019.101328.

Fausey, C. M., Jayaraman, S., and Smith, L. B. (2016). From faces to hands: Changing visual input in the first two years. *Cognition*, 152: 101–107. Doi:10.1016/j.cognition.2016.03.005.

Ferguson, K. T., Kulkofsky, S., Cashon, C. H., and Casasola, M. (2009). The development of specialized processing of own-race faces in infancy. *Infancy*, 14: 263–284. Doi:10.1080/15250000902839369.

Freire, A. and Lee, K. (2001). Face recognition in 4- to 7-year-olds: Processing of configural, featural, and paraphernalia information. *Journal of Experimental Child Psychology*, 80: 347–371. Doi: 10.1006/jecp.2001.2639.

Freire, A., Lee, K., and Symons, L. A. (2000). The face-inversion effect as a deficit in the encoding of configural information: Direct evidence. *Perception*, 29: 159–170. Doi:10.1068/p3012.

Gava, L., Valenza, E., Turati, C., and de Schonen, S. (2008). Effect of partial occlusion on newborns' face preference and recognition. *Developmental Science*, 11(4): 563–574. Doi:10.1111/j.1467-7687.2008.00702.x.

Grady, C. L., Mondloch, C. J., Lewis, T. L., and Maurer, D. (2014). Early visual deprivation from congenital cataracts disrupts activity and functional connectivity in the face network. *Neuropsychologia*, 57, 122–139. Doi:10.1016/j.neuropsychologia.2014.03.005.

Ge, L., Anzures, G., Wang, Z., Kelly, D. J., Pascalis, O., Quinn, P. C., Slater, A. M., Yang, Z., and Lee, K. (2008). An inner face advantage in children's recognition of familiar peers. *Journal of Experimental Child Psychology*, 101: 124–136. Doi: 10.1016/j.jecp.2008.05.006.

Ge, L., Luo, J., Nishimura, M., and Lee, K. (2003). The lasting impression of chairman Mao: Hyperfidelity of familiar-face memory. *Perception*, 32: 601–614. Doi:10.1068/p5022.

Geldart, S., Mondloch, C. J., Maurer, D., de Schonen, S., and Brent, H. P. (2002). The effect of early visual deprivation on the development of face processing. *Developmental Science*, 5(4): 490–501. Doi:10.1111/1467-7687.00242.

Germine, L. T., Duchaine, B., and Nakayama, K. (2011). Where cognitive development and aging meet: Face learning ability peaks after age 30. *Cognition*, 118: 201–210. Doi:10.1016/j.cognition.2010.11.002.

Gilchrist, A. and McKone, E. (2003). Early maturity of face processing in children: Local and relational distinctiveness effects in 7-year-olds. *Visual of Cognition*, 10: 769–793. Doi:10.1080/13506280344000022.

Gliga, T., Elsabbagh, M., Andravizou, A., & Johnson, M. (2009). Faces attract infants' attention in complex displays. *Infancy*, 14, 550–562. Doi: 10.1080/15250000903144199.

Gluckman, M. and Johnson, S. P. (2013). Attentional capture by social stimuli in young infants. *Frontiers in Psychology*, 4: 527. Doi:10.3389/fpsyg.2013.00527.

Goffaux, V. and Rossion, B. (2006). Faces are "spatial"—holistic face perception is supported by low spatial frequencies. *Journal of Experimental Psychology: Human Perception and Performance*, 32: 1023–1039. Doi:10.1037/0096-1523.32.4.1023.

Golarai, G., Ghahremani, D. G., Whitfield-Gabrieli, S., Reiss, A., Eberhardt, J. L., Gabrieli, J. D. E., and Grill-Spector, K. (2007). Differential development of high-level visual cortex correlates with category-specific recognition memory. *Nature Neuroscience*, 10: 512–522. Doi:10.1038/nn1865.

Golarai, G., Hong, S., Haas, B.W., Galaburda, A.M., Mills, D.L., Bellugi, U., Grill-Spector, K., and Reiss, A.L. (2010). The fusiform face area is enlarged in Williams syndrome. *The Journal of Neuroscience*, 30: 6700–6712. Doi:10.1523/JNEUROSCI.4268-09.2010.

Golarai, G., Liberman, A., and Grill-Spector, K. (2017). Experience shapes the development of neural substrates of face processing in human ventral temporal cortex. *Cerebral Cortex*, 27: 1–16. Doi:10.1093/cercor/bhv314.

Golby, A. J., Gabrieli, J. D. E., Chiao, J., and Eberhardt, J. (2001). Differential responses in the fusiform region to same-race and other-race faces. *Nature Neuroscience*, 4: 845–850. Doi:10.1038/90565.

Gottlieb, G. (1976). The roles of experience in the development of behavior and the nervous system. In *Studies on the Development of Behavior and the Nervous System*. G. Gottlieb, ed. vol. 3. Elsevier, pp. 25–54. Doi:10.1016/B978-0-12-609303-2.50008-X.

Grossmann, T., Missana, M., Friederici, A. D., and Ghazanfar, A. A. (2012). Neural correlates of perceptual narrowing in cross-species face-voice matching. *Developmental Science*, 15: 830–839. Doi:10.1111/j.1467-7687.2012.01179.x.

Guellai, B. and Streri, A. (2011). Cues for early social skills: Direct gaze modulates newborns' recognition of talking faces. *PLoS One*, 6(4): e18610. Doi:10.1371/journal.pone.0018610.

Guy, M. W., Zieber, N., and Richards, J. E. (2016). The cortical development of specialized face processing in infancy. *Child Development*, 87: 1581–1600. Doi:10.1111/cdev.12543.

Haith, M. M., Bergman, T., and Moore, M. J. (1977). Eye contact and face scanning in early infancy. *Science*, 198: 853–855. Doi:10.1126/science.918670.

Halit, H., de Haan, M., and Johnson, M. H. (2003). Cortical specialisation for face processing: Face-sensitive event-related potential components in 3- and 12-month-old infants. *Neuroimage*, 19: 1180–1193. Doi:10.1016/S1053-8119(03)00076-4.

Halit, H., Csibra, G., Volein, A., and Johnson, M. H. (2004). Face-sensitive cortical processing in early infancy. *Journal of Child Psychology and Psychiatry*, 45: 1228–1234. Doi:10.1111/j.1469-7610.2004.00321.x.

Hancock, P. J. B., Bruce, V., and Burton, A. M. (2000). Recognition of unfamiliar faces. *Trends in Cognitive Sciences*, 4: 330–377. Doi:10.1016/S1364-6613(00)01519-9.

Hasson, U., Hendler, T., Bashat, D. B., and Malach, R. (2001). Vase or face? A neural correlate of shape-selective grouping processes in the human brain. *Journal of Cognitive Neuroscience*, 13: 744–753. Doi:10.1162/08989290152541412.

Haxby, J. V., Hoffman, E. A., and Gobbini, M. I. (2000). The distributed human neural system for face perception. *Trends in Cognitive Sciences*, 4: 223–233. Doi:10.1016/S1364-6613(00)01482-0.

Hayden, A., Bhatt, R. S., Joseph, J. E., and Tanaka, J. W. (2007a). The other-race effect in infancy: Evidence using a morphing technique. *Infancy*, 12: 95–104. Doi:10.1111/j.1532-7078.2007.tb00235.x.

Hayden, A., Bhatt, R. S., Reed, A., Corbly, C. R., and Joseph, J. E. (2007b). The development of expert face processing: Are infants sensitive to normal differences in second-order relational information? *Journal of Experimental Child Psychology*, 97: 85–98. Doi:10.1016/j.jecp.2007.01.004.

Hayden, A., Bhatt, R. S., Zieber, N., and Kangas, A. (2009). Race-based perceptual asymmetries underlying face processing in infancy. *Psychonomic Bulletin and Review*, 16: 270–275. Doi:10.3758/PBR.16.2.270.

Hayward, W. G., Crookes, K., and Rhodes, G. (2013). The other-race effect: Holistic coding differences and beyond. *Visual Cognition*, 21: 1224–1247. Doi:10.1080/13506285.2013.824530.

Hayward, W. G., Rhodes, G., and Schwaninger, A. (2008). An own-race advantage for components as well as configurations in face recognition. *Cognition*, 106: 1017–1027. Doi:10.1016/j.cognition.2007.04.002.

Heron-Delaney, M., Anzures, G., Herbert, J. S., Quinn, P. C., Slater, A. M., Tanaka, J. W., Lee, K., and Pascalis, O. (2011). Perceptual training prevents the emergence of the other race effect during infancy. *PloS One*, 6(5): e19858. Doi:10.1371/journal.pone.0019858.

Heron-Delaney, M., Damon, F., Quinn, P. C., Méary, D., Xiao, N. G., Lee, K., and Pascalis, O. (2017). An adult face bias in infants that is modulated by face race. *International Journal of Behavioral Development*, 41: 581–587. Doi:10.1177/0165025416651735.

Hills, P. J., Holland, A. M., and Lewis, M. B. (2010). Aftereffects for face attributes with different natural variability: Children are more adaptable than adolescents. *Cognitive Development*, 25: 278–289. Doi:10.1016/j.cogdev.2010.01.002.

Hole, G. J. (1994). Configurational factors in the perception of unfamiliar faces. *Perception*, 23, 65–74. Doi:10.1068/p230065.

Hole, G. J., George, P. A., and Dunsmore, V. (1999). Evidence for holistic processing of faces viewed as photographic negatives. *Perception*, 28: 341–359. Doi:10.1068/p2622.

Hole, G. J., George, P. A, Eaves, K., and Rasek, A. (2002). Effects of geometric distortions on face-recognition performance. *Perception*, 31: 1221–1240. Doi: 10.1068/p3252.

Honda, Y., Nakato, E., Otsuka, Y., Kanazawa, S., Kojima, S., Yamaguchi, M. K., and Kakigi, R. (2010). How do infants perceive scrambled face? A near-infrared spectroscopic study. *Brain Research*, 1308: 137–146. Doi:10.1016/j.brainres.2009.10.046.

Hugenberg, K., Young, S. G., Bernstein, M. J., and Sacco, D. F. (2010). The categorization- individuation model: An integrative account of the other-race recognition deficit. *Psychological Review*, 117; 1168–1187. Doi:10.1037/a0020463.

Hunnius, S. and Geuze, R. H. (2004). Developmental changes in visual scanning of dynamic faces and abstract stimuli in infants: A longitudinal study. *Infancy*, 6: 231–255. Doi:10.1207/s15327078in0602_5.

Ichikawa, H., Nakato, E., Igarashi, Y., Okada, M., Kanazawa, S., Yamaguchi, M. K., and Kakigi, R. (2019). A longitudinal study of infant view-invariant face processing during the first 3-8 months of life. *Neuroimage*, 186: 817–824. Doi:10.1016/j.neuroimage.2018.11.031.

Jaquet, E., Rhodes, G., and Hayward, W. G. (2008). Race-contingent aftereffects suggest distinct perceptual norms for different race faces. *Visual Cognition*, 16: 734–753. Doi:10.1080/13506280701350647.

Jakobsen, K. V., Umstead, L., and Simpson, E. A. (2016). Efficient human face detection in infancy. *Developmental Psychobiology*, 58: 129–136. Doi:10.1002/dev.21338.

Jayaraman, S., Fausey, C. M., and Smith, L. B. (2015). The faces in infant-perspective scenes change over the first year of life. *PLoS One*, 10(5): e0123780. Doi:10.1371/journal.pone.0123780.

Jayaraman, S., Fausey, C. M., and Smith, L. B. (2017). Why are faces denser in the visual experiences of younger than older infants. *Developmental Psychology*, 53: 38–49. Doi:10.1037/dev0000230.

Jayaraman, S. and Smith, L. B. (2018). Faces in early visual environments are persistent not just frequent. *Vision Research*, 157(6): 213–221. Doi:10.1016/j.visres.2018.05.005.

Jeffery, L., McKone, E., Haynes, R., Firth, E., Pellicano, E., and Rhodes, G. (2010). Four- to six-year-old children use norm-based coding in face-space. *Journal of Vision*, 10(5): 1–19. Doi:10.1167/10.5.18.

Jeffery, L., Read, A., and Rhodes, G. (2013). Four-year-olds use norm-based coding for face identity. *Cognition*, 127: 258–263. Doi:10.1016/j.cognition.2013.01.008.

Jeffrey, L. and Rhodes, G. (2011). Insights into the development of face recognition mechanisms revealed by face aftereffects. *British Journal of Psychology*, 102: 799–815. Doi:10.1111/j.2044-8295.2011.02066.x.

Jeffery, L., Rhodes, G., McKone, E., Pellicano, E., Crookes, K., and Taylor, E. (2011). Distinguishing norm-based from exemplar-based coding of identity in children: Evidence from face identity aftereffects. *Journal of Experimental Psychology: Human Perception and Performance*, 37: 1824–1840. Doi:10.1037/a0025643.

Jenkins, R., White, D., Montfort, X. V., and Burton, A. M. (2011). Variability in photos of the same face. *Cognition*, 121: 313–323. Doi:10.1016/j.cognition.2011.08.001.

Johnson, M. H., Griffin, R., Csibra, G., Halit, H., Farroni, T., de Haan, M., Tucker, L. A., Baron-Cohen, S., and Richards, J. (2005). The emergence of the social brain network: Evidence from typical and atypical development. *Developmental Psychopathology*, 17(3): 599–619. Doi:10.1017/S0954579405050297.

Johnson, M. H., Senju, A., and Tomalski, P. (2015). The two-process theory of face processing: Modifications based on two decades of data from infants and adults. *Neuroscience and Biobehavioral Reviews*, 50: 169–179. Doi:10.1016/j.neubiorev.2014.10.009.

Johnston, R. A. and Edmonds, A. J. (2009). Familiar and unfamiliar face recognition: A review. *Memory*, 17: 577–596. Doi:10.1080/09658210902976969.

Kanwisher, N., McDermott, J., and Chun, M.M. (1997). The fusiform face area: A module in human extrastriate cortex specialized for face perception. *The Journal of Neuroscience*, 17: 4302–4311. Doi:10.1523/JNEUROSCI.17-11-04302.1997.

Kelly, D. J., Duarte, S., Meary, D., Bindemann, M., and Pascalis, O. (2019). Infants rapidly detect human faces in complex naturalistic visual scenes. *Developmental Science*, 22(6): e12829. Doi:10.1111/desc.12829.

Kelly, D. J., Liu, S., Ge, L., Quinn, P. C., Slater, A. M., Lee, K., Liu, Q., and Pascalis, O. (2007a). Cross-race preferences for same-race faces extend beyond the African versus Caucasian contrast in 3-month-old infants. *Infancy*, 11: 87–95. Doi:10.1080/15250000709336871.

Kelly, D. J., Liu, S., Lee, K., Quinn, P. C., Pascalis, O., Slater, A. M., and Ge, L. (2009). Development of the other-race effect during infancy: Evidence toward universality? *Journal of Experimental Child Psychology*, 104: 105–114. Doi:10.1016/j.jecp.2009.01.006.

Kelly, D. J., Quinn, P. C., Slater, A. M., Lee, K., Ge, L., and Pascalis, O. (2007b). The other-race effect develops during infancy: Evidence of perceptual narrowing. *Psychological Science*, 18: 1084–1089. Doi:10.1111/j.1467-9280.2007.02029.x.

Kelly, D. J., Quinn, P. C., Slater, A. M., Lee, K., Gibson, A., Smith, M., Ge, L., and Pascalis, O. (2005). Three-month-olds, but not newborns, prefer own-race faces. *Developmental Science*, 8: F31–F36. Doi:10.1111/j.1467-7687.2005.0434a.x.

Kemp, R., Towell, N., and Pike, G. (1997). When seeing should not be believing: Photographs, credit cards and fraud. *Applied Cognitive Psychology*, 11: 211–222. Doi:10.1002/(SICI)1099-0720(199706)11:3<211::AID-ACP430>3.0.CO;2-O.

Kleiner, K. A. (1987). Amplitude and phase spectra as indices of infants' pattern preferences. *Infant Behavior and Development*, 10: 49–59. Doi:10.1016/0163-6383(87)90006-3.

Kobayashi, M., Otsuka, Y., Kanazawa, S., Yamaguchi, M. K., and Kakigi, R. (2014). The processing of faces across non-rigid facial transformation develops at 7 months of age: A fNIRS-adaptation study. *BMC Neuroscience*, 15: 81. Doi:10.1186/1471-2202-15-81.

Kobayashi, M., Otsuka, Y., Nakato, E., Kanazawa, S., Yamaguchi, M. K., and Kakigi, R. (2011). Do infants represent the face in a viewpoint-invariant manner? Neural adaptation study as measured by near-infrared spectroscopy. *Frontiers in Human Neuroscience*, 5(153): 1–12. Doi:10.3389/fnhum.2011.00153.

Kobayashi, M., Macchi Cassia, V., Kanazawa, S., Yamaguchi, M. K., and Kakigi, R. (2018). Perceptual narrowing towards adult faces is a cross-cultural phenomenon in infancy: A behavioral and near-infrared spectroscopy study with Japanese infants. *Developmental Science*, 21(1): e12498. Doi:10.1111/desc.12498.

Komban, S. J., Alonso, J.-M., and Zaidi, Q. (2011). Darks are processed faster than lights. *Journal of Neuroscience*, 31: 8654–8658. Doi:10.1523/JNEUROSCI.0504-11.2011.

Kramer, R., Ritchie, K. L., and Burton, M. A. (2015). Viewers extract the mean from images of the same person: A route to face learning. *Journal of Vision*, 15(4): 1–9. Doi:10.1167/15.4.1.

Langton, S. R. H., Law, A. S., Burton, A. M., and Schweinberger, S. R. (2008). Attention capture by faces. *Cognition*, 107: 330–342. Doi:10.1016/j.cognition.2007.07.012.

Laurence, S. and Mondloch, C. J. (2016). That's my teacher! Children's ability to recognize personally familiar and unfamiliar faces improves with age. *Journal of Experimental Child Psychology*, 143: 123–138. Doi:10.1016/j.jecp.2015.09.030.

Laurence, S., Zhou, X., and Mondloch, C. J. (2016). The flip side of the other-race coin: They all look different to me. *British Journal of Psychology*, 107: 374–388. Doi:10.1111/bjop.12147.

Le Grand, R., Mondloch, C. J., Maurer, D., and Brent, H. P. (2001a). Neuroperception. Early visual experience and face processing. *Nature*, 410(6831): 890. Doi:10.1038/35073749.

Le Grand, R., Mondloch, C. J., Maurer, D., and Brent, H. P. (2001b). Correction: Early visual experience and face processing. *Nature*, 412: 786.

Le Grand, R., Mondloch, C. J., Maurer, D., and Brent, H. P. (2003). Expert face processing requires visual input to the right hemisphere during infancy. *Nature Neuroscience*, 6(10): 1108–1112. Doi:10.1038/nn1121.

Le Grand, R., Mondloch, C. J., Maurer, D., Brent, H. P. (2004). Impairment in holistic face processing following early visual deprivation. *Psychological Science*, 15: 762–768. Doi:10.1111/j.0956-7976.2004.00753.x.

Leleu, A., Rekow, D., Poncet, F., Schaal, B., Durand, K., Rossion, B., and Baudouin, J. Y. (2019). Maternal odor shapes rapid face categorization in the infant brain. *Developmental Science*, e12877. Doi:10.1111/desc.12877.

Leopold, D. A., O'Toole, A. J., Vetter, T., and Blanz, V. (2001). Prototype-referenced shape encoding revealed by high-level aftereffects. *Nature Neuroscience*, 4: 89–94. Doi:10.1038/82947.

Lewis, T. L. and Maurer, D. (1992). The development of the temporal and nasal visual fields during infancy. *Vision Research*, 32, 903–911. Doi:10.1016/0042-6989(92)90033-F.

Liégeois, F., Bentejac, L., and de Schonen, S. (2000). When does inter-hemispheric integration of visual events emerge in infancy? A developmental study on 19- to 28-month-old infants. *Neuropsychologia*, 38: 1382–1389. Doi:10.1016/S0028-3932(00)00041-5.

Little, A. C., DeBruine, L. M., Jones, B. C., and Watt, C. (2008). Category contingent aftereffects for faces of different races, ages and species. *Cognition*, 106: 1537–1547. Doi:10.1016/j.cognition.2007.06.008.

Lui, J., Li, J., Feng, L., Li, L., Tian, J., and Lee, K. (2014). Seeing Jesus in toast: Neural and behavioral correlates of face pareidolia. *Cortex*, 53: 60–77. Doi:10.1016/j.cortex.2014.01.013.

Liu, S., Quinn, P. C., Wheeler, A., Xiao, N., Ge, L., and Lee, K. (2011). Similarity and difference in the processing of same- and other-race faces as revealed by eye tracking in 4- to 9-month-olds. *Journal of Experimental Child Psychology*, 108: 180–189. Doi:10.1016/j.jecp.2010.06.008.

Macchi Cassia, V., Bulf, H., Quadrelli, E., and Proietti, V. (2014). Age-related face processing bias in infancy: Evidence of perceptual narrowing for adult faces. *Developmental Psychobiology*, 58: 238–248. Doi:10.1002/dev.21191.

Macchi Cassia, V., Kuefner, D., Picozzi, M., and Vescovo, E. (2009a). Early experience predicts later plasticity for face processing: Evidence for the reactivation of dormant effects. *Psychological Science*, 20: 853–859. Doi:10.1111/j.1467-9280.2009.02376.x.

Macchi Cassia, V., Kuefner, D., Westerlund, A., and Nelson, C. (2006a). A behavioural and ERP investigation of 3-month-olds' face preferences. *Neuropsychologia*, 44: 2113–2125. Doi:10.1016/j.neuropsychologia.2005.11.014.

Macchi Cassia, V., Kuefner, D., Westerlund, A., and Nelson, C. A. (2006b). Modulation of face-sensitive event-related potentials by canonical and distorted human faces: The role of vertical symmetry and up-down featural arrangement. *Journal of Cognitive Neuroscience*, 18: 1343–1358. Doi:10.1162/jocn.2006.18.8.1343.

Macchi Cassia, V., Pisacane, A., and Gava, L. (2012). No own-age bias in 3-year-old children: More evidence for the role of early experience in building face-processing biases. *Journal of Experimental Child Psychology*, 113: 372–382. Doi:10.1016/j.jecp.2012.06.014.

Macchi Cassia, V., Picozzi, M., Kuerfner, D., Bricolo, E., and Turati, C. (2009b). Holistic processing for faces and cars in preschool-aged children and adults: Evidence from the composite effect. *Developmental Science*, 12: 236–248. Doi:10.1111/j.1467-7687.2008.00765.x.

Macchi Cassia, V., Proietti, V., and Pisacane, A. (2013). Early and later experience with one younger sibling affects face processing abilities of 6-year-old children. *International Journal of Behavioral Development*, 37: 160–165. Doi:10.1177/0165025412469175.

Macchi Cassia, V., Turati, C., and Simion, F. (2004). Can a nonspecific bias toward top-heavy patterns explain newborns' face preference? *Psychological Science*, 15: 379–383. Doi:10.1111/j.0956-7976.2004.00688.x.

Macchi Cassia, V., Valenza, E., Simion, F., and Leo, I. (2008). Congruency as a nonspecific perceptual property contributing to newborns' face preference. *Child Development*, 79: 807–820. Doi:10.1111/j.1467-8624.2008.01160.x.

MacLin, O. H. and Malpass, R. S. (2001). Racial categorization of faces: The ambiguous race face effect. *Psychology, Public Policy, and Law*, 7: 98–118. Doi:10.1037/1076-8971.7.1.98.

Markant, J., Oakes, L. M., and Amso, D. (2016). Visual selective attention biases contribute to the other-race effect among 9-month-old infants. *Developmental Psychobiology*, 58: 355–365. Doi:10.1002/dev.21375.

Matthews, C. M., Davis, E. E., and Mondloch, C. J. (2018). Getting to know you: The development of mechanisms underlying face learning, *Journal of Experimental Child Psychology*, 167: 295–313. Doi:10.1016/j.jecp.2017.10.012.

Matthews, C. M. and Mondloch, C. J. (2018). Finding an unfamiliar face in a line-up: Viewing multiple images of the target is beneficial on target-present trials but costly on target-absent trials. *British Journal of Psychology*, 109: 758–776. Doi:10.1111/bjop.12301.

Matthews, C. M. and Mondloch, C. J. (2022). Learning faces from variability: Four- and five-year-olds differ from older children and adults. *Journal of Experimental Child Psychology*, 213: 105259. Doi:10.1016/j.jecp.2021.105259.

Matthews C. M. and Mondloch, C. J. (2021). Learning and recognizing facial identity in variable images: New insights from older adults. *Visual Cognition*, 29: 708–731. Doi:10.1080/13506285.2021.2002994.

Maurer, D. (1983). The scanning of compound figures by young infants. *Journal of Experimental Child Psychology*, 35: 437–448. Doi:10.1016/0022-0965(83)90019-X.

Maurer, D. (2016). How the baby learns to see: Donald O. Hebb Award Lecture, Canadian Society for Brain, Behaviour, and Cognitive Science. Ottawa, June 2015. *Canadian Journal of Experimental Psychology*, 70: 195–200. Doi:10.1037/cep0000096.

Maurer, D. and Lewis, T. L. (1991). The development of peripheral vision and its physiological underpinnings. In M. J. S. Weiss and P. R. Zelazo (eds.), *Newborn Attention: Biological Constraints and the Influence of Experience*. Norwood, NJ: Ablex Publishing, pp. 218–255. Doi:10.1016/0022-0965(83)90019-X.

Maurer, C. and Maurer, D. (2019). *Pretty Ugly: Why We Like Some Songs, Faces, Foods, Plays, Pictures, Poems, etc., and Dislike Others*. Newcastle: Cambridge Scholars Publishing.

Maurer, D. and Salapatek, P. (1976). Developmental changes in the scanning of faces by young infants. *Child Development* 47, 523–527. Doi: 10.2307/1128813.

Maurer, D. and Werker, J. F. (2014). Perceptual narrowing during infancy: a comparison of language and faces. *Developmental Psychobiology*, 56, 154–178. Doi:10.1002/dev.21177.

McKone, E. and Boyer, B.L. (2006). Sensitivity of 4-year-olds to featural and second-order relational changes in face distinctiveness. *Journal of Experimental Child Psychology*, 94, 134–162. Doi:10.1016/j.jecp.2006.01.001.

McKone, E., Crookes, K., Jeffery, L., and Dilks, D. D. (2012). A critical review of the development of face recognition: Experience is less important than previously believed. *Cognitive Neuropsychology*, 29: 174–212. Doi: 10.1080/02643294.2012.660138.

McCarthy, G., Puce, A., Gore, J. C., and Trent, A. (1997). Face-specific processing in the human fusiform gyrus. *Journal of Cognitive Neuroscience*, 9: 605–610. Doi:10.1162/jocn.1997.9.5.605.

Megreya, A. M. and Bindemann, M. (2015). Developmental improvement and age-related decline in unfamiliar face matching. *Perception*, 44: 5–22. Doi:10.1068/p7825.

Megreya, A. M. and Burton, A. M. (2006). Unfamiliar faces are not faces: Evidence from a matching task. *Memory and Cognition*, 34: 865–876. Doi:10.3758/BF03193433.

Megreya, A. M. and Burton, A. M. (2008). Matching faces to photographs: Poor performance in eyewitness memory (without the memory). *Journal of Experimental Psychology: Applied*, 14: 364–372. Doi:10.1037/a0013464.

Megreya, A. M., Sandford, A., and Burton, A. M. (2013). Matching face images taken on the same day or months apart: The limitations of photo ID. *Applied Cognitive Psychology*, 27: 700–706. Doi:10.1002/acp.2965.

Megreya, A. M., White, D., and Burton, A. M. (2011). The other-race effect does not rely on memory: Evidence from a matching task. *Quarterly Journal of Experimental Psychology*, 64: 1473–1483. Doi:10.1080/17470218.2011.575228.

Meinhardt-Injac, B., Boutet, I., Persike, M., Meinhardt, G., and Imhof, M. (2017). From development to aging: Holistic face perception in children, younger and older adults. *Cognition*, 158: 134–146. Doi:10.1016/j.cognition.2016.10.020.

Meissner, C. A. and Brigham, J.C. (2001). Thirty years of investigating the own-race bias in memory for faces: A meta-analytic review. *Psychology, Public Policy, and Law*, 7: 3–35. Doi:10.1037/1076-8971.7.1.3.

Meissner, C. A., Susa, K. J., and Ross, A. B. (2013). Can I see your passport please? Perceptual discrimination of own- and other-race faces. *Visual Cognition*, 21: 1287–1305. Doi:10.1080/13506285.2013.832451.

Michel, C., Caldara, R., and Rossion, B. (2006a). Same-race faces are perceived more holistically than other-race faces. *Visual Cognition*, 14: 55–73. Doi:10.1080/13506280500158761.

Michel, C., Rossion, B., Han, J., Chung, C. S., and Caldara, R. (2006b). Holistic processing is finely tuned for faces of one's own race. *Psychological Science*, 17: 608–615. Doi:10.1111/j.1467-9280.2006.01752.x.

Mondloch, C. J., Dobson, K. S., Parsons, J., and Maurer, D. (2004). Why 8-year-olds cannot tell the difference between Steve Martin and Paul Newman: Factors contributing to the slow development of sensitivity to the spacing of facial features. *Journal of Experimental Child Psychology*, 89: 159–181. Doi:10.1016/j.jecp.2004.07.002.

Mondloch, C. J., Elms, N., Maurer, D., Rhodes, G., Hayward, W. G., Tanaka, J. W., and Zhou, G. (2010a). Processes underlying the cross-race effect: An investigation of holistic, featural, and relational processing of own-race versus other-race faces. *Perception*, 39: 1065–1085. Doi:10.1068/p6608.

Mondloch, C. J., Geldart, S., Maurer, D., and Le Grand, R. (2003a). Developmental changes in face processing skills. *Journal of Experimental Child Psychology*, 86: 67–84. Doi:10.1016/S0022-0965(03)00102-4.

Mondloch, C. J., Le Grand, R., and Maurer, D. (2002). Configural face processing develops more slowly than featural face processing. *Perception*, 31: 533–566. Doi:10.1068/p3339.

Mondloch, C. J., Le Grand, R., and Maurer, D. (2003b). Early visual experience is necessary for the development of some—but not all—aspects of face processing. In O. Pascalis and A. Slater (eds.), *The Development of Face Processing in Infancy and Early Childhood*. Hauppauge, NY: Nova Science, pp. 99–117.

Mondloch, C. J., Leis, A., and Maurer, D. (2006a). Recognizing the face of Johnny, Suzy, and me: Insensitivity to the spacing among features at 4 years of age. *Child Development*, 77: 234–243. Doi:10.1111/j.1467-8624.2006.00867.x.

Mondloch, C. J., Lewis, T. L., Budreau, D. R., Maurer, D., Dannemiller, J. L., Stephens, B. R., and Kleiner-Gathercoal, K. A. (1999). Face perception during early infancy. *Psychological Science*, 10: 419–422. Doi: 10.1111/1467-9280.00179.

Mondloch, C. J., Lewis, T. L., Levin, A. V., and Maurer, D. (2013a). Infant face preferences after binocular visual deprivation. *International Journal of Behavioral Development*, 37: 148–153. Doi:10.1177/0165025412471221.

Mondloch, C. J., Maurer, D., and Ahola, S. (2006b). Becoming a face expert. *Psychological Science*, 17: 930–934. Doi:10.1111/j.1467-9280.2006.01806.x.

Mondloch, C. J., Pathmam, T., Maurer, D., Le Grand, R., and de Schonen, S. (2007). The composite face effect in six-year-old children: Evidence of adult-like holistic face processing. *Visual Cognition*, 15: 564–577. Doi:10.1080/13506280600859383.

Mondloch, C. J., Robbins, R., and Maurer, D. (2010b). Discrimination of facial features by adults, 10-year-olds, and cataract-reversal patients. *Perception*, 39: 184–194. Doi:10.1068/p6153.

Mondloch, C. J., Segalowitz, S. J., Lewis, T. L., Dywan, J., Le Grand, R., and Maurer, D. (2013b). The effect of early visual deprivation on the development of face detection. *Developmental Science*, 16: 728–742. Doi:10.1111/desc.12065.

Mondloch, C. J. and Thomson, K. (2008). Limitations in 4-year-old children's sensitivity to the spacing among facial features. *Child Development*, 79: 1513–1523. Doi:10.1111/j.1467-8624.2008.01202.x.

Morton, J. and Johnson, M. H. (1991). CONSPEC and CONLERN: A two-process theory of infant face recognition. *Psychological Review*, 98: 164–181. Doi:10.1037/0033-295x.98.2.164.

Moulson, M. C., Shannon, R. W., and Nelson, C. A. (2011). Neural correlates of visual recognition in 3-month-old infants: The role of experience. *Developmental Psychobiology*, 53: 416–424. Doi:10.1002/dev.20532.

Murphy, J., Ipser, A., Gaigg, S. B, and Cook, R. (2015). Exemplar variance supports robust learning of facial identity. *Journal of Experimental Psychology: Human Perception and Performance*, 41: 577–581. Doi:10.1037/xhp0000049.

Natu, V. S., Barmett, M. A., Hartley, J., Gomez, J., Stigliani, A., and Grill-Spector, K. (2016). Development of neural sensitivity to face identity correlates with perceptual discriminability. *Journal of Neuroscience*, 36: 10893–10907. Doi:10.1523/JNEUROSCI.1886-16.2016.

Neil, L., Cappagli, G., Karaminis, T., Jenkins, R., and Pellicano, E. (2016). Recognizing the same face in different contexts: Testing within-person face recognition in typical development and in autism. *Journal of Experimental Child Psychology*, 143: 139–153. Doi:10.1016/j.jecp.2015.09.029.

Newcombe, N. and Lie, N. (1995). Overt and covert recognition of faces in children and adults. *Psychological Science*, 6: 241–245. Doi:10.1111/j.1467-9280.1995.tb00599.x

Nishimura, M., Maurer, D., and Gao, X. (2009). Exploring children's face-space: A multidimensional scaling analysis of the mental representation of facial identity. *Journal of Experimental Child Psychology*, 103: 355–375. Doi:10.1016/j.jecp.2009.02.005.

Nishimura, M., Maurer, D., Jeffrey, L., Pellicano, E., and Rhodes, G. (2008). Fitting the child's mind to the world: Adaptive norm-based coding of facial identity in 8-year-olds. *Developmental Science*, 11: 620–627. Doi:10.1111/j.1467-7687.2008.00706.x.

Nordt, M., Semmelmann, K., Genç, E., and Weigelt, S. (2018). Age-related increase of image-invariance in the fusiform face area. *Developmental Cognitive Neuroscience*, 31, 46–57. Doi:10.1016/j.dcn.2018.04.005.

Nordt, M. and Weigelt, S. (2017). Face recognition is similarly affected by viewpoint in school-aged children and adults. *Peer Journal*, 5(e3253):1–14. Doi:10.7717/peerj.3253.

Nakato, E., Kanazawa, S., and Yamaguchi, M. K. (2018). Holistic processing in mother's face perception for infants. *Infant Behavior and Development*, 50: 257–263. Doi:10.1016/j.infbeh.2018.01.007.

Otsuka, Y., Nakato, E., Kanazawa, S., Yamaguchi, M. K., Watanabe, S., and Kakigi R. (2007). Neural activation to upright and inverted faces in infants measured by near infrared spectroscopy. *Neuroimage*, 34(1): 399–406. Doi:10.1016/j.neuroimage.2006.08.013.

Pascalis, O., De Haan, M., & Nelson, C. A. (2002). Is face processing species-specific during the first year of life? *Science*, 296(5571), 1321–1323.Doi: 10.1126/science.1070223.

Pascalis, O., de Haan, M., Nelson, C. A., and de Schonen, S. (1998). Long-term recognition memory for faces assessed by visual paired comparison in 3- and 6-month-old infants. *Journal of Experimental Psychology: Learning, Memory, and Cognition*, 24: 249–260. Doi:10.1037/0278-7393.24.1.249.

Pascalis, O. and de Schonen, S. (1994). Recognition memory in 3-to 4-day-old human neonates. *NeuroReport*, 5: 1721–1724. Doi:10.1097/00001756-199409080-00008.

Pascalis, O., de Schonen, S., Morton, J., Deruelle, C., and Fabre-Gremet, M. (1995). Mother's face recognition by neonates: A replication and an extension. *Infant Behavior and Development*, 18: 79–85. Doi:10.1016/0163-6383(95)90009-8.

Pascalis, O., Scott, L. S., Kelly, D. J., Shannon, R. W., Nicholson, E., Coleman, M., and Nelson, C. A. (2005). Plasticity of face processing in infancy. *Proceedings of the National Academy of Sciences USA*, 102: 5297–5300. Doi:10.1073/pnas.0406627102.

Peelen, M. V., Glaser, B., Vuilleumier, P., and Eliez, S. (2009). Differential development of selectivity for faces and bodies in the fusiform gyrus. *Developmental Science*, 12: F16–F25. Doi:10.1111/j.1467-7687.2009.00916.x.

Pellicano, E. and Rhodes, G. (2003). Holistic processing of faces in preschool children and adults. *Psychological Science*, 14: 618–622. Doi:10.1046/j.0956-7976.2003.psci_1474.x.

Pellicano, E., Rhodes, G., and Peters, M. (2006). Are preschoolers sensitive to configural information in faces? *Developmental Science*, 9: 270–277. Doi:10.1111/j.1467-7687.2006.00489.x.

Peykarjou, S. and Hoehl, S. (2013). Three-month-olds' brain responses to upright and inverted faces and cars. *Developmental Neuropsychologia*, 38(4): 272–280. Doi:10.1080/87565641.2013.786719.

Pickron, C. B., Fava, E., and Scott, L. S. (2017). Follow my gaze: Face race and sex influence gaze-cued attention in infancy. *Infancy*, 22(5): 626–644. Doi:10.1111/infa.12180.

Pimperton, H., Pellicano, E., Jeffrey, L., and Rhodes, G. (2009). The role of higher-level adaptive coding mechanisms in the development of face recognition. *Journal of Experimental Child Psychology*, 104: 229–238. Doi:10.1016/j.jecp.2009.05.009.

Potter, T. and Corneille, O. (2008). Locating attractiveness in the face space: Faces are more attractive when closer to their group prototype. *Psychonomic Bulletin and Review*, 15: 615–622.
Doi:10.3758/PBR.15.3.615.

Proietti, V., Laurence, S., Matthews, C. M., Zhou, X., and Mondloch, C. J. (2019). Attending
to identity cues reduces the own-age but not the own-race recognition advantage. *Vision
Research*, 157: 184–191. Doi:10.1016/j.visres.2017.11.010.

Putzar, L., Hötting, K., and Röder, B. (2010). Early visual deprivation affects the development of face recognition and of audio-visual speech perception. *Restorative Neurology and
Neuroscience*, 28: 251–257. Doi:10.3233/RNN-2010-0526.

Quinn, P. C., Lee, K., Pascalis, O., and Xiao, N. G. (2020). Emotional expressions reinstate
recognition of other-race faces in infants following perceptual narrowing. *Developmental
Psychology*, 56: 15–27. Doi:10.1037/dev0000858.

Quinn, P. C., Uttley, L., Lee, K., Gibson, A., Smith, M., Slater, A. M., and Pascalis, O. (2008).
Infant preference for female faces occurs for same- but not other-race faces. *Journal of
Neuropsychology*, 2(Pt 1): 15–26. Doi:10.1348/174866407X231029.

Quinn, P. C., Yahr, J., Kuhn, A., Slater, A. M., and Pascalils, O. (2002). Representation of
the gender of human faces by infants: A preference for female. *Perception*, 31: 1109–1121.
Doi:10.1068/p3331.

Reid, V. M., Dunn, K., Young, R. J., Amu, J., Donovan, T., and Reissland, N. (2017). The human
fetus preferentially engages with face-like visual stimuli. *Current Biology*, 27: 1825–1828.e3.
Doi:10.1016/j.cub.2017.05.044.

Rhodes, G., Ewing, L., Jeffery, L., Avard, E., and Taylor, L. (2014). Reduced adaptability, but no
fundamental disruption, of norm-based face-coding mechanisms in cognitively able children and adolescents with autism. *Neuropsychologia*, 62: 262–268. Doi:10.1016/j.neuropsyc
hologia.2014.07.030.

Rhodes, G. and Jeffrey, L. (2006). Adaptive norm-based coding of facial identity. *Vision
Research*, 46: 2977–2987. Doi:10.1016/j.visres.2006.03.002.

Rhodes, G., Jeffrey, R., Taylor, L., Hayward, W. G., and Ewing, L. (2014). Individual
differences in adaptive coding of face identity are linked to individual differences in
face recognition ability. *Psychology: Human Perception and Performance*, 40: 897–903.
Doi:10.1037/a0035939.

Rhodes, G., Jeffrey, L., Watson, T. L., Clifford, C. W., and Nakayama, K. (2003). Fitting the mind
to the world: Face adaptation and attractiveness aftereffects. *Psychological Science*, 14: 558–
566. Doi:10.1046/j.0956-7976.2003.psci_1465.x.

Rhodes, G., Locke, V., Ewing, L., and Evangelista, E. (2009). Race coding and the other-race
effect in face recognition. *Perception*, 38: 232–241. Doi:10.1068/p6110.

Rhodes, G. and Tremewan, T. (1996). Averageness, exaggeration, and facial attractiveness.
Psychological Science, 7: 105–110. Doi:10.1111/j.1467-9280.1996.tb00338.x.

Righi, G., Westerlund, A., Congdon, E. L., Troller-Renfree, S., and Nelson, C. A. (2014). Infants'
experience-dependent processing of male and female faces: Insights from eye tracking and
event-related potentials. *Developmental Cognitive Neuroscience*, 8, 144–152. Doi:10.1016/
j.dcn.2013.09.005.

Ritchie, K. L. and Burton, A. M. (2017). Learning faces from variability. *Quarterly Journal of
Experimental Psychology*, 70: 897–905. Doi:10.1080/17470218.2015.1136656.

Robbins, R. A., Maurer, D., Hatry, A., Anzures, G., and Mondloch, C. J. (2012). Effects of normal
and abnormal visual experience on the development of opposing aftereffects for upright and
inverted faces. *Developmental Science*, 15: 194–203. Doi:10.1111/j.1467-7687.2011.01116.x.

Robbins, R. A., Nishimura, M., Mondloch, C. J., Lewis, T. L., and Maurer, D. (2010). Deficits in sensitivity to spacing after early visual deprivation in humans: A comparison of human faces, monkey faces, and houses. *Developmental Psychobiology*, 52: 775–781. Doi:10.1002/dev.20473.

Robbins, R. A., Shergill, Y., Maurer, D., and Lewis, T. L. (2011). Development of sensitivity to spacing versus feature changes in pictures of houses: Evidence for slow development of a general spacing detection mechanism? *Journal of Experimental Child Psychology*, 109: 371–382. Doi:10.1016/j.jecp.2011.02.004.

Röder, B., Ley, P., Shenoy, B. H., Kekunnaya, R., and Bottari, D. (2013). Sensitive periods for the functional specialization of the neural system for human face processing. *Proceedings of the National Academy of Sciences USA*, 110(42): 16760–16765. Doi:10.1073/pnas.1309963110.

Rossion, B. and Jacques, C. (2011). The N170: Understanding the time-course of face perception in the human brain. In S. Luck and E. Kappnman (eds.), *The Oxford Handbook of Event-Related Potential Components*. Oxford: Oxford University Press, pp. 115–142.

Sai, F. Z. (2005). The role of the mother's voice in developing mother's face preference: Evidence for intermodal perception at birth. *Infant and Child Development*, 14: 29–50. Doi:10.1002/icd.376.

Sangrigoli, S. and De Schonen, S. (2004). Recognition of own-race and other-race faces by three-month-old infants. *Journal of Child Psychology and Psychiatry*, 45(7): 1219–1227. Doi:10.1111/j.1469-7610.2004.00319.x.

Scherf, K. S., Behrmann, M., Humphreys, K., and Luna, B. (2007). Visual category-selectivity for faces, places and objects emerges along different developmental trajectories. *Developmental Science*, 10: F15–F30. Doi:10.1111/j.1467-7687.2007.00595.x.

Schwiedrzik, C. M., Melloni, L., and Schurger, A. (2018). Mooney face stimuli for visual perception research. *PLOS One*, 13(7): 1–11. Doi:10.1371/journal.pone.0200106.

Scott, L. S. and Monesson, A. (2009). The origin of biases in face perception. *Psychological Science*, 20: 676–680. Doi:10.1111/j.1467-9280.2009.02348.x.

Scott, L. S. and Monesson, A. (2010). Experience-dependent neural specialization during infancy. *Neuropsychologia*, 48: 1857–1861. Doi:10.1016/j.neuropsychologia.2010.02.008.

Short, L. A., Balas, B., and Wilson, C. (2017). The effect of educational environment on identity recognition and perceptions of within-person variability. *Visual Cognition*, 25: 940–948. Doi:10.1080/13506285.2017.1360974.

Short, L. A., Hatry, A. J., and Mondloch, C. J. (2011). The development of norm-based coding and race-specific face prototypes: An examination of 5- and 8-year-olds' face space. *Journal of Experimental Child Psychology*, 108: 338–357. Doi:10.1016/j.jecp.2010.07.007.

Short, L. A., Lee, K., Fu, G., and Mondloch, C. J. (2014). Category-specific face prototypes are emerging, but not yet mature, in 5-year-old children. *Journal of Experimental Child Psychology*, 126: 161–177. Doi:10.1016/j.jecp.2014.04.004.

Simion, F., Macchi Cassia, V., Turati, C., and Valenza, E. (2001). The origins of face perception: Specific versus non-specific mechanisms. *Infant and Child Development*, 10: 59–65. Doi:10.1002/icd.247.

Simion, F., Valenza, E., Umiltà, C., and Dalla Barba, B. (1998). Preferential orienting to faces in newborns: a temporal-nasal asymmetry. *Journal of Experimental Psychology: Human Perception and Performance*, 24: 1399–1405. Doi:10.1037/0096-1523.24.5.1399.

Simpson, E. A., Maylott, S. E., Mitsven, S. G., Zeng, G., and Jakobsen, K. V. (2020). Face detection in 2- to 6-month-old infants is influenced by gaze direction and species. *Developmental Science*, 23: e12902. Doi:10.1111/desc.12902.

Simpson, E. A., Varga, K., Frick, J. E., and Fragaszy, D. (2011). Infants experience perceptual narrowing for nonprimate faces. *Infancy*, 16, 318–330. Doi:10.1111/ j.1532-7078.2010.00052.x.

Slater, A., Morison, V., and Rose, D. (1983). Perception of shape by the newborn baby. *British Journal of Developmental Psychology*, 1: 135–142. Doi:10.1111/j.2044-835X.1983.tb00551.x.

Sugden, N. A. and Marquis, A. R. (2017). Meta-analytic review of the development of face discrimination in infancy: Face race, face gender, infant age, and methodology moderate face discrimination. *Psychological Bulletin*, 143: 1201–1244. Doi:10.1037/bul0000116.

Sugden, N. A. and Moulson, M. C. (2017). Hey baby, what's "up"? One- and 3-month-olds experience faces primarily upright but non-upright faces offer the best views. *Quarterly Journal of Experimental Psychology*, 70: 959–969. Doi:10.1080/17470218.2016.1154581.

Sugden, N. A. and Moulson, M. C. (2018). These are the people in your neighbourhood: Consistency and persistence in infants' exposure to caregivers', relatives', and strangers' faces across contexts. *Vision Research*, 157: 230–241. Doi:10.1016/j.visres.2018.09.005.

Sugden, N. A., Mohamed-Ali, M. I., and Moulson, M. C. (2014). I spy with my little eye: Typical, daily exposure to faces documented from a first-person infant perspective. *Developmental Psychobiology*, 56: 249–261. Doi:10.1002/dev.21183.

Sugita, Y. (2008). Face perception in monkeys reared with no exposure to faces. *Proceedings of the National Academy of Sciences USA*, 105: 394–398. Doi:10.1073/pnas.0706079105.

Tanaka, J. W. and Farah, M. J. (1993). Parts and wholes in face recognition. *The Quarterly Journal of Experimental Psychology Section A*, 46: 225–245. Doi:10.1080/14640749308401045.

Tanaka, J. W., Kay, J. B., Grinnell, E., Stansfield, B., and Szechter, L. (1998). Face recognition in young children: When the whole is greater than the sum of its parts. *Visual Cognition*, 5: 479–496. Doi:10.1080/713756795.

Tanaka, J. W., Heptonstall, B., and Hagen, S. (2013). Perceptual expertise and the plasticity of other-race face recognition. *Visual Cognition*, 21: 1183–1201. Doi:10.1080/ 13506285.2013.826315.

Tanaka, J.W., Kiefer, M., and Bukach, C.M. (2004). A holistic account of the own-race effect in face recognition: Evidence from a cross-cultural study. *Cognition*, 93: 81–89. Doi:10.1016/ j.cognition.2003.09.011.

Tanaka, J. W. and Sengco, J. A. (1997). Features and their configuration in face recognition. *Memory and Cognition*, 25: 583–592. Doi:10.3758/BF03211301.

Tham, D. S., Bremner, J. G., and Hay, D. (2015). In infancy the timing of emergence of the other-race effect is dependent on face gender. *Infant Behavior and Development*, 40: 131–138. Doi:10.1016/j.infbeh.2015.05.006.

Tham, D. S. Y., Woo, P. J., and Bremner, J. G. (2019). Development of the other-race effect in Malaysian-Chinese infants. *Developmental Psychobiology*, 61: 107–115. Doi:10.1002/dev.21783.

Timeo, S., Brigadoi, S., and Farroni, T. (2019). Perception of Caucasian and African faces in 5- to 9-month-old Caucasian infants: A functional near-infrared spectroscopy study. *Neuropsychologia*, 126: 3–9. Doi:10.1016/j.neuropsychologia.2017.09.011.

Tong, F., Nakayama, K., Vaughan, J. T., Kanwisher, N. (1998). Binocular rivalry and visual awareness in human extrastriate cortex. *Neuron*, 21(4): 753–759. Doi:10.1016/ S0896-6273(00)80592-9.

Turati, C. (2004). Why faces are not special to newborns: An alternative account of the face preference. *Current Directions in Psychological Science*, 13: 5–8. Doi:10.1111/ j.0963-7214.2004.01301002.x.

Turati, C., Bulf, H., and Simion, F. (2008). Newborns' face recognition over changes in view-point. *Cognition*, 106: 1300–1321. Doi:10.1016/j.cognition.2007.06.005.

Turati, C., Macchi Cassia, V., Simion, F., and Leo, I. (2006). Newborns' face recognition: Role of inner and outer facial features. *Child Development*, 77: 297–311. Doi:10.1111/j.1467-8624.2006.00871.x.

Turati, C., Sangrigoli, S., Ruel, J., and de Schonen, S. (2004). Evidence of the face inversion effect in 4-month-old infants. *Infancy*, 6: 275–297. Doi:10.1207/s15327078in0602_3.

Turati, C., Simion, F., Milani, I., and Umiltà, C. (2002). Newborns' preference for faces: What is crucial? *Developmental Psychology*, 38(6): 875–882. Doi:10.1037/0012-1649.38.6.875.

Turk, M. and Pentland, A. (1991). Eigenfaces for Recognition. *Journal of Cognitive Neuroscience*, 3: 71–86. Doi:10.1162/jocn.1991.3.1.71.

Tzourio-Mazoyer, N., De Schonen, S., Crivello, F., Reutter, B., Aujard, Y., and Mazoyer, B. (2002). Neural correlates of woman face processing by 2-month-old infants. *Neuroimage*, 15: 454–461. Doi:10.1006/nimg.2001.0979.

Valentine, T. (1991). A unified account of the effects of distinctiveness, inversion, and race in face recognition. *The Quarterly Journal of Experimental Psychology Section A: Human Experimental Psychology*, 43: 161–204. Doi:10.1080/14640749108400966.

Valentine, T., Darling, S., and Donnelly, M. (2004). Why are average faces attractive? The effect of view and averageness on the attractiveness of female faces. *Psychonomic Bulletin and Review*, 11: 582–487. Doi:10.3758/BF03196599.

Valentine, T., Lewis, M. B., and Hills, P. J. (2016). Face-space: A unifying concept in face recognition research. *Quarterly Journal of Experimental Psychology*, 69: 1996–2019. Doi:10.1080/17470218.2014.990392.

Valenza, E., Simion, F., Macchi Cassia, V., and Umiltà, C. (1996). Face preference at birth. *Journal of Experimental Psychology: Human Perception and Performance*, 22: 892–903. Doi:10.1037/0096-1523.22.4.892.

Vernetti, A., Ganea, N., Tucker, L., Charman, T., Johnson, M. H., and Senju, A. (2018). Infant neural sensitivity to eye gaze depends on early experience of gaze communication. *Developmental Cognitive Neuroscience*, 34: 1–6. Doi:10.1016/j.dcn.2018.05.007.

Vogel, M., Monesson, A., and Scott, L. S. (2012). Building biases in infancy: The influence of race on face and voice emotion matching. *Developmental Science*, 15: 359–372. Doi:10.1111/j.1467-7687.2012.01138.x.

von Hofsten, O., von Hofsten, C., Sulutvedt, U., Laeng, B., Brennen, T., and Magnussen, S. (2014). Simulating newborn face perception. *Journal of Vision*, 14(13): 16. Doi:10.1167/14.13.16.

Webster, M. A. Maclin, O. H. (1999). Figural aftereffects in the perception of faces. *Psychonomic Bulletin and Review*, 6: 647–653. Doi:10.3758/BF03212974.

Weigelt, S., Koldewyn, K., Dilks, D. D., Balas, B., McKone, E., and Kanwisher, N. (2014). Domain-specific development of face memory but not face perception. *Developmental Science*, 17: 47–58. Doi:10.1111/desc.12089.

Wheeler, A., Anzures, G., Quinn, P. C., Pascalis, O., Omrin, D. S., and Lee, K. (2011). Caucasian infants scan own- and other-race faces differently. *PLOS One*, 6(4): e18621. Doi:10.1371/journal.pone.0018621.

Wilkinson, N., Paikan, A., Gredebäck, G., Rea, F., and Metta, G. (2014). Staring us in the face? An embodied theory of innate face preference. *Developmental Science*, 17: 809–825. Doi:10.1111/desc.12159.

Wilson, R. R., Blades, M., Coleman, M., and Pascalis, O. (2009). Unfamiliar face recognition in children with autistic spectrum disorders. *Infant and Child Development*, 18: 545-555. Doi:10.1002/icd.638.

Wright, D. B., Boyd, C. E., and Tredoux, C. G. (2003). Inter-racial contact and the own-race bias for face recognition in South Africa and England. *Applied Cognitive Psychology*, 17(3): 365-373. Doi:10.1002/acp.898.

Young, A. W. and Burton, A. M. (2017). Recognizing faces. *Current Directions in Psychological Science*, 26: 212-217. Doi:10.1177/0963721416688114.

Young, A. W., Hellawell, D., and Hall, D. C. (1987). Configurational information in face perception. *Perception*, 16: 747-759. Doi:10.1068/p160747.

Young, S. G., Hugenberg, K., Bernstein, M. J., and Sacco, D. F. (2012). Perception and motivation in face recognition: A critical review of theories of the cross-race effect. *Personality and Social Psychology Review*, 16: 116-142. Doi:10.1177/ 1088868311418987.

Zhao, M., Hayward, W. G., and Bülthoff, I. (2014). Holistic processing, contact, and the other-race effect in face recognition. *Vision Research*, 105: 61-69. Doi:10.1016/j.visres.2014.09.006.

Zhou, X., Matthews, C. M., Baker, K. A., and Mondloch, C. J. (2018). Becoming familiar with a newly encountered face: Evidence of an own-race advantage. *Perception*, 47: 807-820. Doi:10.1177/0301006618783915.

Zhou, X. and Mondloch, C. J. (2016). Recognizing "Bella Swan" and "Hermione Granger": No own-race advantage in recognizing photos of famous faces. *Perception*, 45: 1426-1429. Doi:10.1177/0301006616662046.

Zieber, N., Kangas, A., Hock, A., Hayden, A., Collins, R., Bada, H., Joseph, J. E., and Bhatt, R. S. (2013). Perceptual specialization and configural face processing in infancy. *Journal of Experimental Child Psychology*, 116: 625-639. Doi:10.1016/j.jecp.2013.07.007.

CHAPTER 10

···

PREVERBAL CATEGORIZATION AND ITS NEURAL CORRELATES

Methods and Findings

···

SABINA PAUEN AND STEFANIE PEYKARJOU

INTRODUCTION

···

CATEGORIZATION is a basic information-processing capacity of rather diverse animate beings, ranging from insects (e.g., Dolev and Nelson, 2014) to humans (Harnad, 2017). Adults use it to structure their perceptual input, to mentally group things, and to generalize experiences across entities and situations (Smith and Medin, 1981). Interestingly, even the more advanced applications of categorical thinking such as generalizing inter-modal mappings can already be found in the first year of life of human infants (e.g., Dewar and Xu, 2010; Vukatana et al, 2015), thus raising the important question of when and how categorization gets started and which brain processes might underlie this fundamental cognitive ability.

Developmental research soon revealed that even newborns are capable of classifying stimuli on various dimensions, including the auditory and visual domain (e.g., Sinnott and Aslin, 1985; Quinn, 2011). Whereas visual information plays a critical role for forming categorical representations of natural objects (Poulin-Dubois and Pauen, 2017; Rakinson and Yermolayeva, 2010), auditory categorization is required for developing language understanding (Liebenthal and Bernstein, 2017) in the infant mind. Over the last decades, many important questions have been raised and partly answered in categorization research: which type of categories are formed at the beginning of life? What is the contribution of perceptual and conceptual information to categorization? How do visual and auditory information interact in developing categories? Focusing on these and related questions, this chapter provides an overview of the current state of the art in infant categorization research. Differing from most previous reviews we will put special emphasis on explaining how behavioural and neurophysiological approaches are used to study the very beginnings of category formation in the first year of life. Following that, we will summarize empirical work on auditory and visual categorization, as well as their intermodal connection. At the end, we present some concluding remarks and discuss potentially fruitful avenues for future research.

Behavioural Measures of Preverbal Categorization

Even though it is not possible to investigate mental processes directly, empirical work exploring categorical thinking in infants started more than 50 years ago. All tasks introduced so far are based on the general idea to compare infants' interest for stimuli from two different categories. Existing studies vary with respect to the nature of the stimuli presented within the auditory and the visual domain (see section, The nature of the stimuli presented), how researchers measure infants' interest in exemplars from contrasting categories (see section, Dependent measures used to assess states of infant attention and interest), and which basic learning mechanisms are used to study category learning, category activation, and category discrimination (see section, Basic learning mechanisms involved in probing categorization skills).

The Nature of the Stimuli Presented

Within the auditory domain, categorization studies often refer to tones, sounds, chords, melodies, rhythms, or human-made sounds, including language-related input (see section, Limitations of behavioural studies on preverbal infant categorization). In the visual domain, 2-D pictures (cartoons, line-drawings, photographs), 3-D objects (toy-models, artificial objects), point-light displays or video clips serve as stimuli (see section, Limitations of behavioural studies on preverbal infant categorization). In both perceptual domains, stimuli presented in different tasks vary largely with respect to their degree of complexity and whether or not they show dynamic changes over time. Furthermore, some studies focus explicitly on representations that cross modalities, thus providing visual and acoustic input.

Dependent Measures Used to Assess States of Infant Attention and Interest

Although young infants still have a limited behavioural repertoire to express their interest in a given stimulus type, even newborns can already orient their head towards an attractive target (e.g., Zelazo et al., 1984). If the target is auditory and elicits attention, infants may increase their sucking rate which can be measured via a rubber-nipple attached to a lever-action recording device (e.g., Bjjeljac-Babic et al., 1993). If the target is visual, the duration of looking may serve as dependent measure (e.g., Fantz, 1961, 1965). When presenting 3-D objects, 3–4-month-olds might grab them and explore them with their mouth, whereas 5–6-month-olds become able to show more differentiated exploratory play and to express their interest by manipulating objects with their hands while looking at them (Ruff and Saltarelli, 1993). There are multiple ways to combine these behavioural measures of attention with basic learning mechanisms.

Basic Learning Mechanisms Involved in Probing Categorization Skills

With respect to the auditory domain, the most frequently used learning mechanism indicating category discrimination in infancy is *operant conditioning*. As the following examples may illustrate, young infants can learn to modulate their spontaneous behaviour in order to control stimulus exposure by teaching infants to adjust their sucking rate (e.g., DeCasper and Fifer, 1980), or to look in a specific direction in order to hear a specific sound (Trainor, 1996), researchers can not only determine to which degree they are interested in a given stimulus category, but also whether they can discriminate exemplars from two contrasting categories. This approach has been very valuable in probing whether infants can discriminate speech sounds (for a review, see Kuhl, 1987). In the so-called *head-turn paradigm* (e.g., Werker et al., 1997), auditory stimuli are presented over a speaker. Infants learn to turn their head in the direction of a dark plexiglass box whenever they detect a change in category. Whenever the infant shows a correct response, the box is illuminated and an animated toy is presented as a reward. Whenever the infant does not respond correctly, nothing happens. The infants' learning rate thus reveals whether or not a given infant is able to discriminate two contrasting auditory categories. For example, the head-turn paradigm has been used very successfully to demonstrate that younger infants (i.e., below 7 months of age) are able to discriminate sounds from different languages, whereas their speech discrimination abilities narrow down to their mother's tongue during the second half of the first year of life (e.g., Polka and Werker, 1994; Werker and Tees, 1984).

With respect to the visual domain, the simplest way to test category discrimination is to check for *perceptual preferences* due to increased saliency of one category, for instance a preference for animate over inanimate stimuli. If exemplars are presented in a mixed sequence and items from one category are looked at for longer times than items from the contrasting category, this is assumed to indicate category discrimination. At about 2–4 months of age, infants' visual system is developed well enough to compare images and to express visual interest by directing their eye gaze, and by showing a looking preference during simultaneous presentations of two stimuli (see Teller, 1979).

As preference tests are conducted in the auditory as well as in the visual domain, it seems important to note that null results are generally difficult to interpret. A null finding may either indicate that the infant does not like any category more than the other, or it may reveal that the infant is not (yet) capable of detecting the difference between categories.

To overcome this problem, developmental psychologists introduced the so-called habituation and orienting response (Sokolov, 1966): whenever a given stimulus is perceived for the first time, it typically elicits an orienting response. Humans initially respond with increased arousal and increased attention. But, with every repetition of the same presentation, this response will weaken, and at some point, infants lose interest and habituate to the stimulus at hand. Only if a different stimulus is now introduced, a new orienting response can be elicited. This recovery of attention (dishabituation) following a previous habituation phase plays a key role in studies probing infants' perceptual discrimination of individual stimuli. The same mechanism can also be adapted to investigate categorization processes, simply by presenting multiple different exemplars of the same category during a familiarization phase, followed by a test phase, including exemplars from a contrasting category to check for an orienting response towards these out-of-category exemplars. When using the general habituation-dishabituation procedure for probing visual categorization, looking or examination durations typically serve as dependent variables (e.g., Behl-Chadha et al., 1996; Mandler and McDonough, 1993), and when using the same paradigm for probing category discrimination in the acoustic domain, head-turn frequency (e.g., Body et al., 1984), or sucking rate (e.g., Kuhl and Miller, 1975) may serve as dependent behavioural measure.

Behavioural Paradigms Based on Habituation or Familiarization and Dishabituation

Next, we take a closer look at frequently used procedures, which are mainly based on the classical habituation-dishabituation paradigm. As already mentioned, every new stimulus first elicits an orienting response which is reflected by increased infant attention. After repeated or longer exposure this orienting response vanishes—a process called habituation. Only if the infant recognizes a stimulus as new, the orienting response gets reactivated. Researchers can either predefine how long/often the habituation stimulus is presented (fixed-trial procedure), or they define a habituation criterion (e.g., looking time has to decrease to half of its initial level). Within the categorization literature, the term habituation is often replaced by the term "familiarization" and describes the process of becoming familiar with a predefined set of stimuli that all belong to the same category. Typically, there is no fixed criterion to indicate that sufficient familiarity has been gathered. Because every exemplar presented is perceptually new, infants may fail to show any decrease in attention but still become familiar with the category. In that case, the only way to determine the existence of a familiarization effect is to look for a significant increase in attention (dishabituation) to a change in category. Empirical findings obtained with this method will be described in more detail later, whereas the focus of the following paragraphs will be on methodological issues.

In *categorical familiarization-dishabituation paradigms*, by comparing similarities and differences across different stimuli presented during a given familiarization phase, infants are assumed to either form a new category online (Younger and Cohen, 1985), or to activate an already existing category representation (Mandler, 1988). In either case they should lose their interest for exemplars of a given category once the corresponding process is completed. At test, exemplars of a contrasting category are presented to check whether infants respond only to perceptual newness (thus showing a continuation of the familiarization process), or whether they also detect a change in category (revealing an orienting response). Importantly, researchers using this paradigm need to make sure that test-exemplars of categories to be contrasted are equally attractive to the age group tested, otherwise observed dishabituation responses may result from a-priori preferences.

In *auditory habituation tasks (AHT)*, the same auditory stimulus from one category is presented repeatedly before a new sound from a different category is introduced at test. Infants can determine how long they want to hear a given sound by sucking at a specific rate (e.g., Trehub, 1973), or by looking at a given visual cue associated with the sound (e.g., Miller, 1983). Once they have reduced their behavioural response to a pre-defined level, they receive an out-of-category stimulus to check for a potential orienting response. By varying the similarity of contrasting exemplars presented, this technique serves to study category discrimination. Similarly, *visual-habituation tasks (VHT)* present an individual visual stimulus repeatedly before changing to a new stimulus from a different category (Fantz, 1961). *Visual-familiarization tasks (VFT)*, on the other hand, present different exemplars of the same category on every trial. To probe category discrimination, looking time at an out-of-category exemplar at test is compared to looking time at the last familiarization stimulus (e.g., Roberts and Horowitz, 1986).

A very prominent paradigm for probing category discrimination in the visual domain is the *visual familiarization novelty preference (VFNP) task*. This task supports similarity comparisons within each given trial, as different exemplars are presented in pairs (e.g., Behl-Chadha, 1996; Eimas and Quinn, 1994; Quinn and Eimas, 1986, 1996a; Quinn et al., 1993). More specifically, different pairs of exemplars of one category are presented during the familiarization phase, followed by combinations of new exemplars from the familiar category and a contrasting category. A preference-for-novelty response to the out-of-category exemplar(s) at test is taken as evidence for category discrimination. This paradigm combines habituation procedures with preference choices to probe pre-verbal categorization.

When infants of 5–6 months or older are tested with visual material, they may not receive 2-D pictures but rather be given 3-D objects. Categorization tasks using 3-D objects to be explored visually and manually are called *object-examination tasks*. Examination duration is defined as a subset of looking time associated with states of focused attention in the presence or absence of manual exploration (Oakes et al., 1991). Since 3-D presentations are assumed to induce focused attention more easily than 2-D presentations (Ruff, 1984), and since object examining is typically associated with systematic changes in heart rate (Elsner et al., 2006), possibly indicating deeper cognitive processing (Lansink et al., 2001; Oakes and Tellinghuisen, 1994), some authors suggested that object examination tasks not only measure the effects of online category learning, but also activate previously acquired semantic knowledge (e.g., Mandler and McDonough, 1993, 1998). Recent empirical evidence supports this view (Mash and Bornstein, 2012; Pauen, 2002a; Träuble et al., 2008).

Classical habituation/familiarization procedures using object examination tasks may either present a different exemplar on every single trial (e.g., Pauen, 2000), or a limited number of exemplars presented in repeating blocks. For example, Mandler and McDonough (1993) presented the same sequence of four different exemplars twice for familiarization. At test, one new exemplar of the familiar category was presented, followed by one new exemplar of a contrasting category (see also Pauen, 2002a, 2002b). A dishabituation response to the first test exemplar reveals that infants are still attentive and recognize the perceptual newness of the new within-category exemplar. For a full-fledged category discrimination response they also need to show an increase in attention from the first to the second (out-of-category) test exemplar, however. Different procedures (i.e., block-wise presentation vs. single-item presentation) produce fairly similar results for the same categorical contrast—at least in object examination studies. Data obtained with different procedures that take looking time as the dependent measure also produce reliable results, but paired comparisons may support online category formation even more than procedures with single-item trials (see Oakes and Ribar, 2005).

Limitations of Behavioural Studies on Preverbal Infant Categorization

Sucking rate, conditioned responses, looking, and examination duration all provide important indicators of infants' interest in exemplars of different categories. By using any of these methods developmental psychologists discovered astonishing categorization skills even at a preverbal age (see later sections for a summary of most important work). At the same time, many questions referring to the mental processes underlying performance still remain unanswered (Rakinson and Yermolayeva, 2010). For example: does it make any difference whether infants are presented with 2-D or 3-D stimuli (Mash and Bornstein, 2012), and whether they are allowed to look at them or touch them (Younger and Furrer, 2003)? And, to what extent is a given categorical dishabituation response at test grounded on bottom-up processes (i.e., online category formation) and/ or on top-down processes (i.e., the activation of pre-existing conceptual knowledge (Pauen and Träuble, 2008; Rakinson and Oakes, 2003)? Whereas the first two questions can be answered by systematically varying aspects of the experimental procedure using behavioural paradigms, questions regarding the interplay of bottom-up and to-down processes can hardly be answered that way, thus calling for more direct measures of infants' brain activity. Here, electroencephalogram (EEG) methods come into play.

EEG Methods for Studying Infant Categorization at a Preverbal Age

For almost 20 years, EEG and event-related-potentials (ERPs) have become more and more popular to explore cognitive processing and categorization at all ages (de Haan, 2007). This is especially true for the acoustic and the visual domain. In general, EEG techniques exploit the fact that mental processes in the human brain lead to changes in electric voltage on the scalp. These changes can be measured by using a cap with multiple electrodes mounted at defined locations. Even though the reliability of such measures may not be high enough to assess inter-individual difference in any reliable way (Munsters et al., 2017), results at the group level seem fairly robust and thus allow for a systematic investigation of cognitive functioning and category discrimination in young infants (Hoehl and Wahl, 2012). A clear advantage of this method is its high temporal resolution, which allows for assessing brain responses in the order of milliseconds. As a result, presentation times of stimuli are much shorter in EEG than in behavioural paradigms (i.e., between 0.2–1.5 seconds). Therefore, a higher number of individual items can be presented to each participant and the precise time-course of processing a given stimulus category can be measured.

ERP Components Associated with Categorization Processes

When providing enough time for the brain to process a given stimulus (i.e., 1–2 seconds), averaging brain responses to stimuli of a given category over many trials results in characteristic waveforms of voltage change following stimulus onset (Luck, 2005). These voltage changes unfold over time and can either be negative or positive compared to a baseline assessment. ERP components are defined as these characteristic deflections from baseline. The specific timing, location, and waveform of any given component vary with age. These age-related changes may be due to cortical re-organization (particularly regarding location and waveform) and myelination/speed of processing (timing).

If the infant brain detects a rare stimulus in a sequence of familiar ones, a so-called mismatch response (MMR) will be elicited around 200 ms following stimulus onset (Näätänen, 1990). In response to auditory stimulation, this component can be found on fronto-central leads, and in response to visual stimulation, it can be found in occipital regions (Cheour et al., 2000), suggesting that it reflects basic perceptual processing. As conscious attention to the stimulus is not required for eliciting an MMN, one can even measure it during sleep in the case of auditory stimulation.

The *negative central (Nc)* is an ERP component that does require conscious attention and is widely used for studying infant visual categorization. The Nc is observed around 300–800 ms following stimulus onset at fronto-central leads (Reynolds et al., 2010; Reynolds and Richards, 2005). It can vary considerably in strength, depending on the degree of attention devoted to a given stimulus or stimulus category. Because the head bones of the infant are not yet closed, signals received at frontal locations are usually highly robust and strong. The brain activity leading to corresponding voltage changes results mainly from activity in the inferior and superior prefrontal cortex and anterior cingulate area (Reynolds et al., 2010) reflecting cognitive processes associated with sustained attention. The Nc can be elicited by acoustic as well as visual stimulation.

The late *positive slow wave (PSW)* is another component of special interest, typically occurring around 1,000 to 1,800 ms following stimulus onset in visual tasks with infants. It increases in strength when more memory updating is required, e.g., when infants perceive a change in category (Nelson, 1994; Snyder, 2010).

EEG Paradigms for Studying Preverbal Categorization

When researchers began to apply ERPs for studying categorization in human infants, they took their inspiration from behavioural categorization studies. Research along these lines started in the auditory domain and focused on newborns' ability to discriminate acoustic features such as temporal regularities, pitch, and phonemes (for more recent work along these lines see Dehaene-Lambertz and Pena, 2001; Háden et al., 2009; Háden et al., 2015). But, with time, new paradigms have been introduced that are also applicable for probing visual categorization. In the following paragraphs, we describe the most prominent paradigms of this kind.

Comparing ERP means for different categories. As in behavioural research, one can probe category preferences in ERP studies by presenting an equal number of exemplars from two contrasting categories in a mixed sequence, and determine the mean brain wave for all exemplars of each category. In the next step, the strength of the Nc as a brain correlate for infants' attention allocation can be compared between both categories (e.g., Jeschonek et al., 2010). Significant differences indicate that one category elicits more attention than the other, which implies that both categories are discriminated in the infant mind. As in behavioural research, this paradigm suffers from the fact that null findings are always difficult to interpret (see examples to follow in Advantages of using EEG measures over behavioural paradigms).

ERP habituation/familiarization paradigms. Similar to behavioural research, this paradigm familiarizes infants with multiple exemplars from the same category. At test, attention for familiar items, novel exemplars of the familiar category, and novel exemplars of a contrasting category are compared with respect to Nc enhancement. In some tasks, familiarization responses are measured by ERPs but category discrimination is measured via preferential looking (e.g., Quinn, Doran, et al. 2010), and in other tasks, familiarization and category discrimination are both measured at brain level (e.g., Grossmann et al., 2009). If the Nc response is more pronounced for out-of-category exemplars than for new within-category exemplars, this indicates category discrimination. In contrast to behavioural studies ERPs need to be averaged across multiple trials (exemplars of the same category) in order to increase the signal-noise ratio, but the minimum number of valid trials is much lower for infants than for adults (de Haan, 2007) as voltage changes on the infant scalp can be measured better due to thinner skin, a thinner skull, and less hair.

Oddball paradigms. This paradigm again contrasts two distinct categories, one serving as "standard," thus providing the majority of all stimuli (e.g., 80 percent), and the other serving as "oddball," providing the minority of stimuli (e.g., 20 percent). Exemplars are presented in a mixed sequence (Leppänen et al., 1999). If the brain detects the rare category, it typically responds with increased ERP responses (MMR, Nc) to oddball stimuli as compared to standard stimuli. Sometimes, two oddball stimuli are presented together with one standard stimulus (e.g., standard:60 percent, oddball 1:20 percent, oddball 2:20 percent, Elsner et al., 2013). In this case the oddball stimulus of the contrasting category is expected to show a more enhanced Nc than the oddball of the standard category. As in behavioural studies, earlier oddball studies presented the same stimulus repeatedly and varied exemplars between sessions, whereas more recent studies present new stimuli on every single trial, thus covering a large range of exemplars per category in every presentation (e.g., Marinovic et al., 2014). The oddball paradigm deviates from classical habituation or familiarization procedures in that it introduces test stimuli not only at the end of the procedure, but rather inter-mixed with standard stimuli throughout the entire experimental session. This implies that the classical habituation or familiarization response may have less impact on the findings, whereas a-priori category discrimination is more likely to have some impact.

Category repetition paradigms. These focus on the adaptation of the brain to the immediate repetition of a given category. Brain processing of a target stimulus is analysed and compared based on whether or not the preceding exemplar matches the target category. In some versions of this task that have been applied with infants, equal numbers of exemplars from two contrasting categories are presented in a semi-randomized sequence and brain responses are analysed for exemplars preceded by a same-category item and for exemplars preceded by a different-category item (e.g., Jeschonek et al., 2010; Peykarjou, Wissner, and Pauen, 2016). Importantly, ERP components (Nc and PSW) for the target exemplars are analysed and interpreted in relation to category membership of the previously presented exemplar.

Fast periodic visual stimulation (FPVS). A fairly novel EEG approach particularly suited to capture rapid visual categorization processes is the FPVS paradigm. In this paradigm, frequency responses are measured rather than time-locked ERPs. To this aim, images from one category are presented periodically between images of another category at a fixed presentation rate (e.g., six items per second, with the out-of-category exemplar always presented in the fifth position). The cortex responds to the presentation rate with increased activity in exactly the same frequency and—should it detect the periodical change in category—also in the frequency corresponding to category changes (Rossion, 2014; Rossion, Torfs, Jacques, and Liu-Shuang, 2015). Cortical activity will only increase in the predefined frequency if participants sort frequently presented stimuli into one category, and infrequently presented stimuli into another.

Comparing Different EEG Paradigms in Infant Categorization Research

All EEG methods described so far compare brain responses to stimuli from two contrasting categories. In most cases, exemplars from both categories are presented with varying frequencies. Exceptions refer to methods that compare mean ERPs across categories or that use a category repetition paradigm. When both categories are presented with different frequencies, the critical question is whether the infrequent category is presented (a) only at the end of the procedure, as this is the case for the ERP habituation/familiarization paradigms, (b) in a randomly mixed sequence, as this is the case for oddball paradigms, or (c) in a predefined rhythm, as this is the case for the fast visual periodic stimulation method, as well as for certain forms of category repetition priming. This may have an impact on the underlying mental processes, as some of these procedures are more likely to support online category learning (e.g., ERP habituation/ familiarization paradigms), while others are more likely to probe the activation of already existing category representations (e.g., oddball paradigms, category repetition priming, fast visual periodic stimulation). Yet another difference refers to the timing of stimulus presentation: the great majority of existing paradigms works with presentation times of about 0.5–1 sec per stimulus, but the fast visual periodic stimulation presents six stimuli per second, thus covering only very fast processes in the infant brain. This may allow us to discriminate initial perceptual processing from higher semantic processing.

The choice of the EEG paradigm to measure categorization skills partly determines which aspects and underlying mental processes are studied. So far, a systematic comparison of categorization responses obtained with different methods (while keeping the age-group tested and the stimulus selection constant) is still missing.

Advantages of Using EEG Measures Over Behavioural Paradigms

Measuring ERPs with the EEG technique has some important advantages over traditional behavioural measures: (1) EEG has a very high temporal resolution in the area of milliseconds (Luck, 2005), thus allowing us to measure the cascade of cognitive processes induced by the presentation of a given stimulus. For example, it is possible to distinguish attention allocation (via MMN or Nc) from memory updating (via PSW).

Hence, we may learn more about the timing associated with processes of categorization in the infant brain. (2) Given the short presentation time needed to elicit a meaningful brain response (i.e., in the range of 170–1,500 ms) it is possible to present a much higher number of trials and thus a larger variety of different exemplars representing a given category than in behavioural studies. This provides the opportunity to study category learning with new material and it increases the validity and generalizability of studies referring to natural categories. (3) Finally, a great strength of the EEG is that it can be applied in highly similar paradigms across the lifespan. This allows for a systematic exploration of age-related differences in neurophysiological processes associated with categorization. In this context, direct comparisons of changes in specific components may reveal important insights about processes of brain maturation that affect the development of categorization skills.

During the past 50 years, behavioural and neural research has produced surprising insights into the beginnings of categorical thinking in human ontogeny. Now that the reader knows which methods have been applied, we can take a closer look at empirical findings regarding auditory, visual, and intermodal categorization. Please note that behavioural and EEG studies complement each other and will be reported together.

INFANT CATEGORIZATION IN THE AUDITORY DOMAIN

Within the auditory categorization literature, we need to distinguish between (a) studies that present only one single stimulus for habituation and one other stimulus from a different category to check for *stimulus discrimination* (e.g., Friedrich et al., 2009; Stefanics et al, 2007), and (b) studies that present different exemplars for each contrasting category, to check for *categorization responses* (e.g., Trehub and Thorbe, 1989; Virtala et al., 2013). In both cases, researchers are interested in category discrimination, but only in the latter case do they actually habituate or familiarize infants with a category.

First work on early auditory discrimination skills revealed that infants discriminate simple tones (e.g., Sinnott and Aslin, 1985) and pitch categories (Háden et al., 2009). Soon, developmental researchers started to test infants' categorization of more complex stimuli like musical chords (Virtala et al., 2013), or dynamic stimuli representing temporal regularities (Carral et al., 2005; Háden et al. 2015; Stefanics et al., 2007), such as rhythms (e.g., Trehub and Thorbe, 1989), beats (e.g., Honig, Bouwer, and Háden, 2014), or acoustic number categories (e.g., Ruusuvirta et al., 2009). Yet a third line of work investigates how infants categorize human-made sounds, like vocal or action sounds (e.g., Friederici and Thierry, 2008; Friedrich et al., 2009; Geangu et al., 2015). Whereas behavioural paradigms dominated initially, ERP paradigms now provide the gold standard for studying infant categorization because this is the most direct way to probe categorization. In sum, findings obtained with different stimuli, measures, and paradigms suggest that preverbal infants can categorize acoustic stimuli at different levels of abstraction. It should be noted, though, that even more recent studies in the auditory domain often use only one single standard stimulus and only one or two out-of-category stimuli to probe infants' category discrimination. Discrimination thresholds define perceptual categories and are thus highly relevant in the given context. However, representing a given category by only one exemplar clearly limits the generalizability of results.

Distinguishing Basic Language Sounds

So far, the processing of language-input reached the greatest interest among researchers exploring auditory categorization (see Liebenthal and Bernstein, 2017 for a recent overview), thereby revealing some important general developmental trends that can also be found in other perceptual domains: even newborns categorize phonemes, such as /pa/ and /ta/, despite changes in speaker identity (e.g., Dehaene-Lambertz and Pena, 2001). Work comparing language perception across cultures gave rise to the so-called *perceptual narrowing hypothesis* (Lewkowicz, 2014), whereas infants younger than 8–10 months of age can discriminate speech sounds from all languages, they seem to lose this ability by the end of their first year of life (e.g., Ortiz-Mantilla et al., 2016; Werker and Tees, 1984; see Tsuji and Cristia, 2013 for a meta-analysis). But, perceptual narrowing is not restricted to the categorization of speech sounds only; it can also be found for sign language (Palmer et.al, 2012), indicating that it characterizes a general mechanism of perceptual learning (see also section, Categorization of faces). Current approaches to explain this phenomenon highlight the relevance of brain maturation: initially, an infant's brain seems prepared to adapt to different environments before it tunes into the specific conditions encountered by a given individual. Categorization plays a key role in this context. Importantly, a decrease in sensitivity to diverse inputs with age does not imply that older children become incapable of discriminating stimuli of a given kind (Flom, 2014); rather, they only seem to lose their natural proficiency to do so very rapidly.

Distinguishing Other Aspects of Human Vocalization and Sound Production

Another important aspect of language is prosody; it helps us to decode the emotion content of verbal expressions. Comparing mean ERPs, but contrasting average responses to words spoken in either an infant-directed way (ID: high pitch, strong pitch modulation) or an adult-directed way (AD: lower pitch, less modulation), Zangl and Mills (2007) found enhanced ERP amplitudes for familiar words spoken in ID as compared to AD speech in the left hemisphere. Infants discriminated ID and AD speech by 6 months around 600–800 ms after word onset, but a more general enhancement of two negative components called N200–400 and N600–800, was observed at 13 months of age. This discrimination ability points to the fact that categorization can also refer to dynamic processes.

Further aspects of human sound production have also been investigated. For example, Miller (1983) found that 2-month-old infants discriminate male from female voices, but also distinguish between different voices within each category—an ability that was not observed at 6 months of age, thus supporting the perceptual narrowing hypothesis. While this does not indicate that voice discrimination is invariably lost (certainly we can discriminate between different voices even later in development), it seems that sensitivity to individual differences decreases in the first year of life. Geangu et al. (2015) were interested in the processing of human vocalizations (including coughing, sighing, yawning) and compared it to the processing of human action sounds, environmental, and mechanical sounds. The authors found that 7-month-old infants process human vocalizations differently from all other sounds, but partly also similarly to human action sounds.

Auditory Categorization—Summary

Whereas early studies on auditory categorization mainly employed behavioural paradigms, ERP measurements soon came into play, presumably because temporal processing is central to auditory processing, and because behavioural responses (like sucking rate, conditioned head-turn, or looking) are only indirectly linked to auditory categorization.

Even though the number of studies addressing auditory categorization skills in infants below 1 year of age is much larger than the body of literature cited in this chapter, the overall message seems clear: results obtained with a large variety of tasks reveal that newborns organize the auditory world in quite sophisticated ways (Winkler et al. 2003). They discriminate sounds based on simple criteria (e.g., pitch), but also based on structural criteria (e.g., rhythm), thus leading to categories that range from simple tones or phonemes to complex acoustic patterns associated with language, music, or environmental noises. Infants come equipped with the innate ability to make all sorts of fine-grained distinctions within the auditory domain but soon start to shape their categories based on early experiences, as evidenced by data supporting the perceptual-narrowing hypothesis. Whether a similar developmental trend can also be found for the visual domain will be discussed next.

INFANT CATEGORIZATION IN THE VISUAL DOMAIN

Studying categorization in the visual domain has one big advantage: looking time provides a behavioural response that is directly linked to infants' attention. Hence, it does not come as a surprise that the vast majority of studies exploring visual categorization use behavioural paradigms. Only recently, ERP studies started to play a more prominent role, as they promise to help us disentangle the impact of online-category formation and the activation of pre-existing categories in the infant's mind because of their high temporal resolution and the identification of different ERP components associated with different cognitive processes.

Categorization research on visual processing started by exploring infants' discrimination of basic stimulus dimensions like form or colour, using either simple behavioural preference tasks (e.g., Fantz, 1961), familiarization-dishabituation paradigms (e.g., Bomba and Siqueland, 1983; Bornstein et al., 1976; Quinn, Slater, et al. 2010; Turati et al., 2010; Younger and Gotlieb, 1988), or ERP oddball paradigms (e.g., Clifford et al., 2009). Today, infant researchers also study the categorization of more complex units like natural objects (see Poulin-Dubois and Pauen, 2017; Rakinson and Yermolayea, 2010 for recent overviews), or action categories (e.g., Song et al., 2016).

Categorizing Natural Objects

Mainly inspired by the work of Rosch et al. (1976) on the hierarchical structure of semantic memory and its development, infant researchers soon became interested in exploring at which level preverbal infants may start to form object categories. Corresponding studies presented global-level contrasts (e.g., land vs. sea animals (Oakes et al., 1997); animals, vehicles, furniture, plants, kitchen utensils (Mandler and McDonough, 1993, 1998); mammals vs. furniture (Behl-Chadha, 1996; Pauen, 2002a); mammals vs. humans (Pauen, 2000); basic level categories (e.g., cats vs. dogs (Quinn et al., 1993); birds, fish, couches, chairs, tables (Behl-Chadha, 1996); as well as subordinate level contrasts (e.g., cats: Siamese vs. tabby; dogs: beagle vs. Saint Bernard (Quinn, 2004a)). In sum, this work suggests that the development of object categories follows a so-called *global-to-basic level shift*: infants start with rather broad categories, which they gradually refine later, presumably based on everyday experiences. Interestingly, the timing of this shift seems to vary with the method used to assess category discrimination: procedures presenting pictures (especially paired presentation formats) and using looking time as dependent measure revealed a global-to-basic level shift between 2 and 4 months (Quinn and Johnson, 2000), whereas procedures using 3-D stimuli and object examination as dependent variable found the same shift between 6 and 11–12 months (Mandler and McDonough, 1993; Pauen, 2002b). It has been speculated that this discrepancy in findings has to do with the fact that the object examination task is more likely to tap on infants' conceptual understanding and knowledge activation, whereas visual familiarity novelty preference tasks are more likely to assess perceptual learning processes (see also section, Processes Underlying Category Formation and Distinction). Today, most experts in the field would probably agree that both processes overlap (e.g., Oakes et al., 1996).

Leaving aside these methodological issues, the early emergence of global-category discrimination itself was a surprise, as exemplars of the same global category (e.g., animate beings) often look rather different from each other. Whether they can be discriminated based on appearance attributes like facial features (Quinn and Eimas, 1996b), part configuration (Rakinson and Butterworth, 1998), motion patterns (Mandler, 1992), or on a combination of different aspects still awaits further clarification. In this context, it may be useful to take a closer look at the brain processes underlying object categorization and its ERP correlates.

First ERP studies addressing visual object categorization followed the general logic of behavioural research paradigms and adapted the familiarization-preference-for novelty task. One of these pioneering studies used an ERP familiarization paradigm with a categorical contrast that 3–4-month-olds have previously been shown to discriminate in tasks using looking time as dependent measure (Quinn, Eimas, and Rosenkrantz, 1993): Quinn et al. (2006) familiarized 6-month-old infants with cat images and then probed category discrimination with novel cat and dog images at test. While the Nc was indeed increased for dog compared to cat images at test (consistent with basic-level categorization), the choice of channels and statistical analyses in this study made it difficult to draw general conclusions. Grossmann et al. (2009) later built up on this work and investigated basic-level categorization in 6-month-old infants following a familiarization phase. More specifically, they presented four different exemplars from one category (either fish, birds, or cars) repeatedly (presented block-wise) during familiarization. At test, they presented combinations of previously used items, novel items belonging to the (now) familiar category, and novel items from a contrasting category. The authors found an increased Nc only for out-of-category items, regardless of whether the novel category item belonged to the same global-level category (e.g., birds vs. fish) or not (e.g., birds vs. cars). This replicates findings from behavioural research demonstrating that infants of the tested age range can indeed make basic-level distinctions, thus confirming and extending the initial ERP findings provided by Quinn et al. (2006).

While ERP studies including a familiarization phase largely confirm behavioural findings, they do not allow any conclusion about infants' spontaneous categorization or existing category representations. First studies that renounced the familiarization phase used ERP oddball or repetition priming paradigms. They compared different ERP responses for categories such as animals, furniture, and human beings: Elsner et al. (2013) presented one item (i.e., a rabbit or a drawer) as standard stimulus on 60 percent of the trials, and one item from the same superordinate category (i.e., a giraffe or a chair, respectively), as well as one from the other global category as oddballs (each on 20 percent of the trials). The amplitude of the Nc was increased for both oddballs, and the amplitude of the PSW was even more increased for the oddball from the contrasting global level compared to the one from the same basic level. This suggests that 7-month-olds discriminate individual exemplars within a given category at the perceptual level, but also recognize the change at the global level during later processing even without any familiarization phase. A baseline study using ERP mean comparisons insured that none of the exemplars was preferred over the other in infants of the tested age range.

Another oddball study contrasted two categories within the animate domain (i.e., humans vs. animals, Marinovic et al., 2014). A total of eighty different standard exemplars were mixed up with twenty oddball exemplars. The stimulus set for humans included people of all ages and different ethnic backgrounds, while the stimulus set for animals included different basic-level categories such as various types of mammals, fish, or birds. This study found an increased Nc for categorical oddballs in 7- but not in 4-month-olds, suggesting that only older infants discriminate humans from other animate beings.

Studies employing the ERP category repetition paradigm analysed whether processing was influenced by the category of the stimulus presented immediately prior to the target. That is, processing of a target item was compared for targets preceded by a same-category or a different-category item. Jeschonek et al. (2010) and Peykarjou et al. (2016) showed infants different animate (humans or animals) and inanimate (furniture) stimuli in semi-randomized sequence. Both studies found that the amplitude of the PSW was lower for repeated than for non-repeated categories, indicating that items were encoded categorically. As mentioned previously, the PSW is an ERP component typically associated with memory processes (Nelson, 1994; Snyder, 2010) and may thus indicate memory updating in infant categorization tasks.

Together, behavioural studies as well as ERP studies on categorization of natural object categories suggest that basic-level categories are discriminated by 7 months (at the latest). They are also consistent with the idea that categorization at the global level precedes categorization at more fine-grained levels. Importantly, ERP studies allow us to discriminate perceptual processing (MMN) from processes of attention allocation (Nc) and memory updating (PSW). Based on evidence for memory updating, it seems plausible to assume that certain categorical distinctions such as the global animate-inanimate distinction may exist in the infant's mind prior to entering the baby lab (Kovack-Lesh et al., 2008; Mandler, 2000a; Pauen, 2002a), although humans seem to have a special status within the animate domain even to young infants (e.g., Marinovic et al., 2014; Pauen, 2000; Quinn and Eimas, 1998).

Categorization of Faces

Among the different features that might contribute to the early emergence of a global animate-inanimate distinction, and the special status of humans within the animate domain, facial features have probably been studied most extensively. One reason may be that face categorization is also relevant for understanding infants' social and language learning. Corresponding studies (see Simion and Di Girorgio, 2015, for a recent review) investigated whether infants can discriminate human faces from other types of stimuli (de Haan and Nelson, 1999), or from faces of other species. Pascalis et al. (2002) found that 6- but not 10-month-old infants were able to discriminate individuals of their own species as well as monkey faces, whereas the older infants only showed category discrimination within their own species. Such findings support the idea that perceptual narrowing may also exist in the visual domain. Behavioural studies that explore how infants categorize human faces by race (e.g., Anzures et al., 2013; Kelly et al., 2005), and/

or gender (e.g., Quinn, Uttlay, Lee, et al., 2008; Younger and Fearing, 1999) also point to a process of perceptual narrowing but little is known so far about the neural correlates underlying this specialization (but see Scott and Monesson, 2010).

In addition, the categorization of facial emotional expressions has been investigated using behavioural (e.g., de Haan and Nelson, 1998; Hoehl and Striano, 2008) as well as ERP methods (e.g., Jessen and Grossmann, 2015). This work revealed that infants tend to respond with increased Nc amplitude towards negative (angry or fearful) as compared to positive or neutral emotional expressions starting at around 6–7 months of age (Peltola et al., 2009).

Some authors speculate that face processing might be special at the neural level (e.g., Hoehl and Peykarjou, 2012), as it elicits characteristic modulations of ERP components that cannot be found when infants process other types of stimuli (de Haan et al., 2003), but more work is needed before this hypothesis can be fully evaluated.

Categorization of Unfamiliar New Stimuli

If one wants to gain insights into general cognitive mechanisms of categorization, it can also be helpful to study the categorization of new artificial stimuli. Comparably little work along these lines seems to exist so far, but some examples illustrate the potential merit of this approach: an early series of studies along these lines was conducted by Younger and Cohen (1983, 1986) who presented infants with line drawings of artificial animals constructed from different animal parts (e.g., body, tail, and feet of different animals combined) to evaluate infants' ability to respond to correlations of features. They demonstrated that infants reliably base categorization on individual features at 4-months of age and on correlations of features by 10 months of age. Other studies (Rakison, 2004; Welder and Graham, 2006) also employed unfamiliar, artificially created stimuli to demonstrate the role of features for categorization. While this line of work employed artificial stimuli, these items were clearly animal-like and composed of animal parts, thereby inviting infants to treat them similarly to familiar animate stimuli.

Bornstein and Mash (2010) presented 5-month-old infants never-before-seen objects either in the laboratory or in their home environment and found that participants subsequently treated novel exemplars of the same category as familiar in an object-examination task. Again, using an object examination task, Träuble and Pauen (2007) demonstrated that 12-month-old infants categorized 3-D objects according to a non-salient part following the demonstration of the functional relevance of the critical part, but not without watching a corresponding demonstration. Both studies point to the relevance of experience for shaping category formation in a laboratory setting.

Using ERPs as dependent measures, Reynolds and Richards (2005) presented abstract and meaningless black-and-white patterns during familiarization and compared processing of familiar and novel stimuli in infants between 4.5 and 7.5 months. This study found an increased Nc for novel black-and-white patterns, but did not provide infants with a categorical contrast. Hence, the study focused on pattern discrimination at the individual rather than the categorical level. In general, ERP studies probing infant categorization with abstract unfamiliar stimuli are still sparse.

Future research thus needs to clarify how the categorization of unfamiliar stimulus categories develops by using not only behavioural but also EEG methods. One potential reason for the lack of studies along these lines may be that unfamiliar categories are less likely to elicit infants' interest than faces or natural objects. This is even more of a concern when using EEG paradigms that require researchers to present many trials subsequently. However, by using colourful and complex stimuli together with paradigms that put fewer demands on infants' attention, such as the FPVS paradigm, this challenge can be met.

Categorization of Motion and Abstract Relations

Apart from basic features, natural objects, artificial stimuli, and faces, categorization research in the visual domain also investigates the classification of motion types (e.g., animate vs. inanimate: Arterberry and Bornstein, 2002; Hirai and Hiraki, 2005; Kaduk et al., 2013; Marshall and Shipley, 2009), thus extending work to dynamic events. Similarly, the formation of relational categories is an important topic of interest. Corresponding work focuses on spatial relational categories (e.g., Casasola et al., 2003; Casasola and Ahn, 2018; Quinn, 1994; 2004b), but also on complex events that combine information about objects and their relations (e.g., physical events: Baillargeon and Whang, 2002; Konishi et al., 2016; e.g., path relations: Pruden et al., 2013). All studies mentioned here rely on behavioural paradigms and are based on the general habituation-dishabituation procedure. ERP studies are still largely missing in this field.

Processes Underlying Category Formation and Distinction

For quite a long time, researchers have been asking about the nature and structure of categories, as well as the process of forming or activating conceptual representations (Madole and Oakes, 1999; Mandler, 2000b; Pauen and Träuble, 2008). As early as the 1990s, behavioural studies revealed that infants can detect feature correlations and form prototypes during the experimental session (Younger, 1992; Younger and Fearing, 1999). Today, computer simulations help us to probe rather specific ideas about category learning and developmental changes in category structure (e.g., French et al., 2005; Westermann and Mareschal, 2012).

Based on the assumption that category learning takes place within the experimental session, a growing number of behavioural studies focuses on how various aspects of the experimental procedure affect this process (e.g., Arterberry et al., 2013; Junge et al., 2018; Oakes and Ribar, 2005; Younger and Furrer, 2003), or they ask how perceptual, causal, and/or functional experiences impact online categorization (e.g., Rakison and Cohen, 1998; Träuble and Pauen, 2007).

But online learning may not be the only mechanism explaining infants' performance in categorization studies. Bottom-up processes (i.e., analysing the perceptual input) and top-down processes (i.e., applying existing knowledge) are likely to overlap in their contribution to categorization responses (e.g., Kovack-Lesh et al., 2008). ERP studies will soon shed more light on how these two processes interact as they allow us to study temporal changes of ERPs during the experimental session in more detail. Due to the fact that presentation times for individual stimuli are much shorter in a given ERP paradigm than in any behavioural paradigm, a larger number of exemplars can be presented to each infant. Hence, one can study changes in ERPs across the familiarization phase or during an oddball session, to learn more about online-category learning and knowledge activation. The distinction of specific ERP components, varying in their timing and association to mental processes, provides additional information regarding the cognitive underpinnings of infant categorization.

Visual Categorization—Summary

Research on visual categorization initially started with studies focusing on single stimulus dimensions (e.g., colour) and simple stimuli (e.g., forms), then turned to studies targeting complex entities (e.g., objects) to exploring dynamic events (e.g., motion patterns, human actions), but also the formation of abstract representations (e.g., spatial relations) and structural features of categories. The exploration of category development of young infants in the visual and the acoustic domain thus followed a similar path from focusing on single and simple stimuli, to focusing on multidimensional and complex stimulus constellations.

Existing evidence suggests that the brain tunes in to the specific exemplars that infants encounter in their first year of life, thus supporting the perceptual-narrowing hypothesis for the visual domain, too—even though this has only been demonstrated for face categories so far. Furthermore, it revealed that object categorization undergoes a process of gradual refinement with age, as reflected by the global-to-basic level shift. These two developmental trends may complement each other: whereas the perceptual narrowing hypothesis states that infants gradually lose their sensitivity to make more fine-grained distinctions among individual exemplars within a given category, but do so only when lacking exposure to varying exemplars, the global-to-basic level shift states that (at least in the field of object categorization) infants become more sensitive to within-category differences with age, unless they frequently encounter a large variety of different exemplars. Both trends are based on the general assumption that learning experiences and real-world exposure to exemplars play a crucial role, and both mechanisms may help infants to find out which type of similarities and differences are actually relevant in a given environment. Even though we have learned a lot from behavioural studies regarding how visual categorization skills develop at a preverbal age, ERP studies can make important additional contributions—especially when it comes to understanding the interaction of top-down and bottom-up processes.

INTERMODAL CATEGORIZATION AND CROSS-MODAL INFLUENCES ON INFANT CATEGORIZATION

So far, we only mentioned studies that address categorization in one domain. This points to the interesting question of how acoustic and visual categories are combined. Some behavioural work focuses on rather basic combinations such as associations between up- and downward visual movements and corresponding acoustic tone changes (e.g., Jeschonek et al., 2012), or between visual and acoustically displayed rhythms (Spelke, 1979), and the number of visual and acoustic events (e.g., Posid and Cordes, 2019; Starkey et al., 1990). Other studies explore the relation between emotional expressions in voices and faces. For instance, when emotional voice sounds (crying/laughing) were presented before the onset of a face with an emotional expression, 5-month-old infants' Nc was increased for emotionally matching displays (Vogel et al., 2012; Grossmann et al., 2006).

A fairly large body (mostly behavioural research) also explores the impact of auditory language categories (e.g., words) on the formation of visual categories (e.g., object categories). Corresponding work strongly suggests that language input can shape cognition even at a preverbal age (e.g., Althaus and Plunkett, 2015; Balaban and Waxman, 1997; Ferguson and Waxman, 2017; Fulkerson and Waxman, 2007). Findings on the relation between labels and categorization have not been consistent, with some studies indicating improved categorization when corresponding labels are presented (Althaus and Mareschal, 2014; Ferry, Hespos, and Waxman, 2010; Fulkerson and Waxman, 2007; Waxman and Markow, 1995) and others reporting no effect or decreased categorization (Deng and Sloutsky, 2015; Robinson and Sloutsky, 2007). Recent theoretical approaches reconcile these findings by claiming that labels are perceived as features of the input rather than category markers, thus influencing processing of the visual input in specific ways across tasks (Deng and Sloutsky, 2015).

Experience, also implied by age, is crucial for explaining the relation between language and visual categorization (Booth and Waxman, 2002; Ferry et al., 2010; Perszyk and Waxman, 2017). A recent ERP study also shows that auditory and visual information (e.g., human voices and pictures of human beings) may activate common intermodal representations (Peykarjou, Wissner, and Pauen, 2020).

CONCLUDING REMARKS

In this chapter we described main behavioural and EEG paradigms developed to study infant categorization. Furthermore, we provided a brief summary of the current state of infant categorization research in the auditory and visual domain as well as research combining both domains. Thereby, we illustrated the broad scope of questions that developmental psychologists try to answer, as well as the advantages or disadvantages of different methodological approaches to explore category learning and category discrimination at a preverbal age. Based on this review we conclude that behavioural paradigms have been applied extensively to infant categorization and have demonstrated rather sophisticated categorization abilities at a preverbal age. More recently, EEG methods have been introduced as an alternative way to probe infant categorization; they have the potential to greatly contribute to our understanding of the ontogeny of human thinking. Two concrete examples may illustrate this point.

One way to promote the field would be to explore in more detail the process of online category formation. In most behavioural studies, the impact of prior experiences and on-line category formation on categorization performance at test are confounded. To overcome this problem, more studies are needed that use abstract, artificial, and/or never-before-seen stimuli. Alternatively (or in addition), we may compare performance at test between infants who previously received a familiarization phase and a control group who did not participate in any kind of familiarization phase. This should be done at the behavioural as well as the neural level in parallel to compare performance across methods. Ideally, the process of online category learning during the familiarization phase could be analysed continuously. Here, EEG paradigms can be helpful, as they allow us to present many exemplars in sequence and maybe also to conduct time-series analyses.

Another fruitful avenue for future research would be to compare different EEG measures. Most existing EEG work is based on post-perceptual processes reflected by the Nc and PSW, but some paradigms also measure very fast categorization responses, which may reflect initial stages of perceptual processing. By presenting the same set of stimuli to same-aged infants, it would be possible to learn more about how basic and higher aspects of perception interact during processes of categorization.

In general, more research is still needed to better understand (a) at what age infants categorize natural stimuli on different levels of abstraction spontaneously, (b) how the processes of online category learning take place in the laboratory setting, (c) how processes of categorization at different levels of abstraction overlap each other, (d) what we can learn about the timing of categorization in the infant brain, and (e) how information from different modalities is combined in forming categories at the preverbal level. Many of these issues can best be addressed by using EEG measures, but ideally, one would like to see similar effects in behavioural and brain research.

REFERENCES

Althaus, N. and Mareschal, D. (2014). Labels direct infants' attention to commonalities during novel category learning. *PLoS ONE*, 9(7), e99670. doi: 10.1371/journal.pone.0099670.

Althaus, N. and Plunkett, K. (2015). Timing matters: The impact of label synchrony on infant categorization. *Cognition*, 139(Mar), 1–9. doi: 10.1016/j.cognition.2015.02.004.

Anzures, G., Quinn, P., Pascalis, O., Slater, A., and Lee, K. (2013). Development of own-race biases. *Visual Cognition*, 21(9–10) (Sep), 1165–1182. doi: 10.1080/13506285.2013.821428.

Arterberry, M. and Bornstein, M. (2002). Infant perceptual and conceptual categorization: The roles of static and dynamic stimulus attributes. *Cognition*, 86(1), 1–24. doi: 10.1016/s0010-0277(02)00108-7.

Arterberry, M., Bornstein, M., and Blumenstyk, J. (2013). Categorization of two-dimensional and three-dimensional stimuli by 18-month-old infants. *Infant Behavior and Development*, 36(4), 786–795. doi: 10.1016/j.infbeh.2013.09.008.

Baillargeon, R. and Whang, S. (2002). Event categorization in infancy. *Trends in Cognitive Sciences*, 6(2) (March), 85–93. doi: 10.1016/S1364-6613(00)01836-2.

Balaban M. and Waxman, S. (1997). Do words facilitate object categorization in 9-month-olds? *Journal of Experimental Child Psychology*, 64(1) (Jan), 3–26. doi: 10.1006/jecp.1996.2332.

Behl-Chadha, G. (1996). Basic-level and superordinate-like categorical representations in early infancy. *Cognition*, 60(2) (Aug), 105–141. doi: 10.1016/0010-0277(96)00706-8.

Bjjeljac-Babic, R., Bertonicini, J., and Mehler, J. (1993). How do 4-day-old infants categorize multisyllabic utterances? *Developmental Psychology*, 29(4), 711–721.

Bomba, P. and Siqueland. E. (1983). The nature and structure of infant form categories. *Journal of Experimental Child Psychology*, 35(2) (Apr), 294–328. doi: 10.1016/0022-0965(83)90085-1.

Booth, A. E. and Waxman, V. (2002). Object names and object functions serve as cues to categories for infants. *Developmental Psychology*, 38, 948–957.

Bornstein, M., Kessen, W., and Weiskopf. (1976). Color vision and hue categorization in young human infants. *Journal of Experimental Psychology: Human Perception and Performance*, 2(1) (Feb), 115–129. doi: 10.1037//0096-1523.2.1.115.

Bornstein, M. and Mash, C. (2010). Experience-based and on-line categorization of objects in early infancy. *Child Development*, 81(3) (May-Jun), 884–897. doi: 10.1111/j.1467-8624.2010.01440.x.

Brody, L., Polka H., and Zelazo, P. (1984). Habituation and dishabituation to speed in the neonate. *Developmental Psychology*, 20(1), 114–119.

Carral, V., Huotilainen, M., Ruusuvirta, T., Fellman, V., Näätänen, R., and Escera, C. (2005). A kind of auditory "primitive intelligence" already present at birth. *European Journal of Neuroscience*, 21(11) (Jun), 3201–3204. doi: 10.1111/j.1460-9568.2005.04144.x.

Casasola, M. and Ahn Y. (2018). What develops in infants' spatial categorization? Korean infants' categorization of containment and tight-fit relations. *Child Development*, 89(4) (Jul), 382–396. doi: 10.1111/cdev.12903.

Casasola, M., Cohen L., and Chiarello, E. (2003). Six-month-old infants' categorization of containment spatial relations. *Child Development*, 74(3) (Jun), 87–121. doi: 10.1016/bs.acdb.2017.10.007.

Cheour, M., Leppänen, P., and Kraus, N. (2000). Mismatch negativity (MMN) as a tool for investigating auditory discrimination and sensory memory in infants and children. *Clinical Neurophysiology*, 111(1) (Jan), 4–16. doi: 10.1016/s1388-2457(99)00191-1.

Clifford, A., Franklin, A., Davies, I., and Holmes, A. (2009). Electrophysiological markers of categorical perception of color in 7-month old infants. *Brain and Cognition*, 71(2) (Jun), 165–172. doi: 10.1016/j.bandc.2009.05.002.

DeCasper, A. and Fifer, W. (1980). Of human bonding: Newborns prefer their mothers' voices. *Science* 208(4448) (Jun), 1174–1176. doi: 10.1126/science.7375928.

de Haan, M. (2007). *Infant EEG and Event-Related-Potentials*. Hove (East Sussex): Psychology Press.

de Haan, M., Johnson M., and Halit, H. (2003). Development of face-sensitive event-related potentials during infancy: A review. *International Journal of Psychophysiology*, *51*(1) (Dec), 45–58. doi: 10.1016/S0167-8760(03)00152-1.

de Haan, M. and Nelson C. (1998). Discrimination and categorization of facial expression of emotion during infancy. In *Perceptual Development. Visual, Auditory, and Speech Perception in Infancy*, A. Slater (eds.), (pp. 287–309), Hove, UK: Psychology Press

de Haan, M. and Nelson, C. (1999). Brain activity differentiates face and object processing in 6-month-old infants. *Developmental Psychology*, *35*(4) (Jul), 1113–1121. doi: 10.1037/0012-1649.35.4.1113.

Dehaene-Lambertz, G. and Pena, M. (2001). Electrophysiological evidence for automatic phonetic processing in neonates. *NeuroReport*, *12*(14) (Oct), 3155–3158. doi: 10.1097/00001756-200110080-00034.

Deng, W. S. and Sloutsky, V. M. (2015). Selective attention, diffused attention, and the development of categorization. *Cognitive Psychology*, *91* (Dec):, 24–62.

Dewar, K. and Xu F. (2010). Induction, overhypothesis, and the origin of abstract knowledge: Evidence from 9-month-old infants. *Psychological Science*, *21*(12) (Dec), 1871–1877. doi: 10.1177/0956797610388810.

Dolev, Y. and Nelson, X. (2014). Innate pattern recognition and categorization in a jumping spider. *PLoS One* 9(6) (Jun), e97819. doi: 10.1371/journal.pone.0097819.

Eimas, P. and Quinn, P. (1994). Studies on the formation of perceptually based basic-level categories in young infants. *Child Development*, *65*(3) (Jun), 903–917.

Elsner, B., Jeschonekm S., and Pauen, S. (2013). Event-related potentials for 7-month-olds processing of animals and furniture items. *Developmental Cognitive Neuroscience*, 3 (Jan), 53–60. doi: 10.1016/j.dcn.2012.09.002.

Elsner, B., Pauen, S., and Jeschonek, S. (2006). Physiological and behavioural parameters of infant categorization. *Developmental Science*, *9*(6) (Nov), 551–556. doi: 10.1111/j.1467-7687.2006.00532.x.

Fantz, R. (1961). The origin of form perception. *Scienctific Amerikan*. (*Scientific American*), *204*(5), 66–73. doi: 10.1038/scientificamerican0561-66.

Fantz, R. (1965). Visual perception from birth as shown by pattern selectivity. *Annals of the New York Academy of Sciences*, *118*(21) (May), 793–814. doi: 10.1111/j.1749-6632.1965.tb40152.x.

Ferguson, B. and Waxman S. (2017). Linking language and categorization in infancy. *Journal of Child Language*, *44*(3) (May), 527–52. doi: 10.1017/S0305000916000568.

Ferry A., Hespos, S., and Waxman, S. (2010). Words facilitate category formation at 3 months. *Child Development*, 81, 472–479.

Flom, R. (2014). Perceptual narrowing: Retrospect and prospect. *Developmental Psychobiology*, *56*(7) (Nov), 1442–1453. doi: 10.1002/dev.21238.

French, R., Mareshall, D., Mermillond, D., and Quinn, P. (2005). The role of bottom-up processing in perceptual categorization by 3- to 4-month-old infants. *Journal of Experimental Psychology General*, *133*(3) (Sep), 382–397. doi: 10.1037/0096-3445.133.3.382.

Friedrich, M., Herold, B., and Friederici, A. (2009). ERP correlates of processing native and non-native language word stress in infants with different language outcomes. *Cortex*, *45*(5) (May), 662–76. doi: 10.1016/j.cortex.2008.06.014.

Friederici, A. and G. Thierry. (2008). Early languages development: Bridging brain and behaviour. In *Trends in Language Acquisition Research (TiLAR) Series Volume 5*, A. Friederici and G. Thierry (eds.), Amsterdam: John Benjamins.

Fulkerson, A. and Waxman, S. (2007). Words (but not Tones) facilitate object categorization: Evidence from 6- and 12-month-olds. *Cognition*, *105*(1) (Oct), 218–228. doi 10.1016/j.cognition.2006.09.005

Geangu, E., Quadrelli, E., Lewis, J., Cassia, V., and Turati, Ch. (2015). By the sound of it. An ERP investigation of human sound processing in 7-month-old infants. *Developmental Cognitive Neuroscience*, *12*(Apr), 134–144. doi: 10.1016/j.dcn.2015.01.005.

Grossmann, T., Gliga, T., Johnson, M., and Mareschal, D. (2009). The neural basis of perceptual category learning in human infants. *Journal of Cognitive Neuroscience*, *21*(12) (Dec), 2276–2286. doi: 10.1162/jocn.2009.21188.

Grossmann, T., Striano, T., Friederici, A. D. (2006). Crossmodal integration of emotional information from face and voice in the infant brain. *Developmental Science*, *9*(3) (May).

Háden, G., Stefanics, G., Vestergaard, M., Denham, S., Sziller, I., and Winkler, I. (2009). Timbre-independent extraction of pitch in newborn infants. *Psychophysiology*, *46*(1) (Jan), 69–74. doi: 10.1111/j.1469-8986.2008.00749.x.

Háden, G. H., Honing, M., Török, and Winkler, I. (2015). Detecting the temporal structure of sound sequences in newborn infants. *International Journal of Psychophysiology*, *96*(1) (Apr), 23–28. doi: 10.1016/j.ijpsycho.2015.02.024.

Harnad, S. (2017). To cognize is to categorize: Cognition is categorization. In *Handbook of Categorization and Cognitive Science*, H. Cohen and C. Lefebvre (eds.), (pp. 20–42), second edition, Amsterdam: Elsevier.

Hirai, M. and Hiraki, K. (2005). An event-related potentials study of biological motion perception in human infants. *Cognitive Brain Research*, *22*(2) (Feb), 301–304. doi: 10.1016/j.cogbrainres.2004.08.008.

Hoehl, S. and Peykarjou, S. (2012). The early development of face processing: What makes faces special? *Neuroscience Bulletin*, *28*(6) (Dec), 765–788. doi: 10.1007/s12264-012-1280-0.

Hoehl, S. and Striano, T. (2008). Neural processing of eyegaze and threat-related emotional facial expressions in infancy. *Child Development*, *79*(6) (Nov-Dec), 1752–1760. doi: 10.1111/j.1467-8624.2008.01223.x.

Hoehl, S. and Wahl, S. (2012). Recording infant ERP data for cognitive research. *Developmental Neuropsychology*, *37*(3) (Apr), 187–209. doi: 10.1080/87565641.2011.627958.

Honing, H., Bouwer, F. L., and Háden, G. P. (2014). Perceiving temporal regularity in music: The role of auditory event-related potentials (ERPs) in probing beat perception. In *Neurobiology of Interval Timing. Advances in Experimental Medicine and Biology*. H. Merchant and van de Lafuente (eds.) (p. 829). New York (NY): Springer. doi: 10.1007/978-1-4939-1782-2_16.

Jeschonek, S., Marinovic, V., Höhl, S., Elsner, B., and Pauen, S. (2010). Do animals and furniture items elicit different brain responses in human infants? *Brain Development*, *32*(10) (Nov), 863–871. doi: 10.1016/j.braindev.2009.11.010.

Jeschonek, S., Pauen, S., and Babocsai, L. (2012). Cross-modal mapping of visual and acoustic displays in infants: The effect of dynamic and static components. *European Journal of Developmental Psychology*, *10*(3) (May), 337–358. doi: 10.1080/17405629.2012.681590.

Jessen, S. and Grossmann, T. (2015). Neural signatures of conscious and unconscious emotional face processing in human infants. *Cortex*, *64* (Mar), 260–70. doi: 10.1016/j.cortex.2014.11.007.

Junge, C., Rooijen, R., and Raijmarkers, M. (2018). Distributional information shapes infants' categorization of objects. *Infancy*, *23*(6) (Sep), 917–926. doi: 10.1111/infa.12258.

Kaduk, K., Elsner, B., and Reid, V. (2013). Discrimination of animate and inanimate motion in 9-month old infants: An ERP study. *Developmental Cognitive Neuroscience*, *6* (Oct), 14–22. doi: 10.1016/j.dcn.2013.05.003.

Kelly, D., Slater, A., Lee, K., Gibson A., Smith, M., Ge, L., and Pascalis, O. (2005). Three-month-olds, but not newborns, prefer own-race faces. *Developmental Science*, 8(6) (Nov), F31–F36. doi: 10.1111/j.1467-7687.2005.0434a.x.

Konishi, H., Pruden, S., Golinkoff, R., and Hirsh-Pasek, K. (2016). Categorization of dynamic realistic motion events: Infants form categories of path before manner. *Journal of Experimental Child Psychology*, 152 (Dec), 54–70. doi: 10.1016/j.jecp.2016.07.002.

Kovack-Lesh, K., Horst, J., and Oakes, L. (2008). The cat is out of the bag: The joint influence of previous experience and looking behavior on infant categorization. *Infancy*, 13(4) (Feb), 285–307. doi: 10.1080/15250000802189428.

Kuhl, P. K. (1987). Perception of speech sound in early infancy. In *Handbook of Infant Perception*, P. Salapatekand and L. Cohen (eds.), (pp. 275–382), New York: Academic Press.

Kuhl, P. and Miller, J. (1975). Speech perception in early infancy: Discrimination of speech-sound categories. *Journal of the Acoustic Society of America*, 58(S1), 56. doi: 10.1121/1.2002199.

Lansink, J., Mintz, S., and Richards, J. (2001). The distribution of infant attention during object examination. *Developmental Science*, 3(2) (Dec), 163–170. doi: 10.1111/1467-7687.00109.

Leppänen, P., Phiko, E., Eklund, K., and Lyytinen, H. (1999). Cortical responses of infants with and without a genetic risk for dyslexia: II Group effects. *Neuroreport*, 10(5) (Apr), 996–973. doi: 10.1097/00001756-199904060-00014.

Lewkowicz, D. (2014). Early experience and multisensory perceptual narrowing. *Developmental Psychobiology*, 56(2) (Feb), 292–350. doi: 10.1002/dev.21197.

Liebenthal, E. and Bernstein, E. (2017). Neural mechanisms of perceptual categorization as precursors to speech perception. *Frontiers in Neuroscience* [online]. https://lib.ugent.be/en/catalog/ebk01:3800000000216224.

Luck, S. (2005). *An Introduction to the Event-Related Potential Technique*. London: The MIT Press.

Madole, K. and Oakes, L. (1999). Making sense of infant categorization: Stable processes and changing representations. *Developmental Review*, 19(2) (Jun), 263–296. doi: 10.1006/drev.1998.0481.

Mandler, J. (1988). How to build a baby: On the development of an accessible representational system. *Cognitive Development*, 3(2) (Apr), 113–136. doi: 10.1016/0885-2014(88)90015-9.

Mandler, J. (1992). How to build a baby: II. Conceptual primitives. *Psychological Review*, 99(4), 587–604. doi: 10.1037/0033-295X.99.4.587.

Mandler, J. (2000a). Perceptual and conceptual processes in infancy. *Journal of Cognition and Development*, 1(1) (Feb), 3–36. doi: 10.1207/S15327647JCD0101N_2.

Mandler, J. (2000b). The origins of concepts. *Journal of Cognition and Development*, 1(1) (Feb), 37–41. doi: 10.1207/S15327647JCD0101N_3.

Mandler, J. and McDonough, L. (1993). Concept formation in infancy. *Cognitive Development*, 8(3) (Jul–Sep), 291–318. doi: 10.1016/S0885-2014(93)80003-C.

Mandler, J. and McDonough, L. (1998). On developing a knowledge base in infancy. *Developmental Psychology*, 36(4) (Nov), 1274–1288. doi: 10.1037//0012-1649.34.6.1274.

Marinovic, V., Hoehl, S., and Pauen S. (2014). Neural correlates of human-animal distinction: An ERP-study on early categorical differentiation with 4- and 7-month-old infants and adults. *Neuropsychologia*, 60(Jul), 60–76. doi: 10.1016/j.neuropsychologia.2014.05.013.

Mash, C. and Bornstein, M. (2012). 5-Month-olds' categorization of novel objects: Task and measure dependence. *Infancy*, 17(2) (Mar), 179–197. doi: 10.1111/j.1532-7078.2011.00076.x.

Marshall, P. and Shipley, T. (2009). Event-related potentials to point-light displays of human actions in 5-month-old infants. *Developmental Neuropsychology*, 34(3) (May), 358–377. doi: 10.1080/87565640902801866.

Miller, C. (1983). Developmental changes in male/female voice classification by infants. *Infant Behavior and Development*, 6(2–3), 313–330. doi: 10.1016/S0163-6383(83)80040-X.

Munsters, N., van Ravensvaij, H., van den Boomen, C., and Kemner C. (2017). Test-retest reliability of infant event related potentials evoked by faces. *Neuropsychologia*, 126(Apr), 20–26. doi: 10.1016/j.neuropsychologia.2017.03.030.

Näätänen, R. (1990). The role of attention in auditory information processing as revealed by event-related potentials and other brain measures of cognitive function. *Behavioral and Brain Sciences*, 13(2), 201–233. doi: 10.1017/S0140525X00078407.

Nelson, C. (1994). Neural correlates of recognition memory in the first postnatal year of life. In *Human Behavior and the Developing Brain*, G. Dawson and K. Fischer (eds), (pp. 269–313), New York: Guilford Press.

Oakes, L., Coppage, D., and Dingel, A. (1997). By land or by sea: The role of perceptual similarity in infants' categorization of animals. *Developmental Psychology*, 33(3) (May), 396–407. doi: 10.1037//0012-1649.33.3.396.

Oakes, L., Madole K., and Cohen, L. (1991). Infants' object examining: Habituation and categorization. *Cognitive Development*, 6(4) (Oct-Dec), 377–92. doi: 10.1016/0885-2014(91)90045-F.

Oakes, L., Plumert, J., Lansink, J., and Merryman J. (1996). Evidence for task-dependent categorization in infancy. *Infant Behavior and Development*, 19(4) (Oct-Dec). 425–440. doi: 10.1016/S0163-6383(96)90004-1.

Oakes, L. and Ribar, R. (2005). A comparison of infants' categorization in paired and successive presentation familiarization tasks. *Infancy*, 7(1) (Jan), 85–98. doi: 10.1207/s15327078in0701_7.

Oakes, L. and Tellinghuisen, D. (1994). Examining in infancy: Does it reflect active processing? *Developmental Psychology*, 30(5) (Sep), 748–756. doi: 10.1037/0012-1649.30.5.748.

Ortiz-Mantilla, S., Hämäläinen, J., Realpe-Bonilla, T., and Benasich, A. (2016). Oscillatory dynamics underlying perceptual narrowing of native phoneme mapping from 6 to 12 months of age. *Journal of Neuroscience*, 36(48) (Nov), 12095–105. doi: 10.1523/JNEUROSCI.1162-16.2016

Palmer, S., Fais, L., Golinkoff, R., and Werker, J. (2012). Perceptual narrowing of linguistic sign occurs in the first year of life. *Child Development*, 83(2) (Mar–Apr), 543–553. doi: 10.1111/j.1467-8624.2011.01715.x.

Pascalis O., de Haan, M., and Nelson, C. (2002). Is face processing species-specific during the first year of life? *Science*, 296(5571) (May), 1321–1323. doi: 10.1126/science.1070223.

Pauen S. (2000). Early differentiation within the animate domain: Are humans something special? *Journal of Experimental Child Psychology*, 75(2) (Feb), 134–151. doi: 10.1006/jecp.1999.2530.

Pauen S. (2002a). Evidence for knowledge-based category discrimination in infancy. *Child Development*, 73(4) (Jul–Aug), 1016–1033. doi: 10.1111/1467-8624.00454.

Pauen, S. (2002b). The global-to-basic level shift in infants' categorical thinking. Evidence from a longitudinal study. *International Journal of Behavioural Development*, 26(2), 492–499. doi: 10.1080/01650250143000445.

Pauen, S. and Träuble, B. (2008). How to investigate the concept of concepts. *Journal of Anthropological Psychology*, 19(Jan), 32–35.

Peltola, M., Leppänen, J., Mäki, S., and Hietanen, J. (2009). Emergence of enhanced attention to fearful faces between 5 and 7 months of age. *Social Cognitive and Affective Neuroscience*, 4(2) (Jun), 134–142. doi: 10.1093/scan/nsn046.

Perszyk, D. and Waxman, S. (2017). Experience is instrumental in tuning a link between language and cognition: Evidence from 6- to 7- month-old infants' object categorization. *Journal of Visualized Experiment*, 122(Apr), 55435. doi: 10.3791/55435.

Peykarjou, S., Wissner, J., and Pauen, S. (2020). Audio-visual priming in 7-month-old infants: An ERP study. *Infant Behavior & Development, 58* (Feb).

Peykarjou, S., Wissner, J., and Pauen, S. (2016). Categorical ERP repetition effects for human and furniture items in 7-month-old infants. *Infant and Child Development, 26*(5) (Sept–Oct), e2016. doi: 10.1002/icd.2016.

Polka, L. and Werker, J. F. (1994). Developmental changes in perception of nonnative vowel contrasts. *Journal of Experimental Psychology in Human Perceptual Performance, 20,* 421–435. doi: 10.1037/0096-1523.20.2.421.

Posid, T. and Cordes, S. (2019). The effect of multimodal information on children's numerical judgments. *Journal of Experimental Child Psychology, 182*(Jun), 166–86. doi: 10.1016/j.jecp.2019.01.003.

Poulin-Dubois, D. and Pauen, S. (2017). The development of categories. What? When? How? In *Handbook of Categorization in Cognitive Science.* H. Cohen and C. Lefebvre (eds.), (pp. 653–66), second edition, Amsterdam: Elsevier.

Pruden, S., Göksun, T., Roseberry, S., Hirsh-Pasek K., and Golinkoff, R. (2013). Infant categorization of path relations during dynamic events. *Child Development, 84*(1) (Jan), 331–345. doi: 10.1111/j.1467–8624.2012.01843.x.

Quinn, P. (1994). The categorization of above and below spatial relations by young infants. *Child Development, 65*(1) (Feb), 58–69. doi 10.1111/j.1467–8624.1994.tb00734.x.

Quinn, P. (2004a). Development of subordinate-level categorization in 3- to 7-month-old infants. *Child Development, 75*(3) (May-Jun), 886–899. doi: 10.1111/j.1467–8624.2004.00712.x.

Quinn, P. (2004b). Spatial representation by young infants: Categorization of spatial relations or sensitivity to a crossing primitive? *Memory and Cognition, 32*(5) (Jul), 852–861. doi: 10.3758/bf03195874.

Quinn, P. (2011). Born to categorize. In *The Wiley-Blackwell Handbook of Child Cognitive Development.* U. Goswami (eds), (pp. 129–152), second edition, United Kingdom: Blackwell Publishing Ltd.

Quin, P., Doran, M., Reiss, J., and Hoffman J. (2010). Neural markers of subordinate-level categorization in 6- to 7-month-old infants. *Developmental Science, 13*(3)(May), 499–507. doi: 10.1111/j.1467–7687.2009.00903.x.

Quinn, P. and Eimas, P. (1986). On categorization in early infancy. *Merrill-Palmer Quarterly, 32*(4) (Oct), 331–363.

Quinn, P. and Eimas, P. (1996a). Perceptual organization and categorization in young infants. In *Advances in Infancy Research Vol 10.* L. Lipsitt and C. Rovee-Collier (eds.), (pp. 1–36), New Jersey: Ablex.

Quinn, P. and Eimas, P. (1996b). Perceptual cues that permit categorical differentiation of animal species by infants. *Journal of Experimental Child Psychology, 63*(1) (Oct), 189–211. doi: 10.1006/jecp.1996.0047.

Quinn, P. and Eimas, P. (1998). Evidence for a global categorical representation of humans by young infants. *Journal of Experimental Child Psychology, 69*(3) (Jun), 151–174. doi: 10.1006/jecp.1998.2443.

Quinn, P., Eimas, P., and Rosenkrantz, S. (1993). Evidence for representations of perceptually similar natural categories by 3-month-old and 4-month-old infants. *Perception, 22*(4), 463–475. doi: 10.1068/p220463.

Quinn, P. and Johnson, M. (2000). Global-before-basic object categorization in connectionist networks and 2-month-old infants. *Infancy, 1*(1) (Jan), 31–49. doi: 10.1207/S15327078IN0101_04

Quinn, P., Slater, A., Brown, E., and Hayes, R. (2010). Developmental change in form categorization in early infancy. *Developmental Psychology*, 19(2) (Dec), 207–218. doi: 10.1348/026151001166038.

Quinn, P., Uttlay, L., Lee, K., Gibson, A., Smith, M., Slater, A. M., and Pascalis, O. (2008). Infant preference for female faces occurs for same- but not other-race faces. *Journal of Neuropsychology*, 2(1), 15–26. doi: 10.1348/174866407X231029

Quinn, P., Westerlund, A., and Nelson, C. (2006). Neural markers of categorization in 6-month-old infants. *Psychological Science*, 17(1) (Jan), 59–66. doi: 10.1111/j.1467–9280.2005.01665.x.

Rakinson, D. (2004). Infants' sensitivity to correlations among static and dynamic features in a category context. *Journal of Experimental Child Psychology*, 89(1), 1–30. doi: 10.1016/j.jecp.2004.06.001.

Rakinson, D. and G. Butterworth. (1998). Infants' use of object parts in early categorization. *Developmental Psychology*, 34(1) (Jan), 49–62. doi: 10.1037/0012-1649.34.1.49.

Rakinson, D. and Cohen, L. (1998). You've got to roll with it, baby: The effect of functional parts on infants' categorization. *Infant Behavior and Development*, 21, 636.

Rakinson, D. and Oakes, L. (2003). *Early Category and Concept Development*. Oxford: Oxford University Press.

Rakinson, D. and Yermolayea, Y. (2010). Infant categorization. *Cognitive Science*, 1(6) (May), 894–905. doi 10.1002/wcs.81.

Reynolds, G., Courage, M., and Richards, J. (2010). Infant attention and visual preferences: Converging evidence from behavior, event-related potentials, and cortical source localization. *Developmental Psychology*, 46(4) (Jul), 886–904. doi: 10.1037/a0019670.

Roberts, K. and Horowitz, F. (1986). Basic-level categorization in seven- and nine-month-old infants. *Journal of Child Language*, 13(2) (Jun), 191–208. doi: 10.1017/s0305000090000800x.

Reynolds, G. and Richards J. (2005). Familiarization, attention, and recognition memory in infancy: An event-related potential and cortical source localization study. *Developmental Psychology*, 41(4) (Jul), 598–615. doi: 10.1037/0012-1649.41.4.598.

Robinson C. and Sloutsky V. (2007). Linguistic labels and categorization in infancy: Do labels facilitate or hinder? *Infancy*, 11, 233–253.

Rosch, E., Mervis, C., Gray, W., and Boyes-Bream, P. (1976). Basic objects in natural categories. *Cognitive Psychology*, 8(3) (Jul), 382–439. doi: 10.1016/0010-0285(76)90013-X.

Rossion, B. (2014). Understanding individual face discrimination by means of fast periodic visual stimulation. *Experimental Brain Research*, 232(6), 1599–1621. doi: 10.1007/s00221-014-3934-9.

Rossion, B., Torfs, K., Jacques, C., and Liu-Shuang, J. (2015). Fast periodic presentation of natural images reveals a robust face-selective electrophysiological response in the human brain. *Journal of Vision*, 15(1), 18. doi: 10.1167/15.1.18.

Ruff, H. (1984). Infants' manipulative exploration of objects: Effects of age and object characteristics. *Developmental Psychology*, 20(1), 9–20. doi: 10.1037/0012-1649.20.1.9.

Ruff, H. and L. Saltarelli. (1993). Exploratory play with objects: Basic cognitive processes and individual differences. In *New Directions for Child Development Vol. 59*, M. Bomstein and A. O'Reilly (eds.), San Francisco: Jossey-Bass.

Ruusuvirta, T., Huotilainen, M., Fellman, V., and Näätänen. R. (2009). Numerical discrimination in event-related potentials to tone sequences. *The European Journal of Neuroscience*, 30(8) (Oct), 1620–1624. doi: 10.1111/j.1460–9568.2009.06938.x.

Scott L. and Monesson, A. (2010). Experience-dependent neural specialization during infancy. *Neuropsychologia, 48*(6) (May), 1857–1861. doi: 10.1016/j.neuropsychologia.2010.02.008.

Simion, F. and Di Giorgio, E. (2015). Face perception and processing in early infancy: Inborn predispositions and developmental changes. *Frontiers in Psychology, 6*(Jul), 969. doi: 10.3389/fpsyg.2015.00969.

Sinnott, J. and Aslin, R. (1985). Frequency and intensity discrimination in human infants and adults. *The Journal of the Acoustical Society of America, 78*. doi: 10.1121/1.392655.

Smith E. and Medin, D. (1981). *Categories and Concepts*. Cambridge: Harvard University Press.

Snyder, K. (2010). Neural correlates of encoding predict infants' memory in the paired-comparison procedure. *Infancy, 15*(3) (Jan), 270–299. doi: 10.1111/j.1532-7078.2009.00015.x.

Sokolov, E. (1966). Orienting reflex as information regulator. In *Psychological Research in the U.S.S.R.* A. Leontiev, A. Ludia, and A. Smirnov (eds.). Moscow: Progress Publishers.

Song, L., Pruden, S. Golinkoff, R., and Hirsh-Pasek, K. (2016). Prelinguistic foundations of verb learning: Infants discriminate and categorize dynamic human actions. *Journal of Experimental Child Psychology, 151*(Nov), 77–95. doi: 10.1016/j.jecp.2016.01.004.

Spelke, E. (1979). Perceiving bimodally specified events in infancy. *Developmental Psychology, 15*(6), 626–636. doi: 10.1037/0012-1649.15.6.626.

Starkey, P., Spelke, E., and Gelman, R. (1990). Numerical abstraction by human infants. *Cognition, 36*(2) (Aug), 97–127. doi: 10.1016/0010-0277(90)90001-z.

Stefanics, G., Háden, G., Huotilainen, M., Balázs, L., Sziller, I., Beke, A., Fellman, V., and Winkler, I. (2007). Auditory temporal grouping in newborn infants. *Psychophysiology, 44*(5) (Sep), 697–702. doi: 10.1111/j.1469-8986.2007.00540.x.

Teller, D. (1979). The forced-choice preferential looking procedure: A psychophysical technique for use with human infants. *Infant Behavior and Development, 2*(Jan), 135–153. doi: 10.1016/S0163-6383(79)80016-8.

Trainor, L. (1996). Infant preferences for infant-directed versus noninfant-directed playsongs and lullabies. *Infant Behavior and Development, 19*(1) (Mar), 83–92. doi: 10.1016/S0163-6383(96)90046-6.

Träuble, B., Babocsai, L., and Pauen S. (2008). Preverbal categorization. The role of real-world experience. In *Cognitive Psychology Research Developments*. F. Columbus (ed.), Hauppauge (NY): Novascience Publisher.

Träuble, B. and Pauen S. (2007). The role of functional information for infant categorization. *Cognition, 105*(2), 362–379. doi: 10.1016/j.cognition.2006.10.003.

Trehub, S. (1973). Infants' sensitivity to vowel and tonal contrasts. *Developmental Psychology, 9*(1) (Jul), 91–96. doi: 10.1037/h0034999.

Trehub, S. and Thorpe, L. A. (1989). Infants' perception of rhythm: Categorization of auditory sequences by temporal structure. *Canadian Journal of Psychology/Revue Canadienne de Psychologie, 43*(2) (Jul), 217–229. doi: 10.1037/h0084223.

Tsuji, S. and Cristia, A. (2013). Perceptual attunement in vowels: A meta-analysis. *Developmental Psychobiology, 56*(2) (Nov), 179–91. doi: 10.102/201179.

Turati, C., Simion, F., and Zanon, L. (2010). Newborns' perceptual categorization for open and closed forms. *Infancy, 4*(3) (Jun), 309–325. doi: 10.1207/S15327078IN0403_01.

Virtala, P., Huotilainen, Partanen, M., Felllman, V., and Tervaniemi, M. (2013). Newborn infants' auditory system is sensitive to Western music chord categories. *Frontiers in Psychology. Auditory Cognitive Neuroscience, 7*(4): 492. doi: 10.3389/fpsyg.2013.00492.

Vogel, M., Monesson, A., and Scott, L. (2012). Building biases in infancy: The influence of race on face and voice emotion matching. *Developmental Science*, 15(3) (May), 359–372. doi: 10.1111/j.1467-7687.2012.01138.x.

Vukatana, E., Graham, S., Curtin, S., and Zepeda, M. (2015). One is not enough: Multiple exemplars facilitate infants' generalizations of novel properties. *Infancy*, 20(5) (Jun), 548–575. doi: 10.1111/infa.12092.

Waxman, S. and Markow, D. (1995). Words as invitations to form categories: Evidence from 12- to 13-month-old infants. *Cognitive Psychology*, 29, 257–302.

Welder, A. N., and Graham, S. A. (2006). Infants' categorization of novel objects with more or less obvious features. *Cognitive Psychology*, 52, 57–91. http://hdl.handle.net/1880/112083

Werker, J., Polka L., and Pegg, J. (1997). The conditioned head turn procedure as a method for testing infant speech perception. *Early Development and Parenting*, 6(3–4), 171–178.

Werker, J. and Tees, R. (1984). Cross-language speech perception. Evidence for perceptual re-organization in the first year of life. *Infant Behavior and Development*, 7(1) (Jan–Mar), 49–63. doi: 10.1016/S0163-6383(84)80022-3.

Westermann, G. and D. Mareschal. (2012). Mechanisms of developmental change in infant categorization. *Cognitive Development* 27(4) (Nov), 367–382. doi: 10.1016/j.cogdev.2012.08.004.

Winkler I., Kushnerenko, E., Horváth, J., Ceponiene, R., Fellman, V., Huotilainen, M., Näätänen, R., and Sussman, E. (2003). Newborn infants can organize the auditory world. *Proceedings of the National Academy of Sciences*, 100(20) (Oct), 11812–15. doi: 10.1073/pnas.2031891100.

Younger, B., (1992). Developmental change in infant categorization: The perception of correlations among facial features. *Child Development*, 63(6) (Dec), 1526–1535.

Younger, B., and Cohen, L. (1983). Infant perception of correlations among attributes. *Child Development*, 54 (4), 858–867. doi:10.2307/1129890.

Younger, B. and Cohen L. (1985). How infants form categories. In *The Psychology of Learning and Motivation*. G. Bower (eds.), (pp. 211–247), San Diego: Academic.

Younger, B., and Cohen, L. (1986). Developmental change in infants' perception of correlations among attributes. *Child Development*, 57 (3), 803–815. doi:10.2307/1130356.

Younger, B. and Fearing, D. (1999). Parsing items into separate categories: Developmental change in infant categorization. *Child Development*, 70(2) (Apr), 291–303. doi: 10.1111/1467-8624.00022.

Younger, B. and Furrer, S. (2003). A comparison of visual familiarization and object-examining measures of categorization in 9-month-old infants. *Infancy*, 4(3) (Jun), 327–348. doi: 10.1207/S15327078IN0403_02.

Younger, B. and Gotlieb, S. (1988). Development of categorization skills: Changes in the nature or structure of infant form categories? *Developmental Psychology*, 24(5), 611–619.

Zangl, R. and Mills, D. (2007). Increased brain activity to infant-directed speech in 6- and 13-month-old infants. *Infancy*, 11(1) (Dec), 31–62. doi: 10.1207/s15327078in1101_2.

Zelazo, P., Brody, V., and Chaika, H. (1984). Neonatal habituation and dishabituation of head turning to rattle sounds. *Infant Behavior and Development*, 7(3) (Jul–Sep), 311–321. doi: 10.1016/S0163-6383(84)80046-6.

CHAPTER 11

MOTOR DEVELOPMENT IN INFANTS AND CHILDREN

KLAUS LIBERTUS

INTRODUCTION

WHAT are the fundamental ingredients that shape developmental outcomes? Genes offer a blueprint defining development broadly. However, in addition to the role played by genes, experiences stemming from interactions with the environment are another foundation of development. This developmental flexibility is critical and allows the organism to adapt to an ever-changing world. Interactions with the environment rely on and are limited by the constraints of our bodies and physical capabilities. Emphasizing the role of the body and its physical abilities is referred to as the embodiment and emphasizes the role of motor skills during the developmental process. Interest in motor development has increased in recent years, aided by research identifying the importance and long-term consequences of early motor experiences and abilities. This chapter will provide an overview of typical and atypical motor development during the first three years of life, question the standard view on the progression of motor development in early childhood, and discuss the effect of learning new motor skills on brain and behavior from an embodied perspective.

DEVELOPMENT AS A DYNAMIC AND EMBODIED PROCESS

Early conceptions of motor development placed a heavy emphasis on maturation as a driving force behind skill emergence (e.g., Gesell, 1934; Shirley, 1933). While maturation is a significant contributor during development, it is not the only factor influencing motor skills development. Environmental factors such as experiences, opportunities, and nutrition play a significant role as well. The dynamic interplay between genes and environment, opportunities and abilities, physical constraints and neural maturation are emphasized in the *dynamic system* view of infant development (Smith and Thelen, 2003). Starting with the classical epigenetic landscape proposed by Waddington (1947), this perspective views development as a process directed towards an attractor state (i.e., genetic predispositions) but progression towards this state is influenced by multicausal environmental factors. Accordingly, the phenotype, or observable characteristics of a child, are influenced in real time by the dynamic interactions between the environment and genes (Tronick and Hunter, 2016). Within the context of motor development, this view allows for motor skills to dynamically adapt to the environment, past experiences, and the rapidly changing body.

Current conceptions of motor development also acknowledge that motor development is embodied (Needham and Libertus, 2011). An embodied perspective on development realizes two things: First, the body changes during development and a different solution to the same problem may become necessary due to changes in body composition (Van Dam, Hallemans, Truijen, Segers, and Aerts, 2011). Second, the body constricts the problem space by its abilities and limitations. Further, embodied perspectives also go beyond the physical domain and realize that intelligence and cognition are grounded by our bodies and their interactions with the environment (Smith and Gasser, 2005). Consequently, cognition results from sensorimotor interactions with the environment (Gibson, 1988; Piaget, 1970). This perspective emphasizes that the brain is not a computer that functions independently of the body. Rather, the brain is embedded in our bodies and mental processes cannot be understood without acknowledging this mind-body connection (Pfeifer and Bongard, 2007).

Viewing motor development as a dynamic and embodied process emphasizes the importance of motor abilities for cognitive, social, and language development (Smith, 2005). Studying cognitive, social, or language development without considering motor development is at best inefficient and at worst misleading. The interactions between different domains of development need to be considered to gain a fuller, more accurate understanding of infant development (Oakes, 2009). Embodied perspectives on cognition and development acknowledge this focus on movement and further emphasize that we can only experience the world from within the constraints of our bodies (Needham and Libertus, 2011). Constraints include how we can perceive the world (e.g., we can perceive some wavelengths of light but not others), how our bodies are shaped (e.g., having hands with opposable thumbs), and the action abilities available to us (e.g., walking but not flying). Seeing motor development as embodiment changes our perception of the brain, its function, and its development. Cognition was previously thought of as mental computation performed by the brain in isolation. The embodied view acknowledges that cognition is the result of sensory-motor interactions (Smith, 2005). Through this new perspective, it is becoming increasingly clear that actions *construct* thinking (Pfeifer and Bongard, 2007), placing motor development at the core of a holistic approach to understanding infant development. Dynamical systems and embodied approaches acknowledge the interrelations between different developmental domains, systems, or levels of analysis.

MOTOR DEVELOPMENT AND BRAIN DEVELOPMENT

Motor development is a core area of developmental psychology, and most parents would readily agree with the idea that motor skills are a key component of cognitive, social, and perceptual development. Observing a new motor skill in their child is highly memorable for parents, and one of the most consequential events with far-reaching consequences. For example, the onset of crawling or walking is proudly shared with others and results in the sudden need for a baby gate or the re-organization of all shelves in the home. Curiously, for many years, researchers in Developmental Psychology did not seem to share this enthusiasm about infants' motor abilities, and the topic of motor development received only limited attention (D. A. Rosenbaum, 2005). It has been argued that "research on movement and motor control has little more than 'Cinderella' status within the field" (Iverson, 2010, p. 2). The relative neglect of motor development in psychological research is surprising given the foundational role of motor-initiated interactions discussed previously. The apparent lack of research on the impact of motor skills for other domains of development is also surprising given that motor behavior can be readily observed with simple means such as a video camera or check-lists (Adolph and Hoch, 2018). Further, motor development also provides externally observable insights into brain and neural development. Eliciting and controlling movement is perhaps the most dominant function of the brain: large sections of the cortex are dedicated to executing movements (primary motor cortex), planning motor acts (premotor cortex, pre-frontal cortex), and registering the proximal (primary sensory cortex) or distal (visual cortex) outcomes of actions on the environment. Indeed, levels of motor functioning can be used as an indicator of Cerebral Palsy, a condition caused by brain damage (P. L. Rosenbaum, Palisano, Bartlett, Galuppi, and Russell, 2008). Therefore, the absence or inefficient execution of movements may indicate atypical brain development or neural damage (Hadders-Algra, 2008).

Motor development and brain development go hand-in-hand. As the brain matures and new connections are established, new motor skills emerge. At the same time, engaging in physical activities has been found to lead to structural changes in the brain (May, 2011). Of course, any type of learning is affecting brain circuits in some way, but learning-by-doing seems to be particularly effective in shaping neural systems (James, 2010). The close connection between motor skills and neural function is beneficial for clinical and practical applications. Direct assessment of brain activity or neural function is costly and requires specialized equipment such as an electroencephalogram (EEG), functional Magnetic Resonance Imaging (fMRI), or Near-infrared spectroscopy (NIRS). In contrast, motor milestone assessments only require behavioral observations or parent report. Therefore, motor milestones are used in clinical practice to determine a child's overall health and development.

TYPICAL MOTOR DEVELOPMENT

Motor Milestones

Watching their child achieve new motor skills is not only of interest to parents but also of value to help assess a child's neurological functioning. Standardized parent-completed questionnaires such as the Ages and Stages Questionnaire (Squires, Bricker, and Potter, 1997) or the non-standardized assessments such as the Early Motor Questionnaire (Libertus and Landa, 2013) take only minutes to be completed and can be easily implemented in primary care settings as first-level screener for potential motor issues (Kjølbye, Drivsholm, Ertmann, Lykke, and Køster-Rasmussen, 2018). Most of these screening tools ask whether the child has already mastered a specific motor skill and are based on the notion that all children attain different motor milestones around certain times during development. However, while motor milestones and motor norms are helpful in clinical settings, research shows that the notion of norms is incorrect and misrepresents the dynamic nature of motor skill acquisition.

Interest in the establishment of motor norms goes back to the earliest baby diaries. For example, Charles Darwin published a report on the behavior of his infant son noting his observations on the child's reflexes and movements (Darwin, 1877). More deliberate efforts to describe the sequential emergence of motor milestones come from pioneering work by Shirley (1933), McGraw (1935), and Gesell (1946). Gesell offered a highly detailed description of the sequence of motor development in his book *An Atlas of Infant Behaviour* (1934). These early perspectives viewed motor development as proceeding stage-like and in an orderly fashion tightly controlled by the maturation of the central nervous system (CNS). Subsequent research has used larger standardization samples to derive time windows for when certain motor skills are expected to emerge (e.g., Bayley, 1969). An example of the somewhat orderly progression and the stage-like emergence of motor skills can be seen in the approximate age of emergence for 13 motor skills shown in Figure 11.1. However, this view is too simplistic as the age of milestone onset varies significantly between individuals and the different skills are not necessarily attained in this order or attained at all (de Onis, 2006). For example, some infants skip the crawling step altogether; others can stand alone before they even attempt crawling. In fact, the presence of substantial individual differences is a defining feature of early motor development. Individual differences in early motor development also offer an argument against the view of brain and CNS maturation as the driving force behind development.

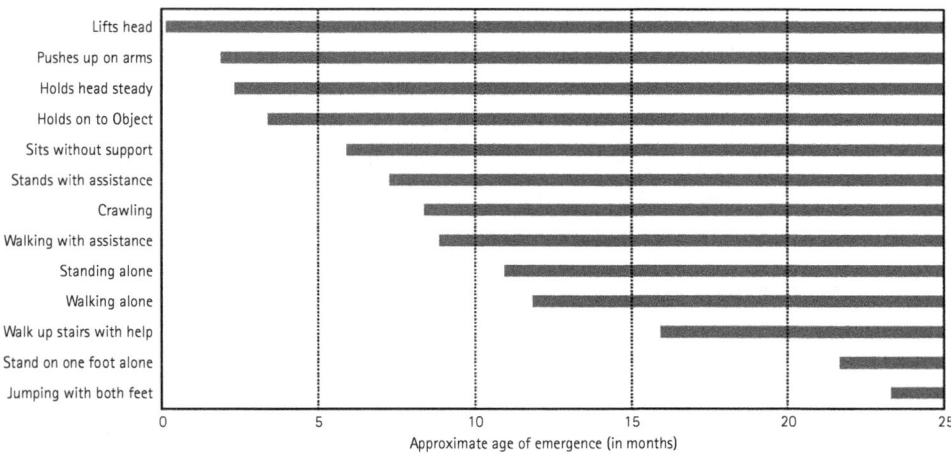

FIGURE 11.1 Approximate age of onset for key motor milestones during the first 2 years of life. Onset patterns suggest a normative progression of motor development, but individual and cultural differences argue against motor "norms."

Approximate onset ages are based on World Health Organization (WHO) norms (de Onis, 2006).

Maturation vs Neonatal Stepping

Maturation constraints play a role in the development of motor skills. When observing the reflexive movements of his son, Darwin (1877) noted that "the perfection of these reflex movements shows that the extreme imperfection of the voluntary ones is not due to the state of the muscles or of the coordinating centers, but to that of the seat of the will" (p. 285). This statement misses the myriad of internal and external factors influencing the developing organism. Each child faces a unique environment and a unique, ever-changing set of constraints. During its development, the motor system meets these constraints and dynamically assembles solutions.

Stepping movements are one example of the influence of multiple factors on motor development. Newborn infants show alternating stepping movements when held upright (Thelen, 1979; Thelen and Fisher, 1982). Early stepping movements subsequently disappear, only to re-emerge slightly later with the onset of walking. This U-shaped developmental trend seems to illustrate how a reflex disappears due to the maturation of inhibitory control and is then replaced by a similar behavior fully under cortical control. However, clever research studies demonstrate that this simple explanation may be incorrect. Neonatal stepping: a) does not disappear if the behavior is practiced repeatedly, b) disappears with weight added to the newborn's feet, and c) re-emerges in older infants when reducing the weight of the legs by placing their legs into water (Thelen, Fisher, and Ridley-Johnson, 1984). Further, the newborn movements may show "alternating" stepping not because the infant is showing an early form of "walking" but rather because it is an easier movement to perform. Changing the context slightly by connecting both legs with a string results in synchronous stepping (Thelen, 1994). This example demonstrates that the child's own experiences, their own bodily constraints (such as weight), the constraints of the current environment, and probably also the child's current mood, and state influence motor development at any given moment. Maturation matters, but it is not the only factor influencing the progression of motor development. The child's body size, individual experiences, the family environment, and cultural context also influence motor development.

Experiences and Culture

Studying and comparing different cultures offers a strong test for the hypothesis that experiences shape motor development in addition to maturational factors. In fact, cultural differences can be so strong that the "norms" proposed in Figure 11.1 would be quite out of place for many countries. For example, infants raised in Kenya seem to show an earlier onset of sitting and walking skills compared to American infants (Super, 1976). More recent research shows similar findings, reporting advanced gross motor skills in infants raised in Cameroon when compared to Germany (Vierhaus et al., 2011). However, the same research shows no difference in infants' fine motor development across the two cultures, suggesting that cultural influences are highly specific and do not generically accelerate development overall. Similar differences in motor development exist between other cultures. For example, walking emerges earlier in American compared to Chinese infants (He, Walle, and Campos, 2015). Differences even exist between children from different Western cultures. For example, significant differences on standardized motor tests have been reported between Flemish and American/Canadian infants (De Kegel et al., 2013; Hoskens, Klingels, and Smits-Engelsman, 2018). Differences persist beyond the infancy period as a comparison between Norwegian, Italian, and Greek children between 6-to 8 years of age shows. Results of this comparison have Norwegian children performing better than Italian or Greek children on a standardized motor assessment (Haga et al., 2018). However, significant differences are also evident between Greek and Italian children even though these two cultures are relatively similar to each other (both are Western, European, and Mediterranean).

Together, cross-cultural research highlights the impact of cultural differences on motor skill emergence and demonstrates that motor development is deeply enculturated (Adolph and Hoch, 2018). Consequently, there is a need to establish culture-specific norms for standardized instruments before use in a given culture or when comparing between cultures (Hoskens et al., 2018; Tso et al., 2018). Motor development should not be studied out of context. Differences between cultures may be the consequence of different parenting practices or parenting beliefs (Atun-Einy, Oudgenoeg-Paz, and van Schaik, 2017). Interaction styles between parent and child may differ across cultures, and toy selections and availability may encourage fine motor skills in one culture but gross motor skills in another (e.g., providing the child with a rattle vs a ball). Parents in different cultures may simply provide children with different experiences and opportunities for learning. If this is the case, one would predict that providing children within one cultural context with different experiences, via training or intervention, should similarly impact early motor development.

Effects of Motor Training

A number of studies demonstrate that training motor skills can result in earlier mastery of motor skills. For example, providing infants with scaffolded reaching experiences at around three months of age leads to an earlier onset of successful grasping in these children (Libertus and Needham, 2010; Needham, Barrett, and Peterman, 2002) Reaching skills in particular seem highly malleable and several research studies show that training reaching skills facilitates infants' independent object exploration and reaching attempts (Guimaraes and Tudella, 2015; Heathcock, Lobo, and Galloway, 2008; Needham, Joh, Wiesen, and Williams, 2014; Needham, Wiesen, Hejazi, Libertus, and Christopher, 2017; Soares, van der Kamp, Savelsbergh, and Tudella, 2013). Similarly, posture and locomotor skills are also receptive to naturalistic (i.e., spontaneously initiated by the caregiver) and experimental training experiences (Donenberg, Fetters, and Johnson, 2018; Hadders-Algra, Brogren, and Forssberg, 1996; Moreira da Silva, Lopes Dos Santos, Righetto Greco, and Tudella, 2017). Finally, long-term follow-up studies on the effects of early motor training show that the effect of training is long-lasting rather than just transient (Libertus, Joh, and Needham, 2016; Wiesen, Watkins, and Needham, 2016). Together, these results demonstrate that early motor skills are malleable and responsive to training or experiences.

Being able to modulate motor skills is important for children with motor delays or disabilities. At the same time, findings that elementary motor skills can be trained also reveal something important about the nature of motor skill learning. As children grow, they learn to reach, sit, crawl, stand, and walk because the dynamics of our limbs attached to our bodies allow for these motor behaviors. However, that does not mean that any of these motor behaviors are "innate" or pre-programmed into our genome. A couple of years later, children learn how to ride a bike or how to balance on a hoverboard. Surely no one would argue that riding a bike is pre-programmed into our genome. But we learn these skills in very similar ways: practice makes perfect and more training leads to faster results. Learning a motor skill such as upright locomotion is one solution to a challenge experienced by the organism given a set of constraints. Motor learning is a dynamic process, and the solution is dynamically acquired. Given a different set of constraints, a different solution would be acquired.

Implications of Early Motor Experiences

Having discussed the progression and dynamic nature of motor development, we will next turn to the implications of acquiring new motor skills. Motor skills emerge very early in development and are children's *gateway* into interactions and exchanges with the world. According to a systems perspective on development, earlier emerging skills lay the foundation for later emerging ones (Thelen and Smith, 1994). Following this argumentation, motor skills are a *prerequisite* for children's learning and growth. Consequently, motor skills should affect children's subsequent development across domains. Several lines of research provide evidence for this hypothesis.

Reaching and sitting skills are two of the earliest emerging motor abilities that have been found to relate to subsequent development. For example, providing 3-month-old infants with scaffolded reaching experiences improves children's object exploration and sustained attention abilities at 15 months of age (Libertus et al., 2016) and individual differences in learning to sit independently between 3 to 6 months of age seem to predict subsequent language learning at both 10 and 14 months (Libertus and Violi, 2016). In fact, several studies suggest that language learning is related to infants' concurrent motor skills. A set of two studies including cross-cultural samples from the US and China show that infants' onset of walking is related to their language development. Walking infants show larger receptive and expressive vocabularies than age-matched non-walking infants (He et al., 2015; Walle and Campos, 2014). This is likely due to the far-reaching changes initiated by walking. Following the onset of independent walking, infants show both an increase in initiation of joint engagement and in responding to joint engagement cues of their parents (Walle, 2016). At the same time, parents' perception of the child changes and they also respond differently to their child (Karasik, Tamis-LeMonda, and Adolph, 2014; Walle, 2016).

Longitudinal studies find similar relations between early motor skills and subsequent language. Earlier onset of sitting and walking skills predict expressive vocabulary growth in 16- to 28-month-olds (Oudgenoeg-Paz, Volman, and Leseman, 2012). The relation between early motor development and subsequent language development seems to decrease over time as children grow older and their language skills increase (Oudgenoeg-Paz, Volman, and Leseman, 2016). However, other findings demonstrate that motor development during infancy has implications for cognitive development as much as 14 years later. A long-term longitudinal study shows that motor exploration ability at five months of age is indirectly related to children's academic achievement at 14 years of age (Bornstein, Hahn, and Suwalsky, 2013). Together, these findings demonstrate that motor skills acquired during infancy can have a long-lasting impact on children's future development.

The importance of motor skills and experiences for children's development across domains continues beyond the infancy period. It is well established that physical activity is beneficial for children's cognitive skills and that recess time at school is important for learning in the classroom (Howie and Pate, 2012). Empirical studies linking motor and cognitive abilities are rare in later childhood but do indicate a link between motor and cognitive skills. For example, motor performance assessed on a standardized measure is related to processing speed as measured on a standardized intelligence test in three- to six-year-old children (Jaščenoka et al., 2018). Fine motor skills, in particular, seem to predict cognitive skills in children, as is evident in 5- to 6-year-olds where fine motor skills have been reported to predict math abilities (Pitchford, Papini, Outhwaite, and Gulliford, 2016).

Similarly, balance skills have been found to predict spatial and proportional reasoning skills in 6-year-olds (Frick and Möhring, 2015). Finally, physical activity outdoors and ball skills have been reported to predict inhibitory efficiency in an intervention design (Pesce et al., 2016). Together, these studies demonstrate that motor skills have important implications for children's future development across domains and may even impact academic performance. This suggests that impaired motor skills may negatively impact development in seemingly unrelated domains. The next section will briefly discuss evidence for this possibility.

ATYPICAL MOTOR DEVELOPMENT

Despite the large variability in typical motor development, motor skills can be delayed or emerge in a form that is not efficient given the constraints faced by the organism. From a distance, fine motor delays may appear as less severe or compromising than gross motor delays. However, given the importance of fine motor behaviors for many daily living activities, the opposite may be true. Further, the preceding review suggests that fine motor skills are often more strongly associated with current and future cognitive development than gross motor skills. Therefore, paying attention to early delays of subtle but important skills such as grasping and object manipulation is important. From a neuroscience perspective, motor delays are significant as they can be used as a behavioral marker for neurological disorders.

Motor delays are a defining feature of cerebral palsy (CP) and developmental co-ordination disorder (DCD). CP is caused by underlying brain abnormalities and affects both gross and fine motor development. Early detection of CP is possible using neuroimaging techniques. Behavioral diagnosis focusing on early motor skills and reflexes can be used as well but is less accurate (Palmer, 2004). Intervention—especially if provided early on—can be effective for children with CP and should focus on foundational motor skills such as postural control and goal-directed reaching or eye-hand coordination (Harbourne, Willett, Kyvelidou, Deffeyes, and Stergiou, 2010). The prevalence of CP is estimated at two to four cases per 1000 births (Stavsky et al., 2017). While CP is considered one of the most common movement disorders in childhood (Stavsky et al., 2017), DCD is even more common. Prevalence numbers very greatly by country with estimates ranging from around 2 percent to 10 percent of children, with a worldwide average of 6 percent (Amador-Ruiz et al., 2018; Tsiotra et al., 2006). Despite the high prevalence, DCD is often not diagnosed, making the health consequences of DCD a neglected problem in pediatrics (Cacola, 2016). DCD is characterized by motor delays and inefficient skill execution that interferes with daily living behaviors. Just as with CP, DCD is often diagnosed relatively late resulting in missed opportunities for effective early intervention strategies. While motor delays are primary in these two disorders, cognitive and social skills may be impaired as well.

Motor Delays in Autism Spectrum Disorders

Autism spectrum disorder (ASD) refers to a group of neurodevelopmental disorders defined by impaired social and communication skills, and by the presence of stereotyped and restricted interests (American Psychiatric Association, 2013). Prevalence of ASD has been estimated at one in 59 children, and most cases are diagnosed after 4 years of age (Baio et al., 2018). Motor delays are not considered a part of ASD and are not included in the diagnostic criteria for ASD. This is surprising, given that motor issues were noted already in the pioneering works of Leo Kanner (1943) and Hans Asperger (1944). The descriptions provided on early ASD patients suggest the presence of poor motor co-ordination, clumsiness, poor postural control and poor fine motor skills (Libertus, Marschik, Einspieler, and Bölte, 2015). More recent findings confirm delayed or at least atypical motor skills in children later diagnosed with ASD (Bhat, Galloway, and Landa, 2012; Libertus, Sheperd, Ross, and Landa, 2014). Recently, a large meta-analysis provided further evidence for the presence of motor delays in ASD and for a relation between motor delays and language delays (West, 2018). These findings suggest that motor skills should be closely monitored in infants and children at familial risk for ASD. Given the important role of motor skill for subsequent development across domains discussed previously, it is likely that motor delays in ASD will have downstream effects on other domains as well. Consequently, it has been suggested that early motor delays in children with ASD may contribute to deficits in the social domain that become evident later in life (Bhat, Landa, and Galloway, 2011). Research does suggest that early motor skills of children at high familial risk for ASD are malleable and respond to intervention (Libertus and Landa, 2014), but the long-term effect of motor interventions on ASD outcomes remains unknown.

While the neural basis for ASD still remains unknown, several studies have identified structural and functional differences in children with and without ASD. For example, white matter connectivity and cortical structure have all been found to differentiate between ASD and typically developing children (Aoki et al., 2017; Fu et al., 2019; Hazlett et al., 2017). Given the close connection between motor and brain development noted earlier, it is not surprising that motor and brain development should be affected in ASD. However, the relation between motor and brain development may even extend into core ASD deficits in the social domain. Social cognition is thought to be supported on the neural level by so-called mirror neurons that allow for mental simulation of an observed action (Caggiano, Fogassi, Rizzolatti, Thier, and Casile, 2009; Del Giudice, Manera, and Keysers, 2009). Decreased or atypical mirror neuron activity has been observed in ASD (Dapretto et al., 2006; Oberman et al., 2005), although this position has been challenged by others (Fan, Decety, Yang, Liu, and

Cheng, 2010). Developmental theories suggest that motor activity and engagement are necessary for neural circuits related to imitation and understanding of others to develop (Meltzoff, 2007). Consequently, impaired motor abilities may result in missed opportunities for learning about the intentions of other's actions. However, this theory remains contested as motor delays in other developmental disorders do not result in the social differences associated with ASD. More research on the connection between motor experiences and social development in typical and atypical development are needed to address this question.

Motor skills can be assessed relatively easy via parent-report measures and during brief observations that may be incorporated into primary care settings. Atypical motor development is present across a range of developmental disorders, including but not limited to CP, DCD, and ASD. Therefore, atypical patterns of motor development are likely to affect a large proportion of all children. Motor assessments and early screening should be a priority and more use of parent-report measures as level 1 screening devices should be implemented (Libertus and Landa, 2013).

FUTURE DIRECTIONS FOR NEUROSCIENTIFIC RESEARCH

Several important questions in the study of motor development remain unanswered—especially in the area of developmental cognitive neuroscience. Motor development has received relatively little attention during the past 20–30 years, a time when neuroscience methods have started to become widely adopted, more accessible, and adapted for infant and child populations. Therefore, developmental cognitive neuroscience is uniquely poised to address some of the open questions on the mechanism, process, and consequences of motor development. At least three broad areas of motor development merit future research.

First, the review presented previously shows that motor development is highly variable and influenced by maturational, experiential, cultural, and other factors. While research has identified some of the factors and experiences that influence motor development, the neural basis for high variability and individual differences remains unknown. For example, scaffolded reaching experiences encourage grasping in 3-month-old children and accelerate the development of a preference for faces (Libertus and Needham, 2011). What underlies these behavioral changes? Does motor training increase neural connectivity? Does it encourage myelination of existing fiber tracts or result in the growth of new connections? Answers to these questions are unknown. Whatever the process, it must act on a fast timescale as brief training experiences of only 3 minutes are already effective in changing behavior (Gerson and Woodward, 2014; Sommerville, Woodward, and Needham, 2005).

Second, this chapter also emphasized that motor development impacts growth across domains with long-lasting implications for cognition and even academic achievement. What are the underlying mechanisms for these connections? Evidence for behavioral developmental cascades (Masten and Cicchetti, 2010) has been described or hypothesized in some studies, but the neural pathways that link motor skills to general cognition have not been explored. Motor skills and physical activity, in general, seem strongly linked to core cognitive abilities such as executive function. The neural basis of executive function has been studied extensively since the beginnings of cognitive neuroscience (Zatorre, 2018). Similarly, motor functions and planning have been carefully localized in the brain (Graziano, 2006). Together, this knowledge provides a solid foundation for future research into how motor skills are linked to cognitive functions and why motor experiences are able to influence cognitive ability.

Finally, the neural basis for atypical motor development in children with developmental disorders needs to be studied further. In cases with no obvious brain or neurological damage, such as children with ASD, the presence of subtle motor delays is surprising. While the consequences of motor delays have been examined, where these motor delays come from remains unclear. Structural and functional neural imaging may prove helpful in this area. Neural re-organization is likely in ASD, and whole brain analyses of fiber tracts suggests differences between children with ASD and typically developing peers (Zhang et al., 2018). Future Developmental Cognitive Neuroscience research should focus on identifying more neural differences in structure and function between children with and without a developmental disorder. But at the same time, it is also important to ask whether neural differences are causing the disorder or are a response or adaptation to the disorder (Johnson, 2017).

Conclusions

Motor development is a core area in Developmental Psychology and needs to be carefully considered in Developmental Cognitive Neuroscience as well. This chapter has discussed the myriad of factors that influence motor development trajectories during childhood. From maturation and genes to experiences and culture, motor development is strongly impacted by external forces and presents as a highly variable developmental process. At the same time, motor development has implications for children's immediate interactions with their surroundings, the information they can access and receive, and their future learning and achievement. The important implications of motor development are increasingly appreciated in education and applied settings, but the neural basis of the downstream effects of motor experiences and activity remain understudied and are an area of future research for Developmental Cognitive Neuroscience. This is especially true for developmental disorders such as Autism Spectrum Disorders, where motor delays are often observed but their neural basis remains unknown.

References

Adolph, K. E. and Hoch, J. E. (2018). Motor development: embodied, embedded, enculturated, and enabling. *Annual Review of Psychology, 70*, 141–164 doi:10.1146/annurev-psych-010418-102836

Amador-Ruiz, S., Gutierrez, D., Martinez-Vizcaino, V., Gulias-Gonzalez, R., Pardo-Guijarro, M. J., and Sanchez-Lopez, M. (2018). Motor competence levels and prevalence of developmental coordination disorder in Spanish children: The MOVI-KIDS Study. *J Sch Health, 88*(7), 538–546. doi:10.1111/josh.12639

American Psychiatric Association. (2013). *Diagnostic and Statistical Manual of mental disorders* (5 ed.). Arlington, VA: American Psychiatric Association.

Aoki, Y., Yoncheva, Y. N., Chen, B., Nath, T., Sharp, D., Lazar, M., ... and Di Martino, A. (2017). Association of white matter structure with autism spectrum disorder and attention-deficit/hyperactivity disorder. *JAMA Psychiatry, 74*(11), 1120–1128. doi:10.1001/jamapsychiatry.2017.2573

Asperger, H. (1944). Die autistischen Psychopathen im Kindesalter. *Archiv für Psychiatrie und Nervenkrankheiten, 117*, 76–136.

Atun-Einy, O., Oudgenoeg-Paz, O., and van Schaik, S. D. M. (2017). Parental beliefs and practices concerning motor development: Testing new tools. *European Journal of Developmental Psychology, 14*(5), 556–604. doi:10.1080/17405629.2016.1263563

Bayley, N. (1969). *The Bayley Scales of Infant Development.* New York: Psychological Corporation.

Baio, J., Wiggins, L., Christensen, D. L., Maenner, M. J., Daniels, J., Warren, Z., ... and Dowling, N. F. (2018). Prevalence of autism spectrum disorder among children aged 8 years—autism and developmental disabilities monitoring network, 11 Sites, United States, 2014. *MMWR Surveill Summ, 67*(6), 1–23. doi:10.15585/mmwr.ss6706a1

Bhat, A. N., Galloway, J. C., and Landa, R. (2012). Relation between early motor delay and later communication delay in infants at risk for autism. *Infant Behavior and Development, 35*, 838–846. doi:10.1016/j.infbeh.2012.07.019

Bhat, A. N., Landa, R. J., and Galloway, J. C. (2011). Current perspectives on motor functioning in infants, children, and adults with autism spectrum disorders. *Physical Therapy, 91*(7), 1116–1129. doi:10.2522/ptj.20100294

Bornstein, M. H., Hahn, C. S., and Suwalsky, J. T. (2013). Physically developed and exploratory young infants contribute to their own long-term academic achievement. *Psychological Science, 24*(10), 1906–1917. doi:10.1177/0956797613479974

Cacola, P. (2016). Physical and mental health of children with developmental coordination disorder. *Front Public Health, 4*, 224. doi:10.3389/fpubh.2016.00224

Caggiano, V., Fogassi, L., Rizzolatti, G., Thier, P., and Casile, A. (2009). Mirror neurons differentially encode the peripersonal and extrapersonal space of monkeys. *Science, 324*(5925), 403–406. doi:10.1126/science.1166818

Dapretto, M., Davies, M. S., Pfeifer, J. H., Scott, A. A., Sigman, M., Bookheimer, S. Y., and Iacoboni, M. (2006). Understanding emotions in others: Mirror neuron dysfunction in children with autism spectrum disorders. *Nat Neurosci, 9*(1), 28–30. doi:10.1038/nn1611

Darwin, C. (1877). A biographical sketch of an infant. *Mind & Language, 2*(7), 285–294. Retrieved from https://www.jstor.org/stable/2246907

De Kegel, A., Peersman, W., Onderbeke, K., Baetens, T., Dhooge, I., and Van Waelvelde, H. (2013). New reference values must be established for the Alberta Infant Motor Scales for accurate identification of infants at risk for motor developmental delay in Flanders. *Child: Care, Health and Development*, 39(2), 260–267. doi:10.1111/j.1365–2214.2012.01384.x

de Onis, M. (2006). WHO Motor Development Study: Windows of achievement for six gross motor development milestones. *Acta Paediatrica*, 95, 86–95.

Del Giudice, M., Manera, V., and Keysers, C. (2009). Programmed to learn? The ontogeny of mirror neurons. *Developmental Science*, 12(2), 350–363. doi:10.1111/j.1467–7687.2008.00783.x

Donenberg, J. G., Fetters, L., and Johnson, R. (2018). The effects of locomotor training in children with spinal cord injury: A systematic review. *Dev Neurorehabil*, 22(4), 1–16. doi:10.1080/17518423.2018.1487474

Fan, Y. T., Decety, J., Yang, C. Y., Liu, J. L., and Cheng, Y. (2010). Unbroken mirror neurons in autism spectrum disorders. *J Child Psychol Psychiatry*, 51(9), 981–988. doi:10.1111/j.1469–7610.2010.02269.x

Frick, A., and Möhring, W. (2015). A matter of balance: Motor control is related to children's spatial and proportional reasoning skills. *Frontiers in Psychology*, 6, 2049. doi:10.3389/fpsyg.2015.02049

Fu, Z., Tu, Y., Di, X., Du, Y., Sui, J., Biswal, B. B., ... and Calhoun, V. D. (2019). Transient increased thalamic-sensory connectivity and decreased whole-brain dynamism in autism. *Neuroimage*, 190, 191–204. doi:10.1016/j.neuroimage.2018.06.003

Gerson, S. A. and Woodward, A. L. (2014). Learning from their own actions: The unique effect of producing actions on infants' action understanding. *Child Development*, 85(1), 264–277. doi:10.1111/cdev.12115

Gesell, A. (1934). *An Atlas of Infant Behavior* (Vol. 1). New Haven, CT: Yale University Press.

Gesell, A. (1946). The Ontogenesis of Infant Behavior. In L. Carmichael (Ed.), *Manual of Child Psychology* (pp. 295–331). New York, NY: John Wiley.

Gibson, E. J. (1988). Exploratory behavior in the development of perceiving, acting and acquiring of knowledge. *Annual Review of Psychology*, 39, 1–41. doi:10.1146/Annurev.Ps.39.020188.000245

Graziano, M. S. A. (2006). The organization of behavioral repertoire in motor cortex. *Annu Rev Neurosci*, 29, 105–134. Retrieved from http://www.ncbi.nlm.nih.gov/entrez/query.fcgi?cmd=Retrieveanddb=PubMedanddopt=Citationandlist_uids=16776581

Guimaraes, E. L. and Tudella, E. (2015). Immediate effect of training at the onset of reaching in preterm infants: Randomized clinical trial. *J Mot Behav*, 47(6), 535–549. doi:10.1080/00222895.2015.1022247

Hadders-Algra, M. (2008). Reduced variability in motor behaviour: An indicator of impaired cerebral connectivity? *Early Hum Dev*, 84(12), 787–789. doi:10.1016/j.earlhumdev.2008.09.002

Hadders-Algra, M., Brogren, E., and Forssberg, H. (1996). Training affects the development of postural adjustments in sitting infants. *Journal of Physiology-London*, 493(1), 289–298. doi:10.1113/jphysiol.1996.sp021383

Haga, M., Tortella, P., Asonitou, K., Charitou, S., Koutsouki, D., Fumagalli, G., and Sigmundsson, H. (2018). Cross-cultural aspects: exploring motor competence among 7- to 8-year-old children from Greece, Italy, and Norway. *Sage Open*, 8(2). doi:10.1177/2158244018768381

Harbourne, R. T., Willett, S., Kyvelidou, A., Deffeyes, J., and Stergiou, N. (2010). A comparison of interventions for children with cerebral palsy to improve sitting postural control: A clinical trial. *Phys Ther*, 90(12), 1881–1898. doi:10.2522/ptj.2010132

Hazlett, H. C., Gu, H., Munsell, B. C., Kim, S. H., Styner, M., Wolff, J. J., ... and Gu, C. H. (2017). Early brain development in infants at high risk for autism spectrum disorder. *Nature*, 542,

348. doi:10.1038/nature21369   https://www.nature.com/articles/nature21369#supplementary-information

He, M., Walle, E. A., and Campos, J. J. (2015). A cross-national investigation of the relationship between infant walking and language development. *Infancy, 20*(3), 283–305. doi:10.1111/infa.12071

Heathcock, J. C., Lobo, M., and Galloway, J. C. (2008). Movement training advances the emergence of reaching in infants born at less than 33 weeks of gestational age: A randomized clinical trial. *Physical Therapy, 88*(3), 310–322.

Hoskens, J., Klingels, K., and Smits-Engelsman, B. (2018). Validity and cross-cultural differences of the Bayley Scales of Infant and Toddler Development, Third Edition in typically developing infants. *Early Hum Dev, 125,* 17–25. doi:10.1016/j.earlhumdev.2018.07.002

Howie, E. K. and Pate, R. R. (2012). Physical activity and academic achievement in children: A historical perspective. *Journal of Sport and Health Science, 1*(3), 160–169. doi:10.1016/j.jshs.2012.09.003

Iverson, J. M. (2010). Developing language in a developing body: The relationship between motor development and language development. *Journal of Child Language, 37*(2), 229–261. doi:10.1017/S0305000909990432

James, K. H. (2010). Sensori-motor experience leads to changes in visual processing in the developing brain. *Dev Sci, 13*(2), 279–288. doi:10.1111/j.1467-7687.2009.00883.x

Jaščenoka, J., Walter, F., Petermann, F., Korsch, F., Fiedler, S., and Daseking, M. (2018). Zum Zusammenhang von motorischer und kognitiver Entwicklung im Vorschulalter. *Kindheit und Entwicklung, 27*(3), 142–152. doi:10.1026/0942-5403/a000254

Johnson, M. H. (2017). Autism as an adaptive common variant pathway for human brain development. *Dev Cogn Neurosci, 25,* 5–11. doi:10.1016/j.dcn.2017.02.004

Kanner, L. (1943). Autistic disturbances of affective contact. *Nervous Child, 2,* 217–250.

Karasik, L. B., Tamis-LeMonda, C. S., and Adolph, K. E. (2014). Crawling and walking infants elicit different verbal responses from mothers. *Developmental Science, 17*(3), 388–395. doi:10.1111/desc.12129

Kjølbye, C. B., Drivsholm, T. B., Ertmann, R. K., Lykke, K., and Køster-Rasmussen, R. (2018). Motor function tests for 0-2-year-old children—A systematic review. *Danish Medical Journal, 65*(6), 21–23.

Libertus, K., Joh, A. S., and Needham, A. W. (2016). Motor training at 3 months affects object exploration 12 months later. *Developmental Science, 19*(6), 1058–1066. doi:10.1111/desc.12370

Libertus, K. and Landa, R. J. (2013). The Early Motor Questionnaire (EMQ): A parental report measure of early motor development. *Infant Behavior and Development, 36*(4), 833–842. doi:10.1016/j.infbeh.2013.09.007

Libertus, K. and Landa, R. J. (2014). Scaffolded reaching experiences encourage grasping activity in infants at high risk for Autism. *Frontiers in Psychology, 5,* 1071. doi:10.3389/fpsyg.2014.01071

Libertus, K., Marschik, P. B., Einspieler, C., and Bölte, S. (2015). Frühe Auffälligkeiten bei Autismus Spektrum Störungen. In S. Sachse (Ed.), *Handbuch Sprachentwicklung und Sprachentwicklungsstörung: Band 3 Frühe Kindheit:* München: Elsevier.

Libertus, K. and Needham, A. (2010). Teach to reach: The effects of active vs. passive reaching experiences on action and perception. *Vision Research, 50*(24), 2750–2757. doi:10.1016/j.visres.2010.09.001

Libertus, K. and Needham, A. (2011). Reaching experience increases face preference in 3-month-old infants. *Developmental Science, 14*(6), 1355–1364. doi:10.1111/j.1467-7687.2011.01084.x

Libertus, K., Sheperd, K. A., Ross, S. W., and Landa, R. J. (2014). Limited fine motor and grasping skills in 6-month-old infants at high risk for Autism. *Child Development*, 85(6), 2218–2231. doi:10.1111/cdev.12262

Libertus, K. and Violi, D. A. (2016). Sit to talk: Relation between motor skills and language development in infancy. *Frontiers in Psychology*, 7(475), 1–8. doi:10.3389/fpsyg.2016.00475

Masten, A. S. and Cicchetti, D. (2010). Developmental cascades. *Development and Psychopathology*, 22(3), 491–495. doi:10.1017/S0954579410000222

May, A. (2011). Experience-dependent structural plasticity in the adult human brain. *Trends Cogn Sci*, 15(10), 475–482. doi:10.1016/j.tics.2011.08.002

McGraw, M. B. (1935). *Growth: A Study of Johnny and Jimmy*. New York, NY: D. Appleton-Century Company.

Meltzoff, A. N. (2007). "Like me": a foundation for social cognition. *Developmental Science*, 10(1), 126–134.

Moreira da Silva, E. S., Lopes Dos Santos, G., Righetto Greco, A. L., and Tudella, E. (2017). Influence of different sitting positions on healthy infants' reaching movements. *J Mot Behav*, 49(6), 603–610. doi:10.1080/00222895.2016.1247034

Needham, A., Barrett, T., and Peterman, K. (2002). A pick-me-up for infants' exploratory skills: Early simulated experiences reaching for objects using "sticky mittens" enhances young infants' object exploration skills. *Infant Behavior and Development*, 25(3), 279–295. doi:10.1016/S0163-6383(02)00097-8

Needham, A., Joh, A. S., Wiesen, S. E., and Williams, N. (2014). Effects of contingent reinforcement of actions on infants' object-directed reaching. *Infancy*, 19(5), 496–517. doi:10.1111/infa.12058

Needham, A. and Libertus, K. (2011). Embodiment in early development. *Wiley Interdisciplinary Reviews: Cognitive Science*, 2(1), 117–123. doi:10.1002/wcs.109

Needham, A., Wiesen, S. E., Hejazi, J. N., Libertus, K., and Christopher, C. (2017). Characteristics of brief sticky mittens training that lead to increases in object exploration. *Journal of Experimental Child Psychology*, 164, 209–224. doi:10.1016/j.jecp.2017.04.009

Oakes, L. M. (2009). The "Humpty Dumpty Problem" in the study of early cognitive development: Putting the infant back together again. *Perspect Psychol Sci*, 4(4), 352–358. doi:10.1111/j.1745-6924.2009.01137.x

Oberman, L. M., Hubbard, E. M., McCleery, J. P., Altschuler, E. L., Ramachandran, V. S., and Pineda, J. A. (2005). EEG evidence for mirror neuron dysfunction in autism spectrum disorders. *Cognitive Brain Research*, 24(2), 190–198. doi:10.1016/j.cogbrainres.2005.01.014

Oudgenoeg-Paz, O., Volman, M. C., and Leseman, P. P. (2012). Attainment of sitting and walking predicts development of productive vocabulary between ages 16 and 28 months. *Infant Behav Dev*, 35(4), 733–736. doi:10.1016/j.infbeh.2012.07.010

Oudgenoeg-Paz, O., Volman, M. C., and Leseman, P. P. (2016). First steps into language? Examining the specific longitudinal relations between walking, exploration and linguistic skills. *Frontiers in Psychology*, 7, 1458. doi:10.3389/fpsyg.2016.01458

Palmer, F. B. (2004). Strategies for the early diagnosis of cerebral palsy. *Journal of Pediatrics*, 145(2), S8-S11. doi:10.1016/j.jpeds.2004.05.016

Pesce, C., Masci, I., Marchetti, R., Vazou, S., Saakslahti, A., and Tomporowski, P. D. (2016). Deliberate play and preparation jointly benefit motor and cognitive development: mediated and moderated effects. *Frontiers in Psychology*, 7, 349. doi:10.3389/fpsyg.2016.00349

Pfeifer, R. and Bongard, J. (2007). *How the Body Shapes the Way We Think—A New View of Intelligence*. Cambridge, MA: The MIT Press.

Piaget, J. (1970). Piaget's Theory. In P. H. Mussen (Ed.), *Carmichael's Manual of Child Psychology* (3rd ed., Vol. 1, pp. 703–732). New York, NY: John Wiley & Sons Inc.

Pitchford, N. J., Papini, C., Outhwaite, L. A., and Gulliford, A. (2016). Fine motor skills predict maths ability better than they predict reading ability in the early primary school years. *Frontiers in Psychology, 7*, 783. doi:10.3389/fpsyg.2016.00783

Rosenbaum, D. A. (2005). The Cinderella of psychology: The neglect of motor control in the science of mental life and behavior. *The American psychologist, 60*(4), 308–317. doi:10.1037/0003-066X.60.4.308

Rosenbaum, P. L., Palisano, R. J., Bartlett, D. J., Galuppi, B. E., and Russell, D. J. (2008). Development of the gross motor function classification system for cerebral palsy. *Dev Med Child Neurol, 50*(4), 249–253. doi:10.1111/j.1469-8749.2008.02045.x

Shirley, M. M. (1933). *The first two years—A study of twenty-five babies* (Vol. 2). Minneapolis, MN: University of Minnesota Press.

Smith, L. B. (2005). Cognition as a dynamic system: Principles from embodiment. *Developmental Review, 25*(3-4), 278–298. doi:10.1016/j.dr.2005.11.001

Smith, L. B. and Gasser, M. (2005). The development of embodied cognition: Six lessons from babies. *Artificial Life, 11*(1-2), 13–29. doi:10.1162/1064546053278973

Smith, L. B. and Thelen, E. (2003). Development as a dynamic system. *Trends in Cognitive Sciences, 7*(8), 343–348. doi: https://doi.org/10.1016/S1364-6613(03)00156-6

Soares, D. de A., van der Kamp, J., Savelsbergh, G. J., and Tudella, E. (2013). The effect of a short bout of practice on reaching behavior in late preterm infants at the onset of reaching: A randomized controlled trial. *Res Dev Disabil, 34*(12), 4546–4558. doi:10.1016/j.ridd.2013.09.028

Sommerville, J. A., Woodward, A. L., and Needham, A. (2005). Action experience alters 3-month-old infants' perception of others' actions. *Cognition, 96*(1), B1–11. doi:10.1016/j.cognition.2004.07.004

Squires, J., Bricker, D., and Potter, L. (1997). Revision of a parent-completed developmental screening tool: Ages and stages questionnaires. *Journal of Pediatric Psychology, 22*(3), 313–328. doi:10.1093/jpepsy/22.3.313

Stavsky, M., Mor, O., Mastrolia, S. A., Greenbaum, S., Than, N. G., and Erez, O. (2017). Cerebral palsy: Trends in epidemiology and recent development in prenatal mechanisms of disease, treatment, and prevention. *Front Pediatr, 5*, 21. doi:10.3389/fped.2017.00021

Super, C. M. (1976). Environmental effects on motor development: the case of "African infant precocity." *Dev Med Child Neurol, 18*(5), 561–567. Retrieved from https://www.ncbi.nlm.nih.gov/pubmed/976610

Thelen, E. (1979). Rhythmical stereotypies in normal human infants. *Animal Behaviour, 27*(Pt 3), 699–715. doi:10.1016/0003-3472(79)90006-X

Thelen, E. (1994). Three-month-old infants can learn task-specific patterns of interlimb coordination. *Psychological Science, 5*(5), 280–284. Retrieved from http://dx.doi.org/10.1111/j.1467-9280.1994.tb00626.x

Thelen, E. and Fisher, D. M. (1982). Newborn stepping: An explanation for a "disappearing" reflex. *Developmental Psychology, 18*(5), 760–775. doi:10.1037/0012-1649.18.5.760

Thelen, E., Fisher, D. M., and Ridley-Johnson, R. (1984). The relationship between physical growth and a newborn reflex. *Infant Behavior and Development, 7*(4), 479–493. Retrieved from http://www.sciencedirect.com/science/article/B6W4K-4F1SFHM-2J/2/247cd84d43221404b40dab05f6695fa3

Thelen, E. and Smith, L. (1994). *A Dynamic Systems Approach to the Development of Cognition and Action.* Cambridge, MA: The MIT Press.

Tronick, E. and Hunter, R. G. (2016). Waddington, dynamic systems, and epigenetics. *Front Behav Neurosci, 10*, 107. doi:10.3389/fnbeh.2016.00107

Tsiotra, G. D., Flouris, A. D., Koutedakis, Y., Faught, B. E., Nevill, A. M., Lane, A. M., and Skenteris, N. (2006). A comparison of developmental coordination disorder prevalence rates in Canadian and Greek children. *Journal of Adolescent Health, 39*(1), 125–127. doi:10.1016/j.jadohealth.2005.07.011

Tso, W. W. Y., Wong, V. C. N., Xia, X., Faragher, B., Li, M., Xu, X., ... and Challis, D. (2018). The Griffiths Development Scales-Chinese (GDS-C): A cross-cultural comparison of developmental trajectories between Chinese and British children. *Child: Care, Health and Development, 44*(3), 378–383. doi:10.1111/cch.12548

Van Dam, M., Hallemans, A., Truijen, S., Segers, V., and Aerts, P. (2011). A cross-sectional study about the relationship between morphology and kinematic parameters in children between 15 and 36 months. *Gait Posture, 34*(2), 159–163. doi:10.1016/j.gaitpost.2011.04.001

Vierhaus, M., Lohaus, A., Kolling, T., Teubert, M., Keller, H., Fassbender, I., ... and Schwarzer, G. (2011). The development of 3- to 9-month-old infants in two cultural contexts: Bayley longitudinal results for Cameroonian and German infants. *European Journal of Developmental Psychology, 8*(3), 349–366. doi:10.1080/17405629.2010.505392

Waddington, C. H. (1947). *Organisers and Genes*. Cambridge, UK: Cambridge University Press.

Walle, E. A. (2016). Infant social development across the transition from crawling to walking. *Frontiers in Psychology, 7*, 960. doi:10.3389/fpsyg.2016.00960

Walle, E. A., and Campos, J. J. (2014). Infant language development is related to the acquisition of walking. *Developmental Psychology, 50*(2), 336–348. doi:10.1037/a0033238

West, K. L. (2018). Infant motor development in autism spectrum disorder: A synthesis and meta-analysis. *Child Development, 90*, 2053–2070. doi:10.1111/cdev.13086

Wiesen, S. E., Watkins, R. M., and Needham, A. W. (2016). Active motor training has long-term effects on infants' object exploration. *Frontiers in Psychology, 7*, 599. doi:10.3389/fpsyg.2016.00599

Zatorre, R. (2018). Brenda Milner and the origins of cognitive neuroscience. *Current Biology, 28*(11), R638-R639. doi:10.1016/j.cub.2018.04.048

Zhang, F., Savadjiev, P., Cai, W., Song, Y., Rathi, Y., Tunc, B., ... and O'Donnell, L. J. (2018). Whole brain white matter connectivity analysis using machine learning: An application to autism. *Neuroimage, 172*, 826–837. doi:10.1016/j.neuroimage.2017.10.029

CHAPTER 12

DEVELOPMENTAL COORDINATION DISORDER

DANIEL BRADY AND HAYLEY LEONARD

WHAT IS DEVELOPMENTAL COORDINATION DISORDER?

DEVELOPMENTAL coordination disorder (DCD) is characterized by motor competence significantly below that expected for an individual's age. The actual presentation of the disorder is somewhat heterogeneous, with different individuals presenting different profiles of impairment. However, common core features include: slower and less accurate motor performance, impaired balance, difficulty in motor learning, and poorer sensorimotor coordination (Geuze, 2003; Wilson et al., 2012). These core features result in more functional difficulties in activities of everyday living, such as: writing and drawing, dressing, using utensils, catching, and running (Summers, Larkin, and Dewey, 2008). A diagnosis of DCD is typically based on the presence of the following four features (American Psychiatric Association, 2013): firstly, the acquisition and execution of coordinated motor skills are below what would be expected at a given chronological age and opportunity for skill learning and use (Criterion A). Secondly, the motor skills deficit significantly or persistently interferes with activities of daily living appropriate to the chronological age and impacts academic productivity, prevocational and vocational activities, leisure, and play (Criterion B). Thirdly, the onset of symptoms is in the early developmental period (Criterion C). Finally, the motor skills deficits cannot be better explained by intellectual disability or visual impairment and are not attributable to a neurological condition affecting movement (Criterion D).

As well as these core symptoms, DCD has also been associated with negative secondary outcomes in a number of domains, including: cognitive (e.g., Leonard and Hill, 2015; Leonard et al., 2015), mental and physical health (e.g., Cairney et al., 2005; Faught et al., 2005; Piek et al., 2007; Pratt and Hill, 2011), social/interpersonal (e.g., Skinner and

Piek, 2001; Miyahara and Piek, 2006), and academic domains (e.g., Miller et al., 2001; Watson and Knott, 2006; Wocadlo and Rieger, 2008).

It is currently estimated that DCD has a prevalence in school-aged children of approximately 5–6 percent (American Psychiatric Association, 2013). This prevalence is also thought to be consistent across races and socio-economic statuses (Blank et al., 2012). Studies indicate that boys are twice as likely as girls to show motor difficulties that can be classified as DCD (Lingam et al., 2009). The estimated cooccurrence of other neurodevelopmental disorders alongside DCD is much higher than would otherwise be expected. The most notable of these cooccurring disorders is attention deficit/hyperactivity disorder (ADHD) with an estimated rate of 35–50 percent (Kaplan et al., 2006), but the rate is also elevated for autism spectrum disorders (ASD; Green et al., 2009) and developmental dyslexia (Nicolson et al., 1999). It is recognized that DCD is not just a childhood disorder, and that a large proportion of children with the disorder will continue to experience problems in adolescence and adulthood (Cousins and Smyth, 2003; Kirby, Sugden, and Purcell, 2014).

THE DEVELOPMENTAL COGNITIVE NEUROSCIENCE OF DCD

While diagnostic criterion D states that DCD is not attributable to a known neurological condition that affects movement, it is generally acknowledged that the root cause of the disorder lies in the brain. Early hypotheses posited that the cause is some form of brain damage (Gubbay et al., 1965); however, the current understanding is centered on the idea that brain development follows an atypical trajectory (Wilson et al., 2012). This section will provide an overview of neuroimaging research that has been conducted in DCD, beginning with investigations of the neural correlates of motor performance in individuals with and without the disorder. Next, we will cover studies examining neural correlates of DCD that were inferred from performance deficits observed in behavioral experiments ("cognitive hypotheses"; Wilson et al., 2012). These hypotheses fit broadly into the following categories: visual perception, visuospatial attention, executive functioning, internal modelling deficits, and imitation (see Wilson et al., 2017, for a recent systematic review). Finally, more recent investigations of structural differences in DCD will be considered.

Neural Correlates of Motor Performance

Given that motor difficulties are core to the disorder, the majority of research investigating the neural correlates of DCD has explored some aspect of motor performance. These include: gross motor control, visually-guided fine motor control, motor learning, and timing.

Gross Motor Control

Pangelinan, Hatfield, and Clark (2013) employed a simple reaching paradigm to explore motor control and its neural correlates in children with DCD. The task required participants to make rapid movements from the center of a graphics tablet to one of two points located diagonally. In order to investigate the neural correlates, electroencephalography (EEG) was recorded throughout the experiment, and analyzed using both event-related potential (ERPs; specifically the movement-related cortical potential, MRCP) and spectral decomposition (focused on alpha (7–13 Hz) and beta (13–30 Hz) bands) approaches. Although no behavioral differences were found between groups, EEG analyses revealed that the DCD group showed a greater mean amplitude in the Fz and Cz electrodes before movement onset, after accounting for age. The spectral analysis showed no changes in alpha or beta activity associated with age, and no differences between groups. Given that there is evidence that localizes the source of the (MCRP) component to motor areas (specifically the sensorimotor and supplementary motor areas), the authors suggest that the DCD group require more cortical resources than their typically developing peers to execute these movements.

Motor overflow is another presentation of poor gross motor control, and is typically defined as "the presence of extraneous movements occurring in parts of the body not actively involved in the performance of a task" (Licari et al., 2015, pp. 3–4). It is fairly common in young children but has also been observed in older children with DCD (Licari, Larkin, and Miyahara, 2006; Licari and Larkin, 2008). Licari and colleagues (2015) explored the neural correlates of motor overflow in DCD using functional magnetic resonance imaging (fMRI), recording the blood oxygenation level dependent (BOLD) signal while the children undertook a task. The BOLD signal represents changes in the blood oxygen level across the brain and is an indirect way of measuring which particular areas are active during tasks (for a more complete explanation see Soares et al., 2016). Children with and without DCD performed two tasks with their dominant hand: a repetitive finger sequencing task and a repetitive hand clenching task. Overflow activity in their non-dominant hand was monitored using a motion-sensitive glove. Children with DCD demonstrated greater overflow than the control group for both tasks, particularly in the finger sequencing task. This task was also associated with greater activation in the left superior and left inferior frontal gyri than the DCD group. Furthermore, the DCD group showed increased activation in the right post-central gyrus. This suggests that over-activation in frontal areas and potential dysfunction in inter-hemispheric inhibition may play a role in motor overflow in DCD.

Visually Guided Fine Motor Control

Thus far, three studies have looked at visually guided fine motor performance in children with DCD, employing tasks where participants used a joystick to either move a cursor to keep pace with a moving target (Kashiwagi et al., 2009), or to follow a trail while remaining within the lines (Zwicker et al., 2010, 2011). Kashiwagi and colleagues (2009) reported lower BOLD activity in the left superior and inferior parietal lobes and the left post-central gyrus for children with DCD compared to controls, and the poorer performance on the task in DCD compared to controls also correlated with BOLD activity in the inferior parietal cortex. This led the authors to suggest that poorer performance on the task could be related to dysfunction in the posterior parietal cortex (PPC). However, the reported poorer performance on the distance measure appears to be driven by a single participant in the DCD group with a particularly poor score, and the correlation with the BOLD response may be driven by the same outlier, meaning these results require cautious interpretation.

A wider range of atypical functional activity was reported by Zwicker and colleagues (2010, 2011), despite demonstrating no behavioral differences in the trail-following task between children with DCD and controls. Specifically, the DCD group showed greater BOLD activity than the control group in nine regions mostly located in the right hemisphere, including: the middle frontal gyrus, supramarginal gyrus, lingual gyrus, parahippocampal gyrus, posterior cingulate gyrus, precentral gyrus, superior temporal gyrus, and cerebellar lobule VI, as well as the left inferior parietal lobule (IPL). In contrast, the control group showed greater activation than the DCD group in five sites primarily located in the left hemisphere, including: the precuneus, superior frontal gyrus, inferior frontal gyrus, and post-central gyrus, as well as the right superior temporal gyrus (Zwicker et al., 2010). After practicing the task for three days outside the scanner (Zwicker et al., 2011), behavioral performance still did not differ between DCD and control groups, but the authors reported statistically significant group by time interactions for a number of cortical areas, including: the IPLs bilaterally, the left fusiform gyrus, the right lingual gyrus, and the right middle frontal gyrus. In addition, three cerebellar areas also showed an interaction: the right crus I, left lobule VI, and left lobule IX. However, no main effects for time or group are reported for these analyses, nor any post-hoc tests, making it difficult to interpret these results for the DCD group.

Fine Motor Sequence Learning

As well as motor control, research into DCD has investigated motor learning. Biotteau and colleagues (2017) used a finger tapping sequence task in order to investigate the neural correlates of motor learning in children with DCD. Two sequences were used: an over-learned sequence, which had been practised for 15 days before the scanning session, and a novel sequence. Motor learning was assessed through accuracy on producing the over-learned sequence when concurrently completing a picture-naming task. Performance during the dual task was poorer than on the over-learned task alone, suggesting that it had not become automatic. Imaging revealed that there was higher activity in the right caudate and right insula during performance of the novel sequence. The authors note that this difference is not surprising given evidence suggesting recruitment of this area during the early stages of motor learning. However, the lack of a control group makes further interpretation difficult.

Timing

Timing is an essential component of efficacious movement control. Individuals with DCD have been reported to have poorer timing ability compared to age-matched controls (Rosenblum and Regev, 2013) but, like most aspects of motor performance in DCD, the neural cause is not clear. Nonetheless, two studies have suggested that frontal areas involved in accurate timing may have reduced activity and connectivity in children with DCD compared to controls (de Castelnau et al., 2008; Debrabant et al., 2013).

In the first of these studies, children responded to a visual stimulus that was presented at high or low frequency by either tapping in synchrony or in syncopation with it (de Castelnau et al., 2008). Coherence in alpha and beta frequency band EEG activity between specific frontal, central, and parietal electrodes was recorded during the task. The DCD group showed more performance variability than the TD group, particularly in the high presentation frequency condition. The younger children with DCD displayed less frontal-central coherence than their TD counterparts; however, this difference diminished with age: the older children demonstrated similar levels of alpha and beta activity compared to age-matched controls.

In the second study, children responded to a stimulus presented on screen with either predictable or unpredictable inter-stimulus intervals (Debrabant et al., 2013). Reaction times decreased and the proportion of anticipatory responses increased in the predictable compared to the unpredictable timing conditions for controls, and greater activation was recorded in the right middle and inferior frontal gyri for the unpredictable timing condition over the predictable timing condition. In contrast, the DCD group demonstrated no differences in either behavioral performance or BOLD activity between the two conditions. When the two groups were compared on activity differences between the two conditions, greater activation of the middle frontal gyrus, temporal-parietal junction (TPJ), and Crus I of the cerebellum was revealed in the control group compared to the DCD group during the unpredictable timing condition. These findings suggest that the reported difficulties in timing in DCD may be due to under-activation in frontal, parietal, and cerebellar areas.

Summary

While each of the six fMRI studies outlined report disparate findings in BOLD responses between children with and without DCD while they undertook motor tasks, there do appear to be some commonalities. Nevertheless, interpretation of these commonalities is problematic due to the difficulties of disentangling task-related activity from broader differences between groups, and because of the small sample sizes in each of the studies. In order to address these issues, Fuelscher and colleagues (2018) used a meta-analytic technique called activation likelihood estimation (ALE) to compare and summarize the overall activation patterns within each of these studies. This technique enables the authors to statistically establish which areas are reliably different between groups and across tasks. The authors found that the middle frontal gyrus, superior frontal gyrus, supramarginal gyrus, IPL, and parts of the cerebellum showed consistently lower activity in the DCD group when compared to controls. They also showed increased activity in thalamic areas in the DCD group. Aside from the cerebellum, none of the areas displaying reduced activity are directly involved in control of movement. They are, however, involved in higher functions related to planning and attention that do indirectly play a role in movement control. By combining the effects of multiple small studies to effectively boost the sample size to approximately eighty participants per group, this study provides the best evidence for the potential neural correlates of motor performance in DCD. However, while these findings are undoubtedly more robust than the findings of the individual studies included in the analysis, they should still be treated cautiously as they are likely to be affected by low power within the individual studies (Button et al., 2013; Nord et al., 2017) and publication bias (Jennings and Van Horn, 2012). Thus, rather than conclusive evidence for involvement of the aforementioned areas, this study should be treated as a robust starting point for future studies of neural correlates of motor performance in DCD.

Cognitive Hypotheses

Visual Perception

Based on evidence collected from behavioral tests of visual perception, this was an early factor thought to underlie the difficulties in DCD (Hulme, Smart, and Moran, 1982; Lord and Hulme, 1987, 1988). Mon-Williams and colleagues (1996) addressed this hypothesis by examining the EEG activity elicited by the presentation of visual stimuli (also known as visually evoked potentials (VEPs)). Although VEP amplitudes of early components for wider stimuli differed between DCD and control groups, the authors suggested that these differences were likely due to the presence of more noise in the recordings from the DCD group (due to inattention and movement). They concluded that low-level visual perception does not appear to play a role in the movement difficulties of DCD. These conclusions have been supported by other work investigating visual perception in DCD (Mon-Williams, Pascal, and Wann, 1994), and the ERP results have been replicated a number of times in subsequent EEG studies (e.g., Tsai et al., 2009). As noted by Mon-Williams and colleagues, however, these results only demonstrate that there are no deficits in low-level visual processing, and it may be that higher levels are affected.

Visuospatial Attention

One suggestion is that individuals with DCD display difficulties in visuospatial attention that may either cause or exacerbate the motor difficulties they experience (Wilson, Maruff, and McKenzie, 1997; Wilson and McKenzie, 1998). Tsai and colleagues have conducted a number of EEG studies to explore the neural basis of these reported attention deficits in DCD (Tsai et al., 2009, 2010; Tsai, Wang, and Tseng, 2012). All of these studies have examined ERPs associated with performance of a visuospatial cuing task. During this task a certain proportion of cues will correctly indicate the position of a subsequent target (valid), a proportion will incorrectly indicate the position of a subsequent target (invalid), and the rest will have no cue. The valid cues reduce the time taken for participants to respond to the target, while there is no difference between target response times for the invalid cue and no cue trials. In all of these studies the behavioral effects of cuing have been stable in both the TD and DCD groups, that is: the DCD group responded slower in all conditions and both groups benefited from the valid cues, but the benefit was smaller for the DCD group.

Tsai and colleagues have examined the mean amplitude and latency of ERP components across all of these studies, namely: N1, N2, and P3 following the target. N1 is an alternate name given to the previously described VEPs and represents early visual processing. N2 and P3 components recorded from these areas are thought to represent higher visual attention/processing and response selection respectively. Across studies, the latency of the P3 component was longer in the invalid condition and the DCD group displayed smaller amplitude and later P3 components, regardless of condition. N2 latency was also longer in the DCD group for two of the studies (Tsai et al., 2009, 2010). After participating in a 10-week exercise program (Tsai, Wang, and Tseng, 2012), children with DCD demonstrated improvements in motor performance and there appeared to be associated changes in P3 amplitude. However, analysis of post-intervention P3 amplitude revealed no differences between the intervention and two control groups. The authors did not report any further post-hoc tests, including comparisons between pre- and post-intervention P3 amplitude, making interpreting the effects of the training intervention on neural signatures of attention difficult.

This research group have also recently looked at oscillatory activity associated with attentional orienting (Wang et al., 2015). The gaze cuing task used in previous studies was repeated, and the behavioral results were largely the same (i.e., both groups benefited from valid cues, but the control group benefited more). Theta-band (4–7 Hz) EEG activity in frontal electrodes was analyzed, given the evidence that activity in this band is associated with spatial orienting. The TD group showed increased power in the theta band for the cued conditions, regardless of cue, compared to the non-cued condition. The DCD group did not show any difference in theta band power for any of the three conditions, suggesting that the source of the theta-band activity produced during spatial orienting may be dysfunctional in DCD. The authors also reported negative correlations between theta power and reaction times; however, these were not corrected for multiple comparisons and should be interpreted cautiously.

Executive Function

Recent evidence has suggested that individuals with DCD may have deficits in executive functioning, which are higher-order cognitive abilities used to regulate and control behavior (Diamond, 2013; see Leonard and Hill, 2014, for a review). Several studies have used neuroimaging to investigate the neural underpinning of this potential executive dysfunction, focusing specifically on visuospatial working memory (VSWM) and inhibitory control.

Two studies have used EEG to assess VSWM in DCD. Both studies used an experimental paradigm where children were asked to compare the positions of stimuli within a grey box (Tsai et al., 2012; Wang et al., 2017). Stimuli were either presented together or after a delay. In the first study, Tsai and colleagues examined the amplitude and latency of the P3 and the positive slow wave ((pSW); a component associated with information retrieval and response selection) components. The P3 amplitude in the response phase was smaller in the DCD group than in controls regardless of the condition, mirroring previously discussed results. Furthermore, the P3 amplitude was larger in the delay conditions than in the immediate condition for all groups, reflecting the reduced accuracy and longer response times after a delay. Finally, there was a significant interaction between group and condition for P3 amplitude, with the DCD group showing significantly smaller amplitudes in the delayed conditions than in the immediate condition, which related to less accurate and longer response times after a delay. The pSW amplitude was also significantly smaller in the DCD group, and in the immediate condition. The authors suggest that these results indicate that the DCD group allocate fewer neural resources during the stimulus evaluation and response selection phase, as indicated by the differences in P3 amplitude. In addition, the group differences in pSW illustrate a more general difficulty in the retrieval process phase for those with DCD.

In the second study, frontal theta band and parietal alpha band activity associated with VSWM was recorded in children with and without DCD (Wang et al., 2017). Frontal theta band activity showed a significant increase in the period directly after stimulus presentation in both conditions. While there were no group differences in the non-delayed condition, the DCD displayed lower activity in the delayed condition, reduced alpha suppression in the posterior region immediately prior to recall, and reduced frontal theta activity for a brief period after recall. These findings provide some support for the previous study, with the DCD children showing atypical activity during the retrieval stage. They also provide some more insight into differences into VSWM processing in DCD, with diminished activity during the encoding and maintenance phases.

Two studies have also used neuroimaging to investigate inhibitory control in children with DCD (Querne et al., 2008; Thornton et al., 2018), but have relied on fMRI to do so. Both studies used a Go-No-Go task to measure inhibition, in which participants must respond by pressing a button when they see one stimulus and withhold this response when they see another stimulus. The number of errors differed between DCD and control children (Querne et al., 2008; Thornton et al., 2018), although this was only the case for those with an additional diagnosis of attention deficit-hyperactivity disorder (ADHD) in the Thornton et al. study (cooccurring diagnoses were not controlled in the other study). Furthermore, Querne and colleagues reported significantly stronger connectivity between the inferior parietal cortex and both the middle frontal cortex and the anterior cingulate cortex in DCD compared to controls, which was bilateral, but more pronounced in the left hemisphere, and decreased connectivity between the right striatum and parietal cortex in children with DCD. However, Thornton and colleagues reported no significant differences between DCD and control groups in

BOLD activity during the task. From these studies, it therefore seems important to control for cooccurring diagnoses when investigating DCD, and this point will be returned to in a later section.

Internal Modelling

One of the main hypotheses grounded in models of motor control is that of the internal modelling deficit. This hypothesis suggests that the motor difficulties experienced by individuals with DCD are due to the inability of the motor system to generate accurate predictions about the consequences of planned actions, which affects its ability to update actions accordingly (Adams et al., 2014). Examining the internal model directly is difficult, but one of the accepted methods is through motor imagery, typically using a hand rotation task. This task asks participants to determine whether the hand presented on screen in a variety of rotations is a right or left hand.

Results have been mixed in relation to studies that have tried to investigate the neural underpinnings of this task. In terms of reaction times to respond to the stimulus, Lust and colleagues (2006) found no significant differences between children with and without DCD. Adult participants with probable DCD also demonstrated this pattern (Kashuk et al., 2017), while adults with a DCD diagnosis did perform worse than the control group overall (Hyde et al., 2018). These three studies used different techniques to assess neural activation during this task. Lust and colleagues reported no significant differences between groups in the amplitude or latency of the rotation-related negativity ERP component (observed in parietal electrodes 300–700 ms after stimulus onset), reflecting their behavioral results. Analysis of task-related fMRI activity by Kashuk and colleagues revealed that the control group showed significantly greater activity in the middle frontal gyrus bilaterally, the left superior parietal lobe, and lobule VI of the cerebellum as the task increased in difficulty, compared to the DCD group. Finally, Hyde and colleagues, using transcranial magnetic stimulation (TMS), revealed differences in the motor-cortical excitability of the two groups during the task. Specifically, the motor-cortical excitability significantly increased for the control group, but there was no change for the DCD group. This suggests that the primary motor cortex (M1) may not be engaged during motor imagery in the DCD group, providing a potential neural correlate for the internal modelling deficit hypothesis. The three studies taken together also highlight the importance of group choice (e.g., children or adults, those with a diagnosis or just screened for motor difficulties), and of the neuroimaging technique used to investigate the task at hand.

Imitation and the Mirror Neuron System (MNS)

Related to the internal modelling deficit is the mirror neuron system, a network of areas in the brain thought to be used when an individual observes, executes, imagines, or imitates an action (Aziz-Zadeh et al., 2006; Iacoboni and Dapretto, 2006). This network is thought to consist of the inferior frontal gyrus (IFG), ventral premotor cortex (vPM), IPL, and the superior temporal sulcus (STS). Alongside the behavioral evidence of motor imagery deficits in DCD, there is evidence that imitation is also impaired (Sinani, Sugden, and Hill, 2011; Elbasan, Kayıhan, and Duzgun, 2012; Reynolds, Kerrigan et al., 2017).

Two studies have been conducted to test this hypothesis, both conducted by Reynolds and colleagues (Reynolds et al., 2015; Reynolds, Billington et al., 2017) with children. The first study included three phases: action observation (viewing a finger sequencing task being executed), action execution (performing the task after viewing a still image of the firsthand stimulus), and action imitation (performing the task while viewing it being executed). Comparison of whole brain activity between groups for each of the conditions revealed more activation in the IFG, bilateral pre-central gyri, and left middle temporal gyrus in the DCD group compared to controls during the observation condition. There were, however, no group differences in the imitation and execution conditions. Region of interest analyses focused on those areas explicitly associated with the MNS revealed condition-specific differences in BOLD activity, and an interaction between condition and group at one site: the pars opercularis. This is located in the IFG and has been shown to be active during action observation and imitation. The control group displayed greater activation than the DCD group in this area during the imitation condition, suggesting that it may not be properly recruited during action imitation in DCD. However, these differences were not statistically significant after correction for multiple comparisons, and it is difficult to interpret the results without the corresponding behavioral data. The authors also note that the overall lack of difference in activity between the imitation and execution conditions for the whole brain analysis precludes a clear interpretation. They speculate that this lack of difference may be due to learning effects from practice of the finger sequencing task and may have masked true between-group differences in activity.

Consequently, Reynolds and colleagues conducted a follow-up study that utilized a modified version of a finger tapping task that has been successfully used in previous research into the MNS (Reynolds, Billington et al., 2017). This simply consisted of tapping the right index finger side-to-side in time with a stimulus, allowing little room for learning effects. As with the previous study, the task was divided into observation, execution, and imitation conditions, but also added a motor imagery condition. As before, BOLD activity was analyzed across the whole brain and in specific regions associated with the MNS, and only weak evidence was found for differences between the groups in the different conditions. Taken together, these studies seem to rule out deficits in the mirror neuron system as a potential explanation for the action imitation difficulties in DCD that have been reported.

Summary

Investigation of the neural correlates of the cognitive hypotheses has provided evidence that rules out the visual perception and mirror neuron hypotheses. The remaining hypotheses have mixed evidence supporting them at best, with the majority requiring more exploration to resolve discrepancies between behavioral and neuroscientific findings. More recent research has therefore turned to considering structural and functional connectivity in DCD, which is outlined further in the next section.

Structural and Functional Connectivity in DCD

White Matter Structure

Diffusion tensor imaging (DTI) uses the diffusion of water molecules in neural tissue to measure the integrity of axonal or white matter tracts in the brain. A number of diffusion indexes can be measured using DTI, but the two most commonly reported are diffusivity and anisotropy (see Soares et al., 2013 for an in-depth description of DTI). Both of these measures quantify how freely the water molecules in the white matter tracts can diffuse. Diffusivity provides a value for the magnitude of diffusion for each axis direction, although it is typically summarized by the mean diffusivity (MD): the mean magnitude of diffusivity across all three axes. Some studies also report axial (AD) and radial diffusivity (RD), which refer to the magnitude of diffusion along the tract and the magnitude of the diffusion in the two directions perpendicular to the tract, respectively. Anisotropy, or more specifically fractional anisotropy (FA), provides a normalized value for diffusivity across all three axis directions. Low values indicate free movement in all directions, while high values indicate that the molecules can only flow in one particular direction.

Three studies have explored the integrity of white matter tracts in DCD using DTI. Zwicker and colleagues (2012) examined the major motor, sensory, and cerebellar white matter pathways in sixteen children, seven with DCD and nine without. No significant differences in FA were reported between the groups for any of the tracts investigated. The DCD group did however show a lower MD than the control group in the corticospinal tract, seemingly driven by lower AD. Furthermore, participants' score on a standardized motor assessment showed a positive correlation with AD in the motor and sensory tracts. Langevin and colleagues (2014) examined different white matter tracts, focusing on the major inter-hemispheric and two intra-hemispheric pathways. Of the eighty-four children recruited for this study, only nine had DCD without any cooccurring disorders. Unlike Zwicker and colleagues, the authors reported no between-group differences in MD but did note lower FA in the DCD group in the posterior portion of the inter-hemispheric tract and the lateral portion of one of the intra-hemispheric tracts. They also reported a significant correlation between scores on a standardized motor assessment and the FA of the posterior portion of the inter-hemispheric tract in the DCD group.

Finally, Debrabrant and colleagues (2016) examined nineteen of the main white matter tracts in children with and without DCD. Of these tracts only the retrolenticular limb of the internal capsule showed differences in the DCD group. Bilaterally, the retrolenticular limb showed greater RD in the DCD group, while the right retrolenticular limb also showed lower FA. Furthermore, in both groups, scores on a standardized test of visuomotor integration were positively correlated with FA in the left retrolenticular limb. Debrabrant and colleagues also constructed a structural connectivity network and examined differences in typical network metrics. The DCD group had lower mean clustering coefficients and global efficiency, suggesting that individuals with DCD have poorer specialization of neural areas and weaker information transfer across the whole brain. Global efficiency was positively correlated with scores on the assessment of visual-motor skill, suggesting that the poorer information transfer across the brain may underlie the core features of DCD.

Together, these studies give some indication that differences in white matter integrity may relate to the core features of DCD, particularly given that all three demonstrate correlations between structural measures and behavioral outcomes. Nonetheless, the lack of consistency in the findings of the three studies, both in terms of areas and measures affected, means that no clear link between DCD symptoms and specific white matter deficits can be drawn.

Grey Matter Structure

Examining grey matter presents an alternative way to explore potential structural correlates of neurodevelopmental disorders such as DCD. This measure gives an indication of the degree of synaptic connectivity in an area; more grey matter suggests greater connectivity and vice versa. The most common way of exploring this is to examine differences in the thickness of the grey matter in different regions of the brain.

Thus far, two studies have used this approach to explore differences in this between DCD and neurotypical individuals. Of the twenty-eight cortical areas explored by Langevin and colleagues (2015), the group of children with DCD only showed thinning in the right temporal pole. The authors also reported several correlations between thickness of grey matter in particular regions and performance on behavioral tasks. These include correlations between the right caudal middle frontal cortex and an inhibition switching task, between the left precentral cortex and a response set task, and between the left entorhinal cortex and an auditory attention task. Strangely, they found no structural-behavioral correlations for a task assessing motor performance, which would be expected for this group. Caeyenberghs and colleagues (2016) reported a difference between children with DCD and controls in the clustering coefficient in the lateral orbitofrontal cortex indicating increased connectivity within that region. However, as the authors make clear, these results should be treated with caution as they were exploratory and not corrected for multiple comparisons. Finally, Reynolds, Licari, and colleagues (2017) measured the volume of grey matter in a particular area rather than just the thickness, thus taking into account differences in surface area and cortical folding. There were no significant differences in overall grey matter, white matter, or total intracranial volumes between groups of children with and without DCD. However, the authors did report that, compared to the control group, the DCD group had smaller relative grey matter volume in the right superior frontal and middle frontal gyri.

Functional Connectivity

Alongside exploring the structural correlates of DCD, it is also possible to explore the functional connectivity using the BOLD signal when the participant is not undertaking a specific task ("resting-state activity"). Resting state activity is examined to see if fluctuations in the BOLD signal in different cortical areas correlate with one another, indicating the degree to which areas are functionally connected with each other.

Only two studies have looked at functional connectivity in children with DCD (McLeod et al., 2014, 2016). The first found that, compared to controls, the DCD group exhibited reduced functional connectivity between the M1 and a number of cortical areas. These included: the right frontal operculum cortex, right supramarginal gyrus, bilateral insular cortices, superior temporal gyri, bilateral caudate, right nucleus accumbens, pallidum, and putamen. The latter study examined which areas showed stronger functional connectivity in the right and left sensorimotor areas, respectively. The DCD group did not show the stronger functional connections between the right sensorimotor areas, and thalamic and cerebellar areas seen in controls. Instead, the opposite was revealed: stronger connections between the left sensorimotor areas and thalamic and cerebellar areas. The authors suggest that these results indicate a lack of hemispheric dominance in DCD, which may explain the bimanual coordination deficits seen in DCD.

Using TMS, He and colleagues (2018) assessed the intra-cortical inhibition and inter-hemispheric inhibition of M1 in young adults with and without DCD. This method provides a direct way of examining functional connectivity within and between the motor cortices. They found no difference in the intra-hemispheric measures. The DCD group did, however, have reduced inter-hemispheric inhibition compared to controls, and there was an association between inter-hemispheric inhibition and manual dexterity in a standardized motor battery. This suggests that the weaker inter-cortical inhibition observed may underlie bimanual coordination difficulties in DCD. However, the authors did not report whether they screened for ADHD in their sample, which presents a potential confounding factor. Cooccurrence of ADHD and DCD is relatively common, and several studies have reported that inter-hemispheric inhibition in individuals with ADHD is reduced (Richter et al., 2007; Schneider et al., 2007).

Summary

Overall, the studies presented exploring structural and functional connectivity in DCD have provided some evidence for broader differences in the structure and connectivity of the brain in DCD. Nonetheless, the support for any one specific cause is weak. Many of the studies presented have small sample sizes, meaning that the research does not have the statistical power to detect anything but the largest differences between groups. Furthermore, many of the studies presented are exploratory, yet they do not explicitly state if and how inflation of Type 1 errors from multiple comparisons was controlled. Further considerations of limitations of the current research and future directions will be discussed in the next section.

CURRENT ISSUES

Within the literature outlined in the previous sections there are a number of issues that should be addressed in order to strengthen research into the neural underpinnings of

DCD. This section will highlight two of the most pressing issues and briefly discuss their ramifications.

Small Sample Sizes

A key issue in the broader field of cognitive neuroscience that has emerged over the last decade is that of the small sample sizes used within imaging studies. In the literature outlined in this chapter, the majority of studies do not exceed a cell size of twenty participants and none exceed forty. This only gives these studies the ability to detect the largest of differences and also renders many of the brain-behavior correlations difficult to interpret (Yarkoni, 2009).

In addition to the more general reasons for small sample sizes in cognitive neuroscience research (i.e., the time and resource cost of recruiting large numbers of participants), DCD-specific recruitment issues can also affect sample size. Despite the fact that the estimated prevalence of DCD in the general population makes it one of the more common developmental disorders (Bishop, 2010), the disorder itself is comparatively unknown amongst parents and teachers (Piek and Edwards, 1997). Consequently, the lack of familiarity with DCD in the public makes recruitment of large samples difficult. Furthermore, the aforementioned high cooccurrence between DCD and ADHD and other disorders can affect recruitment. The majority of studies outlined in this chapter aim to recruit samples without cooccurrences, which can lead to a dramatic reduction in the number of potential participants. Finally, subclinical symptoms of other disorders, such as increased tactile or auditory sensitivity, or inattention/hyperactivity, can affect the amount and quality of data that can be recorded through neuroimaging methods with individuals with DCD, meaning that final sample sizes can be reduced.

Age Range of Samples

A second issue within the DCD literature discussed relates to the age of the samples: currently the range of ages examined is relatively narrow (mainly 8–12 years old). Ensuring that the age range of the sample collected for a particular study is constrained allows for more robust conclusions to be drawn about neural development within DCD at that time point, particularly with small sample sizes. However, the limited range of ages examined within the literature becomes a limiting factor when, as is the case currently, the majority of studies focus on a single relatively narrow age range. This limits the understanding of the broader trajectory of neuromotor development, the atypical trajectory of DCD, where these trajectories begin to diverge, and what happens to these trajectories in adulthood.

For adult samples, the main limiting factor is the lack of reliable tools for identifying DCD in adulthood. Specifically, none of the widely used motor ability assessments are standardized beyond 21 years (Hands, Licari, and Piek, 2015), making fulfilment of diagnostic criterion A difficult. For children under 5 years, there is variability in motor development, rate of acquisition of activities of daily living, motivation, and cooperation of children that makes giving a formal diagnosis unreliable in all but the most severe cases (Blank et al., 2012). This is compounded by the relative difficulty in collecting neural data from children below the age of 7 (Poldrack, Paré-Blagoev, and Grant, 2002; Wilke et al., 2003; Yerys et al., 2009; Raschle et al., 2012). Potential solutions to this and other problems are discussed further in the following section.

POTENTIAL SOLUTIONS AND FUTURE DIRECTIONS

Increasing Sample Sizes

Bringing DCD closer to the fore in public and professional consciousness is ultimately the solution to the problem of recruiting larger sample sizes. However, this will not happen overnight and so short- and medium-term solutions are needed. One such solution is establishing multi-lab studies. This approach would spread the burden of collecting and testing samples across multiple laboratories (potentially across multiple countries), reducing the need for a single lab to recruit many participants alone. Using this approach to establish an OpenfMRI-type database (Poldrack et al., 2013) of imaging data would benefit DCD research in numerous ways.

First, it would allow for the collection and analysis of samples with broader age ranges than typically reported in the literature, without the drawback of blurring findings due to age-related variability. Second, a large sample would allow for subgroups within DCD to be further explored, such as those with different cooccurring disorders, or those with different profiles of motor difficulties. Third, it would allow replication of some of the previously described explorations of the neural correlates of DCD with a much larger dataset. Finally, it would provide a resource from which future hypotheses could be generated and tested without the need for more data collection.

Broadening the Developmental Scope of the Research

As discussed, the suggestions for increasing the sample sizes of DCD research would also allow for the broadening of age ranges included as part of that research. However, this would only extend the age range tested to within the broader area of certainty of diagnosis (from 8–12 up to 7–17). A more comprehensive solution that addresses the neural correlates of DCD in adults and children younger than 7 is a more complicated prospect.

For adults, a medium-term solution lies in a more longitudinal approach, that is: identifying children with DCD (for which the tools are well established), and then following them up in adulthood to see if motor difficulties remain. This would also allow exploration of neural features that may contribute to the persistence of DCD into adulthood. However, this is likely to require considerable time and resources to achieve, especially as a large initial sample will need to be recruited in order to account for drop-outs and resolutions.

Refining the currently available tests to reliably identify children below the age of 5 with motor difficulties would also be of use, but, as noted, the variability in motor ability below 5 may make this task impossible. Once again, a longitudinal approach provides an alternate solution through the use of "at-risk" recruitment. That is, for neurodevelopmental disorders, there are certain factors that put an individual at a higher risk of developing the disorder than others. Two examples specific to DCD include: being related to someone who already has DCD, and premature birth (Martin, Piek, and Hay, 2006; Edwards et al., 2011). Identifying and testing infants and toddlers who are at risk of DCD and following them up at an age when a diagnosis could be properly established would provide a window into the early neural motor development of DCD. Furthermore, given that it is likely that a proportion of these children will not develop DCD, this approach could also potentially provide insight into specific neural factors that influence persistence.

Nonetheless, the difficulty of collecting usable data from younger children with DCD also needs to be addressed, and to accomplish this we have to turn to other techniques for collecting neural data. Functional near-infrared spectroscopy (fNIRS) presents an alternative to EEG and fMRI. This technique is able to record BOLD activity from shallow cortical areas without the need for a magnetic resonance imaging (MRI) scanner. In addition, because it consists of head mounted sensors (much like EEG), it is much less affected by head movements than either fMRI or EEG. Indeed, it has been successfully used in studies that look at the BOLD responses in infants (Nishiyori, 2016). fNIRS is not without its limitations, however, the most notable of which is the poorer spatial resolution compared to fMRI: it is unable to provide the direct link between activity and structure available from fMRI. Despite this, combining a technique such as fNIRS with an at-risk recruitment approach provides an excellent way of examining the neural correlates of DCD in younger children.

Future Techniques

In addition to fNIRS there are a number of other neuroscientific techniques that could be employed to explore DCD further. Here we will discuss two: brain stimulation and magnetic resonance spectroscopic imaging (MRSI).

Brain stimulation techniques are already being used to explore the potential involvement of the motor cortex in DCD (e.g., He et al., 2018; Hyde et al., 2018). There is the scope, however, to expand into using TMS to examine the involvement of other cortical areas, especially as TMS can be used to establish a causal relationship between a specific behavior and an area of the cortex (Walsh and Cowey, 2000). Furthermore, forms of transcranial electric stimulation have been used to alter the excitability of areas of the cortex to modulate behavioral performance in a range of experimental paradigms and with a variety of participants (Kuo and Nitsche, 2012).

Similarly, MRSI is another powerful technique that could be used in the future. MRSI uses the same principles as MRI but, instead of recording the relaxation of water molecules in the body, it can detect signals from a number of other molecules, including neurotransmitters. This potentially allows for a more specific understanding of the neural underpinnings of DCD; rather than simply concluding that a specific brain area is involved, it may be possible to implicate a specific type of neuron based on differences in neurotransmitter concentrations. An excellent example of this is work conducted by Stagg and colleagues looking at the role of GABA in M1 and motor learning (Stagg, Bachtiar, and Johansen-Berg, 2011; Stagg, 2014). However, given that this technique can detect concentrations of multiple molecules in a single scan, it adds another dimension to already complex MRI data sets. Without strong, theory driven hypotheses for the involvement of a particular neurotransmitter, this increases the risk of false positives. Based on the findings outlined here, this theoretical basis does not yet exist for DCD.

Conclusions

While some of the studies presented in this chapter have provided evidence to rule out certain hypotheses, overall, there is no clear consensus about the neural underpinnings of DCD from the current literature. There is, however, some evidence for the involvement of a number of areas, including: the cerebellum, the basal ganglia, the parietal lobe, and parts of the frontal lobe. In addition, there is speculative evidence for the involvement of white matter pathways in the disorder. A clearer picture of the developmental cognitive neuroscience of DCD is unlikely to emerge unless some of the major issues in the field are addressed by future research. Sharing data to increase the size of datasets, taking a developmental perspective, and making use of a range of new methods as part of a theory-led approach will be key to developing our understanding of the underlying mechanisms involved in DCD. Given the prevalence of the disorder and its impact on both motor and non-motor domains, this could have a significant impact on the lives of those with DCD, as well as on the knowledge of researchers and practitioners.

References

Adams, I. L. J. et al. (2014). Compromised motor control in children with DCD: A deficit in the internal model?—A systematic review, *Neuroscience & Biobehavioral Reviews*, 47, 225–244. doi: 10.1016/j.neubiorev.2014.08.011.

American Psychiatric Association (2013). *Diagnostic and Statistical Manual of Mental Disorders, Fifth Edition*. Washington, DC: American Psychiatric Publishing.

Aziz-Zadeh, L. et al. (2006). Lateralization of the human mirror neuron system, *Journal of Neuroscience*, 26(11), 2964–2970. doi: 10.1523/JNEUROSCI.2921–05.2006.

Biotteau, M. et al. (2017). Neural changes associated to procedural learning and automatization process in developmental coordination disorder and/or developmental dyslexia, *European Journal of Paediatric Neurology*, 21(2), 286–299. doi: 10.1016/j.ejpn.2016.07.025.

Bishop, D. V. M. (2010). Which neurodevelopmental disorders get researched and why?, *PLoS ONE*, 5(11), e15112. doi: 10.1371/journal.pone.0015112.

Blank, R. et al. (2012). European Academy for Childhood Disability (EACD): Recommendations on the definition, diagnosis and intervention of developmental coordination disorder (long version), *Developmental Medicine and Child Neurology*, 54(1), 54–93. doi: 10.1111/j.1469–8749.2011.04171.x.

Button, K. S. et al. (2013). Power failure: why small sample size undermines the reliability of neuroscience, *Nature Reviews Neuroscience*, 14(5), 365–376. doi: 10.1038/nrn3475.

Caeyenberghs, K. et al. (2016). Neural signature of developmental coordination disorder in the structural connectome independent of comorbid autism, *Developmental Science*, 19(4), 599–612. doi: 10.1111/desc.12424.

Cairney, J. et al. (2005). Developmental coordination disorder, generalized self-efficacy toward physical activity, and participation in organized and free play activities, *The Journal of Pediatrics*, 147(4), 515–520. doi: 10.1016/j.jpeds.2005.05.013.

de Castelnau, P. et al. (2008). A study of EEG coherence in DCD children during motor synchronization task, *Human Movement Science*, 27(2), 230–241. doi: 10.1016/j.humov.2008.02.006.

Cousins, M. and Smyth, M. M. (2003). Developmental coordination impairments in adulthood, *Human Movement Science*, 22(4–5), 433–459.

Debrabant, J. et al. (2013). Neural underpinnings of impaired predictive motor timing in children with developmental coordination disorder, *Research in Developmental Disabilities*, 34(5), 1478–1487. doi: 10.1016/j.ridd.2013.02.008.

Debrabant, J. et al. (2016). Brain connectomics of visual-motor deficits in children with developmental coordination disorder, *The Journal of Pediatrics*, 169, 21–7.e2. doi: 10.1016/j.jpeds.2015.09.069.

Diamond, A. (2013). Executive functions, *Annual Review of Psychology*, 64, 135–168. doi: 10.1146/annurev-psych-113011-143750.

Edwards, J. et al. (2011). Developmental coordination disorder in school-aged children born very preterm and/or at very low birth weight: A systematic review, *Journal of Developmental & Behavioral Pediatrics*, 32(9), 678–687. doi: 10.1097/DBP.0b013e31822a396a.

Elbasan, B., Kayıhan, H., and Duzgun, I. (2012) Sensory integration and activities of daily living in children with developmental coordination disorder, *Italian Journal of Pediatrics*, 38, 14. doi: 10.1186/1824-7288-38-14.

Faught, B. E. et al. (2005) Increased risk for coronary vascular disease in children with developmental coordination disorder, *Journal of Adolescent Health*, 37(5), 376–380. doi: 10.1016/j.jadohealth.2004.09.021.

Fuelscher, I. et al. (2018). Differential activation of brain areas in children with developmental coordination disorder during tasks of manual dexterity: An ALE meta-analysis, *Neuroscience & Biobehavioral Reviews*, 86, 77–84. doi: 10.1016/j.neubiorev.2018.01.002.

Geuze, R. H. (2003). Static balance and developmental coordination disorder, *Human Movement Science*, 22(4–5), 527–548. doi: 10.1016/j.humov.2003.09.008.

Green, D. et al. (2009). Impairment in movement skills of children with autistic spectrum disorders, *Developmental Medicine and Child Neurology*, 51(4), 311–316. doi: 10.1111/j.1469–8749.2008.03242.x.

Gubbay, S. S. et al. (1965). Clumsy children: A study of apraxia and agnosic deficits in 21 children, *Brain*, 88, 295–312.

Hands, B., Licari, M. and Piek, J. (2015). A review of five tests to identify motor coordination difficulties in young adults, *Research in Developmental Disabilities*, 41–2, 40–51. doi: 10.1016/j.ridd.2015.05.009.

He, J. L. et al. (2018). Interhemispheric cortical inhibition is reduced in young adults with developmental coordination disorder, *Frontiers in Neurology*, 9, 179–191. dci: 10.3389/fneur.2018.00179.

Hulme, C., Smart, A., and Moran, G. (1982). Visual perceptual deficits in clumsy children, *Neuropsychologia*, 20(4), 475–481. doi: 10.1016/0028-3932(82)90046-X.

Hyde, C. et al. (2018). Corticospinal excitability during motor imagery is reduced in young adults with developmental coordination disorder, *Research in Developmental Disabilities*, 72, 214–224. doi: 10.1016/j.ridd.2017.11.009.

Iacoboni, M. and Dapretto, M. (2006). The mirror neuron system and the consequences of its dysfunction, *Nature Reviews Neuroscience*, 7(12), 942–951. doi: 10.1038/nrn2024.

Jennings, R. G. and Van Horn, J. D. (2012). Publication bias in neuroimaging research: Implications for meta-analyses, *Neuroinformatics*, 10(1), 67–80. doi: 10.1007/s12021-011-9125-y.

Kaplan, B. et al. (2006). Comorbidity, co-occurrence, continuum: What's in a name?, *Child: Care, Health and Development*, 32(6), 72331. doi: 10.1111/j.1365–2214.2006.00689.x.

Kashiwagi, M. et al. (2009). Parietal dysfunction in developmental coordination disorder: A functional MRI study, *NeuroReport*, 20(15), 1319–1324. doi: 10.1097/WNR.0b013e32832f4d87.

Kashuk, S. R. et al. (2017). Diminished motor imagery capability in adults with motor impairment: An fMRI mental rotation study, *Behavioural Brain Research*, 334, 86–96. doi: 10.1016/j.bbr.2017.06.042.

Kirby, A., Sugden, D., and Purcell, C. (2014). Diagnosing developmental coordination disorders, *Archives of Disease in Childhood*, 99(3), 292–296. doi: 10.1136/archdischild-2012-303569.

Kuo, M.-F. and Nitsche, M. A. (2012). Effects of transcranial electrical stimulation on cognition, *Clinical EEG and Neuroscience*, 43(3), 192–199. doi: 10.1177/1550059412444975.

Langevin, L. M. et al. (2014). Common white matter microstructure alterations in pediatric motor and attention disorders, *The Journal of Pediatrics*, 164(5), 1157–64.e1. doi: 10.1016/j.jpeds.2014.01.018.

Langevin, L. M., Macmaster, F. P., and Dewey, D. (2015). Distinct patterns of cortical thinning in concurrent motor and attention disorders, *Developmental Medicine and Child Neurology*, 57(3), 257–264. doi: 10.1111/dmcn.12561.

Leonard, H. C. et al. (2015). Executive functioning, motor difficulties, and developmental coordination disorder, *Developmental Neuropsychology*, 40(4), 201–215. do.: 10.1080/87565641.2014.997933.

Leonard, H. C. and Hill, E. L. (2014). Review: The impact of motor development on typical and atypical social cognition and language: A systematic review, *Child and Adolescent Mental Health*, 19(3), 163–170. doi: 10.1111/camh.12055.

Leonard, H. C. and Hill, E. L. (2015). Executive difficulties in developmental coordination disorder: Methodological issues and future directions, *Current Developmental Disorders Reports*, 2(2), 141–149. doi: 10.1007/s40474-015-0044-8.

Licari, M. K. et al. (2015). Cortical functioning in children with developmental coordination disorder: A motor overflow study., *Experimental Brain Research*, 233(6), 1703–1710. doi: 10.1007/s00221-015-4243-7.

Licari, M. and Larkin, D. (2008) Increased associated movements: Influence of attention deficits and movement difficulties, *Human Movement Science*. (Developmental Coordination Disorder), 27(2), 310–324. doi: 10.1016/j.humov.2008.02.013.

Licari, M., Larkin, D., and Miyahara, M. (2006). The influence of developmental coordination disorder and attention deficits on associated movements in children, *Human Movement Science*. (Approaches to Sensory-Motor Development in Infants and Children), 25(1), 90–99. doi: 10.1016/j.humov.2005.10.012.

Lingam, R. et al. (2009). Prevalence of developmental coordination disorder using the DSM-IV at 7 years of age: A UK population-based study, *Pediatrics*, 123(4), e693–e700. doi: 10.1542/peds.2008-1770.

Lord, R. and Hulme, C. (1987). Perceptual judgements of normal and clumsy children, *Developmental Medicine and Child Neurology*, 29(2), 250–257. doi: 10.1111/j.1469-8749.1987.tb02143.x.

Lord, R. and Hulme, C. (1988). Visual perception and drawing ability in clumsy and normal children., *British Journal of Developmental Psychology*, 6(1), 1–9. doi: 10.1111/j.2044-835X.1988.tb01075.x.

Lust, J. M. et al. (2006). An EEG study of mental rotation-related negativity in children with developmental coordination disorder, *Child: Care, Health and Development*, 32(6), 649–663. doi: 10.1111/j.1365-2214.2006.00683.x.

Martin, N. C., Piek, J. P., and Hay, D. (2006). DCD and ADHD: A genetic study of their shared aetiology, *Human Movement Science*. (Approaches to Sensory-Motor Development in Infants and Children), 25(1), 110–124. doi: 10.1016/j.humov.2005.10.006.

McLeod, K. R. et al. (2014) Functional connectivity of neural motor networks is disrupted in children with developmental coordination disorder and attention-deficit/hyperactivity disorder, *NeuroImage: Clinical*, 4, 566–575. doi: 10.1016/j.nicl.2014.03.010.

McLeod, K. R. et al. (2016). Atypical within- and between-hemisphere motor network functional connections in children with developmental coordination disorder and attention-deficit/hyperactivity disorder, *NeuroImage: Clinical*, 12, 157–164. doi: 10.1016/j.nicl.2016.06.019.

Miller, L. T. et al. (2001). Clinical description of children with developmental coordination disorder, *Canadian Journal of Occupational Therapy*, 68(1), 5–15. doi: 10.1177/000841740106800101.

Miyahara, M. and Piek, J. (2006). Self-esteem of children and adolescents with physical disabilities: Quantitative evidence from meta-analysis, *Journal of Developmental and Physical Disabilities*, 18(3), 219–234. doi: 10.1007/s10882-006-9014-8.

Mon-Williams, M. A. et al. (1996). Visual evoked potentials in children with developmental coordination disorder, *Ophthalmic and Physiological Optics*, 16(2), 178–183. doi: 10.1046/j.1475-1313.1996.95000364.x.

Mon-Williams, M., Pascal, E., and Wann, J. P. (1994). Ophthalmic factors in developmental coordination disorder, *Adapted Physical Activity Quarterly*, 11(2), 170–178.

Nicolson, R. I. et al. (1999). Association of abnormal cerebellar activation with motor learning difficulties in dyslexic adults, *Lancet*, 353(9165), 1662–1667. dci: 10.1016/S0140-6736(98)09165-X.

Nishiyori, R. (2016). fNIRS: An emergent method to document functional cortical activity during infant movements, *Frontiers in Psychology*, 7, 533–544. doi: 10.3389/fpsyg.2016.00533.

Nord, C. L. et al. (2017). Power-up: A reanalysis of 'power failure' in neuroscience using mixture modeling, *Journal of Neuroscience*, 37(34), 8051–8061. doi: 10.1523/jneurosci.3592-16.2017.

Pangelinan, M. M., Hatfield, B. D., and Clark, J. E. (2013). Differences in movement-related cortical activation patterns underlying motor performance in children with and without developmental coordination disorder, *Journal of Neurophysiology*, 109(12), 3041–3050. doi: 10.1152/jn.00532.2012.

Piek, J. et al. (2007). Depressive symptomatology in child and adolescent twins with attention-deficit hyperactivity disorder and/or developmental coordination disorder, *Twin Research and Human Genetics*, 10(4), 587–596. doi: 10.1375/twin.10.4.587.

Piek, J. P. and Edwards, K. (1997). The identification of children with developmental coordination disorder by class and physical education teachers, *British Journal of Educational Psychology*, 67(1), 55–67. doi: 10.1111/j.2044–8279.1997.tb01227.x.

Poldrack, R. A. et al. (2013). Toward open sharing of task-based fMRI data: The OpenfMRI project, *Frontiers in Neuroinformatics*, 7, 12–24. doi: 10.3389/fninf.2013.00012.

Poldrack, R. A., Paré-Blagoev, E. J., and Grant, P. E. (2002). Pediatric functional magnetic resonance imaging: Progress and challenges, *Topics in Magnetic Resonance Imaging*, 13(1), 61.

Pratt, M. L. and Hill, E. L. (2011). Anxiety profiles in children with and without developmental coordination disorder, *Research in Developmental Disabilities*, 32(4), 1253–1259. doi: 10.1016/j.ridd.2011.02.006.

Querne, L. et al. (2008). Dysfunction of the attentional brain network in children with developmental coordination disorder: A fMRI study, *Brain Research*, 1244, 89–102. doi: 10.1016/j.brainres.2008.07.066.

Raschle, N. et al. (2012). Pediatric neuroimaging in early childhood and infancy: Challenges and practical guidelines, *Annals of the New York Academy of Sciences*, 1252, 43–50. doi: 10.1111/j.1749–6632.2012.06457.x.

Reynolds, J. E. et al. (2015). Mirror neuron activation in children with developmental coordination disorder: A functional MRI study, *International Journal of Developmental Neuroscience*, 47, 309–319. doi: 10.1016/j.ijdevneu.2015.10.003.

Reynolds, J. E., Billington, J. et al. (2017). Mirror neuron system activation in children with developmental coordination disorder: A replication functional MRI study, *Research in Developmental Disabilities*, 84, 16–27. doi: 10.1016/j.ridd.2017.11.012.

Reynolds, J. E., Kerrigan, S. et al. (2017). Poor imitative performance of unlearned gestures in children with probable developmental coordination disorder, *Journal of Motor Behavior*, 49(4), 378–387. doi: 10.1080/00222895.2016.1219305.

Reynolds, J. E., Licari, M. K. et al. (2017). Reduced relative volume in motor and attention regions in developmental coordination disorder: A voxel-based morphometry study, *International Journal of Developmental Neuroscience*, 58, 59–64. doi: 10.1016/j.ijdevneu.2017.01.008.

Richter, M. M. et al. (2007). Cortical excitability in adult patients with attention-deficit/hyperactivity disorder (ADHD), *Neuroscience Letters*, 419(2), 137–141. doi: 10.1016/j.neulet.2007.04.024.

Rosenblum, S. and Regev, N. (2013). Timing abilities among children with developmental co-ordination disorders (DCD) in comparison to children with typical development, *Research in Developmental Disabilities*, 34(1), 218–227. doi: 10.1016/j.ridd.2012.07.011.

Schneider, M. et al. (2007). Impaired cortical inhibition in adult ADHD patients: A study with transcranial magnetic stimulation, *Journal of Neural Transmission. Supplementum*, (72), 303–309.

Sinani, C., Sugden, D., and Hill, E. L. (2011). Gesture production in school vs. clinical samples of children with developmental coordination disorder (DCD) and typically developing children, *Research in Developmental Disabilities*, 32(4), 1270–1282. doi: 10.1016/j.ridd.2011.01.030.

Skinner, R. A. and Piek, J. (2001). Psychosocial implications of poor motor coordination in children and adolescents, *Human Movement Science*, 20(1–2), 73–94. doi: 10.1016/S0167-9457(01)00029-X.

Soares, J., Marques, P., Alves, V., and Sousa, N. (2013). A hitchhiker's guide to diffusion tensor imaging. *Frontiers in Neuroscience*, 7, 31. doi: 10.3389/fnins.2013.00031

Soares, J. M., et al. (2016). A hitchhiker's guide to functional magnetic resonance imaging. *Frontiers in neuroscience*, 10, 515. doi: 10.3389/fnins.2016.00515

Stagg, C. J. (2014). Magnetic resonance spectroscopy as a tool to study the role of GABA in motor-cortical plasticity, *NeuroImage*, 86, 19–27. doi: 10.1016/j.neuroimage.2013.01.009.

Stagg, C. J., Bachtiar, V., and Johansen-Berg, H. (2011). The role of GABA in human motor learning, *Current Biology*, 21(6), 480–484. doi: 10.1016/j.cub.2011.01.069.

Summers, J., Larkin, D., and Dewey, D. (2008). Activities of daily living in children with developmental coordination disorder: Dressing, personal hygiene, and eating skills, *Human Movement Science*, 27(2), 215–229. doi: 10.1016/j.humov.2008.02.002.

Thornton, S. et al. (2018). Functional brain correlates of motor response inhibition in children with developmental coordination disorder and attention deficit/hyperactivity disorder, *Human Movement Science*, 59, 134–142. doi: 10.1016/j.humov.2018.03.018.

Tsai, C.-L. et al. (2009). Mechanisms of deficit of visuospatial attention shift in children with developmental coordination disorder: A neurophysiological measure of the endogenous Posner paradigm, *Brain and Cognition*, 71(3), 246–258. doi: 10.1016/j.bandc.2009.08.006.

Tsai, C.-L. et al. (2010). Deficits of visuospatial attention with reflexive orienting induced by eye-gazed cues in children with developmental coordination disorder in the lower extremities: An event-related potential study, *Research in Developmental Disabilities*, 31(3), 642–655. doi: 10.1016/j.ridd.2010.01.003.

Tsai, C.-L. et al. (2012). The neurophysiological performance of visuospatial working memory in children with developmental coordination disorder. *Developmental Medicine & Child Neurology*, 54(12), 1114–1120. doi: 10.1111/j.1469-8749.2012.04408.x.

Tsai, C.-L., Wang, C.-H., and Tseng, Y.-T. (2012). Effects of exercise intervention on event-related potential and task performance indices of attention networks in children with developmental coordination disorder, *Brain and Cognition*, 79(1), 12–22. doi: 10.1016/j.bandc.2012.02.004.

Walsh, V. and Cowey, A. (2000). Transcranial magnetic stimulation and cognitive neuroscience. *Nature Reviews Neuroscience*, 1, 73–80. doi: 10.1038/35036239.

Wang, C.-H. et al. (2015). Frontal midline theta as a neurophysiological correlate for deficits of attentional orienting in children with developmental coordination disorder: Theta oscillation, attentional orienting, and DCD, *Psychophysiology*, 52(6), 80112. doi: 10.1111/psyp.12402.

Wang, C.-H. et al. (2017). Neural oscillation reveals deficits in visuospatial working memory in children with developmental coordination disorder, *Child Development*, 88(5), 1716–1726. doi: 10.1111/cdev.12708.

Watson, L. and Knott, F. (2006). Self-esteem and coping in children with developmental coordination disorder, *The British Journal of Occupational Therapy*, 69(10), 450–456. doi: 10.1177/030802260606901003.

Wilke, M. et al. (2003). Functional magnetic resonance imaging in pediatrics, *Neuropediatrics*, 34(5), 225–233. doi: 10.1055/s-2003-43260.

Wilson, P. et al. (2012). Understanding performance deficits in developmental coordination disorder: A meta-analysis of recent research, *Developmental Medicine and Child Neurology*, 55(3), 20–23. doi: 10.1111/j.1469-8749.2012.04436.x.

Wilson, P. H. et al. (2017). Cognitive and neuroimaging findings in developmental coordination disorder: New insights from a systematic review of recent research, *Developmental Medicine and Child Neurology*, 59(11), 1117–1129. doi: 10.1111/dmcn.13530.

Wilson, P. H. and McKenzie, B. E. (1998). Information processing deficits associated with developmental coordination disorder: A meta-analysis of research findings, *Journal of Child Psychology and Psychiatry, and Allied Disciplines*, 39(6), 829–840.

Wilson, P., Maruff, P., and McKenzie, B. E. (1997). Covert orienting of visuospatial attention in children with developmental coordination disorder, *Developmental Medicine and Child Neurology*, 39(11), 736–745. doi: 10.1111/j.1469-8749.1997.tb07375.x.

Wocadlo, C. and Rieger, I. (2008). Motor impairment and low achievement in very preterm children at eight years of age, *Early Human Development*, 84(11), 769–776. doi: 10.1016/j.earlhumdev.2008.06.001.

Yarkoni, T. (2009). Big correlations in little studies: Inflated fMRI correlations reflect low statistical power—commentary on Vul et al. (2009), *Perspectives on Psychological Science*, 4(3), 294–8. doi: 10.1111/j.1745-6924.2009.01127.x.

Yerys, B. E. et al. (2009). The fMRI success rate of children and adolescents: Typical development, epilepsy, attention deficit/hyperactivity disorder, and autism spectrum disorders, *Human Brain Mapping*, 30(10), 3426–3435. doi: 10.1002/hbm.20767.

Zwicker, J. G. et al. (2010). Brain activation of children with developmental coordination disorder is different than peers, *Pediatrics*, 126(3), e678–e686. doi: 10.1542/peds.2010-0059.

Zwicker, J. G. et al. (2011). Brain activation associated with motor skill practice in children with developmental coordination disorder: An fMRI study, *International Journal of Developmental Neuroscience*, 29(2), 145–152. doi: 10.1016/j.ijdevneu.2010.12.002.

Zwicker, J. G. et al. (2012). Developmental coordination disorder: A pilot diffusion tensor imaging study, *Pediatric Neurology*, 46(3), 162–167. doi: 10.1016/j.pediatrneurol.2011.12.007.

CHAPTER 13

SLEEP DEVELOPMENT IN INFANCY AND CHILDHOOD

HANNAH CORCORAN, FALLON COOKE, AND AMANDA L. GRIFFITHS

INTRODUCTION

SLEEP was recently recognized by the American Academy of Sleep Medicine as the third pillar of health alongside nutrition and exercise (Ramar et al. 2021). Sleep needs to be suitably timed, of appropriate duration for age, and adequate quality. Thirty-four percent of children and 75 percent of adolescents fail to get enough sleep (National Center for Chronic Disease Prevention and Health Promotion 2017). Sleep duration for children has been down-trending across recent decades (Iglowstein et al. 2003; Singh and Kenney 2013; Kocevska et al. 2021). Sleep disorders, medical or behavioral, can affect quality of sleep, manifesting as difficulty falling asleep, restless sleep, frequent waking, early morning waking, or daytime tiredness.

Much of our understanding of the function of sleep is the result of studying the impact of sleep deprivation. Sleep is a constant for humans of all ages; however, its exact purpose remains unclear. Sleep allows energy conservation, physical and mental restoration (Adam and Oswald 1977; Shapiro and Flanigan 1993), consolidation of memory (Diekelmann and Born 2010), regulation of growth, and maintenance of immunity (Gamaldo et al. 2012), and it is vital for neurodevelopment (Mignot 2008; Lim and Dinges 2010; Assefa et al. 2015). Short- and long-term deficits in cognitive performance, behavior, mood, and physical health are well-known consequences of sleep deprivation (Pilcher and Huffcutt 1996; Boonstra et al. 2007; Orzeł-Gryglewska 2010; Beattie et al. 2015). The important role of sleep in development, cerebral maturation, and neuroplasticity can be extrapolated from the increased requirements for sleep during significant periods of brain development and disordered development that occurs with significant sleep disruption (Dahl 2007; Peirano and Algarín 2007; Ednick et al. 2009).

Current knowledge of sleep in children recognizes and pays tribute to the pioneers of this field including William C. Dement, Thomas Anders, Nathaniel Kleitman, Eugene Aserisky, Carole Marcus, Christian Guilleminault, Mary Carskadon, and Avi Sadeh.

THE CHANGING NATURE OF SLEEP THROUGH CHILDHOOD

The duration, pattern, and architecture of sleep varies across the life course. From infants sleeping sixteen hours a day with twenty- to fifty-minute sleep cycles with a preponderance of rapid eye movement (REM) sleep, through to adolescents with later sleep onset and wake times. An appreciation of typical sleep in infants, children, and adolescents is integral to our understanding of development (Jenni and Carskadon 2005). Likewise, knowledge of childhood development is fundamental to understanding sleep disorders, both medical and behavioral.

NORMAL SLEEP ARCHITECTURE

The architecture of sleep recognizes two distinct sleep stages with unique physiological and polysomnographic differences. REM sleep is a period of high cortical activity marked by clear saccadic eye movements, irregular heart and respiratory rates, and nearly absent skeletal muscle tone and activity, where the diaphragm acts as the main inspiratory muscle (Tabachnik et al. 1981). Electroencephalogram (EEG) waveforms in REM sleep are low voltage mixed frequency non-alpha rhythms. Conversely relatively low brain activity is seen in Non-Rapid Eye Movement (NREM) sleep, where body movements are preserved. NREM is further divided into stages based on increasing depth of sleep. Stage N1, light sleep, is often brief with reduced alpha activity and slow rolling eye movements. Hypnagogic hallucinations and hypnic jerks can occur. Stage N2 is characterized by bursts of rhythmic EEG activity called sleep spindles and high-amplitude slow wave spikes called K complexes. Slow wave sleep is important for sleep continuity, restoration, and daytime function, and dominates in Stage N3 (Dijk 2009). Growth hormone secretion increases in slow wave sleep, indicating its importance for growth (Sassin et al. 1969; Born et al. 1988). Slow wave activity is also associated with brain and skill maturation (Volk and Huber 2015).

Infant sleep cycles differ from those of older children. Infant sleep is entered through REM sleep, and cycles are short, initially around forty minutes (see Figure 13.1). A sleep cycle in older children and adults consists of the sequence from wakefulness, through

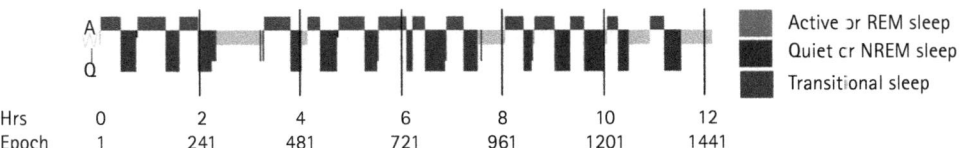

FIGURE 13.1 Sleep architecture of a neonate. Red represents REM or active sleep (56 percent of total sleep time in this recording), blue NREM or quiet sleep (41 percent of total sleep time), and green is indeterminant or transitional sleep (4 percent of total sleep time). Grey is awake periods, with wakes for feeds present.

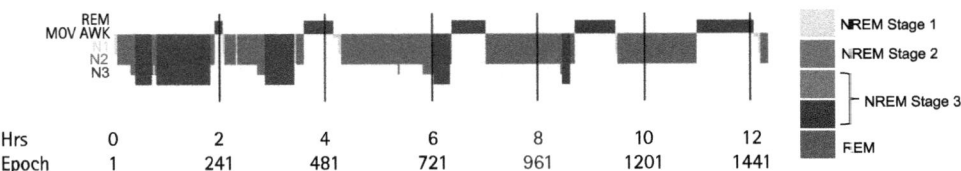

FIGURE 13.2 Sleep architecture of a three-year-old child. Cycles show how NREM stage 3 (29 percent of total sleep time in this recording) dominates in the first third of the night, and later cycles contain more NREM stage 2 (47 percent of total sleep time), alternating with REM sleep (23 percent of total sleep time) until morning. This pattern of cycling from awake, to N1, N2, N3, and REM, continues through childhood and adolescence, with normal adult hypnograms varying only in the duration of each stage.

NREM stages of sleep into deep sleep, then back to REM sleep (see Figure 13.2), with multiple cycles each night (Davis et al. 2004; Crabtree and Williams 2009).

SLEEP ONTOGENY

Sleep architecture and duration changes with age (see Table 13.1) and parallels normal development (Graven and Browne 2008). Adult sleep patterns gradually emerge as children mature, with shorter sleep duration (see Figure 13.3), longer sleep cycles, less daytime sleep, and a reduction in the amount of REM sleep (Galland et al. 2012).

REGULATION OF THE SLEEP-WAKE CYCLE

The regulation of sleep occurs through the interaction of two mechanisms—a homeostatic process known as Process S regulating sleep intensity and the circadian process known as Process C regulating the timing of sleep (Borbély 1982). In Process S, the drive for sleep gradually increases during awake periods due to the build-up of sleep-inducing

Table 13.1 Sleep patterns and developmental sleep–related changes by age

Age groups	Sleep pattern and physiology	Developmental aspects
Newborns	Sleep is polyphasic, lasting 14–17 hours, with shorter sleep cycles of 20–50 minutes (Jenni and Carskadon 2005; Hirshkowitz et al. 2015). Sleep is made up of a "quiet" pattern of bilateral bursts of high-amplitude slow waves alternating with low amplitude, and more "active" sleep where newborns move and make noise and can be easily awoken. Sleep is entered through the REM pattern, and REM sleep dominates (Emde and Metcalf 1970; Mirmiran, Maas, and Ariagno 2003).	Frequent waking and crying as sleep cycles are shorter and sleep is less efficient, making it more easily disturbed (Davis et al. 2004). Sleep-wake cycles are dependent on feeding and hunger (Jenni and Carskadon 2005).
Infants (2–12 months old)	Total sleep duration is 12–16 hours, made up of longer overnight periods with 1–2 awakenings usual overnight, and 1–4 daytime naps (Stern et al. 1969; Sheldon 2005). Sleep-wake transitions and NREM sleep emerge. Sleep is entered through the NREM pattern and REM sleep makes up 40 percent of total sleep (Davis et al. 2004).	The circadian rhythm emerges from three months (Armstrong, Quinn, and Dadds 1994) as infants become increasingly responsive to environmental and social sleep cues (Davis et al. 2004). Infants begin to self-soothe and sleep through the night from three months (St James-Roberts et al. 2015).
Toddlers (1–3 years old)	Total sleep time of 11–14 hours made up of 1–2 daytime naps, and forty-minute sleep cycles with overnight waking becoming less common (Galland et al. 2012). The sleep EEG shows further maturation with REM sleep now approximately 30 percent of total sleep (Kahn et al. 1996).	Sleep consolidation increases and daytime sleep decreases (Kahn et al. 1996).
Pre-schoolers (4–5 years old)	Further gradual reduction in REM sleep, and daytime naps begin to disappear with overnight sleep consolidated into one long period (Kahn et al. 1996).	Bedtimes may become prolonged as desire for autonomy increases (Kahn et al. 1996).
Primary school years (6–12 years old)	Sleep cycles begin to reach adult lengths of ninety minutes (Mindell and Owens 2015) with ideal sleep duration of 9–12 hours (Paruthi et al. 2016b). Sleep is highly efficient as NREM stage 2 sleep increases (Kahn et al. 1996).	A discrepancy between school and non-school night sleep can occur. Daytime sleepiness may be noticeable (Gradisar, Gardner, and Dohnt 2011; Mindell and Owens 2015).
Adolescents (>12 years old)	Sleep physiology approximates adult norms, with requirements of 8–10 hours of sleep. Slow wave sleep declines (Galland et al. 2012; Gradisar et al. 2011; Kahn et al. 1996).	Natural tendency for later bed and wake times. Increasing encroachment of evening activities on sleep.

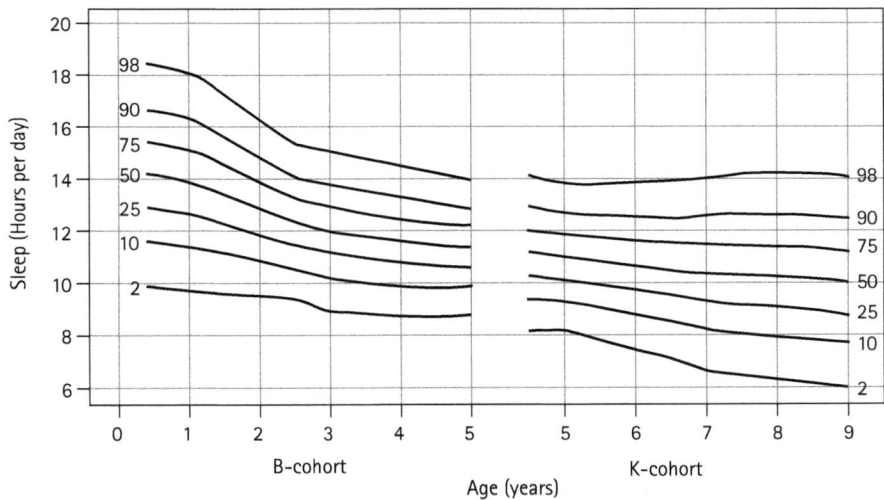

FIGURE 13.3 Total sleep duration from birth to nine years of age. Duration of sleep decreases with increasing age as described in the large longitudinal study by (Price et al. 2014) where "B-cohort" refers to 5,107 0 to 12-month-old infants, and "K-cohort" refers to 4,983 4- to 5-year-old children whose sleep was recorded every two years for four years.

substances in the brain. Slow wave sleep, the amount of which is dependent on the time awake and length of sleep, resets this homeostatic sleep pressure or sleep debt (Borb and Achermann 1999). Circadian rhythms emerge in utero in response to maternal melatonin secretion and are regulated from three months of age (Jenni and Carskadon 2005). The central circadian pacemaker is the suprachiasmatic nucleus (SCN) of the hypothalamus (Borbély and Achermann 1992). The SCN rhythm is driven by clock genes, themselves regulated by cues such as light, behavioral patterns, and the distribution of prior sleep and waking. Body temperature and melatonin excretion are measurable markers of the circadian clock. Melatonin, released from the pineal gland, promotes sleep onset and is suppressed by bright light hitting the retina (Jan et al. 2009).

The ultradian rhythm is the interaction between the homeostatic and circadian rhythms. It signals the alteration of NREM and REM sleep and determines peaks and troughs in alertness and drowsiness that cycle throughout a twenty-four-hour period (Borbély and Achermann 1992). Infants are commonly most alert in the morning hours. From eighteen months of age, peak wakefulness is often from 4.00 to 9.00 pm (Ma et al. 1993), hence the emergence of bedtime resistance and other evening behavioral sleep problems in toddlers (LeBourgeois et al. 2013). Adolescents, with a slight shift in their circadian timing, are more alert in the late morning and evening, influencing later sleep onset and wake times common in teenagers (Valdez et al. 2014).

Circadian phenotypes originate in childhood (Lee et al. 2011; Simpkin et al. 2014). Evening type (ET) individuals, also known as "night owls," prefer to stay awake and rise later, and notice peak performance in late evenings. Morning type (MT) or "morning larks" prefer to rise early, with peak performance occurring in the morning (Roenneberg

2003). Evening type preferences in childhood have been associated with aggression (Schlarb et al. 2014), depression (Haraden et al. 2017), and bedwetting (Wei and Praharaj 2019). Requirements for sleep also vary between individuals as some are better able to tolerate fewer hours of sleep (Brieva et al. 2021).

THE ROLE OF SLEEP

Brain plasticity, the brain's ability to change and adapt because of experience, is most prevalent in childhood (Mundkur 2005). Sleep appears to facilitate such plasticity. Sleep allows new memory formation and consolidation and the maintenance of learning capacity (Ruch et al. 2021). Brain regions undergoing more substantial plasticity are accordingly more affected by sleep deprivation (Wang et al. 2011). Lack of sleep in childhood can have significant impacts on cognitive function, mood, mental, cardiovascular, cerebrovascular, and metabolic health (Jan et al. 2010; Paruthi et al. 2016a; Ordway et al. 2021).

DEVELOPMENTAL CONSEQUENCES OF DISRUPTED SLEEP IN EARLY CHILDHOOD

Early childhood is a period of significant neurodevelopment. Disrupted sleep during infancy, toddlerhood, and the preschool years can impact developmental trajectories across childhood (Winsper and Wolke 2014).

DISRUPTED SLEEP IN EARLY CHILDHOOD AND MENTAL HEALTH

In a longitudinal, community-based cohort study, infants with persistently disrupted sleep at eight months of age experienced increased rates of emotional and behavioral difficulties at age two years (Wake et al. 2006). A similar longitudinall study of sixty-eight infants found those with persistent night waking, bedtime difficulties, and parental reported fussiness experienced increased emotional and behavioral problems at three and a half years of age (Scher et al. 2005).

Large longitudinal studies have examined the association between infant sleep and childhood mental health outcomes ten or more years later. Parent-reported infant sleep difficulties in 1,460 mother-infant dyads were recorded at six, nine, and twelve months of age, and children later evaluated at ages four and ten with parent-reported sleep questionnaires and diagnostic interview (Cook et al. 2020). Infants with persistent and severe sleep difficulties across the first postnatal year were found to have increased odds of experiencing emotional symptoms at age four. At age ten they had over twice the odds of meeting diagnostic criteria for an emotional disorder and over twice the odds of reporting significant symptoms of anxiety, in comparison to infants who slept well.

The presence of persistent and severe sleep disruption during infancy can predict children with heightened susceptibility to mental health problems and attentional difficulties during early to middle childhood. In older age groups, similar findings have been observed. A longitudinal study of 4,983 children followed biannually from age four until twelve to thirteen years (Quach et al. 2018) found that sleep problems were bidirectionally associated with externalizing problems at several time points during childhood. That is, sleep problems appeared to exacerbate externalizing difficulties, and externalizing difficulties exacerbated future sleep problems. The authors suggested that early emotional dysregulation in infancy may be implicated in the development of behavioral problems during childhood.

Research also supports a developmental cascade model of sleep problems and emotional and self-regulation difficulties from infancy through to age eight to nine years (Williams et al. 2017). Examining over 4,000 children, sleep problems were found to have an enduring negative impact on children's emotional and attentional regulation.

Early identification and treatment of disrupted sleep in childhood improves outcomes. Not only is less disrupted sleep in infancy and toddlerhood associated with improved parent mental health (Hiscock and Wake 2001; Piteo et al. 2013), family functioning (Bayer et al. 2007; Cook et al. 2017), and infant development (Tham et al. 2017), it may also play a protective role in establishing strong mental health during childhood.

Evaluating the specific components of infant sleep that are associated with poor outcomes has been difficult due to inconsistencies between study methodologies. Increased sleep efficiency and fewer night waking episodes as measured by actigraphy in twelve-month-old infants was associated with fewer parent-reported behavioral difficulties at age three to four years (Sadeh et al. 2015). Short sleep duration on actigraphy in three- to four-month-old females were associated with higher parent-reported social and emotional problems at eighteen to twenty-four months of age (Saenz et al. 2015). The timing of sleep may be more important than duration. Late bedtimes and wake times and bedtimes that vary significantly across the week have been associated with worse behavior, poorer mental health (Biggs et al. 2011), obesity, lower physical activity levels, more screen time (Golley et al. 2013), poorer child quality of life, and poorer maternal mental health (Quach et al. 2016). Whether sleep is measured via parent report or more objective measures, poor sleep in infancy is consistently associated with poor emotional and behavioral outcomes during childhood.

The impact of poor sleep can be compounded by other environmental and familial factors. For example, mothers who report sleep problems in their infant, report higher rates of postnatal depression, anxiety, fatigue, as well as intimate partner violence (Hiscock et al. 2007; O'Connor et al. 2007; Giallo et al. 2015)—all of which can drastically shape family functioning and subsequently, child development.

When sleep problems co-occur with unsettled or dysregulated behavior (e.g., excessive crying, tantrums, feeding difficulties, and/or difficult temperament) in infancy and the toddler years, risk for later mental health and behavioral problems increases dramatically (DeGangi et al. 2000; Hemmi et al. 2011). Children experiencing a combination of sleeping, crying, and feeding problems at age fifteen to eighteen months had over twelve times the odds of experiencing very highly dysregulated behavior across childhood to age nine and a half years (Winsper and Wolke 2014). When present at twenty-four to thirty months, children had over twenty-nine times the odds of experiencing very highly dysregulated behavior across childhood. Children who had only sleep problems during infancy and no feeding or crying problems had around one and a half times greater odds of having very highly dysregulated behavior across childhood. Severely dysregulated one-year-old infants, who experienced severe problems with sleeping, crying, feeding, and tantrums and had a difficult temperament, were found to have over ten times the risk of clinically significant mental health concerns at age eleven, in comparison to infants who were more settled (Cook et al. 2019). Infants with sleep problems without other regulatory difficulties did experience slightly increased symptoms of mental health difficulties during childhood. However, they were not any more likely to fall into the clinical range than infants who were settled. Rather, it was the combination of sleep problems with other regulation difficulties that identified infants with the greatest risk for poor mental health outcomes. Sleep problems occurring alongside other regulatory difficulties during infancy may mark the beginning of a particularly poor developmental trajectory for some children. Treating sleep difficulties seems to be a logical first step for very dysregulated infants, given that other regulatory difficulties such as crying, temper tantrums, and feeding difficulties for example, may be less amenable to change. Early identification and treatment of sleep difficulties may prevent or reduce poor emotional and behavioral outcomes during childhood.

DISRUPTED SLEEP IN EARLY CHILDHOOD AND COGNITION AND LANGUAGE DEVELOPMENT

Good sleep facilitates the consolidation of memories (Diekelmann and Born 2010). Friedrich et al. found that infants retain and reorganize the meanings of new words when they sleep (Friedrich et al. 2015). Napping has been shown to help fifteen-month-old toddlers remember the grammatical pattern of an artificial language (Hupbach et al. 2009). Conversely, in a longitudinal study of 1,029 six- to forty-month-old twins, children with language delays at sixty months of age tended to have poorer sleep consolidation at six and eighteen months, than those who did not have language delay or who had early, transient language delay (Dionne et al. 2011). In addition, Dearing et al. observed that circadian sleep regulation at seven and nineteen months of age was associated with better language scores at age twenty-four months (Dearing et al. 2001).

Longitudinally, disrupted sleep during infancy is associated with later language outcomes. One-year-old infants who had severe sleep difficulties comorbid with excessive crying and feeding difficulties were five times more likely to have language difficulties at age five and continued to have significantly poorer language scores at age eleven years (Cook et al. 2021). Excessive infant night waking and crying has been associated with less optimal parent-infant relatedness (Räihä et al. 2002; Philbrook and Teti 2016), and as warm, responsive parent-child interactions are important for early language development (Tamis-LeMonda et al. 2004; Landry et al. 2006), infants with sleeping and crying problems may be exposed to fewer rich and stimulating interactions with their parents. More parental resources may be devoted to trying to settle their infant or it may be more difficult for parents to engage in rich, meaningful interactions with an infant who is tired (Conway et al. 2018; Cook et al. 2021). Improving infant sleep may help create a better home environment for language learning to occur.

Sleep quality in infancy is linked to cognitive development (Tham et al. 2017). During infancy, more frequent night waking (Scher et al. 2005), and snoring (Piteo et al. 2011) were associated with poorer scores on the Bayley Scales of Infant and Toddler Development Mental Development Index, and greater sleep efficiency with more favorable score (Scher, 2005). Infants aged eleven to thirteen months were also found to have better problem-solving skills when their sleep efficiency was greater (Gibson et al. 2012). One-year-old infants who achieved more night sleep scored more highly on an executive functioning task at age four (Bernier et al. 2013). Thus, sleep quality plays an important role in cognition, memory, and language acquisition.

A growing body of research demonstrates that disrupted sleep in early childhood is associated with poor developmental outcomes across childhood, including behavior problems, low language skills, and cognitive difficulties. Environmental and genetic factors likely intersect to impact both sleep and developmental outcomes, and more research is required to understand the unique contributions of each of these. Early detection and treatment of sleep difficulties are important avenues for improving developmental outcomes for some children.

CONSEQUENCES OF INSUFFICIENT SLEEP DURING LATE CHILDHOOD AND ADOLESCENCE

Late childhood and adolescence are significant periods of development. Identity exploration and social and emotional development occur during a time of increasing academic and social demands (Hart and Carlo 2005; Sebastian et al. 2008; Steensma et al.

2013; Damon et al. 2019). External changes are mirrored by structural and functional brain changes which are evident on brain imaging studies of adolescents (Giedd 2004; Choudhury et al. 2006; Burnett et al. 2011; Blakemore 2012). As a consequence of significant changes during this sensitive life period, developmental vulnerabilities emerge (Steinberg 2005).

Poor sleep during adolescence is linked to a range of physical and mental problems long term. There is a strong link between poor sleep during adolescence and obesity. Multiple meta-analyses have found that short sleep duration is associated with a considerably higher risk of obesity (Cappuccio et al. 2008; Fatima et al. 2016; Miller et al. 2018; Deng et al. 2021), as high as two times the risk (Fatima et al. 2015). Academically, sleepiness, poor sleep quality, and short duration have been shown to significantly impact school performance (Dewald et al. 2010). Later school times, which fit with adolescent circadian timing of later sleep and wake times, have been raised as a possible solution to combat this (Bowers and Moyer 2017). Additionally, adolescents with chronically poor sleep are at increased risk of sports and musculoskeletal injuries (Gao et al. 2019).

Sleep disturbances in adolescents are common, affecting at least 26 percent of teenagers (Gradisar et al. 2011; Liang et al. 2021). The frequency and impact of electronic technology use in teenagers has a compounding effect—being associated with reduced sleep duration and prolonged sleep latency (Mei et al. 2018). Positively, behavioral sleep interventions remain effective into adolescence (Griggs et al. 2020). Early identification and management of sleep difficulties in older children and adolescents is important in mitigating the negative impacts on educational, social, and physical wellbeing.

SLEEP DISORDERS

Sleep for optimal function requires adequate duration of quality sleep, occurring regularly and with appropriate timing, and the absence of sleep disorders (Ramar et al. 2021). Sleep problems are common in childhood, affecting 25 to 44 percent of children (Fricke-Oerkermann et al. 2007; Ipsiroglu et al. 2002; Mindell et al; 2013; Narasimhan et al. 2020; Owens 2007; Pagel et al. 2007; Schlarb et al. 2015), and 50 to 90 percent of children in high risk groups (Corkum et al. 2011; Richdale et al. 2009; Wang et al. 2022). Sleep problems in children are also often under-reported (Blunden et al. 2004; Richardson et al. 2021).

Sleep disorders are frequently characterized as behavioral or medical in origin (see Table 13.2 for examples).

ASSESSING SLEEP IN CHILDREN

Children's sleep can be evaluated through a variety of methods, including thorough history taking, parent and child completed questionnaires, sleep diaries, and objective

Table 13.2 Common sleep problems by age

Age group	Examples of common sleep problems
Infants (0–12 months old)	• Sleep onset associations (can develop from 4–6 months) • Separation anxiety (from 6 months–3 years)
Toddlers (1–3 years old)	• Separation anxiety (from 6 months–3 years) • Bedtime resistance • Sleep onset associations • Limit-setting disorders • Confusional arousals • Obstructive sleep apnoea and sleep disordered breathing (peaks at 2–8 years of age)
Pre-schoolers (4–5 years old)	• Obstructive sleep apnoea and sleep disordered breathing (peaks at 2–8 years of age) • Bedtime resistance • Nightmares • Sleep walking
Primary school years (6–12 years old)	• Night terrors • Sleep walking • Bruxism • Obstructive sleep apnoea and sleep disordered breathing (peaks at 2–8 years of age) • Rhythmic movement disorders • Poor sleep habits
Adolescents (>12 years old)	• Delayed sleep phase syndrome • Poor sleep habits • Insufficient sleep duration with daytime tiredness • Insomnia • Restless legs syndrome • Narcolepsy

measurement. Standardized questionnaires, such as the Brief Infant Sleep Questionnaire, Children's Sleep Habits Questionnaire, Sleep Disturbance Scale for Children, and Paediatric Sleep Questionnaire (Lewandowski, Toliver-Sokol and Palermo 2011) help evaluate sleep problems. Sleep diaries provide a visual representation of sleep patterns—recording bedtime, sleep time, and wake time as well as associated behaviors—and correlate well, though slightly over-estimate, sleep duration when compared with more objective recordings (Gaina et al. 2004; Asaka and Takada 2011; Short et al. 2012; Mazza et al. 2020). Actigraphy records gross motor movement through a wristwatch-like device to estimate sleep-wake cycles and can be a useful adjunct to a sleep diary (Mazza et al. 2020).

The gold standard measurement of sleep is laboratory-based, attended, multi-channel polysomnography (PSG) with video monitoring. PSG tracks continuous electrophysiological data on eye movement, skeletal muscle activation, oxygen level, respiratory rate, heart rate, and EEG. It is important to note that PSG is not indicated in the clinical assessment of all sleep disorders. In fact, it can make assessment and treatment more problematic in some disorders, such as insomnia. However, it does remain a useful tool providing avenues for investigation and management of certain sleep disorders that require a detailed analysis of sleep characteristics (Aurora et al. 2012).

Home based sleep technology is a rapidly expanding and ever-changing field and provides exciting opportunities to monitor children's sleep in their usual sleep environment. Telehealth-supported pediatric home sleep apnea testing in children aged 5–18 years achieves technical success in almost 90 percent of patients investigated for OSA, with 90 percent achieving equal to or greater than six hours sleep duration and excellent family acceptability (Griffiths et al. 2022). There is also evidence that home studies are feasible in children with neurodevelopmental conditions, and children as young as four months of age (Russo et al. 2021). Furthermore, there is excellent concordance between gold standard lab-based PSG and more limited channel home PSG in children aged 5–18 years (Withers et al. 2022).

Wearables refer to sleep trackers worn usually on the wrist that monitor arm movement and heart rate during sleep, and nearables are mattress-based devices with pressure sensors that measure movement, respiration, and heart rate (Bianchi 2018). Smartphone applications provide this data directly back to the consumer, which can be shared with professionals. These devices are often well-tolerated (Sheth et al. 2018; Mackintosh et al. 2019; Oygür et al. 2021), though sensitivity, reliability, and validity of the data in children are less well known (Toon et al. 2016; Pesonen and Kuula, 2018; Kruizinga et al. 2021; Stražišar 2021).

MEDICAL DISORDERS OF SLEEP

Medical disorders of sleep can cause significant sleep disruption and deficit and have detrimental effects on cognition, attention, memory, mood, and behavior (Colonna et al. 2015).

OBSTRUCTIVE SLEEP APNOEA

Sleep disordered breathing covers the spectrum from simple snoring to obstructive sleep apnoea (OSA). Children with OSA can present with noisy breathing and disturbed sleep as infants, snoring and mouth-breathing in pre-school years, and behavioral and dental problems in childhood, with apnoeas, increased respiratory effort during sleep, restless sleep, and daytime tiredness presenting at any age (Li and Lee 2009). Children with OSA can display significant levels of inattention and hyperactivity that may mimic attention deficit hyperactivity disorder (ADHD) (Sedky et al. 2014). Diagnosis of OSA can be difficult, as not all children who snore have OSA, and

OSA can be present without snoring (Carroll et al. 1995). The general prevalence of OSA in children is 1–4 percent (Lumeng and Chervin 2008; Bixler et al. 2009).

The mechanism of OSA in children is often different to what occurs in adults. Airway narrowing can occur because of adenoid and tonsillar hypertrophy, craniofacial anomalies, neuromuscular diseases, as well as obesity (Katz and D'Ambrosio 2008). There is an association between OSA and enuresis, postulated through effects of apnoea and arousal on antidiuretic hormone secretion (Jeyakumar et al. 2012). Although PSG is the gold-standard diagnostic tool, it is not always recommended due to lack of availability, cost, and lack of consensus regarding interpretation (Chan et al. 2004). Long-term, untreated OSA can impact growth through growth hormone secretion abnormalities and increased energy expenditure (Li and Lee 2009), as well as cardiovascular health, mood (Crabtree et al. 2004), academic ability, attention, and executive function (Kohler et al. 2010; Cardoso et al. 2018).

The mainstay of treatment of OSA is adenotonsillectomy. Two randomized controlled trials of tonsillectomy compared to watchful waiting found improvements in behavior and PSG parameters, but no difference in formally assessed cognition (Marcus et al. 2013; Waters et al. 2020). A systematic review of smaller studies found improvements in intelligence quotient (IQ) scores after adenotonsillectomy (Song et al. 2016). Cure after adenotonsillectomy, however, can be variable with the chance of residual disease, and other treatment options should be discussed (Reckley et al. 2018). Refractory sleep disordered breathing in children with underlying medical conditions may require continuous positive airway pressure (CPAP), repeat or further surgery, and long-term follow-up.

NARCOLEPSY

Narcolepsy is the constellation of sleep attacks, cataplexy, hypnagogic hallucinations, sleep paralysis, and daytime sleepiness (Challamel et al. 1994). It should be suspected in the presence of habitual napping in over five-year-olds and unusual napping in older children. Onset is commonly in childhood and adolescence. Inheritance is strong with likely a multifactorial genetic basis, though the precise cause is unknown. It has been postulated to be autoimmune in origin from a reduction in neurons in the hypothalamus that produce hypocretin—known to help control wakefulness (Miyagawa and Tokunaga 2019). PSG with the multiple sleep latency test is recommended for diagnostic purposes. Diagnostic criteria includes two or more sleep-onset REM sleep periods and a mean sleep latency of less than eight minutes across four to five daytime naps (Aldrich et al. 1997; Bassetti et al. 2019). Narcolepsy causes severely fragmented sleep with repeated REM disruptions with notable impacts on social, mental and physical health, and education (Inocente et al. 2014). Medical therapy to maintain alertness and combat sleepiness, planned napping, education, and support are central to management (Guilleminault and Pelayo 2000).

PARASOMNIAS

Parasomnias are episodic unpleasant recurrent behaviors, experiences, or physiological changes that occur during sleep or sleep transitions (Laberge et al. 2000). They are more common in children than adults and more common in children with underlying neurodevelopmental disorders. Common parasomnias in children include confusional arousals, sleep walking, and sleep terrors (Singh et al. 2018). A vicious cycle of lack of sleep can occur, as parasomnias are a symptom of sleep deprivation and in turn cause poor sleep.

PERIODIC LIMB MOVEMENT DISORDER AND RESTLESS LEGS SYNDROME

Restless legs syndrome (RLS) is the urge to move the legs due to an uncomfortable feeling, usually occurring just prior to sleep onset. Paediatric RLS differs slightly to RLS in adults, with diagnostic criteria including a family history of RLS, and a developmentally appropriate description of the symptoms rather than the simple description of an urge to move the legs usually accompanied or caused by uncomfortable and unpleasant sensations in the legs. RLS can be associated with a deterioration in behavior or academic functioning (Picchietti et al. 2013).

Periodic limb movement disorder (PLMD) comprises brief stereotyped contractions of the legs that occur in NREM sleep and cause arousals which disturb sleep (Natarajan 2010). PLMD is easily recognizable on PSG (Lesage and Hening 2004).

Similar pathophysiology is implicated in both RLS and PLMD, including genetic susceptibility, iron deficiency, and associated dopaminergic dysfunction. Consistent bedtime routines, exercise, and healthy diet can be effective in children. Iron supplementation is the first-line pharmacological treatment, with promising results (Simakajornboon et al. 2003; Gurbani et al. 2019; DelRosso et al. 2021). Other pharmacological options are less well studied in RLS and PLMD in children, but may include dopaminergic agents and melatonin (Dye et al. 2019).

CIRCADIAN RHYTHM DISORDERS

Misalignment of the circadian rhythm cycle can result in sleep disorders. Delayed sleep phase syndrome is characterized by persistent and intractable later phase shifts in sleep and wake times that interfere with general functioning (Wyatt 2004). It commonly arises in adolescence, affecting 3–4 percent of teenagers (Sivertsen et al. 2013; Danielsson et al. 2016). Sufferers report difficulty falling asleep (though with sleep maintenance preserved), difficulty waking, and daytime sleepiness. Management includes melatonin, chronotherapy, sleep hygiene, and light therapy (Zisapel 2001; Kim et al. 2013).

BEHAVIORAL DISORDERS OF SLEEP

Problematic sleep behaviors in children are common, such as bedtime resistance, difficulties falling asleep, frequent overnight waking, and early morning waking (see Table 13.3). Like medical sleep disorders, these can equally impact daytime functioning, mood, and child and family quality of life.

MANAGEMENT OF BEHAVIORAL SLEEP DISORDERS

Initial management of behavioral sleep problems, both in neurotypical and neurodiverse children, usually comprises optimal bedtime routines and introduction of positive sleep habits (Meltzer and Mindell 2004), before embarking on specific behavioral management strategies. Healthy sleep habits (also referred to as sleep hygiene; see Table 13.4) are sleep-related behaviors and environmental cues that stimulate well-timed and effective sleep (Galland and Mitchell 2010).

Behavioral interventions are highly efficacious (Meltzer and Mindell 2014) in behavioral sleep disorders. Behavioral interventions are structured approaches to sleep behaviors that aim to reduce problematic sleep behaviors. Common examples include bedtime fading, the bedtime pass, and graduated extinction methods such as "camping out" and "controlled comforting." The bedtime pass, useful for young children with bedtime resistance, is a notecard that is exchangeable for one trip out of the bedroom for children after bedtime (Freeman 2006). It was found to reduce the rate of children calling out and leaving their room (Moore et al. 2007). Bedtime fading, where bedtime is delayed to approximate when the child naturally falls asleep, then is brought forward by fifteen minutes each night, allows gradual adjustment of the circadian rhythm— helpful for those with delayed sleep onset (Morgenthaler et al. 2006; Cooney et al. 2018; Delemere and Dounavi 2018). Graduated extinction can assist infants over six months of age, toddlers, and preschoolers with sleep-onset associations and frequent waking. In controlled comforting, the parent settles the child in bed until quiet, then leaves the room and responds to crying by returning at set intervals of increasing duration. Studies show significant reductions in child sleep problems and maternal depression (Hiscock and Wake 2002), which persist even after two years (Hiscock et al. 2007), with no detrimental effect on child or parent's mental health, nor child-parent attachment five years

Table 13.3 Examples of common behavioral sleep disorders, precipitants, and management options

Disorder	Presenting symptoms	Underlying precipitants	Commonly affected age groups	Management options
Behavioral insomnia of childhood, sleep-onset association type	• Difficulty settling alone • Crying or fussing at bedtime, or refusal to go to bed • Overnight waking and parent seeking	Objects and people are required for the child to fall asleep at bedtime and return to sleep following normal night-time arousals.	• Infants • Toddlers • Pre-schoolers	• Parent/carer education • Consistent bedtime routines • Graduated extinction (e.g., camping out) • Controlled comforting
Behavioral insomnia of childhood, limit-setting type	• Bedtime refusal • Repeated bedtime requests (e.g., drink, toilet etc.)	Inconsistent limit-setting by parents and caregivers.	• Toddlers • Pre-schoolers • School-aged children	• Healthy sleep habits • Parent/carer education • Bedtime pass
Night-time fears	• Fearful behaviors • Increasing agitation and distress at bedtime • Insistence on light and other comforts	Modelling from others, aversive classical conditioning, and exposure to negative information.	• Pre-schoolers • School-aged children	• Relaxation • Guided imagery • Positive self-statements • Self-control techniques with rewards • Systematic desensitization.
Night terrors	• Sudden arousals from sleep with screaming or yelling and intense fear • Limited recall of event • Child inconsolable	Increased during intercurrent illness, sleep deprivation, and presence of psychosocial stressors. Often co-occur with other parasomnias. Family history of same common. Arise from slow wave sleep.	• Pre-schoolers • School-aged children	• Reassurance and education • Scheduled waking
Inadequate sleep hygiene	• Late sleep onset • Daytime sleepiness • Low mood	Lack of regular habits and practices that are conducive to sleeping well.	• School-aged children • Adolescents	• Sleep education • Consistent sleep-wake schedule • Limit caffeine and other stimulants • Limit evening technology and technology use in the bedroom • Bedtime fading

Compiled references for Table 13.3: Disorder descriptions (Kales et al. 1980; Muris et al. 2001; Schredl 2001; Mindell and Meltzer 2008). Management (Lask 1988; Owens et al. 1999; Hiscock and Wake 2002; Gradisar et al. 2016; van Horn and Street 2018; Hiscock et al. 2021).

> **Table 13.4 Examples of healthy sleep habits**
>
> - Regular sleep and wake times, varying by no more than fifteen minutes night to night.
> - Low light and noise levels prior to bed and whilst asleep.
> - Consistent bedtime routines comprising low-stimulation activities.
> - Minimize evening exposure to bright light, particularly light from electronic screens.
> - Limit caffeine intake to morning hours, noting that caffeine can be hidden in many children's food and drinks (e.g., soda/fizzy drinks, hot chocolate, ice-cream, chocolate, energy and vitamin drinks).

later (Price et al. 2012). In camping out, the parent or carer stays with the child to fall asleep, then gradually moves themselves away from the child each night, eventually withdrawing their presence completely. It has been found to reduce sleep latency and night wakings (Mindell et al. 2006), with no negative impacts found for parent-child attachment, stress response, or child behavior and emotions (Gradisar et al. 2016).

SLEEP AND PSYCHIATRIC DISORDERS

Sleep and mental health problems have a strong bidirectional relationship (Chorney et al. 2008; Alvaro et al. 2013). Changes in sleep patterns are recognized symptoms of mental health disorders: hypersomnia or insomnia form part of the diagnostic criteria of depression; insomnia is common in anxiety; prolonged wakefulness is a feature of mania in bipolar disorder; and nightmares are a symptom in post-traumatic stress disorder (Suni 2020). Etiology of poor sleep in those with mood disorders may be intrinsic to the disorder—for example, with reduced serotonin levels in depression and sleep-arousal regulation variations in bipolar disorder—or extrinsic, as the result of comorbid conditions or medication side effects (Harvey 2011). Changes in mood can also be due to an underlying sleep disorder. Therefore, an understanding of psychiatric conditions when evaluating sleep and recognition of sleep disorders in children presenting with variations in mood and behavior, is vital.

Children and adolescents with diagnosed mental health conditions frequently display more sleep problems than community samples (Ivanenko et al. 2006; Reigstad et al. 2010). Three- to four-year-old children rated as having lower mental health status were more likely to have sleepiness, snoring, anxiety before sleep, and circadian rhythm abnormalities (Horiuchi et al. 2021). Conversely, poor sleep is associated with poor mental health. In school-aged children, poor sleep habits (Shamsaei et al. 2019) and excessive daytime sleepiness (Calhoun et al. 2011) were associated with more symptoms of poor mental health. Sleep-related problems were positively associated with obsessive compulsive disorder (OCD) symptom severity, child-rated anxiety, and parent-rated child internalizing problems (Storch et al. 2008). Examining adolescents, later bedtime, shorter sleep duration, and periods of weekend oversleep were associated with increased odds of suicidality, mood, and anxiety disorders (Zhang et al. 2017). Summative reviews conclude that sleep problems in children and adolescents are associated with anxiety and depression (Brown et al. 2018; Marino et al. 2021).

Poor sleep during childhood is a risk factor for mental health disorders later in life. A systematic review including over 360,000 adolescents found that short sleep duration was associated with a 55 percent increase in the likelihood of subsequent mood deficits (Short et al. 2020). Sleep disturbance is also a significant risk factor for suicide (Fitzgerald et al. 2011; Bernert et al. 2015; Pigeon et al. 2016). Trouble sleeping at twelve to fourteen-years of age significantly predicted suicidal thoughts and self-harm behaviors at fifteen to seventeen years of age (Wong et al. 2011). Longitudinal studies following children have found associations between persistent sleep problems in childhood and anxiety disorders in adulthood (Gregory et al. 2005; Shanahan et al. 2014) and a systematic review of studies in all age groups found insomnia could predict depression (Alvaro et al. 2013). Early sleep problems are an important modifiable factor on the trajectory to mental ill health.

ANXIETY AND SLEEP

Sleep problems in children with anxiety are highly common. Ninety to 92 percent of children with anxiety had one sleep-related problem and 82 percent two or more (Storch et al. 2008; Chase and Pincus 2011). On objective measures, school-aged children with anxiety had later sleep onset times and experienced significantly less sleep compared to non-anxious children (Hudson et al. 2009). Peri-sleep-onset cortisol is higher in children with anxiety disorders than children with depression or controls (Forbes et al. 2006).

Anxiety in children can manifest as excessive night-time worry (Danielsson et al. 2013), bedtime resistance, sleep onset latency, and frequent awakening (Iwadare et al. 2015; Brown et al. 2018). Developmentally appropriate night-time fears, such as fear of the dark or separation anxiety, must be separated from pathological anxiety and behavioral sleep disorders in children, by examining the pervasiveness through developmental stages, degree of impact, duration of symptoms, and family history.

Managing sleep can improve anxiety and vice versa. A systematic review in adolescents with sleep problems and anxiety found cognitive-behavioral sleep interventions improved sleep quality, sleep onset latency, daytime sleepiness, as well as overall anxiety (Blake et al. 2017). Pharmacological management of anxiety can also improve sleep, with children receiving fluvoxamine compared with placebo for anxiety demonstrating improved sleep (Alfano et al. 2007).

DEPRESSION AND SLEEP

Depression is defined as affect dysregulation with clinically significant impairments in mood and motivation (American Psychiatric Association 2013) and there are strong links between depression and sleep. A study found 72.7 percent of children with depression suffer sleep disturbance and 53.5 percent suffer insomnia (Liu et al. 2007). Adolescents with depression experience longer sleep onset latency, more waking after sleep onset, and lower objective sleep efficiency, and high rates of sleep disturbance on subjective report (Lovato and Gradisar 2014). Depression severity in adolescents also correlates with the degree of sleep disturbance (Gupta et al. 2019). Importantly, children with sleep problems are more likely to have unrecognized symptoms of depression (Liu et al. 2007; Lee et al. 2017).

PSG differs in children with depression compared with controls (Emslie et al. 2001). Children with depression demonstrated delayed sleep onset, more awake time in bed, decreased sleep efficiency. Studies have found differences in EEG temporal coherence between depressed children whose mental health recovered and those who did not (Emslie et al. 2001), and sharp waves and polyspikes in frontal regions of 75 percent of children with depressive symptoms (Arana-Lechuga et al. 2008). However, a meta-analysis concluded that the majority of depressed youth had similar architecture to controls and hence it was not an effective individual marker for depression (Augustinavicius et al. 2014).

Parental mental health and infant sleep are intimately linked. Maternal depression is associated with infant sleep problems (O'Connor et al. 2007; Armitage et al. 2009) and maternal report of infant sleep problems predict higher maternal depression scores (Hiscock and Wake 2001; Dennis and Ross 2005). Importantly, improving infant sleep supports maternal mental health, where a randomized trial targeting infant sleep problems was found to improve maternal mental health as well as infant sleep (Hiscock et al. 2007), even two years after the infant sleep intervention (Hiscock et al. 2008).

Poor sleep quality and quantity is recognized as one of the most significant modifiable factors determining risk, severity, and relapse of prevalent mental disorders such as depression (Franzen and Buysse 2008; Peach et al. 2016; Richardson et al. 2019; Blanken et al. 2020). In 116 adolescents with insomnia, cognitive behavioral therapy for insomnia achieved long-term reduction in psychopathology (de Bruin et al. 2018). A systematic review noted that cognitive-behavioral sleep interventions improve sleep quantity and quality, as well as depression symptoms in adolescents (Blake et al. 2017). Management of sleep problems in children with psychiatric disorders requires an integrated approach considering behavioral measures, healthy sleep habits, cognitive-behavioral therapy, and medication.

UNDERSTANDING SLEEP IN NEURODIVERSE CHILDREN

Sleep and ADHD

Sleep problems are common in children with ADHD (Corkum et al. 2001; Accardo et al. 2012; Cortese et al. 2013; Becker et al. 2015), reported in up to 73.3 percent of those diagnosed with the condition (Sung et al. 2008). Bedtime resistance, delayed sleep onset, frequent night awakenings, and daytime sleepiness are commonly reported (Accardo et al. 2012; Scott et al. 2013). Objective measures with actigraphy and polysomnography show lower sleep efficiency (Virring et al. 2016), prolonged sleep onset latency, lower total sleep time (Cortese et al. 2009), and higher number of arousals (Silvestri et al. 2009; Stephens et al. 2013) and movements (Miano et al. 2006) in children with ADHD than in typically developing children. Sleep disturbances in children with ADHD can result in poorer working memory (Sciberras et al. 2015), lower quality of life (Accardo et al. 2012), functional impairment, and aggravation of ADHD symptoms (Cortese et al. 2013).

Dysregulation of arousal in ADHD, with paradoxical hypo-arousal and fatigue, has been postulated to explain the close relationship between sleep, attention, and hyperactivity (Virring et al. 2016). The crossover of central nervous system dysfunction in dopamine signaling in both ADHD and some sleep disorders is also recognized (Cortese et al. 2005). Brain regions involved in the regulation of arousal and impacted by sleep deprivation are also implicated in the pathophysiology of ADHD (Silvestri et al. 2009). Meta-analysis data has postulated links between altered slow wave activity (SWA) and cyclic alternating pattern (CAP) in children with ADHD, and proposed an association with delayed peak cerebral cortical thickness on magnetic resonance imaging (MRI) which might explain differing developmental outcomes (Biancardi et al. 2021).

Disrupted sleep in children with hyperactivity and attention concerns may be intrinsic to ADHD, secondary to a medical or behavioral sleep disorder or comorbid psychiatric disorder, or the result of medication. Delayed sleep onset is a well-known side effect of stimulant medication (Sanabra et al. 2021). Mental health conditions and disturbed sleep are intimately linked, with 33 percent of children with ADHD having a comorbid psychiatric diagnosis (Accardo et al. 2012). Importantly, sleep disorders in children can mimic ADHD (Chervin et al. 2002a; Cortese et al. 2013) where sleep deprivation causes inattention, hyperactivity, disturbed mood, and impulsivity (Paavonen et al. 2009). Similarly, sleep disordered breathing is common in children with ADHD, with snoring reported in 33 percent of children with ADHD, compared with 9 percent of controls (Chervin et al. 1997). Studies have shown surgical treatment of OSA in children improves hyperactivity and impulsivity symptoms (Marcus et al. 2012). RLS and periodic limb movement disorder are the most common medical sleep disorders reported in children with ADHD (Chervin et al. 2002b). Forty-four per cent of children with ADHD have symptoms of RLS compared to baseline incidence of 10–15 percent of children (Silvestri et al. 2009). Iron deficiency, implicated in RLS, is also more common in ADHD (Cortese et al. 2005).

Screening for sleep disorders should occur in any child presenting with concerns regarding concentration and overactivity, and sleep should be evaluated at regular intervals in children with ADHD. Behavioral sleep interventions for children with ADHD are effective in improving ADHD symptomatology, working memory (Sciberras et al. 2015), and quality of life (Accardo et al. 2012; Sciberras et al. 2019). Melatonin can improve sleep latency, total sleep time, and sleep efficiency in children with ADHD, and clonidine has been shown to reduce insomnia (Barrett et al. 2013).

Sleep and Autism Spectrum Disorder

Disturbed sleep in autism spectrum disorder (ASD) is often multifactorial. The neuro-chemical and genetic differences in subsets of people with ASD are increasingly recognized and help explain the high prevalence of sleep difficulties.

Behavioral sleep problems are pervasive in children with ASD, reported in 50–93 percent of those with the condition (Hering et al. 1999; Goldman et al. 2012; Mazurek et al. 2019; Tesfaye et al. 2021)—more prevalent than both neurotypical children (Black 2021) and children with other developmental disorders (Krakowiak et al. 2008; Tzischinsky et al. 2018; Carmassi et al. 2019). Common problems include delayed sleep onset (Mayes and Calhoun 2009; Hodge et al. 2014), lower total sleep time (Humphreys et al. 2014; Elrod and Hood 2015), frequent waking (Hering et al. 1999), restless sleep, early morning awakenings (Carmassi et al. 2019; Black 2021), and parasomnias (Díaz-Román et al. 2018). Ninety-four percent of children with ASD had more than one sleep problem (McLay et al. 2021). Sleep studies are often poorly tolerated in children with ASD, though PSG has demonstrated prolonged sleep latency (Lambert et al. 2013), reduced sleep efficiency (Chen et al. 2020), reduced REM sleep, and changes in the frequency of slow wave sleep (Kotagal and Broomall 2012; Buck et al. 2021). Difficulties with sleep can present as early as thirty months of age in children with ASD (Humphreys et al. 2014) and are seen across childhood and adolescence (Goldman et al. 2012; Hodge et al. 2014). Poor sleep in children with ASD impacts negatively on family functioning (Krakowiak et al. 2008; Veatch et al. 2015; Black 2021), parental sleep (Papadopoulos et al. 2019), and children's quality of life (Delahaye et al. 2014).

ASD severity correlates with sleep problems (Mayes and Calhoun 2009; Tudor et al. 2012). Conversely, sleep deprivation has been shown to correspond with worsening core ASD symptoms and behavior (Veatch et al. 2015; Tesfaye et al. 2021). Tudor et al. noted that sleep onset latency was the strongest predictor of stereotyped behavior, social interaction deficit, and overall ASD severity, with the authors concluding that lack of sleep contributes to the intensification of most ASD symptoms.

Behavioral, biological, and genetic differences in children with ASD underpin the high prevalence and complex nature of sleep disruption (see Table 13.5).

Comorbidities of high frequency with ASD, such as anxiety, depression, and epilepsy, should be considered in those presenting with disrupted sleep. Management often requires a multi-faceted approach, addressing sleep habits, family education, behavioral interventions (Papadopoulos et al. 2019), evaluation and modulation of sensory seeking, and commonly exogenous melatonin (Cuomo et al. 2017; Souders et al. 2017). Melatonin is highly effective in children with ASD, improving sleep onset latency, total sleep time, night waking, bedtime resistance, and sleep efficiency (Gringras et al. 2012; Buckley et al. 2020). Meta-analysis also demonstrated improvements in daytime behavior and social interaction with melatonin (Rossignol and Frye 2011). Weighted blankets have proven ineffective in research (Gringras et al. 2014; Buckley et al. 2020).

Table 13.5 Factors common in autism spectrum disorder that may impact sleep

Behavioral	Neurochemical	Genetic
Sensory and behavioral differences in children with ASD.	Abnormal production, breakdown, and receptor sites of neurotransmitters in individuals with ASD.	Aberrations in gene loci regulating circadian rhythms in individuals with ASD.
• Reduced awareness of social cues that play an important role in the sleep-wake cycle (Richdale and Prior 1995; Veatch, Maxwell-Horn and Malow 2015).	• Depressed melatonin levels (Rossignol and Frye 2011; Tordjman et al. 2013; Carmassi et al. 2019).	• Mutations in CLOCK and CLOCK-related gene expression (Wimpory, Nicholas, and Nash 2002; Bourgeron 2007; Nicholas et al. 2007; Yang et al. 2016; Carmassi et al. 2019).
• Preferences for routine and difficulties with transitions (Reynolds and Malow 2011).	• Lower melatonin metabolites including urinary 6SM (Tordjman et al. 2005; Leu et al. 2011; Carmassi et al. 2019).	• Mutations in melatonin-regulating pathway genes (Veatch et al. 2015; Carmassi et al. 2019; Ballester-Navarro et al. 2021).
• Differences in arousal (Richdale et al. 2014; Souders et al. 2017).	• Abnormal melatonin circadian rhythm (Kulman et al. 2000; Rossignol and Frye 2011).	• Altered levels of circadian gene mRNA levels (Souders et al. 2017; Carmassi et al. 2019).
• Sensory differences, such as hypersensitivity to touch, that may interfere with sleep (Reynolds, Lane, and Thacker 2012; Mazurek and Petroski 2015; Tzischinsky et al. 2018).	• Changes in melatonin receptors and enzymes (Melke et al. 2008; Souders et al. 2017).	
• Higher rates of separation anxiety (Gillott, Furniss, and Walter 2001).	• Abnormalities in GABA-ergic function (Reynolds and Malow 2011; Singh and Zimmerman 2015; Ballester et al. 2020).	
	• Changes in serotonin levels (Mulder et al. 2004; Nakamura et al. 2010; Pagan et al. 2014).	

Sleep in Other Neurodevelopmental Disorders

Sleep problems are more common in children with any neurodevelopmental disability (Dorris et al. 2008; Tietze et al. 2012), due to disruption of the usual progression of sleep development. Sleep issues in pertinent conditions are mentioned below as further in-depth discussion is beyond the scope of this chapter.

- Children with cerebral palsy often have difficulties with sleep-wake transitions, initiating and maintaining sleep, and have increased rates of sleep disordered breathing (Fitzgerald et al. 2009). Possible explanations include differences in tone and higher rates of comorbid conditions, notably epilepsy, that impact sleep (Newman et al. 2006; Simard-Tremblay et al. 2011).
- In Smith-Magenis syndrome—a genetic disorder affecting development—sleep disturbance is nearly universal, with night waking, early morning waking, and short total sleep time common (Smith et al. 1998; Trickett et al. 2018; Shayota and Elsea 2019). REM sleep is often reduced (Greenberg et al. 1996). Etiology relates to

impaired circadian rhythms with paradoxical melatonin secretion (de Leersnyder et al. 2001, 2003; Spruyt et al. 2016).

- Children with neuromuscular disorders are at increased risk of hypoventilation, central and obstructive sleep apnoea, and respiratory and neurological consequences of nocturnal hypoxia (Dhand and Dhand 2006; Alves et al. 2009).
- Children with Down syndrome have higher rates of obstructive sleep apnoea, secondary to hypotonia, and airway abnormalities (Dyken et al. 2003; Shott et al. 2006).
- Children with Rett syndrome often have poor sleep quality with night laughing and behavioral bedtime difficulties (Wong et al. 2015; Boban et al. 2016).

The Impact of Adverse Childhood Experiences on Sleep

Poor sleep is increasingly recognized as having origins in childhood (Palagini et al. 2015). Challenging early life circumstances influence physiology and subsequently sleep and can disturb development with lifelong consequences.

Adverse childhood experiences (ACEs) are stressful events or circumstances occurring in childhood that are potentially traumatic. They include exposure to violence, abuse, neglect, suicide, substance use, and mental health problems (Centers for Disease Control and Prevention 2021). Individual responses to trauma early in life vary, ranging from mild short-term psychological stress, persisting stress with impairments in day-to-day functioning, to chronic post-traumatic stress disorder (Center for Substance Abuse Treatment (US) 2014). ACEs have been linked to a range of poor outcomes in adulthood across many domains, including physical health, wellbeing and social circumstances, with significant costs to society (Felitti et al. 1998; Felitti 2002; Bellis et al. 2018; Schurer et al. 2019; Novais et al. 2021). The influence of adverse childhood experiences on sleep in later life is no different.

Childhood adversity is associated with sleep problems during adulthood according to large-scale studies. These findings persist even when controlling for confounding variables, and demonstrate a strong dose-response relationship between the level of adversity and degree of impact on adult sleep (Koskenvuo et al. 2010; Chapman et al. 2011; Kajeepeta et al. 2015).

The enduring effects of childhood adversity can be understood through the impact of stress on the developing brain (Sanford et al. 2014). The sleep-wake cycle is vulnerable to stress, especially during childhood (Sadeh 1996; Ordway et al. 2021). Childhood trauma, through a lack of feeling of safety during significant periods of development and neuronal plasticity predisposes a child to constant watchfulness (van Someren 2021). This hyper-arousal can result in dysfunctional sleep (Montgomery and Foldspang 2001; Palagini et al. 2015; Riemann et al. 2015).

Adequate sleep, particularly REM sleep, is necessary for the processing of emotions (Charuvastra and Cloitre 2009; Vanderheyden et al. 2015; Fellman 2021). Adequate REM sleep leads to lower levels of emotionality following salient events (van der Helm et al. 2011), and conversely, emotions heighten in the setting of disrupted REM sleep (Wassing et al. 2019). Childhood adversity is known to be associated with fragmented REM sleep (Insana et al. 2012). Where REM sleep is inadequate, stressful experiences are unable to be processed, and the stress-response persists. Ongoing distress results in further sleep problems, notably insomnia, and risk of subsequent mental health disorders (van Someren 2021). Restorative sleep in childhood in the setting of stress is key to recovery from traumatic events (Charuvastra and Cloitre 2009; Spilsbury 2009).

The impact of specific severe adverse childhood experiences on sleep and development is sparsely reported in the literature. Children exposed to interpersonal violence are more likely to have sleep problems, even twelve months post exposure to violence (Spilsbury et al. 2016). Adequate sleep following exposure to violence has been noted to be protective for subsequent development (El-Sheikh and Kelly 2011) and mood disorder severity (Nowakowski et al. 2016). For children in immigration detention who are already at risk in the setting of high rates of trauma, sleep deprivation is commonly found as a result of environmental conditions of constant light, cold temperatures, inadequate bedding, and routine waking (Peeler et al. 2020). Lorek et al. (2009) reported that 90 percent of children in immigration detention had sleep problems and many regressed in sleep-associated behaviors. Regarding sexual assault, abused adolescents had higher rates of sleep disturbance, and disrupted sleep was a risk factor for revictimization (Noll et al. 2006). In the foster care system, fostered children were five times more likely than other children to display problematic behaviors following short sleep duration (Tininenko et al. 2010).

The COVID-19 pandemic provides an opportunity to examine the impact of stress on children's sleep on a global scale. Sleep problems in children doubled during the pandemic according to a systematic review and sleep duration recommendations were not met in 49 percent of children (Sharma et al. 2021). Screen time (Bruni et al. 2021), loneliness (Becker et al. 2021), reporting of death rates (Simor et al. 2021), and fears of contracting COVID-19 (Li et al. 2021) were some factors affecting children's sleep during this time.

Stressful events in childhood, both acute and chronic, negatively impact children's sleep, compounding the detrimental effects of childhood adversity on adult health and prosperity.

CONCLUSION

Adequate sleep quantity and quality is vitally important for children's development. The developing brain is particularly vulnerable to sleep deprivation due to the degree of neuronal plasticity taking place during this time. Despite extensive evidence of the importance of sleep, large numbers of children are not achieving adequate sleep on a regular basis. This may be due to underlying disease, unrecognized sleep disorders, poor sleep habits, or inadequate opportunity. Children with specific developmental disorders, such as ADHD and autism, not only have a unique set of challenges, but may require more intensive interventions to improve their sleep and developmental outcomes.

Insufficient sleep in childhood is now becoming a public health issue. Experts are calling for greater attention to the developmental understanding of sleep in children. The American Sleep Association, recognizing that sleep is essential to health, has appealed for greater sleep health education, clinical focus, and public health promotion.

Disrupted sleep in childhood, with its consequential developmental and emotional impairments, can have devastating and long-lasting effects into adulthood. Given that sleep is modifiable in childhood, this is the critical time to intervene to establish healthy sleep practices and behaviors as a foundation for lifelong health.

ACKNOWLEDGMENT

The authors would like to thank Associate Professor Harriet Hiscock for guidance with this publication.

REFERENCES

Accardo, J. A., Marcus, C. A., Leonard, M. B., Shults, J., Meltzer, L. J., and Elia, J. (2012). Associations between psychiatric comorbidities and sleep disturbances in children with attention-deficit/hyperactivity disorder. *Journal of Developmental and Behavioral Pediatrics*, 33(2), 97–105. doi:10.1097/DBP.0b013e31823f6853

Adam, K., and Oswald, I. (1977). Sleep is for tissue restoration. *Journal of the Royal College of Physicians of London*, 11(4), 376.

Aldrich, M. S., Chervin, R. D., and Malow, B. A. (1997). Value of the multiple sleep latency test (MSLT) for the diagnosis of narcolepsy. *Sleep*, 20(8), 620–629.

Alfano, C. A., Ginsburg, G., and Kingery, J. N. (2007). Sleep-related problems among children and adolescents with anxiety disorders. *Journal of the American Academy of Child & Adolescent Psychiatry*, 46(2), 224–232. doi.org/10.1097/01.chi.0000242233.06011.8e

Alvaro, P. K., Roberts, R. M., and Harris, J. K. (2013). A systematic review assessing bidirectionality between sleep disturbances, anxiety, and depression. *Sleep*, 36(7), 1059–1068. doi:10.5665/sleep.2810

Alves, R. S. C., Resende, M. B. D., Skomro, R. P., Souza, F. J. F. B., and Reed, U. C. (2009). Sleep and neuromuscular disorders in children. *Sleep Medicine Reviews*, 13(2), 133–148.

American Psychiatric Association, DSM-5 Task Force (2013). *Diagnostic and Statistical Manual of Mental Disorders: DSM-5*™ (5th ed.). Arlington, VA: American Psychiatric Association.

Arana-Lechuga, Nuñez-Ortiz R., Terán-Pérez, G., Castillo-Montoya, C., Jiménez-Anguiano, A., Gonzalez-Robles, R. O., Castro-Roman, R., and Velázquez-Moctezuma, J. (2008). Sleep-EEG patterns of school children suffering from symptoms of depression compared to healthy controls. *The World Journal of Biological Psychiatry*, 9(2), 115–120.

Armitage, R., Flynn, H., Hoffmann, R., Vazquez, D., Lopez, J., and Marcus, S. (2009). Early developmental changes in sleep in infants: the impact of maternal depression. *Sleep*, *32*(5), 693–696.

Armstrong, K. L., Quinn, R. A., and Dadds, M. R. (1994). The sleep patterns of normal children. *Medical Journal of Australia*, *161*(3), 202–206. doi:10.5694/j.1326-5377.1994.tb127383.x

Asaka, Y., and Takada, S. (2011). Comparing sleep measures of infants derived from parental reports in sleep diaries and acceleration sensors. *Acta Paediatrica*, *100*(8), 1158–1163.

Assefa, S. Z., Diaz-Abad, M., Wickwire, E. M., and Scharf, S. M. (2015). The functions of sleep. *AIMS Neuroscience*, *2*(3), 155–171.

Augustinavicius, J. L. S., Zanjani, A., Zakzanis, K. K., and Shapiro, C. M. (2014). Polysomnographic features of early-onset depression: A meta-analysis. *Journal of Affective Disorders*, *158*, 11–18. doi:10.1016/j.jad.2013.12.009

Aurora, R. N., Lamm, C. I., Zak, R. S., Kristo, D. A., Bista, S. R., Rowley, J. A., and Casey, K. R. (2012). Practice parameters for the non-respiratory indications for polysomnography and multiple sleep latency testing for children. *Sleep*, *35*(11), 1467–1473.

Ballester, P., Richdale, A. L., Baker, E. K., and Peiró, A. M. (2020). Sleep in autism: A biomolecular approach to aetiology and treatment. *Sleep Medicine Reviews*, *54*, 101357. doi:10.1016/j.smrv.2020.101357

Ballester-Navarro, P., Martínez-Madrid, M. J., Javaloyes-Sanchís, A., Belda-Canto, C., Aguilar, V., Richdale, A. L., Muriel, J., Morales, D., and Peiró, A. M. (2021). Interplay of circadian clock and melatonin pathway gene variants in adults with autism, intellectual disability and sleep problems. *Research in Autism Spectrum Disorders*, *81*, 101715.

Barrett, J. R., Tracy, D. K., and Giaroli, G. (2013). To sleep or not to sleep: A systematic review of the literature of pharmacological treatments of insomnia in children and adolescents with attention-deficit/hyperactivity disorder. *Journal of Child and Adolescent Psychopharmacology*, *23*(10), 640–647. Doi:10.1089/cap.2013.0059

Bassetti, C. L. A., Adamantidis, A., Burdakov, D., Han, F., Gay, S., Kallweit, U., Khatami, R., Koning, F., Kornum, B. R., Lammers, G. J., and Liblau, R. S. (2019). Narcolepsy—clinical spectrum, aetiopathophysiology, diagnosis and treatment. *Nature Reviews Neurology*, *15*(9), 519–539.

Bayer, J. K., Hiscock, H., Hampton, A., and Wake, M. (2007). Sleep problems in young infants and maternal mental and physical health. *Journal of Paediatrics and Child Health*, *43*(1–2), 66–73. Doi:10.1111/j.1440-1754.2007.01005.x

Beattie, L., Kyle, S. D., Espie, C. A., and Biello, S. M. (2015). Social interactions, emotion and sleep: A systematic review and research agenda. *Sleep Medicine Reviews*, *24*, 83–100.

Becker, S. P., Dvorsky, M. R., Breaux, R., Cusick, C. N., Taylor, K. P., and Langberg, J. M. (2021). Prospective examination of adolescent sleep patterns and behaviors before and during COVID-19. *Sleep*, *44*(8), 1–11. Doi:10.1093/sleep/zsab054

Becker, S. P., Langberg, J. M., and Evans, S. W. (2015). Sleep problems predict comorbid externalizing behaviors and depression in young adolescents with attention-deficit/hyperactivity disorder. *European Child and Adolescent Psychiatry*, *24*(8), 897–907. Doi:10.1007/s00787-014-0636-6

Bellis, M. A., Hughes, K., Ford, K., Hardcastle, K. A., Sharp, C. A., Wood, S., Homolova, L., and Davies, A. (2018). Adverse childhood experiences and sources of childhood resilience: A retrospective study of their combined relationships with child health and educational attendance. *BMC Public Health*, *18*(1), 792. doi:10.1186/s12889-018-5699-8

Bernert, R. A., Kim, J. S., Iwata, N. G., and Perlis, M. L. (2015). Sleep disturbances as an evidence-based suicide risk factor. *Current Psychiatry Reports*, 17(3), 554.

Bernier, A., Beauchamp, M. H., Bouvette-Turcot, A.-A., Carlson, S. M., and Carrier, J. (2013). Sleep and cognition in preschool years: Specific links to executive functioning. *Child Development*, 84(5), 1542–1553. doi:10.1111/cdev.12063

Biancardi, C., Sesso, G., Masi, G., Faraguna, U., and Sicca, F. (2021). Sleep EEG microstructure in children and adolescents with attention deficit hyperactivity disorder: A systematic review and meta-analysis. *Sleep*, 44(7), zsab006. doi:10.1093/sleep/zsab006

Bianchi, M. T. (2018). Sleep devices: Wearables and nearables, informational and interventional, consumer and clinical. *Metabolism*, 84, 99–108.

Biggs, S. N., Lushington, K., van den Heuvel, C. J., Martin, A. J., and Kennedy, J. D. (2011). Inconsistent sleep schedules and daytime behavioral difficulties in school-aged children. *Sleep Medicine*, 12(8), 780–786.

Bixler, E. O., Vgontzas, A. N., Lin, H. M., Liao, D., Calhoun, S., Vela-Bueno, A., Fedok, F., Vlasic, V., and Graff, G. (2009). Sleep disordered breathing in children in a general population sample: Prevalence and risk factors. *Sleep*, 32(6), 731–736.

Black, B. (2021). Sleep and autism: An impactful and complex relationship that requires a personalized medicine approach. *Sleep*, 44(9), zsab178. doi: 10.1093/sleep/zsab178

Blake, M. J., Sheeber, L. B., Youssef, G. J., Raniti, M. B., and Allen, N. B. (2017). Systematic review and meta-analysis of adolescent cognitive-behavioral sleep interventions. *Clinical Child and Family Psychology Review*, 20(3), 227–249.

Blakemore, S. J. (2012). Imaging brain development: The adolescent brain. *Neuroimage*, 61(2), 397–406.

Blanken, T. F., Borsboom, D., Penninx, B. W., and Van Someren, E. J. (2020). Network outcome analysis identifies difficulty initiating sleep as a primary target for prevention of depression: A 6-year prospective study. *Sleep*, 43(5), zsz288.

Blunden, S., Lushington, K., Lorenzen, B., Ooi, T., Fung, F., and Kennedy, D. (2004). Are sleep problems under-recognised in general practice? *Archives of Disease in Childhood*, 89(8), 708–712.

Boban, S., Wong, K., Epstein, A., Anderson, B., Murphy, N., Downs, J., and Leonard, H. (2016). Determinants of sleep disturbances in Rett syndrome: Novel findings in relation to genotype. *American Journal of Medical Genetics Part A*, 170(9), 2292–2300.

Boonstra, T. W., Stins, J. F., Daffertshofer, A., and Beek, P. J. (2007). Effects of sleep deprivation on neural functioning: An integrative review. *Cellular and Molecular Life Sciences*, 64(7), 934–946.

Borbély, A. A. (1982). A two process model of sleep regulation. *Human Neurobiology*, 1(3), 195–204.

Borbély, A. A., and Achermann, P. (1992). Concepts and models of sleep regulation: An overview. *Journal of Sleep Research*, 1(2), 63–79.

Borbély, A. A., and Achermann, P. (1999). Sleep homeostasis and models of sleep regulation. *Journal of Biological Rhythms*, 14(6), 557–568.

Born, J., Muth, S., and Fehm, H. L. (1988). The significance of sleep onset and slow wave sleep for nocturnal release of growth hormone (GH) and cortisol. *Psychoneuroendocrinology*, 13(3), 233–243.

Bourgeron, T. (2007). The possible interplay of synaptic and clock genes in autism spectrum disorders. *Cold Spring Harbor Symposia on Quantitative Biology*, 72, 645–654. doi:10.1101/sqb.2007.72.020

Bowers, J. M., and Moyer, A. (2017) Effects of school start time on students' sleep duration, daytime sleepiness, and attendance: a meta-analysis, *Sleep Health*, 3(6), 423–431.

Brieva, T. E., Casale, C. E., Yamazaki, E. M., Antler, C. A., and Goel, N. (2021). Cognitive throughput and working memory raw scores consistently differentiate resilient and vulnerable groups to sleep loss. *Sleep*, 44(12), zsab197. doi:10.1093/sleep/zsab197

Brown, W. J., Wilkerson, A. K., Boyd, S. J., Dewey, D., Mesa, F., and Bunnell, B. E. (2018). A review of sleep disturbance in children and adolescents with anxiety. *Journal of Sleep Research*, 27(3), e12635. doi:10.1111/jsr.12635

Bruni, O., Malorgio, E., Doria, M., Finotti, E., Spruyt, K., Melegari, M. G., Villa, M. P., and Ferri, R. (2021). Changes in sleep patterns and disturbances in children and adolescents in Italy during the Covid-19 outbreak. *Sleep Medicine*, 91, 166–174. doi: 10.1016/j.sleep.2021.02.003

Buck, C., Kawai, M., O' Hara, R., Chick, C., Anker, L., and Schneider, L. (2021). 607 Longer slow wave sleep and exacerbated core symptom severity in autism spectrum disorder. *Sleep*, 44(S 2), A239–A239. doi:10.1093/sleep/zsab072.605

Buckley, A. W., Hirtz, D., Oskoui, M., Armstrong, M. J., Batra, A., Bridgemohan, C., Coury, D., Dawson, G., Donley, D., Findling, R. L., and Gaughan, T. (2020). Practice guideline: Treatment for insomnia and disrupted sleep behavior in children and adolescents with autism spectrum disorder: Report of the Guideline Development, Dissemination, and Implementation Subcommittee of the American Academy of Neurology. *Neurology*, 94(9), 392–404.

Burnett, S., Sebastian, C., Cohen Kadosh, K., and Blakemore, S.-J. (2011). The social brain in adolescence: Evidence from functional magnetic resonance imaging and behavioural studies. *Neuroscience & Biobehavioral Reviews*, 35(8), 1654–1664.

Calhoun, S. L., Vgontzas, A. N., Fernandez-Mendoza, J., Mayes, S. D., Tsaoussoglou, M., Basta, M., and Bixler, E. O. (2011). Prevalence and risk factors of excessive daytime sleepiness in a community sample of young children: The role of obesity, asthma, anxiety/depression, and sleep. *Sleep*, 34(4), 503–507.

Cappuccio, F. P., Taggart, F. M., Kandala, N.-B., Currie, A., Peile, E., Stranges, S., and Miller, M. A. (2008). Meta-analysis of short sleep duration and obesity in children and adults. *Sleep*, 31(5), 619–626.

Cardoso, T. da S. G., Pompéia, S., and Miranda, M. C. (2018). Cognitive and behavioral effects of obstructive sleep apnea syndrome in children: A systematic literature review. *Sleep Medicine*, 46, 46–55.

Carmassi, C., Palagini, L., Caruso, D., Masci, I., Nobili, L., Vita, A., and Dell'Osso, L. (2019). Systematic review of sleep disturbances and circadian sleep desynchronization in autism spectrum disorder: Toward an integrative model of a self-reinforcing loop. *Frontiers in Psychiatry*, 10, 366.

Carroll, J. L., McColley, S. A., Marcus, C. L., Curtis, S., and Loughlin, G. M. (1995). Inability of clinical history to distinguish primary snoring from obstructive sleep apnea syndrome in children. *Chest*, 108(3), 610–618.

Centers for Disease Control and Prevention (CDC) (US) (2021). Preventing adverse childhood experiences. Retrieved from https://www.cdc.gov/violenceprevention/aces/fastfact.html.

Challamel, M.-J., Mazzola, M. E., Nevsimalova, S., Cannard, C., Louis, J., and Revol, M. (1994). Narcolepsy in children. *Sleep*, 17(S_8), S17–S20.

Chan, J., Edman, J. C., and Koltai, P. J. (2004). Obstructive sleep apnea in children. *American Family Physician*, 69(5), 1147–1154.

Chapman, Wheaton A. G., Anda, R. F., Croft, J. B., Edwards, V. J., Liu, Y., Sturgis, S. L., and Perry, G. S. (2011). Adverse childhood experiences and sleep disturbances in adults. *Sleep Medicine*, *12*(8), 773–779. doi.org/10.1016/j.sleep.2011.03.013

Charuvastra, A., and Cloitre, M. (2009). Safe enough to sleep: Sleep disruptions associated with trauma, posttraumatic stress, and anxiety in children and adolescents. *Child and Adolescent Psychiatric Clinics of North America*, *18*(4), 877–891. doi.org/10.1016/j.chc.2009.04.002.

Chase, R. M., and Pincus, D. B. (2011) Sleep-related problems in children and adolescents with anxiety disorders. *Behavioral Sleep Medicine*, *9*(4), 224–236. doi:10.1080/15402002.2011.606768

Chen, X., Liu, H., Wu, Y., Xuan, K., Zhao, T., and Sun, Y. (2020). Characteristics of sleep architecture in autism spectrum disorders: A meta-analysis based on polysomnographic research. *Psychiatry Research*, *296*, 113677.

Chervin, R. D., Archbold, K. H., Dillon, J. E., Panahi, P., Pituch, K. J., Dahl, R. E., and Guilleminault, C. (2002a). Inattention, hyperactivity, and symptoms of sleep-disordered breathing. *Pediatrics*, *109*(3), 449–456. doi:10.1542/peds.109.3.449

Chervin, R. D., Archbold, K. H., Dillon, J. E., Pituch, K. J., Panahi, P., Dahl, R. E., and Guilleminault, C. (2002b). Associations between symptoms of inattention, hyperactivity, restless legs, and periodic leg movements. *Sleep*, *25*(2), 213–218.

Chervin, R. D., Dillon, J. E., Bassetti, C., Ganoczy, D. A., and Pituch, K. J. (1997). Symptoms of sleep disorders, inattention, and hyperactivity in children. *Sleep*, *20*(12), 1185–1192. doi:10.1093/sleep/20.12.1185

Chorney, D. B., Detweiler, M. F., Morris, T. L., and Kuhn, B. R. (2008). The interplay of sleep disturbance, anxiety, and depression in children. *Journal of Pediatric Psychology*, *33*(4), 339–348. doi:10.1093/jpepsy/jsm105

Choudhury, S., Blakemore, S.-J., and Charman, T. (2006). Social cognitive development during adolescence. *Social Cognitive and Affective Neuroscience*, *1*(3), 165–174.

Colonna, A., Smith, A. B., Pal, D. K., and Gringras, P. (2015). Novel mechanisms, treatments, and outcome measures in childhood sleep. *Frontiers in Psychology*, *6*. doi:10.3389/fpsyg.2015.00602

Conway, L. J., Levickis, P. A., Smith, J., Mensah, F., Wake, M., and Reilly, S. (2018). Maternal communicative behaviours and interaction quality as predictors of language development: findings from a community-based study of slow-to-talk toddlers. *International Journal of Language & Communication Disorders*, *53*(2), 339–354. doi:10.1111/1460-6984.12352

Cook, F., Conway, L. J., Giallo, R., Gartland, D., Sciberras, E., and Brown, S. (2020). Infant sleep and child mental health: A longitudinal investigation. *Archives of Disease in Childhood*, *105*(7), 655–660.

Cook, F., Conway, L. J., Omerovic, E., Cahir, P., Giallo, R., Hiscock, H., Mensah, F., Bretherton, L., Bavin, E., Eadie, P., Brown, S., and Reilly, S. (2021). Infant regulation: Associations with child language development in a longitudinal cohort. *The Journal of Pediatrics*, *233*, 90–97.

Cook, F., Giallo, R., Hiscock, H., Mensah, F., Sanchez, K., and Reilly, S. (2019). Infant regulation and child mental health concerns: A longitudinal study. *Pediatrics*, *143*(3):e20180977.

Cook, F., Giallo, R., Petrovic, Z., Coe, A., Seymour, M., Cann, W., and Hiscock, H. (2017). Depression and anger in fathers of unsettled infants: a community cohort study. *Journal of Paediatrics and Child Health*, *53*(2), 131–135.

Cooney, M. R., Short, M. A., and Gradisar, M. (2018). An open trial of bedtime fading for sleep disturbances in preschool children: A parent group education approach. *Sleep Medicine*, *46*, 98–106.

Corkum, P., Davidson, F., and MacPherson, M. (2011). A framework for the assessment and treatment of sleep problems in children with attention-deficit/hyperactivity disorder. *Pediatric Clinics*, 58(3), 667–683.

Corkum, P., Tannock, R., Moldofsky, H., Hogg-Johnson, S., and Humphries, T. (2001). Actigraphy and parental ratings of sleep in children with attention-deficit/hyperactivity disorder (ADHD). *Sleep*, 24(3), 303–312.

Cortese, S., Brown, T. E., Corkum, P., Gruber, R., O'Brien, L. M., Stein, M., Weiss, M., and Owens, J. (2013). Assessment and management of sleep problems in youths with attention-deficit/hyperactivity disorder. *Journal of the American Academy of Child & Adolescent Psychiatry*, 52(8), 784–796.

Cortese, S., Faraone, S. V., Konofal, E., and Lecendreux, M. (2009). Sleep in children with attention-deficit/hyperactivity disorder: Meta-analysis of subjective and objective studies. *Journal of the American Academy of Child & Adolescent Psychiatry*, 48(9), 894–908.

Cortese, S., Konofal, E., Lecendreux, M., Arnulf, I., Mouren, M. C., Darra, F., and Bernardina, B. D. (2005). Restless legs syndrome and attention-deficit/hyperactivity disorder: A review of the literature. *Sleep*, 28(8), 1007–1013.

Crabtree, V. M., Varni, J. W., and Gozal, D. (2004). Health-related quality of life and depressive symptoms in children with suspected sleep-disordered breathing. *Sleep*, 27(6), 1131–1138. doi:10.1093/sleep/27.6.1131

Crabtree, V. M., and Williams, N. A. (2009). Normal sleep in children and adolescents. *Child and Adolescent Psychiatric Clinics*, 18(4), 799–811.

Cuomo, B. M., Vaz, S., Lee, E. A. L., Thompson, C., Rogerson, J. M., and Falkmer, T. (2017). Effectiveness of sleep-based interventions for children with autism spectrum disorder: a meta-synthesis. *Pharmacotherapy: The Journal of Human Pharmacology and Drug Therapy*, 37(5), 555–578.

Dahl, R. E. (2007). Sleep and the developing brain. *Sleep*, 30(9), 1079–1080. doi:10.1093/sleep/30.9.1079

Damon A, W., Menon, J., and Bronk, K. C. (2019). The development of purpose during adolescence. In J. L. Furrow and L. M. Wagener (eds.), *Beyond the Self: Perspectives on Identity and Transcendence Among Youth: A Special Issue of Applied Developmental Science*, 119–128. New York, NY: Routledge.

Danielsson, K., Markström, A., Broman, J. E., von Knorring, L., and Jansson-Fröjmark, M. (2016). Delayed sleep phase disorder in a Swedish cohort of adolescents and young adults: Prevalence and associated factors. *Chronobiology International*, 33(10), 1331–1339.

Danielsson, N. S., Harvey, A. G., MacDonald, S., Jansson-Fröjmark, M., and Linton, S. J. (2013). Sleep disturbance and depressive symptoms in adolescence: The role of catastrophic worry. *Journal of Youth and Adolescence*, 42, 1223–1233.

Davis, K. F., Parker, K. P., and Montgomery, G. L. (2004). Sleep in infants and young children: Part one: Normal sleep. *Journal of Pediatric Health Care*, 18(2), 65–71.

de Bruin, E. J., Bögels, S. M., Oort, F. J., and Meijer, A. M. (2018). Improvements of adolescent psychopathology after insomnia treatment: Results from a randomized controlled trial over 1 year. *Journal of Child Psychology and Psychiatry*, 59(5), 509–522.

de Leersnyder, H., Bresson, J. L., De Blois, M. C., Souberbielle, J. C., Mogenet, A., Delhotal-Landes, B., Salefranque, F., and Munnich, A. (2003). β1-adrenergic antagonists and melatonin reset the clock and restore sleep in a circadian disorder, Smith-Magenis syndrome. *Journal of Medical Genetics*, 40(1), 74–78.

de Leersnyder, H., de Blois, M. C., Claustrat, B., Romana, S., Albrecht, U., von Kleist-Retzow, J. C., Delobel, B., Viot, G., Lyonnet, S., Vekemans, M., and Munnich, A. (2001). Inversion of the circadian rhythm of melatonin in the Smith-Magenis syndrome. *The Journal of Pediatrics*, *139*(1), 111–116.

Dearing, E., McCartney, K., Marshall, N. L., and Warner, R. M. (2001). Parental reports of children's sleep and wakefulness: Longitudinal associations with cognitive and language outcomes. *Infant Behavior and Development*, *24*(2), 151–170.

DeGangi, G. A., Breinbauer, C., Roosevelt, J. D., Porges, S., and Greenspan, S. (2000). Prediction of childhood problems at three years in children experiencing disorders of regulation during infancy. *Infant Mental Health Journal: Official Publication of the World Association for Infant Mental Health*, *21*(3), 156–175.

Delahaye, J., Kovacs, E., Sikora, D., Hall, T. A., Orlich, F., Clemons, T. E., Van Der Weerd, E., Glick, L., and Kuhlthau, K. (2014). The relationship between health-related quality of life and sleep problems in children with autism spectrum disorders. *Research in Autism Spectrum Disorders*, *8*(3), 292–303.

Delemere, E., and Dounavi, K. (2018). Parent-implemented bedtime fading and positive routines for children with autism spectrum disorders. *Journal of Autism and Developmental Disorders*, *48*(4), 1002–1019.

DelRosso, L. M., Ferri, R., Chen, M. L., Kapoor, V., Allen, R. P., Mogavero, M. P., and Picchietti, D. L. (2021). Clinical efficacy and safety of intravenous ferric carboxymaltose treatment of pediatric restless legs syndrome and periodic limb movement disorder. *Sleep Medicine*, *87*, 114–118.

Deng, X., He, M., He, D., Zhu, Y., Zhang, Z., and Niu, W. (2021). Sleep duration and obesity in children and adolescents: Evidence from an updated and dose–response meta-analysis. *Sleep Medicine*, *78*, 169–181.

Dennis, C., and Ross, L. (2005). Relationships among infant sleep patterns, maternal fatigue, and development of depressive symptomatology. *Birth*, *32*(3), 187–193.

Dewald, J. F., Meijer, A. M., Oort, F. J., Kerkhof, G. A., and Bögels, S. M. (2010). The influence of sleep quality, sleep duration and sleepiness on school performance in children and adolescents: A meta-analytic review. *Sleep Medicine Reviews*, *14*(3), 179–189.

Dhand, U. K., and Dhand, R. (2006). Sleep disorders in neuromuscular diseases. *Current Opinion in Pulmonary Medicine*, *12*(6), 402–408.

Díaz-Román, A., Zhang, J., Delorme, R., Beggiato, A., and Cortese, S. (2018). Sleep in youth with autism spectrum disorders: Systematic review and meta-analysis of subjective and objective studies. *Evidence Based Mental Health*, *21*(4), 146–154.

Diekelmann, S., and Born, J. (2010). The memory function of sleep. *Nature Reviews Neuroscience*, *11*(2), 114–126.

Dijk, D. J. (2009). Regulation and functional correlates of slow wave sleep. *Journal of Clinical Sleep Medicine*, *5*(S 2), S6–S15.

Dionne, G., Touchette, E., Forget-Dubois, N., Petit, D., Tremblay, R. E., Montplaisir, J. Y., and Boivin, M. (2011). Associations between sleep-wake consolidation and language development in early childhood: A longitudinal twin study. *Sleep*, *34*(8), 987–995.

Dorris, L., Scott, N., Zuberi, S., Gibson, N., and Espie, C. (2008). Sleep problems in children with neurological disorders. *Developmental Neurorehabilitation*, *11*(2), 95–114.

Dye, T. J., Gurbani, N., and Simakajornboon, N. (2019). How does one choose the correct pharmacotherapy for a pediatric patient with restless legs syndrome and periodic limb movement disorder? Expert guidance. *Expert Opinion on Pharmacotherapy, 20*(13), 1535–1538.

Dyken, M. E., Lin-Dyken, D. C., Poulton, S., Zimmerman, M. B., and Sedars, E. (2003). Prospective polysomnographic analysis of obstructive sleep apnea in Down syndrome. *Archives of Pediatrics & Adolescent Medicine, 157*(7), 655–660.

Ednick, M., Cohen, A. P., McPhail, G. L., Beebe, D., Simakajornboon, N., and Amin, R. S. (2009). A review of the effects of sleep during the first year of life on cognitive, psychomotor, and temperament development. *Sleep, 32*(11), 1449–1458.

Elrod, M. G., and Hood, B. S. (2015). Sleep differences among children with autism spectrum disorders and typically developing peers: A meta-analysis. *Journal of Developmental & Behavioral Pediatrics, 36*(3), 166–177.

El-Sheikh, M., and Kelly, R. J. (2011). Sleep in children: Links with marital conflict and child development. In M. El-Sheikh, R. J. Kelly, and L. E. Philbrook (eds.), *Sleep and Development: Familial and Socio-cultural Considerations*, 3–28. New York, NY: Oxford University Press. doi:10.1093/acprof:oso/9780195395754.003.0001

Emde, R. N., and Metcalf, D. R. (1970). An electroencephalographic study of behavioral rapid eye movement states in the human newborn. *Journal of Nervous and Mental Disease, 150*(5), 376–86.

Emslie, G. J., Armitage, R., Weinberg, W. A., Rush, A. J., Mayes, T. L., and Hoffmann, R. F. (2001). Sleep polysomnography as a predictor of recurrence in children and adolescents with major depressive disorder. *International Journal of Neuropsychopharmacology, 4*(2), 159–168.

Fatima, Y., Doi, S. A. R., and Mamun, A. A. (2015). Longitudinal impact of sleep on overweight and obesity in children and adolescents: A systematic review and bias-adjusted meta-analysis. *Obesity Reviews, 16*(2), 137–149.

Fatima, Y., Doi, S. A. R., and Mamun, A. A. (2016) Sleep quality and obesity in young subjects: A meta-analysis. *Obesity Reviews, 17*(11), 1154–1166.

Felitti, V. J. (2002). The relation between adverse childhood experiences and adult health: Turning gold into lead. *The Permanente Journal, 6*(1), 44–47. Retrieved from https://pub med.ncbi.nlm.nih.gov/30313011

Felitti, V. J., Anda, R. F., Nordenberg, D., Williamson, D. F., Spitz, A. M., Edwards, V., and Marks, J. S. (1998). Relationship of childhood abuse and household dysfunction to many of the leading causes of death in adults: The Adverse Childhood Experiences (ACE) Study. *American Journal of Preventive Medicine, 14*(4), 245–258.

Fellman, V., Heppell, P. J., and Rao, S. (2021). Afraid and awake: The interaction between trauma and sleep in children and adolescents. *Child and Adolescent Psychiatric Clinics, 30*(1), 225–249.

Fitzgerald, C. T., Messias, E., and Buysse, D. J. (2011). Teen sleep and suicidality: Results from the youth risk behavior surveys of 2007 and 2009. *Journal of Clinical Sleep Medicine, 7*(4), 351–356. doi: 10.5664/JCSM.1188

Fitzgerald, D. A., Follett, J., and van Asperen, P. P. (2009). Assessing and managing lung disease and sleep disordered breathing in children with cerebral palsy. *Paediatric Respiratory Reviews, 10*(1), 18–24.

Forbes, E. E., Williamson, D. E., Ryan, N. D., Birmaher, B., Axelson, D. A., and Dahl, R. E. (2006). Peri-sleep-onset cortisol levels in children and adolescents with affective disorders. *Biological Psychiatry, 59*(1), 24–30.

Franzen, P. L., and Buysse, D. J. (2008). Sleep disturbances and depression: Risk relationships for subsequent depression and therapeutic implications. *Dialogues in Clinical Neuroscience, 10*(4), 473.

Freeman, K. A. (2006). Treating bedtime resistance with the bedtime pass: A systematic replication and component analysis with 3-year-olds. *Journal of Applied Behavior Analysis, 39*(4), 423–428.

Fricke-Oerkermann, L., Plück, J., Schredl, M., Heinz, K., Mitschke, A., Wiater, A., and Lehmkuhl, G. (2007). Prevalence and course of sleep problems in childhood. *Sleep, 30*(10), 1371–1377.

Friedrich, M., Wilhelm, I., Born, J., and Friederici, A. D. (2015). Generalization of word meanings during infant sleep. *Nature communications, 6*(1), 6004.

Gaina, A., Sekine, M., Chen, X., Hamanishi, S., and Kagamimori, S. (2004). Validity of child sleep diary questionnaire among junior high school children. *Journal of Epidemiology, 14*(1), 1–4.

Galland, B. C., and Mitchell, E. A. (2010). Helping children sleep. *Archives of Disease in Childhood, 95*(10), 850–853.

Galland, B. C., Taylor, B. J., Elder, D. E., and Herbison, P. (2012). Normal sleep patterns in infants and children: A systematic review of observational studies. *Sleep Medicine Reviews, 16*(3), 213–222.

Gamaldo, C. E., Shaikh, A. K., and McArthur, J. C. (2012).The sleep-immunity relationship. *Neurologic Clinics, 30*(4), 1313–1343. doi:10.1016/j.ncl.2012.08.007

Gao, B., Dwivedi, S., Milewski, M. D., and Cruz, A. I. (2019). Lack of sleep and sports injuries in adolescents: A systematic review and meta-analysis. *Journal of Pediatric Orthopaedics, 39*(5), e324–e333.

Giallo, R., Woolhouse, H., Gartland, D., Hiscock, H., and Brown, S. (2015). The emotional–behavioural functioning of children exposed to maternal depressive symptoms across pregnancy and early childhood: A prospective Australian pregnancy cohort study. *European Child & Adolescent Psychiatry, 24*, 1233–1244.

Gibson, R., Elder, D., and Gander, P. (2012). Actigraphic sleep and developmental progress of one-year-old infants. *Sleep and Biological Rhythms, 10*(2), 77–83. doi:10.1111/j.1479-8425.2011.00525.x

Giedd, J. N. (2004). Structural magnetic resonance imaging of the adolescent brain. *Annals of the New York Academy of Sciences, 1021*(1), 77–85.

Gillott, A., Furniss, F., and Walter, A. (2001). Anxiety in high-functioning children with autism. *Autism, 5*(3), 277–286.

Goldman, S. E., Richdale, A. L., Clemons, T., and Malow, B. A. (2012). Parental sleep concerns in autism spectrum disorders: variations from childhood to adolescence. *Journal of Autism and Developmental Disorders, 42*, 531–538.

Golley, R. K., Maher, C. A., Matricciani, L., and Olds, T. S. (2013). Sleep duration or bedtime? Exploring the association between sleep timing behaviour, diet and BMI in children and adolescents. *International Journal of Obesity, 37*(4), 546–551.

Gradisar, M., Gardner, G., and Dohnt, H. (2011). Recent worldwide sleep patterns and problems during adolescence: A review and meta-analysis of age, region, and sleep. *Sleep Medicine, 12*(2), 110–118. doi:10.1016/j.sleep.2010.11.008

Gradisar, M., Jackson, K., Spurrier, N. J., Gibson, J., Whitham, J., Williams, A. S., Dolby, R., and Kennaway, D. J. (2016). Behavioral interventions for infant sleep problems: A randomized controlled trial. *Pediatrics, 137*(6): e0151486

Graven, S. N., and Browne, J. V. (2008). Sleep and brain development: The critical role of sleep in fetal and early neonatal brain development. *Newborn and Infant Nursing Reviews, 8*(4), 173–179.

Greenberg, F., Lewis, R. A., Potocki, L., Glaze, D., Parke, J., Killian, J. J., Murphy, M. A., Williamson, D., Brown, F., Dutton, R., and Lupski, J. R. (1996). Multi-disciplinary clinical study of Smith-Magenis syndrome (deletion 17p11. 2). *American Journal of Medical Genetics, 62*(3), 247–254.

Gregory, A. M., Caspi, A., Eley, T. C., Moffitt, T. E., O'Connor, T. G., and Poulton, R. (2005). Prospective longitudinal associations between persistent sleep problems in childhood and anxiety and depression disorders in adulthood. *Journal of Abnormal Child Psychology, 33*, 157–163.

Griffiths, A., Mukushi, A., and Adams, A. M. (2022). Telehealth-supported level 2 pediatric home polysomnography. *Journal of Clinical Sleep Medicine, 18*(7), 1815–1821.

Griggs, S., Conley, S., Batten, J., and Grey, M. (2020). A systematic review and meta-analysis of behavioral sleep interventions for adolescents and emerging adults. *Sleep medicine reviews, 54*, 101356.

Gringras, P., Gamble, C., Jones, A. P., Wiggs, L., Williamson, P. R., Sutcliffe, A., Montgomery, P., and Appleton, R. (2012). Melatonin for sleep problems in children with neurodevelopmental disorders: Randomised double masked placebo controlled trial. *British Medical Journal 345* :e6664

Gringras, P., Green, D., Wright, B., Rush, C., Sparrowhawk, M., Pratt, K. K., Allgar, V., Hooke, N., Moore, D., Zaiwalla, Z., and Wiggs, L. (2014). Weighted blankets and sleep in autistic children—A randomized controlled trial. *Pediatrics, 134*(2), 298–306.

Guilleminault, C., and Pelayo, R. (2000). Narcolepsy in children. *Pediatric Drugs, 2*(1), 1–9.

Gupta, P., Sagar, R., and Mehta, M. (2019). Subjective sleep problems and sleep hygiene among adolescents having depression: A case-control study. *Asian Journal of Psychiatry, 44*, 150–155. doi.org/10.1016/j.ajp.2019.07.034

Gurbani, N., Dye, T. J., Dougherty, K., Jain, S., Horn, P. S., and Simakajornboon, N. (2019). Improvement of parasomnias after treatment of restless leg syndrome/periodic limb movement disorder in children. *Journal of Clinical Sleep Medicine, 15*(5), 743–748.

Haraden, D. A., Mullin, B. C., and Hankin, B. L. (2017). The relationship between depression and chronotype: A longitudinal assessment during childhood and adolescence. *Depression and Anxiety, 34*(10), 967–976.

Hart, D., and Carlo, G. (2005). Moral development in adolescence. *Journal of Research on Adolescence, 15*(3), 223–233.

Harvey, A. G. (2011). Sleep and circadian functioning: critical mechanisms in the mood disorders? *Annual Review of Clinical Psychology, 7*, 297–319.

Hemmi, M. H., Wolke, D., and Schneider, S. (2011). Associations between problems with crying, sleeping and/or feeding in infancy and long-term behavioural outcomes in childhood: A meta-analysis. *Archives of Disease in Childhood, 96*(7), 622–9. doi:10.1136/adc.2010.191312

Hering, E., Epstein, R., Elroy, S., Iancu, D. R., and Zelnik, N. (1999). Sleep patterns in autistic children. *Journal of Autism & Developmental Disorders, 29*(2), 143–147.

Hirshkowitz, M., Whiton, K., Albert, S. M., Alessi, C., Bruni, O., DonCarlos, L., Hazen, N., Herman, J., Katz, E. S., Kheirandish-Gozal, L., Neubauer, D. N., O'Donnell, A. E., Ohayon M., Peever, J., Rawding, R., Sachdeva, R. C., Setters, B., Vitiello, M. V., Ware, J. C., and Adams Hillard, P. J. (2015). National Sleep Foundation's sleep time duration recommendations: Methodology and results summary. *Sleep Health, 1*(1), 40–43.

Hiscock, H., Bayer, J., Gold, L., Hampton, A., Ukoumunne, O. C., and Wake, M. (2007). Improving infant sleep and maternal mental health: A cluster randomised trial. *Archives of Disease in Childhood*, 92(11), 952–958.

Hiscock, H., Ng, O., Crossley, L., Chow, J., Rausa, V., and Hearps, S. (2021). Sleep Well Be Well: Pilot of a digital intervention to improve child behavioural sleep problems. *Journal of Paediatrics and Child Health*, 57(1), 33–40.

Hiscock, H., and Wake, M. (2001). Infant sleep problems and postnatal depression: A community-based study. *Pediatrics*, 107(6), 1317–22.

Hiscock, H., and Wake, M. (2002). Randomised controlled trial of behavioural infant sleep intervention to improve infant sleep and maternal mood. *British Medical Journal*, 324(7345), 1062.

Hodge, D., Carollo, T. M., Lewin, M., Hoffman, C. D., and Sweeney, D. P. (2014). Sleep patterns in children with and without autism spectrum disorders: Developmental comparisons. *Research in Developmental Disabilities*, 35(7), 1631–1638.

Horiuchi, F., Kawabe, K., Oka, Y., Nakachi, K., Hosokawa, R., and Ueno, S. I. (2021). Mental health and sleep habits/problems in children aged 3–4 years: A population study. *BioPsychoSocial Medicine*, 15(1), 1–9.

Hudson, J. L., Gradisar, M., Gamble, A., Schniering, C. A., and Rebelo, I. (2009). The sleep patterns and problems of clinically anxious children. *Behaviour Research and Therapy*, 47(4), 339–344.

Humphreys, J. S., Gringras, P., Blair, P. S., Scott, N., Henderson, J., Fleming, P. J., and Emond, A. M. (2014). Sleep patterns in children with autistic spectrum disorders: A prospective cohort study. *Archives of Disease in Childhood*, 99(2), 114–118.

Hupbach, A., Gomez, R. L., Bootzin, R. R., and Nadel, L. (2009). Nap-dependent learning in infants. *Developmental Science*, 12(6), 1007–1012.

Iglowstein, I., Jenni, O. G., Molinari, L., and Largo, R. H. (2003). Sleep duration from infancy to adolescence: Reference values and generational trends. *Pediatrics*, 111(2), 302–307.

Inocente, C. O., Gustin, M. P., Lavault, S., Guignard-Perret, A., Raoux, A., Christol, N., Gerard, D., Dauvilliers, Y., Reimão, R., Bat-Pitault, F., and Lin, J. S., Arnulf, I., Lecendreux, M., and Franco, P. (2014). Quality of life in children with narcolepsy. *CNS Neuroscience & Therapeutics*, 20(8), 763–771.

Insana, S. P., Kolko, D. J., and Germain, A. (2012). Early-life trauma is associated with rapid eye movement sleep fragmentation among military veterans. *Biological Psychology*, 89(3), 570–579.

Ipsiroglu, O. S., Fatemi, A., Werner, I., Paditz, E., and Schwarz, B. (2002). Self-reported organic and nonorganic sleep problems in schoolchildren aged 11 to 15 years in Vienna. *Journal of Adolescent Health*, 31(5), 436–442.

Ivanenko, A., McLaughlin Crabtree, V. M., O'Brien, L. M., Brien, and Gozal, D. (2006). Sleep complaints and psychiatric symptoms in children evaluated at a pediatric mental health clinic. *Journal of Clinical Sleep Medicine*, 2(1), 42–48.

Iwadare, Y., Kamei, Y., Usami, M., Ushijima, H., Tanaka, T., Watanabe, K., Kodaira, M., and Saito, K. (2015). Behavioral symptoms and sleep problems in children with anxiety disorder. *Pediatrics International*, 57(4), 690–693.

Jan, J. E., Reiter, R. J., Bax, M. C., Ribary, U., Freeman, R. D., and Wasdell, M. B. (2010). Long-term sleep disturbances in children: A cause of neuronal loss. *European Journal of Paediatric Neurology*, 14(5), 380–390.

Jan, J. E., Reiter, R. J., Wasdell, M. B., and Bax, M. (2009). The role of the thalamus in sleep, pineal melatonin production, and circadian rhythm sleep disorders. *Journal of Pineal Research*, 46(1), 1–7.

Jenni, O. G., and Carskadon, M. A. (2005). Chapter 1: Normal human sleep at different ages: Infants to adolescents key concepts. *Basics of Sleep Guide*, 11–19. Sleep Research Society. Retrieved from http://scifun.org/conversations/Conversations4Teachers/Chapter 1.pdf

Jeyakumar, A., Rahman, S. I., Armbrecht, E. S., and Mitchell, R. (2012). The association between sleep-disordered breathing and enuresis in children. *The Laryngoscope*, 122(8), 1873–1877.

Kahn, A., Dan, B., Groswasser, J., Franco, P., and Sottiaux, M. (1996). Normal sleep architecture in infants and children. *Journal of Clinical Neurophysiology*, 13(3), 184–197.

Kajeepeta, S., Gelaye, B., Jackson, C. L., and Williams, M. A. (2015). Adverse childhood experiences are associated with adult sleep disorders: A systematic review. *Sleep Medicine*, 16(3), 320–330.

Kales, A., Soldatos, C. R., Bixler, E. O., Ladda, R. L., Charney, D. S., Weber, G., and Schweitzer, P. K. (1980). Hereditary factors in sleepwalking and night terrors. *The British Journal of Psychiatry*, 137(2), 111–118.

Katz, E. S., and D'Ambrosio, C. M. (2008). Pathophysiology of pediatric obstructive sleep apnea. *Proceedings of the American Thoracic Society*, 5(2), 253–262.

Kim, M. J., Lee, J. H., and Duffy, J. F. (2013). Circadian rhythm sleep disorders. *Journal of Clinical Outcomes Management*, 20(11), 513.

Kocevska, D., Lysen, T. S., Dotinga, A., Koopman-Verhoeff, M. E., Luijk, M. P., Antypa, N., Biermasz, N. R., Blokstra, A., Brug, J., Burk, W. J. Comijs, H. C., Corpeleijn, E., Dashti, H. S., de Bruin, E. J., de Graaf, R., Derks, I. P. M., Dewald-Kaufmann, J. F., Elders, P. J. M., Gemke, R. J. B. J., ... & Tiemeier, H. (2021). Sleep characteristics across the lifespan in 1.1 million people from the Netherlands, United Kingdom and United States: A systematic review and meta-analysis. *Nature Human Behaviour*, 5(1), 113–122.

Kohler, M. J., Lushington, K., and Kennedy, J. D. (2010). Neurocognitive performance and behavior before and after treatment for sleep-disordered breathing in children. *Nature and Science of Sleep*, 2, 159.

Koskenvuo, K., Hublin, C., Partinen, M., Paunio, T., and Koskenvuo, M. (2010). Childhood adversities and quality of sleep in adulthood: A population-based study of 26,000 Finns. *Sleep Medicine*, 11(1), 17–22.

Kotagal, S., and Broomall, E. (2012). Sleep in children with autism spectrum disorder. *Pediatric Neurology*, 47(4), 242–251. doi:10.1016/j.pediatrneurol.2012.05.007

Krakowiak, P., Goodlin-Jones, B. E. T. H., Hertz-Picciotto, I. R. V. A., Croen, L. A., and Hansen, R. L. (2008). Sleep problems in children with autism spectrum disorders, developmental delays, and typical development: A population-based study. *Journal of Sleep Research*, 17(2), 197–206.

Kruizinga, M. D., Heide, N. V. D., Moll, A., Zhuparris, A., Yavuz, Y., Kam, M. D., Stuurman, F. E., Cohen, A. F., and Driessen, G. J. A. (2021). Towards remote monitoring in pediatric care and clinical trials—Tolerability, repeatability and reference values of candidate digital endpoints derived from physical activity, heart rate and sleep in healthy children. *PloS One*, 16(1), e0244877.

Kulman, G., Lissoni, P., Rovelli, F., Roselli, M. G., Brivio, F., and Sequeri, P. (2000). Evidence of pineal endocrine hypofunction in autistic children. *Neuroendocrinology Letters*, 21(1), 31–34.

Laberge, L., Tremblay, R. E., Vitaro, F., and Montplaisir, J. (2000). Development of parasomnias from childhood to early adolescence. *Pediatrics*, 106(1), 67–74.

Lambert, A., Tessier, S., Chevrier, É., Scherzer, P., Mottron, L., and Godbout, R. (2013). Sleep in children with high functioning autism: Polysomnography, questionnaires and diaries in a non-complaining sample. *Sleep Medicine, 14*, e137–e138.

Landry, S. H., Smith, K. E., and Swank, P. R. (2006). Responsive parenting: Establishing early foundations for social, communication, and independent problem-solving skills. *Developmental Psychology, 42*(4), 627–642. doi:10.1037/0012-1649.42.4.627

Lask, B. (1988). Novel and non-toxic treatment for night terrors. *British Medical Journal, 297*(6648), 592.

LeBourgeois, M. K., Wright Jr, K. P., LeBourgeois, H. B., and Jenni, O. G. (2013). Dissonance between parent-selected bedtimes and young children's circadian physiology influences nighttime settling difficulties. *Mind, Brain, and Education, 7*(4), 234–242.

Lee, J. H., Kim, I. S., Kim, S. J., Wang, W., and Duffy, J. F. (2011). Change in individual chronotype over a lifetime: A retrospective study. *Sleep Medicine Research, 2*(2), 48–53.

Lee, J., Na, G., Joo, E. Y., Lee, M., and Lee, J. (2017). Clinical and polysomnographic characteristics of excessive daytime sleepiness in children. *Sleep and Breathing, 21*, 967–974.

Lesage, S., and Hening, W. A. (2004). The restless legs syndrome and periodic limb movement disorder: A review of management. *Seminars in Neurology, 24*(3), 249–259.

Leu, R. M., Beyderman, L., Botzolakis, E. J., Surdyka, K., Wang, L., and Malow, B. A. (2011). Relation of melatonin to sleep architecture in children with autism. *Journal of Autism and Developmental Disorders, 41*, 427–433.

Lewandowski, A. S., Toliver-Sokol, M., and Palermo, T. M. (2011). Evidence-based review of subjective pediatric sleep measures. *Journal of Pediatric Psychology, 36*(7), 780–793.

Li, H. Y., and Lee, L. A. (2009). Sleep-disordered breathing in children. *Chang Gung Medical Journal, 32*(3), 247–257. doi:10.1542/pir.2018-0142

Li, Y., Zhou, Y., Ru, T., Niu, J., He, M., and Zhou, G. (2021). How does the COVID-19 affect mental health and sleep among Chinese adolescents: A longitudinal follow-up study. *Sleep Medicine, 85*, 246–258.

Liang, M., Guo, L., Huo, J., and Zhou, G. (2021). Prevalence of sleep disturbances in Chinese adolescents: A systematic review and meta-analysis. *Plos One, 16*(3), e0247333.

Lim, J., and Dinges, D. F. (2010). A meta-analysis of the impact of short-term sleep deprivation on cognitive variables. *Psychological Bulletin, 136*(3), 375.

Liu, X., Buysse, D. J., Gentzler, A. L., Kiss, E., Mayer, L., Kapornai, K., Vetró, A., and Kovacs, M. (2007). Insomnia and hypersomnia associated with depressive phenomenology and comorbidity in childhood depression. *Sleep, 30*(1), 83–90.

Lorek, A., Ehntholt, K., Nesbitt, A., Wey, E., Githinji, C., Rossor, E., and Wickramasinghe, R. (2009). The mental and physical health difficulties of children held within a British immigration detention center: A pilot study. *Child Abuse & Neglect, 33*(9), 573–585.

Lovato, N., and Gradisar, M. (2014). A meta-analysis and model of the relationship between sleep and depression in adolescents: Recommendations for future research and clinical practice. *Sleep Medicine Review, 18*(6), 521–529. doi:10.1016/j.smrv.2014.03.006

Lumeng, J. C., and Chervin, R. D. (2008). Epidemiology of pediatric obstructive sleep apnea. *Proceedings of the American Thoracic Society, 5*(2), 242–252.

Ma, G., Segwa, M., Nomura, Y., Kondo, Y., Yanagitani, M., and Higurashi, M. (1993). The development of sleep-wakefulness rhythm in normal infants and young children. *The Tohoku Journal of Experimental Medicine, 171*(1), 29–41.

Mackintosh, K. A., Chappel, S. E., Salmon, J., Timperio, A., Ball, K., Brown, H., Macfarlane, S., and Ridgers, N. D. (2019). Parental perspectives of a wearable activity tracker for children younger than 13 years: Acceptability and usability study. *JMIR mHealth and uHealth, 7*(11), e13858.

Marcus, C. L., Brooks, L. J., Ward, S. D., Draper, K. A., Gozal, D., Halbower, A. C., Jones, J., Lehmann, C., Schechter, M. S., Sheldon, S., and Shiffman, R. N. (2012). Diagnosis and management of childhood obstructive sleep apnea syndrome. *Pediatrics, 130*(3), e714–e755.

Marcus, C. L., Moore, R. H., Rosen, C. L., Giordani, B., Garetz, S. L., Taylor, H. G., Mitchell, R. B., Amin, R., Katz, E. S., Arens, R., and Paruthi, S. (2013). A randomized trial of adenotonsillectomy for childhood sleep apnea. *New England Journal Medicine, 368*, 2366–2376.

Marino, C., Andrade, B., Campisi, S. C., Wong, M., Zhao, H., Jing, X., Aitken, M., Bonato, S., Haltigan, J., Wang, W., and Szatmari, P. (2021). Association between disturbed sleep and depression in children and youths: A systematic review and meta-analysis of cohort studies. *JAMA Network Open, 4*(3), e212373. doi:10.1001/jamanetworkopen.2021.2373

Mayes, S. D., and Calhoun, S. L. (2009). Variables related to sleep problems in children with autism. *Research in Autism Spectrum Disorders, 3*(4), 931–941.

Mazurek, M. O., Dovgan, K., Neumeyer, A. M., and Malow, B. A. (2019). Course and predictors of sleep and co-occurring problems in children with autism spectrum disorder. *Journal of Autism and Developmental Disorders, 49*(5), 2101–2115. doi:10.1007/s10803-019-03894-5

Mazurek, M. O., and Petroski, G. F. (2015). Sleep problems in children with autism spectrum disorder: Examining the contributions of sensory over-responsivity and anxiety. *Sleep Medicine, 16*(2), 270–279.

Mazza, S., Bastuji, H., and Rey, A. E. (2020). Objective and subjective assessments of sleep in children: comparison of actigraphy, sleep diary completed by children and parents' estimation. *Frontiers in Psychiatry, 11*, 495.

McLay, L., France, K., Blampied, N., van Deurs, J., Hunter, J., Knight, J., Hastie, B., Carnett, A., Woodford, E., Gibbs, R., and Lang, R. (2021). Function-based behavioral interventions for sleep problems in children and adolescents with autism: Summary of 41 clinical cases. *Journal of Autism and Developmental Disorders, 51*, 418–432.

Mei, X., Zhou, Q., Li, X., Jing, P., Wang, X., and Hu, Z. (2018). Sleep problems in excessive technology use among adolescent: A systematic review and meta-analysis. *Sleep Science and Practice, 2*(1), 1–10.

Melke, J., Goubran Botros, H., Chaste, P., Betancur, C., Nygren, G., Anckarsäter, H., Rastam, M., Ståhlberg, O., Gillberg, I. C., Delorme, R., and Chabane, N. (2008). Abnormal melatonin synthesis in autism spectrum disorders. *Molecular Psychiatry, 13*(1), 90–98.

Meltzer, L. J., and Mindell, J. A. (2004). Nonpharmacologic treatments for pediatric sleeplessness. *Pediatric Clinics, 51*(1), 135–151.

Meltzer, L. J., and Mindell, J. A. (2014). Systematic review and meta-analysis of behavioral interventions for pediatric insomnia. *Journal of Pediatric Psychology, 39*(8), 932–948.

Miano, S., Donfrancesco, R., Bruni, O., Ferri, R., Galiffa, S., Pagani, J., Montemitro, E., Kheirandish, L., Gozal, D., and Villa, M. P. (2006). NREM sleep instability is reduced in children with attention-deficit/ hyperactivity disorder. *Sleep, 29*(6), 797–803. doi:10.1093/sleep/29.6.797

Mignot, E. (2008). Why we sleep: The temporal organization of recovery. *pLoS Biology, 6*(4), e106.

Miller, M. A., Kruisbrink, M., Wallace, J., Ji, C., and Cappuccio, F. P. (2018). Sleep duration and incidence of obesity in infants, children, and adolescents: A systematic review and meta-analysis of prospective studies. *Sleep, 41*(4), zsy018.

Mindell, J. A., Kuhn, B., Lewin, D. S., Meltzer, L. J., and Sadeh, A. (2006). Behavioral treatment of bedtime problems and night wakings in infants and young children. *Sleep: Journal of Sleep and Sleep Disorders Research*, 29(10), 1263–1276.

Mindell, J. A., and Meltzer, L. J. (2008). Behavioural sleep disorders in children and adolescents. *Annals Academy of Medicine Singapore*, 37(8), 722–728.

Mindell, J. A., and Owens, J. A. (2015). *A Clinical Guide to Pediatric Sleep: Diagnosis and Management of Sleep Problems*. Philadelphia, PA: Lippincott Williams and Wilkins.

Mindell, J. A., Sadeh, A., Kwon, R., and Goh, D. Y. (2013). Cross-cultural differences in the sleep of preschool children. *Sleep Medicine*, 14(12), 1283–1289.

Mirmiran, M., Maas, Y. G. H., and Ariagno, R. L. (2003). Development of fetal and neonatal sleep and circadian rhythms. *Sleep Medicine Reviews*, 7(4), 321–334.

Miyagawa, T., and Tokunaga, K. (2019). Genetics of narcolepsy. *Human Genome Variation*, 6(1), 1–8.

Montgomery, E., and Foldspang, A. (2001). Traumatic experience and sleep disturbance in refugee children from the Middle East. *The European Journal of Public Health*, 11(1), 18–22.

Moore, B. A., Friman, P. C., Fruzzetti, A. E., and MacAleese, K. (2007). Brief report: Evaluating the bedtime pass program for child resistance to bedtime—a randomized, controlled trial. *Journal of Pediatric Psychology*, 32(3), 283–287.

Morgenthaler, T. I., Owens, J., Alessi, C., Boehlecke, B., Brown, T. M., Coleman Jr, J., Friedman, L., Kapur, V. K., Lee-Chiong, T., Pancer, J., and Swick, T. J. (2006). Practice parameters for behavioral treatment of bedtime problems and night wakings in infants and young children. *Sleep*, 29(10), 1277–1281.

Mulder, E. J., Anderson, G. M., Kema, I. P., De Bildt, A., Van Lang, N. D., Den Boer, J. A., and Minderaa, R. B. (2004). Platelet serotonin levels in pervasive developmental disorders and mental retardation: diagnostic group differences, within-group distribution, and behavioral correlates. *Journal of the American Academy of Child & Adolescent Psychiatry*, 43(4), 491–499.

Mundkur, N. (2005). Neuroplasticity in children. *The Indian Journal of Pediatrics*, 72(10), 855–857.

Muris, P., Merckelbach, H., Ollendick, T. H., King, N. J., and Bogie, N. (2001). Children's nighttime fears: Parent-child ratings of frequency, content, origins, coping behaviors and severity. *Behaviour Research and Therapy*, 39(1), 13–28.

Nakamura, K., Sekine, Y., Ouchi, Y., Tsujii, M., Yoshikawa, E., Futatsubashi, M., Tsuchiya, K. J., Sugihara, G., Iwata, Y., Suzuki, K., and Matsuzaki, H. (2010). Brain serotonin and dopamine transporter bindings in adults with high-functioning autism. *Archives of General Psychiatry*, 67(1), 59–68.

Narasimhan, U., Anitha, F. S., Anbu, C., and Abdul hameed, M. F. (2020). The spectrum of sleep disorders among children: a cross-sectional study at a South Indian tertiary care hospital. *Cureus*, 12(4):: e7535.

Natarajan, R. (2010). Review of periodic limb movement and restless leg syndrome. *Journal of Postgraduate Medicine*, 56(2), 157.

National Center for Chronic Disease Prevention and Health Promotion (2017). Sleep and sleep disorders: Data and statistics. Centers for Disease Control and Prevention. Retrieved from https://www.cdc.gov/sleep/data_statistics.html

Newman, C. J., O'Regan, M., and Hensey, O. (2006). Sleep disorders in children with cerebral palsy. *Developmental Medicine and Child Neurology*, 48(7), 564–568.

Nicholas, B., Rudrasingham, V., Nash, S., Kirov, G., Owen, M. J., and Wimpory, D. C. (2007). Association of Per1 and Npas2 with autistic disorder: Support for the clock genes/social timing hypothesis. *Molecular Psychiatry*, 12(6), 581–592.

Noll, J. G., Trickett, P. K., Susman, E. J., and Putnam, F. W. (2006). Sleep disturbances and childhood sexual abuse. *Journal of Pediatric Psychology*, 31(5), 469–480. doi:10.1093/jpepsy/jsj040

Novais, M., Henriques, T., Vidal-Alves, M. J., and Magalhães, T. (2021). When problems only get bigger: The impact of adverse childhood experience on adult health. *Frontiers in Psychology*, 12. doi.org/10.3389/fpsyg.2021.693420

Nowakowski, S., Choi, H. J., Meers, J., and Temple, J. R. (2016). Inadequate sleep as a mediating variable between exposure to interparental violence and depression severity in adolescents. *Journal of Child & Adolescent Trauma*, 9(2), 109–114. doi:10.1007/s40653-016-0091-2

O'Connor, T. G., Caprariello, P., Blackmore, E. R., Gregory, A. M., Glover, V., Fleming, P., and ALSPAC Study Team (2007). Prenatal mood disturbance predicts sleep problems in infancy and toddlerhood. *Early Human Development*, 83(7), 451–458.

Ordway, M. R., Condon, E. M., Ibrahim, B. B., Abel, E. A., Funaro, M. C., Batten, J., Sadler, L. S., and Redeker, N. S. (2021). A systematic review of the association between sleep health and stress biomarkers in children. *Sleep Medicine Reviews*, 59, 101494. doi: 10.1016/j.smrv.2021.101494

Orzeł-Gryglewska, J. (2010). Consequences of sleep deprivation. *International Journal of Occupational Medicine and Environmental Health*, 23(1), 95–114. doi: 10.2478/v10001-010-0004-9

Owens, J. (2007). Classification and epidemiology of childhood sleep disorders. *Sleep Medicine Clinics*, 2(3), 353–361.

Owens, J. L., France, K. G., and Wiggs, L. (1999). Behavioural and cognitive-behavioural interventions for sleep disorders in infants and children: A review. *Sleep Medicine Reviews*, 3(4), 281–302.

Oygür, I., Su, Z., Epstein, D. A., and Chen, Y. (2021). The lived experience of child-owned wearables: Comparing children's and parents' perspectives on activity tracking. *Proceedings of the 2021 CHI Conference on Human Factors in Computing Systems*, 1–12.

Paavonen, E. J., Raikkonen, K., Lahti, J., Komsi, N., Heinonen, K., Pesonen, A. K., Jarvenpaa, A. L., Strandberg, T., Kajantie, E., and Porkka-Heiskanen, T. (2009). Short sleep duration and behavioral symptoms of attention-deficit/ hyperactivity disorder in healthy 7- to 8-year-old children. *Pediatrics*, 123(5), e857–e864. doi:10.1542/peds.2008-2164

Pagan, C., Delorme, R., Callebert, J., Goubran-Botros, H., Amsellem, F., Drouot, X., Boudebesse, C., Le Dudal, K., Ngo-Nguyen, N., Laouamri, H., and Gillberg, C. (2014). The serotonin-N-acetylserotonin-melatonin pathway as a biomarker for autism spectrum disorders. *Translational Psychiatry*, 4(11), e479–e479.

Pagel, J. F., Forister, N., and Kwiatkowki, C. (2007). Adolescent sleep disturbance and school performance: the confounding variable of socioeconomics. *Journal of Clinical Sleep Medicine*, 3(01), 19–23.

Palagini, L., Drake, C. L., Gehrman, P., Meerlo, P., and Riemann, D. (2015) Early-life origin of adult insomnia: Does prenatal–early-life stress play a role? *Sleep Medicine*, 16(4), 446–456. doi.org/10.1016/j.sleep.2014.10.013

Papadopoulos, N., Sciberras, E., Hiscock, H., Mulraney, M., McGillivray, J., and Rinehart, N. (2019). The efficacy of a brief behavioral sleep intervention in school-aged children with

ADHD and comorbid autism spectrum disorder. *Journal of Attention Disorders, 23*(4), 341–350.

Paruthi, S., Brooks, L. J., D'Ambrosio, C., Hall, W. A., Kotagal, S., Lloyd, R. M., Malow, B. A., Maski, K., Nichols, C., Quan, S. F., and Rosen, C. L. (2016a). Consensus statement of the American Academy of Sleep Medicine on the recommended amount of sleep for healthy children: Methodology and discussion. *Journal of Clinical Sleep Medicine, 12*(11), 1549–1561.

Paruthi, S., Brooks, L. J., D'Ambrosio, C., Hall, W. A., Kotagal, S., Lloyd, R. M., Malow, B. A., Maski, K., Nichols, C., Quan, S. F., and Rosen, C. L. (2016b). Recommended amount of sleep for pediatric populations: A consensus statement of the American Academy of Sleep Medicine. *Journal of Clinical Sleep Medicine, 12*(6), 785–786.

Peach, H., Gaultney, J. F., and Gray, D. D. (2016). Sleep hygiene and sleep quality as predictors of positive and negative dimensions of mental health in college students. *Cogent Psychology, 3*(1), 1168768.

Peeler, K. R., Hampton, K., Lucero, J., and Ijadi-Maghsoodi, R. (2020). Sleep deprivation of detained children: Another reason to end child detention. *Health and Human Rights, 22*(1), 317–320.

Peirano, P. D., and Algarín, C. R. (2007). Sleep in brain development. *Biological Research, 40*(4), 471–478.

Pesonen, A.-K., and Kuula, L. (2018). The validity of a new consumer-targeted wrist device in sleep measurement: An overnight comparison against polysomnography in children and adolescents. *Journal of Clinical Sleep Medicine, 14*(4), 585–591.

Philbrook, L. E., and Teti, D. M. (2016). Bidirectional associations between bedtime parenting and infant sleep: Parenting quality, parenting practices, and their interaction. *Journal of Family Psychology, 30*(4), 431–441. doi:10.1037/fam0000198

Picchietti, D. L., Bruni, O., de Weerd, A., Durmer, J. S., Kotagal, S., Owens, J. A., Simakajornboon, N., and International Restless Legs Syndrome Study Group (2013). Pediatric restless legs syndrome diagnostic criteria: An update by the International Restless Legs Syndrome Study Group. *Sleep Medicine, 14*(12), 1253–1259.

Pigeon, W. R., Bishop, T. M., and Titus, C. E. (2016). The relationship between sleep disturbance, suicidal ideation, suicide attempts, and suicide among adults: A systematic review. *Psychiatric Annals, 46*(3), 177–186.

Pilcher, J. J., and Huffcutt, A. I. (1996). Effects of sleep deprivation on performance: A meta-analysis. *Sleep, 19*(4), 318–326.

Piteo, A. M., Kennedy, J. D., Roberts, R. M., Martin, A. J., Nettelbeck, T., Kohler, M. J., and Lushington, K. (2011). Snoring and cognitive development in infancy. *Sleep Medicine, 12*(10), 981–987.

Piteo, A. M. Roberts, R. M., Nettelbeck, T., Burns, N., Lushington, K., Martin, A. J., and Kennedy, J. D. (2013). Postnatal depression mediates the relationship between infant and maternal sleep disruption and family dysfunction. *Early Human Development, 89*(2), 69–74.

Price, A. M. H., Brown, J. E., Bittman, M., Wake, M., Quach, J., and Hiscock, H. (2014). Children's sleep patterns from 0 to 9 years: Australian population longitudinal study. *Archives of Disease in Childhood, 99*(2), 119–125. doi:10.1136/archdischild-2013-304150

Price, A. M., Wake, M., Ukoumunne, O. C., and Hiscock, H. (2012). Five-year follow-up of harms and benefits of behavioral infant sleep intervention: Randomized trial. *Pediatrics, 130*(4), 643–651.

Quach, J., Price, A. M., Bittman, M., and Hiscock, H. (2016). Sleep timing and child and parent outcomes in Australian 4–9-year-olds: A cross-sectional and longitudinal study. *Sleep Medicine*, *22*, 39–46.

Quach, J. L., Nguyen, C. D., Williams, K. E., and Sciberras, E. (2018). Bidirectional associations between child sleep problems and internalizing and externalizing difficulties from preschool to early adolescence. *JAMA Pediatrics*, *172*(2), e174363. doi:10.1001/jamapediatrics.2017.4363

Räihä, H., Lehtonen, L., Huhtala, V., Saleva, K., and Korvenranta, H. (2002). Excessively crying infant in the family: Mother–infant, father–infant and mother–father interaction. *Child: Care, Health and Development*, *28*(5), 419–429.

Ramar, K., Malhotra, R. K., Carden, K. A., Martin, J. L., Abbasi-Feinberg, F., Aurora, R. N., Kapur, V. K., Olson, E. J., Rosen, C. L., Rowley, J. A., and Shelgikar, A. V. (2021). Sleep is essential to health: An American Academy of Sleep Medicine position statement. *Journal of Clinical Sleep Medicine*, *17*(10), 2115–2119. doi: 10.5664/jcsm.9476

Reckley, L. K., Fernandez-Salvador, C., and Camacho, M. (2018). The effect of tonsillectomy on obstructive sleep apnea: An overview of systematic reviews. *Nature and Science of Sleep*, *10*, 105.

Reigstad, B., Jørgensen, K., Sund, A. M., and Wichstrøm, L. (2010). Prevalences and correlates of sleep problems among adolescents in specialty mental health services and in the community: What differs? *Nordic Journal of Psychiatry*, *64*(3), 172–180.

Reynolds, A. M., and Malow, B. A. (2011). Sleep and autism spectrum disorders. *Pediatric Clinics*, *58*(3), 685–698.

Reynolds, S., Lane, S. J., and Thacker, L. (2012). Sensory processing, physiological stress, and sleep behaviors in children with and without autism spectrum disorders. *OTJR: Occupation, Participation and Health*, *32*(1), 246–257.

Richardson, C., Oar, E., Fardouly, J., Magson, N., Johnco, C., Forbes, M., and Rapee, R. (2019). The moderating role of sleep in the relationship between social isolation and internalising problems in early adolescence. *Child Psychiatry & Human Development*, *50*(6), 1011–1020.

Richardson, C., Ree, M., Bucks, R. S., and Gradisar, M. (2021). Paediatric sleep literacy in Australian health professionals. *Sleep Medicine*, *81*, 327–335.

Richdale, A. L., Baker, E., Short, M., and Gradisar, M. (2014). The role of insomnia, pre-sleep arousal and psychopathology symptoms in daytime impairment in adolescents with high-functioning autism spectrum disorder. *Sleep Medicine*, *15*(9), 1082–1088.

Richdale, A. L., and Prior, M. R. (1995). The sleep/wake rhythm in children with autism. *European Child & Adolescent Psychiatry*, *4*(3), 175–186.

Richdale, A. L., and Schreck, K. A. (2009). Sleep problems in autism spectrum disorders: prevalence, nature, & possible biopsychosocial aetiologies. *Sleep medicine reviews*, *13*(6), 403–411.

Riemann, D., Nissen, C., Palagini, L., Otte, A., Perlis, M. L., and Spiegelhalder, K. (2015). The neurobiology, investigation, and treatment of chronic insomnia. *The Lancet Neurology*, *14*(5), 547–558.

Roenneberg, T., Wirz-Justice, A., and Merrow, M. (2003). Life between clocks: Daily temporal patterns of human chronotypes. *Journal of Biological Rhythms*, *18*(1), 80–90.

Rossignol, D. A., and Frye, R. E. (2011). Melatonin in autism spectrum disorders: A systematic review and meta-analysis. *Developmental Medicine and Child Neurology*, *53*(9), 783–792. doi:10.1111/j.1469-8749.2011.03980

Ruch, S., Valiadis, M., and Gharabaghi, A. (2021). Sleep to learn. *Sleep*, *44*(8), zsab160. doi:10.1093/sleep/zsab160

Russo, K., Greenhill, J., and Burgess, S. (2021). Home (Level 2) polysomnography is feasible in children with suspected sleep disorders. *Sleep Medicine, 88*, 157–161.

Sadeh, A. (1996). Stress, trauma and sleep in children. *Child and Adolescent Psychiatric Clinics of North America, 5*, 685–700.

Sadeh, A., De Marcas, G., Guri, Y., Berger, A., Tikotzky, L., and Bar-Haim, Y. (2015). Infant sleep predicts attention regulation and behavior problems at 3–4 years of age. *Developmental Neuropsychology, 40*(3), 122–137. doi:10.1080/87565641.2014.973498

Saenz, J., Yaugher, A., and Alexander, G. M. (2015). Sleep in infancy predicts gender specific social-emotional problems in toddlers. *Frontiers in Pediatrics, 3*. doi:10.3389/FPED.2015.00042

Sanabra, M., Gómez-Hinojosa, T., Alcover, C., Sans, O., and Alda, J. A. (2021). Effects of stimulant treatment on sleep in attention deficit hyperactivity disorder (ADHD). *Sleep and Biological Rhythms, 19*(1), 69–77.

Sanford, L. D., Suchecki, D., and Meerlo, P. (2014). Stress, arousal, and sleep. In P. Meerlo, R. Benca, and T. Abel (eds.), Sleep, Neuronal Plasticity and Brain Function, 379–410. *Current Topics in Behavioral Neurosciences, Vol. 25.* Berlin and Heidelberg: Springer. doi.org/10.1007/7854_2014_314

Sassin, J. F., Parker, D. C., Mace, J. W., Gotlin, R. W., Johnson, L. C., and Rossman, L. G. (1969). Human growth hormone release: Relation to slow-wave sleep and sleep-waking cycles. *Science, 165*(3892), 513–515.

Scher, A. (2005). Infant sleep at 10 months of age as a window to cognitive development. *Early Human Development, 81*(3), 289–292. Doi:10.1016/J.EARLHUMDEV.2004.07.005

Scher, A., Zukerman, S., and Epstein, R. (2005). Persistent night waking and settling difficulties across the first year: Early precursors of later behavioural problems? *Journal of Reproductive and Infant Psychology, 23*(1), 77–88. Doi:10.1080/02646830512331330929

Schlarb, A. A., Gulewitsch, M. D., Weltzer, V., Ellert, U., and Enck, P. (2015). Sleep duration and sleep problems in a representative sample of German children and adolescents. *Health, 7*(11), 1397.

Schlarb, A. A., Sopp, R., Ambiel, D., and Grünwald, J. (2014). Chronotype-related differences in childhood and adolescent aggression and antisocial behavior: A review of the literature. *Chronobiology International, 31*(1), 1–16.

Schredl, M. (2001). Night terrors in children: Prevalence and influencing factors. *Sleep and Hypnosis, 3*, 68–72.

Schurer, S., Trajkovski, K., and Hariharan, T. (2019). Understanding the mechanisms through which adverse childhood experiences affect lifetime economic outcomes. *Labour Economics, 61*, 101743. doi.org/10.1016/j.labeco.2019.06.007

Sciberras, E., DePetro, A., Mensah, F., and Hiscock, H. (2015). Association between sleep and working memory in children with ADHD: A cross-sectional study. *Sleep Medicine, 16*(10), 1192–1197. doi:10.1016/j.sleep.2015.06.006

Sciberras, E., Mulraney, M., Mensah, F., Oberklaid, F., Efron, D., and Hiscock, H. (2019). Sustained impact of a sleep intervention and moderators of treatment outcome for children with ADHD: A randomised controlled trial. *Psychological Medicine, 50*(2), 210–219. doi:10.1017/S0033291718004063

Scott, N., Blair, P. S., Emond, A. M., Fleming, P. J., Humphreys, J. S., Henderson, J., and Gringras, P. (2013). Sleep patterns in children with ADHD: A population-based cohort study from birth to 11 years. *Journal of Sleep Research, 22*(2), 121–128. doi:10.1111/j.1365-2869.2012.01054.x

Sebastian, C., Burnett, S., and Blakemore, S.-J. (2008). Development of the self-concept during adolescence. *Trends in Cognitive Sciences, 12*(11), 441–446.

Sedky, K., Bennett, D. S., and Carvalho, K. S. (2014). Attention deficit hyperactivity disorder and sleep disordered breathing in pediatric populations: A meta-analysis. *Sleep Medicine Reviews*, 18(4), 349–356.

Shamsaei, F., Daraei, M. M., and Aahmadinia, H. (2019). The relationship between sleep habits and mental health in Iranian elementary school children. *Sleep Science*, 12(2), 94.

Shanahan, L., Copeland, W. E., Angold, A., Bondy, C. L., and Costello, E. J. (2014). Sleep problems predict and are predicted by generalized anxiety/depression and oppositional defiant disorder. *Journal of the American Academy of Child & Adolescent Psychiatry*, 53(5), 550–558.

Shapiro, C. M., and Flanigan, M. J. (1993). ABC of sleep disorders. Function of sleep. *British Medical Journal*, 306(6874), 383.

Sharma, M., Aggarwal, S., Madaan, P., Saini, L., and Bhutani, M. (2021). Impact of COVID-19 pandemic on sleep in children and adolescents: A systematic review and meta-analysis. *Sleep Medicine*, 84, 259–267.

Shayota, B. J., and Elsea, S. H. (2019). Behavior and sleep disturbance in Smith-Magenis syndrome. *Current Opinion in Psychiatry*, 32(2), 73.

Sheldon, S. H. (2005). Sleep in infants and children. In L.-C. Teofilo (ed.), *Sleep: A Comprehensive Handbook*, 507–510. Hoboken, NJ: John Wiley & Sons.

Sheth, A., Yip, H. Y., Jaimini, U., Kadariya, D., Sridharan, V., Venkataramanan, R., Banerjee, T., Thirunarayan, K., and Kalra, M. (2018). Feasibility of recording sleep quality and sleep duration using Fitbit in children with asthma. *Sleep*, 41(1), A297, https://doi.org/10.1093/sleep/zsy061.798

Short, M. A., Gradisar, M., Lack, L. C., Wright, H., and Carskadon, M. A. (2012). The discrepancy between actigraphic and sleep diary measures of sleep in adolescents. *Sleep Medicine*, 13(4), 378–384.

Short, M. A., Booth, S. A., Omar, O., Ostlundh, L., and Arora, T. (2020). The relationship between sleep duration and mood in adolescents: A systematic review and meta-analysis. *Sleep Medicine Reviews*, 52, 101311.

Shott, S. R., Amin, R., Chini, B., Heubi, C., Hotze, S., and Akers, R. (2006). Obstructive sleep apnea: Should all children with Down syndrome be tested? *Archives of Otolaryngology—Head & Neck Surgery*, 132(4), 432–436.

Silvestri, R., Gagliano, A., Aricò, I., Calarese, T., Cedro, C., Bruni, O., Condurso, R., Germanò, E., Gervasi, G., Siracusano, R., and Vita, G. (2009). Sleep disorders in children with attention-deficit/hyperactivity disorder (ADHD) recorded overnight by video-polysomnography. *Sleep Medicine*, 10(10), 1132–1138. doi:10.1016/j.sleep.2009.04.003

Simakajornboon, N., Gozal, D., Vlasic, V., Mack, C., Sharon, D., and McGinley, B. M. (2003). Periodic limb movements in sleep and iron status in children. *Sleep*, 26(6), 735–738.

Simard-Tremblay, E., Constantin, E., Gruber, R., Brouillette, R. T., and Shevell, M. (2011). Sleep in children with cerebral palsy: A review. *Journal of Child Neurology*, 26(10), 1303–1310.

Simor, P., Polner, B., Báthori, N., Sifuentes-Ortega, R., Van Roy, A., Albajara Sáenz, A., Luque González, A., Benkirane, O., Nagy, T., and Peigneux, P. (2021). Home confinement during the COVID-19: Day-to-day associations of sleep quality with rumination, psychotic-like experiences, and somatic symptoms. *Sleep*, 44(7), zsab029.

Simpkin, C. T., Jenni, O. G., Carskadon, M. A., Wright Jr, K. P., Akacem, L. D., Garlo, K. G., and LeBourgeois, M. K. (2014). Chronotype is associated with the timing of the circadian clock and sleep in toddlers. *Journal of Sleep Research*, 23(4), 397–405.

Singh, G. K., and Kenney, M. K. (2013). Rising prevalence and neighborhood, social, and behavioral determinants of sleep problems in US children and adolescents, 2003–2012. *Sleep Disorders*, 2013, 1–15. Doi:10.1155/2013/394320

Singh, K., and Zimmerman, A. W. (2015). Sleep in autism spectrum disorder and attention deficit hyperactivity disorder. *Seminars in Pediatric Neurology*, 22(2), 113–125. Doi:10.1016/j.spen.2015.03.006

Singh, S., Kaur, H., Singh, S., and Khawaja, I. (2018). Parasomnias: A comprehensive review. *Cureus*, 10(12), e3807. doi 10.7759/cureus.3807

Sivertsen, B., Pallesen, S., Stormark, K. M., Bøe, T., Lundervold, A. J., and Hysing. M. (2013). Delayed sleep phase syndrome in adolescents: Prevalence and correlates in a large population based study. *BMC Public Health*, 13(1), 1–10.

Smith, A. C. M., Dykens, E., and Greenberg, F. (1998). Sleep disturbance in Smith-Magenis syndrome (del 17 p11. 2). *American Journal of Medical Genetics*, 81(2), 186–191.

Song, S. A., Tolisano, A. M., Cable, B. B., and Camacho, M. (2016). Neurocognitive outcomes after pediatric adenotonsillectomy for obstructive sleep apnea: A systematic review and meta-analysis. *International Journal of Pediatric Otorhinolaryngology*, 83, 205–210.

Souders, M. C., Zavodny, S., Eriksen, W., Sinko, R., Connell, J., Kerns, C., Schaaf, R., and Pinto-Martin, J. (2017). Sleep in children with autism spectrum disorder. *Current Psychiatry Reports*, 19(6), 34.

Spilsbury, J. C. (2009). Sleep as a mediator in the pathway from violence-induced traumatic stress to poorer health and functioning: A review of the literature and proposed conceptual model. *Behavioral Sleep Medicine*, 7(4), 223–244. doi:10.1080/15402000903190207

Spilsbury, J. C., Frame, J., Magtanong, R., and Rork, K. (2016). Sleep environments of children in an urban US setting exposed to interpersonal violence. *Behavioral Sleep Medicine*, 14(6), 585–601. doi:10.1080/15402002.2015.1048449

Spruyt, K., Braam, W., Smits, M., and Curfs, L. M. (2016). Sleep complaints and the 24-h melatonin level in individuals with Smith–Magenis syndrome: Assessment for effective intervention. *CNS Neuroscience & Therapeutics*, 22(11), 928–935.

St James-Roberts, I., Roberts, M., Hovish, K., and Owen, C. (2015). Video evidence that London infants can resettle themselves back to sleep after waking in the night, as well as sleep for long periods, by 3 months of age. *Journal of Developmental and Behavioral Pediatrics*, 36(5), 324–329. doi:10.1097/DBP.0000000000000166

Steensma, T. D., Kreukels, B. P., de Vries, A. L., and Cohen-Kettenis, P. T. (2013). Gender identity development in adolescence. *Hormones and Behavior*, 64(2), 288–297.

Steinberg, L. (2005) Cognitive and affective development in adolescence, *Trends in Cognitive Sciences*, 9(2), 69–74.

Stephens, R. J., Chung, S. A., Jovanovic, D., Guerra, R., Stephens, B., Sandor, P., and Shapiro, C. M. (2013). Relationship between polysomnographic sleep architecture and behavior in medication-free children with TS, ADHD, TS and ADHD, and controls. *Journal of Developmental and Behavioral Pediatrics*, 34(9), 688–696. doi:10.1097/DBP.0000000000000012

Stern, E., Parmelee, A. H., Akiyama, Y., Schultz, M. A., and Wenner, W. H. (1969). Sleep cycle characteristics in infants. *Pediatrics*, 43(1), 65–70.

Storch, E. A., Murphy, T. K., Lack, C. W., Geffken, G. R., Jacob, M. L., and Goodman, W. K. (2008). Sleep-related problems in pediatric obsessive-compulsive disorder. *Journal of Anxiety Disorders*, 22(5), 877–885.

Stražišar, B. G. (2021) Sleep Measurement in Children—Are We on the Right Track? *Sleep Medicine Clinics*, 16(4), 649–660.

Substance Abuse and Mental Health Services Administration (2014). *Trauma-informed care in behavioral health services* (pp. 59–67). Rockville, MD: Substance Abuse and Mental Health Services Administration. Retrieved from https://www.ncbi.nlm.nih.gov/books/NBK207201/pdf/Bookshelf_NBK207201.pdf.

Sung, V., Hiscock, H., Sciberras, E., and Efron, D. (2008). Sleep problems in children with attention-deficit/hyperactivity disorder: Prevalence and the effect on the child and family. *Archives of Pediatrics and Adolescent Medicine*, 162(4), 336–342. doi:10.1001/archpedi.162.4.336

Suni, E. (2020). Mental Health and Sleep. Sleep Foundation. Retrieved from https://www.sleepfoundation.org/mental-health

Tabachnik, E., Muller, N. L., Bryan, A. C., and Levison, H. (1981). Changes in ventilation and chest wall mechanics during sleep in normal adolescents. *Journal of Applied Physiology*, 51(3), 557–564.

Tamis-LeMonda, C. S., Shannon, J. D., Cabrera, N. J., and Lamb, M. E. (2004). Fathers and mothers at play with their 2- and 3-year-olds: Contributions to language and cognitive development. *Child Development*, 75(6), 1806–1820. doi:10.1111/j.1467-8624.2004.00818.x

Tesfaye, R., Wright, N., Zaidman-Zait, A., Bedford, R., Zwaigenbaum, L., Kerns, C. M., Duku, E., Mirenda, P., Bennett, T., Georgiades, S., and Smith, I. M. (2021). Investigating longitudinal associations between parent reported sleep in early childhood and teacher reported executive functioning in school-aged children with autism. *Sleep*, 44(9), zsab122.

Tham, E. K., Schneider, N., and Broekman, B. F. (2017). Infant sleep and its relation with cognition and growth: A narrative review. *Nature and Science of Sleep*, 9, 135. doi:10.2147/NSS.S125992

Tietze, A.-L., Blankenburg, M., Hechler, T., Michel, E., Koh, M., Schlüter, B., and Zernikow, B. (2012). Sleep disturbances in children with multiple disabilities. *Sleep Medicine Reviews*, 16(2), 117–127.

Tininenko, J. R., Fisher, P. A., Bruce, J., and Pears, K. C. (2010). Sleep disruption in young foster children. *Child Psychiatry & Human Development*, 41(4), 409–424.

Toon, E., Davey, M. J., Hollis, S. L., Nixon, G. M., Horne, R. S., and Biggs, S. N. (2016). Comparison of commercial wrist-based and smartphone accelerometers, actigraphy, and PSG in a clinical cohort of children and adolescents. *Journal of Clinical Sleep Medicine*, 12(3), 343–350.

Tordjman, S. Anderson, G. M., Pichard, N., Charbuy, H., and Touitou, Y. (2005) Nocturnal excretion of 6-sulphatoxymelatonin in children and adolescents with autistic disorder. *Biological Psychiatry*, 57(2), 134–138.

Tordjman, S., Najjar, I., Bellissant, E., Anderson, G. M., Barburoth, M., Cohen, D., Jaafari, N., Schischmanoff, O., Fagard, R., Lagdas, E., and Kermarrec, S. (2013). Advances in the research of melatonin in autism spectrum disorders: Literature review and new perspectives. *International Journal of Molecular Sciences*, 14(10), 20508–20542.

Trickett, J., Heald, M., Oliver, C., and Richards, C. (2018). A cross-syndrome cohort comparison of sleep disturbance in children with Smith-Magenis syndrome, Angelman syndrome, autism spectrum disorder and tuberous sclerosis complex. *Journal of Neurodevelopmental Disorders*, 10(1), 1–14.

Tudor, M. E., Hoffman, C. D., and Sweeney, D. P. (2012). Children with autism: sleep problems and symptom severity. *Focus on Autism and Other Developmental Disabilities*, 27(4), 254–262.

Tzischinsky, O., Meiri, G., Manelis, L., Bar-Sinai, A., Flusser, H., Michaelovski, A., Zivan, O., Ilan, M., Faroy, M., Menashe, I., and Dinstein, I. (2018). Sleep disturbances are associated with specific sensory sensitivities in children with autism. *Molecular Autism, 9*(1), 1–10.

Valdez, P., Ramírez, C., and García, A. (2014). Circadian rhythms in cognitive processes: Implications for school learning. *Mind, Brain, and Education, 8*(4), 161–168.

van der Helm, E., Yao, J., Dutt, S., Rao, V., Saletin, J. M., and Walker, M. P. (2011). REM sleep depotentiates amygdala activity to previous emotional experiences. *Current Biology, 21*(23), 2029–2032.

van Horn, N. L., and Street, M. (2018). *Night Terrors.* Orlando, FL: Statpearls Publishing.

van Someren, E. J. W. (2021). Brain mechanisms of insomnia: New perspectives on causes and consequences. *Physiological Reviews, 101*(3), 995–1046.

Vanderheyden, W. M., George, S. A., Urpa, L., Kehoe, M., Liberzon, I., and Poe, G. R. (2015). Sleep alterations following exposure to stress predict fear-associated memory impairments in a rodent model of PTSD. *Experimental Brain Research, 233*(8), 2335–2346.

Veatch, O. J., Pendergast, J. S., Allen, M. J., Leu, R. M., Johnson, C. H., Elsea, S. H., and Malow, B. A. (2015). Genetic variation in melatonin pathway enzymes in children with autism spectrum disorder and comorbid sleep onset delay. *Journal of Autism and Developmental Disorders, 45*(1), 100–110.

Veatch, O. J., Maxwell-Horn, A. C., and Malow, B. A. (2015). Sleep in autism spectrum disorders. *Current Sleep Medicine Reports, 1*(2), 131–140.

Virring, A., Lambek, R., Thomsen, P. H., Møller, L. R., and Jennum, P. J. (2016). Disturbed sleep in attention-deficit hyperactivity disorder (ADHD) is not a question of psychiatric comorbidity or ADHD presentation. *Journal of Sleep Research, 25*(3), 333–340. Doi:10.1111/jsr.12377

Volk, C., and Huber, R. (2015). Sleep to grow smart? *Archives Italiennes de Biologie, 153*(2–3), 99–109.

Wake, M., Morton-Allen, E., Poulakis, Z., Hiscock, H., Gallagher, S., and Oberklaid, F. (2006). Prevalence, stability, and outcomes of cry-fuss and sleep problems in the first 2 years of life: Prospective community-based study. *Pediatrics, 117*(3), 836–842. doi:10.1542/peds 2005-0775

Wang, G., Grone, B., Colas, D., Appelbaum, L., and Mourrain, P. (2011). Synaptic plasticity in sleep: Learning, homeostasis and disease. *Trends in Neurosciences, 34*(9), 452–463.

Wang, R., Chen, J., Tao, L., Qiang, Y., Yang, Q., and Li, B. (2022). Prevalence of sleep problems and its association with preterm birth among kindergarten children in a rural area of Shanghai, China. *Frontiers in Pediatrics, 10.*

Wassing, R., Benjamins, J. S., Talamini, L. M., Schalkwijk, F., and Van Someren, E. J. (2019). Overnight worsening of emotional distress indicates maladaptive sleep in insomnia. *Sleep, 42*(4), zsy268.

Waters, K. A., Chawla, J., Harris, M. A., Heussler, H., Black, R. J., Cheng, A. T., and Lushington, K. (2020). Cognition after early tonsillectomy for mild OSA. *Pediatrics, 145*(2): e20191450.

Wei, N. S., and Praharaj, S. K. (2019). Chronotypes and its association with psychological morbidity and childhood parasomnias. *Indian Journal of Psychiatry, 61*(6), 598.

Williams, K. E., Berthelsen, D., Walker, S., and Nicholson, J. M. (2017). A developmental cascade model of behavioral sleep problems and emotional and attentional self-regulation across early childhood. *Behavioral Sleep Medicine, 15*(1), 1–21. doi:10.1080/15402002.2015.1065410

Wimpory, D., Nicholas, B., and Nash, S. (2002). Social timing, clock genes and autism: A new hypothesis. *Journal of Intellectual Disability Research, 46*(4), 352–358.

Winsper, C., and Wolke, D. (2014). Infant and toddler crying, sleeping and feeding problems and trajectories of dysregulated behavior across childhood. *Journal of Abnormal Child Psychology*, 42(5), 831–843. doi:10.1007/s10802-013-9813-1

Withers, A., Maul, J., Rosenheim, E., O'Donnell, A., Wilson, A., and Stick, S. (2022). Comparison of home ambulatory type 2 polysomnography with a portable monitoring device and in-laboratory type 1 polysomnography for the diagnosis of obstructive sleep apnea in children. *Journal of Clinical Sleep Medicine*, 18(2), 393–402.

Wong, K., Leonard, H., Jacoby, P., Ellaway, C., and Downs, J. (2015). The trajectories of sleep disturbances in Rett syndrome. *Journal of Sleep Research*, 24(2), 223–233.

Wong, M. M., Brower, K. J., and Zucker, R. A. (2011). Sleep problems, suicidal ideation, and self-harm behaviors in adolescence. *Journal of Psychiatric Research*, 45(4), 505–511.

Wyatt, J. K. (2004). Delayed sleep phase syndrome: pathophysiology and treatment options. *Sleep*, 27(6), 1195–1203.

Yang, Z., Matsumoto, A., Nakayama, K., Jimbo, E. F., Kojima, K., Nagata, K. I., Iwamoto, S., and Yamagata, T. (2016). Circadian-relevant genes are highly polymorphic in autism spectrum disorder patients. *Brain and Development*, 38(1), 91–99.

Zhang, J., Paksarian, D., Lamers, F., Hickie, I. B., He, J., and Merikangas, K. R. (2017). Sleep patterns and mental health correlates in US adolescents. *The Journal of Pediatrics*, 182, 137–143.

Zisapel, N. (2001). Circadian rhythm sleep disorders. *CNS Drugs*, 15(4), 311–328.

CHAPTER 14

DEVELOPMENT OF THE MICROBIOME-GUT-BRAIN AXIS AND ITS EFFECT ON BEHAVIOR

REBECCA C. KNICKMEYER

WHAT IS THE GUT MICROBIOME?

THE human body is home to trillions of microorganisms including bacteria, fungi, protozoa, and viruses (Ursell et al., 2012). The vast majority of these cells inhabit the intestinal tract and are referred to as the "gut microbiota" (Sender et al., 2016). The "gut microbiome" refers to the combined genetic material of these organisms. This is a complex community of symbiotic organisms, many of which can be considered mutualistic with the host; that is they are involved in a mutually beneficial relationship, with some being pathobionts (potentially disease-causing organisms which, under normal circumstances, live as symbionts) (Hornef, 2015). Recent technological advances in genetic sequencing have allowed scientists to probe this complex community in tremendous detail, and have led to a surge of scientific and public interest in the possibility that individual variation in the gut microbiome could affect human health and development, including cognitive development and risk for mental illness.

EFFECTS OF THE GUT MICROBIOME ON BRAIN AND BEHAVIOR IN ANIMALS

Studies conducted in rodents provide compelling evidence that the composition of the gut microbiome influences brain development and function. In particular, multiple laboratories have shown that germ-free (axenic) mice display reduced anxiety-like behavior in tests such as the elevated plus maze, light/dark test, and open field test (e.g., Clarke et al., 2013; Diaz Heijtz et al., 2011; Neufeld et al., 2011). Germ-free mice are generally produced by hysterectomy rederivation (caesarean delivery under aseptic conditions) or embryo transfer using germ-free foster mothers and are maintained in isolators to keep them free of all microorganisms, including those typically found in the gut. Germ-free mice also exhibit impaired sociability (Desbonnet et al., 2014) and recognition memory, increased locomotor activity and self-grooming, and increased corticosterone responses to stress (Sudo et al., 2004). Behavioral changes are accompanied by changes in neurochemistry and gene expression in brain regions implicated in the processing of threat, emotional regulation, and social cognition, including the amygdala, hippocampus, and medial prefrontal cortex (Luczynski et al., 2016; Stilling et al., 2018). Increased hippocampal levels of serotonin have been reported in male, but not female, germ-free mice (Clarke et al., 2013). Reduced levels of brain derived neurotrophic factor have been reported in the cortex, amygdala (Diaz Heijtz et al., 2011), and hippocampus (Diaz Heijtz et al., 2011 ; Sudo et al., 2004; Clarke et al., 2013 ; Gareau et al., 2011) (but, see Neufeld et al. (2011) for an opposing report). There are reports that the dopamine receptor subunit Drd1a is increased in the germ-free hippocampus (Diaz Heijtz et al., 2011), that glutamate receptor subunit Nr1 is decreased in the cortex, and that glutamate receptor subunit Nr2 is reduced in the cortex, hippocampus, and central nucleus of the amygdala (Neufeld et al., 2011; Sudo et al., 2004). In the striatum, expression of the synaptic proteins synaptophysin and postsynaptic density protein-95 is increased (Diaz Heijtz et al., 2011). Furthermore, genes involved in neuronal activity are upregulated in the germ-free amygdala, suggesting potential amygdalar hyperactivity (Stilling et al., 2015), though this seems somewhat paradoxical given the well-replicated reduction in behavioral anxiety exhibited by these animals. One possible explanation is that amygdalar hyperactivity in the absence of threat leads to a blunted transcriptional response in the presence of actual threat. This hypothesis is supported by a recent study of cued fear conditioning, in which germ-free mice demonstrated impaired recall accompanied by reduced numbers of differentially expressed genes and differentially expressed miRNAs in response to fear conditioning (Hoban et al., 2018). Changes in adult hippocampal neurogenesis (Ogbonnaya et al., 2015) have also been reported. Changes in early postnatal neurogenesis have yet to be explored.

Intriguingly, while some of these phenotypes are normalized by the introduction of a typical murine microbiota post weaning, many are not. Phenotypes that persist despite recolonization include increased hippocampal serotonin levels (Clarke et al., 2013), upregulation of neuronal activity genes in the amygdala (Stilling et al., 2015), and increased new neuron survival in the adult hippocampus (Ogbonnaya et al., 2015). Reversal of anxiety-related behaviors may depend on the exact timing of recolonization as well as genetic background (Luczynski et al., 2016). Overall, these results suggest that early development may be a critical period for microbiome effects on the murine brain.

Effects of the Gut Microbiome on Brain and Behavior in Humans

The majority of human studies on the microbiome-gut-brain axis can be grouped into two areas: those that examine the impact of probiotics (and more recently, prebiotics) on behavioral outcomes and those describing variant microbe populations that occur in disease states. Studies of probiotics will be addressed in the Clinical Implications section. The current section will focus on the latter group of studies. Altered gut microbes have been reported in people with autism, depression, schizophrenia, attention deficit hyperactivity disorder (ADHD), and eating disorders (for autism, see Gondalia et al., 2012; Song et al., 2004; Finegold et al., 2002; Parracho et al., 2005; Tomova et al., 2015; Williams et al., 2011; Wang et al., 2013; De Angelis et al., 2013; Finegold et al., 2010; Williams et al., 2012; Kang et al., 2013; Adams et al., 2011; Pulikkan et al., 2018; Kang et al., 2018; Kushak et al., 2017; Strati et al., 2017 ; Inoue et al., 2016, for depression, see Naseribafrouei et al., 2014; Liu et al., 2016; Jiang et al., 2015; Lin et al., 2017; Chen et al., 2018a; Chen et al., 2018b; Aizawa et al., 2016; Kelly et al., 2016; Zheng et al., 2016; Valles-Colomer et al., 2019, for schizophrenia, see Shen et al., 2018, for ADHD, see Aarts et al., 2017, Jiang et al., 2018, and for eating disorders, see Kleiman et al., 2015). Of these, autism and depression represent the most frequently studied disorders with at least 18 relevant publications for autism and at least ten relevant publications for depression. However, there is substantial variability in the specific bacterial groups identified, as well as the direction of effect. Sample sizes are generally small, and many studies fail to correct for multiple comparisons, despite the highly multivariate nature of the data. Consistency might be improved by using shotgun metagenomics or metatranscriptomics, rather than 16S ribosomal RNA (rRNA) sequencing, given studies suggesting different assemblages of bacterial species can exhibit similar functional profiles (Turnbaugh et al., 2009).

It is also important to acknowledge that altered microbial profiles could arise as a consequence of living with a psychiatric condition, rather than reflecting a causal relationship. For example, children with autism are frequently placed on special diets, receive multiple psychiatric medications, and interact with their environment in unique ways, all of which might influence the composition of the microbiome. One way to address this difficulty is to perform longitudinal studies with dense sampling of both the microbiome and symptomatology. A study of anorexia and the gut microbiome carried out by Kleiman et al. (2015) represents an important step in this direction. The authors studied the gut microbiota in patients with anorexia both at inpatient admission (when participants were at <75 percent of their ideal body weight) and at discharge (when weight was at least partially restored). The authors found greater differences in bacterial composition between individuals with anorexia and healthy controls before vs after hospital-based renourishment, suggesting initial differences might have reflected malnutrition. Prospective, longitudinal cohort studies are also needed to document whether variation in microbial composition precedes onset of specific psychiatric disorders. This would strengthen the argument for causality. Evidence for causality could also come from studies of "humanized" germ-free mice. At present, two independent studies have reported that transfer of microbial communities from depressed humans resulted in depression-like behaviors in germ-free mice when compared with transfer of microbial communities from healthy controls, suggesting a causal role for the gut microbiome in the development of depression (Kelly et al., 2016; Zheng et al., 2016).

The Importance of Infancy

Historically, scientists have considered the human fetal environment, including the fetal gut, to be sterile. This viewpoint has recently been challenged by studies documenting bacterial communities in the placenta, amniotic fluid, and meconium using molecular techniques. The possibility that colonization begins in utero continues to be controversial, particularly as supportive studies suffer from some methodological weaknesses including 1) use of molecular approaches with an insufficient detection limit to study microbial populations with very low biomass, 2) failure to include appropriate controls for contamination, and 3) failure to provide evidence of bacterial viability (Perez-Munoz et al., 2017). Whether or not the fetal gut is entirely sterile, pregnancy still represents an important period for microbial effects on human neurodevelopment, as the maternal microbiome may alter maternal physiology in ways that either increase or decrease risk for later psychiatric conditions (Kim et al., 2017; Wang et al., 2019). In addition, variation in the vaginal microbiome may influence the developing microbiome-gut-brain axis, since exposure to the vaginal microbiota during parturition initiates postnatal colonization of the infant gut (Jasarevic et al., 2015).

During infancy and early childhood, the diversity of species within each individual (alpha diversity) increases rapidly as new taxa colonize the gut. Initially the community is dominated by facultative aerobic Enterobacteriaceae. This is followed by a rise in the proportion of anaerobic microbes including *Bifidobacterium*, *Bacteroides*, and *Clostridium*. Between 6 and 24 months, there is a gradual increase in members of the order Clostridiales including Ruminococcaceae, *Faecalibacterium*, and *Lachnospira*, taxa which are characteristic of adult microbiomes (Bokulich et al., 2016; Planer et al., 2016). The first colonizers of the postnatal gut are acquired from the maternal vaginal microbiota (in vaginally born infants) and through incidental exposures to skin bacteria in the hospital environment (in infants delivered via caesarean section) (Dominguez-Bello et al., 2010). Subsequent shifts are intimately tied to changes in infant feeding. In exclusively or predominantly breastfed infants, the second wave of postnatal colonizers represent taxa that facilitate lactate utilization (e.g., *Lactobacillus* species) (Koenig et al., 2011) and those that feed on human milk oligosaccharides (e.g., *Bifidobacterium* species) (Walker and Iyengar, 2015). Early introduction of formula significantly accelerates development of the microbiota during the first 6 to 7 months of life, increasing overall diversity and reducing the abundance of *Bifidobacterium* (Planer et al., 2016), but is also associated with slower increases in diversity during the second year of life (Bokulich et al., 2016). In contrast, early antibiotic exposures delay maturation of the microbiome in the first year of life but do not have long-term effects in the second year of life (Bokulich et al., 2016). The introduction of solid foods is accompanied by increases in bacterial groups involved in carbohydrate utilization, vitamin biosynthesis, and xenobiotic degradation (Koenig et al., 2011). Maturation into an adult-like microbiota occurs between 1 and 4 years of age with the cessation of breastfeeding (Backhed et al., 2015).

While feeding practices, antibiotic exposures, and delivery method have a marked impact on the development of the gut microbiome, it is important to note that substantial inter-individual variability exists even among infants with no major medical or dietary interventions (Eggesbo et al., 2011; Jost et al., 2012). This is also evident in recent unpublished data from our research group (see Figure 14.1). Other factors that may contribute to individual variation include day care attendance (Thompson et al., 2015), maternal prenatal stress (Zijlmans et al., 2015), presence of older siblings in the home (Carlson et al., 2018), and host genotype (Goodrich et al., 2016).

As these dramatic changes take place in the gut, rapid and dramatic changes are also occurring in the human brain. Grey matter volumes double in the first year of life (Knickmeyer et al., 2008), reflecting the proliferation and migration of glia, as well as exuberant synaptogenesis (Stiles and Jernigan, 2010). Growth continues, albeit at a slower pace, in the second and third years of life and by age 4, children have already reached 96 percent of their adult brain volume. Cortical thickness peaks between 1 and 2 years of age and declines thereafter, whereas surface area continues to expand into late

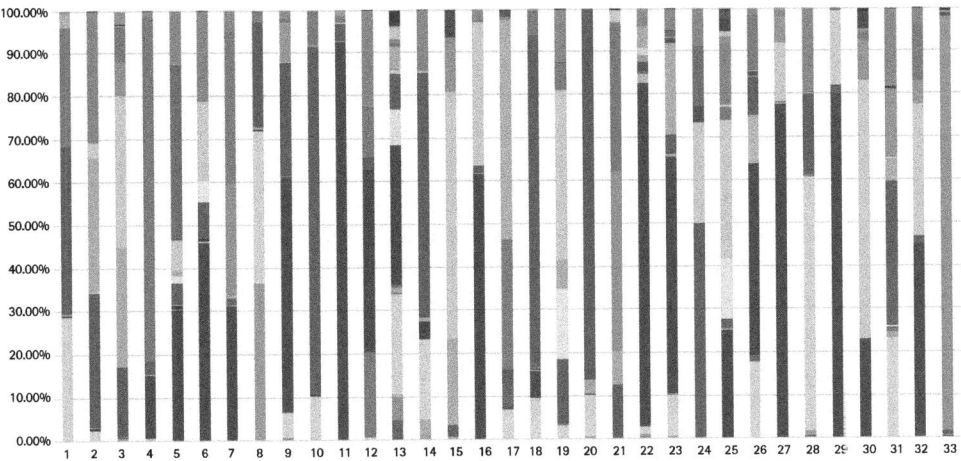

FIGURE 14.1 Substantial microbial diversity in infants with no major medical or dietary interventions. Mean relative abundance of fecal bacteria at the genus level in 33 infants between 2 weeks and 2 months of age. All children were vaginally delivered, exclusively breastfed, and had not been exposed to antibiotics either in infancy or during the last 2 weeks of gestation.

childhood or beyond. As regards white matter maturation, major white-matter tracts such as the corpus callosum, superior and inferior longitudinal fasciculi, arcuate and cingulum are present at the time of birth, but are, for the most part, unmyelinated. The infant will achieve an adult-like pattern of myelination by 2 years of age (Gilmore et al., 2018). Functional networks mature significantly in this period as well, with all canonical adult networks (except the sensorimotor network) increasing in intranetwork connectivity (Gao et al., 2015). Concurrent with the rapid pace of structural and functional brain development is an equally rapid development of motor skills, cognition, social abilities, and language (Flavell et al., 2002).

Relationships between microbial colonization and brain development have just begun to be explored in human infants. In 2015, researchers reported that composition of the gut microbiome showed sex-specific associations with infant temperament at 18–27 months of age (Christian et al., 2015). In 2018, our group reported that microbial composition at 1 year of age is associated with cognitive outcomes at 2 years of age (Carlson et al., 2018). Specifically, a cluster of children characterized by relatively high levels of *Bacteroides* exhibited the highest level of performance (90th percentile) while a cluster of children characterized by relatively high levels of *Faecalibacterium* showed the lowest level of performance (72nd percentile). Notably, *Bacteroides* is more typical of infantile microbiota, whereas *Faecalibacterium* is more typical of adult microbiotas. In addition, high alpha diversity at 1 year of age was associated with poorer cognitive performance at 2 years of age. Both results suggest that accelerated development of the gut microbiota may have negative effects on infant brain development. The neural circuits underlying these findings have not yet been delineated. An exploratory analysis of regional grey matter volumes by Carlson et al. (2018) revealed several significant associations with microbiome measures. However, these findings were anatomically distributed, mostly unilateral, and varied in direction of effect. Therefore, they should be treated with caution until replicated. A subsequent study revealed that alpha diversity at 1 year of age was associated with functional connectivity between the supplementary motor area (SMA, representing the sensorimotor network) and the inferior parietal lobule (IPL) (Gao et al., 2019). SMA-IPL connectivity was also associated with cognitive abilities at 2 years of age, suggesting a potential pathway between microbial diversity and cognitive development in infancy. Alpha diversity was also significantly associated with functional connectivity between the amygdala and thalamus and between the anterior cingulate cortex and anterior insula, key structures in neural circuits for fear-reactivity and anxiety-related behaviors.

Studies incorporating other neuroimaging phenotypes, such as cortical thickness and surface area, diffusion tensor imaging, and functional activation, are expected to provide additional insights into the neural circuits mediating the microbiome's impact on infant cognition. Additional studies are also needed to determine whether microbiome-associated changes in infant fear circuitry have implications for fear behavior and later risk for psychopathology. Fine-grained longitudinal studies are required to delineate temporal relationships between microbial colonization and infant brain development. It is important to note that brain development in this period exhibits significant spatiotemporal complexity. Consequently, variations/disruptions in microbial colonization could have quite specific effects on neuroimaging and behavioral outcomes, depending on their timing.

How Does the Microbiome Communicate with the Brain?

The mechanisms responsible for microbiota effects on brain development remain poorly understood and are likely quite complex. Several potential mechanisms that have attracted research attention include 1) modulation of systemic immunity, 2) production of neuroactive metabolites, and 3) processing of nutrients and environmental chemicals.

Modulation of the Immune System

The microbiota plays a key role in the development and function of the mucosal and systemic immune systems. Excessive contact between the immune systems and the microbiota (for example through a dysfunctional intestinal barrier or "leaky gut") may result in exaggerated proinflammatory responses (Kelly et al., 2015), while inadequate exposure, particularly in early life, can impair immune development and function, resulting in chronic low-grade inflammation (Rook et al., 2014). Notably, many patients with anxiety and depression exhibit an inflammatory phenotype (Miller et al., 2013). Inflammation has also been associated with adult bipolar disorder (Goldstein et al., 2009) and schizophrenia (Potvin et al., 2008). Studies of children with autism spectrum disorders (ASD) also suggest a pro-inflammatory state, though there is significant heterogeneity in the specific cytokine profiles reported (Mitchell and Goldstein, 2014). Inflammatory cytokines, including tumor necrosis factor-α (TNF-α), interferon-γ (IFN-γ) and interleukin-6 (IL-6) influence brain function, development, and neurochemistry in animal models (Bercik et al., 2010; Clarke et al., 2013; Desbonnet et al., 2008). Inflammatory cytokines increase the expression of the transporters responsible for reuptake of dopamine, norepinephrine, and serotonin and reduce the availability of tetrahydrobiopterin (BH4), an essential cofactor for tryptophan hydroxylase and tyrosine hydroxylase, rate-limiting enzymes for the synthesis of serotonin, dopamine, and norepinephrine (Miller et al., 2013). Pro-inflammatory cytokines also enhance the activity of the key enzyme indoleamine 2,3-dioxygenase (IDO) (King and Thomas, 2007). This shifts tryptophan metabolism away from the serotonin pathway and toward the kynurenine (KYN) pathway, resulting in central serotonin depletion and increasing levels of kynurenic acid (KYNA), 3-hydroxykynurenine (3HK), and quinolinic acid (QUIN) (Myint, 2012). KYNA is an antagonist of excitatory amino acid receptors and noncompetitively blocks 7-nicotinic acetylcholine receptors. While generally considered a protective metabolite, its accumulation beyond physiological levels could disrupt brain development. QUIN generates free radicals, and in high concentration excites NMDA receptors, causing excitotoxicity. 3HK can also damage neurons via toxic free radicals, oxidative stress, and lipid peroxidation (Nemeth et al., 2007).

Production of Neuroactive Metabolites

The gut microbiota produces numerous neuroactive molecules. *Lactobacillus* species and *Bifidobacterium* species, key constituents in the infant microbiome, produce gamma-aminobutyric acid (GABA), a major inhibitory neurotransmitter (Barrett et al., 2012). Multiple lactic acid bacteria including *Lactobacillus planatarum* and *Streptococcus thermophiles* produce serotonin (Ozogul, 2011) as does *E. coli* K-12 (Shishov et al., 2009). In addition, spore-forming microbes, such as Clostridial species, promote serotonin production by enterochromaffin cells (Yano et al., 2015). Dopamine is produced by certain *Escherichia, Bacillus, Lactococcus, Lactobacillus*, and *Streptococcus* species (Shishov et al., 2009; Ozogul, 2011; Tsavkelova et al., 2000), and norepinephrine is produced by certain *Escherichia* and *Bacillus* species (Tsavkelova et al., 2000; Shishov et al., 2009). Histamine is produced by a number of fermentative bacteria including *Lactobacillus* (Ozogul, 2011; Landete et al., 2008; Coton et al., 1998). The microbiota may also produce acetylcholine (Horiuchi et al., 2003). It is unlikely that these neurotransmitters act directly on the brain. Instead, they may induce epithelial cells to release molecules that act upon the enteric nervous system or act directly on afferent axons. The vagus nerve may play a key role in transmitting signals from the enteric nervous system to the brain, as vagotomy eliminates some probiotic-induced changes in murine behavior and brain chemistry (Bercik et al., 2011; Bravo et al., 2011). Interestingly, researchers have recently identified a type of gut sensory epithelial cell that synapses with vagal neurons (Kaelberer et al., 2018). These cells, which the researchers have called "neuropod" cells, sense ingested nutrients and microbial metabolites, and would allow rapid communication between the gut and the brain.

Another important group of metabolites are the short chain fatty acids (SCFA), which include butyrate, acetate, and propionate. These are small, organic monocarboxylic acids produced during anaerobic fermentation by multiple bacteria including *Bacteroides, Bifidobacterium, Propionibacterium, Eubacterium, Lactobacillus, Clostridium, Roseburia*, and *Prevotella*. SCFA can influence inflammatory responses through G protein-coupled receptors (Gpr41 and Gpr43) in the small intestine, colon, and adipocytes. They may also act within the brain to alter gene expression and subsequent development. They are capable of crossing the blood-brain barrier, and are taken up by both glia and, to a lesser extent, neurons (Wall et al., 2014). Butyrate and propionate enhance histone acetylation through inhibition of histone deactylase (Steliou et al., 2012; Nguyen et al., 2007), while acetate enhances histone acetylation by increasing acetyl coenzyme A (Ariyannur et al., 2010). High doses of sodium butyrate (1.2 g/kg for 4 weeks) produce antidepressant effects in mice (Schroeder et al., 2007). Administration of propionate to rats alters neurotransmitter systems, and impairs cognition and social behavior (MacFabe et al., 2007; Thomas et al., 2012; Shultz et al., 2015; Shultz et al., 2009; Nankova et al., 2014; Alfawaz et al., 2014). Acetate acts in the hypothalamus to regulate appetite (Frost et al., 2014) and also reduces neuroinflammation (Brissette et al., 2012).

In general, metabolomics analyses suggest that gut microbes exert profound effects on blood metabolites in mammals (Wikoff et al., 2009).

Processing of Nutrients and Environmental Chemicals

Adequate nutrition is necessary to support the explosive brain growth that occurs during prenatal development, infancy, and early childhood. The gut microbiota plays an important role in energy harvest and nutrient metabolism. Consequently, individual variation in the microbiota may influence brain development through differential processing of the diets consumed by children and their pregnant mothers (Goyal et al., 2015). Table 14.1 summarizes how specific dietary factors influence neurodevelopmental

Table 14.1 Relationships between dietary factors, neurodevelopment, and the microbiome

Deficiency	Neuron proliferation (7 weeks GA to 18 months)	Dendrite growth (15 weeks GA to 2 years)	Synapse formation and pruning (23 weeks GA through adolescence)	Myelination (12 weeks GA into adulthood)	Programmed cell death (23 weeks GA through adolescence)	Microbiome
Calories	↓	↓	↓	↓	↑	Caloric salvage
Fatty acids	↓	–	–	↓	–	May modulate uptake
Iron	↓[a]	↓	↓[a]	↓	–	Competition with host
Zinc	↓	↓	–	–	–	Competition with host
Choline	↓	–	–	–	↑	Competition with host
Folic acid[b]	↓	↑	↑	↓	↑	Produced by microbiota
B₆	–	↓	↓	↓	–	Competition with host

See "Processing of nutrients and environmental chemicals" for supporting citations.

[a] Findings restricted to hippocampus.

[b] Gestational deficiency also causes neural tube defects.

processes and how these may be influenced by the microbiome. First, caloric restriction reduces neuron proliferation, dendritogenesis, synaptogenesis, and myelination, while increasing neuronal cell death and synaptic elimination (Prado and Dewey, 2014). One of the major roles of the microbiota is caloric salvage. Microbes in the large intestine provide a variety of enzymes capable of breaking down carbohydrates and protein that would otherwise by indigestible by humans (Krajmalnik-Brown et al., 2012). Second, fatty acids are required to synthesize membrane phospholipids during neurogenesis and are also key structural components of myelin. Gestational deficiency of docosohexanoic acid (DHA) results in reduced neuron proliferation while both prenatal and postnatal deficiency reduces the amount of myelin produced. DHA and arachidonic acid also regulate neurotransmission (Prado and Dewey, 2014). Deficiency of omega-3 polyunsaturated fatty acids has been linked to ADHD (Chang et al., 2018). Administration of *Bifidobacterium breve* strain NCIMB702258 alters fatty acid composition of the murine brain (Wall et al., 2014). The mechanism for this finding is not yet clear, but may be related to fatty acid uptake in the small intestines. Third, gestational iron deficiency decreases the overall size of the hippocampus as well as dendritic complexity and synaptic maturity of this structure. Iron deficiency during gestation and early postnatal development also decreases myelin synthesis (Prado and Dewey, 2014). Iron is an essential nutrient for certain bacteria. Consequently, there is competition between bacteria and their host for this vital nutrient (Kortman et al., 2014). Fourth, gestational zinc deficiency results in decreased neuronal cell number and reduced dendritic arborization. Zinc also modulates synapse function (Prado and Dewey, 2014). Numerous bacterial cells require zinc uptake systems for growth. As with iron, this can result in competition between bacteria and their host (Gielda and DiRita, 2012). Fifth, competition between microbes and the host also occurs for choline, an essential nutrient and methyl donor critical to epigenetic regulation (Romano et al., 2015; Romano et al., 2017). Gestational choline stimulates neurogenesis and reduces neuronal cell death (Prado and Dewey, 2014). Sixth, folate is an essential vitamin which the body needs to make DNA and RNA, metabolize amino acids, and control gene expression via methylation. Deficiency in early gestation is associated with neural tube defects (Prado and Dewey, 2014). Deficiency later in gestation decreases neuroprogenitor proliferation and increases apoptosis (Craciunescu et al., 2004). Gestational and postnatal deficiency may disrupt myelination (Steinfeld et al., 2009; Hirono and Wada, 1978). While most research has focused on deficiencies, elevated folic acid levels may inhibit neurite extension and synaptogenesis (Wiens et al., 2015). Intestinal microbes including *Bifidobacterium bifidum* and *Bifidobacterium longum* produce high concentrations of folate, though it is unclear whether this contributes substantially to folate levels in the host (Biesalski, 2016). Finally, deficiency in Vitamin B_6 during gestation and early postnatal life results in reduced dendritic branching, decreased synaptic density, and reduced myelination (Prado and Dewey, 2014). Vitamin B_6 is required for aminotransferase metabolism in bacteria and may represent yet another incidence of competition between microbes and their host (Biesalski, 2016).

Environmental chemicals are also metabolized by the microbiota with potential implications for their neurotoxicity (Claus et al., 2016). For example, polycyclic aromatic hydrocarbons (PAHs) represent one of the most widespread organic pollutants. PAH exposure disrupts behavior, learning, and memory in rodent models, and is associated with impaired white matter development and externalizing behaviors in human children (Peterson et al., 2015). Microbes present in the human colon are capable of converting non-estrogenic PAHs into estrogenic metabolites (Van de Wiele et al., 2005) suggesting that PAHs may act as endocrine-disrupting chemicals. Methylmercury represents another important developmental neurotoxicant (Grandjean and Herz, 2011). Demethylation of methylmercury by the microbiota helps protect the host from these neurotoxic effects. Both germ-free mice and those treated with antibiotics exhibit reduced fecal secretion of mercury and increased mercury accumulation in the brain, as well as other tissues (Seko et al., 1981; Nakamura et al., 1977). The microbiota also modify arsenic in multiple ways, sometimes increasing toxicity and sometimes decreasing toxicity (Claus et al., 2016). Arsenic exposure is associated with neurological and cognitive dysfunction in both children and adults (Tyler and Allan, 2014). The relevance of microbial metabolism of arsenic to neurodevelopmental toxicity warrants further investigation. Additional substrates derived from the interaction of the gut microbiota with environmental chemicals can be found in Claus et al. (2016).

CLINICAL IMPLICATIONS

The research described thus far suggests that modulation of the gut microbiota may be a tractable strategy for supporting cognitive development and reducing risk for psychiatric disorders. This could be achieved by multiple means including dietary interventions, ingestion of antibiotics or probiotics, and fecal microbiota transplant. This section will focus on the latter two possibilities.

Probiotics are live microorganisms that when consumed (as in a food or a dietary supplement) maintain or restore beneficial bacteria to the digestive tract. They may also exert positive effects by preventing colonization and overgrowth of pathogens (Piewngam et al., 2018). At present, there are at least 25 published clinical trials evaluating the impact of probiotics on symptoms of anxiety and/or depression. Approximately two-thirds of these studies report significant positive effects while others report no significant effects (Allen et al., 2016; Mohammadi et al., 2015; Pinto-Sanchez et al., 2017; Akkasheh et al., 2016; Kazemi et al., 2018; Messaoudi et al., 2011; Raygan et al., 2018; Kato-Kataoka et al.; 2016, Slykerman et al., 2017; Yang et al., 2016; Steenbergen et al., 2015; Takada et al., 2016; Majeed et al., 2018; Nishihira et al., 2014; Lorenzo-Zuniga et al., 2014; Benton et al., 2007; Chung et al., 2014; Kelly et al., 2017; Lyra et al., 2016; Marcos et al., 2004; Ostlund-Lagerstrom et al., 2016; Romijn et al., 2017; Simren et al., 2010; Takada et al., 2017).

Heterogeneity between studies may result from variation in the species, strain, and dosage administered, length of treatment, and the demographic makeup of each sample, among other factors (Sarkar et al., 2016). It is important to note that while the majority of the studies were randomized and double-blinded, there were some common methodological weaknesses. In particular, most studies included multiple outcomes, did not specify which outcome was primary, and did not adjust for multiple comparisons. Furthermore, most studies were carried out on healthy adults. Although such studies are informative, results may not generalize to individuals with clinically significant depression or anxiety. Other studies focused on individuals with inflammatory bowel syndrome (IBS) (Lyra et al., 2016; Simren et al., 2010; Lorenzo-Zuniga et al., 2014; Pinto-Sanchez et al., 2017), where improvements in mood might be secondary to improvements in IBS symptoms. Three studies focused on individuals with clinically diagnosed mental illness (Major Depressive Disorder); including one study of individuals with comorbid IBS and MDD. Encouragingly, all three studies reported reduced depressive symptoms post-treatment (Kazemi et al., 2018; Akkasheh et al., 2016; Majeed et al., 2018).

There are also four small studies evaluating the impact of probiotics on behavior in children with ASD (Sandler et al., 2000; Kaluzna-Czaplinska and Blaszczyk, 2012; Parracho et al., 2010; Shaaban et al., 2018), but only Parracho et al. used a randomized, double-blind design. This was a cross-over study of 15 children and both probiotic and placebo were associated with better scores on the Developmental Behavior Checklist. A randomized controlled trial of the probiotic mixture Vivomixx© in 100 preschoolers with ASD is currently ongoing and will incorporate electroencephalography (EEG) measures as well as measures of symptomatology (Santocchi et al., 2016). Finally, an extremely intriguing study by Partty et al. (2015) reports that infants treated with *Lactobacillus rhamnosus* for the first 6 months of life had lower rates of ADHD and Asperger syndrome at age 13 than those treated with placebo. It should be noted that this particular study was originally designed and statistically powered to evaluate the impact of early probiotic treatment on atopic eczema, not to evaluate more uncommon neuropsychiatric disorders such as ADHD and Asperger syndrome. Only six of 75 children with outcome data received a diagnosis of ADHD or Asperger syndrome. Hence, a change in diagnostic status of even one child could significantly change the pattern of results. In addition, there was a high rate of drop-out during follow-up that could have influenced the results. Nonetheless, this study provides preliminary evidence that manipulating the gut microbiota in early life might have long-term benefits for mental health.

In evaluating the current literature on probiotics and mental health, the following caveats must be kept in mind: 1) currently available probiotics were selected for their influence on gut health, not neuropsychiatric outcomes. 2) Currently available probiotics are primarily aerobic and do not make up a significant portion of the typical human gut microbiome, which is mostly anaerobic, and 3) currently available probiotics seldom produce long-lasting, or even transient, changes in the composition of the gut microbiota (Kristensen et al., 2016). This may be because ingested probiotics must survive inhospitable environments during their passage through the gastrointestinal tract. It may also be because they represent a numeric minority in the context of the existing ecosystem and are out-competed by other members of the community. Intriguingly, recent research suggests that some individuals may be more permissive of probiotic colonization while others may be relatively resistant (Zmora et al., 2018). Although these are serious issues, they are not necessarily insurmountable. Additional preclinical and translational research may identify novel probiotics that are more relevant to neuropsychiatric outcomes (so-called psychobiotics), and advances in pharmaco-engineering may improve delivery of probiotics to the microbiome, even in resistant individuals. As this field moves forward, it will be important to consider the context in which probiotic therapy is recommended and to remain alert for possible negative effects. For example, a recent study reported that post-antibiotic treatment with probiotics actually delayed reestablishment of the gut microbiome and transcriptome (Suez et al., 2018).

In contrast to probiotic interventions, which include only a few bacterial species, fecal microbiota transplant or FMT involves the transfer of an entire microbial community from a healthy donor into the intestinal tract of a recipient. FMT is an established therapy for treatment of recurrent *Clostridium difficile* infections (Bagdasarian et al., 2015) and has shown therapeutic potential for inflammatory bowel disease, obesity, and metabolic syndrome (Gupta et al., 2016). In 2017, the first open-label clinical trial of FMT for ASD was published (Kang et al., 2017). Eighteen children with ASD received 14 days of oral vancomycin treatment followed by 12–24 h fasting with bowel cleansing. The gut was repopulated with microbiota by administering Standardized Human Gut Microbiota (SHGM) either orally or rectally. Children continued to receive a lower dose of SHGM orally for 7–8 weeks. A stomach-acid suppressant was also administered to increase the survival of SHGM through the stomach. ASD symptoms improved significantly, according to parent report, and remained improved 8 weeks after treatment ended. FMT increased overall bacterial diversity as well as the abundance of specific taxa including *Bifidobacterium*, *Prevotella*, and *Desulfovibrio*, changes which persisted for at least 8 weeks. As this was an open-label trial, positive results could represent placebo effects. None-the-less, the results are encouraging and suggest that a randomized, double-blind, placebo-controlled study with lab-based clinical assessments, as well as parent report measures, is warranted.

Finally, it is worth noting that pharmaceuticals as fundamental as tetracycline are microbial products and microbial molecules are routinely used in the clinic as cholesterol-lowering drugs, immune regulators, antibiotics, and antitumor therapies. Ultimately, research into the developing microbiome-gut-brain axis may lead to the identification of microbial metabolites that could one day serve as novel therapeutics for complex psychiatric illnesses, or, alternatively, as drug targets. Preclinical work suggests that this will be a promising area. For example, Hsiao et al. (2013) utilized metabolomic profiling to identify serum metabolites influenced by maternal immune activation (MIA), an experimental animal model for neurodevelopmental disorders, and postnatal treatment with *Bacteroides fragilis*. They identified a specific metabolite, 4-ethylphenylsulfate (4EPS), which increased markedly in the serum of MIA offspring, but returned to normal levels with *B. fragilis* treatment. Administration of 4EPS from 3 to 6 weeks of age produced behavioral abnormalities which were highly similar to those induced by MIA.

CONCLUSIONS

Studies conducted in rodents provide compelling evidence that the gut microbiota influence development and function of the host brain. In particular, multiple reports show that experimental manipulations which alter the intestinal microbiota impact stress-related behaviors highly relevant to psychiatric disorders including anxiety and depression. In humans, infancy represents the foundational period in the establishment of the gut microbiome, with an adult-like microbiome emerging between 1 and 4 years of age. This is also the most rapid and dynamic stage of postnatal brain development. Consequently, early alterations in microbial colonization could have long-term effects on mental health. The mechanisms responsible for microbiota effects on brain development are poorly understood, but activation of the peripheral immune system, production of neuroactive metabolites, and processing of nutrients and environmental chemicals are all likely important. Ultimately, manipulation of the gut microbiota may prove to be a useful tool for treating and/or preventing psychiatric disorders, and for supporting cognitive development.

REFERENCES

Aarts, E., Ederveen, T. H. A., Naaijen, J., Zwiers, M. P., Boekhorst, J., Timmerman, H. M., Smeekens, S. P., Netea, M. G., Buitelaar, J. K., Franke, B., Van Hijum, S., and Arias Vasquez, A. (2017). Gut microbiome in ADHD and its relation to neural reward anticipation. *PLoS One*, 12(9), e0183509.

Adams, J. B., Johansen, L. J., Powell, L. D., Quig, D., and Rubin, R. A. (2011). Gastrointestinal flora and gastrointestinal status in children with autism—comparisons to typical children and correlation with autism severity. *BMC Gastroenterol*, 11, 22.

Aizawa, E., Tsuji, H., Asahara, T., Takahashi, T., Teraishi, T., Yoshida, S., Ota, M., Koga, N., Hattori, K., and Kunugi, H. (2016). Possible association of Bifidobacterium and Lactobacillus in the gut microbiota of patients with major depressive disorder. *J Affect Disord*, 202, 254–257.

Akkasheh, G., Kashani-Poor, Z., Tajabadi-Ebrahimi, M., Jafari, P., Akbari, H., Taghizadeh, M., Memarzadeh, M. R., Asemi, Z., and Esmaillzadeh, A. (2016). Clinical and metabolic response to probiotic administration in patients with major depressive disorder: A randomized, double-blind, placebo-controlled trial. *Nutrition*, 32(3), 315–320.

Alfawaz, H. A., Bhat, R. S., Al-Ayadhi, L., and El-Ansary, A. K. (2014). Protective and restorative potency of Vitamin D on persistent biochemical autistic features induced in propionic acid-intoxicated rat pups. *BMC Complement Altern Med*, 14, 416.

Allen, A. P., Hutch, W., Borre, Y. E., Kennedy, P. J., Temko, A., Boylan, G., Murphy, E., Cryan, J. F., Dinan, T. G., and Clarke, G. (2016). Bifidobacterium longum 1714 as a translational psychobiotic: Modulation of stress, electrophysiology and neurocognition in healthy volunteers. *Transl Psychiatry*, 6(11), e939.

Ariyannur, P. S., Moffett, J. R., Madhavarao, C. N., Arun, P., Vishnu, N., Jacobowitz, D. M., Hallows, W. C., Denu, J. M., and Namboodiri, A. M. A. (2010). Nuclear-cytoplasmic localization of acetyl coenzyme a synthetase-1 in the rat brain. *Journal of Comparative Neurology*, 518(15), 2952–2977.

Backhed, F., Roswall, J., Peng, Y., Feng, Q., Jia, H., Kovatcheva-Datchary, P., Li, Y., Xia, Y., Xie, H., Zhong, H., Khan, M. T., Zhang, J., Li, J., Xiao, L., Al-Aama, J., Zhang, D., Lee, Y. S., Kotowska, D., Colding, C., Tremaroli, V., Yin, Y., Bergman, S., Xu, X., Madsen, L., Kristiansen, K., Dahlgren, J., and Wang, J. (2015). Dynamics and stabilization of the human gut microbiome during the first year of life. *Cell Host Microbe*, 17(6), 852.

Bagdasarian, N., Rao, K., and Malani, P. N. (2015). Diagnosis and treatment of Clostridium difficile in adults: A systematic review. *JAMA*, 313(4), 398–408.

Barrett, E., Ross, R. P., O'toole, P. W., Fitzgerald, G. F., and Stanton, C. (2012). gamma-Aminobutyric acid production by culturable bacteria from the human intestine. *J Appl Microbiol*, 113, 411–417.

Benton, D., Williams, C., and Brown, A. (2007). Impact of consuming a milk drink containing a probiotic on mood and cognition. *Eur J Clin Nutr*, 61(3), 355–361.

Bercik, P., Park, A. J., Sinclair, D., Khoshdel, A., Lu, J., Huang, X., Deng, Y., Blennerhassett, P. A., Fahnestock, M., Moine, D., Berger, B., Huizinga, J. D., Kunze, W., Mclean, P. G., Bergonzelli, G. E., Collins, S. M., and Verdu, E. F. (2011). The anxiolytic effect of Bifidobacterium longum NCC3001 involves vagal pathways for gut-brain communication. *Neurogastroenterology and Motility*, 23(12), 1132–1139.

Bercik, P., Verdu, E. F., Foster, J. A., Macri, J., Potter, M., Huang, X., Malinowski, P., Jackson, W., Blennerhassett, P., Neufeld, K. A., Lu, J., Khan, W. I., Corthesy-Theulaz, I., Cherbut, C., Bergonzelli, G. E., and Collins, S. M. (2010). Chronic gastrointestinal inflammation induces anxiety-like behavior and alters central nervous system biochemistry in mice. *Gastroenterology*, 139(6), 2102–2112 e1.

Biesalski, H. K. (2016). Nutrition meets the microbiome: Micronutrients and the microbiota. *Ann N Y Acad Sci*, 1372(1), 53–64.

Bokulich, N. A., Chung, J., Battaglia, T., Henderson, N., Jay, M., Li, H., A, D. L., Wu, F., Perez-Perez, G. I., Chen, Y., Schweizer, W., Zheng, X., Contreras, M., Dominguez-Bello, M. G., and Blaser, M. J. (2016). Antibiotics, birth mode, and diet shape microbiome maturation during early life. *Sci Transl Med*, 8(343), 343ra82.

Bravo, J. A., Forsythe, P., Chew, M. V., Escaravage, E., Savignac, H. M., Dinan, T. G., Bienenstock, J., and Cryan, J. F. (2011). Ingestion of Lactobacillus strain regulates emotional behavior and central GABA receptor expression in a mouse via the vagus nerve. *Proceedings of the National Academy of Sciences of the United States of America*, 108(38), 16050–16055.

Brissette, C. A., Houdek, H. M., Floden, A. M., and Rosenberger, T. A. (2012). Acetate supplementation reduces microglia activation and brain interleukin-1beta levels in a rat model of Lyme neuroborreliosis. *J Neuroinflammation*, 9, 249.

Carlson, A. L., Xia, K., Azcarate-Peril, M. A., Goldman, B. D., Ahn, M., Styner, M. A., Thompson, A. L., Geng, X., Gilmore, J. H., and Knickmeyer, R. C. (2018). Infant gut microbiome associated with cognitive development. *Biological Psychiatry*, 83(2), 148–159.

Chang, J. P., Su, K. P., Mondelli, V., and Pariante, C. M. (2018). Omega-3 Polyunsaturated fatty acids in youths with attention deficit hyperactivity disorder: A systematic review and meta-analysis of clinical trials and biological studies. *Neuropsychopharmacology*, 43(3), 534–545.

Chen, J. J., Zheng, P., Liu, Y. Y., Zhong, X. G., Wang, H. Y., Guo, Y. J., and Xie, P. (2018a). Sex differences in gut microbiota in patients with major depressive disorder. *Neuropsychiatr Dis Treat*, *14*, 647–655.

Chen, Z., Li, J., Gui, S., Zhou, C., Chen, J., Yang, C., Hu, Z., Wang, H., Zhong, X., Zeng, L., Chen, K., Li, P., and Xie, P. (2018b). Comparative metaproteomics analysis shows altered fecal microbiota signatures in patients with major depressive disorder. *Neuroreport*, *29*(5), 417–425.

Christian, L. M., Galley, J. D., Hade, E. M., Schoppe-Sullivan, S., Kamp Dush, C., and Bailey, M. T. (2015). Gut microbiome composition is associated with temperament during early childhood. *Brain, Behavior, and Immunity*, *45*, 118–127.

Chung, Y. C., Jin, H. M., Cui, Y., Kim, D. S., Jung, J. M., Park, J. I., Jung, E. S., Choi, E. K., and Chae, S. W. (2014). Fermented milk of Lactobacillus helveticus IDCC3801 improves cognitive functioning during cognitive fatigue tests in healthy older adults. *Journal of Functional Foods*, *10*, 465–474.

Clarke, G., Grenham, S., Scully, P., Fitzgerald, P., Moloney, R. D., Shanahan, F., Dinan, T. G., and Cryan, J. F. (2013). The microbiome-gut-brain axis during early life regulates the hippocampal serotonergic system in a sex-dependent manner. *Molecular Psychiatry*, *18*(6), 666–673.

Claus, S. P., Guillou, H., and Ellero-Simatos, S. (2016). The gut microbiota: A major player in the toxicity of environmental pollutants? *NPJ Biofilms Microbiomes*, *2*, 16003.

Coton, E., Rollan, G., Bertrand, A., and Lonvaud-Funel, A. (1998). Histamine-producing lactic acid bacteria in wines: Early detection, frequency, and distribution. *American Journal of Enology and Viticulture*, *49*(2), 199–204.

Craciunescu, C. N., Brown, E. C., Mar, M. H., Albright, C. D., Nadeau, M. R., and Zeisel, S. H. (2004). Folic acid deficiency during late gestation decreases progenitor cell proliferation and increases apoptosis in fetal mouse brain. *J Nutr*, *134*(1), 162–166.

De Angelis, M., Piccolo, M., Vannini, L., Siragusa, S., De Giacomo, A., Serrazzanetti, D. I., Cristofori, F., Guerzoni, M. E., Gobbetti, M., and Francavilla, R. (2013). Fecal microbiota and metabolome of children with autism and pervasive developmental disorder not otherwise specified. *PLoS One*, *8*(10), e76993.

Desbonnet, L., Clarke, G., Shanahan, F., Dinan, T. G., and Cryan, J. F. (2014). Microbiota is essential for social development in the mouse. *Mol Psychiatry*, *19*(2), 146–148.

Desbonnet, L., Garrett, L., Clarke, G., Bienenstock, J., and Dinan, T. G. (2008). The probiotic Bifidobacteria infantis: An assessment of potential antidepressant properties in the rat. *Journal of Psychiatric Research*, *43*(2), 164–174.

Diaz Heijtz, R., Wang, S., Anuar, F., Qian, Y., Bjorkholm, B., Samuelsson, A., Hibberd, M. L., Forssberg, H., and Pettersson, S. (2011). Normal gut microbiota modulates brain development and behavior. *Proceedings of the National Academy of Sciences of the United States of America*, *108*(7), 3047–3052.

Dominguez-Bello, M. G., Costello, E. K., Contreras, M., Magris, M., Hidalgo, G., Fierer, N., and Knight, R. (2010). Delivery mode shapes the acquisition and structure of the initial microbiota across multiple body habitats in newborns. *Proceedings of the National Academy of Sciences of the United States of America*, *107*(26), 11971–11975.

Eggesbo, M., Moen, B., Peddada, S., Baird, D., Rugtveit, J., Midtvedt, T., Bushel, P. R., Sekelja, M., and Rudi, K. (2011). Development of gut microbiota in infants not exposed to medical interventions. *APMIS*, *119*(1), 17–35.

Finegold, S. M., Dowd, S. E., Gontcharova, V., Liu, C., Henley, K. E., Wolcott, R. D., Youn, E., Summanen, P. H., Granpeesheh, D., Dixon, D., Liu, M., Molitoris, D. R., and Green, J. A., 3rd

(2010). Pyrosequencing study of fecal microflora of autistic and control children. *Anaerobe*, *16*(4), 444–453.

Finegold, S. M., Molitoris, D., Song, Y., Liu, C., Vaisanen, M. L., Bolte, E., McTeague, M., Sandler, R., Wexler, H., Marlowe, E. M., Collins, M. D., Lawson, P. A., Summanen, P., Baysallar, M., Tomzynski, T. J., Read, E., Johnson, E., Rolfe, R., Nasir, P., Shah, H., Haake, D. A., Manning, P., and Kaul, A. (2002). Gastrointestinal microflora studies in late-onset autism. *Clin Infect Dis*, *35*(Suppl 1), S6–S16.

Flavell, J. H., Miller, P. H., and Miller, S. A. (2002). *Cognitive Development*. Upper Saddle River, NJ, Prentice Hall.

Frost, G., Sleeth, M. L., Sahuri-Arisoylu, M., Lizarbe, B., Cerdan, S., Brody, L., Anastasovska, J., Ghourab, S., Hankir, M., Zhang, S., Carling, D., Swann, J. R., Gibson, G., Viardot, A., Morrison, D., Thomas, E. L., and Bell, J. D. (2014). The short-chain fatty acid acetate reduces appetite via a central homeostatic mechanism. *Nature Communications*, *5*.

Gao, W., Alcauter, S., Smith, J. K., Gilmore, J. H., and Lin, W. (2015). Development of human brain cortical network architecture during infancy. *Brain Struct Funct*, *220*(2), 1173–1186.

Gao, W., Salzwedel, A. P., Carlson, A. L., Xia, K., Azcarate-Peril, M. A., Styner, M. A., Thompson, A. L., Geng, X., Goldman, B. D., Gilmore, J. H., and Knickmeyer, R. C. (2019). Gut microbiome and brain functional connectivity in infants: A preliminary study focusing on the amygdala. *Psychopharmacology*, (Berl), *236*(5), 1641–1651.

Gareau, M. G., Wine, E., Rodrigues, D. M., Cho, J. H., Whary, M. T., Philpott, D. J., MacQueen, G., and Sherman, P. M. (2011). Bacterial infection causes stress-induced memory dysfunction in mice. *Gut*, *60*(3), 307–317.

Gielda, L. M. and Dirita, V. J. (2012). Zinc competition among the intestinal microbiota. *MBio*, *3*(4), e00171–12.

Gilmore, J. H., Knickmeyer, R. C., and Gao, W. (2018). Imaging structural and functional brain development in early childhood. *Nat Rev Neurosci*, *19*(3), 123–137.

Goldstein, B. I., Kemp, D. E., Soczynska, J. K., and McIntyre, R. S. (2009). Inflammation and the phenomenology, pathophysiology, comorbidity, and treatment of bipolar disorder: A systematic review of the literature. *J Clin Psychiatry*, *70*(8), 1078–1090.

Gondalia, S. V., Palombo, E. A., Knowles, S. R., Cox, S. B., Meyer, D., and Austin, D. W. (2012). Molecular characterisation of gastrointestinal microbiota of children with autism (with and without gastrointestinal dysfunction) and their neurotypical siblings. *Autism Research*, *5*(6), 419–427.

Goodrich, J. K., Davenport, E. R., Beaumont, M., Jackson, M. A., Knight, R., Ober, C., Spector, T. D., Bell, J. T., Clark, A. G., and Ley, R. E. (2016). Genetic Determinants of the Gut Microbiome in UK twins. *Cell Host Microbe*, *19*(5), 731–743.

Goyal, M. S., Venkatesh, S., Milbrandt, J., Gordon, J. I., and Raichle, M. E. (2015). Feeding the brain and nurturing the mind: Linking nutrition and the gut microbiota to brain development. *Proc Natl Acad Sci U S A*, *112*(46), 14105–14112.

Grandjean, P. and Herz, K. T. (2011). Methylmercury and brain development: Imprecision and underestimation of developmental neurotoxicity in humans. *Mt Sinai J Med*, *78*(1), 107–118.

Gupta, S., Allen-Vercoe, E., and Petrof, E. O. (2016). Fecal microbiota transplantation: In perspective. *Therap Adv Gastroenterol*, *9*(2), 229–239.

Hirono, H. and Wada, Y. (1978). Effects of dietary folate deficiency on developmental increase of myelin lipids in rat brain. *J Nutr*, *108*(5), 766–772.

Hoban, A. E., Stilling, R. M., Moloney, G., Shanahan, F., Dinan, T. G., Clarke, G., and Cryan, J. F. (2018). The microbiome regulates amygdala-dependent fear recall. *Mol Psychiatry*, *23*(5), 1134–1144.

Horiuchi, Y., Kimura, R., Kato, N., Fujii, T., Seki, M., Endo, T., Kato, T., and Kawashima, K. (2003). Evolutional study on acetylcholine expression. *Life Sci*, 72(15), 1745–1756.

Hornef, M. (2015). Pathogens, commensal symbionts, and pathobionts: Discovery and functional effects on the host. *ILAR J*, 56(2), 159–162.

Hsiao, E. Y., Mcbride, S. W., Hsien, S., Sharon, G., Hyde, E. R., Mccue, T., Codelli, J. A., Chow, J., Reisman, S. E., Petrosino, J. F., Patterson, P. H., and Mazmanian, S. K. (2013). Microbiota modulate behavioral and physiological abnormalities associated with neurodevelopmental disorders. *Cell*, 155(7), 1451–1463.

Inoue, R., Sakaue, Y., Sawai, C., Sawai, T., Ozeki, M., Romero-Perez, G. A., and Tsukahara, T. (2016). A preliminary investigation on the relationship between gut microbiota and gene expressions in peripheral mononuclear cells of infants with autism spectrum disorders. *Biosci Biotechnol Biochem*, 80(12), 2450–2458.

Jasarevic, E., Howerton, C. L., Howard, C. D., and Bale, T. L. (2015). Alterations in the vaginal microbiome by maternal stress are associated with metabolic reprogramming of the off-spring gut and brain. *Endocrinology*, 156(9), 3265–3276.

Jiang, H., Ling, Z., Zhang, Y., Mao, H., Ma, Z., Yin, Y., Wang, W., Tang, W., Tan, Z., Shi, J., Li, L., and Ruan, B. (2015). Altered fecal microbiota composition in patients with major depressive disorder. *Brain Behav Immun*, 48, 186–194.

Jiang, H. Y., Zhou, Y. Y., Zhou, G. L., Li, Y. C., Yuan, J., Li, X. H., and Ruan, B. (2018). Gut microbiota profiles in treatment-naive children with attention deficit hyperactivity disorder. *Behav Brain Res*, 347, 408–413.

Jost, T., Lacroix, C., Braegger, C. P., and Chassard, C. (2012). New insights in gut microbiota establishment in healthy breast fed neonates. *PLoS One*, 7(8), e44595.

Kaelberer, M. M., Buchanan, K. L., Klein, M. E., Barth, B. B., Montoya, M. M., Shen, X., and Bohorquez, D. V. (2018). A gut-brain neural circuit for nutrient sensory transduction. *Science*, 361(6408), eaat5236.

Kaluzna-Czaplinska, J. and Blaszczyk, S. (2012). The level of arabinitol in autistic children after probiotic therapy. *Nutrition*, 28(2), 124–126.

Kang, D. W., Adams, J. B., Gregory, A. C., Borody, T., Chittick, L., Fasano, A., Khoruts, A., Geis, E., Maldonado, J., Mcdonough-Means, S., Pollard, E. L., Roux, S., Sadowsky, M. J., Lipson, K. S., Sullivan, M. B., Caporaso, J. G., and Krajmalnik-Brown, R. (2017). Microbiota transfer therapy alters gut ecosystem and improves gastrointestinal and autism symptoms: An open-label study. *Microbiome*, 5(1), 10.

Kang, D. W., Ilhan, Z. E., Isern, N. G., Hoyt, D. W., Howsmon, D. P., Shaffer, M., Lozupone, C. A., Hahn, J., Adams, J. B., and Krajmalnik-Brown, R. (2018). Differences in fecal microbial metabolites and microbiota of children with autism spectrum disorders. *Anaerobe*, 49, 121–131.

Kang, D. W., Park, J. G., Ilhan, Z. E., Wallstrom, G., Labaer, J., Adams, J. B., and Krajmalnik-Brown, R. (2013). Reduced incidence of Prevotella and other fermenters in intestinal microflora of autistic children. *PLoS One*, 8(7), e68322.

Kato-Kataoka, A., Nishida, K., Takada, M., Kawai, M., Kikuchi-Hayakawa, H., Suda, K., Ishikawa, H., Gondo, Y., Shimizu, K., Matsuki, T., Kushiro, A., Hoshi, R., Watanabe, O., Igarashi, T., Miyazaki, K., Kuwano, Y., and Rokutan, K. (2016). Fermented milk containing Lactobacillus casei strain shirota preserves the diversity of the gut microbiota and relieves abdominal dysfunction in healthy medical students exposed to academic stress. *Appl Environ Microbiol*, 82(12), 3649–3658.

Kazemi, A., Noorbala, A. A., Azam, K., Eskandari, M. H., and Djafarian, K. (2019). Effect of probiotic and prebiotic vs placebo on psychological outcomes in patients with major depressive disorder: A randomized clinical trial. *Clin Nutr*, *38*(2), 522–528.

Kelly, J. R., Allen, A. P., Temko, A., Hutch, W., Kennedy, P. J., Farid, N., Murphy, E., Boylan, G., Bienenstock, J., Cryan, J. F., Clarke, G., and Dinan, T. G. (2017). Lost in translation? The potential psychobiotic Lactobacillus rhamnosus (JB-1) fails to modulate stress or cognitive performance in healthy male subjects. *Brain Behav Immun*, *61*, 50–59.

Kelly, J. R., Borre, Y., O'Brien, C., Patterson, E., El Aidy, S., Deane, J., Kennedy, P. J., Beers, S., Scott, K., Moloney, G., Hoban, A. E., Scott, L., Fitzgerald, P., Ross, P., Stanton, C., Clarke, G., Cryan, J. F., and Dinan, T. G. (2016). Transferring the blues: Depression-associated gut microbiota induces neurobehavioural changes in the rat. *J Psychiatr Res*, *82*, 109–118.

Kelly, J. R., Kennedy, P. J., Cryan, J. F., Dinan, T. G., Clarke, G., and Hyland, N. P. (2015). Breaking down the barriers: The gut microbiome, intestinal permeability and stress-related psychiatric disorders. *Front Cell Neurosci*, *9*, 392.

Kim, S., Kim, H., Yim, Y. S., Ha, S., Atarashi, K., Tan, T. G., Longman, R. S., Honda, K., Littman, D. R., Choi, G. B., and Huh, J. R. (2017). Maternal gut bacteria promote neurodevelopmental abnormalities in mouse offspring. *Nature*, *549*(7673), 528–532.

King, N. J. and Thomas, S. R. (2007). Molecules in focus: Indoleamine 2,3-dioxygenase. *Int J Biochem Cell Biol*, *39*(12), 2167–2172.

Kleiman, S. C., Watson, H. J., Bulik-Sullivan, E. C., Huh, E. Y., Tarantino, L. M., Bulik, C. M., and Carroll, I. M. (2015). The intestinal microbiota in acute anorexia nervosa and during renourishment: Relationship to depression, anxiety, and eating disorder psychopathology. *Psychosom Med*, *77*(9), 969–981.

Knickmeyer, R., Gouttard, S., Kang, C., Evans, D., Wilber, K., Smith, J. K., Hamer, R. M., Lin, W., Gerig, G. and Gilmore, J. H. (2008). A structural MRI study of human brain development from birth to 2 years. *Journal of Neuroscience*, *28*(47), 12176–12182.

Koenig, J. E., Spor, A., Scalfone, N., Fricker, A. D., Stombaugh, J., Knight, R., Angenent, L. T., and Ley, R. E. (2011). Succession of microbial consortia in the developing infant gut microbiome. *Proceedings of the National Academy of Sciences of the United States of America*, *108* Suppl 1, 4578–4585.

Kortman, G. A., Raffatellu, M., Swinkels, D. W., and Tjalsma, H. (2014). Nutritional iron turned inside out: Intestinal stress from a gut microbial perspective. *FEMS Microbiol Rev*, *38*(6), 1202–1234.

Krajmalnik-Brown, R., Ilhan, Z. E., Kang, D. W., and Dibaise, J. K. (2012). Effects of gut microbes on nutrient absorption and energy regulation. *Nutr Clin Pract*, *27*(2), 201–214.

Kristensen, N. B., Bryrup, T., Allin, K. H., Nielsen, T., Hansen, T. H., and Pedersen, O. (2016). Alterations in fecal microbiota composition by probiotic supplementation in healthy adults: A systematic review of randomized controlled trials. *Genome Med*, *8*(1), 52.

Kushak, R. I., Winter, H. S., Buie, T. M., Cox, S. B., Phillips, C. D., and Ward, N. L. (2017). Analysis of the duodenal microbiome in autistic individuals: Association with carbohydrate digestion. *J Pediatr Gastroenterol Nutr*, *64*(5), e110–e116.

Landete, J. M., De Las Rivas, B., Marcobal, A., and Munoz, R. (2008). Updated molecular knowledge about histamine biosynthesis by bacteria. *Critical Reviews in Food Science and Nutrition*, *48*(8), 697–714.

Lin, P., Ding, B., Feng, C., Yin, S., Zhang, T., Qi, X., Lv, H., Guo, X., Dong, K., Zhu, Y., and Li, Q. (2017). Prevotella and Klebsiella proportions in fecal microbial communities are potential characteristic parameters for patients with major depressive disorder. *J Affect Disord*, *207*, 300–304.

Liu, Y., Zhang, L., Wang, X., Wang, Z., Zhang, J., Jiang, R., Wang, X., Wang, K., Liu, Z., Xia, Z., Xu, Z., Nie, Y., Lv, X., Wu, X., Zhu, H., and Duan, L. (2016). Similar fecal microbiota signatures in patients with diarrhea-predominant irritable bowel syndrome and patients with depression. *Clin Gastroenterol Hepatol*, *14*(11), 1602–1611 e5.

Lorenzo-Zuniga, V., Llop, E., Suarez, C., Alvarez, B., Abreu, L., Espadaler, J., and Serra, J. (2014). I.31, a new combination of probiotics, improves irritable bowel syndrome-related quality of life. *World J Gastroenterol*, *20*(26), 8709–8716.

Luczynski, P., Mcvey Neufeld, K. A., Oriach, C. S., Clarke, G., Dinan, T. G., and Cryan, J. F. (2016). Growing up in a bubble: Using germ-free animals to assess the influence of the gut microbiota on brain and behavior. *Int J Neuropsychopharmacol*, *19*(8), pyw020.

Lyra, A., Hillila, M., Huttunen, T., Mannikko, S., Taalikka, M., Tennila, J., Tarpila, A., Lahtinen, S., Ouwehand, A. C., and Veijola, L. (2016). Irritable bowel syndrome symptom severity improves equally with probiotic and placebo. *World J Gastroenterol*, *22*(48), 10631–10642.

Macfabe, D. F., Cain, D. P., Rodriguez-Capote, K., Franklin, A. E., Hoffman, J. E., Boon, F., Taylor, A. R., Kavaliers, M., and Ossenkopp, K. P. (2007). Neurobiological effects of intraventricular propionic acid in rats: Possible role of short chain fatty acids on the pathogenesis and characteristics of autism spectrum disorders. *Behav Brain Res*, *176*(1), 149–169.

Majeed, M., Nagabhushanam, K., Arumugam, S., Majeed, S., and Ali, F. (2018). Bacillus coagulans MTCC 5856 for the management of major depression with irritable bowel syndrome: A randomised, double-blind, placebo controlled, multi-centre, pilot clinical study. *Food Nutr Res*, *62*.

Marcos, A., Warnberg, J., Nova, E., Gomez, S., Alvarez, A., Alvarez, R., Mateos, J. A., and Cobo, J. M. (2004). The effect of milk fermented by yogurt cultures plus Lactobacillus casei DN-114001 on the immune response of subjects under academic examination stress. *Eur J Nutr*, *43*(6), 381–389.

Messaoudi, M., Lalonde, R., Violle, N., Javelot, H., Desor, D., Nejdi, A., Bisson, J. F., Rougeot, C., Pichelin, M., Cazaubiel, M., and Cazaubiel, J. M. (2011). Assessment of psychotropic-like properties of a probiotic formulation (Lactobacillus helveticus R0052 and Bifidobacterium longum R0175) in rats and human subjects. *British Journal of Nutrition*, *105*(5), 755–764.

Miller, A. H., Haroon, E., Raison, C. L., and Felger, J. C. (2013). Cytokine targets in the brain: Impact on neurotransmitters and neurocircuits. *Depression and Anxiety*, *30*(4), 297–306.

Mitchell, R. H. and Goldstein, B. I. (2014). Inflammation in children and adolescents with neuropsychiatric disorders: A systematic review. *J Am Acad Child Adolesc Psychiatry*, *53*(3), 274–296.

Mohammadi, A. A., Jazayeri, S., Khosravi-Darani, K., Solati, Z., Mohammadpour, N., Asemi, Z., Adab, Z., Djalali, M., Tehrani-Doost, M., Hosseini, M., and Eghtesadi, S. (2015). Effects of probiotics on biomarkers of oxidative stress and inflammatory factors in petrochemical workers: A randomized, double-blind, placebo-controlled trial. *Int J Prev Med*, *6*, 82.

Myint, A. M. (2012). Kynurenines: From the perspective of major psychiatric disorders. *FEBS J*, *279*(8), 1375–1385.

Nakamura, I., Hosokawa, K., Tamura, H., and Miura, T. (1977). Reduced mercury excretion with feces in germfree mice after oral administration of methyl mercury chloride. *Bull Environ Contam Toxicol*, *17*(5), 528–533.

Nankova, B. B., Agarwal, R., Macfabe, D. F., and La Gamma, E. F. (2014). Enteric bacterial metabolites propionic and butyric acid modulate gene expression, including

CREB-dependent catecholaminergic neurotransmission, in PC12 cells—possible relevance to autism spectrum disorders. *PLoS One*, 9(8), e103740.

Naseribafrouei, A., Hestad, K., Avershina, E., Sekelja, M., Linlokken, A., Wilson, R., and Rudi, K. (2014). Correlation between the human fecal microbiota and depression. *Neurogastroenterol Motil*, 26(8), 1155–1162.

Nemeth, H., Robotka, H., Toldi, J., and Vecsei, L. (2007). Kynurenines in the central nervous system. *Central Nervous System Agents in Medicinal Chemistry*, 7, 45–56.

Neufeld, K. M., Kang, N., Bienenstock, J., and Foster, J. A. (2011). Reduced anxiety-like behavior and central neurochemical change in germ-free mice. *Neurogastroenterology and Motility*, 23(3), 255–64, e119.

Nguyen, N. H., Morland, C., Gonzalez, S. V., Rise, F., Storm-Mathisen, J., Gundersen, V., and Hassel, B. (2007). Propionate increases neuronal histone acetylation, but is metabolized oxidatively by glia. Relevance for propionic acidemia. *J Neurochem*, 101(3), 806–814.

Nishihira, J., Kagami-Katsuyama, H., Tanaka, A., Nishimura, M., Kobayashi, T., and Kawasaki, Y. (2014). Elevation of natural killer cell activity and alleviation of mental stress by the consumption of yogurt containing Lactobacillus gasseri SBT2055 and Bifidobacterium longum SBT2928 in a double-blind, placebo-controlled clinical trial. *Journal of Functional Foods*, 11, 261–268.

Ogbonnaya, E. S., Clarke, G., Shanahan, F., Dinan, T. G., Cryan, J. F., and O'Leary, O. F. (2015). Adult hippocampal neurogenesis is regulated by the microbiome. *Biol Psychiatry*, 78(4), e7–9.

Ostlund-Lagerstrom, L., Kihlgren, A., Repsilber, D., Bjorksten, B., Brummer, R. J., and Schoultz, I. (2016). Probiotic administration among free-living older adults: A double blinded, randomized, placebo-controlled clinical trial. *Nutr J*, 15(1), 80.

Ozogul, F. (2011). Effects of specific lactic acid bacteria species on biogenic amine production by foodborne pathogen. *International Journal of Food Science and Technology*, 46(3), 478–484.

Parracho, H. M., Bingham, M. O., Gibson, G. R., and McCartney, A. L. (2005). Differences between the gut microflora of children with autistic spectrum disorders and that of healthy children. *J Med Microbiol*, 54(Pt 10), 987–991.

Parracho, H. M. R. T., Gibson, G. R., Knott, F., Bosscher, D., Kleerebezem, M., and McCartney, A. L. (2010). A double-blind, placebo-controlled, crossover-designed probiotic feeding study in children diagnosed with autistic spectrum disorders. *International Journal of Probiotics and Prebiotics*, 5(2), 69–74.

Partty, A., Kalliomaki, M., Wacklin, P., Salminen, S., and Isolauri, E. (2015). A possible link between early probiotic intervention and the risk of neuropsychiatric disorders later in childhood: A randomized trial. *Pediatr Res*, 77(6), 823–828.

Perez-Munoz, M. E., Arrieta, M. C., Ramer-Tait, A. E., and Walter, J. (2017). A critical assessment of the "sterile womb" and "in utero colonization" hypotheses: Implications for research on the pioneer infant microbiome. *Microbiome*, 5(1), 48.

Peterson, B. S., RAuh, V. A., Bansal, R., Hao, X., Toth, Z., Nati, G., Walsh, K., Miller, R. L., Arias, F., Semanek, D., and Perera, F. (2015). Effects of prenatal exposure to air pollutants (polycyclic aromatic hydrocarbons) on the development of brain white matter, cognition, and behavior in later childhood. *JAMA Psychiatry*, 72(6), 531–540.

Piewngam, P., Zheng, Y., Nguyen, T. H., Dickey, S. W., Joo, H. S., Villaruz, A. E., Glose, K. A., Fisher, E. L., Hunt, R. L., LI, B., Chiou, J., Pharkjaksu, S., Khongthong, S., Cheung, G. Y. C., Kiratisin, P., and Otto, M. (2018). Pathogen elimination by probiotic Bacillus via signalling interference. *Nature*, 562(7728), 532–537.

Pinto-Sanchez, M. I., Hall, G. B., Ghajar, K., Nardelli, A., Bolino, C., Lau, J. T., Martin, F. P., Cominetti, O., Welsh, C., Rieder, A., Traynor, J., Gregory, C., De Palma, G., Pigrau, M., Ford, A. C., Macri, J., Berger, B., Bergonzelli, G., Surette, M. G., Collins, S. M., Moayyedi, P., and Bercik, P. (2017). Probiotic Bifidobacterium longum NCC3001 reduces depression scores and alters brain activity: A pilot study in patients with irritable bowel syndrome. *Gastroenterology*, 153(2), 448–459 e8.

Planer, J. D., Peng, Y., Kau, A. L., Blanton, L. V., Ndao, I. M., Tarr, P. I., Warner, B. B., and Gordon, J. I. (2016). Development of the gut microbiota and mucosal IgA responses in twins and gnotobiotic mice. *Nature*, 534(7606), 263–266.

Potvin, S., Stip, E., Sepehry, A. A., Gendron, A., Bah, R., and Kouassi, E. (2008). Inflammatory cytokine alterations in schizophrenia: A systematic quantitative review. *Biological Psychiatry*, 63(8), 801–808.

Prado, E. L. and Dewey, K. G. (2014). Nutrition and brain development in early life. *Nutr Rev*, 72(4), 267–284.

Pulikkan, J., Maji, A., Dhakan, D. B., Saxena, R., Mohan, B., Anto, M. M., Agarwal, N., Grace, T., and Sharma, V. K. (2018). Gut microbial dysbiosis in Indian children with autism spectrum disorders. *Microb Ecol*, 76(4), 1102–1114.

Raygan, F., Ostadmohammadi, V., Bahmani, F., and Asemi, Z. (2018). The effects of vitamin D and probiotic co-supplementation on mental health parameters and metabolic status in type 2 diabetic patients with coronary heart disease: A randomized, double-blind, placebo-controlled trial. *Prog Neuropsychopharmacol Biol Psychiatry*, 84(Pt A), 50–55.

Romano, K. A., Martinez-Del Campo, A., Kasahara, K., Chittim, C. L., Vivas, E. I., Amador-Noguez, D., Balskus, E. P., and Rey, F. E. (2017). Metabolic, epigenetic, and transgenerational effects of gut bacterial choline consumption. *Cell Host Microbe*, 22(3), 279–290 e7.

Romano, K. A., Vivas, E. I., Amador-Noguez, D., and Rey, F. E. (2015). Intestinal microbiota composition modulates choline bioavailability from diet and accumulation of the proatherogenic metabolite trimethylamine-N-oxide. *MBio*, 6(2), e02481.

Romijn, A. R., Rucklidge, J. J., Kuijer, R. G., and Frampton, C. (2017). A double-blind, randomized, placebo-controlled trial of Lactobacillus helveticus and Bifidobacterium longum for the symptoms of depression. *Aust N Z J Psychiatry*, 51(8), 810–821.

Rook, G. A., Raison, C. L., and Lowry, C. A. (2014). Microbiota, immunoregulatory old friends and psychiatric disorders. *Adv Exp Med Biol*, 817, 319–356.

Sandle, R. R. H., Finegold, S. M., Bolte, E. R., Buchanan, C. P., Maxwell, A. P., Vaisanen, M. L., Nelson, M. N., and Wexler, H. M. (2000). Short-term benefit from oral vancomycin treatment of regressive-onset autism. *J Child Neurol*, 15(7), 429–435.

Santocchi, E., Guiducci, L., Fulceri, F., Billeci, L., Buzzigoli, E., Apicella, F., Calderoni, S., Grossi, E., Morales, M. A., and Muratori, F. (2016). Gut to brain interaction in autism spectrum disorders: A randomized controlled trial on the role of probiotics on clinical, biochemical and neurophysiological parameters. *BMC Psychiatry*, 16, 183.

Sarkar, A., Lehto, S. M., Harty, S., Dinan, T. G., Cryan, J. F., and Burnet, P. W. J. (2016). Psychobiotics and the Manipulation of Bacteria-Gut-Brain Signals. *Trends in Neurosciences*, 39(11), 763–781.

Schroeder, F. A., Lin, C. L., Crusio, W. E., and Akbarian, S. (2007). Antidepressant-like effects of the histone deacetylase inhibitor, sodium butyrate, in the mouse. *Biol Psychiatry*, 62(1), 55–64.

Seko, Y., Miura, T., Takahashi, M., and Koyama, T. (1981). Methyl mercury decomposition in mice treated with antibiotics. *Acta Pharmacol Toxicol* (Copenh), 49(4), 259–265.

Sender, R., Fuchs, S., and Milo, R. (2016). Revised estimates for the number of human and bacteria cells in the body. *PLoS Biol*, *14*(8), e1002533.

Shaaban, S. Y., El Gendy, Y. G., Mehanna, N. S., El-Senousy, W. M., El-Feki, H. S. A., Saad, K., and El-Asheer, O. M. (2018). The role of probiotics in children with autism spectrum disorder: A prospective, open-label study. *Nutr Neurosci*, *21*(9), 676–681

Shen, Y., Xu, J., Li, Z., Huang, Y., Yuan, Y., Wang, J., Zhang, M., Hu, S., and Liang, Y. (2018). Analysis of gut microbiota diversity and auxiliary diagnosis as a biomarker in patients with schizophrenia: A cross-sectional study. *Schizophr Res*, *197*, 470–477.

Shishov, V. A., Kirovskaya, T. A., Kudrin, V. S., and Oleskin, A. V. (2009). Amine Neuromediators, Their precursors, and oxidation products in the culture of Escherichia coli K-12. *Applied Biochemistry and Microbiology*, *45*(5), 494–497.

Shultz, S. R., Aziz, N. A., Yang, L., Sun, M., MacFabe, D. F., and O'Brien, T. J. (2015). Intracerebroventricular injection of propionic acid, an enteric metabolite implicated in autism, induces social abnormalities that do not differ between seizure-prone (FAST) and seizure-resistant (SLOW) rats. *Behav Brain Res*, *278*, 542–548.

Shultz, S. R., MacFabe, D. F., Martin, S., Jackson, J., Taylor, R., Boon, F., Ossenkopp, K. P., and Cain, D. P. (2009). Intracerebroventricular injections of the enteric bacterial metabolic product propionic acid impair cognition and sensorimotor ability in the Long-Evans rat: Further development of a rodent model of autism. *Behav Brain Res*, *200*(1), 33–41.

Simren, M., Ohman, L., Olsson, J., Svensson, U., Ohlson, K., Posserud, I., and Strid, H. (2010). Clinical trial: The effects of a fermented milk containing three probiotic bacteria in patients with irritable bowel syndrome—a randomized, double-blind, controlled study. *Aliment Pharmacol Ther*, *31*(2), 218–227.

Slykerman, R. F., Hood, F., Wickens, K., Thompson, J. M. D., Barthow, C., Murphy, R., Kang, J., Rowden, J., Stone, P., Crane, J., Stanley, T., Abels, P., Purdie, G., Maude, R., Mitchell, E. A., and the Probiotic In Pregnancy Study Group. (2017). Effect of Lactobacillus rhamnosus HN001 in pregnancy on postpartum symptoms of depression and anxiety: A randomised double-blind placebo-controlled trial. *EBioMedicine*, *24*, 159–165.

Song, Y., Liu, C., and Finegold, S. M. (2004). Real-time PCR quantitation of clostridia in feces of autistic children. *Appl Environ Microbiol*, *70*(11), 6459–6465.

Steenbergen, L., Sellaro, R., Van Hemert, S., Bosch, J. A., and Colzato, L. S. (2015). A randomized controlled trial to test the effect of multispecies probiotics on cognitive reactivity to sad mood. *Brain Behav Immun*, *48*, 258–264.

Steinfeld, R., Grapp, M., Kraetzner, R., Dreha-Kulaczewski, S., Helms, G., Dechent, P., Wevers, R., Grosso, S., and Gartner, J. (2009). Folate receptor alpha defect causes cerebral folate transport deficiency: A treatable neurodegenerative disorder associated with disturbed myelin metabolism. *Am J Hum Genet*, *85*(3), 354–363.

Steliou, K., Boosalis, M. S., Perrine, S. P., Sangerman, J., and Faller, D. V. (2012). Butyrate histone deacetylase inhibitors. *Biores Open Access*, *1*(4), 192–198.

Stiles, J. and Jernigan, T. L. 2010. The basics of brain development. *Neuropsychology Review*, *20*(4), 327–348.

Stilling, R. M., Moloney, G. M., Ryan, F. J., Hoban, A. E., Bastiaanssen, T. F., Shanahan, F., Clarke, G., Claesson, M. J., Dinan, T. G., and Cryan, J. F. (2018). Social interaction-induced activation of RNA splicing in the amygdala of microbiome-deficient mice. *Elife*, *7*.

Stilling, R. M., Ryan, F. J., Hoban, A. E., Shanahan, F., Clarke, G., Claesson, M. J., Dinan, T. G., and Cryan, J. F. (2015). Microbes and neurodevelopment—Absence of microbiota during early life increases activity-related transcriptional pathways in the amygdala. *Brain Behav Immun*, 50, 209–220.

Strati, F., Cavalieri, D., Albanese, D., De Felice, C., Donati, C., Hayek, J., Jousson, O., Leoncini, S., Renzi, D., Calabro, A. and De Filippo, C. (2017). New evidences on the altered gut microbiota in autism spectrum disorders. *Microbiome*, 5(1), 24.

Sudo, N., Chida, Y., Aiba, Y., Sonoda, J., Oyama, N., YU, X. N., Kubo, C., and Koga, Y. (2004). Postnatal microbial colonization programs the hypothalamic-pituitary-adrenal system for stress response in mice. *J Physiol*, 558 (Pt 1), 263–275.

Suez, J., Zmora, N., Zilberman-Schapira, G., Mor, U., Dori-Bachash, M., Bashiardes, S., Zur, M., Regev-Lehavi, D., Brik, R. B., Federici, S., Horn, M., Cohen, Y., Moor, A. E., Zeevi, D., Korem, T., Kotler, E., Harmelin, A., Itzkovitz, S., Maharshak, N., Shibolet, O., Pevsner-Fischer, M., Shapiro, H., Sharon, I., Halpern, Z., Segal, E., and Elinav, E. (2018). Post-antibiotic gut mucosal microbiome reconstitution is impaired by probiotics and improved by autologous FMT. *Cell*, 174(6), 1406–1423.

Takada, M., Nishida, K., Gondo, Y., Kikuchi-Hayakawa, H., Ishikawa, H., Suda, K., Kawai, M., Hoshi, R., Kuwano, Y., Miyazaki, K., and Rokutan, K. (2017). Beneficial effects of Lactobacillus casei strain Shirota on academic stress-induced sleep disturbance in healthy adults: A double-blind, randomised, placebo-controlled trial. *Benef Microbes*, 8(2), 153–162.

Takada, M., Nishida, K., Kataoka-Kato, A., Gondo, Y., Ishikawa, H., Suda, K., Kawai, M., Hoshi, R., Watanabe, O., Igarashi, T., Kuwano, Y., Miyazaki, K., and Rokutan, K. (2016). Probiotic Lactobacillus casei strain Shirota relieves stress-associated symptoms by modulating the gut-brain interaction in human and animal models. *Neurogastroenterol Motil*, 28(7), 1027–1036.

Thomas, R. H., Meeking, M. M., Mepham, J. R., Tichenoff, L., Possmayer, F., Liu, S., and MacFabe, D. F. (2012). The enteric bacterial metabolite propionic acid alters brain and plasma phospholipid molecular species: Further development of a rodent model of autism spectrum disorders. *J Neuroinflammation*, 9, 153.

Thompson, A. L., Monteagudo-Mera, A., Cadenas, M. B., Lampl, M. L., and Azcarate-Peril, M. A. (2015). Milk- and solid-feeding practices and daycare attendance are associated with differences in bacterial diversity, predominant communities, and metabolic and immune function of the infant gut microbiome. *Front Cell Infect Microbiol*, 5, 3.

Tomova, A., Husarova, V., Lakatosova, S., Bakos, J., Vlkova, B., Babinska, K., and Ostatnikova, D. (2015). Gastrointestinal microbiota in children with autism in Slovakia. *Physiol Behav*, 138, 179–187.

Tsavkelova, E. A., Botvinko, I. V., Kudrin, V. S., and Oleskin, A. V. (2000). Detection of neurotransmitter amines in microorganisms with the use of high-performance liquid chromatography. *Dokl Biochem*, 372(1–6), 115–117.

Turnbaugh, P. J., Hamady, M., Yatsunenko, T., Cantarel, B. L., Duncan, A., Ley, R. E., Sogin, M. L., Jones, W. J., Roe, B. A., Affourtit, J. P., Egholm, M., Henrissat, B., Heath, A. C., Knight, R., and Gordon, J. I. (2009). A core gut microbiome in obese and lean twins. *Nature*, 457(7228), 480–484.

Tyler, C. R. and Allan, A. M. (2014). The effects of arsenic exposure on neurological and cognitive dysfunction in human and rodent studies: A review. *Curr Environ Health Rep*, 1, 132–147.

Ursell, L. K., Metcalf, J. L., Parfrey, L. W., and Knight, R. (2012). Defining the human microbiome. *Nutr Rev*, 70 Suppl 1, S38–44.

Valles-Colomer, M., Falony, G., Darzi, Y., Tigchelaar, E. F., Wang, J., Tito, R. Y., Schiweck, C., Kurilshikov, A., Joossens, M., Wijmenga, C., Claes, S., van Oudenhove, L., Zhernakova, A., Vieira-Silva, S., and RAES, J. (2019). The neuroactive potential of the human gut microbiota in quality of life and depression. *Nat Microbiol*, 4(4), 623–632.

Van De Wiele, T., Vanhaecke, L., Boeckaert, C., Peru, K., Headley, J., Verstraete, W., and Siciliano, S. (2005). Human colon microbiota transform polycyclic aromatic hydrocarbons to estrogenic metabolites. *Environ Health Perspect*, 113(1), 6–10.

Walker, W. A. and Iyengar, R. S. (2015). Breast milk, microbiota, and intestinal immune homeostasis. *Pediatr Res*, 77(1–2), 220–228.

Wall, R., Cryan, J. F., Ross, R. P., Fitzgerald, G. F., Dinan, T. G., and Stanton, C. (2014). Bacterial Neuroactive Compounds Produced by Psychobiotics. *Microbial Endocrinology: The Microbiota-Gut-Brain Axis in Health and Disease*, 817, 221–239.

Wang, L., Christophersen, C. T., Sorich, M. J., Gerber, J. P., Angley, M. T., and Conlon, M. A. (2013). Increased abundance of Sutterella spp. and Ruminococcus torques in feces of children with autism spectrum disorder. *Mol Autism*, 4(1), 42.

Wang, X., Yang, J., Zhang, H., Yu, J., and Yao, Z. (2019). Oral probiotic administration during pregnancy prevents autism-related behaviors in offspring induced by maternal immune activation via anti-inflammation in mice. *Autism Res*, 12(4), 576–588

Wiens, D., Dewitt, A., Kosar, M., Underriner, C., Finsand, M., and Freese, M. (2015). Influence of folic acid on neural connectivity during dorsal root ganglion neurogenesis. *Cells Tissues Organs*, 201(5), 342–353.

Wikoff, W. R., Anfora, A. T., Liu, J., Schultz, P. G., Lesley, S. A., Peters, E. C., and Siuzdak, G. (2009). Metabolomics analysis reveals large effects of gut microflora on mammalian blood metabolites. *Proceedings of the National Academy of Sciences of the United States of America*, 106(10), 3698–3703.

Williams, B. L., Hornig, M., Buie, T., Bauman, M. L., Cho Paik, M., Wick, I., Bennett, A., Jabado, O., Hirschberg, D. L., and Lipkin, W. I. (2011). Impaired carbohydrate digestion and transport and mucosal dysbiosis in the intestines of children with autism and gastrointestinal disturbances. *PLoS One*, 6(9), e24585.

Williams, B. L., Hornig, M., Parekh, T., and Lipkin, W. I. (2012). Application of novel PCR-based methods for detection, quantitation, and phylogenetic characterization of Sutterella species in intestinal biopsy samples from children with autism and gastrointestinal disturbances. *MBio*, 3(1), e00261-11.

Yang, H., Zhao, X., Tang, S., Huang, H., Zhao, X., Ning, Z., Fu, X., and Zhang, C. (2016). Probiotics reduce psychological stress in patients before laryngeal cancer surgery. *Asia Pac J Clin Oncol*, 12(1), e92-6.

Yano, J. M., Yu, K., Donaldson, G. P., Shastri, G. G., Ann, P., Ma, L., Nagler, C. R., Ismagilov, R. F., Mazmanian, S. K., and Hsiao, E. Y. (2015). Indigenous bacteria from the gut microbiota regulate host serotonin biosynthesis. *Cell*, 161(2), 264–276.

Zheng, P., Zeng, B., Zhou, C., Liu, M., Fang, Z., Xu, X., Zeng, L., Chen, J., Fan, S., Du, X., Zhang, X., Yang, D., Yang, Y., Meng, H., Li, W., Melgiri, N. D., Licinio, J., Wei, H., and Xie, P. (2016). Gut microbiome remodeling induces depressive-like behaviors through a pathway mediated by the host's metabolism. *Mol Psychiatry*, 21(6), 786–796.

Zijlmans, M. A., Korpela, K., Riksen-Walraven, J. M., De Vos, W. M., and De Weerth, C. (2015). Maternal prenatal stress is associated with the infant intestinal microbiota. *Psychoneuroendocrinology*, 53, 233–245.

Zmora, N., Zilberman-Schapira, G., Suez, J., Mor, U., Dori-Bachash, M., Bashiardes, S., Kotler, E., Zur, M., Regev-Lehavi, D., Brik, R. B., Federici, S., Cohen, Y., Linevsky, R., Rothschild, D., Moor, A. E., Ben-Moshe, S., Harmelin, A., Itzkovitz, S., Maharshak, N., Shibolet, O., Shapiro, H., Pevsner-Fischer, M., Sharon, I., Halpern, Z., Segal, E., and Elinav, E. (2018). Personalized gut mucosal colonization resistance to empiric probiotics is associated with unique host and microbiome features. *Cell*, 174(6), 1388–1405.

CHAPTER 15

THE ROLE OF THE MICROBIOTA-GUT-BRAIN AXIS IN NEURODEVELOPMENT AND MENTAL HEALTH IN CHILDHOOD AND ADOLESCENCE

KEVIN CHAMPAGNE-JORGENSEN AND KAREN-ANNE MCVEY NEUFELD

INTRODUCTION

THERE is a growing body of literature recognizing the importance of the intestinal microbiota in the development and function of the central nervous system (CNS). Animal studies ranging from those that use germ-free (GF) mice (animals that are bred and maintained in the total absence of bacteria), mice that have been treated with antibiotics during various stages in development, pre- and probiotic feeding studies, fecal microbial transplantation experiments, as well as work that has examined the impact of pathogenic infection have all indicated that the bacterial status of the intestinal lumen has far-reaching and long-term effects on brain neurochemistry, anatomy and behavior (Bercik et al., 2011b; Diaz Heijtz et al., 2011; Neufeld et al., 2011; Bravo et al., 2011; Collins et al., 2013; Hsaio et al., 2013; Ogbonnaya et al., 2015; Luczynski et al., 2016b; Bastiaanssen et al., 2019). In addition, preliminary evidence in human clinical trials indicates that feeding certain pre- and probiotics may have beneficial effects on brain function (Benton et al., 2007; Messaoudi et al., 2011a; Messaoudi et al., 2011b; Tillisch et al., 2013; Schmidt et al., 2015; Steenbergen et al., 2015; Huang et al., 2016; Cerdo et al., 2017; Reis et al., 2018). There is keen interest in microbiota studies particularly focused on early life, childhood and young adulthood, as this is the time of life when the brain is most plastic and vulnerable to peripheral influence and thus more likely to be sensitive to changes in the intestinal microbiome. Similarly, work focusing on the potential role of the intestinal microbiome in the pathogenesis of a variety of childhood neurodevelopmental illnesses is of high priority. In this chapter we will review the basic concepts of the microbiota-gut-brain axis as well as briefly touch on what is known with regard to potential mechanisms whereby this crosstalk occurs. We will outline current thinking regarding general patterns of intestinal microbial change during early childhood and adolescence and emphasize that dynamic shifts in intestinal colonization occur concurrently with the journey to neurodevelopmental milestones. We will review the body of evidence that supports the hypothesis that bugs can alter neural function, with specific focus on what is known regarding bacterial input on cognitive development in youth. Much of the neurodevelopmental work done to date is preclinical in nature, but there is some evidence of microbial alterations in children with neurodevelopmental disorders that will also be reviewed. Lastly, we will discuss the hypothesis that specifically targeting the intestinal microbiome will provide promising and novel therapeutic opportunities in the treatment of brain-related disease.

MICROBIOTA-GUT-BRAIN AXIS (A PRIMER) AND MECHANISMS FOR THE CROSSTALK

The gut-brain axis is a bidirectional highway of communication linking the peripheral functioning of the gastrointestinal tract with central brain processing, including emotional and cognitive centers. It has been studied both preclinically and clinically for several decades and its existence has been well established in large part due to work focused on the highly comorbid psychiatric and bowel disorders (Mayer, 2011; Al Omran and Aziz, 2014). During roughly the last decade, this axis has been reframed in order to highlight the intestinal microbiota as a key player, as an ever-growing body of research has brought to light the physiological importance of the trillions of bugs residing and working in the intestinal tract (Rhee et al., 2009; Bercik, 2011; Cryan and Dinan, 2012; Mayer et al., 2014; Quigley, 2018). Adult humans house roughly 0.2 kg of intestinal microbiota (Sender et al., 2016a), and the ratio of bacterial cells to human host cells is now thought to be ~ 1:1 (Sender et al., 2016b). Both bacteria and host benefit from a mutualistic relationship; the host gains as bacteria aid in the breakdown of otherwise indigestible food products, and in addition these bacteria provide a protective biofilm against pathogenic bugs. The bacteria benefit from their relationship with the host as they gain a protected environment, rich in life-giving resources (Backhed et al., 2005).

Perhaps even more interesting than the existence of an essentially invisible but highly complex ecosystem flourishing within our guts is the more far-reaching consequences of host-microbe interactions; we now know that these bugs are essential for health and wellbeing, and that this extends to brain function and potentially human mental health. It is now clear that the microbiota influence the expression of host behavior and furthermore it is quite possible that the pathophysiology of psychiatric disease is at least in part dictated by our relationship with intestinal microbiota (Foster and McVey Neufeld, 2013).

While the body of literature surrounding the effects of microbes on the brain is steadily growing, there are far fewer studies examining mechanisms by which the bacteria may exert their influence. We do know that there are a number of systems involved in communication between the bugs, gut, and brain, and that these work in parallel. Neural (both autonomic and enteric), immune, metabolic, and endocrine pathways are all engaged in the constant crosstalk between microbiota and the brain (Cryan and Dinan, 2012; Foster and McVey Neufeld, 2013; Mayer et al., 2014; El Aidy et al., 2015). In addition, the microbiota themselves are known to produce a variety of chemical messengers including short chain fatty acids, oligosaccharides, neurotransmitters, and other neuroactive agents (Puertollano et al., 2014; Wall et al., 2014; Donia and Fischbach, 2015; Levy et al., 2016). While it is currently unclear how these bacteria-derived neuroactive products might act on the brain, or indeed even if their production is being altered in physiologically relevant levels by bugs, this potential mechanism of action whereby intestinal bacteria can influence the host brain warrants careful examination (Forsythe et al., 2016).

Some of the difficulties in teasing apart mechanisms by which the microbiota may exert influence on the brain can be highlighted by an examination of work done in GF animals. First studies indicating a role for commensal bacteria on behavior and brain neurochemistry were initially conducted in GF mice, and these animals have been extremely useful in highlighting the importance of bacterial colonization to host health—particularly with respect to CNS function and behavior. GF mice are raised and maintained in aseptic environments and demonstrate significant alterations in behavior, with deficits observed in anxiety-like behavior (Diaz Heijtz et al., 2011; Neufeld et al., 2011; Clarke et al., 2013), social behavior (Desbonnet et al., 2014; Arentsen et al., 2015), cognition and memory (Gareau et al., 2011), and fear learning behavior (Hoban et al., 2018). In addition, these animals show altered expression of brain mRNA for molecules key to learning and memory and neurotransmission (Sudo et al., 2004; Gareau et al., 2011; Diaz Heijtz et al., 2011; Neufeld et al., 2011; Clarke et al., 2013; Arentsen et al., 2015). GF animals also show impairments in hypothalamic-pituitary-adrenal axis activity (Sudo et al., 2004; Neufeld et al., 2011; Clarke et al., 2013), neurogenesis (Ogbannaya et al., 2015), blood-brain barrier function (Braniste et al., 2014), microglial activation (Erny et al., 2015), and myelination (Hoban et al., 2016b). Anatomically, the brains of GF mice are found to be significantly different from conventionally housed mice, in terms of both increased hippocampal and amygdalar volume, as well as dendritic morphology (Luczynski et al., 2016b). These findings can be, however, difficult to interpret when trying to tease apart mechanisms of action. GF mice are known to be, from an immunological and metabolic perspective, distinctly abnormal. Indeed historically, GF mice were primarily known for their use in immunological research. Defects are clearly observed in the lymphoid tissue development of spleen, thymus, and lymph nodes (Gensollen et al., 2016). The mucosal immune system is significantly undeveloped, and the toll-like receptors therein are decreased or absent in the GF gut (Abrams et al., 1963; Wostmann et al., 1970). In addition these mice have reduced immunoglobulin A (IgA) secretion as well as fewer and smaller lymphoid patches in the gut wall and immune responses are blunted as compared to normally housed "conventional" animals (Clarke et al., 2013). Given these obvious immune impairments, and that immune function clearly impacts that of the brain, it is difficult then to ascertain if CNS alterations in GF mice are a direct result of the complete absence of colonizing bacteria, or are instead due to their significant immune deficiencies (Luczynski et al., 2016a). Interestingly, colonizing GF mice with intestinal bacteria is sufficient to generate an almost entirely normal immune response, particularly if bacteria are introduced in the early postnatal period (Crabbe et al., 1970; Rakoff-Nahoum et al., 2008; Geuking et al., 2011), while it seems only some of the observed CNS abnormalities can be normalized outside of a critical window for early life colonization. Similarly, the metabolic deficits inherent to GF mice may be responsible for observed CNS dysfunction. GF animals need to eat more in order to maintain body weight (Wostmann et al., 1983) and are in fact resistant to diet-induced obesity (Backhed et al., 2004; Backhed et al., 2007). Given that microbiota are responsible for harvesting otherwise inaccessible nutrients, it is also unsurprising that they have reduced short chain fatty acid production (Hoverstad and Midtvedt, 1986) and require vitamins B and K supplementation in order to survive (Gustafsson, 1959; Sumi et al., 1977). Again, it is difficult to ascertain if the metabolic deficiencies observed in GF animals are in fact responsible for some of the alterations observed in CNS functioning. The enteric nervous system (ENS) too is significantly different in GF mice as compared to their conventionally reared counterparts. The ENS is considered immature in GF mice, and it can be normalized after colonization with intestinal bacteria (McVey Neufeld et al., 2013; Collins et al., 2014; De Vadder et al., 2018). Neurons in the ENS control gut motility and other gastrointestinal physiological functions, but also directly communicate with the vagus nerve. Electrophysiological recordings made from the afferent nerve connecting gut to brain have also been demonstrated as abnormal in GF mice compared to controls (McVey Neufeld et al., 2015). Previous work in mice has demonstrated that an intact vagal nerve is necessary in order to transmit some aspects of microbiota-gut-brain communication, but it seems as though this varies depending on the specific bacterial species in question, as both vagal dependence (Lyte et al., 2006; Bercik et al., 2011b; Bravo et al., 2011) and independence (Bercik et al., 2010; Bercik et al., 2011a) have been demonstrated in vagotomy experiments. It is clear that the neural, immune, endocrine, and metabolic systems are all responsive to changes in the intestinal bacterial status, but also that all are able to reciprocally exert influence on the gut microbiota themselves (Hooper et al., 2012; Nicholson et al., 2012; Neuman et al., 2015; Zaneveld et al., 2017). However, due to the overlapping nature of these systems, and that they are all able to independently influence the function of the CNS, this also makes the study of the microbiota-gut-brain axis and the mechanisms of communication particularly challenging. There has thus been a well-reasoned call for an increase in multidisciplinary approaches to the study of the microbiota-gut-brain axis in order to further our understanding of how these systems work in concert to impact human health (Johnstone and Cohen Kadosh, 2019). However for the time being we are certain that cross-communication does occur, but are still in the relative infancy of understanding exactly how this happens.

MICROBIOTA DEVELOPMENT

Our understanding of what makes up the constituent bacteria in the average healthy human gut over the lifespan is under constant revision, in part due to the continual refinement of techniques used to identify and quantitatively analyze resident organisms (Mallick et al., 2017). While it has been challenging to link specific microbial types or combinations to human health outcomes, at least in broad strokes it has been possible through several large, population-scale studies to identify a working definition of what constitutes a "healthy microbiome" (Lloyd-Price et al., 2016).

The controversy and changing opinion with regard to what constitutes a healthy bacterial load begins with the fetus. While this topic is discussed in detail in the chapter of this textbook entitled "Development of the Microbiome-Gut-Brain Axis and its Effect on Behavior," we will briefly say that for more than a century, the developing human was considered to be completely void of bacteria on any surface, including the placenta and womb itself. This is currently contentious, as it has recently been proposed that bacteria are in fact present during the gestational period and that the human placenta, meconium, and amniotic fluid contain their own constituent microbiome (Jimenez et al., 2005; Jimenez et al., 2008; Aagard et al., 2014; Collado et al., 2016; Mishra et al., 2021). However, some studies have failed to find evidence for cultivable bacteria (Perez-Munoz et al., 2017; Kennedy et al., 2021) or bacterial nucleic acids (de Goffau et al., 2019; Kuperman et al., 2020; Kennedy et al., 2021) in fetal meconium or associated tissues. Irrespective of the controversy, it is widely accepted that the majority of colonization of the gut occurs primarily during birth and in the immediate subsequent postnatal days and weeks (Dominguez-Bello et al., 2010; Funkhouser and Bordenstein, 2013; Dominguez-Bello et al., 2019).

The first 3–4 years of life are characterized by a rapid increase in bacterial diversity (Koenig et al., 2011; Yatsunenko et al., 2012; Stewart et al., 2018; Niu et al., 2020) driven in large part by environmental exposure and increasingly complex food intake (Koenig et al., 2011; Stewart et al., 2018) which then stabilizes in early childhood to a relatively constant state (Stewart et al., 2018; Yatsunenko et al., 2012). The childhood microbiome has, as a result, often been considered comparable to that of an adult; nonetheless there appear to be subtle continuing changes that occur with further maturation of the gut communities even into adolescence (Ihekweazu et al., 2018). For example, one study compared the fecal microbiota of adolescent children to those of healthy adults and found that, while the number of species detected was not significantly different between groups, the relative abundances of specific bacterial genera were different (Agans et al., 2011). Higher numbers of *Bifidobacterium* were observed in the younger cohort, with adults presenting with more *Bacteroides* (Agans et al., 2011; Hollister et al., 2015). Subsequent research found that children harbored bacteria linked to growth and development, while adult bacterial communities were associated with inflammation and obesity (Hollister et al., 2015). There is also some evidence that gut microbiota composition may be influenced by hormonal changes at puberty. Yuan and colleagues (Yuan et al., 2020) found that, while no differences in diversity measures were detectable between children pre- or post-puberty, there were distinct changes in some genera which in several cases correlated with testosterone levels. Interestingly, adolescence may also be a period during which microbiome composition is repaired following perturbations earlier in life. While microbiome development is known to be substantially influenced in early life by both delivery mode and feeding type (formula vs breastfed) (Stewart et al., 2018), a recent study found that these effects were largely undetectable by adolescence (Cioffi et al., 2020).

Altogether, research is beginning to demonstrate that childhood and adolescence are important periods of microbiome development, but more studies are needed to examine the normal development and change of intestinal bacteria over the lifespan. Most of the studies that have been carried out to date have dealt with the extremes of age, and we particularly need more information on the gradual change in bacterial communities that occurs during later periods of childhood and through adulthood. Importantly, we do know that intestinal bacteria are susceptible to environmental influence (McVey Neufeld et al., 2016a; McVey Neufeld et al., 2016b), and that this differs from person to person over the lifespan. Once we have established what can be considered normal and healthy gut colonization patterns, studies examining how life experience and environmental challenge may contribute to changes in the intestinal microbiota and thus to human health need follow. What is abundantly clear now is that, from the perinatal period throughout childhood and adolescence, the gut is continuously colonizing with bacterial species, and that this is happening in parallel with ongoing nervous system development (see Figure 15.1). Given that we know the intestinal microbiota have a lasting and significant impact on brain development and function (Foster and McVey Neufeld, 2013), it is crucial that we gain an understanding of how these two systems developmentally mature together.

BRAIN DEVELOPMENT DURING CHILDHOOD
AND ADOLESCENCE

This section focuses on neurodevelopment following the immediate postnatal period, leading up to and including adolescence. Again, for a full discussion of both pre- and

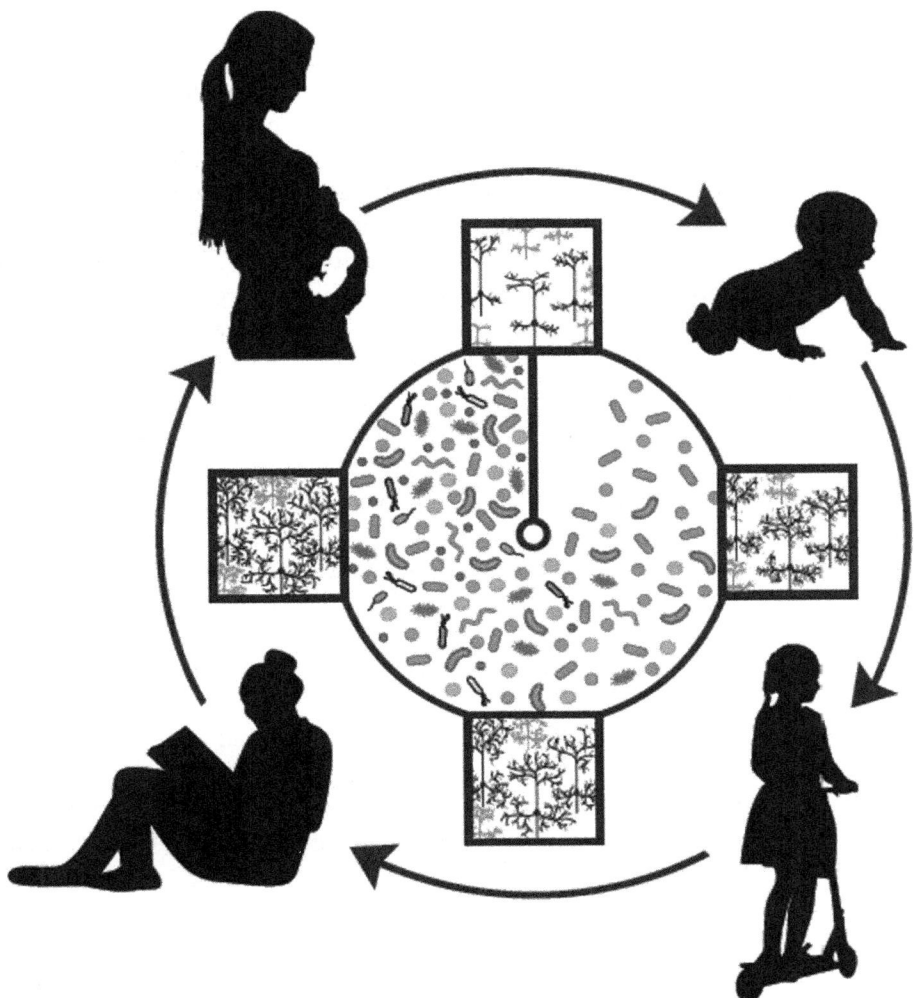

FIGURE 15.1 Microbiota and neural development from infancy to adulthood. The intestinal microbiota and the brain develop in parallel from infancy, through childhood and adolescence into adulthood. Reciprocal interactions occurring between the brain and gut highlight the importance of a normal trajectory of development for the intestinal bacteria, as perturbations may lead to adverse mental health outcomes and brain function later in life.

early postnatal neurodevelopment, we direct the reader to the chapter in this textbook entitled "Development of the Microbiome-Gut-Brain Axis and its Effect on Behavior." The infant brain is approximately four times smaller than it will be in adulthood and given that neurogenesis by this time is largely complete (van Dyck and Morrow, 2017), the growth in brain size over childhood and adolescence must therefore be primarily due to other factors. During the early years of childhood, the rapid increases seen in gray matter are likely due to increased synaptic density, dendritic elaboration, and glia production (van Dyck and Morrow, 2017), while increases in white matter reflect ongoing myelination of axons (Craik and Bialystok, 2006). Research demonstrates continued myelination and increased white matter volume in most brain regions through to early adulthood (Semple et al., 2013), but it is important to emphasize that otherwise much of the continued brain development during childhood and adolescence is region specific (Brenhouse and Andersen, 2011). For example, in regions of the brain involved in sensory processing, early synaptogenesis is quickly followed by a period of increased synaptic pruning with synaptic density reaching adult levels just prior to or around adolescence (Cragg, 1975; Huttenlocher, 1979; Huttenlocher et al., 1982; Bourgeois and Rakic 1993; Huttenlocher and Dabholkar, 1997). In contrast, synaptic development follows a very different time course in the prefrontal cortex. Here, there is a proliferation of synapses during both childhood and adolescence, followed by a period of senescence, then a subsequent elimination of synaptic connections postpuberty (Huttenlocher, 1979; Bourgeois et al., 1993 Woo et al., 1997).

While it was previously thought that the brain was largely fully developed by the time puberty was reached, we now know that this is not the case. Adolescence is a period of profound neuroanatomical change and maturation (Spear, 2000; Blakemore and Choudhury, 2006; Sturman and Moghaddam, 2011) and for a detailed discussion of these changes please see an excellent summary by Brenhouse and Andersen, 2011. Two major processes continue during the adolescent period: synaptic pruning and myelination. Myelination accelerates in adolescence, thus increasing the efficiency of interneuronal communication and resulting in increased speed of information flow between the various brain regions (Benes et al., 1989; Spear, 2013). This is in contrast to synaptic density, which decreases after peaking in childhood. This synaptic refinement is especially prevalent in adolescence, and is thought to increase brain efficiency (Spear, 2013), and results in an overall decrease in synaptic connections moving into adulthood (Huttenlocher, 1979; Woo et al., 1997). Synaptic pruning can be significant, and can result in the loss of up to 50% of synaptic connections in some regions, while other regions can see little decline (Spear, 2013). Pruning is an experience-dependent process and results in a more nuanced wiring, whereby frequently used connections become stronger and less used connections are eliminated (Holtmaat and Svoboda, 2009). While it is still currently unclear exactly how these processes occur, resident microglia in the brain are necessary for both the strengthening and elimination of synapses (Hua and Smith, 2004; Schafer et al., 2012). These processes are especially notable in the prefrontal cortex and other areas associated with cognition, where structural and functional changes in adolescence correlate with improvements in inhibitory control and other adult-typical behaviors (Burnett et al., 2011; Spear, 2013).

Both brain imaging and postmortem studies have provided additional evidence of the ongoing maturation of the frontal cortex during adolescence, and adds to the evidence of delayed maturation in frontal regions thought to reflect immaturities in certain cognitive processes like attention regulation and cognitive control (Huttenlocher, 1979; Huttenlocher et al., 1982; Spear, 2013). Frontal and parietal lobes both show increases in gray matter volume over childhood, which peak in adolescence, and then slowly begin to decrease (Giedd et al., 1999; Sowell et al., 2001). At the same time, white matter shows a fairly constant increase, likely due to the observed increases in myelination (Barnea-Goraly et al., 2005). While further work is necessary examining the exact nature of change in the adolescent brain and the mechanisms by which these occur, what is clear is that the teenage brain is no longer viewed as existing in a static period of development. Increasingly we are recognizing that other physiological systems like the intestinal microbiota play an integral role in how these developmental changes manifest.

CHILDHOOD AND ADOLESCENT MICROBIOTA-GUT-BRAIN AXIS

The ongoing changes occurring in the brain during childhood and adolescence highlight both its plasticity but also potential vulnerability to environmental input. It is vital to stress that the critical window of brain development occurring during childhood and adolescence is occurring simultaneous to ongoing changes in the intestinal microbiome. The composition and diversity of the gut microbiota undergoes a shift in structure and balance during adolescence, the trajectory of which can also be altered by the environment (Borre et al., 2014; Cowan et al., 2021). In addition, a crucial factor that cannot be ignored is the developing youth's exposure to a wide variety of environmental stimuli all of which are known to influence both the gut and brain. A host of environmental stimuli such as alcohol, prescription and nonprescription drug intake, changes to nutritional status and sleep schedule, as well as the occurrence of social stressors are all frequently part of a youth life experience and are all also known to impact the intestinal microbiota and the brain alike (McVey Neufeld et al., 2016a; McVey Neufeld et al., 2016b). It is also important to be cognizant that adolescence is often the first time when psychiatric disorders first become apparent. Given that gut microbiota can influence the development of the brain, changes occurring at the intestinal microbial level may contribute to the manifestation of psychiatric illnesses associated with puberty. It is thus imperative that we investigate the potential harmful effects of such environmental exposures to the ongoing development of the microbiota-gut-brain axis and how these may affect the incidence and manifestation of neurodevelopmental and psychiatric disorders.

Findings from Animal Studies

Much has been gleaned regarding the potential role for intestinal microbiota in neurodevelopment from experiments using GF mice. As stated previously these animals, which have been born and raised in the complete absence of bacteria, show abnormal brain development and functioning (Sudo et al., 2004; Diaz Heijtz et al., 2011; Neufeld et al., 2011; Clarke et al., 2013; Braniste et al., 2014; Desbonnet et al., 2014; Erny et al., 2015; Stilling et al., 2015; Gacias et al., 2016; Chen et al., 2017; Hoban et al., 2018; Thion et al., 2018). Of particular interest is work that has demonstrated both altered social behaviors and cognitive functioning. GF mice typically demonstrate sociability deficits and social memory impairments (Desbonnet et al., 2014; Arentsen et al., 2015; Buffington et al., 2016; Stilling et al., 2018; Sgritta et al., 2019), and these works have strengthened the hypothesis that microbiota-gut-brain axis dysfunction may play an important role in the etiology and development of autism spectrum disorder. Learning and memory is also impaired in GF animals (Gareau et al., 2011; Hoban et al., 2018; Luk et al., 2018; Chu et al., 2019) again emphasizing the importance of examining the role for intestinal microbial disruption in human cognitive disorders.

Given the limitations of GF studies (discussed previously) and their apparent lack of clinical relevance, work examining the impact of induced microbial disruption during early life and up to adolescence has gained importance. While this chapter focuses on studies examining the effects of microbial disruption occurring during periods relevant to childhood and adolescence, we direct the reader to the chapter in this text entitled "Development of the Microbiome-Gut-Brain Axis and its Effect on Behavior" for a discussion of the effects of microbial disruption occurring during the perinatal period. Data gathered from antibiotic feeding studies in adult rodents is increasingly showing resulting alterations to brain and behavior (Frohlich et al., 2016; Hoban et al., 2016a; Wang et al., 2015) indicating central effects following intestinal bacterial disruption. However antibiotic studies conducted in rodents around the period of weaning from their mothers up until what is considered an "early adulthood" developmental timepoint have not been plentiful to date. One such study examined the effects of feeding penicillin V to young mice for one week prior to weaning and then examined behavior and mRNA expression in brain tissue in adulthood. Antibiotic treated male mice showed altered social behavior with concomitant changes to hippocampal gene expression of the arginine vasopressin receptor 1A and 1B as well as changes to immune cell activation. Interestingly, feeding probiotic simultaneous with the antibiotic resulted in normalization of behaviors, brain gene expression and immune cell functioning (Kayyal et al., 2020). A separate antibiotic feeding study demonstrated that antibiotics administered during the adolescent period in mice resulted in significant changes to adult behavior, brain neurochemistry, and intestinal microbial content when compared to control animals (Desbonnet et al., 2015). At weaning, or 3 weeks of age in mice, animals were given an antibiotic treatment cocktail via the drinking water for 4 weeks, and then behavior was measured later in adulthood. These treated mice showed decreased anxiety-like behavior, altered cognitive function in the novel object recognition test, and also deficits in social behaviors. In addition, antibiotic treated mice showed significantly increased serum tryptophan and reduced kynurenine levels. They also had reduced gene expression of hippocampal brain-derived neurotrophic factor (BDNF), and less vasopressin messenger ribonucleic acids (mRNA) in the hypothalamus (Desbonnet et al., 2015). This study was the first to show that antibiotic treatment during adolescence in mice can result in significant changes to brain function and gene expression in adulthood and demonstrates that more work in this area is certainly required.

Another rodent study examined the effects of treating adolescent rats with propionic acid, a short chain fatty acid that is produced by intestinal bacteria (MacFabe et al., 2011). Propionic acid is a normal product of metabolism, but both it and its derivatives have been implicated in disease, particularly autism spectrum disorder. Male rats were treated with propionic acid intracerebroventricularly prior to behavioral testing, and the authors found that these animals showed impairments in social behavior and cognitive tests. When the brains were analyzed via immunohistochemistry they observed a significant upregulation of activated microglia. It should be noted that while the propionic acid was delivered via intracerebroventricular injection in these experiments, this acid does get manufactured in the gut by commensal bacteria and does readily pass the blood-brain barrier in a nonpathological state. The authors argue that their behavioral and immunohistochemical results provide evidence that propionic acid could be linked to autism spectrum disorder.

Findings from Human Studies

While interest is high, there have still been relatively few studies examining the relationship between the intestinal microbiome and behavior in children and adolescents. Again, while not covered in this chapter, there are a few studies examining the relationship between the intestinal bacteria and temperament prior to the first year of life but less at later timepoints in development. One association study measured the constituent microbiota and cognitive performance at later stages of development and showed that the composition of the microbiota at one year of age successfully predicted cognitive performance at age 2. In this study, lower bacterial diversity measures were predictive of better cognitive performance (Carlson et al., 2018). There have also been studies examining the potential ramifications of antibiotic use on children's cognitive performance. In the first of these by Slykerman and colleagues, the authors reported that antibiotic use prior to one year of life was associated with lower IQ and reading ability as well as behavioral disturbances when assessed at early school age (Slykerman et al., 2017b). In a follow-up study these authors went on to examine the relationship between the timing of antibiotic use prior to 2 years of age and neurocognitive outcomes at age 11 through use of a regression model (Slykerman et al., 2019). The authors reported that children who received antibiotics for the first time prior to 6 months of age had significantly lower cognitive abilities at age 11, including difficulties with verbal comprehension and increased problems with executive functioning, impulsivity, hyperactivity, and attention-deficit hyperactivity disorder (ADHD) (Slykerman et al., 2019). It is important to note however that these authors did not control for the reasons behind the antibiotic use in early childhood and thus a cautious interpretation of the results must follow. A more recent study examined premature babies that received perinatal antibiotics either via the mother prior to birth or during their stay at a neonatal hospital unit. In this case the authors did exclude those infants found to have been suffering from infection, and found an association between antibiotic use and attention issues at 4–5 years of age

418   JORGENSEN AND NEUFELD

(Firestein et al., 2019). These children also demonstrated EEG patterns consistent with those observed in children suffering from ADHD (Firestein et al., 2019).

There have been a number of pediatric studies examining the resident microbiota of individuals suffering from ADHD, and many report that these patients have an altered microbial profile as compared to age-matched, healthy control subjects (Aarts et al., 2017; Jiang et al., 2018; Wan et al., 2020). While these studies and others report alterations to the microbial signature of patients with ADHD as compared to controls, findings between studies have been inconsistent, and thus it has not been possible to find a specific microbial signature. For this reason, a larger scale study examining the gut microbiota of treatment-naïve adults suffering from ADHD was conducted, in which authors examined 100 ADHD subjects and 100 sex-matched controls. These authors found differing relative abundances of several microbial taxa in subjects with ADHD, and at the family level lower abundances of Gracilibacteraceae, and higher levels of Selenomonadaceae, and Veillonellaceae (Richarte et al., 2021).

There is currently keen interest in potential associations between altered gut microbiota and autism spectrum disorder (Vuong and Hsiao, 2017). Autism is a neurodevelopmental disorder characterized by impairments in social communication and repetitive, stereotyped behaviors. Causes of autism are unclear, but there is high comorbidity with gastrointestinal dysfunction such as diarrhea/constipation and abdominal pain. Altered intestinal microbiota has also been reported in these patients (Parracho et al., 2005; Strati et al., 2017). There have been several studies reporting an improvement in autism symptoms in patients following a course of antibiotics (Sandler et al., 2000; Ramirez et al., 2013; Urbano et al., 2014), which has been similarly modeled in animal studies (Wellmann et al., 2014; Kumar and Sharma, 2016). In addition there have been a few probiotic studies designed to assess any improvement in behavioral symptoms following a course of treatment. One such study reported an improvement in bowel function and possibly some behavior alterations in a 12-week oral treatment study of 3–16-year-olds (Parracho et al., 2010). Another study reported that feeding a probiotic supplement up until 6 months of age resulted in a reduced risk for autism disorder at 13 years of age (Partty et al., 2015). Again, future well-controlled studies are needed in this area in order to further our understanding of the potential role of the microbiota-gut-brain axis in autism.

FUTURE DIRECTIONS

While research into the microbiota-gut-brain axis has received an enormous amount of attention in the last decade, there is still a paucity of information regarding both mechanism of action of the bugs on brain neurodevelopment, and on human data and clinical trials. This is particularly true for the childhood and adolescent developmental periods. In spite of the lack of information regarding mechanism, interest and study in microbiota-gut-brain communication is expanding rapidly, and it is unsurprising that attention has now been directed toward developing therapeutics targeting this axis in the treatment of brain-related disease. Much work is currently aimed at understanding how pre- and probiotics may potentially treat these often-refractory illnesses. The therapeutic potential of bacteria in the treatment of psychiatric illnesses has indeed led to the development of the term *psychobiotic,* referring to a live organism that can be ingested to produce particular health benefits (Dinan et al., 2013). To date, some studies have been conducted examining the potential for mind altering effects of pre- and probiotics, and most have been conducted in healthy adults with mood and anxiety symptoms as outcome measures (Benton et al., 2007; Messaoudi et al., 2011a; Messaoudi et al., 2011b; Mohammadi et al., 2015; Steenbergen et al., 2015; Takada et al., 2016; Pinto-Sanchez et al., 2017; Slykerman et al., 2017a; Johnstone et al., 2021), although a few of these studies have been conducted in the depressed patient population (Akkasheh et al., 2016; Kazemi et al., 2018). Far fewer studies have focused on youth and their potential response to pre- or probiotic treatment (for a systematic review of these studies see Cohen Kadosh et al., 2021). This promising line of research needs to be expanded in childhood and adolescent clinical trials such as those reviewed in order to determine the potential benefits of pre- and probiotic treatment during neurodevelopment.

CONCLUDING REMARKS

The developing brain is unsurpassed in its potential for growth and plasticity, and yet at the same time incredibly vulnerable to outside environmental influence, including exposure to the changing status of intestinal microbes. From the prenatal period up to and including adolescence, the brain is maturing and changing, simultaneous with changes mirrored in the intestinal bacteria. This can be viewed as an exciting opportunity, as the potential for brain plasticity can be harnessed by therapeutics targeted to the peripheral gut microbiome. We would be wise to turn our attention to furthering our understanding of both the normal healthy progression of intestinal microbiota in the youth, as well as to the impact of abnormal microbial development on the brain. That the microbiome could act as a potential therapeutic target in the improvement of mental resiliency and in the treatment of brain disorders is a challenging but exciting possibility for the future of mental health.

REFERENCES

Aagard, K., Ma, J., Antony, K. M., Ganu, R., Petrosino, J., and Versalovic, J. (2014). The placenta harbors a unique microbiome. *Science Translational Medicine*, 6: 237ra65.

Aarts, E., Ederveen, T. H. A., Naaijen, J., Zwiers, M. P., Boekhorst, J., et al (2017). Gut microbiome in ADHD and its relation to neural reward anticipation. *PLOS One*, 12: e0183509.

Abrams, G. D., Bauer, H., and Sprinz, H. (1963). Influence of the normal flora on mucosal morphology and cellular renewal in the ileum. A comparison of germ-free and conventional mice. *Laboratory Investigations*, 12: 355–364.

Agans, R., Rigsbee, L., Kenche, H., Michail, S., Khamis, H. J., and Paliy, O. (2011). Distal gut microbiota of adolescent children is different from that of adults. *FEMS Microbiology Ecology*, 77: 404–412.

Akkasheh, G., Kashani-Poor, Z., Tajabadi-Ebrahimi, M., Jafari, P., Akbari, H., et al. (2016). Clinical and metabolic response to probiotic administration in patients with major depressive disorder: A randomized, double-blind, placebo-controlled trial. *Nutrition*, 32: 315–320.

Al Omran, Y. and Aziz, Q. (2014). The brain-gut axis in health and disease. *Advances in Experimental Medicine and Biology*, 817: 135–153.

Arentsen, T., Raith, H., Qian, Y., Forssberg, H., and Heijtz, R. D. (2015). Host microbiota modulates development of social preference in mice. *Microbiota in Health and Disease*, 26: 29719.

Backhed, F., Ding, H., Wang, T., Hooper, L. V., Koh, G. Y., Nagy, A., Semenkovich, C. F., and Gordon J. I. (2004). The gut microbiota as an environmental factor that regulates fat storage. *Proceedings of the National Academy of Sciences U S A*, 101: 15718–15723.

Backhed, F., Ley, R. E., Sonnenburg, J. L., Peterson, D. A., and Gordon, J. I. (2005). Host-bacterial mutualism in the human intestine. *Science*, 307: 1915–1920.

Backhed, F., Manchester, J. K., Semenkovich, C. F., and Gordon, J. I. (2007). Mechanisms underlying the resistance to diet-induced obesity in germ-free mice. *Proceedings of the National Academy of Sciences U S A*, 104: 979–984.

Barnea-Goraly, N., Menon, V., Eckert, M., Tamm, L., Bammer, R., Karchemskiy, A., Dant, C. C., and Reiss, A. L. (2005). White matter development during childhood and adolescence: A cross-sectional diffusion tensor imaging study. *Cerebral Cortex*, 15(1848): 54.

Bastiaanssen, T. F. S., Cowan, C. S. M., Claesson, M. J., Dinan, T. G., and Cryan, J. F. (2019). Making sense of ...the microbiome in psychiatry. *International Journal of Neuropsychopharmacology*, 22: 37–52.

Benes, F. M. (1989). Myelination of cortical-hippocampal relays during late adolescence. *Schizophrenia Bulletin*, 15: 585–593.

Benton, D., Williams, C., and Brown, A. (2007). Impact of consuming a milk drink containing a probiotic on mood and cognition. *European Journal of Clinical Nutrition*, 61: 355–361.

Bercik, P. (2011). The microbiota-gut-brain axis: Learning from intestinal bacteria? *Gut*, 60: 288–289.

Bercik, P., Verdu, E. F., Foster, J. A., Macri, J., Potter, M., Huang, X., Malinowski, P., Jackson, W., Blennerhassett, P., Neufeld, K. A., Khan, W. I., Corthesy-Theulaz, I., Cherbut, C., Bergonzelli, G. E., and Collins, S. M. (2010). Chronic gastrointestinal inflammation induces anxiety-like behavior and alters central nervous system biochemistry in mice. *Gastroenterology*, 139: 2102–2112.

Bercik, P., Denou, E., Collins, J., Jackson, W., Lu, J., Jury, J., Deng, Y., Blennerhassett, P., Macri, J., McCoy, K. D., Verdu, E. F., and Collins, S. M. (2011a). The intestinal microbiota affect central levels of brain-derived neurotropic factor and behavior in mice. *Gastroenterology*, 141: 599–609.

Bercik, P., Park, A. J., Sinclair, D., Khoshdel, A., Lu, J., Huang, X., Deng, Y., Blennerhassett, P. A., Fahnestock, M., Moine, D., Berger, B., Huizinga, J. D., Kunze, W., McLean, P. G., Bergonzelli, G. E., Collins, S. M., and Verdu, E. F. (2011b). The anxiolytic effect of Bifidobacterium longum NCC3001 involves vagal pathways for gut-brain communication. *Neurogastroenterology & Motility*, 23: 1132–1139.

Blakemore, S. J. and Choudhury, S. (2006). Brain development during puberty: State of the science. *Developments in Science*, 9: 11–14.

Borre, Y. E., O'Keefe, G. W., Clarke, G., Stanton, C., Dinan, T. G., et al. (2014). Microbiota and neurodevelopmental windows: Implications for brain disorders. *Trends in Molecular Medicine*, 20: 509–518.

Bourgeois, J. P. and Rakic, P. (1993). Changes of synaptic density in the primary visual cortex of the macaque monkey from fetal to adult stage. *Journal of Neuroscience*, 13: 2801–2820.

Braniste, V., Al-Asmakh, M., Kowal, C., Anuar, F., Abbaspour, A., Toth, M., Korecka, A., Bakocevic, N., Ng, L. G., Gulyas, B., Halldin, C., Hultenby, K., Nilsson, H., Hebert, H., Volpe, B. T., Diamond, B., and Pettersson, S. (2014). The gut microbiota influences blood-brain barrier permeability in mice. *Science Translational Medicine*, 6: 263ra158.

Bravo, J. A., Forsythe, P., Chew, M. V., Escaravage, E., Savignac, H. M., Dinan, T. G., Bienenstock, J., and Cryan, J. F. (2011). Ingestion of Lactobacillus strain regulates emotional behavior and central GABA receptor expression in a mouse via the vagus nerve. *Proceedings of the National Academy of Sciences*, U S A, 108: 16050–16055.

Brenhouse, H. C. and Andersen, S. L. (2011). Developmental trajectories during adolescence in males and females: A cross-species understanding of underlying brain changes. *Neuroscience and Biobehavioral Reviews*, 35: 1687–1703.

Buffington, S. A., Di Prisco, G. V., Auchtung, T. A., Ajami, N. J., Petrosino, J. F., and Costa-Mattioli, M. (2016). Microbial reconstitution reverses maternal diet-induced social and synaptic deficits in offspring. *Cell*, 165: 1762–1775.

Burnett, S., Sebastian, C., Cohen Kadosh, K., and Blakemore, S. J. (2011). The social brain in adolescence: Evidence from functional magnetic resonance imaging and behavioural studies. *Neuroscience and Biobehavioral Reviews*, 35: 1654–1664.

Carlson, A. L., Xia, K., Azcarate-Peril, M. A., Goldman, B. D., Ahn, M., et al. (2018). Infant gut microbiome associated with cognitive development. *Biological Psychiatry*, 83: 148–159.

Cerdo, T., Ruiz, A., Suarez, A., and Campoy, C. (2017). Probiotic, prebiotic, and brain development. *Nutrients*, 9: 1247.

Chen, J. J., Zeng, B. H., Li, W. W., Zhou, C. J., Fan, S. H., et al. (2017). Effects of gut microbiota on the microRNA and mRNA expression in the hippocampus of mice. *Behavioural Brain Research*, 322: 34–41.

Chu, C., Murdock, M. H., Jing, D., Won, T. H., Chung, H., et al. (2019). The microbiota regulate neuronal function and fear extinction learning. *Nature*, 574: 543–548.

Cioffi, C. C., Tavalire, H. F., Neiderhiser, J. M., Bohannan, B., and Leve, L. D. (2020). History of breastfeeding but not mode of delivery shapes the gut microbiome in childhood. *PLOS One*, 15: e0235223.

Clarke, G., Grenham, S., Scully, P., Fitzgerald, P., Moloney, R. D., Shanahan, F., Dinan, T. G., and Cryan, J. F. (2013). The microbiome-gut-brain axis during early life regulates the hippocampal serotonergic system in a sex-dependent manner. *Molecular Psychiatry*, 18: 666–673.

Cohen Kadosh, K., Basso, M., Knytl, P., Johnstone, N., Lau, J. Y. F., et al. (2021). Psychobiotic interventions for anxiety in young people: A systematic review and meta-analysis, with youth consultation. *Translational Psychiatry*, 11: 352.

Collado, M. C., Rautava, S., Aakko, J., Isolauri, E., and Salminen, S. (2016). Human gut colonization may be initiated in utero by distinct microbial communities in the placenta and amniotic fluid. *Scientific Reports*, 6: 23129.

Collins, S. M., Kassam, Z., and Bercik, P. (2013). The adoptive transfer of behavioral phenotype via the intestinal microbiota: Experimental evidence and clinical implications. *Current Opinions in Microbiology*, 16: 240–245.

Collins, J., Borojevic, R., Verdu, E. F., Huizinga, J. D., and Ratcliffe, E. M. (2014). Intestinal microbiota influence the early postnatal development of the enteric nervous system. *Neurogastroenterology and Motility*, 26: 98–107.

Cowan, C. S. M., Dinan, T. G., and Cryan, J. F. (2021). Annual Research Review: Critical windows—the microbiota-gut-brain axis in neurocognitive development. *Journal of Child Psychology*, 6: 353–371.

Crabbe, P. A., Nash, D. R., Bazin, H., Eyssen, H., and Heremans, J. F. (1970). Immunohistochemical observations on lymphoid tissues from conventional and germ-free mice. *Laboratory Investigations*, 22: 448–457.

Cragg, B. G. (1975). The development of synapses in the visual system of the cat. *Journal of Comparative Neurology*, 160: 147–166.

Craik, F. and Bialystock, E. (2006). Cognition through the lifespan: Mechanisms of change. *Trends in Cognitive Science*, 10: 131–138.

Cryan, J. F. and Dinan, T. G. (2012). Mind-altering microorganisms: The impact of the gut microbiota on brain and behaviour. *Nature Reviews Neuroscience*, 13: 701–712.

de Goffau, M. C., Lager, S., Sovio, U., Gaccioli, F., Cook E., Peacock, S. J., Parkhill, J., Charnock-Jones, D. S., and Smith, G. C. S. (2019). Human placenta has no microbiome but can contain potential pathogens. *Nature*, 572: 329–334.

Desbonnet, L., Clarke G., Shanahan, F., Dinan, T. G., and Cryan, J. F. (2014). Microbiota is essential for social development in the mouse. *Molecular Psychiatry*, 19: 146–148.

Desbonnet, L., Clarke, G., Traplin, A., O'Sullivan, O., Crispie, F., Moloney, R. D., Cotter, P. D., Dinan, T. G., and Cryan, J. F. (2015). Gut microbiota depletion from early adolescence in mice: Implications for brain and behaviour. *Brain, Behavior, and Immunity*, 48: 165–173.

De Vadder, F., Grasset, E., Manneras Holm, L., Karsenty, G., Macpherson, A. J., Olofsson, L. E., and Backhed, F. (2018). Gut microbiota regulates maturation of the adult enteric nervous system via enteric serotonin networks. *Proceedings of the National Academy of Sciences U S A*, 115: 6458–6463.

Diaz Heijtz, R., Wang, S., Anuar F., Qian, Y., Bjorkholm, B., Samuelsson, A., Hibberd, M. L., Forssberg, H., and Pettersson, S. (2011). Normal gut microbiota modulates brain development and behavior. *Proceedings of the National Academy of Sciences U S A*, 108: 3047–3052.

Dinan, T. G., Stanton, C., and Cryan, J. F. (2013). Psychobiotics: A novel class of psychotropic. *Biological Psychiatry*, 74: 720–726.

Dominguez-Bello, M. G., Costello, E. K., Contreras, M., Magris, M., Hidalgo, G., Fierer, N., and Knight, R. (2010). Delivery mode shapes the acquisition and structure of the initial microbiota across multiple body habitats in newborns. *Proceedings of the National Academy of Sciences U S A*, 107: 11971–11975.

Dominguez-Bello, M. G., Godoy-Vitorino, F., Knight, R., and Blaser, M. J. (2019). Role of the microbiome in human development. *Gut*, 68: 1108–1114.

Donia, M. S. and Fischbach, M. A. (2015). HUMAN MICROBIOTA. Small molecules from the human microbiota. *Science*, 349: 1254766.

El Aidy, S., Dinan, T. G., and Cryan, J. F. (2015). Gut microbiota: The conductor in the orchestra of immune-neuroendocrine communication. *Clinical Theory*, 37: 954–967.

Erny, D., Hrabe de Angelis, A. L., Jaitin, D., Wieghofer, P., Staszewski, O., David, E., Keren-Shaul, H., Mahlakoiv, T., Jakobshagen, K., Buch, T., Schwierzeck, V., Utermohlen, O., Chun, E., Garrett, W. S., McCoy, K. D., Diefenbach, A., Staeheli, P., Stecher, B., Amit, I., and Prinz, M. (2015). Host microbiota constantly control maturation and function of microglia in the CNS. *Nature Neuroscience*, 18: 965–977.

Firestein, M. R., Myers, M. M., Austin, J., Stark, R. I., Barone, J. L., et al. (2019). Perinatal antibiotics alter preterm infant EEG and neurobehaviour in the Family Nurture Intervention trial. *Developmental Psychobiology*, 61: 661–669.

Forsythe, P., Kunze, W., and Bienenstock, J. (2016). Moody microbes or fecal phrenology: What do we know about the microbiota-gut-brain axis? *BMC Medical*, 14: 58.

Foster, J. A. and McVey Neufeld, K. A. (2013). Gut-brain axis: How the microbiome influences anxiety and depression. *Trends in Neuroscience*, 36: 305–312.

Frohlich, E. E., Farzi, A., Mayerhofer, R., Reichmann, F., Jacan, A., et al. (2016). Cognitive impairment by antibiotic-induced gut dysbiosis: Analysis of gut microbiota-brain communication. *Brain, Behavior, and Immunity*, 56: 140–155.

Funkhouser, L. J. and Bordenstein, S. R, (2013). Mom knows best: The universality of maternal microbial transmission. *PLOS Biol*, 11: e1001631.

Gacias, M., Gaspari, S., Santos, P.-M. G., Tamburini, S., Andrade, M. et al. (2016). Microbiota-driven transcriptional changes in the prefrontal cortex override genetic differences in social behavior. *eLife*, 5: e13442.

Gareau, M. G., Wine, E., Rodrigues, D. M., Cho, J. H., Whary, M. T., Philpott, D. J., MacQueen, G., and Sherman, P. M. (2011). Bacterial infection causes stress-induced memory dysfunction in mice. *Gut*, 60: 307–317.

Gensollen, T., Iyer, S. S., Kasper, D. L., and Blumberg, R. S. (2016). How colonization by microbiota in early life shapes the immune system. *Science*, 352: 539–544.

Geuking, M. B., Cahenzli, J., Lawson, M. A., Ng, D. C., Slack, E., Hapfelmeier, S., McCoy, K. D., and Macpherson, A. J. (2011). Intestinal bacterial colonization induces mutualistic regulatory T cell responses. *Immunity*, 34: 794–806.

Giedd, J. N., Blumenthal, J., Jeffries, N. O., Castellanos, F. X., Liu, H., Zijdenbos, A., Paus, T., Evans, A. C., and Rapoport, J. L. (1999). Brain development during childhood and adolescence: A longitudinal MRI study. *Nature Neuroscience*, 2: 861–863.

Gustafsson, B. E. (1959). Vitamin K deficiency in germfree rats. *Annual Report of the New York Academy of Sciences*, 78: 166–174.

Hoban, A. E., Moloney, R. D., Golubeva, A. V., McVey Neufeld, K. A., O'Sullivan, O., et al. (2016a). Behavioural and neurochemical consequences of chronic gut microbiota depletion during adulthood in the rat. *Neuroscience*, 339: 463–477.

Hoban, A. E., Stilling, R. M., Ryan, F. J., Shanahan, F., Dinan, T. G, Claesson, M. J., Clarke G., and Cryan J. F. (2016b). Regulation of prefrontal cortex myelination by the microbiota. *Translational Psychiatry*, 6: e774.

Hoban, A. E., Stilling, R. M., Moloney, G., Shanahan, F., Dinan, T. G., Clarke, G., and Cryan J. F. (2018). The microbiome regulates amygdala-dependent fear recall. *Molecular Psychiatry*, 23: 1134–1144.

Hollister, E. B., Riehle, K., Luna, R. A., Weidler, E. M., Rubio-Gonzales, M., Mistretta, T.-A., Raza, S., Doddapaneni, H. V., Metcalf, G. A., Muzny, D. M., Gibbs, R. A., Petrosino, J. F., Shulman, R. J., and Versalovic, J. (2015). Structure and function of the healthy pre-adolescent pediatric gut microbiome. *Microbiome*, 3: 36.

Holtmaat, A. and Svoboda, K. (2009). Experience-dependent structural synaptic plasticity in the mammalian brain. *Nature Reviews Neuroscience*, 10: 647–658.

Hooper, L. V., Littman, D. R., and Macpherson, A. J. (2012). Interactions between the microbiota and the immune system. *Science*, 336: 1268–1273.

Hoverstadt, T. and Midtvedt, T. (1986). Short-chain fatty acids in germfree mice and rats. *Journal of Nutrition*, 116: 1772–1776.

Hsiao, E. Y., McBride, S. W., Hsien, S., Sharon, G., Hyde, E. R., et al. (2013). Microbiota modulate behavioral and physiological abnormalities associated with neurodevelopmental disorders. *Cell*, 155: 1451–1463.

Hua, J. Y. and Smith, S. J. (2004). Neural activity and the dynamics of central nervous system development. *Nature Neuroscience*, 7: 327–332.

Huang, R., Wang, K., and Hu, J. (2016). Effect of probiotics on depression: A systematic review and meta-analysis of randomized controlled trials. *Nutrients*, 8: 483.

Huttenlocher, P. R. (1979). Synaptic density in human frontal cortex—developmental changes and effects of aging. *Brain Research*, 163: 195–205.

Huttenlocher, P. R. and Dabholkar, A. S. (1997). Regional differences in synaptogenesis in human cerebral cortex. *Journal of Comparative Neurology*, 387: 167–178.

Huttenlocher, P. R., de Courten, C., Garey L. J., and Van der Loos, H. (1982). Synaptogenesis in human visual cortex: Evidence for synaptic elimination during normal development. *Neuroscience Letters*, 33: 247–252.

Ihekweazu, F. D. and Versalovic J. (2018). Development of the pediatric gut microbiome: Impact on health and disease. *American Journal of Medical Science*, 356: 413–423.

Jiang, H.-Y., Zhou, Y.-Y., Zhou, G.-L., Li, Y.-C., Yuan, J., et al. (2018). Gut microbiota profiles in treatment-naïve children with attention deficit hyperactivity disorder. *Behavioural Brain Research*, 347: 408–413.

Jimenez, E., Fernandez, L., Marin, M. L., Martin, R., Odriozola, J. M., Nueno-Palop, C., Narbad, A., Olivares, M., Xaus, J., and Rodriguez, J. M. (2005). Isolation of commensal bacteria from umbilical cord blood of healthy neonates born by cesarean section. *Current Microbiology*, 51: 270–274.

Jimenez, E., Marin, M. L., Martin, R., Odriozola, J. M., Olivares, M., Xaus, J., Fernandez, L., and Rodriguez, J. M, (2008). Is meconium from healthy newborns actually sterile? *Research Microbiology*, 159: 187–193.

Johnstone, N. and Cohen Kadosh, K. (2019). Why a developmental cognitive neuroscience approach may be key for future-proofing microbiota-gut-brain research. *Behavior Brain Science*, 42: e73.

Johnstone, N., Milesi, C., Burn, O., van den Bogert, B., Nauta, A., et al. (2021). Anxiolytic effect of a galacto-oligosaccharides prebiotic in healthy females (18–25 years). with corresponding changes in gut bacterial composition. *Scientific Reports*, 11: 8302.

Kayyal, M., Javkar, T., Mian, M. F., Binyamin, D., Koren, O., et al. (2020). Sex dependent effect of post-natal penicillin on brain, behavior and immune regulation are prevented by concurrent probiotic treatment. *Science Reports*, 10: 10318.

Kazemi, A., Noorbala, A. A., Azm, K., Eskandari, M. H., and Djafarian, K. (2018). Effect of probiotic and prebiotic vs placebo on psychological outcomes in patients with major depressive disorder: A randomized clinical trial. *Clinical Nutrition*, 38: 522–528.

Kennedy, K. M., Gerlach, M. J., Adam, T., Heimesaat, M. M., Rossi, L., Surette, M. G., Sloboda, D. M., and Braun, T. (2021). Fetal meconium does not have a detectable microbiota before birth. *Nature Microbiology*, Doi:10.1038/s41564-021-00904-0 [Online ahead of print].

Koenig, J. E., Spor, A., Scalfone, N., Fricker, A. D., Stombaugh, J., Knight, R., Angenent, L. T., and Ley, R. E. (2011). Succession of microbial consortia in the developing infant gut microbiome. *Proceedings of the National Academy of Sciences* U S A, 108 (Suppl 1): 4578–4585.

Kumar, H. and Sharma, B. (2016). Minocycline ameliorates prenatal valproic acid induced autistic behaviour, biochemistry and blood brain barrier impairments in rats. *Brain Research*, 1630: 83–97.

Kuperman, A. A., Zimmerman, A., Hamadia, S., Ziv, O., Gurevich, V., Fichtman, B., Gavert, N., Straussman, R., Rechnitzer, H., Barzilay, M., Shvalb, S., Bornstein, J., Ben-Shachar, I., Yagel, S., Haviv, I., and Koren, O. (2020). Deep microbial analysis of multiple placentas shows no evidence for a placental microbiome. *British Journal of Obstetrics and Gynaecology*, 127: 159–169.

Levy, M., Thaiss, C. A., and Elinav, E. (2016). Metabolites: Messengers between the microbiota and the immune system. *Genes Developments*, 30: 1589–1597.

Lloyd-Price, J., Abu-Ali, G., and Huttenhower, C. (2016). The healthy human microbiome. *Genome Medicine*, 8: 51.

Luczynski, P., McVey Neufeld, K. A., Oriach, C. S., Clarke, G., Dinan, T. G., and Cryan, J. F. (2016a). Growing up in a bubble: Using germ-free animals to assess the influence of the gut microbiota on brain and behavior. *International Journal of Neuropsychopharmacology*, 19: pyw020.

Luczynski, P., Whelan, S. O., O'Sullivan, C., Clarke, G., Shanahan, F., Dinan, T. G., and Cryan, J. F (2016b). Adult microbiota-deficient mice have distinct dendritic morphological changes: Differential effects in the amygdala and hippocampus. *European Journal of Neuroscience*, 44: 2654–2666.

Luk, B., Veeraragavan, S., Engevik, M., Balderas, M., Major, A., et al. (2018). Postnatal colonization with human "infant-type" Bifidobacterium species alters behavior of adult gnotobiotic mice. *PLOS One*, 13: e0196510.

Lyte, M., Li, W., Opitz, N., Gaykema, R. P. A., and Goehler, L. E. (2006). Induction of anxiety-like behavior in mice during the initial stages of infection with the agent of murine colonic hyperplasia Citrobacter rodentium. *Physiological Behavior*, 89: 350–357.

MacFabe, D. F., Cain, N. E., Boon, F., Ossenkopp, K. P., and Cain, D. P. (2011). Effects of the enteric bacterial metabolic product propionic acid on object-directed behavior, social behavior, cognition, and neuroinflammation in adolescent rats: Relevance to autism spectrum disorder. *Behavior Brain Research*, 217: 47–54.

Mallick, H., Ma, S., Franzosa, E. A., Vatanen, T., Morgan, X. C., and Huttenhower, C. (2017). Experimental design and quantitative analysis of microbial community multiomics. *Genome Biology*, 18: 228.

Mayer, E. A. (2011). Gut feelings: The emerging biology of gut-brain communication. *Nature Reviews Neuroscience*, 12: 453–466.

Mayer, E. A., Knight, R., Mazmanian, S. K., Cryan, J. F., and Tillisch, K. (2014). Gut microbes and the brain: Paradigm shift in neuroscience. *Journal of Neuroscience*, 34: 15490–15496.

McVey Neufeld, K. A., Mao, Y. K., Bienenstock, J., Foster, J. A., and Kunze, W. A. (2013). The microbiome is essential for normal gut intrinsic primary afferent neuron excitability in the mouse. *Neurogastroenterology and Motility*, 25: e183–e188.

McVey Neufeld, K. A., Perez-Burgos, A., Mao, Y. K., Bienenstock, J., and Kunze, W. A. (2015). The gut microbiome restores intrinsic and extrinsic nerve function in germ-free mice accompanied by changes in calbindin. *Neurogastroenterology and Motility*, 27: 627–636.

McVey Neufeld, K. A., Luczynski, P., Dinan, T. G., and Cryan, J. F. (2016a). Reframing the teenage wasteland: Adolescent microbiota-gut-brain axis. *Canadian Journal of Psychiatry*, 61: 214–221.

McVey Neufeld, K. A., Luczynski, P., Seira Oriach, C., Dinan, T. G., and Cryan, J. F. (2016b). What's bugging your teen?—the microbiota and adolescent mental health. *Neuroscience and Biobehavioral Reviews*, 70: 300–312.

Messaoudi, M., Violle, N., Bisson, J. F., Desor, D., Javelot, H., and Rougeot, C. (2011a). Beneficial psychological effects of a probiotic formulation (Lactobacillus helveticus R0052 and Bifidobacterium longum R0175) in healthy human volunteers. *Gut Microbes*, 2: 256–261.

Messaoudi, M., Lalonde, R., Violle N., Javelot H, Desor D., Nejdi A., Bisson, J. F., Fougeot, C., Pichelin, M., Cazaubiel, M., and Cazaubiel, J. M. (2011b). Assessment of psychotropic-like properties of a probiotic formulation (Lactobacillus helveticus R0052 and Bifidobacterium longum R0175) in rats and human subjects. *British Journal of Nutrition*, 105: 755–764.

Mishra, A., Lai, G. C., Yao, L. J., Aung, T. T., Shental, N., et al. (2021). Microbial exposure during early human development primes fetal immune cells. *Cell*, 184: 3394–3409.

Mohammadi, A. A., Jazayeri, S., Khosravi-Darani, K., Solati, Z., Mohammadpour, N., Asemi, Z., Adab, Z., Djalali, M., Tehrani-Doost, M., Hosseini, M., and Eghtesadi, S. (2015). The effects of probiotics on mental health and hypothalamic-pituitary-adrenal axis: A randomized double-blind, placebo-controlled trial in petrochemical workers. *Nutritional Neuroscience*, 19: 387–395.

Neufeld, K. M., Kang, N., Bienenstock, J., and Foster, J. A. (2011). Reduced anxiety-like behaviour and central neurochemical change in germ-free mice. *Neurogastroenterology and Motility*, 23: 255–264.

Neuman, H., Debelius J. W., Knight R., and Koren O. (2015). Microbial endocrinology: The interplay between the microbiota and the endocrine system. *FEMS Microbial Review*, 39: 509–521.

Nicholson, J. K., Holmes, E., Kinross, J., Burcelin, R., Gibson, G., Jia, W., and Pettersson, S. (2012). Host-gut microbiota metabolic interactions. *Science*, 336: 1262–1267.

Niu, J., Xu, L., Qian, Y., Sun, Z., Yu, D., Huang, J., Zhou, X., Wang, Y., Zhang, T., Ren, R., Li, Z., Yu, J., and Gao, X. (2020). Evolution of the gut microbiome in early childhood: A cross-sectional study of Chinese children. *Frontiers in Microbiology*, 11: 439.

Ogbannaya, E. S., Clarke G., Shanahan, F., Dinan, T. G., Cryan, J. F., and O'Leary, O. F. (2015). Adult hippocampal neurogenesis is regulated by the microbiome. *Biological Psychiatry*, 78: e7–9.

Parracho, H. M., Bingham, M. O., Gibson, G. R., and McCartney, A. L. (2005). Differences between the gut microflora of children with autistic spectrum disorders and that of healthy children. *Journal of Medical Microbiology*, 54(Pt 10): 987–991.

Parracho, H. M. R. T., Gibson, G. R., Knott, F., Bosscher, D., Kleerebezem, M., et al. (2010). A double-blind, placebo-controlled, crossover-designed probiotic feeding study in children diagnosed with autistic spectrum disorders. *International Journal of Probiotics and Prebiotics*, 5: 69–74.

Partty, A., Kalliomaki, M., Wacklin, P., Salminen, S., and Isolauri, E. (2015). A possible link between early probiotic intervention and the risk of neuropsychiatric disorders later in childhood: A randomized trial. *Pediatric Research*, 77: 823–828.

Perez-Munoz, M. E., Arrieta, M. C., Ramer-Tait, A. E., and Walter, J. (2017). A critical assessment of the "sterile womb" and "in utero colonization" hypotheses: Implications for research on the pioneer infant microbiome. *Microbiome*, 5: 48.

Pinto-Sanchez, M. I., Hall, G. B., Ghajar, K., Nardelli, A., Bolino, C., et al. (2017). Probiotic Bifidobacterium longum NCC3001 reduces depression scores and alters brain activity: A pilot study in patients with irritable bowel syndrome. *Gastroenterology*, 153: 448–459.

Puertollano, E., Kolida, S., and Yaqoob, P. (2014). Biological significance of short-chain fatty acid metabolism by the intestinal microbiome. *Current Opinion in Clinical Nutrition and Metabolic Care*, 17: 139–144.

Quigley, E. M. M. (2018). The gut-brain-axis and the microbiome: Clues to pathophysiology and opportunities for novel management strategies in irritable bowel syndrome (IBS). *Journal of Clinical Medicine*, 7: 6.

Rakoff-Nahoum, S. and Medzhitov, R. (2008). Innate immune recognition of the indigenous microbial flora. *Mucosal Immunology Supplement*, 1: S10–14.

Ramirez, P. L., Barnhill, K., Gutierrez, A., Schutte, C., and Hewitson, L. (2013). Improvements in behavioral symptoms following antibiotic therapy in a 14-year-old male with autism. *Case Reports in Psychiatry*, 2013: 239034.

Reis, D. J., Ilardi, S. S., and Punt, S. E. W. (2018). The anxiolytic effect of probiotics: A systematic review and meta-analysis of the clinical and preclinical literature. *PLOS One*, 13: e0199041.

Rhee, S. H., Pothoulakis, C., and Mayer, E. A. (2009). Principles and clinical implications of the brain-gut-enteric microbiota axis. *Nature Reviews in Gastroenterology and Hepatology*, 6: 306–314.

Richarte, V., Sanchez-Mora, C., Corrales, M., Fadeuilhe, C., Vilar-Ribo, L., et al. (2021). Gut microbiota signature in treatment-naïve attention-deficit/hyperactivity disorder. *Translational Psychiatry*, 11: 382.

Sandler, R. H., Finegold, S. M., Bolte, E. R., Buchanan, C. P., Maxwell, A. P., et al. (2000). Short-term benefit from oral vancomycin treatment of regressive-onset autism. *Journal of Child Neurology*, 15: 429–435.

Schafer, D. P., Lehrman, E. K., Kautzman, A. G., Koyama, R., Mardinly. A. R., et al. (2012). Microglia sculpt postnatal neural circuits in an activity and complement-dependent manner. *Neuron*, 74: 691–705.

Schmidt, K., Cowen, P. J., Harmer, C. J., Tzortzis, G., Errington, S., and Burnet, P. W. J. (2015). Prebiotic intake reduces the waking cortisol response and alters emotional bias in healthy volunteers. *Psychopharmacology* (Berl), 232: 1793–1801.

Semple, B. D., Blomgren, K., Gimlin, K., Ferriero, D. M., and Noble-Haeusslein, L. J. (2013). Brain development in rodents and humans: Identifying benchmarks of maturation and vulnerability to injury across species. *Progress in Neurobiology*, 106–107: 1–16.

Sender, R., Fuchs, S., and Milo, R. (2016a). Revised estimates for the number of human and bacterial cells in the body. *PLoS Biology*, 14: e1002533.

Sender, R., Fuchs, S., and Milo, R. (2016b). Are we really vastly outnumbered? Revisiting the ratio of bacterial to host cells in humans. *Cell*, 164: 337–340.

Sgritta, M., Dooling, S. W., Buffington, S. A., Momin, E. N., and Francis, M. B. (2019). Mechanisms underlying microbial-mediated changes in social behavior in mouse models of autism spectrum disorder. *Neuron*, 101: 246–259.

Slykerman, R. F., Hood, F., Wickens, K., Thompson, J. M. D., Barthow, C., et al. (2017a). Effect of Lactobacillus rhamnosus HN001 in pregnancy on postpartum symptoms of depression and anxiety: A randomized double-blind placebo-controlled trial. *EBioMedicine*, 24: 159–165.

Slykerman, R. F., Thompson, J, Waldie, K. E., Murphy, R., Wall, C., and Mitchell, E. A. (2017b). Antibiotics in the first year of life and subsequent neurocognitive outcomes. *Acta Pediatrica*, 106: 87–94.

Slykerman, R. F., Coomarasamy, C., Wickens, K., Thompson, J. M. D., Stanley, T. V., et al. (2019). Exposure to antibiotics in the first 24 months of life and neurocognitive outcomes at 11 years of age. *Psychopharmacology* (Berl), 236: 1573–1582.

Sowell, E. R., Thompson, P. M., Tessner, K. D., and Toga, A. W. (2001). Mapping continued brain growth and gray matter density reduction in dorsal frontal cortex: Inverse relationships during postadolescent brain maturation. *Journal of Neuroscience*, 21: 8819–8829.

Spear, L. P. (2000). The adolescent brain and age-related behavioral manifestations. *Neuroscience Biobehavior Review*, 24: 417–463.

Spear, L. P. (2013). Adolescent neurodevelopment. *Journal of Adolescent Health*, 52(2 Suppl 2): S7–13.

Steenbergen, L., Sellaro, R., van Hemert, S., Bosch, J., and Colzato, L. S. (2015). A randomized controlled trial to test the effect of multispecies probiotics on cognitive reactivity to sad mood. *Brain Behavior and Immunity*, 48: 258–264.

Stewart, C. J., Ajami, N. J., O'Brien, J. L., Hutchinson, D. S., Smith, D. P., et al. (2018). Temporal development of the gut microbiome in early childhood from the TEDDY study. *Nature*, 562: 583–588.

Stilling, R. M., Ryan F. J., Hoban, A. E., Shanahan, F., Clarke, G., et al. (2015). Microbes and neurodevelopment—Absence of microbiota during early life increases activity-related transcriptional pathways in the amygdala. *Brain Behavior and Immunity*, 50: 209–220.

Stilling, R. M., Moloney, G. M., Ryan, F. J., Hoban, A. E., Bastiaanssen, T., et al. (2018). Social interaction-induced activation of RNA splicing in the amygdala of microbiome-deficient mice. *ELife*, 7: ee33070.

Strati, F., Cavalieri, D., Albanese, D., De Felice, C., Donati, C., et al. (2017). New evidences on the altered gut microbiota in autism spectrum disorders. *Microbiome*, 5: 24.

Sturman, D. A. and Moghaddam, B. (2011). The neurobiology of adolescence: Changes in brain architecture, functional dynamics, and behavioral tendencies. *Neuroscience and Biobehavioral Reviews*, 35: 1704–1712.

Sudo, N., Chida, Y., Aiba, Y., Sonoda, J., Oyama, N., Yu, X.-N., Kubo, C., and Koga, Y. (2004). Postnatal microbial colonization programs the hypothalamic-pituitary-adrenal system for stress response in mice. *Journal of Physiology*, 558(Pt 1): 263–275.

Sumi, Y., Miyakawa, M., Kanzaki, M., and Kotake, Y. (1977). Vitamin B-6 deficiency in germfree rats. *Journal of Nutrition*, 107: 1707–1714.

Takada, M., Nishida, K., Kataoka-Kato, A., Gondo, Y., Ishikawa, H., et al. (2016). Probiotic Lactobacillus casei strain Shirota relieves stress-associated symptoms by modulating the gut-brain interaction in human and animal models. *Neurogastroenterology and Motility*, 28: 1027–1036.

Thion, M. S., Low, D., Silvin, A., Chen, J., Grisel, P., et al. (2018). Microbiome influences prenatal and adult microglia in a sex-specific manner. *Cell*, 172: 500–516.

Tillisch K., Labus J., Kilpatrick L., Jiang Z., Stains J., et al. (2013). Consumption of fermented milk product with probiotic modulates brain activity. *Gastroenterology*, 144: 1394–1401.

Urbano, M., Okwara, L., Manser, P., Hartmann, K., Herndon, A., and Deutsch, S. I. (2014). A trial of D-cycloserine to treat stereotypies in older adolescents and young adults with autism spectrum disorder. *Journal of Neuropsychiatry and Clinical Neuroscience*, 27: 133–138.

van Dyck, L. I. and Morrow, E. M. (2017). Genetic control of postnatal human brain growth. *Current Opinion in Neurology*, 30: 114–124.

Vuong, H. E., and Hsiao, E. Y. (2017). Emerging roles for the gut microbiome in autism spectrum disorder. *Biological Psychiatry*, 81: 411–423.

Wall, R., Cryan, J. F., Ross, R. P., Fitzgerald, G. F., Dinan, T. G., and Stanton C. (2014). Bacterial neuroactive compounds produced by psychobiotics. *Advances in Experimental Medicine and Biology*, 817: 221–239.

Wan, L., Ge, W.-R., Zhang, S., Sun, Y.-L., Wang, B., et al. (2020). Case-control study of the effects of the gut microbiota composition on neurotransmitter metabolic pathways in children with attention deficit hyperactivity disorder. *Frontiers of Neuroscience*, 14: 127.

Wang, T., Hu, X., Liang, S., Li, W., Wu, X., et al. (2015). Lactobacillus fermentum NS9 restores the antibiotic induced physiological and psychological abnormalities in rats. *Beneficial Microbes*, 6: 707–717.

Wellmann, K. A., Varlinskaya, E. I., and Mooney, S. M. (2014). D-cycloserine ameliorates social alterations that results from prenatal exposure to valproic acid. *Brain Research Bulletin*, 108: 1–9.

Woo, T. U., Pucak, M. L., Kye, C. H., Matus, C. V., and Lewis, D. A. (1997). Peripubertal refinement of the intrinsic and associational circuitry in monkey prefrontal cortex. *Neuroscience*, 80: 1149–1158.

Wostmann, B. S., Pleasants, J. R., Bealmear, P., and Kincade, P. W. (1970). Serum proteins and lymphoid tissues in germ-free mice fed a chemically defined, water soluble, low molecular weight diet. *Immunology*, 19: 443–448.

Wostmann, B. S., Larkin, C., Moriarty, A., and Bruckner-Kardoss, E. (1983). Dietary intake, energy metabolism, and excretory losses of adult male germfree Wistar rats. *Laboratory Animal Science*, 33: 46–50.

Yatsunenko, T., Rey, F. E., Manary, M. J., Trehan, I., Dominguez-Bello, M. G., et al. (2012). Human gut microbiome viewed across age and geography. *Nature*, 486: 222–227.

Yuan, X., Chen, R., Zhang, Y., Lin, X., and Yang, X. (2020). Gut microbiota: Effect of pubertal status. *BMC Microbiology*, 20: 334.

Zaneveld, J. R., McMinds, R., and Vega Thurber, R. (2017). Stress and stability: applying the Anna Karenina principle to animal microbiomes *Nature Microbiology*, 2: 17121.

CHAPTER 16

··

ATTENTION IN INFANCY
AND CHILDHOOD

··

ALEXANDER FRASER AND GAIA SCERIF

IN our highly complex environments, a fundamental question for developmental scientists is how attentive observers select which stimuli are important for learning, and which are not, from early infancy into adulthood. Attention in adults gates what they encode into short-term and long-term memory, and these processes in turn provide the mental workspace to attend most efficiently. Understanding why and how these relationships hold in younger attentive observers is both scientifically necessary and key to translating basic research into effective educational, teaching, and learning strategies. There is very strong evidence for dynamic interaction between attention, memory, and learning, even in adults. This suggests that the developmental cognitive neuroscience of attention needs to move away from valuable but descriptive exercises (e.g., is infants' or children's attention poorer compared to adults?), to understanding the mechanisms that drive change in attention development (i.e., why are there similarities and differences in attentional processes across infants, children, and adults? What drives them?).

In a first section, we therefore provide an overview of how, from very early in infancy, we are equipped with exquisite, though gradually improving, attentional selection skills. Our second section focuses on how data on infancy and childhood highlight mechanisms for the bidirectional interplay between attentional control, learning, and memory, rather than a unidirectional bottleneck or filter on incoming information. Thirdly, we focus on how influences of attention are not equal for all stimuli. These differences are apparent through the effects that highly salient social attentional biases (for eye gaze, faces, and people) have on learning, later recall from memory and further influences on memory-guided attention orienting. This is because they modify the attentional state of learners across development and do so in ways that need to be understood better in neurodiverse populations, such as autistic individuals. Social attention in autism is the focus of our fourth section. In our final section, we discuss how this approach complements work on social attention in autistic individuals, from childhood into adulthood. We conclude with highlighting some useful avenues for future research on the neuroscience of attention development.

TRAJECTORIES OF ATTENTION MECHANISMS FROM INFANCY AND CHILDHOOD INTO ADULTHOOD

Multiple attentional control mechanisms influence online processing by the adult attentive brain, influencing perception, short-term memory, all the way to encoding into and recall from long-term memory. Indeed, diversity of attentional mechanisms is a key feature of all influential attention models for the adult brain. Starting from influences on perception, classic models of adult attention suggest that ongoing task goals bias the immediate visual selection of input in the context of the complex bottom-up flow of incoming stimuli (Desimone and Duncan 1995; Kastner and Ungerleider 2000). Moment-by-moment selection depends on top-down goal representations that play a key role in resolving the competition between the target of visual attention and salient distractors (Bundesen, 1990; Duncan and Humphreys 1989). Other classic models also emphasize the distinction between endogenous goal-driven and exogenous stimulus-driven influences on attention in the adult brain (Corbetta and Shulman 2002) as well as how overlapping but separable attention mechanisms govern behavior (Petersen and Posner 2012; Posner, 1980; Posner and Petersen 1990). Posner and colleagues, in particular, identify interacting but separable neural networks supporting attention deployment in space through spatial orienting and time through alerting processes, and over goals through executive attention (Fan et al. 2002; Fan et al. 2005). Spatial representations play a key role in adult models of attention, both in the context of healthy adult brains and cases of adult brain damage (Bisiach and Luzzatti. 1978; Driver and Vuilleumier, 2001).

Descriptions of differential improvements in attentional functions have been extremely valuable and abound. Authoritative and influential reviews, for example, have described very aptly the onset of distinct attentional processes and their reaching a plateau, in infancy (e.g., Colombo 2001) and in childhood (e.g., Rueda et al. 2004). These influential reviews have highlighted that, in infancy, multiple mechanisms of attention onset at different times over the first three years of life, starting from simple attentional processes, to more complex executive functions, with accompanying changes in their underlying neural mechanisms (Fiske and Holmboe 2019). Later in childhood, cross-sectional data have suggested that simple alerting mechanisms, spatial attentional orienting, and executive skills improve along distinct timeframes over childhood (Rueda et al. 2004), although more recent, again, cross-sectional, work has highlighted how they interact (e.g., see Pozuelos et al. (2014) for replication, extension, and interactions). More recently, bridges across infancy and childhood have been theorized, first over the preschool years and beyond (Garon et al. 2008) and, more recently over the first three years of life (Hendry et al. 2016). Of note, we currently lack long-term longitudinal data across attentional functions, so for now full trajectories are inferred from cross-sectional data. The neural correlates of these differences across ages have been reviewed more extensively elsewhere over adolescence (Crone 2009), childhood (Amso and Scerif 2015), toddlerhood (Fiske and Holmboe 2019), and infancy (Richards 2010; Richards et al. 2010), as well as in the context of broader mechanisms for neurocognitive development (Johnson 2001, 2011), but here we provide a brief overview.

Trajectories for Distinct Attention Processes in Infancy

From the first months of life, changes in attention are indexed by changes in the way in which infants control their eye-movements, beginning by being influenced reflexively by the environment surrounding the infant, and becoming increasingly goal directed. Many aspects of oculomotor control show dramatic but temporally dissociated improvements between birth and four months (Johnson 1994). In young infants, an inability to disengage from salient stimuli, "sticky fixation," is observable only in the first month in typical development (Atkinson et al. 1992). Alongside the control of overt eye-movements, infants between four and six months of age become increasingly able to orient covert attention to stimuli in the environment, as indexed by the benefits that attentional cues (in particular gaze cues, to which we return) accrue to their orienting (Hood 1993; Johnson et al. 1994). In contrast, although the ability to inhibit over orienting towards salient peripheral stimuli emerges from three or four months of age (Johnson 1995), it continues to develop over early childhood and well into adulthood as indexed by the increasing accuracy in producing antisaccades (Luna et al. 2008). However, complexities and interactions abound: infant attention seems driven by a combination of exogenous and endogenous factors. For example, Kannass, Oakes, and colleagues (Kannass, Oakes and Shaddy 2006; Oakes et al. 2002) assessed the role of target familiarity on distraction latency during object exploration at 6.5 and nine-to-ten months of age. They showed that older infants exhibited longer latencies as they investigated novel toys compared to familiar toys, whereas younger infants resisted distraction more poorly, and without differentiation, suggesting that over the first year of life infants develop greater endogenous attention focus, in particular for new objects about which to learn.

In neural terms, these gradual changes in the control of covert and overt saccades have been accounted for by increasing cortical control on subcortical mechanisms (Atkinson et al. 1992; Johnson 1990). The step-like changes characteristic of the first year of age suggest the coming on line of distinct elements of the distributed adult oculomotor control circuit: in adults, attentional control over saccades involves connections between visual areas V1, V2, and V4; parietal cortex; and the frontal eye fields, as well as the superior colliculus and basal ganglia. Sticky fixation depends on the predomin-antly inhibitory input from the basal ganglia to the superior colliculus, in the context of poor cortical control at around one month of age, but increasing frontal and parietal influence on both eye-movements and covert attention unlock this sticky fixation from early in the first year. For example, early electrophysiological evidence pertaining to eye-movements already indicated that infant brains before one year of age deploy fronto-parietal mechanisms when preparing eye-movements (Csibra et al. 1997, 1998). Recent developments in methods that allow clearer localization than electroencephalography (EEG), such as functional Near Infrared Spectroscopy (fNIRS), have highlighted early influences of classic control nodes in frontal and parietal cortex, when young infants direct attention, both to incoming input, and to higher level representations that might guide their actions (Werchan et al. 2016). They have also shown subtle modulation of responses to repeated stimulus presentations (Emberson et al. 2017). Therefore, the an-swer to the question of when attentional mechanisms begin to acquire the adult-like efficiency suggests that the early origins of these active selection processes can be traced all the way back to the first year of life, at least for simple actions and stimuli.

Attention Trajectories in Childhood

Moving from infancy into childhood, beyond the relatively less well charted period of toddler transition, a large body of work has mapped the development of attention mechanisms. This work has primarily been based on the attention model by Posner and colleagues (Petersen and Posner 2012; Posner and Petersen 1990) and has exploited the adaptation of the adult Attentional Network Test (ANT; Fan et al. 2002). This paradigm requires participants to report the direction of a centrally presented arrow head flanked by either directionally congruent, incongruent or neutral arrow heads (an index of ex-ecutive attention deployment), while also measuring the benefits of spatial-orienting cues and temporally alerting cues. The ANT was modified to be used with children as young as four or six years of age (Rueda et al. 2004) and it identified distinct trajectories of spatial attention orienting, alerting of attention, and executive attention in children, although more recent work has highlighted the interactions between these processes, ra-ther than their independence (Pozuelos et al. 2014). Similar findings were also achieved by researchers who instead developed batteries of multiple differentiable attentional tasks, such as the "test of everyday attention for children" (Manly et al. 2001).

The development of a paradigm such as the ANT, aimed at measuring distinguishable attentional mechanisms, has influenced the developmental cognitive neuroscience of attention, as well as psychological theory, because of its uptake by multiple neurosci-ence studies. For example, Konrad and colleagues (2005) used the ANT to study the differences and similarities with which attentional networks for children, adolescents, and adults. They found that children showed significantly reduced brain activation in right-sided frontal-midbrain regions during alerting, in the right-sided temporo-parietal junction during spatial reorienting of attention, and in the dorsolateral prefrontal cortex during executive control of attention. In addition, children activated significantly more brain regions outside those regions of interest, suggesting protracted changes both within and outside the networks that are so efficiently used by adults. Indeed, the more executive nodes of attentional networks and their connectivity patterns continue to develop well into adolescence and into adulthood (Crone 2009; Fair et al., 2007, 2009). Of note, while the Posnerian model of adult attention has been very heavily employed in a developmental context, other neuroscientific models of attention deployment, like for example biased competition (Desimone and Duncan 1995), have only recently been clearly adopted to investigate the interplay between top-down and bottom-up mechanisms on visual attention from first principles, and their developmental trajectories (Kim and Kastner 2019). We shall return to these opportunities in the section on future directions.

THE DEVELOPMENTAL DYNAMICS OF ATTENTION, MEMORY, AND LEARNING

Returning to the adult attention models described briefly earlier, those models operationalize attentional control as the ability to bias the processing of incoming stimuli to enhance dimensions that are relevant to the task at hand, and therefore construe attention as a key gating factor for further cognitive processing. Despite differences across these models, core was the concept of attention as a filter, bias, or bottleneck, constraining further processing. Increasingly, however, views of how adult attentive brains operate incorporate bidirectional interactions between attention and working memory goals or long-term memories (Chun et al. 2011; Gazzaley and Nobre 2012) as well as low-level critical alerting rhythms (Buschman and Kastner 2015; Halassa and Kastner 2017). It is in particular the interface between attention and internally held representations that has increasingly become the focus of developmental cognitive neuroscientists, both for cognitive and their neural reasons.

Converging empirical work has suggested that attentional control constrains the efficiency with which information is encoded and maintained in visual memory (e.g., McNab and Klingberg 2008). However, the relationship between attentional control and memory is bidirectional, because information in short-term and long-term memory influences how attention is deployed (Doherty et al. 2017; Summerfield et al. 2006). However, much of this bidirectional work had been carried out in adults, even though active attentional control in function of memory and learning are challenged to a much higher degree over early childhood. It is also of note that the models of infant and childhood attention development we described earlier treated attentional processes as relatively independently from other developing processes, as they were keenly focused on tracing the onset and maturation of distinct attention in and of itself. In essence, they did not ask about interactions with the short-term and working memory representations that guide them. In contrast, recent work has highlighted how attention interacts with memory in differentiable ways and ways that distinguish infants, children, and adults.

Interactions between Attention and Memory in Infancy and Early Childhood

The deployment of attention in service of short-term memory in infancy shows some exciting developments. Ross-Sheehy and colleagues (Ross-Sheehy et al. 2003) assessed visual short-term memory capacity in four- to thirteen-month-old infants by comparing their looking behaviour to changing and non-changing stimulus streams presented side by side. In each stream, one to six colored squares repeatedly appeared and disappeared. In changing streams, the color of a different randomly chosen square changed each time the display reappeared. The youngest infants preferred changing streams only when the displays contained one object, suggesting a limit to their ability to use change detection to guide their attention, whereas older infants preferred changing streams when the displays contained up to four objects. These findings provided good evidence that visual short-term memory capacity increases significantly across the first year of life. Ross-Sheehy, Oakes, and Luck (2011) built on the change detection paradigm above with five- and ten-month-old infants, who experienced the change detection task above, with arrays of three differently colored squares. On each trial one square changed color and one square was cued; sometimes the cued item was the changing item, and sometimes the changing item was not the cued item. The older infants demonstrated longer looking for the cued item when the cue was a spatial pre-cue, suggesting that they are sensitive to spatial cues. Even younger infants could exhibit this enhanced memory, although the necessary cue needed to be moving and not static. These findings showed that even infants younger than six months can use attention cues to encode information in visual short-term memory (VSTM) about individual items in complex arrays, and that attention-directing cues influence both perceptual and VSTM encoding of stimuli in infants.

There is also growing evidence that basic attentional mechanisms have longer-term influences than short-term memory, from infancy onwards (Markant and Amso, 2013, 2014; Markant et al., 2015). For example, Markant and Amso (2013) found that visual selection mechanisms limit distractor interference during item encoding for infants, a process that is key to retention in long-term memory. In a modified spatial cueing task, nine-month-old infants encoded multiple objects following orienting cues that required them to inhibit distractor information, as opposed to a condition that did not. When their memory was tested, infants in the distractor suppression condition retrieved item-specific information from memory by discriminating items that were old from new items. These data suggested that developing selective attention (and more precisely, suppression of distracting information) enhances efficacy of memory encoding for subsequent retrieval.

This research used eye-movement and behavioral markers, rather than neural mechanisms. It will benefit from exciting developments that increasingly combine infant-friendly neuroscience techniques, such as EEG and fNIRS, as these are necessary to understand the processes by which attentional mechanisms not only mature over the first year of life, but also how they interface efficiently with memory and goal directed behavior at the transition from infancy into childhood. Work on older children has comparable behavioral and neural data.

Interactions between Attention and Memory over Childhood

Interactions between attention and memory continue to develop, in terms of biases influencing both short- and long-term memory. Different traditions differ in whether they use the term "working memory" interchangeably with "short-term memory" or distinguish between the two. However, perhaps one of the most robust findings in developmental science is the fact that, as we saw for infants (Ross-Sheehy et al. 2003), young children's (visual, but also auditory) short-term memory spans are lower than those of older children and adults (Cowan et al. 2005; Cowan et al., 2010; Cowan, 2017; Gathercole et al. 2004). Recent behavioral, eye-tracking, and neural evidence suggests that attentional factors may constrain children's short-term memory span and its development over childhood. Attention may influence how well children and adults remember in different ways: by dynamically preparing to encode information better, or by refreshing it while it is held in memory. As the attentional networks that support adaptive cognitive control are slow to develop, their maturation may also constrain the efficiency with which memories are encoded and maintained.

Let us take, for example, a very simple memory task, like being presented with four items that then disappear, to then be asked whether a memory probe item was part of the initial array or not. Using a version of this task, Astle and colleagues (2015) noted that children, in particular, are highly variable in how they manage to recruit cognitive control in service of memory. Astle and colleagues (2015) recorded oscillatory brain activity using magnetoencephalography (MEG) as nine- to twelve-year-old children and adults performed a VSTM task. They tested whether the strength of fronto-parietal activity correlated with VSTM performance, focusing in particular on fluctuations in the theta frequency range (4–7 Hz) because these have been suggested to reflect attentional influences on short-term memory (Liebe et al. 2012). In children, theta activity within a right lateralized fronto-parietal network in anticipation of the memoranda predicted the accuracy with which those memory items were retrieved later, suggesting that the inconsistent use of anticipatory control mechanism at encoding contributes to trial-to-trial variability in children's VSTM maintenance.

In addition to the general involvement of attention at encoding, spatially selective attention refreshment mechanisms seem to play an even more specific role in maintenance of visual information. Adapting the classic Posner spatial attention-orienting paradigm, multiple studies have shown that spatially directed cues presented during the maintenance period facilitate adults' accurate recall from memory (e.g., Griffin and Nobre 2003). Exploiting this paradigm, Shimi, Nobre, Astle, and Scerif (2014b) asked whether the interactions between spatial attentional cues and memory would differ in children and adults. Although children as young as seven years of age were as capable as adults to draw benefits from spatial attentional cues that are presented prior to encoding (commonly referred to as "pre-cues") to better encode into short-term memory, their ability to use cues presented during the maintenance period (commonly referred to as "retro-cues") was less well developed. Furthermore, EEG data using the same paradigm (Shimi, Kuo, Astle, Nobre, and Scerif 2014a) tested known neural markers of spatial orienting that had previously been measured in response not to memory, but to incoming percepts: the early directing attention negativity (EDAN; associated with attention cue processing), anterior directing attention negativity (ADAN; associated with preparatory attentional filtering), and late directing attention positivity (LDAP; associated with motor preparation) onset at increasing times post-stimulus onset. These have distinct topographies, but are associated with attention-orienting mechanisms (Eimer, 1996; Seiss et al. 2009). Here, they were measured in the context of a VSTM task, and in adults and ten-year-olds. Adults elicited a set of neural markers that were broadly similar in preparation for encoding and during maintenance, consistent with those that have been found for visual search and memory maintenance (Luck and Hillyard, 1994; Luck and Vogel, 2013). In contrast, in children, these processes dissociated, with little evidence of EDAN and ADAN in response to retro cues. Furthermore, in children, individual differences in the amplitude of neural markers of prospective orienting related to individual differences in VSTM capacity, suggesting that children with high capacity are more efficient at selecting information for encoding into VSTM. Extending this work to even younger children, Guillory et al. (2018) found an increasing refinement in short-term memory capacity in four- to seven-year-olds. They found that pre-cues were more effective than retro-cues in improving young children's short-term memory capacity.

As a whole, the emerging developmental literature on attentional cueing in favor of short-term memory suggests that developing spatial attentional control skills contribute to young children's ability to maintain items in visual short-term memory, interacting with other factors (such as load itself and decay of information; Shimi and Scerif 2017). Memory is influenced by attention, but it is also very clear that the contents of memory have a powerful influence on attention itself. Starting from the realm of short-term memory representations, an open question is how attentional biases interact with the nature of the internal memory codes. Recent work has begun to address this. For example, Shimi and Scerif (2015) asked seven-year-olds, eleven-year-olds, and adults to complete a retro-cueing paradigm as described above: spatial cues guided participants' attention to the likely location of the to-be-probed item during maintenance. The memoranda contained either highly familiar items or unfamiliar abstract shapes. Replicating earlier findings, all participants benefited from cues during maintenance, although benefits were smaller for seven-year-olds than for older participants. Critically, attentional benefits interacted with the nature of the memoranda: better VSTM maintenance was obtained for cued familiar items and differentially more so for adults compared to children. These data suggested that, in addition to the efficiency with which spatial attentional biases operate during maintenance, the characteristics of to-be-remembered items influence VSTM and differentially so over childhood. Attentional biases during maintenance seem to operate more efficiently on memory representations that are more familiar and therefore can be retrieved more easily, pointing to the need to consider the influence of memory representations themselves on attention orienting.

Work investigating memory-guided attention orienting tackles most directly the influence of memory traces on attention. Paradigms geared at investigating memory-guided attention and its neural mechanisms were, as discussed earlier, developed for use with adult participants (Stokes et al. 2012; Summerfield et al. 2006). They have been recently adapted to be used in children. Nussenbaum and colleagues (2019) pitted against each other the effects of salient visual cues and of memory traces on attention orienting in children and in adults. Over three complementary experiments, children demonstrated faster reaction times to targets both when they were cued by sudden visual events and by memories, and the effects of memories on attention were surprisingly smaller in adults relative to children. These findings suggest that memories may be a particularly robust source of influence on attention in children.

In their entirety, these findings suggest that the interplay between attentional biases, differential memory traces (short term and long term) and memory-guided attention is complex and modulated by age-related differences from infancy. Complementary methodologies in developmental cognitive neuroscience will be needed to shed light on the mechanisms through which attention and memory interact over development. For example, combining eye-movement methods and child-friendly techniques such as fNIRS, EEG, and/or MEG, may shed light on the neural sources and consequences of attention and memory interplay.

Attention, Learning, and Memory Dynamics: The Special Attentional Role of Social Stimuli and a Case for Interactive Specialization of Social Attention

To adults, people and faces in particular, are a unique stimulus in our environment. We use faces as our primary method of identifying others as well as to interpret the emotional state and intentions of another individual. By investigating how we attend to faces throughout development we can begin to understand how our attention is influenced by others, how this is necessary for development of other behaviors, as well as how this may impact on somebody if these abilities are developing differently.

Famously, Morton and Johnson (1991) pitted against each other two modes of infant face processing: CONSPEC, guiding attention to face-like patterns in newborn infants and CONLEARN, the process of learning information about faces, with the latter initially proposed to only influence infants from the age of two months. However, there is evidence that newborns are able to identify individuals, showing a preference for their mother's face within three days of birth (Bushnell et al. 1989; Pascalis et al. 1995; Walton et al. 1997). There is also evidence that newborns can distinguish and show a preference towards happy faces over neutral faces (Farroni et al. 2007), even if these processes may be guided by very simple perceptual biases, such as a preference for top-heavy stimuli, of which faces are a great example (Macchi Cassia et al. 2006). This would therefore suggest that rudimentary learning about conspecifics appears very early in infant development.

When viewing a face, most individuals will naturally attend to the eyes as the most prominent feature (Kliemann et al. 2010; Or, Peterson and Eckstein 2015; van Belle et al. 2010). This again has been seen in newborns who show a preference to gaze at faces with the eyes open compared to eyes shut (e.g., Batki et al. 2000) and two-to-five-day-old infants being able to discriminate between faces with direct versus averted eye gaze (Farroni et al. 2004a). As children develop, the eyes become a more significant stimulus and the ability to use the directions of the eyes to shift their own attention becomes key in the development of social cognition abilities. Joint attention is the ability of an individual to understand where another person is attending and then shift attention to the same object or location (Butterworth and Jarrett 1991). Other behaviors such as pointing also allow the child to understand where the other person is attending so that they may also shift their attention to the same location or object.

Eye Gaze as an Attentional Cue

To be able to make accurate references based on eye gaze we must be able to accurately interpret the direction of another person's eye gaze. This is not a trivial task, considering that the observer needs to be able to process the gaze direction from the target's point of view to be able to interpret it properly. Therefore, humans have developed complex, yet efficient processes that achieve this (Langton et al. 2000; Vecera and Johnson 1995). This is also aided by the fact that the human eye has evolved to help display gaze direction. That is, a relatively small dark area (the iris and pupil) surrounded by a large white sclera compared to other primates means that we can discriminate gaze direction efficiently (Kobayashi and Kohshima 2001). However, the age at which our ability to follow gaze truly onsets is not easy to pin down. It has been suggested that children's ability to truly determine eye contact and where they are looking may not develop until the age of three (Doherty and Anderson 2000).

However, there are clear examples of children younger than one year, even as young as three months old being able to follow the eye gaze of another person (Hood et al. 1998; Scaife and Bruner 1975), although these abilities may rely more on movement than processing of the eye gaze itself. For example, Farroni, Massaccesi, Pividori, and Johnson (2004b) found that neonates must observe the eyes move and a static image was not enough to establish joint attention. Further evidence found that in nine-month-olds, those that had developed gaze following would also respond to static images, but those who did not display gaze following would not respond. Further to this, infants that were four-to-five-months old would follow the direction that a face moved, even if the eyes were directing the opposite way, whereas an adult would attend towards the direction of the eye gaze regardless of the head direction (Bayliss et al. 2005; Farroni et al. 2000). These abilities appear to occur around the age of twelve-to-eighteen months old. Although these abilities are simple, early joint attention abilities in children have been found to predict other behavioral abilities such as noun acquisition (Morales et al. 2000; Tomasello and Farrar 2016) as well as theory-of-mind abilities (Charman et al. 2000).

A full review of the development of face and gaze processing is well beyond the scope of the current chapter, but we point to some key attentional development considerations. Influential models of face-processing development emphasize the crucial interaction of progressive attention to social stimuli, and the increasing specialization of cortical circuits for face processing (Cohen Kadosh and Johnson 2007; Johnson et al. 2009). These models predict that increasing exposure and experience with faces and eye gaze as attentional cues in a developing child's environment are crucial to the interactive nature of face and gaze processing development. As faces, eyes, and people play a central role, not only as stimuli for processing in and of themselves, but as cues to guide attention, there should be important functional consequences of the development of social processing for the functional consequences of attention.

As we process people, faces, and gaze increasingly well and in a more specialized fashion, will these improvements have an impact on the processing of the stimuli to which we attend? This question has of course been intensely studied in the context of benefits to reaction time and accuracy in perceiving or discriminating attended to stimuli, when they are cued by gaze. However, longer-term consequences on memory and other functional consequences on those objects (for example, their affective nature), have been studied less often (Bayliss et al. 2006). In a recent line of work (Fraser 2019, unpublished doctoral thesis), we studied the functional consequences of the processing of eye gaze as a cue to attention in neurotypical adults and in autistic adults and using both behavioral and eye-tracking measures. We found that in neurotypical adults gaze acts as a powerful attention cue that influences not only simple perception and discrimination, but also memory of (i.e., accurate recall of having previously seen an object) and affective judgments of (e.g., how much an object was liked or preferred) gazed-at objects, compared to objects that are not cued by directed gaze. Over development, these short-term influences may have progressive effects on what children learn from social attention cues, to which they are attracted and for which they become increasingly expert processors, while at the same time their processing of faces and eyes also develops, pointing to a highly interactive interplay between attention, learning, and social cue processing.

We shall return to these progressive interactions and how they may operate in autistic individuals in our final section. We will first review the effects of people as social agents when they are not directly relevant to our task goals and attention focus.

Social Agents Draw Attention to Influence Learning and Memory

While it is very clear that social agents are a powerful attention cue when they direct us to stimuli in our environment, sometimes people in our environments are not relevant to the task at hand. Imagine for example a complex and busy outdoor scene, in which we are trying to learn to locate a small object (a toy or another small object), both familiar people (family, friends) and strangers might not be helpful attention cues to our goal, because people come and go, and they may not always index where an object is hidden/located. In such cases, paying undue attention to people in the scene may in fact be counter-productive.

For example, first in adults, Doherty and colleagues (2017) asked participants to search for targets in scenes containing social or non-social distractors. Subsequent memory precision for target locations was tested. Eye-tracking revealed significantly more attentional capture to social compared to non-social distractors matched for low-level visual salience. Critically, memory precision for target locations was poorer for social scenes, suggesting a role for attentional orienting to and away from distractors on long-term memory. In an extension to younger children, Doherty and colleagues (2019a) found that children directed first looks to the social distractor even more than adults and that memory precision was lower, for both children and adults, when a social distractor was present. The powerful effects of social distractors alert us to the fact that attentional biases do not operate equivalently across stimuli of all types, but that pre-existing preferences for certain stimuli also guide attention.

In addition, Doherty and colleagues (2019b) asked whether the differential effects of social scenes on memory alter subsequent memory-guided attention orienting, and corresponding anticipatory dynamics of 8–12 Hz alpha-band oscillations as measured with EEG. Alpha has long been used as an index of attentional allocation in space (e.g., Gould et al. 2011). Doherty, van Ede and colleagues (2019b) asked young adults to search for targets in scenes that contained either social or non-social distractors and their memory precision tested. Subsequently, reaction time was measured as participants oriented to targets appearing in those scenes at either valid (previously learned) locations or invalid (different) locations. Replicating previous work, memory precision was poorer for target locations in social scenes. In addition, distractor type moderated the validity effect during memory-guided attentional orienting, with a larger cost in reaction time when targets appeared at invalid (different) locations within scenes with social distractors. Poorer memory performance for scenes with social distractors was also marked by reduced anticipatory dynamics of spatially lateralized 8–12 Hz alpha-band oscillations. The functional consequences of a social attention bias therefore extend from memory to memory-guided attention orienting, a bidirectional chain that may further reinforce attentional biases. As we know from previous work that these attentional biases may be even stronger in children compared to adults, ongoing work should be focusing on the neural mechanisms of memory-guided attention for social versus non-social stimuli.

Indeed, the nature of surrounding distractors plays an important role in how attention and memory interact for children in comparison to adults. Earlier we described a study involving six-to-ten-year-old children and young adults, learning to find targets in social and non-social scenes (Doherty et al. 2019a). After the learning and memory phases, these same participants were asked to perform a speeded target detection task, in which targets were either a location that was consistent or inconsistent with their memories (i.e., when their memories validly or invalidly cued attention to the location of upcoming targets). Intriguingly, although both groups were less precise in remembering targets that had appeared in social versus non-social scenes (showing the influence of distractors on memory), children demonstrated overall better memory precision than adults, a finding that was unexpected and one that needs replication. It resonates, however, with the stronger memory-guided attention effects in children compared to adults that we described earlier (Nussenbaum et al. 2019). Furthermore, when participants detected previously learnt targets within visual scenes, adults were slower for targets appearing at unexpected (invalid) locations within social scenes compared to non-social scenes, whereas children did not show this cost, suggesting that social memory traces may play a different role for them, compared to adults.

In summary, then, behavioral and neural evidence points to how age-related differences in attentional processes, from infancy into childhood and adulthood, have implications for the efficiency with which attentive observers encode and maintain information in short-term memory, as well as encode into and retrieve information from long-term memory. A crucial point is that the nature of the items to which attention is directed (social biases or the learning history that surrounds certain stimuli) has an influence on attention itself, with eye gaze as a social cue resulting in powerful functional consequences, not only for perception, but also memory, and affective judgments.

ATYPICAL SOCIAL ATTENTION AND DEVELOPMENT IN AUTISTIC INDIVIDUALS

As in the previous section, we shall consider social stimuli first as the target of attention (e.g., faces as attention cues) and then social stimuli as attention-grabbing stimuli in complex scenes, even when they are not directly relevant to the task at hand. Here, we will provide an overview of how attentional these processes may operate differently and have alternative functional consequences for autistic individuals, whose attention to people may be subtly different over developmental time. Please note that throughout this section we use "identity first" language (e.g., autistic people), rather than "people first" language (people with autism), to align with preferences expressed by the autistic community (Kenny et al. 2016).

Previous research has found that when viewing faces, autistic people will spend less time looking at the eyes of the face compared to neurotypical controls (Dalton et al.

2005; Kliemann et al. 2012; Kliemann et al. 2010). There is even some evidence that time spent attending to eyes when viewing social scenes is able to predict autism (Klin et al. 2002). Further research from Jones and Klin (2013) actually found that attention to eyes actually does occur in children who later went on to develop autism but is lower than expected for those children between the ages of two to six months, an age at which typically developing children begin to attend to eyes more. However, this does not mean that autistic people are unable to use the gaze of others. Leekam and colleagues (1997) found that autistic children are less likely to spontaneously follow the direction of others but were able to say what item a person was looking at when specifically told to do so, suggesting that atypical gaze following is not an issue with geometric analysis, but rather atypical spontaneous monitoring of others' gaze. A follow-up experiment by Leekam and Hunnissett (1998) found that spontaneous gaze cueing was developmentally delayed: children with a verbal mental age of forty-eight months were able to spontaneously follow the gaze of another person, whereas children with lower mental ages had difficulties, although their behavior improved when other cues (pointing and verbal) were added. Further to this, Riby and colleagues (2013), using a more naturalistic stimulus with adult participants, found that if autistic participants were informed that the model in a natural scene was looking at a target object they would spend more time looking at the face and eyes of the model. However, this did not result in more time looking at the target object and their accuracy was lower than that of the control group. Therefore, even though they may be attending to the eyes they still have difficulties extracting information on gaze direction.

Despite these strong suggestions of a functional link between individual differences in eye-gaze processing, broader social cognition and learning over typical development and in autism, it is not fully understood why autistic people show patterns of social attention that are atypical when compared to typically developed individuals. Kliemann and colleagues (2010) found that when autistic participants were presented with a face at eye level with a fixation cross they would shift their attention away from the eyes whereas typical controls would not. This, combined with evidence that suggests that autistic individuals may show a hyperactive reaction in the amygdala when viewing eyes (Dalton et al. 2005; Kliemann et al. 2012) supports the proposal that direct eye contact in autism may result in an aversive sensation, that may in turn delay the development of spontaneous gaze perception. If this aversive reaction results in diminished eye gaze from a young age, autistic individuals may not develop joint attention abilities in a similar way to typically developing children. This may result in different gaze following during development and into adulthood, which in turn may impact on development of other social abilities.

A substantial literature has investigated gaze-cueing effects in autistics (e.g., Nation and Penny 2008). For example, eye-tracking methods have detailed how autistic people attend to faces (Dalton et al. 2005; Kliemann et al. 2010, 2012). It has been noted that autistic individuals tend to spend less time attending to eyes (Dalton et al. 2005; Klin et al. 2002) and show spikes in amygdala activity when they engage with direct eye contact (Kliemann et al. 2010, 2012). One explanation is that autistic individuals show reduced attraction to direct eye contact (Tanaka and Sung 2016), impeding effective gaze cueing. However, if instructed that the gaze cues are predictive, and therefore should be followed, autistic participants will benefit from the gaze cue (Ristic et al. 2005). Other paradigms have been used to investigate how attention to an object can predict preferences towards an object (Schotter et al. 2010): people spend more time attending to an object that they will say they like compared to an item they say they dislike. Schotter and colleagues (2010) argued that it was selective encoding that drove the gaze bias which boosted the liking effect.

Interestingly, this work has tended to focus on how eye-gaze speeds or increases the accuracy of responses to briefly presented stimuli (e.g., by speeding and/or improving discrimination of stimuli), rather than exploring longer-term influences, such as for example affective appraisal of attended stimuli, or their encoding and retrieval from memory. We attempted to investigate longer-term functional consequences of gaze attention cues in both neurotypical adults and autistic adults, by combining eye-tracking with a paradigm initially developed for neurotypical adults (e.g., Bayliss et al. 2006), to investigate what has been termed "the liking effect." The liking effect refers to an observable preference for objects that have been cued by the eyes of a face, but not by the direction of an arrow. Our preliminary data combining eye-tracking with the liking paradigm suggest a gaze cueing bias for both neurotypical and autistic participants for measures of visuospatial attention. However, there were attentional effects on affective judgments of objects for the typical group and not the autistic group. This would suggest that even though attention is cued by eye gaze for autistic participants, they are not utilizing gaze in the same way to influence their judgments of objects. It is plausible that autistic participants were encoding both congruently and incongruently cued items equally, which would eliminate affective differences. In addition, in a direct gaze condition both groups attended to the eye more than the face, although this was attenuated in the autism group, and there was a greater focus towards the middle of the eyes around the crest of the nose for this group. It has been suggested that attending to this space is often used as a coping strategy for many autistic individuals who have difficulties in making direct eye contact (Trevisan et al. 2017). As attention was partially averted from the eyes, again, functional consequences on affective judgments may be reduced. These conclusions are only preliminary and will require replication with larger samples.

A further dimension to the interplay between social processing, attention, and memory in autism is to be considered when faces are not to be the direct target of attention, but rather act as distractors. Doherty and colleagues (2017, 2019b) investigated the distracting influence of social stimuli in scenes while participants searched for targets, remembered the locations of targets, and then used those memories to guide attention subsequently. Most recently, we employed this paradigm, combining behavioral responses with eye-tracking, for children and young people with fragile X syndrome (FXS), a condition of identified genetic origin that is associated with a high prevalence of autistic symptoms (Guy et al. 2020). We found that children and young people with FXS performed well at the experimental task and showed similar accuracy and speed in locating targets in natural scenes to children of equivalent verbal abilities.

They also learned target locations over blocks, but their memory of target locations was not as precise as that of comparison children. In addition, participants with FXS initially directed few first looks to salient social items within the scenes, but these looks increased over blocks, perhaps indicating initial social aversion, that was nonetheless overcome. Like neurotypical children, children with FXS also dwelled gaze upon social items while recalling target locations from memory. As a whole, these data suggest that, although there may be initial differences in processing of social stimuli within complex scenes, when people do not carry task-relevant information, attention to those scenes, memory and memory-guided attention function well. We have employed a similar approach with a large group of autistic children and young people, finding similarly good attention, memory, and memory-guided attention (Guy et al. forthcoming).

In conclusion, we demonstrated that eye-gaze cues affect not only perception of target objects, but potentially also their affective judgments, and do so differentially in autistic and neurotypical people. Furthermore, eye-tracking analysis showed that the preferences for items in the neurotypical group were driven by more time spent looking at the congruently cued items compared to the incongruently cued items. However, this effect was not observed in a sample of autistic individuals, suggesting that they are not influenced by eye gaze in the same way, and this in turn may have consequences on functional consequences of eye gaze: the selective encoding of information and resulting changes in affective judgments. As a whole, these findings point to atypical social attention to faces when they carry task-relevant information, whereas atypicalities are far reduced when people are not central to the task at hand, both in children and adults.

Conclusions and Future Directions at the Interface Between Attention, Memory, and Social Attention Development

In summary, we have overviewed evidence on the developmental trajectories of attentional processes from infancy into childhood and young adulthood. We have emphasized how early goal- and memory-related activity bias attention from very early on in infancy, and therefore how the interactions between attention, memory, and learning are the target of much recent work in the developmental cognitive neuroscience of this area. On the one hand, the attentional state of the system at encoding, even in children, predicts memory, as indexed by the activity of a fronto-parietal oscillatory network at encoding as predictor of memory accuracy. In addition, spatial attentional cues affect encoding into memory and maintenance. While encoding benefits of attentional cues are adult-like from middle childhood, attentional benefits on visual memory maintenance are still developing. Furthermore, we have reviewed a new series of studies geared to experimentally study the interplay between attention and longer-term memory, in children and adults, and special considerations in the context of social stimuli such as faces and gaze.

There exist a number of gaps in the knowledge we have described so far. First of all, a better understanding of the longitudinal relationships between early attentional markers and later controlled processes, is a field that continues to be relatively starkly divided into researchers who study attentional processes and their neural correlates in infancy, and those who begin mapping attentional processes in early childhood (from three years of age onwards). The period between one and two years of age remains relatively uncharted from a cognitive neuroscience point of view, and even rarer are longitudinal studies that investigate the stability, change, and predictive role of early attentional mechanisms for later development in childhood, although at least relevant behavioral data are emerging (Hendry et al. 2016; Fiske and Holmboe 2019). In addition, the extent to which attentional processes can be trained, whether this is more successful early in development, and whether attention training results in transfer to untrained but correlated developing functions, remains uncertain and highly debated (e.g., Wass et al. 2012). Furthermore, it is very clear that the nature of the to-be-attended stimuli also matters: social stimuli can carry task-relevant information, or they can distract. A multi-method approach including developmental data, eye-tracking data, and neural data may shed light on the typical trajectories of these interactions, but also on how social attention and its functional consequences over time operate differently in autistics from early in development.

In conclusion, the evidence thus far suggests that attentional biases during learning influence later memory, but intriguingly we found evidence of greater memory precision in children than in adults, and greater effects of memory traces on attention reorienting, again in children compared to adults. The interplay of attention, short- and long-term memories are bidirectional. Attention influences multiple memory processes in ways that dissociate developmentally. Information held in memory influences attention, again in developmentally informative ways, and ways that are dependent on the prior learning history of the developing system, as well as the nature of to-be-attended stimuli.

References

Amso, D. and Scerif, G. (2015). The attentive brain: Insights from developmental cognitive neuroscience. *Nature Reviews Neuroscience, 16*(10), 606–619. doi:10.1038/nrn4025

Astle, D. E., Luckhoo, H., Woolrich, M., Kuo, B.-C., Nobre, A. C., and Scerif, G. (2015). The neural dynamics of fronto-parietal networks in childhood revealed using magnetoencephalography. *Cerebral Cortex, 25*(10), 3868–3876.

Atkinson, J., Hood, B., Wattambell, J., and Braddick, O. (1992). Changes in infants' ability to switch visual-attention in the 1st 3 months of life. *Perception, 21*(5), 643–653.

Bayliss, A.P., Paul, M.A., Cannon, P.R. et al. (2006). Gaze cuing and affective judgments of objects: I like what you look at. *Psychon Bull Rev 13*, 1061–1066. https://doi.org/10.3758/BF03213926

Batki, A., Baron-Cohen, S., Wheelwright, S., Connellan, J., and Ahluwalia, J. (2000). Is there an innate gaze module? Evidence from human neonates. *Infant Behavior and Development*, 23, 223–229. doi.org/10.1016/S0163-6383(01)00037-6

Bayliss, A. P., di Pellegrino, G., and Tipper, S. P. (2005). Sex differences in eye gaze and symbolic cueing of attention. *The Quarterly Journal of Experimental Psychology*, 58(4), 631–650. doi. org/10.1080/02724980443000124

Bisiach, E. and Luzzatti, C. (1978). Unilateral neglect of representational space. *Cortex*, 14(1), 129–133.

Bundesen, C. (1990). A theory of visual-attention. *Psychological Review*, 97, 523–547. doi:10.1037//0033-295x.97.4.523

Buschman, T. J. and Kastner, S. (2015). From behavior to neural dynamics: An integrated theory of attention. *Neuron*, 88(1), 127–144. doi:10.1016/j.neuron.2015.09.017

Bushnell, I. W. R., Sai, F., and Mullin, J. T. (1989). Neonatal recognition of the mother's face. *British Journal of Developmental Psychology*, 7(1), 3–15. doi.org/10.1111/j.2044-835X.1989.tb00784.x

Butterworth, G. and Jarrett, N. (1991). What minds have in common is space: Spatial mechanisms serving joint visual attention in infancy. *British Journal of Developmental Psychology*, 9(1), 55–72. doi.org/10.1111/j.2044-835X.1991.tb00862.x

Charman, T., Baron-Cohen, S., Swettenham, J., Baird, G., Cox, A., and Drew, A. (2000). Testing joint attention, imitation, and play as infancy precursors to language and theory of mind. *Cognitive Development*, 15(4), 481–498. doi.org/10.1016/S0885-2014(01)00037-5

Chun, M. M., Golomb, J. D., and Turk-Browne, N. B. (2011). A taxonomy of external and internal attention. *Annual Review of Psychology*, 62, 73–101. doi.org/10.1146/annurev.psych.093008.100427

Cohen Kadosh, K., and Johnson, M. H. (2007). Developing a cortex specialized for face perception. *Trends in Cognitive Sciences*, 11(9), 367–369. doi.org/10.1016/j.tics.2007.06.007

Colombo, J. (2001). The development of visual attention in infancy. *Annual Review of Psychology*, 52, 337–367. doi:10.1146/annurev.psych.52.1.337

Corbetta, M. and Shulman, G. L. (2002). Control of goal-directed and stimulus-driven attention in the brain. *Nature Reviews Neuroscience*, 3(3), 201–215. doi:10.1038/nrn755

Cowan, N. (2017). The many faces of working memory and short-term storage. *Psychonomic Bulletin & Review*, 24(4), 1158–1170. doi:10.3758/s13423-016-1191-6

Cowan, N., Elliott, E. M., Saults, J. S., Morey, C. C., Mattox, S., Hismjatullina, A., and Conway, A. R. A. (2005). On the capacity of attention: Its estimation and its role in working memory and cognitive aptitudes. *Cognitive Psychology*, 51(1), 42–100. doi:10.1016/j.cogpsych.2004.12.001

Cowan, N., Morey, C. C., AuBuchon, A. M., Zwilling, C. E., and Gilchrist, A. L. (2010). Seven-year-olds allocate attention like adults unless working memory is overloaded. *Developmental Science*, 13(1), 120–133. doi:10.1111/j.1467-7687.2009.00864.x

Crone, E. A. (2009). Executive functions in adolescence: Inferences from brain and behavior. *Developmental Science*, 12(6), 825–830. doi:10.1111/j.1467-7687.2009.00918.x

Csibra, G., Johnson, M. H., and Tucker, L. A. (1997). Attention and oculomotor control: A high-density ERP study of the gap effect. *Neuropsychologia*, 35(6), 855–865. doi:10.1016/s0028-3932(97)00016-x

Csibra, G., Tucker, L. A., and Johnson, M. H. (1998). Neural correlates of saccade planning in infants: A high-density ERP study. *International Journal of Psychophysiology, 29*(2), 201–215. doi.org/10.1016/S0167-8760(98)00016-6

Dalton, K. M., Nacewicz, B. M., Johnstone, T., Schaefer, H. S., Gernsbacher, M. A., Goldsmith, H. H., et al. (2005). Gaze fixation and the neural circuitry of face processing in autism. *Nature Neuroscience, 8*(4), 519–526. doi.org/10.1038/nn1421

Desimone, R. and Duncan, J. (1995). Neural mechanisms of selective visual-attention. *Annual Review of Neuroscience, 18*, 193–222.

Doherty, B. R., Fraser, A., Nobre, A. C., and Scerif, G. (2019a). The functional consequences of social attention on memory precision and on memory-guided orienting in development. *Developmental Cognitive Neuroscience, 36*, 100625. doi.org/10.1016/j.dcn.2019.100625

Doherty, B. R., Patai, E. Z., Duta, M., Nobre, A. C., and Scerif, G. (2017). The functional consequences of social distraction: Attention and memory for complex scenes. *Cognition, 158*, 215–223. doi:10.1016/j.cognition.2016.10.015

Doherty, B. R., van Ede, F., Fraser, A., Patai, E. Z., Nobre, A. C. N., and Scerif, G. (2019b). The functional consequences of social attention for memory-guided attention orienting and anticipatory neural dynamics. *Journal of Cognitive Neuroscience, 31*(5), 686–698. doi: 10.1162/jocn_a_01379

Doherty, M. J. and Anderson, J. R. (2000). A new look at gaze: Preschool children's understanding of eye-direction. *Cognitive Development, 14*(4), 549–571. doi.crg/10.1016/S0885-2014(99)00019-2

Driver, J. and Vuilleumier, P. (2001). Perceptual awareness and its loss in unilateral neglect and extinction. *Cognition, 79*(1–2), 39–88.

Duncan, J. and Humphreys, G. W. (1989). Visual-search and stimulus similarity. *Psychological Review, 96*(3), 433–458. doi:10.1037/0033-295x.96.3.433

Eimer, M. (1996). The N2pc component as an indicator of attentional selectivity. *Electro-encephalography and Clinical Neurophysiology, 99*, 225–234, doi:10.1016/0013-4694(96)95711-9

Emberson, L. L., Cannon, G., Palmeri, H., Richards, J. E., and Aslin, R. N. (2017). Using fNIRS to examine occipital and temporal responses to stimulus repetition in young infants: Evidence of selective frontal cortex involvement. *Developmental Cognitive Neuroscience, 23*, 26–38. doi.org/10.1016/j.dcn.2016.11.002

Fair, D. A., Cohen, A. L., Power, J. D., Dosenbach, N. U. F., Church, J. A., Miezin, F. M., et al. (2009). Functional brain networks develop from a "local to distributed" organization. *Plos Computational Biology, 5*(5). doi:10.1371/journal.pcbi.1000381

Fair, D. A., Dosenbach, N. U. F., Church, J. A., Cohen, A. L., Brahmbhatt, S., Miezin, F. M., et al. (2007). Development of distinct control networks through segregation and integration. *Proceedings of the National Academy of Sciences of the United States of America, 104*(33), 13507–13512.

Fan, J., McCandliss, B. D., Fossella, J., Flombaum, J. I., and Posner, M. I. (2005). The activation of attentional networks. *NeuroImage, 26*(2), 471–479. doi.org/10.1016/j.neuroimage.2005.02.004

Fan, J., McCandliss, B. D., Sommer, T., Raz, A., and Posner, M. I. (2002). Testing the efficiency and independence of attentional networks. *Journal of Cognitive Neuroscience, 14*(3), 340–347. doi:10.1162/089892902317361886

Farroni, T., Johnson, M. H., Brockbank, M., and Simion, F. (2000). Infants' use of gaze direction to cue attention: The importance of perceived motion. *Visual Cognition, 7*(6), 705–718. doi.org/10.1080/13506280050144399

Farroni, T., Johnson, M. H., and Csibra, G. (2004a). Mechanisms of eye gaze perception during infancy. *Journal of Cognitive Neuroscience*, *16*(8), 1320–1326. doi.org/10.1162/0898929042304787

Farroni, T., Massaccesi, S., Pividori, D., and Johnson, M. H. (2004b). Gaze following in newborns. *Infancy*, *5*(1), 39–60. doi.org/10.1207/s15327078in0501_2

Farroni, T., Menon, E., Rigato, S., and Johnson, M. H. (2007). The perception of facial expressions in newborns. *European Journal of Developmental Psychology*, *4*(1), 2–13. doi.org/10.1080/17405620601046832

Fiske, A. and Holmboe, K. (2019). Neural substrates of early executive function development. *Developmental Review*, *52*, 42–62. doi.org/10.1016/j.dr.2019.100866

Fraser, A. (2019). Evaluating the liking effect paradigm and the impact of realism in face processing among neurotypical and autistic populations (Doctoral dissertation, University of Oxford, Oxford, UK). Retrieved from https://ora.ox.ac.uk/objects/uuid:5f04e1ac-450a-4b9a-ad33-d9be848b2dc8

Garon, N., Bryson, S. E., and Smith, I. M. (2008). Executive function in preschoolers: A review using an integrative framework. *Psychological Bulletin*, *134*(1), 31–60.

Gathercole, S. E., Pickering, S. J., Ambridge, B., and Wearing, H. (2004). The structure of working memory from 4 to 15 years of age. *Developmental Psychology*, *40*(2), 177–190. doi:10.1037/0012-1649.40.2.177

Gazzaley, A. and Nobre, A. C. (2012). Top-down modulation: bridging selective attention and working memory. *Trends in Cognitive Sciences*, *16*(2), 129–135. doi.org/10.1016/j.tics.2011.11.014.

Gould, I. C., Rushworth, M. F., and Nobre, A. C. (2011). Indexing the graded allocation of visuospatial attention using anticipatory alpha oscillations. *Journal of Neurophysiology*, *105*(3), 1318–1326. doi.org/10.1152/jn.00653.2010

Griffin, I. C. and Nobre, A. C. (2003). Orienting attention to locations in internal representations. *Journal of Cognitive Neuroscience*, *15*(8), 1176–1194. doi:10.1162/089892903322598139

Guillory, S. B., Gliga, T., and Kaldy, Z. (2018). Quantifying attentional effects on the fidelity and biases of visual working memory in young children. *Journal of Experimental Child Psychology*, *167*, 146–161. doi: 10.1016/j.jecp.2017.10.005

Guy, J., Ng-Cordell, E., Doherty, B. R., Duta, M., and Scerif, G. (2020). Understanding attention, memory and social biases in fragile X syndrome: Going below the surface with a multi-method approach. *Research in Developmental Disabilities*, *104*, 103693. doi:10.1016/j.ridd.2020.103693

Guy, J., Ng-Cordell, E., Doherty, B. R., Duta, M., and Scerif, G. (forthcoming). Attention, memory and social biases in autism: Not always atypical.

Halassa, M. M. and Kastner, S. (2017). Thalamic functions in distributed cognitive control. *Nature Neuroscience*, *20*(12), 1669–1679. doi:10.1038/s41593-017-0020-1

Hendry, A., Jones, E. H., and Charman, T. (2016). Executive function in the first three years of life: Precursors, predictors and patterns. *Developmental Review*, *42*, 1–33.

Hood, B. M. (1993). Inhibition of return produced by covert shifts of visual-attention in 6-month-old infants. *Infant Behavior & Development*, *16*(2), 245–254. doi:10.1016/0163-6383(93)80020-9

Hood, B. M., Willen, J. D., and Driver, J. (1998). Adult's eyes trigger shifts of visual attention in human infants. *Psychological Science*, *9*(2), 131–134. doi.org/10.1111/1467-9280.00024

Johnson, M. H. (1990). Cortical maturation and the development of visual attention in early infancy. *Journal of Cognitive Neuroscience*, 2(2), 81–95. doi:10.1162/jocn.1990.2.2.81

Johnson, M. H. (1994). Visual attention and the control of eye movements in early infancy. In C. Umeltà and M. Moscovitch (Eds.), *Attention and Performance XV: Conscious and Nonconscious Information Processing, Vol. 15*, 291–310. Cambridge, MA: The MIT Press. doi. org/10.7551/mitpress/1478.003.0018

Johnson, M. H. (1995). The inhibition of automatic saccades in early infancy. *Developmental Psychobiology*, 28(5), 281–291. doi:10.1002/dev.420280504

Johnson, M. H. (2001). Functional brain development in humans. *Nature Reviews Neuroscience*, 2(7), 475–483.

Johnson, M. H. (2011). Interactive specialization: A domain-general framework for human functional brain development? *Developmental Cognitive Neuroscience*, 1(1), 7–21. doi:10.1016/j.dcn.2010.07.003

Johnson, M. H., Grossmann, T., and Cohen Kadosh, K. (2009). Mapping functional brain development: Building a social brain through interactive specialization. *Developmental Psychology*, 45(1), 151–159. doi.org/10.1037/a0014548

Johnson, M. H., Posner, M. I., and Rothbart, M. K. (1994). Facilitation of saccades toward a covertly attended location in early infancy. *Psychological Science*, 5(2), 90–93. doi:10.1111/j.1467-9280.1994.tb00636.x

Jones, W. and Klin, A. (2013). Attention to eyes is present but in decline in 2-6-month-old infants later diagnosed with autism. *Nature*, 504(7480), 427–431. doi.org/10.1038/nature12715

Kannass, K. N., Oakes, L. M., and Shaddy, D. J. (2006). A longitudinal investigation of the development of attention and distractibility. *Journal of Cognition and Development*, 7(3), 381–409. doi:10.1207/s15327647jcd0703_8

Kastner, S. and Ungerleider, L. G. (2000). Mechanisms of visual attention in the human cortex. *Annual Review of Neuroscience*, 23, 315–341. doi:10.1146/annurev.neuro.23.1.315

Kenny, L., Hattersley, C., Molins, B., Buckley, C., Povey, C., and Pellicano, E. (2016). Which terms should be used to describe autism? Perspectives from the UK autism community. *Autism: The International Journal of Research and Practice*, 20(4), 442–462. doi.org/10.1177/1362361315588200

Kim, N. Y. and Kastner, S. (2019). A biased competition theory for the developmental cognitive neuroscience of visuo-spatial attention. *Current Opinion in Psychology*, 29, 219–228. doi.org/10.1016/j.copsyc.2019.03.017

Kliemann, D., Dziobek, I., Hatri, A., Baudewig, J., and Heekeren, H. R. (2012). The role of the amygdala in atypical gaze on emotional faces in autism spectrum disorders. *Journal of Neuroscience*, 32(28), 9469–9476. https://doi.org/10.1523/JNEUROSCI.5294-11.2012

Kliemann, D., Dziobek, I., Hatri, A., Steimke, R., and Heekeren, H. R. (2010). Atypical reflexive gaze patterns on emotional faces in autism spectrum disorders. *Journal of Neuroscience*, 30(37), 12281–12287. doi.org/10.1523/JNEUROSCI.0688-10.2010

Klin, A., Jones, W., Schultz, R., Volkmar, F., and Cohen, D. (2002). Visual fixation patterns during viewing of naturalistic social situations as predictors of social competence in individuals with autism. *Archives of General Psychiatry*, 59(9), 809–816. doi.org/10.1001/archpsyc.59.9.809

Kobayashi, H. and Kohshima, S. (2001). Unique morphology of the human eye and its adaptive meaning: Comparative studies on external morphology of the primate eye. *Journal of Human Evolution*, 40(5), 419–435. doi.org/10.1006/jhev.2001.0468

Konrad, K., Neufang, S., Thiel, C. M., Specht, K., Hanisch, C., Fan, J., Herpertz-Dahlmann, B., and Fink, G. R. (2005). Development of attentional networks: An fMRI study with children and adults. *NeuroImage*, 28(2), 429–439. doi.org/10.1016/j.neuroimage.2005.06.065

Langton, S., Watt, R., and Bruce, I. (2000). Do the eyes have it? Cues to the direction of social attention. *Trends in Cognitive Sciences*, 4(2), 50–59. doi: 10.1016/s1364-6613(99)01436-9

Leekam, S., Baron-Cohen, S., Perrett, D., Milders, M., and Brown, S. (1997). Eye-direction detection: A dissociation between geometric and joint attention skills in autism. *British Journal of Developmental Psychology*, 15(1), 77–95. doi.org/10.1111/j.2044-835X.1997.tb00726.x

Leekam, S. and Hunnissett, E. (1998). Targets and cues: Gaze following in children with autism. *Journal of Child Psychology and Psychiatry*, 39(7), 951–962.

Liebe, S., Hoerzer, G. M., Logothetis, N. K., and Rainer, G. (2012). Theta coupling between V4 and prefrontal cortex predicts visual short-term memory performance. *Nature Neuroscience*, 15(3), 456–462. doi.org/10.1038/nn.3038

Luck, S. J. and Hillyard, S. A. (1994). Electrophysiological correlates of feature analysis during visual-search. *Psychophysiology*, 31, 291–308.

Luck, S. J. and Vogel, E. K. (2013). Visual working memory capacity: From psychophysics and neurobiology to individual differences. *Trends in Cognitive Sciences*, 17, 391–400.

Luna, B., Velanova, K., and Geier, C. F. (2008). Development of eye-movement control. *Brain and Cognition*, 68(3), 293–308. doi:10.1016/j.bandc.2008.08.019

Macchi Cassia, V., Kuefner, D., Westerlund, A., and Nelson, C. A. (2006). Modulation of face-sensitive event-related potentials by canonical and distorted human faces: The role of vertical symmetry and up-down featural arrangement. *Journal of Cognitive Neuroscience*, 18(8), 1343–1358. doi.org/10.1162/jocn.2006.18.8.1343

Manly, T., Anderson, V., Nimmo-Smith, I., Turner, A., Watson, P., and Robertson, I. H. (2001). The differential assessment of children's attention: The test of everyday attention for children (TEA-Ch), normative sample and ADHD performance. *Journal of Child Psychology and Psychiatry*, 42, 1065–1081.

Markant, J. and Amso, D. (2013). Selective memories: Infants' encoding is enhanced in selection via suppression. *Developmental Science*, 16(6), 926–940. doi:10.1111/desc.12084

Markant, J. and Amso, D. (2014). Leveling the playing field: Attention mitigates the effects of intelligence on memory. *Cognition*, 131(2), 195–204. doi:10.1016/j.cognition.2014.01.006

Markant, J., Worden, M. S., and Amso, D. (2015). Not all attention orienting is created equal: Recognition memory is enhanced when attention orienting involves distractor suppression. *Neurobiology of Learning and Memory*, 120, 28–40. doi:10.1016/j.nlm.2015.02.006

McNab, F. and Klingberg, T. (2008). Prefrontal cortex and basal ganglia control access to working memory. *Nature Neuroscience* 11, 103–107.

Morales, M., Mundy, P., Delgado, C. E. F., Yale, M., Messinger, D., Neal, R., and Schwartz, H. K. (2000). Responding to joint attention across the 6-through 24-month age period and early language acquisition. *Journal of Applied Developmental Psychology*, 21(3), 283–298. doi 10.1016/S0193-3973(99)00040-4

Morton, J. and Johnson, M. H. (1991). CONSPEC and CONCERN: A two-process theory of infant face recognition. *Psychological Review*, 98(2), 164–181.

Nation, K. and Penny, S. (2008). Sensitivity to eye gaze in autism: Is it normal? Is it automatic? Is it social? *Development and Psychopathology*, 20(1), 79–97. doi.org/10.1017/S0954579408000047

Nussenbaum, K., Scerif, G., and Nobre, A. C. N. (2019). Differential effects of salient visual events on memory-guided attention in adults and children. *Child Development*, 90(4), 1369–1388. doi: 10.1111/cdev.13149

Oakes, L. M., Kannass, K. N., and Shaddy, D. J. (2002). Developmental changes in endogenous control of attention: The role of target familiarity on infants' distraction latency. *Child Development*, 73(6), 1644–1655. doi:10.1111/1467-8624.00496

Or, C. C.-F., Peterson, M. F., and Eckstein, M. P. (2015). Initial eye movements during face identification are optimal and similar across cultures. *Journal of Vision*, 15(13), 12. https://doi.org/10.1167/15.13.12

Pascalis, O., de Schonen, S., Morton, J., Deruelle, C., and Fabre-Grenet, M. (1995). Mother's face recognition by neonates: A replication and an extension. *Infant Behavior and Development*, 18(1), 79–85. doi.org/10.1016/0163-6383(95)90009-8

Petersen, S. E. and Posner, M. I. (2012). The attention system of the human brain: 20 years after. *Annual Review of Neuroscience*, 35, 73–89. doi:10.1146/annurev-neuro-062111-150525

Posner, M. I. (1980). Orienting of attention. *Quarterly Journal of Experimental Psychology*, 32, 3–25. doi:10.1080/00335558008248231

Posner, M. I. and Petersen, S. E. (1990). The attention system of the human brain. *Annual Review of Neuroscience*, 13, 25–42. doi:10.1146/annurev.neuro.13.1.25

Pozuelos, J. P., Paz-Alonso, P. M., Castillo, A., Fuentes, L. J., and Rueda, M. R. (2014). Development of attention networks and their interactions in childhood. *Developmental Psychology*, 50(10), 2405–2415.

Riby, D. M., Hancock, P. J., Jones, N., and Hanley, M. (2013). Spontaneous and cued gaze-following in autism and Williams syndrome. *Journal of Neurodevelopmental Disorders*, 5(1), 13. doi.org/10.1186/1866-1955-5-13

Richards, J. E. (2010). The development of attention to simple and complex visual stimuli in infants: Behavioral and psychophysiological measures. *Developmental Review*, 30(2), 203–219. doi: 10.1016/j.dr.2010.03.005

Richards, J. E., Reynolds, G. D., and Courage, M. L. (2010). The neural bases of infant attention. *Current Directions in Psychological Science*, 19(1), 41–46. doi:10.1177/0963721409360003

Ristic, J., Mottron, L., Friesen, C. K., Iarocci, G., Burack, J. A., and Kingstone, A. (2005). Eyes are special but not for everyone: the case of autism. *Brain Research. Cognitive Brain Research*, 24(3), 715–718. https://doi.org/10.1016/j.cogbrainres.2005.02.007

Ross-Sheehy, S., Oakes, L. M., and Luck, S. J. (2003). The development of visual short-term memory capacity in infants. *Child Development*, 74(6), 1807–1822. doi:10.1046/j.1467-8624.2003.00639.x

Ross-Sheehy, S., Oakes, L. M., and Luck, S. J. (2011). Exogenous attention influences visual short-term memory in infants. *Developmental Science*, 14(3), 490–501. doi:10.1111/j.1467-7687.2010.00992.x

Rueda, M. R., Fan, J., McCandliss, B. D., Halparin, J. D., Gruber, D. B., Lercari, L. P., and Posner, M. I. (2004). Development of attentional networks in childhood. *Neuropsychologia*, 42(8), 1029–1040. doi:10.1016/j.neuropsychologia.2003.12.012

Scaife, M. and Bruner, J. S. (1975). The capacity for joint visual attention in the infant. *Nature*, 253(5489), 265–266. https://doi.org/10.1038/253265a0

Schotter, E. R., Berry, R. W., McKenzie, C. R., and Rayner, K. (2010). Gaze bias: Selective encoding and liking effects. *Visual Cognition*, 18(8), 1113–132.

Seiss, E., Driver, J., and Eimer, M. (2009). Effects of attentional filtering demands on preparatory ERPs elicited in a spatial cueing task. *Clinical Neurophysiology: Official Journal of the*

International Federation of Clinical Neurophysiology, 120(6), 1087–1095. https://doi.org/10.1016/j.clinph.2009.03.016

Shimi, A., Kuo, B.-C., Astle, D. E., Nobre, A. C., and Scerif, G. (2014a). Age group and individual differences in attentional orienting dissociate neural mechanisms of encoding and maintenance in visual STM. *Journal of Cognitive Neuroscience, 26*(4), 864–877. doi:10.1162/jocn_a_00526

Shimi, A., Nobre, A. C., Astle, D., and Scerif, G. (2014b). Orienting attention within visual short-term memory: Development and mechanisms. *Child Development, 85*(2), 578–592. doi:10.1111/cdev.12150

Shimi, A. and Scerif, G. (2015). The interplay of spatial attentional biases and mental codes in VSTM: Developmentally informed hypotheses. *Developmental Psychology, 51*(6), 731–743. doi:10.1037/a0039057

Shimi, A. and Scerif, G. (2017). Towards an integrative model of visual short-term memory maintenance: Evidence from the effects of attentional control, load, decay, and their interactions in childhood. *Cognition, 169*, 61–83. doi:10.1016/j.cognition.2017.08.005

Stokes, M. G., Atherton, K., Patai, E. Z., and Nobre, A. C. (2012). Long-term memory prepares neural activity for perception. *Proceedings of the National Academy of Sciences of the United States of America, 109*(6), E360–E367. doi:10.1073/pnas.1108555108

Summerfield, J. J., Lepsien, J., Gitelman, D. R., Mesulam, M. M., and Nobre, A. C. (2006). Orienting attention based on long-term memory experience. *Neuron, 49*(6), 905–916. doi:10.1016/j.neuron.2006.01.021

Tanaka, J. W., and Sung, A. (2016). The "Eye Avoidance" Hypothesis of Autism Face Processing. *Journal of Autism and Developmental Disorders, 46*(5), 1538–1552. https://doi.org/10.1007/s10803-013-1976-7

Tomasello, M. and Farrar, M. J. (2016). Joint attention and early language. *Child Development, 57*(6), 1454–1463. doi.org/10.2307/1130423

Trevisan, D. A., Roberts, N., Lin, C., and Birmingham, E. (2017). How do adults and teens with self-declared Autism Spectrum Disorder experience eye contact? A qualitative analysis of first-hand accounts. *PloS One, 12*(11), e0188446.

van Belle, G., Ramon, M., Lefèvre, P., and Rossion, B. (2010). Fixation patterns during recognition of personally familiar and unfamiliar faces. *Frontiers in Psychology, 1*(JUN), 1–8. doi.org/10.3389/fpsyg.2010.00020

Vecera, S. P. and Johnson, M. H. (1995). Gaze detection and the cortical processing of faces: Evidence from infants and adults. *Visual Cognition, 2*(1), 59–87. doi.org/10.1080/13506289508401722

Walton, G. E., Armstrong, E. S., and Bower, T. G. R. (1997). Faces as forms in the world of the newborn. *Infant Behavior and Development, 20*(4), 537–543. doi.org/10.1016/S0163-6383(97)90042-4

Wass, S. V., Scerif, G., and Johnson, M. H. (2012). Training attentional control and working memory—is younger, better? *Developmental Review, 32*(4), 360–387.

Werchan, D. M., Collins, A. G. E., Frank, M. J., and Amso, D. (2016). Role of prefrontal cortex in learning and generalizing hierarchical rules in 8-month-old infants. *Journal of Neuroscience, 36*(40), 10314–10322. doi:10.1523/jneurosci.1351-16.2016

THE EFFECTS OF SOCIOECONOMIC ADVERSITY ON THE DEVELOPMENT OF BRAIN SYSTEMS FOR ATTENTION AND SELF-REGULATION

COURTNEY STEVENS AND ERIC PAKULAK

INTRODUCTION

HUMAN brain development is shaped through the dynamic interplay of genetic and experiential factors. However, across brain systems, there is variability in the relative balance of these two factors, with some systems more genetically constrained and others more malleable as a result of experience (e.g., Stevens and Neville, 2014). This uneven profile of plasticity suggests that some brain systems may be particularly sensitive to aspects of environmental variability associated with early risk and adversity, while others might be more resilient in the face of environmental variation. In this chapter, we review research identifying brain systems that are more sensitive to early risk and adversity, as well as the key pathways linking aspects of early adversity to alterations in neural function. This research suggests that the structure and function of an integrated network underlying self-regulation and attention—including the prefrontal cortex (PFC), amygdala, and hippocampus—is particularly sensitive to the effects of chronic stress associated with early risk and adversity, with implications for functioning across a heterogeneous range of domains.

Although the term "early risk and adversity" is broad, our focus here is on early socioeconomic disadvantage. Socioeconomic disadvantage can be measured in various ways, and the most commonly used measure, socioeconomic status (SES), is typically based on parental education, occupation, and/or income during a child's development (Hollingshead, 1975; McLoyd, 1998). At the same time, SES, however measured, is a proxy variable representing a number of co-occurring factors, including but not limited to availability of high-quality public schools, access to health care, exposure to environmental pollutants and toxins, and differences in nutrition, exercise, and sleep (e.g., Brooks-Gunn et al., 1995; Brooks-Gunn and Duncan, 1997; Evans, 2004; McEwen and McEwen, 2017; Ursache and Noble, 2016). An exhaustive discussion of each of these factors is beyond the scope of this chapter, and we refer the interested reader to other related reviews (e.g., Brito and Noble, 2014; Bruce et al., 2013; Buckhalt, 2011; Lipina and Posner, 2012; McEwen and Gianaros, 2010; Nusslock and Miller, 2016; Propper and Holochwost, 2013; Shonkoff et al., 2012). Here, we focus on aspects of the caregiving environment associated with SES that have implications for stress response systems, as these factors have been linked through correlational and experimental research to a range of developmental outcomes (Evans, 2004; Evans et al., 2005; Farah et al., 2008) and can also be targeted through family-based intervention programs (Blair and Raver, 2016; Neville et al., 2013). As such, we emphasize a specific set of malleable pathways amenable to family-based intervention that link early adversity to brain development.

In the sections that follow, we begin by briefly reviewing the relationship between SES and a range of outcome measures, highlighting attention and self-regulation as foundational neural systems displaying a high degree of neuroplasticity, with differences linked to SES background. Drawing on both experimental data from animal models and correlational evidence from humans, we then turn to evidence linking alterations in these systems to chronic stress associated with early adversity as well as differences in supportive caregiving in response to environmental stressors. The experimental data provide important causal evidence for the effects of both early stress and supportive caregiving on the development of key neural systems, with correlational evidence suggesting parallels in human development. Finally, we examine research identifying potential buffers for these neural systems that can inform early intervention, focusing on the role of family-based programs that include supports for the caregiving environment. In this respect, we frame our understanding of neuroplasticity as incorporating both sensitivity in the face of adverse experiences as well as resilience through buffering and adaptive change.

SOCIOECONOMIC DISPARITIES

Socioeconomic Disparities: Overview

Socioeconomic disadvantage during childhood is a robust predictor of a range of later life outcomes, including poorer physical and mental health (for reviews, see Miller et al., 2011, McEwen and Gianaros, 2010; Shonkoff et al., 2012; Schickedanz et al., 2015), as well as lower levels of cognitive and educational attainment (e.g., Bradbury et al., 2015; Hackman et al., 2010; McFarland et al., 2017; Ursache and Noble, 2016). While broader than socioeconomic adversity, a recent meta-analysis of the relationship between cumulative early adverse experiences and later health found that the strength of these associations for some outcomes is modest (e.g., obesity, diabetes) or moderate (e.g., smoking, cancer, heart and respiratory disease) and is strongest for sexual risk-taking, mental illness, problematic alcohol and drug use, and interpersonal and self-directed violence (Hughes et al., 2017). While these relationships are undoubtedly complex, evidence suggests that many outcomes observed later in life can be traced to the impacts of early adversity on multiple and integrated biological systems (e.g., Nusslock and Miller, 2016; Pakulak et al., 2018). By identifying links between aspects of the environment and the development of neurobiological systems, such research can provide mechanistic insight into how early socioeconomic disadvantage becomes biologically embedded and ultimately impacts a strikingly wide and heterogeneous range of outcome measures (McEwen and McEwen, 2017; McEwen and Gianaros, 2010; Ursache and Noble, 2016; Nusslock and Miller, 2016; Pakulak et al., 2018).

Sensitivity of Attention and Self-Regulation

In this chapter, we focus on systems important for attention and self-regulation given that they are foundational systems for learning across domains (Blair and Raver, 2015; Stevens and Bavelier, 2012) that also display relatively greater plasticity compared to other neural systems (Stevens and Neville, 2014). We also focus on these systems because an integrated neural network underlying attention and self-regulation— including the PFC, hippocampus, and amygdala—is also crucial to many aspects of health and immune system function, and thus may link more broadly to the diverse range of outcomes associated with early socioeconomic adversity (e.g., Nusslock and Miller, 2016; Pakulak et al., 2018). While early adversity also affects other neurocognitive systems, and in particular those for language (e.g., Hackman and Farah, 2003; Noble et al., 2005; Ursache and Noble, 2016), these are beyond the scope of this review.

Before examining evidence for socioeconomic disparities in attention/self-regulation, we first clarify the cognitive constructs of self-regulation and attention that serve as the focus of this chapter. Self-regulation is defined as primarily volitional regulation of attention, emotion, and executive function for the purposes of goal-directed actions (Blair and Raver, 2012; Blair and Raver, 2015). While models of attention differ somewhat in their subdivisions and terminology, all models recognize the importance of a basic level of arousal and the importance of focused selection of specific stimuli for further processing, either transiently or in a sustained manner. Selective attention includes processes of *enhancing* selected signals (signal enhancement) and *suppressing* irrelevant information (distractor suppression). Distractor suppression is a part of early selection and is also considered to be part of executive function. Executive function (EF) subsumes a diverse set of psychological processes, including core cognitive skills such as inhibitory control, working memory, and cognitive flexibility (e.g., Diamond, 2006; Diamond, 2013). As attention and self-regulation skills emerge in infancy and undergo robust changes during childhood (e.g., Carlson, 2005; Diamond, 2006; Ridderinkhof and van der Stelt, 2000), they are potentially sensitive to variability in the environment during the early years of life.

A number of studies document differences in aspects of attention and self-regulation as a function of socioeconomic background. For example, early socioeconomic adversity is associated with poorer performance on specific aspects of attention and self-regulation, including attention shifting, inhibitory control, response inhibition, and working memory (Blair et al., 2011; Farah et al., 2006; Noble et al., 2005; Noble et al., 2007; Sarsour et al., 2011; Mezzacappa, 2004). Socioeconomic differences in aspects of attention emerge as early as infancy (Lipina et al., 2005) and have been documented throughout childhood (Noble et al., 2005; Noble et al., 2007) and into adulthood (Evans and Schamberg, 2009). Moreover, evidence indicates a gradient relationship between the duration of socioeconomic disadvantage during childhood and effects on attentional outcomes, as the amount of time in which family income is below the poverty line predicts performance on executive function tasks at age four (Raver et al., 2013).

While beyond the scope of this review, socioeconomic disadvantage is also related to differences in aspects of socio-emotional processing, mediated by the cortico-amygdala network, and related behavioral outcomes. Children from lower SES backgrounds show higher rates of internalizing and externalizing behaviors and conduct disorders, as well as higher rates of depression and anxiety (e.g., Duncan et al., 1994; Goodman et al., 2003; McLoyd, 1998; Merikangas et al., 2010, Tracy et al., 2008), and self- and parent-reported psychological well-being (Evans and English, 2002). Adolescents from lower SES backgrounds also are more likely to judge ambiguous scenarios as threatening (Chen and Matthews, 2003). There is also evidence for a complex relationship between early development of temperamental reactivity, emotional regulation, and the later development of EF (Ursache et al., 2013).

Electrophysiological studies indicate that lower SES is also associated with reduced effects of selective attention on neural processing, an effect observed in preschool (Giuliano et al., in press; Hampton Wray et al., 2017), early childhood (Stevens et al., 2009), and adolescent (D'Angiulli et al., 2008) samples. Moreover, these electrophysiology studies suggest socioeconomic differences in selective attention are specific to reduced suppression of distracting information in the environment, as opposed to enhancing task-relevant information (Giuliano et al., in press; Hampton Wray et al., 2017; Stevens et al., 2009). In addition, recent evidence suggests that the relationship between socioeconomic adversity and distractor suppression is mediated by sympathetic nervous system function, which may have implications for theories on how relationships between early adversity and physiological regulatory systems may impact health in the longer term (Giuliano et al., in press). As well, a recent electrophysiological study indicates that lower SES is associated with reduced error-related negativity and frontal theta in toddlers, which are believed to index the aspects of executive function involving the PFC and anterior cingulate cortex (Conejero et al., 2018).

Taken together, these results are consistent with the hypothesis that differences in self-regulation and attention associated with early adversity may be one of the primary mechanisms by which poverty affects academic outcomes, as reduced suppression of environmental information might be adaptive in more chaotic environments associated with early adversity, but maladaptive in a classroom environment (Blair and Raver, 2012; Blair and Raver, 2015). For example, while it may be advantageous to attend more broadly to information in the periphery in certain environments, this is less advantageous in a classroom setting where the suppression of distracting activity from classmates may help a student sustain attention to the teacher or task at hand. As will be discussed, a neural network supporting this regulatory behavior is particularly sensitive to early experience.

Sensitivity of Underlying Neural Networks

Neuroimaging studies examining the relationship between SES and brain development highlight the sensitivity of an integrated network of brain areas subserving aspects of attention and self-regulation, including the PFC, amygdala, and hippocampus. These regions, though not the only regions sensitive to socioeconomic disparities, are the focus here given their structural and functional connections to one another, their role in regulatory function (including attention and self-regulation), and their broader involvement in the regulation of stress and immune responses (e.g., McEwen and Gianaros, 2010; Nusslock and Miller, 2016; Pakulak et al., 2018).

The PFC supports many aspects of top-down regulation, including the inhibition of inappropriate responses and the promotion of task-relevant actions (e.g., Arnsten, 2009). The PFC connects to multiple regions, including the amygdala and hippocampus, with different subdivisions associated with different aspects of attention and self-regulation (Diamond, 2013). These functional subdivisions are organized in a topographical manner, with extensive connections to cortical and subcortical areas. More dorsal and lateral regions of the PFC are involved in the regulation of attention, thought, and action and have extensive connections to sensory and motor cortices, while more ventral and medial regions mediate emotional regulation and have extensive connections to subcortical areas including the amygdala, striatum, and hypothalamus. In addition, the PFC has connections to areas in the brainstem that produce catecholamines such as dopamine, epinephrine, and norepinephrine that underlie physiological changes associated with the stress response, and in particular the "fight or flight" response mediated by the sympathetic nervous system.

The amygdala facilitates the detection of biologically relevant emotionally-valenced stimuli, with a general tendency to prioritize negative information known as the "negativity bias" (Cacioppo et al., 1999). The amygdala is hypothesized to serve as a rapid detector of potential threats and mediator of adaptive responses to potential threats through mobilization of the sympathetic nervous system (e.g., Thayer and Lane, 2009). The amygdala and PFC are structurally and functionally connected, with top-down control and inhibition of the amygdala by the ventromedial PFC thought to reflect an integration of the external context (potential threat) with the internal context (perceptions of control over the potential threat) (e.g., Maier et al., 2006; Thayer et al., 2012). When PFC inhibition of the amygdala is reduced, a shift occurs from slower, more thoughtful PFC-regulated action to more reflexive and rapid emotional action (e.g., Arnsten, 2009). However, developmentally, the amygdala matures earlier than the PFC and can exert early influence over PFC development, suggesting that whereas later in development the PFC regulates amygdala activity, this relationship may be reversed early in development, with amygdala activity in early life shaping PFC function (for a review, see Tottenham and Gabard-Durnam, 2018).

Finally, the hippocampus, also closely connected to the amygdala as well as the PFC, is important for many aspects of learning and memory, including the management of spatial and episodic memory of events related to potential threats and other experiences, as well as the control of mood (McEwen and Gianaros, 2010). While less involved in self-regulation and attention directly, the hippocampus plays an important regulatory role in the stress response. The hippocampus contributes to perception of potential threats via contextual memory for the environmental conditions associated with events related to potential threat. This connectivity is adaptive, as events with more emotional salience are better remembered (for review, see Eichenbaum et al., 1992).

Structurally, socioeconomic adversity is associated with differences in both volume and surface area in the PFC (Noble et al., 2012; Noble et al., 2015; Raizada et al., 2008), amygdala (Noble et al., 2012; Luby et al., 2013), and hippocampus (Noble et al., 2012; Noble et al., 2015; Hanson et al., 2011; Jenkins et al., 2011; for extensive review, see Brito and Noble, 2014). Although there is some degree of inconsistency that likely depends both on the timing of the stressor and the time of measurement (for discussion, see Tottenham and Sheridan, 2010), early socioeconomic adversity has been primarily associated with structural atrophy in the PFC and hippocampus and both atrophy and hypertrophy of the amygdala. Given the structural connections among regions, this pattern could be associated with heightened threat sensitivity via increased amygdala reactivity and reduced inhibition of the amygdala by the ventromedial PFC. As well, underscoring the importance of caregiver relationships, detailed in section "Caregiving as a potential buffer," it is noteworthy that parental nurturance at age four has been found to predict hippocampal volume in a sample of adolescents from lower SES backgrounds (Rao et al., 2010) and caregiver support has been shown to mediate the effects of early adversity on hippocampal volume (Luby et al., 2013). These findings suggest that aspects of the caregiving environment associated with socioeconomic adversity may represent a key pathway linking socioeconomic adversity to differences in brain development.

Related functional imaging studies indicate that socioeconomic disadvantage is associated with patterns of hypo- and hyper-activity across the PFC, amygdala, and hippocampus, as well as altered functional connectivity among these regions. For example, a recent study examined resting-state functional connectivity among a number of brain regions including the PFC, hippocampus, and amygdala in a sample of over 100 children aged 7–12 (Barch et al., 2016). Socioeconomic disadvantage was associated with reduced negative connectivity between regions of the frontal lobe and both the amygdala and hippocampus, suggesting reduced top-down inhibition and control of these regions by the PFC. Other studies suggest altered function of this brain network. For example, a functional magnetic resonance imaging (fMRI) study of executive function found that early socioeconomic adversity was associated with poorer performance and greater activation of PFC during a novel rule learning task, which suggested less efficient PFC function in that children from lower SES backgrounds may have to engage this region more strongly to achieve the same level of performance (Sheridan et al., 2012). Early socioeconomic adversity is also associated with aspects of prefrontal regulatory function related to reward processing. For example, adults from lower SES backgrounds show reduced activation in brain areas important for reward and motivation as well as reduced dorsomedial PFC and anterior cingulate cortex activation and decreased functional connectivity between prefrontal and striatal regions in studies of reward activation and monetary gain (Gianaros et al., 2011).

Because SES is a proxy variable, these findings underscore the importance of identifying specific environmental factors associated with early socioeconomic adversity that might mediate the relationship between SES and brain development. While there are many candidate pathways, a consistent set of findings highlights the impact of chronic stress associated with socioeconomic adversity, as well as the critical role of the caregiving environment as both a potential stressor and a buffer to other stressors. The brain is central to this complexity, given that the regulatory systems that are the focus of this chapter are both critical to determining and reacting to potential environmental stressors and also sensitive to the stress hormones associated with the response to stressors. In this respect, the brain is both agent and patient in the stress response system, and as discussed next, this complexity has implications for the wide range of poor outcomes across the lifespan associated with early socioeconomic adversity.

CHRONIC STRESS AND BRAIN DEVELOPMENT

Chronic Stress: Overview

Converging evidence points to the central role of chronic stress as a key pathway linking early adversity to alterations in cognition and brain development (e.g., Blair and Raver, 2016, Champagne and Meaney, 2006; Evans and Kim, 2013; Gunnar and Quevado, 2007; McEwen et al., 2016; Shonkoff et al., 2012). This research highlights the sensitivity of the PFC, hippocampus, and amygdala, which are all rich in glucocorticoid receptors, to stressors. In particular, research on the effects of chronic stress on brain development capitalizes on the use of animal models, where variations in the environment can be controlled in ways not possible with studies of humans. This experimental approach thus allows causal conclusions to be drawn about the effect of the caregiving environment on different outcomes. Here, we begin by reviewing evidence from animal models, focusing on experimental data that allow such causal conclusions to be drawn, before turning to correlational evidence from humans suggesting that similar principles may underlie the SES-related disparities reviewed previously.

Research with Animal Models

Studies with animal models demonstrate the profound effects of stress exposure on the developing brain, and in particular the effects of stressors in the early environment (Coplan et al., 1996; Liu et al., 2000). In initial correlational research, for example, it was observed that rat pups raised by mothers who were high in nurturance (e.g., higher levels of licking, grooming, and arch-backed nursing) performed better on cognitive tasks and also showed differences in hippocampal neurochemistry and neuroarchitecture (Liu et al., 2000; Liu et al., 1997). To determine the causal impact of these maternal nurturance behaviors, subsequent experimental studies cross-fostered rats born to mothers who were higher or lower in nurturance behaviors. In other words, a rat genetically related to a low nurturance rat mother could be cross-fostered to be reared by a mother who was high in nurturance behaviors, and vice versa (Liu et al., 2000; Francis et al., 1999). An important feature of these studies is that the "high" and "low" nurturance conditions represented natural variation in the caregiving of rat mothers, but experimentation allowed rat pups to be randomly assigned to these different rearing conditions. These experimental studies extended the early correlational work, showing that rats reared by (non-genetically related) mothers who were high in nurturance showed subsequent changes in behavior and neural structure, now resembling rat pups raised by their (genetically related) high nurturance mothers (Liu et al., 2000).

Subsequent research further linked the caregiving environment to alterations in gene expression, suggesting epigenetic linkages and a means for the intergenerational transfer of the effects of the early caregiving environment (Weaver et al., 2004; Champagne, 2008; Champagne, 2018; Curley et al., 2017). For example, it has been shown that epigenetic mechanisms underlie the effect of maternal care on hippocampal function via changes in the expression of glucocorticoid receptors. As adults, pups of high nurturance mothers show increased expression of glucocorticoid receptors in the hippocampus, which is in turn associated with enhanced sensitivity to glucocorticoids that improves the negative-feedback function of the hippocampus in stress regulation and results in a more modest HPA response to stressors and lower anxiety (Weaver et al., 2004; Weaver et al., 2007; Murgatroyd et al., 2009).

However, a key question is the extent to which maternal behavior can itself be influenced—in other words, the extent to which this pathway might be malleable. A series of elegant studies suggest this is the case, and in particular demonstrate that stressors can impair caregiving behavior of rat mothers. For example, rat mothers who typically display high nurturance behaviors can be induced to behave in less nurturing ways to future offspring if exposed to a stressor (physical restraint) during pregnancy (Champagne and Meaney, 2006). The effects of this manipulation are apparent in the offspring, who in adulthood show different levels of oxytocin binding, as well as differences in their own maternal caregiving behavior. Moreover, these effects persist to future litters of offspring, even when no stress is applied during that pregnancy (Champagne and Meaney, 2006).

One interesting but often overlooked feature of this research is the specificity of effects for particular rat mothers or pups. For example, in the study by Champagne and Meaney previously described, it was rat mothers initially high in nurturing behaviors who were negatively affected by prenatal stress, while there was no corresponding benefit observed for low nurturing mothers not exposed to stress during pregnancy (Champagne and Meaney, 2006). Likewise, whereas rat pups born to low nurturing mothers show a beneficial effect of subsequent rearing by a high nurturing mother, decrements are less frequently observed for rats born to high-nurturing mothers but cross-fostered to low-nurturing mothers (Liu et al., 2000). The reason for this specificity is unclear but may relate to prenatal experiences and/or genetic factors. However, these added nuances suggest that maternal stressors can be deleterious to what might otherwise be more adaptive, nurturing caregiving behavior, and also that a less nurturing caregiving environment will have more effect on some offspring than others. In the section "Caregiving as a potential buffer," we describe similar specificity emerging from intervention research with humans.

While research with humans is typically done in parallel to the animal literature (see "Research with humans"), one recent cross-species study provides compelling evidence for the convergence of research across species on the caregiving environment as an important mediator between early adversity and behavioral and brain outcomes (Perry et al., 2018). Perry and colleagues used a previously developed manipulation of the rodent caregiving environment originally conceived as representing chronic stress. In this manipulation, one week after giving birth, rat mothers and their pups were randomly assigned to one of two conditions for five days. In the control condition, rat mothers had abundant nest-making materials, whereas in the "scarcity-adversity" condition, rat mothers were provided insufficient materials for nesting. The authors reframed this manipulation as representing a rodent model of some aspects of socioeconomic adversity, and in particular deprivation and attendant stressors. Rat mothers in the scarcity-adversity condition exhibited altered caregiving behaviors, including reduced sensitive caregiving (e.g., presence in nest with pups, nursing pups, maintaining pups in nest) and increased negative caregiving (e.g., rough pup transport, stepping on pups, self- (versus pup-) grooming). As well, their rat pups showed reduced regulation of behavior by maternal cues. For example, rat pups in the scarcity-adversity condition showed reduced preference for maternal odor and were less soothed by maternal presence following brief social isolation. Further, functional imaging indicated that in response to maternal odor, rat pups in the scarcity-adversity condition showed reduced functional connectivity between brain regions processing odor (aPCX) and the PFC, a finding the authors speculate may relate to maternal caregiving modulating PFC development.

A unique feature of Perry and colleagues' research was a cross-species comparison to a human longitudinal dataset analyzing the developmental trajectories from birth to 15 months of 1169 children in two low-income, rural communities. Using data collected when infants were 6 months of age, a continuous composite measure of poverty-related "scarcity-adversity" was created based on family income-to-needs ratio, as well as indicators of economic strain, household density, neighborhood noise/safety, maternal education, and consistent partnership with a co-habiting spouse/partner. Poverty-related scarcity-adversity at 6 months predicted decreased sensitive caregiving by mothers at 15 months, as well as increased negative caregiving. Higher levels of early scarcity-adversity were also associated with poorer measures of infant attention, general cognitive development, and positive affect at 15 months. Further, these relationships were mediated by parenting quality, suggesting important parallels between the rodent models of the effects of stress and adversity on newborn brain development and the correlational data from humans described in more detail in the following.

Research with Humans

Experimental data with animal models have the advantage of identifying cause-effect relationships including experimental manipulation of specific factors associated with socioeconomic adversity, such as the caregiving environment and stress exposure. In contrast, research with humans relies on correlational and observational findings, where single factors may be the focus of inquiry, but all possible confounding factors cannot be removed. However, the available data from human development are strikingly consistent with the animal studies discussed previously and also support a relationship between stress and brain development, and in particular a role of the early caregiving relationship in shaping these processes.

In support of a role for stress regulation in understanding SES-related disparities, a consistent finding reveals dysregulation of aspects of the stress response system following early adversity, and in particular the hypothalamic-pituitary-adrenal (HPA) axis. For example, dysregulation of cortisol levels has been observed under conditions of socioeconomic adversity (Lupien et al., 2000; Lupien et al., 2001; Blair et al., 2011; Evans and English, 2002), as well as more extreme forms of early adversity including institutionalized care (Gunnar and Vazquez, 2001) and maltreatment (Cicchetti and Rogosch, 2001). Moreover, there is evidence to suggest that cortisol, as a measure of stress reactivity, mediates the relationship between socioeconomic adversity and cognitive function, as well as between aspects of parenting and cognitive function (Blair et al., 2011).

Increasingly, evidence also suggests that chronic stress associated with early adversity in humans affects PFC and amygdala function across development. The amygdala shows an earlier developmental time course than the PFC, and it has been posited that caregivers provide an external regulator of amygdala reactivity during early development in ways that can both buffer and amplify affective behavior and amygdala activation in early childhood (Debiec and Sullivan, 2014; Hostinar et al., 2014; for more complete discussion, see Tottenham and Gabard-Durnam, 2018). Consistent with this, adolescents who faced more extreme early adversity in the form of institutionalization show greater amygdala reactivity to emotional stimuli (Gee et al., 2013; Tottenham et al., 2011), and young adults from lower SES backgrounds show greater amygdala activity to threatening facial expressions (Gianaros et al., 2008). Young adults from lower SES backgrounds show both greater amygdala and reduced PFC activity during effortful regulation of negative emotion, and chronic stress exposure across development mediates the relationship between early adversity and PFC activation (Kim et al., 2013).

This increased reactivity may result from insufficient top-down regulation by the PFC, a pattern that may be related to the development of connectivity between the amygdala and PFC. There is evidence suggesting that in individuals who have been exposed to early adversity, more mature connections between the amygdala and the PFC appear earlier in development (Gee et al., 2013; Wolf and Herringa, 2016). This is consistent with the hypothesis that early experience can affect amygdala function but also in turn affect the development of connections with other regulatory systems in ways that can influence interactions with the environment into maturity (Tottenham and Gabard-Durnam, 2018). Thus, evidence from studies of humans is consistent with that from animal studies in revealing the sensitivity of neurobiological systems serving important regulatory functions, including cognitive and emotional, to both chronic stress as well as differences in caregiving. Given this, it is important to consider potential buffers of early socioeconomic adversity and interventions designed in consideration of this evidence.

CAREGIVING AS A BUFFER AND EFFECTS OF INTERVENTION

The evidence from animal models and human studies discussed earlier illustrates the sensitivity of these integrated neurobiological regulatory systems to the environment early in development. Just as the human brain and related systems display sensitivity to adversity that can have profound effects across the lifespan, the exquisite plasticity of the brain also confers opportunity for resilience in the face of early adversity. Indeed, despite the profound effects reviewed here, a substantial proportion of children who experience early adversity avoid many of these poor outcomes. One way this resilience can come about is via experiences that buffer these vulnerable systems in the context of adverse environmental conditions. While this picture is undoubtedly complex, here we focus on what is likely the most powerful buffer: the early caregiving environment.

Caregiving as a Potential Buffer

As previously discussed, abundant evidence from studies of animal models indicates that early caregiving experience influences the development of the structure and function of neural systems important for stress regulation and threat appraisal. Differences in the development of these systems can in turn impact the function of systems important for self-regulation and attention. Evidence from studies of rodents demonstrates how differences in early parental nurturance influence the development of these important regulatory systems in ways that shape how an animal interacts with potential threats in the environment. By highlighting the importance of the establishment of a relationship, early in development, with a consistently responsive caregiver (Loman and Gunnar, 2010), this evidence also points to early caregiving as a potential buffer that provides a target for early interventions with the potential to promote resilience in the face of adversity.

Evidence from studies of humans points to the critical role of parental sensitivity and responsiveness in the development of a secure attachment relationship, which is in turn important for the development of neurobiological systems supporting regulatory and appraisal function (e.g., Gunnar et al., 1996). Retrospective evidence from studies of humans has found that high levels of caregiver nurturance early in development buffer against the long-term health problems associated with early adversity (Miller et al., 2011). However, as discussed, more stressful environments associated with socioeconomic adversity increase the amount of stress experienced by parents, which in turn reduces the likelihood of sensitive maternal child care and the development of secure attachments, increasing the likelihood of stressful interactions with caregivers (e.g., Blair and Raver, 2012; Meaney, 2010).

An important point in the consideration of early caregiving as a buffer is the degree to which development is shaped by both parent caregiving behaviors and also by individual differences in a child (Blair and Raver, 2012). Individual differences in the natural tendency to be more or less reactive can result in differential responses to variation in the caregiving environment in ways that lead to recursive feedback processes. For example, the development of self-regulation is shaped by this recursive feedback between the environment and differences in emotional reactivity and higher-order attention and executive control processes, and in contexts of early adversity this can result in the development of reactivity profiles with consequences for school readiness and broader effects (Blair and Raver, 2012; Blair and Raver, 2015). This has important implications for the development of interventions targeting regulatory systems, as individual differences in reactivity also interact with the degree of environmental adversity and caregiving. Biological sensitivity to context can be adaptive in more supportive caregiving environments but less so in more adverse environments (Ellis and Boyce, 2008), suggesting that interventions targeting the caregiving environment have the potential to capitalize on natural tendencies to be more sensitive to the environment.

Interventions Targeting the Caregiving Environment

As with the human studies discussed previously that reveal our increased understanding of relationships between early adversity and integrated neurobiological regulatory systems, human studies pointing to caregiving as a potential buffer are also correlational and thus limited in the degree to which causation can be inferred. However, experimental studies build on correlational studies in ways that inform both theories regarding causal pathways and also policies that seek to ameliorate these costly effects. Building on much of the evidence already discussed, translational researchers are designing, implementing, and assessing interventions that include consideration of the neurobiological systems affected by socioeconomic adversity. Consistent with the research reviewed earlier, evidence from these studies suggests that interventions addressing the caregiving environment may be a particularly effective approach. In addition, increasingly such studies are assessing the efficacy of interventions by measuring the effects on one or more of the neurobiological systems that are the focus of this chapter, providing valuable evidence on the potential for intervention to alter the adverse developmental trajectory of these vulnerable systems as well as on the mechanisms by which such interventions may be effective. We close this chapter by briefly highlighting several interventions that target early socioeconomic adversity and illustrate the ability for vulnerable systems to be modified for the better through the use of targeted programs addressing the pathways identified earlier. The interventions highlighted here specifically target the early caregiving environment and also include assessment of one or more of the regulatory systems discussed previously.

A large and growing body of evidence documents the efficacy of interventions targeting the early caregiving environment in improving outcomes for children who have experienced early adversity, ranging from more extreme cases such as adoption out of institutional rearing (Tottenham and Sheridan, 2010) to interventions targeting typically developing children from backgrounds of socioeconomic adversity (Neville et al., 2013). For example, in our research we have examined the effects of an eight-week, two-generation intervention for families of preschool children enrolled in Head Start, a national program for families living at or below the poverty line. The intervention combines small group activities for children focused on attention and self-regulation with small-group training for parents, providing tools and strategies for the home focused on increasing parental responsiveness, consistency, and predictability in order to reduce family stress. Children randomly assigned to receive the intervention show increased effects of selective attention on neural processing from before to after the training relative to children in both active and passive control groups, highlighting the plasticity of regulatory neurobiological systems and underscoring the importance of targeting caregiving (Neville et al., 2013). In addition, this study also illustrates the potential of two-generation interventions that simultaneously target attention and self-regulation in children and family stress in parents.

Interventions that target caregiving have also provided evidence for the malleability of the neuroendocrine and autonomic nervous systems that interact with the neural systems discussed to support stress regulation. For example, the number of intervention studies that incorporate cortisol into rigorous experimental designs had doubled in the last ten years (Slopen et al., 2014). This review found that, of 19 studies that took this approach, 18 found significant change in cortisol with intervention, many of which targeted caregiving. Importantly, all eight studies that included a comparison group of children experiencing less socioeconomic adversity found that patterns of cortisol activity in intervention groups of children from lower SES backgrounds changed with intervention to more closely resemble patterns of children from higher SES backgrounds. For example, Fisher and colleagues have shown that interventions targeting stress in foster parents and preschool-aged foster children reduce stress in foster parents and normalize diurnal cortisol patterns in foster children to levels more comparable with community controls (Fisher et al., 2007; Fisher and Stoolmiller, 2008). Other evidence suggests that targeting a classroom setting can also positively impact stress physiology as well as self-regulation. Blair and Raver (2014) showed that a classroom-based program targeting self-regulation in kindergarten children had positive effects on self-regulation, academic outcomes, and cortisol; these effects on cortisol were specific to children from high-poverty schools.

Other intervention work suggests that multiple stress regulatory systems exhibit considerable plasticity and also provides evidence that there may be sensitive periods for interventions targeting certain aspects of caregiving. McLaughlin and colleagues (2015) found that institutionalized children show blunted cortisol and autonomic reactivity in response to psychosocial stressors compared to children randomly assigned to placement into high-quality foster care. In contrast to later age of placement, they also found that placement into high-quality foster care in the first two years of life is associated with normalization of cortisol reactivity as well as more flexible autonomic reactivity during a social task, suggesting sensitive periods underlying the plasticity of these systems.

Given the incorporation of immune system function into recent models of the relationship between early adversity and poor outcomes across the lifespan (Nusslock and Miller, 2016), intervention studies that examine potential effects on immune system function are important. Emerging evidence suggests that the immune system is responsive to early interventions focused on the caregiving environment. Miller and colleagues (2014) demonstrated that African-American adolescents who had experienced socioeconomic adversity and who were randomly assigned as children to receive an intervention designed to strengthen parenting, family relationships, and youth competencies had better immune system function than adolescents from the comparison condition. These effects were mediated by improvements in parenting, again highlighting the importance of targeting the caregiving environment. Interestingly, the authors hypothesize that changes in the caregiving environment may have led children to adopt an adaptive "shift and persist" strategy that involves a combination of acceptance and endurance in the face of adversity. Consistent with this hypothesis, this strategy moderates the relationship between SES and both stress regulatory and immune system function in adolescents and their parents (Chen et al., 2015). The identification of adaptive psychosocial characteristics that are potentially malleable with targeted caregiving interventions that affect neurobiological regulatory systems represents a promising future direction in research in this area.

Consideration of Broader Social Structures

Finally, it is important to note that a comprehensive understanding of the effects of socioeconomic adversity on the development of neurobiological systems requires a consideration of broader social structure (for recent reviews, see McEwen and McEwen, 2017; Ungar et al., 2013). In addition to the potential for nurturing caregiving as a buffer, evidence from sociology documents the importance of social structures and networks that support caregivers who themselves may be dealing with stressful environments (e.g., George, 2013; Pearlin and Bierman, 2013). This evidence extends the concept of resilience to include both individual qualities and experiences as well as social circumstances that include community, neighborhood, cultural contexts, and institutional structures and resources, and non-hierarchical interactions between individuals and socio-ecological environments throughout development (e.g., Ungar et al., 2013).

This in turn underscores the importance of a careful consideration of broader social contexts and their interactions at different levels when designing interventions for caregivers, especially with social changes in the ways people receive and interact with information. For example, recent evidence suggests that parents have stronger preferences for the delivery of evidence-based parenting information via mass media and self-administered formats such as television, online programs, and written materials compared to more traditional formats such as home visits and weekly parenting groups (Metzler et al., 2012). As well, the human caregiving environment extends beyond *maternal* caregiving to include fathers, step-parents, grandparents, and other extended social networks. Thus, while much of the research discussed previously (including all of the research from animal models), emphasized the maternal caregiving environment, future research should more explicitly consider the role of other caregivers in the support of child development. At least some of the intervention research already described explicitly invited all parents/caregivers (including grandparents, step-parents, etc.) to participate (e.g., Neville et al., 2013), but neither participation nor implementation of the intervention was tracked by caregiver status. These broader social structures, including shifting parental preferences for how information is communicated and diverse family structures and shared caregiving responsibility, suggest new opportunities and directions for interventions targeting the caregiving environment.

CONCLUSION

Early socioeconomic adversity strongly predicts a striking range of poor outcomes across the lifespan. The research described here highlights the sensitivity of an integrated network of brain areas supporting attention and self-regulation to early adversity. With striking consistency across animal and human studies, this research has highlighted the deleterious effects of chronic stress on early brain development, and in particular the prefrontal cortex, hippocampus, and amygdala. This research also highlights the role of the caregiving environment in mediating these relationships, with data from animal models providing important causal evidence for the role of stressors in affecting the caregiving environment. Increasingly, evidence from both animal and human studies has also identified the caregiving environment as a malleable pathway that in humans is amenable to family-based interventions. These programs support the potential for sensitive, responsive caregiving to act as a buffer that promotes resilience in the face of some adverse experiences. At the same time, recognizing the multidimensional nature of poverty (e.g., Brooks-Gunn et al., 1995; Brooks-Gunn and Duncan, 1997; Evans, 2004; McEwen and McEwen, 2017; Ursache and Noble, 2016; Ungar et al., 2013), we emphasize that efforts to address the effects of poverty cannot focus exclusively on the caregiving environment but must address the larger set of factors, including those at the community and policy level, that contribute to the negative sequelae of poverty. As research expands across disciplines to address these numerous factors, we will continue to progress in our understanding of the complex interplay between biological and environmental factors that underlie the effects of early socioeconomic adversity as well as the exquisite plasticity of the human brain that provides hope for the amelioration of these effects.

REFERENCES

Arnsten, A. (2009). Stress signalling pathways that impair prefrontal cortex structure and function. *Nature Reviews Neuroscience, 10,* 410–422.

Barch, D., Pagliaccio, D., Belden, A., Harms, M., Gaffrey, M., Sylvester, C., Tillman, R., and Luby, J. (2016). Effect of hippocampal and amydala connectivity on the relationship between preschool poverty and school-age depression. *American Journal of Psychiatry, 173,* 625–634.

Blair, C., Granger, D., Willoughby, M., Mills-Koonce, R., Cox, M., Greenberg, M., Kivlighan, K., Fortunato, C., and FLP Investigators (2011). Salivary cortisol mediates effects of poverty and parenting on executive functions in early childhood. *Child Development, 82,* 1970–1984.

Blair, C. and Raver, C. (2012). Child development in the context of adversity: Experiential canalization of brain and behavior. *American Psychologist, 67,* 309–318.

Blair, C. and Raver, C. (2014). Closing the achievement gap through modification of neurocognitive and neuroendocrine function: Results from a cluster randomized controlled trial of an innovative approach to the education of children in kindergarten. *PLoS ONE, 9,* e112393.

Blair, C. and Raver, C. (2015). School readiness and self-regulation: A developmental psychobiological approach. *Annual Review of Psychology, 66,* 711–731.

Blair, C. and Raver, C. (2016). Poverty, stress, and brain development: New directions for prevention and intervention. *Academic Pediatrics, 16,* S30–S36.

Bradbury, B., Corak, M., Waldfogel, J., and Washbrook, E. (2015). *Too Many Children Left Behind: The US Achievement Gap in Comparative Perspective.* New York: Russell Sage Foundation.

Brito, N. and Noble, K. (2014). Socioeconomic status and structural brain development. *Frontiers in Neuroscience, 8,* 1–12.

Brooks-Gunn, J. and Duncan, G. (1997). The effects of poverty on children. *The Future of Children*, 7, 55–71.

Brooks-Gunn, J., Klebanov, P., and Liaw, F. (1995). The learning, physical, and emotional environment of the home in the context of poverty: The Infant Health and Development Program. *Children and Youth Services Review*, 17, 251–276.

Bruce, J., Gunnar, M., Pears, K., and Fisher, P. (2013). Early adverse care, stress neurobiology, and prevention science: Lessons learned. *Prevention Science*, 14, 247–256.

Buckhalt, J. A. (2011). Insufficient sleep and the socioeconomic status achievement gap. *Child Development Perspectives*, 5, 59–65.

Cacioppo, J., W, G., and Berntson, G. (1999). The affect system has parallel and integrative processing components: Form follows function. *Journal of Personality and Social Psychology*, 76, 839–855.

Carlson, S. (2005). Developmentally sensitive measures of executive function in preschool children. *Developmental Neuropsychology*, 28, 595–616.

Champagne, F. (2008). Epigenetic mechanisms and the transgenerational effects of maternal care. *Frontiers in Neuroendocrinology*, 29, 386–397.

Champagne, F. (2018). Beyond the maternal epigenetic legacy. *Nature Neuroscience*, 21, 773–4.

Champagne, F. and Meaney, M. (2006). Stress during gestation alters postpartum maternal care and the development of the offspring in a rodent model. *Biological Psychiatry*, 59, 1227–1235.

Chen, E. and Matthews, K. (2003). Development of the cognitive appraisal and understanding of social events (CAUSE) videos. *Health Psychology*, 22, 106.

Chen, E., Mclean, K. C., and Miller, G. E. (2015). Shift-and-persist strategies: Associations with socioeconomic status and the regulation of inflammation among adolescents and their parents. *Psychosomatic Medicine*, 77, 371–382.

Cicchetti, D. and Rogosch, F. (2001). The impact of child maltreatment and psychopathology on neuroendocrine functioning. *Development and Psychopathology*, 13, 783–804.

Conejero, Á., Guerra, S., Abundis-Guitiérrez, A., and Rueda, M. R. (2018). Frontal theta activation associated with error detection in toddlers: Influence of familial socioeconomic status. *Developmental Science*, 21, e12494.

Coplan, J., Andrews, M., Rosenblum, L., Owens, M., Friedman, S., Gorman, J., and Nemeroff, C. (1996). Persistent elevations of cerebrospinal fluid concentrations of corticotropin-releasing factor in adult nonhuman primates exposed to early-life stressors: Implications for the pathophysiology of mood and anxiety disorders. *Proceedings of the National Academy of Sciences*, 93, 1619–1623.

Curley, J., Mashoodh, R., and Champagne, F. (2017). Transgenerational Epigenetics. In: Tollefsbol, T. (ed.) *Handbook of Epigenetics: The New Molecular and Medical Genetics*. London: Academic Press.

D'angiulli, A., Herdman, A., Stapells, D., and Hertzman, C. (2008). Children's event-related potentials of auditory selective attention vary with their socioeconomic status. *Neuropsychology*, 22, 293–300.

Debiec, J. and Sullivan, R. (2014). Intergenerational transmission of emotional trauma through amygdala-dependent mother-to-infant transfer of specific fear. *Proceedings of the National Academy of Sciences*, 111, 12222–12227.

Diamond, A. (2006). The early development of executive functions. In: Bialystok, E. and Craik, F. (eds.) *Lifespan Cognition: Mechanisms of Change*. Oxford: Oxford University Press.

Diamond, A. (2013). Executive functions. *Annual Review of Psychology*, 64, 135–168.

472 COURTNEY STEVENS AND ERIC PAKULAK

Duncan, G., Brooks-Gunn, J., and Klebanov, P. (1994). Economic deprivation and early childhood development. *Child Development*, 65, 296–318.

Eichenbaum, H., Otto, T., and Cohen, J. (1992). The hippocampus—what does it do? *Behavioral and Neural Biology*, 57, 2–36.

Ellis, B. and Boyce, W. (2008). Biological sensitivity to context. *Psychological Science*, 17, 183–7.

Evans, G. (2004). The environment of childhood poverty. *American Psychologist*, 59, 77–92.

Evans, G. and English, K. (2002). The environment of poverty: Multiple stressor exposure, psychophysiological stress, and socioemotional adjustment. *Child Development*, 73, 1238–1248.

Evans, G., Gonnella, C., Marcynyszyn, L., Gentile, L., and Salpekar, N. (2005). The role of chaos in poverty and children's socioemotional adjustment. *Psychological Science*, 16, 560–565.

Evans, G. and Kim, P. (2013). Childhood poverty, chronic stress, self-regulation, and coping. *Child Development Perspectives*, 7, 43–48.

Evans, G. and Schamberg, M. (2009). Childhood poverty, chronic stress, and adult working memory. *Proceedings of the National Academy of Sciences*, 106, 6545–6549.

Farah, M., Betancourt, L., Shera, D., Savage, J., Giannetta, J., Brodsky, N., Malmud, E. and Hurt, H. (2008). Environmental stimulation, parental nurturance, and cognitive development in humans. *Developmental Science*, 11, 793–801.

Farah, M., Shera, D., Savage, J., Betancourt, L., Giannetta, J., Brodsky, N., Malmud, E., and Hurt, H. (2006). Childhood poverty: Specific associations with neurocognitive development. *Brain Research*, 1110, 166–174.

Fisher, P. and Stoolmiller, M. (2008). Intervention effects on foster parent stress: Associations with child cortisol levels. *Development and Psychopathology*, 20, 1003–1021.

Fisher, P., Stoolmiller, M., Gunnar, M., and Burraston, B. (2007). Effects of a therapeutic intervention for foster preschoolers on diurnal cortisol activity. *Psychoneuroendocrinology*, 32, 892–905.

Francis, D., Diorio, J., Liu, D., and Meaney, M. (1999). Nongenomic transmission across generations of maternal behavior and stress response in the rat. *Science*, 286, 1155–1158.

Gee, D., Gabard-Durnam, L., Flannery, J., Goff, B., Humphreys, K., Telzer, E., Hare, T., Bookheimer, S., and Tottenham, N. (2013). Early developmental emergence of human amygdala–prefrontal connectivity after maternal deprivation. *Proceedings of the National Academy of Sciences*, 110, 15638–15643.

George, L. (2013). Life-course perspectives on mental health. In: Aneshensel, C., Phelan, J., and Bierman, A. (eds.) *Handbook of the Sociology of Mental Health*. New York: Springer.

Gianaros, P., Horenstein, J., Hariri, A., Sheu, L., Manuck, S., Matthews, K., and Cohen, S. (2008). Potential neural embedding of parental social standing. *Social Cognitive and Affective Neuroscience*, 3, 91–96.

Gianaros, P., Manuck, S., Sheu, L., Votruba-Drzal, E., Craig, A., and Hariri, A. (2011). Parental education predicts corticostriatal functionality in adulthood. *Cerebral Cortex*, 21, 896–910.

Giuliano, R., Karns, C., Roos, L., Bell, T., Petersen, S., Skowron, E., Neville, H., and Pakulak, E. in press. Effects of early adversity on neural mechanisms of distractor suppression are mediated by sympathetic nervous system activity in preschool-aged children. *Developmental Psychology*.

Goodman, E., Slap, G. B., and Huang, B. (2003). The health impact of socioeconomic status on adolescent depression and obesity. *American Journal of Public Health*, 93, 1844–1850.

Gunnar, M., Brodersen, L., Machmias, M., Buss, K., and Rigatuso, J. (1996). Stress reactivity and attachment security. *Developmental Psychobiology*, 29, 191–204.

Gunnar, M. and Quevado, K. (2007). The neurobiology of stress and development. *Annual Review of Psychology*, *58*, 145–173.

Gunnar, M. and Vazquez, D. (2001). Low cortisol and a flattening of expected daytime rhythm: Potential indices of risk in human development. *Development and Psychopathology*, *13*, 515–538.

Hackman, D. and Farah, M. (2008). Socioeconomic status and the developing brain. *Trends in Cognitive Science*, *13*, 65–73.

Hackman, D., Farah, M., and Meaney, M. (2010). Socioeconomic status and the brain: Mechanistic insights from human and animal research. *Nature Reviews Neuroscience*, *11*, 651–659.

Hampton Wray, A., Stevens, C., Pakulak, E., Isbell, E., Bell, T., and Neville, H. (2017). Deveopment of selective attention in preschool-age children from lower socioeconomic status background. *Developmental Cognitive Neuroscience*, *26*, 101–111.

Hanson, J., Chandra, A., Wolfe, B., and Pollak, S. (2011). Association between income and the hippocampus. *PLoS ONE*, *6*, e18712.

Hollingshead, A. B. (1975). *Four factor index of social status*. Unpublished manuscript, Yale University, New Haven, Connecticut.

Hostinar, C., Sullivan, R., and Gunnar, M. (2014). Psychobiological mechanisms underlying the social buffering of the hypothalamic–pituitary–adrenocortical axis: A review of animal models and human studies across development. *Psychological Bulletin*, *140*, 256–282.

Hughes, K., Bellis, M., Hardcastle, K., Sethi, D., Butchart, A., Mikton, C., Jones, L., and Dunne, M. (2017). The effect of multiple adverse childhood experiences on health: A systematic review and meta-analysis. *The Lancet*, *2*, e356–e366.

Jenkins, S. R., Belanger, A., Connally, M. L., Boals, A., and Durón, K. M. (2011). First-generation undergraduate students' social support, depression, and life satisfaction. *Journal of College Counseling*, *16*, 129–142.

Kim, P., Evans, G. W., Angstadt, M., Ho, S. S., Sripada, C. S., Swain, J. E., Liberzon, I., and Phan, K. L. (2013). Effects of childhood poverty and chronic stress on emotion regulatory brain function in adulthood. *Proceedings of the National Academy of Sciences*, *110*, 18442–18447.

Lipina, S., Martelli, M., Vuelta, B., and Colombo, J. (2005). Performance on the A-not-B task of Argentinian infants from unsatisfied and satisfied basic needs homes. *Interamerican Journal of Psychology*, *39*, 49–60.

Lipina, S. and Posner, M. I. (2012). The impact of poverty on the development of brain networks. *Frontiers in Human Neuroscience*, *6*, 1–12.

Liu, D., Diorio, J., Day, J., Francis, D., and Meaney, M. (2000). Maternal care, hippocampal synaptogenesis, and cognitive development in rats. *Nature*, *3*, 798–806.

Liu, D., Diorio, J., Tannenbaum, B., Caldji, C., Francis, D., Freedman, A., Sharma, S., Pearson, D., Plotsky, P., and Meaney, M. (1997). Maternal care, hippocampal glucocorticoid receptors, and hypothalamic-pituatary-adrenal responses to stress. *Science*, *277*, 1659–1662.

Loman, M. M. and Gunnar, M. R. (2010). Early experience and the development of stress reactivity and regulation in children. *Neuroscience and Biobehavioral Reviews*, *34*, 867–876.

Luby, J., Belden, A., Botteron, K., Marrus, N., Harms, M., Babb, C., Nishino, T., and Barch, D. (2013). The effects of poverty on childhood brain development: The mediating effect of caregiving on stressful life events. *JAMA Pediatrics*, *167*, 1135–1142.

Lupien, S. J., King, S., Meaney, M., and Mcewen, B. (2000). Child's stress hormone levels correlate with mother's socioeconomic status and depressive state. *Biological Psychiatry*, *48*, 976–980.

Lupien, S. J., King, S., Meaney, M., and Mcewen, B. (2001). Can poverty get under your skin? Basal cortisol levels and cognitive function in children from low and high socioeconomic status. *Development and Psychopathology*, 13, 653–676.

Maier, S., Amal, J., Baratta, M., Paul, E., and Watkins, L. (2006). Behavioral control, the medial prefrontal cortex, and resilience. *Dialogues in Clinical Neuroscience*, 8, 397–407.

Mcewen, B. and Gianaros, P. (2010). Central role of the brain in stress and adaptation: Links to socioeconomic status, health, and disease. *Annals of the NY Academy of Sciences*, 1186, 190–222.

Mcewen, B., Nasca, C., and Gray, J. (2016). Stress effects on neuronal structure: Hippocampus, amygdala, and prefrontal cortex. *Neuropsychopharmacology Reviews*, 41, 3–23.

Mcewen, C. and McEwen, B. (2017). Social structure, adversity, toxic stress, and intergenerational poverty: An early childhood model. *Annual Review of Sociology*, 43, 445–472.

McFarland, J., Hussar, B., De Brey, C., Snyder, T., Wang, X., Wilkinson-Flicker, S., Gebrekristos, S., Zhang, J., Rathbun, A., Barmer, A., Bullock Mann, F., and Hinz, S. (2017). The Condition of Education 2017 (NCES 2017-144). *U.S. Department of Education, National Center for Education Statistics*. Washington, DC: Washington.

Mclaughlin, K., Sheridan, M., Tibu, F., Fox, N., Zeanah, C., and Nelson, C. (2015). Causal effects of the early caregiving environment on development of stress response systems in children. *Proceedings of the National Academy of Sciences*, 112, 5637–5642.

McLoyd, V. (1998). Socioeconomic disadvantage and child development. *American Psychologist*, 53, 185–204.

Meaney, M. (2010). Epigenetics and the biological definition of gene x environment interactions. *Child Development*, 81, 41–79.

Merikangas, K., He, J., Burstein, M., Swanson, A., Avenevoli, S., Cui, L., Benjet, C., Georgiades, K., and Swendsen, J. (2010). Lifetime prevalence of mental health disorders in US adolescents: Results from the National Comorbidity Survey Replication—Adolescent Supplement (NCS-A). *Journal of the American Academy of Child and Adolescent Psychiatry*, 49, 980–989.

Metzler, C., Sanders, M., Rusby, J., and Crowley, R. (2012). Using consumer preference information to increase the reach and impact of media-based parenting interventions in a public health approach to parenting support. *Behavior Therapy*, 43, 257–270.

Mezzacappa, E. (2004). Alerting, orienting, and executive attention: Developmental properties and sociodemographic correlates in an epidemiological sample of young, urban children. *Child Development*, 75, 1373–1386.

Miller, G., Brody, G., Yu, T., and Chen, E. (2014). A family-oriented psychosocial intervention reduces inflammation in low-SES African American youth. *Proceedings of the National Academy of Sciences*, 111, 11287–11292.

Miller, G., Chen, E., and Parker, K. (2011). Psychological stress in childhood and susceptibility to the chronic diseases of aging: Moving toward a model of behavioral and biological mechanisms. *Psychological Bulletin*, 137, 959.

Murgatroyd, C., Patchev, A., Wu, Y., Micale, V., Bockmühl, Y., Fischer, D., Holsboer, F., Wotjak, C., Almeida, O., and Spengler, D. (2009). Dynamic DNA methylation programs persistent adverse effects of early-life stress. *Nature Neuroscience*, 12, 1559–1566.

Neville, H., Stevens, C., Pakulak, E., Bell, T., Fanning, J., Klein, S., and Isbell, E. (2013). Family-based training program improves brain function, cognition, and behavior in lower socioeconomic status preschoolers. *Proceedings of the National Academy of Sciences, 110,* 12138–12143.

Noble, K., Houston, S., Brito, N., Bartsch, H., Kan, E., Kuperman, J., Akshoomoff, N., Amaral, D., Bloss, C., Libiger, O., Schork, N., Murray, S., Casey, B. J., Chang, L., Ernst, T., Frazier, J., Gruen, J., Kennedy, D., Van Zijl, P., Mostofsky, S., Kaufmann, W., Kenet, T., Dale, A., Jernigan, T., and Sowell, E. (2015). Family income, parental education, and brain structure in children and adolescents. *Nature Neuroscience, 18,* 773–778.

Noble, K., Houston, S., Kan, E., and Sowell, E. (2012). Neural correlates of socioeconomic status in the developing human brain. *Developmental Science, 15,* 516–527.

Noble, K., Mccandliss, B., and Farah, M. (2007). Socioeconomic gradients predict individual differences in neurocognitive abilities. *Developmental Science, 10,* 464–480.

Noble, K., Norman, M. F., and Farah, M. (2005). Neurocognitive correlates of socioeconomic status in kindergarten children. *Developmental Science, 8,* 74–87.

Nusslock, R. and Miller, G. (2016). Early-life adversity and physical and emotional health across the lifespan: A neuroimmune network hypothesis. *Biological Psychiatry, 80,* 23–32.

Pakulak, E., Stevens, C., and Neville, H. (2018). Neuro-, cardio-, and immunoplasticity: Effects of early adversity. *Annual Review of Psychology, 69,* 131–156.

Pearlin, L. and Bierman, A. (2013). Current issues and future directions in research into the stress process. In: Aneshensel, C., Phelan, J., and Bierman, A. (eds.) *Handbook of the Sociology of Mental Health.* New York: Springer.

Perry, R. J., Finegood, E., Braren, S., Dejoseph, M., Putrino, D., Wilson, D., Sullivan, R., Raver, C., Blair, C., and F. L. P. K. I. (2018). Developing a neurobehavioral animal model of poverty: Drawing cross-species connections between environments of scarcity-adversity, parenting quality, and infant outcome. *Development and Psychopathology, 31*(2), 2018, 1–20.

Propper, C. and Holochwost, S. (2013). The influence of proximal risk on the early development of the autonomic nervous system. *Developmental Review, 33,* 151–67.

Raizada, R., Richards, T., Meltzoff, A. and Kuhl, P. (2008). Socioeconomic status predicts hemispheric specialization of the left inferior frontal gyrus in young children. *NeuroImage, 40,* 1392–1401.

Rao, H., Betancourt, L., Giannetta, J., N, B., Korczykowski, M., Avants, B., Gee, J., Wang, J., Hurt, H., Detre, J., and Farah, M. (2010). Early prenatal care is important for hippocampal maturation: Evidence from brain morphology in humans. *NeuroImage, 49,* 1144–1150.

Raver, C., Blair, C., And Willoughby, M. (2013). Poverty as a predictor of 4-year-olds' executive function: New perspectives on models of differential susceptibility. *Developmental Psychology, 49,* 292–304.

Ridderinkhof, K. and Van Der Stelt, O. (2000). Attention and selection in the growing child: Views derived from developmental psychophysiology. *Biological Psychology, 54,* 55–106.

Sarsour, K., Sheridan, M., Jutte, D., Nuru-Jeter, A., Hinshaw, S., and Boyce, W. (2011). Family socioeconomic status and child executive functions: The roles of language, home environment, and single parenthood. *Journal of the International Neuropsychological Society, 17,* 120–132.

Schickedanz, A., Dreyer, B., and Halfon, N. (2015). Childhood poverty: Understanding and preventing the adverse impacts of a most-prevalent risk to pediatric health and well-being. *Pediatric Clinics of North America, 62,* 1111–1135.

Sheridan, M., Sarsour, K., Jutte, D., D'esposito, M., and Boyce, W. (2012). The impact of social disparity on prefrontal function in childhood. *PLoS One, 7,* e35744.

Shonkoff, J., Garner, A., Siegel, B., Dobbins, M., Earls, M., Mcguinn, L., Pascoe, J., and Wood, D. (2012). The lifelong effects of early childhood adversity and toxic stress. *Pediatrics, 129*, e232–e246.

Slopen, N., Mclaughlin, K. A., and Shonkoff, J. P. (2014). Interventions to improve cortisol regulation in children: A systematic review. *Pediatrics, 133*(2), 312–326.

Stevens, C. and Bavelier, D. (2012). The role of selective attention on academic foundations: A cognitive neuroscience perspective. *Developmental Cognitive Neuroscience, 2S*, S30–S48.

Stevens, C., Lauinger, B., and Neville, H. (2009). Differences in the neural mechanisms of selective attention in children from different socioeconomic backgrounds: An event-related brain potential study. *Developmental Science, 12*, 634–646.

Stevens, C. and Neville, H. (2014). Specificity of experiential effects in neurocognitive development. In: Gazzaniga, M. (ed.) *The Cognitive Neurosciences V*. Cambridge, MA: MIT Press.

Thayer, J., Åhs, F., Fredrikson, M., Soller, J., and Wager, T. (2012). A meta-analysis of heart rate variability and neuroimaging studies: Implications for heart rate variability as a marker of stress and health. *Neuroscience and Biobehavioral Reviews, 36*, 747–756.

Thayer, J. and Lane, R. (2009). Claude Bernard and the heart-brain connection: Further elaboration of a model of neurovisceral integration. *Neuroscience and Biobehavioral Reviews, 33*, 81–88.

Tottenham, N. and Gabard-Durnam, J. (2018). The developing amygdala: A student of the world and a teacher of the cortex. *Current Opinion in Psychology, 17*, 55–60.

Tottenham, N., Hare, T., Millner, A., Gilhooly, T., Zevin, J., and Casey, B. (2011). Elevated amygdala response to faces following early deprivation. *Developmental Science, 14*, 190–204.

Tottenham, N. and Sheridan, M. (2010). A review of adversity, the amygdala, and the hippocampus: A consideration of developmental timing. *Frontiers in Human Neuroscience, 3*, 1–18.

Tracy, M., Zimmerman, F., Galea, S., Mccauley, E., and Vander Stoep, A. (2008). What explains the relation between family poverty and childhood depressive symptoms? *Journal of Psychiatric Research, 42*, 1163–1175.

Ungar, M., Ghazinour, M., and Richter, J. (2013). Annual research review: What is resilience within the social ecology of human development? *Journal of Child Psychology and Psychiatry, 54*, 348–366.

Ursache, A., Blair, C., Stifter, C. and Voegtline, K. (2013). Emotional reactivity and regulation in infancy interact to predict executive functioning in early childhood. *Developmental Psychology, 49*, 127–137.

Ursache, A. and Noble, K. (2016). Neurocognitive development in socioeconomic context: Multiple mechanisms and implications for measuring socioeconomic status. *Psychophysiology, 53*, 71–82.

Weaver, I., Cervoni, N., Champagne, F., Alessio, A., Sharma, S., Seckl, J., Dymov, S., Szyf, M., and Meaney, M. (2004). Epigenetic programming by maternal behavior. *Nature Neuroscience, 7*, 847–854.

Weaver, I., D'alessio, A., Brown, S., Hellstrom, I., Dymov, S., Sharma, S., Szyf, M., and Meaney, M. (2007). The transcription factor nerve growth factor-inducible protein mediates epigenetic programming: Altering epigenetic marks by immediate-early genes. *Journal of Neuroscience, 27*, 1756–1768.

Wolf, R. and Herringa, R. (2016). Prefrontal-amygdala dysregulation to threat in pediatric posttraumatic stress disorder. *Neuropsychopharmacology, 41*, 822–831.

CHAPTER 18

THE ROLE OF ATTENTION IN THE DEVELOPMENT OF CREATIVITY

ANNA R. KAUER AND PAUL T. SOWDEN

INTRODUCTION

Creativity: Core Concepts

CREATIVE outputs are commonly regarded as involving the conjunction of at least two key elements. These are that a creative output should be both novel and also effective (Plucker, Beghetto and Dow, 2004; Runco and Jaeger, 2012). Either one of these elements alone is not sufficient for an output to be regarded as creative (Simonton, 2012). However, the extent to which an output is novel has to be judged in relation to the context in which it is produced. For instance, a child's discovery that cheese can be melted to make tasty food, might be novel to them but not to their parents. On the other hand, the discovery of an affordable, scalable, renewable, means of meeting our entire energy needs would be novel to the entire world population. Thus, it has been suggested that creativity can be represented along a continuum (Kaufman and Beghetto, 2009). *Mini-c* creativity represents an output that is novel and effective for an individual and that might change their view of the world. *Little-c* creativity represents outputs that might impress one's friends and family. *Pro-c* creativity would be outputs that are at a level of achievement generally only acquired after many years of mastering domain specific skills and knowledge and that would be recognized as creative by experts working in that domain. Finally, *Big-C* creativity are those outputs that have an enduring influence on humanity, often long after a creator's life is over. Given the time taken to achieve at Pro-c or Big-C levels of creativity, these would generally not be a feature of the creativity seen during childhood (although there are exceptions; cf. Drake and Winner, 2009).

Consequently, the present chapter focuses most closely on mini-c and little-c creativity. In addition to creative outputs or achievement, we also discuss the concept of creative potential, where "potential refers to a latent state which may be considered part of an individual's 'human capital'" (Walberg, 1988); (Lubart, Zenasni, and Barbot, 2013, p. 41), and is typically operationalized through measurement of aptitude at tasks requiring processes such as convergent and divergent thinking and varies across domains. Further, whilst research on creativity typically explores factors that influence creative outputs or products, if we are to understand and influence creativity itself, we must attempt to understand the underlying creative process.

A range of theories propose that creativity might be the product of ordinary cognitive and affective processes (Finke, Ward, and Smith, 1992). This gives rise to a research approach that seeks to understand these constituent processes and, in particular, how their combination gives rise to creativity in a manner that is ubiquitous to humans and enables us to flexibly adapt to everyday situations to which we don't have a predetermined, previously tried, response. These research approaches have begun to make significant strides in recent years. One particularly promising recent body of work (Beaty, Benedek, Silvia, and Schacter, 2016; Ellamil, Dobson, Beeman, and Christoff, 2012; Pringle and Sowden, 2017) suggests that when thinking creatively, cognitive processes involved in associative and analytic thinking, which have often been regarded to work competitively (Evans, 2008), begin to work cooperatively, and that this cooperation is predictive of creative outcomes. These processes and their interaction have been mapped to three underpinning brain networks, namely the default mode network, the executive control network, and the salience network (Beaty et al., 2018).

A key element of the associative processing that contributes to the generation of creative ideas is the retrieval of information from memory and combination of this information with inputs from the external environment. Novel combinations of these information sources can then be analyzed for their potential to produce an effective response to a presented situation. Thus, we might expect both inwardly and externally directed attention to be key elements in the creative process because of their role in the control of memory search and the filtering of external inputs from the environment. Consequently, the focus of the present chapter is to consider how the development of attention might be expected to interact with the development of creativity.

Attention: Core Concepts

Attention is a core cognitive function that controls the selection of relevant information and maintenance of focus (Carrasco, 2011; Driver, 2001). Although the precise way in which attention differentiates between relevant and irrelevant information remains an issue of debate, it is generally accepted that attention is the set of mechanisms resulting from the human brain's limited capacity for processing information, and that its allocation can be controlled (Buschman and Kastner, 2015; Desimone and Duncan, 1995). Although there are many different models of attention, there is largely agreement that there are two main processes: "early" attention, which is exogenous, a bottom-up process, and occurs at the earliest attentional level of sensorial perception; and "late" attention, which follows early attention in the time course, is endogenous, and a top-down process related to attentional control and executive function. More recent work in neuroscience has suggested that these attentional processes are not only cognitively but also anatomically distinct from one another. The processes of sensory perception and the shifting of attention appear broadly to be handled by a ventral frontoparietal system, while top-down and voluntary attentional processes are largely controlled by a dorsal frontoparietal system (Corbetta and Shulman, 2002). These two sets of processes do not run in isolation, but interact flexibly in order to enable rapid and dynamic switching between bottom-up sensory stimuli and top-down goals (Petersen and Posner, 2012; Vossel, Geng, and Fink, 2014).

The development of attentional control, also called "executive attention," has been a particular focus of developmental research, as it determines the extent to which an individual is able to ignore distracting information and choose to maintain an attentional focus on achieving task goals. Attentional control appears primarily to be governed by frontal areas of the brain, particularly the anterior cingulate cortex, as well as to be closely linked to executive functions such as working memory (Astle and Scerif, 2011; Posner and Petersen, 1990).

In this chapter we discuss both the "early" processing of sensory stimuli and the "late" attentional control processes in relation to creativity and its development, as well as considering more recent attentional "network" approaches. We also consider the role of sustained attention, a long-term form of attentional control.

Attention and its Relationship to the Development of Creativity

Creativity in adults has been strongly linked with certain attentional processes including permeable early attention, attentional control, and flexibility in switching between attentional foci. In adults, it has been suggested that these attentional processes may differentially predict both creative potential and creative achievement (Zabelina and Robinson, 2010; Zabelina, Saporta, and Beeman, 2016) but, at present, we have little understanding of whether this same relationship is found in children and what the developmental trajectory of attention from child to creative adult might be. While some attentional differences emerge early, others such as working memory and attentional control continue to develop throughout childhood and adolescence, in association with maturation of the prefrontal cortex (Booth et al., 2003). However, at present we have yet to explore how the combination of attention patterns that appear linked to creative achievement in adults, manifests throughout cognitive development.

Understanding the developmental trajectory of creative attention is now an important research goal, as creativity is increasingly seen as a vital skill for success in the rapidly automating job market of the fourth industrial revolution (Boston Consulting

Group, 2015; Department for Digital, Culture, Media, and Sport, 2017; Frey and Osborne, 2013; Manyika et al., 2017; Roberts, 2006; Robinson, 1999). Creativity was recently listed by the World Economic Forum as one of the top three most important skills for the job market of the future, with the other two being the closely related skills of problem solving and critical thinking (World Economic Forum, 2016). Unsurprisingly therefore, the Programme for International Student Assessment (PISA) has decided to include creative thinking tests in 2021, alongside those for reading, maths, and sciences (Lucas and Spencer, 2017). However, this is set against a background of what is argued to be plummeting levels of creativity among school leavers, together with an increasing inequality of opportunity (Durham Commission, 2019; Johnes, 2017; Kim, 2011; Warwick Commission, 2015). It is clear, therefore, that developing creativity needs to become an integral part of mainstream education as a matter of urgency. However, considerably more research is also needed in order to identify the most important precursors of creativity before targeted teaching strategies to promote creativity can be developed, in conjunction with interventions to support creative children.

In what follows, we first consider the evidence for the relationship between attention and creativity in adults, before proceeding to consider the implications of the development of attention during childhood on creative potential.

ATTENTION AND CREATIVITY IN ADULTS

Early Attention and Creativity

Early visual and auditory attention are the precursors to thought, memory, learning, and action. Therefore, it is not surprising that individual differences in early attention mechanisms appear to influence a person's level of creativity. In particular, broad early attention is often linked to real-world creative achievement. Broad attention was first described by Mendelsohn and Griswold (1964) as being characterized by a wide focus, which simultaneously notices or takes in a large range of stimuli. They contrasted this with narrow attention, which focuses on a smaller range of stimuli and effectively screens irrelevant information from awareness. Broad early attention appears to allow seemingly irrelevant information to leak into awareness, or be noticed, which is believed to keep the mind open to a wider range of possibilities, enabling the generation of more creative ideas (Abraham, 2013).

This notion has been supported by empirical work. For example, a recent eye-tracking study found that the increased attentional processing of apparently irrelevant objects on the periphery of a display resulted in the generation of more original and creative solutions among participants asked to generate original uses for common objects (Agnoli, Franchin, Rubaltelli, and Corazza, 2015). The incorporation of seemingly irrelevant information into the idea generation process may support the formation of remote associations as proposed by Mednick (1962), who argued that creative individuals were characterized by a propensity to synthesize elements into new combinations and that the more remote or unusual the associations between the elements, the more creative the idea or solution. Similarly, Gabora (2010) proposes that when thinking associatively to support creative idea generation, attention is spread over a wider, more diffuse, set of neural cliques[1], thereby supporting the formation of more remote associations between items and therefore greater creativity.

There is considerable anecdotal evidence that this kind of broad attention may be a characteristic shared by many eminent creative achievers (see Kasof, 1997 for a review); Charles Darwin for example noted that he felt himself to be unusual in "noticing things which easily escape attention." Extreme sensitivity to the environment, and to noise in particular, also seems to be a common complaint, with Darwin, Proust, Wagner, Schopenhauer, and Elgar, among others, all having reported a need for intense quiet and solitude as the result of their inability to screen out distracting sounds.

Numerous behavioral studies have provided further evidence of a link between creativity and this type of permeable attention. For example, on dichotic listening tasks, creative participants made more intrusion errors when being asked to repeat the information heard in one ear while also remembering the information presented to the other ear (Rawlings, 1985). Creative individuals were also more likely to incorporate irrelevant information in solutions to anagrams and when recalling phrases (Mendelsohn and Griswold, 1964).

It has been suggested that this tendency for irrelevant information to leak into the awareness of creative individuals is because they do not pre-categorize stimuli as irrelevant. In support of this theory, highly creative individuals have been found to exhibit reduced galvanic skin response habituation rates to auditory stimuli, compared with less creative individuals (Martindale, Anderson, Moore, and West, 1996). Continuing to pay attention to stimuli that have previously been shown to be irrelevant, rather than screening them from awareness, also results in faster learning on Latent Inhibition tasks by creative individuals (Carson, Peterson, and Higgins, 2003). Once again, the mechanism responsible for this appears to be broad versus selective attention. In line with this, individuals high in openness to experience, a trait closely related to creativity and also characterized by "permeable" attention, have been shown to combine visual information more flexibly, even at a low perceptual level (Antinori, Carter, and Smillie, 2017), suggesting that they may experience the world in a qualitatively different way. Recent neuroscientific studies have also provided evidence that these individual differences may be due to genetic variation in sensory gating mechanisms. For example, the P50 event related potential (ERP) is an index of early sensory gating that shows considerable inter-person variation, has been measured in infants as early as 1-month old (de Haan, 2013), and has been directly linked to real-world creative achievement (Zabelina, O'Leary, Pornpattananangkul, Nusslock, and Beeman, 2015).

Permeable, or "leaky" sensory gating has been linked to reduced density of dopamine D2 receptors in the thalamic area (Takahashi, Higuchi, and Suhara, 2006). This same neural architecture has been found in high performers on divergent thinking tasks (a laboratory measure of creative potential) and it has been suggested that this generates a state of "creative bias" (de Manzano, Cervenka, Karabanov, Farde, and Ullén, 2010). This is because a lower density of these receptors appears to have the effect of reducing thalamic gating thresholds, thereby decreasing the automatic regulation and filtering of sensory information (Yasuno et al., 2004). The resultant increase in sensory information flow is likely to widen the range of potential associations that can be combined into new ideas, leading to more original and fluent idea generation (de Manzano et al., 2010). As the researchers note, "thinking outside the box might be facilitated by having a somewhat less intact box" (de Manzano et al., 2010).

Creativity and Disorders of Early Attention

The fact that both a reduced P50 response and reduced density of thalamic D2 receptors are also markers of schizophrenia, may indicate an element of shared neural architecture between the two groups, together with a common pattern of permeable attention or reduced cognitive inhibition (Green and Williams, 1999). Suggestions of a common link are far from new (Kozbelt, Kaufman, Walder, Ospina, and Kim, 2014); indeed there is a long and controversial history of associating professional creativity and mental illness that goes back at least as far as Aristotle (Eysenck, 1993; Richards, Kinney, Lunde, Benet, and Merzel, 1988) and which remains persistent to this day (J.C. Kaufman, 2014). As well as anecdotal evidence that the relatives of creative achievers, such as Albert Einstein and James Joyce, suffered from schizophrenia (Andreasen, 1987, 2011), there has also been evidence from twin studies that both creativity and schizophrenia have a heritable basis (Piffer and Hur, 2014; Vinkhuyzen, van der Sluis, Posthuma, and Boomsma, 2009). This has been further supported by two large-scale studies in Sweden using longitudinal population registry data (Kyaga et al., 2011, 2013). Interestingly, these showed that people in the majority of creative professions were *less* likely to be diagnosed themselves with psychiatric disorders than matched controls, with the exception of bipolar disorder, which was somewhat more common amongst those in creative occupations. However, people in the majority of creative professions were more likely to be first-degree relatives of patients with schizophrenia, bipolar disorder and anorexia nervosa, as well as siblings of individuals with autism. The one exception group was writers, who themselves were moderately more likely than controls to be diagnosed with unipolar depression, schizophrenia, anxiety disorders, substance abuse, and to commit suicide. So far there has been little exploration of what might cause this higher incidence of mental health problems in writers, although Kyaga references Crow's (2008) theory that psychosis has its genetic origin in the human facility for language, which would explain why writers in particular might show greater susceptibility.

In fact, the field of genetics has made several interesting contributions to this idea of a shared link, with comparative genomic studies of several species finding strong evidence for the positive selection during the evolution of modern humans of a number of genes putatively associated with both schizophrenia and creativity (Crespi, Summers, and Dorus, 2007; Kozbelt et al., 2014; Lo et al., 2017). Given that the positive selection of genes that confer such profound cognitive and attentional impairments makes little sense in evolutionary terms, unless they also provide strong benefits, it has been argued that polymorphisms carrying elevated risk of schizophrenia remain in the human gene pool as the byproduct of a shared genetic link with creativity (Carson, 2011; Crespi et al., 2007; DeYoung, Grazioplene, and Peterson, 2012). The hypothesis is not that creativity is enhanced by clinical schizophrenia per se, but that it may be facilitated by some of the milder aspects of schizotypal cognition found at the non-clinical end of the spectrum, such as increased divergent thinking, non-conformity, impulsivity, formation of unusual associations, and reduced cognitive inhibition. It has been suggested, therefore, that the relationship between schizotypal cognition and creativity is not linear but, rather, an inverted U shape (Abraham, 2014; Martindale, 2007), as severe psychiatric disorders such as schizophrenia are likely to be more detrimental than beneficial for meaningful creative achievement.

Thus, whilst historically most research in this area has focused on deficits in clinical populations, there is now emerging interest in the cognitive advantages that this genotype may confer in healthy individuals, particularly when permeable attention is combined with high intelligence. For example, Kéri (2009) found that the highest levels of creative achievement were found in people who carried the T/T genotype (which has previously been related to psychosis risk) when coupled with high intelligence quota (IQ). Carson et al. (2003) found a similar link between low Latent Inhibition (a concept that has also previously been linked to schizophrenia) and high IQ in creative achievers.

Latent Inhibition is a pre-conscious gating mechanism that was first discovered in animal studies of classical conditioning. Human versions of the task date back to the late 1950s (Lubow, 1973), and measure the ability to ignore a non-reinforced stimulus, i.e., the extent to which an individual can screen irrelevant information from conscious awareness (Lubow and Gewirtz, 1995). Typically, participants are exposed to what is apparently an irrelevant stimulus while they complete a masking task. Then, they are given what appears to be a completely new task, but one which can only be solved by paying attention to the apparently irrelevant stimulus from the first phase (which continues to be present). Individuals are said to show low Latent Inhibition when, in the second phase, they do not automatically ignore the "irrelevant" stimulus from the first phase, and therefore, learn the rule for the new task more quickly. Those said to show "intact" Latent Inhibition exhibit delayed learning as they automatically discount the stimulus from the first phase as "irrelevant," and therefore take longer to make the connection.

Carson et al. (2003) found that creative achievers were seven times more likely to show a combination of low Latent Inhibition and high IQ than individuals with low creative achievements. This led Carson (2014) to posit a "shared vulnerability model of creativity and psychopathology" in which certain risk factors for psychiatric disorders, such as leaky early attention and neural hyperconnectivity, are moderated by protective factors such as high IQ, cognitive flexibility and increased working memory capacity, in order to become conducive to creative achievement. Carson's model clearly drew on Latent Inhibition research which had suggested that individuals with low Latent Inhibition were at risk from sensory overload and potential psychosis (Lubow and

Gewirtz, 1995). However, although a robust construct in animal models, the evidence for Latent Inhibition in humans has been inconsistent, with a recent review concluding that Latent Inhibition tasks were essentially too flawed to support any conclusions as to the cause of retardation in learning (Byrom, Msetfi, and Murphy, 2018). Further to this, although there is now convergence on the neural correlates of intelligence (Jung and Haier, 2007; Neubauer and Fink, 2009), there so far appears to have been no investigation of how the potential mechanism suggested by Carson (2014) might function, and therefore these conclusions must be treated with caution. There is, however, evidence in the opposite direction, which suggests that even non-clinical schizotypy may negatively impact working memory and intelligence among other cognitive functions (Matheson and Langdon, 2008; Noguchi et al., 2008).

However, the much-hypothesized link between the attention patterns characteristic of creativity and those of schizophrenia has received support from several neuroimaging studies. For example, in a sample of participants with intact working memory abilities and no history of psychological illness, researchers found that the more creative an individual was, the less they were able to deactivate the precuneus while carrying out a difficult working memory task (Takeuchi et al., 2011). Similar neural behavior has also been observed in individuals with schizophrenia, as well as in their first-degree relatives (Whitfield-Gabrieli et al., 2009). This is interesting because the precuneus is thought to be part of the default network (which will be discussed in more detail later), that typically shows greater activation during rest but deactivates during the performance of tasks that require focused attention (Cavanna and Trimble, 2006; Raichle et al., 2001). This brain region has been associated with attending to environmental stimuli (Corbetta et al., 2008), the unsystematic gathering of information (Raichle et al. 2001), and also mental representations involving the self, including personal memories (Cavanna and Trimble, 2006). Therefore, attenuated activity in this area is thought to suppress task-irrelevant stimuli that could disrupt successful task performance (Raichle et al., 2001) and, indeed, this behavior was observed in the control participants of both studies (Takeuchi et al., 2011; Whitfield-Gabrieli et al., 2009). Fink et al., (2014) investigated this further and compared groups of psychometrically high- and low-schizotypal individuals with controls and found that schizotypy and creativity were both positively associated with reduced deactivation of the right precuneus. Therefore, by "failing to deactivate" their precuneus, as Takeuchi et al. (2011) describe it, more creative individuals and those on the schizotypal spectrum can be perceived as allowing more stimuli to enter their awareness and maintaining more broadly oriented attention, enabling them to generate more remote associations and original ideas.

In summary, permeable early attention is likely to be most relevant to ideation, the aspect of creativity responsible for the generation of novel ideas or making of novel connections. Although this is the element of creativity that has received most emphasis from researchers to date (as will be discussed in the next section), with many tests of creativity being constructed solely on measures of idea generation (e.g., the Torrance Tests of Creative Thinking (Torrance, 2008) and the Alternate Uses Task (Guilford, Christensen, Merrifield, and Wilson, 1978)), this is only one part of the process and does not lead to meaningful creative achievement by itself (Cropley, 2006). Indeed, most recent models of creativity (cf. Sowden, Pringle, and Gabora, 2015) encompass at least two processes: a process of generating ideas, and a goal-directed process of elaborating them into finished creative products. As will now be discussed, attentional control plays a vital role in both of these processes.

Late Attention and Creativity

Flexible Attentional Control and the Generation of New Ideas

Historically, creativity has been operationalized in research as the ability to generate new ideas on divergent thinking tests such as the Alternate Uses Task (Gamble and Kellner, 1968; Torrance, 2008; Guilford et al., 1978; Wallach and Kogan, 1965). The concept of divergent thinking originates from Guilford's theory of creative thought as something that branches off into many different directions (Guilford, 1950). In accordance with this, divergent thinking tests are laboratory tests of creative potential to which there is not one correct answer. In fact, these tests require participants to generate multiple meaningful answers to a question or problem that are as creative as possible within a short time frame.

For example, Guilford's Alternate Uses Test (Guilford et al., 1978) asks participants to generate as many different and creative uses as possible for an everyday object, such as a brick. An obvious and therefore non-creative use for a brick would be to build houses or other objects with it, while a more creative use might be something along the lines of using it as a marker or pumice stone. The responses to these tests are typically evaluated for quantity (fluency) and originality (statistical infrequency) to produce a total divergent thinking score, with more creative responders producing a larger number of original ideas. Guilford's approach has been, and still remains, the dominant one in the psychometric measurement of creativity (J. C. Kaufman et al., 2008).

However, it has been argued that the attentional profile of a creative individual, as operationalized by divergent thinking tests, is very different to that of a creative individual, as operationalized by real-world creative achievement (Zabelina et al., 2015, 2016; Zabelina and Ganis, 2018). In contrast to the broad or leaky attention associated with creative achievement, divergent thinking performance is associated with flexible control of attention (Martindale, 1998; Zabelina and Ganis, 2018; Zabelina and Robinson, 2010). To perform well on these tasks, an individual needs the flexible attentional control to consider a task from multiple angles, generate new ideas at speed, and identify what is valuable and promising, while at the same time engaging top-down control in order to focus, shut out distractions, and suppress or disengage from irrelevant or less creative responses (Gilhooly, Fioratou, Anthony, and Wynn, 2007; Groborz and Necka, 2003; Nijstad, De Dreu, Rietzschel, and Baas, 2010; Nusbaum and Silvia, 2011; Zabelina and Robinson, 2010). The fact that divergent thinking tests are typically administered in timed conditions, requiring the rapid shifting from one generated idea to the next, is also thought to increase the involvement of the executive attention network at the expense of more unfocused attentional states such as day dreaming and the relaxation of mental constraints, thought to be more conducive to real-world creativity (Duncker, 1945; Mok, 2014; Plucker and Makel, 2010). Unsurprisingly therefore, highly creative individuals, as measured by high performance on divergent thinking tasks, show greater attentional flexibility and speed than low divergent thinkers (Vartanian, 2009).

Recent electroencephalogram (EEG) and functional magnetic resonance imaging (fMRI) research has provided additional evidence of the involvement of executive attention in divergent thinking task performance, showing activation in regions such as the inferior frontal gyrus and inferior parietal cortex, regions linked to response selection, interference resolution, and attentional control (Abraham, Beudt, Ott, and von Cramon, 2012; Benedek, Jauk, Sommer, Arendasy, and Neubauer, 2014; Chrysikou and Thompson-Schill, 2011; Fink and Benedek, 2014).

Therefore, the attentional profile associated with creativity may depend to a great extent on how creativity is operationalized (Zabelina, 2018; Zabelina and Ganis, 2018; Zabelina et al., 2015). This is an important point to make because of the extremely large body of research that has regarded measures of divergent thinking as being synonymous with creativity, and extrapolated from these findings that the attentional profile needed to perform well on these tests is the same one that is necessary for real-world creative achievement, despite there being strong indications to the contrary.

Undoubtedly, many of the aspects required for high performance on divergent thinking tests are also relevant to creative achievement such as ideation, the attentional flexibility not to get stuck on an unpromising idea, and the executive control required to focus (Batey and Furnham, 2006; Gilhooly et al., 2007; Nusbaum and Silvia, 2011). However, there are clear indications that, in addition to requiring a different attentional profile to the one needed by high scorers on inventories of creative achievement, divergent thinking tests may also be measuring a different construct of creativity. For example, divergent thinking scores often do not differentiate between modestly and highly creative individuals (Batey and Furnham, 2006; Silvia et al., 2008) and a large number of studies have found no, or only a modest association between divergent thinking and real-world creativity (Barron and Harrington, 1981; Runco and Acar, 2012).

Sustained Attentional Control and the Development of Ideas

Although the "in the moment" process of generating new ideas has received most research attention over the years, the process whereby these ideas are developed into publicly recognized creative achievements is of equal importance. Whereas ideation requires the attentional control to shift and inhibit attention flexibly, once an idea or creative product has been selected as promising, it must then be refined and developed, a process which requires a long-term and persistent form of attentional control. In the real world, as opposed to a laboratory environment, this stage of the creative process can take years (Simonton, 1999).

The ability to maintain an extended and intense attentional focus on task goals over such long periods has been argued not only to be unusual, but also to be one of the defining cognitive characteristics of highly creative individuals (Eysenck, 1993; Feist,

2006; Richards, Kinney, Benet, et al., 1988). Further evidence of sustained attention lies in the fact that most creative fields require many years of intensive study and practice in order to master what is already known in a domain before an individual can begin to formulate their own creative ideas (Simonton, 1999; Wallas, 1926). It has been argued that the extended attentional focus and maintenance of task goals underlying this process of preparation and development, are as important for creative achievement as personality factors such as grit and motivation, and may reflect a core cognitive preference for a persistent attentional focus (Zabelina and Beeman, 2013).

However, the effort of maintaining this type of persistent attentional focus can have costs as well as benefits. Although such immersive thinking has been associated with more creative ideas (Nijstad et al., 2010) it may also reduce attentional flexibility, resulting in real-world creative achievers (as opposed to high performers on divergent thinking tests) making more mistakes on tasks that require them to shift their level of attention (Zabelina and Beeman, 2013).

Shifting Attentional Focus and Creativity

From the discussion so far, it is apparent that creativity involves the recruitment of a range of attentional processes. Neither generative and evaluative, nor associative and analytic thinking are likely to rely on one mode of attention exclusively, but rather to be supported by a range of attentional processes such as broad attention, flexible attention and sustained attention. While leaky attention may support more creative responses by promoting originality, focused top-down attentional control may support fluency and flexibility through enabling shifts between multiple information categories. In addition, sustained attention may support fluency and flexibility through persistent retrieval of information from one category (see Nijstad et al., 2010). Unsurprisingly therefore, Groborz and Necka (2003) have described creative attention as being a balancing act.

As has been extensively argued, it seems certain that creative individuals do not favor just one sort of attention, but are in fact characterized by their ability to shift flexibly between different types (Gabora, 2010; Sowden et al., 2015; Vartanian, 2009; Vartanian, Martindale, and Kwiatkowsky, 2007; Zabelina and Robinson, 2010). As Gabora (2010) has argued, it would be dangerous and impractical to live one's life entirely in a state of defocused or associative attention, and therefore entering into this attentional state only makes adaptive sense if there is also a mechanism to switch to more focused and analytic modes when required. In line with this, it has been argued by several researchers that creativity requires an enhanced ability to shrink or expand the field of attention according to the task or situation (Finke, Ward, and Smith, 1992; Gabora, 2002; Howard-Jones and Murray, 2003; Martindale, 1998; Martindale, Smith, Ward, and Finke, 1995).

There continues to be speculation regarding exactly how this shifting is accomplished, in particular regarding the mechanism that facilitates the switching of attention between internal knowledge and external information for the purposes of idea generation, and to working memory for the purposes of synthesizing, simulating, and evaluating ideas. The frontal pole (Brodman area 10) is one brain area that is thought to be key to this process, together with lateral regions of the prefrontal cortex, as these regions of the brain are thought to manage the division of abstract (i.e., internal) versus concrete information processing (Badre, 2008; Koechlin, 2016; Ramnani and Owen, 2004). It also appears that the salience network, which detects and determines the relevance of environmental stimuli, plays an important role in shifting between externally and internally directed attention (Chand and Dhamala, 2016; Menon and Uddin, 2010; Uddin, 2015).

The salience network is also thought to manage the interaction between other large-scale brain networks such as the default mode network, which is associated with generative and spontaneous processing, and the central executive network, which is linked to more convergent and evaluative thinking (Beaty et al., 2014, 2016, 2018). It has been suggested that, during creative cognition, the salience network may identify promising ideas that are the result of generative processing and forward them to the executive attention system for higher-order processing (Jung, Mead, Carrasco, and Flores 2013). This area of research is relatively new, but the pattern now emerging seems to be that creativity requires the dynamic coupling and increased cooperation of two large-scale brain systems that normally act in opposition to one another (Beaty et al., 2014, 2016, 2018; Fox, Zhang, Snyder, and Raichle, 2009). The extent to which these networks cooperate appears to be related to the task itself, and the presence or absence of task goals: for example, spontaneous idea generation not involving a clear task goal appears to involve less coupling (Liu et al., 2015), while the evaluation of ideas seems to require considerably more interaction between the networks (Ellamil et al., 2012). The most recent work in this area has now identified a pattern of brain network connectivity that is not only characteristic, but also predictive of high-creative thinking ability, and which involves the simultaneous coupling and uncoupling of cortical areas within the default, salience, and executive networks (Beaty et al., 2018). Therefore, according to this approach, the salience network is the mechanism that mediates the way in which the various attentional processes interact. Interestingly, the coactivation of default and executive networks is in line with the finding discussed earlier, that the precuneus, the proposed hub of the default network, remains activated in more creative individuals during executive network dependent working memory tasks.

THE ROLE OF ATTENTION IN THE DEVELOPMENT OF CREATIVITY IN CHILDREN

As noted at the beginning of this chapter, there has to date been very little research into the development of attention in highly creative children. This may, in part, be because there is still much we do not know about the developmental trajectory of attention in the general population, although neuroscience is rapidly advancing our understanding in this area.

The Development of Attention

Attention develops rapidly from birth (Colombo, 2001). However, early and late attention networks appear to mature at different speeds (Johnson, Posner, and Rothbart, 1991). The posterior cortical and subcortical regions associated with early attention appear to mature earlier than the anterior systems involved in late attentional control (Nigg, 2000; Posner, Rothbart, Thomas-Thrapp, 1997; Posner, Rothbart, Sheese, and Voelker, 2014). In line with this, full attentional control takes many years to develop, mirroring the continuing maturation of the cortical regions of the parietal and temporal lobes throughout childhood and adolescence (Amso and Scerif, 2015; Booth et al., 2003; Miller and Cohen, 2001; Plude, Enns, and Brodeur, 1994; Zelazo, Müller, Frye, and Marcovitch, 2003).

As a result, children find it more difficult than adults to focus, divide, switch and sustain attention even into adolescence (Curtindale, Laurie-Rose, Bennet-Murphy, and Hull, 2007; Davidson, Amso, Anderson, and Diamond, 2006; Enns and Akhtar, 1989; Enns and Girgus, 1985; Huizinga, Dolan, and van der Molen 2006; Schul, Townsend, and Stiles, 2003; Wainwright and Bryson, 2005). As their attentional gating is poorer, children also display greater interference effects in both auditory and visual modalities (Hanania and Smith, 2010; Hanauer and Brooks, 2003; MacLeod, 1991; Remington et al., 2014). Behaviorally, this manifests in younger children tending to be slower, more distractible, and more error prone on attentional tasks than older children and adults (Hanauer and Brooks, 2003; Huang-Pollock, Carr, and Nigg, 2002; Remington, Cartwright-Finch, and Lavie, 2014).

In neural terms, the immature attentional system appears to be characterized by a greater overall volume of activation compared with adults performing similar tasks, suggesting that the maturation of these networks results in improved efficiency and focus (Durston et al., 2002; Fair et al., 2007, 2008). A good example of how much more attentional effort is required by immature cognitive control networks is Rueda, Posner, Rothbart, and Davis-Stober's (2004) study of the N2, an ERP associated with executive control (Kopp et al., 1996). The researchers found that 4-year-old children not only take almost double the amount of time of adults to first produce the N2, but then have to maintain it for ten times as long before making their response. In addition, the adults displayed much more focused neural activity than the children, who showed a pattern of much broader activation (Rueda, Posner, Rothbart, and Davis-Stober (2004).

With regard to perceptual capacity, it has been argued that capacity also increases with age, with younger children's attentional capacity for, and awareness of stimuli outside of their attentional focus being exhausted at much lower levels than those of older children and adults (Remington et al., 2014). The combination of reduced perceptual capacity and weaker cognitive control in young children manifests in what can seem a somewhat paradoxical pattern of behavior of them tending to be both more distractible when task load is low but also more focused than older children and adults when task load is high (Carmel, Fairnie, and Lavie, 2012; Remington et al., 2014).

In summary, the general consensus is that overall children, particularly at young ages, are likely to show poorer sensory gating and greater distractor interference than adults, as well as to demonstrate less flexible attention, and more effortful attentional control. However, there is one aspect of attention where very young children often outperform adults, which is on tests of distributed attention. As has been discussed, prior to the onset of the development of selective attention between 4 and 7 years old, infants and young children lack the attentional control needed for selective processing (Hanania and Smith, 2010). Instead of attending to information selectively, they typically show a very wide attentional focus compared to adults, and an equal level of processing of both task-relevant and task-irrelevant information (Posner and Rothbart, 2007; Rueda, Posner, and Rothbart, 2005). Therefore, although younger children display neither the processing speed, nor the efficiency of adults, they can exhibit much more thorough processing of unattended information compared to adults who only encode a small subset. This has been illustrated by empirical work, which has found that younger children outperform older children and adults in this respect on memory, categorization, and inattentional blindness tasks (Best, Yim, and Sloutsky, 2013; Coch, Sanders, and Neville, 2005; Deng and Sloutsky, 2016; Plebanek and Sloutsky, 2017).

Existing Research on Attention, Creativity, and Children

To date, we do not know whether the relationship between leaky early attention, attentional control, and creativity is the same in healthy children as in adults, although there are suggestions in the literature that this may be the case. For example, children who noticed unexpected objects in an inattentional blindness task, produced more original ideas and solutions in divergent thinking tasks (Memmert, 2009). Further, adolescents high in openness to experience, the personality trait strongly linked to creative achievement, were quicker to perceive implicit patterns in probabilistic sequence learning tasks (S. B. Kaufman, DeYoung, Gray, Jiménez, and Brown 2010). Given the indications that sensory gating appears to be genetically determined, at least in part, it seems plausible that permeable early attention is present from birth, as has been discussed earlier.

However, as attentional control is something that continues to develop even into late adolescence, it is likely to be challenging for researchers to isolate exactly what distinguishes the attention pattern of highly creative individuals during childhood. One difficulty is that the pattern of either greater distractibility or perseveration, depending on the circumstances (Enns and Girgus, 1985; Huang-Pollock et al., 2002; Remington et al., 2014), appears on the surface to be not too dissimilar to the pattern of permeable attention and greater perseveration discussed earlier in this chapter as being characteristic of creative achievers. This might suggest that variation in levels of creative achievement amongst children will be reduced relative to adults. Further, as noted at the beginning of this chapter, creative achievement in childhood is unlikely to be at the level of professional or world-renowned creators making it more difficult to identify actual high levels of creative achievement in children. Therefore, realistically, research may have to focus more on the study of creative potential, which makes the identification of creative children potentially even more problematic (Beghetto and Kaufman, 2007).

For these reasons, it is perhaps not surprising that a majority of the studies to have specifically linked creativity and attention in children have focused on attentional disorders, as the impact on creativity of attention patterns that show an extreme divergence from the norm is likely to be more clearly discernible. As has already been discussed, there are additional grounds for believing that there may be a link between creativity and attentional disorders, given the evidence that a certain level of neurodiversity may be beneficial for creativity (Abraham, 2014; Martindale, 2007).

Creativity and Childhood Disorders of Attention

The similarities between creative individuals and those with attention deficit hyperactivity disorder (ADHD) have been a particular research focus. For example, it has been argued that the previously discussed pattern of diffuse early attention and the inability to inhibit irrelevant stimuli is similar to the one found in ADHD (Auerbach, Benjamin, Faroe, Geller, and Ebstein, 2001; Cramond, 1994; Pritchard, Healey, and Neumann 2006). In line with this, individuals with ADHD have also been found to have leaky early attention in the form of low Latent Inhibition (White, 2007). White (2020) has argued that adults with ADHD have a very broad concept of what constitutes relevance in a given context, in line with many of the theories of how broad early attention relates to creativity discussed earlier in this chapter (e.g., Eysenck, 1993; Mednick, 1962; Mendelsohn and Griswold, 1964).

There are also other neural and genetic commonalities between the two groups. Like creative thinkers, adults with ADHD also demonstrate inefficient suppression of default network activity during cognitively demanding tasks (Beaty et al., 2016, 2018; Castellanos and Proal, 2012; Fassbender et al., 2009; Takeuchi et al., 2011; Uddin et al., 2008). Spontaneous activity in the default network is thought to cause fluctuations in focused attention during goal-directed activity, which results in a cognitive style that is more spontaneous, unstructured, and more likely to promote creative and divergent thinking (Acar and Runco, 2012; Eysenck, 1993; Finke, 1996; Sonuga-Barke and Castellanos, 2007). Another shared link appears to be the prevalence in both groups of the DRD4-7R gene which is related to dopaminergic transmission (Auerbach et al., 2001; Dietrich and Kanso, 2010; Munafò, Yalcin, Willis-Owen, and Flint, 2008). Commonly referred to as the "novelty gene," the DRD4-7R has been associated with behaviours commonly associated with both groups such as a preference for novelty, sensation-seeking, and a more dispersed attention pattern in children from 1 year old (Auerbach et al., 2001).

There is some behavioral support for this theory, with several studies having found that children and adolescents with ADHD produce more creatively unusual and original ideas than children without ADHD (Abraham, Windmann, Seifen, Daum, and

Güntürkün, 2006; Cramond, 1994; Fugate, Zentall, and Gentry, 2013; Gonzalez-Carpio, Serrano, and Nieto, 2017; Shaw and Brown, 1990, 1991). Individuals with ADHD, like other distractible individuals with mild executive function deficits, appear to be better at overcoming the constraints of existing knowledge or conceptual boundaries in order to generate more unusual and original ideas (Abraham et al., 2006; Abraham and Windmann, 2007; White, 2020). However, it should also be noted that several studies, including a recent meta-analysis, have found either no, or no conclusive evidence of enhanced creativity in children with ADHD (Aliabadi, Davari-Ashtiani, Khademi and Arabgol, 2016; Healey and (Aliabadi, Davari-Ashtiani, Khademi, and Arabgol, 2016; Healey and Rucklidge, 2005; Paek, Abdulla, and Cramond, 2016).

Further, this body of research focuses chiefly on creative ideation, in line with Simonton's (2004) conception of creativity as being characterized by a thinking style that embraces novelty and unconventionality and is not constrained by existing knowledge and norms. However, as previously discussed, ideation is just one part of creativity and, while ADHD may confer advantages in this respect, it appears to have a detrimental effect in relation to other aspects of the creative process such as persistence and cognitive control. For example, Gonzalez-Carpio et al. (2017), found that children with ADHD displayed enhanced creativity compared with controls on measures of originality and unusualness of perspective, but not on measures relating to developing an idea or persisting with it. White and Shah (2011) found that individuals with ADHD showed a preference for the ideational phase of problem solving and had less interest in the clarification and development phase. Abraham et al. (2006) found that adolescents with ADHD had an enhanced ability for overcoming contextual constraints, but struggled to produce an invention that was functional. Furthermore, divergent thinking in its extreme form without the concomitant cognitive control needed to evaluate and develop these new ideas has been dismissed as only quasi or pseudo creativity (Cropley, 2006), so the results of such studies should be interpreted with caution.

Despite this, there are aspects of this research that help to elucidate the way in which individual differences in attentional permeability and spontaneity, as well as in cognitive control, may strongly influence how creativity develops in healthy children from an early age. It suggests that children with broad and permeable attention and a more spontaneous processing style may show greater originality than those with attention that is narrower and more constrained, and it seems plausible that this may be a precursor to greater creativity in later life (Acar and Runco, 2012; Eysenck, 1993; Simonton, 2004). Also, given the current consensus that ADHD is the result of deficits in self-regulation and cognitive control, rather than an attention deficit per se (Barkley, 1997; Dramsdahl, Westerhausen, Haavik, Hugdahl, and Plessen, 2011; Slobodin, Cassuto, and Berger, 2015), together with the evidence of a shared genetic link, it seems plausible that the early attention processes of creative individuals and those with ADHD may not be dissimilar. However, the poorer performance of children with cognitive control deficits such as ADHD relative to normal controls on the evaluation and development of their original ideas indicates that individual differences in levels of cognitive control may have an early impact on other aspects of creative development. Therefore, it has been suggested that individuals with ADHD may benefit from effective executive function training, and that this is the missing piece of the puzzle that would enable them potentially to become successful creators (Barkley, 1997; Cortese et al., 2015; S. B. Kaufman, 2013; Klingberg et al., 2005).

A more nuanced view of the coexistence of creativity and attention disorders has been provided by Healey and Rucklidge (2006), who found elevated levels of some ADHD traits in 40 percent of their sample of healthy creative children and adolescents, over four times the number that would typically be expected in the general child population. Although none of these children met the full criteria for an ADHD diagnosis, they did show evidence of significant deficits in executive function, such as impairments in processing speed, reaction time, and naming speed. This meant that their cognitive performance fell between that of creative children with ADHD, and creative children who were without ADHD symptoms and appeared to have normal executive functioning. The small sample size of this study means the results should be interpreted with caution, but a longitudinal study of similar groups of children would provide valuable information as to whether it is the mild ADHD cognitive profile or the one with intact executive function skills that is most beneficial for creative achievement in the long term. There is some evidence that ADHD symptoms in adults may be associated with recognized creative achievement (Boot et al., 2017), but a thorough analysis of the prevalence of mild ADHD symptomology among adult creative achievers is also needed in order to clarify this relationship.

SUGGESTIONS FOR FUTURE RESEARCH

To date, there has been considerable research on creativity and giftedness in children, including extensive longitudinal studies, many of which have spanned decades (Holahan, Sears and Cronbach, 1995; Torrance, 1993). Clearly, creative achievement is not only determined by attention, but is a complex mix of elements that also involves motivation and personality factors, as well as opportunity. However, the majority of these individual difference factors have already been identified and explored (Lucas and Spencer, 2017; Mumford and Gustafson, 1988). In this final section of the chapter we briefly discuss some key research directions that we believe would greatly benefit our understanding of the role of attention in the development of creativity:

1) To gain a more detailed understanding of how the cognitive processing skills necessary for creativity develop, and what the potential roadblocks to the development of these skills might be.

2) The creation of a standard behavioral measure of permeable early attention that can be used with adults as well as children.

3) An investigation of whether any reduction in divergent thinking ability in children after 5 years old is the natural result of the development of selective attention and the change from a broader attentional style.

Understanding the Developmental Trajectory of Attention and Creativity

To date there has been little research focus on the *development* of the cognitive processing skills necessary for creativity. Permeable attention and executive control have both been strongly associated with creative achievement in adults and, as has been already discussed in this chapter, there are indications that there may be a similar link in children. However, at present, we have little idea of whether there is a trajectory in terms of the development of attentional processes from child with creative potential to creative adult.

One important question that needs answering is what the roadblocks are to the development of creativity: for instance, if it is found, as argued by Carson (2014), that permeable attention only facilitates creative achievement when coupled with strong executive control, what characterizes those children who maintain the permeable attention patterns of early childhood but also go on to achieve strong executive control compared to those that do not? Further to this, we need to gain a more precise idea of the relationship between different dimensions of ADHD symptomatology in childhood (McLennan, 2016), and the corresponding creative prognosis in adulthood, rather than confining research only to clinical cases. We also need to develop a much clearer understanding of what the developmental *attentional* milestones are that need to be reached in order to ensure the maximum likelihood of a child achieving their creative potential in the fullest sense, as defined at the beginning of this chapter. In addition, this research needs to be combined with the development of new ways of measuring creative potential that can be used right across the lifespan and, as discussed earlier, that encompass more than just divergent thinking ability.

Measuring Permeable Early Attention

The biggest hurdle to further research in this area is the problem of how to assess permeable attention in children. Although there is a long and established track record of testing executive function, together with creative ideation and cognitive flexibility as measured by divergent thinking tests, there is currently no standard behavioral measure of permeable attention. Previous work on this topic has used a wide variety of methods including dichotic listening tasks (Coch et al., 2005; Rawlings, 1985), Stroop tasks (Gamble and Kellner, 1968; Golden, 1975; Wang et al., 2018), P50 ERP response ratios (Zabelina et al., 2015), genotyping (Kéri, 2009), and connectome-based predictive modelling (Beaty et al., 2018). However, many of these do not lend themselves well to large-scale studies of children in relatively naturalistic settings. As discussed earlier in this chapter, a set of behavioral tasks that has been particularly linked to research on creative achievement is Latent Inhibition tasks (e.g., Carson et al., 2003; Peterson and Carson, 2000). Although there have recently been some efforts to modernize these tasks (Evans, Gray, and Snowden, 2007), they remain unreliable, as has been discussed earlier in this chapter, and have no track record of working well with healthy child populations (Kaniel and Lubow, 1986; Lubow, Toren, and Kaplan, 2000). The concept of Latent Inhibition in humans also remains problematic, with there still being no consensus on the psychological mechanisms underlying the construct (Byrom et al., 2018).

Does the Developmental Transition from Distributed to Selective Attention have an Impact on Levels of Creativity in Children?

As has been discussed at length in this chapter, a broad attentional focus that takes in a wide range of stimuli may be beneficial for creative ideation. Children aged 5 tend still to favor a distributed attention style and the majority of children (98 percent) of this age and stage of cognitive development have been found to score very highly on divergent thinking tests (Land and Jarman, 1992), suggesting that creative potential may initially be very widely distributed in the general population indeed. The same study found that by the time these same children were 10 years old, this figure had declined to 30 percent, and to 12 percent by age 15, a steep decline that has often been attributed to the emphasis the education system places on convergent thinking and rote learning rather than on promoting creative thinking and originality (Delis et al., 2007; Robinson, 1999). However, it would merit investigation as to what extent this decline in divergent thinking performance may be the consequence of the typical development of selective attention processes, and associated narrowing of attentional focus, and whether the mild attentional deficits discussed in this chapter might be what differentiates the 12 percent of children who continue to score highly.

CONCLUSION

In summary, this chapter has presented evidence that the various stages of creative thinking from spontaneous idea generation to its evaluation and refinement into a creative product, require the involvement of a large number of attentional processes. These have included broad, flexible, executive, and sustained attention. While leaky attention may support more creative responses by promoting originality, focused top-down attentional control appears to support fluency and flexibility through enabling shifts between multiple information categories. In addition, sustained attention is necessary to carry out the process of refining and developing an idea into a publicly recognized creative achievement.

We have also discussed the growing body of evidence supporting the theory that mild attentional deficits, observable in individuals at the lower end of the ADHD and schizotypy symptomatology spectrum, may confer advantages for creative thinking. In addition, we have considered the promising contribution of brain network approaches which have already been shown to reliably predict creative thinking ability. Given the evidence for the importance of attentional processes to creative thinking, we suggest that the development of attention has an important role to play in terms of whether a child reaches his or her creative potential, and that gaining a thorough understanding of the necessary attentional milestones should now be a research priority. In conclusion, the evidence presented in this chapter indicates that the development of attention is likely to be as integral to the development of creativity in children as it is to all other aspects of their cognition.

NOTE

1. A neural clique is a cell assembly that collectively responds to an experience, with some cliques responding to situation specific elements and others to abstract properties (Lin, Osan, and Tsien, 2006).

REFERENCES

Abraham, A. (2013). The promises and perils of the neuroscience of creativity. *Frontiers in Human Neuroscience*, 7, 246. doi:10.3389/fnhum.2013.00246

Abraham, A. (2014). Is there an inverted-U relationship between creativity and psychopathology? *Frontiers in Psychology*, 5, 750. doi:10.3389/fpsyg.2014.00750.

Abraham, A., Beudt, S., Ott, D. V. M., and von Cramon, D.Y. (2012). Creative cognition and the brain: Dissociations between frontal, parietal–temporal and basal ganglia groups. *Brain Research*, 1482, 55–70. doi:10.1016/j.brainres.2012.09.007.

Abraham, A. and Windmann, S. (2007). Creative cognition: The diverse operations and the prospect of applying a cognitive neuroscience perspective. *Methods*, 42(1), 38–48. doi:10.1016/j.ymeth.2006.12.007.

Abraham, A., Windmann, S., Siefen, R., Daum, I., and Güntürkün, O. (2006). Creative thinking in adolescents with attention deficit hyperactivity disorder (ADHD). *Child Neuropsychology*, 12(2), 111–123. doi:10.1080/09297040500320691.

Acar, S. and Runco, M. A. (2012). Psychoticism and creativity: A meta-analytic review. *Psychology of Aesthetics, Creativity, and the Arts*, 6(4), 341–350. doi:10.1037/a0027497.

Agnoli, S., Franchin, L., Rubaltelli, E., and Corazza, G. E. (2015). An eye-tracking analysis of irrelevance processing as moderator of openness and creative performance. *Creativity Research Journal*, 27(2), 125–132. doi:10.1080/10400419.2015.1030304.

Aliabadi, B., Davari-Ashtiani, R., Khademi, M., and Arabgol, F. (2016). Comparison of creativity between children with and without attention deficit hyperactivity disorder: A case-control study. *Iranian Journal of Psychiatry*, 11(2), 99–103.

Amso, D. and Scerif, G. (2015). The attentive brain: Insights from developmental cognitive neuroscience. *Nature Reviews Neuroscience*, 16(10), 606–619. doi:10.1038/nrn4025.

Andreasen, N. C. (1987). Creativity and mental illness: Prevalence rates in writers and their first-degree relatives. *The American Journal of Psychiatry*, 144(10), 1288–1292. doi:10.1176/ajp.144.10.1288.

Andreasen, N. C. (2011). A journey into chaos: Creativity and the unconscious. *Mens Sana Monographs*, 9(1), 42–53. doi:10.4103/0973-1229.77424.

Antinori, A., Carter, O. L., and Smillie, L. D. (2017). Seeing it both ways: Openness to experience and binocular rivalry suppression. *Journal of Research in Personality*, 68, 15–22. doi:10.1016/j.jrp.2017.03.005.

Astle, D. E. and Scerif, G. (2011). Interactions between attention and visual short-term memory (VSTM): What can be learnt from individual and developmental differences? *Neuropsychologia*, 49(6), 1435–1445. doi:10.1016/j.neuropsychologia.2010.12.001.

Auerbach, J. G., Benjamin, J., Faroy, M., Geller, V., and Ebstein, R. (2001). DRD4 related to infant attention and information processing: A developmental link to ADHD? *Psychiatric Genetics*, 11(1), 31–35. doi:10.1097/00041444-200103000-00006.

Badre, D. (2008). Cognitive control, hierarchy, and the rostro–caudal organization of the frontal lobes. *Trends in Cognitive Sciences*, 12(5), 193–200. doi:10.1016/j.tics.2008.02.004.

Barkley, R. A. (1997). Behavioral inhibition, sustained attention, and executive functions: Constructing a unifying theory of ADHD. *Psychological Bulletin*, 121(1), 65–94. doi:10.1037/0033-2909.121.1.65.

Barron, F. and Harrington, D. M. (1981). Creativity, intelligence and personality. *Annual Review of Psychology*, 32(1), 439–476. doi:10.1146/annurev.ps.32.020181.002255.

Batey, M. and Furnham, A. (2006). Creativity, intelligence, and personality: A critical review of the scattered literature. *Genetic, Social, and General Psychology Monographs*, 132(4), 355–429. doi:10.3200/mono.132.4.355-430.

Beaty, R. E., Benedek, M., Silvia, P. J., and Schacter, D. L. (2016). Creative cognition and brain network dynamics. *Trends in Cognitive Sciences*, 20(2), 87–95. doi:10.1016/j.tics.2015.10.004.

Beaty, R. E., Benedek, M., Wilkins, R. W., Jauk, E., Fink, A., Silvia, P. J., Hodges, D. A., Koschutnig, K., and Neubauer, A. C. (2014). Creativity and the default network: A functional connectivity analysis of the creative brain at rest. *Neuropsychologia*, 64, 92–98. doi:10.1016/j.neuropsychologia.2014.09.019.

Beaty, R. E., Kenett, Y. N., Christensen, A. P., Rosenberg, M. D., Benedek, M., Chen, Q., Fink, A., Qiu, J., Kwapil, T. R., Kane, M. J., and Silvia, P. J. (2018). Robust prediction of individual creative ability from brain functional connectivity. *Proceedings of the National Academy of Sciences*, 115(5), 1087–1092. doi:10.1073/pnas.1713532115.

Beghetto, R. A. and Kaufman, J. C. (2007). Toward a broader conception of creativity: A case for "mini-c" creativity. *Psychology of Aesthetics, Creativity, and the Arts*, 1(2), 73–79. Doi:10.1037/1931-3896.1.2.73.

Benedek, M., Jauk, E., Sommer, M., Arendasy, M., and Neubauer, A. C. (2014). Intelligence, creativity, and cognitive control: The common and differential involvement of executive functions in intelligence and creativity. *Intelligence*, 46, 73–83. doi:10.1016/j.intell.2014.05.007.

Best, C. A., Yim, H., and Sloutsky, V. M. (2013). The cost of selective attention in category learning: Developmental differences between adults and infants. *Journal of Experimental Child Psychology*, 116(2), 105–119. doi:10.1016/j.jecp.2013.05.002.

Boot, N., Nevicka, B., and Baas, M. (2017). Subclinical symptoms of attention-deficit/hyperactivity disorder (ADHD) are associated with specific creative processes. *Personality and Individual Differences*, 114, 73–81. doi:10.1016/j.paid.2017.03.050.

Booth, J. R., Burman, D. D., Meyer, J. R., Lei, Z., Trommer, B. L., Davenport, N. D., Li, W., Parrish, T. B., Gitelman, D. R., and Mesulam, M. M. (2003b). Neural development of selective attention and response inhibition. *NeuroImage*, *20*(2), 737–751. doi:10.1016/s1053-8119(03)00404-x.

Buschman, T. J. and Kastner, S. (2015). From behavior to neural dynamics: An integrated theory of attention. *Neuron*, *88*(1), 127–144. doi:10.1016/j.neuron.2015.09.017.

Byrom, N. C., Msetfi, R. M., and Murphy, R. A. (2018). Human latent inhibition: Problems with the stimulus exposure effect. *Psychonomic Bulletin & Review*, *25*(6), 2102–2118. doi:10.3758/s13423-018-1455-4.

Carmel, D., Fairnie, J., and Lavie, N. (2012). Weight and see: Loading working memory improves incidental identification of irrelevant faces. *Frontiers in Psychology*, *3*, 286. doi:10.3389/fpsyg.2012.00286.

Carrasco, M. (2011). Visual attention: The past 25 years. *Vision Research*, *51*(13), 1484–1525. doi:10.1016/j.visres.2011.04.012.

Carson, S. H. (2011). Creativity and psychopathology: A shared vulnerability model. *The Canadian Journal of Psychiatry*, *56*(3), 144–153. doi:10.1177/070674371105600304.

Carson, S. H. (2014). *Leveraging the "Mad Genius" Debate: Why we Need a Neuroscience of Creativity and Psychopathology*. Harvard: Harvard University Press. https://dash.harvard.edu/handle/1/13347498.

Carson, S. H., Peterson, J., and Higgins, D. M. (2003). Decreased latent inhibition is associated with increased creative achievement in high-functioning individuals. *Journal of Personality and Social Psychology*, *85*(1), 499–506. doi:10.1037/0022-3514.85.3.499.

Castellanos, F. X. and Proal, E. (2012). Large-scale brain systems in ADHD: Beyond the prefrontal–striatal model. *Trends in Cognitive Sciences*, *16*(1), 17–26. doi:10.1016/j.tics.2011.11.007.

Cavanna, A. E. and Trimble, M. R. (2006). The precuneus: A review of its functional anatomy and behavioural correlates. *Brain*, *129*(3), 564–583. doi:10.1093/brain/awl004.

Chand, G. B. and Dhamala, M. (2016). Interactions among the brain default-mode, salience, and central-executive networks during perceptual decision-making of moving dots. *Brain Connectivity; New Rochelle*, *6*(3), 249–254. doi:10.1089/brain.2015.0379.

Chrysikou, E. G. and Thompson-Schill, S. L. (2011). Dissociable brain states linked to common and creative object use. *Human Brain Mapping*, *32*(4), 665–675. doi:10.1002/hbm.21056.

Coch, D., Sanders, L. D., and Neville, H. J. (2005). An event-related potential study of selective auditory attention in children and adults. *Journal of Cognitive Neuroscience*, *17*(4), 605–622. doi:10.1162/0898929053467631.

Colombo, J. (2001). The development of visual attention in infancy. *Annual Review of Psychology; Palo Alto*, *52*, 337–367. doi:10.1146/annurev.psych.52.1.337.

Corbetta, M., Patel, G., and Shulman, G. L. (2008). The reorienting system of the human brain: From environment to theory of mind. *Neuron*, *58*(3), 306–324. doi:10.1016/j.neuron.2008.04.017.

Corbetta, M. and Shulman, G. L. (2002). Control of goal-directed and stimulus-driven attention in the brain. *Nature Reviews Neuroscience*, *3*(3), 201–215. doi:10.1038/nrn755.

Cortese, S., Ferrin, M., Brandeis, D., Buitelaar, J., Daley, D., Dittmann, R. W., Holtmann, M., Santosh, P., Stevenson, J., Stringaris, A., Zuddas, A., and Sonuga-Barke, E. J. S. (2015). Cognitive training for attention-deficit/hyperactivity disorder: Meta-Analysis of clinical and neuropsychological outcomes from randomized controlled trials. *Journal of the American Academy of Child and Adolescent Psychiatry*, *54*(3), 164–174. doi:10.1016/j.jaac.2014.12.010.

Cramond, B. (1994). Attention-deficit hyperactivity disorder and creativity—what is the connection? *The Journal of Creative Behavior, 28*(3), 193–210. doi:10.1002/j.2162-6057.1994.tb01191.x.

Crespi, B., Summers, K., and Dorus S. (2007). Adaptive evolution of genes underlying schizophrenia. *Proceedings: Biological Sciences, 274*(1627), 2801–2810. doi:10.1098/rspb.2007.0876.

Cropley, A. (2006). In praise of convergent thinking. *Creativity Research Journal, 18*(3), 391–404. doi:10.1207/s15326934crj1803_13.

Crow, T. J. (2008). The "Big Bang" theory of the origin of psychosis and the faculty of language. *Schizophrenia Research, 102*(1–3), 31–52. doi:10.1016/j.schres.2008.03.010.

Curtindale, L., Laurie-Rose, C., Bennett-Murphy, L., and Hull, S. (2007). Sensory modality, temperament, and the development of sustained attention: A vigilance study in children and adults. *Developmental Psychology, 43*(3), 576–589. doi:10.1037/0012-1649.43.3.576.

Davidson, M. C., Amso, D., Anderson, L. C., and Diamond, A. (2006). Development of cognitive control and executive functions from 4 to 13 years: Evidence from manipulations of memory, inhibition, and task switching. *Neuropsychologia, 44*(11), 2037–2078. doi:10.1016/j.neuropsychologia.2006.02.006.

de Haan, M. (2013). *Infant EEG and Event-Related Potentials*. London: Psychology Press. doi:10.4324/9780203759660.

de Manzano, Ö., Cervenka, S., Karabanov, A., Farde, L., and Ullén, F. (2010). Thinking outside a less intact box: Thalamic dopamine D2 receptor sensities are negatively related to psychometric creativity in healthy individuals. *PLoS ONE, 5*(5), e10670. doi:10.1371/journal.pone.0010670.

Delis, D. C., Lansing, A., Houston, W. S., Wetter, S., Han, S. D., Jacobson, M., Holdnack, J., and Kramer, J. (2007). Creativity lost—the importance of testing higher-level executive functions in school-age children and adolescents. *Journal of Psychoeducational Assessment, 25*(1), 29–40. doi:10.1177/0734282906292403.

Deng, W. (Sophia) and Sloutsky, V. M. (2016). Selective attention, diffused attention, and the development of categorization. *Cognitive Psychology, 91*, 24–62. doi:10.1016/j.cogpsych.2016.09.002.

Department for Digital, Culture, Media and Sport. (2017). *Creative Industries' Record Contribution to UK Economy*. London: Department for Digital, Culture, Media and Sport. https://www.gov.uk/government/news/creative-industries-record-contribution-to-uk-economy.

Desimone, R. and Duncan, J. (1995). Neural mechanisms of selective visual attention. *Annual Review of Neuroscience, 18*(1), 193–222. doi:10.1146/annurev.ne.18.030195.001205.

DeYoung, C. G., Grazioplene, R. G., and Peterson, J. B. (2012). From madness to genius: The openness/intellect trait domain as a paradoxical simplex. *Journal of Research in Personality, 46*(1), 63–78. doi:10.1016/j.jrp.2011.12.003.

Dietrich, A. and Kanso, R. (2010). A review of EEG, ERP, and neuroimaging studies of creativity and insight. *Psychological Bulletin, 136*(5), 822–848. doi:10.1037/a0019749.

Drake, J. E. and Winner, E. (2009). Precocious realists: perceptual and cognitive characteristics associated with drawing talent in non-autistic children. *Philosophical Transactions of the Royal Society B, 364*, 1149–1158. doi:10.1098/rstb.2008.0295

Dramsdahl, M., Westerhausen, R., Haavik, J., Hugdahl, K., and Plessen, K. J. (2011). Cognitive control in adults with attention-deficit/hyperactivity disorder. *Psychiatry Research, 188*(3), 406–410. doi:10.1016/j.psychres.2011.04.014.

Driver, J. (2001). A selective review of selective attention research from the past century. *British Journal of Psychology*, 92(1), 53–78. doi:10.1348/000712601162103.

Duncker, K. (1945). On problem-solving. *Psychological Monographs*, 58(5), i–113. doi:10.1037/h0093599.

Durham Commission On Creativity and Education. (2019). https://www.dur.ac.uk/resources/creativitycommission/DurhamReport.pdf.

Durston, S., Thomas, K. M., Yang, Y., Uluğ, A. M., Zimmerman, R. D., and Casey, B. J. (2002). A neural basis for the development of inhibitory control. *Developmental Science*, 5(4), F9. doi:10.1111/1467-7687.00235.

Ellamil, M., Dobson, C., Beeman, M., and Christoff, K. (2012). Evaluative and generative modes of thought during the creative process. *NeuroImage*, 59(2), 1783–1794. doi:10.1016/j.neuroimage.2011.08.008.

Enns, J. T. and Akhtar, N. (1989). A developmental study of filtering in visual attention. *Child Development*, 60(5), 1188. doi:10.2307/1130792.

Enns, J. T. and Girgus, J. S. (1985). Developmental changes in selective and integrative visual attention. *Journal of Experimental Child Psychology*, 40(2), 319–337. doi:10.1016/0022-0965(85)90093-1.

Evans, J. St. B. T. (2008). Dual-process accounts of reasoning, judgment and social cognition. *Annual Review of Psychology*, 59, 255–278. doi:10.1146/annurev.psych.59.103006.093629.

Evans, L. H., Gray, N. S., and Snowden, R. J. (2007). A new continuous within-participants latent inhibition task: Examining associations with schizotypy dimensions, smoking status and gender. *Biological Psychology*, 74(3), 365–373. doi:10.1016/j.biopsycho.2006.09.007.

Eysenck, H. J. (1993). Creativity and personality: Suggestions for a theory. *Psychological Inquiry*, 4(3), 147. doi:10.1207/s15327965pli0403_1.

Fair, D. A., Cohen, A. L., Dosenbach, N. U. F., Church, J. A., Miezin, F. M., Barch, D. M., Raichle, M. E., Petersen, S. E., and Schlaggar, B. L. (2008). The maturing architecture of the brain's default network. *Proceedings of the National Academy of Sciences*, 105(10), 4028–4032. doi:10.1073/pnas.0800376105.

Fair, D. A., Dosenbach, N. U. F., Church, J. A., Cohen, A. L., Brahmbhatt, S., Miezin, F. M., Barch, D. M., Raichle, M. E., Petersen, S. E., and Schlaggar, B. L. (2007). Development of distinct control networks through segregation and integration. *Proceedings of the National Academy of Sciences*, 104(33), 13507–13512. doi:10.1073/pnas.0705843104.

Fassbender, C., Zhang, H., Buzy, W. M., Cortes, C. R., Mizuiri, D., Beckett, L., and Schweitzer, J. B. (2009). A lack of default network suppression is linked to increased distractibility in ADHD. *Brain Research*, 1273, 114–128. doi:10.1016/j.brainres.2009.02.070.

Feist, G. J. (2006). How development and personality influence scientific thought, interest, and achievement. *Review of General Psychology*, 10(2), 163–182. doi:10.1037/1089-2680.10.2.163.

Fink, A. and Benedek, M. (2014). EEG alpha power and creative ideation. *Neuroscience & Biobehavioral Reviews*, 44, 111–23. doi:10.1016/j.neubiorev.2012.12.002.

Fink, A., Weber, B., Koschutnig, K., Benedek, M., Reishofer, G., Ebner, F., Papousek, I., and Weiss, E. M. (2014). Creativity and schizotypy from the neuroscience perspective. *Cognitive, Affective, & Behavioral Neuroscience*, 14(1), 378–387. doi:10.3758/s13415-013-0210-6.

Finke, R. A., Ward, T. B., and Smith, S. M. (1992). *Creative Cognition: Theory, Research and Applications*. Cambridge, MA: MIT Press.

Finke, R. A. (1996). Imagery, creativity, and emergent structure. *Consciousness and Cognition*, 5(3), 381–393. doi:10.1006/ccog.1996.0024.

Fox, M. D., Zhang, D., Snyder, A. Z., and Raichle, M. E. (2009). The global signal and observed anticorrelated resting state brain networks. *Journal of Neurophysiology*, 101(6), 3270–3283. doi:10.1152/jn.90777.2008.

Frey, C. B. and Osborne, M. A. (2013). *The Future of Employment: How Susceptible are Jobs to Computerisation?* Oxford: Oxford Martin School, University of Oxford. http://www.oxfordmartin.ox.ac.uk/publications/view/1314.

Fugate, C. M., Zentall, S. S., and Gentry, M. (2013). Creativity and working memory in gifted students with and without characteristics of attention deficit hyperactive disorder: Lifting the mask. *Gifted Child Quarterly*, 57(4), 234–246. doi:10.1177/0016986213500069.

Gabora, L. (2002). Amplifying phenomenal information: Toward a fundamental theory of consciousness. *Journal of Consciousness Studies*, 9(8), 3–29.

Gabora, L. (2010). Revenge of the "neurds": Characterizing creative thought in terms of the structure and dynamics of memory. *Creativity Research Journal*, 22(1), 1–13. doi:10.1080/10400410903579494.

Gamble, K. R. and Kellner, H. (1968). Creative functioning and cognitive regression. *Journal of Personality and Social Psychology*, 9(3), 266–271. doi:10.1037/h0025911.

Gilhooly, K. J., Fioratou, E., Anthony, S. H., and Wynn, V. (2007). Divergent thinking: Strategies and executive involvement in generating novel uses for familiar objects. *British Journal of Psychology*, 98(4), 611–625. doi:10.1111/j.2044-8295.2007.tb00467.x.

Golden, C. J. (1975). The measurement of creativity by the Stroop color and word test. *Journal of Personality Assessment*, 39(5), 502–506. doi:10.1207/s15327752jpa3905_9.

Gonzalez-Carpio, G., Serrano, J. P., and Nieto, M. (2017). Creativity in children with attention deficit hyperactivity disorder (ADHD). *Psychology*, 08(03), 319–334. doi:10.4236/psych.2017.83019.

Green, M. J, and Williams, L. M. (1999). Schizotypy and creativity as effects of reduced cognitive inhibition. *Personality and Individual Differences*, 27(2), 263–275. 10.1016/s0191-8869(98)00238-4.

Groborz, M. and Necka, E. (2003). Creativity and cognitive control: Explorations of generation and evaluation skills. *Creativity Research Journal*, 15(2 & 3), 183–197. 10.1080/10400419.2003.9651411.

Guilford, J. P. (1950). Creativity. *American Psychologist*, 5(9), 444–454. doi:10.1037/h0063487.

Guilford, J. P., Christensen, P., Merrifield, P., and Wilson, R. (1978). *Alternate Uses: Manual of Instructions and Interpretations*. London: Sheridan Psychological Services.

Hanania, R. and Smith, L. B. (2010). Selective attention and attention switching: Towards a unified developmental approach. *Developmental Science*, 13(4), 622–635. doi:10.1111/j.1467-7687.2009.00921.x.

Hanauer, J. B. and Brooks, P. J. (2003). Developmental change in the cross-modal Stroop effect. *Perception & Psychophysics*, 65(3), 359–366. doi:10.3758/bf03194567.

Healey, D. and Rucklidge, J. J. (2005). An exploration into the creative abilities of children with ADHD. *Journal of Attention Disorders*, 8(3), 88–95. doi:10.1177/1087054705277198.

Healey, D. and Rucklidge, J. J. (2006). An investigation into the relationship among ADHD symptomatology, creativity, and neuropsychological functioning in children. *Child Neuropsychology*, 12(6), 421–438. doi:10.1080/09297040600806086.

Holahan, C., K., Sears, R. R., and Cronbach, L. J. (1995). *The Gifted Group in Later Maturity*. Stanford: Stanford University Press.

Howard-Jones, P. A. and Murray, S. (2003). Ideational productivity, focus of attention, and context. *Creativity Research Journal*, 15(2–3), 153166.

Huang-Pollock, C. L., Carr, T. H., and Nigg, J. T. (2002). Development of selective attention: Perceptual load influences early versus late attentional selection in children and adults. *Developmental Psychology, 38*(3), 363–375. doi:10.1037/0012-1649.38.3.363.

Huizinga, M., Dolan, C. V., and van der Molen, M. W. (2006). Age-related change in executive function: Developmental trends and a latent variable analysis. *Neuropsychologia, 44*(11), 2017–036. doi:10.1016/j.neuropsychologia.2006.01.010.

Johnes, R. (2017). *Entries to Arts Subjects at Key Stage 4.* London: Education Policy Institute. https://epi.org.uk/publications-and-research/entries-arts-subjects-key-stage-4/

Johnson, M. H., Posner, M. I., and Rothbart, M. K. (1991). Components of visual orienting in early infancy: Contingency learning, anticipatory looking, and disengaging. *Journal of Cognitive Neuroscience, 3*(4), 335–344. doi:10.1162/jocn.1991.3.4.335.

Jung, R. E. and Haier, R. J. (2007). The parieto-frontal integration theory (P-FIT) of intelligence: Converging neuroimaging evidence. *Behavioral and Brain Sciences, 30*(2), 135–154. doi:10.1017/s0140525x07001185.

Jung, R. E., Mead, B. S., Carrasco, J., and Flores, R. A. (2013). The structure of creative cognition in the human brain. *Frontiers in Human Neuroscience, 7*, 330. doi:10.3389/fnhum.2013.00330.

Kaniel, S. and Lubow, R. E. (1986). Latent inhibition—a developmental study. *British Journal of Developmental Psychology, 4*, 367–375.

Kasof, J. (1997). Creativity and breadth of attention. *Creativity Research Journal, 10*(4), 303–15. doi:10.1207/s15326934crj1004_2.

Kaufman, J. C. (ed.). (2014). *Creativity and Mental Illness.* Cambridge UK: Cambridge University Press.

Kaufman, J. C. and Beghetto, R. A. (2009). Beyond big and little: The four C model of creativity. *Review of General Psychology, 13*, 1–12. doi:10.1037/a0013688

Kaufman, J. C., Plucker, J. A., and Baer, J. (2008). *Essentials of Creativity Assessment.* Hoboken, NJ: John Wiley and Sons.

Kaufman, S. B. (2013). *Ungifted: Intelligence Redefined.* UK: Hachette.

Kaufman, S. B., DeYoung, C. G., Gray, J. R., Jiménez, L., Brown, J., and Mackintosh, N. (2010). Implicit learning as an ability. *Cognition, 116*(3), 321–340. doi:10.1016/j.cognition.2010.05.011.

Kéri, S. (2009). Genes for psychosis and creativity: A promoter polymorphism of the Neuregulin 1 gene is related to creativity in people with high intellectual achievement. *Psychological Science, 20*(9), 1070–1073. doi:10.1111/j.1467-9280.2009.02398.x.

Kim, K. H. (2011). The creativity crisis: The decrease in creative thinking scores on the Torrance tests of creative thinking. *Creativity Research Journal, 23*(4), 285–295. doi:10.1080/10400419.2011.627805.

Klingberg, T., Fernell, E., Olesen, P. J., Johnson, M., Gustafsson, P., Dahlström, K., Gillberg, C. G., Forssberg, H., and Westerberg, H. (2005). Computerized training of working memory in children with ADHD: A randomized, controlled trial. *Journal of the American Academy of Child & Adolescent Psychiatry, 44*(2), 177–186. doi:10.1097/00004583-200502000-00010.

Koechlin, E. (2016). Prefrontal executive function and adaptive behavior in complex environments. *Current Opinion in Neurobiology, 37*, 1–6. doi:10.1016/j.conb.2015.11.004.

Kopp, B., Rist, F., and Mattler, U. (1996). N200 in the flanker task as a neurobehavioral tool for investigating executive control. *Psychophysiology, 33*(3), 282–294. doi:10.1111/j.1469-8986.1996.tb00425.x.

Kozbelt, A., Kaufman, S. B., Walder, D. J., Ospina, L. H., and Kim, J. U. (2014). The evolutionary genetics of the creativity–psychosis connection. In J. C. Kaufman (ed.), *Creativity and Mental Illness* (pp. 102–132). Cambridge UK: Cambridge University Press. doi:10.1017/CBO9781139128902.009.

Kyaga, S., Landén, M., Boman, M., Hultman, C. M., Långström, N., and Lichtenstein, P. (2013). Mental illness, suicide and creativity: 40-Year prospective total population study. *Journal of Psychiatric Research*, 47(1), 83–90. doi:10.1016/j.jpsychires.2012.09.010.

Kyaga, S., Lichtenstein, P., Boman, M., Hultman, C., Långström, N., and Landén, M. (2011). Creativity and mental disorder: Family study of 300 000 people with severe mental disorder. *British Journal of Psychiatry*, 199(05), 373–379. doi:10.1192/bjp.bp.110.085316.

Land, G. and Jarman, B. (1992). *Breakpoint and Beyond: Mastering the Future Today.* Manhattan, NY: Harper Business.

Lin, L., Osan, R. and Tsien, J. Z. (2006). Organizing principles of real-time memory encoding: Neural clique assemblies and universal neural codes. *Trends in Neurosciences*, 29(1), 48–57. doi:10.1016/j.tins.2005.11.004.

Liu, S., Erkkinen, M. G., Healey, M. L., Xu, Y., Swett, K. E., Chow, H. M., and Braun, A. R. (2015). Brain activity and connectivity during poetry composition: Toward a multidimensional model of the creative process. *Human Brain Mapping*, 36(9), 3351–3372. doi:10.1002/hbm.22849.

Lo, M. T., Hinds, D. A., Tung, J. Y., Franz, C., Fan, C. C., Wang, Y., ... and Chen, C H. (2017). Genome-wide analyses for personality traits identify six genomic loci and show correlations with psychiatric disorders. *Nature Genetics*, 49(1), 152–156. doi:10.1038/ng.3736.

Lubart, T., Zenasni, F. and Barbot, B. (2013). Creative potential and its measurement. *International Journal for Talent Development and Creativity*, 1(2), 41–50.

Lubow, R. E. (1973). Latent inhibition. *Psychological Bulletin*, 79(6), 398–407. doi:10.1037/h0034425.

Lubow, R. E. and Gewirtz, J. C. (1995). Latent inhibition in humans: Data, theory, and implications for schizophrenia. *Psychological Bulletin*, 117(1), 87–103. doi:10.1037/0033-2909.117.1.87.

Lubow, R. E., Toren, P., and Kaplan, O. (2000). The effects of target and distractor familiarity on visual search in anxious children: Latent inhibition and novel pop-out. *Journal of Anxiety Disorders*, 14(1), 41–56. doi:10.1016/S0887-6185(99)00038-9

Lucas, B. and Spencer, E. (2017). *Teaching Creative Thinking: Developing Learners Who Generate Ideas and Can Think Critically.* Carmarthen, UK: Crown House Publishing Ltd.

MacLeod, C. M. (1991). Half a century of research on the Stroop effect: An integrative review. *Psychological Bulletin*, 109(2), 163–203. doi:10.1037/0033-2909.109.2.163.

Manyika, J., Lund, S., Chui, M., Bughin, J., Woetzel, J., Batra, P., Ko, R. and Sanghvi, S. (2017). *Jobs Lost, Jobs Gained: Workforce Transitions in a Time of Automation.* San Francisco, CA: McKinsey Global Institute. https://www.mckinsey.com/featured-insights/future-of-work/jobs-lost-jobs-gained-what-the-future-of-work-will-mean-for-jobs-skills-and-wages#

Martindale, C. (1998). Biological bases of creativity. In R. J. Sternberg (ed.), *Handbook of Creativity* (pp. 137–152). Cambridge UK: Cambridge University Press; Cambridge Core. doi:10.1017/CBO9780511807916.009.

Martindale, C. (2007). Creativity, primordial cognition, and personality. *Personality and Individual Differences*, 43(7), 1777–1785. doi:10.1016/j.paid.2007.05.014.

Martindale, C., Anderson, K., Moore, K., and West, A. N. (1996). Creativity, oversensitivity, and rate of habituation. *Personality and Individual Differences*, 20(4), 423–427. doi:10.1016/0191-8869(95)00193-X.

Martindale, C., Smith, S. M., Ward, T. B., and Finke, R. A. (1995). Creativity and connectionism. In *The Creative Cognition Approach* (pp. 249–268). Cambridge, MA: MIT Press.

Matheson, S. and Langdon, R. (2008). Schizotypal traits impact upon executive working memory and aspects of IQ. *Psychiatry Research*, 159(1–2), 207–214. doi:10.1016/j.psychres.2007.04.006.

McLennan, J. D. (2016). Understanding attention deficit hyperactivity disorder as a continuum. *Canadian Family Physician*, 62(12), 979–972. doi: 10.3810/pgm.2010.09.2206.

Mednick, S. (1962). The associative basis of the creative process. *Psychological Review*, 69(3), 220–32. doi:10.1037/h0048850.

Memmert, D. (2009). Noticing unexpected objects improves the creation of creative solutions—inattentional blindness by children influences divergent thinking negatively. *Creativity Research Journal*, 21(2–3), 302–304. doi:10.1080/10400410802633798.

Mendelsohn, G. A. and Griswold, B. B. (1964). Differential use of incidental stimuli in problem solving as a function of creativity. *The Journal of Abnormal and Social Psychology*, 68(4), 431–436. doi:10.1037/h0040166.

Menon, V. and Uddin, L. Q. (2010). Saliency, switching, attention and control: A network model of insula function. *Brain Structure and Function*, 214(5), 655–667. doi:10.1007/s00429-010-0262-0.

Miller, E. K. and Cohen, J. D. (2001). An integrative theory of prefrontal cortex function. *Annual Review of Neuroscience*, 24(1), 167. doi:10.1146/annurev.neuro.24.1.167.

Mok, L. W. (2014). The interplay between spontaneous and controlled processing in creative cognition. *Frontiers in Human Neuroscience*, 8, 663. doi:10.3389/fnhum.2014.00663.

Mumford, M. D. and Gustafson, S. B. (1988). Creativity syndrome: Integration, application, and innovation. *Psychological Bulletin*, 103(1), 27–43. doi:10.1037/0033-2909.103.1.27.

Munafò, M. R., Yalcin, B., Willis-Owen, S. A., and Flint, J. (2008). Association of the dopamine D4 receptor (DRD4) gene and approach-related personality traits: Meta-analysis and new data. *Biological Psychiatry*, 63(2), 197–206. doi:10.1016/j.biopsych.2007.04.006.

Neubauer, A. C. and Fink, A. (2009). Intelligence and neural efficiency: Measures of brain activation versus measures of functional connectivity in the brain. *Intelligence*, 37(2), 223–229. doi:10.1016/j.intell.2008.10.008.

Nigg, J. T. (2000). On inhibition/disinhibition in developmental psychopathology: Views from cognitive and personality psychology and a working inhibition taxonomy. *Psychological Bulletin*, 126(2), 220–246. 10.1037/0033-2909.126.2.220.

Nijstad, B. A., Dreu, C. K. W. D., Rietzschel, E. F., and Baas, M. (2010). The dual pathway to creativity model: Creative ideation as a function of flexibility and persistence. *European Review of Social Psychology*, 21(1), 34–77. doi:10.1080/10463281003765323.

Noguchi, H., Hori, H., and Kunugi, H. (2008). Schizotypal traits and cognitive function in healthy adults. *Psychiatry Research*, 161(2), 162–169. doi:10.1016/j.psychres.2007.07.023.

Nusbaum, E. C. and Silvia, P. J. (2011). Are intelligence and creativity really so different?: Fluid intelligence, executive processes, and strategy use in divergent thinking. *Intelligence*, 39(1), 36–45. doi:10.1016/j.intell.2010.11.002.

Paek, S. H., Abdulla, A. M., and Cramond, B. (2016). A meta-analysis of the relationship between three common psychopathologies—ADHD, anxiety, and depression—and indicators of little-c creativity. *Gifted Child Quarterly*, 60(2), 117–133. doi:10.1177/0016986216630600.

Petersen, S. E. and Posner, M. I. (2012). The attention system of the human brain: 20 years after. *Annual Review of Neuroscience*, 35, 73–89. doi:10.1146/annurev-neuro-062111-150525.

Peterson, J. B. and Carson, S. H. (2000). Latent inhibition and openness to experience in a high-achieving student population. *Personality and Individual Differences*, 28(2), 323–332. doi:10.1146/annurev-neuro-062111-150525.

Piffer, D. and Hur, Y.-M. (2014). Heritability of creative achievement. *Creativity Research Journal*, 26(2), 151–157. doi:10.1080/10400419.2014.901068.

Plebanek, D. J. and Sloutsky, V. M. (2017). Costs of selective attention: When children notice what adults miss. *Psychological Science*, 28(6), 723–732. doi:10.1177/0956797617693005.

Plucker, J. A., Beghetto, R. A., and Dow, G. T. (2004). Why isn't creativity more important to educational psychologists? Potentials, pitfalls, and future directions in creativity research. *Educational Psychologist*, 39, 83–96. doi:10.1207/s15326985ep3902_1.

Plucker, J. A. and Makel, M. C. (2010). Assessment of creativity. In *The Cambridge Handbook of Creativity* (pp. 48–73). Cambridge, UK: Cambridge University Press. doi:10.1017/CBO9780511763205.005.

Plude, D. J., Enns, J. T., and Brodeur, D. (1994). The development of selective attention: A lifespan overview. *Acta Psychologica*, 86(2–3), 227–272. doi:10.1016/0001-6918(94)90004-3.

Posner, M. I. and Petersen, S. E. (1990). The attention system of the human brain. *Annual Review of Neuroscience*, 13, 25–42. doi:10.1146/annurev.ne.13.030190.000325.

Posner, M. I. and Rothbart, M. K. (2007). Research on attention networks as a model for the integration of psychological science. *Annual Review of Psychology*, 58(1), 1–23. doi:10.1146/annurev.psych.58.110405.085516.

Posner, M. I., Rothbart, M. K., Sheese, B. E., and Voelker, P. (2014). Developing attention: Behavioral and brain mechanisms. *Advances in Neuroscience*, 2014, 1–9. doi:10.1155/2014/405094.

Posner, M. I., Rothbart, M. K., and Thomas-Thrapp, L. (1997). Functions of orienting in early infancy. In *Attention and Orienting: Sensory and Motivational Processes.* (pp. 327–345). NJ, USA: Lawrence Erlbaum Associates Publishers.

Pringle, A. and Sowden, P. T. (2017). Unearthing the creative thinking process: Fresh insights from a think aloud study of garden design. *Psychology of Aesthetics, Creativity & the Arts*, 11, 344–358. doi:10.1037/aca0000144.

Pritchard, V. E., Healey, D., and Neumann, E. (2006). Assessing selective attention in ADHD, highly creative, and normal young children via Stroop negative priming effects. In *Cognition and Language: Perspectives from New Zealand.* (pp. 207–224). Bowen Hills: Australian Academic Press.

Raichle, M. E., MacLeod, A. M., Snyder, A. Z., Powers, W. J., Gusnard, D. A., and Shulman, G. L. (2001). A default mode of brain function. *Proceedings of the National Academy of Sciences*, 98(2), 676–682. doi:10.1073/pnas.98.2.676.

Ramnani, N. and Owen, A. M. (2004). Anterior prefrontal cortex: Insights into function from anatomy and neuroimaging. *Nature Reviews Neuroscience*, 5(3), 184–94. doi:10.1038/nrn1343.

Rawlings, D. (1985). Psychoticism, creativity and dichotic shadowing. *Personality and Individual Differences*, 6(6), 737–742. doi:10.1016/0191-8869(85)90084-4.

Remington, A., Cartwright-Finch, U., and Lavie, N. (2014). I can see clearly now: The effects of age and perceptual load on inattentional blindness. *Frontiers in Human Neuroscience*, 8, 229. doi:10.3389/fnhum.2014.00229.

Richards, R., Kinney, D. K., Benet, M., and Merzel, A. P. (1988). Assessing everyday creativity: Characteristics of the Lifetime Creativity Scales and validation with three large samples. *Journal of Personality and Social Psychology*, 54(3), 476–485. doi:10.1037/0022-3514.54.3.476.

Richards, R., Kinney, D. K., Lunde, I., Benet, M., and Merzel, A. P. C. (1988). Creativity in manic-depressives, cyclothymes, their normal relatives, and control subjects. *Journal of Abnormal Psychology*, 97(3), 281–288. doi:10.1037/0021-843X.97.3.281.

Roberts, P. (2006). *Nurturing creativity in young people. A report for Andrew Adonis Parliamentary Under-Secretary of State for Education and Skills, James Purnell Minister for Creative Industries and Tourism and David Lammy Minister for Culture.* London: Department for Culture, Media and Sport. http://www.creativetallis.com/uploads/2/2/8/7/2287089/nurturing-1.pdf

Robinson, K. (1999). *All our Futures: Creativity, Culture and Education. A Report for the Rt. Hon David Blunkett MP Secretary of State for Education and Employment and for the Rt. Hon Chris Smith MP Secretary of State for Culture, Media and Sport.* London: National Advisory Committee on Creative and Cultural Education. http://sirkenrobinson.com/pdf/allourfutures.pdf

Rueda, M. R., Posner, M. I., and Rothbart, M. K. (2005). The development of executive attention: Contributions to the emergence of self-regulation. *Developmental Neuropsychology*, 28(2), 573–594. doi:10.1207/s15326942dn2802_2.

Rueda, M. R., Posner, M. I., Rothbart, M. K., and Davis-Stober, C. P. (2004). Development of the time course for processing conflict: An event-related potentials study with 4 year olds and adults. *BMC Neuroscience*, 5(1), 39. doi:10.1186/1471-2202-5-39.

Runco, M. A. and Acar, S. (2012). Divergent thinking as an indicator of creative potential. *Creativity Research Journal*, 24(1), 66–75. doi:10.1080/10400419.2012.652929.

Runco, M. A. and Jaeger, G. J. (2012). The standard definition of creativity. *Creativity Research Journal*, 24, 92–96. doi:10.1080/10400419.2012.650092

Schul, R., Townsend, J., and Stiles, J. (2003). The development of attentional orienting during the school-age years. *Developmental Science*, 6(3), 262–272. doi:10.1111/1467-7687.00282.

Shaw, G. A. and Brown, G. (1990). Laterality and creativity concomitants of attention problems. *Developmental Neuropsychology*, 6(1), 39–56. doi:10.1080/87565649009540448.

Shaw, G. A. and Brown, G. (1991). Laterality, implicit memory and attention disorder. *Educational Studies*, 17(1), 15–23. doi:10.1080/0305569910170102.

Silvia, P. J., Winterstein, B. P., Willse, J. T., Barona, C. M., Cram, J. T., Hess, K. I., Martinez, J. L., and Richard, C. A. (2008). Assessing creativity with divergent thinking tasks: Exploring the reliability and validity of new subjective scoring methods. *Psychology of Aesthetics, Creativity, and the Arts*, 2(2), 68–85. doi:10.1037/1931-3896.2.2.68.

Simonton, D. K. (1999). Talent and its development: An emergenic and epigenetic model. *Psychological Review*, 106(3), 435–457. doi:10.1037/0033-295X.106.3.435.

Simonton, D. K. (2004). *Creativity in Science: Chance, Logic, Genius, and Zeitgeist.* Cambridge UK: Cambridge University Press.

Simonton, D. K. (2012). Taking the U.S. Patent Office Criteria seriously: A quantitative three-criterion creativity definition and its implications. *Creativity Research Journal*, 24, 97–106. doi:10.1080/10400419.2012.676974

Slobodin, O., Cassuto, H., and Berger, I. (2015). Age-related changes in distractibility: Developmental trajectory of sustained attention in ADHD. *Journal of Attention Disorders*, 22(14), 1333–1343. doi:10.1177/1087054715575066.

Sonuga-Barke, E. J. S. and Castellanos, F. X. (2007). Spontaneous attentional fluctuations in impaired states and pathological conditions: A neurobiological hypothesis. *Neuroscience & Biobehavioral Reviews*, 31(7), 977–986. doi:10.1016/j.neubiorev.2007.02.005.

Sowden, P. T., Pringle, A., and Gabora, L. (2015). The shifting sands of creative thinking: Connections to dual-process theory. *Thinking & Reasoning*, *21*(1), 40–60. doi:10.1080/13546783.2014.885464.

Takahashi, H., Higuchi, M., and Suhara, T. (2006). The role of extrastriatal dopamine D2 receptors in schizophrenia. *Biological Psychiatry*, *59*(10), 919–928. doi:10.1016/j.biopsych.2006.01.022.

Takeoff in Robotics Will Power the Next Productivity Surge in Manufacturing. (2015). Boston: Boston Consulting Group. https://www.bcg.com/press/10feb2015-robotics-power-productivity-surge-manufacturing.

Takeuchi, H., Taki, Y., Hashizume, H., Sassa, Y., Nagase, T., Nouchi, R., and Kawashima, R. (2011). Failing to deactivate: The association between brain activity during a working memory task and creativity. *NeuroImage*, *55*(2), 681–7. doi:10.1016/j.neuroimage.2010.11.052.

The Future of Jobs: Employment, Skills and Workforce Strategy for the Fourth Industrial Revolution (Global Challenge Insight Report). (2016). Geneva: World Economic Forum. http://www3.weforum.org/docs/WEF_Future_of_Jobs.pdf

Torrance, E. P. (1993). The beyonders in a 30 year longitudinal study of creative achievement. *Roeper Review*, *15*, 131–135. doi:10.1080/02783199309553486.

Torrance, E. P. (2008). *Torrance Tests of Creative Thinking: Revised Manual.* Bensenville, IL: Scholastic Testing Service.

Uddin, L. Q. (2015). Salience processing and insular cortical function and dysfunction. *Nature Reviews Neuroscience*, *16*(1), 55–61. doi:10.1038/nrn3857.

Uddin, L. Q., Kelly, A. M. C., Biswal, B. B., Margulies, D. S., Shehzad, Z., Shaw, D., Ghaffari, M., Rotrosen, J., Adler, L. A., Castellanos, F. X., and Milham, M. P. (2008). Network homogeneity reveals decreased integrity of default-mode network in ADHD. *Journal of Neuroscience Methods*, *169*(1), 249–254. doi:10.1016/j.jneumeth.2007.11.031

Vartanian, O. (2009). Variable attention facilitates creative problem solving. *Psychology of Aesthetics, Creativity, and the Arts*, *3*(1), 57–59. doi:10.1037/a0014781.

Vartanian, O., Martindale, C., and Kwiatkowski, J. (2007). Creative potential, attention, and speed of information processing. *Personality and Individual Differences*, *43*(6), 1470–1480. doi:10.1016/j.paid.2007.04.027.

Vinkhuyzen, A. A. E., van der Sluis, S., Posthuma, D., and Boomsma, D. I. (2009). The heritability of aptitude and exceptional talent across different domains in adolescents and young adults. *Behavior Genetics*, *39*(4), 380–392. doi:10.1007/s10519-009-9260-5.

Vossel, S., Geng, J. J., and Fink, G. R. (2014). Dorsal and ventral attention systems. *The Neuroscientist*, *20*(2), 150–159. doi:10.1177/1073858413494269.

Wainwright, A. and Bryson, S. E. (2005). The development of endogenous orienting: Control over the scope of attention and lateral asymmetries. *Developmental Neuropsychology*, *27*(2), 237–255. doi:10.1207/s15326942dn2702_3.

Wallach, M. A. and Kogan, N. (1965). A new look at the creativity-intelligence distinction. *Journal of Personality*, *33*(3), 348. doi:10.1111/j.1467-6494.1965.tb01391.x.

Wallas, G. (1926). *The Art of Thought.* New York, NY: Harcourt Brace.

Walberg, Herbert J. (1988) Creativity and talent as learning. In Sternberg, R. (ed.), *The Nature of Creativity*. New York: Cambridge University Press. pp. 340–361.

Wang, L., Long, H., Plucker, J. A., Wang, Q., Xu, X., and Pang, W. (2018). High schizotypal individuals are more creative? The mediation roles of overinclusive thinking and cognitive inhibition. *Frontiers in Psychology*, *9*, 1766. doi:10.3389/fpsyg.2018.01766.

Warwick Commission. (2015). *Enriching Britain: Culture, Creativity and Growth*. Warwick, UK: Warwick University.

White, H. A. (2007). Inhibitory control of proactive interference in adults with ADHD. *Journal of Attention Disorders, 11*(2), 141–149. doi:10.1177/1087054706295604.

White, H. A. (2020). Thinking "Outside the Box": Unconstrained creative generation in adults with attention deficit hyperactivity disorder. *The Journal of Creative Behavior, 54*(2), 472–483. doi:10.1002/jocb.382.

White, H. A. and Shah, P. (2011). Creative style and achievement in adults with attention-deficit/hyperactivity disorder. *Personality and Individual Differences, 50*(5), 673–677. doi:10.1016/j.paid.2010.12.015.

Whitfield-Gabrieli, S., Thermenos, H. W., Milanovic, S., Tsuang, M. T., Faraone, S. V., McCarley, R. W., Shenton, M. E., Green, A. I., Nieto-Castanon, A., LaViolette, P., Wojcik, J., Gabrieli, J. D. E., and Seidman, L. J. (2009). Hyperactivity and hyperconnectivity of the default network in schizophrenia and in first-degree relatives of persons with schizophrenia. *Proceedings of the National Academy of Sciences, 106*(4), 1279–1284. doi:10.1073/pnas.0809141106.

Yasuno, F., Suhara, T., Okubo, Y., Sudo, Y., Inoue, M., Ichimiya, T., Takano, A., Nakayama, K., Halldin, C., and Farde, L. (2004). Low dopamine D2 receptor binding in subregions of the thalamus in schizophrenia. *American Journal of Psychiatry, 161*(6), 1016–1022. doi:10.1176/appi.ajp.161.6.1016.

Zabelina, D. L. (2018). Attention and creativity. In Rex Eugene Jung and O. Vartanian (eds.), *The Cambridge Handbook of the Neuroscience of Creativity* (pp. 161–179). Cambridge UK: Cambridge University Press.

Zabelina, D. L. and Beeman, M. (2013). Short-term attentional perseveration associated with real-life creative achievement. *Frontiers in Psychology, 4*, 191. doi:10.3389/fpsyg.2013.00191.

Zabelina, D. L. and Ganis, G. (2018). Creativity and cognitive control: Behavioral and ERP evidence that divergent thinking, but not real-life creative achievement, relates to better cognitive control. *Neuropsychologia, 118*, 20–8. doi:10.1016/j.neuropsychologia.2018.02.014.

Zabelina, D. L., O'Leary, D., Pornpattananangkul, N., Nusslock, R., and Beeman, M. (2015). Creativity and sensory gating indexed by the P50: Selective versus leaky sensory gating in divergent thinkers and creative achievers. *Neuropsychologia, 69*, 77–84. doi:10.1016/j.neuropsychologia.2015.01.034.

Zabelina, D. L. and Robinson, M. D. (2010). Creativity as flexible cognitive control. *Psychology of Aesthetics, Creativity, and the Arts, 4*(3), 136–143. doi:10.1037/a0017379.

Zabelina, D. L., Saporta, A., and Beeman, M. (2016). Flexible or leaky attention in creative people? Distinct patterns of attention for different types of creative thinking. *Memory & Cognition, 44*(3), 488–498. doi:10.3758/s13421-015-0569-4.

Zelazo, P. D., Müller, U., Frye, D., and Marcovitch, S. (2003). The development of executive function. *Monographs of the Society for Research in Child Development, 68*(3), 1–27. doi:10.1111/j.0037-976x.2003.00261.x.

CHAPTER 19

TRAINING COGNITION WITH VIDEO GAMES

PEDRO CARDOSO-LEITE, MORTEZA ANSARINIA,
EMMANUEL SCHMÜCK, AND DAPHNE BAVELIER

INTRODUCTION

ACROSS all ages, cognitive abilities play an important role in our quality of life and the type of life we lead. At young ages, executive functions, a cornerstone of cognitive abilities, are thought to determine educational achievement (e.g., Diamond, 2013; Bull, Espy, and Wiebe, 2008; Geary, Berch, and Mann Koepke, 2019; Goldin et al., 2014) and more generally to be "critical for success in school and life" (Diamond, Barnett, Thomas, and Munro, 2007). Longitudinal studies in young children, for example, report that cognitive abilities predict educational achievement attained 2 years later (Bull et al., 2008; Gathercole, Pickering, Knight, and Stegmann, 2004). Among executive functions, attentional control abilities have been of special interest as they mediate a various array of skills, from sustained attention in school to divided attention in team sports. In older adults, for example, attentional abilities correlate with driving accidents—the shrinkage in a person's useful field of view, which is the spatial extent of their visual field to which they effectively pay attention, is strongly associated with a higher incidence of car accidents prior to the attentional test (Ball, Owsley, Sloane, Roenker, and Bruni, 1993). The central role cognitive abilities play in our lives has led to many attempts to devise behavioral training programs to improve cognition, and in particular executive functions (Katz, Shah, and Meyer, 2018). While cognitive enhancement raises ethical concerns (similar to doping in sports), it also holds the promise for broad societal benefits (Bavelier et al., 2019).

Numerous forms of cognitive training exist; yet, their efficiency and the underlying causal mechanisms remain controversial. This is the case, for example, of interventions attempting to improve fluid intelligence by training executive functions (e.g., Jaeggi,

Buschkuehl, Jonides, and Perrig, 2008; Au et al., 2015; Melby-Lervåg and Hulme, 2013). A key concern in the cognitive training literature is that training of specialized cognitive functions may lead to improvements in only those trained functions (i.e., "near transfer") and may not transfer to a broader range of tasks and situations (i.e., "far transfer"). While the necessary conditions for far transfer remain to be firmly established, variety in the training regime and the trained functions appear to be key factors (for an example in the domain of sports, see Güllich, 2018). An alternative perspective on the plasticity of cognitive abilities is not to focus on targeted interventions designed by researchers, but rather to consider the impact of changes in our environment. The Flynn effect, or the rise of IQ scores through the 20th century, is one such example. With the advent of digital media, our lifestyle and cognitive activities—starting at the youngest ages—have radically changed over the past decades. For example, it has been argued that the excessive consumption of multiple media at the same time (e.g., texting while watching TV and browsing the web) may cause an attentional impairment in filtering out distraction (Ophir, Nass, and Wagner, 2009); although more recent data are less clear cut (for reviews, see Wiradhany and Nieuwenstein, 2017; Uncapher and Wagner, 2018). Whether those media-based environmental changes are for the better or for the worse remains highly debated (e.g., Ophir et al., 2009; Bavelier, Green, and Dye, 2010; Sparrow, Liu, and Wegner, 2011). Yet, investigating those effects holds the promise of bringing new insights into human brain plasticity and cognitive training.

Digital media occupy an increasingly large portion of our waking time. In the US, 8–12-year-olds spend close to 6 hours on media each day (Rideout, 2016)—with similar trends being reported all over the world (e.g., Bodson, 2017; Waller, Willemse, Genner, Suter, and Süss, 2016). Digital media affect every aspect of our lives; these effects are complex and not fully understood yet (Bavelier et al., 2010). They depend not only on the specific medium being used but also how they are consumed and what content they deliver (e.g., Cardoso-Leite et al., 2016). Here we limit our scope by focusing on the effects of playing video games on cognition. This choice is motivated by three main points: (i) while the field of media and cognition is quite young, it is already clear that not all media use have the same impact on cognition, implying that different media uses need to be considered separately (as stated earlier, media multitasking may be related to attentional deficits, while playing specific video games can instead be linked to attentional improvements); (ii) video games stand on their own by immersing players in extremely rich and complex experiences with high cognitive demands (a person watching television may spend hours without performing any significant action, while people playing video games may perform multiple meaningful decisions and actions per second); (iii) and finally, the literature concerning the impact of video game play on behavior, including cognition, is arguably one of the best documented today. We will focus only on the relationship between video game play and cognition and will not consider other aspects that might be equally important but are outside the scope of this work, such as the impact of violence, self-image, well-being, creativity, social functioning, or addiction (for reviews on such topics see, Gentile et al., 2017; Király, Tóth, Urbán, Demetrovics, and Maraz, 2017; Stanhope, Owens, and Elliott, 2015).

Almost everyone plays video games now. Although the term video game raises the stereotypical image of the adolescent glued to his screen, there are now as many females, 50 or older, playing video games as there are boys under 18 playing video games. Interestingly, these two groups do not engage with the same genres of video games; older females mostly play puzzle or casual games, while boys play predominantly action-packed, role-playing games. This state of affairs highlights the need to pay close attention to video game genre or the type of experience different video games deliver. In 2015, both "tweens" (8–12 years old) and teens (13–18-year-olds) in the US devoted on average about 1 hour and 20 minutes to playing video games each day; with boys playing substantially more than girls (Rideout and Robb, 2019). The relationship between video game play and cognition has been investigated in various large-scale correlation studies that collect data about children's gaming habits and various measures of interest (Adachi and Willoughby, 2013; Kovess-Masfety et al., 2016; Stanhope et al., 2015). One such study, conducted in Europe reported that video game play was associated with enhanced intellectual, social and academic functioning (as rated by the child's teacher; Kovess-Masfety et al., 2016). Another associated gaming in 7–11 years old with faster response speeds, enhanced sustained attention, and academic performance; but only for intermediate amounts of video game play per day (Pujol et al., 2016). A recent study on 3- to 7-year-old children furthermore documents that casual video game play at this young age may increase fluid intelligence (Fikkers, Piotrowski, and Valkenburg, 2019). It thus appears that, at a macro-level, playing video games *in general* might have beneficial effects on cognition and educational achievement. However, in these studies, researchers typically don't evaluate the effects associated with *specific* genres of video games. Thus, the previous macro-level effects actually represent an average over numerous micro-level effects induced for example by playing different genres of video games. Some of these micro-level effects may be negative and others positive.

The purpose of this chapter is to review the scientific evidence regarding the relationship between playing commercially available video games, as assessed behaviorally or through brain imaging and their potential impact on human cognitive abilities. While many unknowns remain on this topic, it appears clear today that among the many factors to consider is the specific genre of video game being played (e.g., Sala, Tatlidil, and Gobet, 2018; Bediou et al., 2018; Powers and Brooks, 2014; Powers, Brooks, Aldrich, Palladino, and Alfieri, 2013; Toril, Reales, and Ballesteros, 2014; Wang et al., 2016). Following this work, we will review our current understanding of the impact of video games, first in general, and then narrowing in on the specific game genre that appears most effective to improve cognition.

WHICH VIDEO GAMES IMPROVE COGNITION?

Video games come in many different flavors; classifying video games in genres has proven elusive and there is no consensual taxonomy to date. Fifteen years ago when the research on the cognitive effects of video games gained significant traction, researchers seemed to commonly classify video games in a small set of video game genres (see Table 19.1). Since then, video games, video gamers, and gaming have changed considerably, and it seems that the video game classifications that have been used in this field are not adequate to characterize contemporary gaming (Dale, Joessel, Bavelier, and Green, 2020; Dale and Green, 2017). This being said, because the current review focuses on past research that tended to use older games, and to keep in line with the cited literature, we will use the game genres as described in Table 19.1. Note that in this literature "action video games" has been used to refer to first- and third-person shooters, although some authors have made it more inclusive. In this review, "action video games" will strictly refer to first- and third-person shooters.

A wide range of commercial video games have been used in psychological research to evaluate their relationship to or impact on cognition (Bediou et al., 2018; Sala et al., 2018). Video game research has proceeded using a variety of study designs, including cross-sectional and intervention studies ("true experiments"). Among the latter, we find studies looking at short-lived effects on the scale of a few minutes and intervention studies looking at more long-lasting effects, from days, to months, or even years (see Figure 19.1 for the design of such intervention studies). True experiments are necessary to rule out the possibility that the observed group differences pre-date the video gaming activities, and thus assess the causal role of video game play.

An important point to keep in mind in this literature is that all video games are not created equal as to their impact on cognition. Specific genres of video games have been shown to be effective in improving some aspects of cognition while others haven't. Studies that lump together all types of video game play are therefore at risk of blurring existing effects; for this reason, a number of studies adopt a more principled approach and focus on specific genres of video games. A recent meta-analysis evaluated the impact of playing video games on cognition using a rather broad view of what counts as a video game (Sala et al., 2018). Figures 19.2 and 19.3 use data from that meta-analysis and list the video games (or other activities) and their frequency of use in intervention studies aiming to enhance cognitive abilities. Figure 19.2 lists the games that were used for the experimental group, while Figure 19.3 lists games and activities that were used in active control groups. Several points are worth noting here. First, experimental and control activities vary widely. This variety makes it difficult to regroup these studies under one common research question as they each test different hypotheses. For example, when contrasting playing *Unreal Tournament* (FPS) vs. *Tetris* (Puzzle), one asks about the cognitive impact of action, first-person shooter games as compared to other games that also load highly on speed of processing and motor control; yet when contrasting playing *Tetris* (Puzzle) vs. *The Sims* (Life-Simulation Game), one rather asks about the possibility of training mental rotation by contrasting a game that requires such processes and one that does not. Second, many of the activities listed are in fact not video games (e.g., paper-and-pencil games, watching videos). When contrasting, for example, playing a specific video game to playing paper-and-pencil games it is unclear if such studies evaluate the effectiveness of a specific video game,

Table 19.1 Main "classical" video game categories cited in the reviewed literature. These categories are based on the Video Game Questionnaire from the Bavelier lab. We provide in the supplemental materials the current version of the video game questionnaire and the selection criteria used in the Bavelier laboratory (version September 2019). The game categories it lists are motivated by research considerations and not by industry classifications. Yet, examples of games and our labels for game categories have evolved over the years in concert with the changing landscape of video games.

Category	Description	Examples
First and Third-person Shooters	game involving medium to long range weapon-based combat in first/third person perspective, against other players or AI characters.	*Call of Duty, Overwatch, Unreal, Counterstrike*
Real Time Strategy/ Multiplayer Online Battle Arena	game in which the player manoeuver units to take control of the map and/or destroy enemy assets, usually in top-down view.	*StarCraft, League of Legends, Age of Empire, Rise of Nations*
Action Role Playing Game/Adventure Game	game involving varied action gameplay (e.g., shooting, close-combat, driving vehicles) in which the player controls a character that can be customized during the course of the game.	*Uncharted, Mass Effect, Skyrim, Rise of The Tomb Raider*
Sports or Driving Games	game that simulates real-life sports or driving a vehicle in the context of a competition.	*Need for Speed, Mario Kart NBA 2K12*
Non-Action Turn-based Role Playing or Fantasy Games	game in which the player controls a character or party of characters that can be customized during the course of the game. Combat emphasizes decision making over rapid actions (i.e., turn-based or cooldown-based actions).	*World of Warcraft, Final Fantasy, Ultima, Pokemon*
Turn-based Simulation, Strategy or Puzzle Games	turn based game centered around player decisions rather actions, involving strategic thinking and problem solving.	*Solitaire, Bejeweled, Angry Birds, The Sims, Restaurant Empire, Rollercoaster Tycoon*
Music Games	games centered around the interaction with a musical score, often involving rhythm and memory.	*Audiosurf, OSU!, Guitar Hero*
Other	games that don't fit into any other category, or of unspecified type.	Cognitive training games edutainment games, older 2D arcade games such as *Pac-man or Zaxxon*.

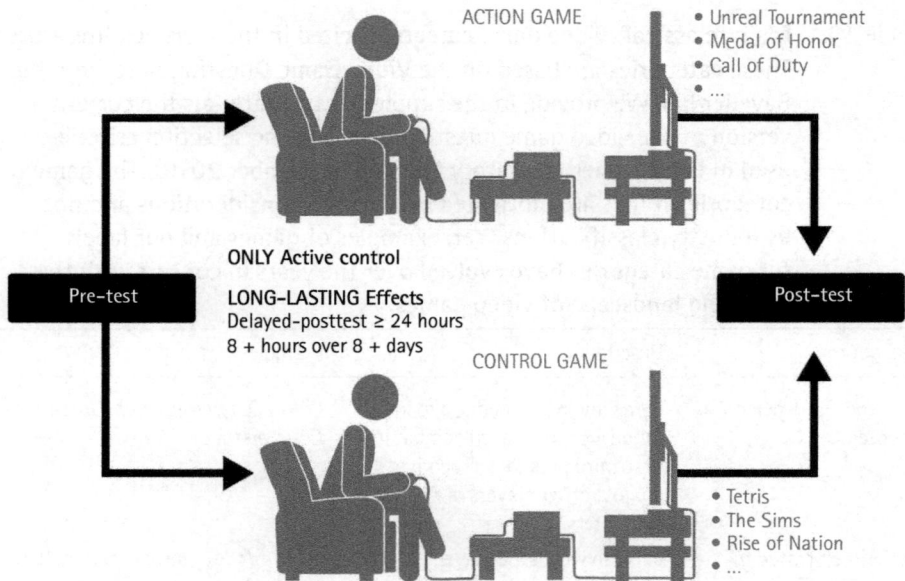

FIGURE 19.1 Intervention design to evaluate the causal impact of playing a specific type of video games on cognition (here termed experimental game). Participants are randomly assigned to play experimental video games or control video games. The training program typically requires at least 8 hours, and typically tens of hours of gameplay, distributed over weeks or months. Participants' cognitive skills are first evaluated on a battery of tests (pre-test) and tested again after completion of their training (post-test). If playing the experimental video games specifically improves the cognitive abilities assessed, then we expect the experimental group to improve more from pre- to post-test than the control group.

the impact of using a console, looking at a screen, or of digital media in general. Given the complexity of interpreting the outcome of grouping together and contrasting such a wide variety of activities, other meta-analyses investigating the impact of video game play on cognition have been more focused. The rationale here has been to group together video game genres that share features hypothesized to enhance cognition and to include only studies using other commercial video games as active controls. Twenty years ago, researchers noticed by chance that study participants playing regularly first- and third-person shooters exhibited outstanding performance in attentional tasks (Bavelier and Green, 2016) and subsequently conducted an experimental study to test and verify the causal impact of playing those types of video games on attention (in contrast to a control group that played a different type of games; Green and Bavelier, 2003). These results led most of the field to focused on the impact of first- and third-person shooter games (e.g., *Unreal Tournament; Medal of Honor* (FPS)), also known as action video games, on cognition. Not surprisingly this is the most represented video game genre in the available literature. This is followed by racing games (e.g., *Mario*

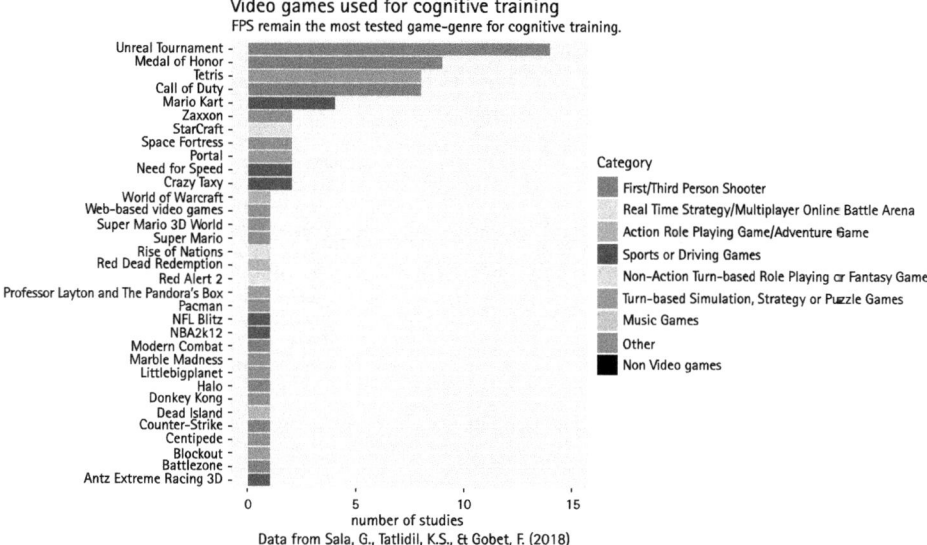

FIGURE 19.2 List of commercial video games used in cognitive training studies from Sala et al. (2018). This list contains a wide range of video game genres that have been used for training in the scientific literature (e.g., first-person shooters, racing games, puzzle games, real-time strategy games, sports games) as well as non-video games (Space Fortress). Large differences in experiences between different game genres (a fast-paced multiplayer FPS is nothing like a slow paced, single player puzzle game) render the interpretation of any such results (positive, negative or null impact on cognition) quite difficult. This figure counts the number of publications cited in Sala et al. (2018) that used a particular video game (out of a total of 63 publications). Note that a publication could involve multiple experiments, each using potentially a different set of video games.

Kart, Crazy Taxi, Need for Speed) with rarer reports on real-time strategy games (e.g., *StarCraft, Rise of Nations*, see Figure 19.2). While we will discuss why these video game genres may be specifically well-tuned to change aspects of cognition, we now turn to the control games used in such studies. As illustrated by Figure 19.3, the video games most commonly used as controls are social simulation games such as *The Sims* (a life simulator game) and puzzle or visuo-motor coordination games (e.g., *Tetris, Ballance, Angry Birds*). This raises the possibility these genres have the least impact on cognition. Yet, it should be clear that different game genres might have different cognitive effects. Thus, depending on the study, the same game may be used for the experimental or for the control group. It appears from these figures that there is minimal overlap between the two lists (Figure 19.2 vs. 19.3; see also Figures 19.4 and 19.5). A notable exception is *Tetris*, which has been frequently used both as a control and as a cognitive training game, especially targeting mental rotation abilities. We will review the literature for the main active game video game genres listed previously.

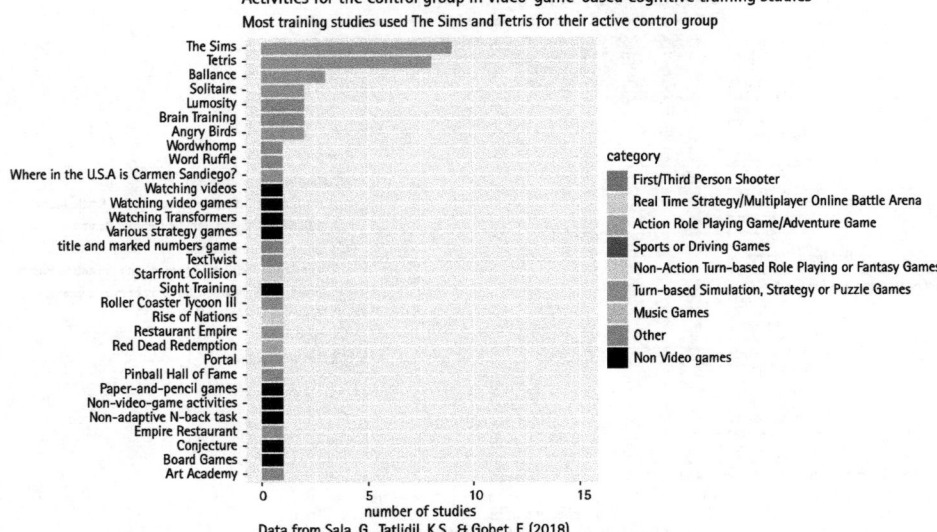

FIGURE 19.3 List of activities used as control treatment in video-game based training studies from Sala et al. (2018). Control treatments vary widely from playing video games to playing paper-and-pencil games; this makes it difficult to abstract the construct measured by such studies. This figure counts the number of publications cited in Sala et al. (2018) that used a particular video game or activity (out of a total of 63 publications). Note that a publication could involve multiple experiments, each using potentially a different set of video games.

Source: Copyright © Pedro Cardoso-Leite. Licensed under CC BY 4.0. Data from Sala, G., Tatlidil, K. S., and Gobet, F. (2018). Video game training does not enhance cognitive ability: A comprehensive meta-analytic investigation. *Psychological Bulletin, 144*(2), 111–139. doi:10.1037/bul0000139.

FIRST- AND THIRD-PERSON SHOOTERS ("ACTION" VIDEO GAMES)

The game genre that has been most studied within the context of cognitive improvement is without a doubt first- and third-person shooters (e.g., Bediou et al. (2018); see also https://www.ncbi.nlm.nih.gov/pubmed/30148383). This category of games has traditionally been called "Action Video Games" (AVG) in the field; however, the changing landscape of video games has made this nomenclature outdated and better classifications are needed (Dale et al., 2020; Dale and Green, 2017). First/third-person shooter games are (1) fast-paced, involving rapidly moving objects (e.g., projectiles), and transient events (e.g., explosion); they (2) require participants to distribute their attention to monitor events from central vision to the visual periphery; they (3) demand a high attentional focus by loading perceptual, cognitive, and motor systems; and finally, they contain (4) temporal and spatial uncertainty preventing full task automatization (Cardoso-Leite, Joessel, and Bavelier, 2020). Games in this category are typically

Action Video games used for cognitive training
Only a small set of video games fit in this category.

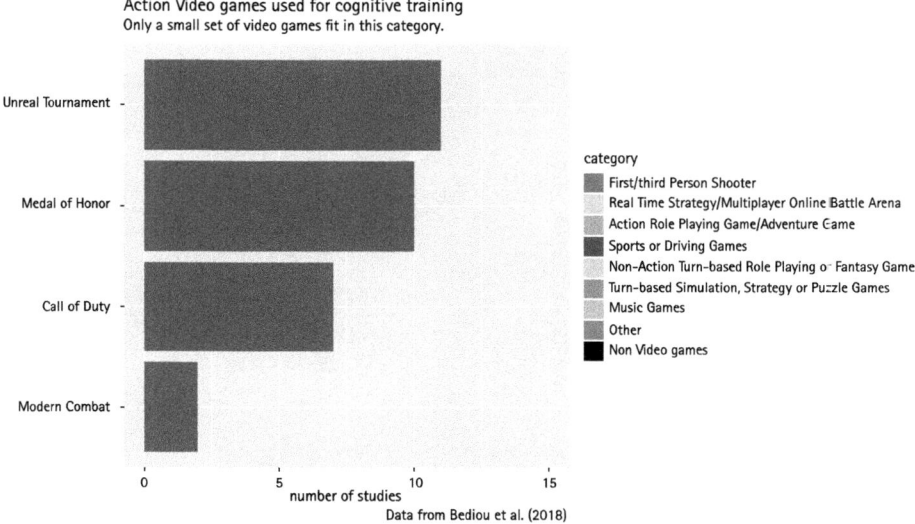

FIGURE 19.4 List of action video games (all first-person shooter games; FPS) used for cognitive training according to Bediou et al. (2018). Focusing on this specific video game genre substantially reduces the number of games titles but still represents a major portion of the scientific literature (contrast this with Figure 19.2). This figure counts the number of publications cited in Bediou et al. (2018) that used a particular video game (out of a total of 23 publications). Note that a publication could involve multiple experiments, each using potentially a different set of video games.

Source: Data from Bediou, B., Adams, D. M., Mayer, R. E., Tipton, E., Green, C. S., and Bavelier, D. (2018). Meta-analysis of action video game impact on perceptual, attentional, and cognitive skills. *Psychological Bulletin*, 144(1), 77–110. doi:10.1037/bul0000130.

violent and include titles like *Medal of Honor* and *Call of Duty*. It is critical to note that, contrary to what some have argued, action video games are not simply "any physically challenging video game in which reaction time plays a crucial role" (Karimpur and Hamburger, 2015, p. 1). There are many games that require fast and accurate responding (e.g., fighting games, games like *The World's Hardest Game*) that do not fulfill the criteria listed previously.

Two types of studies investigated the relationship between action video gaming and cognition: correlation studies—where habitual first-/third-person shooter video game players (AVGP) are contrasted to individuals playing almost no video games at all (i.e., non-video game players (NVGP))—and intervention studies—where individuals with only moderate video game play experience are asked to play either an action video game or a non-action video game for multiple hours distributed over weeks (see Figure 19.1). Correlational studies document significant differences between habitual AVGP and NVGP, leaving unclear the source of the difference. Intervention studies can clarify the causal role of video game play, as they evaluate whether game play changes performance between a baseline time before participants engage in the game play to a time after they have completed their game play training. Research on action video games has matured over the past 20 years and there is now a growing body of correlational and

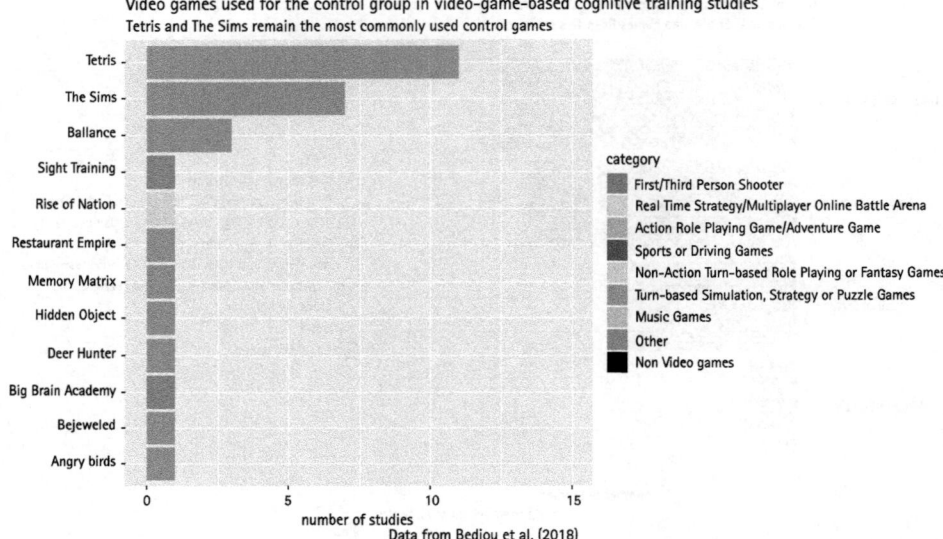

FIGURE 19.5 List of games used in the active control treatment when action video games were tested for cognitive training as tabulated by Bediou et al. (2018). This list includes only commercial video games (with the exception of the Sight Training program; contrast this with 3). This figure counts the number of publications cited in Bediou et al. (2018) that used a particular video game (out of a total of 23 publications). Note that a publication could involve multiple experiments, each using potentially a different set of video games.

Source: Data from Bediou, B., Adams, D. M., Mayer, R. E., Tipton, E., Green, C. S., and Bavelier, D. (2018). Meta-analysis of action video game impact on perceptual, attentional, and cognitive skills. *Psychological Bulletin*, *144*(1), 77–110. doi:10.1037/bul0000130.

intervention studies—almost all of which however focus on healthy young adults. These intervention studies show for example that playing action video games rather than other forms of video games, causes improvements in visual perceptual abilities (Chopin, Bediou, and Bavelier, 2019), spatial cognition (Spence and Feng, 2010), some forms of memory (Pavan et al., 2019; Sungur and Boduroglu, 2012), and perhaps even academic topics such as reading (Franceschini et al., 2015) or mathematical skills (Libertus et al., 2017). A recent meta-analysis has evaluated the impact of action video games on cognition, subdividing outcome variables into one of seven cognitive domains (Bediou et al., 2018): (1) perception, (2) top-down attention, (3) spatial cognition, (4) multitasking, (5) inhibition, (6) problem solving, and (7) verbal cognition. Data from correlational studies show that habitual AVGP outperform NVGP in all of these domains with statistically significant effects for all but the less studied (6) problem-solving category. Data from intervention studies show a similar trend, with AVG training causing numerically improved performance in all domains as compared to training with other commercial games. These effects are however smaller in size and less reliable than those observed in correlational studies, certainly calling for caution. Of all the domains studied, we note that top-down attention and spatial cognition seem most reliably improved by action video gaming interventions. The reduced effect sizes in intervention studies compared to correlational studies may be due to action video game players in the latter having substantially more gaming experience than the tens of hours typical of training studies. The reduced reliability on the other hand is due to both the effect sizes being smaller and to the reduced number of intervention studies per domain. As more studies are conducted, it will become clearer how much each specific domain may be positively impacted by playing action video games.

Most action video game studies focus on healthy young adults. A reason for this is that action video games are not adequate for children because of their violent content and they are not adequate for older adults because of their high difficulty level. While no experimental study would expose children to violent video games, some children do in fact play those age-inappropriate, violent games in their homes. In their meta-analysis, Bediou et al. (2018) list three such cross-sectional studies focusing on the relationship between action video game and children's cognition. One such study tested typically developing children and young adults, with ages ranging from 7 to 22 years, on three attentional tasks: the Useful Field of View (UFOV) (spatial attention), the Attentional Blink (AB) (temporal attention), and the Multiple Object Tracking (MOT) task (sustained, dynamic attention; Dye, Green, and Bavelier, 2009). In addition, these authors collected survey data about each participant's video gaming habits, allowing them to form two subgroups of participants: AVGP and NVGP. This type of data can be used to describe the time course of attentional development and evaluate how these time courses differ between AVGP and NVGP. The results show that AVGP presented a time course of attentional development that was accelerated compared to that of NVGP. The extent and onset of these group differences depended on the specific aspect of attention being considered. AVGP performed better than NVGP on the temporal attention task (i.e., attentional blink) starting at age 7–10, on the spatial attention (i.e., UFOV) at ages 11–13 and on the dynamic attention task (i.e., MOT) at ages 14–17. Such results confirm that various components of attention mature at different speeds and suggest they may be differentially affected by action video game play. Overall, the cross-sectional data collected on children present a pattern of results similar to what is observed in adults and indicate that action video games training may also be effective at younger ages.

To investigate the causal role of action video games on cognition in children, while avoiding exposing them to violent content, a few studies have selected commercial, age-appropriate mini-games that contain features similar to those attributed to action video games. Franceschini et al. (2013) have used this approach in 7–13-year-old dyslexic Italian children to test the hypothesis that enhancing visual attention in Italian readers may in part alleviate reading difficulties. Children trained for 12 hours over two weeks either on action-like mini-games or control mini-games from *Rayman Raving Rabbids*. Note that *Rayman Raving Rabbids* comprises a large set of varied, small party games and thus does not technically fall in the first- or third-person shooter category. However, the authors rated each of the party games from that collection as being action video game-like or not based on game features typically assigned to action video games. Mini games classified as action video game-like were used for the experimental group while the mini games devoid of action mechanics were used for the control group. The training was distributed over about 2 weeks; those children assigned to play the action-like mini-games displayed improvements in attention and in reading abilities, at least as measured by timed tasks of reading, as compared to a control group that played non-action-like games for the same amount of time (Franceschini et al., 2013). These first results were later confirmed in an English-speaking sample of dyslexic children (Franceschini et al., 2017) and supported by a small sample, correlational study on typically reading French adults (Antzaka et al., 2017). Yet, whether action or phonologically based video games may help remediate dyslexia certainly remains controversial as other intervention studies have failed to find a positive impact on reading acquisition (Łuniewska et al., 2018). Moreover, a recent large sample correlational study that contrasted children who report playing video games to those that do not found a negative association between video game play and reading (Seok and DaCosta, 2019). The interpretation of this latter result remains difficult as it did not differentiate between game genres and had an overrepresentation of male children in the video game players group (indeed, if most video game players are boys, it's unclear if the effects relate to playing video games rather than other factors associated with boys being worse readers). Exploiting the proposal that action video game enhanced top-down attention, a recent study documents enhanced ability at performing optimal cue combination in 4–5-year-old children after 7.5 hours of action video game-like mini-games (e.g., *Fruit Ninja*), as compared to playing control mini games (e.g., *Puzzingo*; training was distributed over 2 weeks; Nava, Föcker, and Gori, 2019). While the reviewed evidence points towards action video games having some efficacy in enhancing cognition, and especially attention in children, the empirical data is scarce and further studies are needed to confirm or infirm these results.

The use of action video gaming to train older adults' cognition is also quite rare (for a review, see Toril et al., 2014). One study had 65–91-year-olds play either a first-person shooter (*Medal of Honor*), a puzzle game (*Tetris*) or an attention-training task (UFOV training) for six 90-minute sessions or nothing (no-contact control group). Contrary to what was observed in younger adults, action video play did not improve attentional performance more than playing the puzzle game (Belchior, Marsiske, Sisco, Yam, and Mann, 2012). However, as pointed out by the authors, action video games might be too hard for older adults and training duration not long enough for them to learn how to play the game before the game could train their cognitive abilities. Indeed, players in the action video game group had to receive a step-by-step, PowerPoint-based explanation of the game by an experienced coach to make the difficulty level "manageable." Supporting this view, Boot, Simons, Stothart, and Stutts (2013) reported lower compliance in the action group than the other training groups in a sample of older adults. Because off-the-shelf action video games are designed to be challenging for adolescents and young adults already cognizant of the genre, they are likely too hard to be used with older adults (for a discussion, see section "Does Action Video Game Play Impact All Ages Equally?" in Bediou et al., 2018). Indeed, training with video games obeys the same learning rules as training with any other forms of behavioral interventions (Stafford and Dewar, 2014). In particular, to be efficient, the training difficulty needs to be adapted to the learner's level, a concept pioneered as early as the 1900s by Vygotsky and his proposed "zone of proximal development." Thus, to train cognition in older adults it might be preferable to specifically design video games tailored for this population (Anguera et al., 2013).

RACING GAMES

One of the most promising game genres for cognitive research is racing video games (Belchior et al., 2019; Cherney, 2008; Li, Chen, and Chen, 2016; Wu and Spence, 2013). This is because they are typically less violent than first-person shooter games; they are also easier to grasp by new gamers (Belchior et al., 2013) and easier to create for developers—which makes this genre ideal for cognitive training research (Anguera et al., 2013). Most importantly, this genre of video games can be easily adapted to capture the key mechanics of first- or third-person shooter games, and thus may offer similar cognitive benefits. One study for example had young adults train for 10 hours on either an FPS (i.e., *Medal of Honor*), a racing game (*Need for Speed*), or a puzzle game (*Ballance*) and evaluated the impact of playing those games on visual search performance (Wu and Spence, 2013). Compared to training on the puzzle game, training on either the FPS or the racing game led to improvements in divided attention and top-down attention control. Similarly, training on an FPS (*Unreal Tournament 2004*) or on a racing game (i.e., *Mario Kart*) may both improve visuo-motor control; although the effects might not be strictly identical (Li et al., 2016).

Racing games have also been used to train older adults. One study had 65–86-year-olds train for a total of 60 hours on either a racing game (i.e., *Crazy Taxi*) or a brain-training software (i.e., PositScience *InSight*), while others were part of a no-training control group (Belchior et al., 2019). The results suggest that both forms of training had modest transfer effects which for some were not present at post-test but only in the follow-up, 3 months later. Mental rotation, which was reported to improve with playing a racing game in younger adults (Cherney, 2008), does not seem to be affected in older adults.

While these studies suggest that using racing games might be a viable pathway to cognitive enhancement, more data is needed to fully substantiate such a claim.

REAL-TIME STRATEGY GAMES

A video game genre that has comparatively gained a lot of attention lately is real-time strategy video games. While older generations of strategy video games, not unlike chess, were mainly focused on strategic thinking and slow paced (i.e., "turn-based"), real-time strategy games include fast-paced action game mechanics. For example, in the real-time strategy game *StarCraft*, the player typically has control over multiple units in parallel, each of which requires frequent orders (e.g., move, attack, build) delivered through precise mouse clicks. Optimal play may require over 200 of such actions per minute (Lewis, Trinh, and Kirsh, 2011).

Using participants' self-reported video gaming habits data, Dale and Green (2017) formed four groups of participants and asked them to complete a large battery of cognitive tasks, including simple response time task, choice response time task, a go/no-go task, the AB task, the UFOV, the MOT, and the Operation Span task. The four groups of participants (about n = 14 per group) were habitual AVGP, habitual real-time strategy players, NVGP, and those who play more frequently but a wider range of game genres (i.e., "Tweeners"). Performance on the cognitive tasks differed between these groups. Overall, AVGP tended to perform best on these tasks and NVGP to perform worse with players in the real-time strategy and Tweeners groups performing somewhere in between these two groups. These cross-sectional results suggest that playing action video games but also real-time strategy games may improve performance on a variety of cognitive tasks.

To evaluate the causal effect of playing real-time strategy games on cognition, one study assigned seventy-two 22-year-old (on average) women to play either one of two versions of *StarCraft* (a real-time strategy game) or *The Sims* (a slow pace life-simulator) for a total of 40 hours (completed on average in 43.7 days; Glass, Maddox, and Love, 2013). The alternative versions of *StarCraft* differed in the amount of information players had to simultaneously keep track of and switch between. Before and after playing these video games, participants underwent a battery of cognitive tasks (including for example the Stroop task, Task Switching, and the Operation Span task) selected to represent a latent construct of "cognitive flexibility." The results show that playing *StarCraft* improved cognitive flexibility more than playing *The Sims*. Additionally, the effects were strongest for the game version with higher load on cognitive flexibility.

Real-time strategy games have also been used for cognitive training in older adults (Basak, Boot, Voss, and Kramer, 2008). 70-year-olds were randomly assigned to either play *Rise of Nations* (a slow paced real-time strategy game) for a total of 23.5 hours (distributed over 4 to 5 weeks) or to a no-training, no-contract control group (about 20 persons per group). Before and after the training (or non-training) all participants completed a battery of tasks covering what the authors call "executive control" (which included tasks like task-switching and the N-back) and "visuospatial skills" (e.g., mental rotation, attentional blink). The authors reported that playing *Rise of Nations* led to improved performance in the executive control but not in visuo-spatial skills (but see, Strenziok et al., 2014).

Studies investigating the association between real-time strategy game play and cognitive abilities in children are hard to find. One study had 3rd graders either play a fire-fighting real-time strategy game (*Fire Department 2: Fire Captain*) or read information about fire-fighting on a webpage for 40 minutes before taking a quiz about fire-fighting which included questions requiring to retrieve factual information, compare situations, and solve problems (Chuang and Chen, 2007b, 2007a). Those who played the game performed better than the reading control group on fact-retrieval and problem-solving items. However, it is rather unlikely that these effects are due to the game being a real-time strategy game (rather than say a puzzle game); instead it appears more plausible that learning about fire-fighting is more engaging and effective when that content is learned through active playing rather than by just reading.

TETRIS

Tetris is arguably one of the most used video games in psychological research. It has been used to reduce cravings for food, drugs, and so on (Skorka-Brown, Andrade, Whalley, and May, 2015), reduce intrusions of mental images related to traumatic events (Holmes, James, Coode-Bate, and Deeprose, 2009), and to tone down the negative emotions associated with specific autobiographical memories (Engelhard, van Uijen, and van den Hout, 2010). *Tetris* has also been used within the domain of cognitive training, sometimes as the experimental game and other times as the active control game (for a review, see Sala et al., 2018).

When used for cognitive training, *Tetris* is thought to train visuospatial cognition and more specifically mental rotation abilities as the game heavily relies on mental rotation. One study, for example, had 8–9-year-old children either play *Tetris* (the experimental group) or *Where in the USA is Carmen Sandiego?* (a commercial game focusing on geography with minimal load on mental rotation; the active control group) for eleven 30-minutes sessions distributed over multiple days (De Lisi and Wolford, 2002). The results showed that playing *Tetris*, but not the control game, improved children's 2D mental rotation abilities as measured using a paper-and-pencil mental rotation test.

Studies on young adults suggest that 6 hours of training on *Tetris* (as compared to a no-contact, no-training group) may improve performance on some visuospatial tasks (Okagaki and Frensch, 1994; see also Boot, Kramer, Simons, Fabiani, and Gratton, 2008; Terlecki, Newcombe, and Little, 2008). The effects however seem to be rather specific—training on a 2D *Tetris* version improved 2D mental rotation but not 3D mental rotation, while training on a 3D version of *Tetris* improved both (Moreau, 2013)—and several studies failed to observe improvements on 2D mental rotation after training on *Tetris* (Pilegard and Mayer, 2018; Sims and Mayer, 2002). *Tetris* has also been used for cognitive training in older adults, however not to train mental rotation but rather as a control game. Yet, one study reported that in older adults playing *Tetris* may improve selective attention to the same extent as an action game or training on the attention task itself (Belchior et al., 2013), perhaps because for this age group, *Tetris* is already challenging and action video games are too difficult. The evidence supporting the usefulness of *Tetris* to improve cognition remains, therefore, mixed.

Casual Mobile Games

Casual mobile video game play is among the most common form of video gaming in the general population and it is increasingly popular among older adults (Chesham, Wyss, Müri, Mosimann, and Nef, 2017; Whitbourne, Ellenberg, and Akimoto, 2013). There have been several attempts to evaluate the impact of such video games on cognition; the results however are not always consistent. Note that we restrict here our review to commercial games and do not include the larger literature on computerized experimental psychology tasks, such as those developed by PositScience, Lumosity, or tested by Owen et al. (2010).

One study for example (Oei and Patterson, 2013), had young adults train for a total of 20 hours over 4 weeks in various such games (*Hidden Expedition-Everest, Memory matrix 1.0, Bejewelled 2, Modern Combat: Sandstorm, The Sims 3*) and reported broad benefits (in various attentional and working memory tasks) only for the group playing the first-person shooter video game on mobile (i.e., *Modern Combat: Sandstorm*). Playing other, more casual video games did however improve performance on specific tasks (e.g., *Bejeweled 2*, improving visual search) suggesting that casual video games might be used for targeted cognitive training interventions. However, using partly a different set of games and outcome measures, the same authors (Oei and Patterson, 2014) reported no benefits of training for 20 hours on an FPS (*Modern Combat*), a real-time strategy (*Starfront Collision*), or a fast-paced arcade game (*Fruit Ninja*). Instead, they reported that slow-paced, physics game (*Cut the Rope*) lead to improvements in executive functions as indexed by performance in task switching, Flanker tasks, and a go/no-go task. The authors provide various suggestions as to why their study failed to show improvements in the action video game trained group (e.g., differences in the experimental design). They also offer that the efficacy of the slow-paced physics game may be explained by that game involving cognitive processes that are important for executive functions (e.g., "strategizing, reframing, and planning"). More research is needed to substantiate these claims.

A recurrent issue in this literature is to determine a priori and explain why training on a given game should improve performance on a given cognitive task. An interesting approach, grounded in Thorndike and Woodworth's principle of identical elements (Thorndike and Woodworth, 1901), consists in first evaluating the extent to which performance in various (casual) games correlates with performance on cognitive tasks, which are typically designed to isolate cognitive processes (Baniqued et al., 2013). Correlations between the two sets of measures may be caused by them involving the same set of underlying cognitive processes. Games that correlate with working memory and reasoning tasks may then be used to train those abilities. Using this approach, Baniqued et al. (2014) had participants play various categories of casual video games for 15 hours and measured their cognitive abilities across a large battery of tasks both before and after that training. The authors note that playing video games selected to tap into working memory and reasoning did not improve performance on working memory and reasoning tasks but instead improved performance on divided attention tasks (Baniqued et al., 2014). While this is undoubtedly an interesting and principled approach, more research is needed to solidify these results and gain insights into the differential effects of various game genres.

The literature reviewed previously highlights the need to consider video game genres separately and argues for an empirical approach that contrasts a specific commercial video game based on the mechanics it embodies rather than one that opposes any kind of video game to any kind of non-video game activity (Dale et al., 2020). While most evidence for the efficacy of video games for cognitive training currently rests on the use of action video games, future studies might reveal that other game genres are also (maybe differently) beneficial for cognition. Such studies may help to identify which game mechanics in video games are important to cause various cognitive improvements. An alternative, yet complementary route, consists in evaluating the neural processes involved in various forms of video game play as well as the consequences of video game play on the human brain. We will now review the literature on the neuroscience of video game play.

THE NEUROSCIENCE OF VIDEO GAME PLAY

Understanding what happens in the brain when people play video games, as well as the consequences that significant amounts of video game play has on brain structure and function, may provide new insights to interpret the behavioral results described previously. Playing video games has been associated with extensive neural alterations all over the brain, from sensorimotor regions to higher-order cortices such as prefrontal areas (Gong et al., 2015, 2019). For example, faster motor response times to visual stimuli in AVGP, compared to people who don't play video games, has been linked to increased white matter integrity in visual and motor pathways (Zhang et al., 2015), and AVGP in particular exhibited reduced brain activity during task preparation in the cuneus, middle occipital gyrus, and cerebellum which was interpreted to be indicative of increased neural efficiency (Gorbet and Sergio, 2018).

In the following sections, we briefly review the literature to highlight how video games affect brain organization, and how these functional and structural changes might in turn explain the reported behavioral consequences of playing video games. Yet, as discussed in the previous behavioral section and exemplified in a recent review (Palaus, Marron, Viejo-Sobera, and Redolar-Ripoll, 2017), identifying the impact of video game play, as if it were a homogenous activity, on brain functions may be misguided. Rather, a more fruitful approach appears to focus on the information-processing demands of the game play, and the exact processes engaged by the player. As a first step in that direction we will consider the impact of video game play on the brain systems linked to first reward system, and then spatial navigation, before turning to the special case of action video games and the frontoparietal networks of attention. Other brain systems (e.g., the motor system) may play important roles, but they will not be considered here.

Reward System

The brain's reward system is involved in learning and motivation. All successful video games tap into this system by using complex reward schedules to engage players for long play durations. Differences in the cognitive effects of training with various genres of video games might be related to differences in how these video game genres specifically activate the reward system. Although recent efforts attempt to characterize the specific cognitive effects of action video gaming involving the reward system (for a review, see Bavelier and Green, 2019), much remains to be uncovered as most research so far has focused on the relationship between video games and the reward system without differentiating what exact type of video game is being played. This being said, recent results show that the reward system may be a key player to consider when studying the effects of video games on the brain.

When contrasting playing a first-person tank shooter game to watching a blank screen, Koepp et al. (1998) reported an increase in dopamine release in the ventral striatum (measured indirectly using Positron Emission Tomography) that correlated with the performance in the game (as measured by the highest game level reached by the participant) demonstrating that playing some video games can indeed causally affect the reward system.

Other studies investigated the potential long-term effects of video game play on brain function and structure. Kühn et al. (2011) observed that 14-year-old children who frequently played video games had a larger left striatum than same aged children who played infrequently, suggesting that prolonged video gaming may affect the structure of their reward system. Furthermore, these changes in structure were accompanied by functional changes in that the frequent video players also displayed a larger BOLD activity than infrequent video players in response to losses during a gambling task. Similar studies conducted on adults provide somewhat different results. Kühn et al. (2014a) observed that past video game experience correlated with gray matter volume in various brain areas (e.g., parahippocampal region) but not in the ventral striatum. These results may indicate that the effects of playing video games on the reward system may critically depend on the players' age.

The evidence presented so far in this section is correlational implying that the observed brain differences may actually not be caused by video gaming but rather preexist and partially determine video gaming habits. There are however at least two studies that used an intervention design (contrasting video game training to a passive control group) in order to probe the direction of the causality effect (Kühn et al., 2014a; Lorenz, Gleich, Gallinat, and Kühn, 2015). Each of these studies had adults in the training group play a 3D platformer game (*Super Mario 64*) for 30 minutes per day over a period of 2 months and compared their changes in brain function and structure to those of a passive control group. Both studies reported that playing video games affected the size of various brain structures but did not, contrary to what was observed in the cross-sectional study on children, observe any structural changes in the striatum. The video game training did however affect the responsiveness of the ventral striatum to rewards. Lorenz et al. (2015) had their participants complete a task while under the fMRI scanner both before and after the video game training (for the intervention group) or before and after the waiting period (for the passive control group). The results show that for the participants in the control group the reward responsiveness in the ventral striatum decreased substantially from pre- to post-test sessions while for the participants in the video gaming group this was not the case: participants trained on the 3D platformer video game exhibited similar activation levels in the ventral striatum in the pre- and post-test session. The authors suggest these results may indicate a greater ability in the video game trained participants to maintain high levels of task motivation through the flexible control of the reward responsiveness of the striatum. They further hypothesize that this video-gaming induced effect on the reward system may be exploited for a broad range of uses cases.

Rewards schedules are a key component of all successful video games and it is still unclear how long-term exposure to video games impacts the reward system. Current evidence supports the view that video games may alter the reward system's functioning as well as its structure (although, possibly only during childhood). While the results reported in this section may apply to all types of video games, the behavioral evidence clearly shows that it is necessary to distinguish various video game genres. The reward schedules implemented in different video game genres may have drastically different effects on the reward system, and through the reward system, on learning. There are ongoing efforts to clarify the possible mechanisms relating playing specifically action video games, the reward system and broad cognitive performance improvements (Howard-Jones and Jay, 2016; Miendlarzewska, Bavelier, and Schwartz, 2016). More work is needed to formalize reward mechanisms in video games and assess the impact of different types of video games on the functional and structural properties of the human reward system.

Spatial Cognition and the Hippocampal Formation

Video game play often requires discovering, and thus navigating, new worlds, be they landscapes, buildings, or intergalactic spaces. Such video games are likely to engage the hippocampus whose role in memory and navigation is well-established (for reviews see Lisman et al., 2017; Eichenbaum, 2017).

Frequent video gaming in adolescence and adulthood has been associated with volumetric changes of gray matter in the hippocampal region and its projections. Kühn et al. (2014a) explored the correlation between gray matter volume and frequent gaming in adults, irrespective of the type of game being played. They measured gaming experience in a unit called *joystick years*, which reflects the lifetime amount of video game play, and evaluated to what extent *joystick years* was correlated with gray matter volume across all regions of the brain. Higher numbers of *joystick years* was associated with larger gray matter volume in both the occipital lobe and the hippocampal formation. Different gray matter volume in these two regions was proposed to reflect superior visuospatial expertise in video game players and to suggest that navigational exploration in early visual processing is affected by playing video games. Interestingly, recent findings also suggest a mediating role of the hippocampal formation during visual guidance (see Nau, Julian, and Doeller, 2018). Another correlational study reported a positive correlation between the amount of time spent on video games and gray matter volume in the hippocampus, in particular the entorhinal cortex that surrounds hippocampus (West et al., 2015). The navigation demands of most video games is in line with such changes in entorhinal cortex as this structure acts as a gateway to the hippocampus, and has been associated with spatial navigation, memory, and the perception of time (Bird and Burgess, 2008).

Changes in hippocampal volumes have been recently qualified as dependent on game genre and player strategies. Kühn et al. (2014a) measured gray matter volume of the hippocampus and entorhinal cortex in relation to the lifetime amount of video game playing. Their results show that while playing puzzle and platformer games was associated with increased parahippocampal volume, playing action video games was associated with a decrease in parahippocampal volume (Kühn and Gallinat, 2014). West et al. (2015) further qualified this effect as being related to particular cognitive strategies gamers might use for navigation, strategies that rely on different brain structures. One strategy that can be qualified as "spatial" involves constructing an internal cognitive map of the environment using landmarks and their relationships and then exploiting this map for navigation. The use of this strategy is thought to involve the hippocampus. An alternative, "non-spatial" strategy might instead rely on memorizing a fixed sequence of actions to be completed from a given location to reach a particular endpoint (e.g., when facing the entrance of the building, go left, then right, then left again). This second strategy therefore does not involve building internal representations but merely memorizing stimulus-response mappings. This non-spatial strategy is thought to involve the striatum. West et al. (2015) used a task where players navigated through a maze in the presence of landmarks that could be exploited to create an internal cognitive map. They then tested the same players on the same maze but removed the landmarks. Participants using a "spatial" strategy would be unable to use their internal maps in this situation as the landmarks were necessary to ground their cognitive map. Participants using a "non-spatial" strategy, on the other hand, would not be affected by this manipulation as they could still execute the memorized sequences of actions to reach the target. West et al. (2015) argue that the decrease in hippocampal volume observed in AVGP relative to NVGP may be accounted for by AVGP relying more systematically on a non-spatial navigation strategy; in agreement with their hypothesis, AVGP performed better than NVGP when landmarks were removed, indicating that they exploited more systematically the non-spatial navigation strategy.

To further investigate the impact of spatial strategy during action video game play on hippocampal volume, West et al. (2018) conducted an intervention experiment comparing three groups of participants, one that was trained for 90 hours on action video games (e.g., *Call of Duty: Modern Warfare*), one that was trained for 90 hours on a 3D platformer video game (e.g., *Super Mario 64*) and a no contact group. Before and after the training entorhinal cortex, gray matter volume in the hippocampus was measured. Contrasting video game genres and play strategies shows that gray matter volume was reduced in the hippocampus after action video game training but only in participants using a non-spatial strategy. Yet, when a spatial strategy was used during training, action video game training resulted in increased hippocampal volume. Interestingly, among those trained on the 3D platformer, spatial learning was associated with increased gray matter volume in the hippocampus and non-spatial learning to increased gray matter volume in entorhinal cortex. The authors confirmed the impact of these results in an additional training experiment which entailed training for 90 hours on action video games (e.g., *Call of Duty Modern: Warfare*). They note that it is only when the use of spatial strategy was encouraged during training that participants showed increased hippocampal formation volume. In conclusion, the neural impact of playing video games is mediated not only by the game genre but also by the very game play characteristics the player exhibits. This state of affairs makes it clear that the impact of video game play on brain organization needs to be qualified according to the processes the players engage while playing. As video games span widely different experiences, looking for the neural correlates of video game play in general is likely to remain an ill-posed research question. Finally, while the possibility to increase hippocampal volume through video games is promising to possibly address cognitive decline and in particular memory loss in aging, the directionality of the effects is yet not well understood. For example, reduction in gray matter volume was also observed after 5 days of intense mental calculation training (4 hours per day with two 10-minute breaks), while at the same time, performance being improved by the training (Takeuchi et al., 2011). Such results indicate that reductions in gray matter volume might not always be negative and/or reflect cognitive decline. Taking everything into account, genres and strategies affect how playing video games alters anatomical structures of the brain, calling for careful consideration of the way video games are designed, what content they present, and what strategies must be used to achieve the goals of the game.

Attentional Networks and Action Video Games

The strongest behavioral evidence regarding the impact of action video game training on cognition concerns increases in players' attentional resources over space, time, and objects, as well as enhanced flexibility in the allocation of attention (Bavelier and Green, 2019). In this section we present functional and structural brain modifications that may underlie such attentional improvements.

Attentional functions are mediated by two main neural networks (Buschkuehl, Jaeggi, and Jonides, 2012): a *ventral network of attention*, which encompasses the temporoparietal junction (TPJ) and ventral frontal cortex (VFC) and has been implicated in switching attention (as when redirecting attention towards a novel element in the environment); a *dorsal network of attention*, which consists of the dorso-lateral prefrontal cortex (DLPFC) and intra-parietal cortex and has been associated with strategic, goal-directed, top-down control over attention allocation. Coordination between the bottom-up and top-down networks has been associated with faster and more accurate responses to targets in a variety of cognitive tasks. Interventions targeting the dorso-lateral prefrontal cortex region, at least in children, enhances executive functions performance, including attentional control (e.g., Wang et al., 2018; Siniatchkin, 2017). Furthermore, these brain structures work in concert with the anterior cingulate cortex (ACC) which monitors and resolves conflicts, regulating in part the activity in the frontoparietal systems of attention (Petersen and Posner, 2012). Action video game play has been associated with more efficient neural activities in frontoparietal regions, and enhanced structural and functional connectivities in prefrontal networks, limbic system, as well as more posterior sensorimotor networks (Gong et al., 2017). This enhanced neural resource allocations in dorsal attentional network may contribute to the improved top-down attentional control and more efficient suppression of distracting information documented in AVGP (Bavelier, Green, Pouget, and Schrater, 2012). Attentional control can indeed optimize the selection of sensory information by two different mechanisms: by selecting more relevant signals, or by suppressing irrelevant signals and preventing noise to be transmitted to higher-order processes. Interestingly, AVGP not only benefit from enhanced attention to targets, they also show superior ability to suppress distractors (Bavelier et al., 2012). To track the fate of distractors during an attention-demanding visual task, several studies measured steady state visually evoked potentials (SSVEP), an imaging technique that uses periodic stimuli to frequency-tag neural responses in the visual cortex. Using this technique, both Mishra, Zinni, Bavelier, and Hillyard (2011) and Krishnan, Kang, Sperling, and Srinivasan (2013) documented active suppression of distractors in AVGP, in line with enhanced selective attention. Since the SSVEP have the same frequency as the driving stimulus, it is possible to concurrently record responses to several stimuli if they are presented at different flickering rates. Mishra et al. (2011) measured SSVEP amplitudes, which are affected by selection and filtering processes in attention, in response to peripheral and foveal stimuli in a target detection task. While the SSVEP amplitude in response to attended targets was the same in AVGP and NVGP, SSVEP amplitude to distractors was decreased in AVGP relative to NVGP, suggesting enhanced filtering of irrelevant information. Similarly, Krishnan et al. (2013) compared SSVEP responses to targets and distractors in two groups of video game players, AVGP and role-playing video game players who served as their control group. Measuring signal-to-noise ratios of evoked potentials to both targets and distractors, Krishnan et al. (2013) showed that playing first-person shooters could improve both the selection of targets and the suppression of distractors.

How bottom-up and top-down processes may change to both improve target selection and distractor suppression was assessed in an fMRI correlational study
comparing AVGP relative to NVGP. Föcker, Mortazavi, Khoe, Hillyard, and Bavelier
(2018) recorded fMRI scans while AVGP and NVGP participated in a cross-modal,
endogenous Posner-cueing task. Young adults were presented with an auditory cue
indicating the most likely location of a subsequent target on which participants were to
perform a difficult, near-threshold visual discrimination task. This paradigm, closely
modeled after Corbetta and Shulman (2002), allows one to separate neural responses
to the auditory cues, which direct the attention allocation for the task to come, from
the neural responses during the difficult visual task itself. The frontoparietal network,
which is thought to mediate attention allocation, was more activated in NVGP than in
AVGP when participants processed the cue and thus prepared for the task to come. This
result may suggest that attention allocation is more efficient in AVGP than in NVGP.
Interestingly, a small percentage of trials were in fact catch trials where only visual noise,
but no visual target, was presented. In these catch trials, participants needed to withhold
their response. AVGP outperformed NVGP on such trials exhibiting fewer false alarms.
Moreover, only for AVGP did activation in the temporoparietal junction, middle frontal
gyrus, and superior parietal cortex predicted their reduced false alarm rate, suggesting
that these areas may operate and interact differently in AVGP compared to NVGP.
Overall, these studies suggest that AVGP may benefit from better attentional control, or
more flexibility in allocating attention, perhaps through a reconfiguration of the cross-
talk between the main frontoparietal areas that mediate attention.

Whether these superior attentional skills result from alterations of processing in
the goal-oriented, top-down attentional network, or rather from better filtering of irrelevant, potentially distracting information within early sensory cortices (or both)
remains an open question. Neural markers of early attentional filtering were compared
in EEG-based correlational studies contrasting AVGP and NVGP. Föcker, Mortazavi,
Khoe, Hillyard, and Bavelier (2019), for example, tested if visual event related potentials
(ERP) components differed between AVGP and NVGP in a high-precision visual selective attention task. Faster response times and improved perceptual performance in
AVGP were observed; yet, early markers of attentional selection such as the posterior
N1 and the P1 were identical across groups. Differences between AVGP and NVGP were
only observed in parietal generators such as the P2 and the anterior N1 components. As
the P2 has been previously linked to task demands (Finnigan, O'Connell, Cummins,
Broughton, and Robertson, 2011; Lefebvre, Marchand, Eskes, and Connolly, 2005), these
results may indicate that AVGP are able to more effectively adapt attentional resources
to varying tasks demands. A similar conclusion was reached by another intervention
ERP study (Wu et al., 2012) that recruited 25 adults and recorded ERPs before and after
10 hours of video game training. Participants with no video game experience in the previous 4 years, were randomly assigned to one of two training groups: the action group
played *Medal of Honor: Pacific Assault* (FPS), whereas the control group played *Ballance*,
a 3D puzzle game. Later, during the testing session, participants performed an attentional visual field task which assesses the ability to detect a target among distractors.

As in Föcker et al. (2019), the two training groups exhibited comparable early sensory ERPs, in line with comparable early attentional selection processes across training. Also, as in Föcker et al. (2019), the action trained group showed an increased P2 amplitude. Moreover, the amplitude of the P3 was also increased in the action trained group, possibly indicating enhanced attentional resources being allocated to the task (Kok, 2001). Overall, these results are in line with the proposal that the differences in attentional performance between AVGP and NVPG may reflect a functional reorganization of the goal-oriented, top-down, dorsal attentional network, with distractor suppression being implemented at a central level, rather than through early perceptual filtering.

Furthermore, playing video games, irrespective of the specific game genre, seems to affect structural and functional properties of parts of the frontal cortex. A longitudinal training experiment study for example, evaluated the structural changes in the dorsolateral prefrontal cortex (DLPFC) resulting from 2 months of training with *Super Mario 64*, a 3D platformer, non-action video game that requires navigational skills (Kühn et al., 2014a). The results indicate that playing this video game induced structural changes by increasing the gray matter volume in the right DLPFC. Similarly, a correlational study reported that the self-reported weekly hours adolescents spent playing video games correlates positively with the thickness of their left DLPFC and left frontal eye fields (FEFs; Kühn et al., 2014b)—cortical thickness is similar but not identical to gray matter volume (Winkler et al., 2010).

It has also been reported that relative to NVGP, AVGP have enhanced intra- and inter-network connectivities in the central executive network and salient network (Gong et al., 2016). These two networks are highlighted using fMRI measurements; the central executive network is associated with working memory, planning, and getting prepared to select an appropriate response to a stimulus, whereas the salient network with nodes in the subcortical reward system has been linked to salient stimuli detection as well as integrating emotional, sensory, and interoceptive signals (Menon, 2015). The central executive network typically contains the DLPFC and is engaged during attention-demanding tasks (Fox, 2006). Further analysis of large-scale networks with diffusion tensor imaging, which evaluates how strongly specific areas are connected, shows that those who spend more weekly hours playing action video games display an increased efficiency (as defined in graph theory) in local, global, and nodal levels of prefrontal, limbic, and sensorimotor networks (Gong et al., 2017). The local, global, and nodal efficiencies, respectively, reflect an increased fault tolerance across the network, improved information flow across the whole network, and the importance of a node, respectively. These neural regions are responsible for processing visual information, spatial orientation, motion perception, selective attention, and integrating multimodal stimuli. This finding supports the view that neural efficiency increases by mediating goal-oriented, top-down attentional processes as a consequence of automating visual sensorimotor tasks and delegating them to areas that handle low-level sensory processing.

While our understanding of the effects of playing video games on the human brain has improved considerably over the last decade, it remains nevertheless limited. Most of the literature reviewed is correlational in nature and based exclusively on adult participants. Studying young adults cross-sectionally is a cost-effective strategy to highlight candidate structures and generate and test hypotheses. Indeed, cross-sectional studies only involve a subject selection phase (using surveys for screening) and an assessment phase, while intervention studies require *in addition* multiple training sessions and a second assessment phase (to serve as a post-intervention test to be compared to the pre-intervention test). Intervention studies furthermore involve a high management cost to assure that participants don't drop out and complete the various steps of the study within the planned time frame. Cross-sectional studies are cost-effective to highlight interesting patterns; however, as for behavioral studies, this strategy needs to be complemented with intervention studies to establish causality and rule out the possibility that the neural differences between habitual action video gamers and non-gamers pre-dated the gaming experiences. Furthermore, the studies reviewed previously were mainly conducted on young adults. However, the mechanisms involved may differ with age as the time course of brain plasticity is likely to differ across brain areas. It will thus be important to include pediatric samples in the future.

Concluding Remarks

Research on the cognitive consequences of video game play has boomed over the past 15 years. As the range of video games tested widens, it becomes apparent that not all video games have the same cognitive impact. Rather, studies systematically contrasting specific game genres indicate that the content of the video game, the user interactions it requires, and attentional processes it engages are of paramount importance. This fact has two consequences. First, it makes little sense to ask about the cognitive impact of video game play; rather, it is important to recognize the variety of experiences video game play affords. Here we have reviewed game genres that have been used over the past 15 years using a game classification that might have been relevant for the covered research but is unlikely to stand up to the drastic changes in game types, gamer profiles, and gaming habits that have emerged since (Dale et al., 2020; Dale and Green, 2017); some initial work is being done to better characterize video gaming for cognitive research (Cardoso-Leite et al., 2020; Dale et al., 2020). Second, there is a need to build better theories on why playing certain video games but not others improves cognitive abilities; one route towards building such theories is to contrast commercial video games which differ by specific game mechanics or by specific content (Cardoso-Leite et al., 2020). Following this strategy, past research has focused mainly on contrasting "action video games" (i.e., mostly first- and third-person shooters) to other commercial video games (e.g., puzzle games). A recent meta-analysis supports a causal relationship between playing action video games and improvements in top-down attention and spatial cognition, with effects on other domains requiring further studies (Bediou et al., 2018). This is not to say, however, that this genre of video games is the only genre of interest for cognitive training. More recently, studies have investigated the effectiveness of racing games and real-time strategy games, which may be suitable for a wider audience than action video games. While promising results have been reported, more research is needed to evaluate the efficacy of these alternative game genres and determine the mechanisms by which they may enhance cognition. The strategy of contrasting multiple game genres within the same study may be useful to both evaluate the relative efficacy of different game genres and to unveil the relevant game mechanics.

The study of how video games in general, and action video games in particular, engage and affect the brain has revealed network-wide changes in reward, memory, and attention brain circuits. This variety of effects suggests that the neural mechanisms responsible for the observed cognitive benefits are likely to go beyond the training of a few specific cognitive processes (Bavelier et al., 2012). Rather, aligned changes in memory, reward processing, and mood, as well as attentional networks efficiency may result in faster processing speed, facilitating in turn a variety of cognitive processes. Future work is needed to unravel the link between the behavioral improvements noted after action video game play and their neural bases. Overall, while significant progress has been made over the past 15 years in our understanding of how to leverage video games for cognitive enhancement, there remain many unknowns in this young emerging field. First, the work so far makes it clear that different genres of video games have different effects on cognition, differences in game mechanics have been hypothesized but they remain to be fully tested to firmly document why playing action video games but not social simulation games may improve cognition, for example. Unpacking key game mechanics is central if we are to leverage lessons from action video game research to design therapeutic or educational video games. Second, although the proposal that attention is a main driver of brain plasticity and learning is well accepted in the neuroscience literature, many of the mechanistic details remain to be worked out when it comes to how action video games implement such mechanisms (Bavelier and Green, 2019; Bavelier et al., 2012). Third, our work has focused so far chiefly on cognition; understanding how to best induce plastic changes in other domains, such as emotion or social behavior, is equally important. Finally, most of the literature so far has focused on adults. As we now better understand the game mechanics that promote brain plasticity, the time has come to ask how to best use video games to foster children's development.

Future Perspectives

Research over the past 15 years has focused mainly on establishing and validating the impact of action video games and probing the breadth of their impact on various cognitive constructs (e.g., top-down attention vs. bottom-up attention). Much remains to be done to catalogue and fully describe the impact of different video game genres on various aspects of behavior. Furthermore, our understanding of the taxonomy of video games needs to be improved so that we can move from vague high-level labels (e.g., "action video games") to objective, measurable indices (e.g. type of attention required; exact reward schedule implemented). In the future, we should be able to make quantitative predictions as to which video game to train on in order to enhance performance in one cognitive construct versus another. The challenges that lie ahead of us will require methodological and theoretical innovations as well as multi-lab and interdisciplinary team work.

Acknowledgment

PCL is supported by the Luxembourg National Research Fund (award ATTRACT/2016/ID/11242114/DIGILEARN). This work also benefited from the support by the Swiss National Fund award 100014_159506, the ONR awards N00014-14-1-0512 and N00014-18-1-2633, and the NSF Cyberlearning award 1227168 to DB.

References

Adachi, P. J. C. and Willoughby, T. (2013). More than just fun and games: The longitudinal relationships between strategic video games, self-reported problem solving skills, and academic grades. *Journal of Youth and Adolescence*, 42(7), 1041–1052. doi:10.1007/s10964-013-9913-9

Anguera, J. A., Boccanfuso, J., Rintoul, J. L., Al-Hashimi, O., Faraji, F., Janowich, J., ... and Gazzaley, A. (2013). Video game training enhances cognitive control in older adults. *Nature*, 501(7465), 97–101. doi:10.1038/nature12486

Antzaka, A., Lallier, M., Meyer, S., Diard, J., Carreiras, M., and Valdois, S. (2017). Enhancing reading performance through action video games: The role of visual attention span. *Scientific Reports*, 7(1), 14563. doi:10.1038/s41598-017-15119-9

Au, J., Sheehan, E., Tsai, N., Duncan, G. J., Buschkuehl, M., and Jaeggi, S. M. (2015). Improving fluid intelligence with training on working memory: A meta-analysis. *Psychonomic Bulletin and Review*, 22(2), 366–377. doi:10.3758/s13423-014-0699-x

Ball, K., Owsley, C., Sloane, M. E., Roenker, D. L., and Bruni, J. R. (1993). Visual attention problems as a predictor of vehicle crashes in older drivers. *Investigative Ophthalmology and Visual Science*, 34(11), 3110–3123.

Baniqued, P. L., Kranz, M. B., Voss, M. W., Lee, H., Cosman, J. D., Severson, J., and Kramer, A. F. (2014). Cognitive training with casual video games: Points to consider. *Frontiers in Psychology*, 4, 1010. doi:10.3389/fpsyg.2013.01010

Baniqued, P. L., Lee, H., Voss, M. W., Basak, C., Cosman, J. D., DeSouza, S., ... and Kramer, A. F. (2013). Selling points: What cognitive abilities are tapped by casual video games? *Acta Psychologica*, 142(1), 74–86. doi:10.1016/j.actpsy.2012.11.009

Basak, C., Boot, W. R., Voss, M. W., and Kramer, A. F. (2008). Can training in a real-time strategy video game attenuate cognitive decline in older adults? *Psychology and Aging*, 23(4), 765–777. doi:10.1037/a0013494

Bavelier, D. and Green, C. S. (2019). Enhancing attentional control: Lessons from action video games. *Neuron*, 104(1), 147–163.

Bavelier, D., Green, C. S., and Dye, M. W. (2010). Children, wired: For better and for worse. *Neuron*, 67(5), 692–701. doi:10.1016/j.neuron.2010.08.035

Bavelier, D., Green, C. S., Pouget, A., and Schrater, P. (2012). Brain plasticity through the life span: Learning to learn and action video games. *Annual Review of Neuroscience, 35*(1), 391–416. doi:10.1146/annurev-neuro-060909-152832

Bavelier, D., Savulescu, J., Fried, L. P., Friedmann, T., Lathan, C. E., Schürle, S., and Beard, J. R. (2019). Rethinking human enhancement as collective welfarism. *Nature Human Behaviour, 3*(3), 204. doi:10.1038/s41562-019-0545-2

Bavelier, D., and Green, C. S. (2016). Brain Tune-Up from Action Video Game Play. *Scientific American, 315*(1), 26–31. doi:10.1038/scientificamerican0716-26

Bediou, B., Adams, D. M., Mayer, R. E., Tipton, E., Green, C. S., and Bavelier, D. (2018). Meta-analysis of action video game impact on perceptual, attentional, and cognitive skills. *Psychological Bulletin, 144*(1), 77–110. doi:10.1037/bul0000130

Belchior, P., Marsiske, M., Sisco, S., Yam, A., and Mann, W. (2012). Older adults' engagement with a video game training program. *Activities, Adaptation and Aging, 36*(4), 269–279. doi:10.1080/01924788.2012.702307

Belchior, P., Marsiske, M., Sisco, S. M., Yam, A., Bavelier, D., Ball, K., and Mann, W. C. (2013). Video game training to improve selective visual attention in older adults. *Computers in Human Behavior, 29*(4), 1318–1324.

Belchior, P., Yam, A., Thomas, K. R., Bavelier, D., Ball, K. K., Mann, W. C., and Marsiske, M. (2019). Computer and videogame interventions for older adults' cognitive and everyday functioning. *Games for Health Journal, 8*(2), 129–143. doi:10.1089/g4h.2017.0092

Bird, C. M. and Burgess, N. (2008). The hippocampus and memory: Insights from spatial processing. *Nature Reviews Neuroscience, 9*(3), 182–194. doi:10.1038/nrn2335

Bodson, L. (2017). *Regards sur les activités quotidiennes des jeunes résidents.* Luxembourg: Institut national de la statistique et des études économiques (STATEC).

Boot, W. R., Simons, D. J., Stothart, C., and Stutts, C. (2013). The pervasive problem with placebos in psychology: Why active control groups are not sufficient to rule out placebo effects. *Perspectives on Psychological Science, 8*(4), 445–454. http://pps.sagepub.com/content/8/4/445.short

Boot, W. R., Kramer, A. F., Simons, D. J., Fabiani, M., and Gratton, G. (2008). The effects of video game playing on attention, memory, and executive control. *Acta Psychologica, 129*(3), 387–398. doi:10.1016/j.actpsy.2008.09.005

Bull, R., Espy, K. A., and Wiebe, S. A. (2008). Short-term memory, working memory, and executive functioning in preschoolers: Longitudinal predictors of mathematical achievement at age 7 years. *Developmental Neuropsychology, 33*(3), 205–228. doi:10.1080/87565640801982312

Buschkuehl, M., Jaeggi, S. M., and Jonides, J. (2012). Neuronal effects following working memory training. *Developmental Cognitive Neuroscience, 2*, S167–S179. doi:10.1016/j.dcn.2011.10.001

Cardoso-Leite, P., Joessel, A., and Bavelier, D. (2020). Games for enhancing cognitive abilities. In J. Plass, R. E. Mayer, and B. D. Homer (Eds.), *Handbook of Game-based Learning.* Cambridge, MA: MIT Press. https://mitpress.mit.edu/books/handbook-game-based-learning

Cardoso-Leite, P., Kludt, R., Vignola, G., Ma, W. J., Green, C. S., and Bavelier, D. (2016). Technology consumption and cognitive control: Contrasting action video game experience with media multitasking. *Attention, Perception, and Psychophysics, 78*(1), 218–241. http://link.springer.com/article/10.3758/s13414-015-0988-0

Cherney, I. D. (2008). Mom, let me play more computer games: They improve my mental rotation skills. *Sex Roles, 59*(11-12), 776–786. doi:10.1007/s11199-008-9498-z

Chesham, A., Wyss, P., Müri, R. M., Mosimann, U. P., and Nef, T. (2017). What older people like to play: Genre preferences and acceptance of casual games. *JMIR Serious Games*, 5(2), e8. doi:10.2196/games.7025

Chopin, A., Bediou, B., and Bavelier, D. (2019). Altering perception: The case of action video gaming. *Current Opinion in Psychology*, 29, 168–173. doi:10.1016/j.copsyc.2019.03.004

Chuang, T.-Y. and Chen, W.-F. (2007a). Effect of computer-based video games on children: an experimental study. *2007 First IEEE International Workshop on Digital Game and Intelligent Toy Enhanced Learning (DIGITEL'07)*, Jhongli City, Taiwan, 114–118. doi:10.1109/DIGITEL.2007.24

Chuang, T.-Y. and Chen, W.-F. (2007b). Effect of digital games on children's cognitive achievement. *Journal of Multimedia*, 2(5).

Corbetta, M. and Shulman, G. L. (2002). Control of goal-directed and stimulus-driven attention in the brain. *Nature Reviews Neuroscience*, 3(3), 201–215. doi:10.1038/nrn755

Dale, G. and Green, C. S. (2017). Associations between avid action and real-time strategy game play and cognitive performance: A pilot study. *Journal of Cognitive Enhancement*, 1(3), 295–317. doi:10.1007/s41465-017-0021-8

Dale, G. and Shawn Green, C. (2017). The changing face of video games and video gamers: Future directions in the scientific study of video game play and cognitive performance. *Journal of Cognitive Enhancement*, 1(3), 280–294. doi:10.1007/s41465-017-0015-6

Dale, G., Joessel, A., Bavelier, D., and Green, C. S. (2020). A new look at the cognitive neuroscience of video game play. *Annals of the New York Academy of Sciences*, 1464(1), 192–203.

De Lisi, R. and Wolford, J. L. (2002). Improving children's mental rotation accuracy with computer game playing. *The Journal of Genetic Psychology*, 163(3), 272–282. doi:10.1080/00221320209598683

Diamond, A. (2013). Executive functions. *Annual Review of Psychology*, 64, 135–168. doi:10.1146/annurev-psych-113011-143750

Diamond, A., Barnett, W. S., Thomas, J., and Munro, S. (2007). Preschool program improves cognitive control. *Science (New York, N.Y.)*, 318(5855), 1387–1388. doi:10.1126/science.1151148

Dye, M. W., Green, C. S., and Bavelier, D. (2009). Increasing speed of processing with action video games. *Current Directions in Psychological Science*, 18(6), 321–326. doi:10.1111/j.1467-8721.2009.01660.x

Eichenbaum, H. (2017). The role of the hippocampus in navigation is memory. *Journal of Neurophysiology*, 117(4), 1785–1796. doi:10.1152/jn.00005.2017

Engelhard, I. M., van Uijen, S. L., and van den Hout, M. A. (2010). The impact of taxing working memory on negative and positive memories. *European Journal of Psychotraumatology*, 1(1), 5623. doi:10.3402/ejpt.v1i0.5623

Fikkers, K. M., Piotrowski, J. T., and Valkenburg, P. M. (2019). Child's play? Assessing the bidirectional longitudinal relationship between gaming and intelligence in early childhood. *Journal of Communication*, 69(2), 124–143. doi:10.1093/joc/jqz003

Finnigan, S., O'Connell, R. G., Cummins, T. D. R., Broughton, M., and Robertson, I. H. (2011). ERP measures indicate both attention and working memory encoding decrements in aging: Age effects on attention and memory encoding ERPs. *Psychophysiology*, 48(5), 601–611. doi:10.1111/j.1469-8986.2010.01128.x

Föcker, J., Mortazavi, M., Khoe, W., Hillyard, S. A., and Bavelier, D. (2018). Neural correlates of enhanced visual attentional control in action video game players: An event-related potential study. *Journal of Cognitive Neuroscience*, 31(3), 377–389. doi:10.1162/jocn_a_01230

Föcker, J., Mortazavi, M., Khoe, W., Hillyard, S. A., and Bavelier, D. (2019). Neural correlates of enhanced visual attentional control in action video game players: An event-related potential study. *Journal of Cognitive Neuroscience, 31*(3), 377–389. doi:10.1162/jocn_a_01230

Fox, J. (2006). Teacher's corner: Structural equation modeling with the SEM package in R. *Structural Equation Modeling, 13*(3), 465–486. http://www.tandfonline.com/doi/abs/10.1207/s15328007sem1303_7

Franceschini, S., Bertoni, S., Ronconi, L., Molteni, M., Gori, S., and Facoetti, A. (2015). "Shall we play a game?": Improving reading through action video games in developmental dyslexia. *Current Developmental Disorders Reports, 2*(4), 318–329. doi:10.1007/s40474-015-0064-4

Franceschini, S., Gori, S., Ruffino, M., Viola, S., Molteni, M., and Facoetti, A. (2013). Action video games make dyslexic children read better. *Current Biology, 23*(6), 462–466. doi:10.1016/j.cub.2013.01.044

Franceschini, S., Trevisan, P., Ronconi, L., Bertoni, S., Colmar, S., Double, K., ... and Gori, S. (2017). Action video games improve reading abilities and visual-to-auditory attentional shifting in English-speaking children with dyslexia. *Scientific Reports, 7*(1). doi:10.1038/s41598-017-05826-8

Gathercole, S. E., Pickering, S. J., Knight, C., and Stegmann, Z. (2004). Working memory skills and educational attainment: Evidence from national curriculum assessments at 7 and 14 years of age. *Applied Cognitive Psychology, 18*(1), 1–16. doi:10.1002/acp.934

Geary, D. C., Berch, D. B., and Mann Koepke, K. (2019). Introduction: Cognitive foundations for improving mathematical learning. In D. C. Geary, D. B. Berch, and K. Mann Koepke (Eds.), *Cognitive Foundations for Improving Mathematical Learning* (Vol. 5, pp. 1–36). Cambridge, MA: Academic Press. doi:10.1016/B978-0-12-815952-1.00001-3

Gentile, D. A., Bailey, K., Bavelier, D., Brockmyer, J. F., Cash, H., Coyne, S. M., ... and Young, K. (2017). Internet gaming disorder in children and adolescents. *Pediatrics, 140*(Supplement 2), S81–S85. doi:10.1542/peds.2016-1758H

Glass, B. D., Maddox, W. T., and Love, B. C. (2013). Real-time strategy game training: Emergence of a cognitive flexibility trait. *PLoS ONE, 8*(8), e70350. doi:10.1371/journal.pone.0070350

Goldin, A. P., Hermida, M. J., Shalom, D. E., Elias Costa, M., Lopez-Rosenfeld, M., Segretin, M. S., ... and Sigman, M. (2014). Far transfer to language and math of a short software-based gaming intervention. *Proceedings of the National Academy of Sciences, 111*(17), 6443–6448. doi:10.1073/pnas.1320217111

Gong, D., He, H., Liu, D., Ma, W., Dong, L., Luo, C., and Yao, D. (2015). Enhanced functional connectivity and increased gray matter volume of insula related to action video game playing. *Scientific Reports, 5*(1), 9763. doi:10.1038/srep09763

Gong, D., He, H., Ma, W., Liu, D., Huang, M., Dong, L., ... and Yao, D. (2016). Functional integration between salience and central executive networks: A role for action video game experience. *Neural Plasticity, 2016.* http://www.hindawi.com/journals/np/2016/9803165/abs/

Gong, D., Ma, W., Gong, J., He, H., Dong, L., Zhang, D., ... and Yao, D. (2017). Action Video Game Experience Related to Altered Large-Scale White Matter Networks. *Neural Plasticity, 2017*, 1–7. doi:10.1155/2017/7543686

Gong, D., Yao, Y., Gan, X., Peng, Y., Ma, W., and Yao, D. (2019). A reduction in video gaming time produced a decrease in brain activity. *Frontiers in Human Neuroscience, 13*, 134. doi:10.3389/fnhum.2019.00134

Gorbet, D. J. and Sergio, L. E. (2018). Move faster, think later: Women who play action video games have quicker visually guided responses with later onset visuomotor-related brain activity. *PLoS ONE, 13*(1), e0189110. doi:10.1371/journal.pone.0189110

Green, C. S. and Bavelier, D. (2003). Action video game modifies visual selective attention. *Nature, 423*(6939), 534–537. doi:10.1038/nature01647

Güllich, A. (2018). Sport-specific and non-specific practice of strong and weak responders in junior and senior elite athletics—a matched-pairs analysis. *Journal of Sports Sciences, 36*(19), 2256–2264. doi:10.1080/02640414.2018.1449089

Holmes, E. A., James, E. L., Coode-Bate, T., and Deeprose, C. (2009). Can playing the computer game "Tetris" reduce the build-up of flashbacks for trauma? A proposal from cognitive science. *PLOS ONE, 4*(1), e4153. doi:10.1371/journal.pone.0004153

Howard-Jones, P. A. and Jay, T. (2016). Reward, learning and games. *Current Opinion in Behavioral Sciences, 10,* 65–72. doi:10.1016/j.cobeha.2016.04.015

Jaeggi, S. M., Buschkuehl, M., Jonides, J., and Perrig, W. J. (2008). Improving fluid intelligence with training on working memory. *Proceedings of the National Academy of Sciences, 105*(19), 6829–6833.

Karimpur, H. and Hamburger, K. (2015). The future of action video games in psychological research and application. *Frontiers in Psychology, 6,* 1747. doi:10.3389/fpsyg.2015.01747

Katz, B., Shah, P., and Meyer, D. E. (2018). How to play 20 questions with nature and lose: Reflections on 100 years of brain-training research. *Proceedings of the National Academy of Sciences, 115*(40), 9897–9904. doi:10.1073/pnas.1617102114

Király, O., Tóth, D., Urbán, R., Demetrovics, Z., and Maraz, A. (2017). Intense video gaming is not essentially problematic. *Psychology of Addictive Behaviors, 31*(7), 807–817. doi:10.1037/adb0000316

Koepp, M. J., Gunn, R. N., Lawrence, A. D., Cunningham, V. J., Dagher, A., Jones, T., . . . Grasby, P. M. (1998). Evidence for striatal dopamine release during a video game. *Nature, 393*(6682), 266–268. doi:10.1038/30498

Kok, A. (2001). On the utility of P3 amplitude as a measure of processing capacity. *Psychophysiology, 38*(3), 557–577.

Kovess-Masfety, V., Keyes, K., Hamilton, A., Hanson, G., Bitfoi, A., Golitz, D., . . . Pez, O. (2016). Is time spent playing video games associated with mental health, cognitive and social skills in young children? *Social Psychiatry and Psychiatric Epidemiology, 51*(2), 349–357. doi:10.1007/s00127-016-1179-6

Krishnan, L., Kang, A., Sperling, G., and Srinivasan, R. (2013). Neural strategies for selective attention distinguish fast-action video game players. *Brain Topography, 26*(1), 83–97. doi:10.1007/s10548-012-0232-3

Kühn, S. and Gallinat, J. (2014). Amount of lifetime video gaming is positively associated with entorhinal, hippocampal and occipital volume. *Molecular Psychiatry, 19*(7), 842–847. doi:10.1038/mp.2013.100

Kühn, S., Gleich, T., Lorenz, R. C., Lindenberger, U., and Gallinat, J. (2014a). Playing *Super Mario* induces structural brain plasticity: Gray matter changes resulting from training with a commercial video game. *Molecular Psychiatry, 19*(2), 265–271. doi:10.1038/mp.2013.120

Kühn, S., Lorenz, R., Banaschewski, T., Barker, G. J., Büchel, C., Conrod, P. J., . . . IMAGEN Consortium. (2014b). Positive association of video game playing with left frontal cortical thickness in adolescents. *PloS One, 9*(3), e91506. doi:10.1371/journal.pone.0091506

Kühn, S., Romanowski, A., Schilling, C., Lorenz, R., Mörsen, C., Seiferth, N., ... Gallinat, J. (2011). The neural basis of video gaming. *Translational Psychiatry*, *1*, e53. doi:10.1038/tp.2011.53

Lefebvre, C. D., Marchand, Y., Eskes, G. A., and Connolly, J. F. (2005). Assessment of working memory abilities using an event-related brain potential (ERP)-compatible digit span backward task. *Clinical Neurophysiology*, *116*(7), 1665–1680. doi:10.1016/j.clinph.2005.03.015

Lewis, J., Trinh, P., and Kirsh, D. (2011). A corpus analysis of strategy video game play in Starcraft: Brood War. In *Proceedings of the 33rd Annual Meeting of the Cognitive Science Society*, Austin, TX: Cognitive Science Society.

Li, L., Chen, R., and Chen, J. (2016). Playing action video games improves visuomotor control. *Psychological Science*, *27*(8), 1092–1108. doi:10.1177/0956797616650300

Libertus, M. E., Liu, A., Pikul, O., Jacques, T., Cardoso-Leite, P., Halberda, J., and Bavelier, D. (2017). The impact of action video game training on mathematical abilities in adults. *AERA Open*, *3*(4), 233285841774085. doi:10.1177/2332858417740857

Lisman, J., Buzsáki, G., Eichenbaum, H., Nadel, L., Ranganath, C., and Redish, A. D. (2017). Viewpoints: How the hippocampus contributes to memory, navigation and cognition. *Nature Neuroscience*, *20*(11), 1434–1447. doi:10.1038/nn.4661

Lorenz, R. C., Gleich, T., Gallinat, J., and Kühn, S. (2015). Video game training and the reward system. *Frontiers in Human Neuroscience*, *9*, 40. doi:10.3389/fnhum.2015.00040

Łuniewska, M., Chyl, K., Dębska, A., Kacprzak, A., Plewko, J., Szczerbiński, M., ... Jednoróg, K. (2018). Neither action nor phonological video games make dyslexic children read better. *Scientific Reports*, *8*(1), 549. doi:10.1038/s41598-017-18878-7

Melby-Lervåg, M. and Hulme, C. (2013). Is working memory training effective? A meta-analytic review. *Developmental Psychology*, *49*(2), 270–291. doi:10.1037/a0028228

Menon, V. (2015). Salience Network. In A. W. Toga (Ed.), *Brain Mapping* (pp. 597–611). Cambridge, MA: Academic Press. doi:10.1016/B978-0-12-397025-1.00052-X

Miendlarzewska, E. A., Bavelier, D., and Schwartz, S. (2016). Influence of reward motivation on human declarative memory. *Neuroscience and Biobehavioral Reviews*, *61*, 156–176. doi:10.1016/j.neubiorev.2015.11.015

Mishra, J., Zinni, M., Bavelier, D., and Hillyard, S. A. (2011). Neural basis of superior performance of action videogame players in an attention-demanding task. *The Journal of Neuroscience: The Official Journal of the Society for Neuroscience*, *31*(3), 992–998. doi:10.1523/JNEUROSCI.4834-10.2011

Moreau, D. (2013). Differentiating two- from three-dimensional mental rotation training effects. *Quarterly Journal of Experimental Psychology (2006)*, *66*(7), 1399–1413. doi:10.1080/17470218.2012.744761

Nau, M., Julian, J. B., and Doeller, C. F. (2018). How the brain's navigation system shapes our visual experience. *Trends in Cognitive Sciences*, *22*(9), 810–825. doi:10.1016/j.tics.2018.06.008

Nava, E., Föcker, J., and Gori, M. (2019). Children can optimally integrate multisensory information after a short action-like mini game training. *Developmental Science*, e12840. doi:10.1111/desc.12840

Oei, A. C. and Patterson, M. D. (2013). Enhancing cognition with video games: A multiple game training study. *PLoS ONE*, *8*(3), e58546. doi:10.1371/journal.pone.0058546

Oei, A. C. and Patterson, M. D. (2014). Playing a puzzle video game with changing requirements improves executive functions. *Computers in Human Behavior*, *37*, 216–228. doi:10.1016/j.chb.2014.04.046

Okagaki, L. and Frensch, P. A. (1994). Effects of video game playing on measures of spatial performance: Gender effects in late adolescence. *Journal of Applied Developmental Psychology*, *15*(1), 33–58. doi:10.1016/0193-3973(94)90005-1

Ophir, E., Nass, C., and Wagner, A. D. (2009). Cognitive control in media multitaskers. *Proceedings of the National Academy of Sciences of the United States of America*, *106*(37), 15583–15587. doi:10.1073/pnas.0903620106

Owen, A. M., Hampshire, A., Grahn, J. A., Stenton, R., Dajani, S., Burns, A. S., ... Ballard, C. G. (2010). Putting brain training to the test. *Nature*, *465*(7299), 775–778. doi:10.1038/nature09042

Palaus, M., Marron, E. M., Viejo-Sobera, R., and Redolar-Ripoll, D. (2017). Neural basis of video gaming: A systematic review. *Frontiers in Human Neuroscience*, *11*, 248. doi:10.3389/fnhum.2017.00248

Pavan, A., Hobaek, M., Blurton, S. P., Contillo, A., Ghin, F., and Greenlee, M. W. (2019). Visual short-term memory for coherent motion in video game players: Evidence from a memory-masking paradigm. *Scientific Reports*, *9*(1), 6027. doi:10.1038/s41598-019-42593-0

Petersen, S. E. and Posner, M. I. (2012). The attention system of the human brain: 20 years after. *Annual Review of Neuroscience*, *35*(1), 73–89. doi:10.1146/annurev-neuro-062111-150525

Pilegard, C. and Mayer, R. E. (2018). Game over for *Tetris* as a platform for cognitive skill training. *Contemporary Educational Psychology*, *54*, 29–41. doi:10.1016/j.cedpsych.2018.04.003

Powers, K. L. and Brooks, P. J. (2014). Evaluating the specificity of effects of video game training. In F. C. Blumberg (Ed.), *Learning by Playing* (pp. 302–330). Oxford: Oxford University Press. doi:10.1093/acprof:osobl/9780199896646.003.0021

Powers, K. L., Brooks, P. J., Aldrich, N. J., Palladino, M. A., and Alfieri, L. (2013). Effects of video-game play on information processing: A meta-analytic investigation. *Psychonomic Bulletin and Review*, *20*(6), 1055–1079. doi:10.3758/s13423-013-0418-z

Pujol, J., Fenoll, R., Forns, J., Harrison, B. J., Martínez-Vilavella, G., Macià, D., ... Sunyer, J. (2016). Video gaming in school children: How much is enough? *Annals of Neurology*, *80*(3), 424–433. doi:10.1002/ana.24745

Rideout, V. J., and Robb, M. B. (2019). *The Common Sense Census: Media Use by Tweens and Teens*. Retrieved from Common Sense Media website: https://www.commonsensemedia. org/sites/default/files/uploads/research/census_researchreport.pdf

Rideout, V. (2016). Measuring time spent with media: The common sense census of media use by US 8- to 18-year-olds. *Journal of Children and Media*, *10*(1), 138–144. doi:10.1080/17482798.2016.1129808

Sala, G., Tatlidil, K. S., and Gobet, F. (2018). Video game training does not enhance cognitive ability: A comprehensive meta-analytic investigation. *Psychological Bulletin*, *144*(2), 111–139. doi:10.1037/bul0000139

Seok, S. and DaCosta, B. (2019). *Video Games as a Literacy Tool: A Comparison of Players' and Nonplayers' Grades, Reading Test Scores, and Self-Perceived Digital Reading Ability*. In K. Graziano (Ed.), *Proceedings of Society for Information Technology & Teacher Education International Conference* (pp. 777–781). Las Vegas, NV: Association for the Advancement of Computing in Education. https://www.learntechlib.org/primary/p/207731/

Sims, V. K. and Mayer, R. E. (2002). Domain specificity of spatial expertise: The case of video game players. *Applied Cognitive Psychology*, *16*(1), 97–115. doi:10.1002/acp.759

Siniatchkin, M. (2017). Anodal tDCS over the left DLPFC improved working memory and reduces symptoms in children with ADHD. *Brain Stimulation*, *10*(2), 517. doi:10.1016/j.brs.2017.01.509

Skorka-Brown, J., Andrade, J., Whalley, B., and May, J. (2015). Playing *Tetris* decreases drug and other cravings in real world settings. *Addictive Behaviors*, *51*, 165–170. doi:10.1016/j.addbeh.2015.07.020

Sparrow, B., Liu, J., and Wegner, D. M. (2011). Google effects on memory: Cognitive consequences of having information at our fingertips. *Science*, *333*(6043), 776–778. doi:10.1126/science.1207745

Spence, I. and Feng, J. (2010). Video games and spatial cognition. *Review of General Psychology*, *14*(2), 92–104. doi:10.1037/a0019491

Stafford, T. and Dewar, M. (2014). Tracing the trajectory of skill learning with a very large sample of online game players. *Psychological Science*, *25*(2), 511–518. doi:10.1177/0956797613511466

Stanhope, J. L., Owens, C., and Elliott, L. J. (2015). Stress reduction: Casual gaming versus guided relaxation. *Human Factors and Applied Psychology Student Conference HFAP Conference*. In *Proceedings of the Human Factors and Applied Psychology Student Conference HFAP Conference*. Daytona Beach, FL. http://commons.erau.edu/hfaphttp://commons.erau.edu/hfap/hfap-2015/papers/9

Strenziok, M., Parasuraman, R., Clarke, E., Cisler, D. S., Thompson, J. C., and Greenwood, P. M. (2014). Neurocognitive enhancement in older adults: Comparison of three cognitive training tasks to test a hypothesis of training transfer in brain connectivity. *NeuroImage*, *85*, 1027–1039. doi:10.1016/j.neuroimage.2013.07.069

Sungur, H. and Boduroglu, A. (2012). Action video game players form more detailed representation of objects. *Acta Psychologica*, *139*(2), 327–334. doi:10.1016/j.actpsy.2011.12.002

Takeuchi, H., Taki, Y., Sassa, Y., Hashizume, H., Sekiguchi, A., Fukushima, A., and Kawashima, R. (2011). Working memory training using mental calculation impacts regional gray matter of the frontal and parietal regions. *PLoS ONE*, *6*(8), e23175. doi:10.1371/journal.pone.0023175

Terlecki, M. S., Newcombe, N. S., and Little, M. (2008). Durable and generalized effects of spatial experience on mental rotation: Gender differences in growth patterns. *Applied Cognitive Psychology*, *22*(7), 996–1013. doi:10.1002/acp.1420

Thorndike, E. L. and Woodworth, R. S. (1901). Influence of improvement in one mental function upon the efficiency of other mental functions. *Psychol Rev*, *8*, 247–261.

Toril, P., Reales, J. M., and Ballesteros, S. (2014). Video game training enhances cognition of older adults: A meta-analytic study. *Psychology and Aging*, *29*(3), 706–716. doi:10.1037/a0037507

Uncapher, M. R. and Wagner, A. D. (2018). Minds and brains of media multitaskers: Current findings and future directions. *Proceedings of the National Academy of Sciences*, *115*(40), 9889–9896. doi:10.1073/pnas.1611612115

Waller, G., Willemse, I., Genner, S., Suter, L., and Süss, D. (2016). *JAMES—Jeunes, activités, médias—enquête Suisse*. Zurich: Haute école des sciences appliquées de Zurich.

Wang, P., Liu, H.-H., Zhu, X.-T., Meng, T., Li, H.-J., and Zuo, X.-N. (2016). Action video game training for healthy adults: A meta-analytic study. *Frontiers in Psychology*, *7*. doi:10.3389/fpsyg.2016.00907

Wang, R., Li, M., Zhao, M., Yu, D., Hu, Y., Wiers, C. E., ... Yuan, K. (2018). Internet gaming disorder: Deficits in functional and structural connectivity in the ventral tegmental area-Accumbens pathway. *Brain Imaging and Behavior*, *13*(4), 1172–1181. doi:10.1007/s11682-018-9929-6

West, G. L., Drisdelle, B. L., Konishi, K., Jackson, J., Jolicoeur, P., and Bohbot, V. D. (2015). Habitual action video game playing is associated with caudate nucleus-dependent navigational strategies. *Proceedings. Biological Sciences*, 282(1808), 20142952. doi:10.1098/rspb.2014.2952

West, G. L., Konishi, K., Diarra, M., Benady-Chorney, J., Drisdelle, B. L., Dahmani, L., ... Bohbot, V. D. (2018). Impact of video games on plasticity of the hippocampus. *Molecular Psychiatry*, 23(7), 1566–1574. doi:10.1038/mp.2017.155

Whitbourne, S. K., Ellenberg, S., and Akimoto, K. (2013). Reasons for playing casual video games and perceived benefits among adults 18 to 80 years old. *Cyberpsychology, Behavior, and Social Networking*, 16(12), 892–897. doi:10.1089/cyber.2012.0705

Winkler, A. M., Kochunov, P., Blangero, J., Almasy, L., Zilles, K., Fox, P. T., ... Glahn, D. C. (2010). Cortical thickness or grey matter volume? The importance of selecting the phenotype for imaging genetics studies. *NeuroImage*, 53(3), 1135–1146. doi:10.1016/j.neuroimage.2009.12.028

Wiradhany, W. and Nieuwenstein, M. R. (2017). Cognitive control in media multitaskers: Two replication studies and a meta-Analysis. *Attention, Perception, and Psychophysics*, 79(8), 2620–2641. doi:10.3758/s13414-017-1408-4

Wu, S. and Spence, I. (2013). Playing shooter and driving videogames improves top-down guidance in visual search. *Attention, Perception, and Psychophysics*, 75(4), 673–686. doi:10.3758/s13414-013-0440-2

Wu, S., Cheng, C. K., Feng, J., D'Angelo, L., Alain, C., and Spence, I. (2012). Playing a first-person shooter video game induces neuroplastic change. *Journal of Cognitive Neuroscience*, 24(6), 1286–1293. doi:10.1162/jocn_a_00192

Zhang, Y., Du, G., Yang, Y., Qin, W., Li, X., and Zhang, Q. (2015). Higher integrity of the motor and visual pathways in long-term video game players. *Frontiers in Human Neuroscience*, 9, 695. doi:10.3389/fnhum.2015.00098

CHAPTER 20

ATTENTION BIASES IN CHILDREN AND ADOLESCENTS

OLIVIA M. ELVIN, KATHERINE M. RYAN,
KATHRYN MODECKI, AND ALLISON M. WATERS

THEORETICAL CONSIDERATIONS AND KEY TENETS ON THE ROLE OF ATTENTION BIASES AND PSYCHOPATHOLOGY IN YOUTH

Theoretical and Developmental Considerations

ATTENTION plays a key role in filtering which information is prioritized to facilitate adaptive behavior (Racer and Dishion, 2012). Broadly, attention is defined as a set of processes that allow individuals to select certain information to the (relative) exclusion of other information (see Yantis, 2000). Attention is understood as being "pushed" or deployed in a deliberate manner in order to maintain attention on a task goal (i.e., top-down), or "pulled" as a result of largely automatic or involuntary sensory stimulation (i.e., bottom-up) (Corbetta and Shulman, 2002; Racer and Dishion, 2012). However, in order to selectively attend to the environment and "filter out" irrelevant information, attention requires a balance of interacting top-down and bottom-up mechanisms to engage in adaptive behavior (Racer and Dishion, 2012; Yantis, 2000). Various systems have been suggested as key in understanding maladaptive attention, such as that of the attention network model (e.g., Berger and Posner, 2000; Posner and Petersen, 1990), which defines attention as the coordination of three systems; alerting, orienting, and executive attention. Alerting is posited as the stage at which individuals may have heightened sensitivity to their environment and is considered closely aligned to vigilance. On the other hand, Posner and colleagues described orienting as engaging, shifting, and disengaging attention from stimuli, and proposed that individual differences in these processes may explain attending to irrelevant stimuli when they are no longer relevant and difficulty focusing on relevant information. Lastly, the executive system aligns with the idea of interacting top-down and bottom-up processes. Executive attention comprises many executive functions and overlaps with other broad functions within the executive system, required to complete tasks such as problem solving. These functions include effortful control in order to inhibit the processing of secondary stimuli (Plude, Enns, and Brodeur, 1994), the inhibition of a dominant response to facilitate self-regulation (Eisenberg, Smith, Sadovsky, and Spinrad, 2004; Rothbart, Ahadi, Hersey, and Fisher, 2001), and inhibitory control to maintain control on task-relevant stimuli and actively inhibit irrelevant stimuli (Plude et al., 1994).

Preferential attention allocation toward salient stimuli is a salient cognitive feature across the lifespan, allowing for the detection of danger within the environment and to respond appropriately. In particular, notionally, humans are biologically wired as a result of evolutionary processes to preferentially attend (and respond to) certain stimuli that are threatening to safety and well-being in order to facilitate avoidance of danger (Öhman and Mineka, 2001). Moreover, attention both shapes and is shaped by developmental and environmental contexts. Individual differences in the capacity to selectively attend to important stimuli and signals, to maintain focus if required, and disengage attention when it no longer aligns with task goals, shapes development from the first days of life (Pérez-Edgar et al., 2017; Posner and Rothbart, 2007). Similarly, exposure to harsh and adverse environments, as opposed to those that are warm and supportive, impacts attention processes differently and at different developmental stages (McLaughlin and Lambert, 2017). Yet, the transition from adaptive to maladaptive attention biases and links to maladjustment and psychopathology are not clear, with some studies suggesting that biased attention to interpersonally salient stimuli (e.g., threatening faces) (e.g., Leppänen and Nelson, 2009) and physically threatening stimuli (e.g., snakes) (LoBue, Buss, Taber-Thomas, and Pérez-Edgar, 2017) is normative in infancy and early childhood and becomes more specific to youth experiencing behavioral and emotional problems with increasing age (Kindt, Brosschot, and Everaerd, 1997). That said, what is clear from this literature is that the expression of attention biases varies across development depending on stimulus emotional salience and intensity and the degree of cognitive load required to process them. That is, attention is likely to be captured in all youth by explicit high-threat stimuli to facilitate safety and well-being (Kindt et al., 1997; Waters, Lipp, and Spence, 2004) and to ambiguous and mildly threatening stimuli when cognitive load is high (Morales, Fu, and Pérez-Edgar, 2016; Waters and Craske, 2016). More specifically, youth with behavioral and emotional problems are likely to be more sensitive to perturbations in attention processes in cases where emotional stimuli are of mild intensity (Waters and Craske, 2016). In addition to stimulus factors, individual differences in the development of effortful control appear to negate attention biases in some youth (Morales et al., 2016), thus contributing to a more balanced and adaptive attention system over time for these individuals. Conversely, failure to develop the capacity to control attention appears to maintain attention biases during development (Muris, de Jong, and Engelen, 2004). Therefore, it is important to consider the interaction between stimulus-driven processes and children's capacity for effortful control in influencing children's attention biases and how these may play a role in the development of internalizing or externalizing problems (Waters and Craske, 2016).

Notably, early theoretical and conceptual models of attention processes in relation to internalizing psychopathology, and anxiety in particular, placed more emphasis on perturbations in early orienting or involuntary (i.e., bottom-up) stages of processing than on later (top-down) control processes as underpinning attention biases for threat stimuli in anxious individuals (e.g., Eysenck, Derakshan, Santos, and Calvo, 2007; Mogg and Bradley, 1998; Williams, Watts, MacLeod, and Mathews, 1988). However, informed by more than three decades of research, contemporary models recognize the greater role of top-down influences, suggesting that attention biases for threat stimuli are likely to be underpinned by both bottom-up and top-down processes (e.g., Mogg and Bradley, 2018; Waters and Craske, 2016). Attention models specific to depression are similar to those of anxiety, recognizing the influence of both top-down and bottom-up influences on attention biases that are observable over longer stimulus presentation durations due to processes characteristic of depression such as rumination and brooding. Thus, attention biases in depression may be characterized by difficulties disengaging from processing negative (e.g., sad) stimuli in addition to the absence of a positivity bias, again highlighting the role of biases in top-down influences on attention in depression (Cisler and Koster, 2010; Peckham, McHugh, and Otto, 2010).

On the other hand, cognitive models of externalizing problems in youth have lagged behind internalizing problems, and as such, the field has lacked a clear understanding of how attention biases influence the development of externalizing problems. Rather, approaches to date have tended to focus on discrete components of externalizing problems. For example, Brotman, Kircanski, Stringaris, Pine, and Leibenluft (2017) emphasize the interacting roles of a decreased threshold to attend to both threat and reward stimuli in youth with high levels of irritability. Similar models have been proposed to explain visual attention to threat cues in the context of aggression or anger, although many of these studies have examined these models within adult populations (e.g., Lin et al., 2016; Wilkowski and Robinson, 2008). In terms of psychopathic traits, most research has focused upon deficits in attending to fear stimuli in youth with high levels of callous-unemotional (CU) traits (Dadds et al., 2006). Given the emphasis on biases in attention allocation toward reward, punishment, and affective stimuli, clearer models are required to explain the similarities between disorders within externalizing psychopathology and those that cut across internalizing and externalizing disorders. The progression of research to address these important issues has required considerable advancement in experimental tasks and methodology reviewed next.

METHODOLOGICAL CONSIDERATIONS

Biases in the selective allocation of attention to emotional stimuli in children have most commonly been assessed with the visual probe task in which a threat and neutral stimulus (typically words or faces) are presented simultaneously for between 500 and 1500 ms, followed by a probe in each location over trials. Differences in reaction time to respond to the probe following the emotional compared to the neutral stimulus provides an index of attention bias (e.g., MacLeod and Matthews, 1988). In more recent years, following a body of evidence that questioned the reliability of reaction-time measures using the dot-probe task (e.g., Price et al., 2015; Staugaard, 2009), considerable effort has been devoted to identifying more reliable behavioral indices of attention bias including attention bias variability, dynamic patterns of attention allocation and use of eye-tracking to index attention orienting, engagement and disengagement and sustained attention (Armstrong and Olatunji, 2012; Price et al., 2014; Zvielli, Bernstein, and Koster, 2014). Moreover, in addition to varying the stimulus presentation duration, type of stimuli presented (words vs faces), and the response format required (probe location; probe classification), recent studies have used individually presented stimuli for up to 10 s duration to index differences in the time course of attention vigilance and avoidance (Shechner et al., 2013). Children's biases in selective attention to emotional stimuli have also been assessed using interference-based paradigms including the emotional Stroop task and the Go/No Go task, in which biases in attention to emotional stimuli are indexed by the extent to which emotional content interferes with responding to neutral stimuli and task-relevant behavior.

Visual search tasks (Öhman and Mineka, 2001) have also been used to assess attention bias and include the presentation of a matrix of faces, including distractor stimuli and the presence or absence of a target stimulus, presented on the screen until a response is selected. Stimuli used within the visual search task have varied from emotional faces, words, or pictures of animals (Cisler, Bacon, and Williams, 2009). Faster reaction times to detect an emotional (e.g., angry or sad) face amongst distractors (e.g., neutral faces) in comparison to healthy controls suggest a negative attention bias. Additionally, slower response times to detect a neutral face amongst threat/sad distractor faces suggests a negative attention bias, also suggesting difficulty disengaging from negative stimuli (e.g., through the employment of inhibitory control) (Waters and Lipp, 2008a).

The emotional Stroop task (Kindt, Bierman, and Brosschot, 1997; Reinholdt-Dunne, Mogg, and Bradley, 2009) has likewise been used to examine attention and inhibitory control, primarily in youth with internalizing problems. The emotional Stroop task assesses inhibitory control and the ability to ignore the interference of emotional stimuli (i.e., threat) whereby reaction times and errors are calculated for the identification of color stimuli (e.g., emotional faces or words) and inhibition of responses for neutral stimuli (Brown et al., 2014). Similar to the Stroop task, the emotional Go/No Go task assesses the ability to inhibit responses to threat stimuli whilst participants respond to a particular face (e.g., neutral faces) and withholding a response to other expressions (e.g., threatening/angry faces), and vice versa. The number of errors and reaction times on the tasks are used as an indication of how effectively an individual can inhibit a dominant response and stay on-task, with slower reaction times and more errors suggesting poorer response inhibition to threat stimuli (Waters and Valvoi, 2009). Variations of the Stroop task and the Go/No Go task have also been used throughout the literature to assess these functions, such as response inhibition (i.e., Stop Signal Task).

The Affective Posner task has been modified based on the traditional Posner task (Posner, 1980) for use in populations to assess attention to reward and punishment stimuli. Individuals are presented with a fixation cross, followed by boxes presented on the screen and a cue. Participants are required to respond to the location of the target presented immediately following the cue on the same (congruent) or opposite (incongruent) side of the screen. Across the tasks, feedback is provided in the form of words (e.g., "good job"/ "wrong"), rewards (e.g., "win 10c"), or punishment (e.g., "lose 10c"). Frustration is built across tasks by providing rigged feedback on tasks, resulting in the loss of money. This task allows for the assessment of sustained attention on reward/punishment cues when eye tracking is utilized, and/or the examination of interference of these cues on performance by comparing reaction times and errors across trials. This task provides an indication of the way in which biases in attention toward punishment and reward stimuli may maintain and/or exacerbate symptomology, particularly in children who are prone to experiencing frustration (Deveney et al., 2013; Rich et al., 2007; Tseng et al., 2017).

Early work focused on the inability of some youth to attend to and inhibit relevant task related stimuli due to core executive function deficits. The Attention Network Test was devised to assess differences in Posner and colleagues' suggested processes of attention by measuring reaction times in responding to targets presented following a cue (i.e., an arrow indicating where the target will be). The target is presented in the same (congruent) or opposite (incongruent) direction, providing an indication of an individual's efficiency of their executive attention network (Fan, McCandliss, Sommer, Raz, and Posner, 2002). As a non-experimental alternative, the Behavior Rating Inventory of Executive Functioning (BRIEF; Gioia, Isquith, Guy, and Kenworthy, 2000) is a normed questionnaire used as a parent or teacher report 86-item measure to provide an indication of executive functioning (e.g., metacognitive skills, behavior regulation, and aspects of attention) in those 5 to 18 years of age.

ATTENTION BIASES AND INTERNALIZING PSYCHOPATHOLOGY IN CHILDREN AND ADOLESCENTS

Internalizing problems (e.g., withdrawal, worry, rumination, avoidance) and related disorders including depression and anxiety in youth are common mental health concerns emerging in mid to late childhood and adolescence (Merikangas, Nakamura and Kessler, 2009). The lifetime prevalence of anxiety is 29 percent and major depressive disorder is 17 percent (Kessler et al., 2005), with these disorders being the leading cause of disability claims, adding to individual and societal costs (Lawrence et al 2015; Mathews, Hall, Vos, Patton, and Degenhardt, 2011; Wittchen et al., 2011). Given the prevalence and costs of internalizing psychopathology, advancing knowledge about the underlying mechanisms such as attention biases is important for informing aetiological models and enhancing treatments.

A number of recent meta-analyses and large-scale reviews have been conducted on the role of attention biases toward negative and positive stimuli in youth with anxiety disorders (e.g., Platt, Waters, Schulte-Koerne, Engelmann and Salemink, 2017; Lau and Waters, 2017; Pergamin-Hight, Naim, Bakermans-Kranenburg, van Ijzendoorn, and Bar-Haim, 2015; Dudeney, Sharp, and Hunt, 2015). These meta-analyses have generally concluded that an attention bias towards threat is characteristic of anxious relative to non-anxious children, with the magnitude of this difference increasing with age (Dudeney et al., 2015). However, findings have been mixed when a wider range of studies using various methodological tasks, conceptualizations, and stimuli are considered, suggesting different tasks may be tapping different elements of the attention process.

Anxiety

Early work into attention biases toward threat in anxious children began with the emotional Stroop task, finding that both anxious and control children demonstrated a bias toward threat (Kindt, Bierman, and Brosschot, 1997; Kindt, Brosschot, and Everaerd, 1997). Such different results relative to adult studies and theoretical models led to the suggestion that the emotional Stroop task was unsuitable for youth as it may invoke high levels of cognitive load that affects all children.

Early reviews of studies employing dot probe tasks using word and face stimuli indicated that children with low to moderate levels of anxiety display a bias toward threat cues, while those with self-reported high anxiety level reported attention bias away from threat cues (Ehrenreich and Gross. 2002). Early studies using the dot probe task yielded similar evidence of a common threat bias when anxious and non-anxious children were presented with high threat-related, positive, and neutral pictures (e.g., Waters, Lipp, and Spence, 2004). Building on this work, a revised task using face stimuli found that severely anxious children displayed attention biases toward both angry and happy faces compared to mildly anxious children and a control group (Waters, Mogg, Bradley, and Pine, 2008). Other variants of the dot probe task that included modified stimuli (e.g., faces with cropped hair; oval shapes instead of faces for neutral trials) (Price et al., 2013) and stressful environmental manipulations (functional magnetic resonance imaging (fMRI) scanner) (Monk et al., 2006) similarly yielded mixed results of both vigilance and avoidance of threat in anxious children.

Anxiety disorders are described as reflecting a general internalizing domain that can be further broken into fear and distress disorders. Fear disorders incorporate specific phobia, social phobia, separation anxiety disorder, panic disorder, and agoraphobia whereas distress disorders include generalized anxiety disorder, dysthymia, and depression (Clark and Watson, 2006). In an effort to determine factors that might contribute to differences in bias direction, one large youth study used happy and threat face stimuli in a dot probe task and found vigilance toward threat in youth with distress-related disorders and avoidance of threat in youth with fear-related disorders (Salum et al., 2013). As a result, these researchers tested children with GAD and those with a fear disorder (e.g., specific phobias) using the visual probe task with emotional faces (500 ms) and compared their performance to healthy controls. Children with GAD showed significant bias toward threat relative to neutral faces, whereas those with a fear disorder showed attention bias away from threat relative to neutral stimuli (Waters, Bradley, and Mogg, 2014). However, Abend et al. (2018) subsequently found evidence of attention vigilance toward threat in youth with social phobia relative to other disorders highlighting that this field of research requires more attention.

In addition to studies that have examined biases in initial attention allocation to threat, more recent work has focused on differences in sustained attention on emotional stimuli in anxious youth. One study found both early and intermediate vigilance patterns in response times (i.e., 500–2000 ms) in addition to avoidance gaze patterns during a dot probe task in youth, regardless of anxiety (Price et al., 2013). Moreover, sustained pupil dilation was observed in anxious youth on trials using fearful faces, along with inflexible pattern of pupillary responding in this group, in comparison to the control group (Price et al., 2013). Anxious youth's late pupil alterations suggest sustained effortful processes in response to threats that are inflexible and likely maladaptive.

Attention biases have more recently been assessed using eye-tracking software during picture viewing tasks and findings have again been mixed with several studies reporting biases for emotional stimuli at early stages of processing (Dodd et al., 2015; Shechner et al., 2013), whereas other studies have found mixed results of both vigilance and avoidance of threat stimuli (Gamble and Rapee, 2009). Along similar lines, young children with social phobia displayed hypervigilance (3000 ms) towards threat (angry) faces following induced anxiety, compared to children in a control group (Seefeldt, Kramer, Tusch-Caffier, and Heinrichs, 2014). With the emergence of new research in children and adolescents, there is mixed evidence regarding patterns of vigilance and avoidance in both reaction time and eye-tracking studies (In-Albon, Kossowsky, and Schneider, 2010). Children with social anxiety and healthy controls completed a visual probe task (5000 ms), viewing different picture pairs consisting of faces and houses; both groups displayed avoidance pattern to angry faces; however the total dwell time on angry faces in favor of houses was negatively associated with anxiety levels, that is, the more anxious children avoided the angry face more often than children who were less anxious (Schmidtendorf, Wiedau, Asbrand, Tuschen-Caffier, and Heinrichs, 2018).

One study employing the Go/No Go task with youth participants revealed that participants in the anxious group had significantly slower reaction times to neutral face Go trials embedded in angry face No-Go trials in the low probability condition (Ladouceur et al., 2006). This demonstrated that processing of emotional facial expression interfered with the performance on another task in youth diagnosed with anxiety disorders. Another study found gender differences with anxious girls responding more slowly to neutral faces with embedded angry versus happy face No-Go trials compared to anxious boys and controls, indicating that angry faces selectively interfered with performance on a neutral task (Waters and Valvoi, 2009).

Similarly, visual search interference tasks demonstrate problems with inhibitory control of attention to threat. Waters and Lipp (2008b) demonstrated that pictures of fear relevant spiders and snakes in the background of other animal stimuli captured attention and interfered in the detection of a neutral target in children, and this was slowed further in the presence of a feared fear relevant distractor. In an older sample the results again showed preferential attentional processing of animal fear relevant stimuli, with those who specifically feared one animal but not the other showed enhanced preferential processing of their feared fear relevant animal (Waters, Lipp, and Randwana, 2011).

Consistent with more contemporary theoretical perspectives (Mogg and Bradley, 2018), biases in attention appear to be an amalgamation of both top-down and bottom-up influences, underpinning selective attention. One novel study examined the association between early bottom-up processes and later stage top-down processes and found that early attention bias towards threat (i.e., early threat orienting) was negatively correlated with later attention bias to threat, suggesting that early threat vigilance was associated with later threat avoidance. Moreover, this association was specific to anxiety (Sylvester, Hudziak, Gaffrey, Barch, and Luby, 2016). The parallel associations between attention bias for sad faces and depression were not observed. Taken together, the evidence suggests that attention biases for threat in anxious youth are linked to differences in underlying involuntary and voluntary attention processes and thus the amalgamation of both bottom-up, stimulus-driven and top-down, cognitive control processes.

At Risk Youth

Given that offspring of parents with an anxiety disorder are 3.5 times more likely to develop an anxiety disorder compared to the offspring of non-anxious parents (Merikangas, Avenevoli, Dierker, and Grillon, 1999), attention biases in these youth may be a risk marker for anxiety. For example, attention biases were assessed using a dot probe paradigm with threat, happy and neutral faces in a study in a large sample of 6- to 12-year-olds. It was found that daughters of mothers with an emotional disorder displayed increased attention to threat stimuli compared to daughters of mothers without an emotional disorder (Montagner et al., 2016). These findings were in keeping with those of Waters, Forrest, Peters, Bradley, and Mogg (2015) whereby an attention bias towards threat was observed in high-risk children of mothers with an emotional disorder who did not allocate attention to positive stimuli compared to those mothers who did direct attention to positive stimuli and low-risk offspring (Waters, Forrest, et al., 2015). Moreover, when assessed at follow-up 12 months later, increased child anxiety symptoms were associated with increased maternal threat attention bias in high-risk, but not low-risk dyads, in cross-sectional but not longitudinal outcomes. Such findings suggest that maternal attention to threat stimuli may contribute to increasing child anxiety symptoms via indirect pathways, such as threat information and modeling of distress in response to the threat stimuli to which mothers with emotional disorders attend (Waters, Candy, and Candy, 2018).

Maternal anxiety during pregnancy can also expose a child to greater risk of behavioral and emotional problems later in life (Bock, Wainstock, Braun, and Segal, 2015). A recent study tested 4-year-old children of mothers experiencing high anxiety in their second trimester, compared to 4-year-olds of mothers with low anxiety. Measuring event related brain potentials (ERPs) whilst the children viewed pleasant, neutral and unpleasant pictures, it was found that children of mothers with high compared to low anxiety during pregnancy displayed more attention bias toward neutral rather than unpleasant stimuli (van den Heuvel, Henrichs, Donker, and Van den Bergh, 2018). The researchers interpreted this as the children showing vigilance toward threat that could result in developing behavioral or emotional problems later in life.

Behavioral inhibition (BI) is a biologically based temperament characterized by strong avoidance of novel and unfamiliar situations and people and a key child-related risk factor for developing anxiety, especially social anxiety. Several studies have found that children high on BI exhibit attention biases towards threat stimuli. For example, one study employing the dot probe task while children were undergoing an fMRI scan, found that trials requiring attention orienting away from threat engaged an executive and threat-attention network in children high on BI (Fu, Taber-Thomas, and Perez-Edgar, 2017). Heightened brain activation was related to increased anxiety, and BI levels accounted for the direct relationship between brain activation and anxiety. Behaviorally inhibited children may engage the executive attention system during threat-related processing as a compensatory mechanism (Fu et al., 2017). Relatedly, findings from another study using a dot probe task (angry and happy faces – 500 ms) suggested that youth with a high fear temperament and low attention control (i.e., ability to employ top-down control) were more likely to exhibit an attention bias toward threat stimuli than youth low on fear temperament and attention control (Susa, Pitică, Benga, and Miclea, 2012).

Another broad-based temperament risk factor for anxiety is high levels of negative affect or the propensity to experience strong negative emotion. Examining associations between negative affect and effortful control, Lonigan and Phillips (2001) found that effortful control moderated the relation between negative affectivity and attention bias. That is, only children with low levels of effortful control and high levels of negative affectivity exhibited an attention bias to threat stimuli (Lonigan and Vasey, 2009). Taken together, parental emotional disorders and child temperament characteristics of BI and negative affect may increase risk for attention biases for threat stimuli and effortful control processes may play a compensatory role in downregulating the risk associated with these underlying temperament risk factors.

Treatment

Another area of importance is the extent to which attention biases influence treatment outcomes (i.e., cognitive behavioral therapy). Although in its infancy, several studies suggest that pre-treatment attention bias toward threat in anxious children were associated with greater reductions in anxiety symptoms as compared to those with pre-treatment attention biases away from threat (Waters, Mogg, and Bradley, 2012; Waters, Potter, Jamieson, Bradley, and Mogg, 2015; Manassis, Hum, Lee, Zhang, and Lewis, 2013). Some studies found mixed results, with better initial response to low-intensity CBT associated with pre-treatment avoidance of severe threat, while subsequent response to more intensive CBT was associated with greater pre-treatment threat vigilance (Legerstee et al., 2009; Legerstee et al., 2010).

Depression

A review of cognitive biases in youth depression (Platt et al., 2017) found generally consistent evidence for positive associations between attention biases and youth depression. Variations that were found were likely attributable to methodological factors or the sample population used. Similar to the anxiety literature, dot probe tasks have been commonly used in studies of youth depression, albeit typically finding a negative (i.e., sad) attention bias in comparison to threat. This negative attention bias was found in a sample of 105 adolescents before and after a visual search based cognitive bias medication training (Platt, Murphy, and Lau, 2015). Additionally, Sylvester et al. (2016) indicated that greater biases toward sad faces was found using a dot probe task with adolescents who had a history of depression, comparative to youth with no history of depression. Moreover, attention biases toward negative words also predicted increased symptoms of depression across a semester for youth (Osinsky, Lösch, Hennig, Alexander, and MacLeod, 2012).

In regard to the interfering effects of negative stimuli on attention in depressed youth, studies that have used the emotional Go/No Go task with faces or words have found reasonably consistent results. Specifically, when comparing depressed youth, anxious youth and controls, findings revealed significantly faster reaction times to sad faces on Go trials, embedded in neutral No Go trials, indicating a bias for sad faces (Ladouceur et al., 2006) and sad words compared to happy words (Maalouf et al., 2012; Kyte, Goodyer, and Sahakian, 2005) in depressed youth.

At Risk Youth

Parental depression is also a major risk factor for offspring depression. Attention bias studies find mixed results regarding attention vigilance and avoidance of sad-related stimuli. A dot probe study examined genetic risk factors in the intergenerational transmission of depression, finding that children (8–12 years old) of clinically depressed mothers (compared to those with no history of depression), exhibited larger biases away from sad faces (Gibb, Benas, Grassia, and McGeary, 2009). In contrast, a study of young children from 5 to 7 years old found that daughters of depressed mothers (vs daughters of non-depressed mothers) showed a stronger bias towards sad faces compared to happy faces (Kujawa et al., 2011).

There is growing evidence that the role of parental criticism is associated with off-spring attention biases. In a recent study, children completed a morphed faces task in which reaction times and event related potential responses were assessed. Children of parents who exhibited high compared to low emotional criticism displayed less attention to all face emotions, suggesting an avoidant attention pattern (James, Owens, Woody, Hall, and Gibb, 2018). These findings are in line with an earlier study in which children of depressed mothers (compared to non-depressed mothers) avoided sad faces (Gibb, Pollak, Hajcak, and Owens, 2016).

Treatment

In contrast to studies of anxiety in youth, no studies at the time of preparing this chapter have examined whether attention biases predict treatment outcomes following existing treatments for depression including cognitive-behavioral and interpersonal therapies. This remains an important direction for future research.

Novel Treatments Targeting Attention Biases Across Internalizing Disorders

Attention bias modification (ABM) is a new treatment approach designed specifically to reduce attention biases for emotional stimuli and thereby symptom severity. The original form of ABM is a computer-based training method based on the visual probe task whereby the probe repeatedly follows the neutral instead of the threat stimulus and thereby implicitly trains attention away from threatening stimuli (i.e., ABM-threat avoidance training; MacLeod and Clarke, 2015). ABM is often compared to a control condition whereby participants complete a typical dot-probe task, with the probe replacing the neutral and threatening stimulus on an equal number of trials (Hakamata et al., 2010; Liu, Taber-Thomas, Fu, and Perez-Edgar, 2018). An initial meta-analysis conducted in 2010 of 12 randomized controlled trials in adults highlighted that ABM produced a significantly greater reduction in anxiety than control training (Hakamata et al., 2010). However, several ABM-threat avoidance studies in both adult and youth samples have since found little support for this approach and a recent meta-analysis of youth studies found no effect of ABM-threat avoidance on anxiety or depression symptom reduction (Cristea, Mogoase, David, and Cuijpers, 2015).

Subsequent findings have continued to be mixed with some studies providing some support for ABM-threat avoidance approaches in reducing anxiety symptoms (Liu et al., 2018; White et al., 2017) whereas other have not (Ollendick et al., 2019). Another study found no change in attention bias in youth in the ABM-threat avoidance group or the control group, however they did find a trend (p = .055) towards reduction in anxiety severity in the ABM group compared to the control group (Chang et al., 2018).

Visual search ABM involves the presentation of matrices of negative/threat-related stimuli with positively oriented targets embedded among them. This explicit form of ABM training is designed to engage top-down cognitive control by encouraging youth to search for and respond to positive targets and inhibit the processing of threat distractors (Mogg, Waters, and Bradley, 2017). Some studies conducted with high anxious and non-clinical adolescents have produced mixed results, with high socially anxious youth assigned to visual search ABM exhibiting a decrease in their attention bias for negative information and self-reported social anxiety symptoms (De Voogd, Wiers, Prins, and Salemink, 2014). However, more recent studies with non-selected adolescents found anxiety and depression symptomology reduction following visual search ABM and control conditions and questions whether online ABM is effective as an early intervention in adolescents (De Voogd, Wiers, and Salemink, 2017).

More encouraging results have been found with positive search training (PST), which explicitly trains anxious children in adaptive, goal-directed attention search strategies to search for positive and calm information and inhibit goal irrelevant negative cues. A clinical sample of 6- to 17-year-olds who participated in PST and one treatment session of exposure therapy had greater reductions in danger expectancies during exposure therapy, including reduced clinician rated phobia, and at three months follow-up, compared to a group that only participated in attention training control (Waters, Farrell et al., 2014). Furthermore, PST was trialed in two other studies with clinically anxious children (7–12 years of age), finding that post-treatment clinician- and parent-report measures of anxiety symptoms improved significantly compared to the wait-list group (Waters, Zimmer-Gembeck, Craske, Pine, Bradley and Mogg, 2016). Furthermore, a recent small-scale study on the neural correlates of PST suggest increased activation in a broad attention network potentially suggestive of enhancing cognitive control (Waters, Cao, et al., 2018).

ATTENTION BIASES AND EXTERNALIZING PSYCHOPATHOLOGY IN CHILDREN AND ADOLESCENTS

Despite much of the research focusing on attention biases in youth with internalizing psychopathology, children and adolescents with increased externalizing symptomology (e.g., irritability, anger, aggression), disorders (disruptive mood dysregulation disorder (DMDD), attention deficit hyperactivity disorder (ADHD), conduct disorder (CD) and oppositional defiant disorder (ODD)), and traits (e.g., callous-unemotional, psychopathy) have also demonstrated attention biases for a range of stimuli, including emotional faces (i.e., angry faces), reward stimuli, and punishment stimuli. Externalizing problems are thought of as a spectrum of behaviors that are directed outward (Humphreys et al., 2019), being one of the most prevalent and persistent types of childhood difficulties. These behaviors have been described in the past as attention problems, self-regulation difficulties (Bornstein, Hahn, and Haynes, 2010), aggressive behaviors, rule-breaking behaviors (Achenbach and Rescorla, 2001), impulsivity, and disruptive difficulties. High rates of comorbidity are seen in youth with externalizing problems, likely due to their similar genetic underpinnings and environmental risk factors (American Psychological Association, 2000). Many externalizing disorders (i.e., ADHD, ODD) appear to change with age, with many young children presenting with more aggressive behaviors or physical violence, whereas an increase in externalizing problems during adolescence generally presents in relation to conduct problems, truancy, and drug use (Bongers, Koot, Jan van der, and Verhulst, 2004). On the other hand, some externalizing disorders (e.g., DMDD) have been associated with later internalizing problems in adulthood, such as anxiety and depression (Copeland Shanahan, Egger, Angold, and Costello, 2014). Despite the differences in the presentation of externalizing problems, these behaviors impact not only on the individual's social, emotional, academic, and behavioral functioning, but also their family and peers. Due to the nature of externalizing psychopathology being linked to exuberance and impulsivity, much of the research has focused on the role of threat, reward, and punishment biases in these children and adolescents (Carlson, Pritchard, and Dominelli, 2013) and their difficulty employing cognitive control (i.e., inhibiting responses), leading to increased attention on these stimuli (Morales et al., 2016).

Irritability

The early work examining attention biases and irritability examined youth with severe mood disorder (SMD) and/or bipolar disorder who exhibit heightened irritability. Specifically, Hommer et al. (2014) found that children and adolescents with SMD had greater attentional biases toward threat in comparison to healthy controls using a dot-probe task. Furthermore, this study also found that greater attention biases were evident in youth with greater symptom severity of SMD (i.e., more severe irritability), regardless of the presence of internalizing psychopathology or not. Following this, Salum et al. (2017) examined attention biases using a dot-probe task in children. It was found that after averaging across short (500 ms) and long (a combination of 500 ms and 1250 ms) stimulus durations, children who were often irritable had greater attention biases toward threat (angry faces) relative to neutral faces when compared to non-irritable children. However, neither Hommer et al. (2014) nor Salum et al. (2017) found attention biases toward happy faces using these tasks. Despite this, within attention biases toward reward stimuli, Rich et al. (2007) found that youth with SMD had deficits in their ability to allocate their attention toward targets on the Affective Posner task in comparison to those with bipolar disorder and healthy controls, regardless of the emotion of the stimuli. Moreover, Kessel et al. (2016) found that of 425 participants, children with severe

DMDD symptoms at age three were associated with increased sensitivity to rewards, indexed by their greater electroencephalogram (EEG) positivity response during a reward task at age nine. Despite the limited evidence, these findings suggest that children prone to experiencing irritability may preferentially attend to threat stimuli, and cues indicative of potential reward.

Anger

Similar to irritability, He et al. (2013) assessed a sample of children using the Affective Posner task and determined that children more prone to anger allocated their visual attention toward reward cues faster than punishment cues. However, many studies within the literature examining individuals characterized by anger (e.g., CD) or aggression focus primarily on the interpretation of emotions as more threatening (i.e., hostile attribution) in adults and adolescents (see Mellentin, Dervisevic, Stenager, Pilegaard, and Kirk, 2015) rather than the preferential allocation of attention toward these stimuli in youth.

Aggression

Within a related phenotype to irritability and anger, aggression has also been examined to determine how biases in attention present. However, at odds to the findings within irritability, Horsley, Orobio de Castro, and van der Schoot (2010) found that, in a sample of 10- to 13-year-olds, youth with heightened aggression did not attend more to hostile cues within cartoon visual settings compared to non-hostile cues. Rather, it was found that children with heightened aggression looked for a longer duration at non-hostile cues and attributed more hostile intent to non-aggressive peers, a finding that aligns with research showing links between childhood aggression and hostile attribution biases (De Castro, Veerman, Koops, Bosch, and Monshouwer, 2002). However, these findings differ to the research of Laue et al. (2018) whereby they found that adolescents who displayed more aggressive behaviors demonstrated a hypersensitivity toward hostile cues using the same social scenes as Horsley et al. (2010). Schippell, Vasey, Cravens-Brown, and Bretveld (2003) also found that on a dot-probe task, adolescents with heightened relational aggression responded more slowly to the probe following socially threatening words relative to neutral words, suggestive of avoidance of threat stimuli. A study by Ciucci et al. (2018) suggested that within a sample of 11- to 15-year-olds, youth seen as more aggressive by their peers had greater attention allocation toward angry faces. However, some studies (e.g., Kimonis, Frick, Fazekas, and Loney, 2006) suggest that the presence of CU traits in youth may lead to a variation in findings due to their specific patterns of interpreting and attending to emotion.

Psychopathy

Attention to emotional stimuli in adolescents with psychopathic traits (e.g., CU traits) has been examined using a variety of tasks and stimuli. Much of this research has found evidence to suggest reduced reactivity to punishment cues and affective stimuli compared to youth without CU traits (e.g., Centifanti and Modecki, 2013; Loney, Frick, Clements, Ellis, and Kerlin, 2003). However, some research suggests that psychopathy is made up of a combination of aggression and CU traits. Specifically, in a sample of incarcerated African American boys with high CU traits and high aggression, deficits in their processing of distress stimuli were found comparative to youth high on aggression and low on CU traits, who demonstrated enhanced attention toward distress stimuli (Kimonis et al., 2006; Kimonis, Graham, and Cauffman, 2017). Similarly, age may act as a moderator. Illustratively, Kosson, McBride, Miller, Riser, and Whitman (2018) found that following a frustration exercise and dot-probe task, younger detained adolescents with higher psychopathic traits had greater biases toward positive and negative affective words compared to those without psychopathic traits. However, older detained youth with psychopathic traits had non-significant biases away from affective (sad and happy) stimuli. Conversely, adolescents with CU traits responded slower to negative word stimuli when compared to adolescents without CU traits (Loney et al., 2003). Perhaps variants in the type of CU traits (i.e., primary versus secondary) explain differential patterns of biases in these youth. Kimonis, Frick, Cauffman, Goldweber, and Skeem (2012) found that adolescents with primary variants of CU traits (described as having core deficits in emotional responding) did not demonstrate attention biases toward distressing stimuli (e.g., crying babies). However, youth with secondary variants of CU traits (prone to negative emotionality) displayed greater attention toward distressing emotional stimuli.

The findings in the literature also appear to vary when assessing psychopathy in youth with narcissistic traits as compared to those with CU traits. Centifanti, Kimonis, Frick, and Aucoin (2013) found that 13- to 18-year-olds with heightened psychopathy-associated narcissism had increased attention toward social distress/disapproval stimuli (e.g., ego threat, social rejection) compared to neutral stimuli. Alongside experimental paradigms, some studies have also examined the neural changes in the brain whilst performing these tasks (see Frick and White, 2008; White et al., 2012), suggesting some patterns of biases in both attention and identification of faces may be attributable to underlying neural processing of emotional stimuli.

ADHD

Alongside attentional difficulties, youth with ADHD are often characterized as emotionally dysregulated, reactive, short-tempered and irritable (Cremone, Lugo-Candelas, Harvey, McDermott, and Spencer, 2018). Children and adolescents with ADHD have demonstrated attentional biases towards positive stimuli in comparison to typically developing peers. Cremone et al. (2018) assessed a sample of

4- to 8-year-olds using a dot-probe task and found that youth with symptoms consistent with ADHD had attention biases toward positive stimuli only in comparison to healthy controls. Similarly, Tripp and Alsop (2001) found that using a signal-detection task, children with ADHD had an increased desire for immediate reward, evidenced in an overall mean bias score that was significantly higher than healthy controls. Current evidence suggests that a positive attention bias improves performance, whereas others suggest it inhibits performance (Cremone et al., 2018). Given the varied nature of emotional biases in youth with ADHD throughout the literature, an important consideration is the extent to which these biases are specific to emotional material (i.e., Pishyareh, 2015), or are primarily the result of problems with executive functions, such as working memory, decision-making, and problems with inhibition (e.g., Denney, Rapport, and Chung, 2005; Klein, Wendling, Huettner, Ruder, and Peper, 2006; Kofler et al., 2018). There appears to be overlap in these core executive functions with emotion and reward and punishment-based stimuli and tasks, for example, the role of inhibitory control and its impact on sustained attention (e.g., Morales et al., 2016).

Sustained Attention

Generally, youth with externalizing problems also have difficulty inhibiting responses in order to sustain attention on tasks when faced with both threat and reward stimuli. Poorer performance on emotional inhibition tasks (e.g., emotional Stroop task) are thought to arise due to dysfunction in executive control. Thus, this reduced ability to inhibit responses to emotional distractors has been suggested as a mediating factor in increased externalizing behaviors due to the combination of attentional biases toward stimuli such as reward and threat, and the difficulty redirecting attention back to task demands (Morales et al., 2016). These biases are thought to maintain externalizing psychopathology in youth due to their preferential allocation of attention toward certain stimuli and their difficulty disengaging attention and attending to task goals (e.g., employing top-down attention).

ADHD

Ma et al. (2018) found that in a sample of 9- to 17-year-olds, appetitive words (e.g., good, lucky) in comparison to neutral words (e.g., pen, year), had a negative impact on interference control, leading to an increased error rate and slower reaction times, however this was not exacerbated in youth with ADHD. Conversely, they found that aversive words (e.g., bad, wrong) when matched with neutral words, were facilitative regarding interference control, with faster reaction times with increasing ADHD symptoms. Alongside these findings, Marsh et al. (2008) found that children with ADHD had slower reaction times in making judgements of the gender of fearful, neutral, and angry faces, suggesting difficulties employing top-down cognitive control. Using a response inhibition task, adolescents with ADHD had slower response times when presented with positive distractors, indicating greater interference (Passarotti, Sweeney, and Pavuluri, 2010). However, considering the impaired executive functioning that has been implicated in ADHD (e.g., Hwang et al., 2015), it is important to note that a meta-analysis conducted by Van Mourik, Oosterlaan, and Sergeant (2005) concluded that generally no differences in interference control were found on the traditional Stroop task in comparison to control groups. This suggests that despite some core deficits in ADHD, the emotional content of stimuli appears to play a larger role in attention biases in these youth. Despite the limited evidence of bias toward threat stimuli, increased sensitivity to receiving rewards in these children and adolescents has been linked to increased externalizing difficulties due to their impaired ability to regulate emotions and high approach to reward due to limited impulse control (Morales et al., 2016).

SMD

Using the Emotional Interrupt Task, it was found that youth with SMD were less accurate in comparison to those with bipolar disorder and healthy controls with both emotional and neutral stimuli, suggestive of deficits in sustaining attention on the task (Rich et al., 2010). Adleman et al. (2011) found that using a simple reversal task, youth with SMD made more errors across all trials when compared to those with bipolar disorder and healthy controls. These findings are similar to those of Deveney et al. (2013) whereby they compared chronically irritable children with healthy controls using an Affective Posner task and found that those with SMD were slower to respond (in comparison to healthy controls) on incongruent trials. Overall, these findings are suggestive of difficulties in disengaging attention (i.e., through the use of inhibitory control) from cues of punishment or negative reward feedback in order to sustain attention on task demands.

Conduct Problems and Psychopathy

An adaptation of a visual search task was used in the study by Hodsoll, Lavie, and Viding (2014) whereby participants were required to search for a male emotional or neutral face amongst two neutral or emotional female face distractors and judge the orientation. It was found that children with conduct problems and low CU traits responded similarly on tasks to typically developing youth, suggesting that both of these groups initially orient to salient aspects of an emotional face. However, youth with conduct problems and high CU traits did exhibit a different pattern, in that they responded similarly on tasks of emotional targets and distractors, suggesting they were not distracted by task-irrelevant emotional faces.

ODD and Psychopathy

Using the Emotional Stroop Task, adolescents with abnormally aggressive CD had slower reaction times on interference trials with distressing images compared to healthy controls. This suggests that youth with aggressive CD have impaired cognitive control when presented with distressing images (Euler, Sterzer, and Stadler, 2014). In a further study using a variant of an emotional Go/No Go task using emoticons, inhibition of responses was required on all neutral stimuli and "go" responses were required on all emotional faces. It was found that in comparison to the control group (low CU and low ODD), 8-year-olds with low CU and high ODD and high CU and high ODD committed the most errors when presented with happy, fearful and neutral faces, with the low CU and high ODD also making more errors relative to controls when presented with sad faces. Moreover, it was found that children with low CU and high ODD were faster to respond to all emotions (happy, angry, sad, and fearful) relative to the control group, and the high CU and low ODD were faster relative to controls when presented with fearful faces. No differences were seen in response time for the group with high CU and high ODD. These results may suggest an underlying difficulty recognizing emotional faces in youth with high CU traits and/or ODD. Additionally, when presented with emotional faces, youth with ODD have difficulty employing effortful control and inhibiting a response to these stimuli (Ezpeleta et al., 2017).

Taken together, youth with externalizing problems and disorders demonstrate biases in attention that vary between disorders. It appears that those who present with problems with irritability, aggression, and anger demonstrate preferential attention towards threat stimuli. However, those with psychopathic traits demonstrate inconsistent results, suggestive of either reduced attention toward emotional stimuli (e.g., distress), or increased attention toward these emotions. Alongside biases toward some negative emotional stimuli, some of these children with externalizing psychopathology also present with biases toward rewards or positive stimuli (e.g., irritable youth, ADHD). Overall, it appears that top-down cognitive control processes, i.e., executive attention, also play a role in the development and maintenance of these biases in youth with externalizing psychopathology, in that these youth preferentially allocate their attention toward certain stimuli, and have difficulty disengaging their attention and attending to task goals (e.g., employing top-down attention).

At Risk Youth

Numerous factors play a role in the development and maintenance of externalizing psychopathology, and of these, parental psychopathology is of central importance (e.g., depression) (Kessel et al., 2017). Additional factors such as witnessing inter-parental hostility has also been suggested as a key risk factor (Fosco and Feinberg, 2014) due to the problems associated with increased negative social cognitions within the family context (Davies, Coe, Hentges, Sturge-Apple, and Ripple, 2018). These hostile home environments have been suggested to lead to the development of attention biases toward threat cues due to the likely early adaptive function of these biases to attend to threat when vulnerable within their hostile family home (e.g. Johnston, Roseby, and Kuehnle, 2009). However, in exploring this notion, Davies et al. (2018) found that when using eye-tracking software, children who were exposed to inter-parental hostility predicted a later reduction in attention toward anger, however not toward fear or sadness, thereby increasing their externalizing symptoms. Alongside interparental hostility, some studies (e.g., Briggs-Gowan et al., 2015; Dykas and Cassidy, 2011) have suggested that negative socialization environments (e.g., physical abuse, negative parenting) can increase the risk of psychopathology due to the development of attention avoidance of hostile faces as this too may be adaptive for avoiding harm (Pine et al., 2005; Shackman, Shackman, and Pollak, 2007).

Temperament appears to be another factor influencing the risk for youth to develop externalizing psychopathology. The understanding and influence of temperament on attention biases in the context of externalizing psychopathology appears to be poorly understood (Frick and Morris, 2004). However, the regulation of emotions through effortful control plays a role in the ability to shift attention and inhibit responses to particular stimuli. Thus, developmental differences in the acquisition of effortful control may contribute to the varied evidence of biases in attention across emotional categories of stimuli (e.g., threat, reward), in addition to age and expression of psychopathology (Frick and Morris, 2004).

Treatment

Treatment approaches within externalizing psychopathology have varied within the context of each presenting concern. Some interventions cut across disorders, targeting problem behaviors such as anger, irritability and aggression, employing well regarded interventions such as CBT. However, the implementation of these treatments each have a different emphasis (i.e., social skills, family interactions), and generally have a largely behavioral focus (Sukhodolsky, Smith, McCauley, Ibrahim, and Piasecka, 2016). Despite the existence of these individual evidence-based approaches to treatment (e.g., Dadds and Hawes, 2006; Kimonis, Bagner, Linares, Blake, and Rodriguez, 2014), as of yet there have been no studies that have examined the influence of attention biases per se on treatment outcomes in youth with externalizing problems.

Novel Treatments Targeting Attention Biases Across Externalizing Disorders

Limited novel interventions have been introduced to directly target attention biases to reduce externalizing behaviors and symptoms in youth. Some studies have looked at the efficacy of more basic attention training (e.g., in studies of youth with ADHD; Tamm, Epstein, Peugh, Nakonezny, and Hughes, 2013) or the modification of the interpretation of emotions (e.g. Hiemstra, De Castro, and Thomaes, 2018; Penton-Voak et al., 2013; Stoddard et al., 2016). However, the effectiveness of ABM approaches specifically is yet to be established. Despite the limited evidence thus far, the similar underlying biases in internalizing and externalizing psychopathology suggest a need to explore the effectiveness of ABM approaches in youth with externalizing problems.

ATTENTIONAL BIASES IN INTERNALIZING AND EXTERNALIZING PSYCHOPATHOLOGY

Internalizing and externalizing problems have been seen to influence each other over the course of childhood and adolescence, with early internalizing difficulties often predicting later externalizing problems in adolescence (Bornstein et al., 2010). Children and adolescents with both internalizing and externalizing problems demonstrate poorer outcomes (Dishion, 2000; Racer and Dishion, 2012), calling for examination of the underlying factors contributing to both of these disorders. Although much of the research has examined attentional processes in internalizing and externalizing psychopathy separately, their shared underlying processes warrant exploration to understand their comorbidity (Fraire and Ollendick, 2013).

Short, Adams, Garner, Sonuga-Barke, and Fairchild (2016) examined attention biases within samples of adolescents with anxiety, CD, both anxiety and CD, and control groups using the visual probe task. It was found that on masked trials, youth with anxiety disorders (both anxiety alone and comorbid with CD) were faster to respond to happy faces in comparison to angry faces. However, individuals with CD without anxiety and the control group were slower to respond when presented with happy faces in comparison to angry faces. Those with comorbid anxiety and CD were faster to respond on masked trials across all conditions when compared to those with CD alone, who were the slowest to respond on masked and unmasked trials across all emotional stimuli (anger, fear, and happy). Additionally, the pattern of attention in both the anxiety and CD-only groups demonstrated delayed disengagement and attention away from emotional faces, however the control and comorbid groups did not. These findings suggest that despite both CD and anxiety disorders displaying a pattern of responding suggestive of avoidance and difficulty disengaging their attention from masked emotional stimuli, perhaps the effects of these biases are negated in youth with comorbid anxiety and CD. However, the reasons for biases within CD were proposed to be attributable to a lack of engagement and motivation in attending to emotional stimuli whereas the biases in anxiety disorders are a hypersensitivity to threat stimuli (Short et al., 2016). Perhaps variances in the streams of CD (i.e., emotionally dysregulated as compared to callous-disruptive) (Andrade, Sorge, Na, and Wharton-Shukster, 2015) further explain the different patterns of biases in youth with CD in those with CU traits or comorbid anxiety disorders.

Given that irritability also presents across both internalizing and externalizing disorders (Brotman et al., 2017), it is perhaps unsurprising that attention biases toward threat stimuli on a dot-probe task did not differ significantly between youth with high irritability or anxiety. However, different brain regions implicated in these biases for heightened irritability as compared to heightened anxiety suggest the possibility of different mechanisms at play. Kircanski et al. (2018) found following a latent variable approach, that this may have been attributable to the greater neural processing required for irritable youth to maintain attention and control during threat incongruent trials, heightened arousal when presented with threat, or possible maladaptive approach behaivours toward threat stimuli in irritable youth. Clearly, this small but important body of research highlights the need for further work to elucidate the common and distinct attention mechanisms that cut across internalizing and externalizing psychopathology. The overlap in internalizing and externalizing psychopathology and their attention biases also highlights the value in measuring these disorders dimensionally to facilitate a greater understanding of the common underlying mechanisms of these disorders.

Future Directions

Although this chapter highlights the patterns of attention biases in children and adolescents, it is clear that further research is required to understand the progression of these biases and patterns of comorbidity between internalizing and externalizing disorders across development. For example, attention avoidance of threat appears to characterize fear disorders (Salum et al., 2013; Waters et al., 2014), which tend to have an earlier onset during childhood than other disorders (e.g., specific phobia; separation anxiety disorder) and tend to predate and predict later anxiety and depressive disorders (Lieb et al., 2016). On the other hand, attention vigilance for threat appears to characterize distress disorders (e.g., GAD; MDD), which tend to onset later in mid-late childhood and adolescence, and attention vigilance appears to be more responsive to various treatments compared to threat avoidance. Therefore, attention avoidance of threat early in development may be a stronger predictor of poor developmental outcomes than early vigilance. Along similar lines, given the high rates of comorbidity across internalizing and externalizing disorders, perhaps early biases in attention in the context of internalizing psychopathology are associated with later maladaptive attention deployment and thus the development of externalizing problems (i.e., monitoring angry faces and reward stimuli). Future research is therefore required to elucidate the long-term developmental outcomes, trajectories and pathways of comorbidity in these biases across both internalizing and externalizing psychopathology. Furthermore, given the apparent transdiagnostic nature of attention biases and the negative outcomes associated with comorbid conditions in children and adolescents, the interplay between attention biases and other factors that place young people on pathways to stronger internalizing versus externalizing psychopathology requires further investigation (Dishion, 2000; Fraire and Ollendick, 2013).

Additionally, recent research with adults has begun to consider links between laboratory assessments of attention biases for emotional stimuli, dynamic changes in daily mood assessed via evidence sampling methods, and psychopathology (Iijima, Takano, Tanno, and Pietromonaco, 2018). This novel work with university students highlights overlaps between anxiety symptoms, attention biases and certain emotional ebbs and flows in daily life. Findings suggest that attention biases associated with anxiety (as measured by a dot-probe task) may be reflective of enhanced immediate responses to negative events but are not necessarily reflected in recovery from stressors or in malleability of anxious mood across time. Such preliminary findings are intriguing and point to an important arena for unraveling how attention biases in youth translate to day-to-day emotion functioning "in situ" (Modecki and Mazza, 2017). Further research is warranted, given that different emotion dynamics arguably reflect different aspects of emotion functioning, are differentially tied to psychopathology, and appear to show developmental differences between youth and adults (Houben,Van Den Noortgate, and Kuppens, 2015; Larson, Csikszentmihalyi, and Graef, 1980). In fact, studies point to the salience of real-world affect dynamics for forecasting increases in adolescent anxiety and depression (Maciejewski et al., 2014) and predicting treatment response among youth with internalizing symptoms (Forbes et al., 2012). All told, improvements in treatment approaches targeting these underlying mechanisms may be enhanced by a greater understanding of the varied forms of attention biases that cut across and are unique to internalizing and externalizing psychopathology and related conditions.

CONCLUSIONS

This chapter provided a review of the key findings and considerations regarding attention biases across development. From this review, it is clear that maladaptive attention deployment cuts across both internalizing and externalizing psychopathology and related conditions. However, the nature of these biases and the mechanisms that underpin them are likely to differ and are not yet well understood. From the emerging literature, it is evident that attention biases reflect perturbations in both top-down and bottom-up processes of attention, with these biases observed toward specific stimuli that are salient to each disorder in internalizing disorders (i.e., threat stimuli in anxiety; negative stimuli in depression). In relation to externalizing disorders, the results are more varied, with studies suggesting biases for affective, reward, and punishment-related stimuli. More advanced conceptual frameworks are required in order to advance our understanding of attention processes and these conditions. Overall, future research needs to examine the relationships between attention biases and the development of psychopathology across time and how the findings of attention biases for emotional stimuli derived from laboratory studies relate to attention deployment in everyday life.

REFERENCES

Abend, R., de Voogd, L., Salemink, E., Wiers, R. W., Perez-Edgar, K., Fitzgerald, A., ... and Bar-Haim, Y. (2018). Association between attention bias to threat and anxiety symptoms in children and adolescents. *Depress Anxiety*, 35(3), 229–238. doi: 10.1002/da.22706

Achenbach, T. and Rescorla, L. (2001). *Manual for the ASEBA School-Age Forms and Profiles*. Burlington: University of Vermont Department of Psychiatry.

Adleman, N. E., Kayser, R., Dickstein, D., Blair, R. J. R., Pine, D., and Leibenluft, E. (2011). Neural correlates of reversal learning in severe mood dysregulation and pediatric bipolar disorder. *Journal of the American Academy of Child and Adolescent Psychiatry*, *50*(11), 1173–1185. doi:10.1016/j.jaac.2011.07.011

American Psychological Association. (2000). *Diagnostic and Statistical Manual of Mental Disorders: DSM-5*. Washington, DC: American Psychiatric Publishing.

Andrade, B. F., Sorge, G. B., Na, J. J., and Wharton-Shukster, E. (2015). Clinical profiles of children with disruptive behaviors based on the severity of their conduct problems, callous-unemotional traits and emotional difficulties. *Child Psychiatry & Human Development*, *46*(4), 567–576. doi:10.1007/s10578-014-0497-8

Armstrong, T. and Olatunji, B. O. (2012). Eye tracking of attention in the affective disorders: A meta-analytic review and synthesis. *Clinical Psychology Review*, *32*(8), 704–723. doi: 10.1016/j.cpr.2012.09.004

Berger, A. and Posner, M. I. (2000). Pathologies of brain attentional networks. *Neuroscience & Biobehavioral Reviews*, *24*(1), 3–5. doi: 10.1016/S01497634(99)00046-9

Bock, J., Wainstock, T., Braun, K., and Segal, M. (2015). Stress in utero: Prenatal programming of brain plasticity and cognition. *Biological Psychiatry*, *78*(5), 315–326.

Bongers, I. L., Koot, H. M., Jan van der, E., and Verhulst, F. C. (2004). Developmental trajectories of externalizing behaviors in childhood and adolescence. *Child Development*, *75*(5), 1523–1537. doi:10.1111/j.1467-8624.2004.00755.x

Bornstein, M. H., Hahn, C.-S., and Haynes, O. M. (2010). Social competence, externalizing, and internalizing behavioral adjustment from early childhood through early adolescence: Developmental cascades. *Development and Psychopathology*, *22*(4), 717–735. doi:10.1017/S0954579410000416

Briggs-Gowan, M. J., Pollak, S. D., Grasso, D., Voss, J., Mian, N. D., Zobel, E., ... and Pine, D. S. (2015). Attention bias and anxiety in young children exposed to family violence. *Journal of Child Psychology and Psychiatry*, *56*(11), 1194–1201. doi:10.1111/jcpp.12397

Brotman, M. A., Kircanski, K., Stringaris, A., Pine, D. S., and Leibenluft, E. (2017). Irritability in youths: A translational model. *American Journal of Psychiatry*, *174*(6), 520–532. doi:10.1176/appi.ajp.2016.16070839

Brown, H. M., Eley, T. C., Broeren, S., MacLeod, C., Rinck, M., Hadwin, J. A., and Lester, K. J. (2014). Psychometric properties of reaction time based experimental paradigms measuring anxiety-related information-processing biases in children. *Journal of Anxiety Disorders*, *28*(1), 97–107. doi: 10.1016/j.janxdis.2013.11.004

Carlson, S. R., Pritchard, A. A., and Dominelli, R. M. (2013). Externalizing behavior, the UPPS-P impulsive behavior scale and reward and punishment sensitivity. *Personality and Individual Differences*, *54*(2), 202–207. doi:10.1016/j.paid.2012.08.039

Centifanti, L. C., Kimonis, E. R., Frick, P. J., and Aucoin, K. J. (2013). Emotional reactivity and the association between psychopathy-linked narcissism and aggression in detained adolescent boys. *Development and Psychopathology*, *25*(2), 473–485. doi:10.1017/S0954579412001186

Centifanti, L. C. M. and Modecki, K. (2013). Throwing caution to the wind: Callous-unemotional traits and risk taking in adolescents. *Journal of Clinical Child & Adolescent Psychology*, *42*(1), 106–119. doi: 10.1080/15374416.2012.719460

Chang, S. W., Kuckertz, J. M., Bose, D., Carmona, A. R., Piacentini, J., and Amir, N. (2018). Efficacy of attention bias training for child anxiety disorders: A randomised controlled trial. *Child Psychiatry & Human Development, 50*(2), 198–208. doi:10.1007/s10578-018-0832-6.

Cisler, J. M., Bacon, A. K., and Williams, N. L. (2009). Phenomenological characteristics of attentional biases toward threat: A critical review. *Cognitive Therapy and Research, 33*(2), 221–234. doi:10.1007/s10608-007-9161-y

Cisler, J. M. and Koster, E. H. W. (2010). Mechanisms of attentional biases towards threat in anxiety disorders: An integrative review. *Clinical Psychology Review, 30*(2), 203–216. doi:10.1016/j.cpr.2009.11.003

Ciucci, E., Kimonis, E., Frick, P. J., Righi, S., Baroncelli, A., Tambasco, G., and Facci, C. (2018). Attentional orienting to emotional faces moderates the association between callous-unemotional traits and peer-nominated aggression in young adolescent school children. *Journal of Abnormal Child Psychology, 46*(5), 1011–1019. doi:10.1007/s10802-017-0357-7

Clarke, L. A. and Watson, D. (2006). Distress and fear disorders: An alternative empirically based taxonomy of the "mood" and "anxiety" disorders. *British Journal of Psychiatry, 189*(6), 481–483. doi:10.1192/bjp.bp.106.03825

Copeland, W. E., Shanahan, L., Egger, H., Angold, A., and Costello, J. (2014). Adult diagnostic and functional outcomes of DSM-5 disruptive mood dysregulation disorder. *The American Journal of Psychiatry, 171*(6), 668–674. doi: 10.1176/appi.ajp.2014.13091213

Corbetta, M. and Shulman, G. L. (2002). Control of goal-directed and stimulus-driven attention in the brain. *Nature Reviews Neuroscience, 3*(3), 201–215. doi:10.1038/nrn755

Cremone, A., Lugo-Candelas, C. I., Harvey, E. A., McDermott, J. M., and Spencer, R. M. C. (2018). Positive emotional attention bias in young children with symptoms of ADHD. *Child Neuropsychology: A Journal on Normal and Abnormal Development in Childhood and Adolescence, 24*(8), 1137–1145. doi:10.1080/09297049.2018.1426743

Cristea, I. A., Mogoase, C., David, D., and Cuijpers, P. (2015). Practitioner review: Cognitive bias modification for mental health problems in children and adolescents: A meta-analysis. *The Journal of Child Psychology and Psychiatry, 56*(7), 723–734. doi: 10.1111/jcpp.12383

Dadds, M. R. and Hawes, D. (2006). *Integrated Family Intervention for Child Conduct Problems: A Behaviour-Attachment-Systems Intervention for Parents.* Bowen Hills, Qld: Australian Academic Press.

Dadds, M. R., Perry, Y., Hawes, D. J., Merz, S., Riddell, A. C., Haines, D. J., … and Abeygunawardane, A. I. (2006). Attention to the eyes and fear-recognition deficits in child psychopathy. *The British Journal of Psychiatry, 189*(3), 280–281. doi:10.1192/bjp.bp.105.018150

Davies, P. T., Coe, J. L., Hentges, R. F., Sturge-Apple, M. L., and Ripple, M. T. (2018). Interparental hostility and children's externalizing symptoms: Attention to anger as a mediator. *Developmental Psychology, 54*(7), 1290–1303. doi:10.1037/dev0000520

De Castro, B. O., Veerman, J. W., Koops, W., Bosch, J. D., and Monshouwer, H. J. (2002). Hostile attribution of intent and aggressive behavior: A meta-analysis. *Child development, 73*(3), 916–934. doi: 10.1111/1467-8624.00447

Denney, C. B., Rapport, M. D., and Chung, K. (2005). Interactions of task and subject variables among continuous performance tests. *Journal of Child Psychology and Psychiatry, 46*(4), 420–435. doi:10.1111/j.1469-7610.2004.00362.x

Deveney, C. M., Connolly, M. E., Haring, C. T., Bones, B. L., Reynolds, R. C., Kim, P., … and Leibenluft, E. (2013). Neural mechanisms of frustration in chronically irritable children. *American Journal of Psychiatry, 170*(10), 1186–1194. doi:10.1176/appi.ajp.2013.12070917

De Voogd, E. L., Wiers, R. W., Prins, P. J. M., and Salemink, E. (2014). Visual search attentional bias modification reduced social phobia in adolescents. *Journal of Behaviour Therapy and Experimental Psychiatry*, *45*(2), 252–259. doi: 10.1016/j.jbtep.2013.11.006

De Voogd, E. L., Wiers, R. W., and Salemink, E. (2017). Online visual search attentional bias modification for adolescents with heightened anxiety and depressive symptoms: A randomized controlled trial. *Behaviour Research and Therapy*, *92*, 57–67. doi: 10.1016/j.brat.2017.02.006

Dishion, T. J. (2000). Cross-setting consistency in early adolescent psychopathology: Deviant friendships and problem behavior sequelae. *Journal of Personality*, *68*(6), 1109–1126. doi: 10.1111/1467-6494.00128

Dodd, H. F., Hudson, J. L., Williams, T., Morris, T., Lazarus, R. S., and Byrow, Y. (2015). Anxiety and attentional bias in preschool-aged children: An eye-tracking study. *Journal of Abnormal Child Psychology*, *43*(6), 1055–1065. doi: 10.1007/s10802-014-9962-x

Dudeney, J., Sharpe, L., and Hunt, C. (2015). Attentional bias towards threatening stimuli in children with anxiety: A meta-analysis. *Clinical Psychology Review*, *40*, 66–75. doi: 10.1016/j.cpr.2015.05.007

Dykas, M. J. and Cassidy, J. (2011). Attachment and the processing of social information across the life span: Theory and evidence. *Psychological Bulletin*, *137*(1), 19–46. doi:10.1037/a0021367

Ehrenreich, J. T. and Gross, A. M. (2002). Biased attentional behavior in childhood anxiety. A review of theory and current empirical investigation. *Clinical Psychology Review*, *22*(7), 991–1008. doi: 10.1016/S0272-7358(01)00123-4

Eisenberg, N., Smith, C. L., Sadovsky, A., and Spinrad, T. L. (2004). Effortful control: Relations with emotion regulation, adjustment, and socialization in childhood. In R. F. Baumeister and K. D. Vohs (Eds.), *Handbook of Self-regulation: Research, Theory and Applications* (pp. 259–282). New York, NY: The Guilford Press.

Euler, F., Sterzer, P., and Stadler, C. (2014). Cognitive control under distressing emotional stimulation in adolescents with conduct disorder. *Aggressive Behavior*, *40*(2), 109–119. doi:10.1002/ab.21508

Eysenck, M. W., Derakshan, N., Santos, R., and Calvo, M. G. (2007). Anxiety and cognitive performance: Attentional control theory. *Emotion*, *7*(2), 336–353. doi: 10.1037/1528-3542.7.2.336

Ezpeleta, L., Navarro, J. B., de la Osa, N., Penelo, E., Trepat, E., Martin, V., and Domènech, J. M. (2017). Attention to emotion through a go/no-go task in children with oppositionality and callous–unemotional traits. *Comprehensive Psychiatry*, *75*, 35–45. doi:10.1016/j.comppsych.2017.02.004

Fan, J., McCandliss, B. D., Sommer, T., Raz, A., and Posner, M. I. (2002). Testing the efficiency and independence of attentional networks. *Journal of Cognitive Neuroscience*, *14*(3), 340–347. doi:10.1162/089892902317361886

Forbes, E. E., Stepp, S. D., Dahl, R. E., Ryan, N. D., Whalen, D., Axelson, D. A., … and Silk, J. S. (2012). Real-world affect and social context as predictors of treatment response in child and adolescent depression and anxiety: An ecological momentary assessment study. *Journal of Child and Adolescent Psychopharmacology*, *22*(1), 37–47. doi: 10.1089/cap.2011.0085

Fosco, G. M. and Feinberg, M. E. (2014). Cascading effects of interparental conflict in adolescence: Linking threat appraisals, self-efficacy, and adjustment. *Development and Psychopathology*, *27*(1), 239–252. doi:10.1017/S0954579414000704

Fraire, M. G. and Ollendick, T. H. (2013). Anxiety and oppositional defiant disorder: A transdiagnostic conceptualization. *Clinical Psychology Review*, *33*(2), 229–240. doi:10.1016/j.cpr.2012.11.004

Frick, P. J. and Morris, A. S. (2004). Temperament and developmental pathways to conduct problems. *Journal of Clinical Child and Adolescent Psychology*, 33(1), 54–68. doi: 10.1207/S15374424JCCP3301_6

Frick, P. J. and White, S. F. (2008). Research review: The importance of callous-unemotional traits for developmental models of aggressive and antisocial behavior. *Journal of Child Psychology and Psychiatry*, 49(4), 359–375. doi:10.1111/j.1469-7610.2007.01862.x

Fu, X., Taber-Thomas, B., and Pérez-Edgar, K. (2017). Frontolimbic functioning during threat-related attention: Relations to early behavioral inhibition and anxiety in children, *Biological Psychology*, 122, 98–109. doi: 10.1016/j.biopsycho.2015.08.010

Gamble, A. L. and Rapee, R. M. (2009). The time-course of attentional bias in anxious children and adolescents. *Journal of Anxiety Disorders*, 23(7), 841–847. doi: 10.1016/j.janxdis.2009.04.001

Gibb, B. E., Benas, J. S. Grassia, M., and McGeary, J. (2009). Children's attentional biases and 5-HTTLPR genotype: Potential mechanisms linking mother and child depression. *Journal of Clinical Child & Adolescent Psychology*, 38(3), 415–426. doi: 10.1080/15374410902851705

Gibb, B. E., Pollak, S. D., Hajcak, G. and Owens, M. (2016). Attentional biases in children of depressed mothers: An event related potential (ERP) study. *Journal of Abnormal Psychology*, 125(8), 1166–1178. doi: 10.1037/abn0000216

Gioia, G. A., Isquith, P. K., Guy, S. C., and Kenworthy, L. (2000). *Behavior rating inventory of executive function: BRIEF*. Odessa, FL: Psychological Assessment Resources.

Hakamata, Y., Lissek, S., Bar-Haim, Y., Britton, J. C., Fox, N. A., Leibenluft, E., Ernst, M., and Pine, D. S. (2010). Attention bias modification treatment: A meta-analysis toward the establishment of novel treatment for anxiety. *Biological Psychiatry*, 68(11), 982–990. doi: 10.1016/j.biopsych.2010.07.021

He, J., Jin, X., Zhang, M., Huang, X., Shui, R., and Shen, M. (2013). Anger and selective attention to reward and punishment in children. *Journal of Experimental Child Psychology*, 115(3), 389–404. doi:10.1016/j.jecp.2013.03.004

Hiemstra, W., De Castro, B. O., and Thomaes, S. (2018). Reducing aggressive children's hostile attributions: A cognitive bias modification procedure. *Cognitive Therapy and Research*, 43(2), 387–398. doi:10.1007/s10608-018-9958-x

Hodsoll, S., Lavie, N., and Viding, E. (2014). Emotional attentional capture in children with conduct problems: The role of callous-unemotional traits. *Frontiers in Human Neuroscience*, 8. doi:10.3389/fnhum.2014.00570

Hommer, R. E., Meyer, A., Stoddard, J., Connolly, M. E., Mogg, K., Bradley, B. P., ... and Brotman, M. A. (2014). Attention bias to threat faces in severe mood dysregulation. *Depression and Anxiety*, 31(7), 559–565. doi:10.1002/da.22145

Horsley, T. M., Orobio de Castro, B., and van der Schoot, M. (2010). In the eye of the beholder: Eye-tracking assessment of social information processing in aggressive behavior. *Journal of Abnormal Child Psychology*, 38(5), 587–599. doi:10.1007/s10802-009-9361-x

Houben, M., Van Den Noortgate, W., and Kuppens, P. (2015). The relation between short-term emotion dynamics and psychological well-being: A meta-analysis. *Psychological Bulletin*, 141(4), 901. doi: 10.1037/a0038822

Humphreys, K. L., Schouboe, S. N., Kircanski, K., Leibenluft, E., Stringaris, A., and Gotlib, I. H. (2019). Irritability, externalizing, and internalizing psychopathology in adolescence: Cross-sectional and longitudinal associations and moderation by sex. *Journal of Clinical Child & Adolescent Psychology*, 48(5), 1–9. doi:10.1080/15374416.2018.1460847

Hwang, S., White, S. F., Nolan, Z. T., Williams, W. C., Sinclair, S., and Blair, R. (2015). Executive attention control and emotional responding in attention-deficit/hyperactivity disorder—a functional MRI study. *NeuroImage: Clinical*, *9*(C), 545–554. doi: 10.1016/j.nicl.2015.10.005

Iijima, Y., Takano, K., Tanno, Y. and Pietromonaco, R. P. (2018). Attentional bias and its association with anxious mood dynamics. *Emotion*, *18*(5), 725–735. doi: 10.1037/emo0000338

In-Albon, T., Kossowsky, J., and Schneider, S. (2010). Vigilance and avoidance of threat in the eye movements of children with separation anxiety disorder. *Journal of Child Psychology*, *38*(2), 225–235. doi:10.1007/s10802-009-9359-4

James, K. M., Owens, M., Woody, M. L., Hall N. T., and Gibb, B. E. (2018). Parental expressed emotion-criticism and neural markers of sustained attention to emotional faces in children. *Journal of Clinical Child & Adolescent Psychology*, *47*(1), 1–10. doi: 10.1080/15374416.2018.1453365

Johnston, J., Roseby, V., and Kuehnle, K. (2009). *In the Name of the Child: A Developmental Approach to Understanding and Helping Children of Conflict and Violent Divorce*. New York, NY: Springer.

Kessel, E. M., Dougherty, L. R., Kujawa, A., Hajcak, G., Carlson, G. A., and Klein, D. N. (2016). Longitudinal associations between preschool disruptive mood dysregulation disorder symptoms and neural reactivity to monetary reward during preadolescence. *Journal of Child and Adolescent Psychopharmacology*, *26*(2), 131–137. doi:10.1089/cap.2015.0071

Kessel, E. M., Kujawa, A., Dougherty, L. R., Hajcak, G., Carlson, G. A., and Klein, D. N. (2017). Neurophysiological processing of emotion in children of mothers with a history of depression: The moderating role of preschool persistent irritability. *Journal of Abnormal Child Psychology*, *45*(8), 1599–1608. doi:10.1007/s10802-017-0272-y

Kessler, R. C., Berglund, P., Demler, O., Jin, R., Merikangas, K. R., and Walters, E. E. (2005). Lifetime prevalence and age-of-onset distributions of DSM-IV disorders in the national comorbidity survey replication, *Arch Gen Psychiatry*, *62*(6), 593–768. doi: 10.1001/archpsych.62.6.593

Kimonis, E. R., Bagner, D. M., Linares, D., Blake, C. A., and Rodriguez, G. (2014). Parent training outcomes among young children with callous–unemotional conduct problems with or at risk for developmental delay. *Journal of Child and Family Studies*, *23*(2), 437–448. doi:10.1007/s10826-013-9756-8

Kimonis, E. R., Frick, P. J., Cauffman, E., Goldweber, A., and Skeem, J. (2012). Primary and secondary variants of juvenile psychopathy differ in emotional processing. *Development and Psychopathology*, *24*(3), 1091–1103. doi: 10.1017/S0954579412000557

Kimonis, E. R., Frick, P. J., Fazekas, H., and Loney, B. R. (2006). Psychopathic traits, aggression, and the processing of emotional stimuli in non-referred children. *Behavioral Sciences and the Law*, *24*(1), 21–37. doi: 10.1002/bsl.668

Kimonis, E. R., Graham, N., and Cauffman, E. (2017). Aggressive male juvenile offenders with callous-unemotional traits show aberrant attentional orienting to distress cues. *Journal of Abnormal Child Psychology*, *46*(3), 519–527. doi:10.1007/s10802-017-0295-4

Kindt, M., Bierman, D., and Brosschot, J. F. (1997). Cognitive bias in spider fear and control children: Assessment of emotional interference by a card format and a single-trial format of the Stroop task. *Journal of Experimental Child Psychology*, *66*(2), 163–179. doi: 10.1006/jecp.1997.2376

Kindt, M., Brosschot, J. F., and Everaerd, W. (1997). Cognitive processing bias of children in a real life stress situation and a neutral situation. *Journal of Experimental Child Psychology*, *64*(1), 79–97. doi: 10.1006/jecp.1996.2336

Kircanski, K., White, L. K., Tseng, W.-L., Wiggins, J. L., Frank, H. R., Sequeira, S., ... and Brotman, M. (2018). A latent variable approach to differentiating neural mechanisms of irritability and anxiety in youth. *JAMA Psychiatry* (Chicago, Ill.), *75*(6), 631–639. doi: 10.1001/jamapsychiatry.2018.0468

Klein, C., Wendling, K., Huettner, P., Ruder, H., and Peper, M. (2006). Intra-subject variability in attention-deficit hyperactivity disorder. *Biological Psychiatry, 60*(10), 1088–1097. doi:10.1016/j.biopsych.2006.04.003

Kofler, M. J., Sarver, D. E., Harmon, S. L., Moltisanti, A., Aduen, P. A., Soto, E. F., and Ferretti, N. (2018), Working memory and organizational skills problems in ADHD. *J Child Psychol Psychiatr, 59*(1), 57–67. doi:10.1111/jcpp.12773

Kosson, D. S., McBride, C. K., Miller, S. A., Riser, N. R. E., and Whitman, L. A. (2018). Attentional bias following frustration in youth with psychopathic traits: Emotional deficit versus negative preception. *Journal of Experimental Psychopathology, 9*(2), doi:10.5127/jep.060116

Kujawa, A. J., Torpey, D., Kim, J., Hajcak, G., Rose, S., Gotlib, I. H., and Klein, D. N. (2011). Attentional biases for emotional faces in young children of mothers with chronic or recurrent depression. *Journal of Abnormal Child Psychology, 39*(1), 125–135. doi:10.1007/s10802-010-9438-6

Kyte, A. A., Goodyer, I. M., and Sahakian, B. J. (2005). Selected executive skills in adolescents with recent first episode major depression. *Journal of Child Psychology and Psychiatry, 46*(9), 995–1005. doi:10.1111/j.1469-7610.2004.00400-x

Ladouceur, C. D., Dahl, R. E., Williamson, D. E., Birmaher, B., Axelson, D. A., Ryan, N, D., and Casey, B. J. (2006). Processing emotional facial expressions influences performance on a Go/NoGo task in pediatric anxiety and depression. *Journal of Child Psychology and Psychiatry, 47*(11), 1107–1115. doi: 10.1111/j.1469-7610.2006.01640.x

Larson, R., Csikszentmihalyi, M., and Graef, R. (1980). Mood variability and the psychosocial adjustment of adolescents. *Journal of Youth and Adolescence, 9*(6), 469–490. doi: 10.1007/978-94-017-9094-9_15

Lau, J.Y.F. and Waters, A.M. (2017). Annual research review: An expanded account of information-processing mechanisms in risk for child and adolescent anxiety and depression. *The Journal of Child Psychology and Psychiatry, 58*(4), 387–407. doi:10.1111/jcpp.12653

Laue, C., Griffey, M., Lin, P.-I., Wallace, K., van der Schoot, M., Horn, P., ... and Barzman, D. (2018). Eye gaze patterns associated with aggressive tendencies in adolescence. *Psychiatric Quarterly, 89*(3), 747–756. doi:10.1007/s11126-018-9573-8

Lawrence, D., Johnson, S., Hafekost, J., Boterhoven de Haan, K., Sawyer, M., Ainley, J. and Zubrick, S. R. (2015). *The Mental Health of Children and Adolescents.* Report on the second Australian child and adolescent survey of mental health and wellbeing. Department of Health, Canberra.

Legerstee, J. S., Tulen, J. H. M., Dierckx, B., Treffers, P. D. A., Verhulst, F. C., and Utens, E. M. W. J. (2010). CBT for childhood anxiety disorder: Differential changes in selective attention between treatment responders and non-responders. *The Journal of Child Psychology and Psychiatry, 51*(2), 162–172. doi: 10.1111/j.1469-7610.2009.02143.x

Legerstee, J. S., Tulen, J. H. M., Kallen, V. L., Dieleman, G. C., Treffers, P. D. A., Verhulst, F. C., and Utens, E. M. W. J. (2009). Threat-related selective attention predicts treatment success in childhood anxiety disorders. *Journal of American Academy of Child and Adolescent Psychiatry, 48*(2), 196–205. doi: 10.1097/CHI.0b013e31819176e4

Leppänen, J. M. and Nelson, C. A. (2009). Tuning the developing brain to social signals of emotions. *Nature Reviews Neuroscience, 10*(1), 37–47. doi: 10.1038/nrn2554

Lieb, R., Miché, M., Gloster, A. T., Beesdo-Baum, K., Meyer, A. H., and Wittchen, H. (2016). Impact of specific phobia on the risk of onset of mental disorders: A 10-year prospective study of adolescents and young adults. *Depression and Anxiety, 33*(7), 667–675. doi:10.1002/da.22487

Lin, P.-I., Hsieh, C.-D., Juan, C.-H., Hossain, M. M., Erickson, C. A., Lee, Y.-H., and Su, M.-C. (2016). Predicting aggressive tendencies by visual attention bias associated with hostile emotions. *PloS One, 11*(2), e0149487. doi:10.1371/journal.pone.0149487

Liu, P., Taber-Thomas, B. C., Fu, X., and Pérez-Edgar, K. E. (2018). Biobehavioral markers of attention bias modification in temperamental risk for anxiety: A randomised control trial. *Journal of the American Academy of Child and Adolescent Psychiatry, 57*(2), 103–110. doi: 10.1016/j.jaac.2017.11.016

LoBue, V., Buss, K. A., Taber-Thomas, B. C., and Pérez-Edgar, K. (2017). Developmental differences in infants' attention to social and nonsocial threats. *Infancy, 22*(3), 403–415. doi: 10.1111/infa.12167

Loney, B. R., Frick, P. J., Clements, C. B., Ellis, M. L., and Kerlin, K. (2003). Callous-unemotional traits, impulsivity, and emotional processing in adolescents with antisocial behavior problems. *J Clin Child Adolesc Psychol, 32*(1), 66–80. doi:10.1207/S15374424JCCP3201_07

Lonigan, C. J. and Phillips, B. M. (2001). Temperamental basis of anxiety disorders in children. In M. W. Vasey and M. R. Dadds (Eds.), *The Developmental Psychopathology of Anxiety* (pp. 60–91). New York: Oxford University Press.

Lonigan, C. J., and Vasey, M. W. (2009). Negative affectivity, effortful control and attention to threat-relevant stimuli. *Journal of Abnormal Child Psychology, 37*, 387–399.

Ma, I., Mies, G. W., Lambregts-Rommelse, N. N. J., Buitelaar, J. K., Cillessen, A. H. N., and Scheres, A. (2018). Does an attention bias to appetitive and aversive words modulate interference control in youth with ADHD? *Child Neuropsychology, 24*(4), 541–557. doi:10.1080/09297049.2017.1296940

Maalouf, F. T., Clark, L., Tavitian, L., Sahakian, B. J., Brent, D. and Phillips, M. L. (2012). Bias to negative emotions: A depression state-dependent marker in adolescent major depressive disorder. *Psychiatry Research, 198*(1), 28–33. doi: 10.1016/j.psychres.2012.01.030

Maciejewski, D. F., van Lier, P. A., Neumann, A., Van der Giessen, D., Branje, S. J., Meeus, W. H., and Koot, H. M. (2014). The development of adolescent generalized anxiety and depressive symptoms in the context of adolescent mood variability and parent-adolescent negative interactions. *Journal of Abnormal Child Psychology, 42*(4), 515–526. doi: 10.1007/s10802-013-9797-x

MacLeod, C. and Clarke, P. J. F. (2015). The attentional bias modification approach to anxiety intervention. *Clinical Psychological Science, 3*, 58–78. doi: 10.1177/2167702614560749

MacLeod, C. and Mathews, A. (1988). Anxiety and the allocation of attention to threat. *The Quarterly Journal of Experimental Psychology, 40*(4), 653–670. doi: 10.1080/14640748808402292

Manassis, K., Hum, K., Lee, T. C., Zhang, G., and Lewis, M. D. (2013). Threat perception predicts cognitive behavioral therapy outcomes in anxious children. *Open Journal of Psychiatry, 3*, 141–148. doi: 10.4236/ojpsych.2013.31A009

Marsh, A. A., Finger, E. C., Mitchell, D. G. V., Reid, M. E., Sims, C., Kosson, D. S., . . . and Blair, R. J. R. (2008). Reduced amygdala response to fearful expressions in children and adolescents with callous-unemotional traits and disruptive behavior disorders. *American Journal of Psychiatry*, 165(6), 712–720. doi:10.1176/appi.ajp.2007.07071145

Mathews, R. R. S., Hall, W. D., Vos, T., Patton, G. C. and Degenhardt, L. (2011). What are the major drivers of prevalent disability burden in young Australians? *Medical Journal of Australia*, 194(5), 232–235. doi: 10.5694/j.1326-5377.2011.tb02951.x

McLaughlin, K. A. and Lambert, H. K. (2017). Child trauma exposure and psychopathology: Mechanisms of risk and resilience. *Current Opinion in Psychology*, 14, 29–34. doi: 10.1016/j.copsyc.2016.10.004

Mellentin, A. I., Dervisevic, A., Stenager, E., Pilegaard, M., and Kirk, U. (2015). Seeing enemies? A systematic review of anger bias in the perception of facial expressions among anger-prone and aggressive populations. *Aggression and Violent Behavior*, 25, 373–383. doi:10.1016/j.avb.2015.09.001

Merikangas, K. R., Avenevoli, S., Dierker, L., and Grillon, C. (1999). Vulnerability factors among children at risk for anxiety disorders. *Society of Biological Psychiatry*, 46(11), 1523–1535. doi: 10.1016/S0006-3223(99)00172-9

Merikangas, K. R., Nakamura, E. F., and Kessler, R. C. (2009). Epidemiology of mental disorders in children and adolescents. *Dialogues in Clinical Neuroscience*, 11(1), 7–20.

Modecki, K. L. and Mazza, G. L. (2017). Are we making the most of ecological momentary assessment data? A comment on Richardson, Fuller-Tyszkiewicz, O'Donnell, Ling, & Staiger, 2017. *Health Psychology Review*, 11(3), 295–297. doi: 10.1080/17437199.2017.1347513

Mogg, K. and Bradley, B. P. (1998). A cognitive-motivational analysis of anxiety. *Behaviour Research and Therapy*, 36(9), 809–848. doi: 10.1016/S0005-7967(98)00063-1

Mogg, K. and Bradley, B. P. (2018). Anxiety and threat-related attention: Cognitive-motivational framework and treatment. *Trends in Cognitive Sciences*, 22(3), 225–240. doi: 10.1016/j.tics.2018.01.001

Mogg, K., Waters, A. M., and Bradley, B. P. (2017). Attention bias modification (ABM): Review of effects of multisession ABM training on anxiety and threat-related attention in high anxious individuals. *Clinical Psychological Science*, 5, 698–717. doi: 10.1177/2167702617696359

Monk, C. S., Nelson, E. E., McClure, E. B., Mogg, K., Bradley, B. P., Leibenluft, E., . . . and Pine, D. S. (2006). Ventrolateral prefrontal cortex activation and attentional bias in response to angry faces in adolescents with generalized anxiety disorder. *American Journal of Psychiatry*, 163(6), 1091–1097. doi: 10.1176/ajp.2006.163.6.1091

Montagner, R., Mogg, K., Bradley, B. P., Pine, D. S., Czykiel, M. S., Miguel, E. C., . . . and Salum, G. A. (2016). Attentional bias to threat in children at-risk for emotional disorders: Role of gender and type of maternal emotional disorder. *European Child and Adolescent Psychiatry*, 25(7), 735–742. doi: 10.1007/s00787-015-0792-3

Morales, S., Fu, X., and Pérez-Edgar, K. E. (2016). A developmental neuroscience perspective on affect-biased attention. *Developmental Cognitive Neuroscience*, 21, 26–41. doi:10.1016/j.dcn.2016.08.001

Muris, P., de Jong, P. J., and Engelen, S. (2004). Relationships between neuroticism, attentional control, and anxiety disorders symptoms in non-clinical children. *Personality and Individual Differences*, 37(4), 789–797. doi:10.1016/j.paid.2003.10.007

Öhman, A. and Mineka, S. (2001). Fears, phobias, and preparedness: Toward an evolved module of fear and fear learning. *Psychological Review*, 108(3), 483. doi: 10.1037/0033-295X.108.3.483

Ollendick, T. H., White, S. W., Richey, J., Kim-Spoon, J., Ryan, S. M., Wieckowski, A. T., and Smith, M. (2019). Attention bias modification treatment for adolescents with social anxiety disorder. *Behavior Therapy*, 50(1), 126–139. doi:10.1016/j.beth.2018.04.002

Osinsky, R., Lösch, A., Hennig, J., Alexander, N., and MacLeod, C. (2012). Attentional bias to negative information and 5-HTTLPR genotype interactively predict students' emotional reactivity to first university semester. *Emotion, 12*(3), 460–469. doi: 10.1037/a0026674

Passarotti, A. M., Sweeney, J. A., and Pavuluri, M. N. (2010). Neural correlates of response inhibition in pediatric bipolar disorder and attention deficit hyperactivity disorder. *Psychiatry Research, 181*(1), 36–43. doi:10.1016/j.pscychresns.2009.07.002

Peckham, A. D., McHugh, R. K., and Otto, M. W. (2010). A meta-analysis of the magnitude of biased attention in depression. *Depression and Anxiety, 27*(12), 1135–1142. doi: 10.1002/da.20755

Penton-Voak, I. S., Thomas, J., Gage, S. H., McMurran, M., McDonald, S., and Munafò, M. H. (2013). Increasing recognition of happiness in ambiguous facial expressions reduces anger and aggressive behaviour. *Psychological Science, 24*(5), 688–697. doi:10.1177/0956797612459657

Perez-Edgar, K., Morales, S., LoBue, V., Taber-Thomas, B. C., Allen, E. K., Brown, K. M., and Buss, K. A. (2017). The impact of negative affect on attention patterns to threat across the first 2 years of life. *Developmental Psychology, 53*(12), 2219–2232. doi:10.1037/dev0000408

Pergamin-Hight, L., Naim, R., Bakermans-Kranenburg, M. J., van Ijzendoorn, M. H., and Bar-Haim, Y. (2015). Content specificity of attention bias to threat in anxiety disorders: A meta-analysis. *Clinical Psychology Review, 35*, 10–18

Pine, D. S., Mogg, K., Bradley, B. P., Montgomery, L., Monk, C. S., McClure, E. B., ... and Kaufman, J. (2005). Attention bias to threat in maltreated children: Implications for vulnerability to stress-related psychopathology. *American Journal of Psychiatry, 162*(2), 291–296. doi:10.1176/appi.ajp.162.2.291

Pishyareh, E., Tehrani-Doost, M., Mahmoodi-Gharaie, J., Khorrami, A., and Rahmdar, S. R. (2015). A comparative study of sustained attentional bias on emotional processing in ADHD children to pictures with eye-tracking. *Iranian Journal of Child Neurology, 9*(1), 64–70.

Platt, B., Murphy, S. E., and Lau, J. Y. F. (2015). The association between negative attention biases and symptoms of depression in a community sample of adolescents. *Peer J, 3*: e1372, doi:10.7717/peerj.1372.

Platt, B., Waters, A.M., Schulte-Koerne, G., Engelmann, L., and Salemink, E. (2017) A review of cognitive biases in youth depression: Attention, interpretation and memory. *Cognition and Emotion, 31*(3), 462–483, doi:10.1080/02699931.2015.1127215

Plude, D. J., Enns, J. T., and Brodeur, D. (1994). The development of selective attention: A lifespan overview. *Acta Psychologica, 86*(2), 227–272. doi: 10.1016/0001-6918(94)90004-3

Posner, M. I. (1980). Orienting of attention. *Quarterly Journal of Experimental Psychology, 32*(1), 3–25. doi:10.1080/00335558008248231

Posner, M. I. and Petersen, S. E. (1990). The attention systems of the human brain. *Annual Review of Neuroscience, 13*, 25–42. doi: 10.1146/annurev.ne.13.030190.000325

Posner, M. I. and Rothbart, M. K. (2007). Research on attention networks as a model for the integration of psychological science. *Annual Review of Psychology, 58*(1), 1–23. doi: 10.1146/annurev.psych.58.110405.085516

Price, R. B., Kuckertz, J. M., Siegle, G. J., Ladouceur, C. D., Silk, J. S., Ryan, N. D., ... and Amir, N. (2015). Empirical recommendations for improving the stability of the dot-probe task in clinical research. *Psychological Assessment, 27*(2), 365–376. doi:10.1037/pas0000036

Price, R. B., Siegle, G. J., Silk, J. S., Ladouceur, C., McFarland, A., Dahl, R. E., and Ryan N. D. (2013). Sustained neural alterations in anxious youth performing an attentional bias task: A pupilometry study. *Depression and Anxiety, 30*(1), 22–30. doi: 10.1002/da.21966

Price, R. B., Siegle, G. J., Silk, J. S., Ladouceur, C. D., McFarland, A., Dahl, R. E., and Ryan, N. D. (2014). Looking under the hood of the dot-probe task: An fMRI study in anxious youth. *Depression and Anxiety*, 31(3), 178–187. doi: 10.1002/da.22255

Racer, K. H. and Dishion, T. J. (2012). Disordered attention: Implications for understanding and treating internalizing and externalizing disorders in childhood. *Cognitive and Behavioral Practice*, 19(1), 31–40. doi:10.1016/j.cbpra.2010.06.005

Reinholdt-Dunne, M. L., Mogg, K., and Bradley, B. P. (2009). Effects of anxiety and attention control on processing pictorial and linguistic emotional information. *Behaviour Research and Therapy*, 47(5), 410–417. doi: 10.1016/j.brat.2009.01.012

Rich, B. A., Brotman, M. A., Dickstein, D. P., Mitchell, D. G. V., Blair, R. J. R., and Leibenluft, E. (2010). Deficits in attention to emotional stimuli distinguish youth with severe mood dysregulation from youth with bipolar disorder. *Journal of Abnormal Child Psychology*, 38(5), 695–706. doi:10.1007/s10802-010-9395-0

Rich, B. A., Schmajuk, M., Perez-Edgar, K. E., Fox, N. A., Pine, D. S., and Leibenluft, E. (2007). Different psychophysiological and behavioral responses elicited by frustration in pediatric bipolar disorder and severe mood dysregulation. *Am J Psychiatry*, 164(2), 309–317. doi:10.1176/ajp.2007.164.2.309

Rothbart, M. K., Ahadi, S. A., Hersey, K. L., and Fisher, P. (2001). Investigations of temperament at three to seven years: The children's behavior questionnaire. *Child Development*, 72(5), 1394–1408. doi: 10.1111/1467-8624.00355

Salum, G. A., Mogg, K., Bradley, B. P., Gadelha, A., Pan, P., Tamananha, A. C., . . . and Pine, D. S. (2013). Threat bias in attention orienting: Evidence of specificity in a large community-based study. *Psychological Medicine*, 43(4), 733–745. doi: 10.1017/S0033291712001651

Salum, G. A., Mogg, K., Bradley, B. P., Stringaris, A., Gadelha, A., Pan, P. M., . . . and Leibenluft, E. (2017). Association between irritability and bias in attention orienting to threat in children and adolescents. *Journal of Child Psychology and Psychiatry*, 58(5), 595–602. doi:10.1111/jcpp.12659

Schippell, P. L., Vasey, M. W., Cravens-Brown, L. M., and Bretveld, R. A. (2003). Suppressed attention to rejection, ridicule, and failure cues: A unique correlate of reactive but not proactive aggression in youth. *Journal of Clinical Child and Adolescent Psychology*, 32(1), 40–55. doi: 10.1207/S15374424JCCP3201_05

Schmidtendorf, S., Wiedau, S., Asbrand, J., Tuschen-Caffier, B., and Heinrichs, N. (2018). Attentional bias in children with social anxiety disorder. *Cognitive Therapy and Research*, 42(3), 273–288. doi: 10.1007/s10608-017-9880-7

Seefeldt, W. L., Kramer, M., Tuschen-Caffier, B., and Heinrichs, N. (2014). Hypervigilance and avoidance in visual attention in children with social phobia. *Journal of Behaviour Therapy and Experimental Psychiatry*, 45(1), 105–112. doi: 10.1016/j.jbtep.2013.09.004

Shackman, J. E., Shackman, A. J., and Pollak, S. D. (2007). Physical abuse amplifies attention to threat and increases anxiety in children. *Emotion*, 7(4), 838–852. doi:10.1037/1528-3542.7.4.838

Shechner, T., Jarcho, J. M., Britton, J. C., Leibenluft, E., Pine, D. S., and Nelson, E. E. (2013). Attention bias of anxious youth during extended exposure of emotional face pairs: An eye-tracking study. *Depression and Anxiety*, 30(1), 14–21. doi: 10.1002/da.21986

Short, R. M. L., Adams, W. J., Garner, M., Sonuga-Barke, E. J. S., and Fairchild, G. (2016). Attentional biases to emotional faces in adolescents with conduct disorder, anxiety disorders, and comorbid conduct and anxiety disorders. *Journal of Experimental Psychopathology*, 7(3), 466–483. doi:10.5127/jep.053915

Staugaard, R. S. (2009). Reliability of two versions of the dot-probe task using photographic faces. *Psychology Science Quarterly*, 51(3), 339–350.

Stoddard, J., Sharif-Askary, B., Harkins, E. A., Frank, H. R., Brotman, M. A., Penton-Voak, I. S., ... and Leibenluft, E. (2016). An open pilot study of training hostile interpretation bias to treat disruptive mood dysregulation disorder. *Journal of Child and Adolescent Psychopharmacology*, 26(1), 49–57. doi:10.1089/cap.2015.0100

Sukhodolsky, D. G., Smith, S. D., McCauley, S. A., Ibrahim, K., and Piasecka, J. B. (2016). Behavioral interventions for anger, irritability, and aggression in children and adolescents. *Journal of Child and Adolescent Psychopharmacology*, 26(1), 58–64. doi:10.1089/cap.2015.0120

Susa, G., Pitica, I., Benga, O., and Miclea, M. (2012). The self-regulatory effect of attentional control in modulating the relationship between attentional biases toward threat and anxiety symptoms in children. *Cognition and Emotion*, 26(6), 1069–1083. doi: 10.1080/02699931.2011.638910

Sylvester, C. M., Hudziak, J. J., Gaffrey, M. S., Barch, D. M., and Luby, J. L. (2016) Stimulus-driven attention, threat bias, and sad bias in youth with a history of an anxiety disorder or depression. *Journal of Abnormal Child Psychology*, 44(2), 219–231. doi: 10.1007/s10802-015-9988-8

Tamm, L., Epstein, J. N., Peugh, J. L., Nakonezny, P. A., and Hughes, C. W. (2013). Preliminary data suggesting the efficacy of attention training for school-aged children with ADHD. *Developmental Cognitive Neuroscience*, 4, 16–28. doi:10.1016/j.dcn.2012.11.004

Tripp, G. and Alsop, B. (2001). Sensitivity to reward delay in children with attention deficit hyperactivity disorder (ADHD). *Journal of Child Psychology and Psychiatry*, 42(5), 691–698. doi:10.1111/1469-7610.00764

Tseng, W. L., Moroney, E., Machlin, L., Roberson-Nay, R., Hettema, J. M., Carney, D., ... and Brotman, M. A. (2017). Test-retest reliability and validity of a frustration paradigm and irritability measures. *Journal of Affective Disorders*, 212, 38–45. doi:10.1016/j.jad.2017.01.024

Van den Heuvel, M. I., Heinrichs, J., Donker, F. C. L., and Van den Bergh, B. R. H. (2018). Children prenatally exposed to maternal anxiety devote more additional resources to neutral pictures. *Developmental Science*, 21(4), e12612. doi: 10.1111/desc.12612

Van Mourik, R., Oosterlaan, J., and Sergeant, J. A. (2005). The Stroop revisited: A meta-analysis of interference control in AD/HD. *Journal of Child Psychology and Psychiatry*, 46(2), 150–165. doi: 10.1111/j.1469-7610.2004.00345.x

Waters, A. M., Bradley, B.P., and Mogg, K. (2014). Biased attention to threat in paediatric anxiety disorders (generalised anxiety disorder, social phobia, specific phobia, separation anxiety disorder) as a function of "distress" versus "fear "categorization. *Psychological Medicine*, 44(3), 607–616. doi: 10.1017/S0033291713000779

Waters, A. M., Candy, E. M., and Candy, S. G. (2018). Attention bias to threat in mothers with emotional disorders predicts increased offspring anxiety symptoms: A joint cross-sectional and longitudinal analysis, *Cognition and Emotion*, 32(4), 892–903. doi: 10.1080/02699931.2017.1349650

Waters, A. M., Cao, Y., Kershaw, R., Kerbler, G. M., Shum, D. H. K., Zimmer-Gembeck, M. J., ... and Cunnington, R. (2018). Changes in neural activation underlying attention processing of emotional stimuli following treatment with positive search training in anxious children. *Journal of Anxiety Disorders*, 55, 22–30. doi: 10.1016/j.janxdis.2018.02004

Waters, A. M. and Craske, M. G. (2016). Towards a cognitive-learning formulation of youth anxiety: A narrative review of theory and evidence and implications for treatment. *Clinical Psychology Review*, 50, 50–66. doi:10.1016/j.cpr.2016.09.008

Waters, A. M., Farrell, L. J., Zimmer-Gembeck, M. J., Milliner, E., Tiralongo, E., Donovan, C. L., ... and Ollendick, T. H. (2014). Augmenting one-session treatment of children's specific phobias with attention training to positive stimuli. *Behaviour Research and Therapy, 62,* 107–119. doi: 10.1016/j/.brat.2014.07.020

Waters, A. M., Forrest, K., Peters, R. M., Bradley, B. P., and Mogg, K. (2015). Attention bias to emotional information in children as a function of maternal emotional disorders and maternal attention biases. *Journal of Behavior Therapy and Experimental Psychiatry, 46,* 158–163. doi: 10.1016/j.jbtep.2014.10.002

Waters, A. M. and Lipp, O. V. (2008a). Visual search for emotional faces in children. *Cognition and Emotion, 22*(7), 1306–1326. doi:10.1080/02699930701755530

Waters, A. M. and Lipp, O, V. (2008b). The influence of animal fear on attentional capture by fear-relevant animal stimuli in children. *Behaviour Research and Therapy, 46*(1), 114–121. doi: 10.1016/j.brat.2007.11.002

Waters, A. M., Lipp, O. V., and Randhawa, R. S. (2011). Visual search with animal fear-relevant stimuli: A tale of two procedures. *Motivation and Emotion, 35*(1), 23–32. doi: 10.1007/s11031-010-9191-8

Waters, A. M., Lipp, O. V., and Spence, S. H. (2004). Attentional bias toward fear-related stimuli: An investigation with nonselected children and adults and children with anxiety disorders. *Journal of Experimental Child Psychology, 89*(4), 320–337. doi: 10.1016/j.jecp.2004.06.003

Waters, A. M., Mogg, K., and Bradley, B. P. (2012). Direction of threat attention bias predicts treatment outcome in anxious children receiving cognitive-behavioural therapy. *Behaviour Research and Therapy, 50*(6), 428–434. doi: 10.1016/j.brat.2012.03.006

Waters, A. M., Mogg, K., Bradley, B. P. and Pine, D. S. (2008). Attentional bias for emotional faces in children with generalised anxiety disorder. *Journal of American Academy of Child and Adolescent Psychiatry, 47*(4), 435–442. doi: 10.1097/CHI.0b013e3181642992

Waters, A. M., Potter, A., Jamesion, L., Bradley, B. P., and Mogg, K. (2015). Predictors of treatment outcomes in anxious children receiving group cognitive-behavioural therapy: Pretreatment attention bias to threat and emotional variability during exposure tasks. *Behaviour Change, 32*(3), 143–158. doi: 10.1017/bec.2015.6

Waters, A. M. and Valvoi, J. S. (2009). Attentional bias for emotional faces in paediatric anxiety disorders: An investigation using the emotional go/no go task. *Journal of Behavior Therapy and Experimental Psychiatry, 40*(2), 306–316. doi:10.1016/j.jbtep.2008.12.008

Waters, A. M., Zimmer-Gembeck, M. J., Craske, M. G., Pine, D. S., Bradley, B. P., and Mogg, K. (2016). A preliminary evaluation of a home-based, computer-delivered attention training treatment for anxious children living in regional communities. *Journal of Experimental Psychopathology, 7*(6), 511–527. doi: 10.5127/jep.053315

White, L. K., Sequeria, S., Britton, J. C., Brotman, M. A., Gold, A. L., Berman, E., ... and Pine, D. S. (2017). Complementary features of attention bias modification therapy and cognitive-behavioural therapy in pediatric anxiety disorders. *American Journal of Psychiatry, 174*(8), 775–784. doi: 10.1176/appi.ajp.2017.16070847

White, S. F., Williams, W. C., Brislin, S. J., Sinclair, S., Blair, K. S., Fowler, K. A., ... and Blair, R. J. (2012). Reduced activity within the dorsal endogenous orienting of attention network to fearful expressions in youth with disruptive behavior disorders and psychopathic traits. *Development and Psychopathology, 24*(3), 1105–1116. doi:10.1017/S0954579412000569

Wilkowski, B. M. and Robinson, M. D. (2008). The cognitive basis of trait anger and reactive aggression: An integrative analysis. *Personality and Social Psychology Review, 12*(1), 3–21. doi:10.1177/1088868307309874

Williams, J. M., Watts, F. N., MacLeod, C., and Matthews, A. (1988). *Cognitive Psychology and Emotional Disorders*. Chichester, U.K.: John Wiley & Sons.

Wittchen, H. U., Jacobi, F., Rehm, J., Gustavsson, A., Svensson, M., Jonsson, B., ... and Steinhausen, H. C. (2011). The size and burden of mental disorders and other disorders of the brain in Europe 2010. *European Neuropsychopharmacology*, 21(9), 655–679. doi: 10.1016/j.euroneuro.2011.07.018

Yantis, S. (2000). Goal-directed and stimulus-driven determinants of attentional control. In S. Monsell and J. Driver (Eds.), *Attention and performance*. Cambridge: MIT Press.

Zvielli, A., Bernstein, A., and Koster, E. H. W. (2014). Temporal dynamics of attentional bias. *Clinical Psychological Science*, 3(5), 772–788. doi:10.1177/2167702614551572

Wilson, A., ... and C. ... A [...] ... [...]
... Blackwell, Oxford: Wiley-Blackwell.

Wilson, R. C., and B. W. [...] ... [...] ...
... (2011) [...] ... [...] ... ...
... ... [...] ... [...] ...

... ... [...] ... [...] ...
... and ... [...] ... [...] ...

... ... and Foster, R. J. [...] ... [...] ...
... [...] ... [...] ...

CHAPTER 21

MEMORY DEVELOPMENT

PATRICIA J. BAUER AND JESSICA A. DUGAN

INTRODUCTION

THE title of this chapter—*Memory Development*—creates the impression that a single entity—*memory*—has a single course of development. Instead, there are several different types of memory, each with its own characteristics and course of development. For example, one type of memory is short term, lasting only seconds or minutes before fading away, whereas another type is long term, which may last as long as a lifetime. Some types of memory have a limited capacity, whereas others are for all practical purposes, limitless. Memory sometimes is context specific and other times highly flexible. In fact, it sometimes seems that the only thing all memory has in common is that it is about the past. Yet even this characterization is not universally true, in that remembering to do something *in the future*, also qualifies as "memory," of the prospective type.

One primary goal of this chapter is to characterize the major forms of memory. This step is necessary in order to accomplish the second goal, which is to describe the course of development of different types of memory over the first two decades of life. Because it features some of the most profound changes, the focus will be on long-term memory. As will become apparent, different types of long-term memory have different courses of development. Differences in the timing and course of development shed light on some of the mechanisms of age-related change in childhood, consideration of which is the third major goal of the chapter.

DIFFERENT FORMS OF MEMORY

The suggestion that memory is not a unitary construct is an old one. Maine de Biran (1804/1929) is credited as the first to suggest that there might be different forms of memory (Schacter, Wagner, and Buckner, 2000). The notion was furthered at the beginning of the twentieth century with studies of wounded veterans from World War I. Kleist (1934), a German physician, examined veterans who had received head wounds from gun shots or shrapnel and the behavioral patterns that seemed to result from them. He observed that there were systematic relations between the site of the wound (and resulting brain lesion) and the type of mental impairment experienced by the veteran. The notion that different parts of the brain subserve different cognitive functions, including different types of memory, received especially strong impetus from the famous case of H.M. (Henry Molaison, 1926–2008), who, at the age of 27, underwent experimental surgery to treat intractable seizures. To treat the seizures, his surgeon removed large portions of the temporal lobes on both sides of the brain (Scoville and Milner, 1957). Subsequent to the surgery, H.M. suffered impairment of some forms of memory, yet not all of his memory capacities were disrupted. For example, H.M. was able to learn new motor skills but was challenged to remember the words or images that appeared on a list. His case in particular led to the notion that there are different types or forms of memory. Whereas some distinctions within the domain are readily accepted, others are sources of debate.

Short- and Long-Term Memory

Characterizing memory along a temporal dimension is relatively uncontroversial. Whereas some memories are short term, lasting only seconds, others last much longer, on the order of days, months, years, and perhaps even a lifetime. Short-term memory generally is recognized as capacity limited. Historically it was thought to hold 7 "units" of information—such as digits in a phone number—plus or minus 2 (Miller, 1956). More recent conceptualizations of short-term memory purport a store of only 3–5 units, depending on the contents of each unit and how they are processed (e.g., how effectively they are grouped or "chunked": Cowan, 2010, 2017). In contrast, long-term memory is virtually limitless in its capacity. There seemingly is no ceiling on the number of items, pieces of information, or personal experiences, that one can maintain in long-term memory stores (more on the temporal dimension and capacity limitations of memory to follow).

Declarative and Non-Declarative Memory

It is generally accepted that within long-term memory, there is a division based on the contents of memory, its function, its rules of operation, and the neural substrates that support it. Although the precise distinctions captured by the different labels given to the divisions are not identical, there is substantial overlap in contemporary conceptualizations of declarative or explicit memory versus non-declarative, procedural, or implicit memory. These two major divisions of long-term memory are illustrated schematically in Figure 21.1.

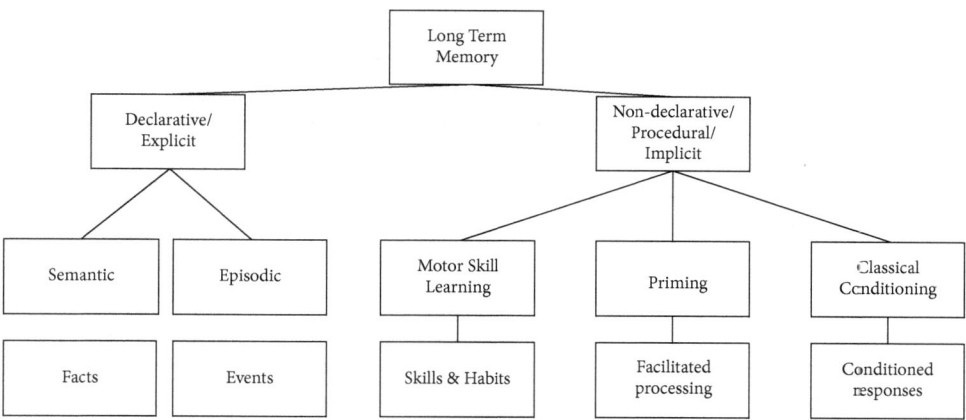

FIGURE 21.1 Schematic representation of major divisions of the long-term memory system, into declarative or explicit memory and non-declarative, procedural, or implicit memory.

Declarative or Explicit Memory

Declarative or explicit memory is devoted to processing of names, dates, places, facts, events, and so forth. These are entities that we think of as being encoded symbolically and that thus can be described with language. In terms of function, declarative memory is specialized for fast processing and learning. New information can be entered into the declarative memory system on the basis of a single trial or experience. In terms of rules of operation, declarative memory is fallible: we forget names, dates, and places, for example. Although there are compelling demonstrations of long-term remembering of lessons learned in high school and college (e.g., foreign language vocabulary: Bahrick, 2000), a great deal of forgetting from declarative memory occurs literally minutes, hours, and days after an experience. Declarative memory also has a specific neural substrate. As reviewed in the following, current conceptualizations suggest that the formation, maintenance, and subsequent retrieval of declarative or explicit memories depends on a multi-component network involving cortical structures as well as medial temporal structures (e.g., Eichenbaum, Sauvage, Fortin, Komorowski, and Lipton, 2012; Squire, 2004; Squire and Wixted, 2011).

Declarative memory is itself subdivided into the categories of semantic memory and episodic memory (e.g., Schacter and Tulving, 1994), with a finer distinction between episodic memory and autobiographical memory. Semantic memory supports general knowledge about the world (Tulving, 1972, 1983). We are consulting semantic memory when we retrieve the facts that the Appalachian Trail is 2190 miles long and crosses 14 states, the southern terminus of the Appalachian Trail is in Georgia, and that the idea for the trail was developed by Benton MacKaye and proposed in 1921. For practical purposes, both the capacity of semantic memory and the longevity of the information stored in it seem infinite. Semantic memory also is not tied to a particular time or place. That is, we know facts and figures, names and dates, yet in most cases, we do not know when and where we learned this information. We might be able to reconstruct how old we were or what grade we were in when we learned some tidbits of information, but unless there was something unique about the experience surrounding the acquisition of this information, we carry it around without address or reference to a specific episode.

In contrast to semantic memory, episodic memory supports retention of information about unique events (Tulving, 1972, 1983), such as hiking a section of the Appalachian Trail or the fact that Georgia was one of the states on a list of state names studied in a memory experiment. Some episodic memories, such as whether a specific state was included in a word list, may not stay with us for very long, and are not especially personally relevant or significant. Yet other episodic memories are personally significant and even self-defining. These so-called *autobiographical* memories are episodic memories that are infused with a sense of personal involvement or ownership (Bauer, 2007, 2015). They are the episodes on which we reflect when we consider who we are and how our previous experiences have shaped us.

Non-Declarative, Procedural, or Implicit Memory

Non-declarative, procedural, or implicit memory is devoted to perceptual and motor skills and procedures. Skilled motor behavior, such as playing piano or jumping hurdles, is not a name, date, place, fact, or event, but a collection of finely tuned motor patterns, behaviors, and perceptual skills that we cannot verbally describe. Most types of non-declarative memory function to support gradual, incremental learning. That is, behavior is modified through practice, experience, or multiple trials. Perhaps as a result of its slow function, a rule of the operation of non-declarative memory is that the learning is relatively infallible. A typical example is riding a bicycle—we may not have ridden a bicycle in many years but when we ride one again, we "just know how." Indeed, breaking an old non-declarative pattern can be quite difficult, as exemplified by the tendency to, in a moment of panic, slam the pedals backward (as we did on our old one-speed bikes) instead of squeezing the hand brakes (as we should on our adult, multi-speed bikes). As might be expected, given the diversity of behaviors supported by non-declarative memory, different types of non-declarative memory depend on different neural structures. For example, motor skill learning, and many types of conditioning seemingly are dependent on cerebellum and subcortical structures such as the basal ganglia, and priming seems dependent on extrastriate cortex.

Just as declarative memory is subdivided, so too is non-declarative memory. The most common categories of non-declarative memory are 1) motor skill learning, 2) priming, and 3) classical conditioning. The range of motor skills that we acquire is limitless. What they have in common is that although there might be a declarative component to them (to stop the bike, put on the brakes), their fluid execution is not accomplished via learning and frequent repetition of the declarative "rules," but motor practice. Behind an exquisite piano performance or a championship winning jump are innumerable hours of practice—repetition after repetition of a passage of music or agility drills. One can learn a simple tune like "*Mary Had a Little Lamb*" or basic jumping movements in an elementary class, but it is not this knowledge on which experts depend. Consider the steps necessary to prepare a vegetable soup from scratch. Potatoes, onions, celery, and carrots must all be chopped before going into the pot, but each requires a different amount of pressure to slice through successfully. Luckily, we do not need to consciously consult our memory to apply the appropriate pressure each and every time. The information is encoded in our muscles and joints; it is not available to conscious access or description.

Another form of non-declarative memory is priming, which is facilitated processing of a stimulus following prior exposure to the stimulus. Perceptual priming occurs when subsequent processing of a stimulus is facilitated by prior perceptual exposure to it. Conceptual priming occurs when there is overlap between related concepts in memory. A person who had recently studied a list that included the word "Alaska" would be more likely to include it in a list of states than an individual who had not studied the item. With regard to the distinction between conscious (declarative) and unconscious (non-declarative) memory, an important point to remember about priming is that it can occur without conscious awareness that the item had been studied earlier, or that subsequent processing was facilitated by earlier processing. That is, facilitated processing occurs even in the absence of recognition or recall of the originally studied item (see Lloyd and Newcombe, 2009; Roediger and McDermott, 1993, for discussions).

Finally, classical conditioning occurs when two stimuli that naturally do not co-occur become associated with one another through repeated pairing. Typically, one of the stimuli—the unconditioned stimulus—evokes a high-probability or even reflexive response termed the unconditioned response. For example, a puff of air to the eye (unconditioned stimulus) produces a blink (unconditioned response). The other stimulus—the conditioned stimulus—is behaviorally neutral, such as a tone. Classical conditioning occurs when, over time and repeated pairing of the two stimuli, the conditioned stimulus (the tone) takes on the same significance as the unconditioned stimulus, such that it alone is sufficient to produce the response: anticipatory eye closure at the sound of the tone (see Woodruff-Pak and Disterhoft, 2008, for a review). This simple form of learning occurs across phyla (e.g., rodents, nonhuman primates, humans), making it one of the clearest examples of learning and memory without awareness.

Relations Between Different Forms of Memory

An important point to keep in mind with regard to the distinction between declarative and non-declarative forms of memory is that, in most cases, one derives both declarative and non-declarative memories from the same experience. Learning to drive a car is a good example. In driver's education class, students rely on declarative memory to learn that to go they step on the accelerator and to stop they step on the brake. Tests to obtain a license to drive probe the student's memory of appropriate following and stopping distance and the steps involved in parallel parking, for example. But one would never become an expert driver based on declarative memory alone. To become a good driver, one must practice executing the motor movements. Although we may be consciously aware that vehicle stopping distance is 15 feet per second, it is not this knowledge that permits us to bring our vehicles to a gentle stop at a stop sign. That skill is acquired through practice, practice, and more practice at driving and braking and otherwise fine-tuning non-declarative memory.

Another important point is that, in most cases, declarative and non-declarative memories not only are acquired in parallel, but they continue to co-exist, even after execution of the behavior no longer seems to require conscious awareness. Many skills, such as driving, start out very demanding of attentional resources yet eventually can be performed almost "automatically," that is, without conscious attention paid to them. Such changes tempt the conclusion that once these skills no longer require conscious attention to execute, they no longer are declarative. Yet such a conclusion is not valid: if declarative memories were to become non-declarative, that would mean that they were no longer accessible to consciousness or declaration. The fact that we can declare that vehicle stopping distance is 15 feet per second, proves that memory of the procedure is accessible to consciousness, even though we do not have to think about feet per second in order to brake. In other words, the declarative memory still exists, even though it is not that memory on which we depend to execute the behavior. In the intact organism, declarative and non-declarative memory co-exist. Many behaviors are executed based on non-declarative memory alone, but that does not mean that one type of memory has "turned into" another. Rather, at the time of learning, both types of memory were acquired; skilled performance of the motor behavior may be supported by one (non-declarative), yet the other (declarative) persists.

The observations that different forms of long-term memory are acquired in parallel and co-exist, possibly throughout the life of the memory, beg the question of why organisms would have multiple types of memory. It is likely that these different forms of memory either evolved in order to deal with competing demands for different kinds of information storage, or that nature took fair advantage of structures that had evolved for others reasons, in order to deal with the demands. Either way, one memory system (declarative) seems specialized for rapid encoding of information that is subject to equally rapid forgetting, whereas the other (non-declarative) seems specialized for acquisition of information at a slower rate, yet seemingly permits more robust retention. Why could not a single system accomplish both tasks? Although this question cannot be answered definitively, computer simulations have revealed that a system that can change rapidly to accept new inputs has difficulty maintaining old inputs. Conversely, a system that is good at maintaining old inputs has difficulty "learning" new things (e.g., McClelland, McNaughton, and O'Reilly, 1995; O'Reilly, Bhattacharyya, Howard, and Ketz, 2014). This analysis suggests that complementary memory systems work in concert in order to avoid interference with existing knowledge, yet still maintain flexibility.

DEVELOPMENTAL CHANGES IN
DECLARATIVE MEMORY

Discussion of developmental changes in all of the different forms of memory is well beyond the scope of this chapter. For this reason, we focus on developmental changes in declarative memory rather than non-declarative memory. We do so for two related reasons. The first reason is that non-declarative memory emerges early in development and is generally considered to undergo little developmental change. Conceptual priming (facilitated processing based on the meaning of the stimulus) has been nominated as an exception to this rule. For example, in response to the instruction to list as many states as they can, subjects are more likely to nominate "Alaska" if they saw the word earlier in the session than if they did not. Conceptual priming is more robust in older relative to younger children (e.g., Perez, Peynircioglu, and Blaxton, 1998). Yet as noted by Lloyd and Newcombe (2009), it is likely that rather than differences in priming, these paradigms may reflect age-related improvements in conceptual knowledge. That is, the mechanism of priming may be age invariant and what accounts for improvements with age is category knowledge. This perspective contributes to the second reason for our focus on declarative rather than non-declarative memory, namely, that there has not been substantial accumulation of new knowledge on the development of non-declarative memory since the last edition of the *Handbook* (see Bauer, 2013b, for a review). For the balance of this chapter, we focus on developmental changes in declarative memory and in episodic and autobiographical memory in particular.

Episodic Memory

As discussed previously, episodic memory supports retention of information about unique events that can be located in a particular place and time (Tulving, 1972, 1983). This form of memory has a protracted course of development, beginning in the first year of life and extending throughout childhood. It is typically researched in two ways: 1) by probing children's recall of naturally occurring events or laboratory events that mimic them, and 2) by testing children's recall or recognition of words on a list or visual images. Because children's memory for events more readily can be studied across both infancy and childhood, our primary focus is on that body of research (see Ghetti and Lee, 2014; Roebers, 2014; for reviews of children's memory for other types of stimuli).

Episodic Memory in the First Years of Life

Until the middle 1980s, it was widely believed that infants and young children were incapable of remembering specific past events. As discussed in Bauer (2006b, 2007), the pessimism as to children's mnemonic abilities stemmed from a number of sources. One salient source of the perspective was the literature on *infantile* or *childhood amnesia*.

Most adults remember few if any memories from the first 3 to 4 years of life and from the ages of 3½ to 7, they have a smaller number of memories than would be expected based on forgetting alone. As reviewed in Bauer (2007, 2008, 2014), the phenomenon is strikingly robust and consistent across time, population, and method (e.g., free recall, response to cue words, questionnaire). Although a variety of explanations for the amnesia have been advanced (see Bauer, 2007, 2008, 2014, for reviews), one of the most common was also the simplest: adults lacked memories from early in life because infants and young children failed to create them.

The perspective on infants' and young children's mnemonic abilities began to change in the middle 1980s as a result of recognition of the importance to memory of meaningful and familiar stimuli and development of a means of assessing event memory in pre- and early verbal children. In an influential series of studies, Nelson and her colleagues (e.g., K. Nelson, 1986) demonstrated that when children were asked to recall "what happens" in the context of everyday events and routines, such as going to McDonald's, their performance was qualitatively similar to that of older children and even adults. Moreover, it became apparent that children did not require multiple experiences of events in order to remember them. For example, Fivush (1984) interviewed kindergarten children after only a single day of school. Although the children had experienced the school-day routine just once, they nevertheless provided well-organized reports of the experience.

The change in perspective on memory ability extended to infants as well, with the development of nonverbal means of assessing memories for specific episodes, namely, elicited and deferred imitation. Elicited and deferred imitation entail use of objects to demonstrate an action or sequence of actions that either immediately (elicited imitation), after some delay (deferred imitation), or both, infants are invited to imitate. As discussed in detail elsewhere (e.g., Bauer, 2006b, 2007; Carver and Bauer, 2001), even though the expression of memory is nonverbal imitation, the task is assumed to depend on declarative memory. We repeat only two components of the argument here. First, once children acquire the requisite language, they talk about events that they experienced as preverbal infants, in the context of imitation tasks (e.g., Bauer, Wenner, and Kroupina, 2002; Cheatham and Bauer, 2005). This is strong evidence that the format in which the memories are encoded is declarative, as opposed to non-declarative or implicit (formats inaccessible to language). Second, the paradigm passes the "amnesia test." Whereas intact adults accurately imitate sequences after a delay, patients with amnesia due to hippocampal lesions perform no better than naïve controls (McDonough, Mandler, McKee, and Squire, 1995). Adolescents and young adults who sustained hippocampal damage early in life also exhibit deficits in performance on the task (Adlam, Vargha-Khadem, Mishkin, and de Haan, 2005). This suggests that the paradigm taps the type of memory that gives rise to verbal report. For these reasons, the task has come to be widely accepted as a nonverbal analogue to verbal report (e.g., Bauer, 2002; Mandler, 1990; Meltzoff, 1990; K. Nelson and Fivush, 2004; Rovee-Collier and Hayne, 2000; Schneider and Bjorklund, 1998; Squire, Knowlton and Musen, 1993), and is widely used to examine developments in memory for specific episodes in the first 2 years of life.

Using an imitation-based technique, Meltzoff (1988) demonstrated that infants as young as 9 months of age were able to defer imitation of an action for 24 hours. Bauer and Shore (1987) found that over a 6-week delay, infants 17 to 23 months of age remembered not only individual actions but temporally ordered sequences of action. That is, even after 6 weeks, they were able to reproduce in the correct temporal order the steps of putting a ball into a cup, covering it with another cup, and shaking the cups to make a rattle. Subsequent research revealed that over the first 2 years of life there are developmental changes in memory for specific episodes along a number of dimensions. Perhaps the most salient change is in the length of time over which memory is apparent. Importantly, because like any complex behavior, the length of time an episode is remembered is determined by multiple factors, there is no "growth chart" function that specifies that children of X age should remember for Y long. Nonetheless, across numerous studies there has emerged evidence that with increasing age, infants tolerate lengthier retention intervals. To illustrate, whereas at 6 months of age, infants remember an average of one action of a three-step sequence for 24 hours (Barr, Dowden, and Hayne, 1996; see also Collie and Hayne, 1999), by 20 months of age, children remember the actions of event sequences over as many as 12 months (Bauer, Wenner, Dropik, and Wewerka, 2000; see Bauer, 2013a; Lukowski and Bauer, 2014, for reviews).

Infants also recall the temporal order of actions in multi-step sequences, though retaining order information presents a cognitive challenge to young infants, in particular, as evidenced by low levels of ordered recall and substantial within-age-group variability in the first year. Although 67 percent of Barr et al.'s (1996) 6-month-olds remembered individual actions over 24 hours, only 25 percent of them remembered actions in the correct temporal order. Among 9-month-olds, approximately 50 percent of infants exhibit ordered reproduction of sequences after a 5-week delay (Bauer, Wiebe, Carver, Waters, and Nelson, 2003; Bauer, Wiebe, Waters, and Bangston, 2001; Carver and Bauer, 1999). By 13 months of age the substantial individual variability in ordered recall has resolved: 78 percent of 13-month-olds exhibit ordered recall after 1 month. Nevertheless, throughout the second year of life, there are age-related differences in children's recall of the order in which actions of multi-step sequences unfolded. The differences are especially apparent under conditions of greater cognitive demand, such as when less support for recall is provided, and after longer delays (Bauer et al., 2000).

The first 2 years of life also are marked by changes in the robustness of memory for specific episodes. For instance, there are changes in the number of experiences that seem to be required in order for infants to remember. In Barr et al. (1996), at 6 months, infants required six exposures to events in order to remember them 24 hours later. If instead they saw the actions demonstrated only three times, they provided no evidence of memory after 24 hours. By the time infants are 14 months of age, a single exposure session is all that is necessary to support recall of multiple different single actions over 4 months (Meltzoff, 1995). Ordered recall of multi-step sequences is apparent after as many as 6 months for infants who received a single exposure to the events at the age of 20 months (Bauer and Leventon, 2013).

Another index of the robustness of memory is the extent to which it is disrupted by interference. One form of interference that has been studied in infancy is changes in context between the time an event is experienced (i.e., encoding) and the time it is tested for persistence (i.e., retrieval). Although there are some suggestions that recall is disrupted if between exposure and test the appearance of the test materials is changed (e.g., Hayne, Boniface, and Barr, 2000; Hayne, MacDonald, and Barr, 1997; Herbert and Hayne, 2000), there also are reports of robust generalization from encoding to test by infants across a wide age range. Infants have been shown to generalize imitative responses across changes in 1) the size, shape, color, and/or material composition of the objects used in demonstration versus test (e.g., Bauer and Dow, 1994; Bauer and Fivush, 1992; Bauer and Lukowski, 2010; Lechuga, Marcos-Ruiz, and Bauer, 2001), 2) the appearance of the room at the time of demonstration of modeled actions and at the time of memory test (e.g., Barnat, Klein, and Meltzoff, 1996; Klein and Meltzoff, 1999), 3) the setting for demonstration of the modeled actions and the test of memory for them (e.g., Hanna and Meltzoff, 1993; Klein and Meltzoff), and 4) the individual who demonstrated the actions and the individual who tested for memory for the actions (e.g., Hanna and Meltzoff). In summary, whereas there is evidence that with age, infants' memories as tested in imitation-based paradigms become more generalizable (e.g., Herbert and Hayne, 2000), there is substantial evidence that from an early age, infants' memories survive changes in context and stimuli.

Developments in the Preschool Years and Beyond

Beginning in the third year of life, verbal assessments become a viable means for testing episodic memory. This opens up new possibilities: children can be tested not only for memory for controlled laboratory events but for events from their lives outside the laboratory as well. This combination of approaches has yielded a wealth of data about children's memories for the routine events that make up their everyday lives, and about their memories for unique events. Some of the events are highly personally significant and contribute to an emerging autobiography or personal past. Major findings from each of these categories are reviewed.

Early studies of young children's memories for the events of their own lives focused on everyday, routine events. The children's reports included actions common to the activities and almost invariably, the actions were mentioned in the temporal order in which they typically occurred. Representative of the findings was the answer provided by a 3-year-old child to the question "What happens when you have a birthday party?": "You cook a cake and eat it" (Nelson and Gruendel, 1986, p. 27). This early research revealed "minimalist," yet nevertheless accurate, reports by children as young as 3 years of age (see also K. Nelson, 1986, 1997). Subsequent studies revealed that with development, children's reports included more information. For example, in addition to mention of cooking a cake and then eating it, 6- and 8-year-old children told of putting up balloons, receiving and then opening presents from party guests, eating birthday cake, and playing games. Relative to younger children, older children also more frequently mentioned alternative actions (e.g., "... and then you have lunch *or whatever you have"*) and optional activities (e.g., "*Sometimes* then they have three games ... then *sometimes* they open up the other presents ...."; K. Nelson and Gruendel, 1986). Whereas some of the differences in younger and older children's reports might be due to the greater number of experiences of events such as birthday parties that older children have, relative to younger children, experience alone does not account for the developmental differences. In laboratory research in which children of different ages are given the same amount of experience with a novel event, older children produce more elaborate reports relative to younger children (e.g., Fivush, Kuebli, and Clubb, 1992; Price and Goodman, 1990).

Young children also form memories of unique events. In an early study, Fivush, Gray, and Fromhoff (1987) found that all of the children in a sample of 2½- to 3-year-olds recalled at least one event that had happened 6 or more months in the past. The children reported the same amount of information about events that had taken place more than 3 months ago as they did about events that had taken place within 3 months. In Hamond and Fivush (1991), 3- and 4-year-old children recalled a trip to Disney World they had taken either 6 months previously or 18 months previously. The amount they remembered did not differ as a function of the delay. Moreover, the older and younger children did not differ in the amount of information they reported about the event. Yet the age groups did differ in how elaborate their reports were. Whereas the younger children tended to provide the minimum required response to a question, the older children tended to provide more elaborate responses.

With development, there are changes in what children include in their reports about events. For example, young children seemingly focus on what is common or routine across experiences, whereas older children and adults focus on what is unique or distinctive. This trend is illustrated in Fivush and Hamond (1990). In response to an interviewer's invitation to talk about going camping, after providing the interviewer with the distinctive information that the family had slept in a tent, a 2½-year-old child went on to report on features that were not unique to the camping experience, such as the fact that they ate dinner and went for a walk. In total, 48 percent of the information that the children reported was judged to be distinctive, implying that 52 percent of it was not. By 4 years of age, children report about three times more distinctive information than typical information (Fivush and Hamond, 1990). One consequence of focus on what is common across experience is that a unique event such as camping gets "fused" into the daily routine of eating and sleeping. In the process, the features that distinguish events from one another may fade into the background and be lost. The result would be fewer memories of episodes that are truly unique. Conversely, with increasing focus on the more distinctive features of events, there is a resulting increase in the number of memories that are truly unique.

With age, children not only include different types of information in their narratives, but they include more information. For example, in research by Fivush and Haden (1997), from 3½ to 6 years of age, the number of propositions children included in the average narrative increased more than two-fold, from 10 to 23. Young children's narratives include basic information about what actions occurred in the event; they feature intensifiers, qualifiers, and internal evaluations; and the actions in the narrative are joined by simple temporal and causal connections (e.g., *then, before, after*; and *because, so, in order to*; respectively). What accounts for the increase in narrative length over this age period is that with age, children provide 1) more information about who was involved and when and where the event occurred, 2) more information about optional or variable actions (e.g., "*When it turned red light*, we stopped"; Fivush and Haden, p. 186), and 3) more elaborations (Fivush and Haden). As a result, relative to younger children's, older children's stories are more complete, easier to follow, and more engaging (see Reese, Haden, Baker-Ward, Bauer, Fivush, and Ornstein, 2011, for a review).

The dramatic increases with age in the amount of information that children *report* tempts the conclusion that there also are age-related increases in the amount of information that children *remember* about events. This is not a "safe" conclusion, however, in light of evidence that perhaps especially for younger children, verbal reports underestimate the richness of memories (e.g., Fivush, Sales, Goldberg, Bahrick, and Parker, 2004; see Bauer, 2015; and Mandler, 1990, for discussions). Indeed, because of the inevitable confounding between increases in age and increases in narrative skills, whereas it is clear that children report more with age, it is not clear whether they also remember more.

Autobiographical Memory

Over the course of the preschool years, autobiographical or personal memory becomes increasingly apparent (see Bauer, 2015, or a review). Autobiographical memories are the memories of events and experiences that make up one's life story or personal past. They are the stories that we tell about ourselves that reveal who we are and how our experiences have shaped our characters. As implied by this description, autobiographical memories differ from "run-of-the-mill" episodic memories in that autobiographical memories are infused with a sense of personal involvement in or ownership of the event. They are memories of events that happened to one's self; in which one participated; and about which one had emotions, thoughts, reactions, and reflections. It is this feature that puts the "auto" in "*auto*biographical."

Throughout the preschool years, children's memories take on more and more autobiographical features (see Bauer, 2007, 2015, for discussions). From a very young age, children include references to themselves in their narratives: "*I* fell down." With age, they increasingly pepper their narratives with the subjective perspective that indicates the significance of the event for the child (Fivush, 2001). For example, they go beyond comment on the objective reality of "falling down" to convey how they felt about the fall: "I fell down and *was so embarrassed* because everybody was watching!" It is this subjective perspective that provides the explanation for why events are funny, or sad, for instance, and thus, of significance to one's self.

There also are changes in the marking of events as having taken place at a specific place and time. For instance, children increasingly include specific references to time, such as "on my birthday," "at Christmas," or "last summer" (K. Nelson and Fivush, 2004). Markings such as these not only establish that an event happened at a time different from the present, but they begin to establish a timeline along which an organized historical record of when events occurred can be constructed. Children also include in their narratives more orienting information, including where events took place and who participated in them (e.g., Fivush and Haden, 1997). These changes serve to distinguish events from one another, thereby making them more distinctive. Children also include more descriptive detail in their reports, suggestive of a sense of re-living of the experience. For example, they include more intensifiers ("Cause she was *very* naughty"), qualifiers ("I *didn't like* her video tape"), elements of suspense ("And *you know what*?"; examples from Fivush and Haden, 1997), and even repetition of the dialogue spoken in the event ("... I said, 'I hope my Nintendo my Super Nintendo is still here.'"; from Ackil, Van Abbema, and Bauer, 2003). The result is a much more elaborate narrative that brings both the storyteller and the listener to the brink of reliving the experience. It is tempting to conclude that these changes account for the finding among adults of a steadily increasing number of memories of events that took place from the ages of 3 to 7 years (Bauer, 2007, 2014).

Relative to those in the preschool years, developments in autobiographical or personal memory in later childhood and adolescence have been relatively neglected. Yet age-related changes in autobiographical reports continue throughout the elementary school years and beyond. An illustration of the types of changes that occur during this period is the breadth or completeness of children's narratives. Like a good newspaper story, a "good" autobiographical narrative includes a number of elements, including information about the *who, what, where, when, why,* and *how* of the experience. The average 7-year-old includes only half the number of these narrative elements than the average 11-year-old (see Bauer, Burch, Scholin, and Güler, 2007).

MECHANISMS OF DEVELOPMENTAL CHANGE

Given that memory is not a single entity or unity construct, it is not surprising that there is not a single answer to the question of "what develops" in the development of memory. Rather, like all complex behaviors, memory is determined by many factors. An adequate explanation of why it develops as it does will entail multiple levels of analysis, ranging from the cellular and molecular events that allow for the storage of information to the cultural influences that shape the expression of memory. Because the focus of this volume is on basic neural and cognitive processes, we focus on these two categories of explanation. The reader is referred to other sources (e.g., Bauer, 2007, 2013a, 2015; Fivush and Zaman, 2014; K. Nelson and Fivush, 2004) for elaboration of the roles of other aspects of development, including conceptual change and social influences on remembering.

Neural Structures and Processes

A thorough review of the neural substrates that support the different types of memory, and the course of development of each, is well beyond the scope of this chapter (see Bachevalier, 2014; Lukowski and Bauer, 2014; C. Nelson, de Haan, and Thomas, 2006, for reviews of subsets of this large literature). Yet a brief review is essential to the goal of identifying possible mechanisms of developmental change. Studies of patients with specific types of lesions and disease and animal models thereof, as well as neuroimaging studies, have made clear that registration of experience and formation of a memory traces to represent it involve multi-stage processes that depend on networks of neural structures. For example encoding, storage, and later retrieval of declarative memories depends on a multi-component network involving temporal (hippocampus, and entorhinal, parahippocampal, and perirhinal cortices) and cortical (including pre-frontal and other association areas) structures (e.g., Dickerson and Eichenbaum, 2010; Eichenbaum and Cohen 2001; Markowitsch 2000; Milner, 2005; Zola and Squire 2000).

The Neural Substrate of Declarative Memory

Formation of a declarative memory begins as the elements that constitute an experience register across primary sensory areas (e.g., visual and auditory). Inputs from primary cortices are projected to unimodal association areas where they are integrated into whole percepts of what objects look, feel, and sound like. Unimodal association areas in turn project to polymodal prefrontal, posterior, and limbic association cortices where inputs from the different sense modalities are integrated and maintained over brief delays (see e.g., Petrides, 1995). Prefrontal structures not only are involved in the initial processing or *encoding* of experiences into long-term traces, but also are implicated in the temporary maintenance of material in short-term and working memory.

For maintenance of traces of experience over delays of longer than a few seconds or minutes, the inputs to the association areas must be stabilized or *consolidated*, a task attributed to medial-temporal structures, in concert with cortical areas (McGaugh, 2000). As illustrated in Figure 21.2, information from association areas converges on perirhinal and parahippocampal structures from which it is projected to lateral and medial entorhinal cortex (respectively) and in turn to hippocampus. Within the hippocampus, conjunctions and relations among the elements of experience are linked into a single event. Association areas share the burden of consolidation, by relating new memories to episodes already in storage: information processed in the hippocampus is projected back through the temporal cortices which in turn project to the association areas that gave rise to their inputs. Eventually, traces are stabilized such that the hippocampus is no longer required to maintain them; consolidated traces are *stored* in neocortex (although, whether memories ever are wholly independent of the hippocampus is debated: see for example, Moscovitch and Nadel, 1998, and Reed and Squire, 1998, for opposing views).

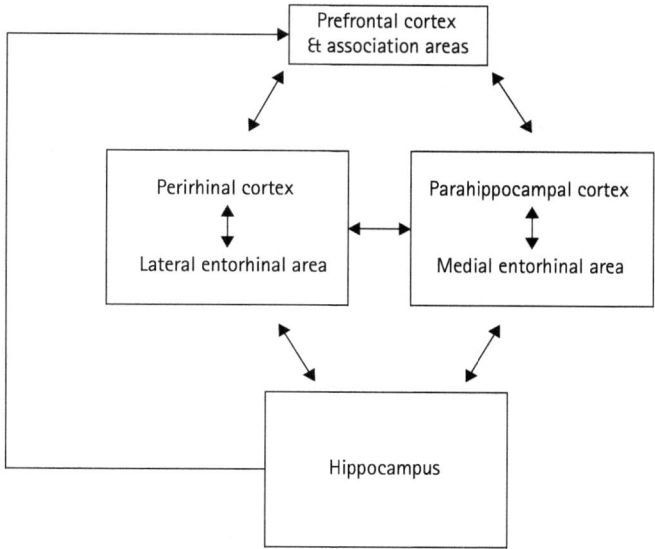

FIGURE 21.2 Schematic representation of the neural structures and interactions involved in consolidation of new memory traces.

Finally, behavioral and neuroimaging data implicate prefrontal cortex in the *retrieval* of memories from long-term storage (e.g., Cabeza, McIntosh, Tulving, Nyberg, and Grady, 1997; Cabeza, Prince, Daselaar, Greenberg, Budde, Dolcos, LaBar, and Rubin, 2004; Maguire, 2001). For example, damage to prefrontal cortex disrupts retrieval of facts and episodes. Deficits are especially apparent 1) in free recall versus recognition, 2) for temporal information versus items, 3) for specific event features, and 4) for source of information. Imaging studies have revealed high levels of activation in the prefrontal cortex during retrieval of episodic memories from long-term stores (reviewed in Gilboa, 2004). Activations in the medial prefrontal cortex are observed during retrieval of internally generated information, such as the thoughts and feelings that put the *auto* in autobiographical memories (Cabeza et al., 2004). Lateral posterior parietal and precuneus also are implicated in retrieval of autobiographical memories. The activations are greater when subjects report retrieving more details about the memory (reviewed in Gilboa, 2004).

Development of the Neural Substrate Supporting Declarative Memory

Developments in the neural substrate supporting declarative memory are summarized in a number of sources (e.g., Bauer, 2007, 2009, 2013a; C. Nelson, 2000; C. Nelson et al., 2006; Richman and Nelson, 2008). In terms of brain development in general, there are changes in both gray matter (the neuronal cell bodies and structures) and white matter (fiber bundles that connect gray matter areas) from infancy well into adolescence (e.g., Caviness, Kennedy, Richelme, Rademacher, and Filipek, 1996; Giedd, Blumenthal, Jeffries, Castellanos, Liu, Zijdenbos, Paus, Evans, and Rapoport, 1999; Gogtay, Giedd, Lusk, Hayashi, Greenstein, Vaituzis, et al., 2004; Sowell, Thompson, Leonard, Welcome, Kan, and Toga, 2004). Reflecting changes in vasculature, glia, neurons, and neuronal processes, gray matter increases until puberty. Beyond puberty, as a result of pruning and other regressive events (i.e., loss of neurons and axonal branches), the thickness of the cortical mantle actually declines (e.g., Giedd et al.; Gogtay et al.; Sowell, Delis, Stiles, and Jernigan, 2001; Van Petten 2004). In contrast to curvilinear change in gray matter volume, white matter volume increases linearly with age (Giedd et al.). Increases in white matter are associated with greater connectivity between brain regions and with myelination processes that continue into young adulthood (e.g., Klingberg, 2008; Schneider, JIl'yasov, Hennig, and Martin, 2004).

In terms of the temporal-cortical episodic memory network, there are a number of indicators that in the human, many components of the medial temporal lobe develop early. For instance, as reviewed by Seress and Abraham (2008), the cells that make up most of the hippocampus are formed in the first half of gestation and virtually all are in their adult locations by the end of the prenatal period. The neurons in most of the hippocampus also begin to connect early in development: synapses are present as early as 15 weeks gestational age. The number and density of synapses both increase rapidly after birth and reach adult levels by approximately 6 postnatal months. Perhaps as a consequence, glucose utilization in the temporal cortex reaches adult levels at the same time (i.e., by about 6 months: Chugani, 1994; Chugani and Phelps, 1986). Thus, there are numerous indices of early maturity of major portions of the medial temporal components of the network.

In contrast to early maturation of most of the hippocampus, development of the dentate gyrus of the hippocampus is protracted (Seress and Abraham, 2008). At birth, the dentate gyrus includes only about 70 percent of the adult number of cells. Thus, roughly 30 percent of the cells are produced postnatally. Indeed, neurogenesis in the dentate gyrus of the hippocampus continues throughout childhood and adulthood (Tanapat, Hastings, and Gould, 2001). It is not until 12 to 15 postnatal months that the morphology of the structure appears adult-like. Maximum density of synaptic connections in the dentate gyrus also is delayed, relative to that in the other regions of the hippocampus. In humans, synaptic density increases dramatically (to well above adult levels) beginning at 8 to 12 postnatal months and reaches its peak at 16 to 20 months. After a period of relative stability, excess synapses are pruned until adult levels are reached at about 4 to 5 years of age (Eckenhoff and Rakic, 1991).

Although the functional significance of later development of the dentate gyrus is not clear, there is reason to speculate that it impacts behavior. As already noted, upon experience of an event, information from distributed regions of cortex converges on the entorhinal cortex. From there, it travels to the hippocampus in one of two ways: via a "long route" or a "short route." The long route involves projections from entorhinal cortex into the hippocampus, by way of the dentate gyrus; the short route bypasses the dentate gyrus. Whereas the functional differences between the routes are not clear, based on data from rodents, it seems that adult-like memory behavior depends on passage of information through the dentate gyrus (Czurkó, Czéh, Seress, Nadel, and Bures, 1997; Nadel and Willner, 1989). This implies that maturation of the dentate gyrus of the hippocampus may be a rate-limiting variable in the development of episodic memory early in life (e.g., Bauer, 2007, 2009; C. Nelson, 1995, 1997, 2000). Finally, hippocampal volume continues to increase gradually throughout childhood and into adolescence (e.g., Gogtay et al., 2004; Pfluger, Weil, Wies, Vollmar, HeissEgger, Scheck, and Hahn, 1999; Utsunomiya, Takano, Okazaki, and Mistudome, 1999). Myelination in the hippocampal region continues throughout adolescence (Arnold and Trojanowski, 1996; Benes, Turtle, Khan, and Farol, 1994; Schneider et al., 2004).

The association areas also undergo a protracted course of development (Bachevalier, 2014). For example, it is not until the seventh prenatal month that all six cortical layers are apparent. The density of synapses in prefrontal cortex increases dramatically at 8 postnatal months and peaks between 15 and 24 months. Pruning to adult levels does not begin until late childhood; adult levels are not reached until late adolescence or early adulthood (Huttenlocher, 1979; Huttenlocher and Dabholkar, 1997). In the years between, in some cortical layers there are changes in the size of cells and the lengths and branching of dendrites (Benes, 2001). Although the maximum density of synapses may be reached as early as 15 postnatal months, it is not until 24 months that synapses develop adult morphology (Huttenlocher, 1979). There also are changes in glucose utilization and blood flow over the second half of the first year and into the second year: blood flow and glucose utilization increase above adult levels by 8 to 12 and 13 to 14 months of age, respectively (Chugani, Phelps, and Mazziotta, 1987). Other maturational changes in prefrontal cortex, such as myelination, continue into adolescence, and adult levels of some neurotransmitters are not seen until the second and third decades of life (Benes, 2001). It is not until adolescence that neurotransmitters such as acetylcholine reach adult levels (discussed in Benes, 2001).

Although much of the attention to developmental changes has focused on the medial-temporal and prefrontal regions, there also are age-related changes in the lateral temporal and parietal cortices. Cortical gray matter changes occur earlier in the frontal and occipital poles, relative to the rest of the cortex, which matures in a parietal-to-frontal direction. The superior temporal cortex is last to mature (though the temporal poles mature early; Gogtay et al., 2004). The late development of this portion of cortex is potentially significant for memory as it is one of the polymodal association areas that plays a role in integration of information across sense modalities.

Functional Consequences of Development of the Temporal-Cortical Network

What are the consequences for behavior of the slow course of development of the neural network that supports declarative memory? At a general level, we may expect concomitant behavioral development: as the neural substrate develops, so does behavior (and vice versa, of course). But precisely how do changes in the medial temporal and cortical structures, and their interconnections, produce changes in behavior? In other words, how do they affect memory representations? To address this question, we must consider the basic processes involved in memory trace formation, storage, and retrieval and how the "recipe" for a memory might be affected by changes in the underlying neural substrate. In other words, we must consider how developmental changes in the substrate for memory relate to changes in the efficacy and efficiency with which information is maintained over the short term, encoded and stabilized for long-term storage, in the reliability and ease with which it is retrieved.

Basic Cognitive and Mnemonic Processes

With developmental changes in the temporal-cortical network, we may expect changes in basic cognitive processes and in behavior. The basic processes involved in memory are encoding, consolidation, and retrieval of memory traces. Although the processes are difficult to cleanly separate from one another (e.g., when encoding ends and consolidation begins is a challenging question to address), they do build on one another and thus are described in the nominal order in which they occur: short-term maintenance and encoding, consolidation and storage, and retrieval.

Encoding

Association cortices are involved in the initial registration and temporary maintenance of experience. Because the prefrontal cortex in particular undergoes considerable postnatal development, it is reasonable to expect that neurodevelopmental changes in it relate to age-related changes in the speed and efficiency with which information is encoded into long-term storage. Consistent with this suggestion, in a longitudinal study, Bauer and her colleagues (Bauer, Wiebe, Carver, Lukowski, Haight, Waters, and Nelson, 2006) found differences in the amplitudes of event-related potential (ERP) responses to familiar stimuli between 9 and 10 months of age that correlated with age-related improvements in recall after a 1-month delay. Behavioral data also indicate developments in encoding throughout the second year of life. For example, relative to 15-month-olds, 12-month-olds require more trials to learn multi-step events to a criterion (learning to a criterion indicates that the material was fully encoded). In turn, 15-month-olds are slower to achieve criterion, relative to 18-month-olds (Howe and Courage, 1997). Indeed, across development, older children learn more rapidly than younger children (Howe and Brainerd, 1989).

Differences in encoding processes continue beyond infancy. For example, Rollins and Riggins (2013) observed that 6-year-olds exhibited a differential pattern of neural responding during encoding, relative to adults. For children, ERP responses to subsequently remembered versus subsequently forgotten items differed at 700–900 milliseconds, whereas for adults, differential responding was observed earlier, around 400–600 milliseconds. Furthermore, the scalp sites at which the neural responses were observed differed between children and adults, suggesting the pattern of engagement of the underlying neural substrates changes with development (Rollins and Riggins, 2013).

Functional magnetic resonance imaging (fMRI) provides insights into the developmental changes in the pattern of engagement of underlying neural substrate. For example, in an encoding task with children 8 years to young adulthood, Ghetti and colleagues (Ghetti, DeMaster, Yonelinas, and Bunge, 2010) observed differences in behavior as well as patterns of functional activations in regions of the medial temporal lobe. The task required that participants encode both items and the colors in which they were presented. Behaviorally, whereas 8-year-olds encoded information about the items, they did not recognize the color in which the item had been presented. In contrast, 14-year-olds and college-age adults correctly recognized the items and their colors. Imaging data revealed differential patterns of activation in the hippocampus during encoding which may have contributed to the failure of the young children to "bind" the item and its color. In addition, among 10-year-olds, 14-year-olds, and adults, prefrontal activation during encoding predicted subsequent memory. This effect was not observed in 8-year-olds. These data illustrate developmental differences in encoding behavior and provide insight into the neural processes that contribute to them.

Consolidation and Storage

As reviewed earlier, medial temporal structures are implicated in the processes by which new memories become "fixed" for long-term storage; cortical association areas are the presumed repositories for long-term memories. In a fully mature adult, neural damage, the changes in synaptic connectivity associated with memory trace consolidation continue for hours, weeks, and even months, after an event. Memory traces are vulnerable throughout this time, as evidenced by the fact that lesions inflicted during the period of consolidation result in deficits in memory whereas lesions inflicted after a trace has been consolidated do not (e.g., Kim and Fanselow, 1992; Takehara, Kawahara, and Kirino, 2003). For the developing organism, the period of consolidation may be one of greater vulnerability for a memory trace, relative to the adult. Not only are some of the implicated neural structures relatively undeveloped (i.e., the dentate gyrus and prefrontal cortex), but the connections between them are still being sculpted and thus are less than fully effective and efficient. As a consequence, even once children have successfully encoded an event, they remain vulnerable to forgetting. Younger children may be more vulnerable to forgetting, relative to older children (Bauer, 2006a).

To examine the role of consolidation and storage processes in long-term memory in 9-month-old infants, Bauer et al. (2003) combined ERP measures of immediate recognition (as an index of encoding), ERP measures of 1-week delayed recognition (as an index of consolidation and storage), and deferred imitation measures of recall after 1 month.

After the delay, 46 percent of the infants evidenced ordered recall of the sequences, and 54 percent did not. At the immediate ERP test, regardless of whether they subsequently recalled the events, the infants evidenced recognition: their ERP responses were different to the old and new stimuli. This strongly implies that the infants had encoded the events. Nevertheless, 1 week later, at the delayed recognition test, the infants who would go on to recall the events recognized the prompts, whereas infants who would not evidence ordered recall did not. Thus, in spite of having encoded the events, a subset of 9-month-olds failed to recognize them after 1 week and subsequently failed to recall them after 1 month. Moreover, the size of the difference in delayed-recognition response predicted recall performance 1 month later. That is, infants who had stronger memory representations after a 1-week delay exhibited higher levels of recall 1 month later (see also Carver, Bauer, and Nelson, 2000). These data strongly imply that at 9 months of age, consolidation and/or storage processes are a source of individual differences in mnemonic performance.

In the second year of life, there are behavioral suggestions of between-age group differences in consolidation and/or storage processes, as well as a replication of the finding among 9-month-olds that intermediate-term consolidation and/or storage failure relates to recall over the long term. In Bauer, Cheatham, Cary, and Van Abbema (2002), 16- and 20-month-olds were exposed to multi-step events and tested for recall immediately (as a measure of encoding) and after 24 hours. Over the delay, the younger children forgot a substantial amount of the information they had encoded: they produced only 65 percent of the target actions and only 57 percent of the ordered pairs of actions that they had learned just 24 hours earlier. For the older children, the amount of forgetting over the delay was not statistically reliable. It is not until 48 hours that children 20 months of age exhibit significant forgetting (Bauer, Van Abbema, and deHaan, 1999). These observations suggest age-related differences in the vulnerability of memory traces during the initial period of consolidation.

The vulnerability of memory traces during the initial period of consolidation is related to the robustness of recall after 1 month. This is apparent from another of the experiments in Bauer et al. (2002), this one involving 20-month-olds only. The children were exposed to multi-step events and then tested for memory for some of the events immediately, some of the events after 48 hours (a delay after which, based on Bauer et al., 1999, some forgetting was expected), and some of the events after 1 month. Although the children exhibited high levels of initial encoding (as measured by immediate recall), they nevertheless exhibited significant forgetting after both 48 hours and 1 month. The robustness of memory after 48 hours predicted 25 percent of the variance in recall 1 month later; variability in level of encoding did not predict significant variance (see also Pathman and Bauer, 2013). This effect is a conceptual replication of that observed with 9-month-olds in Bauer et al. (2003; see Bauer, 2005; and Howe and Courage, 1997, for additional evidence of a role for post-encoding processes in long-term recall). The findings that infants who are "good consolidators" have high levels of long-term recall are reminiscent of Bosshardt, Degonda, Schmidt, Boesiger, Nitsch, Hock, and Henke (2005) with adults: fMRI activations 1 day after learning were predictive of forgetting 1 month later.

Changes in the processes by which memory representations are consolidated and stored can be expected to continue throughout the preschool years. However, although neuroimaging techniques such as ERPs could be brought to bear on the question, as they are in the infancy period, such studies have not been conducted with preschool-age children. Neither is there a plethora of well-controlled behavioral studies to address the question. Yet, studies of preservation of autobiographical memories over long periods of time suggest age-related differences in consolidation (as well as retrieval, see following). As an illustration, in Bauer and Larkina (2015), children 4, 6, and 8 years, and adults, were enrolled in a prospective study of autobiographical memory over a 4-year period. In Year 1, all participants reported on a number of events from the recent past. Different versions of the events then were tested for recall 1 year later, 2 years later, and 3 years later. Although all of the events were well recalled in Year 1 (demonstrating that the events had been encoded into memory), after the delays, children recalled fewer of the events than the adults, and the younger children (4-year-olds) recalled fewer of the events than the older children (8-year-olds). The pattern of data implicates developmental differences in the success of consolidation processes (see also Bauer et al., 2007; Bauer and Larkina, 2014).

Retrieval

The prefrontal cortex is implicated in retrieval of memories from long-term storage sites. A prefrontal cortex undergoes a long period of postnatal development, making it a likely candidate source of age-related differences in long-term recall. Surprisingly, although retrieval processes are a compelling candidate source of developmental differences in long-term recall, there are few data with which to evaluate their contribution. A major reason is that most studies do not allow for assignment of relative roles of the processes that take place before retrieval, namely, encoding and consolidation. As discussed in the section on encoding, older children learn more rapidly than younger children. Yet, age-related differences in encoding effectiveness rarely are taken into account. In fact, in many studies, no measures of encoding or initial learning are obtained. In addition, with standard testing procedures, it is difficult to know whether a memory representation has lost its integrity and become unavailable (consolidation or storage failure) or whether the memory trace remains intact but has become inaccessible with the cues provided (retrieval failure). Implication of retrieval processes as a source of developmental change requires that encoding be controlled, and that memory be tested under conditions of high support for retrieval.

In the infancy period, one of the studies that permits assessment of the contributions of consolidation and/or storage relative to retrieval processes is Bauer et al. (2000; see also Bauer et al., 2003, described earlier). The study provided data on children of multiple ages (13, 16, and 20 months) tested over delays of 1 to 12 months. Immediate recall of half of the events was tested, thus providing a measure of encoding. Because the children were given what amounted to multiple test trials, without intervening study trials, there were multiple opportunities for retrieval. As discussed by Howe and his colleagues (e.g., Howe and Brainerd, 1989; Howe and O'Sullivan, 1997), for intact memory traces, retrieval attempts strengthen the trace and route to retrieval of it, thereby increasing accessibility on each test trial. Conversely, lack of improvement across test trials implies that the trace was no longer available (although see Howe and O'Sullivan, 1997, for multiple nuances of this argument). Third, immediately after the recall tests, re-learning was tested. That is, after the second test trial the experimenter demonstrated each event once and allowed the children to imitate. Since Ebbinghaus (1885), relearning has been used to distinguish between an intact but inaccessible memory trace and a trace that has disintegrated. Specifically, if the number of trials required to relearn a stimulus was smaller than the number required to learn it initially, savings in relearning was said to have occurred. Savings presumably accrues because the products of relearning are integrated with an existing (though not necessarily accessible) memory trace. Conversely, the absence of savings is attributed to storage failure: there is no residual trace upon which to build. In developmental studies, age-related differences in relearning would suggest that the residual memory traces available to children of different ages are differentially intact.

To eliminate encoding processes as a potential source of developmental differences in long-term recall, in a re-analysis of the data from Bauer et al. (2000) subsets of 13- and 16-month-olds and subsets of 16- and 20-month-olds were matched for levels of encoding (as measured by immediate recall; Bauer, 2005). The amount of information the children forgot over the delays then was examined. For both comparisons, even though they were matched for levels of encoding, younger children exhibited more forgetting relative to older children. The age effect was apparent on both test trials. Moreover, in both cases, for older children, levels of performance after the single relearning trial were as high as those at initial learning. In contrast, for younger children, performance after the relearning trial was lower than at initial learning. Together, the findings of age-related differential loss of information over time and of age effects in re-learning strongly implicate storage processes, as opposed to retrieval processes, as the major source of age-related differences in long-term recall.

The conclusions from the infancy literature are consistent with the results of research with older children conducted within the trace-integrity framework (Brainerd, Reyna, Howe, and Kingma, 1990) and conceptually-related fuzzy-trace theory (Brainerd and Reyna, 1990). In this tradition, to eliminate encoding differences as a source of age-related effects, participants are brought to a criterion level of learning prior to imposition of a delay. To permit evaluation of the contributions of storage processes versus retrieval processes, participants are provided multiple test trials, without intervening study trials. In one such study, 4- and 6-year-old children learned and then recalled 8-item picture lists (Howe, 1995). In this study, as in virtually every other study conducted within this tradition (reviewed in Howe and O'Sullivan, 1997), the largest proportion of age-related variance in children's recall was accounted for by memory failure at the level of consolidation and/or storage, as opposed to retrieval. Whereas consolidation and/or storage failure rates decline throughout childhood, retrieval failure rates remain at relatively constant levels (Howe and O'Sullivan, 1997). The apparent lack of change in retrieval failure rates throughout childhood implies that retrieval processes are not a major source of developmental change during this period.

Whereas analyses of behavior implicate storage processes as a major source of developmental change, the neuroimaging literature features suggestions of developmental differences in retrieval processes (though these data must be considered with the caveat that the studies do not explicitly control for possible differences in encoding). For example, Rollins and Riggins (2013) used ERPs to examine item and context memory among children 3 to 6 years of age. They reported increases in levels of retrieval of the conjunction of the item in context in this period (see Bauer, Stewart, White, and Larkina, 2016, for similar behavioral findings in 4- to 8-year-olds). They also noted increases in the amplitude of the neural response to items that were remembered in context.

In an fMRI study, Ofen and colleagues (Ofen, Chai, Schuil, Whitfield-Gabrieli, and Gabrieli, 2012) investigated neural activations as children, adolescents, and young adults engaged in retrieval of pictures of scenes from memory. Across the age groups, successful retrieval was related to activation in medial temporal, frontal, and parietal areas. Retrieval-related activations increased with age in the frontal and parietal areas but not in the medial temporal lobe. Research by Bauer, Pathman, Inman, Campanella, and Hamann (2016) suggests that age-related differences in activation are greatest early in the period of retrieval, as children and adults engage in the process of search for and access of the relevant memory, relative to later in the retrieval period, when memory traces are being elaborated.

Other research suggests that changes in retrieval success are related to differential patterns of activity in the hippocampus. In a study with 8- to 11-year-olds and adults (DeMaster and Ghetti, 2013), among adults, retrieval of items in context was related to activation in the head of the hippocampus and the anterior region of the hippocampal body. This pattern was not observed in children. Instead, when children successfully retrieved items in context, there was greater activation in the tail of the hippocampus. For adults, activation in the hippocampal tail did not vary as a function of retrieval success. This pattern may suggest a shift with development in recruitment of the subregions of the hippocampus during retrieval. Finally, echoing the findings of increased retrieval-related activations in frontal regions with age observed by Ofen and colleagues (2012), DeMaster and Ghetti (2013) reported that adults showed greater activation in the anterior PFC for items judged as "old." For children, anterior PFC activations were high and did not vary as a function of retrieval success.

Overall, there are strong indications of developmental differences in both the amount of information that is retrieved from memory and in the specificity of the information (i.e., specific items in particular contexts). There also are suggestions of differential levels and patterns of neural activation during retrieval as a function of age. Nevertheless, it is important to keep in mind that many studies of retrieval processes do not control for potential differences in encoding (or in consolidation), thus complicating interpretation of the findings.

CONCLUSION

The title of this chapter—*Memory Development*—gives the impression that its subject will be singular: a singular system with a singular course of development. On the contrary, there are many different forms of memory, each with its own characteristics and developmental course. Although some broad generalizations apply, most of what we can be said to know about memory is relevant within a limited frame and for a subset of the types of memory. Continued progress in understanding memory and its development requires that appropriate distinctions be maintained.

Historically, most types of memory were thought to be relatively late to develop. This expectation was perhaps nowhere more apparent than in reference to episodic and autobiographical memory. Research in the last decades of the twentieth century made clear that the assumption was unwarranted. When they are tested with meaningful and personally relevant events and materials, even young children show evidence of mnemonic competence. Thus, in sharp contrast to expectations of developmental discontinuities in event memory, there is ample evidence that the capacity to remember past events develops early. The research also made clear, however, that development is a protracted event, beginning in infancy and continuing into late adolescence. A full accounting of the development of "memory," thus requires a long-term perspective.

Finally, it is a truism to say that complex behaviors are determined by many factors. In the field of the development of memory, this dictum must be embraced, wholeheartedly. Within the same space of time as the mnemonic capabilities of even young children were chronicled, progress in explaining the timing and course of development was made at a variety of different levels. Although much remains to be discovered, understanding of the cellular and molecular events that permit the storage and later retrieval of information now is within reach. Similarly, we are on the verge of understanding how basic memory processes determine the life-course of a memory at different points in developmental time. We may look forward to the day when multiple levels of explanation come together into a comprehensive account of the processes and determinants of the capacities we call *memory*.

REFERENCES

Ackil, J. K., Van Abbema, D. L., and Bauer, P. J. (2003). After the storm: Enduring differences in mother-child recollections of traumatic and nontraumatic events. *Journal of Experimental Child Psychology, 84*, 286–309.

Adlam, A.-L. R., Vargha-Khadem, F., Mishkin, M., and de Haan, M. (2005). Deferred imitation of action sequences in developmental amnesia. *Journal of Cognitive Neuroscience, 17*(2), 240–248.

Arnold, S. E. and Trojanowski, J. Q. (1996). Human fetal hippocampal development: I. Cytoarchitecture, myeloarchitecture, and neuronal morphologic features. *Journal of Comparative Neurology, 367*, 274–292.

Bachevalier, J. (2014). The development of memory from a neurocognitive and comparative perspective. In P. J. Bauer and R. Fivush (Eds.), *The Wiley-Blackwell Handbook on the Development of Children's Memory* (pp. 109–125). West Sussex, UK: Wiley-Blackwell.

Bahrick, H. P. (2000). Long-term maintenance of knowledge. In E. Tulving and F. I. M. Craik (Eds.), *The Oxford Handbook of Memory* (pp. 347–362). New York: Oxford University Press.

Barnat, S. B., Klein, P. J., and Meltzoff, A. N. (1996). Deferred imitation across changes in context and object: Memory and generalization in 14-month-old children. *Infant Behavior and Development, 19*, 241–251.

Barr, R., Dowden, A., and Hayne, H. (1996). Developmental change in deferred imitation by 6- to 24-month-old infants. *Infant Behavior and Development, 19*, 159–170.

Bauer, P. J. (2002). Long-term recall memory: Behavioral and neuro-developmental changes in the first 2 years of life. *Current Directions in Psychological Science, 11*, 137–141.

Bauer, P. J. (2005). Developments in declarative memory: Decreasing susceptibility to storage failure over the second year of life. *Psychological Science, 16*, 41–47.

Bauer, P. J. (2006a). Constructing a past in infancy: A neuro-developmental account. *Trends in Cognitive Sciences, 10*, 175–181.

Bauer, P. J. (2006b). Event memory. In D. Kuhn and R. Siegler (Volume editors: *Volume 2— Cognition, Perception, and Language*), W. Damon and R. M. Lerner (editors-in-chief). *Handbook of Child Psychology, Sixth Edition* (pp. 373–425). Hoboken, NJ: John Wiley & Sons, Inc.

Bauer, P. J. (2007). *Remembering the Times of Our Lives: Memory in Infancy and Beyond*. Mahwah, NJ: Erlbaum.

Bauer, P. J. (2008). Infantile amnesia. In M. M. Haith and J. B. Benson (Eds.), *Encyclopedia of Infant and Early Childhood Development* (pp. 51–61). San Diego, CA: Academic Press.

Bauer, P. J. (2009). The cognitive neuroscience of the development of memory. In M. L. Courage and N. Cowan (Eds.), *The Development of Memory in Infancy and Childhood, Second Edition* (pp. 115–144). New York, NY: Psychology Press.

Bauer, P. J. (2013a). Memory. In P. D. Zelazo (Ed.), *Oxford Handbook of Developmental Psychology, Volume 1: Body and Mind* (pp. 505–541). New York, NY: Oxford University Press.

Bauer, P. J. (2013b). Memory development. To appear in H. Tager-Flusberg (Section Editor: *Cognitive Development*), J. Rubenstein and P. Rakic (Editors-in-Chief). *Developmental Neuroscience—Basic and Clinical Mechanisms, A Comprehensive Reference*. Amsterdam, NE: Elsevier.

Bauer, P. J. (2014). The development of forgetting: Childhood amnesia. In P. J. Bauer and R. Fivush (Eds.), *The Wiley-Blackwell Handbook on the Development of Children's Memory* (pp. 519–544). West Sussex, UK: Wiley-Blackwell.

Bauer, P. J. (2015). A complementary processes account of the development of childhood amnesia and a personal past. *Psychological Review, 2*, 204–231. doi:10.1037/a0038939

Bauer, P. J., Burch, M. M., Scholin, S. E., and Güler, O. E. (2007). Using cue words to investigate the distribution of autobiographical memories in childhood. *Psychological Science, 18*, 910–916.

Bauer, P. J., Cheatham, C. L., Cary, M. S., and Van Abbema, D. L. (2002). Short-term forgetting: Charting its course and its implications for log-term remembering. In S. P. Shohov (Ed.), *Advances in Psychology Research,* Volume 9 (pp. 53–74). Huntington, NY: Nova Science Publishers.

Bauer, P. J. and Dow, G. A. A. (1994). Episodic memory in 16- and 20-month-old children: Specifics are generalized, but not forgotten. *Developmental Psychology, 30*, 403–417.

Bauer, P. J. and Fivush, R. (1992). Constructing event representations: Building on a foundation of variation and enabling relations. *Cognitive Development, 7*, 381–401.

Bauer, P. J. and Larkina, M. (2014). Childhood amnesia in the making: Different distributions of autobiographical memories in children and adults. *Journal of Experimental Psychology: General, 143*(2), 597–611. doi:10.1037/a0033307

Bauer, P. J. and Larkina, M. (2015). Predicting remembering and forgetting of autobiographical memories in children and adults: A 4-year prospective study. *Memory, 24*, 1345–1368. doi:10.1080/09658211.2015.1110595

Bauer, P. J. and Leventon, J. S. (2013). Memory for one-time experiences in the second year of life: Implications for the status of episodic memory. *Infancy, 18*, 755–781.

Bauer, P. J. and Lukowski, A. F. (2010). The memory is in the details: Relations between memory for the specific features of events and long-term recall in infancy. *Journal of Experimental Child Psychology, 107*, 1–14.

Bauer, P. J., Pathman, T., Inman, C., Campanella, C., and Hamann, S. (2016). Neural correlates of autobiographical memory retrieval in children and adults. *Memory.* Advance Online Publication. doi:10.1080/09658211.2016.1186699

Bauer, P. J. and Shore, C. M. (1987). Making a memorable event: Effects of familiarity and organization on young children's recall of action sequences. *Cognitive Development, 2*, 327–338.

Bauer, P. J., Stewart, R., White, E. A., and Larkina, M. (2016). A place for every event and every event in its place: Memory for locations and activities by 4-year-old children. *Journal of Cognition and Development, 17*(2) 244–263. doi:10.1080/15248372.2014.959521

Bauer, P. J., Van Abbema, D. L., and de Haan, M. (1999). In for the short haul: Immediate and short-term remembering and forgetting by 20-month-old children. *Infant Behavior and Development, 22*, 321–343.

Bauer, P. J., Wenner, J. A., Dropik, P. L., and Wewerka, S. S. (2000). Parameters of remembering and forgetting in the transition from infancy to early childhood. *Monographs of the Society for Research in Child Development, 65* (4, Serial No. 263).

Bauer, P. J., Wenner, J. A., and Kroupina, M. G. (2002). Making the past present: Later verbal accessibility of early memories. *Journal of Cognition and Development, 3*, 21–47.

Bauer, P. J., Wiebe, S. A., Carver, L. J., Lukowski, A. F., Haight, J. C., Waters, J. M., and Nelson, C. A. (2006). Electrophysiological indices of encoding and behavioral indices of recall: Examining relations and developmental change late in the first year of life. *Developmental Neuropsychology, 29*, 293–320.

Bauer, P. J., Wiebe, S. A., Carver, L. J., Waters, J. M., and Nelson, C. A. (2003). Developments in long-term explicit memory late in the first year of life: Behavioral and electrophysiological indices. *Psychological Science, 14*, 629–635.

Bauer, P. J., Wiebe, S. A., Waters, J. M., and Bangston, S. K. (2001). Reexposure breeds recall: Effects of experience on 9-month-olds' ordered recall. *Journal of Experimental Child Psychology, 80*, 174–200.

Benes, F. M. (2001). The development of prefrontal cortex: The maturation of neurotransmitter systems and their interaction. In C. A. Nelson and M. Luciana (Eds.), *Handbook of Developmental Cognitive Neuroscience* (pp. 79–92). Cambridge, MA: The MIT Press.

Benes, F. M., Turtle, M., Khan, Y., and Farol, P. (1994). Myelination of a key relay zone in the hippocampal formation occurs in the human brain during childhood, adolescence, and adulthood. *Archives of General Psychiatry, 51*, 477–484.

Bosshardt, S., Degonda, N., Schmidt, C. F., Boesiger, P., Nitsch, R. M., Hock, C., and Henke, K. (2005). One month of human memory consolidation enhances retrieval-related hippocampal activity. *Hippocampus, 15*, 1026–1040.

Brainerd, C. J. and Reyna, V. F. (1990). Gist is the grist: Fuzzy-trace theory and the new intuitionism. *Developmental Review*, *10*, 3–47.

Brainerd, C. J., Reyna, V. F., Howe, M. L., and Kingma, J. (1990). The development of forgetting and reminiscence. *Monographs of the Society for Research in Child Development*, *55* (3–4, Serial No. 222).

Cabeza, R., McIntosh, A. R., Tulving, E., Nyberg, L., and Grady, C. L. (1997). Age-related differences in effective neural connectivity during encoding and recall. *NeuroReport*, *8*, 3479–3483.

Cabeza, R., Prince, S. E., Daselaar, S.M., Greenberg, D. L., Budde, M., Dolcos, F., LaBar, K. S., and Rubin, D. C. (2004). Brain activity during episodic retrieval of autobiographical and laboratory events: An fMRI study using a novel photo paradigm. *Journal of Cognitive Neuroscience*, *16*, 1583–1594.

Carver, L. J. and Bauer, P. J. (1999). When the event is more than the sum of its parts: Nine-month-olds' long-term ordered recall. *Memory*, *7*, 147–174.

Carver, L. J. and Bauer, P. J. (2001). The dawning of a past: The emergence of long-term explicit memory in infancy. *Journal of Experimental Psychology: General*, *130*, 726–745.

Carver, L. J., Bauer, P. J., and Nelson, C. A. (2000). Associations between infant brain activity and recall memory. *Developmental Science*, *3*, 234–246.

Caviness, V. S., Kennedy, D. N., Richelme, C., Rademacher, J., and Filipek, P. A. (1996). The human brain age 7–11 years: A volumetric analysis based on magnetic resonance images. *Cerebral Cortex*, *6*, 726–736.

Cheatham, C. L. and Bauer, P. J. (2005). Construction of a more coherent story: Prior verbal recall predicts later verbal accessibility of early memories. *Memory*, *13*, 516–532.

Chugani, H. T. (1994). Development of regional blood glucose metabolism in relation to behavior and plasticity. In G. Dawson and K. Fischer (Eds.), *Human Behavior and the Developing Brain* (pp. 153–175). New York: Guilford.

Chugani, H. T. and Phelps, M. E. (1986). Maturational changes in cerebral function determined by 18FDG positron emission tomography. *Science*, *231*, 840–843.

Chugani, H. T., Phelps, M., and Mazziotta, J. (1987). Positron emission tomography study of human brain functional development. *Annals of Neurology*, *22*, 487–497.

Cowan, N. (2010). The magical mystery four: How is working memory capacity limited, and why? *Current Directions in Psychological Science*, *19*(1): 51–57.

Cowan, N. (2017). The many faces of working memory and short-term storage. *Psychonomic Bulletin and Review*, *24*(4), 1158–1170.

Collie, R. and Hayne, H. (1999). Deferred imitation by 6- and 9-month-old infants: More evidence of declarative memory. *Developmental Psychobiology*, *35*, 83–90.

Czurkó, A., Czéh, B., Seress, L., Nadel, L., and Bures, J. (1997). Severe spatial navigation deficit in the Morris water maze after single high dose of neonatal X-ray irradiation in the rat. *Proceedings of the National Academy of Sciences*, *94*, 2766–2771.

de Biran, M. (1804/1929). *The Influence of Habit on the Faculty of Thinking*. Baltimore, MD: Williams and Wilkins.

DeMaster, D. M. and Ghetti, S. (2013). Developmental differences in hippocampal and cortical contributions to episodic retrieval. *Cortex*, *49*(6), 1482–1493.

Dickerson, B. C. and Eichenbaum, H. (2010). The episodic memory system: Neurocircuitry and disorders. *Neuropsychopharmacology*, *35*(1), 86–104.

Ebbinghaus, H. (1885). *On Memory* (H. A. Ruger and C. E. Bussenius, Translators). New York: Teachers' College, 1913. Paperback edition, New York: Dover, 1964.

Eckenhoff, M. and Rakic, P. (1991). A quantitative analysis of synaptogenesis in the molecular layer of the dentate gyrus in the rhesus monkey. *Developmental Brain Research, 64*, 129–135.

Eichenbaum, H., Sauvage, M., Fortin, N., Komorowski, R., and Lipton, P. (2012). Towards a functional organization of episodic memory in the medial temporal lobe. *Neuroscience & Biobehavioral Reviews, 36*(7), 1597–1608.

Eichenbaum, H., and Cohen, N. J. (2001). *From Conditioning to Conscious Recollection: Memory Systems of the Brain.* New York: Oxford University Press.

Fivush, R. (1984). Learning about school: The development of kindergarteners' school scripts. *Child Development, 55*, 1697–1709.

Fivush, R. (2001). Owning experience: Developing subjective perspective in autobiographical narratives. In C. Moore and K. Lemmon (Eds.), *The Self in Time: Developmental Perspectives* (pp. 35–52). Mahwah, NJ: Erlbaum.

Fivush, R., Gray, J. T., and Fromhoff, F. A. (1987). Two-year-olds talk about the past. *Cognitive Development, 2*, 393–409.

Fivush, R. and Haden, C. A. (1997). Narrating and representing experience: Preschoolers' developing autobiographical accounts. In P. van den Broek, P. J. Bauer, and T. Bourg (Eds.), *Developmental Spans in Event Representation and Comprehension: Bridging Fictional and Actual Events* (pp. 169–198). Mahwah, NJ: Erlbaum.

Fivush, R. and Hamond, N. R. (1990). Autobiographical memory across the preschool years: Toward reconceptualizing childhood amnesia. In R. Fivush and J. A. Hudson (Eds.), *Knowing and Remembering in Young Children* (pp. 223–248). New York: Cambridge University Press.

Fivush, R., Keubli, J., and Clubb, P. A. (1992). The structure of events and event representations: Developmental analysis. *Child Development, 63*, 188–201.

Fivush, R., Sales, J. M., Goldberg, A., Bahrick, L., and Parker, J. F. (2004). Weathering the storm: Children's long-term recall of Hurricane Andrew. *Memory, 12*, 104–118.

Fivush, R. and Zaman, W. (2014). Gender, subjective perspective, and autobiographical consciousness. In P. J. Bauer and R. Fivush (Eds.), *The Wiley-Blackwell Handbook on the Development of Children's Memory* (pp. 586–604). West Sussex, UK: Wiley-Blackwell.

Ghetti, S., DeMaster, D. M., Yonelinas, A. P., and Bunge, S. A. (2010). Developmental differences in medial temporal lobe function during memory encoding. *Journal of Neuroscience, 30*, 9548–9556.

Ghetti, S. and Lee, J. K. (2014). The development of recollection and familiarity during childhood: Insight from studies of behavior and brain. In P. J. Bauer and R. Fivush (Eds.), *The Wiley-Blackwell Handbook on the Development of Children's Memory* (pp. 309–335). West Sussex, UK: Wiley-Blackwell.

Giedd, J. N., Blumenthal, J., Jeffries, N. O., Castellanos, F. X., Liu, H., and Zijdenbos, A., Paus, T., Evans, A. C., and Rapoport, J. L. (1999). Brain development during childhood and adolescence: A longitudinal MRI study. *Nature Neuroscience, 2*, 861–863.

Gilboa, A. (2004). Autobiographical and episodic memory—one and the same? Evidence from prefrontal activation in neuroimaging studies. *Neuropsychologia, 42*, 1336–1349.

Gogtay, N., Giedd, J. N., Lusk, L., Hayashi, K. M., Greenstein, D., Vaituzis, A. C., et al. (2004). Dynamic mapping of human cortical development during childhood through early adulthood. *PNAS, 101*, 8174–8179.

Hamond, N. R. and Fivush, R. (1991). Memories of Mickey Mouse: Young children recount their trip to Disneyworld. *Cognitive Development, 6*, 433–448.

Hanna, E. and Meltzoff, A. N. (1993). Peer imitation by toddlers in laboratory, home, and day-care contexts: Implications for social learning and memory. *Developmental Psychology*, *29*, 702–710.

Hayne, H., Boniface, J., and Barr, R. (2000). The development of declarative memory in human infants: Age-related changes in deferred imitation. *Behavioral Neuroscience*, *114*, 77–83.

Hayne, H., MacDonald, S., and Barr, R. (1997). Developmental changes in the specificity of memory over the second year of life. *Infant Behavior and Development*, *20*, 233–245.

Herbert, J. and Hayne, H. (2000). Memory retrieval by 18–30-month-olds: Age-related changes in representational flexibility. *Developmental Psychology*, *36*, 473–484.

Howe, M. L. (1995). Interference effects in young children's long-term retention. *Developmental Psychology*, *31*, 579–596.

Howe, M. L. and Brainerd, C. J. (1989). Development of children's long-term retention. *Developmental Review*, *9*, 301–340.

Howe, M. L. and Courage, M. L. (1997). Independent paths in the development of infant learning and forgetting. *Journal of Experimental Child Psychology*, *67*, 131–163.

Howe, M. L. and O'Sullivan, J. T. (1997). What children's memories tell us about recalling our childhoods: A review of storage and retrieval processes in the development of long-term retention. *Developmental Review*, *17*, 148–204.

Huttenlocher, P. R. (1979). Synaptic density in human frontal cortex: Developmental changes and effects of aging. *Brain Research*, *163*, 195–205.

Huttenlocher, P. R. and Dabholkar, A. S. (1997). Regional differences in synaptogenesis in human cerebral cortex. *Journal of Comparative Neurology*, *387*, 167–178.

Kim, J. J. and Fanselow, M. S. (1992). Modality-specific retrograde amnesia of fear. *Science*, *256*, 675–677.

Klein, P. J. and Meltzoff, A. N. (1999). Long-term memory, forgetting, and deferred imitation in 12-month-old infants. *Developmental Science*, *2*, 102–113.

Kleist, K. (1934). Kriegsverletzungen des Gehirns in inhrer Bedeutung fur die Hirnlokalisation and Hirnpathologie. In K. Bonhoeffer (Ed.), *Handbuch der Aerztlichen Erfahrungen im Weltkriege 1914/1918, Volume 4: Geistes- und Nervenkrankheiten* (pp. 343–1360). Leipzig: Barth.

Klingberg, T. (2008). White matter maturation and cognitive development during childhood. In C. A. Nelson and M. Luciana (Eds.), *Handbook of Developmental Cognitive Neuroscience, Second Edition* (pp. 237–243). Cambridge, MA: The MIT Press.

Lechuga, M. T., Marcos-Ruiz, R., and Bauer, P. J. (2001). Episodic recall of specifics and generalisation coexist in 25-month-old children. *Memory*, *9*, 117–132.

Lloyd, M. E. and Newcombe, N. S. (2009). Implicit memory in childhood: Reassessing developmental invariance. In M. L. Courage and N. Cowan (Eds.), *The Development of Memory in Infancy and Childhood* (93–113). New York, NY: Taylor and Francis.

Lukowski, A. F. and Bauer, P. J. (2014). Long-term memory in infancy and early childhood. In P. J. Bauer and R. Fivush (Eds.), *The Wiley-Blackwell Handbook on the Development of Children's Memory* (pp. 230–254). West Sussex, UK: Wiley-Blackwell.

Maguire, E. A. (2001). Neuroimaging studies of autobiographical event memory. *Philosophical Transactions Royal Society of London*, *356*, 1441–1451.

Mandler, J. M. (1990). Recall of events by preverbal children. In A. Diamond (Ed.), *The Development and Neural Bases of Higher Cognitive Functions* (pp. 485–516). New York: New York Academy of Sciences.

Markowitsch, H. J. (2000). Neuroanatomy of memory. In E. Tulving and F. I. M. Craik (Eds.), *The Oxford Handbook of Memory* (pp. 465–484). New York: Oxford University Press.

McClelland, J. L., McNaughton, B. L., and O'Reilly, R. C. (1995). Why there are complementary learning systems in the hippocampus and neocortex: Insights from the successes and failures of connectionist models of learning and memory. *Psychological Review, 102,* 419–457.

McDonough, L., Mandler, J. M., McKee, R. D., and Squire, L. R. (1995). The deferred imitation task as a nonverbal measure of declarative memory. *Proceedings of the National Academy of Sciences, 92,* 7580–7584.

McGaugh, J. L. (2000). Memory—A century of consolidation. *Science, 287,* 248–251.

Meltzoff, A. N. (1988). Infant imitation and memory: Nine-month-olds in immediate and deferred tests. *Child Development, 59,* 217–225.

Meltzoff, A. N. (1990). The implications of cross-modal matching and imitation for the development of representation and memory in infants. In A. Diamond (Ed.), *The Development and Neural Bases of Higher Cognitive Functions* (pp. 1–31). New York: New York Academy of Sciences.

Meltzoff, A. N. (1995). What infant memory tells us about infantile amnesia: Long-term recall and deferred imitation. *Journal of Experimental Child Psychology, 59,* 497–515.

Miller, G. A. (1956). The magical number seven, plus or minus two: Some limits on our capacity for processing information. *Psychological Review, 63,* 81–97.

Milner, B. (2005). The medial temporal-lobe amnesic syndrome. *Psychiatric Clinics, 28*(3), 599–611.

Moscovitch, M. and Nadel, L. (1998). Consolidation and the hippocampal complex revisited: In defense of the multiple-trace model. *Current Opinion in Neurobiology, 8,* 297–300.

Nadel, L. and Willner, J. (1989). Some implications of postnatal maturation of the hippocampus. In V. Chan-Palay and C. Köhler (Eds.), *The Hippocampus—New Vistas* (pp. 17–31). New York: Alan R. Liss.

Nelson, C. A. (1995). The ontogeny of human memory: A cognitive neuroscience perspective. *Developmental Psychology, 31,* 723–738.

Nelson, C. A. (1997). The neurobiological basis of early memory development. In N. Cowan (Ed.), *The Development of Memory in Childhood* (pp. 41–82). Hove East Sussex: Psychology Press.

Nelson, C. A. (2000). Neural plasticity and human development: The role of early experience in sculpting memory systems. *Developmental Science, 3,* 115–136.

Nelson, C. A., de Haan, M., and Thomas, K. (2006). Neural bases of cognitive development. In D. Kuhn and R. Siegler (Volume Editors: *Volume 2—Cognition, Perception, and Language*), W. Damon and R. M. Lerner (Editors-in-Chief). *Handbook of Child Psychology, Sixth Edition* (pp. 3–57). Hoboken, NJ: John Wiley and Sons, Inc.

Nelson, K. (1986). *Event knowledge: Structure and Function in Development.* Hillsdale, NJ: Erlbaum.

Nelson, K. (1997). Event representations then, now, and next. In P. van den Broek, P. J. Bauer, and T. Bourg (Eds.), *Developmental Spans in Event Representation and Comprehension: Bridging Fictional and Actual Events* (pp. 1–26). Mahwah, NJ: Erlbaum.

Nelson, K. and Fivush, R. (2004). The emergence of autobiographical memory: A social cultural developmental theory. *Psychological Review, 111,* 486–511.

Nelson, K. and Gruendel, J. (1986). Children's scripts. In K. Nelson (Ed.), *Event Knowledge: Structure and Function in Development* (pp. 21–46). Hillsdale, NJ: Erlbaum.

Ofen, N., Chai, X. J., Schuil, K. D. I., Whitfield-Gabrieli, S. and Gabrieli, J. D. E. (2012). The development of brain systems for successful memory retrieval of scenes. *Journal of Neuroscience*, *32*(29), 10012–10020.

O' Reilly, R. C., Bhattacharyya, R., Howard, M. D., and Ketz, N. (2014). Complementary learning systems. *Cognitive Science*, *38*, 1229–1248.

Pathman, T. and Bauer, P. J. (2013). Beyond initial encoding: Measures of the post-encoding status of memory traces predict long-term recall in infancy. *Journal of Experimental Child Psychology*, *114*, 321–338.

Perez, L. A., Peynircioglu, Z. F., and Blaxton, T. A. (1998). Developmental differences in implicit and explicit memory performance. *Journal of Experimental Child Psychology*, *70*, 167–185.

Petrides, M. (1995). Impairments on nonspatial self-ordered and externally ordered working memory tasks after lesions of the mid-dorsal part of the lateral frontal cortex in monkeys. *The Journal of Neuroscience*, *15*, 359–375.

Pfluger, T., Weil, S., Wies, S. Vollmar, C., Heiss, D., Egger, J., Scheck, R., and Hahn, K. (1999). Normative volumetric data of the developing hippocampus in children based on magnetic resonance imaging. *Epilepsia*, *40*, 414–423.

Price, D. W. W. and Goodman, G. S. (1990). Visiting the wizard: Children's memory for a recurring event. *Child Development*, *61*, 664–680.

Reed, J. M. and Squire, L. R. (1998). Retrograde amnesia for facts and events: Findings from four new cases. *The Journal of Neuroscience*, *18*, 3943–3954.

Reese, E., Haden, C., Baker-Ward, L., Bauer, P. J., Fivush, R., and Ornstein, P. O. (2011). Coherence of personal narratives across the lifespan: A multidimensional model and coding method. *Journal of Cognition and Development*, *12*, 424–462. doi:10.1080/15248372.2011.587854

Richman, J. and Nelson, C. A. (2008). Mechanisms of change: A cognitive neuroscience approach to declarative memory development. In C. A. Nelson and M. Luciana (Eds.), *Handbook of Developmental Cognitive Neuroscience, Second Edition* (pp. 541–552). Cambridge, MA: The MIT Press.

Roebers, C. M. (2014). Children's deliberate memory development: The contribution of strategies and metacognitive processes. In P. J. Bauer and R. Fivush (Eds.), *The Wiley-Blackwell Handbook on the Development of Children's Memory* (pp. 865–894). West Sussex, UK: Wiley-Blackwell.

Roediger, H. L. and McDermott, K. B. (1993). Implicit memory in normal human subjects. In F. Boller and J. Grafman (Eds.), *Handbook of Neuropsychology* (Vol. 8, pp. 63–131). Amsterdam: Elsevier.

Rollins, L. and Riggins, T. (2013). Developmental changes in memory encoding: insights from event-related potentials. *Developmental Science*, *16*(4), 599–609.

Rovee-Collier, C. and Hayne, H. (2000). Memory in infancy and early childhood. In E. Tulving and F. I. M. Craik (Eds.), *The Oxford Handbook of Memory* (pp. 267–282). New York: Oxford University Press.

Schacter, D. L. and Tulving, E. (1994). What are the memory systems of 1994? In D. L. Schacter and E. Tulving (Eds.), *Memory Systems* (pp. 1–38). Cambridge, MA: MIT Press.

Schacter, D. L., Wagner, A. D., and Buckner, R. L. (2000). Memory systems of 1999. In E. Tulving and F. I. M. Craik (Eds.), *The Oxford Handbook of Memory* (pp. 627–643). New York: Oxford University Press.

Schneider, J. F. L., Il'yasov, K. A., Hennig, J., and Martin, E. (2004). Fast quantitative diffusion-tensor imaging of cerebral white matter from the neonatal period to adolescence. *Neuroradiology*, *46*, 258–266.

Schneider, W. and Bjorklund, D. F. (1998). Memory. In D. Kuhn and R. S. Siegler (Volume Eds.) *Cognition, Perception, and Language, Volume 2*; W. Damon (Editor-in-Chief), *Handbook of Child Psychology, Fifth Edition* (pp. 467–521). New York: John Wiley and Sons.

Scoville, W. B. and Milner, B. (1957). Loss of recent memory after bilateral hippocampal lesions. *Journal of Neurological and Neurosurgical Psychiatry, 20*, 11–12.

Seress, L. and Abraham H. (2008). Pre- and postnatal morphological development of the human hippocampal formation. In C. A. Nelson and M. Luciana (Eds.), *Handbook of Developmental Cognitive Neuroscience, Second Edition* (pp. 187–212). Cambridge, MA: The MIT Press.

Sowell, E. R., Delis, D., Stiles, J., and Jernigan, T. L. (2001). Improved memory functioning and frontal lobe maturation between childhood and adolescence: A structural MRI study. *Journal of International Neuropsychological Society, 7*, 312–322.

Sowell, E. R., Thompson, P. M., Leonard, C. M., Welcome, S. E., Kan, E, and Toga, A. W. (2004). Longitudinal mapping of cortical thickness and brain growth in normal children. *Journal of Neuroscience, 24*, 8223–8231.

Squire, L. R. (2004). Memory systems of the brain: A brief history and current perspective. *Neurobiology of Learning and Memory, 82*(3), 171–177.

Squire, L. R., Knowlton, B., and Musen, G. (1993). The structure and organization of memory. *Annual Review of Psychology, 44*, 453–495.

Squire, L. R. and Wixted, J. T. (2011). The cognitive neuroscience of human memory since HM. *Annual Review of Neuroscience, 34*, 259–288.

Takehara, K., Kawahara, S., and Kirino, Y. (2003). Time-dependent reorganization of the brain components underlying memory retention in trace eyeblink conditioning. *The Journal of Neuroscience, 23*, 9897–9905.

Tanapat, P., Hastings, N. B., and Gould, E. (2001). Adult neurogenesis in the hippocampal formation. In C. A. Nelson and M. Luciana (Eds.), *Handbook of Developmental Cognitive Neuroscience* (pp. 93–105). Cambridge, MA: The MIT Press.

Tulving, E. (1972). Episodic and semantic memory. In E. Tulving and W. Donaldson (Eds.), *Organization of Memory* (pp. 381–403). New York: Academic Press.

Tulving, E. (1983). *Elements of Episodic Memory*. Oxford: Oxford University Press.

Utsunomiya, H., Takano, K., Okazaki, M., and Mistudome, A. (1999). Development of the temporal lobe in infants and children: Analysis by MR-based volumetry. *American Journal of Neuroradiology, 20*, 717–723.

Van Petten, C. (2004). Relationship between hippocampal volume and memory ability in healthy individuals across the lifespan: Review and meta-analysis. *Neuropsychologia, 42*, 1394–1413.

Woodruff-Pak, D. S. and Disterhoft, J. F. (2008). Where is the trace in trace conditioning? *Trends in Neurosciences, 31*, 105–112.

Zola, S. M. and Squire, L. R. (2000). The medial temporal lobe and the hippocampus. In E. Tulving and F. I. M. Craik (Eds.), *The Oxford Handbook of Memory* (pp. 485–500). New York: Oxford University Press.

CHAPTER 22

LANGUAGE DEVELOPMENT IN INFANCY

ENIKŐ LADÁNYI AND JUDIT GERVAIN

INTRODUCTION

ALL neurotypical infants acquire one or several languages during early development with remarkable ease and no formal instruction—an impressive feat unparalleled by adults' explicit, conscious, and effortful learning of a foreign language later in life. While the neural, cognitive, and perceptual mechanisms leading to this developmental achievement still challenge researchers, in the past half century, a considerable amount of empirical knowledge has been gathered shedding light on some of its most important mechanisms. First, behavioral and more recently brain imaging techniques have been developed to test infants' abilities.

The imaging methods most often used with young infants to assess speech and language processing abilities are electroencephalography (EEG; De Haan, 2013), measuring the electrical activity of the brain at the scalp, and near-infrared spectroscopy (Aslin and Mehler, 2005; Gervain et al., 2011; Lloyd-Fox et al., 2009), measuring the metabolic/ hemodynamic correlates of neural activity (see corresponding chapters in the current volume for details about the two methods). These methods are safe, non-invasive, inexpensive, portable, and relatively easy to use; even with very young infants. Less frequently, magnetoencephalography (MEG), measuring the magnetic correlates of the electrophysiological activity of the brain, and magnetic resonance imaging (MRI), measuring the hemodynamic response, are also used, but they are more sensitive to motion artifacts and are generally more constraining and less well tolerated by infants. The electrophysiological and hemodynamic methods are also sometimes combined, as they provide complementary information. Electrophysiological methods provide high temporal resolution, while hemodynamic methods offer good spatial localization. Furthermore, structural MRI can probe brain anatomy, maturation, and myelination (Dubois et al., 2014), which can then be correlated with functional measures provided by a concurrent electrophysiological measure. While there are methodological challenges, near-infrared spectroscopy electroencephalography (NIRS-EEG) co-registration has been successfully implemented even in very young infants (Cabrera and Gervain, 2020; Telkemeyer et al., 2009; Wallois et al., 2012), and studies with older children have used concurrent MEG and MRI (Travis et al., 2011; Roberts et al., 2013).

The aim of the current chapter is to provide an overview of our current knowledge about how language emerges in typically developing children during the first 3 years of life. We will anchor these findings in behavioral research and general theories of language development. We will, however, mainly focus on the underlying neural mechanisms and the methods that allow us to query the infant brain.

PERCEPTUAL ATTUNEMENT TO THE NATIVE LANGUAGE AND CRITICAL PERIODS FOR LANGUAGE ACQUISITION

Language has long been recognized as a critical period phenomenon (Lenneberg, 1967). Critical periods are time windows in an organism's life span, often during early development, when experience has a particularly strong influence, and if relevant experience is missing or altered, development does not unfold in a typical manner. First Penfield and Roberts (1959) and later Lenneberg (1967) suggested that language was such a critical period phenomenon. This suggestion was closely tied to the biological grounding of language, often linked to nativist theories (Chomsky, 2007, 1959). Empirical evidence for this hypothesis comes from case studies of children who missed out on early language input due to social isolation, e.g., Genie (Curtiss et al., 1974). Among these children, those who were found before puberty recovered language, while those who were discovered and introduced to language later, never developed native-like competence. Similar patterns were observed for the L2 English competence of immigrants to the US: those who arrived before the age 7–8 years achieved full native command of English, whereas those who moved to the US later, did not and their level of English negatively correlated with the age at which they arrived (e.g., Johnson and Newport, 1989). Despite such evidence, the existence of a critical period for language was called into question by some (e.g., Chiswick and Miller, 2008; Flynn and Manuel, 1991) and the notion of a sensitive period was sometimes proposed instead. This latter notion is typically used to imply that the onset and offset of the period are less sharp and the underlying biological mechanisms less deterministic than for critical periods and/or that the ability may be acquired outside of the period, just with greater effort than during the sensitive period (Nelson, Zeanah, and Fox, 2019).

Research on the critical period for language gained new momentum in the last decade with advances in cellular and systems-level neuroscience research on well-established critical phenomena in animals (Hensch, 2005, 2003). This work tied the concept of linguistic critical periods, previously mainly assessed at the behavioral level, to mechanisms of brain plasticity and established that language does indeed have (a) critical period(s) in this neurally defined sense (Werker and Hensch, 2015). Other developmental phenomena have also been reinterpreted in this perspective (Maurer and Werker, 2014). Evidence for this comes from a growing body of empirical studies showing that the experiential or chemical processes known to control critical periods in animals also impact language acquisition (and other cognitive domains such as face perception or absolute pitch) in similar ways (Gervain et al., 2013; Maurer et al., 2020; Reh et al., 2018; Weikum et al., 2012). For instance, untreated maternal depression delays the closure of the critical period for phoneme discrimination, while exposure to serotonin reuptake inhibitors (SRIs) (in infants whose mothers are depressed, but medicated), accelerates it (Weikum et al., 2012).

Under this view, the critical period for language acquisition is linked to high neural plasticity during the early years of life, and it closes due to the reduction of this plasticity brought about by changes in the excitatory-inhibitory balance of the brain (Werker and Hensch, 2015). This closure may be different for different aspects of language, and hence it is possible to view language as having multiple critical periods, with a final closure at the onset of puberty.

During this critical period, the brain becomes increasingly specialized for the native language(s), with processing shifting from broad and distributed to more focal and more lateralized areas (Cristia et al., 2014; Minagawa-Kawai et al., 2011; Sato et al., 2010).

This increasing neural specialization or commitment to the native language is believed to be the neural substrate of an analogous phenomenon in perception, assessed behaviorally or using brain imaging: the perceptual narrowing, also called attunement, to the native language(s). During this process, infants' initially broadly based and universal perceptual and discrimination abilities become increasingly finetuned to linguistic contrasts and features used in the native language, while their ability to discriminate nonnative contrasts often declines. For instance, in their classical study, Werker and Tees (1984) showed that 6-month-old English-learning infants are able to discriminate between phonemes from the Hindi and the Salish languages that are not part of the English phoneme repertoire, but they were not able to do so anymore at 10–12 months of age. The sections that follow will describe in detail how this process unfolds for different aspects of speech perception from phoneme discrimination to tone perception.

This neurobiologically anchored perspective of language development emphasizing interactions between maturation and experience departs from the more traditional theoretical dichotomy of nature/nurture, i.e., the more traditional question of whether the language faculty is an innate or a learned ability. As the new perspective emphasizes, this strict dichotomy is ill-posed, as genetically endowed processes act in interaction with experience to give rise to the unfolding of an organism's development (Gervain and Mehler, 2010). Recent epigenetic studies now also confirm this perspective, highlighting synergistic interactions between the genetic endowment and environmental factors (Gelman, 1991; Markman et al., 2011; Roth and Sweatt, 2009).

Another shift in perspective that recent empirical results brought about suggests that the different areas of language develop in closer interaction than previously believed (de la Cruz Pavia et al., 2021; Swingley, 2021; Werker, 2018). Rather than following each other in a sequential manner, attunement to the native phoneme inventory, word learning, and grammar learning start very early, proceed in parallel and mutually support one another, as the next sections will highlight.

The Language Network in the Infant Brain

One central question about the neural substrate of language is whether and how the brain networks involved change during development. Existing results show remarkable similarities between the structures related to language processing in adults and infants.

One important question is the lateralization of speech and language processing in the brain (Friederici, 2012). Processing of different aspects of auditory stimuli have been found to show a functional asymmetry. In right-handed and ambidextrous adults as well as in a certain percentage of left handers, language processing has been argued to be lateralized in the left hemisphere, while the right hemisphere is assumed to be specialized in music processing (e.g., Alho et al., 1998; Tervaniemi and Hugdahl, 2003). Several accounts have been proposed to explain this lateralization. It has been proposed, for instance, that the left hemisphere may contain neurons specialized in processing shorter, rapidly changing, or temporally modulated sounds, while the right hemisphere may preferentially process slowly changing and/or spectrally modulated sounds (Hickok and Poeppel, 2007; Zatorre and Belin, 2001). While there is considerable empirical evidence from behavioral and imaging experiments as well as clinical/lesion studies to establish lateralization, recent findings suggest that lateralization may nevertheless be a less general characteristic of auditory processing than previously thought, with more abstract linguistic functions such as syntax and semantics truly lateralized, while speech processing itself may be more bilateral (Poeppel, 2014).

Whether language is lateralized already in the infant brain has received considerable attention. The behavioral studies about such asymmetries have yielded mixed results. While an early study (Bertoncini et al., 1989) found that neonates showed a right ear preference for speech in a dichotic listening paradigm (auditory stimuli are presented either only to the left or right ear) similarly to adults, suggesting left lateralization of speech processing, other studies did not find such a preference in neonates (e.g., Best et al., 1982).

Functional MRI studies showed a stronger activation in the left language-related areas (superior temporal and angular gyri) to speech than to music or biological nonspeech sounds (monkey vocalizations), at 1–4 months (Dehaene-Lambertz et al., 2002; Dehaene-Lambertz et al., 2010; Shultz et al., 2014) supporting left lateralization of speech processing. Interestingly, Shultz et al. (2014) showed that the response to biological nonspeech in left temporal cortex decreased with age, while the specialization for speech increased from 1–4 months.

A series of NIRS studies also address the question of lateralization and did so in newborn infants. A pioneering study by Peña et al. (2003) found stronger left lateralized activation in newborns in response to their native language, Italian, presented normally, i.e., forward, than to the same material presented backwards or to silence. The response to forward Italian involved the same regions as in adults, mainly the middle and superior temporal areas and the inferior frontal regions including Broca's area. This first study thus established adult-like specialization for speech processing already at birth.

Two subsequent studies then replicated the advantage in the left hemisphere for the native language when it was presented forward as compared to backward, but found no such lateralized advantage for a nonnative language, suggesting that lateralization may be related to (prenatal) experience (native language English and nonnative language Spanish: May et al., 2018; native language Japanese and nonnative language English: Sato et al., 2010).

However, another study found no forward advantage and observed bilateral activation for both the native (English) and the nonnative language (Tagalog), although activation was overall greater for the native language, suggesting that prenatal experience also played a role, even if it did not lead to an observable lateralization difference (May et al., 2011). It needs to be noted, though, that in this study, the speech stimuli were low-pass filtered at 400 Hz to imitate the speech signal heard in utero. It may be the case that, following the hypotheses about different temporal and/or spectral modulations driving hemispheric specialization (Hickok and Poeppel, 2007; Zatorre and Belin, 2001), low-pass filtered stimuli, from which the rapidly changing acoustic features have been removed, failed to trigger a left hemisphere advantage. The NIRS-EEG co-recording studies directly testing this hypothesis in newborns, as well as in 3- and 6-month-old infants, did not provide conclusive evidence (Telkemeyer et al., 2011, 2009). All three groups showed a strong bilateral rather than the predicted left lateralized activation to noise sequences that were rapidly modulated in time (~ 25 msec, i.e., at the (sub)phonemic level), whereas slowly modulated sequences (~ 160 msec and 300 msec, i.e., syllabic level) selectively activated the right temporoparietal cortex, as predicted (Hickok and Poeppel, 2007; Zatorre and Belin, 2001), although this activation was somewhat weaker overall than for the fast stimuli. It needs to be noted that the study used nonspeech stimuli (narrowband noise sequences), which in adults are sufficient to trigger left lateralization for fast-changing stimuli and right lateralization for the slower ones (Boemio et al., 2005), but it remains to be tested whether speech or other ecologically more valid stimuli trigger the predicted lateralization in infants.

In addition, structural studies of the developing brain also indicate that structural asymmetries characterizing the adult brain are present already from the first month of life and may be related to functional asymmetries (e.g., Dubois et al., 2009; Glasel et al., 2011).

Another structural feature which is gaining increasing attention is the connectivity of the language network, and the infant brain has been found surprisingly adult-like in this regard. While the development and pruning of synapses and the myelination of axons is intense during the first years of life, white matter bundles connecting language-related areas are structured in ways similar to those of adults. In adults, two pathways, the dorsal and the ventral pathways can be differentiated which are supporting phonological and semantic processing, respectively (Friederici, 2005). The two pathways are segregated already in 1–4-month-old infants (Dubois et al., 2016). Functional connectivity analyses based on NIRS (Benavides-Varela et al., 2017; Homae et al., 2011, 2010; Molavi et al., 2014), functional magnetic resonance imaging (fMRI) (Cusack et al., 2018) and EEG (Kühn-Popp et al., 2016; Mundy et al., 2003; Righi et al., 2014) data also show considerable functional connectedness between the different language-related areas, with temporo-frontal connections present very early on, and the connections between homologous areas of the two hemispheres increasing over the first few months of development.

How can the brain process all at once the different units of speech, which occur simultaneously, organized into an embedded hierarchy (e.g., the sound /b/ is a phoneme embedded in the syllable /bi/, which itself is embedded in the word *baby*, which is embedded in the phrase *lovely baby*, and so on)? One proposal that has recently received much attention is that a hierarchy of embedded neural oscillations, which match the relevant linguistic units in frequency, allows the brain to process these units simultaneously (Giraud and Poeppel, 2012). Thus, delta oscillations (1–3 Hz) are associated with the processing of prosodic phrases, theta oscillations (4–8 Hz) for syllables, and delta oscillations (> 35 Hz) for (sub)phonemic units. Similarly to linguistic units, the oscillations are organized into a cascading hierarchy, such that the phase of slower oscillations determines the amplitude (or phase) of the faster ones. In nonhuman animals and human adults, considerable electrophysiological evidence supports this model and the oscillations have been localized to the auditory cortex (Giraud et al., 2007; Lakatos et al., 2005; Poeppel, 2014). While the presence and role of such oscillations in language development is only now starting to be investigated (Kalashnikova et al., 2018; Ortiz-Barajas et al., 2021; Telkemeyer et al., 2011), it has been hypothesized (Gervain, 2018) that the chronological sequence of experience with speech during early development, i.e., an initial, low-pass filtered prenatal speech signal followed by the postnatal full-band signal, may underlie the organization of the oscillations. The low-pass filtered intrauterine signal may serve to fine-tune the slower oscillations to the rhythm of the native language already prenatally. Then the postnatal signal, which now also contains the higher frequencies, fine-tunes the faster oscillations, which thus get nested in the already operational slower frequencies. Some promising first results supporting this hypothesis are now emerging (Ortiz-Barajas and Gervain, 2020).

To summarize, the infant brain is language-ready from birth. It shows remarkable similarities to the adult brain in its structure and functions, paving the way for language development and processing.

NEWBORNS' SPEECH PERCEPTION ABILITIES: UNIVERSAL MECHANISMS AND PRENATAL SHAPING

Newborn infants' perceptual and learning abilities have received considerable attention, as they shed light on the "initial state" of the language faculty. Behavioral methods such as high-amplitude sucking (HAS; Floccia et al., 1997), and more recently neuroimaging techniques such as electroencephalography (EEG; De Haan, 2013) and near-infrared spectroscopy (NIRS; Gervain et al., 2011) have been used. These studies have shown that newborns have a wide range of broadly based, universal speech perception abilities, allowing them to learn any language they are exposed to. But speech is also heard in the womb; it is thus not surprising that the existing studies have also revealed abilities that are already modified by prenatal experience. We will now discuss the universal abilities and those shaped by experience in turn.

Newborns show remarkable speech perception abilities despite their immature auditory systems (Eggermont and Moore, 2012). Many of these abilities are universally present, irrespective of what language(s) an infant heard in utero. As a first task, newborns need to identify speech among other sounds in the environment. Newborns and 2-month-old infants can indeed recognize speech, and show a strong preference for it over equally complex sine wave analogues (Vouloumanos and Werker, 2004). However, the category "speech" may be relatively broad at birth roughly corresponding to primate vocalizations, as newborns show equal preference for human speech and rhesus monkey vocalizations (Vouloumanos et al., 2010). It is only by 3 months that infants show a unique preference for speech over both sine wave analogues and monkey calls (Vouloumanos et al., 2010).

Paralleling this behavioral preference for speech, young infants also show a brain specialization for speech processing. Three-month-olds, full-term neonates, and even premature newborns activate approximately the same brain network as adults, i.e., the superior and middle temporal gyri, the inferior parietal cortex, and the inferior frontal gyrus, including Broca's area, in response to language, but not to nonlinguistic controls such as backward speech (Dehaene-Lambertz, Dehaene, and Hertz-Pannier, 2002; Mahmoudzadeh et al., 2013; Peña et al., 2003). As we will discuss, this specialization may already be shaped by prenatal experience.

Newborns are also able to discriminate languages from one another, even if those are unfamiliar to them, on the basis of their different rhythmic properties (Mehler et al., 1988; Nazzi et al., 1998), such as the relative proportion of vowels in the speech signal (Ramus et al., 1999) and other related acoustic measures (Dellwo, 2006; Grabe and Low, 2002; Loukina et al., 2011; Wiget et al., 2010). Rhythmic discrimination does not require familiarity with the languages: newborns prenatally exposed to French are able to discriminate between English and Japanese. So can tamarin monkeys

(Ramus et al., 2000), suggesting that rhythmic discrimination might be a general property of the primate or mammalian auditory system. Using this ability, infants born into a multilingual environment can immediately detect that they are being exposed to different languages, at least if those are rhythmically different. Bilingual newborns have indeed been shown to be able to discriminate their two languages from a third, rhythmically different language (Byers-Heinlein, Burns, and Werker, 2010). The ability to distinguish languages within a rhythmic class emerges by about 3.5–4 months of life and requires familiarity with at least one of the languages (Bosch and Sebastian-Galles, 1997; Molnar et al., 2013). Similarly, discriminating between the native dialect and a nonnative dialect within a language emerges around 5 months (Cristia et al., 2014). These abilities are assumed to rely on the recognition of specific phonemes or phonotactic regularities characteristic of one of the languages and it has been shown to be under brain maturational control—infants born preterm show this ability at the relevant maturational, i.e., corrected, age, not at the relevant chronological age (Peña et al., 2012, 2010).

In addition to the broad prosodic properties of language such as rhythm, newborns can also process smaller units within the speech signal. Behavioral results show, for instance, that they readily detect the acoustic cues correlated with word boundaries (Christophe, Dupoux, Bertoncini, and Mehler, 1994). They have also been found to be sensitive to the prosodic makeup of words at the syllable level (Sansavini et al., 1997), readily discriminating words with different lexical stress patterns, i.e., stress-initial (trochaic) vs stress-final (iambic). Interestingly, however, they cannot tell apart words with different numbers of phonemes, if the number of syllables is the same.

In addition to the previously discussed broad-based abilities, an increasing number of studies suggests that fetuses already learn from experience with speech heard in the womb. Human fetuses can hear from about 24–28 weeks of gestation (Eggermont and Moore, 2012). Using MEG, brain activity can already be measured in fetuses, as MEG is less sensitive to the amniotic fluid and other tissues than other imaging techniques (Chen et al., 2019). MEG studies with fetuses have shown that fetuses show an auditory evoked response from 28 weeks of gestation at the latest, and its latency decreases with gestational age (Draganova et al., 2005, 2007; Govindan et al., 2008; Hartkopf et al., 2016; Holst et al., 2005; Lengle et al., 2001; Lutter et al., 2006; Sheridan et al., 2008). Similarly, fetuses detect changes in sound frequency from the third trimester of pregnancy (Draganova et al., 2007; Muenssinger et al., 2013; Sheridan et al., 2008).

Auditory experience with speech thus starts in the womb. However, the intrauterine speech signal is different from the signal heard outside of the womb. Maternal tissues act as low-pass filters at about ~ 400–800 Hz (Armitage et al., 1980; DeCasper et al., 1994; Gerhardt et al., 1990; Lecanuet and Granier-Deferre, 1993). As a result, prosody, i.e., the melody and rhythm of speech, is preserved, but the fine details necessary to identify individual sounds, especially consonants and words are suppressed. Infants' first experience with speech thus consists mainly of prosodic information (Gervain, 2018, 2015).

This prenatal experience already shapes fetuses' and newborn infants' speech perception abilities. At the most general level, newborns recognize and prefer their mother's voice over other female voices (DeCasper and Fifer, 1980), their native language over other languages (Mehler et al., 1988; Moon, Cooper, and Fifer, 1993) and a story heard frequently in the womb over other stories (DeCasper et al., 1994; Kisilevsky et al., 2009).

But they seem to learn even more specific details about their native language. Since vowels have the most energy in the speech signal and carry prosodic patterns most strongly, some vowels seem to be already learned in part prenatally. Indeed, newborns show a preference for a novel vowel, absent in their prenatal input, over a native one (Moon, Lagercrantz, and Kuhl, 2013). Another study suggests that fetuses can also learn word-level prosodic information (Partanen et al., 2013), readily detecting a change in pitch trained prenatally, which untrained newborns do not recognize.

Infants also show evidence of learning about the prosody of larger linguistic units, such as phrases and utterances, prenatally. Languages vary as a function of what acoustic cues mark prosodic prominence in their phonological phrases. In some languages, such as French or English, prominence is carried by a durational contrast, i.e., the prominent element is lengthened as compared to the non-prominent one (e.g., to Ro:me, note that the vowel of the prominent content word is longer than the vowel of the non-prominent grammatical morpheme). In these languages, the prominent element typically occupies a phrase-final position, so phrases have an iambic prosodic pattern. In other languages, like Japanese or Turkish, prominence is indicated by a pitch/intensity contrast. In these languages, prominence is phrase-initial, i.e., trochaic, so the higher or louder element is at the phrase onset (e.g., Japanese: ^Tokyo kara "Tokyo to"). This alternation of prominent and non-prominent elements creates a rhythmic prosodic pattern readily perceivable even by listeners who are unfamiliar with a given language (Langus et al., 2016). Newborn infants also seem to pick up on this pattern from their prenatal exposure (Abboub et al., 2016). When presented with pairs of pure tones contrasting in duration, pitch or intensity that were consistent with the patterns found in natural languages (iambic pairs for the durational contrast, and trochaic pairs for the pitch and intensity contrasts) compared to pairs that were inconsistent with these patterns (trochees for duration, iambs for pitch/intensity), the newborn brain showed a greater response to the inconsistent patterns, but only for the acoustic cue that marks prosodic prominence in the language the infants were exposed to prenatally.

Prenatal experience also shapes the brain specialization for language processing, as discussed in the section "The Language Network in the Infant Brain." Newborns' brain responses to speech in the native language are different from responses to nonnative languages, be this a difference in lateralization (May et al., 2018; Sato et al., 2012) or in the magnitude of the response (May et al., 2011). This strongly suggests that the brain network for speech processing is being sculpted by prenatal experience. Furthermore, this network is already specialized for processing speech, as a whistled language does not activate it despite being a human communication system (May et al., 2018).

PHONEME PERCEPTION

Phoneme perception is one of the most studied areas of language development. It is also the paradigm example of attunement to the native language. Very young infants, up to about 4–6 months of age, can discriminate almost all phonemes appearing in the world's languages, even those that do not appear in their native language and that adult speakers of the same language are unable to discriminate, as has been shown both behaviorally (Eimas et al., 1971; Werker and Curtin, 2005) and electrophysiologically (using EEG: Dehaene-Lambertz and Baillet, 1998; using MEG: Kujala et al., 2004). Infants' phoneme perception, like that of adults, is categorical, especially for consonants, possibly less so for vowels (Swingley, 2021). Categorical perception means that a given acoustic difference between two sounds is discriminated when it spans a phoneme boundary, but a difference of the same magnitude falling within the boundaries of a phoneme category is not treated as contrastive (even though infants are sensitive to such acoustic differences; McMurray and Aslin, 2005). The electrophysiological correlates of phoneme discrimination have been localized to the left temporal areas in infants, similarly to adults (Dehaene-Lambertz and Gliga, 2004). This universal discrimination repertoire is one of the hallmarks of young infants' broad-based abilities, allowing infants to learn any language they are exposed to. Interestingly, animals can also discriminate phonemes at similar acoustic boundaries (Kuhl, 1981, 1986), suggesting that phoneme perception is a basic perceptual ability rooted in general mammalian auditory mechanisms. It is, therefore, available to young infants prior to experience.

With several months of experience with the native language, nonnative sound discrimination declines (Werker and Tees, 1984), while the discrimination of contrasts found in the native language is maintained or even improves (Kuhl et al., 2006; Narayan et al., 2010). This perceptual attunement or narrowing toward the native sound repertoire takes place around 4–6 months for vowels (Kuhl et al., 1992) and around 10–12 months for consonants (Werker and Tees, 1984). As discussed before, the system nevertheless remains plastic for several years after attunement. It is thus possible to learn the phoneme inventory of another language in a native manner until age 6–8 (or the onset of puberty the latest), as studies with immigrants (Johnson and Newport, 1989) and international adoptees suggest (Pallier et al., 2003; Pierce et al., 2014; Ventureyra et al., 2004). Infants growing up multilingually go through the same perceptual narrowing and can discriminate the contrasts of all of their native languages by the end of the first year of life (Albareda-Castellot et al., 2011; e.g., Bosch and Sebastian-Galles, 2003). In addition to phonemes, similar perceptual attunement has been found for lexical tones (Mattock and Burnham, 2006; Yeung et al., 2013), and sign language signs (Baker et al., 2006). Interestingly, the ability to discriminate some nonnative contrasts does not get lost. Some features, although not all, of Zulu click sounds remain discriminable to nonnative adults (Best et al., 1988). This has been explained by the unusual, almost nonlinguistic nature of these sounds.

Recent results suggest that phoneme discrimination may be facilitated by systematic associations between sounds and objects, implying that the relationship between phoneme perception and word learning is mutual (Werker and Yeung, 2005). Thus 9-month-old infants can successfully discriminate a nonnative sound contrast if each sound is embedded in a nonword that is associated with an object, but fail due to perceptual attunement if the nonwords are presented alone (Yeung and Werker, 2009).

NIRS studies have shown that perceptual attunement is supported by a concomitant neural reorganization. Japanese infants have been found to show differential activation to a vowel length contrast (/a/ vs /a:/) in the left temporal region at 6–7 months and from 13 months onwards, but not at 10–11 months, showing a U-shaped developmental curve, indicating reorganization. Furthermore, discrimination was bilateral at the younger age, but left lateralized at the later ages (Minagawa-Kawai et al., 2007).

In sum, infants start their journey into language as universal listeners, but by the end of the first year of life, they become native language expert listeners, as their perceptual systems and brains reorganize to better perceive those linguistic contrasts that they encounter in the native language.

WORD LEARNING

As an early step of language development, infants have to extract words from the continuous speech they are exposed to, which is a challenging task given that word boundaries do not have reliable acoustic markers in speech (Swingley, 2009). Nonetheless, by 6–9 months, infants recognize a few of the most frequent word forms and can associate them with their referents (Bergelson and Swingley, 2012; Tincoff and Jusczyk, 1999) and by the time they turn 18 months, most of them are able to understand about 150 words and can successfully produce 50 words (Fenson et al., 1993). The task of word learning involves segmenting out potential word forms from the input and linking them to possible referents. Both of these learning tasks have received considerable attention in the literature.

Several cues have been proposed to help infants segment words from speech: the distributions of the cooccurrence statistics of different linguistic units such as syllables, lexical stress, phonotactics, and allophone distributions. Statistical learning is the ability to discover patterns such as frequencies or conditional probabilities in the input (Romberg and Saffran, 2010). Based on this information, we are able to extract chunks in which statistical coherence between units (phonemes or syllables) is high, i.e., assuming a boundary where statistical coherence is low. For instance, if two syllables (e.g., *ba* and *by*) appear very often together (the transitional probability between them is high), then it is very likely that they constitute a word (*baby*), while syllables which appear together only occasionally (e.g., *by* and *sleeps* in the sentence *The baby sleeps*) probably belong to different words—an idea going back to structuralist linguistics (Harris, 1955). Behavioral studies show that infants are able to use statistical information to extract word forms from the input already from 8 months in behavioral experiments (Goodsitt, Morgan, and Kuhl, 1993; Johnson and Tyler, 2010; Saffran, Aslin, and Newport, 1996; Saffran and Kirkham, 2018, for review) and as early as birth when assessed with brain imaging (NIRS: Fló et al., 2019; EEG: Teinonen et al., 2009).

Prosodic cues were also shown to play a role in word segmentation. Languages have their characteristic word stress patterns. In English, for instance, most multisyllabic words have the main stress on their first syllable showing a strong-weak (trochaic) stress pattern. English 7.5-month-old infants can use lexical stress to segment words with a strong-weak (e.g., *doctor*) pattern, but this same heuristic induces them in error when trying to segment a word with a weak-strong pattern (e.g., *guitar*; Höhle et al., 2009; Jusczyk et al., 1999a). Studies in Dutch-learning infants found evidence for using prosodic cues for word segmentation at a slightly later age, at 9 months (Kuijpers et al., 1998). ERP studies, however, have shown that the ability is present at 7 months in Dutch (Kooijman et al., 2005, 2009). In a language like French, in which there is no lexical stress, infants first segment the basic rhythmic unit of their native language, which is the syllable, and the segmentation of multisyllabic words appears only later—at 16 months, when tested behaviorally (Nazzi et al., 2006) and at 12 months, when tested electrophysiologically (Goyet et al., 2010). However, when the processing of disyllabic words with initial and final stress (*"papa* vs *pa"pa*) are compared directly using electrophysiological measures, German and French infants already show preferential processing for the pattern that is most common in their native language at 4 months of age (Friederici et al., 2007). German infants showed a mismatch response to words with final stress, since a large majority of German words are trochaic, i.e., stress-initial, while French infants showed a mismatch response to stress-initial words, since in French, while there is no lexical stress, final lengthening at the end of prosodic phrases creates words with an iambic, i.e., prominence-final pattern. In a language with lexical pitch accent, like Japanese, infants have been shown to be able to behaviorally discriminate high-low and low-high pitch accent patterns at both 4 and 10 months, but simultaneous NIRS recordings revealed that at 4 months, discrimination activated the bilateral temporal cortices and the processing of pitch accented words and tone analog evoked similar activation, whereas at 10 months, there was an advantage for words over tone analogs, and it was left lateralized (Sato et al., 2010). The authors explained this as a signature of neural specialization for language, as processing shifts from acoustic to linguistic and gets increasingly lateralized.

Another possible cue to segmentation is phonotactics. Knowing that the sequence /br/ is frequent in the initial positions of English words/syllables, while /nt/ typically appears at the end can help the learner posit word boundaries. Infants have been shown to be able to rely on phonotactic cues for segmentation starting at 9 months (Mattys et al., 1999; Mattys and Jusczyk, 2001; Saffran and Thiessen, 2003).

The distribution of allophones (phonetically distinct variants of a phoneme that do not contribute to distinctions of meaning) in different positions within words also provides indications about word boundaries. In English, for instance, aspirated stop consonants appear in the initial positions of stressed syllables, their unaspirated allophones appear elsewhere. Consequently, aspirated stops are good cues to word onsets in English. Infants as young as 2 months are able to discriminate the different allophones of a phoneme (Hohne and Jusczyk, 1994), and by 10.5 months, they can use allophonic cues to posit word boundaries (Jusczyk et al., 1999b).

Further cues to segmentation that have been identified in the literature include already known words, such as the infant's name or *Mummy*, and so on (Bortfeld et al., 2005), and boundaries of large prosodic units, which, given the hierarchical organization of prosody, necessarily coincide with word boundaries (Shukla et al., 2011).

As they segment out the first word forms, infants also start to link them to possible meanings. Learning the meaning of a word poses a logical problem: meanings are underdetermined by the situation (formulated as the gavagai-problem by Quine, 1960). It has been suggested that infants have inherent biases to favor some interpretations rather than others, allowing them to reduce the possible space of meanings and converge on the correct one (Markman, 1994, 1991; Merriman, 1991). Thus, infants assume that labels refer to whole objects rather than to object parts, to basic level categories such as "dog" rather than to subordinate or superordinate categories (German Shepherd and mammal, respectively). Also, if infants already know a name for an object, they will follow the mutual exclusivity constraint, and associate a new label with a new object, i.e., they avoid synonyms for the same object (Au and Glusman, 1990; Markman and Wachtel, 1988). Interestingly, this constraint is modulated by experience. Monolingual infants readily honor it, whereas in multilingual infants, who need to be able to learn several labels, one in each language, for the same object, there is a negative correlation between adherence to the principle and the number of languages they are exposed to (Byers-Heinlein and Werker, 2009).

Word learning and lexical access have been investigated in great detail using electrophysiological techniques (see reviews in Friedrich, 2017; Morgan et al., 2020; Kuhl and Rivera-Gaxiola, 2008). One much-studied event related potential component is a negative shift that appears at around 200–500 ms (N200–500) after word onset, if the infant is presented with a familiar word compared to an unfamiliar word. This component is present from 9 months, but undergoes gradual left lateralization between 13 and 20 months as infants' familiarity with specific words increases and their processing becomes more efficient (Molfese 1990, Molfese et al., 1990, 1993; Mills et al., 1993, 1997, 2005; Thierry et al., 2003). A similar signal was found over the outer lateral frontal regions from 12 months of age in priming studies in which infants were presented with a picture and then either with the name of the picture (congruous) or with an unrelated word (incongruous). According to the authors, this component resulted from the pre-activation of the acoustic-phonological structure after the picture was presented, and was interpreted as a word form priming effect (Friedrich and Friederici, 2004, 2005a, 2005b). Another component, a broadly distributed negativity with a centro-parietal maximum from around 400 ms, appeared to incongruous as compared to congruous words and was found at a later age than the word form priming effect. This effect was interpreted as a sematic priming effect, an infant correlate of the adult N400.

N400 in adults is a negative shift 250–600 ms after stimulus onset with a peak at about 400 ms and it is the strongest on right centroparietal areas. It reflects semantic processing and it is reduced if processing is facilitated by the context (e.g., the word is presented together with a picture matching the word, a sentence with which the word is semantically consistent, or is preceded by a semantically related word) and is larger when the context does not support the processing of the word (a non-matching picture or a semantically unrelated word is presented before/together with the target word) (see a summary on N400 in Kutas and Federmeier, 2011). Spatiotemporal features of N400 in infants have been found in some studies to differ from those of the adult component in several respects.

In a series of studies with 12-, 14-, and 19-month-old infants, Friedrich and Friederici (Friedrich and Friederici, 2004, 2005a, 2005b, see summary in Friedrich, 2017) aimed to disentangle two developmental phases of lexical acquisition, based on the previous two ERP signals. In the first phase, infants are only assumed to represent an associative relationship between word forms and objects, while later in development, they are hypothesized to establish a referential relationship. The empirical evidence for this account comes from priming tasks, in which infants were first presented with a picture of an object, and then with a word that was either congruent or incongruent with the picture. Only 14 and 19-month-old infants showed an N400-like effect, while 12-month-olds did not, even if they knew the words according to parental report. According to the authors, these results may mean that while infants can associate word forms with objects earlier, they do not exhibit referential connections between words and their meanings until 14 months of age. With this shift, infants no longer set up links between two simultaneously occurring events (a sound pattern and a specific object) but become able to use language as a way of representing the world with pairing sound patterns with concepts, which also contributes to the vocabulary spurt during the second year of life (Nazzi and Bertoncini, 2003). Interestingly, several studies found that infants with a more adult-like N400 show better language skills behaviorally than infants with missing or less developed N400 (Friedrich and Friederici, 2004; Friedrich and Friederici, 2006).

A subsequent study managed to find the N400 component in younger infants, at least under optimal stimulus conditions. Parise and Csibra (2012) explored 9-month-old infants' brain responses in a similar object-word matching paradigm as did previous studies, but in their study infants were not presented with prerecorded words. Rather either the mother or the experimenter produced the labels and importantly, they were free to use complete utterances to name the objects as well as gestures like pointing. In this modified experimental setting, infants showed an N400 effect in the condition in which their mother produced the words, suggesting that semantic processing of words is present already at 9 months of age. Furthermore, infants as young as 14 months of age also show an N400 when their social partner experiences a semantic incongruity (Forgács et al., 2019), suggesting that infants start tracking others' comprehension of language very early on, and that early word learning ties into theory of mind and other aspects of social cognition essentially from its onset.

The N400 can also be used to investigate the organization of the mental lexicon. In a picture-word matching paradigm (Torkilsden et al., 2006), 20-month-old toddlers were presented with words which were either congruent (picture of a dog—*dog*) or incongruent with the picture and incongruent words were either from the same category (*cat*) or from a different category (*car*). Similarly to adults (Federmeier and Kutas, 1999), toddlers showed a larger and earlier incongruity response for between-category violations than for within category violations, suggesting that categorical organization is already present in the lexicon of 20-month-olds. Interestingly, productive vocabulary was associated with the spatiotemporal features of the ERP response. A subsequent study (Rama et al., 2013) showed a similar effect when words rather than pictures were used as primes in 24-month-olds, but at 18 months, the effect was only present in participants with a high productive vocabulary. According to the authors, these results suggest that words are organized by semantic categories in the mental lexicon at 2 years and already at 18 months in some children.

Furthermore, the N400 was also used as a probe into when infants start processing sentence-level semantics. According to passive listening paradigms infants are sensitive to semantic violations caused by an incongruent word within a sentence from 19 months, as shown by an N400-like component in response to sentences with semantic violations (e.g., *The cat drinks the ball*) vs semantically congruous sentences (Friedrich and Friederici, 2005c; Silva-Pereyra et al., 2005a, 2005b). Spatiotemporal features of this N400 are similar to the adult N400, but its longer latency indicates slower semantic processing in toddlers than in adults.

The Beginnings of Grammar

Increasing evidence suggests that the acquisition of grammar starts earlier than previously believed, in parallel with perceptual narrowing and word learning (Gervain, 2022), supporting a view of language development in which different levels interact and mutually support each other. Artificial grammar learning studies have highlighted infants' ability to represent different abstract relations between linguistic units, and they have also allowed researchers to tap into the earliest knowledge infants have of their native grammar (Gervain et al., 2018; Gomez and Gerken, 2000).

As language unfolds over time, sequential order is one of the most basic properties of the structure of a language. It is thus crucial for infants to be able to encode the order of words in a sequence. Indeed, it has been shown that infants can detect the violation of word order in sequences as early as birth when tested with NIRS (Benavides-Varela and Gervain, 2017) or at 2 months of age when tested behaviorally (Mandel et al., 1996).

While surface order is highly relevant, natural language grammars encode different structural dependencies between abstract linguistic categories. It is thus important to assess when infants first show sensitivity to these. Infants have been shown to be able to discriminate regularities based on adjacent repetitions (e.g., ABB: "mubaba," "penana," and so on) from random controls (ABC: "mubage," "penaku," and so on) already from birth both using high-amplitude sucking (Gervain et al., 2011) as well as NIRS (Gervain et al., 2008, 2012), suggesting that they are able to represent one of the most basic abstract relations, identity. They can also discriminate between sequences with adjacent repetitions in initial vs final positions (AAB vs ABB; Gervain et al., 2012), but cannot discriminate nonadjacent repetitions (ABA) from random sequences (Gervain et al., 2008). Furthermore, using NIRS, the response to such repetition-based patterns was localized to the left temporal and inferior frontal areas, i.e., the language network, including Broca's area (Gervain et al., 2008, 2012). By 7–8 months, infants can also generalize adjacent and nonadjacent repetition patterns to a novel vocabulary, i.e., sequences made up of syllables they were not trained on (Marcus et al., 1999), showing that they encode an abstract structure, not just item-based information about specific syllables. Some authors, however, have suggested that repetitions or identity may be a Gestalt-like pattern, which is automatically detected at the perceptual level and might not require a fully symbolic representation (Endress et al., 2009, 2005). An increasing body of literature has investigated the developmental trajectory of repetition-based regularity learning (for a review and a meta-analysis, see Rabagliati et al., 2019), and results suggest that infants succeed in generalizing the pattern already by 5 months of age, if stimuli are presented multimodally (Frank et al., 2009).

The ability to encode adjacent and nonadjacent dependencies between nonidentical items has also received considerable attention, as these abilities are crucial for morphosyntax, where structural relationships may hold between neighboring (e.g., suffixation in morphology, grammatical morpheme-content word relations in syntax) or distant items (e.g., circumfixes in morphology, or agreement relations in syntax). The detection of adjacent regularities (of the type aX, bY) has been documented behaviorally at 12 months (Gómez and Lakusta, 2004). Nonadjacent dependencies within words, i.e., at the morphological level, have been found to appear between 7 and 12 months (Marchetto and Bonatti, 2015). The learning of nonadjacent regularities between words, i.e., at the syntactic level, has been observed behaviorally at 15 and 17 months and even at 12 months with longer exposure to stimuli (Gómez and Maye, 2005; Gomez, 2002). Electrophysiological studies, however, suggest that these abilities may be present earlier than behaviorally observable. The EEG responses of 4-month-old German infants exposed to a nonadjacent morphological dependency in Italian between an auxiliary and a verb form (_sta cantando_ "is singing" vs _puo cantare_ "can sing") differed upon hearing grammatical forms as compared to violations (e.g., _sta cantare_) (Mueller et al., 2012; Friederici et al., 2011). Similarly, in a paradigm resembling that of Marchetto and Bonatti (2015), infants showed an EEG signature of learning the nonadjacent dependency already at 8 months (Kabdebon et al., 2015).

Infants' emerging knowledge of the specific properties of their native grammar has also been investigated. Infants have been shown to establish morphosyntactic categories very early on. Babies are sensitive to the phonological minimality of function words (grammatical words encoding structure, such as articles, prepositions, pronouns, and so on) and can use this feature to distinguish them from content words (words that carry lexical meaning, such as nouns and verbs) already at birth (Shi et al., 1999). They can detect functors (i.e., grammatical morphemes, e.g., prepositions, articles, pronouns) in continuous speech by 7–9 months (Höhle and Weissenborn, 2003) and can later use functors to segment out and categorize the content words that appear with them (in German: Hohle et al., 2006; in French: Shi and Melançon, 2010; Shi and Werker, 2003; in English: Shi and Werker, 2001). By their first birthday, infants show different brain responses to continuous speech as compared with the same stimuli in which a tone was superimposed on the functors (Shafer et al., 1998).

By 8 months of age, infants are also sensitive to the relative order of functors and content words in their native language and use word frequency as a cue to the two categories. Infants acquiring languages with opposite word orders, such as functor-initial Italian or French and functor-final Japanese, parse a structurally ambiguous artificial grammar in which frequent functors and infrequent content words strictly alternate the opposite way: Italian and French infants prefer sequences starting with a frequent word, mirroring the word order of these languages, while Japanese infants prefer sequences ending in frequent words, as is the case in Japanese (Bernard and Gervain, 2012). Bilingual infants growing up simultaneously with a frequent-initial and a frequent-final language use the different phrasal prosodies of these two language types as an additional cue to establish word order preference (Gervain et al., 2013). Furthermore, in this paradigm, 8-month-old infants readily accept substitutions for infrequent words, i.e., they already treat content words as open classes, but not frequent words, i.e., they treat functors as closed classes (Marino et al., 2020). By 17 months, they map infrequent words, but not frequent ones onto objects, suggesting that they know that content words, but not functors, have referents (Hochmann et al., 2010). Taken together, these results suggest that by 8 months, infants establish the morphosyntactic categories of functors and content words, the universal building blocks of all natural language grammars and can use them to build a rudimentary representation of the basic word order of the native language.

Functors may be free, i.e., independent words, or bound morphemes, i.e., prefixes or suffixes attached to stems. Different languages encode the same structural relationship with free vs bound functors to differing degrees. In English, for instance, prepositions, i.e., free functors, are used to signal relations such as location (e.g., *in the house*), while Turkish, Japanese, or Basque, these relations are expressed using found morphemes (e.g., Basque: *etxean*, where *etxe* is "house" and the suffix *–an* indicates location). In languages using bound morphemes extensively, infants need to learn to decompose these morphologically complex forms to acquire grammar as well as to learn the lexicon. It has indeed been shown that by the time infants enter the vocabulary spurt by the middle of the second year of life, they have sufficient knowledge of at least the most frequent suffixes of their native language to decompose morphologically complex words, both in morphologically poorer (French: Marquis and Shi, 2012; English: Mintz, 2013) and richer (Hungarian: Ladányi et al., 2020) languages.

During the second year of life, children start to produce words and word combinations. At this time, their knowledge of syntax is advanced enough to allow them to use the main syntactic frame of a sentence to determine the argument structure and thus (some aspects of) the meaning of novel verbs. Specifically, they can track the number and the position of noun phrases in sentences, and use this information to categorize novel verbs as intransitive, those appearing with a single noun phrase (e.g., sleep), transitive, appearing with two noun phrases (e.g., hit, kick) or ditransitive, having three noun phrases (e.g., give). This syntactic bootstrapping (Christophe et al., 2008; Gleitman and Gleitman, 1994) of verb argument structure and meaning is another illustration of the interdependence of different levels of language.

Toddlers' growing syntactic knowledge has also been investigated electrophysiologically by assessing when and how ERP components known from adult syntactic processing emerge. The most frequently investigated components in relation to syntactic processing in adults are the Left Anterior Negativity (LAN) and the P600. The LAN peaks at around 200 ms after the word violating the syntactic structure of a sentence and it reflects the violation of syntactic expectations based on the preceding context. The P600, which is a positive shift peaking at around 600 ms, reflects a controlled process and appears when a sustained period is needed for the integration of syntactic with other information or reanalysis is required (Friederici, 1995, 2002; Hahne and Friederici, 1999). Studies show that a late positive shift, which can be interpreted as the toddler variant of the P600, appears earlier than the LAN, at 24 months (Oberecker and Friederici, 2006), suggesting that the rule-based analysis of sentences is already present at that age. This toddler P600, has a longer latency than in adults and its spatial distribution is also different, suggesting that the process becomes adult-like later (Silva-Pereyra et al., 2005b). The earliest age in which a LAN-like component was observed for syntactic violation was 32 months (Oberecker et al., 2005), suggesting that automatic syntactic processing appears later than controlled processes.

Although the previous studies indicate that toddlers are sensitive to syntactic violations, and they are able to analyze the syntactic structure of the sentence, it was also pointed out (Brusini et al., 2016) that in these paradigms errors could be detected without syntactic analysis, based on surface word strings stored in memory. Grammatical word sequences should be familiar to children, while violations are never or rarely heard in the input. To exclude this possibility, the authors conducted a similar passive listening experiment as in the studies presented previously but they used pseudowords in the sentences. Toddlers learnt the words 1 week before the ERP experiment. An early left anterior negative wave at about 100–400 ms and a late posterior positive shift at about 700–900 ms were found after word onset to grammatically incorrect words, which were very similar to the LAN and P600 components shown by adults for syntactic violations.

Taken together, the previous studies suggest that infants start acquiring grammar alongside words and phoneme categories during the second half of the first year of life, and by toddlerhood, they exhibit sophisticated morphosyntactic knowledge of their native language.

CONCLUSIONS

We have reviewed the most important empirical findings exploring how language develops from prenatal experience to the toddler years. Our knowledge of how the infant mind and brain perceive, process, and learn language have increased considerably in the last decades, due to improved developmental and imaging techniques, suitable to work even with the youngest infants.

These results suggest that some of our previous concepts of language development, such as the sequential view whereby infants first learn sounds, then words and then grammar, need to be revised in favor of a more integrative view emphasizing the mutual relationships between different aspects of language.

Furthermore, language development is viewed in an increasingly biological manner, supported by genetic, epigenetic, and imaging studies that allow us to start to understand how biologically endowed abilities and experience synergistically interact during many possible developmental trajectories to language.

ACKNOWLEDGMENTS

The authors were supported by the ERC Consolidator Grant (773202 ERC-2017-COG "BabyRhythm"; https://erc.europa.eu/funding/consolidator-grants) and the ECOS-Sud grant nr. C20S02 to J. G. and the European Union's Horizon 2020 research and innovation program under the Marie Sklodowska-Curie grant agreement No. 641858 (PredictAble project). E. L. was also supported by funding from the National Institutes of Health: the NIH Common Fund through the Office of NIH Director, and the National Institute on Deafness and Other Communication Disorders, under Award Numbers DP2HD098859 and R01DC016977. The content is solely the responsibility of the authors and does not necessarily represent the official views of the NIH.

REFERENCES

Abboub, N., Nazzi, T., and Gervain, J. (2016). Prosodic grouping at birth. *Brain and Language*, 162, 46–59.

Albareda-Castellot, B., Pons, F., and Sebastian-Galles, N. (2011). Acquisition of phonetic categories in bilingual infants: new data from an anticipatory eye movement paradigm. *Developmental Science*, 14(2), 395–401.

Alho, K., Connolly, J. F., Cheour, M., Lehtokoski, A., Huotilainen, M., Virtanen, J., Aulanko, R., and Ilmoniemi, R. J. (1998). Hemispheric lateralization in preattentive processing of speech sounds. *Neuroscience Letters*, 258, 9–12.

Armitage, S. E., Baldwin, B. A., and Vince, M. A. (1980). The fetal sound environment of sheep. *Science*, 208, 1173–1174.

Aslin, R. N. and Mehler, J. (2005). Near-infrared spectroscopy for functional studies of brain activity in human infants: Promise, prospects, and challenges. *Journal of Biomedical Optics*, 10, 11009.

Au, T. K.-F. and Glusman, M. (1990). The principle of mutual exclusivity in word learning: To honor or not to honor? *Child Development*, 61, 1474–1490.

Baker, S. A., Golinkoff, R. M., and Petitto, L. A. (2006). New insights into old puzzles from infants' categorical discrimination of silent phonetic units. *Language Learning and Development*, 2, 147–162.

Benavides-Varela, S. and Gervain, J. (2017). Learning word order at birth: A NIRS study. *Developmental Cognitive Neuroscience*, 25, 198–208.

Benavides-Varela, S., Siugzdaite, R., Gómez, D. M., Macagno, F., Cattarossi, L., and Mehler, J. (2017). Brain regions and functional interactions supporting early word recognition in the face of input variability. *Proceedings of the National Academy of Sciences*, 114(29), 7588–7593.

Bergelson, E. and Swingley, D. (2012). At 6–9 months, human infants know the meanings of many common nouns. *Proceedings of the National Academy of Sciences*, 109, 3253–3258.

Bernard, C. and Gervain, J. (2012). Prosodic cues to word order: What level of representation? *Frontiers in Psychology*, 3, 451.

Bertoncini, J., Morais, J., Bijeljac-Babic, R., McAdams, S., Peretz, I., and Mehler, J. (1989). Dichotic perception and laterality in neonates. *Brain and Language*, 37(4), 591–605.

Best, C. T., Hoffman, H., and Glanville, B. B. (1982). Development of infant ear asymmetries for speech and music. *Perception and Psychophysics*, 31(1), 75–85.

Best, C. T., McRoberts, G. W., and Sithole, N. M. (1988). Examination of perceptual reorganization for nonnative speech contrasts: Zulu click discrimination by English-speaking adults and infants. *The Journal of Experimental Psychology: Human Perception and Performance*, 14, 345–360.

Boemio, A., Fromm, S., Braun, A., and Poeppel, D. (2005). Hierarchical and asymmetric temporal sensitivity in human auditory cortices. *Nature Neuroscience*, 8, 389–395.

Bortfeld, H., Morgan, J. L., Golinkoff, R. M., and Rathbun, K. (2005). Mommy and me: Familiar names help launch babies into speech-stream segmentation. *Psychological Science*, 16, 298–304.

Bosch, L. and Sebastian-Galles, N. (1997). Native-language recognition abilities in 4-month-old infants from monolingual and bilingual environments. *Cognition*, 65, 33–69.

Bosch, L. and Sebastian-Galles, N. (2003). Simultaneous bilingualism and the perception of a language-specific vowel contrast in the first year of life. *Language and Speech*, 46, 217–243.

Brusini, P., Dehaene-Lambertz, G., Dutat, M., Goffinet, F., and Christophe, A. (2016). ERP evidence for on-line syntactic computations in 2-year-olds. *Developmental Cognitive Neuroscience*, 19, 164–173.

Byers-Heinlein, K., Burns, T. C., and Werker, J. F. (2010). The roots of bilingualism in newborns. *Psychological Science*, 21(3), 343–348.

Byers-Heinlein, K. and Werker, J. F. (2009). Monolingual, bilingual, trilingual: Infants' language experience influences the development of a word learning heuristic. *Developmental Science*, 12, 815–823.

Cabrera, L. and Gervain, J. (2020). Speech perception at birth: The brain encodes fast and slow temporal information. *Science Advances*, 6(30), eaba7830.

Chen, Y. H., Saby, J., Kuschner, E., Gaetz, W., Edgar, J. C., and Roberts, T. (2019). Magnetoencephalography and the infant brain. *NeuroImage*, 189, 445–458.

Chiswick, P. B. R. and Miller, P. W. (2008). A test of the critical period hypothesis for language learning. *Journal of Multilingual and Multicultural Development*, 29, 16–29.

Chomsky, N. (1959). A review of B. F. Skinner's verbal behavior. *Language*, 35, 26–58.

Chomsky, N. (2007). Of minds and language. *Biolinguistics*, 1, 9–27.

Christophe, A., Dupoux, E., Bertoncini, J., and Mehler, J. (1994). Do infants perceive word boundaries? An empirical study of the bootstrapping of lexical acquisition. *Journal of Acoustic Society of America*, 95, 1570–1580.

Christophe, A., Millotte, S., Bernal, S., and Lidz, J. (2008). Bootstrapping lexical and syntactic acquisition. *Language and Speech*, 51, 61–75.

Cristia, A., Minagawa-Kawai, Y., Egorova, N., Gervain, J., Filippin, L., Cabrol, D., and Dupoux, E. (2014). Neural correlates of infant accent discrimination: An fNIRS study. *Developmental Science*, 17, 628–635.

Curtiss, S., Fromkin, V. A., Krashen, S., Rigler, D., and Rigler, M. (1974). The linguistic development of Genie. *Language*, 50, 528–554.

Cusack, R., Wild, C. J., Zubiaurre-Elorza, L., and Linke, A. C. (2018). Why does language not emerge until the second year? *Hearing Research*, 366, 75–81.

De Haan, M. (2013). *Infant EEG and Event-Related Potentials*. London: Psychology Press.

de la Cruz Pavia, I., Marino, C., and Gervain, J. (2021). Learning Word Order: Early Beginnings. *Trends in Cognitive Sciences*, 25(9), 802–812.

DeCasper, A. J. and Fifer, W. P. (1980). Of human bonding: Newborns prefer their mothers' voices. *Science*, 208, 1174–1176.

DeCasper, A. J., Lecanuet, J.-P., Busnel, M. C., and Granier-Deferre, C. (1994). Fetal reactions to recurrent maternal speech. *Infant Behavior & Development*, 17, 159–164.

Dehaene-Lambertz, G. and Baillet, S. (1998). A phonological representation in the infant brain. *Neuroreport*, 9, 1885–1888.

Dehaene-Lambertz, G., Dehaene, S., and Hertz-Pannier, L. (2002). Functional neuroimaging of speech perception in infants. *Science*, 298, 2013–2015.

Dehaene-Lambertz, G. and Gliga, T. (2004). Common neural basis for phoneme processing in infants and adults. *Journal of Cognitive Neuroscience*, 16(8), 1375–1387.

Dehaene-Lambertz, G., Montavont, A., Jobert, A., Allirol, L., Dubois, J., Hertz-Pannier, L., and Dehaene, S. (2010). Language or music, mother or Mozart? Structural and environmental influences on infants' language networks. *Brain and Language*, 114, 53–65.

Dellwo, V. (2006). Rhythm and speech rate: A variation coefficient for Δ C. In P. Karnowski and I. Szigeti (Eds.), *Language and Language Processing: Proceedings of the 38th Linguistic Colloquium, Piliscsaba 2003* (pp. 231–241). Frankfurt: Peter Lang.

Draganova, R., Eswaran, H., Murphy, P., Huotilainen, M., Lowery, C., and Preissl, H. (2005). Sound frequency change detection in fetuses and newborns, a magnetoencephalographic study. *Neuroimage*, 28(2), 354–361.

Draganova, R., Eswaran, H., Murphy, P., Lowery, C., and Preissl, H. (2007). Serial magnetoencephalographic study of fetal and newborn auditory discriminative evoked responses. *Early Human Development*, 83(3), 199–207.

Dubois, J., Dehaene-Lambertz, G., Kulikova, S., Poupon, C., Hüppi, P. S., and Hertz-Pannier, L. (2014). The early development of brain white matter: A review of imaging studies in fetuses, newborns and infants. *Neuroscience*, 276, 48–71.

Dubois, J., Hertz-Pannier, L., Cachia, A., Mangin, J. F., Le Bihan, D., and Dehaene-Lambertz, G. (2009). Structural asymmetries in the infant language and sensori-motor networks. *Cerebral Cortex*, 19(2), 414–423.

Dubois, J., Poupon, C., Thirion, B., Simonnet, H., Kulikova, S., Leroy, F., Hertz-Pannier, L., and Dehaene-Lambertz, G. (2016). Exploring the early organization and maturation of linguistic pathways in the human infant brain. *Cerebral Cortex*, 26(5), 2283–2298.

Eggermont, J. J. and Moore, J. K. (2012). Morphological and Functional Development of the Auditory Nervous System. In L. Werner, R. R. Fay, and A. N. Popper (Eds.), *Human Auditory Development* (pp. 61–105). New York: Springer.

Eimas, P. D., Siqueland, E. R., Jusczyk, P. W., and Vigorito, J. (1971). Speech perception in infants. *Science*, 171, 303–306.

Endress, A. D., Nespor, M., and Mehler, J. (2009). Perceptual and memory constraints on language acquisition. *Trends in Cognitive Sciences*, 13, 348–353.

Endress, A. D., Scholl, B. J., and Mehler, J. (2005). The role of salience in the extraction of algebraic rules. *The Journal of Experimental Psychology*, 134, 406–419.

Federmeier, K. D. and Kutas, M. (1999). A rose by any other name: Long-term memory structure and sentence processing. *Journal of Memory and Language*, 41(4), 469–495.

Fenson, L., Dale, P., Reznick, S., Thal, D., Bates, E., Hartung, J., et al. (1993). *The MacArthur Communicative Development Inventories: User's guide and technical manual*. San Diego, CA: Singular Publishing Group.

Fló, A., Brusini, P., Macagno, F., Nespor, M., Mehler, J., and Ferry, A. L. (2019). Newborns are sensitive to multiple cues for word segmentation in continuous speech. *Developmental Science*, 22, e12802.

Floccia, C., Christophe, A., and Bertoncini, J. (1997). High-amplitude sucking and newborns: The quest for underlying mechanisms. *Journal of Experimental Child Psychology*, 64(2), 175–198.

Flynn, S. and Manuel, S. (1991). Age-dependent effects in language acquisition: An evaluation of critical period hypotheses. In L. Eubank (Ed.), *Point Counterpoint: Universal Grammar in the Second Language* (pp. 117–145). Amsterdam, Philadelphia: John Benjamins Press.

Forgács, B., Parise, E., Csibra, G., Gergely, G., Jacquey, L., and Gervain, J. (2019). Fourteen-month-old infants track the language comprehension of communicative partners. *Developmental Science*, 22, e12751.

Frank, M. C., Slemmer, J. A., Marcus, G. F., and Johnson, S. P. (2009). Information from multiple modalities helps five-month-olds learn abstract rules. *Developmental Science*, 12, 504–509.

Friederici, A. D. (1995). The time course of syntactic activation during language processing: A model based on neuropsychological and neurophysiological data. *Brain and Language*, 50(3), 259–281.

Friederici, A. D. (2002). Towards a neural basis of auditory sentence processing. *Trends in Cognitive Sciences*, 6(2), 78–84.

Friederici, A. D. (2005). Neurophysiological markers of early language acquisition: From syllables to sentences. *Trends in Cognitive Sciences*, 9, 481–488.

Friederici, A. D. (2012). The cortical language circuit: From auditory perception to sentence comprehension. *Trends in Cognitive Sciences*, 16(5), 262–268.

Friederici, A. D., Friedrich, M., and Christophe, A. (2007). Brain responses in 4-month-old infants are already language specific. *Current Biology*, 17, 1208–1211.

Friederici, A. D., Mueller, J. L., and Oberecker, R. (2011). Precursors to natural grammar learning: Preliminary evidence from 4-month-old infants. *PLoS One*, 6(3), e17920.

Friedrich, M. and Friederici, A. D. (2006). Early N400 development and later language acquisition. *Psychophysiology*, 43(1), 1–12.

Friedrich, M. and Friederici, A. D. (2004). N400-like semantic incongruity effect in 19-month-olds: Processing known words in picture contexts. *Journal of Cognitive Neuroscience*, 16(8), 1465–1477.

Friedrich, M. and Friederici, A. D. (2005a). Phonotactic knowledge and lexical-semantic processing in one-year-olds: Brain responses to words and nonsense words in picture contexts. *Journal of Cognitive Neuroscience*, 17(11), 1785–1802.

Friedrich, M. and Friederici, A. D. (2005b). Lexical priming and semantic integration reflected in the event-related potential of 14-month-olds. *Neuroreport*, 16(6), 653–656.

Friedrich, M. and Friederici, A. D. (2005c). Semantic sentence processing reflected in the event-related potentials of one-and two-year-old children. *Neuroreport*, 16(16), 1801–1804.

Friedrich, M. (2017). ERP indices of word learning: What do they reflect and what do they tell us about the neural representations of early words? In N. Mani and G. Westermann (Eds.), *Early Word Learning* (pp. 123–137). New York: Routledge.

Gelman, R. (1991). Epigenetic foundations of knowledge structures: Initial and transcendent constructions. In S. Carey and R. Gelman (Eds.), *The Epigenesis of Mind: Essays on Biology and Cognition* (pp. 293–322). New York: Lawrence Erlbaum Associates.

Gerhardt, K. J., Abrams, R. M., and Oliver, C. C. (1990). Sound environment of the fetal sheep. *American Journal of Obstetrics and Gynecology*, 162, 282–287.

Gervain, J. (2015). Plasticity in early language acquisition: The effects of prenatal and early childhood experience. *Current Opinion in Neurobiology*, 35, 13–20.

Gervain, J. (2018). The role of prenatal experience in language development. *Current Opinions in Behavioral Science*, 21, 62–67.

Gervain, J. (2022). The early acquisition of morphology in agglutinating languages: The case of Hungarian. In G. Csibra, J. Gervain, and K. Kovács (Eds.), *A Life in Cognition* (pp. 109–123). Cham: Springer.

Gervain, J., Berent, I., and Werker, J. F. (2012). Binding at birth: The newborn brain detects identity relations and sequential position in speech. *Journal of Cognitive Neuroscience*, 24(3), 564–574.

Gervain, J., de la Cruz Pavia, I., and Gerken, L. (2018). Behavioural and imaging studies of infant artificial grammar learning. *Topics in Cognitive Science*, 12(3), 815–827.

Gervain, J., Macagno, F., Cogoi, S., Pena, M., and Mehler, J. (2008). The neonate brain detects speech structure. *Proceedings of the National Academy of Sciences*, 105, 14222–14227.

Gervain, J. and Mehler, J. (2010). Speech perception and language acquisition in the first year of life. *Annual Review of Psychology*, 61, 191–218.

Gervain, J., Mehler, J., Werker, J. F., Nelson, C. A., Csibra, G., Lloyd-Fox, S., Shukla, M., and Aslin, R. N. (2011). Near-infrared spectroscopy: A report from the McDonnell infant methodology consortium. *Developmental Cognitive Neuroscience*, 1(1), 22–46.

Gervain, J., Vines, B. W., Chen, L. M., Seo, R. J., Hensch, T. K., Werker, J. F., and Young, A. H., (2013). Valproate reopens critical-period learning of absolute pitch. *Frontiers in Systems in Neuroscience*, 7, 102.

Giraud, A.-L., Kleinschmidt, A., Poeppel, D., Lund, T. E., Frackowiak, R. S., and Laufs, H. (2007). Endogenous cortical rhythms determine cerebral specialization for speech perception and production. *Neuron*, 56, 1127–1134.

Giraud, A. L. and Poeppel, D. (2012). Cortical oscillations and speech processing: Emerging computational principles and operations. *Nature Neuroscience*, 15, 511–517.

Glasel, H., Leroy, F., Dubois, J., Hertz-Pannier, L., Mangin, J. F., and Dehaene-Lambertz, G. (2011). A robust cerebral asymmetry in the infant brain: The rightward superior temporal sulcus. *NeuroImage*, 58(3), 716–723.

Gleitman, L. R. and Gleitman, H. (1994). A picture is worth a thousand words, but that's the problem: The role of syntax in vocabulary acquisition. In B. Lust, M. Suñer, and J. Whitman

(Eds.), *Syntactic Theory and First Language Acquisition: Cross Linguistic Perspectives, Vol. 1. Heads, Projections, and Learnability* (pp. 291–299). Hillsdale, NJ, USA: Lawrence Erlbaum Associates, Inc.

Gómez, R. and Maye, J. (2005). The developmental trajectory of nonadjacent dependency learning. *Infancy*, 7(2), 183–206.

Gomez, R. L. (2002). Variability and detection of invariant structure. *Psychological Science*, 13(5), 431–436.

Gomez, R. L. and Gerken, L. (2000). Infant artificial language learning and language acquisition. *Trends in Cognitive Sciences*, 4, 178–186.

Gómez, R. L. and Lakusta, L. (2004). A first step in form-based category abstraction by 12-month-old infants. *Developmental Science*, 7, 567–580.

Goodsitt, J. V., Morgan, J. L., and Kuhl, P. K. (1993). Perceptual strategies in prelingual speech segmentation. *Journal of Child Language*, 20(2), 229–252.

Govindan, R. B., Wilson, J. D., Preissl, H., Murphy, P., Lowery, C. L., and Eswaran, H. (2008). An objective assessment of fetal and neonatal auditory evoked responses. *Neuroimage*, 43(3), 521–527.

Goyet, L., de Schonen, S., and Nazzi, T. (2010). Words and syllables in fluent speech segmentation by French-learning infants: An ERP study. *Brain Research*, 1332, 75–89.

Grabe, E. and Low, E. L. (2002). Durational Variability in Speech and the Rhythm Class Hypothesis. In C. Gussenhoven and N. Warner (Eds.), *Papers in Laboratory Phonology 7* (pp. 515–546). Berlin: Mouton de Gruyter.

Hahne, A. and Friederici, A. D. (1999). Electrophysiological evidence for two steps in syntactic analysis: Early automatic and late controlled processes. *Journal of Cognitive Neuroscience*, 11(2), 194–205.

Harris, Z. (1955). From phoneme to morpheme. *Language*, 31, 190–222.

Hartkopf, J., Schleger, F., Weiss, M., Hertrich, I., Kiefer-Schmidt, I., Preissl, H., and Muenssinger, J. (2016). Neuromagnetic signatures of syllable processing in fetuses and infants provide no evidence for habituation. *Early Human Development*, 100, 61–66.

Hensch, T. K. (2003). Controlling the critical period. *Neuroscience Research*, 47, 17–22.

Hensch, T. K. (2005). Critical period plasticity in local cortical circuits. *Nature Reviews Neuroscience*, 6, 877–888.

Hickok, G. and Poeppel, D. (2007). The cortical organization of speech processing. *Nature Reviews Neuroscience*, 8, 393–402.

Hochmann, J.-R., Endress, A. D., and Mehler, J. (2010). Word Frequency as a Cue for Identifying Function Words in Infancy. *Cognition*, 115(3), 444–457.

Höhle, B., Bijeljac-Babic, R., Herold, B., Weissenborn, J., and Nazzi, T. (2009). Language specific prosodic preferences during the first half year of life: Evidence from German and French infants. *Infant Behavior and Development*, 32(3), 262–274.

Hohle, B., Schmitz, M., Santelmann, L. M., and Weissenborn, J. (2006). The recognition of discontinuous verbal dependencies by German 19-month-olds: Evidence for lexical and structural influences on children's early processing capacities. *Language Learning and Development*, 2, 277–300.

Höhle, B. and Weissenborn, J. (2003). German-learning infants' ability to detect unstressed closed-class elements in continuous speech. *Developmental Science*, 6, 122–127.

Hohne, E. A. and Jusczyk, P. W. (1994). Two-month-old infants' sensitivity to allophonic differences. *Perception & Psychophysics*, 56, 613–623.

Holst, M., Eswaran, H., Lowery, C., Murphy, P., Norton, J., and Preissl, H. (2005). Development of auditory evoked fields in human fetuses and newborns: A longitudinal MEG study. *Clinical Neurophysiology*, 116(8), 1949–1955.

Homae, F., Watanabe, H., Nakano, T., and Taga, G. (2011). Large-scale brain networks underlying language acquisition in early infancy. *Frontiers in Psychology*, 2, 93.

Homae, F., Watanabe, H., Otobe, T., Nakano, T., Go, T., Konishi, Y., and Taga, G. (2010). Development of global cortical networks in early infancy. *Journal of Neuroscience*, 30, 4877–4882.

Johnson, E. K. and Tyler, M. D. (2010). Testing the limits of statistical learning for word segmentation. *Developmental Science*, 13(2), 339–345.

Johnson, J. S. and Newport, E. L. (1989). Critical period effects in second language learning: The influence of maturational state on the acquisition of English as a second language. *Cognitive Psychology*, 21, 60–99.

Jusczyk, P. W., Houston, D. M., and Newsome, M. (1999a). The beginnings of word segmentation in English-learning infants. *Cognitive Psychology*, 39(3–4), 159–207.

Jusczyk, P. W., Hohne, E. A., and Bauman, A. (1999b). Infants' sensitivity to allophonic cues to word segmentation. *Perception & Psychophysics*, 61, 1465–1476.

Kabdebon, C., Pena, M., Buiatti, M., and Dehaene-Lambertz, G. (2015). Electrophysiological evidence of statistical learning of long-distance dependencies in 8-month-old preterm and full-term infants. *Brain and Language*, 148, 25–36.

Kalashnikova, M., Peter, V., Di Liberto, G., Lalor, E., and Burnham, D. (2018). Infant-directed speech facilitates seven-month-old infants' cortical tracking of speech. *Scientific Reports*, 8, 13745.

Kisilevsky, B. S., Hains, S. M. J., Brown, C. A., Lee, C. T., Cowperthwaite, B., Stutzman, S. S., Swansburg, M. L., Lee, K., Xie, X., Huang, H., et al. (2009). Fetal sensitivity to properties of maternal speech and language. *Infant Behavior & Development*, 32, 59–71.

Kooijman, V., Hagoort, P., and Cutler, A. (2005). Electrophysiological evidence for prelinguistic infants' word recognition in continuous speech. *Cognitive Brain Research*, 24(1), 109–116.

Kooijman, V., Hagoort, P., and Cutler, A. (2009). Prosodic structure in early word segmentation: ERP evidence from Dutch ten-month-olds. *Infancy*, 14(6), 591–612.

Kuhl, P. and Rivera-Gaxiola, M. (2008). Neural substrates of language acquisition. *Annual Review of Neuroscience*, 31, 511–534.

Kuhl, P. K. (1981). Discrimination of speech by nonhuman animals: Basic auditory sensitivities conducive to the perception of speech-sound categories. *Journal of the Acoustical Society of America*, 70, 340–349.

Kuhl, P. K. (1986). Theoretical contributions of tests on animals to the special-mechanisms debate in speech. *Experimental Biology*, 45, 233–265.

Kuhl, P. K., Stevens, E., Hayashi, A., Deguchi, T., Kiritani, S., and Iverson, P. (2006). Infants show a facilitation effect for native language phonetic perception between 6 and 12 months. *Developmental Science*, 9, F13–F21.

Kuhl, P. K., Williams, K. A., Lacerda, F., Stevens, K. N., and Lindblom, B. (1992). Linguistic experience alters phonetic perception in infants by 6 months of age. *Science*, 255, 606–608.

Kühn-Popp, N., Kristen, S., Paulus, M., Meinhardt, J., and Sodian, B. (2016). Left hemisphere EEG coherence in infancy predicts infant declarative pointing and preschool epistemic language. *Social Neuroscience*, 11(1), 49–59.

Kuijpers, C. T. L., Coolen, R., Houston, D., and Cutler, A. (1998). Using the headturning technique to explore cross-linguistic performance differences. In C. Rovee-Collier, L. Lipsitt, and H. Hayne (Eds.), *Advances in Infancy Research, vol. 12* (pp. 205–220). Stanford, CT: Ablex Publishers.

Kujala, A., Huotilainen, M., Hotakainen, M., Lennes, M., Parkkonen, L., Fellman, V., and Näätänen, R. (2004). Speech-sound discrimination in neonates as measured with MEG. *NeuroReport*, 15, 2089–2092.

Kutas, M. and Federmeier, K. D. (2011). Thirty years and counting: Finding meaning in the N400 component of the event-related brain potential (ERP). *Annual Review of Psychology*, 62, 621–647.

Ladányi, E., Kovács, Á. M., and Gervain, J. (2020). How 15-month-old infants process morphologically complex forms in an agglutinative language? *Infancy*, 25, 190–204.

Lakatos, P., Shah, A. S., Knuth, K. H., Ulbert, I., Karmos, G., Schroeder, and C. E. (2005). An oscillatory hierarchy controlling neuronal excitability and stimulus processing in the auditory cortex. *Journal of Neurophysiology*, 94, 1904–1911.

Langus, A., Mehler, J., and Nespor, M. (2016). Rhythm in language acquisition. *Neuroscience & Biobehavioral Reviews*, 81, 158–166.

Lecanuet, J. P. and Granier-Deferre, C. (1993). Speech Stimuli in the Fetal Environment. In B. de Boysson-Bardies et al. (Eds.), *Developmental Neurocognition: Speech and Face Processing in the First Year of Life* (pp. 237–248). Dordrecht: Springer.

Lengle, J. M., Chen, M., and Wakai, R. T. (2001). Improved neuromagnetic detection of fetal and neonatal auditory evoked responses. *Clinical Neurophysiology*, 112(5), 785–792.

Lenneberg, E. H. (1967). *The Biological Foundations of Language*. New York: Wiley Press.

Lloyd-Fox, S., Blasi, A., and Elwell, C. E. (2009). Illuminating the developing brain: The past, present and future of functional near infrared spectroscopy. *Neuroscience & Biobehavioral Reviews*, 34(3), 269–284.

Loukina, A., Kochanski, G., Rosner, B., Keane, E., and Shih, C. (2011). Rhythm measures and dimensions of durational variation in speech. *Journal of the Acoustical Society of America*, 129, 3258–3270.

Lutter, W. J., Maier, M., and Wakai, R. T. (2006). Development of MEG sleep patterns and magnetic auditory evoked responses during early infancy. *Clinical Neurophysiology*, 117(3), 522–530.

Mahmoudzadeh, M., Dehaene-Lambertz, G., Fournier, M., Kongolo, G., Goudjil, S., Dubois, J., Grebe, R., and Wallois, F. (2013). Syllabic discrimination in premature human infants prior to complete formation of cortical layers. *Proceedings of the National Academy of Sciences*, 110, 4846–4851.

Mandel, D. R., Kemler Nelson, D. G., and Jusczyk, P. W. (1996). Infants remember the order of words in a spoken sentence. *Cognitive Development*, 11, 181–196.

Marchetto, E. and Bonatti, L. L. (2015). Finding words and word structure in artificial speech: The development of infants' sensitivity to morphosyntactic regularities. *Journal of Child Language*, 42(4), 873–902.

Marcus, G. F., Vijayan, S., Rao, S. B., and Vishton, P. M. (1999). Rule learning by seven-month-old infants. *Science*, 283, 77–80.

Marino, C., Bernard, C., and Gervain, J. (2020). Word Frequency Is a Cue to Lexical Category for 8-Month-Old Infants. *Current Biology*, 30(8), 1380–1386.

Markman, E. M. (1991). The whole-object, taxonomic, and mutual exclusivity assumptions as initial constraints on word meanings. In S. A. Gelman and J. P. Byrnes (Eds.), *Perspectives on*

Language and Thought: Interrelations in Development (pp. 72–106). New York: Cambridge University Press.

Markman, E. M. (1994). Constraints on word meaning in early language acquisition. *Lingua*, 92, 199–227.

Markman, E. M. and Wachtel, G. F. (1988). Children's use of mutual exclusivity to constrain the meanings of words. *Cognitive Psychology*, 20, 121–157.

Markman, T. M., Quittner, A. L., Eisenberg, L. S., Tobey, E. A., Thal, D., Niparko, J. K., Wang, N.-Y., and Team, Cd. I. (2011). Language development after cochlear implantation: An epigenetic model. *Journal of Neurodevelopmental Disorders*, 3(4), 388–404.

Marquis, A. and Shi, R. (2012). Initial morphological learning in preverbal infants. *Cognition*, 122, 61–66.

Mattock, K. and Burnham, D. (2006). Chinese and English Infants' tone perception: Evidence for perceptual reorganization. *Infancy*, 10, 241–265.

Mattys, S. L. and Jusczyk, P. W. (2001). Phonotactic cues for segmentation of fluent speech by infants. *Cognition*, 78, 91–121.

Mattys, S. L., Jusczyk, P. W., Luce, P. A., and Morgan, J. L. (1999). Phonotactic and prosodic effects on word segmentation in infants. *Cognitive Psychology*, 38, 465–494.

Maurer, D., Ghloum, J. K., Gibson, L. C., Watson, M. R., Chen, L. M., Akins, K., Enns, J. T., Hensch, T. K., and Werker, J. F. (2020). Reduced perceptual narrowing in synesthesia. *Proceedings of the National Academy of Sciences*, 117, 10089–10096.

Maurer, D. and Werker, J. F. (2014). Perceptual narrowing during infancy: A comparison of language and faces. *Developmental Psychobiology*, 56, 154–178.

May, L., Byers-Heinlein, K., Gervain, J., and Werker, J. F. (2011). Language and the newborn brain: Does prenatal language experience shape the neonate neural response to speech? *Frontiers in Psychology*, 2, 222.

May, L., Gervain, J., Carreiras, M., and Werker, J. F. (2018). The specificity of the neural response to speech at birth. *Developmental Science*, 21, e12564.

McMurray, B. and Aslin, R. N. (2005). Infants are sensitive to within-category variation in speech perception. *Cognition*, 95, B15–B26.

Mehler, J., Jusczyk, P. W., Lambertz, G., Halsted, N., Bertoncini, J., and Amiel-Tison, C. (1988). A precursor of language acquisition in young infants. *Cognition*, 29, 143–178.

Merriman, W. E. (1991). The mutual exclusivity bias in children's word learning: A reply to Woodward and Markman. *Developmental Review*, 11, 164–191.

Mills, D. L., Coffey-Corina S., and Neville, H. J. (1997). Language comprehension and cerebral specialization from 13–20 months. *Developmental Neuropsychology*, 13, 233–237.

Mills, D. L., Coffey-Corina, S. A., and Neville, H. J. (1993). Language acquisition and cerebral specialization in 20-month-old infants. *Journal of Cognitive Neuroscience*, 5, 317–334.

Mills, D. L., Plunkett K., Prat, C., and Schafer, G. (2005). Watching the infant brain learn words: Effects of vocabulary size and experience. *Cognitive Development*, 20, 19–31.

Minagawa-Kawai, Y., Cristià, A., and Dupoux, E. (2011). Cerebral lateralization and early speech acquisition: A developmental scenario. *Developmental Cognitive Neuroscience*, 1(3), 217–232.

Minagawa-Kawai, Y., Mori, K., Naoi, N., and Kojima, S. (2007). Neural attunement processes in infants during the acquisition of a language-specific phonemic contrast. *Journal of Neuroscience*, 27, 315–321.

Mintz, T. H. (2013). The segmentation of sub-lexical morphemes in English-learning 15-month-olds. *Frontiers in Psychology*, 4, 24.

Molavi, B., May, L., Gervain, J., Carreiras, M., Werker, J. F., and Dumont, G. A. (2014). Analyzing the resting state functional connectivity in the human language system using near infrared spectroscopy. *Frontiers in Human Neuroscience*, 7, 921.

Molfese, D. L. (1990). Auditory evoked responses recorded from 16-month-old human infants to words they did and did not know. *Brain and Language*, 38(3), 345–363.

Molfese, D. L., Morse, P. A., and Peters, C. J. (1990). Auditory evoked responses to names for different objects: Cross-modal processing as a basis for infant language acquisition. *Developmental Psychology*, 26(5), 780.

Molfese, D. L., Wetzel, W. F., and Gill, L. A. (1993). Known versus unknown word discriminations in 12-month-old human infants: Electrophysiological correlates. *Developmental Neuropsychology*, 9(3–4), 241–258.

Molnar, M., Gervain, J., and Carreiras, M. (2013). Within-rhythm class native language discrimination abilities of Basque-Spanish monolingual and bilingual infants at 3.5 months of age. *Infancy*, 19(3), 326–337.

Moon, C., Cooper, R. P., Fifer, and W. P. (1993). Two-day-olds prefer their native language. *Infant Behavior and Development*, 16, 495–500.

Moon, C., Lagercrantz, H., and Kuhl, P. K. (2013). Language experienced in utero affects vowel perception after birth: A two-country study. *Acta Paediatrica*, 102(2), 156–160.

Morgan, E. U., van der Meer, A., Vulchanova, M., Blasi, D. E., and Baggio, G. (2020). Meaning before grammar: A review of ERP experiments on the neurodevelopmental origins of semantic processing. *Psychonomic Bulletin & Review*, 27(3), 441–464.

Mueller, J. L., Friederici, A. D., and Männel, C. (2012). Auditory perception at the root of language learning. *Proceedings of the National Academy of Sciences*, 109(39), 15953–15958.

Muenssinger, J., Matuz, T., Schleger, F., Kiefer-Schmidt, I., Goelz, R., Wacker-Gussmann, A., Birbaumer, N., and Preissl, H. (2013). Auditory habituation in the fetus and neonate: An fMEG study. *Developmental Science*, 16(2), 287–295.

Mundy, P., Fox, N., and Card, J. (2003). EEG coherence, joint attention and language development in the second year. *Developmental Science*, 6(1), 48–54.

Narayan, C. R., Werker, J. F., and Beddor, P. S. (2010). The interaction between acoustic salience and language experience in developmental speech perception: Evidence from nasal place discrimination. *Developmental Science*, 13, 407–420.

Nazzi, T. and Bertoncini, J. (2003). Before and after the vocabulary spurt: Two modes of word acquisition? *Developmental Science*, 6, 136–142.

Nazzi, T., Bertoncini, J., and Mehler, J. (1998). Language discrimination by newborns: Toward an understanding of the role of rhythm. *The Journal of Experimental Psychology. Human Perception Perform*, 24, 756–766.

Nazzi, T., Iakimova, G., Bertoncini, J., Frédonie, S., and Alcantara, C. (2006). Early segmentation of fluent speech by infants acquiring French: Emerging evidence for crosslinguistic differences. *Journal of Memory and Language*, 54(3), 283–299.

Nelson, C. A., 3rd, Zeanah, C. H., and Fox, N. A. (2019). How early experience shapes human development: The case of psychosocial deprivation. *Neural Plasticity*, 2019, 1676285.

Oberecker, R. and Friederici, A. D. (2006). Syntactic event-related potential components in 24-month-olds' sentence comprehension. *Neuroreport*, 17(10), 1017–1021.

Oberecker, R., Friedrich, M., and Friederici, A. D. (2005). Neural correlates of syntactic processing in two-year-olds. *Journal of Cognitive Neuroscience*, 17(10), 1667–1678.

Ortiz-Barajas, M. C. and Gervain, J. (2020). *Neural Oscillations and Speech Processing During the First Months of Life*. Presented at the vICIS Biennial Congress, Glasgow, Scotland (virtual).

Ortiz-Barajas, M. C., Guevara Erra, R., and Gervain, J. (2021). The origins and development of speech envelope tracking during the first months of life. *Developmental Cognitive Neuroscience*, 48, 100915.

Pallier, C., Dehaene, S., Poline, J.-B., LeBihan, D., Argenti, A.-M., Dupoux, E., and Mehler, J. (2003). Brain imaging of language plasticity in adopted adults: Can a second language replace the first? *Cerebral Cortex*, 13, 155–161.

Parise, E. and Csibra, G. (2012). Electrophysiological evidence for the understanding of maternal speech by 9-month-old infants. *Psychological Science*, 23(7), 728–733.

Partanen, E., Kujala, T., Näätänen, R., Liitola, A., Sambeth, A., and Huotilainen, M. (2013). Learning-induced neural plasticity of speech processing before birth. *Proceedings of the National Academy of Sciences*, 110, 15145–15150.

Pena, M., Maki, A., Kovacic, D., Dehaene-Lambertz, G., Koizumi, H., Bouquet, F., and Mehler, J. (2003). Sounds and silence: An optical topography study of language recognition at birth. *Proceedings of the National Academy of Sciences*, 100, 11702–11705.

Peña, M., Pittaluga, E., and Mehler, J. (2010). Language acquisition in premature and full-term infants. *Proceedings of the National Academy of Sciences*, 107, 3823–3828.

Peña, M., Werker, J. F., and Dehaene-Lambertz, G. (2012). Earlier speech exposure does not accelerate speech acquisition. *Journal of Neuroscience*, 32, 11159–11163.

Penfield, W. and Roberts, L. (1959). *Speech and Brain Mechanisms*. Princeton, NJ: Princeton University Press.

Pierce, L. J., Klein, D., Chen, J.-K., Delcenserie, A., and Genesee, F. (2014). Mapping the unconscious maintenance of a lost first language. *Proceedings of the National Academy of Sciences*, 111, 17314–17319.

Poeppel, D. (2014). The neuroanatomic and neurophysiological infrastructure for speech and language. *Current Opinion in Neurobiology*, 28, 142–149.

Quine, W. V. (1960). *Word and Object*. Cambridge, MA: MIT Press.

Rabagliati, H., Ferguson, B., and Lew-Williams, C. (2019). The profile of abstract rule learning in infancy: Meta-analytic and experimental evidence. *Developmental Science*, 22, e12704.

Rämä, P., Sirri, L., and Serres, J. (2013). Development of lexical-semantic language system: N400 priming effect for spoken words in 18- and 24-month old children. *Brain and Language*, 125(1), 1–10.

Ramus, F., Hauser, M. D., Miller, C., Morris, D., and Mehler, J. (2000). Language discrimination by human newborns and by cotton-top tamarin monkeys. *Science*, 288, 349–351.

Ramus, F., Nespor, M., and Mehler, J. (1999). Correlates of linguistic rhythm in the speech signal. *Cognition*, 73, 265–292.

Reh, R., Arredondo, M., and Werker, J. F. (2018). Understanding individual variation in levels of second language attainment through the lens of critical period mechanisms. *Bilingualism: Language Cognition*, 21(5), 930–931.

Righi, G., Tierney, A. L., Tager-Flusberg, H., and Nelson, C. A. (2014). Functional connectivity in the first year of life in infants at risk for autism spectrum disorder: An EEG study. *PLoS ONE*, 9(8), e105176.

Roberts, T. P., Lanza M. R., Dell J., Qasmieh S., Hines K., Blaskey L., Zarnow D. M., Levy S. E., Edgar J. C., and Berman J. I. (2013). Maturational differences in thalamocortical white matter microstructure and auditory evoked response latencies in autism spectrum disorders. *Brain Research*, 1537, 79–85.

Romberg, A. R. and Saffran, J. R. (2010). Statistical learning and language acquisition. *Wiley Interdisciplinary Reviews: Cognitive Science*, 1(6), 906–914.

Roth, T. L. and Sweatt, J. D. (2009). Regulation of chromatin structure in memory formation. *Current Opinion in Neurobiology*, 19, 336–342.

Saffran, J. R. and Kirkham, N. Z. (2018). Infant statistical learning. *Annual Review of Psychology*, 69, 181–203.

Saffran, J. R., Aslin, R. N., and Newport, E. L. (1996). Statistical learning by 8-month-old infants. *Science*, 274(5294), 1926–1928.

Saffran, J. R. and Thiessen, E. D. (2003). Pattern induction by infant language learners. *Developmental Psychology*, 39, 484–494.

Sansavini, A., Bertoncini, J., Giovanelli, G. (1997). Newborns discriminate the rhythm of multisyllabic stressed words. *Developmental Psychology*, 33, 3–11.

Sato, H., Hirabayashi, Y., Tsubokura, H., Kanai, M., Ashida, T., Konishi, I., Uchida-Ota, M., Konishi, Y., and Maki, A. (2012). Cerebral hemodynamics in newborn infants exposed to speech sounds: A whole-head optical topography study. *Human Brain Mapping*, 33, 2092–2103.

Sato, Y., Sogabe, Y., and Mazuka, R. (2010). Development of hemispheric specialization for lexical pitch-accent in Japanese infants. *Journal of Cognitive Neuroscience*, 22(11), 2503–2513.

Shafer, V. L., Shucard, D. W., Shucard, J. L., and Gerken, L. (1998). An electrophysiological study of infants' sensitivity to the sound patterns of English speech. *Journal of Speech and Language Hearing Research*, 41, 874–886.

Sheridan, C. J., Preissl, H., Siegel, E. R., Murphy, P., Ware, M., Lowery, C. L., and Eswaran, H. (2008). Neonatal and fetal response decrement of evoked responses: A MEG study. *Clinical Neurophysiology*, 119(4), 796–804.

Shi, R. and Melançon, A. (2010). Syntactic categorization in French-learning infants. *Infancy*, 15, 517–533.

Shi, R. and Werker, J. F. (2001). Six-month-old infants' preference for lexical words. *Psychological Science*, 12, 71–76.

Shi, R. and Werker, J. F. (2003). The basis of preference for lexical words in 6-month-old infants. *Developmental Science*, 6, 484–488.

Shi, R., Werker, J. F., and Morgan, J. L. (1999). Newborn infants' sensitivity to perceptual cues to lexical and grammatical words. *Cognition*, 72, B11–B21.

Shukla, M., White, K. S., and Aslin, R. N. (2011). Prosody guides the rapid mapping of auditory word forms onto visual objects in 6-mo-old infants. *Proceedings of the National Academy of Sciences*, 108, 6038.

Shultz, S., Vouloumanos, A., Bennett, R. H., and Pelphrey, K. (2014). Neural specialization for speech in the first months of life. *Developmental Science*, 17, 766–774.

Silva-Pereyra, J., Klarman, L., Lin, L. J. F., and Kuhl, P. K. (2005a). Sentence processing in 30-month-old children: An event-related potential study. *Neuroreport*, 16(6), 645–648.

Silva-Pereyra, J., Rivera-Gaxiola, M., and Kuhl, P. K. (2005b). An event-related brain potential study of sentence comprehension in preschoolers: Semantic and morphosyntactic processing. *Cognitive Brain Research*, 23(2–3), 247–258.

Swingley, D. (2009). Contributions of infant word learning to language development. *Philosophical Transactions of the Royal Society B: Biological Sciences*, 364(1536), 3617–3632.

Swingley, D. (2021). Infants' learning of speech sounds and word forms. In A. Papafragou, J. C. Trueswell, and L. R. Gleitman (Eds.), *Oxford Handbook of the Mental Lexicon* (pp. 267–291). Oxford: Oxford University Press.

Teinonen, T., Fellman, V., Naatanen, R., Alku, P., and Huotilainen, M. (2009). Statistical language learning in neonates revealed by event-related brain potentials. *BMC Neuroscience*, 10, 21.

Telkemeyer, S., Rossi, S., Koch, S. P., Nierhaus, T., Steinbrink, J., Poeppel, D., Obrig, H., and Wartenburger, I. (2009). Sensitivity of newborn auditory cortex to the temporal structure of sounds. *Journal of Neuroscience*, 29, 14726–14733.

Telkemeyer, S., Rossi, S., Nierhaus, T., Steinbrink, J., Obrig, H., and Wartenburger, I. (2011). Acoustic processing of temporally modulated sounds in infants: Evidence from a combined near-infrared spectroscopy and EEG study. *Frontiers in Psychology*, 1, 62.

Tervaniemi, M. and Hugdahl, K. (2003). Lateralization of auditory-cortex functions. *Brain Research Reviews*, 43, 231–246.

Thierry, G., Vihman, M., and Roberts, M. (2003). Familiar words capture the attention of 11-month-olds in less than 250 ms. *NeuroReport*, 14, 2307–2310.

Tincoff, R. and Jusczyk, P. W. (1999). Some beginnings of word comprehension in 6-month-olds. *Psychological Science*, 10, 172–175.

Travis, K. E., Leonard, M. K., Brown, T. T., Hagler Jr, D. J., Curran, M., Dale, A. M., Elman, J. L., and Halgren, E. (2011). Spatiotemporal neural dynamics of word understanding in 12- to 18-month-old-infants. *Cerebral Cortex*, 21(8), 1832–1839.

Ventureyra, V. A., Pallier, C., and Yoo, H.-Y. (2004). The loss of first language phonetic perception in adopted Koreans. *Journal of Neurolinguistics*, 17, 79–91.

von Koss Torkildsen, J., Sannerud, T., Syversen, G., Thormodsen, R., Simonsen, H. G., Moen, I., Smith, L., and Lindgren, M. (2006). Semantic organization of basic-level words in 20-month-olds: An ERP study. *Journal of Neurolinguistics*, 19(6), 431–454.

Vouloumanos, A., Hauser, M. D., Werker, J. F., and Martin, A. (2010). The tuning of human neonates' preference for speech. *Child Development*, 81, 517–527.

Vouloumanos, A. and Werker, J. F. (2004). Tuned to the signal: The privileged status of speech for young infants. *Developmental Science*, 7, 270.

Wallois, F., Mahmoudzadeh, M., Patil, A., and Grebe, R. (2012). Usefulness of simultaneous EEG–NIRS recording in language studies. *Brain and Language*, 121, 110–123.

Weikum, W. M., Oberlander, T. F., Hensch, T. K., and Werker, J. F. (2012). Prenatal exposure to antidepressants and depressed maternal mood alter trajectory of infant speech perception. *Proceedings of the National Academy of Sciences*, 109, 17221–17227.

Werker, J. F. (2018). Perceptual beginnings to language acquisition. *Applied Psycholinguistics*, 39, 703–728.

Werker, J. F. and Curtin, S. (2005). PRIMIR: A developmental model of speech processing. *Language Learning and Development*, 1, 197–234.

Werker, J. F. and Hensch, T. K. (2015). Critical periods in speech perception: New directions. *Annual Review of Psychology*, 66, 173–196.

Werker, J. F. and Tees, R. C. (1984). Cross-language speech perception: Evidence for perceptual reorganization during the first year of life. *Infant Behavior & Development*, 7, 49–63.

Werker, J. F. and Yeung, H. H. (2005). Infant speech perception bootstraps word learning. *Trends in Cognitive Sciences*, 9, 519–527.

Wiget, L., White, L., Schuppler, B., Grenon, I., Rauch, O., and Mattys, S. L. (2010). How stable are acoustic metrics of contrastive speech rhythm? *Journal of the Acoustical Society of America*, 127, 1559–1569.

Yeung, H. H., Chen, K. H., and Werker, J. F. (2013). When does native language input affect phonetic perception? The precocious case of lexical tone. *The Journal of Memory and Language*, 68, 123–139.

Yeung, H. H. and Werker, J. F. (2009). Learning words' sounds before learning how words sound: 9-Month-olds use distinct objects as cues to categorize speech information. *Cognition*, 113, 234–243.

Zatorre, R. J. and Belin, P. (2001). Spectral and temporal processing in human auditory cortex. *Cerebral Cortex*, 11, 946.

CHAPTER 23

NEURAL BASIS OF SPEECH AND LANGUAGE IMPAIRMENTS IN DEVELOPMENT

The case of developmental language disorder

SALOMI S. ASARIDOU AND KATE E. WATKINS

INTRODUCTION

DEVELOPMENTAL disorders of speech and language are relatively common (affecting 8–12.6% of children at school age; Law et al. 2000) yet poorly understood in terms of their neural basis or etiology. There is considerably less awareness among the public of these disorders especially in comparison with dyslexia, autism, and attention deficit hyperactivity disorder (ADHD) (Bishop et al. 2012). Research into these speech and language disorders also receives considerably less funding than those that are better known (Bishop 2010).

The causes of developmental disorders of speech and language are unknown. Environmental factors, including socio-economic disadvantage, fail to predict these disorders (Reilly et al. 2010). Biological factors, such as known hearing impairment or frank neurological findings that might explain the occurrence of speech or language acquisition problems are ruled out when diagnosing these disorders. With regard to potential neurological causes of these disorders, it is also worth noting that on the whole, speech and language learning difficulties are absent or mild in children with either congenital or early acquired brain injuries that are unilateral (Woods and Teuber 1978; Thal et al. 1991; Bates and Dick 2002; Newport et al. 2017). Even when the whole of the left hemisphere is absent due to an early prenatal disruption, language can be typically acquired and any impairments are relatively subtle (Asaridou et al.

2020). In cases of severe epilepsy due to disease or malformations, hemispherectomy is sometimes performed to isolate or remove a whole hemisphere. Even in cases of left hemispherectomy, language development is mildly affected and can recover to within the normal range (Curtiss, de Bode, and Mathern 2001).

An interesting question arises from these observations: why is plasticity and reorganization or recovery of language function available to children with focal unilateral brain injury but not to children who DLD to have no obvious neural disruptions? One explanation is that the disruptions that occur in children with developmental disorders of speech and language are diffuse and bilateral. There has been considerable focus on the genetic causes of speech and language disorders since the discovery of a large family— the KE family—with a mutation in the gene *FOXP2*, which results in verbal dyspraxia in family members carrying the mutation (Lai et al. 2001; Vargha-Khadem et al. 2005). The neural correlates of this disorder were investigated using brain imaging to reveal subtle differences in the size and function of several brain regions, including those not typically considered to be "language" areas, such as the striatum and cerebellum (Watkins et al. 2002; Belton et al. 2003; Liégeois et al. 2003; Argyropoulos et al. 2019).

Disruptions arising from genetic causes of speech and language disorders are likely to affect neuronal and glial size and number, and the development of axonal projections that connect brain areas and support communication between them. Advances in brain imaging techniques means we are able to study the neural correlates of speech and language disorders noninvasively and in vivo. Studies of brain structure in these disorders use image acquisition protocols and analysis techniques that are sensitive to differences in the size and shape of cortical and subcortical brain regions and the white matter pathways that connect them. Studies of brain function can shed light on the underlying functional organization of brain networks supporting speech and language and other behavior and determine patterns of adaptive and maladaptive reorganization.

We will focus on developmental language disorder, a poorly recognized yet common disability affecting up to 10% of children (Tomblin et al. 1997; Norbury et al. 2016b). Here, we add to the existing literature (Liégeois, Mayes, and Morgan 2014; Morgan, Bonthrone, and Liegeois 2016) by reviewing the most recent findings from studies using structural and functional brain imaging, in order to further our understanding of the neural correlates of these language learning difficulties.

DEVELOPMENTAL LANGUAGE DISORDER

Children with Developmental Language Disorder (DLD; formerly known as Specific Language Impairment) struggle to acquire their native language for no apparent reason. More specifically, children with DLD perform significantly lower than their peers in receptive or expressive language skills or both, while their non-verbal intelligence is usually within the normal range for their age (although this is not a criterion for language disorder according to the Diagnostic and Statistical Manual of Mental Disorders

(DSM-5); American Psychiatric Association 2013). Importantly, their language performance cannot be attributed to acquired brain lesions, deprivation, or another known neurological or developmental disorder such as Down's syndrome or autism spectrum disorder. This impairment of language skills is "severe enough to hinder academic progress" in about 7% of children at school entry (Norbury et al. 2016a). The etiology of the disorder is still poorly understood. Heritability studies and genome screening of DLD cohorts offer initial evidence that the disorder may be linked to genetic disruptions (Bishop 2014). These disruptions most likely affect the maturation and plasticity of neural circuits crucial for speech and language at pre- and post-natal developmental stages (Newbury, Fisher, and Monaco 2010; den Hoed and Fisher 2020).

Neuropsychological Profile and Comorbidity

Children with DLD can present with speech sound discrimination and expressive phonological problems, for example, in nonword repetition (Conti-Ramsden, Botting, and Faragher 2001); they have difficulty learning new words, and have poor receptive and expressive vocabulary for their age; they have problems learning grammar and exhibit grammatical difficulties in both receptive and expressive language (Conti-Ramsden, Botting, and Faragher 2001); lastly, they sometimes have poor ability to understand discourse and can show impaired social communication (Bishop 2014). An example of the language produced by a 13-year-old child with DLD when asked to narrate a cartoon story from the Expression, Reception, and Recall of Narrative Instrument (ERRNI; Bishop 2004) is illustrated in Figure 23.1. The child's language production is limited and more typical of a much younger child; the narrative contains over-regularization errors and some lexical substitutions. DLD frequently cooccurs with dyslexia (Snowling et al. 2020); ADHD (Margari et al. 2013), developmental coordination disorder (Margari et al. 2013), and speech sound disorder (Eadie et al. 2015). Even though it is a developmental disorder, which means that its symptoms first appear in childhood, this does not mean that, as children develop, they outgrow the problem. While the profile might change, the impairment is likely to continue through childhood, adolescence, and eventually adulthood, with an increase in social problems (St Clair et al. 2011), and overall lower academic and employment achievements (Conti-Ramsden et al. 2018).

THE NEURAL BASIS OF DEVELOPMENTAL
LANGUAGE DISORDER

Since the 1990s, magnetic resonance imaging (MRI) was used to investigate the neural correlates of developmental disorders of speech and language. We will review first

"He saw two girls.
And they wanted to get ice-cream with him.
They did.
And then they had a good day.
But the little girl put the fish in a different bag.
And the doll in his bag.
Then they setted off.
Then the fish gave the mom the bag.
She looked at it and it was just a baby doll.
She ran to the phone.
Tried to call.
But she saw two little girls with the yellow bag.
And took out the fish.
And put it in the fish where the mummy lived."

FIGURE 23.1 Example of Narrative Recall produced by a 13-year-old with DLD elicited with a cartoon story. The example illustrates limited language production below the level expected for a 13-year-old; sentences are short and simple, there is an over-regularization error, and other mistakes that could be due to grammatical difficulties or are omissions or lexical substitutions.

Adapted from *Expression, Reception and Recall of Narrative Instrument*, Dorothy Bishop, Pearson (2004).

the studies that focused on brain function followed by those examining structural differences in grey and white matter.

Brain Function

The cognitive neuroscience approach to developmental language disorder ultimately aims to provide a mechanistic understanding of the disorder. A better mechanistic understanding of how language is implemented in the brain of individuals with language impairment can in turn inform cognitive theories and provide possible targets for intervention.

Over the past two decades, a relatively small number of studies used functional magnetic resonance imaging (fMRI) to investigate brain activity evoked by language processing in DLD. The studies differed in the experimental paradigms they used and the underlying processes they assessed. The sample sizes in these studies were also small, which is problematic especially considering the diversity in the impairments exhibited by children with DLD and their developmental trajectories.

The initial fMRI studies of children with DLD indicated a pattern of reduced, hypo- or underactivity (relative to control groups of typically developing children (TD) during task performance. Passive listening tasks evoked weaker activity in children with DLD

(N = 5) in auditory speech perception areas in the posterior superior temporal cortex in both hemispheres (Hugdahl et al. 2004). During a more engaging task involving listening and responding to questions, children with DLD (N = 8) showed a pattern of underactivity in left hemisphere "language" areas, including motor and parietal cortex as well as the inferior frontal regions (i.e., Broca's area) (Weismer et al. 2005). Similarly, during a task involving listening to phrases and covertly retrieving a word, children with DLD (N = 8) showed underactivity in both left and right posterior superior temporal cortex, the left inferior frontal cortex, and, subcortically, in the right putamen (Badcock et al. 2012).

One of the larger studies of children with DLD (N = 21) used a small battery of four language tasks involving both expressive and receptive language processing of visual (pictures) and auditory stimuli (De Guibert et al. 2011). The children with DLD showed differences in task-related activity on only two tasks: a definition task similar to that used by Badcock et al. (2012) and a covert visual naming task that tapped phonological processing. They showed significantly reduced activity during the definition task in the left posterior superior temporal and inferior parietal cortex. Conversely, when performing the naming task (covertly producing rhyming triplets e.g., "poule, boule, moule"), the DLD group showed higher activation in the right inferior frontal cortex extending to the anterior insula and, subcortically, to the head of the caudate nucleus.

From these few preliminary studies, a pattern emerged to indicate generally reduced activity in the classic language areas—posterior superior temporal bilaterally and left inferior frontal cortex—during language processing in children with DLD (Figure 23.2). However, the use of covert and passive listening tasks precludes an understanding as to whether this is due to poor performance, or reduced attention and engagement with the task in these language impaired groups. More recent studies have attempted to address this issue by measuring and considering task performance.

In a study designed to test implicit word learning, Plante and colleagues (2017) did not observe any difference in learning performance between young adults with DLD and controls; however, the DLD group engaged the superior temporal gyrus, inferior frontal gyrus, and supramarginal gyrus to a larger extent than the control group to achieve the same level of performance. On the other hand, using a nonword repetition task, Pigdon and colleagues (2020) found significant differences in performance between children with DLD and typically developing children, but there were no differences between the two groups in brain activity patterns while performing the task. In a very large sample of children with DLD (N = 50), activity evoked during overt verb generation revealed that when performance differences between groups were eliminated, so were differences in task-related activity (Krishnan et al. 2020). Only when a subset of children with DLD who performed the task poorly (N = 14) were compared with a group of typically developing controls who performed the task at ceiling, was a trend observed towards reduced activity in the DLD group in left inferior frontal cortex, right motor cortex, and the caudate nucleus bilaterally.

A possible explanation for patterns of reduced activity during language task when comparing groups is that the activity patterns of individuals in the different groups

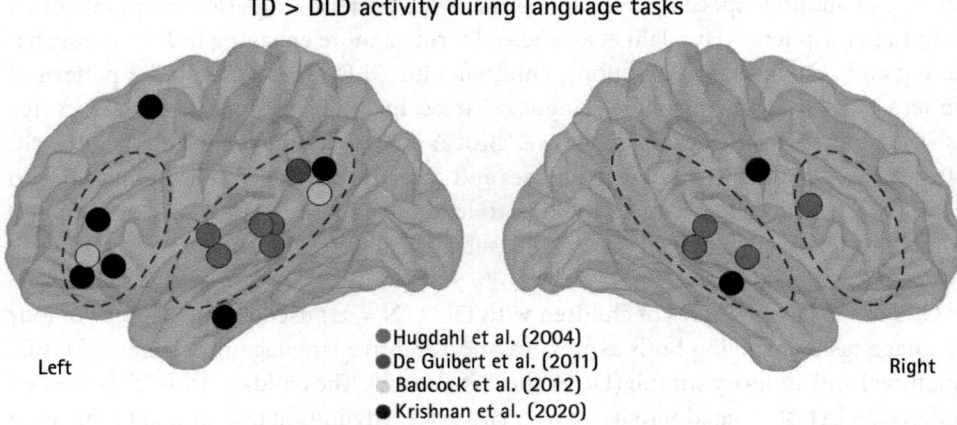

FIGURE 23.2 Cortical activation peaks from studies comparing activity during language processing (e.g., speech vs. reversed speech or baseline) in TD > DLD. Peaks are mapped onto the ICBM-152 brain template mesh. Peak activations across studies are clustered in the inferior frontal cortex (especially on the left) and the right and left superior temporal cortex (dashed areas). NB: Only studies reporting MNI peak coordinates are included. Peaks based on N = 5, Hugdahl et al. (2004); N = 21, De Guibert et al. (2011); N = 10, Badcock et al. (2012); N = 14, Krishnan et al. (2020).

might be poorly aligned. That is, the children with DLD might vary in terms of the location of focal brain activity or show more broadly distributed processing during task, whereas a typically developing group activate the same brain regions consistently. The averaging of activity within these groups would then give the appearance that the DLD group had lower activity in the areas robustly activated by task in the control group. This possibility was explored by Krishnan and colleagues (2020) by mapping the overlap of areas activated in individual children rather than averaging. The analysis revealed that the areas activated in children with DLD who could perform the overt verb generation task were remarkably similar to those activated in the typically developing group. There was little indication that children with DLD had more diffuse activity or distributed processing or recruited areas outside those typically activated by controls (see Figure 23.3).

As far back as the early 1900s, it was proposed that developmental disorders of speech and language were caused by a failure to develop language dominance, specifically a pattern of left-hemisphere specialization for language (Travis 1978; Bishop 1990, 2013). Anecdotal observations that children with speech and language disorders were more likely to be left-handed lent support to this view. Consequently, a few studies used fMRI to evaluate language lateralization during task performance in children with DLD. There was tentative support for an overall lack of leftward lateralization in DLD in the early studies (De Guibert et al. 2011; Badcock et al. 2012). However, these findings are difficult to evaluate in light of the overall reduction in patterns of activity in these studies. Recruitment of areas in the right hemisphere (as seen in the de Guibert

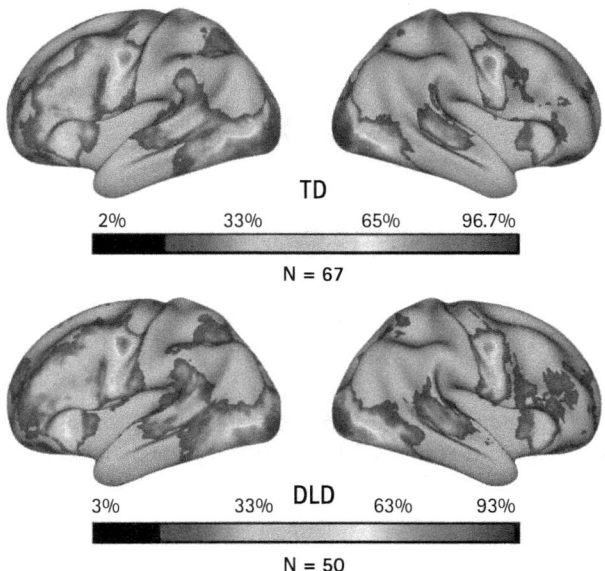

FIGURE 23.3 Activity evoked during verb generation versus baseline in typically developing (TD) and DLD children from the study by Krishnan et al. (2020). Colored overlays on left (left side of image) and right (right side of image) hemisphere lateral surfaces represent the proportion of children in each group who showed robust activation at these locations. Warm colors indicate areas that were commonly activated during verb generation (e.g., generate the verb "sit" to a picture of a chair). Both groups of children activated very similar areas.

task involving covert naming of phonologically related words), would contribute to a pattern of reduced left lateralization also. In the large sample of children with DLD studied by Krishnan and colleagues (2020) there were no differences in the distributions of language laterality indices in the DLD group compared with the typically developing group; all showed a group average of left-lateralized activity during verb generation (Figure 23.4).

A more recent neural explanation of DLD is the Procedural Deficit Hypothesis proposed by Ullman and Pierpont (Ullman and Pierpont 2005; see also updated hypothesis in Ullman et al. 2020) and recently updated to the Procedural *Circuit* Deficit Hypothesis to emphasize its focus on dysfunction in the neural circuits involved in procedural memory rather than procedural memory itself (Ullman, Earle, Walenski, and Janacseck 2020). Multiple memory systems in the brain were first proposed by Milner and colleagues based on the classic observations that implicit learning is generally spared in patients with amnesia due to medial temporal lobe damage (Scoville and Milner 1957). As much of language learning is implicit and could be considered similar to some aspects of motor skill learning, which depend on the integrity of subcortical circuits including the basal ganglia and cerebellum, Ullman proposes that children with speech and language impairments would show deficits in these brain areas (Ullman

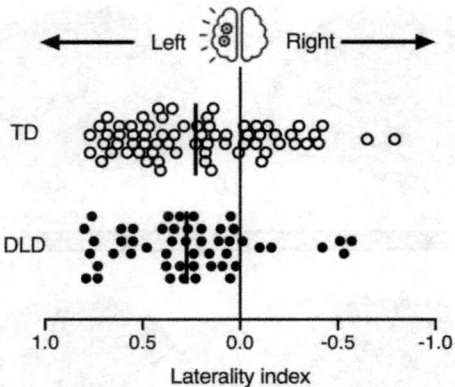

FIGURE 23.4 Language lateralization in children with DLD. Data plotted show individual language lateralization estimates from an fMRI study of verb generation in children with DLD (filled circles) compared with TD children (open circles) from the Krishnan et al. 2020 study. The LI-tool box (Wilke and Lidzba 2007) was used to produce a weighted mean of lateralization measurements across a range of statistical thresholds. The solid line represents the group median. The majority of children in each group were left lateralized and the groups were not significantly different.

and Pierpont 2005; Lum et al. 2012; Ullman et al. 2020). Findings in the KE family, who have a form of DLD that primarily affects their verbal praxis, lent support to this view; affected members of the KE family showed overactivity in the caudate nucleus during word repetition (Vargha-Khadem et al. 1998), and reduced activity in the putamen during verb generation (Liégeois et al. 2003). The fMRI studies reviewed thus far offer similar tentative support for the procedural circuit deficit hypothesis. Both overactivity of the right caudate nucleus (De Guibert et al. 2011) and underactivity of the right putamen (Badcock et al. 2012) were reported in children with DLD. However, when performance was matched between groups, the large sample of children with DLD studied by Krishnan and colleagues (2020) failed to show predicted differences in putamen activity evoked by verb generation. Reduced activity was observed in the caudate nucleus bilaterally when the analysis was restricted to the small subgroup of children who had difficulty performing the verb generation task (Krishnan et al. 2020). Further work is needed to resolve the discrepant claims among studies and cohorts of both under- and over-activation of subcortical nuclei and to understand the relationship with brain structural differences, which will be described, and with task performance.

The mechanistic insight provided by these fMRI findings is limited primarily due to differences between studies in their statistical power to detect group effects—with the exception of the study by Krishnan et al. (2020), the average DLD group sample size was 16 and ranged between ten and 21 participants. The use of different paradigms that posed distinct demands on performance also hinders the interpretation of the findings. Performance is an important issue for consideration in fMRI studies of developmental disorders. In order to make appropriate inferences, experimenters have to ask whether their fMRI paradigms test speech and language processing or whether they rely heavily on other cognitive abilities such as executive functions (for example, when task rules are complicated: perform A when screen is white and B when screen is black) or meta-linguistic skills (for example, when the task requires awareness of language rules: judge whether two words rhyme or not). While testing "pure" speech and language processing might be impossible, one way to attribute activation differences between DLD and control individuals to language impairment rather than performance is to ensure that the two groups do not differ in their ability to execute the task as in the Krishnan et al. (2020) study. Task-free, resting-state fMRI could be an alternative option to task fMRI, offering insight into the brain's "spontaneous" rather than evoked activation patterns. These patterns of activation are not random but follow an intrinsic functional organization that resembles that seen in task fMRI. To the best of our knowledge, there are as of yet no published resting-state fMRI studies on DLD, potentially due to the fact that resting-state data are difficult to acquire in younger ages. We anticipate that as it becomes more common in studies of typical development—especially in large cohort imaging studies—resting-state fMRI will become a method of choice in studying the neural underpinnings of DLD.

Brain Structure—Grey Matter in DLD

As per definition, the language difficulties observed in children with DLD cannot be attributed to congenital or acquired neurological impairments, such as epilepsy, micro-cephaly, stroke, or traumatic brain injury. While gross anatomical differences between children with DLD and typically developing children are therefore unlikely, subtle differences that can be detected at a group level are not. In the late 1990s and early 2000s, several studies investigated whether children with DLD showed abnormalities in brain morphology, grey matter and white matter volume, using manual or semi-automated methods of MRI brain images. In this chapter we will be discussing studies that were conducted within the last decade (2010–20) that used automatic and reproducible methods to detect anatomical correlates of DLD. Three studies performed whole brain voxel-based morphometry (Badcock et al. 2012; Girbau-Massana et al. 2014; Pigdon et al. 2019), while the rest focused on a priori defined regions of interest (Hodge et al. 2010; Kurth et al. 2018; Lee, Dick, and Tomblin 2020). Perisylvian cortical brain regions as well as subcortical structures, including the basal ganglia and the cerebellum, were examined.

Three studies focused on a set of perisylvian regions motivated by their involvement in speech and language processing (Kurth et al. 2018; Pigdon et al. 2019; Lee, Dick, and Tomblin 2020). Out of these, two studies found larger grey mater volume in individuals with DLD in the ventral part of the right inferior frontal gyrus, but in different parts: one in pars triangularis (Kurth et al. 2018) and the other in pars orbitalis (Lee, Dick, and Tomblin 2020). Given the unexpected laterality of the effect, these results were interpreted in the context of broader memory and learning processes that might be affected in DLD (Kurth et al. 2018; Lee, Dick, and Tomblin 2020). It should be noted that participants in these two studies belonged to different age groups (10-year-old children vs. adolescents and young adults), and that the studies used different methodologies, and focused on different set of regions of interest, all of which may explain the discrepancy in their findings. Interestingly, whole brain analysis findings revealed more grey matter in the left rather than right inferior frontal gyrus (the orbital part) in children with DLD (Badcock et al. 2012). This structural abnormality was accompanied by reduced functional activation in this area during a speech task (Badcock et al. 2012), a pattern that had been observed in affected members of the KE family who are phenotypically similar to DLD (Belton et al. 2003; Liégeois et al. 2003). With respect to other perisylvian brain regions, adolescents and young adults with DLD showed more grey matter in the left transverse temporal gyrus, the location of primary auditory cortex, (Lee, Dick, and Tomblin 2020) whereas that was not the case in younger children with DLD (Badcock et al. 2012; Kurth et al. 2018; Pigdon et al. 2019); instead, they exhibited less grey matter in the posterior superior temporal sulcus bilaterally, extending to the superior temporal gyrus in the right hemisphere (Badcock et al. 2012). Differences were also reported in grey matter in the temporal poles bilaterally (DLD > TD; Lee, Dick, and Tomblin 2020), right insula, left intraparietal sulcus (DLD > TD; Badcock et al. 2012), right superior occipital gyrus (DLD > TD; Girbau-Massana et al. 2014), the medial frontal polar cortex, right medial superior parietal cortex, left occipital pole (DLD < TD; Badcock et al. 2012), right postcentral gyrus, and right and left medial occipital gyri (DLD < TD; Girbau-Massana, Garcia-Marti, Marti-Bonmati et al. 2014).

The lack of consensus seen in cortical findings also characterizes subcortical findings. In their whole brain analysis, Badcock and colleagues (2012) reported less grey matter in children with DLD in subcortical structures, specifically in the right caudate nucleus and right side of the midbrain at the level of the substantia nigra. Siblings of children with DLD also showed less grey matter in the right caudate nucleus which led to the speculation that this abnormality might constitute a heritable risk factor for DLD (Badcock et al. 2012). In a study focusing on grey matter volume in the corticostriatal system as well as the hippocampus and cerebral lobes, Lee and colleagues (2013) did not find any abnormality in the caudate nucleus. They did, however, find larger grey matter volumes in the putamen, right globus pallidus, nucleus accumbens, and hippocampus in individuals with DLD. In accordance with Badcock et al. (2012) but in contrast with Lee et al. (2013), the affected members of the KE family showed reduced caudate nucleus volume bilaterally (Watkins et al. 2002; Belton et al. 2003), as did a single case of a boy with a de novo *FOXP2* deletion (Liégeois et al. 2016). This case also had reduced volume in the hippocampus bilaterally unlike a group of young adults with DLD who showed increased hippocampal volume (Lee, Nopoulos, and Tomblin 2013). Lastly, a whole brain analysis approach by Pigdon et al. (2019) revealed a significant difference in the cerebellum such that children with DLD had significantly more grey matter in the right cerebellar lobule VI/V compared with controls. Interestingly, using a region-of-interest approach to investigate grey and white matter abnormalities in the cerebellum, Hodge and colleagues (2010) found that grey matter volume in the vermis was smaller in children with DLD. They also found a significant asymmetry such that DLD showed more leftward, and the control group more rightward asymmetry in cerebellar lobule VIIIA (Hodge et al. 2010). In the KE family, using cerebellum-specific voxel-based morphometry (VBM) and volumetry, affected members showed reduced grey matter volume bilaterally in the cerebellar lobule VIIa Crus I compared with unaffected members and unrelated controls (Argyropoulos et al. 2019).

What do these differences mean in terms of function? Are they pathological or compensatory? One way that this has been investigated is by testing whether grey matter volume was related to behavioral performance. The outcomes of such investigations are against the idea that the observed abnormalities are compensatory. Volume in the right inferior frontal gyrus (pars triangularis), which was higher in DLD, correlated negatively with core, receptive and expressive language scores (Kurth et al. 2018). Grey matter in the right caudate nucleus correlated negatively with nonword repetition in children with DLD (Badcock *et al.* 2012) and in affected members of the KE family (Watkins et al. 2002) while grey matter in other subcortical structures including the putamen, right globus pallidus, nucleus accumbens, and hippocampus was negatively correlated with composite language scores (Lee, Nopoulos, and Tomblin 2013). Lastly, language scores were positively correlated with increasing rightward asymmetry in the cerebellar lobule VIIIA, which was lower in DLD, and the volume of the anterior vermis, again lower in DLD, correlated positively with nonword repetition, vocabulary, and non-verbal intelligence quotient (IQ) (Hodge et al. 2010). In the affected members of the KE family, the cerebellar Crus I volume reductions correlated with the volume reductions in the caudate nucleus and with performance on tasks that revealed core impairments in non-word repetition and non-verbal orofacial praxis (Argyropoulos et al. 2019). The heterogeneity that characterizes the functional and structural imaging findings discussed so far in DLD is seen here in the results of analyses that correlate measures of brain structure with behavioral scores. The measures and scores analyzed vary across studies and the relationships between them are often in unexpected directions, e.g., abnormally larger volumes associated with poor task performance or abnormally smaller volumes associated with better task performance. Such patterns suggest maladaptive rather than compensatory changes. These could reflect a failure to develop at a normal growth rate or a failure to undergo normal maturational processes that result in cortical thinning and volume reduction. Longitudinal analyses could be helpful here. Studies would benefit also from methods aimed at reducing the number of variables included in analyses and preregistered hypotheses with appropriate corrections and sample sizes to guard against reporting of false positives and false negatives.

Grey matter volume characteristics in DLD tell us only part of the story. Grey matter changes dynamically over the course of development, with reductions in the frontal, temporal and parietal lobes (Tamnes et al. 2017) most likely due to "synaptic pruning" (Huttenlocher 1979). These changes occur in tandem with changes in white matter, which increases during development enabling efficient long-distance connectivity between different regions of the brain (Mills et al. 2016). In the following section we will discuss white matter connectivity in DLD and how it may play an important role in abnormal neurodevelopment.

Brain Structure—White Matter Connectivity in DLD

All complex, higher-order skills, such as language, rely on contributions from several interconnected brain areas. The idea that language problems can result from disruption in neural pathways connecting cortical "language" areas dates back to the Wernicke–Lichtheim Model of Aphasia (Lichtheim 1885). The advent of the diffusion-weighted MRI (dMRI) technique has revolutionized the field by enabling imaging of white matter connectivity in vivo. The technique measures the diffusion of water molecules in the brain and capitalizes on the fact that water diffusion in white matter regions varies depending on the degree of myelination and coherence of fiber tract directions (Pierpaoli and Basser 1996). In the presence of myelinated and coherent fiber tracts, water tends to diffuse anisotropically, along the fiber direction, thereby offering an indirect measure of white matter connectivity. Equipped with dMRI and sophisticated analysis tools, researchers have made progress in identifying fiber pathways in the human brain and updating the 19th century models of language connectivity. White matter pathways for language feature prominently in current models of speech and language processing (Hickok and Poeppel 2007; Friederici 2009; Rauschecker and Scott 2009; Ueno et al. 2011; Catani and Bambini 2014). It is therefore not surprising that a number of studies have focused their investigations on the characteristics of white matter pathways in DLD.

The fiber tracts that run dorsally and connect posterior superior temporal and inferior parietal areas to inferior frontal regions, are by far the most investigated white matter pathways in DLD (see Figure 23.5). They are part of the dorsal stream for language processing (Hickok and Poeppel 2007; Saur et al. 2008; Rauschecker and Scott 2009) and consist of the superior longitudinal fasciculus and the arcuate fasciculus—sometimes viewed as synonymous tracts in the literature (see Dick and Tremblay 2012 for an in-depth discussion). It has been suggested that the direct segment of the arcuate fasciculus connects temporal to frontal regions while the indirect segments connect parietal to frontal (anterior part) and temporal to parietal (posterior part) regions (Catani and Mesulam 2008). Others have identified the fiber tract connecting parietal to frontal areas as the superior longitudinal fasciculus, specifically the third branch which originates in the supramarginal gyrus and can also be traced in monkeys (Schmahmann et al. 2007; Barbeau, Descoteaux, and Petrides 2020). Although the terminology and the exact anatomical origins and terminations of these tracts are still debated (Dick and Tremblay 2012; Bajada, Lambon Ralph, and Cloutman 2015), here we will use the term arcuate fasciculus to refer to the temporo-frontal pathway and superior longitudinal fasciculus (third branch) to refer to the parieto-frontal pathway (see Figure 23.5).

With respect to its functional role, the dorsal stream for language enables the mapping of speech sounds to articulation (Hickok and Poeppel 2007; Saur et al. 2008)

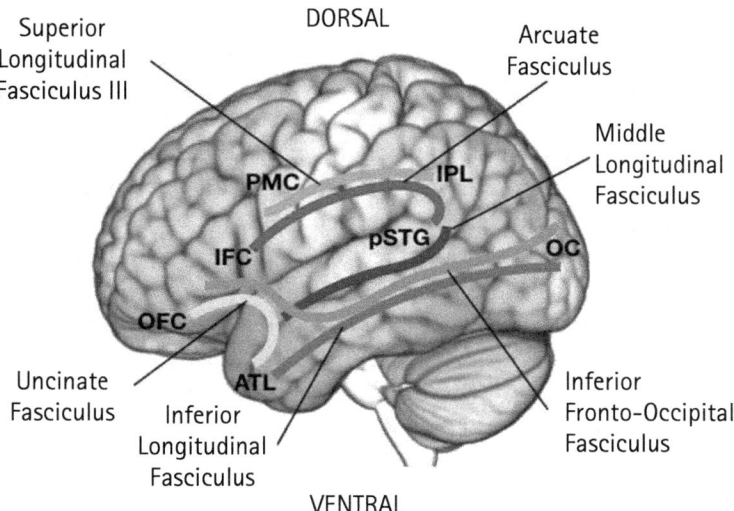

FIGURE 23.5 Schematic representation of white matter tracts for dorsal and ventral language processing streams reported in studies of DLD. Each line represents a tract with the arcuate fasciculus in red and the superior longitudinal fasciculus in cyan comprising the dorsal processing stream, and the inferior longitudinal fasciculus in brown, the inferior fronto-occipital fasciculus in green, the middle longitudinal fasciculus in blue and the uncinate fasciculus in yellow comprising the ventral processing stream. ATL: anterior temporal lobe; IFC: inferior frontal cortex; IPL: inferior parietal lobule; OC: occipital cortex; OFC: orbitofrontal cortex; PMC: premotor cortex; pSTL: posterior superior temporal lobe.

and is therefore important for speech perception and production. Some researchers have argued for a role of the superior longitudinal fasciculus in speech perception and for the arcuate fasciculus in syntactic processing (Papoutsi et al. 2011; Wilson et al. 2011; Friederici 2012). Given that children with DLD present with expressive language difficulties as well as difficulties with grammar, a number of studies have tried to identify the dorsal tracts in this group and to quantify their microstructural properties with dMRI. These studies suggest that fractional anisotropy (FA) along the left and right arcuate fasciculus (Verhoeven et al. 2012; Vydrova et al. 2015) and superior longitudinal fasciculus (Lee, Dick, and Tomblin 2020) is reduced in children and young adults with DLD (see Morgan et al. 2018, however, who reported no differences in dorsal language tracts in DLD). While there is no longitudinal data on white matter development in children with DLD, cross-sectional findings indicate that while FA along the dorsal tracts increases with age in typically developing children that is not the case for individuals with DLD (Lee, Dick, and Tomblin 2020).

Fractional anisotropy is sensitive but not specific to a number of tissue properties, including myelination, axonal tortuosity, axonal size and diameter, and axonal density (Jones, Knösche and Turner 2013). The interpretation of differences in FA between children with DLD and typically developing children is therefore not entirely straightforward. At least one study demonstrated that group differences in FA and mean diffusivity were driven by increase in radial diffusivity (i.e. diffusion perpendicular to the fiber axis) in children with DLD (Roberts et al. 2014; Vydrova et al. 2015). As myelin wrapping of axons impedes the diffusion of water in the direction perpendicular to the long axis of the axon, increased diffusivity in the radial direction is thought to indicate reduced myelination of axons. Thus, this finding points to a difference in myelination of dorsal tracts in the DLD group compared with controls.

Whether the characteristics of the dorsal tracts relate to language performance is still unclear. In typically developing children, diffusion properties of the left arcuate fasciculus have been associated with the rate of vocabulary growth (Su et al. 2018, 2020), phonological skills (Yeatman et al. 2011), phonological awareness (Saygin et al. 2013; Travis et al. 2017), word reading development (Wang et al. 2017), and overall receptive language and language content (Broce et al. 2015). Similarly, in adults, FA in the left arcuate fasciculus has been positively related to vocabulary size (Teubner-Rhodes et al. 2016) and phonological vocabulary learning (López-Barroso et al. 2013). Interestingly, early damage due to stroke in left dorsal language tracts can have long-lasting effects, in particular on speech repetition abilities (Northam et al. 2018), which highlights the important role of the dorsal stream in mapping sound to articulation and the possibility that these functions may not reorganize as successfully as functions that depend on the ventral stream (see also Asaridou et al. 2020). Findings in studies comparing DLD with typically developing children have been inconsistent: one study reported positive associations between FA in the left and right arcuate fasciculus and *receptive* language performance in children with DLD but not in typically developing children (Verhoeven et al. 2012); others found no associations (Roberts et al. 2014; Verly et al. 2019).

Fiber tracts that are part of the ventral stream for language processing (Hickok and Poeppel 2007; Saur et al. 2008; Rauschecker and Scott 2009) have also been investigated in individuals with DLD, although less so than the dorsal tracts. The ventral stream for language is considered to be largely bilateral and involved in speech recognition, specifically in mapping sound to meaning (Hickok and Poeppel 2007; Saur et al. 2008), with some researchers arguing that it also supports semantic processing (Friederici and Gierhan 2013; Bajada, Lambon Ralph, and Cloutman 2015). It consists of several fiber pathways: the inferior fronto-occipital fasciculus, which runs ventrally through the extreme capsule and connects posterior temporal and lateral occipital lobes to the frontal lobe (also referred to as the extreme capsular fasciculus); the inferior longitudinal fasciculus, which connects the lateral occipital to the anterior temporal lobe; the uncinate fasciculus, a hook-shaped tract which connects the temporal pole to the frontal lobe; and the middle longitudinal fasciculus which connects the parietal to the anterior temporal lobe (see Forkel et al. 2014; Bajada, Lambon Ralph, and Cloutman 2015 for detailed descriptions) (see Figure 23.5).

Children and young adults with DLD show decreased FA in the inferior fronto-occipital fasciculus bilaterally (Vydrova et al. 2015; Lee, Dick, and Tomblin 2020), in the inferior longitudinal fasciculus in the left (Vydrova et al. 2015; Verly et al. 2019) and right hemispheres (Lee, Dick, and Tomblin 2020), in the uncinate fasciculus bilaterally (Vydrova et al. 2015), and in the left middle longitudinal fasciculus (Verly et al. 2019) (but note the negative findings reported by Morgan et al. 2018). With respect to the underlying tissue properties, reduced FA in these studies is driven by both increased radial diffusivity, reflecting lower myelin content, and decreased axonal diffusivity (diffusion along the fiber axon), reflecting reduced fiber coherence or increased tortuosity (Song et al. 2002), in children with DLD (Vydrova et al. 2015). Adolescents and young adults with DLD did not show age-related increases in FA in the left inferior longitudinal fasciculus and right uncinate fasciculus that were observed in the control group (Lee, Dick, and Tomblin 2020). Whether these findings can be interpreted as differences in the developmental trajectory of white matter tracts or the consequence of poor language performance in DLD is an empirical question that only longitudinal studies can address. Interestingly, no correlation was found between FA in ventral tracts and language performance in children with DLD (Verly et al. 2019). In typically developing children, fractional anisotropy in ventral tracts has been associated with higher phonological skills (Vandermosten et al. 2015; Travis et al. 2017; Walton, Dewey, and Lebel 2018), and lower radial diffusivity with better word learning performance, when learning derived from semantic context (Ripollés et al. 2017). The lateralization patterns of the dorsal and ventral fiber tracts have also been investigated in DLD. In typically developing children, the arcuate fasciculus and the inferior fronto-occipital fasciculus are left lateralized (Johnson et al. 2014; Budisavljevic et al. 2015; Zhao et al. 2016). Higher FA in the left compared with the right homologue of these tracts have been detected as early as one year of age (Stephens et al. 2020), an effect attributed to genetic as well as environmental factors (Budisavljevic et al. 2015). Bearing in mind the evidence for atypical or inconsistent functional lateralization in children with DLD, it was hypothesized that this group might show different lateralization patterns in the white matter tracts that are important for language. There has been some support for this hypothesis from one study reporting leftward asymmetry in FA in typically developing children but no hemispheric asymmetry in dorsal and ventral white matter tracts in children with DLD, except for the middle longitudinal fasciculus (Verly et al. 2019). Other findings, however, report normal leftward hemispheric asymmetry in FA in children with DLD in the arcuate fasciculus, inferior fronto-occipital fasciculus, and inferior longitudinal fasciculus, no different from age-matched controls (Vydrova et al. 2015). When tract volume, rather than FA, was the measure of interest, only the inferior fronto-occipital fasciculus showed rightward volumetric inter-hemispheric asymmetry in children with DLD, which was significantly different from controls (Vydrova et al. 2015). Thus, the evidence that individuals with DLD show inconsistent hemispheric white matter asymmetry so far is weak and studies with larger samples are needed in order to satisfactorily test this hypothesis.

Very few studies have investigated white matter pathways beyond the dorsal-ventral streams in DLD. Based on the hypothesis that domain-general procedural learning mechanisms are impaired in DLD (Ullman and Pierpont 2005; Krishnan, Watkins, and Bishop 2016; Lee, Nopoulos, and Tomblin 2020), white matter pathways in the corticostriatal and corticocerebellar systems, thought to support procedural memory, were investigated in adolescents and young adults with DLD (Lee, Nopoulos, and Tomblin 2020). These were compared with pathways in the medial temporal lobe regions involved in declarative memory, which is considered relatively spared in DLD (although see Lum and Conti-Ramsden 2013). They found significantly lower FA in adolescents and young adults with DLD in the corticostriatal tracts (anterior limb of internal capsule bilaterally and right posterior limb); in the corticocerebellar tracts (superior and middle cerebellar peduncles); and in the medial temporal region tracts (fornix bilaterally and the right hippocampal part of the cingulum bundle). In addition, they observed that while FA in the left anterior and posterior limbs of internal capsule, the cerebellar peduncles, and the fornix increased with age in the control participants, it did not in participants with DLD (Lee, Nopoulos, and Tomblin 2020). Another white matter tract that was examined in children with DLD is the corticobulbar tract (which contributes to fibers in the posterior limb of the internal capsule). Morgan et al. (2018) hypothesized that atypical development of this speech-motor pathway might be contributing to DLD, however, they found significantly lower FA only in children with pure developmental speech sound disorder, and not in those with DLD.

In summary, there is evidence to support the idea that white matter connections important for speech and language, as well as memory and learning, are "mis-wired" in DLD. The most robust finding in the literature thus far has been that children with DLD show lower FA in the left arcuate fasciculus compared with age-matched typically developing children. That being said, not all studies that set out to test white matter connectivity reported differences between DLD and typically developing individuals (see for example Roberts et al. 2014; Morgan et al. 2018; Verly et al. 2019). The evidence is fragmented, characterized by a notable lack of agreement between dMRI studies which limits meaningful interpretations. Sample sizes and characteristics, acquisition protocols, quality control procedures, data analysis methods/pipelines can vary widely between studies, often leading to discrepant or even conflicting findings. Furthermore, the high phenotypic heterogeneity in children with DLD might indicate that there is a one-to-many rather than a one-to-one mapping between DLD and white matter pathways in the brain. Larger studies that examine white matter connectivity as well as the phenotypic variation longitudinally are necessary for an in-depth understanding of how one relates to the other.

FUTURE DIRECTIONS

Having presented past findings on the neural correlates of a developmental speech and language disorder, we now turn to the future and outline ways the field can leverage cognitive neuroscience methods to understand developmental speech and language impairments.

The need for future studies with larger sample sizes cannot be emphasized enough. Studies with small sample sizes have low statistical power. Underpowered studies not only have reduced likelihood of detecting a true effect but also reduced likelihood of finding a statistically significant result that reflects a true effect (Button et al. 2013). The field is already shifting towards large cohort studies such as the Adolescent Brain Cognitive Development Study (ABCD) (Casey et al. 2018) which follows more than 10,000 children longitudinally and investigates brain development and child health during adolescence with a focus on risk, which aims to identify potential predictors of later drug abuse. We anticipate that the success of such studies in typically developing children will have a positive effect on research on atypical language development, in particular by encouraging data sharing practices as well as research collaborations across institutions. Initiatives such as the Autism Brain Imaging Data Exchange (ABIDE) project (Di Martino et al. 2014, 2017) that aggregated MRI data on autism spectrum disorder across sites and made them publicly available are setting an example that we hope researchers working on developmental language disorder will follow.

Future studies not only need to recruit larger samples but also well-characterized samples. Detailed phenotyping is essential when examining brain-behavior relationships in highly heterogeneous developmental disorders such as DLD. If the phenotype is poorly defined, findings can be hard to interpret or even meaningless. For example, a child with DLD might present with primarily expressive language problems while another with receptive language problems. Assigning a diagnostic label to these children is useful for clinical purposes, however, it might make little sense from a cognitive neuroscience research perspective to treat them the same. A correlational rather than categorical approach to linking behavior to brain structure and function might be fruitful (Krishnan et al. 2020). A transdiagnostic approach might yield better insights into the intricate ways in which brain structure and function are associated with atypical language development by reflecting the dimensionality and complexity within a disorder (Bishop 1997; Siugzdaite et al. 2020). Brain imaging findings could in turn help to further map out the complexity in behavior by revealing subtypes of functional and anatomical differences within a population. In addition to the complexity at the level of behavior, there is complexity at the level of brain organization that needs to be considered. Different neural correlates can underlie the same disorder and the same neural correlate can characterize different disorders. Future studies will therefore need to move away from the idea of one-to-one brain region-to-behavior mapping and towards approaches that take into consideration the whole-brain network organization (Bassett and Sporns 2017; Siugzdaite et al. 2020).

Lastly, developmental disorders can only be understood by taking into account the dynamic interactions between genes, brain, cognition, and environment over time (Karmiloff-Smith 2009; Johnson 2011) using longitudinal observations. We have already emphasized the importance of longitudinal designs for developmental cognitive neuroscientific studies of speech and language disorders. Such studies are under way in the investigation of developmental stuttering for example (Chang et al. 2019). Longitudinal studies will allow us to examine patterns of recovery, compensation, as well as potentially lasting impairments. More importantly, they will help disentangle atypical brain structure and function that is causally related to the disorder from atypical brain structure and function that is a consequence of growing up and living with the disorder.

ACKNOWLEDGEMENTS

Our work using brain imaging in children with developmental language disorder is supported by the Medical Research Council UK project grant MR/P024149/1. We would like to thank our colleagues Dr. Saloni Krishnan and Dr. Gabe Cler for their support and work on this project and informal discussions. We also thank them and Professor Dorothy Bishop for advice and useful discussions about DLD. We thank Ellie Rushworth for the artwork used in Figure 23.1.

REFERENCES

American Psychiatric Association. (2013). *Diagnostic and Statistical Manual of Mental Disorders (DSM-5)*. 5th ed. Washington, DC: American Psychiatric Association.

Argyropoulos, G. P. D., Watkins, K. E., Belton-Pagnamenta, E., et al. (2019). Neocerebellar Crus I abnormalities associated with a speech and language disorder due to a mutation in FOXP2. *Cerebellum* 18: 309–319.

Asaridou, S. S., Demir-Lira, E., Goldin-Meadow, S., et al. (2020). Language development and brain reorganization in a child born without the left hemisphere. *Cortex* 127: 290–312.

Badcock, N. A., Bishop, D. V. M., Hardiman, M. J., et al. (2012). Co-Localisation of abnormal brain structure and function in specific language impairment. *Brain Lang* 120: 310–320.

Bajada, C. J., Lambon Ralph, M. A., and Cloutman, L. L. (2015). Transport for language south of the Sylvian fissure: The routes and history of the main tracts and stations in the ventral language network. *Cortex* 69: 141–151.

Barbeau, E. B., Descoteaux, M., and Petrides, M. (2020). Dissociating the white matter tracts connecting the temporo-parietal cortical region with frontal cortex using diffusion tractography. *Sci Rep* 10: 1–13.

Bassett, D. S. and Sporns, O. (2017). Network neuroscience. *Nat Neurosci* 20: 353–364.

Bates, E., and Dick, F. 2002. Language, gesture, and the developing brain. *Dev Psychobiol* 40: 293–310.

Belton, E., Salmond, C. H., Watkins, K. E., et al. (2003). Bilateral brain abnormalities associated with dominantly inherited verbal and orofacial dyspraxia. *Hum Brain Mapp* 18: 194–200.

Bishop, D. V. and Clark, B., Conti-Ramsden, G., et al. (2012). RALLI: An internet campaign for raising awareness of language learning impairments. *Child Lang Teach Ther* 28: 259–262.

Bishop, D. V. B. (1990). *Handedness and Developmental Disorder*. Oxford: Blackwell.

Bishop, D. V. M. (1997). Cognitive neuropsychology and developmental disorders: Uncomfortable bedfellows. *Q J Exp Psychol Sect A* 50: 899–923.

Bishop, D. V. M. (2004). *Expression, Reception and Recall of Narrative Instrument: ERRNI*. London: Harcourt Assessment.

Bishop, D. V. M. (2010). Which neurodevelopmental disorders get researched and why? Morty R. E. (ed.). *PLoS One* 5: e15112.

Bishop, D. V. M. (2013). Cerebral asymmetry and language development: Cause, correlate, or consequence? *Science* (80): 340. Doi: 10.1126/science.1230531.

Bishop, D. V. M. (2014). *Uncommon Understanding: Development and Disorders of Language Comprehension in Children, Classic Edition*. Psychology Press.

Broce, I., Bernal, B., Altman, N., et al. (2015). Fiber tracking of the frontal aslant tract and subcomponents of the arcuate fasciculus in 5–8-year-olds: Relation to speech and language function. *Brain Lang* 149: 66–76.

Budisavljevic, S., Dell'Acqua, F., Rijsdijk, F. V., et al. (2015). Age-related differences and heritability of the perisylvian language networks. *J Neurosci* 35: 12625–12634.

Button, K. S., Ioannidis, J. P. A., Mokrysz, C., et al. (2013). Power failure: Why small sample size undermines the reliability of neuroscience. *Nat Rev Neurosci* 14: 365–376.

Casey, B. J., Cannonier, T., Conley, M. I., et al. (2018). The Adolescent Brain Cognitive Development (ABCD) study: Imaging acquisition across 21 sites. *Dev Cogn Neurosci* 32: 43–54.

Catani, M. and Bambini, V. A. (2014). Model for Social Communication And Language Evolution and Development (SCALED). *Curr Opin Neurobiol* 28: 165–171.

Catani, M. and Mesulam, M. (2008). The arcuate fasciculus and the disconnection theme in language and aphasia: History and current state. *Cortex* 44: 953–961.

Chang, S. E., Garnett, E. O., Etchell, A., et al. (2019). Functional and neuroanatomical bases of developmental stuttering: Current insights. *Neuroscientist* 25: 566–582.

Conti-Ramsden, G., Botting, N. and Faragher, B. (2001). Psycholinguistic markers for specific language impairment (SLI). *J Child Psychol Psychiatry Allied Discip* 42: 741–748.

Conti-Ramsden, G. and Durkin, K., Toseeb, U., et al. (2018). Education and employment outcomes of young adults with a history of developmental language disorder. *Int J Lang Commun Disord* 53: 237–255.

Curtiss, S., de Bode, S., and Mathern, G. W. (2001). Spoken language outcomes after hemispherectomy: Factoring in etiology. *Brain Lang* 79: 379–396.

Dick, A. S. and Tremblay, P. (2012). Beyond the arcuate fasciculus: Consensus and controversy in the connectional anatomy of language. *Brain* 135: 3529–3550.

Eadie, P., Morgan, A., Ukoumunne, O. C., et al. (2015). Speech sound disorder at 4 years: Prevalence, comorbidities, and predictors in a community cohort of children. *Dev Med Child Neurol* 57: 578–584.

Forkel, S. J., Thiebaut de Schotten, M., Kawadler, J. M., et al. (2014). The anatomy of fronto-occipital connections from early blunt dissections to contemporary tractography. *Cortex* 56: 73–84.

Friederici, A. D. (2009). Pathways to language: Fiber tracts in the human brain. *Trends Cogn Sci* 13: 175–181.

Friederici, A. D. (2012). The cortical language circuit: From auditory perception to sentence comprehension. *Trends Cogn Sci* 16: 262–268.

Friederici, A. D. and Gierhan, S. M. (2013). The language network. *Curr Opin Neurobiol* 23: 250–254.

Girbau-Massana, D., Garcia-Marti, G., Marti-Bonmati, L., et al. (2014). Gray-white matter and cerebrospinal fluid volume differences in children with specific language impairment and/or reading disability. *Neuropsychologia* 56: 90–100.

De Guibert, C., Maumet, C., Jannin, P., et al. (2011). Abnormal functional lateralization and activity of language brain areas in typical specific language impairment (developmental dysphasia). *Brain* 134: 3044–3058.

Hickok, G. and Poeppel, D. (2007). The cortical organization of speech processing. *Nat Rev Neurosci* 8: 393–402.

Hodge, S. M, Makris, N., Kennedy, D. N., et al. (2010). Cerebellum, language, and cognition in autism and specific language impairment. *J Autism Dev Disord* 40: 300–316.

den Hoed, J. and Fisher, S. E. (2020). Genetic pathways involved in human speech disorders. *Curr Opin Genet Dev* 65: 103–111.

Hugdahl, K., Gundersen, H., Brekke, C., et al. (2004). fMRI brain activation in a Finnish family with specific language impairment compared with a normal control group. *J Speech, Lang Hear Res* 47: 162–172.

Huttenlocher, P. R. (1979). Synaptic density in human frontal cortex—developmental changes and effects of aging. *Brain Res* 163: 195–205.

Johnson, M. H. (2011). Interactive specialization: A domain-general framework for human functional brain development? *Dev Cogn Neurosci* 1: 7–21.

Johnson, R. T, Yeatman, J. D, Wandell, B. A., et al. (2014). Diffusion properties of major white matter tracts in young, typically developing children. *Neuroimage* 88: 143–154.

Jones, D. K., Knösche, T. R., and Turner, R. (2013). White matter integrity, fiber count, and other fallacies: The do's and don'ts of diffusion MRI. *Neuroimage* 73: 239–254.

Karmiloff-Smith, A. (2009). Nativism versus neuroconstructivism: Rethinking the study of developmental disorders. *Dev Psychol* 45: 56–63.

Krishnan, S., Asaridou, S. S., Cler, G. J., et al. (2020). Functional organisation for verb generation in children with language disorders. *Neuroimage* 226: 117599.

Krishnan, S.,Watkins, K. E., and Bishop D. V. M. (2016). Neurobiological basis of language learning difficulties. *Trends Cogn Sci* 20: 701–714.

Kurth, F., Luders, E., Pigdon, L., et al. (2018). Altered gray matter volumes in language-associated regions in children with developmental language disorder and speech sound disorder. *Dev Psychobiol* 60: 814–824.

Lai, C. S. L., Fisher, S. E., Hurst, J. A., et al. (2001). A forkhead-domain gene is mutated in a severe speech and language disorder. *Nature* 413: 519–523.

Law, J., Boyle, J., Harris, F., et al. (2000). Prevalence and natural history of primary speech and language delay: Findings from a systematic review of the literature. *Int J Lang Commun Disord* 35: 165–188.

Lee, J. C., Dick, A. S., and Tomblin, J. B. (2020). Altered brain structures in the dorsal and ventral language pathways in individuals with and without developmental language disorder (DLD). *Brain Imaging Behav* 14: 2569–2586.

Lee, J. C., Nopoulos, P. C., and Tomblin, J. B. (2013). Abnormal subcortical components of the corticostriatal system in young adults with DLI: A combined structural MRI and DTI study. *Neuropsychologia* 51: 2154–2161.

Lee, J. C., Nopoulos, P. C., and Tomblin, J. B. (2020). Procedural and declarative memory brain systems in developmental language disorder (DLD). *Brain Lang* 205: 104789.

Lichtheim, L. (1885). On aphasia. *Brain* 7: 433–484.

Liégeois, F., Baldeweg, T., Connelly, A., et al. (2003). Language fMRI abnormalities associated with FOXP2 gene mutation. *Nat Neurosci* 6: 1230–1237.

Liégeois, F., Mayes, A., and Morgan, A. (2014). Neural correlates of developmental speech and language disorders: Evidence from neuroimaging. *Curr Dev Disord Reports* 1: 215–227.

Liégeois, F. J., Hildebrand, M. S., Bonthrone, A., et al. (2016). Early neuroimaging markers of FOXP2 intragenic deletion. *Sci Rep* 6: 1–9.

López-Barroso, D., Catani, M., Ripollés, P., et al. (2013). Word learning is mediated by the left arcuate fasciculus. *Proc Natl Acad Sci U S A* Doi: 10.1073/pnas.1301696110.

Lum, J. A. G. and Conti-Ramsden, G. (2013). Long-term memory: A review and meta-analysis of studies of declarative and procedural memory in specific language impairment. *Top Lang Disord* 33: 282–297.

Lum, J. A. G., Conti-Ramsden, G., Page, D., et al. (2012). Working, declarative and procedural memory in specific language impairment. *Cortex* 48: 1138–1154.

Margari, L., Buttiglione, M., Craig, F., et al. (2013). Neuropsychopathological comorbidities in learning disorders. *BMC Neurol* 13: 198.

Di Martino, A., O'Connor, D., Chen, B., et al. (2017). Enhancing studies of the connectome in autism using the autism brain imaging data exchange II. *Sci Data* 4: 1–15.

Di Martino, A., Yan, C. G., Li, Q., et al. (2014). The autism brain imaging data exchange: Towards a large-scale evaluation of the intrinsic brain architecture in autism. *Mol Psychiatry* 19: 659–667.

Mills, K. L., Goddings, A. L., Herting, M. M., et al. (2016). Structural brain development between childhood and adulthood: Convergence across four longitudinal samples. *Neuroimage* 141: 273–281.

Morgan, A., Bonthrone, A., and Liegeois, F. J. (2016). Brain basis of childhood speech and language disorders: Are we closer to clinically meaningful MRI markers? *Curr Opin Pediatr* 28: 725–730.

Morgan, A. T., Su, M., Reilly, S., et al. (2018). A brain marker for developmental speech disorders. *J Pediatr* 198: 234–239.e1.

Newbury, D. F., Fisher, S. E., and Monaco, A. P. (2010). Recent advances in the genetics of language impairment. *Genome Med* 2: 1–8.

Newport, E. L., Landau, B., Seydell-Greenwald, A., et al. (2017). Revisiting Lenneberg's hypotheses about early developmental plasticity: Language organization after left-hemisphere perinatal stroke. *Biolinguistics* 11: 407–422.

Norbury, C. F., Gooch, D., Baird, G., et al. (2016a). Younger children experience lower levels of language competence and academic progress in the first year of school: Evidence from a population study. *J Child Psychol Psychiatry* 57: 65–73.

Norbury, C. F., Gooch, D., Wray, C., et al. (2016b). The impact of nonverbal ability on prevalence and clinical presentation of language disorder: Evidence from a population study. *J Child Psychol Psychiatry Allied Discip* 57: 1247–1257.

Northam G. B., Adler S., Eschmann, K. C. J., et al. (2018). Developmental conduction aphasia after neonatal stroke. *Ann Neurol* 83: 664–675.

Papoutsi, M., Stamatakis, E. A., Griffiths, J., et al. (2011). Is left fronto-temporal connectivity essential for syntax? Effective connectivity, tractography and performance in left-hemisphere damaged patients. *Neuroimage* 58: 656–664.

Pierpaoli, C. and Basser, P. J. (1996). Toward a quantitative assessment of diffusion anisotropy. *Magn Reson Med* 36: 893–906.

Pigdon, L., Willmott, C., Reilly, S., et al. (2019). Grey matter volume in developmental speech and language disorder. *Brain Struct Funct* 224(9): 3387–3398.

Pigdon, L., Willmott, C., Reilly, S., et al. (2020). The neural basis of nonword repetition in children with developmental speech or language disorder: An fMRI study. *Neuropsychologia* 138: 107312.

Plante, E., Patterson, D., Sandoval, M., et al. (2017). An fMRI study of implicit language learning in developmental language impairment. *NeuroImage Clin* 14: 277–285.

Rauschecker, J. P. and Scott, S. K. (2009). Maps and streams in the auditory cortex: Nonhuman primates illuminate human speech processing. *Nat Neurosci* 12: 718–724.

Reilly, S., Wake, M., Ukoumunne, O. C., et al. (2010). Predicting language outcomes at 4 years of age: findings from Early Language in Victoria Study. *Pediatrics* 126: e1530–e1537.

Ripollés, P., Biel, D., Peñaloza, C., et al. (2017). Strength of temporal white matter pathways predicts semantic learning. *J Neurosci* 37: 11101–11113.

Roberts, T. P. L., Heiken, K., Zarnow, D., et al. (2014). Left hemisphere diffusivity of the arcuate fasciculus: Influences of autism spectrum disorder and language impairment. *Am J Neuroradiol* 35: 587–592.

Saur, D., Kreher, B. W., Schnell, S., et al. (2008). Ventral and dorsal pathways for language. *Proc Natl Acad Sci U S A* 105: 18035–18040.

Saygin, Z. M., Norton, E. S., Osher, D. E., et al. (2013). Tracking the roots of reading ability: White matter volume and integrity correlate with phonological awareness in prereading and early-reading kindergarten children. *J Neurosci* 33: 13251–13258.

Schmahmann, J. D., Pandya, D. N., Wang, R., et al. (2007). Association fibre pathways of the brain: Parallel observations from diffusion spectrum imaging and autoradiography. *Brain* 130: 630–653.

Scoville, W. B. and Milner, B. (1957). Loss of recent memory after bilateral hippocampal lesions. *J Neurol Neurosurg Psychiatry* 20: 11–21.

Siugzdaite, R., Bathelt, J., Holmes, J., et al. (2020). Transdiagnostic brain mapping in developmental disorders. *Curr Biol* 30: 1245–1257.e4.

Snowling, M. J., Hayiou-Thomas, M. E., Nash, H. M., et al. (2020). Dyslexia and developmental language disorder: Comorbid disorders with distinct effects on reading comprehension. *J Child Psychol Psychiatry* 61: 672–680.

Song, S. K., Sun, S. W., Ramsbottom, M. J., et al. (2002). Dysmyelination revealed through MRI as increased radial (but unchanged axial) diffusion of water. *Neuroimage* 17: 1429–36.

St Clair, M. C., Pickles, A., Durkin, K., et al. (2011). A longitudinal study of behavioral, emotional and social difficulties in individuals with a history of specific language impairment (SLI). *J Commun Disord* 44: 186–199.

Stephens, R. L., Langworthy, B. W., Short, S. J., et al. (2020). White matter development from birth to 6 years of age: A longitudinal study. *Cereb Cortex* 30(12): 6152–6168.

Su, M., Thiebaut de Schotten, M., Zhao, J., et al. (2018). Vocabulary growth rate from preschool to school-age years is reflected in the connectivity of the arcuate fasciculus in 14-year-old children. *Dev Sci* 21(5): e12647.

Su, M., Thiebaut de Schotten, M., Zhao, J., et al. (2020). Influences of the early family environment and long-term vocabulary development on the structure of white matter pathways: A longitudinal investigation. *Dev Cogn Neurosci* 42: 100767.

Tamnes, C. K., Herting, M. M., Goddings, A-L., et al. (2017). Development of the cerebral cortex across adolescence: A multisample study of inter-related longitudinal changes in cortical volume, surface area, and thickness. *J Neurosci* 37: 3402–3412.

Teubner-Rhodes, S., Vaden, K. I., Cute, S. L., et al. (2016). Aging-resilient associations between the arcuate fasciculus and vocabulary knowledge: Microstructure or morphology? *J Neurosci* 36: 7210–7222.

Thal, D. J., Marchman, V., Stiles, J., et al. (1991). Early lexical development in children with focal brain injury. *Brain Lang* 40: 491–527.

Tomblin, J. B., Records, N. L., Buckwalter, P., et al. (1997). Prevalence of specific language impairment in kindergarten children. *J Speech, Lang Hear Res* 40: 1245–1260.

Travis, K. E., Adams, J. N., Kovachy, V. N., et al. (2017). White matter properties differ in 6-year old readers and pre-readers. *Brain Struct Funct* 222: 1685–1703.

Travis, L. E. (1978). The cerebral dominance theory of stuttering: 1931–1978. *J Speech Hear Disord* 43: 278–281.

Ueno T., Saito, S., Rogers, T. T., et al. (2011). Lichtheim 2: Synthesizing aphasia and the neural basis of language in a neurocomputational model of the dual dorsal-ventral language pathways. *Neuron* 72: 385–396.

Ullman, M. T., Earle, F. S., Walenski, M., et al. (2020). The neurocognition of developmental disorders of language. *Annu Rev Psychol* 71: 389–417.

Ullman, M. T. and Pierpont, E. I. (2005). Specific language impairment is not specific to language: The procedural deficit hypothesis. *Cortex* 41: 399–433.

Vandermosten, M., Vanderauwera, J., Theys, C., et al. (2015). A DTI tractography study in pre-readers at risk for dyslexia. *Dev Cogn Neurosci* 14: 8–15.

Vargha-Khadem, F., Gadian, D. G., Copp, A., et al. (2005). FOXP2 and the neuroanatomy of speech and language. *Nat Rev Neurosci* 6: 131–138.

Vargha-Khadem, F., Watkins, K. E., Price, C. J., et al. (1998). Neural basis of an inherited speech and language disorder. *Proc Natl Acad Sci U S A* 95: 12695–12700.

Verhoeven, J. S., Rommel, N., Prodi, E., et al. (2012). Is there a common neuroanatomical substrate of language deficit between autism spectrum disorder and specific language impairment? *Cereb Cortex* 22: 2263–2271.

Verly, M., Gerrits, R., Sleurs, C., et al. (2019). The mis-wired language network in children with developmental language disorder: Insights from DTI tractography. *Brain Imaging Behav* 13: 973–984.

Vydrova, R., Komarek, V., Sanda, J., et al. (2015). Structural alterations of the language connectome in children with specific language impairment. *Brain Lang* 151: 35–41.

Walton, M., Dewey, D., and Lebel, C. (2018). Brain white matter structure and language ability in preschool-aged children. *Brain Lang* 176: 19–25.

Wang, Y., Mauer, M. V., Raney, T., et al. (2017). Development of tract-specific white matter pathways during early reading development in at-risk children and typical controls. *Cereb Cortex* 27: 2469–2485.

Watkins, K. E., Vargha-Khadem, F., Ashburner, J., et al. (2002). MRI analysis of an inherited speech and language disorder: Structural brain abnormalities. *Brain* 125: 465–478.

Weismer, S. E., Plante, E., Jones, M., et al. (2005). A functional magnetic resonance imaging investigation of verbal working memory in adolescents with specific language impairment. *J Speech, Lang Hear Res* 48: 405–425.

Wilke, M. and Lidzba, K. (2007). LI-tool: A new toolbox to assess lateralization in functional MR-data. *J Neurosci Methods* 163: 128–136.

Wilson, S. M., Galantucci, S., Tartaglia, M. C., et al. (2011). Syntactic processing depends on dorsal language tracts. *Neuron* 72: 397–403.

Woods, B. T. and Teuber, H-L. (1978). Changing patterns of childhood aphasia. *Ann Neurol* 3: 273–280.

Yeatman, J. D., Dougherty, R. F., Rykhlevskaia, E., et al. (2011). Anatomical properties of the arcuate fasciculus predict phonological and reading skills in children. *J Cogn Neurosci* 23: 3304–3317.

Zhao, J., Thiebaut de Schotten, M., Altarelli, I., et al. (2016). Altered hemispheric lateralization of white matter pathways in developmental dyslexia: Evidence from spherical deconvolution tractography. *Cortex* 76: 51–62.

CHAPTER 24

MULTI-LANGUAGE ACQUISITION AND COGNITIVE DEVELOPMENT

ROBERTO FILIPPI, EVA PERICHE-TOMAS,
ANDRIANI PAPAGEORGIOU,
ORIANNA BAIRAKTARI, AND PETER BRIGHT

A MULTILINGUAL WORLD

In an increasingly culturally diverse world, knowledge of more than one language has become a prominent phenomenon of modern societies. It is estimated that over half of the world's population is bilingual due to factors such as educational requirements and migration (Grosjean, 2010; Grosjean, 2012). In Europe, 56 percent of the population can converse fluently in more than one language (European Commission, 2012). Likewise, in the US, 51 percent of students have been reported to use English as their second language (National Clearinghouse for English Language Acquisition, 2006).

The impact of second language learning on cognitive ability is vigorously debated in the literature. Early studies with children with different language backgrounds demonstrated a negative impact on cognitive development (Saer, 1922, 1923) and this led to changes in educational policy, which, in turn, dissuaded parents from raising their children in a bilingual environment due to the potential risk of incurring delays in intellectual development. This negative view of bilingualism was later challenged by reports of bilingual cognitive advantages on similar tests of general ability to those employed in the earlier studies (Peal and Lambert, 1962; Ben-Zeev, 1977). A candidate explanation for this discrepancy across studies concerns the level of methodological control, particularly with respect to matching across monolingual and bilingual groups. Socio-economic status (SES) has emerged as an important potential confound in these studies (Hakuta and Diaz, 1985), and recent research has more systematically factored in and/or controlled for group differences on this variable and other related measures.

Some of these studies have demonstrated that, even though bilinguals may have weaker verbal skills (Bialystok and Luk, 2012; Gollan et al., 2005), the acquisition of a second language promotes better cognitive ability for inhibiting irrelevant non-verbal information (Bialystok and Martin, 2004; Costa, Hernandez, and Sebastian-Galles, 2008).

One of the most influential explanatory frameworks for the observed benefits associated with bilingualism is the Inhibitory Control Model (ICM) proposed by Green (1986, 1998). This model suggests that general attentional resources are recruited in the voluntary control of language selection and production, while inhibition of the non-target language occurs during communication (Hilchey and Klein, 2011). The ICM is the theoretical framework which laid the foundations for the development of the *Bilingual Advantage Hypothesis* (Bialystok, 1999), which suggests that the continuous use of two or more languages leads to the reinforcement of general non-linguistic abilities (e.g., selective attention) resulting from continuous switching during communication (Bialystok, 2017). More recently, the ICM has been developed to incorporate the *Adaptive Control Hypothesis* (ACH). This model predicts that cognitive control is closely related to the use of language in context (Abutalebi and Green, 2016; Green and Abutalebi, 2013). For bilingual speakers the interactional context is important in order to adapt their cognitive control processes and to tune the networks of control. The relevant literature has shown that bilinguals, in comparison to monolinguals, are more efficient at background monitoring which helps to detect cues more frequently and elicit the appropriate response. Early empirical evidence for a bilingual advantage was demonstrated by Bialystok and Majumber (1998) in a study of two bilingual groups (English-French and Bengali-French) and a matched monolingual group of 7–10-year-old children on tasks requiring non-linguistic attentional control. Subsequent studies demonstrated a bilingual advantage on executive tasks requiring inhibition of irrelevant information and the ability to shift between targets (see Bialystok, 2009 for a review). Further support for this advantage has been obtained through studies that use visual (e.g., Bialystok, 1992; Bialystok et al., 2004) and auditory tasks (e.g., Filippi et al., 2012; Filippi et al., 2015) in children and adults. Nonetheless, as with the early studies demonstrating a bilingual disadvantage, these recent claims of a bilingual advantage have been scrutinized for having small size samples and poor experimental control over confounding variables such as SES and immigration status (Anton et al., 2014; Duñabeitia et al., 2014; Filippi, Ceccolini et al., 2020; Gathercole et al., 2014; Paap and Greenberg, 2013; Paap, Johnson, and Sawi, 2015; Morton and Harper, 2007).

CROSS-EXAMINING THE FOUNDATIONS OF "THE BILINGUAL ADVANTAGE"

The cumulative counter-evidence of this so-called bilingual advantage has raised the prospect that educational, cultural, social, and cognitive factors are so complex and heterogeneous that the required levels of matching between monolingual and multilingual groups may be impossible to achieve (for discussion see Filippi, D'Souza, and Bright, 2018).

In addition, controversies associated with the definition of bilingualism and monolingualism, specification of second language fluency in bilinguals, and heterogeneity in second language abilities in those classified as monolingual add an additional layer of complexity in this field of study (Hoffman, 2014). One of the most overlooked issues in bilingual research is the composition of monolingual groups, within which participants may vary considerably with respect to regional dialects and in their experience with other languages. Yet, these groups are typically treated as homogenous and naïve with respect to non-native forms of communication. This ongoing controversy in regard to the existence of a bilingual advantage has deeper roots than any of the previously mentioned studies have considered. Bilingualism as a concept is very complicated as it can interact with a variety of variables and produce unique results depending on specific circumstances and specific populations. Within the literature the term bilingualism has been defined in a variety of ways, from very restrictive (i.e., native-like mastery of two languages; Bloomfield, 1935) to more liberal and inclusive such as the alternate use of the two languages, irrespective of fluency (Weinreich, 1953). Perhaps the most popular description applied in this field has been that a bilingual is a person who uses two languages during their daily life (Grosjean, 1994). However, the fact that there is no incorporation of fluency or precise measurement in this conceptualization may, in part, sustain poor progress towards resolution of debate. Such concerns have provoked some researchers to question whether the bilingual advantage hypothesis is a viable subject for investigation. To promote progress in this field, we suggest that more sophisticated approaches will be required which incorporate current and long-standing issues raised on both sides of the debate. In particular, application of a more coherent developmental framework might be required in order to determine whether and when—in the developmental trajectory—multilingual environments might encourage cognitive change over and above that associated with monolingual contexts. Such a developmentally informed approach should, for example, provide the basis for identifying points of divergence in executive functions between monolinguals and bilinguals, how this differentiation evolves during development, and under what conditions bilingualism may offset age-related cognitive deterioration.

One theoretical framework that may provide a more integrated and holistic approach to these issues is encapsulated by the *neuroconstructivism* hypothesis (Karmiloff Smith, 1998). A fundamental claim is that acquired cognitive abilities emerge from context-dependent interactions that occur both within the child (e.g., neural assemblies) and between the child and the environment (e.g., object exploration; Karmiloff-Smith, 1998). Relevant research has mainly focused on understanding cognitive development in early childhood. This strategy takes into account the specialization of neurons that occurs during early development (Johnson, 2011) which may lead to "neural commitment" (Kuhl et al., 2006), suggesting that the ability of these neurons to respond to different or new forms of stimulation later in development will be constrained. Recent data have suggested that bilingual children are more efficient in processing information compared to their monolingual counterparts, and that this may result from enhanced sensitivity to novelty (D' Souza and D'Souza, 2016). We might therefore assume that bilingual infants will show reduced familiarity and increased novelty preference compared to monolingual infants (Filippi, D'Souza, and Bright, 2018). This has been demonstrated through studies with bilingual infants who have shown an improved ability to discriminate between familiar and novel languages, in comparison to their monolingual counterparts (Ramus et al., 2000; Mehler et al., 1988).

Research on the developmental impact of bilingualism in preschool children has employed measures of response inhibition as indices of cognitive control. In these tasks, the participants are required to inhibit an automatic or prepotent response tendency (the key experimental feature in the Simon, Flanker, and Stroop tasks). Bilingual children tend to outperform monolinguals on these tasks, even when controlling for alternative explanatory factors such as culture and social demographics at ages 4 (Yang, Yang and Lust, 2011), 6 (Barac and Bialystok, 2011), and 8 (Bialystok and Viswananthan, 2009). Moreover, 3-year-old bilingual children have been found to perform better on tests assessing Theory of Mind (ToM) (Kovacs, 2009; Bialystok and Senman, 2004), an ability which, like response inhibition, is closely associated with cognitive control and executive function (Austin, Groppe, and Elsner, 2014).

In summary, cognitive benefits associated with multi-language acquisition remain fiercely debated, and this impasse may, at least in part, relate to insufficient methodological rigor and the absence of a comprehensive, developmentally informed theoretical framework. New approaches are required in which these limitations can be properly addressed. Adoption of a consistent and broad developmental approach will ensure that like is compared with like, and will encourage a more fine-grained and plausible conception of the trajectory of cognitive change associated with multi-language processing across the lifespan.

BILINGUAL DEVELOPMENT IN LINGUISTIC AND NON-LINGUISTIC DOMAINS

A straightforward definition of bilingualism is the ability to use two languages in daily life, yet defining bilingualism for research purposes has proved problematic, not least due to the varying characteristics (e.g., level of proficiency and age of acquisition) that can influence classification as bilingual. The most accepted levels of dual language acquisition include native (from birth), early (preschool children with regular and continued exposure to more than one language) and late (learning a second language later in life).

Language skills in young bilingual children are highly varied as a result of the diversity in their language experience. The two main categories of dual language learning in children are native and early, an important distinction that requires careful consideration because the rates of development for each may vary considerably (Genesee, 2008).

Extensive research in native bilingual children has shown that when compared to their monolingual counterparts they are better able to 1) distinguish between familiar and unfamiliar languages at the age of 4 months (Bosch and Sebastián-Gallés, 1997), 2) comprehend segmented words from continuous speech in both languages at the age of 7 (Polka and Sundara, 2003), and 3) engage in canonical and variegated babbling at around 7–10 and 10–12 months of age, respectively (Maneva and Genesee, 2002; Oller, Eilers, Urbano, and Cobo-Lewis, 1997). Furthermore, they produce their first words in both languages at about the same age as monolingual children (12–13-months-old; Patterson and Pearson, 2004). In terms of their total conceptual vocabulary, when both languages were considered for bilinguals, the rate of vocabulary growth and overall vocabulary size were comparable to monolinguals (Pearson et al., 1993). Differences in vocabulary size, however, can be observed if only one language is considered for bilinguals (Ben-Zeev, 1977; Rosenblum and Pinker, 1983).

In contrast, research evidence for early bilinguals is limited and as a consequence they have been aggregated with the native bilingual population. For those few studies that have separately assessed early bilinguals, evidence indicates that they require 2 to 5 years to perform at the same level as a native speaker in their second language (see Callan, 2008, for a critical review). Taken together, these findings indicate that native bilinguals' language acquisition is systematic and exhibits the same critical milestones as the monolingual children, while early bilinguals' language acquisition seems to follow a similar process, however in a less timely fashion.

Other studies have suggested that bilingualism confers attentional advantages in language processing (Bialystok, 1987, 1988, 1999, 2001a, 2001b, 2007; Bialystok et al., 2004; Bialystok, Craik and Ryan, 2006), non-linguistic problem solving (Bialystok, 1999, 2006a, 2006b, 2007, 2009) and metalinguistic awareness (Bialystok, Majumber, and Martin, 2003). The latter has received most attention among the reported effects of bilingualism. Researchers in the field have subdivided metalinguistic awareness into two aspects, *representation* and *control of attention*. Representation "in this context" has been defined as the ability to think and analyse formal aspects of language (Dillon, 2009). Furthermore, it has been suggested that representation is regarded as a process, that develops with the general language ability as well as the improvement of proficiency and acquisition of literacy (Bialystok, 1988, 1999). Conversely, control of attention, which—in this context—is the ability to selectively attend to specific aspects of language and ignore irrelevant information, has consistently been reported as a bilingual-specific advantage in linguistic and non-linguistic tasks (Bialystok and Viswanathan, 2009; Bialystok, Craik and Luk, 2008).

Further research on non-linguistic domains has shown that bilingual cognitive benefits are dependent on language proficiency. Indeed, it has been demonstrated that in order to promote cognitive benefits through bilingualism, the person has to have reached age-specific proficiency in both respective languages (Antoniou et al., 2016; Crivello et al., 2016; Weber et al., 2016).

Numerous studies within the cognitive field of bilingualism have focused on the executive control system. Executive function is an umbrella term that encompasses the set of higher-order processes (such as inhibitory control, working memory, and attentional flexibility) that govern goal-directed actions and adaptive responses to novel, complex, or ambiguous situations (Hughes et al., 2006). Bilingual research in executive function has demonstrated advantages in cognitive flexibility and some aspects of inhibition, but there is no consistent evidence of an advantageous effect on performance related to working memory (Grundy, Anderson, and Bialystok, 2017).

Of these reported effects, cognitive flexibility, the ability to switch or "shift" between two different goal-relevant thoughts or actions (Davidson et al., 2006), has been identified as a function that is enhanced through the process of multi-language acquisition. In particular, young bilinguals have been found to outperform monolinguals on measures of task-switching. One task typically used to measure shifting ability is the Dimensional Change Sort task (DCCS; Zelazo, Fryes, and Rapus, 1996), in which participants are required to arrange cards on the basis of one dimension (e.g., color) and then alternate and use another dimension (e.g., shape). Shifting ability is determined by level of performance in the post-switch phase, with poor performance indicative of cognitive inflexibility. Bilingual children have been found to outperform monolinguals at age 4 to 5 (Carlson and Meltzoff, 2008). Bilingual adults have also shown reduced switching costs (difference in response time between switch and non-switch trials) compared to their monolingual counterparts in task switching, indicating superior shifting ability (Prior and MacWhinney, 2010).

Interference suppression or inhibitory control refers to the ability to resolve conflict by attending to relevant stimuli and ignoring distracting information (Friedman and Miyake, 2004). The literature has demonstrated bilingual advantages in this ability in studies with children (e.g., Kappa and Colombo, 2013; Martin-Rhee and Bialystok, 2008; Diamond, Kirkham, and Amso, 2002) and adults (e.g., Costa et al., 2009; Bialystok and DePape, 2009; Bialystok, Craik and Luk, 2008; Costa, Hernández, and Sebastián-Gallés, 2008; Bialystok et al., 2005). However, others have not replicated these results (e.g., Gathercole et al., 2014; Paap and Greenberg, 2013; Kousaie and Phillips, 2012a; Humphrey and Valian, 2012). There is evidence that the bilingual advantage may also operate in older age (Gold, Johson, and Powell., 2013; Emmorey et al., 2008; Bialystok, Craik, and Luk, 2008; Bialystok et al., 2005; Bialystok et al., 2004), underpinning claims that bilingualism may help offset age-related cognitive deterioration (Bialystok, Craik, and Luk, 2008). Nevertheless, bilingual advantages in ageing samples have not been found in classic motor response inhibition tasks such as the standard antisaccade task (Bialystok, Craik, and Ryan, 2006) or on other motor inhibition tasks with comparatively low working memory demands (Billig and Scholl, 2011).

Inhibitory control has predominantly been measured in two non-linguistic paradigms, the Simon and Flanker tasks. Much of the early evidence on the bilingual advantage however, has been based on Simon performance (Bialystok and Martin, 2004; Bialystok et al., 2005; Bialystok, Craik, and Ryan, 2006; Martin-Rhee and Bialystok, 2008). This task has been widely used due to its applicability in all age groups. Nonetheless, the results obtained from the Simon task should be interpreted with caution because performance among non-linguistic interference tasks has been shown to be poorly correlated (Stins et al., 2005; Kousaie and Phillips, 2012b; Paap and Greenberg, 2013). Thus, rather than reflecting a bilingual advantage in general inhibition, reported effects may, instead, reflect some aspect of performance which is specifically engaged by that particular task (and might therefore be of limited theoretical value (Paap and Greenberg, 2013).

Most studies of inhibitory control have focused on visual-based response inhibition, while auditory inhibition has received less attention. Research has shown that bilinguals outperform monolinguals in dichotic listening tasks with non-verbal (Bak, Vega-Mendoza, and Sorace, 2014; Soveri et al., 2011) and verbal stimuli (Bak, Vega-Mendoza, and Sorace, 2014). Moreover, other research has shown a bilingual advantage in auditory processing in a sentence interpretation task incorporating verbal interference, but only when target sentences were presented in the dominant language—an effect observed both in children (Filippi et al., 2015) and adults (Filippi et al., 2012). Moreover, the bilingual advantage was greatest when processing syntactically complex relative to simple sentences, a finding consistent with another reports highlighting a disproportionate bilingual advantage on the Flanker task in conditions incorporating high relative to low monitoring demands (Costa et al., 2009). These studies, therefore, indicate that higher-level cognitive processes associated with the bilingual advantage may be disproportionately engaged in more cognitively demanding or complex tasks.

The concept of *working memory* refers to the ability to store and manipulate information in the service of ongoing goal-directed demands (Zelazo et al., 2003), and to the extent that bilingualism confers a genuine advantage in cognitive control, we might expect bilinguals to outperform monolinguals on related tasks. Studies employing classic tests of working memory capacity (e.g., Digit Span and Corsi Block) report equivalent monolingual/bilingual performance (e.g., Bialystok and Martin, 2004; Bialystok, Craik, and Luk, 2008; Engel de Abreu et al., 2012; Engel de Abreu, 2011) but advantages typically emerge on tasks with high cognitive load conditions which place stronger monitoring demands on task performance (Bialystok et al., 2004; Grundy and Timmer, 2017; Blom et al., 2014; Morales, Calvo, and Bialystok, 2013; Biedroń and Szczepaniak, 2012; Hernandez, Costa, and Humphreys, 2012; Bialystok, 2010; Bialystok, Craik, and Luk, 2008; Feng, Diamond, and Bialystok, 2007). Typically, demanding tests of working memory which employ non-verbal stimuli are most likely to result in an observed advantage in bilinguals (Calvo, Ibanez, and Garcia, 2016), perhaps due to the generally poorer performance on verbal, relative to non-verbal, tasks observed in bilingual participants (Bialystok, 2009; Kerrigan et al., 2017). Whether working memory tasks place relatively passive/bottom-up vs active/top-down demands on performance also appears to predict the likelihood of observing a bilingual advantage (Hernandez, Costa, and Humphreys, 2012). For example, a bilingual advantage has been reported on a visual search task condition in which visual distractors held in working memory must be intentionally ignored (top-down performance), but not in a condition in which unexpected/task irrelevant visual distractors must be ignored (bottom-up performance), indicating equivalent attentional "capture" in bilinguals and monolinguals.

Bilingual cognitive costs have also been reported, on verbal fluency (Rosselli et al., 2000; Golan, Montoya, and Werner, 2002; Sandoval et al., 2010), lexical access (Roberts et al., 2002; Gollan, et al., 2005), receptive vocabulary (Bialystok, 2007, 2009; Ivanova and Costa, 2008) and in the solving of mathematical word problems with and without additional numerical distractors (Kempert, Saalbach, and Hardy, 2011). Of these observed disadvantages, lexical access deficits have received the most attention. The two most prevalent explanatory theories include: (a) possible "weaker links" in brain connections responsible for effortless and coherent speech production caused by limited use of both languages (Michael and Gollan, 2005) and (b) conflict caused by competition from the non-target language which requires inhibition via executive control (Green, 1998).

BILINGUALISM PROTECTING AGAINST COGNITIVE DECLINE

Increased demands on cognition associated with knowledge and effective manipulation of two different languages provides the context for the bilingual advantage hypothesis. There is some evidence that ongoing disproportionate demands in the bilingual mind may alleviate deleterious effects of ageing on cognition (Adesope et al., 2010; Bialystok et al., 2004; Salvatierra and Rosselli, 2011). The main evidence base for this contentious view comprises retrospective reports that, among patients with a dementia diagnosis, monolinguals receive their diagnosis significantly earlier (by approximately 4 years) than bilinguals (Bialystok, Craik, and Freedman, 2007; Craik, Bialystok, and Freedman, 2010; Schweizer et al., 2012).

Bak, Vega-Mendoza, and Sorace (2014) exploited the Lothian Birth Cohort (Deary et al., 2007) to assess cognition in 853 participants first tested in 1947 (age 11) and again in 2008–10, reporting more significant deterioration in intellectual ability and reading in monolinguals relative to those classified as bilingual (a significant minority within this cohort). Nevertheless, these findings have been challenged by null results that demonstrate no such effect of bilingualism (Crane et al., 2009; Sanders et al., 2012; Lawton, Gasquoine, and Weimer, 2015). A very recent study in which monolingual and bilingual older adults were rigorously matched by age, gender, and socio-economic status has shown no difference between the two groups in non-verbal reasoning, verbal working memory, visuo-spatial memory, response inhibition, or problem solving/decision making (Papageorgiou et al., 2019).

At present, therefore, the issue of whether bilingualism offers a genuine protective effect against age-related cognitive deterioration remains contested and unresolved.

The Multilingual Brain

Neuroimaging has provided the opportunity to investigate the neural correlates of executive control in the bilingual brain. Advocates of the bilingual advantage hypothesis propose that language experience shapes domain-general cognition and this process underpins the greater development of higher-level cognitive control associated with bilingualism (Costa and Sebastian-Galles, 2014), through inhibitory control mechanisms in the dorsolateral prefrontal cortex (DLPFC) as suggested by the Inhibitory Control Model (ICM; Green, 1998). This theory has been supported by studies of picture naming performance (Hernandez, 2009; Hernandez et al., 2001) and has been developed further to incorporate brain regions associated with multi-language processing and control (Abutalebi and Green, 2008; Green and Abutalebi, 2013).

Studies have reported increased grey matter volume and density in the left inferior parietal structures (areas involved in verbal fluency processing; Mechelli et al., 2004;), the left putamen (articulatory and phonological processes; Abutalebi et al., 2013), left caudate nucleus (phonological detections; Li, Legault, and Litcofsky, 2014), and Heschl's gyrus (auditory processing; Ressel et al., 2012) in bilinguals relative to age-matched monolinguals. Another study in bilinguals revealed a significant positive correlation between tissue density in the right posterior paravermis in the cerebellum and the ability to control verbal interference from the dominant language while processing heard sentences in the weaker language (Filippi et al., 2011; 2020). This was the first study to identify and isolate the importance of cerebellar areas in language processing. Many neuroimaging studies have excluded the cerebellum from analysis in order to, for example, increase sensitivity by limiting the field of view, and therefore may have missed important structural/functional relationships.

Structural differences have also been reported in the white matter of specific brain areas. Studies involving older participants have shown stronger white matter integrity in bilinguals' corpus callosum, occipital fasciculus, uncinate fasciculus and bilateral superior longitudinal fasciculi (Luk et al., 2011). Functional magnetic resonance imaging (fMRI) studies have demonstrated stronger activation in the DLPFC and left inferior frontal cortex (IFC) in native bilinguals relative to monolinguals during the completion of a syntactic sentence judgement task including plausible and non-plausible sentences (Kovelman, Baker, and Petitto, 2008; Klein et al., 2014). Further evidence for enhanced recruitment of language processing areas in the bilingual brain was presented by Parker Jones et al. (2012). These authors identified stronger activation of five left-hemisphere language areas (dorsal precentral gyrus, pars triangularis, pars opercularis, superior temporal gyrus, and planum temporale) in bilinguals during word retrieval and articulation tasks. Other studies have demonstrated that the left caudate and the left anterior cingulate cortex are disproportionately recruited during bilingual language processing in highly proficient early bilinguals (Abutalebi, 2008; Abutalebi et al., 2008; Abutalebi et al., 2007; Garbin et al., 2011).

Neuroimaging studies have also examined the effect of bilingualism on executive control in children aged 8–11 years old. In an fMRI study of Simon task performance, Mohades et al. (2014) reported stronger and more extensive activation in incongruent trials compared to congruent trials in bilingual children when compared to their monolingual counterparts. The main areas associated with this group difference included the superior temporal gyrus, caudate nucleus, middle frontal gyrus, posterior cingulate gyrus, and precuneus. These regions are typically recruited for language conflict situations, however the authors concluded that the stronger and more extensive activation in these areas could have effects on non-verbal conflict processing in bilingual children. This could explain bilinguals' lower non-verbal performance in comparison to monolinguals.

A study of Simon performance in older adults has identified a different distributed network associated with conflict resolution, within which the left inferior parietal cortex was disproportionately recruited in the bilingual participants and the right middle frontal gyrus disproportionately recruited in the monolinguals (Ansaldo, Ghazi-Saidi, and Adrover-Roig, 2015).

In a study with young adults in which the Flanker task was used to examine inhibition control, the activated network associated with bilingual performance included the bilateral inferior frontal and temporal cortices and subcortical regions thought to be involved in cognitive control (Luk et al., 2010). Since both the Simon and Flanker tasks measure inhibition control, it might have been logical to predict that they would recruit the same or similar brain networks. However, a comparison of these two studies shows that this was not the case. Candidate explanations include the possibility that adults recruit different networks than children, that the task taps different forms of inhibition, or that they engage both shared and distinct cognitive mechanisms. Relevant research comparing switching to non-switching tasks in children have reported increased activation of the left inferior frontal gyrus for bilinguals (Garbin et al., 2010; Rodriguez-Pujadas et al., 2013).

With regard to the possible contribution of lifelong use of two or more languages to protect the brain from the effects of ageing, neuroimaging research has shown that bilingualism may contribute to the *neural reserve*, which involves neurostructural changes in grey and white matter and *neural compensation*, which involves neuro-functional changes associated with brain connectivity (Abutalebi et al., 2007). These phenomena have been studied both in neurologically healthy participants and in patients with progressive neurological conditions. Bilingual research with healthy older adults has indicated neuroplastic changes in the grey matter of left anterior inferior temporal gyrus (Abutalebi et al., 2014), left and right inferior parietal lobe (Abutalebi et al., 2015a) and left and right anterior cingulate cortex (Abutalebi et al., 2015b). Additionally, Luk et al. (2011) found increased white matter connectivity in bilinguals relative to age-matched monolinguals in the corpus callosum extending anteriorly to frontal white matter.

In a series of fMRI experiments involving monolingual and bilingual adults, with an age range between 60–70 years old, Gold et al. (2013) presented evidence that bilinguals had significantly decreased activation than monolinguals of left lateral frontal cortex and cingulate cortex. Despite this, their cognitive performance was comparable to their monolingual counterparts (i.e., same performance with less neuronal effort), suggesting that lifelong bilingualism may indicate a more efficient use of resources in crucial brain areas that could potentially offset age-related brain degeneration.

Bilingualism research in patients with Alzheimer's disease (AD) and mild cognitive impairment (MCI) has not been as extensive as that conducted in neurologically healthy ageing participants. One study with Alzheimer's patients indicated that bilinguals and monolinguals matched on age and education showed similar cognitive performance (on memory, language, attention, visuospatial function, and executive function tasks) even though the bilinguals demonstrated a more extensive medial temporal lobe atrophy (Schweizer et al., 2012). These results were also supported by Duncan et al. (2018) who found that multilingual MCI and Alzheimer patients had increased grey matter density in language and cognitive control areas compared to matched within-diagnosis monolingual patients. The relationship between a lifelong multilingual experience and possible beneficial effects on neurodegenerative disorders is still poorly understood and requires further systematic inquiry, perhaps by adopting a developmental trajectory approach under a neuroconstructivist framework (see Filippi, D'Souza, and Bright, 2018, for discussion).

In summary, we have outlined the ongoing debate around the existence and characteristics of the bilingual advantage. Much current research activity is directed towards resolving the binary debate on whether there is a genuine cognitive advantage, which is driven by the process of acquiring a second language, and it remains fiercely contested on both sides (e.g., Bialystok, 2009; Paap and Greenberg, 2013). The fact that the issue remains unresolved has naturally raised significant concerns and argument regarding group allocations, terminology, implemented tasks, and the adequacy of control over alternative explanatory variables. There is currently very limited, if any, evidence of a trend towards consensus in this debate. However, we suggest that systematic implementation of a developmental, neuroconstructivist framework will offer the opportunity to work towards a comprehensive model within which the inconsistencies in the literature might be resolved. We hope that such a framework will provide better insight on the cognitive impact of second language learning beyond a straightforward advantage or disadvantage, allowing researchers to identify more qualitative differences in cognitive processing associated with different forms and levels of ongoing linguistic experience throughout the developmental trajectory.

BILINGUALISM IN ATYPICAL CIRCUMSTANCES

Research on the effects of bilingualism in the developing brain has predominantly focused on typical development. However, it may also be instructive to consider bilingual development from an atypical perspective. Exploration of both typical and atypical populations, in which developmental trends in cognition are assessed in monolingual and bilingual contexts, provides the opportunity to identify theoretically and clinically important behavioral and neurocognitive outcomes (Filippi and Karmiloff-Smith, 2013). In particular, parents, educators, and practitioners may gain insight on whether and how bilingualism interacts with developmental disorders and adapt parental, educational and/or clinical practices accordingly. Research relevant to this issue is sparse, but in the following sections we review some current perspectives.

Autism Spectrum Disorder (ASD)

Parents of children with autism spectrum disorder (ASD) believe (Yu, 2013) or are commonly advised (Kay-Raining Bird Lamond, and Holden, 2012) that their children should not be raised bilingual, as this would put an additional strain on already pronounced communication difficulties. However, research is inconsistent with this position, and provides limited if any evidence that multi-language acquisition places additional strain on children with ASD in the social domain (Iarocci, Hutchinson, and O'Toole, 2017; Reetze, Zou, and Sheng, 2015; Hambly and Fombonne, 2014; Valicenti-McDermott et al., 2013; Ohashi et al., 2012; Hambly and Fombonne, 2012), language development (Beauchamp and McLeod, 2017; Kay-Raining Bird, Lamond, and Holden, 2012), or vocabulary, once non-verbal intelligence (Gonzalez-Barrero and Nadig, 2016; Petersen, Marinova-Todd, and Mirenda, 2012) and EF (Iarocci, Hutchinson, and O'Toole, 2017; Hambly and Fombonne, 2014; see also Drysdale, van der Meer, and Kagohara, 2015 for a review) has been accounted for. Moreover, timing of acquisition does not differentiate native and early bilinguals; children acquiring a second language before and after 12 months of age do not differ in their social or cognitive outcomes (Hambly and Fombonne, 2014). Most importantly, evidence suggests that bilingual children show comparable functional communication and executive function outcomes to monolinguals even when they were not tested in their primary language (Iarocci, Hutchinson, and O'Toole, 2017). Indeed, some have even reported a bilingual advantage. For example, Valicenti-McDermott et al. (2013) reported advanced pretend play, cooing, and gestures in bilingual toddlers with ASD compared to matched monolingual ASD toddlers. Furthermore, bilinguals with ASD aged from 3 to 7 years old display larger total vocabulary than monolinguals (Petersen, Marinova-Todd, and Mirenda, 2012) and a shifting advantage has been reported for bilinguals with ASD compared to their monolingual counterparts (Gonzalez-Barrero and Nadig, 2019).

Despite this weight of evidence, a smaller number of studies report a bilingual *disadvantage* in this population. For example, there is evidence that higher exposure to a second language is associated with poorer receptive and expressive language (Chaidez, Hansen, and Hertz-Picciotto, 2012). Furthermore, an argument has been made regarding the applicability of a bilingual benefit; for example, while a bilingual shifting advantage has been found in experimental settings (i.e., by performing the DCCS task), autistic bilingual individuals showed no difference in everyday set-shifting when compared with their monolingual peers (Gonzalez-Barrero and Nadig, 2019). Social constraints have also been found in parent-based interventions, with individualized behavioral instructions. Monolingual 3-month-olds have been reported to display more gestures than bilinguals, indicative of superior communication, but bilinguals catch up by the age of 1 or 2 years, therefore possibly benefiting more from the intervention (Zhou et al., 2019).

In summary, despite some evidence for disadvantages of raising autistic children in bilingual environments, there may be important potential advantages, both cognitive and social. A bilingual environment may offer greater opportunities for communication and development of social skills (Iarocci. Hutchinson, and O'Toole, 2017). Parents also report greater satisfaction in immersing their child in both the majority and minority language, because this allows them to better express themselves with their child in their language of preference and create an additional cultural connection with their child (Beauchamp and McLeod, 2017; Yu, 2013).

Attention Deficit Hyperactivity Disorder (ADHD)

Bilingualism research in attention deficit hyperactivity disorder (ADHD) is less developed than in ASD. This developmental disorder is characterized by deficits in interference suppression (Boonstra et al., 2010), and, given that bilingualism is frequently associated with enhanced interference suppression, we might expect bilingual environments to be particularly beneficial for children with the condition. Indeed, early evidence showed levels of second language proficiency in children diagnosed with ADHD was negatively correlated with ADHD symptoms (Toppelberg et al., 2002).

However, more recent studies have provided conflicting results. For example, Bialystok et al. (2017) investigated the interaction between ADHD and executive function in four groups of young adults (bilingual/monolingual—with ADHD/without ADHD). Results suggested that ADHD might confer a greater burden in response inhibition in bilinguals than in monolinguals (Bialystok et al., 2017). Another study reports that bilingualism may disadvantage those with ADHD. Mor, Yitzhaki-Amsalem, and Prior (2015) provided evidence for poor interference suppression in university students with ADHD, with the lowest scores produced by bilinguals.

Down Syndrome (DS)

Children with Down syndrome (DS) typically present with evidence of poor language acquisition (Cleave et al., 2014). Therefore, parents are typically advised to refrain from raising their children as bilingual (Paradis, Genesse, and Grago, 2011). However, there is no reliable evidence that bilingualism adds additional strain in children with DS when matched on mental age with typically developed children (Cleave et al., 2014; Feltmate and Kay-Raining Bird, 2008; Kay- Raining Bird et al., 2005). Some research has identified more constrained linguistic expression in bilinguals with DS (Kay-Raining Bird et al., 2005), but bilingual DS children surpass their monolingual counterparts in their ability to learn new words and even resemble typically developing children in this domain (Cleave et al., 2014).

Dyslexia, Specific Language Impairment (SLI), and Reading Difficulties

Even in the area of developmental language disorders there is a dearth of work investigating the effects of being raised in a multilingual environment. For example, dyslexia is sensitive to the morphological and phonological characteristics of a given language. Therefore, the level of difficulty in acquiring a second language will be dependent on the language characteristics. Chinese, for example, has logograms (i.e., a sign or character representing a word or phrase) which means that dyslexia is less apparent in Chinese speakers in general (Ho and Fong, 2005). Moreover, Chinese dyslexics may exhibit difficulties learning English but they are less likely to suffer from dyslexia in both languages (Ho and Fong, 2005). Nevertheless, there are some studies showing the dyslexic effect across languages regardless of their differences (Siegel, 2016; Gupta and Jamal, 2007). While typically developing individuals with Hindi as their native language (L1) show greater reading accuracy in Hindi than in English, dyslexic participants had difficulties with reading in both languages (Gupta and Jamal, 2007). Furthermore, dyslexics seem to improve their phonological literacy skills when the language learned is characterized by phonemic orthography (Siegel, 2016).

The effects of bilingualism on those with reading difficulties has received scant attention in the literature. Perhaps the only study conducted to date demonstrated that bilingual children with reading difficulties elicit the slowest responses on a grammatical production task in comparison to monolingual children (Jacobson and Schwartz, 2005). Further research is clearly warranted.

Another language learning disability that has been studied in conjunction with bilingualism is SLI. This impairment is characterized by difficulties in language development that cannot be accounted for by other cognitive/physical delays or difference in socio-economic status. While some parents appear reluctant to raise children with SLI as bilingual (Paradis, Crago, and Genesee, 2006), research does not indicate additional difficulties associated with bilingualism (Paradis. Crago, and Genesee, 2006). In this line of research, native bilinguals with extensive exposure to both languages have shown comparable understanding of morphology and syntax to monolinguals in their dominant language (Morgan, Restrepo, and Auza, 2013), or both (Illuz-Cohen and Walters, 2012).

NEW CHALLENGES IN BILINGUAL RESEARCH

As previously discussed, the claimed benefits acquired through bilingualism have been robustly challenged in the recent literature (e.g., Morton and Harper, 2007; de Bruin, Treccani, and Della Sala, 2015; Paap and Greenberg, 2013). Across multiple measures of executive functioning, equivalent performance in monolinguals and bilinguals has been reported (Duñabeitia et al., 2014; Paap, Johnson, and Sawi, 2014). A systematic review also demonstrated that the bilingual advantage in relation to conflict resolution (i.e., inhibitory abilities of bilinguals) is sporadic, that the research is based on weak findings produced by small studies and that they inadequately account for potential confounding variables (Hilchey and Klein, 2011). Publication bias has been raised as particularly problematic in this field (de Bruin, Treccani, and Della Sala, 2015) and there is an on-going debate concerning convergent and ecological validity of tasks routinely employed (e.g., Paap, Johnson, and Sawi, 2014).

Research will benefit from an informed consensus on who can be considered a bilingual, what the appropriate criteria for monolingual controls are, and how best to categorize levels of bilingualism. Native and early bilinguals may present different behavioral (Aguilar-Mediavilla et al., 2014) and neurological (Mohades et al., 2012) outcomes, and level of biliteracy (the ability to read and write in more than one language) might also be an important factor associated with cognitive development. These concerns have received scant attention in the literature. We suggest that the distinction between basic and formal knowledge of a second language is an important consideration, but it is not routinely assessed or controlled for, in studies of bilingual cognition.

Additionally, the literature lacks a clear and well-grounded theory that could explain how and why the management of two languages would impact executive functioning (Hartsuiker, 2015; Treccani and Mulatti, 2015). The only model that provides a prediction about the effect of the bilingual advantage on inhibitory control is the Inhibitory Control Model (ICM; Green, 1986). Nevertheless, even this model has not been successful in predicting or explaining the infrequently reported advantages in inhibitory control nor why these advantages often extend to congruent (i.e., conditions with no obvious element of response inhibition) in addition to incongruent trials. Moreover, further research would need to be conducted on the direction of causality between bilingualism and executive functioning given the level of inconsistency in the field (Cox et al., 2016; Li and Grant, 2015).

We have outlined some of the concerns regarding tasks employed in the literature. In particular, the Simon task has been criticized as ineffective in capturing the specific executive mechanism(s) that drive observed cognitive advantages associated with isolating an interference effect from a wider EF effect (Borgmann et al., 2007). Moreover, performance on typical laboratory tests of executive function may not reflect real world executive/goal directed abilities (Gonzalez-Barrero and Nadig, 2019).

Finally, while bilingualism refers to language development and linguistic processing, most tasks of executive function used in bilingual research place visual rather than verbal or auditory demands on participants, and are arguably, therefore, tapping at least partially independent functional networks than those exercised via the process of acquiring an additional language (Filippi, D'Souza, and Bright, 2018). Future research should address these concerns by employing tasks that are more reflective of the actual bilingual experience.

The contribution of alternative explanatory variables for the bilingual advantage, such as socio-economic status, cultural considerations, and genetics have received considerable recent attention (e.g., Paap and Greenberg, 2013; Filippi, D'Souza, and Bright, 2018) but more work is needed given the ongoing unresolved balance of evidence. In parallel, research into developmental disorders might provide a window on how bilingual experience may interact in the presentation of symptoms and trajectory of these conditions (Beauchamp and McLeod, 2017; Kay-Raining Bird, Lamond, and Holden, 2012) as well as executive function (Drysdale, van der Meer, and Kagohara, 2015). Such observations may have implications not only with respect to our understanding of developmental conditions themselves but also the wider impact that bilingualism has on typical development.

A Neuroconstructivist Developmental Approach to Bilingual Research

No single experience renders someone bilingual. It is, rather, a combination of the amount of exposure (Chaidez, Hansen, and Hertz-Picciotto, 2012), type of exposure (Iarocci, Hutchison and O'Toole, 2017), and even personal attributions to language status (Verhoeven and Aarts, 1998). Whether bilinguals excel or fall behind in given cognitive domains compared to monolinguals is currently unclear. What *is* clear, however, is that bilingualism offers a different developmental experience.

It may seem absurd to state that development always needs to be studied with the aim to understand development! Yet, as in many other research areas also, bilingual research is not immune to study infants, children, and young adults in a non-developmental way. The vast majority of published work to date focuses on single age groups in single domains. Neuroconstructivism, by contrast, stresses the need to study *change over developmental time* and to focus on the interaction of different domains across time.

Indeed, bilingualism is a continuous experience and, therefore, we suggest that it is important to adopt a neuroconstructivist approach, monitoring progression and assessing possible factors that interact with and impact on other forms of cognition at particular points in the developmental trajectory (Filippi, D'Souza, and Bright, 2018).

CONCLUSIONS

In this article we have encouraged the adoption of a developmental approach that may ultimately resolve intractable positions currently espoused in the literature and lead to a broad and comprehensive understanding of multi-language processing starting from early life to older age. Future bilingualism research should also incorporate genetics, sociocultural, and environmental factors, and extend theoretical frameworks to include atypical development (e.g., autism, ADHD, and Down Syndrome). Research would benefit from longitudinal models and focus on younger and older age groups for a more holistic view of the bilingual experience (Filippi, D'Souza, and Bright, 2018).

Monolingual/bilingual group comparisons of overall task performance are certainly necessary to identify possible points of divergence in cognitive ability, but other approaches may also be informative in identifying *qualitatively* different approaches to thinking and operating within complex environments (which may or may not underpin quantitative measures of performance). Neurological research, for example, shows that it is not a matter of better or worse processing, there are different brain areas (Mohades et al., 2012; Luk et al., 2010) and event-related potential latencies (Oren and Brezniz, 2005: Moreno et al., 2014; Fernandez et al., 2013) involved in how bilinguals and monolinguals process cognitive tasks. The importance of this field of research is not restricted to straightforward determination of the existence (or nature) of the bilingual advantage, but also has potential ramifications for educational practices in the service of educational attainment. In this context, it may be instructive to explore potential differences in the approaches adopted by monolingual and bilingual participants independently of actual performance outcomes in order to identify optimal/individualized educational practices to promote subsequent learning.

The aim of this chapter is to offer the reader a comprehensive overview of the history of bilingualism and its developmental impact on cognition by providing behavioral and imaging evidence to date from typical and atypical samples. We also reviewed current claims in the literature ranging from the pernicious to the advantageous effects of second language experience, as well as covering the controversies in the field concerning the methodological approaches and the replicability, accuracy, and interpretation of findings. The chapter concludes by recommending that relevant future work should incorporate language experience and its assessment if it is to capture how multi-language acquisition and processing influences high-level cognition.

ACKNOWLEDGMENTS

This work was supported by the Leverhulme Trust UK [RPG-2015-024] and the British Academy [SG162171]. A special thought goes to Prof. Annette Karmiloff-Smith who inspired our research.

References

Abutalebi, J. (2008). Neural aspects of second language representation and language control. *Acta Psychologica*, 128(3), 466–478.

Abutalebi, J. and Green, D. (2008). Control mechanisms in bilingual language production: Neural evidence from language switching studies. *Language and Cognitive Processes*, 23(4), 557–582.

Abutalebi, J. and Green, D. (2016). Neuroimaging of language control in bilinguals: Neural adaptation and reserve. *Bilingualism: Language and Cognition*, 19(4), 689–698.

Abutalebi, J., Annoni, J., Zimine, I., Pegna, A., Seghier, M., Lee-Jahnke, H., Lazeyras, F., Cappa, S., and Khateb, A. (2008). Language control and lexical competition in bilinguals: An event-related fMRI study. *Cerebral Cortex*, 18(7), 1496–1505.

Abutalebi J., Brambati S., Annoni J., Moro A., Cappa S., and Perani D. (2007). The neural cost of the auditory perception of language switches: An event-related functional magnetic resonance imaging study in bilinguals. *The Journal of Neuroscience*, 27, 13762–13769.

Abutalebi, J., Canini, M., Della Rosa, P., Green, D., and Weekes, B. (2015a). The neuroprotective effects of bilingualism upon the inferior parietal lobule: A structural neuroimaging study in aging Chinese bilinguals. *Journal of Neurolinguistics*, 33, 3–13.

Abutalebi, J., Canini, M., Della Rosa, P., Sheung, L., Green, D., and Weekes, B. (2014). Bilingualism protects anterior temporal lobe integrity in aging. *Neurobiology of Aging*, 35(9), 2126–2133.

Abutalebi, J., Della Rosa, P., Gonzaga, A., Keim, R., Costa, A., and Perani, D. (2013). The role of the left putamen in multilingual language production. *Brain and Language*, 125(3), 307–315.

Abutalebi, J., Guidi, L., Borsa, V., Canini, M., Della Rosa, P., Parris, B., and Weekes, B. (2015b). Bilingualism provides a neural reserve for aging populations. *Neuropsychologia*, 69, 201–210.

Adesope, O., Lavin, T., Thompson, T., and Ungerleider, C. (2010). A systematic review and meta-analysis of the cognitive correlates of bilingualism. *Review of Educational Research*, 80(2), 207–245.

Aguilar-Mediavilla, E., Buil-Legaz, L., Perez-Castello, J. A., Rigo-Carratala, E., and Adrover-Roig, D. (2014). Early preschool processing abilities predict subsequent reading outcomes in bilingual Spanish–Catalan children with specific language impairment (SLI). *Journal of Communication Disorders*, 50, 19–35.

Ansaldo, A., Ghazi-Saidi, L., and Adrover-Roig, D. (2015). Interference control in elderly bilinguals: Appearances can be misleading. *Journal of Clinical and Experimental Neuropsychology*, 37(5), 455–470.

Antón, E., Duñabeitia, J., Estévez, A., Hernández, J., Castillo, A., Fuentes, L., Davidson, D., and Carreiras, M. (2014). Is there a bilingual advantage in the ANT task? Evidence from children. *Frontiers in Psychology*, 5, 398.

Antoniou K., Grohmann K., Kambanaros M., and Katsos N. (2016). The effect of childhood bilectalism and multilingualism on executive control. *Cognition*, 149, 18–30.

Austin, G., Groppe, K., and Elsner, B. (2014). The reciprocal relationship between executive function and theory of mind in middle childhood: A 1-year longitudinal perspective. *Frontiers in Psychology*, 5, 655.

Bak, T., Vega-Mendoza, M., and Sorace, A. (2014). Never too late? An advantage on tests of auditory attention extends to late bilinguals. *Frontiers in Psychology*, 5, 485.

Barac, R. and Bialystok, E. (2011). Cognitive development of bilingual children. *Language Teaching*, 44(1), 36–54.

Beauchamp, M. and MacLeod, A. (2017). Bilingualism in children with autism spectrum disorder: Making evidence based recommendations. *Canadian Psychology/Psychologie Canadienne*, *58*(3), 250.

Ben-Zeev, S. (1977). The influence of bilingualism on cognitive strategy and cognitive development. *Child Development*, *48*(3), 1009–1018.

Bialystok, E. (1987). Influences of bilingualism on metalinguistic development. *Interlanguage Studies Bulletin* (Utrecht), *3*(2), 154–166.

Bialystok, E. (1988). Levels of bilingualism and levels of linguistic awareness. *Developmental psychology*, *24*(4), 560.

Bialystok, E. (1992). Attentional control in children's metalinguistic performance and measures of field independence. *Developmental Psychology*, *28*(4), 654.

Bialystok, E. (1999). Cognitive complexity and attentional control in the bilingual mind. *Child Development*, *70*(3), 636–644.

Bialystok, E. (2001a). *Bilingualism in Development: Language, Literacy, and Cognition*. Cambridge: Cambridge University Press.

Bialystok, E. (2001b). Metalinguistic aspects of bilingual processing. *Annual Review of Applied Linguistics*, *21*, 169–181.

Bialystok, E. (2006a). Effect of bilingualism and computer video game experience on the Simon task. *Canadian Journal of Experimental Psychology/Revue Canadienne de Psychologie Expérimentale*, *60*(1), 68.

Bialystok, E. (2007). Cognitive effects of bilingualism: How linguistic experience leads to cognitive change. *International Journal of Bilingual Education and Bilingualism*, *10*(3), 210–223.

Bialystok, E. (2009). Bilingualism: The good, the bad, and the indifferent. *Bilingualism: Language and Cognition*, *12*(1), 3–11.

Bialystok, E. (2010). Global–local and trail-making tasks by monolingual and bilingual children: Beyond inhibition. *Developmental Psychology*, *46*(1), 93.

Bialystok, E. (2017). The bilingual adaptation: How minds accommodate experience. *Psychological Bulletin*, *143*(3), 233–262.

Bialystok, E. and DePape, A. (2009). Musical expertise, bilingualism, and executive functioning. *Journal of Experimental Psychology: Human Perception and Performance*, *35*(2), 565.

Bialystok, E. and Luk, G. (2012). Receptive vocabulary differences in monolingual and bilingual adults. *Bilingualism: Language and Cognition*, *15*(2), 397–401.

Bialystok, E. and Majumder, S. (1998). The relationship between bilingualism and the development of cognitive processes in problem solving. *Applied Psycholinguistics*, *19*(1), 69–85.

Bialystok, E. and Martin, M. (2004). Attention and inhibition in bilingual children: Evidence from the dimensional change card sort task. *Developmental Science*, *7*(3), 325–339.

Bialystok, E. and Senman, L. (2004). Executive processes in appearance–reality tasks: The role of inhibition of attention and symbolic representation. *Child Development*, *75*(2), 562–579.

Bialystok, E. and Viswanathan, M. (2009). Components of executive control with advantages for bilingual children in two cultures. *Cognition*, *112*(3), 494–500.

Bialystok, E., Craik, F., and Freedman, M. (2007). Bilingualism as a protection against the onset of symptoms of dementia. *Neuropsychologia*, *45*(2), 459–464.

Bialystok, E., Craik, F., and Luk, G. (2008). Cognitive control and lexical access in younger and older bilinguals. *Journal of Experimental Psychology: Learning, Memory, and Cognition*, *34*(4), 859.

Bialystok, E., Craik, F., and Ryan, J. (2006). Executive control in a modified antisaccade task: Effects of aging and bilingualism. *Journal of experimental psychology: Learning, Memory, and Cognition*, 32(6), 1341.

Bialystok, E., Craik, F., Grady, C., Chau, W., Ishii, R., Gunji, A., and Pantev, C. (2005). Effect of bilingualism on cognitive control in the Simon task: Evidence from MEG. *NeuroImage*, 24(1), 40–49.

Bialystok, E., Craik, F., Klein, R., and Viswanathan, M. (2004). Bilingualism, aging, and cognitive control: Evidence from the Simon task. *Psychology and Aging*, 19(2), 290.

Bialystok, E., Hawrylewicz, K., Wiseheart, M., and Toplak, M. (2017). Interaction of bilingualism and attention-deficit/hyperactivity disorder in young adults. *Bilingualism: Language and Cognition*, 20(3), 588–601.

Bialystok, E., Majumder, S., and Martin, M. (2003). Developing phonological awareness: Is there a bilingual advantage? *Applied Psycholinguistics*, 24(1), 27–44.

Biedroń, A. and Szczepaniak, A. (2012). Working memory and short-term memory abilities in accomplished multilinguals. *The Modern Language Journal*, 96(2), 290–306.

Billig, J. and Scholl, A. (2011). The impact of bilingualism and aging on inhibitory control and working memory. *Organon*, 26(51).

Blom, E., Küntay, A., Messer, M., Verhagen, J., and Leseman, P. (2014). The benefits of being bilingual: Working memory in bilingual Turkish–Dutch children. *Journal of Experimental Child Psychology*, 128, 105–119.

Bloomfield, L., (1935). Linguistic aspects of science. *Philosophy of Science*, 2(4), 499–517.

Boonstra, A., Kooij, J., Oosterlaan, J., Sergeant, J., and Buitelaar, J. (2010). To act or not to act, that's the problem: Primarily inhibition difficulties in adult ADHD. *Neuropsychology*, 24(2), 209.

Borgmann, K., Risko, E., Stolz, J., and Besner, D. (2007). Simon says: Reliability and the role of working memory and attentional control in the Simon task. *Psychonomic Bulletin and Review*, 14(2), 313–319.

Bosch, L., and Sebastián-Gallés, N. (1997). Native-language recognition abilities in 4-month-old infants from monolingual and bilingual environments. *Cognition*, 65(1), 33–69.

Callan, E. (2008). Critical Review: Does bilingualism slow language development in children? 20 (2012, November), 2007–2008.

Calvo, N., Ibáñez, A., and García, A. (2016). The impact of bilingualism on working memory: A null effect on the whole may not be so on the parts. *Frontiers in Psychology*, 7, 265.

Carlson, S. and Meltzoff, A. (2008). Bilingual experience and executive functioning in young children. *Developmental Science*, 11, 282–298.

Chaidez, V., Hansen, R., and Hertz-Picciotto, I. (2012). Autism spectrum disorders in Hispanics and non-Hispanics. *Autism*, 16(4), 381–397.

Cleave, P., Bird, E., Trudeau, N., and Sutton, A. (2014). Syntactic bootstrapping in children with Down syndrome: The impact of bilingualism. *Journal of Communication Disorders*, 49, May–June, 42–54.

Costa, A. and Sebastián-Gallés, N. (2014). How does the bilingual experience sculpt the brain? *Nature Reviews Neuroscience*, 15(5), 336.

Costa, A., Hernández, M., and Sebastián-Gallés, N. (2008). Bilingualism aids conflict resolution: Evidence from the ANT task. *Cognition*, 106(1), 59–86.

Costa, A., Hernández, M., Costa-Faidella, J., and Sebastián-Gallés, N. (2009). On the bilingual advantage in conflict processing: Now you see it, now you don't. *Cognition*, 113(2), 135–149.

Cox, S., Bak, T., Allerhand, M., Redmond, P., Starr, J., Deary, I., and MacPherson, S. (2016). Bilingualism, social cognition and executive functions: A tale of chickens and eggs. *Neuropsychologia*, *91*, October, 299–306.

Craik, F., Bialystok, E., and Freedman, M. (2010). Delaying the onset of Alzheimer disease Bilingualism as a form of cognitive reserve. *Neurology*, *75*(19), 1726–1729.

Crane, P., Gibbons, L., Arani, K., Nguyen, V., Rhoads, K., McCurry, S., Launer, L., Masaki, K., and White, L. (2009). Midlife use of written Japanese and protection from late life dementia. *Epidemiology* (Cambridge, Mass.), *20*(5), 766.

Crivello C., Kuzyk O., Rodrigues M., Friend M., Zesiger P., and Poulin-Dubois D. (2016). The effects of bilingual growth on toddlers' executive function. *J. Exp. Child Psychol.* 141, 121–132.

D'Souza, D. and D'Souza, H. (2016). Bilingual language control mechanisms in anterior cingulate cortex and dorsolateral prefrontal cortex: A developmental perspective. *Journal of Neuroscience*, *36*(20), 5434–5436.

Davidson, M., Amso, D., Anderson, L., and Diamond, A. (2006). Development of cognitive control and executive functions from 4 to 13 years: Evidence from manipulations of memory, inhibition, and task switching. *Neuropsychologia*, *44*(11), 2037–2078.

Deary, I., Gow, A., Taylor, M., Corley, J., Brett, C., Wilson, V., Campbell, H., Whalley, L., Visscher, P., Porteous, D., and Starr, J. (2007). The Lothian birth cohort 1936: A study to examine influences on cognitive ageing from age 11 to age 70 and beyond. *BMC Geriatrics*, *7*(1), 28.

De Bruin, A., Treccani, B., and Della Sala, S. (2015). Cognitive advantage in bilingualism: An example of publication bias? *Psychological Science*, *26*(1), 99–107.

Diamond, A., Kirkham, N., and Amso, D. (2002). Conditions under which young children CAN hold two rules in mind and inhibit a prepotent response. *Developmental Psychology*, *38*, 352–362.

Dillon, A. (2009). Metalinguistic awareness and evidence of cross-linguistic influence among bilingual learners in Irish primary schools. *Language Awareness*, *18*(2), 182–197.

Directorate-General for Communication European Commission. (2012). Europeans and their languages. *Special eurobarometer*, *386*.

Drysdale, H., van der Meer, L., and Kagohara, D. (2015). Children with autism spectrum disorder from bilingual families: A systematic review. *Review Journal of Autism and Developmental Disorders*, *2*(1), 26–38.

Duñabeitia, J., Hernández, J., Antón, E., Macizo, P., Estévez, A., Fuentes, L., and Carreiras, M. (2014). The inhibitory advantage in bilingual children revisited: Myth or reality? *Experimental Psychology*, *61*(3), 234.

Duncan, H., Nikelski, J., Pilon, R., Steffener, J., Chertkow, H., and Phillips, N. (2018). Structural brain differences between monolingual and multilingual patients with mild cognitive impairment and Alzheimer disease: Evidence for cognitive reserve. *Neuropsychologia*, *109*, 270–282.

Emmorey, K., Luk, G., Pyers, J., and Bialystok, E. (2008). The source of enhanced cognitive control in bilinguals: Evidence from bimodal bilinguals. *Psychological Science*, *19*. 1201–1206.

Engel de Abreu, P. (2011). Working memory in multilingual children: Is there a bilingual effect? *Memory*, *19*(5), 529–537.

Engel de Abreu, P., Cruz-Santos, A., Tourinho, C., Martin, R., and Bialystok, E. (2012). Bilingualism enriches the poor: Enhanced cognitive control in low-income minority children. *Psychological Science*, *23*(11), 1364–1371.

Feltmate, K. and Kay-Raining Bird, E. (2008). Language Learning In Four Bilingual Children with Down syndrome: A detailed analysis of vocabulary and morphosyntax. *Revue Canadienne d'Orthophonie et d'Audiologie*, *32*(1), 7.

Feng, X., Diamond, A., and Bialystok, E. (2007). Manipulating information in working memory: An advantage for bilinguals. In *Poster presented at the biennial meeting of the Society for Research in Child Development*, Boston, MA (vol. 29).

Fernandez, M., Tartar, J., Padron, D., and Acosta, J. (2013). Neurophysiological marker of inhibition distinguishes language groups on a non-linguistic executive function test. *Brain and Cognition*, *83*(3), 330–336.

Filippi, R., Ceccolini, A., Periche-Tomas, E., Papageorgiou, A., and Bright, P. (2020). Developmental trajectories of control of verbal and non-verbal interference in speech comprehension in monolingual and multilingual children. *Cognition*, *200*, 104252.

Filippi, R., D'Souza, D., and Bright, P. (2018). *International Journal of Bilingualism*.

Filippi, R., Karmiloff-Smith, A. (2013). What can developmental disorders teach us about typical development?, in Marshall, C. R. (Ed.) *Current Issues in Developmental Disorders*. London, UK: Psychology Press.

Filippi, R., Leech, R., Thomas, M., Green, D., and Dick, F. (2012). A bilingual advantage in controlling language interference during sentence comprehension. *Bilingualism: Language and Cognition*, *15*(04), 858–872.

Filippi, R., Morris, J., Richardson, F., Bright, P., Thomas, M., Karmiloff-Smith, A., and Marian, V. (2015). Bilingual children show an advantage in controlling verbal interference during spoken language comprehension. *Bilingualism: Language and Cognition*, *18*(3), 490–501.

Filippi, R., Periche Tomas, E., Papageorgiou, A., and Bright, P. (2020). A role for the cerebellum in the control of verbal interference: Comparison of bilingual and monolingual adults. *PloS one*, *15*(4), e0231288.

Filippi, R., Richardson, F., Dick, F., Leech, R., Green, D., Thomas, M., and Price, C. (2011). The right posterior paravermis and the control of language interference. *Journal of Neuroscience*, *31*(29), 10732–10740.

Friedman, N. and Miyake, A. (2004). The relations among inhibition and interference control functions: A latent-variable analysis. *Journal of Experimental Psychology: General*, *133*(1), 101.

Garbin, G., Costa, A., Sanjuan, A., Forn, C., Rodriguez-Pujadas, A., Ventura, N., Belloch, V., Hernandez, M., and Avila, C. (2011). Neural bases of language switching in high and early proficient bilinguals. *Brain and Language*, *119*(3), 129–135.

Garbin, G., Sanjuan, A., Forn, C., Bustamante, J., Rodríguez-Pujadas, A., Belloch, V., Hernández, M., Costa, A., and Ávila, C. (2010). Bridging language and attention: Brain basis of the impact of bilingualism on cognitive control. *Neuroimage*, *53*(4), 1272–1278.

Gathercole, V., Thomas, E., Kennedy, I., Prys, C., Young, N., Viñas-Guasch, N., Roberts, E., Hughes, E., and Jones, L. (2014). Does language dominance affect cognitive performance in bilinguals? Lifespan evidence from preschoolers through older adults on card sorting, Simon, and metalinguistic tasks. *Frontiers in Psychology*, *5*, 11.

Genesee, F. (2008). Early dual language learning. *Zero to Three*, *29*(1), 17–23.

Gold, B., Johnson, N., and Powell, D. (2013). Lifelong bilingualism contributes to cognitive reserve against white matter integrity declines in aging. *Neuropsychologia*, *51*(13), 2841–2846.

Gold, B., Kim, C., Johnson, N., Kryscio, R., and Smith, C. (2013). Lifelong bilingualism maintains neural efficiency for cognitive control in aging. *Journal of Neuroscience*, *33*(2), 387–396.

Gollan, T.H., Montoya, R.I., and Werner, G., (2002). Semantic and letter fluency in Spanish-English bilinguals. *Neuropsychology*, 16(41), 562–576.

Gollan, T., Montoya, R., Fennema-Notestine, C., and Morris, S. (2005). Bilingualism affects picture naming but not picture classification. *Memory & Cognition*, 33(7), 1220–1234.

Gonzalez-Barrero, A. and Nadig, A. (2016). Verbal fluency in bilingual children with autism spectrum disorders. *Linguistic Approaches to Bilingualism*, 7(3), 460–475.

Gonzalez-Barrero, A. and Nadig, A. (2019). Can bilingualism mitigate set-shifting difficulties in children with autism spectrum disorders? *Child Development*, 90(4), 1043–1060.

Green, D. (1986). Control, activation, and resource: A framework and a model for the control of speech in bilinguals. *Brain and Language*, 27(2), 210–223.

Green, D. (1998). Mental control of the bilingual lexico-semantic system. *Bilingualism: Language and Cognition*, 1, 67–81.

Green, D. and Abutalebi, J. (2013). Language control in bilinguals: The adaptive control hypothesis. *Journal of Cognitive Psychology*, 25(5), 515–530.

Grosjean, F. (1994). Individual bilingualism. In *The Encyclopedia of language and Linguistics*, Oxford: Pergamon Press, 3, 1656–1660.

Grosjean, F. (2010). *Bilingual*. Harvard University Press.

Grosjean, F. (2012). An attempt to isolate, and then differentiate, transfer and interference. *International Journal of Bilingualism*, 16(1), 11–21.

Grundy, J. and Timmer, K. (2017). Bilingualism and working memory capacity: A comprehensive meta-analysis. *Second Language Research*, 33(3), 325–340.

Grundy, J., Anderson, J., and Bialystok, E. (2017). Neural correlates of cognitive processing in monolinguals and bilinguals. *Annals of the New York Academy of Sciences*, 1396(1), 183–201.

Gupta, A. and Jamal, G. (2007). Reading strategies of bilingual normally progressing and dyslexic readers in Hindi and English. *Applied Psycholinguistics*, 28(1), 47–68.

Hakuta, K. and Diaz, R. (1985). The relationship between degree of bilingualism and cognitive ability: A critical discussion and some new longitudinal data. In Nelson, K. E., (Ed.) *Children's Language* (pp. 319–344). Hillside, NJ: Lawrence Erlbaum.

Hambly, C. and Fombonne, E. (2012). The impact of bilingual environments on language development in children with autism spectrum disorders. *Journal of Autism and Developmental Disorders*, 42, 1342–1352.

Hambly, C. and Fombonne, E. (2014). Factors influencing bilingual expressive vocabulary size in children with autism spectrum disorders. *Research in Autism Spectrum Disorders*, 8(9), 1079–1089.

Hartsuiker, R. (2015). Why it is pointless to ask under which specific circumstances the bilingual advantage occurs. *Cortex*, 73, 336–337.

Hernandez, A. (2009). Language switching in the bilingual brain: What's next? *Brain and Language*, 109(2–3), 133–140.

Hernández, M., Costa, A., and Humphreys, G. (2012). Escaping capture: Bilingualism modulates distraction from working memory. *Cognition*, 122(1), 37–50.

Hernandez, A., Dapretto, M., Mazziotta, J., and Bookheimer, S. (2001). Language switching and language representation in Spanish–English bilinguals: An fMRI study. *NeuroImage*, 14(2), 510–520.

Hilchey, M. and Klein, R. (2011). Are there bilingual advantages on nonlinguistic interference tasks? Implications for the plasticity of executive control processes. *Psychonomic Bulletin and Review*, 18, 625–658.

Ho, C. and Fong, K. (2005). Do Chinese dyslexic children have difficulties learning English as a second language? *Journal of Psycholinguistic Research, 34*(6), 603–618.

Hoffmann, C. (2014). *Introduction to bilingualism.* London and New York: Routledge.

Hughes, C., Shaunessy, E., Brice, A., Ratliff, M., and McHatton, P. (2006). Code switching among bilingual and limited English proficient students: Possible indicators of giftedness. *Journal for the Education of the Gifted, 30*(1), 7–28.

Humphrey, A. and Valian, V. (2012). Multilingualism and cognitive control: Simon and flanker task performance in monolingual and multilingual young adults. In *53rd Annual Meeting of the Psychonomic Society*, Minneapolis, MN.

Iarocci, G., Hutchison, S., and O'Toole, G. (2017). Second language exposure, functional communication, and executive function in children with and without autism spectrum disorder (ASD). *Journal of autism and developmental disorders, 47*(6), 1818–1829.

Iluz-Cohen, P. and Walters, J. (2012). Telling stories in two languages: Narratives of bilingual preschool children with typical and impaired language. *Bilingualism: Language and Cognition, 15*(1), 58–74.

Ivanova, I. and Costa, A. (2008). Does bilingualism hamper lexical access in speech production? *Acta Psychologica, 127*(2), 277–288.

Jacobson, P. and Schwartz, R. (2005). English past tense use in bilingual children with language impairment. *American Journal of Speech-Language Pathology, 14*(4), 313–323.

Johnson, M. (2011). Interactive specialization: A domain-general framework for human functional brain development? *Developmental Cognitive Neuroscience, 1*(1), 7–21.

Kapa, L. and Colombo, J. (2013). Attentional control in early and later bilingual children. *Cognitive Development, 28*(3), 233–246.

Karmiloff-Smith, A. (1998). Development itself is the key to understanding developmental disorders. *Trends in Cognitive Sciences, 2*(10), 389–398.

Kay-Raining Bird, E., Cleave, P., Trudeau, N., Thordardottir, E., Sutton, A., and Thorpe, A. (2005). The language abilities of bilingual children with Down syndrome. *American Journal of Speech-Language Pathology, 14*, 187–199.

Kay-Raining Bird, E., Lamond, E., and Holden, J. (2012). Survey of bilingualism in autism spectrum disorders. *International Journal of Language and Communication Disorders, 47*(1), 52–64.

Kempert, S., Saalbach, H., and Hardy, I. (2011). Cognitive benefits and costs of bilingualism in elementary school students: The case of mathematical word problems. *Journal of Educational Psychology, 103*(3), 547.

Kerrigan, L., Thomas, M., Bright, P., and Filippi, R. (2017). Evidence of an advantage in visuospatial memory for bilingual compared to monolingual speakers. *Bilingualism: Language and Cognition, 20*(3), 602–612.

Klein, D., Mok, K., Chen, J., and Watkins, K. (2014). Age of language learning shapes brain structure: A cortical thickness study of bilingual and monolingual individuals. *Brain and language, 131*, 20–24.

Kousaie, S. and Phillips, N. (2012a). Ageing and bilingualism: Absence of a "bilingual advantage" in Stroop interference in a nonimmigrant sample. *Quarterly Journal of Experimental Psychology, 65*(2), 356–369.

Kousaie, S. and Phillips, N. (2012b). Conflict monitoring and resolution: Are two languages better than one? Evidence from reaction time and event-related brain potentials. *Brain Research, 1446*, 71–90.

Kovács, Á. (2009). Early bilingualism enhances mechanisms of false-belief reasoning. *Developmental Science*, 12(1), 48–54.

Kovelman, I., Baker, S., and Petitto, L. (2008). Bilingual and monolingual brains compared: A functional magnetic resonance imaging investigation of syntactic processing and a possible "neural signature" of bilingualism. *Journal of Cognitive Neuroscience*, 20(1), 153–169.

Kuhl, P., Stevens, E., Hayashi, A., Deguchi, T., Kiritani, S., and Iverson, P. (2006). Infants show a facilitation effect for native language phonetic perception between 6 and 12 months. *Developmental Science*, 9(2).

Lawton, D., Gasquoine, P., and Weimer, A. (2015). Age of dementia diagnosis in community dwelling bilingual and monolingual Hispanic Americans. *Cortex*, 66, 141–145.

Li, P. and Grant, A. (2015). Identifying the causal link: Two approaches toward understanding the relationship between bilingualism and cognitive control. *Cortex*, 73, 358–360.

Li, P., Legault, J., and Litcofsky, K. (2014). Neuroplasticity as a function of second language learning: Anatomical changes in the human brain. *Cortex*, 58, 301–324.

Luk, G., Anderson, J., Craik, F., Grady, C., and Bialystok, E. (2010). Distinct neural correlates for two types of inhibition in bilinguals: Response inhibition versus interference suppression. *Brain and Cognition*, 74(3), 347–357.

Luk, G., Bialystok, E., Craik, F., and Grady, C. (2011). Lifelong bilingualism maintains white matter integrity in older adults. *Journal of Neuroscience*, 31(46), 16808–16813.

Maneva, B., and Genesee, F. (2002). Bilingual babbling: Evidence for language differentiation in dual language acquisition. In *Proceedings of the Annual Boston University Conference on Language Development* (vol. 26, no. 1, pp. 383–392).

Martin-Rhee, M. and Bialystok, E. (2008). The development of two types of inhibitory control in monolingual and bilingual children. *Bilingualism: Language and Cognition*, 11(1), 81–93.

Mechelli, A., Crinion, J., Noppeney, U., O'doherty, J., Ashburner, J., Frackowiak, R., and Price, C. (2004). Neurolinguistics: Structural plasticity in the bilingual brain. *Nature*, 431(7010), 757.

Mehler, J., Jusczyk, P., Lambertz, G., Halsted, N., Bertoncini, J., and Amiel-Tison, C. (1988). A precursor of language acquisition in young infants. *Cognition*, 29(2), 143–178.

Michael, E. and Gollan, T. (2005). Being and becoming bilingual. Individual differences and consequences for language production. In J. R. Kroll and A. de Groot (Eds.), *Handbook of Bilingualism* (pp. 389–407), New York: Oxford University.

Mohades, S., Struys, E., Van Schuerbeek, P., Mondt, K., Van De Craen, P., and Luypaert, R. (2012). DTI reveals structural differences in white matter tracts between bilingual and monolingual children. *Brain Research*, 1435, 72–80.

Mohades, S., Struys, E., Van Schuerbeek, P., Baeken, C., Van De Craen, P., and Luypaert, R. (2014). Age of second language acquisition affects nonverbal conflict processing in children: An fMRI study. *Brain and Behavior*, 4(5), 626–642.

Mor, B., Yitzhaki-Amsalem, S., and Prior, A. (2015). The joint effect of bilingualism and ADHD on executive functions. *Journal of Attention Disorders*, 19(6), 527–541.

Morales, J., Calvo, A., and Bialystok, E. (2013). Working memory development in monolingual and bilingual children. *Journal of Experimental Child Psychology*, 114(2), 187–202.

Moreno, S., Wodniecka, Z., Tays, W., Alain, C., and Bialystok, E. (2014). Inhibitory control in bilinguals and musicians: Event related potential (ERP) evidence for experience-specific effects. *PloS One*, 9(4), e94169.

Morgan, G., Restrepo, M., and Auza, A. (2013). Comparison of Spanish morphology in monolingual and Spanish–English bilingual children with and without language impairment. *Bilingualism: Language and Cognition*, 16(3), 578–596.

Morton, J. and Harper, S. (2007). What did Simon say? Revisiting the bilingual advantage. *Developmental Science*, 10(6), 719–726.

National Clearinghouse for English Language Acquisition and Language Instruction Educational Programs (NCELA). (2006). The growing numbers of limited English proficient students. Retrieved from http://www.ncela.gwu.edu/files/uploads/4/GrowingLEP_0 506.pdf

Ohashi, J., Mirenda, P., Marinova-Todd, S., Hambly, C., Fombonne, E., Szatmari, P., Bryson, S., Roberts, W., Smith, I., Vaillancourt, T., and Volden, J. (2012). Comparing early language development in monolingual- and bilingual-exposed young children with autism spectrum disorders. *Research in Autism Spectrum Disorders*, 6(2), 890–897.

Oller, D. K., Eilers, R. E., Urbano, R., and Cobo-Lewis, A. B. (1997). Development of precursors to speech in infants exposed to two languages. *Journal of child language*, 24(2), 407–425.

Oren, R. and Breznitz, Z. (2005). Reading processes in L1 and L2 among dyslexic as compared to regular bilingual readers: Behavioral and electrophysiological evidence. *Journal of Neurolinguistics*, 18(2), 127–151.

Paap, K. and Greenberg, Z. (2013). There is no coherent evidence for a bilingual advantage in executive processing. *Cognitive Psychology*, 66, 232–258.

Paap, K., Johnson, H., and Sawi, O. (2014). Are bilingual advantages dependent upon specific tasks or specific bilingual experiences? *Journal of Cognitive Psychology*, 26(6), 615–639.

Paap, K., Johnson, H., and Sawi, O. (2015). Bilingual advantages in executive functioning either do not exist or are restricted to very specific and undetermined circumstances. *Cortex*, 69, August, 265–278.

Papageorgiou, A., Bright, P., Periche Tomas, E., and Filippi, R. (2019). Evidence against a cognitive advantage in the older bilingual population. *Quarterly Journal of Experimental Psychology*, 72(6), 1354–1363.

Paradis, J., Crago, M., and Genesee, F. (2006). Domain-general versus domain-specific accounts of specific language impairment: Evidence from bilingual children's acquisition of object pronouns. *Language Acquisition*, 13(1), 33–62.

Paradis, J., Genesee, F., and Crago, M. B. (2011). *Dual Language Development & Disorder* (pp. 88–108). Baltimore: Paul H. Brookes Publishing.

Parker Jones, Ō., Green, D., Grogan, A., Pliatsikas, C., Filippopolitis, K., Ali, N., Lee, H., Ramsden, S., Gazarian, K., Prejawa, S., and Seghier, M. (2012). Where, when and why brain activation differs for bilinguals and monolinguals during picture naming and reading aloud. *Cerebral Cortex*, 22(4), 892–902.

Patterson, J. L., and Pearson, B. Z. (2004). Bilingual lexical development: Influences, contexts, and processes. In B. A. Goldstein (Ed.), *Bilingual Language Development and Disorders in Spanish-English Speakers* (pp. 77–104). Paul H Brookes Publishing.

Peal, E. and Lambert, W. (1962). The relation of bilingualism to intelligence. *Psychological Monographs: General and Applied*, 76(27), 1.

Pearson, B. Z., Fernández, S. C., and Oller, D. K. (1993). Lexical development in bilingual infants and toddlers: Comparison to monolingual norms. *Language learning*, 43(1), 93–120.

Petersen, J., Marinova-Todd, S., and Mirenda, P. (2012). Brief report: An exploratory study of lexical skills in bilingual children with autism spectrum disorder. *Journal of Autism and Developmental Disorders*, 42(7), 1499–1503.

Polka, L., and Sundara, M. (2003, August). Word segmentation in monolingual and bilingual infant learners of English and French. In *Proceedings of the 15th International Congress of Phonetic Sciences* (pp. 1021–1024).

Prior, A. and MacWhinney, B. (2010). A bilingual advantage in task switching. *Bilingualism: Language and Cognition*, 13(2), 253–262.

Ramus, F., Hauser, M., Miller, C., Morris, D., and Mehler, J. (2000). Language discrimination by human newborns and by cotton-top tamarin monkeys. *Science*, 288(5464), 349–351.

Reetze, R., Zou, X., and Sheng, L. (2015). Communicative development in bilingually exposed Chinese children with autism spectrum disorders. *Journal of Speech, Language, and Hearing Research*, 58(3), 813–825.

Ressel, V., Pallier, C., Ventura-Campos, N., Díaz, B., Roessler, A., Ávila, C., and Sebastián-Gallés, N. (2012). An effect of bilingualism on the auditory cortex. *Journal of Neuroscience*, 32(47), 16597–16601.

Roberts, P., Garcia, L.J., Desrochers, A., and Hernandez, D. (2002). English performance of proficient bilingual adults on the Boston Naming Test. *Aphasiology*, 16(4–6), 635–645.

Rodríguez-Pujadas, A., Sanjuán, A., Ventura-Campos, N., Román, P., Martin, C., Barceló, F., Costa, A., and Ávila, C. (2013). Bilinguals use language-control brain areas more than monolinguals to perform non-linguistic switching tasks. *PLoS One*, 8(9), 73028.

Rosenblum, T., and Pinker, S. A. (1983). Word magic revisited: Monolingual and bilingual children's understanding of the word-object relationship. *Child Development*, 54(3), 773–780.

Rosselli, M., Ardila, A., Araujo, K., Weekes, V., Caracciolo, V., Padilla, M., and Ostrosky-Solí, F. (2000). Verbal fluency and repetition skills in healthy older Spanish-English bilinguals. *Applied Neuropsychology*, 7(1), 17–24.

Saer, D. (1922). An inquiry into the effect of bilingualism upon the intelligence of young children. *Journal of Experimental Pedagogy & Training College Record*, 6, 232–274.

Saer, D. (1923). The effect of bilingualism on intelligence. *British Journal of Psychology*, 14(1), 25–38.

Salvatierra, J. and Rosselli, M. (2011). The effect of bilingualism and age on inhibitory control. *International Journal of Bilingualism*, 15(1), 26–37.

Sanders, A., Hall, C., Katz, M., and Lipton, R. (2012). Non-native language use and risk of incident dementia in the elderly. *Journal of Alzheimer's Disease*, 29(1), 99–108.

Sandoval, T., Gollan, T., Ferreira, V. and Salmon, D. (2010). What causes the bilingual disadvantage in verbal fluency? The dual-task analogy. *Bilingualism: Language and Cognition*, 13(2), 231–252.

Schweizer, T., Ware, J., Fischer, C., Craik, F., and Bialystok, E. (2012). Bilingualism as a contributor to cognitive reserve: Evidence from brain atrophy in Alzheimer's disease. *Cortex*, 48(8), 991–996.

Siegel, L. (2016). Bilingualism and Dyslexia: The case of children learning English as an additional language. In *Multilingualism, Literacy and Dyslexia* (pp. 137–147). New York: Routledge.

Soveri, A., Laine, M., Hämäläinen, H., and Hugdahl, K. (2011). Bilingual advantage in attentional control: Evidence from the forced-attention dichotic listening paradigm. *Bilingualism: Language and Cognition*, 14(3), 371–378.

Stins, J., Polderman, J., Boomsma, D., and de Geus, E. (2005). Response interference and working memory in 12-year-old children. *Child Neuropsychology*, 11(2), 191–201.

Toppelberg, C., Medrano, L., Morgens, L., and Nieto-Castanon, A. (2002). Bilingual children referred for psychiatric services: Associations of language disorders, language skills, and psychopathology. *Journal of the American Academy of Child & Adolescent Psychiatry*, 41(6), 712–722.

Treccani, B. and Mulatti, C. (2015). No matter who, no matter how . . . and no matter whether the white matter matters. Why theories of bilingual advantage in executive functioning are so difficult to falsify. *Cortex, 73*, 349–351.

Valicenti-McDermott, M., Tarshis, N., Schouls, M., Galdston, M., Hottinger, K., Seijo, R., Shulman, L., and Shinnar, S. (2013). Language differences between monolingual English and bilingual English-Spanish young children with autism spectrum disorders. *Journal of Child Neurology, 28*(7), 945–948.

Verhoeven, L. and Aarts, R. (1998). Attaining functional biliteracy in the Netherlands. In A. Y. Durgunoğlu and L. Verhoeven (Eds.), *Literacy Development in a Multilingual Context: Cross-Cultural Perspectives* (pp. 111–133). Mahwah, NJ, US: Lawrence Erlbaum Associates Publishers.

Weber R., Johnson A., Riccio C., and Liew J. (2016). Balanced bilingualism and executive functioning in children. *Bilingualism, 19*, 425–431

Weinreich, U. (1953). *Languages in Contact: Findings and Problems.* NY: Linguistic Circle.

Yang, S., Yang, H., and Lust, B. (2011). Early childhood bilingualism leads to advances in executive attention: Dissociating culture and language. *Bilingualism, Language and Cognition, 14*(3), 412–422.

Yu, B. (2013). Issues in bilingualism and heritage language maintenance: Perspectives of minority-language mothers of children with autism spectrum disorders. *American Journal of Speech-Language Pathology, 22*(1), 10–24.

Zelazo, P., Frye, D., and Rapus, T. (1996). An age-related dissociation between knowing rules and using them. *Cognitive Development, 11*(1), 37–63.

Zelazo, P., Müller, U., Frye, D., Marcovitch, S., Argitis, G., Boseovski, J., Chiang, J., Hongwanishkul, D., Schuster, B., Sutherland, A., and Carlson, S. (2003). The development of executive function in early childhood. *Monographs of the Society for Research in Child Development*, i–151.

Zhou, V., Munson, J. A., Greenson, J., Hou, Y., Rogers, S., and Estes, A. M. (2019). An exploratory longitudinal study of social and language outcomes in children with autism in bilingual home environments. *Autism, 23*(2), 394–404.

CHAPTER 25

NUMERICAL AND MATHEMATICAL ABILITIES IN CHILDREN

Behavioral and neural correlates

ROI COHEN KADOSH AND TERESA IUCULANO

INTRODUCTION

THE ability to compare and reason about numerical attributes of sets develops very early in life. Infants and even newborns are able to perceptually discriminate which of two sets is larger in numerosity for 1:3 ratios (Izard et al. 2009), with a steep developmental curve in acuity within the first year of life (i.e., by nine to ten months they are able to "*solve for*" 2:3 ratios) (Lipton and Spelke 2003; Xu and Arriaga 2007). This ability seems to be present also in several mammalian and non-mammalian species (Agrillo et al. 2008; Brannon and Terrace 2000; Dacke and Srinivasan 2008; Ditz and Nieder 2015; Meck and Church 1983; Nieder et al. 2002; Shumaker et al. 2001; Tudusciuc and Nieder 2007) and has been referred to as the "number sense". Inasmuch as rudimentary numerical skills such as these are ontogenetically and phylogenetically ancient, the ability to reason symbolically about these attributes, which is the essence of modern mathematics, is an extremely recent acquisition in our lineage and requires the coordinated support of several brain systems over development.

When a child first learns mathematical symbols such as 1, 2, 3, and 4, these are usually associated with meaningless counting sequences (Sarnecka and Carey 2008). Only later, and with more formal teaching, these symbols are associated with their corresponding semantic numerical meanings and their intrinsic properties (Gelman and Gallistel 1978; Sarnecka 2015). These include simple rules of cardinality and ordinality that precede, and are foundational, to the ability to reason about symbolic numerosities and eventually being able to perform formal arithmetical operations on them (i.e., to add, to subtract, etc.). Cardinality includes all principles derived from bijective and injective functions such as: (i) when the elements of one set can be put in one-to-one correspondence with the elements of the other, then the two sets have the same numerosity (i.e., they are exactly equal)—bijection; and (ii) the set denoted by the greater number includes the set denoted by the smaller one (e.g., a set of four items will include a set of three items)—injection. This means that adding (or subtracting) an item to (from) a set, will affect its numerosity.

Formal principles of ordinality emerge from cardinal principles: the final number of a counting sequence represents the cardinality of that sequence; and the next word of the counting sequence can be obtained by adding one unit to the current cardinality (Giaquinto 1995; see also Gelman and Gallistel 1978). As previously mentioned, the approximate, non-symbolic equivalent of these types of abilities is present already in infants (for a review, see de Hevia 2016). Habituation type of paradigms have shown that infants as young as four to six months are able to succeed—with adequate accuracy levels—on tasks that require to compare, add, or subtract non-symbolic numerosities from sets: they are aware when changes in numerosities occur (McCrink and Wynn 2007), and they seem to have an accurate arithmetic expectancy regarding numerical transformations on them (Wynn 1992, 1998). Some rudimentary ordinality properties also seem to emerge very early during development. A recent study has shown that pre-verbal infants as young as four months old can detect ordinal relationships of size-based and numerical-based sequences (de Hevia, Addabbo, et al. 2017; see also Brannon 2002).

Yet, within our exact (and symbolic) number system, both these cardinality and ordinality properties are highly complex to acquire, require many hours of formal education and are susceptible to many heuristic-based and confusion errors (Gelman and Gallistel 1978; Houdé and Tzourio-Mazoyer 2003). The question is, how—during their first years of schooling—do most children become efficient "symbolic calculators"? How do they successfully step away from wrong counting procedures, inaccurate applications of one-to-one principles, to finally (and successfully) learn arithmetic rules and procedures on Arabic numerals?

The recent advent of several non-invasive, in vivo neuroimaging techniques, such as electroencephalography (EEG) and functional magnetic resonance imaging (fMRI), has started to shed light into how the brain supports this knowledge acquisition such that, at some point during development, a child "just knows" that $3 + 4 = 7$. These functional brain systems include visual and auditory association cortices—and their interactions. These brain regions decode the visual and phonological features of the stimuli (i.e., the symbol "3" or the spoken word "three"), and have their hubs in the higher level visual cortex, particularly in and around the fusiform gyrus (FG) (Abboud et al. 2015; Ansari 2008; Cantlon et al. 2006, 2009; Cantlon et al. 2011; Grotheer et al. 2016; Holloway et al. 2013; Shum et al., 2013); and in phonological and language processing regions anchored in the superior temporal sulcus (STS) and adjacent middle temporal gyrus (MTG) (Thompson et al. 2004). A semantic meaning to these originally arbitrary visual and verbal symbols arises through interactions with quantity processing systems residing in the posterior aspect of the parietal lobe, particularly the intraparietal sulcus (IPS). Developing such association provides the most sophisticated form of numerical reasoning: two sets of *"four apples"* and *"four ideas"* denote the same quantity, and this "shared property" can be referred to by the symbol (or word) "4." The IPS has been systematically implicated in studies of number processing, number representation, and number reasoning (Cantlon et al. 2009, 2011; Castaldi et al. 2016; Castelli et al. 2006; Cohen Kadosh et al. 2007, 2008, 2011; Cutini et al. 2014; Dehaene and Cohen 1995; Dehaene et al. 2003; Hyde et al. 2010; Piazza et al. 2004, 2007; Vogel et al. 2015), possibly due to its topological proximity to cortical regions implicated in other cognitive tasks from which the ability to count and manipulate numbers/quantities may derive, phylogenetically. These include regions of the frontal eye fields and posterior and superior parietal cortices supporting selective spatial attention (Corbetta and Shulman 2002; Schaffelhofer and Scherberger 2016), fine motor and eye movements (i e., visuomotor; visual saccades; Cohen Kadosh et al. 2011; Simon et al. 2002), visual short-term memory (Knops et al. 2014; Luck and Vogel 1997, 2013; Todd and Marois 2004), as well as space perception (i.e., right-left discrimination in space; Gold et al. 1995; Hirnstein et al. 2011; Mayer et al. 1999), finger representations (Krinzinger et al., 2011), and allocentric and egocentric space representations (di Pellegrino and Làdavas 2015; Liu et al. 2017; Zaehle et al. 2007). These studies, which have been based mostly on adult participants, have suggested that once a child learns to form a systematic pairing between the number "3" and the quantity of *"three items,"* they can start performing formal arithmetical operations.

Learning rules and principles (i.e., "+" and "–" signs; or formal bijection/injection properties of sets), while starting to inhibit intuitive (and wrong) heuristics that do not meet the formal rule(s) learned, is supported by cognitive control inhibitory processes within the prefrontal cortex (Houdé and Tzourio-Mazoyer 2003; Hung et al. 2018; Supekar and Menon 2012). The prefrontal cortex is responsible for different types of cognitive learning (Chein and Schneider 2012). One of the key roles of this brain region is to support cognitive control processes, such as inhibition. In the case of numerical processing, this includes mastering the concept that irrelevant, incongruent perceptual properties of sets, such as size, density, or length, do not influence the numerosity of that set. Concomitantly, interactions between the prefrontal cortex and posterior parietal regions—that form the working memory executive network—help to maintain and dynamically update intermediate results and procedures to facilitate calculation processes (Halgren et al. 2002; Zago et al. 2008). Finally, the hippocampus, within the medial temporal lobe, and its functional interactions with cortical regions, predominantly in the dorsolateral prefrontal and posterior parietal cortices, is involved in forming stable declarative memory associations (Davachi et al. 2003) between addends (i.e., "3" and "4") and results (i.e., "7"), ultimately facilitating the automatic retrieval of arithmetical facts from memory (Qin et al. 2014; Rosenberg-Lee et al. 2018).

In this chapter our goal is to provide a comprehensive and timely overview on the aforementioned neurocognitive processes.

NUMERICAL ABILITIES: DOMAIN-SPECIFIC AND DOMAIN-GENERAL COGNITIVE SCAFFOLDS

Developing Cardinal Skills

The numerosity of a set (i.e., its cardinality) is originally detected—yet processed approximately even for small numerosities—from a very early age (de Hevia 2016; Izard et al. 2009; Wynn 1992, 1998). Gradually, its acuity improves: within the first years of schooling, typically developing children can successfully discriminate sets of non-symbolic quantities with a 5:6 ratio (Halberda and Feigenson 2008). Such ability then steadily improves across development (Halberda et al. 2012) (Figure 25.1a), and it has been shown to be moderately associated with mathematical skills later in life (Halberda et al. 2008, 2012; see also Libertus et al. 2012) (Figure 25.1a). A stronger—and more reliable—domain-specific predictor of mathematical skills seems to be symbolic numerical magnitude processing (De Smedt et al. 2013; Schneider et al. 2017).

Critically, in a recent study, Vanbinst and colleagues report that the strength of the prediction of symbolic number comparison to arithmetical skills is comparable to that of phonological awareness to reading (Vanbinst et al. 2016; see also Gersten and Chard 1999).

The ability to discriminate between numerosities has been attributed to the appropriate functioning of the posterior aspect of the parietal lobes, particularly the IPS. Using near-infrared spectroscopy (NIRS) in a group of six-month-olds, Hyde and colleagues demonstrated that the right IPS is sensitive to changes (1:2 ratio) of non-symbolic numerosities—but not shape—(Figure 25.1b,c); while right occipital areas are sensitive to changes in shape—but not numerosities (Hyde et al. 2010). A "cortical bias" of the posterior parietal cortex towards numerosity-specific aspects of the visual world has been demonstrated in infants as young as three months. Using EEG during a dishabituation paradigm, Izard and colleagues have shown that—similar to four-year-old children and adults—three-month-old infants' electrical brain activity displays sensitivity to the cardinal properties of sets in regions of the posterior parietal lobe. Sensitivity to changes in other visual properties of the sets (i.e., object identity) was instead detected in regions of the occipital cortex—again, similar to four-year-olds and adults (Izard et al. 2008).

Notably, activity of the IPS has been shown to linearly scale with numerosities up to four or five (Xu and Chun 2006; i.e., the so-called subitizing range—or *visual short-term memory*—see Vogel and Machizawa 2004) which suggests the existence of an exact-tracking processing system for a small number of items, supported by the posterior parietal cortex (but see Piazza et al. 2002). Critically, subitizing skills are impaired in children with developmental dyscalculia (DD; a severe form of mathematical

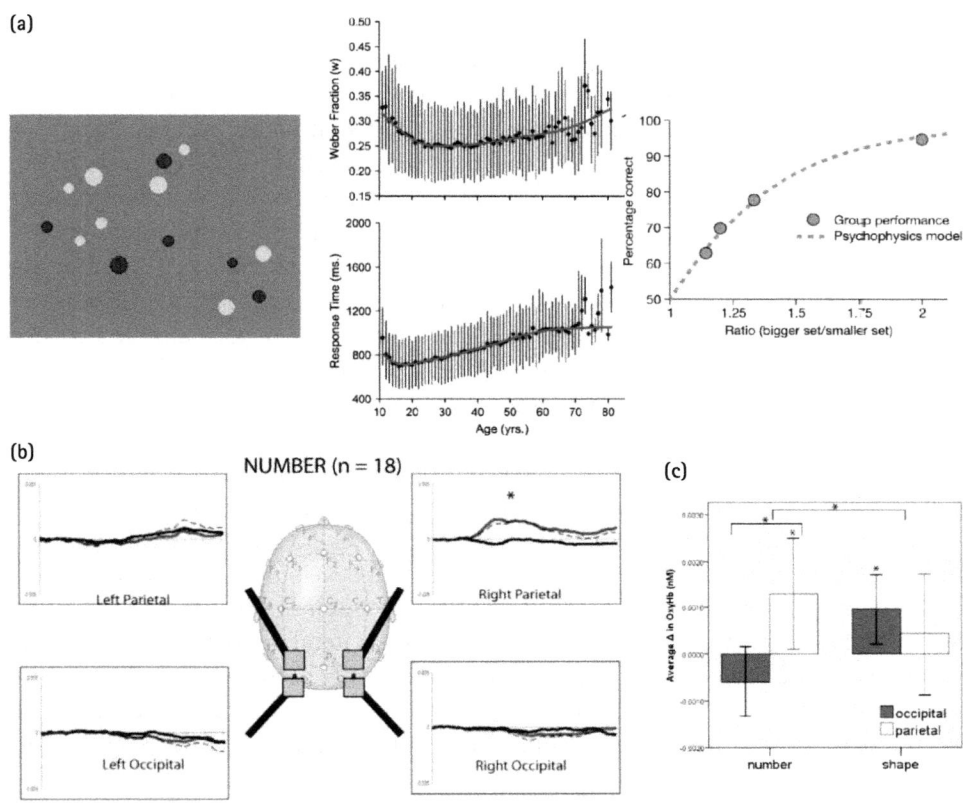

FIGURE 25.1 Development of cardinal skills—behavioral and neural correlates.

(a) *Left.* Non-symbolic numerosity judgement task to assess numerical acuity—used with permission from Halberda et al. (2008). *Middle.* Numerical acuity peaks at about the age of thirty years old—used with permission from Halberda et al. (2012). Two behavioral indices of numerical/numerosity proficiency are represented: on the *Top*: Weber-fraction. *Bottom*: Reaction times; *Right*: Numerical acuity significantly predicts mathematical skills later in life—used with permission from Halberda et al. (2008).

(b) At six months of age, the right IPS is uniquely sensitive to changes in numerosity—used with permission from Hyde et al. (2010).

(c) This effect is not evident for changes in shape (Hyde et al., 2010).

Note: Y-axes reflect relative change in oxygenated hemoglobin concentration (micromol) measured by Near-Infrared Spectroscopy.

learning disability) (Koontz 1996; Landerl et al. 2004), and subitizing deficits have been associated with aberrancies in posterior parietal regions, particularly the IPS (Arp and Fagard 2005; Arp et al. 2006; Bruandet et al. 2004; Cipolotti et al. 1991; Paterson et al. 2006; Simon et al. 2005, 2008). More generally, it has been suggested/proposed that subitizing constitutes a fundamental building block for the appropriate development of arithmetical skills (Landerl et al. 2004; Reeve et al. 2012).

For numerosities >4 the acuity of the system decreases, and its pace of processing increases—even in typical populations (Chi and Klahr 1975). Yet, the cognitive processes associated with these (more difficult) numerosity discriminations are still supported, with increased specialization/modulation throughout development, by the IPS (Piazza et al. 2002). For instance, with larger numerosities (i.e., >4), a well reported finding is that the sensitivity to *distance/ratio* is reflected by faster and more accurate responses for larger *distance/ratio* (Moyer and Landauer 1967). Notably, at the brain level, the "neural distance/ratio effect" is marked by greater functional activity for numbers that are closer in numerosity (e.g., 4 versus 5). These are harder to discriminate compared to two numbers that are further apart from each other (e.g., 4 versus 9). This effect has been observed in several brain regions, including the IPS in adults (Cohen Kadosh et al. 2011; Piazza et al. 2004). At younger ages, prefrontal cortex regions display a greater *neural distance/ratio effect* (Ansari and Dhital 2006; Price et al. 2007) for both symbolic and non-symbolic stimuli. Yet, other studies did show parietal modulation as a function of the "*distance/ratio effect*" already in children cohorts (Cantlon et al. 2006). We suggest that these mixed findings likely reflect the aid of working memory and executive processes to numerosity discriminations in children, and the greater degree of variability at younger ages—occurring before efficient numerical skills are consolidated. This suggestion fits well with the view that throughout development, the need of high-level (domain-general) processes for online tracking of items in space and their active storage and manipulation—heavily supported by dorsolateral prefrontal cortex regions and their connectivity with the parietal cortex—gradually becomes expendable when representations for approximate (and later exact) numerosities start to form within the IPS (i.e., anterior-to-posterior shift; see also later section "Developing Formal Arithmetic").

Format-dependent symbolic representations (i.e., Arabic numerals) have also been shown to be modulated by the IPS (Vogel et al. 2017), as well as by high-level visual areas in the ventral temporal occipital cortex, particularly those adjacent to the FG (Grotheer et al. 2016). This brain region has been systematically reported to be highly sensitive to complex visual forms/stimuli (i.e., letters, faces, objects) (Gauthier et al. 1999, 2000). Interestingly, the FG seems also to subtend semantic representations of numbers: a *distance/ratio effect* has been reported in adults during numerical comparison tasks (Ansari 2008; Cantlon et al. 2009; Holloway et al. 2013; but see Cohen Kadosh et al. 2007). Such higher-order semantic specialization of the FG may be the result of the strengthening of the coupling within the dorsal visual stream that directly connects the visual and association cortices (Jung et al. 2017). Critically, the intrinsic functional connectivity between the FG and the IPS has been shown to significantly predict gains in mathematical abilities over development (Evans et al. 2015), suggesting a critical role of this pathway in children's mathematical skill acquisition.

Together, this evidence suggests that both the IPS and the FG—and their functional and structural interactions—help to create neural representations of numbers in multiple formats (i.e., symbolic, non-symbolic). With the support of these two core brain systems, the human brain learns more complex mathematical computations through development and schooling, and with the support of other—less domain-specific—brain systems anchored in the prefrontal cortex and in the medial aspects of the temporal lobe.

Developing Ordinal Skills and Processing Spatio-Temporal Information

Most of the early work on numerical and mathematical cognitive skills has mainly focused on cardinality properties and their representations (Butterworth 1999, 2011; Dehaene et al. 2003; Piazza et al. 2006; Reeve et al. 2012), while ordinality aspects have largely been neglected.

Only quite recently, behavioral studies have begun to report a significant association between mathematical performance and semantic (i.e., magnitude-relevant) tasks of ordinality, such as: (i) ordinality judgements of triplets of symbolic numbers (Goffin and Ansari 2016; Kucian et al. 2011; Lyons et al. 2014); (ii) counting of forward and backward sequences/lists (O'Connor et al. 2018); as well as (iii) order working memory (WM) tasks (Attout et al. 2014; O'Connor et al. 2018). Critically, the ability to assess ordinal properties of numerical symbols seems to increase with development and schooling: in grades 1 to 2, performance on tasks tapping on cardinal properties of symbolic numbers has been shown to be the most predictive of arithmetical skills, while by grade 6, it is the performance on numerical ordinal tasks (with symbolic numbers) that emerges as the best predictor of arithmetic performance (Lyons et al. 2014; but see Attout et al. 2014).

Ordinality sequences—without a semantic magnitude connotation—are properties that emerge very early in life, when children start reciting counting sequences (i.e., number words) and even non-counting (i.e., alphabet, days of the week, months) lists (Gelman and Gallistel 1978; Sarnecka and Carey 2008). Ordering sequences without a semantic magnitude connotation (i.e., non-numerical ordering abilities) can also include remembering and processing spatio-temporal relationships such as daily events. Critically, a recent longitudinal study has demonstrated that the ability to correctly identify/report spatio-temporal patterns of daily events in first grade was the strongest longitudinal predictor of math skills at the end of grade 2. Notably, these non-numerical ordering skills in this age range were a stronger predictor than numerical ordering skills (i.e., with a semantic magnitude connotation), including order WM, and number ordering tasks. This finding highlights a novel spatio-temporal component of ordinality skills that can be regarded more as domain-general, rather than magnitude-specific, and which seems to be crucial for the appropriate development of math skills at the beginning of formal education. The ability to understand and remember order relationships (e.g., daily events as they occurred) has a strong episodic component, and may be linked to the processing and representations of other spatial and temporal relationships, which are mostly supported by medial temporal lobe brain systems (Buzsáki and Moser 2013; Moser et al. 2008; O'Keefe and Burgess 1996). Specifically, the hippocampus, in the medial temporal lobe, has been systematically associated with successful encoding and retrieval of spatio-temporal contexts (Davachi et al. 2003; Nadel et al. 2000; Rajah et al. 2010; Ross and Slotnick 2008). Hence, the ability to build a strong relationship between spatio-temporal contexts/information—supported by subcortical regions in the hippocampus—may be a precursor/foundation to mathematical learning.

The existence of a spatial (left-to-right ordered) component in the representation of numerical magnitude/numerosities has been supported by the so-called Spatial Numerical Association of Response Codes (SNARC) effect (Dehaene et al. 1993), whereas in right-handed subjects responses to numerical judgements for small numbers are faster left-than-right hand, while numerical judgements for large numbers are faster right-than-left hand. This effect has supported the notion of the "mental number line" which postulates that, at least in Western cultures, numbers are spatially represented from left to right (Dehaene et al. 1993). Critically, this association occurs when ordinal information is relevant (i.e., numerical judgements) or irrelevant (i.e., parity judgements on numbers) to the task, suggesting that such spatial component is activated when numbers are presented, independently of the task. Moreover, this spatial coding seems to occur also when subjects are asked to judge non-numerical ordinal information/ stimuli such as months and letters, suggesting a similar—or shared—neurocognitive system that processes ordinal information, that is spatially organized (Gevers et al. 2004). Even more critically, a recent study of seven-month olds suggests that mapping ordinal information into a directional space is an early predisposition that precedes language and symbolic knowledge acquisition (Bulf et al. 2017). This datum hints at the idea that a left-to-right spatially organized mental representation of ordered dimensions may be rooted in biologically determined constraints of human brain development (Bulf et al. 2017; de Hevia et al. 2017) and is also observed in newborn chicks (Rugani et al. 2015).

Regardless of the spatial bias that seems to characterize both numerically-relevant and numerically-irrelevant sequences/stimuli, and which likely taps onto a phylogenetically and ontogenetically ancient—and shared (across numerical and non-numerical dimensions)—neurocognitive system, a clear dissociation seems to exist between cardinality and ordinality processes later in development. This may suggest a critical role of experience and schooling on *numerosity coding*. A brain-based dissociation between cardinal and ordinal properties of numbers dates back to the neuropsychological literature whereas a double dissociation has been reported for cardinal and ordinal aspects of numerosities: patient MAR (Dehaene and Cohen 1997) and SE (Delazer and Butterworth 1997) showed intact ordinal but impaired numerical processing, whereas patient BOO (Dehaene and Cohen 1997) and CO (Turconi and Seron 2002) showed the reverse pattern with preserved numerical but impaired ordinal sequence processing. Moreover, in healthy adults cardinality and ordinality judgement tasks have been shown not to correlate with one another—even though both, yet uniquely, predict math performance (Goffin and Ansari 2016). Further challenging the notion of a common neural substrate for cardinality and ordinality magnitude processing—particularly when stimuli are presented in symbolic format—is the reverse *distance effect*, which has been systematically reported in various studies. The reverse *distance effect* stipulates that judging whether 1 and 2 are in order (either ascending or descending) is faster and less error-prone than judging whether 1 and 3 are in order (Turconi et al. 2006; Goffin and Ansari 2016; Lyons and Beilock 2013). This can be interpreted—yet it is only speculative—as a potential failure to inhibit automatically processed counting sequences that are acquired a-semantically—and with a recital component—during early language acquisition (Gelman and Gallistel 1978; Sarnecka and Carey 2008). However, it was also found that ordinality and cardinality predictability accounts of math scores are independent of cognitive control tasks (Goffin and Ansari 2016).

Furthermore, fMRI studies report dissociable networks of activation between symbolic cardinality and ordinality judgements, while their non-symbolic neural correlates were very highly overlapping. Specifically, during non-symbolic judgements, a cohort of typically-developing adult participants showed prominent activation of a fronto-parietal network for both cardinality judgements (i.e., select the larger numerosity among two alternatives) and ordinality judgements (i.e., select whether a given sequence of dots is presented in order or not). Yet, when the symbolic equivalent of such task was presented, participants activated a canonical fronto-parietal network during the cardinality task, while regions of the Supplementary Motor Area (SMA) and pre-SMA were recruited during the ordinality task (Lyons and Beilock 2013). Dissociable neural representations of number symbols have also been reported in perceptual and association cortices. Specifically, it has been shown that the neural similarity between number symbols was positively predicted by the lexical frequency of numerals in parietal and occipital brain areas. By contrast, the similarity space of non-symbolic analogue quantities was positively predicted by the *distance/ratio effect* in prefrontal, parietal, and occipital brain areas (Lyons and Beilock 2018). This seems to support the idea that a failure to inhibit early acquired counting lists may influence similarities and differences of *numerosity codes,* particularly for symbolic stimuli. Hence, while sometimes it is important that the incremental properties of numbers get associated with one other, in other circumstances, and with certain tasks, it may be beneficial to inhibit the phonological/recital aspects of numerical sequencing in order to better access and manipulate its associated cardinality. This notion has important implications for cognitive control theories of numerical and mathematical cognition and development.

Learning to Inhibit Irrelevant Information

So far, we have detailed the basic properties of numbers (i.e., cardinality, ordinality) and their behavioral and neural correlates, as well as described the potential involvement of some executive functions—such as working memory—in the development of these numerosity processes. In this section we will focus on other general neurocognitive processes, which are involved during development and learning of math-related concepts.

For example, learning rules and principles (i.e., the "+" and "-" signs; knowing the bijection and injection functions/properties), and starting to inhibit wrong/intuitive heuristics that do not meet or do not comply with the formal rule learned, both represent a fundamental milestone of mathematical learning over development (Brookman-Byrne et al. 2018; Houdé and Tzourio-Mazoyer 2003). The highest level of proficiency with numbers lies in (i) the ability to apply different rules given the context (i.e., "+" and "-" signs; flexibility), and (ii) adequately mastering the rules of abstraction. For instance, inasmuch as the perceptual properties of the numerical stimuli—such as size, density, or length—are salient, they do not influence the numerosity of the set and become therefore irrelevant (i.e., they need to be inhibited; Clayton and Gilmore 2015; Gilmore et al. 2016; Houdé and Tzourio-Mazoyer 2003).

Over development, we learn to break away from perception—which relies heavily on automatic and heuristic processes—and to apply learned rules to solve a specific (numerical) problem. This is a process that is supported by inhibitory control regions within the prefrontal cortex, particularly in its ventrolateral aspect, encompassing the anterior insular cortex (AIC) and the anterior cingulate cortex (ACC; Houdé and Tzourio-Mazoyer 2003).

As previously mentioned, certain types of information (i.e., spatial information, but also other magnitude's information such as luminance) can sometimes aid numerical estimation, yet they can also be detrimental when they are yielding conflicting information (Cohen Kadosh et al. 2008) to the numerical task. In the latter case, these types of (redundant) information can interfere with numerical judgements. For example, in a numerical Stroop paradigm, responses are faster when the numerical and size dimensions are congruent with each other (i.e., facilitation); while responses are slower and more error-prone when the numerical and size dimensions are incongruent with each other (i.e., interference; Girelli et al. 2000; Henik and Tzelgov 1982; Rubinsten et al. 2002).

A recent prominent behavioral theory of mathematical cognition has postulated that the relationship between "number sense" and mathematical skills is only significant when the numerical and non-numerical information vary incongruently with each other—e.g., when of two to-be-compared sets the one that has the smaller number of items also has the larger item size or total occupied area and vice versa. This *datum* suggests that inhibitory/cognitive control processes can greatly influence our ability to perform numerical tasks (Fuhs and Mcneil 2013; Gilmore et al. 2013; Viarouge et al. 2023) and, in turn, poor inhibitory control skills may significantly jeopardize adequate mathematical learning. Some studies report that, during number comparison tasks, children with DD are selectively impaired only on incongruent trials and perform adequately on congruent trials (Gilmore et al. 2013; Szucs et al. 2013). Furthermore, in a series of EEG/event-related potential (ERP) studies, Szucs and colleagues demonstrated that, as children progress through education and become more and more proficient with numerical stimuli, brain responses during task-irrelevant trials/stimuli show a signature of weakening interference and strengthening facilitation, suggesting a parallel development of general cognitive control skills and automatic number processing (Soltész et al. 2011; Szucs et al. 2007, 2009).

Critically, over development, other forms of interference emerge that require the aid of neurocognitive inhibitory control mechanisms. This is the case of numerical judgements of decimals-, negative number-, and fraction-like stimuli. School-aged children are likely to think that 1.45 is larger than 1.5 because 45 is larger than 5. In a recent study, it was shown that the ability to compare decimal numbers in which the smallest number has the greatest number of digits, is rooted in the ability to inhibit the learned misconception that the greater the number of digits, the greater the magnitude; and in the ability to inhibit the perceptual input of the length of the decimal number per se (Roell et al. 2017).

The results of these studies converge on the idea that prefrontal cortex mechanisms are essential to mathematical learning at multiple stages of math skills development. Namely, inhibitory control processes—anchored in the prefrontal cortex—are required to keep suppressing irrelevant information, biases, and rules, as a function of the context and the task at hand, prior to number processing becoming the most relevant dimension and starts being automatized.

Furthermore, prefrontal cortex mechanisms may be essential to achieve, through development and training, the ability for numerical abstraction (Cohen Kadosh and Walsh 2009).

Together, the data presented here support the notion of a complex neurocognitive interaction, mediated by prefrontal cortex brain regions that aids the successful development of basic numerical skills.

MATHEMATICAL ABILITIES: DOMAIN-SPECIFIC AND DOMAIN-GENERAL COGNITIVE SCAFFOLDS

Developing Formal Arithmetic

Thus far, we have shown that some of the foundational building blocks of mathematical knowledge lie in (i) the ability to adequately represent quantities in various connotations and formats, as well as in (ii) being able to "extract" the unique numerosity property of an items set (or of a number) among other perceptual and semantic *biases*. Yet, and as mentioned earlier, when we need to perform arithmetical computations on numbers, other and more sophisticated skills need to be developed. For example, (i) we have to know and understand that adding a number to a set or subtracting an item from a set will affect the numerosity of that set; and (ii) we have to keep track of intermediate results, etc. Notably, approximate forms of these types of arithmetical computations are evident already in infants and toddlers (Barth et al. 2005, 2006; Gilmore et al. 2007; Wynn 1992).

However, complex manipulations of symbolic and abstract quantities (starting from formal counting) require the aid of higher-order cognitive functions that support maintenance of numerical representations in working memory over several seconds. This is achieved through the dynamic interaction of lateral fronto-parietal brain circuits. Critically, the involvement of these working memory circuits changes dramatically with development (Menon 2014). A consistent profile of diminished reliance on lateral prefrontal cortex regions and increased reliance on posterior parietal regions (i.e., the so-called anterior-to-posterior shift) has been documented in a wide range of studies involving numerical comparisons as well as arithmetical computations (Ansari and Dhital 2006; Ansari et al. 2005; Cantlon et al. 2009). In a seminal study, Rivera and colleagues found that relative to adults, children tend to engage the posterior parietal cortex less, and the prefrontal cortex more when solving arithmetic problems (Rivera et al. 2005), likely reflecting increased specialization and automatization (possibly even through processes of inhibitory control over development; see next section "Learning to Inhibit Irrelevant Strategies for Arithmetic Problem-Solving while Strengthening Arithmetical Representations for Fact Retrieval"). This idea has been supported by arithmetic learning paradigms in adults, which have shown initial prefrontal activation and a shift towards posterior parietal cortex regions as a function of learning and experience (Delazer et al. 2005; Ischebeck et al. 2007). When looking on neurotransmitters in the parietal and prefrontal cortices, it has been shown that neurotransmitters (γ-aminobutyric acid and glutamate) in the parietal cortex explain current mathematical performance and predict future mathematical achievement during childhood, adolescence, and young adulthood (Zacharopoulos et al. 2021b), while mathematical learning, or the lack of it, is related to a neurotransmitter (GABA) in the prefrontal cortex (Zacharopoulos et al. 2021a).

Other studies have more directly addressed the link between working memory abilities and arithmetic skills. For example, Dumontheil and Klingberg found that IPS activity during a visuospatial working memory task predicted arithmetic performance two years later in a sample of six- to sixteen-year-olds (Dumontheil and Klingberg 2012). Further analyses on the neural correlates of individual components of working memory have provided evidence for the fractionation of neurofunctional systems associated with distinct working memory components during arithmetic problem-solving (Arsalidou and Taylor 2011; Metcalfe et al. 2013). Specifically, analysis of the relation between the central executive, phonological, and visuospatial components of working memory, with brain activation during an arithmetic task, revealed that visuospatial working memory is a strong predictor of mathematical skills in seven- to nine-year-old children (Metcalfe et al. 2013). This study also demonstrated that visual working memory is associated with increased problem complexity-related responses in left dorsolateral and right ventrolateral prefrontal cortex as well as in the bilateral IPS and supramarginal gyrus (SMG). Notably, visuospatial working memory and the central executive component were associated with largely distinct patterns of brain responses during arithmetic problem-solving and overlap was only observed in the ventral aspects of the left supramarginal gyrus, suggesting that this region is an important locus for the integration of information in working memory during arithmetic problem-solving in children (Ansari 2008; Dehaene et al. 2003; Kawashima et al. 2004; Kucian et al. 2008; Menon et al. 2000; Rivera et al. 2005).

Finally, analysis of intrinsic functional connectivity suggests that a network of prefrontal and parietal cortical areas supports the longitudinal development of numerical abilities from seven to fourteen years of age (Evans et al. 2015).

These findings further confirm the pivotal role of overlapping parietal-frontal working memory circuits in children's mathematical skill development.

Learning to Inhibit Irrelevant Strategies for Arithmetic Problem-Solving while Strengthening Arithmetical Representations for Fact Retrieval

Over development, children's gains in arithmetical skills and mathematical problem-solving are characterized by the gradual replacement of inefficient procedural strategies with more efficient ones. Namely, various counting strategies (i.e., verbal/finger counting) give way to direct retrieval of arithmetic facts (Cho et al. 2011; Geary 2011; Geary and Brown 1991; Qin et al. 2014). This is a process that has been attributed to an initial engagement of the hippocampal subcortical system within the medial temporal lobe and, later, its functional connectivity with lateral fronto-parietal cortical regions (Qin et al. 2014; Rosenberg-Lee et al. 2018). Within this framework, during an addition problem, the association of addend-pairs with their correct solution is first strengthened within the associative learning system residing in the hippocampus. Once such association is made, the so-formed "arithmetical facts" are "transferred" to cortical areas where representations can then "reside" for handy/easy retrieval (Davachi et al. 2003; Qin et al. 2014).

This notion has been validated by a study that implemented a hybrid cross-sectional/longitudinal design, and showed that children's transition from counting to memory-based retrieval strategies over a 1- to 2-year interval period (i.e., from seven-to-nine to ten-to-eleven-year-olds) was mediated by increased hippocampal activation and increased hippocampal-neocortical functional connectivity (Qin et al. 2014). Critically, following an initial increase in hippocampal engagement during middle childhood, this hippocampal dependency decreased during adolescence and adulthood despite further improvements in memory-based problem-solving strategies. This pattern of initial increase and subsequent decrease in activation provides support for the notion that the hippocampus plays a time-limited role in the early phase of knowledge acquisition (McClelland et al. 1995; Tse et al. 2007). The critical, and timely, involvement of the hippocampus in the development of optimal retrieval-based skills was further demonstrated in a training study whereas seven-to-nine-year-old children showed significant increases in hippocampal engagement and hippocampal-cortical interactions following a eight-week arithmetic training (Rosenberg-Lee et al. 2018). Critically, in both studies, changes in subcortico-cortical interactions were paired with increases in fact retrieval strategy-use.

As the representations for arithmetical facts are formed through appropriate cross-talk between hippocampal and cortical systems, those older, inefficient strategies need to be inhibited. This is supported mainly by ventrolateral and ventromedial regions of the prefrontal cortex encompassing the AIC and the ACC.

Critically, weak prefrontal control signals have been associated with worse performance on a arithmetic problem-solving task, both in adults and children (Supekar and Menon 2012). The strength of causal signals from the AIC to ventral parietal and prefrontal regions in the SMG and the ventrolateral prefrontal cortex (VLPFC) respectively, significantly predicted better performance. Furthermore, in adults, an additional link from the AIC to the ACC has been reported—which significantly and uniquely predicted their performance on the same arithmetic problem-solving task.

Collectively, these results suggest that, over development, together with an anterior-to-posterior functional brain activity shift within the dorsal aspect of the fronto-parietal network (see previous section "Developing Formal Arithmetic"), a posterior-to-anterior functional brain activity shift in the ventral aspect of the same network may support the development of inhibitory control processes that help to suppress perceptual, or other types of, cognitive heuristics/strategies in favor of more optimal ones. In parallel, subcortical-cortical interactions support the fine-tuning of arithmetical representations in the form of "learned facts." Together, these processes subtend the development of successful math problem-solving skills.

INTERVENTION

This chapter presents converging evidence for the involvement of multiple brain systems—and the cognitive processes/functions they subtend—to the appropriate acquisition of numerical and mathematical abilities. Thus, it is critical to consider all of them when designing intervention approaches for children that are at the end of the distribution for mathematical and numerical abilities (e.g., children with mathematical learning disabilities (MLD), children with DD). The following section delineates/describes the work that has been done so far on exercising/training/boosting these processes, both at the domain-specific, as well as at the domain-general level. These studies' benefits are two fold: they allow (i) to test whether the above-mentioned neurocognitive mechanisms are causally involved in the development of mathematical and numerical skills, and (ii) to pave the way towards effective interventions in cases of atypical development.

Domain-Specific

Numerosity-Based Interventions

As discussed, numerosity manipulation has been suggested as one of the processes at the core of successful math-knowledge acquisition. Studies on healthy adults, typically developing children and even preschoolers, have shown that numerosity-based training can improve arithmetic performance (Hyde, Khanum, and Spelke, 2014; Park, Bermudez, Roberts, and Brannon, 2016; Park and Brannon, 2013, 2014 – but see Kim, Jang, and Cho, 2018). Hence, numerosity-based interventions have naturally been advocated for the remediation of MLD and DD cases (Butterworth and Yeo 2004). Notably, these interventions have capitalized on symbolic and non-symbolic material and on the use of "numerical manipulatives" (i.e., Cuisenaire rods, number tracks, number cards; Butterworth and Yeo 2004) to aid the learner in forming an accurate representation of (each) numerosity, through active visual manipulations.

Thus far, several training studies have also looked at the effects of software-based learning programs designed to foster active manipulations on numerosities in children with mathematical difficulties. Wilson and colleagues implemented a computer program (The Number Race) specifically designed to train symbolic and non-symbolic number sense through a series of computer-based activities involving counting, and number/magnitude comparisons (Wilson et al. 2006). This study showed only a selected "within-domain" improvement whereas children's basic numerical (i.e., number sense) skills got better after training, while no significant gains were observed in their arithmetical performance. Similarly, in a study that employed a software program focused on strengthening the mapping between numerosities and their symbolic representations (Graphogame-Math) performance gains were minimal, and so were transfer effects to arithmetical competences (Räsänen et al. 2009).

Critically—and similarly to school-age children with DD—a study on preschoolers at risk for developmental learning disabilities showed that an approximate arithmetic training benefited performance only on early informal, but not formal, arithmetic (Szkudlarek and Brannon 2018).

By contrast, interventions capitulating on visual number-to-space training have shown greater promise. Kucian and colleagues used a custom-made computer-based number-line estimation task (rescue calcularis) to train a population of eight-to-ten-year-old children with DD (Kucian et al. 2011). After five weeks' training, improvements were evident on the number-line task as well as on an arithmetic task (featuring addition and subtraction problems). Yet, after the training, children with DD still differed in performance compared to their (matched) typically developing peers. Critically, after training, both groups—albeit to a different extent—showed decreases in functional brain activity when performing the number-line estimation task. Specifically, decreases in functional brain activity were evident in regions of the posterior parietal and prefrontal cortices, both critical for numerical manipulations at this age (Ansari and Dhital 2006; Ansari et al. 2005; see also previous section "Developing Cardinal Skills"). Notably, the results of this study suggest that even though this type of training could not fully rescue performance in the DD group, it could still elicit transfer effects. Most importantly, number-line training induced brain plasticity effects even in the DD group, further highlighting the potential of this approach to treat children with math difficulties. In the wake of these results, other number-based digitalized training programs have been developed that capitalize on the link between non-symbolic and symbolic numerosities onto number lines/tracks (Butterworth et al. 2011; for a review, see also Dowker 2017).

Together, the training studies reviewed here suggest that a number sense training alone may not be sufficient to enhance arithmetical performance, particularly in children at the lower end of the distribution of abilities. In contrast, number-line training was able to elicit some transfer effects, even in children with DD. This indicates that a cognitive training that integrates cardinal as well as visuo-spatial (i.e., ordinal) properties of numbers may aid the developing brain with a more solid numerical representation critical for successful arithmetical performance. In line with this, a recent number sense training study in DD has shown that gains in arithmetic were fully mediated by performance on a visual perception task (i.e., figure matching; Cheng et al. 2020). Indeed, visuospatial processing has long been posited critical for successful mathematical performance (Gilligan et al. 2020; Sella et al. 2016; Verdine et al. 2014).

Yet, it is also possible that other domain-general processes could come to play in mediating this relationship, such as inhibitory processes, or mechanisms of visual attention—which could facilitate performance during visual-form perception and, as such, mediate the relationship between number sense training and arithmetic transfer. More generally, this suggests that to propel significant transfer effects to arithmetical performance, particularly at early stages of development, and in populations who are struggling with math, a number sense training may be accompanied (or should embed) the training of other domain-general processes that could stimulate multiple neurocognitive systems for numerical and mathematical functioning.

Arithmetic-Based Interventions

A large bulk of studies on the remediation of arithmetical skills are classroom-based intervention studies. These often combine conceptual and procedural activities with speeded practice to strengthen problem-solving skills and promote efficient retrieval of arithmetical facts (Fuchs et al. 2008, 2009; Powell et al. 2009; for a comprehensive review, see Dowker 2005, 2017).

In a recent study, Iuculano and colleagues have looked at the effects—at the brain level—of a comprehensive arithmetic intervention similar to these (Fuchs et al. 2008, 2010; Powell et al. 2009)—yet in a 1:1 setting—on a selected group of children with MLD. The authors showed that, in parallel with performance normalization, eight weeks of 1:1 arithmetic training elicited extensive functional brain changes in eight-to-nine-year-old children with MLD (Iuculano et al. 2015). This comprehensive arithmetic training normalized aberrant functional responses in the MLD group, to the level of neurotypical peers. Brain plasticity effects were evident in a distributed network of prefrontal, parietal, and ventral temporal-occipital brain regions. Remarkably, machine learning algorithm techniques revealed that brain activity patterns in children with MLD were significantly discriminable from neurotypical peers before training, but statistically indistinguishable after. Notably, changes in brain activity after training in MLD were characterized by significant reduction of widespread over-activation in multiple neurocognitive systems in prefrontal, parietal, and ventral temporal occipital cortices (Figure 25.2). This suggests that this type of training can induce global changes across distributed brain systems that encompass multiple stages of the information processing hierarchy necessary for successful arithmetic problem-solving (Iuculano and Menon 2018; Iuculano et al. 2018). Specifically, by

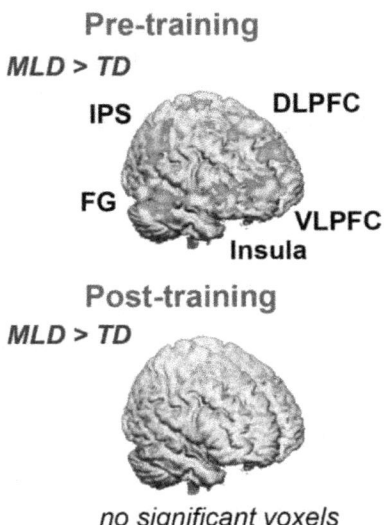

Pre-training

MLD > TD

IPS   DLPFC

FG

VLPFC
Insula

Post-training

MLD > TD

no significant voxels

FIGURE 25.2 Neural correlates of arithmetic-based intervention in mathematical learning disabilities.

Top. Before training, children with Mathematical Learning Disabilities (MLD) showed greater functional engagement—during arithmetic problem solving—of multiple neurocognitive systems encompassing prefrontal, parietal, and ventral temporal occipital cortices.

Bottom. Following training, no differences in functional brain activation—during arithmetic problem-solving—were detected between children with MLD and their Typically Developing (TD) peers—adapted from: Iuculano et al., (2015).

Abbreviations: DLPFC = Dorsolateral Prefrontal Cortex; VLPFC = Ventrolateral Prefrontal Cortex; IPS = Intraparietal Sulcus; FG = Fusiform Gyrus

facilitating the development of quantity representations and the use of sophisticated counting procedures, this type of training might place fewer demands on quantitative and higher-order visual form processes supported by the posterior parietal and ventral temporal-occipital cortices. Concurrently, it might facilitate efficient processing strategies by decreasing load on cognitive neural resources (e.g., working memory, non-verbal reasoning, attention) supported by the prefrontal cortex. More generally, and similarly to what has been discussed before, these findings suggest that a comprehensive training, one which integrates conceptual as well as procedural aspects of arithmetic learning—rather than focusing on isolated components of number knowledge and arithmetic—might be more effective in remediating deficits in multiple neurocognitive systems in children with math difficulties. Critically, it needs be noted that not all children improved after the training. This result calls for a more sophisticated, fine-grained approach for low attainers. Recently, Dowker and colleagues have developed a behavioral protocol for an ad hoc individualized training program—that capitalizes on numerical and arithmetical principles as well as retrieval strategies—and which unfolds as a function of the baseline profile of the child (i.e., Catch Up Numeracy). Notably, behavioral results show significant improvements following this protocol (Holmes and Dowker 2013). The next step will be to implement these types of adaptive arithmetic training programs digitally, as well as test their validity and benefits at the brain level (similar to Iuculano et al. 2015; Kucian et al. 2011). Furthermore, longevity effects are another crucial aspect that needs to be systematically addressed in future studies (see Koponen et al. 2018).

Domain-General

Executive Functions-Based Interventions

Executive functions are defined as a class of cognitive processes critical to perform goal-directed and purposeful behaviors. They include WM, planning, and cognitive/inhibitory control. We have already discussed the involvement of executive processes in numerical and arithmetic skills acquisition over development. For example, we have seen that functional activity of the IPS during a visuo-spatial WM task significantly predicts subsequent arithmetic performance in a sample of six-to-sixteen-year-olds (see previous section "Developing Formal Arithmetic;" Dumontheil and Klingberg 2012; see also Metcalfe et al. 2013). We have also analyzed the importance of cognitive control processes in fostering appropriate numerosity and arithmetic-strategy "tuning" (Soltész et al. 2011; Supekar and Menon 2012; Szucs et al. 2007). Given the involvement of these processes in successful numerical and arithmetical acquisition, we can ask whether cognitive trainings that tap into these processes can be useful in remediating (or enhancing) math performance over development.

The majority of WM training studies have focused on improving its visuo-spatial component with promising "within domain" results. The most consistently implicated brain correlates of these trainings have been found across regions of the prefrontal and posterior parietal cortices—i.e., the canonical WM network (Constantinidis and Klingberg 2016). Given the involvement of these brain regions (and processes) to math learning and cognition, it has been hypothesized that WM training could lead to significant transfer effects to math performance.

In a cognitive WM training study, Bergman-Nutley and Klingberg tested the effects of a five-week adaptive WM training (Cogmed Working Memory Training Program; Klingberg et al. 2005) on arithmetic fluency in a group of seven-to-fifteen-year-olds (the majority of whom suffered from attention deficit hyperactivity disorder (ADHD)). Their results showed improvements on the math fluency test—yet, effect sizes of the improvement were very small (d = 0.2), particularly when compared to other cognitive abilities assessed: visual-WM, and the ability to follow instructions (0.67 < d < 0.90) (Bergman-Nutley and Klingberg 2014). As previously discussed, given that math competency is the result of the interaction of multiple neurocognitive systems and abilities, minimal transfer effects following a visuo-spatial WM training alone are not surprising.

A better approach may be to train multiple executive functions (EF) simultaneously. A recent study has indeed combined a comprehensive training of EFs in a group of primary school children and assessed its effects on academic outcome measures. Training occurred for ten weeks and, similar to Klingberg and colleagues (2005), was also adaptive. The authors report changes in multiple facets of EFs, including inhibitory control, flexibility, as well as on school grades of reading and math (Goldin et al. 2014). Even though this study had no imaging component, it provides preliminary (and corroborating) evidence for the unique benefits of comprehensive training protocols for strengthening math competencies over development.

Embodied-Based Interventions—A Proxy to Foster Ordinal Skills and Spatial Information

This line of research capitalizes on the fact that the development of accurate numerical representations could be fostered by bodily sensory (and motor) experiences. In a recent series of studies, Nuerk and colleagues showed that kindergarteners that were trained to solve a magnitude comparison task on a digital dance mat demonstrated significant improvements on a number-line estimation task, as well as on a subtest of a standardized mathematical achievement battery (TEDI-MATH) (Cress et al. 2010). Notably, improvements of the "dance-mat group" were greater than those of a control group trained on the same magnitude comparison task—and for the same amount of time—but using a personal computer. Similar results were found when training first-graders to indicate the position of a given number by walking to the estimated location on a number line on the floor (Link et al. 2013).

At the brain level, the dorsolateral prefrontal cortex (DLPFC) seems to be a key player in this process. The DLPFC has a pivotal role of integrating multiple forms of information in WM, allowing for flexible (and accurate) mental representations (Prabhakaran et al. 2000). Notably, transcranial direct current stimulation (tDCS) of the DLPFC during a number-line training involving body movements along a physical line was shown to improve performance in a group of healthy adult participants. Improvements were seen on the number-line task (trained), as well as on a WM task (untrained). Critically, training effects were specific to the experimental group that received DLPFC stimulation combined with bodily mental number-line training, and lasted up to two months after training (Looi et al. 2016). Similarly, transcranial random noise stimulation (tRNS) over the bilateral DLPFCs combined with a similar mental number-line bodily training improved learning and performance of children with MLD (Looi et al. 2017). Notably, training gains correlated positively with improvement on a standardized math test, and this effect was strengthened by tRNS.

We would like to emphasize that we do not aim to provide a comprehensive review of the intervention literature in this field (for a more comprehensive review on this topic, see Dowker 2005, 2017). However, all in all, the studies reviewed here suggest that cognitive trainings designed to "stimulate" multiple aspects of math knowledge (and the neurocognitive systems subtending it) may yield the best outcomes to significantly improve (or remediate) arithmetical skills—particularly in individuals falling at the lower end of the distribution of abilities.

CONCLUSIONS AND FUTURE DIRECTIONS

In this chapter, we report on how numerical and mathematical cognition is developed from a neurocognitive perspective. Such development is associated with quantitative and qualitative changes in multiple brain networks, as well as the involvement of a myriad of cognitive processes that can be considered both domain-specific and domain-general. We reviewed these multiple distributed neurocognitive systems, which are involved in quantity manipulation, perceptual (coding and decoding, inhibitory control, and working) and associative declarative (memories), as well as the corresponding brain regions and networks that support and orchestrate the development of numerical and mathematical cognitive skills. The involvement of these neurocognitive processes and systems is complex, dynamic, and, most importantly, a function of development, proficiency of skills, and the task at hand. While we have focused on positive findings, it is important to note that not all the studies find a support that domain-specific (e.g., numerosity) or domain-general (e.g., working memory) are involved in mathematical ability (van Bueren et al. 2022), and further studies should examine the effects due to experimental settings, educational systems, and culture. In addition, while we focused mainly on fMRI, it is important to acknowledge other approaches that could allow more comprehensive understanding of brain development and its link to mathematical development, such as neurotransmitters, metabolic changes, or aperiodic brain activity (van Bueren et al. 2022). Further research in developmental cognitive neuroscience can foster our understanding of the involved neurocognitive mechanisms as a function of age in typical and atypical populations, providing the required knowledge to design and promote effective interventions at the cognitive and neural levels.

REFERENCES

Abboud, S., Maidenbaum, S., Dehaene, S., and Amedi, A. (2015). A number-form area in the blind. *Nature Communications*. doi.org/10.1038/ncomms7026

Agrillo, C., Dadda, M., Serena, G., and Bisazza, A. (2008). Do fish count? Spontaneous discrimination of quantity in female mosquitofish. *Animal Cognition*. doi.org/10.1007/s10071-008-0140-9

Ansari, D. (2008). Effects of development and enculturation on number representation in the brain. *Nature Reviews Neuroscience*. doi.org/10.1038/nrn2334

Ansari, D. and Dhital, B. (2006). Age-related changes in the activation of the intraparietal sulcus during nonsymbolic magnitude processing: An event-related functional magnetic resonance imaging study. *Journal of Cognitive Neuroscience*. doi.org/10.1162/jocn.2006.18.11.1820

Ansari, D., Garcia, N., Lucas, E., Hamon, K., and Dhitalm, B. (2005). Neural correlates of symbolic number processing in children and adults. *Neuroreport*. doi.org/10.1097/01.wnr.0000183905.23396.f1

Arp, S. and Fagard, J. (2005). What impairs subitizing in cerebral palsied children? *Developmental Psychobiology*. doi.org/10.1002/dev.20069

Arp, S., Taranne, P., and Fagard, J. (2006). Global perception of small numerosities (subitizing) in cerebral-palsied children. *Journal of Clinical and Experimental Neuropsychology*. doi.org/10.1080/13803390590935426

Arsalidou, M. and Taylor, M. J. (2011). Is 2+2=4? Meta-analyses of brain areas needed for numbers and calculations. *NeuroImage*. doi.org/10.1016/j.neuroimage.2010.10.009

Attout, L., Noël, M.-P., and Majerus, S. (2014). The relationship between working memory for serial order and numerical development: A longitudinal study. *Developmental Psychology*. doi.org/10.1037/a0036496

Barth, H., La Mont, K., Lipton, J., Dehaene, S., Kanwisher, N., and Spelke, E. (2006). Non-symbolic arithmetic in adults and young children. *Cognition*. doi.org/10.1016/j.cognition.2004.09.011

Barth, H., La Mont, K., Lipton, J., and Spelke, E. S. (2005). Abstract number and arithmetic in preschool children. *Proceedings of the National Academy of Sciences*. doi.org/10.1073/pnas.0505512102

Bergman-Nutley, S. and Klingberg, T. (2014). Effect of working memory training on working memory, arithmetic and following instructions. *Psychological Research*. doi.org/10.1007/s00426-014-0614-0

Brannon, E. M. (2002). The development of ordinal numerical knowledge in infancy. *Cognition*. doi.org/10.1016/S0010-0277(02)00005-7

Brannon, E. M. and Terrace, H. S. (2000). Representation of the numerosities 1–9 by rhesus macaques (Macaca mulatta). *Journal of Experimental Psychology: Animal Behavior Processes*. doi.org/10.1037/0097-7403.26.1.31

Brookman-Byrne, A., Mareschal, D., Tolmie, A., and Dumontheil, I. (2018). Inhibitory control and counterintuitive science and maths reasoning in adolescence. *PLoS ONE*, 13(6), e0198973.

Bruandet, M., Molko, N., Cohen, L., and Dehaene, S. (2004). A cognitive characterization of dyscalculia in Turner syndrome. *Neuropsychologia*. doi.org/10.1016/j.neuropsychologia.2003.08.007

Bulf, H., De Hevia, M. D., Gariboldi, V., and Cassia, V. M. (2017). Infants learn better from left to right: A directional bias in infants' sequence learning. *Scientific Reports*. doi.org/10.1038/s41598-017-02466-w

Butterworth, B. (1999). *The Mathematical Brain*. London: Macmillan.

Butterworth, B. (2011). Foundational numerical capacities and the origins of dyscalculia. In S. Dehaene and E. Brannon (eds), *Space, Time and Number in the Brain: Searching for the Foundations of Medical Thought*, 249–265. London: Academic Press. doi.org/10.1016/B978-0-12-385948-8.00016-5

Butterworth, B., Sashank, V., and Laurillard, D. (2011). Dyscalculia: From brain to education. *Science*, 332, 1049–1053. doi.org/10.1126/science.1201536

Butterworth, B. and Yeo, D. (2004). *Dyscalculia Guidance: Helping Pupils with Specific Learning Difficulties in Maths*. London: NferNelson Publishing.

Buzsáki, G. and Moser, E. I. (2013). Memory, navigation and theta rhythm in the hippocampal-entorhinal system. *Nature Neuroscience*. doi.org/10.1038/nn.3304

Cantlon, J. F., Brannon, E. M., Carter, E. J., and Pelphrey, K. A. (2006). Functional imaging of numerical processing in adults and 4-y-old children. *PLoS Biology*. doi.org/10.1371/journal.pbio.0040125

Cantlon, J. F., Libertus, M. E., Pinel, P., Dehaene, S., Brannon, E. M., and Pelphrey, K. A. (2009). The neural development of an abstract concept of number. *Journal of Cognitive Neuroscience*. doi.org/10.1162/jocn.2008.21159

Cantlon, J. F., Pinel, P., Dehaene, S., and Pelphrey, K. A. (2011). Cortical representations of symbols, objects, and faces are pruned back during early childhood. *Cerebral Cortex*. doi. org/10.1093/cercor/bhq078

Castaldi, E., Aagten-Murphy, D., Tosetti, M., Burr, D., and Morrone, M. C. (2016). Effects of adaptation on numerosity decoding in the human brain. *NeuroImage*. doi.org/10.1016/ j.neuroimage.2016.09.020

Castelli, F., Glaser, D. E., and Butterworth, B. (2006). Discrete and analogue quantity processing in the parietal lobe: A functional MRI study. *Proceedings of the National Academy of Sciences*. doi.org/10.1073/pnas.0600444103

Chein, J. M. and Schneider, W. (2012). The brain's learning and control architecture. *Current Directions in Psychological Science*. doi.org/10.1177/0963721411434977

Cheng, D., Xiao, Q., Cui, J., Chen, C., Zeng, J., Chen, Q., et al. (2020). Short-term numerosity training promotes symbolic arithmetic in children with developmental dyscalculia: The mediating role of visual form perception. *Developmental Science*, 23(4), e12910. doi.org/ 10.1111/desc.12910

Chi, M. T. H. and Klahr, D. (1975). Span and rate of apprehension in children and adults. *Journal of Experimental Child Psychology*. doi.org/10.1016/0022-0965(75)90072-7

Cho, S., Ryali, S., Geary, D. C., and Menon, V. (2011). How does a child solve 7 + 8? Decoding brain activity patterns associated with counting and retrieval strategies. *Developmental Science*. doi.org/10.1111/j.1467-7687.2011.01055.x

Cipolotti, L., Butterworth, B., and Denes, G. (1991). A specific deficit for numbers in a case of dense acalculia. *Brain*. doi.org/10.1093/brain/114.6.2619

Clayton, S. and Gilmore, C. (2015). Inhibition in dot comparison tasks. *ZDM*. doi.org/10.1007/ s11858-014-0655-2

Cohen Kadosh, R., Bahrami, B., Walsh, V., Butterworth, B., Popescu, T., and Price, C. (2011). Specialization in the human brain: The case of numbers. *Frontiers in Human Neuroscience*, 5:62. doi: 10.3389/fnhum.2011.00062.eCollection 2011

Cohen Kadosh, R., Cohen Kadosh, K., Kaas, A., Henik, A., and Goebel, R. (2007). Notation-dependent and -independent representations of numbers in the parietal lobes. *Neuron*. doi. org/10.1016/j.neuron.2006.12.025

Cohen Kadosh, R., Lammertyn, J., and Izard, V. (2008). Are numbers special? An overview of chronometric, neuroimaging, developmental and comparative studies of magnitude representation. *Progress in Neurobiology*. doi.org/10.1016/j.pneurobio.2007.11.001

Constantinidis, C. and Klingberg, T. (2016). The neuroscience of working memory capacity and training. *Nature Reviews Neuroscience*. doi.org/10.1038/nrn.2016.43

Corbetta, M. and Shulman, G. L. (2002). Control of goal-directed and stimulus-driven attention in the brain. *Nature Reviews Neuroscience*. doi.org/10.1038/nrn755

Cress, U., Fischer, U., Moeller, K., Sauter, C., and Nuerk, H. C. (2010). The use of a digital dance mat for training kindergarten children in a magnitude comparison task. Learning in the Disciplines. In ICLS 2010: *Conference Proceedings, 9th International Conference of the Learning Sciences*, Vol. 1, 105–112. Chicago, IL: International Society of Learning Sciences.

Cutini, S., Scatturin, P., Basso Moro, S., and Zorzi, M. (2014). Are the neural correlates of subitizing and estimation dissociable? An fNIRS investigation. *NeuroImage*. doi.org/ 10.1016/j.neuroimage.2013.08.027

Dacke, M. and Srinivasan, M. V. (2008). Evidence for counting in insects. *Animal Cognition*. doi.org/10.1007/s10071-008-0159-y

Davachi, L., Mitchell, J. P., and Wagner, A. D. (2003). Multiple routes to memory: Distinct medial temporal lobe processes build item and source memories. *Proceedings of the National Academy of Sciences*. doi.org/10.1073/pnas.0337195100

de Hevia, M. D. (2016). Core mathematical abilities in infants: Number and much more. In M. Cappelletti and W. Fias (eds), *The Mathematical Brain Across the Lifespan*, Progress in Brain Research series, Vol. 227, 53–74. Amsterdam: Elsevier. doi.org/10.1016/bs.pbr.2016.04.014

de Hevia, M. D., Addabbo, M., Nava, E., Croci, E., Girelli, L., and Macchi Cassia, V. (2017). Infants' detection of increasing numerical order comes before detection of decreasing number. *Cognition*. doi.org/10.1016/j.cognition.2016.10.022

de Hevia, M. D., Veggiotti, L., Streri, A., and Bonn, C. D. (2017). At birth, humans associate "few" with left and "many" with right. *Current Biology*. doi.org/10.1016/j.cub.2017.11.024

De Smedt, B., Noël, M.-P., Gilmore, C., and Ansari, D. (2013). The relationship between symbolic and non-symbolic numerical magnitude processing skills and the typical and atypical development of mathematics: A review of evidence from brain and behavior. *Trends in Neuroscience and Education*, 2(2), 48–55.

Dehaene, S., Bossini, S., and Giraux, P. (1993). The mental representation of parity and number magnitude. *Journal of Experimental Psychology: General*. doi.org/10.1037/0096-3445.122.3.371

Dehaene, S. and Cohen, L. (1995). Towards an anatomical and functional model of number processing. *Mathematical Cognition*, 1(1), 83–120.

Dehaene, S. and Cohen, L. (1997). Cerebral pathways for calculation: Double dissociation between rote verbal and quantitative knowledge of arithmetic. *Cortex*. doi.org/10.1016/S0010-9452(08)70002-9

Dehaene, S., Piazza, M., Pinel, P., and Cohen, L. (2003). Three parietal circuits for number processing. *Cognitive Neuropsychology*. doi.org/10.1080/02643290244000239

Delazer, M. and Butterworth, B. (1997). A dissociation of number meanings. *Cognitive Neuropsychology*. doi.org/10.1080/026432997381501

Delazer, M., Ischebeck, A., Domahs, F., Zamarian, L., Koppelstaetter, F., Siedentopf, C. M., et al. (2005). Learning by strategies and learning by drill—evidence from an fMRI study. *NeuroImage*. doi.org/10.1016/j.neuroimage.2004.12.009

di Pellegrino, G. and Làdavas, E. (2015). Peripersonal space in the brain. *Neuropsychologia*. doi.org/10.1016/j.neuropsychologia.2014.11.011

Ditz, H. M., and Nieder, A. (2015). Neurons selective to the number of visual items in the corvid songbird endbrain. *Proceedings of the National Academy of Sciences*. doi.org/10.1073/pnas.1504245112

Dowker, A. (2005). Early identification and intervention for students with mathematical difficulties. *Journal of Learning Disabilities*, 38(4), 324–332. doi.org/10.1177/00222194050380040801

Dowker, A. (2017). Interventions for primary school children with difficulties in mathematics. *Advances in Child Development and Behavior*, 53, 255–287. doi.org/10.1016/bs.acdb.2017.04.004

Dumontheil, I. and Klingberg, T. (2012). Brain activity during a visuospatial working memory task predicts arithmetical performance 2 years later. *Cerebral Cortex*. doi.org/10.1093/cercor/bhr175

Evans, T. M., Kochalka, J., Ngoon, T. J., Wu, S. S., Qin, S., Battista, C., et al. (2015). Brain structural integrity and intrinsic functional connectivity forecast 6 year longitudinal growth in children's numerical abilities. *The Journal of Neuroscience*. doi.org/10.1523/JNEUROSCI.0216-15.2015

Fuchs, L. S., Powell, S. R., Hamlett, C. L., Fuchs, D., Cirino, P. T., and Fletcher, J. M. (2008). Remediating computational deficits at third grade: A randomized field trial. *Journal of Research on Educational Effectiveness, 1*(1), 2–32. doi.org/10.1080/19345740701692449

Fuchs, L. S., Powell, S. R., Seethaler, P. M., Cirino, P. T., Fletcher, J. M., Fuchs, D., et al. (2009). Remediating number combination and word problem deficits among students with mathematics difficulties: a randomized control trial. *Journal of Educational Psychology, 101*, 561–576.

Fuchs, L. S., Powell, S. R., Seethaler, P. M., Fuchs, D., Hamlett, C. L., Cirino, P. T., et al. (2010). A framework for remediating number combination deficits. *Exceptional Children*. doi.org/10.1177/001440291007600201

Fuhs, M. W. and Mcneil, N. M. (2013). ANS acuity and mathematics ability in preschoolers from low-income homes: Contributions of inhibitory control. *Developmental Science*. doi.org/10.1111/desc.12013

Gauthier, I., Skudlarski, P., Gore, J. C., and Anderson, A. W. (2000). Expertise for cars and birds recruits brain areas involved in face recognition. *Nature Neuroscience*. doi.org/10.1038/72140

Gauthier, I., Tarr, M. J., Anderson, A. W., Skudlarski, P., and Gore, J. C. (1999). Activation of the middle fusiform "face area" increases with expertise in recognizing novel objects. *Nature Neuroscience*. doi.org/10.1038/9224

Geary, D. C. (2011). Cognitive predictors of achievement growth in mathematics: A 5-year longitudinal study. *Developmental Psychology*. doi.org/10.1037/a0025510

Geary, D. C. and Brown, S. C. (1991). Cognitive addition: Strategy choice and speed-of-processing differences in gifted, normal, and mathematically disabled children. *Developmental Psychology*. doi.org/10.1037/0012-1649.27.3.398

Gelman, R. and Gallistel, C. R. (1978). *The Child's Understanding of Number*. Cambridge, MA: Harvard University Press.

Gersten, R. and Chard, D. (1999). Number sense: Rethinking arithmetic instruction for students with mathematical disabilities. *Journal of Special Education*. doi.org/10.1177/002246699903300102

Gevers, W., Reynvoet, B., and Fias, W. (2004). The mental representation of ordinal sequences is spatially organized: Evidence from days of the week. *Cortex; A Journal Devoted to the Study of the Nervous System and Behavior*. doi.org/10.1016/S0010-9452(08)70938-9

Giaquinto, M. (1995). Concepts and calculation. *Mathematical Cognition, 1*, 61–81.

Gilligan, K. A., Thomas, M. S. C., and Farran, E. K. (2020). First demonstration of effective spatial training for near transfer to spatial performance and far transfer to a range of mathematics skills at 8 years. *Developmental Science, 23*(4), e12909. doi.org/doi.org/10.1111/desc.12909

Gilmore, C., Attridge, N., Clayton, S., Cragg, L., Johnson, S., Marlow, N., et al. (2013). Individual differences in inhibitory control, not non-verbal number acuity, correlate with mathematics achievement. *PLoS ONE*. doi.org/10.1371/journal.pone.0067374

Gilmore, C., Cragg, L., Hogan, G., and Inglis, M. (2016). Congruency effects in dot comparison tasks: Convex hull is more important than dot area. *Journal of Cognitive Psychology*. doi.org/10.1080/20445911.2016.1221828

Gilmore, C. K., McCarthy, S. E., and Spelke, E. S. (2007). Symbolic arithmetic knowledge without instruction. *Nature*. doi.org/10.1038/nature05850

Girelli, L., Lucangeli, D., and Butterworth, B. (2000). The development of automaticity in accessing number magnitude. *Journal of Experimental Child Psychology*, *76*(2), 104–122. doi.org/10.1006/jecp.2000.2564

Goffin, C. and Ansari, D. (2016). Beyond magnitude: Judging ordinality of symbolic number is unrelated to magnitude comparison and independently relates to individual differences in arithmetic. *Cognition*. doi.org/10.1016/j.cognition.2016.01.018

Gold, M., Adair, J. C., Jacobs, D. H., and Heilman, K. M. (1995). Right-left confusion in Gerstmann's syndrome: A model of body centered spatial orientation. *Cortex*, *31*(2), 267–283. doi.org/10.1016/s0010-9452(13)80362-0.

Goldin, A. P., Hermida, M. J., Shalom, D. E., Costa, M. E., Lopez-Rosenfeld, M., Segretin, et al. (2014). Far transfer to language and math of a short software-based gaming intervention. *Proceedings of the National Academy of Sciences of the United States of America*. doi.org/10.1073/pnas.1320217111

Grotheer, M., Herrmann, K.-H., and Kovacs, G. (2016). Neuroimaging evidence of a bilateral representation for visually presented numbers. *Journal of Neuroscience*. doi.org/10.1523/JNEUROSCI.2129-15.2016

Halberda, J. and Feigenson, L. (2008). Developmental change in the acuity of the "number sense": The approximate number system in 3-, 4-, 5-, and 6-year-olds and adults. *Developmental Psychology*. doi.org/10.1037/a0012682

Halberda, J., Ly, R., Wilmer, J. B., Naiman, D. Q., and Germine, L. (2012). Number sense across the lifespan as revealed by a massive Internet-based sample. *Proceedings of the National Academy of Sciences*. doi.org/10.1073/pnas.1200196109

Halberda, J., Mazzocco, M. M. M., and Feigenson, L. (2008). Individual differences in non-verbal number acuity correlate with maths achievement. *Nature*. doi.org/10.1038/nature07246

Halgren, E., Boujon, C., Clarke, J., Wang, C., and Chauvel, P. (2002). Rapid distributed fronto-parieto-occipital processing stages during working memory in humans. *Cerebral Cortex*. doi.org/10.1093/cercor/12.7.710

Henik, A. and Tzelgov, J. (1982). Is three greater than five: The relation between physical and semantic size in comparison tasks. *Memory and Cognition*. doi.org/10.3758/BF03202431

Hirnstein, M., Bayer, U., Ellison, A., and Hausmann, M. (2011). TMS over the left angular gyrus impairs the ability to discriminate left from right. *Neuropsychologia*. doi.org/10.1016/j.neuropsychologia.2010.10.028

Holloway, I. D., Battista, C., Vogel, S. E., and Ansari, D. (2013). Semantic and perceptual processing of number symbols: Evidence from a cross-linguistic fMRI adaptation study. *Journal of Cognitive Neuroscience*. doi.org/10.1162/jocn_a_00323

Holmes, W. and. Dowker, A. (2013). Catch up numeracy: A targeted intervention for children who are low-attaining in mathematics. *Research in Mathematics Education*, *15*(3), 249–265. doi.org/10.1080/14794802.2013.803779

Houdé, O. and Tzourio-Mazoyer, N. (2003). Neural foundations of logical and mathematical cognition. *Nature Reviews Neuroscience*. doi.org/10.1038/nrn1117

Hung, Y., Gaillard, S. L., Yarmak, P., and Arsalidou, M. (2018). Dissociations of cognitive inhibition, response inhibition, and emotional interference: Voxelwise ALE meta-analyses of fMRI studies. *Human Brain Mapping*. doi.org/doi: 10.1002/hbm.24232

Hyde, D. C., Boas, D. A., Blair, C., and Carey, S. (2010). Near-infrared spectroscopy shows right parietal specialization for number in pre-verbal infants. *NeuroImage*. doi.org/10.1016/j.neuroimage.2010.06.030

Hyde, D. C., Khanum, S., and Spelke, E. S. (2014). Brief non-symbolic, approximate number practice enhances subsequent exact symbolic arithmetic in children. *Cognition*. doi.org/ 10.1016/j.cognition.2013.12.007

Ischebeck, A., Zamarian, L., Egger, K., Schocke, M., and Delazer, M. (2007). Imaging early practice effects in arithmetic. *NeuroImage*. doi.org/10.1016/j.neuroimage.2007.03.051

Iuculano, T., Rosenberg-Lee, M., Richardson, J., Tenison, C., Fuchs, L., Supekar, K., and Menon, V. (2015). Cognitive tutoring induces widespread neuroplasticity and remediates brain function in children with mathematical learning disabilities. *Nature Communications*, 6(1), 8453. https://doi.org/10.1038/ncomms9453

Iuculano, T. and Menon, V. (2018). Development of mathematical reasoning. In J. T. Wixted and S. Ghetti (eds), *Stevens' Handbook of Experimental Psychology: Developmental and Social Psychology*, Vol. 4 (4th ed.), 183–222. Chichester: John Wiley and Sons Inc.

Iuculano, T., Padmanabhan, A., and Menon, V. (2018). Systems neuroscience of mathematical cognition and learning: Basic organization and neural sources of heterogeneity in typical and atypical development. In A. Henik and W. Fias (eds), *Heterogeneity of Function in Numerical Cognition*, 287–336. Cambridge, MA: Academic Press. doi.org/10.1016/ B978-0-12-811529-9.00015-7

Iuculano, T,. Rosenberg-Lee, M., Richardson, J., Tenison, C., Fuchs, L., Supekar, et al. (2015). Cognitive tutoring induces widespread neuroplasticity and remediates brain function in children with mathematical learning disabilities. *Nature Communications*, 6, 8453. doi.org/ 10.1038/ncomms.9453

Izard, V., Dehaene-Lambertz, G., and Dehaene, S. (2008). Distinct cerebral pathways for object identity and number in human infants. *PLoS Biology*. doi.org/10.1371/journal.pbio.0060011

Izard, V., Sann, C., Spelke, E. S., and Streri, A. (2009). Newborn infants perceive abstract numbers. *Proceedings of the National Academy of Sciences*. doi.org/10.1073/pnas.0812142106

Jung, J. Y., Cloutman, L. L., Binney, R. J., and Lambon Ralph, M. A. (2017). The structural connectivity of higher order association cortices reflects human functional brain networks. *Cortex*. doi.org/10.1016/j.cortex.2016.08.011

Kadosh, R. C. and Walsh, V. (2009). Numerical representation in the parietal lobes: Abstract or not abstract? *Behavioral and Brain Sciences*. doi.org/10.1017/S0140525X09990938

Kawashima, R., Taira, M., Okita, K., Inoue, K., Tajima, N., Yoshida, H., et al. (2004). A functional MRI study of simple arithmetic—a comparison between children and adults. *Cognitive Brain Research*. doi.org/10.1016/j.cogbrainres.2003.10.009

Kim, N., Jang, S., and Cho, S. (2018). Testing the efficacy of training basic numerical cognition and transfer effects to improvement in children's math ability. *Frontiers in Psychology*. doi. org/10.3389/fpsyg.2018.01775

Klingberg, T., Fernell, E., Olesen, P. J., Johnson, M., Gustafsson, P., Dahlström, K., et al. (2005). Computerized training of working memory in children with ADHD—a randomized, controlled trial. *Journal of the American Academy of Child and Adolescent Psychiatry*. doi. org/10.1097/00004583-200502000-00010

Knops, A., Piazza, M., Sengupta, R., Eger, E., and Melcher, D. (2014). A shared, flexible neural map architecture reflects capacity limits in both visual short-term memory and enumeration. *Journal of Neuroscience*. doi.org/10.1523/JNEUROSCI.2758-13.2014

Koontz, K. L. (1996). Identifying simple numerical stimuli: Processing inefficiencies exhibited by arithmetic learning disabled children. *Mathematical Cognition*. doi.org/10.1080/ 135467996387525

Koponen, T., K., Sorvo, R., Dowker, A., Raikkonen, E., Viholainen, H., Aro, M., et al. (2018). Does multi-component strategy training improve calculation fluency among poor performing elementary school children? *Frontiers in Psychology*, 9, 1187. 10.3389/fpsvg.2018.01187

Krinzinger, H., Koten, J. W., Horoufchin, H., Kohn, N., Arndt, D., Sahr, K., et al. (2011). The role of finger representations and saccades for number processing: An fMRI study in children. *Frontiers in Psychology*. doi.org/10.3389/fpsyg.2011.00373

Kucian, K., Grond, U., Rotzer, S., Henzi, B., Schönmann, C., Plangger, F., et al. (2011). Mental number line training in children with developmental dyscalculia. *NeuroImage*. doi.org/10.1016/j.neuroimage.2011.01.070

Kucian, K., Von Aster, M., Loenneker, T., Dietrich, T., and Martin, E. (2008). Development of neural networks for exact and approximate calculation: A fMRI study. *Developmental Neuropsychology*. doi.org/10.1080/87565640802101474

Landerl, K., Bevan, A., and Butterworth, B. (2004). Developmental dyscalculia and basic numerical capacities: A study of 8-9-year-old students. *Cognition*. doi.org/10.1016/j.cognition.2003.11.004

Libertus, M. E., Odic, D., and Halberda, J. (2012). Intuitive sense of number correlates with math scores on college-entrance examination. *Acta Psychologica*. doi.org/10.1016/j.actpsy.2012.09.009

Link, T., Moeller, K., Huber, S., Fischer, U., and Nuerk, H. C. (2013). Walk the number line—an embodied training of numerical concepts. *Trends in Neuroscience and Education*. doi.org/10.1016/j.tine.2013.06.005

Lipton, J. S. and Spelke, E. S. (2003). Origins of number sense: Large-number discrimination in human infants. *Psychological Science*. doi.org/10.1111/1467-9280.01453

Liu, N., Li, H., Su, W., and Chen, Q. (2017). Common and specific neural correlates underlying the spatial congruency effect induced by the egocentric and allocentric reference frame. *Human Brain Mapping*. doi.org/10.1002/hbm.23508

Looi, C. Y., Duta, M., Brem, A. K., Huber, S., Nuerk, H. C., and Cohen Kadosh, R. (2016). Combining brain stimulation and video game to promote long-term transfer of learning and cognitive enhancement. *Scientific Reports*. doi.org/10.1038/srep22003

Looi, C. Y., Lim, J., Sella, F., Lolliot, S., Duta, M., Avramenko, A. A., et al. (2017). Transcranial random noise stimulation and cognitive training to improve learning and cognition of the atypically developing brain: A pilot study. *Scientific Reports*, 7(1), 4633. doi.org/10.1038/s41598-017-04649-x

Luck, S. J., and Vogel, E. K. (1997). The capacity of visual working memory for features and conjunctions. *Nature*. doi.org/10.1038/36846

Luck, S. J. and Vogel, E. K. (2013). Visual working memory capacity: From psychophysics and neurobiology to individual differences. *Trends in Cognitive Sciences*. doi.org/10.1016/j.tics.2013.06.006

Lyons, I. M. and Beilock, S. L. (2013). Ordinality and the nature of symbolic numbers. *Journal of Neuroscience*. doi.org/10.1523/JNEUROSCI.1775-13.2013

Lyons, I. M. and Beilock, S. L. (2018). Characterizing the neural coding of symbolic quantities. *NeuroImage*. doi.org/10.1016/j.neuroimage.2018.05.062

Lyons, I. M., Price, G. R., Vaessen, A., Blomert, L., and Ansari, D. (2014). Numerical predictors of arithmetic success in grades 1-6. *Developmental Science*. doi.org/10.1111/desc.12152

Mayer, E., Martory, M. D., Pegna, A. J., Landis, T., Delavelle, J., and Annoni, J. M. (1999). A pure case of Gerstmann syndrome with a subangular lesion. *Brain*. doi.org/10.1093/brain/122.6.1107

McClelland, J. L., McNaughton, B. L., and O'Reilly, R. C. (1995). Why there are complementary learning systems in the hippocampus and neocortex: Insights from the successes and failures of connectionist models of learning and memory. *Psychological Review.* doi.org/10.1037/0033-295X.102.3.419

McCrink, K. and Wynn, K. (2007). Ratio abstraction by 6-month-old infants. *Psychological Science.* doi.org/10.1111/j.1467-9280.2007.01969.x

Meck, W. H. and Church, R. M. (1983). A mode control model of counting and timing processes. *Journal of Experimental Psychology: Animal Behavior Processes.* doi.org/10.1037/0097-7403.9.3.320

Menon, V. (2014). Arithmetic in the child and adult brain. In R. Cohen Kadosh and A. Dowker (eds), *The Oxford Handbook of Mathematical Cognition*, Oxford Library of Psychology series, 502–530. Oxford: Oxford University Press. doi.org/10.1093/oxfordhb/9780199642342.013.041

Menon, V., Rivera, S. M., White, C. D., Eliez, S., Glover, G. H., and Reiss, A. L. (2000). Functional optimization of arithmetic processing in perfect performers. *Cognitive Brain Research.* doi.org/10.1016/S0926-6410(00)00010-0

Metcalfe, A. W. S., Ashkenazi, S., Rosenberg-Lee, M., and Menon, V. (2013). Fractionating the neural correlates of individual working memory components underlying arithmetic problem solving skills in children. *Developmental Cognitive Neuroscience.* doi.org/10.1016/j.dcn.2013.10.001

Moser, E. I., Kropff, E., and Moser, M.-B. (2008). Place cells, grid cells, and the brain's spatial representation system. *Annual Review of Neuroscience.* doi.org/10.1146/annurev.neuro.31.061307.090723

Moyer, R. S. and Landauer, T. K. (1967). Time required for judgements of numerical inequality [47]. *Nature.* doi.org/10.1038/2151519a0

Nadel, L., Samsonovich, A., Ryan, L., and Moscovitch, M. (2000). Multiple trace theory of human memory: Computational, neuroimaging, and neuropsychological results. *Hippocampus.* doi.org/10.1002/1098-1063(2000)10:4<352::AID-HIPO2>3.0.CO;2-D

Nieder, A., Freedman, D. J., and Miller, E. K. (2002). Representation of the quantity of visual items in the primate prefrontal cortex. *Science.* doi.org/10.1126/science.1072493

O'Connor, P. A., Morsanyi, K., and McCormack, T. (2018). Young children's non-numerical ordering ability at the start of formal education longitudinally predicts their symbolic number skills and academic achievement in maths. *Developmental Science.* doi.org/10.1111/desc.12645

O' Keefe, J. and Burgess, N. (1996). Geometric determinants of the place fields of hippocampal neurons. *Nature.* doi.org/10.1038/381425a0

Park, J., Bermudez, V., Roberts, R. C., and Brannon, E. M. (2016). Non-symbolic approximate arithmetic training improves math performance in preschoolers. *Journal of Experimental Child Psychology.* doi.org/10.1016/j.jecp.2016.07.011

Park, J. and Brannon, E. M. (2013). Training the approximate number system improves math proficiency. *Psychological Science.* doi.org/10.1177/0956797613482944

Park, J. and Brannon, E. M. (2014). Improving arithmetic performance with number sense training: An investigation of underlying mechanism. *Cognition.* doi.org/10.1016/j.cognition.2014.06.011

Paterson, S. J., Girelli, L., Butterworth, B., and Karmiloff-Smith, A. (2006). Are numerical impairments syndrome specific? Evidence from Williams syndrome and Down's syndrome. *Journal of Child Psychology and Psychiatry and Allied Disciplines.* doi.org/10.1111/j.1469-7610.2005.01460.x

Piazza, M., Izard, V., Pinel, P., Le Bihan, D., and Dehaene, S. (2004). Tuning curves for approximate numerosity in the human intraparietal sulcus. *Neuron.* doi.org/10.1016/j.neuron.2004.10.014

Piazza, M., Mechelli, A., Butterworth, B., and Price, C. J. (2002). Are subitizing and counting implemented as separate or functionally overlapping processes? *NeuroImage.* doi.org/10.1006/nimg.2001.0980

Piazza, M., Mechelli, A., Price, C. J., and Butterworth, B. (2006). Exact and approximate judgements of visual and auditory numerosity: An fMRI study. *Brain Research.* doi.org/10.1016/j.brainres.2006.05.104

Piazza, M., Pinel, P., Le Bihan, D., and Dehaene, S. (2007). A magnitude code common to numerosities and number symbols in human intraparietal cortex. *Neuron.* doi.org/10.1016/j.neuron.2006.11.022

Powell, S. R., Fuchs, L. S., Fuchs, D., Cirino, P. T., and Fletcher, J. M. (2009). Effects of fact retrieval tutoring on third-grade students with math difficulties with and without reading difficulties. *Learning Disabilities Research and Practice.* doi.org/10.1111/j.1540-5826.2008.01272.x

Prabhakaran, V., Narayanan, K., Zhao, Z., and Gabriel, J. D. E. (2000). Integration of diverse information in working memory within the frontal lobe. *Nature Neuroscience.* doi.org/10.1038/71156

Price, G. R., Holloway, I., Räsänen, P., Vesterinen, M., and Ansari, D. (2007). Impaired parietal magnitude processing in developmental dyscalculia. *Current Biology.* doi.org/10.1016/j.cub.2007.10.013

Qin, S., Cho, S., Chen, T., Rosenberg-Lee, M., Geary, D. C., and Menon, V. (2014). Hippocampal-neocortical functional reorganization underlies children's cognitive development. *Nature Neuroscience.* doi.org/10.1038/nn.3788

Rajah, M. N., Kromas, M., Han, J. E., and Pruessner, J. C. (2010). Group differences in anterior hippocampal volume and in the retrieval of spatial and temporal context memory in healthy young versus older adults. *Neuropsychologia.* doi.org/10.1016/j.neuropsychologia.2010.10.010

Räsänen, P., Salminen, J., Wilson, A. J., Aunio, P., and Dehaene, S. (2009). Computer-assisted intervention for children with low numeracy skills. *Cognitive Development.* doi.org/10.1016/j.cogdev.2009.09.003

Reeve, R., Reynolds, F., Humberstone, J., and Butterworth, B. (2012). Stability and change in markers of core numerical competencies. *Journal of Experimental Psychology: General.* doi.org/10.1037/a0027520

Rivera, S. M., Reiss, A. L., Eckert, M. A., and Menon, V. (2005). Developmental changes in mental arithmetic: Evidence for increased functional specialization in the left inferior parietal cortex. *Cerebral Cortex.* doi.org/10.1093/cercor/bhi055

Roell, M., Viarouge, A., Houdé, O., and Borst, G. (2017). Inhibitory control and decimal number comparison in school-aged children. *PLoS ONE.* doi.org/10.1371/journal.pone.0188276

Rosenberg-Lee, M., Iuculano, T., Bae, S. R., Richardson, J., Qin, S., Jolles, D., et al. (2018). Short-term cognitive training recapitulates hippocampal functional changes associated with one year of longitudinal skill development. *Trends in Neuroscience and Education, 10.* 19–29. doi.org/10.1016/j.tine.2017.12.001

Ross, R. S. and Slotnick, S. D. (2008). The hippocampus is preferentially associated with memory for spatial context. *Journal of Cognitive Neuroscience.* doi.org/10.1162/jocn.2008.20035

Rubinsten, O., Henik, A., Berger, A., and Shahar-Shalev, S. (2002). The development of internal representations of magnitude and their association with Arabic numerals. *Journal of Experimental Child Psychology, 81*(1), 74–92. doi.org/10.1006/jecp.2001.2645

Rugani, R., Vallortigara, G., Priftis, K., and Regolin, L. (2015). Number-space mappings in the newborn chick resembles humans' mental number line. *Science, 347*(6221), 534–536. doi.org/10.1126/science.aaa1379.

Sarnecka, B. W. (2015). Learning to represent exact numbers. *Synthese, 198*(Suppl 5), 1001–1018. doi.org/10.1007/s11229-015-0854-6

Sarnecka, B. W. and Carey, S. (2008). How counting represents number: What children must learn and when they learn it. *Cognition.* doi.org/10.1016/j.cognition.2008.05.007

Schaffelhofer, S. and Scherberger, H. (2016). Object vision to hand action in macaque parietal, premotor, and motor cortices. *ELife.* doi.org/10.7554/eLife.15278

Schneider, M., Beeres, K., Coban, L., Merz, S., Susan Schmidt, S., Stricker, J., et al. (2017). Associations of non-symbolic and symbolic numerical magnitude processing with mathematical competence: a meta-analysis. *Developmental Science.* doi.org/10.1111/desc.12372

Sella, F., Sader, E., Lolliot, S., and Cohen Kadosh, R. (2016). Basic and advanced numerical performances relate to mathematical expertise but are fully mediated by visuospatial skills. *Journal of Experimental Psychology: Learning Memory and Cognition.* doi.org/10.1037/xlm0000249

Shum, J., Hermes, D., Foster, B. L., Dastjerdi, M., Rangarajan, V., Winawer, J., et al. (2013). A Brain Area for Visual Numerals. *Journal of Neuroscience.* doi.org/10.1523/JNEUROSCI.4558-12.2013

Shumaker, R. W., Beck, B. B., Palkovich, A. M., Guagnano, G. A., and Morowitz, H. (2001). Spontaneous use of magnitude discrimination and ordination by the orangutan (Pongo pygmaeus). *Journal of Comparative Psychology.* doi.org/10.1037/0735-7036.115.4.385

Simon, O., Mangin, J. F., Cohen, L., Le Bihan, D., and Dehaene, S. (2002). Topographical layout of hand, eye, calculation, and language-related areas in the human parietal lobe. *Neuron.* doi.org/10.1016/S0896-6273(02)00575-5

Simon, T. J., Bearden, C. E., McDonald Mc-Ginn, D., and Zackai, E. (2005). Visuospatial and numerical cognitive deficits in children with chromosome 22q11.2 deletion syndrome. *Cortex.* doi.org/10.1016/S0010-9452(08)70889-X

Simon, T. J., Takarae, Y., DeBoer, T., McDonald-McGinn, D. M., Zackai, E. H., and Ross, J. L. (2008). Overlapping numerical cognition impairments in children with chromosome 22q11.2 deletion or Turner syndromes. *Neuropsychologia.* doi.org/10.1016/j.neuropsychologia.2007.08.016

Soltész, F., White, S., and Szucs, D. (2011). Event-related brain potentials dissociate the developmental time-course of automatic numerical magnitude analysis and cognitive control functions during the first three years of primary school. *Developmental Neuropsychology.* doi.org/10.1080/87565641.2010.549982

Supekar, K. and Menon, V. (2012). Developmental maturation of dynamic causal control signals in higher-order cognition: A neurocognitive network model. *PLoS Computational Biology.* doi.org/10.1371/journal.pcbi.1002374

Szucs, D., Soltész, F., Jármi, É., and Csépe, V. (2007). The speed of magnitude processing and executive functions in controlled and automatic number comparison in children: An electroencephalography study. *Behavioral and Brain Functions.* doi.org/10.1186/1744-9081-3-23

Szkudlarek, E. and Brannon, E. M. (2018). Approximate arithmetic training improves informal math performance in low achieving preschoolers. *Frontiers in Psychology*. doi.org/10.3389/fpsyg.2018.00606

Szucs, D., Devine, A., Soltesz, F., Nobes, A., and Gabriel, F. (2013). Developmental dyscalculia is related to visuo-spatial memory and inhibition impairment. *Cortex*. doi.org/10.1016/j.cortex.2013.06.007

Szűcs, D., Soltész, F., Jármi, É. et al. (2007). The speed of magnitude processing and executive functions in controlled and automatic number comparison in children: an electro-encephalography study. *Behavioral and Brain Functions*, 3, 23.

Szucs, D., Soltész, F., and White, S. (2009). Motor conflict in Stroop tasks: Direct evidence from single-trial electro-myography and electro-encephalography. *NeuroImage*. doi.org/10.1016/j.neuroimage.2009.05.048

Thompson, J. C., Abbott, D. F., Wheaton, K. J., Syngeniotis, A., and Puce, A. (2004). Digit representation is more than just hand waving. *Cognitive Brain Research*. doi.org/10.1016/j.cogbrainres.2004.07.001

Todd, J. J. and Marois, R. (2004). Capacity limit of visual short-term memory in human posterior parietal cortex. *Nature*. doi.org/10.1038/nature02466

Tse, D., Langston, R. F., Kakeyama, M., Bethus, I., Spooner, P. A., Wood, E. R., ... Morris, R. G. M. (2007). Schemas and memory consolidation. *Science*. doi.org/10.1126/science.1135935

Tudusciuc, O. and Nieder, A. (2007). Neuronal population coding of continuous and discrete quantity in the primate posterior parietal cortex. *Proceedings of the National Academy of Sciences of the United States of America*. doi.org/10.1073/pnas.0705495104

Turconi, E., Campbell, J. I., and Seron, X. (2006). Numerical order and quantity processing in number comparison. *Cognition*, 98(3), 273–285. https://doi.org/10.1016/j.cognition.2004.12.002

Turconi, E. and Seron, X. (2002). Dissociation between order and quantity meanings in a patient with Gerstmann syndrome. *Cortex*. doi.org/10.1016/S0010-9452(08)70069-8

van Bueren, N. E. R., van der Ven, S. H. G., Roelofs, K., Cohen Kadosh, R., and Kroesbergen, E. H. (2022). Predicting math ability using working memory, number sense, and neurophysiology in children and adults. Brain Sciences, 12(5), 550. www.mdpi.com/2076-3425/12/5/550

Vanbinst, K., Ansari, D., Ghesquière, P., and Smedt, B. De. (2016). Symbolic numerical magnitude processing is as important to arithmetic as phonological awareness is to reading. *PLoS ONE*. doi.org/10.1371/journal.pone.0151045

Verdine, B. N., Golinkoff, R. M., Hirsh-Pasek, K., Newcombe, N. S., Filipowicz, A. T., and Chang, A. (2014). Deconstructing building blocks: Preschoolers' spatial assembly performance relates to early mathematical skills. *Child Development*. doi.org/10.1111/cdev.12165

Viarouge, A., Lee, H., and Borst, G. (2023). Attention to number requires magnitude-specific inhibition. *Cognition*, 230, 105285. doi.org/doi.org/10.1016/j.cognition.2022.105285

Vogel, E. K. and Machizawa, M. G. (2004). Neural activity predicts individual differences in visual working memory capacity. *Nature*. doi.org/10.1038/nature02447

Vogel, S. E., Goffin, C., and Ansari, D. (2015). Developmental specialization of the left parietal cortex for the semantic representation of Arabic numerals: An fMR-adaptation study. *Developmental Cognitive Neuroscience*. doi.org/10.1016/j.dcn.2014.12.001

Vogel, S. E., Goffin, C., Bohnenberger, J., Koschutnig, K., Reishofer, G., Grabner, R. H., et al. (2017). The left intraparietal sulcus adapts to symbolic number in both the visual and auditory modalities: Evidence from fMRI. *NeuroImage*. doi.org/10.1016/j.neuroimage.2017.03.048

Wilson, A. J., Revkin, S. K., Cohen, D., Cohen, L., and Dehaene, S. (2006). An open trial assessment of "the number race," an adaptive computer game for remediation of dyscalculia. *Behavioral and Brain Functions*. doi.org/10.1186/1744-9081-2-20

Wynn, K. (1992). Addition and subtraction by human infants. *Nature, 358*(6389), 749–750. doi.org/10.1038/358749a0

Wynn, K. (1998). Psychological foundations of number: Numerical competence in human infants. *Trends in Cognitive Sciences*. doi.org/10.1016/S1364-6613(98)01203-0

Xu, F. and Arriaga, R. I. (2007). Number discrimination in 10-month-old infants. *British Journal of Developmental Psychology*. doi.org/10.1348/026151005X90704

Xu, Y. and Chun, M. M. (2006). Dissociable neural mechanisms supporting visual short-term memory for objects. *Nature*. doi.org/10.1038/nature04262

Zacharopoulos, G., Sella, F., and Cohen Kadosh, R. (2021a). The impact of a lack of mathematical education on brain development and future attainment. *Proceedings of the National Academy of Sciences, 118*(24), e2013155118.

Zacharopoulos, G., Sella, F., Cohen Kadosh, K., Hartwright, C., Emir, U., and Cohen Kadosh, R. (2021b). Predicting learning and achievement using GABA and glutamate concentrations in human development. *PLOS Biology, 19*(7), e3001325.

Zaehle, T., Jordan, K., Wüstenberg, T., Baudewig, J., Dechent, P., and Mast, F. W. (2007). The neural basis of the egocentric and allocentric spatial frame of reference. *Brain Research*. doi.org/10.1016/j.brainres.2006.12.044

Zago, L., Petit, L., Turbelin, M. R., Andersson, F., Vigneau, M., and Tzourio-Mazoyer, N. (2008). How verbal and spatial manipulation networks contribute to calculation: An fMRI study. *Neuropsychologia*. doi.org/10.1016/j.neuropsychologia.2008.03.001

CHAPTER 26

COGNITIVE NEUROSCIENCE OF DYSCALCULIA AND MATH LEARNING DISABILITIES

VINOD MENON, AARTHI PADMANABHAN, AND FLORA SCHWARTZ

INTRODUCTION

DYSCALCULIA or mathematical disability (MD) is a specific learning disability characterized by difficulties in processing numerical and mathematical information despite normal IQ (Butterworth, Varma et al., 2011; Kaufmann, Mazzocco et al., 2013; Price and Ansari, 2013). Developmental dyscalculia was originally described as a deficit solely limited to the understanding of quantity and numerical sets (Butterworth, 2005), but cases of pure developmental dyscalculia are relatively rare as numerical processing deficits are often accompanied by difficulties in other visuospatial and verbal domains (von Aster and Shalev, 2007; Devine, Soltész et al., 2013). More generally, children and adults who are in the bottom quartile of mathematics achievement are at risk for poor long-term outcomes in mathematics learning and in real-world endeavors that require mathematics (Parsons and Bynner, 1997; Parsons and Bynner, 2005). The definition of dyscalculia has therefore been revised to include a broader group of individuals with MD, and increasingly the focus is on characterization of multicomponent deficits and impairments of functional circuits that impact acquisition of numerosity and performance on numerical problem-solving skills. Individuals with MD show poor performance on a broad range of basic numerical tasks including magnitude judgment (Geary, Hamson et al., 2000; Ashkenazi, Rubinsten et al., 2009; Mussolin, Mejias et al., 2010) and enumeration (Geary and Brown 1991; Geary, Bow-Thomas et al.,1992; Knootz and Berch 1996; Landerl, Bevan et al., 2004; Schleifer and Landerl, 2011). They also lag behind their typically developing (TD) peers in basic arithmetic problem-solving skills

(Geary, Bow-Thomas et al., 1992; Shalev, Auerbach et al., 2000; Shalev, Manor et al., 2005). This review thus necessarily takes a systems neuroscience approach to the investigation of MD.

This chapter summarizes current knowledge about the neurocognitive basis of MD. We also discuss how recent advances in functional and structural neuroimaging provide a more comprehensive understanding of neurocognitive impairments in affected individuals. Considerable progress has been made over the past several decades in identifying the cognitive deficits that contribute to MD (Geary, Hoard et al., 2007; De Smedt, Noël et al., 2013; De Visscher and Noël, 2014; Attout and Majerus, 2015), and in recent years, the identification of the underlying brain systems (De Smedt, Holloway et al., 2011; Kucian, Loenneker et al., 2011; Supekar, Swigart et al., 2013; Demir, Prado et al., 2014; Attout, Salmon et al., 2015; Rosenberg-Lee, Ashkenazi et al., 2015; McCaskey, von Aster et al., 2017; McCaskey, von Aster et al., 2018; Schwartz, Epinat-Duclos et al., 2018). We first provide an overview of central cognitive theories and models of MD, including impairments in non-symbolic and symbolic representations of quantity and impairments in key "domain-general" cognitive processes including memory and cognitive control. We then describe key aspects of brain areas and functional circuits that are impacted by the disability, focusing on basic number sense and arithmetic problem-solving—the two domains of numerical cognition most widely investigated in children and adults with MD. Next, we consider the specific role of brain systems underlying working memory (WM) and cognitive control deficits in MD. Finally, we review evidence for deficits in intrinsic structural and functional circuits, which provide a convergent view of neural systems that underlie MD.

DIAGNOSIS AND PREVALENCE OF MD

The revised Diagnostic and Statistical Manual for Mental Disorders, 5th Edition (DSM-5) defines MD as a specific learning disorder with persistent difficulties in acquiring academically relevant mathematical skills, which cannot be attributed to intellectual disabilities or neurological disorders (American Psychiatric Association, 2013). MD is typically manifested in the early school years and is characterized by deficits that persist over at least six months (Chu, Vanmarle et al. 2013; Kaufmann, Mazzocco et al., 2013). Children with MD typically perform one to two years below grade level and show difficulties in one or more of the four domains identified in DSM-5: number sense, memorization of arithmetic facts, accurate and fluent calculation, and mathematical reasoning. However, the DSM-5 is limited in definition and scope of MD and does not provide specific diagnostic measures to assess individual domains of information processing deficits in affected individuals.

A number of standardized neuropsychological assessments have been developed over the past decades to overcome limitations of DSM-based criteria for defining MD.

The Woodcock-Johnson (Woodcock, 1977), Wechsler Individual Achievement Test (Wechsler, 1992), and the Wide Range Achievement Test (Wilkinson and Robertson, 2006) are among the more widely used measures of mathematical skills and academic achievement in children and adults in the United States and similar measures have been developed for use in other countries, such as the Zareki (von Aster, Weinhold Zulauf et al., 2006). Behavioral and neurocognitive studies tend to vary considerably in the specific criteria used to define MD, with diagnostic cut-offs on standardized neuropsychological assessments ranging typically from the 3rd percentile to the 25th percentile (Devine, Soltész et al., 2013). Discrepancy between standard math scores and IQ have also been used in many previous studies but the validity of this approach has been questioned (Mazzocco and Myers, 2003; Alloway and Alloway, 2010). Accordingly, IQ-discrepancy criteria are now neither included in the DSM-5 (2013) as a defining characteristic of MD nor are they widely used in the cognitive neuroscience literature.

In the past decade, developmental and behavioral studies have converged on more precise cutoff criteria and distinctions between subtypes of children with MD at or below the bottom quartile of mathematics achievement. Children who score at or below the 10th percentile on standardized mathematics achievement tests for at least two consecutive academic years are typically categorized as having a core math learning disorder or developmental dyscalculia and those scoring between the 11th and the 25th percentiles at any time point are characterized as low achieving (Geary, Hoard et al., 2007; Murphy, Mazzocco et al., 2007). Even though children composing these two groups represent different cut-offs on a normally distributed continuum of mathematical abilities, they often differ in their underlying cognitive strengths and weaknesses and profiles of comorbid deficits (De Smedt and Gilmore, 2011; Geary, Hoard et al., 2012). Children with MD, as a group, tend to score low average in intelligence tests and often have co-occurring deficits in reading and WM (Murphy, Mazzocco et al., 2007; Geary, Hoard et al., 2012). It is important to note here that the use of different diagnostic criteria to classify MD can impact the consistency of findings across studies, and their neurocognitive interpretations, because of variation in the precise profile of cognitive deficits in study participants, a view that has been increasingly noted in recent years (Rubinsten and Henik, 2009; Fias, Menon et al., 2013; Kaufmann, Mazzocco et al., 2013; Henik and Fias, 2018).

Neurocognitive Theories and a Multicomponent View of MD

In this section we present an overview of multicomponent cognitive models of MD that have emerged from cognitive and neuroscience studies of numerical cognition in MD and TD individuals. There are two major "domain-specific" aspects of information processing deficits in MD: (i) number sense and quantity manipulation, and (ii) arithmetic fact retrieval and problem-solving (Figures 26.1, 26.2).

The first is a core deficit in number sense, the ability to make judgments about quantity, and to reason with symbolic representations of quantity (Wilson and Dehaene, 2007; Piazza, Facoetti et al., 2010; Butterworth, Varma et al., 2011). Consistent with this hypothesis, it has been demonstrated that, compared to TD children, children with MD have lower than expected abilities in quantity estimation (Piazza, Facoetti et al., 2010; Mazzocco, Feigenson et al., 2011a) and abnormal magnitude representations (Ashkenazi, Mark-Zigdon et al., 2009; Mussolin, Mejias et al., 2010). Deficits in number sense also arise from weakness in automatically mapping symbols to their internal magnitude representations, reflecting a relatively greater impairment in symbolic, compared to non-symbolic processing (Rubinsten and Henik, 2005; Rousselle and Noel, 2007). A second essential aspect of MD relates to weakness in arithmetic problem-solving abilities and difficulties in committing arithmetic facts to memory. Importantly, difficulties in these core building blocks manifest in the earliest years of elementary schooling and, if not remediated, can have a lifelong impact on mathematical learning and skill acquisition in subsequent stages where more complex mathematical procedures need to be learnt. Unsurprisingly, therefore, these two categories of deficits constitute the most widely investigated neurocognitive domains in MD.

In recent years, researchers have also identified "domain-general" cognitive factors underlying information processing deficits in MD. The two most well studied of these factors are WM and cognitive control. Individuals with MD show prominent deficits in the use of developmentally appropriate arithmetic procedures, and these impairments have been attributed to weaknesses in manipulation of quantity in WM independent of "domain-specific" deficits in number sense. Meta-analyses of behavioral studies in children with MD have pointed to a consistent pattern of visuospatial and verbal WM deficits in affected children (Swanson, Howard et al., 2006). Deficits in cognitive control, including interference resolution, also influence problem-solving and decision making across multiple domains of mathematics, including comparison of numerical magnitude (Gilmore, Attridge et al., 2013; Bugden and Ansari, 2016), arithmetic problem-solving (De Visscher and Noël, 2014; De Visscher, Szmalec et al., 2015), and logical reasoning (Morsanyi, Devine et al., 2013). Finally, difficulties with word-based reasoning problems (Jordan, Hanich et al., 2003) are an important source of vulnerability and a significant number of children with MD have comorbid reading disability including dyslexia (Lewis, Hitch et al., 1994; Knopik, Alarcon et al., 1997; Willcutt, Petrill et al., 2013; Wilson, Andrewes et al., 2015). MD arises from a complex interplay between "domain-specific" and "domain-general" deficits. In this view, number processing and mathematical problem-solving are built on multiple neurocognitive components that are implemented by distinct and overlapping brain systems. Impairments in any of these components can compromise efficiency of numerical problem-solving skills and constitute risk factors for MD (Figures 26.1, 26.2).

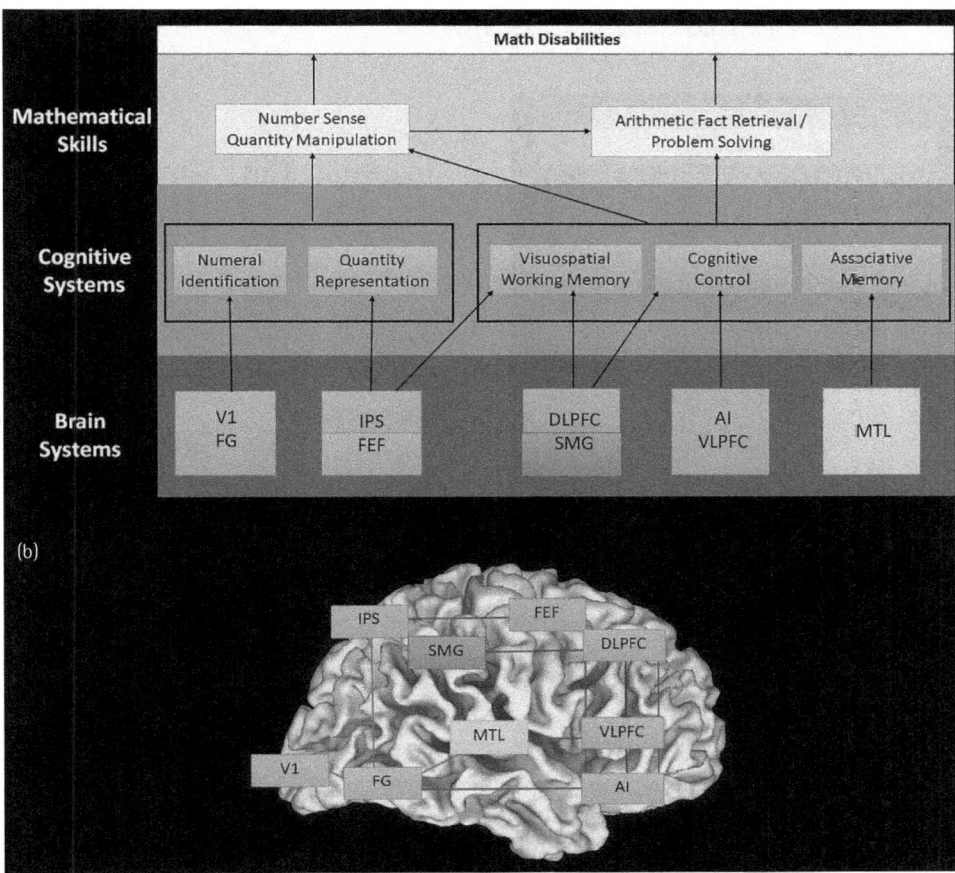

FIGURE 26.1 (a) Cognitive and brain systems that underpin MD. Mathematical difficulties arise from impairments in two core domains of mathematical cognition: (i) number sense and quantity manipulation, and (ii) arithmetic fact retrieval and problem-solving. Impairments in number sense and quantity manipulation arise from weak symbolic and non-symbolic representations of quantity as well as "domain-general" deficits in visuospatial working memory capacity and cognitive control. Impairments in arithmetic fact retrieval and problem-solving arise from deficits in the ability to manipulate internal representations of quantity as well as "domain-general" deficits in visuospatial working memory, cognitive control, and associative memory encoding and retrieval. Impairments in any of these components can compromise efficiency of numerical problem-solving skills and constitute risk factors for MD. (2) Schematic diagram of number, arithmetic, memory, and cognitive control circuits impaired in MD. The fusiform gyrus (FG) in the inferior temporal cortex decodes number form and together with the intra-parietal sulcus (IPS) in the parietal cortex which helps builds visuo-spatial representations of numerical quantity. Distinct parietal-frontal circuits differentially link the IPS and supramarginal gyrus (SMG) with the frontal eye field (FEF) and dorsolateral prefrontal cortex (DLPFC), respectively. These circuits facilitate visuospatial working memory for objects in space and create a hierarchy of short-term representations that allow manipulation of multiple discrete quantities over several seconds. The declarative memory system anchored in the medial temporal cortex (MTL)—the hippocampus, specifically, plays an important role in long-term memory formation and generalization beyond individual problem attributes. Finally, prefrontal control circuits anchored in the anterior insula (AI), ventrolateral prefrontal cortex (VLPFC) and DLPFC serve as flexible hubs for integrating information across attentional and memory systems, thereby facilitating goal-specific problem-solving and decision making.

Source: Part b: Reprinted from *Trends in Neuroscience and Education*, 2(2): 43–47, Wim Fias, Vinod Menon, and Denes Szucs, Multiple Components of Developmental Dyscalculia. https://doi.org/10.1016/j.tine.2013.06.006. Copyright © 2013 Elsevier GmbH. All rights reserved.

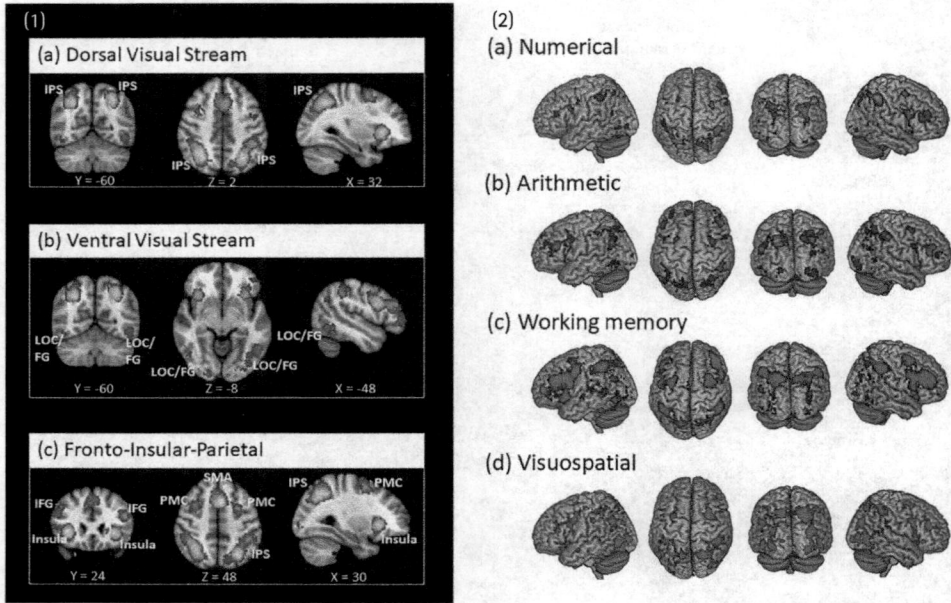

FIGURE 26.2 (1) Canonical brain areas involved in arithmetic problem-solving. (a) Dorsal visual stream anchored in the intraparietal sulcus within the posterior parietal cortex. (b) Ventral visual stream anchored in the lateral occipital cortex (LOC) and FG. (c) Prefrontal cortex control system anchored in the anterior insula, inferior frontal gyrus (IFG) and premotor cortex (PMC). Maps are based on a meta-analysis of "arithmetic" studies in neurosynth.org (Yarkoni et al., 2011). (2) Common patterns of fronto-parietal network activations elicited by numerical, arithmetic, working memory and visuospatial processing tasks. Results from a meta-analysis conducted using Neurosynth (www.neurosynth.org) with the corresponding search terms.

NUMBER SENSE DEFICITS IN MD

Behavioral Findings

Humans, as well as other species, are endowed with a core capacity to discriminate between sets of objects that differ on numerosity (Dehaene, 1997; Butterworth, 1999; Carey, 2004). This core capacity is present in infants (Izard, Dehaene-Lambertz et al., 2008; Izard, Sann et al., 2009). Infants can discriminate between displays of small numerosity—e.g., they respond when the display changes from two to three objects, or from three objects to two (Starkey and Cooper, 1980; Starkey, Spelke et al., 1990; van Loosbroek, 1990). An evolutionarily endowed capacity for numerosity is grounded in evidence that such capacity has adaptive value, but importantly also serves as the foundation for number sense and manipulating numerical quantity. Weaknesses in representing, accessing, and manipulating numerical quantity—number sense—have been hypothesized to be a core deficit in MD (Butterworth, Varma et al., 2011). Children with MD show deficits in magnitude judgment in both symbolic (e.g., Arabic numerals, number words) (Rousselle and Noel, 2007; Mussolin, Mejias et al., 2010) and non-symbolic formats (e.g., dots patterns and random sticks patterns) (Price, Holloway et al., 2007).

An important feature of mathematical development is that symbolic problem-solving skills are built upon, and activate, non-symbolic quantity representations (Gilmore, McCarthy et al., 2007). When children learn to count and acquire a symbolic system for representing numbers, they map these symbols onto a preexisting system involving approximate visuospatial representations of quantity (Mundy and Gilmore, 2009). Children with MD have difficulties with both non-symbolic quantity representation, as assessed using the approximate number system (ANS) (Halberda, Mazzocco et al., 2008; Mazzocco, Feigenson et al., 2011a), and symbolic quantity representation, as assessed by standard achievement tests and cognitive tasks involving Arabic numerals (Geary, 1993; Jordan, Hanich et al., 2003; Rousselle and Noel, 2007; Butterworth, Varma et al., 2011; Mazzocco, Feigenson et al., 2011a; Geary, Hoard et al., 2012). Moreover, they are particularly impaired at mapping between symbolic and non-symbolic representations.

Although MD is typically identified after the first years of formal schooling, there are hints that number sense deficits may be apparent early in development in individuals who later go on to develop MD (Butterworth, 2005; Butterworth, 2010). Number sense involves two distinct processes: one for subitizing for smaller numbers and the other an ANS for larger numbers and both processes have early developmental origins. Subitizing, the ability to spontaneously recognize small numbers up to three or four develops as early as 6-months of age (Wynn, 1992). ANS involves the ability to discriminate between relatively large numerosities without counting. Performance of the ANS follows Fechner's psychophysical principles (Fechner, 1966) and is a function of the ratio between two sets of objects. This skill is also evident as early as 6-months of age (Libertus, Brannon et al., 2011) and continues to develop until late childhood (Mazzocco, Feigenson et al., 2011b). An intermediate stage of number sense development relates to knowledge of the cardinal properties of numbers, for example, *threeness* and *fourness* of sets of objects, and their mapping to visual symbols such as "3" and "4" (Carey, 2004).

Lower subitizing performance has been reported in MD as compared to TD (Landerl, Bevan et al., 2004; Schleifer and Landerl, 2011), and spontaneous discrimination of large numerical magnitudes is also more difficult for individuals with MD (Price, Holloway et al., 2007; Piazza, Facoetti et al., 2010; Mazzocco, Feigenson et al., 2011a). Lastly, there is evidence that individuals with MD might have difficulties with the mapping of exact numerical quantities onto their symbol, which might delay refinement of the ANS in later childhood (Noël and Rousselle, 2011). There has been considerable debate regarding the relative contributions of symbolic and non-symbolic magnitude judgment to overall math abilities in TD children and adults (De Smedt, Noël et al., 2013). While some studies have pointed to the primacy of non-symbolic capacities (Halberda, Mazzocco et al., 2008), others have suggested a stronger relation between symbolic skills and math abilities (Rousselle and Noel, 2007; Holloway and Ansari, 2009; Kolkman, Kroesbergen et al., 2013). The emerging consensus is that symbolic skills may be a more robust predictor of math skills (Lyons, Price et al., 2014). Crucially, among symbolic skills, counting (the ability to correctly sequence numbers orally) is a good predictor of the subsequent development of arithmetic skills (Geary, 1993) and an early source of vulnerability in children who subsequently go on to develop MD (Geary, 2004).

Neurobiological Correlates

Neuroimaging investigations of number sense deficits in MD have used symbolic number and non-symbolic dot comparison tasks with different types of control tasks. While some functional neuroimaging studies use fine-grained manipulations involving small and large distance between pairs of numbers to investigate neural distance effects similar to the behavioral distance effect (Ashkenazi, Mark-Zigdon et al., 2009), others have used non-numerical control tasks with the goal of elucidating number-specific differences in brain activation. The design of these tasks is based on the landmark behavioral research of Moyer and Landauer demonstrating a numerical distance effect: adults are significantly slower and more error prone when they compare numbers with a smaller distance (e.g., 1 and 2) compared to a larger distance (e.g., 1 and 7) (Moyer and Landauer, 1967). Notably, the type of control task used (e.g., numerical distance effect vs. non-numerical) can influence the pattern of aberrancies reported in MD. We first focus on studies of the neural distance effect because they provide greater specificity arising from closely matched perceptual, cognitive, and response factors.

Studies of number sense deficits in MD have focused on the intra-parietal sulcus (IPS), a brain region important for numerical magnitude judgment at all stages of development including infancy (Hyde, Boas et al., 2010), pre-school children (Cantlon, Brannon et al., 2006) as well as older children and adults (Pinel, Dehaene et al., 2001). Specifically, in neurotypical individuals, the IPS is modulated by numerical distance (Dehaene, Piazza et al., 2003). Using a non-symbolic quantity comparison task, Price and colleagues demonstrated weak modulation of IPS activity in a group of 12-year-old children with MD (Figure 26.3a) (Price, Holloway et al., 2007). Aberrant activation of the IPS and adjoining posterior parietal cortex has also been reported in younger children for both symbolic (Mussolin, De Volder et al., 2010) and non-symbolic number comparison tasks (Kucian, Loenneker et al., 2006; Kaufmann, Vogel et al., 2009; Kucian, Loenneker et al., 2011). Lastly, reduced modulation of the left IPS by numerical distance has also been reported in children with MD for symbolic tasks (Figure 26.3b) (Mussolin, De Volder et al., 2010).

More generally, the causal role of the right parietal cortex in numerical processing is supported by brain stimulation studies in adults, who showed that transcranial magnetic stimulation (TMS) over the right parietal lobe, which likely encompassed the IPS, disrupted numerical magnitude judgment and discrimination in healthy adults (Cohen Kadosh, Cohen Kadosh et al., 2007). Convergent on these findings, children with MD also show structural anomalies in the right IPS as discussed in a later section of this

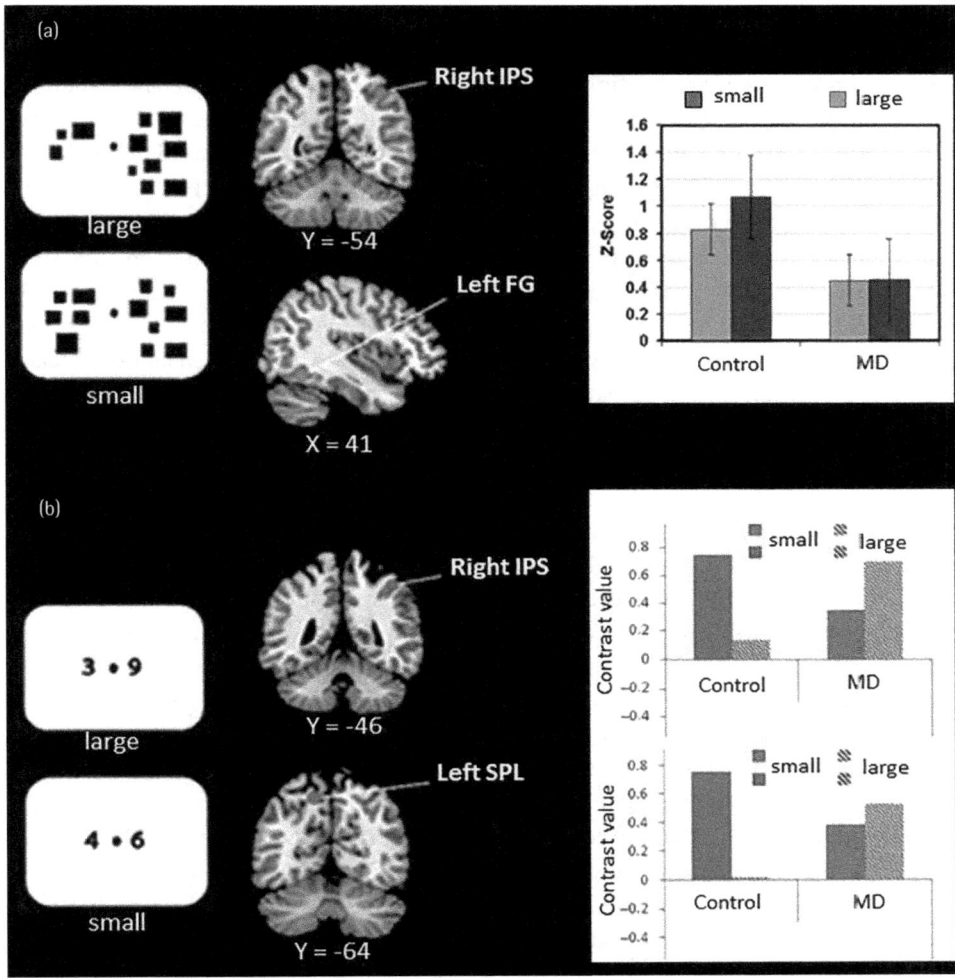

FIGURE 26.3 Aberrant parietal and ventro-temporal cortex responses during magnitude judgment. (a) Reduced modulation of brain responses by numerical distance during non-symbolic quantity comparison in children with MD, relative to control children. Results from an interaction analysis of group (MD, Control) x distance (small, large) in the right intraparietal sulcus (IPS) and left FG. The left panel shows stimuli used in the large and small distance conditions. The middle panel depicts brain areas that showed a neural distance effect. The right panel shows signal levels in each condition by group. (b) Reduced modulation by numerical distance during symbolic quantity comparison in children with MD relative to Control. Results from an interaction analysis of group x distance in the right IPS and left superior parietal lobule (SPL)

Source: Part a: Adapted from *Current Biology*, 17 (24), Gavin R. Price, Ian Holloway, Pekka Räsänen, Manu Vesterinen, and Daniel Ansari, Impaired parietal magnitude processing in developmental dyscalculia, PR1042–R1043, https://doi.org/10.1016/j.cub.2007.10.013. Copyright © 2007 Elsevier Ltd. All rights reserved.

Part b: adapted from Christophe Mussolin, Anne De Volder, Cécile Grandin, Xavier Schlögel, Marie-Cécile Nassogne, and Marie-Pascale Noël (2010). Neural correlates of symbolic number comparison in developmental dyscalculia. *Journal of Cognitive Neuroscience*, 22(5): 860–874. © 2010 by the Massachusetts Institute of Technology.

chapter. Taken together, these results suggest that atypical activity of the IPS may be a proximal cause of impairments in processing numerical magnitudes in MD.

In addition to weak modulation of the IPS, individuals with MD also show reduced activity in the ventral temporal occipital cortex (VTOC) and multiple areas of the prefrontal cortex (PFC) during basic number comparison tasks. Decreased activation in relation to numerical distance has been detected in the dorsolateral PFC during symbolic (Mussolin, De Volder et al., 2010) and non-symbolic number comparison tasks (Price, Holloway et al., 2007), but the laterality of PFC findings has not been consistent across these studies. Aberrant engagement of the fusiform gyrus (FG) within the VTOC in MD has also been reported in both symbolic and non-symbolic tasks with evidence of deficits (Price, Holloway et al., 2007) as well as over-compensation (Kucian, Loenneker et al., 2011). Another study that parametrically varied the numerical distance during a non-symbolic comparison task found evidence for compensatory engagement of bilateral supplementary motor area and the right FG in children with MD (Kucian, Loenneker et al., 2011).

Studies using low-level non-numerical control tasks have provided further insights into the neurocognitive basis of MD. A careful examination of the differential contrasts reported in various studies suggests that when activation during magnitude judgment are compared against low-level non-numerical control tasks, children with MD often show hyper-activation when compared to TD children, relative to hypo-activation during magnitude comparison as described previously. A conflicting pattern of hypo- and hyper-activation in MD reported in studies to date may therefore be directly related to the control task used. For example, in the study by Mussolin and colleagues, despite weak modulation by numerical distance, when activations were compared to passive fixation, children with MD showed greater activation than TD children in several brain regions, including right supramarginal gyrus and right postcentral gyrus (Mussolin, De Volder et al., 2010). Similarly, when magnitude judgment was contrasted against judgment of palm rotation, children with MD showed greater activation in left and right IPS, supramarginal gyrus, paracentral lobule and right superior frontal gyrus (Kaufmann, Vogel et al., 2009). Similar to children, adults with MD have also been reported to show compensatory hyper-activation in multiple PFC regions, including right superior frontal gyrus and left inferior frontal gyrus (IFG) (Cappelletti and Price, 2014). These findings suggest that individuals with MD can engage compensatory mechanisms in task-relevant brain regions during numerical information processing.

Despite the lack of consistency in the use of control conditions between studies, the general pattern that emerges from these studies is one of abnormal activity in distributed brain areas that extend beyond the IPS into multiple regions of the PFC, posterior parietal cortex (PPC) and VTOC regions bilaterally. Comparisons with low level baseline tasks reveal compensatory engagement of PPC, PFC, and VTOC in ways that may allow individuals with MD to achieve similar levels of performance as their TD peers on simpler numerical quantity processing tasks. At the same time, evidence from well-controlled tasks involving the numerical distance effects suggests weaker magnitude comparison-related modulation of brain responses in children with MD.

ARITHMETIC PROBLEM-SOLVING
DEFICITS IN MD

Behavioral Findings

Weak arithmetic problem-solving abilities, including poor fluency in retrieval of math facts, are a major area of difficulty in MD (Geary, 2004), even with solving simple arithmetic problems (Svenson and Broquist, 1975; Geary, 1990; Geary and Brown, 1991; Gross-Tsur, Manor et al., 1996; Barrouillet, Fayol et al., 1997; Jordan and Montani, 1997; Ostad, 1997; Ostad, 1998; Hanich, Jordan et al., 2001; Jordan, Hanich et al., 2003). In fact, poor fluency in retrieval of math facts in children with MD is typically the domain in which the math difficulties are first noticed (Geary, 2004). Most of the behavioral research on children with MD has focused on the strategies they use to solve single- or double-digit arithmetic problems (Geary, 1990; Jordan, Levine et al., 1995; Hanich, Jordan et al., 2001). A shift from immature to mature problem-solving strategies, such as counting to numerical fact retrieval, is a cardinal feature of mathematical development in children (Geary, 1994; Siegler, 1996). Children with MD use immature procedures and, unlike their TD peers, they do not show a developmental shift from calculation to efficient memory-based problem-solving (Jordan and Montani, 1997; Geary, Hamson et al., 2000), suggesting difficulties in storing and accessing facts from long-term memory (Garnett and Fleischner, 1983; Geary and Brown, 1991; Jordan and Montani, 1997; Ostad, 1997; Ostad, 1998). Moreover, the speed and accuracy with which individual strategies are executed improves with development and experience. Children with MD continue to use less mature strategies such as finger counting, to solve arithmetic problems (Geary, 1993; Shalev, Auerbach et al., 2000), commit more counting-procedure errors than do their TD peers (Jordan and Montani, 1997; Geary, Hamson et al., 2000), and rely on finger counting and use the sum procedure for many more years than TD children do (Geary and Brown, 1991; Geary, Hamson et al., 2000; Ostad, 2000).

Neurobiological Correlates

Most neuroimaging studies of aberrant brain response and representations during arithmetic problem-solving in MD have focused on addition operations, drawing on an extensive body of behavioral research. These studies have found evidence for atypical functional activation in the IPS as well as multiple PFC regions during problem-solving tasks involving mental addition. Crucially, differences in multiple functional systems can be found even when TD and MD groups are matched on performance. However, the direction of effects between MD and TD has not always been consistent with individuals with MD showing both hyper- and hypo-activation of relevant brain regions relative to TD individuals. For example, Kucian and colleagues (2006) using a task that measured approximate and exact addition problem-solving as well as magnitude estimation, found reduced brain activity in the right IPS, ventrolateral and dorsolateral PFC in children with MD (3rd to 6th graders), but only for addition problems that participants were asked to solve approximately (Kucian, Loenneker et al., 2006). In contrast, Davis and colleagues (2009), using a similar addition task, found hyper-activation in many regions of the PPC and PFC during both exact and approximate addition trials in 3rd grade children with MD (Davis, Cannistraci et al., 2009). These discrepancies in hyper- vs. hypo-activation in MD might be related to differences in the criteria used to define MD, the age ranges studied, and developmental changes that occur during key periods of skill acquisition between grades 3 and 6.

Moreover, more targeted studies of 7–9-year-old (grades 2 and 3) children with MD, who were matched on intelligence, WM, and reading with their TD peers, suggest that profiles of activation differences between MD and TD also depends crucially on control tasks used (Ashkenazi, Rosenberg-Lee et al., 2012; Rosenberg-Lee, Ashkenazi et al., 2015). When compared to passive fixation baseline, children with MD typically engage regions of the PPC and PFC at similar, or even higher, levels than those seen in their well-matched TD peers (Rosenberg-Lee, Ashkenazi et al., 2015). However, when "complex" "x + y" problems are used relative to a "simple" "x + 1" control task (Barrouillet and Lépine, 2005), children with MD fail to show appropriate increases in brain responses with increasing arithmetic complexity (Figure 26.4a) in the IPS but also the adjoining superior parietal lobule in the dorsal PPC, supramarginal gyrus in the ventral PPC, and bilateral dorsolateral PFC.

There is also evidence of weak modulation of neural representations for distinct arithmetic problems in children with MD, which has been probed using multivariate pattern analyses (MVPA) (Ashkenazi, Rosenberg-Lee et al., 2012). MVPA examines the spatial pattern of multi-voxel brain activity in a specific region of interest between task conditions, which are independent of overall differences in signal level (Formisano and Kriegeskorte, 2012). While TD children showed high differentiation between complex and simple problems in the bilateral IPS, children with MD showed less differentiated activation patterns independent of overall differences in signal level (Figure 26.4b) (Ashkenazi, Rosenberg-Lee et al., 2012). Taken together, these results suggest that children with MD not only have weak modulation of response in key brain regions but also fail to generate distinct neural representations for distinct arithmetic problems.

OPERATION-SPECIFIC DEFICITS AND FUNCTIONAL HYPER-CONNECTIVITY IN MD

Although most behavioral studies of children with MD have focused on addition problems, there is also evidence of impairments in subtraction (Ostad, 2000; Jordan, Hanich et al., 2003). Children with MD are particularly impaired at solving subtraction,

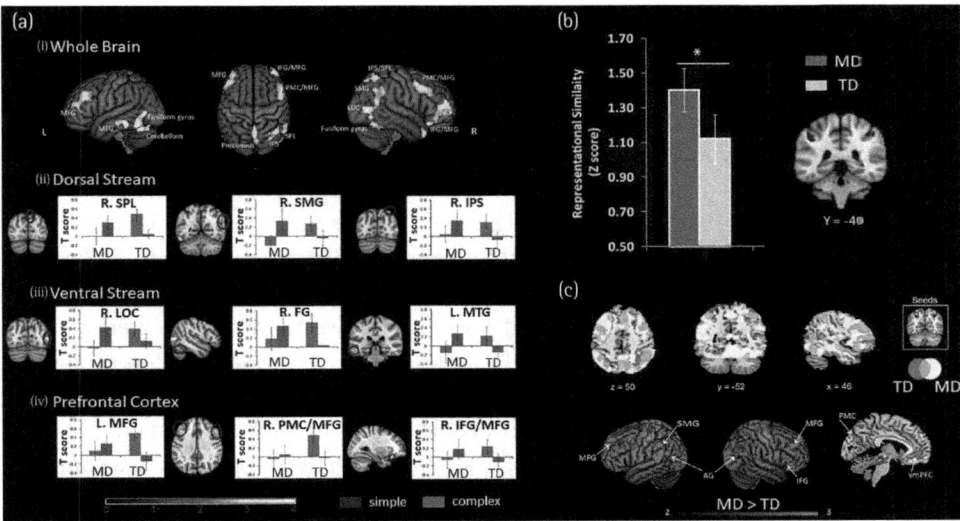

FIGURE 26.4 Aberrant parietal, ventro-temporal and prefrontal cortex responses during arithmetic problem-solving. (a) Reduced modulation of brain responses during arithmetic problem-solving in children with MD, compared to TD children. (*i*) Surface rendering of brain areas activated in the MD and TD groups by "complex" (x + y) vs. "simple" (x + 1) and arithmetic problems. (*ii,iii,iv*) Signal levels elicited by each problem type. FG = fusiform gyrus; IPS = intraparietal sulcus; LOC = lateral occipital cortex; MFG= middle frontal gyrus; MTG = middle temporal gyrus; PMC= premotor cortex; SMG = supramarginal gyrus; SPL = superior parietal lobule. (b) Representational similarity analysis demonstrating greater similarity of multi-voxel brain responses to "complex" and "simple" problems in the anterior IPS (hIP2: −44, −40, 48) of the left and right IPS; *p < .05. (c) IPS connectivity during arithmetic problem-solving in the MD (shown in yellow) and TD (shown in red) groups. Note the more extensive connectivity in the MD group. Brain regions that showed greater IPS connectivity in the MD group included multiple frontal, parietal and occipital regions: bilateral angular gyrus (AG), left supramarginal gyrus (SMG), right middle frontal gyrus (MFG), right inferior frontal gyrus (IFG), posterior-medial cortex (PMC) and ventral medial prefrontal cortex (vmPFC).

Sources: Parts 1 and 2: Reprinted from *Developmental Cognitive Neuroscience*, 2 (Supplement 1), Sarit Ashkenazi, Miriam Rosenberg-Lee, Caitlin Tenison, and Vinod Menon, Weak task-related modulation and stimulus representations during arithmetic problem solving in children with developmental dyscalculia, S152–S166, https://doi.org/10.1016/j.dcn.2011.09.006 Copyright © 2011 Elsevier Ltd. Published by Elsevier Ltd. All rights reserved.

Part 3: Adapted from *Developmental Science*, 18(3), Miriam Rosenberg-Lee, Sarit Ashkenazi, Tianwen Chen, Christina B. Young, David C. Geary, and Vinod Menon, Brain hyper-connectivity and operation-specific deficits during arithmetic problem solving in children with developmental dyscalculia, 351–372, https://doi.org/10.1111/desc.12216. Copyright © 2014 John Wiley and Sons Ltd.

relative to addition problems and one report has suggested that they continue to rely extensively on finger counting to solve subtraction problems even in the 7th grade (Ostad, 1999). Despite their surface similarity, subtraction problems rely more on quantity-based calculation procedures and are less well rehearsed using memory retrieval strategies relative to addition (Campbell and Xue, 2001; Barrouillet, Mignon et al., 2008). Contrasting brain responses to these two distinct but related operations has provided further insights into problem-solving deficits arising from weaknesses in executing calculation procedures in MD.

Relative to TD children, children with MD demonstrate aberrant brain responses during the solving of subtraction problems (Rosenberg-Lee, Ashkenazi et al., 2015). Remarkably, despite poorer performance on subtraction problems, children with MD show hyper-activation in multiple IPS and superior parietal lobule subdivisions in the dorsal PPC as well as the FG in VTOC. They also engage the right anterior insula and dorsolateral PFC, right supplementary motor area and bilateral superior frontal gyrus, and PFC regions involved in cognitive and motor control more than TD children (Figure 26.4c). Together, these results suggest that children with MD show more extensive dysfunction in multiple PPC, VTOC, and PFC areas while solving subtraction problems. Critically, hyper-activation of multiple brain areas may reflect the need for more processing resources during calculation-based problem-solving, such as subtraction, rather than an inability to activate task-relevant brain areas. Furthermore, effective connectivity analyses revealed hyper-connectivity, rather than hypo-connectivity, between the IPS and multiple brain systems including regions within the lateral fronto-parietal and default mode networks in children with MD during both addition and subtraction (Rosenberg-Lee, Ashkenazi et al., 2015) (Figure 26.4c). These findings suggest the IPS and its functional circuits are a major locus of dysfunction during both addition and subtraction problem-solving in MD. Moreover, recent findings suggest that inappropriate task modulation and hyper-connectivity, rather than under-engagement and under-connectivity, are also key neural mechanisms underlying problem-solving difficulties in children with MD.

WM Deficits in MD

Behavioral Findings

WM deficits also contribute to weak math problem-solving skills in MD (Geary, Hoard et al., 2007). There is growing evidence that WM deficits contribute to multiple aspects of MD, encompassing not only complex arithmetic problem-solving, which requires online manipulation of information in WM, but also basic quantity representation (Chu, Vanmarle et al., 2013; Fias, Menon et al., 2013), number ordering (Attout and Majerus, 2017) and word problem-solving (Fung and Swanson, 2017). Among the three main components of the Baddeley WM model—the central executive, visuospatial sketchpad, and phonological loop (Baddeley, 1996; Baddeley, Emslie et al., 1998), visuospatial WM is a particular source of numerical problem-solving deficits in MD (Wilson and Swanson, 2001; Swanson, Howard et al. 2006; Rotzer, Loenneker et al. 2009; Ashkenazi, Rosenberg-Lee et al., 2013; Szucs, Devine et al., 2013; Attout and Majerus, 2015). Furthermore, visuospatial WM is a specific deficit in children with MD when they are compared to children with reading disabilities (Swanson, Howard et al., 2006) and is also a strong predictor of math learning over development (Geary, 2011).

Neurobiological Findings

The importance of visuospatial WM and associated fronto-parietal processing during arithmetic problem-solving is further highlighted by neuroimaging studies of children with MD. Rotzer and colleagues (2009) found that children with MD had lower scores on a Corsi Block Tapping test, a standard measure of visuospatial WM ability, than TD children. This study also found that children with MD had lower activity levels in the right IFG, right IPS, and right insula during a visuospatial WM task, and that right IPS activity was positively correlated with visuospatial WM ability.

More direct evidence for a neurocognitive link between WM deficits and MD was provided by Ashkenazi and colleagues using standardized measures of all three components of WM, central executive, visuospatial, and phonological loop (Ashkenazi, Rosenberg-Lee et al., 2013) and examined their role in modulating brain responses to numerical problem-solving. Children with MD demonstrated lower visuospatial WM, even with normal IQ and preserved abilities on the phonological and central executive components of WM. Crucially, activations in IPS, and dorsolateral and ventrolateral PFC were positively correlated with visuospatial WM ability in TD children, but no such relation was seen in children with MD. These results suggest that visuospatial WM is a specific source of vulnerability in MD and thus needs to be seriously considered as a key component in cognitive, neurobiological, and developmental models of the disability. Notably, extant findings point to the IPS as a common locus of visuospatial WM and arithmetic problem-solving deficits, storing and manipulating quantity representations in MD (Metcalfe, Rosenberg-Lee et al., 2013; Iuculano, Rosenberg-Lee et al., 2015; Jolles, Ashkenazi et al., 2016).

COGNITIVE CONTROL DEFICITS IN MD

Behavioral Findings

A core deficit underlying MD and persistent low achievement in mathematics is difficulty committing basic arithmetic facts to long-term memory or retrieving them from memory (Temple, 1991; Geary, Hoard et al., 2012). This deficit is thought to arise from proactive interference, that is, confusion of related, previously learned information with current learning (De Visscher and Noël, 2013; De Visscher and Noël, 2014). Proactive control is specially required when the learning of arithmetic facts is highly prone to interference. For example, when asked to solve "6 x 3 = ?," participants might answer "24"

(6 x 4) or "12" (6 x 2) instead of "18" (Campbell and Graham, 1985). One possible reason for these relatively common errors is that arithmetic facts may be stored in memory in the form of a "network," whose size increases over the course of development, and whose nodes consist in operands (Ashcraft, 1992). Arithmetic problem-solving (production or verification of a result) is thought to activate relevant nodes as well as adjacent nodes, leading irrelevant facts to compete with the problem to solve. For example, when asked to solve "6 x 3," the competitors "6 x 4" and "6 x 2" can also be activated in WM and interfere with problem-solving. Proactive control is thus required to inhibit the competitors and select the correct answer. In this context, individual differences in the learning of arithmetic facts might come from differences in resisting interference in WM (Barrouillet, Fayol et al., 1997; Barrouillet and Lépine, 2005; De Visscher and Noël, 2014).

Barrouillet and Lépine (2005) found that adolescents with higher proactive control were better at retrieving arithmetic facts, while individuals who were more sensitive to interference showed lower arithmetic fluency (Barrouillet and Lépine, 2005). Similarly, adolescents with MD also show fact retrieval deficits arising from a lower ability to resist interference (generated by competitors), despite intact encoding in memory (Barrouillet, Fayol et al., 1997). Consistent with this initial finding, recent work has confirmed that difficulties with arithmetic problem-solving in MD are associated with hypersensitivity to interference (De Visscher and Noël, 2013; De Visscher and Noël, 2014; De Visscher, Szmalec et al., 2015). Thus, for example, children ages 8 to 10 with low arithmetic fluency showed poorer WM performance relative to TD children under high-interference, but not under low-interference conditions (De Visscher and Noël, 2014). Finally, adults with a history of MD, and in particular with low arithmetic fluency, showed the same type of WM impairments under high interference conditions (De Visscher, Szmalec et al., 2015). Taken together, these results suggest that hypersensitivity to high interference in WM contributes to numerical problem-solving deficits in MD.

Neurobiological Correlates

Two neuroimaging studies in adults have investigated the neural correlates of the interference effects during arithmetic problem-solving. They revealed increased brain activity under high interference (as compared to low interference) in the left IFG (De Visscher, Vogel et al., 2018) and in the right IFG, bilateral insula, medial PFC, and premotor and motor cortex (De Visscher, Berens et al., 2015). Interestingly, brain activity in the left IFG during high interference was negatively correlated with arithmetic skills (De Visscher, Vogel et al., 2018), pointing to greater need for cognitive control resources in individuals with low arithmetic fluency. These findings are consistent with reports of hyperactivity and over-engagement of multiple cognitive control regions of the PFC reported in MD (Rosenberg-Lee, Ashkenazi et al., 2015), especially when neural activations are compared using low-level task baselines, as summarized previously.

INTRINSIC BRAIN CONNECTIVITY IN MD

As described previously, numerical cognition relies on interactions between multiple functional brain systems, including those subserving quantity processing, WM, and cognitive control (Fias, Menon et al., 2013; Arsalidou, Pawliw-Levac et al., 2018) (Figure 26.1). An important question is whether aberrations in brain activation and connectivity are a task/state dependent with variable profiles of differences arising from the type of numerical tasks used (e.g., non-symbolic number comparison, symbolic number comparison, addition, subtraction) and baseline ("control") conditions (e.g., fixation, mental rotation, manipulation of task complexity) used in various studies, or whether they arise from differences in task performance between MD and TD groups. Intrinsic functional connectivity using task-free resting state fMRI has the potential to address questions related to functional brain organization in MD, provide insights into impairments in functional brain circuity and how they might influence the ability of individuals with MD to perform numerical tasks, and circumvent methodological issues arising from differences in the type of numerical task and control conditions used, as well as confounding effects of task performance and strategy use (Fair, Schlaggar et al., 2007; Church, Wenger et al., 2009; Koyama, Di Martino et al., 2011; Supekar, Uddin et al., 2013; Uddin, Supekar et al., 2013; Finn, Shen et al., 2014).

Using this approach, Jolles, Ashkenazi et al. (2016) characterized intrinsic functional connectivity of the IPS in children with MD, relative to a group of TD children who were matched on age, gender, IQ, WM, and reading abilities (Jolles, Ashkenazi et al., 2016) using a multilevel analytical approach (Figure 26.5). First, compared to TD children, children with MD showed hyper-connectivity of the IPS with multiple prefrontal, parietal and basal ganglia regions. Hyperconnectivity of the IPS with prefrontal and parietal cortex was also associated with poorer math ability in children with MD (Figure 26.5a). Furthermore, machine learning algorithms revealed that aberrant IPS connectivity patterns accurately discriminated children with MD and TD children. Lastly, children with MD showed higher levels of spontaneous low-frequency fluctuations in multiple frontal and parietal areas, including IPS, superior parietal lobule, supramarginal gyrus, and hippocampus (Figure 26.5b). Notably, children with MD showed higher low-frequency fluctuations in multiple fronto-parietal areas that overlapped with brain regions that exhibited hyper-connectivity with the IPS. These results suggest that intrinsic hyper-connectivity and enhanced low-frequency fluctuations may limit flexible resource allocation, and contribute to aberrant recruitment of task-related brain regions during numerical problem-solving in MD.

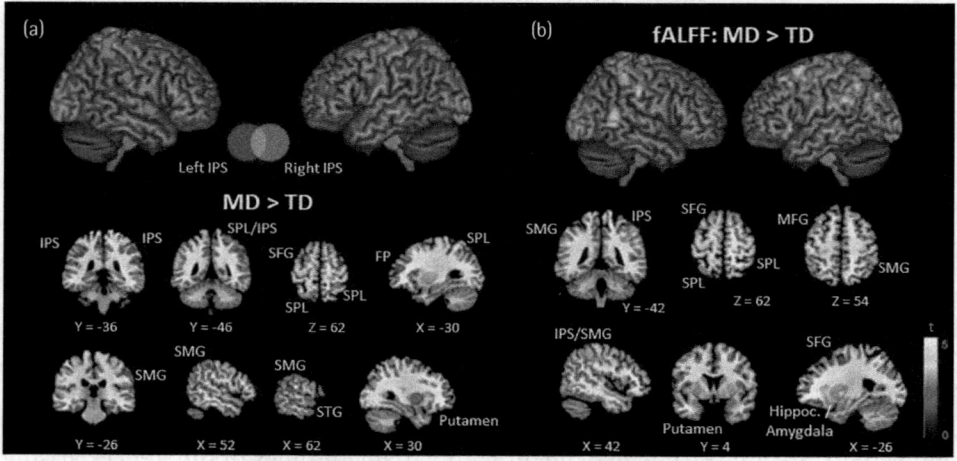

FIGURE 26.5 Aberrant intrinsic functional connectivity in MD. (a) Brain areas that showed greater IPS connectivity in children with mathematical disabilities (MD) compared to typically developing (TD) children. Children with MD showed hyper-connectivity between bilateral IPS and multiple dorsal frontal and parietal cortical regions, between bilateral IPS and right hemisphere SMG and STG, and between left IPS and right putamen. Greater connectivity for MD > TD in red (left IPS), blue (right IPS), and green (both left and right IPS). Coordinates are in MNI space. FP = frontal pole; IPS = intraparietal sulcus; SFG = superior frontal gyrus; SMG = supramarginal gyrus; SPL = superior parietal lobe; STG = superior temporal gyrus. (b) Greater fALFF in children with MD compared to TD children. fALFF = Fractional Amplitude of Low Frequency Fluctuations; MFG = middle frontal gyrus

Source: Adapted from *Developmental Science*, 19(4): 613–631, Dietsje Jolles, Sarit Ashkenazi, John Kochalka, Tanya Evans, Jennifer Richardson, Miriam Rosenberg-Lee, Hui Zhao, Kaustubh Supekar, Tianwen Chen, and Vinod Menon, *Parietal Hyper-Connectivity, Aberrant Brain Organization, and Circuit-Based Biomarkers in Children with Mathematical Disabilities*. https://doi.org/10.1111/desc.12399. Copyright © 2016 John Wiley and Sons Ltd.

NEUROANATOMICAL BASIS OF MD

A more general view of neurobiological deficits in MD, without the confounding effects of task performance, is provided by neuroanatomical investigations of gray and white matter (Rotzer, Kucian et al., 2008; Rykhlevskaia, Uddin et al., 2009; Kucian, Ashkenazi et al., 2014). As reviewed in the previous sections, the profile of aberrant functional activations varies considerably with the level of task difficulty, the type of operation performed, and the control task against which activations are compared. Gray and white matter integrity are crucial for all cognitive operations and systematic identification of anatomical deficits can provide concrete and convergent evidence for core areas of deficits in MD.

Decreased gray matter volume in multiple parietal, temporal, and occipital areas, including bilateral IPS, superior parietal lobule, FG, para hippocampal gyrus, hippocampus, and anterior temporal cortex areas that have been implicated in numerical problem-solving has been consistently reported in children and adolescents with

MD (Rotzer, Kucian et al., 2008, Rykhlevskaia, Uddin et al., 2009; Ranpura, Isaacs et al., 2013) (Figure 26.6a). One study reported reduced gray matter volume in superior parietal lobule, IPS, FG, para hippocampal gyrus, and right anterior temporal cortex as well as reduced structural connectivity in long-range white matter projection fibers linking the right FG with temporo-parietal cortex in children with MD (Figure 26.6b) (Rykhlevskaia, Uddin et al., 2009). DTI studies also suggest a link between arithmetic skills and white matter integrity in left fronto-parietal tracts (Tsang, Dougherty et al., 2009), and in several tracts that directly pass through the IPS (Rykhlevskaia, Uddin et al., 2009; Li, Hu et al., 2013; Kucian, Ashkenazi et al., 2014). Additionally, deficits in the inferior fronto-occipital fasciculus and inferior and superior longitudinal fasciculus, tracts that link parietal and occipital cortex with the PFC have also been detected (Figure 26.6c). Taken together, findings to date suggest that both gray matter reductions in posterior brain areas and weak integrity of white matter tracts linking them with the PFC are a potential locus of information processing deficits and learning in MD.

INTERVENTIONS AND REMEDIATION OF COGNITIVE DEFICITS AND MATH DIFFICULTIES IN MD

Behavioral Findings

Cognitive interventions have been shown to be highly effective in improving math performance in children (Fuchs, Fuchs et al., 2013). Here, we review interventions that have targeted distinct math difficulties as well as general cognitive deficits in MD. Previous behavioral research has shown that interventions using speeded retrieval of number combinations together with conceptual number knowledge can significantly improve math skills in elementary school children (Christensen and Gerber, 1990; Okolo, 1992; Fuchs, Fuchs et al. 2002; Fuchs, Fuchs et al. 2004; Fuchs, Fuchs, Hamlet et al., 2006; Fuchs, Fuchs, Compton et al., 2006; Fuchs, Fuchs, Compton et al. 2007; Fuchs, Fuchs and Luther, 2007). Specifically, extensive research by Fuchs and colleagues has shown that interventions that directly target specific areas of math, such as the memorization of arithmetic facts (Fuchs, Powell et al., 2009; Powell, Fuchs et al., 2009), conceptual understanding of arithmetic procedures (Powell, Fuchs et al., 2009), and word problem-solving (Fuchs, Fuchs et al. 2004), can significantly improve numerical skills in children with MD. The most consistent evidence of training-related math gains comes from a series of studies using computer-assisted tutoring that taught children to recall simple math facts, with immediate feedback for improving problem-solving strategies. Training sessions typically consisted of 30 to 45 min sessions several times a week for about 3 months (Fuchs, Fuchs et al., 2013). This line of research has reported high rates of success as assessed by the number of children with MD who showed prolonged effects of intervention.

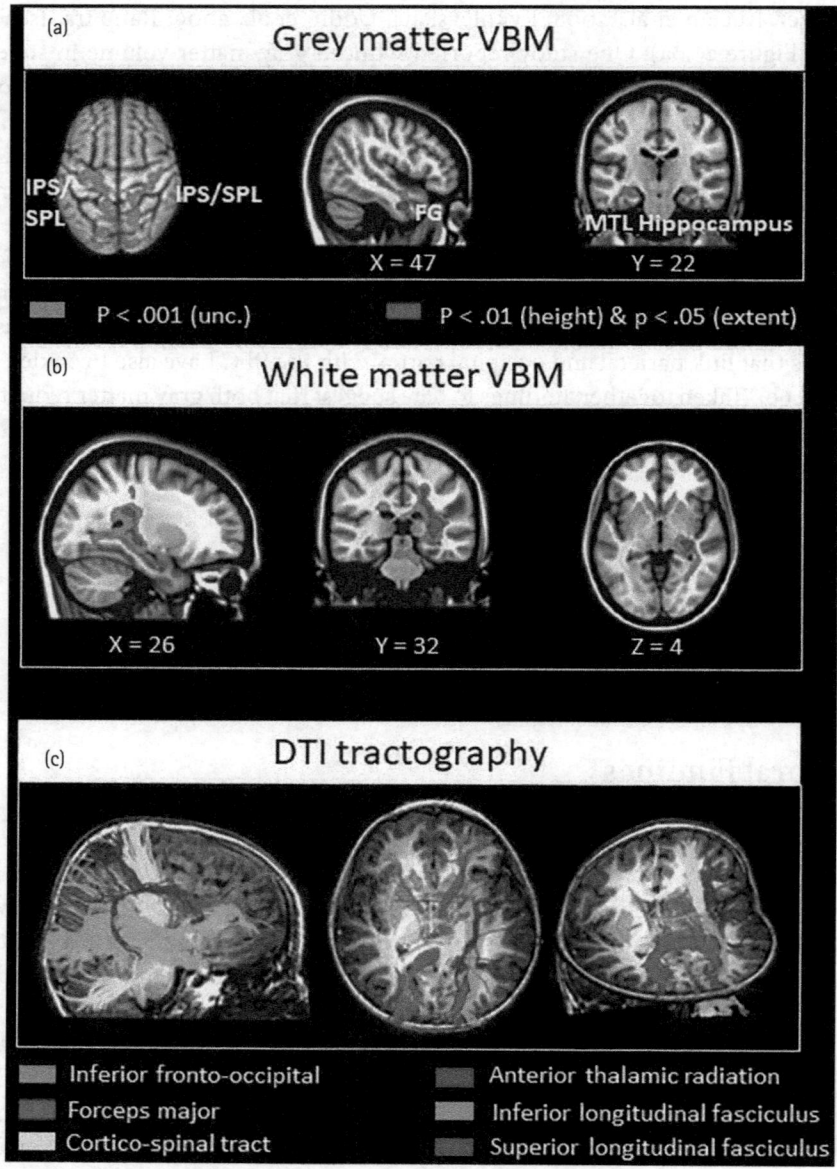

FIGURE 26.6 Gray and white matter deficits in MD. (a) Brain areas that show reduced gray matter volume in the MD group: bilateral superior parietal lobule (SPL), bilateral intra-parietal sulcus (IPS), lingual and fusiform gyri (FG), and hippocampus in the medial temporal cortex (MTL). (b) Reduced white matter volume in right temporal-parietal cortex in the MD group. (c) Lower fiber density in inferior longitudinal fasciculus, inferior fronto-occipital fasciculus, and caudal forceps major compared to typically developing children. The right panel depicts reduced connectivity in children with MD for long-range white matter projection fibers that link the right FG with temporal-parietal areas.

Source: Adapted from Rykhlevskaia E, Uddin LQ, Kondos L, and Menon V (2009). Neuroanatomical correlates of developmental dyscalculia: Combined evidence from morphometry and tractography. *Frontal Human Neuroscience* 3:51. https://doi.org/10.3389/neuro.09.051.2009. © 2009 Rykhlevskaia, Uddin, Kondos and Menon.

Studies focusing on improving number sense (Wilson, Revkin et al., 2006) and their spatial representation (Kucian, Grond et al., 2011) have also been shown to moderately improve math skills in MD. Children with MD who show deficits in elementary number knowledge may benefit from focused training in numerical magnitudes and mapping with number symbols during early school years, as suggested by a software-based training in 7–9-year-old children with MD (Wilson, Revkin et al., 2006). After training, children with MD also showed improvement in subtraction and symbolic as well as nonsymbolic number comparison (Wilson, Revkin et al., 2006). Similarly, in a sample of 8- to 10-year-old children with MD, programs designed to remediate immature mental number line improved spatial representation of numbers, and most importantly improved arithmetic skills (Kucian, Grond et al., 2011). Similarly, a computerized training focused on spatial number representations, arithmetic, and word problem-solving found improvement on number representations as well as arithmetic problem-solving (Rauscher, Kohn et al., 2016). These findings suggest that tapping into basic numerical and spatial skills can have generalizable effects on math performance.

Other remediation and training studies aimed at improving math learning in MD, or in TD individuals, have targeted general cognitive abilities such as attention (Ashkenazi and Henik, 2012), spatial skills (Uttal, Meadow et al., 2013), and finger representations, with mixed results. Using a video-game targeting attention orienting skills in adults with MD, Ashkenazi and Henik (2012) showed that although participants' attention improved after training, no transfer occurred on math skills. Cheng and Mix (2014) found that mental rotation training improved arithmetic problem-solving in TD children (Cheng and Mix, 2014). Another study that trained finger representation in children in 1st grade who showed poor finger gnosia resulted in improved symbolic and nonsymbolic skills (Gracia-Bafalluy and Noël, 2008), suggesting that improved finger gnosia can be a pathway for improving math skills in children with MD with visuomotor deficits and profiles (Kinsbourne and Warrington, 1963).

Taken together, these studies also suggest that the remediation of poor math skills in children with MD might require focused training on different cognitive mechanisms. Cognitive factors influencing individual intervention gains may vary depending on overall cognitive abilities, age range targeted, the difficulty and modality of the training, and affective factors (Chodura, Kuhn et al., 2015; Powell, Cirino et al., 2017). Behavioral studies indicate that children with MD can benefit from structured interventions, but how these improvements occur and why some children are more responsive than others is not known (Fuchs, Fuchs et al. 2004; Griffin, 2007; Fuchs, Powell et al., 2008; Chodura, Kuhn et al., 2015). Neuroimaging studies might be useful to better characterize individual differences in response to intervention and tailor cognitive intervention to individual profiles of cognitive, numerical, as well as affective vulnerabilities. Interestingly, intervention studies focusing on arithmetic problem-solving in children have also reported improvement in cognitive and affective domains.

A noteworthy finding to emerge from these cognitive intervention studies is that tutoring designed to improve numerical problem-solving skills may also decrease math anxiety levels (Supekar, Swigart et al., 2013; Berkowitz, Schaeffer et al., 2015).

Math anxiety involves negative emotions and stress in situations involving manipulation and reasoning about numerical problems in a wide variety of academic and life situations (Richardson and Suinn, 1972). The neural correlates of math anxiety have been investigated in adults (Lyons and Beilock, 2012a, 2012b) and in children (Young, Wu et al., 2012; Kucian, McCaskey et al., 2018). These studies point to increased activity in brain regions associated with emotion regulation in high math anxious individuals, in addition to a decrease in brain activity in fronto-parietal regions that support WM processes during problem-solving (Young, Wu et al., 2012). Moreover, right amygdala volume has been reported to be reduced in high math anxious children (Kucian, McCaskey et al., 2018). These findings suggest that affective factors contribute to problem-solving deficits in children with MD, and should be taken into account while designing interventions to remediate math skills in MD.

Neurobiological Correlates

Several fMRI studies have now begun to examine the extent to which cognitive training alters aberrant functional activity and connectivity in relevant neurocognitive systems in MD. The effects of training can work either by normalizing deficits or by engagement of compensatory mechanisms to facilitate improved task performance.

Cognitive tutoring studies of speeded practice on math fact retrieval have shown that 8 weeks of 1:1 tutoring in 7–9-year-old children with MD can normalize aberrant functional brain responses to the level of TD peers (Iuculano, Rosenberg-Lee et al., 2015). Brain plasticity was evident not just in the IPS, but in a distributed network of prefrontal, parietal, and ventral temporal-occipital brain areas important for numerical problem-solving (Figure 26.7a). Remarkably, machine learning algorithms revealed that brain activity patterns in children with MD were significantly different from TD peers before tutoring, but statistically indistinguishable after tutoring (Figure 26.7b). Crucially, findings did not provide evidence for a "compensatory" model of plasticity, which would posit that after training, children with MD would recruit additional and distinct brain systems compared to TD peers. Instead, these results indicate that intervention benefits were related to decreased load on prefrontal WM and dorsal attentional systems during arithmetic problem-solving.

Similarly, intervention studies involving mental number line training found increased recruitment of bilateral parietal areas, in parallel with decreased PFC activity, in children with MD after training as compared to before training (Kucian, Grond et al., 2011). Prefrontal and parietal activations in MD were more similar to TD peers after training. This type of training was additionally associated with normalization of brain hyperconnectivity between the IPS and other cortical areas to levels seen in TD individuals (Michels, O'Gorman et al., 2018).

Training-related studies have also demonstrated that plasticity of white-matter tracts as well as intrinsic functional circuits is associated with individual differences in math learning (Jolles, Supekar et al., 2016; Jolles, Wassermann et al., 2016). Using novel fiber

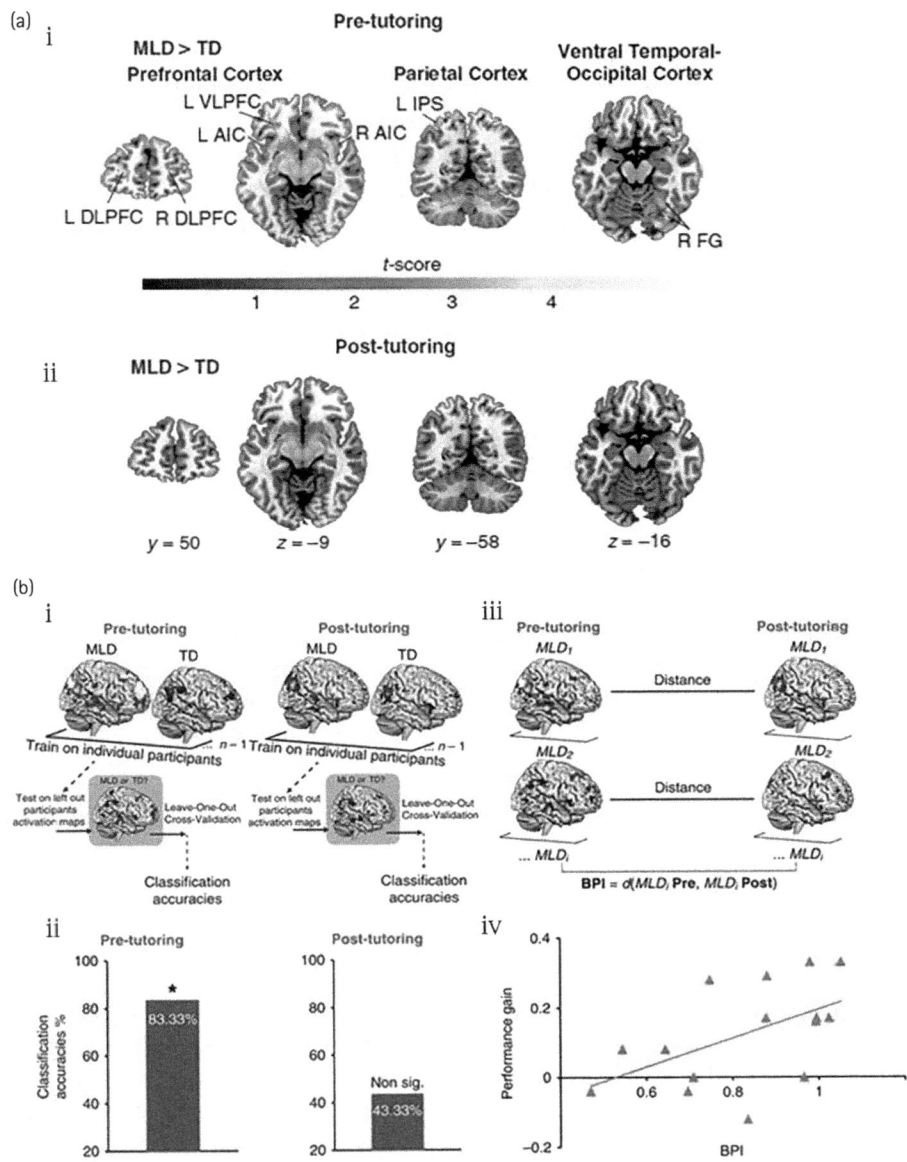

FIGURE 26.7 Cognitive training and neuroplasticity in MD. (a) Normalization of aberrant functional brain responses in children with MD after 8 weeks of math tutoring. (i) Before tutoring, children with MD showed significant differences in brain activation levels compared with TD children. Significant group differences were evident in multiple cortical areas in the Prefrontal Cortex, including the bilateral Dorsolateral Prefrontal Cortices (DLPFC), and the left Ventrolateral Prefrontal Cortex (VLPFC), as well as the bilateral Anterior Insular Cortices (AIC); in the Parietal Cortex encompassing the left Intraparietal Sulcus (IPS); and in the Ventral Temporal–Occipital Cortex including the right FG. (ii) After 8 weeks of tutoring, functional brain responses in MD children (n=15) normalized to the levels of TD children. (b) Multivariate brain

tracking algorithms Jolles, Wassermann et al., 2016 identified sections of the superior longitudinal fasciculus linking frontal and parietal, parietal and temporal and frontal and temporal cortices. They found that individual differences in behavioral gains in math after two months of tutoring were specifically correlated with plasticity in the left frontotemporal tract that connects posterior temporal and lateral prefrontal cortices, coursing through the PPC. This tract is well positioned to integrate symbolic, numerical, and control processing carried out by distributed brain regions (Arsalidou and Taylor, 2011). Notably, additional analyses showed that brain-behavior correlations were present for multiple behavioral measures associated with numerical cognition. Analysis of intrinsic functional connectivity data further showed that training also strengthened IPS connectivity with the lateral PFC, VTOC, and hippocampus. Crucially, increased IPS connectivity was associated with individual performance gains, highlighting the behavioral significance of plasticity in IPS circuits. Tutoring-related changes in IPS connectivity were distinct from those of the adjacent angular gyrus, which did not predict performance gains (Jolles, Supekar et al., 2016).

Overall, the results of cognitive training studies in children point to plasticity of multiple brain circuits associated with the development of more or less specialized math and cognitive skills. It is noteworthy that quantitative measures of brain function and intrinsic brain organization can provide a more sensitive endophenotypic marker of skill acquisition than behavioral measures. The development of robust brain-based biomarkers will be a significant step towards understanding children's predicted learning trajectories and may facilitate improved training and intervention programs.

FIGURE 26.7 Continued
activity patterns-based classification of MD children and association with performance gains. (i) Classification Analysis flowchart. A linear classifier built using support vector machines (SVM) with Leave-One-Out Cross-Validation (LOOCV) was used to classify children with MD from TD children based on patterns of brain activation during arithmetic problem-solving, before and after tutoring. (ii) Classification accuracies pre- and post-tutoring. Brain activation patterns between MD and TD children during arithmetic problem-solving were significantly and highly discriminable before tutoring, while the groups were no longer discriminable by their patterns of brain activity after tutoring. (iii) Brain Plasticity Index (BPI) in children with MD. A distance metric d was computed to quantify tutoring-induced functional brain plasticity effects pre-tutoring versus post-tutoring in children with MD. d was calculated individually for each MD child by computing a multivariate spatial correlation between pre- and post-tutoring patterns of brain activity, and subtracting it from 1. (iv) Relation between tutoring-induced functional brain plasticity and performance gain. A significant positive correlation was observed between BPI and individual performance gains associated with tutoring in children with MD. Performance gain represents change in arithmetic problem-solving accuracy from pre- to post-tutoring

CONCLUSIONS AND FUTURE DIRECTIONS

As with behavioral studies, our understanding of the neurobiology of MD is inherently constrained by criteria used to define the disorder. Despite this limitation, neurocognitive studies are revealing important new insights into brain and cognitive processes disrupted in MD.

We described multicomponent cognitive models of MD that have emerged from cognitive and neuroscience studies of numerical cognition in MD and TD individuals, emphasizing number sense and arithmetic problem-solving, the two major "domain-specific" aspects of information processing deficits in MD. This review has emphasized the role of "domain-general" deficits, most importantly visuospatial WM and cognitive control, which impact the ability to manipulate quantity, retrieve facts and resolve intrusion errors. In this view MD arises from a complex interplay between "domain-specific" and "domain-general" deficits that are implemented by overlapping brain circuits. While functional neuroimaging studies have overtly focused on the IPS as the locus of numerical information processing deficits in children with MD, it is now clear that individuals with MD show deficits in a distributed inter-connected set of brain regions that include the IPS and other parietal cortex regions, inferior temporal cortex, medial temporal lobe, and multiple PFC areas. Impairments in one or more of these brain regions and their associated brain circuits can compromise efficiency of numerical problem-solving skills and constitute risk factors for MD.

An important future research direction is the characterization of MD subtypes, including investigations of how impairments to different neurocognitive systems are related to weaknesses in math abilities. In this context, it will be crucial to investigate heterogeneity arising from co-occurring reading and attention deficits. Another important topic of ongoing research is to determine the extent to which brain networks supporting math learning are malleable, and the type of instruction that can target these networks at different developmental periods. Due to the challenges associated with neuroimaging studies of heterogeneous learning disabilities, these efforts will benefit from future multi-site studies. Finally, stronger interdisciplinary collaborations between psychology, education and neuroscience are needed to advance diagnosis and remediation of MD in affected children.

ACKNOWLEDGMENTS

Supported by grants from the National Institutes of Health (HD047520 and HD059205).

REFERENCES

Alloway, T. P. and R. G. Alloway (2010). Investigating the predictive roles of working memory and IQ in academic attainment. *Journal of Experimental Child Psychology, 106*(1): 20–29.

American Psychiatric Association (2013). *Diagnostic and Statistical Manual of Mental Disorders*. Arlington, VA: American Psychiatric Association.

Arsalidou, M., M. Pawliw-Levac, M. Sadeghi, and J. Pascual-Leone (2018). Brain areas associated with numbers and calculations in children: Meta-analyses of fMRI studies. *Developmental Cognitive Neuroscience, 30*: 239–250.

Arsalidou, M. and M. J. Taylor (2011). Is 2+2=4? Meta-analyses of brain areas needed for numbers and calculations. *NeuroImage, 54*(3): 2382–2393.

Ashcraft, M. H. (1992). Cognitive arithmetic: A review of data and theory. *Numerical Cognition, 44*(1): 75–106.

Ashkenazi, S. and A. Henik (2012). Does attentional training improve numerical processing in developmental dyscalculia? *Neuropsychology, 26*(1): 45–56.

Ashkenazi, S., N. Mark-Zigdon, and A. Henik (2009). Numerical distance effect in developmental dyscalculia. *Cognitive Development, 24*(4): 387–400.

Ashkenazi, S., M. Rosenberg-Lee, A. W. Metcalfe, A. G. Swigart, and V. Menon (2013). Visuospatial working memory is an important source of domain-general vulnerability in the development of arithmetic cognition. *Neuropsychologia, 51*(11): 2305–2317.

Ashkenazi, S., M. Rosenberg-Lee, C. Tenison, and V. Menon (2012). Weak task-related modulation and stimulus representations during arithmetic problem solving in children with developmental dyscalculia. *Developmental Cognitive Neuroscience, 2*(Suppl 1): S152–S166.

Ashkenazi, S., O. Rubinsten, and A. Henik (2009). Attention, automaticity, and developmental dyscalculia. *Neuropsychology, 23*(4): 535–540.

Attout, L. and S. Majerus (2015). Working memory deficits in developmental dyscalculia: The importance of serial order. *Child Neuropsychology, 21*(4): 432–450.

Attout, L. and S. Majerus (2017). Serial order working memory and numerical ordinal processing share common processes and predict arithmetic abilities. *British Journal of Developmental Psychology, 36*(2): 285–298.

Attout, L., E. Salmon, and S. Majerus (2015). Working memory for serial order is dysfunctional in adults with a history of developmental dyscalculia: Evidence from behavioral and neuroimaging data. *Developmental Neuropsychology, 40*(4): 230–247.

Baddeley, A. (1996). The fractionation of working memory. *Proceedings of the National Academy of Sciences U S A, 93*(24): 13468–13472.

Baddeley, A., H. Emslie, J. Kolodny, and J. Duncan (1998). Random generation and the executive control of working memory. *Quarterly Journal of Experimental Psychology, 51*(4): 819–852.

Barrouillet, P., M. Fayol, and E. Lathuliere (1997). Selecting between competitors in multiplication tasks: An explanation of the errors produced by adolescents with learning difficulties. *International Journal of Behavioral Development, 21*(2): 253–275.

Barrouillet, P. and R. Lépine (2005). Working memory and children's use of retrieval to solve addition problems. *Journal of Experimental Child Psychology, 91*(3): 183–204.

Barrouillet, P., M. Mignon, and C. Thevenot (2008). Strategies in subtraction problem solving in children. *Journal of Experimental Child Psychology, 99*(4): 233–251.

Berkowitz, T., M. W. Schaeffer, E. A. Maloney, L. Peterson, C. Gregor, S. C. Levine, and S. L. Beilock (2015). Math at home adds up to achievement in school. *Science, 350*(6257): 196–198.

Bugden, S. and D. Ansari (2016). Probing the nature of deficits in the "Approximate Number System" in children with persistent developmental dyscalculia. *Developmental Science, 19*(5): 817–833.

Butterworth, B. (1999). *The Mathematical Brain*. London: Macmillan.

Butterworth, B. (2005). Developmental dyscalculia. In *Handbook of Mathematical Cognition*, pp. 455–467. Hove, UK: J. I. D. Campbell.

Butterworth, B. (2010). Foundational numerical capacities and the origins of dyscalculia. *Trends in Cognitive Sciences*, 14(12): 534–541.

Butterworth, B., S. Varma, and D. Laurillard (2011). Dyscalculia: From brain to education. *Science*, 332(6033): 1049–1053.

Campbell, J. I. and Q. Xue (2001). Cognitive arithmetic across cultures. *Journal of Experimental Psychology: General*, 130(2): 299–315.

Campbell, J. I. D. and D. J. Graham (1985). Mental multiplication skill: Structure, process, and acquisition. *Canadian Journal of Psychology/Revue canadienne de psychologie*, 39(2): 338–366.

Cantlon, J. F., E. M. Brannon, E. J. Carter, and K. A. Pelphrey (2006). Functional Imaging of Numerical Processing in Adults and 4-y-Old Children. *PLoS Biology*, 4(5): e125.

Cappelletti, M. and C. J. Price (2014). Residual number processing in dyscalculia. *NeuroImage: Clinical*, 4: 18–28.

Carey, S. (2004). *On the Origin of Concepts*. New York: Oxford University Press.

Cheng, Y.-L. and K. S. Mix (2014). Spatial training improves children's mathematics ability. *Journal of Cognition and Development*, 15(1): 2–11.

Chodura, S., J.-T. Kuhn and H. Holling (2015). Interventions for children with mathematical difficulties: A meta-analysis. *Zeitschrift für Psychologie*, 223(2): 129–144.

Christensen, C. A. and M. M. Gerber (1990). Effectiveness of computerized drill and practice games in teaching basic math facts. *Exceptionality*, 1: 149–165.

Chu, F. W., K. Vanmarle, and D. C. Geary (2013). Quantitative deficits of preschool children at risk for mathematical learning disability. *Frontal Psychology*, 4: 195.

Church, J. A., K. K. Wenger, N. U. Dosenbach, F. M. Miezin, S. E. Petersen, and B. L. Schlaggar (2009). Task control signals in pediatric Tourette syndrome show evidence of immature and anomalous functional activity. *Frontiers in Human Neuroscience*, 3: 38.

Cohen Kadosh, R., K. Cohen Kadosh, T. Schuhmann, A. Kaas, R. Goebel, A. Henik, and A. T. Sack (2007). Virtual dyscalculia induced by parietal-lobe TMS impairs automatic magnitude processing. *Current Biology*, 17(8): 689–693.

Davis, N., C. J. Cannistraci, B. P. Rogers, J. C. Gatenby, L. S. Fuchs, A. W. Anderson, and J. C. Gore (2009). Aberrant functional activation in school age children at-risk for mathematical disability: A functional imaging study of simple arithmetic skill. *Neuropsychologia*, 47(12): 2470–2479.

De Smedt, B. and C. K. Gilmore (2011). Defective number module or impaired access? Numerical magnitude processing in first graders with mathematical difficulties. *Journal of Experimental Child Psychology*, 108(2): 278–292.

De Smedt, B., I. D. Holloway, and D. Ansari (2011). Effects of problem size and arithmetic operation on brain activation during calculation in children with varying levels of arithmetical fluency. *NeuroImage*, 57(3): 771–781.

De Smedt, B., M.-P. Noël, C. Gilmore, and D. Ansari (2013). How do symbolic and non-symbolic numerical magnitude processing skills relate to individual differences in children's mathematical skills? A review of evidence from brain and behavior. *Trends in Neuroscience and Education*, 2(2): 48–55.

De Visscher, A., S. C. Berens, J. L. Keidel, M.-P. Noël, and C. M. Bird (2015). The interference effect in arithmetic fact solving: An fMRI study. *NeuroImage*, 116: 92–101.

De Visscher, A. and M.-P. Noël (2013). A case study of arithmetic facts dyscalculia caused by a hypersensitivity-to-interference in memory. *Cortex*, 49(1): 50–70.

De Visscher, A. and M.-P. Noël (2014). Arithmetic facts storage deficit: The hypersensitivity-to-interference in memory hypothesis. *Developmental Science*, 17(3): 434–442.

De Visscher, A., A. Szmalec, L. Van Der Linden, and M.-P. Noël (2015). Serial-order learning impairment and hypersensitivity-to-interference in dyscalculia. *Cognition*, 144: 38–48.

De Visscher, A., S. E. Vogel, G. Reishofer, E. Hassler, K. Koschutnig, B. De Smedt, and R. H. Grabner (2018). Interference and problem size effect in multiplication fact solving: Individual differences in brain activations and arithmetic performance. *NeuroImage*, 172: 718–727.

Dehaene, S. (1997). *The Number Sense: How the Mind Creates Mathematics*. New York: Oxford University Press.

Dehaene, S., M. Piazza, P. Pinel, and L. Cohen (2003). Three parietal circuits for number processing. *Cognitive Neuropsychology*, 20(3): 487–506.

Demir, Ö. E., J. Prado, and J. R. Booth (2014). The differential role of verbal and spatial working memory in the neural basis of arithmetic. *Developmental Neuropsychology*, 39(6): 440–458.

Devine, A., F. Soltész, A. Nobes, U. Goswami, and D. Szűcs (2013). Gender differences in developmental dyscalculia depend on diagnostic criteria. *Learning and Instruction*, 27: 31–39.

Fair, D. A., B. L. Schlaggar, A. L. Cohen, F. M. Miezin, N. U. Dosenbach, K. K. Wenger, M. D. Fox, A. Z. Snyder, M. E. Raichle, and S. E. Petersen (2007). A method for using blocked and event-related fMRI data to study "resting state" functional connectivity. *Neuroimage*, 35(1): 396–405.

Fechner, G. (1966). *Elements of Psychophysics*. Vol. 1. New York: Holt, Rinehart and Winston.

Fias, W., V. Menon, and D. Szucs (2013). Multiple components of developmental dyscalculia. *Trends in Neuroscience and Education*, 2(2): 43–47.

Finn, E. S., X. Shen, J. M. Holahan, D. Scheinost, C. Lacadie, X. Papademetris, S. E. Shaywitz, B. A. Shaywitz, and R. T. Constable (2014). Disruption of functional networks in dyslexia: A whole-brain, data-driven analysis of connectivity. *Biological Psychiatry*, 76(5): 397–404.

Formisano, E. and N. Kriegeskorte (2012). Seeing patterns through the hemodynamic veil--the future of pattern-information fMRI. *NeuroImage*, 62(2): 1249–1256.

Fuchs, L., D. Fuchs, C. Hamlet, S. Powell, A. Capizzi, and P. Seethaler (2006). The effects of computer-assisted instruction on number combination skill in at-risk first graders. *Journal of Learning Disabilities*, 39(5): 467–475.

Fuchs, L. S., D. Fuchs, and D. L. Compton (2013). Intervention effects for students with comorbid forms of learning disability: Understanding the needs of nonresponders. *Journal of Learning Disabilities*, 46(6): 534–548.

Fuchs, L. S., Fuchs, D., Compton, D. L., Bryant, J. D., Hamlett, C. L., and Seethaler, P. M. (2007). Mathematics screening and progress monitoring at first grade: Implications for responsiveness-to-intervention. *Exceptional Children*, 73(1): 311–330.

Fuchs, L. S., D. Fuchs, D. L. Compton, S. R. Powell, P. M. Seethaler, A. M. Capizzi, C. Schatschneider, and J. M. Fletcher (2006). The cognitive correlates of third-grade skill in arithmetic, algorithmic computation, and arithmetic word problems. *Journal of Educational Psychology*, 98(1): 29–43.

Fuchs, L. S., Fuchs, D., and Luther, K.H. (2007). Extending responsiveness-to-intervention to mathematics at first and third grades. *Learning Disabilities Research and Practice*, 22(13–24).

Fuchs, L. S., Fuchs, D., Prentice, K., Hamlett, C.L., Finelli, R., and Courey, S.J. (2004). Enhancing mathematical problem solving among third-grade students with schema-based instruction. *Journal of Educational Psychology*, 96: 635–647.

Fuchs, L. S., Fuchs, D., Yazdian, L., and Powell, S.R. (2002). Enhancing first-grade children's mathematical development with peer-assisted learning strategies. *School Psychology Review*, *31*: 569–584.

Fuchs, L. S., S. R. Powell, C. L. Hamlett, D. Fuchs, P. T. Cirino, and J. M. Fletcher (2008). Remediating computational deficits at third grade: A randomized field trial. *Journal of Research on Educational Effectiveness*, *1*(1): 2–32.

Fuchs, L. S., S. R. Powell, P. M. Seethaler, P. T. Cirino, J. M. Fletcher, D. Fuchs, C. L. Hamlett, and R. O. Zumeta (2009). Remediating number combination and word problem deficits among students with mathematics difficulties: A randomized control trial. *Journal of Educational Psychology*, *101*(3): 561–576.

Fung, W. and H. L. Swanson (2017). Working memory components that predict word problem solving: Is it merely a function of reading, calculation, and fluid intelligence? *Memory & Cognition*, *45*(5): 804–823.

Garnett, K. and J. E. Fleischner (1983). Automatization and basic fact performance of normal and learning disabled children. *Learning Disability Quarterly*, *6*(2): 223–230.

Geary, D. C. (1990). A componential analysis of an early learning deficit in mathematics. *Journal of Experimental Child Psychology*, *49*(3): 363–383.

Geary, D. C. (1993). Mathematical disabilities: Cognitive, neuropsychological, and genetic components. *Psychological Bulletin*, *114*(2): 345–362.

Geary, D. C. (1994). *Children's Mathematical Development: Research and Practical Applications*. Washington, DC: American Psychological Association.

Geary, D. C. (2004). Mathematics and Learning Disabilities. *Journal of Learning Disabilities*, *37*(1): 4–15.

Geary, D. C. (2011). Cognitive predictors of achievement growth in mathematics: A 5-year longitudinal study. *Developmental Psychology*, *47*(6): 1539–1552.

Geary, D. C., C. C. Bow-Thomas, and Y. Yao (1992). Counting knowledge and skill in cognitive addition: A comparison of normal and mathematically disabled children. *Journal of Experimental Child Psychology*, *54*(3): 372–391.

Geary, D. C. and S. C. Brown (1991). Cognitive addition: Strategy choice and speed-of-processing differences in gifted, normal, and mathematically disabled children. *Developmental Psychology*, *27*(3): 398–406.

Geary, D. C., C. O. Hamson, and M. K. Hoard (2000). Numerical and arithmetical cognition: A longitudinal study of process and concept deficits in children with learning disability. *Journal of Experimental Child Psychology*, *77*(3): 236–263.

Geary, D. C., M. K. Hoard, J. Byrd-Craven, L. Nugent and C. Numtee (2007). Cognitive mechanisms underlying achievement deficits in children with mathematical learning disability. *Child Development*, *78*(4): 1343–1359.

Geary, D. C., M. K. Hoard, L. Nugent, and D. H. Bailey (2012). Mathematical cognition deficits in children with learning disabilities and persistent low achievement: A five-year prospective study. *Journal of Educational Psychology*, *104*(1): 206–223.

Gilmore, C., N. Attridge, S. Clayton, L. Cragg, S. Johnson, N. Marlow, V. Simms, and M. Inglis (2013). Individual differences in inhibitory control, not non-verbal number acuity, correlate with mathematics achievement. *PLoS One*, *8*(6): e67374.

Gilmore, C. K., S. E. McCarthy, and E. S. Spelke (2007). Symbolic arithmetic knowledge without instruction. *Nature*, *447*(7144): 589–591.

Gracia-Bafalluy, M. and M.-P. Noël (2008). Does finger training increase young children's numerical performance? *Cortex*, *44*(4): 368–375.

Griffin, S. (2007). Early intervention for children at risk of developing mathematical learning difficulties. In D. B. Berch and M. M. Mazzocco (Eds.) *Why is math so hard for some children?* 83–105. Baltimore, MD, Paul H. Brookes Publishing Co.

Gross-Tsur, V., O. Manor and R. S. Shalev (1996). Developmental dyscalculia: Prevalence and demographic features. *Developmental Medical Child Neurology*, 38(1): 25–33.

Halberda, J., M. M. Mazzocco, and L. Feigenson (2008). Individual differences in non-verbal number acuity correlate with maths achievement. *Nature*, 455(7213): 665–668.

Hanich, L. B., N. C. Jordan, D. Kaplan, and J. Dick (2001). Performance across different areas of mathematical cognition in children with learning difficulties. *Journal of Educational Psychology*, 93(3): 615–626.

Henik, A. and W. Fias (2018). *Heterogeneity of Function in Numerical Cognition*. Cambridge, MA: Academic Press.

Holloway, I. D. and D. Ansari (2009). Mapping numerical magnitudes onto symbols: The numerical distance effect and individual differences in children's mathematics achievement. *Journal of Experimental Child Psychology*, 103(1): 17–29.

Hyde, D. C., D. A. Boas, C. Blair, and S. Carey (2010). Near-infrared spectroscopy shows right parietal specialization for number in pre-verbal infants. *NeuroImage*, 53(2): 647–652.

Iuculano, T., M. Rosenberg-Lee, J. Richardson, C. Tenison, L. Fuchs, K. Supekar, and V. Menon (2015). Cognitive tutoring induces widespread neuroplasticity and remediates brain function in children with mathematical learning disabilities. *Nature Communications*, 6(1): 8453.

Izard, V., G. Dehaene-Lambertz, and S. Dehaene (2008). Distinct cerebral pathways for object identity and number in human infants. *PLoS Biol*, 6(2): e11.

Izard, V., C. Sann, E. S. Spelke, and A. Streri (2009). Newborn infants perceive abstract numbers. *Proceedings of the National Academy of Sciences*, 106(25): 10382–10385.

Jolles, D., S. Ashkenazi, J. Kochalka, T. Evans, J. Richardson, M. Rosenberg-Lee, H. Zhao, K. Supekar, T. Chen, and V. Menon (2016). Parietal hyper-connectivity, aberrant brain organization, and circuit-based biomarkers in children with mathematical disabilities. *Developmental Science*, 19(4): 613–631.

Jolles, D., K. Supekar, J. Richardson, C. Tenison, S. Ashkenazi, M. Rosenberg-Lee, L. Fuchs, and V. Menon (2016). Reconfiguration of parietal circuits with cognitive tutoring in elementary school children. *Cerebral Cortex*, 83: 231–245.

Jolles, D., D. Wassermann, R. Chokhani, J. Richardson, C. Tenison, R. Bammer, L. Fuchs, K. Supekar, and V. Menon (2016). Plasticity of left perisylvian white-matter tracts is associated with individual differences in math learning. *Brain Structure and Function*, 221(3): 1337–1351.

Jordan, N. C., L. B. Hanich, and D. Kaplan (2003). Arithmetic fact mastery in young children: A longitudinal investigation. *Journal of Experimental Child Psychology*, 85(2): 103–119.

Jordan, N. C., S. C. Levine, and J. Huttenlocher (1995). Calculation abilities in young children with different patterns of cognitive functioning. *Journal of Learning Disability*, 28(1): 53–64.

Jordan, N. C. and T. O. Montani (1997). Cognitive arithmetic and problem solving: A comparison of children with specific and general mathematics difficulties. *Journal of Learning Disabilities*, 30(6): 624–634.

Kaufmann, L., M. M. Mazzocco, A. Dowker, M. von Aster, S. M. Gobel, R. H. Grabner, A. Henik, N. C. Jordan, A. D. Karmiloff-Smith, K. Kucian, O. Rubinsten, D. Szucs, R. Shalev, and H. C. Nuerk (2013). Dyscalculia from a developmental and differential perspective. *Frontiers in Psychology*, 4: 516.

Kaufmann, L., S. E. Vogel, M. Starke, C. Kremser, M. Schocke, and G. Wood (2009). Developmental dyscalculia: Compensatory mechanisms in left intraparietal regions in response to nonsymbolic magnitudes. *Behavior Brain Function*, 5: 35.

Kinsbourne, M. and E. K. Warrington (1963). The developmental Gerstmann syndrome. *Archives of Neurology*, 8(5): 490–501.

Knootz, K. L. and D. B. Berch (1996). Identifying simple numerical stimuli: Processing inefficiencies exhibited by arithmetic learning disabled children. *Mathematical Cognition*, 2(1): 23.

Knopik, V. S., M. Alarcon, and J. C. DeFries (1997). Comorbidity of mathematics and reading deficits: Evidence for a genetic etiology. *Behavioral Genetics*, 27(5): 447–453.

Kolkman, M. E., E. H. Kroesbergen, and P. P. M. Leseman (2013). Early numerical development and the role of non-symbolic and symbolic skills. *Learning and Instruction*, 25: 95–103.

Koyama, M. S., A. Di Martino, X. N. Zuo, C. Kelly, M. Mennes, D. R. Jutagir, F. X. Castellanos, and M. P. Milham (2011). Resting-state functional connectivity indexes reading competence in children and adults. *Journal of Neuroscience*, 31(23): 8617–8624.

Kucian, K., S. S. Ashkenazi, J. Hanggi, S. Rotzer, L. Jancke, E. Martin, and M. von Aster (2014). Developmental dyscalculia: A dysconnection syndrome? *Brain Structure and Function*, 219(5): 1721–1733.

Kucian, K., U. Grond, S. Rotzer, B. Henzi, C. Schonmann, F. Plangger, M. Galli, E. Martin, and M. von Aster (2011). Mental number line training in children with developmental dyscalculia. *Neuroimage*, 57(3): 782–795.

Kucian, K., T. Loenneker, T. Dietrich, M. Dosch, E. Martin, and M. von Aster (2006). Impaired neural networks for approximate calculation in dyscalculic children: A functional MRI study. *Behavioural Brain Functioning*, 2: 31.

Kucian, K., T. Loenneker, E. Martin, and M. von Aster (2011). Non-symbolic numerical distance effect in children with and without developmental dyscalculia: A parametric fMRI study. *Developmental Neuropsychology*, 36(6): 741–762.

Kucian, K., U. McCaskey, R. O'Gorman Tuura, and M. von Aster (2018). Neurostructural correlate of math anxiety in the brain of children. *Translational Psychiatry*, 8(1): 273.

Landerl, K., A. Bevan, and B. Butterworth (2004). Developmental dyscalculia and basic numerical capacities: A study of 8-9-year-old students. *Cognition*, 93(2): 99–125.

Lewis, C., G. J. Hitch, and P. Walker (1994). The prevalence of specific arithmetic difficulties and specific reading difficulties in 9- to 10-year-old boys and girls. *Journal of Child Psychology and Psychiatry*, 35(2): 283–292.

Li, Y., Y. Hu, Y. Wang, J. Weng, and F. Chen (2013). Individual structural differences in left inferior parietal area are associated with schoolchildrens' arithmetic scores. *Frontiers in Human Neuroscience*, 7: e844.

Libertus, M. E., E. M. Brannon, and M. G. Woldorff (2011). Parallels in stimulus-driven oscillatory brain responses to numerosity changes in adults and seven-month-old infants. *Developmental Neuropsychology*, 36(6): 651–667.

Lyons, I. M. and S. L. Beilock (2012a). Mathematics anxiety: Separating the math from the anxiety. *Cerebral Cortex*, 22(9): 2102–2110.

Lyons, I. M. and S. L. Beilock (2012b). When math hurts: Math anxiety predicts pain network activation in anticipation of doing math. *PLoS One*, 7(10): e48076.

Lyons, I. M., G. R. Price, A. Vaessen, L. Blomert, and D. Ansari (2014). Numerical predictors of arithmetic success in grades 1–6. *Developmental Science*, 17(5): 714–726.

Mazzocco, M. M., L. Feigenson, and J. Halberda (2011a). Impaired acuity of the approximate number system underlies mathematical learning disability (dyscalculia). *Child Development*, 82(4): 1224–1237.

Mazzocco, M. M., L. Feigenson, and J. Halberda (2011b). Preschoolers' precision of the approximate number system predicts later school mathematics performance. *PLoS One*, 6(9): e23749.

Mazzocco, M. M. M. and G. F. Myers (2003). Complexities in identifying and defining mathematics learning disability in the primary school-age years. *Annals of Dyslexia*, 53(1): 218–253.

McCaskey, U., M. von Aster, U. Maurer, E. Martin, R. O'Gorman Tuura, and K. Kucian (2018). Longitudinal brain development of numerical skills in typically developing children and children with developmental dyscalculia. *Frontiers in Human Neuroscience*, 11: 629.

McCaskey, U., M. von Aster, R. O'Gorman Tuura, and K. Kucian (2017). Adolescents with developmental dyscalculia do not have a generalized magnitude deficit – processing of discrete and continuous magnitudes. *Frontiers in Human Neuroscience*, 11: 102.

Menon, V. (2016). Memory and cognitive control circuits in mathematical cognition and learning. In Cappelletti M. and Fias W. (Eds.), The Mathematical Brain Across the Lifespan, Vol. 227, Chapter 7, 159–186, *Progress in Brain Research*. Amsterdam: Elsevier Ltd.

Metcalfe, A. W. S., M. Rosenberg-Lee, S. Ashkenazi, and V. Menon (2013). Fractionating the neural correlates of individual working memory components underlying problem solving skills in young children. *Developmental Cognitive Neuroscience*, 6: 162–175.

Michels, L., R. O'Gorman, and K. Kucian (2018). Functional hyperconnectivity vanishes in children with developmental dyscalculia after numerical intervention. *Developmental Cognitive Neuroscience*, 30: 291–303.

Morsanyi, K., A. Devine, A. Nobes, and D. Szűcs (2013). The link between logic, mathematics and imagination: Evidence from children with developmental dyscalculia and mathematically gifted children. *Developmental Science*, 16(4): 542–553.

Moyer, R. S. and T. K. Landauer (1967). Time required for judgements of numerical inequality. *Nature*, 215(5109): 1519.

Mundy, E. and C. K. Gilmore (2009). Children's mapping between symbolic and nonsymbolic representations of number. *Journal of Experimental Child Psychology*, 103(4): 490–502.

Murphy, M. M., M. M. M. Mazzocco, L. B. Hanich, and M. C. Early (2007). Cognitive characteristics of children with mathematics learning disability (MLD) vary as a function of the cutoff criterion used to define MLD. *Journal of Learning Disabilities*, 40(5): 458–478.

Mussolin, C., A. De Volder, C. Grandin, X. Schlogel, M. C. Nassogne, and M. P. Noel (2010). Neural correlates of symbolic number comparison in developmental dyscalculia. *Journal Cognitive Neuroscience*, 22(5): 860–874.

Mussolin, C., S. Mejias, and M. P. Noel (2010). Symbolic and nonsymbolic number comparison in children with and without dyscalculia. *Cognition*, 115(1): 10–25.

Noël, M.-P. and L. Rousselle (2011). Developmental changes in the profiles of dyscalculia: An explanation based on a double exact-and-approximate number representation model. *Frontiers in Human Neuroscience*, 5: 165.

Okolo, C. M. (1992). The effect of computer-assisted instruction format and initial attitude on the arithmetic facts proficiency and continuing motivation of students with learning disabilities. *Exceptionality*, 3: 195–211.

Ostad, S. A. (1997). Developmental differences in addition strategies: A comparison of mathematically disabled and mathematically normal children. *British Journal of Educational Psychology*, 67(3): 345–357.

Ostad, S. A. (1998). Developmental differences in solving simple arithmetic word problems and simple number-fact problems: A comparison of mathematically normal and mathematically disabled children. *Mathematical Cognition*, 4(1): 1–19.

Ostad, S. A. (1999). Developmental progression of subtraction strategies: A comparison of mathematically normal and mathematically disabled children. *European Journal of Special Needs Education*, 14: 21–36.

Ostad, S. A. (2000). Cognitive subtraction in a developmental perspective: Accuracy, speed-of-processing and strategy-use differences in normal and mathematically disabled children. *Focus on Learning Problems in Mathematics*, 22: 18–31.

Parsons, S. and J. Bynner (1997). Numeracy and employment. *Education and Training*, 39(2): 43–51.

Parsons, S. and J. Bynner (2005). *Does numeracy matter more?* London: National Research and Development Centre for Adult Literacy and Numeracy.

Piazza, M., A. Facoetti, A. N. Trussardi, I. Berteletti, S. Conte, D. Lucangeli, S. Dehaene, and M. Zorzi (2010). Developmental trajectory of number acuity reveals a severe impairment in developmental dyscalculia. *Cognition*, 116(1): 33–41.

Pinel, P., S. Dehaene, D. Rivière, and D. LeBihan (2001). Modulation of parietal activation by semantic distance in a number comparison task. *NeuroImage*, 14(5): 1013–1026.

Powell, S. R., P. T. Cirino, and A. S. Malone (2017). Child-level predictors of responsiveness to evidence-based mathematics intervention. *Exceptional Children*, 83(4): 359–377.

Powell, S. R., L. S. Fuchs, D. Fuchs, P. T. Cirino, and J. M. Fletcher (2009). Effects of fact retrieval tutoring on third-grade students with math difficulties with and without reading difficulties. *Learning Disabilities Research and Practice*, 24(1): 1–11.

Price, G. and D. Ansari (2013). Dyscalculia: Characteristics, causes, and treatments. *Numeracy*, 6(1).

Price, G. R., I. Holloway, P. Rasanen, M. Vesterinen, and D. Ansari (2007). Impaired parietal magnitude processing in developmental dyscalculia. *Current Biology*, 17(24): R1042–R1043.

Ranpura, A., E. Isaacs, C. Edmonds, M. Rogers, J. Lanigan, A. Singhal, J. Clayden, C. Clark and B. Butterworth (2013). Developmental trajectories of grey and white matter in dyscalculia. *Trends in Neuroscience and Education*, 2(2): 56–64.

Rauscher, L., J. Kohn, T. Käser, V. Mayer, K. Kucian, U. McCaskey, G. Esser, and M. von Aster (2016). Evaluation of a computer-based training program for enhancing arithmetic skills and spatial number representation in primary school children. *Frontiers in Psychology*, 7: 913.

Richardson, F. C. and R. M. Suinn (1972). The mathematics anxiety rating scale: Psychometric data. *Journal of Counseling Psychology*, 19(6): 551–554.

Rosenberg-Lee, M., S. Ashkenazi, T. Chen, C. B. Young, D. C. Geary, and V. Menon (2015). Brain hyper-connectivity and operation-specific deficits during arithmetic problem solving in children with developmental dyscalculia. *Developmental Science*, 18(3): 351–372.

Rotzer, S., K. Kucian, E. Martin, M. v. Aster, P. Klaver, and T. Loenneker (2008). Optimized voxel-based morphometry in children with developmental dyscalculia. *NeuroImage*, 39(1): 417–422.

Rotzer, S., T. Loenneker, K. Kucian, E. Martin, P. Klaver, and M. von Aster (2009). Dysfunctional neural network of spatial working memory contributes to developmental dyscalculia. *Neuropsychologia*, 47(13): 2859–2865.

Rousselle, L. and M. P. Noel (2007). Basic numerical skills in children with mathematics learning disabilities: a comparison of symbolic vs non-symbolic number magnitude processing. *Cognition*, 102(3): 361–395.

Rubinsten, O. and A. Henik (2005). Automatic activation of internal magnitudes: A study of developmental dyscalculia. *Neuropsychology*, 19(5): 641–648.

Rubinsten, O. and A. Henik (2009). Developmental dyscalculia: Heterogeneity might not mean different mechanisms. *Trends in Cognitive Sciences*, 13(2): 92–99.

Rykhlevskaia, E., L. Q. Uddin, L. Kondos, and V. Menon (2009). Neuroanatomical correlates of developmental dyscalculia: Combined evidence from morphometry and tractography. *Frontiers in Human Neuroscience*, 3: 51.

Schleifer, P. and K. Landerl (2011). Subitizing and counting in typical and atypical development. *Developmental Science* 14(2): 280–291.

Schwartz, F., J. Epinat-Duclos, J. Léone, A. Poisson, and J. Prado (2018). Impaired neural processing of transitive relations in children with math learning difficulty. *NeuroImage: Clinical*, 20: 1255–1265.

Shalev, R. S., J. Auerbach, O. Manor, and V. Gross-Tsur (2000). Developmental dyscalculia: prevalence and prognosis. *European Child Adolescent Psychiatry*, 9(Suppl 2): II58–II64.

Shalev, R. S., O. Manor, and V. Gross-Tsur (2005). Developmental dyscalculia: a prospective six-year follow-up. *Developmental Medicine & Child Neurology*, 47(2): 121–125.

Siegler, R. S. (1996). *Emerging Minds: The Process of Change in Children's Thinking.* New York: Oxford University Press.

Starkey, P. and R. G. Cooper, Jr. (1980). Perception of numbers by human infants. *Science*, 210(4473): 1033–1035.

Starkey, P., E. S. Spelke and R. Gelman (1990). Numerical abstraction by human infants. *Cognition*, 36(2): 97–127.

Supekar, K., A. G. Swigart, C. Tenison, D. D. Jolles, M. Rosenberg-Lee, L. Fuchs, and V. Menon (2013). Neural predictors of individual differences in response to math tutoring in primary-grade school children. *Proceedings of the National Academy of Sciences*, 110(20): 8230–8235.

Supekar, K., L. Q. Uddin, A. Khouzam, J. Phillips, W. D. Gaillard, L. E. Kenworthy, B. E. Yerys, C. J. Vaidya, and V. Menon (2013). Brain hyperconnectivity in children with autism and its links to social deficits. *Cell Reports*, 5(3): 738–747.

Svenson, O. and S. Broquist (1975). Strategies for solving simple addition problems: A comparison of normal and subnormal children. *Scandinavian Journal of Psychology*, 16(2): 143–148.

Swanson, H. L., C. B. Howard, and L. Saez (2006). Do different components of working memory underlie different subgroups of reading disabilities? *Journal of Learning Disability*, 39(3): 252–269.

Szucs, D., A. Devine, F. Soltesz, A. Nobes, and F. Gabriel (2013). Developmental dyscalculia is related to visuo-spatial memory and inhibition impairment. *Cortex*, 49(10): 2674–2688.

Temple, C. M. (1991). Procedural dyscalculia and number fact dyscalculia: Double dissociation in developmental dyscalculia. *Cognitive Neuropsychology*, 8(2): 155–176.

Tsang, J. M., R. F. Dougherty, G. K. Deutsch, B. A. Wandell, and M. Ben-Shachar (2009). Frontoparietal white matter diffusion properties predict mental arithmetic skills in children. *Proceedings of the National Academy of Sciences of the United States of America*, 106(52): 22546–22551.

Uddin, L. Q., K. Supekar and V. Menon (2013). Reconceptualizing functional brain connectivity in autism from a developmental perspective. *Frontiers in Human Neuroscience*, 7: 458.

Uttal, D. H., N. G. Meadow, E. Tipton, L. L. Hand, A. R. Alden, C. Warren and N. S. Newcombe (2013). The malleability of spatial skills: A meta-analysis of training studies. *Psychological Bulletin*, 139(2): 352–402.

van Loosbroek, E., and Smitsman, A. (1990). Visual perception of numerosity in infancy. *Developmental Psychology*, 26(6): 916–922.

von Aster, M., M. Weinhold Zulauf, and R. Horn (2006). *ZAREKI-R: Die neuropsychologische Testbatterie für Zahlenverarbeitung und Rechnen bei Kindern*, revidierte Version. Frankfurt: Hartcourt.

von Aster, M. G. and R. S. Shalev (2007). Number development and developmental dyscalculia. *Developmental Medicine and Child Neurology*, 49(11): 868–873.

Wechsler, D. (1992). *Wechsler Individual Achievement Test*. San Antonio, TX: Psychological Corporation.

Wilkinson, G. S. and G. J. Robertson (2006). *Wide Range Achievement Test (WRAT4)*. Lutz, FL: Psychological Assessment Resources.

Willcutt, E. G., S. A. Petrill, S. Wu, R. Boada, J. C. Defries, R. K. Olson, and B. F. Pennington (2013). Comorbidity between reading disability and math disability: Concurrent psychopathology, functional impairment, and neuropsychological functioning. *Journal of Learning Disabilities*, 46(6): 500–516.

Wilson, A. J., S. G. Andrewes, H. Struthers, V. M. Rowe, R. Bogdanovic, and K. E. Waldie (2015). Dyscalculia and dyslexia in adults: Cognitive bases of comorbidity. *Learning and Individual Differences* 37: 118–132.

Wilson, A. J. and S. Dehaene (2007). Number sense and developmental dyscalculia. In D. Coch, G. Dawson, and K. W. Fischer, *Human Behavior, Learning, and the Developing Brain: Atypical Development*, 212–238. New York, NY, US: Guilford Press.

Wilson, A. J., S. K. Revkin, D. Cohen, L. Cohen, and S. Dehaene (2006). An open trial assessment of "The Number Race", an adaptive computer game for remediation of dyscalculia. *Behavioral and Brain Functions* 2: 20.

Wilson, K. M. and H. L. Swanson (2001). Are mathematics disabilities due to a domain-general or a domain-specific working memory deficit? *Journal of Learning Disability*, 34(3): 237–248.

Woodcock, R. W. (1977). *Woodcock-Johnson Psycho-Educational Battery*. Technical Report.

Wynn, K. (1992). Addition and subtraction by human infants. *Nature*, 358(6389): 749.

Yarkoni, T., R. A. Poldrack, T. E. Nichols, D. C. Van Essen, and T. D. Wager (2011). Large-scale automated synthesis of human functional neuroimaging data. *Nature Methods*, 8: 665–670.

Young, C. B., S. S. Wu, and V. Menon (2012). The neurodevelopmental basis of math anxiety. *Psychology Sciences*, 23(5): 492–501.

CHAPTER 27

......

COGNITIVE NEUROSCIENCE OF DEVELOPMENTAL DYSLEXIA

......

ANNA A. MATEJKO, CAMERON C. MCKAY, SIKOYA M. ASHBURN, GABRIELLE-ANN TORRE, AND GUINEVERE F. EDEN

INTRODUCTION

LEARNING to read is a complex process that takes years to master. For example, in alphabetic languages, children learn the names of letters and letter-sound correspondences at the onset of formal instruction, allowing them to then go on to blend sounds together to decode new words. At the same time, they increasingly recognize frequently occurring parts of words and whole words by sight. While most children become fluent readers, between 5 and 10 percent of children struggle to learn to read due to the reading disability developmental dyslexia (referred to as "dyslexia" throughout this chapter). Children with dyslexia are characterized by poor reading accuracy and fluency, with their ability to decode words significantly below what would be expected based on age and opportunity (American Psychiatric Association 2013). They often never catch up to their typically developing peers, with the gap between poor readers and typical readers in first grade tending to persist into adolescence and early adulthood (Ferrer et al. 2015).

Because dyslexia has life-long social, personal, and professional consequences (Gerber 2012), it has been of great interest to better understand its causes. Behavioral research has characterized the cognitive manifestations of dyslexia, and neuroimaging and genetic research has probed the underlying biological bases. Much of this work has converged on an underlying phonological processing deficit as the cause of impaired reading in dyslexia (Snowling 2001; Vellutino et al. 2004). Phonological processing is "the ability to quickly and correctly hear, store, recall and make different speech sounds"

(National Center on Improving Literacy). The exact mechanism by which the phonological processing deficit occurs is still unclear, however, it is not the only cause of dyslexia, with other oral language skills and domain-general cognitive abilities also playing a significant role (McCardle et al. 2001). Further, there are likely to be multiple genetic and environmental origins of dyslexia (Peterson and Pennington 2012, 2015).

Brain imaging research has revealed the neural signature of reading in typical learners and how it differs in those with dyslexia. It has also contributed to theories on the etiology of dyslexia, characterization of subtypes, and shed light on how biological factors (genetics, sex) and environmental/experiential factors (socioeconomic status and language background) affect this reading disability. Brain imaging research has also provided some insights into the mechanisms of successful intervention and has advanced our understanding of comorbidity with other disabilities. In this chapter, we review these topics, after a review of the behavioral work on reading acquisition and reading disability, which provided the foundation for brain imaging studies.

BEHAVIORAL STUDIES OF TYPICAL AND ATYPICAL READING

Skilled readers are able to extract meaning from text by integrating many domain-specific skills, such as linguistic and print knowledge (Scarborough 2003), as well as domain-general skills such as attention and working memory (Cain et al. 2004; Franceschini et al. 2012; Peng and Fuchs 2014). This research has arrived at a consensus on the importance of several key skills for reading, including phonological awareness (PA), orthographic awareness, verbal short-term/working memory, and speeded lexical retrieval. PA is a phonological processing skill that refers to one's ability to recognize how words are made up of sounds, and to discriminate and manipulate those sounds (e.g., isolating the sounds of a word or blending sounds) (Anthony and Francis 2008; Scarborough and Brady 2002). Together with the ability to associate sounds with letters (alphabetic principle), PA facilitates the blending of sounds to successfully decode words. Not only is strong PA associated with better performance on reading measures (Swanson et al. 2003), but studies focused on improving children's PA through intervention have shown that PA is causally linked to reading ability (Ehri et al. 2001).

While PA is clearly a strong predictor of reading acquisition, other skills, such as orthographic awareness, also play an important role (Badian 1994; Richards et al. 2006). Specifically, letter/print knowledge and visual word form recognition are critical to becoming a fluent reader, because they allow children to directly map the orthography of familiar words onto their semantic representations and thereby also provide an avenue for reading irregular words that cannot be sounded out. Orthographic and phonological processing are used in tandem and training both skills improves reading more than training PA alone (Lee and Burkam 2002). While PA is important for beginning readers when they need to sound out unfamiliar words through decoding, orthographic processing has been found to be a stronger predictor later, when children/adolescents rely on sight-word reading (Badian 1995).

In addition to phonological and orthographic awareness, other skills are also important for and predictive of reading acquisition (Alloway and Alloway 2010; Norton and Wolf 2012). Verbal short-term memory and working memory are important to retain and manipulate information for text comprehension and semantic processing (Daneman and Carpenter 1980; Daneman and Merikle 1996), especially when children are learning to read (Baddeley 1979). Speeded lexical retrieval (i.e., rapid automatized naming of letters and numbers), which is the speed of access to the visual-verbal relationships necessary for reading (Denckla and Cutting 1999) is thought to be important for fluent reading (Badian 1995). These two skills, together with PA, are described under the umbrella term of phonological processing, because all three involve the retrieval of phonological codes. However, each of these skills makes a significant independent contribution to predicting later reading outcome in typically developing children (Wagner and Torgesen 1987).

The same measures that predict reading in general are found to be impaired in children/adolescents with dyslexia. Those with dyslexia often have poor phonological awareness (Peterson and Pennington 2015; Vellutino et al. 2004), orthographic awareness (Badian 2005), verbal short-term/working memory (Peng and Fuchs 2014), and speeded lexical retrieval (Denckla and Rudel 1976; Norton and Wolf 2012) in isolation or in combination. Different combinations of these deficits have led to the notion of subtypes of dyslexia. For example, a double deficit in phonological awareness and speeded lexical retrieval may lead to more severe reading problems and greater resistance to treatment (Denckla and Cutting 1999; Lovett et al. 2000; Ozernov-Palchik et al. 2017).

Together, this behavioral literature points to multiple cognitive deficits in dyslexia, with an emphasis on impaired phonological processing. It has also provided a framework by which to uncover functional and anatomical differences in dyslexia.

FUNCTIONAL AND ANATOMICAL BASES OF READING

Functional Neural Networks Underlying Typical Reading

Functional magnetic resonance imaging (fMRI) studies of children/adolescents and adults have identified a set of left-lateralized regions that are involved in reading (Figure 27.1), including the occipito-temporal, temporo-parietal, and inferior-frontal cortices (Martin et al. 2015; Sandak et al. 2012; Turkeltaub et al. 2003). Beginning with the back of the brain, the occipito-temporal cortex (OTC) is recruited for orthographic processing, with the visual word form area supporting direct lexical access (Dehaene and Cohen 2011; Glezer et al. 2016; Turkeltaub et al. 2008). These ventral regions utilize visual object-recognition areas to support fast, fluent, and automatic reading, especially in more proficient, older readers (Pugh et al. 2001). However, these regions may also be involved in some aspects of phonological processing (Dietz et al. 2005; Price and Devlin 2011; Twomey et al. 2011). The temporo-parietal cortex (TPC), including the supramarginal, angular, and superior temporal gyri, is activated during phonological processing and cross-modal integration (Glezer et al. 2016; Pugh et al. 2013; Simos et al. 2002; Turkeltaub et al. 2003; Weiss et al. 2018). These dorsal regions utilize language areas for grapheme-phoneme mapping (print-to-sound correspondences) to decode unfamiliar words, especially in novice readers. The left inferior frontal cortex (IFC) is thought to facilitate articulation of the words (Pugh et al. 2001). However, its role is more complex, with IFC having been shown to be involved in phonological, semantic, and orthographic processing (Glezer et al. 2016; Poldrack et al. 1999; Sandak et al. 2004; Wu et al. 2012). The engagement of all of these regions changes over development, where dorsal regions are more engaged in younger readers as they learn to decode new words, and ventral regions serve more proficient readers who recognize more words by sight (Booth et al. 2001; Dehaene et al. 2010; Martin et al. 2015; Schlaggar et al. 2002). Yet, connections between these regions are important during reading acquisition (Wise Younger et al. 2017). Figure 27.1(a) illustrates these age-specific differences in meta-analyses of reading in children/adolescents and adults (Martin et al. 2015). The age of the "children" in the original studies ranged from six to fifteen years, with study-specific averages ranging from seven to fourteen years. Like other studies described in this chapter, we will use the term "children," but note that some of these studies do in fact include adolescents.

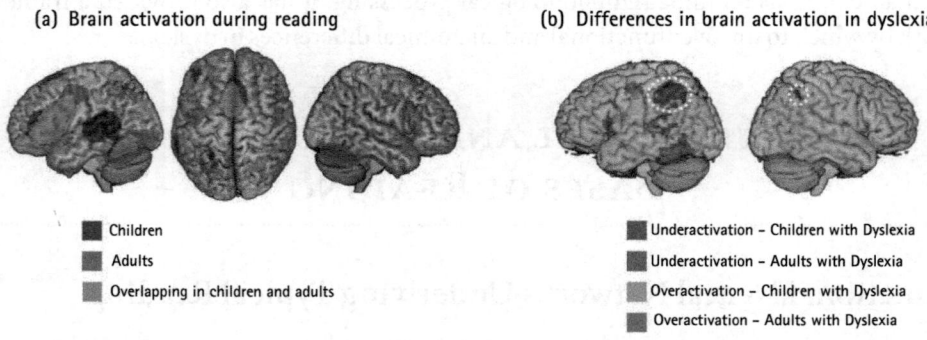

FIGURE 27.1 (a) Regions recruited for reading in typical children and adults (from Martin et al. 2015).
(b) Differences in brain function observed in studies examining children and adults with dyslexia compared to controls.

(Reproduced from Richlan et al. (2011) with permission from Elsevier.)

Functional Differences in Children with Dyslexia

Studies of dyslexia have overall found less activation of the left OTC, TPC, and IFC compared to controls across a wide range of reading tasks (real words, pseudowords, implicit, and explicit tasks) (Maisog et al. 2008; Paulesu et al. 2014; Richlan et al. 2009, 2011). One popular theory is that a primary deficit in phonological processing within the TPC leads to a secondary (later) impairment of sight-word reading in the OTC (Sandak et al. 2012). However, meta-analyses indicate that both adults and children with dyslexia under-activate the OTC (Richlan et al. 2011), suggesting that the deficit in the occipito-temporal cortex may be primary in nature (Kronbichler and Kronbichler 2018; Richlan 2012). Most studies find under-activation in the left IFC in dyslexia (Maisog et al. 2008; Paulesu et al. 2014; Richlan et al. 2009, 2011), indicating that all three regions known to be active during reading in typical readers are under-activated in those with dyslexia. There have also been findings of over-activation (e.g., in the left precentral cortex for children/adolescents and adults; Richlan et al. 2009, 2011). Over-activation has often been interpreted as a possible compensatory mechanism in dyslexic readers (Hancock et al. 2017; Shaywitz et al. 1998). More fine-grained techniques such as sensitivity gradients (e.g., Olulade et al. 2015) or multivoxel pattern analysis (Boets et al. 2013) continue to shed light on the role of each of these regions and how they differ in dyslexia.

Dyslexia is also characterized by poorer connectivity among these regions (Paulesu et al. 1996). Task-based and resting-state studies have found reduced functional connectivity of left occipito-temporal cortex with left frontal areas (Finn et al. 2014; Schurz et al. 2015), of bilateral auditory cortex with the left inferior frontal gyrus (IFG; Boets et al. 2013), and left intraparietal sulcus with left middle frontal gyrus (Koyama et al. 2013). Some consider abnormal connectivity to be the primary deficit in dyslexia (Boets et al. 2013), highlighting the importance of examining both functional connectivity and regional brain activity in understanding the etiology of dyslexia.

An important consideration is whether these differences in brain activity and functional connectivity predate the onset of reading and are therefore the cause of dyslexia, rather than the consequence of an impoverished reading experience. Several of the brain regions observed to differ in dyslexia relative to age-matched controls (OTC, TPC, IFC) are also found when comparing those with dyslexia to reading-matched controls, suggesting that these are not driven by a lack of reading experience (Hoeft et al. 2007). Also, studies of children with a family history of dyslexia compared to those with no family history, indicate under-activation in bilateral occipito-temporal and left temporo-parietal brain regions (Raschle et al. 2012), as well as the left inferior frontal cortex (Raschle et al. 2014), suggesting that early functional anomalies in these "at-risk" children pre-date the onset of reading. Interestingly, these studies indicate that both dorsal and ventral pathways may be affected in dyslexia prior to reading instruction. Research has begun to examine how brain function differs in children at risk for dyslexia who then go on to have dyslexia, versus those who are at risk, but ultimately do not develop dyslexia (Ozernov-Palchik and Gaab 2016). Regarding the latter, a recent multi-voxel pattern analysis study by Yu et al. (2020) found that children who have a family risk of dyslexia, but do not go on to develop the learning disability, have over-activation in the right IFC compared to controls during a phonological processing task and that patterns of brain activity in the right IFC were correlated with future reading achievement. This suggests that the engagement of contralateral right hemisphere regions may serve as a "protective" mechanism to shield against the reading disability that has affected other members of their family (Yu et al. 2020). Taken together the OTC, TPC, and IFC have been shown to be altered in dyslexia, likely causing the difficulties in learning to read.

Neuroanatomical Differences in Dyslexia

Decades of research have focused on the neuroanatomical bases of dyslexia and structural MRI (sMRI) studies continue to be heavily utilized to this day. In addition to providing various measures of brain anatomy, a significant advantage of anatomical scans is that they can be administered with relative ease in very young participants and do not involve the complexities involved in fMRI, such as designing child-friendly tasks. The use of sMRI in younger children provides critical insights into the early neurobiological manifestations of dyslexia, prior to the onset of reading, therefore disambiguating cause from consequence of poor reading.

Gray Matter

Meta-analyses have found lower grey matter volumes (GMV) in the left OTC, TPC, IFC as well cerebellum (Eckert et al. 2016; Linkersdörfer et al. 2012; Richlan et al. 2013). Some of these differences are further modulated by the profile of impairments (Jednoróg et al. 2014) and are unique to dyslexia rather than a function of being at the lower end of the distribution of reading abilities (Torre and Eden 2019). Notably, they are observed in children at risk for dyslexia prior to receiving literacy instructions (Raschle et al. 2011; Black et al. 2012; Hosseini et al. 2013), although longitudinal research is still required to examine GMV in those at risk for dyslexia who then go on to develop dyslexia.

Surface-based morphometry studies on dyslexia have also found less cortical thickness (CT) in the left OTC (fusiform gyrus) and right superior temporal gyrus compared to controls (Ma et al. 2015). Again, valuable insights are provided by studies conducted with young children identified to be at risk for dyslexia who go on to have dyslexia. Here there have been findings in bilateral OTC (fusiform gyrus), but not the TPC regions known to be involved in reading (Beelen et al. 2019). It has been proposed from some longitudinal studies that there are differences in surface-based morphometry that predate reading instruction in areas subserving auditory, visual, and executive functions, and that differences in regions involved in reading only emerge after reading instruction (Clark et al. 2014; Kuhl et al. 2020).

In sum, GMV differences in dyslexia are largely in the same regions that differ in activation during reading or phonological tasks, even in children who have not yet started to read. However, surface-based morphometry findings suggest that anatomical abnormalities in dyslexia are more dynamic, with some manifesting prior to, and others following reading instruction. Further studies are needed for both approaches in order to get a firm grasp of the developmental trajectory of gray matter anatomy in children with dyslexia, and as has been noted, results of these studies will be more compelling if they use more rigorous methodologies and larger sample sizes (Ramus et al. 2018).

White Matter

Diffusion tensor imaging (DTI) has revealed relationships between white matter pathways and individual differences in reading, as well as atypical white matter in children with dyslexia (Vandermosten, Boets, Wouters, et al. 2012). Five major white matter tracts that support reading are the inferior longitudinal fasciculus (ILF), inferior fronto-occipital fasciculus (IFO), arcuate fasciculus (AF), corona radiata (CR), and corpus callosum (CC; see Figure 27.2).

The ILF connects the occipito-temporal cortex to the lateral and anterior temporal regions (Catani and Thiebaut de Schotten 2008), and is thought to support orthographic processing. The IFO connects the occipito-temporal cortex with the inferior frontal cortex and is thought to support associations between orthography and semantics (Epelbaum et al. 2008; Vandermosten, Boets, Poelmans, et al. 2012; Yeatman et al. 2013). Several studies have found atypical white matter metrics in dyslexia in the ILF (Steinbrink et al. 2008; Su et al. 2018; Wang et al. 2016), while findings in the IFO are mixed, with some studies showing differences in children with reading difficulties, or with a family history of dyslexia (Odegard et al. 2009; Vandermosten et al. 2015), but others have not (Su et al. 2018; Vandermosten, Boets, Poelmans, et al. 2012).

The AF has projections from the superior temporal cortex and inferior parietal cortex to the insula and IFG (Catani and Thiebaut de Schotten 2008) and is part of the larger

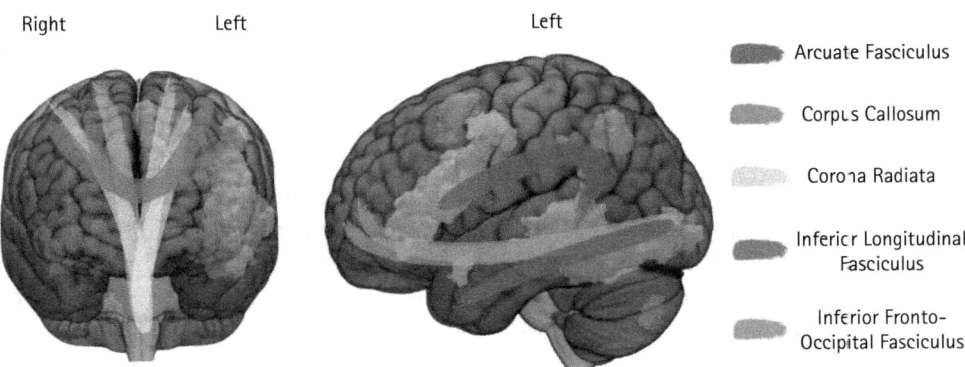

FIGURE 27.2 Schematic showing white matter tracts related to reading and dyslexia. Corpus callosum is not visualized in the sagittal view.

(Figure is adapted from Vandermosten, Boets, Wouters, et al. 2012.)

superior longitudinal fasciculus (Kamali et al. 2014). It is thought to be involved in language processing (Romeo et al. 2018; Torre et al. 2019) and not surprisingly, therefore, the AF has the largest body of evidence of all tracts demonstrating its role in reading ability and dyslexia. Specifically, white matter metrics of the AF are correlated with reading ability and phonological processing (Catani et al. 2005; Saygin et al. 2013; Yeatman et al. 2011), and differ in children with dyslexia (Hoeft et al. 2011; Vanderauwera et al. 2017; Vandermosten, Boets, Poelmans, et al. 2012; Wang et al. 2016). Pre-readers (Wang et al. 2016) and even infants (Langer et al. 2017) with a family history of dyslexia have been reported to have atypical white matter in the left AF compared to those with no family history (however, see Vandermosten et al. 2015). Longitudinal studies have confirmed that white matter anomalies in the left AF are present prior to reading instruction in those who go on to develop dyslexia (Van Der Auwera et al. 2021; Wang et al. 2016). Interestingly, similar to the right inferior frontal activation in children who are at risk for dyslexia, but do not develop the learning disability as described in the fMRI study above (Yu et al. 2020), the right superior longitudinal fasciculus may also play a "protective" role in reading outcomes for children with dyslexia, or for those who are at risk of developing dyslexia (Hoeft et al. 2011; Wang et al. 2016). This suggests that compensatory mechanisms in right hemisphere tracts may offset compromised white matter in the left hemisphere, allowing children who are at risk for dyslexia to develop typical reading skills.

The CR has ascending fibers from the thalamus to the cortex, and descending fibers from the fronto-parietal cortex to the brain stem and subcortical nuclei (Catani and Thiebaut de Schotten 2008), suggesting its role in reading could be related to perceptual and motor functions. Some (Beaulieu et al. 2005; Niogi et al. 2010) but not all studies (Yeatman et al. 2011) have found that individual differences in reading performance are correlated with CR microstructure. Mixed findings have also been reported in studies of children with dyslexia, where some find atypical white matter in the CR (Niogi and McCandliss 2006; Odegard et al. 2009) while others have not (Vandermosten, Boets, Poelmans, et al. 2012).

Finally, the CC is a commissural pathway that connects homologous regions of the left and right hemispheres. The CC may play a role in the lateralization of language- and reading-related processes, such that "stronger" white matter microstructure in the CC is paradoxically related to poorer reading (Dougherty et al. 2007; Odegard et al. 2009). Children with dyslexia have been shown to have stronger cross-hemispheric white-matter connections in the CC compared to controls (Frye et al. 2008), which could contribute to weaker left-hemisphere lateralization of language-related skills (Illingworth and Bishop 2009; Xu et al. 2015).

Together, studies of structural connectivity suggest impaired anatomical connections in those with dyslexia between posterior and frontal regions known to be engaged during reading, converging with some of the functional connectivity findings discussed above. However, for diffusion imaging studies the methodological details are quite variable as the field is evolving and the need for more studies with greater consistency is especially urgent (Ramus et al. 2018).

Contributions of Biological and Environmental Factors to the Neural Underpinnings of Dyslexia

The above-described observations in dyslexia are further shaped by biological and experiential influences, such as genetics, sex, socioeconomic status (SES), and language background. Next, we will briefly consider how some of these risk factors (Mascheretti et al. 2018; Thompson et al. 2015) have been addressed in neuroimaging studies of dyslexia.

Biological Factors

Genetics

Dyslexia is familial and heritable (Castles et al. 1999; DeFries et al. 1987; Gallagher et al. 2000; Gilger et al. 1991; Pennington 1991), with 30 to 50 percent of children with a family history of dyslexia exhibiting symptoms of poor reading (Gilger et al. 1991). Genetic risk is a powerful predictor of reading outcome, adding significantly to other predictive measures such as phonological processing skills (Pennington and Lefly 2001). While no one gene is causally linked to dyslexia (Galaburda et al. 2006), several candidate genes have been identified based on family studies, twin studies, and genetic linkage studies (Fisher and DeFries 2002; Gialluisi et al. 2020; Peterson and Pennington 2015). Neuroimaging research has implicated the candidate gene *KIAA0319* in abnormal brain activation responses to sound in the primary auditory cortex in children with dyslexia (Centanni et al. 2018). The candidate genes *DCDC2, DYX1C1, KIAA0319*, shown in animal work to partially control neuronal migration (Kere 2011), are related to the integrity of several white matter structures found to be affected in dyslexia (left superior longitudinal fasciculus and corpus callosum; Darki et al. 2012).

Sex

Males are at a much higher risk of having dyslexia than females. Once thought to be an artifact of referral bias to studies, this preponderance of males with dyslexia (Katusic et al. 2001; Liederman et al. 2005) has been confirmed in epidemiological studies of representative English-speaking populations (Flannery et al. 2000; Rutter et al. 2004). This raises the question of whether the brain-based anomalies associated with dyslexia are the same in males and females. One study found that dyslexia in males and females is not characterized by the same GMV abnormalities in left hemisphere language regions (Evans, Flowers, Napoliello, and Eden 2014), while another found that females with dyslexia drive the differences observed in brain anatomy (CT) and function in the visual-word form area (Altarelli et al. 2013). Since sex-specific differences in brain structure are found in the general population of children (Herting et al. 2015; Lenroot and Giedd

2010; Peper et al. 2009; Sowell et al. 2007; Witte et al. 2009), further research into sex-specific anatomical as well as functional differences in dyslexia is warranted (Krafnick and Evans 2019).

Environmental Factors

Socioeconomic Status

Because socioeconomic status (SES) determines quantity and quality of the home language environment (Hoff 2003) and school instruction (Burkam and Lee 2002), it is one of the most pervasive factors associated with reading outcome (Peterson and Pennington 2015). Childhood SES is measured by factors such as parental education, occupation, or household income. The role of SES on the neural bases of dyslexia has not yet been fully explored since participants are usually of higher SES (Monzalvo et al. 2012). However, the need for such studies is motivated by the finding that SES is related to brain anatomy in typically developing children (Noble et al. 2015). Following a reading intervention, children with dyslexia are more likely to make gains in reading and demonstrate changes in CT (in bilateral occipito-temporal and temporo-parietal regions) if they are from lower-SES backgrounds (or if they have more severe reading difficulties) (Christodoulou et al. 2017; Romeo et al. 2017). Inadequate instruction or little opportunity for reading create a situation where interventions in children with dyslexia become more impactful, leading to greater behavioral and brain-based changes in low-relative-to-high SES students. It has been suggested that environmental influences are stronger in children with dyslexia of low SES, while genetic influences are stronger in children of high SES (Friend et al. 2008). Together, these findings emphasize the importance of bringing about environmental changes (early enrichment and instruction) in order to raise reading levels among children with dyslexia, especially for those children from low SES. They also highlight the strong moderating factor of SES in neuroimaging studies of dyslexia and reading intervention.

Language

The behavioral research described above has been Anglo- and alphabet-centric, with ongoing efforts to expand this work to other languages and writing systems. From this it has been shown that there are weaker relationships between phonological awareness and reading in logographic (e.g., Chinese) compared to alphabetic languages (e.g., English; Goswami 2002), while orthographic awareness, semantics, and the establishment of motor programs may be more important for reading in logographic languages (Goswami 2002; Tan et al. 2005). Neuroimaging studies of typical reading demonstrate some differences attributed to different scripts (Bolger et al. 2005), while generally showing universal patterns of brain activity across languages and writing systems (Rueckl et al. 2015). Likewise, there are findings for script-specific (Siok et al. 2008) as well as universal (Hu et al. 2010) manifestations of dyslexia (Richlan 2020).

Amongst alphabetic languages, the prevalence of dyslexia is influenced by a language's orthography. The incidence of dyslexia is lower in populations that have a more "shallow" orthography (direct grapheme-to-phoneme mapping) such as German, compared to more "deep" orthography (indirect grapheme-to-phoneme mapping) such as English, because the deep orthographies exacerbate slow and inaccurate reading (Bonifacci and Snowling 2008; Lallier et al. 2014; Landerl et al. 1997, 2013). Yet, functional neuroimaging studies of dyslexia demonstrate some consistent differences in activation for reading across alphabetic languages of different orthographic depth (Martin et al. 2016; Paulesu et al. 2001).

THE BRAIN-BASES OF TREATMENT OF DEVELOPMENTAL DYSLEXIA

Behavioral studies have provided evidence for the ability of language-based interventions to improve reading skills in children with dyslexia (Lee and Burkam 2002; Melby-Lervåg et al. 2012; Torgesen et al. 2001). The neural mechanisms underlying such successful interventions can be elucidated through neuroimaging.

A meta-analysis of studies using fMRI before and after a range of interventions found increased activation in the bilateral IFG, left thalamus, left middle occipital gyrus, and right posterior cingulate cortex (Barquero et al. 2014). When regions that differ in dyslexia (e.g., left IFG) change following an intervention, they tend to be characterized as "normalization," while changes in regions not associated with dyslexia (e.g., thalamus) are thought to represent "compensation," either because they involve domain-general processes (e.g., attention or memory; Barquero et al. 2014) or because they signify the adoption of reading-related processes by brain regions that are not usually utilized for that purpose (Eden et al. 2004). It has been suggested, however, that "normalization" is not likely to occur in brain regions that are the underlying cause of deficits in dyslexia (Xia et al. 2017). Changes in activation following reading intervention have been observed in a variety of age ranges, from pre-literate children (Yamada et al. 2011) to adults (Eden et al. 2004). Research has also examined whether changes in brain activity differ between children who respond well to an intervention compared to those that do not, finding "responders" have significantly greater activation in the left inferior parietal cortex (Odegard et al. 2008) and right fusiform gyrus (Nugiel et al. 2019) compared to "non-responders." However, not all studies report brain-based changes in dyslexia following their improved reading skills (Krafnick et al. 2022). This could be because the location of changes vary between individuals to the point where they cannot be captured by a group analysis; and for studies where they are present, variations in location between studies may result in little convergence in the literature. Indeed, a more recent meta-analysis concluded that brain-based changes following interventions in dyslexia do not actually converge across neuroimaging studies (Perdue et al. 2022). Compensatory changes are likely to be more variable in their location between participants than normalization changes, since they represent an alternative, lesser used path. As such, care has to be taken when using functional neuroimaging (and structural neuroimaging as discussed next) to arbitrate between mechanisms like normalization versus compensation, if one of these two mechanisms is less likely to be observed in group maps due to greater variability amongst participants. Lastly, specifics of the intervention programs also vary in terms of the skills they target, their efficacy, and their short- and long-term outcomes. These differences may be another reason for the lack of convergence across studies.

Studies of brain structure have also reported intervention-induced changes in regions associated with reading, such as GMV in the left fusiform (Krafnick et al. 2011) and CT in the left inferior/middle temporal and supramarginal gyri (Romeo et al. 2017). Changes in GMV in bilateral hippocampal regions (Krafnick et al. 2011) and CT in right hemisphere areas (Romeo et al. 2017) could be interpreted as compensation since they are outside of regions typically associated with reading. As noted above, increases in CT were found in "responders" relative to "non-responders," with non-responders representing about half the sample receiving the intervention (Romeo et al. 2017). As with brain activation, some anatomical differences in dyslexia remain unaltered by interventions (Ma et al. 2015), emphasizing the enduring challenges in treating this reading disability.

Several studies have also measured intervention-induced changes in white matter microstructure in children with dyslexia. Huber et al. (2018) found changes in the left ILF and AF, which as noted above, have often been found to be compromised in dyslexia. Changes in these tracts did not eliminate pre-intervention differences between controls and those with dyslexia, rather, the intervention made these differences larger. On the other hand, another study found intervention-induced changes in a tract near the CR, which was interpreted as normalization because white matter integrity in this region was lower in the group with dyslexia (as compared to the control group) prior to but not following the intervention (Keller and Just 2009).

In sum, successful intervention can result in changes to both brain function and structure, but the specific results are mixed; and there is not a one-to-one mapping of intervention-induced changes in activation and brain structure. Further, the mechanisms of these changes ("normalization" and "compensation") are poorly understood and there is no single mechanism at work. Future research needs to address these issues and should also strive to include long-term outcomes to gauge the brain mechanisms that support long-term success.

DEFICITS OUTSIDE OF LANGUAGE AND COGNITION

Dyslexia is commonly characterized as a language-based learning disability. However, there are reports of a range of sensory and motor deficits, some of which are considered in context of the language-based and cognitive deficits, and others separate from them. Two examples will be briefly discussed here.

Some behavioral studies of dyslexia have reported relatively poor motor sequence learning (Nicolson et al. 1999), balance (Nicolson and Fawcett 1990; Rochelle and Talcott 2006; Stoodley et al. 2005), handwriting (Hassid 1995; Holmes 1939), timing (Overy et al. 2003), automaticity (Nicolson and Fawcett 1990), and aberrant eye blink conditioning (Nicolson et al. 2002), giving rise to the cerebellar deficit theory (Nicolson et al. 2001). The cerebellar deficit hypothesis proposes deleterious effects on motor function in children with dyslexia (manifested in poor handwriting and balance), while faulty connections between the cerebellum and cortical frontal regions supporting articulatory processing lead to difficulties in phonological processing and reading (Nicolson et al. 2001). However, behavioral deficits associated with the cerebellum in dyslexia are observed far less often than phonological deficits (Ramus 2003). Functional brain imaging studies of motor tasks have found adults with dyslexia have atypical activation in the right cerebellum during motor sequence tasks (Menghini et al. 2006; Nicolson et al. 1999). However, the cerebellum is not active during reading in typically developing children (Martin et al. 2015). Meta-analyses (Maisog et al. 2008; Richlan et al. 2011) and a targeted study (Ashburn et al. 2020) comparing typical readers to those with dyslexia during word processing report no differences in cerebellar activity, while one meta-analysis found greater activity in the cerebellum in dyslexia (Linkersdörfer et al. 2012). Structural differences have been found in the cerebellum in dyslexia (Eckert et al. 2003; Pernet et al. 2009; Stoodley 2014), and have been reported in some (Eckert et al. 2016; Linkersdörfer et al. 2012), but not all (Richlan et al. 2013) meta-analyses. It should also be noted that avenues for intervention under this cerebellar hypothesis framework have not been successful (Rack et al. 2007). This, together with the mixed support for behavioral, functional and anatomical differences of the cerebellum in dyslexia means this theory is not widely accepted.

Turning to another theory, visual psychophysical studies indicate relatively poor performance in dyslexia when processing images with low spatial contrast and high temporal frequency (Lovegrove et al. 1980), as well as when engaging in visual motion perception (Cornelissen and Stein 1995). These processes are supported by the visual magnocellular system (Shapley 1990), giving rise to the magnocellular deficit theory of dyslexia (Stein 1997; Talcott et al. 1998). Neuroimaging studies have revealed less activation in area V5/MT (part of the magnocellular visual stream) during a visual motion task in those with dyslexia when compared to age-matched controls (Demb et al. 1998; Eden et al. 1996). However, these findings do not persist when participants are matched on reading level (rather than age), suggesting that the visual motion deficit is not causal to reading (Olulade et al. 2013). Further, activity in V5/MT increased in children following gains in reading, indicating that this part of the visual system may be modified by learning to read (Boets et al. 2011; Olulade et al. 2013).

These studies highlight how neuroimaging has contributed to distinguishing between the manifestations of dyslexia that are causal to the reading difficulties from those that are an epiphenomenon of the disorder, or those that are the consequence of less reading experience.

COMORBIDITY

Dyslexia frequently co-occurs with other learning difficulties (Margari et al. 2013) such as attention-deficit/hyperactivity disorder (ADHD) and dyscalculia (also referred to as maths disability), raising the question of whether they have shared etiologies. Brain imaging can shed important insights into this question, but the number of studies to date is limited because they require the inclusion of multiple, carefully defined groups (i.e., those with and without comorbidity), which comes with higher costs.

Attention-Deficit/Hyperactivity Disorder

ADHD is neurodevelopmental disorder characterized by pervasive symptoms of inattention and/or hyperactivity-impulsivity (American Psychiatric Association 2013). Neuroimaging studies of ADHD have found functional (Cortese et al. 2012; Norman et al. 2016) and structural (Frodl and Skokauskas 2012; McGrath and Stoodley 2019; Norman et al. 2016) differences in the basal ganglia and frontal cortex.

There is a high co-occurrence of dyslexia and ADHD, with comorbidity rates ranging from 9 to 44 percent (Capano et al. 2008; Semrud-Clikeman et al. 1992; Sexton et al. 2012; Willcutt et al. 2010; Willcutt and Pennington 2000). Several models have been proposed to explain this comorbidity, the most prominent of these being the cognitive subtype/independent disorders model, the phenocopy model, and the multiple deficit/common etiology model. The cognitive subtype model/independent disorders model argues that comorbid dyslexia and ADHD is a unique disorder that is distinct from either dyslexia or ADHD in isolation (Rucklidge and Tannock 2002). There is support for this model from behavioral work demonstrating that children with comorbid dyslexia and ADHD have greater impairments on tasks such as rapid naming and processing speed, than children with either dyslexia or ADHD alone (Nigg et al. 1998; Rucklidge and Tannock 2002; Willcutt et al. 2010). While there is no functional neuroimaging data in support of this model, there is evidence that those with comorbid dyslexia and ADHD have uniquely less CT in the left middle temporal gyrus (Langer et al. 2019).

The phenocopy model hypothesizes that one disorder is a secondary consequence of the other. For example, ADHD can distract a child from reading, causing a child with ADHD to appear to have reading impairments and hence be diagnosed with both disorders (Rabiner and Coie 2000; Rapport et al. 1999). Conversely, the frustrations that come with dyslexia can cause a child to appear inattentive and/or hyperactive (Pennington et al. 1993). This model would be supported by the observation that brain regions that reliably differ in ADHD or reliably differ in dyslexia are not both altered in children who have behavioral manifestations for both ADHD and dyslexia. Langer et al. (2019) found less activity (on both phonological processing and reading fluency tasks) and less CT in the left fusiform gyrus in the group with dyslexia, but not in the comorbid group, which may indicate that the problems with reading in the comorbid group stemmed from their ADHD (as left fusiform gyrus was intact). But at the same time, the authors found lower GMV in the basal ganglia in the ADHD group relative to controls as expected, but no such basal ganglia difference in the comorbid group, similarly suggesting the comorbid group's inattentive and/or hyperactive behaviors were driven by their reading disability (as their basal ganglia were intact).

Finally, the multiple deficit/common etiology model stipulates that there are multiple probabilistic risk factors for each disorder, and that comorbidity occurs in children who exhibit shared risk factors associated with dyslexia and ADHD (McGrath et al. 2011; Pennington 2006). This model is supported by shared genetic risk (genetic correlations between dyslexia and ADHD) and shared behavioral deficits in processing speed, working memory, inhibition, and sustained attention (McGrath et al. 2011; Pennington 2006). Neuroimaging evidence in support of this model would constitute overlapping differences in all three clinical groups (dyslexia, ADHD, and dyslexia combined with ADHD) relative to controls. Such evidence was reported in a multimodal imaging study that found less activity during a language task and less CT in the left supplementary motor area/anterior cingulate cortex (Langer et al. 2019). Shared differences in GMV across all three disorders have also been reported, this time in the right caudate nucleus and superior frontal gyrus (Jagger-Rickels et al. 2018), however, a meta-analysis of GMV in dyslexia and another of ADHD found no shared differences at all (McGrath and Stoodley 2019).

In sum, efforts to understand comorbidity of dyslexia and ADHD have not resulted in clear support for any one particular model and data from a single study provides support for different models.

Dyscalculia

Dyscalculia is a learning disability characterized by difficulties in processing numerical information, learning arithmetic facts, maths reasoning, and calculation (American Psychiatric Association 2013), as described by Menon et al. in this volume. Atypical brain function and structure have been found in the posterior parietal (including the intraparietal sulci), prefrontal, and occipito-temporal cortices in those with dyscalculia (Fias et al. 2013; Matejko and Ansari 2018; Peters and De Smedt 2018). While studies of structural and functional connectivity are mixed, they have revealed atypical connectivity between intraparietal, prefrontal, and medial or inferior temporal brain regions (Matejko and Ansari 2015; Moeller et al. 2015).

While dyslexia and dyscalculia are seemingly distinct disorders, dyslexia co-occurs with the dyscalculia in 30 to 70 percent of children (Landerl and Moll 2010; Lewis et al. 1994; Moll et al. 2018; Willcutt et al. 2013, 2019). Three models have been proposed by (Ashkenazi et al. 2013) to explain comorbidity between these disabilities: the additive, verbally mediated, and the domain-general model (see Figure 27.3). The additive model ascribes comorbidity to two independent and etiologically distinct learning disabilities that happen to co-occur; basic number processing deficits lead to their dyscalculia and phonological processing impairments lead to their dyslexia (Landerl et al. 2009). Yet, an additive account is in conflict with the rates of comorbidity between dyslexia and

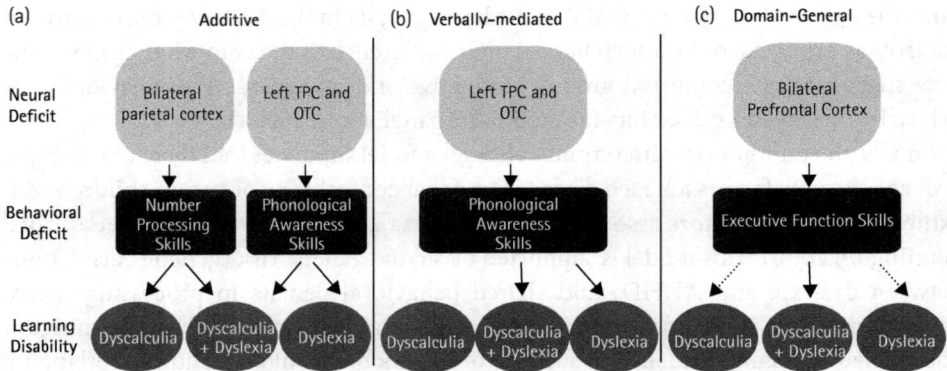

FIGURE 27.3 Theoretical models of comorbidity between dyslexia and dyscalculia.

(Adapted from Ashkenazi et al. 2013.)

dyscalculia, which are higher than would be expected by chance (Landerl and Moll 2010). Further, a multivariate neuroimaging study found that while brain activity during arithmetic could discriminate between controls and children with either dyscalculia, dyslexia, or combined dyslexia and dyscalculia, there were no differences amongst these three groups with learning disabilities (Peters et al. 2018). Similarly, anatomical studies in children (Skeide et al. 2018) and adults (Moreau et al. 2019) have found no differences in GMV or CT between groups with dyscalculia, dyslexia, and dyslexia combined with dyscalculia. These findings do not support the additive model.

The verbally mediated model proposes that compromised phonological processing also impairs maths skills. Evidence for this model is found in behavioral studies that report weaknesses in arithmetic fact retrieval (solving arithmetic problems from memory) in dyslexia (Boets and De Smedt 2010; De Smedt and Boets 2010), and in neuroimaging research reporting altered activity in left temporo-parietal regions during maths tasks in children with dyslexia (Evans, Flowers, Napoliello, Olulade, et al. 2014). The results from Peters et al. (2018) and Skeide et al. (2018) noted above do not support this model as dyscalculia would be expected to differ from dyslexia and from dyslexia with dyscalculia. However, of all models, this model will be especially dependent on the criteria used for determining dyscalculia and the measures employed to gauge math performance. This is because measures involving fact retrieval (but not procedural computation) are related to phonemic awareness in children with (Matejko et al. 2022) and without (De Smedt et al. 2010) learning disabilities, meaning that if dyscalculia is defined by low performance on fact retrieval, then dyslexia is more likely to co-occur.

The domain-general model posits that a weakness in domain-general skills impact both reading and maths, leading to comorbidity. This is supported by behavioral research reporting that children with combined reading and maths disability have poor domain-general skills (working memory and processing speed; Willcutt et al. 2013), which often exceed those found in children with dyslexia only or dyscalculia only (Peng and Fuchs 2014). This model is also supported by findings of atypical cortical surface complexity and resting-state connectivity within systems involved in memory encoding and retrieval (specifically, the right parahippocampal gyrus) in children with combined reading and maths impairments compared to typically developing children and to children with reading or math disability only (Skeide et al. 2018). This model is similar to the multiple deficit/common etiology model described in the prior section on ADHD (while the other two models are not analogous).

In sum, neuroimaging studies do not yield a unified account that converges on one model that explains comorbidity of dyslexia and dyscalculia. It is likely that more than these mechanism are at play and that there is also significant heterogeneity within those with dyscalculia (Peters and Ansari 2019).

Taken together, comorbidity with dyslexia is likely to arise in a variety of ways and neuroimaging studies have not been able to unequivocally point to one model over another for any kind of comorbidity. Future studies on comorbidity with dyslexia will need to be addressed with larger numbers of participants and take a dimensional individual differences approach to fully understand dyslexia and its co-occurrence with other disorders.

CONCLUSIONS AND FUTURE DIRECTIONS

Brain imaging studies characterizing typical reading acquisition, dyslexia, and intervention-related changes in dyslexia have contributed to our understanding of dyslexia. Brain imaging studies have revealed that dyslexia is associated with impairments at multiple levels, and is characterized by functional and structural impairments within the left OTC, TPC, and IFC, as well as atypical connections between these regions. The brain bases of dyslexia are related to factors such as genetics and sex, as well as socioeconomic status and language context, indicating that the neural underpinnings of dyslexia are modulated by both biological and environmental factors. As such, these factors will need to be taken into consideration during participant selection and characterization in future studies. Targeted reading intervention has been shown to lead to changes in brain function and structure, but consistency of results across studies is lacking. This could be due to the mechanistic aspects of brain changes following treatment (individual variability in the characteristics of plasticity) and/or differences in methodological approaches taken (interventions used, participant selection, etc.). Brain imaging research has supported the notion that dyslexia is attributed to a phonological deficit, together with deficits in the domains of orthographic processing and verbal short-term/working memory, while sensory or motor systems are less likely to be causally related to dyslexia. When considered in the context of other frequently co-occurring conditions (ADHD or dyscalculia), the complexities of dyslexia become augmented. Specifically, those with dyslexia and accompanying disorders do not necessarily differ in brain function or anatomy, raising the question of whether there are common mechanisms across these disorders, or whether these similarities could be related to arbitrary diagnostic boundaries, or heterogeneity. On the other hand, studies that do report findings of unique mechanisms for dyslexia with and without accompanying disorders do not converge.

Taken together, while there have been significant advances in the neurobiology of dyslexia, there still remain several important unresolved questions that should be addressed in future research. First, longitudinal work has begun to use brain imaging measures as a biomarker for later outcomes in dyslexia (Hoeft et al. 2011; Wang et al. 2016; Yu et al. 2020), and offers the opportunity to disambiguate causal differences from those due to less reading experience (Clark et al. 2014; Kuhl et al. 2020; Van Der Auwera et al. 2021; Wang et al. 2016). Some aspects of early brain function and structure appear to facilitate typical reading development in children who are at risk for dyslexia (due to a family history of dyslexia), suggesting they provide "protective" mechanisms (Wang et al. 2016; Yu et al. 2020). Therefore, longitudinal research is needed to not only help us understand the different developmental trajectories in reading acquisition, but also to decipher brain-based indicators of resiliency. Secondly, while research has begun to examine comorbidity in dyslexia, much more remains to be learned about the neural bases of dyslexia in the presence of conditions that affect learning in other domains. But this can only be achieved with larger samples, and can benefit from longitudinal or intervention designs. Finally, larger samples have been called for to also improve reliability of future research, together with careful consideration of participant selection and rigorous analyses methods (Ramus 2003). Against the body of existing research, these future directions will provide a more comprehensive understanding of the neurodevelopmental trajectories of dyslexia and, ultimately, inform better identification and treatment.

ABBREVIATIONS

ADHD	Attention deficit/hyperactivity disorder
AF	arcuate fasciculus
CC	corpus callosum
CR	corona radiata
CT	cortical thickness
DTI	diffusion tensor imaging
GMV	grey matter volume
IFC	inferior frontal cortex
IFG	inferior frontal gyrus
IFO	inferior fronto-occipital fasciculus
ILF	inferior longitudinal fasciculus
fMRI	functional magnetic resonance imaging
sMRI	structural magnetic resonance imaging
OTC	occipito-temporal cortex
PA	phonological awareness
SES	socioeconomic status
TPC	temporo-parietal cortex

REFERENCES

Alloway, T. P. and Alloway, R. G. (2010). Investigating the predictive roles of working memory and IQ in academic attainment. *Journal of Experimental Child Psychology*, 106(1), 20–29. doi. org/10.1016/j.jecp.2009.11.003

Altarelli, I. Monzalvo, K. Iannuzzi, S. Fluss, J. Billard, C. Ramus, F., et al. (2013). A functionally guided approach to the morphometry of occipitotemporal regions in developmental dyslexia: Evidence for differential effects in boys and girls. *Journal of Neuroscience*, 33(27), 11296–11301. doi.org/10.1523/JNEUROSCI.5854-12.2013

American Psychiatric Association (APA) (2013). *Diagnostic and Statistical Manual of Mental Disorders* (5th ed.)—DSM 5. Washington, DC: APA Publishing. doi.org/10.1176/appi. books.9780890425596.744053

Anthony, J. L. and Francis, D. J. (2008). Development of phonological awareness skills. *Current Directions in Psychological Science*, 14(5), 255–259. doi.org/10.1007/s10936-008-9085-z

Ashburn, S. M., Flowers, D. L., Napoliello, E. M., and Eden, G. F. (2020). Cerebellar function in children with and without dyslexia during single word processing. *Human Brain Mapping*, 41(1), 120–138. doi.org/10.1002/hbm.24792

Ashkenazi, S., Black, J. M., Abrams, D. A., Hoeft, F., and Menon, V. (2013). Neurobiological underpinnings of math and reading learning disabilities. *Journal of Learning Disabilities*, 46(6), 549–569. doi.org/10.1177/0022219413483174

Baddeley, A. D. (1979). Working memory and reading. In P. A. Kolers, M. E. Wrolstad, and H. Bouma (eds), *Processing of Visible Language*, Vol. 13, Nato Conference series, 355–370. Boston, MA: Springer.

Badian, N. A. (1994). Preschool prediction: Orthographic and phonological skills, and reading. *Annals of Dyslexia*, 44(1), 1–25. doi.org/10.1007/BF02648153

Badian, N. A. (1995). Predicting reading ability over the long term: The changing roles of letter naming, phonological awareness and orthographic processing. *Annals of Dyslexia*, 45, 79–96.

Badian, N. A. (2005). Does a visual-orthographic deficit contribute to reading disability? *Annals of Dyslexia*, 55(1). doi.org/10.1007/s11881-005-0003-x

Barquero, L. A. Davis, N., and Cutting, L. E. (2014). Neuroimaging of reading intervention: A systematic review and activation likelihood estimate meta-analysis. *PLoS ONE*, 9(1), e83668. doi.org/10.1371/journal.pone.0083668

Beaulieu, C., Plewes, C., Paulson, L. A., Roy, D., Snook, L., Concha, L., et al. (2005). Imaging brain connectivity in children with diverse reading ability. *NeuroImage*, 25(4), 1266–1271. doi.org/10.1016/j.neuroimage.2004.12.053

Beelen, C., Vanderauwera, J., Wouters, Vandermosten, M., and Ghesquière, P. (2019). Atypical gray matter in children with dyslexia before the onset of reading instruction. *Cortex*, 121, 399–413. doi.org/10.1016/j.cortex.2019.09.010

Black, J. M., Tanaka, H., Stanley, L., Nagamine, M., Zakerani, N., Thurston, A., et al. (2012). Maternal history of reading difficulty is associated with reduced language-related gray matter in beginning readers. *Neuroimage*, 59(3), 3021e3032. doi.org/10.1016/ j.neuroimage.2011.10.024

Boets, B. and De Smedt, B. (2010). Single-digit arithmetic in children with dyslexia. *Dyslexia*, 16(2), 183–191. doi.org/10.1002/dys.403

Boets, B., Op De Beeck, H. P., Vandermosten, M., Scott, S. K., Gillebert, C. R., Mantini, D., et al. (2013). Intact but less accessible phonetic representations in adults with dyslexia. *Science*, 342(6163), 1251–1254. doi.org/10.1126/science.1244333

Boets, B., Vandermosten, M., Cornelissen, P., Wouters, J., and Ghesquière, P. (2011). Coherent motion sensitivity and reading development in the transition from prereading to reading stage. *Child Development*, 82(3). doi.org/10.1111/j.1467-8624.2010.01527.x

Bolger, D. J., Perfetti, C. A., and Schneider, W. (2005). Cross-cultural effect on the brain revisited: Universal structures plus writing system variation. *Human Brain Mapping*, 25(1). doi.org/10.1002/hbm.20124

Bonifacci, P. and Snowling, M. J. (2008). Speed of processing and reading disability: A cross-linguistic investigation of dyslexia and borderline intellectual functioning. *Cognition*, 107(3), 999–1017. doi.org/10.1016/J.COGNITION.2007.12.006

Booth, J. R., Burman, D. D., Van Santen, F. W., Harasaki, Y., Gitelman, D. R., Parrish, T. B., et al. (2001). The development of specialized brain systems in reading and oral-language. *Child Neuropsychology*, 7(3). doi.org/10.1076/chin.7.3.119.8740

Burkam, D. T. and Lee, V. E. (2002). Inequality at the Starting Gate: Social Background Differences in Achievement as Children Begin School. Economic Policy Institute, Washington, DC.

Cain, K., Oakhill, J., and Bryant, P. (2004). Children's reading comprehension ability: Concurrent prediction by working memory, verbal ability, and component skills. *Journal of Educational Psychology*, 96(1), 31–42. doi.org/10.1037/0022-0663.96.1.31

Capano, L., Minden, D., Chen, S. X., Schachar, R. J., and Ickowicz, A. (2008). Mathematical learning disorder in school-age children with attention-deficit hyperactivity disorder. *Canadian Journal of Psychiatry*, 53(6), 392–399. doi.org/10.1177/070674370805300609

Castles, A., Datta, H., Gayan, J., and Olson, R. K. (1999). Varieties of developmental reading disorder: Genetic and environmental influences. *Journal of Experimental Child Psychology*, 72(2). doi.org/10.1006/jecp.1998.2482

Catani, M., Jones, D. K., and Ffytche, D. H. (2005). Perisylvian language networks of the human brain. *Annals of Neurology*, 57(1), 8–16.

Catani, M. and Thiebaut de Schotten, M. (2008). A diffusion tensor imaging tractography atlas for virtual in vivo dissections. *Cortex*, 44(8), 1105–1132. doi.org/10.1016/j.cortex.2008.05.004

Centanni, T. M., Pantazis, D., Truong, D. T., Gruen, J. R., Gabrieli, J. D. E., and Hogan, T. P. (2018). Increased variability of stimulus-driven cortical responses is associated with genetic variability in children with and without dyslexia. *Developmental Cognitive Neuroscience*, 34, 7–17. doi.org/10.1016/j.dcn.2018.05.008

Christodoulou, J. A., Cyr, A., Murtagh, J., Chang, P., Lin, J., Guarino, A. J. et al. (2017). Impact of intensive summer reading intervention for children with reading disabilities and difficulties in early elementary school. *Journal of Learning Disabilities*, 50(2), 115–127. doi.org/10.1177/0022219415617163

Clark, K. A., Helland, T., Specht, K., Narr, K. L., Manis, F. R., Toga, A. W., et al. (2014). Neuroanatomical precursors of dyslexia identified from pre-reading through to age 11. *Brain*, 137(12), 3136–3141. doi.org/10.1093/brain/awu229

Cornelissen, P. and Stein, J. (1995). Contrast sensitivity and coherent motion detection measured at photopic luminance levels in dyslexics and controls. *Vision*, 35(10), 1482–1494.

Cortese, S., Kelly, C., Chabernaud, C., Proal, E., Di Martino, A., Milham, M. P., et al. (2012). Toward systems neuroscience of ADHD: A meta-analysis of 55 fMRI studies. *American Journal of Psychiatry*, 169(10), 1038–1055.

Daneman, M. and Carpenter, P. A. (1980). Individual differences in working memory during reading. *Journal of Verbal Learning and Verbal Behavior*, 19(4), 450–466. doi.org/10.1016/S0022-5371(80)90312-6

Daneman, M. and Merikle, P. M. (1996). Working memory and language comprehension: A meta-analysis. *Psychonomic Bulletin and Review*, 3(4), 422–433. doi.org/10.3758/BF03214546

Darki, F., Peyrard-Janvid, M., Matsson, H., Kere, J., and Klingberg, T. (2012). Three dyslexia susceptibility genes, DYX1C1, DCDC2, and KIAA0319, affect temporo-parietal white matter structure. *Biological Psychiatry*, 72(8), 671–676. doi.org/10.1016/j.biopsych.2012.05.008

De Smedt, B. and Boets, B. (2010). Phonological processing and arithmetic fact retrieval: Evidence from developmental dyslexia. *Neuropsychologia*, 48(14), 3973–3981. doi.org/10.1016/j.neuropsychologia.2010.10.018

De Smedt, B., Taylor, J., Archibald, L., and Ansari, D. (2010). How is phonological processing related to individual differences in children's arithmetic skills? *Developmental Science*, 13(3), 508–520. doi.org/10.1111/j.1467-7687.2009.00897

DeFries, J. C., Fulker, D. W., and LaBuda, M. C. (1987). Evidence for a genetic aetiology in reading disability of twins. *Nature*, 329(6139), 537–539. doi.org/10.1038/329537a0

Dehaene, S. and Cohen, L. (2011). The unique role of the visual word form area in reading. *Trends in Cognitive Sciences*, 15(6), 254–262. doi.org/10.1016/j.tics.2011.04.003

Dehaene, S., Pegado, F., Braga, L. W., Ventura, P., Nunes Filho, G., Jobert, A., et al. (2010). How learning to read changes the cortical networks for vision and language. *Science*, 330(6009), 1359–1364. doi.org/10.1126/science.1194140

Demb, J. B., Boynton, G. M., and Heeger, D. J. (1998). Functional magnetic resonance imaging of early visual pathways in dyslexia. *The Journal of Neuroscience*, 18(17), 13.

Denckla, M. B. and Cutting, L. E. (1999). History and significance of rapid automatized naming. *Annals of Dyslexia*, 49, 29–42. doi.org/10.1007/s11881-999-0018-9

Denckla, M. B. and Rudel, R. G. (1976). Rapid "automatized" naming (R.A.N): Dyslexia differentiated from other learning disabilities. *Neuropsychologia*, 14(4), 471–479.

Dietz, N. A. E., Jones, K. M., Gareau, L., Zeffiro, T. A., and Eden, G. F. (2005). Phonological decoding involves left posterior fusiform gyrus. *Human Brain Mapping*, 26(2), 81–93. doi.org/10.1002/hbm.20122

Dougherty, R. F., Ben-Shachar, M., Deutsch, G. K., Hernandez, A., Fox, G. R., and Wandell, B. A. (2007). Temporal-callosal pathway diffusivity predicts phonological skills in children. *Proceedings of the National Academy of Sciences of the United States of America*, 104(20), 8556–8561. doi.org/10.1073/pnas.0608961104

Eckert, M. A., Berninger, V. W., Vaden, K. I., Gebregziabher, M., and Tsu, L. (2016). Gray matter features of reading disability: A combined meta-analytic and direct analysis approach. *ENeuro*, 3(1). doi.org/10.1523/ENEURO.0103-15.2015

Eckert, M. A., Leonard, C. M., Richards, T. L., Aylward, E. H., Thomson, J., and Berninger, V. W. (2003). Anatomical correlates of dyslexia: Frontal and cerebellar findings. *Brain*, 126(2), 482–494. doi.org/10.1093/brain/awg026

Eden, G. F., Jones, K. M., Cappell, K., Gareau, L., Wood, F. B., Zeffiro, T. A., et al. (2004). Neural changes following remediation in adult developmental dyslexia. *Neuron*, 44(3), 411–422.

Eden, G. F., VanMeter, J. W., Rumsey, J. M., Maisog, J. M., Woods, R. P., and Zeffiro, T. A. (1996). Abnormal processing of visual motion in dyslexia revealed by functional brain imaging. *Nature*, 382(6586), 66–69. doi.org/10.1038/382066a0

Ehri, L. C., Nunes, S. R., Stahl, S. A., and Willows, D. M. (2001). Systematic phonics instruction helps students learn to read: Evidence from the national reading panel's meta-analysis. *Review of Educational Research*, 71(3), 393–447. doi.org/10.3102/00346543071003393

Epelbaum, S., Pinel, P., Gaillard, R., Delmaire, C., Perrin, M., Dupont, S., et al. (2008). Pure alexia as a disconnection syndrome: New diffusion imaging evidence for an old concept. *Cortex*, 44(8), 962–974. doi.org/10.1016/j.cortex.2008.05.003

Evans, T. M., Flowers, D. L., Napoliello, E. M., and Eden, G. F. (2014). Sex-specific gray matter volume differences in females with developmental dyslexia. *Brain Structure and Function*, 219(3), 1041–1054. doi.org/10.1007/s00429-013-0552-4

Evans, T. M., Flowers, D. L., Napoliello, E. M., Olulade, O. A., Eden, G. F., Napoliello, E. M., et al. (2014). The functional anatomy of single-digit arithmetic in children with developmental dyslexia. *NeuroImage*, 101(7), 644–652. doi.org/10.1016/j.neuroimage.2014.07.028

Ferrer, E., Shaywitz, B. A., Holahan, J. M., Marchione, K. E., Michaels, R., and Shaywitz, S. E. (2015). Achievement gap in reading is present as early as first grade and persists through adolescence. *Journal of Pediatrics*, 167(5), 1121–1125. doi.org/10.1016/j.jpeds.2015.07.045

Fias, W., Menon, V., and Szucs, D. (2013). Multiple components of developmental dyscalculia. *Trends in Neuroscience and Education*, 2(2), 43–47. doi.org/10.1016/j.tine.2013.06.006

Finn, E. S., Shen, X., Holahan, J. M., Scheinost, D., Lacadie, C., Papademetris, X., et al. (2014). Disruption of functional networks in dyslexia: A whole-brain, data-driven analysis of connectivity. *Biological Psychiatry*, 76(5), 397–404. doi.org/10.1016/j.biopsych.2013.08.031

Fisher, S. E. and DeFries, J. C. (2002). Developmental dyslexia: Genetic dissection of a complex cognitive trait. *Nature Reviews Neuroscience*, 3(10), 767–780. doi.org/10.1038/nrn936

Flannery, K. A., Liederman, J., Daly, L., and Schultz, J. (2000). Male prevalence for reading disability is found in a large sample of black and white children free from ascertainment bias. *Journal of the International Neuropsychological Society*, 6(4), 433–442.

Franceschini, S., Gori, S., Ruffino, M., Pedrolli, K., and Facoetti, A. (2012). A causal link between visual spatial attention and reading acquisition. *Current Biology*, 22(9), 814–819. doi.org/10.1016/j.cub.2012.03.013

Friend, A., DeFries, J. C., and Olson, R. K. (2008). Parental education moderates genetic influences on reading disability. *Psychological Science*, 19(11), 1124–1130. doi.org/10.1111/j.1467-9280.2008.02213.x

Frodl, T. and Skokauskas, N. (2012). Meta-analysis of structural MRI studies in children and adults with attention deficit hyperactivity disorder indicates treatment effects. *Acta Psychiatrica Scandinavica*, 125(2), 114–126. doi.org/10.1111/j.1600-0447.2011.01786.x

Frye, R. E., Hasan, K., Xue, L., Strickland, D., Malmberg, B., Liederman, J., et al. (2008). Splenium microstructure is related to two dimensions of reading skill. *NeuroReport*. doi.org/10.1097/WNR.0b013e328314b8ee

Galaburda, A. M., LoTurco, J., Ramus, F., Fitch, R. H., and Rosen, G. D. (2006). From genes to behavior in developmental dyslexia. *Nature Neuroscience*, 9(10), 1213–1217. doi.org/10.1038/nn1772

Gallagher, A., Frith, U., and Snowling, M. J. (2000). Precursors of literacy delay among children at genetic risk of dyslexia. *Journal of Child Psychology and Psychiatry and Allied Disciplines*, 41(2), 203–213. doi.org/10.1111/1469-7610.00601

Gerber, P. J. (2012). The impact of learning disabilities on adulthood: A review of the evidenced-based literature for research and practice in adult education. *Journal of Learning Disabilities*, 45(1), 31–46. doi.org/10.1177/0022219411426858

Gialluisi, A., Andlauer, T. F. M., Mirza-Schreiber, N., Moll, K., Becker, J., Hoffmann, P., et al. (2020). Genome-wide association study reveals new insights into the heritability and genetic correlates of developmental dyslexia. *Molecular Psychiatry*, *26*, 3004–3017. doi.org/10.1038/s41380-020-00898-x

Gilger, J. W., Pennington, B. F., and Defries, J. C. (1991). Risk for reading disability as a function of parental history in three family studies. *Reading and Writing*, *3*(3–4), 205–217. doi.org/10.1007/BF00354958

Glezer, L. S., Eden, G., Jiang, X., Luetje, M., Napoliello, E., Kim, J., et al. (2016). Uncovering phonological and orthographic selectivity across the reading network using fMRI-RA. *NeuroImage*, *138*, 248–256. doi.org/10.1016/j.neuroimage.2016.05.072

Goswami, U. (2002). Phonology, reading development, and dyslexia: A cross-linguistic perspective. *Annals of Dyslexia*, *52*, 139–163. doi.org/10.1007/s11881-002-0010-0

Hancock, R., Richlan, F., Hoeft, F., Francisco, S., Francisco, S., States, U., et al. (2017). Possible roles for fronto-striatal circuits in reading disorder. *Neuroscience and Biobehavioral Reviews*, *72*, 243–260. doi.org/10.1016/j.neubiorev.2016.10.025.Possible

Hassid, E. I. (1995). A case of language dysfunction associated with cerebellar infarction. *Journal of Neurologic Rehabilitation*, *9*(3), 157–160. doi.org/10.1177/154596839500900304

Herting, M. M., Gautam, P., Spielberg, J. M., Dahl, R. E., and Sowell, E. R. (2015). A longitudinal study: Changes in cortical thickness and surface area during pubertal maturation. *PLOS ONE*, *10*(3), e0119774. doi.org/10.1371/journal.pone.0119774

Hoeft, F., McCandliss, B. D., Black, J. M., Gantman, A., Zakerani, N., Hulme, C., et al. (2011). Neural systems predicting long-term outcome in dyslexia. *Proceedings of the National Academy of Sciences*, *108*(1), 361. doi.org/10.1073/pnas.1008950108

Hoeft, F., Meyler, A., Hernandez, A., Juel, C., Taylor-Hill, H., Martindale, J. L., et al. (2007). Functional and morphometric brain dissociation between dyslexia and reading ability. *Proceedings of the National Academy of Sciences of the United States of America*, *104*(10), 4234–4239. doi.org/10.1073/pnas.0609399104

Hoff, E. (2003). The specificity of environmental influence: Socioeconomic status affects early vocabulary development via maternal speech. *Child Development*, *74*(5), 1368–1378. doi.org/10.1111/1467-8624.00612

Holmes, G. (1939). The cerebellum of man. *Brain*, *62*(1), 1–30.

Hosseini, S. H., Black, J. M., Soriano, T., Bugescu, N., Martinez, R., Raman, M. M., et al. (2013). Topological properties of large-scale structural brain networks in children with familial risk for reading difficulties. *Neuroimage*, *71*, 260e274. doi.org/10.1016/j.neuroimage.2013.01.013

Hu, W., Lee, H. L., Zhang, Q., Liu, T., Geng, L. B., Seghier, M. L., et al. (2010). Developmental dyslexia in Chinese and English populations: Dissociating the effect of dyslexia from language differences. *Brain*, *133*(6), 1694–1706. doi.org/10.1093/brain/awq106

Huber, E., Donnelly, P. M., Rokem, A., and Yeatman, J. D. (2018). Rapid and widespread white matter plasticity during an intensive reading intervention. *Nature Communications*, *9*(1), 2260. doi.org/10.1038/s41467-018-04627-5

Illingworth, S. and Bishop, D. V. M. (2009). Atypical cerebral lateralisation in adults with compensated developmental dyslexia demonstrated using functional transcranial Doppler ultrasound. *Brain and Language*, *111*(1), 61–65. doi.org/10.1016/j.bandl.2009.05.002

Jagger-Rickels, A. C., Kibby, M. Y., and Constance, J. M. (2018). Global gray matter morphometry differences between children with reading disability, ADHD, and comorbid reading disability/ADHD. *Brain and Language*, *185*, 54–66. doi.org/10.1016/j.bandl.2018.08.004

Jednoróg, K., Gawron, N., Marchewka, A., Heim, S., and Grabowska, A. (2014). Cognitive subtypes of dyslexia are characterized by distinct patterns of grey matter volume. *Brain Structure and Function*, 219(5). doi.org/10.1007/s00429-013-0595-6

Kamali, A., Flanders, A. E., Brody, J., Hunter, J. V., and Hasan, K. M. (2014). Tracing superior longitudinal fasciculus connectivity in the human brain using high resolution diffusion tensor tractography. *Brain Structure and Function*, 219(1), 269–281. doi.org/10.1007/s00429-012-0498-y

Katusic, S. K., Colligan, R. C., Barbaresi, W. J., Schaid, D. J., and Jacobsen, S. J. (2001). Incidence of reading disability in a population-based birth cohort, 1976–1982, Rochester, Minn. *Mayo Clinic Proceedings*, 76(11), 1081–1092.

Keller, T. A. and Just, M. A. (2009). Altering cortical connectivity: Remediation-induced changes in the white matter of poor readers. *Neuron*, 64(5), 624–631. doi.org/10.1016/j.neuron.2009.10.018

Kere, J. (2011). Molecular genetics and molecular biology of dyslexia. *Wiley Interdisciplinary Reviews: Cognitive Science*, 2(4), 441–448. doi.org/10.1002/wcs.138

Koyama, M. S., Di Martino, A., Kelly, C., Jutagir, D. R., Sunshine, J., Schwartz, S. J., et al. (2013). Cortical signatures of dyslexia and remediation: An intrinsic functional connectivity approach. *PLoS ONE*, 8(2). doi.org/10.1371/journal.pone.0055454

Krafnick, A. J. and Evans, T. M. (2019). Neurobiological sex differences in developmental dyslexia. *Frontiers in Psychology*, 9. www.frontiersin.org/articles/10.3389/fpsyg.2018.02669

Krafnick, A. J., Flowers, D. L., Napoliello, E. M., and Eden, G. F. (2011). Gray matter volume changes following reading intervention in dyslexic children. *NeuroImage*, 57(3), 733–741. doi.org/10.1016/j.neuroimage.2010.10.062

Krafnick, A. J., Napoliello, E. M., Flowers, D. L., and Eden, G. F. (2022). The role of brain activity in characterizing successful reading intervention in children with dyslexia. *Frontiers in Neuroscience*, 16. www.frontiersin.org/articles/10.3389/fnins.2022.898661

Kronbichler, L. and Kronbichler, M. (2018). The importance of the left occipitotemporal cortex in developmental dyslexia. *Current Developmental Disorders Reports*, 5(1), 1–8. doi.org/10.1007/s40474-018-0135-4

Kuhl, U., Neef, N. E., Kraft, I., Schaadt, G., Dörr, L., Brauer, J., et al. (2020). The emergence of dyslexia in the developing brain. *NeuroImage*, 211(January), 116633. doi.org/10.1016/j.neuroimage.2020.116633

Lallier, M., Valdois, S., Lassus-Sangosse, D., Prado, C., and Kandel, S. (2014). Impact of orthographic transparency on typical and atypical reading development: Evidence in French-Spanish bilingual children. *Research in Developmental Disabilities*, 35(5), 1177–1190. doi.org/10.1016/J.RIDD.2014.01.021

Landerl, K., Fussenegger, B., Moll, K., and Willburger, E. (2009). Dyslexia and dyscalculia: Two learning disorders with different cognitive profiles. *Journal of Experimental Child Psychology*, 103(3), 309–324. doi.org/10.1016/j.jecp.2009.03.006

Landerl, K. and Moll, K. (2010). Comorbidity of learning disorders: Prevalence and familial transmission. *Journal of Child Psychology and Psychiatry and Allied Disciplines*, 51(3), 287–294. doi.org/10.1111/j.1469-7610.2009.02164.x

Landerl, K., Ramus, F., Moll, K., Lyytinen, H., Leppänen, P. H. T., Lohvansuu, K., et al. (2013). Predictors of developmental dyslexia in European orthographies with varying complexity. *Journal of Child Psychology and Psychiatry*, 54(6), 686–694. doi.org/10.1111/jcpp.12029

Landerl, K., Wimmer, H., and Frith, U. (1997). The impact of orthographic consistency on dyslexia: A German-English comparison. *Cognition*, *63*(3), 315–334. doi.org/10.1016/S0010-0277(97)00005-X

Langer, N., Benjamin, C., Becker, B. L. C., and Gaab, N. (2019). Comorbidity of reading disabilities and ADHD: Structural and functional brain characteristics. *Human Brain Mapping*. doi.org/10.1002/hbm.24552

Langer, N., Peysakhovich, B., Zuk, J., Drottar, M., Sliva, D. D., Smith, S., et al. (2017). White matter alterations in infants at risk for developmental dyslexia. *Cerebral Cortex*, *27*(2), 1027–1036. doi.org/10.1093/cercor/bhv281

Lee, V. E. and Burkam, D. T. (2002). *Inequality at the Starting Gate: Social Background Differences in Achievement as Children Begin School*. Economic Policy Institute.

Lenroot, R. K. and Giedd, J. N. (2010). Sex differences in the adolescent brain. *Brain and Cognition*, *72*(1), 46–55. doi.org/10.1016/j.bandc.2009.10.008

Lewis, C., Hitch, G. J., and Walker, P. (1994). The prevalence of specific arithmetic difficulties and specific reading difficulties in 9- to 10-year-old boys and girls. *Journal of Child Psychology and Psychiatry*, *35*(2), 283–292. doi.org/10.1111/j.1469-7610.1994.tb01162.x

Liederman, J., Kantrowitz, L., and Flannery, K. (2005). Male vulnerability to reading disability is not likely to be a myth: A call for new data. *Journal of Learning Disabilities*, *38*(2). doi.org/10.1177/00222194050380020201

Linkersdörfer, J., Lonnemann, J., Lindberg, S., Hasselhorn, M., and Fiebach, C. J. (2012). Grey matter alterations co-localize with functional abnormalities in developmental dyslexia: An ALE meta-analysis. *PLoS ONE*, *7*(8), e43122. doi.org/10.1371/journal.pone.0043122

Lovegrove, W., Bowling, A., Badcock, D., and Blackwood, M. (1980). Specific reading disability: Differences in contrast sensitivity as a function of spatial frequency. *Science*, *210*(4468), 439–440. doi.org/10.1126/science.7433985

Lovett, M. W., Steinbach, K. A., and Frijters, J. C. (2000). Remediating the core deficits of developmental reading disability: A double-deficit perspective. *Journal of Learning Disabilities*, *33*(4), 334–358. doi.org/10.1177/002221940003300406

Ma, Y., Koyama, M. S., Milham, M. P., Castellanos, F. X., Quinn, B. T., Pardoe, H., et al. (2015). Cortical thickness abnormalities associated with dyslexia, independent of remediation status. *NeuroImage: Clinical*, *7*, 177–186. doi.org/10.1016/j.nicl.2014.11.005

Maisog, J. M., Einbinder, E. R., Flowers, D. L., Turkeltaub, P. E., and Eden, G. F. (2003). A meta-analysis of functional neuroimaging studies of dyslexia. *Annals of the New York Academy of Sciences*, *1145*, 237–259. doi.org/10.1196/annals.1416.024

Margari, L., Buttiglione, M., Craig, F., Cristella, A., de Giambattista, C., Matera, E., et al. (2013). Neuropsychopathological comorbidities in learning disorders. *BMC Neurology*, *13*(1), 198. doi.org/10.1186/1471-2377-13-198

Martin, A., Kronbichler, M., and Richlan, F. (2016). Dyslexic brain activation abnormalities in deep and shallow orthographies: A meta-analysis of 28 functional neuroimaging studies. *Human Brain Mapping*, *37*(7), 2676–2699. doi.org/10.1002/hbm.23202

Martin, A., Schurz, M., Kronbichler, M., and Richlan, F. (2015). Reading in the brain of children and adults: A meta-analysis of 40 functional magnetic resonance imaging studies. *Human Brain Mapping*, *36*(5), 1963–1981. doi.org/10.1002/hbm.22749

Mascheretti, S., Andreola, C., Scaini, S., and Sulpizio, S. (2018). Beyond genes: A systematic review of environmental risk factors in specific reading disorder. *Research in Developmental Disabilities*, *82*, 147–152. doi.org/10.1016/j.ridd.2018.03.005

Matejko, A. A. and Ansari, D. (2015). Drawing connections between white matter and numerical and mathematical cognition: A literature review. *Neuroscience and Biobehavioral Reviews, 48*, 35–52. doi.org/10.1016/j.neubiorev.2014.11.006

Matejko, A. A. and Ansari, D. (2018). Contributions of functional magnetic resonance imaging (fMRI) to the study of numerical cognition. *Journal of Numerical Cognition, 4*(3), 505–525. doi.org/10.5964/jnc.v4i3.136

Matejko, A. A., Lozano, M. C., Schlosberg, N., McKay, C. C., Core, L., Resvine, C., et al. (2022). The relationship between phonological processing and arithmetic in children with learning disabilities, *Developmental Science, 26*(2), e13294. doi:10.1111/desc.13294

McCardle, P., Scarborough, H. S., and Catts, H. W. (2001). Predicting, explaining, and preventing children's reading difficulties. *Learning Disabilities Research and Practice, 16*(4). doi.org/10.1111/0938-8982.00023

McGrath, L. M., Pennington, B. F., Shanahan, M. A., Santerre-Lemmon, L. E., Barnard, H. D., Willcutt, E. G., et al. (2011). A multiple deficit model of reading disability and attention-deficit/ hyperactivity disorder: Searching for shared cognitive deficits. *Journal of Child Psychology and Psychiatry and Allied Disciplines, 52*(5). doi.org/10.1111/j.1469-7610.2010.02346.x

McGrath, L. M. and Stoodley, C. J. (2019). Are there shared neural correlates between dyslexia and ADHD? A meta-analysis of voxel-based morphometry studies. *Journal of Neurodevelopmental Disorders, 11*(1). doi.org/10.1186/s11689-019-9287-8

Melby-Lervåg, M., Lyster, S.-A. H., and Hulme, C. (2012). Phonological skills and their role in learning to read: A meta-analytic review. *Psychological Bulletin, 138*(2), 322–352. doi.org/10.1037/a0026744

Menghini, D., Hagberg, G. E., Caltagirone, C., Petrosini, L., and Vicari, S. (2006). Implicit learning deficits in dyslexic adults: An fMRI study. *NeuroImage, 33*(4), 1218–1226. doi.org/10.1016/j.neuroimage.2006.08.024

Moeller, K., Willmes, K., and Klein, E. (2015). A review on functional and structural brain connectivity in numerical cognition. *Frontiers in Human Neuroscience, 9*(May), 1–14. doi.org/10.3389/fnhum.2015.00227

Moll, K., Landerl, K., Snowling, M. J., and Schulte-Körne, G. (2018). Understanding comorbidity of learning disorders: Task-dependent estimates of prevalence. *Journal of Child Psychology and Psychiatry and Allied Disciplines, 60*(3), 286–294. doi.org/10.1111/jcpp.12965

Monzalvo, K., Fluss, J., Billard, C., Dehaene, S., and Dehaene-Lambertz, G. (2012). Cortical networks for vision and language in dyslexic and normal children of variable socioeconomic status. *NeuroImage, 61*(1), 258–274. doi.org/10.1016/j.neuroimage.2012.02.035

Moreau, D., Wiebels, K., Wilson, A. J., and Waldie, K. E. (2019). Volumetric and surface characteristics of gray matter in adult dyslexia and dyscalculia. *Neuropsychologia, 127*, 204–210. doi.org/10.1016/j.neuropsychologia.2019.02.002

Nicolson, R. I., Daum, I., Schugens, M. M., Fawcett, A. J., and Schulz, A. (2002). Eyeblink conditioning indicates cerebellar abnormality in dyslexia. *Experimental Brain Research, 143*(1), 42–50. doi.org/10.1007/s00221-001-0969-5

Nicolson, R. I. and Fawcett, A. J. (1990). Automaticity: A new framework for dyslexia research? *Cognition, 35*(2), 159–182. doi.org/10.1016/0010-0277(90)90013-A

Nicolson, R. I., Fawcett, A. J., Berry, E. L., Jenkins, I. H., Dean, P., and Brooks, D. J. (1999). Association of abnormal cerebellar activation with motor learning difficulties in dyslexic adults. *The Lancet, 353*(9165), 1662–1667.

Nicolson, R. I., Fawcett, A. J., and Dean, P. (2001). Developmental dyslexia: The cerebellar deficit hypothesis. *Trends in Neurosciences*, 24(9), 508–511.

Nigg, J. T., Carte, E. T., Hinshaw, S. P., and Treuting, J. J. (1998). Neuropsychological correlates of childhood attention-deficit/hyperactivity disorder: Explainable by comorbid disruptive behavior or reading problems? *Journal of Abnormal Psychology*, 107(3). doi.org/10.1037/0021-843X.107.3.468

Niogi, S., Mukherjee, P., Ghajar, J., and McCandliss, B. D. (2010). Individual differences in distinct components of attention are linked to anatomical variations in distinct white matter tracts. *Frontiers in Neuroanatomy*, 4(2). doi.org/10.3389/neuro.05.002.2010

Niogi, S. N. and McCandliss, B. D. (2006). Left lateralized white matter microstructure accounts for individual differences in reading ability and disability. *Neuropsychologia*, 44(11), 2178–2188. doi.org/10.1016/j.neuropsychologia.2006.01.011

Noble, K. G., Houston, S. M., Brito, N. H., Bartsch, H., Kan, E., Kuperman, J. M., et al. (2015). Family income, parental education and brain structure in children and adolescents. *Nature Neuroscience*, 18(5), 773–778. doi.org/10.1038/nn.3983

Norman, L. J., Carlisi, C., Lukito, S., Hart, H., Mataix-Cols, D., Radua, J., et al. (2016). Structural and functional brain abnormalities in attention-deficit/hyperactivity disorder and obsessive-compulsive disorder: A comparative meta-analysis. *JAMA Psychiatry*, 73(8), 815–825. doi.org/10.1001/jamapsychiatry.2016.0700

Norton, E. S. and Wolf, M. (2012). Rapid automatized naming (RAN) and reading fluency: Implications for understanding and treatment of reading disabilities. *Annual Review of Psychology*, 63(1), 427–452. doi.org/10.1146/annurev-psych-120710-100431

Nugiel, T., Roe, M. A., Taylor, W. P., Cirino, P. T., Vaughn, S. R., Fletcher, J. M., et al. (2019). Brain activity in struggling readers before intervention relates to future reading gains. *Cortex*, 111, 286–302. doi.org/10.1016/j.cortex.2018.11.009

Odegard, T. N., Farris, E. A., Ring, J., McColl, R., and Black, J. (2009). Brain connectivity in non-reading impaired children and children diagnosed with developmental dyslexia. *Neuropsychologia*, 47, 1972–1977. doi.org/10.1016/j.neuropsychologia.2009.03.009

Odegard, T. N., Ring, J., Smith, S., Biggan, J., and Black, J. (2008). Differentiating the neural response to intervention in children with developmental dyslexia. *Annals of Dyslexia*, 58(1), 1–14. doi.org/10.1007/s11881-008-0014-5

Olulade, O. A., Flowers, D. L., Napoliello, E. M., and Eden, G. F. (2015). Dyslexic children lack word selectivity gradients in occipito-temporal and inferior frontal cortex. *NeuroImage: Clinical*, 7, 742–754. doi.org/10.1016/j.nicl.2015.02.013

Olulade, O. A., Napoliello, E. M., and Eden, G. F. (2013). Abnormal visual motion processing is not a cause of dyslexia. *Neuron*, 79(1), 180–190. doi.org/10.1016/j.neuron.2013.05.002

Overy, K., Nicolson, R. I., Fawcett, A. J., and Clarke, E. F. (2003). Dyslexia and music: Measuring musical timing skills. *Dyslexia*, 9(1), 18–36. doi.org/10.1002/dys.233

Ozernov-Palchik, O. and Gaab, N. (2016). Tackling the "dyslexia paradox": Reading brain and behavior for early markers of developmental dyslexiax. *Wiley Interdisciplinary Reviews: Cognitive Science*, 7(2), 156–176. doi.org/10.1002/wcs.1383

Ozernov-Palchik, O., Norton, E. S., Sideridis, G., Beach, S. D., Wolf, M., Gabrieli, J. D. E., et al. (2017). Longitudinal stability of pre-reading skill profiles of kindergarten children: Implications for early screening and theories of reading. *Developmental Science*, 20(5), 1–18. doi.org/10.1111/desc.12471

Paulesu, E., Danelli, L., and Berlingeri, M. (2014). Reading the dyslexic brain: Multiple dysfunctional routes revealed by a new meta-analysis of PET and fMRI activation studies. *Frontiers in Human Neuroscience, 8*(November), 1–20. doi.org/10.3389/fnhum.2014.00830

Paulesu, E., Démonet, J.-F., Fazio, F., McCrory, E., Chanoine, V., Brunswick, N., et al. (2001). Dyslexia: Cultural diversity and biological unity. *Science, 291*(5511), 2165–2167. doi.org/10.1126/science.1057179

Paulesu, E., Frith, U., Snowling, M., Gallagher, A., Morton, J., Frackowiak, R. S. J., et al. (1996). Is developmental dyslexia a disconnection syndrome? doi.org/10.1093/brain/119.1.143

Peng, P. and Fuchs, D. (2014). A meta-analysis of working memory deficits in children with learning difficulties: Is there a difference between verbal domain and numerical domain? *Journal of Learning Disabilities, 49*(1). doi.org/10.1177/0022219414521667

Pennington, B. F. (1991). Evidence for major gene transmission of developmental dyslexia. *JAMA, 266*(11), 1527–1534. doi.org/10.1001/jama.266.11.1527

Pennington, B. F. (2006). From single to multiple deficit models of developmental disorders. *Cognition, 101*(2). doi.org/10.1016/j.cognition.2006.04.008

Pennington, B. F., Groisser, D., and Welsh, M. C. (1993). Contrasting cognitive deficits in attention deficit hyperactivity disorder versus reading disability. *Developmental Psychology, 29*(3). doi.org/10.1037/0012-1649.29.3.511

Pennington, B. F. and Lefly, D. L. (2001). Early reading development in children at family risk for dyslexia. *Child Development, 72*(3). doi.org/10.1111/1467-8624.00317

Peper, J. S., Brouwer, R. M., Schnack, H. G., van Baal, G. C., van Leeuwen, M., van den Berg, S. M., et al. (2009). Sex steroids and brain structure in pubertal boys and girls. *Psychoneuroendocrinology, 34*(3), 332–342. doi.org/10.1016/j.psyneuen.2008.09.012

Perdue, M. V., Mahaffy, K., Vlahcevic, K., Wolfman, E., Erbeli, F., Richlan, F., et al. (2022). Reading intervention and neuroplasticity: A systematic review and meta-analysis of brain changes associated with reading intervention. *Neuroscience and Biobehavioral Reviews, 132,* 465–494. doi.org/10.1016/j.neubiorev.2021.11.011

Pernet, C. R., Poline, J., Demonet, J., and Rousselet, G. A. (2009). Brain classification reveals the right cerebellum as the best biomarker of dyslexia. *BMC Neuroscience, 10*(1), 67. doi.org/10.1186/1471-2202-10-67

Peters, L. and Ansari, D. (2019). Are specific learning disorders truly specific, and are they disorders? *Trends in Neuroscience and Education, 17,* 100115. doi.org/10.1016/j.tine.2019.100115

Peters, L., Bulthé, J., Daniels, N., Op de Beeck, H., and De Smedt, B. (2018). Dyscalculia and dyslexia: Different behavioral, yet similar brain activity profiles during arithmetic. *NeuroImage: Clinical, 18,* 663–674. doi.org/10.1016/j.nicl.2018.03.003

Peters, L. and De Smedt, B. (2018). Arithmetic in the developing brain: A review of brain imaging studies. *Developmental Cognitive Neuroscience, 30,* 265–279. doi.org/10.1016/j.dcn.2017.05.002

Peterson, R. L. and Pennington, B. (2012). Developmental dyslexia. *Lancet, 379,* 1997–2007. doi.org/10.1016/S0140-6736(12)60198-6

Peterson, R. L. and Pennington, B. F. (2015). Developmental dyslexia. *Annual Review of Clinical Psychology, 11*(1), 283–307. doi.org/10.1146/annurev-clinpsy-032814-112842

Poldrack, R. A., Wagner, A. D., Prull, M. W., Desmond, J. E., Glover, G. H., and Gabrieli, J. D. E. (1999). Functional specialization for semantic and phonological processing in the left inferior prefrontal cortex. *NeuroImage, 10*(1). doi.org/10.1006/nimg.1999.0441

Price, C. J. and Devlin, J. T. (2011). The interactive account of ventral occipitotemporal contributions to reading. *Trends in Cognitive Sciences, 15*(6), 246–253. doi.org/10.1016/j.tics.2011.04.001

Pugh, K. R., Landi, N., Preston, J. L., Mencl, W. E., Austin, A. C., Sibley, D., et al. (2013). The relationship between phonological and auditory processing and brain organization in beginning readers. *Brain and Language, 125*(2), 173–183. doi.org/10.1016/j.bandl.2012.04.004

Pugh, K. R., Mencl, W. E., Jenner, A. R., Katz, L., Frost, S. J., Lee, J. R., et al. (2001). Neurobiological studies of reading and reading disability. *Journal of Communication Disorders, 34*(6), 479–492. doi.org/10.1016/S0021-9924(01)00060-0

Rabiner, D. and Coie, J. D. (2000). Early attention problems and children's reading achievement: A longitudinal investigation. The Conduct Problems Prevention Research Group. *Journal of the American Academy of Child and Adolescent Psychiatry, 39*(7), 859–867.

Rack, J. P., Snowling, M. J., Hulme, C., and Gibbs, S. (2007). No evidence that an exercise-based treatment programme (DDAT) has specific benefits for children with reading difficulties. *Dyslexia, 13*(2), 97–104. doi.org/10.1002/dys.335

Ramus, F. (2003). Developmental dyslexia: Specific phonological deficit or general sensori-motor dysfunction? *Current Opinion in Neurobiology, 13*(2), 212–218. doi.org/10.1016/S0959-4388(03)00035-7

Ramus, F., Altarelli, I., Jednoróg, K., Zhao, J., and Scotto di Covella, L. (2018). Neuroanatomy of developmental dyslexia: Pitfalls and promise. *Neuroscience and Biobehavioral Reviews, 84*, 434–452. doi.org/10.1016/j.neubiorev.2017.08.001

Rapport, M. D., Scanlan, S. W., and Denney, C. B. (1999). Attention-deficit/hyperactivity disorder and scholastic achievement: A model of dual developmental pathways. *Journal of Child Psychology and Psychiatry and Allied Disciplines, 40*(8), 1169–1183. doi.org/10.1017/S0021963099004618

Raschle, N. M., Chang, M., and Gaab, N. (2011). Structural brain alterations associated with dyslexia predate reading onset. *NeuroImage, 57*(3), 742–749. doi.org/10.1016/j.neuroimage.2010.09.055

Raschle, N. M., Stering, P. L., Meissner, S. N., and Gaab, N. (2014). Altered neuronal response during rapid auditory processing and its relation to phonological processing in prereading children at familial risk for dyslexia. *Cerebral Cortex, 24*(9), 2489–2501. doi.org/10.1093/cercor/bht104

Raschle, N., Zuk, J., Ortiz-Mantilla, S., Sliva, D. D., Franceschi, A., Grant, P. E., Benasich, A. A., and Gaab, N. (2012). Pediatric neuroimaging in early childhood and infancy: Challenges and practical guidelines. *Annals of the New York Academy of Sciences, 1252*(1), 43–50. doi.org/10.1111/j.1749-6632.2012.06457.x

Richards, T. L., Aylward, E. H., Berninger, V. W., Field, K. M., Grimme, A. C., Richards, A. L., and Nagy, W. (2006). Individual fMRI activation in orthographic mapping and morpheme mapping after orthographic or morphological spelling treatment in child dyslexics. *Journal of Neurolinguistics, 19*(1), 56–86. doi.org/10.1016/j.jneuroling.2005.07.003

Richlan, F. (2012). Developmental dyslexia: Dysfunction of a left hemisphere reading network. *Frontiers in Human Neuroscience, 6*(May), 1–5. doi.org/10.3389/fnhum.2012.00120

Richlan, F. (2020). The functional neuroanatomy of developmental dyslexia across languages and writing systems. *Frontiers in Psychology, 11.* doi.org/10.3389/fpsyg.2020.00155

Richlan, F., Kronbichler, M., and Wimmer, H. (2009). Functional abnormalities in the dyslexic brain: A quantitative meta-analysis of neuroimaging studies. *Human Brain Mapping, 30*(10), 3299–3308. doi.org/10.1002/hbm.20752

Richlan, F., Kronbichler, M., and Wimmer, H. (2011). Meta-analyzing brain dysfunctions in dyslexic children and adults. *NeuroImage, 56*(3), 1735–1742. doi.org/10.1016/j.neuroimage.2011.02.040

Richlan, F., Kronbichler, M., and Wimmer, H. (2013). Structural abnormalities in the dyslexic brain: A meta-analysis of voxel-based morphometry studies. *Human Brain Mapping*, 34(11), 3055–3065. doi.org/10.1002/hbm.22127

Rochelle, K. S. H. and Talcott, J. B. (2006). Impaired balance in developmental dyslexia? A meta-analysis of the contending evidence. *Journal of Child Psychology and Psychiatry*, 47(11), 1159–1166. doi.org/10.1111/j.1469-7610.2006.01641.x

Romeo, R. R., Christodoulou, J. A., Halverson, K. K., Murtagh, J., Cyr, A. B., Schimmel, C., et al. (2017). Socioeconomic status and reading disability: Neuroanatomy and plasticity in response to intervention. *Cerebral Cortex*, 28(7), 2297–2312. doi.org/10.1093/cercor/bhx131

Romeo, R. R., Segaran, J., Leonard, J. A., Robinson, S. T., West, M. R., Mackey, et al. (2018). Language exposure relates to structural neural connectivity in childhood. *The Journal of Neuroscience*, 38(36), 7870–7877. doi.org/10.1523/JNEUROSCI.0484-18.2018

Rucklidge, J. J. and Tannock, R. (2002). Neuropsychological profiles of adolescents with ADHD: Effects of reading difficulties and gender. *Journal of Child Psychology and Psychiatry and Allied Disciplines*, 43(8). doi.org/10.1111/1469-7610.00227

Rueckl, J. G., Paz-Alonso, P. M., Molfese, P. J., Kuo, W. J., Bick, A., Frost, S. J., et al. (2015). Universal brain signature of proficient reading: Evidence from four contrasting languages. *Proceedings of the National Academy of Sciences of the United States of America*, 112(50), 15510–15515. doi.org/10.1073/pnas.1509321112

Rutter, M., Caspi, A., Fergusson, D., Horwood, L. J., Goodman, R., Maughan, B., et al. (2004). Sex differences in developmental reading disability. *JAMA*, 291(16), 2007. doi.org/10.1001/jama.291.16.2007

Sandak, R., Frost, S. J., Rueckl, J. G., Landi, N., Mencl, W. E., Katz, L., et al. (2012). How does the brain read words? In M. Spivey, K. McRae, and M. F. Joanisse (eds), *The Cambridge Handbook of Psycholinguistics*, Cambridge Handbooks in Psychology series, 218–236. New York, NY: Cambridge University Press.

Sandak, R., Mencl, W. E., Frost, S. J., and Pugh, K. R. (2004). The neurobiological basis of skilled and impaired reading: Recent findings and new directions. *Scientific Studies of Reading*, 8, 273–292. DOI: 10.1207/S1532799Xssr0803_6

Saygin, Z. M., Norton, E. S., Osher, D. E., Beach, S. D., Cyr, A. B., Ozernov-Palchik, O., et al. (2013). Tracking the roots of reading ability: White matter volume and integrity correlate with phonological awareness in prereading and early-reading kindergarten children. *Journal of Neuroscience*, 33(33), 13251–13258. doi.org/10.1523/JNEUROSCI.4383-12.2013

Scarborough, H. S. (2003). Connecting early language and literacy to later reading (dis) abilities: Evidence, theory, and practice. In S. Neuman and D. Dickinson (eds), *Handbook for Research in Early Literacy* (pp. 97–110). New York: Guilford Press.

Scarborough, H. S. and Brady, S. A. (2002). Toward a common terminology for talking about speech and reading: A glossary of the "phon" words and some related Terms. *Journal of Literacy Research*, 34(3), 299–336. doi.org/10.1207/s15548430jlr3403_3

Schlaggar, B. L., Brown, T. T., Lugar, H. M., Visscher, K. M., Miezin, F. M., and Petersen, S. E. (2002). Functional neuroanatomical differences between adults and school-age children in the processing of single words. *Science*, 296(5572), 1476–1479. doi.org/10.1126/science.1069464

Schurz, M., Wimmer, H., Richlan, F., Ludersdorfer, P., Klackl, J., and Kronbichler, M. (2015). Resting-state and task-based functional brain connectivity in developmental dyslexia. *Cerebral Cortex*, 25(10), 3502–3514. doi.org/10.1093/cercor/bhu184

Semrud-Clikeman, M., Biederman, J., Sprich-Buckminster, S., Lehman, B. K., Faraone, S. V., and Norman, D. (1992). Comorbidity between ADDH and learning disability: A review and report in a clinically referred sample. *Journal of the American Academy of Child and Adolescent Psychiatry*, 31(3), 439–448. doi.org/10.1097/00004583-199205000-00009

Sexton, C. C., Gelhorn, H. L., Bell, J. A., and Classi, P. M. (2012). The co-occurrence of reading disorder and ADHD: Epidemiology, treatment, psychosocial impact, and economic burden. *Journal of Learning Disabilities*, 45(6), 538–564. doi.org/10.1177/0022219411407772

Shapley, R. (1990). Visual sensitivity and parallel retinocortical channels. *Annual Review of Psychology*, 41(1), 635–658. doi.org/10.1146/annurev.ps.41.020190.003223

Shaywitz, S. E., Shaywitz, B. A., Pugh, K. R., Fulbright, R. K., Constable, R. T., Mencl, W. E., et al. (1998). Functional disruption in the organization of the brain for reading in dyslexia. *Proceedings of the National Academy of Sciences of the United States of America*, 95(5), 2636–2641.

Simos, P. G., Breier, J. I., Fletcher, J. M., Foorman, B. R., Castillo, E. M., and Papanicolaou, A. C. (2002). Brain mechanisms for reading words and pseudowords: An integrated approach. *Cerebral Cortex*, 12(3), 297–305. doi.org/10.1093/cercor/12.3.297

Siok, W. T., Niu, Z., Jin, Z., Perfetti, C. A., and Tan, L. H. (2008). A structural-functional basis for dyslexia in the cortex of Chinese readers. *Proceedings of the National Academy of Sciences*, 105(14), 5561–5566. doi.org/10.1073/pnas.0801750105

Skeide, M. A., Evans, T. M., Mei, E. Z., Abrams, D. A., and Menon, V. (2018). Neural signatures of co-occurring reading and mathematical difficulties. *Developmental Science*, 21(6), e12680. doi.org/10.1111/desc.12680

Snowling, M. J. (2001). From language to reading and dyslexia. *Dyslexia*, 7(1), 37–45. doi.org/10.1002/dys.185

Sowell, E. R., Peterson, B. S., Kan, E., Woods, R. P., Yoshii, J., Bansal, R., et al. (2007). Sex differences in cortical thickness mapped in 176 healthy individuals between 7 and 87 years of age. *Cerebral Cortex*, 17(7), 1550–1560. doi.org/10.1093/cercor/bhl066

Stein, J. (1997). To see but not to read; the magnocellular theory of dyslexia. *Trends in Neurosciences*, 20(4), 147–152. doi.org/10.1016/S0166-2236(96)01005-3

Steinbrink, C., Vogt, K., Kastrup, A., Müller, H. P., Juengling, F. D., Kassubek, J., et al. (2008). The contribution of white and gray matter differences to developmental dyslexia: Insights from DTI and VBM at 3.0 T. *Neuropsychologia*, 46(13), 3170–3178. doi.org/10.1016/j.neuropsychologia.2008.07.015

Stoodley, C. J. (2014). Distinct regions of the cerebellum show gray matter decreases in autism, ADHD, and developmental dyslexia. *Frontiers in Systems Neuroscience*, 8. doi.org/10.3389/fnsys.2014.00092

Stoodley, C. J., Fawcett, A. J., Nicolson, R. I., and Stein, J. F. (2005). Impaired balancing ability in dyslexic children. *Experimental Brain Research*, 167(3), 370–380. doi.org/10.1007/s00221-005-0042-x

Su, M., Zhao, J., Thiebaut de Schotten, M., Zhou, W., Gong, G., Ramus, F., et al. (2018). Alterations in white matter pathways underlying phonological and morphological processing in Chinese developmental dyslexia. *Developmental Cognitive Neuroscience*, 31, 11–19. doi.org/10.1016/j.dcn.2018.04.002

Swanson, H. L., Trainin, G., Necoechea, D. M., and Hammill, D. D. (2003). Rapid naming, phonological awareness, and reading: A meta-analysis of the correlation evidence. *Review of Educational Research*, 73(4), 407–440. doi.org/10.3102/00346543073004407

Talcott, J. B., Hansen, P. C., Willis-Owen, C., McKinnell, I. W., Richardson, A. J., and Stein, J. F. (1998). Visual magnocellular impairment in adult developmental dyslexics. *Neuro-Ophthalmology*, 20(4), 187–201. doi.org/10.1076/noph.20.4.187.3931

Tan, L. H., Spinks, J. A., Eden, G. F., Perfetti, C. A., and Siok, W. T. (2005). Reading depends on writing, in Chinese. *Proceedings of the National Academy of Sciences*, 102(24), 8781–8785. doi.org/10.1073/pnas.0503523102

Thompson, P. A., Hulme, C., Nash, H. M., Gooch, D., Hayiou-Thomas, E., and Snowling, M. J. (2015). Developmental dyslexia: Predicting individual risk. *Journal of Child Psychology and Psychiatry and Allied Disciplines*, 56(9), 976–987. doi.org/10.1111/jcpp.12412

Torgesen, J. K., Alexander, A. W., Wagner, R. K., Rashotte, C. A., Voeller, K. K., Conway, T., et al. (2001). Intensive remedial instruction for children with severe reading disabilities: Immediate and long-term outcomes from two instructional approaches. *Journal of Learning Disabilities*, 34(1), 33–58. doi.org/10.1177/002221940103400104

Torre, G. A. A. and Eden, G. F. (2019). Relationships between gray matter volume and reading ability in typically developing children, adolescents, and young adults. *Developmental Cognitive Neuroscience*, 36, 100636. doi.org/10.1016/j.dcn.2019.100636

Torre, G.-A. A., McKay, C. C., and Matejko, A. A. (2019). The early language environment and the neuroanatomical foundations for reading. *The Journal of Neuroscience*, 39(7), 1136–1138. doi.org/10.1523/JNEUROSCI.2895-18.2018

Turkeltaub, P. E., Flowers, D. L., Lyon, L. G., and Eden, G. F. (2008). Development of ventral stream representations for single letters. *Annals of the New York Academy of Sciences*, 1145, 13–29. doi.org/10.1196/annals.1416.026

Turkeltaub, P. E., Gareau, L., Flowers, D. L., Zeffiro, T. A., and Eden, G. F. (2003). Development of neural mechanisms for reading. *Nature Neuroscience*, 6(7), 767–773. doi.org/10.1038/nn1065

Twomey, T., Kawabata Duncan, K. J., Price, C. J., and Devlin, J. T. (2011). Top-down modulation of ventral occipito-temporal responses during visual word recognition. *NeuroImage*, 55(3), 1242–1251. doi.org/10.1016/j.neuroimage.2011.01.001

Van Der Auwera, S., Vandermosten, M., Wouters, J., Ghesquière, P., and Vanderauwera, J. (2021). A three-time point longitudinal investigation of the arcuate fasciculus throughout reading acquisition in children developing dyslexia. *NeuroImage*, 237, 118087. doi.org/10.1016/j.neuroimage.2021.118087

Vanderauwera, J., Wouters, J., Vandermosten, M., and Ghesquière, P. (2017). Early dynamics of white matter deficits in children developing dyslexia. *Developmental Cognitive Neuroscience*, 27, 69–77. doi.org/10.1016/j.dcn.2017.08.003

Vandermosten, M., Boets, B., Poelmans, H., Sunaert, S., Wouters, J., and Ghesquiere, P. (2012). A tractography study in dyslexia: Neuroanatomic correlates of orthographic, phonological and speech processing. *Brain*, 135(3), 935–948. doi.org/10.1093/brain/awr363

Vandermosten, M., Boets, B., Wouters, J., and Ghesquière, P. (2012). A qualitative and quantitative review of diffusion tensor imaging studies in reading and dyslexia. *Neuroscience and Biobehavioral Reviews*, 36(6), 1532–1552. doi.org/10.1016/j.neubiorev.2012.04.002

Vandermosten, M., Vanderauwera, J., Theys, C., De Vos, A., Vanvooren, S., Sunaert, S., et al. (2015). A DTI tractography study in pre-readers at risk for dyslexia. *Developmental Cognitive Neuroscience*, 14, 8–15. doi.org/10.1016/j.dcn.2015.05.006

Vellutino, F. R., Fletcher, J. M., Snowling, M. J., and Scanlon, D. M. (2004). Specific reading disability (dyslexia): What have we learned in the past four decades? *Journal of Child Psychology and Psychiatry*, 45(1), 2–40.

Wagner, R. K. and Torgesen, J. K. (1987). The nature of phonological processing and its causal role in the acquisition of reading skills. *Psychological Bulletin*, *101*(2), 192–212. doi.org/10.1037/0033-2909.101.2.192

Wang, Y., Mauer, M. V., Raney, T., Peysakhovich, B., Becker, B. L. C., Sliva, D. D., et al. (2016). Development of tract-specific white matter pathways during early reading development in at-risk children and typical controls. *Cerebral Cortex*, *27*(4), 2469–2485. doi.org/10.1093/cercor/bhw095

Weiss, Y., Cweigenberg, H. G., and Booth, J. R. (2018). Neural specialization of phonological and semantic processing in young children. *Human Brain Mapping*, *39*(11), 4334–4348. doi.org/10.1002/hbm.24274

Willcutt, E. G., Betjemann, R. S., McGrath, L. M., Chhabildas, N. A., Olson, R. K., DeFries, J. C., et al. (2010). Etiology and neuropsychology of comorbidity between RD and ADHD: The case for multiple-deficit models. *Cortex*, *46*, 1345–1361. doi.org/10.1016/j.cortex.2010.06.009

Willcutt, E. G., McGrath, L. M., Pennington, B. F., Keenan, J. M., DeFries, J. C., Olson, R. K., et al. (2019). Understanding comorbidity between specific learning disabilities. *New Directions for Child and Adolescent Development*, *165*, 91–109. doi.org/10.1002/cad.20291

Willcutt, E. G. and Pennington, B. F. (2000). Comorbidity of reading disability and attention-deficit/hyperactivity disorder: Differences by gender and subtype. *Journal of Learning Disabilities*, *33*(2), 179–191. doi.org/10.1177/002221940003300206

Willcutt, E. G., Petrill, S. A., Wu, S., Boada, R., Defries, J. C., Olson, R. K., et al. (2013). Comorbidity between reading disability and math disability: Concurrent psychopathology, functional impairment, and neuropsychological functioning. *Journal of Learning Disabilities*, *46*, 500–516. doi.org/10.1177/0022219413477476

Wise Younger, J., Tucker-Drob, E., Booth, J. R., Wise, J., Tucker-drob, E., and Booth, J. R. (2017). Longitudinal changes in reading network connectivity related to skill improvement. *NeuroImage*, *158*(February), 90–98. doi.org/10.1016/j.neuroimage.2017.06.044

Witte, A. V., Savli, M., Holik, A., Kasper, S., and Lanzenberger, R. (2009). Regional sex differences in grey matter volume are associated with sex hormones in the young adult human brain. *NeuroImage*, *49*, 1205–1212. doi.org/10.1016/j.neuroimage.2009.09.046

Wu, J., Wang, B., Yan, T., Li, X., Bao, X., and Guo, Q. (2012). Different roles of the posterior inferior frontal gyrus in Chinese character form judgment differences between literate and illiterate individuals. *Brain Research*, *1431*, 69–76. doi.org/10.1016/j.brainres.2011.10.052

Xia, Z., Hancock, R., and Hoeft, F. (2017). Neurobiological bases of reading disorder Part I: Etiological investigations. *Linguistics and Language Compass*, *11*(4), 1–18. doi.org/10.1111/lnc3.12239

Xu, M., Yang, J., Siok, W. T., and Tan, L. H. (2015). Atypical lateralization of phonological working memory in developmental dyslexia. *Journal of Neurolinguistics*, *33*, 67–77. doi.org/10.1016/j.jneuroling.2014.07.004

Yamada, Y., Stevens, C., Dow, M., Harn, B. A., Chard, D. J., and Neville, H. J. (2011). Emergence of the neural network for reading in five-year-old beginning readers of different levels of pre-literacy abilities: An fMRI study. *NeuroImage*, *57*(3), 704–713. doi.org/10.1016/j.neuroimage.2010.10.057

Yeatman, J. D., Dougherty, R. F., Rykhlevskaia, E., Sherbondy, A. J., Deutsch, G. K., Wandell, B. A., et al. (2011). Anatomical properties of the arcuate fasciculus predict phonological and reading skills in children. *Journal of Cognitive Neuroscience*, *23*(11), 3304–3317. doi.org/10.1162/jocn_a_00061

Yeatman, J. D., Rauschecker, A. M., and Wandell, B. A. (2013). Anatomy of the visual word form area: Adjacent cortical circuits and long-range white matter connections. *Brain and Language*, 125(2), 146–155. doi.org/10.1016/j.bandl.2012.04.010

Yu, X., Zuk, J., Perdue, M. V., Ozernov-Palchik, O., Raney, T., Beach, S. D., et al. (2020). Putative protective neural mechanisms in prereaders with a family history of dyslexia who subsequently develop typical reading skills. *Human Brain Mapping*, 41(10), 2827–2845. doi.org/10.1002/hbm.24980

....

CHAPTER 28

NEUROCOGNITIVE UNDERPINNINGS OF SOCIAL DEVELOPMENT DURING CHILDHOOD

....

NIKOLAUS STEINBEIS

INTRODUCTION

....

HUMANS are an inordinately social species. We are capable of profound emotions in response to others' joys and sorrows and hugely adept at predicting what others are likely to do. How human adults arrive at such fully fledged experiences and abilities and what the developmental mechanisms are which lead to their emergence and change constitutes a burgeoning field of scientific inquiry. This chapter focusses on developmental mechanisms during childhood, namely a set of regulatory processes that is intricately related to age-related changes in social behaviors and abilities.

I begin by arguing that prosocial behavior carries a cost. Sharing resources means less for oneself and helping another requires time and physical or psychological effort. At the very least prosocial behavior implies opportunity costs with regards to both resource and recipient. Studies have shown that children are aware of these costs, since their prosocial behavior is modulated by both the potentially incurred cost (Eisenberg-Berg et al., 1979; Svetlova et al., 2010) and the value of the resource (Blake and Rand, 2010). I propose that the development of prosociality can be viewed through a value-based decision-making framework (Ruff and Fehr, 2014), whereby the potential costs of prosociality are weighed up against the possible benefits derived from being prosocial, both in the short- and the long-term. I argue that for decisions to be swayed in favor of prosociality, the costs of being prosocial need to be regulated. These regulatory processes undergo protracted developmental change due to the maturation of the underlying neural circuitry.

The first part of this chapter reviews studies showing that distinct forms of prosociality demand distinct regulatory processes (Steinbeis, 2018a). Thus, sharing requires behavioral control and helping requires emotion regulating. These regulatory processes draw on distinct neural mechanisms, which might also explain the observed lack of positive relationships between the two types of prosociality in development (Dunfield et al., 2011; Paulus et al., 2012).

I then discuss social abilities, namely understanding others' minds and feelings (Steinbeis, 2016). Humans face countless social interactions with friends and strangers on a daily basis. In spite of vast differences between these encounters, the contexts within which they occur and the social knowledge available to us, we appear to achieve our social goals with apparent ease. There is much evidence that understanding what others think and feel requires us to rely at least partly on our own projections of what we would think and feel in comparable situations (Nickerson, 1999; O'Brien and Ellsworth, 2012). Such claims are buttressed by findings of shared neural representations when thinking about oneself and others or sharing others' feelings (Jenkins et al., 2008; Lamm et al., 2011). However, to draw on one's own thoughts and feelings only makes sense when these are aligned with those of others. Therefore, to enable smooth social interactions a distinction has to be made between mental representations and emotional experiences of self and other. Known as self-other distinction, this ability has been related to representing others' mental states (Spengler et al., 2009) as well as the ability to empathize (Silani et al., 2013). I will argue that self-other distinction is a crucial mechanism that allows for accurate social understanding. This mechanism undergoes protracted development, which helps to account for age-related changes in understanding others.

As becomes evident, particularly in the section on the role of empathy in helping, there are strong relationships between understanding others' thoughts and feelings and prosocial behavior. Thus, theory of mind predicts both sharing (Takagishi et al., 2010; Cowell et al., 2015) and helping (Imuta et al., 2016) in children; equally empathy is strongly related to sharing (Eisenberg, 2000). In spite of these relationships, for the sake of this chapter, we will offer a separate discussion for each of these behaviors—abilities and their underlying mechanisms. Importantly, work in adults represents a necessary preamble in terms of understanding neural mechanisms. While the focus of this chapter is on the developmental mechanisms during childhood, studies on adults are discussed throughout.[1]

Prior work has mostly focused on the emergence of social behavior and abilities in early childhood around the preschool years (ages 3–5 years; Warneken and Tomasello, 2006; Eisenberg, 2000; Saxe et al., 2004). How this changes during middle childhood (ages 6–12) remains poorly understood (Steinbeis, 2018a). This is particularly noteworthy as middle childhood has been identified as a critical developmental period predicting later mental health and problem behaviors (Feinstein and Bynner, 2004; Mah and Ford-Jones, 2012; Pedersen et al., 2007). Well-adjusted social behavior is crucial for positive developmental trajectories and as such understanding these behaviors and abilities during a significant developmental stage is of particular importance.

PROSOCIAL BEHAVIOR

Prosocial behavior can take multiple forms, such as sharing (i.e., food or money), helping, or comforting others in distress. These forms of prosociality have been documented already in infancy (Dunfield et al., 2011), suggesting early and deep-seated ontogenetic roots (Warneken, 2016). While considerable attention has been dedicated to understanding the origins of prosociality, much less is known about how these behaviors develop throughout childhood. The first part of this chapter reviews studies on the neurocognitive mechanisms of prosocial development, specifically sharing and helping, during middle and older childhood.[2]

The Neurocognitive Mechanisms of Sharing

Sharing valuable resources can be observed in toddlers as young as 15 months (Schmidt and Sommerville, 2011). At this age, sensitivity towards equal distributions and fairness norms also emerges (Geraci and Surian, 2011; McAuliffe et al., 2017). Around 3 years, children state that sharing equally is the norm and from then on, they increasingly follow such sharing norms with their actual behavior (Smith et al., 2013). Complying with social norms constitutes a long-term goal, which conflicts with the more immediate satisfaction of reward maximization (Buckholtz, 2015). Resolving such conflict in favor of sharing according to the norm requires behavioral control. There is by now increasing evidence that behavioral control is positively correlated to sharing in both preschoolers (Aguilar-Pardo et al., 2013; Paulus et al., 2015) and school children (Blake et al., 2015; Smith et al., 2013) using a variety of measures of behavioral control (i.e., day-night tasks, questionnaire measures). More recently, these correlative findings have been extended to show that explicit experimental manipulations of behavioral control impact sharing directly. Thus, in one study, it was shown that children aged 6–9 years shared less after having engaged in a behavioral motor control task compared to sharing after a speeded reaction time task (Steinbeis, 2018b). In another study, 6–9-year-olds shared more after having listened to stories priming behavioral control compared to stories that did not (Steinbeis and Over, 2017). In sum, there is a sizeable body of literature suggesting that behavioral control aids sharing in accordance with the prevailing social norms during middle childhood.

In adults, it has been shown that social norm compliance relies on lateral prefrontal cortical (PFC) brain regions (Spitzer et al., 2007; Ruff et al., 2013). Activity in left and right dorsolateral prefrontal cortex (DLPFC) was positively correlated with sharing under threat of punishment (Spitzer et al., 2007), while disrupting activity in right DLPFC reduced such sharing (Ruff et al., 2013). It has been argued that the top-down modulation from DLPFC of subjective value-signals in the ventromedial prefrontal cortex (VMPFC) is key for both implementing and complying with social norms (Baumgartner et al., 2011). Lateral PFC areas are among the brain regions undergoing the most protracted age-related loss of grey matter volume throughout childhood and adolescence (Lenroot and Giedd, 2006). Further, linear age-related increases in structural connectivity are most delayed in white matter bordering prefrontal cortex (Lebel et al., 2012), which in turn impacts the extent of functional connectivity (Hagmann et al., 2010). Lateral portions of the PFC are involved in actively maintaining task goals, biasing attention, and implementing behaviors (Miller and Cohen, 2001). The maturation of the lateral PFC also underpins the development of these functions in children (Rubia et al., 2007; Bunge et al., 2002), which makes it a suitable candidate region supporting the emergence of behavioral control mediated social norm compliance.

A recent study showed that social norm compliant sharing in children increased between the ages of 6–13 years and this correlated directly with an independent measure of motor control (Steinbeis et al., 2012). Simultaneously recorded brain activity showed that this age-related increase in social norm compliance correlated positively with activity in the left DLPFC. Further, the age-dependent increase in activity in this region mediated the developmental increase in behavioral motor control, which in turn predicted the increase in social norm compliance throughout childhood. While connectivity analyses were not conducted in this study, it is likely that decisions to share are supported by increased functional coupling between DLPFC and brain areas that compute the value of decisions, such as the VMPFC. Such a mechanism has been shown to support decisions in favor of long-term goals in similar scenarios in both adults (Hare et al., 2009; Baumgartner et al., 2011; Hare et al., 2014) as well as during middle childhood (Steinbeis et al., 2016). This interpretation is buttressed by findings from a recent study measuring event-related potentials (ERPs), which show an increase of regulatory processes in bringing about sharing during childhood (Meidenbauer et al., 2016). Thus, in older children, the P3, a component reflecting behavioral control mechanisms predicted equal sharing of resources, whereas in younger children this was predicted by the EPN, an early component reflecting affective evaluation.

In sum, the development of behavioral control, supported by the maturation of function and connectivity of prefrontal cortical circuitry can account for the observed changes in sharing during middle and late childhood. Such a mechanism helps to shift decisions away from the immediate desires of reward maximization to complying to social norms of equal sharing.

The Neurocognitive Mechanisms of Helping

Studies on the ontogeny of prosociality have largely focused on helping, showing that this emerges as early as 14 months (Warneken and Tomasello, 2007). Helping others also implies costs in terms of time, effort, and opportunities. Two key motivations have been argued to underlie helping. A selfish desire to *reduce one's own distress*, also known as personal distress; and an altruistic desire to *reduce the other's distress*, also known as empathic concern (Batson, 1991; Eisenberg, 2000). Whereas personal distress leads to helping only when there is no other recourse of stopping one's own distress (i.e., fleeing the situation), empathic concern leads to helping across a range of situations. Especially in children, only indicators of empathic concern were shown to predict helping (for a review see Eisenberg, 2000). It has been argued that empathic concern arises out of the interplay of an emotional response to the need of another and a sufficiently strong regulation of this emotional response (Eisenberg, 2000). Thus, the literature on the development of helping during childhood suggests that those children both high in emotional responding and emotion regulation are the ones most likely to help (Eisenberg et al., 1998). Support for this comes from studies using parental questionnaires, parent and teacher ratings, as well as psychophysiological indicators suggesting that empathic concern is a good predictor of helping behavior in childhood.

In adults, it has been shown that observing painful or unpleasant experiences of others activates circuitry that is also recruited when undergoing this experience oneself (Singer et al., 2004; Wicker et al., 2003). This circuitry comprises the bilateral anterior insula and medial/anterior cingulate cortex (ACC) (Lamm et al., 2011). Activity in the anterior insula was shown to correlate positively with empathic concern ratings (Singer et al., 2004) and activity of this region when watching others in pain was positively related to helping behavior, albeit only to in-group members (Hein et al., 2010). It has been argued that the anterior insula performs value computations related to prosocial behavior (Hare et al., 2010). More recently, it was shown that especially connectivity between the ACC and the anterior insula was positively related to prosocial behavior following an empathy induction (Hein et al., 2016). The ACC has been implicated in top-down regulation and control of negative emotions and processing of emotional conflict (Etkin et al., 2011) and in this case might function as affective regulation mechanism to produce empathic concern in turn leading to prosocial behavior.

Several studies on the neurocognitive development of empathic concern and helping have been performed in children. These studies show that children aged 7–12 years activate anterior insula as well as anterior mid-cingulate when observing the pain of another (Decety et al., 2008). Similar activation patterns could already be seen from 4 years of age (Michalska et al., 2013). A study testing 7–40-year-old participants showed that activity in the amygdala in response to seeing others' pain decreased with age, while activity in lateral prefrontal cortex increased, suggesting a potential decrease in distress-related and an increase in emotion regulation related brain functions (Decety and Michalska, 2010). A recent longitudinal study in children from 10 to 13 years of age showed that empathic concern at age 10 predicted activation of lateral prefrontal regions, namely the inferior frontal gyrus (IFG), which in turn was linked to helping (Flournoy et al., 2016). Importantly, the IFG activation overlapped with coordinates typically found for cognitively effortful processing. A most recent study also shows that age-related changes in costly helping (i.e., taking on experiences of disgust for another while having the option to also escape) were strongly linked to increased connectivity between the right anterior insula and lateral and medial PFC areas (Hoffmann et al., in preparation).

In sum, the neurocognitive mechanisms of helping during childhood comprise affective responding to the emotional state of another, as coded in the anterior insula, in combination with regulatory mechanisms of the experienced affect instantiated in pre-frontal cortical brain regions.

INTERIM SUMMARY

This first part summarizes the recent literature on the neurocognitive underpinnings of prosocial behavior in childhood. One key finding is that neurocognitive mechanisms differ depending on the prosocial behavior in question. Whereas sharing requires be-havioral control, helping needs the regulation of emotions in response to another's dis-tress. These distinct mechanisms draw on distinct neural circuitry, which in turn could explain why different types of prosocial behavior correlate so poorly during develop-ment (Dunfield et al., 2011).

These findings can be contextualized within a value-based framework. It is important to note that value in this case entails the monetary reward at stake as well as the in-trinsic motivation to affiliate with others (Over, 2016). It has been shown that volun-tary giving is associated with activity in reward-related brain regions (Harbaugh et al., 2007), a supposed correlate of the experience of "warm glow" of giving. Aknin et al. (2012) show that giving leads to happiness in 2-year-olds and Hepach et al. (2012) argue that children of the same age are intrinsically motivated to see others helped. Further, learning to incur rewards for others is directly related to trait empathy (Lockwood et al., 2015). Acting prosocially will be a function of how much one values rewards for oneself on the one side as well as the value of rewards for others and potentially associated regu-latory mechanisms. There is strong evidence that the maturation of prefrontal cortically mediated regulation supports the development of prosocial behavior during childhood. A value-based framework provides a mechanistic account of developmental change in prosocial behavior. Such a framework can also simultaneously account for maturational changes in observable behavior and accommodate for the effect of contextual variables likely to influence the computation of costs and benefits associated with the various available social behaviors.

SOCIAL UNDERSTANDING

We can understand others' behavior in different ways, focusing for instance on their mental states and intentions or on their emotional states (Kanske et al., 2015). These two routes to understanding others are typically referred to as theory of mind and empathy (Singer, 2006). Both have been studied in terms of their emergence and development (Eisenberg, 2000; Eisenberg et al., 1994; Bischof-Köhler, 1988; Apperly and Butterfill, 2009; Wimmer and Perner, 1983).

Understanding Other Minds

One of the most studied phenomena in social development is when children can reliably attribute mental states to others. This pertains to their ability to appreciate that someone looking at the same object but from a different angle might see it differently (i.e., level 2 perspective-taking) as well as attributing beliefs to others. The classic measure for assessing belief attribution is the so-called false-belief test (Wimmer and Perner, 1983; Baron-Cohen et al., 1985). Proposed as the best assessment of a child's theory of mind (Dennett, 1978) are tests tapping the ability to attribute a belief accurately to another agent even when this conflicts with reality and the child's own beliefs. It has been shown that whereas 3-year-olds typically fail such tasks, most children pass by 4 years (Wellman et al., 2001). Passing this task is seen as a fundamental shift in children's understanding of others between the ages of 3 and 4 years (Flavell, 1999; Perner, 1991) in that they build a representation of others' mental states which can differ from their own.[3]

Others' thoughts, beliefs, and intentions, in short mental states, rarely align with one's own (Brass et al., 2009; Santiesteban et al., 2012b). False-belief tests perfectly exemplify the kind of self-other distinction required for accurate perspective-taking to take place in that children need to inhibit attributing their own belief, which is incongruent to the agent's whose belief they in turn need to predict.[4]

The best evidence to date in support of a link between the attribution of mental states and a mechanism of self-other distinction comes from a recent study looking at the effects of training either imitation, imitation-control, or motor control on a subsequent test of visual perspective-taking in adults (Santiesteban et al., 2012b).[5] The visual perspective-taking task (also known as the Director's task; Keysar et al., 2003) requires participants to take the viewpoint of a "director" who gives instructions to move objects on a shelf. Experimental trials entail a conflict between the director's and the participant's perspective, explicitly requiring self-other distinction to avoid making errors. Only training imitation inhibition increased participants' performance on experimental trials, an effect that was absent when training either imitation or motor control more generally. This study thus demonstrates that processes of self-other distinction (and not executive functions per se, as these would have been targeted by the motor control training) underlie perspective-taking especially when the other's perspective potentially conflicts with our own and mediates our tendency to egocentrically attribute our own states.

Evidence that self-other distinction plays a role in the attribution of mental states also comes from the field of neuroimaging. Meta-analyses in adults have shown that the ability to distinguish self-generated actions from externally produced ones relies on some of the same brain region required for mental state attribution, such as the temporo-parietal junction TPJ; Decety and Lamm, 2007). The TPJ is at the intersection of the posterior end of the superior temporal sulcus, the inferior parietal lobule and the lateral occipital cortex. This region is large and has been shown to have heterogeneous projections depending on sub-regions within the TPJ (Mars et al., 2012). Thus, parcellating the TPJ based on diffusion-weighted imaging tractography yielded three distinct regions of TPJ, each with unique resting state functional connectivity profiles. Specifically, a posterior TPJ cluster showed greatest connectivity with other brain regions involved in the attribution of mental states, such as the posterior cingulate, the temporal poles and the medial prefrontal cortex (Frith and Frith, 2006). Consistent with the functional specificity of a posterior cluster of TPJ in cognitive self-other distinction, it has been shown that even within the same set of subjects the same part of posterior TPJ is activated for both inhibiting imitative tendencies and attributing mental states (Spengler et al., 2009).

Evidence for a causal role of TPJ supporting self-other distinction to make accurate mental state attributions in adults comes from a recent study using transcranial direct current stimulation (tDCS; Santiesteban et al., 2012a). Using excitatory (anodal) as well as inhibitory (cathodal) stimulation, it was shown that anodal stimulation improved both the control of imitative tendencies (i.e., distinguishing in favor of self-representations) and also improved visual perspective-taking (i.e., distinguishing in favor of other representations). At the same time, social judgments that did not lead to a conflict of self and other representations were left unchanged by tDCS stimulation. While these findings do not speak for a selective involvement of a specific sub-region of the TPJ, they do provide compelling evidence that TPJ is critically involved in self-other distinction, a key process for reducing the effects of egocentricity in social judgments conflicting with one's own mental state.

So far, imaging studies that capture this crucial developmental period are lacking, with the exception of two recent studies. The first is a cross-sectional study of children aged 3 to 4 years. Children were given classic false-belief tasks and measures of structural brain development were also obtained. In a tract-based statistics analysis it was found that white matter development in the right TPJ correlated with false-belief understanding (Grosse Wiesmann et al., 2017). The importance of right TPJ for passing false-belief tests was underscored by a more recent study demonstrating a relationship between cortical thickness and surface area in TPJ and performance on explicit false-belief tests in 3–4-year-olds (Wiesmann et al., 2020). Interestingly, cortical structure in a separate network of brain regions including the supramarginal gyrus was related to performance on implicit false-belief tests in the same sample. This is in line with a previous study by Hyde et al. (2015) suggesting that anterior portions of TPJ might already play a role in implicit belief ascription in 7-month-old infants using fNIRS. Whereas it is still unclear whether TPJ is also relevant during actual task-performance, these data demonstrate a crucial relevance and early functional specialization of right TPJ for the observed developmental transition between 3–4 years in successfully passing explicit false-belief tests. Earlier studies provide indirect evidence which corroborate our conclusion (Sabbagh et al., 2009). Thus, Sabbagh et al. (2009) measured amplitude and coherence from resting state electroencephalogram (EEG) and linked this with performance on a false-belief task in 4-year-olds. They showed that current density of alpha waves in right TPJ (and dorsomedial prefrontal cortex) correlated best with individual performance in false-belief tasks. Both studies suggest that the structural maturation of rTPJ is critical for self-other distinction in perspective-taking, which leads to a leap in reduced cognitive egocentricity in children.

That brain regions for understanding other minds develop with great specificity early in ontogeny is confirmed by a more recent study (Richardson et al., 2018), which looked at the development of functional brain networks underlying theory of mind abilities in 122 children aged 3–12 years. Children watched short films depicting animate characters highlighting their mental states and physical sensations (usually pain). Activity in two cortical networks known to be involved in the processing of mental states (i.e., bilateral TPJ, precuneus, dorsal, and vental medial prefrontal cortex) and pain (i.e., bilateral medial frontal gyrus, insula, and dorsal anterior cingulate cortex) respectively. It was found that the two networks were functionally distinct by age 3 years and that functional specialization in each network increased throughout childhood.

Understanding Others' Emotions

As the preceding review illustrates, there is compelling evidence that the *simultaneous* representation of one's own and others' mental states relies on processes of self-other distinction, which in turn draws on posterior parts of the TPJ. It has been argued that similar processes underlie the understanding of others' emotions specifically empathic responses (Decety and Lamm, 2007; Lamm et al., 2007; Lamm et al., 2011). As a result, for the experience of empathy, a distinction needs to be made that the *primary* source of one's feeling is the perception of someone else's experience (Singer, 2006), critically differentiating empathy from emotional contagion. Further, failing to uphold a boundary between self and other when seeing another in pain can lead to feelings of personal distress. Such a self-centered response hinders orienting towards the other and responding empathically, in turn negatively affecting prosocial behavior (Batson et al., 1987). More tentative evidence in support of a role for self-other distinction in empathy comes from a meta-analysis showing that empathy, perspective-taking, and distinguishing between one's own and others' actions activates the same areas of TPJ (Decety and Lamm, 2007).

Whereas paradigms on empathic responses typically measure the extent to which we share the feelings of another when we are in an otherwise neutral state, processes of self-other distinction might not necessarily be required. A better way of testing for a potential role of self-other distinction in understanding others' feelings would be through the induction of feelings congruent or incongruent to another. Precisely such a design was deployed by means of affective visuo-tactile stimulation (Silani et al., 2013). Adult participants were invited in pairs and underwent separate but simultaneous stimulation through two experimenters. Stimulation was performed such that participants were touched with material of either positive or negative valence while seeing, at the same time, a depiction of what they were stimulated with as well as what stimulation the other participant was currently undergoing. This way it was possible to create affectively congruent (i.e., both participants received positive or negative touch) or incongruent experiences (i.e., one participant received negative touch while the other received positive touch and vice versa). After each trial, participants were asked to either rate how they themselves felt or how they thought the other felt. Using this procedure, it was possible to assess the extent to which participants might be egocentrically biased in their attribution of emotional states. Such a bias should manifest itself through stronger attributions of emotions to others when these are congruent with one's own states compared to when they are incongruent. Exactly such an emotional egocentricity bias was found. Thus, in a paradigm analogous to perspective-taking studies it was shown that egocentricity occurs also in the affective domain.

Do processes of self-other distinction required to overcome emotional egocentricity rely on the same brain regions as those relied on to overcome cognitive egocentricity? In two fMRI studies in adults, it was found that temporo-parietal areas were activated (Silani et al., 2013). The activations were not part of the more posterior cluster typically found for the attribution of mental states however but rather located more anteriorly, comprising a brain region also known as the supramarginal gyrus (SMG). An overlap of brain regions involved in perspective-taking showed that these were anatomically distinct clusters. Evidence in favor of rSMG subserving mechanisms of self-other distinction was indicated by findings of increased emotional egocentricity after performing low-frequency repetitive transcranial magnetic stimulation (rTMS) over temporo-parietal regions compared to sham stimulation.[6] This suggests that whereas temporo-parietal regions might overall be functionally involved in self-other distinction, precisely which part is recruited depends on the nature of the distinction made (i.e., motor, cognitive or affective). One reason for this could be distinct connectivity profiles of subregions of TPJ, which make posterior regions more suited for self-other distinction in the cognitive and anterior regions more suited for such computations in the affective domain. The findings by Mars et al. (2012) already indicate that posterior rTPJ is intrinsically more connected to brain regions involved in mental state attribution and perspective-taking (i.e., posterior cingulate, temporal poles, and medial prefrontal cortex; Frith and Frith, 2006), whereas anterior portions, corresponding to the rSMG show stronger intrinsic connectivity to brain regions implicated in empathic responses (i.e., anterior insula; anterior cingulate cortex; Singer et al., 2004; Lamm et al., 2011). Such a functional distinction was confirmed more recently using seeds derived from activations in the context of overcoming emotional egocentricity in adults and children (Steinbeis et al., 2015). This suggests a functional differentiation in temporo-parietal areas in terms of their role of self-other distinction in different social domains.

As outlined previously, it has been argued that understanding others' emotional states, also known as empathy, requires a form of self-other distinction (Bird and Viding, 2014; Lamm et al., 2011; Decety and Lamm, 2007). The best empirical evidence for this actually comes from developmental psychology. In a seminal paper, Bischof-Köhler (1988) made such a link explicit by giving children aged 16–24 months a mirror self-recognition test and observing their emotional and behavioral responses to the plight of a confederate adult (confederate cries in response to her teddy's arm breaking off; see also Bischof-Kohler, 1994). She shows that the children passing the mirror self-recognition test (and thus with a mature concept of self and presumably capable of making the distinction between self and others) were also significantly more likely to help the confederate (i.e., comforting; attempts to repair teddy). To date, this study is one of the few pieces of evidence which actually demonstrates the crucial role of self-other distinction in bringing about empathic responses.

Apart from a potential role of self-other distinction in producing the shift from emotional contagion to empathic responses, such a process would be even more necessary when bringing about accurate empathic judgments when the other's state differs from one's own (like in the study by Silani et al., 2013). To test for developmental changes in this ability a study was conducted with children ranging from 6 to 13 years of age as well as adults using monetary rewards and punishments to induce positive and negative affective states (Steinbeis et al., 2015). As in the previous study by Silani et al. (2013), participants were made to feel either congruent or incongruent to one another and asked to rate how they thought the other felt. Two particular age-trends emerged: while overall affective egocentricity declined throughout childhood, adults were also less egocentric than children overall. Importantly both of these findings could be independently replicated using a new paradigm inducing emotions by means of pleasant and unpleasant gustatory stimulation (i.e., sweetened water/juice and saline/quinine solutions respectively; Hoffmann et al., 2015). Crucially, the study by Hoffmann et al. (2015) could show that the developmental decrease in affective egocentricity was critically mediated by resolving conflict between two concurrent emotional representations, which is precisely the sort of computation subserved by self-other distinction. Thus, the self-other distinction required for overcoming affective egocentricity increased significantly both throughout childhood as well as into adulthood. Importantly, both studies included several measures of cognitive perspective-taking. While these were also shown to change significantly with age there were no correlations between emotional egocentricity and cognitive perspective-taking, suggesting that at the behavioral level self-other distinction in the cognitive and the affective domains are dissociable in development.

Using neuroimaging Steinbeis et al. (2015) investigated the neurocognitive mechanisms underlying this developmental effect. Children showed significantly reduced activity of the rSMG compared to adults and this activation difference overlapped with the rSMG activation in the study by Silani et al. (2013). Further, the smaller the affective egocentricity the stronger the functional connectivity between rSMG and left DLPFC, but only in adults and not in children. These data suggest that children are affectively more egocentric than adults as a result of poorer affective self-other distinction in turn due to the prolonged maturation of the relevant brain regions such as rSMG. We also established by means of performing a functional resting-state analysis of seeds in the rTPJ and rSMG that in children the distinct networks of increased connectivity with brain regions involved in mental state and affective state attribution respectively are already present and comparable to adults. This suggests a functional specificity of both rTPJ and rSMG in children as young as 6 years that makes each of these brain regions particularly suited to performing computations of self-other distinction in the cognitive and the affective domain respectively. At the same time however, it appears to be the functional recruitment of these regions when actually having to perform the computations requires further functional maturation (Gweon et al., 2012; Steinbeis et al., 2015).

Interim Summary

In sum, mechanisms of self-other distinction appear to play a crucial role both in the context of understanding others' mental as well as affective states. This is supported by evidence of improvements in perspective-taking following training the inhibition of imitative tendencies on the one hand as well as findings of a strong effect of one's own emotional experience on judgments of others' emotional states. While such mechanisms recruit temporo-parietal brain areas both when needed to overcome cognitive and affective egocentricity, in each case different subregions are activated. Thus, more posterior parts of TPJ are relevant for overcoming cognitive egocentricity, anterior parts of TPJ, specifically rSMG performs such a function for overcoming affective egocentricity. Presumably each of these subregions are uniquely suited to overcome egocentricity in these respective domains as a function of their intrinsic connectivity with brain regions involved in attributing mental and affective states respectively.[7] Given that these temporo-parietal regions are amongst the latest of the cortex to mature (Gogtay et al., 2004), such a protracted developmental process might also relate to the development of self-other distinction and its role in understanding others' mental and emotional states.

Conclusion

This chapter summarizes the most recent literature on the neurocognitive mechanisms that support the development of prosocial behavior and social abilities in childhood. Each of these domains is marked by developmental changes, such as increased sharing and helping as well as improved theory of mind and reduced emotional egocentricity. Each of these changes is supported by specific neurocognitive mechanisms related to the development of a regulatory process, such as behavioral control, self-other distinction and emotion regulation respectively. Crucially, this set of studies demonstrates that these developmental changes are supported by maturational changes in neural circuits to a highly specific set of socio-cognitive and affective processes. Interestingly, the brain regions identified to be involved in sharing, helping, as well as theory of mind and empathy during childhood are remarkably similar to the ones found in adults. This suggests that these brain regions specialize early in terms of their function, and undergo further functional and structural maturation throughout childhood to support socio-cognitive and affective changes.

There is strong evidence that the maturation of prefrontal cortically mediated regulation supports the development of prosocial behavior during childhood. One key finding is that neurocognitive mechanisms differ depending on the prosocial behavior in question. Whereas sharing requires behavioral control, helping needs the regulation of emotions in response to another's distress. These mechanisms draw on distinct neural circuitry, which in turn could explain why different types of prosocial behavior correlate so poorly during development (Dunfield et al., 2011).

This chapter also presents evidence suggesting a crucial role of self-other distinction in understanding others' mental and emotional states in adults as well as in children, particularly when these states are in conflict with our own. Further, self-other distinction appears to be differentiated between cognitive and affective domains. This is supported by a lack of correlations in tasks requiring self-other distinction at the behavioral level as well as distinct neural correlates for each. Both in adults, as well as in children, there is good evidence to suggest that even though cognitive and affective self-other distinctions rely on temporo-parietal brain regions, they rely on distinct sub-regions, namely posterior TPJ and SMG respectively. One reason for this differential recruitment of distinct sub-regions of temporo-parietal cortex could be the intrinsic functional connectivity of TPJ and SMG. Thus, TPJ is more connected to brain regions involved in attributing mental states, whereas SMG is more connected to regions involved in attributing affective states, a pattern present already in childhood.

In sum, the development of social behavior and abilities is crucially subserved by age-related changes in regulatory mechanisms subserved by specific late maturing neural circuitry. Virtually all mental health problems are also characterized by atypical social behavior. Given the importance of social behaviors for mental health and well-being, identifying underlying developmental mechanisms will assist in devising interventions targeting these mechanisms to lead to improved social skills. Future research will need to focus on how social skills can be improved, and how this can be directly titrated to individual differences and developmental stage (i.e., toddlerhood, childhood, adolescence). There is much work to be done in this field, but the policy implications are considerable. The mechanistic work outlined previously is a first step in this direction.

Acknowledgements

This research is funded by an ERC Starting Grant for Innovative Ideas (ERC-2016-StG-715282) and a Jacobs Foundation Early Career Research Fellowship.

Notes

1. There is a plethora of work on social behavior and social processes in adolescence. I do not review this here, but point to this excellent review: Crone, E. A. and Fuligni, A. J. (2020). Self and others in adolescence. *Annual Review of Psychology*, 71(71), 447–69.

2. The literature on neurocognitive mechanisms of comforting in children is virtually non-existent.

3. Interestingly, a study on the attribution of desires shows that children can reliably attribute a desire to others incongruent to their own from around 18 months (Repacholi, B. M. and Gopnik, A. 1997. Early reasoning about desires: Evidence from 14- and 18-month-olds. *Developmental Psychology*, 33, 12–21). Some see this as evidence for the fundamentally different nature of belief attribution compared to other states (Saxe, R., Carey, S., and Kanwisher, N. 2004. Understanding other minds: Linking developmental psychology and functional neuroimaging. *Annual Review of Psychology*, 55, 87–124).

4. One crucial question is however if children may not have acquired a theory of mind prior to passing explicit false-belief tests. Explicit false-belief tests have been criticized partly because they rely on well-developed language skills as well as executive functions, abilities which also develop between the ages of 3 and 4 years (Bloom, P. and German, T. P. 2000. Two reasons to abandon the false belief task as a test of theory of mind. *Cognition*, 77, B25–B31). Such task-related demands may mask the underlying abilities that for instance infants might possess (Onishi, K. H. and Baillargeon, R. 2005. Do 15-month-old infants understand false beliefs? *Science*, 308, 255–8). Using false-belief tasks that do not explicitly refer to the belief context (and thus avoid confounding this with mastery of language and level of executive functions) when asking for responses it has been shown that infants already in their second year of life pass non-verbal false-belief tests (ibid.) (Southgate, V., Senju, A., and Csibra, G. 2007. Action anticipation through attribution of false belief by 2-year-olds. *Psychol Sci*, 18, 587–92). This would mean that passing a false-belief test at around 4 years is more a test of how well mechanisms of self-other distinction are in place as opposed to whether the child can represent another's mental state. Findings of an early theory of mind have been questioned however as to whether they really reflect an infant's access to another's belief as opposed to relying on basic behavioral cues (Perner, J. and Ruffman, T. 2005. Infants' insight into the mind: How deep? *Science*, 308, 214–16) or effects of novelty biasing responses (Heyes, C. 2014. False belief in infancy: A fresh look. *Developmental Science*, 17, 647–59). Thus, depending on whether evidence in favor of early forms of theory of mind holds up against current criticisms, passing false-belief tests at around the age of 4 years reflects a critical developmental step in terms of self-other distinction and perhaps also the acquisition of a theory of mind.

5. The control or inhibition of automatic imitation of others' actions can be used as an index for self-other distinction. Typically participants are asked to lift their index or middle finger in response to a number, while watching either congruent (i.e., the same) or incongruent (i.e., the opposite) finger movements of a videotaped hand (Brass, M., Bekkering, H., Wohlschlager, A., and Prinz, W. 2000. Compatibility between observed and executed finger movements: comparing symbolic, spatial, and imitative cues. *Brain Cognition*, 44, 124–43). Whereas in congruent trials the videotaped hand and the instructed movement are identical, in incongruent trials the instructed and observed movements differ. Movements in congruent trials are quasi-imitative leading to faster reaction times, while in incongruent trials such imitative tendencies have to be controlled, leading to slower reaction times. The resulting difference provides a measure for the extent of the interference and makes the degree of self-other distinction in the domain of perception and action measurable.

6. Note that even though the regional specificity of rTMS is too imprecise to say with any certainty that only rSMG was affected and in turn leading to increased emotional egocentricity, the fact that only rSMG (and not rTPJ) was activated in the fMRI studies makes

it reasonable to assume that only the disruption of brain regions functionally implicated would lead to such an increase in emotional egocentricity.

7. Interestingly a recent study has shown that psychosocial stress can lead to comparable detriments across motor, cognitive and affective domains as indicated by a reduction in imitation control as well as greater cognitive and affective egocentricity (Tomova, L., Von Dawans, B., Heinrichs, M., Silani, G., and Lamm, C. 2014. Is stress affecting our ability to tune into others? Evidence for gender differences in the effects of stress on self-other distinction. *Psychoneuroendocrinology*, 43, 95–104). While this tells us little about the underlying mechanisms, it shows that social contexts can have a global effect on self-other distinction across domains.

REFERENCES

Aguilar-Pardo, D., Martinez-Arias, R., and Colmenares, F. (2013). The role of inhibition in young children's altruistic behaviour. *Cognitive Procession*, 14, 301–307.

Aknin, L. B., Hamlin, J. K., and Dunn, E. W. (2012). Giving leads to happiness in young children. *PLoS One*, 7, e39211.

Apperly, I. A. and Butterfill, S. A. (2009). Do humans have two systems to track beliefs and belief-like states? *Psychological Review*, 116, 953–970.

Baron-Cohen, S., Leslie, A. M., and Frith, U. (1985). Does the autistic child have a "theory of mind"? *Cognition*, 21, 37–46.

Batson, C. D. (1991). *The Altruism Question: Toward a Social-Psychological Answer*, Hillsdale, NJ: Erlbaum.

Batson, C. D., Fultz, J., and Schoenrade, P. A. (1987). Distress and empathy – 2 qualitatively distinct vicarious emotions with different motivational consequences. *Journal of Personality*, 55, 19–39.

Baumgartner, T., Knoch, D., Hotz, P., Eisenegger, C., and Fehr, E. (2011). Dorsolateral and ventromedial prefrontal cortex orchestrate normative choice. *Nature Neuroscience*, 14, 1468–U149.

Bird, G. and Viding, E. (2014). The self to other model of empathy: Providing a new framework for understanding empathy impairments in psychopathy, autism, and alexithymia. *Neuroscience Biobehavioral Review*, 47, 520–532.

Bischof-Kohler, D. (1994). Self-object and interpersonal emotions. Identification of own mirror image, empathy and prosocial behavior in the 2nd year of life. *Z Psychol Z Angew Psychol*, 202, 349–377.

Bischof-Köhler, D. (1988). Ueber den Zusammenhang von Empathie und der Faehigkeit, sich im Spiegel zu erkennen. *Schweizerische Zeitschrift fuer Psychologie*, 47, 147–159.

Blake, P. R., Piovesan, M., Montinari, N., Warneken, F., and Gino, F. (2015). Prosocial norms in the classroom: The role of self-regulation in following norms of giving. *Journal of Economic Behavior and Organization*, 115, 18–29.

Blake, P. R. and Rand, D. G. (2010). Currency value moderates equity preference among young children. *Evolution and Human Behavior*, 31, 210–218.

Bloom, P. and German, T. P. (2000). Two reasons to abandon the false belief task as a test of theory of mind. *Cognition*, 77, B25–B31.

Brass, M., Bekkering, H., Wohlschlager, A., and Prinz, W. (2000). Compatibility between observed and executed finger movements: Comparing symbolic, spatial, and imitative cues. *Brain Cognition*, 44, 124–143.

Brass, M., Ruby, P., and Spengler, S. (2009). Inhibition of imitative behaviour and social cognition. *Philosophical Transactions of the Royal Society B-Biological Sciences*, 364, 2359–67.

Buckholtz, J. W. (2015). Social norms, self-control, and the value of antisocial behavior. *Current Opinion in Behavioral Sciences*, 3, 122–129.

Bunge, S. A., Dudukovic, N. M., Thomason, M. E., Vaidya, C. J., and Gabrieli, J. D. (2002). Immature frontal lobe contributions to cognitive control in children: Evidence from fMRI. *Neuron*, 33, 301–111.

Cowell, J. M., Samek, A., List, J., and Decety, J. (2015). The curious relation between theory of mind and sharing in preschool age children. *Plos One*, 10, 1–8.

Crone, E. A. and Fuligni, A. J. (2020). Self and others in adolescence. *Annual Review of Psychology*, 71(71), 447–469.

Decety, J. and Lamm, C. (2007). The role of the right temporoparietal junction in social interaction: How low-level computational processes contribute to meta-cognition. *Neuroscientist*, 13, 580–593.

Decety, J. and Michalska, K. J. (2010). Neurodevelopmental changes in the circuits underlying empathy and sympathy from childhood to adulthood. *Developmental Science*, 13, 886–899.

Decety, J., Michalska, K. J., and Akitsuki, Y. (2008). Who caused the pain? An fMRI investigation of empathy and intentionality in children. *Neuropsychologia*, 46, 2607–2614.

Dennett, D. (1978). Beliefs about beliefs. *Behavioral and Brain Sciences*, 1, 568–570.

Dunfield, K., Kuhlmeier, V. A., O'Connell, L., and Kelley, E. (2011). Examining the diversity of prosocial behavior: Helping, sharing, and comforting in infancy. *Infancy*, 16, 227–247.

Eisenberg-Berg, N., Haake, R., Hand, M., and Sadalla, E. (1979). Effects of instructions concerning ownership of a toy on preschoolers' sharing and defensive behaviors. *Developmental Psychology*, 15, 460–461.

Eisenberg, N. (2000). Emotion, regulation, and moral development. *Annual Review of Psychology*, 51, 665–697.

Eisenberg, N., Fabes, R. A., Murphy, B., Karbon, M., Maszk, P., Smith, M., O'Boyle, C., and Suh, K. (1994). The relations of emotionality and regulation to dispositional and situational empathy-related responding. *Journal of Perceptive Social Psychology*, 66, 776–797.

Eisenberg, N., Fabes, R. A., Shepard, S. A., Murphy, B. C., Jones, S., and Guthrie, I. K. (1998). Contemporaneous and longitudinal prediction of children's sympathy from dispositional regulation and emotionality. *Developmental Psychology*, 34, 910–924.

Etkin, A., Egner, T., and Kalisch, R. (2011). Emotional processing in anterior cingulate and medial prefrontal cortex. *Trends Cognitive Science*, 15, 85–93.

Feinstein, L. and Bynner, J. (2004). The importance of cognitive development in middle childhood for adulthood socioeconomic status, mental health, and problem behavior. *Child Development*, 75, 1329–1339.

Flavell, J. H. (1999). Cognitive development: Children's knowledge about the mind. *Annual Review of Psychology*, 50, 21–45.

Flournoy, J. C., Pfeifer, J. H., Moore, W. E., Tackman, A. M., Masten, C. L., Mazziotta, J. C., Iacoboni, M., and Dapretto, M. (2016). Neural reactivity to emotional faces may mediate the relationship between childhood empathy and adolescent prosocial behavior. *Child Development*, 87, 1691–1702.

Frith, C. D. and Frith, U. (2006). The neural basis of mentalizing. *Neuron*, 50, 531–534.

Geraci, A. and Surian, L. (2011). The developmental roots of fairness: Infants' reactions to equal and unequal distributions of resources. *Developmental Science*, 14, 1012–1020.

Gogtay, N., Giedd, J. N., Lusk, L., Hayashi, K. M., Greenstein, D., Vaituzis, A. C., Nugent, T. F., Herman, D. H., Clasen, L. S., Toga, A. W., Rapoport, J. L., and Thompson, P. M. (2004). Dynamic mapping of human cortical development during childhood through early adulthood. *Proceedings of the National Academy of Sciences U S A*, 101, 8174–8179.

Grosse Wiesmann, C., Schreiber, J., Singer, T., Steinbeis, N., and Friederici, A. D. (2017). White matter maturation is associated with the emergence of Theory of Mind in early childhood. *National Communication*, 8, 14692.

Gweon, H., Dodell-Feder, D., Bedny, M., and Saxe, R. (2012). Theory of mind performance in children correlates with functional specialization of a brain region for thinking about thoughts. *Child Development*, 83, 1853–1868.

Hagmann, P., Sporns, O., Madan, N., Cammoun, L., Pienaar, R., Wedeen, V. J., Meuli, R., Thiran, J. P., and Grant, P. E. (2010). White matter maturation reshapes structural connectivity in the late developing human brain. *Proceedings of the National Academy of Sciences U S A*, 107, 19067–19072.

Harbaugh, W. T., Mayr, U., and Burghart, D. R. (2007). Neural responses to taxation and voluntary giving reveal motives for charitable donations. *Science*, 316, 1622–1625.

Hare, T. A., Camerer, C. F., Knoepfle, D. T., and Rangel, A. (2010). Value computations in ventral medial prefrontal cortex during charitable decision-making incorporate input from regions involved in social cognition. *Journal of Neuroscience*, 30, 583–590.

Hare, T. A., Camerer, C. F., and Rangel, A. (2009). Self-control in decision-making involves modulation of the vmPFC valuation system. *Science*, 324, 646–648.

Hare, T. A., HAKIMI, S., and Rangel, A. (2014). Activity in dlPFC and its effective connectivity to vmPFC are associated with temporal discounting. *Frontiers of Neuroscience*, 8, 50.

Hein, G., Morishima, Y., Leiberg, S., Sul, S., and Fehr, E. (2016). The brain's functional network architecture reveals human motives. *Science*, 351, 1074–1078.

Hein, G., Silani, G., Preuschoff, K., Batson, C. D., and Singer, T. (2010). Neural responses to ingroup and outgroup members' suffering predict individual differences in costly helping. *Neuron*, 68, 149–160.

Hepach, R., Vaish, A. and Tomasello, M. (2012). Young children are intrinsically motivated to see others helped. *Psychological Science*, 23, 967–972.

Heyes, C. (2014). False belief in infancy: a fresh look. *Developmental Science*, 17, 647–659.

Hoffmann, F., Grosse Wiesmann, C., singer, T. and Steinbeis, N. In preparation. Maturation of prefrontal-insular connectivity predicts developmental shift towards altruistic helping in childhood.

Hoffmann, F., Singer, T. and Steinbeis, N. (2015). Children's increased emotional egocentricity compared to adults is mediated by age-related differences in conflict processing. *Child Devopment*, 86, 765–780.

Hyde, D. C., Aparicio Betancourt, M., and Simon, C. E. (2015). Human temporal-parietal junction spontaneously tracks others' beliefs: A functional near-infrared spectroscopy study. *Human Brain Mapping*, 36, 4831–4846.

Imuta, K., Henry, J. D., Slaughter, V., Selcuk, B., and Ruffman, T. (2016). Theory of mind and prosocial behavior in childhood: A meta-analytic review. *Developmental Psychology*, 52, 1192–1205.

Jenkins, A. C., Macrae, C. N. and Mitchell, J. P. (2008). Repetition suppression of ventromedial prefrontal activity during judgments of self and others. *Proceedings of the National Academy of Sciences U S A*, 105, 4507–4512.

Kanske, P., Bockler, A., Trautwein, F. M., and Singer, T. (2015). Dissecting the social brain: Introducing the EmpaToM to reveal distinct neural networks and brain-behavior relations for empathy and theory of mind. *Neuroimage*, 122, 6–19.

Keysar, B., Lin, S. H., and Barr, D. J. (2003). Limits on theory of mind use in adults. *Cognition*, 89, 25–41.

Lamm, C., Batson, C. D., and Decety, J. (2007). The neural substrate of human empathy: Effects of perspective-taking and cognitive appraisal. *Journal of Cognitive Neuroscience*, 19, 42–58.

Lamm, C., Decety, J., and Singer, T. (2011). Meta-analytic evidence for common and distinct neural networks associated with directly experienced pain and empathy for pain. *Neuroimage*, 54, 2492–2502.

Lebel, C., Gee, M., Camicioli, R., Wieler, M., Martin, W., and Beaulieu, C. (2012). Diffusion tensor imaging of white matter tract evolution over the lifespan. *Neuroimage*, 60, 340–352.

Lenroot, R. K. and Giedd, J. N. (2006). Brain development in children and adolescents: Insights from anatomical magnetic resonance imaging. *Neuroscience Biobehavioural Review*, 30, 718–729.

Lockwood, P. L., Apps, M. A. J., Roiser, J. P. and Viding, E. (2015). Encoding of vicarious reward prediction in anterior cingulate cortex and relationship with trait empathy. *Journal of Neuroscience*, 35, 13720–13727.

Mah, V. K. and Ford-Jones, E. L. (2012). Spotlight on middle childhood: Rejuvenating the 'forgotten years'. *Paediatrics & Child Health*, 17, 81–83.

Mars, R. B., Sallet, J., Schuffelgen, U., Jbabdi, S., Toni, I., and Rushworth, M. F. (2012). Connectivity-based subdivisions of the human right "temporoparietal junction area": Evidence for different areas participating in different cortical networks. *Cerebral Cortex*, 22, 1894–1903.

McAuliffe, K., Blake, P. R., Steinbeis, N., and Warneken, F. (2017). The developmental foundations of human fairness. *Nature Human Behaviour*, 1, 0042.

Meidenbauer, K. L., Cowell, J. M., Killen, M., and Decety, J. (2016). A developmental neuroscience study of moral decision making regarding resource allocation. *Child Development*, 89, 1177–1192.

Michalska, K. J., Kinzler, K. D. and Decety, J. (2013). Age-related sex differences in explicit measures of empathy do not predict brain responses across childhood and adolescence. *Developmental Cognitive Neuroscience*, 3, 22–32.

Miller, E. K. and Cohen, J. D. (2001). An integrative theory of prefrontal cortex function. *Annual Review of Neuroscience*, 24, 167–202.

Nickerson, R. S. (1999). How we know – and sometimes misjudge – what others know: Imputing one's own knowledge to others. *Psychological Bulletin*, 125, 737–759.

O'Brien, E. and Ellsworth, P. C. (2012). More than skin deep: Visceral states are not projected onto dissimilar others. *Psychological Science*, 23, 391–396.

Onishi, K. H. and Baillargeon, R. (2005). Do 15-month-old infants understand false beliefs? *Science*, 308, 255–258.

Over, H. (2016). The origins of belonging: Social motivation in infants and young children. *Philosophical Translation Royal Society London B Biological Sciences*, 371, 20150072.

Paulus, M., Kuhn-Popp, N., Licata, M., Sodian, B., and Meinhardt, J. (2012). Neural correlates of prosocial behavior in infancy: Different neurophysiological mechanisms support the emergence of helping and comforting. *Neuroimage*, 66C, 522–530.

Paulus, M., Licata, M., Kristen, S., Thoermer, C., Woodward, A., and Sodian, B. (2015). Social understanding and self-regulation predict preschoolers' sharing with friends and disliked peers: A longitudinal study. *International Journal of Behavioral Development*, 39, 53–64.

Pedersen, S., Vitaro, F., Barker, E. D., and Borge, A. I. (2007). The timing of middle-childhood peer rejection and friendship: Linking early behavior to early-adolescent adjustment. *Child Development*, 78, 1037–1051.

Perner, J. (1991). *Understanding the Representational Mind. Learning, Development and Conceptual Change*. MIT Press.

Perner, J. and Ruffman, T. (2005). Infants' insight into the mind: How deep? *Science*, 308, 214–216.

Repacholi, B. M. and Gopnik, A. (1997). Early reasoning about desires: Evidence from 14- and 18-month-olds. *Developmental Psychology*, 33, 12–21.

Richardson, H., Lisandrelli, G., Riobueno-Naylor, A., and Saxe, R. (2018). Development of the social brain from age three to twelve years. *National Communication*, 9, 1027.

Rubia, K., Smith, A. B., Taylor, E., and Brammer, M. (2007). Linear age-correlated functional development of right inferior fronto-striato-cerebellar networks during response inhibition and anterior cingulate during error-related processes. *Human Brain Mapping*, 28, 1163–1177.

Ruff, C. C. and Fehr, E. (2014). The neurobiology of rewards and values in social decision making. *National Review of Neuroscience*, 15, 549–562.

Ruff, C. C., Ugazio, G., and Fehr, E. (2013). Changing social norm compliance with noninvasive brain stimulation. *Science*, 342, 482–484.

Sabbagh, M. A., Bowman, L. C., Evraire, L. E., and Ito, J. M. (2009). Neurodevelopmental correlates of theory of mind in preschool children. *Child Development*, 80, 1147–1162.

Santiesteban, I., Banissy, M. J., Catmur, C., and Bird, G. (2012a). Enhancing social ability by stimulating right temporoparietal junction. *Current Biology*, 22, 2274–2277.

Santiesteban, I., White, S., Cook, J., Gilbert, S. J., Heyes, C., and Bird, G. (2012b). Training social cognition: From imitation to Theory of Mind. *Cognition*, 122, 228–235.

Saxe, R., Carey, S., and Kanwisher, N. (2004). Understanding other minds: Linking developmental psychology and functional neuroimaging. *Annual Review of Psychology*, 55, 87–124.

Schmidt, M. F. H. and Sommerville, J. A. (2011). Fairness expectations and altruistic sharing in 15-month-old human infants. *Plos One*, 6, e23223. https://doi.org/10.1371/journal.pone.0023223.

Silani, G., Lamm, C., Ruff, C., and Singer, T. (2013). Right supramarginal gyrus is crucial to overcome emotional egocentricity bias in social judgments. *Journal of Neuroscience*, 33, 15466–15476.

Singer, T. (2006). The neuronal basis and ontogeny of empathy and mind reading: Review of literature and implications for future research. *Neuroscience and Biobehavioral Reviews*, 30, 855–863.

Singer, T., Seymour, B., O'Doherty, J., Kaube, H., Dolan, R. J., and Frith, C. D. (2004). Empathy for pain involves the affective but not sensory components of pain. *Science*, 303, 1157–1162.

Smith, C. E., Blake, P. R., and Harris, P. L. (2013). I should but I won't: Why young children endorse norms of fair sharing but do not follow them. *PLoS One*, 8, e59510.

Southgate, V., Senju, A., and Csibra, G. (2007). Action anticipation through attribution of false belief by 2-year-olds. *Psychological Science*, 18, 587–592.

Spengler, S., Von Cramon, D. Y., and Brass, M. (2009). Control of shared representations relies on key processes involved in mental state attribution. *Human Brain Mapping*, 30, 3704–18.

Spitzer, M., Fischbacher, U., Herrnberger, B., Gron, G., and Fehr, E. (2007). The neural signature of social norm compliance. *Neuron*, 56, 185–196.

Steinbeis, N. (2016). The role of self-other distinction in understanding others' mental and emotional states: Neurocognitive mechanisms in children and adults. *Philosophical Translations Royal Society of London B Biological Science*, 371, 20150074.

Steinbeis, N. (2018a). Neurocognitive mechanisms of prosociality in childhood. *Current Opinion in Psychology*, 20, 30–34.

Steinbeis, N. (2018b). Taxing behavioral control diminishes sharing and costly punishment in childhood. *Developmental Science*, 21, e12492.

Steinbeis, N., Bernhardt, B. C., and Singer, T. (2012). Impulse control and underlying functions of the left DLPFC mediate age-related and age-independent individual differences in strategic social behavior. *Neuron*, 73, 1040–1051.

Steinbeis, N., Bernhardt, B. C., and Singer, T. (2015). Age-related differences in function and structure of rSMG and reduced functional connectivity with DLPFC explains heightened emotional egocentricity bias in childhood. *Social Cognitive Affective Neuroscience*, 10, 302–310.

Steinbeis, N., Haushofer, J., Fehr, E. and Singer, T. (2016). Development of behavioral control and associated vmPFC-DLPFC connectivity explains children's increased resistance to temptation in intertemporal choice. *Cerebral Cortex*, 26, 32–42.

Steinbeis, N. and Over, H. (2017). Enhancing behavioral control increases sharing in children. *Journal of Experimental Child Psychology*, 159, 310–318.

Svetlova, M., Nichols, S. R., and Brownell, C. A. (2010). Toddlers' prosocial behavior: from instrumental to empathic to altruistic helping. *Child Development*, 81, 1814–1827.

Takagishi, H., Kameshima, S., Schug, J., Koizumi, M., and Yamagishi, T. (2010). Theory of mind enhances preference for fairness. *Journal of Experimental Child Psychology*, 105, 130–137.

Tomova, L., Von Dawans, B., Heinrichs, M., Silani, G. and Lamm, C. (2014). Is stress affecting our ability to tune into others? Evidence for gender differences in the effects of stress on self-other distinction. *Psychoneuroendocrinology*, 43, 95–104.

Warneken, F. (2016). Insight into the biological foundation of human altruistic sentiments. *Current Opinion in Psychology*, 7, 51–56.

Warneken, F. and Tomasello, M. (2006). Altruistic helping in human infants and young chimpanzees. *Science*, 311, 1301–1303.

Warneken, F. and Tomasello, M. (2007). Helping and cooperation at 14 months of age. *Infancy*, 11, 271–294.

Wellman, H. M., Cross, D., and Watson, J. (2001). Meta-analysis of theory-of-mind development: The truth about false belief. *Child Development*, 72, 655–684.

Wicker, B., Keysers, C., Plailly, J., Royet, J. P., Gallese, V., and Rizzolatti, G. (2003). Both of us disgusted in My insula: The common neural basis of seeing and feeling disgust. *Neuron*, 40, 655–664.

Wiesmann, C. G., Friederici, A. D., Singer, T., and Steinbeis, N. (2020). Two systems for thinking about others' thoughts in the developing brain. *Proceedings of the National Academy of Sciences of the U S A*, 117, 6928–6935.

Wimmer, H. and Perner, J. (1983). Beliefs about beliefs – representation and constraining function of wrong beliefs in young children's understanding of deception. *Cognition*, 13, 103–128.

CHAPTER 29

PUBERTY AND SOCIAL BRAIN DEVELOPMENT

MARJOLEIN E. A. BARENDSE AND
JENNIFER H. PFEIFER

INTRODUCTION

PUBERTY can be defined as the transitional period marking the start of adolescence. It is characterized by substantial change in many areas of development, including hormonal, physical, neuronal, psychological, and social domains. A rapidly expanding body of work shows how development in each of these areas is related to one another. In this chapter, we outline how puberty, and its underlying hormonal and physical changes, might elicit a sensitive period for the development of the social brain. The ultimate aim of puberty is to achieve sexual maturation. However, in humans, a complex set of social skills required to manage adult relationships develops alongside this. We start by giving an overview of the hormonal processes underlying puberty and the mechanisms through which they might impact the structure and functioning of the brain. Next, we lay out the knowledge gained so far about how pubertal development relates to the structural and functional development of the social brain. We suggest ways forward to understand the relevance of individual differences in pubertal processes to social brain development, and to unravel the mechanisms via which pubertal processes impact social behavior through social brain functioning.

PUBERTAL PROCESSES

Puberty can generally be divided into two phases. The earliest phase is adrenarche, which starts around age 7–8 (Campbell, 2006; Reiter, Fuldauer, and Root, 1977). The zona reticularis of the adrenal glands forms during adrenarche, which produces increasing levels of the adrenal hormones dehydroepiandrosterone (DHEA) and its sulphate (DHEAS). The levels of these hormones continue to increase throughout adolescence and reach their peak in early adulthood (Søeborg et al., 2014; Auchus and Rainey, 2004). Both DHEA and DHEAS are neuroactive steroids that can alter neuronal excitability in partially overlapping and partially unique ways (Maninger et al., 2009). They can both be metabolized in the brain and converted into each other. Testosterone also increases during adrenarche, in equivalent levels in boys and girls. During this time, most testosterone is of adrenal origin, metabolized via the conversion of DHEA in the adrenal cortex. DHEA and testosterone show a diurnal pattern with a peak in the early morning (Matchock, Dorn, and Susman, 2007). The rise in testosterone during adrenarche is associated with the appearance of some body hair and skin oiliness, though these changes occur later than the hormonal changes and are much more subtle than the physical changes linked to gonadarche.

The more well-known phase of puberty, gonadarche, starts 2–3 years after the start of adrenarche, about a year earlier in girls than in boys (Mouritsen et al., 2013). It is triggered by the re-activation of the hypothalamic-pituitary-gonadal (HPG)-axis, which has been quiescent since infancy. This leads to increases in the levels of luteinizing hormone (LH) and follicle-stimulating hormone (FSH), which later on results in steep increases in the levels of the sex hormones testosterone (most strongly for boys) and estradiol and progesterone (most strongly for girls), as the gonads develop (Søeborg et al., 2014). LH and FSH are released from the pituitary in pulses, mostly during the night, whereas estradiol (like testosterone) shows a diurnal variation that peaks in the morning (Matchock, Dorn, and Susman, 2007; Styne and Grumbach, 2011). A monthly cycle in estradiol and progesterone levels develops in girls prior to or around menarche (Hansen, Hoffman, and Ross, 1975). Menarche is one of the later steps/milestones in pubertal development for girls and occurs at median age of 12.25 to 12.75 years (Biro et al., 2018; Lazzeri et al., 2018; Villamor et al., 2017). However, this varies by race/ethnicity within the US, with black and Latina girls experiencing menarche earlier than white girls (Wu et al., 2002; Biro et al., 2018), and it is likely different outside of the western world. The increasing levels of sex hormones trigger the prominent bodily changes (secondary sex characteristics) visible during gonadarche, including breast or testicular growth, and pubic and facial hair. Although most bodily changes occur in early-to-mid adolescence, both gonadal and adrenal hormones continue to increase into the third decade of life (Søeborg et al., 2014). The two phases of puberty are suggested to be independent of each other, as children with a clinical disturbance in adrenarche mostly show a normal progression through gonadarche, and vice versa (Styne and Grumbach, 2011). For more information on the measurement of pubertal processes, see Box 29.1.

Box 29.1 Measuring puberty

Several aspects of puberty can be measured and related to other domains of development. To examine normative development, researchers have commonly assessed pubertal status, which refers to the stage of puberty that an individual is currently in. This can be assessed through self-reports or physical examination of secondary sexual characteristics. The most commonly used self-assessment instruments to measure pubertal stage are the Pubertal Development Scale (Petersen et al., 1988) and the Tanner stage line-drawings (Morris and Udry, 1980), which can both be converted into five stages: pre-, early, mid-, late, and post-puberty. Researchers have also indexed normative pubertal development by examining levels of adrenal and gonadal hormones. As discussed in "Pubertal processes", levels of these hormones strongly increase during puberty, thus higher hormone levels are considered an indicator of more advanced pubertal development. Age and pubertal status tap into different but partially overlapping developmental processes that contribute to brain development. Therefore, some of the studies examining associations between pubertal stage or hormone levels and brain development have controlled for age in their models to capture variance uniquely related to pubertal development. The downside is that shared variance between age and puberty is filtered out, which is a larger component when the age range is wide. Therefore, in the subsequent sections we mention whether studies controlled for age, and specify the age range of study when it is either really narrow or really broad.

Individual variation in when and how rapidly pubertal development occurs is large. This variance results in other measurable aspects of puberty, timing, and tempo. Pubertal timing is a person's pubertal status relative to same-age peers; in other words, someone with early timing shows a more advanced development than most of his/her peers. Puberty is considered to be precocious, i.e., clinically early, if gonadarche starts before age 8 for girls or 9 for boys (Latronico, Brito, and Carel, 2016). So, above these ages, there is a large normative variation, with ranges for the onset of gonadarche found to be 4 to 5 years in many studies (Biro et al., 2018; Semiz et al., 2008; Aksglaede et al., 2009). Timing of puberty can be indexed by taking any of the measures of pubertal status described in the previous paragraph and varying out age, either statistically (taking the residuals from regressing puberty on age) or by taking a group of participants who are all the same age. Further, timing of gonadarche in girls is sometimes conceptualized as the age at menarche.

Finally, tempo refers to the speed of progression through puberty. This also shows substantial variation, for example, the time between onset of breast development and age at menarche ranged from 2 to 3.5 years (interquartile range) in a large sample of US girls (Biro et al., 2018). Measuring tempo requires multiple measurements in the same individual (ideally three or more), and this methodological complexity is probably the reason that tempo is the least well-studied aspect of puberty in relation to brain development or health outcomes.

The timing and tempo of adrenarche have gathered much less attention but are thought to also show considerable and relevant variation between children (Barendse et al., 2020b; Byrne et al., 2017). The status, timing, and tempo of adrenarche and gonadarche can all be examined separately, of course depending on the age range studied and the hormonal and physical development indicators collected.

Considering the multitude of processes that puberty encompasses, as well as the many indicators of pubertal development as described above, one can imagine that measuring puberty has been a challenge for the field. One of the ways researchers have dealt with this, is to measure pubertal development in multiple ways within the same study, which has often led to discrepant findings between measures, as detailed in "Puberty and brain structure" and "Puberty and social brain function." The differences between the sexes in secondary sexual characteristics, hormone levels, and timing of gonadarche further complicates this. Researchers should at least determine interactions between sex and their puberty predictor variable(s), if not examining each sex separately.

MECHANISMS LINKING PUBERTY AND BRAIN DEVELOPMENT

Pubertal changes are likely to shape brain structure and functioning through several mechanisms, including a direct impact of hormones on the brain and more indirect pathways, for example through changes in the social environment. The direct impact of hormones on the brain has been studied most extensively in animal research. This has led to an influential theory on the link between hormones and brain structure and function: the organizational-activational hypothesis (adapted to puberty by Schulz, Molenda-Figueira, and Sisk, 2009; originally posited by Phoenix et al., 1959). This theory states that steroid hormones influence behavior both through long-lasting effects on the organization of neural circuits and through faster activational effects on neural function, although the activational effects can also be long lasting. Although fetal development is the most crucial phase for brain organization, both organizational and activational processes are thought to take place during puberty (Schulz, Molenda-Figueira, and Sisk, 2009).

The theory is supported by findings of animal research that testosterone, estradiol, DHEA, and DHEAS have effects on neuronal growth, axon growth and survival of neurons (Arevalo, Azcoitia, and Garcia-Segura, 2015; Maninger et al., 2009; Ahmed et al., 2008; McCarthy, 2008) and can have a long-lasting impact on brain organization (Schulz, Molenda-Figueira, and Sisk, 2009). These effects are often timing dependent, such that administering the hormone to adult animals has no or a different effect. The effects are also regionally specific, depending on (among other things) the distribution of androgen and estrogen receptors. For example, gonadal hormones affect neurogenesis and cellular proliferation in the medial amygdala, leading to bigger volumes in males (in Syrian hamsters (De Lorme et al., 2012); and rats (Ahmed et al., 2008)). Castration of male animals blocks the effects of these hormones, removing the sex difference that normally develops in the volume of the medial amygdala (Ahmed et al., 2008). In the medial prefrontal cortex on the other hand, pubertal hormones are thought to influence cell death, leading to a reduction in volume (in female rats only; Markham, Morris, and Juraska, 2007; Koss et al., 2015). Ovariectomy of female animals prevents this sex difference from occurring, suggesting that it is driven by estrogens or other ovarian hormones.

Apart from organizational effects, adrenal and gonadal hormones can also modulate actions of neurons, having an activational effect. DHEA and DHEAS impact neuronal activation by influencing neurotransmitters, e.g., DHEAS can act as a $GABA_A$ receptor antagonist and both DHEAS and DHEA administration can change the concentration of several neurotransmitters in the hypothalamus (Maninger et al., 2009; Prough, Clark, and Klinge, 2016). Testosterone and estradiol do this more indirectly by affecting gene expression after attaching to androgen or estrogen receptors, or by activating other proteins/peptides; estradiol can stimulate the expression of oxytocin for example (Bos et al., 2012). On top of that, adrenal and gonadal hormones can be converted into each other: testosterone is converted to estradiol through aromatization, DHEA and DHEAS can be converted into each other through (de)sulfation, and testosterone is synthesized from DHEA through androstenedione. It should be noted that the mechanisms described previously are based on animal and in vitro research and it is unknown how well these proposed mechanisms of action of pubertal hormones on neurobiology translate to humans; even within the animal literature the full extent of these processes is still being determined (Prough, Clark, and Klinge, 2016).

THE SOCIAL BRAIN

The social brain is a broad term to describe a network of regions involved in the processing of social information, understanding the minds of others, and social decision making (Nelson et al., 2005; Burnett et al., 2011; Alcalá-López et al., 2018). The core of this network consists of the temporoparietal junction, superior temporal sulcus, medial prefrontal cortex, and anterior temporal cortex/temporal pole. More broadly, the network encompasses areas for detection of social information, such as from faces; affective or salience areas; regions important for social cognition; and regions involved in regulation, integration of information, and decision making. The areas for social information detection include the fusiform face area, superior temporal sulcus (STS), and anterior temporal cortex (ATC) (Nelson et al., 2005; Burnett et al., 2011). Affective areas include the limbic regions, such as the insula and amygdala, as well as the ventral striatum and, in some situations, anterior cingulate cortex (ACC) and medial prefrontal cortex (mPFC) (Nelson et al., 2005; Burnett et al., 2011; Alcalá-López et al., 2018). The temporoparietal junction (TPJ), posterior STS, (dorso)medial PFC, and ATC/temporal pole are involved in social cognition (Nelson et al., 2005; Burnett et al., 2011). Regulatory/integration regions include the ACC, mPFC, and lateral prefrontal regions (Nelson et al., 2005; Burnett et al., 2011). See Figure 29.1 for a visualization of these regions. It should be noted that the function of many of these regions is not exclusive to social brain processes (Alcalá-López et al., 2018), the lateral prefrontal regions for example also have a role in regulating behavior in non-social situations and the ventral striatum is also an indicator of the salience or reward value of non-social stimuli.

A large set of white matter tracts are involved in connecting all of these regions in the social brain. This includes the uncinate fasciculus, connecting amygdala and ventral PFC; the genu of the corpus callosum, connecting left and right PFC; the splenium of the corpus callosum, connecting left with right temporal cortex (and visual cortex);

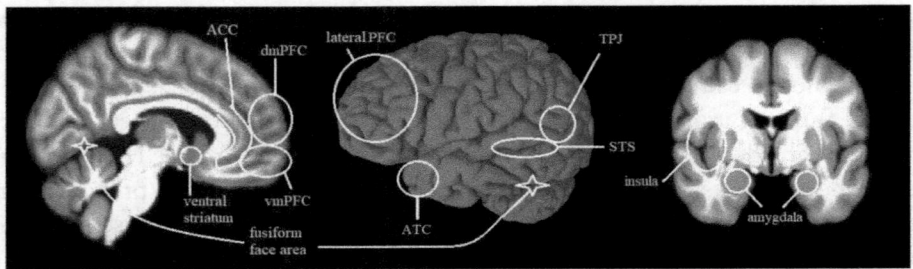

FIGURE 29.1 Regions implicated in social brain processes.
ACC = anterior cingulate cortex, ATC = anterior temporal cortex, dmPFC = dorsomedial prefrontal cortex, PFC = prefrontal cortex, STS = superior temporal sulcus, TPJ = temporoparietal
junction, vmPFC = ventromedial prefrontal cortex. Image is author's own.

the frontostriatal tracts, connecting the striatum with the PFC; and the cingulum, connecting the ACC to several limbic regions.

PUBERTY AND BRAIN STRUCTURE

Over the course of normal brain development, white matter grows gradually in volume; during adolescence this is likely due to a combination of myelination and increase in axon diameter (Paus, 2010). There is also continued development of white matter micro-structure, with decreases in MD, and increases in FA, fiber density, and fiber cross-section (Genc et al., 2018; Herting et al., 2017). Cortical gray matter, however, expands early in life, but reduces in volume during adolescence through cortical thinning and, after early adolescence, reductions in surface area (Lenroot et al., 2007; Mills et al., 2016; Wierenga et al., 2014). See Box 29.2 for an explanation of how these indices of gray and white matter are measured. The speed and timing of these normal structural changes vary by region of the brain, with basic sensory regions maturing earliest and prefrontal and temporal areas and connected white matter tracts showing much more extended development. Since the social brain contains many frontal and temporal regions, as described in "The social brain," it is not surprising that they continue to develop structurally during adolescence. This was examined in a large longitudinal study of both sexes (Mills et al., 2014). The findings showed that volume and cortical thickness of the mPFC, TPJ, and posterior STS decrease from late childhood to early adulthood, whereas surface area starts to decrease after early adolescence (around age 13). The ATC showed a different trajectory, with increases in cortical thickness during adolescence and decreases in volume and surface area. In the following sections, we will discuss associations between pubertal processes and development of these regions as well as the broader social brain network.

Box 29.2 Measuring brain structure and function

Adolescent brain development is most commonly measured using magnetic resonance imaging (MRI). Different types of MRI scans tap into different aspects of brain development: structural MRI obviously provides information about brain structure, diffusion-weighted imaging (DWI) about white matter structure and connectivity, and functional MRI (fMRI) about brain activation during specific tasks or during resting state. For more details of the physics behind these different scans, see Weishaupt, Köchli, and Marincek (2006).

Structural scans were initially used to obtain the volumes of cortical and subcortical regions. More recently, researchers have distinguished between cortical thickness and surface area of the cortex, which both underlie cortical volume, but develop in unique ways (Wierenga et al., 2014). Structural MRI scans can additionally be used to measure white matter volume.

DWI is used to obtain more refined properties of the white matter than its overall volume. DWI measures diffusion of water molecules in the brain. In white matter fibers, molecules can easily diffuse along the direction of the fiber, but not in a perpendicular direction. Therefore, much stronger diffusion in one direction compared to other directions indicates the presence of fibers. This diffusion information can be used to extract white matter fiber tracts with tractography, or to create an index of microstructure such as fractional anisotropy (FA), mean diffusivity (MD), radial diffusivity (RD), or axial diffusivity (AD). FA and MD are commonly used, broad summary measures of white matter microstructure, with higher FA indicating stronger diffusion in one direction compared to other directions. AD is diffusivity on the main axis (along the axon), whereas RD is thought of as diffusivity perpendicular to the axon. Recently, more advanced DWI protocols have led to the development of a range of additional indices of white matter (micro)structure, including neurite density index, fiber density, and fiber cross-section.

Finally, functional MRI provides an indirect index of the activation of brain regions by measuring blood-oxygenation-level-dependent (BOLD) signals, based on the fact that active neurons require and receive more oxygen from the blood than inactive neurons. A wide variety of fMRI tasks exists; some popular paradigms relevant to social brain functioning include the presentation of affective faces, providing social feedback, excluding the participants during a ball-tossing game, presenting scripts describing social events, or letting the participants reflect on what they or others think of them. Both functional activation and connectivity can be examined with fMRI, the latter looking at the extent to which brain regions activate 'in synchrony', either during a specific task context or during resting state. In the next sections, we will discuss the current state of the literature on the associations between pubertal processes and each of the aspects of brain development explained above (gray matter structure, white matter structure, and function).

Cortical Brain Development

Pubertal development appears to support the normal developmental process of cortical thinning and cortical volume reductions over the course of adolescence. Most studies have reported negative associations between gonadarcheal development (either pubertal stage based on secondary sexual characteristics or gonadal hormone levels) and cortical thickness and volume (Vijayakumar et al., 2018). However, part of these associations are dependent on sex and the specific brain regions involved are not always replicated across studies. We will now highlight some examples, emphasizing social brain areas. Higher testosterone in males and higher estradiol levels in both sexes have been associated with reduced gray matter volume in the ACC (in a sample spanning age 8 to 25; Koolschijn, Peper, and Crone, 2014). Similarly, a longitudinal study found higher testosterone levels to be associated with lower cortical thickness in the ACC, posterior cingulate, and dorsolateral PFC of mid-to-late adolescent males but not females (Nguyen, McCracken, et al., 2013a). In studies examining secondary sexual characteristics, increasing pubertal stage has been related to decreasing surface area of the STS in males (Herting et al., 2015) and decreasing cortical thickness in the precuneus and lateral PFC (Vijayakumar et al., 2021); and, in a sample of 9-year-old children, males and females who had started puberty had decreased mPFC and lateral PFC gray matter density compared to those who were prepubertal (Peper et al., 2009). The adrenarcheal hormone DHEA seems to have an opposite effect, at least in mid-to-late childhood, being associated with *increasing* cortical thickness in the dorsolateral PFC, TPJ, premotor cortex, and medial temporal lobe of both sexes (Nguyen, McCracken, et al., 2013b). In this age range, cortical volumes and thickness are typically still increasing, including in the frontal and temporal lobes (Lenroot et al., 2007), thus DHEA might contribute to these increases.

Subcortical Brain Development

Similar to cortical brain development, sex differences have been found in the associations between pubertal development and subcortical brain volumes. The most consistent finding is an association between pubertal development and amygdala volume in males. In male adolescents, increasing pubertal development (in the form of secondary sexual characteristics and testosterone levels) has been repeatedly linked with increasing volume of the amygdala (Goddings et al., 2014; Herting et al., 2014; Wierenga et al., 2018; Bramen et al., 2011). The picture is less clear for females, with either weaker increases than males (Goddings et al., 2014), no significant differences (Wierenga et al. 2018), or decreases (Herting et al., 2014; Bramen et al., 2011; Blanton et al., 2012) in amygdala volume with pubertal development.

In the striatum, the findings are also mixed. More developed secondary sexual characteristics (Goddings et al., 2014) and higher testosterone levels (Wierenga et al., 2018) are related to smaller nucleus accumbens in males. In females, no significant association between puberty and nucleus accumbens volume was found in one study (Wierenga et al., 2018) and a negative association in another study (Goddings et al., 2014). Further pubertal development was related to a larger putamen in males but not females (Wierenga et al., 2018). However, another study found a quadratic effect in both sexes: putamen volume did not change or slightly increased from prepuberty to mid-puberty, but then decreased from mid- to post-puberty (Goddings et al., 2014).

For the hippocampus, effects of pubertal development also seem to be sex-specific. In females, negative associations have been found between pubertal development based on secondary sexual characteristics and hippocampus volume (Bramen et al., 2011; Blanton et al., 2012), whereas in males positive associations with secondary sexual characteristics or testosterone and hippocampus volume have been observed (Bramen et al., 2011; Goddings et al., 2014; Wierenga et al., 2018). However, one study found no significant associations between either secondary sexual characteristics or testosterone levels and hippocampal volumes (Herting et al., 2014).

White Matter Development

Levels of pubertal hormones as well as secondary sexual characteristics are associated with measures of brain connectivity (Perrin et al., 2008; Herting et al., 2012; Peper et al., 2008; Menzies et al., 2015; Herting et al., 2017). Most of these associations have been found after controlling for age. More advanced pubertal stage appears to be related to larger volumes of white matter, but findings are more mixed for microstructure.

Several studies have examined white matter microstructure in relation to secondary sexual characteristics (Asato et al., 2010; Menzies et al., 2015; Herting et al., 2017; Genc et al., 2017). In the first study, with a wide age range (8–28 years old), males and females with more developed secondary sexual characteristics had more developed white matter microstructure (Asato et al., 2010). Almost all investigated tracts continued to develop throughout the whole course of puberty, including tracts relevant to the social brain like the uncinate fasciculus, frontostriatal connections, and the corpus callosum. In line with this, two more recent studies have demonstrated lower mean diffusivity (MD) across many major tracts in males (Menzies et al., 2015) and higher fiber density in the corpus callosum in both sexes (Genc et al., 2017) with more advanced pubertal maturation. However, the two longitudinal studies on changes in secondary sexual characteristics in relation to white matter microstructure development have shown a more complex pattern, with no associations of pubertal development with change in fiber density and sex-specific (both positive and negative) associations with change in fractional anisotropy (FA) (Herting et al., 2017; Genc et al., 2018).

Studies using levels of sex hormones to represent pubertal development have shown varied associations with white matter microstructure. This includes associations in the direction of normal white matter development, such as a positive association between morning serum testosterone and whole-brain white matter volume (controlling for age; Perrin et al., 2008). However, several studies have reported sex-specific findings, such as positive relations of morning serum testosterone and estradiol levels (controlling for age) with FA values in many white matter tracts, which was more pronounced in males than females (Herting et al., 2012). Similarly, a longitudinal study reported positive associations between change in testosterone and in FA in the cingulum and corpus

callosum in females, but not males (Ho et al., 2020). Negative associations for both maturation of secondary sexual characteristics and testosterone levels with MD values have
also been found in males in many white matter tracts, including tracts connecting social
brain areas such as the uncinate fasciculus, cingulum, and corpus callosum (Menzies
et al. 2015). However, there are also some opposite and null findings. For example, one
study in a sample of 8–25-year-olds found a positive association between morning saliva
testosterone levels and MD, most strongly in females, and no significant results for estradiol (Peper et al., 2015). A study in 9-year-old boys and girls showed a positive association between morning saliva DHEA levels and MD in a wide range of tracts (Barendse
et al., 2018). Since the study included a very limited age range and controlled for age, this
finding was interpreted as an effect of relatively early timing of adrenarche.

Summary

The research so far indicates that associations of secondary sexual characteristics and of
adrenal and gonadal hormones with white matter development might be widespread.
Effects on cortical and subcortical brain development appear less widespread and less
consistent across the brain. They still include much of the social brain, however, such as
the medial and lateral PFC, ACC, STS, and amygdala. The mechanisms through which
pubertal hormones can influence brain structure might differ between the gray and
white matter because of different receptor distributions or distributions of glial cells.
Androgen and estrogen receptors have been found in several types of glia, including
astrocytes and oligodendrocytes (Finley and Kritzer, 1999; Jung-Testas and Baulieu,
1998), which have different functions in the gray and white matter. Furthermore, actions
of pubertal hormones on cell bodies and dendrites (i.e., gray matter) can have down-
stream effects on the associated white matter. Pubertal influences are further likely to
vary between gray matter regions because of varying levels of androgen and estrogen
receptors.

   However, findings are also not completely consistent within regions/tracts, with some
studies reporting effects in opposite directions compared to other studies. As illustrated
by the previous findings reviewed, associations between hormone levels and subcortical
and cortical brain volumes depend on sex. Findings might have additionally differed
between studies because of the varying age ranges studied and analysis techniques
applied. Animal research has shown that mechanisms of gonadal hormone action can
differ between subregions of subcortical areas (De Lorme et al., 2012; Meyer, Ferres-
Torres, and Mas, 1978), which could also contribute to inconsistencies in how levels
of these hormones relate to volume of an entire area. Further longitudinal research in
large samples is needed, as well as direct comparisons of analysis techniques and analysis of subregions, to shed light on the source of inconsistencies in the current literature. Longitudinal research is particularly powerful because of its ability to characterize
intra-individual development and to observe the temporal order of events.

PUBERTY AND SOCIAL BRAIN FUNCTION

A popular idea in the field of adolescent brain development is that puberty renders adolescents more sensitive to social information by altering the way in which their brain processes this information (Nelson et al., 2005). This could in turn affect their behavior during social situations. Recent research has provided some support for this idea, using a range of functional MRI paradigms, including viewing affective faces, rejection/exclusion, and social cognition paradigms (see Box 29.2).

Affective Face Processing

Facial expressions are a prevalent source of social information, and presenting participants with affective faces is a commonly used paradigm in research on socio-affective brain function. Facial expression recognition skills are still improving in early adolescence, and improve with pubertal development specifically for complex expressions (Motta-Mena and Scherf, 2017). Scherf and colleagues (2012) proposed that pubertal hormones instigate changes in face processing by influencing motivation for new social behaviors and by reorganizing the neural network underlying face processing. Several studies have empirically tested the second part of this proposal by examining associations between pubertal development and neural correlates of affective face processing, often focusing on the amygdala because of its well-known role in emotion processing. Two studies using region of interest (ROI) analyses reported decreased amygdala activation to neutral faces in those with more developed secondary sexual characteristics, but no difference in amygdala activation to fearful faces (Ferri et al., 2014; Forbes et al., 2011). This suggests that pubertal development is related to decreased salience of neutral faces, rather than increased salience of affective faces.

The two longitudinal studies on this topic, however, have shown inconsistent results. Spielberg and colleagues (2014) reported that increasing testosterone levels across early adolescence were positively linked to increasing amygdala reactivity to angry and fearful faces in both sexes, with a follow-up study in the same sample showing an additional link with reduced connectivity between the amygdala and the orbitofrontal cortex (Spielberg et al., 2015). Vijayakumar and colleagues (2019) on the other hand, showed a quadratic (U-shaped) association between testosterone and amygdala activation in females, but not males. Further, Spielberg and colleagues (2014) found positive associations between nucleus accumbens activity and changes in testosterone levels, whereas Vijayakumar and colleagues (2019) reported no association with testosterone levels and a negative association with secondary sexual characteristics. These discrepancies might be partially due to the wider age range in Vijayakumar and colleagues (2019) and partly to the fact that Spielberg and colleagues (2014) focused on the amount of intra-individual change in hormone level and activation. The findings also suggest that hormonal changes (or specifically changes in testosterone) might have different effects than pubertal development as indexed by physical characteristics. Finally, a study on pubertal timing reported increased amygdala, visual cortex, and thalamus activation to both neutral and affective faces in those with earlier pubertal timing (using the first two time points of the sample of Vijayakumar et al. (2019), age 10, and age 13; Moore et al., 2012), suggesting individual variation in pubertal development is relevant to affective face processing.

The studies above have all focused on gonadarcheal development. Adrenarcheal development has also been linked to neural activation to affective faces: males and females with higher DHEA levels for their age, showed lower midcingulate activation to affective faces (but no association between DHEA levels and amygdala activation was observed; Whittle et al., 2015). The authors interpreted this as timing of adrenarche moderating affective face processing.

Social Feedback and Ostracism

Another common way to examine sensitivity to social information is to provide participants with peer feedback, in the form of acceptance and rejection or inclusion and exclusion. Being included or accepted elicits activation in reward-related areas of the striatum and in the medial prefrontal cortex (Dalgleish et al., 2017; Sherman et al., 2016). Rejection and exclusion on the other hand activate the subgenual ACC and precuneus/posterior cingulate, as well as the regulatory region of the ventrolateral prefrontal cortex (Vijayakumar, Cheng, and Pfeifer, 2017). Surprisingly, only one study has examined the association between pubertal development and neural activation during peer rejection. In this study of adolescents of both sexes, having more developed secondary sexual characteristics was related to more activation in the bilateral amygdala/parahippocampal gyrus and the caudate/subgenual ACC in response to rejection by peers (Silk et al., 2014). These effects remained significant when controlling for age. From this, it seems that part of the normal neural responses to rejection become strengthened during puberty. The neural response to acceptance feedback, however, was not associated with pubertal development. Replication of this finding is warranted, particularly using social feedback paradigms that reflect (online or offline) social interactions that adolescents have in their everyday life (e.g., Sherman et al., 2016).

Social Cognition

More advanced social-cognitive skills, including mentalizing and perspective taking, are still developing in adolescence (Dumontheil, Apperly, and Blakemore, 2010). Several studies have examined how neural processes underlying these skills develop and relate to puberty. In a small sample of early adolescents of both sexes, more developed secondary sexual characteristics at age 13, and greater longitudinal increases in development of these characteristics from age 10 to 13, related to increased activity in social cognition related regions while witnessing peer rejection (including dorsomedial PFC, posterior cingulate, TPJ, and temporal pole) (Masten et al., 2013). This suggests increasing engagement of social brain areas with pubertal development during a perspective taking/empathy task. Also, activity in many of these same brain areas was related to higher self-reported empathic abilities, and increasing empathic abilities over time, suggesting that these areas may play a role in the development of social-cognitive skills. Building on studies focused on basic emotion processing, a study in females only looked at associations between puberty and processing of social emotions (embarrassment and guilt), which require more perspective taking than basic emotions (Goddings et al., 2012). Adolescents with higher pubertal hormone levels (testosterone, estradiol, and DHEA), controlling for age, had increased temporal pole activation while processing social emotions. However, there were no differences in activation between those at lower versus higher pubertal stage based on secondary sexual characteristics.

Self-Perception

Another set of studies have begun investigating how puberty relates to the neural correlates of self-evaluation. Although thinking about oneself is not a social behavior in itself, people extensively use social information to inform their self-evaluations. One study looked explicitly at this by using reflected self-evaluations (evaluating oneself in the eyes of a peer) and demonstrated increased activation in a ventral striatum ROI for males and females with more developed secondary sexual characteristics (Jankowski et al., 2014). The first such study examining neural activation in relation to direct self-evaluation observed that males and females with greater change in self-reported pubertal development over 3 years had larger increases in vmPFC activation during self-evaluation of social traits, in particular (Pfeifer et al., 2013). These preliminary investigations led to a subsequent cross-sectional investigation of early adolescent girls (ages 10–13 years) who completed a similar self-evaluation fMRI task and also provided data to compute several metrics of pubertal development (self-report and basal salivary hormone levels over the previous month). Contrary to expectations, more advanced pubertal development was not associated with greater activity in anterior cortical midline structures during self-evaluation of social traits (Barendse et al., 2020a), despite having 95% power to detect an f^2 of 0.12 or larger. It is possible that pubertal effects are very small in cross-sectional data and are more evident when examining longitudinal changes within individuals, or pubertal effects are dependent on the domains of self-concept assessed. For example, a large fMRI study of self-evaluations across social, academic, and physical appearance domains did not report pubertal associations, but did find that mPFC activity increased with age for physical appearance self-evaluations only (van der Cruijsen et al., 2018).

Summary

Pubertal development has been associated with the neural processing of various forms of social information. The salience of neutral facial expressions might be lower at more advanced pubertal stages, but the evidence is mixed on whether (negative) affective faces are more salient for those at more advanced pubertal development. Further, pubertal development might increase adolescents' sensitivity to rejection feedback (in terms of higher amygdala and subgenual ACC activation), but not to acceptance. Pubertal development is also linked to increased engagement of the social brain during social cognition tasks. This might support further development of social-cognitive skills (Masten et al., 2013), although not many studies have directly tested associations with social behavior.

However, the strength of all of these conclusions is limited by the small sample sizes typically used (the average N of reviewed studies was 57, median N was 48, ranging from 16 to 143), the lack of replications and the low number of longitudinal studies. Longitudinal research is important given the large between-person variation in brain function, which can be separated from within-person (intra-individual) development by doing repeated measurements. Also, there is limited research on individual variation in the timing or tempo of puberty in relation to social brain development, even though this variation is large and has been consistently linked to mental health outcomes (Ullsperger and Nikolas, 2017).

FUTURE DIRECTIONS AND CONCLUSION

Overall, both the animal and human literature to date suggests that pubertal development is relevant for the development of both structure and function of the social brain. Pubertal status has been associated with the structural development of many social brain regions and their connections, partly in interaction with sex. Functionally, puberty might increase the sensitivity of the brain to social information. Pubertal timing, although based on a much more limited set of studies, has also been associated with brain structure and function in adolescence. However, conclusions on the direction and localization of associations between puberty and brain development are limited by a range of methodological issues that have led to inconsistencies between studies. These issues include small sample sizes, variation in the age range of studies, in the measures of pubertal development, and in the techniques used to analyze brain structure. To determine the origin of the inconsistencies and resolve them, larger longitudinal projects are needed that focus on the social brain. In the largest longitudinal brain development studies to date, the Adolescent Brain and Cognitive Development (ABCD) study and the Human Connectome Project Development (HCP-D) study, social brain development is not a core focus, and the measurement of pubertal development in ABCD is relatively limited, but the HCP-D assesses puberty in greater depth. Further, studies that apply and compare multiple analysis techniques for structural and functional brain development are warranted to examine methodological sources of inconsistencies between studies.

Studies that examined pubertal development using both hormone levels and self-report of physical changes (secondary sexual characteristics) have often found differential effects of both. It remains important to examine both of these indicators of pubertal development, especially when focusing on the social brain. This is because visible physical changes directly impact both how other people perceive you and how you perceive yourself, which in turn can influence social interactions that shape the social brain. This more indirect, psychosocial mechanism is in contrast to the direct mechanisms that are often suggested of hormonal impact on brain development and lacks empirical investigation even though it has been suggested as a mechanism linking pubertal processes to risk for mental health disorders such as depression in girls (Cyranowski et al., 2000; Mendle, Turkheimer, and Emery, 2007).

A related point is that studies on the links between pubertal development and social brain function have often neglected to examine real-world social behavior or functioning. This could include self-reported or sociometric measures of social status, measures of friendship quality, but also experimental tasks on prosocial behavior or empathy outside of the scanner, such as those measuring prosocial behavior towards victims after observation of social exclusion in the scanner (Masten et al., 2011; Will et al., 2013). Examining mechanisms from (variation in) pubertal development through brain development to social functioning is an important avenue for future research because it will help us understand the origins of changes in adolescents' social behavior.

Further, as described in "Measuring puberty," interindividual differences in the timing and tempo of pubertal processes are large. These differences are also relevant for mental health (Barendse et al., 2020b; Belsky et al., 2015; Ullsperger and Nikolas, 2017). However, most of the research to date examined pubertal stage and thus normal developmental changes. One of the reasons might be that it is complex to acquire a dataset with enough participants and time points to examine variation in change over time between individuals, especially when including measures of brain development. Nonetheless, it is important for future research to examine this variation because of the relevance for identifying adolescents at risk for mental illness or health-risking behavior and eventually creating more targeted prevention or intervention efforts.

In conclusion, there is evidence demonstrating that pubertal development is relevant for the development of both structure and function of the social brain. However, methodological issues such as small sample size, and variation in study design, and analysis limit the possibility of drawing more specific conclusions. Apart from overcoming these methodological problems, future research is needed that focuses on individual differences in pubertal processes and their relevance to development of the social brain, as well as research examining the mechanisms from pubertal processes through social brain functioning to social behavior.

References

Ahmed, E. I., Zehr, J. L., Schulz, K. M., Lorenz, B. H., DonCarlos, L. L., and Sisk, C. L. (2008), Pubertal hormones modulate the addition of new cells to sexually dimorphic brain regions, *Nature Neuroscience*, 11(9), 995–997.

Aksglaede, L., Sorensen, K., Petersen, J. H., Skakkebaek, N. E., and Juul, A. (2009). Recent decline in age at breast development: The Copenhagen Puberty Study, *Pediatrics*, 123(5), e932–e939.

Alcalá-López, D., Smallwood, J., Jefferies, E., Van Overwalle, F., Vogeley, K., Mars, R. B., Turetsky, B. I., Laird, A. R., Fox, P. T., Eickhoff, S. B., and Bzdok, D. (2018). Computing the social brain connectome across systems and states, *Cerebral Cortex*, 28(7), 2207–2232.

Arevalo, M-.A., Azcoitia, I., and Garcia-Segura, L. M. (2015). The neuroprotective actions of oestradiol and oestrogen receptors, *Nature Reviews. Neuroscience*, 16(1), 17–29.

Asato, M. R., Terwilliger, R., Woo, J., and Luna, B. (2010). White matter development in adolescence: A DTI study, *Cerebral Cortex*, 20(9), 2122–2131.

Auchus, R. J. and Rainey, W. E. (2004). Adrenarche – physiology, biochemistry and human disease, *Clinical Endocrinology*, 60(3), 288–296.

Barendse, M. E. A., Cosme, D., Flournoy, J. C., Vijayakumar, N., Cheng, T. W., Allen, N. B., and Pfeifer, J. H. (2020a). Neural correlates of self-evaluation in relation to age and pubertal development in early adolescent girls'. *Developmental Cognitive Neuroscience*, 44, 100799.

Barendse, M. E. A., Simmons, J. G., Byrne, M. L., Seal, M. L., Patton, G., Mundy, L., Wood, S. J., Olsson, C. A., Allen, N. B., and Whittle, S. (2018). Brain structural connectivity during adrenarche: Associations between hormone levels and white matter microstructure, *Psychoneuroendocrinology*, 88, 70–77.

Barendse, M. E. A., Simmons, J. G., Byrne, M. L., Seal, M. L., Patton, G., Mundy, L., Wood, S. J., Olsson, C. A., Allen, N. B., and Whittle, S. (2020b). Adrenarcheal timing longitudinally predicts anxiety symptoms via amygdala connectivity during emotion processing, *Journal of the American Academy of Child & Adolescent Psychiatry* , 59(6), 739–748.

Belsky, J., Ruttle, P. L., Boyce, W. T., Armstrong, J. M., and Essex, M. J. (2015). Early adversity, elevated stress physiology, accelerated sexual maturation, and poor health in females, *Developmental Psychology*, 51(6), 816–22.

Biro, F. M., Pajak, A., Wolff, M. S., Pinney, S. M., Windham, G. C., Galvez, M. P., Greenspan, L. C., Kushi, L. H., and Teitelbaum, S. L. (2018). Age of menarche in a longitudinal US cohort, *Journal of Pediatric and Adolescent Gynecology*, 31(4), 339–345.

Blanton, R. E., Cooney, R. E., Joormann, J., Eugène, F., Glover, G. H., and Gotlib, I. H. (2012). Pubertal stage and brain anatomy in girls, *Neuroscience*, 217, 105–12.

Bos, P A., Panksepp, J., Bluthé, R. M., and Honk, J. Van. (2012). Acute effects of steroid hormones and neuropeptides on human social-emotional behavior: A review of single administration studies, *Frontiers in Neuroendocrinology*, 33(1), 17–35.

Bramen, J. E., Hranilovich, J. A., Dahl, R. E., Forbes, E. E., Chen, J., Toga, A. W., Dinov, I. D., Worthman, C. M., and Sowell, E. R. (2011). Puberty influences medial temporal lobe and cortical gray matter maturation differently in boys than girls matched for sexual maturity, *Cerebral Cortex*, 21(3), 636–646.

Burnett, S., Sebastian, C., Cohen Kadosh, K., and Blakemore, S-. J. (2011). The social brain in adolescence: Evidence from functional magnetic resonance imaging and behavioural studies, *Neuroscience and Biobehavioral Reviews*, 35(8), 1654–1664.

Byrne, M. L., Whittle, S., Vijayakumar, N., Dennison, M., Simmons, J. G., and Allen, N. B. (2017). A systematic review of adrenarche as a sensitive period in neurobiological development and mental health, *Developmental Cognitive Neuroscience*, 25, 12–28.

Campbell, B. (2006). Adrenarche and the evolution of human life history, *American Journal of Human Biology: The Official Journal of the Human Biology Council*, 18(5), 569–589.

van der Cruijsen, R., Peters, S., van der Aar, L. P. E., and Crone, E. A. (2018). The neural signature of self-concept development in adolescence: The role of domain and valence distinctions, *Developmental Cognitive Neuroscience*, 30, 1–12.

Cyranowski, J. M., Frank, E., Young, E., and Shear, M. K. (2000). Adolescent onset of the gender difference in lifetime rates of major depression: A theoretical model, *Archives of General Psychiatry*, 57(1), 21–7.

Dalgleish, T., Walsh, N. D., Mobbs, D., Schweizer, S., van Harmelen, A-. L., Dunn, B., Dunn, V., Goodyer, I., and Stretton, J. (2017). Social pain and social gain in the adolescent brain: A common neural circuitry underlying both positive and negative social evaluation. *Scientific Reports*, 7(1), 42010.

Dumontheil, I., Apperly, I. A., and Blakemore, S-. J. (2010). Online usage of theory of mind continues to develop in late adolescence, *Developmental Science*, 13(2), 331–338.

Ferri, J., Bress, J. N., Eaton, N. R., and Proudfit, G. H. (2014). The impact of puberty and social anxiety on amygdala activation to faces in adolescence, *Developmental Neuroscience*, 36(3–4), 239–249.

Finley, S. K. and Kritzer, M. F. (1999). Immunoreactivity for intracellular androgen receptors in identified subpopulations of neurons, astrocytes and oligodendrocytes in primate prefrontal cortex, *Journal of Neurobiology*, 40(4), 446–457.

Forbes, E. E., Phillips, M. L., Silk, J. S., Ryan, N. D., and Dahl, R. E. (2011). Neural systems of threat processing in adolescents: Role of pubertal maturation and relation to measures of negative affect, *Developmental Neuropsychology*, 36(4), 429–452.

Genc, S., Seal, M. L., Dhollander, T., Malpas, C. B., Hazell, P., and Silk, T. J. (2017). White matter alterations at pubertal onset, *NeuroImage*, 156, 286–292.

Genc, S., Smith, R. E., Malpas, C. B., Anderson, V., Nicholson, J. M., Efron, D., Sciberras, E., Seal, M. L., and Silk, T. J. (2018). Development of white matter fibre density and morphology over childhood: A longitudinal fixel-based analysis, *NeuroImage*, 183, 666–676.

Goddings, A- L., Burnett Heyes, S., Bird, G., Viner, R. M., and Blakemore, S-. J. (2012). The relationship between puberty and social emotion processing, *Developmental Science*, 15(6), 801–811.

Goddings, A-. L., Mills, K. L., Clasen, L. S., Giedd, J. N., Viner, R. M., and Blakemore, S-. J. (2014). The influence of puberty on subcortical brain development, *NeuroImage*, 88, 242–251.

Hansen, J. W., Hoffman, H. J., and Ross, G. T. (1975). Monthly gonadotropin cycles in premenarcheal girls, *Science*, 190(4210), 161–163.

Herting, M. M., Gautam, P, Spielberg, J. M., Dahl, R. E., and Sowell, E. R. (2015). A longitudinal study: Changes in cortical thickness and surface area during pubertal maturation, *PloS one*, 10(3), e0119774.

Herting, M. M., Gautam, P., Spielberg, J. M., Kan, E., Dahl, R. E., and Sowell, E. R. (2014). The role of testosterone and estradiol in brain volume changes across adolescence: A longitudinal structural MRI study, *Human Brain Mapping*, 35(11), 5633–5645.

Herting, M. M., Kim, R., Uban, K. A., Kan, E., Binley, A., and Sowell, E. R. (2017). Longitudinal changes in pubertal maturation and white matter microstructure, *Psychoneuroendocrinology*, 81, 70–79.

Herting, M. M., Maxwell, E. C., Irvine, C., and Nagel, B. J. (2012). The impact of sex, puberty, and hormones on white matter microstructure in adolescents, *Cerebral Cortex*, 22(9), 1979–1992.

Ho, T. C., Colich, N. L., Sisk, L. M., Oskirko, K., Jo, B., and Gotlib, I. H. (2020). Sex differences in the effects of gonadal hormones on white matter microstructure development in adolescence, *Developmental Cognitive Neuroscience*, 42, 100773.

Jankowski, K. F., Moore, W. E., Merchant, J. S., Kahn, L. E., and Pfeifer, J. H. (2014). But do you think I'm cool? Developmental differences in striatal recruitment during direct and reflected social self-evaluations, *Developmental Cognitive Neuroscience*, 8, 40–54.

Jung-Testas, I. and Baulieu, E. E. (1998). Steroid hormone receptors and steroid action in rat glial cells of the central and peripheral nervous system, *The Journal of Steroid Biochemistry and Molecular Biology*, 65(1–6), 243–251.

Koolschijn, P. C. M. P., Peper, J. S., and Crone, E. A. (2014). The influence of sex steroids on structural brain maturation in adolescence, *PloS one*, 9(1). e83929.

Koss, W. A., Lloyd, M. M., Sadowski, R. N., Wise, L. M., and Juraska, J. M. (2015). Gonadectomy before puberty increases the number of neurons and glia in the medial prefrontal cortex of female, but not male, rats, *Developmental Psychobiology*, 57(3), 305–312.

Latronico, A. C., Brito, V. N., and Carel, J-. C. (2016). Causes, diagnosis, and treatment of central precocious puberty, *The Lancet Diabetes & Endocrinology*, 4(3), 265–274.

Lazzeri, G., Tosti, C., Pammolli, A., Troiano, G., Vieno, A., Canale, N., et al. (2018). Overweight and lower age at menarche: Evidence from the Italian HBSC cross-sectional survey, *BMC Women's Health*, 18(1), 168–175.

Lenroot, R. K., Gogtay, N., Greenstein, D. K., Wells, E. M., Wallace, G. L., Clasen, L. S., Blumenthal, J. D., Lerch, J., Zijdenbos, A. P., Evans, A. C., Thompson, P. M., and Giedd, J. N. (2007). Sexual dimorphism of brain developmental trajectories during childhood and adolescence, *NeuroImage*, 36(4), 1065–1073.

De Lorme, K. C., Schulz, K. M., Salas-Ramirez, K. Y., and Sisk, C. L. (2012). Pubertal testosterone organizes regional volume and neuronal number within the medial amygdala of adult male Syrian hamsters, *Brain Research*, 1460, 33–40.

Maninger, N., Wolkowitz, O. M., Reus, V. I., Epel, E. S., and Mellon, S. H. (2009), Neurobiological and neuropsychiatric effects of dehydroepiandrosterone (DHEA) and DHEA sulfate (DHEAS), *Frontiers in Neuroendocrinology*, 30(1), 65–91.

Markham, J. A., Morris, J. R., and Juraska, J. M. (2007). Neuron number decreases in the rat ventral, but not dorsal, medial prefrontal cortex between adolescence and adulthood, *Neuroscience*, 144(3), 961–968.

Masten, C. L., Eisenberger, N. I., Pfeifer, J. H., Colich, N. L., and Dapretto, M. (2013). Associations among pubertal development, empathic ability, and neural responses while witnessing peer rejection in adolescence, *Child Development*, 84(4), 1338–1354.

Masten, C. L., Morelli, S. A., and Eisenberger, N. I. (2011). An fMRI investigation of empathy for 'social pain' and subsequent prosocial behavior, *NeuroImage*, 55, 381–388.

Matchock, R. L., Dorn, L. D., and Susman, E. J. (2007). Diurnal and seasonal cortisol, testosterone, and DHEA rhythms in boys and girls during puberty, *Chronobiology International*, 24(5), 969–990.

McCarthy, M. M. (2008). Estradiol and the developing brain, *Physiological Reviews*, 88(1), 91–124.

Mendle, J., Turkheimer, E., and Emery, R. E. (2007). Detrimental psychological outcomes associated with early pubertal timing in adolescent girls, *Developmental Review*, 27(2), 151–171.

Menzies, L., Goddings, A-. L., Whitaker, K. J., Blakemore, S-. J., and Viner, R. M. (2015). The effects of puberty on white matter development in boys, *Developmental Cognitive Neuroscience*, *11*, 116–128.

Meyer, G., Ferres-Torres, R., and Mas, M. (1978). The effects of puberty and castration on hippocampal dendritic spines of mice. A Golgi study, *Brain Research*, *155*(1), 108–112.

Mills, K. L., Goddings, A-. L., Herting, M. M., Meuwese, R., Blakemore, S-. J., Crone, E. A., Dahl, R. E., Güroğlu, B., Raznahan, A., Sowell, E. R., and Tamnes, C. K. (2016). Structural brain development between childhood and adulthood: Convergence across four longitudinal samples, *NeuroImage*, *141*, 273–281.

Mills, K. L., Lalonde, F., Clasen, L. S., Giedd, J. N., and Blakemore, S-. J. (2014). Developmental changes in the structure of the social brain in late childhood and adolescence, *Social Cognitive and Affective Neuroscience*, *9*(1), 123–131.

Moore, W. E., Pfeifer, J. H., Masten, C. L., Mazziotta, J. C., Iacoboni, M., and Dapretto, M. (2012). Facing puberty: Associations between pubertal development and neural responses to affective facial displays, *Social Cognitive and Affective Neuroscience*, *7*(1), 35–43.

Morris, N. M. and Udry, J. R. (1980). Validation of a self-administered instrument to assess stage of adolescent development, *Journal of Youth and Adolescence*, *9*(3), 271–280.

Motta-Mena, N. V. and Scherf, K. S. (2017), Pubertal development shapes perception of complex facial expressions, *Developmental Science*, *20*(4), 1–10.

Mouritsen, A., Aksglaede, L., Soerensen, K., Hagen, C. P., Petersen, J. H., Main, K. M., and Juul, A. (2013). The pubertal transition in 179 healthy Danish children: Associations between pubarche, adrenarche, gonadarche, and body composition, *European Journal of Endocrinology*, *168*, 129–136.

Nelson, E. E., Leibenluft, E., McClure, E. B., and Pine, D. S. (2005). The social re-orientation of adolescence: A neuroscience perspective on the process and its relation to psychopathology, *Psychological Medicine*, *35*(2), 163–174.

Nguyen, T-. V., McCracken, J, Ducharme, S., Botteron, K. N., Mahabir, M., Johnson, W., Israel, M., Evans, A. C., and Karama, S. (2013a). Testosterone-related cortical maturation across childhood and adolescence, *Cerebral Cortex*, *23*(6), 1424–1432.

Nguyen, T-. V., McCracken, J. T., Ducharme, S., Cropp, B. F., Botteron, K. N., Evans, A. C., and Karama, S. (2013b). Interactive effects of dehydroepiandrosterone and testosterone on cortical thickness during early brain development, *The Journal of Neuroscience*, *33*(26), 10840–108408.

Paus, T. (2010). Growth of white matter in the adolescent brain: Myelin or axon? *Brain and Cognition*, *72*(1), 26–35.

Peper, J. S., Brouwer, R. M., Schnack, H. G., van Baal, G. C. M., van Leeuwen, M., van den Berg, S. M., Delemarre-Van de Waal, H. A., Janke, A. L., Collins, D. L., Evans, A. C., Boomsma, D. I., Kahn, R. S., and Hulshoff Pol, H. E. (2008), Cerebral white matter in early puberty is associated with luteinizing hormone concentrations, *Psychoneuroendocrinology*, *33*(7), 909–915.

Peper, J. S., de Reus, M. A., van den Heuvel, M. P., and Schutter, D. J. L. G. (2015). Short fused? Associations between white matter connections, sex steroids, and aggression across adolescence, *Human Brain Mapping*, *36*(3), 1043–1052.

Peper, J. S., Schnack, H. G., Brouwer, R. M., Van Baal, G. C. M., Pjetri, E., Székely, E., van Leeuwen, M., van den Berg, S. M., Collins, D. L., Evans, A. C., Boomsma, D. I., Kahn, R. S., and Hulshoff Pol, H. E. (2009). Heritability of regional and global brain structure at the onset of puberty: A magnetic resonance imaging study in 9-year-old twin pairs, *Human Brain Mapping*, *30*(7), 2184–2196.

Perrin, J. S., Hervé, P-. Y., Leonard, G., Perron, M., Pike, G. B., Pitiot, A., Richer, L., Veillette, S., Pausova, Z., and Paus, T. (2008). Growth of white matter in the adolescent brain: Role of testosterone and androgen receptor, *The Journal of Neuroscience*, 28(38), 9519–9524.

Petersen, A. C., Crockett, L., Richards, M., and Boxer, A. (1988). A self-report measure of pubertal status: Reliability, validity, and initial norms, *Journal of Youth and Adolescence*, 17(2), 117–133.

Pfeifer, J. H., Kahn, L. E., Merchant, J. S., Peake, S. J., Veroude, K., Masten, C. L., Lieberman, M. D., Mazziotta, J. C., and Dapretto, M. (2013). Longitudinal change in the neural bases of adolescent social self-evaluations: Effects of age and pubertal development, *The Journal of Neuroscience*, 33(17), 7415–7419.

Phoenix, C. H., Goy, R. W., Gerall, A. A., and Young, W. C. (1959). Organizing action of prenatally administered testosterone propionate on the tissues mediating mating behavior in the female guinea pig, *Endocrinology*, 65, 369–382.

Prough, R. A., Clark, B. J., and Klinge, C. M. (2016). Novel mechanisms for DHEA action, *Journal of Molecular Endocrinology*, 56(3), R139–R155.

Reiter, E. O., Fuldauer, V. G., and Root, A. W. (1977). Secretion of the adrenal androgen, dehydroepiandrosterone sulfate, during normal infancy, childhood, and adolescence, in sick infants, and in children with endocrinologic abnormalities, *The Journal of Pediatrics*, 90(5), 766–770.

Scherf, K. S., Behrmann, M., and Dahl, R. E. (2012). Facing changes and changing faces in adolescence: A new model for investigating adolescent-specific interactions between pubertal, brain and behavioral development, *Developmental Cognitive Neuroscience*, 2(2), 199–219.

Schulz, K. M., Molenda-Figueira, H. A., and Sisk, C. L. (2009). Back to the future: The organizational-activational hypothesis adapted to puberty and adolescence, *Hormones and Behavior*, 55(5), 597–604.

Semiz, S., Kurt, F., Kurt, D. T., Zencir, M., and Sevinç, O. (2008). Pubertal development of Turkish children, *Journal of Pediatric Endocrinology & Metabolism*, 21(10), 951–961.

Sherman, L. E., Payton, A. A., Hernandez, L. M., Greenfield, P. M., and Dapretto, M. (2016). The power of the like in adolescence: Effects of peer influence on neural and behavioral responses to social media, *Psychological Science*, 27(7), 1027–1035.

Silk, J. S., Siegle, G. J., Lee, K. H., Nelson, E. E., Stroud, L. R., and Dahl, R. E. (2014). Increased neural response to peer rejection associated with adolescent depression and pubertal development, *Social Cognitive and Affective Neuroscience*, 9(11), 1798–1807.

Søeborg, T., Frederiksen, H., Mouritsen, A., Johannsen, T. H., Main, K. M., Jørgensen, N., Petersen, J. H., Andersson, A-. M., and Juul, A. (2014). Sex, age, pubertal development and use of oral contraceptives in relation to serum concentrations of DHEA, DHEAS, 17α-hydroxyprogesterone, Δ4-androstenedione, testosterone and their ratios in children, adolescents and young adults, *Clinica chimica acta; International Journal of Clinical Chemistry*, 437, 6–13.

Spielberg, J. M., Forbes, E. E., Ladouceur, C. D., Worthman, C. M., Olino, T. M., Ryan, N. D., and Dahl, R. E. (2015). Pubertal testosterone influences threat-related amygdala-orbitofrontal cortex coupling, *Social Cognitive and Affective Neuroscience*, 10(3), 408–415.

Spielberg, J. M., Olino, T. M., Forbes, E. E., and Dahl, R. E. (2014). Exciting fear in adolescence: Does pubertal development alter threat processing? *Developmental Cognitive Neuroscience*, 8, 86–95.

Styne, D. M. and Grumbach, M. M. (2011). Puberty: Ontogeny, neuroendocrinology, physiology, and disorders, in S. Melmed, K. S. Polonsky, P. R. Larsen, and H. M. Kronenberg (Eds.), *Williams Textbook of Endocrinology* (pp. 1054–1201). Elsevier: Philadelphia.

Ullsperger, J. M. and Nikolas, M. A. (2017). A meta-analytic review of the association between pubertal timing and psychopathology in adolescence: Are there sex differences in risk? *Psychological Bulletin*, *143*(9), 903–938.

Vijayakumar, N., Cheng, T. W., and Pfeifer, J. H. (2017). Neural correlates of social exclusion across ages: A coordinate-based meta-analysis of functional MRI studies, *NeuroImage*, *153*, 359–368.

Vijayakumar, N., Op de Macks, Z., Shirtcliff, E. A., and Pfeifer, J. H. (2018). Puberty and the human brain: Insights into adolescent development, *Neuroscience & Biobehavioral Reviews*, *92*, 417–436.

Vijayakumar, N., Pfeifer, J. H., Flournoy, J. C., Hernandez, L. M., and Dapretto, M. (2019). Affective reactivity during adolescence: Associations with age, puberty and testosterone' *Cortex*, *117*, 336–350.

Vijayakumar, N., Youssef, G. J., Allen, N. B., Anderson, V., Efron, D., Hazell, P., ... and Silk, T. (2021). A longitudinal analysis of puberty-related cortical development. *NeuroImage*, *228*, 117684.

Villamor, E., Marín, C., Mora-Plazas, M., and Oliveros, H. (2017). Micronutrient status in middle childhood and age at menarche: Results from the Bogotá School Children Cohort, *British Journal of Nutrition*, *118*(12), 1097–1105.

Weishaupt, D., Köchli, V. D., and Marincek, B. (2006). *How Does MRI Work?* Springer: Berlin, Heidelberg.

Whittle, S., Simmons, J. G., Byrne, M. L., Strikwerda-Brown, C., Kerestes, R., Seal, M. L., Olsson, C. A., Dudgeon, P., Mundy, L. K., Patton, G. C., and Allen, N. B. (2015), Associations between early adrenarche, affective brain function and mental health in children, *Social Cognitive and Affective Neuroscience*, *10*(9), 1282–1290.

Wierenga, L. M., Bos, M. G. N., Schreuders, E., Kamp, F. van der, Peper, J. S., Tamnes, C. K., and Crone, E. A. (2018). Unraveling age, puberty and testosterone effects on subcortical brain development across adolescence, *Psychoneuroendocrinology*, *91*, 105–114.

Wierenga, L. M., Langen, M., Oranje, B., and Durston, S. (2014), Unique developmental trajectories of cortical thickness and surface area, *NeuroImage*, *87*, 120–126.

Will, G-. J, Crone, E. A., van den Bos, W., and Güroğlu, B. (2013). Acting on observed social exclusion: Developmental perspectives on punishment of excluders and compensation of victims, *Developmental Psychology*, *49*, 2236–2244.

Wu, T., Mendola, P., and Buck, G. M. (2002). Ethnic differences in the presence of secondary sex characteristics and menarche among US girls: The Third National Health and Nutrition Examination Survey, 1988–1994, *Pediatrics*, *110*(4), 752–757.

CHAPTER 30

NEUROCOGNITIVE DEVELOPMENTAL CHANGES IN TRUST AND RECIPROCITY ACROSS ADOLESCENCE

SARAH M. BURKE, SUZANNE VAN DE GROEP, PHILIP BRANDNER, AND EVELINE A. CRONE

INTRODUCTION

ADOLESCENCE is an important transition period between childhood and adulthood with pronounced developmental changes in cognitive and social-affective abilities (Crone and Dahl, 2012). The onset of adolescence is defined by the start of puberty around the age of 9–11 years, although this is typically somewhat earlier for girls than for boys and differs depending on culture (Shirtcliff, Dahl, and Pollak, 2009; Blakemore, Burnett, and Dahl, 2010). Pubertal development is associated with extensive gonadal hormone changes, which have massive effects on the appearance of adolescents, such as facial features and body characteristics; but also affects brain development and social sensitivities (Peper and Dahl, 2013). Pubertal development lasts until approximately ages 15–16-years (Braams et al., 2015), after which adolescents enter mid-to-late adolescence, a transition period into adulthood. The end point of adolescence is more difficult to define and depends on when adolescents reach the stage of mature social goals. The end phase of adolescence is therefore partly culturally defined and occurs around the early twenties (Steinberg, 2008; Crone and Dahl, 2012).

One of the most prominent changes in adolescence concerns the way adolescents experience and develop social relationships (Cillessen and Rose, 2005). Adolescents spend less time with their parents and more time with their peers, and they also develop more intimate friendships (Newcomb, Bukowski, and Pattee, 1993). Social acceptance and social rejection are highly salient events in the lives of adolescents (Peters et al., 2011; Sebastian et al., 2011), and these have become even more prominent with the involvement of social media (Crone and Konijn, 2018). Two processes that are highly important for developing secure and intimate social relationships are trust and reciprocity (Rilling and Sanfey, 2011). Trust helps individuals to develop social relationships, whereas reciprocity is crucial for maintaining social relationships (Lahno, 1995). Intriguing questions concern the potential developmental changes in trust and reciprocity. What processes might be driving these changes? And are the transitions in trust and reciprocity related to general changes in cognitive and social functions or may these be specific for adolescence?

The transition to more intimate social relations and friendships has been associated with the development of social perspective taking (Selman, 1980; Güroğlu, van den Bos, and Crone, 2014). Perspective taking allows adolescents to have a better understanding of motives, intentions, and goals of others, and therefore are important for social relations that are based upon trust and reciprocity. In earlier research it was suggested that perspective taking was already mature in early childhood, because at that time children are able to use Theory of Mind (ToM) and understand that others have different perspectives than their own (Carlson, Moses, and Claxton, 2004). However, more recent experimental studies demonstrated that more advanced forms of perspective taking, such as measured with the Director Game where individuals need to make choices based on the perspectives of others, developed until late adolescence (Dumontheil, Apperly, and Blakemore, 2010). These findings fit well with research on executive functions such as working memory and cognitive flexibility, for which it was also shown that most pronounced developmental improvements are observed before the age of 10 years, but continuing refinement of executive functions is observed until late adolescence (Huizinga, Dolan, and van der Molen, 2006). Therefore, one of the reasons that adolescents develop more advanced trust and reciprocity may be because of increasing social-cognitive perspective taking skills.

Recent insights based on brain research have shown that many of the changes in social behavior are accompanied by continuing maturation of brain regions critically involved in perspective taking and social behavior. Neuroimaging methods may allow us to gain further insight in the processes that change during adolescence. Social reasoning and perspective taking consistently engage a network in the brain that is referred to as the "social brain" (Frith and Frith, 2003; Amodio and Frith, 2006). This network involves the medial prefrontal cortex, temporal-parietal junction, superior temporal sulcus, and anterior temporal lobe (Blakemore, 2008; Rilling and Sanfey, 2011). Interestingly, these regions show pronounced morphological changes especially during the teen-age years, after which there is a period of less change (Mills et al., 2014). Possibly, these changes in brain structure coincide with developmental changes in social-cognitive processes, although research relating structural development to individual trajectories is still in its early stages (Foulkes and Blakemore, 2018). Functional magnetic resonance imaging (fMRI) has been informative in understanding which brain regions are engaged during different types of social decision making (Rilling and Sanfey, 2011). Several meta-analyses have confirmed that children, adolescents, and adults engage the same network when performing social perspective taking or decision-making tasks, but the extent of activation differs between age groups (Burnett et al., 2011; Crone and Dahl, 2012). Therefore, the study of brain activity during social tasks is a promising method to study developmental changes in trust and reciprocity; the social processes that are important for forming and maintaining social relationships.

In this chapter, we investigate the development of trust and reciprocity from a behavioral and neuroscience perspective. In the first section, we will show that economic games that manipulate trust and reciprocity conditions can be informative for understanding motivations for trusting others, including the developmental changes in these processes. Next, we will describe neuroimaging studies that examined trust and reciprocity in adults and adolescents. Special emphasis will be given to possible individual differences relating to gender, perspective taking skills and social context, factors that all may influence trust and reciprocity behavior. Finally, the chapter will end with several compelling questions that are important for relating lab-based experimental designs and neuroscience studies to understanding trust and reciprocity in the complexity of adolescents' daily lives.

Economic Games and Their Value in Studying Social Behavior

In recent years, psychologists and neuroscientists have increasingly turned to economic games to study social decision making (Rilling and Sanfey, 2011; Will and Güroğlu, 2016). Economic games are game theoretical paradigms derived from behavioral economics, which are used to investigate social decisions that have actual consequences for the players involved, for example monetary gains or losses (Gummerum, Hanoch, and Keller, 2008; Rilling and Sanfey, 2011; Will and Güroğlu, 2016). In the field of behavioral economics, it has been repeatedly shown that human behavior in economic games deviates from behavior as predicted by game theory. For example, even when given full power over a pool of resources players will not keep everything to themselves. As such, these games have been interpreted to serve as an indication of people's social preferences (Gummerum, Hanoch, and Keller, 2008). As has been put forward in previous research, there are several reasons why economic games are suited for studying social decisions over the course of development, including adolescence (Will and Güroğlu, 2016). First, economic games can be used across cultures and across the lifespan, which enables researchers to compare social behaviors, such as trust and reciprocity, between different cultures and age groups (Gummerum, Hanoch, and Keller, 2008; Will and Güroğlu, 2016). Second, economic games are very structured in nature, making them suited for fMRI paradigms to study real-time social interactions in controlled environments (Rilling and Sanfey, 2011; Tzieropoulos, 2013). Third, as was mentioned previously, certain aspects of economic games can be manipulated such that mechanisms underlying social decisions, such as perspective taking, can be studied and be disentangled from other mechanisms (Will and Güroğlu, 2016).

One economic game that is of particular interest to study trust and reciprocity is the Trust Game (Berg, Dickhaut, and McCabe, 1995). In this game, two players are involved in dividing a certain set of resources or tokens (Berg, Dickhaut, and McCabe, 1995; Malhotra, 2004). The first player (the trustor or investor) can choose how to divide the initial stake between themselves and a second player (the trustee). The amount of resources the trustor sends to the second player is multiplied (usually tripled) and given to the trustee. The second player then decides how much they want to give back to the first player (see Figure 30.1A). Since the first publication on the Trust Game in 1995, various versions have emerged. For example, the amount of resources sent to the second player is doubled or quadrupled instead of tripled (Gummerum, Hanoch, and Keller, 2008), or the amount that the second player gives back to the first player may be tripled as well (Gummerum, Hanoch, and Keller, 2008). In another version of the task the trustor and trustee have dichotomous choices: either to trust (give some of the stake away) or not to trust (to keep most of the stake to themselves) in case of the trustor, and either to reciprocate (give back a lot of the multiplied stake) or defect (give back little of the multiplied stake) in case of the trustee (Figure 30.1B) (Malhotra, 2004; van den Bos et al., 2010; van den Bos, van Dijk, and Crone, 2012). Another distinction can be made in the way the Trust Game is administered: either as a "one-shot" game, meaning that after a single investment exchange the game over, or as a "multi-round" game, in which the same set of players interact over several trials and during which they can form certain opinions on the other player's behavior and adapt the own behavior in a next trial accordingly. Multi-round paradigms, employing dichotomous trust and reciprocity decision options have especially been used in neuroimaging studies of the Trust Game (van den Bos et al., 2009, 2011).

Different stages of the Trust Game can be distinguished, measuring complex behavior that relies on different aspects of social cognition. In the *trust stage*, the investor needs to calculate the risk associated with trust decisions, needs to consider the likelihood of a reciprocal response from the other player, and his/her trust decisions may be based on the anticipation and valuing of both financial self-interest and social reward. In the *reciprocity stage*, the trustee will evaluate the offer as either generous or not, will consider social norm values and expectations of the investor, and will prepare a response accordingly. Both the trustee during the reciprocity stage and the investor during the *feedback stage*, when it is revealed how much money the trustee returned, engage in outcome monitoring that is necessary for building a reputation in case of repeated interactions and on which behavior in following trials is based.

Studies examining trust and reciprocity behavior in Trust Games, using both the traditional (Sutter and Kocher, 2007; Fett, Shergill et al., 2014) and dichotomous (van den Bos et al., 2010; Güroğlu, van den Bos, and Crone, 2014; Van de Groep et al., 2018) designs with anonymous others as targets have been inconsistent with regard to the development of trust decisions over the course of adolescence. The first

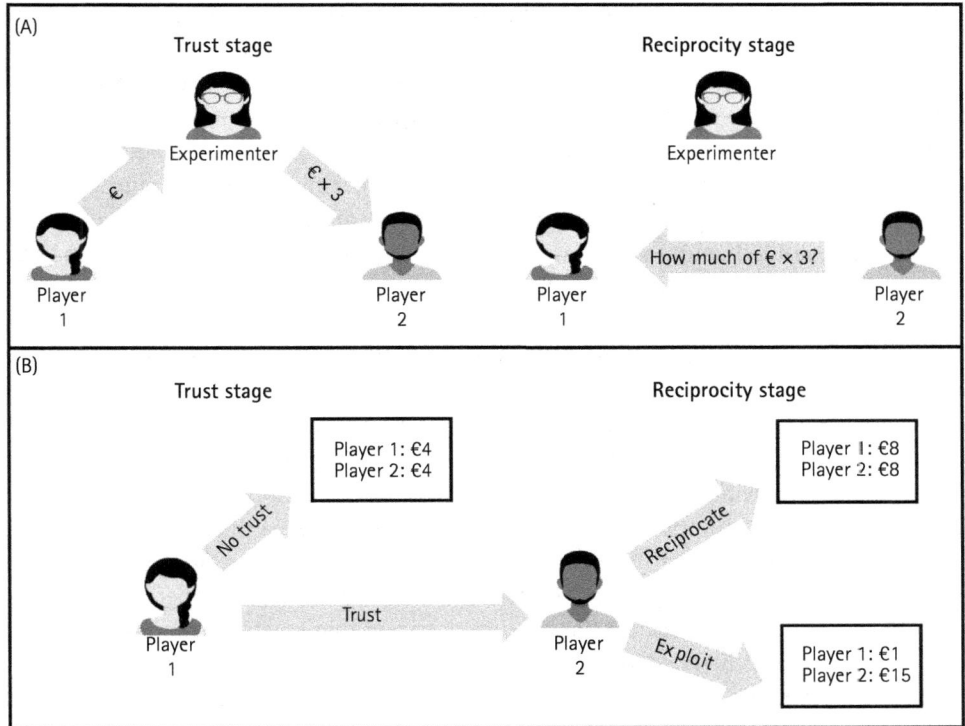

FIGURE 30.1 A) Depiction of a classic Trust Game. In round 1, the Trust stage, Player 1 (the trustor or investor) decides how much of a stake will be given to the second player (the trustee). The amount she gives away to the second player (depicted here as "€") is multiplied by the experimenter, in this case tripled, depicted as "€ x 3," which is a measure of trust. In round two of a classic Trust Game, the Reciprocity stage, Player 2 received "€ x 3" from Player 1 and can now decide how much of "€ x 3" he wants to give back to the Player 1, which is a measure of reciprocity. Note that in some versions of the Trust Game, the amount that is given back to Player 1 is multiplied (e.g. tripled) by the experimenter as well. B) Depiction of a dichotomous Trust Game. In the first round, Player 1 can either decide to trust (i.e., to transfer the stake to the Player 2) or not to trust (i.e., to select the outcome depicted at the top). If Player 1 decides to trust, the stake is multiplied, in this case doubled. In the second round, Player 2 can either decide to reciprocate (give back a large proportion of the multiplied stake) or exploit (give back a small proportion of the multiplied stake). Note that such labels (i.e., trust, no trust, reciprocate, exploit) would not be visible to participants but are shown here for clarification.

developmental studies that examined trust and reciprocity using the Trust Game have found age-related increases in both trust and reciprocity between childhood and adolescence (Sutter and Kocher, 2007; van den Bos et al., 2010; van den Bos, van Dijk, and Crone, 2012), with some reporting further increases in mid-adolescence (van den Bos et al., 2010), while others reported no further changes during early and mid-adolescence (Fett, Shergill et al., 2014; Güroğlu, van den Bos, and Crone, 2014) or between late adolescence and adulthood (Lemmers-Jansen et al., 2017). Recently, a study that included a large sample of 496 adolescents, aged 12–18 years, found a general stability, and thus no age-related changes in trust, but, in contrast to previous studies, a *decrease* in reciprocity behavior over the course of adolescence (Van de Groep et al., 2018). It should however be noted that many of these studies cross-sectionally compared different age groups (e.g., comparing 9-, 12-, 16-, and 22-year-olds) instead of investigating development across a continuous age range, and only few studies included young age groups (children and young adolescents), which may have hampered the detection of specific age-related changes across adolescence. Taken together, trust seems to increase between childhood and mid-adolescence and then stabilizes, whereas for reciprocity behavior different developmental patterns across adolescence have been observed.

Overall, the different studies do not provide a clear pattern of trust and reciprocity development in adolescence, which may have to do with the complexity of social decision making. As will be outlined, various studies have shown that males and females differ in their trust (but not reciprocity) decisions (Derks, Lee, and Krabbendam, 2014; Lemmers-Jansen et al., 2017; Van de Groep et al., 2018). In addition, adolescents, when growing older, get increasingly sensitive for the context in which social decision making takes place, and in recent years it has become apparent that trust, reciprocity, and their development are influenced by individual differences, for example in perspective-taking skills (van den Bos et al., 2010; Fett, Shergill et al., 2014; Güroğlu, van den Bos, and Crone, 2014; Van de Groep et al., 2018). Therefore, varying situational contexts and individual difference variables, that critically interact with age, all influence social decision making and may explain the inconsistent developmental patterns of trust and reciprocity behavior during adolescence.

Sex differences in prosocial behavior have been well-described, with women showing e.g., more generosity and altruism compared to men (Zak et al., 2005; Innocenti and Pazienza, 2006). Also, men and women have been found to differ in mind-reading abilities and empathy (Schulte-Rüther et al., 2008; Krach et al., 2009; Singer and Lamm, 2009; Frank, Baron-Cohen, and Ganzel, 2015; Adenzato et al., 2017). These sex differences in social cognition and prosocial behavior have been linked to sex hormones. For instance, testosterone levels, which are multiple times higher in men than women, were reported to be negatively associated with prosocial tendencies, but positively correlated with aggression and avoidance behavior, and testosterone furthermore was shown to promote status-seeking and social dominance motives (Harris et al., 1996; Zak et al., 2009; Eisenegger, Haushofer, and Fehr, 2011; Riedl and Javor, 2012). However, notably, these findings do not necessarily translate to females showing more trust than males. Instead, various studies in adolescents have shown that boys show *higher* levels of trust than girls (see Figure 30.2a), both in single and repeated Trust Game interactions (Derks, Lee and Krabbendam, 2014; Lemmers-Jansen et al., 2017; Van de Groep et al., 2018), a finding that matches several reports in adults across a wide age range of 18–84 years (Chaudhuri and Gangadharan, 2007; Migheli, 2007; Croson and Gneezy, 2009; Garbarino and Slonim, 2009; Dittrich, 2015).

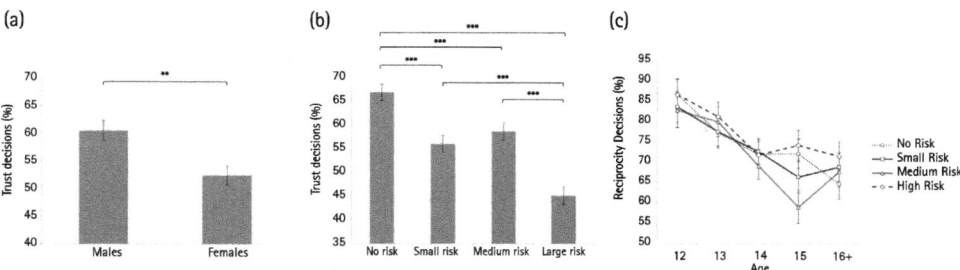

FIGURE 30.2 a) Main effect of risk on trust decisions (error bars denote S.E.); b) Main effect of sex on trust decisions (error bars denote S.E.); c) Percentages of reciprocity decisions in each condition across the five age groups (error bars denote S.E.).

Adapted with permission from van de Groep et al. (2018), "Developmental Changes and Individual Differences in Trust and Reciprocity in Adolescence," *Journal of Research on Adolescence*. John Wiley & Sons. Published online October 16th, 2018, DOI 10.1111/jora.12459 Copyright © 2018 Society for Research on Adolescence.

This paradoxical association between prosocial behavior, in particular trust and sex, might be explained by the fact that females' trust levels are more context-sensitive than those of men (Croson and Gneezy, 2009; Balliet et al., 2011). For example, when their trust was violated, females appeared to stay more trusting than males and were more likely to restore trust, as has been shown in both adults (Haselhuhn et al., 2014) and adolescents (Lemmers-Jansen et al., 2017). Some studies have used more simple forms of economic exchange to understand the dynamics of context. Using 2-choice allocation games, Meuwese et al. (2015), in a large sample of 1216 children and adolescents (8–18 years of age), found an age-related increase in the preference for efficient outcomes (i.e., maximization of total available resources), which here would be trust, that was stronger for boys than for girls. This finding is in line with a sociocultural perspective on sex differences in trust decisions, in which male gender roles promote agentic (i.e., outcome-based and efficient) behavior (Eagly, 2009; Derks, Lee, and Krabbendam, 2014; Lemmers-Jansen et al., 2017). Interestingly, Derks et al. (2014) noted that prosocial orientation and male sex were two independent predictors of trust behavior, and proposed that sex differences in trust might be explained by higher tendencies for risk taking, sensation seeking, and optimism in men (Ben-Ner and Halldorsson, 2010) and greater risk aversion in women (Byrnes, Miller, and Schafer, 1999; Croson and Gneezy, 2009). Thus, trust decisions may be defined as a form of prosocial risk-taking behavior (Do, Guassi Moreira, and Telzer, 2017), and males seem to engage in this behavior more often than females.

For reciprocity, it has been suggested that adolescence represents an important period of transition from general reciprocity to context-specific reciprocity, that is similar for males and females (van den Bos et al., 2009, 2011; van den Bos, van Dijk, and Crone, 2012; Van de Groep et al., 2018).

As an example of the influence of social context, Güroğlu et al. (2014) found that trust and reciprocity behavior, and its developmental course across adolescence, were dependent on the beneficiary. Adolescents showed more trust and reciprocity to friends than to neutral peers, antagonists, or anonymous others, and reciprocity decreased in mid- and late adolescence towards anonymous peers (in 15- and 18-year-olds) and towards antagonists (only in 18-year-olds), but reciprocity towards friends and neutral peers remained stable across age groups.

Another study by Fett, Shergill et al. (2014) manipulated the interaction context in a multi-round Trust Game. In one set of twenty rounds the adolescent participants (13–18 years old), in the role of the investor, learned that their (anonymous) interaction partner behaved in a cooperative manner (first repayment of either 100 percent, 150 percent, or 200 percent of the invested amount and subsequent increases of the repaid amount if investors increased their trust), whereas in another set of twenty rounds reciprocal decisions of another interaction partner were rather unfair (i.e., repayment was 50 percent, 75 percent, or 100 percent of the invested amount and subsequent decreases in repaid amount, if participants had decided to trust more). As expected, participants adaptively changed their trust decisions during the course of the Trust Game rounds; participants increasingly trusted in the cooperative context and decreased the amounts invested in the unfair context. However, in contrast to previous studies, no age effects were observed, which might have been related to their relatively small sample size of fifty adolescents.

Studies in which levels of risk were manipulated, thus in which investors had to take into account relatively higher or lower chances for betrayal of trust, have revealed that reciprocal decisions by the trustee were highly dependent on such risk context (see Figure 30.2b). Van den Bos et al. (2010, 2011) found that adolescents reciprocated more often when the risk for the investor to trust was relatively high. Interestingly, this differentiation between levels of risk was only observed in the mid-adolescent (mean age 16 years) and late adolescent (mean age 22 years) age groups, whereas reciprocity rates were similar for both risk-conditions (high vs. low risk) in the 9- and 12-year-olds (van den Bos et al., 2010).

This age-dependent effect of risk differentiation on reciprocal behavior has recently been replicated by van de Groep et al. (2018). In particular, the group of 15-year-olds tended to differentiate more between four different risk levels compared to the younger (12–14 years old) participants (see Figure 30.2c); they showed relatively higher reciprocity rates when trust decisions were either not risky or when the risk taken to be trusted was high, but relatively less reciprocity when investors had taken medium or small risks.

Previously, perspective taking skills have been found to improve with age and have been associated with increased pro-social behavior (Johnson, 1975; Güroglu, Bos, and Crone, 2014). In the study by van den Bos et al. (2010) not only risk levels were manipulated in their Trust Game design, but it was also distinguished between low versus high benefits for the trustee when being trusted by the investor. The authors found age-related changes in sensitivity to this benefit manipulation. Only the oldest age group (22-year-olds) trusted more often when the benefits for the trustee were relatively high, thereby valuing outcomes for the other player that may eventually also benefit themselves. Also, 12-, 16-, and 22-year-old adolescents, but not the 9-year-old children showed increased reciprocity when benefits for the trustee were high. This suggests that advanced forms of perspective taking, required for the differentiation between risks and benefits involved in trust and reciprocity decisions, develop with age (van den Bos et al., 2010, 2011). Furthermore, the reported different patterns of trust and reciprocity development across adolescence may be explained by age-dependent changes in the sensitivity for social context.

Although, in the study by Fett, Shergill et al. (2014) no age effects were found, the authors were able to relate individual differences in perspective taking skills, measured with a Theory of Mind task, to the adaptive changes in trust behavior that adolescents showed in different contexts. Those participants that scored higher on the perspective-taking task, were better able to adapt their trust behavior: in the cooperative context, they showed higher investments, and were quicker to detect and respond to unfair behavior by decreasing investments (Fett, Shergill et al., 2014). In line with this, individual differences in certain forms of empathy, in interaction with age have been shown to modulate social decision making. For example, Sutter (2007) noted that the relative importance of intentions rather than outcomes of economic decisions increases with age across adolescence. In addition, van de Groep et al. (2018) found that in particular the intention to comfort others, mediated the relation between a decrease of reciprocity with increasing adolescent age.

Thus, age-related changes and individual differences in perspective taking skills and empathy may explain the increased sensitivity for social context in trust and reciprocity decisions during mid-adolescence (van den Bos et al., 2010, 2011; Güroglu, Bos, and Crone, 2014; Van de Groep et al., 2018).

TRUST AND RECIPROCITY IN THE BRAIN

Within the last decades cognitive neuroscience has advanced our understanding of how the brain guides social behavior. Trust and reciprocity decisions encompass several complex processes that engage different interconnected brain regions. Two neural networks within the context of trust and reciprocity, which will be discussed in more detail soon, may be distinguished (see Figure 30.3). First, the so-called social brain network (Van Overwalle, 2009), which sub-serves socio-cognitive functions, such as social perspective taking and cognitive control. Social brain regions are involved e.g., during considerations of the other player's expectations and inhibition of selfish decisions (exploiting trust). As trust decisions usually are not purely rational, but rather intuitive, gut-feeling-based decisions, an important second neural network involved controls social-affective functions during value-based decision making. This neural network is active during e.g., experiencing reward when trust is reciprocated or when evaluating chances for betrayal.

Though many distributed areas seem to be involved in social cognition, the central hub of the social brain network lies, most likely, within the medial prefrontal cortex. In

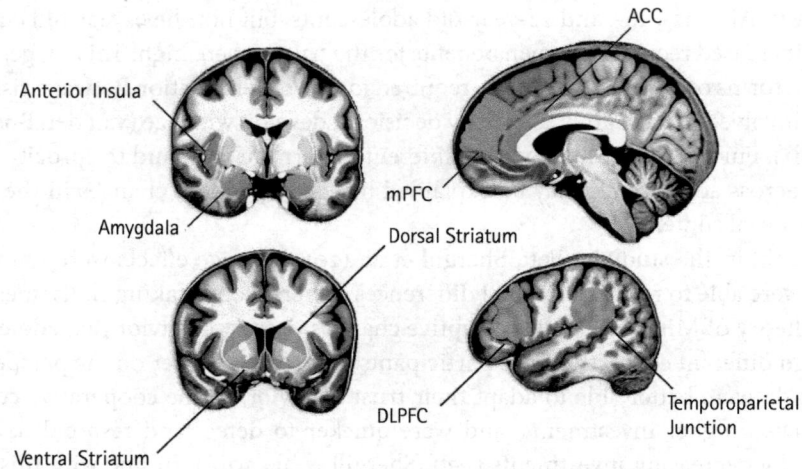

FIGURE 30.3 Overview of the brain regions recruited in social situations involving trust and reciprocity decisions. These networks consist of brain regions playing a role in mentalizing and social perspective taking (temporo-parietal junction, medial prefrontal cortex), reward learning (ventral and dorsal striatum, and anterior cingulate cortex), emotional processing (amygdala, anterior insula, anterior cingulate cortex), as well as impulse control (dorsolateral prefrontal cortex). For simplicity and clarity, we categorized these regions as either more social-cognitive in their primary mechanism (blue) or social-affective (orange).

particular, the *anterior medial prefrontal cortex*, with its extensive interconnections to limbic and associative regions, has been linked to social perspective taking, integrating the perspectives of self and other, and mentalizing feelings, thoughts, and beliefs of oneself and those of others (Rilling et al., 2004; Amodio and Frith, 2006). In the Trust Game, this region is critically involved when the interaction partners are inferring each other's intention to trust and to reciprocate trust (Amodio and Frith, 2006; Krueger et al., 2007; King-Casas et al., 2008) and when, in the role of the trustee, choosing to exploit trust rather than to reciprocate, thus when keeping most of the invested amount of money for her/himself (van den Bos et al., 2009). The anterior medial prefrontal cortex is thus involved during mentalizing and is sensitive to increases in personal outcomes.

The *dorso-medial prefrontal subregion* has been found to encode cognitive processes related to initial learning whether somebody is considered trustworthy and the subsequent evaluation of one's reciprocity options (Behrens et al., 2008; Rilling and Sanfey, 2011), and the *dorsal striatum*, during the course of a repeated Trust Game, is involved in reinforcement learning and outcome monitoring, and was found to support the process of learning an interaction partner's reputation (Delgado, Frank, and Phelps, 2005; Fareri et al., 2015).

Trust and reciprocity behaviors do not only require cognitive functions such as mentalizing and outcome monitoring, but in order to display cooperative behavior, self-oriented impulses need to be regulated. Therefore, brain regions that have been associated with cognitive control, such as the *anterior cingulate cortex* and *dorso-lateral prefrontal cortex*, have been found to be active during rejections of unfair monetary offers, when suppressing selfish responses, and when there is conflict between social norms and personal interests (Sanfey et al., 2003; Knoch et al., 2009; van den Bos et al., 2009).

Another key brain region in social cognition is the *temporo-parietal junction*, which is implicated in mental state representation, and was found to be sensitive to risk-manipulation when reciprocating trust. Brain activation in this region was found to be higher during reciprocal choices, when the risk the investor had taken to be trusted was high. Thus, the temporo-parietal junction, found to be important for shifting attention from self to other, is involved during consideration of the consequences of one's own and the partner's decisions (Güroğlu, Van Den Bos, and Crone, 2009; van den Bos et al., 2009, 2011; Abu-Akel and Shamay-Tsoory, 2011; Eddy, 2016). As will be outlined in more detail, the region of the temporo-parietal junction has been found to show age-dependent changes in functioning, getting more engaged during reciprocal decision-making processes with older adolescent age.

Besides the more cognitive aspects of decision making during the Trust Game, several additional brain regions are involved during social-affective functioning and value-based decision making. The first impression of a person's trustworthiness is gained through facial cues. The decision to trust an unknown partner is highly predictable based on that person's facial trustworthiness (van 't Wout and Sanfey, 2008). The *amygdala* has been shown to play a central role in judging trustworthiness of people's faces and coding fear of betrayal (Adolphs, Tranel, and Damasio, 1998; Adolphs, 2002; Winston et al., 2002; Rilling and Sanfey, 2011), thereby guiding social decisions in whom to engage with and whom to avoid. Interestingly, and even though more recent neuroimaging studies have not reported any amygdala involvement during the Trust Game, a lesion study showed that the amygdala plays a crucial role in extracting social value from the behavior of others, even in the absence of facial cues (Koscik and Tranel, 2011). In contrast to the healthy control participants, who used a "tit-for-tat" reciprocity strategy, patients with amygdala damage were less likely to adapt to betrayal behavior of the other player (Koscik and Tranel, 2011).

The *ventro-medial prefrontal cortex* was found to encode valuing of (future) social reward signals, received from reciprocal decisions, and therefore this region was suggested to be important for long-term benefit calculations that arise out of collaborative relationships (Rilling and Sanfey, 2011; Fareri, Chang, and Delgado, 2015). In addition, recently Haas et al. (2015) reported that individuals characterized as being more trusting (in terms of self-reported and behavioral trust tendencies) had greater gray matter volume in the bilateral ventro-medial prefrontal cortex (and in bilateral anterior insula) compared to participants tending to be less trusting (Haas et al., 2015).

In close connection with the ventro-medial prefrontal cortex, the *anterior insula* represents another key player in trust decisions. Inherent to initial trust decisions is a certain component of risk or uncertainty, such that when trust is given, it might be betrayed. This uncertainty component is greater in one-shot Trust Games, compared to multi-round interactions, during which partners build a reputation and get able to predict the partner's behavior. Activation in the right anterior insula has been linked to recognition of such risk in initial trust decisions and was associated with negative emotions related to betrayal aversion in the investor (Aimone, Houser, and Weber, 2014; Belfi, Koscik, and Tranel, 2015). Similarly, anticipated aversive feelings such as guilt reflected in right anterior insula activations may be elicited in the trustee when considering the social norms one should comply with (e.g., repaying trust) (Fehr, 2009; Chang et al., 2011; Belfi, Koscik, and Tranel, 2015; Montague, Lohrenz, and Dayan, 2015). Notably, the perception of social norms is not independent of the social context, the target (unknown other or a friend), and individual differences in prosocial preference. For example, right anterior insula activation was observed when *personal* social norms were violated, as in pro-social individuals who exploited trust, and in pro-selfs when reciprocating (King-Casas et al., 2008; van den Bos et al., 2009; Fareri, Chang, and Delgado, 2015; Bellucci et al., 2017).

Finally, the striatum, in particular the *ventral striatum*, is critically involved in encoding rewarding decision outcomes, for instance, when trust is being reciprocated (Delgado, Frank, and Phelps, 2005; Fareri et al., 2015). Of note, the outcome's reward value may be composed of both self-interested value and social-reward value (Fareri et al., 2015). Furthermore, predicted outcomes, which thus reward anticipation when the investor intends to trust during the later stages of a multi-round Trust Game, have been linked to striatal activations (King-Casas et al., 2005). The striatum is thus essential for integrating action outcomes and associated future decisions in trust and reciprocity interactions.

Thus, during social interaction in the Trust Game, brain regions important for social perspective taking, reward learning, emotional processing, and impulse control are recruited (see Figure 30.3). However, these social affective and socio-cognitive functions are not yet mature and continue to develop throughout childhood and adolescence. Also, many of the regions of the social brain network show protracted development until adulthood. Therefore, in the Trust Game, adolescents as compared to adults may recruit these brain areas to varying degrees, which may explain and provide further information on the neuro-biological mechanisms of the different developmental patterns of trust and reciprocity behavior across adolescence.

Neuro-Biological Development of Trust and Reciprocity

It has been argued that adolescents, when growing older show a shift from self-oriented behavior towards other-oriented behavior, which becomes increasingly complex and aids adolescents to develop and maintain stable, close relationships (Crone and Dahl, 2012). This change in social behavior is accompanied by significant changes in brain functioning. In recent years, a growing number of studies focused on the developmental neuro-cognitive aspects of trust and reciprocity behavior and on the corresponding changes in brain functioning during the Trust Game (van den Bos et al., 2011; Fett, Gromann, et al., 2014; Lemmers-Jansen et al., 2017). Overall, these studies showed that adolescents during trust and reciprocity decisions recruit similar brain regions as adults, but that the pattern of activation critically changes with age.

The first study that investigated trust and reciprocity behavior from a neuro-developmental perspective was conducted by van den Bos et al. (2011). Sixty-two participants (thirty-two male) were divided into three age groups: 12–14-year-olds, 15–17-year-olds, and 18–22-year-olds, thus comparing early, mid-, and late adolescence. Participants performed an fMRI Trust Game in the role of the trustee, in order to examine behavioral responses and corresponding brain activation when receiving trust and when deciding to either reciprocate or exploit trust. In line with their previous behavioral study (van den Bos et al., 2010), the authors found that older participants were better able to identify the intentions of others, indicated by larger risk-level related reciprocity differentiation with older age, which, in addition, positively correlated with an increase in brain activation in the temporo-parietal junction. Thus, the better participants were able to consider the risk (high/low) the investor had taken to trust them, the stronger the temporo-parietal junction was recruited after receiving trust. Furthermore, only the group of 18–22-year-olds showed activation in the dorso-lateral prefrontal cortex when receiving trust. In contrast, activation in the anterior medial prefrontal cortex during reciprocal choices was observed only in the youngest age group, indeed indicating a shift from more self-oriented processing in early adolescence towards more other-oriented social decision making in late adolescence. The authors concluded that mid-adolescence thus marks an important transition period during which intention consideration is emerging (van den Bos et al., 2011). In addition, when choosing to exploit trust (i.e., maximizing self-gain), van den Bos et al. (2011) found that, across age groups, individual differences in average reciprocity rates were predictive for brain activation in the anterior insula, dorsal anterior cingulate cortex, and dorso-lateral prefrontal cortex. In line with studies in adults (van den Bos et al., 2009; Fareri, Chang and Delgado, 2015), activation in these regions was greater in the "pro-socials" (relatively higher reciprocity rates) than in the "pro-selfs" (relatively lower reciprocity rates), indicating conflict between personal and social norms, and violation of personal norms for reciprocal behavior (King-Casas et al., 2008; van den Bos et al., 2009; Fareri, Chang, and Delgado, 2015; Bellucci et al., 2017). Interestingly, this finding was independent of the participant's age.

Another neuro-developmental study that examined the neural correlates of trust decisions using the Trust Game (Fett, Gromann, et al., 2014), included a sample of fifty-four participants, with a wide age range from adolescence to mid-adulthood (13–49 years of age). Using a similar task design as in their behavioral study (Fett, Shergill, et al., 2014), participants performed two sets of an iterated Trust Game in the role of the investor in the fMRI scanner. In one of the sets the trustee was acting cooperatively, in the other the interaction partner acted unfairly. Thus, neural responses to trust behavior, and their interaction with age were compared between conditions (unfair, cooperative). In line with previous reports, it was found that age was positively associated with initial trust, suggesting that with older age investors anticipate reciprocal responses by the trustee and increasingly expect others to act benevolently. Also, Fett, Shergill et al.

(2014) found that older age predicted quicker adjustment to the interaction partner's be-havior (i.e., keeping trust levels high when partner is cooperative, and showing a steeper decline in trust decisions when partner acted unfairly).

During trust decisions, irrespective of condition, older age was associated with increasing activation in mentalizing regions, such as the temporo-parietal junction, medial prefrontal cortex, precuneus (one of the cortical mid-line structures that, simi-larly as the medial prefrontal cortex, is involved in mentalizing and self-other distinc-tion [Northoff and Bermpohl, 2004]). Age-related decreases in activation were observed in the dorso-medial prefrontal cortex, orbito-frontal cortex, and caudate nucleus. This suggests, in line with the findings of van den Bos et al., (2011) that increasing mentalizing abilities, from adolescence to adulthood, are accompanied by stronger engagement of regions important for social perspective taking, and that brain areas involved in reward learning and reward anticipation are less recruited in trust decisions with age, possibly due to reduced prediction error signaling (Fett, Gromann, et al., 2014). Interestingly, in the unfair context, age was positively associated with activation in the anterior cingulate cortex, reflecting conflict signaling in situations when expectations of benevolence were not met, which older participants seemed to experience more strongly than younger participants.

The most recent study investigating trust and reciprocity behavior and their neuro-developmental correlates included a sample of forty-three participants (twenty-two male), between 16–27 years of age (Lemmers-Jansen et al., 2017). Though participants' age range did not include the youngest, pre- and early adolescent groups, the study is the first to compare male and female adolescents, thereby providing initial evidence for the neuro-developmental correlates of sex-specific trust behavior between mid-adolescence and early adulthood. With a similar experimental design of two different social contexts (cooperative vs. unfair) as in Fett, Gromann, et al., 2014), the authors replicated the be-havioral finding of males showing higher initial trust rates compared to females (Derks, Lee, and Krabbendam, 2014; Dittrich, 2015; Van de Groep et al. (2018). In addition, sex differences in trust behavior became especially apparent in the unfair context, with males showing a much steeper decline in investments compared to females. Also, their finding of a general stability of trust, thus no age-related changes in first investments, between mid-adolescence and adulthood, is in line with recent literature suggesting that major developmental changes in trust and reciprocity behavior occur in the transition from childhood to adolescence (Fett, Shergill, et al., 2014; Güroğlu, van den Bos, and Crone, 2014; Van de Groep et al., 2018).

Interestingly, despite the absence of any changes in trust behavior with age, Lemmers-Jansen et al. (2017) found age-related increases of brain activation during the investment phase in the temporo-parietal junction and dorso-lateral prefrontal cortex, which was similar for males and females, and comparable to the Trust Game study by van den Bos et al. (2011) that included a slightly younger sample. Therefore, age-related increases in brain activation in these regions during trust decisions may suggest a gen-eral increase in mentalizing, and optimized goal-directed decision making with age. However, how exactly the changes in brain functioning interact with e.g., different aspects of perspective taking and refinement of social decision making remains to be determined.

Contrary to the findings by Fett, Gromann, et al. (2014) the authors found an age-related increase in activation in the caudate nucleus during the repayment phase, i.e., when feedback about outcomes for the participant and the trustee was revealed, irrespective of the interaction context (unfair or cooperative interaction partner). In addition, females showed stronger activation of the caudate than males, specifically in cooperative interactions, whereas males more strongly engaged the temporo-parietal junction. This suggests that caudate activation not simply reflects social reward signaling (which would not be expected from unfair interactions), but rather that the neuro-biological mechanisms underlying reward learning change with age, and in a sex-specific manner, such that males and females use differential cognitive strategies when processing repayment outcomes (Lemmers-Jansen et al., 2017).

Together, these findings suggest that pro-social, rather than selfish motives for reciprocity behavior continue to develop during adolescence, accompanying the shift from more self-directed towards other-directed cognitive functioning. This refinement of social decision making, which is critically dependent on factors such as social context, sex, and individual difference variables, is reflected in age-related changes in distinct neural systems (Blakemore, 2008).

CONCLUSIONS AND FUTURE DIRECTIONS

In this chapter we have reviewed evidence for the notion that adolescence forms a crucial period of socio-cognitive and social-affective development that parallels the social re-orientation from a family/parental context to a social context more focused on friends and peers. Thereby, adolescents get able to form close and stable relationships, that are crucially based on trust and reciprocity. Economic games, such as the Trust Game have proven to be valuable in studying the development of specific aspects of social decision making, distinguishing moments of consideration to trust another person, and to weigh the options for repaying that trust (reciprocity).

Behavioral studies employing the Trust Game have shown that trust and reciprocity development across adolescence are highly complex and that the different developmental patterns observed [e.g., linearly increasing trust and reciprocity with age (Sutter and Kocher, 2007; van den Bos et al., 2010), linearly decreasing reciprocity with age (Van de Groep et al., 2018), no age-related changes of trust (Fett, Shergill, et al., 2014; Van de Groep et al., 2018)] may very well be explained by interactions of age with factors such as sex, risk-level associated with decision to trust, and social context. In recent years, neuroimaging studies on trust and reciprocity development have started to emerge, supporting the idea that during adolescence more advanced forms of social perspective taking develop, which is accompanied by functional brain changes in regions of the social brain network (e.g., temporo-parietal junction), and in regions important for impulse control (e.g., dorso-lateral prefrontal cortex), and computation of value (e.g., caudate nucleus). However, replication of these initial findings is greatly needed and many questions concerning the neuro-biological mechanisms underlying trust and reciprocity development remain to be answered.

In order to pin-point the exact nature of developmental changes in social decision-making and their underlying neural correlates, studies need to include larger samples, covering a broader age range from pre-adolescence to early adulthood. In addition, even though cross-sectional comparisons in the current literature provide valuable information on differences in e.g., trust behavior between age groups, future studies should use longitudinal designs in order to better characterize age-related changes in brain and behavior.

Furthermore, it will be important to adopt a multi-modal approach in the characterization of neural changes that accompany age-related changes in trust and reciprocity behavior. First, evidence from studies in adults, showing that volume differences in the medial prefrontal cortex and anterior insula were predictive for trust behavior (Haas et al., 2015), might be inspirational for future neuro-developmental studies in linking the vast morphological brain changes that occur during adolescence to specific functional changes in social decision making (Mills et al., 2014). Moreover, besides understanding which brain regions are involved in trust and reciprocity decisions, it is important to better characterize their interaction during different (e.g., risk) contexts and in relation to individual difference variables (e.g., different forms of empathy). This could be done by functional and structural connectivity analyses between brain regions engaged during fMRI Trust Game paradigms.

Another question for future neuro-developmental studies to elucidate concerns the neuro-biological basis for the robust finding of sex differences in trust behavior. Lemmers-Jansen et al. (2017) have provided initial evidence for sex-specific strategies and corresponding brain activations during trust decisions. However, their sample included only post-pubertal adolescents and adults. It would therefore be interesting to investigate whether sex differences in the age-dependent changes of trust emerge as a function of pubertal status and rise in sex hormone levels. In addition, it would be interesting to examine how behavioral sex differences in terms of risk taking and risk aversion tendencies interact with pubertal status in relation to brain activation patterns during the Trust Game.

Finally, it will be an important endeavor for future studies to consider the ecological validity of their study designs and findings. For example, studies may use socio-metric procedures to identify participants' real-life friends and disliked peers, or include age-matched confederates in task paradigms of social decision making (Sharp, Burton and Ha, 2011; Güroğlu, van den Bos, and Crone, 2014; Hoorn et al., 2016; Schreuders et al., 2018). This would increase ecological validity of both the neural and behavioral measures of trust and reciprocity decisions. Investigation of their changes across adolescence in relation to lab-setting based findings will greatly extend our knowledge on the development of trust and reciprocity behavior during adolescence.

REFERENCES

Abu-Akel, A. and Shamay-Tsoory, S. (2011). Neuroanatomical and neurochemical bases of theory of mind. *Neuropsychologia, 49*(11), 2971–2984. doi:10.1016/j.neuropsychologia.2011.07.012.

Adenzato, M. et al. (2017). Gender differences in cognitive theory of mind revealed by transcranial direct current stimulation on medial prefrontal cortex. *Scientific Reports, 7*, 41219. doi:10.1038/srep41219.

Adolphs, R. (2002). Trust in the brain. *Nature Neuroscience, 5*(3), 192–193. doi:10.1038/nn0302-192.

Adolphs, R., Tranel, D., and Damasio, A. R. (1998). The human amygdala in social judgment. *Nature, 393*(6684), 470–474. doi:10.1038/30982.

Aimone, J. A., Houser, D., and Weber, B. (2014). Neural signatures of betrayal aversion: An fMRI study of trust. *Proceedings of the Royal Society B: Biological Sciences, 281*(1782), 20132127. doi:10.1098/rspb.2013.2127.

Amodio, D. M. and Frith, C. D. (2006). Meeting of minds: The medial frontal cortex and social cognition. *Nature Reviews Neuroscience, 7*(4), 268–277. doi:10.1038/nrn1884.

Balliet, D., et al. (2011). Sex differences in cooperation: A meta-analytic review of social dilemmas. *Psychological Bulletin, 137*(6), 881–909. doi:10.1037/a0025354.

Behrens, T. E. J. et al. (2008) Associative learning of social value. *Nature, 456*(7219), 245–249. doi:10.1038/nature07538.

Belfi, A. M., Koscik, T. R., and Tranel, D. (2015). Damage to the insula is associated with abnormal interpersonal trust. *Neuropsychologia, 71*, 165–172. doi:10.1016/j.neuropsychologia.2015.04.003.

Bellucci, G., et al. (2017). Neural signatures of trust in reciprocity: A coordinate-based meta-analysis. *Human Brain Mapping, 38*(3), 1233–1248. doi:10.1002/hbm.23451.

Ben-Ner, A. and Halldorsson, F. (2010). Trusting and trustworthiness: What are they, how to measure them, and what affects them. *Journal of Economic Psychology, 31*(1), 64–79.

Berg, J., Dickhaut, J., and McCabe, K. (1995). Trust, reciprocity, and social history. *Games and Economic Behavior, 10*(1), 122–142. doi:10.1006/GAME.1995.1027.

Blakemore, S.-J. (2008). The social brain in adolescence. *Nature Reviews, Neuroscience, 9*(4), 267–277. doi:10.1038/nrn2353.

Blakemore, S.-J., Burnett, S., and Dahl, R. E. (2010). The role of puberty in the developing adolescent brain. *Human Brain Mapping, 31*(6), 926–933. doi:10.1002/hbm.21052.

van den Bos, W., et al. (2009). What motivates repayment? Neural correlates of reciprocity in the Trust Game. *Social Cognitive and Affective Neuroscience, 4*(3), 294–304. doi:10.1093/scan/nsp009.

van den Bos, W. et al. (2010). Development of trust and reciprocity in adolescence. *Cognitive Development, 25*(1), 90–102. doi:10.1016/j.cogdev.2009.07.004.

van den Bos, W., et al. (2011). Changing brains, changing perspectives: The neurocognitive development of reciprocity. *Psychological Science, 22*(1), 60–70. doi:10.1177/0956797610391102.

van den Bos, W., van Dijk, E., and Crone, E. A. (2012). Learning whom to trust in repeated social interactions: A developmental perspective. *Group Processes & Intergroup Relations, 15*(2), 243–256. doi:10.1177/1368430211418698.

Braams, B. R., et al. (2015). Longitudinal changes in adolescent risk-taking: A comprehensive study of neural responses to rewards, pubertal development, and risk-taking behavior. *Journal of Neuroscience, 35*(18), 7226–7238. doi:10.1523/JNEUROSCI.4764-14.2015.

Burnett, S. et al. (2011). The social brain in adolescence: Evidence from functional magnetic resonance imaging and behavioural studies. *Neuroscience and Biobehavioral Reviews*, 35(8), 1654–1664. doi:10.1016/j.neubiorev.2010.10.011.

Byrnes, J. P., Miller, D. C., and Schafer, W. D. (1999). Gender differences in risk taking: A meta-analysis. *Psychological Bulletin*, 125(3), 367–383. doi:10.1037/0033-2909.125.3.367.

Carlson, S. M., Moses, L. J., and Claxton, L. J. (2004). Individual differences in executive functioning and theory of mind: An investigation of inhibitory control and planning ability. *Journal of Experimental Child Psychology*, 87(4), 299–319. doi:10.1016/j.jecp.2004.01.002.

Chang, L. J., et al. (2011). Triangulating the neural, psychological, and economic bases of guilt aversion. *Neuron*, 70(3), 560–572. doi:10.1016/j.neuron.2011.02.056.

Chaudhuri, A. and Gangadharan, L. (2007). An experimental analysis of trust and trustworthiness. *Southern Economic Journal*, 73(4).

Cillessen, A. H. N. and Rose, A. J. (2005). Understanding popularity in the peer system. *Current Directions in Psychological Science*, 14(2), 102–105. doi:10.1111/j.0963-7214.2005.00343.x.

Crone, E. A. and Dahl, R. E. (2012). Understanding adolescence as a period of social–affective engagement and goal flexibility. *Nature Reviews Neuroscience*, 13(9), 636–650. doi:10.1038/nrn3313.

Crone, E. A. and Konijn, E. A. (2018). Media use and brain development during adolescence. *Nature Communications*, 9(1), 588. doi:10.1038/s41467-018-03126-x.

Croson, R. and Gneezy, U. (2009). Gender differences in preferences. *Journal of Economic Literature*, 47(2), 448–474. doi:10.1257/jel.47.2.448.

Delgado, M. R., Frank, R. H., and Phelps, E. A. (2005). Perceptions of moral character modulate the neural systems of reward during the trust game. *Nature Neuroscience*, 8(11), 1611–1618. doi:10.1038/nn1575.

Derks, J., Lee, N. C., and Krabbendam, L. (2014). Adolescent trust and trustworthiness: Role of gender and social value orientation. *Journal of Adolescence*, 37(8), 1379–1386. doi:10.1016/j.adolescence.2014.09.014.

Dittrich, M. (2015). Gender differences in trust and reciprocity: Evidence from a large-scale experiment with heterogeneous subjects. *Applied Economics*, 47(36), 3825–3838. doi:10.1080/00036846.2015.1019036.

Do, K. T., Guassi Moreira, J. F., and Telzer, E. H. (2017). But is helping you worth the risk? Defining prosocial risk taking in adolescence. *Developmental Cognitive Neuroscience*, 25, 260–271. doi:10.1016/j.dcn.2016.11.008.

Dumontheil, I., Apperly, I. A., and Blakemore, S.-J. (2010). Online usage of theory of mind continues to develop in late adolescence. *Developmental Science*, 13(2), 331–338. doi:10.1111/j.1467-7687.2009.00888.x.

Eagly, A. H. (2009). The his and hers of prosocial behavior: An examination of the social psychology of gender. *American Psychologist*, 64(8), 644–658. doi:10.1037/0003-066X.64.8.644.

Eddy, C. M. (2016). The junction between self and other? Temporo-parietal dysfunction in neuropsychiatry. *Neuropsychologia*, 89, 465–477. doi:10.1016/j.neuropsychologia.2016.07.030.

Eisenegger, C., Haushofer, J., and Fehr, E. (2011). The role of testosterone in social interaction. *Trends in Cognitive Sciences*, 15(6), 263–271. doi:10.1016/j.tics.2011.04.008.

Fareri, D. S., Chang, L. J., and Delgado, M. R. (2015). Computational substrates of social value in interpersonal collaboration. *Journal of Neuroscience*, 35(21), 8170–8180. doi:10.1523/JNEUROSCI.4775-14.2015.

Fehr, E. (2009). On the economics and biology of trust. *Journal of the European Economic Association*, 7(2–3), 235–266. doi:10.1162/JEEA.2009.7.2–3.235.

Fett, A.-K. J., Gromann, P. M., et al. (2014). Default distrust? An fMRI investigation of the neural development of trust and cooperation. *Social Cognitive and Affective Neuroscience*, 9(4), 395–402. doi:10.1093/scan/nss144.

Fett, A.-K. J., Shergill, S. S., et al. (2014). Trust and social reciprocity in adolescence— a matter of perspective-taking. *Journal of Adolescence*, 37(2), 175–184. doi:10.1016/j.adolescence.2013.11.011.

Foulkes, L. and Blakemore, S. J. (2018). Studying individual differences in human adolescent brain development, *Nature Neuroscience*, 21(3), 315–323. doi:10.1038/s41593-018-0078-4.

Frank, C. K., Baron-Cohen, S., and Ganzel, B. L. (2015). Sex differences in the neural basis of false-belief and pragmatic language comprehension. *NeuroImage*, 105, 300–311. doi:10.1016/j.neuroimage.2014.09.041.

Frith, U. and Frith, C. D. (2003). Development and neurophysiology of mentalizing. *Philosophical Transactions of the Royal Society of London. Series B, Biological sciences*, 358(1431), 459–473. doi:10.1098/rstb.2002.1218.

Garbarino, E. and Slonim, R. (2009). The robustness of trust and reciprocity across a heterogeneous U.S. population. *Journal of Economic Behavior and Organization*, 69(3), 226–240. doi:10.1016/j.jebo.2007.06.010.

Gummerum, M., Hanoch, Y., and Keller, M. (2008). When child development meets economic game theory: An interdisciplinary approach to investigating social development. *Human Development*, 51(4), 235–261. doi:10.1159/000151494.

Güroglu, B., Bos, W., van den and Crone, E. A. (2014). Sharing and giving across adolescence: An experimental study examining the development of prosocial behavior. *Frontiers in Psychology*, 5(APR), 1–13. doi:10.3389/fpsyg.2014.00291.

Güroğlu, B., van den Bos, W., and Crone, E. A. (2014). Sharing and giving across adolescence: An experimental study examining the development of prosocial behavior. *Frontiers in psychology*, 5, 291. doi:10.3389/fpsyg.2014.00291.

Güroğlu, B., Van Den Bos, W., and Crone, E. A. (2009). Neural correlates of social decision making and relationships: A developmental perspective. *Annals of the New York Academy of Sciences*, 1167(1), 197–206. doi:10.1111/j.1749–6632.2009.04502.x.

Haas, B. W., et al. (2015). The tendency to trust is reflected in human brain structure. *NeuroImage*, 107, 175–181. doi:10.1016/j.neuroimage.2014.11.060.

Harris, J. A., et al. (1996). Salivary testosterone and self-report aggressive and pro-social personality characteristics in men and women. *Aggressive Behavior*, 22(5), 321–31. doi:10.1002/(SICI)1098-2337(1996)22:5<321::AID-AB1>3.0.CO;2-M.

Haselhuhn, M. P. et al. (2014). Gender differences in trust dynamics: Women trust more than men following a trust violation. *Journal of Experimental Social Psychology*, 56, 104–109.

Hoorn, J. Van et al. (2016). Neural correlates of prosocial peer influence on public goods game donations during adolescence. *Social Cognitive and Affective Neuroscience*, 11(6), 923–933. doi:10.1093/scan/nsw013.

Huizinga, M., Dolan, C. V., and van der Molen, M. W. (2006). Age-related change in executive function: Developmental trends and a latent variable analysis. *Neuropsychologia*, 44(11), 2017–2036. doi:10.1016/j.neuropsychologia.2006.01.010.

Innocenti, A. and Pazienza, M. G. (2006). Altruism and Gender in the Trust Game. Labsi Working Paper No. 5/2006. *SSRN Electronic Journal*. doi:10.2139/ssrn.884378.

Johnson, D. W. (1975). Cooperativeness and social perspective taking. *Journal of Personality and Social Psychology*, 31(2), 241–244. doi:10.1037/h0076285.

King-Casas, B. et al. (2005). Getting to know you: reputation and trust in a two-person economic exchange. *Science, 308*(5718), 78–83. doi:10.1126/science.1108062.

King-Casas, B., et al. (2008). The rupture and repair of cooperation in borderline personality disorder. *Science, 321*(5890), 806–810. doi:10.1126/science.1156902.

Knoch, D. et al. (2009). Disrupting the prefrontal cortex diminishes the human ability to build a good reputation. *Proceedings of the National Academy of Sciences, 106*(49), 20895–20899. doi:10.1073/pnas.0911619106.

Koscik, T. R. and Tranel, D. (2011). The human amygdala is necessary for developing and expressing normal interpersonal trust. *Neuropsychologia, 49*(4), 602–611. doi:10.1016/j.neuropsychologia.2010.09.023.

Krach, S., et al. (2009). Are women better mindreaders? Sex differences in neural correlates of mentalizing detected with functional MRI. *BMC.Neurosci, 10*, 9. doi:10.1186/1471-2202-10-9.

Krueger, F. et al. (2007). Neural correlates of trust. *Proceedings of the National Academy of Sciences, 104*(50), 20084–20089. doi:10.1073/pnas.0710103104.

Lahno, B. (1995). Trust, reputation, and exit in exchange relationships. *Journal of Conflict Resolution, 39*(3), 495–510. doi:10.1177/0022002795039003005.

Lemmers-Jansen, I. L. J., et al. (2017). Boys vs. girls: Gender differences in the neural development of trust and reciprocity depend on social context. *Developmental Cognitive Neuroscience, 25*, 235–245. doi:10.1016/j.dcn.2017.02.001.

Malhotra, D. (2004). Trust and reciprocity decisions: The differing perspectives of trustors and trusted parties. *Organizational Behavior and Human Decision Processes, 94*(2), 61–73. doi:10.1016/j.obhdp.2004.03.001.

Meuwese, R., et al. (2015). Development of equity preferences in boys and girls across adolescence. *Child Development, 86*(1), 145–158. doi:10.1111/cdev.12290.

Migheli, M. (2007). Trust, gender and social capital: Experimental evidence from three western European countries. *SSRN Electronic Journal*. doi:10.2139/ssrn.976380.

Mills, K. L., et al. (2014). Developmental changes in the structure of the social brain in late childhood and adolescence. *Social Cognitive and Affective Neuroscience, 9*(1), 123–131. doi:10.1093/scan/nss113.

Montague, P. R., Lohrenz, T., and Dayan, P. (2015). The three R's of trust. *Current Opinion in Behavioral Sciences, 3*, 102–106. doi:10.1016/j.cobeha.2015.02.009.

Newcomb, A. F., Bukowski, W. M., and Pattee, L. (1993). Children's peer relations: A meta-analytic review of popular, rejected, neglected, controversial, and average sociometric status. *Psychological Bulletin, 113*(1), 99–128.

Northoff, G. and Bermpohl, F. (2004). Cortical midline structures and the self. *Trends in Cognitive Sciences, 8*(3), 102–107. doi:10.1016/j.tics.2004.01.004.

Van Overwalle, F. (2009). Social cognition and the brain: A meta-analysis. *Human Brain Mapping, 30*(3), 829–858. doi:10.1002/hbm.20547.

Peper, J. S. and Dahl, R. E. (2013). Surging hormones: Brain-behavior interactions during puberty. *Current Directions in Psychological Science, 22*(2), 134–139. doi:10.1177/0963721412473755.

Peters, E., et al. (2011). Peer rejection and HPA activity in middle childhood: Friendship makes a difference. *Child Development, 82*(6), 1906–1920. doi:10.1111/j.1467-8624.2011.01647.x.

Riedl, R. and Javor, A. (2012). The biology of trust: Integrating evidence from genetics, endocrinology, and functional brain imaging. *Journal of Neuroscience, Psychology, and Economics, 5*(2), 63–91. doi:10.1037/a0026318.

Rilling, J. K., et al. (2004). The neural correlates of theory of mind within interpersonal interactions. *NeuroImage, 22*(4), 1694–1703. doi:10.1016/j.neuroimage.2004.04.015.

Rilling, J. K. and Sanfey, A. G. (2011). The neuroscience of social decision making. *Annual Review of Psychology, 62*, 23–48. doi:10.1146/annurev.psych.121208.131647.

Sanfey, A. G., et al. (2003). The neural basis of economic decision-making in the ultimatum game. *Science, 300*(5626), 1755–1758. doi:10.1126/science.1082976.

Schreuders, E., et al. (2018). Friend versus foe: Neural correlates of prosocial decisions for liked and disliked peers. *Cognitive, Affective, & Behavioral Neuroscience, 18*(1), 127–142. doi:10.3758/s13415-017-0557-1.

Schulte-Rüther, M., et al. (2008). Gender differences in brain networks supporting empathy. *NeuroImage, 42*(1), 393–403. doi:10.1016/J.NEUROIMAGE.2008.04.180.

Sebastian, C. L., et al. (2011). Developmental influences on the neural bases of responses to social rejection: Implications of social neuroscience for education. *NeuroImage, 57*(3), 686–694. doi:10.1016/j.neuroimage.2010.09.063.

Selman, R. L. (1980) *The Growth of Interpersonal Understanding: Developmental and Clinical Analyses.* New York, NY: Academic Press.

Sharp, C., Burton, P. C., and Ha, C. (2011). "Better the devil you know": a preliminary study of the differential modulating effects of reputation on reward processing for boys with and without externalizing behavior problems. *European Child & Adolescent Psychiatry, 20*(11–12), 581–592. doi:10.1007/s00787-011-0225-x.

Shirtcliff, E. A., Dahl, R. E., and Pollak, S. D. (2009). Pubertal development: Correspondence between hormonal and physical development. *Child Development, 80*(2), 327–337. doi:10.1111/j.1467-8624.2009.01263.x.

Singer, T. and Lamm, C. (2009). The social neuroscience of empathy. *Annals of the New York Academy of Sciences, 1156*(1), 81–96. doi:10.1111/J.1749-6632.2009.04418.X.

Steinberg, L. (2008). A social neuroscience perspective on adolescent risk-taking. *Developmental Review, 28*(1), 78–106. doi:10.1016/j.dr.2007.08.002.

Sutter, M. (2007). Outcomes versus intentions: On the nature of fair behavior and its development with age. *Journal of Economic Psychology, 28*(1), 69–78. doi:10.1016/j.joep.2006.09.001.

Sutter, M. and Kocher, M. G. (2007). Age and the development of trust and reciprocity. *Games and Economic Behavior, 59*, 364–82. doi:10.2139/ssrn.480184.

Tzieropoulos, H. (2013) The Trust Game in neuroscience: A short review. *Social Neuroscience, 8*(5), 407–416. doi:10.1080/17470919.2013.832375.

Van de Groep, S., et al. (2018). Developmental changes and individual differences in trust and reciprocity in adolescence. *Journal of Research on Adolescence, 30*(S1), 192–208. doi:10.1111/jora.12459.

van 't Wout, M. and Sanfey, A. G. (2008). Friend or foe: The effect of implicit trustworthiness judgments in social decision-making. *Cognition, 108*(3), 796–803. doi:10.1016/j.cognition.2008.07.002.

Will, G.-J. and Güroğlu, B. (2016). A neurocognitive perspective on the development of social decision-making. In Reuter, M. and Montag, C. (eds.) *Neuroeconomics, Studies in Neuroscience, Psychology and Behavioral Economics*, pp. 293–309. Berlin, Heidelberg: Springer. doi:10.1007/978-3-642-35923-1_15.

Winston, J. S. S., et al. (2002). Automatic and intentional brain responses during evaluation of trustworthiness of faces. *Nature Neuroscience, 5*(3), 277–283. doi:10.1038/nn816.

Zak, P. J., et al. (2005). The neuroeconomics of distrust: Sex differences in behavior and physiology. *American Economic Review, 95*(2), 360–363. doi:10.1257/000282805774669709.

Zak, P. J., et al. (2009). Testosterone administration decreases generosity in the ultimatum game. *PLoS ONE, 4*(12), e8330. doi:10.1371/journal.pone.0008330.

CHAPTER 31

NEUROBIOLOGICAL SUSCEPTIBILITY TO PEER INFLUENCE IN ADOLESCENCE

KATHY T. DO, MITCHELL J. PRINSTEIN, AND EVA H. TELZER

INTRODUCTION

ACROSS rodents, non-human primates, and humans, adolescence is marked by a universal "social restructuring" characterized by increasingly complex social development (e.g., perspective taking) and inhibitory control and a reorientation from parental to peer contexts (Blakemore and Mills, 2014; Nelson, Leibenluft, McClure, and Pine, 2005). Compared with children, adolescents spend more time with peers, form more sophisticated and hierarchical social relationships, are more sensitive to peer acceptance, and become increasingly self-conscious (Blakemore and Mills, 2014; Brown, 2004; Steinberg and Morris, 2001). This social reorientation and heightened sensitivity to peer contexts may render adolescents particularly susceptible to social influence. Indeed, compared to children and adults, adolescents engage in more risk taking in the presence of peers (e.g., Gardner and Steinberg, 2005), tend to conform to the attitudes of their peers about risky (Knoll, Magis-Weinberg, Speekenbrink, and Blakemore, 2015) and prosocial (Foulkes, Leung, Fuhrmann, Knoll, and Blakemore, 2018) behaviors, show heightened embarrassment when being observed by their peers (Somerville et al., 2013), and show compromised emotion regulation in the presence of socially appetitive cues (Perino, Miernicki, and Telzer, 2016; Somerville, Hare, and Casey, 2011). Heightened social influence susceptibility may reflect maturational changes in how the brain responds to social information. Indeed, functional reorganization and maturation of the developing brain continues well into young adulthood (Sowell et al., 2003, 2004). Such neural plasticity, in conjunction with dynamic social environments, may render adolescents particularly sensitive to social influences in their environment and play a critical role in shaping both positive and negative developmental trajectories.

The current chapter presents emerging findings from developmental science, social psychology, and cognitive neuroscience that begin to address the construct of susceptibility to peer influence, particularly in moderating adolescents' conformity to their peers' risky behaviors. First, we provide a brief overview of the increasingly salient role of peers in socializing risk-related attitudes and behaviors during adolescence. Using the differential susceptibility theory as a framework, we consider how an individual's social influence susceptibility emerges and interacts with various peer environments in ways that provoke shifts in adolescent behavior. We then explore candidate biomarkers of social influence susceptibility that may index an individual's level of susceptibility to social contexts, with a particular focus on neurobiological biomarkers. Finally, we conclude with future directions examining how a heightened susceptibility to peer influence can facilitate both adaptive and maladaptive behavior, depending on the type of peer environment.

PEER INFLUENCES ON ADOLESCENT RISK TAKING

A widely held belief shared by many parents, educators, and policymakers is that peers steer youth toward engaging in negative, health-compromising behaviors that they otherwise would not have participated in alone. Indeed, decades of research have shown that peers' engagement in risk-taking behaviors is a consistent and robust predictor of adolescents' own risk-taking behaviors (Prinstein and Dodge, 2008). In fact, peer interactions in infancy and toddlerhood start to facilitate the early ontogeny of group reputation concerns (Engelmann, Herrmann, and Tomasello, 2018; Engelmann, Over, Herrmann, and Tomasello, 2013) and norm-based behavior, such as learning and reinforcing contextual and societal norms (Schmidt, Butler, Heinz, and Tomasello, 2016; Schmidt, Rakoczy, and Tomasello, 2012; reviewed in Tomasello, 2014), both of which contribute to individuals' susceptibility to social influence. However, often overlooked is that conformity to peer influences can be a positive social skill throughout human ontogeny, and perhaps especially in adolescence, because it enables individuals to develop stronger affiliations with others and gain entry into diverse social groups (Tomasello, 2014, 2016; Tomasello, Kruger, and Ratner, 1993). Compared to those with a lower susceptibility to peer influence, individuals with a greater susceptibility to peer influence may be more likely to leverage this developing social skill, thus benefiting from the many important and meaningful social connections inherent in conforming to others. Despite the social advantages conferred by peer conformity, researchers have primarily examined how an increased susceptibility to peer influence can exacerbate associations between peer influence and disproportionate rates of risk taking during adolescence (Allen et al., 2006; Prinstein et al., 2011; Widman et al., 2016). Studying individual differences in susceptibility to peer influence represents a promising opportunity to disentangle the person- versus environment-specific factors that contribute to both the positive and negative effects of peer conformity during adolescence.

Peer environments afford a unique opportunity for adolescents to develop and update their own risk-related attitudes and behaviors over time. Several theories from social and developmental psychology have suggested that conformity is motivated by a desire to align oneself with positive role models and consequently develop a more favorable sense of self-identity (Baldwin, 1895; Berger, 2008; Cooley, 1902; Gibbons, Gerrard, Blanton, and Russell, 1998; Mead, 1934). Being especially attuned to social evaluation as peer relationships become increasingly salient may lead some adolescents to conform to the norms and behaviors of their peer group in an effort to enhance social belonging (i.e., "fitting in") or establish a more favorable self-identity. Adolescents often prioritize enhancing their reputational status to a greater extent than children and adults—even beyond their pursuit of friendships, romantic interests, or personal achievements (e.g., academic) (LaFontana and Cillessen, 2010). Adolescents who have, or value attaining, high social status are particularly susceptible to peer influence, which has long-term consequences on health outcomes (Cillessen and Mayeux, 2004). For example, peers' deviant behaviors predict adolescents' future deviant behaviors, especially among adolescents who are highly susceptible to high-status peers (Choukas-Bradley, Giletta, Widman, Cohen, and Prinstein, 2014; Prinstein, Brechwald, and Cohen, 2011).

Adolescents' engagement in health risk behaviors is not only shaped by the type of peers that an individual selectively affiliates with (peer selection), but also by the actual or perceived norms of the peer group (peer socialization). Shifts in risk-related attitudes or behaviors can occur in response to social influence from one peer, small peer groups, or broader social networks, which also vary in magnitude based on the strength and quality of those social ties (reviewed in Telzer, Rogers, and Van Hoorn, 2017). The most common method of understanding risk-taking trajectories is investigating the degree to which perceived and actual peer norms socialize adolescents' engagement in health risk behaviors over time (reviewed in Albert et al., 2013; Brechwald and Prinstein, 2011; Simons-Morton and Farhat, 2010), as (mis) perceptions of peer norms about risk taking have been shown to redirect adolescents' antecedent attitudes and behaviors toward riskier *and* healthier outcomes. Peers can socialize adolescents' risky decision making in both implicit and explicit ways. Implicit forms of peer influence include positive reinforcement of deviant behavior during peer interactions (i.e., deviancy training) (Dishion, Spracklen, Andrews, and Patterson, 1996), being physically present (e.g., Gardner and Steinberg, 2005), or modeling specific attitudes or behaviors that are then internalized by adolescents as valued in the peer context (e.g., Cascio et al., 2015; Knoll, Leung, Foulkes, and Blakemore, 2017; Knoll et al., 2015; Welborn et al., 2015). In some cases, adolescents over- or under-estimate peers' attitudes and behaviors around health risk behaviors and use these erroneous peer norms to guide their own decisions, a phenomenon known as the false consensus effect (Helms et al., 2014; Prinstein and Wang, 2005). Importantly, peer norms need only be perceived to influence behavior, as perceptions of peers' risk behaviors can be an even stronger predictor of adolescents' own risk behavior compared to actual peer behavior (Fromme and Ruela, 1994; Iannotti and Bush, 1992; Slagt, Dubas, Deković, Haselager, and van Aken, 2015). Peers can also influence adolescent behavior explicitly, by providing feedback (e.g., likes) (e.g., van Hoorn, van Dijk, Güroğlu, and Crone, 2016; van Hoorn, van Dijk, Meuwese, Rieffe, and Crone, 2014) or verbally reinforcing group norms (e.g., Chein, Albert, O'Brien, Uckert, and Steinberg, 2011). Although both forms of peer influence are associated with changes in adolescents' health-related risk taking (e.g., Cascio et al., 2015; Chein, Albert, O'Brien, Uckert, and Steinberg, 2011; Dishion, Spracklen, Andrews, and Patterson, 1996; Falk et al., 2014), previous work suggests that explicit peer influence may be more effective at altering adolescent risk behavior than implicit forms of peer influence (Engelmann, Moore, Monica Capra, and Berns, 2012; Munoz Centifanti, Modecki, Maclellan, and Gowling, 2014; Somerville et al., 2018). However, implicit or indirect peer pressures likely occur more frequently in real life (Simons-Morton and Farhat, 2010).

Increasingly, researchers are showing that peer effects on adolescents' risky decisions are more nuanced than previously reported. Peers can redirect adolescent risk behavior toward fundamentally different health outcomes based on the extent of peer involvement (e.g., mere presence versus peer monitoring), type of decision context (e.g., "hot" or "cold" decisions), and whether risk outcomes additionally impact others (Do, Guassi Moreira, and Telzer, 2016; Powers et al., 2018; Somerville et al., 2018). Furthermore, the association between peers' and adolescents' risk-taking behaviors can be moderated by several psychosocial factors at the individual level (reviewed in Brechwald and Prinstein, 2011). For example, individual differences in self-regulation (e.g., Gardner, Dishion, and Connell, 2008), social reputation (e.g., Prinstein, Meade, and Cohen, 2003), and family conflict (e.g., Prinstein, Boergers, and Spirito, 2001) can exacerbate or buffer the degree to which peer pressures alter health-related attitudes or behaviors. Taken together, these findings suggest that although peers strongly influence adolescent risk taking, several environmental and psychosocial conditions can amplify or attenuate these relations.

Although peer influence is characteristically associated with monotonic increases in adolescent risk taking, there is emerging support for the role of peers in promoting risk-averse preferences, which can buffer against the normative uptick in health risk behaviors during adolescence. For instance, adolescents actually engage in less risky behavior when their peers explicitly endorse risk-averse preferences (Cascio et al., 2015; Engelmann et al., 2012), an effect that is stronger among early and late adolescents relative to adults (Engelmann et al., 2012). The social status of the influencing peers can further moderate adolescents' conformity to peers' health risk behaviors. Whereas adolescents exhibit higher rates of conformity when risk-related attitudes and behaviors are endorsed by high status (e.g., popular) peers, adolescents are more likely to diverge from—and act against—these norms if endorsed by less desirable (e.g., low status) peers (i.e., anticonformity) (Brechwald and Prinstein, 2011; Cohen and Prinstein, 2006; Teunissen et al., 2012; Widman et al., 2016). In fact, high status peers may be more effective at discouraging rather than encouraging health-compromising risk taking during adolescence (Teunissen et al., 2012), underscoring the equally powerful effect of peers for influencing positive health outcomes during adolescence. Indeed, peers can promote prosocial, other-focused behaviors, such as sharing, cooperating, helping, intentions to volunteer, as well as learning and exploration (Barry and Wentzel, 2006; Choukas-Bradley, Giletta, Cohen, and Prinstein, 2015; Foulkes et al., 2018; Silva, Shulman, Chein, and Steinberg, 2016; van Hoorn, van Dijk, Güroğlu, and Crone, 2016; Wentzel, Filisetti, and Looney, 2007). If peers can also re-direct adolescents toward positive behavior change (Telzer, van Hoorn, Rogers, and Do, 2018; van Hoorn, Fuligni, Crone, and Galván, 2016), then interventions that cap-italize on adolescents' peer influence susceptibility in positive environments may be particularly promising for improving youth adjustment.

Although adolescents are, on average, more susceptible to peer influences compared to children and adults, not all adolescents may be equally influenced by their peers (Allen et al., 2006; Brittain, 1963; Prinstein et al., 2011; Steinberg and Monahan, 2007; Widman et al., 2016). Studies rarely have examined individual variability in, or sus-ceptibility to, peer influence. However, extant research using both self-reported and performance-based approaches to study susceptibility have demonstrated that this is a normally distributed construct (Prinstein et al., 2011; Widman et al., 2016), suggesting approximately equal proportions of those extremely high or low in susceptibility as compared to other age mates. Individual differences in gender, pubertal develop-ment, social anxiety, and sensitivity to peer status have been shown to render some adolescents more at risk to negative peer pressures than others (Prinstein et al., 2011; Widman et al., 2016). Furthermore, adolescents who show a higher sensitivity to peer influences tend to experience distinct psychosocial outcomes compared to those who are relatively resistant to peer pressures (Allen et al., 2006; Prinstein et al., 2011). For example, one study reported longitudinal associations between best friends' and adolescents' own deviant behavior only among adolescents who demonstrated high levels of peer influence susceptibility to high-status classmates on a laboratory-based task (Prinstein et al., 2011). Notably, adolescents with a low susceptibility to high-status peers or a general (i.e., high or low) susceptibility to low-status peers did not show peer-related changes in deviant behavior over time (Prinstein et al., 2011), suggesting that those with a high susceptibility to (high-status) peers may be particularly more vulnerable to negative peer socialization effects over time. Moreover, growing evi-dence suggests that adolescents are also highly susceptible to prosocial peer influences (Foulkes et al., 2018; van Hoorn, Fuligni, et al., 2016; van Hoorn, van Dijk, et al., 2016; van Hoorn et al., 2014), yet little is known about the long-term consequences of having a high sensitivity to peer influence in more positive social contexts. Taken together, these data underscore the need to examine between-person variation in the level of susceptibility to peer influence in future research, particularly in response to different types of peers and social contexts.

Differential Susceptibility to Peer Influence: A Biological Sensitivity Perspective

Research on adolescent development has increasingly shifted its focus toward understanding person-by-environment interactions. This approach diverges from a long history of studies suggesting that environmental effects apply equally to all youth, which fail to consider that an adolescent's personal characteristics may determine if, and how much, the environment influences developmental outcomes. Several biological sensitivity models have been proposed to describe how person- and environment-level characteristics interact to predict developmental outcomes. The differential susceptibility theory (Belsky et al., 2007; Belsky and Pluess, 2009; Schriber and Guyer, 2016) and biological sensitivity to context theory (Boyce and Ellis, 2005; Ellis et al., 2011) propose that some individuals are more at risk than others if a personal vulnerability and negative environmental stressor interact beyond a certain threshold. Those who have an increased susceptibility to the environment are considered to be at a developmental disadvantage, as they are disproportionately impacted by environmental adversity compared to those who may not be as susceptible to their social environment. Although highly susceptible individuals are adversely affected by negative environmental contexts, susceptible youth are also more likely to benefit disproportionately from positive environmental contexts. In other words, heightened susceptibility to the environment can influence adolescent adjustment in a for-better *and* for-worse manner, resulting in more adaptive outcomes in positive contexts (for better) and more maladaptive outcomes in negative contexts (for worse). In contrast, the adjustment of low susceptible individuals is less affected by changes in the environment, regardless of how enriching or impoverished the context may be.

The differential susceptibility model is particularly relevant for investigating the various ways in which peers influence shifts in adolescent behavior, and the role of individual susceptibility in moderating these associations. In particular, adolescents' biological systems will shape the way they perceive and attend to their environments, resulting in youth reacting to their environments to match its demands. Thus, adolescents may be more or less sensitive to peer contexts as indexed by biomarkers of peer influence susceptibility. For example, adolescents who are more sensitive (e.g., greater neurobiological sensitivity to peer-related cues) *and* are in a negative peer environment (e.g., around deviant peers) may be pushed to engage in risk taking, whereas in a positive peer environment (e.g., around prosocial peers) may be pushed to engage in prosocial behaviors. In contrast, adolescents who are less sensitive (e.g., blunted neurobiological sensitivity to peer contexts) will be less affected by the type of environment they are in and overall less susceptible to peer influence (Figure 31.1). A closer examination of potential biomarkers of susceptibility to social influence may provide

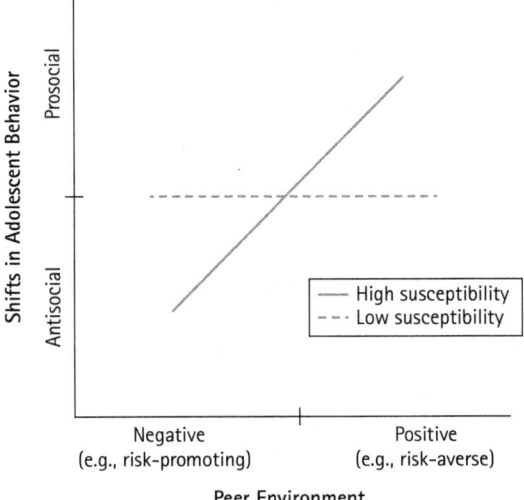

FIGURE 31.1 Understanding neurobiological susceptibility to peer influence from a differential susceptibility framework. According to the differential susceptibility theory, adolescents with a higher susceptibility to social contexts (solid blue line) may be differentially impacted by their environment in a for better *and* for worse manner; not only do they show a higher vulnerability to engage in greater antisocial behavior in negative peer environments, but they also show a higher sensitivity to engage in greater prosocial behavior in positive peer environments. In contrast, adolescents with a lower susceptibility to social contexts (dashed gray line) may show average peer-related shifts in behavior that may not be disproportionately more enhanced or vulnerable in positive or negative peer environments, respectively.

new insight into the mechanisms that push some adolescents towards or away from increased risk behaviors in peer contexts.

NEURAL BIOMARKERS OF ADOLESCENT SUSCEPTIBILITY TO PEER INFLUENCE

Prior work suggests that an individual's level of susceptibility to their environment can be indexed by three types of biomarkers: 1) phenotypic markers, or observable traits (e.g., temperament); 2) genetic markers (e.g., presence of dopamine receptor D4 polymorphism); and 3) endophenotypic markers, or traits that fall somewhere between phenotypic and genetic markers (e.g., neurobiology). While extensive literature from developmental psychology has examined various phenotypic and genetic markers of differential susceptibility to peer influence (e.g., reviewed in Belsky and Pluess, 2009), less attention has been paid to understanding how endophenotypic markers—beyond physiological responses—interact with peer environments to affect adolescent behavior.

In particular, little is known about which neurobiological markers relate to and explain differences in susceptibility (see Schriber and Guyer, 2016; for exceptions see Whittle et al., 2011; Yap et al., 2008). Given extensive reorganization in brain structure and function during adolescence, coupled with a social restructuring of the brain that confers social contexts highly salient (Nelson, Jarcho, and Guyer, 2016), considering how neural networks serve as moderators of social experience can greatly inform theories of adolescent development (Schriber and Guyer, 2016). While one or several susceptibility biomarkers may cumulatively enhance an individual's sensitivity to their social environments over time, these behavioral and neurobiological predispositions can manifest in critically different ways based on the type of environments that individuals are exposed to. Thus, while person- and environment-specific factors independently affect susceptibility to social influence in adolescence, examining interactions between these characteristics is critical for identifying those who may be most susceptible not only to maladaptive outcomes, but also adaptive outcomes (as in differential susceptibility theory).

Normative ontogenetic changes in the adolescent brain confer heightened neurobiological responses to peer contexts that are distinct from other developmental periods. Insight from developmental cognitive neuroscience has identified that brain regions involved in social, cognitive, and affective processing are involved in encoding social information from peer contexts that later inform adolescent behavior (reviewed in Telzer, van Hoorn, et al., 2018). Using functional magnetic resonance imaging (fMRI), this research has started to delineate the neural correlates of different aspects of peer socialization (e.g., peer presence or explicit feedback), social belonging (e.g., peer exclusion or acceptance), and affective processing of peer-related stimuli (e.g., happy versus angry peer faces) (Chein et al., 2011; Guyer et al., 2009; Pfeifer et al., 2011), with the goal of identifying the basic mechanisms that explain the role of peers in adolescent decision making and behavior. Of course, a major caveat of fMRI research is that a single brain region can be coopted for multiple psychological functions, limiting the specificity of future conclusions about the neural mechanisms of peer influence. For example, while the ventral striatum (VS) is classically implicated in reward processing during adolescence (reviewed in Telzer, 2016), it has also been associated with emotion regulation (Pfeifer et al., 2011) and learning (Davidow et al., 2016). In other words, while a psychological process might show consistent and robust relations with a specific brain region, it is improbable that there is a one-to-one mapping of a single psychological function on to a single brain region (Poldrack, 2011).

In spite of these methodological limitations, an intriguing possibility is that differential activation in social, cognitive, and affective brain systems reflects youths' susceptibility to peers at a neurobiological level. Affective regions, such as the VS, ventromedial prefrontal cortex (vmPFC), orbitofrontal cortex (OFC), and the amygdala, are implicated in processing social rewards and punishments, as well as detecting socially and motivationally salient information (Guyer, McClure-Tone, Shiffrin, Pine, and Nelson, 2009; Pfeifer et al., 2011; Telzer, 2016; Telzer et al., 2018). In addition, the social brain network, which includes the temporoparietal junction (TPJ), posterior superior temporal sulcus (pSTS), and dorsal and medial prefrontal cortex (dmPFC, mPFC), is involved in higher-order social cognition, such as self-other processing or mentalizing about others' thoughts and feelings to inform one's own behavior (Blakemore, 2008; Blakemore and Mills, 2014; Crone and Dahl, 2012; Telzer, 2016; Telzer et al., 2018). Finally, cognitive control regions, such as the medial and ventral/dorsal lateral prefrontal cortex (mPFC, vlPFC, dlPFC) and anterior cingulate cortex (ACC), play an important role in regulating reactivity within, and connectivity between, regions that encode social information from the environment (Falk et al., 2014; Haber and Knutson, 2010; Masten et al., 2009; Telzer et al., 2018).

Across diverse social contexts (e.g., being observed by peers, making risky decisions, watching static and dynamic faces), adolescents exhibit greater activation than children and/or adults in affective and social cognitive regions, as well as altered (as evidenced by both greater and lower) activation in regulatory regions (Do and Galván, 2016; Galvan et al., 2006; McCormick, Perino, and Telzer, 2018), an effect that is modulated by peer contexts (e.g., when a peer is present) (Ambrosia et al., 2018; Chein et al., 2011; Falk et al., 2014; Peake, Dishion, Stormshak, Moore, and Pfeifer, 2013; Pfeifer et al., 2013; Telzer, Miernicki, et al., 2018). While a wide range of peer influences are linked to changes in adolescent risk taking at the behavioral and neural level (Albert, Chein, and Steinberg, 2013; Telzer et al., 2017; van Hoorn et al., 2016), the role of neurobiological susceptibility markers in moderating these relations has received considerably less attention. An approach gaining traction in recent work explores how potential markers of neurobiological susceptibility can predict changes in adolescent risk-taking behavior.

Of particular interest to this investigation is the dopaminergic mesolimbic circuitry, which is thought to sensitize the adolescent brain to overweigh the sense of reward associated with engaging in risk behaviors in the presence of peers (e.g., Chein et al., 2011). Compared to adults, adolescents who evince greater VS and OFC activation when taking risks in the presence of peers report lower resistance to peer influence and show greater increases in risk-taking behavior following peer influence (Chein et al., 2011). In addition, adolescents who change their attitudes to align more with their peers show increased activation in the vmPFC, suggesting that conforming to their peers may be intrinsically valued, thereby increasing the efficacy of social influence pressures over time (Welborn et al., 2015). Finally, conforming to peers' behaviors is instrumental for gaining social acceptance and establishing stronger peer connections, an effect that may be exacerbated in youth who are rejected by their peers. Indeed, adolescents experiencing peer adversity (e.g., chronic peer victimization (Telzer, Miernicki, et al., 2018); chronic peer conflict (Telzer, Fuligni, Lieberman, Miernicki, and Galván, 2015)) show enhanced activation in the VS, insula, and amygdala during risk taking, potentially reflecting the anticipation of social rewards (e.g., peer approval) by engaging in risky behavior.

Prior studies from behavioral genetics have suggested that negative peer pressures influence real-world risk-taking behaviors (e.g., substance use), an effect only found among individuals with an enhanced genetic sensitivity to environmental experiences as indexed by different polymorphisms of the dopamine D4 receptor (DRD4) gene (Griffin, Harrington Cleveland, Schlomer, Vandenbergh, and

Feinberg, 2015; Larsen et al., 2010). Specifically, the 7-repeat DRD4 allele is consistently identified as a putative vulnerability factor for physical and mental health outcomes because it appears to decrease dopamine receptor efficiency through its effect on dopaminergic mesolimbic systems (Asghari et al., 1995; Hutchison, Lachance, Niaura, Bryan, and Smolen, 2002). Blunted dopamine activity in these neural circuits is often associated with an increased sensitivity to social input from negative peer environments, thereby reinforcing the rewarding nature of engaging in motivated and appetitive behaviors. Indeed, prior work suggests that the 7-repeat DRD4 allele not only impacts reactivity to risk-related cues (Hutchison et al., 2002), but also moderates the role of peer influence processes on health risk behavior (Griffin et al., 2015; Larsen et al., 2010; Mrug and Windle, 2014). Although speculative, one interpretation is that individuals with a genetic susceptibility to peers' deviant behaviors may experience a greater sense of reward from emulating peers' behavior, perhaps out of a purported desire to gain peer acceptance or increase social connection (Griffin et al., 2015; Larsen et al., 2010). Collectively, these findings implicate high dopamine reactivity to risky decisions as a potential biomarker that modulates the perceived reward value associated with capitulating to peer influences, which may subsequently shape both the inability to resist negative peer pressures, as well as a general propensity to engage in risk taking.

In addition to dopamine-associated biomarkers, neural circuits involved in higher-order social cognitive processing (e.g., mentalizing) moderate the effect of peer influence on adolescent decision making. For instance, adolescents who change their own attitudes more towards those of their parents or peers show increased activation in the dmPFC and TPJ, suggesting that when shifting their attitudes, adolescents incorporate the thoughts and intentions of their parents and peers to inform their decision making (Cascio et al., 2015; Welborn et al., 2015). Further supporting the link between social sensitivity and adolescent risk taking, adolescents who show greater activation in social brain regions (e.g., mPFC, precuneus, PCC) when viewing their own peers relative to parents report greater risk-taking behavior and affiliation with risky peers (Saxbe, Del Piero, Immordino-Yang, Kaplan, and Margolin, 2015). Other studies have similarly shown that high reactivity within social brain regions (dmPFC, TPJ, PCC) following peer exclusion prompts increased risk taking in peer contexts (Falk et al., 2014; Peake et al., 2013; Telzer, Miernicki, et al., 2018), perhaps in an attempt to promote social affiliation. For example, after being excluded by a peer, adolescents who are more susceptible to peer influence, as indexed by self-reported resistance to peer influence (RPI), show greater activation in the TPJ when making risky decisions, with TPJ activation predicting heightened risky behavior (Peake et al., 2013). These findings suggest that neural differences observed in peer-evaluative contexts may implicitly upweight the value of peers' risky attitudes and behaviors, thus pushing youth to adjust their behavior to fit in with expected peer group norms (i.e., riskier decisions).

Finally, impaired recruitment of brain regions supporting self-regulation and impulse control may also serve as a biomarker of social influence susceptibility. Highly salient, affectively-arousing peer contexts are hypothesized to compromise self-regulation abilities during adolescence, conferring an increased risk for maladaptive decision making in the presence of peers. For example, increases in dangerous driving behavior (i.e., risk taking that results in car crashes) following explicit peer feedback is only found among adolescents who not only report higher behavioral approach tendencies (e.g., oriented toward social rewards), but also exhibit reduced activation in the mPFC, a region implicated in the regulation of emotional states (Segalowitz et al., 2012). Similarly, adolescents reporting higher self-perceived susceptibility to peer influence exhibit greater functional connectivity between brain regions involved in action observation (frontoparietal and temporo-occipital regions) and executive control and decision making (prefrontal cortex) when passively processing others' movements that are socially salient (Grosbras et al., 2007). These findings suggest that, even when passively encoding salient stimuli from the environment, early adolescents with a higher susceptibility to peer influence may already recruit less regulatory resources when they are reasoning about their own behavioral responses, which could result in maladaptive decision making compared to those with a lower susceptibility to peer influence. Relatedly, longitudinal increases in VS activation to emotional expressions are associated with a stronger ability to resist peer influence from childhood to early adolescence (Pfeifer et al., 2011). Although counterintuitive at first glance, the VS has been implicated in emotion regulation (Masten et al., 2009; Wager, Davidson, Hughes, Lindquist, and Ochsner, 2008), suggesting that the VS is serving a regulatory process that may be compromised in adolescents with higher susceptibility to peer influence.

Taken together, these findings suggest that neurobiological processes that encode affective, social-cognitive, and regulatory cues from the environment contribute to peer influence susceptibility. Thus, peer contexts may elicit a reduced regulation of reward-related processing during risky decision making and an intensification of social-affective reactivity. In order to better understand how these neural networks interact and work in tandem to influence adolescent behavior, it is key to examine functional connectivity. Adolescent behavior is determined by dynamically interacting functional brain systems. Increasingly, researchers are implementing connectivity techniques including functional and effective connectivity, which measure the temporal correlation between neurophysiological events (Sporns and Honey, 2007) and the influence one neural network exerts over another (Friston, 2009), respectively. Although still a nascent area of inquiry, studies have emerged showing differences in connectivity patterns are associated with differences in adolescent decision making. For example, adolescents showed heightened mPFC-striatum connectivity when being observed by a peer compared to children and adults, which may represent the integration of social signals with motivational systems that ultimately result in increased risk taking in the presence of peers (Somerville et al., 2013). Indeed, we have shown that longitudinal increases in mPFC-striatum connectivity underlies longitudinal increases in risk taking during adolescence (Qu, Galvan, Fuligni, Lieberman, and Telzer, 2015). Moreover, during rest, VS-mPFC connectivity shows linear increases from childhood to adolescence and correlates with pubertal hormones (Fareri et al., 2015), as well as reward sensitivity (van Duijvenvoorde, Peters, Braams, and Crone, 2016), cognitive control, and substance use (Lee and Telzer, 2016), which may reflect a developmental shift in the way social and motivational systems interact during adolescence. By examining the coupling of these dynamically interacting neural systems across different social contexts and across development, we will gain a deeper understanding of how neural systems work in tandem to inform adolescent behavior.

NEUROBIOLOGICAL SUSCEPTIBILITY TO PEER INFLUENCE: FOR BETTER *AND* FOR WORSE?

Most studies to date have examined how peer influence susceptibility relates to negative, deviant behaviors, with far less known about adolescents' susceptibility to positive peer influences. Prior research suggests that greater activation in still-maturing regulatory brain systems can facilitate adolescents' susceptibility to positive peer influences. For instance, adolescents who exhibit greater activation in the response inhibition network (e.g., IFG) are more likely to drive safely in the presence of a peer promoting cautious driving practices, an effect not observed in response to negative peer influences (Cascio et al., 2015). Thus, neural circuits involved in response inhibition may facilitate an increased capacity to resist maladaptive forms of peer pressures during risky decision making. These results highlight the potentially adaptive benefits of peer influence susceptibility, such that ongoing changes in cognitive regulation can be leveraged to redirect adolescents with an increased susceptibility to peer contexts toward more positive relative to negative decisions.

Previous work from behavioral genetics has also identified genetic biomarkers that serve to buffer negative peer environments. Similar to the role of DRD4 polymorphisms in altering sensitivity to negative environmental influences (Belsky and Pluess, 2009), researchers have shown that the 7-repeat DRD4 allele can also increase susceptibility to positive peer experiences, which ultimately buffers against adolescent delinquency over time (e.g., Kretschmer, Dijkstra, Ormel, Verhulst, and Veenstra, 2013; reviewed in Belsky and Pluess, 2009; Ellis, Boyce, Belsky, Bakermans-Kranenburg, and Van Ijzendoorn, 2011). These results highlight genetic factors as another way to identify individuals who might be most responsive to different social influences, which can be leveraged to foster more positive peer effects on adolescent risk taking.

Moving beyond risk-taking and negative behaviors, researchers have also begun to examine how social contexts interact to promote positive behaviors in adolescents. For example, the presence of peers impacts adolescents' prosocial behaviors (e.g., contributing coins to a group), via enhanced activity in several social brain regions including the mPFC, TPJ, precuneus, and STS (van Hoorn et al., 2016). Such peer effects in social brain regions are largest for younger (12–13-year-old) than older (15–16-year-old) adolescents, suggesting that younger adolescents may be more sensitive to positive social influence from peers. Indeed, behavioral work examining risk perceptions has shown that 12–14-year-olds are more sensitive to social influence from peers than are adults, whereas older adolescents (age 15–18 years) show similar sensitivity to influence as adults (Knoll et al., 2015). These findings suggest that heightened social brain activation may underlie this increased sensitivity to social influence in young adolescents, and expands the scope of social influence work to positive, prosocial behaviors. Together, these studies underscore the increasing salience of peers in adolescence, insofar that they become especially attuned to their peers' evaluations, and highlight both the positive and negative effects of peer conformity.

To our knowledge, no study to date has explicitly tested a neurobiological susceptibility framework for social influence, which posits that neurobiological biomarkers can influence susceptibility in a for-better and for-worse fashion depending on the environmental context. Some initial evidence suggests there may be largely overlapping neural biomarkers for positive and negative social influence susceptibility. Indeed, heightened neural activation in socio-affective brain regions respond to both positive and negative cues. For instance, heightened VS activation, while traditionally discussed as a vulnerability for risk taking, can also be protective against risk taking depending on the social context. When adolescents show heightened VS activation in risk-taking tasks (Galvan, Hare, Voss, Glover, and Casey, 2007; Qu et al., 2015; Telzer, Fuligni, Lieberman, and Galván, 2013), particularly in the presence of peers (Chein et al., 2011), they show increases in real-life risk taking, suggesting that individual differences in VS activation may be a neurobiological susceptibility marker, and in a risky context will promote risky behavior. However, adolescents who show heightened VS activation in positive, prosocial contexts show decreases in risk taking (Telzer et al., 2013), suggesting that the same susceptibility biomarker can promote positive adjustment depending on the social context. Despite the promise of a neurobiological susceptibility to peer influence model, by which biomarkers of susceptibility interact with the social environment, much work is needed to further support this model.

CONCLUSIONS AND FUTURE DIRECTIONS

Given a significant rise in mortality rates from preventable health risk behaviors during adolescence, researchers have directed considerable attention toward identifying innovative approaches to the study of social influence susceptibility. Here, we argue that taking a neurobiological susceptibility to peer influence framework will help us understand potential biomarkers that serve as both vulnerabilities and opportunities for adolescent risk taking, depending on the peer context. Although researchers have only begun to unpack the inherently complex construct of peer influence susceptibility, exciting advances from developmental psychology and cognitive neuroscience suggest that not all adolescents are equally influenced by their peers. Using a diverse set of methods ranging from questionnaires to in-vivo experiments during fMRI, this burgeoning literature seeks to delineate how adolescents' level of behavioral and neurobiological susceptibility to peers can shape whether, how, and how much their own attitudes and behaviors fluctuate across different peer environments. Exploring individual variability in, or susceptibility to, peer influence represents a promising approach for identifying adolescents who may be most at risk or resilient to negative peer pressures, which has important implications for tailoring interventions that prevent risk-taking behavior directly. Yet, despite encouraging evidence from behavioral, genetic, and brain studies, the construct of susceptibility to peer influence has largely remained elusive and understudied, particularly in more positive social contexts.

Importantly, further research is needed to address several methodological and conceptual issues that still remain. Methodologically, there is considerable variability not only in the use of self-reported versus performance-based measures of susceptibility to peer influence, but also in the statistical analyses that have been proposed to test susceptibility effects (i.e., moderation analyses). In addition, the effect of individual susceptibility to peers on adolescent conformity to peer pressures is likely going to vary based on whether the behavior is positive (e.g., prosocial) or negative (e.g., antisocial), if it is a behavior that adolescents are already experienced with (e.g., cigarette use, but not cocaine use), and the perceived severity and consequences of engaging in that behavior (e.g., drug use versus cutting class). The nature of peer interactions as well as the latency between exposure to peers' behavior and adolescents' own decisions may also be relevant for determining the extent to which peer influence susceptibility pushes some adolescents toward (or away from) their peers' behaviors more than others. Based on these overall considerations, future empirical work is necessary for developing a more comprehensive account of defining, measuring, and testing the effect of peer influence susceptibility during adolescence.

Guided by a rich literature from developmental science, cognitive neuroscience, and social psychology, this review explores the use of neurobiological biomarkers to index an individual's susceptibility to their social environment. Collectively, this research underscores the need to consider a wider range of peer contexts when studying how adolescent susceptibility and peer contexts independently and mutually influence risk-related attitudes and behaviors. Neurobiological indices of susceptibility to peer influence may be particularly useful for identifying adolescents who are extremely oriented toward peers, which can be leveraged to target peer socialization efforts that encourage healthier attitudes and behaviors during adolescence.

Acknowledgments

The authors would like to thank Michael Tomasello for comments on an early version of this manuscript.

REFERENCES

Albert, D., Chein, J., and Steinberg, L. (2013). Peer influences on adolescent decision making. *Current Directions in Psychological Science*, 22(2), 114–20. doi:10.1177/0963721412471347.

Allen, J. P., Porter, M. R., and Mcfarland, F. C. (2006). Leaders and followers in adolescent close friendships: Susceptibility to peer influence as a predictor of risky behavior, friendship instability, and depression. *Developmental Psychopathology*, 18(1), 155–172. doi:10.1017/S0954579406060093.

Ambrosia, M., Eckstrand, K. L., Morgan, J. K., Allen, N. B., Jones, N. P., Sheeber, L., ... and Forbes, E. E. (2018). Temptations of friends: Adolescents' neural and behavioral responses to best friends predict risky behavior. *Social Cognitive and Affective Neuroscience*, 13 (5), 483–491. doi:10.1093/scan/nsy028.

Asghari, V., Sanyal, S., Buchwaldt, S., Paterson, A., Jovanovic, V., and Van Tol, H. H. (1995). Modulation of intracellular cyclic AMP levels by different human dopamine D4 receptor variants. *Journal of Neurochemistry*, 65(3), 1157–1165. doi:10.1046/j.1471-4159.1995.65031157.x.

Baldwin, J. M. (1895). *Mental Development in the Child and the Race: Method and Processes.* New York: Macmillan Publishing. doi:10.1037/10003-000.

Barry, C. M. and Wentzel, K. R. (2006). Friend influence on prosocial behavior: The role of motivational factors and friendship characteristics. *Developmental Psychology*, 42(1), 153–163. doi:10.1037/0012-1649.42.1.153.

Belsky, J., Bakermans-Kranenburg, M. J., and van Ijzendoorn, M. H. (2007). For better and for worse: Differential susceptibility to environmental influences. *Current Directions in Psychological Science*, 16(6), 300–304. doi:10.1111/j.1467-8721.2007.00525.x.

Belsky, J. and Pluess, M. (2009). Beyond diathesis stress: Differential susceptibility to environmental influences. *Psychological Bulletin*, 135(6), 885–908. doi:10.1037/a0017376.

Berger, J. (2008). Identity signaling, social influence, and social contagion. In M. J. Prinstein and K. A. Dodge (eds.), *Understanding Peer Influence in Children and Adolescents* (pp. 181–199). New York, NY: The Guilford Press.

Blakemore, S.-J. (2008). The social brain in adolescence. *Nature Reviews Neuroscience*, 9(4), 267–277. doi:10.1038/nrn2353.

Blakemore, S.-J. and Mills, K. L. (2014). Is adolescence a sensitive period for socio-cultural processing? *Annual Review of Psychology*, 65, 187–207. doi: 10.1146/annurev-psych-010213-115202.

Boyce, W. T. and Ellis, B. J. (2005). Biological sensitivity to context: I. An evolutionary-developmental theory of the origins and functions of stress reactivity. *Development and Psychopathology*, 17(2), 271–301. doi:10.1017/S0954579405050145.

Brechwald, W. A., and Prinstein, M. J. (2011). Beyond homophily: A decade of advances in understanding peer influence processes. *Journal of Research on Adolescence*, 21(1), 166–179. doi:10.1111/j.1532-7795.2010.00721.x

Brittain, C. V. (1963). Adolescent choices and parent-peer cross-pressures. *American Sociological Review*, 28(3), 385–391.

Brown, B. B. (2004). Adolescents' relationships with peers. In R. Lerner and L. Steinberg (eds.), *Handbook of Adolescent Psychology* (2nd ed., pp. 363–394). New York: Wiley.

Cascio, C. N., Carp, J., O'Donnell, M. B., Tinney, F. J., Bingham, C. R., Shope, J. T., ...and Falk, E. B. (2015). Buffering social influence: Neural correlates of response inhibition predict driving safety in the presence of a peer. *Journal of Cognitive Neuroscience*, 27(1), 83–95. doi:10.1162/jocn_a_00693.

Chein, J., Albert, D., O'Brien, L., Uckert, K., and Steinberg, L. (2011). Peers increase adolescent risk taking by enhancing activity in the brain's reward circuitry. *Developmental Science*, 14(2), F1–F10. doi:10.1111/j.1467-7687.2010.01035.x.

Choukas-Bradley, S., Giletta, M., Cohen, G. L., and Prinstein, M. J. (2015). Peer influence, peer status, and prosocial behavior: An experimental investigation of peer socialization of adolescents' intentions to volunteer. *Journal of Youth and Adolescence*, 44, 2197–2210. doi:10.1007/s10964-015-0373-2.

Choukas-Bradley, S., Giletta, M., Widman, L., Cohen, G. L., and Prinstein, M. J. (2014). Experimentally-measured susceptibility to peer influence and adolescent sexual behavior trajectories: A preliminary study. *Developmental Psychology*, 50(9), 2221–2227. doi:10.1037/a0037300.

Cillessen, A. H. N. and Mayeux, L. (2004). Sociometric status and peer group behavior: Previous findings and current directions. In J. B. Kupersmidt and K. A. Dodge (eds.), *Children's Peer Relations: From Development to Intervention* (pp. 3–20). Washington DC, US: American Psychological Association. doi:10.1037/10653-001.

Cohen, G. L. and Prinstein, M. J. (2006). Peer contagion of aggression and health risk behavior among adolescent males: An experimental investigation of effects on public conduct and private attitudes. *Child Development*, 77(4), 967–983. doi:10.1111/j.1467-8624.2006.00913.x.

Cooley, C. H. (1902). *Human Nature and the Social Order*. New York, NY: Charles Scribner's Sons.

Crone, E. A. and Dahl, R. E. (2012). Understanding adolescence as a period of social-affective engagement and goal flexibility. *Nature Reviews Neuroscience*, 13, 636–650. doi:10.1038/nrn3313.

Davidow, J. Y., Foerde, K., Galván, A., Shohamy, D., Dahl, R. E., Crone, E. A., and Gluck, M. A. (2016). An upside to reward sensitivity: The hippocampus supports enhanced reinforcement learning in adolescence. *Neuron*, 92(1), 93–99. doi:10.1016/j.neuron.2016.08.031.

Dishion, T. J., Spracklen, K. M., Andrews, D. W., and Patterson, G. R. (1996). Deviancy training in male adolescent friendships. *Behavior Therapy*, 27(3), 373–390. doi:10.1016/S0005-7894(96)80023-2.

Do, K. T. and Galván, A. (2016). Neural sensitivity to smoking stimuli is associated with cigarette craving in adolescent smokers. *Journal of Adolescent Health*, 58(2), 186–194. doi:10.1016/j.jadohealth.2015.10.004.

Do, K. T., Guassi Moreira, J. F., and Telzer, E. H. (2016). But is helping you worth the risk? Defining prosocial risk taking in adolescence. *Developmental Cognitive Neuroscience*, 25, 260–271. doi:10.1016/j.dcn.2016.11.008.

Ellis, B. J., Boyce, W. T., Belsky, J., Bakermans-Kranenburg, M. J., and Van Ijzendoorn, M. H. (2011). Differential susceptibility to the environment: An evolutionary–neurodevelopmental theory. *Development and Psychopathology*, 23(1), 7–28. doi:10.1017/S0954579410000611.

Engelmann, J. B., Moore, S., Monica Capra, C., and Berns, G. S. (2012). Differential neurobiological effects of expert advice on risky choice in adolescents and adults. *Social Cognitive and Affective Neuroscience*, 7(5), 557–567. doi:10.1093/scan/nss050.

Engelmann, J. M., Herrmann, E., and Tomasello, M. (2018). Concern for group reputation increases prosociality in young children. *Psychological Science*, 29(2), 181–190. doi:10.1177/0956797617733830.

Engelmann, J. M., Over, H., Herrmann, E., and Tomasello, M. (2013). Young children care more about their reputation with ingroup members and potential reciprocators. *Developmental Science*, 16(6), 952–958. doi:10.1111/desc.12086.

Falk, E. B., Cascio, C. N., Brook O'Donnell, M., Carp, J., Tinney, F. J., Bingham, C. R., . . . and Simons-Morton, B. G. (2014). Neural responses to exclusion predict susceptibility to social influence. *Journal of Adolescent Health*, 54(5), S22–S31. doi:10.1016/j.jadohealth.2013.12.035.

Fareri, D. S., Gabard-Durnam, L., Goff, B., Flannery, J., Gee, D. G., Lumian, D. S., . . . and Tottenham, N. (2015). Normative development of ventral striatal resting state connectivity in humans. *Neuroimage*, 118, 422–437. doi:10.1016/j.neuroimage.2015.06.022.

Foulkes, L., Leung, J. T., Fuhrmann, D., Knoll, L. J., and Blakemore, S.-J. (2018). Age differences in the prosocial influence effect. *Developmental Science*, 21(6), 1–9. doi:10.1111/desc.12666.

Friston, K. (2009). Causal modelling and brain connectivity in functional magnetic resonance imaging. *PLoS Biology*, 7(2), e1000033. doi:10.1371/journal.pbio.1000033.

Fromme, K. and Ruela, A. (1994). Mediators and moderators of young adults' drinking. *Addiction*, 89(1), 63–71. doi:10.1111/j.1360-0443.1994.tb00850.x.

Galvan, A., Hare, T., Voss, H., Glover, G., and Casey, B. J. (2007). Risk-taking and the adolescent brain: Who is at risk? *Developmental Science*, 10(2), F8–F14. doi:10.1111/j.1467-7687.2006.00579.x.

Galvan, A., Hare, T. A., Parra, C. E., Penn, J., Voss, H., Glover, G., and Casey, B. J. (2006). Earlier development of the accumbens relative to orbitofrontal cortex might underlie risk-taking behavior in adolescents. *Journal of Neuroscience*, 26(25), 6885–6892. doi:10.1523/JNEUROSCI.1062-06.2006.

Gardner, M. and Steinberg, L. (2005). Peer influence on risk taking, risk preference, and risky decision making in adolescence and adulthood: An experimental study. *Developmental Psychology*, 41(4), 625–635. doi:10.1037/0012-1649.41.4.625.

Gardner, T. W., Dishion, T. J., and Connell, A. M. (2008). Adolescent self-regulation as resilience: Resistance to antisocial behavior within the deviant peer context. *Journal of Abnormal Child Psychology*, 36, 273–284. doi:10.1007/s10802-007-9176-6.

Gibbons, F. X., Gerrard, M., Blanton, H., and Russell, D. W. (1998). Reasoned action and social reaction: Willingness and intention as independent predictors of health risk. *Journal of Personality and Social Psychology*, 74(5), 1164–1180. doi:10.1037/0022-3514.74.5.1164.

Griffin, A. M., Harrington Cleveland, H., Schlomer, G. L., Vandenbergh, D. J., and Feinberg, M. E. (2015). Differential susceptibility: The genetic moderation of peer pressure on alcohol use. *Journal of Youth and Adolescence*, 44, 1841–1853. doi:10.1007/s10964-015-0344-7.

Grosbras, M.-H., Jansen, M., Leonard, G., Mcintosh, A., Osswald, K., Poulsen, C., . . . and Paus, T. (2007). Neural mechanisms of resistance to peer influence in early adolescence. *Journal of Neuroscience*, 27(30), 8040–8045. doi:10.1523/JNEUROSCI.1360-07.2007.

Guyer, A. E., McClure-Tone, E. B., Shiffrin, N. D., Pine, D. S., and Nelson, E. E. (2009). Probing the neural correlates of anticipated peer evaluation in adolescence. *Child Development*, 80(4), 1000–1015. doi:10.1111/j.1467-8624.2009.01313.x.

Haber, S. N., and Knutson, B. (2010). The reward circuit: Linking primate anatomy and human imaging. *Neuropsychopharmacology Reviews*, 35, 4–26. doi:10.1038/npp.2009.129.

Helms, S. W., Choukas-Bradley, S., Widman, L., Giletta, M., Cohen, G. L., and Prinstein, M. J. (2014). Adolescents misperceive and are influenced by high status peers' health risk, deviant, and adaptive behavior. *Developmental Psychology*, 50(12), 2697–2714. doi:10.1037/a0038178.

Hutchison, K. E., Lachance, H., Niaura, R., Bryan, A., and Smolen, A. (2002). The DRD4 VNTR polymorphism influences reactivity to smoking cues. *Journal of Abnormal Psychology*, 111(1), 134–143. doi:10.1037//0021-843X.111.1.134.

Iannotti, R. J. and Bush, P. J. (1992). Perceived vs. actual friends' use of alcohol, cigarettes, marijuana, and cocaine: Which has the most influence? *Journal of Youth and Adolescence*, 21, 375–389. doi:10.1007/BF01537024.

Knoll, L. J., Leung, J. T., Foulkes, L., and Blakemore, S.-J. (2017). Age-related differences in social influence on risk perception depend on the direction of influence. *Journal of Adolescence, 60*, 53–63. doi:10.1016/j.adolescence.2017.07.002.

Knoll, L. J., Magis-Weinberg, L., Speekenbrink, M., and Blakemore, S.-J. (2015). Social influence on risk perception during adolescence. *Psychological Science, 26*(5), 583–592. doi:10.1177/0956797615569578.

Kretschmer, T., Dijkstra, J. K., Ormel, J., Verhulst, F. C., and Veenstra, R. (2013). Dopamine receptor D4 gene moderates the effect of positive and negative peer experiences on later delinquency: The Tracking Adolescents' Individual Lives survey study. *Development and Psychopathology, 25*(4pt1), 1107–1117. doi:10.1017/S0954579413000400.

LaFontana, K. and Cillessen, A. H. N. (2010). Developmental changes in the priority of perceived status in childhood and adolescence. *Social Development, 19*(1), 130–147. doi:10.1111/j.1467-9507.2008.00522.x.

Larsen, H., van der Zwaluw, C. S., Overbeek, G., Granic, I., Franke, B., and Engels, R. C. M. E. (2010). A variable-number-of-tandem-repeats polymorphism in the dopamine D4 receptor gene affects social adaptation of alcohol use. *Psychological Science, 21*(8), 1064–1068. doi:10.1177/0956797610376654.

Lee, T.-H. and Telzer, E. H. (2016). Negative functional coupling between the right fronto-parietal and limbic resting state networks predicts increased self-control and later substance use onset in adolescence. *Developmental Cognitive Neuroscience, 20*, 35–42. doi:10.1016/j.dcn.2016.06.002.

Masten, C. L., Eisenberger, N. I., Borofsky, L. A., Pfeifer, J. H., McNealy, K., Mazziotta, J. C., and Dapretto, M. (2009). Neural correlates of social exclusion during adolescence: Understanding the distress of peer rejection. *Social Cognitive and Affective Neuroscience, 4*(2), 143–157. doi:10.1093/scan/nsp007.

McCormick, E. M., Perino, M. T., and Telzer, E. H. (2018). Not just social sensitivity: Adolescent neural suppression of social feedback during risk taking. *Developmental Cognitive Neuroscience, 30*, 134–141. doi:10.1016/j.dcn.2018.01.012.

Mead, G. H. (1934). *Mind, Self and Society from the Standpoint of a Social Behaviorist*. Chicago, IL: University of Chicago Press.

Mrug, S. and Windle, M. (2014). DRD4 and susceptibility to peer influence on alcohol use from adolescence to adulthood. *Drug Alcohol Dependence, 145*, 168–173. doi:10.1016/j.drugalcdep.2014.10.009.

Munoz Centifanti, L. C., Modecki, K. L., Maclellan, S., and Gowling, H. (2014). Driving under the influence of risky peers: An experimental study of adolescent risk taking. *Journal of Research on Adolescence, 26*(1), 207–222. doi:10.1111/jora.12187.

Nelson, E. E., Jarcho, J. M., and Guyer, A. E. (2016). Social re-orientation and brain development: An expanded and updated view. *Developmental Cognitive Neuroscience, 17*, 118–127. doi:10.1016/j.dcn.2015.12.008.

Nelson, E. E., Leibenluft, E., McClure, E. B., and Pine, D. S. (2005). The social re-orientation of adolescence: A neuroscience perspective on the process and its relation to psychopathology. *Psychological Medicine, 35*(2), 163–174. doi:10.1017/s0033291704003915.

Peake, S. J., Dishion, T. J., Stormshak, E. A., Moore, W. E., and Pfeifer, J. H. (2013). Risk-taking and social exclusion in adolescence: Neural mechanisms underlying peer influences on decision-making. *Neuroimage, 82*, 23–34. doi:10.1016/j.neuroimage.2013.05.061.

Perino, M. T., Miernicki, M. E., and Telzer, E. H. (2016). Letting the good times roll: Adolescence as a period of reduced inhibition to appetitive social cues. *Social Cognitive and Affective Neuroscience, 11*(11), 1762–1771. doi:10.1093/scan/nsw096.

Pfeifer, J. H., Kahn, L. E., Merchant, J. S., Peake, S. J., Veroude, K., Masten, C. L., ... and Dapretto, M. (2013). Longitudinal change in the neural bases of adolescent social self-evaluations: Effects of age and pubertal development. *Journal of Neuroscience, 33*(17), 7415–7419. doi:10.1523/JNEUROSCI.4074-12.2013.

Pfeifer, J. H., Masten, C. L., Moore, W. E., Oswald, T. M., Mazziotta, J. C., Iacoboni, M., and Dapretto, M. (2011). Entering adolescence: Resistance to peer influence, risky behavior, and neural changes in emotion reactivity. *Neuron, 69*(5), 1029–1036. doi:10.1016/j.neuron.2011.02.019.

Poldrack, R. A. (2011). Inferring mental states from neuroimaging data: From reverse inference to large-scale decoding. *Neuron, 72*(5), 692–697. doi:10.1016/j.neuron.2011.11.001.

Powers, K. E., Yaffe, G., Hartley, C. A., Davidow, J. Y., Kober, H., and Somerville, L. H. (2018). Consequences for peers differentially bias computations about risk across development. *Journal of Experimental Psychology: General, 147*(5), 671–682. doi:10.1037/xge0000389.

Prinstein, M. J., Boergers, J., and Spirito, A. (2001). Adolescents' and their friends' health-risk behavior: Factors that alter or add to peer influence. *Journal of Pediatric Psychology, 26*(5), 287–298. doi:10.1093/jpepsy/26.5.287.

Prinstein, M. J., Brechwald, W. A., and Cohen, G. L. (2011). Susceptibility to peer influence: Using a performance-based measure to identify adolescent males at heightened risk for deviant peer socialization. *Developmental Psychology, 47*(4), 1167–1172. doi:10.1037/a0023274.

Prinstein, M. J. and Dodge, K. A. (eds.). (2008). *Understanding Peer Influence in Children and Adolescents*. New York, NY: Guilford Press.

Prinstein, M. J., Meade, C. S., and Cohen, G. L. (2003). Adolescent oral sex, peer popularity, and perceptions of best friends' sexual behavior. *Journal of Pediatric Psychology, 28*(4), 243–249. doi:10.1093/jpepsy/jsg012.

Prinstein, M. J. and Wang, S. S. (2005). False consensus and adolescent peer contagion: Examining discrepancies between perceptions and actual reported levels of friends' deviant and health risk behaviors. *Journal of Abnormal Child Psychology, 33*, 293–306. doi:10.1007/s10802-005-3566-4.

Qu, Y., Galvan, A., Fuligni, A. J., Lieberman, M. D., and Telzer, E. H. (2015). Longitudinal changes in prefrontal cortex activation underlie declines in adolescent risk taking. *Journal of Neuroscience, 35*(32), 11308–11314. doi:10.1523/JNEUROSCI.1553-15.2015.

Saxbe, D., Del Piero, L., Immordino-Yang, M. H., Kaplan, J., and Margolin, G. (2015). Neural correlates of adolescents' viewing of parents' and peers' emotions: Associations with risk-taking behavior and risky peer affiliations. *Social Neuroscience, 10*(6), 592–604. doi:10.1080/17470919.2015.1022216.

Schmidt, M. F. H., Butler, L. P., Heinz, J., and Tomasello, M. (2016). Young children see a single action and infer a social norm: Promiscuous normativity in 3-year-olds. *Psychological Science, 27*(10), 1360–1370. doi:10.1177/0956797616661182.

Schmidt, M. F. H., Rakoczy, H., and Tomasello, M. (2012). Young children enforce social norms selectively depending on the violator's group affiliation. *Cognition, 124*(3), 325–333. doi:10.1016/j.cognition.2012.06.004.

Schriber, R. A. and Guyer, A. E. (2016). Adolescent neurobiological susceptibility to social context. *Developmental Cognitive Neuroscience, 19*, 1–18. doi:10.1016/j.dcn.2015.12.009.

Segalowitz, S. J., Santesso, D. L., Willoughby, T., Reker, D. L., Campbell, K., Chalmers, H., and Rose-Krasnor, L. (2012). Adolescent peer interaction and trait surgency weaken medial prefrontal cortex responses to failure. *Social Cognitive and Affective Neuroscience, 7*, 115–124. doi:10.1093/scan/nsq090.

Silva, K., Shulman, E. P., Chein, J., and Steinberg, L. (2016). Peers increase late adolescents' exploratory behavior and sensitivity to positive and negative feedback. *Journal of Research on Adolescence, 26*(4), 696–705. doi:10.1111/jora.12219.

Simons-Morton, B. G. and Farhat, T. (2010). Recent findings on peer group influences on adolescent smoking. *Journal of Primary Prevention, 31*, 191–208. doi:10.1007/s10935-010-0220-x.

Slagt, M., Dubas, J. S., Deković, M., Haselager, G. J. T., and van Aken, M. A. G. (2015). Longitudinal associations between delinquent behaviour of friends and delinquent behaviour of adolescents: Moderation by adolescent personality traits. *European Journal of Personality, 29*(4), 468–477. doi:10.1002/per.2001.

Somerville, L. H., Haddara, N., Sasse, S. F., Skwara, A. C., Moran, J. M., and Figner, B. (2019). Dissecting "peer presence" and "decisions" to deepen understanding of peer influence on adolescent risky choice. *Child Development, 90*(6), 2086–2103. doi:10.1111/cdev.13081.

Somerville, L. H., Hare, T., and Casey, B. J. (2011). Frontostriatal maturation predicts cognitive control failure to appetitive cues in adolescents. *Journal of Cognitive Neuroscience, 23*(9), 2123–2124. doi:10.1162/jocn.2010.21572.

Somerville, L. H., Jones, R. M., Ruberry, E. J., Dyke, J. P., Glover, G., and Casey, B. J. (2013). The medial prefrontal cortex and the emergence of self-conscious emotion in adolescence. *Psychological Science, 24*(8), 1554–1562. doi:10.1177/0956797613475633.

Sowell, E. R., Peterson, B. S., Thompson, P. M., Welcome, S. E., Henkenius, A. L., and Toga, A. W. (2003). Mapping cortical change across the human life span. *Nature Neuroscience, 6*, 309–315. doi:10.1038/nn1008.

Sowell, E. R., Thompson, P. M., Leonard, C. M., Welcome, S. E., Kan, E., and Toga, A. W. (2004). Longitudinal mapping of cortical thickness and brain growth in normal children. *Journal of Neuroscience, 24*(38), 8223–8231. doi:10.1523/JNEUROSCI.1798-04.2004.

Sporns, O. and Honey, C. J. (2007). Identification and classification of hubs in brain networks. *PLoS ONE, 2*(10), e1049.doi:10.1371/journal.pone.0001049.

Steinberg, L. and Monahan, K. C. (2007). Age differences in resistance to peer influence. *Developmental Psychology, 43*(6), 1531–1543. doi:10.1037/0012-1649.43.6.1531.

Steinberg, L. and Morris, A. S. (2001). Adolescent development. *Annual Review of Psychology, 52*, 83–110. doi:10.1146/annurev.psych.52.1.83.

Telzer, E. H. (2016). Dopaminergic reward sensitivity can promote adolescent health: A new perspective on the mechanism of ventral striatum activation. *Developmental Cognitive Neuroscience, 17*, 57–67. doi:10.1016/j.dcn.2015.10.010.

Telzer, E. H., Fuligni, A. J., Lieberman, M. D., and Galván, A. (2013). Ventral striatum activation to prosocial rewards predicts longitudinal declines in adolescent risk taking. *Developmental Cognitive Neuroscience, 3*, 45–52. doi:10.1016/j.dcn.2012.08.004.

Telzer, E. H., Fuligni, A. J., Lieberman, M. D., Miernicki, M. E., and Galván, A. (2015). The quality of adolescents' peer relationships modulates neural sensitivity to risk taking. *Social Cognitive and Affective Neuroscience, 10*(3), 389–398. doi:10.1093/scan/nsu064.

Telzer, E. H., Miernicki, M. E., and Rudolph, K. D. (2018). Chronic peer victimization heightens neural sensitivity to risk taking. *Development and Psychopathology, 30*(1), 13–26. doi:10.1017/S0954579417000438.

Telzer, E. H., Rogers, C. R., and Van Hoorn, J. (2017). Neural correlates of social influence on risk taking and substance use in adolescents. *Addiction Reports, 4*, 333–341. doi:10.1007/s40429-017-0164-9.

Telzer, E. H., van Hoorn, J., Rogers, C. R., and Do, K. T. (2018). Social influence on positive youth development: A developmental neuroscience perspective. In *Advances in*

Child Development and Behavior (pp. 215–258). Cambridge, UK: Elsevier. doi:10.1016/bs.acdb.2017.10.003.

Teunissen, H. A., Spijkerman, R., Prinstein, M. J., Cohen, G. L., Engels, R. C. M. E., and Scholte, R. H. J. (2012). Adolescents' conformity to their peers' pro-alcohol and anti-alcohol norms: The power of popularity. *Alcoholism: Clinical & Experimental Research, 36*(7), 1257–1267. doi:10.1111/j.1530-0277.2011.01728.x.

Tomasello, M. (2014). The ultra-social animal. *European Journal of Social Psychology, 44*(3), 187–194. doi:10.1002/ejsp.2015.

Tomasello, M. (2016). Cultural learning redux. *Child Development, 87*(3), 643–653. doi:10.1111/cdev.12499.

Tomasello, M., Kruger, A. C., and Ratner, H. H. (1993). Cultural learning. *Behavioral and Brain Sciences, 16*(03), 495–511. doi:10.1017/S0140525X0003123X.

van Duijvenvoorde, A. C. K., Peters, S., Braams, B. R., and Crone, E. A. (2016). What motivates adolescents? Neural responses to rewards and their influence on adolescents' risk taking, learning, and cognitive control. *Neuroscience & Biobehavioral Reviews, 70*, 135–147. doi:10.1016/j.neubiorev.2016.06.037.

van Hoorn, J., Fuligni, A. J., Crone, E. A., and Galván, A. (2016). Peer influence effects on risk-taking and prosocial decision-making in adolescence: Insights from neuroimaging studies. *Current Opinion in Behavioral Sciences, 10*, 59–64. doi:10.1016/j.cobeha.2016.05.007.

van Hoorn, J., van Dijk, E., Güroğlu, B., and Crone, E. A. (2016). Neural correlates of prosocial peer influence on public goods game donations during adolescence. *Social Cognitive and Affective Neuroscience, 11*(6), 923–933. doi:10.1093/scan/nsw013.

van Hoorn, J., van Dijk, E., Meuwese, R., Rieffe, C., and Crone, E. A. (2014). Peer influence on prosocial behavior in adolescence. *Journal of Research on Adolescence, 26*(1), 90–100. doi:10.1111/jora.12173.

Wager, T. D., Davidson, M. L., Hughes, B. L., Lindquist, M. A., and Ochsner, K. N. (2008). Prefrontal-subcortical pathways mediating successful emotion regulation. *Neuron, 59*(6), 1037–1050. doi:10.1016/j.neuron.2008.09.006.

Welborn, B. L., Lieberman, M. D., Goldenberg, D., Fuligni, A. J., Galvan, A., and Telzer, E. H. (2015). Neural mechanisms of social influence in adolescence. *Social Cognitive and Affective Neuroscience, 11*(1), 100–109. doi:10.1093/scan/nsv095

Wentzel, K. R., Filisetti, L., and Looney, L. (2007). Adolescent prosocial behavior: The role of self-processes and contextual cues. *Child Development, 78*(3), 895–910. doi:10.1111/j.1467-8624.2007.01039.x.

Whittle, S., Yap, M. B. H., Sheeber, L., Dudgeon, P., Yücel, M., Pantelis, C., ... and Allen, N. B. (2011). Hippocampal volume and sensitivity to maternal aggressive behavior: A prospective study of adolescent depressive symptoms. *Development and Psychopathology, 23*(1), 115–129. doi:10.1017/S0954579410000684.

Widman, L., Choukas-Bradley, S., Helms, S. W., and Prinstein, M. J. (2016). Adolescent susceptibility to peer influence in sexual situations. *Journal of Adolescent Health, 58*(3), 323–329. doi:10.1016/j.jadohealth.2015.10.253.

Yap, M. B. H., Whittle, S., Yücel, M., Sheeber, L., Pantelis, C., Simmons, J. G., and Allen, N. B. (2008). Interaction of parenting experiences and brain structure in the prediction of depressive symptoms in adolescents. *Archives of General Psychiatry, 65*(12), 1377. doi:10.1001/archpsyc.65.12.1377.

CHAPTER 32

··

EARLY COGNITIVE AND BRAIN DEVELOPMENT IN INFANTS AND CHILDREN WITH ASD

··

ANNA KOLESNIK-TAYLOR AND E. J. H. JONES

INTRODUCTION

··

AUTISM Spectrum Disorder (ASD) is an overarching term for a group of neurodevelopmental conditions that are characterized by social communication and interaction difficulties as well as restricted, repetitive patterns of behavior, interests, or activities (DSM-5; American Psychiatric Association 2013). The condition can be reliably diagnosed from around 2–3 years, although the current UK median is around 4.5 years (Brett et al. 2016; Petrou et al. 2018). Sensory atypicalities have been reported in over 90% of individuals with ASD (Leekam et al. 2007) and have recently been included as part of the diagnostic criteria. Individual symptom profiles are highly heterogeneous and the complexity is increased by the co-occurrence of cognitive and psychiatric conditions such as Attention-Deficit/Hyperactivity Disorder (ADHD), Social Anxiety, Intellectual Disability (ID), and Opposition Defiance Disorder (ODD) (Simonoff et al. 2008). Further, the etiology is also highly heterogeneous with a range of identified environmental and genetic factors that each typically contribute a small amount of variance in explaining diagnosis (Chaste and Leboyer 2012; Grove et al. 2019; Karimi et al. 2017). Up to 10% of individuals with ASD do have a known genetic disorder, including Fragile X, Tuberous Sclerosis Complex, and Neurofibromatosis Type 1, or a medical condition such as epilepsy, hydrocephalus, fetal alcohol syndrome, and cerebral palsy (Heil and Schaaf 2012; Huguet et al. 2013; Kielinen et al. 2004).

Identifying causal paths that link genetic and environmental risk factors to later behavior is a critical target for the field, because of the potential to yield new routes to intervention. Symptoms of ASD likely emerge through a complex developmental cascade of interactions between genetics, the brain, cognition, behavior, and the child's interaction with their environment (Jones et al. 2014). Further, vulnerability factors act predominantly prenatally (Hornig et al. 2017; Sanders 2015). A focus on early brain development is key to understanding different causal paths to ASD as well as addressing the need to reduce the gap between initial reports of parental concerns (12–18 months) and the average age of diagnosis (4–6 years) (Ozonoff et al. 2009). Associating these paths with particular biological systems could also help the development of pharmacological therapies to complement existing behavioral interventions (Brian et al. 2015; Loth et al. 2014). Indeed, one key advantage of studying ASD in infancy is the possibility of early detection and intervention during key stages of brain development and enhanced plasticity (Jones et al. 2014).

Prospective longitudinal studies of infants with an older sibling with ASD provide a way to study early brain and cognitive development before ASD symptoms emerge (Bryson et al. 2007; Jones et al. 2014; Ozonoff et al. 2011). Such studies are feasible because of the high heritability of ASD (up to 90%; Faraone and Larsson (2019); Tick et al. (2016)). Infants with an older sibling or first-degree relative with a diagnosis of ASD, also known as *infant siblings* have an elevated likelihood of receiving the diagnosis themselves, with an incidence rate of up to 20% (Ozonoff et al. 2011). Using this strategy means that a much higher percentage of infants who will go on to receive a diagnosis of ASD can be identified and recruited, relative to the 1–3% incidence in the general population (Russell et al. 2014).

This chapter summarizes and evaluates the recent evidence on the cognitive development of infants and children with ASD. We organize the review based on domains most commonly associated with ASD symptomatology, including social communication and interaction, language, and repetitive behaviors. We assess both behavioral findings (reports from observational, parent-report and standardized behavioral assessments) as well as cognitive and brain data (e.g., eye-tracking, electroencephalography (EEG), or functional near-infrared spectroscopy (fNIRS), see Figure 32.1). We focus particularly on domains that may provide insight into causal paths to later symptoms, because they represent foundational skills that are likely to influence later developmental progression. Lastly, we consider how these recent findings fit into whole-brain theories of the emergence of ASD.

SOCIAL COMMUNICATION AND INTERACTION

Behavioral Findings

Difficulties in social communication are a core feature of ASD. Although it was historically believed that up to 50% of individuals with ASD were non-verbal (Lord and Paul 1997), reports of earlier identification and interventions in children with ASD diagnosed

- **Eye tracking** - a technique where a light is used to illuminate the eye and causes visible reflections that are picked up by a camera. This image is then reconstructed as a vector of an angle between the light source of the cornea and the pupil as well as other geometric aspects of the reflection, which can then be used to calculate gaze direction. Most trackers are tolerant of dynamic head movement which allows for minimal restrictions on their natural actions. This makes them ideal for infant and child studies, as well as atypical populations.

- **Electroencephalography (EEG)** - is a non-invasive method of recording electrical activity of the brain, which is picked up via electrodes placed on the scalp. The EEG signal contains task relevant activity as well as background noise. Task relevant activity is time-locked to presentation of the stimulus, which is the general principle behind Event-Related Potentials technique and evoked oscillations (see Luck 2014 for comprehensive overview).

- **Functional Magnetic Resonance Imaging (fMRI)** – is used to measure brain activity with high spatial precision by detecting blood flow changes. This technique is extremely challenging in babies and children due to excessive movement. Successful studies have recorded functional and structural activity during natural sleep, although acquisition rates remain lower than other techniques.

- **Functional Near Infrared Spectroscopy (fNIRS)** – is a non-invasive optical imaging technique that measures changes in oxygenated and deoxygenated hemoglobin (Oxy-Hb and deoxy-Hb) using two wavelengths of infrared light. fNIRS is particularly effective in babies and young children as they have thinner skulls and very little hair that allows for close contact of the sensors.

- **Magnetic spectroscopy (MRS)** – a novel technique that detects electromagnetic signals produced within the nuclei of any cell. Can be used to obtain concentrations of chemicals in the brain.

FIGURE 32.1   (Box 1) Common experimental techniques used in developmental research.

at 2 years found that only 14–20% of the sample had little or no functional language in adolescence (Lord et al. 2004; Sigman and McGovern 2005). A range of other communication difficulties that include lack of gestures and reduced eye-contact is common. There is great variability reported in verbal communication of children with ASD, as well as an overall reduction of verbal-to-non-verbal discrepancies with age and improvement in language functioning (Charman and Stone 2008). Recent reports suggest that girls with ASD and children who receive high levels of verbal input from parents show better language and communication outcomes (Eriksson et al. 2012; Nadig and Bang 2017). Sex differences and environmental factors are likely to be important early on in shaping trajectories. In order to understand later individual differences in language and communication outcomes, it is critical to look at early differences in the prelinguistic skills that provide a foundation for later language development. In the current chapter, we overview a set of preverbal skills that emerge early and are fundamental to later communication and interaction abilities; in particular, joint attention, symbol use, and social affiliation.

Joint attention is defined as coordinated attention between two people and a third object or event, including following a non-verbal prompt such as a point or gaze shift. Aspects of joint attention typically emerge around 3–6 months of age (Marshall and Fox 2005). Behaviors such as matching head direction to mother's head turn are visible as early as 6 months of age (Morales et al. 1998), and initiation of joint attention through behaviors like pointing are demonstrated by most infants by 12 months of age (Carpenter et al. 1998). Longitudinal studies have revealed positive associations between amount of joint attention in a parent-child dyad and later vocabulary development in the first year of life (Morales et al. 1998; Tomasello and Farrar 1986), but not the second year (Morales et al. 1998), which suggests there may be sensitive periods in which joint attention facilitates infant development. Initiation of joint attention by the child emerges over the second year of life, with several studies suggesting this should be treated as a separate construct which is not strongly correlated with subsequent language development (Akhtar and Gernsbacher 2007).

A range of reports have shown that children with ASD experience difficulties with joint attention. These include difficulties in following gaze, less positive affect during interaction, poorer coordination of gaze and action, reduced frequency of initiation of joint attention, and reduced gaze and/or cue following during a clinical observation or from interactions with an unfamiliar adult (Charman et al. 1997; Loveland and Landry 1986; Mundy et al. 1990). These difficulties may be present from infancy; for example, Sullivan and colleagues (2007) showed that less responding to joint attention cues (e.g., following a gaze shift or head turn without a point) at 14 months predicted later ASD outcome. Since then, a number of prospective studies with *infant siblings*, i.e., children with an older sibling with a diagnosis of ASD have shown early difficulties in joint attention (Macari et al. 2012; Nyström et al. 2019). For example, Nyström and colleagues (2019) reported findings from a live eye-tracking study of 112 infants. Eighty-one infants with elevated familial likelihood of ASD showed an atypical developmental trajectory of joint attention between 10–18 months of age, including reduced frequency of response to initiation of joint attention as well as reduced number of instances where infants initiated episodes of joint attention (Nyström et al. 2019). Trajectories for initiation of joint attention further distinguished between infants who went onto receive a diagnosis of ASD versus elevated likelihood infants who were typically developing (see Franchini et al. (2019) for a recent review of the literature).

Other studies found no differences in initiation of joint attention and later ASD outcome in infants younger than 14 months (Landa et al. 2007; Macari et al. 2012; Rozga et al. 2011). Rozga and colleagues (2011) reported findings from 137 infants that at 12 months of age episodes of joint attention were initiated equally, irrespective of ASD outcome. There were also no differences in the number of joint attention episodes in a smaller sample of infants with later ASD (*n* = 13) at 12 months in the Macari and colleagues (2012) cohort. These inconsistencies may be attributed to differences in methodologies and heterogeneity in the sample, although initial findings raise the question whether "gaze alternation" or initiation of joint attention may be atypical across populations with later ASD and other developmental delays. Further research is needed to support whether this can be considered a cumulative risk factor (Jones et al. 2014).

Differences in joint attention are thought to have cascading effects on language development, as language learning occurs in the context of the caregiver referring to objects and providing shared meanings of words (Charman and Stone 2008). Joint attention allows for development of reference within a social context, and has been found to be a necessary and sufficient precursor to word learning (Akhtar and Gernsbacher 2007). At 18 months and above, a set of behaviors including joint attention, imitation, and object play were associated with language development and intellectual functioning in ASD (Charman et al. 2005; Shumway and Wetherby 2009; Wetherby and Prizant 2002). This has been supported by several longitudinal studies (Poon et al. 2012; Sullivan et al. 2007). In a retrospective study of home videos of 29 children with ASD by Poon and colleagues (2012), joint attention at 9–12 and 15–18 months was associated with communication and intellectual function ability at 3–7 years. It should be noted that no control group was present in this investigation and there is some participant selection bias present when recruiting parents of children with ASD diagnoses for retrospective data analysis. In a prospective study, Sullivan and colleagues (2007) examined associations between joint attention responses in 51 infants with family history of ASD. Researchers found that response to joint attention was predictive of language performance in the 16 infants with later ASD and 8 infants later rated to be on the broader autism spectrum (BAP), which suggests that joint attention abilities in infancy and toddlerhood may be used as an early marker for language impairment risk, even if the criteria for a diagnosis of ASD are not met.

Symbol processing and motor control are other domains of difficulty that may contribute to delays in communication difficulties in ASD. Symbol use is important for social communication as one must understand that an object has a shared verbal meaning with others in your environment. Further, symbols are used in play in typical developing children (DeLoache et al. 1999; Orr and Geva 2015), whilst reports suggest that infants with later ASD tend to engage in functional play with objects (Koul et al. 2001; Rutherford and Rogers 2003; Williams et al. 2001). Closer examination of functional play in children with ASD, Down's Syndrome, and typical development by Williams and colleagues (2001) revealed qualitative differences in play behaviors, including increased rigidity, reduced elaboration, and presence of actions that are not appropriate to the situational context. Due to the rarity of the genetic conditions, however, the comparison samples were relatively small ($n = 15$ per group) and have limited generalizability.

Beyond the movements involved in speech control and articulation, the increasing proficiency in motor skills including reaching, crawling, and walking allows the infant to explore objects and people in their environment (Iverson 2010). It is therefore unsurprising that a strong relationship has been found between early development of motor abilities, language and communication in infants with and without later ASD (Bedford et al. 2016; Sheinkopf et al. 2000). For example, Sheinkopf and colleagues (2000) examined children with ASD ($n = 15$) and developmental delay ($n = 11$) and suggested that deficits in verbal communication arise from poor motor control of the speech mechanism and difficulties in capacity for symbolic representation. In this study, vocal atypicalities were not associated with joint attention abilities in children with ASD, which suggested that the two deficits arise from distinct mechanisms; although no study to date has examined this relationship in early infancy (Sheinkopf et al. 2000). Instead, some children who do learn to speak use echolalia (imitation of speech of others; Roberts 2014), where a whole utterance is repeated verbatim and is used as the symbol for a situation or event. Successful language acquisition can occur when children begin to break down the utterances into single words and give them meaning, after which some aspects of grammar and complex vocabulary can be acquired (Roberts 2014). Although these studies have relatively small sample sizes, the rich datasets analyzed have allowed important insights into the quality and variability of speech in ASD.

Children with ASD show reduced social affiliation or affect relative to their neurotypical peers, including atypical face-to-face gaze, poor comprehension of social and emotional cues, and perspective-taking abilities. This has been associated with difficulty in forming and maintaining of social and affective relationships, which are essential for further social and communicative skill development (Warlaumont et al. 2014; Yirmiya et al. 2006). Children with ASD may show less social smiling in response to their mothers relative to typically developing children (Dawson et al. 1990). One study found that higher levels of maternal responsiveness (i.e., behaviors that are sensitive, contingent, and support and follow the lead of the infant in social, emotional, communicative, and play behaviors) at 9 months of age were associated with higher social smiling at 18 months in infants with and without familial likelihood of ASD (Harker et al. 2016). Higher levels of maternal directiveness (i.e., behaviors including prompting, instructing, and/or requesting, putting the focus on infant's attention or behavior) predicted an overall slower growth of social smiling abilities between 9 and 18 months. However, in the infants with increased likelihood of ASD, maternal directiveness at 9 months predicted a positive growth of social smiling between 9 and 12 months. The study also found both responsiveness and directiveness within both samples of mothers and infants demonstrating that the two styles can occur within a single play session. The research suggests that there are complex ways in which parental behavior may relate to or influence children's social interaction outcomes.

Nichols and colleagues (2014) examined the precursors of reduced social affect in 15-month-old infants, and found reduced eye-contact and non-social smiling (general affect) in infants with later ASD ($n = 15$) relative to infant siblings with later typical development ($n = 27$) or neurotypical controls ($n = 25$). Importantly, both infant sibling groups showed reduced social smiling in comparison to the neurotypical infants irrespective of later developmental outcome; which suggested that early behavioral differences do not translate directly into later clinical difficulties, and other contributing factors are likely to be at play (Nichols et al. 2014).

Overall, differences in joint attention, symbol use, and social affect are three aspects of social development that have been found to be associated with later ASD in infants. While these non-verbal abilities could represent precursors of more advanced aspects of social communication, future research is necessary to understand whether interventions to address these low-level behaviors in infancy and childhood (e.g., Kourassanis-Velasquez and Jones (2018)) will have long-lasting positive effects on social skills.

Cognitive Paradigms

Looking is one of the first behaviors that emerges in infancy and provides an important modality through which young infants learn about the world. Recent technological developments have allowed us to capture infant gaze behavior with unprecedented precision through eye tracking (Oakes 2012). Measures of gaze direction and pupil dilation are believed to tap into active cognitive processes. For example, looking direction towards a particular object is driven by underlying neural mechanisms that regulate detection, identification and discrimination of the presented visual stimulus (Aslin 2012). Different looking behaviors have been attributed to specific mechanisms: for example, longer looking times are associated with preference and faster saccadic reaction times are associated with learning/age-related changes in stimulus processing (Hepach and Westermann 2016; Oakes 2010; Yeung et al. 2016).

Relative to observational and standardized behavioral measures of social attention, eye-tracking paradigms offer greater resolution on what the infant/child is looking at and for how long. Eye tracking does not require a baseline level of verbal or cognitive ability, and thus can be used with preverbal children and infants. From eye-tracking data, we are able to detect subtle variations in viewing behavior from early infancy, which has been utilized in prospective longitudinal study designs. Thus, eye tracking is a useful methodology for studying the early factors that may contribute to or index the emergence of ASD symptoms.

Eye-tracking literature on the early predictors of ASD has primarily focused on attention to people and social interaction. Social attention refers to the way attention is affected by the presence of other people and is synonymous with non-verbal communication. It is often operationalized as looking towards faces, eyes, or people (Freeth et al. 2013; Salley and Colombo 2016). A systematic review of visual social attention paradigms by Guillon and colleagues (2014) concluded that the majority of eye-tracking studies report reduced social attention in individuals with ASD relative to neurotypical controls, including in young children (Bedford et al. 2014; Chawarska et al. 2012, 2013; Dawson et al. 2012). The extent to which toddlers with ASD show atypical patterns of social attention in ASD may be context dependent; specifically, the presence of child-directed speech and a larger number of speakers may elicit bigger differences in looking towards faces in children with ASD relative to controls (Chawarska et al. 2012).

The pattern of data for looking to the eye region of faces is more mixed (Guillon et al. 2014; Jones and Klin 2013; Merin et al. 2007). For example, Merin and colleagues (2007) found that 6-month-old infants demonstrated similar patterns of behavior, or the established Still Face effect[1] irrespective of developmental outcome. Closer inspection of the pattern of looking behavior across the sample revealed a subgroup within an infant sibling sample; infants with reduced gaze to the eyes (10 out of 11 infants in the subgroup) were classified as "at-risk of ASD". However, 16 further "at-risk" infants maintained a high degree of gaze at the eyes during the experiment. These subgroups did not differ in their social/affective measures, although it is difficult to interpret these findings in the context of ASD as the majority of the infant siblings will not go on to receive a diagnosis in childhood. A more recent longitudinal study of 103 infants over 10 time points found a decline in looking towards the eyes between 2 and 6 months of age in infants with a later diagnosis of ASD, whilst no such pattern was observed in infants who were typically developing at age 3 (Jones and Klin 2013). In interpreting this pattern of data, we may need to distinguish between the different attentional mechanisms that may underlie a profile of looking. Infant models of attention distinguish between alertness (the ability to *attain* an alert state), spatial orienting (the shifting of attention to a specific spatial locus, the "where" system), attention to object features (analysis of input and processing of features that lead to identification or the "what" system) and endogenous attention (direction of attention to the stimulus voluntarily); which are assumed to have different neural substrates and developmental time courses (see Colombo (2001) and Salley and Colombo (2016) for review). A range of evidence suggests that orienting to social stimuli may be relatively intact in ASD (Elsabbagh et al. 2013; Guillon et al. 2014; Jones and Klin 2013). Specifically, basic social orienting at 2 months was greater in infants who went on to receive a diagnosis of ASD in the cohort described by Jones and Klin (2013). Based on this and other evidence, Johnson (2014) proposes that social orienting is intact in ASD and thus models that had proposed that an early failure of "innate" social orienting would reduce long-term self-directed exposure to social information are not tenable (Johnson 2014). Rather, it may be that differences in social motivation or alertness contribute to alterations in endogenous attention and impact development of executive function systems (Chevallier et al. 2012; Hendry et al. 2016; Kapp et al. 2019).

Another early emerging bias in typically developing infants is the tendency to look at biological motion (Falck-Ytter et al. 2011; Simion et al. 2008, 2011). This is thought to be an important evolutionary mechanism to identify and preferentially attend to own species. Although orienting to faces may be intact, it has been suggested that a reduction in the bias towards biological motion from early infancy causes cascading social difficulties in later life (Falck-Ytter et al. (2013), Klin et al. (2009), Yang et al. (2016); see Todorova et al. (2019) for recent meta-analysis). Several research groups have reported that toddlers and children with ASD may have difficulties in perceiving global, coherent motion when viewing random-dot cinematograms, which may suggest broader difficulties with integrating information (Milne et al. 2002; Spencer et al. 2000). Others have argued that the reductions in biological motion preference in a sample of children with ASD ($n = 21$) may actually reflect interference from an increased interest in audiovisual synchrony (AVS), which could draw attention to the non-social environment (Klin et al. 2009). Specifically, researchers found that visual preference for biological motion displays had a strong positive association with the degree of preference for AVS in toddlers with ASD, but not in neurotypical controls. In contrast, a study of ten 3-year-olds with ASD found that unlike typically developing controls ($n = 14$), they did not orient towards either AVS or biological motion, and this response did not correlate with chronological or developmental age (Falck-Ytter et al. 2013, 2018). There remains mixed evidence as to whether atypical biases in visual attention could contribute to the later development of ASD. Larger sample sizes as well as longitudinal assessments are needed to map the inter- and intra-group variability of biological motion perception.

When considering differences in social attention, it is important to consider whether they are specific to the social domain or whether they represent more domain-general aspects of attention. One potential non-social marker of ASD is "sticky attention," or difficulties in attentional disengagement. Early studies indicated that toddlers with

ASD take longer to disengage from a central stimulus to shift their attention to a peripheral stimulus (Landry and Bryson 2004; Sabatos-DeVito et al. 2016; Sacrey et al. 2014). This may extend to social stimuli (Franchini et al. 2017; Katarzyna et al. 2010); although Vernetti and colleagues (2018) reported atypical orienting only to less predictable social stimuli (i.e., gaze-contingent stimuli that produced a randomized social response). Similar effects have been found in infants with a later diagnosis with ASD. Elsabbagh and colleagues (2013) examined 106 infants longitudinally and reported that while there were no group differences at 6–9 months, 10–15-month-old infants with later ASD showed prolonged disengagement times relative to neurotypical infants (see also Elsabbagh et al. 2009; Zwaigenbaum et al. 2005). Sacrey and colleagues (2014) reviewed social orienting and disengagement in ASD studies throughout the lifespan and suggested that whilst some atypicalities are established by 12 months, a third of studies examined found null or counter results in ASD groups (i.e., faster disengagement in ASD groups), which were attributed to differences in stimulus type and inter-stimulus-interval duration. Studies that do not find significant differences between children and infants with later ASD and tardive dyskinesia (Fischer et al. 2016; Landry and Parker 2013; Shah et al. 2013) have argued that "sticky" attention may not be a core feature of ASD phenotype and thus does not play a key role in its etiology.

Neuroimaging

Technological advancements in brain imaging techniques have allowed us to visualize neural activation in infancy through functional electroencephalography (EEG) and magnetic resonance imaging (fMRI; Figure 32.1). In the developing brain, the main features of social processing and communication that have been examined in association with ASD outcome are: (1) atypical cortical face processing, (2) neural activation of "social brain" networks, and (3) voice processing.

Several EEG studies have shown differences in perceptual processing of social information in children and infants with ASD. Of particular interest in these studies are the N290 and P400 event-related potential (ERP) components. Recent studies have shown that the infant N290 and P400 (a negativity that occurs around 290 ms and a positivity around 400 ms after stimulus onset recorded over the posterior regions) are analogous to the N170 component in older children and adults (de Haan and Nelson 1997; Leppänen et al. 2007; Scott et al. 2006). The N170 is a "face" sensitive component that reflects experience-related expertise with faces (Allison et al. 2000; Haxby et al. 2002). A recent meta-analysis indicates that the N170 peak appears later in ASD (Kang et al. 2018), an effect that may reflect reduced expertise and is also seen in younger children (Webb et al. 2006, 2011, 2012). Delayed ERP responses to faces may also indicate an overall delay in processing of social stimuli, which could underlie differences in social processing at a behavioral level. Some support for this hypothesis comes from the observation of the relationship between speed of brain response to faces and adaptive social behavior (socialization subscale on the Vineland Adaptive Behavior Scales) (Webb et al. 2011).

Similar effects have been examined in studies of infants with older siblings with ASD (Key and Stone 2012; Luyster et al. 2014). Key and Stone (2012) reported similar ERP responses in infants with later ASD and neurotypical controls when viewing pictures of infants' mothers, whilst a delayed P400 was reported only in the neurotypicals when viewing strangers' faces. A longitudinal study of 131 infants, around 20–30 infants with good EEG at each time point, by Luyster and colleagues (2014) looked at age-related change between 6 and 36 months in the pattern of N290 and P400 responses to familiar and novel faces. No group differences were reported in the strength or topography of the N290 or P400 components, although neither peak amplitude nor latency were examined. Two other studies have observed faster P400 latencies to faces in infant siblings group (Elsabbagh et al. 2012 ($n = 40$); Jones et al. 2018 ($n = 27$)) indicating that alterations in early face processing may be age-dependent in early infancy. Finally, Elsabbagh et al. (2012) report diminished P400 responses to differences in gaze direction in young infants with later ASD, which may be related to individual differences in joint attention reviewed above.

Another component of interest in social processing is the Negative Central component, also known as the Nc, which is characterized as a gradual negative deflection (400–850 ms) recorded over fronto-central regions that is thought to reflect attentional engagement (de Haan et al. 2003; Nelson and Monk 2001). A study of thirty-four 3–4-year-old children with ASD showed a robust difference in the Nc response to familiar and unfamiliar objects but not to familiar versus unfamiliar faces, as was reported in a chronological age-matched neurotypical group ($n = 19$) and developmentally delayed group without autism ($n = 16$; Dawson et al. 2002). In a smaller subsample of 16 children with ASD aged 18–30 months, Webb and colleagues (2011) found that the modulation of the Nc component to familiar and unfamiliar faces in older children with ASD resembled the component observed in younger typically-developing children matched on social developmental age. The researchers suggested that there is evidence for delayed rather than atypical development of face recognition. In the infant literature, Luyster and colleagues (2014) showed an absence of maturation in the amplitude of the Nc response to familiar and unfamiliar faces between 6 months and 2 years in infants with an older sibling with ASD relative to typically developing controls; however, Key and Stone (2012) did not report differences between outcome groups in the amplitude of Nc responses to faces at 9 months, although suggested that differences appear in response to alterations in facial features (i.e., eye or mouth change). It was reported that the size and latency of ERP responses was related to parent-reported levels of expressive and receptive communication abilities irrespective of "risk status." Interestingly, higher sensory hypersensitivity at 2 years predicted a larger Nc amplitude to faces at age 4 in a cohort with an ASD diagnosis (Jones et al. 2018), suggesting there may be interrelations between different ASD-relevant domains over early development.

A range of studies has examined specific brain structures that may be involved in social processing in ASD. Studies with adults have reported a lack/reduced activation of the fusiform gyrus (or the "fusiform face area," FFA) and the amygdala, structures associated with face and social processing in typically developing adults (Bilalić et al.

2011; Kanwisher and Yovel 2006). Other studies found that children with ASD display comparable activation in the FFA when viewing images of their parents and other children but not unfamiliar adults (Pierce and Redcay 2008), as well as activation outside the FFA in adults with ASD (Pierce et al. 2001). Ensuring individuals with ASD look at the eye region of a face with a fixation cross can also produce neurotypical neural responses in the FFA (Perlman et al. 2011). These findings suggest that alterations in FFA activity may be related to patterns of social attention and developmental history of interest in faces. Indeed, in one child with ASD, Grelotti and colleagues (2005) reported activation in the FFA and the amygdala during discriminating between Pokémon characters (of special interest to the child scanned) rather than familiar and unfamiliar faces. A follow-up study with a larger group of 20 adults and children obtained similar results (Foss-Feig et al. 2016). This may indicate that face-processing systems become specialized for areas of expertise, which in ASD may not necessarily be faces. Infant data will be required to resolve whether atypicalities in face processing systems precede or follow the declines in attention to the face that are observed in the first year of life (Ozonoff et al. 2010).

There are currently no reported fMRI studies of visual functional brain activation in awake infants with later ASD because the methodology is very difficult to use with awake infants (though see Blasi et al. 2011; Deen et al. 2017). Functional near Infrared Spectroscopy (fNIRS) is a related method that enables the study of activation in different brain areas in awake infants by measuring changes in concentration of oxygenated and deoxygenated blood (see Wilcox and Biondi (2015) for review). An early study of 4–6-month-old infants showed diminished hemodynamic responses in temporo-parietal brain regions in response to auditory and visual social stimuli in infants with older siblings with ASD ($n = 18$) relative to typically developing infants ($n = 16$) (Figure 32.2; Lloyd-Fox et al. 2013), and these changes were related to later ASD outcome (Lloyd-Fox et al. 2018).These results are consistent with altered early specialization of the social brain, especially since greater responses to social vs non-social stimuli within this brain region appear soon after birth in neurotypical infants (Farroni et al. 2004). The findings with respect to sibling status were later replicated in an independent cohort of 29 infant siblings (Braukmann et al. 2018). Researchers showed reduced activation to social stimuli in the right posterior temporal cortex relative to neurotypical controls. This early data does suggest that alterations in social brain functioning may emerge before clear differences in social behavior.

RESTRICTIVE-REPETITIVE BEHAVIORS

Behavioral Findings

Restricted and repetitive behaviors (RRBs) in ASD can include simple actions such as rocking, hand-flapping, or echolalia. While some repetitive behaviors are noted in typically developing children in the first years of life and are believed to be a normal feature

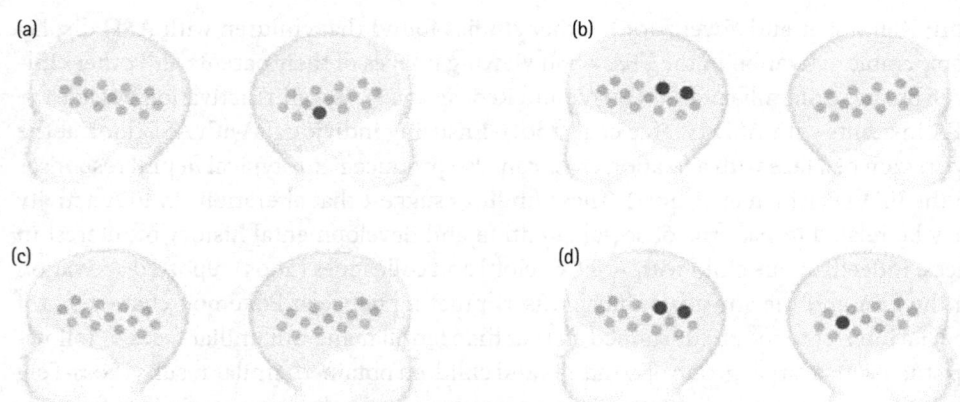

FIGURE 32.2 Neural activation to vocal and non-vocal selective responses in neurotypical infants (a,b) and infant siblings (c,d). The statistically significant effects (two-tailed, p, 0.05) are displayed for (a,c) the vocal. non-vocal and (b,d) the non-vocal. vocal selective responses. The channels that revealed a significantly greater response during the specified time window of activation are plotted in dark gray (increase in HbO_2 concentration) and light gray (decrease in HHb concentration).

of development (Berkson and Tupa 2000), the amount and frequency of RRBs is significantly higher in autism (Bodfish et al. 2000). Individuals with ASD may struggle with changes to established routines or schedules. RRBs in autism appear more consistently with age, and may have detrimental effects for early learning and social development (Leekam et al. 2011), while other literature has suggested the soothing effects of RRBs (Johnson et al. 2014; Scahill and Challa 2016). Moore and Goodson (2003) conducted behavioral observations of 20 children with an early diagnosis of with ASD at 2–3 years and followed them up again between 4 and 5. It was found that while social communication symptoms were persistent, RRBs became more apparent over developmental time. A review of behavioral data suggested that the reduction in complex mannerisms in infants and toddlers was accompanied by an increase in restricted interests and occupations in older children and adolescents (Leekam et al. 2011). These behaviors may be further related to cognitive abilities, as IQ was found to be negatively associated with the number of reported RRBs (Bishop et al. 2006; Chen et al. 2009).

The majority of literature on the emergence of RRBs has involved retrospective parent report measures (e.g., through "gold standard diagnostic tools" such as the autism diagnostic inventory (ADI-R; Rutter et al. 2003). For example, Charman and colleagues (2005) compared trajectories in 134 children with ASD from age 2–3 years to age 7. Over time, scores on the RRBs domain on the ADI-R remained relatively steady, relative to age-related decreases in social interaction and non-verbal communication;

yet individual differences in reported RRBs at 2 or 3 were not significant predictors of symptom severity in childhood. Researchers concluded that it is essential to look at developmental changes in behaviors over time, as there are distinct trajectories across domains as well as individual differences. Due to adaptation and change in the nature of RRBs, it may be difficult to make associations between infant repetitive behaviors and their impact on later cognitive functioning. For instance, while increases in RRBs were recorded in children with ASD between 24–48 months based on the standardized ADI scoring, expanding the assessment scale to include a broader range of RRBs showed a decrease in total repetitive behaviors over time as well as an association with reduced impact on child and family activities (Honey et al. 2008).

Within a prospective design, RRBs can be measured through observational captures of behavior. Ozonoff, Macari and colleagues (2008) reported atypical visual and tactile object exploration in nine 12-month-old infants with later ASD (total $n = 35$), suggesting that RRBs may be detected earlier than previously considered. Other studies reported some overlap in stereotyped behavior in 12–18-month-old infant siblings relative to the typically developing controls (Loh et al. 2007). One reason that there may not be group differences observed in early development is because young typically developing children also show repetitive behavior patterns (Bodfish et al. 2000; Evans et al. 1997). It may be that particular types of RRBs such as those shown during play, unusual visual regard, and motor stereotypies are more specific to a diagnosis of ASD (Honey et al. 2007; Melo et al. 2019). On the other hand, Loh and colleagues (2007) conducted retrospective analyses on videos of 9 children with family history of ASD vs 15 control participants. The actual number of children with a diagnosis in this group was unknown at the time of analysis, and the majority of the "high risk" sample actually went on to develop typically.

Cognitive Paradigms

The mechanisms underlying RRBs have been less researched than the social communication domain. However, a range of explanations have been offered at present that include differences in (1) sensory sensitivity, (2) cognitive flexibility, and (3) executive functioning, which are discussed in the following sections. Firstly, up to 90% of individuals with ASD report hyper- and hypo-sensitivity, and sensory overload (Leekam et al. 2007; Rogers et al. 2003; Rogers and Ozonoff 2005). Within the DSM-V, sensory difficulties are now grouped with the RRB subdomain because of a range of evidence that individual differences are associated (Gabriels et al. 2008; Mandy et al. 2012). Chen and colleagues (2009) reported positive associations between sensory processing symptoms and RRBs in children with ASD. In addition, several reports have found higher levels of sensory symptoms in infants with later ASD through parent report (Ben-Sasson and Carter 2013; Germani et al. 2014), which was interpreted as a possible precursor to emergence of RRBs.

A second hypothesis about difficulties relating to RRBs centers around cognitive functioning (Lopez et al. 2005). Executive function (EF) is defined as a set of higher-order cognitive processes that are necessary for cognitive control of thought and action. Dysregulation of EF has been attributed to the emergence of RRBs, as failing to control attention and inhibit behavioral responses can lead individuals with ASD to become "locked into" a specific behavior and/or thought (Lopez et al. 2005; South et al. 2007; Turner 1997). Older children and adolescents with high-functioning ASD showed associations between RRBs and EF (Wisconsin Card Sorting task; South et al. (2007)). Impaired cognitive flexibility, i.e., the ability to learn and adapt to changing feedback from the environment has been implicated in ASD and has been found to be associated with RRBs. Crawley and colleagues (2020) observed a large sample of 321 individuals with ASD and found that children with a diagnosis showed poorer cognitive flexibility than controls in a probabilistic reversal learning paradigm. Their perform-ance improved with age, although reduced cognitive flexibility was associated with RRB in adults. The study suggests that persistent cognitive *in*flexibility may be indicative of more severe ASD outcomes, although further study is needed to understand the specifi-city of this task in longitudinal paradigms.

Difficulties in EF and cognitive control have been studied in pre-school children. Two early studies found poorer performance, such as increased number of pervasive errors in children with ASD, not all tasks may have been developmentally appropriate for the sample (Griffith et al. 1999; Hughes 2002; McEvoy et al. 1993). EF atypicalities have been linked to development of Theory of Mind (ToM), as reduced EF performance (set shifting, planning and inhibition) was associated with poorer ToM in 5-year-olds with ASD (Pellicano 2007). In infancy research, limited attempts have been made to study emergent EF abilities, due to an overall difficulty in identifying the developmental pathways, although it has been suggested that attentional orienting and disengagement, as discussed previously are important in the development of higher-order cognitive processing in the first year of life (Holmboe et al. 2008; Johnson et al. 1991; Posner and Cohen 1984). In a comprehensive review, Hendry and colleagues (2016) argued that im-pulse control and cognitive flexibility are present by 3 years of age in typical develop-ment; this may be an important target for future work in ASD.

Neuroimaging

Recent studies have begun to address the possible neural substrates that may explain the emergence of repetitive behaviors in infant siblings. One study examined EEG connect-ivity, where hyper-connectivity was observed in the alpha range in infants with familial likelihood of ASD. Elevated functional connectivity at 14 months was associated with RRBs in the nine children that went on to receive a diagnosis of ASD at 3 years (Orekhova et al. 2014). Replication of this methodological analysis with an independent sample of 13 individuals with later ASD who provided good data by Haartsen and colleagues (2019) did not show significant associations between global EEG-connectivity and ASD outcome. Within a combined sample of 90 infants from the "high-risk" subgroups (HR-atypical $n = 21$, HR-TD $n = 47$), the association between higher functional connectivity at 14 months and RRBs (particularly restricted interests) at 36 months was replicated (Haartsen et al. 2019), which suggested that connectivity may be useful as a candidate predictor of variation in ASD symptom severity.

Magnetic resonance imaging (MRI) has also been used to study the brain mechanisms that may underlie RRBs. In a longitudinal study of 49 children with ASD at 9.5 and 11.5 years, ritualistic behavior was further associated with increased rate of growth of the striatum in ASD (Langen et al. 2014). Further, McKinnon and colleagues (2019) used fMRI in 168 sleeping infants to map out trajectories of connectivity at 12 and 24 months. In the 20 infants with later ASD who provided good data, stereotyped and repeated behavior was associated with functional connectivity in the default mode network, temporal, and visual areas. The combination of longitudinal and cross-sectional data in this study as well as a lack of direct measures of RRBs in infancy, however limits our interpretation of the emergence of these behaviors. This is a more general limitation within the literature, where parental reports of early behavior are attributed to alterations in brain function or morphology.

Alternatively, we can consider early sensory processing alterations as a precursor of RRBs. Given the links between atypical sensory processing and repetitive behavior in children with ASD (Chen et al. 2009; Joosten and Bundy 2010), the early maturation of sensory processing networks in the brain are important to consider prospectively (Emberson et al. 2015). Kolesnik and colleagues (2019) addressed the possible neural underpinnings of sensory atypicalities by reporting heightened cortical reactivity to auditory tones (characterized by reduced habituation and increased inter-trial coherence) in 8-month-old infants with later ASD. Although an association was reported between cortical reactivity at 8 months and the Social Responsiveness Scale scores at 3 years, the relationship with the RRBs subscale was not specifically examined (Kolesnik et al. 2019). Additionally, it was reported that stronger pupillary light responses at 9–10 months were associated with more severe ASD outcome at 3 years (Nyström et al. 2018). Whilst atypicalities in the pupillary light reflex have been attributed to elevated sensory sensitivity in older children with ASD (Daluwatte et al. 2015), early relative constriction of the pupil was not associated with later RRBs in a dimensional analysis irrespective of developmental outcome (Nyström et al. 2018). In summary, there is a clear need for a more extensive evaluation of the early developmental roots of later RRBs in prospective longitudinal studies.

Mechanisms

Theoretical Perspectives

Probabilistic epigenesis (Gottlieb 2007) proposes that there are ongoing bidirectional interactions between genes, brain structure, the environment, and cognition. Activity-dependent development is an integral component of this approach, where infants are

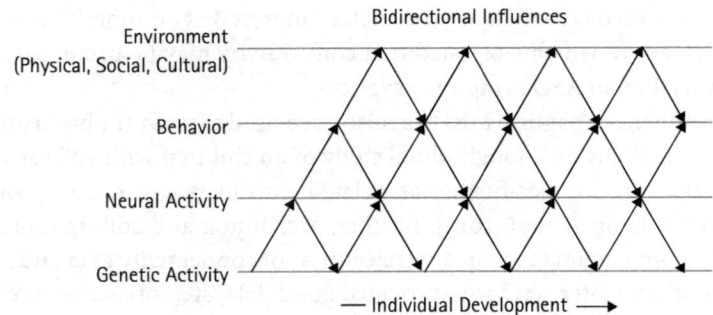

FIGURE 32.3 Metatheoretical model of probabilistic epigenesis. Note the relationships between different systems and their bidirectional nature. This aims to more accurately represent the dynamics of early human development.

Reproduced with permission from Gottlieb, G. Probabilistic epigenesis. *Developmental Science*, 10(1): 1–11. © 2007 The Author. Journal compilation © 2007 Blackwell Publishing Ltd

actively choosing the appropriate input that guides further specialization (Figure 32.3). Elements of the theory have also been integrated into the Interactive Specialization (IS) framework (Johnson 2000, 2011). The interactive component of the framework predicts that neural change will occur across multiple regions as networks will adjust to optimize overall function; activation of regions will become more fine-tuned to given stimuli or task processing as specialization extends from specific regions to networks; and the specialization process will be impacted by the environment across different functional networks. Additionally, specialization of a single region occurs within the context of interactions with other regions. Under such theories, an understanding of autism requires an integrated view of data from multiple levels of analysis.

Sampling from the environment may be one key process that drives alterations in neurodevelopmental paths. For example, the lack of social engagement in infants with an elevated likelihood of ASD may prompt different indirect caregiver behaviors including scaffolding and increased parental stress, which may indirectly provide fewer opportunities for social learning. Such a model has received some support through observational studies of parent-child interactions with infant siblings (Adamson et al. 2012; Beurkens et al. 2013) including associations with later developmental outcome (Wan et al. 2013). Fear of danger/injury might also mean that caregivers limit exploration behaviors, and the infant receives less experience interacting with their environment (Karmiloff-Smith 2009).

The IS framework does not consider atypical developmental trajectories to be purely based on deficits and argues there are many abilities that fall within the typical or even superior range in children and adults with ASD. Others have also proposed that individuals with neurodevelopmental disorders such as ASD and ADHD do not demonstrate deficits specific only to 'social' and 'attention' abilities respectively, but instead show a wide range of strengths and weaknesses across different modalities (Donkers et al. 2015; Glod et al. 2016). Rejection of the deficit model has been widely recognized in the literature on visual search and attention tasks (Joseph et al. 2009; Keehn et al. 2008; Milne et al. 2013). For example, infants with later ASD may be more accurate at detecting a distinctive visual stimulus within an array (Cheung et al. 2018).

The framework can be further used to interpret findings from an fMRI study by Pierce and colleagues (2001), which reported that individuals with ASD are able to process social stimuli through unique brain circuitry. Superior abilities in speech discrimination in a small sample of seven adults with ASD relative to eight neurotypical controls were associated with better receptive language skills, and suggested that this may be an important mechanism in ensuring better language outcomes (Heaton et al. 2008). Others found responses to language stimuli beyond the language regions of the left hemisphere, specifically in the right- and right-medial frontal regions (Dawson et al. 1986; Redcay and Courchesne, 2008). In this way, "atypical" lateralization of speech processing in ASD may actually indicate an adaptive pathway that facilitates language development (Jones et al. 2014), although the sample sizes were comparably small across these studies. As such, the need for longitudinal methods needs to be recognized in future studies, including models that span multiple domains and consider the role of the environment and the bidirectional influence of the infant on their own development.

Neural Accounts

In this chapter, we discussed early behavioral differences in infancy and childhood that are associated with a later ASD outcome and that might shine a light on underlying mechanisms. Past theoretical assumptions of ASD symptomatology have often focused on trying to isolate selective cortical regions whose impairment might give rise to the phenotype. It has been proposed that ASD emergence is associated with altered structure of the cerebellum (Courchesne et al. 1988), the frontal cortex (Pennington and Welsh 1995; Shalom 2009), and excess cerebrospinal fluid in early development (Shen 2018). Yet, infants with familial likelihood of ASD have shown atypicalities across a range of functional, structural, and task-related measures prior to the onset of behavioral symptoms (Haartsen et al. 2016; Hazlett et al. 2017; Jones et al. 2014). Consistent with theories including interactive specialization and neuro-constructivism, it has been proposed that the early developmental roots of autism may be related to a broader network of brain regions and their connectivity.

Structurally, the overgrowth of the corpus callosum has been reported in infants with later ASD, with largest differences with neurotypical controls recorded at 6 months. Further, size and thickness of the corpus callosum at 6 and 12 months of age were associated with repetitive behaviors at 2 years in 213 children with familial likelihood of ASD (Wolff et al. 2015). There is consistent evidence for general brain volume overgrowth (Courchesne et al. 2003; Hazlett et al. 2011), which is associated with altered neural patterning and wiring. Neural connections between regions show developmental change in early development, and atypical long and short distance cortical connections have been proposed as a mechanism in infants with later ASD. For example, Subramanian and colleagues (2015) have suggested that an imbalance in growth pathways in organ, network, and cellular mechanisms (both over and under-growth) may lead to ASD outcome.

Brain overgrowth may not be the only contributing factor to the increasing head circumference associated with ASD (Webb et al. 2007). Shen (2018) reviewed the function of cerebrospinal fluid (CSF) and suggested that infants with later ASD showed increased levels of CSF at 6 months, which has been supported by several studies (Hazlett et al. 2011; Shen et al. 2013, 2017). Further, excessive amount of CSF fluid was associated with motor delays and deficits in social processing (Hazlett et al. 2017; Shen et al. 2017). Identification of these differences prior to the onset of behavioral symptoms led researchers to argue that conventional structural MRI scans can be used to identify altered CSF flow and be used as an early indicator of atypical neurodevelopment. Future research would benefit from tracking of the effects of concentrations of the fluid in the brain and spinal cord on behavioral predictors of ASD.

A further proposed neural mechanism is atypical brain lateralization. Lateralization, i.e., functional dominance of one hemisphere over the other on a particular function, is thought to occur early in development. Several studies have reported higher right hemisphere activation during a speech perception task in toddlers and infants with an elevated likelihood of ASD (Eyler et al. 2012; Seery et al. 2013), relative to the expected left hemispheric lateralization observed in typical development. Right hemisphere lateralization of speech processing at 12 months further predicted ASD outcome at 3 years of age, suggesting that this may be an early diagnostic marker (Finch et al. 2017). 'Atypical' lateralization, however, has also been seen in neurotypical children and adults that are left-handed or ambidextrous. Children and adolescents with ASD who did show the typical pattern of left hemisphere lateralization were more likely to be right handed across several independent cohorts (Finch et al. 2017; Groen et al. 2008; Knaus et al. 2008; Szaflarski et al. 2006); which suggests that lateralization research in ASD requires further study.

EXCITATION/INHIBITION BALANCE

In an influential paper, Rubenstein and Merzenich (2003) proposed that an imbalance of excitatory and inhibitory neurotransmitter function reduces signal-to-noise ratio and leads to the emergence of hyper-excitable brain networks, which gives rise to the ASD phenotype. The proposed excitation/inhibition (E/I) balance theory argues that over-reactivity at the synapse impairs the basic mechanisms of information processing and learning and as such, adequate communication and language ability cannot develop (Lee et al. 2016; Rubenstein 2010). In support of this theory, several magnetic resonance spectroscopy (MRS) studies also reported enhanced glutamatergic (Uzunova et al. 2014) and reduced GABAergic signaling (Hussman 2001; Pizzarelli and Cherubini 2011) in individuals with ASD. Rippon and colleagues published a review of the literature, which summarized that GABAergic activity diminishes inhibitory control in signaling and prevents efficient processing of incoming information (Rippon et al. 2007). It has been argued that GABAergic hypofunction alone could elicit the E/I imbalance by compromising connectivity and increasing cortical noise at baseline. A longitudinal MRS study in children with ASD found region-specific alterations in GABA concentration, including lower GABA in the motor and auditory regions of interest but not the visual system, again highlighting the importance of looking at the whole brain and interactions between different areas (Gaetz et al. 2014). Although both motor and auditory system deficits are reported in ASD, no significant associations were found with the behavioral measures collected during the study. This is, perhaps, not surprising due to the complex neurophysiological processes between neurotransmitter activity and behavior that are not captured by the correlational study design. Atypical GABAergic signaling in ASD has been corroborated by evidence from a larger sample MRS investigation with children with and without the condition, which found reduced GABA levels in sensorimotor areas but not in the visual cortex (Puts et al. 2017).

Neural activity at different frequencies has been used as a predictor of cognitive function and as a putative marker of E/I balance in several studies in children with ASD and their first-degree relatives with small to moderate sample sizes (Orekhova et al. 2007; Port et al. 2017; Rojas et al. 2008). Specifically, activity in the high-frequency gamma band is thought to reflect the orchestrated stability between populations of excitatory and inhibitory neurons (Buzsáki and Wang 2012). As such, observing the strength and change in activation of gamma power may offer a suitable alternative for the empirical study of neurotransmitter activity in the brain of individuals with a high likelihood of ASD (see E/I theory by Rubenstein and Merzenich 2003). Several studies discussed previously have shown evidence of atypical gamma activity during rest and auditory processing in infants with later ASD (Kolesnik et al. 2019; Tierney et al. 2012). Dimensional associations between gamma activity, social processing, and language ability further support the role of early neurotransmitter activity on cognition and behavior. Taken together, these findings suggest region-specific reductions in GABA within sensory areas in individuals with ASD, which is supported by atypical gamma activity measured by EEG and behavioral findings of atypical and sensory responsivity to sounds and failure to habituate to sensory aspects of the environment (Glod et al. 2016; Rojas et al. 2008; Vivanti et al. 2018).

Finally, it is important to note that there is evidence for both increases and decreases in neural excitability in ASD, as well as several studies which found no difference between individuals with autism and neurotypical controls (see Dickinson et al. (2016) for review). Further, alterations within neurotransmitter balance are not specific to ASD, but are observed in many neurodevelopmental, psychiatric, and genetic disorders (Spencer et al. 2014). The latter point is particularly important to consider in order to understand the common and divergent pathways to ASD and how molecular paths to the ASD phenotype may be shared across different conditions (Gandal et al. 2018; Gao and Penzes 2015). Further work examining GABA concentrations, as well as refinement of the present analysis methods using MRS (Rojas et al. 2015; Stagg et al. 2011), are needed to understand the distribution and timing of E/I balance and its implications for brain development in ASD.

FUNCTIONAL CONNECTIVITY

Another prominent theory of the neural mechanisms that may underpin ASD is alterations in whole-brain connectivity. Several studies have suggested that dysregulated communication between brain areas could emerge early in infancy and contribute to later ASD diagnosis. Indeed, a longitudinal fNIRS study showed reduced functional connectivity in 3-month-old and 12-month-old infant siblings (i.e., children with a first-degree relative with a confirmed ASD diagnosis) compared to neurotypical controls (Keehn et al. 2013). It further appears that whilst neurotypical infants showed increased functional connectivity with age, a decrease is observed in infant siblings. No differences were observed between groups during 6- and 9-month observation, which warrants further investigation. Further, other studies in 14-month-old infants with later ASD have shown that *over*-connectivity in the alpha frequency range predicts later ASD symptoms (Orekhova et al. 2014; Haartsen et al. 2019). These apparently opposing findings of connectivity at 12 and 14 months may be due to differences in brain imaging methods, stimulus, and sample size used. Further, these studies may be addressing different underlying mechanisms (i.e., blood oxygenation level vs. electrophysiological activity in the brain). Further, outcome data was not available for the fNIRS cohort at the time of publication, and thus follow up studies to examine the subgroup of infants with later ASD is needed to better understand the differences in results across these studies.

Protective Factors

Decades of genetic and epidemiological research showed that a combination of genetic and environmental risk factors can be associated with increased ASD likelihood (Chaste and Leboyer 2012; Gardener et al. 2011). At present, there is insufficient evidence to implicate a single pre- or postnatal factor in etiology of ASD, although it appears that complex bidirectional interactions between several genetic, epigenetic, and environmental factors may affect likelihood and severity of the condition. Protective factors are individual, relational, and contextual variables that improve long-term outcomes of neurodevelopmental conditions despite early developmental risk (Szatmari 2018).

Several protective factors have been associated with less severe ASD phenotypes, including right-hemisphere lateralization and processing of faces outside the FFA, which are discussed above (Finch et al. 2017; Pierce et al. 2001; Seery et al. 2013). Enhanced social attention has been proposed as a protective factor in infant siblings (Chawarska et al. 2016). Between 6 and 12 months, females with elevated familial likelihood of ASD ($n = 35$) showed enhanced social attention to speaking faces, which were related to less severe social impairments at age 2 (Social Attention subscale of the ADOS[2]) relative to males in the same group and typically developing female and male controls ($n = 61$). The authors interpret this data as showing that social attention is a protective factor, as it is associated with less severe manifestation of socio-communicative abilities.

Resilience factors may also be external to the child. For example, parental coping strategies and positive sibling relationships appear to reduce the overall reported difficulty associated with ASD symptoms (Rivers and Stoneman 2003), which suggests that parent-targeted interventions may be highly appropriate to support development in core areas (Hatamzadeh et al. 2010; Masse et al. 2016; Solomon et al. 2008) and to partially prevent the emergence of later associated mental health conditions (Hastings and Brown 2002; Walton and Ingersol 2015). This is supported by some early promising data from parent-mediated intervention studies in infant siblings (Green et al. 2015, 2017). Lastly, studies of toddlers and children with ASD found that enhanced EF abilities were associated with less severe symptomatology, assessed through the ADOS and standardized parent-report questionnaires (McCrimmon et al. 2016; Pugliese et al. 2015).

Taking a systems neuroscience approach, Johnson and colleagues (2021) conceptualized this system in the AMEND framework, where early stage processing or "predictors" are dissociated from later neurocognitive modifiers, i.e., brain systems that push individual developmental trajectories towards a particular behavioral outcome. Early atypicalities in emerging sensory motor/posterior regions are associated with risk for later developing autism, which is consistent with existing literature. Anterior systems, on the other hand, are later developing, and may be able to compensate or compound the detrimental effects of atypical sensory/motor processing. As a result, we may see dysregulation in executive attention or intellectual functioning as they are working to modify low-level processing difficulties. It was argued that separating early predictors from modifiers is highly useful for understanding trajectories, mitigating risk, and understanding of the dimensional nature of ASD and co-occurring conditions.

Conclusion/Summary

In this chapter, we highlighted the importance of studying early markers of cognition longitudinally. We explored early cognitive profiles of infants and children with ASD and considered the underlying cellular and whole-brain mechanisms that may be involved. Advances in brain imaging technology have allowed us to identify several candidate markers predictive of later developmental outcome prior to emergence of behavioral symptoms, yet there are several gaps in the literature that need to be addressed. Differences in participant ages, diagnostic criteria, sample sizes, and experimental methodologies used are largely varied across studies, which has resulted in somewhat poor replicability in the literature. Therefore, it is important to reflect on the theoretical and conceptual models available, and whether they are suitable to address the complexity and dynamic nature of human development and cognition. In future research, it is necessary to address the specificity of putative behavioral, cognitive, and neural markers to ASD relative to other pervasive developmental conditions and general developmental delay as well as generate multi-methodological and computational modelling studies that integrate across different levels of analysis to better understand the gene-brain-behavior relationships, while embracing complexity.

NOTES

1. Still Face Effect—a behavioral paradigm which examines reciprocal social interaction. During the task, the mother engages with the child for a predetermined time, then is asked to keep a still face for 1 minute, which is followed by another period of reciprocal interaction. These pre- and post-Still Face interaction episodes have been found to elicit an established pattern of behavior, including smiling, gaze aversion, and negative *affect* (Gusella et al. 1988; Tarabulsy et al. 2003; Toda and Fogel 1993).
2. *Autism Diagnostic Observation Schedule, Second Edn: ADOS-2*, Pearson Assessment (2012); Gotham et al. (2009).

REFERENCES

Adamson, L. B., Bakeman, R., Deckner, D. F., and Nelson, P. B. 2012. Rating parent–child interactions: Joint engagement, communication dynamics, and shared topics in autism, Down syndrome, and typical development. *Journal of Autism and Developmental Disorders*, 42(12): 2622–2635. https://doi.org/10.1007/s10803-012-1520-1

Akhtar, N. and Gernsbacher, M. A. 2007. Joint attention and vocabulary development: A critical look. *Language and Linguistics Compass*, 1(3): 195–207. https://doi.org/10.1111/j.1749-818X.2007.00014.x

Allison, T., Puce, A., and McCarthy, G. 2000. Social perception from visual cues: Role of the STS region. *Trends in Cognitive Sciences*, 4(7): 267–278. https://doi.org/10.1016/S1364-6613(00)01501-1

American Psychiatric Association. 2013. *Diagnostic and Statistical Manual of Mental Disorders (DSM-5®)*. Arlington, VA: American Psychiatric Association.

Aslin, R. N. 2012. Infant eyes: A window on cognitive development. *Infancy: The Official Journal of the International Society on Infant Studies*, 17(1): 126–140. https://doi.org/10.1111/j.1532-7078.2011.00097.x

Autism Diagnostic Observation Schedule, Second Edn: ADOS-2 Pearson Assessment. n.d. Retrieved 4 December 2020, from https://www.pearsonclinical.co.uk/Psychology/ChildMentalHealth/ChildMentalHealth/ADOS-2/autism-diagnostic-observation-schedule-second-edition.aspx

Bedford, R., Pickles, A., Gliga, T., Elsabbagh, M., Charman, T., and Johnson, M. H. 2014. Additive effects of social and non-social attention during infancy relate to later autism spectrum disorder. *Developmental Science*, 17(4): 612–620. https://doi.org/10.1111/desc.12139

Bedford, R., Pickles, A., and Lord, C. 2016. Early gross motor skills predict the subsequent development of language in children with autism spectrum disorder. *Autism Research: Official*

Journal of the International Society for Autism Research, 9(9): 993–1001. https://doi.org/10.1002/aur.1587

Ben-Sasson, A. and Carter, A. S. 2013. The contribution of sensory–regulatory markers to the accuracy of ASD screening at 12 months. *Research in Autism Spectrum Disorders*, 7(7): 879–888. https://doi.org/10.1016/j.rasd.2013.03.006

Berkson, G. and Tupa, M. 2000. Early development of stereotyped and self-injurious behaviors. *Journal of Early Intervention*, 23(1): 1–19. https://doi.org/10.1177/105381510002300010401

Beurkens, N. M., Hobson, J. A., and Hobson, R. P. 2013. Autism severity and qualities of parent-child relations. *Journal of Autism and Developmental Disorders*, 43(1): 168–178. https://doi.org/10.1007/s10803-012-1562-4

Bilalić, M., Langner, R., Ulrich, R., and Grodd, W. 2011. Many faces of expertise: Fusiform face area in chess experts and novices. *Journal of Neuroscience*, 31(28): 10206–10214. https://doi.org/10.1523/JNEUROSCI.5727-10.2011

Bishop, S. L., Richler, J., and Lord, C. 2006. Association between restricted and repetitive behaviors and nonverbal IQ in children with autism spectrum disorders. *Child Neuropsychology*, 12(4–5): 247–267. https://doi.org/10.1080/09297040600630288

Blasi, A., Mercure, E., Lloyd-Fox, S., Thomson, A., Brammer, M., Sauter, D., Deeley, Q., Barker, G. J., Renvall, V., Deoni, S., Gasston, D., Williams, S. C. R., Johnson, M. H., Simmons, A., and Murphy, D. G. M. 2011. Early specialization for voice and emotion processing in the infant brain. *Current Biology*, 21(14): 1220–1224. https://doi.org/10.1016/j.cub.2011.06.009

Bodfish, J. W., Symons, F. J., Parker, D. E., and Lewis, M. H. 2000. Varieties of repetitive behavior in autism: Comparisons to mental retardation. *Journal of Autism and Developmental Disorders*, 30(3): 237–243. https://doi.org/10.1023/A:1005596502855

Braukmann, R., Lloyd-Fox, S., Blasi, A., Johnson, M. H., Bekkering, H., Buitelaar, J. K., and Hunnius, S. 2018. Diminished socially selective neural processing in 5-month-old infants at high familial risk of autism. *European Journal of Neuroscience*, 47(6): 720–728. https://doi.org/10.1111/ejn.13751

Brett, D., Warnell, F., McConachie, H., and Parr, J. R. 2016. Factors affecting age at ASD diagnosis in UK: No evidence that diagnosis age has decreased between 2004 and 2014. *Journal of Autism and Developmental Disorders*, 46: 1974–1984. https://doi.org/10.1007/s10803-016-2716-6

Brian, J. A., Bryson, S. E., and Zwaigenbaum, L. 2015. Autism spectrum disorder in infancy: Developmental considerations in treatment targets. *Current Opinion in Neurology*, 28(2): 117–123. https://doi.org/10.1097/WCO.0000000000000182

Bryson, S. E., Zwaigenbaum, L., Brian, J., Roberts, W., Szatmari, P., Rombough, V., and McDermott, C. 2007. A prospective case series of high-risk infants who developed autism. *Journal of Autism and Developmental Disorders*, 37(1): 12–24. https://doi.org/10.1007/s10803-006-0328-2

Buzsáki, G. and Wang, X.-J. 2012. Mechanisms of gamma oscillations. *Annual Review of Neuroscience*, 35: 203–225. https://doi.org/10.1146/annurev-neuro-062111-150444

Carpenter, M., Nagell, K., and Tomasello, M. 1998. Social cognition, joint attention, and communicative competence from 9 to 15 months of age. *Monographs of the Society for Research in Child Development*, 63(4): i–vi, 1–143.

Charman, T. and Stone, W. 2008. *Social and Communication Development in Autism Spectrum Disorders: Early Identification, Diagnosis, and Intervention*. New York: Guilford Press.

Charman, T., Swettenham, J., Baron-Cohen, S., Cox, A., Baird, G., and Drew, A. 1997. Infants with autism: An investigation of empathy, pretend play, joint attention, and imitation. *Developmental Psychology*, 33(5): 781–789. https://doi.org/10.1037/0012-1649.33.5.781

Charman, T., Taylor, E., Drew, A., Cockerill, H., Brown, J.-A., and Baird, G. 2005. Outcome at 7 years of children diagnosed with autism at age 2: Predictive validity of assessments conducted at 2 and 3 years of age and pattern of symptom change over time. *Journal of Child Psychology and Psychiatry*, 46(5): 500–513. https://doi.org/10.1111/j.1469-7610.2004.00377.x

Chaste, P. and Leboyer, M. 2012. Autism risk factors: Genes, environment, and gene-environment interactions. *Dialogues in Clinical Neuroscience*, 14(3): 281–292.

Chawarska, K., Macari, S., Powell, K., DiNicola, L., and Shic, F. 2016. Enhanced social attention in female infant siblings at risk for autism. *Journal of the American Academy of Child & Adolescent Psychiatry*, 55(3): 188–195. https://doi.org/10.1016/j.jaac.2015.11.016

Chawarska, K., Macari, S., and Shic, F. 2012. Context modulates attention to social scenes in toddlers with autism. *Journal of Child Psychology and Psychiatry*, 53(8): 903–913. https://doi.org/10.1111/j.1469-7610.2012.02538.x

Chawarska, K., Macari, S., and Shic, F. 2013. Decreased spontaneous attention to social scenes in 6-month-old infants later diagnosed with autism spectrum disorders. *Biological Psychiatry*, 74(3): 195–203. https://doi.org/10.1016/j.biopsych.2012.11.022

Chen, Y.-H., Rodgers, J., and McConachie, H. 2009. Restricted and repetitive behaviours, sensory processing and cognitive style in children with autism spectrum disorders. *Journal of Autism and Developmental Disorders*, 39(4): 635–642. https://doi.org/10.1007/s10803-008-0663-6

Cheung, C. H. M., Bedford, R., Johnson, M. H., Charman, T., and Gliga, T. 2018. Visual search performance in infants associates with later ASD diagnosis. *Developmental Cognitive Neuroscience*, 29: 4–10. https://doi.org/10.1016/j.dcn.2016.09.003

Chevallier, C., Kohls, G., Troiani, V., Brodkin, E. S., and Schultz, R. T. 2012. The social motivation theory of autism. *Trends in Cognitive Sciences*, 16(4): 231–239. https://doi.org/10.1016/j.tics.2012.02.007

Colombo, J. 2001. The development of visual attention in infancy. *Annual Review of Psychology*, 52(1): 337–367. https://doi.org/10.1146/annurev.psych.52.1.337

Courchesne, E., Yeung-Courchesne, R., Hesselink, J. R., and Jernigan, T. L. 1988. Hypoplasia of cerebellar vermal lobules VI and VII in autism. *New England Journal of Medicine*, 318(21): 1349–1354. https://doi.org/10.1056/NEJM198805263182102

Courchesne, Eric, Carper, R., and Akshoomoff, N. 2003. Evidence of brain overgrowth in the first year of life in autism. *JAMA*, 290(3): 337–344. https://doi.org/10.1001/jama.290.3.337

Crawley, D., Zhang, L., Jones, E. J. H., Ahmad, J., Oakley, B., Cáceres, A. S. J., Charman, T., Buitelaar, J. K., Murphy, D. G. M., Chatham, C., Ouden, H. den, Loth, E., and Group, the E.-A. L. 2020. Modeling flexible behavior in childhood to adulthood shows age-dependent learning mechanisms and less optimal learning in autism in each age group. *PLOS Biology*, 18(10): e3000908. https://doi.org/10.1371/journal.pbio.3000908

Daluwatte, C., Miles, J. H., Sun, J., and Yao, G. 2015. Association between pupillary light reflex and sensory behaviors in children with autism spectrum disorders. *Research in Developmental Disabilities*, 37: 209–215. https://doi.org/10.1016/j.ridd.2014.11.019

Dawson, G., Bernier, R., and Ring, R. H. 2012. Social attention: A possible early indicator of efficacy in autism clinical trials. *Journal of Neurodevelopmental Disorders*, 4(1): 11. https://doi.org/10.1186/1866-1955-4-11

Dawson, G., Carver, L., Meltzoff, A. N., Panagiotides, H., McPartland, J., and Webb, S. J. 2002. Neural correlates of face and object recognition in young children with autism spectrum disorder, developmental delay, and typical development. *Child Development*, 73(3): 700–717. https://doi.org/10.1111/1467-8624.00433

Dawson, G., Finley, C., Phillips, S., and Galpert, L. 1986. Hemispheric specialization and the language abilities of autistic children. *Child Development*, 57(6): 1440–1453. JSTOR. https://doi.org/10.2307/1130422

Dawson, G., Hill, D., Spencer, A., Galpert, L., and Watson, L. 1990. Affective exchanges between young autistic children and their mothers. *Journal of Abnormal Child Psychology*, 18(3): 335–345. https://doi.org/10.1007/BF00916569

de Haan, M., Johnson, M. H., and Halit, H. 2003. Development of face-sensitive event-related potentials during infancy: A review. *International Journal of Psychophysiology*, 51(1): 45–58. https://doi.org/10.1016/S0167-8760(03)00152-1

de Haan, M. and Nelson, C. A. 1997. Recognition of the mother's face by six-month-old infants: A neurobehavioral study. *Child Development*, 68(2): 187–210. https://doi.org/10.2307/1131845

Deen, B., Richardson, H., Dilks, D. D., Takahashi, A., Keil, B., Wald, L. L., Kanwisher, N., and Saxe, R. 2017. Organization of high-level visual cortex in human infants. *Nature Communications*, 8(1): 13995. https://doi.org/10.1038/ncomms13995

DeLoache, J. S., de Mendoza, O. A. P., and Anderson, K. N. 1999. Multiple factors in early symbol use: Instructions, similarity, and age in understanding a symbol-referent relation. *Cognitive Development*, 14(2): 299–312. https://doi.org/10.1016/S0885-2014(99)00006-4

Dickinson, A., Jones, M., and Milne, E. 2016. Measuring neural excitation and inhibition in autism: Different approaches, different findings and different interpretations. *Brain Research*, 1648: 277–289. https://doi.org/10.1016/j.brainres.2016.07.011

Donkers, F. C. L., Schipul, S. E., Baranek, G. T., Cleary, K. M., Willoughby, M. T., Evans, A. M., Bulluck, J. C., Lovmo, J. E., and Belger, A. 2015. Attenuated auditory event-related potentials and associations with atypical sensory response patterns in children with autism. *Journal of Autism and Developmental Disorders*, 45(2): 506–523. https://doi.org/10.1007/s10803-013-1948-y

Elsabbagh, M., Fernandes, J., Webb, S. J., Dawson, G., Charman, T., and Johnson, M. H. 2013. Disengagement of visual attention in infancy is associated with emerging autism in toddlerhood. *Biological Psychiatry*, 74(3): 189–194. https://doi.org/10.1016/j.biopsych.2012.11.030

Elsabbagh, M., Mercure, E., Hudry, K., Chandler, S., Pasco, G., Charman, T., Pickles, A., Baron-Cohen, S., Bolton, P., and Johnson, M. H. 2012. Infant neural sensitivity to dynamic eye gaze is associated with later emerging autism. *Current Biology*, 22(4): 338–342. https://doi.org/10.1016/j.cub.2011.12.056

Elsabbagh, M., Volein, A., Holmboe, K., Tucker, L., Csibra, G., Baron-Cohen, S., Bolton, P., Charman, T., Baird, G., and Johnson, M. H. 2009. Visual orienting in the early broader autism phenotype: Disengagement and facilitation. *Journal of Child Psychology and Psychiatry*, 50(5): 637–642. https://doi.org/10.1111/j.1469-7610.2008.02051.x

Emberson, L. L., Richards, J. E., and Aslin, R. N. 2015. Top-down modulation in the infant brain: Learning-induced expectations rapidly affect the sensory cortex at 6 months. *Proceedings of the National Academy of Sciences*, 112(31): 9585–9590. https://doi.org/10.1073/pnas.1510343112

Eriksson, M., Marschik, P. B., Tulviste, T., Almgren, M., Pereira, M. P., Wehberg, S., Marjanovič-Umek, L., Gayraud, F., Kovacevic, M., and Gallego, C. 2012. Differences between girls and boys in emerging language skills: Evidence from 10 language communities. *British Journal of Developmental Psychology*, 30(2): 326–343. https://doi.org/10.1111/j.2044-835X.2011.02042.x

Evans, D. W., Leckman, J. F., Carter, A., Reznick, J. S., Henshaw, D., King, R. A., and Pauls, D. 1997. Ritual, habit, and perfectionism: The prevalence and development of compulsive-like behavior in normal young children. *Child Development*, 68(1): 58–68. https://doi.org/10.1111/j.1467-8624.1997.tb01925.x

Eyler, L. T., Pierce, K., and Courchesne, E. 2012. A failure of left temporal cortex to specialize for language is an early emerging and fundamental property of autism. *Brain*, 135(3): 949–960. https://doi.org/10.1093/brain/awr364

Falck-Ytter, T., Bakker, M., and von Hofsten, C. 2011. Human infants orient to biological motion rather than audiovisual synchrony. *Neuropsychologia*, 49(7): 2131–2135. https://doi.org/10.1016/j.neuropsychologia.2011.03.040

Falck-Ytter, T., Nyström, P., Gredebäck, G., Gliga, T., Bölte, S., and EASE team. 2018. Reduced orienting to audiovisual synchrony in infancy predicts autism diagnosis at 3 years of age. *Journal of Child Psychology and Psychiatry, and Allied Disciplines*, 59(8): 872–880. https://doi.org/10.1111/jcpp.12863

Falck-Ytter, T., Rehnberg, E., and Bölte, S. 2013. Lack of visual orienting to biological motion and audiovisual synchrony in 3-year-olds with autism. *PLoS ONE*, 8(7): e68816. https://doi.org/10.1371/journal.pone.0068816

Faraone, S. V. and Larsson, H. 2019. Genetics of attention deficit hyperactivity disorder. *Molecular Psychiatry*, 24(4): 562–575. https://doi.org/10.1038/s41380-018-0070-0

Farroni, T., Massaccesi, S., Pividori, D., and Johnson, M. H. 2004. Gaze following in newborns. *Infancy*, 5(1): 39–60. https://doi.org/10.1207/s15327078in0501_2

Finch, K. H., Seery, A. M., Talbott, M. R., Nelson, C. A., and Tager-Flusberg, H. 2017. Lateralization of ERPs to speech and handedness in the early development of Autism Spectrum Disorder. *Journal of Neurodevelopmental Disorders*, 9: 4. https://doi.org/10.1186/s11689-017-9185-x

Fischer, J., Smith, H., Martinez Pedraza, F., Carter, A. S., Kanwisher, N., and Kaldy, Z. 2016. Unimpaired attentional disengagement in toddlers with autism spectrum disorder. *Developmental Science*, 19(6): 1095–1103. https://doi.org/10.1111/desc.12386

Foss-Feig, J. H., McGugin, R. W., Gauthier, I., Mash, L. E., Ventola, P., and Cascio, C. J. 2016. A functional neuroimaging study of fusiform response to restricted interests in children and adolescents with autism spectrum disorder. *Journal of Neurodevelopmental Disorders*, 8(1): 1–12. https://doi.org/10.1186/s11689-016-9149-6

Franchini, M., Armstrong, V. L., Schaer, M., and Smith, I. M. 2019. Initiation of joint attention and related visual attention processes in infants with autism spectrum disorder: Literature review. *Child Neuropsychology: A Journal on Normal and Abnormal Development in Childhood and Adolescence*, 25(3): 287–317. https://doi.org/10.1080/09297049.2018.1490706

Franchini, M., Glaser, B., Wood de Wilde, H., Gentaz, E., Eliez, S., and Schaer, M. 2017. Social orienting and joint attention in preschoolers with autism spectrum disorders. *PLoS ONE*, 12(6): e0178859. https://doi.org/10.1371/journal.pone.0178859

Freeth, M., Foulsham, T., and Kingstone, A. 2013. What affects social attention? Social presence, eye contact and autistic traits. *PLoS ONE*, 8(1). https://doi.org/10.1371/journal.pone.0053286

Gabriels, R. L., Agnew, J. A., Miller, L. J., Gralla, J., Pan, Z., Goldson, E., Ledbetter, J. C., Dinkins, J. P., and Hooks, E. 2008. Is there a relationship between restricted, repetitive, stereotyped behaviors and interests and abnormal sensory response in children with autism spectrum disorders? *Research in Autism Spectrum Disorders*, 2(4): 660–670. https://doi.org/10.1016/j.rasd.2008.02.002

Gaetz, W., Bloy, L., Wang, D. J., Port, R. G., Blaskey, L., Levy, S. E., and Roberts, T. P. L. 2014. GABA estimation in the brains of children on the autism spectrum: Measurement precision and regional cortical variation. *NeuroImage*, 86: 1–9. https://doi.org/10.1016/j.neuroimage.2013.05.068

Gandal, M. J., Haney, J. R., Parikshak, N. N., Leppa, V., Ramaswami, G., Hartl, C., Schork, A. J., Appadurai, V., Buil, A., Werge, T. M., Liu, C., White, K. P., CommonMind

Consortium, PsychENCODE Consortium, iPSYCH-BROAD Working Group, Horvath, S., and Geschwind, D. H. 2018. Shared molecular neuropathology across major psychiatric disorders parallels polygenic overlap. *Science* (New York, N.Y.), 359(6376): 693–697. https://doi.org/10.1126/science.aad6469

Gao, R. and Penzes, P. 2015. Common mechanisms of excitatory and inhibitory imbalance in schizophrenia and autism spectrum disorders. *Current Molecular Medicine*, 15(2): 146–167.

Gardener, H., Spiegelman, D., and Buka, S. L. 2011. Perinatal and neonatal risk factors for autism: A comprehensive meta-analysis. *Pediatrics*, 128(2): 344–355. https://doi.org/10.1542/peds.2010-1036

Germani, T., Zwaigenbaum, L., Bryson, S., Brian, J., Smith, I., Roberts, W., Szatmari, P., Roncadin, C., Sacrey, L. A. R., Garon, N., and Vaillancourt, T. 2014. Brief report: Assessment of early sensory processing in infants at high-risk of autism spectrum disorder. *Journal of Autism and Developmental Disorders*, 44(12): 3264–3270. https://doi.org/10.1007/s10803-014-2175-x

Glod, M., Riby Deborah M., Honey Emma, and Rodgers Jacqui. 2016. Sensory atypicalities in dyads of children with autism spectrum disorder (ASD) and their parents. *Autism Research*, 10(3): 531–538. https://doi.org/10.1002/aur.1680

Gotham, K., Pickles, A., and Lord, C. 2009. Standardizing ADOS scores for a measure of severity in autism spectrum disorders. *Journal of Autism and Developmental Disorders*, 39(5): 693–705. https://doi.org/10.1007/s10803-008-0674-3

Gottlieb, G. 2007. Probabilistic epigenesis. *Developmental Science*, 10(1): 1–11. https://doi.org/10.1111/j.1467-7687.2007.00556.x

Green, J., Charman, T., Pickles, A., Wan, M. W., Elsabbagh, M., Slonims, V., Taylor, C., McNally, J., Booth, R., Gliga, T., Jones, E. J. H., Harrop, C., Bedford, R., and Johnson, M. H. 2015. Parent-mediated intervention versus no intervention for infants at high risk of autism: A parallel, single-blind, randomised trial. *The Lancet Psychiatry*, 2(2): 133–140. https://doi.org/10.1016/S2215-0366(14)00091-1

Green, J., Pickles, A., Pasco, G., Bedford, R., Wan, M. W., Elsabbagh, M., Slonims, V., Gliga, T., Jones, E., Cheung, C., Charman, T., and Johnson, M. 2017. Randomised trial of a parent-mediated intervention for infants at high risk for autism: Longitudinal outcomes to age 3 years. *Journal of Child Psychology and Psychiatry*, 58(12): 1330–1340. https://doi.org/10.1111/jcpp.12728

Grelotti, D. J., Klin, A. J., Gauthier, I., Skudlarski, P., Cohen, D. J., Gore, J. C., Volkmar, F. R., and Schultz, R. T. 2005. FMRI activation of the fusiform gyrus and amygdala to cartoon characters but not to faces in a boy with autism. *Neuropsychologia*, 43(3): 373–385. https://doi.org/10.1016/j.neuropsychologia.2004.06.015

Griffith, E. M., Pennington, B. F., Wehner, E. A., and Rogers, S. J. 1999. Executive functions in young children with autism. *Child Development*, 70(4): 817–832. https://doi.org/10.1111/1467-8624.00059

Groen, W. B., Zwiers, M. P., van der Gaag, R.-J., and Buitelaar, J. K. 2008. The phenotype and neural correlates of language in autism: An integrative review. *Neuroscience & Biobehavioral Reviews*, 32(8): 1416–1425. https://doi.org/10.1016/j.neubiorev.2008.05.008

Grove, J., Ripke, S., Als, T. D., Mattheisen, M., Walters, R. K., Won, H., Pallesen, J., Agerbo, E., Andreassen, O. A., Anney, R., Awashti, S., Belliveau, R., Bettella, F., Buxbaum, J. D., Bybjerg-Grauholm, J., Bækvad-Hansen, M., Cerrato, F., Chambert, K., Christensen, J. H., … and Børglum, A. D. 2019. Identification of common genetic risk variants for autism spectrum disorder. *Nature Genetics*, 51(3): 431–444. https://doi.org/10.1038/s41588-019-0344-8

Guillon, Q., Hadjikhani, N., Baduel, S., and Rogé, B. 2014. Visual social attention in autism spectrum disorder: Insights from eye tracking studies. *Neuroscience & Biobehavioral Reviews*, 42: 279–297. https://doi.org/10.1016/j.neubiorev.2014.03.013

Gusella, J. L., Muir, D., and Tronick, E. Z. 1988. The effect of manipulating maternal behavior during an interaction on three- and six-month-olds' affect and attention. *Child Development*, 59(4): 1111–1124. https://doi.org/10.1111/j.1467-8624.1988.tb03264.x

Haartsen, R., Jones, E. J., and Johnson, M. H. 2016. Human brain development over the early years. *Current Opinion in Behavioral Sciences*, 10: 149–154. https://doi.org/10.1016/j.cob eha.2016.05.015

Haartsen, R., Jones, E. J. H., Orekhova, E. V., Charman, T., and Johnson, M. H. 2019. Functional EEG connectivity in infants associates with later restricted and repetitive behaviours in autism; a replication study. *Translational Psychiatry*, 9(1): 1–14. https://doi.org/10.1038/s41 398-019-0380-2

Harker, C. M., Ibañez, L. V., Nguyen, T. P., Messinger, D. S., and Stone, W. L. 2016. The effect of parenting style on social smiling in infants at high and low risk for ASD. *Journal of Autism and Developmental Disorders*, 46(7): 2399–2407. https://doi.org/10.1007/s10 803-016-2772-y

Hastings, R. P. and Brown, T. 2002. Behavior problems of children with autism, parental self-efficacy, and mental health. *American Journal on Mental Retardation*, 107(3): 222–232. https://doi.org/10.1352/0895-8017(2002)107<0222:BPOCWA>2.0.CO;2

Hatamzadeh, A., Pouretemad, H., and Hassanabadi, H. 2010. The effectiveness of parent–child interaction therapy for children with high functioning autism. *Procedia – Social and Behavioral Sciences*, 5: 994–997. https://doi.org/10.1016/j.sbspro.2010.07.224

Haxby, J. V., Hoffman, E. A., and Gobbini, M. I. 2002. Human neural systems for face recognition and social communication. *Biological Psychiatry*, 51(1): 59–67. https://doi.org/10.1016/S0006-3223(01)01330-0

Hazlett, H. C., Gu, H., Munsell, B. C., Kim, S. H., Styner, M., Wolff, J. J., Elison, J. T., Swanson, M. R., Zhu, H., Botteron, K. N., Collins, D. L., Constantino, J. N., Dager, S. R., Estes, A. M., Evans, A. C., Fonov, V. S., Gerig, G., Kostopoulos, P., McKinstry, R. C., ... and The IBIS Network. 2017. Early brain development in infants at high risk for autism spectrum disorder. *Nature*, 542(7641): 348–351. https://doi.org/10.1038/nature21369

Hazlett, H. C., Poe, M., Gerig, G., Styner, M., Chappell, C., Smith, R. G., Vachet, C., and Piven, J. 2011. Early brain overgrowth in autism associated with an increase in cortical surface area before age 2. *Archives of General Psychiatry*, 68(5): 467–476. https://doi.org/10.1001/archge npsychiatry.2011.39

Heaton, P., Hudry, K., Ludlow, A., and Hill, E. 2008. Superior discrimination of speech pitch and its relationship to verbal ability in autism spectrum disorders. *Cognitive Neuropsychology*, 25(6): 771–782. https://doi.org/10.1080/02643290802336277

Heil, K. M. and Schaaf, C. P. 2012. The genetics of autism spectrum disorders – a guide for clinicians. *Current Psychiatry Reports*, 15(1): 334. https://doi.org/10.1007/s11920-012-0334-3

Hendry, A., Jones, E. J. H., and Charman, T. 2016. Executive function in the first three years of life: Precursors, predictors and patterns. *Developmental Review*, 42: 1–33. https://doi.org/10.1016/j.dr.2016.06.005

Hepach, R. and Westermann, G. 2016. Pupillometry in infancy research. *Journal of Cognition and Development*, 17(3): 359–377. https://doi.org/10.1080/15248372.2015.1135801

Holmboe, K., Pasco Fearon, R. M., Csibra, G., Tucker, L. A., and Johnson, M. H. 2008. Freeze-Frame: A new infant inhibition task and its relation to frontal cortex tasks during infancy and early childhood. *Journal of Experimental Child Psychology*, 100(2): 89–114. https://doi. org/10.1016/j.jecp.2007.09.004

Honey, E., Leekam, S., Turner, M., and McConachie, H. 2007. Repetitive behaviour and play in typically developing children and children with autism spectrum disorders. *Journal of Autism and Developmental Disorders*, 37(6): 1107–1115. https://doi.org/10.1007/s10 803-006-0253-4

Honey, E., McConachie, H., Randle, V., Shearer, H., and Couteur, A. S. L. 2008. One-year change in repetitive behaviours in young children with communication disorders including autism. *Journal of Autism and Developmental Disorders*, 38(8): 1439–1450. https://doi.org/ 10.1007/s10803-006-0191-1

Hornig, M., Bresnahan, M. A., Che, X., Schultz, A. F., Ukaigwe, J. E., Eddy, M. L., Hirtz, D., Gunnes, N., Lie, K. K., Magnus, P., Mjaaland, S., Reichborn-Kjennerud, T., Schjolberg, S., Oyen, A.-S., Levin, B., Susser, E. S., Stoltenberg, C., and Lipkin, W. I. 2017. Prenatal fever and autism risk. *Molecular Psychiatry*. http://dx.doi.org/10.1038/mp.2017.119

Hughes, C. 2002. Executive functions and development: Emerging themes. *Infant and Child Development*, 11(2): 201–209. https://doi.org/10.1002/icd.297

Huguet, G., Ey, E., and Bourgeron, T. 2013. The genetic landscapes of autism spectrum disorders. *Annual Review of Genomics and Human Genetics*, 14(1): 191–213. https://doi.org/ 10.1146/annurev-genom-091212-153431

Hussman, J. P. 2001. Letters to the editor: Suppressed GABAergic Inhibition as a common factor in suspected etiologies of autism. *Journal of Autism and Developmental Disorders*, 31(2): 247–248. https://doi.org/10.1023/A:1010715619091

Iverson, J. M. 2010. Developing language in a developing body: The relationship between motor development and language development. *Journal of Child Language*, 37(2): 229–261. https://doi.org/10.1017/S0305000909990432

Johnson, Mark H. 2000. Functional brain development in infants: Elements of an interactive specialization framework. *Child Development*, 71(1): 75–81. https://doi.org/10.1111/ 1467-8624.00120

Johnson, Mark H. 2011 Interactive Specialization: A domain-general framework for human functional brain development? *Developmental Cognitive Neuroscience*, 1(1): 7–21. https://doi. org/10.1016/j.dcn.2010.07.003

Johnson, Mark H. 2014. Autism: Demise of the innate social orienting hypothesis. *Current Biology*, 24(1): R30–R31. https://doi.org/10.1016/j.cub.2013.11.021

Johnson, Mark H., Posner, M. I., and Rothbart, M. K. 1991. Components of visual orienting in early infancy: Contingency learning, anticipatory looking, and disengaging. *Journal of Cognitive Neuroscience*, 3(4): 335–344. https://doi.org/10.1162/jocn.1991.3.4.335

Johnson, M. H., Charman, T., Pickles, A., and Jones, E. J. H. 2021. Annual research review: Anterior Modifiers in the Emergence of Neurodevelopmental Disorders (AMEND)—a systems neuroscience approach to common developmental disorders. *Journal of Child Psychology and Psychiatry*, 62(5): 610–630. https://doi.org/10.1111/jcpp.13372

Johnson, N. L., Bekhet, A., Robinson, K., and Rodriguez, D. 2014. Attributed meanings and strategies to prevent challenging behaviors of hospitalized children with autism: Two perspectives. *Journal of Pediatric Health Care*, 28(5): 386–393. https://doi.org/10.1016/ j.pedhc.2013.10.001

Jones, E. J. H, Dawson, G., and Webb, S. J. 2018. Sensory hypersensitivity predicts enhanced attention capture by faces in the early development of ASD. *Developmental Cognitive Neuroscience*. https://doi.org/10.1016/j.dcn.2017.04.001

Jones, Emily J. H., Gliga, T., Bedford, R., Charman, T., and Johnson, M. H. 2014. Developmental pathways to autism: A review of prospective studies of infants at risk. *Neuroscience & Biobehavioral Reviews*, 39: 1–33. https://doi.org/10.1016/j.neubiorev.2013.12.001

Jones, W. and Klin, A. 2013. Attention to eyes is present but in decline in 2-6-month-old infants later diagnosed with autism. *Nature*, 504(7480): 427–431. https://doi.org/10.1038/nature12715

Joosten, A. V. and Bundy, A. C. 2010. Sensory processing and stereotypical and repetitive behaviour in children with autism and intellectual disability. *Australian Occupational Therapy Journal*, 57(6): 366–372. https://doi.org/10.1111/j.1440-1630.2009.00835.x

Joseph, R. M., Keehn, B., Connolly, C., Wolfe, J. M., and Horowitz, T. S. 2009. Why is visual search superior in autism spectrum disorder? *Developmental Science*, 12(6): 1083–1096. https://doi.org/10.1111/j.1467-7687.2009.00855.x

Kang, E., Keifer, C. M., Levy, E. J., Foss-Feig, J. H., McPartland, J. C., and Lerner, M. D. 2018. Atypicality of the N170 event-related potential in autism spectrum disorder: A meta-analysis. *Biological Psychiatry: Cognitive Neuroscience and Neuroimaging*, 3(8): 657–666. https://doi.org/10.1016/j.bpsc.2017.11.003

Kanwisher, N. and Yovel, G. 2006. The fusiform face area: A cortical region specialized for the perception of faces. *Philosophical Transactions of the Royal Society B: Biological Sciences*, 361(1476): 2109–2128. https://doi.org/10.1098/rstb.2006.1934

Kapp, S. K., Goldknopf, E., Brooks, P. J., Kofner, B., and Hossain, M. 2019. Expanding the critique of the social motivation theory of autism with participatory and developmental research. *Behavioral and Brain Sciences*, 42: e94. https://doi.org/10.1017/S0140525X18002479

Karimi, P., Kamali, E., Mousavi, S. M., and Karahmadi, M. 2017. Environmental factors influencing the risk of autism. *Journal of Research in Medical Sciences: The Official Journal of Isfahan University of Medical Sciences*, 22: 27. https://doi.org/10.4103/1735-1995.200272

Karmiloff-Smith, A. 2009. Nativism versus neuroconstructivism: Rethinking the study of developmental disorders. *Developmental Psychology*, 45(1): 56–63. https://doi.org/10.1037/a0014506

Katarzyna, C., Fred, V., and Ami, K. 2010. Limited attentional bias for faces in toddlers with autism spectrum disorders. *Archives of General Psychiatry*, 67(2): 178–185. https://doi.org/10.1001/archgenpsychiatry.2009.194

Keehn, B., Brenner, L., Palmer, E., Lincoln, A. J., and Müller, R.-A. 2008. Functional brain organization for visual search in ASD. *Journal of the International Neuropsychological Society*, 14(6): 990–1003. https://doi.org/10.1017/S1355617708081356

Keehn, B., Wagner, J., Tager-Flusberg, H., and Nelson, C. A. 2013. Functional connectivity in the first year of life in infants at-risk for autism: A preliminary near-infrared spectroscopy study. *Frontiers in Human Neuroscience*, 7: 444. https://doi.org/10.3389/fnhum.2013.00444

Key, A. P. F. and Stone, W. L. 2012. Processing of novel and familiar faces in infants at average and high risk for autism. *Developmental Cognitive Neuroscience*, 2(2): 244–255. https://doi.org/10.1016/j.dcn.2011.12.003

Kielinen, M., Rantala, H., Timonen, E., Linna, S.-L., and Moilanen, I. 2004. Associated medical disorders and disabilities in children with autistic disorder: A population-based study. *Autism*, 8(1): 49–60. https://doi.org/10.1177/1362361304040638

Klin, A., Lin, D. J., Gorrindo, P., Ramsay, G., and Jones, W. 2009. Two-year-olds with autism orient to non-social contingencies rather than biological motion. *Nature*, 459(7244): 257–261. https://doi.org/10.1038/nature07868

Knaus, T. A., Silver, A. M., Lindgren, K. A., Hadjikhani, N., and Tager-Flusberg, H. 2008. FMRI activation during a language task in adolescents with ASD. *Journal of the International Neuropsychological Society*, 14(6): 967–979. https://doi.org/10.1017/S1355617708081216

Kolesnik, A., Ali, J. B., Gliga, T., Guiraud, J., Charman, T., Johnson, M. H., and Jones, E. J. H. 2019. Increased cortical reactivity to repeated tones at 8 months in infants with later ASD. *Translational Psychiatry*, 9(1): 46. https://doi.org/10.1038/s41398-019-0393-x

Koul, R. K., Schlosser, R. W., and Sancibrian, S. 2001. Effects of symbol, referent, and instructional variables on the acquisition of aided and unaided symbols by individuals with autism spectrum disorders. *Focus on Autism and Other Developmental Disabilities*, 16(3): 162–169. https://doi.org/10.1177/108835760101600304

Kourassanis-Velasquez, J. and Jones, E. A. 2018. Increasing joint attention in children with autism and their peers. *Behavior Analysis in Practice*, 12(1): 78–94. https://doi.org/10.1007/s40617-018-0228-x

Landa, R. J., Holman, K. C., and Garrett-Mayer, E. 2007. Social and communication development in toddlers with early and later diagnosis of autism spectrum disorders. *Archives of General Psychiatry*, 64(7): 853–864. https://doi.org/10.1001/archpsyc.64.7.853

Landry, O. and Parker, A. 2013. A meta-analysis of visual orienting in autism. *Frontiers in Human Neuroscience*, 7: 833. https://doi.org/10.3389/fnhum.2013.00833

Landry, R. and Bryson, S. E. 2004. Impaired disengagement of attention in young children with autism. *Journal of Child Psychology and Psychiatry*, 45(6): 1115–1122. https://doi.org/10.1111/j.1469-7610.2004.00304.x

Langen, M., Bos, D., Noordermeer, S. D. S., Nederveen, H., van Engeland, H., and Durston, S. 2014. Changes in the development of striatum are involved in repetitive behavior in autism. *Biological Psychiatry*, 76(5): 405–411. https://doi.org/10.1016/j.biopsych.2013.08.013

Lee, E., Lee, J., and Kim, E. 2016. Excitation/inhibition imbalance in animal models of autism spectrum disorders. *Biological Psychiatry*, 81(10): 838–847. https://doi.org/10.1016/j.biopsych.2016.05.011

Leekam, S. R., Nieto, C., Libby, S. J., Wing, L., and Gould, J. 2007. Describing the sensory abnormalities of children and adults with autism. *Journal of Autism and Developmental Disorders*, 37(5): 894–910. https://doi.org/10.1007/s10803-006-0218-7

Leekam, S. R., Prior, M. R., and Uljarevic, M. 2011. Restricted and repetitive behaviors in autism spectrum disorders: A review of research in the last decade. *Psychological Bulletin*, 137(4): 562–593. https://doi.org/10.1037/a0023341

Leppänen, J. M., Moulson, M. C., Vogel-Farley, V. K., and Nelson, C. A. 2007. An ERP study of emotional face processing in the adult and infant brain. *Child Development*, 78(1): 232–245. https://doi.org/10.1111/j.1467-8624.2007.00994.x

Lloyd-Fox, S., Blasi, A., Elwell, C. E., Charman, T., Murphy, D. G., and Johnson, M. H. 2013. Reduced neural sensitivity to social stimuli in infants at risk for autism. *Proceedings of the Royal Society B*, 280(1758): 20123026. https://doi.org/10.1098/rspb.2012.3026

Lloyd-Fox, S., Blasi, A., Pasco, G., Gliga, T., Jones, E. J. H., Murphy, D. G. M., Elwell, C. E., Charman, T., and Johnson, M. H. 2018. Cortical responses before 6 months of life associate with later autism. *European Journal of Neuroscience*, 47(6): 736–749. https://doi.org/10.1111/ejn.13757

Loh, A., Soman, T., Brian, J., Bryson, S. E., Roberts, W., Szatmari, P., Smith, I. M., and Zwaigenbaum, L. 2007. Stereotyped motor behaviors associated with autism in high-risk infants: A pilot videotape analysis of a sibling sample. *Journal of Autism and Developmental Disorders*, 37(1): 25–36. https://doi.org/10.1007/s10803-006-0333-5

Lopez, B. R., Lincoln, A. J., Ozonoff, S., and Lai, Z. 2005. Examining the relationship between executive functions and restricted, repetitive symptoms of autistic disorder. *Journal of Autism and Developmental Disorders*, 35(4): 445–460. https://doi.org/10.1007/s10803-005-5035-x

Lord, C., and Paul, R. 1997. Language and communication in autism. In D. J. Cohen and F. R. Volkmar (eds.), *Handbook of Autism and Pervasive Developmental Disorders*, 2nd ed., pp. 195–225. New York: Wiley.

Lord, C., Risi, S., and Pickles, A. 2004. Trajectory of Language Development in Autistic Spectrum Disorders. In M. L. Rice and S. F. Warren (eds.), *Developmental language disorders: From phenotypes to etiologies*, pp. 7–29. Mahwah, NJ: Lawrence Erlbaum Associates Publishers.

Loth, E., Spooren, W., and Murphy, D. G. 2014. New treatment targets for autism spectrum disorders: EU-AIMS. *The Lancet Psychiatry*, 1(6): 413–415. https://doi.org/10.1016/S2215-0366(14)00004-2

Loveland, K. A. and Landry, S. H. 1986. Joint attention and language in autism and developmental language delay. *Journal of Autism and Developmental Disorders*, 16(3): 335–349. https://doi.org/10.1007/BF01531663

Luyster, R. J., Powell, C., Tager-Flusberg, H., and Nelson, C. A. 2014. Neural measures of social attention across the first years of life: Characterizing typical development and markers of autism risk. *Developmental Cognitive Neuroscience*, 8: 131–143. https://doi.org/10.1016/j.dcn.2013.09.006

Macari, S. L., Campbell, D., Gengoux, G. W., Saulnier, C. A., Klin, A. J., and Chawarska, K. 2012. Predicting developmental status from 12 to 24 months in infants at risk for autism spectrum disorder: A preliminary report. *Journal of Autism and Developmental Disorders*, 42(12): 2636–2647. https://doi.org/10.1007/s10803-012-1521-0

Mandy, W. P. L., Charman, T., and Skuse, D. H. 2012. Testing the construct validity of proposed criteria for DSM-5 autism spectrum disorder. *Journal of the American Academy of Child & Adolescent Psychiatry*, 51(1): 41–50. https://doi.org/10.1016/j.jaac.2011.10.013

Marshall, P. J. and Fox, N. A. 2005. *The Development of Social Engagement: Neurobiological Perspectives*. Oxford: Oxford University Press.

Masse, J. J., McNeil, C. B., Wagner, S., and Quetsch, L. B. 2016. Examining the efficacy of parent–child interaction therapy with children on the autism spectrum. *Journal of Child and Family Studies*, 25(8): 2508–2525. https://doi.org/10.1007/s10826-016-0424-7

McCrimmon, A. W., Matchullis, R. L., and Altomare, A. A. 2016. Resilience and emotional intelligence in children with high-functioning autism spectrum disorder. *Developmental Neurorehabilitation*, 19(3): 154–161. https://doi.org/10.3109/17518423.2014.927017

McEvoy, R. E., Rogers, S. J., and Pennington, B. F. 1993. Executive function and social communication deficits in young autistic children. *Journal of Child Psychology and Psychiatry, and Allied Disciplines*, 34(4): 563–578. https://doi.org/10.1111/j.1469-7610.1993.tb01036.x

McKinnon, C. J., Eggebrecht, A. T., Todorov, A., Wolff, J. J., Elison, J. T., Adams, C. M., Snyder, A. Z., Estes, A. M., Zwaigenbaum, L., Botteron, K. N., McKinstry, R. C., Marrus, N., Evans, A., Hazlett, H. C., Dager, S. R., Paterson, S. J., Pandey, J., Schultz, R. T., Styner, M. A., ... Pruett, J. R. 2019. Restricted and repetitive behavior and brain functional connectivity in infants at risk for developing autism spectrum disorder. *Biological Psychiatry: Cognitive Neuroscience and Neuroimaging*, 4(1): 50–61. https://doi.org/10.1016/j.bpsc.2018.09.008

Melo, C., Ruano, L., Jorge, J., Ribeiro, T. P., Oliveira, G., Azevedo, L., and Temudo, T. 2019. Prevalence and determinants of motor stereotypies in autism spectrum disorder: A systematic review and meta-analysis. *Autism*, 24(3): 569–590. https://doi.org/10.1177/1362361319869118

Merin, N., Young, G. S., Ozonoff, S., and Rogers, S. J. 2007. Visual fixation patterns during reciprocal social interaction distinguish a subgroup of 6-month-old infants at-risk for autism from comparison infants. *Journal of Autism and Developmental Disorders*, 37(1): 108–121. https://doi.org/10.1007/s10803-006-0342-4

Milne, E., Dunn, S. A., Freeth, M., and Rosas-Martinez, L. 2013. Visual search performance is predicted by the degree to which selective attention to features modulates the ERP between 350 and 600 ms. *Neuropsychologia*, 51(6): 1109–1118. https://doi.org/10.1016/j.neuropsychologia.2013.03.002

Milne, E., Swettenham, J., Hansen, P., Campbell, R., Jeffries, H., and Plaisted, K. 2002. High motion coherence thresholds in children with autism. *Journal of Child Psychology and Psychiatry*, 43(2): 255–263. https://doi.org/10.1111/1469-7610.00018

Moore, V. and Goodson, S. 2003. How well does early diagnosis of autism stand the test of time? Follow-up study of children assessed for autism at age 2 and development of an early diagnostic service. *Autism: The International Journal of Research and Practice*, 7(1): 47–63. https://doi.org/10.1177/1362361303007001005

Morales, M., Mundy, P., and Rojas, J. 1998. Following the direction of gaze and language development in 6-month-olds. *Infant Behavior and Development*, 21(2): 373–377. https://doi.org/10.1016/S0163-6383(98)90014-5

Mundy, P., Sigman, M., and Kasari, C. 1990. A longitudinal study of joint attention and language development in autistic children. *Journal of Autism and Developmental Disorders*, 20(1): 115–128. https://doi.org/10.1007/BF02206861

Nadig, A. and Bang, J. 2017. Parental input to children with ASD and its influence on later language. In L. R. Naigles (ed.), *Innovative Investigations of Language in Autism Spectrum Disorder*, pp. 89–113. Washington, DC: American Psychological Association; Berlin, Germany: Walter de Gruyter GmbH. https://doi.org/10.1037/15964-006

Nelson, C. A. and Monk, C. S. 2001. The use of event-related potentials in the study of cognitive development. In C. A. Nelson and M. Luciana (eds.), *Handbook of Developmental Cognitive Neuroscience*, pp. 125–135. Cambridge, MA: MIT Press.

Nichols, C. M., Ibañez, L. V., Foss-Feig, J. H., and Stone, W. L. 2014. Social smiling and its components in high-risk infant siblings without later ASD symptomatology. *Journal of Autism and Developmental Disorders*, 44(4): 894–902. https://doi.org/10.1007/s10803-013-1944-2

Nyström, P., Gliga, T., Nilsson Jobs, E., Gredebäck, G., Charman, T., Johnson, M. H., Bölte, S., and Falck-Ytter, T. 2018. Enhanced pupillary light reflex in infancy is associated with autism diagnosis in toddlerhood. *Nature Communications*, 9(1): 1–5. https://doi.org/10.1038/s41467-018-03985-4

Nyström, P., Thorup, E., Bölte, S., and Falck-Ytter, T. 2019. Joint attention in infancy and the emergence of autism. *Biological Psychiatry*, 86(8): 631–638. https://doi.org/10.1016/j.biopsych.2019.05.006

Oakes, L. M. 2010. Using habituation of looking time to assess mental processes in infancy. *Journal of Cognition and Development: Official Journal of the Cognitive Development Society*, 11(3): 255–268. https://doi.org/10.1080/15248371003699977

Oakes, L. M. 2012. Advances in eye tracking in infancy research. *Infancy*, 17(1): 1–8. https://doi.org/10.1111/j.1532-7078.2011.00101.x

Orekhova, E. V., Elsabbagh, M., Jones, E. J., Dawson, G., Charman, T., and Johnson, M. H. 2014. EEG hyper-connectivity in high-risk infants is associated with later autism. *Journal of Neurodevelopmental Disorders*, 6: 40. https://doi.org/10.1186/1866-1955-6-40

Orekhova, E. V., Stroganova, T. A., Nygren, G., Tsetlin, M. M., Posikera, I. N., Gillberg, C., and Elam, M. 2007. Excess of high frequency electroencephalogram oscillations in boys with autism. *Biological Psychiatry*, 62(9): 1022–1029. https://doi.org/10.1016/j.biops ych.2006.12.029

Orr, E. and Geva, R. 2015. Symbolic play and language development. *Infant Behavior and Development*, 38: 147–161. https://doi.org/10.1016/j.infbeh.2015.01.002

Ozonoff, S., Iosif, A.-M., Baguio, F., Cook, I. C., Hill, M. M., Hutman, T., Rogers, S. J., Rozga, A., Sangha, S., Sigman, M., Steinfeld, M. B., and Young, G. S. 2010. A prospective study of the emergence of early behavioral signs of autism. *Journal of the American Academy of Child & Adolescent Psychiatry*, 49(3): 256–266.e2. https://doi.org/10.1016/j.jaac.2009.11.009

Ozonoff, S., Macari, S., Young, G. S., Goldring, S., Thompson, M., and Rogers, S. J. 2008. Atypical object exploration at 12 months of age is associated with autism in a prospective sample. *Autism: The International Journal of Research and Practice*, 12(5): 457–472. https://doi.org/10.1177/1362361308096402

Ozonoff, S., Young, G. S., Carter, A., Messinger, D., Yirmiya, N., Zwaigenbaum, L., Bryson, S., Carver, L. J., Constantino, J. N., Dobkins, K., Hutman, T., Iverson, J. M., Landa, R., Rogers, S. J., Sigman, M., and Stone, W. L. 2011. Recurrence risk for autism spectrum disorders: A baby siblings research consortium study. *Pediatrics*, 128(3): e488–e495. https://doi.org/10.1542/peds.2010-2825

Ozonoff, S., Young, G. S., Steinfeld, M. B., Hill, M. M., Cook, I., Hutman, T., Macari, S., Rogers, S. J., and Sigman, M. 2009. How early do parent concerns predict later autism diagnosis? *Journal of Developmental and Behavioral Pediatrics*, 30(5): 367–375.

Pellicano, E. 2007. Links between theory of mind and executive function in young children with autism: Clues to developmental primacy. *Developmental Psychology*, 43(4): 974–990. https://doi.org/10.1037/0012-1649.43.4.974

Pennington, B. F. and Welsh, M. 1995. Neuropsychology and developmental psychopathology. In D. Cicchetti and D. J. Cohen (eds.), *Developmental Psychopathology*, Vol. 1: *Theory and Methods*, pp. 254–290. New York: John Wiley & Sons.

Perlman, S. B., Hudac, C. M., Pegors, T., Minshew, N. J., and Pelphrey, K. A. 2011. Experimental manipulation of face-evoked activity in the fusiform gyrus of individuals with autism. *Social Neuroscience*, 6(1): 22–30. https://doi.org/10.1080/17470911003683185

Petrou, A. M., Parr, J. R., and McConachie, H. 2018. Gender differences in parent-reported age at diagnosis of children with autism spectrum disorder. *Research in Autism Spectrum Disorders*, 50: 32–42. https://doi.org/10.1016/j.rasd.2018.02.003

Pierce, K., Müller, R.-A., Ambrose, J., Allen, G., and Courchesne, E. 2001. Face processing occurs outside the fusiform "face area" in autism: Evidence from functional MRI. *Brain*, 124(10): 2059–2073. https://doi.org/10.1093/brain/124.10.2059

Pierce, K. and Redcay, E. 2008. Fusiform function in children with an autism spectrum disorder is a matter of "Who". *Biological Psychiatry*, 64(7): 552–560. https://doi.org/10.1016/j.biopsych.2008.05.013

Pizzarelli, R. and Cherubini, E. 2011. Alterations of GABAergic signaling in autism spectrum disorders. *Neural Plasticity*, 2011: e297153. https://doi.org/10.1155/2011/297153

Poon, K. K., Watson, L. R., Baranek, G. T., and Poe, M. D. 2012. To what extent do joint attention, imitation, and object play behaviors in infancy predict later communication and intellectual functioning in ASD? *Journal of Autism and Developmental Disorders*, 42(6): 1064–1074. https://doi.org/10.1007/s10803-011-1349-z

Port, R. G., Gaetz, W., Bloy, L., Wang, D.-J., Blaskey, L., Kuschner, E. S., Levy, S. E., Brodkin, E. S., and Roberts, T. P. L. 2017. Exploring the relationship between cortical GABA concentrations, auditory gamma-band responses and development in ASD: Evidence for an altered maturational trajectory in ASD. *Autism Research: Official Journal of the International Society for Autism Research*, 10(4): 593–607. https://doi.org/10.1002/aur.1686

Posner, M. I. and Cohen, Y. 1984. Components of visual orienting. *Attention and Performance X*, 32: 531–556.

Pugliese, C. E., Anthony, L., Strang, J. F., Dudley, K., Wallace, G. L., and Kenworthy, L. 2015. Increasing adaptive behavior skill deficits from childhood to adolescence in autism spectrum disorder: Role of executive function. *Journal of Autism and Developmental Disorders*, 45(6): 1579–1587. https://doi.org/10.1007/s10803-014-2309-1

Puts, N. A. J., Wodka, E. L., Harris, A. D., Crocetti, D., Tommerdahl, M., Mostofsky, S. H., and Edden, R. A. E. 2017. Reduced GABA and altered somatosensory function in children with autism spectrum disorder. *Autism Research*, 10(4): 608–619. https://doi.org/10.1002/aur.1691

Redcay, E. and Courchesne, E. 2008. Deviant functional magnetic resonance imaging patterns of brain activity to speech in 2–3-year-old children with autism spectrum disorder. *Biological Psychiatry*, 64(7): 589–598. https://doi.org/10.1016/j.biopsych.2008.05.020

Rippon, G., Brock, J., Brown, C., and Boucher, J. 2007. Disordered connectivity in the autistic brain: Challenges for the "new psychophysiology." *International Journal of Psychophysiology*, 63(2): 164–172. https://doi.org/10.1016/j.ijpsycho.2006.03.012

Rivers, J. W. and Stoneman, Z. 2003. Sibling relationships when a child has autism: Marital stress and support coping. *Journal of Autism and Developmental Disorders*, 33(4): 383–394. https://doi.org/10.1023/A:1025006727395

Roberts, Jacqueline M. A. 2014. Echolalia and language development in children with autism. In J. Arciuli and J. Brock (eds.), *Communication in Autism*, pp. 53–74. Amsterdam, NL: John Benjamins Publishing Company.

Rogers, S. J., Hepburn, S., and Wehner, E. 2003. Parent reports of sensory symptoms in toddlers with autism and those with other developmental disorders. *Journal of Autism and Developmental Disorders*, 33(6): 631–642. https://doi.org/10.1023/B:JADD.0000006000.38991.a7

Rogers, S. J. and Ozonoff, S. 2005. Annotation: What do we know about sensory dysfunction in autism? A critical review of the empirical evidence. *Journal of Child Psychology and Psychiatry, and Allied Disciplines*, 46(12): 1255–1268. https://doi.org/10.1111/j.1469-7610.2005.01431.x

Rojas, D. C., Becker, K. M., and Wilson, L. B. 2015. Magnetic resonance spectroscopy studies of glutamate and GABA in autism: Implications for excitation-inhibition imbalance theory. *Current Developmental Disorders Reports*, 2(1): 46–57. https://doi.org/10.1007/s40474-014-0032-4

Rojas, D. C., Maharajh, K., Teale, P., and Rogers, S. J. 2008. Reduced neural synchronization of gamma-band MEG oscillations in first-degree relatives of children with autism. *BMC Psychiatry*, 8: 66. https://doi.org/10.1186/1471-244X-8-66

Rozga, A., Hutman, T., Young, G. S., Rogers, S. J., Ozonoff, S., Dapretto, M., and Sigman, M. 2011. Behavioral profiles of affected and unaffected siblings of children with autism: Contribution of measures of mother–infant interaction and nonverbal communication. *Journal of Autism and Developmental Disorders*, 41(3): 287–301. https://doi.org/10.1007/s10803-010-1051-6

Rubenstein, J. L. R. 2010. Three hypotheses for developmental defects that may underlie some forms of autism spectrum disorder. *Current Opinion in Neurology*, 23(2): 118–123. https://doi.org/10.1097/WCO.0b013e328336eb13

Rubenstein, J. L. R. and Merzenich, M. M. 2003. Model of autism: Increased ratio of excitation/inhibition in key neural systems. *Genes, Brain and Behavior*, 2(5): 255–267. https://doi.org/10.1034/j.1601-183X.2003.00037.x

Russell, G., Rodgers, L. R., Ukoumunne, O. C., and Ford, T. 2014. Prevalence of parent-reported ASD and ADHD in the UK: Findings from the Millennium Cohort Study. *Journal of Autism and Developmental Disorders*, 44(1): 31–40. https://doi.org/10.1007/s10803-013-1849-0

Rutherford, M. D. and Rogers, S. J. 2003. Cognitive underpinnings of pretend play in autism. *Journal of Autism and Developmental Disorders*, 33(3): 289–302. https://doi.org/10.1023/A:1024406601334

Rutter, M., Le Couteur, A., and Lord, C. 2003. *Autism Diagnostic Interview-Revised (ADI-R)*. Pearson Clinical. https://www.pearsonclinical.com.au/products/view/290

Sabatos-DeVito, M., Schipul, S. E., Bulluck, J. C., Belger, A., and Baranek, G. T. 2016. Eye tracking reveals impaired attentional disengagement associated with sensory response patterns in children with autism. *Journal of Autism and Developmental Disorders*, 46(4): 1319–1333. https://doi.org/10.1007/s10803-015-2681-5

Sacrey, L.-A. R., Armstrong, V. L., Bryson, S. E., and Zwaigenbaum, L. 2014. Impairments to visual disengagement in autism spectrum disorder: A review of experimental studies from infancy to adulthood. *Neuroscience & Biobehavioral Reviews*, 47: 559–577. https://doi.org/10.1016/j.neubiorev.2014.10.011

Salley, B. and Colombo, J. 2016. Conceptualizing social attention in developmental research. *Social Development* (Oxford, England), 25(4): 687–703. https://doi.org/10.1111/sode.12174

Sanders, S. J. 2015. First glimpses of the neurobiology of autism spectrum disorder. *Current Opinion in Genetics & Development*, 33: 80–92. https://doi.org/10.1016/j.gde.2015.10.002

Scahill, L. and Challa, S. A. 2016. Repetitive behavior in children with autism spectrum disorder: Similarities and differences with obsessive-compulsive disorder. In L. Mazzone and B. Vitiello (eds.), *Psychiatric Symptoms and Comorbidities in Autism Spectrum Disorder*, pp. 39–50. New York: Springer International Publishing. https://doi.org/10.1007/978-3-319-29695-1_3

Scott, L. S., Shannon, R. W., and Nelson, C. A. 2006. Neural correlates of human and monkey face processing in 9-month-old infants. *Infancy*, 10(2): 171–186. https://doi.org/10.1207/s15327078in1002_4

Seery, A. M., Vogel-Farley, V., Tager-Flusberg, H., and Nelson, C. A. 2013. Atypical lateralization of ERP response to native and non-native speech in infants at risk for autism spectrum disorder. *Developmental Cognitive Neuroscience*, 5: 10–24. https://doi.org/10.1016/j.dcn.2012.11.007

Shah, P., Gaule, A., Bird, G., and Cook, R. 2013. Robust orienting to protofacial stimuli in autism. *Current Biology*, 23(24): R1087–R1088. https://doi.org/10.1016/j.cub.2013.10.034

Shalom, D. B. 2009. The medial prefrontal cortex and integration in autism. *The Neuroscientist*, 15(6): 589–598. https://doi.org/10.1177/1073858409336371

Sheinkopf, S. J., Mundy, P. C., Oller, D. K., and Steffens, M. 2000. Vocal atypicalities of pre-verbal autistic children. *Journal of Autism and Developmental Disorders*, 30(4): 345–354. https://doi.org/10.1023/A:1005531501155

Shen, M. D. 2018. Cerebrospinal fluid and the early brain development of autism. *Journal of Neurodevelopmental Disorders*, 10(1): 39. https://doi.org/10.1186/s11689-018-9256-7

Shen, M. D., Kim, S. H., McKinstry, R. C., Gu, H., Hazlett, H. C., Nordahl, C. W., Emerson, R. W., Shaw, D., Elison, J. T., Swanson, M. R., Fonov, V. S., Gerig, G., Dager, S. R., Botteron, K. N., Paterson, S., Schultz, R. T., Evans, A. C., Estes, A. M., Zwaigenbaum, L., ... Piven, J.

2017. Increased extra-axial cerebrospinal fluid in high-risk infants who later develop autism. *Biological Psychiatry*, 82(3): 186–193. https://doi.org/10.1016/j.biopsych.2017.02.1095

Shen, M. D., Nordahl, C. W., Young, G. S., Wootton-Gorges, S. L., Lee, A., Liston, S. E., Harrington, K. R., Ozonoff, S., and Amaral, D. G. 2013. Early brain enlargement and elevated extra-axial fluid in infants who develop autism spectrum disorder. *Brain*, 136(9): 2825–2835. https://doi.org/10.1093/brain/awt166

Shumway, S. and Wetherby, A. M. 2009. Communicative acts of children with autism spectrum disorders in the second year of life. *Journal of Speech, Language, and Hearing Research*, 52(5): 1139–1156. https://doi.org/10.1044/1092-4388(2009/07-0280)

Sigman, M. and McGovern, C. W. 2005. Improvement in cognitive and language skills from preschool to adolescence in autism. *Journal of Autism and Developmental Disorders*, 35(1): 15–23. https://doi.org/10.1007/s10803-004-1027-5

Simion, F., Di Giorgio, E., Leo, I., and Bardi, L. 2011. The processing of social stimuli in early infancy: From faces to biological motion perception. In O. Braddick, J. Atkinson, and G. M. Innocenti (eds.), *Progress in Brain Research* (Vol. 189, pp. 173–193). Amsterdam, NL: Elsevier. https://doi.org/10.1016/B978-0-444-53884-0.00024-5

Simion, F., Regolin, L., and Bulf, H. 2008. A predisposition for biological motion in the newborn baby. *Proceedings of the National Academy of Sciences*, 105(2): 809–813. https://doi.org/10.1073/pnas.0707021105

Simonoff, E., Pickles, A., Charman, T., Chandler, S., Loucas, T., and Baird, G. 2008. Psychiatric disorders in children with autism spectrum disorders: Prevalence, comorbidity, and associated factors in a population-derived sample. *Journal of the American Academy of Child & Adolescent Psychiatry*, 47(8): 921–929. https://doi.org/10.1097/CHI.0b013e318179964f

Solomon, M., Ono, M., Timmer, S., and Goodlin-Jones, B. 2008. The effectiveness of parent–child interaction therapy for families of children on the autism spectrum. *Journal of Autism and Developmental Disorders*, 38(9): 1767–1776. https://doi.org/10.1007/s10803-008-0567-5

South, M., Ozonoff, S., and McMahon, W. M. 2007. The relationship between executive functioning, central coherence, and repetitive behaviors in the high-functioning autism spectrum. *Autism: The International Journal of Research and Practice*, 11(5): 437–451. https://doi.org/10.1177/1362361307079606

Spencer, J., O'Brien, J., Riggs, K., Braddick, O., Atkinson, J., and Wattam-Bell, J. 2000. Motion processing in autism: Evidence for a dorsal stream deficiency. *NeuroReport*, 11(12): 2765–2767.

Spencer, A. E., Uchida, M., Kenworthy, T., Keary, C. J., and Biederman, J. 2014. Glutamatergic dysregulation in pediatric psychiatric disorders: A systematic review of the magnetic resonance spectroscopy Literature. *Journal of Clinical Psychiatry*, 75(11): 1226–1241. https://doi.org/10.4088/JCP.13r08767

Stagg, C. J., Bachtiar, V., and Johansen-Berg, H. 2011. What are we measuring with GABA magnetic resonance spectroscopy? *Communicative & Integrative Biology*, 4(5): 573–575. https://doi.org/10.4161/cib.4.5.16213

Subramanian, M., Timmerman, C. K., Schwartz, J. L., Pham, D. L., and Meffert, M. K. 2015. Characterizing autism spectrum disorders by key biochemical pathways. *Frontiers in Neuroscience*, 9: 313. https://doi.org/10.3389/fnins.2015.00313

Sullivan, M., Finelli, J., Marvin, A., Garrett-Mayer, E., Bauman, M., and Landa, R. 2007. Response to joint attention in toddlers at risk for autism spectrum disorder: A prospective study. *Journal of Autism and Developmental Disorders*, 37(1): 37–48. https://doi.org/10.1007/s10803-006-0335-3

Szaflarski, J. P., Holland, S. K., Schmithorst, V. J., and Byars, A. W. 2006. FMRI study of language lateralization in children and adults. *Human Brain Mapping*, 27(3): 202–212. https://doi.org/10.1002/hbm.20177

Szatmari, P. 2018. Risk and resilience in autism spectrum disorder: A missed translational opportunity? *Developmental Medicine & Child Neurology*, 60(3): 225–229. https://doi.org/10.1111/dmcn.13588

Tarabulsy, G., Deslandes, J., St-Laurent, D., Moss, E., Lemelin, J.-P., Bernier, A., and Dassylva, J.-F. 2003. Individual differences in infant still-face response. *Infant Behavior and Development*, 26: 421–438. https://doi.org/10.1016/S0163-6383(03)00039-0

Tick, B., Bolton, P., Happé, F., Rutter, M., and Rijsdijk, F. 2016. Heritability of autism spectrum disorders: A meta-analysis of twin studies. *Journal of Child Psychology and Psychiatry*, 57(5): 585–595. https://doi.org/10.1111/jcpp.12499

Tierney, A. L., Gabard-Durnam, L., Vogel-Farley, V., Tager-Flusberg, H., and Nelson, C. A. 2012. Developmental trajectories of resting EEG power: An endophenotype of autism spectrum disorder. *PLOS ONE*, 7(6): e39127. https://doi.org/10.1371/journal.pone.0039127

Toda, S. and Fogel, A. 1993. Infant response to the still-face situation at 3 and 6 months. *Developmental Psychology*, 29(3): 532–538. https://doi.org/10.1037/0012-1649.29.3.532

Todorova, G. K., Hatton, R. E. M., and Pollick, F. E. 2019. Biological motion perception in autism spectrum disorder: A meta-analysis. *Molecular Autism*, 10(1): 49. https://doi.org/10.1186/s13229-019-0299-8

Tomasello, M., and Farrar, M. J. 1986. Joint Attention and Early Language. *Child Development*, 57(6): 1454–1463. JSTOR. https://doi.org/10.2307/1130423

Turner, M. 1997. Towards an executive dysfunction account of repetitive behaviour in autism. In J. Russell (ed.), *Autism as an Executive Disorder*, pp. 57–100. Oxford: Oxford University Press.

Uzunova, G., Hollander, E., and Shepherd, J. 2014. The role of ionotropic glutamate receptors in childhood neurodevelopmental disorders: Autism spectrum disorders and fragile x syndrome. *Current Neuropharmacology*, 12(1): 71–98.

Vernetti, A., Senju, A., Charman, T., Johnson, M. H., and Gliga, T. 2018. Simulating interaction: Using gaze-contingent eye-tracking to measure the reward value of social signals in toddlers with and without autism. *Developmental Cognitive Neuroscience*, 29: 21–29. https://doi.org/10.1016/j.dcn.2017.08.004

Vivanti, G., Hocking, D. R., Fanning, P. A. J., Uljarevic, M., Postorino, V., Mazzone, L., and Dissanayake, C. 2018. Attention to novelty versus repetition: Contrasting habituation profiles in autism and Williams syndrome. *Developmental Cognitive Neuroscience*, 29: 54–60. https://doi.org/10.1016/j.dcn.2017.01.006

Walton, K. M., and Ingersoll, B. R. 2015. Psychosocial adjustment and sibling relationships in siblings of children with autism spectrum disorder: Risk and protective factors. *Journal of Autism and Developmental Disorders*, 45(9): 2764–2778. https://doi.org/10.1007/s10803-015-2440-7

Wan, M. W., Green, J., Elsabbagh, M., Johnson, M., Charman, T., and Plummer, F. 2013. Quality of interaction between at-risk infants and caregiver at 12–15 months is associated with 3-year autism outcome. *Journal of Child Psychology and Psychiatry*, 54(7): 763–771. https://doi.org/10.1111/jcpp.12032

Warlaumont, A. S., Richards, J. A., Gilkerson, J., and Oller, D. K. 2014. A social feedback loop for speech development and its reduction in autism. *Psychological Science*, 25(7): 1314–1324. https://doi.org/10.1177/0956797614531023

Webb, S. J., Dawson, G., Bernier, R., and Panagiotides, H. 2006. ERP evidence of atypical face processing in young children with autism. *Journal of Autism and Developmental Disorders*, 36(7): 881. https://doi.org/10.1007/s10803-006-0126-x

Webb, S. J., Jones, E. J. H., Merkle, K., Venema, K., Greenson, J., Murias, M., and Dawson, G. 2011. Developmental change in the ERP responses to familiar faces in toddlers with autism spectrum disorders versus typical development. *Child Development*, 82(6): 1868–1886. https://doi.org/10.1111/j.1467-8624.2011.01656.x

Webb, S. J., Merkle, K., Murias, M., Richards, T., Aylward, E., and Dawson, G. 2012. ERP responses differentiate inverted but not upright face processing in adults with ASD. *Social Cognitive and Affective Neuroscience*, 7(5): 578–587. https://doi.org/10.1093/scan/nsp002

Webb, S. J., Nalty, T., Munson, J., Brock, C., Abbott, R., and Dawson, G. 2007. Rate of head circumference growth as a function of autism diagnosis and history of autistic regression. *Journal of Child Neurology*, 22(10): 1182–1190. https://doi.org/10.1177/0883073807306263

Wetherby, A. M. and Prizant, B. M. 2002. *Communication and Symbolic Behavior Scales: Developmental Profile*. Baltimore, MD: Paul H Brookes Publishing Co.

Wilcox, T. and Biondi, M. 2015. FNIRS in the developmental sciences. *WIREs Cognitive Science*, 6(3): 263–283. https://doi.org/10.1002/wcs.1343

Williams, E., Reddy, V., and Costall, A. 2001. Taking a closer look at functional play in children with autism. *Journal of Autism and Developmental Disorders*, 31(1): 67–77. https://doi.org/10.1023/A:1005665714197

Wolff, J. J., Gerig, G., Lewis, J. D., Soda, T., Styner, M. A., Vachet, C., Botteron, K. N., Elison, J. T., Dager, S. R., Estes, A. M., Hazlett, H. C., Schultz, R. T., Zwaigenbaum, L., and Piven, J. 2015. Altered corpus callosum morphology associated with autism over the first 2 years of life. *Brain*, 138(7): 2046–2058. https://doi.org/10.1093/brain/awv118

Yang, D., Pelphrey, K. A., Sukhodolsky, D. G., Crowley, M. J., Dayan, E., Dvornek, N. C., Venkataraman, A., Duncan, J., Staib, L., and Ventola, P. 2016. Brain responses to biological motion predict treatment outcome in young children with autism. *Translational Psychiatry*, 6(11): e948–e948. https://doi.org/10.1038/tp.2016.213

Yeung, H. H., Denison, S., and Johnson, S. P. 2016. Infants' looking to surprising events: When eye-tracking reveals more than looking time. *PLOS ONE*, 11(12): e0164277. https://doi.org/10.1371/journal.pone.0164277

Yirmiya, N., Gamliel, I., Pilowsky, T., Feldman, R., Baron-Cohen, S., and Sigman, M. 2006. The development of siblings of children with autism at 4 and 14 months: Social engagement, communication, and cognition. *Journal of Child Psychology and Psychiatry*, 47(5): 511–523. https://doi.org/10.1111/j.1469-7610.2005.01528.x

Zwaigenbaum, L., Bryson, S., Rogers, T., Roberts, W., Brian, J., and Szatmari, P. 2005. Behavioral manifestations of autism in the first year of life. *International Journal of Developmental Neuroscience: The Official Journal of the International Society for Developmental Neuroscience*, 23(2–3): 143–152. https://doi.org/10.1016/j.ijdevneu.2004.05.001

CHAPTER 33

SOCIAL COGNITIVE AND INTERACTIVE ABILITIES IN AUTISM

HANNA THALER, LAURA ALBANTAKIS,
AND LEONHARD SCHILBACH

INTRODUCTION

AUTISM spectrum disorder, henceforth referred to as autism in this chapter, is a neurodevelopmental condition. One of its hallmarks are difficulties in social cognition and interaction. Social cognitive abilities can be described as a set of skills that help us process other people's intentions, emotions, and more generally, mental states in order to understand their behavior. A wealth of experimental research has documented social processing difficulties in autism. Early on in development, autistic children lag behind age-matched peers in mastering abilities such as inferring others' mental states (Baron-Cohen, Leslie, and Frith, 1985; Senju et al., 2010) or following others' eye gaze (Charman, 2003). These differences can also be related to differences in brain structure and function (Hazlett et al., 2017), and may partially be compensated for across development (e.g., White et al., 2009). Despite these compensatory developments, autism is known as a lifelong condition. Many aspects of social functioning continue to be affected throughout adulthood and detrimentally impact quality of life in important ways (Kim and Bottema-Beutel, 2019). It has been questioned whether traditional ways of studying social cognitive abilities can succeed in explaining the difficulties autistic people face in everyday life. Crucially, real life requires the use of social abilities in an interactive context, in which we do not only have to process, but also quickly adapt to the social information available (Schilbach et al., 2012; Schilbach et al., 2013). This has brought about an *interactive turn*, shifting focus from individual cognition to interacting minds (cf. Frith and Frith, 1999; Schilbach, 2016; 2019). Resulting two-person accounts of autism research have strived to capture what happens when autistic people interact with other, autistic and non-autistic, people (Redcay and Schilbach, 2019). Another challenge in autism research has been to explain social difficulties within a more unified framework of the autistic brain. Here, the Bayesian brain account represents a novel attempt of relating social and nonsocial characteristics of autism to more general aspects of autistic perception (e.g., Bolis et al., 2017).

This chapter gives an overview of the theoretical frameworks and research findings that explain social interactive aspects of autism. Its focus lies on autism with no accompanying intellectual disability, often referred to as so-called *high-functioning* autism. First, we give a brief introduction to longstanding research on theory of mind. In the next section, we focus on different aspects of empathy. The third section of this chapter describes how social abilities can be studied in a more interactive context. The last section introduces the Bayesian brain account of autism. Finally, we outline several more recent developments in autism research that may shape the following decades.

THEORY OF MIND IN AUTISM

Early attempts at explaining social cognitive difficulties in autism have focused on the question whether autistic children have a theory of mind (TOM). TOM, also referred to as *mentalizing*, describes the ability to cognize about own and others' mental states (cf. Premack and Woodruff, 1978). In acquiring a TOM, children gradually become aware how behavior is shaped by beliefs, intentions, and emotions. They learn to infer these states in others, and to predict how others behave on the basis of their inferences. An early indicator of TOM is children's awareness of others' false beliefs. Typically, developing children gradually improve their false belief understanding between 2.5 and 5 years of age, and are able to recognize that someone's thoughts deviate from the actual state of the world when they are 4–5 years old (see meta-analysis including more than 4000 children by Wellman, Cross, and Watson, 2001).

Autism and TOM: Evidence from Behavioral Studies

The first study with autistic children pointed to profound differences in their TOM development. Baron-Cohen and colleagues (1985) observed that 80% of autistic children (n = 20, mean age: 12 years) failed the so-called Sally-Anne-task, an explicit test of false belief understanding that at least 85% of children with Down's syndrome (n = 14, mean age: 11 years) and non-autistic children (n = 27, mean age: 4.5 years) could solve. Their findings inspired the mindblindness hypothesis, which proposes that autistic people show social and communicative difficulties due to a failure of attributing mental states to others (Baron-Cohen, 1995; Frith, 2001). Since then a number of findings has emerged confirming that autistic children are delayed in their development of a TOM (discussed in Baron-Cohen 2000; Happé, 1995). Although these delays are also seen in children and adolescents with high-functioning autism, verbal intelligence seems to counteract mentalizing difficulties. When autistic children reach a certain level of verbal skills, many of them manage to solve first-order false belief tasks (e.g., Happé, 1995; White et al., 2009). This may indicate an important confound, in that even children with preserved TOM abilities could fail at explicit tasks because they do not understand the words used.

There is mixed evidence on whether TOM abilities of autistic people continue to be affected in adulthood. One hypothesis is that autistic adults do struggle with more complex, or advanced, forms of TOM. Examples of complex TOM include the understanding of ambiguous, nonliteral remarks, sarcasm, and second-order TOM. Second-order false beliefs require inferences about mental states that address the mental state of a third person, e.g., *he falsely believes that she thinks*. A study employing the Strange Stories task observed that autistic adults (n = 18) performed worse than non-autistic children and adults (n = 49) in finding appropriate explanations for other people's ambiguous statements (Happé, 1994; replicated by Jolliffe and Baron-Cohen, 1999). This was contradicted in a later study, finding no differences between autistic and non-autistic adults (N = 38) in a short form of the Strange Stories task, despite group differences at inferring thoughts and feelings in dynamic videotaped interactions (Ponnet et al. 2004). In a study with 63 children and 80 adults, autistic children performed worse at understanding stories conveying first- and second-order false beliefs, whereas autistic adults only showed difficulties with second-order false beliefs (White et al., 2009). However, a large study on children and adolescents (N = 254) did not observe any differences between autistic and non-autistic groups in the understanding of social stories requiring second-order TOM, comprehension of sarcasm and other advanced forms of TOM (Scheeren et al., 2013). In line with previous findings, Scheeren and colleagues observed that advanced TOM skills were associated with verbal abilities. This indicates that also the assessment of complex TOM may be confounded by differences in verbal abilities.

It has also been proposed that autistic people may struggle more with implicit rather than explicit mentalizing (e.g., Senju et al., 2009; for a detailed discussion of evidence see Sodian et al., 2015). This hypothesis argues that autistic people can learn to compensate for their explicit mentalizing difficulties, and that using compensatory strategies later allows them to infer others' mental states in a conscious and deliberate manner similar to how non-autistic persons may solve a math problem. However, compensatory strategies still fall short of permitting implicit inferences during spontaneous communication and interaction. This account is interesting in that it might explain why autistic people often succeed in TOM tasks despite experiencing difficulties in understanding others' intentions in real life. In a first, seminal study testing this account, Senju and colleagues (2009) tracked eye gaze of autistic (n = 17) and non-autistic (n = 19) adults during a nonverbal false belief task. They observed that non-autistic adults engaged in unprompted behavior that is typical for these situations: they spent more time looking at the correct solution. Autistic participants, on the other hand, showed no looking bias toward either solution. Intriguingly, no group differences were observed in an explicit version of this false belief task. These findings support the notion that autistic adults do not spontaneously infer others' mental states to anticipate their behavior. In another study, Senju and colleagues (2010) found evidence suggesting that this difference is already present at an early age. In contrast to their non-autistic peers (n = 17), autistic children (n = 12) with a mean age of 8 years did not display a looking bias indicative of spontaneous false belief attribution, and also performed worse in an explicit false belief task. However, studies have raised doubts on the methods used to measure implicit TOM. For instance, a recent replication study could not replicate original results of four anticipatory-looking tasks (Kulke et al., 2018). Crucially, this failed replication also concerned the task used by Senju and colleagues (2009, 2010). Thus, answering the question of whether autistic people have greater difficulties with implicit TOM, will first require further advances in capturing implicit TOM.

Although not all studies discussed previously suggest universal TOM difficulties, a recent meta-analysis on social and nonsocial cognitive functioning in autism does indicate that TOM difficulties count as one of the most consistent research findings (Velikonja, Fett, and Velthorst, 2019). According to this analysis, autistic adults display the largest difficulties in studies of TOM (n = 39) and emotion processing (n = 18), with respective Hedges g estimates of -1.09 and -0.80.

Mentalizing and the Brain: The TOM Network

Neuroimaging studies assessing brain activation during TOM tasks have observed a pattern of involved regions that are typically referred to as the TOM, or mentalizing, network. The main hubs of this network are considered the medial prefrontal cortex (MPFC), bilateral superior temporal sulcus (STS), bilateral temporoparietal junction (TPJ), as well as a region spanning precuneus and posterior cingulate cortex (see reviews by Carrington and Bailey, 2009; Abu-Akel and Shamay-Tsoory, 2011). The relative involvement of these regions depends on what types and components of TOM are assessed. For instance, the TPJ seems to be particularly important for processing false beliefs (e.g., Saxe and Kanwisher, 2003) and is believed to have a role in distinguishing own and others' mental states. Abu-Akel and Shamay-Tsoory (2011) further contrast affective and cognitive TOM, considering dorsolateral prefrontal cortex and striatum as two contributing areas involved in executive and affective components of TOM, respectively. There is evidence that also processes considered "precursors" of TOM, such as pupil mimicry of an interaction partner (Prochazkova et al., 2018) and participation in gaze-based social interaction engage the TOM network (Redcay and Schilbach. 2019).

TOM and Brain Development

Development of TOM is connected to continuous functional and structural development of the brain. A recent study presented children (n = 122) and adults (n = 33) with movie scenes depicting others' mental versus bodily states (Richardson et al., 2018). Already, at the age of 3 years, mental and bodily state scenes distinctly engaged the TOM network and the pain network. Interestingly, this functional specialization of the two networks could also be seen in children who still failed false-belief tasks. From ages 3–12, and with continuous TOM maturation, the two networks became increasingly anticorrelated.

Brain development in autism is altered early on, with studies detecting differences to typically developing peers within the first 2 years of life (for review of evidence see Lainhart, 2015). A compelling study of 148 infants with high and low familial risk of autism observed that those diagnosed with autism at 24 months showed higher brain volume overgrowth between 12 and 24 months (Hazlett et al., 2017). Brain volume overgrowth at that age was also associated with higher scores in the autism diagnostic observation schedule (ADOS) and in the communication and symbolic behavior scales (CSBS). A deep-learning algorithm, mostly stemming from measurements made as early as 6 months after birth, could predict autism diagnosis on the basis of hyperexpansive cortical surface area.

Autism and TOM: Evidence from Brain Imaging Studies

In one of the first studies on TOM-related brain activity, five male autistic adults underwent positron emission tomography (PET) imaging while reading stories requiring attributions of mental or physical states (Happé et al., 1996). Compared to a sample of six male non-autistic adults (Fletcher et al., 1995), autistic participants performed worse in answering questions about the TOM condition and did not show activation in the left medial frontal cortex during TOM processing. They also showed less pronounced activation differences between conditions in bilateral temporal poles and the left superior temporal gyrus. Another early PET study of 20 autistic and non-autistic adults used animated triangles to represent "interactions" requiring mental state attribution (Castelli et al., 2002). Compared to the control group, autistic participants performed worse at interpreting mental states and exhibited reduced activation during TOM-related processing in the basal temporal area, STS, and superior frontal gyrus. They also showed reduced connectivity between extrastriate regions, involved in early visual processing, and the STS. In a similar study using the Frith-Happé animations (White et al., 2011), autistic children (n = 13) performed worse than non-autistic peers (n = 13) at attributing mental and nonmental states to triangles, and exhibited reduced activation in the TOM network, the so-called mirror neuron system and the cerebellum (Kana et al., 2015).

A wealth of neuroimaging studies has since emerged that has rendered a more complex picture of the TOM network in autism. In line with the findings discussed previously, several studies suggest that autistic individuals may recruit the TOM network in a less discriminate manner, distinguishing less between mental and nonmental states. For instance, contrasting inferences on intentions vs physical and emotional content, one study found less pronounced TOM-related activation differences in 18 autistic adults compared to 18 non-autistic controls (Mason et al., 2008). Autistic participants also exhibited reduced functional connectivity within the TOM network—associated with structural differences in the corpus callosum—and between TOM and language processing regions. Although no behavioral results were reported, the authors interpreted this overall picture of neural differences as evidence for a less efficient use of the TOM network, suggesting that it might be related to connectivity differences spanning the whole brain. Using the animated triangle task, another study with autistic and non-autistic adults (N = 24) observed lower functional connectivity within the frontal and posterior parts of the TOM network in autistic participants, although there was no difference in behavioral performance (Kana et al., 2009). Less discriminate TOM-related activation was also observed in a study on the processing of opinion sentences (Lombardo et al., 2011). In contrast to controls (n = 33), autistic adults (n = 29) recruited the right TPJ in a comparable manner in response to mental and physical states. This lack of specificity was related to social difficulties as assessed by the Autism Diagnostic Interview-Revised (ADI-R). A recent magnetoencephalography (MEG) study indicates that reduced TOM-related activation can already be observed during childhood (Yuk et al., 2018). In that study, autistic and non-autistic children (N = 41) performed equally well in attributing true versus false beliefs. However, autistic children exhibited reduced activation of left TPJ, which coincided with increased activation of right inferior frontal gyrus. The authors suggest that inferior frontal activation may represent increased use of executive functioning as a compensatory strategy. Conversely, there are some studies that have observed greater involvement of the TOM network in autism. Wang and colleagues (2006) found that autistic children and adolescents (n = 18) were less accurate than their peers (n = 18) in an explicit task on irony comprehension, but showed greater activation in STS and inferior frontal gyrus during the processing of different cues. In a longitudinal study, White and colleagues (2014) compared autistic children who passed false belief tasks to those who did not. Both autistic groups (n = 22) showed increased TOM-related activation relative to their non-autistic peers (n = 11). These differences were observed at two time points 5 years apart, despite a large overall improvement in mentalizing performance.

Studies have found potential explanations for both hyper- and hypoactivation of the TOM network in autism. Increased TOM-related activation has typically been interpreted as more effortful processing, whereas reduced activation and intra- or interregional functional connectivity have been seen as indications for a lower propensity to engage in mentalizing, or lower efficiency. Finally, it has been suggested that greater involvement of regions outside the core TOM network, such as the inferior frontal gyrus, may point to alternative strategies used by autistic people to compensate for mentalizing difficulties. These interpretations are partially supported by evidence linking brain activation to lower task performance or scores of social functioning, although several of the described studies did not find behavioral differences. Differences in neurocognitive functioning could also be seen when task performance was comparable, or in autistic people whose TOM seemed preserved. One potential explanation could be that brain imaging is sensitive to fine-grained differences in social processing that the tasks used in those studies do not capture.

Importantly, there are several studies observing no group differences in brain activation during TOM-related processing. Dufour and colleagues (2013) did not detect any differences in cortical activation between autistic (n = 31) and non-autistic (n = 462) adults during the processing of an explicit, verbal false belief task. Furthermore, a large recent study did not observe any group differences in brain activation between autistic (n = 205) and non-autistic (n = 189) children and adults during the processing of an animated shapes task (Moessnang et al., 2020). Considering the large sample size of the latter study, this casts doubt on a general hyper- or hypoactivation of the TOM network in autism. To determine whether there are reliable differences in TOM-related brain connectivity, additional studies with larger sample sizes will be needed.

In sum, research on TOM highlights difficulties, but also points to heterogeneity within the autistic population. It has been suggested that autistic people can learn to compensate for TOM difficulties but struggle to do so implicitly. Difficulties with implicit mentalizing would likely show in real-time interactions and impact one's ability to generate adequate responses, such as empathic behavior. The upcoming sections of this chapter will focus on how empathy and interactive behaviors present in autism.

Empathy in Autism

Some studies seem to suggest that empathy is impaired in autism (e.g., Baron-Cohen and Wheelwright, 2004; Lombardo et al., 2007). Indeed, autistic people often report to be less empathetic in psychometric instruments such as the empathy quotient (EQ: Baron-Cohen and Wheelwright, 2004) (e.g., Wheelwright et al., 2006; Lombardo et al., 2007; Johnson, Filliter, and Murphy, 2009) or the interpersonal reactivity index (IRI: Davis, 1983) (e.g., Lombardo et al., 2007; Demurie, De Corel, and Roeyers, 2011; however, for findings of comparable scores in self-reported emotional empathy see Rogers et al., 2007; Dziobek et al., 2008, and Thaler et al., 2018). Experimental research on empathy in autism has produced more diverging results. An important limitation of self-reported empathy may be that individuals' answers depend on their introspective abilities (cf. Dziobek et al., 2008), a domain that autistic people tend to have difficulties with (Uddin, 2011). Disagreements on whether autistic people experience difficulties with empathy may also be related to the use of different conceptualizations of empathy (Decety and Jackson, 2004).

Theoretical Accounts of Empathy

Several definitions of empathy distinguish between cognitive and emotional, or affective, empathy (Davis, 1983; Shamay-Tsoory, Aharon-Peretz, and Perry, 2009). Cognitive empathy typically means to adopt the perspective of the other person and be aware of his or her emotions, while affective empathy is about producing an adequate emotional response. One hypothesis states that autistic people only show difficulties in the cognitive component, while affective empathy is preserved (e.g., Dziobek et al., 2008; Bird and Viding, 2014). The empathy imbalance hypothesis (Smith, 2009) and the intense world theory (Markram and Markram, 2010) propose that autistic people may even display heightened affective empathy, along with a general tendency for hyperarousal. Others have argued that difficulties experienced by autistic people may lie in precursors of empathy, such as emotion recognition and social information processing (Bird et al., 2013; Bird and Viding, 2014). Differences in self-other distinction, the process by which own emotional experiences are isolated from those of others (Decety and Lamm, 2006), have also been suggested to impact autistic people's understanding of others' emotional states (cf. Lombardo and Baron-Cohen, 2011).

Cognitive vs Affective Empathy

Support for a distinction between cognitive and affective empathy stems from studies contrasting autism with psychopathy. Individuals with psychopathic tendencies tend to have preserved TOM but show reduced affective empathy. The reverse seems to apply to autistic individuals. For instance, a study by Jones and colleagues (2010) compared boys with autism, psychopathic tendencies, conduct disorder, or no diagnosis (N = 96) in their empathic responses to hypothetical scenarios about other people's dismay. Boys with psychopathic tendencies were the only group reporting reduced fear responses, but performed well in a task assessing TOM. Conversely, autistic boys reported a typical level of affective responses, but experienced difficulties in the TOM task. Another study (Schwenck et al., 2012) observed lower cognitive empathy in autistic boys (n = 55) as opposed to lower affective empathy in peers with conduct disorder and a high number of callous-unemotional traits (n = 36). Findings in that study also indicated that empathic abilities improved with age within all comparison groups (N = 192), pointing to similar development of empathy regardless of the presence of a diagnosis.

Hadjikhani and colleagues (2014) presented autistic adults and adolescents (n = 38) and matched controls (n = 35) with dynamic facial expressions of patients with shoulder pain. Although groups differed in self-reported empathy as assessed with the EQ, there were no group differences in brain activation during the video viewing. In the autistic group, self-reported emotional empathy was associated with activation in regions typically involved in emotion processing. Although no behavioral responses were assessed, the authors interpreted findings as support for preserved emotional contagion in autism. Other neuroimaging studies depicting others' pain also point to preserved emotional empathy (Krach et al., 2015) with potentially heightened affective arousal (Fan et al., 2014; Gu et al., 2015). Presenting videos of body parts getting hurt in physical accidents, Fan and colleagues (2014) observed elevated somatosensory area (SI/SII) activation and blunted activation in anterior insula (AI) in autistic male young adults (n = 24) compared to non-autistic peers (n = 21). In a second study with a similar sample (n = 40), they did not detect any group differences in *mu* suppression, indicating that sensorimotor resonance was preserved. Autistic participants showed heightened responses in N2 amplitude, indicating greater early affective arousal during empathic processing. Cortical and hemodynamic responses were modulated differentially by social complexity and individual pain sensitivity, highlighting modulatory roles of social understanding and sensory processing. Social complexity also modulated empathic responses of autistic adults in the study by Krach and colleagues (2015). When presented with more complex social scenarios, autistic male participants (n = 16) rated emotion intensity of others lower than non-autistic participants (n = 16), showed blunted pupil dilation, and reduced brain activation in AI and anterior cingulate cortex (ACC).

In sum, behavioral and neuroimaging data support the notion that autistic individuals show preserved, if not elevated, affective empathy. Findings showing a moderating effect of social-cognitive demands are in line with the proposal that it is difficulties in social information processing, rather than in empathy per se, that can impact empathic responses (Bird and Viding, 2014).

The Roles of Alexithymia and Social Information Processing

Bird and colleagues (2013) have argued that any difficulties in empathy seen in autism may stem from elevated levels of alexithymia. Alexithymia is characterized by difficulties in identifying and describing own emotions. Research indicates that alexithymic individuals also have difficulties judging others' emotions (Parker, Taylor, and Bagby, 1993; Lane et al., 1996; Prkachin, Casey, and Prkachin, 2009; Uljarevic and Hamilton, 2013), and display reduced empathic responses (Moriguchi et al., 2007; Grynberg et al., 2010). While approximately 10% of the general population possess this trait (Salminen et al., 1999), the percentage of autistic people with alexithymia is presumed to be considerably higher. A recent meta-analysis of 15 studies with relatively small autistic samples (N ≤ 88) has estimated prevalence to be at 49.93% (Kinnaird, Stewart, and Tchanturia, 2019).

First evidence for a role of alexithymia in empathy stems from a study that compared 36 autistic and non-autistic men with matched alexithymia levels. Bird and colleagues (2010) examined neural responses to watching a close companion receive electrical pain stimulation. Brain activation in an a-priori defined region of interest in AI was negatively associated with alexithymic traits. There was no difference in this correlation between autistic and non-autistic participants. The authors interpreted findings as supporting their hypothesis that autism does not impact affective empathy. In line with a previous study (Silani et al., 2008), affective empathy—in the form of self-reported unpleasantness—was neither affected by alexithymia nor autism. Another recent study assessed how 32 autistic and non-autistic adults with comparable alexithymia levels evaluated others' facial expressions of pain (Thaler et al., 2018). Viewing videos depicting patients' shoulder pain, both groups reported similar levels of unpleasantness, indicating comparable affective empathy. However, autistic participants underestimated actual pain intensities experienced by patients more than the control group. Results highlight that autistic people may experience difficulties in precursors of empathy that cannot be explained by alexithymia, such as the intensity of dynamic facial expressions.

It has been argued that difficulties in social information processing—including deficits in emotion identification and TOM—could impact empathic behavior even when empathy per se is preserved (cf. Bird and Viding, 2014). Compelling evidence for how empathy and TOM are linked in real life stems from a study evaluating an 8-week training of TOM for autistic children (Holopainen et al., 2018). Empathic responsiveness was assessed via structural observations of video-recorded interactions between autistic children and experimenters. Compared to a waitlist control group (n = 63), those children who took part in the training (n = 72) showed increased empathic responsiveness after the intervention. Coll and colleagues (2017) have recently proposed a measurement framework that distinguishes problems with emotion identification from empathizing difficulties. Quantifying empathy as the extent to which a person shares the emotional state they *believe* the other person to be in, this framework accounts for the possibility that the empathizing individual misinterprets another's emotions. This argument is interesting because it may redefine what has been interpreted as reduced empathic behavior in autism as more general difficulties in social information processing.

The Double Empathy Problem

When studying empathy and TOM in autism, it is important to consider the idea that social misunderstandings do not just happen *within* one person, but concern all interaction partners. Referring to this issue as the "double empathy problem," Milton (2012) has argued that non-autistic people also struggle with taking the perspective and understanding the experiences of autistic people. This phenomenon has also been described as a "social interaction mismatch" (Bolis et al., 2017; see section "Autism as a Problem of Interpersonal Misattunement" for further details). Indeed, there is experimental evidence supporting this notion. For instance, in a study involving 27 autistic and non-autistic rating participants, autistic people's posed facial expressions of emotions were recognized less well than those of non-autistic people (Brewer et al., 2016). Across a set of three studies (respective samples of 214, 37, and 98), non-autistic adults evaluated autistic children and adults appearing in videos of social interactions less favorably than they evaluated non-autistic peers (Sasson et al., 2017). Participants perceived autistic protagonists as less likeable and approachable, more socially awkward, and reported they would be less likely to socially engage or develop a friendship with them. Finally, in a study with 361 adolescents and adults, Cage, Di Monaco, and Newell (2018) observed that participants exhibited dehumanizing attitudes toward autistic people, expressed in a lower attribution of uniquely human character traits. The double empathy problem highlights the importance of investigating and reframing social-cognitive difficulties in autism as difficulties concerning all interaction partners, including autistic and non-autistic people.

Autism and Social Interaction

There is increasing interest in studying social abilities in settings that are interactive, and thus, more closely resemble real life (cf. Schilbach, 2016). The premise of this *interactive turn* is that being social is less about observing others and more about interacting with them. As discussed in this section, it seems that it is precisely this interaction that may often prove challenging.

Joint Attention in Autism

Research increasingly suggests that autism may best be described as a social inter-action disorder that manifests in forming adequate behavioral responses during so-cial encounters (cf. Schilbach, 2016). One important element of interactions is joint attention (JA), defined as the process of coordinating attention with interaction part-ners to send and receive information about a third entity (e.g., other people, events, or objects). In contrast to dyadic interactions such as mutual gaze, JA has a triadic char-acter, structured as a "referential triangle" between two agents and that third entity (Carpenter, Nagell, and Tomasello, 1998; Pfeiffer, Vogeley, and Schilbach, 2013). Self-initiated or initiating JA (IJA) describes the process of making another individual follow one's own gaze, whereas other-initiated or responding JA (RJA) entails following an-other person's gaze to share a common referential target (Mundy and Newell, 2007). JA is vital for navigating spontaneous and unstructured settings. It is considered an im-portant developmental precursor for other social and cognitive abilities.

Typically developing children can follow other people's gaze direction to a location outside of their visual field by 9 months of age, allowing them to communicate and interact with their environment before language development (Mundy and Newell, 2007). Up to the age of 18 months, toddlers will usually have developed a large reper-toire of JA behaviors (Mundy et al., 2007). They then continue to refine these skills up to adolescence (e.g., Oberwelland et al., 2016; also see review by Mundy, 2018). Social and communicative difficulties, including eye contact, gaze behavior, and response to name, are the most conclusive discriminators for autism around the age of 12–18 months (Stone, Ousley, and Littleford, 1997; Charman et al., 2000; Charman, 2003; Gotham et al., 2007). Early delays can also be observed in more domain-general areas of visual attention. For instance, a study with 104 infants with and without familial risk for autism observed that by 14 months, those later diagnosed with autism were less efficient in disengaging from visual stimuli (Elsabbagh et al., 2013). They also showed a less consistent development of speed and flexibility of visual orienting between 7 and 14 months of age. As these abilities are considered crucial in JA, it is unsurprising that autistic children show difficulties in both IJA and RJA (for reviewed evidence see Mundy and Newell, 2007; Mundy, Sullivan, and Mastergeorge, 2009). Studies indicate that deficits in IJA are more stable and universal, while RJA difficulties can disappear later in childhood. Importantly, early delays in IJA and RJA development correlate with later difficulties in language and social abilities (Sigman and McGovern, 2005; Toth et al., 2006; Anderson et al., 2007; Delinicolas and Young, 2007; Redcay et al., 2013). This highlights that it may be beneficial to include training in JA in early intervention programs for autistic children (Charman, 2003).

Autism and JA: Evidence from Brain Imaging Studies

During the last decade, research on JA has undergone remarkable developments in replacing purely *observational* tasks with socially *interactive* ones. Real-time gaze-contingent experimental setups during functional magnetic resonance imaging (fMRI), for instance, have been used to account for the reciprocal nature of JA while exploring associated neural mechanisms (Schilbach et al., 2010, 2013; Oberwelland et al., 2016). A series of studies employing socially interactive tasks indicate that the JA network in typically developed adults involves the MPFC extending into the medial orbitofrontal cortex, the right posterior STS, and the right TPJ (Redcay, Dodell-Feder, and Pearrow, 2010; Schilbach et al., 2010; Pfeiffer et al., 2014; Bilek et al., 2015; Oberwelland et al., 2016). Aside from this general JA network, the ACC, ventral striatum, and amygdala, all of which are associated with motivation and emotion processing, seem to contribute to JA processes (Schilbach et al., 2010; Pfeiffer et al., 2014). These findings indicate that JA goes beyond the mere exchange of information and intentions. It appears that sharing attention with other people is associated with feelings of reward, and thus has a socio-communicative purpose in itself (Schilbach et al., 2010; Schilbach, 2015; Oberwelland et al., 2016). Findings of increased reward-based brain activation during social interaction have also been observed in children (Alkire et al., 2018). When they believed to be interacting with a peer rather than an artificial agent during mental- and nonmental state reasoning, they responded more quickly and reported greater enjoyment. Interactive context was associated with brain activation in the TOM network, cortical midline regions, bilateral insula, striatum, and amygdala.

In another study, Oberwelland and colleagues (2016) looked at the developmental trajectories of JA in 32 children and adolescents using an interactive task during a magnetic resonance imaging (MRI) scan. Overall, JA recruited a similar set of brain regions than in adults. There was, however, an interesting discrepancy between different types of JA, pointing to a more reward-associated role of IJA as compared to RJA. While activation during RJA was confined to regions involved in visual attention and social processing, IJA involved activation in a wider range of brain regions typically attributed to social cognition, decision-making, emotions, and motivational/reward processing. Only activation during IJA was significantly enhanced when participants interacted with a familiar rather than an unfamiliar partner. Regarding developmental effects, the authors found some evidence for a greater differentiation in the recruitment of TPJ with increasing age. Adolescents showed greater activation in bilateral TPJ during IJA, whereas in children, no difference between IJA and RJA could be observed.

Oberwelland and colleagues (2017) used the same task to investigate neural correlates of JA in a sample of 32 male children and adolescents with and without autism. Although there were no group differences in saccade behavior or overall accuracy during JA, non-autistic participants showed greater brain activation during JA in two clusters associated with the so-called social brain, including temporal pole, STS, TPJ, and precuneus. Only within the non-autistic group, IJA was associated with frontal activation involving the MPFC. In autistic participants, self-related activation in that same frontal cluster was negatively associated with self-reported difficulties in social communication. Integrating this finding with previous results, the authors discuss that a lower involvement of frontal regions during IJA might reflect difficulties in self-other distinction (Schulte-Rüther et al., 2014) or in the disengagement of visual attention (Keehn, Müller, and Townsend, 2013). There were no group differences in the modulation of brain activation by familiarity of the interaction partner. Results indicate that, even in a simple and relatively controlled interaction, autistic children and adolescents show differences in their neural processing of JA.

Sensorimotor Interaction in Autism

One factor that may contribute to social interaction difficulties is sensorimotor processing. For instance, Dowd, Rinehart, and McGinley (2010) have argued that motor impairments seen in autism do not just happen in parallel, but exacerbate social interaction problems, and may be represented in overlapping neural circuits. Attempts have been made to go beyond visuospatial interaction and eye gaze, to study interacting bodies (Schilbach, 2016; Bolis et al., 2017; Zapata-Fonseca et al., 2018).

In one of the first studies to investigate sensorimotor interaction in autism, Zapata-Fonseca and colleagues (2018) implemented a minimalist experimental paradigm known as perceptual crossing (PC), modified from Lenay (2008). Ten dyads of autistic and non-autistic adults engaged in tactile-based interactions in a shared one-dimensional virtual space. Participants were connected to each other via a sensorimotor loop: they appeared as avatars in the virtual space, navigated a computer mouse with their dominant hand, and a Braille stimulator with the index finger of the other hand. Moving around in virtual space, both participants received sensory feedback via the Braille stimulator when they crossed each other, or when they interacted with objects serving as distractors. They were asked to click the computer mouse whenever they thought they had crossed the other person's avatar. With this setup, researchers could compare participants' abilities in social contingency detection. Interestingly, autistic participants performed equally well as control participants, showing comparable sensitivity to the other person's tactile properties. This indicates that, at least in a reductionist virtual environment as given in this paradigm, autistic individuals are able to detect an agent that responds to their actions and can discriminate between this agent-based movement and nonreactive, object-based movement. In discussing their unexpected finding, Zapata-Fonseca and colleagues (2018) have raised important issues pertaining to computer-mediated interactions. Despite their dyadic and real-time features, virtual setups are typically far more reductionist than real-life encounters. They convey limited sensory input, and focus on a goal-oriented task that typically does not rely too much on social information. Conversely, real-time social interaction requires people to adapt to unexpected changes in environment as well as changing intentions of interaction partners. The authors argue that autistic people might benefit from a virtual setup, as it provides fewer potentially distracting or confusing sources of information.

Despite these issues, examining real-time social interaction within sensorimotor tasks appears to be a promising approach for gaining a better understanding of autism. Two-person psychophysiology and multilevel analysis of intersubjectivity could increase ecological validity (Bolis and Schilbach, 2018). It will also be important to compare other variations of dyads within larger samples, including autistic-autistic dyads.

Autism as a Problem of Interpersonal Misattunement

The hypothesis of *interpersonal misattunement* redefines autism and other psychiatric conditions not merely as disorders of the individual brain, but as cumulative misattunement between people (Bolis et al., 2017). Already at an early age, the level of synchronicity between an infant and their caregiver correlates with the development of subsequent communicative forms such as JA and language (Siller and Sigman, 2002; Bolis and Schilbach, 2018). Bolis and Schilbach (2018) distinguish two types of cumulative attunement that may be important for synchronizing interactions: propositional attunement (knowing what to do) and pragmatic attunement (knowing how to do something). They argue that autistic people, irrespective of verbal skills, may struggle with spontaneous interactions because they show less pragmatic attunement with others. Based on this hypothesis, interactions within homogeneous dyads, such as two autistic people, would be expected to appear smoother compared to heterogeneous dyads. As such, describing autism as a cumulative mismatch might potentially alleviate social stigma (Bolis et al., 2017). Clinical settings subscribing to this hypothesis could do this by prioritizing the coupling dynamics and "tuning" of interaction *between* people over the tuning of individual behavior. A preliminary evaluation of an autism-specific group psychotherapy focusing on these interactive elements suggests that autistic people benefit from a setting in which they interact with other autistic people (Parpart et al., 2018).

Studies on Autistic-Autistic Interaction

Some recent studies have started to investigate social interaction in autistic-autistic dyads. In a study with 52 autistic and non-autistic adults, Wadge and colleagues (2019) compared different dyad combinations in a game focused on reconstructing a geometric target using communication. They observed that autistic pairs and mixed pairs performed worse than non-autistic pairs at solving these problems. The difference was driven by a lower use of shared signals, pointing to less communicative alignment. Another study focused on the hypothesis that autistic people develop shared understanding in other ways than non-autistic people. To examine how this process might come about, Heasman and Gillespie (2018) analyzed dialogues in video-recorded interactions of 30 autistic adults playing video games. The authors identified two outstanding characteristics in these pair-wise interactions. First, participants made generous assumptions of common ground, in that they frequently shifted topics without introducing them to the other person. Second, participants had a low demand for coordination, in that they showed a high tolerance for misunderstandings or ignored turns. Interestingly, both factors seemed to have large beneficial effects in improving rapport and reciprocity. More support for the idea that autistic people exhibit differences, rather than deficits, in their interactional style stems from a recent study of 172 adults performing semi-structured interactions within varying dyad constellations (Crompton et al., 2020). Compared to non-autistic pairs and mixed pairs, external (non-autistic and autistic) raters perceived interactional rapport as highest in autistic pairs. In sum, these studies demonstrate how moving away from a deficit-oriented perspective may further enhance our understanding of autistic social interaction.

THE BAYESIAN BRAIN ACCOUNT OF AUTISM

Autism is characterized by a diverse set of characteristics. Yet the most prominent autism theories, such as accounts of weak central coherence (Happé and Frith, 2006) and mindblindness (Baron-Cohen, 1995), are typically more well-suited to explaining specific features rather than the full phenotype. The Bayesian brain account may offer a more unifying theoretical framework of autism. More generally, this account can describe psychiatric and neurodevelopmental disorders in terms of how individuals adapt to changing environments using different sources of information (for a detailed discussion of clinical applications of Bayesian accounts, see Haker, Schneebeli, and Stephan, 2016). As such, it could also be useful in explaining strengths seen in autism, such as enhanced perceptual functioning (e.g., Bonnel et al., 2003; reviewed in Mottron et al., 2006). The Bayesian brain account views social difficulties as rooted in more domain-general inferential processes and provides plausible hypotheses for how these differences are expressed on a biological level.

Bayesian Inference and Predictive Coding

According to Bayesian accounts of cognition and perception the human brain makes sense of the external world by continuously inferring what states of the environment are causing sensory inputs (Friston, 2010). In other words, the brain does not passively perceive the world, but actively strives to explain and predict it. The brain makes these inferences by relating current sensory input (observed data or likelihood) to what was previously learned about the environment (prior belief or prior). This process is formalized using Bayes' theorem: in graphical terms, the prior, likelihood, and posterior, can be represented as probability distributions. Applying this Bayesian brain account to autism, Pellicano and Burr (2012) have proposed that autistic individuals favor immediate sensory evidence over prior beliefs. They suggested that autistic prior probability distributions are broader, thus exerting a weaker influence on perception. Building on this hypothesis, autistic people may be less influenced by previous information because they *expect* their priors to be less precise (Friston, Lawson, and Frith, 2013; Lawson, Rees, and Friston, 2014).

To understand how Bayesian inference is instantiated in physiological processes in body and brain, it can be explained within a predictive coding framework (cf. Clark, 2013; Hohwy, 2014; Hohwy, 2017). In predictive coding terms, the means of prior probability distributions represent predictions, and the difference between predicted sensory input and actual sensory input is expressed in terms of prediction errors. Following the free energy principle, the brain attempts to minimize surprise by minimizing unexplained information, i.e., prediction errors. It can do so via two mechanisms—by acting in accordance with predictions (active inference), or by updating the internal model (perceptual inference) to improve future predictions. An example of perceptual inference would be learning to identify the position of crickets based on their auditory signals (cf. Skewes and Gebauer, 2016). An example of active inference would be directing eyes toward aspects that the brain predicts, resulting in more frequent saccades toward predicted content. Predictions and prediction errors are generated at different layers in the cortical hierarchy. Prediction errors are passed up in this hierarchy, while predictions are passed down. Biologically, these processes are instantiated by modulating postsynaptic gain in superficial pyramidal neurons (e.g., Feldman and Friston, 2010). Applying this framework to autistic perception, Van de Cruys and colleagues (2014) have proposed that prediction errors in autism have a higher than expected precision and are less flexible.

As Palmer, Lawson, and Hohwy (2017) have pointed out, these Bayesian accounts of autism share the assumption that autistic people have a higher learning rate. Importantly, learning rates fluctuate depending on the weighting balance of predictions and prediction errors, a balance that responds to different sources of uncertainty at different hierarchical levels. For instance, learning rates decline when precisions of our sensory environment are more stable over time, and are augmented when conditions are more volatile (e.g., Behrens et al., 2007). Tracking these fluctuations in learning rates may provide important clues to how autistic people adapt to changing environments (e.g., see the following for a study by Lawson, Mathys, and Rees, 2017).

Autism and the Bayesian Brain

There is now emerging evidence from experimental studies putting the Bayesian brain account of autism to a test (also see discussion by Palmer, Lawson, and Hohwy, 2017). To assess how individuals weight priors, studies typically manipulate the expectations of participants using, for instance, contextual cues or illusions. The rubber-hand illusion is brought about by applying synchronous sensory stimulation to an individual's concealed hand and a rubber hand situated next to it. Watching the rubber hand being stimulated and feeling that stimulation on one's own hand helps create a sense of ownership for the rubber hand. When the rubber hand is moved, participants experience a drift in the estimated position of their own hand, and this effect can influence subsequent motor movements. Palmer and colleagues (2013) examined the effects of this illusion in a sample of 24 adults with high- and low-levels of autistic traits. They observed that participants with fewer autistic traits were more influenced by the illusion, displaying differences in proprioceptive drift and smoothness of grasping movements between conditions. Participants with more autistic traits exhibited no such differences between conditions, despite reporting subjective effects of the illusion. This indicates that they integrated contextual variation less into their perceptual decisions and actions, which could be explained by a lower expected prior precision. In a second study, Palmer and colleagues (2015) extended these findings by comparing the effects of this illusion in 15 autistic adults and 30 control participants with a high vs low number of autistic traits. All three groups reported a subjective experience of the illusion and showed comparable magnitudes of perceptual integration, including proprioceptive drift. However, an effect on the smoothness of grasp movements was only observed in control participants with a low number of autistic traits. Results indicate that both autism and autistic traits are associated with a lower integration of context, but highlight that this integration might particularly concern active inference in the form of motor movements.

One way of directly contrasting individuals' prior precisions and sensory precisions is to manipulate response criteria and discriminability in signal detection experiments. In a study with 29 adults, Skewes, Jegindø, and Gebauer (2015) introduced statistical contexts and sensory noise to assess how they influence perceptual decisions on visual orientation. They observed that participants with more autistic traits integrated statistical contexts less into their response criteria, indicating lower prior precisions. Conversely, there was no difference between people with high- and low-levels of autistic traits in the influence of sensory noise on discriminability, pointing to comparable sensory precisions. Skewes and Gebauer (2016) extended these findings in a signal detection study with 20 autistic and 23 non-autistic adults, focusing on auditory localization. Autistic participants responded less optimally to manipulations of statistical context and sensory noise, indicating a less efficient integration of prior information with sensory information.

Manning and colleagues (2017) were the first to test whether autistic children (n = 34) have higher learning rates than non-autistic children (n = 32) and adults (n = 19) under volatile circumstances. Contrasting stable and volatile reward probabilities in a learning task, they estimated learning rates with a reinforcement learning mechanism, applying a delta learning rule. They found no group differences in overall learning rates, nor in increases of learning rates in the volatile condition. Results indicate that proposed differences in precision weighting in autism might be more constrained than previously assumed. As the authors argue, differences may be confined to more complex forms of perceptual learning, such as a more continuous tracking of reward probabilities, or taking into account several types of information. As such, models that implement hierarchical layers may be better suited for capturing autism-related differences. Lawson, Mathys, and Rees (2017) used the hierarchical Gaussian filter (HGF: cf. Mathys et al., 2011) to examine expectations of volatility in 49 autistic and non-autistic adults. They observed that autistic participants showed less surprise in response to volatility changes, represented in less modulation of reaction times, error rates, and pupil dilation by expectedness. Autistic participants also showed a greater weighting of expectations about volatility, represented in increased updating of learning rates on volatility. Findings indicate that autistic people may overestimate how unstable their environment is, resulting in a less flexible weighting of prediction errors under uncertainty.

Despite these promising first findings, it should be noted that evidence for the Bayesian brain account of autism is still limited. Studies comparing Bayesian inference in different psychiatric and neurodevelopmental conditions will be important to determine in more detail in what parameters they differ from each other and how these differences relate to causes and phenotypes (e.g., see Lanillos et al., 2020 for a comparison of Bayesian brain and other neural network models in schizophrenia vs autism).

Autism and Social Bayes

Some recent studies have tested the Bayesian brain account of autism in a social-cognitive context. One example for how weighting of social priors can be assessed is the direct gaze bias. It describes the phenomenon that people tend to perceive eye gaze as more direct when depicted eyes contain more perceptual noise. Under a Bayesian framework this can be explained by a greater reliance on priors during conditions of high uncertainty, and it would be expected to be reduced in the case of autism. Contrary to these predictions, a recent study indicates that the effect can also be seen in autistic individuals (Pell et al., 2016). Applying a more complex Bayesian hierarchical model, Sevgi and colleagues (2020) examined how autistic traits are associated with the weighting of social cues. In a study with 36 adults, the authors assessed participants' propensity to integrate eye gaze in a reward-based learning task. The task varied reward probabilities of eye gaze and a nonsocial cue, as well as the volatility of reward probabilities. Participants' responses were modeled using the HGF (Mathys et al., 2011). In line with a Bayesian brain account of autism, autistic traits were associated with poorer task performance, explained by a suboptimal integration of social cues during volatile phases. More specifically, people with more autistic traits tended to over-rely on gaze context when its accuracy was low. Chambon and colleagues (2017) built on signal detection experiments by Skewes, Jegindø, and Gebauer (2015), assessing prior and sensory precisions in inferences on social versus non-social motor intentions. The non-autistic control group (n = 20) responded more to manipulations of priors for social inferences. No such difference was seen in autistic participants (n = 18), supporting a Bayesian interpretation of attenuated social priors. Another recent study examined how 36 male autistic and non-autistic adults coded social prediction errors in the brain (Balsters et al., 2017). In a new false belief task, the researchers assessed social prediction errors by contrasting actual and expected outcomes of others' decisions. Autistic participants were less good at using reward probabilities to identify likely outcomes, and this difference was stronger for outcomes of another person or a computer-generated agent. As in previous studies, social prediction error was associated with activation of the gyral surface of the anterior cingulate cortex (ACCg), represented in a negative blood-oxygenation level dependent (BOLD) response to others' unexpected rewards. Importantly, however, this ACCg response to social prediction error could not be observed in the autistic group. Results indicate that the magnitude of brain responses to social prediction errors may be attenuated in autism. Findings on social aspects of Bayesian inference may explain how social communicative features such as difficulty responding to others or a reduced tendency to use gestures and other nonverbal signals, develop in autism. These communicative skills are typically acquired in interaction with others. Here, less efficient forms of inferential processing—in the form of attenuated social priors or attenuated neural processing of social prediction errors—would be expected to influence skill development. As can be seen from the discussed studies, the Bayesian brain account of autism is a promising theoretical framework that will need further testing and refinement.

General Outlook and Perspectives

This chapter gives an overview of theory and empirical findings on how social cognitive and interactive abilities develop differently in autism. To conclude, we would like to point out developments in this field that are expected to shape the next decades of autism research. First, it appears that the difficulties autistic people experience in real life happen primarily during interactions with others. Innovations in experimental settings increasingly allow researchers to study social functioning through a second-person rather than a third-person perspective (cf. Schilbach, 2015; Schilbach, 2016). One example of how these interactions can be targeted more directly is the use of mobile phone apps in research and interventions (Koumpouros and Kafazis, 2019). These apps could already prove useful in early childhood, facilitating early detection and intervention. Examples could be first screening tools helping parents recognize features warranting further diagnostic assessment, as well as self-guided trainings for parents of children with a confirmed diagnosis. Second, there has been some development in explaining autistic features within an overarching theory. The Bayesian brain account of autism describes these traits in terms of Bayesian inference, making plausible assumptions for underlying neurobiological mechanisms. A step forward will be to further integrate this framework with the view of autism as a social interaction disorder (cf. Bolis and Schilbach, 2018). Third, it will be important that theoretical advances are visible in the development of clinical and social interventions. A stronger focus will need to be put on social functioning in everyday life, including employment, access to support systems, and interpersonal relationships. A fourth issue apparent in the discussed studies is that sample sizes are often quite small. This may be an important reason for mixed and inconsistent evidence. To tackle this problem, open science methods will likely gain more relevance, as well as multicenter studies and databases with wider geographical and temporal scopes. One interesting example of a longitudinal, nationwide database is the Nederlands Autisme Register (NAR) (Begeer et al., 2017). The NAR was created together with autistic stakeholders and is designed to provide feedback to its autistic members that is useful to them. One fifth and final issue concerns the question of how autism studies can benefit autistic people. An important study in the UK showed that autistic people and their families did not perceive research outcomes as leading to concrete improvements in their lives (Pellicano, Dinsmore, and Charman, 2014). To address this issue, attempts have been made to develop research processes in which autistic people take a more active part (e.g., Fletcher-Watson et al., 2018). These initiatives also seem highly relevant for studying social abilities, considering that they are tightly linked to autistic people's quality of life (Tobin et al., 2014; Kim and Bottema-Beutel, 2019) and tend to be one of their main priorities in psychotherapy interventions (e.g., Gawronski et al., 2011). As is also highlighted in emerging accounts such as the double empathy problem, autism researchers need to critically reflect on how they can best form an understanding of the perspectives of autistic people. To establish and improve this understanding, the lived experience of the autism community seems vital.

REFERENCES

Abu-Akel, A. and Shamay-Tsoory, S. (2011). Neuroanatomical and neurochemical bases of theory of mind. *Neuropsychologia*, 49(11): 2971–2984. Doi: 10.1016/j.neuropsychologia.2011.07.012.

Alkire, D. et al. (2018). Social interaction recruits mentalizing and reward systems in middle childhood. *Human Brain Mapping*, 39(10): 3928–3942. Doi: 10.1002/hbm.24221.

Anderson, D. K. et al. (2007). Patterns of growth in verbal abilities among children with autism spectrum disorder. *Journal of Consulting and Clinical Psychology*, 75(4): 594–604. Doi: 10.1037/0022-006X.75.4.594.

Balsters, J. H. et al. (2017). Disrupted prediction errors index social deficits in autism spectrum disorder. *Brain*, 140(1): 235–246. Doi: 10.1093/brain/aww287.

Baron-Cohen, S. (1995). *Mindblindness: An Essay on Autism and Theory of Mind*. Cambridge, Massachusetts: MIT press.

Baron-Cohen, S. (2000). Theory of mind and autism: A fifteen-year review. In Baron-Cohen, S., Tager-Flusberg, H., and Cohen, D. J. (eds.) *Understanding Other Minds: Perspectives from Developmental Cognitive Neuroscience*. 2nd edn. New York, NY, US: Oxford University Press, pp. 3–20.

Baron-Cohen, S., Leslie, A. M., and Frith, U. (1985). Does the autistic child have a "theory of mind"? *Cognition*, 21(1): 37–46. Doi: 10.1016/0010-0277(85)90022-8.

Baron-Cohen, S. and Wheelwright, S. (2004). The empathy quotient: An investigation of adults with Asperger syndrome or high functioning autism, and normal sex differences. *Journal of Autism and Developmental Disorders*, 34(2): 163–175. Doi: 10.1023/B:JADD.0000022607.19833.00.

Begeer, S. et al. (2017). Brief report: Influence of gender and age on parent reported subjective well-being in children with and without autism. *Research in Autism Spectrum Disorders*, 35: 86–91. Doi: 10.1016/j.rasd.2016.11.004.

Behrens, T. E. J. et al. (2007). Learning the value of information in an uncertain world. *Nature Neuroscience*, 10(9): 1214–1221. Doi: 10.1038/nn1954.

Bilek, E. et al. (2015). Information flow between interacting human brains: Identification, validation, and relationship to social expertise. *Proceedings of the National Academy of Sciences*, 112(16): 5207–5212. Doi: 10.1073/pnas.1421831112.

Bird, G. et al. (2010). Empathic brain responses in insula are modulated by levels of alexithymia but not autism. *Brain*, 133(5): 1515–1525. Doi: 10.1093/brain/awq060.

Bird, G. and Cook, R. (2013). Mixed emotions: The contribution of alexithymia to the emotional symptoms of autism. *Translational Psychiatry*, 3(7): e285–e285. Doi: 10.1038/tp.2013.61.

Bird, G. and Viding, E. (2014). The self to other model of empathy: Providing a new framework for understanding empathy impairments in psychopathy, autism, and alexithymia. *Neuroscience & Biobehavioral Reviews*, 47: 520–532. Doi: 10.1016/j.neubiorev.2014.09.021.

Bolis, D. et al. (2017). Beyond autism: Introducing the dialectical misattunement hypothesis and a Bayesian account of intersubjectivity. *Psychopathology*, 50(6): 355–372. Doi: 10.1159/000484353.

Bolis, D. and Schilbach, L. (2018). Observing and participating in social interactions: Action perception and action control across the autistic spectrum. *Developmental Cognitive Neuroscience*, 29: 168–175. Doi: 10.1016/j.dcn.2017.01.009.

Bonnel, A. et al. (2003). Enhanced pitch sensitivity in individuals with autism: A signal detection analysis. *Journal of Cognitive Neuroscience*, 15(2): 226–235. Doi: 10.1162/089892903321208169.

Brewer, R. et al. (2016). Can neurotypical individuals read autistic facial expressions? Atypical production of emotional facial expressions in autism spectrum disorders: Expression production in autism. *Autism Research*, 9(2): 262–271. Doi: 10.1002/aur.1508.

Cage, E., Di Monaco, J., and Newell, V. (2018). Understanding, attitudes and dehumanisation towards autistic people. *Autism*, p. 136236131881129. Doi: 10.1177/1362361318811290.

Carpenter, M. et al. (1998). Social cognition, joint attention, and communicative competence from 9 to 15 months of age. *Monographs of the Society for Research in Child Development*, 63(4): i. Doi: 10.2307/1166214.

Carrington, S. J. and Bailey, A. J. (2009). Are there theory of mind regions in the brain? A review of the neuroimaging literature. *Human Brain Mapping*, 30(8): 2313–2335. Doi: 10.1002/hbm.20671.

Castelli, F. et al. (2002). Autism, Asperger syndrome and brain mechanisms for the attribution of mental states to animated shapes. *Brain*, 125(8): 1839–1849. Doi: 10.1093/brain/awf189.

Chambon, V. et al. (2017). Reduced sensitivity to social priors during action prediction in adults with autism spectrum disorders. *Cognition*, 160: 17–26. Doi: 10.1016/j.cognition.2016.12.005.

Charman, T. et al. (2000). Testing joint attention, imitation, and play as infancy precursors to language and theory of mind. *Cognitive Development*, 15(4): 481–498. Doi: 10.1016/S0885-2014(01)00037-5.

Charman, T. (2003). Why is joint attention a pivotal skill in autism? *Philosophical Transactions of the Royal Society of London. Series B: Biological Sciences*, 358(1430): 315–324. Doi: 10.1098/rstb.2002.1199.

Clark, A. (2013). Whatever next? Predictive brains, situated agents, and the future of cognitive science. *Behavioral and Brain Sciences*, 36(3): 181–204. Doi: 10.1017/S0140525X12000477.

Coll, M. -P. et al. (2017). Are we really measuring empathy? Proposal for a new measurement framework. *Neuroscience & Biobehavioral Reviews*, 83: 132–139. Doi: 10.1016/j.neubiorev.2017.10.009.

Crompton, C. J. et al. (2020). Neurotype-matching, but not being autistic, influences self and observer ratings of interpersonal rapport. *Frontiers in Psychology*, 11: 2961. Doi: 10.3389/fpsyg.2020.586171.

Davis, M. H. (1983). Measuring individual differences in empathy: Evidence for a multidimensional approach. *Journal of personality and social psychology*, 44(1): 113–126.

Decety, J. and Jackson, P. L. (2004). The functional architecture of human empathy. *Behavioral and Cognitive Neuroscience Reviews*, 3(2): 71–100. Doi: 10.1177/1534582304267187.

Decety, J. and Lamm, C. (2006). Human empathy through the lens of social neuroscience. *The Scientific World Journal*, 6: 1146–1163. Doi: 10.1100/tsw.2006.221.

Delinicolas, E. K. and Young, R. L. (2007). Joint attention, language, social relating, and stereotypical behaviours in children with autistic disorder. *Autism*, 11(5): 425–436. Doi: 10.1177/1362361307079595.

Demurie, E., De Corel, M., and Roeyers, H. (2011). Empathic accuracy in adolescents with autism spectrum disorders and adolescents with attention-deficit/hyperactivity disorder. *Research in Autism Spectrum Disorders*, 5(1): 126–134. Doi: 10.1016/j.rasd.2010.03.002.

Dowd, A., Rinehart, N., and McGinley, J. (2010). Motor function in children with autism: Why is this relevant to psychologists? *Clinical Psychologist*, 14(3): 90–96. Doi: 10.1080/13284207.2010.525532.

Dufour, N. et al. (2013). Similar brain activation during false belief tasks in a large sample of adults with and without autism. *PLOS One*. S. Gilbert (ed.) 8(9): e75468. Doi: 10.1371/journal.pone.0075468.

Dziobek, I. et al. (2008). Dissociation of cognitive and emotional empathy in adults with Asperger syndrome using the Multifaceted Empathy Test (MET). *Journal of Autism and Developmental Disorders*, 38(3): 464–473. Doi: 10.1007/s10803-007-0486-x.

Elsabbagh, M. et al. (2013). Disengagement of visual attention in infancy is associated with emerging autism in toddlerhood. *Biological Psychiatry*, 74(3): 189–194. Doi: 10.1016/j.biopsych.2012.11.030.

Fan, Y.-T. et al. (2014). Empathic arousal and social understanding in individuals with autism: Evidence from fMRI and ERP measurements. *Social Cognitive and Affective Neuroscience*, 9(8): 1203–1213. Doi: 10.1093/scan/nst101.

Feldman, H. and Friston, K. J. (2010). Attention, uncertainty, and free-energy. *Frontiers in Human Neuroscience*, 4. Doi: 10.3389/fnhum.2010.00215.

Fletcher, P. (1995). Other minds in the brain: A functional imaging study of "theory of mind" in story comprehension. *Cognition*, 57(2): 109–128. Doi: 10.1016/0010-0277(95)00692-R.

Fletcher-Watson, S. et al. (2018). Making the future together: Shaping autism research through meaningful participation. *Autism*, 943–953, 136236131878672. Doi: 10.1177/1362361318786721.

Friston, K. (2010). The free-energy principle: A unified brain theory? *Nature Reviews Neuroscience*, 11(2): 127–138. Doi: 10.1038/nrn2787.

Friston, K. J., Lawson, R., and Frith, C. D. (2013). On hyperpriors and hypopriors: Comment on Pellicano and Burr. *Trends in Cognitive Sciences*, 17(1): 1. Doi: 10.1016/j.tics.2012.11.c03.

Frith, C. D. and Frith, U. (1999). Interacting minds—a biological basis. *Science*, 286(5445): 1692–1695. Doi: 10.1126/science.286.5445.1692.

Frith, U. (2001). Mind blindness and the brain in autism. *Neuron*, 32(6): 969–979. Do:: 10.1016/S0896-6273(01)00552-9.

Gawronski, A. et al. (2011). Erwartungen an eine Psychotherapie von hochfunktionalen erwachsenen Personen mit einer Autismus-Spektrum-Störung. *Fortschritte der Neurologie · Psychiatrie*, 79(11): 647–654. Doi: 10.1055/s-0031-1281734.

Gotham, K. et al. (2007). The autism diagnostic observation schedule: Revised algorithms for improved diagnostic validity. *Journal of Autism and Developmental Disorders*, 37(4): 613–627. Doi: 10.1007/s10803-006-0280-1.

Grynberg, D. et al. (2010). Alexithymia in the interpersonal domain: A general deficit of empathy? *Personality and Individual Differences*, 49(8): 845–850. Doi: 10.1016/j.paid.2010.07.013.

Gu, X. et al. (2015). Autonomic and brain responses associated with empathy deficits in autism spectrum disorder: Autonomic and brain responses in ASD. *Human Brain Mapping*, 36(9): 3323–3338. Doi: 10.1002/hbm.22840.

Hadjikhani, N. et al. (2014). Emotional contagion for pain is intact in autism spectrum disorders. *Translational Psychiatry*, 4(1): e343–e343. Doi: 10.1038/tp.2013.113.

Haker, H., Schneebeli, M., and Stephan, K. E. (2016). Can Bayesian theories of autism spectrum disorder help improve clinical practice? *Frontiers in Psychiatry*, 7. Doi: 10.3389/fpsyt.2016.00107.

Happé, F. and Frith, U. (2006). The weak coherence account: Detail-focused cognitive style in autism spectrum disorders. *Journal of Autism and Developmental Disorders*, 36(1): 5–25. Doi: 10.1007/s10803-005-0039-0.

Happé, F. G. E. (1994). An advanced test of theory of mind: Understanding of story characters' thoughts and feelings by able autistic, mentally handicapped, and normal children and adults. *Journal of Autism and Developmental Disorders*, 24(2): 129–154. Doi: 10.1007/BF02172093.

Happé, F. G. E. (1995). The role of age and verbal ability in the theory of mind task performance of subjects with autism. *Child Development*, 66(3): 843. Doi: 10.2307/1131954.

Happé, F. G. E. et al. (1996). "Theory of mind" in the brain. Evidence from a PET scan study of Asperger syndrome. *Neuroreport*, 8(1): 197–201.

Hazlett, H. C. et al. (2017). Early brain development in infants at high risk for autism spectrum disorder. *Nature*, 542(7641): 348–351. Doi: 10.1038/nature21369.

Heasman, B. and Gillespie, A. (2018). Neurodivergent intersubjectivity: Distinctive features of how autistic people create shared understanding. *Autism*, 910–921. Doi: 10.1177/1362361318785172.

Hohwy, J. (2014). The self-evidencing brain. *Noûs*, 50(2): 259–285. Doi: 10.1111/nous.12062.

Hohwy, J. (2017). Priors in perception: Top-down modulation, Bayesian perceptual learning rate, and prediction error minimization. *Consciousness and Cognition*, 47: 75–85. Doi: 10.1016/j.concog.2016.09.004.

Holopainen, A. et al. (2018). Does theory of mind training enhance empathy in Autism? *Journal of Autism and Developmental Disorders*. Doi: 10.1007/s10803-018-3671-1.

Johnson, S. A., Filliter, J. H., and Murphy, R. R. (2009). Discrepancies between self- and parent-perceptions of autistic traits and empathy in high functioning children and adolescents on the Autism spectrum. *Journal of Autism and Developmental Disorders*, 39(12): 1706–1714. Doi: 10.1007/s10803-009-0809-1.

Jolliffe, T. and Baron-Cohen, S. (1999). The strange stories test: A replication with high-functioning adults with autism or Asperger syndrome. *Journal of Autism and Developmental Disorders*, 29(5): 395–406.

Jones, A. P. et al. (2010). Feeling, caring, knowing: Different types of empathy deficit in boys with psychopathic tendencies and autism spectrum disorder. *Journal of Child Psychology and Psychiatry*, 51(11): 1188–1197. Doi: 10.1111/j.1469-7610.2010.02280.x.

Kana, R. K. et al. (2009). Atypical frontal-posterior synchronization of Theory of Mind regions in autism during mental state attribution. *Social Neuroscience*, 4(2): 135–152. Doi: 10.1080/17470910802198510.

Kana, R. K. et al. (2015). Aberrant functioning of the theory-of-mind network in children and adolescents with autism. *Molecular Autism*, 6(1): 1–12. Doi: 10.1186/s13229-015-0052-x.

Keehn, B., Müller, R.-A., and Townsend, J. (2013). Atypical attentional networks and the emergence of autism. *Neuroscience & Biobehavioral Reviews*, 37(2): 164–183. Doi: 10.1016/j.neubiorev.2012.11.014.

Kinnaird, E., Stewart, C., and Tchanturia, K. (2019). Investigating alexithymia in autism: A systematic review and meta-analysis. *European Psychiatry*, 55: 80–89. Doi:10.1016/j.eurpsy.2018.09.004.

Kim, S. Y. and Bottema-Beutel, K. (2019). A meta regression analysis of quality of life correlates in adults with ASD. *Research in Autism Spectrum Disorders*, 63, 23–33. Doi: 10.1016/j.rasd.2018.11.004

Koumpouros, Y. and Kafazis, T. (2019). Wearables and mobile technologies in autism spectrum disorder interventions: A systematic literature review. *Research in Autism Spectrum Disorders*, 66: 101405. Doi: 10.1016/j.rasd.2019.05.005.

Krach, S. et al. (2015). Evidence from pupillometry and fMRI indicates reduced neural response during vicarious social pain but not physical pain in autism: Vicarious physical and social pain in ASD. *Human Brain Mapping*, 36(11): 4730–4744. Doi: 10.1002/hbm.22949.

Kulke, L. et al. (2018). Is implicit theory of mind a real and robust phenomenon? Results from a systematic replication study. *Psychological Science*, 29(6): 888–900. Doi: 10.1177/0956797617747090.

Lainhart, J. E. (2015). Brain imaging research in autism spectrum disorders: In search of neuropathology and health across the lifespan. *Current Opinion in Psychiatry*, 28(2): 76. Doi: 10.1097/YCO.0000000000000130.

Lane, R. D. et al. (1996). Impaired verbal and nonverbal emotion recognition in alexithymia. *Psychosomatic Medicine*, 58(3): 203–210. Doi: 10.1097/00006842-199605000-00002.

Lanillos, P., Oliva, D., Philippsen, A., Yamashita, Y., Nagai, Y., and Cheng, G. (2020). A review on neural network models of schizophrenia and autism spectrum disorder. *Neural Networks*, 122, 338–363. Doi: 10.1016/j.neunet.2019.10.014

Lawson, R. P., Mathys, C., and Rees, G. (2017). Adults with autism overestimate the volatility of the sensory environment. *Nature Neuroscience*, 20(9): 1293–1299. Doi: 10.1038/nn.4615.

Lawson, R. P., Rees, G., and Friston, K. J. (2014). An aberrant precision account of autism. *Frontiers in Human Neuroscience*, 8. Doi: 10.3389/fnhum.2014.00302.

Lenay, C. (2008). Médiations techniques des interactions perceptives: rencontres tactiles dans les environnements numériques partagés. *Social Science Information*, 47(3): 331–352. Doi: 10.1177/0539018408092576.

Lombardo, M. V. et al. (2007). Self-referential cognition and empathy in autism. *PLOS One*. P. Zak, (ed.) 2(9): e883. Doi: 10.1371/journal.pone.0000883.

Lombardo, M. V. et al. (2011). Specialization of right temporo-parietal junction for mentalizing and its relation to social impairments in autism. *NeuroImage*, 56(3): 1832–1838. Doi: 10.1016/j.neuroimage.2011.02.067.

Lombardo, M. V. and Baron-Cohen, S. (2011). The role of the self in mindblindness in autism. *Consciousness and Cognition*, 20(1): 130–140. Doi: 10.1016/j.concog.2010.09.006.

Manning, C. et al. (2017). Children on the autism spectrum update their behaviour in response to a volatile environment. *Developmental Science*, 20(5): e12435. Doi: 10.1111/desc.12435.

Markram, K. and Markram, H. (2010). The intense world theory—a unifying theory of the neurobiology of autism. *Frontiers in Human Neuroscience*, 4: 224. Doi: 10.3389/fnhum.2010.00224.

Mason, R. A. et al. (2008). Theory of mind disruption and recruitment of the right hemisphere during narrative comprehension in autism. *Neuropsychologia*, 46(1): 269–280. Doi: 10.1016/j.neuropsychologia.2007.07.018.

Mathys, C. et al. (2011). A Bayesian foundation for individual learning under uncertainty. *Frontiers in Human Neuroscience*, 5: 39. Doi: 10.3389/fnhum.2011.00039.

Milton, D. E. M. (2012). On the ontological status of autism: The "double empathy problem". *Disability & Society*, 27(6): 883–887. Doi: 10.1080/09687599.2012.710008.

Moriguchi, Y. et al. (2007). Empathy and judging other's pain: An fMRI study of alexithymia. *Cerebral Cortex*, 17(9): 2223–2234. Doi: 10.1093/cercor/bhl130.

Moessnang, C. et al. (2020). Social brain activation during mentalizing in a large autism cohort: The Longitudinal European Autism Project. *Molecular Autism*, 11(1): 17. Doi: 10.1186/s13229-020-0317-x.

Mottron, L. et al. (2006). Enhanced perceptual functioning in autism: An update, and eight principles of autistic perception. *Journal of Autism and Developmental Disorders*, 36(1): 27–43. Doi: 10.1007/s10803-005-0040-7.

Mundy, P. et al. (2007). Individual differences and the development of joint attention in infancy. *Child Development*, 78(3): 938–954. Doi: 10.1111/j.1467-8624.2007.01042.x.

Mundy, P. (2018). A review of joint attention and social-cognitive brain systems in typical development and autism spectrum disorder. *European Journal of Neuroscience*, 47(6): 497–514. Doi: 10.1111/ejn.13720.

Mundy, P. and Newell, L. (2007). Attention, joint attention, and social cognition. *Current Directions in Psychological Science*, 16(5): 269–274. Doi: 10.1111/j.1467-8721.2007.00518.x.

Mundy, P., Sullivan, L., and Mastergeorge, A. M. (2009). A parallel and distributed-processing model of joint attention, social cognition and autism. *Autism Research*, 2(1): 2–21. Doi: 10.1002/aur.61.

Oberwelland, E. et al. (2016). Look into my eyes: Investigating joint attention using interactive eye-tracking and fMRI in a developmental sample. *NeuroImage*, 130: 248–260. Doi: 10.1016/j.neuroimage.2016.02.026.

Oberwelland, E. et al. (2017). Young adolescents with autism show abnormal joint attention network: A gaze contingent fMRI study. *NeuroImage: Clinical*, 14: 112–121. Doi: 10.1016/j.nicl.2017.01.006.

Palmer, C. J. et al. (2013). Movement under uncertainty: The effects of the rubber-hand illusion vary along the nonclinical autism spectrum. *Neuropsychologia*, 51(10): 1942–1951. Doi: 10.1016/j.neuropsychologia.2013.06.020.

Palmer, C. J. et al. (2015). Context sensitivity in action decreases along the autism spectrum: A predictive processing perspective. *Proceedings of the Royal Society B: Biological Sciences*, 282(1802): 20141557–20141557. Doi: 10.1098/rspb.2014.1557.

Palmer, C. J., Lawson, R. P., and Hohwy, J. (2017). Bayesian approaches to autism: Towards volatility, action, and behavior. *Psychological Bulletin*, 143(5): 521–542. Doi: 10.1037/bul0000097.

Parpart, H., Krankenhagen, M., Albantakis, L., Henco, L., Friess, E., and Schilbach, L. (2018). Schematherapie-informiertes, soziales Interaktionstraining. *Psychotherapeut*, 63(3), 235–242. Doi: 10.1007/s00278-018-0271-7.

Parker, J. D., Taylor, G. J., and Bagby, M. (1993). Alexithymia and the recognition of facial expressions of emotion. *Psychotherapy and psychosomatics*, 59(3-4): 197–202.

Pell, P. J. et al. (2016). Intact priors for gaze direction in adults with high-functioning autism spectrum conditions. *Molecular Autism*, 7(1). Doi: 10.1186/s13229-016-0085-9.

Pellicano, E. and Burr, D. (2012). When the world becomes "too real": a Bayesian explanation of autistic perception. *Trends in Cognitive Sciences*, 16(10): 504–510. Doi: 10.1016/j.tics.2012.08.009.

Pellicano, E., Dinsmore, A., and Charman, T. (2014). What should autism research focus upon? Community views and priorities from the United Kingdom. *Autism*, 18(7): 756–770. Doi: 10.1177/1362361314529627.

Pfeiffer, U. J. et al. (2014). Why we interact: On the functional role of the striatum in the subjective experience of social interaction. *NeuroImage*, 101: 124–137. Doi: 10.1016/j.neuroimage.2014.06.061.

Pfeiffer, U. J., Vogeley, K., and Schilbach, L. (2013). From gaze cueing to dual eye-tracking: Novel approaches to investigate the neural correlates of gaze in social interaction. *Neuroscience & Biobehavioral Reviews*, 37(10): 2516–2528. Doi: 10.1016/j.neubiorev.2013.07.017.

Ponnet, K. S. et al. (2004). Advanced mind-reading in adults with Asperger syndrome. *Autism*, 8(3): 249–266. Doi: 10.1177/1362361304045214.

Premack, D. and Woodruff, G. (1978). Does the chimpanzee have a theory of mind? *Behavioral and brain sciences*, 1(4): 515–526. Doi:10.1017/S0140525X00076512.

Prkachin, G. C., Casey, C., and Prkachin, K. M. (2009). Alexithymia and perception of facial expressions of emotion. *Personality and Individual Differences*, 46(4): 412–417. Doi: 10.1016/j.paid.2008.11.010.

Prochazkova, E. et al. (2018). Pupil mimicry promotes trust through the theory-of-mind network. *Proceedings of the National Academy of Sciences*, 115(31): E7265–E7274. Doi: 10.1073/pnas.1803916115.

Redcay, E. et al. (2010). Live face-to-face interaction during fMRI: A new tool for social cognitive neuroscience. *NeuroImage*, 50(4): 1639–1647. Doi: 10.1016/j.neuroimage.2010.01.052.

Redcay, E. et al. (2013). Intrinsic functional network organization in high-functioning adolescents with autism spectrum disorder. *Frontiers in Human Neuroscience*, 7. Doi: 10.3389/fnhum.2013.00573.

Redcay, E. and Schilbach, L. (2019). Using second-person neuroscience to elucidate the mechanisms of social interaction. *Nature Reviews Neuroscience*, 20(8): 495–505. Doi: 10.1038/s41583-019-0179-4.

Richardson, H. et al. (2018). Development of the social brain from age three to twelve years. *Nature Communications*, 9(1): 1–12. Doi: 10.1038/s41467-018-03399-2.

Rogers, K. et al. (2007). Who cares? Revisiting empathy in Asperger syndrome. *Journal of Autism and Developmental Disorders*, 37(4): 709–715. Doi: 10.1007/s10803-006-0197-8.

Salminen, J. K. et al. (1999). Prevalence of alexithymia and its association with sociodemographic variables in the general population of Finland. *Journal of Psychosomatic Research*, 46(1): 75–82. Doi: 10.1016/S0022-3999(98)00053-1.

Sasson, N. J. et al. (2017). Neurotypical peers are less willing to interact with those with autism based on thin slice judgments. *Scientific Reports*, 7(1): 1–10. Doi: 10.1038/srep40700.

Saxe, R. and Kanwisher, N. (2003). People thinking about thinking people. The role of the temporo-parietal junction in "theory of mind." *NeuroImage*, 19(4): 1835–1842. Doi: 10.1016/S1053-8119(03)00230-1.

Scheeren, A. M. et al. (2013). Rethinking theory of mind in high-functioning autism spectrum disorder: Advanced theory of mind in autism. *Journal of Child Psychology and Psychiatry*, 54(6): 628–635. Doi: 10.1111/jcpp.12007.

Schilbach, L. et al. (2010). Minds made for sharing: Initiating joint attention recruits reward-related neurocircuitry. *Journal of Cognitive Neuroscience*, 22(12): 2702–2715. Doi: 10.1162/jocn.2009.21401.

Schilbach, L. et al. (2012). Shall we do this together? Social gaze influences action control in a comparison group, but not in individuals with high-functioning autism. *Autism*, 16(2): 151–162. Doi: 10.1177/1362361311409258.

Schilbach, L. et al. (2013). Toward a second-person neuroscience. *Behavioral and Brain Sciences*, 36(4): 393–414. Doi: 10.1017/S0140525X12000660.

Schilbach, L. (2015). Eye to eye, face to face and brain to brain: Novel approaches to study the behavioral dynamics and neural mechanisms of social interactions. *Current Opinion in Behavioral Sciences*, 3: 130–135. Doi: 10.1016/j.cobeha.2015.03.006.

Schilbach, L. (2016). Towards a second-person neuropsychiatry. *Philosophical Transactions of the Royal Society B: Biological Sciences*, 371(1686): 20150081. Doi: 10.1098/rstb.2015.0081.

Schilbach, L. (2019). Using interaction-based phenotyping to assess the behavioral and neural mechanisms of transdiagnostic social impairments in psychiatry. *European Archives of Psychiatry and Clinical Neuroscience*, 269(3): 273–274. Doi: 10.1007/s00406-019-00998-y.

Schulte-Rüther, M. et al. (2014). Age-dependent changes in the neural substrates of empathy in autism spectrum disorder. *Social Cognitive and Affective Neuroscience*, 9(8): 1118–1126. Doi: 10.1093/scan/nst088.

Schwenck, C. et al. (2012). Empathy in children with autism and conduct disorder: Group-specific profiles and developmental aspects: Empathy in children with autism and CD. *Journal of Child Psychology and Psychiatry*, 53(6): 651–659. Doi: 10.1111/j.1469-7610.2011.02499.x.

Senju, A. et al. (2009). Mindblind eyes: An absence of spontaneous theory of mind in Asperger syndrome. *Science*, 325(5942): 883–885. Doi: 10.1126/science.1176170.

Senju, A. et al. (2010). Absence of spontaneous action anticipation by false belief attribution in children with autism spectrum disorder. *Development and Psychopathology*, 22(2): 353–360. Doi: 10.1017/S0954579410000106.

Sevgi, M. et al. (2020). Social Bayes: Using Bayesian modeling to study autistic trait–related differences in social cognition *Biological Psychiatry*, 87(2): 185–193. Doi: 10.1016/j.biopsych.2019.09.032.

Shamay-Tsoory S.G., Aharon-Peretz J., and Perry D. (2009). Two systems for empathy: a double dissociation between emotional and cognitive empathy in inferior frontal gyrus versus ventromedial prefrontal lesions. *Brain*, 132(3): 617–27. Doi: 10.1093/brain/awn279.

Sigman, M. and McGovern, C. W. (2005). Improvement in cognitive and language skills from preschool to adolescence in autism. *Journal of Autism and Developmental Disorders*, 35(1): 15–23. Doi: 10.1007/s10803-004-1027-5.

Silani, G. et al. (2008). Levels of emotional awareness and autism: An fMRI study. *Social Neuroscience*, 3(2): 97–112. Doi: 10.1080/17470910701577020.

Siller, M. and Sigman, M. (2002). The behaviors of parents of children with autism predict the subsequent development of their children's communication. *Journal of Autism and Developmental Disorders*, 32(2): 77–89. Doi: 10.1023/A:1014884404276.

Skewes, J. C. and Gebauer, L. (2016). Brief report: Suboptimal auditory localization in autism spectrum disorder: Support for the Bayesian account of sensory symptoms. *Journal of Autism and Developmental Disorders*, 46(7): 2539–2547. Doi: 10.1007/s10803-016-2774-9.

Skewes, J. C., Jegindø, E. -M., and Gebauer, L. (2015). Perceptual inference and autistic traits. *Autism*, 19(3): 301–307. Doi: 10.1177/1362361313519872.

Smith, A. (2009). The empathy imbalance hypothesis of autism: A theoretical approach to cognitive and emotional empathy in autistic development. *The Psychological Record*, 59(3): 489–510.

Sodian, B., Schuwerk, T., and Kristen, S. (2015). Implicit and spontaneous theory of mind reasoning in autism spectrum disorders. In Fitzgerald, M. (ed.) *Autism Spectrum Disorder–Recent Advances*. InTech. Doi: 10.5772/59393.

Stone, W. L., Ousley, O. Y., and Littleford, C. D. (1997). Motor imitation in young children with autism: What's the object? *Journal of abnormal child psychology*, 25(6): 475–485. Available at: http://www.ncbi.nlm.nih.gov/pubmed/9468108.

Thaler, H. et al. (2018). Typical pain experience but underestimation of others' pain: Emotion perception in self and others in autism spectrum disorder. *Autism*, 22(6): 751–762. Doi: 10.1177/1362361317701269.

Tobin, M. C., Drager, K. D. R., and Richardson, L. F. (2014). A systematic review of social participation for adults with autism spectrum disorders: Support, social functioning, and quality of life. *Research in Autism Spectrum Disorders*, 8(3), 214–229. Doi: 10.1016/j.rasd.2013.12.002.

Toth, K., et al. (2006). Early predictors of communication development in young children with autism spectrum disorder: Joint attention, imitation, and toy play. *Journal of Autism and Developmental Disorders*, 36(8): 993–1005. Doi: 10.1007/s10803-006-0137-7.

Uddin, L. Q. (2011). The self in autism: An emerging view from neuroimaging. *Neurocase*, 17(3): 201–208. Doi: 10.1080/13554794.2010.509320.

Uljarevic, M. and Hamilton, A. (2013). Recognition of emotions in autism: A formal meta-analysis. *Journal of Autism and Developmental Disorders*, 43(7): 1517–1526. Doi: 10.1007/s10803-012-1695-5.

Van de Cruys, S., et al. (2014). Precise minds in uncertain worlds: Predictive coding in autism. *Psychological Review*, 121(4): 649–675. Doi: 10.1037/a0037665.

Velikonja, T., Fett, A. -K., and Velthorst, E. (2019). Patterns of nonsocial and social cognitive functioning in adults with autism spectrum disorder: A systematic review and meta-analysis. *JAMA Psychiatry*, 76(2): 135. Doi: 10.1001/jamapsychiatry.2018.3645.

Wadge, H. et al. (2019). Communicative misalignment in autism spectrum disorder. *Cortex*, 115: 15–26. Doi: 10.1016/j.cortex.2019.01.003.

Wang, A. T. et al. (2006). Neural basis of irony comprehension in children with autism: The role of prosody and context. *Brain*, 129(4): 932–943. Doi: 10.1093/brain/awl032.

Wellman, H. M., Cross, D., and Watson, J. (2001). Meta-analysis of theory-of-mind development: The truth about false belief. *Child Development*, 72(3): 655–684. Available at: http://www.jstor.org/stable/1132444.

Wheelwright, S. et al. (2006). Predicting autism spectrum quotient (AQ) from the systemizing quotient-revised (SQ-R) and empathy quotient (EQ). *Brain Research*, 1079(1): 47–56. Doi: 10.1016/j.brainres.2006.01.012.

White, S. et al. (2009). Revisiting the strange stories: Revealing mentalizing impairments in autism. *Child Development*, 80(4): 1097–1117. Doi: 10.1111/j.1467-8624.2009.01319.x.

White, S. J. et al. (2011). Developing the Frith-Happé animations: A quick and objective test of theory of mind for adults with autism. *Autism Research*, 4(2): 149–154. Doi: 10.1002/aur.174.

White, S. J. et al. (2014). Autistic adolescents show atypical activation of the brain's mentalizing system even without a prior history of mentalizing problems. *Neuropsychologia*, 56: 17–25. Doi: 10.1016/j.neuropsychologia.2013.12.013.

Yuk, V. et al. (2018). Do you know what I'm thinking? Temporal and spatial brain activity during a theory-of-mind task in children with autism. *Developmental Cognitive Neuroscience*, 34: 139–147. Doi: 10.1016/j.dcn.2018.08.001.

Zapata-Fonseca, L. et al. (2018). Sensitivity to social contingency in adults with high-functioning autism during computer-mediated embodied interaction. *Behavioral Sciences*, 8(2): 22. Doi: 10.3390/bs8020022.

CHAPTER 34

AFFECTIVE DISORDERS IN DEVELOPMENT

MICHELE A. MORNINGSTAR, WHITNEY I. MATTSON, AND ERIC E. NELSON

INTRODUCTION

ANXIETY and depression are among the most commonly experienced psychiatric disorders. Prevalence rates derived from large-scale epidemiology studies suggest that 16–20 percent of adults in the United States will experience depression during their lifetime (Blazer et al., 1994; Kessler et al., 2003; Kessler et al., 1993), and 29 percent will have an anxiety disorder (Kessler et al., 2005). These affective disorders are associated with role impairments in many life domains (Kessler et al., 2003) and place a significant burden on individuals, their families, and society (Greenberg et al., 2015).

Much effort has been directed towards understanding the etiology of affective conditions, in order to prevent their onset or provide effective therapeutic interventions. One approach has been to investigate the development of these disorders longitudinally to identify mechanistic factors involved in their ontogeny. Chronic affective disorders often first emerge in childhood or adolescence (Birmaher et al., 1996; Kessler and Wang, 2008). Early-onset anxiety and depression are sometimes conceptualized as a more severe form of these disorders than adult-onset versions (Andersen and Teicher, 2008; Kaufman et al., 2001; Zisook et al., 2007), given their chronicity (Weissman et al., 1999) and the recursive detrimental impact of these conditions on individuals' social, affective, and cognitive development. Not all individuals who develop affective disorders in their youth go on to experience chronic internalizing problems. However, diagnoses of depression or anxiety during adolescence are robustly related to increased risk of recurrence, or of developing another psychiatric disorder, in early or late adulthood (Beesdo et al., 2007; Colman et al., 2007; Costello et al., 2003), making it crucial to identify and temper factors contributing to these conditions in youth.

Though depression and anxiety have been noted to occur in preschoolers (e.g., Luby, 2010), the onset of affective disorders in early childhood is relatively uncommon. During mid-childhood, some forms of anxiety—like separation anxiety or specific phobias—become much more common (American Psychiatric Association, 2013; Roza et al., 2003). However, most internalizing disorders do not demonstrate much increase in prevalence rates until the early teenage years (Angold et al., 1998; Kessler et al., 2005). Adolescence is considered a sensitive period for the emergence of a number of psychiatric disorders (Kessler et al., 2007; Paus et al., 2008), and of affective disorders in particular (Kessler et al., 2005). For instance, the incidence of depression increases markedly in early adolescence (Costello et al., 2003), possibly related to pubertal development (Angold and Costello, 2006). While 6–8 percent of the general population report depression in the last 12 months (American Psychiatric Association, 2013; Kessler et al., 2003), estimates of prevalence range from 14 percent to 27 percent during early adolescence and emerging adulthood (Beesdo et al., 2007; Merikangas et al., 2010). Anxiety disorders are even more prevalent in childhood and adolescence, affecting 32 percent of 13- to 18-year-olds in an American national survey (Merikangas et al., 2010)—compared to a 12-month prevalence rate of 10 percent in the adult population (Somers et al., 2006). In particular, the rates of panic disorder (Costello et al., 2003) and social anxiety (Beesdo et al., 2007) increase dramatically during the teenage years.

The surge in incidence of affective disorders during adolescence coincides with a myriad of changes in body, brain, and behavior. Dramatic changes to the physical body's appearance and function are paralleled by extensive maturation of the brain's structure, function, and organization. At the interpersonal level, there is a broad restructuring of social support networks brought on by schooling changes, increased demands of extracurricular activities, and evolving relationships with family members, peers, and emerging romantic interests (Larson and Richards, 1991; Steinberg and Morris, 2001). Adolescents begin spending less time with their family and more time with their peers, with whom they develop strong and complex relationships (Hartup, 1996).

Social re-orientation away from caregivers and towards peers facilitates the development of social and romantic relationships, yet also represents a challenge in adaptation for youth (Masten, 2005). Adolescents are tasked with managing multifaceted emotions and cognitions about their peers, their social network, and their place within it. For instance, they must learn to navigate their emotions and those of others, and to integrate their emerging sense of self with others' perceptions of them, in order to achieve competent social integration. At a neural level, the acquisition and interpretation of this complex socio-emotional information is thought to be guided by three processing nodes: detection, affect, and cognitive regulation. As described by the Social Information Processing Network model (SIPN; Nelson et al., 2005), these nodes are thought to work both serially and in conjunction with one another to integrate the perception of relevant social cues in the environment, determine their emotional significance, and regulate cognitive interpretation and behavioral responses (see Figure 34.1).

Because of the confluence of novel contexts and behaviors, emergent neural networks, and the highly motivated aspect of social behavior, adolescence may be an

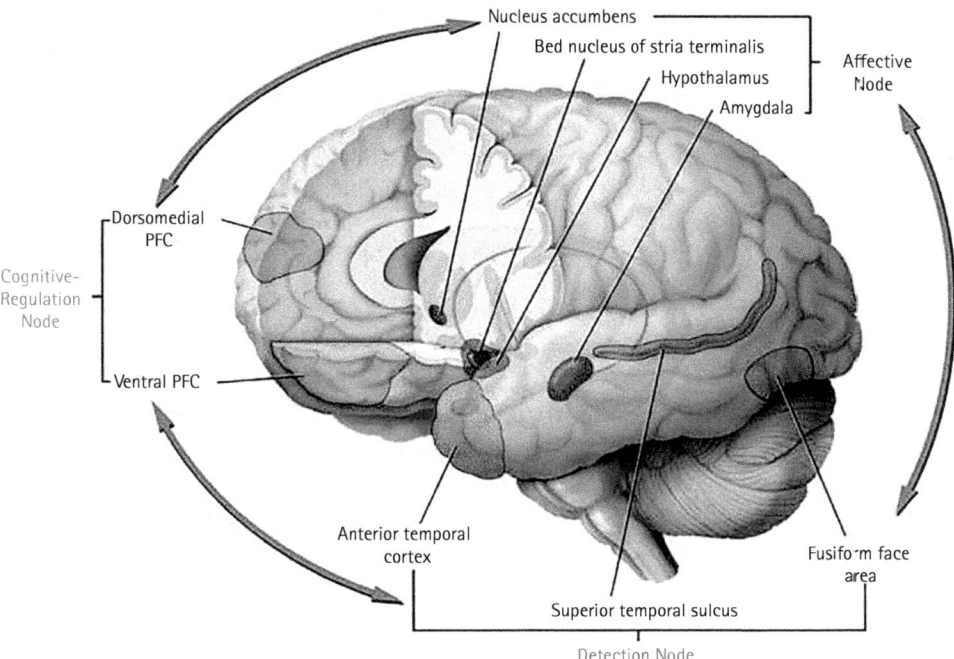

FIGURE 34.1 The Social Information Processing Network (SIPN) model.

Source: Reproduced from Nelson et al., 2005.
Reprinted with permission of Cambridge University Press.

especially vulnerable period for disruptions to social and emotional maturation (Crone and Dahl, 2012; Dahl et al., 2018; Paus et al., 2008; Steinberg, 2005). In particular, adolescence may be a sensitive period for the emergence of disorders that feature problematic cognitive and emotional appraisals of the environment, such as depression and anxiety. Information processing is likely disrupted at multiple levels in these affective disorders. For instance, heightened emotional responding in the affect node can alter cognitive processes like attention, learning, and the selection of behavioral responses in ways that perpetuate difficulties with psychosocial adjustment. Further, the emotional salience with which stimuli are imbued can alter basic sensation and perception processes in the detection node, which can in turn influence the type of information that is gleaned from the environment. Thus, abnormal brain response in one or many processing nodes may accompany or even promote faulty appraisals of environmental cues that then pave the way for disordered affect and internalizing problems.

The goal of this chapter is to characterize social information processing in pediatric anxiety and depression. We limit our scope to major depressive disorder (excluding bipolar disorder) and the anxiety disorders of social anxiety and generalized anxiety, which are common and well-researched diagnoses in adolescents. We utilize the SIPN framework to guide our review of cognitive neuroscience evidence for disrupted neural response to socio-emotional information in adolescents with affective disorders. Because adolescence itself has also been characterized as a series of disruptions to neural networks, we begin with an overview of expected maturational changes in brain structure and function within the SIPN nodes. We then review evidence for disturbances to these developmental processes in anxiety and depression.

NORMATIVE DEVELOPMENTAL CHANGES OF ADOLESCENCE

Neuroimaging studies have demonstrated that the brain undergoes regulated changes in both structural and functional organization throughout the teenage years, and in some instances well into the third decade of life (Brain Development Cooperative, 2012; Gogtay et al., 2004; Golarai et al., 2015; Lebel et al., 2012). These findings have had important implications for how we understand the factors that influence adolescent development and potential mediators of behavioral and emotional disorders that emerge during brain maturation (Giedd and Rapoport, 2010; Paus et al., 2008).

Structural Brain Changes

Structural brain maturation involves alterations to both gray matter (reflecting alterations to cell bodies, dendritic arbors, and synaptic contacts), and white matter (reflecting maturation of the myelin sheath that insulates axons). In most brain regions, gray matter undergoes non-linear patterns of maturation in which an initial increase in local gray matter volume is followed by a protracted period of gray matter reduction (Brain Development Cooperative, 2012; Cao et al., 2015; Gogtay et al., 2004). This process is thought to reflect a relatively nonspecific proliferation of cells and synaptic contacts followed by a targeted "pruning" of processes and connections that are not protected by use (Hensch, 2005; Petanjek et al., 2011; Stiles, 2008). In contrast, most studies of white matter development have demonstrated a linear increase in both the volume and effectiveness of white matter tracts in the brain (Lebel et al., 2012; Paus, 2010).

The timing of these processes differs markedly across brain areas. Regions of the neocortex that are phylogenetically older, such as primary sensory regions, tend to mature sooner than those with more recent evolutionary appearance, such as high-level association cortices (Gogtay et al., 2004). The teenage years are marked by age-related volume reductions in brain cortex regions that play an important role in high-level social cognitive functions, like mentalizing, perceptual integration, and executive functions (Blakemore and Mills, 2014; Mills et al., 2014). Subcortical regions such as the amygdala, striatum, and thalamus also undergo detectable changes in gray matter throughout adolescence, and many of these are thought to be specifically impacted by puberty

(Goddings et al., 2014). Lastly, a number of recent neuroimaging studies of white matter development have also reported developmental changes linked to puberty (Menzies et al., 2015; Peper et al., 2011a; Peper et al., 2011b). However, the development of white matter tracts is even more protracted than that of gray matter, with peak metrics of organization occurring in the late twenties or early thirties for a number of tracts (Lebel et al., 2012; Olson et al., 2015).

The extensive growth in efficiency and organization of functional circuits during adolescence are thought to encourage the shaping of substrates and neural connections engaged during social behavior (Blakemore and Mills, 2014). From the standpoint of the SIPN, the most protracted structural organization of the adolescent period coincides with executive functions in the cognitive regulation node.

Functional Brain Changes

Functional imaging studies have also revealed age-related changes in patterns of response to provocative stimuli across adolescence. Enhanced responding has been found within regions such as the amygdala, striatum, anterior insula, and anterior cingulate cortex, which are associated with motivational aspects of affective experience (Braams et al., 2016; Casey and Jones, 2010; Guyer et al., 2016; Moore et al., 2012; Smith et al., 2014). These findings are consistent with self-reports of more frequent, more intense, and more variable emotionality in daily life during the adolescent period than earlier or later in development (Arnett, 1999; Larson et al., 2002; Steinberg et al., 2008; Steinberg and Morris, 2001). Adolescent hyper-emotionality at the experiential and neural level is strongly modulated by social context (Blakemore and Mills, 2014; Crone and Dahl, 2012; Guyer et al., 2016; Pfeifer and Blakemore, 2012; Smith et al., 2013). For example, teenagers are more responsive to peer acceptance and rejection, more likely to take risks in the presence of peers, and more attentive to peer than non-peer faces than are children or adults (Guyer et al., 2009; Moor et al., 2010; Picci and Scherf, 2016; Smith et al., 2015). These changes, which are associated primarily with the affect node of the SIPN model, may be directly linked to the impact of gonadal hormones on subcortical brain function (Forbes and Dahl, 2010; Goddings et al., 2012; Scherf et al., 2013), particularly within affect-related regions (Chein et al., 2011; Forbes and Dahl, 2010; Jones et al., 2014; Nelson et al., 2014; Scherf et al., 2013; Smith et al., 2014). From an evolutionary standpoint, this enhanced motivational response to salient environmental cues may promote behavioral engagement with developmentally relevant stimuli, and thereby facilitate the crafting of adaptive brain circuits (Crone and Dahl, 2012; Nelson, 2017; Nelson et al., 2016; Nelson et al., 2014).

A second major change in brain development is increased activity within the prefrontal cortex (PFC), particularly when engaging executive functions such as working memory, inhibitory control, flexible response selection, and long-term behavioral planning (Casey and Jones, 2010; Crone, 2009; Nelson and Guyer, 2011; Steinberg, 2005). These functions are associated with the cognitive regulation node of the SIPN model. Age-related increases in executive function have been consistently found across adolescence (Crone, 2009), and a number of functional imaging correlates within the PFC have been linked to this enhanced performance (Crone, 2009; Ordaz et al., 2013; Satterthwaite et al., 2013). In general, improvement on executive function tasks is associated with greater (or more efficient) engagement of prefrontal regions, combined with inhibitory control over brain activity which is task irrelevant (such as within the limbic system and default mode network). Both behavioral and brain-based studies have revealed that executive functioning is particularly degraded in adolescence by the presence of either emotional or social distractors (Chein et al., 2011; Crone, 2009); this degradation suggests that the heightened salience of peers requires an additional level of control within cognitive-regulatory regions.

Several influential theories of adolescent development posit that the behavioral hyper-emotionality manifested by teenagers is due to a mismatch in the timing of maturation between early-developing subcortical structures and later-developing neocortical structures across adolescence. The motivation-relevant regions in the limbic system are sensitized at puberty by increased levels of circulating hormones, whereas prefrontal regions that downregulate emotional activation mature in a more protracted manner that is largely independent of pubertal hormones. This differential maturation pattern between the affective and executive brain systems has also been advanced as a factor contributing to adolescent vulnerability to affective disorders (Casey et al., 2011; Ernst, 2014; Steinberg, 2005).

While there clearly is support for differential timing of cortical and subcortical development, these "dual systems" models have been criticized for not adequately accounting for the complexity of emotions, psychopathology, or complex neuronal communication (Crone and Dahl, 2012; Pfeifer and Allen, 2012). One deficiency in these models is that the contributions of limbic activation *to* prefrontal functioning are not adequately accounted for (Crone and Dahl, 2012; Nelson et al., 2014). Most dual systems models of brain development frame the functional significance of emerging connections between prefrontal and limbic systems as unidirectional, where prefrontal circuits exert inhibitory control on emotional responsiveness. However, communication between limbic and PFC is *bidirectional*: while inhibitory inputs from the PFC can in fact dampen emotional hyper-excitability, it is also likely that activations from limbic structures play an important role in shaping neocortical circuits in the PFC and elsewhere. As such, emotionality can play an important role in guiding behavioral trajectories (Crone and Dahl, 2012; Jarcho et al., 2015; Nelson, 2017; Nelson et al., 2016; Nelson et al., 2014).

Development within the SIPN Network

From a functional standpoint, much of the evidence reviewed previously is consistent with a dual systems model. Namely, puberty-related increases in activation within limbic structures precede the development of prefrontal systems that exert some inhibitory influence over limbic activity. However, the development of social cognitive circuitry involved in socio-emotional processing is likely much more complicated, because interactions among the regions influence both activity among each node and its developmental trajectory.

How can the SIPN framework be interpreted in terms of an elaborated network approach to development? The SIPN has a loosely defined modular structure that consists of perceptual, affective, and regulatory components (Nelson et al., 2005). The "detection" node deciphers the basic properties of a stimulus and categorizes its social characteristics. The "affect" node assesses the emotional significance of a stimulus and oversees approach/avoidance orientations towards environmental cues. Lastly, the "cognitive regulation" node engages skills such as theory of mind, inhibition of responses via impulse control, goal-directed behavior, understanding of emotion and regulation of emotional impulses, and cognitive representations of others in the social world.

The evidence has largely supported the contention that these nodes have unique developmental profiles. The most dramatic maturational changes in functionality occur for perceptual processing in early childhood, affective processing in early adolescence, and executive functioning in late adolescence. However, it is important to recognize that these nodes are all interconnected. The emergence of puberty has a profound effect on the functioning of the amygdala and striatum; further, because of interconnections between these structures and others, puberty also impacts the functioning and structural organization of areas involved in perceptual processing, like the fusiform face area and the temporal voice area (Fruhholz et al., 2012; Scherf et al., 2013). The same can be said of emergent executive functions, in that heightened activity in the amygdala and striatum impacts the structure, function, and development of prefrontal regions (Jarcho et al., 2015; Jones et al., 2014; Nelson, 2017; Nelson et al., 2014; Scherf et al., 2013; Somerville et al., 2013).

Within the framework of the SIPN, adolescent development can be characterized as puberty-induced enhanced volatility of subcortical structures within the affect node, and more slowly emerging functionality and refinement of cognitive regulation functions instantiated by the maturation of the PFC. An important aspect of this model, however, is that the developmental progression of the entire system is impacted by changes in any one component. The impact of specific deviations on the functioning of the entire system is important to consider not only for general developmental models, but also to help understand individual differences that emerge in socio-emotional processing across development.

Deviations to Typical Development

Affective disorders emerging in adolescence can represent heightened disruptions in or between SIPN nodes that go beyond expected changes in adolescence. For example, within the affect node, attentional biases towards threatening stimuli are stronger in adolescents than in adults (Pine, 2007). These biases are not pathological on their own but may be deployed ineffectively or inappropriately in anxiety. Similarly, though healthy adolescents show increased reactivity to rejection in neural networks encompassing the amygdala, subgenual anterior cingulate cortex (ACC), insula, and nucleus accumbens, this pattern is enhanced in adolescents who are experiencing depression (Silk et al., 2014). Moreover, affective disorders may be characterized by function that runs counter to the expected developmental trajectory within a node or network. For instance, the structural and functional connectivity between the amygdala and PFC decreases from childhood to early adulthood. However, *increased* connectivity between these regions is associated with greater anxiety and depression in children and adults (Jalbrzikowski et al., 2017).

We will now examine abnormalities in social information processing, within each of the SIPN nodes, in pediatric anxiety and depression. Comorbidity is common between these two disorders, with 25–50 percent of depressed youth also being diagnosed with an anxiety disorder, and 10–15 percent of anxious youth also experiencing depression (Axelson and Birmaher, 2001). This correlation between anxiety and depression may stem from shared negative affectivity and biases in information processing (Axelson and Birmaher, 2001). Nonetheless, as most research conducted thus far has considered these disorders unitarily, we will review evidence for social information processing deficits in anxiety and depression separately.

Pediatric Anxiety

Epidemiology

Anxiety disorders are the most prevalent of childhood psychiatric diagnoses, with a third of children being affected during childhood or adolescence (Merikangas et al., 2010). Anxiety disorders commonly emerge in childhood or early adolescence, particularly in females (Roza et al., 2003). The average age of onset depends on the particular form of anxiety: separation anxiety is most common in young children (Copeland et al., 2014), social anxiety is particularly likely to emerge in youth, and generalized anxiety disorder (GAD) rates increase during the teenage years with a further rise in prevalence around age 30.

Across several forms of anxiety, common features are excessive fear, hypervigilance, and worry that are persistent (for more than 6 months) and out of proportion to any actual threat. The content of the fear delineates the various types of anxiety disorders: for example, social anxiety is characterized by fear about being negatively evaluated by others in social situations, and GAD involves persistent worry across many domains of life. Anxious individuals will experience highly aversive emotions when faced with the feared stimulus and will typically avoid anxiety-provoking situations if possible. This highly maladaptive response temporarily reduces negative affect, but also serves to increase the potency of the threat on subsequent encounters.

The current chapter focuses on GAD and social anxiety, given that these forms of anxiety are often diagnosed and well-researched in adolescents. Pediatric versions of these diagnoses follow similar diagnostic criteria as in adults, with the following youth-specific variants: in youth with social anxiety, individuals must be fearful of interactions that occur with peers—and not just interactions with adults. Fear or anxiety can be expressed via crying, tantrums, freezing, clinging to others, or not speaking. In children with GAD, worry tends to center on competence or performance quality (e.g., in school or sports), even when no formal evaluation will occur. Affected youth typically seek excessive amounts of reassurance and approval from others. Further, the physical symptoms typical of GAD in adults (e.g., fatigue, restlessness, irritability, tension, or sleep disturbances) are less prevalent and not necessary to diagnose GAD in children.

Pediatric anxiety disorders negatively impact youths' functioning with peers, at school, during leisure activities, and within their families (Essau et al., 2014; Ezpeleta et al., 2001). Childhood anxiety significantly increases risk of both depression (Rapee et al., 2009) and continued anxiety disorders in adulthood (Pine et al., 1999). Besides increased risk for persistent internalizing problems, adolescent anxiety is also associated with a number of non-specific adverse outcomes in adulthood, including poor academic and career achievement, increased stress, strained family relationships, substance dependence, and overall reduced life satisfaction (Essau et al., 2014; Woodward and Fergusson, 2001).

Etiology

Several factors have been implicated in the onset of childhood anxiety, including parenting style, modeling and learning experiences, life events, and cognitive factors such as information-processing biases (Rapee et al., 2009). Genetics and temperament-related factors have also been linked to the emergence of pediatric anxiety. Notably, behavioral inhibition in infancy predicts two- to four-fold increased risk for anxiety disorders later in childhood (Perez-Edgar and Fox, 2005), and social anxiety in particular (Thai et al., 2016). However, the typical course for the emergence of anxiety disorders is thought to involve a complex interplay of many factors across development (Pine and Fox, 2015).

Broadly speaking, anxiety disorders are conceptualized as a disruption in the processing of threat-related information. A normative "fear system" is thought to develop early in infancy, manifesting as separation anxiety or fearfulness around strangers (Gullone, 2000). Rodent and primate models suggest that the amygdala is an important component of the brain circuits involved in the behavioral and physiological responses to threat (Blackford and Pine, 2012). There is no clear or simple relationship between amygdala activation and the experience of fear (LeDoux, 2014; LeDoux and Pine, 2016). However, the amygdala is thought to be an important neural component in orchestrating emotional behaviors towards salient stimuli in a variety of contexts, including attention and reinforcement learning related to both positive and negative stimuli (Blackford and Pine, 2012). Aberrant functional and structural properties of the amygdala have repeatedly been found in anxiety disorders in both children and adults (Blackford and Pine, 2012; Bruhl et al., 2014; Fox et al., 2015; Qin et al., 2014). In addition, anxious individuals show atypical activation in a number of regions with interconnections to the amygdala, including the ventral PFC, insula, anterior cingulate cortex, and striatum (Bruhl et al., 2014; Fox and Kalin, 2014; Ghashghaei and Barbas, 2002; Pine, 2007; Ray and Zald, 2012). Amygdala circuits incorporating PFC regions (medial PFC, ventrolateral PFC, and orbitofrontal cortex) and the ACC appear particularly relevant for anxiety. Afferent inputs from the amygdala to the ACC and PFC may drive cognitive function; in addition, efferent inputs from higher cortical regions may inhibit and regulate amygdala activity. Broadly, dysfunction within these networks may result in alterations in attention orienting, threat appraisal, and memory and learning processes like fear conditioning and performance monitoring (Pine, 2007; Shechner et al., 2012). In the next section, we review evidence for altered social information processing in pediatric anxiety, within each node of the SIPN. Due to the paucity of research on abnormalities in perceptual processes ("detection") in anxiety, we will concentrate on attentional and cognitive biases towards emotional or threatening stimuli ("affect" and "cognitive regulation").

Disrupted Social Information Processing

Affect

Processing of threat-related information

Adolescents' processing of emotional cues is typically inferred from their neural response to emotional stimuli, such as facial expressions (Kerestes et al., 2014). Typically developing teenagers show heightened amygdala response to emotional cues (Casey et al., 2010; Guyer et al., 2008b; Somerville et al., 2011), with fearful faces specifically eliciting greater amygdala, orbitofrontal cortex, and ACC activation in adolescents than adults (Monk et al., 2003).

Anxious children consistently show aberrant amygdala activation when processing negative facial expressions (review in Blackford and Pine, 2012). When rating fearful faces, children with GAD showed greater functional connectivity between the amygdala and insula than did controls, which was correlated with the severity of their anxiety symptoms (McClure et al., 2007). Similarly, adolescents with social anxiety had greater activation in both the amygdala and rostral ACC while viewing fearful faces (Blair et al., 2011), with amygdala activity positively correlating with social anxiety symptoms (Killgore and Yurgelun-Todd, 2005). Aberrant amygdala function during the processing of social information may be a precursor or risk factor for the emergence of anxiety disorders, as it was also noted in behaviorally inhibited adolescents viewing fearful faces (Pérez-Edgar et al., 2007).

Activation in the amygdala and ventrolateral PFC (vlPFC) are inversely correlated when healthy adolescents are asked to rate fearful faces. However, this pattern is less pronounced in anxious children (Monk et al., 2008), who instead seem to demonstrate hyper-activation in both the amygdala and PFC. Children with GAD showed heightened activation in the vlPFC when rating masked angry faces (Monk et al., 2006). Similarly, socially anxious teenagers' ventromedial PFC (vmPFC) activation when viewing fearful faces was positively correlated with the severity of their anxiety (Blair et al., 2011). Such patterns in anxious groups have been hypothesized to reflect temporally extended processing in the amygdala-PFC loop. While non-anxious youth's PFC may inhibit amygdala signals quickly when mandated, anxious children's PFC may take longer to act (Blackford and Pine, 2012). This prolonged co-activation between executive and limbic structures may be a marker of disruption within neural networks engaged in emotional processing, as greater functional connectivity between the amygdala and PFC has been associated with heightened anxiety (Guyer et al., 2008a).

Bias in attention orienting

Attentional biases have often been assessed with dot-probe tasks, in which participants simultaneously view two stimuli of presumably different valence and/or salience. Participants are then asked to identify or detect a target that appears in lieu of one of the two stimuli. Faster responses to a target that followed the salient stimulus are thought to index attentional bias towards that cue. Research on attention biases in anxiety have typically contrasted response latency to neutral vs. angry faces, with rapid detection of anger interpreted as enhanced attention bias to threat.

Healthy individuals typically show no attention orientation towards mild threats (and possibly a tendency to avoid them); conversely, anxious individuals automatically attend to threatening stimuli (Pine, 2007). Like anxious adults, children with anxiety disorders or youth who score highly on measures of anxiety orient their attention more towards threat-related information, and the threshold for what constitutes threat is lower (meta-analysis by Dudeney et al., 2015; Heim-Dreger et al., 2006; Shechner et al., 2012; Waters et al., 2010). This bias coincides with heightened activation in the amygdala (Pine, 2007) and increased engagement of the vlPFC for the regulation of attention (Shechner et al., 2012). During an attention-orienting task with masked emotional (happy, angry) and neutral faces, youth with GAD had greater right amygdala activation (which correlated with anxiety severity) and less negative coupling between the right amygdala and right vlPFC when viewing masked angry faces, compared to controls (Monk et al., 2008). Further, trait anxiety in healthy youth was associated with greater attentional bias towards angry faces in a dot-probe task, and greater associated right dorsolateral PFC activation (Telzer et al., 2008). Thus, aberrant function in the amygdala-PFC network may underlie attentional biases towards negative emotional cues.

Threat orientation differences between anxious and non-anxious groups may emerge later in childhood, when the normative biased attention to threat has subsided for typically developing but not anxious youth (Dudeney et al., 2015; Kindt et al., 2003; Vasey et al., 1995). Indeed, it is perhaps adaptive for children to orient towards threats, though this tendency disappears with increasing age. With maturity, the threshold for threat information to influence orienting may increase (Pine, 2007). Anxiety may represent a disruption to expected age-related changes in attentional processes.

It is important to note that enhanced sensitivity to threat information can sometimes manifest as avoidance, or bias away from the threat cue, in anxious populations (Bar-Haim et al., 2010; Bar-Haim et al., 2007). As the intensity of the threat increases, healthy individuals attend to it more; anxious individuals instead begin to avoid it (Pine, 2007). Avoidance of threat information may be a compensatory response. For instance, adolescents with GAD, compared to controls, showed greater attention bias *away* from angry faces in a dot-probe task and accompanying increased right vlPFC activation, which here correlated with less severe anxiety symptoms (Monk et al., 2006). Alternatively, this pattern may be reconciled in a two-stage model of information processing, whereby anxious individuals quickly orient their attention to a threat and quickly shift away from it (Shechner et al., 2012). This hypothesis may explain some contradictory findings in the literature on attention orienting, as longer stimulus exposure times during the dot-probe task may be capturing both threat vigilance and avoidance responses in anxious populations. However, it should be noted that recent work has called into question the validity of the dot-probe test as a measure of attentional bias (e.g., Chapman et al., 2019; Schmukle, 2005).

Sensitivity to reward

The coding of a rewarding stimulus' salience sculpts modulation of behavior during stimulus-response learning (Ernst et al., 2011; Frank and Fossella, 2011). Adolescents show enhanced sensitization to reward, a heightened approach motivation, and an immature avoidance system; together, these tendencies may lead to increased reward-seeking and risk-taking behavior (Somerville et al., 2010). Developmental changes in the reward system are thought to be mediated by the neural dopamine system, particularly in the midbrain and connected regions with high concentrations of dopamine neurons—like the medial PFC (mPFC) and striatum (Haber and Knutson, 2010). Healthy adolescents show strong striatum response to rewards (Luking et al., 2016), a developmental shift in sensitization that is thought to be related to the levels of gonadal hormones circulating in puberty (Braams et al., 2015). This enhanced striatal activity during reward anticipation and outcome was correlated with higher positive affect in daily life (Forbes et al., 2009), as well as greater reported happiness when winning large rewards (Ernst et al., 2005).

Unlike anxious adults, children with anxiety do not necessarily demonstrate attenuated response to positive stimuli (Frewen et al., 2008; Shechner et al., 2012). Children with anxiety and behavioral inhibition show increased sensitivity to reward and heightened striatum response during monetary incentive tasks, but only when the reward is dependent on their performance (Bar-Haim et al., 2009; Blackford and Pine, 2012; Guyer et al., 2012). These findings highlight the involvement of cognitive regulation systems in reward processing in anxiety.

Cognitive regulation

Deviations in functioning within the cognitive regulation node have most often been investigated via tasks that tap into error monitoring, the conscious appraisal of threat, and the capacity to regulate emotional responses. Electroencephalography (EEG) reveals that event-related potentials (ERPs) in response to performance-based tasks distinguish anxious from non-anxious children.

First, there is evidence that anxious youth show enhanced error-related negativity (ERN) signals, which peak after an error has been committed, in tasks that require the monitoring of performance. For instance, anxious 8- to 14-year-olds had greater ERN amplitude in the ACC during a Flanker task than controls (Ladouceur et al., 2006). Clinically anxious youth of a similar age also showed greater ERN and correct-response negativities than healthy youth in a Go-No-Go task (Hum et al., 2013). Further, the anxious group showed greater posterior P1 and frontal N2 amplitudes during the task, which are thought to index heightened attention/arousal and cognitive control, respectively.

Larger ERNs, which may signal enhanced processing of mistakes, may constitute a risk factor for the development of anxiety. Indeed, beyond the effect of baseline anxiety and familial risk, enhanced ERNs during a Go-No-Go task at age 6 predicted the onset of anxiety disorders by age 9 (Meyer et al., 2015). Similarly, children who exhibited both behavioral inhibition and enhanced ERNs on a Flanker task were at greater risk for anxiety diagnoses in adolescence (McDermott et al., 2009). However, in considering the role of ERN responses to errors, it is important to note that the relationship between ERN amplitude and anxiety may change with age. Indeed, older children showed larger ERNs to errors, which were related to increased parent-report of anxiety; however, younger children showed opposite patterns, whereby smaller ERNs were related to increased anxiety (Meyer et al., 2012). The relationship between ERP signals and anxious symptomatology is not yet well-understood, though results regarding error monitoring in anxiety do seem to converge with findings about performance-based reward sensitivity.

Second, both children and adults with anxiety disorders show biases in their appraisal of threat information. In contrast with the implicit attention orienting reviewed previously, the conscious interpretation of threat occurs later in the processing stream and engages higher-order cognitive capacities. Anxious groups show a lower threshold for classifying something as dangerous than non-anxious groups (Pine, 2007). This may be coupled with neural engagement of the amygdala and PFC circuit, as adolescents with GAD (compared to controls) showed more amygdala and vmPFC activation when appraising threatening faces than when passively viewing them (McClure et al., 2007).

Lastly, anxious youth may show differential emotional regulation capabilities than non-anxious peers, though research on emotional control is scarce. Healthy adolescents typically have greater amygdala activity relative to adults during tasks that require emotional regulation (e.g., emotional Go-No-Go), but this activation attenuates with repeated exposure to the emotional cues. However, self-reported trait anxiety was linked to reduced habituation of the amygdala despite ongoing exposure. This failure to habituate was linked to reduced functional connectivity between the amygdala and PFC

(Hare et al., 2008), suggesting that this network may be aberrant in both the aforementioned processing of threat-related information and the regulation of associated emotional responses.

Summary

Taken together, the reviewed evidence suggests that pediatric anxiety is associated with a confluence of factors that impact social information processing. Normative development generates enhanced engagement of motivation-related circuits during adolescence, particularly in regard to peer-related stimuli. This enhanced emotionality occurs when executive systems are still undergoing maturation. Anxiety in youth is characterized by the further potentiation of social threat cues, evidenced by enhanced amygdala *and* PFC response to fearful stimuli. The same neural pattern also underlies anxious children's tendency to orient towards, and subsequently avoid, threatening information in dot-probe tasks. The enhanced response in the amygdala-PFC network does not follow the expected habituation to emotional stimuli in anxious children compared to non-anxious youth. Further, the threshold for appraising cues as threatening is lower in anxious groups than in typically-developing youth.

Moreover, anxious adolescents also demonstrate dysfunctional monitoring of performance and errors, which may alter their reward sensitivity and ability to make behavioral adjustments. Anxiety is linked to increased responding to reward within the striatum when the reward is contingent upon task performance. Findings in EEG studies also point to concern with performance in anxiety, as anxious children demonstrate greater neural responses to errors on inhibitory or cognitive control tasks than their non-anxious peers. In sum, these findings suggest that anxiety is linked to enhanced performance monitoring and heightened sensitivity to threatening or emotional information at a neural level, which may underlie phenotypic concerns with evaluation and worry about negative outcomes in pediatric social anxiety or GAD.

PEDIATRIC DEPRESSION

Epidemiology

The incidence of depression rises steeply after puberty, especially in girls (Andersen and Teicher, 2008; Angold and Costello, 2006; Roza et al., 2003). It is estimated that 4 percent of teenagers report experiencing depression in the last year (Costello et al., 2006; Thapar et al., 2012). The most commonly noted risk factors for depression are a family history of the disorder and exposure to psychosocial stress (Thapar et al., 2012), such as maltreatment or problematic family relationships (Restifo and Bögels, 2009; Rueter et al., 1999), or peer victimization (Hawker and Boulton, 2000). Symptoms of depression in childhood or adolescence robustly predict depression in adulthood (Pine et al., 1999; Wilcox and Anthony, 2004).

As in adults, pediatric depression is defined as having persistent depressed mood or a loss of pleasure or interest over several weeks. These symptoms are accompanied by other concerns, including changes in weight or sleep, fatigue, attention and concentration deficits, feelings of worthlessness or excessive guilt, psychomotor disturbances, or recurrent thoughts of death or suicide. In addition to excessive sadness, irritability is increasingly recognized as a symptom of depression, and may be especially pronounced in pediatric populations (Emslie et al., 2005). In school-aged children, depression has been associated with more intense affect globally, and less effective emotion regulation capacities (Silk et al., 2003). In contrast, during adolescence, depression is more closely associated with emotional blunting (particularly anhedonia), as well as disturbances in sleep, concentration problems, and suicidality (Sørensen et al., 2005; Teicher et al., 1993). Perhaps relatedly, 10- to 17-year-olds with depression reported fewer peer interactions and less overall positive affect (Axelson et al., 2003), at a time when typically developing individuals demonstrate increases in both these experiences (Steinberg and Morris, 2001).

Etiology

Theories of pediatric depression suggest that hormonal, social, and neural changes contribute to the increased prevalence of this disorder during adolescence (Casey et al., 2011; Forbes and Dahl, 2005; Kerestes et al., 2014). Typically developing adolescents tend to show enhanced reward sensitivity and appetitive behavior, which is supported by increased activity within reward-related structures such as the ventral striatum (Braams et al., 2015; Casey et al., 2008). In contrast, hypoactivation within these circuits may yield low levels of positive affect and appetitive behavior, which are hallmarks of depression (Andersen and Teicher, 2008). Indeed, blunted dopaminergic substrates in the ventral tegmental area and nucleus accumbens have been implicated in depression (Nestler and Carlezon, 2006). These findings suggest that alterations in the function of reward-related neural structures may be responsible for depressive deficits in reward sensitivity.

In addition to showing attenuated responses to reward, adolescents with depression have been found to show heightened neural response to threat and aversive stimuli (Beesdo et al., 2009b). Furthermore, a more reactive glucocorticoid system during adolescence may make teenagers particularly vulnerable to stressful experiences around the time of puberty. Adolescents have heightened glucocorticoid receptors in the hippocampus, increased glucocorticoid receptor expression in the cortex, and prolonged corticosterone response after stress (Pryce, 2008; Romeo and McEwen, 2006). The hippocampus in particular is known to be susceptible to the effects of stress: for instance, maladaptive secretion of stress-related hormones such as glucocorticoids and mineralocorticoids can reduce neurogenesis and synaptogenesis in the hippocampus, a process that may be related to psychopathology (Andersen and Teicher, 2008). Indeed, the hippocampus has been involved in mood regulation, and responds to anti-depressants in non-human primates (McEwen, 2000; Perera et al., 2007). It has been suggested that dysfunction within this region disrupts the context provided to emotional stimuli in depression (Andersen and Teicher, 2008).

Lastly, abnormal functional connectivity between the limbic and frontal regions of the brain during resting state has been linked to adolescent depression (Chattopadhyay et al., 2017; Connolly et al., 2017; Zhang, 2017). One theory proposes that the development of the PFC in the context of unachieved rewards may suppress the (limbic) reward system and lead to subsequent depressive symptoms (Davey et al., 2008). We will now review evidence for disrupted social information processing and associated activation within the affect (primarily limbic) and cognitive regulation (primarily prefrontal) nodes of the SIPN model.

Disrupted Social Information Processing

Affect

Response to social and emotional stimuli

The emotional profile of depression can be characterized as enhanced response to negatively valenced stimuli, combined with a blunted response to positively valenced cues (Beesdo et al., 2009a; Forbes and Dahl, 2005; Joiner et al., 2002; Kujawa and Burkhouse, 2017; Lonigan et al., 1994; Nandrino et al., 2004). This neural and behavioral pattern has been observed in children diagnosed with depression and in children who are at familial risk for the disorder (high-risk [HR] youth), suggesting that an endophenotype of depression may be detectable (Olino et al., 2015; Sharp et al., 2014; Silk et al., 2006; Silk et al., 2014).

Because of the social domain's importance during adolescence, interpersonal experiences may be particularly sensitive to motivational perturbations. Indeed, both depressed and HR adolescents demonstrate blunted activation in motivational circuitry in response to happy faces (Monk et al., 2008) or social acceptance cues (Olino et al., 2015; c.f., Silk et al., 2014). Conversely, they also show heightened amygdala response to *aversive* social cues, such as threatening (i.e., fearful and angry) or sad faces (Gaffrey et al., 2011; Kerestes et al., 2014; c.f., Thomas et al., 2001). Further, amygdala response fails to habituate to threatening stimuli in depression, and persistent hyperactivity to social stimuli within this structure has been noted in HR adolescents (Swartz et al., 2015). The anti-depressant fluoxetine can reduce this aberrant amygdala activation in depressed adolescents (Tao et al., 2012), reinforcing the hypothesis that excessive and prolonged amygdala response to threat may play a role in the occurrence of this disorder.

Moreover, when receiving negative social feedback (e.g., rejection in simulated peer interactions), depressed adolescents also showed increased activation in a number of other limbic systems beyond the amygdala, including the subgenual ACC, anterior insula, and nucleus accumbens (Silk et al., 2014). Greater subgenual ACC activation to rejection in particular was linked to depressed symptoms 1 year later (Masten et al., 2011). This finding reinforces the notion that aberrant limbic response to negative socio-emotional cues may lead to long-term brain changes that maintain depression in youth.

Reward processing

Anhedonic symptoms may represent a failure in approach systems that regulate motivation for, sensitivity to, and enjoyment of reward (Forbes and Dahl, 2005). Indeed, deficits in approach motivation and reward processing are consistently noted in HR and depressed youth (Kujawa and Burkhouse, 2017). Though typically developing adolescents show increased reward-seeking during this developmental stage, depressed teenagers show *less* reward sensitivity (Forbes and Dahl, 2005), particularly in response to monetary (Kerestes et al., 2014) or social rewards (Forbes and Dahl, 2012). For instance, depressed 11-year-old boys did not distinguish between choices yielding small or large rewards in a decision-making task. Neglecting to choose the option associated with a high-probability large reward was linked to risk for depression (and anxiety) 1 year later (Forbes et al., 2007).

Youth with depression and HR children also show blunted *neural* responses to reward, which predict symptom severity or illness onset in adolescence beyond the impact of baseline symptoms (Luking et al., 2016). For example, blunted reward positivity (RewP) and feedback negativity (FN) ERP responses to reward have been noted in high-risk youth (Kujawa and Burkhouse, 2017), with smaller FNs being linked to greater depressive symptoms and illness onset 2 years later (Bress et al., 2013). This abnormal reward response in depression is typically denoted by reduced activation in the striatum and greater activation in the mPFC during anticipation and receipt of reward (Forbes and Dahl, 2005; Forbes and Dahl, 2012; Forbes et al., 2006; Kujawa and Burkhouse, 2017), as well as after the receipt of a reward (Gotlib et al., 2010; Luking et al., 2016; Olino et al., 2014; Sharp et al., 2014). Forbes and Dahl (2012) propose that typical reward responses may occur in the striatum of depressed youth, but be quickly overridden by inhibitory signals from an overactive mPFC, resulting in shorter and less intense positive affect in response to reward. Alternatively, mPFC activation may represent an attempt to elicit greater dopamine release from an underactive striatum. Both reduced caudate activation during anticipation of reward and enhanced activity in the mPFC during reward outcome have been linked to increased depressive symptoms 2 years later (Morgan et al., 2013). This link supports the potentially causal role of this neural response pattern in depressive symptomatology.

Interestingly, while striatal response to rewards are blunted, enhanced sensitivity in the striatum, anterior insula, and parahippocampus has been noted in HR children who either lose a reward (Luking et al., 2016) or receive a punishing outcome (Gotlib et al., 2010). In social realms, greater reactivity to negative social feedback could lead to withdrawal from the environment, with failure to adjust behavior to create positive reinforcing social experiences (Kujawa and Burkhouse, 2017). When combined with attenuated response to positive or rewarding cues, such a pattern could be involved in the maintenance of depression, since hypersensitivity to rejection by peers and difficulties in social relationships both predict psychiatric difficulties during adolescence (Gould et al., 2003; Nelson et al., 2005).

Cognitive regulation

In addition to altered patterns of subcortical response to affective information, there is evidence that executive functions and higher-level cognitive engagement may also be impaired in adolescent depression. The timing of disorder onset may be particularly important in predicting executive control deficits: depression emerging in early adolescence has a significantly greater impact on the developmental trajectory of cognitive control than depression with onset in late adolescence or young adulthood (Vijayakumar et al., 2016).

As in adults, the anterior cingulate is an important contributor to executive dysfunction in adolescent depression (Drevets et al., 2008; Lai, 2013). Depressed youth have been found to show enhanced fronto-cingulate activation in cognitive control tasks (Kerestes et al., 2014), a pattern that may reflect abnormal connectivity within frontal networks. For instance, while the subgenual ACC is typically functionally coupled with frontal regions in healthy children, this area is instead increasingly connected with the thalamus and parietal regions in children with a history of preschool-onset depression (Gaffrey et al., 2010). Similarly, HR children show hypoconnectivity between the subgenual ACC and the left dorsolateral PFC, but hyperconnectivity between the right amygdala and right inferior frontal gyrus (Chai et al., 2016). White matter tracts connecting the right subgenual ACC to the amygdala were also less effective in depressed adolescents (Cullen et al., 2010). Thus, alterations in the connections between the cingulate, limbic system, and PFC may be representative of pediatric depression.

Youth depression has also been associated with altered cognitive processing of affective information, manifested as deficits in emotion recognition and regulation. For instance, depression in children and adolescents has been associated with impaired recognition of facial (Demenescu et al., 2010) and vocal (Morningstar et al., 2019) expressions of emotion. At a neural level, these deficits may be associated with aberrant limbic activation, since depressed adolescents showed greater left amygdala activation than controls during a face-emotion matching task (Yang et al., 2010). More research is needed to understand what neural patterns underlie the incorrect interpretation of emotional stimuli. In regard to emotion regulation, depressed youth showed greater amygdala activation when asked to maintain an emotion in response to a particular stimulus, but decreased connectivity between the amygdala and mPFC or insula (Perlman et al., 2012). Similarly, in a negative mood induction and subsequent emotion regulation task, children with a history of preschool depression showed enhanced amygdala and reduced PFC activity. Heightened amygdala response corresponded to increased severity of negative mood (Pagliaccio et al., 2012). Echoing findings regarding the *automatic* processing of emotion cues in depression, the amygdala-PFC loop may also be impaired during the controlled interpretation of emotional information.

Summary

In sum, pediatric depression is characterized by blunted experiential and neural response to positive stimuli and rewards, a pattern that is thought to represent vulnerability for depression onset and maintenance (Kujawa and Burkhouse, 2017). Additionally, depressed and HR youth tend to show enhanced and prolonged limbic activation in response to negative stimuli. These neural patterns have been linked to the severity of symptoms or the prospective emergence of depression in youth, underscoring their involvement in the onset or maintenance of depressive symptomatology. In conjunction with aberrant amygdalar response, the PFC is overly active during reward processing. The combination of altered responsiveness in the amygdala and striatal regions ("affect" node) and aberrant function in the cingulate and prefrontal areas ("cognitive regulation" node) may be related to the anhedonia and blunted affect that is typical of depression.

CONCLUSION

Sensitive periods of organization for the developing brain have been highlighted in early childhood (e.g., Andersen, 2003; Lewis and Maurer, 2005; Stanwood and Levitt, 2004), but the teenage years may present a similar opportunity for reorganization and growth of processes related to social and emotional development (e.g., Blakemore and Mills, 2014; Burke et al., 2017). The onset of puberty triggers a number of dramatic changes in the body and behavior of adolescents, which increases teenagers' vulnerability to affective disorders. Indeed, the prevalence of anxiety and depression surges markedly during adolescence. The emergence of these disorders during this developmental stage poses a significant obstacle to normative development and has been associated with ongoing psychiatric difficulties in adulthood.

Adolescence is marked by physical and emotional detachment from the family structure, increased reliance on peer networks for social support, the emergence of entirely new physical and emotional dynamics associated with intimacy, and an emerging sense of self and agency in the world. Many of these developmental shifts are initiated by changes in motivation-related circuits encompassing the limbic system. Puberty-induced changes in the amygdala and striatum are particularly noteworthy in this context, as they have a clear association with affective disorders in adults (Nelson et al., 2014; Scherf et al., 2013). Emerging dynamics between these limbic regions and the PFC also appear to be particularly important for both developmental changes in behavior and vulnerability to affective disorders in adolescence. From an evolutionary standpoint, regulated shifts in stimulus salience (e.g., peers vs. parents) and the associated sculpting of new circuits in the neocortex is an adaptive response permitting integration within novel social groups. However, from the standpoint of individual adjustment, these changes result in a period of emotional intensity and vulnerability for long-term maladaptive modes of behavioral and neural response to social stimuli.

Affective disorders in youth are likely a product of a combination of factors related to genetic risk, temperament, early life experiences, family environment, social learning, and neural correlates underlying information processing. Anxiety and depression are both characterized by problematic cognitive and emotional appraisals of environmental cues, including enhanced processing of threat and attentional biases towards negative stimuli. Shifting interactions between limbic and executive networks during adolescence may underlie the maladaptive processing of both positive and negative stimuli. Indeed, the relative engagement of the amygdala and prefrontal circuits in response to salient stimuli can differentiate youth with affective disorders from their typically developing peers. Healthy adolescents typically show enhanced amygdala and reduced prefrontal engagement to emotional cues, regardless of their valence. Conversely, threatening or negative stimuli elicit high activation in *both* the amygdala and PFC in anxious and depressed youth. This neural response translates to automatic attention to threat and enhanced processing of negative cues, both of which are hallmarks of anxiety and depression. Additionally, positive stimuli recruit less amygdala activation in depressed youth than in healthy adolescents, a pattern that is associated with reduced pursuit and enjoyment of reward. Thus, anxiety and depression are characterized by aberrant neural response in limbic and frontal control regions, which alters affective and cognitive processing of emotional cues.

Altered social information processing may have recursive effects on psychosocial well-being via its impact on behavior. For instance, heightened processing of negative cues in social realms may lead to social withdrawal and the reinforcing of negative beliefs about the self and others. Experience-based learning from such repeated occurrences can then influence the acquisition of function within relevant SIPN nodes, thus contributing to either the onset or maintenance of affective disorders. Early intervention to rectify or ameliorate deficits in social information processing and consequent behavior patterns may be most effective in preventing the long-term consequences of pediatric anxiety and depression.

REFERENCES

American Psychiatric Association (2013). *Diagnostic and Statistical Manual of Mental Disorders (5th edition)*, Arlington, VA, APA.

Andersen, S. L. (2003). Trajectories of brain development: Point of vulnerability or window of opportunity? *Neuroscience and Biobehavioral Reviews, 27*, 3–18.

Andersen, S. L., and Teicher, M. H. (2008). Stress, sensitive periods and maturational events in adolescent depression. *Trends in Neurosciences, 31*, 183–91.

Angold, A., and Costello, E. J. (2006). Puberty and depression. *Child and Adolescent Psychiatric Clinics of North America, 15*, 919–937.

Angold, A., Costello, E. J., and Worthman, C. M. (1998). Puberty and depression: The roles of age, pubertal status and pubertal timing. *Psychological Medicine, 28*, 51–61.

Arnett, J. J. (1999). Adolescent storm and stress, reconsidered. *American Psychologist, 54*, 317–326.

Axelson, D. A., Bertocci, M. A., Lewin, D. S., Trubnick, L. S., Birmaher, B., Williamson, D. E., Ryan, N. D., and Dahl, R. E. (2003). Measuring mood and complex behavior in natural environments: Use of ecological momentary assessment in pediatric affective disorders. *Journal Of Child And Adolescent Psychopharmacology, 13*, 253–266.

Axelson, D. A., and Birmaher, B. (2001). Relation between anxiety and depressive disorders in childhood and adolescence. *Depression and Anxiety, 14*, 67–78.

Bar-Haim, Y., Fox, N. A., Benson, B., Guyer, A. E., Williams, A., Nelson, E. E., Perez-Edgar, K., Pine, D. S., and Ernst, M. (2009). Neural correlates of reward processing in adolescents with a history of inhibited temperament. *Psychological Science, 20*, 1009–1018.

Bar-Haim, Y., Holoshitz, Y., Eldar, S., Frenkel, T. I., Muller, D., Charney, D. S., Pine, D. S., Fox, N. A., and Wald, I. (2010). Life-threatening danger and suppression of attention bias to threat. *American Journal of Psychiatry, 167*, 694–698.

Bar-Haim, Y., Lamy, D., Pergamin, L., Bakermans-Kranenburg, M. J., and Van, I. M. H. (2007). Threat-related attentional bias in anxious and nonanxious individuals: A meta-analytic study. *Psychological Bulletin, 133*, 1–24.

Beesdo, K., Bittner, A., Pine, D. S., Stein, M. B., Hofler, M., Lieb, R., and Wittchen, H. U. (2007). Incidence of social anxiety disorder and the consistent risk for secondary depression in the first three decades of life. *Archives of General Psychiatry, 64*, 903–912.

Beesdo, K., Knappe, S., and Pine, D. S. (2009a). Anxiety and anxiety disorders in children and adolescents: Developmental issues and implications for DSM-V. *Psychiatric Clinics of North America, 32*, 483–524.

Beesdo, K., Lau, J. Y., Guyer, A. E., Mcclure-Tone, E. B., Monk, C. S., Nelson, E. E., Fromm, S. J., Goldwin, M. A., Wittchen, H. U., Leibenluft, E., Ernst, M., and Pine, D. S. (2009b). Common and distinct amygdala-function perturbations in depressed vs anxious adolescents. *Archives of General Psychiatry, 66*, 275–285.

Birmaher, B., Ryan, N. D., Williamson, D. E., Brent, D. A., and Kaufman, J. (1996). Childhood and adolescent depression: A review of the past 10 years. Part II. *Journal of the American Academy of Child and Adolescent Psychiatry, 35*, 1575–1583.

Blackford, J. U., and Pine, D. S. (2012). Neural substrates of childhood anxiety disorders: A review of neuroimaging findings. *Child and Adolescent Psychiatric Clinics of North America, 21*, 501–525.

Blair, K. S., Geraci, M., Korelitz, K., Otero, M., Towbin, K., Ernst, M., Leibenluft, E., Blair, R. J. R., and Pine, D. S. (2011). The pathology of social phobia is independent of developmental changes in face processing. *The American Journal of Psychiatry, 168*, 1202–1209.

Blakemore, S. J., and Mills, K. L. (2014). Is adolescence a sensitive period for sociocultural processing? *Annual Review of Psychology, 65*, 187–207.

Blazer, D. G., Kessler, R. C., Mcgonagle, K. A., and Swartz, M. S. (1994). The prevalence and distribution of major depression in a national community sample: The National Comorbidity Survey. *American Journal of Psychiatry, 151*, 979–986.

Braams, B. R., Peper, J. S., Van Der Heide, D., Peters, S., and Crone, E. A. (2016). Nucleus accumbens response to rewards and testosterone levels are related to alcohol use in adolescents and young adults. *Developmental Cognitive Neuroscience, 17*, 83–93.

Braams, B. R., Van Duijvenvoorde, A. C., Peper, J. S., and Crone, E. A. (2015). Longitudinal changes in adolescent risk-taking: A comprehensive study of neural responses to rewards, pubertal development, and risk-taking behavior. *Journal of Neuroscience, 35*, 7226–7238.

Brain Development Cooperative, G. (2012). Total and regional brain volumes in a population-based normative sample from 4 to 18 years: The NIH MRI Study of Normal Brain Development. *Cerebral Cortex, 22*, 1–12.

Bress, J. N., Foti, D., Kotov, R., Klein, D. N., and Hajcak, G. (2013). Blunted neural response to rewards prospectively predicts depression in adolescent girls. *Psychophysiology, 50*, 74–81.

Bruhl, A. B., Delsignore, A., Komossa, K., and Weidt, S. (2014). Neuroimaging in social anxiety disorder: A meta-analytic review resulting in a new neurofunctional model. *Neuroscience and Biobehavioral Reviews, 47*, 260–280.

Burke, A. R., Mccormick, C. M., Pellis, S. M., and Lukkes, J. L. (2017). Impact of adolescent social experiences on behavior and neural circuits implicated in mental illnesses. *Neuroscience and Biobehavioral Reviews, 76*, 280–300.

Cao, B., Mwangi, B., Hasan, K. M., Selvaraj, S., Zeni, C. P., Zunta-Soares, G. B., and Soares, J. C. (2015). Development and validation of a brain maturation index using longitudinal neuroanatomical scans. *NeuroImage, 117*, 311–318.

Casey, B. J., Getz, S., and Galvan, A. (2008). The adolescent brain. *Development Revisited, 28*, 62–77.

Casey, B. J., and Jones, R. M. (2010). Neurobiology of the adolescent brain and behavior: Implications for substance use disorders. *Journal of the American Academy of Child and Adolescent Psychiatry, 49*, 1189–1201; quiz 1285.

Casey, B. J., Jones, R. M., Levita, L., Libby, V., Pattwell, S. S., Ruberry, E. J., Soliman, F., and Somerville, L. H. (2010). The storm and stress of adolescence: Insights from human imaging and mouse genetics. *Developmental Psychobiology, 52*, 225–235.

Casey, B. J., Jones, R. M., and Somerville, L. H. (2011). Braking and Accelerating of the Adolescent Brain. *Journal of Research on Adolescence, 21*, 21–33.

Chai, X. J., Hirshfeld-Becker, D., Biederman, J., Uchida, M., Doehrmann, O., Leonard, J. A., Salvatore, J., Kenworthy, T., Brown, A., Kagan, E., De Los Angeles, C., Gabrieli, J. D. E., and Whitfield-Gabrieli, S. (2016). Altered intrinsic functional brain architecture in children at familial risk of major depression. *Biological Psychiatry, 80*, 849–858.

Chapman, A., Devue, C., and Grimshaw, G. M. (2019). Fleeting reliability in the dot-probe task. *Psychological Res, 83*, 308–320.

Chattopadhyay, S., Tait, R., Simas, T., Van Nieuwenhuizen, A., Hagan, C. C., Holt, R. J., Graham, J., Sahakian, B. J., Wilkinson, P. O., Goodyer, I. M., and Suckling, J. (2017). Cognitive behavioral therapy lowers elevated functional connectivity in depressed adolescents. *EBioMedicine, 17*, 216–222.

Chein, J., Albert, D., O'brien, L., Uckert, K., and Steinberg, L. (2011). Peers increase adolescent risk taking by enhancing activity in the brain's reward circuitry. *Developmental Science, 14*, F1–F10.

Colman, I., Ploubidis, G. B., Wadsworth, M. E., Jones, P. B., and Croudace, T. J. (2007). A longitudinal typology of symptoms of depression and anxiety over the life course. *Biological Psychiatry, 62*, 1265–1271.

Connolly, C. G., Ho, T. C., Blom, E. H., Lewinn, K. Z., Sacchet, M. D., Tymofiyeva, O., Simmons, A. N., and Yang, T. T. (2017). Resting-state functional connectivity of the amygdala and longitudinal changes in depression severity in adolescent depression. *Journal of Affective Disorders, 207*, 86–94.

Copeland, W. E., Angold, A., Shanahan, L., and Costello, E. J. (2014). Longitudinal patterns of anxiety from childhood to adulthood: The Great Smoky Mountains Study. *Journal of the American Academy of Child and Adolescent Psychiatry, 53*, 21–33.

Costello, E. J., Mustillo, S., Erkanli, A., Keeler, G., and Angold, A. (2003). Prevalence and development of psychiatric disorders in childhood and adolescence. *Archives of General Psychiatry*, 60, 837–844.

Costello, J. E., Erkanli, A., and Angold, A. (2006). Is there an epidemic of child or adolescent depression? *Journal of Child Psychology and Psychiatry*, 47, 1263–71.

Crone, E. A. (2009). Executive functions in adolescence: Inferences from brain and behavior. *Developmental Science*, 12, 825–830.

Crone, E. A., and Dahl, R. E. (2012). Understanding adolescence as a period of social–affective engagement and goal flexibility. *Nature Reviews Neuroscience*, 13, 636–650.

Cullen, K. R., Klimes-Dougan, B., Muetzel, R., Mueller, B. A., Camchong, J., Houri, A., Kurma, S., and Lim, K. O. (2010). Altered white matter microstructure in adolescents with major depression: A preliminary study. *Journal of the American Academy of Child and Adolescent Psychiatry*, 49, 173–83.e1.

Dahl, R. E., Allen, N. B., Wilbrecht, L., and Suleiman, A. B. (2018). Importance of investing in adolescence from a developmental science perspective. *Nature*, 554, 441–450.

Davey, C. G., Yucel, M., and Allen, N. B. (2008). The emergence of depression in adolescence: development of the prefrontal cortex and the representation of reward. *Neuroscience and Biobehavioral Reviews*, 32, 1–19.

Demenescu, L. R., Kortekaas, R., Den Boer, J. A., and Aleman, A. (2010). Impaired attribution of emotion to facial expressions in anxiety and major depression. *PLoS One*, 5, e15058.

Drevets, W. C., Savitz, J., and Trimble, M. (2008). The subgenual anterior cingulate cortex in mood disorders. *CNS Spectrums*, 13, 663–681.

Dudeney, J., Sharpe, L., and Hunt, C. (2015). Attentional bias towards threatening stimuli in children with anxiety: A meta-analysis. *Clinical Psychology Review*, 40, 66–75.

Emslie, G. J., Mayes, T. L., and Ruberu, M. (2005). Continuation and maintenance therapy of early-onset major depressive disorder. *Pediatric Drugs*, 7, 203–217.

Ernst, M. (2014). The triadic model perspective for the study of adolescent motivated behavior. *Brain and Cognition*, 89, 104–111.

Ernst, M., Daniele, T., and Frantz, K. (2011). New perspectives on adolescent motivated behavior: Attention and conditioning. *Developmental Cognitive Neuroscience*, 1, 377–389.

Ernst, M., Nelson, E. E., Jazbec, S., Mcclure, E. B., Monk, C. S., Leibenluft, E., Blair, J., and Pine, D. S. (2005). Amygdala and nucleus accumbens in responses to receipt and omission of gains in adults and adolescents. *NeuroImage*, 25, 1279–1291.

Essau, C. A., Lewinsohn, P. M., Olaya, B., and Seeley, J. R. (2014). Anxiety disorders in adolescents and psychosocial outcomes at age 30. *Journal of Affective Disorders*, 163, 125–132.

Ezpeleta, L., Keeler, G., Erkanli, A., Costello, E. J., and Angold, A. (2001). Epidemiology of psychiatric disability in childhood and adolescence. *Journal of Child Psychology and Psychiatry*, 42, 901–914.

Forbes, E. E., and Dahl, R. E. (2005). Neural systems of positive affect: Relevance to understanding child and adolescent depression? *Developmental Psychopathology*, 17, 827–850.

Forbes, E. E., and Dahl, R. E. (2010). Pubertal development and behavior: Hormonal activation of social and motivational tendencies. *Brain and Cognition*, 72, 66–72.

Forbes, E. E., and Dahl, R. E. (2012). Research review: Altered reward function in adolescent depression: What, when and how? *Journal of Child Psychology and Psychiatry*, 53, 3–15.

Forbes, E. E., Hariri, A. R., Martin, S. L., Silk, J. S., Moyles, D. L., Fisher, P. M., Brown, S. M., Ryan, N. D., Birmaher, B., Axelson, D. A., and Dahl, R. E. (2009). Altered striatal activation predicting real-world positive affect in adolescent major depressive disorder. *American Journal of Psychiatry*, *166*, 64–73.

Forbes, E. E., May, J. C., Siegle, G. J., Ladouceur, C. D., Ryan, N. D., Carter, C. S., Birmaher, B., Axelson, D. A., and Dahl, R. E. (2006). Reward-related decision-making in pediatric major depressive disorder: An fMRI study. *Journal of Child Psychology and Psychiatry*, *47*, 1031–1040.

Forbes, E. E., Shaw, D. S., and Dahl, R. E. (2007). Alterations in reward-related decision making in boys with recent and future depression. *Biological Psychiatry*, *61*, 633–639.

Fox, A. S., and Kalin, N. H. (2014). A translational neuroscience approach to understanding the development of social anxiety disorder and its pathophysiology. *American Journal of Psychiatry*, *171*, 1162–1173.

Fox, A. S., Oler, J. A., Tromp Do, P. M., Fudge, J. L., and Kalin, N. H. (2015). Extending the amygdala in theories of threat processing. *Trends in Neurosciences*, *38*, 319–329.

Frank, M. J., and Fossella, J. A. (2011). Neurogenetics and pharmacology of learning, motivation, and cognition. *Neuropsychopharmacology*, *36*, 133–152.

Frewen, P. A., Dozois, D. J., Joanisse, M. F., and Neufeld, R. W. (2008). Selective attention to threat versus reward: Meta-analysis and neural-network modeling of the dot-probe task. *Clinical Psychology Review*, *28*, 307–337.

Fruhholz, S., Ceravolo, L., and Grandjean, D. (2012). Specific brain networks during explicit and implicit decoding of emotional prosody. *Cerebral Cortex*, *22*, 1107–1117.

Gaffrey, M. S., Luby, J. L., Belden, A. C., Hirshberg, J. S., Volsch, J., and Barch, D. M. (2011). Association between depression severity and amygdala reactivity during sad face viewing in depressed preschoolers: An fMRI study. *Journal of Affective Disorders*, *129*, 364–370.

Gaffrey, M. S., Luby, J. L., Repovš, G., Belden, A. C., Botteron, K. N., Luking, K. R., and Barch, D. M. (2010). Subgenual cingulate connectivity in children with a history of preschool-depression. *NeuroReport*, *21*, 1182–1188.

Ghashghaei, H. T., and Barbas, H. (2002). Pathways for emotion: Interactions of prefrontal and anterior temporal pathways in the amygdala of the rhesus monkey. *Neuroscience*, *115*, 1261–1279.

Giedd, J. N., and Rapoport, J. L. (2010). Structural MRI of pediatric brain development: What have we learned and where are we going? *Neuron*, *67*, 728–734.

Goddings, A. L., Burnett Heyes, S., Bird, G., Viner, R. M., and Blakemore, S. J. (2012). The relationship between puberty and social emotion processing. *Developmental Science*, *15*, 801–811.

Goddings, A. L., Mills, K. L., Clasen, L. S., Giedd, J. N., Viner, R. M., and Blakemore, S. J. (2014). The influence of puberty on subcortical brain development. *NeuroImage*, *88*, 242–251.

Gogtay, N., Giedd, J. N., Lusk, L., Hayashi, K. M., Greenstein, D., Vaituzis, A. C., Nugent, T. F., Herman, D. H., Clasen, L. S., and Toga, A. W. (2004). Dynamic mapping of human cortical development during childhood through early adulthood. *Proceedings of the National Academy of Sciences of the United States of America*, *101*, 8174–819.

Golarai, G., Liberman, A., and Grill-Spector, K. (2015). Experience shapes the development of neural substrates of face processing in human ventral temporal cortex. *Cerebral Cortex*, *27*, bhv314.

Gotlib, I. H., Hamilton, J. P., Cooney, R. E., Singh, M. K., Henry, M. L., and Joormann, J. (2010). Neural processing of reward and loss in girls at risk for major depression. *Archives of General Psychiatry*, *67*, 380–387.

Gould, M. S., Greenberg, T., Velting, D. M., and Shaffer, D. (2003). Youth suicide risk and preventive interventions: A review of the past 10 years. *Journal of the American Academy of Child and Adolescent Psychiatry*, *42*, 386–405.

Greenberg, P. E., Fournier, A.-A., Sisitsky, T., Pike, C. T., and Kessler, R. C. (2015). The economic burden of adults with major depressive disorder in the United States (2005 and 2010). *The Journal of Clinical Psychiatry, 76*, 155–162.

Gullone, E. (2000). The development of normal fear: A century of research. *Clinical Psychology Review, 20*, 429–451.

Guyer, A. E., Choate, V. R., Detloff, A., Benson, B., Nelson, E. E., Perez-Edgar, K., Fox, N. A., Pine, D. S., and Ernst, M. (2012). Striatal functional alteration during incentive anticipation in pediatric anxiety disorders. *American Journal of Psychiatry, 169*, 205–212.

Guyer, A. E., Lau, J. Y., Mcclure-Tone, E. B., Parrish, J., Shiffrin, N. D., Reynolds, R. C., Chen, G., Blair, R. J., Leibenluft, E., Fox, N. A., Ernst, M., Pine, D. S., and Nelson, E. E. (2008a). Amygdala and ventrolateral prefrontal cortex function during anticipated peer evaluation in pediatric social anxiety. *Archives of General Psychiatry, 65*, 1303–1312.

Guyer, A. E., Mcclure-Tone, E. B., Shiffrin, N. D., Pine, D. S., and Nelson, E. E. (2009). Probing the neural correlates of anticipated peer evaluation in adolescence. *Child Development, 80*, 1000–1015.

Guyer, A. E., Monk, C. S., Mcclure-Tone, E. B., Nelson, E. E., Roberson-Nay, R., Adler, A. D., Fromm, S. J., Leibenluft, E., Pine, D. S., and Ernst, M. (2008b). A developmental examination of amygdala response to facial expressions. *Journal of Cognitive Neuroscience, 20*, 1565–1582.

Guyer, A. E., Silk, J. S., and Nelson, E. E. (2016). The neurobiology of the emotional adolescent: From the inside out. *Neuroscience and Biobehavioral Reviews, 70*, 74–85.

Haber, S. N., and Knutson, B. (2010). The reward circuit: linking primate anatomy and human imaging. *Neuropsychopharmacology, 35*, 4–26.

Hare, T. A., Tottenham, N., Galvan, A., Voss, H. U., Glover, G. H., and Casey, B. (2008). Biological substrates of emotional reactivity and regulation in adolescence during an emotional Go-NoGo task. *Biological Psychiatry, 63*, 927–934.

Hartup, W. W. (1996). The company they keep: Friendships and their developmental significance. *Child Development, 67*, 1–13.

Hawker, D. S., and Boulton, M. J. (2000). Twenty years' research on peer victimization and psychosocial maladjustment: A meta-analytic review of cross-sectional studies. *Journal of Child Psychology and Psychiatry, 41*, 441–455.

Heim-Dreger, U., Kohlmann, C. W., Eschenbeck, H., and Burkhardt, U. (2006). Attentional biases for threatening faces in children: Vigilant and avoidant processes. *Emotion, 5*, 320–325.

Hensch, T. K. (2005). Critical period plasticity in local cortical circuits. *Nature Reviews Neuroscience, 6*, 877–888.

Hum, K. M., Manassis, K., and Lewis, M. D. (2013). Neural mechanisms of emotion regulation in childhood anxiety. *Journal of Child Psychology and Psychiatry, 54*, 552–564.

Jalbrzikowski, M., Larsen, B., Hallquist, M. N., Foran, W., Calabro, F., and Luna, B. (2017). Development of white matter microstructure and intrinsic functional connectivity between the amygdala and ventromedial prefrontal cortex: Associations with anxiety and depression. *Biological Psychiatry, 82*, 511–521.

Jarcho, J. M., Romer, A. L., Shechner, T., Galvan, A., Guyer, A. E., Leibenluft, E., Pine, D. S., and Nelson, E. E. (2015). Forgetting the best when predicting the worst: Preliminary observations on neural circuit function in adolescent social anxiety. *Developmental Cognitive Neuroscience, 13*, 21–31.

Joiner, T. E., Lewinsohn, P. M., and Seeley, J. R. (2002). The core of loneliness: Lack of pleasurable engagement—more so than painful disconnection—predicts social impairment, depression onset, and recovery from depressive disorders among adolescents. *Journal of Personality Assessment, 79*, 472–91.

Jones, R. M., Somerville, L. H., Li, J., Ruberry, E. J., Powers, A., Mehta, N., Dyke, J., and Casey, B. J. (2014). Adolescent-specific patterns of behavior and neural activity during social reinforcement learning. *Cognitive, Affective, and Behavioral Neuroscience, 14*, 683–697.

Kaufman, J., Martin, A., King, R. A., and Charney, D. (2001). Are child-, adolescent-, and adult-onset depression one and the same disorder? *Biological Psychiatry, 49*, 980–1001.

Kerestes, R., Davey, C. G., Stephanou, K., Whittle, S., and Harrison, B. J. (2014). Functional brain imaging studies of youth depression: a systematic review. *NeuroImage: Clinical, 4*, 209–231.

Kessler, R. C., Angermeyer, M., Anthony, J. C., de Graaf, R., Demyttenaere, K., Gasquet, I., G, D. E. G., Gluzman, S., Gureje, O., Haro, J. M., Kawakami, N., Karam, A., Levinson, D., Medina Mora, M. E., Oakley Browne, M. A., Posada-Villa, J., Stein, D. J., Adley Tsang, C. H., Aguilar-Gaxiola, S., Alonso, J., Lee, S., Heeringa, S., Pennell, B. E., Berglund, P., Gruber, M. J., Petukhova, M., Chatterji, S., and Ustun, T. B. (2007). Lifetime prevalence and age-of-onset distributions of mental disorders in the World Health Organization's World Mental Health Survey Initiative. *World Psychiatry, 6*, 168–176.

Kessler, R. C., Berglund, P., Demler, O., Jin, R., Koretz, D., Merikangas, K. R., Rush, A. J., Walters, E. E., Wang, P. S., and National Comorbidity Survey, R. (2003). The epidemiology of major depressive disorder: Results from the National Comorbidity Survey Replication (NCS-R). *JAMA, 289*, 3095–3105.

Kessler, R. C., Berglund, P., Demler, O., Jin, R., Merikangas, K. R., and Walters, E. E. (2005). Lifetime prevalence and age-of-onset distributions of DSM-IV disorders in the National Comorbidity Survey Replication. *Archives of General Psychiatry, 62*, 593–602.

Kessler, R. C., Mcgonagle, K. A., Swartz, M., Blazer, D. G., and Nelson, C. B. (1993). Sex and depression in the National Comorbidity Survey. I: Lifetime prevalence, chronicity and recurrence. *Journal of Affective Disorders, 29*, 85–96.

Kessler, R. C., and Wang, P. S. (2008). The descriptive epidemiology of commonly occurring mental disorders in the United States. *Annual Review of Public Health, 29*, 115–129.

Killgore, W. D. S., and Yurgelun-Todd, D. A. (2005). Social anxiety predicts amygdala activation in adolescents viewing fearful faces. *NeuroReport, 16*, 1671–1675.

Kindt, M., Bögels, S., and Morren, M. (2003). Processing bias in children with separation anxiety disorder, social phobia and generalised anxiety disorder. *Behaviour Change, 20*, 143–150.

Kujawa, A., and Burkhouse, K. L. (2017). Vulnerability to depression in youth: Advances from affective neuroscience. *Biological Psychiatry: Cognitive Neuroscience and Neuroimaging, 2*, 28–37.

Ladouceur, C. D., Dahl, R. E., Birmaher, B., Axelson, D. A., and Ryan, N. D. (2006). Increased error-related negativity (ERN) in childhood anxiety disorders: ERP and source localization. *Journal of Child Psychology and Psychiatry, 47*, 1073–1082.

Lai, C. H. (2013). Gray matter volume in major depressive disorder: A meta-analysis of voxel-based morphometry studies. *Psychiatry Research: Neuroimaging, 211*, 37–46.

Larson, R., and Richards, M. H. (1991). Daily companionship in late childhood and early adolescence: Changing developmental contexts. *Child Development, 62*, 284–300.

Larson, R. W., Moneta, G., Richards, M. H., and Wilson, S. (2002). Continuity, stability, and change in daily emotional experience across adolescence. *Child Development, 73*, 1151–1165.

Lebel, C., Gee, M., Camicioli, R., Wieler, M., Martin, W., and Beaulieu, C. (2012). Diffusion tensor imaging of white matter tract evolution over the lifespan. *NeuroImage, 60*, 340–352.

Ledoux, J. E. (2014). Coming to terms with fear. *Proceedings of the National Academy of Sciences of the United States of America*, 111, 2871–288.

Ledoux, J. E., and Pine, D. S. (2016). Using neuroscience to help understand fear and anxiety: A two-system framework. *American Journal of Psychiatry*, 173, 1083–1093.

Lewis, T. L., and Maurer, D. (2005). Multiple sensitive periods in human visual development: Evidence from visually deprived children. *Developmental Psychobiology*, 46, 163–183.

Lonigan, C. J., Carey, M. P., and Finch, A. J., Jr. (1994). Anxiety and depression in children and adolescents: Negative affectivity and the utility of self-reports. *Journal of Consulting and Clinical Psychology*, 62, 1000–1008.

Luby, J. L. (2010). Preschool depression: The importance of identification of depression early in development. *Current Directions in Psychological Science*, 19, 91–95.

Luking, K. R., Pagliaccio, D., Luby, J. L., and Barch, D. M. (2016). Reward processing and risk for depression across development. *Trends in Cognitive Sciences*, 20, 456–468.

Masten, A. S. (2005). Peer relationships and psychopathology in developmental perspective: Reflections on progress and promise. *Journal of Clinical Child and Adolescent Psychology*, 34, 87–92.

Masten, C. L., Eisenberger, N. I., Borofsky, L. A., Mcnealy, K., Pfeifer, J. H., and Dapretto, M. (2011). Subgenual anterior cingulate responses to peer rejection: A marker of adolescents' risk for depression. *Developmental Psychopathology*, 23, 283–292.

Mcclure, E. B., Monk, C. S., Nelson, E. E., Parrish, J. M., Adler, A., Blair, R. J., Fromm, S., Charney, D. S., Leibenluft, E., Ernst, M., and Pine, D. S. (2007). Abnormal attention modulation of fear circuit function in pediatric generalized anxiety disorder. *Archives of General Psychiatry*, 64, 97–106.

Mcdermott, J. M., Perez-Edgar, K., Henderson, H. A., Chronis-Tuscano, A., Pine, D. S., and Fox, N. A. (2009). A history of childhood behavioral inhibition and enhanced response monitoring in adolescence are linked to clinical anxiety. *Biological Psychiatry*, 65, 445–448.

Mcewen, B. S. (2000). Effects of adverse experiences for brain structure and function. *Biological Psychiatry*, 48, 721–731.

Menzies, L., Goddings, A. L., Whitaker, K. J., Blakemore, S. J., and Viner, R. M. (2015). The effects of puberty on white matter development in boys. *Developmental Cognitive Neuroscience*, 11, 116–128.

Merikangas, K. R., He, J. P., Burstein, M., Swanson, S. A., Avenevoli, S., Cui, L., Benjet, C., Georgiades, K., and Swendsen, J. (2010). Lifetime prevalence of mental disorders in U.S. adolescents: Results from the National Comorbidity Survey Replication—Adolescent Supplement (NCS-A). *Journal of the American Academy of Child and Adolescent Psychiatry*, 49, 980–989.

Meyer, A., Proudfit, G. H., Torpey-Newman, D. C., Kujawa, A., and Klein, D. N. (2015). Enhanced error-related brain activity in children predicts the onset of anxiety disorders between the ages of 6 and 9. *Journal of Abnormal Psychology*, 124, 266–274.

Meyer, A., Weinberg, A., Klein, D. N., and Hajcak, G. (2012). The development of the error-related negativity (ERN) and its relationship with anxiety: Evidence from 8- to 13-year-olds. *Developmental Cognitive Neuroscience*, 2, 152–161.

Mills, K. L., Lalonde, F., Clasen, L. S., Giedd, J. N., and Blakemore, S. J. (2014). Developmental changes in the structure of the social brain in late childhood and adolescence. *Social Cognitive and Affective Neuroscience*, 9, 123–131.

Monk, C. S., Mcclure, E. B., Nelson, E. E., Zarahn, E., Bilder, R. M., Leibenluft, E., Charney, D. S., Ernst, M., and Pine, D. S. (2003). Adolescent immaturity in attention-related brain engagement to emotional facial expressions. *NeuroImage*, 20, 420–428.

Monk, C. S., Nelson, E. E., Mcclure, E. B., Mogg, K., Bradley, B. P., Leibenluft, E., Blair, R. J., Chen, G., Charney, D. S., Ernst, M., and Pine, D. S. (2006). Ventrolateral prefrontal cortex activation and attentional bias in response to angry faces in adolescents with generalized anxiety disorder. *American Journal of Psychiatry*, *163*, 1091–1097.

Monk, C. S., Telzer, E. H., Mogg, K., Bradley, B. P., Mai, X., Louro, H. M. C., Chen, G., Mcclure-Tone, E. B., Ernst, M., and Pine, D. S. (2008). Amygdala and ventrolateral prefrontal cortex activation to masked angry faces in children and adolescents with generalized anxiety disorder. *Archives of General Psychiatry*, *65*, 568–576.

Moor, B. G., Leijenhorst, L. V., Rombouts, S. A. R. B., Crone, E. A., and Van Der Molen, M. W. (2010). Do you like me? Neural correlates of social evaluation and developmental trajectories. *Social Neuroscience*, *1*, 1–22.

Moore, W. E., 3rd, Pfeifer, J. H., Masten, C. L., Mazziotta, J. C., Iacoboni, M., and Dapretto, M. (2012). Facing puberty: Associations between pubertal development and neural responses to affective facial displays. *Social Cognitive and Affective Neuroscience*, *7*, 35–43.

Morgan, J. K., Olino, T. M., Mcmakin, D. L., Ryan, N. D., and Forbes, E. E. (2013). Neural response to reward as a predictor of increases in depressive symptoms in adolescence. *Neurobiological Disorders*, *52*, 66–74.

Morningstar, M., Dirks, M. A., Rappaport, B. I., Pine, D. S., and Nelson, E. E. (2019). Associations between anxious and depressive symptoms and the recognition of vocal socioemotional expressions in youth. *Journal of Clinical Child and Adolescent Psychology*, *48*, 491–500.

Nandrino, J. L., Dodin, V., Martin, P., and Henniaux, M. (2004). Emotional information processing in first and recurrent major depressive episodes. *Journal of Psychiatric Research*, *38*, 475–484.

Nelson, E. E. (2017). Learning through the ages: How the brain adapts to the social world across development. *Cognitive Development*, *42*, 84–94.

Nelson, E. E., and Guyer, A. E. (2011). The development of the ventral prefrontal cortex and social flexibility. *Developmental Cognitive Neuroscience*, *1*, 233–245.

Nelson, E. E., Jarcho, J. M., and Guyer, A. E. (2016). Social re-orientation and brain development: An expanded and updated view. *Developmental Cognitive Neuroscience*, *17*, 118–127.

Nelson, E. E., Lau, J. Y., and Jarcho, J. M. (2014). Growing pains and pleasures: How emotional learning guides development. *Trends in Cognitive Sciences*, *18*, 99–108.

Nelson, E. E., Leibenluft, E., Mcclure, E. B., and Pine, D. S. (2005). The social re-orientation of adolescence: A neuroscience perspective on the process and its relation to psychopathology. *Psychological Medicine*, *35*, 163–174.

Nestler, E. J., and Carlezon, W. A., Jr. (2006). The mesolimbic dopamine reward circuit in depression. *Biological Psychiatry*, *59*, 1151–1159.

Olino, T. M., Mcmakin, D. L., Morgan, J. K., Silk, J. S., Birmaher, B., Axelson, D. A., Williamson, D. E., Dahl, R. E., Ryan, N. D., and Forbes, E. E. (2014). Reduced reward anticipation in youth at high-risk for unipolar depression: A preliminary study. *Developmental Cognitive Neuroscience*, *8*, 55–64.

Olino, T. M., Silk, J. S., Osterritter, C., and Forbes, E. E. (2015). Social reward in youth at risk for depression: A preliminary investigation of subjective and neural differences. *Journal of Child and Adolescent Psychopharmacology*, *25*, 711–721.

Olson, I. R., Von Der Heide, R. J., Alm, K. H., and Vyas, G. (2015). Development of the uncinate fasciculus: Implications for theory and developmental disorders. *Developmental Cognitive Neuroscience*, *14*, 50–61.

Ordaz, S. J., Foran, W., Velanova, K., and Luna, B. (2013). Longitudinal growth curves of brain function underlying inhibitory control through adolescence. *Journal of Neuroscience, 33*, 18109–18124.

Pagliaccio, D., Luby, J., Gaffrey, M., Belden, A., Botteron, K., Gotlib, I. H., and Barch, D. M. (2012). Anomalous functional brain activation following negative mood induction in children with pre-school onset major depression. *Developmental Cognitive Neuroscience, 2*, 256–267.

Paus, T. (2010). Growth of white matter in the adolescent brain: Myelin or axon? *Brain and Cognition, 72*, 26–35.

Paus, T., Keshavan, M., and Giedd, J. N. (2008). Why do many psychiatric disorders emerge during adolescence? *Nature Reviews Neuroscience, 9*, 947–957.

Peper, J. S., Hulshoff Pol, H. E., Crone, E. A., and Van Honk, J. (2011a). Sex steroids and brain structure in pubertal boys and girls: A mini review of neuroimaging studies. *Neuroscience, 191*, 28–37.

Peper, J. S., Van Den Heuvel, M. P., Mandl, R. C., Hulshoff Pol, H. E., and Van Honk, J. (2011b). Sex steroids and connectivity in the human brain: A review of neuroimaging studies. *Psychoneuroendocrinology, 36*, 1101–1113.

Perera, T. D., Coplan, J. D., Lisanby, S. H., Lipira, C. M., Arif, M., Carpio, C., Spitzer, G., Santarelli, L., Scharf, B., Hen, R., Rosoklija, G., Sackeim, H. A., and Dwork, A. J. (2007). Antidepressant-induced neurogenesis in the hippocampus of adult nonhuman primates. *Journal of Neuroscience, 27*, 4894–4901.

Perez-Edgar, K., and Fox, N. A. (2005). Temperament and anxiety disorders. *Child and Adolescent Psychiatric Clinics of North America, 14*, 681–706, viii.

Pérez-Edgar, K., Roberson-Nay, R., Hardin, M. G., Poeth, K., Guyer, A. E., Nelson, E. E., Mcclure, E. B., Henderson, H. A., Fox, N. A., Pine, D. S., and Ernst, M. (2007). Attention alters neural responses to evocative faces in behaviorally inhibited adolescents. *NeuroImage, 35*, 1538–1546.

Perlman, G., Simmons, A. N., Wu, J., Hahn, K. S., Tapert, S. F., Max, J. E., Paulus, M. P., Brown, G. G., Frank, G. K., Campbell-Sills, L., and Yang, T. T. (2012). Amygdala response and functional connectivity during emotion regulation: A study of 14 depressed adolescents. *Journal of Affective Disorders, 139*, 75–84.

Petanjek, Z., Judas, M., Simic, G., Rasin, M. R., Uylings, H. B., Rakic, P., and Kostovic, I. (2011). Extraordinary neoteny of synaptic spines in the human prefrontal cortex. *Proceedings of the National Academy of Sciences of the United States of America, 108*, 13281–13286.

Pfeifer, J. H., and Allen, N. B. (2012). Arrested development? Reconsidering dual-systems models of brain function in adolescence and disorders. *Trends in Cognitive Sciences, 16*, 322–329.

Pfeifer, J. H., and Blakemore, S. J. (2012). Adolescent social cognitive and affective neuroscience: Past, present, and future. *Social Cognitive and Affective Neuroscience, 7*, 1–10.

Picci, G., and Scherf, K. S. (2016). From caregivers to peers: Puberty shapes human face perception. *Psychological Science, 27*, 1461–1473.

Pine, D. S. (2007). Research review: A neuroscience framework for pediatric anxiety disorders. *Journal of Child Psychology and Psychiatry, 48*, 631–648.

Pine, D. S., Cohen, E., Cohen, P., and Brook, J. (1999). Adolescent depressive symptoms as predictors of adult depression: Moodiness or mood disorder? *American Journal of Psychiatry, 156*, 133–135.

Pine, D. S., and Fox, N. A. (2015). Childhood antecedents and risk for adult mental disorders. *Annual Review of Psychology, 66*, 459–485.

Pryce, C. R. (2008). Postnatal ontogeny of expression of the corticosteroid receptor genes in mammalian brains: Inter-species and intra-species differences. *Brain Research Reviews*, *57*, 596–605.

Qin, S., Young, C. B., Duan, X., Chen, T., Supekar, K., and Menon, V. (2014). Amygdala sub-regional structure and intrinsic functional connectivity predicts individual differences in anxiety during early childhood. *Biological Psychiatry*, *75*, 892–900.

Rapee, R. M., Schniering, C. A., and Hudson, J. L. (2009). Anxiety disorders during childhood and adolescence: Origins and treatment. *Annual Review of Clinical Psychology*, *5*, 311–341.

Ray, R. D., and Zald, D. H. (2012). Anatomical insights into the interaction of emotion and cognition in the prefrontal cortex. *Neuroscience and Biobehavioral Reviews*, *36*, 479–501.

Restifo, K., and Bögels, S. (2009). Family processes in the development of youth depression: Translating the evidence to treatment. *Clinical Psychology Review*, *29*, 294–316.

Romeo, R. D., and Mcewen, B. S. (2006). Stress and the adolescent brain. *Annals of the New York Academy of Sciences*, *1094*, 202–214.

Roza, S. J., Hofstra, M. B., Van Der Ende, J., and Verhulst, F. C. (2003). Stable prediction of mood and anxiety disorders based on behavioral and emotional problems in childhood: A 14-year follow-up during childhood, adolescence, and young adulthood. *American Journal of Psychiatry*, *160*, 2116–2121.

Rueter, M. A., Scaramella, L., Wallace, L., and Conger, R. D. (1999). First onset of depressive or anxiety disorders predicted by the longitudinal course of internalizing symptoms and parent-adolescent disagreements. *Archives of General Psychiatry*, *56*, 726–732.

Satterthwaite, T. D., Wolf, D. H., Erus, G., Ruparel, K., Elliott, M. A., Gennatas, E. D., Hopson, R., Jackson, C., Prabhakaran, K., Bilker, W. B., Calkins, M. E., Loughead, J., Smith, A., Roalf, D. R., Hakonarson, H., Verma, R., Davatzikos, C., Gur, R. C., and Gur, R. E. (2013). Functional maturation of the executive system during adolescence. *Journal of Neuroscience*, *33*, 16249–16261.

Scherf, K. S., Smyth, J. M., and Delgado, M. R. (2013). The amygdala: An agent of change in adolescent neural networks. *Hormones and Behavior*, *64*, 298–313.

Schmukle, S. C. (2005). Unreliability of the dot probe task. *European Journal of Personality*, *19*, 595–605.

Sharp, C., Kim, S., Herman, L., Pane, H., Reuter, T., and Stratharn, L. (2014). Major depression in mothers predicts reduced ventral striatum activation in adolescent female offspring with and without depression. *Journal of Abnormal Psychology*, *123*, 298–309.

Shechner, T., Britton, J. C., Perez-Edgar, K., Bar-Haim, Y., Ernst, M., Fox, N. A., Leibenluft, E., and Pine, D. S. (2012). Attention biases, anxiety, and development: Toward or away from threats or rewards? *Depression and Anxiety*, *29*, 282–94.

Silk, J. S., Shaw, D. S., Forbes, E. E., Lane, T. L., and Kovacs, M. (2006). Maternal depression and child internalizing: The moderating role of child emotion regulation. *Journal of Clinical Child and Adolescent Psychology*, *35*, 116–126.

Silk, J. S., Siegle, G. J., Lee, K. H., Nelson, E. E., Stroud, L. R., and Dahl, R. E. (2014). Increased neural response to peer rejection associated with adolescent depression and pubertal development. *Social Cognitive and Affective Neuroscience*, *9*, 1798–1807.

Silk, J. S., Steinberg, L., and Sheffield Morris, A. S. (2003). Adolescents' emotion regulation in daily life: Links to depressive symptoms and problem behavior. *Child Development*, *74*, 1869–1880.

Smith, A. R., Chein, J., and Steinberg, L. (2013). Impact of socio-emotional context, brain development, and pubertal maturation on adolescent risk-taking. *Hormones and Behavior*, *64*, 323–32.

Smith, A. R., Steinberg, L., and Chein, J. (2014). The role of the anterior insula in adolescent decision making. *Developmental Neuroscience, 36*, 196–209.

Smith, A. R., Steinberg, L., Strang, N., and Chein, J. (2015). Age differences in the impact of peers on adolescents' and adults' neural response to reward. *Developmental Cognitive Neuroscience, 11*, 75–82.

Somers, J. M., Goldner, E. M., Waraich, P., and Hsu, L. (2006). Prevalence and incidence studies of anxiety disorders: A systematic review of the literature. *The Canadian Journal of Psychiatry, 51*, 100–113.

Somerville, L. H., Fani, N., and Mcclure-Tone, E. B. (2011). Behavioral and neural representation of emotional facial expressions across the lifespan. *Developmental Neuropsychology, 36*, 408–428.

Somerville, L. H., Jones, R. M., and Casey, B. J. (2010). A time of change: Behavioral and neural correlates of adolescent sensitivity to appetitive and aversive environmental cues. *Brain and Cognition, 72*, 124–133.

Somerville, L. H., Jones, R. M., Ruberry, E. J., Dyke, J. P., Glover, G., and Casey, B. J. (2013). The medial prefrontal cortex and the emergence of self-conscious emotion in adolescence. *Psychological Science, 24*, 1554–1562.

Sørensen, M. J., Nissen, J. B., Mors, O., and Thomsen, P. H. (2005). Age and gender differences in depressive symptomatology and comorbidity: An incident sample of psychiatrically admitted children. *Journal of Affective Disorders, 84*, 85–91.

Stanwood, G. D., and Levitt, P. (2004). Drug exposure early in life: functional repercussions of changing neuropharmacology during sensitive periods of brain development. *Current Opinion in Pharmacology, 4*, 65–71.

Steinberg, L. (2005). Cognitive and affective development in adolescence. *Trends in Cognitive Sciences, 9*, 69–74.

Steinberg, L., Albert, D., Cauffman, E., Banich, M., Graham, S., and Woolard, J. (2008). Age differences in sensation seeking and impulsivity as indexed by behavior and self-report: Evidence for a dual systems model. *Developmental Psychology, 44*, 1764–1778.

Steinberg, L., and Morris, A. S. (2001). Adolescent development. *Annual Review of Psychology, 52*, 83–110.

Stiles, J. (2008). *The Fundamentals of Brain Development: Integrating Nature and Nurture.* Cambridge, MA, Harvard University Press.

Swartz, J. R., Williamson, D. E., and Hariri, A. R. (2015). Developmental change in amygdala reactivity during adolescence: Effects of family history of depression and stressful life events. *American Journal of Psychiatry, 172*, 276–283.

Tao, R., Calley, C. S., Hart, J., Mayes, T. L., Nakonezny, P. A., Lu, H., Kennard, B. D., Tamminga, C. A., and Emslie, G. J. (2012). Brain activity in adolescent major depressive disorder before and after fluoxetine treatment. *American Journal of Psychiatry, 169*, 381–388.

Teicher, M. H., Glod, C. A., Harper, D., Magnus, E., Brasher, C., Wren, F., and Pahlavan, K. (1993). Locomotor activity in depressed children and adolescents: I. Circadian dysregulation. *Journal of the American Academy of Child and Adolescent Psychiatry, 32*, 760–769.

Telzer, E. H., Mogg, K., Bradley, B. P., Mai, X., Ernst, M., Pine, D. S., and Monk, C. S. (2008). Relationship between trait anxiety, prefrontal cortex, and attention bias to angry faces in children and adolescents. *Biological Psychiatry, 79*, 216–222.

Thai, N., Taber-Thomas, B. C., and Perez-Edgar, K. E. (2016). Neural correlates of attention biases, behavioral inhibition, and social anxiety in children: An ERP study. *Developmental Cognitive Neuroscience, 19*, 200–210.

Thapar, A., Collishaw, S., Pine, D. S., and Thapar, A. K. (2012). Depression in adolescence. *Lancet, 379*, 1056–1067.

Thomas, K. M., Drevets, W. C., Dahl, R. E., Ryan, N. D., Birmaher, B., Eccard, C. H., Axelson, D., Whalen, P. J., and Casey, B. J. (2001). Amygdala response to fearful faces in anxious and depressed children. *Archives of General Psychiatry, 58*, 1057–1063.

Vasey, M. W., Daleiden, E. L., Williams, L. L., and Brown, L. M. (1995). Biased attention in childhood anxiety disorders: A preliminary study. *Journal of Abnormal Child Psychology, 23*, 267–279.

Vijayakumar, N., Whittle, S., Yucel, M., Byrne, M. L., Schwartz, O., Simmons, J. G., and Allen, N. B. (2016). Impaired maturation of cognitive control in adolescents who develop major depressive disorder. *Journal of Clinical Child and Adolescent Psychology, 45*, 31–43.

Waters, A. M., Henry, J., Mogg, K., Bradley, B. P., and Pine, D. S. (2010). Attentional bias towards angry faces in childhood anxiety disorders. *Journal of Behavior Therapy and Experimental Psychiatry, 41*, 158–164.

Weissman, M. M., Wolk, S., Wickramaratne, P., Goldstein, R. B., Adams, P., Greenwald, S., Ryan, N. D., Dahl, R. E., and Steinberg, D. (1999). Children with prepubertal-onset major depressive disorder and anxiety grown up. *Archives of General Psychiatry, 56*, 794–801.

Wilcox, H. C., and Anthony, J. C. (2004). Child and adolescent clinical features as forerunners of adult-onset major depressive disorder: Retrospective evidence from an epidemiological sample. *Journal of Affective Disorders, 82*, 9–20.

Woodward, L. J., and Fergusson, D. M. (2001). Life course outcomes of young people with anxiety disorders in adolescence. *Journal of the American Academy of Child and Adolescent Psychiatry, 40*, 1086–1093.

Yang, T. T., Simmons, A. N., Matthews, S. C., Tapert, S. F., Frank, G. K., Max, J. E., Bischoff-Grethe, A., Lansing, A. E., Brown, G., Strigo, I. A., Wu, J., and Paulus, M. P. (2010). Adolescents with major depression demonstrate increased amygdala activation. *Journal of the American Academy of Child and Adolescent Psychiatry, 49*, 42–51.

Zhang, F. (2017). Resting-state functional connectivity abnormalities in adolescent depression. *EBioMedicine, 17*, 20–1.

Zisook, S., Rush, A. J., Lesser, I., Wisniewski, S. R., Trivedi, M., Husain, M. M., Balasubramani, G. K., Alpert, J. E., and Fava, M. (2007). Preadult onset vs. adult onset of major depressive disorder: A replication study. *Acta Psychiatrica Scandinavica, 115*, 196–205.

CHAPTER 35

DEPRESSION IN YOUNG PEOPLE

Cognitive biases and targets for intervention

VICTORIA PILE AND JENNIFER Y. F. LAU

INTRODUCTION

DEPRESSION refers to a group of symptoms and behaviors characterized by changes in emotion, cognition and behavior. Changes in mood include sadness, irritability and anhedonia (a reduction or loss of capacity to experience pleasure). Changes in cognition can include effects on cognitive functioning (e.g., concentration, decision-making) but also the presence of mood-congruent thoughts such as decreased self-esteem. Physically, people with depression often become less active and may experience changes in appetite, weight and sleep (National Institute of Clinical Excellence, 2015). To receive a diagnosis according to the Diagnostic and Statistical Manual of Mental Disorders (American Psychiatric Association, 2013), a set of symptoms of depression must be present for most of the day for two weeks and significantly impact on that individual's functioning. Whilst depression is often considered to present similarly in children and young people as in adults, there are some important developmental differences. Children are perhaps more likely to present with somatic features, such as aches and pains, whilst adolescents may show more cognitive features, such as poor concentration, worthlessness and self-criticism (Luby et al., 2003). In youth, irritability is a cardinal criterion for diagnosis (and can be present rather than low mood) which is not the case for adults (Stringaris, Maughan, Copeland, Costello, and Angold, 2013; Stringaris, Vidal-Ribas, Brotman, and Leibenluft, 2018).

Depression is a common condition, with 20 per cent of people estimated to experience an episode at some point in their life (Kessler and Wang, 2009). Epidemiological studies indicate that females are more than twice as likely as males to experience depression (Mccrone, Dhanasiri, Patel, Knapp, and Lawton-Smith, 2008). Depression is a leading cause of disability worldwide (Whiteford et al., 2013) with estimated UK costs reaching £12.15 billion by 2026 (Mccrone et al., 2008). It is highly recurrent with 75 per cent of affected individuals experiencing a second episode, usually relapsing within two years of recovery from a previous episode (Boland and Keller 2009). The peak age of first onset of depression is in adolescence. Adolescent depression is associated with psychological and social difficulties, including higher social dysfunction, poorer academic performance, more physical ill health complaints and is a leading risk factor for suicide in youth (Andersen and Teicher, 2008; Fombonne, Wostear, Cooper, Harrington, and Rutter, 2001; Zisook et al., 2007). Moreover, compared to adult-onset, adolescent-onset depression is associated with an increased risk of chronicity and reduced functioning (de Girolamo, Dagani, Purcell, Cocchi, and McGorry, 2012; Richards, 2011).

In the UK, the first line treatment for 12- to 18-year-olds with moderate to severe depression is individual cognitive behavioral therapy (CBT) for at least three months (National Institute for Health and Care Excellence, 2019). If CBT is not appropriate then another specific psychological intervention can be offered: interpersonal psychotherapy, family therapy, brief psychosocial intervention or psychodynamic psychotherapy. The National Institute for Health and Care Excellence (NICE) identified that, compared to control groups, individual CBT produced greater gains during treatment and that these are maintained at follow-up. However, other research has found only modest effect sizes for psychological therapy in youth and insufficient evidence to recommend one intervention in particular (Cox, Callahan, et al., 2014; Goodyer et al., 2016). For 5- to 11-year-olds with moderate to severe depression, family based interpersonal therapy, family therapy, individual CBT, or psychodynamic psychotherapy can be considered (National Institute for Health and Care Excellence, 2019). However, there has been very limited research for this age group (National Institute for Health and Care Excellence, 2019).

The use of antidepressant medication in children and young people with depression emerged from its evidence base in adults. However, medication alone is not recommended for initial treatment due to concerns over unpleasant side effects, poor efficacy, addictive potential and increased suicide risk (Healy, 2003; National Institute for Clinical Excellence, 2015; Springer, Rubin, and Beevers, 2011; Wright et al., 2014). There is some evidence that combining medication and CBT may be most effective in reducing depressive symptoms and preventing relapse in youth, although evidence is limited (Cox, Callahan, et al., 2014; Cox, Fisher, et al., 2014). Novel interventions that can enhance effectiveness are therefore needed.

As well as improving effectiveness, innovation in treatments for depression in children and adolescents should also address current challenges in under-detection and under-treatment, which are also associated with increased healthcare costs (Wright et al., 2017). It is estimated that 75 per cent of youth with mood disorders in the US go undetected (Coyle et al., 2003) and of those referred to child and adolescent mental health services (CAMHS) in the United Kingdom (UK) between 28 per cent and 75 per cent do not receive a service (Children's Commissioner, 2016; Pile, Shammas, and Smith, 2019).

Research underlines the importance of accessing services early, with young people who have not had contact with services at age 14 being seven times more likely to report clinical depression at age 17, compared to peers who were similarly depressed but had contact with services (Neufeld et al., 2017).

Adolescence is a key period of opportunity to prevent persistent and recurrent depression from developing. Findings suggest that negative experiences (such as depression) occurring in adolescence could alter developmental trajectories, resulting in long-lasting maladaptive neurobiological and cognitive changes that persist into adulthood (Lau and Pile, 2015; Lau, 2013; Post, 1992; Sokolov and Kutcher, 2001). Importantly, as neurodevelopmental and physiological changes as well as increased flexibility and learning potential characterize adolescence (Blakemore, 2012), interventions administered at this juncture could be more effective. Indeed, recent government initiatives have recognized the importance of administering early preventative interventions to young people who show early signs of distress to prevent problems deteriorating (Secretary of State for Health and Secretary of State for Education, 2017). Yet, as the present range of early interventions are not always effective (Calear and Christensen, 2010; Werner-seidler, Perry, Calear, Newby, and Christensen, 2017) and difficult to access, a fuller understanding of the risk mechanisms contributing to adolescent depression is needed. There is little doubt that the aetiology of depression is complex and multifactorial, drawing on biological, social, and psychological factors. However, cognitive factors have increasingly been a focus of research attention given clear theoretical models, impetus from service users to focus on psychological factors and the relatively ready translation of cognitive factors into intervention.

Cognitive theories of depression were first suggested approximately fifty years ago. They propose that risk for depression is increased by a person's thoughts, attitudes, and interpretations as well as the way in which they attend to and remember information. According to cognitive theory, appraisals determine whether an emotion is experienced, and which one is experienced. Cognitive biases therefore impact on emotional regulation capabilities and may confer risk for emotional disorders. Biases may also act to alter how easily emotion regulation strategies are used, rather than directly acting on emotional experience. Cognitive biases are proposed to not only feature in depressive episodes (or as a response to depression) but also play a crucial role in enhancing a person's vulnerability for developing depression, increasing severity and risk for relapse following recovery. Often, these are proposed as part of a vulnerability-stress hypothesis, notably that the onset of depression is associated with a precipitating stressor (e.g., a negative life event or environmental factor) acting alongside existing psychological vulnerability (e.g., cognitive information-processing biases). It is suggested that the cognitive vulnerability factor magnifies the effect of the stressor on symptomatology.

In this chapter, we will focus on one set of risk factors associated with depression: biases in information-processing. Because biased information-processing factors may maintain symptoms but also mediate the effects of distal risk factors such as genetics or early life adversity on their presentation, they may be more amenable to change through therapeutic techniques that are more accessible. First, we review evidence for cognitive biases in the processing of emotional material, followed by a brief review of the role of general executive functioning deficits in depression. We focus on three cognitive biases: those in attention, interpretation and memory. In each section, we will define the bias and describe methods of assessment; briefly summarize adult findings as well as review research in youth with depression and in community samples; and then summarize research investigating whether the bias is a cognitive vulnerability factor (e.g., in youth at risk of depression). Second, we review current treatments for depression and advances in these treatments based on work with cognitive biases.

COGNITIVE BIASES IN DEPRESSION

Attention Biases

It has been proposed that depression is associated with enhanced, and perhaps automatic, perception of emotion-relevant cues in the environment. Attention bias to negative information refers to the differential allocation of attention towards negative over non-negative stimuli (Bar-Haim, Lamy, Pergamin, Bakermans-Kranenburg, and van IJzendoorn, 2007). Individuals differ in the extent that they allocate attention to negative stimuli, which can magnify the person's perceived negativity in the environment in a maladaptive way.

A commonly used task for measuring attention bias in young people is the dot probe task. In this task, two stimuli—one negative and one positive/neutral—are presented together for around 500–1500 ms on a screen (e.g., a sad face alongside a happy face). The stimuli disappear and one is replaced by a probe (e.g., a letter p or q) and the aim of the task is to respond to the probe (e.g., press the correct letter key) as soon as it appears. Attention bias is derived from comparing the response times to probes that replace negative stimuli (congruent trials) and those that replace non-negative stimuli (incongruent trials). If an individual preferentially allocates their attention to negative stimuli, then their response times will be shorter for congruent compared to incongruent trials. Inhibitory control tasks (e.g., Go/No-go task) have also been used to assess attention bias. Targets fall into two categories: negative or positive words. Participants are presented with the target word among non-targets words for around 300 ms (Hare, Tottenham, Davidson, Glover, and Casey, 2005). The target category changes for each block with distractors as non-targets words. Faster reaction time (RT)s to negative (compared to positive or neutral) targets indicate attention bias. Early studies used the emotional version of the Stroop task, where words are presented in different colored ink and the participant is asked to say the color out loud (rather than the word). In the emotional version of this task, words that are depression-related, threat-related, and trauma-related can be included. The emotional Stroop task is now considered to be more a measure of inhibition of attention to negative stimuli rather than spatial orienting of visual attention—and therefore assessing a different component of attention bias to that of the dot-probe task. Indeed, it seems likely that attention bias has several components and the various tasks developed to measure it may tap different aspects.

A more recent paradigm is the Emotional Visual Search Task (EVST) (De Voogd, Wiers, Prins, and Salemink, 2014). There are several variations of this task. In the one used for measuring attentional biases in depression, participants are presented with a 4 x 4 matrix of fifteen distractor faces and one target face. The matrix is displayed until the participant selects their response. For some blocks, the target will be positive with negative distractors whilst this is reversed for other blocks. Negative attention bias is assessed by faster reaction times to detect the negative over positive target. These tasks have a longer stimulus presentation time and are thought to measure overt rather than covert attentional processes (e.g., Weierich, Treat, and Hollingworth, 2008). In contrast, Posner's spatial cueing paradigm (Posner, 1980) measures covert attention biases by presenting cue stimuli for a shorter time (e.g., 150 ms). Participants are asked to respond to an emotionally neutral target which appears on the left- or right-hand side of a central fixation cross. A cue (emotionally valenced or neutral) appears either on the same or opposite side of the screen as the target. A covert engagement bias is indicated when reaction times are quicker for targets that follow negative cues (compared to neutral) on the same side of the target. A covert disengagement bias for negative information is indicated when reaction times are slower for targets that follow negative cues (compared to neutral) on the opposite side of the screen.

A key advantage of many of these tasks is that attention bias is indexed as a within subject difference in responses to congruent and incongruent probes. This means that individual factors (such as impulsivity or difficulties with attention) are minimized as explanations for the bias, as it would be assumed that this is equivalent across trials. More recently, eye tracking has also been included in these paradigms to provide a more thorough assessment of attention allocation (Harrison and Gibb, 2015).

The method of measuring this bias (i.e., which component is being measured) is important; a meta-analysis in adults found that studies using the dot probe showed significant differences between depressed and non-depressed groups (d = 0.52), whilst studies using the emotional Stroop task showed only marginally significant effects (Peckham, McHugh, and Otto, 2010). Depression, as opposed to anxiety, appears to be characterized by difficulties disengaging attention from negative stimuli (once the stimulus has become the focus of their attention) rather than the involuntary orientation to negative information in their environment (Gotlib and Joormann, 2010; Joormann and Stanton, 2016). Indeed, research has now shown that clinically depressed individuals and those at risk for depression (e.g., subclinical symptoms of depression) show difficulties in disengaging attention from negative information and do not demonstrate an attention bias for positive information (which is found in non-depressed individuals) (Peckham et al., 2010). Similarly, the content of the stimuli is important. In adults, depression has been associated with a selective bias for negative and/or unpleasant stimuli (e.g., a bias for sad over happy faces; Everaert, Koster, and Derakshan, 2012). This is in contrast to anxiety disorders, where attention is preferably allocated to threat-related stimuli (e.g., social threat words in social anxiety) (Cisler and Koster, 2010).

For young people, initial studies suggested no differences in attentional biases for negative information between depressed and non-depressed peers (Dalgleish et al., 2003; Neshat-Doost, Moradi, Taghavi, Yule, and Dalgleish, 2000; Neshat Doost, Taghavi, Moradi, Yule, and Dalgleish, 1997). However, this may be due to the relatively small sample sizes in these studies and use of the emotional Stroop task in two of the studies (Dalgleish et al., 2003; Neshat Doost et al., 1997). More recent studies have found differences in attention biases for negative stimuli between depressed and non-depressed groups using the visual probe task (Hankin, Gibb, Abela, and Flory, 2010). Similar to the adult literature, these biases are observed at longer stimulus durations (500 to 1250 ms) and therefore are likely to reflect a bias in voluntary attentional processes.

Inhibitory control tasks (e.g., Go/No-go task) have mostly shown that young people with depression are more likely to allocate their attention towards negative stimuli, compared to young people without depression (Kyte, Goodyer, and Sahakian, 2005; Ladouceur et al., 2006; Maalouf et al., 2012), however, one study found no relationship (Micco, Henin, and Hirshfeld-Becker, 2014). Further to these findings, one study which tracked participant eye movement whilst passively viewing positive and negative faces, found that young people with depression spent longer attending to happy compared to sad faces (Harrison and Gibb, 2015). This finding contrasts with the previous studies and the authors suggest that at longer stimulus durations (e.g., 20s), young people with depression engage in emotional regulation strategies (i.e., allocate their attention away from negative stimuli). [This finding is similar to adult studies, where a meta-analysis indicated that people with depression spend less time attending to sad stimuli than those without depression (Armstrong and Olatunji, 2012).]

For young people, studies in community samples have been more mixed. One study found a relationship between negative attention bias for negative faces and depressive symptoms (Platt, Murphy, and Lau, 2015), whereas others have found no relationship (De Voogd et al., 2014) or that the relationship is better explained by anxiety (Reid, Salmon, and Lovibond, 2006). Interestingly, one study found that the relationship between attention bias and negative affectivity (a temperamental factor thought to confer vulnerability for depression) was only present for children with low (and not high) effortful control (indicator of poor executive control) (Lonigan and Vasey, 2009). This implies that those with better executive control are able to modify the impact that attention bias has on their mood.

Studies investigating whether attention biases are a vulnerability factor for depression have found conflicting results (Lau and Waters, 2017). Some studies have found attention bias towards sad faces, whilst others have found attention bias away from sad faces in non-depressed children of depressed mothers (compared to never depressed mothers) (Connell, Patton, Klostermann, and Hughes-Scalise, 2013; Gibb, Benas, Grassia, and McGeary, 2009; Joormann, Talbot, and Gotlib, 2007; Kujawa et al., 2011). These differences may be explained by differences in methodology and sample characteristics (although stimulus durations were similar). Further to this, a study assessing attention biases in young people in remission from depression found no evidence for attention bias to negative emotion stimuli (Maalouf et al., 2012).

Interpretation Biases

Negative interpretations are a core constituent of cognitive models of depression. Interpretation bias refers to a tendency to interpret ambiguous events or information in a negative way. They are most commonly assessed by presenting, in words or pictures, an ambiguous scenario where both negative and positive (or benign) interpretations are possible. Paradigms assessing interpretation bias can be divided into offline and online tasks (Hirsch, Meeten, Krahé, and Reeder, 2016). Offline tasks (e.g., self-report questionnaires or tasks requiring the ordering of potential interpretations) allow participants to think about ambiguous material without having to report the first interpretation that comes to mind. These tasks are vulnerable to various biases including demand effects (participants understand the purpose of the task and respond to meet the experimenter's expectations) and selection bias (selecting from a range of interpretations rather than responding with the first inference that comes to mind). The recognition task aims to make the purpose of the task less obvious and avoids some of these challenges. The recognition task measures interpretation bias in two parts. First, participants read a set of ambiguous scenarios, which include a non-emotional title (e.g., First Birthday Party), the scenario where the last word is a fragment (e.g., "You are organizing your first real party for your birthday at your parents' basement. At the party, you see some people in the corner and hear them t_lk_ng)". Participants are asked to complete this fragment (which does not emotionally disambiguate the scenario) and complete a comprehension question (e.g., "Did you organize a party for your birthday?"). In the second part, after reading all the scenarios, participants are presented with the title of the scenario and four sentences, which participants are asked to rate based on their similarity to the scenario. Two of these sentences are targets (positive and negative interpretations of the scenario) and the other two are "foils" (positive and negative sentences not interpreting the scenario). The advantage of the recognition task is its ecological validity, given that realistic events are described. A disadvantage is that it does not differentiate between interpretations generated when the participant first encounters the ambiguous material and interpretations that occur later, at retrieval, when the participant is able to reflect.

In contrast to offline tasks, online tasks use ambiguous information to prime target words and then response times are used to indicate how the ambiguous information has been interpreted. So, faster reaction times will be observed when the target matches the inferred meaning. An example of ambiguous stimuli used in these tasks are homophones and homographs, words that have the same sound but different meanings (for example "pain/pane") and words with multiple meanings (for example, "mug"), as well as ambiguous scenarios. For example, participants are asked to generate sentences from a given word (e.g., "hit") and this sentence is then coded as negative/neutral/positive (e.g., "the boy hit him" versus "the song was a big hit") (Eley et al., 2008). Online tasks are less subject to biases. However, the usefulness of homophones and homographs is limited by the low number of these words that have clear negative/positive meanings (and that these meanings are well matched).

Research in adults with depression has been inconclusive. Some studies have reported that people with depression are more likely to endorse negative interpretations (e.g., Hindash and Amir, 2012; Lawson, MacLeod, and Hammond, 2002), whilst others have found no relationship (Lawson and MacLeod, 1999), even following a negative mood induction (Bisson and Sears, 2007). However, a recent meta-analysis in adults concluded that depression is associated with interpretation bias but that methodological factors shape conclusions (as the effect was only found in studies using offline measures) (Everaert, Podina, and Koster, 2017). There were similar effect sizes for both the presence of negative interpretation bias and a lack of positive interpretation bias.

There have only been a small number of studies assessing interpretation bias in young people with depression, using either questionnaires or a version of the recognition task. Depressed youth endorse more negative interpretations and less positive interpretations, relative to healthy controls (Micco, Henin, and Hirshfeld-Becker, 2014; Orchard, Pass, and Reynolds, 2016). Interpretation bias also shows a positive correlation with depressive symptom severity in those with diagnoses of depression (Micco et al., 2014) as well as in community samples (Dineen and Hadwin, 2004; Eley et al., 2008; Reid, Salmon, and Lovibond, 2006). Some community studies have also used online tasks to measure interpretation bias, indicating associations between depressive symptoms and interpretation bias (Eley et al., 2008; Reid et al., 2006). However, it has been suggested that anxiety may account for the findings given the high comorbidity of anxiety with depression and findings that biased interpretations also characterize youth anxiety. One study found that the relationship was independent of anxiety (Eley et al., 2008), whereas another study suggested the opposite (Reid et al., 2006). Interestingly, anxiety (compared to depression) was particularly associated with interpretation bias in a study that compared multiple cognitive biases and their links with depression and anxiety (Smith, Reynolds, Orchard, Whalley, and Chan, 2018).

In terms of whether interpretation bias is a vulnerability factor for depression, there has been limited research. One study found increased negative interpretation bias in never-depressed daughters of mothers with depression, compared to daughters of mothers with no psychopathology (Dearing and Gotlib, 2009). A second found that, in a sample of young adults who had experienced child abuse, those with greater negative interpretation bias were more likely to experience symptoms of depression (Wells, Vanderlind, Selby, and Beevers, 2014). There is also initial evidence that interpretation bias may play a role in increasing risk for depression. One study found that interpretation bias predicted increases in depression symptoms 4–6 weeks later in a student sample (Rude, Wenzlaff, Gibbs, Vane, and Whitney, 2002).

A related bias is attribution bias, where the cause of events is attributed to internal factors if negative and external factors if positive. Initial research suggests that, compared to healthy controls, young people with depression attribute negative events to internal causes (e.g., failing an exam because I'm not clever) and positive events to external causes (e.g., passing an exam because the teacher set easy questions; Lau, Rijsdijk, and Eley, 2006; Lau, Belli, Gregory, Napolitano, and Eley, 2012).

Memory Biases

There are several ways in which memories are processed differently in depression, including biases in memory recall, autobiographical memory and in response to autobiographical memories.

Memory Recall

It is suggested that depression is characterized by preferential recall of negative over positive material (Joormann and Stanton, 2016; Mathews and MacLeod, 2005). To assess this most studies use free-recall tasks from words that have been previously encoded. One method to encode stimuli is the self-referent encoding task (SRET; Hammen and Zupan, 1984). Here, participants are given a set of positive and negative adjectives and asked to rate how much these adjectives describe them. Following this task, they are asked to recall as many words as possible with memory bias classically measured as the proportion of words that the person is able to recall and has endorsed as self-referent. Adults with depression (or healthy participants who have undergone a negative mood induction) recall and recognize more negative words and fewer positive words compared to controls (Gotlib and Joormann, 2010). A meta-analysis found that people with depression remember 10 per cent more negative words than positive, whereas controls show a bias for positive information (Matt et al., 1992). This bias is more consistently found for explicit memory than for studies investigating implicit or recognition memory. Indeed, biases may be eradicated when encoding or recall of emotional material depends solely on perceptual processing (Watkins, 2002). For example, implicit memory biases are reliably found when at encoding the person is asked to imagine themselves in a scene related to the word but are not as consistently obtained when participants are asked at encoding to count the number of letters (Watkins, Martin, and Stern, 2000). These differences suggest that these biases could result from differences in the elaboration of emotional material.

In youth, research is more limited (Platt, Waters, Schulte-Koerne, Engelmann, and Salemink, 2017). One study found group differences, between depressed and non-depressed youth, in enhanced recall of negative compared to positive self-referent adjectives (Zupan, Hammen, and Jaenicke, 1987). A further study found similar group differences in recall (but not recognition) of negative material and, interestingly, that this association was enhanced with age (Neshat-Doost, Taghavi, Moradi, Yule, and Dalgleish, 1998). However, other studies have not replicated this finding. This includes a study comparing clinically depressed with non-depressed youth for a memory bias for negative words (Dalgleish et al., 2003); a study using a mood induction to investigate free recall of self-referent words in currently, previously and never depressed youth (Timbremont, Braet, Bosmans, and Van Vlierberghe, 2008); and a study assessing memory biases in psychiatric inpatients (Gençöz, Voelz, Gençöz, Pettit, and Joiner, 2001). Although the final study did not find enhanced recall of negative adjectives, they did find that lower recall of positive adjectives predicted depressive symptomatology. In community studies using the SRET, one study found evidence of an association between negative words and symptoms of depression, anxiety or aggression (Reid et al., 2006) and another large study found that depression was associated with higher negative and lower positive recall (Goldstein, Hayden, and Klein, 2014). A cross-sectional study comparing various cognitive biases and their associations with depression and anxiety, found that depressive symptoms were particularly related to self-referential memory bias (Smith et al., 2018).

In terms of whether memory biases for self-referent words are a cognitive-vulnerability factor for depression, one study assessed this bias in daughters of depressed mothers (Asarnow, Thompson, Joormann, and Gotlib, 2014). They found no negative memory bias in daughters of depressed vs non-depressed mothers, however, they found that genotype modified the effect. Daughters of mothers homozygous for the Met allele of the COMT Val158Met genotype recalled less positive self-referent words than those homozygous for the Val allele. This implies that the Val/Val genotype may serve a protective role. A large community sample found that reduced recall of positive words (but not enhanced recall of negative words) at age 6 predicted depression at age 9, implying that less positive processing may indicate vulnerability for later depression (Goldstein et al., 2014).

Biases in Autobiographical Memory

Autobiographical memory relates to the ability to recollect facts and personal events from one's life (Conway and Pleydell-Pearce, 2000) and is important for the individual's sense of self and daily functioning, even in psychiatrically healthy individuals. There are a number of ways in which biases in autobiographical memory have been associated with depression, including people with depression having less specific autobiographical memories (overgeneral memory (OGM)), having more negative and intrusive memories and also having fewer and less powerful positive memories (Dalgleish and Werner-Seidler, 2014; Gotlib and Joormann, 2010).

OGM is a phenomenon where individuals have difficulty retrieving specific autobiographical memories and instead generate categorical or extended memories. This applies to both positive and negative memories. OGM is most commonly assessed using the autobiographical memory task (AMT) where participants are asked to respond to negative, positive, and/or neutral cue words with an autobiographical memory. The responses are then coded by the experimenter. Specific memories identify unique events, occurring at a particular time and place, and are in contrast with memories that are of repeated events (categorical memories) or events that last longer than a day (extended memories). OGM has been strongly implicated in adult depression in meta-analyses, being not only associated with current symptoms but also with the onset and course of depression (Sumner, Griffith, and Mineka, 2010). Importantly, research with adults has demonstrated that OGM is specific to depression (rather than anxiety or general distress); does not reflect a general defect of memory functioning (Williams et al., 2007); and is not purely a correlate of current low mood state (Spinhoven et al., 2006).

The dominant model for OGM proposes that OGM develops in childhood (Williams and Broadbent, 1986). A review of seventeen articles concluded that childhood onset of OGM is supported and that OGM predicts depression in youth (Hitchcock, Nixon, and Weber, 2014), reporting large to moderate effect sizes (e.g., Park, Goodyer, and Teasdale, 2004; Rawal and Rice, 2012). This contrasts with young people with anxiety, where there has been very little research (Lau and Waters, 2017). There has also been some investigation of whether OGM predicts development of depression and/or maintenance of depression. In 10- to 18-year-old girls without current symptoms of depression but at familial risk for depression, OGM was found to predict development of depression one year later (Rawal and Rice, 2012). Similarly, in females with elevated symptoms of depression, OGM predicted depression one year later (Hipwell, Sapotichne, Klostermann, Battista, and Keenan, 2011). In terms of relapse, OGM predicts recurrence of symptoms sixteen months later in adolescents with major depressive disorder (MDD) (Sumner et al., 2011). OGM has also been found to be elevated in both adolescents with current depression and those with remitted depression (relative to those with no history of depression), suggesting that it may represent a trait-like vulnerability and is not only a state marker of current depression (Champagne et al., 2016).

Four longitudinal studies with community samples have found varying results. One study found that OGM interacted with familial emotional abuse to predict depression after 8 months (Stange, Hamlat, Hamilton, Abramson, and Alloy, 2013). Hamlat et al. (2015) identified that females with OGM in combination with higher levels of rumination were most vulnerable to developing depressive symptoms following negative life events. They also found that having more specific autobiographical memories reduced the impact of negative life events on depression symptoms in males and females, and so may serve a protective function. However, other studies have found no relationship (Crane et al., 2016) or that OGM to negative words was associated with increases in anxiety (and not depression) for those with high rumination (Gutenbrunner, Salmon, and Jose, 2018). Initial evidence, therefore, is mixed but suggests that OGM could be associated with the maintenance and relapse of depression in youth, especially if at elevated risk of psychopathology. Some suggest that the role OGM plays in the development and maintenance of psychopathology (e.g., anxiety vs depression) varies across the levels and nature of concurrent cognitive vulnerabilities (Gutenbrunner et al., 2018).

Intrusive memories are also very common in depression across the age range (44–87 per cent prevalence) (Meiser-Stedman, Dalgleish, Yule, and Smith, 2012; Williams et al., 2007). These are memories of past events, entering awareness unbidden and producing strong psychophysiological stress responses. They represent a form of facilitated negative retrieval, perhaps lying in the encoding and representation of the memory (Brewin, Gregory, Lipton, and Burgess, 2010). Intrusion frequency and a person's response to intrusions (Williams and Moulds, 2008) predict depression severity in adults (Williams and Moulds, 2007), with emerging evidence indicating a similarly powerful association in adolescence (Meiser-Stedman et al., 2012). As well as remembering the past, experiencing negative intrusive images of the future (for example of suicide) could be central in a person's experience of depression (Holmes, Blackwell, Burnett Heyes, Renner, and Raes, 2016) and is beginning to be investigated in youth (Pile and Lau, 2020).

Recalling positive autobiographical memories can repair mood following a negative mood induction (Joormann and Siemer, 2004) and mood-incongruent recall is often used as a mood-repair strategy (e.g., Rusting and DeHart, 2000). However, evidence suggests that people with depression are unable to use positive autobiographical memories to repair negative mood (Joormann and Siemer, 2004; Joormann, Siemer, and Gotlib, 2007) and that the recall of positive autobiographical memories may even be detrimental (Joormann, Siemer, et al., 2007). This may be for several reasons. It may be that, in remembering positive past events, comparisons are made with current negative state and the discrepancy between the two (perhaps also leading to rumination). Importantly, if rumination is prevented by instruction, adults with depression can utilize positive memories to improve their mood (Werner-Seidler and Moulds, 2012b). Alternatively, the quality of the positive memories has also been shown to be less vivid and emotionally intense in adults with depression (compared to non-depressed adults) (Werner-Seidler and Moulds, 2011, 2012a). A recent study found that adolescents with depression had difficulty recalling rich positive autobiographical memories, compared to healthy controls (Begovic et al., 2017). Interestingly, these deficits were also found in unaffected siblings and those in remission, suggesting that they are a vulnerability factor for depression and persist in remission.

An extension to this work on positive imagery for past events is the investigation of positive future imagery, emanating from clinical observations that hopelessness (an inability to see anything positive in the future) and disturbances in the anticipation of events (i.e., positive emotional experience when considering a goal) are core features of depression. Research in nonclinical adult samples has demonstrated that depression is associated with less vivid *positive* future imagery whilst anxiety is associated with more vivid *negative* future imagery (for review, see Holmes, Blackwell, Heyes, Renner, and Raes, 2016). One community study has replicated these findings in young people (Pile and Lau, 2018).

Responses to Negative Autobiographical Memories

The recollection of negative past events can be distressing and there are a number of likely responses to these memories in order to reduce their emotional impact, including avoidance and suppression. Avoidance and/or suppression of these memories has been shown to be more common in people with depression (Beblo et al., 2012; Sumner, 2012). Indeed, Williams et al.'s (1996) affect-regulation hypothesis identifies that a function of OGM is to minimize negative affect associated with negative memories. This theory was elaborated in the CaR-FA-X model, in which three processes influence OGM: capture and rumination, functional avoidance, and impaired executive control. A review of thirty-eight studies in adults (Sumner, 2012) concluded that there is robust evidence for associations between OGM and rumination and impaired executive control, and that OGM is likely to be a cognitive avoidance strategy. Experimental studies demonstrate that a negative event leads to more subjective stress in high compared to low (memory) specific adults (Raes, Hermans, de Decker, Eelen, and Williams, 2003). So, whilst this strategy may be beneficial in the short term it is problematic in the long term. For example, increased suppression increases intrusion frequency in people with depression and may actually enhance access to other distressing memories (Dalgleish and Yiend, 2006).

Support for the CaR-FA-X model in young people is more limited. Some studies have demonstrated that negative self-beliefs and rumination (capture and rumination) are associated with greater OGM (Park et al., 2004; Valentino, Toth, and Cicchetti, 2009) and a longitudinal study (mentioned previously) found that females with greater OGM combined with higher levels of rumination were most likely to experience increased depressive symptoms (Hamlat et al., 2015). Studies investigating executive control have yielded mixed results; this includes studies examining the relationship between OGM and inhibition (Raes, Verstraeten, Bijttebier, Vasey, and Dalgleish, 2010; Valentino, Bridgett, Hayden, and Nuttall, 2012) as well as OGM and working memory (e.g., Kuyken, Howell, and Dalgleish, 2006; Nixon, Ball, Sterk, Best, and Beatty, 2013; Valentino et al., 2012). The majority of studies also support a relationship between trauma exposure and OGM in young people with moderate to large effect sizes, across a range of trauma types (Brennen et al., 2010; Decker et al., 2003; Nixon et al., 2013).

Developmental factors may play an important role in the mechanisms underpinning the CaR-FA-X model, with some of the components being more or less implicated at different developmental stages. Preadolescents may ruminate less than early adolescents and, especially for females, rumination may increase with age (Hampel and Petermann, 2005; Jose and Brown, 2008). Therefore, age may moderate the relationship between rumination and OGM. Executive functioning continues to develop throughout childhood and adolescence (Blakemore and Choudhury, 2006) and so executive control errors could be more common at an earlier age. Functional avoidance may be more common in young children due to reduced emotional regulation strategies (Sumner, 2012) with the literature indicating that avoidant coping has a negative relationship with age (Phipps, Fairclough, and Mulhern, 1995). Therefore, the literature suggests that whilst the capture and rumination mechanism may be more influential in older children and adolescents, functional avoidance and executive control mechanisms might be more important in younger children.

A GENERAL DEFICIT IN EXECUTIVE FUNCTIONING

It has been suggested that depression is characterized by a general deficit in executive functioning, rather than specific biases in the processing of emotional material and so could explain observed cognitive biases. Executive functioning includes higher order neurocognitive processes, such as working memory, attention, and inhibitory control that are important for goal directed behavior (Blakemore and Choudhury, 2006). People with depression report general cognitive difficulties, for example in memory and concentration (Burt, Zembar, and Niederehe, 1995). However, there is limited evidence to suggest a pervasive deficit in general cognitive functioning that is specific to depression (Gotlib and Joormann, 2010). For example, a meta-analysis found more consistent evidence for memory impairments in inpatients than outpatients and these impairments are reported across a range of psychopathology (Burt et al., 1995). The affective interference hypothesis (Siegle, Ingram, and Matt, 2002) proposes that (as people with depression are focused on processing emotional material) performance on tasks requiring them to process emotional stimuli will be intact whilst they will struggle on tasks where emotional aspects of the stimuli need to be ignored in preference to other aspects. In adults, there is evidence that depression is associated with difficulties in controlling attention, supporting the affective interference hypothesis. For example, studies suggest that with careful instruction and without the opportunity to ruminate, deficits are decreased or eliminated (Gotlib and Joormann, 2010).

Executive functioning develops throughout childhood and adolescence and continues to early adulthood, mirroring protracted development of the prefrontal cortex (Best and Miller, 2010; Gogtay et al., 2004; Luna, Padmanabhan, and O'Hearn, 2010). However, very little research has investigated executive functioning in children and adolescents with depression. A recent review concluded that there is little support for general deficits in executive functioning in depression (Vilgis, Silk, and Vance, 2015). It highlighted that there was no evidence for deficits in response inhibition, attentional set shifting, selective attention, verbal working memory, and verbal fluency and that more studies are needed to investigate potential deficits in spatial working memory processing, sustaining attention, planning, negative attentional bias, reward processing, and decision-making.

INTERVENTIONS TARGETING COGNITIVE BIASES

CBT is one of the most effective interventions for depression in adults. This focuses on recognizing and modifying biased interpretations and dysfunctional automatic thoughts in order to ameliorate the symptoms of depression. The research reviewed above has, in part, aimed to develop a more comprehensive account of how cognitive biases contribute to depression to advance cognitive theory of depression and so improve intervention. However, there has been little change in front line treatments for adult depression from those first conceptualized by Beck (1976). In Beck's cognitive model of emotional disorders (Beck, 1976), he proposed a "negative cognitive triad" where negative views are held about the self, the world and the future. Beck proposed that schemas or existing memory representations lead the individual to filter environmental stimuli, such that their attention may be focused on stimuli that are consistent with their existing schema. Depression—proposed to be associated with themes such as failure, loss, and worthlessness—will have information processing biases that direct processing to environmental stimuli that are congruent with these themes.

Exploration of cognitive biases has led to suggested innovations in intervention, including cognitive bias modification (CBM) training and memory specificity training. These interventions also test causality as they tend to target one cognitive process assessing whether modification of this bias is possible and whether modification reduces symptoms. Targeting these biases has also been incorporated into complex interventions, for example as third wave cognitive behavioural therapies.

TECHNIQUES TO MODIFY ATTENTION AND INTERPRETATION BIASES

CBM paradigms have investigated the possibility of training either attention or interpretation biases. These paradigms attempt to shift these cognitive biases from negative material to favor neutral or positive material, through systematic, computerized training. CBM of attention (CBM-A) training paradigms aim to train attention away from negative or towards positive information, typically using a dot probe task where the probe is presented more frequently behind the positive (or neutral) stimuli than behind the negative stimuli. CBM of interpretation (CBM-I) training paradigms usually use ambiguous scenarios and resolve them in a positive or neutral way. For the purposes of comparing effects of modifying biases on mood, some CBM studies also train attention/interpretation in a negative/neutral way. In general, CBM has generated a great amount of interest, both as an experimental paradigm but also as an intervention. Unfortunately, results have been mixed, especially for effects on depressive (rather than anxiety) symptoms. A meta-analysis in adults that investigated anxiety and depression studies together found that CBM demonstrated a medium effect on biases (g = 0.49) and that this was stronger for interpretation (g = 0.81) than for attention (g = 0.29) biases (Hallion and Ruscio, 2011). CBM was found to significantly modify symptoms of anxiety but not depression (when they are examined separately).

In studies focused on depressive symptoms in young people, initial evidence suggested that attention biases can be modified if the same task is used for assessment and intervention (De Voogd et al., 2014) but that this does not transfer onto other measures of attention bias (Platt et al., 2015). Both studies did not show any effect on negative mood and both were in healthy populations. For CBM-I, results are mixed with one study showing changes in cognitive biases and mood (Lothmann, Holmes, Chan, and Lau, 2011) but this was not replicated in a second study (Micco et al., 2014). Results for anxiety in youth have been more promising (e.g., Lau and Pile, 2015) with a recent meta-analysis showing that CBM-I had moderate effects on negative and positive interpretations (g = -0.70 and g = -0.52 respectively) and a small but significant effect on anxiety (Krebs et al., 2017). However, a meta-analysis that considered CBM on both anxiety and depression in youth found no significant effects of CBM interventions for mental health, anxiety, and depression (Cristea, David, and Cuijpers, 2015). Inconsistency in the results may, in part, be due to the huge variation in delivery of CBM (e.g., single or multiple sessions).

An extension to CBM paradigms has been the inclusion of imagery in an attempt to enhance effects. A study in adults compared training positive interpretation biases using imagery with training biases verbally, found that the imagery condition had a greater impact on positive mood, interpretation bias and protected against induced negative mood state more than the verbal condition (Holmes, Lang, and Shah, 2009). Studies have begun to investigate techniques to enhance positive imagery, including imagery CBM which involves repeated practice in generating positive mental imagery to ambiguous stimuli. In adults with depression, one week of generating daily imagery of positive scenarios (compared to just listening to them) decreased depressive symptoms and negative interpretive bias (Torkan et al., 2014). Other work suggests that positive imagery-based CBM may not improve depressive symptoms overall, but may reduce anhedonia in an exploratory analysis (Blackwell et al., 2015). The use of imagery CBM has been recently extended to healthy adolescents, with an experimental investigation indicating that generation of positive images can increase positive affect and reduce negative interpretation bias (Burnett Heyes et al., 2017).

TECHNIQUES TO TARGET MEMORY BIASES

Targeting Autobiographical Memory Biases

Memory specificity training (MEST) aims to teach people to be more specific in their recall of autobiographical memories. MEST is usually delivered in small groups with the main focus being on generating specific autobiographical memories to cue words. Two early-stage trials indicate that MEST can improve memory specificity, reduce depressive symptomatology and improve day-to-day cognition in adolescents with depression (Neshat-Doost et al., 2012) and female adults with depression (Raes, Williams, and Hermans, 2009). However, a recent cluster-randomized controlled pilot trial with adults found no difference in depressive symptoms at three-month follow-up between the group receiving MEST and an active control group (Werner-Seidler et al., 2018). In this trial, MEST did improve memory specificity relative to the control group but both groups showed improvement in self-reported depressive symptoms. Further to this, researchers have suggested that it may be important to train flexibility in autobiographical memory (e.g., being able to flexibly shift between general and specific memories), with investigation of this type of training just beginning (Hitchcock et al., 2015, 2016).

CBM techniques have also been applied to memory biases for negative material. One method applied in adults has been to teach individuals, through repeated mental exercises, to process emotional events in a concrete and specific way (rather than an abstract or ruminative way). Initial studies indicated promising results, showing both differences in depressive symptoms and rumination between the intervention and control groups (Watkins, Baeyens, and Read, 2009; Watkins and Moberly, 2009). However, a larger randomized controlled trial did not replicate the effects on group differences in depressive symptoms (Watkins et al., 2012).

Another approach to target the distressing impact of negative memories is imagery rescripting. Imagery rescripting aims to modify the content of negative imagery and re-script the negative beliefs associated to be something more benign or positive. Imagery rescripting has mostly been applied in anxiety disorders but initial studies suggest it can be effectively applied in depression (Brewin et al., 2009; Wheatley et al., 2007). A recent meta-analysis on imagery receipting in adults, indicated promising results from nineteen studies across different disorders (Morina, Lancee, and Arntz, 2017). This meta-analysis suggested that imagery rescripting was effective in reducing symptoms from pre-treatment to post-treatment ($g = 1.22$) and follow-up ($g = 1.79$), even when symptom change was compared to a control condition (pre to post-treatment: $g = 0.90$). Importantly, imagery rescripting also indicated large effects on comorbid depression, aversive imagery, and encapsulated beliefs. Experimental studies using the trauma film paradigm also suggest that techniques used in imagery rescripting (verbally updating negative images) reduce negative intrusions relative to exposure and other control groups (Pile, Barnhofer, and Wild, 2015). As yet, imagery rescripting has not been researched in adolescents, although studies are forthcoming (Pile et al., 2018).

As discussed previously, recalling positive autobiographical memories is a powerful emotional regulation strategy that can be used to reduce negative affect. Elaborating positive material with sensory, affective, and visual detail is proposed to enhance the affective impact of the positive memory and so could potentially allow people with depression to harness their positive memories. An experimental study with adults with depression demonstrated that the process of recall for positive memories may determine their emotional impact (Werner-Seidler and Moulds, 2012b). This study showed that the positive emotional impact of memories was enhanced for those participants who focused on the detailed concrete aspects of memory rather than focused on the memory in an abstract way (e.g., thinking about causes and consequences of the memory). Similarly, when positive material is elaborated using imagery its emotional impact is enhanced (Holmes et al., 2009; Holmes, Mathews, Dalgleish, and Mackintosh, 2006). These studies offer initial experimental evidence that adults with depression can benefit from the elaboration of positive material; these techniques have not yet been researched in youth.

THIRD WAVE CBT

Developments of CBT that attempt to change the relationship that people have with their memories and thoughts (rather than changing the content of those thoughts) are known as "third wave" therapies. This includes rumination-focused CBT (RF-CBT) and mindfulness-based cognitive therapy (MBCT). RF-CBT attempts to enhance CBT by targeting ruminative processes in depression and very early results are promising in adults (Watkins et al., 2011). MBCT has mostly been applied to adults with recurrent depression to reduce relapse and has good evidence for relapse prevention (Kuyken et al., 2016; Piet and Hougaard, 2011). Studies are also beginning to apply MBCT to adolescents (Ames, Richardson, Payne, Smith, and Leigh, 2014) with a large school-based RCT planned (Kuyken et al., 2017). However, both therapies are complex and require highly trained therapists to administer them. Some studies have therefore attempted to distil the core ingredients of MBCT, in particular, in an attempt to make them simpler and more readily available. For example, experimental studies teaching self-distancing (Kross, Gard, Deldin, Clifton, and Ayduck, 2012) and perspective broadening show initial promise (Schartau, Dalgleish, and Dunn, 2009).

CONCLUDING REMARKS

Originally, it was proposed that both anxiety and depression are characterized by biases in all aspects of information processing. This has received little empirical support. Instead, particular biases have been strongly implicated in our cognitive understanding of depression whilst others have not. The majority of studies show that certain cognitive biases are present during depressive episodes (for example comparing people with depression to those without) but evidence for these biases outside of depressive episodes has been less forthcoming.

Young people with depression may show attention biases. It is suggested that young people with depression show a tendency to allocate their attention towards negative stimuli and then towards positive stimuli at later voluntary stages of processing. Studies of children of depressed mothers suggest that attention biases could be a cognitive vulnerability factor for depression in youth but more studies with consistent methodology are needed to establish a clear relationship. Biases in interpretation are implicated in youth depression. There is also some (limited) evidence that these relationships are present in young people at risk for depression and so not simply a by-product of depression. However, these results could be accounted for by comorbid anxiety, with more research required to establish specificity. Memory biases for emotional material are perhaps the most robustly replicated cognitive bias in adult depression (Joormann and Stanton, 2016; Mathews and MacLeod, 2005). Importantly, they also appear to be disorder specific, with data on memory biases in other disorders being contradictory. Biases in memory recall are consistently found in adults with evidence in young people more mixed. Overgeneral memory is linked to depression across the age range and seems likely to play a role in the maintenance, relapse and possibly onset of depression. Intrusive memories are common in depression and deficits in positive autobiographical memories may limit emotional regulation capabilities. Cognitive responses to negative memories also appear important to understand these biases (and how to target them) in depression. Given that depression appears to be characterized by a systematic and pervasive bias for negative material, it is perhaps unsurprising that the disorder is characterized by a deeply negative sense of the self, the world, and the future.

Despite limited effectiveness and availability of mainstream psychological therapies for adolescent depression, there is not yet support for alternative approaches. Initial studies targeting attention and interpretation biases in depression appeared promising but there is little evidence to suggest that modifying the biases transfer into gains in relieving distressing symptoms. Techniques to target memory biases and the development of CBT techniques in third wave therapies are beginning to emerge but so far there has been insufficient research to test them, especially in young people.

As we have presented here, research investigating cognitive biases has typically examined them in isolation and has focused on adults. Cognitive biases may play a greater role in the maintenance of psychopathology when they act in combination rather than alone (Hirsch, Clark, and Mathews, 2006). In adults, evidence suggests that cognitive biases are inter-correlated and that training participants in one bias can impact on another (Blaut, Paulewicz, Szastok, Prochwicz, and Koster, 2013; Ellis, Wells, Vanderlind, and Beevers, 2014; Everaert, Duyck, and Koster, 2014; Everaert et al., 2012; Everaert, Tierens, Uzieblo, and Koster, 2013). This work is beginning to be extended to young people (Klein, de Voogd, Wiers, and Salemink, 2017; Orchard and Reynolds, 2018). In general, less research has been conducted with young people compared to adults. Some of the challenges of conducting work with adults with depression are the high levels of comorbidity, medication use and chronicity of the disorder. These factors are all likely to contribute to variability in the findings of which cognitive biases are driving the disorders. Adolescence may, therefore, be a time when these biases can be investigated without so many (or such severe) confounding factors. However, adolescence is characterized by neurodevelopmental and physiological changes. Although these are likely to establish cognitive functions and shape emotional regulation (Yurgelun-Todd, 2007), they may also mean that the relationships between depression and cognitive biases change throughout childhood and stabilize in adolescence (Hankin et al., 2009). Age and stage of development may therefore significantly influence the relationship between these information processing biases and depressive symptoms and is not often studied in these investigations. There is some implication that with age, associations between the biases and symptoms grow stronger or are more consistently observed (Waite, Codd, and Creswell, 2015). There are several possible explanations to this including increased stability in neurocognitive factors with age, appropriateness of the tasks for younger children or that biases are linked to certain experiences occurring during childhood and adolescence. Further investigation of how cognitive biases may interact both with each other and with ongoing cognitive development would be valuable. The majority of the reviewed research is cross-sectional and would benefit from longitudinal or experimental studies that can tease apart the temporal nature of the biases and symptomatology to determine causation.

REFERENCES

American Psychiatric Association. (2013). *Diagnostic and Statistical Manual of Mental Disorders* (5th ed.). Arlington, VA: Author.

Ames, C. S., Richardson, J., Payne, S., Smith, P., and Leigh, E. (2014). Mindfulness-based cognitive therapy for depression in adolescents. *Child and Adolescent Mental Health, 19*(1), 74–78. doi:10.1111/camh.12034

Andersen, S. L. and Teicher, M. H. (2008). Stress, sensitive periods and maturational events in adolescent depression. *Trends in Neurosciences, 31*(4), 183–191.

Armstrong, T., and Olatunji, B. O. (2012). Eye tracking of attention in the affective disorders: A meta-analytic review and synthesis. *Clinical Psychology Review, 32*(8), 704–723. doi:10.1016/j.cpr.2012.09.004

Asarnow, L. D., Thompson, R. J., Joormann, J., and Gotlib, I. H. (2014). Children at risk for depression: Memory biases, self-schemas, and genotypic variation. *Journal of Affective Disorders, 159*, 66–72. doi:10.1016/j.jad.2014.02.020

Bar-Haim, Y., Lamy, D., Pergamin, L., Bakermans-Kranenburg, M. J., and van IJzendoorn, M. H. (2007). Threat-related attentional bias in anxious and nonanxious individuals: A meta-analytic study. *Psychological Bulletin, 133*(1), 1–24. doi:10.1037/0033-2909.133.1.1

Beblo, T., Fernando, S., Klocke, S., Griepenstroh, J., Aschenbrenner, S., and Driessen, M. (2012). Increased suppression of negative and positive emotions in major depression. *Journal of Affective Disorders, 141*(2–3), 474–479. doi:10.1016/j.jad.2012.03.019

Beck, A. T. (1976). *Cognitive Therapy and the Emotional Disorders.* New York: International Universities Press.

Begovic, E., Panaite, V., Bylsma, L. M., George, C., Kovacs, M., Yaroslavsky, I., ... Rottenberg, J. (2017). Positive autobiographical memory deficits in youth with depression histories and their never-depressed siblings. *British Journal of Clinical Psychology, 56*(3), 329–346. doi:10.1111/bjc.12141

Best, J. R. and Miller, P. M. (2010). A developmental perspective on executive function. *Child Development, 81*(6), 1641–1660.

Bisson, M. A. S. and Sears, C. R. (2007). The effect of depressed mood on the interpretation of ambiguity, with and without negative mode induction. *Cognition and Emotion, 21*(3), 614–645. doi:10.1080/02699930600750715

Blackwell, S. E., Browning, M., Mathews, A., Pictet, A., Welch, J., Davies, J., ... Holmes, E. A. (2015). Positive imagery-based cognitive bias modification as a web-based treatment tool for depressed adults: A randomized controlled trial. *Clinical Psychological Science, 3*(1), 91–111. doi:10.1177/2167702614560746

Blakemore, S.-J. and Choudhury, S. (2006). Development of the adolescent brain: Implications for executive function and social cognition. *Journal of Child Psychology and Psychiatry, 47*(3–4), 296–312. doi:10.1111/j.1469-7610.2006.01611.x

Blakemore, S. J. (2012). Imaging brain development: The adolescent brain. *NeuroImage*, *61*(2), 397–406. doi:10.1016/j.neuroimage.2011.11.080

Blaut, A., Paulewicz, B., Szastok, M., Prochwicz, K., and Koster, E. (2013). Are attentional bias and memory bias for negative words causally related? *Journal of Behavior Therapy and Experimental Psychiatry*, *44*(3), 293–299. doi:10.1016/j.jbtep.2013.01.002

Boland, Robert J., and Keller, M. B. (2009). Course and outcome of depression. In *Handbook of Depression*, (eds.) Gotlib I. H., and Hammen C. L., pp. 23–43. New York, NY: Guilford.

Brennen, T., Hasanović, M., Zotović, M., Blix, I., Skar, A. S., Prelić, N. K., . . . Gavrilov-Jerković, V. (2010). Trauma exposure in childhood impairs the ability to recall specific autobiographical memories in late adolescence. *Journal of Traumatic Stress*, *23*(2), 240–247. doi:10.1002/jts.

Brewin, C. R., Gregory, J. D., Lipton, M., and Burgess, N. (2010). Intrusive images in psychological disorders: Characteristics, neural mechanisms, and treatment implications. *Psychological Review*, *117*(1), 210–232.

Brewin, C. R., Wheatley, J., Patel, T., Fearon, P., Hackmann, A., Wells, A., . . . Myers, S. (2009). Imagery rescripting as a brief stand-alone treatment for depressed patients with intrusive memories. *Behaviour Research and Therapy*, *47*(7), 569–576. doi:10.1016/j.brat.2009.03.008

Burnett Heyes, S., Pictet, A., Mitchell, H., Raeder, S. M., Lau, J. Y. F., Holmes, E. A., and Blackwell, S. E. (2017). Mental imagery-based training to modify mood and cognitive bias in adolescents: Effects of valence and perspective. *Cognitive Therapy and Research*, *41*(1), 73–88. doi:10.1007/s10608-016-9795-8

Burt, D. B., Zembar, M. J., and Niederehe, G. (1995). Depression and memory impairment: A meta-analysis of the association, its pattern, and specificity. *Psychological Bulletin*, *117*(2), 285.

Calear, A. L. and Christensen, H. (2010). Systematic review of school-based prevention and early intervention programs for depression. *Journal of Adolescence*, *33*(3), 429–438. doi:10.1016/j.adolescence.2009.07.004

Champagne, K., Burkhouse, K. L., Woody, M. L., Feurer, C., Sosoo, E., and Gibb, B. E. (2016). Brief report: Overgeneral autobiographical memory in adolescent major depressive disorder. *Journal of Adolescence*, *52*, 72–75. doi:10.1016/j.adolescence.2016.07.008

Children's Commissioner. (2016). *Lightning Review: Access to Child and Adolescent Mental Health Services*. doi:10.1017/CBO9781107415324.004

Cisler, J. M. and Koster, E. H. W. (2010). Mechanisms of attentional biases towards threat in anxiety disorders: An integrative review. *Clinical Psychology Review*, *30*(2), 203–216. doi:10.1016/j.cpr.2009.11.003

Connell, A. M., Patton, E., Klostermann, S., and Hughes-Scalise, A. (2013). Attention bias in youth: Associations with youth and mother's depressive symptoms moderated by emotion regulation and affective dynamics during family interactions. *Cognition and Emotion*, *27*(8), 1522–1534. doi:10.1080/02699931.2013.803459

Conway, M. A. and Pleydell-Pearce, C. W. (2000). The Construction of Autobiographical Memories in the Self-Memory System. *Psychological Review*, *107*(2), 261–288.

Cox, G., Callahan, P., Churchill, R., Hunot, V., Merry, S., Parker, A., and Hetrick, S. (2014). Psychological therapies versus antidepressant medication, alone and in combination for depression in children and adolescents (Review). *The Cochrane Library*, (11).

Cox, G., Fisher, C., De Silva, S., Phelan, M., Akinwale, O., Simmons, M., and Hetrick, S. (2014). Interventions for preventing relapse and recurrence of a depressive disorder in children and adolescents (Review). *The Cochrane Library*, (11).

Coyle, J. T., Pine, D. S., Charney, D. S., Lewis, L., Nemeroff, C. B., Carlson, G. a., ... Hellander, M. (2003). Depression and bipolar support alliance consensus statement on the unmet needs in diagnosis and treatment of mood disorders in children and adolescents. *Journal of the American Academy of Child and Adolescent Psychiatry*, 42(12), 1494–1503. doi:10.1097/00004583-200312000-00017

Crane, C., Heron, J., Gunnell, D., Lewis, G., Evans, J., Williams, M. G., ... Evans, J. (2016). Adolescent over-general memory, life events and mental health outcomes: Findings from a UK cohort study. *Memory*, 0(0), 1–16. doi:10.1080/09658211.2015.1008014

Cristea, I. A., David, D., and Cuijpers, P. (2015). Practitioner review: Cognitive bias modification for mental health problems in children and adolescents a meta-analysis. *Journal of Child Psychology and Psychiatry*, 56(7), 723–734. doi:10.1111/jcpp.12383

Dalgleish, T., Taghavi, R., Neshat-Doost, H., Moradi, A., Canterbury, R., and Yule, W. (2003). Patterns of processing bias for emotional information across clinical disorders: a comparison of attention, memory, and prospective cognition in children and adolescents with depression, generalized anxiety, and posttraumatic stress disorder. *Journal of Clinical Child and Adolescent Psychology*, 32(1), 10–21. doi:10.1207/S15374424JCCP3201

Dalgleish, T. and Werner-Seidler, A. (2014). Disruptions in autobiographical memory processing in depression and the emergence of memory therapeutics. *Trends in Cognitive Sciences*, 18(11), 596–604. doi:10.1016/j.tics.2014.06.010

Dalgleish, T. and Yiend, J. (2006). The effects of suppressing a negative autobiographical memory on concurrent intrusions and subsequent autobiographical recall in dysphoria. *Journal of Abnormal Psychology*, 115(3), 467–473.

de Girolamo, G., Dagani, J., Purcell, R., Cocchi, A., and McGorry, P. D. (2012). Age of onset of mental disorders and use of mental health services: Needs, opportunities and obstacles. *Epidemiology and Psychiatric Sciences*, 21(01), 47–57. doi:10.1017/S2045796011000746

De Voogd, E. L., Wiers, R. W., Prins, P. J. M., and Salemink, E. (2014). Visual search attentional bias modification reduced social phobia in adolescents. *Journal of Behavior Therapy and Experimental Psychiatry*, 45(2), 252–259. doi:10.1016/j.jbtep.2013.11.006

Dearing, K. F. and Gotlib, I. H. (2009). Interpretation of ambiguous information in girls at risk for depression. *Journal of Abnormal Child Psychology*, 37(1), 79–91. doi:10.1007/s10802-008-9259-z

Decker, A. De, Hermans, D., Raes, F., Eelen, P., Decker, A. De, Hermans, D., ... Eelen, P. (2003). Autobiographical memory specificity and trauma in inpatient adolescents and trauma in inpatient adolescents. *Journal of Clinical Child and Adolescent Psychology ISSN:*, 32(1), 22–31.

Dineen, K. A. and Hadwin, J. A. (2004). Anxious and depressive symptoms and children's judgements of their own and others' interpretation of ambiguous social scenarios. *Journal of Anxiety Disorders*, 18(4), 499–513. doi:10.1016/S0887-6185(03)00030-6

Eley, T. C., Gregory, A. M., Lau, J. Y. F., McGuffin, P., Napolitano, M., Rijsdijk, F. V., and Clark, D. M. (2008). In the face of uncertainty: A twin study of ambiguous information, anxiety and depression in children. *Journal of Abnormal Child Psychology*, 36(1), 55–65. doi:10.1007/s10802-007-9159-7

Ellis, A. J., Wells, T. T., Vanderlind, W. M., and Beevers, C. G. (2014). The role of controlled attention on recall in major depression. *Cognition and Emotion*, 28(3), 520–529. doi:10.1080/02699931.2013.832153

Everaert, J., Duyck, W., and Koster, E. H. W. (2014). Attention, interpretation, and memory biases in subclinical depression: A proof-of-principle test of the combined cognitive biases hypothesis. *Emotion*, 14(311), 1–34.

Everaert, J., Koster, E. H. W., and Derakshan, N. (2012). The combined cognitive bias hypothesis in depression. *Clinical Psychology Review*, 32(5), 413–424. doi:10.1016/j.cpr.2012.04.003

Everaert, J., Podina, I. R., and Koster, E. H. W. (2017). A comprehensive meta-analysis of interpretation biases in depression. *Clinical Psychology Review*, 58, 33–48. doi:10.1016/j.cpr.2017.09.005

Everaert, J., Tierens, M., Uzieblo, K., and Koster, E. H. W. (2013). The indirect effect of attention bias on memory via interpretation bias: Evidence for the combined cognitive bias hypothesis in subclinical depression. *Cognition and Emotion*, 27(8), 1450–1459. doi:10.1080/02699931.2013.787972

Fombonne, E., Wostear, G., Cooper, V., Harrington, R., and Rutter, M. (2001). The Maudsley long-term follow-up of child and adolescent depression: 2. Suicidality, criminality and social dysfunction in adulthood. *The British Journal of Psychiatry*, 179(3), 218–223. doi:10.1192/bjp.179.3.218

Gençöz, T., Voelz, Z. R., Gençöz, F., Pettit, J. W., and Joiner, T. E. (2001). Specificity of information processing styles to depressive symptoms in youth psychiatric inpatients. *Journal of Abnormal Child Psychology*, 29(3), 255–262. doi:10.1023/A:1010385832566

Gibb, B. E., Benas, J. S., Grassia, M., and McGeary, J. (2009). Children's attentional biases and 5-HTTLPR genotype: Potential mechanisms linking mother and child depression. *Journal of Clinical Child and Adolescent Psychology*, 38(3), 415–426. doi:10.1080/15374410902851705

Gogtay, N., Giedd, J. N., Lusk, L., Hayashi, K. M., Greenstein, D., Vaituzis, A. C., ... Thompson, P. M. (2004). Dynamic mapping of human cortical development during childhood through early adulthood. *Proceedings of the National Academy of Sciences of the United States of America*, 101(21), 8174–8179. doi:10.1073/pnas.0402680101

Goldstein, B. L., Hayden, E. P., and Klein, D. N. (2014). Stability of self-referent encoding task performance and associations with change in depressive symptoms from early to middle childhood. *Cognition and Emotion*, 29(8), 1445–1455. doi:10.1080/02699931.2014.990358

Goodyer, I. M., Reynolds, S., Barrett, B., Byford, S., Dubicka, B., Hill, J., ... Midgley, N. (2016). Cognitive behavioural therapy and short-term psychoanalytical psychotherapy versus a brief psychosocial intervention in adolescents with unipolar major depressive disorder (IMPACT): A multicentre, pragmatic, observer-blind, randomised controlled superiority trial. *The Lancet Psychiatry*, 0366(16), 1–11. doi:10.1016/S2215-0366(16)30378-9

Gotlib, I. H. and Joormann, J. (2010). Cognition and depression: current status and future directions. *Annual Review of Clinical Psychology*, 6(1), 285–312. doi:10.1146/annurev.clinpsy.121208.131305

Gutenbrunner, C., Salmon, K., and Jose, P. E. (2018). Do overgeneral autobiographical memories predict increased psychopathological symptoms in community youth? A 3-year longitudinal investigation. *Journal of Abnormal Child Psychology*, 46, 197–208. doi:10.1007/s10802-017-0278-5

Hallion, L. S. and Ruscio, A. M. (2011). A meta-analysis of the effect of cognitive bias modification on anxiety and depression. *Psychological Bulletin*, 137(6), 940–958. doi:10.1037/a0024355

Hamlat, E. J., Connolly, S. L., Hamilton, J. L., Stange, J. P., Abramson, L. Y., and Alloy, L. B. (2015). Rumination and overgeneral autobiographical memory in adolescents: An integration of cognitive vulnerabilities to depression. *Journal of Youth and Adolescence*, 44(4), 806–818. doi:10.1007/s10964-014-0090-2

Hammen, C. and Zupan, B. A. (1984). Self-schemas, depression, and the processing of personal information in children. *Journal of Experimental Child Psychology*, 37(3), 598–608. doi:10.1016/0022-0965(84)90079-1

Hampel, P. and Petermann, F. (2005). Age and gender effects on coping in children and adolescents. *Journal Of Youth and Adolescence*, 34(2), 73–83. doi:10.1007/s10964-005-3207-9

Hankin, B. L., Gibb, B. E., Abela, J. R. Z., and Flory, K. (2010). Selective attention to affective stimuli and clinical depression among youths: Role of anxiety and specificity of emotion. *Journal of Abnormal Psychology*, 119(3), 491–501. doi:10.1037/a0019609

Hankin, B. L., Oppenheimer, C., Jenness, J., Barrocas, A., Shapero, B. G., and Goldband, J. (2009). Developmental origins of cognitive vulnerabilities to depression: Review of processes contributing to stability and change across time. *Journal of Clinical Psychology*, 65(12), 1327–1338. doi:10.1002/jclp

Hare, T. A., Tottenham, N., Davidson, M. C., Glover, G. H., and Casey, B. J. (2005). Contributions of amygdala and striatal activity in emotion regulation. *Biological Psychiatry*, 57(6), 624–632. doi:10.1016/j.biopsych.2004.12.038

Harrison, A. J. and Gibb, B. E. (2015). Attentional biases in currently depressed children: An eye-tracking study of biases in sustained attention to emotional stimuli. *Journal of Clinical Child and Adolescent Psychology*, 44(6), 1008–1014. doi:10.1080/15374416.2014.930688

Healy, D. (2003). Lines of evidence on the risks of suicide with selective serotonin reuptake inhibitors. *Psychotherapy and Psychosomatics*, 72(2), 71–79.

Hindash, A. H. C. and Amir, N. (2012). Negative interpretation bias in individuals with depressive symptoms. *Cognitive Therapy and Research*, 36(5), 502–511. doi:10.1007/s10608-011-9397-4

Hipwell, A. E., Sapotichne, B., Klostermann, S., Battista, D., and Keenan, K. (2011). Autobiographical memory as a predictor of depression vulnerability in girls. *Journal of Clinical Child and Adolescent Psychology*, 40(2), 254–265. doi:10.1080/15374416.2011.546037

Hirsch, C. R., Clark, D. M., and Mathews, A. (2006). Imagery and interpretations in social phobia: Support for the combined cognitive biases hypothesis. *Behavior Therapy*, 37(3), 223–236. doi:10.1016/j.beth.2006.02.001

Hirsch, C. R., Meeten, F., Krahé, C., and Reeder, C. (2016). Resolving ambiguity in emotional disorders: The nature and role of interpretation biases. *Annual Review of Clinical Psychology*, 12(1), 281–305. doi:10.1146/annurev-clinpsy-021815-093436

Hitchcock, C., Hammond, E., Rees, C., Panesar, I., Watson, P., Werner-Seidler, A., and Dalgleish, T. (2015). Memory Flexibility Training (MemFlex) to reduce depressive symptomatology in individuals with major depressive disorder: Study protocol for a randomised controlled trial. *Trials*, 16(1), 1–8. doi:10.1186/s13063-015-1029-y

Hitchcock, C., Mueller, V., Hammond, E., Rees, C., Werner-Seidler, A., and Dalgleish, T. (2016). The effects of autobiographical memory flexibility (MemFlex) training: An uncontrolled trial in individuals in remission from depression. *Journal of Behavior Therapy and Experimental Psychiatry*, 52, 92–98. doi:10.1016/j.jbtep.2016.03.012

Hitchcock, C., Nixon, R. D. V, and Weber, N. (2014). A review of overgeneral memory in child psychopathology. *British Journal of Clinical Psychology*, 53(2), 170–193. doi:10.1111/bjc.12034

Holmes, E. A., Blackwell, S. E., Burnett Heyes, S., Renner, F., and Raes, F. (2016). Mental imagery in depression: Phenomenology, potential mechanisms, and treatment implications. *Annual Review of Clinical Psychology*, 12, 249–280. doi:10.1146/annurev-clinpsy-021815-092925

Holmes, E. A., Lang, T. J., and Shah, D. M. (2009). Developing interpretation bias modification as a "cognitive vaccine" for depressed mood: Imagining positive events makes you feel better than thinking about them verbally. *Journal of Abnormal Psychology*, 118(1), 76–88. doi:10.1037/a0012590

Holmes, E. A., Mathews, A., Dalgleish, T., and Mackintosh, B. (2006). Positive interpret-ation training: Effects of mental imagery versus verbal training on positive mood. *Behavior Therapy*, 37(3), 237–247. doi:10.1016/j.beth.2006.02.002

Joormann, J. and Siemer, M. (2004). Memory accessibility, mood regulation, and dysphoria: Difficulties in repairing sad mood with happy memories? *Journal of Abnormal Psychology*, 113(2), 179–188. doi:10.1037/0021-843X.113.2.179

Joormann, J., Siemer, M., and Gotlib, I. H. (2007). Mood regulation in depression: Differential effects of distraction and recall of happy memories on sad mood. *Journal of Abnormal Psychology*, 116(3), 484–490. doi:10.1037/0021-843X.116.3.484

Joormann, J. and Stanton, C. H. (2016). Examining emotion regulation in depression: A review and future directions. *Behaviour Research and Therapy*, 86, 35–49. doi:10.1016/j.brat.2016.07.007

Joormann, J., Talbot, L., and Gotlib, I. H. (2007). Biased processing of emotional information in girls at risk for depression. *Journal of Abnormal Psychology*, 116(1), 135–143. doi:10.1037/0021-843X.116.1.135

Jose, P. E. and Brown, I. (2008). When does the gender difference in rumination begin? Gender and age differences in the use of rumination by adolescents. *Journal of Youth and Adolescence*, 37, 180–192. doi:10.1007/s10964-006-9166-y

Kessler, R. C. and Wang, P. S. (2009). The epidemiology of depression. In *Handbook of Depression*, 2nd edition (eds.) I. H. Gotlib and C. L. Hammen, pp. 5–22. New York: Guilford.

Klein, A. M., de Voogd, L., Wiers, R. W., and Salemink, E. (2017). Biases in attention and in-terpretation in adolescents with varying levels of anxiety and depression. *Cognition and Emotion*, 9931, 1–9. doi:10.1080/02699931.2017.1304359

Krebs, G., Pile, V., Grant, S., Degli Esposti, M., Montgomery, P., and Lau, Y. F. (2017). Cognitive bias modification of interpretations for anxiety in children and adolescents: A system-atic review and meta-analysis. *Journal of Child Psychology and Psychiatry*, 59(8), 831–844. doi:10.1002/APP.38652.

Kross, E., Gard, D., Deldin, P., Clifton, J., and Ayduck, O. (2012). "Asking Why" From a dis-tance: Its cognitive and emotional consequences for people with major depressive disorder. *Journal of Abnormal Psychology*, 121(3), 559–569. doi:10.1037/a0028808

Kujawa, A. J., Torpey, D., Kim, J., Hajcak, G., Rose, S., Gotlib, I. H., and Klein, D. N. (2011). Attentional biases for emotional faces in young children of mothers with chronic or re-current depression. *Journal of Abnormal Child Psychology*, 39(1), 125–135. doi:10.1007/s10802-010-9438-6

Kuyken, W., Howell, R., and Dalgleish, T. (2006). Overgeneral autobiographical memory in depressed adolescents with, versus without, a reported history of trauma. *Journal of Abnormal Psychology*, 115(3), 387–396. doi:10.1037/0021-843X.115.3.387

Kuyken, W., Nuthall, E., Byford, S., Crane, C., Dalgleish, T., Ford, T., ... Williams, J. M. G. (2017). The effectiveness and cost-effectiveness of a mindfulness training programme in schools compared with normal school provision (MYRIAD): Study protocol for a randomised controlled trial. *Trials*, 18(194), 1–17. doi:10.1186/s13063-017-1917-4

Kuyken, W., Warren, F. C., Taylor, R. S., Whalley, B., Crane, C., Bondolfi, G., ... Teasdale, J. D. (2016). Efficacy of mindfulness-based cognitive therapy in prevention of depressive relapse: An individual patient data meta-analysis from randomized trials. *JAMA Psychiatry*, 73(6), 565–574. doi:10.1001/jamapsychiatry.2016.0076

Kyte, Z. A., Goodyer, I. M., and Sahakian, B. J. (2005). Selected executive skills in adolescents with recent first episode major depression. *Journal of Child Psychology and Psychiatry*, 46(9), 995–1005. doi:10.1111/j.1469-7610.2004.00400.x

Ladouceur, C. D., Dahl, R. E., Williamson, D. E., Birmaher, B., Axelson, D. A., Ryan, N. D., and Casey, B. J. (2006). Processing emotional facial expressions influences performance on a Go/NoGo task in pediatric anxiety and depression. *Journal of Child Psychology and Psychiatry*, 47(11), 1107–1115. doi:10.1111/j.1469-7610.2006.01640.x

Lau, J. Y. F. (2013). Cognitive bias modification of interpretations: A viable treatment for child and adolescent anxiety? *Behaviour Research and Therapy*, 51(10), 614–622. doi:10.1016/j.brat.2013.07.001

Lau, J. Y. F., Belli, S. D., Gregory, A. M., Napolitano, M., and Eley, T. C. (2012). The role of children's negative attributions on depressive symptoms: An inherited characteristic or a product of the early environment? *Developmental Science*, 15(4), 569–578. doi:10.1111/j.1467-7687.2012.01152.x

Lau, J. Y. F. and Pile, V. (2015). Can cognitive bias modification of interpretations training alter mood states in children and adolescents? A reanalysis of data from six studies. *Clinical Psychological Science*, 3(1), 112–125. doi:10.1177/2167702614549596

Lau, J. Y. F., Rijsdijk, F., and Eley, T. C. (2006). I think, therefore I am: A twin study of attributional style in adolescents. *Journal of Child Psychology and Psychiatry*, 47(7), 696–703. doi:10.1111/j.1469-7610.2005.01532.x

Lau, J. Y. F. and Waters, A. M. (2017). Annual research review: An expanded account of information-processing mechanisms in risk for child and adolescent anxiety and depression. *Journal of Child Psychology and Psychiatry*, 58(4), 387–407. doi:10.1111/jcpp.12653

Lawson, C. and MacLeod, C. (1999). Depression and the interpretation of ambiguity. *Behaviour Research and Therapy*, 37(5), 463–474. doi:10.1016/S0005-7967(98)00131-4

Lawson, C., MacLeod, C., and Hammond, G. (2002). Interpretation revealed in the blink of an eye: Depressive bias in the resolution of ambiguity. *Journal of Abnormal Psychology*, 111(2), 321–328. doi:10.1037/0021-843X.111.2.321

Lonigan, C. J. and Vasey, M. W. (2009). Negative affectivity, effortful control, and attention to threat-relevant stimuli. *Journal of Abnormal Child Psychology*, 37(3), 387–399. doi:10.1007/s10802-008-9284-y

Lothmann, C., Holmes, E. A., Chan, S. W. Y., and Lau, J. Y. F. (2011). Cognitive bias modification training in adolescents: Effects on interpretation biases and mood. *Journal of Child Psychology and Psychiatry*, 52(1), 24–32. doi:10.1111/j.1469-7610.2010.02286.x

Luby, J. L., Heffelfinger, A. K., Mrakotsky, C., Brown, K. M., Hessler, M. J., Wallis, J. M., and Spitznagel, E. L. (2003). The clinical picture of depression in preschool children. *Journal of the American Academy of Child and Adolescent Psychiatry*, 42(3), 340–348. doi:10.1097/00004583-200303000-00015

Luna, B., Padmanabhan, A., and O'Hearn, K. (2010). What has fMRI told us about the Development of Cognitive Control through Adolescence? *Brain and Cognition*, 72(1), 101–113. doi:10.1016/j.bandc.2009.08.005

Maalouf, F. T., Clark, L., Tavitian, L., Sahakian, B. J., Brent, D., and Phillips, M. L. (2012). Bias to negative emotions: A depression state-dependent marker in adolescent major depressive disorder. *Psychiatry Research*, 198(1), 28–33. doi:10.1016/j.psychres.2012.01.030

Mathews, A. and MacLeod, C. (2005). Cognitive vulnerability to emotional disorders. *Annual Review of Clinical Psychology*, 1, 167–195. doi:10.1146/annurev.clinpsy.1.102803.143916

Matt, G. E., Vázquez, C., Campbell, and W. K. (1992). Mood-congruent recall of affectively toned stimuli: A meta-analytic review article. *Clinical Psychology Review*, 12(2), 227–255.

Mccrone, P., Dhanasiri, S., Patel, A., Knapp, M., and Lawton-Smith, S. (2008). *The Kings Fund: Paying the Price: The Cost of Mental Health Care in England to 2026*. London: The Kings Fund.

Meiser-Stedman, R., Dalgleish, T., Yule, W., and Smith, P. (2012). Intrusive memories and depression following recent non-traumatic negative life events in adolescents. *Journal of Affective Disorders*, 137(1–3), 70–78. doi:10.1016/j.jad.2011.12.020

Micco, J. A., Henin, A., and Hirshfeld-Becker, D. R. (2014). Efficacy of interpretation bias modification in depressed adolescents and young adults. *Cognitive Therapy and Research*, 38(2), 89–102. doi:10.1007/s10608-013-9578-4

Morina, N., Lancee, J., and Arntz, A. (2017). Imagery rescripting as a clinical intervention for aversive memories: A meta-analysis. *Journal of Behavior Therapy and Experimental Psychiatry*, 55, 6–15. doi:10.1016/j.jbtep.2016.11.003

National Institute for Clinical Excellence. (2015). *CG28 Depression in Children and Young People: Full Guidance*, available at http://guidance.nice.org.uk/CG28/Guidance. (28).

National Institute for Health and Care Excellence. (2019). *Depression in Children and Young People: Identification and Management*, available at doi:10.1211/CP.2018.20204575

Neshat-Doost, H. T., Dalgleish, T., Yule, W., Kalantari, M., Ahmadi, S. J., Dyregrov, A., and Jobson, L. (2012). Enhancing autobiographical memory specificity through cognitive training: An intervention for depression translated from basic science. *Clinical Psychological Science*, 1(1), 84–92. doi:10.1177/2167702612454613

Neshat-Doost, H. T., Taghavi, M. R., Moradi, A. R., Yule, W., and Dalgleish, T. (1998). Memory for emotional trait adjectives in clinically depressed youth. *Journal of Abnormal Psychology*, 107(4), 642.

Neshat-Doost, Hamid T., Moradi, A. R., Taghavi, M. R., Yule, W., and Dalgleish, T. (2000). Lack of attentional bias for emotional information in clinically depressed children and adolescents on the dot probe task. *Journal of Child Psychology and Psychiatry*, 41(3), 363–368. doi:10.1017/S0021963099005429

Neshat Doost, H. T., Taghavi, M. R., Moradi, A. R., Yule, W., and Dalgleish, T. (1997). The performance of clinically depressed children and adolescents on the modified Stroop paradigm. *Personality and Individual Differences*, 23(5), 753–759. doi:10.1016/S0191-8869(97)00097-4

Neufeld, S. A. S., Dunn, V. J., Jones, P. B., Croudace, T. J., and Goodyer, I. M. (2017). Reduction in adolescent depression after contact with mental health services: A longitudinal cohort study in the UK. *The Lancet Psychiatry*, 4(2), 120–127. doi:10.1016/S2215-0366(17)30002-0

Nixon, R. D. V., Ball, S.-A., Sterk, J., Best, T., and Beatty, L. (2013). Autobiographical memory in children and adolescents with acute stress and chronic posttraumatic stress disorder. *Behaviour Change*, 30(03), 180–198. doi:10.1017/bec.2013.17

Orchard, F., Pass, L., and Reynolds, S. (2016). 'It was all my fault'; Negative interpretation bias in depressed adolescents. *Journal of Abnormal Child Psychology*, 44(5), 991–998. doi:10.1007/s10802-015-0092-x

Orchard, F. and Reynolds, S. (2018). The combined influence of cognitions in adolescent depression: Biases of interpretation, self-evaluation, and memory. *British Journal of Clinical Psychology*, 57(4), 420–435. doi:10.1111/bjc.12184

Park, R. J., Goodyer, I. M., and Teasdale, J. D. (2004). Effects of induced rumination and distraction on mood and overgeneral autobiographical memory in adolescent major depressive disorder and controls. *Journal of Child Psychology and Psychiatry*, 45(5), 996–1006. doi:10.1111/j.1469-7610.2004.t01-1-00291.x

Peckham, A. D., McHugh, R. K., and Otto, M. W. (2010). A meta-analysis of the magnitude of biased attention in depression. *Depression and Anxiety*, 27(12), 1135–1142. doi:10.1002/da.20755

Phipps, S., Fairclough, D., and Mulhern, R. K. (1995). Avoidant coping in children with cancer. *Journal of Pediatric Psychology*, *20*(2), 217–232. doi:10.1093/jpepsy/20.2.217

Piet, J. and Hougaard, E. (2011). The effect of mindfulness-based cognitive therapy for prevention of relapse in recurrent major depressive disorder: A systematic review and meta-analysis. *Clinical Psychology Review*, *31*(6), 1032–1040. doi:10.1016/j.cpr.2011.05.002

Pile, V., Barnhofer, T., and Wild, J. (2015). Updating versus exposure to prevent consolidation of conditioned fear. *PloS One*, *10*(4), e0122971.

Pile, V. and Lau, J. Y. F. (2018). Looking forward to the future: Impoverished vividness for positive prospective events characterises low mood in adolescence. *Journal of Affective Disorders*, *238*, 269–276. doi:10.1016/j.jad.2018.05.032

Pile, V. and Lau, J. Y. F. (2020). Intrusive images of a distressing future: Links between prospective mental imagery, generalized anxiety and a tendency to suppress emotional experience in youth. *Behaviour Research and Therapy*, *124*, 103508. doi:10.1016/j.brat.2019.103508

Pile, V., Shammas, D., and Smith, P. (2019). Assessment and treatment of depression in children and young people in the United Kingdom: Comparison of access to services and provision at two time points. *Clinical Child Psychology and Psychiatry*, *25*(1), 119–132. doi:10.1177/1359104519858112

Pile, V., Smith, P., Leamy, M., Blackwell, S. E., Meiser-Stedman, R., Stringer, D., ... Lau, J. Y. F. (2018). A brief early intervention for adolescent depression that targets emotional mental images and memories: Protocol for a feasibility randomised controlled trial (IMAGINE trial). *Pilot and Feasibility Studies*, *4*(1), 97.

Platt, J. D., Brantut, N., Rice, J. R. (2015). Strain localization driven by thermal decomposition during seismic shear. *JGR Solid Earth*, *120*(6), 4405–4433. doi:10.1002/2014JB011493

Platt, B., Murphy, S. E., and Lau, J. Y. (2015). The association between negative attention biases and symptoms of depression in a community sample of adolescents. *PeerJ*, *3*, e1372.

Platt, B., Waters, A. M., Schulte-Koerne, G., Engelmann, L., and Salemink, E. (2017). A review of cognitive biases in youth depression: Attention, interpretation and memory. *Cognition and Emotion*, *31*(3), 462–483. doi:10.1080/02699931.2015.1127215

Posner, M. I. (1980). Orienting of attention. *The Quarterly Journal of Experimental Psychology*, *32*(1), 3–25. doi:10.1080/00335558008248231

Post, M. (1992). Transduction of psychosocial stress into the neurobiology of recurrent affective disorder. *American Journal of Psychitary*, *149*(8), 999–1010.

Raes, F., Hermans, D., de Decker, A., Eelen, P., and Williams, J. M. G. (2003). Autobiographical memory specificity and affect regulation: An experimental approach. *Emotion*, *3*(2), 201–206. doi:10.1037/1528-3542.3.2.201

Raes, F., Verstraeten, K., Bijttebier, P., Vasey, M. W., and Dalgleish, T. (2010). Inhibitory control mediates the relationship between depressed mood and overgeneral memory recall in children. *Journal of Clinical Child and Adolescent Psychology*, *39*(2), 276–281. doi:10.1080/15374410903532684

Raes, F., Williams, J. M. G., and Hermans, D. (2009). Reducing cognitive vulnerability to depression: A preliminary investigation of memory specificity training (MEST) in inpatients with depressive symptomatology. *Journal of Behavior Therapy and Experimental Psychiatry*, *40*(1), 24–38. doi:10.1016/j.jbtep.2008.03.001

Rawal, A. and Rice, F. (2012). Examining overgeneral autobiographical memory as a risk factor for adolescent depression. *Journal of the American Academy of Child and Adolescent Psychiatry*, *51*(5), 518–527. doi:10.1016/j.jaac.2012.02.025

Reid, S. C., Salmon, K., and Lovibond, P. F. (2006). Cognitive biases in childhood anxiety, depression, and aggression: Are they pervasive or specific? *Cognitive Therapy and Research*, *30*(5), 531–549. doi:10.1007/s10608-006-9077-y

Richards, D. (2011). Prevalence and clinical course of depression: A review. *Clinical Psychology Review*, *31*(7), 1117–1125. doi:10.1016/j.cpr.2011.07.004

Rude, S. S., Wenzlaff, R. M., Gibbs, B., Vane, J., and Whitney, T. (2002). Negative processing biases predict subsequent depressive symptoms. *Cognition and Emotion*, *16*(3), 423–440. doi:10.1080/02699930143000554

Rusting, C. L. and DeHart, T. (2000). Retrieving positive memories to regulate negative mood: Consequences for mood-congruent memory. *Journal of Personality and Social Psychology*, *78*(4), 737–752. doi:10.1037//0022-3514.78.4.737

Schartau, P. E. S., Dalgleish, T., and Dunn, B. D. (2009). Seeing the bigger picture: Training in perspective broadening reduces self-reported affect and psychophysiological response to distressing films and autobiographical memories. *Journal of Abnormal Psychology*, *118*(1), 15–27. doi:10.1037/a0012906

Secretary of State for Health and Secretary of State for Education. (2017). *Transforming Children and Young People's Mental Health Provision: A Green Paper*.

Siegle, G. J., Ingram, R. E., and Matt, G. E. (2002). Affective interference: An explanation for negative attention biases in dysphoria? *Cognitive Therapy and Research*, *26*(1), 73–87. doi:10.1023/A:1013893705009

Smith, E., Reynolds, S., Orchard, F., Whalley, H. C., and Chan, S. W. (2018). Cognitive biases predict symptoms of depression, anxiety and wellbeing above and beyond neuroticism in adolescence. *Journal of Affective Disorders*, *241*, 446–453. doi:10.1016/j.jad.2018.08.051

Sokolov, S. and Kutcher, S. (2001). Adolescent depression: Neuroendocrine aspects. In I. M. Goodyer (ed.), *The Depressed Child and Adolescent. Cambridge Child and Adolescent Psychiatry Series*, pp. 233–266. Cambridge: Cambridge University Press.

Spinhoven, P., Bockting, C. L. H., Schene, A. H., Koeter, M. W. J., Wekking, E. M., and Williams, J. M. G. (2006). Autobiographical memory in the euthymic phase of recurrent depression. *Journal of Abnormal Psychology*, *115*(3), 590–600. doi:10.1037/0021-843X.115.3.590

Springer, D. W., Rubin, A., and Beevers, C. G. (2011). *Treatment of Depression in Adolescents and Adults: Clinician's Guide to Evidence-Based Practice*. New York: John Wiley and Sons.

Stange, J. P., Hamlat, E. J., Hamilton, J. L., Abramson, L. Y., and Alloy, L. B. (2013). Overgeneral autobiographical memory, emotional maltreatment, and depressive symptoms in adolescence: Evidence of a cognitive vulnerability–stress interaction. *Journal of Adolescence*, *36*(1), 201–208. doi:10.1016/j.adolescence.2012.11.001

Stringaris, A., Maughan, B., Copeland, W. S., Costello, E. J., and Angold, A. (2013). Irritable mood as a symptom of depression in youth: Prevalence, developmental, and clinical correlates in the Great Smoky Mountains study. *Journal of the American Academy of Child and Adolescent Psychiatry*, *52*(8), 831–840. doi:10.1016/j.jaac.2013.05.017

Stringaris, A., Vidal-Ribas, P., Brotman, M. A., and Leibenluft, E. (2018). Practitioner review: Definition, recognition, and treatment challenges of irritability in young people. *Journal of Child Psychology and Psychiatry*, *59*(7), 721–739. doi:10.1111/jcpp.12823

Sumner, J. A. (2012). The mechanisms underlying overgeneral autobiographical memory: An evaluative review of evidence for the CaR-FA-X model. *Clinical Psychology Review*, *32*(1), 34–48. doi:10.1016/j.cpr.2011.10.003

Sumner, J. A., Griffith, J. W., and Mineka, S. (2010). Overgeneral autobiographical memory as a predictor of the course of depression: A meta-analysis. *Behaviour Research and Therapy*, *48*(7), 614–625. doi:10.1016/j.brat.2010.03.013

Sumner, J. A., Griffith, J. W., Mineka, S., Rekart, K. N., Zinbarg, R. E., and Craske, M. G. (2011). Overgeneral autobiographical memory and chronic interpersonal stress as predictors of the course of depression in adolescents. *Cognition and Emotion*, *25*, 183–192. doi:10.1080/02699931003741566

Timbremont, B., Braet, C., Bosmans, G., and Van Vlierberghe, L. (2008). Cognitive biases in depressed and non-depressed referred youth. *Clinical Psychology and Psychotherapy*, *15*(5), 329–339. doi:10.1002/cpp.579

Torkan, H., Blackwell, S. E., Holmes, E. A., Kalantari, M., Neshat-Doost, H. T., Maroufi, M., and Talebi, H. (2014). Positive imagery cognitive bias modification in treatment-seeking patients with major depression in Iran: A pilot study. *Cognitive Therapy and Research*, *38*(2), 132–145. doi:10.1007/s10608-014-9598-8

Valentino, K., Bridgett, D. J., Hayden, L. C., and Nuttall, A. K. (2012). Abuse, depressive symptoms, executive functioning, and overgeneral memory among a psychiatric sample of children and adolescents. *Journal of Clinical Child and Adolescent Psychology*, *41*(4), 491–498. doi:10.1080/15374416.2012.660689

Valentino, K., Toth, S. L., and Cicchetti, D. (2009). Autobiographical memory functioning among abused, neglected, and nonmaltreated children: The overgeneral memory effect. *Journal of Child Psychology and Psychiatry*, *50*(8), 1029–1038. doi:10.1111/j.1469-7610.2009.02072.x

Vilgis, V., Silk, T. J., and Vance, A. (2015). Executive function and attention in children and adolescents with depressive disorders: A systematic review. *European Child and Adolescent Psychiatry*, *24*(4), 365–384. doi:10.1007/s00787-015-0675-7

Waite, P., Codd, J., and Creswell, C. (2015). Interpretation of ambiguity: Differences between children and adolescents with and without an anxiety disorder. *Journal of Affective Disorders*, *188*, 194–201. doi:10.1016/j.jad.2015.08.022

Watkins, E. R., Baeyens, C. B., and Read, R. (2009). Concreteness training reduces dysphoria: Proof-of-principle for repeated cognitive bias modification in depression. *Journal of Abnormal Psychology*, *118*(1), 55–64. doi:10.1037/a0013642

Watkins, E. R., Taylor, R. S., Byng, R., Baeyens, C., Read, R., Pearson, K., and Watson, L. (2012). Guided self-help concreteness training as an intervention for major depression in primary care: A Phase II randomized controlled trial. *Psychological Medicine*, *42*(7), 1359–1371. doi:10.1017/S0033291711002480

Watkins, Edward R., Mullan, E., Wingrove, J., Rimes, K., Steiner, H., Bathurst, N., ... Scott, J. (2011). Rumination-focused cognitive-behavioural therapy for residual depression: Phase II randomised controlled trial. *British Journal of Psychiatry*, *199*(4), 317–322. doi:10.1192/bjp.bp.110.090282

Watkins, Edward R. and Moberly, N. J. (2009). Concreteness training reduces dysphoria: A pilot proof-of-principle study. *Behaviour Research and Therapy*, *47*(1), 48–53. doi:10.1016/j.brat.2008.10.014

Watkins, P. C. (2002). Implicit memory bias in depression. *Cognition and Emotion*, *16*(3), 381–402. doi:10.1080/02699930143000536

Watkins, P. C., Martin, C. K., and Stern, L. D. (2000). Unconscious memory bias in depression: Perceptual and conceptual processes. *Journal of Abnormal Psychology*, *109*(2), 282–289. doi:10.1037/0021-843X.109.2.282

Weierich, M. R., Treat, T. A., and Hollingworth, A. (2008). Theories and measurement of visual attentional processing in anxiety. *Cognition and Emotion*, 22(6), 985–1018. doi:10.1080/02699930701597601

Wells, T. T., Vanderlind, W. M., Selby, E. A., and Beevers, C. G. (2014). Childhood abuse and vulnerability to depression: Cognitive scars in otherwise healthy young adults. *Cognition and Emotion*, 28(5), 821–833. doi:10.1080/02699931.2013.864258

Werner-Seidler, A., Hitchcock, C., Bevan, A., McKinnon, A., Gillard, J., Dahm, T., ... Dalgleish, T. (2018). A cluster randomized controlled platform trial comparing group memory specificity training (MEST) to group psychoeducation and supportive counselling (PSC) in the treatment of recurrent depression. *Behaviour Research and Therapy*, 105, 1–9. doi:10.1016/j.brat.2018.03.004

Werner-Seidler, A. and Moulds, M. L. (2011). Autobiographical memory characteristics in depression vulnerability: Formerly depressed individuals recall less vivid positive memories. *Cognition and Emotion*, 25(6), 1087–1103. doi:10.1080/02699931.2010.531007

Werner-Seidler, A., and Moulds, M. L. (2012a). Characteristics of self-defining memory in depression vulnerability. *Memory*, 20(8), 935–948. doi:10.1080/09658211.2012.712702

Werner-Seidler, A. and Moulds, M. L. (2012b). Mood repair and processing mode in depression. *Emotion*, 12(3), 470–478. doi:10.1037/a0025984

Werner-seidler, A., Perry, Y., Calear, A. L., Newby, J. M., and Christensen, H. (2017). School-based depression and anxiety prevention programs for young people: A systematic review and meta-analysis. *Clinical Psychology Review*, 51, 30–47. doi:10.1016/j.cpr.2016.10.005

Wheatley, J., Brewin, C. R., Patel, T., Hackmann, A., Wells, A., Fisher, P., and Myers, S. (2007). "I'll believe it when I can see it": Imagery rescripting of intrusive sensory memories in depression. *Journal of Behavior Therapy and Experimental Psychiatry*, 38(4), 371–385. doi:10.1016/j.jbtep.2007.08.005

Whiteford, H. A., Degenhardt, L., Rehm, J., Baxter, A. J., Ferrari, A. J., Erskine, H. E., ... Vos, T. (2013). Global burden of disease attributable to mental and substance use disorders: Findings from the Global Burden of Disease Study 2010. *Lancet*, 382(9904), 1575–1586. doi:10.1016/S0140-6736(13)61611-6

Williams, A. D. and Moulds, M. L. (2007). An investigation of the cognitive and experiential features of intrusive memories in depression. *Memory*, 15(8), 912–920. doi:10.1080/09658210701508369

Williams, A. D. and Moulds, M. L. (2008). Negative appraisals and cognitive avoidance of intrusive memories in depression: A replication and extension. *Depression and Anxiety*, 25(7), E26–33. doi:10.1002/da.20409

Williams, J. M. G., Barnhofer, T., Crane, C., Herman, D., Raes, F., Watkins, E., and Dalgleish, T. (2007). Autobiographical memory specificity and emotional disorder. *Psychological Bulletin*, 133(1), 122–148. doi:10.1037/0033-2909.133.1.122

Williams, J. M. G. and Broadbent, K. (1986). Autobiographical memory in suicide attempters. *Journal of Abnormal Psychology*, 95(2), 144–149. doi:10.1037/0021-843X.95.2.144

Williams, J. M. G., Ellis, N. C., Tyers, C., Healy, H., Rose, G., and Macleod, A. K. (1996). The specificity of autobiographical memory and imageability of the future. *Memory and Cognition*, 24(1), 116–125. doi:10.3758/BF03197278

Wright, B., Tindall, L., Littlewood, E., Adamson, J., Allgar, V., Bennett, S., ... Ali, S. (2014). Computerised cognitive behaviour therapy for depression in adolescents: Study protocol for a feasibility randomised controlled trial. *BMJ Open*, 4(10). doi:10.1136/bmjopen-2014-006488

Wright, B., Tindall, L., Littlewood, E., Allgar, V., Abeles, P., Trépel, D., and Ali, S. (2017). Computerised cognitive-behavioural therapy for depression in adolescents: Feasibility results and 4-month outcomes of a UK randomised controlled trial. *BMJ Open*, *7*(1), 1–11. doi:10.1136/bmjopen-2016-012834

Yurgelun-Todd, D. (2007). Emotional and cognitive changes during adolescence. *Current Opinion in Neurobiology*, *17*(2), 251–257.

Zisook, S., Rush, A. J., Lesser, I., Wisniewski, S. R., Trivedi, M., Husain, M. M., ... Fava, M. (2007). Preadult onset vs. adult onset of major depressive disorder: A replication study. *Acta Psychiatrica Scandinavica*, *115*(3), 196–205. doi:10.1111/j.1600-0447.2006.00868.x

Zupan, B. A., Hammen, C., and Jaenicke, C. (1987). The effects of current mood and prior depressive history on self-schematic processing in children. *Journal of Experimental Child Psychology*, *43*(1), 149–158. doi:10.1016/0022-0965(87)90056-7

CHAPTER 36

···

SLEEP, ANXIETY, AND DEPRESSION IN ADOLESCENCE

A developmental cognitive neuroscience approach

···

MONIKA B. RANITI, JOHN TRINDER,
AND NICHOLAS B. ALLEN

INTRODUCTION

···

SLEEP is a ubiquitous, active, and distinct neurological process that is essential for human development and functioning (Colrain and Baker 2011). It is thought to be involved in growth, tissue repair, restoration, learning, and memory consolidation (Krueger et al. 2016), and appears to be particularly important during periods of brain maturation such as adolescence (Dahl and Lewin 2002). Of the various functions affected by the neurobiological and psychosocial changes occurring in adolescence, patterns of sleep and wakefulness are among the most substantially altered (Colrain and Baker 2011). Consequently, adolescents are susceptible to experiencing sleep disturbances (Carskadon 2011). Sleep shares considerable overlap with arousal and emotion regulation systems, such that disruptions to sleep can contribute to affective dysregulation and vice versa, and thus increase adolescents' vulnerability for internalizing problems (Dahl and Harvey 2007; Dahl 1996). Indeed, sleep problems commonly co-occur with depression and anxiety in adolescence and are increasingly being recognized as a potential transdiagnostic mechanism underlying the comorbidity between these disorders (Harvey et al. 2011; Blake et al. 2018). Notably, accumulating evidence suggests that sleep disturbances commonly precipitate the onset of depression during adolescence (Lovato and Gradisar 2014) and are also intimately related to anxiety (Alfano 2018). Sleep problems, therefore, may represent a promising and novel target for therapeutic intervention in the prevention of adolescent onset mental disorders.

ADOLESCENT SLEEP PROBLEMS

Sleep problems, including insufficient and poor-quality sleep, are common during adolescence in what is increasingly being regarded as an international public health issue (Becker et al. 2015; Adolescent Sleep Working Group et al. 2014). Current consensus guidelines from the American Academy of Sleep Medicine recommend that adolescents aged 13 to 18 years regularly obtain 8 to 10 hours sleep within a 24-hour period for optimal health and functioning (e.g., cognitive, emotional, and physical; Paruthi et al. 2016). Guidelines from the National Sleep Foundation in the United States add that fewer than 7 hours or more than 11 hours of sleep is not recommended (Hirshkowitz et al. 2015). These recommendations are supported by evidence from experimental and longitudinal studies which indicate that adolescents sleep around 9 to 9.25 hours per night under unrestricted sleep opportunities (Carskadon and Acebo 2002; Carskadon et al. 1980; Warner et al. 2008), and that this is required for optimal cognitive/attention functioning (Short et al. 2018) and emotional regulation (Fuligni et al. 2019). Indeed, adolescents themselves report needing about 9.25 hours sleep per night on average "to feel their best every day" (Short et al. 2013d: p. 105).

In addition to insufficient sleep duration, adolescents also experience poor-quality sleep including difficulties initiating and maintaining sleep, disrupted or uncomfortable sleep, and feeling unrefreshed upon awakening. Around 30% of community-based adolescents report difficulties falling asleep (Short et al. 2013d) and up to 65% report prolonged sleep onset latency (> 30 minutes) which may be indicative of a clinical sleep disorder (Hysing et al. 2013; Short et al. 2013d; Raniti et al. 2017; Alexandru et al. 2006). Although less common, around 12% of adolescents experience difficulties staying asleep (Short et al. 2013d). Nearly half of adolescents report non-refreshing sleep (Short et al. 2013d) and at least a third of adolescents report excessive daytime sleepiness (Andrade et al. 1993). Disturbingly, it is estimated that two-thirds of adolescents meet criteria for at least one symptom of a clinical sleep disorder (Short et al. 2013a). Two of the most common sleep disorders in adolescence are insomnia disorder (de Zambotti et al. 2018) and delayed sleep-wake phase disorder (DSWPD; Nesbitt 2018).

Insomnia disorder is characterized by frequent complaints of dissatisfaction with sleep quality or quantity associated with difficulty initiating, difficulty maintaining, and/or early-morning awakenings in conjunction with significant daytime distress or impairment (American Psychiatric Association 2013). The median age of onset for adolescent insomnia is 11 years (Johnson et al. 2006b), and the lifetime prevalence rate of adolescent insomnia disorder in adolescents has been estimated to range from 7.9% to 13.6% in community-based populations (Chung et al. 2014; Johnson et al. 2006b; Hysing et al. 2013), which is similar to the prevalence in adults (Hillman and Lack 2013;

Buysse et al. 2008). However, insomnia symptoms have been reported in up to 40% of adolescents (Chung et al. 2014; Johnson et al. 2006b; Short et al. 2013a; Hysing et al. 2013). In adolescents, the most common symptoms of insomnia disorder are difficulty falling asleep (68.5%) and reports of non-restorative sleep (48.1%; Johnson et al. 2006b).

DSWPD is a circadian rhythm disorder that is most prevalent during adolescence (Nesbitt 2018). It is characterized by a significant delay in sleep timing (usually > 2 hours) such that sleep is misaligned with societal norms (American Psychiatric Association 2013). The pattern of extremely delayed sleep onset can cause difficulties with falling asleep, waking up, and daytime sleepiness because an individual's desired sleep-wake schedule conflicts with comparatively early, socially imposed wake times (e.g., for school, work, social commitments; Johnson et al. 2006b). The prevalence of DSWPD in adolescents has been reported to range from 1% to 16% (Gradisar and Crowley 2013; Nesbitt 2018; Gradisar et al. 2011b), although over half of adolescents meet criteria for least one diagnostic symptom (Lovato et al. 2013).

Factors Contributing to Adolescent Sleep Problems

Adolescents are vulnerable to similar factors known to contribute to the development of sleep problems and disorders, particularly insomnia, in adults. Such factors can include: genetic vulnerability (Taheri and Mignot 2002); female gender (Klink et al. 1992; Zhang et al. 2016); negative and stressful life events such as ill health or interpersonal conflict and loss (Bastien et al. 2004); a tendency toward hyperarousal including cognitive (e.g., worry and rumination) hyperarousal (Carney et al. 2010; Bootzin and Epstein 2011; Riemann et al. 2010); eveningness chronotype preference (Schmidt and Van der Linden 2015); no or low levels of regular physical activity (Kredlow et al. 2015; Bartel et al. 2015); and perceived neighborhood safety (Hill et al. 2016; Hawkins and Takeuchi 2016). Of the many developmental changes to occur during adolescence, the pattern of sleep and wakefulness is among the most dramatically altered (Colrain and Baker 2011; Carskadon 2011). These changes in adolescent sleep patterns are generally attributed to both biological and psychosocial factors that uniquely converge during adolescence, in what has been described as a "perfect storm" of factors (Carskadon 2011; Crowley et al. 2018). As a consequence, adolescents as a cohort are particularly susceptible to experiencing disruptions to the quantity, quality, and timing of sleep (Carskadon 2011; Crowley et al. 2018).

Neurodevelopmental Factors

One of the most prominent developmental changes to adolescent sleep is a habitual delay in nocturnal sleep onset (Carskadon and Acebo 2002). This delay has been attributed, in part, to maturational changes to the intrinsic homeostatic and circadian sleep processes that determine sleep propensity, and influence the duration, intensity, and timing of sleep (Borbély 1982; Carskadon and Acebo 2002; Colrain and Baker 2011). A homeostatic process, known as Process S, regulates the distribution of sleep in response to prior sleep and wakefulness, and reflects the "pressure" or drive for sleep (Borbély and Achermann 1999). Sleep pressure accumulates during wakefulness to increase sleep propensity and dissipates during sleep to reduce sleep propensity (Borbély and Achermann 1999). Following sleep loss, recovery sleep may be partially extended but the relationship between sleep loss and recovery sleep is not linear (Borbély 1982). That is, hours of "sleep debt" cannot be "paid off" in equivalent hours of compensatory sleep. Rather, Process S compensates for sleep loss by regulating sleep propensity and intensity of subsequent sleep (Borbély and Achermann 2017). The amount of slow-wave sleep or "deep" sleep a person receives (the degree of slow-wave activity, power density in the delta frequencies ~ 0.5–4.5Hz, or "depth" of sleep), measured during electroencephalography recording (EEG), is considered to be a physiological marker of Process S (Dijk et al. 1990). However, Processes S can also be inferred using behavioral markers including cues for sleepiness (e.g., yawning, droopy eye lids, heavy limbs) or propensity to fall asleep rapidly (e.g., < 5 minutes; Samson et al. 2013).

Beginning at around 11 years of age and approximately aligned with the onset of puberty, the processes of brain reorganization and synaptic pruning occurring during adolescence result in decreases in synaptic density (Colrain and Baker 2011; Feinberg and Campbell 2010). This leads to a progressive reduction in the accumulation of homeostatic sleep pressure during wakefulness such that adolescents experience a reduction in sleep drive (Feinberg et al. 2006). This reduction in homeostatic sleep drive is reflected in marked decreases in slow-wave sleep over adolescence (Campbell and Feinberg 2009). Consistent with continued brain changes as adolescents develop, older adolescents have been shown to have a slower build-up of homeostatic sleep pressure than younger adolescents (Jenni et al. 2005).

A circadian process, known as Process C, represents an internal daily rhythm that regulates sleep onset (Borbély 1982; Cajochen et al. 2003). It is thought to be controlled by an endogenous circadian oscillator (a "clock-like" mechanism) that originates in the suprachiasmatic nucleus of the anterior hypothalamus and has an intrinsic period averaging 24.2 hours (Borbély et al. 2016). In healthy adults, the circadian signal for sleep tends to be highest in the very early morning (during the normal sleep period) and the circadian signal for wakefulness increases in strength through the day until the early to middle afternoon (Borbély 1982). Melatonin secretion is one process that follows the circadian rhythm and influences sleep timing, and thus can be used as a biological marker of Process C (Cajochen et al. 2003). Melatonin begins to be secreted approximately 1 hour before normal sleep onset, assuming the absence of light, to signal sleep and melatonin offset towards morning promotes wakefulness (Cajochen et al. 2003). Although predominately an endogenous process, Process C is also influenced by rhythmically occurring external cues (i.e., zeitgebers; Borbély 1982). The most potent of these cues is environmental light as the suprachiasmatic nucleus is connected to the retina via the retinohypothalamic tract (Borbély et al. 2016). Light exposure suppresses melatonin secretion and therefore allows sleep/wake rhythms to be synchronized with the 24-hour day/night cycle (Cajochen et al. 2003).

Although individuals can differ in the phase position of their circadian rhythm, during adolescence there is a shift in circadian phase which delays the optimal time for nocturnal sleep onset (Carskadon et al. 1997). As a result, adolescents experience a shift towards a later, "eveningness" type circadian preference or "chronotype" (e.g., Carskadon et al. 1993). It has been hypothesized that this shift occurs via maturational changes related to Process C including a lengthening of the intrinsic period of the oscillator, and an increased sensitivity to the phase-delaying effects of evening light and/ or a decreased sensitivity to phase-advancing morning light (Carskadon 2011; Colrain and Baker 2011; Carskadon et al. 1999; Crowley et al. 2007). For example, evidence indicates that compared to children, adolescents release melatonin later in the evening and are more sensitive to evening light (Carskadon et al. 1997). Further, melatonin offset has been shown to be later in more mature, compared to less mature, adolescents (Carskadon and Acebo 2002).

When taken together, sleep onset is most likely to occur when homeostatic drive for sleep is relatively high and aligns with an appropriate time in the circadian rhythm (Borbély et al. 2016; Borbély 1982). Through normal maturation, adolescents experience a reduction in sleep drive and shift to a later circadian phase, both of which become amplified as adolescents get older. This manifests as a progressive delay in nocturnal sleep onset that plateaus around the age of 20 years old (Roenneberg et al. 2004). Thus, the combined effect of these changes to homeostatic and circadian sleep processes mean that adolescents are biologically capable of later bedtimes compared to children. They also indicate that older adolescents are capable of even later bedtimes than younger adolescents.

Psychosocial Factors

In addition to biological factors, a number of psychosocial factors salient during adolescence have also been implicated in delaying adolescents' bedtime and nocturnal sleep onset (Carskadon 2011; Bartel et al. 2015; Becker et al. 2015). These factors can result in continued waking activity in the evening displacing time for sleep and/or being physiologically, emotionally, and cognitively arousing such that sleep onset becomes delayed (Carskadon 2011; Crowley et al. 2018). For example, compared to children, adolescents have increased academic demands, extra-curricular activities, and for many, the emergence of part-time employment (Carskadon and Acebo 2002; Carskadon 1990; Bachman and Schulenberg 1993; Dorofaeff and Denny 2006; Adam et al. 2007). Also, social interactions are vital during adolescence and adolescents typically have more freedom to socialize with peers outside of school hours, in person and electronically. Both the amount of time spent socializing in addition to the quality of social interaction have been shown to impact sleep (Becker et al. 2015). The more time spent with friends has been associated with less sleep duration on school nights and greater variability of sleep duration (Fuligni and Hardway 2006; Kuo et al. 2015) as social interactions occur in place of sleep and with an irregular schedule. Loneliness, problems within peer relationships, and bullying and victimization, have also been related to shorter sleep and irregular sleep patterns (Mahon 1994; Kubiszewski et al. 2014; Harris et al. 2013; Roberts et al. 2002; Xu et al. 2012). However, positive, supportive relationships are related to less sleep disruption and longer sleep (Maume 2013).

During adolescence, parental influence over bedtimes is also reduced as adolescents develop increased autonomy and control over their routines and schedules. Parent set bedtimes are associated with earlier bedtimes (Short et al. 2011), longer sleep duration (Bartel et al. 2015; Short et al. 2013c), better sleep quality (Buxton et al. 2015), and improved daytime wakefulness and less fatigue (Short et al. 2011), although they are not associated with improved sleep onset latency (Short et al. 2011; Bartel et al. 2015).

Adolescents frequently use electronic devices, most commonly smart phones, computers, and gaming consoles (Gamble et al. 2014) and adolescents' use of technology has increased since the beginning of this century (Bucksch et al. 2016; Rideout et al. 2010). Over 90% of adolescents have access to at least one electronic device in their bedroom at night (Hysing et al. 2015b; Gradisar et al. 2013) and over 70% report two or more (Gamble et al. 2014). These electronic devices can have a detrimental impact on sleep (Gradisar et al. 2013). They have been associated with decreased sleep duration (Bartel et al. 2015; Hale and Guan 2015), later bedtimes (Bartel et al. 2015; Shochat et al. 2010), later sleep onset and rise times (Gamble et al. 2014), increased sleep deficiency (Hysing et al. 2015b; Buxton et al. 2015), and to a lesser extent longer sleep onset latency (Shochat et al. 2010; Hysing et al. 2015b). Nearly half of adolescents report using an electronic device when they should be sleeping (Gamble et al. 2014). The use of electronic devices might be disruptive to sleep because it displaces sleep and/or because it promotes alertness and physiological, emotional, and cognitive arousal during the pre-sleep period (Cain and Gradisar 2010; Bartel and Gradisar 2017; Harbard et al. 2016). The blue bright light emitted by a number of these devices also contributes to the suppression of melatonin secretion and to delayed circadian phase (Carskadon et al. 1997), to which adolescents may be particularly sensitive (Crowley et al. 2015).

Effect of Delayed Sleep Onset on Adolescent Sleep Patterns

As described, both biological and psychosocial factors contribute to delaying adolescents' sleep onset and encourage later bedtimes, particularly as adolescents become older. In 13-year-olds, average weekday bedtime is around 10 pm and by age 18 it shifts to around 11:30 pm (Zhang et al. 2017). While delayed sleep onset occurs in the course of normal development, it poses challenges to obtaining adequate sleep when coupled with early and typically inflexible school start times which are as early as 7:10 am in the United States (Wolfson and Carskadon 1998) and 8:00 am in Australia (Warner et al. 2008). Further, wake times shift earlier throughout adolescence due to earlier school start times (Crowley et al. 2014) and adolescents will not necessarily go to bed earlier to compensate (Wolfson et al. 2007). Thus, on school nights, adolescents have a restricted opportunity for sleep and this contributes to many adolescents habitually failing to obtain a sufficient amount of sleep, particularly as they get older (Sadeh et al. 2009).

It is well-established that insufficient sleep in adolescents is associated with daytime sleepiness to levels that impact daily functioning (Owens et al. 2014). For example, in a National Sleep Foundation poll in the United States, 28% of adolescents reported falling asleep in class at least once a week and 14% reported oversleeping and missing or arriving late to school (National Sleep Foundation 2006). Many adolescents also report needing assistance waking in the morning (Yang et al. 2005). Therefore, in order to cope with accumulated sleep debt over the school week and immediate daytime consequences (Samson et al. 2013), adolescents engage in compensatory "catch-up" or recovery sleep on the weekends and school holidays (Hansen et al. 2005). This recovery sleep is characterized by even later bedtimes and longer sleep duration than on school nights (Hansen et al. 2005).

A number of studies have demonstrated that weekend bedtimes are nearly 2 hours later than on school nights and total sleep time is typically 1 to 2 hours longer (Warner et al. 2008; Zhang et al. 2017; Olds et al. 2010, Short et al. 2013b). The discrepancy between school day and weekend sleep duration generally becomes larger as adolescents get older, with average duration around 30 minutes at age 13 and 90 minutes around age 17, with some differences reported as large as 2 hours (Short et al. 2013d; Zhang et al. 2017; Olds et al. 2010; Paavonen et al. 2016; Wolfson and Carskadon 1998).

While weekend catch-up sleep likely offsets some of the sleep debt accumulated by adolescents, it is unlikely to be sufficient to rectify chronic sleep restriction experienced by adolescents (Kim et al. 2011). Experimental studies (Lo et al. 2017; Lo et al. 2016) have demonstrated that weekend catch-up sleep, even when combined with napping on school days, for sleep restricted adolescents is "inferior to obtaining sufficient and regular school-night sleep" (Crowley et al. 2018: p. 62). Further, frequent shifts in sleep timing can contribute to what has been termed "social jetlag" (Wittmann et al. 2006; Hansen et al. 2005; Touitou 2013). Because weekend recovery sleep tends to occur at a less appropriate circadian phase, this reduces its restorative value (Carskadon et al. 2004) and can further perpetuate sleep problems such as difficulties falling asleep (Gradisar et al. 2011a).

CONSEQUENCES OF ADOLESCENT SLEEP PROBLEMS

Adolescent sleep problems are associated with a range of immediate and long-term consequences to health and functioning (Shochat et al. 2014). Sleep restriction and disturbances in adolescence have been associated with: poor school performance

(Dewald et al. 2010; Gillen-O'Neel et al. 2013) and attendance (Hysing et al. 2015a); impaired cognitive performance (Carskadon et al. 1981; Randazzo et al. 1998; Astill et al. 2012); drowsy driving and motor vehicle accidents (Pizza et al. 2010; Danner and Phillips 2008); self-harm (Lundh et al. 2012; Hysing et al. 2015c); risk-taking behaviors (Telzer et al. 2013; O'Brien and Mindell 2005; Shochat et al. 2014, e.g., risky sexual behaviors, substance use and misuse; McKnight-Eily et al. 2011; Sivertsen et al. 2015); delinquency (Clinkinbeard et al. 2011; Meldrum et al. 2015); behavioral (Meijer et al. 2010; Zhang et al. 2017) and interpersonal problems (Roberts et al. 2008); somatic symptoms such as gastrointestinal problems (Zhou et al. 2012) and headaches (Bruni et al. 2008); lower health-related quality of life (Hart et al. 2005) and poorer perceptions of health (Zhang et al. 2017); cardiovascular (Narang et al. 2012; Javaheri et al. 2008) and metabolic disease risk (Martinez-Gomez et al. 2011; Park et al. 2016); and poor-quality diet, sedentary behavior, and being overweight or obese (Golley et al. 2013; Garaulet et al. 2011; Seicean et al. 2007).

Adolescent Sleep Problems and Affective Functioning

Of the numerous impairments associated with adolescence sleep problems, deficits to affective functioning are some of the most striking (Dahl 1999). In adolescents, sleep restriction has been associated with an increase in negative emotions (McMakin et al. 2016), a decrease in positive emotions (Dagys et al. 2012; McGlinchey et al. 2011), and increases in stress/anxiety, irritability/anger, and confusion (Talbot et al. 2010; Baum et al. 2014; Short and Louca 2015). Perceptions of poor sleep quality have been associated with more negative and less positive affect the following day (van Zundert et al. 2015). In one study, late school term (> 11:15 pm) and summer vacation (> 1:30 am) bedtimes predicted later emotional distress around 7 years later (Asarnow et al. 2014). Sleep loss in adolescents is also related to: poorer emotion regulation (Baum et al. 2014; Vriend et al. 2013) and strategy use (e.g., less problem solving and greater rumination and avoidance; Palmer et al. 2018); greater levels of negative social evaluative emotions (e.g., feeling rejection, embarrassment; Tavernier et al. 2016); negative affective behavior when resolving conflict in real world social contexts (McMakin et al. 2016); and may contribute to difficulties performing emotional and cognitive processing simultaneously, which is crucial for social competence tasks (Dahl and Lewin 2002). Further, sleep and arousal are antithetical processes which also share considerable overlap with affective functioning (Dahl 1996).

Inadequate sleep, therefore, is related to changes in affective functioning in what is considered to be a complex and bidirectional relationship (Kahn et al. 2013). Not only can sleep loss result in problems with emotions, emotional states of both arousal and distress can in turn contribute to difficulties falling asleep or having disturbed sleep, and lead to a vicious deleterious cycle over time (Kahn et al. 2013; Dahl 1999). These cycles may be particularly damaging during the sensitive developmental window of adolescence, a time of susceptibility to affective dysregulation, and may place additional strains on adolescents' self-regulatory capacities (Dahl and Lewin 2002). While the precise mechanisms underlying the sleep-affect/emotion relationship, particularly during adolescence are unclear (Palmer et al. 2018), they are likely related to several neurobiological (e.g., energy-related alertness; arousal), behavioral (e.g., social withdrawal due to fatigue), and cognitive (e.g., coping style) processes (Kahn et al. 2013; Dahl and Lewin 2002; Palmer and Alfano 2017). For example, neural pathways identified to be important in the production and regulation of emotions and behavior (e.g., the prefrontal cortex and limbic structures such as the amygdala) are also those activated in sleep states (see Palmer and Alfano 2017). Nonetheless, adolescents seem to be particularly susceptible to emotional dysregulation following insufficient sleep (Kahn et al. 2013). For example, compared to sleep deprived adults, adolescents have been found to experience significantly less positive emotion following sleep deprivation (McGlinchey et al. 2011). Sleep deprived younger adolescents have also been shown to perceive catastrophizing worries as more threatening (i.e., greater anxiety) relative to older adolescents and adults (Talbot et al. 2010).

Given that compromised emotion processing and regulation is understood to be an underlying mechanism for the development and maintenance of psychopathology (Sloan et al. 2017), particularly internalizing problems (Gregory and Sadeh 2016; Palmer et al. 2018), several researchers have suggested that the relationship between sleep disturbance and affective processes may, in part, underlie the risk for psychopathology that develops during adolescence (Palmer et al. 2018; Palmer and Alfano 2017; Harvey et al. 2011; Dahl and Lewin 2002). Indeed, recent models of adolescent sleep disturbance acknowledge mental health as a contributing factor to the emergence of sleep problems in adolescence (Becker et al. 2015; Crowley et al. 2018).

ADOLESCENT SLEEP PROBLEMS AND MENTAL HEALTH

Adolescent sleep problems have been linked to various forms of psychopathology (Dahl and Harvey 2007; Gregory and Sadeh 2016). Insufficient and/or disrupted sleep commonly co-occurs in adolescents with psychiatric and neurodevelopmental disorders including depressive (Johnson et al. 2006a), anxiety (Forbes et al. 2008), bipolar (Harvey et al. 2006), attention deficit hyperactivity (Hysing et al. 2016), oppositional defiant (Shanahan et al. 2014), and autism spectrum (Øyane and Bjorvatn 2005) disorders. Over half of adolescents with insomnia disorder have a comorbid psychiatric disorder (Johnson et al. 2006b). In large non-clinical adolescent populations, the rate of co-occurrence between short or disturbed sleep and symptoms of poor mental health (e.g., anxiety, depressed mood, mood disturbance, suicidal ideation) is high (Roberts et al. 2001; Johnson et al. 2000; Morrison et al. 1992; Coulombe et al. 2010; Kaneita et al. 2009). In light of these associations, sleep disturbances have been conceptualized as transdiagnostic as they are a feature of multiple, and often comorbid, mental health problems (Harvey et al. 2011).

In addition to sleep and mental health problems commonly co-occurring, accumulating evidence from longitudinal studies indicates that adolescent sleep disturbances may precipitate the onset of internalizing and externalizing problems (Pieters et al. 2015), poor mental health status (Kaneita et al. 2009), depression (Lovato and Gradisar 2014), anxiety (Roberts and Duong 2017), suicide (Wong and Brower 2012), bipolar disorder (Ritter et al. 2015), and substance use (Pieters et al. 2015). In general, there is evidence to support a bidirectional relationship between sleep and mental health over time, in which poor sleep increases risk for future mental health problems, and vice versa (Shochat et al. 2014; Alvaro et al. 2013). Therefore, disturbed sleep is more than just a symptom of psychopathology—sleep problems contribute to, and may even be a cause of, perpetuating cycles of vulnerability that increase risk for mental ill health during adolescence (Harvey 2016). However, despite the rise in depression appearing to parallel the rise in sleep problems during adolescence, it is only relatively recently that sleep disturbances have been explored in the etiology of adolescent onset internalizing problems.

Adolescent Sleep Problems
and Depression

One of the most established correlates of sleep disturbance across the lifespan is depression. Indeed, insomnia and/or hypersomnia are a formal diagnostic criterion for major depressive episode (American Psychiatric Association 2013). Sleep complaints, most commonly insomnia symptoms, are present in up to 90% of individuals with major depressive disorder (MDD) (Franzen and Buysse 2008; Tsuno et al. 2005; Ohayon et al. 1998) and are the most common residual symptoms following remission from a depressive episode (Carney et al. 2007; Nierenberg et al. 1999). Insomnia is associated with reduced effectiveness of depression treatment and higher rates of depression recurrence (Ohayon and Roth 2003; Sunderajan et al. 2010; Perlis et al. 1997). Rates of depression are higher in individuals with insomnia compared to those without sleep problems (Breslau et al. 1996; Ford and Kamerow 1989). Further, some of the most reliable physiological changes occurring in depressed individuals are sleep EEG abnormalities (American Psychiatric Association 2013) including prolonged sleep onset latency and greater periods of wake after sleep onset; reduced slow-wave sleep; shortened rapid eye movement (REM) latency; an extended first REM period; and increased REM density (Riemann et al. 2001; i.e., number of eye movements; Baglioni et al. 2014).

There is a substantial body of evidence to suggest that the relationship between sleep problems and depressive symptoms is particularly marked in adolescence. Findings from several cross-sectional studies conducted with community samples of adolescents have linked self-reported reduced sleep quality (Raniti et al. 2017), quantity (Pasch et al. 2014; Wang et al. 2017), difficulty initiating sleep (Hayley et al. 2015), symptoms of insomnia (Alvaro et al. 2014; Roberts et al. 2001), greater bedtime variability (Wolfson and Carskadon 1998); and an evening chronotype (Pabst et al. 2009) to greater levels of depressive symptoms (Franic et al. 2014). Similar associations between sleep problems and depression have been demonstrated in clinical populations. Approximately 80% of clinically depressed adolescents report sleeping problems (Patton et al. 2000), with the majority (75%) reporting insomnia and fewer reporting hypersomnia (25%; Dahl 1999). In a meta-analysis of relationships between sleep and depression in adolescents, Lovato and Gradisar (2014) found that compared to non-depressed adolescents, adolescents with MDD were more likely to report more sleep disturbances, insomnia, hypersomnia, and poor sleep quality. These sleep complaints were consistent with findings from objective data in which depressed adolescents had significantly more wakefulness in bed, including longer sleep onset latency, more night-time awakenings, and lower sleep efficiency (Lovato and Gradisar 2014). A number of studies have demonstrated EEG changes in depressed adolescents that are congruent with those found in adults (Rao et al. 2009; Kutcher et al. 1992; Lopez et al. 2010; Robert et al. 2006) such as higher REM density (Lovato and Gradisar 2014). However, it should be acknowledged that EEG alterations have not been consistently demonstrated (Goetz et al. 1987; Khan and Todd 1990; Dahl et al. 1990), which has been attributed to small sample sizes and study heterogeneity (e.g., gender, age, clinical features of depression such as inpatient/outpatient status, and medication use; Santangeli et al. 2017).

Relative to other developmental stages, the relationship between sleep and depression in adolescence appears to be particularly strong (Gregory and Sadeh 2016). In cross-sectional analyses, Johnson et al. (2000) found a stronger association between sleep problems and depression at age 11 compared to age 6. Similarly, in a study of nearly 500 participants, Gregory and O'Connor (2002) found that childhood sleep disturbances predicted affective problems in mid-adolescence and that strength of the relationship increased with age. Adolescents with depression have also been shown to have significantly higher levels of cortisol (indicative of changes to the hypothalamic-pituitary-adrenal (HPA) axis implicated in affective disorders) around the sleep onset period than depressed children, which may be a consequence of HPA and sleep systems becoming vulnerable to dysregulation following puberty and creating an opportunity for depression to influence cortisol levels at sleep onset (Forbes et al. 2006). Further, the association between eveningness and depressive symptoms has been shown to be significant in individuals younger than 21 years old, but not in individuals in their 30s and 40s (Kim et al. 2010).

In addition to depression, insufficient (Liu 2004) and poor-quality sleep or non-restorative sleep (Park et al. 2013) are also associated with and may predict suicide ideation, attempts, and death by suicide in adolescents (Perlis et al. 2016). In a sample of over 6,500 adolescents, Wong and Brower (2012) found trouble falling or staying asleep to be a robust predictor of suicidal thoughts and attempts 3 years later. In a large population-based survey conducted in Korea of over 8,000 adolescents in grades 7 to 11, shorter weekday sleep duration and longer weekend oversleep was significantly associated with increased suicidal ideation (Lee et al. 2012). The relationship with weekend over-sleep remained significant even after controlling for depressive symptoms, insomnia symptoms, snoring, and daytime sleepiness (Lee et al. 2012).

Sleep Problems as a Predictor of Adolescent Onset Depression

There is evidence to suggest that sleep problems and depression likely have bidirectional prospective relationships across the lifespan (Alvaro et al. 2013). That is, sleep problems, particularly insomnia symptoms and disorder, have been shown to predict later de-pressive symptoms and disorders and vice versa (Hayley et al. 2015; e.g., Alvaro et al. 2013; Alvaro et al. 2014; Lovato et al. 2017; Breslau et al. 1996; Chang et al. 1997; Ford and Kamerow 1989). However, findings from several reviews and meta-analyses indi-cate that sleep problems precede the onset of depression, including in adolescence, more consistently than the reverse (Baglioni et al. 2011a; Baglioni et al. 2011b; Pigeon et al. 2017; Lovato and Gradisar 2014).

In a review and meta-analysis of 21 longitudinal epidemiological studies, Baglioni and colleagues (2011a) examined the predictive relationship between insomnia symptoms or disorder and the onset of clinical depression. The results showed that non-depressed individuals with insomnia, including adolescents (Roane and Taylor 2008; Roberts et al. 2002) had a two-fold risk for developing depression compared to individuals with no sleep problems (Baglioni et al. 2011a). Specifically, in individuals with insomnia and no depression at baseline, the incidence of depression at follow-up (an average of 6 years later) was 11.3%. In contrast, incident depression was 2.4% in individuals without in-somnia at baseline. The results also indicated that chronic insomnia can exist for years prior to the onset of a depressive episode. The authors concluded that while insomnia can be a symptom of depression, in some cases may represent a prodrome of depression, or may simply be the first or most noticeable symptom of an imminent depressive epi-sode, it can also be conceptualized as a separate disorder that independently increases the risk for future depression (Baglioni et al. 2011a). This finding is consistent with a sys-tematic review of 11 studies (published after those included in the Baglioni et al. (2011a) review), in which insomnia was found to prospectively predict new onset (not relapsed or recurrent) depression (Pigeon et al. 2017). Even when a bidirectional relationship between insomnia and depression has been demonstrated, insomnia might predict de-pression more consistently than the reverse (Alvaro et al. 2013).

The prospective relationship between sleep problems and depression has also been demonstrated within primarily adolescent samples. In a meta-analysis of 23 studies conducted with individuals aged 12 to 20 years, Lovato and Gradisar (2014) examined the directionality of relationships between sleep and depression. Using data from lon-gitudinal and treatment studies with a mean follow-up of 5.64 years, they found that sleep problems, particularly variables related to wakefulness in bed including sleep onset latency, wake after sleep onset, and sleep efficiency, but not total sleep time, act as a precursor to depression in adolescents. They found little support for the reverse associations, although noted that fewer studies have investigated the predictive value of depression on subsequent sleep problems. These findings have been supported by subsequent longitudinal studies (e.g., Alvaro et al. 2017), including for example, a large prospective study of over 1,000 adolescents and young adults aged 14 to 24 years and without a major mental health disorder at baseline who were assessed three times over 10 years (Ritter et al. 2015). Difficulty falling asleep and restless or disturbed sleep, but not early morning awakening, increased the risk for the incidence of major depression (odds ratio = 1.41) after adjusting for key covariates (e.g., age, sex, parental affective disorders; Ritter et al., 2015). Together these findings also suggest that different types of sleep problems may be differentially associated with depression risk in young people (Walker and Harvey 2010), highlighting the need for studies to use specific definitions of sleep variables (e.g., sleep onset latency) rather than overly inclusive terms (e.g., sleep problems; Alvaro et al. 2013).

Thus, there is sufficient evidence to indicate that sleep disturbances, particularly insomnia and problems related to wakefulness in bed, can be a precursor to the onset of depression during adolescence. Therefore, the early identification and treatment of sleep problems during adolescence has the potential to be an effective depression prevention strategy (Baglioni et al. 2011b; Ford and Kamerow 1989).

ADOLESCENT SLEEP PROBLEMS AND ANXIETY

In addition to the relationship between adolescent sleep problems and depression, there is robust evidence to support a relationship between self-reported sleep problems and anxiety in adolescents (Willis and Gregory 2015). Cross-sectional studies conducted with community samples have demonstrated that reduced sleep duration (Sarchiapone et al. 2014), higher levels of sleep disturbances (Nielsen et al. 2000, Alfano et al. 2009), poorer sleep quality (Raniti et al. 2018, Nielsen et al. 2000), longer sleep onset latency (Forbes et al. 2008), feeling unrefreshed upon awakening (Alfano et al. 2010), and daytime sleepiness (Alfano et al. 2010), greater intra-individual variability of sleep duration and onset latency (Becker et al. 2017), and evening chronotype (Pabst et al. 2009) are all associated with greater levels of anxiety symptoms. In clinical samples, nearly 90% of children and adolescents with an anxiety disorder experience at least one sleep problem and over half report three or more (Alfano et al. 2007; Alfano et al. 2010), with the number of sleep problems being positively associated with anxiety severity (Alfano et al. 2007). The most common sleep disturbances related to anxiety disorders in adolescents are difficulty initiating sleep and restless or unrefreshing sleep (Shanahan et al. 2014). Bedtime resistance, refusal to sleep alone, and night-time fears are also common in younger children (Alfano 2018). Around a quarter of adolescents with an anxiety disorder also meet criteria for insomnia disorder (Johnson et al. 2006a).

Sleep disturbances are a formal diagnostic criterion for GAD and separation anxiety disorder but have been associated with all anxiety subtypes (American Psychiatric Association 2013; Johnson et al. 2006a; Alfano 2018). The odds of having comorbid insomnia are also increased across all anxiety disorder diagnoses by a factor of 4 (Johnson et al. 2006a). However, a number of studies have shown that sleep problems are highest in youth with GAD, and this association is more consistently demonstrated compared to other anxiety disorders (Alfano et al. 2006; Alfano et al. 2007; Chase and Pincus 2011; Alfano et al. 2010). Around 90% of individuals with GAD report sleeping problems, most commonly difficulty falling asleep (Alfano et al. 2010; Alfano et al. 2007). Alfano et al. (2010) demonstrated that adolescents with GAD report greater levels of sleeping problems than adolescents with obsessive-compulsive disorder (OCD), separation anxiety disorder, or social phobia.

Comparatively few studies have used objective measures to assess sleep in anxious adolescents and results are more mixed. Forbes et al. (2008) found that children and adolescents with anxiety disorders have less slow-wave sleep and longer sleep onset latency than healthy controls when measured with polysomnography. This finding is congruent with the findings of other studies conducted in youth with generalized anxiety (Alfano et al. 2013; Fuller et al. 1997). In contrast, Wilhelm and colleagues (2017) found no sleep EEG differences in the amount of slow-wave sleep or slow-wave activity (EEG frequency 0.75–4.5 Hz) in children and adolescents with social anxiety disorder compared to controls. They did, however find a widespread reduction in sleep spindle activity (9–16 Hz) which has been associated with emotion regulation, stress response, and use of coping strategies in children (Wilhelm et al. 2017; Mikoteit et al. 2013; Mikoteit et al. 2012).

When sleep is measured with actigraphy (typically over 1 week), some studies have found that anxious individuals have increased sleep onset latency which corroborates subjective reports. Adolescents with anxiety, particularly GAD, have been found to take up to twice as long to fall asleep compared to healthy controls (Alfano et al. 2015; Cousins et al. 2011; Mullin et al. 2017; Forbes et al. 2008). However, several studies have failed to find differences in the sleep parameters of anxious youth compared to healthy controls (Brown et al. 2018; Baddam et al. 2018; Alfano 2018), even when differences have been shown on these parameters when sleep is measured via self-report. One possibility for this discrepancy may be due to measurement. Actigraphy assesses movement rather than sleep per se, therefore lying motionless but awake in bed may be coded as sleep (Martin and Hakim 2011; Tang and Harvey 2004). Self-report measures are often retrospective and can be prone to bias, therefore anxious individuals may misperceive their sleep as wake or interpret brief awakenings as longer than they are as they may be more attuned to sleep-related threats (Harvey and Tang 2012). Thus, actigraphy may lead to an underestimation of sleep onset latency and overestimation of sleep duration, whereas self-report measures may lead to the opposite (Harvey and Tang 2012).

When findings from studies measuring sleep subjectively and objectively are considered together, evidence suggests that problems with sleep initiation and perceptions of poor sleep are the primary concerns of anxious adolescents (Cowie et al. 2014). They also highlight that a multi-method assessment is preferable for assessing sleep in anxious adolescents (Alfano 2018), although all clinically significant sleep complaints should be taken seriously, regardless of whether they are corroborated by objective evidence (McMakin and Alfano 2015).

Prospective Relationships Between Sleep Problems and Anxiety

Relative to the literature examining prospective relations between sleep problems and depression, the evidence for anxiety is more limited, particularly in adolescent samples (Roberts and Duong 2017; Pigeon et al. 2017). Recent reviews have synthesized current evidence to suggest that sleep problems precede the onset of later anxiety (McMakin and Alfano 2015; Cowie et al. 2014; Brown et al. 2018; Pigeon et al. 2017; Alvaro et al. 2013; Willis and Gregory 2015) and vice versa (Alvaro et al. 2013) across the lifespan. The majority of studies examining the relationship between adolescent sleep problems and anxiety have used cross-sectional designs, but there have been some longitudinal studies that enable the direction of temporal associations to be examined (e.g., Willis and Gregory 2015; Alfano et al. 2013; Comer et al. 2012). Again, evidence appears to support a bidirectional relationship between sleep problems and anxiety, but prospective relationships may differ according to the type of sleep problem, the anxiety subtype (Alvaro et al. 2017), and the duration of follow-up. One particularly important longitudinal study was recently reported by Orchard and colleagues (2020). They found that although sleep difficulties were not cross-sectionally associated with anxiety disorders at 15 years, they were prospectively associated with the onset of both anxiety and depression. Specifically, less total sleep time on school nights at age 15, but not weekend total sleep time or chronotype, was associated with anxiety and depression diagnoses at age 17 and 24. Also, anxiety diagnoses at age 24 were predicted by sleep onset latency, again specifically on school nights. These findings suggest that features of sleep during school nights (when level of stress is presumably higher) may indicate greater risk for the subsequent development of case level anxiety disorders.

Potential Mechanisms Underlying Associations Between Sleep Problems, Depression, and Anxiety

There are several possible mechanisms that are likely to account for how adolescent sleep disturbances may contribute to the development of depression, particularly in anxious individuals. Overall, however, these mechanisms remain poorly understood and evidence derived from adolescent samples is particularly limited (Blake et al. 2018). Broadly speaking, sleep problems may contribute to general psychosocial distress that exacerbates an adolescents' pre-existing vulnerabilities (Pigeon et al. 2017) to compound the risk for depression over time (Lovato and Gradisar 2014), consistent with the diathesis-stress model of mental illness (Pigeon et al. 2017). For example, disturbed sleep may increase physiological stress reactivity and impair an individual's ability to adaptively respond to stressors (Li et al. 2018), which contributes to negative emotionality and affective instability (Li et al. 2018; Ehring et al. 2010; Erk et al. 2010; Dinges et al. 1997). Compared to adults, adolescents, especially those with anxiety, may be particularly susceptible to these effects of sleep disturbance as they may be less likely to have the skills to self-regulate and self-soothe during periods of heightened distress (Alfano et al. 2007).

More specifically, mechanisms that likely operate in parallel, sequential, and interacting patterns over time have also been proposed to underlie associations between sleep problems, depression, and anxiety (Blake et al. 2018). In a recent review, Blake, Trinder, and Allen (2018) synthesized evidence from a variety of study types (cross-sectional, longitudinal, twin studies, experimental, intervention) and identified several of these mechanisms within biological (e.g., dysregulation in genetic, brain, cortisol, inflammatory cytokine, memory, and sleep architecture processes), psychological (e.g., cognitive inflexibility and attention biases and dysfunctional beliefs and attitudes about sleep), and social (e.g., reduced and impaired social interactions, unhelpful parenting behaviors, and family stress) domains. The extent to which insomnia, depression, and anxiety are independent and distinct diagnostic entities that mutually influence each other or whether they represent different manifestations of a shared underlying process or phenotype is yet to be established (Blake et al. 2018).

For example, disrupted sleep may contribute to symptoms of depression by negatively influencing processes related to attention and memory (Blake et al. 2018). Sleep deprived adolescents may selectively attend to more negative information (e.g., Hallion and Ruscio 2011; Harris et al. 2015; Milkins et al. 2016; Nota and Coles 2018; Price et al. 2016) and be more likely to encode negative memories compared to other memories (Yoo et al. 2007). Moreover, such negative evaluations may extend to perceptions of sleep quality which is detrimental to both sleep and mood (Lovato and Gradisar 2014). In support of this, a longitudinal study investigating adolescent sleep in a naturalistic setting conducted by Bei et al. (2017) found that the relationship between low mood and intra-individual variability in sleep onset latency was accelerated (i.e., a curvilinear association) via perceived sleep quality.

Further, wakefulness in bed reinforces cycles of rumination and negative repetitive thinking (risk factors for depression) which contribute to continued sleep disturbances. Indeed, many of the direct consequences of sleep loss are also symptoms of a depressive episode including low or irritable mood, loss of energy and fatigue, changes in appetite, and diminished ability to think or concentrate, which may contribute to further sleep disruptions and negatively reinforcing depressionogenic cycles (Lovato and Gradisar 2014; Cousins et al. 2011).

In addition, sleepiness and fatigue associated with poor sleep may lead to the amotivation, decreased engagement in pleasurable activities, and withdrawal from social interactions associated with depression. This may also be related to other impairments such as to reward processing in which sleep loss blunts positive responses to rewarding experiences (Zohar et al. 2005), or to emotion expressivity and recognition that are important social interactions and affiliative behaviors (Palmer and Alfano 2017; Beattie et al. 2015).

Therefore, the exact mechanisms that contribute to associations between sleep, depression, and anxiety in adolescents are only beginning to be understood. However, to the extent shared mechanisms exist, it underscores the likelihood that adolescent sleep disturbances play a role in the etiology of both depression and anxiety. Further, it suggests that sleep treatment interventions, many of which are thought to modify sleep via these mechanisms (e.g., decreasing arousal and avoidance, improving emotion regulation) are likely to influence both depression and anxiety.

Conclusions and Future Directions

Adolescent sleep problems including insufficient sleep, poor-quality sleep, and related daytime impairments are common in adolescents, and this is increasingly being recognized as a key public health issue. In fact, adolescents are particularly susceptible to experiencing sleep problems owing to a "perfect storm" of biological and psycho-social factors that are unique to this stage of life. The negative consequences of sleep problems, particularly to mental health, are becoming increasingly recognized, and are likely driven in part by disruptions to affective functioning. Adolescent sleep problems commonly co-occur with depression and anxiety, particularly generalized anxiety, and likely share complex bidirectional relationships over time. Notably, evidence indicates that sleep problems, particularly insomnia symptoms, typically precede the emergence of and are an independent risk factor for the onset of depression in adolescents, and sleep problems have been proposed as a mediating link in the sequential comorbidity between anxiety and depression.

There are a number of important future directions to explore. First, it is important to extend work showing that sleep improvement interventions can improve mental health by exploring whether sleep treatment interventions tailored to the unique developmental needs of adolescents may represent a novel approach to prevent adolescent depression, anxiety, and potentially to reduce mental health crises such as suicide (Blake and Allen, 2020). Second, the investigation of both objective and subjective measures of sleep change is also important given the findings from previous studies that these measures can often provide divergent outcomes. Finally, using intervention studies to understand the mechanisms by which the sleep intervention may result in positive outcomes, including sleep improvement and immunological and affective mechanisms (with assessment across several levels of analysis, including self-report, behavior, and molecules), will provide a significantly deeper understanding of the processes involved in any observed benefits. Of particular relevance to this latter issue, the use of intensive longitudinal data collection approaches, whereby both sleep and mental health risk factor mechanisms are assessed on a daily or weekly basis is now more feasible by using mobile and wearable computing methods that allow these data to be collected without significant participant burden. Such studies can provide a rich dataset for modeling the action of therapeutic mechanisms that operate during interventions using time series analysis techniques and will also allow such studies to not only address questions about the efficacy of interventions, but to also make a signification contribution to our fundamental understanding of the associations between sleep, neurobehavioral risk factors, and mental health outcomes.

REFERENCES

Adam, E. K., Snell, E. K., and Pendry, P. 2007. Sleep timing and quantity in ecological and family context: A nationally representative time-diary study. *Journal of Family Psychology*, 21(1): 4–19.

Adolescent Sleep Working Group, Committee on Adolescence and Council on School Health 2014. School start times for adolescents. *Pediatrics*, 134(3): 642–649.

Alexandru, G. et al. 2006. Epidemiological aspects of self-reported sleep onset latency in Japanese junior high school children. *Journal of Sleep Research*, 15(3): 266–275.

Alfano, C. A. 2018. (Re) conceptualizing sleep among children with anxiety disorders: Where to next? *Clinical Child and Family Psychology Review*, 21(4); 482–499.

Alfano, C. A., Ginsburg, G. S., and Kingery, J. N. 2007. Sleep-related problems among children and adolescents with anxiety disorders. *Journal of the American Academy of Child & Adolescent Psychiatry*, 46(2): 224–232.

Alfano, C. A., Patriquin, M. A., and De Los Reyes, A. 2015. Subjective–objective sleep comparisons and discrepancies among clinically-anxious and healthy children. *Journal of Abnormal Child Psychology*, 43(7): 1343–1353.

Alfano, C. A. et al. 2006. Preliminary evidence for sleep complaints among children referred for anxiety. *Sleep Medicine*, 7(6): 467–473.

Alfano, C. A. et al. 2009. Sleep problems and their relation to cognitive factors, anxiety, and depressive symptoms in children and adolescents. *Depression and Anxiety*, 26(6); 503–512.

Alfano, C. A. et al. 2010. Pre-sleep arousal and sleep problems of anxiety-disordered youth. *Child Psychiatry and Human Development*, 41(2): 156–167.

Alfano, C. A. et al. 2013. Polysomnographic sleep patterns of non-depressed, non-medicated children with generalized anxiety disorder. *Journal of Affective Disorders*, 147(1–3): 379–384.

Alvaro, P. K., Roberts, R. M., and Harris, J. K. 2013. A systematic review assessing bidirectionality between sleep disturbances, anxiety, and depression. *Sleep*, 36(7): 1059–1068.

Alvaro, P. K., Roberts, R. M., and Harris, J. K. 2014. The independent relationships between insomnia, depression, subtypes of anxiety, and chronotype during adolescence. *Sleep Medicine*, 15(8): 934–941.

Alvaro, P. K. et al. 2017. The direction of the relationship between symptoms of insomnia and psychiatric disorders in adolescents. *Journal of Affective Disorders*, 207: 167–174.

American Psychiatric Association. 2013. *Diagnostic and Statistical Manual of Mental Disorders*. Arlington, VA: American Psychiatric Association Publishing.

Andrade, M. M. et al. 1993. Sleep characteristics of adolescents: A longitudinal study. *Journal of Adolescent Health*, 14(5): 401–406.

Asarnow, L. D., McGlinchey, E., and Harvey, A. G. 2014. The effects of bedtime and sleep duration on academic and emotional outcomes in a nationally representative sample of adolescents. *Journal of Adolescent Health*, 54(3): 350–356.

Astill, R. G. et al. 2012. Sleep, cognition, and behavioral problems in school-age children: A century of research meta-analyzed. *Psychological Bulletin*, 138(6): 1109–1147.

Bachman, J. G. and Schulenberg, J. 1993. How part-time work intensity relates to drug use, problem behavior, time use, and satisfaction among high school seniors: Are these consequences or merely correlates? *Developmental Psychology*, 29(2); 220–235.

Baddam, S. et al. 2018. Sleep disturbances in child and adolescent mental health disorders: A review of the variability of objective sleep markers. *Medical Sciences*, 6(2): 46–69.

Baglioni, C. et al. 2011a. Insomnia as a predictor of depression: A meta-analytic evaluation of longitudinal epidemiological studies. *Journal of Affective Disorders*, 135(1–3): 10–19.

Baglioni, C. et al. 2011b. Clinical implications of the causal relationship between insomnia and depression: How individually tailored treatment of sleeping difficulties could prevent the onset of depression. *EPMA Journal*, 2(3): 287–293.

Baglioni, C. et al. 2014. Sleep changes in the disorder of insomnia: a meta-analysis of polysomnographic studies. *Sleep Medicine Reviews*, 18(3): 195–213.

Bartel, K. and Gradisar, M. 2017. New directions in the link between technology use and sleep in young people. In S. Nevšímalová and O. Bruni (eds.), *Sleep Disorders in Children*, 69–80. Switzerland: Springer International Publishing. https://doi.org/10.1007/978-3-319-28640-2_4.

Bartel, K. A., Gradisar, M., and Williamson, P. 2015. Protective and risk factors for adolescent sleep: A meta-analytic review. *Sleep Medicine Reviews*, 21: 72–85.

Bastien, C. H., Vallieres, A., and Morin, C. M. 2004. Precipitating factors of insomnia. *Behavioral sleep medicine*, 2(1): 50–62.

Baum, K. T. et al. 2014. Sleep restriction worsens mood and emotion regulation in adolescents. *Journal of Child Psychology and Psychiatry*, 55(2): 180–190.

Beattie, L. et al. 2015. Social interactions, emotion and sleep: A systematic review and research agenda. *Sleep Medicine Reviews*, 24: 83–100.

Becker, S. P., Langberg, J. M., and Byars, K. C., 2015. *Advancing a Biopsychosocial and Contextual Model of Sleep in Adolescence: A review and Introduction to the Special Issue.* Springer.

Becker, S. P. et al. 2017. Intraindividual variability of sleep/wake patterns in relation to child and adolescent functioning: A systematic review. *Sleep Medicine Reviews*, 34: 94–121.

Bei, B. et al. 2017. Too long, too short, or too variable? Sleep intraindividual variability and its associations with perceived sleep quality and mood in adolescents during naturalistically unconstrained sleep. *Sleep*, 40(2): 1–8.

Blake, M. J., and Allen, N. B. 2020. Prevention of internalizing disorders and suicide via adolescent sleep interventions. *Current Opinion in Psychology*, 34: 37–42.

Blake, M., Trinder, J. A., and Allen, N. B. 2018. Mechanisms underlying the association between insomnia, anxiety, and depression in adolescence: Implications for behavioral sleep interventions. *Clinical Psychology Review*, 63: 25–40.

Bootzin, R. R. and Epstein, D. R. 2011. Understanding and treating insomnia. *Annual Review of Clinical Psychology*, 7: 435–458.

Borbély, A. A. 1982. A two process model of sleep regulation. *Human Neurobiology*, 1(3): 195–204.

Borbély, A. A. and Achermann, P. 1999. Sleep homeostasis and models of sleep regulation. *Journal of Biological Rhythms*, 14(6): 559–570.

Borbély, A. A. and Achermann, P. 2017. Sleep homeostasis and models of sleep regulation. In Meir H. Kryger, T. Roth, and William C. Dement (eds.), *Principles and Practice of Sleep Medicine* (Sixth Edition), pp. 377–387. Philadelphia: Elsevier.

Borbély, A. A. et al. 2016. The two-process model of sleep regulation: A reappraisal. *Journal of Sleep Research*, 25(2): 131–143.

Breslau, N. et al. 1996. Sleep disturbance and psychiatric disorders: A longitudinal epidemiological study of young adults. *Biological Psychiatry*, 39(6): 411–418.

Brown, W. J. et al. 2018. A review of sleep disturbance in children and adolescents with anxiety. *Journal of Sleep Research*, 27(3): e12635.

Bruni, O. et al. 2008. Relationships between headache and sleep in a non-clinical population of children and adolescents. *Sleep Medicine*, 9(5): 542–548.

Bucksch, J. et al. 2016. International trends in adolescent screen-time behaviors from 2002 to 2010. *Journal of Adolescent Health*, 58(4): 417–425.

Buxton, O. M. et al. 2015. Sleep in the modern family: Protective family routines for child and adolescent sleep. *Sleep Health*, 1(1): 15–27.

Buysse, D. J. et al. 2008. Prevalence, course, and comorbidity of insomnia and depression in young adults. *Sleep*, 31(4): 473–480.

Cain, N. and Gradisar, M. 2010. Electronic media use and sleep in school-aged children and adolescents: A review. *Sleep Medicine*, 11(8): 735–742.

Cajochen, C., Kräuchi, K., and Wirz-Justice, A. 2003. Role of melatonin in the regulation of human circadian rhythms and sleep. *Journal of Neuroendocrinology*, 15(4): 432–437.

Campbell, I. G. and Feinberg, I. 2009. Longitudinal trajectories of non-rapid eye movement delta and theta EEG as indicators of adolescent brain maturation. *Proceedings of the National Academy of Sciences*, 106(13): 5177–5180.

Carney, C. E. et al. 2007. A comparison of rates of residual insomnia symptoms following pharmacotherapy or cognitive-behavioral therapy for major depressive disorder. *The Journal of Clinical Psychiatry*, 68(2): 254–260.

Carney, C. E. et al. 2010. Distinguishing rumination from worry in clinical insomnia. *Behaviour Research and Therapy*, 48(6): 540–546.

Carskadon, M. A. 1990. Patterns of sleep and sleepiness in adolescents. *Pediatrician*, 17(1): 5–12.

Carskadon, M. A. 2011. Sleep in adolescents: The perfect storm. *Pediatric Clinics*, 58(3): 637–647.

Carskadon, M. A. and Acebo, C. 2002. Regulation of sleepiness in adolescents: Update, insights, and speculation. *Sleep*, 25(6): 606–614.

Carskadon, M. A., Acebo, C., and Jenni, O. G. 2004. Regulation of adolescent sleep: implications for behavior. *Annals of the New York Academy of Sciences*, 1021(1): 276–291.

Carskadon, M. A., Kim, H. and Dement, W. C. 1981. Sleep loss in young adolescents. *Sleep*, 4(3): 299–312.

Carskadon, M. A., Vieira, C., and Acebo, C. 1993. Association between puberty and delayed phase preference. *Sleep*, 16(3): 258–262.

Carskadon, M. A. et al. 1980. Pubertal changes in daytime sleepiness. *Sleep*, 2(4): 453–460.

Carskadon, M. A. et al. 1997. An approach to studying circadian rhythms of adolescent humans. *Journal of Biological Rhythms*, 12(3): 278–289.

Carskadon, M. A. et al. 1999. Intrinsic circadian period of adolescent humans measured in conditions of forced desynchrony. *Neuroscience Letters*, 260(2): 129–132.

Chang, P. P. et al. 1997. Insomnia in young men and subsequent depression: The Johns Hopkins Precursors Study. *American Journal of Epidemiology*, 146(2): 105–114.

Chase, R. M. and Pincus, D. B. 2011. Sleep-related problems in children and adolescents with anxiety disorders. *Behavioral Sleep Medicine*, 9(4): 224–236.

Chung, K. F., Kan, K. K. K., and Yeung, W. F. 2014. Insomnia in adolescents: Prevalence, help-seeking behaviors, and types of interventions. *Child and Adolescent Mental Health*, 19(1): 57–63.

Clinkinbeard, S. S. et al. 2011. Sleep and delinquency: Does the amount of sleep matter? *Journal of Youth and Adolescence*, 40(7): 916–930.

Colrain, I. M. and Baker, F. C. 2011. Changes in sleep as a function of adolescent development. *Neuropsychology Review*, 21(1): 5–21.

Comer, J. S., Pincus, D. B., and Hofmann, S. G. 2012. Generalized anxiety disorder and the proposed associated symptoms criterion change for DSM-5 in a treatment-seeking sample of anxious youth. *Depression and Anxiety*, 29(12): 994–1003.

Coulombe, J. A. et al. 2010. Sleep problems, tiredness, and psychological symptoms among healthy adolescents. *Journal of Pediatric Psychology*, 36(1): 25–35.

Cousins, J. C. et al. 2011. The bidirectional association between daytime affect and nighttime sleep in youth with anxiety and depression. *Journal of Pediatric Psychology*, 36(9): 969–979.

Cowie, J. et al. 2014. Addressing sleep in children with anxiety disorders. *Sleep Medicine Clinics*, 9(2): 137–148.

Crowley, S. J., Acebo, C., and Carskadon, M. A. 2007. Sleep, circadian rhythms, and delayed phase in adolescence. *Sleep Medicine*, 8(6): 602–612.

Crowley, S. J. et al. 2014. A longitudinal assessment of sleep timing, circadian phase, and phase angle of entrainment across human adolescence. *PLoS One*, 9(11): e112199.

Crowley, S. J. et al. 2015. Increased sensitivity of the circadian system to light in early/mid-puberty. *The Journal of Clinical Endocrinology & Metabolism*, 100(11): 4067–4073.

Crowley, S. J. et al. 2018. An update on adolescent sleep: New evidence informing the perfect storm model. *Journal of Adolescence*, 67: 55–65.

Dagys, N. et al. 2012. Double trouble? The effects of sleep deprivation and chronotype on adolescent affect. *Journal of Child Psychology and Psychiatry*, 53(6): 660–667.

Dahl, R. E. 1996. The regulation of sleep and arousal: Development and psychopathology. *Development and Psychopathology*, 8(1): 3–27.

Dahl, R. E. 1999. The consequences of insufficient sleep for adolescents. *Phi Delta Kappan*, 80(5): 354–359.

Dahl, R. E. and Harvey, A. G. 2007. Sleep in children and adolescents with behavioral and emotional disorders. *Sleep Medicine Clinics*, 2(3): 501–511.

Dahl, R. E. and Lewin, D. S. 2002. Pathways to adolescent health sleep regulation and behavior. *Journal of Adolescent Health*, 31(6): 175–184.

Dahl, R. E. et al. 1990. EEG sleep in adolescents with major depression: The role of suicidality and inpatient status. *Journal of Affective Disorders*, 19(1): 63–75.

Danner, F. and Phillips, B. 2008. Adolescent sleep, school start times, and teen motor vehicle crashes. *Journal of Clinical Sleep Medicine*, 4(06): 533–535.

de Zambotti, M. et al. 2018. Insomnia disorder in adolescence: Diagnosis, impact, and treatment. *Sleep Medicine Reviews*, 39: 12–24.

Dewald, J. F. et al. 2010. The influence of sleep quality, sleep duration and sleepiness on school performance in children and adolescents: A meta-analytic review. *Sleep Medicine Reviews*, 14(3): 179–189.

Dijk, D.-J. et al. 1990. Electroencephalogram power density and slow wave sleep as a function of prior waking and circadian phase. *Sleep*, 13(5): 430–440.

Dinges, D. F. et al. 1997. Cumulative sleepiness, mood disturbance, and psychomotor vigilance performance decrements during a week of sleep restricted to 4–5 hours per night. *Sleep*, 20(4): 267–277.

Dorofaeff, T. F. and Denny, S. 2006. Sleep and adolescence. Do New Zealand teenagers get enough? *Journal of Paediatrics and Child Health*, 42(9): 515–520.

Ehring, T. et al. 2010. Emotion regulation and vulnerability to depression: Spontaneous versus instructed use of emotion suppression and reappraisal. *Emotion*, 10(4): 563–572.

Erk, S. et al. 2010. Acute and sustained effects of cognitive emotion regulation in major depression. *Journal of Neuroscience*, 30(47): 15726–15734.

Feinberg, I. and Campbell, I. G. 2010. Sleep EEG changes during adolescence: An index of a fundamental brain reorganization. *Brain and Cognition*, 72(1): 56–65.

Feinberg, I. et al. 2006. The adolescent decline of NREM delta, an indicator of brain maturation, is linked to age and sex but not to pubertal stage. *American Journal of Physiology-Regulatory, Integrative and Comparative Physiology*, 291(6): R1724–R1729.

Forbes, E. E. et al. 2006. Peri-sleep-onset cortisol levels in children and adolescents with affective disorders. *Biological Psychiatry*, 59(1): 24–30.

Forbes, E. E. et al. 2008. Objective sleep in pediatric anxiety disorders and major depressive disorder. *Journal of the American Academy of Child and Adolescent Psychiatry*, 47(2): 148–155.

Ford, D. E. and Kamerow, D. B. 1989. Epidemiologic study of sleep disturbances and psychiatric disorders: An opportunity for prevention? *JAMA*, 262(11): 1479–1484.

Franic, T. et al. 2014. Suicidal ideations and sleep-related problems in early adolescence. *Early Intervention in Psychiatry*, 8(2): 155–162.

Franzen, P. L. and Buysse, D. J. 2008. Sleep disturbances and depression: Risk relationships for subsequent depression and therapeutic implications. *Dialogues in Clinical Neuroscience*, 10(4): 473–481.

Fuligni, A. J. and Hardway, C. 2006. Daily variation in adolescents' sleep, activities, and psychological well-being. *Journal of Research on Adolescence*, 16(3): 353–378.

Fuligni, A. J., Bai, S., Krull, J. L., and Gonzales, N. A. (2019). Individual differences in optimum sleep for daily mood during adolescence. *Journal of Clinical Child & Adolescent Psychology*, 48(3), 469–479.

Fuller, K. H. et al. 1997. Generalized anxiety and sleep architecture: A polysomnographic investigation. *Sleep*, 20(5): 370–376.

Gamble, A. L. et al. 2014. Adolescent sleep patterns and night-time technology use: Results of the Australian Broadcasting Corporation's Big Sleep Survey. *PLoS One*, 9(11): e111700.

Garaulet, M. et al. 2011. Short sleep duration is associated with increased obesity markers in European adolescents: Effect of physical activity and dietary habits. The HELENA study. *International Journal of Obesity*, 35(10): 1308.

Gillen-O'Neel, C., Huynh, V. W., and Fuligni, A. J. 2013. To study or to sleep? The academic costs of extra studying at the expense of sleep. *Child Development*, 84(1): 133–142.

Goetz, R. R. et al. 1987. Electroencephalographic sleep of adolescents with major depression and normal controls. *Archives of General Psychiatry*, 44(1): 61–68.

Golley, R. K. et al. 2013. Sleep duration or bedtime? Exploring the association between sleep timing behaviour, diet and BMI in children and adolescents. *International Journal of Obesity*, 37(4): 546–551.

Gradisar, M. and Crowley, S. J. 2013. Delayed sleep phase disorder in youth. *Current Opinion in Psychiatry*, 26(6): 580–585.

Gradisar, M., Gardner, G., and Dohnt, H. 2011b. Recent worldwide sleep patterns and problems during adolescence: A review and meta-analysis of age, region, and sleep. *Sleep Medicine*, 12(2): 110–118.

Gradisar, M. et al. 2011a. A randomized controlled trial of cognitive-behavior therapy plus bright light therapy for adolescent delayed sleep phase disorder. *Sleep*, 34(12): 1671–1680.

Gradisar, M. et al. 2013. The sleep and technology use of Americans: Findings from the National Sleep Foundation's 2011 Sleep in America poll. *Journal of Clinical Sleep Medicine*, 9(12): 1291–1299.

Gregory, A. M. and O'Connor, T. 2002. Sleep problems in childhood: A longitudinal study of developmental change and association with behavior. *Journal of the American Academy of Child & Adolescent Psychiatry*, 41(8): 964–971.

Gregory, A. M. and Sadeh, A. 2016. Annual Research Review: Sleep problems in childhood psychiatric disorders—A review of the latest science. *Journal of Child Psychology and Psychiatry*, 57(3): 296–317.

Hale, L. and Guan, S. 2015. Screen time and sleep among school-aged children and adolescents: A systematic literature review. *Sleep Medicine Reviews*, 21: 50–58.

Hallion, L. S. and Ruscio, A. M. 2011. A meta-analysis of the effect of cognitive bias modification on anxiety and depression. *Psychological Bulletin*, 137(6): 940.

Hansen, M. et al. 2005. The impact of school daily schedule on adolescent sleep. *Pediatrics*, 115(6): 1555–1561.

Harbard, E. et al. 2016. What's keeping teenagers up? Prebedtime behaviors and actigraphy-assessed sleep over school and vacation. *Journal of Adolescent Health*, 58(4): 426–432.

Harris, R. A., Qualter, P., and Robinson, S. J. 2013. Loneliness trajectories from middle childhood to pre-adolescence: Impact on perceived health and sleep disturbance. *Journal of Adolescence*, 36(6): 1295–1304.

Harris, K. et al. 2015. Sleep-related attentional bias in insomnia: A state-of-the-science review. *Clinical Psychology Review*, 42: 16–27.

Hart, C. N., Palermo, T. M., and Rosen, C. L. 2005. Health-related quality of life among children presenting to a pediatric sleep disorders clinic. *Behavioral Sleep Medicine*, 3(1): 4–17.

Harvey, A. G. 2016. A transdiagnostic intervention for youth sleep and circadian problems. *Cognitive and Behavioral Practice*, 23(3): 341–355.

Harvey, A. G., Mullin, B. C., and Hinshaw, S. P. 2006. Sleep and circadian rhythms in children and adolescents with bipolar disorder. *Development and Psychopathology*, 18(4): 1147–1168.

Harvey, A. G. and Tang, N. K. 2012. (Mis) perception of sleep in insomnia: A puzzle and a resolution. *Psychological Bulletin*, 138(1): 77–101.

Harvey, A. G. et al. 2011. Sleep disturbance as transdiagnostic: Consideration of neurobiological mechanisms. *Clinical Psychology Review*, 31(2): 225–235.

Hawkins, S. S. and Takeuchi, D. T. 2016. Social determinants of inadequate sleep in US children and adolescents. *Public Health*, 138: 119–126.

Hayley, A. C. et al. 2015. Symptoms of depression and difficulty initiating sleep from early adolescence to early adulthood: A longitudinal study. *Sleep*, 38(10): 1599–1606.

Hill, T. D. et al. 2016. Perceived neighborhood safety and sleep quality: A global analysis of six countries. *Sleep Medicine*, 18: 56–60.

Hillman, D. R. and Lack, L. C. 2013. Public health implications of sleep loss: the community burden. *Medical Journal of Australia*, 199(8): S7–S10.

Hirshkowitz, M. et al. 2015. National Sleep Foundation's updated sleep duration recommendations. *Sleep Health: Journal of the National Sleep Foundation*, 1(4): 233–243.

Hysing, M., Stormark, K. M., and O'Connor, R. C. 2015c. Sleep problems and self-harm in adolescence. *The British Journal of Psychiatry*, 207(4): 306–312.

Hysing, M. et al. 2013. Sleep patterns and insomnia among adolescents: A population-based study. *Journal of Sleep Research*, 22(5): 549–556.

Hysing, M. et al. 2015a. Sleep and school attendance in adolescence: Results from a large population-based study. *Scandinavian Journal of Public Health*, 43(1): 2–9.

Hysing, M. et al. 2015b. Sleep and use of electronic devices in adolescence: Results from a large population-based study. *BMJ Open*, 5(1): e006748.

Hysing, M. et al. 2016. Association between sleep problems and symptoms of attention deficit hyperactivity disorder in adolescence: Results from a large population-based study. *Behavioral Sleep Medicine*, 14(5): 550–564.

Javaheri, S. et al. 2008. Sleep quality and elevated blood pressure in adolescents. *Circulation*, 118(10): 1034–1040.

Jenni, O. G., Achermann, P., and Carskadon, M. A. 2005. Homeostatic sleep regulation in adolescents. *Sleep*, 28(11): 1446–1454.

Johnson, E. O., Chilcoat, H. D., and Breslau, N. 2000. Trouble sleeping and anxiety-depression in childhood. *Psychiatry Research*, 94(2): 93–102.

Johnson, E. O., Roth, T., and Breslau, N. 2006a. The association of insomnia with anxiety disorders and depression: Exploration of the direction of risk. *Journal of Psychiatric Research*, 40(8): 700–708.

Johnson, E. O. et al. 2006b. Epidemiology of DSM-IV insomnia in adolescence: Lifetime prevalence, chronicity, and an emergent gender difference. *Pediatrics*, 117(2): e247–e256.

Kahn, M., Sheppes, G., and Sadeh, A. 2013. Sleep and emotions: Bidirectional links and underlying mechanisms. *International Journal of Psychophysiology*, 89(2): 218–228.

Kaneita, Y. et al. 2009. Associations between sleep disturbance and mental health status: A longitudinal study of Japanese junior high school students. *Sleep Medicine*, 10(7): 780–786.

Khan, A. U. and Todd, S. 1990. Polysomnographic findings in adolescents with major depression. *Psychiatry Research*, 33(3): 313–320.

Kim, S. J. et al. 2010. Age as a moderator of the association between depressive symptoms and morningness-eveningness. *Journal of Psychosomatic Research*, 68(2); 159–164.

Kim, S. J. et al. 2011. Relationship between weekend catch-up sleep and poor performance on attention tasks in Korean adolescents. *Archives of Pediatrics and Adolescent Medicine*, 165(9): 806–812.

Klink, M. E. et al. 1992. Risk factors associated with complaints of insomnia in a general adult population: Influence of previous complaints of insomnia. *Archives of Internal Medicine*, 152(8): 1634–1637.

Kredlow, M. A. et al. 2015. The effects of physical activity on sleep: A meta-analytic review. *Journal of Behavioral Medicine*, 38(3): 427–449.

Krueger, J. M. et al. 2016. Sleep function: Toward elucidating an enigma. *Sleep Medicine Reviews*, 28: 46–54.

Kubiszewski, V. et al. 2014. Bullying, sleep/wake patterns and subjective sleep disorders: Findings from a cross-sectional survey. *Chronobiology International*, 31(4): 542–553.

Kuo, S. I.-C. et al. 2015. Mexican American adolescents' sleep patterns: Contextual correlates and implications for health and adjustment in young adulthood. *Journal of Youth and Adolescence*, 44(2): 346–361.

Kutcher, S. et al. 1992. REM latency in endogenously depressed adolescents. *The British Journal of Psychiatry*, 161(3): 399–402.

Lee, Y. J. et al. 2012. Insufficient sleep and suicidality in adolescents. *Sleep*, 35(4): 455–460.

Li, Y., Starr, L. and Wray-Lake, L. 2018. Insomnia mediates the longitudinal relationship between anxiety and depressive symptoms in a nationally representative sample of adolescents. *Depression and Anxiety*, 35(6): 583–591.

Liu. 2004. Sleep and adolescent suicidal behavior. *Sleep*, 27(7): 1351–1358.

Lo, J. C. et al. 2016. Cognitive performance, sleepiness, and mood in partially sleep deprived adolescents: The need for sleep study. *Sleep*, 39(3): 687–698.

Lo, J. C. et al. 2017. Neurobehavioral impact of successive cycles of sleep restriction with and without naps in adolescents. *Sleep*, 40(2): 1–9.

Lopez, J., Hoffmann, R., and Armitage, R. 2010. Reduced sleep spindle activity in early-onset and elevated risk for depression. *Journal of the American Academy of Child & Adolescent Psychiatry*, 49(9): 934–943.

Lovato, N. and Gradisar, M. 2014. A meta-analysis and model of the relationship between sleep and depression in adolescents: Recommendations for future research and clinical practice. *Sleep Medicine Reviews*, 18(6): 521–529.

Lovato, N. et al. 2013. Delayed sleep phase disorder in an Australian school-based sample of adolescents. *Journal of Clinical Sleep Medicine*, 9(09): 939–944.

Lovato, N. et al. 2017. An investigation of the longitudinal relationship between sleep and depressed mood in developing teens. *Nature and Science of Sleep*, 9: 3–10.

Lundh, L.-G., Bjärehed, J., and Wångby-Lundh, M. 2012. Poor sleep as a risk factor for nonsuicidal self-injury in adolescent girls. *Journal of Psychopathology and Behavioral Assessment*, 35(1): 85–92.

Mahon, N. E. 1994. Loneliness and sleep during adolescence. *Perceptual and Motor Skills*, 78(1): 227–231.

Martin, J. L. and Hakim, A. D. 2011. Wrist actigraphy. *Chest*, 139(6): 1514–1527.

Martinez-Gomez, D. et al. 2011. Sleep duration and emerging cardiometabolic risk markers in adolescents. The AFINOS study. *Sleep Medicine*, 12(10): 997–1002.

Maume, D. J. 2013. Social ties and adolescent sleep disruption. *Journal of Health and Social Behavior*, 54(4): 498–515.

McGlinchey, E. L. et al. 2011. The effect of sleep deprivation on vocal expression of emotion in adolescents and adults. *Sleep*, 34(9): 1233–1241.

McKnight-Eily, L. R. et al. 2011. Relationships between hours of sleep and health-risk behaviors in US adolescent students. *Preventive Medicine*, 53(4–5): 271–273.

McMakin, D. L. and Alfano, C. A. 2015. Sleep and anxiety in late childhood and early adolescence. *Current Opinion in Psychiatry*, 28(6): 483–489.

McMakin, D. L. et al. 2016. The impact of experimental sleep restriction on affective functioning in social and nonsocial contexts among adolescents. *Journal of Child Psychology and Psychiatry*, 57(9): 1027–1037.

Meijer, A. M. et al. 2010. Longitudinal relations between sleep quality, time in bed and adolescent problem behaviour. *Journal of Child Psychology and Psychiatry*, 51(11): 1278–1286.

Meldrum, R. C., Barnes, J., and Hay, C. 2015. Sleep deprivation, low self-control, and delinquency: A test of the strength model of self-control. *Journal of Youth and Adolescence*, 44(2): 465–477.

Mikoteit, T. et al. 2012. Visually detected NREM Stage 2 sleep spindles in kindergarten children are associated with stress challenge and coping strategies. *The World Journal of Biological Psychiatry*, 13(4): 259–268.

Mikoteit, T. et al. 2013. Visually detected NREM Stage 2 sleep spindles in kindergarten children are associated with current and future emotional and behavioural characteristics. *Journal of Sleep Research*, 22(2): 129–136.

Milkins, B. et al. 2016. The potential benefits of targeted attentional bias modification on cognitive arousal and sleep quality in worry-related sleep disturbance. *Clinical Psychological Science*, 4(6): 1015–1027.

Morrison, D. N., McGee, R., and Stanton, W. R. 1992. Sleep problems in adolescence. *Journal of the American Academy of Child & Adolescent Psychiatry*, 31(1): 94–99.

Mullin, B. C. et al. 2017. A preliminary multimethod comparison of sleep among adolescents with and without generalized anxiety disorder. *Journal of Clinical Child and Adolescent Psychology*, 46(2): 198–210.

Narang, I. et al. 2012. Sleep disturbance and cardiovascular risk in adolescents. *CMAJ: Canadian Medical Association Journal*, 184(17): E913–E920.

National Sleep Foundation. 2006. *Sleep in America Poll* [online]. Available from: http://slee pfoundation.org/sites/default/files/2006_summary_of_findings.pdf.

Nesbitt, A. D. 2018. Delayed sleep-wake phase disorder. *Journal of Thoracic Disease*, 10(Suppl 1): S103.

Nielsen, T. A. et al. 2000. Development of disturbing dreams during adolescence and their relation to anxiety symptoms. *Sleep*, 23(6): 1–10.

Nierenberg, A. A. et al. 1999. Residual symptoms in depressed patients who respond acutely to fluoxetine. *The Journal of Clinical Psychiatry*, 60(4): 221–225.

Nota, J. A. and Coles, M. E. 2018. Shorter sleep duration and longer sleep onset latency are related to difficulty disengaging attention from negative emotional images in individuals with elevated transdiagnostic repetitive negative thinking. *Journal of Behavior Therapy and Experimental Psychiatry*, 58: 114–122.

O'Brien, E. M. and Mindell, J. A. 2005. Sleep and risk-taking behavior in adolescents. *Behavioral Sleep Medicine*, 3(3): 113–133.

Ohayon, M. M., Caulet, M., and Lemoine, P. 1998. Comorbidity of mental and insomnia disorders in the general population. *Comprehensive Psychiatry*, 39(4): 185–197.

Ohayon, M. M. and Roth, T. 2003. Place of chronic insomnia in the course of depressive and anxiety disorders. *Journal of Psychiatric Research*, 37(1): 9–15.

Olds, T. et al. 2010. Normative data on the sleep habits of Australian children and adolescents. *Sleep*, 33(10): 1381–1388.

Orchard, F. et al. 2020. Self-reported sleep patterns and quality amongst adolescents: Cross-sectional and prospective associations with anxiety and depression. *Journal of Child Psychology and Psychiatry*, 61(10): 1126–1137.

Owens, J. et al. Adolescent Sleep Working Group and Committee on Adolescence. 2014. Insufficient sleep in adolescents and young adults: An update on causes and consequences. *Pediatrics*, 134(3): e921–e932.

Øyane, N. M. and Bjorvatn, B. 2005. Sleep disturbances in adolescents and young adults with autism and Asperger syndrome. *Autism*, 9(1): 83–94.

Paavonen, E. J. et al. 2016. Brief behavioral sleep intervention for adolescents: An effectiveness study. *Behavioral Sleep Medicine*, 14(4): 351–366.

Pabst, S. R. et al. 2009. Depression and anxiety in adolescent females: The impact of sleep preference and body mass index. *Journal of Adolescent Health*, 44(6): 554–560.

Palmer, C. A. and Alfano, C. A. 2017. Sleep and emotion regulation: An organizing, integrative review. *Sleep Medicine Reviews*, 31: 6–16.

Palmer, C. A. et al. 2018. Associations among adolescent sleep problems, emotion regulation, and affective disorders: Findings from a nationally representative sample. *Journal of Psychiatric Research*, 96: 1–8.

Park, J. H., Yoo, J. H., and Kim, S. H. 2013. Associations between non-restorative sleep, short sleep duration and suicidality: Findings from a representative sample of Korean adolescents. *Psychiatry and Clinical Neuroscience*, 67(1): 28–34.

Park, H. et al. 2016. Sleep and inflammation during adolescence. *Psychosomatic Medicine*, 78(6): 677–685.

Paruthi, S. et al. 2016. Recommended amount of sleep for pediatric populations: A consensus statement of the American Academy of Sleep Medicine. *Journal of Clinical Sleep Medicine*, 12(6): 785–786.

Pasch, K. E. et al. 2014. Adolescent sleep, risk behaviors, and depressive symptoms: Are they linked? *American Journal of Health Behaviour*, 34(2): 237–248.

Patton, G. C. et al. 2000. Adolescent depressive disorder: A population-based study of ICD-10 symptoms. *Australian & New Zealand Journal of Psychiatry*, 34(5): 741–747.

Perlis, M. L. et al. 1997. Self-reported sleep disturbance as a prodromal symptom in recurrent depression. *Journal of Affective Disorders*, 42(2–3): 209–212.

Perlis, M. L. et al. 2016. Suicide and sleep: Is it a bad thing to be awake when reason sleeps? *Sleep Medicine Reviews*, 29: 101–107.

Pieters, S. et al. 2015. Prospective relationships between sleep problems and substance use, internalizing and externalizing problems. *Journal of Youth and Adolescence*, 44(2): 379–388.

Pigeon, W. R., Bishop, T. M., and Krueger, K. M. 2017. Insomnia as a precipitating factor in new onset mental illness: A systematic review of recent findings. *Current Psychiatry Reports*, 19(8): 44.

Pizza, F. et al. 2010. Sleep quality and motor vehicle crashes in adolescents. *Journal of Clinical Sleep Medicine*, 6(1): 41–45.

Price, R. B. et al. 2016. From anxious youth to depressed adolescents: Prospective prediction of 2-year depression symptoms via attentional bias measures. *Journal of Abnormal Psychology*, 125(2): 267–278.

Randazzo, A. C. et al. 1998. Cognitive function following acute sleep restriction in children ages 10–14. *Sleep*, 21(8): 861–868.

Raniti, M. B. et al. 2017. Sleep duration and sleep quality: Associations with depressive symptoms across adolescence. *Behavioral Sleep Medicine*, 15(3): 198–215.

Raniti, M. B. et al. 2018. Factor structure and psychometric properties of the Pittsburgh Sleep Quality Index in community-based adolescents. *Sleep*, 41(6): zsy066.

Rao, U., Hammen, C. L., and Poland, R. E. 2009. Risk markers for depression in adolescents: Sleep and HPA measures. *Neuropsychopharmacology*, 34(8): 1936.

Rideout, V. J., Foehr, U. G., and Roberts, D. F. 2010. Generation M 2: Media in the Lives of 8- to 18-Year-Olds. *Henry J. Kaiser Family Foundation*. Retrieved from https://files.eric.ed.gov/fulltext/ED527859.pdf.

Riemann, D., Berger, M., and Voderholzer, U. 2001. Sleep and depression—results from psychobiological studies: An overview. *Biological Psychology*, 57(1–3): 67–103.

Riemann, D. et al. 2010. The hyperarousal model of insomnia: A review of the concept and its evidence. *Sleep Medicine Reviews*, 14(1): 19–31.

Ritter, P. S. et al. 2015. Disturbed sleep as risk factor for the subsequent onset of bipolar disorder–Data from a 10-year prospective-longitudinal study among adolescents and young adults. *Journal of Psychiatric Research*, 68: 76–82.

Roane, B. M. and Taylor, D. J. 2008. Adolescent insomnia as a risk factor for early adult depression and substance abuse. *Sleep*, 31(10): 1351–1356.

Robert, J. J. et al. 2006. Sex and age differences in sleep macroarchitecture in childhood and adolescent depression. *Sleep*, 29(3): 351–358.

Roberts, R. E. and Duong, H. T. 2017. Is there an association between short sleep duration and adolescent anxiety disorders? *Sleep Medicine*, 30: 82–87.

Roberts, R. E., Roberts, C. R., and Chen, I. G. 2001. Functioning of adolescents with symptoms of disturbed sleep. *Journal of Youth and Adolescence*, 30(1): 1–18.

Roberts, R. E., Roberts, C. R., and Chen, I. G. 2002. Impact of insomnia on future functioning of adolescents. *Journal of Psychosomatic Research*, 53(1): 561–569.

Roberts, R. E., Roberts, C. R., and Duong, H. T. 2008. Chronic insomnia and its negative consequences for health and functioning of adolescents: A 12-month prospective study. *Journal of Adolescent Health*, 42(3): 294–302.

Roenneberg, T. et al. 2004. A marker for the end of adolescence. *Current Biology*, 14(24): R1038–R1039.

Sadeh, A. et al. 2009. Sleep and the transition to adolescence: A longitudinal study. *Sleep*, 32(12): 1602–1609.

Samson, R., Blunden, S., and Banks, S. 2013. The characteristics of sleep and sleep loss in adolescence: A review. *International Review of Social Science and Humanities*, 4(2): 90–107.

Santangeli, O. et al. 2017. Sleep and slow-wave activity in depressed adolescent boys: A preliminary study. *Sleep Medicine*, 38: 24–30.

Sarchiapone, M. et al. 2014. Hours of sleep in adolescents and its association with anxiety, emotional concerns, and suicidal ideation. *Sleep Medicine*, 15(2): 248–254.

Schmidt, R. E. and Van der Linden, M. 2015. The relations between sleep, personality, behavioral problems, and school performance in adolescents. *Sleep Medicine Clinics*, 10(2): 117–123.

Seicean, A. et al. 2007. Association between short sleeping hours and overweight in adolescents: Results from a US Suburban High School survey. *Sleep Breath*, 11(4): 285–293.

Shanahan, L. et al. 2014. Sleep problems predict and are predicted by generalized anxiety/depression and oppositional defiant disorder. *Journal of the American Academy of Child and Adolescent Psychiatry*, 53(5): 550–558.

Shochat, T., Cohen-Zion, M., and Tzischinsky, O. 2014. Functional consequences of inadequate sleep in adolescents: A systematic review. *Sleep Medicine Reviews*, 18(1): 75–87.

Shochat, T., Flint-Bretler, O., and Tzischinsky, O. 2010. Sleep patterns, electronic media exposure and daytime sleep-related behaviours among Israeli adolescents. *Acta Paediatrica*, 99(9): 1396–1400.

Short, M. A. et al. 2013a. Identifying adolescent sleep problems. *PLoS One*, 8(9): e75301.

Short, M. A. et al. 2013b. Estimating adolescent sleep patterns: Parent reports versus adolescent self-report surveys, sleep diaries, and actigraphy. *Nature and Science of Sleep*, 5: 23–26.

Short, M. A. et al. 2013c. A cross-cultural comparison of sleep duration between US and Australian adolescents: The effect of school start time, parent-set bedtimes, and extracurricular load. *Health Education and Behavior*, 40(3): 323–330.

Short, M. A. et al. 2013d. The sleep patterns and well-being of Australian adolescents. *Journal of Adolescence*, 36(1): 103–110.

Short, M. A. et al. 2011. Time for bed: Parent-set bedtimes associated with improved sleep and daytime functioning in adolescents. *Sleep*, 34(6): 797–800.

Short, M. A. and Louca, M. 2015. Sleep deprivation leads to mood deficits in healthy adolescents. *Sleep Medicine*, 16(8): 987–993.

Short, M. A. et al. 2018. Estimating adolescent sleep need using dose-response modeling. *Sleep*, 41(4): zsy011.

Sivertsen, B. et al. 2015. Sleep and use of alcohol and drug in adolescence. A large population-based study of Norwegian adolescents aged 16 to 19 years. *Drug and Alcohol Dependence*, 149: 180–186.

Sloan, E. et al. 2017. Emotion regulation as a transdiagnostic treatment construct across anxiety, depression, substance, eating and borderline personality disorders: A systematic review. *Clinical Psychology Review*, 57: 141–163.

Sunderajan, P. et al. 2010. Insomnia in patients with depression: A STAR* D report. *CNS spectrums*, 15(6): 394–406.

Taheri, S. and Mignot, E. 2002. The genetics of sleep disorders. *The Lancet Neurology*, 1(4) 242–250.

Talbot, L. S. et al. 2010. Sleep deprivation in adolescents and adults: Changes in affect *Emotion*, 10(6): 831.

Tang, N. K. and Harvey, A. G. 2004. Effects of cognitive arousal and physiological arousal on sleep perception. *Sleep*, 27(1): 69–78.

Tavernier, R. et al. 2016. Daily affective experiences predict objective sleep outcomes among adolescents. *Journal of sleep research*, 25(1): 62–69.

Telzer, E. H. et al. 2013. The effects of poor quality sleep on brain function and risk taking in adolescence. *Neuroimage*, 71: 275–283.

Touitou, Y. 2013. Adolescent sleep misalignment: A chronic jet lag and a matter of public health. *Journal of Physiology - Paris*, 107(4): 323–326.

Tsuno, N., Besset, A., and Ritchie, K. 2005. Sleep and depression. *The Journal of clinical psychiatry*, 66: 1254–1269.

van Zundert, R. M. et al. 2015. Reciprocal associations between adolescents' night-time sleep and daytime affect and the role of gender and depressive symptoms. *Journal of Youth and Adolescence*, 44(2): 556–569.

Vriend, J. L. et al. 2013. Manipulating sleep duration alters emotional functioning and cognitive performance in children. *Journal of Pediatric Psychology*, 38(10): 1058–1069.

Walker, M. P. and Harvey, A. G. 2010. Obligate symbiosis: Sleep and affect. *Sleep Medicine Reviews*, 14(4): 215–217.

Wang, Y. et al. 2017. Sleep patterns and their association with depression and behavior problems among Chinese adolescents in different grades. *PsyCh journal*, 6(4): 253–262.

Warner, S., Murray, G., and Meyer, D. 2008. Holiday and school-term sleep patterns of Australian adolescents. *Journal of Adolescence*, 31(5): 595–608.

Wilhelm, I. et al. 2017. Widespread reduction in sleep spindle activity in socially anxious children and adolescents. *Journal of Psychiatric Research*, 88: 47–55.

Willis, T. A. and Gregory, A. M. 2015. Anxiety disorders and sleep in children and adolescents. *Sleep Medicine Clinics*, 10(2): 125–131.

Wittmann, M. et al. 2006. Social jetlag: Misalignment of biological and social time. *Chronobiology International*, 23(1–2): 497–509.

Wolfson, A. R. and Carskadon, M. A. 1998. Sleep schedules and daytime functioning in adolescents. *Child Development*, 69(4): 875–887.

Wolfson, A. R. et al. 2007. Middle school start times: The importance of a good night's sleep for young adolescents. *Behavioral Sleep Medicine*, 5(3): 194–209.

Wong, M. M. and Brower, K. J. 2012. The prospective relationship between sleep problems and suicidal behavior in the National Longitudinal Study of Adolescent Health. *Journal of Psychiatric Research*, 46(7): 953–959.

Xu, Z. et al. 2012. Sleep quality of Chinese adolescents: Distribution and its associated factors. *Journal of Paediatrics and Child Health*, 48(2): 138–145.

Yang, C.-K. et al. 2005. Age-related changes in sleep/wake patterns among Korean teenagers. *Pediatrics*, 115(Supplement 1): 250–256.

Yoo, S.-S. et al. 2007. The human emotional brain without sleep—a prefrontal amygdala disconnect. *Current Biology*, 17(20): R877–R878.

Zhang, J. et al. 2016. Emergence of sex differences in insomnia symptoms in adolescents: A large-scale school-based study. *Sleep*, 39(8): 1563–1570.

Zhang, J. et al. 2017. Sleep patterns and mental health correlates in US adolescents. *The Journal of Pediatrics*, 182: 137–143.

Zhou, H. Q. et al. 2012. Functional gastrointestinal disorders among adolescents with poor sleep: A school-based study in Shanghai, China. *Sleep and Breathing*, 16(4): 1211–1218.

Zohar, D. et al. 2005. The effects of sleep loss on medical residents' emotional reactions to work events: a cognitive-energy model. *Sleep*, 28(1): 47–54.

....

CHAPTER 37

BORDERLINE PERSONALITY DISORDER IN DEVELOPMENT

....

MAURIZIO SICORELLO AND CHRISTIAN SCHMAHL

BORDERLINE personality disorder (BPD) is an enduring, trait-like syndrome characterized by impulsive behavior, as well as instability in affect, and interpersonal problems (American Psychiatric Association, 2013). Even though its estimated prevalence of around 3 percent (Trull et al., 2010) is relatively low compared to disorders like depression and general anxiety disorder, it is accompanied by a high psychological burden and mortality, making it an important target for health interventions (Zimmermann and Iwanski, 2014).

Previously, the BPD diagnosis was primarily given to adults, mainly due to the stigma associated with BPD, the unfinished consolidation of personality before entering adulthood, and the difficulties in distinguishing between normative and pathological developmental trajectories (Sharp and Tackett, 2014). Over the last 20 years, however, there has been a shift towards exploring BPD at earlier ages. This shift is based on accumulating empirical findings that BPD has identifiable precursors in childhood, can be reliably diagnosed even in early adolescence, and is generally less stable than previously thought (Sharp and Tackett, 2014). For example, two longitudinal studies reported remission rates above 80 percent over a 10-year follow-up period (Zanarini, Frankenburg, Hennen, et al., 2006; Gunderson et al., 2011). These findings underline Lenzenweger and Cicchetti's (2005) position that BPD research should take place in a developmental context, investigating BPD features over the full lifespan. Only from a developmental perspective can we identify both etiological and maintaining factors to guide evidence-based clinical interventions. Following these arguments, more longitudinal studies have been published in recent years (Stepp et al., 2015).

Complementary to the increasing developmental emphasis, behavioral genetic studies have shown that BPD is highly influenced by genetic factors, with heritability estimates ranging between 42 and 70 percent (De Clercq et al., 2014). Establishing the link between gene–environment interactions and developmental trajectories towards psychopathology is a key challenge to improving early detection and prevention, as well as our biological understanding of the disorder (Beauchaine et al., 2008). However, not unlike most other mental disorders, the search for predictive genes has been relatively unsuccessful (Calati et al., 2013). Here, endophenotypes have been proposed as a viable solution (Gottesman and Gould, 2003). Endophenotypes are objectively measurable, even though not directly observable, dimensional characteristics, which bridge the gap between genetic predisposition, environmental influences, and the eventual onset of a disorder (Caspi and Moffitt, 2006).

Research on neurobiological endophenotypes of BPD has increased rapidly, paralleling the trend of expanding BPD research to include more prospective studies. Nevertheless, even though most developmental theories strongly incorporate biological perspectives (Linehan, 1993; Crowell et al., 2009; Bandelow et al., 2010; Hughes et al., 2012), there are virtually no longitudinal studies which target the prospective neurobiological development of BPD. This might be due to its relatively low prevalence, which necessitates the screening of large populations, making cost- and work-intensive longitudinal neuroimaging studies almost impossible. For this reason, it is crucial that early biomarkers of BPD are carefully identified. A first step towards this goal is to integrate cross-sectionally attained neurobiological findings of BPD with developmental theories, etiological research, and insights into normative neurodevelopment, which is the purpose of this chapter.

DEVELOPMENTAL PERSPECTIVES ON BPD

The most influential developmental theory of BPD is the biosocial developmental model of borderline personality (Linehan, 1993). In its most recent formulation (Crowell et al., 2009), it posits that extreme genetically predisposed trait impulsivity is expressed early in life and increases the likelihood to develop emotional and behavioral disturbances later on. Indeed, according to some estimates, 80 percent of trait impulsivity is genetically inherited (Beauchaine and Neuhaus, 2008) and BPD symptoms in early adolescence are predicted by impulsivity and externalizing symptoms in childhood (Belsky et al., 2012; De Clercq et al., 2014). Impulsivity of course represents a risk factor for a variety of externalizing and internalizing disorders and may account for high comorbidities with BPD symptoms. Therefore, impulsivity in BPD is the first phenomenon we will review from a biological perspective in this chapter.

According to the biosocial model, high trait impulsivity, together with predisposed negative affectivity, and emotional sensitivity, transact with an invalidating family environment to create persistent emotional lability, which is the core feature of BPD (Linehan, 1993). An invalidating environment consists in downplaying the child's emotions as excessive and inappropriate. Consequently, children do not learn to adequately identify and thus regulate their emotions. At the same time, extreme emotional outbursts are intermittently reinforced as they represent one of the few effective behaviors the child has at its disposal to gain the caregiver's attention (Crowell et al.,

2009). From a biological perspective, this is often viewed as an imbalance of heightened bottom-up emotional intensity and lowered top-down regulatory control. The psychological result, broadly called emotional dysregulation, is what we will turn to after our review of impulsivity.

Even though adverse childhood experiences are highly prevalent in almost every mental disorder, numbers are particularly high for BPD with up to 92 percent of patients reporting childhood trauma (Frias et al., 2016). Nevertheless, it is widely accepted that traumatic events are neither necessary nor sufficient for the development of BPD (De Aquino Ferreira et al., 2018). Childhood trauma is related to insecure attachment style (Baer and Martinez, 2006), which is also highly prevalent in BPD (Agrawal et al., 2004). Together with impulsive behavior and emotional lability, insecure attachment disrupts interpersonal relationships later in life, further exacerbating symptoms (e.g., Baryshnikov et al., 2017). As a result, BPD patients are extremely sensitive to social rejection and have fragile social relationships (American Psychiatric Association, 2013; Gao et al., 2017). In the last subsection of this chapter, we will focus on the biological findings elucidating these interpersonal factors.

Taken together, theory and empirical studies imply that BPD symptoms are largely inherited, but are shaped by adverse childhood experiences, which results in a downward spiral of impulsivity, emotional lability, and interpersonal problems. To deal with extreme emotional tension, BPD patients often resort to maladaptive coping strategies such as substance abuse and non-suicidal self-injury. The biosocial model postulates the latter as a likely precursor to BPD, because a majority of BPD patients engage in self-injury and report having first done so in early adolescence (Zanarini, Frankenburg, Ridolfi, et al., 2006). Adhering to the chronology proposed by the biosocial model, we will review the neurobiology of impulsivity, emotion dysregulation, and interpersonal problems in BPD (Figure 37.1 illustrates brain regions of interest for these domains).

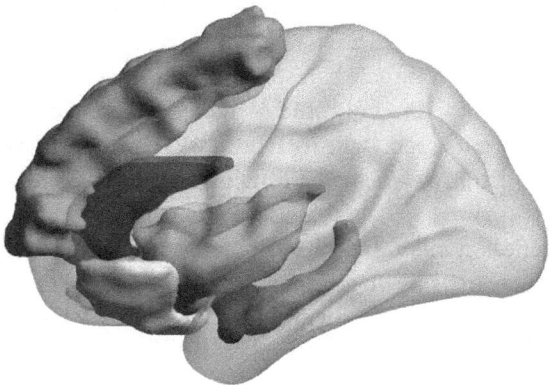

dlPFC:
Impulsivity and emotion regulation

ACC:
Impulsivity and interpersonal problems

OFC:
Impulsivity and emotion regulation

Insula:
Emotion dysregulation and interpersonal problems

Amygdala:
Emotion dysregulation

Hippocampus:
Emotional intensity

FIGURE 37.1 Regions of interest for symptom domains of borderline personality disorder.

Sections of this chapter which review the empirical evidence on these regions are indicated in Figure 37.1 in italics underneath the anatomical labels. Regions were visualized using the BrainNet Viewer (Xia et al., 2013).

IMPULSIVITY

Impulsivity is a primary clinical feature of BPD and can be defined as the relative inability to inhibit one's own initial response to external or internal stimuli, with potentially negative effects on the self or others (Chamberlain and Sahakian, 2007). Concrete impulsive behaviors comprise high-risk behaviors such as substance abuse, binge eating, aggressive outbursts, and sudden relationship breakups (American Psychiatric Association, 2013). BPD patients have higher scores on self-report scales for impulsivity compared to healthy controls and patients with unipolar or bipolar depression (Sebastian et al. 2013). Furthermore, prospective longitudinal studies have shown that low self-control at age 5 predicts BPD symptoms throughout adolescence (Stepp et al., 2015).

One behavioral genetic study found that BPD and antisocial personality disorder share genetic risk, most likely due to impulsive aggression (Kendler et al., 2008). Among the specific BPD symptoms, impulsive self-harm and suicidal behavior are most heritable, which is for the most part explained by genetic variance not shared with other symptoms (Reichborn-Kjennerud et al., 2013). This underlines the biosocial model's perspective that inherited impulsivity is a core factor in BPD etiology, that is genetically distinct from other symptoms (Crowell et al., 2009). Several studies have attempted to explain such heritability by focusing on genes related to the serotonin system, implicated in impulse control and affective processes (Krakowski, 2003). So far, these studies have investigated the tryptophan hydroxylase gene (TPH; the rate-limiting enzyme in the serotonin synthesis), serotonin receptor genes, and genes related to the degradation of serotonin such as the serotonin transporter (5-HTTLPR) and monoaminoxidase A. However, meta-analytical aggregation did not reveal any significant associations between these genes and the BPD diagnosis (Calati et al., 2013; Amad et al., 2014). Notably, most studies on TPH investigated TPH-1, which is mainly responsible for the synthesis of peripheral serotonin (Savelieva et al., 2008). The only study which investigated the centrally active TPH-2 did report a significant association with both BPD on a categorical as well as aggression and emotional lability scores on a dimensional level (Perez-Rodriguez et al., 2010).

In general, main effects of specific genes on mental disorders are rarely detected and have mostly been surpassed by an approach focusing on gene–environment interactions, which mirror diathesis–stress models more closely (Caspi and Moffitt, 2006). A large-scale behavioral genetic study found that genetic contributions to BPD shrink when individuals have been victims of sexual assault, demonstrating that genetic effects can be overwritten by relevant experiences (Distel et al., 2011). Two studies reported that

TPH-1 (Wilson et al., 2012) and 5-HTTLPR (Wagner et al., 2009) moderated the impact of serious life events on BPD. A study on children and adolescents reported a significant interaction between the 5-HTTLPR and chronic stress on the occurrence of self-injurious thoughts and behaviors, which was replicated in a separate sample (Hankin et al., 2015). Still, these results are not specific to impulsivity as all studies used the BPD diagnosis as the outcome. Also, the relative success of gene–environment interaction studies does not mean that genetic main effects are inconceivable. Categorical diagnoses are a distal criterion, which are influenced by clinical conventions and a loss of information in comparison to dimensional measures. A recent meta-analysis demonstrated that the relationship between 5-HTTLPR and impulsivity depends on the measurement paradigm, with the smallest effect for dichotomous diagnoses and increasing effect sizes for trait-dimensional, neuropsychological, and neurobiological assessments, the latter being by far the strongest (Jonas and Markon, 2014). Pointing in the same direction, Zetzsche et al. (2007) did not find differential distributions of the A1 receptor between a BPD and a healthy group; they did however find that patients carrying the G-allele had smaller amygdala volumes, a structure we will turn to in more detail in the section on emotion dysregulation.

A person's ability to cognitively inhibit impulsive responses typically increases monotonically from childhood through adulthood (Tottenham and Galván, 2016), with particularly strong increases in early childhood (Best and Miller, 2010). Electroencephalographic (EEG) as well as functional magnetic resonance imaging (fMRI) studies have shown that brain regions involved in cognitive control—especially the dorsolateral prefrontal cortex (dlPFC), the orbitofrontal cortex (OFC), and the anterior cingulate cortex (ACC)—are already recruited during early childhood, but evolve throughout development, marked by longitudinal decreases in dlPFC and increases in ACC activity during response inhibition (Luna et al., 2010; Hoyniak, 2017). These regions are likely neural substrates of impulsive symptoms in BPD, due to their involvement in the processing of cognitive conflicts, monitoring of behavioral inhibition, and delay of gratification (Rosario, 2012). Nevertheless, studies on metabolism in these brain regions have been relatively inconsistent, as we will review next.

An early positron emission tomography (PET) study reported lower glucose metabolism in prefrontal areas and the ACC for BPD patients, while later studies also reported hypermetabolism (Juengling et al., 2003) and no differences (Lange et al., 2005). A more recent study reported both hypometabolism in prefrontal areas and hypermetabolism in the ACC (Salavert et al., 2011). Wolf et al. (2012) found diminished cerebral blood flow in the medial OFC and increased blood flow in the lateral OFC of BPD patients. Surprisingly, increased blood flow in both areas was related to higher impulsivity scores on a self-report measure, painting a rather complicated picture with unclear clinical implications. Another study found that cerebral blood flow in prefrontal areas and the ACC were correlated with higher scores on a hostility measure, but not with a BPD severity measure (Schulz et al., 2013). This again emphasizes the importance of dimensional assessments and was incorporated in later studies, which focused on specific neurotransmitters such as glutamate and gamma aminobutyric acid (GABA).

Proton magnetic resonance spectroscopy imaging is the only non-invasive neuroimaging method which can account for the concentration of specific neuro-metabolites and therefore has the potential to add nuance to the inconsistent litera-ture. Still, as for PET studies, the evidence on specific neurotransmitters gathered with this method appears to be mixed. One study found both that BPD patients had higher glutamate concentrations in the ACC and that higher glutamate levels were positively related to impulsivity scores (Hoerst et al., 2010). A second study reported similar group differences for the ACC (Rüsch et al., 2010), while a third study reported *lower* glutamate concentrations in the ACC of patients with cluster B personality disorders and higher concentrations in the dlPFC (Smesny et al., 2018). Neither of the latter two studies found significant correlations with dimensional measures of psychopathology. Contrarily, two other studies did not find any group differences, but did find relationships with dimen-sional measures of impulsivity: Coccaro et al. (2013) reported a positive relationship be-tween cerebrospinal glutamate concentrations and impulsivity/aggression (Coccaro et al., 2013). Ende et al. (2016) reported that impulsivity had a positive relationship with ACC glutamate and a negative relationship with ACC GABA concentrations, in both BPD patients and healthy individuals. Taken together, transmitter concentrations in the ACC appear to be a central substrate for impulsivity, but do not consistently distinguish BPD patients from healthy controls.

The inconsistent group differences reported previously might be explained by the fact that BPD patients do not score higher on all facets of impulsivity. Most studies measuring response inhibition with established cognitive paradigms (such as the go-no-go task and the Stroop test) did not find any behavioral differences between BPD patients and healthy controls (Sebastian et al., 2013), while studies using more motiv-ational paradigms such as the Iowa Gambling task did report higher behavioral im-pulsivity in BPD and deficits in gratification postponement (Paret et al., 2017). Studies which distinguished between trait facets of impulsivity found that BPD features are most strongly correlated with negative urgency (i.e., impulsive behavior during distress), while sensation seeking in particular had a weak or even negative correlation with BPD (Peters et al., 2013; Hahn et al., 2016). Consequently, some studies have shown that cog-nitive inhibition deteriorates after stress induction in BPD patients, but not in healthy controls, even though several other studies did not replicate this effect (Cackowski et al., 2014).

Similarly, several fMRI studies investigated the effect of emotional distractors on cognitive control. However, here results are yet again mixed. While one study reported that only healthy controls showed increased frontal and ACC activity during negative emotional stimuli (Wingenfeld et al., 2010), another study found this effect only in BPD patients (Holtmann et al., 2013). Two other studies employed the go/no-go task with emotional distractors. The first study found that negative emotional stimuli decreased activation in the medial OFC and the subgenual ACC and increased activation in the lateral OFC, insula, and dorsal ACC during response inhibition trials in the BPD group, while opposite patterns were found in the control group (Silbersweig et al., 2007). The OFC activation pattern fits the results concerning resting state blood flow reviewed previously (Wolf et al., 2012). Similarly, the second study reported decreased activity in lateral inferior frontal regions (Jacob et al., 2013). Importantly, neither of the previous studies found greater levels of behavioral performance deterioration in the BPD group during emotional trials. Nevertheless, at least two studies did find a stronger effect of emotional distractors on working memory performance in BPD patients (Krause-Utz et al., 2012; Prehn et al., 2013). On a neural level, the first study reported larger amygdala reactivity in BPD independent of experimental condition (Krause-Utz et al., 2012). Furthermore, hippocampus activity increased only in the BPD group between emotional and neutral trials. In this group, amygdala activity was positively related to reaction times, whilst dlPFC activity related negatively to reaction times. In the other study, amygdala activity was generally higher in trials with negative compared to neutral background images (Prehn et al., 2013). This effect was significantly stronger in the BPD group. Hence, emotional stimuli seem to have an effect on some cognitive processes in BPD, but the effect of negative emotions on impulsive symptoms specifically remains rather unclear (Herman et al., 2018). Therefore, some argue that impulsivity in BPD should be attributed to comorbid attention deficit hyperactivity disorder (ADHD) (still compatible with the biosocial model), or is a consequence of emotion dysregulation (Sebastian et al., 2013).

EMOTION DYSREGULATION

Emotion dysregulation refers to maladaptive abnormalities in the generation and regulation of emotions (Sheppes et al., 2015). It is generally considered to be the most central aspect of BPD and is strongly related to behavioral and interpersonal problems (Linehan, 1993; Carpenter and Trull, 2013). Additionally, it is the most consistent indicator of BPD, with 90 percent of patients showing emotion dysregulation, as well as being the BPD feature most predictive of treatment utilization (Zanarini et al., 2004; Bagge et al., 2005). In the biosocial model (Crowell et al., 2009), emotional lability plays a major role as an early risk factor for BPD. This has been confirmed by several longitudinal studies which, just as for impulsivity, successfully predicted adolescent BPD from emotional instability as early as age 5 and adult BPD from emotional instability at age 12 (Stepp et al., 2016).

Theoretical accounts differ markedly in the number and labeling of emotion dysregulation components (D'Agostino et al., 2017). In the original formulation of the biosocial theory, Linehan (1993) divided emotion generation in BPD into three components: *emotional sensitivity* refers to a lowered threshold for an emotion to be elicited, *emotional intensity* posits that these triggered emotions are more intense, and *slow return to baseline* means that, once elicited, negative emotions are more enduring in individuals with BPD. Abnormalities in BPD regarding these three components have recently been confirmed using daily diary studies. These studies demonstrated a higher baseline of negative affect for BPD patients, with greater instability, and a slower return to their individual set point (Ebner-Priemer et al., 2015). More recent studies revealed similar patterns for various other mental disorders as well (Santangelo et al., 2014).

Emotional Sensitivity

In psychobiological research, it is particularly challenging to disentangle heightened emotional sensitivity and emotional intensity, as both components can explain differences in emotional responses between BPD patients and healthy controls across varying experimental settings. A strong indicator of emotional sensitivity is a tendency to interpret neutral or ambiguous stimuli as emotional, which would also lead to earlier detection of genuinely emotional content. Meta-analyses on facial emotion recognition accuracy revealed that BPD patients were less accurate in correctly recognizing facial emotions. Interestingly, the most pronounced effect was found for neutral faces, indicating that BPD patients are biased to falsely interpret neutral faces as emotional (Daros et al., 2013; Mitchell et al., 2014). Similar deficits in facial emotion recognition are also observed in schizophrenia; the question thus remains whether this effect is specific to BPD. To answer this question, two studies compared BPD patients to schizophrenia patients and healthy controls (Catalan et al., 2016; Van Dijke et al., 2016). Both studies replicated the finding that BPD patients are significantly less accurate than healthy controls in identifying neutral facial expressions. The same trend was found in the schizophrenia patients, but the effects were smaller and did not reach statistical significance in either study. This could mean that the incorrect attribution of emotion is a general feature of psychopathology but is pronounced in BPD.

One EEG study showed pictures of facial expressions, digitally manipulated to vary along a continuum between angry and happy (Izurieta Hidalgo et al., 2016). They found a general increase of the P100 amplitude in the BPD group. This component reflects early processing of visual stimuli in the occipital cortex, which is independent of facial recognition and can be modulated by emotional input via the amygdala (Vuilleumier and Driver, 2007). Strikingly, the P100 component was altered by the emotional valence of the facial expressions in healthy controls, while remaining unchanged in the BPD group. Furthermore, while the emotion categorization related P300 component was sensitive to both positive and negative faces in healthy controls, BPD patients showed sensitivity exclusively towards angry faces. Taken together, these findings demonstrate deficits in the neural processing of emotional faces at several stages. In BPD patients, early processing is less affected by emotional valence and they exhibit reduced activation in response to faces with positive valences in later stages, which is related to deficient emotion recognition on a behavioral level.

Several fMRI studies reported stronger amygdala activity while viewing neutral pictures in a BPD group compared to healthy controls (Donegan et al., 2003; Niedtfeld et al., 2010; Krause-Utz et al., 2012). The amygdala is an important structure for attentional guidance in fear recognition (Adolphs and Tranel, 2003; Calder et al., 1996), but has recently been argued to contain face-sensitive neurons, which respond in the absence of emotional expression (Todorov, 2012). However, meta-analyses show that amygdala activity is elevated more strongly by emotional faces (regardless of valence), reaffirming that increased amygdala activation in BPD patients, in the presence of neutral faces, might be due to their heightened emotional sensitivity (Costafreda et al., 2008; Sergerie et al., 2008).

Emotional Intensity

Once a relevant emotional cue is detected, BPD patients tend to experience increased emotional intensity, a phenomenon often termed *emotional* or *stress reactivity*. A meta-analysis of 19 fMRI studies comparing the neural response to negative stimuli between BPD patients and healthy controls revealed heightened amygdala activity in the BPD group (Schulze et al., 2016). Importantly, this effect was found in studies where BPD patients were unmedicated, granting further evidence that amygdala reactivity is a core feature underlying BPD symptoms which could be alleviated through pharmacological treatment. Even though virtually all emotion sensitive brain regions respond to emotional stimuli of both positive and negative valence, the left amygdala and the left anterior insula are the only regions where such "valence general neurons" have been identified to have an additional preference for negative valence (Lindquist et al., 2016). An updated meta-analysis confirmed the increased reactivity of the left amygdala in BPD (Schulze et al., 2019). These results were further compared with studies on two other mental disorders which often cooccur with BPD: post-traumatic stress disorder and major depression. Reactivity was increased only in the *right* amygdala for patients with post-traumatic stress disorder, whereas amygdala reactivity was unchanged in patients with major depression. This might indicate some specificity of left amygdala reactivity to processes involved in BPD.

The amygdala usually reaches its developmental peak volume between the 10th and the 12th life year, while its general reactivity declines between childhood and adolescence (Uematsu et al., 2012; Silvers et al., 2017). Increased volumes are related to disorganized attachment in early childhood, which in turn is an important predictor for BPD (Lyons-Ruth et al., 2016). It is therefore reasonable to expect that amygdala abnormalities develop in childhood, but so far these abnormalities have been restricted to function, with no volume differences found in an adolescent BPD sample (Chanen et al., 2008). In adults, a multimodal meta-analysis found that amygdala and hippocampal areas that were more reactive in the presence of negative stimuli also had smaller volumes (Schulze et al., 2016). This multimodal approach can be viewed as more meaningful than structural analyses alone because it incorporates the functional abnormalities of specific brain regions. A meta-analysis corroborated these results, reporting amygdala volume declines in BPD patients with age, indicating that smaller volumes develop over time, rather than representing an early risk-factor (Kimmel et al., 2016). Importantly, amygdala hyperreactivity was already found in an adolescent sample with self-injurious behavior, a diagnostic indicator of BPD (Plener et al., 2012). Taken together, heightened amygdala activity appears to be more robust than structural differences. Nevertheless, the evidence suggests that differences in amygdala function are related to structural changes, but more research on child and adolescent samples is needed.

In addition to amygdala reactivity, reactivity of the HPA-stress-axis has been thoroughly investigated in BPD. The HPA-axis comprises the expression of corticotropin releasing hormone (CRH) in the hypothalamus, subordinately also in the amygdala and the bed nucleus of the stria terminalis (Lebow and Chen, 2016), which triggers the release of adrenocorticotropin releasing hormone (ACTH) from the pituitary. ACTH then travels through the blood stream and triggers the release of cortisol from the cortices of the adrenal glands, which has a variety of effects on the body and the brain. Several studies used the Trier Social Stress Test to induce psychosocial stress in a laboratory setting (Kirschbaum et al., 1993). Most studies found lower baseline cortisol levels and a blunted cortisol response in BPD (Simeon et al., 2007; Nater et al., 2010), even in comparison to another patient group with cluster C personality disorders, which speaks against a general effect of psychopathology (Aleknaviciute et al., 2016; Deckers et al., 2015), which might again indicate a certain degree of specificity. These effects were also found in adolescent samples with non-suicidal self-injury, a diagnostic criterion of BPD (Kaess et al., 2012; Plener et al., 2017). In the study by Simeon et al. (2007), lower average plasma cortisol was related to more dissociative symptoms, which is thought to be indicative of disorder severity. Even though hypocortisolism appears to contradict the hypothesis of increased emotional reactivity, a meta-analysis showed that daily cortisol output initially increases in individuals immediately after they experienced severe stress and then permanently decreases later on (Miller et al., 2007). They also found a decrease of cortisol output in individuals with post-traumatic stress disorder (PTSD), again emphasizing the biological similarities between the two disorders.

Traumatic experiences play an important role in BPD, but as they are neither a necessary nor sufficient antecedent of BPD the results described might indicate that HPA abnormalities relate more to adverse childhood experiences than BPD per se. This idea has also received support from brain research. The hippocampus is a brain structure involved in contextual fear learning and possesses a large number of cortisol receptors that inhibit the activity of the HPA-axis as a part of a negative feedback loop (Sapolsky, 2015). Sustained high cortisol levels lead to reductions in hippocampal volume, as observed in BPD patients (Nunes et al., 2009; Ruocco et al., 2012; Schulze et al., 2016) as well as in patients with PTSD (Karl et al., 2006), depression (Wise et al., 2016), and healthy subjects with a history of childhood maltreatment (Dannlowski et al., 2012). Another meta-analysis of studies comparing BPD patients with and without comorbid PTSD found smaller volumes in the comorbid PTSD sample (Rodrigues et al., 2011). It therefore seems more likely that HPA-axis abnormalities are not a specific feature of BPD, but rather a consequence of severe adverse experiences (Steudte et al., 2013), which may or may not lead to BPD.

Slower Return to Baseline

After a stressor is removed, Linehan (1993) argued that BPD patients experience a slower return to their individual affective baseline. Studies on self-reports from daily life, as well as experimental studies, support this hypothesis (Reitz et al., 2012; Ebner-Priemer et al., 2015), but physiological studies are less clear. Walter and colleagues (2008) monitored the cortisol response of adolescents with BPD after a conflict discussion with their mothers and observed a delayed rise of cortisol in the BPD group, but not a slowed decrease. The aforementioned cortisol studies did not find a slowed return to baseline either, and few studies have investigated this phenomenon using other biomarkers. A larger number of studies has examined emotional habituation, a potentially related phenomenon, which might provide some insight.

Three fMRI studies repeatedly exposed their participants to the same set of emotional pictures (Hazlett et al., 2012; Koenigsberg et al., 2014; Dudas et al., 2017). The first study used pictures of different emotional categories. Healthy controls showed a reduction of striatal activity over time, while BPD patients continuously maintained low striatal activity (Dudas et al., 2017). The second study only used pictures with negative valence. Instead of habituation effects, they found that dorsal ACC activity was higher during the second picture presentation in the control group and lower in the BPD group (Koenigsberg et al., 2014). The dorsal ACC (dACC) is a brain region engaged in pain processing, social rejection, and the allocation of attentional resources. However, the dACC also seems to play a role in top-down emotion regulation (Botvinick et al., 2004; Ochsner and Gross, 2014; Woo et al., 2014). A greater decrease of dACC activity between repetitions was correlated with affective instability in both the BPD as well as the control group. Similarly, they found that connectivity between amygdala and insula increased from first to second picture presentation in the control group, but not in the BPD group. The third study found not only that amygdala activity returned to baseline more slowly in BPD patients at initial picture exposure, but also that the amygdala was activated to a greater degree by repeated exposure, compared to both healthy individuals and patients with schizotypal personality disorder (Hazlett et al., 2012). A later reanalysis revealed that the effect during repeated exposure was stronger for BPD patients carrying the Met-allele on the BDNF Val66Met gene (Perez-Rodriguez et al., 2017). Another study showed that a serotonergic multi-marker score was associated with the degree of habituation in the amygdala (Piel et al., 2018), again demonstrating that serotonergic genes are relevant for functional BPD biomarkers, but not exclusively for the impulsivity component.

One study investigating fear conditioning reported that BPD patients did not differ from healthy controls on self-reports of valence and arousal during fear acquisition. Instead, negative ratings of the pain-coupled stimulus decreased less in the BPD group during extinction and ratings of the safe CS- increased in the same phase (Krause-Utz et al., 2016). These findings suggest that BPD is less about the initial response to threat, but more about the stability of this response and its generalization to save contexts. On a neural level, healthy controls exhibited a decrease over time of amygdala reactivity during shock-coupled trials, while there was no decrease over time for BPD patients. Together with the findings by Hazlett et al. (2012), these outcomes present the clearest evidence for reduced habituation in BPD. However, the restriction of this effect to shock-coupled trials could mean that this effect was due to habituation of pain responses and not due to habituation of fear. The study also did not find any habituation in the skin conductance response, which fits with the results of Kamphausen et al. (2013), who reported a decrease in amygdala reactivity only in healthy controls. Interestingly, amygdala activity and skin conductance responses were positively correlated only in the control group and negatively in the BPD group. This implies an uncoupling of central and peripheral processes in BPD, as amygdala activity normally triggers a skin conductance response (Laine et al., 2009), and might explain the disparate findings of emotional intensity in these two domains.

Emotion Regulation

Emotion regulation refers to the process of appraising and purposefully changing an emotion and therefore determines the speed of return to an affective baseline (Ebner-Priemer et al., 2015). Emotion regulation strategies can be applied before and after an emotion is elicited, and focus either on controlling the antecedence of an emotion (e.g., changing objective situations or subjective appraisals) or the emotional response, (e.g., controlling one's facial expression; Sheppes et al., 2015). Deficits in emotion regulation play a central role in most BPD theories (Linehan, 1993; Crowell et al., 2009; Hughes et al., 2012), but also represent a transdiagnostic feature of affective disorders in general (Aldao et al., 2010); Hofmann et al., 2012).

Neurobiological models of emotion regulation largely focus on the regulatory effect of prefrontal regions on limbic structures (Banks et al., 2007; Wager et al., 2008), but have become increasingly complex in distinguishing different components of regulation (Kohn et al., 2014; Ochsner and Gross, 2014). The most popular fMRI task involves the presentation of aversive pictures, either with the instruction to reappraise or to passively view the content (Ochsner et al., 2002). Reappraisal refers to the reinterpretation of the content, for example by imagining benign outcomes for a threatening scene, or by distancing oneself from the content, such as when taking the perspective of an uninvolved observer. Most consistently, reappraisal results in an upregulation of prefrontal regions generally involved in cognitive tasks (e.g., the dlPFC, dorsomedial PFC [dmPFC], ventrolateral PFC [vlPFC], and the ACC) and a concurrent downregulation of the amygdala (Buhle et al., 2014; Kohn et al., 2014).

Several studies have investigated the neural correlates of deficient emotion regulation in BPD and found decreased activation in prefrontal regions during reappraisal (Koenigsberg et al., 2009; Schulze et al., 2011; Lang et al., 2012). The first study found that although recruitment of the dlPFC and vlPFC during reappraisal is common in BPD patients and healthy controls, patients show a lower activation of the ACC and amygdala activation instead of deactivation (Koenigsberg et al., 2009). Lowered activation of the ACC was also observed in the second study (Lang et al., 2012). They also observed similar activations in a trauma-exposed control group without BPD, again questioning the specificity of these findings. The third study reported less activation of the OFC and increased insula activation, while both groups equally recruited the dlPFC (Schulze et al., 2011). Complementing these results, a PET study found a disconnect between the metabolic rates of the OFC and the amygdala in BPD patients, while they were positively correlated in healthy controls (New et al., 2007). Taken together, these studies indicate that BPD patients have deficits in downregulating the amygdala and insula, which are sensitive indicators of emotional reactivity (Paret et al., 2014). Nevertheless, these deficits were not accompanied by a lower recruitment of all primary prefrontal regions related to reappraisal, but only the ACC and OFC. Especially the role of the OFC in emotion regulation is unclear, even though its general role in affective learning (e.g., fear extinction) makes it an interesting target to illuminate the specific neural emotion regulation deficiencies in BPD (Ochsner and Gross, 2014).

In contrast to the relatively early maturation of subcortical structures, cortical grey matter structures exhibit a cubic developmental trajectory with a peak volume during early adolescence and a consecutive volume decline (Shaw et al., 2008). This decline is often attributed to neural pruning accompanied by increased myelination (Paus, 2005). The prefrontal cortex has the longest maturation phase, losing mass until the mid-to-late 20s (Gogtay et al., 2004). The developmental mismatch between prefrontal and limbic structures is particularly large during mid-adolescence and is therefore thought to play a role in adolescent risk-taking and affective instability (Somerville et al., 2010; Shulman et al., 2016). This developmental gap might be particularly large in BPD, as these are two core aspects of the disorder. Volumetric studies point towards decreased volume in the amygdala and increased frontal grey matter volume in BPD (Schulze et al., 2016). The latter could imply that neurodevelopmental interference occurs in adolescence, when individual frontal volumes are at their greatest, before pruning takes place. In fact, some studies indicate that emotion dysregulation emerges only after childhood (Crowell et al., 2009). The reduced amygdala volume in patients should theoretically alleviate the mismatch in fronto-limbic development, although BPD patients still exhibit larger amygdala activity, despite reduced volumes (Schulze et al., 2016).

Generally, adolescents use less adaptive emotion regulation strategies than young adults (Zimmermann and Iwanski, 2014). McRae et al. (2012) investigated the neural correlates of emotion regulation between childhood and adulthood and found that emotion regulation capabilities increased with age together with enhanced vlPFC activity during cognitive reappraisal. Nevertheless, the functional BPD studies previously reported did not find differences in the vlPFC, hence it is unclear whether emotion regulation deficits in BPD are merely a disruption of normative brain development.

Dialectic behavioral therapy (DBT) is an intervention specifically tailored to emotion regulation deficits in BPD (Linehan, 1993). Combined neuroimaging/DBT studies revealed decreased amygdala and prefrontal activity during a reappraisal task after therapy, with therapy-non-responders showing no change in most regions (Schmitt et al., 2016). The prefrontal decrease is harder to interpret, as basic studies have so far related reappraisal success to *increased* frontal activity (Wager et al., 2008; Ochsner et al., 2002). DBT studies also found decreased amygdala activity after therapy (Niedtfeld et al., 2017), which in one study was related to self-reported changes in emotion regulation capabilities (Goodman et al., 2014). As amygdala reactivity is an indicator of emotional intensity, rather than regulation, this finding indicates that the two processes are most likely interrelated. Even though basic cognitive emotion regulation skills might be present, the extreme emotions BPD patients undergo might affect their choice and implementation success of emotion regulation strategies.

The relationship between emotional intensity and emotion regulation has been a particular focus in recent years. Against the classic distinction between adaptive and maladaptive strategies Sheppes and Levin (2013) emphasized the adaptiveness of flexible strategy choice. This choice generally depends on the type of emotion and its intensity (Dixon-Gordon et al., 2015). Specifically, people are more likely to use distraction strategies during high intensity stimuli and cognitive reappraisal during low intensity stimuli. This was confirmed with neural measures in an event-related potential (ERP) study which also provided evidence that distraction was the more effective strategy (Shafir et al., 2016). This could possibly explain the high tendency of BPD patients to injure themselves (American Psychiatric Association, 2013). Non-suicidal self-injury is usually employed as a means of emotion regulation when negative emotions become unbearably intense (Andover and Morris, 2014). This behavior is very common, with a lifetime prevalence around 18 percent (Muehlenkamp et al., 2012), and can be interpreted as a strategy of extreme distraction when emotions can otherwise not be avoided. The behavior usually begins in early adolescence and was therefore proposed as a precursor of BPD by the biosocial model (Crowell et al., 2009). Painful stimulation can lead to amygdala deactivation in BPD patients who resort to self-injurious behavior (Schmahl et al., 2006). This effect was also observed in the presence of negative stimuli and could be reversed through successful DBT (Niedtfeld et al., 2017).

In sum, emotional processes play a central role in BPD and neurobiological studies have been able to shed light on different processes involved, which were already proposed by Linehan's (1993) theory, namely emotional sensitivity, emotional intensity, and a slowed return to baseline. The developmental neurocognitive basis of emotion regulation deficits in BPD is a promising area, but needs more research to be conclusive. Importantly, although deficits are particularly pronounced in BPD, they also play a prominent role in other affective disorders and therefore lack specificity (Aldao et al., 2010). For this reason, it is crucial to consider the interplay between emotion dysregulation and interpersonal difficulties.

INTERPERSONAL PROBLEMS

Theories of BPD highlight the etiological relevance of the early caregiving environment. They argue for the importance of invalidating environments, which fail to acknowledge the legitimacy of a child's emotion (Linehan, 1993). Consequently, the child does not learn to appropriately recognize, label, and regulate its emotions (e.g., Deborde et al., 2012). At the same time, extreme emotional outbursts are negatively reinforced, because this is an effective way to gain the caregiver's attention, giving way to emotional intensity and interpersonal conflicts (Crowell et al., 2009). Social baseline theory, a successor of the biosocial model, posits that emotion dysregulation is better understood as both an individual and interpersonal process (Hughes et al., 2012). It assumes that social proximity is a baseline affect regulation strategy. This coregulation of emotions may conserve regulatory resources throughout the lifespan and has only recently gained more attention of researchers under the term *interpersonal emotion regulation* (Williams et al., 2018).

Early life is a period when humans are particularly reliant on coregulation from their primary caregiver. Early attachments promote emotion regulation skills and the ability to form stable adult relationships, which BPD patients do not sufficiently have (American Psychiatric Association, 2013). This period is also characterized by rapid neural development (Hughes et al., 2012) and environmental susceptibility, where parenting behavior creates the foundation for later attachment styles (Hughes et al., 2012). Early parenting affects neural development of cortical and limbic structures through epigenetic pathways, though most studies looked at extreme cases of childhood maltreatment (Belsky and De Haan, 2011; Roth and Sweatt, 2011).

Most BPD patients have an insecure attachment style and experienced abuse or neglect during childhood (Agrawal et al., 2004; Winsper et al., 2016; De Aquino Ferreira et al., 2018). These experiences most likely represent prospective risk factors (Stepp et al., 2015). In one possible scenario, the relationship between adverse childhood experiences and BPD is mediated by insecure attachment in a first step and emotion regulation in a second step (Kuo et al., 2015; Fossati et al., 2016; Frias et al., 2016; Baryshnikov et al., 2017). Hippocampal atrophy, as found in BPD (Schulze et al., 2016), has consistently been associated with early adversity (Teicher and Samson, 2016), including hostile parenting (Luby et al., 2013). These effects are exacerbated in carriers of a risk polymorphism on the 5-HTTLPR (Everaerd et al., 2012), which also contributes to BPD risk. Studies on amygdala grey matter in maltreated individuals point towards increased volume during childhood, followed by decreased volume in adulthood (Teicher and Samson, 2016). This might reflect an early sensitization of the amygdala (reflected by enlargement) to later stressful experiences (leading to shrinking). This would generally fit the pattern that adult BPD patients have a smaller and more reactive amygdala (Schulze et al., 2016) and the meta-analytic finding that amygdala volumes decrease with age in BPD patients (Kimmel et al., 2016). One study found that both BPD patients and healthy controls with unresolved attachment had a stronger amygdala reactivity to attachment-related material (Buchheim et al., 2016). Another study found that negative fronto-limbic connectivity, which has been associated with emotion regulation success (Lee et al., 2012), mediated the effect of early adversity on BPD symptoms such that negative connectivity was dampened in individuals who experience more adversity (Vai et al., 2018).

Interestingly, it is not necessarily the early experiences that have the strongest effect on neurobiological development. One study demonstrated that hippocampal volume is most susceptible to maltreatment between the ages 7 and 14 (Pechtel et al., 2014). Similarly, a prospective study found that attachment measured in infancy was not predictive of adult BPD, while studies measuring attachment in late childhood and adolescence did find a prospective relationship (Stepp et al., 2015). Hence, attachment problems relevant to BPD might develop at a later stage than initially presumed. This could explain why two prospective studies found that insecure attachment in very young children (≤2 years) was predictive of *increased* (instead of decreased) amygdala volume in adulthood (Moutsiana et al., 2015; Lyons-Ruth et al., 2016). These children also had deficits in upregulating positive emotion, but not in down-regulating negative emotions (Moutsiana et al., 2014). These differences between neurobiological correlates of early attachment and BPD might again suggest that attachment becomes a factor in the neurodevelopment of BPD, but at a later stage. Still, we need more proximal neurobiological data of younger samples to draw a firm conclusion.

Underlining the centrality of later interpersonal problems, one prospective study demonstrated that bullying between age 8–10 contributed more to BPD symptoms at age 11 than both maladaptive parenting at the same age and dysregulated behavior at a younger age (Winsper et al., 2017). Dysregulated behavior was only predictive as long as bully victimization was not controlled for. This might indicate a mediational pathway where dysregulation increases the likelihood to be bullied, which contributes to BPD. The centrality of peer relationships highlights the importance of counselling services and educational programs in schools. Two other prospective studies found that emotion dysregulation mediates the relationship between BPD symptoms and aggression perpetration, as well as victimization (Scott et al., 2014; Newhill et al., 2012). This indicates a vicious circle between interpersonal problems and emotion dysregulation. Daily diary studies corroborate this idea by demonstrating a general reciprocal relationship between interpersonal problems and affect, which is particularly strong in BPD, even compared to other mental disorder (Hepp et al., 2017, 2018). Moreover, induced stress intensifies state anger in BPD, but not ADHD or healthy controls, granting some experimental support for causality (Cackowski et al., 2017).

BPD patients are more sensitive to social rejection than healthy controls (Gao et al., 2017). They have a lower expectation to be accepted, adjust their expectations more strongly to negative than to positive feedback, and respond with anger not only to negative, but also to positive feedback (Liebke et al., 2018). The sensitivity to social rejection is frequently investigated with the Cyberball paradigm (Williams and Jarvis, 2006); a virtual ball tossing game where participants either receive a fair or an unfair amount of ball tosses by the other players. This task most consistently activates ventral and dorsal portions of the ACC, but also the anterior insula and the OFC (Cacioppo et al., 2013; Rotge et al., 2015).

Adults experience diminishing sensitivity to ostracism across their lifespan (Hawkley et al., 2011), being less sensitive than adolescents (Sebastian et al., 2010), while differences during the transition between childhood and adolescence are less conclusive

(Wölfer and Scheithauer, 2012). On a neural level, decrease as well as increase of ACC activity during adolescence have been observed (Bolling et al., 2011; Gunther Moor et al., 2012). A recent meta-analysis did not find an effect of age, however, adult samples were strongly overrepresented and it could be that neural activity may reach a plateau during emerging adulthood. Therefore, the neurodevelopment of rejection sensitivity remains unclear.

Neuroimaging studies using cyberball in BPD samples point with relative consistency in a rather surprising direction: BPD patients do not necessarily differ in their neural activity during social exclusion, but most markedly exhibit heightened activity in the *inclusion* condition (Domsalla et al., 2014; Gutz et al., 2015; Brown et al., 2017). More specifically, one study found that dACC activity and feelings of exclusion were similarly increased in BPD patients and healthy controls, but both measures were also increased during the inclusion condition in the BPD sample (Domsalla et al., 2014). An EEG study reported the same results for the P3b component, an ERP thought to be reflective of dACC activity, even compared to another control group of patients with social anxiety disorder (Gutz et al., 2015). Brown et al. (2017) compared adult BPD patients to adolescents with non-suicidal self-injury, as well as an adolescent and a healthy adult control group. Both patient groups had stronger dACC activity during exclusion than their healthy counterparts, but only the adult BPD group had larger activity in the anterior insula during social *inclusion*. It appears that a strong response to social exclusion is a sensitive indicator of psychopathology: a recent meta-analysis confirmed that BPD patients, but also patients with depressive, anxiety, and body dysmorphic disorders, are more sensitive to exclusion (Gao et al., 2017). Nevertheless, the finding that BPD patients behave as if they were rejected during social *inclusion* appears to be a promising candidate for a *specific* indicator of the disorder, on both a subjective and a neurobiological level.

Research on the biological foundations of these interpersonal deficits has recently started to expand on two fronts: the oxytocin system and hyperscanning. Oxytocin is a neuropeptide involved in pair bonding, parturition, and lactation (Donaldson and Young, 2008) which has received increasing attention due to its role in social cognition (Lopatina et al., 2018). Lower plasma oxytocin has been observed in BPD patients as well as people who experienced emotional neglect or abuse during childhood (Bertsch, Schmidinger, et al., 2013; Müller et al., 2019). There is also evidence that intranasally administered oxytocin can alleviate amygdala hyperreactivity and the bias towards threatening facial cues in BPD patients (Bertsch, Gamer, et al., 2013), but it also led to a reduction of trust, cooperation, and prosocial behavior in other studies (Servan et al., 2018). Hyperscanning refers to simultaneous neuroimaging of two participants, usually with the goal to assess the synchrony between brain signals. A recent study employed a joint attention task where participants communicated via eye gaze during an fMRI session (Bilek et al., 2017). The activity in the temporoparietal junction, which is related to theory of mind, was more strongly correlated between actual dyads than between arbitrarily matched brain signals. Most strikingly, when healthy controls were paired with current BPD patients, synchrony was lower compared to pairings with remitted BPD patients or other healthy controls. There are still many unknowns, as the oxytocin system and hyperscanning are relatively new areas within BPD research, even though both bear great potential.

Conclusions and Future Directions

In this chapter, we aimed to describe the neurobiological endophenotype of BPD and synthesize it with the available developmental evidence, along the lines of the biosocial model. The model positions early, genetically inherited impulsivity as an important risk factor, but the genes which explain the large heritability of BPD remain largely unknown. The best candidates appear to be genes related to the serotonergic system. Nevertheless, genetic effects are highly dependent on environmental influences and appear to be related not only to impulsivity, but also to emotion processing. Moreover, a recent study has cast doubt on the general idea that diagnostic categories of mental disorders can be satisfyingly explained with a candidate gene approach (Border et al., 2019). If prediction is the main goal, polygenic risk scores derived from large samples might alleviate some shortcomings of the candidate gene approach. Polygenic risk scores have already been created for other mental disorders and can accommodate gene-environment interactions, even though effect sizes often remain relatively modest (e.g., Peyrot et al., 2014). At the same time, theoretical models should be advanced not by searching for "BPD genes," but through understanding how specific genes are related to the neurobiological components of relevant mental processes like impulsivity and emotion generation.

Concerning the construct of impulsivity, target brain areas such as the ACC do not show unequivocal differences between BPD patients and healthy controls. In contrast, more studies do suggest that impulsivity in BPD has to be understood ina context of emotion processing. In this area, neurobiological results are more conclusive and support that BPD patients have greater emotional sensitivity, experience emotions more strongly, and do not sufficiently habituate to negative emotional material. The amygdala is at the core of these functional findings and patients who exhibit amygdala hyperreactivity also appear to have smaller amygdala volumes. Research should disentangle central moderators and confounders of this effect, especially type and timing of adverse experiences, and further investigate its clinical significance and malleability from a process-oriented perspective.

Emotion regulation has a particularly prominent role in most BPD theories. fMRI studies support that patients are less successful in reappraising negative content, as indexed by heightened amygdala and insula activation. Importantly, most frontal regions associated with top-down control of emotions do not distinguish patients and healthy controls. The neurobiology of emotion regulation is still an evolving research area and BPD research would particularly profit from more basic research to understand the neural complexities of emotion regulation. Also, problems with emotion regulation can be understood from both angles of top-down regulatory deficits and bottom-up regulatory demands posed by heightened emotional intensity. These two are not mutually exclusive, but could be combined in an index of imbalances between demands and resources, analogous to the theory of developmental mismatches between fronto-limbic structures in adolescence (Somerville et al., 2010).

Importantly, none of these phenomena appear to be specific to BPD. Most are found in other disorders or healthy individuals with traumatic experiences and should be regarded as transdiagnostic features, even though some have been reported to be particularly pronounced in BPD. Hence, young people who have the circumscribed endophenotype are most likely at risk for a variety of mood and stress-related disorders, while the similarity to traumatized but resilient individuals could even suggest that certain biomarkers are a mere by-product of stressful experiences and not related to any negative outcome in itself.

A potentially specific feature of BPD, which might distinguish it from many other conditions, is the constant feeling of being socially rejected, which is reflected in a stronger activation of the ACC. Longitudinal studies suggest that interpersonal experiences do not only affect the risk for BPD when they occur during a confined period at a very early age. Instead, interpersonal experiences contribute to symptoms throughout childhood and adolescence. Hence, whether ACC reactivity during social *inclusion* represents a risk factor or is acquired relatively late, remains an open empirical question.

To move forward, there is no way around conducting longitudinal studies. The most promising path might be to screen young children who experienced trauma, neglect, or abuse, learn about their neural top-down control capabilities, limbic structures, HPA-axis functioning, serotonin, and oxytocin system, to distinguish resilient from vulnerable individuals. Moreover, we cannot assume that BPD as a categorical diagnosis will neatly map onto one biological phenotype. Heterogeneity within the BPD group and surprising similarities among different mental disorders make this seem unlikely. To advance our knowledge, we need to align dimensional traits and biological systems connected to specific processes—a perspective increasingly embraced by current research.

References

Adolphs, Ralph and Daniel Tranel. (2003). Amygdala damage impairs emotion recognition from scenes only when they contain facial expressions. *Neuropsychologia, 41*(10), 1281–1289. doi:10.1016/S0028-3932(03)00064-2.

Agrawal, Hans R., John Gunderson, Bjarne M. Holmes, and Karlen Lyons-Ruth. (2004). Attachment studies with borderline patients: A review. *Harvard Review of Psychiatry, 12*(2), 94–104. doi:10.1080/10673220490447218.

Aldao, Amelia, Susan Nolen-Hoeksema, and Susanne Schweizer. (2010). Emotion-regulation strategies across psychopathology: A meta-analytic review. *Clinical Psychology Review, 30*(2), 217–237. doi:10.1016/j.cpr.2009.11.004.

Aleknaviciute, Jurate, Joke H. M. Tulen, Astrid M. Kamperman, Yolanda B. de Rijke, Cornelis G. Kooiman, and Steven A. Kushner. (2016). Borderline and cluster C personality disorders manifest distinct physiological responses to psychosocial stress. *Psychoneuroendocrinology, 72*(October), 131–138. doi:10.1016/j.psyneuen.2016.06.010.

Amad, Ali, Nicolas Ramoz, Pierre Thomas, Renaud Jardri, and Philip Gorwood. (2014). Genetics of borderline personality disorder: Systematic review and proposal of an integrative model. *Neuroscience and Biobehavioral Review*, 40(March), 6–19. doi:10.1016/j.neubiorev.2014.01.003.

American Psychiatric Association. 2013. *Diagnostic and Statistical Manual of Mental Disorders*. 5th ed. Arlington: American Psychiatric Publishing.

Andover, Margaret S. and Blair W. Morris. (2014). Expanding and clarifying the role of emotion regulation in nonsuicidal self-injury. *Canadian Journal of Psychiatry*, 59(11), 569–575. doi:10.1177/070674371405901102.

Aquino Ferreira, Lucas F. De, Fábio H. Queiroz Pereira, Ana M. L. Neri Benevides, and Matias C. Aguiar Melo. (2018). Borderline personality disorder and sexual abuse: A systematic review. *Psychiatry Research*, 262(December), 70–77. doi:10.1016/j.psychres.2018.01.043.

Baer, Judith, and Colleen D. Martinez. (2006). Child maltreatment and insecure attachment: A meta-analysis. *Journal of Reproductive and Infant Psychology*, 24(3), 187–197. doi:10.1080/02646830600821231.

Bagge, Courtney L., Stephanie D. Stepp, and Timothy J. Trull. (2005). Borderline personality disorder features and utilization of treatment over two years. *Journal of Personality Disorders*, 19(4), 420–439. doi:10.1521/pedi.2005.19.4.420.

Bandelow, Borwin, Christian Schmahl, Peter Falkai, and Dirk Wedekind. (2010). Borderline personality disorder: A dysregulation of the endogenous opioid system? *Psychological Review*, 117(2), 623–636. doi:10.1037/a0018095.

Banks, Sarah J., Kamryn T. Eddy, Mike Angstadt, Pradeep J. Nathan, and K. Luan Phan. 2007. Amygdala-frontal connectivity during emotion regulation. *Social Cognitive and Affective Neuroscience*, 2(4), 303–312. doi:10.1093/scan/nsm029.

Baryshnikov, Ilya, Grigori Joffe, Maaria Koivisto, Tarja Melartin, Kari Aaltonen, Kirsi Suominen, Tom Rosenström, Petri Näätänen, Boris Karpov, Martti Heikkinen, and Erkki Isometsä. (2017). Relationships between self-reported childhood traumatic experiences, attachment style, neuroticism and features of borderline personality disorders in patients with mood disorders. *Journal of Affective Disorders*, 210(December), 82–88. doi:10.1016/j.jad.2016.12.004.

Beauchaine, Theodore P. and Emily Neuhaus. (2008). Impulsivity and vulnerability to psychopathology. In: *Child and Adolescent Psychopathology*, (pp. 129–156), Theodore P. Beauchaine and Stephen P. Hinshaw (eds.). Hoboken: Wiley.

Beauchaine, Theodore P., Emily Neuhaus, Sharon L. Brenner, and Lisa Gatzke-Kopp. (2008). Ten good reasons to consider biological processes in prevention and intervention research. *Development and Psychopathology*, 20(3), 745–774. doi:10.1017/S0954579408000369.

Belsky, Daniel W., Avshalom Caspi, Louise Arseneault, Wiebke Bleidorn, Peter Fonagy, Marianne Goodman, Renate Houts, and Terrie E. Moffitt. (2012). Etiological features of borderline personality related characteristics in a birth cohort of 12-year-old children. *Development and Psychopathology*, 24(1), 251–265. doi:10.1017/S0954579411000812.

Belsky, Jay and Michelle De Haan. (2011). Annual research review: Parenting and children's brain development: The end of the beginning. *Journal of Child Psychology and Psychiatry and Allied Disciplines*, 52(4), 409–428. doi:10.1111/j.1469-7610.2010.02281.x.B

Bertsch, Katja, Matthias Gamer, Brigitte Schmidt, Ilinca Schmidinger, Stephan Walther, Thorsten Kästel, Knut Schnell, Christian Büchel, Gregor Domes, and Sabine C. Herpertz. (2013). Oxytocin and reduction of social threat hypersensitivity in women with borderline personality disorder. *American Journal of Psychiatry, 170*(10), 1169–1177. doi:10.1176/appi. ajp.2013.13020263.

Bertsch, Katja, Ilinca Schmidinger, Inga D. Neumann, and Sabine C. Herpertz. (2013). Reduced plasma oxytocin levels in female patients with borderline personality disorder. *Hormones and Behavior, 63*(3), 424–429. doi:10.1016/j.yhbeh.2012.11.013.

Best, John R. and Patricia H. Miller. (2010). A developmental perspective on executive function. *Child Development, 81*(6), 1641–1660. doi:10.1111/j.1467-8624.2010.01499.x.

Bilek, Edda, Gabriela Stößel, Axel Schäfer, Laura Clement, Matthias Ruf, Lydia Robnik, Corinne Neukel, Heike Tost, Peter Kirsch, and Andreas Meyer-Lindenberg. (2017). State-dependent cross-brain information flow in borderline personality disorder. *JAMA Psychiatry, 74* (9), 949–957. doi:10.1001/jamapsychiatry.2017.1682.

Bolling, Danielle Z., Naomi B. Pitskel, Ben Deen, Michael J. Crowley, Linda C. Mayes, and Kevin A. Pelphrey. (2011). Development of neural systems for processing social exclusion from childhood to adolescence. *Developmental Science, 14*(6), 1431–1444. doi:10.1111/ j.1467-7687.2011.01087.x.

Border, Richard., Johnson, Emma C., Evans, Luke M., Smolen, Andrew, Berley, Noah, Sullivan, Patrick F., and Keller, Matthew C. (2019). No support for historical candidate gene or candidate gene-by-interaction hypotheses for major depression across multiple large samples. *American Journal of Psychiatry, 176*(5), 376–387. doi:10.1176/appi.ajp.2018.18070881

Botvinick, Matthew M., Jonathan D. Cohen, and Cameron S. Carter. (2004). Conflict monitoring and anterior cingulate cortex: An update. *Trends in Cognitive Sciences, 8*(12), 539–546. doi:10.1016/j.tics.2004.10.003.

Brown, Rebecca C., Paul L. Plener, Georg Groen, Dominik Neff, Martina Bonenberger, and Birgit Abler. (2017). Differential neural processing of social exclusion and inclusion in adolescents with non-suicidal self-injury and young adults with borderline personality disorder. *Frontiers in Psychiatry, 8*(November), 267. doi:10.3389/fpsyt.2017.00267.

Buchheim, Anna, Susanne Erk, Carol George, Horst Kächele, Philipp Martius, Dan Pokorny, Manfred Spitzer, and Henrik Walter. (2016). Neural response during the activation of the attachment system in patients with borderline personality disorder: An FMRI Study. *Frontiers in Human Neuroscience, 10*(August), 1–13. doi:10.3389/fnhum.2016.00389.

Buhle, Jason T., Jennifer A. Silvers, Tor D. Wage, Richard Lopez, Chukwudi Onyemekwu, Hedy Kober, Jochen Webe, and Kevin N. Ochsner. (2014). Cognitive reappraisal of emotion: A meta-analysis of human neuroimaging studies. *Cerebral Cortex, 24*(11), 2981–2990. doi:10.1093/cercor/bht154.

Cacioppo, Stephanie, Chris Frum, Erik Asp, Robin M. Weiss, James W. Lewis, and John T. Cacioppo. (2013). A quantitative meta-analysis of functional imaging studies of social rejection. *Scientific Reports, 3*(1), 2027. doi:10.1038/srep02027.

Cackowski, Sylvia, Annegret Krause-Utz, Julia Van Eijk, Katrin Klohr, Stephanie Daffner, Esther Sobanski, and Gabriele Ende. (2017). Anger and aggression in borderline personality disorder and attention deficit hyperactivity disorder – does stress matter? *Borderline Personality Disorder and Emotion Dysregulation, 4*(March), 6. doi:10.1186/ s40479-017-0057-5.

Cackowski, Sylvia., A. C. Reitz, Gabriele Ende, Nikolaus Kleindienst, Martin Bohus, Christian Schmahl, and Annegret Krause-Utz. (2014). Impact of stress on different components of impulsivity in borderline personality disorder. *Psychological Medicine, 44*(15), 3329–3340. doi:10.1017/S0033291714000427.

Calati, Raffaella, Florence Gressier, Martina Balestri, and Alessandro Serretti. (2013). Genetic modulation of borderline personality disorder: Systematic review and meta-analysis. *Journal of Psychiatric Research*, 47(10), 1275–1287. doi:10.1016/j.jpsychires.2013.06.002.

Calder, Andrew J., Andrew W. Young, Duncan Rowland, David I. Perrett, John R. Hodges, and Nancy L. Etcoff. (1996). Facial emotion recognition after bilateral amygdala damage: Differentially severe impairment of fear. *Cognitive Neuropsychology*, 13(5), 699–745. doi:10.1080/026432996381890.

Carpenter, Ryan W., and Timothy J. Trull. (2013). Components of emotion dysregulation in borderline personality disorder: A review. *Current Psychiatry Reports*, 15(1), 335. doi:10.1007/s11920-012-0335-2.

Caspi, Avshalom and Terrie E. Moffitt. (2006). Gene-environment interactions in psychiatry: Joining forces with neuroscience. *Nature Reviews. Neuroscience*, 7(7), 583–590. doi:10.1038/nrn1925.

Catalan, Ana, Maider Gonzalez De Artaza, Sonia Bustamante, Pablo Orgaz, Luis Osa, Virxinia Angosto, Cristina Valverde, Amaia Bilbao, Arantza Madrazo, Jim Van Os, and Miguel Angel Gonzalez-Torres. (2016). Differences in facial emotion recognition between first episode psychosis, borderline personality disorder and healthy controls. *PLoS ONE*, 11(7), p. e0160056. doi:10.1371/journal.pone.0160056.

Chamberlain, Samuel R., and Barbara J. Sahakian. (2007). The neuropsychiatry of impulsivity. *Current Opinion in Psychiatry*, 20(3), 255–261. doi:10.1097/YCO.0b013e3280ba4989.

Chanen, Andrew M, Dennis Velakoulis, Kate Carison, Karen Gaunson, Stephen J. Wood, Hok Pan Yuen, Murat Yücel, Henry J. Jackson, Patrick D. McGorry, and Christos Pantelis. (2008). Orbitofrontal, amygdala and hippocampal volumes in teenagers with first-presentation borderline personality disorder. *Psychiatry Research: Neuroimaging*, 163(2), 116–125. doi:10.1016/j.pscychresns.2007.08.007.

Clercq, Barbara De, Mieke Decuyper, and Elien de Caluwé. (2014). Developmental manifestations of borderline personality pathology from an age-specific dimensional personality disorder trait framework. In *Handbook of Borderline Personality Disorder in Children and Adolescents*, pp. 81–94, Carla Sharp and Jennifer L. Tackett (eds.). New York: Springer.

Coccaro, Emil F., Royce Lee, and Paul Vezina. (2013). Cerebrospinal fluid glutamate concentration correlates with impulsive aggression in human subjects. *Journal of Psychiatric Research*, 47(9), 1247–1253. doi:10.1016/j.jpsychires.2013.05.001.

Costafreda, Sergi G., Michael J. Brammer, Anthony S. David, and Cynthia H. Y. Fu. (2008). Predictors of amygdala activation during the processing of emotional stimuli: A meta-analysis of 385 PET and FMRI studies. *Brain Research Reviews*, 58(1), 57–70. doi:10.1016/j.brainresrev.2007.10.012.

Crowell, Sheila E., Theodore P. Beauchaine, and Marsha M. Linehan. (2009). A biosocial developmental model of borderline personality: Elaborating and extending Linehan's theory. *Psychological Bulletin*, 135(3), 495–510. doi:10.1037/a0015616.

D'Agostino, Alessandra, Serena Covanti, Mario Rossi Monti, and Vladan Starcevic. (2017). Reconsidering emotion dysregulation. *Psychiatric Quarterly*, 88(4), 807–825. doi:10.1007/s11126-017-9499-6.

Dannlowski, Udo, Anja Stuhrmann, Victoria Beutelmann, Peter Zwanzger, Thomas Lenzen, Dominik Grotegerd, Katharina Domschke, Christa Hohoff, Patricia Ohrmann, Jochen Bauer, Christian Lindner, Christian Postert, Carsten Konrad, Volker Arolt, Walter Heindel, Thomas Suslow, and Harald Kugel. (2012). Limbic scars: Long-term consequences of childhood maltreatment revealed by functional and structural magnetic resonance imaging. *Biological Psychiatry*, *71*(4), 286–293. doi:10.1016/j.biopsych.2011.10.021.

Daros, Alexander R., Konstantine K. Zakzanis, and Anthony. C. Ruocco. (2013). Facial emotion recognition in borderline personality disorder. *Psychological Medicine*, *4* (9), 1953–1963. doi:10.1017/S0033291712002607.

Deborde, Anne-Sophie, Raphaële Miljkovitch, Caroline Roy, Corinne Dugré-Le Bigre, Alexandra Pham-Scottez, Mario Speranza, and Maurice Corcos. (2012). Alexithymia as a mediator between attachment and the development of borderline personality disorder in adolescence. *Journal of Personality Disorders*, *26*(5), 676–688. doi:10.1521/pedi.2012.26.5.676.

Deckers, Janneke W. M., Jill Lobbestael, Guido A. Van Wingen, Roy P. C. Kessels, Arnoud Arntz, and Jos I. M. Egger. (2015). The influence of stress on social cognition in patients with borderline personality disorder. *Psychoneuroendocrinology*, *52*(1), 119–129. doi:10.1016/j.psyneuen.2014.11.003.

Dijke, Annemiek Van, Mascha T. Van Wout, Julian D. Ford, and André Aleman. (2016). Deficits in degraded facial affect labeling in schizophrenia and borderline personality disorder. *PLoS ONE*, *11*(6), p. e0154145. doi:10.1371/journal.pone.0154145.

Distel, Marijn A., Christel M. Middeldorp, Timothy J. Trull, Catherine A. Derom, Gonneke Willemsen, and Dorret I. Boomsma. (2011). Life events and borderline personality features: The influence of gene-environment interaction and gene-environment correlation. *Psychological Medicine*, *41*(4), 849–860. doi:10.1017/S0033291710001297.

Dixon-Gordon, Katherine L., Amelia Aldao, and Andres De Los Reyes. (2015). Emotion regulation in context: Examining the spontaneous use of strategies across emotional intensity and type of emotion. *Personality and Individual Differences*, *86*: 271–276. doi:10.1016/j.paid.2015.06.011.

Domsalla, Melanie, Georgia Koppe, Inga Niedtfeld, Sabine Vollstädt-Klein, Christian Schmahl, Martin Bohus, and Stefanie Lis. (2014). Cerebral processing of social rejection in patients with borderline personality disorder. *Social Cognitive and Affective Neuroscience*, *9*(11), 1789–1797. doi:10.1093/scan/nst176.

Donaldson, Zoe R., and Larry J. Young. (2008). Oxytocin, vasopressin, and the neurogenetics of sociality. *Science*, *322*(5903), 900–904. doi:10.1126/science.1158668.

Donegan, Nelson H., Charles A. Sanislow, Hilary P. Blumberg, Robert K. Fulbright, Cheryl Lacadie, Pawel Skudlarski, John C. Gore, Ingrid R. Olson, Thomas H. McGlashan, and Bruce E. Wexler. (2003). Amygdala hyperreactivity in borderline personality disorder: Implications for emotional dysregulation. *Biological Psychiatry*, *54*(11), 1284–1293. doi:10.1016/S0006-3223(03)00636-X.

Dudas, Robert B., Tom B. Mole, Laurel S. Morris, Chess Denman, Emma Hill, Bence Szalma, Davy Evans, Barnaby Dunn, Paul Fletcher, and Valerie Voon. (2017). Amygdala and DlPFC abnormalities, with aberrant connectivity and habituation in response to emotional stimuli in females with BPD. *Journal of Affective Disorders*, *208*: 460–466. doi:10.1016/j.jad.2016.10.043.

Ebner-Priemer, Ulrich W., Marlies Houben, Philip Santangelo, Nikolaus Kleindienst, Francis Tuerlinckx, Zita Oravecz, Gregory Verleysen, Katrijn Van Deun, Martin Bohus, and Peter Kuppens. (2015). Unraveling affective dysregulation in borderline personality disorder: A theoretical model and empirical evidence. *Journal of Abnormal Psychology*, *124*(1), 186–198. doi:10.1037/abn0000021.

Ende, Gabriele, Sylvia Cackowski, Julia Van Eijk, Markus Sack, Traute Demirakca, Nikolaus Kleindienst, Martin Bohus, Esther Sobanski, Annegret Krause-Utz, and Christian Schmahl.

(2016). Impulsivity and aggression in female BPD and ADHD patients: Association with ACC glutamate and GABA concentrations. *Neuropsychopharmacology*, 41(2), 410–418. doi:10.1038/npp.2015.153.

Everaerd, Daphne, Lotte Gerritsen, Mark Rijpkema, Thomas Frodl, Iris Van Oostrom, Barbara Franke, Guillén Fernández, and Indira Tendolkar. (2012). Sex modulates the interactive effect of the serotonin transporter gene polymorphism and childhood adversity on hippocampal volume. *Neuropsychopharmacology*, 37(8), 1848–1855. doi:10.1038/npp.2012.32.

Fossati, Andrea, Kim L. Gratz, Antonella Somma, Cesare Maffei, and Serena Borroni. (2016). The mediating role of emotion dysregulation in the relations between childhood trauma history and adult attachment and borderline personality disorder features: A study of Italian nonclinical participants. *Journal of Personality Disorders*, 30(5), 653–676. doi:10.1521/pedi_2015_29_222.

Frias, Alvaro, Carol Palma, Núria Farriols, Laura Gonzalez, and Anna Horta. (2016). Anxious adult attachment may mediate the relationship between childhood emotional abuse and borderline personality disorder. *Personality and Mental Health*, 10(4), 274–284. doi:10.1002/pmh.1348.

Gao, Shuling, Mark Assink, Andrea Cipriani, and Kangguang Lin. (2017). Associations between rejection sensitivity and mental health outcomes: A meta-analytic review. *Clinical Psychology Review*, 57(August), 59–74. doi:10.1016/j.cpr.2017.08.007.

Gogtay, Nitin, Jay N. Giedd, Leslie Lusk, Kiralee M. Hayashi, Deanna Greenstein, A. Catherine Vaituzis, Tom F. Nugent, David H. Herman, Liv S. Clasen, Arthur W. Toga, Judith L. Rapoport, and Paul M. Thompson. (2004). Dynamic mapping of human cortical development during childhood through early adulthood. *Proceedings of the National Academy of Sciences of the United States of America*, 101(21), 8174–8179. doi:10.1073/pnas.0402680101.

Goodman, Marianne, David Carpenter, Cheuk Y. Tang, Kim E. Goldstein, Jennifer Avedon, Nicolas Fernandez, Kathryn A. Mascitelli, Nicholas J. Blair, Antonia S. New, Joseph Triebwasser, Larry J. Siever, and Erin A. Hazlett. (2014). Dialectical behavior therapy alters emotion regulation and amygdala activity in patients with borderline personality disorder. *Journal of Psychiatric Research*, 57(1), 108–116. doi:10.1016/j.jpsychires.2014.06.020.

Gottesman, Irving I. and Todd D. Gould. (2003). The endophenotype concept in psychiatry: etymology and strategic intentions. *American Journal of Psychiatry*, 160(4), 636–645. doi:10.1176/appi.ajp.160.4.636.

Gunderson, John G., Robert L. Stout, Thomas H. McGlashan, M. Tracie Shea, Leslie C. Morey, Carlos M. Grilo, Mary C. Zanarini, Shirley Yen, John C. Markowitz, Charles Sanislow, Emily Ansell, Anthony Pinto, and Andrew E. Skodol. (2011). Ten-year course of borderline personality disorder: Psychopathology and function from the collaborative longitudinal personality disorders study. *Archives of General Psychiatry*, 68 (8), 827–837. doi:10.1001/archgenpsychiatry.2011.37.

Gunther Moor, Bregtje, Berna Güroğlu, Zdeňa A. Op de Macks, Serge A. R. B. Rombouts, Maurits W. Van der Molen, and Eveline A. Crone. (2012). Social exclusion and punishment of excluders: Neural correlates and developmental trajectories. *NeuroImage*, 59(1), 708–717. doi:10.1016/j.neuroimage.2011.07.028.

Gutz, Lea, Babette Renneberg, Stefan Roepke, and Michael Niedeggen. (2015). Neural processing of social participation in borderline personality disorder and social anxiety disorder. *Journal of Abnormal Psychology* 124 (2), 421–431. doi:10.1037/a0038614.

Hahn, Austin M., Raluca M. Simons, and Christine K. Hahn. (2016). Five factors of impulsivity: Unique pathways to borderline and antisocial personality features and subsequent alcohol problems. *Personality and Individual Differences*, *99*(September), 313–319. doi:10.1016/j.paid.2016.05.035.

Hankin, Benjamin L., Andrea L. Barrocas, Jami F. Young, Brett Haberstick, and Andrew Smolen. (2015). 5-HTTLPR x interpersonal stress interaction and nonsuicidal self-injury in general community sample of youth. *Psychiatry Research*, *225*(3), 609–612. doi:10.1016/j.psychres.2014.11.037.

Hawkley, Louise C., Kipling D. Williams, and John T. Cacioppo. (2011). Responses to ostracism across adulthood. *Social Cognitive and Affective Neuroscience*, *6*(2), 234–243. doi:10.1093/scan/nsq045.

Hazlett, Erin A., Jing Zhang, Antonia S. New, Yuliya Zelmanova, Kim E. Goldstein, M. Mehmet Haznedar, David Meyerson, Marianne Goodman, Larry J. Siever, and King Wai Chu. (2012). Potentiated amygdala response to repeated emotional pictures in borderline personality disorder. *Biological Psychiatry*, *72*(6), 448–456. doi:10.1016/j.biopsych.2012.03.027.

Hepp, Johanna, Sean P. Lane, Ryan W. Carpenter, Inga Niedtfeld, Whitney C. Brown, and Timothy J. Trull. (2017). Interpersonal problems and negative affect in borderline personality and depressive disorders in daily life. *Clinical Psychological Science*, *5*(3), 470–484. doi:10.1177/2167702616677312.

Hepp, Johanna, Sean P. Lane, Andrea M. Wycoff, Ryan W. Carpenter, and Timothy J. Trull. (2018). Interpersonal stressors and negative affect in individuals with borderline personality disorder and community adults in daily life: A replication and extension. *Journal of Abnormal Psychology*, *127*(2), 183–189. doi:10.1037/abn0000318.

Herman, Aleksandra M., Hugo D. Critchley, and Theodora Duka. (2018). The role of emotions and physiological arousal in modulating impulsive behaviour. *Biological Psychology*, *133*(March), 30–43. doi:10.1016/j.biopsycho.2018.01.014.

Hoerst, Mareen, Wolfgang Weber-Fahr, Nuran Tunc-Skarka, Matthias Ruf, Martin Bohus, Christian Schmahl, and Gabriele Ende. (2010). Correlation of glutamate levels in the anterior cingulate cortex with self-reported impulsivity in patients with borderline personality disorder and healthy controls. *Arch Gen Psychiatry*, *67*(9), 946–954. doi:10.1001/archgenpsychiatry.2010.93.

Hofmann, Stefan G., Alice T. Sawyer, Angela Fang, and Anu Asnaani. (2012). Emotion dysregulation model of mood and anxiety disorders. *Depression and Anxiety*, *29*(5), 409–416. doi:10.1002/da.21888.

Holtmann, Jana, Maike C. Herbort, Torsten Wüstenberg, Joram Soch, Sylvia Richter, Henrik Walter, Stefan Roepke, and Björn H. Schott. (2013). Trait anxiety modulates fronto-limbic processing of emotional interference in borderline personality disorder. *Frontiers in Human Neuroscience*, *7*(March), 1–21. doi:10.3389/fnhum.2013.00054.

Hoyniak, Caroline. (2017). Changes in the NoGo N2 event-related potential component across childhood: A systematic review and meta-analysis. *Developmental Neuropsychology*, *42*(1), 1–24. doi:10.1080/87565641.2016.1247162.

Hughes, Amy E., Sheila E. Crowell, Lauren Uyeji, and James A. Coan. (2012). A developmental neuroscience of borderline pathology: Emotion dysregulation and social baseline theory. *Journal of Abnormal Child Psychology*, *40*(1), 21–33. doi:10.1007/s10802-011-9555-x.

Izurieta Hidalgo, Natalie A., Rieke Oelkers-Ax, Krisztina Nagy, Falk Mancke, Martin Bohus, Sabine C. Herpertz, and Katja Bertsch. (2016). Time course of facial emotion processing in women with borderline personality disorder: An ERP study. *Journal of Psychiatry and Neuroscience*, *41*(1), 16–26. doi:10.1503/jpn.140215.

Jacob, Gitta A., Kerstin Zvonik, Susanne Kamphausen, Alexandra Sebastian, Simon Maier, Alexandra Philipsen, Ludger Tebartz Van Elst, Klaus Lieb, and Oliver Tüscher. (2013). Emotional modulation of motor response inhibition in women with borderline personality disorder: An FMRI study. *Journal of Psychiatry and Neuroscience, 38*(3), 164–172. doi:10.1503/jpn.120029.

Jonas, Katherine G. and Kristian E. Markon. (2014). A meta-analytic evaluation of the endophenotype hypothesis: Effects of measurement paradigm in the psychiatric genetics of impulsivity. *Journal of Abnormal Psychology, 123*(3), 660–675. doi:10.1037/a0037094.

Juengling, Freimut D., Christian Schmahl, Bernd Heßlinger, Dieter Ebert, James D. Bremner, Julia Gostomzyk, Martin Bohus, and Klaus Lieb. (2003). Positron emission tomography in female patients with borderline personality disorder. *Journal of Psychiatric Research, 37*(2), 109–115. doi:10.1016/S0022-3956(02)00084-5.

Kaess, Michael, Markus Hille, Peter Parzer, Christiane Maser-Gluth, Franz Resch, and Romuald Brunner. (2012). Alterations in the neuroendocrinological stress response to acute psychosocial stress in adolescents engaging in nonsuicidal self-injury. *Psychoneuroendocrinology, 37*(1), 157–161. doi:10.1016/j.psyneuen.2011.05.009.

Kamphausen, Susanne, Patricia Schröder, Simon Maier, Kerstin Bader, Bernd Feige, Christoph P. Kaller, Volkmar Glauche, Sabine Ohlendorf, Ludger Tebartz Van Elst, Stefan Klöppel, Gitta A. Jacob, David Silbersweig, Klaus Lieb, and Oliver Tüscher. (2013). Medial prefrontal dysfunction and prolonged amygdala response during instructed fear processing in borderline personality disorder. *World Journal of Biological Psychiatry 14* (4), 307–318. doi:10.3109/15622975.2012.665174.

Karl, Anke, Michael Schaefer, Loretta S. Malta, Denise Dörfel, Nicolas Rohleder, and Annett Werner. (2006). A meta-analysis of structural brain abnormalities in PTSD. *Neuroscience & Biobehavioral Reviews, 30* (7), 1004–1031. doi:10.1016/j.neubiorev.2006.03.004.

Kendler, Kenneth S., Steven H. Aggen, Nikolai Czajkowski, Espen Røysamb, Kristian Tambs, Svenn Torgersen, Michael C. Neale, and Ted Reichborn-Kjennerud. (2008). The structure of genetic and environmental risk factors for DSM-IV personality disorders. *Archives of General Psychiatry, 65*(12), 1438–1446. doi:10.1001/archpsyc.65.12.1438.

Kimmel, Christine L., Omar M. Alhassoon, Scott C. Wollman, Mark J. Stern, Adlyn Perez-Figueroa, Matthew G. Hall, Joscelyn Rompogren, and Joaquim Radua. (2016). Age-related parieto-occipital and other gray matter changes in borderline personality disorder: A meta-analysis of cortical and subcortical structures. *Psychiatry Research: Neuroimaging, 251*(May), 15–25. doi:10.1016/j.pscychresns.2016.04.005.

Kirschbaum, Clemens, Karl M. Pirke, and Dirk Hellhammer. (1993). The 'Trier Social Stress Test' – A tool for investigating psychobiological stress responses in a laboratory setting. *Neuropsychobiology, 28*(1–2), 76–81. doi:10.1159/000119004.

Koenigsberg, Harold W., Bryan T. Denny, Jin Fan, Xun Liu, Stephanie Guerreri, Sarah J. Mayson, Liza Rimsky, Antonia S. New, Marianne Goodman, and Larry J. Siever. (2014). The neural correlates of anomalous habituation to negative emotional pictures in borderline and avoidant personality disorder patients. *American Journal of Psychiatry, 171*(1), 82–90. doi:10.1176/appi.ajp.2013.13070852.

Koenigsberg, Harold W., Larry J. Siever, Hedok Lee, Scott Pizzarello, Antonia S. New, Marianne Goodman, Hu Cheng, Janine Flory, and Isak Prohovnik. (2009). Neural correlates of emotion processing in borderline personality disorder. *Psychiatry Research: Neuroimaging 172* (3), 192–199. doi:10.1016/j.pscychresns.2008.07.010.

Kohn, Nils, Simon B. Eickhoff, Maryse Scheller, Angela R. Laird, Peter T. Fox, and Ute Habel. (2014). Neural network of cognitive emotion regulation – an ALE meta-analysis and MACM analysis. *NeuroImage, 87*(February), 345–355. doi:10.1016/j.neuroimage.2013.11.001.

Krakowski, Menaham. (2003). Violence and serotonin: Influence of impulse control, affect regulation, and social functioning. *Journal of Neuropsychiatry and Clinical Neurosciences, 15*(3), 294–305. doi:10.1176/appi.neuropsych.15.3.294.

Krause-Utz, Annegret, Jana Keibel-Mauchnik, Ulrich Ebner-Priemer, Martin Bohus, and Christian Schmahl. (2016). Classical conditioning in borderline personality disorder: An FMRI study. *European Archives of Psychiatry and Clinical Neuroscience, 266*(4), 291–305. doi:10.1007/s00406-015-0593-1.

Krause-Utz, Annegret, Nicole Y. L. Oei, Inga Niedtfeld, Martin Bohus, Philip Spinhoven, Christian Schmahl, and Bernet M. Elzinga. (2012). Influence of emotional distraction on working memory performance in borderline personality disorder. *Psychological Medicine, 42* (10), 2181–2192. doi:10.1017/S0033291712000153.

Kuo, Janice R., Jennifer E. Khoury, Rebecca Metcalfe, Skye Fitzpatrick, and Alasdair Goodwill. (2015). An examination of the relationship between childhood emotional abuse and borderline personality disorder features: The role of difficulties with emotion regulation. *Child Abuse and Neglect, 39*(January), 147–155. doi:10.1016/j.chiabu.2014.08.008.

Laine, Christopher M., Kevin M. Spitler, Clayton P. Mosher, and Katalin M. Gothard. (2009). Behavioral triggers of skin conductance responses and their neural correlates in the primate amygdala. *Journal of Neurophysiology, 101*(4), 1749–1754. doi:10.1152/jn.91110.2008.

Lang, Simone, Boris Kotchoubey, Carina Frick, Carsten Spitzer, Hans Jörgen Grabe, and Sven Barnow. (2012). Cognitive reappraisal in trauma-exposed women with borderline personality disorder. *NeuroImage, 59*(2), 1727–1734. doi:10.1016/j.neuroimage.2011.08.061.

Lange, Claudia, Lutz Kracht, Karl Herholz, Ulrich Sachsse, and Eva Irle. (2005). Reduced glucose metabolism in temporo-parietal cortices of women with borderline personality disorder. *Psychiatry Research: Neuroimaging, 139*(2), 115–126. doi:10.1016/j.pscychresns.2005.05.003.

Lebow, Maya A. and Aiguo Chen. (2016). Overshadowed by the amygdala: The bed nucleus of the stria terminalis emerges as key to psychiatric disorders. *Molecular Psychiatry, 21*(4), 450–463. doi:10.1038/mp.2016.1.

Lee, Hyejeen, Aaron S. Heller, Carien M. Van Reekum, Brady Nelson, and Richard J. Davidson. (2012). Amygdala-prefrontal coupling underlies individual differences in emotion regulation. *NeuroImage, 62*(3), 1575–1581. doi:10.1016/j.neuroimage.2012.05.044.

Lenzenweger, Mark F. and Cicchetti, Dante. (2005). Toward a developmental psychopathology approach to borderline personality disorder. *Development and Psychopathology, 17*(4), 893–898. doi:10.1017/S095457940505042X.

Liebke, Lisa, Georgia Koppe, Melanie Bungert, Janine Thome, Sophie Hauschild, Nadine Defiebre, Natalie A. Izurieta Hidalgo, Christian Schmahl, Martin Bohus, and Stefanie Lis. (2018). Difficulties with being socially accepted: An experimental study in borderline personality disorder. *Journal of Abnormal Psychology, 127*(7), 670–682. doi:10.1037/abn0000373.

Lindquist, Kristen A., Ajay B. Satpute, Tor D. Wager, Jochen Weber, and Lisa Feldman Barrett. (2016). The brain basis of positive and negative affect: Evidence from a meta-analysis of the human neuroimaging literature. *Cerebral Cortex, 26*(5), 1910–1922. doi:10.1093/cercor/bhv001.

Linehan, Marsha M. (1993). *Cognitive-Behavioral Treatment of Borderline Personality Disorder.* New York: Guilford Press.

Lopatina, Olga L., Yulia K. Komleva, Yana V. Gorina, Haruhiro Higashida, and Alla B. Salmina. (2018). Neurobiological aspects of face recognition: The role of oxytocin. *Frontiers in Behavioral Neuroscience*, 12(August), 1–11. doi:10.3389/fnbeh.2018.00195.

Luby, Joan, Andy Belden, Kelly N Botteron, Natasha Marrus, Michael P. Harms, Casey Babb, Tomoyuki Nishino, and Deanna Barch. (2013). The effects of poverty on childhood brain development: The mediating effect of caregiving and stressful life events. *Journal of American Pediatrics*, 167(12), 1135–1142. doi:10.1001/jamapediatrics.2013.3139.

Luna, Beatriz, Aarthi Padmanabhan, and Kirsten O'Hearn. (2010). What has FMRI told us about the development of cognitive control through adolescence? *Brain and Cognition*, 72(1), 101–113. doi:10.1016/j.bandc.2009.08.005.

Lyons-Ruth, Karlen, Pia Pechtel, Annie Seungyeon Yoon, Carl Anderson, and Martin Teicher. (2016). Disorganized attachment in infancy predicts greater amygdala volume in adulthood. *Behavioural Brain Research*, 308(July), 83–93. doi:10.1016/j.bbr.2016.03.050.

McRae, Kateri, James J. Gross, Jochen Weber, Elaine R. Robertson, Peter Sokol-Hessner, Rebecca D. Ray, John D. E. Gabrieli, and Kevin N. Ochsner. (2012). The development of emotion regulation: An FMRI study of cognitive reappraisal in children, adolescents and young adults. *Social Cognitive and Affective Neuroscience*, 7(1), 11–22. doi:10.1093/scan/nsr093.

Miller, Gregory E., Edith Chen, and Eric S. Zhou. (2007). If it goes up, must it come down? Chronic stress and the hypothalamic-pituitary-adrenocortical axis in humans. *Psychological Bulletin*, 133 (1), 25–45. doi:10.1037/0033-2909.133.1.25.

Mitchell, Amy E., Geoffrey L. Dickens, and Marco M. Picchioni. (2014). Facial emotion processing in borderline personality disorder: A Systematic review and meta-analysis. *Neuropsychology Review*, 24(4), 166–184. doi:10.1007/s11065-014-9254-9.

Moutsiana, Christina, Pasco Fearon, Lynne Murray, Peter Cooper, Ian Goodyer, Tom Johnstone, and Sarah Halligan. (2014). Making an effort to feel positive: insecure attachment in infancy predicts the neural underpinnings of emotion regulation in adulthood. *Journal of Child Psychology and Psychiatry and Allied Disciplines*, 55(9), 999–1008. doi:10.1111/jcpp.12198.

Moutsiana, Christina, Tom Johnstone, Lynne Murray, Pasco Fearon, Peter J. Cooper, Christos Pliatsikas, Ian Goodyer, and Sarah L. Halligan. (2015). Insecure attachment during infancy predicts greater amygdala volumes in early adulthood. *Journal of Child Psychology and Psychiatry and Allied Disciplines*, 56(5), 540–548. doi:10.1111/jcpp.12317.

Muehlenkamp, Jennifer J., Laurence Claes, Lindsey Havertape, and Paul L. Plener. (2012). International prevalence of adolescent non-suicidal self-injury and deliberate self-harm. *Child and Adolescent Psychiatry and Mental Health*, 6(March), p. 10. doi:10.1186/1753-2000-6-10.

Müller, Laura E., Katja Bertsch, Konstatin Bülau, Sabine C. Herpertz, and Anna Buchheim. (2019). Emotional neglect in childhood shapes social dysfunctioning in adults by influencing the oxytocin and the attachment system: Results from a population-based study. *International Journal of Psychophysiology*, 136 (February). doi:10.1016/j.ijpsycho.2018.05.011.

Nater, Urs M., Martin Bohus, Elvira Abbruzzese, Beate Ditzen, Jens Gaab, Nikolaus Kleindienst, Ulrich Ebner-Priemer, Jana Mauchnik, and Ulrike Ehlert. (2010). Increased psychological and attenuated cortisol and alpha-amylase responses to acute psychosocial stress in female patients with borderline personality disorder. *Psychoneuroendocrinology*, 35(10), 1565–1572. doi:10.1016/j.psyneuen.2010.06.002.

New, Antonia S., Erin A. Hazlett, Monte S. Buchsbaum, Marianne Goodman, Serge A. Mitelman, Randall Newmark, Roanna Trisdorfer, M. Mehmet Haznedar, Harold W.

Koenigsberg, Janine Flory, and Larry J. Siever. (2007). Amygdala–prefrontal disconnection in borderline personality disorder. *Neuropsychopharmacology*, 32(7), 1629–1640. doi:10.1038/sj.npp.1301283.

Newhill, Christina E., Shaun M. Eack, and Edward P. Mulvey. (2012). A growth curve analysis of emotion dysregulation as a mediator for violence in individuals with and without borderline personality disorder. *Journal of Personality Disorders*, 26(3), 452–467. doi:10.1521/pedi.2012.26.3.452.

Niedtfeld, Inga, Ruth Schmitt, Dorina Winter, Martin Bohus, Christian Schmahl, and Sabine C. Herpertz. (2017). Pain-mediated affect regulation is reduced after dialectical behavior therapy in borderline personality disorder: A longitudinal FMRI study. *Social Cognitive and Affective Neuroscience*, 12(5), 739–747. doi:10.1093/scan/nsw183.

Niedtfeld, Inga, Lars Schulze, Peter Kirsch, Sabine C. Herpertz, Martin Bohus, and Christian Schmahl. (2010). Affect regulation and pain in borderline personality disorder: A possible link to the understanding of self-injury. *Biological Psychiatry* 68 (4), 383–391. doi:10.1016/j.biopsych.2010.04.015.

Nunes, Paulo Menezes, Amy Wenzel, Karinne Tavares Borges, Cristianne Ribeiro Porto, Renato Maiato Caminha, and Irismar Reis De Oliveira. (2009). Volumes of the hippocampus and amygdala in patients with borderline personality disorder: A meta-analysis. *Journal of Personality Disorders*, 23(4), 333–345. doi:10.1521/pedi.2009.23.4.333.

Ochsner, Kevin N., Silvia A. Bunge, James J. Gross, and John D. E. Gabrieli. (2002). Rethinking feelings: An FMRI study of the cognitive regulation of emotion. *Journal of Cognitive Neuroscience*, 14(8), 1215–1229. doi:10.1162/089892902760807212.

Ochsner, Kevin N. and James J. Gross. (2014). The neural bases of emotion and emotion regulation: A valuation perspective. In *Handbook of Emotion Regulation*, (pp. 23–24)2, James J. Gross (eds.). New York: Guilford Press.

Paret, Christian, Christine Jennen-Steinmetz, and Christian Schmahl. (2017). Disadvantageous decision-making in borderline personality disorder: Partial support from a meta-analytic review. *Neuroscience and Biobehavioral Reviews*, 72(January), 301–309. doi:10.1016/j.neubiorev.2016.11.019.

Paret, Christian, Rosemarie Kluetsch, Matthias Ruf, Traute Demirakca, Raffael Kalisch, Christian Schmahl, and Gabriele Ende. (2014). Transient and sustained BOLD signal time courses affect the detection of emotion-related brain activation in FMRI. *NeuroImage*, 103(December), 522–532. doi:10.1016/j.neuroimage.2014.08.054.

Paus, Tomáš. (2005). Mapping brain maturation and cognitive development during adolescence. *Trends in Cognitive Sciences* 9 (2), 60–68. doi:10.1016/j.tics.2004.12.008.

Pechtel, Pia, Karlen Lyons-Ruth, Carl M. Anderson, and Martin H. Teicher. 2014. Sensitive periods of amygdala development: The role of maltreatment in preadolescence. *NeuroImage*, 97(August), 236–244. doi:10.1016/j.neuroimage.2014.04.025.

Perez-Rodriguez, M. Mercedes, Antonia S. New, Kim E. Goldstein, Daniel Rosell, Qiaoping Yuan, Zhifeng Zhou, Colin Hodgkinson, David Goldman, Larry J. Siever, and Erin A. Hazlett. (2017). Brain-derived neurotrophic factor Val66Met genotype modulates amygdala habituation. *Psychiatry Research: Neuroimaging*, 263(February), 85–92. doi:10.1016/j.pscychresns.2017.03.008.

Perez-Rodriguez, M. Mercedes, Shauna Weinstein, Antonia S. New, Laura Bevilacqua, Qiaoping Yuan, Zhifeng Zhou, Colin Hodgkinson, Marianne Goodman, Harold W. Koenigsberg, David Goldman, and Larry J. Siever. (2010). Tryptophan-hydroxylase 2 haplotype association with borderline personality disorder and aggression in a sample of patients with personality disorders and healthy controls. *Journal of Psychiatric Research*, 44(15), 1075–1081. doi:10.1016/j.jpsychires.2010.03.014.

Peters, Jessica R., Brian T. Upton, and Ruth A. Baer. (2013). Brief report: Relationships between facets of impulsivity and borderline personality features. *Journal of Personality Disorders*, 27(4), 547–552. doi:10.1521/pedi_2012_26_044.

Peyrot, Wouter J., Milaneschi, Yuri, Abdellaoui, Abdel, Sullivan, Patrick F., Hottenga, Jouke J., Boomsma, Dorret I., and Penninx, Brenda W. J. H. (2014). Effect of polygenic risk scores on depression in childhood trauma. *British Journal of Psychiatry*, 205(2), 113–119. doi:10.1192/bjp.bp.113.143081

Piel, J. H., Tristram A. Lett, Carolin Wackerhagen, Michael M. Plichta, Sebastian Mohnke, Oliver Grimm, N. Romanczuk-Seiferth, Franz J. Degenhardt, Heike Tost, Stephanie Witt, Markus Nöthen, Marcella Rietschel, Andreas Heinz, Andreas Meyer-Lindenberg, Henrik Walter, and Susanne Erk. (2018). The effect of 5-HTTLPR and a serotonergic multi-marker score on amygdala, prefrontal and anterior cingulate cortex reactivity and habituation in a large, healthy FMRI cohort. *European Neuropsychopharmacology*, 28 (3), 415–427. doi:10.1016/j.euroneuro.2017.12.014.

Plener, Paul L., Nikola Bubalo, Anne K. Fladung, Andrea G. Ludolph, and Dorothée Lulé. (2012). Prone to excitement: Adolescent females with non-suicidal self-injury (NSSI) show altered cortical pattern to emotional and NSS-related material. *Psychiatry Research: Neuroimaging*, 203(2–3), 146–152. doi:10.1016/j.pscychresns.2011.12.012.

Plener, Paul L., Katrin Zohsel, Erika Hohm, Arlette F. Buchmann, T. Banaschewski, Ulrich S. Zimmermann, and Manfred Laucht. (2017). Lower cortisol level in response to a psycho-social stressor in young females with self-harm. *Psychoneuroendocrinology*, 76(February), 84–87. doi:10.1016/j.psyneuen.2016.11.009.

Prehn, Kristin, Lars Schulze, Sabine Rossmann, Christoph Berger, Knut Vohs, Monika Fleischer, Karlheinz Hauenstein, Peter Keiper, Gregor Domes, and Sabine C. Herpertz. (2013). Effects of emotional stimuli on working memory processes in male criminal offenders with borderline and antisocial personality disorder. *World Journal of Biological Psychiatry*, 14(1), 71–78. doi:10.3109/15622975.2011.584906.

Reichborn-Kjennerud, Ted, Eivind Ystrom, Michael C. Neale, Steven H. Aggen, Suzanne E. Mazzeo, Gun Peggy Knudsen, Kristian Tambs, Nikolai O. Czajkowski, and Kenneth S. Kendler. (2013). Structure of genetic and environmental risk factors for symptoms of DSM-IV borderline personality disorder. *JAMA Psychiatry*, 70(11), 1206–1214. doi:10.1001/jamapsychiatry.2013.1944.

Reitz, Sarah, Annegret Krause-Utz, Esther M. Pogatzki-Zahn, Ulrich Ebner-Priemer, Martin Bohus, and Christian Schmahl. (2012). Stress regulation and incision in borderline person-ality disorder—a pilot study modeling cutting behavior. *Journal of Personality Disorders*, 26 (4), 605–615. doi:10.1521/pedi.2012.26.4.605.

Rodrigues, Elisabete, A. E. Wenzel, M. P. Ribeiro, L. C. Quarantini, A. Miranda-Scippa, E. de Sena, and I. R. de Oliveira. (2011). Hippocampal volume in borderline personality dis-order with and without comorbid posttraumatic stress disorder: A meta-analysis. *European Psychiatry*, 26 (7), 452–456. doi:10.1016/j.eurpsy.2010.07.005.

Rosario, Maria R. (2012). Effortful control. In *Handbook of Temperament*, (pp. 145–67), Marcel Zentner and Rebecca L. Shiner (eds.). New York: Guilford Press.

Rotge, Jean Y., Cedric Lemogne, Sophie Hinfray, Pascal Huguet, Ouriel Grynszpan, Eric Tartour, Nathalie George, and Philippe Fossati. (2015). A meta-analysis of the anterior cingulate contribution to social pain. *Social Cognitive and Affective Neuroscience*, *10*(1), 19–27. doi:10.1093/scan/nsu110.

Roth, Tania L. and John D. Sweatt. (2011). Annual research review: Epigenetic mechanisms and environmental shaping of the brain during sensitive periods of development. *Journal of Child Psychology and Psychiatry and Allied Disciplines*, *52*(4), 398–408. doi:10.1111/j.1469-7610.2010.02282.x.

Ruocco, Anthony C., Sathya Amirthavasagam, and Konstantine K. Zakzanis. (2012). Amygdala and hippocampal volume reductions as candidate endophenotypes for borderline personality disorder: A meta-analysis of magnetic resonance imaging studies. *Psychiatry Research: Neuroimaging*, *201*(3), 245–252. doi:10.1016/j.pscychresns.2012.02.012.

Rüsch, Nicolas, Miriam Boeker, Martin Büchert, Volkmar Glauche, Carl Bohrmann, Dieter Ebert, Klaus Lieb, Jürgen Hennig, and Ludger Tebartz Van Elst. (2010). Neurochemical alterations in women with borderline personality disorder and comorbid attention-deficit hyperactivity disorder. *The World Journal of Biological Psychiatry*, *11*(2), 372–381. doi:10.3109/15622970801958331.

Salavert, José, Miquel Gasol, Eduard Vieta, Ana Cervantes, Carlos Trampal, and Juan Domingo Gispert. (2011). Fronto-limbic dysfunction in borderline personality disorder: A 18F-FDG positron emission tomography study. *Journal of Affective Disorders*, *131*(1–3), 260–267. doi:10.1016/j.jad.2011.01.001.

Santangelo, Philip, Iris Reinhard, Lutz Mussgay, Regina Steil, Günther Sawitzki, Christoph Klein, Timothy J. Trull, Martin Bohus, and Ulrich W. Ebner-Priemer. (2014). Specificity of affective instability in patients with borderline personality disorder compared to posttraumatic stress disorder, bulimia nervosa, and healthy controls. *Journal of Abnormal Psychology*, *123*(1), 258–272. doi:10.1037/a0035619.

Sapolsky, Robert M. (2015). Stress and the brain: Individual variability and the inverted-U. *Nature Neuroscience*, *18* (10), 1344–1346. doi:10.1038/nn.4109.

Savelieva, Katerina V., Shulei Zhao, Vladimir M. Pogorelov, Indrani Rajan, Qi Yang, Emily Cullinan, and Thomas H. Lanthorn. (2008). Genetic disruption of both tryptophan hydroxylase genes dramatically reduces serotonin and affects behavior in models sensitive to antidepressants. *PloS One*, *3*(10), p. e3301. doi:10.1371/journal.pone.0003301.

Schmahl, Christian, Martin Bohus, Fabrizio Esposito, Rolf-Detlef Treede, Francesco Di Salle, Wolfgang Greffrath, Petra Ludaescher, Anja Jochims, Klaus Lieb, Klaus Scheffler, Juergen Hennig, and Erich Seifritz. (2006). Neural correlates of antinociception in borderline personality disorder. *Archives of General Psychiatry*, *63*(6), 659–667. doi:10.1001/archpsyc.63.6.659.

Schmitt, Ruth, Dorina Winter, Inga Niedtfeld, Sabine C. Herpertz, and Christian Schmahl. (2016). Effects of psychotherapy on neuronal correlates of reappraisal in female patients with borderline personality disorder. *Biological Psychiatry: Cognitive Neuroscience and Neuroimaging*, *1*(6), 548–557. doi:10.1016/j.bpsc.2016.07.003.

Schulz, S. Charles, Jazmin Camchong, Ann Romine, Amanda Schlesinger, Michael Kuskowski, Jose V. Pardo, Kathryn R. Cullen, and Kelvin O. Lim. (2013). An exploratory study of the relationship of symptom domains and diagnostic severity to PET scan imaging in borderline personality disorder. *Psychiatry Research: Neuroimaging*, *214*(2), 161–168. doi:10.1016/j.pscychresns.2013.05.007.

Schulze, Lars, Gregor Domes, Alexander Krüger, Christoph Berger, Monika Fleischer, Kristin Prehn, Christian Schmahl, Annette Grossmann, Karlheinz Hauenstein, and Sabine C.

Herpertz. (2011). Neuronal correlates of cognitive reappraisal in borderline patients with affective instability. *Biological Psychiatry, 69*(6), 564–573. doi:10.1016/j.biopsych.2010.10.025.

Schulze, Lars, Christian Schmahl, and Inga Niedtfeld. (2016). Neural correlates of disturbed emotion processing in borderline personality disorder: A multimodal meta-analysis. *Biological Psychiatry, 79*(2), 97–106. doi:10.1016/j.biopsych.2015.03.027.

Schulze, Lars, Andreas Schulze, Babette Renneberg, Christian Schmahl, and Inga Niedtfeld. (2019). Neural correlates of affective disturbances. A comparative meta-analysis of negative affect processing in borderline personality disorder, major depression, and posttraumatic stress disorder. *Biological Psychiatry: Cognitive Neuroscience and Neuroimaging, 4*(3), 220–232. doi:10.1016/j.bpsc.2018.11.004

Scott, Lori N., Stephanie D. Stepp, and Paul A. Pilkonis. (2014). Prospective associations between features of borderline personality disorder, emotion dysregulation, and aggression. *Personality Disorders: Theory, Research, and Treatment, 5*(3), 278–288. doi:10.1037/per0000070.

Sebastian, Alexandra, Gitta Jacob, Klaus Lieb, and Oliver Tüscher. (2013). Impulsivity in borderline personality disorder: A matter of disturbed impulse control or a facet of emotional dysregulation? *Current Psychiatry Reports, 15*(2), 339. doi:10.1007/s11920-012-0339-y.

Sebastian, Catherine, Essi Viding, Kipling D. Williams, and Sarah Jayne Blakemore. (2010). Social brain development and the affective consequences of ostracism in adolescence. *Brain and Cognition, 72*(1), 134–145. doi:10.1016/j.bandc.2009.06.008.

Sergerie, Karine, Caroline Chochol, and Jorge L. Armony. (2008). The role of the amygdala in emotional processing: A quantitative meta-analysis of functional neuroimaging studies. *Neuroscience and Biobehavioral Reviews, 32*(4), 811–830. doi:10.1016/j.neubiorev.2007.12.002.

Servan, A., Jerome Brunelin, and Emmanuel Poulet. (2018). The effects of oxytocin on social cognition in borderline personality disorder. *Encephale, 44*(1), 46–51. doi:10.1016/j.encep.2017.11.001.

Shafir, Roni, Ravi Thiruchselvam, Gaurav Suri, James J. Gross, and Gal Sheppes. (2016). Neural processing of emotional-intensity predicts emotion regulation choice. *Social Cognitive and Affective Neuroscience, 11*(12), 1863–1871. doi:10.1093/scan/nsw114.

Sharp, Carla and Jennifer L. Tackett. (2014). Introduction: An idea whose time has come. In: *Handbook of Borderline Personality Disorder in Children and Adolescents*, (pp. 3–8), Carla Sharp and Jennifer L. Tackett (eds.). New York: Springer.

Shaw, Philip, Noor J. Kabani, Jason. P. Lerch, Kristen Eckstrand, Rhoshel Lenroot, Nitin Gogtay, Deanna Greenstein, Liv Clasen, Alan Evans, Judith L. Rapoport, Jay N. Giedd, and Steve P. Wise. (2008). Neurodevelopmental trajectories of the human cerebral cortex. *Journal of Neuroscience, 28*(14), 3586–3594. doi:10.1523/JNEUROSCI.5309-07.2008.

Sheppes, Gal, and Ziv Levin. (2013). Emotion regulation choice: Selecting between cognitive regulation strategies to control emotion. *Frontiers in Human Neuroscience, 7*(May), 1–4. doi:10.3389/fnhum.2013.00179.

Sheppes, Gal, Gaurav Suri, and James J. Gross. (2015). Emotion regulation and psychopathology. *Annual Review of Clinical Psychology, 11*(1), 379–405. doi:10.1146/annurev-clinpsy-032814-112739.

Shulman, Elizabeth P., Ashley R. Smith, Karol Silva, Grace Icenogle, Natasha Duell, Jason Chein, and Laurence Steinberg. (2016). The dual systems model: Review, reappraisal, and reaffirmation. *Developmental Cognitive Neuroscience, 17*(February), 103–117. doi:10.1016/j.dcn.2015.12.010.

Silbersweig, David, John F. Clarkin, Martin Goldstein, Otto F. Kernberg, Oliver Tuescher, Kenneth N. Levy, Gary Brendel, Hong Pan, Manfred Beutel, Michelle T. Pavony, Jane

Epstein, Mark F. Lenzenweger, Kathleen M. Thomas, Michael I. Posner, and Emily Stern. (2007). Failure of frontolimbic inhibitory function in the context of negative emotion in borderline personality disorder. *American Journal of Psychiatry*, 164(12), 1832–1841. doi:10.1176/appi.ajp.2007.06010126.

Silvers, Jennifer A., Catherine Insel, Alisa Powers, Peter Franz, Chelsea Helion, Rebecca Martin, Jochen Weber, Walter Mischel, B. J. Casey, and Kevin N. Ochsner. (2017). The transition from childhood to adolescence is marked by a general decrease in amygdala reactivity and an affect-specific ventral-to-dorsal shift in medial prefrontal recruitment. *Developmental Cognitive Neuroscience*, 25: 128–137. doi:10.1016/j.dcn.2016.06.005.

Simeon, Daphne, Margaret Knutelska, Lisa Smith, Bryann R. Baker, and Eric Hollander. (2007). A preliminary study of cortisol and norepinephrine reactivity to psychosocial stress in borderline personality disorder with high and low dissociation. *Psychiatry Research*, 149(1–3), 177–184. doi:10.1016/j.psychres.2005.11.014.

Smesny, Stefan, Johanna Große, Alexander Gussew, Kerstin Langbein, Nils Schönfeld, Gerd Wagner, Matias Valente, and Jürgen R. Reichenbach. (2018). Prefrontal glutamatergic emotion regulation is disturbed in cluster B and C personality disorders – A combined 1 H/ 31 P-MR spectroscopic study. *Journal of Affective Disorders* 227 (October), 688–697. doi:10.1016/j.jad.2017.10.044.

Somerville, Leah H., Rebecca M. Jones, and B. J. Casey. (2010). A time of change: Behavioral and neural correlates of adolescent sensitivity to appetitive and aversive environmental cues. *Brain and Cognition*, 72(1), 124–133. doi:10.1016/j.bandc.2009.07.003.

Stepp, Stephanie D., Sophie A. Lazarus, and Amy L. Byrd. (2015). A systematic review of risk factors prospectively associated with borderline personality disorder: Taking stock and moving forward. *Personality Disorders: Theory, Research, and Treatment*, 7(4), 316–323. doi:10.1037/per0000186.

Stepp, Stephanie D., Lori N. Scott, Neil P. Jones, Diana J. Whalen, and Alison E. Hipwell. (2016). Negative emotional reactivity as a marker of vulnerability in the development of borderline personality disorder symptoms. *Developmental Psychopathology*, 28(1), 213–224. doi:10.1017/S0954579415000395.

Steudte, Susann, Clemens Kirschbaum, Wei Gao, Nina Alexander, Sabine Schönfeld, Jürgen Hoyer, and Tobias Stalder. (2013). Hair cortisol as a biomarker of traumatization in healthy individuals and posttraumatic stress disorder patients. *Biological Psychiatry*, 74(9), 639–646. doi:10.1016/j.biopsych.2013.03.011.

Teicher, Martin H. and Jacqueline A. Samson. (2016). Annual research review: Enduring neurobiological effects of childhood abuse and neglect. *Journal of Child Psychology and Psychiatry and Allied Disciplines*, 57(3), 241–266. doi:10.1111/jcpp.12507.

Todorov, Alexander. (2012). The role of the amygdala in face perception and evaluation. *Motivation and Emotion*, 36(1), 16–26. doi:10.1007/s11031-011-9238-5.

Tottenham, Nim, and Adriana Galván. (2016). Stress and the adolescent brain: Amygdala-prefrontal cortex circuitry and ventral striatum as developmental targets. *Neuroscience and Biobehavioral Reviews*, 70: 2017–2027. doi:10.1016/j.neubiorev.2016.07.030.

Trull, Timothy J., Seungmin Jahng, Rachel L. Tomko, Phillip K. Wood, and Kenneth J. Sher. (2010). Revised NESARC personality disorder diagnoses: Gender, prevalence, and comorbidity with substance dependence disorders. *Journal of Personality Disorders*, 24(4), 412–426. doi:10.1521/pedi.2010.24.4.412.

Uematsu, Akiko, Mie Matsui, Chiaki Tanaka, Tsutomu Takahashi, Kyo Noguchi, Michio Suzuki, and Hisao Nishijo. (2012). Developmental trajectories of amygdala and hippocampus from infancy to early adulthood in healthy individuals. *PLoS ONE, 7*(10), p. e46970. doi:10.1371/journal.pone.0046970.

Vai, Benedetta, Laura Sforzini, Raffaele Visintini, Martina Riberto, Chiara Bulgarelli, Davide Ghiglino, Elisa Melloni, Irene Bollettini, Sara Poletti, Cesare Maffei, and Francesco Benedetti. (2018). Corticolimbic connectivity mediates the relationship between adverse childhood experiences and symptom severity in borderline personality disorder. *Neuropsychobiology, 76*(2), 105–115. doi:10.1159/000487961.

Vuilleumier, Patrik and Jon Driver. (2007). Modulation of visual processing by attention and emotion: Windows on causal interactions between human brain regions. *Philosophical Transactions of the Royal Society B: Biological Sciences, 362*(1481), 837–855. doi:10.1098/rstb.2007.2092.

Wager, Tor D., Matthew L. Davidson, Brent L. Hughes, Martin A. Lindquist, and Kevin N. Ochsner. (2008). Prefrontal-subcortical pathways mediating successful emotion regulation. *Neuron, 59*(6), 1037–1050. doi:10.1016/j.neuron.2008.09.006.

Wagner, Stefanie, Ömür Baskaya, Klaus Lieb, Norbert Dahmen, and André Tadić. (2009). The 5-HTTLPR polymorphism modulates the association of serious life events (SLE) and impulsivity in patients with borderline personality disorder. *Journal of Psychiatric Research, 43*(13), 1067–1072. doi:10.1016/j.jpsychires.2009.03.004.

Walter, Marc, Jean F. Bureau, Bjarne M. Holmes, Eszter A. Bertha, Michael Hollander, Joan Wheelis, Nancy H. Brooks, and Karlen Lyons-Ruth. (2008). Cortisol response to interpersonal stress in young adults with borderline personality disorder: A pilot study. *European Psychiatry, 23*(3), 201–204. doi:10.1016/j.eurpsy.2007.12.003.

Williams, Kipling D. and Blair Jarvis. (2006). Cyberball: A program for use in research on interpersonal ostracism and acceptance. *Behavior Research Methods, 38*(1), 174–180. doi:10.3758/BF03192765.

Williams, W. Craig, Sylvia A. Morelli, Desmond C. Ong, and Jamil Zaki. (2018). Interpersonal emotion regulation: Implications for affiliation, perceived support, relationships, and well-being: Correction to Williams et al. (2018). *Journal of Personality and Social Psychology, 115*(4), 656. doi:10.1037/pspi0000161.

Wilson, Scott T., Barbara Stanley, David A. Brent, Maria A. Oquendo, Yung-yu Huang, Fatemeh Haghighi, Colin A. Hodgkinson, and J. John Mann. (2012). Interaction between tryptophan hydroxylase I polymorphisms and childhood abuse is associated with increased risk for borderline personality disorder in adulthood. *Psychiatric Genetics, 22* (1), 15–24. doi:10.1097/YPG.0b013e32834c0c4c.

Wingenfeld, Katja, Carsten Spitzer, Nina Rullkötter, and Bernd Löwe. (2010). Borderline personality disorder: Hypothalamus pituitary adrenal axis and findings from neuroimaging studies. *Psychoneuroendocrinology, 35*(1), 154–170. doi:10.1016/j.psyneuen.2009.09.014.

Winsper, Catherine, James Hall, Vicky Y. Strauss, and Dieter Wolke. (2017). Aetiological pathways to borderline personality disorder symptoms in early adolescence: Childhood dysregulated behaviour, maladaptive parenting and bully victimisation. *Borderline Personality Disorder and Emotion Dysregulation, 4*(1), 10. doi:10.1186/s40479-017-0060-x.

Winsper, Catherine, Suzet Tanya Lereya, Steven Marwaha, Andrew Thompson, Julie Eyden, and Swaran P. Singh. (2016). The aetiological and psychopathological validity of borderline personality disorder in youth: A systematic review and meta-analysis. *Clinical Psychology Review, 44*(March), 13–24. doi:10.1016/j.cpr.2015.12.001.

Wise, Toby, J. Radua, E. Via, N. Cardoner, O. Abe, T. M. Adams, F. Amico, Y. Cheng, J. H. Cole, C. De Azevedo Marques Périco, D. P. Dickstein, T. F.D. Farrow, T. Frodl, G. Wagner,

I. H. Gotlib, O. Gruber, B. J. Ham, D. E. Job, M. J. Kempton, et al. (2016). Common and distinct patterns of grey-matter volume alteration in major depression and bipolar disorder: Evidence from voxel-based meta-analysis. *Molecular Psychiatry, 22*(10), 1455–1463. doi:10.1038/mp.2016.72.

Wolf, Robert C., Philipp A. Thomann, Fabio Sambataro, Nenad Vasic, Markus Schmid, and Nadine D. Wolf. (2012). Orbitofrontal cortex and impulsivity in borderline personality disorder: An MRI study of baseline brain perfusion. *European Archives of Psychiatry and Clinical Neuroscience, 262*(8), 677–685. doi:10.1007/s00406-012-0303-1.

Wölfer, Ralf, and Herbert Scheithauer. (2012). Ostracism in childhood and adolescence: Emotional, cognitive, and behavioral effects of social exclusion. *Social Influence, 8*(4), 217–236. doi:10.1080/15534510.2012.706233.

Woo, Choong-Wan, Leonie Koban, Ethan Kross, Martin A. Lindquist, Marie T. Banich, Luka Ruzic, Jessica R. Andrews-Hanna, and Tor D. Wager. (2014). Separate neural representations for physical pain and social rejection. *Nature Communications, 5*(November), 5380. doi:10.1038/ncomms6380.

Xia, Mingrui, Jinhui Wang, Yong He. (2013). BrainNet Viewer: A network visualization tool for human brain connectomics. *PLoS ONE, 8*: e68910. doi:10.1371/journal.pone.0068910

Zanarini, Mary C., Frances R. Frankenburg, John Hennen, D. Bradford Reich, and Kenneth R. Silk. (2006). Prediction of the 10-year course of borderline personality disorder. *American Journal of Psychiatry, 163*(5), 827–832. doi:10.1176/ajp.2006.163.5.827.

Zanarini, Mary C., Frances R. Frankenburg, John Hennen, Donald Bradford Reich, and Kenneth R. Silk. (2004). Axis I comorbidity in patients with borderline personality disorder: 6-year follow-up and prediction of time to remission. *The American Journal of Psychiatry, 161*(11), 2108–2114. doi:10.1176/appi.ajp.161.11.2108.

Zanarini, Mary C., Frances R. Frankenburg, Maria E. Ridolfi, Shari Jager-Hyman, John Hennen, and John G. Gunderson. (2006). Reported childhood onset of self-mutilation among borderline patients. *Journal of Personality Disorders, 20*(1), 9–15. doi:10.1521/pedi.2006.20.1.9.

Zetzsche, T., U. W. Preuss, B. Bondy, T. Frodl, P. Zill, G. Schmitt, N. Koutsouleris, D. Rujescu, C. Born, M. Reiser, H.-J. Möller, and E. M. Meisenzahl. (2007). 5-HT1A receptor gene C –1019 G polymorphism and amygdala volume in borderline personality disorder. *Genes, Brain and Behavior, 7*(3), 306–313. doi:10.1111/j.1601-183X.2007.00353.x.

Zimmermann, Peter, and Alexandra Iwanski. (2014). Emotion regulation from early adolescence to emerging adulthood and middle adulthood: Age differences, gender differences, and emotion-specific developmental variations. *International Journal of Behavioral Development, 38*(2), 182–194. doi:10.1177/0165025413515405.

CHAPTER 38

PSYCHOTIC EXPERIENCES, COGNITION, AND NEURODEVELOPMENT

LEON FONVILLE AND JOSEPHINE MOLLON

INTRODUCTION

PSYCHOTIC symptoms such as hallucinations and delusions, whereby perceptions and beliefs are disconnected from reality, are traditionally conceptualized as core features of psychotic disorders. However, while these symptoms are indeed an indication of severe mental illness in a small minority of the population, they are also expressed at levels that do not reach clinical significance in a larger minority of the general population than clinically diagnosed psychotic disorders (van Os, 2003; van Os et al., 2000). These so-called subclinical symptoms—referred to herein as psychotic experiences (PEs)—have led to a reconsideration of the psychosis phenotype where symptoms are thought to be expressed along a continuum rather than as a simple dichotomy of either present or absent (Strauss, 1969, van Os et al., 2000).[1] This reappraisal has generated an elegant account of individual differences in the severity and specificity of psychotic traits, as well as valuable new insights into the pathophysiology and aetiology of psychotic disorders.

Psychotic symptoms typically emerge during adolescence and early adulthood as precursors to clinical psychotic disorders such as schizophrenia—and this is captured in the operational criteria used to identify individuals considered at high risk of developing a psychotic disorder (Yung et al., 1998). However, while the emergence of psychotic symptoms substantially increases risk of developing a clinical psychotic disorder, in the majority of cases these symptoms dissipate over time and most individuals do not transition to the full clinical disorder (Zammit et al., 2013). The risk of transitioning to a clinical psychotic disorder has been associated with several indices of the severity of PEs—such as the number and frequency of symptoms, associated levels of distress and help-seeking, their persistence over time, as well as comorbid symptoms of depression and anxiety—with a dose-response increase in risk over time (Kaymaz et al., 2012). The presence of PEs may also be an important risk factor for non-psychotic disorders (Kelleher et al., 2012a), even in the absence of a transition to a clinical psychotic disorder. There is further evidence that PEs are associated with increased risk of suicide (Yates et al., 2018), poorer general functioning (Healy et al., 2018), and a dose-response relationship with several disability measures (Navarro-Mateu et al., 2017) that is greater than what can be explained by a comorbid mental disorder. Therefore, characterizing the profile of PEs may help improve multiple health outcomes.

Phenomenological and temporal continuity between PEs and psychotic disorders suggests some similarity in their underlying pathways and this is reflected to some degree by overlapping environmental risk factors, as well as familial clustering of symptoms (van Os et al., 2009). However, efforts to identify genetic factors underlying the extended psychosis phenotype have, at best, produced weak findings (Sieradzka et al., 2014), and, at worst, none (Jones et al., 2016; Zammit et al., 2013, 2008). Examination of putative endophenotypes, such as cognitive and brain abnormalities, may be more fruitful in advancing knowledge about the genetic and biological pathways underlying psychosis. Moreover, since clinical and subclinical psychotic symptoms typically manifest around adolescence (Paus et al., 2008)—a period of substantial social, cognitive, and physiological changes—and neurodevelopmental disturbances during this period may play a particularly important role in the emergence and modulation of these symptoms.

Nevertheless, perturbations in brain and behavior are not limited to the adolescent period. In this chapter, we will discuss both early and late deviations from typical neurodevelopment. We will further examine similarities, as well as differences, between neurodevelopmental trajectories of individuals who transition to clinical psychotic disorders versus those who do not. Finally, we will end with an account of some potential underlying mechanisms, as well as suggestions for future research.

PSYCHOTIC EXPERIENCES AND COGNITIVE FUNCTION

Lending support to the hypothesized psychosis continuum is evidence that the cognitive profile associated with subclinical psychotic experiences bears similarities to that of first episode and chronic schizophrenia patients. Most of this evidence derives from studies of childhood, adolescence, and young adulthood which demonstrated increased risk for conversion to clinical psychosis in those presenting with PEs during these developmental periods (Cannon et al., 2002; Horwood et al., 2008; Kelleher et al., 2013; Polanczyk et al., 2010). However, some studies examined the association between PEs and cognitive functioning across adulthood (Johns et al., 2004; Simons et al., 2007), including late adulthood (65+ years of age) (Henderson et al., 1998; Östling et al., 2004). Studies aimed to disentangle the age-associated cognitive trajectories associated with

PEs have utilized both cross-sectional (Gur et al., 2014; Mollon et al., 2018) and longitudinal designs (Carey et al., 2018; Kremen et al., 1998; Mollon et al., 2018; Niarchou et al., 2013), again focusing mostly on the period between childhood and early adulthood.

Studies examining the association between intelligence quotient (IQ) and psychotic episode (PE)s have focused exclusively on childhood. The largest of these studies utilized data from the Avon Longitudinal Study of Parents and Children (ALSPAC) to examine the association between IQ at age 8 and PEs at age 12 (Horwood et al., 2008). Out of a sample of 6751 children, 13.7 percent, 5.6 percent and 2.6 percent were rated as having "broad," "narrow" and "frequent" PEs at age 12, respectively. Interestingly, a non-linear association between IQ and risk for PEs was reported, with both below and above average IQ significantly increasing risk for PEs, regardless of how PEs were defined i.e., "broad," "narrow" or "frequent." This is in contrast with the broadly linear relationship that has been reported between IQ and risk for psychotic disorders, where individuals with below average IQ are at increased risk compared to those with average IQ, who are in turn at increased risk compared to those with above average IQ. A second study utilized data from the Environmental Risk (E-Risk) Longitudinal Twin Study (Polanczyk et al., 2010). Out of a sample of 2127 children, 5.9 percent reported at least one definite PE at age 12, in almost perfect agreement with the 5.6 percent from the ALSPAC cohort. At age 5, children with PEs already showed an IQ deficit of 7.5 points below those without PEs. A very similar effect size was reported between IQ at ages 3–11 and PEs at age 11 in the Dunedin Multidisciplinary Health and Development Study (Cannon et al., 2002), despite a lower prevalence of PEs (1.6 percent). While somewhat mixed, current findings suggest a medium IQ deficit in children who report PEs, and the magnitude of IQ deficit in children who later develop schizophrenia has been reported to be identical (Woodberry et al., 2008), suggesting the importance of PEs as a risk factor for later clinical psychotic disorders. However, the breadth of the cognitive deficits in individuals with PEs remains to be seen, as well as the age-associated changes in these deficits.

Studies examining the association between PEs and cognitive functioning have also looked beyond IQ, to specific cognitive functions. For example, Kelleher and colleagues (2013) administered the comprehensive Measurement and Treatment Research to Improve Cognition in Schizophrenia (MATRICS) cognitive battery to a sample of 165 11- to 13-year-olds, who were also assessed for psychotic experiences using the Schedule for Affective Disorders and Schizophrenia for School-Age Children (K-SADS). Children with PEs showed cognitive deficits across a range of cognitive measures, but most notably in those tapping executive and working memory functions as well as processing speed. A follow-up study provided mixed evidence for an ongoing association between the presence of PEs—either during childhood, adolescence, or early adulthood—and processing speed deficits during early adulthood (Carey et al., 2018). Since executive functions and processing speed deficits have been reported to be among the largest in schizophrenia patients, these findings lend support to the notion of PEs as an extended phenotype of clinical psychotic symptoms. Similar findings have also been reported in the Avon Longitudinal Study of Pregnancy and Childhood (ALSPAC) cohort (Niarchou et al., 2013), where processing speed deficits at age 8, but not age 11, were associated with increased risk for PEs at age 12. Moreover, when adjusting for performance on other cognitive measures, only processing speed remained associated with increased risk. Working memory deficits at both ages 8 and 11, as well as attention deficits at ages 8 and 10, were also associated with elevated risk for PEs. However, improving processing speed performance over time between ages 8 and 11 was also reported to be associated with increased risk for PEs, although the cognitive measure used to assess processing speed at age 11 may be more traditionally conceptualized as a measure of attention than processing speed. Nevertheless, the association between processing speed and PEs remains unclear due to the fact that another study on cognitive functioning and PEs in adults found no association between processing speed and PEs, but did report working memory and memory deficits (Mollon et al., 2016). The only other study, to our knowledge, to have administered a comprehensive cognitive battery to individuals also assessed for PEs is the Philadelphia Neurodevelopmental Cohort (PNC; Gur et al., 2014), where deficits were reported across measures of executive function, memory, complex cognition, social cognition, and sensorimotor speed. The largest deficits, however, were on measures of complex and social cognition. While studies examining the association between specific cognitive abilities and PEs remain few, their findings suggest that PEs are associated with some degree of impairment across cognitive domains that are also impaired in individuals with clinical psychotic disorders.

Studies of the cognitive trajectories associated with PEs have utilized both cross-sectional and longitudinal designs and again have focused on the first two decades of life, with one exception (Mollon et al., 2016). In the PNC, fluctuating cognitive lags, whereby age-associated growth was reduced in individuals with PEs compared to their peers, were reported between ages 8 and 21 on measures of complex and social cognition (Gur et al., 2014). An increase in cognitive impairment with increasing age has also been reported in the South East London Community Health (SELCoH) study, a general population based sample of adults (Mollon et al., 2016). In this study, adults with PEs aged 50 to 90 years old showed larger deficits across general cognitive ability, working memory, and memory than those aged 35 to 49, who in turn showed larger deficits than those aged 25 to 34. However, the association between age and magnitude of deficit was non-linear, with those with PEs aged 16 to 24 years old showing an overall magnitude of deficits similar to 35 to 49-year-olds. Longitudinal studies have also reported some evidence for age-associated increases in cognitive deficits in young people with PEs. For example, Kremen and colleagues (1998) reported that IQ decline between ages 4 and 7 predicted presence of PEs at age 23 and that IQ decline was a more important predictor of PEs than age 7 IQ alone. In the ALSPAC cohort, on the other hand, change between ages 8 and 10/11 on measures of attention and working memory were not associated with PEs at age 12. Moreover, as mentioned previously, improvement on measures of processing speed between ages 8 and 11 was associated with PEs at age 12. These findings have recently been extended to incorporate subsequent cognitive assessments at age 20, as well as ascertainment of PEs at age 18 (Mollon et al., 2018). In line with the previous findings reported by Niarchou and colleagues (2013), individuals with PEs at age 18 showed small, but static, cognitive deficits between the ages of 8 and 20. Moreover, a stable IQ deficit was also found between the ages of 18 months and 8 years (Mollon et al., 2018). The discrepancy between findings from the National Collaborative Perinatal Project (NCPP) (Kremen et al., 1998) and ALSPAC (Mollon et al., 2018; Niarchou et al., 2013) may be due to differences in the age at which PEs were ascertained. Since the majority of PEs are transitory, it may be that individuals with PEs from the NCPP, who were assessed for PEs 5 years later than individuals from the ALSPAC cohort, represent a more severe subgroup.

While the studies of PEs and cognitive function in older adults are few, they are worth mentioning due to the fact that the prevalence of PEs remains substantial throughout adulthood. First, Henderson and colleagues (1998) examined the association between PEs and cognitive functioning (processing speed and episodic memory) in a sample of adults aged 70 years and over. Out of a sample of 935, 7 percent were rated as having experienced PEs and these individuals showed large and significant deficits on measures of processing speed and episodic memory. When individuals considered to be cognitively impaired according to the Mini Mental State Examination (MMSE) were excluded, the episodic memory deficit no longer reached significance, but the processing speed deficit remained significant and of moderate effect size (0.51 SD). Second, Östling and colleagues (2004) examined a sample of 347 adults aged 85 years of age, from which individuals with dementia had been excluded. Out of a sample of 261, 14.7 percent were rated as having PEs using psychiatric examinations, informant interviews and medical records. Individuals with PEs showed significant impairments across most cognitive tests, including measures of verbal, reasoning, spatial, perceptual speed, and memory ability. Impairments ranged from small (e.g., memory: 0.34 SD), to medium (e.g., spatial ability: 0.57 SD), to large (e.g., reasoning ability: 0.79 SD).

Overall, there is substantial evidence that subclinical PEs throughout the life course are associated with some degree of cognitive impairment. The quantitative and qualitative nature of this impairment, however, remains unclear, with several reports of medium deficits in general cognitive ability or IQ, but mixed findings in terms of specific cognitive abilities. Cross-sectional and longitudinal evidence suggests some moderation of cognitive deficits by age, such that impairments increase with age. However, the cross-sectional evidence suggests that not all deficits are moderated by age (Gur et al., 2014), and also that the effect of age may be non-linear (Mollon et al., 2016). Longitudinal evidence is also mixed, with evidence for IQ decline (Kremen et al., 1998), improvements in processing speed (Carey et al., 2018; Niarchou et al., 2013), and stable deficits (Mollon et al., 2018). In studies of older adults with PEs—though limited—the available evidence suggests that PEs in this population may be associated with particularly large deficits (Henderson et al., 1998; Östling et al., 2004). While PEs are one of the largest risk factors for clinical psychotic disorders, the majority are transitory. Therefore, future studies that examine the cognitive correlates of PEs that remit, persist, and convert to clinical psychosis will improve their predictive power. Finally, no study to date has been able to examine the duration of PEs and whether persisting PEs are associated with greater cognitive deficits than those with recent onset. While most studies to date have focused on the first two decades of life in order to better predict and prevent the developmental sequelae of psychotic disorders, adults with PEs may represent a substantial minority of individuals who experience substantial and burdensome cognitive impairments.

Characterizing the Brain in PEs Using Magnetic Resonance Imaging

The study of the brain in relation to PEs is fairly recent and has so far been limited to cross-sectional studies comparing those with PEs to typically developing peers. Though it is easy to highlight how findings overlap with the general psychosis literature, it is important to note that there is a widespread pattern of disturbances in psychotic disorders and subtle differences in the same regions may or may not reflect a vulnerability to psychosis. In addition, variation in definition and assessment of PEs could play a role in explaining some of the mixed findings (Fonville et al., 2017; Lee et al., 2016) and it should be pointed out that many of the early studies had limited sample sizes (n ≈ 60). In recent years an increasing number of large-scale studies have examined PEs in representative populations[2] and we will place more emphasis on these studies.

Early functional MRI studies have often found no difference in behavioral performance with some degree of dysfunction involving prefrontal areas. Increased blood oxygenation level-dependent (BOLD) signal has been reported during paradigms involving Theory of Mind (Meer et al., 2013; Modinos et al., 2010c, 2011), emotion processing (Modinos et al., 2010b, 2012), and response inhibition in children (Jacobson et al., 2010), but not in adults (Meer et al., 2013). More recent large-scale studies have provided mixed support for these initial findings. Wolf and colleagues (2015) examined PEs in the PNC study—ages 11 to 22—and found an increased response to threatening stimuli in the prefrontal lobe and amygdala. Bourque and colleagues (2017) instead found increased signal in the hippocampus and amygdala in a sample of 14-year-olds with PEs that were part of the IMAGEN study. This study also reported a reduced prefrontal BOLD signal in PEs during response inhibition (Bourque et al., 2017). A follow-up longitudinal study of this sample examined prefrontal function during reward learning at ages 14 and 19 (Papanastasiou et al., 2018). This study appeared to show a relative normalization of their BOLD response at age 19. Importantly, individuals with and without PEs were defined based on their rating at age 19 and did not take their rating at age 14 into account. Groups had also been matched for IQ at age 14, which makes it unlikely that initial differences were driven by a lag in cognitive development in those with PEs. Two population-based studies have looked at working memory function in individuals with PEs; ALSPAC (Fonville et al., 2015) and PNC (Wolf et al., 2015). While both studies showed some degree of impairment in performance, only Wolf and colleagues (2015) reported a lower functional response within a fronto-parietal network (FPN). An exploratory subanalysis of PEs in this study further showed a correlation between bilateral prefrontal activation and general cognitive ability. The fact that both studies indicated impairment in working memory is important, given that very different samples were used. The study by Wolf and colleagues (2015) used a wide age range from late childhood to early adulthood whereas the study by Fonville and colleagues (2015) used age-matched young adults and further differentiated between those who had PEs and those who continue to have PEs 2 years after their initial assessment.

A few studies have examined the functional networks of the brain in those with PEs and identified several affected networks. Orr and colleagues (2014) obtained a small sample of young adults and reported that PEs showed greater connectivity within FPN and cingulo-opercular (CON) control networks but reduced connectivity between these networks and the default mode network (DMN). Additionally, severity of PEs positively correlated with the connectivity strength between the CON and the visual cortices. Satterthwaite and colleagues (2015) examined PEs in the PNC and described hypoconnectivity within the CON and hyperconnectivity between frontal regions and this network. Amico and colleagues (2017) reported reduced connectivity within the CON as well as reduced fronto-opercular connectivity in children aged 11 to 13 in the ABDS sample. Sheffield and colleagues (2016) examined the functional connectome in a subsample from the Human Connectome Project selected for PEs and found reduced efficiency of the CON in association with PEs. Furthermore, efficiency of this network was positively correlated with cognitive ability and was found to play a mediating role between PEs and cognitive ability. Studies that focused solely on subclinical auditory hallucinations found increased connectivity between left and right temporal regions (Amico et al., 2017; Diederen et al., 2013) as well as greater connectedness of temporal regions (Lutterveld et al., 2014). Both states of hypo- and hyperconnectivity of the DMN have been described but—with one exception (Satterthwaite et al., 2015)—these seem to point towards reduced DMN connectivity (Amico et al., 2017; Barber et al., 2017) and some further show reduced frontal-DMN coupling (Orr et al., 2014).

Structural MRI studies have not been able to consistently link PEs to abnormalities in the brain. Early studies of grey matter volume (GMV) have found both increased (Jacobson et al., 2010; Modinos et al., 2010a) and decreased (Jacobson et al., 2010; Pelletier-Baldelli et al., 2014) local GMV but there is little agreement across results. More recently, work from the PNC has highlighted abnormalities in the temporal lobe. Satterthwaithe and colleagues (2016) reported reduced global GMV along with additional local reductions that include the prefrontal, orbitofrontal, and medial temporal lobe. Interestingly, they highlighted a group-by-age interaction in the medial temporal lobe: children with PEs show greater GMV whereas adolescents show lower GMV relative to their typically developing peers. An in-depth analysis of the medial temporal lobe further highlighted reduced GMV in the entorhinal cortex and this reduction was strongly correlated to cognitive function (Roalf et al., 2016). Results from the ALSPAC cohort have been mixed. Fonville and colleagues (2015) did not find any differences in GMV but this study considered both transient and persisting symptoms. A second study that looked at cortical surface morphometry—using the same distinction within PEs—did find reduced gyrification (local gyrification index: lGI) in association with PEs (Fonville et al., 2018). Lower lGI in the middle temporal gyrus was associated with persistence of PEs and an interaction effect highlighted relatively lower lGI with increasing brain volume in PEs compared to their peers in prefrontal and occipital regions. Drakesmith and colleagues (2016b)—looking at the mere presence of PEs—instead found a reduction in GMV in the supramarginal gyrus. This study also found reduced cortical myelination in temporal-parietal and prefrontal cortical areas when further considering the severity of PEs, and reduction in temporal-parietal areas also correlated with maternal education.

Studies of white matter structure have been limited and many of these studies did not carry out segmentation of the actual tracts. For example, Jacobson and colleagues (2010) found reduced fractional anisotropy (FA) in temporal and occipital areas of the brain and the relevant fiber tracts running through the identified regions were the cingulum bundle, inferior fronto-occipital fasciculus (IFOF), and superior longitudinal fasciculus (SLF). Two studies have found that severity of PEs is associated with reduced FA (DeRosse et al., 2017; Drakesmith et al., 2016a). The study by Drakesmith and colleagues (2016a) compared PEs to their typical peers and saw the effect—reduced FA proximal to the corpus callosum and corona radiata—become more pronounced when considering severity of PEs. DeRosse and colleagues (2017) instead examined the correlation between severity of PEs and FA and found a negative correlation proximal to the corticospinal tract as well as the IFOF and inferior longitudinal fasciculus. Drakesmith and colleagues (2016a) also found that frontomedial FA had a mediating effect on the relationship between childhood IQ and PEs during adolescence. Only one study has reported increased FA in PEs—a study on children aged 11 to 13 that were part of the ABDS—and found bilateral increased FA in striatal regions (O'Hanlon et al., 2015). This was considered to be part of the uncinate fasciculus (UF). This study did segment the relevant tracts and confirmed increased FA in the UF. It also highlighted increased FA in the anterior section of the IFOF. Finally, there have been two studies that examined the topology of the structural connectome in PEs (Drakesmith et al., 2015; van Dellen et al., 2016). Both highlighted altered topology in parietal, temporal, and cingulate areas that point towards reduced connectedness of key network "hubs"[3] in PEs similar to schizophrenia. The study by van Dellen and colleagues (2016) also involved individuals with a psychotic disorder and the disturbances seen in PEs were present in more pronounced form in those with a psychotic disorder.

It is challenging to make any kind of definitive statement about the state of the brain in individuals with PEs given the relatively small—but growing—body of work to draw from, but it is clear from the literature that 1) *the vast majority of studies on PEs show differences in either brain function, structure, or organization* and 2) *any reported differences appear to show quite subtle effects*. Additionally, we do see spatial convergence within and across imaging modalities. Functional imaging studies have found prefrontal dysfunction associated with PEs across a range of tasks as well as disturbances in network connectivity. Studies of structural connectivity similarly implicate tracts and areas that branch to and from the frontal lobes. However, in terms of grey matter volume we instead see disturbances primarily in temporal-parietal areas of the brain and with little evidence of structural abnormalities in the prefrontal cortex. These are in agreement with the network disturbances that primarily seem to affect the highly connected hub areas of the structural connectome in temporal, parietal, and cingulate regions.

LINKING COGNITIVE AND NEURAL ABNORMALITIES IN PEs

As with any novel area of research, findings on brain abnormalities and cognitive impairments associated with PEs are mixed. Moreover, very few studies have been able to characterize deviations in brain structure and function as well as cognition simultaneously in the same sample. However, findings from these two modalities do show some degree of convergence and in this section, we will attempt to integrate the neuropsychological and neuroimaging evidence regarding PEs.

There is substantial evidence for lower IQ in individuals with PEs, suggesting that—just as with psychotic disorders—lower childhood IQ may be a risk factor for the manifestation of PEs. Additionally, studies on the ALSPAC cohort have been able to examine these deficits in childhood IQ in relation to brain imaging data. First, Fonville and colleagues (2015) reported lower IQ at age 8 in those with persistent, but not transient, PEs compared to their typically developing peers at age 20. However, the persistent PEs group did not show significant differences on imaging metrics compared to their typically developing peers and childhood IQ was not correlated with grey matter volume. In a second study of cortical morphometry in the ALSPAC sample, Fonville and colleagues (2018) reported differences in temporal gyrification between individuals with and without PEs when childhood IQ was taken into consideration. Drakesmith and colleagues (2016a), on the other hand, reported that frontomedial white matter measures at age 20 mediated the association between childhood IQ and PEs at age 18. Here, the authors proposed that longstanding neurodevelopmental aberrations to white matter circuitry first present as IQ deficits in childhood that are still observable during adulthood, when they may contribute to the manifestation of PEs. Though we do not see a clear spatial overlap, it is worth noting the similarity with findings from Sheffield and colleagues (2016) where global efficiency of the CON partially mediated the relationship between PEs and cognitive ability. These findings do seem to lend support to the notion of local disturbances in neurodevelopment that precede the manifestation of PEs. Additionally, Drakesmith and colleagues (2015) had previously reported on reduced efficiency of nodes that are part of the CCN, though did not find network-level differences. On the other hand, PEs were associated with dynamic neurocognitive delays in the PNC (Gur et al., 2014) but showed no relationship with the reduced connectivity within the CON (Satterthwaite et al., 2015). However, it is important to note that individuals with PEs in the ALSPAC cohort showed an average childhood IQ that was within the normal range. Thus, findings from this cohort should be considered in this context, especially when compared to findings from other cohorts. The longstanding notion that individuals with larger brains perform better on tests of general intelligence and that this effect may generalize across age groups and cognitive domains (Pietschnig et al., 2015) may play a role in the differences in findings between these population-based samples. This discrepancy in measures of general cognitive ability may account for the findings that total brain and grey matter volumes do not differ between individuals with and without PEs in the ALSPAC cohort (Fonville et al., 2018), but smaller intracranial volumes were observed in individuals with PEs in the PNC (Satterthwaite et al., 2016).

While the evidence for an association between PEs and cognitive deficits is most consistent regarding IQ or general cognitive ability, there is also evidence for disturbances in specific cognitive domains. Deficits in working memory function have been reported in both children and young adults with PEs (Gur et al., 2014; Kelleher et al., 2013; Mollon et al., 2016) and poorer working memory performance has also been reported in fMRI studies (Fonville et al., 2015; Wolf et al., 2015).[4] Findings from both of these fMRI studies showed the typical involvement of FPN areas, as well as increasing signal with increasing task difficulty. However, reduced BOLD signal in dorsolateral prefrontal (DLPFC) areas during task performance in individuals with PEs compared to controls was reported only in the Wolf and colleagues (2015) study. Interestingly, signal strength in the DLPFC was also positively correlated with general cognitive ability in this study, suggesting the possibility that the observed differences in prefrontal function may not reflect a specific impairment in working memory, but rather a general cognitive deficit. Similarly, in another study from the PNC, disturbances in network connectivity—which focused on the role of the frontal regions in relation to the DMN and the CON—were reported in individuals with PEs (Satterthwaite et al., 2015). Specifically, decoupling of frontal regions and the DMN was reported, as well as hyperconnectivity within the DMN. Moreover, the authors reported a greater coupling of frontal regions and the CON, as well as reduced connectivity within the CON. Interestingly, hyperconnectivity of the DMN was associated with poorer general cognitive ability, but also with specific components of the cognitive test battery, such as executive function, complex reasoning, episodic memory, and social cognition. These findings may reflect disturbances in top-down regulation by frontal cognitive control regions in individuals with PEs, which may, in turn, underlie the reported cognitive deficits in both specific and general cognitive domains (Gur et al., 2014). One such specific domain may be social cognition, since fMRI studies have implicated aberrations in the prefrontal cortex during tasks of theory of mind (Meer et al., 2013; Modinos et al., 2011, 2010b), as well as emotion processing (Modinos et al., 2012, 2010b; Wolf et al., 2015). As mentioned previously, several lines of complementary evidence suggest disturbances in the prefrontal lobe, for example altered FA in association tracts projecting to prefrontal areas (DeRosse et al., 2017; Drakesmith et al., 2016a) and anterior sections of the IFOF (O'Hanlon et al., 2015) as well as reduced prefrontal cortical myelination (Drakesmith et al., 2016b). However, it is also worth reiterating that convincing evidence for prefrontal disturbances in grey matter volume is lacking.

PEs and Conversion to Clinical
Psychotic Disorders

So far, we have presented substantial evidence to suggest that individuals with PEs show at least some abnormalities in cognition, as well as brain structure and function. However, if PEs are to be a useful predictor of future psychiatric outcomes, particularly psychotic disorders, an important question remains: how do the neurodevelopmental abnormalities associated with PEs match those observed in individuals who develop a clinical psychotic disorder? Psychotic disorders, such as schizophrenia, are generally ascribed a strong neuropathological basis, with both early and late neurodevelopmental processes influencing aetiology and pathogenesis (Bearden et al., 2015). For example, disturbances in cognitive functioning are already evident in very early childhood, long before the manifestation of overt psychotic symptoms (Cannon et al., 2002; Polanczyk et al., 2010; Reichenberg et al., 2010). Prospective data on neurodevelopmental trajectories in schizophrenia are scarce but suggest some degree of cognitive decline prior to conversion to clinical psychosis (Mollon and Reichenberg, 2018). Around the time of transition to a clinical psychotic disorder—typically during late adolescence and/or early adulthood—evidence suggests substantial deterioration in many aspects of general functioning, as well as cognitive deficits, and finally the emergence of psychotic symptoms (Bearden et al., 2015). In terms of neurodevelopmental events after the onset of clinical psychosis, however, longitudinal studies of first-episode patients with schizophrenia have failed to find evidence for a continuing decline in cognitive functions, at least within the first 5 years after onset (Bozikas and Andreou, 2011). This apparent stabilization of cognitive dysfunction has been thought to be indicative of deficits that result from neurodevelopmental abnormalities occurring prior to transition to clinical psychosis (Bora and Murray, 2014).

Evidence regarding the trajectory of brain abnormalities prior to conversion to psychosis has mostly come from studies of clinical high-risk populations, i.e., individuals experiencing psychotic symptoms at a frequency or severity below that required for a clinical diagnosis. Similarly to patients with schizophrenia, these individuals at high risk for psychosis show cognitive deficits across a range of domains and, importantly, these deficits are greater in individuals who later transition to the full psychotic disorder compared to those who do not (Fusar-Poli et al., 2012a). A similar pattern of deficits has also been reported in imaging studies of HR. Specifically, functional and structural MRI studies have highlighted abnormalities primarily in prefrontal, temporal, and cingulate regions (Fusar-Poli, 2012; Fusar-Poli et al., 2011, 2007), with differences being somewhat more pronounced in individuals who convert to clinical psychosis versus those whose symptoms remit (Fusar-Poli et al., 2011; Smieskova et al., 2010). While evidence regarding white matter abnormalities in HR individuals is scarce, findings thus far have pointed to reduced FA primarily in the corpus callosum, as well as the major association fibers that pass through frontal and temporal regions of the brain (Karlsgodt et al., 2012).

If we consider PEs as part of the prodromal phase of clinical psychosis, do we see evidence of similar cognitive deficits and agreement in the location of functional and anatomical brain disturbances? We have presented evidence of substantial cognitive impairments in PEs during both adolescence and adulthood, but no evidence that this impairment is part of an ongoing decline towards clinical psychotic disorders. Similarly, while imaging studies of individuals with PEs have reported functional and anatomical disturbances in prefrontal and temporal regions, which overlap with those seen in individuals at high risk for psychosis (Fusar-Poli et al., 2011), it is uncertain whether these anomalies coincide with the emergence of clinical symptoms.

Moreover, there is mixed evidence for poorer outcomes with persistence of PEs, but there have been findings linking temporal lobe disturbances to the severity of PEs (Drakesmith et al., 2016b; Fonville et al., 2018; Roalf et al., 2016). A similar association has been reported in first-episode psychosis (Fusar-Poli et al., 2012b), and high risk for psychosis populations, with greater abnormalities associated with an increased risk of transitioning to clinical psychosis (Fusar-Poli et al., 2011). However, without longitudinal imaging data on individuals with PEs it remains uncertain whether these abnormalities indicate developmental anomalies or, rather, progressive deterioration.

PEs AND NEURODEVELOPMENT

We have discussed dynamic and stable cognitive deficits across the lifespan and differences in brain function, structure, and organization across childhood, adolescence, and adulthood in relation to PEs. What can these findings tell us about neurodevelopment in PEs?

Figure 38.1 shows a conceptual representation of cognitive development and changes in brain network topology from childhood to adulthood.[5] In Figure 38.1a, we show differences between healthy controls and individuals with PEs in cognitive developmental trajectories and these findings reflect those of Mollon and colleagues (2018), as well as of other studies discussed above. The shift towards more distributed brain networks and the refinement of relevant networks represented in Figures 38.1b and 38.1c is based on the work of Fair and colleagues (2007, 2009).

As shown in Figure 38.1a, disturbances in cognitive function emerge at an early age and can be present prior to the manifestation of PEs (Kremen et al., 1998; Mollon et al., 2018; Niarchou et al., 2013). The magnitude of this effect—and the changes over time—have been attributed to small cognitive lags and static deficits, with no sign of a progressive deterioration of function in relation to PEs (Mollon et al., 2018). Evidence of disturbances in brain function—notably prefrontal—that coincide with cognitive difficulties suggest an early deviation from typical development in those that will exhibit PEs. Below, we will briefly discuss findings from cross-sectional imaging studies of individuals with PEs in relation to typical neurodevelopment shown in Figures 38.1b and 38.1c.

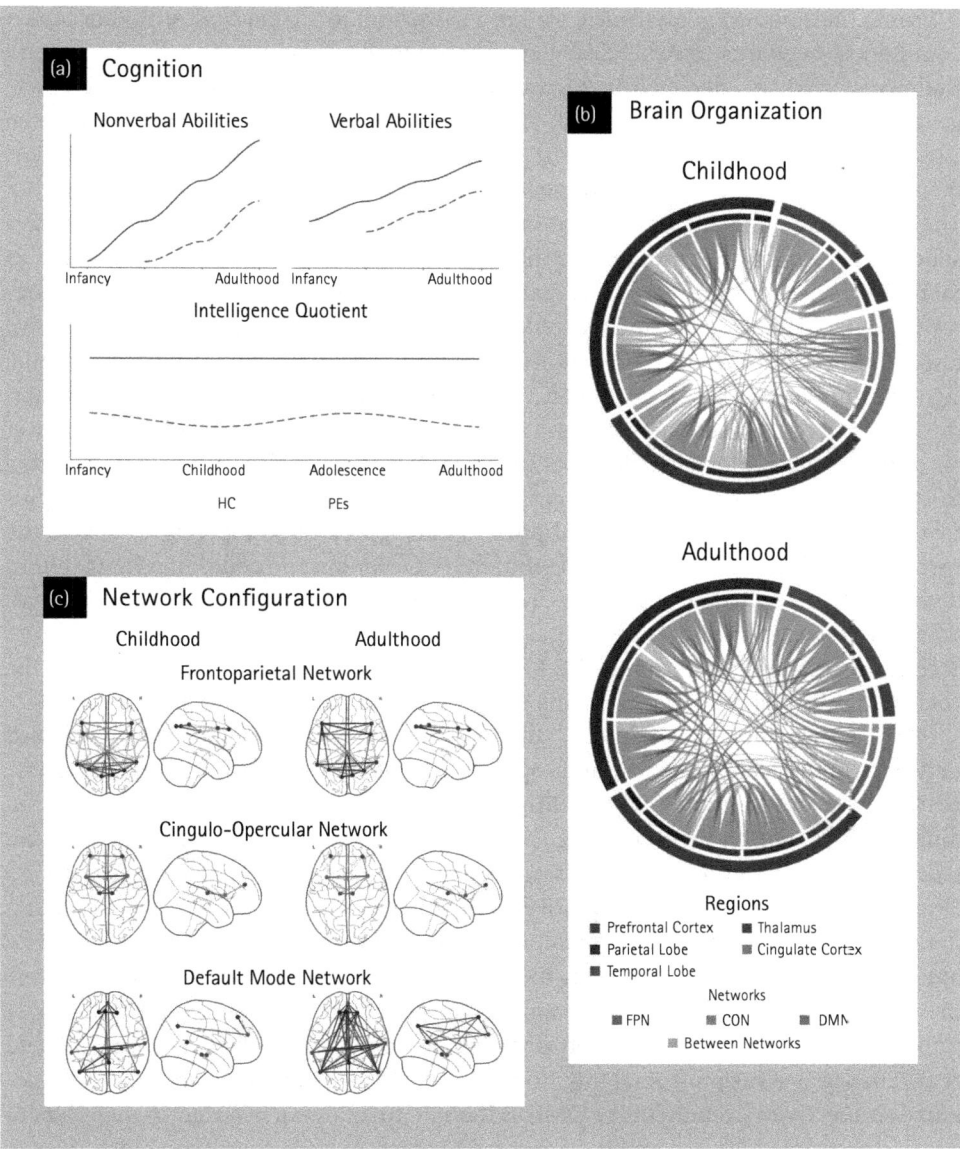

FIGURE 38.1 Cognitive trajectories and development of the frontoparietal (FPN), cingulo-opercular (CON), and default mode networks (DMN). Panel (a) Developmental trajectories for cognitive function from infancy to adulthood in healthy controls (HC) and individuals with psychotic experiences (PEs). Panel (b) Functional reorganization of brain regions associated with FPN, CON, and DMN from childhood to adulthood. Panel (c) Corresponding network configurations during childhood and adulthood.

From childhood to adolescence, we see a shift from local anatomically constrained community networks towards distributed functionally defined networks (Vértes and Bullmore, 2015). As depicted in Figure 38.1b–c, this segregation leads to the emergence of specialized networks—such as the FPN, CON, and DMN (Fair et al., 2007; Sato et al., 2014)—and it has been suggested that the prefrontal cortex plays a prominent role in the development of these networks (Johnson et al., 2015; Sato et al., 2014). The number and strength of connections between frontal hub regions and cortical and subcortical regions increases during this period (Hwang et al., 2013). There is also an increased recruitment of prefrontal areas during this period on tasks that require cognitive control (Luna et al., 2010). A reduction in prefrontal BOLD signal on cognitive tasks during childhood and early adolescence could be interpreted as a developmental lag (Bourque et al., 2017; Jacobson et al., 2010; Papanastasiou et al., 2018). In this regard, the associations between cognition and hyperconnectivity of the DMN (Satterthwaite et al., 2015) as well as global efficiency of the CON (Sheffield et al., 2016) could be related to early disturbances in top-down regulation by prefrontal regions involved in cognitive control. Myelination is a crucial process during neurodevelopment to enable the increased modularity of the brain (Menon, 2013). O'Hanlon and colleagues (2015) found increased FA in the IFOF and UF during childhood, both tracts that are slower to mature and project to and from prefrontal regions (Lebel et al., 2008). This could indicate accelerated maturation in those with PEs, but it could also reflect an adaptation to the aberrant prefrontal functioning during this period.

From adolescence to adulthood, there is an increasing reliance on specialized networks and an emphasis towards optimizing network configuration (Luna et al., 2010; Vértes and Bullmore, 2015). This could explain why there are fewer reports on differences in prefrontal BOLD signal. Papanastasiou and colleagues (2018) found reduced prefrontal BOLD during reward learning at age 14, but at age 19 there was no clear difference in BOLD or cognitive performance. Fonville and colleagues (2015) found no evidence of a working memory deficit or aberrant BOLD response at age 20 but did report a greater reliance on top-down frontal signaling within a FPN. While there is less evidence of focal disturbances in the prefrontal lobe, we still see evidence of disturbances in aforementioned networks (Barber et al., 2017; Orr et al., 2014; Sheffield et al., 2016) as well as prefrontal functional coupling (Modinos et al., 2010b). The functional connectivity strength of the DMN is thought to follow an inverted U-shape where it is strongest in adulthood and is considered to correlate with cognitive ability (Mak et al., 2017). Adults with PEs have been found to spend more time in a state characterized by hypoconnectivity within the DMN (Barber et al., 2017). This could mean that DMN perturbations do not normalize with maturation. Correspondingly, network analyses of white matter pathways have found global reductions in connectivity and reduced local efficiency involving parietal hubs that are part of the DMN (Drakesmith et al., 2015; van Dellen et al., 2016).

A key question still unanswered is how the severity of PEs affects these reported brain abnormalities (DeRosse et al., 2017; Drakesmith et al., 2016b, 2016a; Fonville et al., 2018; Orr et al., 2014; Satterthwaite et al., 2016; van Dellen et al., 2016). However, without longitudinal imaging studies it will be difficult to distinguish the aetiology of PEs from psychotic disorders and typical development.

POTENTIAL MECHANISMS

The notion that psychotic experiences and psychotic disorders lie on a continuum partly originates from evidence that these phenomena share risk factors. In this chapter, we have summarized evidence that subclinical PEs are associated with similar, but not identical, neurocognitive dysfunction and brain abnormalities as clinical psychotic disorders. The limited information that is available does not point towards a single "hit" or a "lag" in neurodevelopment but suggests dynamic changes from childhood through adulthood are more likely. There is also evidence that subclinical and clinical psychosis share a multitude of other risk factors. It follows, therefore, that these phenomena may reflect at least partly overlapping mechanisms. In this section we will summarize evidence regarding several of these potential mechanisms.

Psychotic disorders, particularly schizophrenia, are widely regarded to be of neurodevelopmental origin. Nevertheless, many environmental factors show an association with psychosis risk. Specifically, cannabis use, prenatal and perinatal complications, and childhood adversity are among those risk factors most robustly associated with psychotic disorders. Several studies have sought to establish whether these same putative environmental mechanisms might also underlie risk for PEs. Cannabis use, which has long been known to increase risk for schizophrenia and other psychotic disorders, has also been shown to increase risk for subclinical psychotic experiences. For example, Kuepper and colleagues (2011) showed, using a general population sample of 14- to 24-year-olds from Germany, that incident cannabis use between baseline and first follow-up increased risk of reporting PEs at second and third follow-up. Moreover, continued cannabis use increased the risk of these PEs persisting between the second and third follow-ups. The longitudinal nature of these data lends credence to the causal nature of the association between cannabis use and psychotic experiences. Similarly, Morgan and colleagues (2014) reported an association between cannabis use and PEs in a sample of adults from the South East London Community Health study (SELCoH). Interestingly, this association only held true for cannabis use in the previous year, with no association between PEs and cannabis use 2 years before. Finally, Zammit and colleagues (2011) used data from ALSPAC to show that individuals who had used cannabis by age 14 were twice as likely to report PEs at age 16 than those who had not used cannabis. There was also evidence of a dose-response relationship between cannabis use and PEs, lending credence to the causality of the association between cannabis and PEs.

Prenatal and perinatal complications, which have long been considered to increase risk for psychosis, have also been examined as a potential mechanism in the development of PEs. The Avon Longitudinal Study of Parents and Children (ALSPAC) has published extensively on this topic, finding evidence for increased risk of PEs in offspring following a range of adverse prenatal and perinatal events. Specifically, Zammit and colleagues (2009) reported that maternal infection, maternal diabetes, resuscitation, and 5-min Apgar score were all associated with increased risk for PEs at age 12. Interestingly, none of these associations were mediated by IQ at age 8, possibly suggesting mediation by more subtle neurodevelopmental deficits than low IQ. Similarly, Betts and colleagues (2014) reported an association between maternal vaginal infection and risk of PEs at age 21 in an Australian birth cohort. However, further analyses revealed that the association between maternal vaginal infection and PEs was not a direct one, but rather, mediated by risk of infant illness susceptibility. Indeed, there is further evidence of an association between early life infection and later risk for PEs, for example inflammation, atopic disorders (Khandaker et al., 2014b), and Epstein–Barr Virus (EBV) (Khandaker et al., 2014a).

Childhood adversity is also associated with increased risk for psychotic disorders and has also been examined in relation to subclinical PEs. For example, Morgan and colleagues (2014) reported increased risk for PEs in adults who had experienced either physical or sexual abuse as children, with the greatest risk being in those who had experienced both. Moreover, there was evidence for synergistic effects of abuse and life events, or, in other words, the combined effect of abuse and life events on PEs was greater than the sum of these individual effects. Thus, similarly to the way in which prenatal events increase risk for later PEs through increased susceptibility to childhood illness, childhood adversity may increase vulnerability to PEs through exposure to later adversity. Similarly, Janssen and colleagues (2004) reported that childhood abuse reported at baseline predicted incidence of PEs at 2-year follow-up in a sample of 4045 adults aged 18 to 64 years old. Finally, Kelleher and colleagues (2008), using a sample of 211 adolescents aged 12 to 15, reported an increased risk of PEs in individuals who had experienced physical or sexual abuse as children. Moreover, risk of PEs was even greater in those who had witnessed domestic violence as children, as well as those with a history of being a perpetrator of bullying.

Several studies have also sought to examine whether the genetic factors underlying psychotic disorders, particularly schizophrenia, are also associated with risk for subclinical psychotic experiences. However, evidence to date suggests that this is not the case. For example, Zammit and colleagues (2014) used data from the Avon Longitudinal Study of Parents and Children (ALSPAC) to examine whether schizophrenia polygenic risk score and variants showing genome-wide association with schizophrenia were also associated with psychotic experiences, concluding that no evidence for such an association was found. Indeed, individuals who had a higher number of risk alleles for genome-wide hits for schizophrenia actually showed a decreased risk of psychotic experiences. The authors extended this finding more recently, reporting an association between schizophrenia polygenic risk scores and negative symptoms and anxiety disorder, but not with psychotic experiences or depression (Jones et al., 2016). Similarly, Simons and colleagues (2007) reported that a phenotypic correlation between negative, but not positive, subclinical psychotic symptoms and processing speed was explained by genetic factors. Thus, there has been no evidence to suggest that positive PEs in adolescents from general population samples share a genetic architecture with schizophrenia. Instead, subclinical negative symptoms may be a more fruitful approach to increasing our understanding of the genetic aetiology of schizophrenia. Correspondingly, while psychotic experiences are somewhat more common during adolescence, the peak age at onset for schizophrenia is in the 20s, suggesting that diverging mechanisms underlie these phenomena.

Future Directions

Experiencing subclinical psychotic experiences in early life clearly increases risk for developing psychotic disorders later in life. However, PEs also increase risk for a multitude of other poor psychiatric outcomes, and many individuals who experience PEs do not go on to experience any further psychiatric symptoms. Nevertheless, clinical and subclinical psychotic symptoms are associated with similar, although not identical, neurocognitive dysfunction and brain abnormalities, suggesting at least partly overlapping aetiologies. While environmental risk factors show some commonality between clinical and subclinical psychosis, the genetic underpinnings have so far shown less of an overlap. Future studies should aim to delineate the neurocognitive correlates of PEs that remit, persist or convert to clinical psychosis to improve their predictive power. Moreover, we have presented evidence of substantial neurocognitive abnormalities in association with the manifestation of PEs across different stages of development, but there is little evidence to suggest that these abnormalities represent an ongoing decline towards clinical psychotic disorders. Future longitudinal studies are needed to establish whether these anomalies coincide with the emergence of clinical symptoms, as well as whether they indicate processes relating to developmental deviations or progressive neurodegeneration.

Notes

1. Debate about the validity and clinical utility of an extended psychosis phenotype is still ongoing (David, 2010; David and Ajnakina, 2016; Kaymaz and van Os, 2010; Lawrie et al., 2010; McGorry and Nelson, 2016; Tandon, 2016; Yung and Lin, 2016).
2. There are four population studies that incorporate research on PEs using MRI; the Philadelphia Neurodevelopmental Cohort (PNC) (Satterthwaite et al., 2014), the Avon Longitudinal Study of Parents and Children (ALSPAC) (Boyd et al., 2013), the IMAGEN Consortium (Schumann et al., 2010), and the Adolescent Brain Development Study (ABDS) (Kelleher et al., 2012b). The PNC and IMAGEN studies used self-report measures whereas ALSPAC and ABDS used semi-structured interviews to define PEs.
3. Network topology of the brain has typically been studied using graph theory. This approach models brain regions as network nodes with connections of varying strength. The brain

connectome is thought to exhibit small-world architecture, where networks communicate with each other through highly connected nodes that function as "hubs" (Bullmore and Sporns, 2009). In this framework, hubs are crucial for efficient information transmission among specialized networks.

4. It should be noted that, with the exception of the community-based study by Kelleher and colleagues (2013) on the ABDS, the fMRI, and neurocognitive studies were part of the same cohorts; ALSPAC and PNC.

5. The changes in brain network topology were based on a paper from Fair and colleagues (2009) and childhood refers to those 7–9 years old and adulthood is those 19–31 years old. This paper used resting-state fMRI to characterize the connectivity changes with age.

References

Amico, F., O'Hanlon, E., Kraft, D., Oertel-Knöchel, V., Clarke, M., Kelleher, I., Higgins, N., Coughlan, H., Creegan, D., Heneghan, M., Power, E., Power, L., Ryan, J., Frodl, T. and Cannon, M. (2017). Functional connectivity anomalies in adolescents with psychotic symptoms. *PloS One*, 12(1), e0169364. doi:10.1371/journal.pone.0169364

Barber, A. D., Lindquist, M. A., DeRosse, P., Karlsgodt, K. H. (2017). Dynamic functional connectivity states reflecting psychotic-like experiences. *Biol. Psychiatry Cogn. Neurosci. Neuroimaging*, 3(5), 443–453. doi:10.1016/j.bpsc.2017.09.008

Bearden, C. E., Meyer, S. E., Loewy, R. L., Niendam, T. A., Cannon, T. D. (2015). The neurodevelopmental model of schizophrenia: Updated, in: Cicchetti, D., Cohen, D. J. (Eds.), *Developmental Psychopathology*. John Wiley & Sons, Inc., Hoboken, NJ, USA, pp. 542–569. doi:10.1002/9780470939406.ch14

Betts, K. S., Williams, G. M., Najman, J. M., Scott, J., Alati, R. (2014). Maternal prenatal infection, early susceptibility to illness and adult psychotic experiences: A birth cohort study. *Schizophr. Res.*, 156, 161–167. doi:10.1016/j.schres.2014.04.013

Bora, E. and Murray, R. M. (2014). Meta-analysis of cognitive deficits in ultra-high risk to psychosis and first-episode psychosis: Do the cognitive deficits progress over, or after, the onset of psychosis? *Schizophr. Bull.*, 40, 744–755. doi:10.1093/schbul/sbt085

Bozikas, V. P. and Andreou, C. (2011). Longitudinal studies of cognition in first episode psychosis: A systematic review of the literature. *Aust. N. Z. J. Psychiatry*, 45, 93–108. doi:10.3109/00048674.2010.541418

Bourque, J., Spechler, P. A., Potvin, S., Whelan, R., Banaschewski, T., Bokde, A. L. W., Bromberg, U., Büchel, C., Quinlan, E. B., Desrivières, S., Flor, H., Frouin, V., Gowland, P., Heinz, A., Ittermann, B., Martinot, J.-L., Paillère-Martinot, M.-L., McEwen, S. C., Nees, F., Orfanos, D. P., Paus, T., Poustka, L., Smolka, M. N., Vetter, N. C., Walter, H., Schumann, G., Garavan, H. and Conrod, P. J. (2017). Functional neuroimaging predictors of self-reported psychotic symptoms in adolescents. *Am. J. Psychiatry*, 174, 566–575. doi:10.1176/appi.ajp.2017.16080897

Boyd, A., Golding, J., Macleod, J., Lawlor, D. A., Fraser, A., Henderson, J., Molloy, L., Ness, A., Ring, S. and Davey Smith, G. (2013). Cohort Profile: The "Children of the 90s"—the index offspring of the Avon Longitudinal Study of Parents and Children. *Int. J. Epidemiol.*, 42, 111–127. doi:10.1093/ije/dys064

Bullmore, E. and Sporns, O. (2009). Complex brain networks: Graph theoretical analysis of structural and functional systems. *Nat. Rev. Neurosci.*, 10, 186–198. doi:10.1038/nrn2575

Cannon, M., Caspi, A., Moffitt, T. E., Harrington, H., Taylor, A., Murray, R. M. and Poulton, R. (2002). Evidence for early-childhood, pan-developmental impairment specific to schizophreniform disorder: Results from a longitudinal birth cohort. *Arch. Gen. Psychiatry*, *59*, 449–456. doi:10.1001/archpsyc.59.5.449

Carey, E., Dooley, N., Gillan, D., Healy, C., Coughlan, H., Clarke, M., Kelleher, I., Cannon, M. (2018). Fine motor skill and processing speed deficits in young people with psychotic experiences: A longitudinal study. *Schizophr. Res.*, *204*, 127–132. doi:10.1016/j.schres.2018.08.014

David, A. S. (2010). Why we need more debate on whether psychotic symptoms lie on a continuum with normality. *Psychol. Med.*, *40*, 1935–1942. doi:10.1017/S0033291710000188

David, A. S. and Ajnakina, O. (2016). Psychosis as a continuous phenotype in the general population: The thin line between normality and pathology. *World Psychiatry*, *15*, 129–130. doi:10.1002/wps.20327

DeRosse, P., Ikuta, T., Karlsgodt, K. H., Peters, B. D., Gopin, C. B., Szeszko, P. R., Malhotra, A. K. (2017). White matter abnormalities associated with subsyndromal psychotic-like symptoms predict later social competence in children and adolescents. *Schizophr. Bull.*, *43*, 152–159. doi:10.1093/schbul/sbw062

Diederen, K. M. J., Neggers, S. F. W., Weijer, A. D. de, Lutterveld, R. van, Daalman, K., Eickhoff, S. B., Clos, M., Kahn, R. S. and Sommer, I. E. C. (2013). Aberrant resting-state connectivity in non-psychotic individuals with auditory hallucinations. *Psychol. Med.*, *43*, 1685–1696. doi:10.1017/S0033291712002541

Drakesmith, M., Caeyenberghs, K., Dutt, A., Zammit, S., Evans, C. J., Reichenberg, A., Lewis, G., David, A. S. and Jones, D. K. (2015). Schizophrenia-like topological changes in the structural connectome of individuals with subclinical psychotic experiences. *Hum. Brain Mapp.*, *36*, 2629–2643. doi:10.1002/hbm.22796

Drakesmith, M., Dutt, A., Fonville, L., Zammit, S., Reichenberg, A., Evans, C. J., Lewis, G., Jones, D. K. and David, A. S. (2016a). Mediation of developmental risk factors for psychosis by white matter microstructure in young adults with psychotic experiences. *JAMA Psychiatry*, *73*(4), 396–406. doi:10.1001/jamapsychiatry.2015.3375

Drakesmith, M., Dutt, A., Fonville, L., Zammit, S., Reichenberg, A., Evans, C. J., McGuire, P., Lewis, G., Jones, D. K. and David, A. S. (2016b). Volumetric, relaxometric and diffusometric correlates of psychotic experiences in a non-clinical sample of young adults. *NeuroImage Clin.*, *12*, 550–558. doi:10.1016/j.nicl.2016.09.002

Fair, D. A., Cohen, A. L., Power, J. D., Dosenbach, N. U. F., Church, J. A., Miezin, F. M., Schlaggar, B. L. and Petersen, S. E. (2009). Functional brain networks develop from a "local to distributed" organization. *PLoS Comput. Biol.*, *5*, e1000381. doi:10.1371/journal.pcbi.1000381

Fair, D. A., Dosenbach, N. U. F., Church, J. A., Cohen, A. L., Brahmbhatt, S., Miezin, F. M., Barch, D. M., Raichle, M. E., Petersen, S. E. and Schlaggar, B. L. (2007). Development of distinct control networks through segregation and integration. *Proc. Natl. Acad. Sci.*, *104*, 13507–13512. doi:10.1073/pnas.0705843104

Fonville, L., Drakesmith, M., Zammit, S., Lewis, G., Jones, D. K., David, A. S. (2018). MRI indices of cortical development in young people with psychotic experiences: Influence of genetic risk and persistence of symptoms. *Schizophr. Bull.*, *45*(1), 169–179. doi:10.1093/schbul/sbx195

Fonville, L., Kadosh, K. C., Drakesmith, M., Dutt, A., Zammit, S., Mollon, J., Reichenberg, A., Lewis, G., Jones, D. K. and David, A. S. (2015). Psychotic experiences, working memory, and the developing brain: A multimodal neuroimaging study. *Cereb. Cortex*, 25(12), 4828–4838. doi:10.1093/cercor/bhv181

Fonville, L., Zammit, S., Lewis, G. and David, A. S. (2017). Review of assessment tools for psychotic-like experiences misses the psychosis-like symptoms semi-structured interview (PLIKSi) in ALSPAC. *Early Interv. Psychiatry*, 11, 269–270. doi:10.1111/eip.12254

Fusar-Poli, P. (2012). Voxel-wise meta-analysis of fMRI studies in patients at clinical high risk for psychosis. *J. Psychiatry Neurosci. JPN*, 37, 106–112. doi:10.1503/jpn.110021

Fusar-Poli, P., Borgwardt, S., Crescini, A., Deste, G., Kempton, M. J., Lawrie, S., Mc Guire, P., Sacchetti, E. (2011). Neuroanatomy of vulnerability to psychosis: A voxel-based meta-analysis. *Neurosci. Biobehav. Rev.*, 35, 1175–1185. doi:10.1016/j.neubiorev.2010.12.005

Fusar-Poli, P., Deste, G., Smieskova, R., Barlati, S., Yung, A. R., Howes, O., Stieglitz, R.-D., Vita, A. and McGuire, P. and Borgwardt, S. (2012a). Cognitive functioning in prodromal psychosis: A meta-analysis. *Arch. Gen. Psychiatry*, 69, 562–571. doi:10.1001/archgenpsychiatry.2011.1592

Fusar-Poli, P., Perez, J., Broome, M., Borgwardt, S., Placentino, A., Caverzasi, E., Cortesi, M., Veggiotti, P., Politi, P., Barale, F., McGuire, P. (2007). Neurofunctional correlates of vulnerability to psychosis: A systematic review and meta-analysis. *Neurosci. Biobehav. Rev.*, 31, 465–484. doi:10.1016/j.neubiorev.2006.11.006

Fusar-Poli, P., Radua, J., McGuire, P. and Borgwardt, S. (2012b). Neuroanatomical maps of psychosis onset: Voxel-wise meta-analysis of antipsychotic-naive VBM Studies. *Schizophr. Bull.*, 38, 1297–1307. doi:10.1093/schbul/sbr134

Gur, R. C., Calkins, M. E., Satterthwaite, T. D., Ruparel, K., Bilker, W. B., Moore, T. M., Savitt, A. P., Hakonarson, H. and Gur, R. E. (2014). Neurocognitive growth charting in psychosis spectrum youths. *JAMA Psychiatry*, 71, 366–374. doi:10.1001/jamapsychiatry.2013.4190

Healy, C., Campbell, D., Coughlan, H., Clarke, M., Kelleher, I. and Cannon, M. (2018). Childhood psychotic experiences are associated with poorer global functioning throughout adolescence and into early adulthood. *Acta Psychiatr. Scand.*, 138, 26–34. doi:10.1111/acps.12907

Henderson, A. S., Korten, A.E., Levings, C., Jorm, A. F., Christensen, H., Jacomb, P. A. and Rodgers, B. (1998). Psychotic symptoms in the elderly: A prospective study in a population sample. *Int. J. Geriatr. Psychiatry*, 13, 484–492. doi:10.1002/(SICI)1099-1166(199807)13:7<484::AID-GPS808>3.0.CO;2-7

Horwood, J., Salvi, G., Thomas, K., Duffy, L., Gunnell, D., Hollis, C., Lewis, G., Menezes, P., Thompson, A., Wolke, D., Zammit, S. and Harrison, G. (2008). IQ and non-clinical psychotic symptoms in 12-year-olds: Results from the ALSPAC birth cohort. *Br. J. Psychiatry*, 193, 185–191. doi:10.1192/bjp.bp.108.051904

Hwang, K., Hallquist, M. N. and Luna, B. (2013). The development of hub architecture in the human functional brain network. *Cereb. Cortex*, 23, 2380–2393. doi:10.1093/cercor/bhs227

Jacobson, S., Kelleher, I., Harley, M., Murtagh, A., Clarke, M., Blanchard, M., Connolly, C., O'Hanlon, E., Garavan, H. and Cannon, M. (2010). Structural and functional brain correlates of subclinical psychotic symptoms in 11–13-year-old schoolchildren. *NeuroImage*, 49, 1875–1885. doi:10.1016/j.neuroimage.2009.09.015

Janssen, I., Krabbendam, L., Bak, M., Hanssen, M., Vollebergh, W., Graaf, R. de and van Os, J. (2004). Childhood abuse as a risk factor for psychotic experiences. *Acta Psychiatr. Scand*, 109, 38–45. doi:10.1046/j.0001-690X.2003.00217.x

Johns, L.C., Cannon, M., Singleton, N., Murray, R. M., Farrell, M., Brugha, T., Bebbington, P., Jenkins, R. and Meltzer, H. (2004). Prevalence and correlates of self-reported psychotic symptoms in the British population. *Br. J. Psychiatry*, 185, 298–305. doi:10.1192/bjp.185.4.298

Johnson, M. H., Jones, E. J. H. and Gliga, T. (2015). Brain adaptation and alternative developmental trajectories. *Dev. Psychopathol.*, *27*, 425–442. doi:10.1017/S0954579415000073

Jones, H. J., Stergiakouli, E., Tansey, K. E., Hubbard, L., Heron, J., Cannon, M., Holmans, P., Lewis, G., Linden, D. E. J., Jones, P. B., Davey Smith, G., O'Donovan, M. C., Owen, M. J., Walters, J. T. and Zammit, S. (2016). Phenotypic manifestation of genetic risk for schizophrenia during adolescence in the general population. *JAMA Psychiatry*, *73*, 221–228. doi:10.1001/jamapsychiatry.2015.3058

Karlsgodt, K., Jacobson, S., Seal, M. and Fusar-Poli, P. (2012). The relationship of developmental changes in white matter to the onset of psychosis. *Curr. Pharm. Des.*, *18*, 422–433. doi:10.2174/138161212799316073

Kaymaz, N., Drukker, M., Lieb, R., Wittchen, H. U., Werbeloff, N., Weiser, M., Lataster, T. and van Os, J. (2012). Do subthreshold psychotic experiences predict clinical outcomes in unselected non-help-seeking population-based samples? A systematic review and meta-analysis, enriched with new results. *Psychol. Med.*, *42*, 2239–2253. doi:Doi 10.1017/S0033291711002911

Kaymaz, N. and van Os, J. (2010). Extended psychosis phenotype—yes: Single continuum—unlikely. *Psychol. Med.*, *40*, 1963–1966. doi:10.1017/S0033291710000358

Kelleher, I., Clarke, M. C., Rawdon, C., Murphy, J. and Cannon, M. (2013). Neurocognition in the extended psychosis phenotype: Performance of a community sample of adolescents with psychotic symptoms on the MATRICS Neurocognitive Battery. *Schizophr. Bull.*, *39*, 1018–1026. doi:10.1093/schbul/sbs086

Kelleher, I., Harley, M., Lynch, F., Arseneault, L., Fitzpatrick, C. and Cannon, M. (2008). Associations between childhood trauma, bullying and psychotic symptoms among a school-based adolescent sample. *Br. J. Psychiatry*, *193*, 378–382. doi:10.1192/bjp.bp.108.049536

Kelleher, I., Keeley, H., Corcoran, P., Lynch, F., Fitzpatrick, C., Devlin, N., Molloy, C., Roddy, S., Clarke, M. C., Harley, M., Arseneault, L., Wasserman, C., Carli, V., Sarchiapone, M., Hoven, C., Wasserman, D. and Cannon, M. (2012a). Clinicopathological significance of psychotic experiences in non-psychotic young people: Evidence from four population-based studies. *Br. J. Psychiatry*, *201*, 26–32. doi:10.1192/bjp.bp.111.101543

Kelleher, I., Murtagh, A., Molloy, C., Roddy, S., Clarke, M. C., Harley, M. and Cannon, M. (2012b). Identification and characterization of prodromal risk syndromes in young adolescents in the community: A population-based clinical interview study. *Schizophr. Bull.*, *38*, 239–246. doi:10.1093/schbul/sbr164

Khandaker, G. M., Stochl, J., Zammit, S., Lewis, G. and Jones, P. B. (2014a). Childhood Epstein-Barr Virus infection and subsequent risk of psychotic experiences in adolescence: A population-based prospective serological study. *Schizophr. Res.*, *158*, 19–24. doi:10.1016/j.schres.2014.05.019

Khandaker, G. M., Zammit, S., Lewis, G. and Jones, P. B. (2014b). A population-based study of atopic disorders and inflammatory markers in childhood before psychotic experiences in adolescence. *Schizophr. Res.*, *152*, 139–145. doi:10.1016/j.schres.2013.09.021

Kremen, W. S., Buka, S. L., Seidman, L. J., Goldstein, J. M., Koren, D. and Tsuang, M. T. (1998). IQ decline during childhood and adult psychotic symptoms in a community sample: A 19-year longitudinal study. *Am. J. Psychiatry*, *155*, 672–677. doi:10.1176/ajp.155.5.672

Kuepper, R., Os, J. van, Lieb, R., Wittchen, H.-U., Höfler, M. and Henquet, C. (2011). Continued cannabis use and risk of incidence and persistence of psychotic symptoms: 10-year follow-up cohort study. *BMJ*, *342*, d738. doi:10.1136/bmj.d738

Lawrie, S. M., Hall, J., McIntosh, A. M., Owens, D. G. C. and Johnstone, E. C. (2010). The "continuum of psychosis": Scientifically unproven and clinically impractical. *Br. J. Psychiatry*, 197, 423–425. doi:10.1192/bjp.bp.109.072827

Lebel, C., Walker, L., Leemans, A., Phillips, L. and Beaulieu, C. (2008). Microstructural maturation of the human brain from childhood to adulthood. *NeuroImage*, 40, 1044–1055. doi:10.1016/j.neuroimage.2007.12.053

Lee, K.-W., Chan, K.-W., Chang, W.-C., Lee, E. H.-M., Hui, C. L.-M. and Chen, E. Y.-H. (2016). A systematic review on definitions and assessments of psychotic-like experiences. *Early Interv. Psychiatry*, 10, 3–16. doi:10.1111/eip.12228

Luna, B., Padmanabhan, A. and O'Hearn, K. (2010). What has fMRI told us about the development of cognitive control through adolescence? *Brain Cogn.*, 72, 101–113. doi:10.1016/j.bandc.2009.08.005

Lutterveld, R. van, Diederen, K. M. J., Otte, W. M. and Sommer, I. E. (2014). Network analysis of auditory hallucinations in nonpsychotic individuals. *Hum. Brain Mapp.*, 35, 1436–1445. doi:10.1002/hbm.22264

Mak, L. E., Minuzzi, L., MacQueen, G., Hall, G., Kennedy, S. H. and Milev, R. (2017). The default mode network in healthy individuals: A systematic review and meta-analysis. *Brain Connect.*, 7, 25–33.doi:10.1089/brain.2016.0438

McGorry, P. and Nelson, B. (2016). Why we need a transdiagnostic staging approach to emerging psychopathology, early diagnosis, and treatment. *JAMA Psychiatry*, 73, 191–192. doi:10.1001/jamapsychiatry.2015.2868

Meer, L. van der, Groenewold, N. A., Pijnenborg, M. and Aleman, A. (2013). Psychosis-proneness and neural correlates of self-inhibition in theory of mind. *PLoS ONE*, 8, e67774. doi:10.1371/journal.pone.0067774

Menon, V. (2013). Developmental pathways to functional brain networks: Emerging principles. *Trends Cogn. Sci.*, 17, 627–640. doi:10.1016/j.tics.2013.09.015

Modinos, G., Mechelli, A., Ormel, J., Groenewold, N. A., Aleman, A. and McGuire, P. K. (2010a). Schizotypy and brain structure: A voxel-based morphometry study. *Psychol. Med.*, 40, 1423–1431. doi:10.1017/S0033291709991875

Modinos, G., Ormel, J. and Aleman, A. (2010b). Altered activation and functional connectivity of neural systems supporting cognitive control of emotion in psychosis proneness. *Schizophr. Res.*, 118, 88–97. doi:10.1016/j.schres.2010.01.030

Modinos, G., Pettersson-Yeo, W., Allen, P., McGuire, P. K., Aleman, A. and Mechelli, A. (2012). Multivariate pattern classification reveals differential brain activation during emotional processing in individuals with psychosis proneness. *NeuroImage*, 59, 3033–3041. doi:10.1016/j.neuroimage.2011.10.048

Modinos, G., Renken, R., Ormel, J. and Aleman, A. (2011). Self-reflection and the psychosis-prone brain: An fMRI study. *Neuropsychology*, 25, 295–305. doi:10.1037/a0021747

Modinos, G., Renken, R., Shamay-Tsoory, S. G., Ormel, J. and Aleman, A. (2010c). Neurobiological correlates of theory of mind in psychosis proneness. *Neuropsychologia*, 48, 3715–3724. doi:10.1016/j.neuropsychologia.2010.09.030

Mollon, J., David, A. S., Morgan, C., Frissa, S., Glahn, D., Pilecka, I., Hatch, S. L. and Hotopf, M. and Reichenberg, A. (2016). Psychotic experiences and neuropsychological functioning in a population-based sample. *JAMA Psychiatry*, 73, 129–138. doi:10.1001/jamapsychiatry.2015.2551

Mollon, J., David, A. S., Zammit, S., Lewis, G. and Reichenberg, A. (2018). Course of cognitive development from infancy to early adulthood in the psychosis spectrum. *JAMA Psychiatry*, 75(3), 270–279. doi:10.1001/jamapsychiatry.2017.4327

Mollon, J. and Reichenberg, A. (2018). Cognitive development prior to onset of psychosis. *Psychol. Med.*, *48*, 392–403. doi: doi:10.1017/S0033291717001970

Morgan, C., Reininghaus, U., Reichenberg, A., Frissa, S., SELCoH study team, Hotopf, M., Hatch, S. L, (2014). Adversity, cannabis use and psychotic experiences: Evidence of cumulative and synergistic effects. *Br. J. Psychiatry*, *204*(5), 346–353. doi:10.1192/bjp.bp.113.134452

Navarro-Mateu, F., Alonso, J., Lim, C. C. W., Saha, S., Aguilar-Gaxiola, S., Al-Hamzawi, A., Andrade, L. H., Bromet, E. J., Bruffaerts, R., Chatterji, S., Degenhardt, L., Girolamo, G. de, Jonge, P. de, Fayyad, J., Florescu, S., Gureje, O., Haro, J. M., Hu, C., Karam, E. G., Kovess-Masfety, V., Lee, S., Medina-Mora, M. E., Ojagbemi, A., Pennell, B.-E., Piazza, M., Posada-Villa, J., Scott, K. M., Stagnaro, J. C., Xavier, M., Kendler, K. S., Kessler, R. C. and McGrath, J. J. (2017). The association between psychotic experiences and disability: Results from the WHO World Mental Health Surveys. *Acta Psychiatr. Scand.*, *136*, 74–84. doi:10.1111/acps.12749

Niarchou, M., Zammit, S., Walters, J., Lewis, G., Owen, M.J. and Bree, M. B. van den, (2013). Defective processing speed and nonclinical psychotic experiences in children: Longitudinal analyses in a large birth cohort. *Am. J. Psychiatry*. *170*(5), 550–557. doi:10 1176/appi. ajp.2012.12060792

O'Hanlon, E., Leemans, A., Kelleher, I., Clarke, M. C., Roddy, S., Coughlan, H., Harley, M., Amico, F., Hoscheit, M. J., Tiedt, L., Tabish, J., McGettigan, A., Frodl, T. and Cannon, M. (2015). White matter differences among adolescents reporting psychotic experiences: A population-based diffusion magnetic resonance imaging study. *JAMA Psychiatry*, *72*, 668–677. doi:10.1001/jamapsychiatry.2015.0137

Orr, J. M., Turner, J. A. and Mittal, V. A. (2014). Widespread brain dysconnectivity associated with psychotic-like experiences in the general population. *NeuroImage Clin.*, *4*, 343–351. doi:10.1016/j.nicl.2014.01.006

Östling, S., Johansson, B. and Skoog, I. (2004). Cognitive test performance in relation to psychotic symptoms and paranoid ideation in non-demented 85-year-olds. *Psychol. Med.* 34, 443–450. doi:10.1017/S0033291703001144

Papanastasiou, E., Mouchlianitis, E., Joyce, D. W., McGuire, P., Banaschewski, T., Bokde, A. L. W., Bromberg, U., Büchel, C., Quinlan, E. B., Desrivières, S., Flor, H., Frouin, V., Garavan, H., Spechler, P., Gowland, P., Heinz, A., Ittermann, B., Martinot, J.-L., Paillère Martinot, M.-L., Artiges, E., Nees, F., Papadopoulos Orfanos, D., Poustka, L., Millenet, S., Fröhner, J. H., Smolka, M. N., Walter, H., Whelan, R., Schumann, G. and Shergill, S., IMAGEN Consortium, (2018). Examination of the neural basis of psychotic-like experiences in adolescence during reward processing. *JAMA Psychiatry*, *75*(10), 1043–1051.doi:10.1001/jamapsychiatry.2018.1973

Paus, T., Keshavan, M. and, Giedd, J. N. (2008). Why do many psychiatric disorders emerge during adolescence? *Nat. Rev. Neurosci.*, *9*, 947–957. doi:10.1038/nrn2513

Pelletier-Baldelli, A., Dean, D. J., Lunsford-Avery, J. R., Smith Watts, A. K., Orr, J. M., Gupta, T., Millman, Z. B. and Mittal, V. A. (2014). Orbitofrontal cortex volume and intrinsic religiosity in non-clinical psychosis. *Psychiatry Res. Neuroimaging*, *222*, 124–130. doi:10.1016/j.pscychresns.2014.03.010

Pietschnig, J., Penke, L., Wicherts, J.M., Zeiler, M. and Voracek, M. (2015). Meta-analysis of associations between human brain volume and intelligence differences: How strong are they and what do they mean? *Neurosci. Biobehav. Rev.*, *57*, 411–432. doi:10.1016/j.neubiorev.2015.09.017

Polanczyk, G., Moffitt, T.E., Arseneault, L., Cannon, M., Ambler, A., Keefe, R. S. E., Houts, R., Odgers, C. L. and Caspi, A. (2010). Etiological and clinical features of childhood psychotic symptoms: Results from a birth cohort. *Arch. Gen. Psychiatry*, *67*, 328–338. doi:10.1001/archgenpsychiatry.2010.14

Reichenberg, A., Caspi, A., Harrington, H., Houts, R., Keefe, R. S. E., Murray, R. M., Poulton, R. and Moffitt, T. E. (2010). Static and dynamic cognitive deficits in childhood preceding adult schizophrenia: A 30-year study. *Am. J. Psychiatry*, *167*, 160–169. doi:10.1176/appi. ajp.2009.09040574

Roalf, D. R., Quarmley, M., Calkins, M. E., Satterthwaite, T. D., Ruparel, K., Elliott, M. A., Moore, T. M., Gur, R. C., Gur, R. E., Moberg, P. J. and Turetsky, B. I. (2016). Temporal lobe volume decrements in psychosis spectrum youths. *Schizophr. Bull.*, *43*(3), 601–610. doi:10.1093/schbul/sbw112

Sato, J. R., Salum, G.A., Gadelha, A., Picon, F. A., Pan, P. M., Vieira, G., Zugman, A., Hoexter, M. Q., Anés, M., Moura, L. M., Gomes Del'Aquilla, M. A., Amaro, E., McGuire, P., Crossley, N., Lacerda, A., Rohde, L. A., Miguel, E. C., Bressan, R. A. and Jackowski, A. P. (2014). Age effects on the default mode and control networks in typically developing children. *J. Psychiatr. Res.*, *58*, 89–95. doi:10.1016/j.jpsychires.2014.07.004

Satterthwaite, T. D., Elliott, M. A., Ruparel, K., Loughead, J., Prabhakaran, K., Calkins, M. E., Hopson, R., Jackson, C., Keefe, J., Riley, M., Mentch, F. D., Sleiman, P., Verma, R., Davatzikos, C., Hakonarson, H., Gur, R. C. and Gur, R. E. (2014). Neuroimaging of the Philadelphia neurodevelopmental cohort. *NeuroImage*. *86*, 544–553. doi:10.1016/j.neuroimage.2013.07.064

Satterthwaite, T. D., Vandekar, S. N., Wolf, D. H., Bassett, D. S., Ruparel, K., Shehzad, Z., Craddock, R. C., Shinohara, R. T., Moore, T. M., Gennatas, E. D., Jackson, C., Roalf, D. R., Milham, M. P., Calkins, M. E., Hakonarson, H., Gur, R. C. and Gur, R. E. (2015). Connectome-wide network analysis of youth with psychosis-spectrum symptoms. *Mol. Psychiatry*, *20*, 1508–1515. doi:10.1038/mp.2015.66

Satterthwaite, T. D., Wolf, D. H., Calkins, M. E., Vandekar, S. N., Erus, G., Ruparel, K., Roalf, D. R., Linn, K. A., Elliott, M. A., Moore, T. M., Hakonarson, H., Shinohara, R.T., Davatzikos, C., Gur, R. E. and Gur, R. C. (2016). Structural brain abnormalities in youth with psychosis spectrum symptoms. *JAMA Psychiatry*, *73*, 515–524. doi:10.1001/jamapsychiatry.2015.3463

Schumann, G., Loth, E., Banaschewski, T., Barbot, A., Barker, G., Büchel, C., Conrod, P. J., Dalley, J. W., Flor, H., Gallinat, J., Garavan, H., Heinz, A., Itterman, B., Lathrop, M., Mallik, C., Mann, K., Martinot, J.-L., Paus, T., Poline, J.-B., Robbins, T. W., Rietschel, M., Reed, L., Smolka, M., Spanagel, R., Speiser, C., Stephens, D. N. and Ströhle, A., Struve, M. (2010). The IMAGEN study: Reinforcement-related behaviour in normal brain function and psychopathology. *Mol. Psychiatry*, *15*, 1128–1139. doi:10.1038/mp.2010.4

Sheffield, J.M., Kandala, S., Burgess, G. C., Harms, M. P. and Barch, D. M. (2016). Cingulo-opercular network efficiency mediates the association between psychotic-like experiences and cognitive ability in the general population. *Biol. Psychiatry Cogn. Neurosci. Neuroimaging*, *1*, 498–506. doi:10.1016/j.bpsc.2016.03.009

Sieradzka, D., Power, R. A., Freeman, D., Cardno, A. G., McGuire, P., Plomin, R., Meaburn, E. L., Dudbridge, F. and Ronald, A. (2014). Are genetic risk factors for psychosis also associated with dimension-specific psychotic experiences in adolescence? *PLoS ONE*. *9*, e94398. doi:10.1371/journal.pone.0094398

Simons, C. J. P., Jacobs, N., Jolles, J., Os and J. van, Krabbendam, L. (2007). Subclinical psychotic experiences and cognitive functioning as a bivariate phenotype for genetic studies in the general population. *Schizophr. Res.*, *92*, 24–31. doi:10.1016/j.schres.2007.01.008

Smieskova, R., Fusar-Poli, P., Allen, P., Bendfeldt, K., Stieglitz, R. D., Drewe, J., Radue, E. W., McGuire, P. K., Riecher-Rössler, A. and Borgwardt, S. J. (2010). Neuroimaging predictors of transition to psychosis—a systematic review and meta-analysis. *Neurosci. Biobehav. Rev., 34*, 1207–1222. doi:10.1016/j.neubiorev.2010.01.016

Strauss, J. S. (1969). Hallucinations and delusions as points on continua function: Rating scale evidence. *Arch. Gen. Psychiatry, 21*, 581–586. doi:10.1001/archpsyc.1969.01740230069010

Tandon, R. (2016). Conceptualizing psychotic disorders: Don't throw the baby out with the bathwater. *World Psychiatry, 15*, 133–134. doi:10.1002/wps.20329

van Dellen, E., Bohlken, M. M., Draaisma, L., Tewarie, P. K., van Lutterveld, R., Mandl, R., Stam, C. J., Sommer, I. E. (2016). Structural brain network disturbances in the psychosis spectrum. *Schizophr. Bull., 42*, 782–789. doi:10.1093/schbul/sbv178

van Os, J. (2003). Is there a continuum of psychotic experiences in the general population? *Epidemiol. Psychiatr. Soc., 12*, 242–252. doi:10.1017/S1121189X00003067

van Os, J., Hanssen, M., Bijl, R.V. and Ravelli, A. (2000). Strauss (1969) revisited: A psychosis continuum in the general population? *Schizophr. Res., 45*, 11–20. doi:10.1016/S0920-9964(99)00224-8

van Os, J., Linscott, R. J., Myin-Germeys, I., Delespaul, P. and Krabbendam, L. (2009). A systematic review and meta-analysis of the psychosis continuum: Evidence for a psychosis proneness-persistence-impairment model of psychotic disorder. *Psychol. Med., 39*, 179–195. doi:10.1017/S0033291708003814

Vértes, P. E. and Bullmore, E. T. (2015). Annual Research Review: Growth connectomics—the organization and reorganization of brain networks during normal and abnormal development. *J. Child Psychol. Psychiatry, 56*, 299–320. doi:10.1111/jcpp.12365

Wolf, D. H., Satterthwaite, T. D., Calkins, M. E., Ruparel, K., Elliott, M. A., Hopson, R. D., Jackson, C. T., Prabhakaran, K., Bilker, W. B., Hakonarson, H., Gur, R. C., Gur, R. E. (2015). Functional neuroimaging abnormalities in youth with psychosis spectrum symptoms. *JAMA Psychiatry, 72*, 456–465. doi:10.1001/jamapsychiatry.2014.3169

Woodberry, K. A., Giuliano, A. J., Seidman, L. J. (2008). Premorbid IQ in schizophrenia: A meta-analytic review. *Am. J. Psychiatry, 165*(5), 579–587. doi:10.1176/appi.ajp.2008.07081242

Yates, K., Lång, U., Cederlöf, M., Boland, F., Taylor, P., Cannon, M., McNicholas, F., DeVylder, J. and Kelleher, I. (2018). Association of psychotic experiences with subsequent risk of suicidal ideation, suicide attempts, and suicide deaths: A systematic review and meta-analysis of longitudinal population studies. *JAMA Psychiatry, 76*(2), 180–189. doi:10.1001/jamapsychiatry.2018.3514

Yung, A. R. and Lin, A. (2016). Psychotic experiences and their significance. *World Psychiatry, 15*(2), 130. doi:10.1002/wps.20327

Yung, A. R., Phillips, L. J., McGorry, P. D., McFarlane, C. A., Francey, S., Harrigan, S., Patton, G. C. and Jackson, H. J. (1998). Prediction of psychosis: A step towards indicated prevention of schizophrenia. *Br. J. Psychiatry. Suppl., 172*, 14–20.

Zammit, S., Hamshere, M., Dwyer, S., Georgiva, L., Timpson, N., Moskvina, V., Richards, A., Evans, D.M., Lewis, G., Jones, P., Owen, M. J. and O'Donovan, M. C. (2014). A population-based study of genetic variation and psychotic experiences in adolescents. *Schizophr. Bull., 40*, 1254–1262. doi:10.1093/schbul/sbt146

Zammit, S., Horwood, J., Thompson, A., Thomas, K., Menezes, P., Gunnell, D., Hollis, C., Wolke, D., Lewis, G. and Harrison, G. (2008). Investigating if psychosis-like symptoms (PLIKS) are associated with family history of schizophrenia or paternal age in the ALSPAC birth cohort. *Schizophr. Res., 104*, 279–286. doi:10.1016/j.schres.2008.04.036

Zammit, S., Kounali, D., Cannon, M., David, A. S., Gunnell, D., Heron, J., Jones, P. B., Lewis, S., Sullivan, S., Wolke, D. and Lewis, G. (2013). Psychotic experiences and psychotic disorders at age 18 in relation to psychotic experiences at age 12 in a longitudinal population-based cohort study. *Am. J. Psychiatry*, *170*, 742–750. doi:10.1176/appi.ajp.2013.12060768

Zammit, S., Odd, D., Horwood, J., Thompson, A., Thomas, K., Menezes, P., Gunnell, D., Hollis, C., Wolke, D., Lewis, G. and Harrison, G. (2009). Investigating whether adverse prenatal and perinatal events are associated with non-clinical psychotic symptoms at age 12 years in the ALSPAC birth cohort. *Psychol. Med.*, *39*, 1457–1467. doi:10.1017/S0033291708005126

Zammit, S., Owen, M. J., Evans, J., Heron, J. and Lewis, G. (2011). Cannabis, COMT and psychotic experiences. *Br. J. Psychiatry*, *199*, 380–385. doi:10.1192/bjp.bp.111.091421

CHAPTER 39

BRAIN DEVELOPMENT IN DEAF CHILDREN OR CHILDREN OF DEAF PARENTS

EVELYNE MERCURE AND VICTORIA MOUSLEY

INTRODUCTION

THE importance of early experience for later language development has been discussed for decades (Lenneberg 1967; Werker and Hensch 2015), but this "critical" or "sensitive period" hypothesis is difficult to test given the universality of human language. Cases of severe language deprivation suggest that language acquisition is limited and atypical following a period of early deprivation (Fromkin et al. 1974). However, cases like these are also associated with a severe limitation of human interaction, parental care, and normal childhood experiences, making it difficult to evaluate the specific, detrimental effects of language deprivation.

Investigating language development among populations whose speech, language, and communication experience differ from each other allows for a better understanding of experience-dependent plasticity in language development. This chapter will present a brief summary of recent literature on the early development of functional specialization for language in the infant brain, before comparing and contrasting brain activation for a familiar and an unfamiliar language. Then we will address language development and brain activation for language in populations of infants with various experiences of communication. First, we will address the natural variability in language input associated with differences in socioeconomic status, as well as the impact of this variability on brain development and language learning. Next, we will compare monolingual and bilingual infants to clarify the impact of increased language diversity on language development and the brain structures that support it. Then our attention will shift to differences in language experience observed in the context of deafness. Hearing infants of deaf mothers are likely to experience less prenatal language if the mother uses mainly sign language and to experience two languages in different modalities. Finally, we will discuss deaf infants' experience of various degrees of language deprivation depending on their family's linguistic context, the level of their deafness, and their access to technology such as cochlear implants. Deaf infants represent a unique population to test hypotheses derived from the notion of sensitive periods. Taken together, this chapter aims to clarify the different neural paths to language development.

EARLY DEVELOPMENT OF FUNCTIONAL SPECIALIZATION FOR LANGUAGE

The negative effects of language deprivation are often attributed to a lack of accessible language input during the brain's sensitive developmental period. Evaluating this explanation requires a clear understanding of typical developmental trajectories of the neural substrates of language. From the first days and months of life, babies activate a network of specialized brain areas mainly in the frontal and temporal lobes, to process speech and non-speech vocalizations, which resembles the pattern observed in response to the same stimuli in adulthood (Altvater-Mackensen et al. 2016; Dehaene-Lambertz et al. 2002, 2006, 2010; May et al. 2018; Minagawa-Kawai et al. 2011; Peña et al. 2003; Perani et al. 2011; Sato et al. 2012; Shultz et al. 2014; Vannasing et al. 2016). As in adults, activation to speech is often found to be greater in amplitude in the left hemisphere than in the right hemisphere in infants (Altvater-Mackensen et al. 2016; Dehaene-Lambertz et al. 2002, 2010; Peña et al. 2003; Vannasing et al. 2016), although this pattern of lateralization is not always reliably observed in infancy (Dehaene-Lambertz et al. 2006; May et al. 2011; May et al. 2018; Mercure, Evans, et al. 2019; Perani et al. 2011). Patterns of brain activation may also be influenced by the familiarity of the language presented to the infants, as will be discussed in the next section.

The presence of these fronto-temporal patterns of functional specialization early in infancy does not necessarily imply that they are "innate" or experience independent. Indeed, functional specialization for speech may develop very early, even prenatally, while receiving the sounds and vibrations of the mother's voice (DeCasper and Fifer 1980). A fetus can hear from the twenty-fourth to twenty-fifth gestational week (Birnholz and Benacerraf 1983). Neonates begin to learn voice and language processing in utero (for review, see Gervain 2015), as evidenced by newborns' preference for their mother's voice over other females' voices (DeCasper and Fifer 1980; Moon et al. 1993), and their ability to recognize the prosodic patterns of their mother's language(s) (Abboub et al. 2016; Partanen et al. 2013).

Interestingly, activation of the temporal, frontal, and parietal areas of the brain has also been observed in response to speech in preterm newborns, who were born at twenty-eight to thirty-two gestational weeks, and therefore have very limited experience of hearing their mother's voice prenatally (Mahmoudzadeh et al. 2013). In these infants, the right inferior frontal cortex appears sensitive to phoneme changes (/ba/ versus /a/) and voice changes (male versus female), while the left inferior frontal cortex is only sensitive to phoneme changes. These results suggest an early bias for language processing in the left inferior frontal cortex. However, studies of children with perinatal brain injury reveal that early anatomical biases are relatively plastic in infancy and that the right hemisphere can be as crucial as the left for language processing and development (Stiles 2000). Despite huge variability between individual cases, children who experience brain injury around the time of birth are likely to later demonstrate delays in the acquisition of language (babbling, early comprehension, first words, and first sentences). These delays are observed at the group level, regardless of which hemisphere suffered the lesion, suggesting a high degree of plasticity in the early development of the brain substrate for language.

LANGUAGE FAMILIARITY

Recent brain imaging studies with infants have addressed the question of language-specific experience-dependent plasticity in the brain representation for language. In other words, how does experience of a specific language influence the neural activation associated with this particular language? Brain imaging studies with newborns have shown that brain activation is of larger amplitude (May et al. 2011; Sato et al. 2012) and more left lateralized (Sato et al. 2012; Vannasing et al. 2016) in response to the language heard in utero compared to an unfamiliar language, which suggests an impact of prenatal or perinatal experience of the familiar language on brain activation for language. In the left temporal and temporo-parietal cortex, forward speech elicits more activation than backward speech for the familiar language, while this difference is not observed for an unfamiliar language (May et al. 2018; Sato et al. 2012, but see May et al. 2011). In older infants, a familiar language can elicit larger activation compared to an unfamiliar language (Fava, Hull, and Bortfeld 2014; Minagawa-Kawai et al. 2011), although this difference is not always reliably observed (Mercure, Evans, et al. 2019). In a group of infants from a wide age range (three to twelve months), increased activation for the familiar language was found in the left temporal region, while increased activation for the unfamiliar language was found in the right temporal region (Fava, Hull, Baumbauer, and Bortfeld 2014). This pattern of results appeared to be driven mainly by older infants, suggesting dynamic modifications of the patterns of functional specialization for the language learned in the first year of life compared to an unfamiliar language.

SOCIOECONOMIC STATUS AND
LANGUAGE EXPERIENCE

While the process of language learning is resilient to many environmental factors, some environmental elements may put children at disproportionate risk for poorer language attainment compared with their peers (for review, see Hoff 2006). Some studies have shown that socioeconomic status (SES) is negatively correlated with language outcomes (Betancourt et al. 2015; Whitehurst 1997). In a classic study, Hart and Risley (1995) showed that children from low SES backgrounds heard approximately thirty million fewer words than more advantaged children (though lively debate ensues, see Sperry et al. 2018a, resulting commentary by Golinkoff et al. 2018, and response by Sperry et al. 2018b). Overall, poorer quality and quantity of language input may be linked to poorer language skills (Hoff 2003; Noble et al. 2007) and academic achievement (Bowey 1995; Field 2010; Walker et al. 1994) among low SES (compared to high SES) children (for review, see Hackman and Farah 2009; Merz et al. 2019).

However, more recent research has found that quantity and quality of input are hugely variable, even within low SES settings (e.g., Gilkerson et al. 2017; Weisleder and Fernald 2013). It is likely to be the case that SES serves as a proxy for many additional environmental factors that may have a variety of effects on specific components of infants' linguistic growth across development. For example, parents of low SES strata are likely to report high rates of parental stress, less support, lower levels of formal education (Noble et al. 2015; Romeo et al. 2018) as well as less free time to play (Evans and English 2002; Rowe 2012). Each of these variables could have extricable effects on children's linguistic growth. Additional work is necessary to continue to disentangle and quantify the impact of environmental factors on infants' language growth.

The emergence of the neural systems that underlie language development may also be linked with childhood SES. Raizada et al. (2008) demonstrated a link between five-year-olds' SES and the degree of hemispheric lateralization in the left inferior frontal gyrus (IFG) during a rhyming task. Importantly, hemispheric lateralization during the rhyming task was not linked with behavioral outcomes of intelligence, language, cognition, or online task performance. Also in the IFG, total volume of grey and white matter revealed marginally significant negative correlations with SES. These findings suggest that childhood SES may be linked with the brain's emerging structure and function, though the reason(s) for the link remain(s) unknown (Raizada et al. 2008). A recent study bolsters the notion that emerging neuroanatomical structures (e.g., connectivity) may be sensitive to environmental factors such as childhood poverty (Kim et al. 2019). Functional MRI studies have reported that low SES is linked with atypical grey matter development in childhood (Hair et al. 2015; Jednorog et al. 2015). Specifically, Hair et al. (2015) found that, for children living below one and a half times the federal poverty line, regional grey matter volumes were three to four percentage points below age- and sex-matched developmental norms (Hair et al. 2015; Jednorog et al. 2015). Interestingly,

Hanson et al. (2013) do not find differences in temporal grey matter volume in early infancy, though they did show that low SES is linked with slower growth rates in the frontal and parietal development in the first months of life compared to high SES counterparts (Hanson et al. 2013). Taken together, neuroimaging studies seem to bolster conclusions drawn in the behavioral literature: links appear to exist between SES and infants' structure and function of regions related to processing language. However, it is important to note that the extent to which brain plasticity may mediate the behavioral relationships shown between SES and language has yet to be fully delineated.

Finally, it is crucial to note that studies on the links between SES and language typically include homogenous samples. The majority of the literature on brain and behavioral development in infancy is concerned with western, educated, industrialized, rich, and democratic populations (a concept called "WEIRD research," Rad et al. 2018). Research on WEIRD samples does not accurately reflect the maturation of infants who come from diverse and variable backgrounds. Alongside a continued attempt to do justice to the nuance of SES effects on the developing human brain and behavior, future research may aim to counter the WEIRD bias that historically categorizes the field itself.

SPOKEN BILINGUALISM

Studying infants who experience two spoken languages from birth offers another angle from which to examine the impact of language experience on the development of the brain's functional organization for language. It is remarkable to see that infants exposed to two or more languages from birth acquire these two linguistic codes (two phonological systems, two lexicons, and two sets of grammatical rules) and learn to keep them apart, while experiencing a reduced exposure to each of these codes compared to monolinguals (Costa and Sebastián-Gallés 2014; Werker 2012). Even more remarkable is the fact that young bilinguals usually follow the same milestones of early language development as monolinguals, including canonical babbling, first-word production, and first-word combinations (Costa and Sebastián-Gallés 2014; Werker 2012). A bilingual experience may elicit some adaptations in speech and language processing, which probably allow bilinguals to acquire two languages at the same time as monolinguals acquire a single language. For example, some studies suggest that bilingual infants appear to rely more on visual cues of articulation in speech processing than monolinguals. Indeed, bilinguals may be better at distinguishing two foreign languages based on silent articulation (Sebastián-Gallés et al. 2012). Bilingual infants may also show increased visual attention to the mouth of talking faces (Pons et al. 2015, but see Morin-Lessard et al. 2019) and increased orientation and sustained attention to face stimuli than monolinguals (Mercure et al. 2018; Mousley et al. 2023).

Crucially, there exists wide variability in bilingual experiences (even early in infancy) which may complicate apparent group differences between monolinguals and bilinguals. Perhaps most obvious is the fact that the two languages a bilingual infant hears can have huge variety in the similarities or differences between them in a variety of domains, including but not limited to speech sound inventories, transitional probabilities, use of lexical stress, rhythm, and word order. The experience of learning two languages which are in the same language family, and are therefore quite similar to each other (i.e., Catalan and Spanish: Bosch and Sebastián-Gallés 1997), may differ dramatically from learning two languages of different families, such as Chinese and Arabic (Floccia et al. 2018). Further, bilingual infants may have only one primary caretaker who is a native bilingual, while the other could be a monolingual or a sequential bilingual, meaning they learned the second language later in childhood (Place and Hoff 2016). Bilingual parents may vary in their fluency, their accent, and their mixing of their two languages while speaking to the infant (Byers-Heinlein 2012). Some bilingual infants may spend time with a number of different speakers of their second language, providing more robust experience with speakers of that language, while others may only have one adult that uses two languages with them (Gollan et al. 2015). It is crucial to consider and to quantify sources of intra-group variability in bilingualism research.

One such example of monolingual and bilingual difference which may also be sensitive to intra-group bilingual factors is that of development trajectories for perceptual attunement. Behavioral studies show that adults successfully discriminate phonetic contrasts in their native language (Rivera-Gaxiola et al. 2000), but struggle to discriminate speech sounds that are not contrastive in their native language (Best and McRoberts 2003; Best et al. 2001). For example, English-speaking adults struggle to perceive contrasts present in Hindi, as it is not an informative phonetic contrast in English (Werker and Tees 1984). In contrast, infants between zero and six months can discriminate phonetic contrasts in almost any language (Best and McRoberts 2003; Eimas et al. 1971; Kuhl et al. 2006; Werker and Tees 1984). An infant born and raised learning English can perceive the Hindi contrast before six months of age (Werker and Tees 1984). With time, infants use their early language experiences to "tune" their perceptual skills for discrimination of their native language's meaningful phonetic contrasts. By twelve months, monolingual infants become "experts" at discriminating native phonetic contrasts, while their perceptual ability for phonetic contrasts that are not contrastive in their native language has declined. Researchers often use an event-related potential component called mismatch negativity (MMN), an electrophysiological brain response elicited by the violation of an auditory "rule" (Näätänen et al. 2007). For example, Rivera-Gaxiola et al. (2005) presented English-learning infants with a string of consistent phonemes (perceived as /da/ in English) and intermittent non-native phonemes (voiced /da/ which is phonemic in Spanish, but not English). They found that seven-month-old infants showed an MMN response to the non-native phonetic contrast, indicating neural discrimination. As a group, eleven-month-olds did not demonstrate an MMN response to the non-native contrast, likely indicating a lack of neural sensitivity to the unfamiliar contrast. In contrast, both the seven- and eleven-month-olds showed MMN responses for native

(English) contrasts, suggesting a maintained neural sensitivity for contrasts with which infants had experience. The concurrent development of maintained sensitivity for native contrasts and declined sensitivity for non-native contrasts over the first year of life is often referred to as "perceptual attunement."

While the behavioral and neural trajectories of perceptual attunement for monolinguals are relatively clear, far less work examines these trends among bilingual infants who have a fundamentally different language experience from monolingual infants. Byers-Heinlein and Fennell (2014) suggest that bilingual infants' trajectories for perceptual attunement may differ from those of monolingual infants, as bilinguals have less exposure to each of their native languages than monolingual infants and have more native phonetic contrasts to learn (Byers-Heinlein and Fennell 2014). Of course, the extent to which a bilingual infants' two languages share particular phonetic properties could modulate a potential monolingual-bilingual difference. Indeed, neuroimaging evidence has suggested that learning Spanish and English from birth (rather than English only) is linked to differences in the neurobiological systems underpinning perceptual attunement in the first year of life (Garcia-Sierra et al. 2016). Using the same double oddball paradigm as Rivera-Gaxiola et al. (2005) but with magnetoencephalography, Ferjan Ramírez et al. (2017) found that English monolingual and Spanish-English bilinguals did not differ in their neural activation for English contrasts, suggesting the groups are equally sensitive to native phonetic contrasts. Spanish-English bilinguals, however, showed a stronger response than English monolinguals to a Spanish phonetic contrast in the right anterior superior and middle temporal gyrus, inferior frontal cortex, sensorimotor speech area, frontal pole, dorsolateral and ventrolateral prefrontal cortex, and orbitofrontal cortex (Ferjan Ramírez et al. 2017). This study suggests that Spanish-English bilingual and English-only monolingual infants' neural systems for perceptual attunement differ based on early language experience. However, behavioral research does not yet concur on this point, and it remains to be known whether the electrophysiological effect is modulated by intra-bilingual factors, such as distance between language pairs.

Generally speaking, bilingual infants perform remarkably well while learning two spoken languages, following similar early milestones of language development to monolingual infants, despite the complexity of their task. This may be achieved by perceptual and attentional adaptations that remain the subject of many investigations into infants' bilingual experiences. So far, research has suggested that infants' sensitivity to visual cues of articulation, visual attention to faces, and electrophysiological sensitivity to phonetic contrasts may be sensitive to experience-dependent variability in early language environments. The following section will provide more evidence for an impact of bilingualism on the development of functional specialization for language by comparing brain activation for infant-directed language in bilingual and monolingual infants, as well as infants learning a spoken and sign language from their deaf mother.

SIGN LANGUAGE EXPERIENCE IN HEARING INFANTS

Hearing infants with deaf mothers are a very interesting population to study from an experience-dependent plasticity angle. Despite having full hearing, they have a very different experience of speech, language, and communication compared to their counterparts with hearing mothers. This difference begins even before their birth. If a deaf mother uses a signed language, such as British Sign Language (BSL), as her preferred mode of communication, her fetus is likely to have reduced prenatal language experience compared to the fetus of a mother using spoken language. After birth, hearing infants with deaf mothers may learn a sign language (e.g., BSL) and/or a spoken language (e.g., English), both of which may be used to different degrees by the mother and/or other family members. For this reason, hearing infants with deaf mothers can be referred to as "bimodal bilinguals," as opposed to "unimodal bilinguals" who are exposed to two languages in the same modality, such as two spoken languages. The languages bimodal bilinguals experience differ in terms of modality—one language is visual and one is auditory—which is a difference that unimodal bilinguals do not experience.

The language produced by a deaf mother to and around her infant is likely to include less audiovisual spoken language than that of a hearing mother. Many deaf signing individuals use speech to communicate with hearing people, but the extent to which they actually "voice" their speech and produce sound, as opposed to silently mouth, is extremely variable (Bishop and Hicks 2005). When addressing her infant, a deaf mother may use sign and/or spoken language, though spoken utterances by deaf mothers tend to be reduced in length and frequency compared to that of hearing mothers (Woll and Kyle 1989). Also, the phonetic properties of a deaf mother's spoken language may differ from those of a hearing mother. The execution of phonological distinctions has been observed to be less precise when produced by deaf than hearing adults and changes in fine pitch control have also been noted (Lane and Webster 1991; Waldstein 1990).

Adult neuroimaging studies robustly demonstrate that sign language is processed in a similar brain network as spoken language in fluent hearing and deaf adults (Capek et al. 2008; Emmorey 2001; Hickok et al. 1996; MacSweeney et al. 2004; Petitto et al. 2000). This is a strong argument for the idea that classical language areas are specialized for the processing of natural languages, regardless of their modality. However, a systematic review of brain imaging studies that compares unimodal and bimodal bilinguals reveals that despite these strong similarities in the neural activation for signed and spoken language, bimodal bilinguals show less neural overlap between their two languages than unimodal bilinguals (Emmorey et al. 2016). This probably reflects the difference in sensory processing of two languages in different modalities. Moreover, during in a picture-naming task with adult bimodal bilinguals, increased activation was found in the left posterior temporal region during a bilingual context as opposed to a monolingual context (Kovelman et al. 2009). This result suggests that the increased effort involved in keeping language representations apart can increase the neural activation of classical language areas in bimodal bilinguals.

Studies of language development in bimodal bilingual children are usually based on case studies or small groups. Some research suggests that bimodal bilingual children show reduced performance compared to monolingual norms in phonology, comprehension, vocabulary, and grammar (Hofmann and Chilla 2015). However, since bimodal bilinguals learn two languages, unimodal bilinguals represent a more appropriate contrast group than monolinguals. Such contrasts with unimodal bilinguals reveal that bimodal bilinguals achieve the early linguistic milestones in each of their languages at the same time as children learning two spoken languages (Hofmann and Chilla 2015; Petitto, Katerelos, et al. 2001). In other words, not only do bimodal bilinguals develop spoken language at a similar rate as other bilingual children, but they also acquire their parents' sign language following the development milestones of spoken language acquisition in children of hearing parents, including sign babbling, the first sign production, and the first sign combinations (Bonvillian et al. 1983; Emmorey 2001; Meier 2016; Petitto, Holowka, et al. 2001; Petitto, Katerelos, et al. 2001; Petitto and Marentette 1991).

We recently completed a large-scale study investigating communicative development and brain activation for language in infants with different language experience. This study included nearly one hundred infants from three different groups: monolingual infants exclusively exposed to English, unimodal bilingual infants exposed to two or more spoken languages, and bimodal bilingual infants exposed to English and BSL from their deaf mothers. Interestingly, this study revealed that bimodal bilinguals differed from the other two groups in terms of their scanning patterns to talking faces (Mercure, Kushnerenko, et al. 2019). Indeed, while both monolinguals and unimodal bilinguals increased their looking time to the mouth of talking faces between four and eight months, no such relationship was found in bimodal bilingual infants. This suggests that the shift in attention towards the mouth which is observed in monolinguals and in unimodal bilinguals at this age requires sufficient experience with audiovisual spoken language. Moreover, when the audio and visual cues of articulation did not match, monolinguals increased their looking time to the mouth, while both groups of bilinguals did not. This result may be interpreted as an indication that unimodal and bimodal bilinguals are less sensitive to audiovisual incongruence in speech as they are used to increased variability in articulation due to foreign accents, or potentially imprecise articulations by deaf adults.

In the same sample of infants, we used functional near infrared spectroscopy (fNIRS) to analyze the impact of language experience on the patterns of brain activation for language. This study presented monolingual, unimodal bilingual, and bimodal bilingual infants with short stories in infant-direct spoken or signed language. Results revealed that language experience had an impact on the neural substrate of language, and these effects of experience appeared to be consistent across different language modalities (Mercure, Evans, et al. 2019). Indeed, unimodal bilinguals presented increased right hemisphere participation for both spoken and signed language, while bimodal bilinguals tended to demonstrate activation of smaller amplitude compared to the other groups, and a reduced number of active channels. Interestingly, this effect was observed in the absence of a language deficit in this group. To the contrary, when the three groups of infants were evaluated via the Mullen Scales of Early Learning (Mullen 1995), bimodal bilinguals outperformed the other two groups in receptive language. This finding is interesting given that this language scale was designed with hearing infants of hearing parents in mind. It included items that may have disadvantaged bimodal bilinguals such as calling the infant's name from behind their back, which is less likely to be used as a means of attracting attention by deaf parents. Moreover, our fNIRS study also used multivariate pattern analysis (MVPA) to compare brain activation to both language modalities. Our original prediction was that language modalities would be more effi-ciently distinguished based on distributed patterns of activation in infants without ex-perience of sign language. Bimodal bilinguals were expected to present more similar patterns of activation for spoken and signed language given that they have experience of each of these language modalities in a communicative setting. MVPA could be used successfully to categorize distributed patterns of activation associated with spoken and signed language in monolinguals. The signal from the left hemisphere was key to this successful categorization. Moreover, this categorization of patterns of activation for spoken and signed language was not significantly better than chance in both groups of bilingual infants. Increased variability in input associated with a bilingual experience (whether unimodal or bimodal) could influence the process of neural specialization for language, leading to more variable brain representations.

LANGUAGE DEPRIVATION IN THE CONTEXT OF DEAFNESS

From a scientific point of view, it is interesting to note that deaf infants vary enormously in terms of their language experience (for review, see Hall et al. 2019). Firstly, deafness varies tremendously. It can be characterized by levels ranging from mild to profound, by symmetry (i.e., in one or both ears), and by etiology. Deaf individuals' access to audi-tory information varies greatly depending on these factors but is likely reduced overall compared to hearing counterparts.

Many deaf people also use sign language. Deaf infants with deaf (DoD) parents who use sign language are immersed in a rich and natural language that they can easily access through the visual modality: their parents' fluent use of visual language (e.g., BSL). On the other hand, deaf infants of hearing parents are mainly surrounded by users of spoken language (e.g., English), a signal that they cannot fully access. Deaf infants with hearing (DoH) parents are therefore likely to be deprived of a fully accessible native language (Hall 2017), while DoD infants can develop a native language in the visual modality. Some hearing parents of deaf infants learn a sign language, though they are unlikely to provide as fluent a model as deaf parents, in the few months after their child has been identified as deaf. Some research suggests that positive language outcomes are still linked to hearing parents teaching their infants sign language while the parent themselves is simultaneously learning it (Caselli et al. 2021). Some deaf infants will also receive aiding technologies, such as cochlear implants, which are biomedical devices surgically implanted in the cochlea, that may increase their access to spoken language. There is incredible variability in the efficacy of implantation and the age at which it is performed. In the United Kingdom, 74 percent of eligible children receive their cochlear implant sometime between zero and three years of age (Raine 2013). There is incredible variability in deafness and language use with deaf infants; however, DoH parents are likely to experience some degree of language deprivation during a hypothesized sensitive period for language acquisition, while DoD parents experience a native fluent language in the visual modality.

Furthermore, studies of mother-child interaction suggest that interactive difficulties are more prevalent in DoH parents than in DoD parents due to the increased difficulty of interacting with someone who has different communicative needs than the parents' own. Hearing mothers use less visual or non-vocal strategies to obtain their deaf infants' attention than deaf mothers (Koester and Lahti-Harper 2010). DoH dyads spent less of their interacting time in coordinated joint attention than DoD dyads (Spencer 2000). Infants with deaf mothers are more likely to maintain their gaze on their mothers, which allows access to the mother's use of sign language, gesture, and lip movements (Spencer et al. 2004). DoD infants demonstrate more developmentally advanced play behavior than DoH infants (Meadow-Orlans et al. 2004), and DoD also outperform DoH in language development (Spencer et al. 2004).

In adulthood, there is convincing evidence that language learning is highly dependent on early language experience in infancy. In deaf adults who use sign language as their primary and preferred mode of communication in adulthood, the age at which they were first exposed to this sign language greatly predicts their proficiency. Phonological processing seems to suffer most from late acquisition. Indeed, late signers are more likely to make phonological substitutions in sign shadowing and repetition tasks than adults who learned sign language from younger (for a review, see Emmorey 2001). Moreover, it has been observed that deaf adults with reduced early language experience perform poorly compared to deaf and hearing adults without language deprivation when learning a new language in adulthood regardless of its modality (sign language or spoken language; Mayberry 2002). Native signers also have better reading and literacy skills in English than deaf late learners of sign language (Chamberlain and Mayberry 2008).

At the brain level, late acquisition of a first language leads to increased activation in the left inferior frontal gyrus in a phonological judgment task compared to deaf individuals who learned sign language from birth. Interestingly, this group difference was observed when the task was performed in both language modalities, which suggests that a delayed first-language acquisition does not only influence the brain activation for this language, but also for languages learned later in life (MacSweeney et al. 2008).

Late acquisition of a first language in deaf adults is also related to reduced activation in frontal and temporal language areas, but increased activation in visual areas of the occipital cortex during grammatical and phonemic judgment tasks (Mayberry et al. 2011). This finding suggests that the classic neural network for language requires early exposure to language in synchrony with brain growth in order to reach its full potential. Anatomical changes were also observed in the same group of deaf adults, with delayed exposure to language associated with lower grey matter concentration in the occipital cortex (Pénicaud et al. 2013).

These adult data suggest that if a first language is not acquired in the period of maximal brain plasticity, its acquisition is limited in its success, and changes in anatomical and functional brain properties are associated. It is unclear, however, how this developmental trajectory begins in infancy. Studying deaf infants and children experiencing various degrees of language deprivation could help draw these differential developmental trajectories, better understand the mechanisms underlying them, and shed light on early interventions and communication patterns that are associated with optimal outcomes for deaf individuals.

CONCLUSIONS

Language development has traditionally been studied among typically hearing infants learning one spoken language. The conclusions we can draw from this body of research, while important, do not necessarily generalize to children born into language-diverse families. Yet, monolingual children represent less than 50 percent of the world population. Given the huge variety of infants' and children's language experiences, research on language outcomes should aim to reflect this diversity. Moreover, from a scientific point of view, comparing populations of infants with different early experience of speech, language, and communication can shed light on the mechanisms of functional specialization in the brain, and especially on the impact of early experience in shaping this process. This strategy will allow a better understanding of the neurobiological mechanisms that underlie language acquisition in all its natural variability.

REFERENCES

Abboub, N., Nazzi, T., and Gervain, J. (2016). Prosodic grouping at birth. *Brain and Language*, *162*, 46–59.

Altvater-Mackensen, N., Mani, N., and Grossmann, T. (2016). Audiovisual speech perception in infancy: The influence of vowel identity and infants' productive abilities on sensitivity to (mis)matches between auditory and visual speech cues. *Developmental Psychology*, *52*, 191–204.

Best, C. C. and McRoberts, G. W. (2003). Infant perception of non-native consonant contrasts that adults assimilate in different ways. *Language and Speech*, *46*, 183–216.

Best, C. T., McRoberts, G. W., and Goodell, E. (2001). Discrimination of non-native consonant contrasts varying in perceptual assimilation to the listener's native phonological system. *The Journal of the Acoustical Society of America*, 109, 775–794.

Betancourt, L. M., Brodsky, N. L., and Hurt, H. (2015). Socioeconomic (SES) differences in language are evident in female infants at 7 months of age. *Early Human Development*, 91, 719–724.

Birnholz, J. C. and Benacerraf, B. R. (1983). The development of human fetal hearing. *Science*, 222, 516–518.

Bishop, M. and Hicks, S. (2005). Orange eyes: Bimodal bilingualism in hearing adults from deaf families. *Sign Language Studies*, 5, 188–230.

Bonvillian, J. D., Orlansky, M. D., Novack, L. L., and Folven, R. J. (1983). Early sign language acquisition and cognitive development. In D. Rogers and J. A. Sloboda (eds.), *The Acquisition of Symbolic Skills*, 207–214. Boston, MA: Springer.

Bosch, L. and Sebastián-Gallés, N. (1997, September 22). The role of prosody in infants' native-language discrimination abilities: The case of two phonologically close languages (Eurospeech 1997). Fifth European Conference on Speech Communication and Technology. Rhodes, Greece.

Bowey, J. A. (1995). Socioeconomic status differences in preschool phonological sensitivity and first-grade reading achievement. *Journal of Educational Psychology*, 87, 476–487.

Byers-Heinlein, K. (2012). Parental language mixing: Its measurement and the relation of mixed input to young bilingual children's vocabulary size. *Bilingualism: Language and Cognition*, 16, 32–48.

Byers-Heinlein, K. and Fennell, C. T. (2014). Perceptual narrowing in the context of increased variation: Insights from bilingual infants. *Developmental Psychobiology*, 56, 274–291.

Capek, C. M., Waters, D., Woll, B., MacSweeney, M., Brammer, M. J., McGuire, P. K., et al. (2008). Hand and mouth: Cortical correlates of lexical processing in British Sign Language and speechreading English. *Journal of Cognitive Neuroscience*, 20, 1220–1234.

Caselli, N., Pyers, J., and Lieberman, A. M. (2021). Deaf children of hearing parents have age-level vocabulary growth when exposed to American Sign Language by 6 months of age. *The Journal of Pediatrics*, 232, 232–236.

Chamberlain, C. and Mayberry, R. I. (2008). American Sign Language syntactic and narrative comprehension in skilled and less skilled readers: Bilingual and bimodal evidence for the linguistic basis of reading. *Applied Psycholinguistics*, 29, 367–388.

Costa, A. and Sebastián-Gallés, N. (2014). How does the bilingual experience sculpt the brain? *Nature Reviews. Neuroscience*, 15, 336–345.

DeCasper, A. J. and Fifer, W. P. (1980). Of human bonding: Newborns prefer their mothers' voices. *Science*, 208, 1174–1176.

Dehaene-Lambertz, G., Dehaene, S., and Hertz-Pannier, L. (2002). Functional neuroimaging of speech perception in infants. *Science*, 298, 2013–2015.

Dehaene-Lambertz, G., Hertz-Pannier, L., and Dubois, J. (2006). Nature and nurture in language acquisition: anatomical and functional brain-imaging studies in infants. *Trends in Neurosciences*, 29, 367–373.

Dehaene-Lambertz, G., Montavont, A., Jobert, A., Allirol, L., Dubois, J., Hertz-Pannier, L., et al. (2010). Language or music, mother or Mozart? Structural and environmental influences on infants' language networks. *Brain and Language*, 114, 53–65.

Eimas, P. D., Siqueland, E. R., Jusczyk, P., and Vigorito, J. (1971). Speech perception in infants. *Science*, 171, 303–306.

Emmorey, K. (2001). *Language, Cognition, and the Brain: Insights from Sign Language Research.* New York, NY: Psychology Press.

Emmorey, K., Giezen, M. R., and Gollan, T. H. (2016). Psycholinguistic, cognitive, and neural implications of bimodal bilingualism. *Bilingualism, 19*, 223–242.

Evans, G. W. and English, K. (2002). The environment of poverty: Multiple stressor exposure, psychophysiological stress, and socioemotional adjustment. *Child Development, 73*, 1238–1248.

Fava, E., Hull, R., Baumbauer, K., and Bortfeld, H. (2014). Hemodynamic responses to speech and music in preverbal infants. *Child Neuropsychology, 20*, 430–448.

Fava, E., Hull, R., and Bortfeld, H. (2014). Dissociating cortical activity during processing of native and non-native audiovisual speech from early to late infancy. *Brain Sciences, 4*, 471–487.

Ferjan Ramírez, N., Ramírez, R. R., Clarke, M., Taulu, S., and Kuhl, P. K. (2017). Speech discrimination in 11-month-old bilingual and monolingual infants: A magnetoencephalography study. *Developmental Science, 20*, e12427.

Field, J. (2010). Listening in the language classroom. *ELT Journal, 64*, 331–333.

Floccia, C., Sambrook, T. D., Luche, C. D., Kwok, R., Goslin, J., White, L., et al. (2018). Vocabulary of 2-year-olds learning English and an additional language: Norms and effects of language distance. *Monographs of the Society for Research in Child Development, 83*, 7–29.

Fromkin, V., Krashen, S., Curtiss, S., Rigler, D., and Rigler, M. (1974). The development of language in Genie: A case of language acquisition beyond the "critical period." *Brain and Language, 1*, 81–107.

Garcia-Sierra, A., Ramírez-Esparza, N., and Kuhl, P. K. (2016). Relationships between quantity of language input and brain responses in bilingual and monolingual infants. *International Journal of Psychophysiology: Official Journal of the International Organization of Psychophysiology, 110*, 1–17.

Gervain, J. (2015). Plasticity in early language acquisition: The effects of prenatal and early childhood experience. *Current Opinion in Neurobiology, 35*, 13–20.

Gilkerson, J., Richards, J. A., and Topping, K. J. (2017). The impact of book reading in the early years on parent–child language interaction. *Journal of Early Childhood Literacy, 17*, 92–110.

Golinkoff, R. M., Hoff, E., Rowe, M. L., Tamis-LeMonda, C. S., and Hirsh-Pasek, K. (2018). Language matters: Denying the existence of the 30-million-word-gap has serious consequences. *Child Development, 90*, 985–992.

Gollan, T. H., Starr, J., and Ferreira, V. S. (2015). More than use it or lose it: The number of speakers effect on heritage language proficiency. *Psychonomic Bulletin & Review, 22*, 147–155.

Hackman, D. A. and Farah, M. J. (2009). Socioeconomic status and the developing brain. *Trends in Cognitive Sciences, 13*, 65–73.

Hair, N. L., Hanson, J. L., Wolfe, B. L., and Pollak, S. D. (2015). Association of child poverty, brain development, and academic achievement. *JAMA Pediatrics, 169*, 822–829.

Hall, W. C. (2017). What you don't know can hurt you: The risk of language deprivation by impairing sign language development in deaf children. *Maternal and Child Health Journal, 21*, 961–965.

Hall, M. L., Hall, W. C., and Caselli, N. K. (2019). Deaf children need language, not (just) speech. *First Language, 39*, 367–395.

Hanson, J. L., Hair, N., Shen, D. G., Shi, F., Gilmore, J. H., Wolfe, B. L., et al. (2013). Family poverty affects the rate of human infant brain growth. *PloS One, 8*, e80954.

Hart, B. and Risley, T. R. (1995). *Meaningful Differences in the Everyday Experience of Young American Children*. Towson, MD: Brookes Publishing.

Hickok, G., Bellugi, U., and Klima, E. S. (1996). The neurobiology of sign language and its implications for the neural basis of language. *Nature*, *381*, 699–702.

Hoff, E. (2003). The specificity of environmental influence: Socioeconomic status affects early vocabulary development via maternal speech. *Child Development*, *74*, 1368–1378.

Hoff, E. (2006). How social contexts support and shape language development. *Developmental Review: DR*, *26*, 55–88.

Hofmann, K. and Chilla, S. (2015). Bimodal bilingual language development of hearing children of deaf parents. *European Journal of Special Needs Education*, *30*, 30–46.

Jednorog, K., Marchewka, A., Altarelli, I., Monzalvo Lopez, A. K., van Ermingen-Marbach, M., Grande, M., et al. (2015). How reliable are gray matter disruptions in specific reading disability across multiple countries and languages? Insights from a large-scale voxel-based morphometry study. *Human Brain Mapping*, *36*, 1741–1754.

Kim, D. J., Davis, E. P., Sandman, C. A., Glynn, L., Sporns, O., O'Donnell, B. F., et al. (2019). Childhood poverty and the organization of structural brain connectome. *NeuroImage*, *184*, 409–416.

Koester, L. S. and Lahti-Harper, E. (2010). Mother-infant hearing status and intuitive parenting behaviors during the first 18 months. *American Annals of the Deaf*, *155*, 5–18.

Kovelman, I., Shalinsky, M. H., White, K. S., Schmitt, S. N., Berens, M. S., Paymer, N., et al. (2009). Dual language use in sign-speech bimodal bilinguals: fNIRS brain-imaging evidence. *Brain and Language*, *109*, 112–123.

Kuhl, P. K., Stevens, E., Hayashi, A., Deguchi, T., Kiritani, S., and Iverson, P. (2006). Infants show a facilitation effect for native language phonetic perception between 6 and 12 months. *Developmental Science*, *9*, F13–F21.

Lane, H. and Webster, J. W. (1991). Speech deterioration in postlingually deafened adults. *The Journal of the Acoustical Society of America*, *89*, 859–866.

Lenneberg, E. H. (1967). The biological foundations of language. *Hospital Practice*, *2*, 59–67.

MacSweeney, M., Campbell, R., Woll, B., Giampietro, V., David, A. S., McGuire, P. K., et al. (2004). Dissociating linguistic and nonlinguistic gestural communication in the brain. *NeuroImage*, *22*, 1605–1618.

MacSweeney, M., Waters, D., Brammer, M. J., Woll, B., and Goswami, U. (2008). Phonological processing in deaf signers and the impact of age of first language acquisition. *NeuroImage*, *40*, 1369–1379.

Mahmoudzadeh, M., Dehaene-Lambertz, G., Fournier, M., Kongolo, G., Goudjil, S., Dubois, J., et al. (2013). Syllabic discrimination in premature human infants prior to complete formation of cortical layers. *Proceedings of the National Academy of Sciences*, *110*, 4846–4851.

May, L., Byers-Heinlein, K., Gervain, J., and Werker, J. F. (2011). Language and the newborn brain: Does prenatal language experience shape the neonate neural response to speech? *Frontiers in Psychology*, *2*, 222.

May, L., Gervain, J., Carreiras, M., and Werker, J. F. (2018). The specificity of the neural response to speech at birth. *Developmental Science*, *21*, e12564.

Mayberry, R. I. (2002). Cognitive development in deaf children: The interface of language and perception in neuropsychology. *Handbook of Neuropsychology*, *8*, 71–107.

Mayberry, R. I., Chen, J. K., Witcher, P., and Klein, D. (2011). Age of acquisition effects on the functional organization of language in the adult brain. *Brain and Language*, *119*, 16–29.

Meadow-Orlans, K. P., Spencer, P. E., and Koester, L. S. (2004). *The World of Deaf Infants*. Oxford: Oxford University Press.

Meier, R. P. (2016). *Sign Language Acquisition*. Oxford: Oxford University Press.

Mercure, E., Evans, S., Pirazzoli, L., Goldberg, L., Bowden-Howl, H., Coulson-Thaker, K., et al. (2019). Language experience impacts brain activation for spoken and signed language in infancy: Insights from unimodal and bimodal bilinguals. *Neurobiology of Language*, 1, 9–32.

Mercure, E., Kushnerenko, E., Goldberg, L., Bowden-Howl, H., Coulson, K., Johnson, M. H., et al. (2019). Language experience influences audiovisual speech integration in unimodal and bimodal bilingual infants. *Developmental Science*, 22, e12701.

Mercure, E., Quiroz, I., Goldberg, L., Bowden-Howl, H., Coulson, K., Gliga, T., et al. (2018). Impact of language experience on attention to faces in infancy: Evidence from unimodal and bimodal bilingual infants. *Frontiers in Psychology*, 9, 1943.

Merz, E. C., Wiltshire, C. A., and Noble, K. G. (2019). Socioeconomic inequality and the developing brain: Spotlight on language and executive function. *Child Development Perspectives*, 13, 15–20.

Minagawa-Kawai, Y., van der Lely, H., Ramus, F., Sato, Y., Mazuka, R., and Dupoux, E. (2011). Optical brain imaging reveals general auditory and language-specific processing in early infant development. *Cerebral Cortex*, 21, 254–261.

Moon, C., Cooper, R. P., and Fifer, W. P. (1993). Two-day-olds prefer their native language. *Infant Behavior and Development*, 16, 495–500.

Morin-Lessard, E., Poulin-Dubois, D., Segalowitz, N., and Byers-Heinlein, K. (2019). Selective attention to the mouth of talking faces in monolinguals and bilinguals aged 5 months to 5 years. *Developmental Psychology*, 55, 1640–1655.

Mousley, V. L., MacSweeney, M., and Mercure, E. M. (2023). Bilingual toddlers show increased attention capture by static faces compared to monolinguals. *Bilingualism: Language and Cognition*, 1–10.

Mullen, E. M. (1995). *Mullen Scales of Early Learning*. Circle Pines, MN: American Guidance Service.

Näätänen, R., Paavilainen, P., Rinne, T., and Alho, K. (2007). The mismatch negativity (MMN) in basic research of central auditory processing: A review. *Clinical Neurophysiology: Official Journal of the International Federation of Clinical Neurophysiology*, 118, 2544–2590.

Noble, K. G., Houston, S. M., Brito, N. H., Bartsch, H., Kan, E., Kuperman, J. M., et al. (2015). Family income, parental education and brain structure in children and adolescents. *Nature Neuroscience*, 18, 773–778.

Noble, K. G., McCandliss, B. D., and Farah, M. J. (2007). Socioeconomic gradients predict individual differences in neurocognitive abilities. *Developmental Science*, 10, 464–480.

Partanen, E., Kujala, T., Näätänen, R., Liitola, A., Sambeth, A., and Huotilainen, M. (2013). Learning-induced neural plasticity of speech processing before birth. *Proceedings of the National Academy of Sciences of the United States of America*, 110, 15145–15150.

Peña, M., Maki, A., Kovacić, D., Dehaene-Lambertz, G., Koizumi, H., Bouquet, F., et al. (2003). Sounds and silence: An optical topography study of language recognition at birth. *Proceedings of the National Academy of Sciences of the United States of America*, 100, 11702–11705.

Pénicaud, S., Klein, D., Zatorre, R. J., Chen, J.-K., Witcher, P., Hyde, K., et al. (2013). Structural brain changes linked to delayed first language acquisition in congenitally deaf individuals. *NeuroImage*, 66, 42–49.

Perani, D., Saccuman, M. C., Scifo, P., Anwander, A., Spada, D., Baldoli, C., et al. (2011). Neural language networks at birth. *Proceedings of the National Academy of Sciences of the United States of America, 108,* 16056–16061.

Petitto, L. A., Holowka, S., Sergio, L. E., and Ostry, D. (2001). Language rhythms in baby hand movements. *Nature, 413,* 35–36.

Petitto, L. A., Katerelos, M., Levy, B. G., Gauna, K., Tétreault, K., and Ferraro, V. (2001). Bilingual signed and spoken language acquisition from birth: implications for the mechanisms underlying early bilingual language acquisition. *Journal of Child Language, 28,* 453–496.

Petitto, L. A. and Marentette, P. F. (1991). Babbling in the manual mode: Evidence for the ontogeny of language. *Science, 251,* 1493–1496.

Petitto, L. A., Zatorre, R. J., Gauna, K., Nikelski, E. J., Dostie, D., and Evans, A. C. (2000). Speech-like cerebral activity in profoundly deaf people processing signed languages: Implications for the neural basis of human language. *Proceedings of the National Academy of Sciences of the United States of America, 97,* 13961–13966.

Place, S. and Hoff, E. (2016). Effects and noneffects of input in bilingual environments on dual language skills in 21/2-year-olds. *Bilingualism, 19,* 1023.

Pons, F., Bosch, L., and Lewkowicz, D. J. (2015). Bilingualism modulates infants' selective attention to the mouth of a talking face. *Psychological Science, 26,* 490–498.

Rad, M. S., Martingano, A. J., and Ginges, J. (2018). Toward a psychology of Homo sapiens: Making psychological science more representative of the human population. *Proceedings of the National Academy of Sciences of the United States of America, 115,* 11401–11405.

Raine, C. (2013). Cochlear implants in the United Kingdom: Awareness and utilization. *Cochlear Implants International, 14,* S32–S37.

Raizada, R. D. S., Richards, T. L., Meltzoff, A., and Kuhl, P. K. (2008). Socioeconomic status predicts hemispheric specialisation of the left inferior frontal gyrus in young children. *NeuroImage, 40,* 1392–1401.

Rivera-Gaxiola, M., Johnson, M. H., Csibra, G., and Karmiloff-Smith, A. (2000). Electrophysiological correlates of category goodness. *Behavioural Brain Research, 112,* 1–11.

Rivera-Gaxiola, M., Silva-Pereyra, J., and Kuhl, P. K. (2005). Brain potentials to native and non-native speech contrasts in 7- and 11-month-old American infants. *Developmental Science, 8,* 162–172.

Romeo, R. R., Leonard, J. A., Robinson, S. T., West, M. R., Mackey, A. P., Rowe, M. L., et al. (2018). Beyond the 30-million-word gap: Children's conversational exposure is associated with language-related brain function. *Psychological Science, 29,* 700–710.

Rowe, M. L. (2012). A longitudinal investigation of the role of quantity and quality of child-directed speech in vocabulary development. *Child Development, 83,* 1762–1774.

Sato, H., Hirabayashi, Y., Tsubokura, H., Kanai, M., Ashida, T., Konishi, I., et al. (2012). Cerebral hemodynamics in newborn infants exposed to speech sounds: A whole-head optical topography study. *Human Brain Mapping, 33,* 2092–2103.

Sebastián-Gallés, N., Albareda-Castellot, B., Weikum, W. M., and Werker, J. F. (2012). A bilingual advantage in visual language discrimination in infancy. *Psychological Science, 23,* 994–999.

Shultz, S., Vouloumanos, A., Bennett, R. H., and Pelphrey, K. (2014). Neural specialization for speech in the first months of life. *Developmental Science, 17,* 766–774.

Spencer, P. E. (2000). Looking without listening: Is audition a prerequisite for normal development of visual attention during infancy? *Journal of Deaf Studies and Deaf Education*, 5, 291–302.

Spencer, P. E., Virginia Swisher, M., and Waxman, R. P. (2004). Visual attention. In K. Meadow-Orlans, P. Spencer, and L. Koester (eds.), *The World of Deaf Infants*, 168–188. New York, NY: Oxford University Press.

Sperry, D. E., Sperry, L. L., and Miller, P. J. (2018a). Reexamining the verbal environments of children from different socioeconomic backgrounds. *Child Development*, 90, 1303–1318.

Sperry, D. E., Sperry, L. L., and Miller, P. J. (2018b). Language *does* matter: But there is more to language than vocabulary and directed speech. *Child Development*, 90, 993–997.

Stiles, J. (2000). Neural plasticity and cognitive development. *Developmental Neuropsychology*, 18, 237–272.

Vannasing, P., Florea, O., González-Frankenberger, B., Tremblay, J., Paquette, N., Safi, D., et al. (2016). Distinct hemispheric specializations for native and non-native languages in one-day-old newborns identified by fNIRS. *Neuropsychologia*, 84, 63–69.

Waldstein, R. S. (1990). Effects of postlingual deafness on speech production: Implications for the role of auditory feedback. *The Journal of the Acoustical Society of America*, 88, 2099–2114.

Walker, D., Greenwood, C. R., Hart, B., and Carta, J. J. (1994). Improving the prediction of early school academic outcomes using socioeconomic status and early language production. *Child Development*, 65, 606–621.

Weisleder, A. and Fernald, A. (2013). Talking to children matters: Early language experience strengthens processing and builds vocabulary. *Psychological Science*, 24, 2143–2152.

Werker, J. (2012). Perceptual foundations of bilingual acquisition in infancy. *Annals of the New York Academy of Sciences*, 1251, 50–61.

Werker, J. F. and Hensch, T. K. (2015). Critical periods in speech perception: New directions. *Annual Review of Psychology*, 66, 173–196.

Werker, J. F. and Tees, R. C. (1984). Cross-language speech perception: Evidence for perceptual reorganization during the first year of life. *Infant Behavior & Development*, 7, 49–63.

Whitehurst, G. J. (1997). Language processes in context: Language learning in children reared in poverty. In L. B. Adamson and M. A. Romski (eds.), *Research on Communication and Language Disorders: Contribution to Theories of Language Development*, 233–266.

Woll, B. and Kyle, J. G. (1989). Communication and language development in children of deaf parents. In S. von Tetzchner, L. S. Siegel, and L. Smith (eds.), *The Social and Cognitive Aspects of Normal and Atypical Language Development*, 129–144. Springer Series in Cognitive Development. New York, NY: Springer. doi: 10.1007/978-1-4612-3580-4_7.

INDEX